# PERIODIC TABLE OF THE ELEMENTS

Legend:

| | | Atomic number |
|---|---|---|
| 1 | | |
| **H** | | Symbol |
| Hydrogen | | Name |
| 1.0079 | | Average atomic mass |

Key:
- Metals
- Metalloids
- Nonmetals

| 1 1A | 2 2A | 3 3B | 4 4B | 5 5B | 6 6B | 7 7B | 8 | 9 8B | 10 | 11 1B | 12 2B | 13 3A | 14 4A | 15 5A | 16 6A | 17 7A | 18 8A |
|---|---|---|---|---|---|---|---|---|---|---|---|---|---|---|---|---|---|
| 1 **H** Hydrogen 1.0079 | | | | | | | | | | | | | | | | | 2 **He** Helium 4.0026 |
| 3 **Li** Lithium 6.941 | 4 **Be** Beryllium 9.0122 | | | | | | | | | | | 5 **B** Boron 10.811 | 6 **C** Carbon 12.011 | 7 **N** Nitrogen 14.007 | 8 **O** Oxygen 15.999 | 9 **F** Fluorine 18.998 | 10 **Ne** Neon 20.180 |
| 11 **Na** Sodium 22.990 | 12 **Mg** Magnesium 24.305 | | | | | | | | | | | 13 **Al** Aluminum 26.982 | 14 **Si** Silicon 28.086 | 15 **P** Phosphorus 30.974 | 16 **S** Sulfur 32.065 | 17 **Cl** Chlorine 35.453 | 18 **Ar** Argon 39.948 |
| 19 **K** Potassium 39.098 | 20 **Ca** Calcium 40.078 | 21 **Sc** Scandium 44.956 | 22 **Ti** Titanium 47.867 | 23 **V** Vanadium 50.942 | 24 **Cr** Chromium 51.996 | 25 **Mn** Manganese 54.938 | 26 **Fe** Iron 55.845 | 27 **Co** Cobalt 58.933 | 28 **Ni** Nickel 58.693 | 29 **Cu** Copper 63.546 | 30 **Zn** Zinc 65.38 | 31 **Ga** Gallium 69.723 | 32 **Ge** Germanium 72.63 | 33 **As** Arsenic 74.922 | 34 **Se** Selenium 78.96 | 35 **Br** Bromine 79.904 | 36 **Kr** Krypton 83.798 |
| 37 **Rb** Rubidium 85.468 | 38 **Sr** Strontium 87.62 | 39 **Y** Yttrium 88.906 | 40 **Zr** Zirconium 91.224 | 41 **Nb** Niobium 92.906 | 42 **Mo** Molybdenum 95.96 | 43 **Tc** Technetium [98] | 44 **Ru** Ruthenium 101.07 | 45 **Rh** Rhodium 102.91 | 46 **Pd** Palladium 106.42 | 47 **Ag** Silver 107.87 | 48 **Cd** Cadmium 112.41 | 49 **In** Indium 114.82 | 50 **Sn** Tin 118.71 | 51 **Sb** Antimony 121.76 | 52 **Te** Tellurium 127.60 | 53 **I** Iodine 126.90 | 54 **Xe** Xenon 131.29 |
| 55 **Cs** Cesium 132.91 | 56 **Ba** Barium 137.33 | 57 **La** Lanthanum 138.91 | 72 **Hf** Hafnium 178.49 | 73 **Ta** Tantalum 180.95 | 74 **W** Tungsten 183.84 | 75 **Re** Rhenium 186.21 | 76 **Os** Osmium 190.23 | 77 **Ir** Iridium 192.22 | 78 **Pt** Platinum 195.08 | 79 **Au** Gold 196.97 | 80 **Hg** Mercury 200.59 | 81 **Tl** Thallium 204.38 | 82 **Pb** Lead 207.2 | 83 **Bi** Bismuth 208.98 | 84 **Po** Polonium [209] | 85 **At** Astatine [210] | 86 **Rn** Radon [222] |
| 87 **Fr** Francium [223] | 88 **Ra** Radium [226] | 89 **Ac** Actinium [227] | 104 **Rf** Rutherfordium [265] | 105 **Db** Dubnium [268] | 106 **Sg** Seaborgium [271] | 107 **Bh** Bohrium [270] | 108 **Hs** Hassium [277] | 109 **Mt** Meitnerium [276] | 110 **Ds** Darmstadtium [281] | 111 **Rg** Roentgenium [280] | 112 **Cn** Copernicium [285] | 113 **Uut** Ununtrium [284] | 114 **Fl** Flerovium [289] | 115 **Uup** Ununpentium [288] | 116 **Lv** Livermorium [293] | 117 **Uus** Ununseptium [294] | 118 **Uuo** Ununoctium [294] |

**6 Lanthanides**

| 58 **Ce** Cerium 140.12 | 59 **Pr** Praseodymium 140.91 | 60 **Nd** Neodymium 144.24 | 61 **Pm** Promethium [145] | 62 **Sm** Samarium 150.36 | 63 **Eu** Europium 151.96 | 64 **Gd** Gadolinium 157.25 | 65 **Tb** Terbium 158.93 | 66 **Dy** Dysprosium 162.50 | 67 **Ho** Holmium 164.93 | 68 **Er** Erbium 167.26 | 69 **Tm** Thulium 168.93 | 70 **Yb** Ytterbium 173.05 | 71 **Lu** Lutetium 174.97 |
|---|---|---|---|---|---|---|---|---|---|---|---|---|---|

**7 Actinides**

| 90 **Th** Thorium 232.04 | 91 **Pa** Protactinium 231.04 | 92 **U** Uranium 238.03 | 93 **Np** Neptunium [237] | 94 **Pu** Plutonium [244] | 95 **Am** Americium [243] | 96 **Cm** Curium [247] | 97 **Bk** Berkelium [247] | 98 **Cf** Californium [251] | 99 **Es** Einsteinium [252] | 100 **Fm** Fermium [257] | 101 **Md** Mendelevium [258] | 102 **No** Nobelium [259] | 103 **Lr** Lawrencium [262] |
|---|---|---|---|---|---|---|---|---|---|---|---|---|---|

We have used the U.S. system as well as the system recommended by the International Union of Pure and Applied Chemistry (IUPAC) to label the groups in this periodic table. The system used in the United States includes a letter and a number (1A, 2A, 3B, 4B, etc.), which is close to the system developed by Mendeleev. The IUPAC system uses numbers 1–18 and has been recommended by the American Chemical Society (ACS). While we show both numbering systems here, we use the IUPAC system exclusively in the book.

| Element | Symbol | Atomic Number | Average Atomic Mass[a] |
|---|---|---|---|
| Actinium | Ac | 89 | [227] |
| Aluminum | Al | 13 | 26.982 |
| Americium | Am | 95 | [243] |
| Antimony | Sb | 51 | 121.76 |
| Argon | Ar | 18 | 39.948 |
| Arsenic | As | 33 | 74.922 |
| Astatine | At | 85 | [210] |
| Barium | Ba | 56 | 137.33 |
| Berkelium | Bk | 97 | [247] |
| Beryllium | Be | 4 | 9.0122 |
| Bismuth | Bi | 83 | 208.98 |
| Bohrium | Bh | 107 | [270] |
| Boron | B | 5 | 10.811 |
| Bromine | Br | 35 | 79.904 |
| Cadmium | Cd | 48 | 112.41 |
| Calcium | Ca | 20 | 40.078 |
| Californium | Cf | 98 | [251] |
| Carbon | C | 6 | 12.011 |
| Cerium | Ce | 58 | 140.12 |
| Cesium | Cs | 55 | 132.91 |
| Chlorine | Cl | 17 | 35.453 |
| Chromium | Cr | 24 | 51.996 |
| Cobalt | Co | 27 | 58.933 |
| Copernicium | Cn | 112 | [285] |
| Copper | Cu | 29 | 63.546 |
| Curium | Cm | 96 | [247] |
| Darmstadtium | Ds | 110 | [281] |
| Dubnium | Db | 105 | [268] |
| Dysprosium | Dy | 66 | 162.50 |
| Einsteinium | Es | 99 | [252] |
| Erbium | Er | 68 | 167.26 |
| Europium | Eu | 63 | 151.96 |
| Fermium | Fm | 100 | [257] |
| Flerovium | Fl | 114 | [289] |
| Fluorine | F | 9 | 18.998 |
| Francium | Fr | 87 | [223] |
| Gadolinium | Gd | 64 | 157.25 |
| Gallium | Ga | 31 | 69.723 |
| Germanium | Ge | 32 | 72.63 |
| Gold | Au | 79 | 196.97 |
| Hafnium | Hf | 72 | 178.49 |
| Hassium | Hs | 108 | [277] |
| Helium | He | 2 | 4.0026 |
| Holmium | Ho | 67 | 164.93 |
| Hydrogen | H | 1 | 1.0079 |
| Indium | In | 49 | 114.82 |
| Iodine | I | 53 | 126.90 |
| Iridium | Ir | 77 | 192.22 |
| Iron | Fe | 26 | 55.845 |
| Krypton | Kr | 36 | 83.798 |
| Lanthanum | La | 57 | 138.91 |
| Lawrencium | Lr | 103 | [262] |
| Lead | Pb | 82 | 207.2 |
| Lithium | Li | 3 | 6.941 |
| Livermorium | Lv | 116 | [293] |
| Lutetium | Lu | 71 | 174.97 |
| Magnesium | Mg | 12 | 24.305 |
| Manganese | Mn | 25 | 54.938 |
| Meitnerium | Mt | 109 | [276] |
| Mendelevium | Md | 101 | [258] |
| Mercury | Hg | 80 | 200.59 |
| Molybdenum | Mo | 42 | 95.96 |
| Neodymium | Nd | 60 | 144.24 |
| Neon | Ne | 10 | 20.180 |
| Neptunium | Np | 93 | [237] |
| Nickel | Ni | 28 | 58.693 |
| Niobium | Nb | 41 | 92.906 |
| Nitrogen | N | 7 | 14.007 |
| Nobelium | No | 102 | [259] |
| Osmium | Os | 76 | 190.23 |
| Oxygen | O | 8 | 15.999 |
| Palladium | Pd | 46 | 106.42 |
| Phosphorus | P | 15 | 30.974 |
| Platinum | Pt | 78 | 195.08 |
| Plutonium | Pu | 94 | [244] |
| Polonium | Po | 84 | [209] |
| Potassium | K | 19 | 39.098 |
| Praseodymium | Pr | 59 | 140.91 |
| Promethium | Pm | 61 | [145] |
| Protactinium | Pa | 91 | 231.04 |
| Radium | Ra | 88 | [226] |
| Radon | Rn | 86 | [222] |
| Rhenium | Re | 75 | 186.21 |
| Rhodium | Rh | 45 | 102.91 |
| Roentgenium | Rg | 111 | [280] |
| Rubidium | Rb | 37 | 85.468 |
| Ruthenium | Ru | 44 | 101.07 |
| Rutherfordium | Rf | 104 | [265] |
| Samarium | Sm | 62 | 150.36 |
| Scandium | Sc | 21 | 44.956 |
| Seaborgium | Sg | 106 | [271] |
| Selenium | Se | 34 | 78.96 |
| Silicon | Si | 14 | 28.086 |
| Silver | Ag | 47 | 107.87 |
| Sodium | Na | 11 | 22.990 |
| Strontium | Sr | 38 | 87.62 |
| Sulfur | S | 16 | 32.065 |
| Tantalum | Ta | 73 | 180.95 |
| Technetium | Tc | 43 | [98] |
| Tellurium | Te | 52 | 127.60 |
| Terbium | Tb | 65 | 158.93 |
| Thallium | Tl | 81 | 204.38 |
| Thorium | Th | 90 | 232.04 |
| Thulium | Tm | 69 | 168.93 |
| Tin | Sn | 50 | 118.71 |
| Titanium | Ti | 22 | 47.867 |
| Tungsten | W | 74 | 183.84 |
| Ununoctium | Uuo | 118 | [294] |
| Ununpentium | Uup | 115 | [288] |
| Ununseptium | Uus | 117 | [294] |
| Ununtrium | Uut | 113 | [284] |
| Uranium | U | 92 | 238.03 |
| Vanadium | V | 23 | 50.942 |
| Xenon | Xe | 54 | 131.29 |
| Ytterbium | Yb | 70 | 173.05 |
| Yttrium | Y | 39 | 88.906 |
| Zinc | Zn | 30 | 65.38 |
| Zirconium | Zr | 40 | 91.224 |

[a] Average atomic mass values for most elements are from *Pure Appl. Chem.* (2011) **83**, 359. Those for B, C, Cl, H, Li, N, O, Si, S, and Tl are from *Pure Appl. Chem.* (2009) **81**, 2131 and are within the ranges cited in the first reference. Atomic masses in brackets are the mass numbers of the longest-lived isotopes of elements with no stable isotopes.

# CHEMISTRY

## THE SCIENCE IN CONTEXT

### FOURTH EDITION

**Thomas R. Gilbert**

NORTHEASTERN UNIVERSITY

**Rein V. Kirss**

NORTHEASTERN UNIVERSITY

**Natalie Foster**

LEHIGH UNIVERSITY

**Geoffrey Davies**

NORTHEASTERN UNIVERSITY

W. W. Norton & Company

NEW YORK · LONDON

W. W. Norton & Company has been independent since its founding in 1923, when William Warder Norton and Mary D. Herter Norton first published lectures delivered at the People's Institute, the adult education division of New York City's Cooper Union. The Nortons soon expanded their program beyond the Institute, publishing books by celebrated academics from America and abroad. By mid-century, the two major pillars of Norton's publishing program—trade books and college texts—were firmly established. In the 1950s, the Norton family transferred control of the company to its employees, and today—with a staff of four hundred and a comparable number of trade, college, and professional titles published each year—W. W. Norton & Company stands as the largest and oldest publishing house owned wholly by its employees.

Editor: Erik Fahlgren

Project Editors: Amy Weintraub, Carla L. Talmadge

Assistant Editor: Renee Cotton

Production Manager: Sean Mintus

Developmental Editor: David Chelton

Marketing Manager: Stacy Loyal

Managing Editor, College: Marian Johnson

Science Media Editor: Robert Bellinger

Associate Media Editor: Jennifer Barnhardt

Assistant Media Editor: Paula Iborra

Design Director: Rubina Yeh

Book Designer: Lissi Sigillo

Photo Editor: Nelson Colon

Photo Researcher: Rona Tuccillo

Composition: Precision Graphics

Illustrations: Precision Graphics; Electrostatic Potential Surfaces: Daniel Zeroka

Manufacturing: Courier—Kendallville

Library of Congress Cataloging-in-Publication Data

Gilbert, Thomas R., author.

Chemistry. The science in context. — Fourth edition / Thomas R. Gilbert, Northeastern University, Rein V. Kirss, Northeastern University, Natalie Foster, Lehigh University, Geoffrey Davies, Northeastern University.

pages cm

Includes index.

**ISBN 978-0-393-91937-0 (hardcover)**

1. Chemistry—Textbooks. I. Kirss, Rein V., author. II. Foster, Natalie, author. III. Davies, Geoffrey, 1942– author. IV. Title.

QD33.2.G55 2014

540—dc23

2013039155

W. W. Norton & Company, Inc., 500 Fifth Avenue, New York, NY 10110

www.wwnorton.com

W. W. Norton & Company Ltd., Castle House, 75/76 Wells Street, London W1T 3QT

4 5 6 7 8 9 0

# Brief Contents

1 Matter and Energy: The Origin of the Universe   2

2 Atoms, Ions, and Molecules: Matter Starts Here   40

3 Stoichiometry: Mass, Formulas, and Reactions   74

4 Solution Chemistry: The Hydrosphere   132

5 Thermochemistry: Energy Changes in Reactions   196

6 Properties of Gases: The Air We Breathe   258

7 A Quantum Model of Atoms: Waves and Particles   316

8 Chemical Bonds: What Makes a Gas a Greenhouse Gas?   372

9 Molecular Geometry: Shape Determines Function   424

10 Intermolecular Forces: The Uniqueness of Water   474

11 Solutions: Properties and Behavior   512

12 Solids: Structures and Applications   560

13 Organic Chemistry: Fuels, Pharmaceuticals, Materials, and Life   602

14 Chemical Kinetics: Reactions in the Air We Breathe   666

15 Chemical Equilibrium: How Much Product Does a Reaction Really Make?   732

16 Acid–Base and Solubility Equilibria: Reactions in Soil and Water   778

17 Metal Ions: Colorful and Essential   844

18 Thermodynamics: Spontaneous and Nonspontaneous Reactions and Processes   882

19 Electrochemistry: The Quest for Clean Energy   926

20 Biochemistry: The Compounds of Life   968

21 Nuclear Chemistry: Applications to Energy and Medicine   1008

22 Life and the Periodic Table   1046

# Contents

List of Descriptive Chemistry Boxes    xvi

List of Applications    xvii

List of ChemTours    xix

About the Authors    xx

Preface    xxi

**1 Matter and Energy: The Origin of the Universe    2**

Explaining How and Why    2

**1.1** Classes of Matter    4

**1.2** Matter: An Atomic View    6

**1.3** Mixtures and How to Separate Them    8

**1.4** A Framework for Solving Problems    10

**1.5** Properties of Matter    12

**1.6** States of Matter    13

**1.7** The Scientific Method: Starting Off with a Bang    16

**1.8** Making Measurements and Expressing the Results    18
SI Units    18
Significant Figures    19
Significant Figures in Calculations    21
Precision and Accuracy    25

**1.9** Unit Conversions and Dimensional Analysis    26

**1.10** Testing a Theory: The Big Bang Revisited    28
Temperature Scales    29
An Echo of the Big Bang    31

Summary    34 • Problem-Solving Summary    34 • Visual Problems    35 •
Questions and Problems    36

How do we measure heat lost from buildings? *(Chapter 1)*

**2 Atoms, Ions, and Molecules: Matter Starts Here    40**

From Atoms to Stars and Back    40

**2.1** The Nuclear Model of Atomic Structure    42
Electrons    42
Radioactivity and the Nuclear Atom    44
Protons and Neutrons    45

**2.2** Isotopes    46

**2.3** Average Atomic Mass    48

**2.4** The Periodic Table of the Elements    50
Navigating the Modern Periodic Table    51

**2.5** Trends in Compound Formation    53
Molecular Compounds    54
Ionic Compounds    55

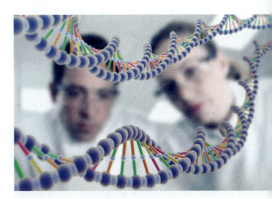

What composes DNA? *(Chapter 2)*

How are the contents of an inhaler balanced for optimal medicine delivery? *(Chapter 3)*

**2.6** Naming Compounds and Writing Formulas  57
Binary Molecular Compounds  57
Binary Ionic Compounds  58
Binary Compounds of Transition Metals  59
Polyatomic Ions  60
Acids  62

**2.7** Nucleosynthesis  62
Primordial Nucleosynthesis  63
Stellar Nucleosynthesis  64

Summary 66 • Problem-Solving Summary 67 • Visual Problems 67 •
Questions and Problems  68

**3** **Stoichiometry: Mass, Formulas, and Reactions**  **74**

The Origins of Life on Earth  74
**3.1** Chemical Reactions and Earth's Early Atmosphere  76
**3.2** The Mole  77
Molar Mass  80
Molecular Masses and Formula Masses  82
Moles and Chemical Equations  85
**3.3** Writing Balanced Chemical Equations  87
**3.4** Combustion Reactions  92
**3.5** Stoichiometric Calculations and the Carbon Cycle  94
**3.6** Determining Empirical Formulas from Percent Composition  98
**3.7** Empirical and Molecular Formulas Compared  103
**3.8** Combustion Analysis  106
**3.9** Limiting Reactants and Percent Yield  110
Calculations Involving Limiting Reactants  110
Actual Yields versus Theoretical Yields  114

*Hydrogen and Helium: The Bulk of the Universe*  118

Summary 120 • Problem-Solving Summary 120 • Visual Problems 122 •
Questions and Problems  124

**4** **Solution Chemistry: The Hydrosphere**  **132**

Oceans and Solutions  132
**4.1** Solutions on Earth and Other Places  134
**4.2** Concentration Units  136
**4.3** Dilutions  142
Determining Concentration  144
**4.4** Electrolytes and Nonelectrolytes  146
**4.5** Acid–Base Reactions: Proton Transfer  148
**4.6** Titrations  153
**4.7** Precipitation Reactions  157
Making Insoluble Salts  157
Using Precipitation in Analysis  161
Saturated Solutions and Supersaturation  164
**4.8** Ion Exchange  165
**4.9** Oxidation–Reduction Reactions: Electron Transfer  167
Oxidation Numbers  168
Considering Electron Transfer in Redox Reactions  170
Balancing Redox Reactions Using Half-Reactions  172
The Activity Series for Metals  175

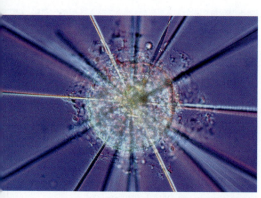

How does this single-cell plankton form its exoskeleton? *(Chapter 4)*

Redox in Nature   177

*Calcium: In the Limelight*   182

Summary 184 • Problem-Solving Summary 184 • Visual Problems 186 • Questions and Problems 187

**5**   **Thermochemistry: Energy Changes in Reactions**   **196**

Sunlight Unwinding   196

**5.1**   Energy: Basic Concepts and Definitions   198
Work, Potential Energy, and Kinetic Energy   198
Kinetic Energy and Potential Energy at the Molecular Level   201

**5.2**   Systems, Surroundings, and Energy Transfer   204
Isolated, Closed, and Open Systems   204
Exothermic and Endothermic Processes   205
Energy Units and *P−V* Work   208

**5.3**   Enthalpy and Enthalpy Changes   211

**5.4**   Heating Curves and Heat Capacity   214
Hot Soup on a Cold Day   214
Cold Drinks on a Hot Day   219

**5.5**   Calorimetry: Measuring Heat Capacity and Enthalpies of Reaction   222
Determining Molar Heat Capacity and Specific Heat   222
Enthalpies of Reaction   225
Determining Calorimeter Constants   227

**5.6**   Hess's Law   229

**5.7**   Standard Enthalpies of Formation and Reaction   233
A Practical Application of Thermochemistry   238

**5.8**   Fuel Values and Food Values   240
Fuel Value   241
Food Value   242

*Carbon: Diamonds, Graphite, and the Molecules of Life*   246

Summary 247 • Problem-Solving Summary 247 • Visual Problems 248 • Questions and Problems 249

How is heat transferred to chill soda cans? *(Chapter 5)*

**6**   **Properties of Gases: The Air We Breathe**   **258**

An Invisible Necessity   258

**6.1**   The Gas Phase   260

**6.2**   Atmospheric Pressure   260

**6.3**   The Gas Laws   265
Boyle's Law: Relating Pressure and Volume   265
Charles's Law: Relating Volume and Temperature   268
Avogadro's Law: Relating Volume and Quantity of Gas   270
Amontons's Law: Relating Pressure and Temperature   271

**6.4**   The Ideal Gas Law   273

**6.5**   Gases in Chemical Reactions   277

**6.6**   Gas Density   280

**6.7**   Dalton's Law and Mixtures of Gases   283

**6.8**   The Kinetic Molecular Theory of Gases   288
Explaining Boyle's, Dalton's, and Avogadro's Laws   288
Explaining Amontons's and Charles's Laws   289
Molecular Speeds and Kinetic Energy   290
Graham's Law: Effusion and Diffusion   293

**6.9**   Real Gases   295
Deviations from Ideality   295

The van der Waals Equation for Real Gases 297

*Nitrogen: Feeding Plants and Inflating Air Bags* 300

Summary 302 • Problem-Solving Summary 302 • Visual Problems 303 • Questions and Problems 306

**7** **A Quantum Model of Atoms: Waves and Particles** 316

Can Nature Be as Absurd as It Seems? 316

**7.1** Light Waves 318
Properties of Waves 319
The Behavior of Waves 320

**7.2** Atomic Spectra 321

**7.3** Particles of Light and Quantum Theory 323
Quantum Theory 324
The Photoelectric Effect 326
Wave–Particle Duality 327

**7.4** The Hydrogen Spectrum and the Bohr Model 328
The Hydrogen Emission Spectrum 328
The Bohr Model of Hydrogen 329

**7.5** Electron Waves 332
De Broglie Wavelengths 332
The Heisenberg Uncertainty Principle 335

**7.6** Quantum Numbers and Electron Spin 337

**7.7** The Sizes and Shapes of Atomic Orbitals 343
*s* Orbitals 343
*p* and *d* Orbitals 344

**7.8** The Periodic Table and Filling the Orbitals of Multielectron Atoms 345

**7.9** Electron Configurations of Ions 351
Ions of the Main Group Elements 351
Transition Metal Cations 352

**7.10** The Sizes of Atoms and Ions 354
Trends in Atomic and Ionic Sizes 356

**7.11** Ionization Energies 357

**7.12** Electron Affinities 360
*A Noble Family: Special Status for Special Behavior* 362

Summary 364 • Problem-Solving Summary 364 • Visual Problems 365 • Questions and Problems 366

**8** **Chemical Bonds: What Makes a Gas a Greenhouse Gas?** 372

The Greenhouse Effect: Good News and Bad 372

**8.1** Types of Chemical Bonds 374

**8.2** Lewis Structures 375
Lewis Symbols 376
Lewis Structures 377
Steps to Follow when Drawing Lewis Structures 377
Lewis Structures of Molecules with Double and Triple Bonds 379
Lewis Structures of Ionic Compounds 381

**8.3** Polar Covalent Bonds 382
Polarity and Type of Bond 384

**8.4** Vibrating Bonds and the Greenhouse Effect 386

**8.5** Resonance 387

**8.6** Formal Charge: Choosing among Lewis Structures 391
Calculating Formal Charge of an Atom in a Resonance Structure 392

Why do mountaineers at high altitudes need supplemental oxygen? *(Chapter 6)*

What physical process is responsible for the brilliant red in fireworks? *(Chapter 7)*

**8.7** Exceptions to the Octet Rule   395
Odd-Electron Molecules   395
Atoms with More than an Octet   397
Atoms with Less than an Octet   400
The Limits of Bonding Models   403

**8.8** The Lengths and Strengths of Covalent Bonds   403
Bond Length   404
Bond Energies   404

*Fluorine and Oxygen: Location, Location, Location*   410

Summary   412 • Problem-Solving Summary   412 • Visual Problems   413 •
Questions and Problems   415

## **9** Molecular Geometry: Shape Determines Function   424

Biological Activity and Molecular Shape   424
**9.1** Molecular Shape   426
**9.2** Valence-Shell Electron-Pair Repulsion Theory (VSEPR)   427
Central Atoms with No Lone Pairs   427
Central Atoms with Lone Pairs   431

**9.3** Polar Bonds and Polar Molecules   437
**9.4** Valence Bond Theory   440
Bonds from Orbital Overlap   440
Hybridization   441
Tetrahedral Geometry: $sp^3$ Hybrid Orbitals   442
Trigonal Planar Geometry: $sp^2$ Hybrid Orbitals   443
Linear Geometry: $sp$ Hybrid Orbitals   445
Octahedral and Trigonal Bipyramidal Geometries: $sp^3d^2$ and $sp^3d$ Hybrid Orbitals   447

**9.5** Shape and Interactions with Large Molecules   449
**9.6** Chirality and Molecular Recognition   452
**9.7** Molecular Orbital Theory   453
Molecular Orbitals of Hydrogen and Helium   455
Molecular Orbitals of Homonuclear Diatomic Molecules   456
Molecular Orbitals of Heteronuclear Diatomic Molecules   460
Molecular Orbitals of $N_2^+$ and Spectra of Auroras   462

*Chalcogens: From Alcohol to Asparagus, the Nose Knows*   464

Summary   466 • Problem-Solving Summary   466 • Visual Problems   467 •
Questions and Problems   467

How are tomatoes ripened? *(Chapter 9)*

## **10** Intermolecular Forces: The Uniqueness of Water   474

Ubiquitous, Essential, and Remarkable   474
**10.1** Interactions between Ions   476
Ion–Ion Interactions   477

**10.2** Interactions Involving Polar Molecules   479
Ion–Dipole Interactions   480
Dipole–Dipole Interactions   481
Hydrogen Bonds   481

**10.3** Dispersion Forces   484
**10.4** Polarity and Solubility   488
Combinations of Intermolecular Forces   490

**10.5** Vapor Pressure of Pure Liquids   492
Vapor Pressure and Temperature   492
Volatility and the Clausius–Clapeyron Equation   493

**10.6** Phase Diagrams: Intermolecular Forces at Work   495
Phases and Phase Transformations   495

**10.7** Some Remarkable Properties of Water   498
Surface Tension and Viscosity   498
Water and Aquatic Life   502

*The Halogens: The Salt of the Earth*   504

Summary  506 · Problem-Solving Summary  506 · Visual Problems  507 ·
Questions and Problems  508

**11** **Solutions: Properties and Behavior**   512

A World of Solutions   512

**11.1** Vapor Pressure of Solutions   514
Vapor Pressure of Solutions: Raoult's Law   515

**11.2** Solubility of Gases in Water   517

**11.3** Energy Changes during Formation and Dissolution of Ionic Compounds   520
Calculating Lattice Energies Using the Born–Haber Cycle   522
Enthalpies of Hydration   525

**11.4** Mixtures of Volatile Solutes   527
Vapor Pressures of Mixtures of Volatile Solutes   527

**11.5** Colligative Properties of Solutions   532
Molality   533
Boiling Point Elevation   534
Freezing Point Depression   535
The van 't Hoff Factor   537
Osmosis and Osmotic Pressure   541
Reverse Osmosis   546

**11.6** Measuring the Molar Mass of a Solute Using Colligative Properties   548

Summary  552 · Problem-Solving Summary  552 · Visual Problems  554 ·
Questions and Problems  555

**12** **Solids: Structures and Applications**   560

Stronger, Tougher, Harder   560

**12.1** The Solid State   562

**12.2** Structures of Metals   563
Stacking Patterns   563
Stacking Spheres and Unit Cells   564
Unit Cell Dimensions   567

**12.3** Alloys   571
Substitutional Alloys   571
Interstitial Alloys   572

**12.4** Metallic Bonds and Conduction Bands   574

**12.5** Semiconductors   575

**12.6** Salt Crystals: Ionic Solids   577

**12.7** Structures of Nonmetals   580

**12.8** Ceramics: Insulators to Superconductors   582
Polymorphs of Silica   583
Ionic Silicates   583
From Clay to Ceramic   584
Superconductors   584

**12.9** X-ray Diffraction: How We Know Crystal Structures   586

*Silicon, Silica, Silicates, Silicone: What's in a Name?*   590

Summary  592 · Problem-Solving Summary  593 · Visual Problems  593 ·
Questions and Problems  595

How have microorganisms adapted to
arsenic in Mono Lake? *(Chapter 9)*

What property enables this water strider
to walk on the water's surface?
*(Chapter 10)*

**13** **Organic Chemistry: Fuels, Pharmaceuticals, Materials, and Life   602**

Carbon Everywhere   602

**13.1**   Carbon: The Scope of Organic Chemistry   604
Families Based on Functional Groups   604
Monomers and Polymers   605
Small Molecules versus Polymers: Physical Properties   606

**13.2**   Alkanes   606
Physical Properties and Structures of Alkanes   608
Drawing Organic Molecules   609
Constitutional (Structural) Isomers   611
Naming Alkanes   613
Cycloalkanes   616
Sources and Uses of Alkanes   617

**13.3**   Alkenes and Alkynes   618
Chemical Reactivities of Alkenes and Alkynes   619
Isomers of Alkenes and Alkynes   620
Naming Alkenes and Alkynes   622
Polymers of Alkenes   623

**13.4**   Aromatic Compounds   626
Isomers of Aromatic Compounds   627
Polymers Containing Aromatic Rings   628

**13.5**   Amines   629

**13.6**   Alcohols, Ethers, and Reformulated Gasoline   630
Alcohols: Methanol and Ethanol   631
Ethers: Diethyl Ether   634
Polymers of Alcohols and Ethers   635

**13.7**   Carbonyl-Containing Compounds   637
Aldehydes and Ketones   638
Carboxylic Acids   638
Esters and Amides   639
Polyesters and Polyamides   641

**13.8**   Chirality   647
Optical Isomerism   647
Chirality in Nature   650

Summary 654 • Problem-Solving Summary 654 • Visual Problems 655 • Questions and Problems 657

What properties allow this sheet of carbon graphene to behave as a semiconductor? *(Chapter 12)*

**14** **Chemical Kinetics: Reactions in the Air We Breathe   666**

Clearing the Air   666

**14.1**   Cars, Trucks, and Air Quality   668

**14.2**   Reaction Rates   669
Experimentally Determined Rates: Actual Values   673
Average and Instantaneous Rates of Formation of NO   674

**14.3**   Effect of Concentration on Reaction Rate   677
Reaction Order and Rate Constants   678
Integrated Rate Laws: First-Order Reactions   683
Reaction Half-Lives   686
Integrated Rate Laws: Second-Order Reactions   688
Pseudo-First-Order Reactions   690
Zero-Order Reactions   693

**14.4**   Reaction Rates, Temperature, and the Arrhenius Equation   694

**14.5** Reaction Mechanisms  699
Elementary Steps  700
Rate Laws and Reaction Mechanisms  702
Mechanisms and Zero-Order Reactions  706

**14.6** Catalysis  706
Catalysts and the Ozone Layer  706
Catalysts and Catalytic Converters  710
Enzymes: Biological Catalysts  712

*The Platinum Group: Catalysts, Jewelry, and Investment*  715

Summary  716 • Problem-Solving Summary  716 • Visual Problems  717 •
Questions and Problems  720

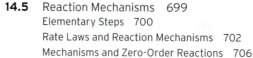

**15** **Chemical Equilibrium: How Much Product Does a Reaction Really Make?**  732

Making Commercial Goods Economically  732
**15.1** The Dynamics of Chemical Equilibrium  734
**15.2** Writing Equilibrium Constant Expressions  737
**15.3** Relationships between $K_c$ and $K_p$ Values  742
**15.4** Manipulating Equilibrium Constant Expressions  745
$K$ for Reverse Reactions  745
$K$ for an Equation Multiplied by a Number  746
Combining $K$ Values  748

**15.5** Equilibrium Constants and Reaction Quotients  750
**15.6** Heterogeneous Equilibria  752
**15.7** Le Châtelier's Principle  754
Effects of Adding or Removing Reactants or Products  755
Effects of Pressure and Volume Changes  756
Effect of Temperature Changes  759
Catalysts and Equilibrium  760

**15.8** Calculations Based on $K$  761

Summary  769 • Problem-Solving Summary  769 • Visual Problems  770 •
Questions and Problems  771

What produced this "brown LA haze"?
*(Chapter 15)*

What is responsible for the color of hydrangeas? *(Chapter 16)*

**16** **Acid–Base and Solubility Equilibria: Reactions in Soil and Water**  778

A Balancing Act  778
**16.1** Acids and Bases: The Brønsted–Lowry Model  780
Strong and Weak Acids  780
Conjugate Acid–Base Pairs  782
Strong and Weak Bases  783
Relative Strengths of Acids and Bases  784

**16.2** pH and the Autoionization of Water  785
The pH Scale  786
pOH  788

**16.3** Calculations Involving pH, $K_a$, and $K_b$  790
Weak Acids  790
Weak Bases  793
pH of Very Dilute Solutions  795

**16.4** Polyprotic Acids  796
Acid Precipitation  796
Acidification of the Ocean  797

**16.5**   Acid Strength and Molecular Structure   800

**16.6**   pH of Salt Solutions   801

**16.7**   The Common-Ion Effect   806

**16.8**   pH Buffers   810
An Environmental Buffer   810
A Physiological Buffer   814
Buffer Range and Capacity   815

**16.9**   Indicators and Acid–Base Titrations   818
Acid–Base Titrations   818
Alkalinity Titrations   822

**16.10**   Solubility Equilibria   825
$K_{sp}$ and $Q$   829
*The Chemistry of Two Strong Acids: Sulfuric and Nitric Acids*   833
Summary 834 • Problem-Solving Summary 834 • Visual Problems 836 •
Questions and Problems 837

**17**   **Metal Ions: Colorful and Essential   844**

The Company They Keep   844

**17.1**   Lewis Acids and Bases   846

**17.2**   Complex Ions   848

**17.3**   Complex-Ion Equilibria   851

**17.4**   Naming Complex Ions and Coordination Compounds   853
Complex Ions with a Positive Charge   853
Complex Ions with a Negative Charge   855
Coordination Compounds   855

**17.5**   Hydrated Metal Ions as Acids   856

**17.6**   Polydentate Ligands   859

**17.7**   Ligand Strength and the Chelate Effect   861

**17.8**   Crystal Field Theory   862

**17.9**   Magnetism and Spin States   867

**17.10**   Isomerism in Coordination Compounds   869
Stereoisomers of Coordination Compounds   870
Enantiomers   871

**17.11**   Coordination Compounds in Biochemistry   872
Summary 876 • Problem-Solving Summary 877 • Visual Problems 877 •
Questions and Problems 879

What role do metals play in the resplendent color change of autumn leaves? *(Chapter 17)*

**18**   **Thermodynamics: Spontaneous and Nonspontaneous Reactions and Processes   882**

The Game of Energy Conversion   882

**18.1**   Spontaneous Processes   884

**18.2**   Thermodynamic Entropy   886

**18.3**   Absolute Entropy and the Third Law of Thermodynamics   890
Entropy and Structure   892

**18.4**   Calculating Entropy Changes   894

**18.5**   Free Energy   895

**18.6**   Temperature and Spontaneity   900

**18.7**   Free Energy and Chemical Equilibrium   902

**18.8**   Influence of Temperature on Equilibrium Constants   907

**18.9** Driving the Human Engine: Coupled Reactions   909

**18.10** Microstates: A Quantized View of Entropy   913

Summary 916 • Problem-Solving Summary 917 • Visual Problems 918 • Questions and Problems 919

How do we fuel energy-efficient cars?
*(Chapter 19)*

## 19 Electrochemistry: The Quest for Clean Energy   926

Efficient Locomotion—Batteries Included   926

**19.1** Redox Chemistry Revisited   928

**19.2** Electrochemical Cells   930
Voltaic and Electrolytic Cells   931
Cell Diagrams   931

**19.3** Standard Potentials   934

**19.4** Chemical Energy and Electrical Work   937

**19.5** A Reference Point: The Standard Hydrogen Electrode   940

**19.6** The Effect of Concentration on $E_{cell}$   942
The Nernst Equation   942
$E°$ and $K$   944

**19.7** Relating Battery Capacity to Quantities of Reactants   946
Nickel–Metal Hydride Batteries   947
Lithium–Ion Batteries   948

**19.8** Corrosion: Unwanted Electrochemical Reactions   950

**19.9** Electrolytic Cells and Rechargeable Batteries   953

**19.10** Fuel Cells   956

Summary 960 • Problem-Solving Summary 961 • Visual Problems 961 • Questions and Problems 963

Did life on Earth begin with self-replicating molecules in places like this?
*(Chapter 20)*

## 20 Biochemistry: The Compounds of Life   968

Function Follows Form   968

**20.1** The Composition of Proteins   970
Amino Acids   970
Chirality   970
Zwitterions   972
Peptides   975

**20.2** Protein Structure and Function   977
Primary Structure   977
Secondary Structure   979
Tertiary and Quaternary Structure   981
Enzymes: Proteins as Catalysts   982

**20.3** Carbohydrates   984
Molecular Structures of Glucose and Fructose   984
Disaccharides and Polysaccharides   985
Glycolysis Revisited   988

**20.4** Lipids   988
Function and Metabolism of Lipids   990
Other Types of Lipids   991

**20.5** Nucleotides and Nucleic Acids   993
From DNA to New Proteins   995

**20.6** From Biomolecules to Living Cells   997

Summary 999 • Problem-Solving Summary 999 • Visual Problems 1000 • Questions and Problems 1002

**21  Nuclear Chemistry: Applications to Energy and Medicine  1008**

The Risks and Benefits of Nuclear Radiation  1008
**21.1**  Binding Energy and Nuclear Stability  1010
**21.2**  Unstable Nuclei and Radioactive Decay  1012
**21.3**  Rates of Radioactive Decay  1017
**21.4**  Radiometric Dating  1018
**21.5**  Measuring Radioactivity  1021
**21.6**  Biological Effects of Radioactivity  1024
Radiation Dosage  1024
Evaluating the Risks of Radiation  1026
**21.7**  Medical Applications of Radionuclides  1028
Therapeutic Radiology  1028
Diagnostic Radiology  1029
**21.8**  Nuclear Fission  1029
**21.9**  Nuclear Fusion and the Quest for Clean Energy  1032

Summary 1037 • Problem-Solving Summary 1038 • Visual Problems 1038 •
Questions and Problems 1039

**22  Life and the Periodic Table  1046**

Elements in Our Bodies  1046
**22.1**  The Periodic Table of Life  1048
**22.2**  Major Essential Elements  1051
Sodium and Potassium  1051
Magnesium and Calcium  1053
Chlorine  1055
Nitrogen  1056
**22.3**  Trace and Ultratrace Essential Elements  1058
Trace Essential Main Group Elements  1058
Trace Essential Transition Elements  1059
Ultratrace Essential Elements  1062
**22.4**  Nonessential Elements  1065
Rubidium and Cesium  1065
Strontium and Barium  1065
Germanium  1065
Antimony  1065
**22.5**  Elements for Diagnosis and Therapy  1066
Diagnostic Applications  1067
Therapeutic Applications  1072
Medical Devices and Materials  1076

Summary 1079 • Problem-Solving Summary 1080 • Visual Problems 1080 •
Questions and Problems 1082

Appendices  APP-1

Glossary  G-1

Answers to Concept Tests and Practice Exercises  ANS-1

Answers to Selected End-of-Chapter Questions and Problems  ANS-11

Credits  C-1

Index  I-1

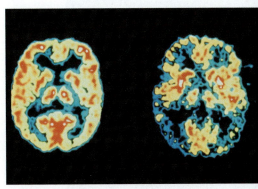

How can positron emission help monitor brain activity? *(Chapter 21)*

# Descriptive Chemistry Boxes

Hydrogen and Helium: The Bulk of the Universe  118

Calcium: In the Limelight  182

Carbon: Diamonds, Graphite, and the Molecules of Life  246

Nitrogen: Feeding Plants and Inflating Air Bags  300

A Noble Family: Special Status for Special Behavior  362

Fluorine and Oxygen: Location, Location, Location  410

Chalcogens: From Alcohol to Asparagus, the Nose Knows  464

The Halogens: The Salt of the Earth  504

Silicon, Silica, Silicates, Silicone: What's in a Name?  590

The Platinum Group: Catalysts, Jewelry, and Investment  715

The Chemistry of Two Strong Acids: Sulfuric and Nitric Acids  833

# Applications

Silicon wafers   7

Seawater distillation and algae filtration   8

Chromatography   9

Driving the Mars rover *Curiosity*   33

Big Bang and primordial nucleosynthesis   63

Star formation and stellar nucleosynthesis   64

Volcanic eruptions   76

Natural gas stoves   92

Atmospheric carbon dioxide   92

Photosynthesis, respiration, and the carbon cycle   94

Power plant emissions   96

Antiasthma drugs   99

Composition of pheromones   106

Hydrogen-powered vehicles   115

Anticancer drugs (Taxol)   116

Evidence for water on Mars   134

Polyvinyl chloride (PVC) pipes   139

Great Salt Lake   140

Saline intravenous infusion   144

Stalactites and stalagmites   151

Chemical weathering   151

Drainage from abandoned coal mines   153

Antacids   156

Barium sulfate for gastrointestinal imaging   160

Rock candy   164

Zeolites for water filtration and water softeners   166

Iron oxides in rocks and soils   177

Drug stability calculations   181

Smokestack scrubbers   183

Rockets   203

Diesel engines   209

Resurfacing an ice rink   212

Instant cold packs   213

Heat sinks and car radiators   220

Recycling aluminum   239

Comparing fuels   241

Flameless heat source   244

Diamond and graphite   246

Earth's atmosphere   260

Barometers and manometers   260

Superstorm Sandy   262

Aerosol cans   272

Bicycle tire pressure   272

Weather balloon pressure   276

Air bag inflation   279

Dieng Plateau gas poisoning disaster   280

Gas mixtures for scuba diving and nitrogen narcosis   284

Compressed oxygen for mountaineering   286

Rainbows   320

Lasers   342

Why fireworks are red   361

Blimps and helium   362

Greenhouse effect   372

Oxyacetylene torches   381

Atmospheric ozone   387

Chlorofluorocarbons (CFCs) and ozone holes   411

Ripening tomatoes   424

Polycyclic aromatic hydrocarbon (PAH) intercalation in DNA   451

Spearmint and caraway aromas   452

Auroras   453

Hydrogen bonds in DNA   482

Supercritical carbon dioxide and dry ice   497

Water striders   499

Aquatic life in frozen lakes   502

Drug efficacy   502

Splenda   504

Antifreeze in car batteries   512

Fractional distillation of crude oil   528

Radiator fluid   536

Osmosis in red blood cells   541

Saline and dextrose intravenous solutions   545

Desalination of seawater via reverse osmosis   546

Purifying water   547

Aluminum alloys   560

Metallurgy of copper and discovery of bronze   571

Stainless steel and reduction of iron ore in blast furnaces   572

Carbon steel—stronger than pure steel 573

Semiconductors in bar-code readers, CD players, and indicator lights 576

Graphene: a versatile material 582

Porcelain and glossy paper 584

Maglev trains 586

Silica and silicates in solids and technology, Silicon Valley 590

Gasoline, kerosene, diesel fuel, and mineral oil 618

Polyethylene: LDPE and HDPE plastics 623

Teflon for surgical procedures 625

Styrofoam and aromatic rings 629

Amphetamine, Benadryl, and adrenaline 629

Ethanol as grain alcohol 632

Polypropylene and vinyl polymers 635

Gasoline additives: ethanol and MTBE 635

Fuel production via methanogenic bacteria 639

Aspirin, ibuprofen, and naproxen 640

Artificial skin and dissolving sutures 642

Plastic soda bottles 642

Synthetic fabrics: Dacron, nylon, and Kevlar 645

Antiasthma drugs and antidepressants 651

Lotions and creams; Vaseline 652

Photochemical smog and catalytic converters 666

Digestion of aspirin 693

Ozone in the stratosphere 706

Smog simulations 713

Platinum in catalysts, jewelry and everyday devices 715

Manufacturing sulfuric acid 732

Lime kilns 752

Acid rain 788

Respiratory acidosis and alkalosis 814

Swimming pool test kits for pH 818

Milk of magnesia—a suspension 825

Food preservatives 860

Anticancer drugs (cisplatin) 870

Chlorophyll, hemoglobin, and myoglobin 872

Autumn leaf colors and cytochromes 873

Instant cold packs 885

Engine efficiency 900

Energy from food; glycolysis 910

Hybrid and electric vehicles 926

Alkaline, nicad, and zinc-air batteries 936

Lead−acid car batteries 943

Rusted iron via oxidation 950

Rechargeable batteries 953

Electrolysis of salt 955

Proton−exchange membrane (PEM) fuel cells 956

Alloys and corrosion at sea 959

Perfect foods and complete proteins 970

Aspartame 975

Sickle-cell anemia and malaria 979

Silk and β-pleated sheets 980

Alzheimer's disease 980

Lactose intolerance 982

Blood types and glycoproteins 984

Ethanol production from cellulose 987

Unsaturated fats, saturated fats, and trans fats 990

Olestra, a modified fat substitute 991

Cholesterol and plaque 992

DNA and RNA 993

Origin of life on Earth and RNA world hypothesis 997

Liquid oil to solid fat 998

Radiometric dating 1018

Scintillation counters and Geiger counters 1022

Fukushima nuclear disaster 1023

Biological effects of radioactivity; Chernobyl; radon gas 1026

Therapeutic and diagnostic radiology 1028

Nuclear weapons and nuclear power 1030

Nuclear fusion in the sun 1032

Tokamak reactors and ITER 1034

Radium paint and the Radium Girls 1036

Dietary reference intake (DRI) for essential elements 1048

The Mad Hatter and mercury poisoning 1050

Neutralizing stomach acid; acid reflux 1055

Diabetes and vanadium 1073

# ▶❚❚ ChemTours

Big Bang   16

Significant Figures   20

Scientific Notation   20

Dimensional Analysis   26

Temperature Conversion   29

Cathode-Ray Tube   42

Millikan Oil-Drop
     Experiment   42

Rutherford Experiment   45

The Periodic Table   51

NaCl Reaction   55

Synthesis of Elements   64

Avogadro's Number   78

Balancing Equations   88

Carbon Cycle   96

Percent Composition   98

Limiting Reactant   111

Molarity   138

Dilution   142

Migration of Ions in
     Solution   146

Saturated Solutions   164

State Functions and Path
     Functions   199

Internal Energy   208

Pressure−Volume Work   210

Heating Curves   214

Calorimetry   224

Hess's Law   229

The Ideal Gas Law   274

Dalton's Law   283

Molecular Speed   291

Molecular Motion   291

Electromagnetic Radiation   319

Light Diffraction   320

Light Emission and
     Absorption   322

Bohr Model of the Atom   329

De Broglie Wavelength   333

Quantum Numbers   337

Electron Configuration   346

Bonding   374

Lewis Dot Structures   377

Partial Charges and Bond
     Dipoles   383

Greenhouse Effect   386

Vibrational Modes   387

Resonance   389

Expanded Valence Shells   398

Estimating Enthalpy
     Changes   406

VSEPR Model   427

Hybridization   448

Chemistry of the Upper
     Atmosphere   453

Molecular Orbitals   457

Intermolecular Forces   481

Phase Diagrams   497

Hydrogen Bonding in Water   498

Capillary Action   500

Raoult's Law   515

Henry's Law   518

Lattice Energy   521

Fractional Distillation   529

Boiling and Freezing
     Points   537

Osmotic Pressure   542

Crystal Packing   564

Unit Cell   564

Allotropes of Carbon   580

Superconductors   584

X-ray Diffraction   586

Structure of Cyclohexane   616

Cyclohexane in 3-D   617

Structure of Benzene   627

Polymers   642

Chirality   648

Chiral Centers   649

Reaction Rate   671

Reaction Order   678

Arrhenius Equation   696

Collision Theory   696

Reaction Mechanisms   700

Equilibrium   734

Equilibrium in the Gas
     Phase   739

Le Châtelier's Principle   754

Solving Equilibrium
     Problems   761

Acid Rain   780

Acid−Base Ionization   780

Autoionization of Water   785

pH Scale   786

Acid Strength and Molecular
     Structure   800

Buffers   810

Acid/Base Titrations   820

Titrations of Weak Acids   820

Crystal Field Splitting   863

Dissolution of Ammonium
     Nitrate   884

Entropy   886

Gibbs Free Energy   895

Equilibrium and
     Thermodynamics   903

Zinc−Copper Cell   931

Cell Potential   934

Alkaline Battery   936

Free Energy   937

Fuel Cell   956

Condensation of Biological
     Polymers   975

Fiber Strength and
     Elasticity   980

Formation of Sucrose   986

Balancing Nuclear
     Equations   1012

Radioactive Decay Modes   1016

Half-Life   1017

Fusion of Hydrogen   1032

# About the Authors

**Thomas R. Gilbert** has a BS in chemistry from Clarkson and a PhD in analytical chemistry from MIT. After 10 years with the Research Department of the New England Aquarium in Boston, he joined the faculty of Northeastern University, where he is currently associate professor of chemistry and chemical biology. His research interests are in chemical and science education. He teaches general chemistry and science education courses and conducts professional development workshops for K–12 teachers. He has won Northeastern's Excellence in Teaching Award and Outstanding Teacher of First-Year Engineering Students Award. He is a fellow of the American Chemical Society and in 2012 was elected to the ACS Board of Directors.

**Rein V. Kirss** received both a BS in chemistry and a BA in history as well as an MA in chemistry from SUNY Buffalo. He received his PhD in inorganic chemistry from the University of Wisconsin, Madison, where the seeds for this textbook were undoubtedly planted. After two years of postdoctoral study at the University of Rochester, he spent a year at Advanced Technology Materials, Inc., before returning to academics at Northeastern University in 1989. He is an associate professor of chemistry with an active research interest in organometallic chemistry.

**Natalie Foster** is emeritus professor of chemistry at Lehigh University in Bethlehem, Pennsylvania. She received a BS in chemistry from Muhlenberg College and MS, DA, and PhD degrees from Lehigh University. Her research interests included studying poly(vinyl alcohol) gels by NMR as part of a larger interest in porphyrins and phthalocyanines as candidate contrast enhancement agents for MRI. She taught both semesters of the introductory chemistry class to engineering, biology, and other nonchemistry majors and a spectral analysis course at the graduate level. She is the recipient of the Christian R. and Mary F. Lindback Foundation Award for distinguished teaching.

**Geoffrey Davies** holds BSc, PhD, and DSc degrees in chemistry from Birmingham University, England. He joined the faculty at Northeastern University in 1971 after postdoctoral research on the kinetics of very rapid reactions at Brandeis University, Brookhaven National Laboratory, and the University of Kent at Canterbury. He is now a Matthews Distinguished University Professor at Northeastern University. His research group has explored experimental and theoretical redox chemistry, alternative fuels, transmetalation reactions, tunable metal–zeolite catalysts and, most recently, the chemistry of humic substances, the essential brown animal and plant metabolites in sediments, soils, and water. He edits a column on experiential and study-abroad education in the *Journal of Chemical Education* and a book series on humic substances. He is a Fellow of the Royal Society of Chemistry and was awarded Northeastern's Excellence in Teaching Award in 1981, 1993, and 1999, and its first Lifetime Achievement in Teaching Award in 2004.

# Preface

Dear Student,

We wrote this book with three overarching goals in mind: to make chemistry interesting, relevant, and memorable; to enable you to see the world from a molecular point of view; and to help you become an expert problem-solver. You have a number of resources available to help you succeed in your general chemistry course. This textbook will be a valuable resource, and we have written it with you, and the different ways you may use the book, in mind.

If you are someone who reads a chapter from the first page to the last, you will see that *Chemistry: The Science in Context*, Fourth Edition, introduces the chemical principles within a chapter using contexts drawn from daily life as well as from other disciplines including biology, environmental science, materials science, astronomy, geology, and medicine. We believe that these contexts make chemistry more interesting, relevant, understandable, and memorable.

We begin each chapter with an introduction that is meant to pique your interest but also to set the stage for the concepts to come. These sections provide glimpses of how the chemistry in the chapter connects to the world. We have used topics that should be familiar to you, but we place them in chemical contexts that we hope will intrigue you.

If you want a quick summary of what is most important in a chapter to direct your studying on selected topics, check the **Learning Outcomes** listed on the first page of each chapter. Whether you are reading the chapter from first page to last, moving from topic to topic in an order you select, or reviewing material for an exam, the Learning Outcomes can help you focus on the key information you need to know and the skills you should acquire.

In every section, you will find **key terms** in boldface in the text and in a **running glossary** in the margin. We have inserted the definitions throughout the text, so you can continue reading without interruption but quickly find key terms when doing homework or reviewing for a test. All key terms are also defined in the Glossary in the back of the book.

## Learning Outcomes

**LO1** Explain kinetic and potential energies at the molecular level
**Sample Exercise 5.1**

**LO2** Identify familiar endothermic and exothermic processes
**Sample Exercise 5.2**

**LO3** Calculate changes in the internal energy of a system
**Sample Exercises 5.3, 5.4**

**LO4** Calculate the amount of heat transferred in physical or chemical processes
**Sample Exercises 5.5, 5.6, 5.7, 5.8**

**LO5** Calculate thermochemical values using data from calorimetry experiments
**Sample Exercises 5.9, 5.10, 5.11**

**LO6** Calculate enthalpies of reaction
**Sample Exercises 5.12, 5.13, 5.15**

**LO7** Recognize and write equations for formation reactions
**Sample Exercises 5.14, 5.16**

**LO8** Calculate and compare fuel and food values and fuel densities
**Sample Exercises 5.17, 5.18**

Approximately once per section, you will find a **Concept Test**. These short, conceptual questions provide a self-check opportunity by asking you to stop and answer a question relating to what you just read. We designed them to help you see for yourself whether you have grasped a key concept and can apply it. You will find answers to Concept Tests in the back of the book.

**CONCEPT TEST**

The energy lost by the beverages inside the 72 cans in the preceding discussion was more than 100 times the heat lost by the cans. What factors contributed to this large difference between the energy lost by the cans and the energy lost by their contents?

*(Answers to Concept Tests are in the back of the book.)*

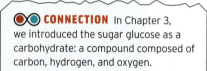

**CONNECTION** In Chapter 3, we introduced the sugar glucose as a carbohydrate: a compound composed of carbon, hydrogen, and oxygen.

New concepts naturally build on previous information, and you will find that many concepts are related to others described earlier in the book. We point out these relationships with **Connection** icons in the margins. These reminders will help you see the big picture and draw your own connections between the major themes covered in the book.

Chemists' unique perspective of natural processes and insights into the properties of substances, from high-performance alloys to the products of biotechnology, are based on understanding these processes and substances at the atomic and molecular levels. A major goal of this book is to help you develop this microscale perspective and link it to macroscopic properties. To help you develop this perspective, we use molecular art to enhance photos and figures and to illustrate what is happening at the atomic and molecular levels. All of the molecular art has been updated, and the fourth edition has more molecular art than previous editions of this book.

**CHEMTOUR** Hess's Law

If you're looking for additional help visualizing a concept, we have more than 100 **ChemTours**, denoted by the ChemTour icon. The ChemTours, available at wwnpag.es/chemtours, demonstrate dynamic processes and help you visualize events at the molecular level. Many ChemTours are interactive, allowing you to manipulate variables and observe resulting changes in a graph or a process. Questions at the end of the ChemTour tutorials offer step-by-step assistance in solving problems and provide useful feedback.

Whereas the biochemical properties of elements and their roles in medicine are the focus of Chapter 22, near the end of many chapters is a **Descriptive Chemistry** box. Descriptive Chemistry boxes throughout the textbook summarize the properties and uses of individual elements or groups of elements that are highlighted in a particular chapter. We discuss where the substances occur in nature and how they are used in ways that touch our lives and shape our world.

Another goal of the book is to help you improve your problem-solving skills. Sometimes the hardest parts of solving a problem are knowing where to start and distinguishing between information that is relevant and information that is not. Once you are clear on where you are starting and where you are going, planning for and carrying out a solution become much easier.

To help you hone your problem-solving skills, we have developed a framework that is introduced in Chapter 1 and used consistently throughout the book. It is a four-step approach we call **COAST**, which is our acronym for (1) **C**ollect and **O**rganize, (2) **A**nalyze, (3) **S**olve, and (4) **T**hink About It. We use these four steps in *every* Sample Exercise and in the solutions to *odd-numbered* problems in the Student's Solutions Manual. They are also used in the hints and feedback embedded in the Smart-Work online homework program. To summarize the four steps:

**Collect and Organize** helps you understand where to begin. In this step we often point out what is given and what you must find, and identify the relevant information that is provided in the problem statement or available elsewhere in the book.

**Analyze** is where we map out a strategy for solving the problem. As part of that strategy we often estimate what a reasonable answer might be.

**Solve** applies our analysis of the problem from the second step to the information and relations from the first step to actually solve the problem. We walk you through each step in the solution so that you can follow the logic as well as the math.

**Think About It** reminds us that an answer is not the last step in solving a problem. Checking if the solution is reasonable in light of an estimate is imperative. Is the answer realistic? Are the units correct? Is the number of significant figures appropriate? Does it make sense with our estimate from the Analyze step?

Many students use the **Sample Exercises** more than any other part of the book. Sample Exercises take the concept being discussed and illustrate how to apply it to solve a problem. We hope that repeated application of COAST will help you refine your problem-solving skills and become an expert problem-solver. When you finish a Sample Exercise, you'll find a **Practice Exercise** to try on your own. If you have the ebook, the Practice Exercises are "live," meaning that you can solve them and receive hints and answer-specific feedback in SmartWork to guide you. Notice that the Sample Exercises and the Learning Objectives are connected. We think this will help you focus efficiently on the main ideas in the chapter.

Students sometimes comment that the questions on an exam are more challenging than the Sample Exercises in a book. To address this, we have added an Integrating Concepts Sample Exercise near the end of each chapter. These exercises require you to use more than one concept from the chapter and may expect you to use concepts from earlier chapters to solve a problem. Please invest your time working through these problems because we think they will further enhance your problem-solving skills and give you an increased appreciation of how chemistry is used in the world.

If you use the book mostly as a reference and problem-solving guide, we have a learning path for you as well. It starts with the **Summary** and a **Problem-Solving Summary** at the end of each chapter. The first is a brief synopsis of the chapter, organized by Learning Outcomes. Key figures have been added to this Summary to provide visual cues as you review. The Problem-Solving Summary

---

**SAMPLE EXERCISE 5.6** The Sign of $\Delta H$ in a Chemical Reaction **LO4**

A chemical cold pack (Figure 5.24) contains a small pouch of water inside a bag of solid ammonium nitrate. To activate the pack, you press on it to rupture the pouch, allowing the ammonium nitrate to dissolve in the water. The result is a cold, aqueous solution of ammonium nitrate. Write a balanced chemical equation describing the dissolution of solid $NH_4NO_3$. If we define the system as the $NH_4NO_3$ and the surroundings as the water, what are the signs of $\Delta H_{sys}$ and $q_{surr}$?

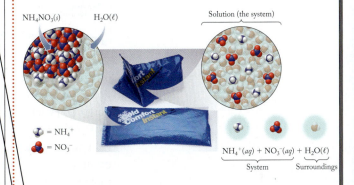

**Collect and Organize** The solution's temperature drops when we dissolve $NH_4NO_3$ in water. We want to write a balanced chemical equation for the dissolution of ammonium nitrate and determine the sign of $\Delta H$ for the solution (the system) and $q_{surr}$.

**Analyze** The first law of thermodynamics says energy must be conserved, so all the energy must be accounted for between system and surroundings.

**Solve** The solubility rules in Table 4.5 tell us that all ammonium salts are soluble. We can write a balanced chemical equation for the dissolution of $NH_4NO_3$ (the system) as

$$NH_4NO_3(s) \rightarrow NH_4^+(aq) + NO_3^-(aq)$$

The low temperature of the cold pack (the system) means that heat flows into it from its warmer surroundings. Therefore, the sign of $\Delta H_{sys}$ is positive ($\Delta H_{sys} > 0$), and the sign of $q_{surr}$ is negative ($q_{surr} < 0$).

**Think About It** The first law of thermodynamics tells us that the signs of $\Delta H_{sys}$ and $q_{surr}$ must be opposite.

**Practice Exercise** Potassium hydroxide can be used to unclog sink drains. The reaction between potassium hydroxide and water is quite exothermic. What are the signs of $\Delta H_{sys}$ and $q_{surr}$?

*(Answers to Practice Exercises are in the back of the book.)*

---

**SUMMARY**

**Learning Outcome 1** **Potential energy (PE)** is the energy of position or composition and is a **state function**. **Kinetic energy (KE)** is the energy of motion. Heating a sample increases the average kinetic energy of the atoms in the sample. Energy is stored in compounds, and energy is absorbed or released when they are transformed into different compounds or when a change of state occurs. (Section 5.1)

**Learning Outcome 2** In an **exothermic process** the system loses energy by heating its surroundings ($q < 0$); in an **endothermic process** the system absorbs energy ($q > 0$) from its surroundings. (Section 5.2)

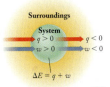

| PROBLEM-SOLVING SUMMARY | | |
|---|---|---|
| **TYPE OF PROBLEM** | **CONCEPTS AND EQUATIONS** | **SAMPLE EXERCISES** |
| Kinetic and potential energy | $\text{KE} = \frac{1}{2}mu^2 \qquad E_{el} \propto \dfrac{Q_1 \times Q_2}{d} \qquad \qquad (5.3)$ | 5.1 |
| Identifying endothermic and exothermic processes, and calculating internal energy change ($\Delta E$) and $P$–$V$ work | For the system: $$\Delta E = q + w \qquad (5.5)$$ where $w = -P\Delta V$. | 5.2, 5.3, 5.4 |
| Predicting the sign of $\Delta H_{sys}$ for physical and chemical changes | Exothermic, $\Delta H_{sys} < 0$ <br> Endothermic, $\Delta H_{sys} > 0$ | 5.5, 5.6 |

organizes the chapter by problem type and summarizes relevant concepts and equations you need to solve each type of problem. The Problem-Solving Summary also points you back to the Sample Exercise that models how to solve each problem and cross-references the Learning Outcomes at the beginning of the chapter.

Following the summaries are groups of questions and problems. The first group consists of **Visual Problems**. In many of them, you are asked to interpret a molecular view of a sample or a graph of experimental data.

**Concept Review Questions and Problems** come next, arranged by topic in the same order as they appear in the chapter. Concept Reviews are qualitative and often ask you to explain why or how something happens. Problems are paired and can be quantitative, conceptual, or a combination of both. **Contextual problems** have a title that describes the context in which the problem is placed. **Additional Problems** can come from any section or combination of sections in the chapter. Some of them incorporate concepts from previous chapters. Problems marked with an asterisk (*) are more challenging and often require multiple steps to solve.

We want you to have confidence in using the answers in the back of the book as well as the Student's Solutions Manual, so we used a rigorous triple-check accuracy program for the fourth edition. Each end-of-chapter question and problem has been solved independently by at least three separate PhD chemists. For the fourth edition the team included Solutions Manual author Bradley Wile and two additional chemical educators. Brad compared his solutions to those from the two reviewers and resolved any discrepancies. This process is designed to ensure clearly written problems and accurate answers in the appendices and Solutions Manual.

No matter how you use this book, we hope it becomes a valuable tool for you and helps you not only understand the principles of chemistry but also apply them to solving global problems, such as diagnosing and treating disease or making more efficient use of Earth's natural resources.

## Changes to the Fourth Edition

Dear Instructor,

Whether you used the third edition of this book or not, we want you to know how the fourth edition compares to its predecessor. Here are some of the general changes we made throughout this edition:

> ➤ We have closely connected the Learning Outcomes, Sample Exercises, and Summary in each chapter, and problems in SmartWork can now be filtered by Learning Objective. This should make studying from the book easier for students and assessing students easier for you.

➤ Most Sample Exercises were revised to improve clarity and continuity. We also added 50 Sample Exercises to the fourth edition, based on reviewer feedback.

➤ Each chapter has a new type of Sample Exercise called Integrating Concepts. These are placed near the end of the chapter and require students to use more than one concept from the chapter, and occasionally concepts from preceding chapters, to solve real-world problems.

➤ We have significantly revised and expanded the art program by updating all of the molecular art, adding more molecular views, and making people, glassware, and equipment more photorealistic.

➤ We have revised or replaced approximately 20% of the end-of-chapter problems. We used feedback from users and reviewers to address areas where we needed more problems or additional problems of varying difficulty.

➤ More than 1000 new problems have been added to SmartWork to support the fourth edition, including more math review problems, visual problems, and more multi-step tutorials. In response to user feedback, most of these problems are written *for* the book but are not *in* the book.

➤ Chapter 14 in the third edition (Thermodynamics) is now Chapter 18 and follows kinetics and equilibrium. This organization more closely matches the order taught at most colleges and universities.

Other changes include moving most of the nuclear chemistry from Chapter 2 to Chapter 21 and rebalancing the material in Chapters 10 and 11. Chapter 10, on intermolecular forces, now concentrates primarily on the behavior and properties of pure materials, whereas Chapter 11 deals with solutions and interactions among particles of different substances. The unique behavior of water is a unifying feature of both chapters.

Because a discussion of the ways chemistry touches our daily lives is part of the fabric of the text, we have routinely incorporated new information in the story lines of the chapters. From the landing of *Curiosity* on Mars to the most recent results on the role of the survival advantage conveyed by sickle-cell anemia in regions where malaria is endemic, current events have supplied us with new contextual examples to enrich our discussions of chemical concepts.

## Teaching and Learning Resources

### SMARTWORK ONLINE HOMEWORK FOR GENERAL CHEMISTRY

wwnorton.com/smartwork/chemistry

Created by chemistry educators, SmartWork is the most intuitive online tutorial and homework management system available for general chemistry. The many question types, including graded molecule drawing, math and chemical equations, ranking tasks, and interactive figures, help students develop and apply their understanding of fundamental concepts in chemistry.

Every problem in SmartWork includes response-specific feedback and general hints using the steps in COAST. Links to the ebook version of *Chemistry: The Science in Context*, Fourth Edition, take students to the specific place in the text where the concept is explained. All problems in SmartWork use the same language and notation as the textbook.

SmartWork also features Tutorial Problems. If students ask for help in a Tutorial Problem, SmartWork breaks the problem down into smaller steps, coaching them with hints, answer-specific feedback, and probing questions within each step. At any point in a Tutorial, a student can return to and answer the original problem.

Assigning, editing, and administering homework within SmartWork is easy. SmartWork allows the instructor to search for problems using both the text's Learning Objectives and Bloom's taxonomy. Instructors can use pre-made assignment sets provided by Norton authors, modify those assignments, or create their own. Instructors can also make changes in the problems at the question level. All instructors have access to our WYSIWYG (What You See Is What You Get) authoring tools—the same ones Norton authors use. Those intuitive tools make it easy to modify existing problems or to develop new content that meets the specific needs of your course.

Wherever possible, SmartWork makes use of algorithmic variables so that students see slightly different versions of the same problem. Assignments are graded automatically, and SmartWork includes sophisticated yet flexible tools for managing class data. Instructors can use the Item Analysis features to assess how students have done on specific problems within an assignment. Instructors can also review individual students' work on problems.

SmartWork for *Chemistry: The Science in Context*, Fourth Edition, features the following problem types:

➤ End-of-Chapter Problems. These problems, which use algorithmic variables when appropriate, all have hints and answer-specific feedback to coach students through mastering single- and multi-concept problems based on chapter content. They make use of all of SmartWork's answer-entry tools.

➤ Multistep Tutorials. These problems offer students who demonstrate a need for help a series of linked, step-by-step subproblems to work. They are based on the Concept Review problems at the end of each chapter. Tutorials make use of student-focused artwork from the Student's Solutions Manual.

➤ Math Review Problems. These problems can be used by students for practice or by instructors to diagnose the mathematical ability of their students.

➤ Ranking Task Problems. These problems ask students to make comparative judgments between items in a set.

➤ Visual and Graphing Problems. These problems challenge students to identify chemical phenomena and to interpret graphs. They use SmartWork's Drag-and-Drop and Hotspot functionality.

➤ Reaction Visualization Problems. Based on both static art and videos of simulated reactions, these problems are designed to help students visualize what happens at the atomic level—and why it happens.

➤ Nomenclature Problems. New matching and multiple-choice problems help students master course vocabulary.

➤ ChemTour Problems. These are based on Norton's popular animations.

 www.nortonebooks.com

### EBOOK

An affordable and convenient alternative to the print text, the ebook retains the content and design of the print book. Students can highlight and take notes with ease, print chapters as needed, and search the text.

The online version of *Chemistry*, Fourth Edition provides students with stand-alone Interactive Practice Exercises. These are self-grading SmartWork problems that allow students to practice solving problems and receive hints and feedback without being penalized for making multiple attempts. The online ebook also allows students one-click access to the 100 ChemTour animations.

The online ebook is available bundled with the print text and SmartWork at no extra cost, or it may be purchased bundled with SmartWork access.

Norton also offers a downloadable PDF version of the ebook as well as a format optimized for delivery to any device—laptop, tablet, or smartphone.

### STUDENT'S SOLUTIONS MANUAL by Bradley Wile, Ohio Northern University

The Student's Solutions Manual provides students with fully worked solutions to select end-of-chapter problems using the **COAST** four-step method (**C**ollect and **O**rganize, **A**nalyze, **S**olve, and **T**hink About It). The Student Solutions Manual contains several pieces of art for each chapter, designed to help students visualize ways to approach problems. This artwork is also used in the hints and feedback within SmartWork.

### CLICKERS IN ACTION: INCREASING STUDENT PARTICIPATION IN GENERAL CHEMISTRY by Margaret Asirvatham, University of Colorado, Boulder

An instructor-oriented resource providing information on implementing clickers in general chemistry courses. *Clickers in Action* contains more than 250 class-tested, lecture-ready questions, with histograms showing student responses, as well as insights and suggestions for implementation. Question types include macroscopic observation, symbolic representation, and atomic/molecular views of processes.

### INSTRUCTOR'S SOLUTIONS MANUAL by Bradley Wile, Ohio Northern University

The Instructor's Solutions Manual provides instructors with fully worked solutions to every end-of-chapter Concept Review and Problem. Each solution uses the **COAST** four-step method (**C**ollect and **O**rganize, **A**nalyze, **S**olve, and **T**hink About It).

### INSTRUCTOR'S RESOURCE MANUAL by Timothy Zauche, University of Wisconsin-Platteville

Each chapter in this complete resource manual for instructors begins with a brief overview of the text chapter, suggestions for integrating the contexts featured in the book into a lecture, and alternate contexts. Each chapter also contains an overview of using selected *Clickers in Action* questions with the chapter and instructor notes for suggested activities from the *ChemConnections* and *Calculations in Chemistry* workbooks. Summaries of the ChemTours and suggested laboratory exercises round out each chapter.

### TEST BANK by David Hanson, Stony Brook University

Norton uses an innovative, evidence-based model to deliver high-quality and pedagogically effective quizzes and testing materials. Each chapter of the Test Bank is structured around an expanded list of student learning objectives and evaluates student knowledge on six distinct levels based on Bloom's Taxonomy: Remembering, Understanding, Applying, Analyzing, Evaluating, and Creating.

Questions are further classified by section and difficulty, making it easy to construct tests and quizzes that are meaningful and diagnostic, according to each instructor's needs. More than 2200 questions are divided into multiple-choice and short answer.

Beginning in the fall of 2015, there will also be an annual update to the digital Test Bank files found on the downloadable instructor's resource site. Every year, for the life of the fourth edition, new questions will be added, and edits to existing questions will be made based on the author use of the Test Bank questions.

The Test Bank is available in ExamView Assessment Suite, Word RTF, and PDF formats.

### INSTRUCTOR'S RESOURCE DISC

This helpful classroom presentation tool features:

➤ Stepwise animations and classroom response questions. Developed by Jeffrey Macedone of Brigham Young University and his team, these animations, which use native PowerPoint functionality and textbook art, help instructors to "walk" students through more than 100 chemical concepts and processes. Where appropriate, the slides contain two types of questions for students to answer in class: questions that ask them to predict what will happen next and why, and questions that ask them to apply knowledge gained from watching the animation. Self-contained notes help instructors adapt these materials to their own classrooms.

➤ Lecture PowerPoint (David Ballantine, Northern Illinois University) slides that include integrated figures from the text, ChemTours, and stick-or-switch clicker questions. These are particularly helpful to first-time teachers of the introductory course.

➤ All ChemTours and multi-level visualizations.

➤ *Clickers in Action* clicker questions for each chapter provide instructors with class-tested questions they can integrate into their course.

➤ Photographs, drawn figures, and tables from the text, available in PowerPoint and JPEG.

### DOWNLOADABLE INSTRUCTOR'S RESOURCES

wwnorton.com/instructors

This password-protected site for instructors includes:

➤ Stepwise animations and classroom response questions. Developed by Jeffrey Macedone of Brigham Young University and his team, these animations, which use native PowerPoint functionality and textbook art, help instructors to "walk" students through more than 100 chemical concepts and processes. Where appropriate, the slides contain two types of questions for students to answer in class: questions that ask them to predict what will happen next and why, and questions that ask them to apply knowledge gained from watching the animation. Self-contained notes help instructors adapt these materials to their own classrooms.

➤ Lecture PowerPoints with stick-or-switch clicker questions.

➤ All ChemTours and multi-level visualizations.

➤ Test bank in PDF, Word RTF, and *ExamView* Assessment Suite formats.

➤ Solutions Manual in PDF and Word, so that instructors may edit solutions.

➤ All of the end-of-chapter questions and problems, available in Word along with the key equations.

➤ Photographs, drawn figures, and tables from the text, available in PowerPoint and JPEG.

➤ *Clickers in Action* clicker questions.

➤ BlackBoard and WebCT materials.

**BLACKBOARD AND WEBCT COURSE CARTRIDGES**

Course cartridges for BlackBoard and WebCT include access to the ChemTours, a Study Plan for each chapter, multiple-choice tests (Stephen Wuerz, Highland Community College), and links to premium content in the ebook and Smart-Work.

## Acknowledgments

As authors of a textbook we are very often asked: "Why is a fourth edition necessary? Has the science changed that much since the third edition?" Although chemistry is a vigorous and dynamic field, most basic concepts presented in an introductory course have not changed dramatically. However, two areas tightly intertwined in this text—pedagogy and context—have experienced significant changes, and those areas are the drivers of this new edition. The review process that involves you, the users of this text, plays a major role in triggering the perfect storm of activity required to produce a new version of a book. The suggestions, comments, critiques, and general feedback you have provided based on previous editions of this book encouraged us to commit to a revision with an eye on making the content work better for you and for all your students. Our deepest thanks and gratitude go to you, the users, for sharing your experiences with us. Your comments at meetings, at focus groups, in emails, and during office visits with the Norton travelers about what works well, what needs to be improved pedagogically, and the new stories and current real-world examples of chemistry that capture your students' interest are the mainstays of this revision. We begin our acknowledgments by thanking all of you.

Our colleagues at W. W. Norton remain a constant source of inspiration and guidance. Their vision of the role of publishers as providers of vetted content and of the importance of accuracy and reliability of information is central to everything we do to create a good and useful package of resources for students. Our highest order of thanks must go to W. W. Norton for having enough confidence in the idea behind the first three editions to commit to the massive labor of the fourth. The people at W. W. Norton with whom we work most closely deserve much more praise than we can possibly express here. Our editor, Erik Fahlgren, continues to be an indefatigable source of guidance, help, creativity, and the occasional threats of "time-outs" when discussions become heated. Erik's involvement in the project is the single greatest reason for its completion, and our greatest thanks are too small an offering for his unwavering focus; he is the consummate professional and a valued friend.

We are pleased to acknowledge the contributions of developmental editor David Chelton. "Just do what David suggests" became a mantra early in this project for good reason. David provided wise and reliable advice based on a diverse collection of comments and suggestions from instructors and his own considerable experience with producing science books. He was always available for discussions of the grandest scheme or the tiniest detail, always with an unflinching eye toward making our presentation clear, cogent, and accessible to students. Our project editor Amy Weintraub was a font of patience as we wrestled with changes in format and design, always striving on our behalf to make a good book even better. Assistant editor Renee Cotton kept our punch lists current, our paperwork flowing, and our collection of Zen quotations appropriate for whatever situations arose. Debra Morton Hoyt took our inchoate ideas and produced a spectacular new cover; Nelson Colon found just the right photo again and again; production manager Sean Mintus worked tirelessly behind the scenes; Jennifer Barnhardt managed the print supplements skillfully; Rob Bellinger brought his high level of creativity and diligence to the media package; and Stacy Loyal was both innovative and strategic in marketing the book and working in the field.

This book has benefited greatly from the care and thought that many reviewers, listed here, gave to their readings of earlier drafts. We owe an extra-special thanks to Brad Wile for his dedicated and precise work on the Solutions Manual. He, along with Steve Trail and Tim Brewer, are the triple-check accuracy team who solved each problem and reviewed each solution for accuracy. We are deeply grateful to David Hanson for his help with the Learning Objectives and for working with us to clarify both our language and our thoughts about thermochemistry. Finally, we greatly appreciate Tim Brewer, Diep Ca, Claire Cohen-Schmidt, Patricia Coleman, Mark Cybulski, Theodore Duello, Nancy Gerber, Sam Glazier, Andrew Ho, Tracy Knowles, Larry Kolopajlo, Rebecca Miller, Robbie Montgomery, Bob Shelton, Steve Trail, Rory Waterman, and Karen Wesenberg-Ward for checking the accuracy of the myriad facts that form the framework of our science.

Thomas R. Gilbert
Rein V. Kirss
Natalie Foster
Geoffrey Davies

## Fourth Edition Reviewers

Ian Balcom, Lyndon State College of Vermont
Shuhsien Batamo, Houston Community College Central Campus
Erin Battin, West Virginia University
Lawrence Berliner, University of Denver
Mark Berry, Brandon University
Narayan Bhat, University of Texas, Pan American
Simon Bott, University of Houston
Ivana Bozidarevic, DeAnza College
Chris Bradley, Mount St. Mary's University
Richard Bretz, Miami University
Timothy Brewer, Eastern Michigan University
Ted Bryan, Briar Cliff University
Donna Budzynski, San Diego Community College
Diep Ca, Shenandoah University
Dean Campbell, Bradley University
David Cedeno, Illinois State University
Stephen Cheng, University of Regina, Main Campus
Tabitha Chigwada, West Virginia University
Allen Clabo, Francis Marion University
Claire Cohen-Schmidt, University of Toledo
Patricia Coleman, Eastern Michigan University
Shara Compton, Widener University
Elisa Cooper, Austin Community College
Mark Cybulski, Miami University
Shadi Dalili, University of Toronto

William Davis, Texas Lutheran University
Milagros Delgado, Florida International University
Theodore Duello, Tennessee State University
Bill Durham, University of Arkansas
Gavin Edwards, Eastern Michigan University
Alegra Eroy-Reveles, San Francisco State University
Andrew Frazer, University of Central Florida
Rachel Garcia, San Jacinto College
Simon Garrett, California State University, Northridge
Nancy Gerber, San Francisco State University
Samantha Glazier, St. Lawrence University
Pete Golden, Sandhills Community College
John Goodwin, Coastal Carolina University
Nathaniel Grove, University of North Carolina, Wilmington
Tammy Gummersheimer, Schenectady County Community College
Kim Gunnerson, University of Washington
Christopher Hamaker, Illinois State University
David Hanson, Stony Brook University
Sara Hein, Winona College
Andy Ho, Lehigh University
Joseph Ho, University of New Mexico
Angela Hoffman, University of Portland
Zhaoyang Huang, Jacksonville University
Donna Ianotti, Brevard Community College
Eugenio Jaramillo, Texas A&M International University

David Johnson, University of Dayton
Vance Kennedy, Eastern Michigan University
Michael Kenney, Case Western Reserve University
Mark Keranen, University of Tennessee, Martin
Edith Kippenhan, University of Toledo
Sushilla Knottenbelt, University of New Mexico
Tracy Knowles, Bluegrass Community and Technical College
Lawrence Kolopajlo, Eastern Michigan University
Stephen Kuebler, University of Central Florida
Liina Ladon, Towson University
Timothy Lash, Illinois State University
Edward Lee, Texas Tech University
Alistair Lees, SUNY Binghamton
Scott Lewis, Kennesaw State University
Joanne Lin, Houston Community College, Eastside Campus
Matthew Linford, Brigham Young University
Karen Lou, Union College
Leslie Lyons, Grinnell College
Susan Marine, Miami University, Middletown Campus
Brian Martinelli, Nevada State College
Laura McCunn, Marshall University
Ryan McDonnell, Cape Fear Community College
Lauren McMills, Ohio University
Rebecca Miller, Lehigh University
Robbie Montgomery, University of Tennessee, Martin
Joshua Moore, Tennessee State University
Steven Neshyba, University of Puget Sound
Mya Norman, University of Arkansas
Joshua Ojwang, Brevard Community College
Maria Pacheco, Buffalo State College
Pedro Patino, University of Central Florida
Vicki Paulissen, Eastern Michigan University
Jessica Parr, University of Southern California
Pete Poston, Western Oregon University

Robert Quandt, Illinois State University
Orlando Raola, Santa Rosa College
Haley Redmond, Texas Tech University
James Reeves, University of North Carolina, Wilmington
Scott Reid, Marquette University
Paul Richardson, Coastal Carolina University
Albert Rives, Wake Forest University
Mark Rockley, Oklahoma State University
Alan Rowe, Norfolk State University
Christopher Roy, Duke University
Erik Ruggles, University of Vermont
Joel Russell, Oakland University
Mark Schraf, West Virginia University
Louis Scudiero, Washington State University
George Shelton, Austin Peay State University
Brock Spencer, Beloit College
Clayton Spencer, Illinois College
Wesley Stites, University of Arkansas
Meredith Storms, University of North Carolina, Pembroke
Katherine Stumpo, University of Tennessee, Martin
Luyi Sun, Texas State University
Charles Thomas, University of Tennessee, Martin
Matthew Thompson, Trent University
Edmund Tisko, University of Nebraska, Omaha
Steve Trail, Elgin Community College
Kris Varazo, Francis Marion University
Andrew Vreugdenhil, Trent University
Haobin Wang, New Mexico State University
Erik Wasinger, California State University, Chico
Rory Waterman, University of Vermont
Karen Wesenberg-Ward, Montana Tech University
Karla Wohlers, North Dakota State University
Mingming Xu, West Virginia University
Corbin Zea, Grand View University

## Previous Editions' Reviewers

William Acree, Jr., University of North Texas
R. Allendoefer, State University of New York, Buffalo
Thomas J. Anderson, Francis Marion University
Sharon Anthony, The Evergreen State College
Jeffrey Appling, Clemson University
Marsi Archer, Missouri Southern State University
Margaret Asirvatham, University of Colorado, Boulder
Robert Balahura, University of Guelph
Anil Banerjee, Texas A&M University, Commerce
Sandra Banks, Mills College
Mikhail V. Barybin, University of Kansas
Mufeed Basti, North Carolina Agricultural & Technical State
    University
Robert Bateman, University of Southern Mississippi
Kevin Bennett, Hood College
H. Laine Berghout, Weber State University
Eric Bittner, University of Houston

David Blauch, Davidson College
Robert Boggess, Radford University
Simon Bott, University of Houston
Michael Bradley, Valparaiso University
Karen Brewer, Hamilton College
Timothy Brewer, Eastern Michigan University
Julia Burdge, Florida Atlantic University
Robert Burk, Carleton University
Andrew Burns, Kent State University
Sharmaine Cady, East Stroudsburg University
Chris Cahill, George Washington University
Kevin Cantrell, University of Portland
Nancy Carpenter, University of Minnesota, Morris
Patrick Caruana, State University of New York, Cortland
David Cedeno, Illinois State University
Tim Champion, Johnson C. Smith University
William Cleaver, University of Vermont

Penelope Codding, University of Victoria
Jeffery Coffer, Texas Christian University
Renee Cole, Central Missouri State University
Andrew L. Cooksy, San Diego State University
Brian Coppola, University of Michigan
Richard Cordell, Heidelberg College
Charles Cornett, University of Wisconsin, Platteville
Mitchel Cottenoir, South Plains College
Robert Cozzens, George Mason University
Margaret Czrew, Raritan Valley Community College
John Davison, Irvine Valley College
Laura Deakin, University of Alberta
Anthony Diaz, Central Washington University
Klaus Dichmann, Vanier College
Mauro Di Renzo, Vanier College
Kelley J. Donaghy, State University of New York, College of
    Environmental Science and Forestry
Michelle Driessen, University of Minnesota
Dan Durfey, Naval Academy Preparatory School
Bill Durham, University of Arkansas
Stefka Eddins, Gardner-Webb University
Dwaine Eubanks, Clemson University
Lucy Eubanks, Clemson University
Jordan L. Fantini, Denison University
Nancy Faulk, Blinn College, Bryan
Tricia Ferrett, Carleton College
Matt Fisher, St. Vincent College
Amy Flanagan-Johnson, Eastern Michigan University
George Flowers, Darton College
Richard Foust, Northern Arizona University
David Frank, California State University, Fresno
Cynthia Friend, Harvard University
Karen Frindell, Santa Rosa Junior College
Barbara Gage, Prince George's Community College
Brian Gilbert, Linfield College
Jack Gill, Texas Woman's University
Arthur Glasfeld, Reed College
Stephen Z. Goldberg, Adelphi University
Frank Gomez, California State University, Los Angeles
John Goodwin, Coastal Carolina University
Steve Gravelle, Saint Vincent College
Tom Greenbowe, Iowa State University
Stan Grenda, University of Nevada
Margaret Haak, Oregon State University
Todd Hamilton, Adrian College
Robert Hanson, St. Olaf College
David Harris, University of California, Santa Barbara
Holly Ann Harris, Creighton University
Donald Harriss, University of Minnesota, Duluth
C. Alton Hassell, Baylor University
Dale Hawley, Kansas State University
Brad Herrick, Colorado School of Mines
Vicki Hess, Indiana Wesleyan University
Paul Higgs, University of Tennessee, Martin
Kimberly Hill Edwards, Oakland University

Donna Hobbs, Augusta State University
Angela Hoffman, University of Portland
Matthew Horn, Utah Valley University
Tim Jackson, University of Kansas
Tamera Jahnke, Southern Missouri State University
Shahid Jalil, John Abbot College
Kevin Johnson, Pacific University
Lori Jones, University of Guelph
Martha Joseph, Westminster College
David Katz, Pima Community College
Jason Kautz, University of Nebraska, Lincoln
Phillip Keller, University of Arizona
Resa Kelly, San Jose State University
Robert Kerber, Stony Brook University
Elizabeth Kershisnik, Oakton Community College
Angela King, Wake Forest University
Larry Kolopajlo, Eastern Michigan University
John Krenos, Rutgers University
C. Krishnan, State University of New York, Stony Brook
Richard Langley, Stephen F. Austin State University
Sandra Laursen, University of Colorado, Boulder
Richard Lavrich, College of Charleston
David Laws, The Lawrenceville School
Willem Leenstra, University of Vermont
Jailson de Lima, Vanier College
George Lisensky, Beloit College
Jerry Lokensgard, Lawrence University
Boon Loo, Towson University
Laura MacManus-Spencer, Union College
Roderick M. Macrae, Marian College
John Maguire, Southern Methodist University
Susan Marine, Miami University, Ohio
Albert Martin, Moravian College
Diana Mason, University of North Texas
Garrett McGowan, Alfred University
Thomas McGrath, Baylor University
Craig McLauchlan, Illinois State University
Heather Mernitz, Tufts University
Stephen Mezyk, California State University, Long Beach
Rebecca Miller, Lehigh University
John Milligan, Los Angeles Valley College
Timothy Minger, Mesa Community College
Ellen Mitchell, Bridgewater College
Dan Moriarty, Siena College
Stephanie Myers, Augusta State College
Richard Nafshun, Oregon State University
Melanie Nilsson, McDaniel College
Mya Norman, University of Arkansas
Sue Nurrenbern, Purdue University
Gerard Nyssen, University of Tennessee, Knoxville
Ken O'Connor, Marshall University
Jodi O'Donnell, Siena College
Jung Oh, Kansas State University, Salinas
MaryKay Orgill, University of Nevada, Las Vegas
Robert Orwoll, College of William and Mary

Gregory Oswald, North Dakota State University
Jason Overby, College of Charleston
Greg Owens, University of Utah
Stephen R. Parker, Montana Tech
Jessica Parr, University of Southern California
Trilisa Perrine, Ohio Northern University
Giuseppe Petrucci, University of Vermont
Julie Peyton, Portland State University
David E. Phippen, Shoreline Community College
Alexander Pines, University of California, Berkeley
Prasad Polavarapu, Vanderbilt University
John Pollard, University of Arizona
Lisa Ponton, Elon University
Gretchen Potts, University of Tennessee, Chattanooga
Robert Pribush, Butler University
Gordon Purser, University of Tulsa
Robert Quandt, Illinois State University
Karla Radke, North Dakota State University
Casey Raymond, State University of New York, Oswego
Alan Richardson, Oregon State University
Mark Rockley, Oklahoma State University
Beatriz Ruiz Silva, University of California, Los Angeles
Pam Runnels, Germanna Community College
Joel W. Russell, Oakland University
Jerry Sarquis, Miami University, Ohio
Nancy Savage, University of New Haven
Barbara Sawrey, University of California, Santa Barbara
Mark Schraf, West Virginia University
Truman Schwartz, Macalester College
Fatma Selampinar, University of Connecticut
Shawn Sendlinger, North Carolina Central University
Susan Shadle, Boise State University
Peter Sheridan, Colgate University
Edwin Sibert, University of Wisconsin, Madison

Ernest Siew, Hudson Valley Community College
Roberta Silerova, John Abbot College
Virginia Smith, United States Naval Academy
Sally Solomon, Drexel University
Xianzhi Song, University of Pittsburgh, Johnstown
Estel Sprague, University of Cincinnati
Steven Strauss, Colorado State University
William Steel, York College of Pennsylvania
Mark Sulkes, Tulane University
Duane Swank, Pacific Lutheran University
Keith Symcox, University of Tulsa
Agnes Tenney, University of Portland
Matthew G. K. Thompson, Trent University
Craig Thulin, Utah Valley University
Edmund Tisko, University of Nebraska, Omaha
Brian Tissue, Virginia Polytechnic Institute and State University
Steve S. Trail, Elgin Community College
Laurie Tyler, Union College
Mike van Stipdonk, Wichita State University
Kris Varazo, Francis Marion University
William Vining, University of Massachusetts
Andrew Vruegdenhill, Trent University
Ed Walton, California State Polytechnic University, Pomona
Wayne Wesolowski, University of Arizona
Charles Wilkie, Marquette University
Ed Witten, Northeastern University
Stephen Wood, Brigham Young University
Cynthia Woodbridge
Mingming Xu, West Virginia University
Tim Zauche, University of Wisconsin, Platteville
Noel Zaugg, Brigham Young University, Idaho
James Zimmerman, Missouri State University
Martin Zysmilich, George Washington University

# CHEMISTRY

## THE SCIENCE IN CONTEXT

# 1

# Matter and Energy: The Origin of the Universe

**1.1** Classes of Matter

**1.2** Matter: An Atomic View

**1.3** Mixtures and How to Separate Them

**1.4** A Framework for Solving Problems

**1.5** Properties of Matter

**1.6** States of Matter

**1.7** The Scientific Method: Starting Off with a Bang

**1.8** Making Measurements and Expressing the Results

**1.9** Unit Conversions and Dimensional Analysis

**1.10** Testing a Theory: The Big Bang Revisited

## Explaining How and Why

For thousands of years people have asked how and why natural events take place: our tendency to ask *how* and *why* is part of human nature. However, only in the last thousand years or so have people moved from speculation toward a rational approach based on science and the scientific method to answer these questions.

Long ago, humans thought the Sun moved across the sky in a chariot pulled by winged horses and driven by a Sun-god. This belief lasted for several thousand years, through the rise and fall of the ancient Egyptian, Greek, and Roman civilizations. Not until 400 years ago did our observations of the sky become dependable enough that people could look for other explanations that they could test through observations and measurements.

Over the centuries, our study of substances on Earth has led to great advances in our ability to make things such as buildings, roads, and ships. The study of how and why things move led to an understanding of energy, which enabled people to see how things work and change in nature. Our forebears' studies of the properties of matter led to today's atomic view of matter. They also asked how and why matter formed the way it did, and how and why the universe itself came to be the way we see it now.

The study of the universe and its origins, called cosmology, is the ultimate example of looking for evidence about what happened a long time ago—in this case billions of years. When finding, examining, and analyzing the light from stars and galaxies, we use the scientific method to understand how they form, grow, and die. Studying the explosions that mark the deaths of stars has revealed clear evidence that stars are huge manufacturing centers of matter, producing heavy solid materials from the lightest gases. One of the current frontiers of science involves sending out space probes to gather samples of cosmic dust and return

**Crab Nebula** This nebula is the remnant of a supernova explosion first observed ▶ on Earth in 1054 AD by Chinese and Japanese astronomers as well as by the Native American Anasazi tribe, who recorded the appearance of a new star in the heavens. The explosion scattered heavy elements throughout its region of the Milky Way Galaxy. The colors in this image correspond to the different elements that were expelled during the explosion.

### Learning Outcomes

**LO1** Describe different forms of matter

**LO2** Relate chemical formulas to molecular structures and vice versa

**LO3** Distinguish between physical processes and chemical reactions and between physical and chemical properties
Sample Exercise 1.1

**LO4** Use a systematic approach (COAST) to solve problems

**LO5** Describe the three states of matter and the transitions between them at the macroscopic and atomic levels
Sample Exercise 1.2

**LO6** Describe the scientific method

**LO7** Express values with the appropriate number of significant figures
Sample Exercises 1.3, 1.7, 1.9

**LO8** Distinguish between exact and uncertain values
Sample Exercises 1.4, 1.5, 1.9

**LO9** Convert values from one system of units to another
Sample Exercises 1.6, 1.7, 1.8, 1.9

**matter** anything that has mass and occupies space.

**mass** the property that defines the quantity of matter in an object.

**energy** the capacity to do work.

**chemistry** the study of the composition, structure, and properties of matter and of the energy consumed or given off when matter undergoes a change.

**substance** matter that has a constant composition and cannot be broken down to simpler matter by any physical process; also called *pure substance*.

them to Earth for analysis (Figure 1.1). Scientists need great imagination and creativity to carry out their search for evidence and understanding, as we will see throughout this book. ■

# 1.1 Classes of Matter

We see, hear, taste, smell, and feel the world around us. All things in it that are physically real—from the air we breathe to the ground we walk on to the Sun's rays that warm us—are forms of either matter or energy. **Matter** is everything in the universe that has **mass** and occupies space. **Energy** is a fundamental component of the universe commonly described as the capacity to do work. The study of the composition, structure, and properties of matter is called **chemistry**. Chemists observe the changes that matter undergoes and measure the amount of energy produced or consumed during those changes. The science of chemistry has led to the synthesis of many new forms of matter that affect the way we live and the planet on which we live. Modern life is difficult to imagine without plastics, computers, cell phones, aspirin, and the countless other synthetic materials and technological innovations made possible through the study of chemistry.

The different forms of matter can be organized according to the classification scheme shown in Figure 1.2. The principal classes are pure substances and mixtures (Figure 1.3). A pure **substance** has a constant composition that does not vary from one sample to another. For example, the composition of pure water does not vary, no matter what its source or how much of it there is. Water cannot be separated into simpler substances by any physical process. By **physical process**, we mean a transformation of a sample of matter, such as a change in its physical state, that does not alter the chemical identities of any of the substances in the sample.

A **mixture** is matter composed of two or more substances that can be separated from one another by a physical process. The pure substances in a mixture are not present in definite proportions, and they retain their chemical identities. In a

(a)

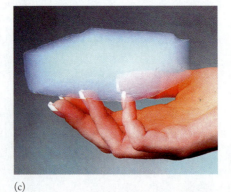

(b)

(c)

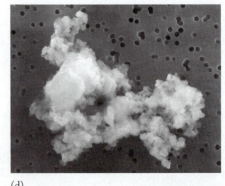

(d)

**FIGURE 1.1** What do scientists do? Among many other things, they have learned about the composition and chemical history of the universe by analyzing particles of interstellar dust, which most often contain nickel, iron, calcium, and sulfur. NASA's *Stardust* probe returned to Earth in 2006 after a 7-year voyage in which it collected interstellar dust and sampled the tail of a comet. Comets are believed to contain material left over from when the solar system formed over 4.5 billion years ago. This sequence shows (a) an artist's rendition of *Stardust* approaching the tail of the comet; (b) Jet Propulsion Laboratory scientist Dr. Peter Tsou holding a *Stardust* sample tray; (c) a sample of the aerogel (a material somewhat like dried Jell-O with very tiny pores) used to trap dust particles; (d) a photomicrograph of a cosmic dust fragment on the aerogel.

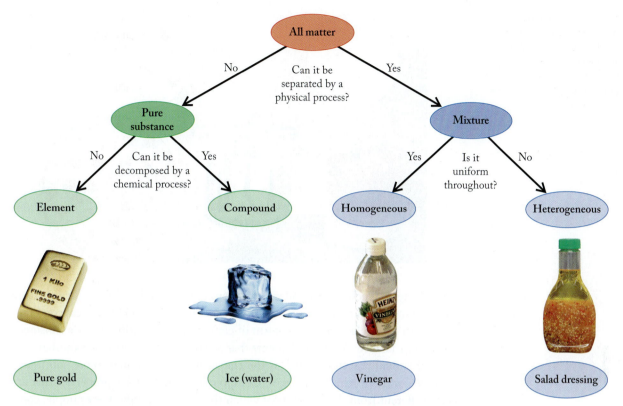

**FIGURE 1.2** The two principal categories of matter are pure substances and mixtures. A pure substance may be a compound (such as water) or an element (such as gold). When the substances making up a mixture are distributed uniformly, as they are in vinegar (a mixture of acetic acid and water), the mixture is homogeneous. When the substances making up a mixture are not distributed uniformly—as when solids are suspended in a liquid and may settle to the bottom of the container, as they do in some salad dressings—the mixture is heterogeneous.

**homogeneous mixture**, the substances making up the mixture are distributed uniformly, and the composition and appearance of the mixture are uniform throughout. Homogeneous mixtures are also called **solutions**, a term that scientists apply to homogeneous mixtures of gases and solids as well as liquids. In contrast, the substances in a **heterogeneous mixture** are not distributed uniformly.

Substances are subdivided into two groups: elements and compounds. An **element** is a pure substance that cannot be broken down into simpler substances. The periodic table inside the front cover shows all the known elements. Only a

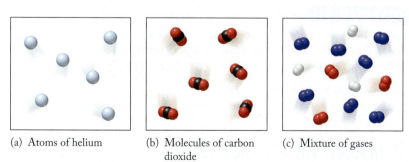

(a) Atoms of helium  (b) Molecules of carbon dioxide  (c) Mixture of gases

**FIGURE 1.3** All matter is made up of either pure substances (of which there are relatively few in nature) or mixtures. (a) The element helium (He), the second most abundant element in the universe, is one example of a pure substance. (b) The compound carbon dioxide ($CO_2$), the gas used in many fire extinguishers, is also a pure substance. (c) This homogeneous mixture contains three substances: nitrogen ($N_2$, blue), hydrogen ($H_2$, white), and oxygen ($O_2$, red).

**physical process** a transformation of a sample of matter, such as a change in its physical state, that does not alter the chemical identity of any substance in the sample.

**mixture** a combination of pure substances in variable proportions in which the individual substances retain their chemical identities and can be separated from one another by a physical process.

**homogeneous mixture** a mixture in which the components are distributed uniformly throughout and have no visible boundaries or regions.

**solution** another name for homogeneous mixture. Solutions are often liquids, but they may also be solids or gases.

**heterogeneous mixture** a mixture in which the components are not distributed uniformly, so that the mixture contains distinct regions of different compositions.

**element** a pure substance that cannot be separated into simpler substances by any chemical process.

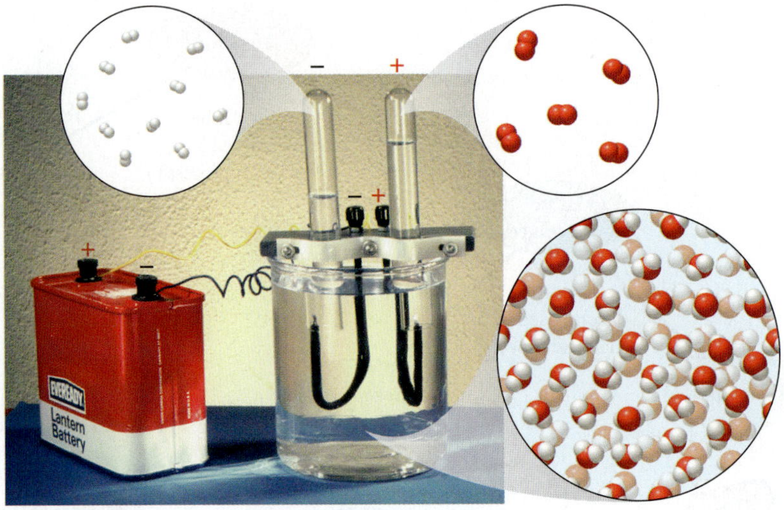

**FIGURE 1.4** An electric current passed through water decomposes the water into oxygen gas and hydrogen gas. The volume ratio of the gases produced is always two volumes of hydrogen for every one volume of oxygen. A quantitative observation like this illustrates the law of constant composition.

**compound** a pure substance that is composed of two or more elements bonded together in fixed proportions and that can be broken down into those elements by some chemical process.

**chemical reaction** the transformation of one or more substances into different substances.

**law of constant composition** all samples of a particular compound contain the same elements combined in the same proportions.

**atom** the smallest particle of an element that retains the chemical characteristics of the element.

**molecule** a collection of atoms chemically bonded together in characteristic proportions.

**chemical formula** a notation for representing elements and compounds; consists of the symbols of the constituent elements and subscripts identifying the number of atoms of each element in one molecule.

**chemical equation** notation in which chemical formulas express the identities and their coefficients express the quantities of substances involved in a chemical reaction.

few elements (such as gold, silver, nitrogen, oxygen, and sulfur) occur in nature uncombined with other elements. Most elements are found in chemical combination with other elements in the form of compounds. A **compound** is a substance that consists of two or more elements that can be separated from one another only in a **chemical reaction**, which is the transformation of one or more substances into one or more other substances. Compounds typically have properties that are very different from the elements of which they are composed. For example, common table salt (sodium chloride), a white solid, is composed of sodium, which is a silvery-gray metal, and chlorine, which is a green gas. Both of these elements are extremely hazardous in their pure forms.

The elements in compounds are present in characteristic and definite proportions. Water, for example, is described by the formula $H_2O$. These symbols mean that pure water always has two particles of the element hydrogen, H, combined with one particle of the element oxygen, O. When hydrogen and oxygen react with each other, this chemical process results in the formation of water, and the combining ratio is *always* two volumes of hydrogen gas for every one volume of oxygen gas. If we reverse the process and decompose water into hydrogen and oxygen (Figure 1.4), which is another chemical process, we always obtain two volumes of hydrogen gas for every one volume of oxygen. This consistency illustrates the **law of constant composition**: every sample of a particular compound always contains the same elements combined in the same proportions.

**CONCEPT TEST**

A compound with the formula NO is present in the exhaust gases leaving a car's engine. As NO travels through the car's exhaust system, some of it decomposes into nitrogen and oxygen gas. What is the volume ratio of nitrogen to oxygen formed from NO?

*(Answers to Concept Tests are in the back of the book.)*

## 1.2 Matter: An Atomic View

An **atom** is the smallest representative particle of an element. If, for example, you were to grind a sample made up solely of the element silicon into the finest dust imaginable, there would be a limit to how tiny a particle of the dust could be and still be silicon. That limit is one atom of silicon.

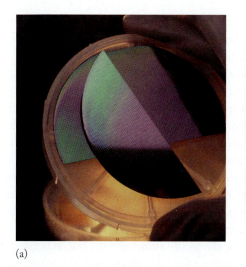

(a)

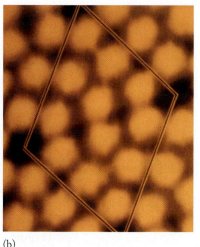

(b)

**FIGURE 1.5** (a) Silicon is used to make computer chips and solar cells. (b) Since the 1980s, scientists have been able to image individual atoms using an instrument called a scanning tunneling microscope (STM). In this STM image, the fuzzy spheres are individual silicon atoms. The radius of each atom is 117 picometers (pm), or 117 trillionths of a meter. These atoms are the tiniest particles of silicon that still retain the chemical characteristics of silicon.

Chemists view matter and its properties on the atomic level, which is also called the microscopic or particulate level. Scientists have even developed instruments to produce images of matter at that level. For example, if the surface of a silicon wafer (Figure 1.5a) used to make computer chips or solar cells is magnified over 100 million times using a device called a scanning tunneling microscope (STM), the result is an image of individual silicon atoms (Figure 1.5b).

Just as elements are made of particles called atoms, many compounds are made of multiatom particles called molecules. A **molecule** is a collection of atoms chemically bonded together in a characteristic pattern and proportion.

Let's take an atomic view of the combination of hydrogen and oxygen to form water. In Figure 1.6 we represent the reaction using models depicting atoms of hydrogen (the white spheres) and oxygen (the red spheres). Hydrogen and oxygen, like several other elements that are gases at room temperature, exist as *diatomic* (two-atom) molecules and therefore are represented by the chemical formulas $H_2$ and $O_2$. A **chemical formula** consists of the symbols of elements and subscripts that indicate the number of atoms of each element in the molecule. For example, the chemical formula $H_2O$ tells us that every molecule of water contains two hydrogen atoms and one oxygen atom.

Whenever hydrogen and oxygen combine to form water, two molecules of $H_2$ combine with one molecule of $O_2$ to form two molecules of $H_2O$. This relation is represented in Figure 1.6 by the molecular models and by the chemical equation beneath them. In a **chemical equation**, chemical formulas represent the identities of

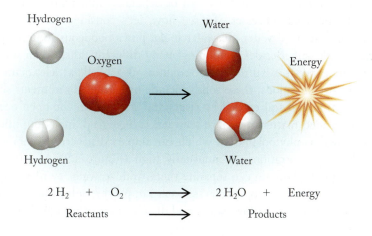

Hydrogen

Oxygen

Hydrogen

Water

Energy

Water

$$2\,H_2 \ + \ O_2 \ \longrightarrow \ 2\,H_2O \ + \ Energy$$

Reactants  $\longrightarrow$  Products

**FIGURE 1.6** The reaction between hydrogen and oxygen is depicted here with space-filling molecular models (white and red spheres) and in the form of a chemical equation. Note that energy is also a product of the reaction.

(a) Chemical formulas:      $H_2O$        $C_2H_6O$

(b) Structural formulas:

(c) Ball-and-stick models:

(d) Space-filling models:

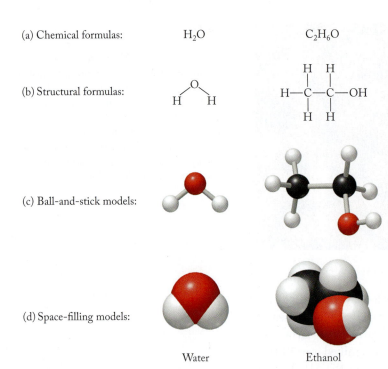

Water            Ethanol

**FIGURE 1.7** Four ways to represent the arrangement of atoms in molecules of water and ethanol: (a) chemical formulas; (b) structural formulas; (c) ball-and-stick models, where white spheres represent hydrogen atoms, black spheres represent carbon atoms, and red spheres represent oxygen atoms; (d) space-filling models. We often use colors in models to distinguish different atoms, but atoms themselves have no color.

**FIGURE 1.8**

----

**chemical bond** the energy that holds two atoms in a molecule together.

**filtration** a process for separating particles suspended in a liquid or a gas from that liquid or gas by passing the mixture through a medium that retains the particles.

**distillation** a separation technique in which the more *volatile* (more easily vaporized) components of a mixture are vaporized and then condensed, thereby separating them from the less volatile components.

----

substances involved in a chemical reaction. The arrow in the equation separates the initial materials (the reactants) from the final materials (the products).

Chemical formulas, such as those shown in Figure 1.7(a), provide information about the proportions of the elements in a compound, but they do not tell us how the atoms of those elements are connected in a molecule of the compound, nor do they tell us anything about the shape of the molecules. We use structural formulas and molecular models to show atom-to-atom connections and molecular shapes. The atoms in a molecule are linked together by **chemical bonds** in a fixed arrangement, and a *structural formula* (Figure 1.7b) shows the atoms and the bonds between them. A structural formula shows how the atoms are connected but does not necessarily indicate the correct angles between bonds or the three-dimensional shape of the molecule.

*Ball-and-stick models* (Figure 1.7c) use spheres to represent atoms and sticks to represent chemical bonds. The advantage of ball-and-stick models is that they show the correct angles between the bonds. However, the relative sizes of the spheres do not always match the relative sizes of the atoms they represent. Another disadvantage of these models is that the atoms must be spaced far apart to accommodate the stick bonds. In real molecules, however, the atoms touch each other. Both of these disadvantages are overcome with *space-filling models* (Figure 1.7d), in which the spheres are drawn to scale and are next to one another as atoms are in real molecules. The disadvantage of space-filling models is that the bond angles between atoms may be hard to see.

**CONCEPT TEST**

Figure 1.8 is the space-filling model of formaldehyde. What is its chemical formula?

*(Answers to Concept Tests are in the back of the book.)*

## 1.3 Mixtures and How to Separate Them

As noted in Figure 1.2, mixtures can be separated into their component substances by physical processes. To see how, let's consider the most abundant mixture on Earth: seawater. Seawater is not just an aqueous (water) solution of common table salt. Besides containing many different salts, ocean water is not even a homogeneous mixture—it also contains suspended particles of undissolved solids. In coastal regions, these solids include fine-grained sediments suspended by wave action and soil eroded by rivers and streams. In the open ocean most of the suspended matter in surface seawater is biological, including microscopic plants known as phytoplankton, which can impart distinctive colors to the sea when their concentrations are unusually high (Figure 1.9). Marine scientists who study phytoplankton can separate them from seawater by **filtration**, which involves passing the heterogeneous seawater sample through a filter that traps the phytoplankton cells but allows water and dissolved salts through (Figure 1.10).

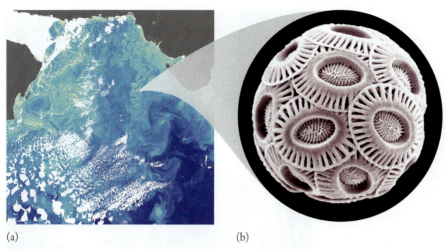

(a)                                        (b)

**FIGURE 1.9** (a) Very high concentrations (called "blooms") of phytoplankton known as *coccolithophores* in the Arabian Sea were photographed by a NASA satellite in July 2005. During intense blooms, coccolithophores turn the color of the ocean a milky aquamarine. The color comes from the chlorophyll in the phytoplankton, the milkiness from sunlight scattering off the organisms. (b) Intricate plates (or *coccoliths*) surround each cell of the organism. The coccoliths are so tiny that 500 of them placed end-to-end would fit in a length of 1 mm.

The particles caught on the filter can be further treated by a technique called solvent extraction. This process may involve soaking the filter in a vial of a liquid that dissolves many of the compounds present inside the phytoplankton cells (Figure 1.11a). These compounds include chlorophyll and other pigments that have distinctive colors. Once dissolved in a solvent, the pigments can be separated from each other using a technique called thin layer chromatography (TLC). In TLC, a small volume of the pigment extract is applied to the porous surface coating a thin plate. Then the edge of the plate nearest the sample is immersed in a shallow bath of solvent. The solvent is drawn up the face of the plate in the same way liquids wick up a paper towel. Different pigments migrate upward with the rising flow of solvent at different rates, producing a distinctive pattern (Figure 1.11b). TLC is a rapid, inexpensive, and widely used method for separating mixtures of compounds of chemical or biological interest.

Filtration is based on the principle that an object cannot pass through a pore that is smaller than the object. Air may also be filtered to remove particles suspended in it. Air filters range in size from the screens used to prevent insects from entering buildings to HEPA (high-efficiency particulate air) filters used in clean rooms to prevent dust, bacteria, or viruses from contacting samples such as high-purity silicon chips or cells in culture that must be protected from contamination.

Seawater is unfit to drink because the concentrations of the salts in it are too high. One way to make seawater drinkable is through another physical separation method—**distillation**—whereby seawater is heated to a temperature at which water is vaporized (Figure 1.12). The water vapor contacts a cool surface, where it condenses back into liquid water and is collected as purified *distillate*. Any dissolved salts and suspended particles remain behind because they are much less volatile than water.

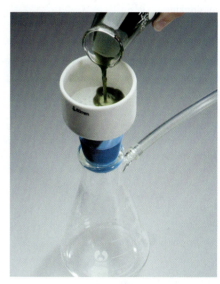

**FIGURE 1.10** Particles in suspension, such as a culture of phytoplankton in seawater, can be separated by filtration.

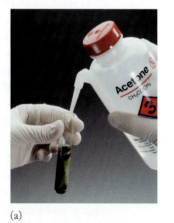

(a)                                        (b)

**FIGURE 1.11** (a) Scientists may extract the material trapped on a filter with a solvent such as acetone that dissolves the chlorophyll and other pigments. (b) Dissolved pigments can be separated from one another using thin layer chromatography, producing a distinctive pattern.

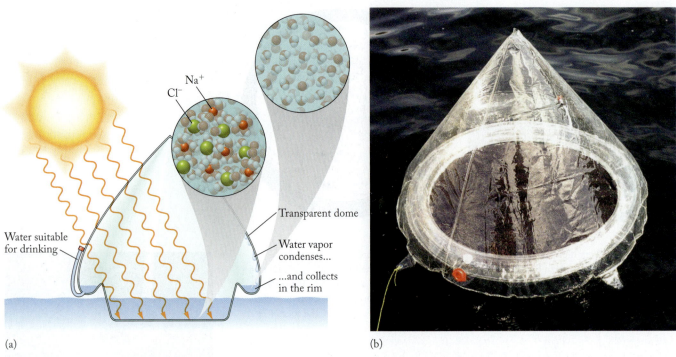

Na$^+$

Cl$^-$

Transparent dome

Water vapor
condenses...

...and collects
in the rim

Water suitable
for drinking

(a)

(b)

**FIGURE 1.12** (a) In a solar still, used in survival gear to provide freshwater from seawater, sunlight passes through the transparent dome and heats a pool of seawater. Water vapor rises from the pool, contacts the inside of the transparent dome, which is relatively cool, and condenses. The distilled water collects in the depression around the rim and then passes into the attached tube, from which one may drink it. (b) A solar still in use.

**CONCEPT TEST**  • • • • • • • • • • • • • • • • • • • • • • • • • • • • • • • • • • • • • • • • • • • • • • • • • • • • • • •

Is gasoline a solution? Explain why you think it is or is not.

*(Answers to Concept Tests are in the back of the book.)*

• • • • • • • • • • • • • • • • • • • • • • • • • • • • • • • • • • • • • • • • • • • • • • • • • • • • • • • • • • • • • • •

# 1.4 A Framework for Solving Problems

Throughout this book we include questions and problems that test your comprehension of the material. Your success in this course depends, in part, on your ability to solve these problems. Solving chemistry problems is like playing a musical instrument: the more you practice, the better you become.

In this section we present a framework for solving problems that we follow in the Sample Exercises throughout the book. Each Sample Exercise is followed by a Practice Exercise involving the same type of problem, which can be solved using a similar approach. We strongly encourage you to hone your problem-solving skills by working all the Practice Exercises as you read the chapters. You should find the approach described here useful in solving the exercises in this book as well as problems you encounter in other courses and other contexts.

We use the acronym COAST (**C**ollect and **O**rganize, **A**nalyze, **S**olve, and **T**hink about the answer) to represent the four steps in this approach. As you read about it here and use it later, keep in mind that COAST is merely a *framework* for solving problems, not a recipe. You can use it as a guide to developing your own approach to solving problems.

**Collect and Organize** The first step in solving a problem is to decide how to use the information given to answer the problem. This step is based on your understanding of the problem, including its fundamental chemical and/or physical principles. In assessing given information, you need to accomplish these tasks:

➤ *Visualize* the problem. For example, close your eyes and see a mixture of gases in a rigid container or a solution in a beaker on a lab bench. Drawing a sketch based on molecular models or an experimental setup may help you visualize how the starting points and final answer are connected.
➤ *Identify* the key concept of the problem. Identify key terms used to express that concept. You may find it useful to restate the problem in your own words.
➤ *Sort* through the information given, separating what is relevant from what is not.
➤ *Assemble* needed key information, including equations, definitions, and constants. To solve problems that require use of an equation, it is often helpful to make a list of the quantities that refer to each symbol in the equation. If you have chosen the correct equation, only one quantity for a symbol should be left as an unknown. This quantity is what you need to find.

**Analyze** The next step is to analyze the information you have to determine how to connect it to the answer you seek. Sometimes you can work backward to create these links: Consider the answer first and think about how you might get to it from the information provided and other sources. This might mean finding some intermediate quantity to use in a later step. If the problem requires a numerical answer, frequently the units of the initial values and the final answer can help you identify how they are connected and which equation(s) may be useful. This step may include rearranging equations to solve for an unknown or setting up conversion factors. You should also look at the numbers involved and estimate your answer.

**Solve** The solutions to most qualitative problems that test your understanding of a concept flow directly from your analysis of the problem. To solve quantitative problems, you need to insert the starting values and appropriate constants into the equations or conversion factors and calculate the answer. In this step, make sure that units are consistent and cancel out as needed and that the answer contains the appropriate number of significant figures. (We discuss significant figures in Section 1.8 and conversion factors in Section 1.9.)

**Think About It** Ask yourself, "Does this answer make sense based on my experience and what I have just learned? Is the numerical answer reasonable? Are the units correct and the number of significant figures appropriate?" Then ask yourself if you could solve another problem, perhaps drawn from another context but based on the same concept. You can also think about how this problem relates to your daily life.

The COAST approach should help you solve problems in a logical way and avoid pitfalls like grabbing an equation that seems to have the right variables and plugging in numbers or resorting to trial and error. As you study the steps in each Sample Exercise, try to answer these questions about each step:

➤ **What** is done in this step?
➤ **How** is it done?
➤ **Why** is it done?

With answers to these questions, you will be ready to solve the Practice Exercises and end-of-chapter Questions and Problems in a systematic way.

## 1.5 Properties of Matter

Every day we use information from our senses to describe the world around us, and in doing so we categorize matter in a variety of ways. We breathe a gaseous mixture of matter called air to bring oxygen into our bodies to metabolize food. Oxygen is odorless, as is nitrogen, the main component of air on Earth, so when we smell something, we know we are inhaling some other gas along with the oxygen and nitrogen. This example illustrates categorizing gases by smell.

Pure substances have distinctive properties. Some substances burn, others put out fires. Some substances are shiny, others are dull. Some are brittle, others are malleable. These and other properties do not vary from one sample of a given pure substance to the next. Consider gold, for example: its color (very distinctive), its hardness (soft for a metal), its malleability (it can be hammered into very thin sheets called gold leaf), and its melting temperature (1064°C) all apply to any sample of pure gold. These characteristics are examples of **intensive properties**: properties that characterize matter independent of the quantity of the material present. On the other hand, an object made of gold, such as an ingot, has a particular length, width, mass, and volume. These properties of a particular sample of a pure substance, which depend on how much of the substance is present, are **extensive properties**.

### CONCEPT TEST

Which of these properties of a sample of pure iron is intensive? (a) mass; (b) density; (c) volume

*(Answers to Concept Tests are in the back of the book.)*

Properties fall into two other general categories: physical and chemical. **Physical properties** are the properties of a pure substance that can be observed or measured without changing the substance into another substance. The intensive properties of gold described above are all physical properties as is its **density (d)**—the ratio of the mass (m) of an object to its volume (V).

$$d = \frac{m}{V} \qquad (1.1)$$

As noted in Section 1.1, gold is one of the few elements found in nature uncombined with other elements. Such *free elements* may exist as single atoms, like He, or as molecules containing atoms of only the element, like $O_2$ and $S_8$. Most elements are not found free in nature but rather are combined with other elements in compounds, like hydrogen in $H_2O$ or sodium in NaCl. As already noted, hydrogen combines with oxygen in a chemical reaction that produces water and energy. This reaction is an example of combustion (burning), and $H_2$ is the fuel. High flammability is a **chemical property** of hydrogen. Like any other chemical property, flammability can be determined only by reacting one substance with another substance and finding that a different material is produced. A substance's chemical reactivity, including the rates of its reactions, the identity of other substances with which it reacts, and the identity of the products formed, all define the chemical properties of the substance.

The physical and chemical properties of a compound can be very different from those of the elements that combine to form it. Water is a liquid at room temperature, for example, whereas hydrogen and oxygen are gases at room temperature. Water expands when it freezes at 0°C; hydrogen (freezing point,

**intensive property** a property that is independent of the amount of substance present.

**extensive property** a property that varies with the quantity of the substance present.

**physical property** a property of a substance that can be observed without changing it into another substance.

**density (d)** the ratio of the mass (m) of an object to its volume (V).

**chemical property** a property of a substance that can be observed only by reacting it to form another substance.

−259°C) and oxygen (−219°C) do not. Oxygen supports combustion reactions and hydrogen is a highly flammable fuel, but water does not support combustion and is not flammable. Because of its physical and chemical properties, water is widely used to put out fires.

---

**SAMPLE EXERCISE 1.1** **Distinguishing Physical and Chemical Properties** **LO3**

Which of the following properties of gold are chemical and which are physical?
a. Gold metal, which is insoluble in water, can be made soluble by reacting it with a mixture of nitric and hydrochloric acids known as aqua regia.
b. Gold melts at 1064°C.
c. Gold can be hammered into sheets so thin that light passes through them.
d. Gold metal can be recovered from gold ore by treating the ore with a solution containing cyanide, which reacts with and dissolves gold.

**Collect and Organize** We are given several properties of gold that are either chemical or physical and are asked to decide which is which.

**Analyze** We can check the relevant definitions from the text: chemical properties describe how a substance reacts with other substances; physical properties can be observed or measured without changing one substance into another.

Properties a and d involve reactions that chemically change gold metal (which does not dissolve in water) into compounds of gold that do dissolve in water. Properties b and c describe processes in which elemental gold remains elemental gold. When it melts, gold changes its physical state, but not its chemical identity. Gold leaf is still solid, elemental gold.

**Solve** Properties a and d are chemical properties, and b and c are physical properties.

**Think About It** When possible, we fall back on our experiences and observations. We know gold jewelry does not dissolve in water. Therefore, dissolving gold metal requires a change in its chemical identity: it can no longer be elemental gold. On the other hand, physical changes such as melting do not alter the chemical identity of the gold. Gold can be melted and then cooled to produce solid gold again.

⚙ **Practice Exercise** Which of the following properties of water are chemical and which are physical?
a. It normally freezes at 0.0°C.
b. It is useful for putting out most fires.
c. A cork floats on it, but a piece of copper sinks.
d. During digestion, starch reacts with water to form sugar.

*(Answers to Practice Exercises are in the back of the book.)*

---

## 1.6 States of Matter

Matter exists in one of three phases or physical states: solid, liquid, or gas. You are probably familiar with the characteristic properties of these states:

➤ A **solid** has a definite volume and shape.
➤ A **liquid** has a definite volume but not a definite shape.
➤ A **gas** (or *vapor*) has neither a definite volume nor a definite shape. Rather, it expands to occupy the entire volume and shape of its container.

Unlike a solid or a liquid, a gas is compressible, which means it can be squeezed into a smaller volume if its container is not rigid and pressure is applied to it.

The differences between solids, liquids, and gases can be understood if we view the three states on the atomic level. As shown in Figure 1.13, each $H_2O$ molecule

**solid** a form of matter that has a definite shape and volume.

**liquid** a form of matter that occupies a definite volume but flows to assume the shape of its container.

**gas** a form of matter that has neither definite volume nor shape and that expands to fill its container; also called *vapor*.

(a) Solid

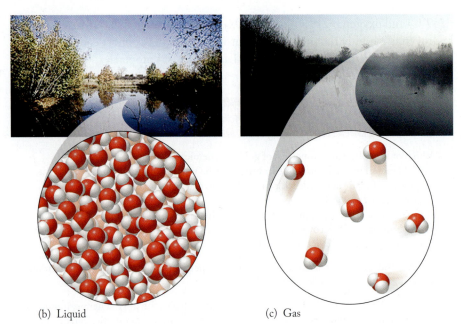

(b) Liquid

(c) Gas

**FIGURE 1.13** Water can exist in three states: (a) solid (ice), (b) liquid (water), or (c) gas (water vapor). (a) In the solid state, each $H_2O$ molecule in ice is held in place in a rigid, three-dimensional array by strong interactions between adjacent molecules. (b) In the liquid state, the $H_2O$ molecules are close together and still interact with each other, but they are free to flow past one another. (c) In the gas state, the $H_2O$ molecules are far apart, largely independent of one another, and move freely. Water vapor is invisible, but we can see clouds and fog that form when atmospheric water vapor condenses to liquid $H_2O$.

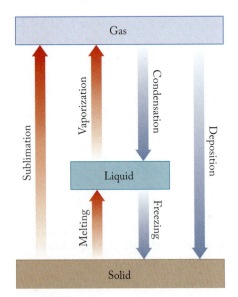

**FIGURE 1.14** Matter changes from one state to another when energy is either added or removed. Arrows pointing upward represent transformations that require the addition of energy; arrows pointing downward represent transformations that release energy.

in ice is mostly surrounded by other $H_2O$ molecules and is locked in place in a three-dimensional array. Molecules in this structure may vibrate, but they are not free to move past the molecules that surround them; they have the same nearest neighbors over time. The $H_2O$ molecules in liquid water are free to flow past one another, but they are still in close proximity; their nearest neighbors change over time. In water vapor the $H_2O$ molecules are widely separated, and the volume they occupy is much smaller than the volume occupied by the vapor. This separation accounts for the compressibility of gases. The molecules in the gas phase are relatively independent and move rapidly throughout the space occupied by the vapor.

We can transform water from one physical state to another by raising or lowering its temperature. Ice forms on a pond when the temperature drops in the winter and the water freezes. This process is reversed when warmer temperatures return in the spring and the ice melts. The heat of the sun may vaporize liquid water during the daytime, but colder temperatures at night may cause the water vapor to condense as dew.

Some solids change directly into gases with no intervening liquid phase. For example, snow may be converted to water vapor on a cold, sunny winter day even though the air temperature remains well below freezing. This transformation of solid directly to vapor is called **sublimation**. The reverse process—in which water vapor forms a layer of frost on a cold night—is an example of a gas being transformed directly into a solid without ever being a liquid, a process called **deposition**.

The physical changes matter undergoes on conversion from one physical state to another are illustrated in Figure 1.14. The transformations represented by upward-pointing arrows require the addition of heat energy; those represented by downward-pointing arrows release heat.

Energy must be added to convert solids into liquids and to convert liquids into gases, as shown in Figure 1.14. In the case of water, energy is needed to overcome the attractions between water molecules that are close together. Based on these images, should it require (a) more than, (b) less than, or (c) about the same amount of energy to melt a gram of ice at its melting point as it does to boil a gram of water that has already been heated to its boiling point? Explain your selection.

*(Answers to Concept Tests are in the back of the book.)*

**sublimation** transformation of a solid directly into a vapor (gas).

**deposition** transformation of a vapor (gas) directly into a solid.

---

**SAMPLE EXERCISE 1.2** **Recognizing the Physical States of Matter** **LO5**

Which physical state is represented in each box of Figure 1.15? (The particles could be atoms or molecules.) What changes of state are indicated by the two arrows? What would the changes of state be if both arrows pointed in the opposite direction?

**Collect and Organize** The particles in each box represent atoms or molecules in a solid, liquid, or gas.

**Analyze** The first step is to recognize the patterns of atomic-level particles in these different states. Once we have identified the initial and final states represented by each pair of boxes, we can name the transition using the information in Figure 1.14. We look for the following patterns among the particles:
- An ordered arrangement of particles that does not fill the box or match the shape of the box represents a solid.
- A less-ordered arrangement of particles that partially fills a box and conforms to its shape represents a liquid.
- A dispersed array of particles distributed throughout a box represents a gas.

**Solve** The particles in the left box in Figure 1.15(a) partially fill the box and adopt the shape of the container; therefore they represent a liquid. The particles in the box on the right are ordered and form a shape different from that of the box, so they represent a solid. The arrow represents a liquid turning into a solid, and thus the transition is freezing. An arrow in the opposite direction would represent melting. The particles in the box on the left in Figure 1.15(b) represent a solid because they are ordered and do not adopt the shape of the box. Those to the right are dispersed throughout the box, representing a gas. The arrow represents a solid turning into a gas and the state change is sublimation. The reverse process—a gas becoming a solid—is called deposition.

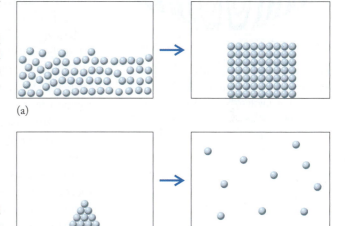

(a)

(b)

**FIGURE 1.15** Changes in the physical state of matter.

**Think About It** Notice that the particles represented in Figure 1.15(a) are very close together, which is consistent with liquids and solids being classified as condensed phases. On the other hand, the particles in gases, as shown in the right panel of Figure 1.15(b), are widely separated.

**Practice Exercise** (a) What physical state is represented by the particles in each box of Figure 1.16, and which change of state is represented? (b) Which change of state would be represented if the arrow pointed in the opposite direction?

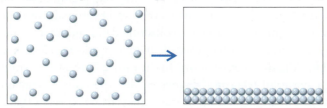

**FIGURE 1.16**

*(Answers to Practice Exercises are in the back of the book.)*

**FIGURE 1.17** For centuries, cultures have passed on creation stories in which a supreme being is responsible for creating the Sun, Moon, Earth, and its inhabitants. According to a creation myth of Native Americans of the Pacific Northwest, a deity in the form of a raven was responsible for releasing the Sun into the sky.

▶❙❙ **CHEMTOUR** Big Bang

........................................................

**scientific method** an approach to acquiring knowledge based on observation of phenomena, development of a testable hypothesis, and additional experiments that test the validity of the hypothesis.

**hypothesis** a tentative and testable explanation for an observation or a series of observations.

**scientific theory (model)** a general explanation of a widely observed phenomenon that has been extensively tested and validated.

........................................................

# 1.7 The Scientific Method: Starting Off with a Bang

In Sections 1.5 and 1.6, we examined the properties and states of matter. In this section and in Section 1.10, we examine the origins of matter and a theory of how the universe came into being.

The ancient Greeks believed that at the beginning of time there was no matter, only a vast emptiness they called Chaos, and that from that emptiness emerged the first supreme being, Gaia (also known as Mother Earth), who gave birth to Uranus (Father Sky). Other cultures and religions have described creation in similar terms of supernatural beings creating ordered worlds out of vast emptiness. The opening verses of the book of Genesis, for example, describe darkness "without form and void" from which God created the heavens and Earth. Similar stories are part of Asian, African, and Native American cultures (Figure 1.17). Today we have the technological capacity to take a different approach to explaining how the universe was formed and what physical forces control it. Our approach is based on the **scientific method** of inquiry (Figure 1.18).

This approach evolved in the late Renaissance, during a time when economic and social stability gave people an opportunity to study nature and question old beliefs. By the early 17th century, the English philosopher Francis Bacon (1561–1626) had published his *Novum Organum* (*New Organ* or *New Instrument*) in which he described how humans can acquire knowledge and understanding of the natural world through observation, experimentation, and reflection. In the process he described, observations of a natural phenomenon lead to a tentative explanation, or **hypothesis**, of what causes the phenomenon. Making a hypothesis and devising experiments to test it in a decisive way is a very creative aspect of science. One measure of its validity is that it enables scientists to predict accurately the results of future experiments and observations. Further testing and observation might support a hypothesis or disprove it, or perhaps require that it be modified so that it explains all of the experimental results. A hypothesis that withstands the tests of many experiments over time and explains the results of further observation and experimentation may be elevated to the rank of a **scientific theory**, or **model**.

Let's examine how the scientific method works using a simple situation from daily life. Most gas stations sell gasoline with three different octane ratings. You hypothesize that the rating of the gasoline matters in determining how far you can drive on a tank of gas. You then test that hypothesis by filling your gas tank on three successive trips with a different grade of gas. Suppose there is no difference in your car's driving range, but then you realize that all the driving you did during this test consisted of short hops within a city. You hypothesize that the kind of driving might make a difference, and that on a longer highway trip your car might do better with premium. On your next long drive you test that hypothesis. Depending on the results, you may decide to use cheaper gas with lower octane rating when you drive in town and only buy premium on long trips. The point is, when applying the scientific method you continually test and refine your hypotheses by running experiments, collecting data, and interpreting the results.

Let's now examine how the scientific method works by addressing a profound question: How did the universe form? In the early 20th century, astronomers using increasingly powerful telescopes discovered that the universe contains billions of galaxies, each containing billions of stars. They also discovered that (1) the other galaxies in the universe are moving away from our own Milky Way and from one another and (2) the speeds with which the galaxies are receding are proportional to the distance between them and the Milky Way—the farthest galaxies are moving away from us the fastest.

Based on this observation of galaxies moving farther and farther away, Georges-Henri Lemaître (1894–1966) proposed in 1927 that the universe we know today formed with an enormous release of energy that quickly transformed into a rapidly expanding, unimaginably hot cloud of matter in accordance with Einstein's famous formula:

$$E = mc^2$$

where $E$ is energy, $m$ is the equivalent mass of matter, and $c$ is the speed of light. To understand Lemaître's explanation of our expanding universe, consider what we would see if the motion of the matter in the universe had somehow been recorded since the beginning of time and we were able to play the recording in reverse. The other galaxies would be moving toward ours (and toward one another), with the ones farthest away closing in the fastest, catching up with those galaxies that were closer to start with but moving more slowly. Eventually, near the beginning, all the matter in the universe would approach the same point at the same time. During this compression, the density of the universe would become enormous and it would get very hot. At the very beginning, the universe would be squeezed into an infinitesimally small space of unimaginably high temperature. At such a temperature the distinction between matter and energy becomes blurred. Indeed, at the very beginning there is no matter, only energy. Now, if we play our recording in the forward direction, we see an instantaneous release of an enormous quantity of energy in an event that has come to be known as the Big Bang.

Lemaître's explanation of how the universe began has been the subject of extensive testing—and controversy—since it was first proposed. Indeed, the term Big Bang was first used by British astronomer Sir Fred Hoyle (1915–2001) to poke fun at the idea. However, the results of many experiments and observations conducted since the 1920s support what is now described as the Big Bang theory. We use the term *theory* to describe an explanation of natural phenomena that is comprehensive, validated by the results of extensive experimentation and observation, and can be used to accurately predict the results of other experiments and observations. In Section 1.10, we examine a few of the experiments that have supported the validity of the Big Bang theory.

**CONCEPT TEST** ..........................................................

If the volume of the universe is expanding and if the mass of the universe is not changing, is the density of the universe (a) increasing, (b) decreasing, or (c) constant?

*(Answers to Concept Tests are in the back of the book.)*
..........................................................

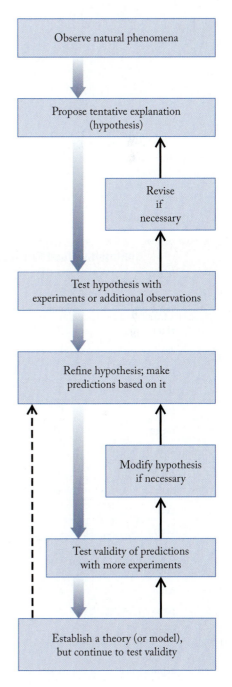

**FIGURE 1.18** In the scientific method, observations lead to a tentative explanation, or hypothesis, which leads to more observations and testing, which may lead to the formulation of a succinct, comprehensive explanation called a theory. The process, which repeats in a self-correcting fashion, represents a highly reasoned way to understand nature. The steps in the process are flexible. Theories like the Big Bang started out with a hypothesis and then progressed by validating its predictions. Scientists do not "prove" a theory, they demonstrate or validate it. Just one counter-example is enough to make a theory yesterday's news. So far there is no scientifically conclusive evidence to invalidate the Big Bang.

# 1.8 Making Measurements and Expressing the Results

Before we examine some of the experiments designed to test the Big Bang theory, we need to comment on the critical importance of accurate measurements in science. Whether we are exploring the nature of the cosmos or determining whether a medication contains the right quantity of a particular drug, accurate measurements are essential to our ability to characterize many of the physical and chemical properties of matter. The rise of scientific inquiry in the 17th and 18th centuries brought about a heightened awareness of the need for accurate measurements as well as the need for expressing those measurements in ways that were understandable to others. Standardization of the units of measurement was essential.

## SI Units

In 1791 French scientists proposed a standard unit of length, which they called the **meter** (m), after the Greek *metron,* which means "measure." They based the length of the meter on 1/10,000,000 of the distance along an imaginary line running from the North Pole to the equator. By 1794, hard work by teams of surveyors had established the length of the meter that is still in use today.

These French scientists also settled on a decimal-based system for designating lengths that are multiples or fractions of a meter (Table 1.1). They chose Greek prefixes for lengths much greater than 1 meter, such as *deka-* and *kilo-* for lengths of 10 meters (1 dekameter) and 1000 meters (1 kilometer). Then they chose Latin

**TABLE 1.1    Commonly Used Prefixes for SI Units**

| PREFIX | | VALUE | |
|---|---|---|---|
| Name | Symbol | Numerical | Exponential |
| zetta | Z | 1,000,000,000,000,000,000,000 | $10^{21}$ |
| exa | E | 1,000,000,000,000,000,000 | $10^{18}$ |
| peta | P | 1,000,000,000,000,000 | $10^{15}$ |
| tera | T | 1,000,000,000,000 | $10^{12}$ |
| giga | G | 1,000,000,000 | $10^{9}$ |
| mega | M | 1,000,000 | $10^{6}$ |
| kilo | k | 1000 | $10^{3}$ |
| hecto | h | 100 | $10^{2}$ |
| deka | da | 10 | $10^{1}$ |
| deci | d | 0.1 | $10^{-1}$ |
| centi | c | 0.01 | $10^{-2}$ |
| milli | m | 0.001 | $10^{-3}$ |
| micro | μ | 0.000001 | $10^{-6}$ |
| nano | n | 0.000000001 | $10^{-9}$ |
| pico | p | 0.000000000001 | $10^{-12}$ |
| femto | f | 0.000000000000001 | $10^{-15}$ |
| atto | a | 0.000000000000000001 | $10^{-18}$ |
| zepto | z | 0.000000000000000000001 | $10^{-21}$ |

prefixes for lengths much smaller than a meter, such as *centi-* and *milli-* for lengths of 1/100 of a meter (1 centimeter) and 1/1000 of a meter (1 millimeter). (See Appendix 1 for a description of scientific notation used in Table 1.1.)

Eventually these decimal prefixes carried over to the names of standard units for other dimensions. In July 1799 a platinum (Pt) rod 1 meter long and a platinum block having a mass of 1 kilogram (1000 grams) were placed in the French National Archives to serve as the legal standards for length and mass. These objects served as references for the metric system for expressing measured quantities.

Since 1960, scientists have, by international agreement, used a modern version of the French metric system: the *Système International d'Unités*, commonly abbreviated SI. There are seven *SI base units* in Table 1.2, and many other SI units are derived from them. For example, a common SI unit for volume, the cubic meter ($m^3$), is derived from the length base unit, the meter. The common SI unit for speed, meters per second (m/s), is derived from the length and time base units. Table 1.3 contains some of these derived units and their equivalents in the U.S. Customary System of units. They include the volume corresponding to 1 cubic decimeter (a cube 1/10 meter on each side), which we call a liter (L).

Modern science requires that the length of the meter, as well as the dimensions of other SI units, be defined by quantities that are much more constant than the length of a platinum rod in Paris. Two such quantities are the speed of light (*c*) and time. In 1983, 1 m was redefined as the distance traveled in 1/299,792,458 of a second by the light emitted from a helium–neon laser. This modern definition of the meter is consistent with the one adopted in France in 1794.

| TABLE 1.2 | Seven SI Base Units | |
|---|---|---|
| **Quantity or Dimension** | **Unit Name** | **Unit Abbreviation** |
| Mass | kilogram | kg |
| Length | meter | m |
| Temperature | kelvin | K |
| Time | second | s |
| Electrical current | ampere | A |
| Amount of a substance | mole | mol |

## Significant Figures

All scientific measurements have one thing in common: there is a limit to how accurate they can be. Nobody is perfect, and no analytical method is perfect either. Every method has an inherent limit in its capacity to produce accurate results, and we need ways to express experimental results that reflect these limits. We do so by expressing numerical data from experiments to the appropriate number of **significant figures**.

| TABLE 1.3 | Conversion Factors for SI and Other Commonly Used Units |
|---|---|
| **Quantity or Dimension** | **Equivalent Units** |
| Mass | 1 kg = 2.205 pounds (lb); 1 lb = 0.4536 kg = 453.6 g |
| | 1 g = 0.03527 ounce (oz); 1 oz = 28.35 g |
| Length (distance) | 1 m = 1.094 yards (yd); 1 yd = 0.9144 m (exactly) |
| | 1 m = 39.37 inches (in); 1 foot (ft) = 0.3048 m (exactly) |
| | 1 in = 2.54 cm (exactly) |
| | 1 km = 0.6214 miles (mi); 1 mi = 1.609 km |
| Volume | 1 $m^3$ = 35.31 $ft^3$; 1 $ft^3$ = 0.02832 $m^3$ |
| | 1 $m^3$ = 1000 liters (L) (exactly) |
| | 1 L = 0.2642 gallon (gal); 1 gal = 3.785 L |
| | 1 L = 1.057 quarts (qt); 1 qt = 0.9464 L |

**meter** the standard unit of length, named after the Greek *metron*, which means "measure," and equivalent to 39.37 inches.

**significant figures** all the certain digits in a measured value plus one estimated digit. The greater the number of significant figures, the greater the certainty with which the value is known.

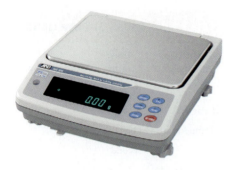

**FIGURE 1.19** The mass of a penny can be measured to the nearest 0.01 gram with the top balance and to the nearest 0.0001 gram with the balance below.

▶‖ **CHEMTOUR** Significant Figures

▶‖ **CHEMTOUR** Scientific Notation

The number of significant figures reported in a measured value, or in a calculation based on one or more measured values, indicates how certain we are of the measured value or values. For example, suppose we determine the mass of a penny on the two balances shown in Figure 1.19, with both balances working properly. The balance at the bottom determines the mass of objects to the nearest 0.0001 gram (g); the balance at the top determines the mass of objects only to the nearest 0.01 g. According to the balance at the top, our penny has a mass of 2.53 g; according to the balance at the bottom, it has a mass of 2.5271 g. The mass obtained from the top balance has three significant figures: the 2, 5, and 3 are considered *significant*, which means we are confident in their values. The mass obtained using the bottom balance has five significant figures (2, 5, 2, 7, and 1). We may conclude that the mass of the penny can be determined with greater certainty with the balance at the bottom.

Now suppose someone bumps the bench beneath both balances. The value displayed at the top balance will probably not be affected, but the balance at the bottom is so sensitive that there will likely be a change in the last digit of the displayed value. This change illustrates an important point: for many measured values there is some uncertainty in the rightmost digit. However, the last measurable digit is considered significant even though we are less certain of its value than we are of the value of the other digits. The significant figures in a number include all the digits we know with certainty plus one digit that is uncertain.

Now imagine that an aspirin tablet is placed on the balance at the bottom in Figure 1.19, and the display reads 0.0810 g. How many significant figures are there in this value? You might be tempted to say five because that is the number of digits displayed. However, the first two zeros are not considered significant because they serve only to set the location of the decimal point. These two zeros function the way exponents do when we express values using scientific notation. Expressing 0.0180 g using scientific notation gives us $1.80 \times 10^{-2}$ g. Only the three digits in the decimal part indicate how precisely we know the value; the "−2" in the exponent does not. (See Appendix 1 for a review of how to express values using scientific notation.)

Why is the rightmost zero in 0.0810 g significant? The answer is related to the ability of the balance to measure masses to the nearest 0.0001 g. If the balance is operating correctly, we can assume that the mass of the tablet is 0.0810 g, not 0.0811 g or 0.0809 g. We may assume that the last digit really is zero, and we need a way to express that. If we dropped the zero and recorded a value of only 0.081 g, we would be implying that we knew the value to only the nearest 0.001 g, which is not the case. The following guidelines will help you handle zeros (highlighted in green) in deciding the number of significant figures in a value:

1. Zeros at the beginning of a value, as in **0.0**592, are never significant. In this example, they just set the decimal place.
2. Zeros at the end of a value and after a decimal point, as in 3.**00** $\times 10^8$, are always significant.
3. Zeros at the end of a value that contains no decimal point, as in 96,5**00**, may or may not be significant. They may be there only to set the decimal place. We should use scientific notation to avoid this ambiguity. Here, $9.65 \times 10^4$ (three significant figures) and $9.6500 \times 10^4$ (five significant figures) are two of the possible interpretations of 96,500.
4. Zeros between nonzero digits, as in 1**0**1.3, are always significant.

How many significant figures are there in the values used as examples in guidelines 1, 2, and 4 on the previous page?

*(Answers to Concept Tests are in the back of the book.)*

## Significant Figures in Calculations

Now let's see how significant figures are used to express the results of calculations involving measured quantities. An important rule to remember is that significant-figure rules should be used only at *the end of a calculation*, never on intermediate results. A reasonable guideline to follow is that at least one digit to the right of the last significant digit should be carried forward in all intermediate steps.

Suppose we suspect that a small nugget of yellow metal is pure gold. We could test our suspicion by determining the mass and volume of the nugget and then calculating its density. If its density matches that of gold (19.3 g/mL), chances are good that the nugget is pure gold because few minerals are that dense. We find that the mass of the nugget is 4.72 g and its volume is 0.25 mL. What is the density of the nugget, expressed in the appropriate number of significant figures?

Using Equation 1.1 to calculate the density ($d$) from the mass ($m$) and volume ($V$) produces the following result:

$$d = \frac{m}{V} = \frac{4.72 \text{ g}}{0.25 \text{ mL}} = 18.88 \text{ g/mL}$$

This density value appears to be slightly less than that of pure gold. However, we need to answer the question, "How well do we know the result?" The mass value is known to three significant figures, but the volume value is known only to two. At this point we need to apply the *weak-link principle*, which is based on the idea that a chain is only as strong as its weakest link. In calculations involving measured values, this principle means that we can know the answer to a calculation only as well as we know the least well-known value used. In calculations involving multiplication or division, the weak link has the fewest significant figures. In this example the weak link is the volume, 0.25 mL. Because it has only two significant figures, our final answer must have only two significant figures. We must convert 18.88 to a number with two significant figures. We do this by a process called rounding off.

*Rounding off* a value means dropping the *insignificant digits* (all digits to the right of the first uncertain digit) and then rounding that first uncertain digit either up or down. If the first digit in the string of insignificant digits dropped is greater than 5, we round up; if it is less than 5, we round down. The question remains of how to handle cases in which the first dropped digit is 5. If there are nonzero digits to the right of the 5, then round up. For example, rounding 45.450001 to three significant figures makes it 45.5 (we rounded up because there was a nonzero digit to the right of the second 5). If there are no nonzero digits to the right of the 5, then a good rule to follow is to round to the nearest even number. Rounding 45.45 or 45.450 to three significant figures makes either value 45.4 because the 4 in the tenths place is the nearest even number. However, we round off 45.55 or 45.550 to 45.6 because the 6 in the tenths place is the nearest even digit.

| TABLE 1.4 | Densities of Common Materials |
|---|---|
| **ITEM** | **DENSITY (AT 20°C)** |
| **Solids** | **g/cm³** |
| Aluminum | 2.7 |
| Copper | 8.92 |
| Lead | 11.3 |
| **Liquids** | **g/mL** |
| Seawater | 1.025 |
| Blood | 1.060 |
| Classic soda | 1.02 |
| Diet soda | 0.982 |
| Gasoline | 0.737 |
| Olive oil | 0.80–0.92 |
| Milk | 1.028–1.035 |
| Turpentine | 0.847 |
| **Gases** | **g/L (at 20°C and 1 atm pressure)** |
| Air | 1.205 |
| Natural gas | 0.7–0.9 |
| Ammonia | 0.717 |
| Carbon dioxide | 1.842 |
| Water vapor | 0.804 |

**FIGURE 1.20** A nugget believed to be pure gold is placed in a graduated cylinder containing 50.0 mL of water. The volume rises to 58.5 mL, which means that the volume of the nugget is 8.5 mL.

In the case of our gold nugget, the density value resulting from our calculation, 18.88 g/mL, must be rounded to two significant figures. Because the first insignificant digit to be dropped (shown in red) is greater than 5, we round up the blue 8 to 9:

$$18.88 \text{ g/mL} = 19 \text{ g/mL}$$

The density of pure gold expressed to two significant figures is also 19 g/mL, so based on how well we know the measured values, we conclude that the nugget could be pure gold. For comparison, the densities of some common materials are given in Table 1.4.

The weak-link principle for significant figures also applies to calculations requiring addition and subtraction. To illustrate, consider how the volume of a gold nugget might be measured. One approach is to measure the volume of water the nugget displaces. Suppose we add 50.0 mL of water to a 100-mL graduated cylinder, as shown in Figure 1.20. Note that the cylinder is graduated in milliliters. However, the space between the graduations allows us to estimate the volumes of samples to the nearest tenth of a milliliter. For example, the meniscus of the water in Figure 1.20 is aligned exactly with the 50-mL graduation. We can correctly record the volume of water as 50.0 mL because we can read the "50" part of this value directly from the cylinder, and we can estimate that the next digit is "0."

Then we place the nugget in the graduated cylinder, being careful not to splash any water out. The level of the water is now about halfway between the 58 and 59 mL graduations and so we estimate the volume of the combined sample is 58.5 mL. Taking the difference of the two estimated values, we have the volume of our nugget:

$$
\begin{array}{r}
58.5 \text{ mL} \\
- 50.0 \text{ mL} \\
\hline
8.5 \text{ mL}
\end{array}
$$

The result has only two significant figures because the initial and final volumes are known to the nearest tenth of a milliliter. Therefore, we can know the difference between them to only the nearest tenth of a milliliter. When measured numbers are added or subtracted, the result has the same number of significant digits to the right of the decimal as the measured number with the fewest digits to the right of the decimal.

Consider one more example. Suppose the average mass of a penny is 2.53 g and a roll of pennies contains 50 of them. We use the top balance in Figure 1.19 to determine the mass of a roll of pennies. Using the tare feature to correct for the mass of the wrapper, we find that the mass of pennies in the roll is 124.01 g. Dividing this value by the average mass of a penny and canceling out the grams units in the numerator and denominator, we get

$$\frac{124.01 \text{ g}}{2.53 \text{ g/penny}} = 49.0 \text{ pennies}$$

If we were measuring uncountable substances, rounding off to 49.0 would be appropriate (because under these circumstances, the measured value of 2.53 g/penny would be the weak link in the calculation). However, because we use only whole numbers to count items, we round to 49 and conclude that the roll contains 49 pennies. We must add one penny to make the roll complete.

Suppose the penny we add is one we know has a mass of 2.5271 g because we determined its mass on the bottom balance in Figure 1.19. What is the mass of all

50 pennies to the appropriate number of significant figures? Adding the mass of the 50th penny to the mass of the other 49, we have

$$
\begin{array}{r}
124.01 \text{ g} \\
+ \quad 2.5271 \text{ g} \\
\hline
126.5371 \text{ g}
\end{array}
$$

The mass of the first 49 pennies is known to two decimal places, and the mass of the 50th is known to four decimal places. Therefore, by the weak-link principle we can know the sum of the two numbers to only two decimal places. This makes the last two digits in the sum, shown in red, not significant, so we must round off the mass to 126.54 g.

---

**SAMPLE EXERCISE 1.3** **Using Significant Figures in Calculations** **LO7**

A nugget of a shiny yellow mineral has a mass of 30.01 g. Its volume is determined by placing it in a 100-mL graduated cylinder containing 56.3 mL of water. The volume after the nugget is added is 62.6 mL. Is the nugget made of gold?

**Collect and Organize** We are given the mass of a shiny yellow nugget and water displacement data that can be used to calculate its volume. Density is mass divided by volume. We know from data given in this section that the density of gold is 19.3 g/mL.

**Analyze** We use Equation 1.1 to calculate the density of the nugget from its mass (30.01 g) and its volume, indicated by the amount of water displaced (the difference between 62.6 and 56.3 mL). We express the density using the weak-link rule to determine the number of significant figures in the value calculated from experimental data. Then we compare the result with the known density of gold to decide whether the nugget could be gold.

**Solve**

$$
d = \frac{m}{V}
$$

$$
= \frac{30.01 \text{ g}}{(62.6 - 56.3) \text{ mL}} = \frac{30.01 \text{ g}}{6.3 \text{ mL}} = \frac{4.7635 \text{ g}}{\text{mL}}
$$

Because we know the volume of the nugget (6.3 mL) to two significant figures, we can know its density to only two significant figures. So we must round off the result to 4.8 g/mL. The density of gold is 19.3 g/mL, so the nugget is not pure gold.

**Think About It** The weak link in this exercise is the volume of the sample calculated from water displacement. Balances such as the bottom one in Figure 1.19 are capable of determining masses of objects weighing tens of grams to the nearest tenth of a milligram. Such mass measurements can have as many as six significant figures, which is many more than in estimated volumes of irregularly shaped samples.

**Practice Exercise** Express the result of the following calculation to the appropriate number of significant figures:

$$
\frac{0.391 \times 0.0821 \times (273 + 25)}{8.401}
$$

Note: None of the starting values are exact numbers.

*(Answers to Practice Exercises are in the back of the book.)*

---

**CONCEPT TEST**

Lap speed in stock car racing is calculated by dividing the distance around a track (Figure 1.21) by the time it takes a car to go around the track. If lap time is measured to the nearest hundredth of a second, which dimension—distance or time—is probably the weak link in the calculation? Explain your answer.

*(Answers to Concept Tests are in the back of the book.)*

**FIGURE 1.21** Dirt track for stock car racing.

**precision** the extent to which repeated measurements of the same variable agree.

**accuracy** agreement between an experimental value and the true value.

Measurements always have some degree of uncertainty, which limits the number of significant figures we can use to report any measurement. On the other hand, some values are known exactly, such as 12 eggs in a dozen and 60 seconds in a minute. Because these values are definitions, there is no uncertainty in them and they are not considered when determining significant figures in the answer to a mathematical calculation in which they appear.

---

**SAMPLE EXERCISE 1.4** **Distinguishing Exact from Uncertain Values** **LO8**

Which of the following data for the Washington Monument in Washington, DC, are exact numbers and which are not?
a. The monument is made of 36,491 white marble blocks.
b. The monument is 169 m tall.
c. There are 893 steps to the top.
d. The mass of the aluminum apex is 2.8 kg.
e. The area of the foundation is 1487 m$^2$.

**Collect and Organize** This exercise presents us with values based on an exact number, such as 12 eggs in a dozen, and those that are not exact numbers, such as the mass of an egg.

**Analyze** One way to distinguish exact from inexact values is to answer the question, "Which quantities can be counted?" Values that can be counted are exact.

**Solve** (a) The number of marble blocks and (c) the number of stairs are quantities we can count, so they are exact numbers. The other three quantities are based on measurements of (b) length, (d) mass, and (e) area and therefore are not exact.

**Think About It** The "you can count them" property of exact numbers works in this exercise and in others even when it takes a long time (the hairs on a dog) or if help is needed in visualizing the objects (borrow a microscope). Actually, these examples bring up a good point: Is there uncertainty in the counting process itself? Can you think of a famous example in American politics of an uncertain counting process?

⚙ **Practice Exercise** Which of the following statistics associated with the Golden Gate Bridge in San Francisco, CA, are exact numbers and which have some uncertainty?
a. The roadway is six lanes wide.
b. The width of the bridge is 27.4 m.
c. The bridge has a mass of $3.808 \times 10^8$ kg.
d. The length of the bridge is 2740 m.
e. The toll for a car traveling south increased to $6.00 on September 1, 2008.

*(Answers to Practice Exercises are in the back of the book.)*

---

**SAMPLE EXERCISE 1.5** **Ranking Uncertainty in Numbers** **LO8**

Rank these inexact values in order of decreasing certainty: $6.45 \times 10^6$, 5.5010, 0.0052.

**Collect and Organize** We have three numbers of very different magnitude and we need to rank them.

**Analyze** The question has to do with numerical uncertainty, not magnitude. Converting the numbers to exponential notation will lead us to the answer. (See Appendix 1 for a description of exponential notation.)

**Solve** The values $6.45 \times 10^6$, 5.5010, 0.0052 when written in exponential notation are $6.45 \times 10^6$, $5.5010 \times 10^0$, and $5.2 \times 10^{-3}$, respectively. Certainty is expressed by the number of significant figures in the numbers irrespective of the exponent. On this basis, the order of decreasing certainty is $5.5010 \times 10^0 > 6.45 \times 10^6 > 5.2 \times 10^{-3}$.

**Think About It** An exponential number has two parts: a quantity whose number of significant figures expresses its degree of certainty, followed by a power of ten, which most often is an exact number.

☼ **Practice Exercise** Give the order of increasing certainty for the inverses of the following values, none of which are exact (the inverse of $x$ is $1/x$): $6.45 \times 10^6$, 5.5010, 0.0052.

*(Answers to Practice Exercises are in the back of the book.)* ■

## Precision and Accuracy

Two terms—precision and accuracy—are used to describe how well a measured quantity or a value calculated from a measured quantity is known. **Precision** indicates how repeatable a measurement is. Suppose we used the top balance in Figure 1.19 to determine the mass of a penny over and over again, and suppose the reading on the balance was always 2.53 g. These results tell us that our determinations of the mass of the penny were precisely 2.53 g. Said another way, the results were precise to the nearest 0.01 g.

Now suppose we use the bottom balance in Figure 1.19 to determine the mass of the same penny five more times and obtain these values:

(a)

| Measurement | Mass (g) |
|:-----------:|:--------:|
| 1 | 2.5270 |
| 2 | 2.5271 |
| 3 | 2.5272 |
| 4 | 2.5271 |
| 5 | 2.5271 |

Note that there is a small variability in the last decimal place. Such variability is common when using a balance that can report masses to the nearest 0.0001 g. These results are consistent with one another, so the balance is precise. Many factors including dust landing on the balance, vibration of the laboratory bench, and transfer of moisture from our fingers as we handle the penny, could all produce a change in mass of 0.0001 g or more.

One way to express the precision of these results is to report the range between the highest and lowest values. In this example it is 2.5270 to 2.5272. Range can also be expressed using the average value (2.5271 g) and the range above (0.0001) and below (0.0001) the average that includes all the observed results. A convenient way to express the observed range in this case is 2.5271 ± 0.0001, where the symbol ± means "plus or minus" the value that follows it.

Precision indicates agreement among repeated measurements, whereas **accuracy** reflects how close the measured value is to the true value. Suppose the true mass of our penny is 2.5267 g. That means the average result obtained with the bottom balance in Figure 1.19—2.5271 g—is 0.0004 g too high. Thus, the measurements made on this balance may be precise to within 0.0001 g, but they are not accurate to within 0.0001 g of the true value. Figure 1.22 shows a way to visualize the difference between accuracy and precision.

The accuracy of a balance can be checked by measuring the mass of objects of known mass. A thermometer can be calibrated by measuring the temperature at which a substance changes state. Ice melts to liquid water at 0.0°C,

(b)

(c)

**FIGURE 1.22** (a) The three dart throws meant to hit the center of the target are both accurate and precise. (b) The three throws meant to hit the center of the target are precise but not accurate. (c) This set of throws is neither precise nor accurate.

**conversion factor** a fraction in which the numerator is equivalent to the denominator but is expressed in different units, making the value of the fraction one.

and an accurate thermometer dipped into a mixture of ice and liquid water reads 0.0°C. At sea level, liquid water boils and changes to water vapor at 100.0°C, which means an accurate thermometer dipped into the boiling water reads that temperature. A measurement that is validated by calibration with an accepted standard material is considered accurate.

## 1.9 Unit Conversions and Dimensional Analysis

Throughout this book we often need to convert quantities from one unit to another. Sometimes it is a matter of using the appropriate prefix, such as expressing the distance between Toronto and Montreal in kilometers instead of meters. Other times we need to convert a value from one unit system to another, such as converting a gasoline price per liter to an equivalent price per U.S. gallon.

To do these calculations and many others, we use an approach sometimes called the *unit factor method* but more often called *dimensional analysis*. The approach makes use of **conversion factors**, which are fractions in which the numerators and denominators have different units but represent equivalent quantities. This equivalency means that multiplying a quantity by a conversion factor is like multiplying by 1: The intrinsic value that the quantity represents does not change; it is simply expressed in different units.

The key to using conversion factors is to set them up correctly with the appropriate units in the numerator and denominator. For example, the odometer of a Canadian car traveling between the airports serving Vancouver, BC, and Seattle, WA, records the distance as 241 km. What is that distance in miles? To find out we need to translate the following equivalency from Table 1.3:

$$1 \text{ km} = 0.6214 \text{ mi}$$

into this conversion factor:

$$\frac{0.6214 \text{ mi}}{1 \text{ km}}$$

When we multiply the distance in kilometers by this conversion factor, the km in the initial value and in the denominator of the conversion factor cancel each other out, and the answer has the desired units of miles:

$$241 \text{ \sout{km}} \times \frac{0.6214 \text{ mi}}{1 \text{ \sout{km}}} = 1.50 \times 10^2 \text{ mi}$$

The general form of the equation for converting a value from one unit to another is

$$\text{\sout{initial units}} \times \frac{\text{desired units}}{\text{\sout{initial units}}} = \text{desired units}$$

▶▮ **CHEMTOUR** Dimensional Analysis

Most conversion factors, such as the one for converting kilometers to miles, also have a "1" in the numerator or denominator. All of these ones are *exactly* one; there is no uncertainty in their value.

---

**SAMPLE EXERCISE 1.6**  **Converting One Unit to Another**  **LO9**

The Star of Africa (Figure 1.23) is one of the world's largest diamonds, having a mass of 106.04 g. What is its mass in milligrams and in kilograms?

**Collect and Organize** We have a mass expressed in grams (g). From Table 1.1, we know that the prefixes *milli-* and *kilo-* represent units that are 1/1000 and 1000 times the value of the given unit "grams."

**Analyze** We seek answers in units of milligrams (mg) and kilograms (kg). We need two conversion factors: one for converting grams to milligrams and one for converting grams to kilograms. Table 1.1 tells us that 1 mg = 0.001 g and 1 kg = 1000 g. Therefore our possibilities for conversion factors are

$$\frac{1 \text{ mg}}{0.001 \text{ g}} \qquad \frac{0.001 \text{ g}}{1 \text{ mg}} \qquad \frac{1 \text{ kg}}{1000 \text{ g}} \qquad \frac{1000 \text{ g}}{1 \text{ kg}}$$

Because we want factors showing our desired units in the numerator, we use the first and third possibilities.

**Solve** The patterns of unit conversion are

$$\boxed{g} \rightarrow \boxed{mg} \text{ and } \boxed{g} \rightarrow \boxed{kg}$$

We multiply the given mass by each conversion factor to obtain the desired results:

$$106.04 \text{ g} \times \frac{1 \text{ mg}}{0.001 \text{ g}} = 106{,}040 \text{ mg} \qquad 106.04 \text{ g} \times \frac{1 \text{ kg}}{1000 \text{ g}} = 0.10604 \text{ kg}$$

These answers are best expressed in scientific notation: $1.0604 \times 10^5$ mg and $1.0604 \times 10^{-1}$ kg, respectively, to convey how many significant figures they contain.

**Think About It** The results make sense because there should be many more of the smaller units (106,040 mg versus 106.04 g) in a given mass and many fewer of the larger units (0.10604 kg versus 106.04 g).

⚙ **Practice Exercise** The Eiffel Tower in Paris is 324 m tall, including a 24-m television antenna that was not there when the tower was built in 1889. What is the height of the Eiffel Tower in kilometers and in centimeters?

*(Answers to Practice Exercises are in the back of the book.)*

**FIGURE 1.23** The Star of Africa is one of the largest cut diamonds in the world. It is mounted in the handle of the Royal Sceptre in the British Crown Jewels displayed in the Tower of London.

---

| SAMPLE EXERCISE 1.7 | **Converting U.S. Units to SI Units** | LO9 |

The summit of Mount Washington in New Hampshire is famous for its bad weather: on average it experiences hurricane-force winds 110 days per year (Figure 1.24). The summit is 6288 feet above sea level. What is this altitude in meters?

**Collect and Organize** We need to convert the height from feet to meters. The fact that the summit is windy is irrelevant to the altitude calculation. Table 1.3 contains conversion factors for converting lengths from U.S. units to SI units.

**Analyze** The pattern of unit conversion is

$$\boxed{ft} \rightarrow \boxed{m}$$

The conversion factor from Table 1.3 with the initial unit in the denominator is

$$\frac{0.3048 \text{ m}}{1 \text{ ft}}$$

Because 1 meter is larger than 1 foot, we expect our final answer to be smaller than the number of feet. Because 1 foot is about one-third of a meter, we can estimate that our answer should be about one-third of 6000, or 2000 m.

**Solve** Multiplying the summit height in feet by this conversion factor, we get

$$6288 \text{ ft} \times \frac{0.3048 \text{ m}}{1 \text{ ft}} = 1917 \text{ m}$$

**Think About It** Our answer is reasonable based on our estimate. It is rounded to four figures because the initial value has four figures.

**FIGURE 1.24** A warning to Mount Washington hikers from the U.S. Forest Service.

⚙ **Practice Exercise** Perhaps the most famous horse race in U.S. history was the 1938 match race between War Admiral and Seabiscuit, both of which stood about 15 hands high. The length unit *hand*, used to measure horses, is exactly 4 inches. How tall were both horses in centimeters?

*(Answers to Practice Exercises are in the back of the book.)*

■ ● ● ● ● ● ● ● ● ● ● ●

---

**SAMPLE EXERCISE 1.8** **Multi-Step Conversions of U.S. Customary to SI Units** LO9

On April 12, 1934, a wind gust of 231 miles per hour was recorded at the summit of Mount Washington in New Hampshire. What is this wind speed in meters per second?

**Collect and Organize** Speed is expressed as distance divided by time. In this conversion, both the given distance unit and the given time unit must change. Table 1.3 contains the relation 1 km = 0.6214 mi. The desired distance unit is meters, so we also need to convert kilometers to meters (1 km = 1000 m, Table 1.1). Finally, we need to convert hours to seconds.

**Analyze** The unit conversion patterns are

$$\boxed{\text{mi}} \rightarrow \boxed{\text{km}} \rightarrow \boxed{\text{m}}$$

and

$$\boxed{\text{hr}} \rightarrow \boxed{\text{min}} \rightarrow \boxed{\text{s}}$$

The conversion factors we need are

$$\frac{1 \text{ km}}{0.6214 \text{ mi}} \qquad \frac{1000 \text{ m}}{1 \text{ km}} \qquad \frac{1 \text{ hr}}{60 \text{ min}} \qquad \frac{1 \text{ min}}{60 \text{ s}}$$

The first conversion factor for distance is written with the given unit in the denominator; in the second conversion factor, the unit in the denominator is the result of the first conversion, and the desired unit is in the numerator. Hour is the unit in the denominator of the initial value. This means that for hour to cancel it must appear in the numerator of a conversion factor for time. To estimate our answer, note that 1000 m in the numerator is divided by about $\frac{2}{3} \times 60 \times 60$ or about 2400, so we expect our answer in m/s to be a little less than half of the value of the speed in mi/hr, or about 100 m/s.

**Solve**

$$231 \, \frac{\text{mi}}{\text{hr}} \times \frac{1 \text{ km}}{0.6214 \text{ mi}} \times \frac{1000 \text{ m}}{1 \text{ km}} \times \frac{1 \text{ hr}}{60 \text{ min}} \times \frac{1 \text{ min}}{60 \text{ s}} = 103 \, \frac{\text{m}}{\text{s}}$$

**Think About It** The answer seems reasonable based on our estimate. We rounded the result to three figures because that is the number of figures in 231 mi/hr.

⚙ **Practice Exercise** If light travels exactly 1 meter in 1/299,792,458 of a second, how many kilometers does it travel in one year?

*(Answers to Practice Exercises are in the back of the book.)*

■ ● ● ● ● ● ● ● ● ● ● ●

---

## 1.10 Testing a Theory: The Big Bang Revisited

Let's return to our discussion of experiments testing the validity of the Big Bang theory. If all the matter in the universe started as a dense cloud of very hot gas that formed following an enormous release of energy, and if the universe has been expanding ever since, then the universe must have been cooling throughout time because gases cool as they expand. This is the principle behind the operation of refrigerators and air conditioners.

## Temperature Scales

If the universe is still expanding and cooling, some warmth must be left over from the Big Bang. By the 1950s, some scientists predicted how much leftover warmth there should be: enough to give interstellar space a temperature of 2.73 K, where K indicates a temperature value on the Kelvin scale.

Several temperature scales are in use today. In the United States the Fahrenheit scale is still the most popular. In the rest of the world and in science, temperatures are most often expressed in degrees Celsius or on the Kelvin scale. The Fahrenheit and Celsius scales differ in two ways, as shown in Figure 1.25. First, their zero points are different. Zero degrees Celsius (0°C) is the temperature at which water freezes under normal conditions, but that temperature is 32 degrees on the Fahrenheit scale (32°F). The other difference is the size of the temperature change corresponding to 1 degree. The difference between the freezing and boiling points of water is 212 − 32 = 180 degrees on the Fahrenheit scale but only 100 − 0 = 100 degrees on the Celsius scale. This difference means that a Fahrenheit degree is 100/180, or 5/9, as large as a Celsius degree.

▶❙❙ CHEMTOUR Temperature Conversion

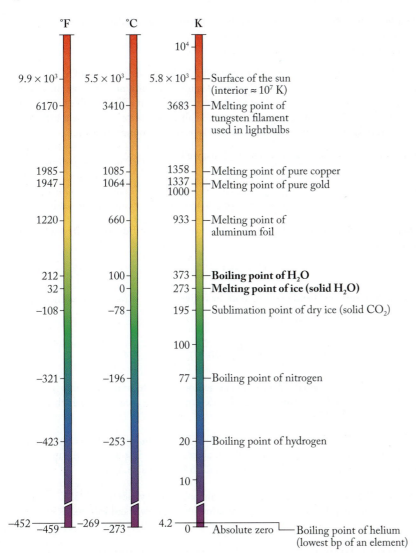

**FIGURE 1.25** Three temperature scales are commonly used today, although the Fahrenheit scale is rarely used in scientific work. The freezing and boiling points of liquid water are the defining temperatures for the scales and degree sizes.

**kelvin (K)** the SI unit of temperature.

**absolute zero (0 K)** the zero point on the Kelvin temperature scale; theoretically the lowest temperature possible.

To convert temperatures from Fahrenheit into Celsius, we need to account for the differences in zero point and in degree size. The following equation does both:

$$°C = \frac{5}{9}(°F - 32) \tag{1.2}$$

The SI unit of temperature (Table 1.2) is the **kelvin (K)**. The zero point on the Kelvin scale is not related to the freezing of a particular substance; rather, it is the coldest temperature—called **absolute zero (0 K)**—that can theoretically exist. It is equivalent to −273.15°C. No one has ever been able to chill matter to absolute zero, but scientists have come very close, cooling samples to less than $10^{-9}$ K.

The zero point on the Kelvin scale differs from that on the Celsius scale, but the size of 1 degree is the same on the two scales. For this reason, the conversion from a Celsius temperature to a Kelvin temperature is simply a matter of adding 273.15 to the Celsius value:

$$K = °C + 273.15 \tag{1.3}$$

### SAMPLE EXERCISE 1.9 Temperature Conversions LO7–LO9

The temperature of interstellar space is 2.73 K. What is this temperature on the Celsius scale and on the Fahrenheit scale?

**Collect and Organize** We are given the temperature of interstellar space and want to convert it to other scales. Equation 1.2 relates Celsius and Fahrenheit temperatures; Equation 1.3 relates Kelvin and Celsius temperatures.

**Analyze** We can first use Equation 1.3 to convert 2.73 K into an equivalent Celsius temperature and then use Equation 1.2 to calculate an equivalent Fahrenheit temperature. The value in degrees Celsius should be close to absolute zero (about −273°C). Because 1°F is about half the size of 1°C, the temperature on the Fahrenheit scale should be a little less than twice the value of the temperature on the Celsius scale (around −500°F).

**Solve** To convert from kelvins to degrees Celsius, we have

$$K = °C + 273.15$$
$$°C = K - 273.15$$
$$= 2.73 - 273.15 = -270.42°C$$

To convert degrees Celsius to degrees Fahrenheit, we rearrange Equation 1.2 to solve for degrees Fahrenheit:

$$°C = \frac{5}{9}(°F - 32)$$

Multiplying both sides by 9 and dividing both by 5 gives us

$$\frac{9}{5}°C = °F - 32$$

$$°F = \frac{9}{5}°C + 32 = \frac{9}{5}(-270.42) + 32 = -454.76°F$$

The value 32°F is considered a definition and so does not determine the number of significant figures in the answer. The number that determines the accuracy to which we can know this value is −270.42°C.

**Think About It** The calculated Celsius value of −270.42°C makes sense because it represents a temperature only a few degrees above absolute zero, just as we estimated. The Fahrenheit value is within 10% of our estimate and so is reasonable, too.

**Practice Exercise** The temperature of the moon's surface varies from −233°C at night to 123°C during the day. What are these temperatures on the Kelvin and Fahrenheit scales?

*(Answers to Practice Exercises are in the back of the book.)*

## An Echo of the Big Bang

In the early 1960s, Princeton University physicist Robert Dicke (Figure 1.26) had suggested the presence of residual energy left over from the Big Bang. He was anxious to test this hypothesis. He proposed building an antenna that could detect microwave energy reaching Earth from outer space. Why did he pick microwaves? Even matter as cold as 2.73 K emits a "glow" (an energy signature), but not a glow you can see or feel, like the infrared rays emitted by any warm object (Figure 1.27). Instead, the glow from a 2.73 K object is in the form of microwave energy.

Dicke's microwave detector was never built because of events occurring a short distance from Princeton. By the early 1960s, the United States had launched *Echo* and *Telstar*, the first communication satellites. These satellites were reflective spheres designed to bounce microwave signals to receivers on Earth. An antenna designed to receive the microwave signals had been built at Bell Laboratories in Holmdel, NJ (Figure 1.28). Two Bell Labs scientists, Robert W. Wilson and Arno A. Penzias, were working to improve the antenna's reception when they encountered a problem. They found that no matter where they directed their antenna, it picked up a background microwave signal much like the hissing sound radios make when tuned between stations. They concluded that the signal was due to a flaw in the antenna or in one of the instruments connected to it. At one point they came up with another explanation: that the source of the background signal was a pair of pigeons roosting on the antenna and coating parts of it with their droppings. However, the problem persisted when the droppings were cleaned up. More testing discounted the flawed-instrument hypothesis but left unanswered the question of where the signal was coming from.

The nuisance signal picked up by the Wilson–Penzias antenna matched the microwave echo of the Big Bang that Dicke had predicted. When the scientists at Bell Labs learned of Dicke's prediction, they realized the significance of the strange background signal, and others did, too. Wilson and Penzias shared the Nobel Prize in Physics in 1978 for discovering the cosmic microwave background radiation of the universe. Dicke did not share in the prize, even though he had accurately predicted what Wilson and Penzias discovered by accident.

A major discovery in science usually leads to new questions. The discovery of cosmic microwaves reinforced the Big Bang theory and also raised questions

**FIGURE 1.26** Robert Dicke (1916–1997) predicted the existence of cosmic microwave background radiation. His prediction was confirmed by the serendipitous discovery of this radiation by Robert Wilson and Arno Penzias.

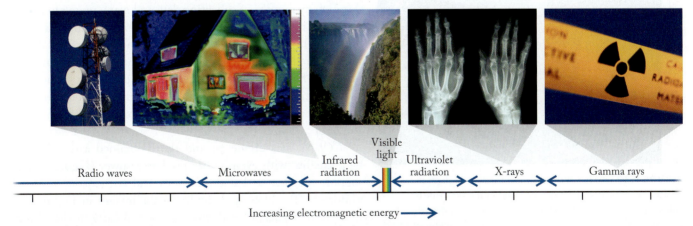

**FIGURE 1.27** The electromagnetic spectrum includes (but is not limited to) microwaves used for cell phone communications, infrared radiation emitted by warm objects (and thus used to evaluate how much heat is lost from buildings), visible light, X-rays used to image the interior of objects, and high-energy gamma rays.

**FIGURE 1.28** In 1965 Robert Wilson (left) and Arno Penzias discovered the microwave echo of the Big Bang while tuning this highly sensitive "horn" antenna in Holmdel, NJ.

about it. The Bell Labs antenna picked up the same microwave signal no matter where in the sky it was pointed. In other words, the cosmic microwaves appeared to be uniformly distributed throughout the universe. If the afterglow from the Big Bang really was uniform, how could galaxies have formed? Some heterogeneity had to arise in the expanding universe—some clustering of the matter in it—to allow galaxies to form.

Scientists doing work related to Dicke's proposed that, if such clustering had occurred, a record of it should exist as subtle differences in the cosmic background radiation. Unfortunately, a microwave antenna in New Jersey or any place on Earth cannot detect such slight variations in signals because too many other sources of microwaves interfere with the measurements. One way for scientists to determine if these heterogeneities exist would be to take readings from a radio antenna in space.

In late 1989, the United States launched such an antenna in the form of the *Cosmic Background Explorer* (*COBE*) satellite. After many months of collecting data and many more months of analyzing it, the results were released in 1992. The image in Figure 1.29(a) appeared on the front pages of newspapers and magazines around the world. This was a major news story because the predicted heterogeneity had been found, providing support for the Big Bang theory of the origin of the universe. At a news conference, the lead scientist on the *COBE* project called the map a "fossil of creation."

*COBE*'s measurements, and those obtained a decade later by a satellite with even higher resolving power (Figure 1.29b), support the theory that the universe did not expand and cool uniformly. The blobs and ripples in the images in Figure 1.29 indicate that galaxy "seed clusters" formed early in the history of the universe. Cosmologists believe these ripples are a record of the next stage after the Big Bang in the creation of matter. This stage is examined in Chapter 2, where we discuss how

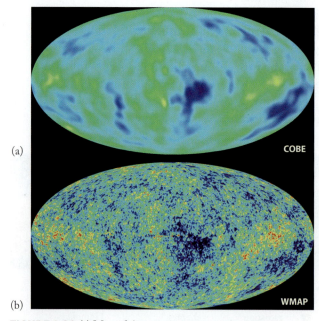

**FIGURE 1.29** (a) Map of the cosmic microwave background released in 1992. It is a 360-degree image of the sky made by collecting microwave signals for a year from the microwave telescopes of the *COBE* satellite. (b) Higher resolution image based on measurements made in 2002 by the *WMAP* satellite. Red regions are up to 200 μK warmer than the average interstellar temperature of 2.73 K, and blue regions are up to 200 μK colder than 2.73 K.

some elements may have formed just after the Big Bang and others continue to be formed by the nuclear reactions that fuel our sun and all the stars in the universe.

**CONCEPT TEST** ••••••••••••••••••••••••••••••••••••••••••••••••••••••••••••••••••••••••

When Lemaître first proposed how the universe may have formed following a cosmic release of energy, would it have been more appropriate to call his explanation a hypothesis or a theory? Why?

*(Answers to Concept Tests are in the back of the book.)*

••••••••••••••••••••••••••••••••••••••••••••••••••••••••••••••••••••••••

**SAMPLE EXERCISE 1.10  Integrating Concepts: Driving around Mars**

In early press conferences about the *Curiosity* rover on Mars (Figure 1.30), scientists and engineers expressed distances in either the English or the metric system and temperatures in K, °C, and °F in response to questions from journalists from different countries. After an early drive, *Curiosity* stopped 8.0 ft away from an interesting football-shaped rock. *Curiosity*'s wheels are 50 cm in diameter, and the rover moves at a speed of 200 m/sol, where 1 sol = 1 Martian day = 24.65 hr.

a. How many minutes would it take the rover to move to within 1.0 ft of the rock?

b. How many rotations of the wheels would be required to move that distance?

c. The rover landed in a valley about 15 miles away from the base of Mt. Sharp. How far is that in km?

d. From the base to the peak of Mt. Sharp is 5.5 km; from the base to the peak of Mt. Everest on Earth is 15,000 ft. Which mountain is taller?

e. One night *Curiosity* recorded a temperature reported as "−132°" on the Fahrenheit scale. If the daytime temperature was 263 K, what was the day/night temperature range in °C?

**Collect and Organize**  We are given a series of measurements and are asked questions that require us to convert units of time, distance, speed, and temperature.

**Analyze**  We can use conversion factors from the chapter and from the table on the inside back cover of the book to help us estimate the answers. For part (a), the rover has to move 7 feet. The speed of the rover is 200 m/sol, which is about 600 ft/24 hr. The distance moved is about 1/100 of 600 ft, so we estimate it would take about

1/100 of a day, or about 0.24 hr, which is about $\frac{1}{4}$ hr or 15 min. For (b), the wheels are 50 cm in diameter or about 20 in. The formula for the circumference of a circle is $\pi \times d \approx 3 \times 20 = 60$ in, or about 5 ft. We estimate it would take more than one turn of the wheels to travel to the rock. For part (c), a mile is about 1.6 km, so we estimate that Mt. Sharp is midway between 15 and 30 km away or about 23 km distant. For part (d), we can estimate that 5.5 km is a bit more than 3 miles and 15,000 ft is a bit less than 3 miles, so we estimate that the heights are similar but Mt. Sharp might be taller. Finally for part (e), −132°F is (−132 − 32)°F = −164°F below the freezing point of water. One degree Celsius is about the size of 2°F, so we estimate the temperature at night was about −84°C. The temperature in Kelvin during the day was 263 K or −10°C, so the range of temperatures on the surface was about [−10°C − (−84°C)], or about 74°C.

**Solve**  To work out the answers more exactly:

a.
$$(8.0 - 1.0) \, \text{ft} \times \frac{12 \, \text{in}}{1 \, \text{ft}} \times \frac{2.54 \, \text{cm}}{1 \, \text{in}} \times \frac{1 \, \text{m}}{100 \, \text{cm}} \times \frac{1 \, \text{sol}}{200 \, \text{m}}$$
$$\times \frac{24.65 \, \text{hr}}{1 \, \text{sol}} \times \frac{60 \, \text{min}}{1 \, \text{hr}} = 16 \, \text{min}$$

b. The diameter of the wheel is 50 cm or
$$50 \, \text{cm} \times \frac{1 \, \text{in}}{2.54 \, \text{cm}} \times \frac{1 \, \text{ft}}{12 \, \text{in}} = 1.64 \, \text{ft}$$

and its circumference is
$$\pi \times d = 3.1416 \times 1.64 \, \text{ft} = 5.15 \, \text{ft}$$

The wheel has 7.0 ft to travel, so it requires
$$7.0 \, \text{ft} \times \frac{1 \, \text{revolution}}{5.15 \, \text{ft}} = 1.4 \, \text{revolutions}$$

c.
$$15 \, \text{mi} \times \frac{1.6093 \, \text{km}}{1 \, \text{mi}} = 24.14 = 24 \, \text{km}$$

d. Mt. Sharp:
$$5.5 \, \text{km} \times \frac{1 \, \text{mi}}{1.6093 \, \text{km}} = 3.4 \, \text{mi}$$

Mt. Everest:
$$15{,}000 \, \text{ft} \times \frac{1 \, \text{mi}}{5280 \, \text{ft}} = 2.8 \, \text{mi}$$

Mt. Sharp is (3.4 − 2.8) = 0.6 miles higher than Mt. Everest.

e.
$$°C = \frac{5}{9}(-132°F - 32°F) = -91.1°C$$

The high temperature that day was 263 K:
$$°C = K - 273.15 = 263 - 273.15 = -10.15 = -10°C$$

(a)

(b)

**FIGURE 1.30** (a) NASA scientists and models of three Martian rovers in the Mars Yard at the Jet Propulsion Laboratory in Pasadena, CA. *Curiosity* (2012) is on the right; *Sojourner* (1997) in the front; *Spirit* (2004) on the left. (b) Tracks of the *Curiosity* rover (bluish circle on the right) on Mars as seen from the Mars Reconnaissance Orbiter.

The temperature differential on the surface that day was

$$-10°C - (-91.1°C) = 81°C$$

**Think About It** Although none of our estimates matched the calculated values exactly, each of them was in the right ball park and served as a useful accuracy check. Two of the starting values: the diameter (50 cm) of *Curiosity's* wheels and its speed (200 m/sol), ended with zeros and had no decimal points. In both cases we assumed the zeros were significant. This is a reasonable assumption given (1) the precision with which the components of *Curiosity* must have been fabricated and (2) the ability of the JPL engineers to control its speed. Moreover, the results of the calculation involving the rover's speed were eventually rounded to only two significant figures.

## SUMMARY

**Learning Outcome 1** **Matter** exists as pure **substances**, which may be either **elements** or **compounds**, and **mixtures**. Mixtures may be **homogeneous** (these mixtures are also called **solutions**) or **heterogeneous**. (Section 1.1)

**Learning Outcome 2** All matter consists of particles, and we use **chemical formulas** consisting of atomic symbols to express the elemental composition of a compound or a polyatomic form of an element. **Chemical equations** describe the proportions of the substances involved in a **chemical reaction**. Space-filling and ball-and-stick models are used to show molecular structure—the three-dimensional arrangement of atoms in a molecule. (Section 1.2)

**Learning Outcome 3** Matter is described and defined in terms of its **physical properties** and by the chemical reactions it undergoes. These physical properties may be used to separate mixtures into pure substances. (Sections 1.3 and 1.5)

**Learning Outcome 4** The COAST framework used in this book to solve problems has four components: **Collect and Organize** information and ideas; **Analyze** the information to determine how it can be used to obtain the answer; **Solve** the answer to the problem (often the math-intensive step); and **Think** about the answer. (Section 1.4)

**Learning Outcome 5** The differences in the states of matter and the transitions between them can be understood by viewing them at the atomic level. (Section 1.6)

**Learning Outcome 6** The **scientific method** is the approach we use to acquire knowledge through observation, testable hypotheses, and experimentation. (Sections 1.7 and 1.10)

**Learning Outcome 7** Accurate measurements expressed in units understandable to others are crucial in science. The limit to how accurate a measurement can be is expressed by the number of **significant figures** in the number. (Sections 1.8 and 1.10)

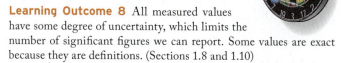

**Learning Outcome 8** All measured values have some degree of uncertainty, which limits the number of significant figures we can report. Some values are exact because they are definitions. (Sections 1.8 and 1.10)

**Learning Outcome 9** Dimensional analysis uses **conversion factors** (fractions in which the numerators and denominators have different units but represent the same quantity) to convert a value from one unit into another unit. (Sections 1.9 and 1.10)

## PROBLEM-SOLVING SUMMARY

| TYPE OF PROBLEM | CONCEPTS AND EQUATIONS | | SAMPLE EXERCISES |
|---|---|---|---|
| **Distinguishing physical properties from chemical properties** | The chemical properties of a substance can be determined only by reacting it with another substance; physical properties can be determined without altering the substance's composition. | | 1.1 |
| **Recognizing physical states of matter** | Particles in a solid are ordered; particles in a liquid are randomly arranged but close together; particles in a gas are separated by space and entirely fill the volume of their container. | | 1.2 |
| **Using significant figures in calculations** | Apply the weak-link rule: the number of significant figures allowed in a calculated quantity involving multiplication or division can be no greater than the number of significant figures in the least-certain value used to calculate it. | | 1.3 |
| **Calculating density from mass and volume** | $d = \dfrac{m}{V}$ | (1.1) | 1.3 |

| TYPE OF PROBLEM | CONCEPTS AND EQUATIONS | SAMPLE EXERCISES |
|---|---|---|
| **Distinguishing exact from uncertain values** | Quantities that can be counted are exact. Measured quantities or conversion factors that are not exact values are inherently uncertain. | 1.4, 1.5 |
| **Doing dimensional analysis and converting units** | Convert values from one set of units to another by multiplying by conversion factors set up so that the original units cancel. | 1.6–1.8 |
| **Converting temperatures** | $$°C = \frac{5}{9}(°F - 32) \qquad (1.2)$$ $$K = °C + 273.15 \qquad (1.3)$$ | 1.9 |

## VISUAL PROBLEMS

*(Answers to boldface end-of-chapter questions and problems are in the back of the book.)*

**1.1.** For each image in Figure P1.1, identify what class of pure substance is depicted (element or compound) and identify the physical state(s).

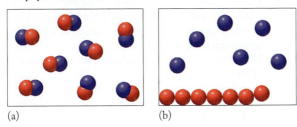

(a)      (b)

**FIGURE P1.1**

**1.2.** For each image in Figure P1.2, identify what class of matter is depicted (an element, a compound, a mixture of elements, or a mixture of compounds) and identify the physical state.

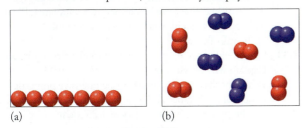

(a)      (b)

**FIGURE P1.2**

**1.3.** How would you describe the change depicted in Figure P1.3?

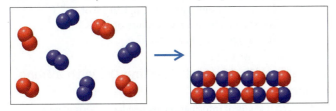

**FIGURE P1.3**

   a. A mixture of two gaseous elements undergoes a chemical reaction, forming a gaseous compound.
   b. A mixture of two gaseous elements undergoes a chemical reaction, forming a solid compound.
   c. A mixture of two gaseous elements undergoes deposition.
   d. A mixture of two gaseous elements condenses.

**1.4.** How would you describe the change depicted in Figure P1.4?

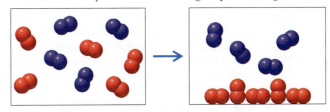

**FIGURE P1.4**

   a. A mixture of two gaseous elements is cooled to a temperature at which one of them condenses.
   b. A mixture of two gaseous compounds is heated to a temperature at which one of them decomposes.
   c. A mixture of two gaseous elements undergoes deposition.
   d. A mixture of two gaseous elements reacts together to form two compounds, one of which is a liquid.

**1.5.** A space-filling model of methanol is shown in Figure P1.5. What is the chemical formula of methanol?

**FIGURE P1.5**

**1.6.** A ball-and-stick model of acetone is shown in Figure P1.6. What is the chemical formula of acetone?

**FIGURE P1.6**

## QUESTIONS AND PROBLEMS ...........................■

### Matter

#### CONCEPT REVIEW

**1.7.** Two students get into an argument over whether the sun is an example of matter or energy. Which point of view is correct? Why?

**1.8.** List three differences and three similarities between a compound and an element.

**1.9.** List one chemical and four physical properties of gold.

**1.10.** Describe three physical properties that gold and silver have in common, and three physical properties that distinguish them.

**1.11.** How might you use filtration to separate a mixture of salt and sand?

**1.12.** How can distillation be used to desalinate seawater?

**1.13.** Which of the following processes is a chemical reaction? (a) distillation; (b) combustion; (c) filtration; (d) condensation

**1.14.** Gasohol is a fuel that contains ethanol dissolved in gasoline. Is gasohol a heterogeneous mixture or a homogeneous one?

**1.15.** Which of the following foods is a heterogeneous mixture? (a) solid butter; (b) a Snickers bar; (c) grape juice; (d) an uncooked hamburger

**1.16.** Which of the following foods are homogeneous mixtures? (a) freshly brewed coffee; (b) vinegar; (c) a slice of white bread; (d) a slice of ham

**1.17.** Which of the following foods are heterogeneous mixtures? (a) apple juice; (b) cooking oil; (c) solid butter; (d) orange juice; (e) tomato juice

**1.18.** Which of the following is a homogeneous mixture? (a) a wedding ring; (b) sweat; (c) Nile River water; (d) human blood; (e) compressed air in a scuba tank.

**1.19.** Give three properties that enable a person to distinguish between table sugar, water, and oxygen.

**1.20.** Give three properties that enable a person to distinguish between table salt, sand, and copper.

**1.21.** Indicate whether each of the following properties is a physical or a chemical property of sodium (Na):
  a. Its density is greater than that of kerosene and less than that of water.
  b. It has a lower melting point than most metals.
  c. It is an excellent conductor of heat and electricity.
  d. It is soft and can be easily cut with a knife.
  e. Freshly cut sodium is shiny, but it rapidly tarnishes in contact with air.
  f. It reacts very vigorously with water to form hydrogen gas ($H_2$) and sodium hydroxide (NaOH).

**1.22.** Indicate whether each of the following is a physical or chemical property of hydrogen gas ($H_2$):
  a. At room temperature, its density is less than that of any other gas.
  b. It reacts vigorously with oxygen ($O_2$) to form water.
  c. Liquefied $H_2$ boils at a very low temperature (−253°C).
  d. $H_2$ gas does not conduct electricity.

**1.23.** Which of the following is not a pure substance? (a) air; (b) nitrogen gas; (c) oxygen gas; (d) argon gas; (e) table salt (sodium chloride)

**1.24.** Which of the following is a pure substance? (a) mineral water; (b) blood; (c) brass (an alloy of copper and zinc); (d) sucrose (table sugar); (e) beer

**1.25.** Which of the following is an element? (a) $Cl_2$; (b) $H_2O$; (c) HCl; (d) NaCl

**1.26.** Which of the following is not an element? (a) $I_2$; (b) Ce; (c) ClF; (d) $S_8$

**1.27.** Which of the following is a homogeneous mixture? (a) filtered water; (b) chicken noodle soup; (c) clouds; (d) trail mix snack; (e) fruit salad

**1.28.** Which of the following is a heterogeneous mixture? (a) air; (b) sugar dissolved in water; (c) muddy river water; (d) brass; (e) table salt (sodium chloride)

**1.29.** Which of the following can be separated by filtration? (a) sugar dissolved in coffee; (b) sand and water; (c) gasoline; (d) alcohol dissolved in water; (e) damp air

*\*1.30.* Enzymes are proteins. Proteins are constituents of egg whites. Assume we have a sample of an enzyme dissolved in water. Would filtration or distillation be a suitable way of separating the enzyme from the water?

**1.31.** Which of the following is an example of a chemical property of formaldehyde ($CH_2O$)?
  a. It has a characteristic, acrid smell.
  b. It is soluble in water.
  c. It burns in air.
  d. It is a gas at room temperature.
  e. It is colorless.

**1.32.** Which of the following is an example of a physical property of silver?
  a. It tarnishes over time.
  b. Tarnished silver can be cleaned to a shiny metallic finish.
  c. It reacts with chlorine to make a white solid.
  d. It sinks in water.

**1.33.** Can an extensive property be used to identify a substance? Explain why or why not.

**1.34.** Which of these statements describe intensive properties of a sample of a substance and which describe extensive properties?
  a. The substance has a density of 1.25 g/cm$^3$.
  b. The substance conducts electricity.
  c. The substance has a water solubility of 0.379 g in 10 mL of water.
  d. The material sublimes at a temperature of −47°C.
  e. The solid melts at a temperature of 157°C.
  f. The sample has a mass of 6.77 g.

### The Scientific Method: Starting Off with a Bang

#### CONCEPT REVIEW

**1.35.** What kinds of information are needed to formulate a hypothesis?

**1.36.** How does a hypothesis become a theory?

**1.37.** Is it possible to disprove a scientific hypothesis?

**1.38.** Why is the theory that matter consists of atoms universally accepted?

**1.39.** How do people use the word *theory* in normal conversation?

**1.40.** Can a theory be proven?

## Making Measurements and Expressing the Results; Unit Conversions and Dimensional Analysis

### CONCEPT REVIEW

**1.41.** Describe in general terms how the SI and U.S. customary systems of units differ.

**1.42.** Suggest two reasons why SI units are not more widely used in the United States.

### PROBLEMS

NOTE: The physical properties of the elements are in Appendix 3.

**1.43.** Which of these uncertain values has the smallest number of significant figures? (a) 545; (b) $6.4 \times 10^{-3}$; (c) 6.50; (d) $1.346 \times 10^2$

**1.44.** Which of these uncertain values has the largest number of significant figures? (a) 545; (b) $6.4 \times 10^{-3}$; (c) 6.50; (d) $1.346 \times 10^2$

**1.45.** Which of these uncertain values has the smallest number of significant figures? (a) 1/545; (b) $1/6.4 \times 10^{-3}$; (c) 1/6.50; (d) $1/1.346 \times 10^2$

**1.46.** Which of these uncertain values has the largest number of significant figures? (a) 1/545; (b) $1/6.4 \times 10^{-3}$; (c) 1/6.50; (d) $1/1.346 \times 10^2$

**1.47.** Which of these uncertain values have four significant figures? (a) 0.0592; (b) 0.08206; (c) 8.314; (d) 5420; (e) $5.4 \times 10^3$

**1.48.** Which of these uncertain values have only three significant figures? (a) 7.02; (b) 6.452; (c) $6.02 \times 10^{23}$; (d) 302; (e) 12.77

**1.49.** Perform each of the following calculations and express the answer with the correct number of significant figures (only the highlighted values are exact):
a. $0.6274 \times 1.00 \times 10^3/[2.205 \times (2.54)^3] =$
b. $6 \times 10^{-18} \times (1.00 \times 10^3) \times 12 =$
c. $(4.00 \times 58.69)/(6.02 \times 10^{23} \times 6.84) =$
d. $[(26.0 \times 60.0)/43.53]/(1.000 \times 10^4) =$

**1.50.** Perform each of the following calculations, and express the answer with the correct number of significant figures (only the highlighted values are exact):
a. $[(12 \times 60.0) + 55.3]/(5.000 \times 10^3) =$
b. $(2.00 \times 183.9)/[(6.02 \times 10^{23}) \times (1.61 \times 10^{-8})^3] =$
c. $0.8161/[2.205 \times (2.54)^3] =$
d. $(9.00 \times 60.0) + (50.0 \times 60.0) + (3.00 \times 10^1) =$

**1.51.** **Boston Marathon** To qualify to run in the 2009 Boston Marathon, a distance of 26.2 miles, an 18-year-old woman had to have completed another marathon in 3 hours and 40 minutes or less. To qualify, what must a woman's average speed have been (a) in miles per hour and (b) in meters per second?

**1.52.** **DVDs** If a DVD is 11.95 cm in diameter and spins at a rate of 10,000 revolutions per minute, how fast does it spin in miles per hour?

**1.53.** An imperial gallon is 4.546 liters. How many U.S. gallons are in an imperial gallon?

**1.54.** **Olympic Mile** An Olympic "mile" is actually 1500 m. What percentage is an Olympic mile of a U.S. mile (5280 ft)?

**1.55.** How many grams are in 1.65 lb (1 lb is 453.6 g)?

**1.56.** How many pounds are in 765.4 g?

**1.57.** How many milliliters are in 2.44 gal?

**1.58.** How many gallons are in 108 mL?

**1.59.** Peter is 5 ft, 11 in tall and Paul is 176 cm tall. Who is taller?

**1.60.** We measured the depth of the water in a swimming pool. At end A it is 1.1 m and at end B it is 72 in. Which is the deep end of the pool?

**\*1.61.** If a wheelchair-marathon racer moving at 13.1 miles per hour expends energy at a rate of 665 Calories per hour, how much energy in Calories would be required to complete a marathon race (26.2 miles) at this pace?

**1.62.** An American sport-utility vehicle has an average mileage rating of 18 miles per gallon. How many gallons of gasoline are needed for a 389-mile trip?

**1.63.** A single strand of natural silk may be as long as $4.0 \times 10^3$ m. Convert this length into miles.

**\*1.64.** **Automotive Engineering** The original (1955) Ford Thunderbird (Figure P1.64, left) was powered by a 292-cubic-inch V-8 engine. The 2005 Thunderbird (Figure P1.64, right) was powered by a 3.9-liter V-8 engine. Which engine was bigger?

**FIGURE P1.64**

**1.65.** What is the mass of a magnesium block that measures 2.5 cm × 3.5 cm × 1.5 cm?

**1.66.** What is the mass of an osmium block that measures 6.5 cm × 9.0 cm × 3.25 cm? Do you think you could lift it with one hand?

**1.67.** What volume of gold would be equal in mass to a piece of copper with a volume of 125 cm$^3$?

**\*1.68.** A small hot-air balloon is filled with $1.00 \times 10^6$ L of air ($d = 1.20$ g/L). As the air in the balloon is heated, it expands to $1.09 \times 10^6$ L. What is the density of the heated air in the balloon?

**1.69.** What is the volume of 1.00 kg of mercury?

**1.70.** A student wonders whether a piece of jewelry is made of pure silver. She determines that its mass is 3.17 g. Then she drops it into a 10 mL graduated cylinder partially filled with water and determines that its volume is 0.3 mL. Could the jewelry be made of pure silver?

*1.71. **Utility Boats for the Navy** A plastic material called HDPE or high-density polyethylene was once evaluated for use in impact-resistant hulls of small utility boats for the Navy. A cube of this material measures $1.20 \times 10^{-2}$ m on a side and has a mass of $1.70 \times 10^{-3}$ kg. Seawater at the surface of the ocean has a density of 1.03 g/cm$^3$. Will this cube float on water?

1.72. **The Sun** The Sun is a sphere with an estimated mass of $2 \times 10^{30}$ kg. If the radius of the sun is $7.0 \times 10^5$ km, what is the average density of the sun in units of grams per cubic centimeter? The volume of a sphere is $\frac{4}{3}\pi r^3$.

1.73. Diamonds are measured in carats, where 1 carat = 0.200 g. The density of diamond is 3.51 g/cm$^3$. What is the volume of a 5.0-carat diamond?

*1.74. If the concentration of mercury in the water of a polluted lake is 0.33 μg (micrograms) per liter of water, what is the total mass of mercury in the lake, in kilograms, if the lake has a surface area of 10.0 km$^2$ and an average depth of 15 m?

*1.75. The widths of copper lines in printed circuit boards must be close to a specified value. Three manufacturers were asked to prepare circuit boards with copper lines that are 0.500 μm (micrometers) wide (1 μm = $1 \times 10^{-6}$ m). Each manufacturer's quality control department reported the following line widths on five sample circuit boards (given in micrometers):

| Manufacturer 1 | Manufacturer 2 | Manufacturer 3 |
|---|---|---|
| 0.512 | 0.514 | 0.500 |
| 0.508 | 0.513 | 0.501 |
| 0.516 | 0.514 | 0.502 |
| 0.504 | 0.514 | 0.502 |
| 0.513 | 0.512 | 0.501 |

a. What is the range of the data provided by each manufacturer?
b. Can any of the manufacturers justifiably advertise that they produce circuit boards with "high precision"?
c. Is there a data set for which this claim is misleading?

*1.76. **Patient Data** Measurements of a patient's temperature are routinely done several times a day in hospitals. Digital thermometers are used, and it is important to evaluate new thermometers and select the best ones. The accuracy of these thermometers is checked by immersing them in liquids of known temperature. Such liquids include an ice–water mixture at 0.0°C and boiling water at 100.0°C at exactly 1 atmosphere pressure (boiling point varies with atmospheric pressure). Suppose the data shown in the following table were obtained on three available thermometers and you were asked to select the "best" one of the three.

| Thermometer | Measured Temperature of Ice Water, °C | Measured Temperature of Boiling Water, °C |
|---|---|---|
| A | −0.8 | 99.9 |
| B | 0.3 | 99.8 |
| C | 0.3 | 100.3 |

Explain your choice of the "best" thermometer for use in the hospital.

## Testing a Theory: The Big Bang Revisited

### CONCEPT REVIEW

1.77. Can a temperature in °C ever have the same value in °F?
1.78. What is meant by an *absolute* temperature scale?

### PROBLEMS

1.79. Liquid helium boils at 4.2 K. What is the boiling point of helium in °C?
1.80. Liquid hydrogen boils at −253°C. What is the boiling point of $H_2$ on the Kelvin scale?

1.81. A person has a fever of 102.5°F. What is this temperature in °C?
1.82. Physiological temperature, or body temperature, is considered to be 37.0°C. What is this temperature in °F?

1.83. **Record Low** The lowest temperature measured on Earth is −128.6°F, recorded at Vostok, Antarctica, in July 1983. What is this temperature on the Celsius and Kelvin scales?
1.84. **Record High** The highest temperature ever recorded in the United States is 134°F at Greenland Ranch, Death Valley, CA, on July 13, 1913. What is this temperature on the Celsius and Kelvin scales?

1.85. The coolant in an automobile radiator freezes at −39°C and boils at 110°C. What are these temperatures on the Fahrenheit scale?
1.86. Silver and gold melt at 962°C and 1064°C, respectively. Convert these two temperatures to the Kelvin scale.

1.87. **Critical Temperature** The discovery of new "high temperature" superconducting materials in the mid-1980s spurred a race to prepare the material with the highest superconducting temperature. The *critical temperatures* ($T_c$)—the temperatures at which the material becomes superconducting—of $YBa_2Cu_3O_7$, $Nb_3Ge$, and $HgBa_2CaCu_2O_6$ are 93.0 K, −250.0°C, and −231.1°F, respectively. Convert these temperatures into a single temperature scale, and determine which superconductor has the highest $T_c$ value.
1.88. As air is cooled, which gas condenses first: $N_2$, $O_2$, or Ar?

## Additional Problems

*1.89. **Agricultural Runoff** A farmer applies 1500 kg of a fertilizer that contains 10% nitrogen to his fields each year. Fifteen percent of the fertilizer washes into a stream that runs through the farm. If the stream flows at an average rate of 1.4 cubic meters per minute, what is the additional concentration of nitrogen (expressed in milligrams of nitrogen per liter) in the stream water due to the farmer's yearly application of fertilizer?
1.90. Your laboratory instructor has given you two shiny, light gray metal cylinders. Your assignment is to determine which one is made of aluminum ($d = 2.699$ g/mL) and which one is made of titanium ($d = 4.54$ g/mL). The mass of each cylinder was determined on a balance to five significant figures. The volume was determined by immersing the cylinders in a graduated cylinder as shown in Figure P1.90. The initial

volume of water was 25.0 mL in each graduated cylinder. The following data were collected:

|  | Mass (g) | Height (cm) | Diameter (cm) |
| --- | --- | --- | --- |
| **Cylinder A** | 15.560 | 5.1 | 1.2 |
| **Cylinder B** | 35.536 | 5.9 | 1.3 |

a. Calculate the volume of each cylinder using the dimensions of the cylinder only.
b. Calculate the volume from the water displacement method.
c. Which volume measurement allows for the greater number of significant figures in the calculated densities?
d. Express the density of each cylinder to the appropriate number of significant figures.

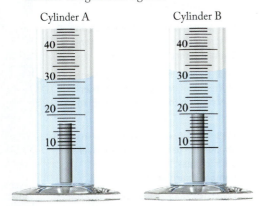

**FIGURE P1.90**

*1.91. Sodium chloride (NaCl) contains 1.54 g Cl for every 1.00 g Na. Which of the following mixtures would react to produce sodium chloride with no Na or Cl left over?
 a. 11.0 g Na and 17.0 g Cl   c. 6.5 g Na and 12.0 g Cl
 b. 6.5 g Na and 10.0 g Cl   d. 6.5 g Na and 8.0 g Cl

*1.92. **Toothpaste Chemistry** Most of the toothpaste sold in the United States contains about 1.00 mg of fluoride per gram of toothpaste. The fluoride compound that is most often used in toothpaste is sodium fluoride, NaF, which is 45% fluoride by mass. How many milligrams of NaF are in a typical 8.2-ounce tube of toothpaste?

*1.93. **Test for HIV** Tests called ELISAs (enzyme-linked immunosorbent assays) detect and quantify substances such as HIV antibodies in biological samples. A "sandwich" assay traps the HIV antibody between two other molecules. The trapping event causes a detector molecule to change color. To make a sandwich assay for HIV, you need the following components: one plate to which the molecules are attached; a 0.550 mg sample of the recognition molecule that "recognizes" the HIV antibody; 1.200 mg of the capture molecule that "captures" the HIV antibody in a sandwich; and 0.450 mg of the detector molecule that produces a visible color when the HIV antibody is captured. You need to make 96 plates for an assay. You are given the following quantities of material: 100.00 mg of the recognition molecule; 100.00 mg of the capture molecule; 50.00 mg of the detector molecule.
a. Do you have sufficient material to make 96 plates?
b. If you do, how much of each material is left after 96 sandwich assays are assembled? If you do not have sufficient material to make 96 assays, how many assays can you assemble?

1.94. Some people believe that large doses of vitamin C can cure the common cold. One commercial over-the-counter product consists of 500.0 mg tablets that are 20% by mass vitamin C. How many tablets are needed for a 1.00 g dose of vitamin C?

1.95. We are building bicycles from separate parts. Each bicycle needs a frame, a front wheel, a rear wheel, two pedals, a set of handlebars, a bike chain, and a set each of front and rear brakes. How many complete bicycles can we make from 111 frames, 81 front wheels, 95 rear wheels, 112 pedals, 47 sets of handlebars, 38 bike chains, 17 front brakes, and 35 rear brakes?

1.96. Each Thursday the 11 kindergarten students in Ms. Goodson's class are each allowed one slice of pie, one cup of orange juice, and two "doughnut holes." The leftovers will be given to the custodian on the night shift. This Thursday the caterer has left two pies that each can be cut into 8 slices, 18 cups of orange juice, and 24 doughnut holes. How many slices of pie, cups of orange juice, and doughnut holes are left for the custodian?

1.97. Manufacturers of trail mix have to control the distribution of items in their products. Deviations of more than 2% outside specifications cause supply problems and downtime in the factory. A favorite trail mix is designed to contain 67% peanuts and 33% raisins. Bags of trail mix were sampled from the assembly line on different days. The bags were opened and the contents counted, with the following results:

| Day | Peanuts | Raisins |
| --- | --- | --- |
| 1 | 50 | 32 |
| 11 | 56 | 26 |
| 21 | 48 | 34 |
| 31 | 52 | 30 |

On which day(s) did the product meet the specification of 65% to 69% peanuts in the bag?

*1.98. Gasoline and water do not mix. Regular grade (87 octane) gasoline has a lower density (0.73 g/mL) than water (1.00 g/mL). A 100 mL graduated cylinder with an inside diameter of 3.2 cm contains 34.0 g of gasoline and 34.0 g of water. What is the combined height of the two liquid layers in the cylinder? The volume of a cylinder is $\pi r^2 h$, where $r$ is the radius and $h$ is the height.

1.99. **Stretchy Springs** Metal springs come in many shapes and sizes. The same force is used to stretch each of two springs A and B. Spring A stretches from its natural length of 4.0 cm to a length of 5.4 cm; spring B's length increases by 15%. Which is the stronger spring, A or B?

1.100. Ms. Goodson's geology classes are popular because of their end-of-the-year field trips. Some last several days, but all involve exactly eight hours of hiking per day. On one three-day trip the class's average hiking speeds were 1.6 mi/hr, 1.4 mi/hr, and 1.7 mi/hr each day, respectively. What was the length of their trip in miles and kilometers?

If your instructor assigns problems in **smartwork**, log in at **smartwork.wwnorton.com.**

# 2

# Atoms, Ions, and Molecules: Matter Starts Here

2.1 The Nuclear Model of Atomic Structure

2.2 Isotopes

2.3 Average Atomic Mass

2.4 The Periodic Table of the Elements

2.5 Trends in Compound Formation

2.6 Naming Compounds and Writing Formulas

2.7 Nucleosynthesis

## Learning Outcomes

**LO1** Write symbols of nuclides and describe the composition of atoms
**Sample Exercises 2.1, 2.2**

**LO2** Explain how the experiments of Thomson, Millikan, and Rutherford contributed to our understanding of atomic structure

**LO3** Identify isotopes and use natural abundance data to calculate average atomic mass
**Sample Exercise 2.3**

**LO4** Use the periodic table to predict the chemical properties of elements
**Sample Exercises 2.4, 2.6**

**LO5** Describe the general differences in the properties of metals, nonmetals, and metalloids

**LO6** Name common molecular and ionic compounds and write and interpret their formulas
**Sample Exercises 2.5, 2.7, 2.8, 2.9, 2.10, 2.11, 2.12**

**LO7** Describe how elements are synthesized in the cores of giant stars

## From Atoms to Stars and Back

We saw in Chapter 1 that the atomic model of matter explains many of the physical and chemical properties of substances. However, as late as the mid-19th century, scientists did not know why one atom differs from another or how atoms interact when they form molecules. In fact, many leading scientists were not convinced that atoms even existed.

By the mid-19th century chemists had begun to categorize elements according to their properties by grouping them into families of related elements. This eventually led to what we now call the periodic table—one of the most useful tools in all of chemistry. The periodic table is based on the properties of the elements and on our understanding of atomic structure.

Understanding the structure of an atom is essential to understanding how and why different atoms combine and whether they form molecules or ions. In this chapter we present basic atomic structure first and then discuss simple compounds that form when atoms bond together. Frequently two elements can form more than one compound, so we need a clear and unambiguous way to refer to these compounds by name. The systems used by chemists for naming various types of compounds are also described in this chapter.

These basic ideas are important because the identities of all forms of matter depend on the identities of their atoms. Whether we are trying to understand how blood carries oxygen to our cells, why a car rusts, why leaves turn colors in the fall, or why aspirin alleviates pain, we need to know the identities of the atoms that make up the materials involved. Concepts of atomic structure have led to methods for determining the overall structures of molecules: elemental composition and molecular structure go hand-in-hand in determining function.

**The Essential Parts of Everything** All matter, from the simplest molecules of hydrogen gas to complex structures like DNA, consists of atoms. Ideas about the structure and composition of atoms described in this chapter form the basis of our understanding of the material world. ▶

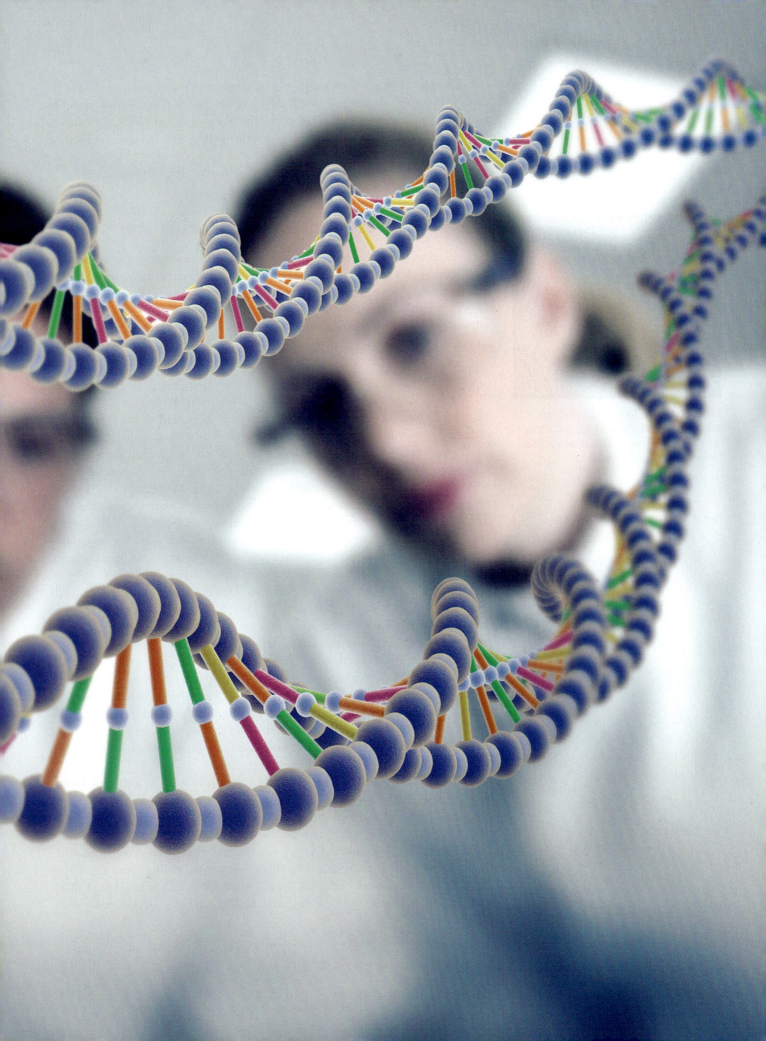

This chapter closes with another basic question: Where do the atoms come from that shape our material world? All elements are produced within stars. As astronomer and author Carl Sagan said: "The nitrogen in our DNA, the calcium in our blood, the carbon in our apple pies were made in the interiors of collapsing stars. We are made of starstuff."[1] ∎

## 2.1 The Nuclear Model of Atomic Structure

By the end of the 19th century, many scientists realized that atoms are not the smallest particles of matter, but rather consist of even smaller *subatomic* particles. This realization came in part from the seminal research of British scientist Joseph John (J. J.) Thomson (1856–1940; Figure 2.1). However, not even Thomson himself recognized the fundamental importance of his work for the future of science and technology.

### Electrons

Figure 2.2 shows the apparatus Thomson used in his experiments. It is called a cathode-ray tube (CRT), and it consists of a glass tube from which most of the air has been removed. Electrodes within the tube are attached to the poles of a high-voltage power supply. The electrode called the cathode is connected to the negative terminal of the power supply, and the anode is connected to the positive terminal. When these connections are made, electricity passes through the glass tube in the form of a beam of **cathode rays** emitted by the cathode. Cathode rays are invisible to the naked eye, but when the end of the CRT opposite the cathode is coated with a phosphorescent material, a glowing spot appears where the beam hits it.

Thomson found that cathode-ray beams are deflected by magnetic fields (Figure 2.2a) and by electric fields (Figure 2.2b). The directions of these deflections established that cathode rays are negatively charged particles. By adjusting the strengths of the electric and magnetic fields, Thomson was able to balance out the deflections (Figure 2.2c). From the strengths of the two opposing fields, he calculated the mass-to-charge ($m/e$) ratio of the particles. Thomson and others observed that these particles always behave the same way and always have the same mass-to-charge ratio, no matter what cathode material is used. This observation established that the particles in the cathode rays, which are now known as **electrons**, are fundamental particles present in all forms of matter.

In 1909, American physicist Robert Millikan (1868–1953) advanced Thomson's work by quantifying the charge on the electron. Thomson had calculated the mass-to-charge ratio of the electron, and if Millikan could measure the charge, then the mass would also be known. Figure 2.3 illustrates Millikan's experiment: Highly energetic X-rays remove electrons from molecules of air in the lower of two connected chambers. Oil drops, falling from the upper chamber into the lower one, pick up these electrons and their charge. Millikan measured the mass of the drops in the absence of an electric field, when

**FIGURE 2.1** J. J. Thomson (1856–1940) discovered electrons in 1897 using a cathode-ray tube, but he was not sure where electrons fit into the structure of atoms.

▶❚❚ **CHEMTOUR** Cathode-Ray Tube

▶❚❚ **CHEMTOUR** Millikan Oil-Drop Experiment

---

[1]Carl Sagan, *Cosmos*, Ballantine Books, 1985, p. 190.

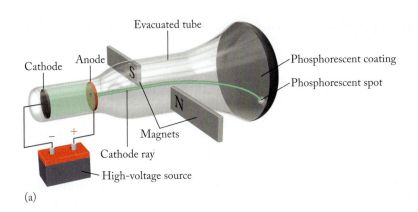

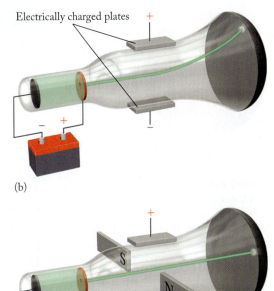

**FIGURE 2.2** A cathode ray is generated when electricity is passed through a tube from which most of the air has been removed. Though invisible, the path of the ray can be inferred by the bright spot it makes in a phosphorescent material coated on the end of the tube. (a) Cathode ray deflected in one direction by a magnetic field; (b) cathode ray deflected in the opposite direction by an electric field; (c) electric and magnetic fields tuned to balance out the deflections.

their rate of fall was governed by gravity. He then turned on and adjusted the electric field to make the drops fall at different rates and even suspended some of them in midair. From the strength of the electric field and the rate of fall, he calculated the charge on a drop. By measuring the charges on hundreds of drops, Millikan determined that the charge on each drop was a whole-number multiple of a minimum charge. He concluded that this minimum charge had to be the charge on one electron. Millikan's value was within 1% of the modern value: $-1.602 \times 10^{-19}$ C. (The coulomb, abbreviated C, is the SI unit for electric charge.) Knowing Thomson's value of $m/e$ for the electron, Millikan calculated the mass of the electron: $9.109 \times 10^{-28}$ g.

The discovery of the electron raised the possibility of there being other subatomic particles. Scientists in the 1890s knew that matter was electrically neutral, but they did not know how the electrons and the positive charges were arranged at the atomic level. Thomson proposed a *plum-pudding model* in which the atom is a diffuse sphere of positive charge with negatively charged electrons embedded in the sphere, like raisins in a plum pudding (Figure 2.4). Thomson's plum-pudding model did not last long. Its demise was linked to another scientific discovery in the 1890s: radioactivity.

**FIGURE 2.3** Millikan's oil-drop experiment.

**cathode rays** streams of electrons emitted by the cathode in a partially evacuated tube.

**electron** a subatomic particle that has a negative charge and essentially zero mass.

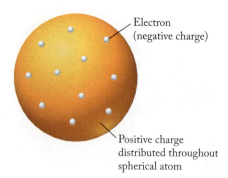

Electron
(negative charge)

Positive charge
distributed throughout
spherical atom

**FIGURE 2.4** In J. J. Thomson's plum-pudding model, atoms consist of electrons distributed throughout a massive, positively charged but very diffuse sphere. The plum-pudding model lasted only a few years before it was replaced by a model based on experiments carried out under the direction of Thomson's former student, Ernest Rutherford.

**FIGURE 2.5** Ernest Rutherford (1871–1937) was born in New Zealand and was awarded a scholarship in 1894 that enabled him to go to Trinity College in Cambridge, England. There he was a research assistant in the laboratory of J. J. Thomson. His contributions included characterizing the properties of α and β particles. By 1907, he was a professor at the University of Manchester, where his famous gold-foil experiments led to our modern view of atomic structure. He received the Nobel Prize in Chemistry in 1908.

**radioactivity** the spontaneous emission of high-energy radiation and particles by materials.

**beta (β) particle** a radioactive emission that is a high-energy electron.

**alpha (α) particle** a radioactive emission with a charge of 2+ and a mass equivalent to that of a helium nucleus.

**CONCEPT TEST**

Why did Thomson's discovery of the electron lead to proposals that a positive particle might exist within the atom?

*(Answers to Concept Tests are in the back of the book.)*

## Radioactivity and the Nuclear Atom

In 1896, French physicist Henri Becquerel (1852–1908) discovered that *pitchblende*, a brownish-black mineral that is the principal source of uranium, produces radiation that can be detected on photographic plates. Becquerel and his contemporaries initially thought that this radiation consisted of the X-rays that had just been discovered by German scientist Wilhelm Conrad Röntgen (1845–1923).[2] However, additional experiments by Becquerel, by the French wife-and-husband team of Marie (1867–1934) and Pierre (1859–1906) Curie, and by British scientist Ernest Rutherford (Figure 2.5) showed that this radiation contained particles as well as rays. Today we use the term **radioactivity** for the spontaneous emission of high-energy radiation and particles by materials such as pitchblende.

In studying the particles emitted by pitchblende, Rutherford found that one type, which he named **beta (β) particles**, penetrate materials better than the second type, which he named **alpha (α) particles**. The degree of deflection of β particles in a magnetic field allowed their mass-to-charge ratio to be calculated, and the results exactly matched the electron mass-to-charge ratio determined by J. J. Thomson. These data established that β particles are simply high-energy electrons.

Rutherford discovered that α particles are deflected by an electric field in the opposite direction from β particles; the same is true for the two particles in a magnetic field. Therefore, he concluded that α particles are positively charged. The β particle (electron) was assigned a relative charge of 1−. The corresponding charge of an α particle is 2+. In addition, α particles have nearly the same mass as an atom of helium, meaning they are over $10^3$ times more massive than β particles.

The experiment that disproved Thomson's plum-pudding model was directed by Rutherford and carried out by two of his students at Manchester University: Hans Geiger (1882–1945; for whom the Geiger counter was named) and Ernest Marsden (1889–1970). Geiger and Marsden bombarded a thin foil of gold with a beam of α particles emitted from a radioactive source (Figure 2.6a). They then measured how many particles were deflected and to what extent they were deflected. Rutherford's hypothesis was that if Thomson's model were correct, most of the α particles would pass straight through the diffuse positive spheres of gold atoms (each atom a "pudding" like the one shown in Figure 2.4), but a few of the particles would interact with the electrons (the "raisins") embedded in these spheres and be deflected slightly (Figure 2.6b).

Geiger and Marsden observed something quite unexpected. For the most part, the α particles did indeed pass directly through the gold. However, about 1 in every 8000 particles was deflected from the foil through an average angle of

[2]Röntgen discovered X-rays in experiments with a cathode-ray tube much like the apparatus used by J. J. Thomson. After completely encasing the tube in a black carton, Röntgen discovered that invisible rays escaped the carton and were detected by a photographic plate. Because he knew so little about these rays, he called them X-rays.

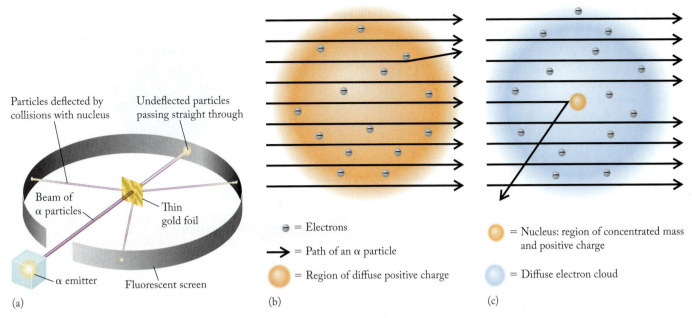

**FIGURE 2.6** (a) In the Rutherford–Geiger–Marsden experiment, α particles from a radioactive source were allowed to strike a thin gold foil. A fluorescent screen surrounded the foil to detect any deflected particles. (b) If Thomson's plum-pudding model were correct, most of the α particles would pass through the gold foil and a few would be deflected slightly. (c) In fact, most particles passed straight through, but a few were scattered widely. This unexpected result led to the theory that an atom has a small, positively charged nucleus that contains most of the mass of the atom.

90 degrees, and a very few (about 1 out of 100,000) bounced back in the direction from which the particles came (Figure 2.6c). Rutherford described his amazement at the result: "It is about as incredible as if you had fired a 15-inch shell at a piece of tissue paper and it came back and hit you." (A cannon shell with a diameter of 15 inches was the largest projectile that could be fired by a British battleship in 1909.)

The results of the gold-foil experiments ended the short life of the plum-pudding model because the model could not account for the large angles of deflection. Rutherford concluded that those deflections occurred because the α particles occasionally encountered small regions of high positive charge and large mass. Rutherford determined that the region of positive charge is only about 1/10,000 of the overall size of a gold atom. His model of the atom became the basis for our current understanding of atomic structure. It assumes that an atom consists of a massive but tiny, positively charged **nucleus** surrounded by a diffuse cloud of negatively charged electrons.

## Protons and Neutrons

In the decade following the gold-foil experiments, Rutherford and others observed that bombarding elements with α particles could change, or *transmute*, these elements into other elements. They also discovered that hydrogen nuclei were sometimes produced during transmutation reactions. By 1920, agreement was growing that hydrogen nuclei, which Rutherford called **protons** (from the Greek *protos*, meaning "first"), were part of all nuclei. For example, to account for the mass and charge of an α particle, Rutherford assumed it was made of four protons, two of which had combined with two electrons to form two electrically neutral particles that he called **neutrons**. Repeated attempts to produce neutrons by neutralizing

▶❚❚ **CHEMTOUR** Rutherford Experiment

**nucleus** (of an atom) the positively charged center of an atom that contains nearly all the atom's mass.

**proton** a positively charged subatomic particle present in the nucleus of an atom.

**neutron** an electrically neutral (uncharged) subatomic particle found in the nucleus of an atom.

**FIGURE 2.7** The modern view of Rutherford's model of the gold atom includes a nucleus that is about 1/10,000 the overall size of the atom. Note that the scales are given in picometers (pm = $10^{-12}$ m); the nucleus would be too small to see if drawn to scale in the left drawing. If an atom were the size of the Rose Bowl (an oval stadium about 210 m across in the narrow direction), the nucleus would be the size of a dime at mid-field.

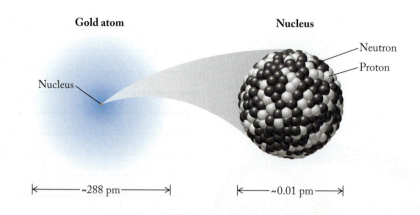

| TABLE 2.1 | Properties of Subatomic Particles | | | | |
|---|---|---|---|---|---|
| | | **MASS** | | **CHARGE** | |
| **Particle** | **Symbol** | **In Atomic Mass Units (amu)** | **In Grams (g)** | **Relative Value** | **Charge (Cª)** |
| Neutron | $^{1}_{0}n$ | $1.00867 \approx 1$ | $1.67493 \times 10^{-24}$ | 0 | 0 |
| Proton | $^{1}_{1}p$ | $1.00728 \approx 1$ | $1.67262 \times 10^{-24}$ | 1+ | $+1.602 \times 10^{-19}$ |
| Electron | $^{0}_{-1}e$ | $5.485799 \times 10^{-4} \approx 0$ | $9.10939 \times 10^{-28}$ | 1− | $-1.602 \times 10^{-19}$ |

ªThe *coulomb* (C) is the SI unit of electric charge. When a current of 1 ampere (see Table 1.2) passes through a conductor for 1 second, the quantity of electric charge that moves past any point in the conductor is 1 C.

protons with electrons were unsuccessful. However, in 1932 one of Rutherford's students, James Chadwick (1891–1974), discovered and characterized free neutrons. With the discovery of neutrons, the current model of atomic structure was complete, as illustrated with the gold atom in Figure 2.7.

Table 2.1 summarizes the properties of neutrons, protons, and electrons. For convenience, the masses of these tiny particles are expressed in **atomic mass units (amu)**. These units are also called **daltons (Da)**, or *unified atomic mass units* (u). One amu is exactly 1/12 the mass of a carbon atom that has six protons and six neutrons in its nucleus. The dalton honors English chemist John Dalton (1766–1844), who published the first table of atomic masses in 1803. As you can see from the data in Table 2.1, the masses of the neutron and proton are approximately the same, and both are assigned a mass of 1 amu. If you compare the amu values in the table with the masses of the particles in grams, you will see that these particles are tiny indeed: 1 amu is only $1.66054 \times 10^{-24}$ g.

## 2.2 Isotopes

Even as Thomson investigated the properties of cathode rays in 1897, other scientists designed and built devices to produce beams of positively charged particles. One of Thomson's former students, Francis W. Aston (1877–1945), built modified cathode-ray tubes that were evacuated except for small quantities of *fill gases*, such as neon. With these tubes he detected conventional beams of cathode rays, but he also detected secondary beams of positively charged particles. Charge was not the only thing different about the particles in these secondary beams. Whereas cathode rays are electrons that all have the same mass and charge no matter what

**atomic mass unit (amu)** unit used to express the relative masses of atoms and subatomic particles; it is exactly 1/12 the mass of one atom of carbon with six protons and six neutrons in its nucleus.

**dalton (Da)** a unit of mass identical to 1 atomic mass unit.

the cathode material or the fill gas is, the masses of Aston's positive rays depended on the identity of the fill gas. The particles in the positive rays were not just individual protons, but rather atoms of the fill gas that had lost electrons and formed positively charged **ions**.

Aston used his positive-ray analyzer (Figure 2.8) to pass positively charged beams of particles through a magnetic field. Each particle in the beam was deflected along a path determined by the particle's mass: the greater the mass, the smaller the deflection. Using the purest sample of neon gas available, Aston determined that most particles had a mass of 20 amu, but about 1 in 10 had a mass of 22 amu. To explain his data, Aston proposed that neon consists of two kinds of atoms, or **isotopes**. Both isotopes of neon have the same number of protons (10) in the nucleus, but one isotope has 10 neutrons in its nucleus, giving it a mass of 20 amu, whereas the other isotope has 12 neutrons in its nucleus, giving it a mass of 22 amu. The nuclei of any element with their particular combination of neutrons and protons are called **nuclides**.

Since the time of John Dalton, scientists had believed that each element *is composed of identical atoms, all of which have the same mass*. Aston's work showed that each element *is composed of atoms all having the same number of protons in their nuclei*. The number of protons is called the **atomic number (Z)** of the element. The total number of **nucleons** (neutrons and protons) in the nucleus of an atom defines its **mass number (A)**. The isotopes of a given element thus all have the same atomic number, Z, but different mass numbers, A. The modern **periodic table of the elements** (inside the front cover of this book) displays the elements in order of atomic number.

The general format for identifying a particular nuclide is

$$_Z^A X$$

where X represents the one- or two-letter symbol for the element. For example, two isotopes of oxygen (O) and lead (Pb) are written:

$$_8^{16}O \qquad _8^{18}O \qquad\qquad _{82}^{206}Pb \qquad _{82}^{208}Pb$$

Because Z and X provide the same information—each by itself identifies the element—the subscript Z is frequently omitted: often the isotope symbol is written $^AX$. This same information—mass number and element name—may also be spelled out. For example, the names of the two isotopes of neon that Aston discovered may be written neon-20 and neon-22.

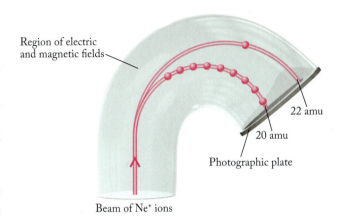

**FIGURE 2.8** Aston's positive-ray analyzer. A beam of positively charged ions of neon gas is passed through a focusing slit into a region of electric and magnetic fields. The ions are separated according to mass: those with a mass of 20 amu—90% of the sample—hit the detector at one spot, and those with a mass of 22 amu—the remaining 10%—hit the detector at a different spot. Aston's positive-ray analyzer was the forerunner of the modern mass spectrometer.

---

**ion** an atom or group of atoms that has a net positive or negative charge.

**isotopes** atoms of an element containing different numbers of neutrons.

**nuclide** the nucleus of a specific isotope of an element.

**atomic number (Z)** the number of protons in the nucleus of an atom.

**nucleon** either a proton or a neutron in a nucleus.

**mass number (A)** the number of nucleons in an atom.

**periodic table of the elements** a chart of the elements in order of their atomic numbers and in a pattern based on their physical and chemical properties.

---

**SAMPLE EXERCISE 2.1    Writing Symbols of Nuclides        LO1**

Write symbols in the form $_Z^A X$ for the nuclides that have (a) 6 protons and 6 neutrons, (b) 11 protons and 12 neutrons, and (c) 92 protons and 143 neutrons.

**Collect and Organize** We know the number of protons and neutrons in the nuclei of three nuclides and are to write symbols in the $_Z^A X$ form where Z is the atomic number, A is the mass number, and X is the symbol of the element.

**Analyze** The number of protons in the nucleus of an atom defines its atomic number (Z) and defines which element it is (X). The sum of the nucleons (protons plus neutrons) is the mass number (A). Summarizing the sequence of steps:

$$\boxed{Z \text{ (given)}} \rightarrow \boxed{X \text{ (from periodic table)}} \rightarrow \boxed{A \text{ (sum of protons and neutrons)}}$$

**Solve**

a. This nuclide has six protons and therefore $Z = 6$; it must be an isotope of carbon. Six protons plus six neutrons gives the isotope a mass number of 12. This isotope of carbon is carbon-12, $^{12}_{6}C$.

b. This nuclide has 11 protons and therefore $Z = 11$; it must be an isotope of sodium. Eleven protons and 12 neutrons give the isotope a mass number of 23. This isotope of sodium is sodium-23, $^{23}_{11}Na$.

c. This nuclide has 92 protons and therefore $Z = 92$; it must be an isotope of uranium. The mass number is $92 + 143 = 235$. This isotope of uranium is uranium-235, $^{235}_{92}U$.

**Think About It** In working through this exercise, did you use the periodic table of the elements? Once you identify the number of protons in a nucleus (its atomic number), finding a symbol and identifying the element it represents is easy because the elements in the periodic table are arranged in order of increasing atomic number.

⚙ **Practice Exercise** Use the format $^{A}X$ to write the symbols of the nuclides having (a) 26 protons and 30 neutrons, (b) 7 protons and 8 neutrons, (c) 17 protons and 20 neutrons, and (d) 19 protons and 20 neutrons.

*(Answers to Practice Exercises are in the back of the book.)*

---

**SAMPLE EXERCISE 2.2** **Counting Nuclide Constituents** **LO1**

What are the number of neutrons in the following nuclides: (a) $^{14}N$; (b) $^{32}P$; (c) $^{157}Gd$?

**Collect and Organize** We are given the symbols of three nuclides and asked to determine the number of neutrons in each of their atoms.

**Analyze** The elements in the periodic table are arranged in order of increasing atomic number, $Z$, which is the number of protons their nuclei contain. We know $Z$ from the element's symbol. Subtracting $Z$ from $A$ will give us the number of neutrons. Summarizing the steps:

| Element symbol $^{A}X$ (given) | → | $Z$ (from periodic table) | → | number of neutrons $= A - Z$ |
|---|---|---|---|---|

**Solve**

a. $^{14}N$ is a nuclide of nitrogen, whose atoms have seven protons. The number of neutrons is $A - Z = 14 - 7 = 7$. $^{14}N$ contains 7 neutrons.

b. From the periodic table, we see that $^{32}P$ is a nuclide of phosphorus with $A = 32$. Since $Z = 15$, the number of neutrons in $^{32}P$ is $32 - 15 = 17$.

c. The periodic table indicates that $^{157}Gd$ is a nuclide of gadolinium, with $Z = 64$. Thus the number of neutrons in this nuclide is $157 - 64 = 93$.

**Think About It** No two elements in the periodic table have nuclei that contain the same number of protons. In fact, the number of protons is unique to any element and is a reason why elements differ.

⚙ **Practice Exercise** Give the number of protons and neutrons in each of these radioactive nuclides: (a) $^{60}Co$, used in cancer therapy; (b) $^{131}I$, used in thyroid therapy; (c) $^{192}Ir$, used to treat coronary disease.

*(Answers to Practice Exercises are in the back of the book.)*

---

# 2.3 Average Atomic Mass

Each of the cells in the periodic table contains the symbol of an element. The number above the symbol is the element's atomic number ($Z$) and the number below the symbol is the element's **average atomic mass**. More precisely, the number below the symbol is the *weighted average* of the masses of all the isotopes of the element.

To understand the meaning of a weighted average, consider the masses and **natural abundances** of the three isotopes of neon in the table to the right. Natural abundances are usually expressed in percentages. Thus, 90.4838% of all neon atoms are neon-20, 9.2465% are neon-22, and only 0.2696% are neon-21. The abundance of neon-21 is so small Aston could not detect it with his positive-ray analyzer. Modern mass spectrometers, which are the source of natural abundance data such as these, are vastly more sensitive and more precise than Aston's prototype.

| Isotope | Mass (amu) | Natural Abundance (%) |
|---------|-----------|----------------------|
| Neon-20 | 19.9924 | 90.4838 |
| Neon-21 | 20.9940 | 0.2696 |
| Neon-22 | 21.9914 | 9.2465 |

To determine the average atomic mass, we multiply the mass of each isotope by its natural abundance (in the language of mathematics, we *weight* the isotope's mass using natural abundance as the *weighting factor*) and then sum the three weighted masses. To simplify the calculation, we first convert the percent abundance values into their decimal equivalents:

$$\text{Average atomic mass of neon} = \begin{array}{l} (19.9924 \text{ amu} \times 0.904838) \\ + (20.9940 \text{ amu} \times 0.002696) \\ + (21.9914 \text{ amu} \times 0.092465) \\ \hline 20.1799 \text{ amu} \end{array}$$

It is important to note that no atom of neon has the average atomic mass; every atom of neon in the universe must have a mass equal to one of the three neon isotopes. The value we have calculated is simply the weighted average of these three isotopic masses.

This method of calculating average atomic mass works for all elements. The general equation for doing these calculations is

$$m_X = a_1 m_1 + a_2 m_2 + a_3 m_3 + \cdots \qquad (2.1)$$

where $m_X$ is the average atomic mass of element X, which has isotopes with masses $m_1, m_2, m_3, \ldots$, the natural abundances of which, expressed in decimal form, are $a_1, a_2, a_3, \ldots$.

---

**SAMPLE EXERCISE 2.3** **Calculating an Average Atomic Mass** **LO3**

The precious metal platinum (Z = 78) has six isotopes with these natural abundances:

| Symbol | Mass (amu) | Natural Abundance (%) |
|--------|-----------|----------------------|
| $^{190}$Pt | 189.96 | 0.014 |
| $^{192}$Pt | 191.96 | 0.782 |
| $^{194}$Pt | 193.96 | 32.967 |
| $^{195}$Pt | 194.97 | 33.832 |
| $^{196}$Pt | 195.97 | 25.242 |
| $^{198}$Pt | 197.97 | 7.163 |

Use these data to calculate the average atomic mass of platinum.

**Collect and Organize** We know the masses and natural abundances of each of the six isotopes of platinum.

**Analyze** Using Equation 2.1, we multiply the mass of each isotope by its natural abundance, expressed as a decimal, and then add the products together.

**average atomic mass** a weighted average of masses of all isotopes of an element, calculated by multiplying the natural abundance of each isotope by its mass in atomic mass units and then summing these products.

**natural abundance** the proportion of a particular isotope, usually expressed as a percentage, relative to all the isotopes of that element in a natural sample.

**period** a horizontal row in the periodic table.

**group** all elements in the same column of the periodic table; also called *family*.

**Solve**

$$
\begin{aligned}
\text{Average atomic mass} = \quad & (189.96 \text{ amu})(0.00014) \\
+ & (191.96 \text{ amu})(0.00782) \\
+ & (193.96 \text{ amu})(0.32967) \\
+ & (194.97 \text{ amu})(0.33832) \\
+ & (195.97 \text{ amu})(0.25242) \\
+ & (197.97 \text{ amu})(0.07163) \\
\hline
& 195.08 \text{ amu}
\end{aligned}
$$

**Think About It** Note that the six values of natural abundances expressed as decimals should add up to 1.00000, and they do. Sometimes this is not the case (check the neon abundances above). Uncertainties in the last decimal place may be due to uncertainties in measured values or in rounding them off. The calculated average atomic mass of platinum is consistent with the value given inside the front cover.

**Practice Exercise** Silver (Ag) has two stable isotopes: $^{107}$Ag, 106.90 amu, and $^{109}$Ag, 108.90 amu. If the average atomic mass of silver is 107.87 amu, what is the natural abundance of each isotope? *Hint*: Let $x$ be the natural abundance of one of the isotopes. Then $1 - x$ is the natural abundance of the other.

*(Answers to Practice Exercises are in the back of the book.)*

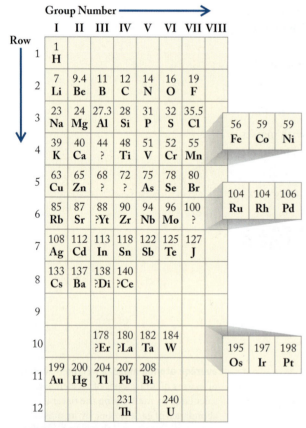

**FIGURE 2.9** Mendeleev organized his periodic table on the basis of chemical and physical properties and atomic masses. He assigned three elements with similar properties to group VIII in rows 4, 6, and 10. Because he did this, the elements in the rows that followed lined up in appropriate groups. In this way, rows 4 and 5 together contain spaces for 18 elements, corresponding to the 18 groups in the modern periodic table.

## 2.4 The Periodic Table of the Elements

Long before chemists knew about electrons, protons, neutrons, and the concept of atomic numbers, they knew that groups of elements, such as Li, Na, and K, or F, Cl, and Br, have similar chemical (and sometimes physical) properties. When the elements are arranged by increasing atomic mass, repeating patterns of similar properties appear among the elements. This *periodicity* in the chemical properties of the elements inspired several 19th-century scientists to create tables of the elements in which the elements were arranged in patterns based on similarities in their chemical properties.

The most successful of these scientists was Russian chemist Dmitri Mendeleev (1834–1907). In 1872 he published a table (Figure 2.9) that was the forerunner of the modern periodic table (Figure 2.10). In addition to organizing all the elements that were known at the time, Mendeleev realized that there might be elements in nature that were yet to be discovered, so he left empty cells in his table for those unknown elements. This allowed him to align the known elements so that those in each column had similar chemical properties. Based on the locations of the empty cells, Mendeleev predicted the chemical properties of the missing elements that ultimately were discovered. Mendeleev arranged the elements in his periodic table in order of increasing atomic mass. In modern periodic tables the elements appear in order of their atomic numbers.

**CONCEPT TEST**

Why was Mendeleev's version of the periodic table accepted so readily?

*(Answers to Concept Tests are in the back of the book.)*

| 1 | | | | | | | | | | | | | | | | | 18 |
|---|---|---|---|---|---|---|---|---|---|---|---|---|---|---|---|---|---|
| 1 H | 2 | | | | | | | | | | | 13 | 14 | 15 | 16 | 17 | 2 He |
| 3 Li | 4 Be | | | | | | | | | | | 5 B | 6 C | 7 N | 8 O | 9 F | 10 Ne |
| 11 Na | 12 Mg | 3 | 4 | 5 | 6 | 7 | 8 | 9 | 10 | 11 | 12 | 13 Al | 14 Si | 15 P | 16 S | 17 Cl | 18 Ar |
| 19 K | 20 Ca | 21 Sc | 22 Ti | 23 V | 24 Cr | 25 Mn | 26 Fe | 27 Co | 28 Ni | 29 Cu | 30 Zn | 31 Ga | 32 Ge | 33 As | 34 Se | 35 Br | 36 Kr |
| 37 Rb | 38 Sr | 39 Y | 40 Zr | 41 Nb | 42 Mo | 43 Tc | 44 Ru | 45 Rh | 46 Pd | 47 Ag | 48 Cd | 49 In | 50 Sn | 51 Sb | 52 Te | 53 I | 54 Xe |
| 55 Cs | 56 Ba | 57 La | 72 Hf | 73 Ta | 74 W | 75 Re | 76 Os | 77 Ir | 78 Pt | 79 Au | 80 Hg | 81 Tl | 82 Pb | 83 Bi | 84 Po | 85 At | 86 Rn |
| 87 Fr | 88 Ra | 89 Ac | 104 Rf | 105 Db | 106 Sg | 107 Bh | 108 Hs | 109 Mt | 110 Ds | 111 Rg | 112 Cn | 113 Uut | 114 Fl | 115 Uup | 116 Lv | 117 Uus | 118 Uuo |

Atomic number / Symbol for element

Period

| 6 Lanthanides | 58 Ce | 59 Pr | 60 Nd | 61 Pm | 62 Sm | 63 Eu | 64 Gd | 65 Tb | 66 Dy | 67 Ho | 68 Er | 69 Tm | 70 Yb | 71 Lu |
|---|---|---|---|---|---|---|---|---|---|---|---|---|---|---|
| 7 Actinides | 90 Th | 91 Pa | 92 U | 93 Np | 94 Pu | 95 Am | 96 Cm | 97 Bk | 98 Cf | 99 Es | 100 Fm | 101 Md | 102 No | 103 Lr |

**FIGURE 2.10** In the modern periodic table, the elements are arranged in order of atomic number ($Z$) and in a pattern related to their physical and chemical properties. The elements shown in tan are classified as metals; those shown in blue, nonmetals; and those shown in green, metalloids (also called semimetals).

## Navigating the Modern Periodic Table

The modern periodic table (Figure 2.10) contains seven horizontal rows (called **periods**) and 18 columns (that each contain a **group** or *family*) of elements (Figure 2.11a). The periods are numbered at the far left of each row, and the group numbers appear at the top of each column. The periodic table inside the front cover shows a second set of column headings containing numbers followed by the letter A or B. These secondary headings were widely used in earlier versions of the table, and many scientists still find them useful.

The first row contains only two elements—hydrogen and helium—and the second and third rows contain only eight. Starting with the fourth row, all 18

▶❚❚ CHEMTOUR The Periodic Table

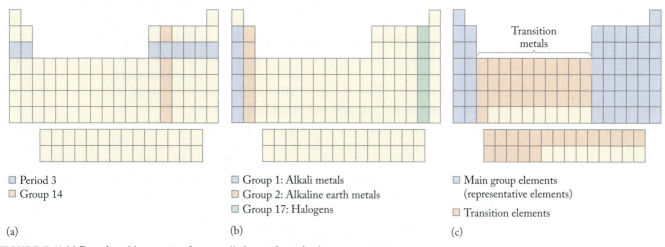

Transition metals

(a)
☐ Period 3
☐ Group 14

(b)
☐ Group 1: Alkali metals
☐ Group 2: Alkaline earth metals
☐ Group 17: Halogens

(c)
☐ Main group elements (representative elements)
☐ Transition elements

**FIGURE 2.11** (a) Periodic tables consist of rows, called periods, and columns containing groups of elements. (b) The commonly used names of groups 1, 2, and 17. (c) The *main group* (or *representative*) elements are in groups 1, 2, and 13–18. They are separated by the *transition metals* in groups 3–12.

**halogens** the elements in group 17 of the periodic table.

**alkali metals** the elements in group 1 of the periodic table.

**alkaline earth metals** the elements in group 2 of the periodic table.

**metals** the elements on the left side of the periodic table that are typically shiny solids that conduct heat and electricity well and are malleable and ductile.

**nonmetals** elements with properties opposite those of metals, including poor conductivity of heat and electricity.

**metalloids** (also called **semimetals**) elements along the border between metals and nonmetals in the periodic table; they have some metallic and some nonmetallic properties.

**main group elements** (also called **representative elements**) the elements in groups 1, 2, and 13 through 18 of the periodic table.

**transition metals** the elements in groups 3 through 12 of the periodic table.

**noble gases** the elements in group 18 of the periodic table.

columns are full. Actually, the sixth and seventh rows contain more elements than there is space for in an 18-column array. These additional elements appear in the two separate rows at the bottom of the main table. Elements in the row with atomic numbers from 58 to 71 are called the lanthanides (after element 57, lanthanum), and those with atomic numbers between 90 and 103 are called actinides. Most of the nuclides of the actinide elements are radioactive, and none of those with atomic numbers above 94 occur in nature. Therefore, they have no *natural* abundance.

Several of the groups have a name in addition to a number. The names are typically based on properties common to all the elements in that group (Figure 2.11b). For example, the elements in group 17 are called **halogens**. The word *halogen* is derived from the Greek for "salt former." Chlorine is a typical halogen: it forms a 1:1 compound (for example, NaCl, table salt) with the elements in group 1, and 2:1 compounds (for example, $CaCl_2$) with the elements in group 2. For their part, group 1 elements, which are called **alkali metals**, form 1:1 compounds with all the group 17 elements and 2:1 compounds, such as $Na_2S$, with the group 16 elements. The group 2 elements, which are called **alkaline earth metals**, form 1:1 compounds, such as MgO, with the group 16 elements and 1:2 compounds with all the group 17 elements. These and other reactivity patterns were the basis for Mendeleev's arrangement of the elements in his periodic table, and they illustrate what we mean by "similar chemical properties."

The elements in the periodic table are also divided into three broad categories highlighted by the three different cell colors in Figure 2.10. Elements in the tan cells are **metals**. They tend to conduct heat and electricity well; they tend to be malleable (capable of being shaped by hammering) or ductile (capable of being drawn out in a wire), and all but mercury (Hg) are shiny solids at room temperature. Elements in the blue cells are **nonmetals**. They are poor conductors of heat and electricity; most are gases at room temperature, the solids among them tend to be brittle, and bromine is a liquid with a low boiling point. The elements in the green cells are called **metalloids** or **semimetals**, so named because they tend to have the physical properties of metals but the chemical properties of nonmetals.

In the modern periodic table, groups 1, 2, and 13 through 18 are referred to collectively as **main group elements**, or **representative elements** (Figure 2.11c). They include the most abundant elements in the solar system and many of the most abundant elements on Earth. Note that these are the "A" elements in the older group labeling system shown on the table inside the front cover. The elements in groups 3 through 12 are called **transition metals**; these are the "B" elements. They are nearly all classic metals: hard, shiny, ductile, malleable, and excellent conductors of heat and electricity.

Take a moment to compare Figures 2.9 and 2.10. First note the similarity in the arrangements of the lighter (smaller atomic number) elements through calcium ($Z = 20$). All of the elements in groups 1, 2, and 13 through 17 of the modern table appear in the same order in Mendeleev's table, but group 18 (the **noble gases**) is missing from Mendeleev's table. There is a good reason for this: Helium was the first noble gas to be discovered, and that was not until 1895, many years after Mendeleev published his table. Noble gases are chemically unreactive, and so were elusive substances for early chemists to isolate and identify. Because Mendeleev arranged his table largely on the basis of reactivity, he had no reason to predict the existence of the noble gases.

## SAMPLE EXERCISE 2.4 Navigating the Periodic Table LO4

Using Figure 2.12, give the symbol and name of each of the following elements:
a. The fourth row alkali metal
b. The halogen with fewer than 16 protons in its nucleus
c. The third row element in group 14
d. The metal (X) in the second row that forms a compound with the chemical formula $XBr_2$

**Collect and Organize** In a, c, and d we are given the row number; in b we know the element has fewer than 16 protons in its nucleus. Information about the group locations comes from group names in a and b, group number in c, and a chemical property in d.

**Analyze** (a) The alkali metals are group 1 elements, so the cell address is group 1, row 4. (b) The halogens are group 17 elements, and the only group 17 element with fewer than 16 protons in its nucleus is fluorine (F, $Z = 9$). (c) The cell address is group 14, row 3. (d) The group 2 elements form 1:2 compounds with Br and the other halogens, so the cell address is group 2, row 2.

**Solve** (a) K, potassium; (b) F, fluorine; (c) Si, silicon; (d) Be, beryllium (Figure 2.12).

**Think About It** Each element has a unique location in the periodic table, determined by its atomic number, which defines the row it is in, and its reactivity with other elements, which defines the group it is in. We assumed that beryllium was the only metal in the second row that could form a 1:2 compound with bromine. This is a valid assumption because the only other metal in the second row is Li, which is a group 1 element whose compound with Br has the formula LiBr.

⚙ **Practice Exercise** Write the symbol and name of each element (or elements):
a. The metalloid in group 15 closest in mass to the noble gas krypton
b. A representative element in the fourth row that is an alkaline earth metal
c. A transition metal in the sixth row with chemical properties similar to zinc ($Z = 30$)
d. A nonmetal in the third row with chemical properties similar to oxygen

*(Answers to Practice Exercises are in the back of the book.)*

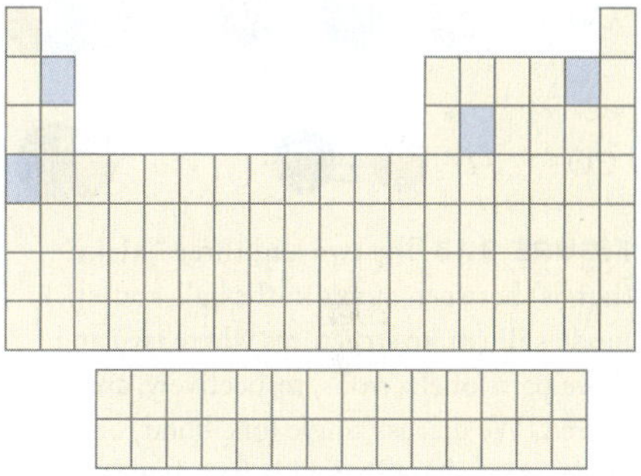

**FIGURE 2.12**

# 2.5 Trends in Compound Formation

In Section 2.1 we noted John Dalton's pioneering work in determining the atomic masses of the elements. Dalton also played a key role in documenting and predicting patterns in how elements combine to form compounds. In Section 2.4 we discussed how Mendeleev used this idea to place elements in different groups in his early periodic table. Both Dalton and Mendeleev knew that when elements combine to form compounds, they do so in characteristic proportions. These proportions are reflected in the chemical formulas of compounds. For example, the formula of carbon dioxide, $CO_2$, tells us that in every molecule of $CO_2$, one atom of carbon is combined with two atoms of oxygen.

Dalton's atomic view of compounds also explains another property of some elements: if the same two elements (for example, S and O) can form more than one compound (as in $SO_2$ and $SO_3$), the ratio of the different masses of O that react with a given mass of S to form the two compounds can be expressed as the ratio of two small whole numbers. This principle was observed experimentally and is known as Dalton's **law of multiple proportions**.

To see what this principle means, consider $SO_2$ and $SO_3$. We determine in an experiment that under one set of conditions, 10 g of sulfur reacts with 10 g of oxygen to form $SO_2$; however, under different conditions, the 10 g of sulfur reacts with 15 g of oxygen to form $SO_3$. The ratio of the two masses of oxygen is 10:15,

**law of multiple proportions** the ratio of the two masses of one element that react with a given mass of another element to form two different compounds is the ratio of two small whole numbers.

**molecular compound** a compound composed of atoms held together in molecules by covalent bonds.

**covalent bond** a bond between two atoms created by sharing one or more pairs of electrons.

**molecular formula** a notation showing the number and type of atoms present in one molecule of a molecular compound.

**ionic compound** a compound composed of positively and negatively charged ions held together by electrostatic attraction.

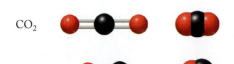

CO₂

CO

**FIGURE 2.13** The two and three "sticks" (bonds) between atoms in the ball-and-stick models illustrate structures where two and three pairs of electrons, respectively, are shared. We discuss single (one bond, one pair of electrons shared), double (two bonds, two pairs), and triple (three bonds, three pairs) bonds in Chapter 8.

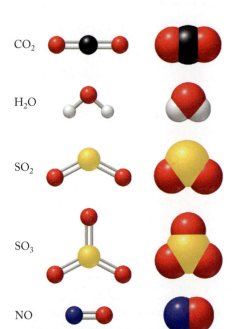

CO₂

H₂O

SO₂

SO₃

NO

**FIGURE 2.15** Ball-and-stick and space-filling models of some molecular compounds that occur in the atmosphere, particularly near sources of automobile and industrial emissions.

or 2:3, which is a ratio of two small whole numbers. This example illustrates the law of multiple proportions.

Similarly, we can confirm experimentally that the mass of oxygen that reacts with a given mass of nitrogen to form $NO_2$ (22.8 g O for every 10.0 g N) is twice as much as the mass of oxygen that reacts with the same mass of nitrogen to form NO (11.4 g O for every 10.0 g N). The ratio of the two oxygen masses is 22.8:11.4, or 2:1, again a ratio of small whole numbers. The law of multiple proportions and the law of constant composition (Section 1.1) were key ideas that formed the basis for Dalton's atomic theory.

**SAMPLE EXERCISE 2.5** **Using Chemical Formulas and the Law of Multiple Proportions** **LO6**

Carbon combines with oxygen to form either CO or $CO_2$ (Figure 2.13), depending on reaction conditions. If 26.6 g of oxygen reacts with 10.0 g of carbon to make $CO_2$, how many grams of oxygen react with 10.0 g of carbon to make CO?

**Collect and Organize** We have formulas for both compounds, CO and $CO_2$, and are told that both reactions involve 10.0 g of carbon.

**Analyze** The ratio of the O atoms to C atoms in CO is 1:1. The ratio of O atoms to C atoms in $CO_2$ is 2:1. Therefore, half as much oxygen reacts with 10.0 g of carbon to make CO as reacts with 10.0 g of carbon to make $CO_2$.

**Solve** (26.6 g of oxygen) $\times \frac{1}{2}$ = 13.3 g of oxygen

**Think About It** Dalton's atomic view of these compounds is expressed by their chemical formulas. We used chemical formulas to calculate the different masses of oxygen required to react with a given mass of carbon to form each compound. In practice, the reverse is done: chemists analyze the masses of the elements in a compound and use that information to determine its chemical formula.

**Practice Exercise** Predict the mass of oxygen required to react with 14.0 g of nitrogen to make $N_2O_5$ if 16.0 g of oxygen reacts with 14.0 g of nitrogen to make $N_2O_2$ (Figure 2.14).

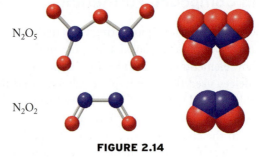

N₂O₅

N₂O₂

**FIGURE 2.14**

*(Answers to Practice Exercises are in the back of the book.)*

## Molecular Compounds

The compounds we have examined so far in this section have been binary (two-element) **molecular compounds** (as may be seen by the molecular structures of several of them in Figure 2.15). The building blocks of these compounds are molecules that contain atoms of two nonmetals. The atoms are held together by shared pairs of electrons called **covalent bonds** (the sticks in the ball-and-stick models in Figure 2.15).

All the compounds in Figure 2.15 may be present in the air we breathe, especially if we live in an area where air quality is affected by exhaust from vehicles, power plants, and factories. All the compounds shown contain oxygen and another nonmetal, so they are examples of *nonmetal oxides*. Each of their chemical formulas specifies the number of atoms of each element in one molecule of the compound. Therefore, these chemical formulas are **molecular formulas**. The fact that the same two elements can form compounds with different molecular formulas means that there are different ways to form covalent bonds between atoms of the same two elements. In later chapters of this book, we demonstrate the impact of different bonding patterns on the chemical properties of molecular compounds.

## Ionic Compounds

Now we shift our focus to binary **ionic compounds**, which all contain a metallic element (shown in tan in the periodic table inside the front cover) combined with a nonmetal (shown in blue). In an ionic compound, each atom of the metal has lost one or more electrons, forming a positively charged ion called a **cation**. Also, each atom of the nonmetal has gained one or more electrons, forming a negatively charged **anion**. Because these ions start out as single atoms, they are called *monatomic* ions. The cations and anions in an ionic compound are held together by the strong electrostatic attraction that ions of opposite charge have for each other.

Consider the formation of a common ionic compound, sodium chloride (table salt). We can use single atoms of sodium (a metal) and chlorine (a nonmetal) to visualize how these elements combine. Each sodium atom loses an electron and forms a sodium ion:

$$Na \rightarrow Na^+ + e^-$$

Correspondingly, each chlorine atom gains an electron and forms a chloride ion:

$$Cl + e^- \rightarrow Cl^-$$

Figure 2.16(a) illustrates the loss and gain of electrons by these atoms. Notice that when the sodium atom loses an electron and forms the sodium ion, the cation is smaller than the neutral atom. When a chlorine atom gains an electron, the anion is larger than the neutral atom. We discuss in later chapters the details of how and why atoms gain or lose electrons and why the relative sizes of ions vary this way.

Figure 2.16(b) shows several crystals of sodium chloride and an atomic view of part of a crystal, revealing the three-dimensional array of sodium ions and chloride ions. Within the crystal, each sodium ion is surrounded by six chloride ions, and each chloride ion is surrounded by six sodium ions. However, the smallest whole-number ratio of sodium ions to chloride ions in the crystal is simply 1:1. Formulas based on the lowest whole-number ratio of the elements in a compound are called **empirical formulas**. The chemical formulas of ionic compounds, such as NaCl for sodium chloride, are examples of empirical formulas. The empirical formula of an ionic compound describes a **formula unit**, the smallest electrically neutral unit within the crystal.

The periodic table helps us predict the charges on the monatomic ions that elements form and thereby helps us predict the chemical formulas of ionic compounds. For example, atoms of group 1 elements each lose one electron and form 1+ ions; atoms of group 2 elements each lose two electrons and form 2+ ions (Figure 2.17). Note that the charges on these monatomic cations match the group numbers. Unfortunately, no strong correlation exists between group number and

**cation** a positively charged particle created when an atom or molecule loses one or more electrons.

**anion** a negatively charged particle created when an atom or molecule gains one or more electrons.

**empirical formula** a formula showing the smallest whole-number ratio of elements in a compound.

**formula unit** the smallest electrically neutral unit of an ionic compound.

▶❚❚ **CHEMTOUR** NaCl Reaction

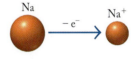
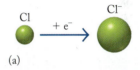

(a)

(b)                                    One formula unit

**FIGURE 2.16** (a) A sodium atom forms a $Na^+$ cation by losing one electron. A chlorine atom forms a $Cl^-$ anion by gaining one electron. (b) Crystals of sodium chloride. The cubic shape of the crystals mirrors the cubic array of $Na^+$ and $Cl^-$ ions that make up its structure. The empirical formula NaCl describes the smallest whole-number ratio of cations to anions in the structure, which is electrically neutral.

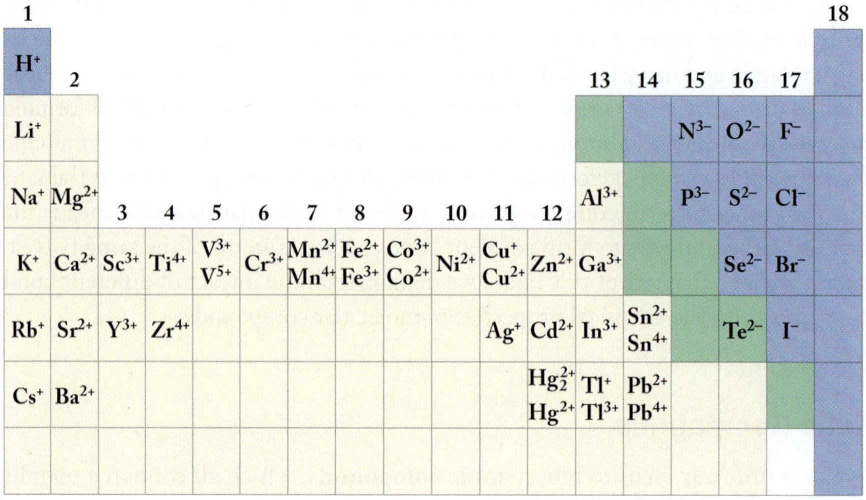

**FIGURE 2.17** The most common charges on the ions of some common elements. With the representative elements (groups 1, 2, and 13–18), all the ions in the group typically have the same charge. All the ions shown are made of only one atom except for the mercury ion $Hg_2^{2+}$.

| 1 | | | | | | | | | | | | | | | | | 18 |
|---|---|---|---|---|---|---|---|---|---|---|---|---|---|---|---|---|---|
| $H^+$ | 2 | | | | | | | | | | | 13 | 14 | 15 | 16 | 17 | |
| $Li^+$ | | | | | | | | | | | | | | $N^{3-}$ | $O^{2-}$ | $F^-$ | |
| $Na^+$ | $Mg^{2+}$ | 3 | 4 | 5 | 6 | 7 | 8 | 9 | 10 | 11 | 12 | $Al^{3+}$ | | $P^{3-}$ | $S^{2-}$ | $Cl^-$ | |
| $K^+$ | $Ca^{2+}$ | $Sc^{3+}$ | $Ti^{4+}$ | $V^{3+}$ $V^{5+}$ | $Cr^{3+}$ | $Mn^{2+}$ $Mn^{4+}$ | $Fe^{2+}$ $Fe^{3+}$ | $Co^{3+}$ $Co^{2+}$ | $Ni^{2+}$ | $Cu^+$ $Cu^{2+}$ | $Zn^{2+}$ | $Ga^{3+}$ | | | $Se^{2-}$ | $Br^-$ | |
| $Rb^+$ | $Sr^{2+}$ | $Y^{3+}$ | $Zr^{4+}$ | | | | | | | $Ag^+$ | $Cd^{2+}$ | $In^{3+}$ | $Sn^{2+}$ $Sn^{4+}$ | | $Te^{2-}$ | $I^-$ | |
| $Cs^+$ | $Ba^{2+}$ | | | | | | | | | | | $Hg_2^{2+}$ $Hg^{2+}$ | $Tl^+$ $Tl^{3+}$ | $Pb^{2+}$ $Pb^{4+}$ | | | |
| | | | | | | | | | | | | | | | | | |

cation charge among the transition metals and the metallic elements on the right side of the periodic table, as you can see in Figure 2.17. Still, similarities are apparent within groups. For example, the most common charge of the group 13 monatomic ions is 3+.

As metallic elements lose electrons in forming ionic compounds, their nonmetallic partners gain them so that the overall charge on the resultant compound is zero. As Figure 2.17 shows, the charge on the monatomic anions formed by the group 17 elements is 1−; the charge on the monatomic anions formed by the group 16 nonmetals is 2−; and the charge is 3− for the nonmetals in group 15. Note that the charge of each of these anions is their group number minus 18.

> **CONCEPT TEST** ...................................................
>
> Which of the following empirical formulas does not represent an electrically neutral compound? *Hint:* Base your selection on the charges of the common ions in Figure 2.17. (a) KBr; (b) $MgF_2$; (c) CsN; (d) $TiO_2$; (e) AgCl
>
> *(Answers to Concept Tests are in the back of the book.)*
> ...................................................

**FIGURE 2.18**

> **CONCEPT TEST** ...................................................
>
> Figure 2.18 shows a space-filling model of hydrogen peroxide. What are the molecular formula and empirical formula of hydrogen peroxide?
>
> *(Answers to Concept Tests are in the back of the book.)*
> ...................................................

---

**SAMPLE EXERCISE 2.6**   **Classifying Compounds as Molecular or Ionic**   **LO4**

Identify each of the following compounds as ionic or molecular: (a) sodium bromide (NaBr); (b) carbon dioxide ($CO_2$); (c) lithium iodide (LiI); (d) magnesium fluoride ($MgF_2$); (e) calcium chloride ($CaCl_2$).

**Collect and Organize**  We are given the formulas of compounds. We can use the periodic table to determine which elements in the compounds are metallic and which are nonmetallic.

**Analyze**  NaBr, LiI, $MgF_2$, and $CaCl_2$ all contain a group 1 or group 2 metal and a group 17 nonmetal. Only $CO_2$ is composed of two nonmetals.

**Solve** (a) NaBr, (c) LiI, (d) $MgF_2$, and (e) $CaCl_2$ are ionic; (b) $CO_2$ is molecular.

**Think About It** In later chapters we will study more complicated compounds than addressed in this exercise. Some covalent bonds have a degree of ionic "character," and we will describe a way based on the elements' positions in the periodic table to determine how much ionic character covalent bonds have.

⚙ **Practice Exercise** Identify the following binary compounds as molecular or ionic: (a) carbon disulfide ($CS_2$); (b) carbon monoxide (CO); (c) ammonia ($NH_3$); (d) water ($H_2O$); (e) sodium iodide (NaI).

*(Answers to Practice Exercises are in the back of the book.)*

## 2.6 Naming Compounds and Writing Formulas

At this point we need to establish some rules for naming compounds and writing their chemical formulas. These names and formulas are a foundation of the language of chemistry. The periodic table is a valuable resource for naming simple compounds, for translating those names into chemical formulas, and for translating chemical formulas into names.

### Binary Molecular Compounds

Translating the molecular formula of a binary molecular compound into the compound name is straightforward:

1. Start with the name of the first element in the formula. This is usually the element farther to the left in the periodic table (lower group number). If both elements are in the same group—for example, sulfur and oxygen—the element with the higher atomic number (larger period number) comes first.
2. Change the ending of the name of the second element to *-ide*.
3. Add prefixes (Table 2.2) to the first and second names to indicate the number of atoms of each type in the molecule. (Exception: do not use the prefix *mono-* with the first element in a name.)

For example, NO is nitrogen monoxide (not *mono*nitrogen monoxide), $NO_2$ is nitrogen dioxide, $SO_2$ is sulfur dioxide, and $SO_3$ is sulfur trioxide. When prefixes ending in *o-* or *a-* (like *mono-* and *tetra-*) precede a name that begins with a vowel (such as *oxide*), the o or a at the end of the prefix is deleted to make the combination of prefix and name easier to pronounce: monoxide, not *mono*oxide.

| TABLE 2.2 | Naming Prefixes for Molecular Compounds |
|---|---|
| one | *mono-* |
| two | *di-* |
| three | *tri-* |
| four | *tetra-* |
| five | *penta-* |
| six | *hexa-* |
| seven | *hepta-* |
| eight | *octa-* |
| nine | *nona-* |
| ten | *deca-* |
| eleven | *undeca-* |
| twelve | *dodeca-* |

**SAMPLE EXERCISE 2.7** **Naming Binary Molecular Compounds** **LO6**

What are the names of the compounds with these chemical formulas: (a) $N_2O$; (b) $N_2O_4$; (c) $N_2O_5$? What are the chemical formulas of these compounds: (d) chlorine dioxide; (e) dinitrogen monoxide; (f) diphosphorus pentoxide?

**Collect and Organize** All six compounds are binary nonmetal oxides and hence are molecular compounds.

**Analyze** We use prefixes from Table 2.2 in the names to indicate the number of atoms of each element present in one molecule.

In the first question, the first element in all three compounds is nitrogen, so the first word in each name is *nitrogen* with the appropriate prefix. The second element in all three compounds is oxygen, so the second word in each name is *oxide* with the appropriate prefixes from Table 2.2.

In the second question, all the molecules contain oxygen, and their names end in oxide. Note that in (d), chlorine is in a higher period than oxygen, so its symbol comes first. The prefixes indicate the numbers of atoms in each molecule.

**Solve** (a) dinitrogen monoxide; (b) dinitrogen tetroxide; (c) dinitrogen pentoxide; (d) $ClO_2$; (e) $N_2O$; (f) $P_2O_5$.

**Think About It** Using the correct prefixes is essential to distinguish among compounds that have different numbers of the same constituent atoms.

⚙ **Practice Exercise** Name these compounds: (a) $P_4O_{10}$; (b) CO; (c) $NCl_3$. What are the formulas for these compounds: (d) sulfur hexafluoride; (e) iodine monochloride; (f) dibromine monoxide?

*(Answers to Practice Exercises are in the back of the book.)*

## Binary Ionic Compounds

To name a binary ionic compound:

1. Start with the name of the cation, which is simply the name of the parent element.
2. Add the name of the anion, which is the name of the element except that the ending is changed to *-ide*.

Prefixes are not used in naming binary ionic compounds of representative elements because group 1 and 2 metals and aluminum in group 13 all have characteristic positive charges, as do the monatomic anions formed by the group 16 and 17 elements. Ionic compounds are electrically neutral, so the negative and positive charges in an ionic compound must balance, which dictates the number of each of the ions in the chemical formula. Therefore, the name *magnesium fluoride*, for example, is unambiguous: it can mean only $MgF_2$.

---

**SAMPLE EXERCISE 2.8** | **Writing Formulas of Binary Ionic Compounds** | **LO6**

---

Write the chemical formula of (a) potassium bromide, (b) calcium oxide, (c) sodium sulfide, (d) magnesium chloride, and (e) aluminum oxide.

**Collect and Organize** The name of each compound consists of the name of one main group metal and one main group nonmetal, which tells us that these are binary ionic compounds.

**Analyze** To write formulas of ionic compounds, assign the charges to the ions based on the group numbers of the parent elements. Locate each element in the periodic table and predict the charge of its most common ion based on location and group number: $K^+$, $Br^-$, $Ca^{2+}$, $O^{2-}$, $Na^+$, $S^{2-}$, $Mg^{2+}$, $Cl^-$, and $Al^{3+}$. If you have difficulty predicting ionic charge, refer to Figure 2.17. Writing chemical formulas of the compounds is an exercise in balancing positive and negative charges.

**Solve** We must balance the positive and negative charges in each compound:
a. In potassium bromide, the ionic charges are 1+ and 1− ($K^+$ and $Br^-$). A 1:1 ratio of the ions is required for electrical neutrality, making the formula KBr.
b. In calcium oxide, the ionic charges are 2+ and 2−. A 1:1 ratio of $Ca^{2+}$ to $O^{2-}$ ions balances their charges, making the formula CaO.
c. In sodium sulfide, the ionic charges are 1+ and 2−. A 2:1 ratio of $Na^+$ to $S^{2-}$ ions is needed: $Na_2S$.
d. In magnesium chloride, the ionic charges are 2+ and 1−. A 1:2 ratio of $Mg^{2+}$ to $Cl^-$ ions is needed: $MgCl_2$.
e. In aluminum oxide, the ionic charges are 3+ and 2−. If we use two $Al^{3+}$ ions for every three $O^{2-}$ ions, the charges will balance. The formula is $Al_2O_3$.

**Think About It** Different approaches may be used to work out the formulas of ionic compounds. The basic principle is that the sum of the total positive and negative charges must balance to give a net charge of zero. If you had difficulty writing the formula of aluminum oxide, try this shortcut: use the charge on each ion as the subscript for the other ion. Thus the 3+ charge on $Al^{3+}$ becomes a subscript 3 of O, and the 2− charge on the oxide ion becomes a subscript 2 of Al, resulting in $Al_2O_3$.

$$Al^{3+}O^{2-} \rightarrow Al_2O_3$$

⚙ **Practice Exercise** Write the chemical formulas of (a) strontium chloride, (b) magnesium oxide, (c) sodium fluoride, and (d) calcium bromide.

*(Answers to Practice Exercises are in the back of the book.)*

## Binary Compounds of Transition Metals

Some metallic elements, including many of the transition metals, form several cations carrying different charges. For example, most of the copper found in nature is present as $Cu^{2+}$; however, some copper compounds contain $Cu^+$ ions. Because the name *copper chloride* could apply to either $CuCl_2$ or CuCl, systematic names are needed to distinguish between the two compounds.

For many years, chemists used different names to identify different cations of the same element. For instance, $Cu^+$ was called the *cuprous* ion and $Cu^{2+}$ was called *cupric*. Similarly, $Fe^{2+}$ and $Fe^{3+}$ have long been called ferrous and ferric ions, respectively. Note that, in both pairs of ions, the name with the lower charge ends in *-ous* and the name of the ion with the higher charge ends in *-ic*.

However, throughout this book we use the more modern system of adding a Roman numeral after the name of the transition metal in a compound to define the charge on the metal ion. This system is often called the Stock system, after the German chemist Alfred Stock (1876–1946). Thus, copper(II) chloride is the chloride of $Cu^{2+}$ ($CuCl_2$) (pronounced "copper two chloride"), and copper(I) chloride (CuCl; "copper one chloride") is the chloride of $Cu^+$.

Roman numerals are used to indicate the charge of nearly all transition metals except for $Ag^+$, $Cd^{2+}$, and $Zn^{2+}$ because these ions are the only ones that these metals normally form.

**SAMPLE EXERCISE 2.9** **Writing Formulas of Transition Metal Compounds** **LO6**

a. Write the chemical formulas of iron(II) sulfide and iron(III) oxide.
b. Write alternative names for these compounds that do not use Roman numerals to indicate the charge on the iron ions.

**Collect and Organize** Because iron is a transition metal, Roman numerals are used to indicate the charges on the iron cations.

**Analyze** The Roman numerals (II) and (III) indicate that the charges on the iron cations are 2+ and 3+, respectively. Oxygen and sulfur are both in group 16. Therefore the charge on both the sulfide ion and oxide ion is 2−. In the old naming system, $Fe^{2+}$ is the *ferrous* ion and $Fe^{3+}$ is the *ferric* ion.

**Solve**
a. A charge balance in iron(II) sulfide is achieved with equal numbers of $Fe^{2+}$ and $S^{2-}$ ions, so the chemical formula is FeS. To balance the different charges on the $Fe^{3+}$ and $O^{2-}$ ions in iron(III) oxide, we need three $O^{2-}$ ions for every two $Fe^{3+}$ ions. Thus the formula of iron(III) oxide is $Fe_2O_3$.
b. The old names of FeS and $Fe_2O_3$ are ferrous sulfide and ferric oxide, respectively.

**polyatomic ions** charged groups of two or more atoms joined together by covalent bonds.

**oxoanions** polyatomic ions that contain oxygen in combination with one or more other elements.

**Think About It** We use the Stock system for designating the charges on transition metal ions, but you may encounter *-ous/-ic* nomenclature in older books and articles.

⚙ **Practice Exercise** Write the formulas of manganese(II) chloride and manganese(IV) oxide.

*(Answers to Practice Exercises are in the back of the book.)*

## Polyatomic Ions

Table 2.3 lists some commonly encountered ions. Most of them are **polyatomic ions**, which means they consist of more than one kind of atom joined by covalent bonds. The ammonium ion ($NH_4^+$) is the only common polyatomic cation; all the others are anions. Polyatomic ions containing oxygen and one or more other elements are called **oxoanions**. Most oxoanions have a name based on the name of the element that appears first in the formula, but the ending is changed to either *-ite* or *-ate*, depending on the number of oxygen atoms in the formula (the *-ate* ions have more O atoms than the corresponding *-ite* ions). When writing the formula of a compound with two or more of the same oxoanion per formula unit, we put parentheses around the formula of the oxoanion to make it clear that the subscript that follows applies to the entire ion. For example, there are three sulfate ions in $Al_2(SO_4)_3$.

If an element forms more than two kinds of oxoanions, as chlorine and the other halogens do, prefixes are used to distinguish among them (Table 2.4). The oxoanion with the largest number of oxygen atoms may have the prefix *per-*, and the one with the smallest number of oxygen atoms may have the prefix *hypo-* in its name. Because these rules may not enable you to predict the chemical formula

### TABLE 2.3 Names and Charges of Some Common Ions

| Name | Chemical Formula | Name | Chemical Formula |
|---|---|---|---|
| Acetate | $CH_3COO^-$ | Hydrogen phosphate | $HPO_4^{2-}$ |
| Ammonium | $NH_4^+$ | Hydrogen sulfite or bisulfite | $HSO_3^-$ |
| Azide | $N_3^-$ | Hydroxide | $OH^-$ |
| Bromide | $Br^-$ | Nitrate | $NO_3^-$ |
| Carbonate | $CO_3^{2-}$ | Nitride | $N^{3-}$ |
| Chlorate | $ClO_3^-$ | Nitrite | $NO_2^-$ |
| Chloride | $Cl^-$ | Oxide | $O^{2-}$ |
| Chromate | $CrO_4^{2-}$ | Perchlorate | $ClO_4^-$ |
| Cyanide | $CN^-$ | Permanganate | $MnO_4^-$ |
| Dichromate | $Cr_2O_7^{2-}$ | Peroxide | $O_2^{2-}$ |
| Dihydrogen phosphate | $H_2PO_4^-$ | Phosphate | $PO_4^{3-}$ |
| Disulfide | $S_2^{2-}$ | Sulfate | $SO_4^{2-}$ |
| Fluoride | $F^-$ | Sulfide | $S^{2-}$ |
| Hydride | $H^-$ | Sulfite | $SO_3^{2-}$ |
| Hydrogen carbonate or bicarbonate | $HCO_3^-$ | Thiocyanate | $SCN^-$ |

| TABLE 2.4 | Oxoanions of Chlorine and Their Corresponding Acids | | |
|---|---|---|---|
| **Ions** | | **Acids** | |
| $ClO^-$ | hypochlorite | $HClO$ | hypochlorous acid |
| $ClO_2^-$ | chlorite | $HClO_2$ | chlorous acid |
| $ClO_3^-$ | chlorate | $HClO_3$ | chloric acid |
| $ClO_4^-$ | perchlorate | $HClO_4$ | perchloric acid |

of an oxoanion from its name or its name from its formula, you need to memorize the formulas, charges, and names of the polyatomic ions in Tables 2.3 and 2.4 that your instructor feels are most important.

---

**SAMPLE EXERCISE 2.10**  **Writing the Formulas of Compounds Containing Oxoanions**  **LO6**

Write the chemical formulas of (a) sodium sulfate and (b) magnesium phosphate.

**Collect and Organize** We are given the names of two compounds containing oxoanions and are to write their chemical formulas. The cations in these compounds are those formed by Na and Mg atoms.

**Analyze** To write the formulas of these ionic compounds, we need to know the formulas and charges of the ions. Sodium is in group 1 and magnesium is in group 2, so the charges on their ions are 1+ and 2+, respectively. The sulfate ion is $SO_4^{2-}$, and phosphate is $PO_4^{3-}$.

**Solve**
a. To balance the charges on $Na^+$ and $SO_4^{2-}$, we need twice as many $Na^+$ ions as $SO_4^{2-}$ ions. Therefore the formula is $Na_2SO_4$.
b. To balance the charges on $Mg^{2+}$ and $PO_4^{3-}$, we need three $Mg^{2+}$ ions for every two $PO_4^{3-}$ ions, which gives us $Mg_3(PO_4)_2$.

**Think About It** To complete this exercise we had to know the formulas and charges of the sulfate and phosphate oxoanions. The charges on the cations could be inferred from the positions of the elements in the periodic table.

⚙ **Practice Exercise** Write the chemical formulas of (a) strontium nitrate and (b) potassium sulfite.

*(Answers to Practice Exercises are in the back of the book.)*

---

**SAMPLE EXERCISE 2.11**  **Naming Compounds Containing Oxoanions**  **LO6**

Name the following compounds: (a) $CaCO_3$; (b) $LiNO_3$; (c) $MgSO_3$; (d) $RbNO_2$; (e) $KClO_3$; and (f) $NaHCO_3$.

**Collect and Organize** The names of ionic compounds begin with the names of the parent elements of the cations followed by the names of the oxoanions.

**Analyze** The cations in these compounds are those formed by atoms of the elements (a) calcium, (b) lithium, (c) magnesium, (d) rubidium, (e) potassium, and (f) sodium. The names of the oxoanions, as listed in Table 2.3, are (a) carbonate, (b) nitrate, (c) sulfite, (d) nitrite, (e) chlorate, and (f) hydrogen carbonate.

**Solve** Combining the names of these cations and oxoanions, we get (a) calcium carbonate, (b) lithium nitrate, (c) magnesium sulfite, (d) rubidium nitrite, (e) potassium chlorate, and (f) sodium hydrogen carbonate.

**Think About It** Sodium hydrogen carbonate is often called sodium bicarbonate. The prefix *bi-* is sometimes used to indicate that there is a hydrogen ion ($H^+$) attached to an oxoanion.

⚙ **Practice Exercise** Name the following compounds: (a) $Ca_3(PO_4)_2$; (b) $Mg(ClO_4)_2$; (c) $LiNO_2$; (d) $NaClO$; (e) $KMnO_4$.

*(Answers to Practice Exercises are in the back of the book.)*

## Acids

Some compounds have special names that highlight particular chemical properties. Among these are acids. We discuss acids in greater detail in later chapters, but for now it is sufficient to say that acids are compounds that release hydrogen ions ($H^+$) when they dissolve in water. Their chemical formulas begin with H. For example, the binary compound HCl is hydrogen chloride. When HCl dissolves in water, it produces the acidic solution we call hydrochloric acid. To name binary acids:

1. Affix the prefix *hydro-* to the name of the element other than hydrogen.
2. Replace the last syllable in the element name from step 1 with the suffix *-ic*, and add *acid*.

The most common binary acids are compounds of hydrogen and the halogens. Their aqueous solutions are hydrofluoric, hydrochloric, hydrobromic, and hydroiodic acid.

The scheme for naming the acids of oxoanions, called *oxoacids*, is illustrated in Table 2.4. If the oxoanion name ends in *-ate*, the name of the corresponding oxoacid ends in *-ic*; if the oxoanion name ends in *-ite*, the name of the oxoacid ends in *-ous*. Thus, the acid of the sulf*ate* ($SO_4^{2-}$) ion is sulfur*ic* acid ($H_2SO_4$) and the acid of the nitr*ite* ($NO_2^-$) ion is nitr*ous* acid ($HNO_2$).

**SAMPLE EXERCISE 2.12  Naming Oxoacids  LO6**

Name the oxoacids formed by the following oxoanions: (a) $SO_3^{2-}$; (b) $ClO_4^-$; (c) $NO_3^-$.

**Collect and Organize** We are given the formulas of three oxoanions and are to name the oxoacids formed when they combine with $H^+$ ions.

**Analyze** According to Table 2.3, the names of the oxoanions are (a) sulfite, (b) perchlorate, and (c) nitrate. When the oxoanion name ends in *-ite*, the corresponding oxoacid name ends in *-ous*. When the anion name ends in *-ate*, the oxoacid name ends in *-ic*.

**Solve** Making the appropriate changes to the endings of the oxoanion names and adding the word *acid*, we get (a) sulfurous acid, (b) perchloric acid, and (c) nitric acid.

**Think About It** Note that you cannot tell the names of the oxoanions just by looking at them. You have to remember the names associated with each family of polyatomic ions.

⚙ **Practice Exercise** Name these acids: (a) $HClO$; (b) $HClO_2$; (c) $H_2CO_3$.

*(Answers to Practice Exercises are in the back of the book.)*

## 2.7 Nucleosynthesis

We began this chapter by discussing remarkable advances in the early 20th century that provided science with a clearer view of atomic structure and of why elements react the way they do. Later in the century, scientists increased our understanding of how the elements may have formed. We explored these concepts and the Big Bang theory in Chapter 1, discussing how the two most abundant elements in the universe—hydrogen and helium—formed shortly after the Big Bang. The question is, starting with hydrogen and helium, where and how did all the other elements form?

Today, theoretical physicists believe that much of the energy released at the instant of the Big Bang had transformed into matter within a few microseconds of its creation. This matter consisted of the smallest of subatomic particles: electrons and **quarks** (Figure 2.19). Less than a millisecond later, the universe had expanded and "cooled" to a mere $10^{12}$ K, and quarks combined with one another to form neutrons and protons. Thus in less than a second, the matter in the universe consisted of the three types of subatomic particles that would eventually make up all atoms.

## Primordial Nucleosynthesis

By about four minutes after the Big Bang, the universe had expanded and cooled to $10^9$ K. In this hot, dense, subatomic "soup," neutrons and protons that collided with one another began to fuse together in a process called primordial **nucleosynthesis**. (We discuss fusion in more detail in Chapter 21.) In one step, protons (p) and neutrons (n) fused to form *deuterons* (d), which are nuclei of the deuterium ($_1^2$H) isotope of hydrogen:

$$_1^1\text{p} + {_0^1}\text{n} \rightarrow {_1^2}\text{d} \qquad (2.2)$$

In writing this equation, we follow the rules described in Section 2.2 for writing nuclide symbols: the superscript is a mass number and the subscript is an atomic number. We use a similar convention for writing the symbols of subatomic particles, except that in this case a subscript represents the charge on the particle. For example, the symbol of the neutron is $_0^1$n because a neutron has a mass number of 1 and a charge of 0.

Deuteron formation proceeded rapidly, consuming most of the neutrons in the universe in a matter of seconds. No sooner did deuterons form than they too were rapidly consumed by colliding with one another and fusing to make $_2^4$He nuclei, which are the same as α particles:

$$2\,_1^2\text{d} \rightarrow {_2^4}\alpha \qquad (2.3)$$

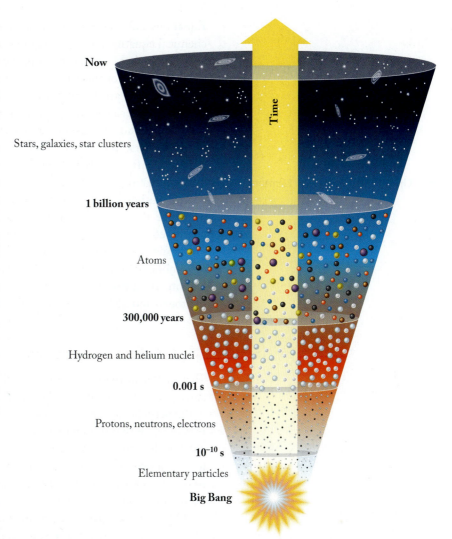

**FIGURE 2.19** Timeline for energy and matter transformations believed to have occurred since the universe began. In this model, protons and neutrons were formed from quarks in the first millisecond after the Big Bang, followed by hydrogen and helium nuclei. Whole atoms of H and He did not form until after 300,000 years of expansion and cooling, and other elements did not form until the first galaxies appeared, around 1 billion years after the Big Bang. According to this model, our solar system, our planet, and all life forms on it are composed of elements synthesized in stars that were born, burned brightly, and then disappeared millions to billions of years after the Big Bang.

**quarks** elementary particles that combine to form neutrons and protons.

**nucleosynthesis** the natural formation of nuclei as a result of fusion and other nuclear processes.

▶❙❙ **CHEMTOUR** Synthesis of Elements

Equations 2.2 and 2.3 are called *nuclear equations*, and they are related to the chemical equations we will begin writing in Chapter 3 in that they are *balanced*. This means that in each equation, the sum of the masses (superscripts) of the particles to the left of the arrow is equal to the sum of the masses of the particles to the right of the arrow. Similarly, the sum of the charges (subscripts) of the particles on the left side is equal to the sum of the charges of the particles on the right.

As a result of the fusion reactions shown in Equations 2.2 and 2.3, the matter of the universe was about 75% (by mass) protons and 25% α particles by about 5 minutes after the Big Bang. Primordial nucleosynthesis then came to a halt. Why didn't α particles and protons fuse together to make heavier particles? Nucleosynthesis requires extreme conditions of high temperature or pressure to overcome the mutual repulsion of positively charged particles. As the universe continued to expand and cool, its temperature dropped below $10^8$ K, which is too low to sustain nuclear fusion. Instead, these cooler temperatures allowed electrons to combine with protons and α particles, forming neutral atoms of hydrogen and helium. As a result, the elemental composition of the universe remained 75% hydrogen and 25% helium for millions of years.

## Stellar Nucleosynthesis

That matter in the universe today still consists mostly of atoms of $^1$H and $^4$He is strong evidence supporting the Big Bang theory. But how did the other elements in the periodic table form, including those that make up most of our planet? Scientists theorize that the synthesis of elements more massive than helium had to wait until nuclear fusion resumed in the first generation of stars. Inside the coalescing masses of hydrogen and helium that would turn into the first stars, the gases underwent enormous compression heating. The nuclear furnaces that are the source of the energy in all stars were ignited as protons began to fuse, making more α particles.

Once nuclear fusion begins in stars, it continues in stages. When the hydrogen at the star's core has all fused into helium, gravity causes the core to contract, causing even higher temperatures and pressures. This is enough for helium to start to fuse, producing carbon, while the star outside the core expands, becoming what astronomers call a giant or supergiant. Over time, the star's core forms shells of heavier elements produced from the fusion of lighter elements. As these shells compress, even the heavier elements begin to fuse.

For the past 13 billion years, fusion reactions of the sort described above have simultaneously fueled the nuclear furnaces of stars and produced isotopes as heavy as $^{56}$Fe (Figure 2.20). However, once the core of a star turns into iron, the star is in trouble because fusion reactions involving iron nuclei do not release energy; instead they *consume* it. For example, fusing an α particle with $^{56}$Fe to make $^{60}$Ni requires the *addition* of energy:

$$^{56}_{26}\text{Fe} + {}^4_2\alpha \rightarrow {}^{60}_{28}\text{Ni}$$

A star with an iron core has essentially run out of fuel. Its nuclear furnace goes out, and the star begins to cool and collapse into itself.

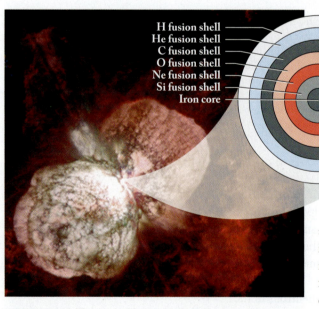

H fusion shell
He fusion shell
C fusion shell
O fusion shell
Ne fusion shell
Si fusion shell
Iron core

**FIGURE 2.20** The star η-Carinae is believed to be evolving toward an explosion. The outer regions are still fueled by energy released as hydrogen isotopes fuse, but the star is increasingly hotter and denser closer to the center. This central heating allows the fusion of larger nuclei and results in the production of $^{56}$Fe in the core.

As the star collapses, compression reheats its core to above $10^9$ K. At such temperatures, nuclei begin to disintegrate into free protons and neutrons. Free neutrons readily fuse with atomic nuclei, producing heavier elements than iron. Repeated fusion in the cores of collapsing stars produces nuclei of the most massive elements in the periodic table.

However, one more chapter in the element-manufacturing story needs to be told: distribution of the products. The enormous heating that occurs when a dying star collapses also produces a gigantic explosion. Cosmologists call such an event a *supernova*. In addition to finishing the job of synthesizing the elemental building blocks found in the universe, a supernova serves as its own element-distribution system, blasting its inventory of elements throughout its galaxy (Figure 2.21). The legacies of supernovas are found in the elemental composition of later-generation stars, like our Sun, and in the planets that orbit these stars. Indeed, everything that exists in our solar system, including all life, is made from the residues of exploding stars.

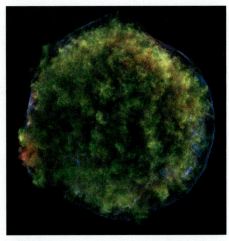

**FIGURE 2.21** This colorized picture from the Chandra X-ray Observatory shows the remains of the supernova of the star Tycho in the constellation Cassiopeia. The expanding bubble of debris colored red and green is a cloud of hot ionized gas inside a more rapidly moving shell of extremely high-energy electrons in blue.

---

**SAMPLE EXERCISE 2.13 Integrating Concepts: Radioactive Isotopes in Practice**

Dmitri Mendeleev predicted the existence of one of the elements that was unknown in his time—technetium ($Z = 43$). Technetium has over 40 isotopes, ranging in mass from 85 to 118 amu. Every isotope of technetium is radioactive. Most technetium on Earth is produced from $^{235}$U in nuclear reactors: 1.00 g of $^{235}$U yields 27 mg of $^{99}$Tc. Very small amounts of technetium have been found in Africa in the ore pitchblende. It is estimated that $1.22 \times 10^{12}$ atoms of $^{99}$Tc are contained in 1.00 kg of pitchblende. Technetium compounds are used most commonly in diagnostic medicine to image organ function (Figure 2.22). Very small quantities of technetium may be added to steel as a corrosion inhibitor, but its use in this fashion is severely limited because of its radioactivity. Given: the atomic mass of $^{99}$Tc is 98.906 amu.

a. What are the complete symbols for the lightest and the heaviest isotopes of technetium?

b. How many protons, neutrons, and electrons are in a neutral atom of $^{99}$Tc?

c. Compare the amount of technetium produced in a nuclear reactor from 1.00 g of $^{235}$U to the amount in 1.00 kg of African pitchblende.

**Collect and Organize** We are given the values of $A$ and $Z$ for Tc as well as information about the amount of $^{99}$Tc available from two different sources.

**Analyze** We can use the values given to write the symbols and to determine the number of protons and neutrons in the nucleus of different isotopes. The number of electrons in the neutral atom is equal to the number of protons in the nucleus. We can use dimensional analysis to convert the number of atoms of Tc to milligrams of Tc in the ore and then mathematically compare the two amounts.

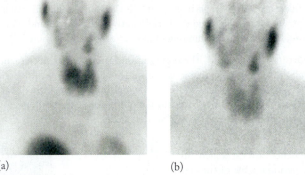

(a)                                    (b)

**FIGURE 2.22** Technetium is used to study the function of organs such as the thyroid gland. Typically (a) an image is taken soon (in this case, 10 minutes) after injection of a radioactive Tc compound. (b) After a period of time (here: 90 minutes), the thyroid is re-imaged to see how much Tc remains. The distribution of Tc should be symmetrical, producing a radiation image with two equal lobes that look like butterfly wings. Here the lobes are not the same, indicating a thyroid function deficiency.

**Solve**

a. For the lightest isotope of Tc, $Z = 43$ and $A = 85$; for the heaviest isotope, $Z = 43$ and $A = 118$. Therefore the two symbols are

lightest isotope: $^{85}_{43}$Tc    heaviest isotope: $^{118}_{43}$Tc

b. For $^{99}$Tc:

| | |
|---|---|
| protons | 43 |
| neutrons | 56 |
| electrons | 43 |

c. 1.00 g of $^{235}$U produces 27 mg $^{99}$Tc; 1.00 kg of pitchblende contains $1.22 \times 10^{12}$ atoms of $^{99}$Tc. We can convert the atoms of Tc per kilogram ore into milligrams of Tc per gram of ore to make a direct comparison.

$$\frac{1.22 \times 10^{12} \text{ atoms } ^{99}\text{Tc}}{1.00 \text{ kg ore}} \times \frac{98.906 \text{ amu}}{\text{atom}} \times \frac{1.66054 \times 10^{-24} \text{ g}}{\text{amu}}$$

$$\times \frac{1000 \text{ mg}}{1 \text{ g}} \times \frac{1 \text{ kg}}{1000 \text{ g}} = \frac{2.00 \times 10^{-10} \text{ mg } ^{99}\text{Tc}}{\text{g ore}}$$

Therefore, the 27 mg $^{99}$Tc produced by 1.00 g $^{235}$U in a reaction is:

$$\frac{27 \text{ mg } ^{99}\text{Tc}}{2.0 \times 10^{-10} \text{ mg } ^{99}\text{Tc}} = 1.34 \times 10^{11}$$

times as much $^{99}$Tc as in 1.00 kg pitchblende ore.

**Think About It** Because one of its isotopes emits gamma radiation that can be easily detected, technetium incorporated into soluble molecules can be used to observe the flow of fluids in organisms. The thyroid is a symmetrical, butterfly-shaped gland that surrounds the trachea. It appears in Figure 2.22(a) as the two large, dark lobes in the upper chest of a patient, imaged 10 minutes after the administration of a technetium-containing agent. The gland should be symmetrical, but one lobe is clearly smaller than the other. The irregular uptake of the agent indicates a malfunctioning thyroid. In Figure 2.22(b) the image taken 90 minutes later is fainter, indicating the clearance of the agent from the gland over time.

## SUMMARY

**Learning Outcome 1** Atoms consist of a nucleus containing **protons** and **neutrons** that accounts for nearly all the mass of the atom and is surrounded by **electrons**. Unique symbols for each **isotope** of every element indicate the identity of an atom by showing its **atomic number** and **mass number**. (Sections 2.1 and 2.2)

**Learning Outcome 2** Thomson's experiments with cathode rays, Millikan's oil-drop experiment, and Rutherford's gold foil experiment, in addition to studies of **radioactivity** by Pierre and Marie Curie, led to our understanding of atomic structure. (Section 2.1)

**Learning Outcome 3** The **average atomic mass** below an element's symbol on the periodic table is the weighted average of all the isotopes of that element. It is calculated by multiplying the mass of each of its stable isotopes by the **natural abundance** of that isotope, and then summing these products. (Section 2.3)

**Learning Outcome 4**
Elements are arranged in the periodic table of the elements in order of increasing atomic number and in a pattern based on their physical and chemical properties. Elements in the same vertical column are said to be in the same **group**, or *family*. (Section 2.4)

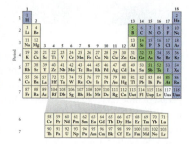

**Learning Outcome 5** The table is further divided several different ways into **main group** (or **representative**) **elements** and **transition metals, metals, metalloids,** and **nonmetals**. The location of an element on the chart gives information about its chemical and physical properties. (Section 2.4)

**Learning Outcome 6** The **law of multiple proportions** describes the ratio of masses of elements in simple compounds in terms of small whole numbers. Compounds may be either **ionic compounds** or **molecular compounds**, and different naming conventions are used for both types. The **empirical formula** of a molecular or ionic compound gives the smallest whole-number ratio of the atoms (or ions) in it. (Sections 2.5 and 2.6)

**Learning Outcome 7** During primordial **nucleosynthesis**, protons and neutrons fused to produce nuclei of helium. The nuclei of atoms as massive as $^{56}$Fe formed when the nuclei of lighter elements fused together in the cores of giant stars in a process called stellar nucleosynthesis, which continues today. Even more massive nuclei formed by a combination of other nuclear reactions, leading to supernovas (explosions of giant stars). (Section 2.7)

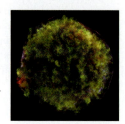

## PROBLEM-SOLVING SUMMARY

| TYPE OF PROBLEM | CONCEPTS AND EQUATIONS | SAMPLE EXERCISES |
|---|---|---|
| **Writing symbols of isotopes** | To the left of the element symbol, place a superscript for the mass number ($A$) and a subscript for the atomic number ($Z$). | 2.1, 2.2 |
| **Calculating the average atomic mass of an element** | Multiply the mass ($m$) of each stable isotope of the element times the natural abundance ($a$) of that isotope; then sum these products: $$m_X = a_1m_1 + a_2m_2 + a_3m_3 + \cdots \qquad (2.1)$$ | 2.3 |
| **Locating elements on the periodic table** | Use the row number, group designation, and chemical properties to locate elements. | 2.4 |
| **Calculating the quantity of an element in one compound based on the quantity of the same element in another compound** | Use the chemical formulas of the two compounds and the law of multiple proportions. | 2.5 |
| **Classifying compounds as ionic or molecular** | Ionic compounds contain metallic and nonmetallic elements; molecular compounds contain nonmetals and/or metalloids. | 2.6 |
| **Naming binary compounds and writing their formulas** | Apply the naming rules in Section 2.6. | 2.7, 2.8 |
| **Naming transition metal compounds and writing their formulas** | Use a Roman numeral to indicate the charge on the transition metal cation. | 2.9 |
| **Naming oxoacids and compounds containing oxoanions and writing their formulas** | Apply the naming rules in Section 2.6. | 2.10, 2.11, 2.12 |

## VISUAL PROBLEMS

*(Answers to boldface end-of-chapter questions and problems are in the back of the book.)*

**2.1.** In Figure P2.1 the blue spheres represent nitrogen atoms and the red spheres represent oxygen atoms. The figure as a whole represents which of the following gases? (a) $N_2O_3$; (b) $N_7O_{11}$; (c) a mixture of $NO_2$ and NO; (d) a mixture of $N_2$ and $O_3$

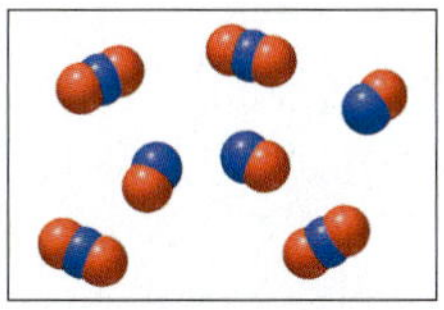

**FIGURE P2.1**

**2.2.** In Figure P2.2 the black spheres represent carbon atoms and the red spheres represent oxygen atoms. Which of the following statements about the two equal-volume compartments is or are true?

**FIGURE P2.2**

  a. The compartment on the left contains $CO_2$; the one on the right contains CO.

  b. The compartments contain the same mass of carbon.

  c. The ratio of the oxygen to carbon in the gas in the left compartment is twice that of the gas in the right compartment.

  d. The pressures inside the two compartments are equal. (Assume that the pressure of a gas is proportional to the number of particles in a given volume.)

**2.3.** Which particle-level diagram is the best representation of an $^{13}_{6}C^+$ ion?

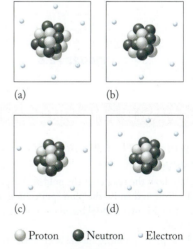

(a)              (b)

(c)              (d)

○ Proton    ● Neutron    ◡ Electron

**FIGURE P2.3**

2.4. Which of the highlighted elements in Figure P2.4 is *not* formed by fusion of lighter elements in the cores of giant stars?

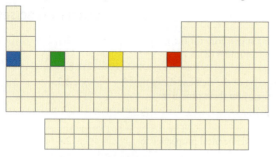

**FIGURE P2.4**

2.5. Which of the highlighted elements in Figure P2.5 is (a) a reactive nonmetal, (b) a chemically inert gas, (c) a reactive metal?

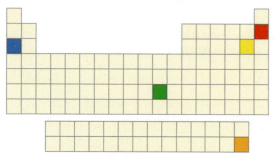

**FIGURE P2.5**

2.6. Which of the highlighted elements in Figure P2.6 forms monatomic ions with a charge of (a) 1+, (b) 2+, (c) 3+, (d) 1−, (e) 2−?

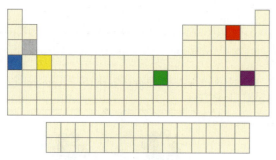

**FIGURE P2.6**

2.7. Which of the highlighted elements in Figure P2.7 forms an oxide with the following formula: (a) XO; (b) $X_2O$; (c) $XO_2$; (d) $X_2O_3$?

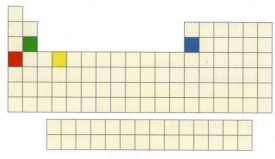

**FIGURE P2.7**

2.8. Which of the highlighted elements in Figure P2.8 forms an oxoanion with the following generic formula: (a) $XO_4^-$; (b) $XO_4^{2-}$; (c) $XO_4^{3-}$; (d) $XO_3^-$?

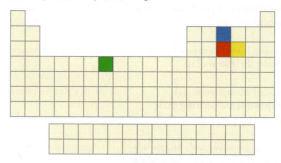

**FIGURE P2.8**

---

## QUESTIONS AND PROBLEMS

### The Nuclear Model of Atomic Structure

**CONCEPT REVIEW**

2.9. Explain how the results of the gold-foil experiment led Rutherford to dismiss the plum-pudding model of the atom and create his own model based on a nucleus surrounded by electrons.

2.10. Had the plum-pudding model been valid, how would the results of the gold-foil experiment have differed from what Geiger and Marsden actually observed?

2.11. What properties of cathode rays led Thomson to conclude that they were not pure energy, but rather particles with an electric charge?

*2.12. What would be observed if the charges on the plates of Millikan's apparatus (Figure 2.3) were reversed?

### Isotopes and Average Atomic Mass

**CONCEPT REVIEW**

2.13. What is meant by a *weighted average*?

2.14. Explain how percent natural abundances are related to average atomic masses.

## PROBLEMS

**2.15.** If the mass number of an isotope is more than twice the atomic number, is the neutron-to-proton ratio less than, greater than, or equal to 1?

**2.16.** In each of the following pairs of isotopes, which isotope has more protons and which has more neutrons? (a) $^{127}$I or $^{131}$I; (b) $^{188}$Re or $^{188}$W; (c) $^{14}$N or $^{14}$C

**2.17.** Boron, lithium, nitrogen, and neon each have two stable isotopes. In which of the following pairs of isotopes is the heavier isotope more abundant? (a) $^{10}$B or $^{11}$B (average atomic mass, 10.81 amu); (b) $^{6}$Li or $^{7}$Li (average atomic mass, 6.941 amu); (c) $^{14}$N or $^{15}$N (average atomic mass, 14.01 amu); (d) $^{20}$Ne or $^{22}$Ne (average atomic mass, 20.18 amu)

**2.18.** The average atomic mass of copper is 63.546 amu. It is composed of copper-63 and copper-65 isotopes. The natural abundance of Cu-63 (62.9296 amu) is 69.17%. What is the natural abundance of Cu-65 (64.9278 amu)?

**2.19.** Naturally occurring chlorine consists of two isotopes: 75.78% $^{35}$Cl and 24.22% $^{37}$Cl. Calculate the average atomic mass of chlorine.

**2.20.** Naturally occurring sulfur consists of four isotopes: $^{32}$S (31.97207 amu, 95.04%); $^{33}$S (32.97146 amu, 0.75%); $^{34}$S (33.96787 amu, 4.20%); and $^{36}$S (35.96708 amu, 0.01%). Calculate the average atomic mass of sulfur in atomic mass units.

**2.21. Chemistry of Mars** The 1997 mission to Mars included a small robot, the *Sojourner*, that analyzed the composition of Martian rocks. Magnesium oxide from a boulder dubbed "Barnacle Bill" was analyzed and found to have the following isotopic composition:

| Mass (amu) | Natural Abundance (%) |
|---|---|
| 39.9872 | 78.70 |
| 40.9886 | 10.13 |
| 41.9846 | 11.17 |

If essentially all of the oxygen in the Martian MgO sample is oxygen-16 (which has an exact mass of 15.9948 amu), is the average atomic mass of magnesium on Mars the same as on Earth (24.31 amu)?

**2.22.** Use the following table of abundances and masses of the three naturally occurring argon isotopes to calculate the mass of $^{40}$Ar.

| Symbol | Mass (amu) | Natural Abundance (%) |
|---|---|---|
| $^{36}$Ar | 35.96755 | 0.337 |
| $^{38}$Ar | 37.96272 | 0.063 |
| $^{40}$Ar | ? | 99.60 |
| Average | 39.948 | |

**2.23.** Use the following table of abundances and masses of five naturally occurring titanium isotopes to calculate the mass of $^{48}$Ti.

| Symbol | Mass (amu) | Natural Abundance (%) |
|---|---|---|
| $^{46}$Ti | 45.95263 | 8.25 |
| $^{47}$Ti | 46.9518 | 7.44 |
| $^{48}$Ti | ? | 73.72 |
| $^{49}$Ti | 48.94787 | 5.41 |
| $^{50}$Ti | 49.9448 | 5.18 |
| Average | 47.87 | |

**2.24.** Strontium has four isotopes: $^{84}$Sr, $^{86}$Sr, $^{87}$Sr, and $^{88}$Sr.
a. How many neutrons are there in each isotope?
b. The natural abundances of the four isotopes are 0.56% $^{84}$Sr (83.9134 amu); 9.86% $^{86}$Sr (85.9094 amu); 7.00% $^{87}$Sr (86.9089 amu); and 82.58% $^{88}$Sr (87.9056 amu). Calculate the average atomic mass of strontium and compare it to the value in the periodic table on the inside front cover.

## The Periodic Table of the Elements
### CONCEPT REVIEW

**2.25.** Mendeleev ordered the elements in his version of the periodic table based on their atomic masses instead of their atomic numbers. Why?

**2.26.** Why did Mendeleev not include the noble gases in his version of the periodic table?

**2.27.** A few elements in Mendeleev's chart are out of order when arranged by mass compared to the modern chart where they are arranged by atomic number. Why does ordering elements in terms of their mass not always work?

*__2.28.__ Tellurium has an average atomic mass of 127.60 amu, which is heavier than iodine, whose mass is 126.90447 amu, yet iodine comes after tellurium on the modern periodic table. Why does this occur? Find other instances where heavier atoms precede lighter ones on the periodic table.

### PROBLEMS

**2.29.** How many protons, neutrons, and electrons are there in the following atoms? (a) $^{14}$C; (b) $^{59}$Fe; (c) $^{90}$Sr; (d) $^{210}$Pb

**2.30.** How many protons, neutrons, and electrons are there in the following atoms? (a) $^{11}$B; (b) $^{19}$F; (c) $^{131}$I; (d) $^{222}$Rn

**2.31.** Fill in the missing information in the following table of four neutral atoms:

| Symbol | $^{23}$Na | ? | ? | ? |
|---|---|---|---|---|
| Number of Protons | ? | 39 | ? | 79 |
| Number of Neutrons | ? | 50 | ? | ? |
| Number of Electrons | ? | ? | 50 | ? |
| Mass Number | ? | ? | 118 | 197 |

**2.32.** Fill in the missing information in the following table of four neutral atoms:

| Symbol | $^{27}Al$ | ? | ? | ? |
|---|---|---|---|---|
| Number of Protons | ? | 42 | ? | 92 |
| Number of Neutrons | ? | 56 | ? | ? |
| Number of Electrons | ? | ? | 60 | ? |
| Mass Number | ? | ? | 143 | 238 |

**2.33.** Fill in the missing information in the following table of ions:

| Symbol | $^{37}Cl^-$ | ? | ? | ? |
|---|---|---|---|---|
| Number of Protons | ? | 11 | ? | 88 |
| Number of Neutrons | ? | 12 | 46 | ? |
| Number of Electrons | ? | 10 | 36 | 86 |
| Mass Number | ? | ? | 81 | 226 |

**2.34.** Fill in the missing information in the following table of ions:

| Symbol | $^{137}Ba^{2+}$ | ? | ? | ? |
|---|---|---|---|---|
| Number of Protons | ? | 30 | ? | 40 |
| Number of Neutrons | ? | 34 | 16 | ? |
| Number of Electrons | ? | 28 | 18 | 36 |
| Mass Number | ? | ? | 32 | 90 |

**2.35.** Which element is most likely to form a cation with a 2+ charge? (a) S; (b) P; (c) Be; (d) Al

**2.36.** Which element is most likely to form an anion with a 2− charge? (a) S; (b) P; (c) Be; (d) Al

**2.37.** Which species contains the greatest number of electrons? (a) F; (b) $O^{2-}$; (c) $S^{2-}$; (d) Cl

**2.38.** Which species contains the smallest number of electrons? (a) F; (b) $O^{2-}$; (c) $S^{2-}$; (d) Cl

**2.39.** Which ions have the same number of electrons as an atom of argon? (a) $S^{2-}$; (b) $P^{3-}$; (c) $Be^{2+}$; (d) $Ca^{2+}$

**2.40.** Which ions have the same number of electrons as an atom of krypton? (a) $Se^{2-}$; (b) $As^{3-}$; (c) $Ca^{2+}$; (d) $K^+$

**2.41.** Which element is a nonmetal? (a) Si; (b) Br; (c) Ca; (d) Ru

**2.42.** Which element is a metalloid? (a) Si; (b) Br; (c) Ca; (d) Ru

**2.43.** Which species contains 10 electrons? (a) $F^-$; (b) $Cl^-$; (c) $Br^-$; (d) $I^-$

**2.44.** Which species contains 18 electrons? (a) $O^{2-}$; (b) $S^{2-}$; (c) $Se^{2-}$; (d) $Te^{2-}$

**2.45.** Which element is a halogen? (a) $N_2$; (b) $Cl_2$; (c) Xe; (d) $H_2$

**2.46.** Which element is a noble gas? (a) $N_2$; (b) $Cl_2$; (c) Xe; (d) $I_2$

**2.47.** Which element is an alkali metal? (a) Na; (b) $Cl_2$; (c) Xe; (d) $Br_2$

**2.48.** Which element is an alkaline earth metal? (a) Na; (b) Ca; (c) Xe; (d) $H_2$

## Trends in Compound Formation

### CONCEPT REVIEW

**2.49.** **Cations in Blood and Urine** Reports from standard blood and urine tests indicate the amounts of sodium, potassium, calcium, and magnesium cations present. Chloride ion is the most abundant anion in both blood and urine; urine also contains some sulfate ion. Write formulas for the chlorides and sulfates of the four cations.

**2.50.** Explain why the law of constant composition is considered a scientific *law*, whereas Dalton's view of the atomic structure of matter is considered a scientific *theory*.

**2.51.** How does Dalton's atomic theory of matter explain the fact that when water is decomposed into hydrogen and oxygen gas, the volume of hydrogen is always twice that of oxygen?

**2.52.** **Pollutants in Automobile Exhaust** In the internal combustion engines that power most automobiles, nitrogen and oxygen may combine to form NO. When NO in automobile exhaust is released into the atmosphere, it reacts with more oxygen, forming $NO_2$, a key ingredient in smog. How do these reactions illustrate Dalton's law of multiple proportions?

**2.53.** Describe the types of elements that combine to form molecular compounds and the types that combine to form ionic compounds.

**2.54.** How do the properties of ionic compounds differ from those of covalent compounds?

### PROBLEMS

**2.55.** Cobalt forms two sulfides: CoS and $Co_2S_3$. Predict the ratio of the two masses of sulfur that combine with a fixed mass of cobalt to form CoS and $Co_2S_3$.

**2.56.** Lead forms two oxides: PbO and $PbO_2$. Predict the ratio of the two masses of oxygen that combine with a fixed mass of lead to form PbO and $PbO_2$.

**2.57.** When 5.0 g of sulfur is combined with 5.0 g of oxygen, 10.0 g of sulfur dioxide is formed. What mass of oxygen would be required to convert 5.0 g of sulfur into sulfur trioxide?

*2.58.** Nitrogen monoxide (NO) is 46.7% nitrogen by mass. Use the law of multiple proportions to calculate the mass percentage of nitrogen in nitrogen dioxide ($NO_2$).

**2.59.** **Seawater** The most abundant anion in seawater is the chloride ion. Write the formulas for the chlorides and sulfates of the most abundant cations in seawater: sodium, magnesium, calcium, potassium, and strontium.

**2.60.** The most abundant cation in seawater is the sodium ion. The evaporation of seawater gives a mixture of ionic compounds containing sodium combined with chloride, sulfate, carbonate, bicarbonate, bromide, fluoride, and tetrahydroborate, $B(OH)_4^-$. Write the chemical formulas of all these compounds.

**2.61.** *Milk of magnesia* is a slurry of $Mg(OH)_2$ in water. What is the chemical name of this compound?

2.62. *Smelling salts*, an antidote for fainting, are made of $(NH_4)_2CO_3$, which smells of ammonia.
   a. Give the chemical name of $(NH_4)_2CO_3$
   b. What is the chemical formula of ammonia?

**2.63.** Write the names of these molecular compounds: (a) HBr; (b) $CS_2$; (c) $NF_3$; (d) $PF_5$.

**2.64.** Write the names of these ionic compounds: (a) $(NH_4)_2S$; (b) CsCl; (c) $TeF_4$; (d) $Rb_2O$.

**2.65.** Which of these compounds consist of molecules and which consist of ions? (a) $P_4O_{10}$; (b) $SrCl_2$; (c) $MgCO_3$; (d) $H_2SO_4$

2.66. Which of these compounds consist of molecules, and which consist of ions? (a) $Mg(OH)_2$; (b) $Ba(NO_3)_2$; (c) $HNO_3$; (d) $NCl_3$

**2.67.** Give the total number of atoms in a formula unit of these compounds: (a) $LaF_3$; (b) $In_2S_3$; (c) $Na_2HPO_4$; (d) $Ca(H_2PO_4)_2$.

2.68. Give the total number of atoms in a formula unit of these compounds: (a) $In_2O_3$; (b) $Ce(SO_4)_2$; (c) $TlF_3$; (d) $Ca_{10}(PO_4)_6(OH)_2$.

## Naming Compounds and Writing Formulas

### CONCEPT REVIEW

**2.69.** Consider a mythical element X, which forms only two oxoanions: $XO_2^{2-}$ and $XO_3^{2-}$. Which of the two has a name that ends in -*ite*?

2.70. Concerning the oxoanions in Problem 2.69, would the name of either of them require a prefix such as *hypo-* or *per-*? Explain why or why not.

**2.71.** What is the role of Roman numerals in the names of the compounds formed by transition metals?

2.72. Why do the names of the ionic compounds formed by the alkali metals and by the alkaline earth metals not include Roman numerals?

### PROBLEMS

**2.73. Toxicity of Nitrogen Oxides** Nitrogen oxides form naturally during the combustion of nitrogen-containing compounds such as coal, diesel fuel, and green plants. Small quantities of all except $N_2O$ (laughing gas) are irritating to eyes, skin, and the respiratory tract. Name the binary compounds of nitrogen and oxygen: (a) $NO_3$; (b) $N_2O_5$; (c) $N_2O_4$; (d) $NO_2$; (e) $N_2O_3$; (f) NO; (g) $N_2O$; (h) $N_4O$.

2.74. More than a dozen binary compounds containing sulfur and oxygen have been identified. Give the chemical formulas for the following six: (a) sulfur monoxide; (b) sulfur dioxide; (c) sulfur trioxide; (d) disulfur monoxide; (e) hexasulfur monoxide (f) heptasulfur dioxide.

**2.75.** Predict the formula and give the name of the binary ionic compound containing the following: (a) sodium and sulfur; (b) strontium and chlorine; (c) aluminum and oxygen; (d) lithium and hydrogen.

2.76. Predict the formula and give the name of the binary ionic compound containing the following: (a) potassium and bromine; (b) calcium and hydrogen; (c) lithium and nitrogen; (d) aluminum and chlorine.

**2.77.** Which of these chemical names is followed by an incorrect chemical formula? (a) calcium oxide, CaO; (b) lithium sulfate, $LiSO_4$; (c) barium sulfide, BaS; (d) potassium oxide, $K_2O$

2.78. Which of these chemical names is followed by an incorrect chemical formula? (a) aluminum nitride, AlN; (b) aluminum sulfate dodecahydrate, $Al_2(SO_4)_3 \cdot 12H_2O$; (c) potassium chloride, $KCl_2$; (d) cesium sulfate, $Cs_2SO_4$

**2.79.** Give the formula and charge of the oxoanion in each of the following compounds: (a) sodium hypobromite; (b) potassium sulfate; (c) lithium iodate; (d) magnesium nitrite.

*2.80. Give the formula and charge of the oxoanion in each of the following compounds: (a) potassium tellurite; (b) sodium arsenate; (c) calcium selenite; (d) potassium chlorate.

**2.81.** Give chemical names of the following ionic compounds: (a) $K_2CO_3$; (b) NaCN; (c) $LiHCO_3$; (d) $Ca(ClO)_2$.

2.82. Give chemical names of the following ionic compounds: (a) $Mg(ClO_4)_2$; (b) $NH_4NO_3$; (c) $Cu(CH_3COO)_2$; (d) $K_2SO_4$.

**2.83.** Give the name or chemical formula of each of the following acids: (a) HF; (b) $HBrO_3$; (c) phosphoric acid; (d) nitrous acid.

2.84. Give the name or chemical formula of each of the following acids: (a) HBr; (b) $HIO_4$; (c) selenous acid; (d) hydrocyanic acid.

**2.85.** Name these compounds: (a) $Na_2O$; (b) $Na_2S$; (c) $Na_2SO_4$; (d) $NaNO_3$; (e) $NaNO_2$.

2.86. Name these compounds: (a) $K_3PO_4$; (b) $K_2O$; (c) $K_2SO_3$; (d) $KNO_3$; (e) $KNO_2$.

**2.87.** Write the chemical formulas of these compounds: (a) potassium sulfide; (b) potassium selenide; (c) rubidium sulfate; (d) rubidium nitrite; (e) magnesium sulfate.

2.88. Write the chemical formulas of these compounds: (a) rubidium nitride; (b) potassium selenite; (c) rubidium sulfite; (d) rubidium nitrate; (e) magnesium sulfite.

**2.89.** Which compound is sodium sulfite? (a) $Na_2S$; (b) $Na_2SO_3$; (c) $Na_2SO_4$; (d) NaHS

2.90. Which compound is calcium nitrate? (a) $Ca_3N_2$; (b) $Ca_2NO_3$; (c) $Ca_2(NO_3)_2$; (d) $Ca(NO_3)_2$

**2.91.** Give the chemical names of the cobalt oxides that have the following formulas: (a) CoO; (b) $Co_2O_3$; (c) $CoO_2$.

2.92. Give the formula of each of the following copper minerals: (a) cuprite, copper(I) oxide; (b) chalcocite, copper(I) sulfide; (c) covellite, copper(II) sulfide.

**2.93.** Name these compounds: (a) MnS; (b) $V_3N_2$; (c) $Cr_2(SO_4)_3$; (d) $Co(NO_3)_2$; (e) $Fe_2O_3$.

2.94. Name these compounds: (a) RuS; (b) $PdCl_2$; (c) $Ag_2O$; (d) $WO_3$; (e) $PtO_2$.

**2.95.** Which of these chemical names is followed by an incorrect chemical formula? (a) iron(II) oxide, FeO; (b) titanium(IV) sulfate, $Ti(SO_4)_2$; (c) cobalt(II) chloride, $CoCl_2$; (d) vanadium(IV) oxide, VO.

2.96. Which of these chemical names is followed by an incorrect chemical formula? (a) manganese(II) phosphate, $Mn_3(PO_4)_3$; (b) nickel(II) bromide, $NiBr_2$; (c) chromium(III) sulfide, $Cr_2S_3$; (d) copper(II) sulfate hexahydrate, $CuSO_4 \cdot 6H_2O$.

## Nucleosynthesis

### CONCEPT REVIEW

**2.97.** Write brief (one-sentence) definitions of *chemistry* and *cosmology*, and then give as many examples as you can of how the two sciences are related.

2.98. In the history of the universe, which of these particles formed first, and which formed last? (a) deuteron; (b) neutron; (c) proton; (d) quark

**2.99.** Chemists do not include quarks in the category of subatomic particles. Why?

2.100. Why did early nucleosynthesis last such a short time?

**2.101.** In the current cosmological model, the volume of the universe is increasing with time. How might this expansion affect the density of the universe?

2.102. **Components of Solar Wind** Most of the ions that flow out from the Sun in the solar wind are hydrogen ions. The ions of which element should be next most abundant?

*2.103. It takes nearly twice the energy to remove an electron from a helium atom as it does to remove an electron from a hydrogen atom. Propose an explanation for this.

## Additional Problems

2.104. One reaction in the process of carbon fusion in massive stars involves the combination of two carbon-12 nuclei to form the nucleus of a new element and an alpha particle. Write an equation that describes this process.

**2.105.** A process called neon fusion takes place in massive stars. In one of the reactions in this process, an alpha particle combines with a neon-21 nucleus to produce another element and a neutron. Write an equation that describes this process.

2.106. In April 1897, J. J. Thomson presented the results of his experiment with cathode-ray tubes (Figure P2.106) in which he proposed that the rays were actually beams of negatively charged particles, which he called "corpuscles."
   a. What is the name we use for these particles today?
   b. Why did the beam deflect when passed between electrically charged plates, as shown in Figure P2.106?
   c. If the polarity of the plates were switched, how would the position of the light spot on the phosphorescent screen change?
   d. If the voltage on the plates were reduced by half, how would the position of the light spot change?

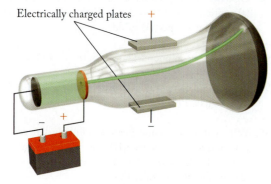

**FIGURE P2.106**

*2.107. Suppose the electrically charged discs at the end of the cathode-ray tube were replaced with a radioactive source, as shown in Figure P2.107. Also suppose the radioactive material inside the source emits α and β particles, as well as rays of energy with no charge. The only way for any of the three kinds of particles or rays to escape the source is through a narrow channel drilled through a block of lead.
   a. How many light spots do you expect to see on the phosphorescent screen?
   b. What are their positions relative to the electrical plates, and which particle produces which spot?

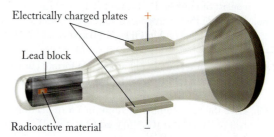

**FIGURE P2.107**

*2.108. Suppose the radioactive material inside the source in the apparatus shown in Figure P2.107 emits protons and α particles, and suppose both kinds of particles have the same velocities.
   a. How many light spots do you expect to see on the phosphorescent screen?
   b. What are their positions on the screen (above, at, or below the center)? Which particle produces which spot?

2.109. Cosmologists estimate that the matter in the early universe was 75% by mass hydrogen-1 and 25% helium-4 when atoms first formed.
   a. Assuming these proportions are correct, what was the ratio of hydrogen to helium *atoms* in the early universe?
   b. The ratio of hydrogen to helium atoms in our solar system is slightly less than 10:1. Compare this value with the value you calculated in part a.
   c. Propose a hypothesis that accounts for the difference in composition between the solar system and the early universe.
   d. Describe an experiment that would test your hypothesis.

2.110. **Sources of Breathable Air** Potassium forms three compounds with oxygen: $K_2O$ (potassium oxide), $K_2O_2$ (potassium peroxide), and $KO_2$ (potassium superoxide). Elemental potassium is rarely encountered; it reacts violently with water and is very corrosive to human tissue. Potassium superoxide is used in self-contained breathing apparatus as a source of oxygen for use in mines, submarines, and spacecraft. Potassium peroxide binds carbon dioxide and is used to scrub (remove) toxic $CO_2$ from the air in submarines. Predict the ratio of the masses of oxygen that combine with a fixed mass of potassium in $K_2O$, $K_2O_2$, and $KO_2$.

2.111. **Stainless Steel** The gleaming metallic appearance of the Gateway Arch (Figure P2.111) in St. Louis, MO, comes from the stainless steel used in its construction. This steel is made mostly of iron,

**FIGURE P2.111**

but it also contains 19% by mass chromium and 9% by mass nickel.

   a. Stainless steel maintains its metallic sheen because the chromium and nickel in it combine with oxygen from the atmosphere, forming a layer of $Cr_2O_3$ and $NiO$ that is too thin to detract from the luster of the steel but that protects the metal beneath from further corrosion. What are the names of these two ionic compounds?

   b. What are the charges of the cations in $Cr_2O_3$ and $NiO$?

**2.112. Bronze Age** Historians and archaeologists often apply the term "Bronze Age" to the period in Mediterranean and Middle Eastern history when bronze was the preferred material for making weapons, tools, and other metal objects. Ancient bronze was an alloy prepared by blending molten copper (90%) and tin (10%) by mass. What is the ratio of copper to tin atoms in a piece of bronze with this composition?

**\*2.113.** In his version of the periodic table, Mendeleev arranged elements based on the formulas of the compounds they formed with hydrogen and oxygen. The elements in one of his eight groups formed compounds with these generic formulas: $MH_3$ and $M_2O_5$, where M is the symbol of an element in the group. Which Roman numeral did Mendeleev assign to this group?

**2.114.** In the Mendeleev table in Figure 2.9, there are no symbols for elements with predicted atomic masses of 44, 68, and 72.

   a. Which elements are these?

   b. Mendeleev anticipated the later discovery of these three elements and gave them tentative names: ekaaluminum, ekaboron, and ekasilicon, reflecting the probability that their properties would resemble those of aluminum, boron, and silicon, respectively. What are the modern names of ekaaluminum, ekaboron, and ekasilicon?

   c. When were these elements finally discovered? To answer this question you may wish to consult a reference such as webelements.com.

**2.115. Medical and Commercial Compounds** Many common compounds have old-fashioned, nonsystematic names or newer commercial names that are widely used. Search for the systematic names and chemical formulas of compounds with these nonsystematic names: (a) magnesia; (b) Epsom salt; (c) K-Dur; (d) lime; (e) baking soda; (f) caustic soda; (g) muriatic acid; (h) zirconia.

**\*2.116.** The ruby shown in Figure P2.116 has a mass of 12.04 carats (1 carat = 200.0 mg). Rubies are mostly $Al_2O_3$.

   a. What percentage of the mass of the ruby is aluminum?

   b. The density of rubies is 4.02 $g/cm^3$. What is the volume of the ruby?

**FIGURE P2.116**

**2.117.** In chemical nomenclature, the prefix *thio-* is used to indicate that a sulfur atom has replaced an oxygen atom in the structure of a molecule or a polyatomic ion.

   a. With this rule in mind, write the formula for the thiosulfate ion.

   b. What is the formula of sodium thiosulfate?

**2.118.** There are two stable isotopes of gallium. Their masses are 68.92558 and 70.9247050 amu. If the average atomic mass of gallium is 69.7231 amu, what is the natural abundance of the lighter isotope?

**2.119.** There are two stable isotopes of bromine. Their masses are 78.9183 and 80.9163 amu. If the average atomic mass of bromine is 79.9091 amu, what is the natural abundance of the heavier isotope?

**\*2.120.** Using the information in the previous question:

   a. Predict the possible masses of individual molecules of $Br_2$.

   b. Calculate the natural abundance of molecules with each of the masses predicted in part a in a sample of $Br_2$.

**\*2.121.** There are three stable isotopes of magnesium. Their masses are 23.9850, 24.9858, and 25.9826 amu. If the average atomic mass of magnesium is 24.3050 amu and the natural abundance of the lightest isotope is 78.99%, what are the natural abundances of the other two isotopes?

**2.122.** Write the names and formulas of the following compounds from this list of elements: Li, Fe, Al, O, C, and N.

   a. A molecular substance $AB_2$, where A is a group 14 element and B is a group 16 element

   b. An ionic compound $C_3D$, where C is a group 1 element and D is a group 15 element

**2.123.** Write the chemical symbol of each of the following species: (a) a cation with a mass number of 24, an atomic number of 12, and a charge of 2+; (b) a member of group 15 that has a charge of 3+, 48 electrons, and 70 neutrons; (c) a noble gas atom with 48 neutrons.

**2.124.** How many periods make up the periodic table? Give the names and symbols of the elements in period 3.

**\*2.125.** Predict some physical and chemical properties of radium (Ra). Predict the melting points of $RaCl_2$ and $RaO$. (*Hint:* find the melting points of the other alkaline earth element chlorides and oxides on the web.)

**2.126.** Based on their positions in the periodic table, predict which of the following elements would be good electrical conductors: Ti, Ne, N, Ag, Tb, Br, and Mo.

**2.127.** Argon has a larger average atomic mass than potassium, yet it is placed before potassium in the modern periodic table. Explain.

If your instructor assigns problems in **smartwork**, log in at **smartwork.wwnorton.com**.

# 3

# Stoichiometry: Mass, Formulas, and Reactions

## The Origins of Life on Earth

In Chapter 2 we described how elements are synthesized in the cores of giant stars and dispersed throughout galaxies when the stars explode as supernovas. The dispersed elements become the building blocks of other stars and planets. As our solar system formed, the inner planets—Mercury, Venus, Earth, and Mars—were rich in nonvolatile elements such as iron, silicon, magnesium, and aluminum. Earth was also rich in oxygen—in the form of stable compounds with these and other elements. These compounds formed the rocks and minerals of Earth's crust and provided an early atmosphere, containing mostly water vapor and carbon dioxide. As Earth cooled, atmospheric water vapor condensed into torrential rains that filled up depressions in the crust, forming the first oceans.

How did the substances in lifeless rocks and air or dissolved in primordial seas become the organic building blocks of life—compounds like simple sugars, amino acids, and the molecules that form DNA? No one knows for sure, but in the early 1950s Nobel Prize winner Harold Urey and his student Stanley Miller showed that subjecting a mixture of gases believed to have been present in Earth's early atmosphere to an electric current (simulating lightning) could produce several amino acids. Samples from the original experiments were analyzed in 2007 and more than 20 amino acids were detected. Amino acids also have been found in meteorites, which have bombarded Earth since it formed. Extensive analyses of the Murchison meteorite, which landed in Australia in 1969, showed that it contains many of the same amino acids synthesized in the Miller/Urey experiments.

The study of the origins of life serves more purposes than satisfying a curiosity. The emerging field of astrobiology explores the possibility of extraterrestrial life and what forms that life might have. Evolutionary biologists study the oldest and most basic reactions of biochemistry to find ways to combat disease-causing microbes around the world. Many fields of study rely on an understanding of the chemical reactions that take place in living things and in the environment.

We begin this chapter with an atomic and molecular level view of the physical processes and chemical reactions that may have occurred

**3.1** Chemical Reactions and Earth's Early Atmosphere

**3.2** The Mole

**3.3** Writing Balanced Chemical Equations

**3.4** Combustion Reactions

**3.5** Stoichiometric Calculations and the Carbon Cycle

**3.6** Determining Empirical Formulas from Percent Composition

**3.7** Empirical and Molecular Formulas Compared

**3.8** Combustion Analysis

**3.9** Limiting Reactants and Percent Yield

### Learning Outcomes

**LO1** Use Avogadro's number and the definition of the mole in calculations
**Sample Exercises 3.1, 3.2, 3.3, 3.4, 3.5, 3.6, 3.7**

**LO2** Write balanced chemical equations that describe chemical reactions
**Sample Exercises 3.8, 3.9, 3.10, 3.11**

**LO3** Use balanced chemical equations to relate the mass of a reactant to the mass of a product
**Sample Exercises 3.12, 3.13**

**LO4** Determine an empirical formula from the percent composition of a substance
**Sample Exercises 3.14, 3.15, 3.16**

**LO5** Determine a molecular formula from the empirical formula and molar mass of a substance
**Sample Exercise 3.17**

**LO6** Use data from combustion reactions in determining empirical formulas of substances
**Sample Exercises 3.18, 3.19**

**LO7** Determine the limiting reactant in a chemical reaction
**Sample Exercises 3.20, 3.21**

**LO8** Calculate the theoretical and percent yields in a chemical reaction
**Sample Exercises 3.22, 3.23**

**Methane Bubbles** An ecologist pierces a large methane bubble trapped in the ▶ ice of an Alaskan lake. Burning methane is an example of a chemical reaction.

**74**

on early Earth. We explore how the quantities of substances produced and consumed in chemical reactions are related. Then we examine the analytical methods that reveal the composition of pure substances. Finally, we consider a class of chemical reactions that consume oxygen and a fuel, liberating energy—including the energy necessary for life. ■

## 3.1 Chemical Reactions and Earth's Early Atmosphere

The Earth that formed 4.6 billion years ago was a hot, molten sphere that gradually separated into distinct regions based on differences in density and melting point. The densest elements, notably iron and nickel, sank to the center of the planet. A less dense mantle, rich in compounds containing aluminum, magnesium, silicon, and oxygen, formed around the core. As time passed and Earth cooled, the mantle fractionated further, allowing a solid crust to form from the components of the mantle that were the least dense and had the highest melting points. The core also separated into a solid inner core and a molten outer core. Figure 3.1 shows the elemental compositions of these layers.

Earth's early crust was torn by the impact of asteroids and widespread volcanic activity. The gases released by these impacts and eruptions generated a primitive atmosphere with a chemical composition very different from the air we breathe now. Earth's early atmosphere was nearly devoid of molecular oxygen but was rich in oxygen-containing compounds, including carbon dioxide ($CO_2$), water vapor ($H_2O$), and other volatile oxides, including sulfur dioxide ($SO_2$), sulfur trioxide ($SO_3$), nitrogen monoxide ($NO$), and nitrogen dioxide ($NO_2$). The most abundant gases released today by volcanic systems like Mount St. Helens are water vapor, $CO_2$, and $SO_2$ (Figure 3.2).

**FIGURE 3.1** Earth is composed of a solid inner core, consisting mostly of nickel and iron, surrounded by a molten outer core of similar composition. A liquid mantle, composed mostly of oxygen, silicon, magnesium, and iron, lies between the outer core and a relatively thin solid crust.

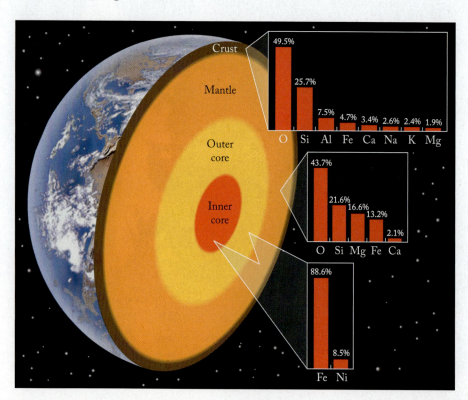

Sometimes these compounds combined to make substances with more elaborate molecular structures. For example, sulfur trioxide gas and water vapor are the **reactants** that combine to produce liquid sulfuric acid, $H_2SO_4$, as a **product** (Figure 3.3). We use the formulas of these substances in an expression called a chemical equation to describe the reaction:

$$SO_3(g) \; + \; H_2O(g) \; \rightarrow \; H_2SO_4(\ell) \tag{3.1}$$

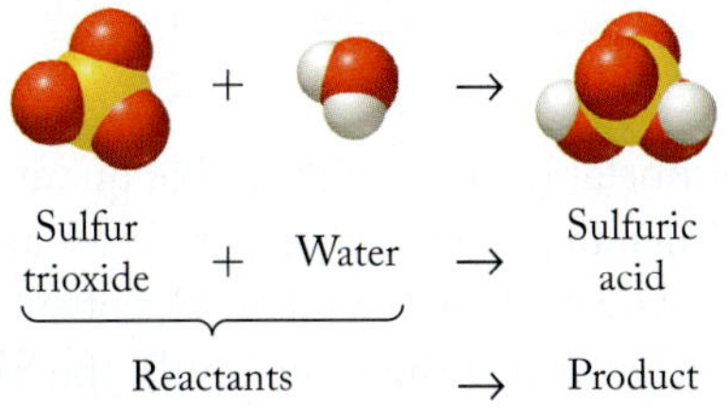

Sulfur trioxide $\;+\;$ Water $\;\rightarrow\;$ Sulfuric acid

Reactants $\;\;\rightarrow\;\;$ Product

**FIGURE 3.3** In this combination reaction, a molecule of $SO_3$ and a molecule of $H_2O$ form a molecule of $H_2SO_4$.

The reaction between $SO_3$ and $H_2O$ is an example of a **combination reaction**— two (or more) substances combine to form one product. We use a reaction arrow to link the reactants and product and show the direction of the reaction. The symbols in parentheses after the chemical formulas indicate the physical states of the reactants and product: $g$ for gas and $\ell$ for liquid. An important feature of any chemical equation is that it is balanced: every atom that is present in the reactants is also present in the products. This conservation of atoms means that there is also a conservation of mass: the sum of the masses of the reactants always equals the sum of the masses of the products.

The sulfuric acid that formed in Earth's early atmosphere eventually fell to the planet's surface as highly acidic rain. This rain would have landed on a crust made up mostly of metal and metalloid oxides, including the mineral hematite, $Fe_2O_3$. When that sulfuric acid encountered hematite, another chemical reaction took place—one that produced water-soluble iron(III) sulfate, $Fe_2(SO_4)_3$, and liquid water. This reaction is described by the following chemical equation:

$$Fe_2O_3(s) + 3\,H_2SO_4(aq) \rightarrow 3\,H_2O(\ell) + Fe_2(SO_4)_3(aq) \tag{3.2}$$

In Equation 3.2 the symbol $s$ means solid and $aq$ stands for aqueous solution, meaning that the substance is dissolved in water. The "3"s in front of both $H_2SO_4$ and $H_2O$ are coefficients that indicate that three molecules of each compound are involved in a complete reaction. These coefficients balance the number of atoms of each element on both sides of the reaction arrow. The absence of a coefficient in front of $Fe_2O_3$ and $Fe_2(SO_4)_3$ is the same as having a coefficient of "1" in front of them. The chemical formula itself stands for one molecule (or one formula unit) of the substance; the chemical formula also indicates the number of moles of each element in one mole of a compound.

## 3.2 The Mole

Equation 3.1 describes a chemical reaction between a molecule of sulfur trioxide and a molecule of water. In our macroscopic (visible) world, chemists rarely work with individual atoms or molecules: they are too small to manipulate easily. Instead, they usually deal with quantities of reactants and products large enough

**FIGURE 3.2** Volcanic eruption of Mount St. Helens in Washington State on May 18, 1980. The most abundant gas released in this eruption was water vapor.

**reactant** a substance consumed during a chemical reaction

**product** a substance formed during a chemical reaction

**combination reaction** a reaction in which two (or more) substances combine to form one product.

**FIGURE 3.4** Converting between a number of particles and an equivalent number of moles (or vice versa) is a matter of dividing (or multiplying) by Avogadro's number. Note how the units cancel, leaving only the units we sought.

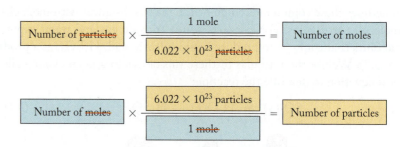

▶‖ **CHEMTOUR** Avogadro's Number

to see and work with comfortably. Such measurable quantities contain enormous numbers of particles—either atoms, molecules, or ions—and consequently chemists need a unit that can relate measurable quantities of substances to the number of particles they contain. That unit is the **mole (mol)**, the SI base unit for expressing quantities of substances (see Table 1.2).

Moles allow us to relate the mass of a pure substance to the number of particles it contains. One mole of any substance is defined as the quantity of the substance that contains the same number of particles as the number of carbon atoms in exactly 12 g of the isotope carbon-12. This number is $6.022 \times 10^{23}$ carbon-12 atoms. This very large value is called **Avogadro's number ($N_A$)** after the Italian scientist Amedeo Avogadro (1776–1856), whose research enabled other scientists to accurately determine the atomic masses of the elements.[1] One mole of water also contains $6.022 \times 10^{23}$ particles—in this case molecules of water. To put a number of this magnitude in perspective, consider a sphere the size of Earth. It would require about 1 mole of basketballs to fill that sphere.

Avogadro's number is often used as part of a conversion factor between the number of particles and the number of moles of a substance. Dividing the number of particles in a sample by Avogadro's number yields the number of moles of those particles. For example, a jet airplane flying at an altitude of 11,300 m (37,000 ft) is flying through air that contains about $7.0 \times 10^{21}$ particles per liter. This number is equivalent to $1.2 \times 10^{-2}$ moles of particles per liter of air:

$$\frac{7.0 \times 10^{21} \text{ particles}}{1 \text{ L}} \times \frac{1 \text{ mol}}{6.022 \times 10^{23} \text{ particles}} = \frac{1.2 \times 10^{-2} \text{ mol}}{L}$$

On the other hand, multiplying a number of moles by Avogadro's number gives us the number of particles in that many moles. For example, a liter bottle of seltzer water contains 55 moles of $H_2O$. The equivalent number of molecules of water in that liter bottle is a very large number:

$$\frac{55 \text{ mol } H_2O}{1 \text{ L}} \times \frac{6.022 \times 10^{23} \text{ molecules } H_2O}{1 \text{ mol } H_2O} = \frac{3.3 \times 10^{25} \text{ molecules } H_2O}{1 \text{ L}}$$

The use of this conversion factor is illustrated in Figure 3.4, and the quantities of some common elements equivalent to 1 mole are illustrated in Figure 3.5. Although it is sometimes useful to know the number of molecules of a substance in a sample, we will mostly be concerned about knowing how many *moles* are in a particular quantity of a substance.

**FIGURE 3.5** The quantities shown are equivalent to 1 mole of each material: 4.003 g of helium gas in the balloon and, left to right in front of the balloon, 32.06 g of solid sulfur, 63.55 g of copper metal, and 200.59 g of liquid mercury.

[1]Amedeo Avogadro did not actually discover the number that bears his name (see Bodner, G. M. *Scientific American* February 16, 2004). Avogadro's number can be calculated using the results of Robert Millikan's experiment described in Chapter 2. Dividing the charge on one mole of electrons (96,485.3383 C/mol) by the charge per electron ($1.60217653 \times 10^{-19}$ coulombs per electron) yields $N_A = 6.02214154 \times 10^{23}$ particles per mole. We round it off to four significant figures throughout this book.

**mole (mol)** an amount of material (atoms, ions, or molecules) that contains Avogadro's number ($N_A = 6.022 \times 10^{23}$) of particles.

**Avogadro's number ($N_A$)** the number of carbon atoms in exactly 12 grams of the carbon-12 isotope; $N_A = 6.022 \times 10^{23}$. It is the number of particles in one mole.

**SAMPLE EXERCISE 3.1**    **Converting Number of Moles into Number of Particles**    **LO1**

It's not unusual for the polluted air above a large metropolitan area to contain as much as $5 \times 10^{-10}$ moles of $SO_2$ per liter of air. What is this concentration of $SO_2$ in molecules per liter?

**Collect and Organize** The problem gives the number of moles of $SO_2$ per liter of air. We want to find the number of molecules of $SO_2$ in 1 L. There are $6.022 \times 10^{23}$ particles in 1 mole of anything.

**Analyze** We convert the number of moles/liter into the number of molecules/liter by multiplying by $6.022 \times 10^{23}$ molecules/mole:

$$\boxed{\dfrac{\text{mole}}{\text{L}}} \xrightarrow{\dfrac{6.022 \times 10^{23}\ \text{molecules}}{1\ \text{mole}}} \boxed{\dfrac{\text{molecules}}{\text{L}}}$$

We can estimate the answer by considering the approximate value of Avogadro's number (about $6 \times 10^{23}$) and the number of moles of $SO_2$ ($5 \times 10^{-10}$). The product of the two is $3 \times 10^{14}$.

**Solve** The number of $SO_2$ molecules per liter is:

$$\frac{5 \times 10^{-10}\ \cancel{\text{mol } SO_2}}{1\ \text{L air}} \times \frac{6.022 \times 10^{23}\ \text{molecules } SO_2}{1\ \cancel{\text{mol } SO_2}} = \frac{3 \times 10^{14}\ \text{molecules } SO_2}{1\ \text{L air}}$$

**Think About It** The result tells us that a tiny fraction of 1 mole of a molecular substance ($10^{-10}$) is equivalent to a very large number of molecules ($10^{14}$), which makes sense given the immensity of Avogadro's number ($10^{23}$).

**Practice Exercise** If 1.0 mL of seawater contains about $2.5 \times 10^{-14}$ moles of dissolved gold, how many atoms of gold are in that volume of seawater?

*(Answers to Practice Exercises are in the back of the book.)*

**SAMPLE EXERCISE 3.2**    **Relating Formulas, Moles, and Particles**    **LO1**

Dental fillings contain compounds with the formulas $Ag_2Hg_3$, $Ag_3Sn$, and $Sn_8Hg$ (Figure 3.6). Which compound contains the most atoms per mole? How many atoms of silver are there in one mole of $Ag_3Sn$?

**Collect and Organize** We are given the formulas of three compounds: $Ag_2Hg_3$, $Ag_3Sn$, and $Sn_8Hg$. Only two of them contain silver.

**Analyze** A chemical formula indicates the moles of each element in one mole of a compound. The number of atoms in a mole of *any* element is defined by Avogadro's number: $N_A = 6.022 \times 10^{23}$ atoms/mol.

**Solve** From the formulas of the three compounds we have:
- 1 mole $Ag_2Hg_3$ contains 2 moles of silver and 3 moles of mercury or $2 + 3 = 5$ total moles of atoms.
- 1 mole $Ag_3Sn$ contains 3 moles of silver and 1 mole of tin or $3 + 1 = 4$ total moles of atoms.
- 1 mole $Sn_8Hg$ contains 8 moles of tin and 1 mole of mercury or $8 + 1 = 9$ total moles of atoms.

Thus, $Sn_8Hg$ contains the most total moles of atoms and the greatest number of atoms:

$$\frac{9\ \cancel{\text{total mol}}}{\text{mol } Sn_8Hg} \times \frac{6.022 \times 10^{23}\ \text{atoms}}{1\ \cancel{\text{mol}}} = \frac{5.420 \times 10^{24}\ \text{atoms}}{\text{mol } Sn_8Hg}$$

The number of atoms of Ag per mol $Ag_3Sn$ is:

$$\frac{3\ \cancel{\text{mol Ag}}}{\text{mol } Ag_3Sn} \times \frac{6.022 \times 10^{23}\ \text{atoms Ag}}{1\ \cancel{\text{mol Ag}}} = \frac{1.807 \times 10^{24}\ \text{atoms Ag}}{\text{mol } Ag_3Sn}$$

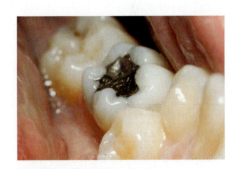

**FIGURE 3.6** One of the materials used to fill dental caries is a mixture of three compounds: $Ag_2Hg_3$, $Ag_3Sn$, and $Sn_8Hg$.

(a)

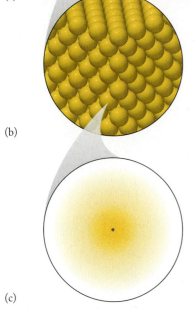

(b)

(c)

**FIGURE 3.7** (a) Five 1-ounce gold coins, each containing about 0.143 mole of pure gold. (b) A lattice of gold atoms. (c) The nucleus of one gold atom, surrounded by electrons.

| 2 |
| He |
| 4.003 |

| Atomic mass of He | 4.003 amu/atom |
|---|---|
| Mass of 1 mol of He | 4.003 g |
| Molar mass of He | 4.003 g/mol |

**FIGURE 3.8** The atomic mass (in amu/atom) and the molar mass (in g/mol) of helium have the same numerical value.

**Think About It** Chemical formulas reveal the number of moles of each element per mole of a compound as well as the total number of atoms per molecule of a compound. When comparing two compounds, their chemical formulas allow us to quickly determine the relative numbers of each type of atom.

⚙ **Practice Exercise** How many electrons are in a one-ounce gold coin (Figure 3.7) that contains 0.143 mol of pure gold?

*(Answers to Practice Exercises are in the back of the book.)*

■

**CONCEPT TEST**

How does a unit of measure such as a gross (144 units) relate to the concept of the mole?

*(Answers to Concept Tests are in the back of the book.)*

## Molar Mass

The mole provides an important link from the values of atomic mass in the periodic table to masses of macroscopic quantities of elements and compounds. Recall from Section 2.3 that the mass value shown in the cell of any element in the periodic table is the average atomic mass of the atoms of the element, expressed in atomic mass units (amu) or daltons (Da). That same value is also the mass of 1 mole of atoms of that element expressed in grams. Thus, the average mass of one atom of helium is 4.003 amu, and the mass of 1 mole of helium is 4.003 g. The mass in grams of 1 mole of any substance is called the **molar mass (*M*)** of the substance, and we say that the molar mass of helium is 4.003 g/mol (Figure 3.8). Note that because the mass of electrons is insignificant compared with that of protons and neutrons (Table 2.1), the mass of an ion is essentially the same as the mass of an atom, even though cations have fewer electrons than neutral atoms and anions have more.

As mentioned earlier, one of the reasons we use moles in chemistry is to allow us to compare the number of particles in differing masses of two different substances. Molar mass can be used to convert between the mass of a sample of any substance and the number of moles in that sample (Sample Exercises 3.3 and 3.4). Remember that chemical reactions take place between particles (e.g., atoms, molecules), but we measure the masses of products and reactants.

The mole enables us to calculate the number of particles in any sample of a given substance from the known mass of the sample. We can do this because the mole represents both a fixed number of particles (Avogadro's number) and the specific mass (the molar mass) of the substance. One useful analogy is to compare golf balls with baseballs. A dozen golf balls and a dozen baseballs have very different masses, but each contains 12 balls (Figure 3.9). Indeed, the mole is sometimes referred to as "the chemist's dozen."

**FIGURE 3.9** A dozen golf balls weigh less than a dozen baseballs, but both contain the same number of balls.

## SAMPLE EXERCISE 3.3 Converting Mass into Number of Moles  LO1

Some antacid tablets contain 425 mg of calcium (as $Ca^{2+}$ ions). How many moles of calcium are in each tablet?

**Collect and Organize** We are given the mass of $Ca^{2+}$ ions in each tablet. We are asked to convert this mass into an equivalent number of moles.

**Analyze** The number of moles of a substance is related to the mass of the substance by its molar mass. We begin by converting milligrams to grams, and then we convert grams of $Ca^{2+}$ into moles of $Ca^{2+}$ using the molar mass of $Ca^{2+}$.

$$\boxed{mg\ Ca^{2+}} \xrightarrow{\frac{1\ g}{10^3\ mg}} \boxed{g\ Ca^{2+}} \xrightarrow{\frac{1}{molar\ mass}} \boxed{mol\ Ca^{2+}}$$

The average atomic mass of a calcium atom is 40.078 amu, which means the molar mass of calcium is 40.08 g/mol when rounded to four significant figures.[2] We can estimate the answer to be about 0.01 mol because we are dividing about 0.4 g by about 40 g/mol.

**Solve**

$$425\ \cancel{mg\ Ca^{2+}} \times \frac{1\ \cancel{g}}{10^3\ \cancel{mg}} \times \frac{1\ mol\ Ca^{2+}}{40.08\ \cancel{g\ Ca^{2+}}} = 0.0106\ mol\ Ca^{2+} = 1.06 \times 10^{-2}\ mol\ Ca^{2+}$$

**Think About It** People seem most comfortable dealing with numbers between 1 and 1000, which is why the mass is given in milligrams in the antacid tablet. The answer here seems appropriate based on the small number of grams of calcium in the tablet.

⚙ **Practice Exercise** The mass of the diamond in Figure 3.10 is 3.25 carats (1 carat = 0.200 g). Diamonds are nearly pure carbon. How many moles of carbon are in the diamond?

*(Answers to Practice Exercises are in the back of the book.)*

**FIGURE 3.10** A 3.25-carat diamond.

## SAMPLE EXERCISE 3.4 Converting Number of Moles into Mass in Grams  LO1

A helium balloon sold at an amusement park contains 0.462 mole of He. How many grams of He are in the balloon?

**Collect and Organize** We are asked to convert 0.462 mol He to grams of He. We will need the molar mass of helium, 4.003 g/mol.

**Analyze** We use molar mass to convert moles of He into grams of He:

$$\boxed{mol\ He} \xrightarrow{molar\ mass} \boxed{g\ He}$$

Because the balloon contains approximately $\frac{1}{2}$ mole of helium, we estimate the answer should be around 2 g.

**Solve**

$$0.462\ \cancel{mol\ He} \times \frac{4.003\ g\ He}{1\ \cancel{mol\ He}} = 1.85\ g\ He$$

**Think About It** Two grams is very little mass, which seems reasonable for a balloon that is lighter than air.

⚙ **Practice Exercise** How many grams of gold are there in 0.250 mole of gold?

*(Answers to Practice Exercises are in the back of the book.)*

**molar mass ($\mathcal{M}$)** the mass of 1 mole of a substance. The molar mass of an element in grams per mole is numerically equal to that element's average atomic mass in atomic mass units.

---

[2]In Sample Exercises involving molar masses, we will usually round off the values given in the periodic table to four significant figures, except for elements with molar masses over 100.

| Molecular mass of $SO_3$ | 80.06 amu/molecule |
|---|---|
| Mass of 1 mol of $SO_3$ | 80.06 g |
| Molar mass of $SO_3$ | 80.06 g/mol |

**FIGURE 3.11** The molecular mass (in amu/molecule) and the molar mass (in g/mol) of sulfur trioxide have the same numerical value.

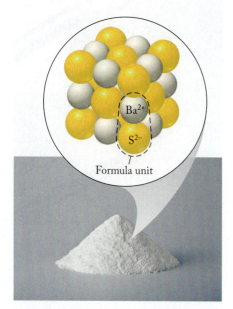

**FIGURE 3.12** Barium sulfide is an ionic compound used in white paint. A crystal of barium sulfide consists of a three-dimensional array of $Ba^{2+}$ ions and $S^{2-}$ ions. The 1:1 ratio of $Ba^{2+}$ ions to $S^{2-}$ ions is represented in the formula unit of the crystal (enclosed in a dashed oval) and in the empirical formula of the compound (BaS).

**CONNECTION** In Section 2.5 we defined the composition of an ionic compound in terms of the formula unit.

**CONCEPT TEST**

Which contains more atoms: 1 gram of gold (Au) or 1 gram of silver (Ag)?

*(Answers to Concept Tests are in the back of the book.)*

## Molecular Masses and Formula Masses

The **molecular mass** of a molecular compound is the sum of the atomic masses of the atoms in that molecule. Because atomic masses are given in atomic mass units, molecular masses are also reported in amu. Thus the molecular mass of sulfur trioxide, $SO_3$, is the sum of the masses of 1 sulfur atom and 3 oxygen atoms:

$$32.06 \text{ amu} + (3 \times 16.00 \text{ amu}) = 80.06 \text{ amu}$$

Just as the molar mass of an element in grams per mole is numerically the same as the atomic mass of one atom of the element in atomic mass units (Figure 3.8), the molar mass ($\mathcal{M}$) of a molecular compound in grams per mole is numerically the same as its molecular mass in atomic mass units. Thus the molar mass of $SO_3$ is 80.06 g/mol (Figure 3.11).

As with atoms and atomic mass, we can use the concept of moles to scale up molecular masses to molar masses, which are quantities we can measure and manipulate. The result is that atomic mass units become grams, atoms and molecules become moles, and we finally have the molar mass of the compound. For example, the molar mass of $SO_3$ in grams per mole is the sum of the masses of 1 mole of sulfur atoms and 3 moles of oxygen atoms:

$$\mathcal{M}_{SO_3} = 32.06 \text{ g/mol} + 3(16.00 \text{ g/mol}) = 80.06 \text{ g/mol } SO_3$$

The concept of molar mass applies to ionic as well as molecular compounds. For example, 1 mole of BaS contains 1 mole (137.33 g) of $Ba^{2+}$ ions and 1 mole (32.06 g) of $S^{2-}$ ions. Therefore, the molar mass of BaS is

$$\mathcal{M}_{BaS} = 137.33 \text{ g/mol} + 32.06 \text{ g/mol} = 169.39 \text{ g/mol BaS}$$

Keep in mind that there are no discrete molecules of BaS. This compound is ionic, and its crystals consist of ordered arrays of cations and anions (Figure 3.12). The formula unit defines the smallest integer ratio of positive and negative ions, 1:1 in this case, that describes the composition of an ionic compound. The mass of one formula unit of an ionic compound is called its **formula mass**.

Figure 3.13 summarizes how to use Avogadro's number and molar masses to convert between the number of moles, the mass, and the number of particles in a given quantity of an element or compound. The calculation of molar mass is further illustrated in Sample Exercise 3.5.

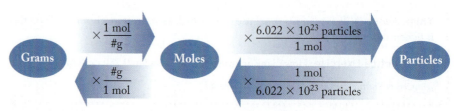

**FIGURE 3.13** The number of moles, number of particles, and the mass of a substance are related by Avogadro's number ($N_A$) and the molar mass ($\mathcal{M}$) of the substance.

**SAMPLE EXERCISE 3.5** **Calculating the Molar Mass of a Compound** **LO1**

Calculate the molar masses of (a) $H_2O$ and (b) $H_2SO_4$.

**Collect and Organize** We are given the formulas of two compounds and want to find their molar masses. The molar mass of a molecular compound is the sum of the molar masses of the elements in the molecular formula of the compound, each multiplied by the number of atoms of that element in the molecular formula. The molar mass of an element in grams per mole can be found in the periodic table.

**Analyze**

a. One mole of $H_2O$ contains 2 moles of H atoms and 1 mole of O atoms. Therefore,

$$\mathcal{M}_{H_2O} = 2 \times \mathcal{M}_H + \mathcal{M}_O$$

b. One mole of $H_2SO_4$ contains 2 moles of H atoms, 1 mole of S atoms, and 4 moles of O atoms. Therefore,

$$\mathcal{M}_{H_2SO_4} = 2 \times \mathcal{M}_H + \mathcal{M}_S + 4 \times \mathcal{M}_O$$

**Solve**

a. The molar mass of $H_2O$ is

$$2(1.008 \text{ g/mol}) + 16.00 \text{ g/mol} = 18.02 \text{ g/mol } H_2O$$

b. The molar mass of $H_2SO_4$ is

$$2(1.008 \text{ g/mol}) + 32.06 \text{ g/mol} + 4(16.00 \text{ g/mol}) = 98.08 \text{ g/mol } H_2SO_4$$

**Think About It** The molar mass of a compound is the sum of the masses of the elements in that compound. Because $H_2SO_4$ contains more atoms and heavier atoms than $H_2O$, we expect the molar mass of sulfuric acid to be larger than that of water.

⚙ **Practice Exercise** Green plants take in water ($H_2O$) and carbon dioxide ($CO_2$) and produce glucose ($C_6H_{12}O_6$) and oxygen ($O_2$). Calculate the molar masses of carbon dioxide, oxygen, and glucose.

*(Answers to Practice Exercises are in the back of the book.)*

**SAMPLE EXERCISE 3.6** **Interconverting Grams, Moles, and Molecules** **LO1**

Lithium carbonate is used in the treatment of depression. Calculate the number of moles and the number of formula units of lithium carbonate in a tablet containing 0.350 g of $Li_2CO_3$.

**Collect and Organize** We are given the mass of $Li_2CO_3$. We are asked to express that mass in terms of the number of moles of $Li_2CO_3$ and the number of formula units. Lithium carbonate is an ionic compound with the formula unit $Li_2CO_3$. We need to determine the formula mass of $Li_2CO_3$ and use Avogadro's number.

**Analyze** Figure 3.13 illustrates the general method we can apply. To convert a mass of $Li_2CO_3$ in grams into the equivalent number of formula units, we divide by its molar mass and multiply by Avogadro's number:

$$\text{g Li}_2\text{CO}_3 \xrightarrow{\frac{1}{\text{molar mass}}} \text{mol Li}_2\text{CO}_3 \xrightarrow{N_A} \text{formula units Li}_2\text{CO}_3$$

Because we have a relatively small mass of $Li_2CO_3$, we expect the number of moles of $Li_2CO_3$ to be less than one. Avogadro's number, however, is very large, so the number of formula units of $Li_2CO_3$ should be very large.

**Solve** First we need to calculate the molar mass of $Li_2CO_3$ by adding twice the molar mass of Li, the molar mass of C, and three times the molar mass of O:

$$2(6.941 \text{ g/mol}) + 12.01 \text{ g/mol} + 3(16.00 \text{ g/mol}) = 73.89 \text{ g/mol } Li_2CO_3$$

**molecular mass** the mass of one molecule of a molecular compound.

**formula mass** the mass of one formula unit of an ionic compound.

Then convert from grams to moles:

$$0.350 \ \cancel{\text{g Li}_2\text{CO}_3} \times \frac{1 \ \text{mol Li}_2\text{CO}_3}{73.89 \ \cancel{\text{g Li}_2\text{CO}_3}} = 0.004737 \ \text{mol Li}_2\text{CO}_3 = 4.737 \times 10^{-3} \ \text{mol Li}_2\text{CO}_3$$

We report the answer with three significant figures, $4.74 \times 10^{-3}$ moles, because the mass of $\text{Li}_2\text{CO}_3$ has only three significant figures.

Carrying on the calculation with the intermediate value and multiplying by Avogadro's number gives us

$$4.737 \times 10^{-3} \ \cancel{\text{mol Li}_2\text{CO}_3} \times \frac{6.022 \times 10^{23} \ \text{formula units Li}_2\text{CO}_3}{1 \ \cancel{\text{mol Li}_2\text{CO}_3}}$$

$$= 2.852 \times 10^{21} \ \text{formula units Li}_2\text{CO}_3$$

The final answer must have only three significant figures, so the number of formula units of $\text{Li}_2\text{CO}_3$ is $2.85 \times 10^{21}$.

**Think About It** Because atoms and ions are tiny, we expect a large number of formula units in a relatively small mass of $\text{Li}_2\text{CO}_3$. We cannot count individual formula units of an ionic compound, but we can convert the number of formula units to mass, a quantity we can readily measure.

⚙ **Practice Exercise** How many moles of $\text{CaCO}_3$ are there in an antacid tablet containing 0.500 g of $\text{CaCO}_3$? How many formula units of $\text{CaCO}_3$ are in this tablet?

*(Answers to Practice Exercises are in the back of the book.)*

---

**SAMPLE EXERCISE 3.7** **Calculating the Mass of an Element in a Mixture of Its Compounds** **LO1**

The discovery that iron metal could be extracted from iron minerals such as $\text{Fe}_2\text{O}_3$ and $\text{Fe}_3\text{O}_4$ ushered in the Iron Age around 2000 BCE. If an ancient metalworker converts all the iron in a mixture of 341 g $\text{Fe}_2\text{O}_3$ and 113 g $\text{Fe}_3\text{O}_4$ to iron metal, how many iron atoms would there be? What mass of iron does this represent?

**Collect and Organize** We are given the masses of two iron oxides and asked to calculate the combined mass of iron in the two oxides. The chemical formulas of the oxides tell us the number of moles of Fe per mole of each oxide. The molar mass of a compound or element can be used to interconvert its mass and the equivalent number of moles as shown in Figure 3.13.

**Analyze** To convert the masses of two iron oxides into equivalent masses of iron we need to calculate the number of moles of each oxide in the mixture and then use the formulas of the two oxides to convert the number of moles of each oxide to moles of iron. That value multiplied by the molar mass of Fe will give us the mass of Fe in the oxide. So, first we need to calculate the molar masses of $\text{Fe}_2\text{O}_3$ and $\text{Fe}_3\text{O}_4$, which will serve as conversion factors for calculating the number of moles of $\text{Fe}_2\text{O}_3$ and $\text{Fe}_3\text{O}_4$ (see Figure 3.13). Next we convert these values into moles of Fe. There are 2 mol Fe per mol $\text{Fe}_2\text{O}_3$ and 3 mol Fe per mol $\text{Fe}_3\text{O}_4$, so the conversion factors in this step are 2 and 3. Then we convert mol Fe values to masses in grams by multiplying by the molar mass of Fe. Finally we sum the two masses of Fe. Summarizing all but the last of these steps for $\text{Fe}_2\text{O}_3$:

$$\boxed{\text{g Fe}_2\text{O}_3} \xrightarrow[\text{molar mass}]{1} \boxed{\text{mol Fe}_2\text{O}_3} \xrightarrow[\text{mol Fe}_2\text{O}_3]{2 \ \text{mol Fe}} \boxed{\text{mol Fe}} \xrightarrow{\text{molar mass}} \boxed{\text{g Fe}}$$

**Solve** First we need to determine the formula masses of $\text{Fe}_2\text{O}_3$ and $\text{Fe}_3\text{O}_4$ using the molar mass of Fe and the molar mass of O:

$$\text{Fe}_2\text{O}_3: 2(55.847 \ \text{g/mol}) + 3(16.00 \ \text{g/mol}) = 159.69 \ \text{g/mol Fe}_2\text{O}_3$$

$$\text{Fe}_3\text{O}_4: 3(55.847 \ \text{g/mol}) + 4(16.00 \ \text{g/mol}) = 231.54 \ \text{g/mol Fe}_3\text{O}_4$$

Converting from grams $Fe_2O_3$ to moles $Fe_2O_3$ and then to number of ions gives us

$$341 \text{ g Fe}_2\text{O}_3 \times \frac{1 \text{ mol Fe}_2\text{O}_3}{159.69 \text{ g Fe}_2\text{O}_3} \times \frac{2 \text{ mol Fe}}{1 \text{ mol Fe}_2\text{O}_3} \times \frac{55.847 \text{ g Fe}}{1 \text{ mol Fe}} = 238.5 \text{ g Fe}$$

We repeat the calculation for $Fe_3O_4$:

$$113 \text{ g Fe}_3\text{O}_4 \times \frac{1 \text{ mol Fe}_3\text{O}_4}{231.54 \text{ g Fe}_3\text{O}_4} \times \frac{3 \text{ mol Fe}}{1 \text{ mol Fe}_3\text{O}_4} \times \frac{55.847 \text{ g Fe}}{1 \text{ mol Fe}} = 81.8 \text{ g Fe}$$

The total mass of iron in the two samples is $238.5 + 81.8 = 320.3$ g.

Since the masses of $Fe_2O_3$ and $Fe_3O_4$ were given to three significant figures, our final answer must also have three significant figures: 320 g Fe or $3.20 \times 10^2$ g Fe.

**Think About It** The mass of iron recovered from a total of 454 g of iron oxides ($\approx$1 pound) represents

$$\frac{320 \text{ g Fe}}{454 \text{ g Fe}_2\text{O}_3 \text{ and Fe}_3\text{O}_4} \times 100\% = 70.5\%$$

of the mass of the oxides. This high percentage makes sense because the molar mass of Fe is more than three times greater than the molar mass of oxygen. These iron oxides were valuable commodities in the Iron Age, as they are today.

**Practice Exercise** How many grams of Pb are contained in a mixture 0.100 kg each of $PbCl(OH)$ and $Pb_2Cl_2CO_3$?

*(Answers to Practice Exercises are in the back of the book.)*

## Moles and Chemical Equations

The concepts of mole and molar mass provide three more interpretations of Equation 3.1:

$$SO_3(g) + H_2O(g) \rightarrow H_2SO_4(\ell)$$

1. The coefficients in the chemical equation tell us that 1 *mole* of $SO_3$ reacts with 1 *mole* of $H_2O$, producing 1 *mole* of $H_2SO_4$.
2. The molar masses of the reactants and products allow us to say that 80.06 *grams* of $SO_3$ react with 18.02 *grams* of $H_2O$, producing 98.08 *grams* of $H_2SO_4$.
3. Avogadro's number tells us that $6.022 \times 10^{23}$ *molecules* of $SO_3$ react with $6.022 \times 10^{23}$ *molecules* of $H_2O$, forming $6.022 \times 10^{23}$ *molecules* of $H_2SO_4$. In terms of lowest whole-number ratios, 1 *molecule* of $SO_3$ reacts with 1 *molecule* of $H_2O$, forming 1 *molecule* of $H_2SO_4$.

The several interpretations of this equation are not limited to 1 mole of $SO_3$ reacting with 1 mole of $H_2O$. In general, the equation tells us that *any* number of moles ($x$ mol) of $SO_3$ react with an equal number of moles ($x$ mol) of water to produce the same quantity ($x$ mol) of sulfuric acid. This interpretation is valid because the mole ratio (the ratio of the coefficients) of $SO_3$ to $H_2O$ to $H_2SO_4$ in the equation is 1:1:1. The mole ratio is specific for a given equation and is different when different substances combine. For example, for the reaction in Equation 3.2 between sulfuric acid and $Fe_2O_3$ (hematite):

$$Fe_2O_3(s) + 3 H_2SO_4(aq) \rightarrow 3 H_2O(\ell) + Fe_2(SO_4)_3(aq)$$

the mole ratio of $Fe_2O_3$ to $H_2SO_4$ is 1:3.

From Equation 3.1, we can determine how much of one reactant is needed to completely react with any quantity of the other reactant. Or we can calculate how

**stoichiometry** the quantitative relation between reactants and products in a chemical reaction.

**law of conservation of mass** the sum of the masses of the reactants in a chemical reaction is equal to the sum of the masses of the products.

much product can be made from any quantity of reactants by multiplying $x$ by the appropriate molar mass.

|  | $SO_3(g)$ | + | $H_2O(g)$ | → | $H_2SO_4(\ell)$ |
|---|---|---|---|---|---|
| Particle ratios: | 1 molecule | + | 1 molecule | → | 1 molecule |
| Mole ratios: | 1 mol | + | 1 mol | → | 1 mol |
| Mass ratios: | 80.06 g | + | 18.02 g | → | 98.08 g |
| General case (moles): | $x$ mol | + | $x$ mol | → | $x$ mol |
| General case (masses): | $x$(80.06 g) | + | $x$(18.02 g) | → | $x$(98.08 g) |

The quantitative relation between the reactants and products involved in a chemical reaction is called the **stoichiometry** of the reaction. In a reaction where all reactants are completely converted into products, the sum of the masses of the reactants equals the sum of the masses of the products (Figure 3.14). This fact illustrates a fundamental relation known as the **law of conservation of mass**, which applies to all chemical reactions. This law works for two reasons: (1) the total number of atoms (and hence moles) of each element to the left of the reaction arrow in the equation representing any chemical reaction must match the total number of atoms (and moles) of that element to the right of the reaction arrow, and (2) the identity of atoms does not change in a chemical reaction.

**FIGURE 3.14** The law of conservation of mass states that the total mass of reactants consumed in a chemical reaction equals the total mass of products formed in the reaction. In the reaction shown here, the combined mass of $SO_3$ and $H_2O$ consumed equals the mass of the $H_2SO_4$ that forms. The number of each kind of atom is the same on the left and right sides of the balanced equation.

$$SO_3 + H_2O \longrightarrow H_2SO_4$$

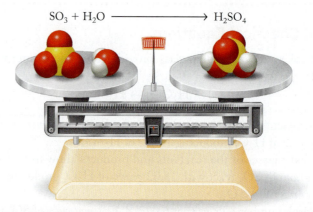

Using two different colored spheres to represent nitrogen and oxygen atoms, sketch a picture describing the reaction between two molecules of $N_2$ and four molecules of $O_2$ to form $NO_2$.

**Collect and Organize** We are given the chemical formulas for two reactants and will use colored spheres to sketch their conversion to a product.

**Analyze** We choose blue spheres to represent nitrogen atoms and red spheres to represent oxygen atoms. We use the atoms contained in two molecules of nitrogen and four molecules of oxygen to construct the appropriate number of $NO_2$ molecules.

**Solve** From a total of 4 nitrogen and 8 oxygen atoms, we can make four molecules of $NO_2$.

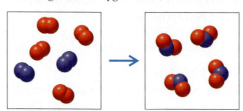

**Think About It** A chemical reaction involves rearrangements of atoms and molecules, resulting in new molecules. The number of each type of atom present before the reaction equals the number of each type of atom after the reaction.

⚙ **Practice Exercise** Using two different colored spheres to represent nitrogen and hydrogen atoms, sketch a picture describing the reaction between two molecules of $N_2$ and six molecules of $H_2$ to form ammonia, $NH_3$.

*(Answers to Practice Exercises are in the back of the book.)*

# 3.3 Writing Balanced Chemical Equations

A *balanced* chemical equation is one that has the same number of atoms of each type on both sides of the equation. In this section we practice writing balanced chemical equations describing chemical reactions that may account for the products observed by Miller and Urey in their famous 1953 experiment on the origins of life on Earth mentioned at the beginning of this chapter and shown in Figure 3.15. A key molecule in the synthesis of glycine, one of the amino acids produced in the Miller/Urey experiment, is hydrogen cyanide, HCN. Hydrogen cyanide can be formed from methane, $CH_4$, and ammonia, $NH_3$, in a reaction that also produces hydrogen, $H_2$. Let's write a balanced chemical equation describing this reaction. We know the formulas of the reactants and products, but we don't know the stoichiometry of the reaction, that is, the coefficients that precede their formulas in a balanced chemical equation. There are several approaches to writing chemical equations. Let's apply the following four-step method, which works well if we have correctly identified all the reactants and products.

1. *Write a preliminary expression containing a single particle (atom, molecule, or formula unit) of each reactant and product with a reaction arrow separating reactants from products. Include symbols indicating physical states.*

Below is such an expression for the reaction of interest with models of the molecules involved.

$$CH_4(g) + NH_3(g) \rightarrow HCN(g) + H_2(g)$$
reactants → products

2. *Check whether the expression is balanced by summing the different types of atoms on each side of the reaction arrow.*

Adding up the atoms of each element in the reactants and products, we find that the expression above is balanced in terms of the numbers of C and N atoms, but not in terms of H atoms: there are 7 on the reactant side and only 3 on the product side.

| Element | Reactant Side | Product Side | Balanced? |
|---|---|---|---|
| C | 1 | 1 | ✔ |
| N | 1 | 1 | ✔ |
| H | 4 + 3 = 7 | 1 + 2 = 3 | ✗ |

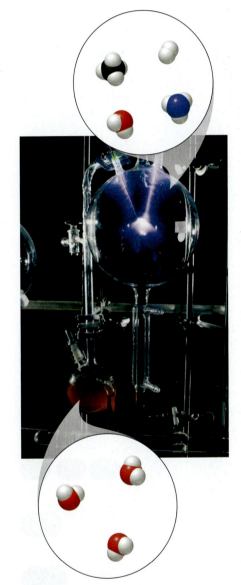

**FIGURE 3.15** This is the type of apparatus that Stanley L. Miller and Harold C. Urey used for their 1953 experiment. The large flask contained water, ammonia, hydrogen, and methane in ratios believed to represent the composition of Earth's early atmosphere. The smaller flask was heated to add water vapor to the gas mixture in the large flask. An electrical current was applied to the tungsten rods attached to the flask to simulate the frequent lightning storms thought to occur on early Earth.

3. *Choose an element that appears in only one reactant and product and balance it first.*

In this example, the two elements that occur only once on each side of the reaction arrow are C and N, but they are already balanced, so we can skip this step.

4. *Choose coefficients for the other substances so that the number of atoms for each element is the same on both sides of the reaction arrow.*

▶❙❙ **CHEMTOUR** Balancing Equations

We need 4 more H atoms on the product side to have a balanced equation. Changing the coefficient in front of either HCN or $H_2$ could add the needed H atoms, but changing the coefficient of HCN would also increase the numbers of C and N atoms, which are already balanced. Therefore, the correct approach is to increase the coefficient of $H_2$ from 1 to 3, which increases the total number of H atoms from $H_2$ to 6.

$$CH_4(g) + NH_3(g) \rightarrow HCN(g) + \underline{3}\ H_2(g)$$

A final check of the number of atoms of each element on both sides of the reaction arrow confirms that the reaction is balanced:

| Element | Reactant Side | Product Side | Balanced? |
|---------|---------------|--------------|-----------|
| C | 1 | 1 | ✔ |
| N | 1 | 1 | ✔ |
| H | 4 + 3 = 7 | 1 + (3 × 2) = 7 | ✔ |

Note that balance was achieved by changing a coefficient, *not* by changing any subscripts in chemical formulas. Changing subscripts would change the identities of substances and that would change the nature of the reaction. Only coefficients may be changed in writing balanced chemical equations.

Let's get more practice writing balanced chemical equations by using another important reaction in the Miller/Urey experiment. Analysis of the vapor above the solution in the large flask revealed the presence of both carbon dioxide and carbon monoxide gas. It has been shown that the reaction between methane and water yields carbon dioxide and hydrogen.

1. *Write a preliminary expression containing a single particle of each reactant and product.* Methane has the formula $CH_4$, water is $H_2O$, carbon dioxide is $CO_2$, and hydrogen is $H_2$. Note that all of the gaseous, nonmetallic elements—except for the noble gases—exist in nature as diatomic molecules: $H_2$, $N_2$, $O_2$, $F_2$, and $Cl_2$. Bromine, a liquid at room temperature, and $I_2$, a solid, are also diatomic (Figure 3.16). Whenever these substances are involved in chemical reactions, they must be written as diatomic molecules.

$$CH_4(g) + H_2O(g) \rightarrow CO_2(g) + H_2(g)$$

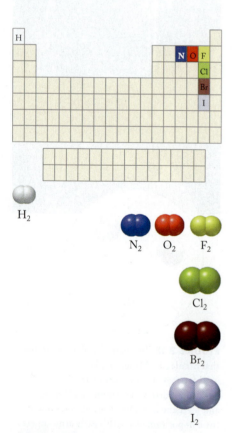

**FIGURE 3.16** The diatomic elements are hydrogen, nitrogen, oxygen, and the group 17 halogens. The halogen astatine (At) at the bottom of group 17 is the rarest terrestrial element, and virtually none of its bulk physical properties are known.

2. *Check whether the expression is balanced.*

| Element | Reactant Side | Product Side | Balanced? |
|---------|---------------|--------------|-----------|
| C | 1 | 1 | ✔ |
| O | 1 | 2 | ✗ |
| H | 4 + 2 = 6 | 2 | ✗ |

The equation as written is not balanced.

3. *Choose an element that appears in only one reactant and product to balance first.* In this case we could choose carbon or oxygen. Carbon is already balanced, so we choose oxygen. With 1 O atom on the reactant side and 2 O atoms on the product side, we can make the number of O atoms equal on both sides of the equation by placing a coefficient of 2 in front of $H_2O$.

$$\_\_ CH_4(g) + \underline{2}\ H_2O(g) \rightarrow \_\_ CO_2(g) + \_\_ H_2(g)$$

Assigning this coefficient balances the number of oxygen atoms but not the number of hydrogen atoms.

| Element | Reactant Side | Product Side | Balanced? |
|---------|---------------|--------------|-----------|
| C | 1 | 1 | ✔ |
| O | 2 × 1 = 2 | 2 | ✔ |
| H | 4 + (2 × 2) = 8 | 2 | ✗ |

4. *Choose coefficients for the other substances so that the numbers of atoms of each element are the same on both sides of the equation.*

We can balance the H atoms by placing a coefficient of 4 in front of $H_2$, which finishes the balancing process:

$$CH_4(g)\ + \underline{2}\ H_2O(g) \rightarrow\ CO_2(g)\ + \underline{4}\ H_2(g)$$

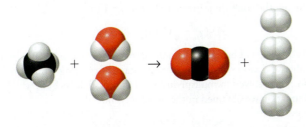

| Element | Reactant Side | Product Side | Balanced? |
|---------|---------------|--------------|-----------|
| C | 1 | 1 | ✔ |
| O | 2 × 1 = 2 | 2 | ✔ |
| H | 4 + (2 × 2) = 8 | 4 × 2 = 8 | ✔ |

**SAMPLE EXERCISE 3.9** **Writing a Balanced Chemical Equation** **LO2**

Write a balanced chemical equation for the gas-phase reaction between $SO_2$ and $O_2$ that forms $SO_3$. This reaction, one of the causes of acid rain, occurs when fuels containing sulfur are burned. Acid rain is harmful to aquatic life and damages terrestrial plants.

**Collect and Organize** We are given the chemical formulas and physical states of the reactants and products and are to write a balanced chemical equation.

**Analyze** In a balanced chemical equation, we must have the same number of atoms of each type on both sides of the equation. To balance the equation, we assign coefficients as needed to make the numbers of atoms of each element on the reactant (left) and product (right) sides of the reaction arrow equal.

**Solve**
1. Write a reaction expression based on single molecules of the reactants and product:

$$SO_2(g) + O_2(g) \rightarrow SO_3(g)$$

2. A check of whether the expression is balanced:

| Element | Reactant Side | Product Side | Balanced? |
|---------|---------------|--------------|-----------|
| S | 1 | 1 | ✔ |
| O | 2 + 2 = 4 | 3 | ✗ |

discloses that it is not.

3. Balancing the O atoms presents a challenge because increasing their number on the product side by increasing the $SO_3$ coefficient from 1 to 2:

$$SO_2(g) + O_2(g) \rightarrow \underline{2}\ SO_3(g)$$

means that there are now two more O atoms and one more S atom on the product side than on the reactant side. Fortunately, both of these imbalances can be solved simultaneously by increasing the $SO_2$ coefficient from 1 to 2:

$$\underline{2}\ SO_2(g) + O_2(g) \rightarrow \underline{2}\ SO_3(g)$$

A check of the atom inventory on both sides confirms that the equation is now balanced:

| Element | Reactant Side | Product Side | Balanced? |
|---------|---------------|--------------|-----------|
| S | **2** × 1 = 2 | **2** × 1 = 2 | ✔ |
| O | (**2** × 2) + 2 = 6 | **2** × 3 = 6 | ✔ |

**Think About It** Rather than increasing the number of O atoms on the product side to bring the equation into balance, we might have tried *decreasing* their number on the reactant side by reducing the coefficient in front of $O_2$ from 1 to $\frac{1}{2}$:

$$SO_2(g) + \tfrac{1}{2} O_2(g) \rightarrow SO_3(g)$$

This fix has the advantage of bringing the number of O atoms into balance without imbalancing the number of S atoms. It has the disadvantage of creating a fractional coefficient, but this problem can be solved by multiplying all of the coefficients by 2, which gives us the same balanced equation we obtained in the Solve section:

$$2\ SO_2(g) + O_2(g) \rightarrow 2\ SO_3(g)$$

We will use this strategy again in Sample Exercise 3.10 and in writing chemical equations for some of the combustion reactions in Section 3.4.

⚙ **Practice Exercise** Balance the chemical equations for (a) the reaction between elemental phosphorus $P_4(s)$ and oxygen to make $P_4O_{10}(s)$ and (b) the reactions of water with $P_4O_{10}(s)$ to produce phosphoric acid, $H_3PO_4(\ell)$.

*(Answers to Practice Exercises are in the back of the book.)*

(Answers to Concept Tests are in the back of the book.)

**CONCEPT TEST**

Why can't we write a balanced equation for the reaction between $SO_2$ and $O_2$ in Sample Exercise 3.9 this way?

$$SO_2(g) + O(g) \rightarrow SO_3(g)$$

Note that a chemical equation is analogous to a mathematical equation. The equation in Sample Exercise 3.9 is still balanced if we multiply all the coefficients by the same number. For example, if we multiply all the coefficients by 2:

$$4\,SO_2(g) + 2\,O_2(g) \rightarrow 4\,SO_3(g)$$

It is also still balanced if we multiply all the coefficients by a fraction, such as $\frac{1}{2}$. However, it is conventional to report a balanced equation with the smallest whole-number coefficients.

**SAMPLE EXERCISE 3.10** **Balancing Chemical Equations Using Fractional Coefficients** **LO2**

Write a balanced chemical equation for the gas-phase reaction in which dinitrogen pentoxide is formed from nitrogen and oxygen.

**Collect and Organize** We are asked to write a balanced chemical equation given the names and physical states of the reactants and products.

**Analyze** Nitrogen and oxygen gases exist in nature as diatomic molecules, so their molecular formulas are $N_2$ and $O_2$. The formula for dinitrogen pentoxide is $N_2O_5$. To write a balanced chemical equation, we must assign coefficients as needed to make the numbers of atoms of each element on the reactant (left) and product (right) sides of the reaction arrow equal.

**Solve**
1. Write an equation using correct formulas.

$$N_2(g) + O_2(g) \rightarrow N_2O_5(g)$$

| Element | Reactant Side | Product Side | Balanced? |
|---|---|---|---|
| N | 2 | 2 | ✔ |
| O | 2 | 5 | ✘ |

The equation is not balanced as written.

2. Two N atoms appear on each side of the equation, so no adjustment of its coefficients is necessary.
3. Oxygen is the only other element to balance in the equation. With 2 O atoms on the reactant side and 5 O atoms on the product side, there is no whole-number coefficient that we can place in front of the $O_2$ to make the number of O atoms equal on both sides of the equation. A coefficient of $\frac{5}{2}$ in front of $O_2$ would balance the number of O atoms, but we need to use whole numbers. Since $\frac{5}{2}$ times 2 is the whole number 5, we can place a coefficient of 5 in front of $O_2$ and a coefficient of 2 in front of $N_2O_5$:

$$\underline{\phantom{2}}\,N_2(g) + \underline{5}\,O_2(g) \rightarrow \underline{2}\,N_2O_5(g)$$

However, the number of nitrogen atoms is no longer equal on both sides of the equation:

| Element | Reactant Side | Product Side | Balanced? |
|---|---|---|---|
| N | 2 | 2 × 2 = 4 | ✘ |
| O | 5 × 2 = 10 | 2 × 5 = 10 | ✔ |

To balance the number of nitrogen atoms, we add a coefficient of 2 in front of $N_2$:

$$\underline{2}\ N_2(g) + \underline{5}\ O_2(g) \rightarrow \underline{2}\ N_2O_5(g)$$

The equation is now balanced.

**Think About It** In balancing a chemical equation, always check the equation after you are done to make sure that an element balanced in one step is still balanced at the end.

⚙ **Practice Exercise** Balance the chemical equation for the reaction between carbon monoxide, $CO(g)$, and oxygen to form carbon dioxide, $CO_2(g)$.

*(Answers to Practice Exercises are in the back of the book.)*

# 3.4 Combustion Reactions

Both methane and carbon monoxide are still present in our atmosphere, although at much lower concentrations than on early Earth. Atmospheric methane (natural gas) can be traced to several sources including wetlands, cattle ranching, rice production, and oil drilling operations. Carbon dioxide in the atmosphere arises from burning carbon-containing compounds. Natural sources of $CO_2$ include volcanic activity and forest fires. In our industrialized world, the greatest use of methane and petroleum-based fuels is energy production. Fuels like methane ($CH_4$) are members of a class of organic compounds known as **hydrocarbons** because they are composed of only hydrogen and carbon.

Natural gas and other hydrocarbon fuels are burned to heat buildings and to warm water. Burning $CH_4$ is an example of an important class of chemical reactions known as **combustion reactions**, or simply *combustion* (Figure 3.17). Combustion refers to the reaction of a fuel with oxygen, such as the burning of a substance in air, which contains 21% oxygen.

When the combustion of a hydrocarbon is complete, the only products are carbon dioxide and water. In the presence of $O_2$, the most stable form of carbon is $CO_2$, not CO, so carbon monoxide also reacts with oxygen to form carbon dioxide. In the combustion reaction between methane and oxygen, the carbon becomes carbon dioxide when a sufficient supply of oxygen is available. The hydrogen in the fuel combines with oxygen to form water vapor. Let's see how to balance the equation describing the combustion of methane.

1. Our first step is to write the formulas of the reactants on the left and products on the right.

$$CH_4(g) + O_2(g) \rightarrow H_2O(g) + CO_2(g)$$

| Element | Reactant Side | Product Side | Balanced? |
|---------|---------------|--------------|-----------|
| C | 1 | 1 | ✔ |
| H | 4 | 2 | ✘ |
| O | 2 | 1 + 2 = 3 | ✘ |

The carbon atoms are balanced but the hydrogen and oxygen atoms are not.

2. Like carbon, H appears in only one reactant and one product. Adding a coefficient of 2 in front of the $H_2O$ makes the number of H atoms equal on both sides of the equation but increases the number of O atoms on the product side:

**FIGURE 3.17** The combustion of methane in a gas stove produces carbon dioxide and water.

**hydrocarbons** a class of organic compounds composed of only hydrogen and carbon.

**combustion reaction** a heat-producing reaction between oxygen and another element or compound.

$$CH_4(g) + \_\_ O_2(g) \rightarrow \underline{2} H_2O(g) + CO_2(g)$$

| Element | Reactant Side | Product Side | Balanced? |
|---------|---------------|--------------|-----------|
| C | 1 | 1 | ✔ |
| H | 4 | 2 × 2 = 4 | ✔ |
| O | 2 | (2 × 1) + 2 = 4 | ✘ |

3. This leaves only the O atoms to balance. A coefficient of 2 in front of $O_2$ makes the atoms of O equal on both sides of the equation:

$$CH_4(g) + \underline{2} O_2(g) \rightarrow \underline{2} H_2O(g) + CO_2(g)$$

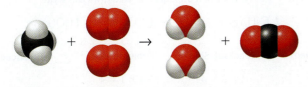

| Element | Reactant Side | Product Side | Balanced? |
|---------|---------------|--------------|-----------|
| C | 1 | 1 | ✔ |
| H | 4 | 2 × 2 = 4 | ✔ |
| O | 2 × 2 = 4 | (2 × 1) + 2 = 4 | ✔ |

Balancing first C, then H, then O is useful in writing the equations for many combustion reactions involving hydrocarbons.

---

**SAMPLE EXERCISE 3.11**  **Writing a Balanced Chemical**  **LO2**
**Equation for a Combustion Reaction**

Methane ($CH_4$) is the principal ingredient in natural gas, but significant concentrations of the gases ethane ($C_2H_6$) and propane ($C_3H_8$) are also present in most natural gas samples. Write a balanced chemical equation describing the complete combustion of $C_2H_6$.

**Collect and Organize**  We are to write a balanced chemical equation for the combustion of $C_2H_6$. The products of complete combustion of hydrocarbons are carbon dioxide and water.

**Analyze**
1. The chemical equation describing the combustion of ethane is

$$C_2H_6(g) + O_2(g) \rightarrow CO_2(g) + H_2O(g)$$

| Element | Reactant Side | Product Side | Balanced? |
|---------|---------------|--------------|-----------|
| C | 2 | 1 | ✘ |
| H | 6 | 2 | ✘ |
| O | 2 | 2 + 1 = 3 | ✘ |

**Solve**
2. Because this reaction is between oxygen and a hydrocarbon, we balance C first, then H, then O. Balance the carbon atoms first by giving $CO_2$ in the product a coefficient of 2:

$$\_\_ C_2H_6(g) + \_\_ O_2(g) \rightarrow \underline{2} CO_2(g) + \_\_ H_2O(g)$$

| Element | Reactant Side | Product Side | Balanced? |
|---------|---------------|--------------|-----------|
| C | 2 | 2 × 1 = 2 | ✔ |
| H | 6 | 2 | ✘ |
| O | 2 | (2 × 2) + 1 = 5 | ✘ |

3. Then balance the hydrogen atoms by giving $H_2O$ a coefficient of 3:

$$\_\ C_2H_6(g) + \_\ O_2(g) \rightarrow \underline{2}\ CO_2(g) + \underline{3}\ H_2O(g)$$

| Element | Reactant Side | Product Side | Balanced? |
|---------|---------------|--------------|-----------|
| C | 2 | $2 \times 1 = 2$ | ✔ |
| H | 6 | $3 \times 2 = 6$ | ✔ |
| O | 2 | $(2 \times 2) + (3 \times 1) = 7$ | ✗ |

4. At this stage, the oxygen atoms cannot be balanced with a simple whole-number coefficient for $O_2$; we need $\frac{7}{2} O_2$. However, if we give $O_2$ a coefficient of 7 and double the coefficients for ethane, carbon dioxide, and water, we can write

$$\underline{2}\ C_2H_6(g) + \underline{7}\ O_2(g) \rightarrow \underline{4}\ CO_2(g) + \underline{6}\ H_2O(g)$$

| Element | Reactant Side | Product Side | Balanced? |
|---------|---------------|--------------|-----------|
| C | $2 \times 2 = 4$ | $4 \times 1 = 4$ | ✔ |
| H | $2 \times 6 = 12$ | $6 \times 2 = 12$ | ✔ |
| O | $7 \times 2 = 14$ | $(4 \times 2) + (6 \times 1) = 14$ | ✔ |

This equation is balanced.

**Think About It** Balancing chemical equations for combustion reactions of hydrocarbons often requires several iterations of the coefficients before all of the elements are balanced.

⚙ **Practice Exercise** Write a balanced chemical equation describing the complete combustion of propane ($C_3H_8$).

*(Answers to Practice Exercises are in the back of the book.)*

# 3.5 Stoichiometric Calculations and the Carbon Cycle

Earth's atmosphere underwent a major change beginning about 2.5 billion years ago with the evolution of green plants and the onset of *photosynthesis*. Photosynthesis is driven by the energy in sunlight and involves several steps, but the overall reaction is

$$6\ CO_2(g) + 6\ H_2O(\ell) \rightarrow C_6H_{12}O_6(aq) + 6\ O_2(g)$$
$$\text{Glucose}$$

The reverse reaction is called *respiration*.

$$C_6H_{12}O_6(aq) + 6\ O_2(g) \rightarrow 6\ CO_2(g) + 6\ H_2O(\ell)$$
$$\text{Glucose}$$

Respiration is the major source of energy for all living things on Earth. The appearance of bacteria capable of photosynthesis is believed to have been the initial source of oxygen in our atmosphere, but today $O_2$ comes mostly from green plants.

Photosynthesis and respiration are key reactions in the *carbon cycle* (Figure 3.18). The two processes are nearly, but not exactly, in balance in Earth's biosphere. If they were exactly in balance, no net change would have taken place in the concentrations of atmospheric carbon dioxide or oxygen in the past 2.5 billion years. However, about 0.01% of the decaying mass of plants and animals (called *detritus*) is incorporated into sediments and soil when organisms die. Shielded in

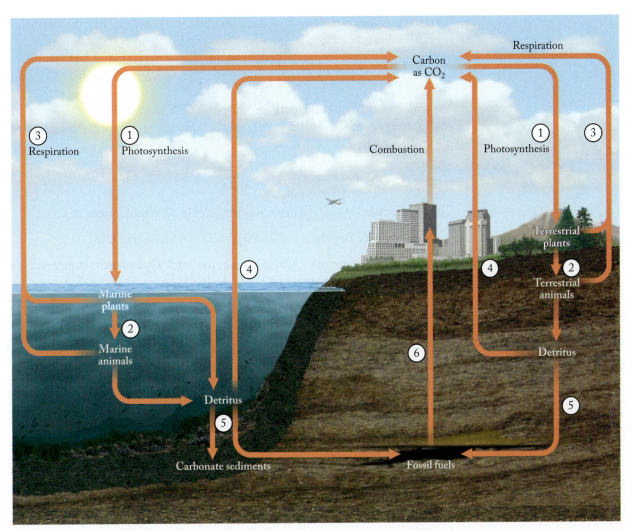

**FIGURE 3.18** The carbon cycle. ① Green plants and marine plants incorporate $CO_2$ into their biomass. ② Some of the plant biomass becomes the biomass of animals. ③ As plants and animals respire, they release $CO_2$ back into the environment. ④ When they die, the decay of their tissues releases most of their carbon content as $CO_2$, but about 0.01% is incorporated into carbonate minerals and deposits of coal, petroleum, and natural gas (fossil fuels, ⑤). ⑥ Mining and the combustion of fossil fuels for human use are shifting the natural equilibrium that has controlled the concentration of $CO_2$ in the atmosphere.

this way from exposure to oxygen, the carbon in this mass is not converted back into $CO_2$. Although 0.01% may not seem like much, over hundreds of millions of years it has added up to the removal of about $10^{20}$ kg of carbon dioxide from the atmosphere. About $10^{15}$ kg of this buried carbon is in the form of fossil fuels: coal, petroleum, and natural gas.

As a result of human activity and the combustion of fossil fuels, the natural balance that limited the concentration of $CO_2$ in the atmosphere is being altered. Annually about 6.8 trillion ($6.8 \times 10^{12}$) kg of carbon is reintroduced into the atmosphere as $CO_2$ as a result of the combustion of fossil fuels, and deforestation adds another $2 \times 10^{12}$ kg each year. The effects of these additions on global climate have been the subject of considerable debate, and we will examine them again in Chapter 8. Here, though, we use a typical reaction involving $CO_2$ to learn how to use a balanced chemical equation to calculate the mass of products in a chemical reaction.

If combustion of fossil fuels adds $6.8 \times 10^{12}$ kg of carbon to the atmosphere each year as $CO_2$, what is the mass of the carbon dioxide added? The mass must be more than $6.8 \times 10^{12}$ kg, because this amount is only the mass due to carbon. The balanced chemical equation for the combustion of carbon to carbon dioxide in excess $O_2$ is

$$C(s) + O_2(g) \rightarrow CO_2(g)$$

To use this equation to determine the mass of $CO_2$ released, we first use the molar mass of carbon, 12.01 g/mol, to convert mass into moles:

$$6.8 \times 10^{12} \text{ kg C} \times \frac{10^3 \text{ g}}{1 \text{ kg}} \times \frac{1 \text{ mol C}}{12.01 \text{ g C}} = 5.7 \times 10^{14} \text{ mol C}$$

We know from the coefficients for $CO_2$ and C in the balanced equation that the mole ratio of $CO_2$ to C is 1:1, so the amount of $CO_2$ produced is also $5.7 \times 10^{14}$ mol:

$$5.7 \times 10^{14} \text{ mol C} \times \frac{1 \text{ mol CO}_2}{1 \text{ mol C}} = 5.7 \times 10^{14} \text{ mol CO}_2$$

To convert moles of $CO_2$ to mass, we first calculate the molar mass of $CO_2$:

$$12.01 \text{ g/mol} + 2(16.00 \text{ g/mol}) = 44.01 \text{ g/mol CO}_2$$

Multiplying the moles of $CO_2$ by the molar mass of $CO_2$ and converting grams to kilograms gives us the mass of $CO_2$:

▶❚ CHEMTOUR Carbon Cycle

$$5.7 \times 10^{14} \text{ mol CO}_2 \times \frac{44.01 \text{ g CO}_2}{1 \text{ mol CO}_2} \times \frac{1 \text{ kg}}{10^3 \text{ g}} = 2.5 \times 10^{13} \text{ kg CO}_2$$

Notice that the answer in each of these steps is the starting point for the next step. Therefore we can combine the three separate calculations into a single calculation:

$$6.8 \times 10^{12} \text{ kg C} \times \frac{10^3 \text{ g}}{1 \text{ kg}} \times \frac{1 \text{ mol C}}{12.01 \text{ g C}} \times \frac{1 \text{ mol CO}_2}{1 \text{ mol C}} \times \frac{44.01 \text{ g CO}_2}{1 \text{ mol CO}_2} \times \frac{1 \text{ kg}}{10^3 \text{ g}}$$

$$= 2.5 \times 10^{13} \text{ kg CO}_2$$

This procedure can be applied to determining the mass of any substance (reactant or product) involved in any chemical reaction if we know (1) the mass of another substance in the reaction and (2) the *stoichiometric relation* between the two substances, that is, their mole ratio in the balanced chemical equation.

---

**SAMPLE EXERCISE 3.12** **Calculating a Product Mass from a Reactant Mass** **LO3**

In 2012 electric power plants in the United States consumed about $1.22 \times 10^{11}$ kg of natural gas. Natural gas is mostly methane, $CH_4$, so we can approximate the combustion reaction generating the energy by the equation:

$$CH_4(g) + 2\,O_2(g) \rightarrow 2\,H_2O(g) + CO_2(g)$$

How many kilograms of $CO_2$ were released into the atmosphere from these power plants in 2012?

**Collect and Organize** We are asked to calculate the mass of $CO_2$ produced from combustion of $1.22 \times 10^{11}$ kg $CH_4$.

**Analyze** The balanced equation tells us that 1 mole of carbon dioxide is produced for every 1 mole of methane consumed. We can change the mass of $CH_4$ in kg to g and then convert into moles of $CH_4$ by multiplying by the conversion factor that defines the mass of $CH_4$ that is equal to the mass of one mole of $CH_4$. In this reaction the moles of $CH_4$ consumed is equal to the number of moles of $CO_2$ produced; the mole ratio is 1:1. Finally we convert moles of $CO_2$ into kilograms of $CO_2$. We can combine all of these steps into a single calculation:

$$\boxed{\text{kg CH}_4} \xrightarrow{\frac{10^3 \text{ g}}{1 \text{ kg}}} \boxed{\text{g CH}_4} \xrightarrow{\frac{1}{\text{molar mass}}} \boxed{\text{mol CH}_4} \xrightarrow{\frac{1 \text{ mol CO}_2}{1 \text{ mol CH}_4}} \boxed{\text{mol CO}_2} \xrightarrow{\text{molar mass}} \boxed{\text{g CO}_2} \xrightarrow{\frac{1 \text{ kg}}{10^3 \text{ g}}} \boxed{\text{kg CO}_2}$$

**Solve** The chemical equation is balanced as written (you should always check to be sure). To convert between grams and moles of $CH_4$ and $CO_2$, we need to calculate the molar masses of these compounds. The molar mass of $CH_4$ is

$$12.01 \text{ g/mol} + 4(1.008 \text{ g/mol}) = 16.04 \text{ g/mol } CH_4$$

and the molar mass of $CO_2$ is

$$12.01 \text{ g/mol} + 2(16.00 \text{ g/mol}) = 44.01 \text{ g/mol } CO_2$$

1. Convert the mass of $CH_4$ into moles:

$$1.22 \times 10^{11} \text{ kg } CH_4 \times \frac{10^3 \text{ g}}{1 \text{ kg}} \times \frac{1 \text{ mol } CH_4}{16.04 \text{ g } CH_4} = 7.606 \times 10^{12} \text{ mol } CH_4$$

2. Convert moles of $CH_4$ into moles of $CO_2$ using the coefficients from the balanced chemical equation:

$$7.606 \times 10^{12} \text{ mol } CH_4 \times \frac{1 \text{ mol } CO_2}{1 \text{ mol } CH_4} = 7.606 \times 10^{12} \text{ mol } CO_2$$

3. Convert moles of $CO_2$ into mass of $CO_2$:

$$7.606 \times 10^{12} \text{ mol } CO_2 \times \frac{44.01 \text{ g } CO_2}{1 \text{ mol } CO_2} \times \frac{1 \text{ kg}}{10^3 \text{ g}} = 3.35 \times 10^{11} \text{ kg } CO_2$$

We can combine the three separate calculations into a single calculation, and in subsequent problems we may not show the individual steps:

$$1.22 \times 10^{11} \text{ kg } CH_4 \times \frac{10^3 \text{ g}}{1 \text{ kg}} \times \frac{1 \text{ mol } CH_4}{16.04 \text{ g } CH_4} \times \frac{1 \text{ mol } CO_2}{1 \text{ mol } CH_4} \times \frac{44.01 \text{ g } CO_2}{1 \text{ mol } CO_2} \times \frac{1 \text{ kg}}{10^3 \text{ g}}$$

$$= 3.35 \times 10^{11} \text{ kg } CO_2$$

**Think About It** The answer $3.35 \times 10^{11}$ kg $CO_2$ is about 1.3% of the mass of $CO_2$ we calculated for total annual $CO_2$ production by combustion of fossil fuels.

⚙ **Practice Exercise** Disposable lighters burn butane ($C_4H_{10}$) and produce $CO_2$ and $H_2O$. Balance the chemical equation for this combustion reaction, and determine how many grams of $CO_2$ are produced by burning 1.00 g $C_4H_{10}$.

*(Answers to Practice Exercises are in the back of the book.)*

---

**SAMPLE EXERCISE 3.13** **Calculating the Masses of Reactants** **LO3**

Copper was among the first metals to be refined from minerals collected by ancient metalworkers. The production of copper was already an industry by 3500 BCE. Cuprite is a copper mineral commonly found near Earth's surface, making it a likely resource for Bronze Age coppersmiths. Cuprite has the formula $Cu_2O$ and can be converted to copper metal by reacting it with charcoal:

$$2 \text{ Cu}_2O(s) + C(s) \rightarrow 4 \text{ Cu}(s) + CO_2(g)$$

How much cuprite and how much carbon are needed to prepare a 256 g copper bracelet such as the one shown in Figure 3.19?

**Collect and Organize** We are given a balanced chemical equation and the mass of product. We are asked to find the masses of the reactants. Because the chemical equation relates moles, not masses, of reactants and products, we need to find the molar mass of each substance in the reaction.

**Analyze** The chemical equation tells us that for every 4 moles of copper we produce, 2 moles of $Cu_2O$ and 1 mole of C are needed. To use these relationships, we first need to convert the mass of copper to the equivalent number of moles of Cu using the molar mass of Cu. We then work two separate problems: one uses the ratio of Cu to $Cu_2O$; the second uses the ratio of Cu to C from the balanced equation to covert to moles of the two reactants. Finally, we use the molar masses of $Cu_2O$ and C to find the mass of each required for the reaction.

**FIGURE 3.19** Copper bracelet.

**percent composition** the composition of a compound expressed in terms of the percentage by mass of each element in the compound.

$$g\ Cu \xrightarrow[\text{molar mass}]{1} mol\ Cu \xrightarrow[\text{4 mol Cu}]{2\ mol\ Cu_2O} mol\ Cu_2O \xrightarrow{\text{molar mass}} g\ Cu_2O$$

$$g\ Cu \xrightarrow[\text{molar mass}]{1} mol\ Cu \xrightarrow[\text{4 mol Cu}]{1\ mol\ C} mol\ C \xrightarrow{\text{molar mass}} g\ C$$

**Solve** Carrying out the steps described in the analysis of the problem:

$$256\ \text{g Cu} \times \frac{1\ \text{mol Cu}}{63.55\ \text{g Cu}} \times \frac{2\ \text{mol Cu}_2\text{O}}{4\ \text{mol Cu}} \times \frac{143.09\ \text{g Cu}_2\text{O}}{1\ \text{mol Cu}_2\text{O}} = 288\ \text{g Cu}_2\text{O}$$

$$256\ \text{g Cu} \times \frac{1\ \text{mol Cu}}{63.55\ \text{g Cu}} \times \frac{1\ \text{mol C}}{4\ \text{mol Cu}} \times \frac{12.01\ \text{g C}}{1\ \text{mol C}} = 12.1\ \text{g C}$$

We are allowed only three significant figures in our answer, so the final masses of the reactants we need are 288 g $Cu_2O$ and 12.1 g C.

**Think About It** To prepare a given mass of copper, we must have sufficient amounts of both copper ore and charcoal. Usually there is a limited supply of ore and more than enough charcoal.

⚙ **Practice Exercise** The copper mineral chalcocite, $Cu_2S$, can be converted to copper simply by heating in air: $Cu_2S(s) + O_2(g) \rightarrow 2\ Cu(s) + SO_2(g)$. How much $Cu_2S$ is needed to make 256 g Cu? How much $SO_2$ is produced?

*(Answers to Practice Exercises are in the back of the book.)*

▶‖ CHEMTOUR Percent Composition

**FIGURE 3.20** The odor of the longleaf pine (*Pinus palustris*) comes from the volatile hydrocarbon pinene.

## 3.6 Determining Empirical Formulas from Percent Composition

Natural sources of methane (e.g. swamps, bogs, rice paddies, and animals) introduce 0.570 teragrams ($5.70 \times 10^{11}$ g) of $CH_4$ into the atmosphere every day. Conifers, such as the longleaf pine shown in Figure 3.20, emit a range of hydrocarbons into the air, including pinene, which has the chemical formula $C_{10}H_{16}$. One way of distinguishing between $CH_4$ and $C_{10}H_{16}$ is in terms of **percent composition**: the composition of a compound expressed in terms of the percentages of the masses of the constituent elements with respect to the total mass of the compound.

To define the percent composition, we must calculate the carbon content of both $CH_4$ and $C_{10}H_{16}$. Suppose we have 1 mole of $CH_4$, which has a molar mass of

$$4(1.008\ \text{g H/mol}) + 12.01\ \text{g C/mol} = 16.04\ \text{g/mol}$$

Of this 16.04 g, C accounts for 12.01 g. Therefore, the C content of $CH_4$ is

$$\frac{\text{mass of C}}{\text{total mass}} = \frac{12.01\ \text{g C}}{16.04\ \text{g CH}_4} = 0.7488$$

$$\text{or} \quad 0.7488 \times 100\% = 74.88\%\ \text{C}$$

It follows that the hydrogen content is

$$\frac{\text{mass of H}}{\text{total mass}} = \frac{4(1.008\ \text{g H})}{16.04\ \text{g CH}_4} = 0.2512 \quad \text{or} \quad 25.12\%\ \text{H}$$

Because $CH_4$ contains only C and H, we could also determine the percent H by subtracting the percent C from 100%:

$$100.00\% - 74.88\% = 25.12\%$$

Thus the percent composition of $CH_4$ is 74.88% C and 25.12% H.

How does this compare with the % C in pinene? If we repeat the calculation for $C_{10}H_{16}$, the molar mass is

$$16(1.008 \text{ g H/mol}) + 10(12.01 \text{ g C/mol}) = 136.23 \text{ g/mol}$$

The C content of $C_{10}H_{16}$ is

$$\frac{\text{mass of C}}{\text{total mass}} = \frac{10(12.01 \text{ g C})}{136.23 \text{ g } C_{10}H_{16}} = 0.8816 = 88.16\%$$

and the hydrogen content is

$$100.00\% - 88.16\% = 11.84\%$$

Pinene contains a higher percent carbon by mass, and a correspondingly lower percent hydrogen, than methane. Actually, methane has the highest percent hydrogen of all hydrocarbons.

---

**SAMPLE EXERCISE 3.14** **Calculating Percent Composition from a Chemical Formula** **LO4**

The compound $C_2F_4H_2$ (sold as HFC-134a) containing C, F, and H is a propellant in the inhalers used by asthma sufferers (Figure 3.21). What is the percent composition of $C_2F_4H_2$?

**Collect and Organize** We are asked to calculate the percent composition of $C_2F_4H_2$.

**Analyze** We need the molar masses of each of the elements to calculate the molar mass of $C_2F_4H_2$ and the contribution of each element to it. To calculate the percent composition of $C_2F_4H_2$, we must determine the percentage by mass of each element in one mole of $C_2F_4H_2$. These percentages can be calculated by dividing the mass of each element in 1 mole of $C_2F_4H_2$ by the molar mass of $C_2F_4H_2$.

**Solve** The molar mass of $C_2F_4H_2$ is

$$2(12.01 \text{ g/mol}) + 4(19.00 \text{ g/mol}) + 2(1.008 \text{ g/mol}) = 102.04 \text{ g/mol}$$

Thus the percent composition of this compound is

$$\%C = \frac{24.02 \text{ g C}}{102.04 \text{ g}} \times 100\% = 23.54\% \text{ C}$$

$$\%F = \frac{76.00 \text{ g F}}{102.04 \text{ g}} \times 100\% = 74.48\% \text{ F}$$

$$\%H = \frac{2.016 \text{ g H}}{102.04 \text{ g}} \times 100\% = 1.98\% \text{ H}$$

**Think About It** Although C and F have similar molar masses, the observation that the percentage of F is nearly three times the percentage of C makes sense because there are two moles of F for every 1 mole of C in $C_2F_4H_2$. It also makes sense that the percentage of the lightest element, hydrogen, is very small.

⚙ **Practice Exercise** Determine the percent composition of Fluosol-DA, $C_{10}F_{18}$, the only FDA-approved synthetic blood substitute (Figure 3.22).

*(Answers to Practice Exercises are in the back of the book.)*

■

**FIGURE 3.21** Patients suffering from asthma and other respiratory ailments use inhalers to relieve their discomfort. The medication is delivered using compounds such as $C_2F_4H_2$ as the propellant.

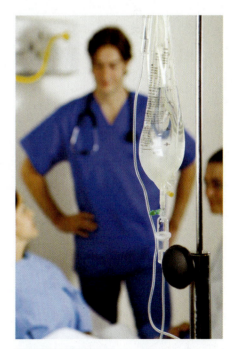

**FIGURE 3.22** The fluorocarbon $C_{10}F_{18}$ is the only FDA-approved synthetic blood substitute.

Note in Sample Exercise 3.14 that the three percentages sum to 100.00%. The percent composition values should always sum to 100%, or very close to 100%, if we have accounted for all the elements that make up the total mass of the compound. The total may deviate slightly from 100% because of rounding.

We typically determine the percent composition of a substance in the laboratory by measuring the amount of each element in a given mass of the substance. We can use these data to determine the formula of the substance. The formula determined from percent composition is the *empirical formula* of the material. There are four steps to deriving an empirical formula from percent composition data:

1. Assume you have exactly 100 g of the substance, so that the percent composition values are equivalent to the values of the masses of the elements expressed in grams. As an initial check, add the mass percentages given in the problem to make sure they total 100%.
2. Convert the mass of each element into moles.
3. Compute the mole ratio by reducing one of the mole values to 1.
4. If necessary, convert the ratio from step 3 into a ratio of whole numbers.

Let's look at an example. Carbon tetrachloride was once widely used as a fire extinguishing agent and as a solvent in dry cleaning. It is 7.808% C and 92.19% Cl. These two values sum to 99.998%, so we know all the elements are accounted for and the compound contains only carbon and chlorine. In exactly 100 grams of this substance, there are

$$7.808\% \text{ C} = 7.808 \text{ g C in } 100 \text{ g}$$

$$92.19\% \text{ Cl} = 92.19 \text{ g Cl in } 100 \text{ g}$$

Now we determine the number of moles of C and Cl in this sample:

$$7.808 \text{ g C} \times \frac{1 \text{ mol C}}{12.01 \text{ g C}} = 0.6501 \text{ mol C} \quad 92.19 \text{ g Cl} \times \frac{1 \text{ mol Cl}}{35.45 \text{ g Cl}} = 2.600 \text{ mol Cl}$$

The mole ratio of carbon to chlorine is 0.6501:2.600, but we need to express this in a chemical formula in terms of whole numbers. We divide both numbers by the smaller of the two; this ensures that one of the numbers in the ratio has a value of 1 and the other will be larger than 1:

$$\frac{0.6501 \text{ mol C}}{0.6501} = 1.000 \text{ or } 1 \text{ mol C} \quad \frac{2.600 \text{ mol Cl}}{0.6501} = 3.999 \text{ or } 4 \text{ mol Cl}$$

That means the empirical formula of carbon tetrachloride is $CCl_4$.

We can use the same techniques for substances that contain more than two elements. Let's see how this procedure works with a commonly used refrigeration gas, HFC-143a. Elemental analysis reveals that the percent composition of a sample is 28.58% C, 67.82% F, and 3.60% H. The sum of these percentages totals 100%, so all elements are accounted for in our analysis. Our calculations are then

$$28.58\% \text{ C} = 28.58 \text{ g C in } 100 \text{ g}$$

$$67.82\% \text{ F} = 67.82 \text{ g F in } 100 \text{ g}$$

$$3.60\% \text{ H} = 3.60 \text{ g H in } 100 \text{ g}$$

and the numbers of moles are

$$28.58 \text{ g C} \times \frac{\text{mol C}}{12.01 \text{ g C}} = 2.380 \text{ mol C} \quad 67.82 \text{ g F} \times \frac{\text{mol F}}{19.00 \text{ g F}} = 3.569 \text{ mol F}$$

$$3.60 \text{ g H} \times \frac{\text{mol H}}{1.008 \text{ g H}} = 3.571 \text{ mol H}$$

The C:F:H mole ratio is 2.378:3.569:3.571. To convert this ratio to small whole numbers, we divide all values by the smallest value:

$$\frac{2.380 \text{ mol C}}{2.380} = 1.000 \text{ or 1 mol C} \qquad \frac{3.569 \text{ mol F}}{2.380} = 1.500 \text{ or 1.5 mol F}$$

$$\frac{3.571 \text{ mol H}}{2.380} = 1.500 \text{ or 1.5 mol H}$$

To change the mole ratio 1:1.5:1.5 to a ratio of whole numbers, we multiply the numbers by a factor that results in whole numbers. Multiplying 1.5 (or $\frac{3}{2}$) by 2 yields 3, so we multiply all three terms in this ratio by 2 to obtain a C:F:H ratio of 2:3:3. Thus the chemical formula is $C_2F_3H_3$.

The formulas $CCl_4$ and $C_2F_3H_3$ are empirical formulas, which we discussed in Chapter 2 as the typical formulas of ionic compounds. *Empirical*, which is related to the word *experiment*, means "derived from experimental data" such as the data from experiments to determine percent composition. Both $CCl_4$ and $C_2F_3H_3$ are molecular compounds, as shown in Figure 3.23(a,b), whose empirical formulas match their molecular formulas. This is not always the case. Consider another hydrocarbon found in natural gas and used as a fuel: butane. Butane has a percent composition of 82.66% C and 17.34% H. We can calculate its empirical formula:

$$82.66\% \text{ C} = 82.66 \text{ g C} \times \frac{1 \text{ mol C}}{12.01 \text{ g C}} = 6.883 \text{ mol C}$$

$$17.34\% \text{ H} = 17.34 \text{ g H} \times \frac{1 \text{ mol H}}{1.008 \text{ g H}} = 17.20 \text{ mol H}$$

Converting the mole ratio 6.883:17.20 gives us

$$\frac{6.883}{6.883} : \frac{17.20}{6.883} = 1:2.499$$

When we multiply both numbers by 2, the whole-number ratio becomes 2:5 and the empirical formula for butane is $C_2H_5$. The structure of a molecule of butane is shown in Figure 3.23(c); it has the molecular formula $C_4H_{10}$. The percent composition data gives us the empirical formula—the lowest whole-number ratio of atoms in the molecule—but the molecular formula shows us the actual numbers of atoms in the molecule.

For ionic compounds such as FeO and $Fe_2O_3$, the chemical formulas represent empirical formulas. However, FeO is made up of a three-dimensional array of $Fe^{2+}$ and $O^{2-}$ ions, and $Fe_2O_3$ is made up of a three-dimensional array of $Fe^{3+}$ and $O^{2-}$ ions. Empirical formulas represent the relative proportions of the ions in the formula unit of an ionic compound. In FeO, the $Fe^{2+}$ and $O^{2-}$ ions are arranged as shown in Figure 3.24, often called the *halite structure* because it is the structure seen in the mineral halite (NaCl). In an ionic compound the chemical formula, the empirical formula, and the formula unit are all the same.

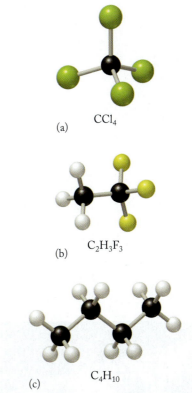

(a) $CCl_4$

(b) $C_2H_3F_3$

(c) $C_4H_{10}$

**FIGURE 3.23** The molecular structures of (a) carbon tetrachloride, $CCl_4$; (b) the refrigerant gas, $C_2F_3H_3$; and (c) the fuel gas butane, $C_4H_{10}$.

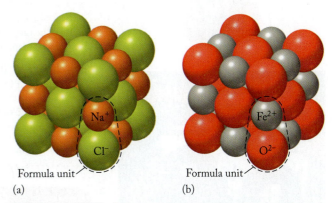

Formula unit (a) $Na^+$ $Cl^-$

Formula unit (b) $Fe^{2+}$ $O^{2-}$

**FIGURE 3.24** (a) The halite structure is the three-dimensional pattern of the $Na^+$ and $Cl^-$ ions in NaCl; (b) the same pattern is found in the $Fe^{2+}$ and $O^{2-}$ ions in FeO.

**CONCEPT TEST** ••••••••••••••••••••••••••••••••••••••••••••••••••••••

Which of the following compounds have the same empirical formula and same percent composition?

    a. Ethylene ($C_2H_4$), a gas used to ripen bananas
    b. Eicosene ($C_{20}H_{40}$), a compound used to attract Japanese beetles to traps
    c. Acetylene ($C_2H_2$), a gas used in welding
    d. Benzene ($C_6H_6$), a known carcinogen found in gasoline

*(Answers to Concept Tests are in the back of the book.)*

**SAMPLE EXERCISE 3.15**   **Deriving an Empirical Formula**   **LO4**
                                  **from Percent Composition**

Nitrous oxide (laughing gas), used in dentists' offices, is a nitrogen oxide that is 63.65% N. What is the empirical formula of nitrous oxide?

**Collect and Organize** We are given the percent composition by mass of N (63.65%) and asked to determine the simplest whole-number ratio of nitrogen and oxygen in the chemical formula.

**Analyze** We can obtain the percent composition by mass of oxygen in nitrous oxide by subtracting the percent N from 100%. We can use the four steps described earlier to determine the empirical formula.

**Solve** The percent O in the sample is:

$$\%O = 100.00\% - \%N = 100.00\% - 63.65\% = 36.35\%$$

Now we have the data we need to determine the empirical formula of nitrous oxide.

1. Convert percent composition values into masses by assuming a sample size of exactly 100 g:

$$63.65\% \text{ N} = 63.65 \text{ g N in } 100 \text{ g}$$
$$36.35\% \text{ O} = 36.35 \text{ g O in } 100 \text{ g}$$

2. Convert masses into moles:

$$63.65 \text{ g N} \times \frac{1 \text{ mol N}}{14.01 \text{ g N}} = 4.543 \text{ mol N}$$

$$36.35 \text{ g O} \times \frac{1 \text{ mol O}}{16.00 \text{ g O}} = 2.272 \text{ mol O}$$

3. Simplify the mole ratio 4.543:2.272 by dividing by the smaller value:

$$\frac{4.543 \text{ mol N}}{2.272} = 2.000 \text{ mol N} \qquad \frac{2.272 \text{ mol O}}{2.272} = 1.000 \text{ mol O}$$

    The mole ratio N:O is 2:1.
4. Since the mole ratio of N to O is already a ratio of whole numbers, we have the empirical formula of nitrous oxide: $N_2O$.

**Think About It** The molar masses of N and O are similar, so a ≈2:1 ratio of the mass of N to the mass of O is consistent with an empirical formula that is 2:1 nitrogen to oxygen.

⚙ **Practice Exercise** A different nitrogen oxide, containing 53.84% N, acts as a vasodilator, lowering blood pressure in the human body. What is its empirical formula?

*(Answers to Practice Exercises are in the back of the book.)*

■••••••••••••••••••••••••••••••••••••••••••••••••••••••••••••••••••

**SAMPLE EXERCISE 3.16**   **Deriving an Empirical Formula**   **LO4**
                                    **of a Compound Containing**
                                    **More than Two Elements**

Rechargeable lithium batteries are essential to portable electronic devices and are increasingly important to the automotive industry in gas–electric hybrid and all-electric vehicles (Figure 3.25). Analysis of the most common lithium-containing mineral, spodumene,

**FIGURE 3.25** (a) Spodumene is the most important source of lithium metal for use in (b) lithium batteries used in portable electronic devices and (c) in automobiles.

shows it is composed of 3.73% Li, 14.50% Al, 30.18% Si, and 51.59% O. What is the empirical formula of spodumene?

**Collect and Organize** We are given the percent composition by mass of spodumene and asked to determine the simplest whole-number ratio of the four elements (Li, Al, Si, and O), which defines the empirical formula of spodumene.

**Analyze** We follow the same steps as in Sample Exercise 3.15, except that here we must calculate more than one mole ratio.

**Solve**

1. Assuming an exactly 100 g sample of spodumene, we have 3.73 g of Li, 14.50 g of Al, 30.18 g of Si, and 51.59 g of O.
2. Converting the masses into moles, we have

$$3.73 \ \text{g Li} \times \frac{1 \ \text{mol Li}}{6.941 \ \text{g Li}} = 0.5374 \ \text{mol Li}$$

$$14.50 \ \text{g Al} \times \frac{1 \ \text{mol Al}}{26.98 \ \text{g Al}} = 0.5374 \ \text{mol Al}$$

$$30.18 \ \text{g Si} \times \frac{1 \ \text{mol Si}}{28.09 \ \text{g Si}} = 1.074 \ \text{mol Si}$$

$$51.59 \ \text{g O} \times \frac{1 \ \text{mol O}}{16.00 \ \text{g O}} = 3.224 \ \text{mol O}$$

3. We divide each mole value by the smallest value to obtain a simple ratio of the four elements:

$$\frac{0.5374 \ \text{mol Li}}{0.5374} = 1.000 \ \text{or} \ 1 \ \text{mol Li} \qquad \frac{0.5374 \ \text{mol Al}}{0.5374} = 1.000 \ \text{or} \ 1 \ \text{mol Al}$$

$$\frac{1.074 \ \text{mol Si}}{0.5374} = 1.999 \ \text{or} \ 2 \ \text{mol Si} \qquad \frac{3.224 \ \text{mol O}}{0.5374} = 5.999 \ \text{or} \ 6 \ \text{mol O}$$

4. All these results are either whole numbers or very close to whole numbers, so no further simplification of terms is needed. Our mole ratio is Li:Al:Si:O = 1:1:2:6, which means the empirical formula of spodumene is $LiAlSi_2O_6$.

**Think About It** The sum of the mass percentages should be very close to 100%; in this case, 3.73% + 14.50% + 30.18% + 51.59% = 100.00% so we know we have accounted for all the elements. Notice that lithium is present in the lowest percentage by mass of all four elements in spodumene. Nevertheless, spodumene remains the most commercially important source of lithium for batteries.

⚙ **Practice Exercise** Spodumene is processed chemically to convert it to a material having the percent composition 18.79% Li, 16.25% C, and 64.96% O. What is the empirical formula of this lithium compound?

*(Answers to Practice Exercises are in the back of the book.)*

■

# 3.7 Empirical and Molecular Formulas Compared

An empirical formula tells us the simplest whole-number ratio of the elements contained in a compound, but as we have already seen, it is not necessarily the molecular formula of a molecular compound. To illustrate the difference between empirical and molecular formulas, let's look at the organic molecule glycolaldehyde (Figure 3.26). About one in 20 of the meteorites that fall to Earth today contains a variety of organic compounds, including glycolaldehyde. Some scientists view the presence of these compounds as evidence that the molecular building blocks of life on Earth may have come from space. Glycolaldehyde is also present in our bodies as a product of the metabolism of sugars and proteins.

**FIGURE 3.26** Glycolaldehyde is one of over 100 organic compounds detected in interstellar gases. It is also the smallest molecule among those identified as sugars.

Elemental analysis of glycolaldehyde gives the following percent composition data: 40.00% C, 6.71% H, and 53.28% O. We can use this information to calculate the empirical formula of the compound and then compare our result with the molecular structure in Figure 3.26. Using the method developed in Section 3.6, we obtain these results:

$$40.00 \text{ g C} \times \frac{1 \text{ mol C}}{12.01 \text{ g C}} = 3.331 \text{ mol C}$$

$$6.71 \text{ g H} \times \frac{1 \text{ mol H}}{1.008 \text{ g H}} = 6.66 \text{ mol H}$$

$$53.28 \text{ g O} \times \frac{1 \text{ mol O}}{16.00 \text{ g O}} = 3.330 \text{ mol O}$$

The mole ratio of carbon to hydrogen to oxygen is 3.331:6.66:3.330 = 1:2:1, which defines an empirical formula of $CH_2O$.

Comparing this empirical formula with the molecule in Figure 3.26 reveals that the two are related but not equivalent. Counting up the atoms in the model, we arrive at the molecular formula $C_2H_4O_2$, which represents the actual numbers of C, H, and O atoms in one molecule of glycolaldehyde. In this case, we must multiply each subscript in the empirical formula by 2 to obtain the molecular formula.

---

**CONCEPT TEST** ...............................................................

Which of the following chemical formulas must represent empirical formulas and which ones must represent molecular formulas?

$C_2H_6O_2$, ethylene glycol, used in antifreeze; $C_3H_8O$, isopropanol, also known as rubbing alcohol; and $C_6H_{12}O_6$, glucose, also known as "blood sugar" and the primary source of energy for our body's cells.

*(Answers to Concept Tests are in the back of the book.)*

.........................................................................................................

**FIGURE 3.27** The empirical formula of the molecular compound formaldehyde is identical to its molecular formula.

Occasionally, empirical and molecular formulas are identical. For example, formaldehyde—which like glycolaldehyde has the percent composition 40.00% C, 6.71% H, 53.28% O—has the molecular formula $CH_2O$, the same as its empirical formula (Figure 3.27). Many other compounds, such as glucose, $C_6H_{12}O_6$, also have the empirical formula $CH_2O$. Each subscript in these molecular formulas is a multiple of the corresponding subscript in the empirical formula. For example, a multiplier (*n*) of 6 converts the empirical formula for glucose and fructose, $CH_2O$, into their common molecular formula, $C_6H_{12}O_6$:

$$(CH_2O)_n = (CH_2O)_6 = C_6H_{12}O_6$$
$$\text{Glucose}$$

The key to translating empirical formulas into molecular formulas is to determine the value of *n*, and the key to determining the value of *n* is knowing the molecular mass. For example, suppose elemental analysis tells us that the empirical formula of a liquid hydrocarbon is $C_4H_5$. Using the stoichiometric relation given by this formula and the average atomic masses of the two elements, we can calculate the mass corresponding to the empirical formula unit:

$$(4 \times 12.01 \text{ amu}) + (5 \times 1.008 \text{ amu}) = 53.08 \text{ amu}$$

Each molecule of this hydrocarbon is made up of some number of empirical formula units, and each of these units contains 4 carbon atoms and 5 hydrogen atoms. Additional analysis of the compound reveals its molecular mass to be 106 amu. How many $C_4H_5$ formula units are needed to make one molecule that has a molecular mass of 106 amu? The following calculation tells us:

$$\frac{\text{molecular mass}}{\text{mass of empirical formula unit}} = \frac{106 \; \text{amu}}{53 \; \text{amu}} = 2 \text{ formula units/molecule}$$

Thus 2 is the value of the multiplier, *n*, and the molecular formula of the compound is

$$(C_4H_5)_n = (C_4H_5)_2 = C_8H_{10}$$

---

**SAMPLE EXERCISE 3.17**   **Deriving a Molecular Formula**   **LO5**
                                  **from an Empirical Formula**

Lycopene (molar mass 536.88 g/mol) and cymene (134.22 g/mol) are natural products found in tomatoes and cumin (a common spice), respectively. Lycopene in the diet has been shown to reduce the risk of prostate cancer, whereas cumin is a home remedy for stomach ailments. Both compounds are 89.49% C and 10.51% H by mass. What are the empirical and molecular formulas of lycopene and cymene?

**Collect and Organize** The problem gives the percent composition of two compounds and their molar masses. We want to determine both the empirical and molecular formulas of lycopene and cymene.

**Analyze** The data on percent composition lead directly to the empirical formula using the procedure developed in Section 3.6. The molar masses of lycopene and cymene in g/mol are numerically the same as their molecular masses in amu and can be used to determine the value of the multipliers needed to convert their empirical formulas into molecular formulas.

**Solve** Assuming a 100 g sample, we have 89.49 g of C and 10.51 g of H, which we convert to moles:

$$89.49 \; \text{g C} \times \frac{1 \; \text{mol C}}{12.01 \; \text{g C}} = 7.451 \; \text{mol C}$$

$$10.51 \; \text{g H} \times \frac{1 \; \text{mol H}}{1.008 \; \text{g H}} = 10.43 \; \text{mol H}$$

The mole ratio of C to H is 7.451:10.43, or 1:1.4. We can convert the fractional mole ratio to a ratio of whole numbers by multiplying by 5:

$$C{:}H = 5 \times (1{:}1.4) = 5{:}7$$

which means the empirical formula is $C_5H_7$.

To determine the molecular formula, we first determine the empirical formula mass of $C_5H_7$:

$$5(12.01 \; \text{amu}) + 7(1.008 \; \text{amu}) = 67.11 \; \text{amu}$$

Next we divide the molecular masses for lycopene and cymene by the empirical formula mass to determine the values of multiplier *n*:

$$n = \frac{\text{molecular mass lycopene}}{\text{empirical formula mass}} = \frac{536.88 \; \text{amu}}{67.11 \; \text{amu}} = 8$$

$$n = \frac{\text{molecular mass lycopene}}{\text{empirical formula mass}} = \frac{134.22 \; \text{amu}}{67.11 \; \text{amu}} = 2$$

The molecular formula of lycopene is

$$(C_5H_7)_n = (C_5H_7)_8 = C_{40}H_{56}$$

and the molecular formula of cymene is

$$(C_5H_7)_n = (C_5H_7)_2 = C_{10}H_{14}$$

**Think About It** The percent composition of a compound reveals only the empirical formula. All compounds with the same empirical formula *must* have the same percent composition.

⚙️ **Practice Exercise** *Pheromones* are chemical substances secreted by members of a species to stimulate a response in other individuals of the same species. For example, certain pheromones are secreted by females of a species to attract males for mating. The percent composition of eicosene (280 g/mol), a compound similar to the Japanese beetle mating pheromone, is 85.63% C and 14.37% H. Determine its molecular formula.

*(Answers to Practice Exercises are in the back of the book.)*

■ ⬤•••••••••••••••••••••••••••••••••••••••••••••••••••

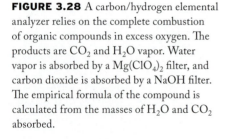

👁️ **CONNECTION** In Section 2.2 we learned about Aston's positive ray analyzer.

How do we determine the molecular mass of a pure substance? We will examine several methods in Chapters 6 and 11; however, for now we apply what we learned in Chapter 2. Francis Aston's work with isotopes of elements using his positive ray analyzer led to the development of **mass spectrometers**, which are instruments used to determine the masses of molecules and to characterize their structures. All mass spectrometers separate and count ions. Many convert neutral atoms or molecules into cations and then separate those cations based on the ratio of their mass ($m$) to their electric charge ($z$). The symbol lower-case $z$ is used in mass spectrometry for the charge on an ion. Data produced by a mass spectrometer are displayed in a graph, called a **mass spectrum** (plural, *spectra*), in which the $m/z$ values of the ions reaching the detector are plotted on the horizontal axis against the intensity (number of particles) on the vertical axis. If the charge on every ion is 1+, the $m/z$ value of each peak is $m/1$, that is, the mass of the ion producing the peak. Among these ions is the **molecular ion ($M^{+\cdot}$)**, whose $m/z$ value represents molecular mass.

## 3.8 Combustion Analysis

In **combustion analysis**, the complete combustion (defined in Section 3.4) of a compound followed by an analysis of the products enable chemists to determine the chemical composition of that compound. To ensure that combustion is complete, the process is carried out in excess oxygen, which means more oxygen is present than the stoichiometric amount. Combusting an organic compound means converting all of the carbon in the compound to $CO_2$ and all of the hydrogen to $H_2O$:

$$C_aH_b + \text{excess } O_2(g) \rightarrow a\,CO_2(g) + b/2\,H_2O(g)$$

Consider burning a hydrocarbon of unknown composition in a chamber through which a stream of pure oxygen flows (Figure 3.28). The $CO_2(g)$ and $H_2O(g)$ produced flow first through a tube packed with $Mg(ClO_4)_2(s)$, which

**FIGURE 3.28** A carbon/hydrogen elemental analyzer relies on the complete combustion of organic compounds in excess oxygen. The products are $CO_2$ and $H_2O$ vapor. Water vapor is absorbed by a $Mg(ClO_4)_2$ filter, and carbon dioxide is absorbed by a NaOH filter. The empirical formula of the compound is calculated from the masses of $H_2O$ and $CO_2$ absorbed.

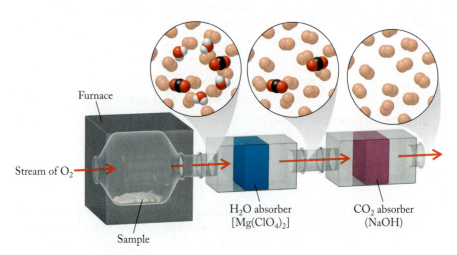

Furnace

Stream of $O_2$

Sample

$H_2O$ absorber $[Mg(ClO_4)_2]$

$CO_2$ absorber (NaOH)

selectively absorbs the $H_2O(g)$, and then through a tube containing $NaOH(s)$, which absorbs the $CO_2(g)$. The masses of these tubes are measured before and after combustion. Suppose the mass of the tube that traps $CO_2(g)$ increases by 1.320 g, and the mass of the tube that traps $H_2O(g)$ increases by 0.540 g. How can we use these results to determine the empirical formula of the hydrocarbon?

First, let's establish what we know about the hydrocarbon in this example:

1. Being a hydrocarbon, it contains only carbon and hydrogen.
2. Complete conversion of its carbon into $CO_2$ produces 1.320 g of $CO_2$.
3. Complete conversion of its hydrogen into $H_2O$ produces 0.540 g of $H_2O$.

To derive an empirical formula for the hydrocarbon, we must determine the number of moles of carbon in 1.320 g of carbon dioxide and the number of moles of hydrogen in 0.540 g of water vapor. These quantities are directly related to the number of moles of carbon and hydrogen in the sample we burned. Converting the mass of carbon dioxide to moles of carbon and the mass of water to moles of hydrogen:

$$1.320 \ \cancel{g \ CO_2} \times \frac{1 \ \cancel{mol \ CO_2}}{44.01 \ \cancel{g \ CO_2}} \times \frac{1 \ mol \ C}{1 \ \cancel{mol \ CO_2}} = 0.02999 \ mol \ C \approx 0.0300 \ mol \ C$$

$$0.540 \ \cancel{g \ H_2O} \times \frac{1 \ \cancel{mol \ H_2O}}{18.02 \ \cancel{g \ H_2O}} \times \frac{2 \ mol \ H}{1 \ \cancel{mol \ H_2O}} = 0.05993 \ mol \ H \approx 0.0600 \ mol \ H$$

The empirical formula is based on these proportions of C and H. If we divide both molar amounts by the smaller one:

$$\frac{0.0300 \ mol \ C}{0.0300} = 1 \ mol \ C \qquad \frac{0.0600 \ mol \ H}{0.0300} = 2 \ mol \ H$$

we find that the empirical formula of the hydrocarbon is $CH_2$.

If we want to extend this analysis to determine a molecular formula, we need to know the molecular mass of the hydrocarbon. Suppose its mass spectrum has a molecular ion with a mass of 84 amu. The formula mass of $CH_2$ is 14 amu. Dividing the molecular mass by the formula mass, we get the multiplier, $n$:

$$n = \frac{84 \ amu}{14 \ amu} = 6$$

The molecular formula is therefore

$$(CH_2)_6 = C_6H_{12}$$

Note that in this problem we did not need to know the initial mass of the sample to determine its empirical formula. We only needed to know that the sample was a hydrocarbon and that it was completely converted into the stated amounts of $CO_2$ and $H_2O$.

What if we did not know our sample was a hydrocarbon? What if it were a pharmacologically promising compound isolated from a tropical plant and we had no idea which elements it contained? Many such compounds are made of carbon, hydrogen, oxygen, and sometimes nitrogen. To calculate the empirical formula of a compound made of C, H, and O, for example, we need to know the proportion of oxygen in it, but there is no simple way of measuring that directly when the compound is burned in excess oxygen. Combustion analyses yield the percentages by mass of all atoms in a sample *except* oxygen. When given data from a combustion analysis, always check to see whether the percentages add up to 100%. If they do, all the elements in the sample are accounted for in the results. If they do not, the missing mass is probably due to oxygen.

**mass spectrometer** an instrument that separates and counts ions according to their mass.

**mass spectrum** a graph of the data from a mass spectrometer, where $m/z$ ratios of the deflected particles are plotted against the number of particles with a particular mass.

**molecular ion ($M^{+\cdot}$)** the peak of highest mass in a mass spectrum; it has the same mass as the molecule from which it came.

**combustion analysis** a laboratory procedure for determining the composition of a substance by burning it completely in oxygen to produce known compounds whose masses are used to determine the composition of the original material.

**SAMPLE EXERCISE 3.18**   **Combustion Analysis of a Hydrocarbon**   **LO6**

Limonene is a hydrocarbon that contributes to the odor of citrus fruits, including lemons. Combustion of 0.671 g of limonene yielded 2.168 g $CO_2$ and 0.710 g $H_2O$. What is the empirical formula of limonene? The molecular mass of limonene is 136.24 amu; what is the molecular formula of limonene?

**Collect and Organize**  We know the masses of $CO_2$ and $H_2O$ produced during the combustion of a hydrocarbon sample, and are asked to determine its empirical formula and then, from its molecular mass, its molecular formula.

**Analyze**  First we determine the number of moles of C and H in the $CO_2$ and $H_2O$ produced during combustion. These values are equal to the number of moles of C and H in the combusted sample. Then we calculate the C:H mole ratio and convert it into a ratio of small whole numbers to obtain the empirical formula. Next we calculate the mass of one empirical formula unit and divide this mass into the average mass of a molecule to obtain the multiplier, $n$, that allows us to convert the empirical formula to a molecular formula.

$$\boxed{\text{g } CO_2} \xrightarrow{\dfrac{1}{\text{molar mass}}} \boxed{\text{mol } CO_2} \xrightarrow{\dfrac{1 \text{ mol C}}{1 \text{ mol } CO_2}} \boxed{\text{mol C}}$$

$$\boxed{\text{g } H_2O} \xrightarrow{\dfrac{1}{\text{molar mass}}} \boxed{\text{mol } H_2O} \xrightarrow{\dfrac{2 \text{ mol H}}{1 \text{ mol } H_2O}} \boxed{\text{mol H}}$$

**Solve**  The moles of C and H in the $CO_2$ and $H_2O$ collected during combustion are

$$2.168 \text{ g } CO_2 \times \frac{1 \text{ mol } CO_2}{44.01 \text{ g } CO_2} \times \frac{1 \text{ mol C}}{1 \text{ mol } CO_2} = 0.04926 \text{ mol C}$$

$$0.710 \text{ g } H_2O \times \frac{1 \text{ mol } H_2O}{18.02 \text{ g } H_2O} \times \frac{2 \text{ mol H}}{1 \text{ mol } H_2O} = 0.0788 \text{ mol H}$$

The mole ratio of the two elements in the sample is

$$0.04926 \text{ mol C} : 0.0788 \text{ mol H}$$

Dividing through by the smallest value (0.04926 mol) gives a mole ratio of 1:1.6. We can convert this ratio to whole numbers by multiplying by 5, making the empirical formula of the sample $C_5H_8$.

The molecular mass of limonene is 136.24 amu. The mass on the empirical formula unit of limonene is 5(12.01 amu) + 8(1.008 amu) = 68.11 amu, so the multiplier $n$ is

$$\frac{\text{average molecular mass}}{\text{empirical formula mass}} = \frac{136.24 \text{ amu}}{68.11 \text{ amu}} = 2$$

and the molecular formula is

$$(C_5H_8)_n = (C_5H_8)_2 = C_{10}H_{16}$$

**Think About It**  The subscripts in our answer, 5 for C and 8 for H, make sense because the moles of C $\approx 5 \times 10^{-2}$ and moles H $\approx 8.00 \times 10^{-2}$ are essentially at a 5:8 ratio.

⚙ **Practice Exercise**  Cembrene A is another naturally occurring hydrocarbon, extracted from coral. Combustion of 0.0341 g of cembrene A yields 0.1101 g $CO_2$ and 0.0360 g $H_2O$. What is the empirical formula of cembrene A?

*(Answers to Practice Exercises are in the back of the book.)*

SAMPLE EXERCISE 3.19    **Combustion Analysis for**    **L06**
**Compounds Containing Oxygen**

Eugenol is an ingredient in several spices, including bay leaves and cloves. Eugenol contains carbon, hydrogen, and oxygen. Combustion of 21.8 mg of eugenol yields 58.5 mg of $CO_2$ and 14.4 mg of $H_2O$. What is the empirical formula of eugenol? The mass spectrum of eugenol shows a molecular ion at 164 amu; what is the molecular formula of eugenol?

**Collect and Organize** We are given the masses of $CO_2$ (58.5 mg) and $H_2O$ (14.4 mg) produced by the combustion of 21.8 mg of eugenol and are asked to determine its empirical formula. We also know its molecular mass and are to determine its molecular formula.

**Analyze** First we determine the number of moles of C and H in the $CO_2$ and $H_2O$ produced during combustion. These values are equal to the number of moles of C and H in the combusted sample. Multiplying these mole values by the atomic masses of C and H yields the mass of these elements in the original sample, which we then subtract from the total mass of the sample to obtain the mass of O in the sample. This mass can then be converted into moles of O in the sample:

$$\boxed{\text{mg sample}} - \boxed{\text{mg C}} - \boxed{\text{mg H}} = \boxed{\text{mg O}} \xrightarrow{\frac{1\,g}{10^3\,mg}} \boxed{g\,O} \xrightarrow{\frac{1}{\text{molar mass}}} \boxed{\text{mol O}}$$

The mole ratio of C:H:O is then calculated and reduced to a ratio of small whole numbers to obtain the empirical formula. We divide the mass of one empirical formula unit into the mass of the molecular ion to obtain the multiplier, $n$, that allows us to convert the empirical formula to a molecular formula.

**Solve** The moles of C and H in the $CO_2$ and $H_2O$ collected during combustion are

$$58.5\ \text{mg}\ CO_2 \times \frac{1\ g\ CO_2}{10^3\ mg\ CO_2} \times \frac{1\ mol\ CO_2}{44.01\ g\ CO_2} \times \frac{1\ mol\ C}{1\ mol\ CO_2} = 1.329 \times 10^{-3}\ mol\ C$$

$$14.4\ \text{mg}\ H_2O \times \frac{1\ g\ H_2O}{10^3\ mg\ H_2O} \times \frac{1\ mol\ H_2O}{18.02\ g\ H_2O} \times \frac{2\ mol\ H}{1\ mol\ H_2O} = 1.598 \times 10^{-3}\ mol\ H$$

The masses of C and H are

$$1.329 \times 10^{-3}\ mol\ C \times \frac{12.01\ g\ C}{1\ mol\ C} = 1.596 \times 10^{-2}\ g\ C$$

$$1.598 \times 10^{-3}\ mol\ H \times \frac{1.008\ g\ H}{1\ mol\ H} = 1.611 \times 10^{-3}\ g\ H$$

The sum of these two masses ($1.596 \times 10^{-2}$ g C + $1.611 \times 10^{-3}$ g H = $1.757 \times 10^{-2}$ g) is less than the mass of the sample ($2.18 \times 10^{-2}$ g). The difference must be the mass of oxygen in the sample:

$$\text{Mass of oxygen} = 2.180 \times 10^{-2}\ g - 1.757 \times 10^{-2}\ g = 4.228 \times 10^{-3}\ g\ O$$

The number of moles of O atoms in the sample is

$$4.228 \times 10^{-3}\ g\ O \times \frac{1\ mol\ O}{16.00\ g\ O} = 2.642 \times 10^{-4}\ mol\ O$$

The mole ratio of the three elements in the sample is

$$1.329 \times 10^{-3}\ mol\ C : 1.598 \times 10^{-3}\ mol\ H : 2.642 \times 10^{-4}\ mol\ O$$

Dividing through by the smallest value ($2.642 \times 10^{-4}$ mol) gives a mole ratio of 5:6:1, making the empirical formula of the sample $C_5H_6O$. The empirical formula mass is

$$(5 \times 12.01\ \text{amu} + 6 \times 1.008\ \text{amu} + 16.00\ \text{amu})n = 82.10\ \text{amu}$$

The molecular ion identifies the molecular mass of eugenol as 164 amu.

Therefore, the multiplier, $n$, to convert the empirical formula into a molecular formula is

$$n = \frac{\text{molecular mass}}{\text{empirical formula mass}} = \frac{164 \text{ amu}}{82.10 \text{ amu}} = 2$$

and the molecular formula is

$$(C_5H_4O)_n = (C_5H_6O)_2 = C_{10}H_{12}O_2$$

**Think About It** Our answer makes sense because we have a simple whole number ratio of the elements.

⚙ **Practice Exercise** Vanillin is a compound containing carbon, hydrogen, and oxygen that gives vanilla beans their distinctive flavor. The combustion of 30.4 mg of vanillin produces 70.4 mg of $CO_2$ and 14.4 mg of $H_2O$. The mass spectrum of vanillin shows a molecular ion at 152 amu. Use this information to determine the molecular formula of vanillin.

*(Answers to Practice Exercises are in the back of the book.)*

# 3.9 Limiting Reactants and Percent Yield

Let's return to photosynthesis, the process responsible for the $O_2$ in our present-day atmosphere and for the energy that sustains life on Earth's surface. The stoichiometry of the reaction calls for equal parts on a mole basis, referred to as *equimolar amounts*, of $CO_2$ and $H_2O$:

$$6\,CO_2(g) + 6\,H_2O(\ell) \rightarrow C_6H_{12}O_6(aq) + 6\,O_2(g)$$
Glucose

Because in nature there is little likelihood of having exactly equal proportions of reactants at a reaction site, let's consider what happens when more than six molecules of water are available for every six molecules of $CO_2$; having water in excess is a common occurrence in biological systems. The photosynthetic production of glucose continues only until all the $CO_2$ is consumed, leaving the extra molecules of water unreacted. In this example, carbon dioxide is the **limiting reactant**, meaning that the extent to which the reaction proceeds is determined by the quantity of $CO_2$ available and not by the quantity of $H_2O$, which is in excess. Figure 3.29 illustrates the concept of a limiting reactant.

Another way of thinking about limiting reactants is to consider the following task. You are asked to assemble units containing one bolt, one nut, and two washers, but you have 100 bolts, 100 nuts, and 100 washers. The washers are the limiting reactant in this scenario.

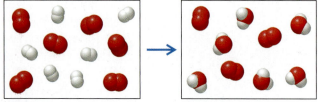

**FIGURE 3.29** The reaction of hydrogen ($H_2$, white molecules) with oxygen ($O_2$, red molecules) produces water ($H_2O$). A mixture containing equal numbers of hydrogen and oxygen molecules produces only as many water molecules as the number of $H_2$ molecules available. In this case, $H_2$ is the limiting reactant and $O_2$ is in excess.

## Calculations Involving Limiting Reactants

Suppose you are asked to calculate the mass of a product formed from given masses of reactants. How do you know if one reactant is limiting? And how do you know which one? It is tempting to select the reactant present in the smaller amount by mass. Avoid this temptation and instead take a systematic approach based on the stoichiometry of the reaction. Several approaches are possible, two of which we present here—one that uses stoichiometric calculations and one that uses mole ratios.

In our first method, we calculate how much product would be formed if reactant A were completely consumed. Then we repeat the calculation for reactant B and any other reactants. Let's try this approach with the reaction between sulfur trioxide gas and water vapor to form sulfuric acid:

$$SO_3(g) + H_2O(g) \rightarrow H_2SO_4(\ell) \tag{3.1}$$

**limiting reactant** a reactant that is consumed completely in a chemical reaction. The amount of product formed depends on the amount of the limiting reactant available.

This reaction was described in Section 3.1 as contributing to the acidity of Earth's early atmosphere. Suppose we carry out this reaction in the laboratory using 20.00 g of $SO_3(g)$ and 10.00 g of $H_2O(g)$. Which is the limiting reactant, and how many grams of $H_2SO_4$ are produced from these masses of reactants?

We first calculate the molar masses and then the numbers of moles of $SO_3$ and $H_2O$ from the two masses:

▶❚❚ **CHEMTOUR** Limiting Reactant

$$\mathcal{M}_{SO_3} = 32.06 \text{ g/mol} + 3(16.00 \text{ g/mol}) = 80.06 \text{ g/mol}$$

$$\mathcal{M}_{H_2O} = 2(1.008 \text{ g/mol}) + 16.00 \text{ g/mol} = 18.02 \text{ g/mol}$$

$$20.00 \text{ g SO}_3 \times \frac{1 \text{ mol SO}_3}{80.06 \text{ g SO}_3} = 0.2498 \text{ mol SO}_3$$

$$10.00 \text{ g H}_2\text{O} \times \frac{1 \text{ mol H}_2\text{O}}{18.02 \text{ g H}_2\text{O}} = 0.5549 \text{ mol H}_2\text{O}$$

Then we calculate the mass of sulfuric acid produced if all of the $SO_3$ is consumed:

$$0.2498 \text{ mol SO}_3 \times \frac{1 \text{ mol H}_2\text{SO}_4}{1 \text{ mol SO}_3} \times \frac{98.08 \text{ g H}_2\text{SO}_4}{1 \text{ mol H}_2\text{SO}_4} = 24.50 \text{ g H}_2\text{SO}_4$$

Next we carry out the same calculation to determine how much sulfuric acid is produced assuming 10.00 g of water reacts completely:

$$0.5549 \text{ mol H}_2\text{O} \times \frac{1 \text{ mol H}_2\text{SO}_4}{1 \text{ mol H}_2\text{O}} \times \frac{98.08 \text{ g H}_2\text{SO}_4}{1 \text{ mol H}_2\text{SO}_4} = 54.42 \text{ g H}_2\text{SO}_4$$

The $SO_3$ is the limiting reactant because, when it is completely consumed, a smaller amount of product is formed. Thus when 20.00 g of $SO_3$ and 10.00 g of $H_2O$ are combined, the maximum amount of product that can form is 24.50 g of $H_2SO_4$. Making that amount of sulfuric acid consumes all the available $SO_3$ but not all the available $H_2O$.

The second method compares the mole ratio of the reactants to the mole ratio required by the balanced chemical equation. Using this approach, we

1. Convert masses of reactants A and B into moles
2. Calculate the mole ratio of A to B
3. Compare this mole ratio with the stoichiometric mole ratio from the balanced chemical equation. If

$$\left(\frac{\text{mol A}}{\text{mol B}}\right)_{\text{given}} > \left(\frac{\text{mol A}}{\text{mol B}}\right)_{\text{stoichiometric}} \tag{3.4}$$

then B is the limiting reactant. If

$$\left(\frac{\text{mol A}}{\text{mol B}}\right)_{\text{given}} < \left(\frac{\text{mol A}}{\text{mol B}}\right)_{\text{stoichiometric}} \tag{3.5}$$

then A is the limiting reactant. If the two ratios are equal, the masses are the stoichiometric amounts and both reactants are consumed completely.

For the same example using 20.00 g of $SO_3$ and 10.00 g of $H_2O$, we calculate the mole ratio of $SO_3$ to $H_2O$ after calculating the moles of $SO_3$ and $H_2O$ available:

$$\frac{\text{mol SO}_3}{\text{mol H}_2\text{O}} = \frac{20.00 \text{ g SO}_3 \times \dfrac{1 \text{ mol SO}_3}{80.06 \text{ g SO}_3}}{10.00 \text{ g H}_2\text{O} \times \dfrac{1 \text{ mol H}_2\text{O}}{18.02 \text{ g H}_2\text{O}}} = \frac{0.2498 \text{ mol SO}_3}{0.5549 \text{ mol H}_2\text{O}} = 0.4502$$

**theoretical yield** the maximum amount of product possible in a chemical reaction for given quantities of reactants; also called *stoichiometric yield.*

Equation 3.1 tells us that the stoichiometric ratio of $SO_3$ to $H_2O$ is 1/1 = 1. The $SO_3/H_2O$ mole ratio for our reaction conditions is 0.2498/0.5549 = 0.4502. Expressing this outcome as a mathematical relationship, we get

$$\left(\frac{\text{mol } SO_3}{\text{mol } H_2O}\right)_{\text{given}} < \left(\frac{\text{mol } SO_3}{\text{mol } H_2O}\right)_{\text{stoichiometric}}$$

This inequality matches Equation 3.5, which tells us that $SO_3$ (reactant A in Equation 3.5) must be the limiting reactant.

These two approaches to determining a limiting reactant lead to the same conclusion and are of similar complexity. In both cases, because the masses of the reactants are given in grams, we must convert those masses into moles. The advantage of the first method is that it not only determines the identity of the limiting reactant, it also determines the **theoretical yield**, the maximum amount of product that can be produced by the given mixture. The theoretical yield is sometimes called the *stoichiometric yield* or *100% yield.*

---

**SAMPLE EXERCISE 3.20    Identifying Limiting Reactants    L07**

During strenuous exercise, respiration (the reaction between glucose and oxygen, Section 3.5) is limited by the availability of oxygen to the muscles. Under *anaerobic* conditions, our bodies produce lactic acid, which accumulates in muscles and leads to the "burning" sensation you may have felt. We can model this situation by considering the reaction between 1.32 g $C_6H_{12}O_6$ and 1.32 g $O_2$ to produce $CO_2$ and water. Is one of these reactants a limiting reactant, or is the mixture stoichiometric? If one reactant is limiting, which one is it?

**Collect and Organize** We are given quantities of two reactants ($C_6H_{12}O_6$ and $O_2$) and want to determine which, if either, is limiting.

**Analyze** We can determine the limiting reactant by calculating the ratio of $C_6H_{12}O_6$ to $O_2$ in the balanced chemical equation for the reaction. That means we will use the second method described in the preceding discussion. An unbalanced equation for this combustion reaction is

$$C_6H_{12}O_6(s) + O_2(g) \rightarrow CO_2(g) + H_2O(g)$$

After balancing this equation, we compare the ratio of the amounts of reactants with the stoichiometric ratio in the balanced equation to determine whether or not one of the reactants is limiting.

**Solve** First we need to balance the combustion equation:

$$\underline{\quad} C_6H_{12}O_6(s) + \underline{\quad} O_2(g) \rightarrow \underline{\quad} CO_2(g) + \underline{\quad} H_2O(g)$$

$$C_6H_{12}O_6(s) + 6 O_2(g) \rightarrow 6 CO_2(g) + 6 H_2O(g)$$

This balanced equation tells us that the stoichiometric ratio of $C_6H_{12}O_6$ to $O_2$ is 1:6. Next we convert the given masses of reactants to moles:

$$1.32 \text{ g } O_2 \times \frac{1 \text{ mol } O_2}{32.00 \text{ g } O_2} = 4.13 \times 10^{-2} \text{ mol } O_2$$

$$1.32 \text{ g } C_6H_{12}O_6 \times \frac{1 \text{ mol } C_6H_{12}O_6}{180.16 \text{ g } C_6H_{12}O_6} = 7.33 \times 10^{-3} \text{ mol } C_6H_{12}O_6$$

Then we calculate the mole ratio

$$\frac{4.13 \times 10^{-2} \text{ mol } O_2}{7.33 \times 10^{-3} \text{ mol } C_6H_{12}O_6} = 5.6$$

This ratio is less than the stoichiometric $O_2$:$C_6H_{12}O_6$ ratio of 6:1, so oxygen is the limiting reactant, and there is a slight excess of glucose.

**Think About It** The calculation indicates that when equal masses of glucose and oxygen react, oxygen is the limiting reactant.

⚙ **Practice Exercise** Any fuel–oxygen mixture that contains more oxygen than is needed to burn the fuel completely is called a *lean mixture*, whereas a mixture containing too little oxygen to allow complete combustion of the fuel is called a *rich mixture*. A high-performance heater that burns propane, $C_3H_8(g)$, is adjusted so that 100.0 g of $O_2(g)$ enters the system for every 100.0 g of propane. Is this mixture rich or lean?

*(Answers to Practice Exercises are in the back of the book.)*

---

| SAMPLE EXERCISE 3.21 | **Limiting Reactants in Chemical Reactions** | L07 |
|---|---|---|

The results of the Miller/Urey experiment require a chemical reaction to account for the formation of glycine. The reaction between carbon dioxide, ammonia, and methane that produces glycine, water, and carbon monoxide is described by the following chemical equation:

$$2\,CO_2(g) + NH_3(g) + CH_4(g) \rightarrow C_2H_5NO_2(s) + H_2O(\ell) + CO(g)$$

How much glycine could be expected from the reaction of 29.3 g $CO_2$, 4.53 g $NH_3$, and 4.27 g $CH_4$?

**Collect and Organize** We are given a balanced chemical equation and the quantities of three reactants ($CH_4$, $NH_3$, and $CO_2$). We are asked how much glycine can be produced, which will require us to identify which one of the reactants is the limiting reactant. Since there are three reactants, we must use the first method discussed above.

**Analyze** First we calculate the number of moles of each reactant we have available. Then, we calculate the maximum amount of glycine that would be produced from each reactant. The reactant yielding the smallest amount of glycine is the limiting reactant and determines the maximum amount of glycine that we can expect from the reaction.

**Solve**
1. Using the molar masses of $CO_2$, $NH_3$, and $CH_4$, we calculate the number of moles of each that are available:

$$29.3 \text{ g CO}_2 \times \frac{1 \text{ mol CO}_2}{44.01 \text{ g CO}_2} = 0.6658 \text{ mol CO}_2$$

$$4.53 \text{ g NH}_3 \times \frac{1 \text{ mol NH}_3}{17.03 \text{ g NH}_3} = 0.2660 \text{ mol NH}_3$$

$$4.27 \text{ g CH}_4 \times \frac{1 \text{ mol CH}_4}{16.04 \text{ g CH}_4} = 0.2662 \text{ mol CH}_4$$

2. Using the mole ratios of $C_2H_5NO_2$ for each reactant and the molar mass of $C_2H_5NO_2$, we calculate the amount of glycine that could be produced from each reactant:

$$0.6658 \text{ mol CO}_2 \times \frac{1 \text{ mol C}_2H_5NO_2}{2 \text{ mol CO}_2} \times \frac{75.07 \text{ g C}_2H_5NO_2}{1 \text{ mol C}_2H_5NO_2} = 25.0 \text{ g C}_2H_5NO_2$$

$$0.2660 \text{ mol NH}_3 \times \frac{1 \text{ mol C}_2H_5NO_2}{1 \text{ mol NH}_3} \times \frac{75.07 \text{ g C}_2H_5NO_2}{1 \text{ mol C}_2H_5NO_2} = 20.0 \text{ g C}_2H_5NO_2$$

$$0.2662 \text{ mol CH}_4 \times \frac{1 \text{ mol C}_2H_5NO_2}{1 \text{ mol CH}_4} \times \frac{75.07 \text{ g C}_2H_5NO_2}{1 \text{ mol C}_2H_5NO_2} = 20.0 \text{ g C}_2H_5NO_2$$

Of these three possible yields of $C_2H_5NO_2$, we can only make the smallest amount, 20.0 g $C_2H_5NO_2$, because both ammonia and methane are limiting reactants in this reaction.

**Think About It** Although the mass of methane available for the reaction is the lowest of the three reactants, there is an equal number of moles of methane and ammonia. Since carbon dioxide is present in excess, both methane and ammonia are limiting reactants. With three or more reactants, one or more may turn out to be in excess.

**actual yield** the amount of product obtained from a chemical reaction, which is often less than the theoretical yield.

**percent yield** the ratio, expressed as a percentage, of the actual yield of a chemical reaction to the theoretical yield.

⚙ **Practice Exercise** One of the intermediates in the synthesis of glycine from ammonia, carbon dioxide, and methane is $C_2H_4N_2$, produced by the balanced chemical equation: $CH_4(g) + CO_2(g) + 2\,NH_3(g) \rightarrow C_2H_4N_2(g) + 3\,H_2O(g)$. How much $C_2H_4N_2$ could be expected from reaction of 14.2 g $CO_2$, 2.27 g $NH_3$, and 2.14 g $CH_4$?

*(Answers to Practice Exercises are in the back of the book.)*

## Actual Yields versus Theoretical Yields

At the beginning of this section, when we calculated the mass of sulfuric acid formed by 10.00 g of water and 20.00 g of sulfur trioxide, we defined the value as a theoretical yield: the maximum amount of product possible from the given quantities of reactants. In nature, industry, or the laboratory, the **actual yield** is often less than the theoretical yield for several reasons. In addition to the reaction we want, the reactants may also undergo other reactions, yielding products in addition to the ones we expect. Sometimes the rate of a reaction is so slow that some reactants remain unreacted even after an extended time. Other reactions do not go to completion no matter how long they are allowed to run, yielding a mixture of reactants and products, the composition of which does not change with time. For these and other reasons, it is useful to distinguish between the theoretical and the actual yields of a chemical reaction and to calculate the **percent yield**:

$$\text{Percent yield} = \frac{\text{actual yield}}{\text{theoretical yield}} \times 100\% \tag{3.6}$$

**SAMPLE EXERCISE 3.22    Calculating Percent Yield    LO8**

The industrial process for making the ammonia used in fertilizer, explosives, and many other products is based on the reaction between nitrogen and hydrogen at high temperature and pressure:

$$N_2(g) + 3\,H_2(g) \rightarrow 2\,NH_3(g)$$

If 18.20 kg of $NH_3$ is produced by a reaction mixture that initially contains 6.00 kg of $H_2$ and an excess of $N_2$, what is the percent yield of the reaction?

**Collect and Organize** We know that the actual yield of $NH_3$ is 18.20 kg. We also know that $H_2$ must be the limiting reactant because the problem specifies the presence of excess $N_2$ and that the reaction mixture initially contained 6.00 kg of $H_2$.

**Analyze** We need to use the mass of $H_2$ to calculate how much $NH_3$ could have been produced—the theoretical yield. We then use the theoretical yield and the actual yield to calculate percent yield. We need the molar masses of $H_2$ and $NH_3$ and the stoichiometry of the reaction, which tells us that 2 moles of $NH_3$ are produced for every 3 moles of $H_2$ consumed.

$$\boxed{\text{g }H_2} \xrightarrow[\text{molar mass}]{1} \boxed{\text{mol }H_2} \xrightarrow[\text{3 mol }H_2]{\text{2 mol }NH_3} \boxed{\text{mol }NH_3} \xrightarrow[\text{molar mass}]{} \boxed{\text{g }NH_3} = \boxed{\text{theoretical yield}}$$

$$\boxed{\text{\% yield}} = \boxed{\frac{\text{actual yield}}{\text{theoretical yield}}} \times 100\%$$

**Solve** The molar masses we need are

$$\mathcal{M}_{H_2} = 2(1.008\text{ g/mol}) = 2.016\text{ g/mol}$$

$$\mathcal{M}_{NH_3} = 14.01\text{ g/mol} + 3(1.008\text{ g/mol}) = 17.03\text{ g/mol}$$

We calculate the theoretical yield of $NH_3$:

$$6.00 \text{ kg } H_2 \times \frac{10^3 \text{ g } H_2}{1 \text{ kg } H_2} \times \frac{1 \text{ mol } H_2}{2.016 \text{ g } H_2} \times \frac{2 \text{ mol } NH_3}{3 \text{ mol } H_2} \times \frac{17.03 \text{ g } NH_3}{1 \text{ mol } NH_3} \times \frac{1 \text{ kg } NH_3}{10^3 \text{ g } NH_3}$$

$$= 33.8 \text{ kg } NH_3$$

Then we divide the actual yield by that to determine the percent yield:

$$\frac{18.20 \text{ kg } NH_3}{33.8 \text{ kg } NH_3} \times 100\% = 53.8\%$$

**Think About It** A yield of about 54% may seem low, but it may be the best that can be achieved for a particular process. A great deal of chemical research goes into trying to improve the percent yield of industrial chemical reactions.

⚙ **Practice Exercise** The combustion of 58.0 g of butane ($C_4H_{10}$) produces 158 g of $CO_2$. What is the percent yield of the reaction?

*(Answers to Practice Exercises are in the back of the book.)*

---

**SAMPLE EXERCISE 3.23** **Effect of Percent Yield on Calculating the Mass of Reactant Needed** **LO8**

One way of producing hydrogen for use in hydrogen-powered vehicles is the water–gas shift reaction:

$$H_2O(g) + CO(g) \rightarrow H_2(g) + CO_2(g)$$

At 200°C, the reaction produces a 96% yield. How many grams of $H_2O$ and CO are needed to generate 176 g $H_2$ under these conditions?

**Collect and Organize** This is a stoichiometry problem like Sample Exercises 3.12 and 3.13 except that the yield of reaction is only 96%.

**Analyze** We need to convert the mass of $H_2$ we wish to make into moles of $H_2$, and then use the mole ratios of $H_2$:CO and $H_2$:$H_2O$ to find the amount of reactants we need. To produce 176 g $H_2$ using a reaction that produces 96% of the theoretical yield, we consider the theoretical yield to be 96% of the total yield we need. Then by analogy to Sample Exercise 3.13, we can calculate the amount of both reactants needed.

**Solve**
1. The moles of $H_2$ we wish to produce is:

$$176 \text{ g } H_2 \times \frac{1 \text{ mol } H_2}{2.016 \text{ g } H_2} = 87.30 \text{ mol } H_2$$

We carry one more figure along than is significant until we get to our final answer, which we will report in the correct number of significant figures.

However, since the reaction only produces 96% of the theoretical yield, we need to base our calculation on what 100% yield would be if 87.30 moles is 96% of the theoretical yield:

$$87.30 \text{ mol} = 0.96x \qquad x = 91 \text{ mol}$$

2. Using the mole ratios of $H_2$:$H_2O$ and $H_2$:CO, we find the moles of $H_2O$ and CO:

$$91 \text{ mol } H_2 \times \frac{1 \text{ mol } H_2O}{1 \text{ mol } H_2} = 91 \text{ mol } H_2O \qquad 91 \text{ mol } H_2 \times \frac{1 \text{ mol CO}}{1 \text{ mol } H_2} = 91 \text{ mol CO}$$

3. We convert the moles of the reactants to the mass of the reactants by multiplying by the molar mass.

$$91 \text{ mol } H_2O \times \frac{18.02 \text{ g } H_2O}{1 \text{ mol } H_2O} = 1.6 \times 10^3 \text{ g } H_2O$$

$$91 \text{ mol CO} \times \frac{28.01 \text{ g CO}}{1 \text{ mol CO}} = 2.5 \times 10^3 \text{ g CO}$$

We are only allowed three significant figures in our answer, so the masses of the reactants we need are: $1.64 \times 10^3$ g $H_2O$ and $2.55 \times 10^3$ g CO.

As a check of our answers, let's calculate how much $H_2$ would be produced from 91 moles of CO if the reaction proceeded in 100% yield:

$$91 \text{ mol CO} \times \frac{1 \text{ mol } H_2}{1 \text{ mol CO}} \times \frac{2.016 \text{ g } H_2}{1 \text{ mol } H_2} = 183 \text{ g } H_2$$

The percent yield is

$$\frac{176 \text{ g } H_2}{183 \text{ g } H_2} \times 100\% = 96\%$$

**Think About It** On a large scale, any yield less than 100% could be a significant cost for rare and expensive reactants.

**Practice Exercise** How much aluminum oxide and how much carbon are needed to prepare 454 g (1 pound) of aluminum by the balanced chemical equation:

$$2 \text{ Al}_2O_3(s) + 3C(s) \rightarrow 4 \text{ Al}(s) + 3 \text{ CO}_2(g)$$

if the reaction proceeds to 78% yield?

*(Answers to Practice Exercises are in the back of the book.)*

---

## SAMPLE EXERCISE 3.24 Integrating Concepts: Taxol™

Taxol™ is a molecular compound used in the treatment of cancer. It was originally isolated from the bark of the Pacific yew tree and subsequently has been synthesized in the laboratory.

a. The yew tree provides about 100 mg Taxol for every 1.00 kg bark. If 1 tree has about 3 kg bark and 9000 trees were needed to isolate enough Taxol for its first clinical trial, how many grams of Taxol were used for those initial tests?

b. One of the laboratory syntheses of Taxol begins with camphor (78.89% C; 10.59% H; $\mathcal{M} = 152.23$ g/mol) and requires 23 different chemical reactions in sequence to convert it into Taxol (66.11% C; 6.02% H; 1.64% N; $\mathcal{M} = 853.88$ g/mol). The overall yield of the process to produce 1 mole of Taxol, starting with 1 mole camphor, is 0.10%. (i) What are the molecular formulas of camphor and Taxol? (ii) How much camphor must one use to make the amount of Taxol provided by 9000 trees?

**Collect and Organize** We are given information about the amount of Taxol in bark, the percent composition of Taxol, and the yield of the laboratory synthesis. Using dimensional analysis and other methods from this chapter, we can answer the questions posed.

**Analyze** The percent compositions given for both substances do not add to 100%. We may assume that both compounds contain oxygen.

**Solve**

a. We begin by calculating the amount of Taxol in 9000 trees:

$$9000 \text{ trees} \times \frac{3 \text{ kg bark}}{1 \text{ tree}} \times \frac{0.100 \text{ g Taxol}}{1.00 \text{ kg bark}} = 3 \times 10^3 \text{ g Taxol}$$

b. (i) The percent carbon in camphor added to the percent hydrogen is less than 100%:

$$78.89\% + 10.59\% = 89.48\%$$

Because the reported percentages do not add to 100%, the camphor most likely contains oxygen, the percent of which we cannot find in a combustion analysis. We can assume the percent oxygen in camphor is 100% − 89.48% = 10.52%. We can now determine the empirical formula of camphor by considering that 100.00 g of camphor would contain 78.89 g C, 10.59 g H, and 10.52 g O:

$$78.89 \text{ g C} \times \frac{1 \text{ mol C}}{12.01 \text{ g C}} = 6.569 \text{ mol C}$$

$$10.59 \text{ g H} \times \frac{1 \text{ mol H}}{1.008 \text{ g H}} = 10.51 \text{ mol H}$$

$$10.52 \text{ g O} \times \frac{1 \text{ mol O}}{16.00 \text{ g O}} = 0.6575 \text{ mol O}$$

Dividing by the smallest number of moles:

$$\frac{6.569 \text{ mol}}{0.6575} = 10 \text{ mol C}$$

$$\frac{10.51 \text{ mol}}{0.6575} = 16 \text{ mol H}$$

$$\frac{0.6575 \text{ mol}}{0.6575} = 1 \text{ mol O}$$

The empirical formula of camphor is $C_{10}H_{16}O$. Its mass is $[(10 \times 12.01 \text{ amu}) + (16 \times 1.008 \text{ amu}) + (16.00 \text{ amu})] = 153.23$ amu, which matches the molecular mass of camphor. Therefore, the molecular formula is the same as the empirical formula.

Carrying out the same procedure for Taxol:

$$66.11\% + 6.02\% + 1.64\% = 73.77\%$$

$$100\% - 73.77\% = 26.23\% \text{ oxygen}$$

Calculating the empirical formula:

$$66.11 \text{ g C} \times \frac{1 \text{ mol C}}{12.01 \text{ g C}} = 5.505 \text{ mol C}$$

$$6.02 \text{ g H} \times \frac{1 \text{ mol H}}{1.008 \text{ g H}} = 5.97 \text{ mol H}$$

$$1.64 \text{ g N} \times \frac{1 \text{ mol N}}{14.01 \text{ g N}} = 0.117 \text{ mol N}$$

$$26.23 \text{ g O} \times \frac{1 \text{ mol O}}{16.00 \text{ g O}} = 1.639 \text{ mol O}$$

Dividing by the smallest number of moles:

$$\frac{5.505 \text{ mol C}}{0.117} = 47 \text{ mol C}$$

$$\frac{5.97 \text{ mol H}}{0.117} = 51 \text{ mol H}$$

$$\frac{0.117 \text{ mol N}}{0.117} = 1 \text{ mol N}$$

$$\frac{1.639 \text{ mol O}}{0.117} = 14 \text{ mol O}$$

The empirical formula of Taxol is $C_{47}H_{51}NO_{14}$, which has a mass of $[(47 \times 12.01 \text{ amu}) + (51 \times 1.008 \text{ amu}) + (1 \times 14.01 \text{ amu}) + (14 \times 16.00 \text{ amu})] = 853.88$ amu. This value is the same as the molecular mass of Taxol, so $C_{47}H_{51}NO_{14}$ is also the molecular formula.

(ii) We need $3 \times 10^3$ g Taxol from a process that has a yield of 0.10%. The molar ratio in the process is 1 mole camphor produces 1 mole Taxol.

$$3 \times 10^3 \text{ g Taxol} = 0.10\% \text{ of the theoretical yield}$$

The theoretical yield for the process is

$$3 \times 10^3 \text{ g} = 0.0010x \qquad x = 3 \times 10^3 \text{ g}/0.0010 = 3 \times 10^6 \text{ g}$$

Starting with that amount of Taxol as the theoretical yield:

$$3 \times 10^6 \text{ g Taxol} \times \frac{1 \text{ mol Taxol}}{853.88 \text{ g Taxol}} \times \frac{1 \text{ mol camphor}}{1 \text{ mol Taxol}}$$

$$\times \frac{152.23 \text{ g camphor}}{1 \text{ mol camphor}} = 5 \times 10^5 \text{ g camphor}$$

Thus, we need half a million grams of camphor to synthesize as much Taxol as there is in 9000 Pacific yew trees.

**Think About It** Camphor is quite inexpensive, but this example clearly illustrates how much material can be required to make a valuable product if the process has a low overall yield.

At the beginning of this chapter we asked several questions, including: How do we describe chemical changes involving different compounds using chemical reactions? How do we determine the composition of a compound? How do we determine the chemical formula of a compound? Chemical reactions are described by balanced chemical equations that contain the information needed to calculate the amounts of reactants required to prepare certain amounts of products. Data from chemical reactions can also be used to determine the percentage composition of compounds, as well as empirical and molecular formulas. Although the examples we used in this chapter relate to processes that have changed Earth and may have given rise to life on Earth, the concepts have broad applicability in all areas of chemistry—from the extraction of metallic elements from minerals to the analysis of molecules of medical and biological importance. The ideas of the mole and stoichiometry are fundamental to understanding chemistry on the microscopic level. The calculations presented in this chapter are part of your growing "toolbox" of skills that we will apply in subsequent chapters.

# Hydrogen and Helium: The Bulk of the Universe

Hydrogen and helium, the only two elements in the first row of the periodic table, are the least dense and the most abundant elements in the universe, accounting for over 99% of all atoms. They are much less abundant on Earth as free elements: Hydrogen gas ($H_2$) constitutes only about 0.00005% of the atmosphere, by volume. The atmospheric concentration of helium is a factor of 10 higher—0.0005% by volume—which still makes He a very minor component.

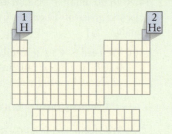

These two elements have similar physical properties, but their chemical properties are remarkably different. Helium is chemically inert, while hydrogen is extremely reactive and occurs in more compounds than any other element. Compounds containing carbon and hydrogen are the major components of all living matter, and water ($H_2O$) is ubiquitous and essential for life on Earth.

Hydrogen is currently being hailed as the fuel of the future, with the potential for large-scale use in internal combustion engines and fuel cells (Figure 3.30). The principal advantage of hydrogen over fossil fuels is its high fuel value—the amount of energy it releases per unit of mass. Hydrogen is also a clean fuel, producing only water and energy and not oxides of carbon, nitrogen, or sulfur, which are products of fossil-fuel combustion and contribute to atmospheric pollution and climate change.

Because hydrogen, as the free element, is rare in nature, it must be extracted from compounds like methane ($CH_4$), the principal component of natural gas. Hydrogen is generated from methane in a process called *steam–methane reforming*:

$$CH_4(g) + H_2O(g) \xrightarrow{1000°C} CO(g) + 3\,H_2(g)$$

Additional $H_2$ is generated when the $CO(g)$ is reacted with more steam in the *water–gas shift reaction*:

$$CO(g) + H_2O(g) \xrightarrow{200°C} CO_2(g) + H_2(g)$$

Another way to generate hydrogen is to split $H_2O$ molecules into $H_2$ and $O_2$ by *electrolysis*:

$$2\,H_2O(\ell) \xrightarrow{\text{electric current}} 2\,H_2(g) + O_2(g)$$

Developing hydrogen as a widely used fuel is hampered by the fact that it cannot be generated in bulk quantities by any efficient means that do not also consume fossil fuels.

Probably the most important use of hydrogen is in producing ammonia in the *Haber–Bosch process*:

$$N_2(g) + 3\,H_2(g) \rightarrow 2\,NH_3(g)$$

The hydrogen used to make ammonia comes from steam–methane reforming, which means the production of hydrogen and ammonia are tightly linked commercially. Ammonia is used as a fertilizer and as a source of nitrogen for manufacturing other nitrogen-containing compounds, including explosives.

Another major use of hydrogen is as a reactant with liquid vegetable oils to make solid edible fats such as margarine. This reaction is called *hydrogenation*, and the products are called *hydrogenated fats*. Some of the oils from partial hydrogenation are *trans fats*, which differ in structure from natural vegetable oils. Trans fats have been used in many processed foods, and they are currently the object of much scrutiny because of their negative impact on human health.

The *deuterium* and *tritium* isotopes of hydrogen are also useful. Deuterium ($^2$H or D) accounts for about 0.015% of the hydrogen in nature. Heavy water, $D_2O$, is an essential component in nuclear reactors where plutonium is generated from uranium and is also crucial to the production of nuclear weapons. Heavy water is so important to this technology that its production is monitored and the material kept

**FIGURE 3.30** A hydrogen fueling station for vehicles.

under strict export controls. Heavy water has a slightly higher boiling point (101.4°C) than $H_2O$ (100.0°C), and in principle $D_2O$ can be separated from $H_2O$ by distillation. In practice, $D_2$ is separated from liquefied $H_2$ and then used to make deuterium compounds that can be purified by subsequent processing. For example, $D_2$ is reacted to form $D_2S$, which is treated with $H_2O$ and converted to $D_2O$. The production of 1 L of $D_2O$ requires about 340,000 L of $H_2O$.

Because tritium ($^3H$) is radioactive, it can be detected easily, which gives rise to its utility. It is used in devices where dim light is needed but no source of electricity is available, such as emergency exit signs in buildings that must function in times of power failure, dials in aircraft, rifle sights, and even in the luminous dials of some watches. In these applications, the radiation from tritium interacts with a thin coating of a luminescent material, causing it to emit light. The level of radioactivity in these devices is not harmful.

On early Earth, hydrogen was retained because it formed compounds with heavier elements, but helium forms no compounds and was too light to be held by Earth's gravitational field. Therefore any helium now on Earth is the product of radioactive decay. Most helium in the world is obtained from natural gas deposits in Texas, Oklahoma, Kansas, the Middle East, and Russia. These locations have significant deposits of uranium ore, and the decay of uranium produces helium, which mixes with underground deposits of natural gas. Helium is separated from the natural gas, in which its concentration may be as high as 0.3% by mass.

Helium has several uses beyond keeping blimps, weather balloons, and decorative balloons aloft. It is mixed with oxygen in air tanks used in deep-sea diving to create nitrogen-free gas mixtures. "Regular" air is about four-fifths nitrogen. Because nitrogen is soluble in blood, divers breathing regular air, with its high nitrogen content, run the risk of developing nitrogen narcosis, a disorienting condition in which the nervous system becomes saturated with nitrogen. Helium does not have the same narcotic properties as nitrogen. Use of helium in air tanks protects divers from decompression sickness (the bends), a painful and life-threatening condition caused by nitrogen bubbles forming in the blood and joints as divers ascend to the surface, because it is less soluble in blood than nitrogen.

Increasingly, helium is used as a *cryogen*, which is a gas that has been liquefied by cooling to a very low temperature. Helium has the lowest boiling point of any element (−268.9°C), and it is used to create and maintain superconducting magnets in instruments ranging from equipment in chemistry and physics laboratories to magnetic resonance imaging (MRI) units in hospitals and clinics (Figure 3.31).

In its liquid phase, helium has some extraordinary physical properties. It exhibits superfluidity, which means it flows so freely that it can flow up the walls of containers. Liquid helium is also a nearly perfect conductor of heat as a superfluid.

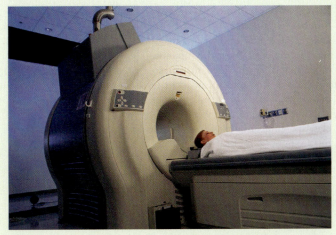

(a)

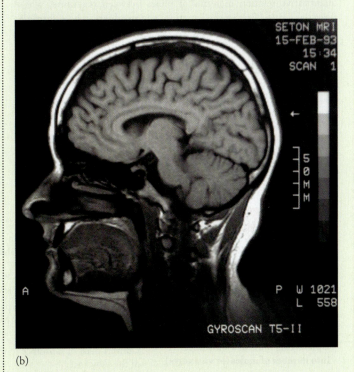

(b)

**FIGURE 3.31** (a) The magnet in a magnetic-resonance-imaging (MRI) unit maintains its superconductivity by being submerged in liquid helium inside the large white cylinder shown here. (b) An MRI image.

## SUMMARY

**Learning Outcome 1** **Avogadro's number** ($N_A = 6.022 \times 10^{23}$ particles) and the **mole** (the amount of a material that contains Avogadro's number of particles) can be used to convert grams of a substance to moles or to number of particles, moles of a substance to grams or to number of particles, and number of particles to moles or to number of grams. One mole of any substance has a mass equal to the sum of the masses of the constituent particles in its formula. (Sections 3.1 and 3.2)

**Learning Outcome 2** Balanced **chemical equations** are essential tools to describe chemical reactions. Correct equations indicate the quantitative relation in terms of moles between **reactants**, whose formulas appear first in the equation, and **products**, whose formulas appear after the reaction arrow. In a balanced chemical equation, the number of atoms of each element is the same on the reactant side and on the product side. (Sections 3.1–3.3)

**Learning Outcome 3** The mole ratio from balanced chemical equations can be used to determine amounts of reactants used and products formed in a chemical reaction. (Section 3.5)

**Learning Outcome 4** Writing equations requires knowing correct formulas for substances. Empirical formulas for substances can be determined from their percent composition by mass. (Section 3.6)

**Learning Outcome 5** The molecular formula of a compound may or may not be the same as the empirical formula. It can be determined from the empirical formula and the molecular mass of the compound. (Section 3.7)

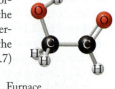

**Learning Outcome 6** Empirical formulas for substances can be determined from the **combustion analysis** (the burning of a substance in oxygen) of compounds. In the combustion analysis of organic compounds, the production of $CO_2$, $H_2O$, and other oxides can provide the percent by mass of all elements in the compound except oxygen, which we can determine by applying the law of conservation of mass. (Sections 3.7 and 3.8)

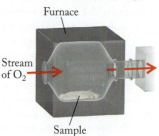

**Learning Outcome 7** Chemical reactions are not always run with exact stoichiometric amounts of reactants. Balanced equations and stoichiometric calculations can be used to determine which substances are **limiting reactants** and, from that information, the **theoretical yield** (the maximum amount of product) of the reaction. (Section 3.9)

**Learning Outcome 8** The **actual yield** (often expressed as **percent yield**) of a chemical reaction is frequently less than the theoretical yield predicted by stoichiometry. If we know the amount of material isolated as a result of a reaction, we can calculate the percent yield as the ratio of actual yield to theoretical yield expressed as a percent. (Section 3.9)

## PROBLEM-SOLVING SUMMARY

| TYPE OF PROBLEM | CONCEPTS AND EQUATIONS | SAMPLE EXERCISES |
|---|---|---|
| **Converting number of particles into number of moles (or vice versa)** | Convert number of particles to moles by dividing by Avogadro's number ($N_A = 6.022 \times 10^{23}$ particles/mol). | 3.1, 3.2 |
| | Convert number of moles to particles by multiplying by Avogadro's number ($N_A = 6.022 \times 10^{23}$ particles/mol). | |
| **Converting mass of a substance into number of moles (or vice versa)** | Convert mass of substance to moles by dividing by the molar mass ($\mathcal{M}$) of the substance. | 3.3, 3.4, 3.6, 3.7 |
| | Convert moles of substance to mass by multiplying by the molar mass ($\mathcal{M}$) of the substance. | |
| **Calculating molar mass of a compound** | Multiply molar mass of each element by its subscript in the compound's formula; then add the products. | 3.5, 3.6, 3.7 |
| **Balancing a chemical reaction** | Change coefficients in the equation so that the number of atoms of each element are the same on the two sides of the reaction arrow. | 3.8, 3.9, 3.10 |

| TYPE OF PROBLEM | CONCEPTS AND EQUATIONS | SAMPLE EXERCISES |
|---|---|---|
| **Writing and balancing a chemical equation for a combustion reaction** | The C and H in organic compounds react with $O_2$ to form $CO_2$ and $H_2O$, for example: $$CH_4(g) + 2\,O_2(g) \rightarrow 2\,H_2O(g) + CO_2(g)$$ Balance moles of C first, then H, then O. | 3.11 |
| **Calculating mass of a product from mass of a reactant** | Use $\mathcal{M}_{reactant}$ to convert mass of reactant into moles of reactant; use reaction stoichiometry to calculate moles of product and then use $\mathcal{M}_{product}$ to calculate mass of product. | 3.12, 3.13 |
| **Calculating percent composition from a chemical formula** | Calculate molar mass of the compound represented by the chemical formula. Determine mass of each element in grams from the chemical formula. Divide each element mass by the molar mass of the compound. Express each result as a percentage; percentages should add up to 100%. | 3.14 |
| **Determining empirical formula from percent composition** | Assuming a 100.00 g sample, assign the mass (g) of each element to equal its percentage. Divide the mass of each element by its molar mass to get moles. Simplify mole ratios to lowest whole numbers, and use those numbers as subscripts in the empirical formula of the compound. | 3.15, 3.16 |
| **Relating empirical and molecular formulas** | Calculate conversion factor $n$ by dividing the compound's molecular mass by its empirical formula mass. Multiply subscripts in the empirical formula by this conversion factor. | 3.17 |
| **Deriving empirical formula from combustion analysis** | For hydrocarbons, convert given masses of $CO_2$ and $H_2O$ into moles of $CO_2$ and $H_2O$, and then to moles of C and H. For compounds containing O, convert moles of C and H into masses of C and H and subtract the sum of these values from the sample mass to calculate mass of O. Convert mass of O to moles of O. Simplify the mole ratio of C to H to O. | 3.18, 3.19 |
| **Identifying limiting reactant** | Method 1: Calculate how much product each reactant could make; reactant making the least amount of product is limiting reactant. <br><br> Method 2: Convert given masses of reactants into moles. Compare mole ratio of reactants to corresponding mole ratio in stoichiometric equation (Equations 3.4 and 3.5). If $$\left(\frac{\text{mol A}}{\text{mol B}}\right)_{given} > \left(\frac{\text{mol A}}{\text{mol B}}\right)_{stoichiometric} \qquad (3.4)$$ then B is the limiting reactant. If $$\left(\frac{\text{mol A}}{\text{mol B}}\right)_{given} < \left(\frac{\text{mol A}}{\text{mol B}}\right)_{stoichiometric} \qquad (3.5)$$ then A is the limiting reactant. | 3.20, 3.21 |
| **Calculating percent yield** | Calculate theoretical yield of product using mass of limiting reactant. Divide actual yield (given) by theoretical yield (Equation 3.6): $$\text{Percent yield} = \frac{\text{actual yield}}{\text{theoretical yield}} \times 100\% \qquad (3.6)$$ | 3.22, 3.23 |

**VISUAL PROBLEMS** ●●●●●●●●●●●●●●●●●●●●●●●●●●●●●●●●●●●●●●●●●●●●●●●●●●●●●●●●●●●●●●●●●

*(Answers to boldface end-of-chapter questions and problems are in the back of the book.)*

**3.1.** Each of the pairs of containers pictured in Figure P3.1 contains substances composed of elements X (red balls) and Y (blue balls). For each pair, write a balanced chemical equation describing the reaction that takes place. Be sure to indicate the physical states of the reactants and products, using the appropriate symbols in parentheses.

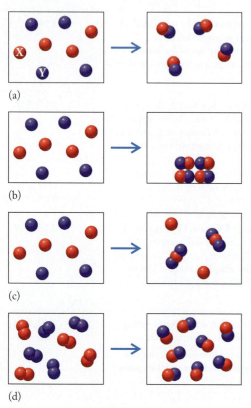

(a)

(b)

(c)

(d)

**FIGURE P3.1**

3.2. Identify the limiting reactant in each of the pairs of containers pictured in Figure P3.2. The red balls represent atoms of element X, the blue balls are atoms of element Y. Each question mark means that there is unreacted reactant left over.

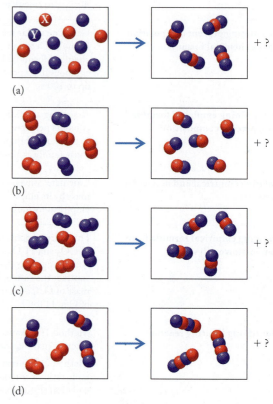

(a)

(b)

(c)

(d)

**FIGURE P3.2**

**3.3.** Which of the following drawings best illustrates the 100% reaction between $N_2$ and $O_2$ to produce $N_2O$? The red balls represent atoms of oxygen and the blue balls are atoms of nitrogen.

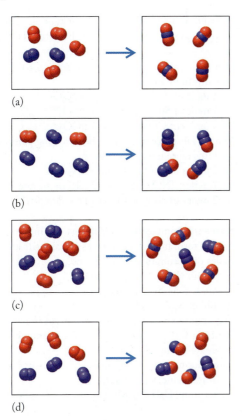

FIGURE P3.3

**3.4.** Is there a limiting reactant in any of the reactions depicted in Figure P3.3?

**3.5.** Which of the following molecules have the same empirical formulas?

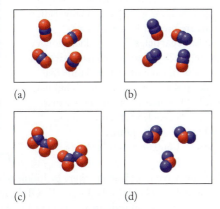

FIGURE P3.5

**3.6.** Which of the following drawings represent balanced chemical equations?

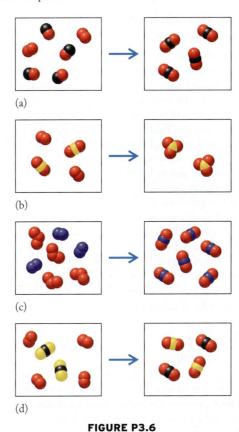

FIGURE P3.6

**3.7.** If the red spheres represent oxygen, black is carbon, blue is nitrogen, and yellow is sulfur, write balanced chemical equations for any unbalanced equations in Figure P3.6.

**3.8.** What is the percent yield of $NH_3$ for the reaction depicted in Figure P3.8 if the blue spheres represent nitrogen and the white spheres hydrogen?

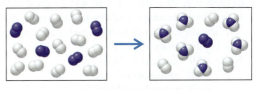

FIGURE P3.8

## QUESTIONS AND PROBLEMS

### Chemical Reactions and Earth's Early Atmosphere

#### CONCEPT REVIEW

**3.9.** On the basis of the distribution of the elements in Earth's layers (see Figure 3.1), which of the following substances should be the most dense? (a) $SiO_2(s)$; (b) $Al_2O_3(s)$; (c) $Fe(\ell)$

**3.10.** On the basis of the compositions and physical states of Earth's various layers (Figure 3.1), which of the following substances has the highest melting point? (a) $Al_2O_3$; (b) Fe; (c) Ni; (d) S

**3.11.** In a combination reaction, is the number of different products equal to, less than, or greater than the number of different reactants?

**\*3.12.** The proportions of the elements that make up the asteroid 433 Eros are similar to those that make up Earth. Scientists believe that this similarity means that the asteroid and Earth formed around the same time. If the asteroid formed after Earth formed a solid crust, and if the asteroid was the product of a collision between Earth and an even larger asteroid, how would the composition of 433 Eros be different from what it actually is?

### The Mole

#### CONCEPT REVIEW

**3.13.** In principle we could use the more familiar unit *dozen* in place of mole when expressing the quantities of particles (atoms, ions, or molecules) in chemical reactions. What would be the disadvantage in doing so?

**3.14.** In what way is the molar mass of an ionic compound and its formula mass the same, and in what ways are they different?

**3.15.** Do molecular compounds containing three atoms per molecule always have a molar mass greater than that of molecular compounds containing two atoms per molecule? Explain.

**3.16.** Without calculating their molar masses (though you may consult the periodic table), predict which of the following oxides of nitrogen has the larger molar mass: $NO_2$ or $N_2O$.

#### PROBLEMS

**3.17.** Earth's atmosphere contains many volatile substances that are present in trace amounts. The following quantities of these trace gases were found in a 1.0 mL sample of air. Calculate the number of moles of each gas in the sample.
   a. $4.4 \times 10^{14}$ atoms of $Ne(g)$
   b. $4.2 \times 10^{13}$ molecules of $CH_4(g)$
   c. $2.5 \times 10^{12}$ molecules of $O_3(g)$
   d. $4.9 \times 10^9$ molecules of $NO_2(g)$

**3.18.** The following quantities of trace gases were found in a 1.0 mL sample of air. Calculate the number of moles of each compound in the sample.
   a. $1.4 \times 10^{13}$ molecules of $H_2(g)$
   b. $1.5 \times 10^{14}$ atoms of $He(g)$
   c. $7.7 \times 10^{12}$ molecules of $N_2O(g)$
   d. $3.0 \times 10^{12}$ molecules of $CO(g)$

**3.19.** How many atoms of titanium are there in 0.125 mole of each of the following?
   a. ilmenite, $FeTiO_3$
   b. titanium(IV) chloride
   c. $Ti_2O_3$
   d. $Ti_3O_5$

**3.20.** How many atoms of iron are there in 2.5 moles of each of the following?
   a. wolframite, $FeWO_4$
   b. pyrite, $FeS_2$
   c. magnetite, $Fe_3O_4$
   d. hematite, $Fe_2O_3$

**3.21.** Which substance in each of the following pairs of quantities contains more moles of oxygen?
   a. 1 mole of $Al_2O_3$ or 1 mole of $Fe_2O_3$
   b. 1 mole of $SiO_2$ or 1 mole of $N_2O_4$
   c. 3 moles of CO or 2 moles of $CO_2$

**3.22.** Which substance in each of the following pairs of quantities contains more moles of oxygen?
   a. 2 moles of $N_2O$ or 1 mole of $N_2O_5$
   b. 1 mole of NO or 1 mole of calcium nitrate
   c. 2 moles of $NO_2$ or 1 mole of sodium nitrite

**3.23.** **Elemental Composition of Minerals** Aluminum, silicon, and oxygen form minerals known as aluminosilicates. How many moles of aluminum are in 1.50 mol of
   a. pyrophyllite, $Al_2Si_4O_{10}(OH)_2$?
   b. mica, $KAl_3Si_3O_{10}(OH)_2$?
   c. albite, $NaAlSi_3O_8$?

**3.24.** The uranium used for nuclear fuel exists in nature in several minerals. Calculate how many moles of uranium are in 1 mole of the following:
   a. carnotite, $K_2(UO_2)_2(VO_4)_2$
   b. uranophane, $CaU_2Si_2O_{11}$
   c. autunite, $Ca(UO_2)_2(PO_4)_2$

**3.25.** How many moles of carbon are there in 500.0 g of carbon?

**3.26.** How many moles of gold are there in 2.00 ounces of gold?

**3.27.** **Cancer Therapy with Iridium Metal** When iridium-192 is used in cancer treatment, a small cylindrical piece of $^{192}Ir$, 0.6 mm in diameter and 3.5 mm long, is surgically inserted into the tumor. If the density of iridium is 22.42 $g/cm^3$, how many iridium atoms are present in the sample?

**3.28.** **Gold Nanoparticles** The product shown in Figure P3.28 contains gold nanoparticles in water. The manufacturer claims that drinking this beverage improves human health. How many gold atoms are in a gold nanoparticle with a diameter of 2.00 nm ($d = 19.3$ g/mL, 1 amu $= 1.66054 \times 10^{-24}$ g)?

**FIGURE P3.28**

**3.29.** How many moles of iron are there in 1 mole of the following compounds? (a) FeO; (b) $Fe_2O_3$; (c) $Fe(OH)_3$; (d) $Fe_3O_4$

**3.30.** How many moles of sodium are there in 1 mole of the following compounds? (a) NaCl; (b) $Na_2SO_4$; (c) $Na_3PO_4$; (d) $NaNO_3$

**3.31.** Calculate the molar masses of the following atmospheric molecules: (a) $SO_2$; (b) $O_3$; (c) $CO_2$; (d) $N_2O_5$.

3.32. Determine the molar masses of the following minerals:
 a. rhodonite, $MnSiO_3$    c. ilmenite, $FeTiO_3$
 b. scheelite, $CaWO_4$    d. magnesite, $MgCO_3$

**3.33.** **Flavoring Additives** Calculate the molar masses of the following common flavors in food:
 a. vanillin, $C_8H_8O_3$    c. anise oil, $C_{10}H_{12}O$
 b. oil of cloves, $C_{10}H_{12}O_2$    d. oil of cinnamon, $C_9H_8O$

3.34. **Sweeteners** Calculate the molar masses of the following common sweeteners:
 a. sucrose, $C_{12}H_{22}O_{11}$    c. aspartame, $C_{14}H_{18}N_2O_5$
 b. saccharin, $C_7H_5O_3NS$    d. fructose, $C_6H_{12}O_6$

**3.35.** Suppose pairs of balloons are filled with 10.0 g of the following pairs of gases. Which balloon in each pair has the greater number of particles? (a) $CO_2$ or NO; (b) $CO_2$ or $SO_2$; (c) $O_2$ or Ar

3.36. If you had equal masses of the substances in the following pairs of compounds, which of the two would contain the greater number of ions? (a) NaBr or KCl; (b) NaCl or $MgCl_2$; (c) $BaCl_2$ or $Li_2CO_3$

**3.37.** How many moles of $SiO_2$ are there in a quartz crystal ($SiO_2$) that has a mass of 45.2 g?

3.38. How many moles of NaCl are there in a crystal of halite that has a mass of 6.82 g?

**3.39.** What is the mass of 0.122 mol $MgCO_3$?

3.40. What is the volume of 1.00 mol benzene ($C_6H_6$) at 20°C? The density of benzene is 0.879 g/mL.

**\*3.41.** The density of uranium (U; 19.05 g/cm³) is more than five times as great as that of diamond (C; 3.514 g/cm³). If you have a cube (1 cm on a side) of each element, which cube contains more atoms?

\*3.42. Aluminum ($d$ = 2.70 g/mL) and strontium ($d$ = 2.64 g/mL) have nearly the same density. If we manufacture two cubes, each containing 1 mole of one element or the other, which cube will be smaller? What are the dimensions of this cube?

## Writing Balanced Chemical Equations

### CONCEPT REVIEW

**3.43.** In a balanced chemical equation, does the number of moles of reactants always equal the number of moles of products?

3.44. In a balanced chemical equation, does the sum of the coefficients for the reactants always equal the sum of the coefficients for the products?

**3.45.** In a balanced chemical equation, must the sum of the masses of all the gaseous reactants always equal the sum of the masses of the gaseous products?

3.46. In a balanced chemical equation, must the sum of the volumes occupied by the gaseous reactants always equal the sum of the volumes occupied by the gaseous products?

### PROBLEMS

**3.47.** Using different colored spheres to represent C and O, sketch the reaction between five C atoms and the necessary number of $O_2$ molecules to produce a 50% mixture of CO and $CO_2$.

3.48. Using different colored spheres to represent N and O, sketch the reaction between three molecules of $N_2$ and sufficient $O_2$ to produce a mixture containing 50% $NO_2$ and 50% $N_2O_4$.

**\*3.49.** **Chemical Weathering of Rocks and Minerals** Balance the following chemical reactions, which contribute to weathering of the iron–silicate minerals ferrosilite ($FeSiO_3$), fayalite ($Fe_2SiO_4$), and greenalite [$Fe_3Si_2O_5(OH)_4$]:
 a. $FeSiO_3(s) + H_2O(\ell) \rightarrow Fe_3Si_2O_5(OH)_4(s) + H_4SiO_4(aq)$
 b. $Fe_2SiO_4(s) + CO_2(g) + H_2O(\ell) \rightarrow$
   $FeCO_3(s) + H_4SiO_4(aq)$
 c. $Fe_3Si_2O_5(OH)_4(s) + CO_2(g) + H_2O(\ell) \rightarrow$
   $FeCO_3(s) + H_4SiO_4(aq)$

\*3.50. Copper was one of the first metals used by humans because it can be refined from a wide variety of copper minerals, such as cuprite ($Cu_2O$), chalcocite ($Cu_2S$), and malachite [$Cu_2CO_3(OH)_2$]. Balance the following reactions for converting these minerals into copper metal:
 a. $Cu_2O(s) + C(s) \rightarrow Cu(s) + CO_2(g)$
 b. $Cu_2O(s) + Cu_2S(s) \rightarrow Cu(s) + SO_2(g)$
 c. $Cu_2CO_3(OH)_2(s) + C(s) \rightarrow Cu(s) + CO_2(g) + H_2O(g)$

3.51. **Physiologically Active Nitrogen Oxides** The oxides of nitrogen are biologically reactive substances now known to be formed endogenously in the human lung: NO is a powerful agent for dilating blood vessels; $N_2O$ is the anesthetic known as laughing gas; $NO_2$ has an acrid odor and is corrosive to lung tissue. Balance the following reactions for the formation of nitrogen oxides:
 a. $N_2(g) + O_2(g) \rightarrow NO(g)$
 b. $NO(g) + O_2(g) \rightarrow NO_2(g)$
 c. $NO(g) + NO_3(g) \rightarrow NO_2(g)$
 d. $N_2(g) + O_2(g) \rightarrow N_2O(g)$

3.52. **Chemistry of Geothermal Vents** Some scientists believe that life on Earth may have originated near deep-ocean vents. Balance the following reactions, which are among those taking place near such vents:
 a. $CH_3SH(aq) + CO(aq) \rightarrow CH_3COSCH_3(aq) + H_2S(aq)$
 b. $H_2S(aq) + CO(aq) \rightarrow CH_3CO_2H(aq) + S_8(s)$

**\*3.53.** Write a balanced chemical equation for each of the following reactions:
 a. Dinitrogen pentoxide reacts with sodium metal to produce sodium nitrate and nitrogen dioxide.
 b. A mixture of nitric acid and nitrous acid is formed when water reacts with dinitrogen tetroxide.
 c. At high pressure, nitrogen monoxide decomposes to dinitrogen monoxide and nitrogen dioxide.
 d. Acetylene, $C_2H_2$, burns and becomes carbon dioxide and water vapor.

3.54. Write a balanced chemical equation for each of the following reactions:
 a. Carbon dioxide reacts with carbon to form carbon monoxide.
 b. Potassium reacts with water to give potassium hydroxide and the element hydrogen.
 c. Phosphorus, $P_4$, burns in air to give diphosphorus pentoxide.
 d. Octane, $C_8H_{18}$, burns and becomes carbon dioxide and water vapor.

## Combustion Reactions; Stoichiometric Calculations and the Carbon Cycle

### CONCEPT REVIEW

**3.55.** If the sum of the masses of the reactants in a chemical equation equals the sum of the masses of the products, must the equation be balanced?

*3.56. There are two ways to write the equation for the combustion of ethane:

$$C_2H_6(g) + \tfrac{7}{2}O_2(g) \rightarrow 3\,H_2O(g) + 2\,CO_2(g)$$
$$2\,C_2H_6(g) + 7\,O_2(g) \rightarrow 6\,H_2O(g) + 4\,CO_2(g)$$

Do these two different ways of writing the equation affect the calculation of how much $CO_2$ is produced from a known quantity of $C_2H_6$?

### PROBLEMS

**3.57. Land Management** The United Nations Intergovernmental Panel on Climate Change reported in June 2000 that better management of cropland, grazing land, and forests would reduce the amount of carbon dioxide in the atmosphere by $5.4 \times 10^9$ kg of carbon per year.
   a. How many moles of carbon are present in $5.4 \times 10^9$ kg of carbon?
   b. How many kilograms of carbon dioxide does this quantity of carbon represent?

**3.58.** Energy generation results in the addition of an estimated $2.7 \times 10^{12}$ kg of $CO_2$ to the atmosphere each year. How many moles of $CO_2$ does this mass represent?

**3.59.** When $NaHCO_3$ is heated above 270°C, it decomposes to $Na_2CO_3(s)$, $H_2O(g)$, and $CO_2(g)$.
   a. Write a balanced chemical equation for the decomposition reaction.
   b. Calculate the mass of $CO_2$ produced from the decomposition of 25.0 g of $NaHCO_3$.

**3.60. Egyptian Cosmetics** Pb(OH)Cl, one of the lead compounds used in ancient Egyptian cosmetics, was prepared from PbO according to the following recipe:

$$PbO(s) + NaCl(aq) + H_2O(\ell) \rightarrow Pb(OH)Cl(s) + NaOH(aq)$$

How many grams of PbO and how many grams of NaCl would be required to produce 10.0 g of Pb(OH)Cl?

**3.61.** The manufacture of aluminum includes the production of cryolite ($Na_3AlF_6$) from the following reaction:

$$6\,HF(g) + 3\,NaAlO_2(s) \rightarrow Na_3AlF_6(s) + 3\,H_2O(\ell) + Al_2O_3(s)$$

How much $NaAlO_2$ (sodium aluminate) is required to produce 1.00 kg of $Na_3AlF_6$?

**3.62.** Chromium metal can be produced from the high-temperature reaction of $Cr_2O_3$ [chromium(III) oxide] with silicon or aluminum by each of the following reactions:

$$Cr_2O_3(s) + 2\,Al(\ell) \rightarrow 2\,Cr(\ell) + Al_2O_3(s)$$
$$2\,Cr_2O_3(s) + 3\,Si(\ell) \rightarrow 4\,Cr(\ell) + 3\,SiO_2(s)$$

   a. Calculate the number of grams of aluminum required to prepare 400.0 g of chromium metal by the first reaction.
   b. Calculate the number of grams of silicon required to prepare 400.0 g of chromium metal by the second reaction.

*3.63. Suppose 25 metric tons of coal that is 3.0% sulfur by mass is burned at an electric power plant (1 metric ton = $10^3$ kg). During combustion, the sulfur is converted into sulfur dioxide. How many tons of sulfur dioxide are produced?

*3.64. The uranium minerals found in nature must be refined and enriched in $^{235}U$ before the uranium can be used as a fuel in nuclear reactors. One procedure for enriching uranium relies on the reaction of $UO_2$ with HF to form $UF_4$, which is then converted into $UF_6$ by reaction with fluorine:

$$UO_2(g) + 4\,HF(aq) \rightarrow UF_4(g) + 2\,H_2O(\ell)$$
$$UF_4(g) + F_2(g) \rightarrow UF_6(g)$$

   a. How many kilograms of HF are needed to completely react with 5.00 kg of $UO_2$?
   b. How much $UF_6$ can be produced from 850.0 g of $UO_2$?

**3.65.** In Brazil automobiles use ethanol, $C_2H_6O$, as fuel whereas in the United States we rely on gasoline. Using $C_8H_{18}$ (octane) to represent gasoline, write balanced chemical equations for the complete combustion of ethanol and octane. Which fuel produces more $CO_2$ per gram of fuel?

**3.66.** Driving 1000 miles a month is not unusual for a short-distance commuter. If your vehicle gets 25 mpg, you would use 40 gallons ($\approx 150$ L) of gasoline every month. If gasoline is approximated as $C_8H_{18}$ ($d = 0.703$ g/mL), how much carbon dioxide does your vehicle emit every month? The *unbalanced* chemical equation for the reaction is

$$C_8H_{18}(\ell) + O_2(g) \rightarrow CO_2(g) + H_2O(g)$$

*3.67. Chalcopyrite ($CuFeS_2$) is an abundant copper mineral that can be converted into elemental copper. How much Cu could be produced from 1.00 kg of $CuFeS_2$?

*3.68. **Mining for Gold** Unlike most metals, gold is found in nature as the pure element. Miners in California in 1849 searched for gold nuggets and gold dust in streambeds, where the denser gold could be easily separated from sand and gravel. However, larger deposits of gold are found in veins of rock and can be separated chemically in a two-step process:

$$(1)\ 4\,Au(s) + 8\,NaCN(aq) + O_2(g) + 2\,H_2O(\ell) \rightarrow$$
$$4\,NaAu(CN)_2(aq) + 4\,NaOH(aq)$$
$$(2)\ 2\,NaAu(CN)_2(aq) + Zn(s) \rightarrow 2\,Au(s) + Na_2[Zn(CN)_4](aq)$$

If a $1.0 \times 10^3$ kg sample of rock is 0.019% gold by mass, how much Zn is needed to react with the gold extracted from the rock? Assume that reactions (1) and (2) are 100% efficient.

## Determining Empirical Formulas from Percent Composition; Empirical and Molecular Formulas Compared

### CONCEPT REVIEW

**3.69.** What is the difference between an empirical formula and a molecular formula?

**3.70.** Do the empirical and molecular formulas of a compound have the same percent composition values?

**3.71.** Is the element with the largest atomic mass always the element present in the highest percentage by mass in a compound?

**3.72.** Sometimes the composition of a compound is expressed as a mole percent or atom percent. Are the values of these parameters likely to be the same for a given compound, or different?

**3.73.** Among the naturally occurring hydrocarbons emitted by plants are three compounds named camphene, carene, and thujene. If all three of these compounds have the same percent composition and the same molar mass, do they have the same empirical formula? Do they have the same molecular formula?

*3.74. How might the compounds in Problem 3.73 differ from each other if they have the same molar mass and percent composition?

## PROBLEMS

**3.75.** Gasoline consists primarily of a mixture of the hydrocarbons $C_6H_{14}$, $C_7H_{16}$, $C_8H_{18}$, and $C_9H_{20}$. Do these four compounds have the same empirical formula?

3.76. The biosynthesis of carbohydrate includes the molecules shown below. Do all three compounds shown in Figure P3.76 have the same empirical formula?

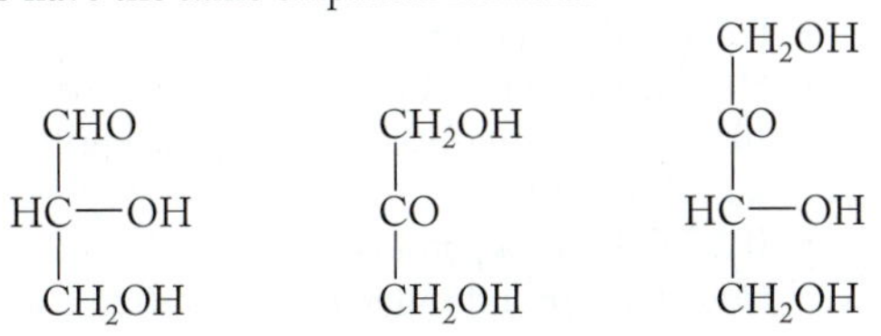

Glyceraldehyde    Dihydroxyacetone    Erythrulose

**FIGURE P3.76**

**3.77.** Calculate the percent composition of (a) $Na_2O$, (b) NaOH, (c) $NaHCO_3$, and (d) $Na_2CO_3$.

3.78. Calculate the percent composition of (a) sodium sulfate, (b) dinitrogen tetroxide, (c) strontium nitrate, and (d) aluminum sulfide.

**3.79. Organic Compounds in Space** The following compounds have been detected in space. Which of them contains the greatest percentage of carbon by mass? Do any two of the following compounds have the same empirical formula?
 a. naphthalene, $C_{10}H_8$
 b. chrysene, $C_{18}H_{12}$
 c. pentacene, $C_{22}H_{14}$
 d. pyrene, $C_{16}H_{10}$

3.80. Of the nitrogen oxides—$N_2O$, NO, $N_2O_3$, $N_2O_2$, $NO_2$, and $N_2O_4$—which are more than 50% oxygen by mass? Which, if any, have the same empirical formula?

**3.81.** Methane ($CH_4$) and tetrafluoromethane ($CF_4$) both contain 20% carbon per mole. Which one has the greater %C by mass?

3.82. Silane ($SiH_4$) is used in the electronics industry to manufacture thin films of silicon. What is the %Si by mass in $SiH_4$? Does silane have the same %Si by mass as disilane, $Si_2H_6$?

**3.83. Surgical Grade Titanium** Medical implants and high-quality jewelry items for body piercings are frequently made of a material known as G23Ti, or surgical-grade titanium. The percent composition of the material is 64.39% titanium, 24.19% aluminum, and 11.42% vanadium. What is the empirical formula for surgical-grade titanium?

3.84. A sample of an iron-containing compound is 22.0% iron, 50.2% oxygen, and 27.8% chlorine by mass. What is the empirical formula of this compound?

**3.85.** In an experiment, 2.43 g of magnesium reacts with 1.60 g of oxygen, forming 4.03 g of magnesium oxide.
 a. Use these data to calculate the empirical formula of magnesium oxide.
 b. Write a balanced chemical equation for this reaction.

3.86. Ferrophosphorus ($Fe_2P$) reacts with pyrite ($FeS_2$), producing iron(II) sulfide and a compound that is 27.87% P and 72.13% S by mass and has a molar mass of 444.56 g/mol.
 a. Determine the empirical and molecular formulas of this compound.
 b. Write a balanced chemical equation for this reaction.

**3.87. Asbestosis** Asbestosis is a lung disease caused by inhaling asbestos fibers. In addition, fiber from a form of asbestos called chrysotile is considered to be a human carcinogen by the U.S. Department of Health and Human Services. Chrysotile's composition is 26.31% magnesium, 20.20% silicon, and 1.45% hydrogen with the remainder of the mass as oxygen. Determine the empirical formula of chrysotile.

3.88. **Chemistry of Soot** A candle flame produces easily seen specks of soot near the edges of the flame, especially when the candle is moved. A piece of glass held over a candle flame will become coated with soot, which is the result of the incomplete combustion of candle wax. Elemental analysis of a compound extracted from a sample of this soot gave these results: 7.74% H and 92.26% C by mass. Calculate the empirical formula of the compound.

**3.89. Making the DNA Bases** Adenine (135.14 g/mol; 44.44% C, 3.73% H, and 51.84% N) was detected in mixtures of HCN, ammonia, and water under conditions that simulate early Earth. This observation suggests a possible origin for one of the bases found in DNA. What are the empirical and molecular formulas for adenine?

3.90. **Making Sugars for RNA** Ribose, the sugar found in RNA, has been detected in experiments designed to mimic the conditions of early Earth. If ribose contains 40.00% C, 6.71% H, and 53.28% O, with a molar mass of 150.13 g/mol, what are the empirical and molecular formulas for ribose?

## Combustion Analysis

### CONCEPT REVIEW

**3.91.** Explain why it is important for combustion analysis to be carried out in an excess of oxygen.

3.92. Why is the quantity of $CO_2$ obtained in a combustion analysis not a direct measure of the oxygen content of the starting compound?

**3.93.** Can the results of a combustion analysis ever give the true molecular formula of a compound?

3.94. What additional information is needed to determine a molecular formula from the results of an elemental analysis of an organic compound?

*3.95. If a compound containing nitrogen is subjected to combustion analysis in excess oxygen, what is the most likely molecular formula for the nitrogen-containing product?

*3.96. Suppose an insufficient amount of oxygen is used for a combustion analysis of a hydrocarbon, and some of the carbon is converted to CO rather than to $CO_2$. Will the empirical formula determined from the results of this analysis be too low or too high in carbon?

**PROBLEMS**

**3.97.** The combustion of 135.0 mg of a hydrocarbon produces 440.0 mg of $CO_2$ and 135.0 mg $H_2O$. The molar mass of the hydrocarbon is 270 g/mol. Determine the empirical and molecular formulas of this compound.

**3.98.** A 0.100 g sample of a compound containing C, H, and O is burned in oxygen, producing 0.1783 g of $CO_2$ and 0.0734 g of $H_2O$. Determine the empirical formula of the compound.

**3.99. GRAS List for Food Additives** The compound geraniol is on the GRAS (generally recognized as safe) list and can be used in foods and personal care products. By itself, geraniol smells like roses but it is frequently blended with other fragrances on the GRAS list and then added to products to produce a pleasant peach- or lemon-like aroma. In an analysis, the complete combustion of 175 mg of geraniol produced 499 mg $CO_2$ and 184 mg $H_2O$. What is the empirical formula for geraniol?

**\*3.100.** The combustion of 40.5 mg of a compound containing C, H, and O, and extracted from the bark of the sassafras tree, produces 110.0 mg of $CO_2$ and 22.5 mg of $H_2O$. The molar mass of the compound is 162 g/mol. Determine its empirical and molecular formulas.

**3.101. Ethnobotany** One of the ingredients in the Native American stomach-ache remedy derived from common chokecherry is caffeic acid. Combustion of 100 mg of caffeic acid yielded 220 mg $CO_2$ and 40.3 mg $H_2O$. Determine the empirical formula of caffeic acid.

**3.102.** Coniine, a substance isolated from poison hemlock, contains only carbon, hydrogen, and nitrogen. Combustion of 5.024 mg of coniine yields 13.90 mg $CO_2$ and 6.048 mg of $H_2O$. What is the empirical formula of coniine?

## Limiting Reactants and Percent Yield

**CONCEPT REVIEW**

**3.103.** If a reaction vessel contains equal masses of Fe and S, a mass of FeS corresponding to which of the following could theoretically be produced?
a. the sum of the masses of Fe and S
b. more than the sum of the masses of Fe and S
c. less than the sum of the masses of Fe and S

**3.104.** Can the percent yield of a chemical reaction ever exceed 100%?

**3.105.** Give two reasons why the actual yield from a chemical reaction is usually less than the theoretical yield.

**3.106.** A chemical reaction produces less than the expected amount of product. Is this result a violation of the law of conservation of mass?

**PROBLEMS**

**3.107. Making Hollandaise Sauce** A recipe for 1 cup of hollandaise sauce calls for $\frac{1}{2}$ cup of butter, $\frac{1}{4}$ cup of hot water, 4 egg yolks, and the juice of a medium-sized lemon. How many cups of this sauce can be made from a pound (2 cups) of butter, a dozen eggs, 4 medium lemons, and an unlimited supply of hot water?

**3.108.** A factory making toy wagons has 13,466 wheels, 3360 handles, and 2400 wagon beds in stock. What is the maximum number of wagons the factory can make?

**3.109.** Ammonia rapidly reacts with hydrogen chloride, making ammonium chloride. Write a balanced chemical equation for the reaction, and calculate the number of grams of excess reactant when 3.0 g of $NH_3$ reacts with 5.0 g of HCl.

**3.110.** Sulfur trioxide dissolves in water, producing $H_2SO_4$. How much sulfuric acid can be produced from 10.0 mL of water ($d = 1.00$ g/mL) and 25.6 g of $SO_3$?

**\*3.111.** Phosgenite, a lead compound with the formula $Pb_2Cl_2CO_3$, is found in ancient Egyptian cosmetics. Phosgenite was prepared by the reaction of PbO, NaCl, $H_2O$, and $CO_2$. An unbalanced equation of the reaction is

$$PbO(s) + NaCl(aq) + H_2O(\ell) + CO_2(g) \rightarrow$$
$$Pb_2Cl_2CO_3(s) + NaOH(aq)$$

a. Balance the equation.
b. How many grams of phosgenite can be obtained from 10.0 g of PbO and 10.0 g NaCl in the presence of excess water and $CO_2$?
c. If 2.72 grams of phosgenite are produced in the laboratory from the amounts of starting materials stated in (b), what is the percent yield of the reaction?

**3.112.** Potassium superoxide, $KO_2$, reacts with carbon dioxide to form potassium carbonate and oxygen:

$$4\,KO_2(s) + 2\,CO_2(g) \rightarrow 2\,K_2CO_3(s) + 3\,O_2(g)$$

This reaction makes potassium superoxide useful in a self-contained breathing apparatus. How much $O_2$ could be produced from 2.50 g of $KO_2$ and 4.50 g of $CO_2$?

**3.113.** The reaction of 3.0 g of carbon with excess $O_2$ yields 6.5 g of $CO_2$. What is the percent yield of this reaction?

**3.114.** Baking soda ($NaHCO_3$) can be made in large quantities by the following reaction:

$$NaCl(aq) + NH_3(aq) + CO_2(aq) + H_2O(\ell) \rightarrow$$
$$NaHCO_3(s) + NH_4Cl(aq)$$

If 10.0 g of NaCl reacts with excesses of the other reactants and 4.2 g of $NaHCO_3$ is isolated, what is the percent yield of the reaction?

**3.115. Chemistry of Fermentation** Yeast converts glucose ($C_6H_{12}O_6$) into ethanol ($d = 0.789$ g/mL) in a process called fermentation. An equation for the reaction can be written as follows:

$$C_6H_{12}O_6(aq) \rightarrow C_2H_5OH(\ell) + CO_2(g)$$

a. Write a balanced chemical equation for this fermentation reaction.
b. If 100.0 g of glucose yields 50.0 mL of ethanol, what is the percent yield for the reaction?

**\*3.116. Composition of Seawater** A 1-liter sample of seawater contains 19.4 g of $Cl^-$, 10.8 g of $Na^+$, and 1.29 g of $Mg^{2+}$.
a. How many moles of each ion are present?
b. If we evaporated the seawater, would there be enough $Cl^-$ present to form the chloride salts of all the sodium and magnesium present?

## Additional Problems

**\*3.117.** As a solution of copper sulfate slowly evaporates, beautiful blue crystals made of Cu(II) and sulfate ions form such that water molecules are trapped inside the crystals. The overall formula of the compound is $CuSO_4 \cdot 5H_2O$.
  a. What is the percent water in this compound?
  b. At high temperatures, the water in the compound is driven off as steam. What mass percentage of the original sample of the blue solid is lost as a result?

3.118. Aluminum is mined as the mineral bauxite, which consists primarily of $Al_2O_3$ (alumina).
  a. How much aluminum is produced from 1 metric ton (1 metric ton = $10^3$ kg) of $Al_2O_3$?

$$2\, Al_2O_3(s) \rightarrow 4\, Al(s) + 3\, O_2(g)$$

  b. The oxygen produced in part a is allowed to react with carbon to produce carbon monoxide:

$$O_2(g) + 2\, C(s) \rightarrow 2\, CO(g)$$

  Balance the following equation describing the reaction of alumina with carbon:

$$Al_2O_3(s) + C(s) \rightarrow Al(s) + CO(g)$$

  c. How much CO is produced from the $O_2$ made in part a?

**\*3.119. Chemistry of Copper Production** "Native," or elemental, copper can be found in nature, but most copper is mined as oxide or sulfide minerals. Chalcopyrite ($CuFeS_2$) is one copper mineral that can be converted to elemental copper in a series of chemical steps. Reacting chalcopyrite with oxygen at high temperature produces a mixture of copper sulfide and iron oxide. The iron oxide is separated from CuS by reaction with sand. CuS is converted to $Cu_2S$ in the process, and the $Cu_2S$ is burned in air to produce Cu and $SO_2$:
  (1) $2\, CuFeS_2(s) + 3\, O_2(g) \rightarrow 2\, CuS(s) + 2\, FeO(s) + 2\, SO_2(g)$
  (2) $FeO(s) + SiO_2(s) \rightarrow FeSiO_3(s)$
  (3) $2\, CuS(s) \rightarrow Cu_2S(s) + \frac{1}{8} S_8(s)$
  (4) $Cu_2S(s) + O_2(g) \rightarrow 2\, Cu(s) + SO_2(g)$
  An average copper penny minted in the 1960s has a mass of about 3.0 g.
  a. How much chalcopyrite had to be mined to produce one dollar's worth of pennies?
  b. How much chalcopyrite had to be mined to produce one dollar's worth of pennies if reaction 1 above had a percent yield of 85% and reactions 2, 3, and 4 had percent yields of essentially 100%?
  c. How much chalcopyrite had to be mined to produce one dollar's worth of pennies if each reaction involving copper proceeded with an 85% yield?

**\*3.120. Mining for Gold** Gold can be extracted from the surrounding rock using a solution of sodium cyanide. While effective for isolating gold, toxic cyanide finds its way into watersheds, causing environmental damage and harming human health.

$$4\, Au(s) + 8\, NaCN(aq) + O_2(g) + 2\, H_2O(\ell) \rightarrow$$
$$4\, NaAu(CN)_2(aq) + 4\, NaOH(aq)$$

$$2\, NaAu(CN)_2(aq) + Zn(s) \rightarrow 2\, Au(s) + Na_2[Zn(CN)_4](aq)$$

  a. If a sample of rock contains 0.009% gold by mass, how much NaCN is needed to extract the gold from 1 metric ton (1 metric ton = $10^3$ kg) of rock as $NaAu(CN)_2$?
  b. How much zinc is needed to convert the $NaAu(CN)_2$ from part a to metallic gold?
  c. The gold recovered in part b is manufactured into a gold ingot in the shape of a cube. The density of gold is 19.3 g/cm$^3$. How big is the cube of gold?

**\*3.121.** Uranium oxides used in the preparation of fuel for nuclear reactors are separated from other metals in minerals by converting the uranium to $UO_x(NO_3)_y(H_2O)_z$, where uranium has a positive charge ranging from 3+ to 6+.
  a. Roasting $UO_x(NO_3)_y(H_2O)_z$ at 400°C leads to loss of water and decomposition of the nitrate ion to nitrogen oxides, leaving behind a product with the formula $U_aO_b$ that is 83.22% U by mass. What are the values of $a$ and $b$? What is the charge on U in $U_aO_b$?
  b. Higher temperatures produce a different uranium oxide, $U_cO_d$, with a higher uranium content, 84.8% U. What are the values of $c$ and $d$? What is the charge on U in $U_cO_b$?
  c. The values of $x$, $y$, and $z$ in $UO_x(NO_3)_y(H_2O)_z$ are found by gently heating the compound to remove all of the water. In a laboratory experiment, 1.328 g of $UO_x(NO_3)_y(H_2O)_z$ produced 1.042 g of $UO_x(NO_3)_y$. Continued heating generated 0.742 g of $U_nO_m$. Using the information in parts a and b, calculate $x$, $y$, and $z$.

**\*3.122.** Large quantities of fertilizer are washed into the Mississippi River from agricultural land in the Midwest. The excess nutrients collect in the Gulf of Mexico, promoting the growth of algae and endangering other aquatic life.
  a. One commonly used fertilizer is ammonium nitrate. What is the chemical formula of ammonium nitrate?
  b. Corn farmers typically use $5.0 \times 10^3$ kg of ammonium nitrate per square kilometer of cornfield per year. Ammonium nitrate can be prepared by the following reaction:

$$NH_3(aq) + HNO_3(aq) \rightarrow NH_4NO_3(aq)$$

  How much nitric acid would be required to make the fertilizer needed for 1 km$^2$ of cornfield per year?
  c. The ammonium ions can be converted into $NO_3^-$ by bacterial action.

$$NH_4^+(aq) + 2\, O_2(g) \rightarrow NO_3^-(aq) + H_2O(\ell) + 2\, H^-(aq)$$

  If 10% of the ammonium component of $5.0 \times 10^2$ kg of fertilizer ends up as nitrate, how much oxygen would be consumed?

3.123. **Composition of Over-the-Counter Medicines** Calculate the number of molecules or formula units of compound in each of the following common, over-the-counter medications:
  a. ibuprofen, a pain reliever and fever reducer that contains 200.0 mg of the active ingredient, $C_{13}H_{18}O_2$
  b. an antacid containing 500.0 mg of calcium carbonate
  c. an allergy tablet containing 4 mg chlorpheniramine ($C_{16}H_{19}N_2Cl$)

3.124. **Chemistry of Pain Relievers** The common pain relievers aspirin ($C_9H_8O_4$), acetaminophen ($C_8H_9NO_2$), and naproxen sodium ($C_{14}H_{13}O_3Na$) are all available in tablets containing 200.0 mg of the active ingredient. Which compound contains the greatest number of molecules per tablet? How many molecules of the active ingredient are present in each tablet?

**3.125.** Some catalytic converters in automobiles contain the manganese oxides $Mn_2O_3$ and $MnO_2$.
   a. Give the names of $Mn_2O_3$ and $MnO_2$.
   b. Calculate the percent manganese by mass in $Mn_2O_3$ and $MnO_2$.
   c. Explain how $Mn_2O_3$ and $MnO_2$ are consistent with the law of multiple proportions.

*\*3.126.* A number of chemical reactions have been proposed for the formation of organic compounds from inorganic precursors. Here is one of them:

$$H_2S(g) + FeS(s) + CO_2(g) \rightarrow FeS_2(s) + HCO_2H(\ell)$$

   a. Identify the ions in FeS and $FeS_2$. Give correct names for each compound.
   *\*b.* How much $HCO_2H$ is obtained by reacting 1.00 g of FeS, 0.50 g of $H_2S$, and 0.50 g of $CO_2$ if the reaction results in a 50.0% yield?

*\*3.127.* The formation of organic compounds by the reaction of iron(II) sulfide with carbonic acid is described by the following chemical equation:

$$2\,FeS + H_2CO_3 \rightarrow 2\,FeO + 1/n\,(CH_2O)_n + 2\,S$$

   a. How much FeO is produced starting with 1.50 g FeS and 0.525 mol of $H_2CO_3$ if the reaction results in a 78.5% yield?
   *\*b.* If the carbon-containing product has a molar mass of $3.00 \times 10^2$ g/mol, what is the chemical formula of the product?

*\*3.128.* **Marine Chemistry of Iron** On the seafloor, iron(II) oxide reacts with water to form $Fe_3O_4$ and hydrogen in a process called serpentization.
   a. Balance the following equation for serpentization:

$$FeO(s) + H_2O(\ell) \rightarrow Fe_3O_4(s) + H_2(g)$$

   b. When $CO_2$ is present, the product is methane, not hydrogen. Balance the following chemical equation:

$$FeO(s) + H_2O(\ell) + CO_2(g) \rightarrow Fe_3O_4(s) + CH_4(g)$$

**3.129.** The solar wind is made up of ions, mostly protons, flowing out from the sun at about 400 km/s. Near Earth, each cubic kilometer of interplanetary space contains, on average, $6 \times 10^{15}$ solar-wind ions. How many moles of ions are in a cubic kilometer of near-Earth space?

**3.130.** The famous Hope Diamond at the Smithsonian National Museum of Natural History has a mass of 45.52 carats (Figure P3.130). Diamond is a crystalline form of carbon.
   a. How many moles of carbon are in the Hope Diamond (1 carat = 200.0 mg)?
   b. How many carbon atoms are in the diamond?

**FIGURE P3.130**

*\*3.131.* E-85 is an alternative fuel for automobiles and light trucks that consists of 85% (by volume) ethanol, $C_2H_5OH$, and 15% gasoline. The density of ethanol is 0.79 g/mL. How many moles of ethanol are in a gallon of E-85?

*\*3.132.* A 100.00 g sample of white powder A is heated to 550°C. At that temperature the powder decomposes, giving off colorless gas B, which is denser than air and is neither flammable nor does it support combustion. The products also include 56 g of a second white powder C. When gas B is bubbled through a solution of calcium hydroxide, substance A re-forms. What are the identities of substances A, B, and C?

*\*3.133.* You are given a 0.6240 g sample of a substance with the generic formula $MCl_2(H_2O)_2$. After completely drying the sample (which means removing the 2 mol of $H_2O$ per mole of $MCl_2$), the sample has a mass of 0.5471 g. What is the identity of element M?

**3.134.** A compound found in crude oil consists of 93.71% C and 6.29% H by mass. The molar mass of the compound is 128 g/mol. What is its molecular formula?

**3.135.** A reaction vessel for synthesizing ammonia by reacting nitrogen and hydrogen is charged with 6.04 kg of $H_2$ and excess $N_2$. A total of 28.0 kg of $NH_3$ is produced. What is the percent yield of the reaction?

**3.136.** If a cube of table sugar, which is made of sucrose, $C_{12}H_{22}O_{11}$, is added to concentrated sulfuric acid, the acid "dehydrates" the sugar, removing the hydrogen and oxygen from it and leaving behind a lump of carbon. What percentage of the initial mass of sugar is carbon?

*\*3.137.* A power plant burns $1.0 \times 10^2$ metric tons of coal that contains 3.0% (by mass) sulfur (1 metric ton = $10^3$ kg). The sulfur is converted to $SO_2$ during combustion.
   a. How many metric tons of $SO_2$ are produced?
   b. When $SO_2$ escapes into the atmosphere it may combine with $O_2$ and $H_2O$, forming sulfuric acid, $H_2SO_4$. Write a balanced chemical equation describing this reaction.
   c. How many metric tons of sulfuric acid, a component of acid rain, could be produced from the quantity of $SO_2$ calculated in part a?

**3.138.** **Reducing SO₂ Emissions** With respect to the previous question, one way to reduce the formation of acid rain involves trapping the $SO_2$ by passing smokestack gases through a spray of calcium oxide and $O_2$. The product of this reaction is calcium sulfate.
   a. Write a balanced chemical equation describing this reaction.
   b. How many metric tons of calcium sulfate would be produced from each ton of $SO_2$ that is trapped?

**3.139.** In the early 20th century, Londoners suffered from severe air pollution caused by burning high-sulfur coal. The sulfur dioxide that was emitted into the air mixed with London fog, forming sulfuric acid. For every gram of sulfur that was burned, how many grams of sulfuric acid could have formed?

*\*3.140.* **Gas Grill Reaction** The burner in a gas grill mixes 24 volumes of air for every one volume of propane ($C_3H_8$) fuel. Like all gases, the volume that propane occupies is directly proportional to the number of moles of it at a given temperature and pressure. Air is 21% (by volume) $O_2$. Is the flame produced by the burner fuel-rich (excess propane in

the reaction mixture), fuel-lean (not enough propane), or stoichiometric (just right)?

**3.141.** A common mineral in Earth's crust has the chemical composition 34.55% Mg, 19.96% Si, and 45.49% O. What is its empirical formula?

*__3.142.__ **Ozone Generators** Some indoor air-purification systems work by converting a little of the oxygen in the air to ozone, which oxidizes mold and mildew spores and other biological air pollutants. The chemical equation for the ozone generation reaction is

$$3\ O_2(g) \rightarrow 2\ O_3(g)$$

It is claimed that one such system generates 4.0 g of $O_3$ per hour from dry air passing through the purifier at a flow of 5.0 L/min. If 1 liter of indoor air contains 0.28 g of $O_2$,
   a. what fraction of the molecules of $O_2$ is converted to $O_3$ by the air purifier?
   b. what is the percent yield of the ozone generation reaction?

*__3.143.__ In the first episode of George Lucas' Star Wars series, Qui-Gon Jinn and Obi-Wan Kenobi can only visit the underwater world of the Gungans by using A99 Aquata Breathers, which allow them to survive underwater for up to two hours. While the tiny devices may be from the far-fetched world of science fiction, current technology exists for transforming carbon dioxide to oxygen. These self-contained re-breathers are used by a select group of underwater cave explorers and can act as self-rescue devices. The chemistry is based on the following chemical reactions, using either potassium superoxide or sodium peroxide:

$$4\ KO_2(s) + 2\ CO_2(g) \rightarrow 2\ K_2CO_3(s) + 3\ O_2(g)$$
$$2\ Na_2O_2(s) + 2\ CO_2(g) \rightarrow 2\ Na_2CO_3(s) + O_2(g)$$

   a. The respiratory rate at rest for an average, healthy adult is 12 breaths per minute. If the average breath takes in 0.500 L of $O_2$ ($d = 1.429$ g/L) into the lungs, how many grams of $KO_2$ are needed to produce enough of oxygen for two hours underwater?
   b. Would you need more or less $Na_2O_2$ to produce an equivalent amount of oxygen?
   c. Given the densities of $KO_2$ (2.14 g/mL), and $Na_2O_2$ (2.805 g/mL), which solid material would occupy less volume in a re-breather device?

**3.144.** Socks containing silver nanoparticles embedded in the fabric are currently marketed as an antidote to smelly socks. Silver is known to have antimicrobial properties, and silver ions are toxic to aquatic life. A study at Arizona State University found that much of the silver particles are lost upon laundering the socks in mild acid.
   a. Each sock in the study began with 1360 µg of silver. How many moles of silver are contained in each sock?
   b. As much as 650 µg of silver was lost after four washings. What percent of the silver was lost?

If your instructor assigns problems in **smartwork**, log in at **smartwork.wwnorton.com**.

# 4

# Solution Chemistry: The Hydrosphere

## Oceans and Solutions

Life on Earth may have begun in the sea. According to one theory, life originated deep beneath the surface of the oceans in hydrothermal vents, where hot gases mixed with water, forming mineral-rich deposits. Others have argued that only shallow ponds, not the marine environment, provide the appropriate concentrations of ions for cells to arise. Whether either theory is correct or not, water is essential for the chemistry of all biological systems because it provides the medium in which the transport of molecules and ions can occur during chemical reactions. Because water is necessary for all life, scientists look for evidence of liquid water wherever they search for life, whether it is somewhere on Earth, somewhere else in our solar system, or beyond.

Since their original formation, Earth's oceans have contained essentially the same volume of seawater, and their chemical content has changed very little over time. How can this be, considering the continuous influx of soluble and suspended material from rivers and runoff from rain? The answer is that dissolved compounds and suspended matter in the sea eventually are incorporated into the mud on the ocean floor, balancing the material entering the oceans. Over time, this material becomes sedimentary rock, which forms from the mud, and is then exposed to Earth's surface during shifts in Earth's crust. The exposed rock is then subject to weathering, and the cycle continues.

The fluids within living cells are saline solutions, although they are not as salty as seawater. The concentration of dissolved matter within our cells also stays remarkably constant in healthy individuals. Cells contain sodium, potassium, chloride, and hydrogen carbonate ions, and when blood tests are done to assess a patient's well-being, the concentrations of these ions are a crucial part of the picture. A shift in their concentrations indicates problems, ranging from kidney disease to drug abuse.

Many other solutions are part of daily life. Vinegar in salad dressing is a solution of acetic acid in water. People with contact lenses probably wash them in a solution containing sodium chloride and cleaning agents. The liquid in an automobile battery is a solution of

**Earth, Water, and Life** Fish, like these salmon, that migrate between seawater and freshwater are exposed to water containing very different concentrations of dissolved substances. ▶

4.1   Solutions on Earth and Other Places

4.2   Concentration Units

4.3   Dilutions

4.4   Electrolytes and Nonelectrolytes

4.5   Acid–Base Reactions: Proton Transfer

4.6   Titrations

4.7   Precipitation Reactions

4.8   Ion Exchange

4.9   Oxidation–Reduction Reactions: Electron Transfer

## Learning Outcomes

**LO1** Express the concentrations of solutions in different units and convert from one set of units to another
**Sample Exercise 4.1**

**LO2** Calculate the molar concentration (molarity) of a solution, the mass of a solute, or the volume of a stock solution required to make a solution of specified molarity
**Sample Exercises 4.2, 4.3, 4.4, 4.5**

**LO3** Calculate the concentration of a solute from stoichiometry and titration data, and from absorbance data and Beer's Law
**Sample Exercises 4.6, 4.9, 4.10**

**LO4** Identify strong electrolytes, weak electrolytes, and nonelectrolytes from conductivity experiments
**Sample Exercise 4.7**

**LO5** Write molecular, overall ionic, and net ionic equations for reactions
**Sample Exercises 4.8, 4.11, 4.12**

**LO6** Predict precipitation reactions using solubility rules, quantify results from precipitation titrations, and explain ion exchange
**Sample Exercises 4.12, 4.13, 4.14, 4.15**

**LO7** Identify redox reactions, oxidizing agents, and reducing agents, and balance redox reactions
**Sample Exercises 4.16, 4.17, 4.18, 4.19, 4.20, 4.21**

(a)

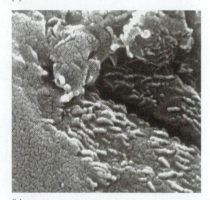

(b)

(c)

**FIGURE 4.1** (a) Meteorite ALH84001, thought to have originated from Mars, contains microscopic features (b) that some scientists believe to be fossilized bacteria. (c) An Earth-grown colony of the bacterium *Escherichia coli*, for comparison.

sulfuric acid in water. Most beverages are solutions—even bottled water contains dissolved ions that give it a refreshing flavor.

What kinds of chemical reactions take place in aqueous solution? In this chapter we focus on four major types of reactions: acid–base, precipitation, ion exchange, and electron transfer reactions. Whether dealing with seawater, freshwater, the fluids within cells, or the myriad solutions we come in contact with in daily life, issues of determining the identity and amount of materials dissolved in aqueous solutions are important in countless situations. Thus, we will also discuss ways of expressing and measuring the concentration of solutions in important environmental, medical, and commercial applications. ■

# 4.1 Solutions on Earth and Other Places

Earth is widely referred to as the "water planet." Life exists here because liquid water is abundant: depressions in Earth's crust contain about $1.5 \times 10^{21}$ L of liquid $H_2O$. Debate over the existence of life elsewhere in the universe hinges on the prospect of liquid water existing on other planets. The controversial belief that Martian meteorites collected in Antarctica contain fossilized life-forms (Figure 4.1) is consistent with evidence that there may have been water on the surface of Mars in the distant past. The Grand Canyon in the southwestern United States was formed by the action of water on rocks, and similar rock formations have been observed on Mars. A series of images from the *Curiosity* rover indicates that water may have flowed on the surface of Mars in the past (Figure 4.2). These and other observations support the idea that liquid water may now exist just under the surface of Mars, and where there is water, there could be life.

Biochemical reactions in all living cells, from single-celled organisms to human beings, require the presence of liquid water, and understanding the chemistry of the biosphere—just like the hydrosphere—requires understanding the principles

**FIGURE 4.2** (a) Rocks on the Martian surface are similar in size and shape to (b) rocks on Earth in an area where a fast-moving stream of water once flowed.

(a)

(b)

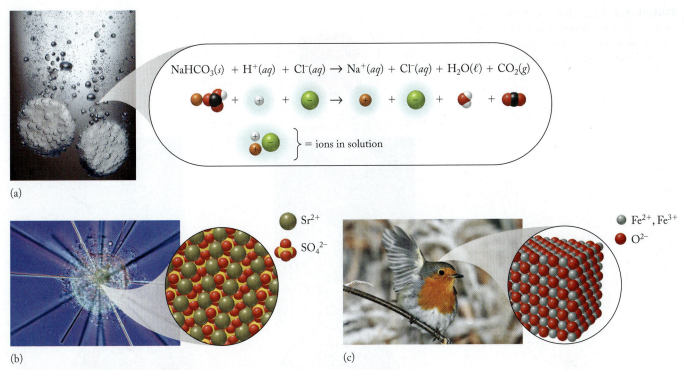

$$NaHCO_3(s) + H^+(aq) + Cl^-(aq) \rightarrow Na^+(aq) + Cl^-(aq) + H_2O(\ell) + CO_2(g)$$

= ions in solution

(a)

(b)

$Sr^{2+}$
$SO_4^{2-}$

(c)

$Fe^{2+}, Fe^{3+}$
$O^{2-}$

**FIGURE 4.3** (a) The reaction between a base (sodium bicarbonate) and stomach acid generates bubbles of carbon dioxide that provide relief for an upset stomach. (b) *Acantharia*, a single-cell species of plankton, has an exoskeleton formed by precipitation of strontium sulfate. (c) The European robin migrates thousands of kilometers, guided by magnetite in its head.

of chemical reactions between substances dissolved in water. For example, when we take an antacid tablet to settle an upset stomach, we are taking advantage of the reaction between acids and bases (Figure 4.3a). The exoskeletons of the microorganism *Acantharia* are composed of single crystals of strontium sulfate, produced by the precipitation of insoluble $SrSO_4$ from solution (Figure 4.3b). Finally, magnetite ($Fe_3O_4$), formed by cellular electron transfer reactions of dissolved iron ions, appears to play a role in the ability of migratory birds and fish to find their way over long distances (Figure 4.3c).

All natural waters, whether saltwater or fresh, contain ionic and molecular compounds dissolved in them. When one element or compound dissolves in another, a solution forms. As defined in Section 1.1, a *solution* is a homogeneous mixture of two or more substances (Figure 4.4). The substance present in the greatest number of moles is called the **solvent**, and all the other substances in the solution are called **solutes**. When the solvent is water, the solution is an *aqueous solution*, and in this chapter you may assume all of the solutions we discuss are aqueous. In the most general case, the solvent does not even have to be a liquid. Many materials in Earth's crust, for example, are *solid solutions*, which are uniform mixtures of solid substances with variable composition.

**CONCEPT TEST**

Which, if any, of the following are solutions? (a) muddy river water; (b) helium gas; (c) clear cough syrup; (d) filtered dry air

*(Answers to Concept Tests are in the back of the book.)*

**solvent** the most abundant component of a solution in terms of moles.

**solute** any component in a solution other than the solvent. A solution may contain one or more solutes.

**FIGURE 4.4** Adding table sugar (the solute), represented by red spheres, to water (the solvent) produces a solution of sugar molecules evenly distributed among water molecules.

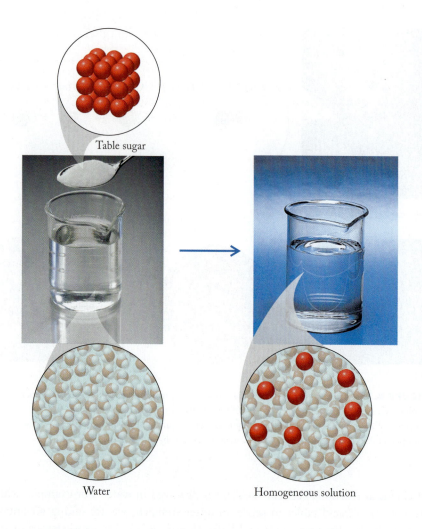

Table sugar

Water                    Homogeneous solution

# 4.2 Concentration Units

The **concentration** of any solution, or the amount of solute in a given amount of solvent or solution, can be expressed in different ways. Qualitatively, a solution can be described as being *concentrated* if it contains a much larger ratio of solute to solvent than does a *dilute* solution. A dilute solution contains a very small amount of solute compared to the amount of solvent. These are relative terms—what constitutes a large amount of solute versus a small amount?—but they are used constantly to compare solutions containing different quantities of solutes.

Many quantitative measures of concentration are based on mass-to-mass ratios, such as milligrams of solute per kilogram of solution. Others are based on mass-to-volume ratios, such as milligrams of solute per liter of solution. For example, when environmental regulatory agencies establish limits[1] on the concentrations of contaminants permitted in drinking water, the limits may be expressed as milligrams of contaminant (solute) per liter of drinking water (solution). When these agencies express the maximum recommended human exposure to these contaminants, a common unit is milligrams of contaminant per kilogram of body mass.

**concentration** the amount of a solute in a particular amount of solvent or solution.

[1]The U.S. Environmental Protection Agency (EPA) sets contaminant concentration limits called *maximum contaminant levels* (MCLs) for air and water, including drinking water. Similar limits have been established by Environment Canada, the World Health Organization, and other environmental agencies.

Two convenient units for expressing very small concentrations are *parts per million* (ppm) and *parts per billion* (ppb). A solution with a concentration of 1 ppm contains 1 part solute for every million parts of solution, which means 1 g of solute for every million ($10^6$) grams of solution. Put another way, each gram of a 1 ppm solution contains one-millionth of a gram ($10^{-6}$ g = 1 µg) of solute. A solution of 1 ppm is also the same as 1 mg of solute per kilogram of solution:

$$1 \text{ ppm} = \frac{1 \text{ µg solute}}{1 \text{ g solution}} \times \frac{1 \text{ mg solute}}{10^3 \text{ µg solute}} \times \frac{10^3 \text{ g solution}}{1 \text{ kg solution}} = \frac{1 \text{ mg solute}}{1 \text{ kg solution}}$$

A concentration of 1 ppb is 1/1000 as concentrated as 1 ppm. Because 1 ppm is the same as 1 mg of solute per kilogram of solution, 1 ppb is equivalent to 1 µg of solute per kilogram of solution:

$$1 \text{ ppb} = \frac{1 \text{ ppm}}{1000} = \frac{0.001 \text{ µg solute}}{1 \text{ g solution}} \times \frac{10^3 \text{ g solution}}{1 \text{ kg solution}} = \frac{1 \text{ µg solute}}{1 \text{ kg solution}}$$

Table 4.1 lists the major ions in seawater and their average concentrations, using three different units. Let's express the smallest concentration in the table (0.00130 g $F^-$/kg seawater) in parts per million. To do so, we convert grams of $F^-$ into milligrams of $F^-$ because 1 mg of solute per kilogram of solution is the same as 1 ppm:

$$\frac{0.00130 \text{ g } F^-}{1 \text{ kg seawater}} \times \frac{1 \text{ mg } F^-}{0.001 \text{ g } F^-} = \frac{1.30 \text{ mg } F^-}{1 \text{ kg seawater}} = 1.30 \text{ ppm } F^-$$

As you can see, parts per million is a particularly convenient unit for expressing the concentration of $F^-$ ion in seawater because it avoids the use of exponents or lots of zeroes to set the decimal place.

| TABLE 4.1 | Average Concentrations of the 11 Major Ions in Seawater | | |
|---|---|---|---|
| Constituent | g/kg[a] | mmol/kg[b] | mmol/L[c] |
| $Na^+$ | 10.781 | 468.96 | 480.57 |
| $K^+$ | 0.399 | 10.21 | 10.46 |
| $Mg^{2+}$ | 1.284 | 52.83 | 54.14 |
| $Ca^{2+}$ | 0.4119 | 10.28 | 10.53 |
| $Sr^{2+}$ | 0.00794 | 0.0906 | 0.0928 |
| $Cl^-$ | 19.353 | 545.88 | 559.40 |
| $SO_4^{2-}$ | 2.712 | 28.23 | 28.93 |
| $HCO_3^-$ | 0.126 | 2.06 | 2.11 |
| $Br^-$ | 0.0673 | 0.844 | 0.865 |
| $B(OH)_3$ | 0.0257 | 0.416 | 0.426 |
| $F^-$ | 0.00130 | 0.068 | 0.070 |
| Total | 35.169 | 1119.87 | 1147.59 |

[a]g/kg = grams of solute per kilogram of solution.
[b]mmol/kg = millimoles of solute per kilogram of solution.
[c]mmol/L = millimoles of solute per liter of solution.

**SAMPLE EXERCISE 4.1**  **Comparing Ion Concentrations in Aqueous Solutions**  LO1

The average concentration of chloride ion in seawater is 19.353 g Cl⁻/kg solution. The World Health Organization recommends that the concentration of Cl⁻ ions in drinking water not exceed 250 ppm ($2.50 \times 10^2$ ppm). How many times as much chloride ion is there in seawater than in the maximum concentration allowed in drinking water?

**Collect and Organize** We are given the concentrations of Cl⁻ ions in two different units, g Cl⁻/kg and ppm, and asked to determine the ratio of Cl⁻ ions in seawater to the upper limit of Cl⁻ ions allowed in drinking water.

**Analyze** We know how to create conversion factors from equalities. We can convert the seawater concentration to milligrams of Cl⁻ per kilogram of seawater using the conversion factor $10^3$ mg/1 g and then use the fact that 1 mg solute/kg solution = 1 ppm. We expect the value in mg/kg to be much larger than the value given in g/kg.

$$\boxed{\frac{\text{g}}{\text{kg}}\,\text{Cl}^-} \xrightarrow{\frac{10^3\ \text{mg}}{\text{g}}} \boxed{\frac{\text{mg}}{\text{kg}}\,\text{Cl}^-} = \boxed{\text{ppm Cl}^-}$$

**Solve**

$$19.353\frac{\text{g Cl}^-}{\text{kg seawater}} \times \frac{10^3\ \text{mg}}{\text{g}} = 19{,}353\frac{\text{mg Cl}^-}{\text{kg seawater}} = 19{,}353\ \text{ppm Cl}^-$$

Next we take the ratio of the two concentrations:

$$\frac{19{,}353\ \text{ppm Cl}^-\ \text{in seawater}}{250\ \text{ppm Cl}^-\ \text{in drinking water}} = 77.4$$

The concentration of Cl⁻ ions in seawater is 77.4 times greater than would be acceptable in drinking water.

**Think About It** Common sense says that seawater is much saltier than drinking water. Drinking seawater induces nausea and vomiting, and may even cause death. We could also convert the drinking water concentration to grams of Cl⁻ per kilogram of drinking water and compare that number with 19.353 g Cl⁻/kg seawater. It does not matter which units you choose to make the comparison as long as the units of the two numbers are the same.

⚙ **Practice Exercise** The World Health Organization (WHO) drinking water standard for arsenic is 10.0 µg/L. Water from some wells in Bangladesh was found to contain as much as 1.2 mg of arsenic per liter of water. How many times above the WHO standard is this level? Assume that all the samples have a density of 1 g/mL.

*(Answers to Practice Exercises are in the back of the book.)*

◉◉ **CONNECTION** In Chapter 1 we saw an example of a solar still used in survival gear to generate freshwater from saltwater.

▶❙❙ **CHEMTOUR** Molarity

Most scientists interested in studying chemical processes in natural waters or laboratory solutions prefer to work with concentrations based on moles of solute rather than mass. For these scientists, the preferred concentration unit is moles of solute per liter of solution. This ratio is called **molarity (M):**[2]

$$\text{molarity } (M) = \frac{\text{moles } (n) \text{ of solute}}{\text{volume } (V) \text{ of solution in liters}}$$

or,

$$M = \frac{n}{V} \tag{4.1}$$

If we know the volume and molarity for any solution, we can readily calculate the mass of solute in the solution. First we rearrange Equation 4.1 to

$$n = V \times M$$

**molarity (M)** the concentration of a solution expressed in moles of solute per liter of solution ($M = n/V$).

[2]Note that molarity is symbolized by $M$ and molar mass is $\mathcal{M}$ throughout this text.

Next, we convert the number of moles of solute, $n$, into mass in grams by multiplying by the molar mass ($\mathcal{M}$):

$$\text{Mass of solute (g)} = \text{moles of solute} \times \mathcal{M}$$

$$\text{g} = \cancel{\text{mol}} \times \frac{\text{g}}{\cancel{\text{mol}}}$$

Substituting ($V \times M$) for moles of solute:

$$m_{\text{solute}} = (V \times M) \times \mathcal{M} \tag{4.2}$$

Note how the units cancel out in Equation 4.2:

$$\text{g} = \left(\cancel{\text{L}} \times \frac{\cancel{\text{mol}}}{\cancel{\text{L}}}\right) \times \frac{\text{g}}{\cancel{\text{mol}}}$$

Equation 4.2 is useful when we want to calculate the mass of a solute needed to prepare a solution of a desired volume and molarity, or when we have a solution of known molarity and want to know the mass of solute in a given volume of the solution.

In many environmental and biological systems, solute concentrations are often much less than 1.0 $M$. In Table 4.1, for instance, the concentration units in column 4 are not moles per liter but *millimoles* per liter, which means we are describing concentration in terms of *millimolarity* (m$M$; 1 m$M$ = $10^{-3}$ $M$) rather than molarity. On an even smaller scale, the concentrations of minor and trace elements in seawater are often expressed in terms of *micromolarity* (μ$M$; 1 μ$M$ = $10^{-6}$ $M$), *nanomolarity* (n$M$; 1 n$M$ = $10^{-9}$ $M$), and even *picomolarity* (p$M$; 1 p$M$ = $10^{-12}$ $M$). The concentration ranges of many biologically active substances in blood, urine, and other biological liquids are also so small that they are often expressed in units such as these.

Column 3 in Table 4.1 shows concentrations as millimoles of solute per kilogram of seawater. Oceanographers prefer this unit to one based on solution volume because the volume of a given mass of water varies with changing temperature and pressure, whereas its mass remains constant.

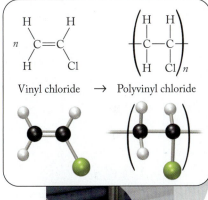

Vinyl chloride → Polyvinyl chloride

**CONCEPT TEST**

Which of the following aqueous sodium chloride solutions is the least concentrated?
(a) 0.0053 $M$ NaCl; (b) 54 m$M$ NaCl; (c) 550 μ$M$ NaCl; (d) 56,000 n$M$ NaCl

*(Answers to Concept Tests are in the back of the book.)*

**SAMPLE EXERCISE 4.2** **Calculating Molarity from Mass and Volume** **LO2**

Pipes made of polyvinyl chloride (PVC; Figure 4.5) are widely used in homes and office buildings, though their use is restricted because vinyl chloride (VC) may leach from them. The maximum concentration of vinyl chloride ($CH_2CHCl$) allowed in drinking water in the United States is 0.002 mg $CH_2CHCl$/L solution. What molarity is this?

**Collect and Organize** We are given a concentration in milligrams of solute per liter of solution and asked to convert it to units of molarity (moles of solute per liter of solution).

**Analyze** The molar mass of a substance relates the mass and number of moles of a given quantity of the substance. Because the solute mass is given in milligrams, we need to convert this mass first to grams and then to moles. Because the number of grams of vinyl chloride is very small, we can expect that the molarity will be smaller still.

$$\boxed{\frac{\text{mg VC}}{\text{L}}} \xrightarrow{\frac{\text{g}}{10^3 \text{ mg}}} \boxed{\frac{\text{g VC}}{\text{L}}} \xrightarrow{\frac{1}{\text{molar mass}}} \boxed{\frac{\text{mol VC}}{\text{L}}}$$

**FIGURE 4.5** Vinyl chloride, a suspected carcinogen, enters drinking water by leaching from pipes made of polyvinyl chloride (PVC). Vinyl chloride is the common name of a small molecule, called a monomer (Greek for "one unit"), from which the very large molecule called a polymer ("many units") is formed. The polymer PVC usually consists of hundreds of molecules of vinyl chloride bonded together.

**Solve** The conversion factors we need are the molar mass of $CH_2CHCl$ and the equality $1\ g = 10^3\ mg$. The molar mass of vinyl chloride is

$$\mathcal{M} = 2(12.01\ g\ C/mol) + 3(1.008\ g\ H/mol) + 35.45\ g\ Cl/mol = 62.49\ g/mol$$

and the molarity is

$$\frac{0.002\ \text{mg}}{L} \times \frac{1\ \text{g}}{10^3\ \text{mg}} \times \frac{1\ mol}{62.49\ \text{g}} = \frac{3.2 \times 10^{-8}\ mol}{L} = 3 \times 10^{-8}\ M \quad \text{or} \quad 0.03\ \mu M$$

**Think About It** This concentration may seem very low, but remember that the amount of vinyl chloride (0.002 mg, or $2 \times 10^{-6}$ g) in 1 L of water is very small. The extremely low limit of only 0.002 mg/L or $3 \times 10^{-8}$ $M$ reflects recognition of the danger vinyl chloride poses to human health.

⚙ **Practice Exercise** When 1.00 L of water from the surface of the Dead Sea is evaporated, 179 g of $MgCl_2$ is recovered. What is the molarity of $MgCl_2$ in the original sample?

*(Answers to Practice Exercises are in the back of the book.)*

---

**SAMPLE EXERCISE 4.3**    **Calculating Molarity from Density**    LO2

A water sample from the Great Salt Lake in Utah contains 83.6 mg $Na^+$/g. What is the molarity of $Na^+$ if the density of the water is 1.160 g/mL?

**Collect and Organize** Our task is to convert concentration units from mg $Na^+$/g of solution to mol $Na^+$/L of solution.

**Analyze** To convert mg $Na^+$/g into mol $Na^+$/L, we need to convert milligrams of solute to grams and then to moles, and we need to convert grams of solution to milliliters and then to liters using the conversion steps shown below.

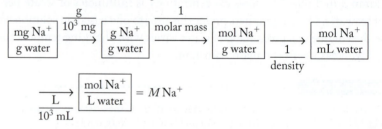

**Solve** The moles of solute are

$$83.6\ \text{mg}\ Na^+ \times \frac{1\ \text{g}}{10^3\ \text{mg}} \times \frac{1\ mol\ Na^+}{22.99\ \text{g}\ Na^+} = 3.64 \times 10^{-3}\ mol\ Na^+$$

The volume of solution is

$$1.000\ \text{g} \times \frac{1\ \text{mL}}{1.160\ \text{g}} \times \frac{1\ L}{10^3\ \text{mL}} = 8.621 \times 10^{-4}\ L$$

The molarity of sodium ion in the water from the Great Salt Lake is

$$\frac{mol}{L} = \frac{3.64 \times 10^{-3}\ mol\ Na^+}{8.621 \times 10^{-4}\ L} = 4.22\ M$$

**Think About It** According to Table 4.1, second column, the $Na^+$ concentration in seawater is about 10 g of $Na^+$ per kilogram of seawater, or about 10 mg of $Na^+$ per gram of seawater, which is approximately 10 mg/mL assuming the density of the water is about 1.0 g/mL. The concentration given for $Na^+$ in the Great Salt Lake, 83.6 mg/g lake water, is more than eight times the seawater concentration, so the molarity should be more than eight times the millimolar concentration given in column 4 of Table 4.1 (480.57 m$M \approx 0.48\ M$). Our answer, 4.22 $M$, is reasonable in light of that comparison. The Great Salt Lake is the result of evaporation of ocean water over thousands of years. Evaporation of the solvent

---

🔴🔵 **CONNECTION** In Chapter 1 we introduced density, the ratio of mass of a quantity of material to its volume or $d = m/V$, as an intensive physical property.

from any solution reduces the volume of solution and increases its concentration because the amount of solute (*n*) is unchanged.

⚙ **Practice Exercise** If the density of ocean water at a depth of 10,000 m is 1.071 g/mL and if 25.0 g of water at that depth contains 190 mg of potassium chloride, what is the molarity of potassium chloride in the sample?

*(Answers to Practice Exercises are in the back of the book.)*

---

**SAMPLE EXERCISE 4.4** **Calculating the Quantity of Solute Needed to Prepare a Solution of Known Concentration** **LO2**

An aqueous solution called *Kalkwasser* [0.0225 *M* Ca(OH)$_2$] may be added to saltwater aquaria to reduce acidity and add Ca$^{2+}$ ions. How many grams of Ca(OH)$_2$ do you need to make 500.0 mL of Kalkwasser?

**Collect and Organize** We know the volume and molarity of the solution we must prepare. We also know the identity of the solute, which enables us to calculate its molar mass.

**Analyze** We use Equation 4.2 to calculate the mass of Ca(OH)$_2$ needed. We also must convert the volume from milliliters to liters to obtain the mass, in grams, of solute required.

$$\boxed{\text{mL solution}} \xrightarrow{\dfrac{\text{L}}{10^3\ \text{mL}}} \boxed{\text{L solution}} \xrightarrow{\dfrac{\text{mol solute}}{\text{L solution}}} \boxed{\text{mol solute}} \xrightarrow{\dfrac{1}{\text{molar mass}}} \boxed{\text{g solute}}$$

The molar mass of Ca(OH)$_2$ is

$$40.08\ \text{g/mol} + 2(16.00\ \text{g/mol}) + 2(1.008\ \text{g/mol}) = 74.10\ \text{g/mol}$$

**Solve** Equation 4.2 can be used to calculate the mass of Ca(OH)$_2$ needed:

$$m\,\text{Ca(OH)}_2 = V \times M_{\text{Ca(OH)}_2} \times \mathcal{M}$$

$$= 500.0\ \cancel{\text{mL}} \times \frac{1\ \cancel{\text{L}}}{10^3\ \cancel{\text{mL}}} \times \frac{0.0225\ \cancel{\text{mol}}}{1\ \cancel{\text{L}}} \times \frac{74.10\ \text{g}}{1\ \cancel{\text{mol}}} = 0.834\ \text{g}$$

Figure 4.6 shows the technique for making this solution. A key feature is that we suspend 0.834 g of solute in 200–300 mL of water and then add additional water until the final solution has the volume specified.

**FIGURE 4.6** Preparing 500.0 mL of 0.0225 *M* Ca(OH)$_2$. (a) Weigh out the desired quantity of solid Ca(OH)$_2$; (b) transfer the Ca(OH)$_2$ to a 500 mL volumetric flask and add 200 to 300 mL of water; (c) swirl to suspend the solute; (d) dilute to exactly 500 mL. Stopper the flask and invert it several times to thoroughly mix the solution.

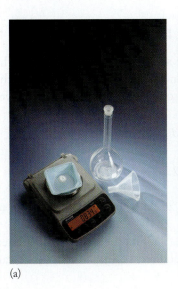
(a)

(b)

(c)

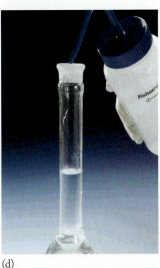
(d)

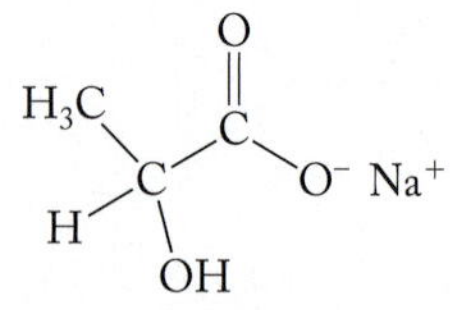

Sodium lactate

**FIGURE 4.7** Structure of sodium lactate, $NaC_3H_5O_3$.

**Think About It** To solve a problem like this, all we need to remember is the definition of molarity: moles of solute per liter of solution. The solution we want has a concentration of 0.0225 *M*, or 0.0225 mol/L. Because we need only 0.5000 L of this solution, we do not need 0.0225 mol of $Ca(OH)_2$, but only half that amount (0.0112 mol). Finally, we calculate the number of grams of calcium hydroxide in 0.0112 mol of calcium hydroxide (0.834 g). The amount of $Ca(OH)_2$ needed is less than a teaspoon.

⚙ **Practice Exercise** An aqueous solution known as Ringer's lactate is administered intravenously to trauma victims suffering from blood loss or severe burns. The solution contains the chloride salts of sodium, potassium, and calcium and is 4.00 m*M* in sodium lactate ($NaC_3H_5O_3$; Figure 4.7). How many grams of sodium lactate are needed to prepare 10.0 L of Ringer's lactate?

*(Answers to Practice Exercises are in the back of the book.)*

■ ●●●●●●●●●●●●●●●●●●●●●●●●●●●●●●●●●●●●●

# 4.3 Dilutions

▶Ⅱ **CHEMTOUR** Dilution

In laboratories such as those that test drinking water quality, **stock solutions** of substances, such as pesticides and toxic metals, are available commercially in concentrations too high to be used directly in analytical procedures. A stock solution is diluted to prepare a solution with a solute concentration of the substance close to its concentration in the sample being tested. **Dilution** is the process of lowering the concentration of a solution by adding solvent to a known volume of the initial solution. The stock solution is concentrated—it contains a much larger solute-to-solvent ratio than does the dilute solution.

We can use the definition of molarity given in Equation 4.1

$$M = \frac{n}{V}$$

to determine how to dilute a stock solution to make a solution of a desired concentration. Suppose we need to prepare 250.0 mL of an aqueous solution that is 0.0100 *M* $Cu^{2+}$ by starting with a stock solution that is 0.1000 *M* $Cu^{2+}$. What volume of the stock solution do we need?

Rearranging Equation 4.1 to solve for the number of moles, *n*, gives the moles of $Cu^{2+}$ needed in the dilute solution we want to make:

$$\text{Moles of solute} = n_{diluted} = V_{diluted} \times M_{diluted}$$

$$= (0.2500 \text{ L}) (0.0100 \text{ mol/L}) = 2.50 \times 10^{-3} \text{ mol } Cu^{2+}$$

The subscript "diluted" refers to the solution we are preparing and *V* refers to the final volume of the solution in L. Using the same equation, we can calculate the *initial* quantities; that is, the volume of the stock solution we need to deliver $2.50 \times 10^{-3}$ mol of $Cu^{2+}$:

$$n_{initial} = V_{initial} \times M_{initial}$$

$$V_{initial} = \frac{n_{initial}}{M_{initial}} = \frac{2.50 \times 10^{-3} \text{ mol}}{0.1000 \text{ mol/L}} = 2.50 \times 10^{-2} \text{ L} = 25.0 \text{ mL}$$

**stock solution** a concentrated solution of a substance used to prepare solutions of lower concentration.

**dilution** the process of lowering the concentration of a solution by adding more solvent.

To make the diluted solution, we add water to 25.0 mL of the stock solution to make a final volume of 250.0 mL (Figure 4.8).

Because the number of moles of solute ($Cu^{2+}$ ions) is the same in the diluted solution and in the portion of stock solution required ($n_{diluted} = n_{initial}$), we can combine these two equations into one that applies to all situations in which we know any three of these four variables: the initial volume ($V_{initial}$) of stock solution, the initial solute concentration ($M_{initial}$), the volume ($V_{diluted}$) of the diluted solution, and the solute concentration ($M_{diluted}$) in the diluted solution:

$$V_{initial} \times M_{initial} = n_{initial} = n_{diluted} = V_{diluted} \times M_{diluted}$$

or simply

$$V_{initial} \times M_{initial} = V_{diluted} \times M_{diluted} \qquad (4.3)$$

Equation 4.3 works because each side of the equation represents a quantity of solute that does not change because of dilution. Furthermore, this equation can be used for *any* units of volume and concentration, as long as the units used to express the initial and diluted volumes are the same and the units for the initial and diluted concentrations are the same.

Equation 4.3 applies to all systems, even those much larger than a laboratory stock solution to be diluted. For instance, in estuaries, where rivers flow into

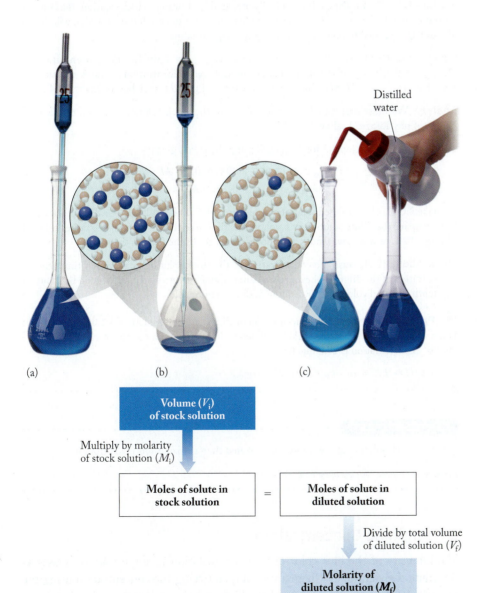

(a)        (b)        (c)

Distilled water

**Volume ($V_i$) of stock solution**

Multiply by molarity of stock solution ($M_i$)

**Moles of solute in stock solution** = **Moles of solute in diluted solution**

Divide by total volume of diluted solution ($V_f$)

**Molarity of diluted solution ($M_f$)**

**FIGURE 4.8** To prepare 250.0 mL of a solution that is 0.0100 $M$ in $Cu^{2+}$, (a) a pipet is used to withdraw 25.0 mL of a 0.1000 $M$ stock solution. (b) This volume of stock solution is transferred to a 250.0 mL volumetric flask. (c) Distilled water is added to bring the volume of the diluted solution to 250.0 mL. Note that the color of the diluted solution is lighter than that of the stock solution.

the sea, salinity is reduced because the seawater is diluted by the freshwater that enters it. Suppose a coastal bay contains $3.8 \times 10^{12}$ L of ocean water mixed with $1.2 \times 10^{12}$ L of river water for a total volume of $5.0 \times 10^{12}$ L. We can calculate the $Na^+$ concentration in the bay if we know that the $Na^+$ concentration in the open ocean is 0.48 $M$ and if we assume that the river water does not contribute significantly to the $Na^+$ content in the bay. Inserting the given values and solving Equation 4.3 for $M_{diluted}$, the $Na^+$ concentration in the bay, we have

$$M_{diluted} = \frac{V_{initial} \times M_{initial}}{V_{diluted}} = \frac{(3.8 \times 10^{12} \text{ L}) \times 0.48 \ M}{5.0 \times 10^{12} \text{ L}} = 0.36 \ M$$

The bay is less salty than the ocean because it is diluted by the freshwater from the river.

---

**SAMPLE EXERCISE 4.5    Calculating Dilutions**                              LO2

The solution used in hospitals for intravenous infusion—called *physiological saline* or *saline solution*—is 0.155 $M$ in NaCl. It is typically prepared by diluting a stock solution, the concentration of which is 1.76 $M$, with water. What volume of stock solution and what volume of water are required to prepare 2.00 L of physiological saline?

**Collect and Organize** We know the volume ($V_{diluted}$ = 2.00 L) and concentration ($M_{diluted}$ = 0.155 $M$) of the diluted solution and the concentration of stock solution available ($M_{initial}$ = 1.76 $M$). That is, we know three of the four variables in Equation 4.3.

**Analyze** We must first find $V_{initial}$, the volume of the stock solution required to make 2.00 L of physiological saline.

**Solve** Solving Equation 4.3 for $V_{initial}$ and using the values given:

$$V_{initial} = \frac{V_{diluted} \times M_{diluted}}{M_{initial}} = \frac{2.00 \text{ L} \times 0.155 \ M}{1.76 \ M} = 0.176 \text{ L}$$

To make 2.00 L of the diluted solution we need to add: 2.00 L – 0.176 L = 1.82 L of water. If we prepare the dilute solution in a volumetric flask like that in Figure 4.8, rather than adding 1.82 L of water, we would add enough water to fill the flask to the mark.

**Think About It** Because the concentration of the stock solution is about ten times the concentration of the diluted solution, the volume of stock solution required should be about one-tenth the final volume. Our result of 5.28 L is reasonable.

⚙ **Practice Exercise** The concentration of $Pb^{2+}$ in a stock solution is 1.000 mg/mL. What volume of this solution should be diluted to 500.0 mL to produce a solution in which the $Pb^{2+}$ concentration is 25.0 mg/L?

*(Answers to Practice Exercises are in the back of the book.)*

---

**CONCEPT TEST**

Why is a stock solution always more concentrated than the solutions made from it?

*(Answers to Concept Tests are in the back of the book.)*

---

## Determining Concentration

The intensity of the blue color of the $Cu^{2+}$ solution in Figure 4.8 decreases as we dilute the solution. Is there a way of quantifying the concentration using the intensity of the color? The color of the $Cu^{2+}$ solution arises from the absorption

**absorbance** a measure of the quantity of light absorbed by a sample.

**Beer's law** relates the absorbance of a solution ($A$) to concentration ($c$), path length ($b$), and the molar absorptivity ($\varepsilon$) by the equation $A = \varepsilon bc$.

**molar absorptivity** a measure of how well a compound or ion absorbs light.

**calibration curve** compares a measurable property, such as absorbance, of a solution of unknown concentration to a set of standard samples of known concentration.

**spectrophotometry** an analytical method that relies on the absorption of light and Beer's law to measure the concentration of a solution.

of visible light, a topic we will explore in more detail in Chapters 7 and 18. For now, it is sufficient to know that the **absorbance** ($A$) of a solution is a measure of the intensity of the color. **Beer's law** (Equation 4.4) relates absorbance to three quantities: concentration ($c$), the path length that the light travels through the solution ($b$), and a constant (called the **molar absorptivity**, or $\varepsilon$) characteristic of the dissolved solute.

$$A = \varepsilon bc \qquad (4.4)$$

Absorbance is measured with a spectrophotometer like that shown in Figure 4.9, in which a solution is placed in a sample cell of fixed length ($b$), typically 1.00 cm. The cell is placed in a beam of light. The value of $A$ reflects how much light is absorbed by the solution. If we express the concentration of the solution, $c$, in molarity ($M$), we can calculate $\varepsilon$. A graph of $A$ versus $c$ for solutions of different $Cu^{2+}$ concentration yields a straight line. Such a graph (Figure 4.9b) is called a **calibration curve**. Once a calibration curve is generated, the concentration of any solution of $Cu^{2+}$ can be determined by measuring its absorbance, finding that value on the $y$-axis, reading across to where it intersects the calibration curve, and then reading the value on the $x$-axis below, which is the concentration. We can also use a calibration curve to determine molar absorptivity. If $b = 1.00$ cm, then the slope of the graph is equal to $\varepsilon$ in units of $M^{-1}cm^{-1}$. This method of analysis is called **spectrophotometry**. Let's see how we can apply Beer's law to the determination of concentration of a solution of iron ions in Sample Exercise 4.6.

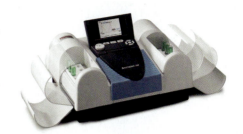

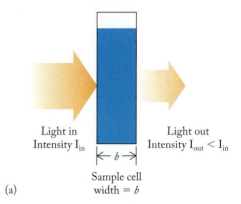

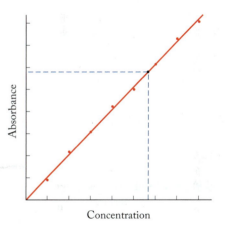

(a) Sample cell width = $b$

Light in Intensity $I_{in}$

Light out Intensity $I_{out} < I_{in}$

(b) Absorbance vs Concentration

**FIGURE 4.9** (a) A spectrophotometer, used to measure the absorbance of solutions. The intensity of light leaving the solution ($I_{out}$) is less than the intensity of light entering the solution ($I_{in}$). (b) The red line represents a calibration curve for Beer's law, plotting absorbance versus concentration for a series of standard samples. The horizontal dashed blue line represents the absorbance of a solution of unknown concentration. By extending the line to where it meets the calibration curve and reading down, the concentration of the unknown solution can be determined.

---

**SAMPLE EXERCISE 4.6    Applying Beer's Law    LO3**

One way of determining the concentration of $Fe^{2+}$ in an environmental sample is to first mix it with phenanthroline (phen) to give an intense orange-red colored solution (Figure 4.10). The absorbance of a $5.00 \times 10^{-5}$ $M$ solution is measured as 0.55 in a 1.00 cm cell. (a) Calculate the molar absorptivity, $\varepsilon$, for the $Fe^{2+}$-phen complex, and (b) determine the concentration of a solution whose absorbance is 0.36.

**Collect and Organize** Given the concentration and absorbance of a solution, we are asked to calculate the molar absorptivity ($\varepsilon$).

**Analyze** (a) We can rearrange Beer's law to solve for $\varepsilon$ and then use this value to find the concentration of the unknown solution. (b) Concentration and absorbance are linearly related by Beer's law, so we predict that a smaller absorbance for the unknown solution should correspond to a lower concentration.

$Fe^{2+}(aq) + 3\ Phen(aq) \rightarrow Fe(Phen)_3^{2+}(aq)$
colorless       colorless         orange-red

**FIGURE 4.10** Adding phenanthroline to a solution of $Fe^{2+}$ forms an orange-red complex ion consisting of an $Fe^{2+}$ ion surrounded by 3 molecules of phenanthroline.

**Solve**

a. Rearranging Equation 4.4 and substituting for $A$, $b$, and $c$:

$$\varepsilon = \frac{A}{bc} = \frac{0.55}{(1.00 \text{ cm})(5.00 \times 10^{-5} \text{ } M)} = 1.1 \times 10^4 \text{ } M^{-1} \text{ cm}^{-1}$$

b. Solving Beer's law for concentration and substituting for $A$, $b$, and $\varepsilon$:

$$c = \frac{A}{\varepsilon b} = \frac{0.36}{(1.1 \times 10^4 \text{ } M^{-1} \cdot \text{cm}^{-1})(1.00 \text{ cm})} = 3.3 \times 10^{-5} \text{ } M$$

**Think About It** As predicted, the concentration of the unknown solution is lower than $5.00 \times 10^{-5} \text{ } M$.

⚙ **Practice Exercise** One analytical method for determining the concentration of copper ion uses a reagent called cuproine to form a complex ion similar to that formed between $Fe^{2+}$ and phenanthroline. The following data were collected for four standard solutions of $Cu^{2+}$:

| Concentration of Copper (M) | Absorbance |
|---|---|
| $1.00 \times 10^{-4}$ | 0.64 |
| $1.25 \times 10^{-4}$ | 0.80 |
| $1.50 \times 10^{-4}$ | 0.96 |
| $1.75 \times 10^{-4}$ | 1.12 |

Construct a calibration curve based on these data and use it to determine the concentration of a copper solution with an absorbance $A = 0.74$.

*(Answers to Practice Exercises are in the back of the book.)*

# 4.4 Electrolytes and Nonelectrolytes

▶❚❚ CHEMTOUR  Migration of Ions in Solution

**electrolyte** a substance that dissociates into ions when it dissolves, enhancing the conductivity of the solvent.

**strong electrolyte** a substance that dissociates completely into ions when it dissolves in water.

**nonelectrolyte** a substance that does not dissociate into ions and therefore does not enhance the conductivity of water when dissolved.

**weak electrolyte** a substance that only partly dissociates into ions when it dissolves in water.

The high concentrations of NaCl and other salts in seawater (Table 4.1) make it a good conductor of electricity. Suppose we immerse two *electrodes* (solid conductors) in distilled water (Figure 4.11a) and connect them to a battery and a lightbulb. For electricity to flow and the bulb to light up, the circuit must be completed by mobile charge carriers (ions) in the solution. The lightbulb does not light when the electrodes are placed in distilled water because there are very few ions in distilled water. When the electrodes are pressed into solid NaCl (Figure 4.11b), no electric current flows and the bulb does not light because the $Na^+$ and $Cl^-$ ions in solid NaCl are fixed in position. However, when the electrodes are immersed in aqueous $0.50 \text{ } M$ NaCl (Figure 4.11c), the bulb lights because the many mobile ions in the solution act as charge carriers between the two electrodes. The saline solution conducts electricity because $Na^+$ ions are attracted to and migrate toward the electrode connected to the negative terminal of the battery and $Cl^-$ ions are attracted to and migrate toward the positive electrode, completing an electric circuit.

Any solute that imparts electrical conductivity to an aqueous solution is called an **electrolyte**. Sodium chloride is considered a **strong electrolyte** because it *dissociates* completely in water, which means it completely breaks up into its component ions when it dissolves, releasing charge-carrying $Na^+$ cations and $Cl^-$ anions.

Many substances do not dissociate when dissolved in water. For instance, a $0.50 \text{ } M$ solution of ethanol ($CH_3CH_2OH$) conducts electricity no better than pure water because each $CH_3CH_2OH$ molecule stays intact in the solution and

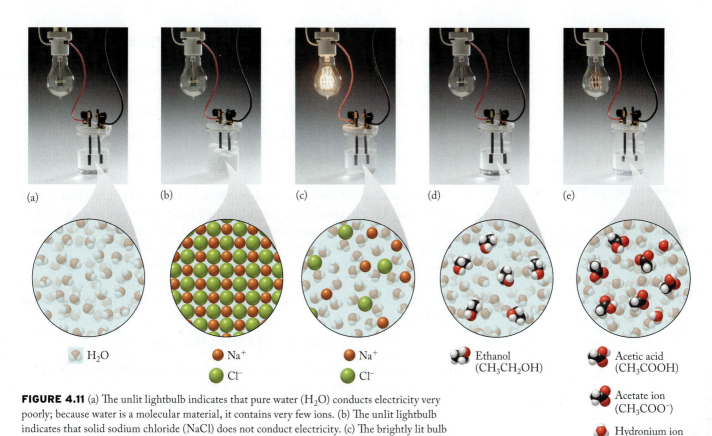

**FIGURE 4.11** (a) The unlit lightbulb indicates that pure water ($H_2O$) conducts electricity very poorly; because water is a molecular material, it contains very few ions. (b) The unlit lightbulb indicates that solid sodium chloride (NaCl) does not conduct electricity. (c) The brightly lit bulb indicates that a 0.50 $M$ solution of NaCl conducts electricity very well. (d) The unlit bulb indicates that a 0.50 $M$ solution of ethanol ($CH_3CH_2OH$) does not conduct electricity any better than pure water. (e) The dimly lit bulb indicates that a 0.50 $M$ solution of acetic acid ($CH_3COOH$) conducts electricity better than pure water or the ethanol solution but not as well as the NaCl solution.

no ions are formed (Figure 4.11d). Solutes that do not form ions when dissolved in water are called **nonelectrolytes**.

Besides strong electrolytes (complete dissociation into ions) and nonelectrolytes (no dissociation), **weak electrolytes** also exist. Weak electrolytes dissociate partially when dissolved in water. One example of a weak electrolyte is acetic acid, $CH_3COOH$. In Figure 4.11(e), the beaker contains a 0.50 $M$ aqueous solution of acetic acid; the acetic acid molarity is equivalent to that of the solutes in parts c and d of the figure. Note that the bulb connected to the acetic acid solution glows but not very brightly. The acetic acid molecules dissociate to some extent, forming hydrogen ions and acetate ions, but the process does not go to completion:

$$CH_3COOH(aq) \rightleftharpoons CH_3COO^-(aq) + H^+(aq)$$

In this case, the arrow with the top half pointing right and the bottom half pointing left indicates that all species are present in solution: undissociated $CH_3COOH$, hydrogen ion, and acetate ion. This type of arrow is used to indicate that the dissociation process is not complete.

**CONCEPT TEST** • • • • • • • • • • • • • • • • • • • • • • • • • • • • • • • • • • • • •

Why were equivalent molar concentrations used in the solutions in Figure 4.11?

*(Answers to Concept Tests are in the back of the book.)*

# 4.5 Acid–Base Reactions: Proton Transfer

When oxides of sulfur, nitrogen, or other nonmetals dissolve in water, they produce hydrogen ions in solution. Examples of this process are

$$N_2O_5(g) + H_2O(\ell) \rightarrow 2\,H^+(aq) + 2\,NO_3^-(aq)$$

$$SO_3(g) + H_2O(\ell) \rightarrow H^+(aq) + HSO_4^-(aq)$$

In this section we focus on the processes that occur in solution when hydrogen ions are produced.

Let's look at the behavior of a common binary acid: hydrochloric acid (HCl). Pure HCl is a molecular compound and a gas. However, when HCl dissolves in water it ionizes completely, producing $H^+(aq)$ and $Cl^-(aq)$. Generally, we use $H^+(aq)$ to describe the cation produced in an aqueous medium by an acid. However, hydrogen ions ($H^+$, which actually are protons, the extremely small nuclei of hydrogen atoms) do not have an independent existence in water because each proton combines with a water molecule forming a **hydronium ion ($H_3O^+$)**:

$$H^+(aq) + H_2O(\ell) \rightarrow H_3O^+(aq)$$

In aqueous acid–base chemistry, the terms hydrogen ion, proton, and hydronium ion all refer to the same species.

When a molecule of HCl dissolves, it donates a proton to a water molecule:

$$\underset{\substack{\text{proton donor}\\\text{(acid)}}}{HCl(g)} \quad + \quad \underset{\substack{\text{proton acceptor}\\\text{(base)}}}{H_2O(\ell)} \quad \rightarrow \quad H_3O^+(aq) + Cl^-(aq) \qquad (4.5)$$

This behavior fits our definition of an **acid** as a *proton donor*. Because the water molecule accepts the proton, water in this case is a *proton acceptor*, which is the corresponding definition of a **base**. These definitions were originally proposed by chemists Johannes Brønsted (1879–1947) and Thomas Lowry (1874–1936), so they are called **Brønsted–Lowry acids** and **Brønsted–Lowry bases**. We will develop this concept more completely later, but for now it provides a convenient way to define acids and bases in aqueous solutions.

A base may also be defined as a substance that produces hydroxide ions in an aqueous solution. With the Brønsted–Lowry definition of a base as a proton acceptor, the hydroxide ion is still classified as a base, but other substances can be categorized as bases as well.

In the reaction between aqueous solutions of HCl and NaOH,

$$HCl(aq) + NaOH(aq) \rightarrow NaCl(aq) + H_2O(\ell) \qquad (4.6)$$

the HCl functions as an acid by donating a proton to the hydroxide ion, and the hydroxide ion functions as a base by accepting the proton; the $H^+$ and $OH^-$ ions combine to form a molecule of water. The remaining two ions—the anion $Cl^-$ and the cation $Na^+$—form a salt, sodium chloride, which remains dissolved in the aqueous solution. This reaction is called a **neutralization reaction** because the acid and the base have been eliminated as a result of the reaction. The products of the neutralization of an acid like HCl with a base like NaOH in solution are always water and a salt. This provides us with the definition of a **salt** as a substance formed along with water as a product of a neutralization reaction.

---

**CONNECTION** In Chapter 3, we described the hydrolysis reactions of nonmetal oxides that produce acidic solutions.

**CONNECTION** We discussed the naming of binary acids in Section 2.6.

---

**hydronium ion ($H_3O^+$)** a $H^+$ ion bonded to a molecule of water, $H_2O$; the form in which the hydrogen ion is found in an aqueous solution.

**acid (Brønsted–Lowry acid)** a proton donor.

**base (Brønsted–Lowry base)** a proton acceptor.

**neutralization reaction** a reaction that takes place when an acid reacts with a base and produces a solution of a salt in water.

**salt** the product of a neutralization reaction; it is made up of the cation of the base in the reaction plus the anion of the acid.

Which drawing in Figure 4.12 describes the particles in an aqueous solution of an acid that is a weak electrolyte and which describes the particles in an aqueous solution of an acid that is a strong electrolyte?

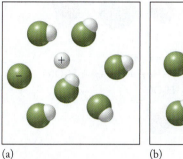

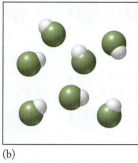

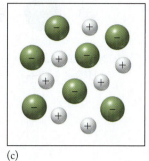

(a)                                      (b)                                      (c)

Unionized acid

Hydrogen ion

Anion

**FIGURE 4.12** Aqueous solutions of binary compounds.

*(Answers to Concept Tests are in the back of the book.)*

---

**SAMPLE EXERCISE 4.7** **Comparing Electrolytes, Acids, and Bases**    **LO4**

Classify each of the following compounds in aqueous solution as a strong electrolyte, weak electrolyte, nonelectrolyte, acid, or base: (a) NaCl; (b) HCl; (c) NaOH; (d) CH$_3$COOH.

**Collect and Organize** We are given the formulas of four compounds and asked to classify them into one or more of five categories. We will work with the definitions of these terms.

**Analyze** Strong electrolytes are substances that dissociate completely into ions when dissolved in water. Weak electrolytes dissociate only partially in water. Nonelectrolytes do not dissociate at all. A Brønsted–Lowry acid is defined as a proton donor and a Brønsted–Lowry base is a proton acceptor.

**Solve**
a. NaCl is an ionic compound that dissociates completely into ions in aqueous solution, as shown in Figure 4.11. Therefore NaCl is a strong electrolyte. Sodium chloride has no protons to donate, and neither Na$^+$ nor Cl$^-$ ions readily accept protons. Thus NaCl is neither an acid nor a base.
b. HCl dissociates completely into H$^+$ and Cl$^-$ ions when dissolved in water, making HCl a strong electrolyte. As a proton donor, aqueous HCl is also a Brønsted–Lowry acid.
c. NaOH is an ionic compound that dissociates completely into Na$^+$ and OH$^-$ ions in aqueous solution. NaOH is a strong electrolyte. The OH$^-$ ion formed upon dissolution of NaOH readily accepts a proton to form water, making NaOH a Brønsted–Lowry base.
d. Some molecules of acetic acid, CH$_3$COOH, when dissolved in water, dissociate into H$^+$ and CH$_3$COO$^-$ ions (Figure 4.11e), creating a weakly conducting solution; thus it is a weak electrolyte. Since acetic acid is a proton donor, it is considered a Brønsted–Lowry acid.

**Think About It** You should not be surprised that some strong electrolytes are acids or bases. Acidity and basicity have different definitions from those for electrolytes, and substances can be classified as both acid or base and strong or weak electrolyte. A weak electrolyte that donates a proton is a weak acid.

**Practice Exercise** Classify each of the following aqueous compounds as a strong electrolyte, weak electrolyte, nonelectrolyte, acid, or base: (a) potassium sulfate; (b) HBr; (c) NH$_3$; (d) ethanol (CH$_3$CH$_2$OH).

*(Answers to Practice Exercises are in the back of the book.)*

**FIGURE 4.13** The neutralization reaction between aqueous hydrochloric acid and sodium hydroxide that produces sodium chloride and water can be written in thee ways: (a) a molecular equation, (b) an overall ionic equation, and (c) a net ionic equation.

(a) Molecular equation  $HCl(aq) + NaOH(aq) \rightarrow NaCl(aq) + H_2O(\ell)$

(b) Overall ionic equation

$H^+(aq) + Cl^-(aq) + Na^+(aq) + OH^-(aq) \rightarrow Na^+(aq) + Cl^-(aq) + H_2O(\ell)$

(c) Net ionic equation = overall ionic equation − spectator ions

$H^+(aq) + \cancel{Cl^-(aq)} + \cancel{Na^+(aq)} + OH^-(aq) \rightarrow \cancel{Na^+(aq)} + \cancel{Cl^-(aq)} + H_2O(\ell)$

$H^+(aq) + OH^-(aq) \rightarrow H_2O(\ell)$

Equation 4.6 is written as a **molecular equation**, meaning each reactant and product is written as a neutral compound. Molecular equations are sometimes the easiest equations to balance, so reactions involving substances that dissociate in solution are often written first as molecular equations to indicate each reactant and product and to simplify balancing the equation.

Another equation representing the reaction in Equation 4.6 is one that shows each reactant and product the way it is present in solution, which means any species that dissociates completely is written as a cation/anion pair:

$$\underbrace{HCl(aq)}_{H^+(aq) + Cl^-(aq)} + \underbrace{NaOH(aq)}_{Na^+(aq) + OH^-(aq)} \rightarrow \underbrace{NaCl(aq)}_{Na^+(aq) + Cl^-(aq)} + H_2O(\ell) \quad (4.7)$$

This form, called an **overall ionic equation**, distinguishes ionic substances from molecular substances in a chemical reaction taking place in solution. Any non-electrolytes or weak electrolytes are written as neutral molecules.

Notice that in Equation 4.7 chloride ions and sodium ions appear on both sides of the reaction arrow. If this were an algebraic equation, we would remove the terms that are the same on both sides. When we do the same thing in chemistry, the result is a **net ionic equation**, as written in Equation 4.8. The ions that are removed (in this case, sodium ion and chloride ion) are **spectator ions**, and the resulting equation shows only the species taking part in the reaction:

$$H^+(aq) + OH^-(aq) \rightarrow H_2O(\ell) \quad (4.8)$$

The spectator ions are unchanged by the reaction and remain in solution. The chemical change in a neutralization reaction occurs when a hydrogen ion reacts with a hydroxide ion to produce a molecule of water. The net ionic equation enables us to focus on those species that actually participate in the reaction. The three ways of depicting the reaction between HCl and NaOH—molecular, overall ionic, and net ionic equations—are summarized in Figure 4.13.

**molecular equation** a balanced equation describing a reaction in solution in which the reactants and products are written as undissociated molecules.

**overall ionic equation** a balanced equation that shows all the species, both ionic and molecular, present in a reaction occurring in aqueous solution.

**net ionic equation** a balanced equation that describes the actual reaction taking place in aqueous solution; it is obtained by eliminating the spectator ions from the overall ionic equation.

**spectator ion** an ion that is present in a reaction vessel when a chemical reaction takes place but is unchanged by the reaction; spectator ions appear in an overall ionic equation but not in a net ionic equation.

**hydrolysis** the reaction of water with another material. The hydrolysis of nonmetal oxides produces acids.

**SAMPLE EXERCISE 4.8** **Writing Neutralization Reaction Equations** LO5

Write the balanced (a) molecular, (b) overall ionic, and (c) net ionic equations that describe the reaction that takes place when an aqueous solution of sulfuric acid is neutralized by an aqueous solution of potassium hydroxide.

**Collect and Organize** Solutions of sulfuric acid and potassium hydroxide take part in a neutralization reaction. Our task is to write three different equations describing the neutralization reaction.

**Analyze** We know that potassium hydroxide (KOH) is the base and sulfuric acid ($H_2SO_4$) is the acid. The products of a neutralization reaction are water and a salt. (a) All species are written as neutral compounds in the molecular equation. The products are water and potassium sulfate, the salt made from the cation of the base and the anion of the acid. (b) We can write each substance in the balanced molecular equation in ionic form to generate the overall ionic equation. (c) Removing spectator ions from the overall ionic equation gives us the net ionic equation.

**Solve**

a. The molecular equation is

$$H_2SO_4(aq) + KOH(aq) \rightarrow K_2SO_4(aq) + H_2O(\ell) \qquad \text{(unbalanced)}$$
$$H_2SO_4(aq) + 2\,KOH(aq) \rightarrow K_2SO_4(aq) + 2\,H_2O(\ell) \qquad \text{(balanced)}$$

b. The overall ionic equation is

$$H^+(aq) + HSO_4^-(aq) + 2\,K^+(aq) + 2\,OH^-(aq) \rightarrow 2\,K^+(aq) + SO_4^{2-}(aq) + 2\,H_2O(\ell)$$

c. The spectator ions are $K^+$ and $SO_4^{2-}$. Removing them gives us the net ionic equation:

$$2\,H^+(aq) + 2\,OH^-(aq) \rightarrow 2\,H_2O(\ell)$$

or

$$H^+(aq) + OH^-(aq) \rightarrow H_2O(\ell)$$

**Think About It** We treat chemical equations just like algebraic equations, cancelling out spectator ions that are unchanged in the reaction and appear on both sides of the reaction arrow in the *overall* ionic equation. After cancelling the spectator ions, we are left with the net ionic equation, which focuses on the species that were changed as a result of the reaction. It is also important to remember that sulfate is a polyatomic ion (see Table 2.3) that retains its identity in solution and does not separate into atoms and ions. Note that the net ionic equation is the same as in the reaction of an aqueous solution of HCl with aqueous NaOH.

⚙ **Practice Exercise** Write balanced (a) molecular, (b) overall ionic, and (c) net ionic equations for the reaction between an aqueous solution of phosphoric acid, $H_3PO_4(aq)$, and an aqueous solution of sodium hydroxide. The products are sodium phosphate and water. Note: phosphoric acid is a weak electrolyte.

*(Answers to Practice Exercises are in the back of the book.)*

For billions of years, neutralization reactions have played key roles in the chemical transformations of Earth's crust that geologists call *chemical weathering*. One type of chemical weathering occurs when carbon dioxide in the atmosphere dissolves in rainwater to create a weakly acidic solution of carbonic acid, $H_2CO_3(aq)$. This solution dissolves calcium carbonate, which occurs as chalk, limestone, and marble and is insoluble in pure water. The reaction between carbonic acid and calcium carbonate is responsible for the formation of stalactites and stalagmites (Figure 4.14). Chemical weathering caused by even more acidic rain is responsible for the degradation of statues and the exteriors of buildings made of marble (Figure 4.15). Table 4.2 lists some common nonmetal oxides that are classified as atmospheric pollutants because they dissolve in water to produce acid rain.

Nonmetal oxides other than carbon dioxide form solutions that are much more acidic than carbonic acid. Dinitrogen pentoxide, for example, forms nitric acid in water in a **hydrolysis** reaction.

$$N_2O_5(g) + H_2O(\ell) \rightarrow 2\,HNO_3(aq) \qquad HNO_3(aq) \rightarrow H^+(aq) + NO_3^-(aq)$$

Similarly, sulfur trioxide forms sulfuric acid:

$$SO_3(g) + H_2O(\ell) \rightarrow H_2SO_4(aq) \qquad H_2SO_4(aq) \rightarrow H^+(aq) + HSO_4^-(aq)$$

**FIGURE 4.14** In Carlsbad Caverns in New Mexico, stalagmites of limestone grow up from the cavern floor and stalactites grow downward from the ceiling.

**FIGURE 4.15** Atmospheric sulfuric acid, made when $SO_3(g)$ dissolves in rainwater, attacks marble statues by converting the calcium carbonate to calcium sulfate, which is more soluble and is slowly washed away by rain and melting snow. The picture on the left was taken in 1908; the one on the right, in 1968.

| TABLE 4.2 | Volatile Nonmetal Oxides and Their Acids | |
|---|---|
| **Oxide** | **Acid** |
| $SO_2$ | $H_2SO_3$ |
| $SO_3$ | $H_2SO_4$ |
| $NO_2$ | $HNO_2$, $HNO_3$ |
| $N_2O_5$ | $HNO_3$ |
| $CO_2$ | $H_2CO_3$ |

| TABLE 4.3 | Strong Acids |
|---|---|
| **Acid** | **Molecular Formula** |
| Hydrochloric acid | HCl |
| Hydrobromic acid | HBr |
| Hydroiodic acid | HI |
| Nitric acid | $HNO_3$ |
| Sulfuric acid | $H_2SO_4$ |
| Perchloric acid | $HClO_4$ |

Nitric acid and sulfuric acid are **strong acids** because they dissociate completely in water. They are listed in Table 4.3 along with other common strong acids. Because one molecule of nitric acid donates one proton, nitric acid is called a *monoprotic acid*. Hydrochloric acid is another strong monoprotic acid. Sulfuric acid can donate up to two protons per molecule, which makes it a *diprotic acid*. However, in some $H_2SO_4$ solutions, only one of the two H atoms in each molecule ionizes. This behavior means that sulfuric acid is a strong acid in terms of donating the first proton. This dissociation results in the formation of the $HSO_4^-$ ion (hydrogen sulfate), which is also an acid because some $HSO_4^-$ ions in solution dissociate into $H^+$ and $SO_4^{2-}$ ions. However, because not all $HSO_4^-$ ions dissociate, the hydrogen sulfate anion is a **weak acid**, which is defined as one that dissociates only partially in water.

The two equations for the dissociation of the two protons of sulfuric acid are written to reflect this behavior. First, the dissociation of the strong acid $H_2SO_4$:

$$H_2SO_4(aq) \rightarrow H^+(aq) + HSO_4^-(aq) \qquad (\rightarrow \text{ means complete ionization})$$

Second, the dissociation of the weak acid $HSO_4^-$:

$$HSO_4^-(aq) \rightleftharpoons H^+(aq) + SO_4^{2-}(aq) \qquad (\rightleftharpoons \text{ means incomplete ionization})$$

For any diprotic acid, the $HX^-$ anion formed when the first proton dissociates is a weaker acid than the original acid $H_2X$. The pattern continues for triprotic acids ($H_3X$) and other *polyprotic acids* as well: in terms of their strength as acids, the proton donors in a triprotic acid are in the order $H_3X > H_2X^- > HX^{2-}$.

**CONCEPT TEST** ••••••••••••••••••••••••••••••••••••••••••••••••••••••

A molecule called EDTA has four acidic protons. We can symbolize it as $H_4E$ to highlight this property. Pick the more acidic member of each pair listed: (a) $H_2E^{2-}$ or $HE^{3-}$; (b) $H_4E$ or $H_3E^-$.

*(Answers to Concept Tests are in the back of the book.)*

••••••••••••••••••••••••••••••••••••••••••••••••••••••••••••••••••••••••••

Strong acids are strong electrolytes, and weak acids are weak electrolytes. A balance, called a *dynamic equilibrium*, is achieved in solutions of weak electrolytes, at which point the concentrations of the reactants and the products in the dissociation reaction do not change. Solutions of weak electrolytes, such as weak acids, are characterized by dissociated ions and undissociated molecules existing together in dynamic equilibrium.

In a classification scheme similar to that used for acids, bases are classified as **strong bases** or **weak bases**, depending on the extent to which they dissociate in aqueous solution. Strong bases include the hydroxides of groups 1 and 2 metals, which dissociate completely when dissolved in water. Table 4.4 lists some common basic minerals that are weak bases. Another example of a weak base is ammonia ($NH_3$). Ammonia in its molecular form is a gas at room temperature. When $NH_3$ dissolves in water it produces a solution that conducts electricity weakly, which means that ammonia is a weak electrolyte:

$$NH_3(aq) \quad + \quad H_2O(\ell) \;\rightleftharpoons\; NH_4^+(aq) + OH^-(aq) \qquad (4.9)$$

$$\underset{\substack{\text{base} \\ \text{(proton acceptor)}}}{} \qquad \underset{\substack{\text{acid} \\ \text{(proton donor)}}}{}$$

Note that water behaves as a base in Equation 4.5 and as an acid in Equation 4.9. Is water an acid or a base? The answer depends on what is dissolved in it. Water is called an **amphiprotic** substance because it can function as a proton acceptor (Brønsted–Lowry base) in solutions of acid solutes or as a proton donor (Brønsted–Lowry acid) in solutions of basic solutes. In the hydrolysis of HCl, water functions as a base; in the hydrolysis of $NH_3$, it functions as an acid.

**CONCEPT TEST** ● ● ● ● ● ● ● ● ● ● ● ● ● ● ● ● ● ● ● ● ● ● ● ●

Which, if any, of these species is amphiprotic? (a) $H_2SO_4$; (b) $HSO_4^-$; (c) $SO_4^{2-}$

*(Answers to Concept Tests are in the back of the book.)*

# 4.6 Titrations

We can use our understanding of the reactions of acids and bases to analyze solutions and determine the concentration of dissolved substances. An approach called **titration** is a common analytical method based on measured volumes of reactants. Acid–base neutralization reactions are frequently the basis of analyses done by titration.

One use of titrations is to analyze water draining from abandoned coal mines. Sulfide-containing minerals called pyrites may be exposed when coal is removed from a site. The ensuing reaction between the minerals and microorganisms in the presence of oxygen produces very acidic solutions of sulfuric acid. To manage these hazardous wastes appropriately, we must know the concentrations of acid present in them. Titrations give us this information.

Suppose we have 100.0 mL of drainage water containing an unknown concentration of sulfuric acid. We can use an acid–base titration in which the sulfuric acid in the water sample is neutralized by reaction with a standard solution of 0.00100 *M* NaOH. The NaOH solution in this case is called the **titrant**; it is a **standard solution**, meaning its concentration is known accurately. The neutralization reaction is

$$H_2SO_4(aq) + 2\,NaOH(aq) \rightarrow Na_2SO_4(aq) + 2\,H_2O(\ell)$$

| TABLE 4.4 | Names and Formulas of Some Basic Minerals |
|---|---|
| **Name**[a] | **Formula** |
| Calcite | $CaCO_3$ |
| Gibbsite | $Al(OH)_3$ |
| Dolomite | $MgCa(CO_3)_2$ |

[a]All these minerals are considered weak bases because of their limited solubilities in water. Strong bases include all the hydroxides of the group 1 and group 2 elements except Be and Mg.

**strong acid** an acid that completely dissociates into ions in aqueous solution.

**weak acid** an acid that only partially dissociates in aqueous solution and so has a limited capacity to donate protons to the medium.

**strong base** a base that completely dissociates into ions in aqueous solution.

**weak base** a base that only partially dissociates in aqueous solution and so has a limited capacity to accept protons.

**amphiprotic** a substance that can behave as either a proton acceptor or a proton donor.

**titration** an analytical method for determining the concentration of a solute in a sample by reacting the solute with a standard solution of known concentration.

**titrant** the standard solution added to the sample in a titration.

**standard solution** a solution of known concentration used in titrations.

To determine the concentration of sulfuric acid in the sample, we determine the volume of titrant needed to neutralize a known volume of the sample.

Suppose 22.4 mL of the NaOH solution is required to react completely with the $H_2SO_4$ in the sample. Because we know the volume and molarity of the NaOH solution, we can calculate the number of moles of NaOH consumed:

$$n = V \times M$$

$$= 22.4 \text{ mL} \times \frac{1 \text{ L}}{10^3 \text{ mL}} \times \frac{1.00 \times 10^{-3} \text{ mol NaOH}}{\text{L}}$$

$$= 2.24 \times 10^{-5} \text{ mol NaOH}$$

We know from the stoichiometry of the reaction that 2 moles of NaOH are required to neutralize 1 mole of $H_2SO_4$, so the number of moles of $H_2SO_4$ in the 100.0 mL sample must be

$$2.24 \times 10^{-5} \text{ mol NaOH} \times \frac{1 \text{ mol } H_2SO_4}{2 \text{ mol NaOH}} = 1.12 \times 10^{-5} \text{ mol } H_2SO_4$$

We can combine these two steps into one mathematical expression and will do so throughout the text:

$$0.02240 \text{ L} \times \frac{1.00 \times 10^{-3} \text{ mol NaOH}}{\text{L}} \times \frac{1 \text{ mol } H_2SO_4}{2 \text{ mol NaOH}}$$

$$= 1.12 \times 10^{-5} \text{ mol } H_2SO_4$$

Therefore the concentration of $H_2SO_4$ is

$$\frac{1.12 \times 10^{-5} \text{ mol } H_2SO_4}{0.1000 \text{ L}} = \frac{1.12 \times 10^{-4} \text{ mol } H_2SO_4}{\text{L}} = 1.12 \times 10^{-4} \, M \, H_2SO_4$$

How did we find out that exactly 22.40 mL of the NaOH solution was needed to react with the sulfuric acid in the sample? A setup for doing an acid–base titration is illustrated in Figure 4.16. The standard solution of NaOH is poured into a *buret*, a narrow glass cylinder with volume markings. The solution is gradually added to the sample until an indicator that changes color in response to $H^+(aq)$ concentration signals that the reaction is complete. The point in the titration when just enough standard solution has been added to completely react with all the solute in the sample is called the **equivalence point** of the titration. If the correct indicator has been chosen, the equivalence point is very close to the **end point**, the point at which the indicator changes color.

To detect the end point in our sulfuric acid–sodium hydroxide titration, we might use the indicator *phenolphthalein*, which is colorless in acidic solutions but pink in basic solutions. The part of the titration that requires the most skill is adding just enough NaOH solution to reach the end point, the point at which a pink color first persists in the solution being titrated. To catch this end point, the standard solution must be added no faster than one drop at a time with thorough mixing between drops.

Titration is an effective method for determining solution concentrations; however, it is not applicable in all situations. We have already seen in Section 4.3 how spectrophotometry can be used to determine concentration. We explore another method in Section 4.7.

(a)

(b)

**FIGURE 4.16** Determining a sulfuric acid concentration. (a) A known volume of the $H_2SO_4$ solution is placed in the flask. The buret is filled with an aqueous NaOH solution of known concentration. A few drops of phenolphthalein indicator solution are added to the flask. (b) Sodium hydroxide is carefully added to the flask until the indicator changes from colorless to pink, signaling that the acid has been neutralized.

SAMPLE EXERCISE 4.9 **Calculating Molarity from Titration Data** **LO3**

Vinegar is an aqueous solution of acetic acid ($CH_3COOH$) that can be made from any source containing starch or sugar. Apple cider vinegar is made from apple juice that is fermented to produce alcohol, which then reacts with oxygen from the air in the presence of certain bacteria to produce vinegar. Commercial vinegar must contain no less than 4 grams of acetic acid per 100 mL of vinegar. Suppose the titration of a 25.00 mL sample of vinegar requires 11.20 mL of a 5.95 *M* solution of NaOH. What is the molarity of the vinegar? Could this be a commercial sample of vinegar?

**Collect and Organize** We are given the volume and concentration of the titrant and the volume of the sample. We are asked to calculate the concentration of the sample.

**Analyze** We start with a balanced chemical equation for the neutralization reaction. We then calculate the molarity of the vinegar so we can determine the number of grams of acetic acid in the sample and thereby determine if the sample is commercial grade.

$$\boxed{\text{L NaOH}} \xrightarrow{\dfrac{\text{mol NaOH}}{\text{L}}} \boxed{\text{mol NaOH}} \xrightarrow{\dfrac{1 \text{ mol } CH_3COOH}{1 \text{ mol NaOH}}} \boxed{\text{mol } CH_3COOH} \xrightarrow{\dfrac{1}{\text{L sample}}} \boxed{\dfrac{\text{mol } CH_3COOH}{\text{L}}}$$

**Solve** The balanced chemical equation for the neutralization reaction is

$$CH_3COOH(aq) + NaOH(aq) \rightarrow H_2O(\ell) + NaCH_3COO(aq)$$

The stoichiometry of the reaction shows that 1 mole of base reacts with 1 mole of acid. The number of moles of acid titrated is

$$0.01120 \text{ L NaOH} \times \frac{5.95 \text{ mol NaOH}}{1 \text{ L NaOH}} \times \frac{1 \text{ mol } CH_3COOH}{1 \text{ mol NaOH}} = 0.0666 \text{ mol } CH_3COOH$$

The molarity of the vinegar is

$$\frac{0.0666 \text{ mol } CH_3COOH}{0.02500 \text{ L vinegar}} = 2.66 \text{ } M$$

The number of grams of acetic acid in the sample is

$$0.0666 \text{ mol } CH_3COOH \times \frac{60.05 \text{ g } CH_3COOH}{1 \text{ mol } CH_3COOH} = 4.00 \text{ g } CH_3COOH$$

The original sample had a volume of 25.00 mL, so the number of grams of acetic acid per 100 mL of sample is

$$\frac{4.00 \text{ g } CH_3COOH}{25.00 \text{ mL vinegar}} \times 100 \text{ mL vinegar} = 16.0 \text{ g}$$

The sample has 16.0 g in 100 mL of solution, so it could certainly be a commercial vinegar.

**Think About It** About 11 mL of titrant was needed to neutralize the sample; that is about half the volume of the sample, so the molarity of the vinegar should be about half that of the titrant. The answer for the concentration of the vinegar seems reasonable. Vinegar this concentrated is typically used for pickling. Vinegar with 4 to 8% acetic acid is table vinegar, used in salad dressings.

⚙ **Practice Exercise** Citric acid ($C_6H_8O_7$; Figure 4.17) is a triprotic acid found naturally in lemon juice, and it is widely used as a flavoring in beverages. What is the molarity of $C_6H_8O_7$ in commercially available lemon juice if 14.26 mL of 1.751 *M* NaOH is required in a titration to neutralize 20.00 mL of the juice? How many grams of citric acid are in 100 mL of the juice?

*(Answers to Practice Exercises are in the back of the book.)*

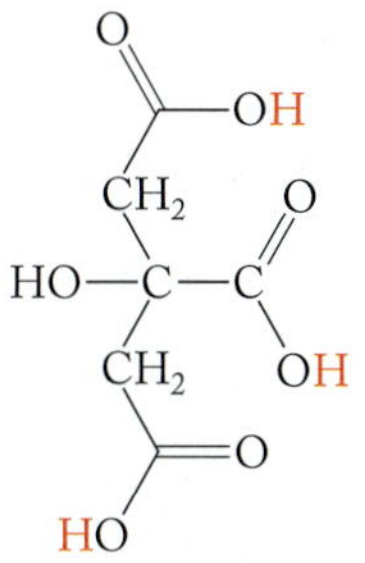

**equivalence point** the point in a titration at which the number of moles of titrant added is stoichiometrically equal to the number of moles of the substance being analyzed.

**end point** the point in a titration that is reached when just enough standard solution has been added to cause the indicator to change color.

Citric acid

**FIGURE 4.17** Structure of citric acid; acidic protons are shown in red.

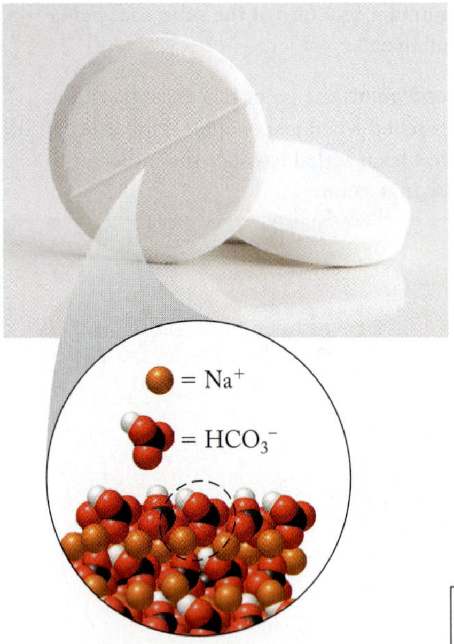


**FIGURE 4.18** Sodium bicarbonate antacid tablets.

## SAMPLE EXERCISE 4.10 Calculating Molarity of a Bicarbonate Solution LO3

Many over-the-counter antacids, like the one shown in Figure 4.18, contain sodium bicarbonate. Sodium bicarbonate reacts with excess stomach acid (aqueous HCl) in a neutralization reaction. A 650 mg tablet is dissolved in 25.00 mL of water. Titration of this solution requires 25.27 mL of a 0.3000 *M* solution of HCl. (a) Write a net ionic equation for the reaction between $NaHCO_3$ and HCl. (b) What is the molarity of the antacid solution? (c) Is the tablet pure $NaHCO_3$?

**Collect and Organize** We are assigned three tasks in this sample exercise. We are given the names of the reactants and asked to write a balanced net ionic equation. We are given the volume and concentration of the titrant and the volume of the sample and asked to calculate the concentration of the solution. Finally, using the results of our calculations, we are asked to assess the purity of the $NaHCO_3$ present in the original sample.

**Analyze** (a) The products of the reaction between $NaHCO_3$ and HCl are sodium chloride, water, and carbon dioxide. We will first balance the molecular equation, and then generate the overall ionic equation and the net ionic equation. (b) The steps involved in calculating the molarity of the sodium bicarbonate solution (not including liter–milliliter conversions) are

$$\boxed{\text{L HCl}} \xrightarrow{\frac{\text{mol HCl}}{\text{L}}} \boxed{\text{mol HCl}} \xrightarrow{\frac{1\text{ mol NaHCO}_3}{1\text{ mol HCl}}} \boxed{\text{mol NaHCO}_3} \xrightarrow{\frac{1}{\text{L sample}}} \boxed{\frac{\text{mol NaHCO}_3}{\text{L}}}$$

(c) The purity of the tablet can be determined by converting the number of moles of $NaHCO_3$ calculated in part b into an equivalent number of grams:

$$\boxed{\text{mol NaHCO}_3} \xrightarrow{\text{molar mass}} \boxed{\text{g NaHCO}_3}$$

and comparing the result to the mass of the tablet (650 mg).

**Solve**

a. The balanced molecular equation for the neutralization reaction is

$$NaHCO_3(aq) + HCl(aq) \rightarrow H_2O(\ell) + CO_2(g) + NaCl(aq)$$

The balanced ionic equation is

$$Na^+(aq) + HCO_3^-(aq) + H^+(aq) + Cl^-(aq) \rightarrow H_2O(\ell) + CO_2(g) + Na^+(aq) + Cl^-(aq)$$

and the net ionic equation is

$$\cancel{Na^+(aq)} + HCO_3^-(aq) + H^+(aq) + \cancel{Cl^-(aq)} \rightarrow H_2O(\ell) + CO_2(g) + \cancel{Na^+(aq)} + \cancel{Cl^-(aq)}$$

$$HCO_3^-(aq) + H^+(aq) \rightarrow H_2O(\ell) + CO_2(g)$$

b. The stoichiometry of the reaction shows that 1 mole of $NaHCO_3$ reacts with 1 mole of HCl. The number of moles of $NaHCO_3$ titrated is

$$0.02527\ \cancel{\text{L HCl}} \times \frac{0.3000\ \cancel{\text{mol HCl}}}{1\ \cancel{\text{L HCl}}} \times \frac{1\text{ mol NaHCO}_3}{1\ \cancel{\text{mol HCl}}} = 0.007581\text{ mol NaHCO}_3$$

The molarity of the $HCO_3^-$ is

$$\frac{0.007581\text{ mol NaHCO}_3}{0.02500\text{ L solution}} = 0.3032\ M\text{ NaHCO}_3$$

c. The number of grams of sodium bicarbonate in the sample is

$$0.007581\ \cancel{\text{mol NaHCO}_3^-} \times \frac{84.01\text{ g NaHCO}_3}{1\ \cancel{\text{mol NaHCO}_3}} = 0.6369\text{ g NaHCO}_3$$

The mass of the tablet was 650 mg. Based on the titration, the sample contained only 637 mg of $NaHCO_3$ and hence is not pure $NaHCO_3$.

**Think About It** The stoichiometry of the net ionic equation indicates a 1:1 ratio of acid to base. If a nearly equivalent volume of aqueous HCl was needed to titrate 25 mL of antacid solution, then we expect the concentration of the base to be similar to that of the acid.

⚙ **Practice Exercise** What is the molarity of calcium bicarbonate if 9.870 mL of 1.000 $M$ HNO$_3$ is required in a titration to neutralize 50.00 mL of a solution of Ca(HCO$_3$)$_2$?

*(Answers to Practice Exercises are in the back of the book.)*

# 4.7 Precipitation Reactions

Some of the most abundant elements in Earth's crust, including silicon (Si), aluminum (Al), and iron (Fe), are not abundant in seawater for the simple reason that most compounds containing these elements are insoluble in water. Earth's crust must be made of compounds with limited water solubility, or else they would never survive as solids in the presence of all the water on our planet. The reasons why some compounds are insoluble in water while others are soluble involve many factors. For now we will rely on the solubility rules in Table 4.5 to predict the solubility of common ionic compounds.

## Making Insoluble Salts

What happens when we add aqueous sulfuric acid to a solution of barium hydroxide? From Section 4.5 we recognize that this is a neutralization reaction; however, as Figure 4.19 illustrates, a white solid, called a **precipitate**, is observed at the bottom of the beaker. Such reactions are called **precipitation reactions**.

We can use Table 4.5 to predict whether a precipitate will form when solutions of ionic solutes are mixed together. For example, does a precipitate form when aqueous solutions of potassium nitrate (KNO$_3$) and sodium iodide (NaI) are mixed? To answer this question we need to:

1. *Recognize that both salts are in solution, which indicates that they are both soluble in water.* Table 4.5 specifies that all compounds containing cations

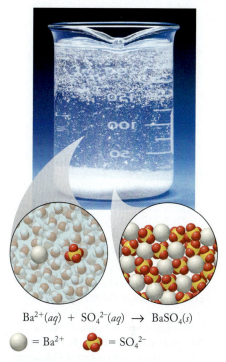

$$Ba^{2+}(aq) + SO_4^{2-}(aq) \rightarrow BaSO_4(s)$$

⚪ = Ba$^{2+}$    🔴 = SO$_4^{2-}$

**FIGURE 4.19** Addition of aqueous sulfuric acid to a solution of barium hydroxide precipitates white barium sulfate.

| TABLE 4.5 | Solubility Rules for Common Ionic Compounds in Water |
| --- | --- |

All compounds containing the following ions are soluble:
- Cations: Group 1 ions (alkali metals) and NH$_4^+$
- Anions: NO$_3^-$ and CH$_3$COO$^-$ (acetate)

Compounds containing the following anions are soluble except as noted:
- Group 17 ions (halides), except the halides of Ag$^+$, Cu$^+$, Hg$_2^{2+}$, and Pb$^{2+}$
- SO$_4^{2-}$, except the sulfates of Ba$^{2+}$, Ca$^{2+}$, Hg$_2^{2+}$, Pb$^{2+}$, and Sr$^{2+}$

Insoluble compounds include the following:
- All hydroxides except those of group 1 cations and Ca(OH)$_2$, Sr(OH)$_2$, and Ba(OH)$_2$
- All sulfides except those of group 1 cations and NH$_4^+$, CaS, SrS, and BaS
- All carbonates except those of group 1 cations and NH$_4^+$
- All phosphates except those of group 1 cations and NH$_4^+$
- Most fluorides, though not those of group 1 cations and NH$_4^+$

**precipitate** a solid product formed from a reaction in solution.

**precipitation reaction** a reaction that produces an insoluble product upon mixing two solutions.

from group 1 in the periodic table are soluble in water. Potassium and sodium are in group 1, so salts containing them are soluble:

Solution 1: $\qquad KNO_3(aq) \rightarrow K^+(aq) + NO_3^-(aq)$

Solution 2: $\qquad NaI(aq) \rightarrow Na^+(aq) + I^-(aq)$

From this information we can set up the reactant side of the equation:

$$K^+(aq) + NO_3^-(aq) + Na^+(aq) + I^-(aq) \rightarrow ?$$

2. *Determine whether any of the possible combinations of ions produces an insoluble product.* $KNO_3$, KI, NaI, and $NaNO_3$ are all soluble according to Table 4.5, so no precipitate forms when the two solutions are mixed:

$$K^+(aq) + NO_3^-(aq) + Na^+(aq) + I^-(aq) \rightarrow \text{no reaction}$$

Does a precipitate form when aqueous solutions of lead(II) nitrate and sodium iodide are mixed? All nitrate salts are soluble, so $Pb(NO_3)_2$ dissolves in water and forms ions. The ions present in the mixed solutions are

$$Pb^{2+}(aq) + 2\,NO_3^-(aq) + Na^+(aq) + I^-(aq) \rightarrow ?$$

The solubility rules in Table 4.5 indicate that halide ions such as $I^-$ of group 17 of the periodic table form insoluble compounds with $Pb^{2+}(aq)$, so we predict that $PbI_2(s)$ will precipitate (Figure 4.20).

Because a reaction takes place, let's write a balanced equation that describes the process, starting with the molecular equation:

Unbalanced: $\qquad Pb(NO_3)_2(aq) + NaI(aq) \rightarrow PbI_2(s) + NaNO_3(aq)$

Balanced: $\qquad Pb(NO_3)_2(aq) + 2\,NaI(aq) \rightarrow PbI_2(s) + 2\,NaNO_3(aq)$

**FIGURE 4.20** (a) One beaker contains a 0.1 $M$ solution of $Pb(NO_3)_2$, and the other contains a 0.1 $M$ solution of NaI. Both solutions are colorless. (b) As the NaI solution is poured into the $Pb(NO_3)_2$ solution, a yellow precipitate of $PbI_2$ forms.

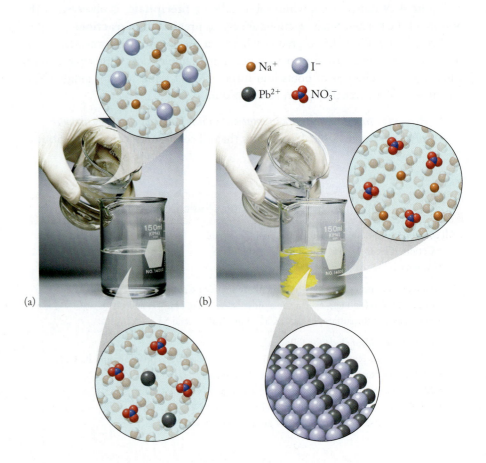

From the balanced molecular equation we get the overall ionic equation:

$$Pb^{2+}(aq) + 2\,NO_3^-(aq) + 2\,Na^+(aq) + 2\,I^-(aq)$$
$$\rightarrow PbI_2(s) + 2\,Na^+(aq) + 2\,NO_3^-(aq)$$

Removing the spectator ions $NO_3^-$ and $Na^+$ gives us the net ionic equation:

$$Pb^{2+}(aq) + 2\,I^-(aq) \rightarrow PbI_2(s)$$

*Soluble* and *insoluble* are qualitative terms. In principle, all ionic compounds dissolve in water to some extent. In practice, we consider a compound to be insoluble in water if the maximum amount that dissolves gives a concentration of less than 0.01 *M*. At this level, a solid appears to be insoluble to the naked eye; the tiny amount that dissolves is negligible compared with the amount that remains in the solid state and in contact with the solvent.

---

**SAMPLE EXERCISE 4.11**   **Writing Net Ionic Equations for Precipitation Reactions**   **LO5, LO6**

Write a balanced net ionic equation for the reaction between aqueous sulfuric acid and barium hydroxide in Figure 4.19.

**Collect and Organize** We are asked to write a balanced net ionic equation given the names of the reactants.

**Analyze** Figure 4.19 shows that the reaction produces a white precipitate. Table 4.5 tells us that this precipitate is barium sulfate, $BaSO_4$. First, we write a balanced molecular equation for the reaction. Next we identify which reactants and products are soluble in water. Write any soluble ionic compounds as ions, leaving any molecular compounds unchanged. Finally, we simplify the overall ionic equation by removing any ions that appear on both sides of the equation to obtain the net ionic equation.

**Solve**
1. Molecular equation: $Ba(OH)_2(aq) + H_2SO_4(aq) \rightarrow BaSO_4(s) + 2\,H_2O(\ell)$
2. Overall ionic equation:

$$Ba^{2+}(aq) + 2\,OH^-(aq) + H^+(aq) + HSO_4^-(aq) \rightarrow BaSO_4(s) + 2\,H_2O(\ell)$$

3. The equation can not be simplified any further, so the net ionic equation is the same as the overall ionic equation in this case.

**Think About It** The reaction between barium hydroxide and sulfuric acid is *both* a neutralization and a precipitation reaction. The neutralization is indicated by the formation of a salt ($BaSO_4$) and water; because the salt is insoluble, the reaction is also a precipitation reaction.

⚙ **Practice Exercise** Write a balanced net ionic equation for the reaction between barium hydroxide and phosphoric acid ($H_3PO_4$) in water.

*(Answers to Practice Exercises are in the back of the book.)*

---

**SAMPLE EXERCISE 4.12**   **Writing Equations for Precipitation Reactions**   **LO5, LO6**

A precipitate forms when aqueous solutions of ammonium sulfate and barium chloride are mixed. Write the net ionic equation for the reaction. Is this the same net ionic equation as in Sample Exercise 4.11?

**Collect and Organize** Mixing two solutions precipitates an insoluble compound. Using the names of the reactants, we are asked to identify the precipitate.

**Analyze** The two salts are in solution, so they both must be soluble in water. This is consistent with the solubility rules in Table 4.5. We need to determine which combinations of the ions produce insoluble solids. First, we need to write correct formulas of the salts. Then we can check Table 4.5 to see which of the possible combinations of ions are insoluble in water.

**Solve**

Solution 1: $(NH_4)_2SO_4(aq) \rightarrow 2\,NH_4^+(aq) + SO_4^{2-}(aq)$

Solution 2: $BaCl_2(aq) \rightarrow Ba^{2+}(aq) + 2\,Cl^-(aq)$

The new combinations are $BaSO_4$ and $NH_4Cl$. Table 4.5 indicates that all ammonium salts are soluble, so the ammonium and chloride ions remain dissolved. Barium sulfate is insoluble, however, and that salt precipitates. The overall ionic equation describes the species in solution:

$$\cancel{2\,NH_4^+(aq)} + SO_4^{2-}(aq) + Ba^{2+}(aq) + \cancel{2\,Cl^-(aq)} \rightarrow \cancel{2\,NH_4^+(aq)} + \cancel{2\,Cl^-(aq)} + BaSO_4(s)$$

We can simplify the equation by eliminating the spectator ions to yield the net ionic equation, which describes the formation of the precipitate:

$$Ba^{2+}(aq) + SO_4^{2-}(aq) \rightarrow BaSO_4(s)$$

This net ionic equation is not the same as for the reaction between barium hydroxide and sulfuric acid in Sample Exercise 4.11. The net ionic equation in Sample Exercise 4.11 is the sum of two balanced net ionic equations:

$$Ba^{2+}(aq) + SO_4^{2-}(aq) \rightarrow BaSO_4(s)$$

and

$$H^+(aq) + OH^-(aq) \rightarrow H_2O(\ell)$$

**Think About It** To identify whether or not a precipitation reaction occurs when you mix two solutions containing ions, check whether an insoluble compound is formed when the ions change partners. In this exercise, sulfate ions originally in the salt ammonium sulfate partner with barium ions originally in the salt barium chloride, forming insoluble $BaSO_4$. The other new combination of ions, $NH_4^+$ and $Cl^-$, remains in solution.

⚙ **Practice Exercise** Does a precipitate form when you mix aqueous solutions of (a) sodium acetate and ammonium sulfate; (b) calcium chloride and mercury(I) nitrate? (c) If you answered yes in either case, write the net ionic equation for the reaction.

*(Answers to Practice Exercises are in the back of the book.)*

Precipitation reactions can be used to synthesize water-insoluble salts. For example, barium sulfate, used in medical procedures to image the gastrointestinal tract, can be made by mixing an aqueous solution of any water-soluble barium salt with a solution of a soluble sulfate salt. The precipitate from such a reaction can be collected by filtration, dried, and used for whatever purpose we may have. The soluble compound remaining in solution can also be collected. The ammonium chloride left in solution in Sample Exercise 4.12, for instance, can be isolated by boiling off the water to leave a residue of solid $NH_4Cl$.

**CONCEPT TEST**

Lead(II) dichromate ($PbCr_2O_7$; the dichromate ion is $Cr_2O_7^{2-}$) is a water-insoluble pigment called school bus yellow that is used to paint lines on highways. Design a synthesis of school bus yellow that uses a precipitation reaction.

*(Answers to Concept Tests are in the back of the book.)*

## Using Precipitation in Analysis

Chemists use the insolubility of ionic compounds to determine concentrations of ions in solution. For example, NaCl (in the form of *rock salt*) is used to melt ice and snow on roads during the winter. As the resulting NaCl solutions run off into nearby water supplies, that water becomes contaminated with high levels of sodium and chloride ions. To determine if sources of drinking water have become contaminated with such runoff, analytical chemists can determine the concentration of chloride ion by reacting a sample of the water with a solution of silver nitrate, $AgNO_3(aq)$. Any chloride ions in the water combine with $Ag^+$ ions to form a precipitate of AgCl:

$$NaCl(aq) + AgNO_3(aq) \rightarrow AgCl(s) + NaNO_3(aq) \qquad (4.10)$$

This precipitate can be filtered and dried, and its mass determined. From its mass, we can calculate the concentration of chloride in the water sample.

Equation 4.10 is the molecular equation describing the formation of AgCl. Because both reactants are soluble ionic compounds and completely dissociate in solution, the overall ionic equation is

$$Na^+(aq) + Cl^-(aq) + Ag^+(aq) + NO_3^-(aq) \rightarrow AgCl(s) + Na^+(aq) + NO_3^-(aq) \quad (4.11)$$

Eliminating the spectator ions $Na^+(aq)$ and $NO_3^-(aq)$ yields the net ionic equation:

$$Ag^+(aq) + Cl^-(aq) \rightarrow AgCl(s) \qquad (4.12)$$

When performing this analysis, we must add enough $Ag^+$ ions to ensure that all the $Cl^-$ ions precipitate. Specifically, the molar ratio $Ag^+(aq)/Cl^-(aq)$ must be greater than one. In this situation, the $Ag^+(aq)$ is *in excess*. Precipitation reactions such as this may also include indicators that produce a visual signal (like a color change) that defines when the reaction is complete.

**CONNECTION** In Chapter 3 we introduced the idea of limiting reactants and reactants in excess when calculating the yield of reactions.

---

**SAMPLE EXERCISE 4.13** **Calculating the Mass of a Precipitate** **LO6**

When hydrofluoric acid gets on the skin, it migrates away from the site of exposure and shuts down the microcapillaries that carry blood. The clinical management of burns on the skin from exposure to hydrofluoric acid, $HF(aq)$, may include the injection of a soluble calcium salt in the skin near the site of contact to stop its migration. As $F^-$ ions from the acid move through the skin, they combine with $Ca^{2+}$ ions and form deposits of insoluble $CaF_2$. If 1.00 mL of a 2.24 $M$ aqueous solution of $Ca^{2+}$ is injected, what is the mass of $CaF_2$ produced if all the calcium ions react with fluoride ions?

**Collect and Organize** We are given the volume and concentration of a solution of $Ca^{2+}$ ions and asked to determine the maximum amount of $CaF_2$ precipitate that could be formed on reaction with $F^-$ ions.

**Analyze** The net ionic reaction for the process involved is

$$Ca^{2+}(aq) + 2\,F^-(aq) \rightarrow CaF_2(s)$$

The problem states that all the calcium ions react, so we calculate the mass of product assuming $Ca^{2+}$ ions are the limiting reactant.

$$\boxed{\text{L } Ca^{2+} \text{ solution}} \xrightarrow{\frac{\text{mol } Ca^{2+}}{\text{L}}} \boxed{\text{mol } Ca^{2+}} \xrightarrow{\frac{\text{mol } CaF_2}{\text{mol } Ca^{2+}}} \boxed{\text{mol } CaF_2} \xrightarrow{\text{molar mass}} \boxed{\text{g } CaF_2}$$

**Solve** We can calculate the number of moles of $Ca^{2+}(aq)$ and use the stoichiometric relationships in the net ionic equation to determine the mass of product:

$$1.00 \times 10^{-3}\ \cancel{\text{L}} \times \frac{2.24\ \cancel{\text{mol } Ca^{2+}}}{1\ \cancel{\text{L}}} \times \frac{1\ \cancel{\text{mol } CaF_2}}{1\ \cancel{\text{mol } Ca^{2+}}} \times \frac{78.08\ \text{g } CaF_2}{1\ \cancel{\text{mol } CaF_2}} = 0.175\ \text{g } CaF_2$$

**Think About It** This problem asks for the mass of $CaF_2$ that could be formed, assuming $Ca^{2+}$ ions were the limiting reactant. However, when this treatment is actually used, it is more likely that the $Ca^{2+}$ ions will be in excess to make sure all the fluoride ions are consumed.

⚙ **Practice Exercise** Vermillion is a very rare and expensive solid natural pigment. Known as Chinese red, it is the pigment used to print the author's signature on works of art. It consists of mercury(II) sulfide (HgS) and is very insoluble in water. What is the maximum number of grams of vermillion you can make from the reaction of 50.00 mL of a 0.0150 $M$ aqueous solution of mercury(II) nitrate $[Hg(NO_3)_2]$ with an excess of a concentrated aqueous solution of sodium sulfide ($Na_2S$)?

*(Answers to Practice Exercises are in the back of the book.)*

---

**SAMPLE EXERCISE 4.14  Calculating a Solute Concentration    LO6
from Mass of a Precipitate**

To determine the concentration of chloride ion in a 100.0 mL sample of groundwater, a chemist adds a large enough volume of a solution of $AgNO_3$ to the sample to precipitate all the $Cl^-$ as AgCl. The mass of the resulting AgCl precipitate is 71.7 mg. What is the chloride concentration in milligrams of $Cl^-$ per liter of groundwater?

**Collect and Organize** We are given the sample volume, 100.0 mL, and the mass of AgCl formed, 71.7 mg. Our task is to determine the chloride concentration in milligrams of $Cl^-$ per liter of groundwater.

**Analyze** We need a balanced chemical equation for the reaction. Although we can use the molecular, overall ionic, or net ionic equation, we choose the net ionic equation because it contains only those species involved in the precipitation reaction. The net ionic equation is

$$Ag^+(aq) + Cl^-(aq) \rightarrow AgCl(s)$$

It shows that 1 mole of $Cl^-$ ions reacts with 1 mole of $Ag^+$ ions to produce 1 mole of AgCl. The molar mass of $Cl^-$ is 35.45 g/mol, and that of AgCl is 143.32 g/mol. We use the methods developed in Chapter 3 to determine the mass of chloride ion in 100.0 mL of groundwater and then use this mass to calculate the concentration we need.

$$\boxed{\text{g AgCl}} \xrightarrow[\text{molar mass}]{1} \boxed{\text{mol AgCl}} \xrightarrow[\text{1 mol AgCl}]{\text{1 mol Cl}^-} \boxed{\text{mol Cl}^-} \xrightarrow[\text{molar mass}]{} \boxed{\text{g Cl}^-} \xrightarrow[\text{g}]{10^3 \text{ mg}} \boxed{\text{mg Cl}^-} \xrightarrow[\text{mL}]{\dfrac{1}{100.0 \text{ mL}}} \boxed{\dfrac{\text{mg Cl}^-}{\text{mL}}} \xrightarrow[\dfrac{L}{10^3 \text{ mL}}]{} \boxed{\dfrac{\text{mg Cl}^-}{\text{L}}}$$

**Solve** Because our molar masses are in grams per mole, a logical first step is to convert the AgCl(s) mass to grams:

$$71.7 \text{ mg} \times \frac{10^{-3} \text{ g}}{1 \text{ mg}} = 0.0717 \text{ g}$$

The mass of $Cl^-$ ions in the 100.0 mL sample of groundwater is therefore

$$0.0717 \text{ g AgCl}(s) \times \frac{1 \text{ mol AgCl}(s)}{143.32 \text{ g AgCl}(s)} \times \frac{1 \text{ mol Cl}^-(aq)}{1 \text{ mol AgCl}(s)}$$

$$\times \frac{35.45 \text{ g Cl}^-(aq)}{1 \text{ mol Cl}^-(aq)} = 0.0177 \text{ g Cl}^-(aq) = 17.7 \text{ mg Cl}^-(aq)$$

The $Cl^-$ ion concentration in milligrams per liter of water is

$$\frac{17.7 \text{ mg Cl}^-}{100.0 \text{ mL}} \times \frac{1000 \text{ mL}}{1 \text{ L}} = 177 \text{ mg Cl}^-/\text{L groundwater}$$

An alternative way to solve this problem is to determine the mass of $Cl^-$ ions in 1 milligram of AgCl(s):

$$\frac{1 \text{ mol Cl}^-}{1 \text{ mol AgCl}} \times \frac{35.45 \text{ g Cl}^-/\text{mol Cl}^-}{143.32 \text{ g AgCl/mol AgCl}} = \frac{0.2473 \text{ g Cl}^-}{\text{g AgCl}} = \frac{0.2473 \text{ mg Cl}}{\text{mg AgCl}}$$

Using this factor with the mass of AgCl(s) precipitated, we get

$$71.7 \; \text{mg AgCl} \times \frac{0.2743 \; \text{mg Cl}^-}{1 \; \text{mg AgCl}} = 17.7 \; \text{mg Cl}^- \text{ precipitated as AgCl}(s)$$

From here, the problem proceeds as before.

**Think About It** In Sample Exercise 4.1 we noted that the chloride concentration of drinking water should not exceed 250 ppm. Since mg/L is equivalent to ppm for dilute solutions, the concentration of $Cl^-$ in the sample for this exercise is 177 ppm, which meets the guidelines for drinking water.

**Practice Exercise** The concentration of $SO_4^{2-}(aq)$ in a 50.0 mL sample of river water is determined using a precipitation titration in which the titrant cation is $Ba^{2+}(aq)$ and $BaSO_4(s)$ is the precipitate. What is the $SO_4^{2-}(aq)$ concentration in molarity if 6.55 mL of a $0.00100 \; M \; Ba^{2+}(aq)$ solution is consumed?

*(Answers to Practice Exercises are in the back of the book.)*

---

**SAMPLE EXERCISE 4.15**  **Limiting Reactants in Solution Reactions**  **LO6**

Acid rain in the form of aqueous $H_2SO_4$ slowly dissolves marble statues, as shown in Figure 4.15. If we add 16.75 mL of $1.00 \times 10^{-2} \; M \; H_2SO_4$ to 25.00 mg of $CaCO_3$, will all of the $CaCO_3$ dissolve?

**Collect and Organize** Given the volume and molarity of the $H_2SO_4$ solution, we can calculate the mass of $CaCO_3$ that reacts.

**Analyze** We need the balanced chemical equation for the reaction to determine the stoichiometry of the reaction between $CaCO_3$ and $H_2SO_4$. The molecular equation is

$$CaCO_3(s) + H_2SO_4(aq) \rightarrow CaSO_4(aq) + H_2O(\ell) + CO_2(g)$$

One mole of $H_2SO_4(aq)$ reacts with 1 mole of $CaCO_3(s)$. The steps involved in calculating the moles of $H_2SO_4$ available and the mass (mg) of calcium carbonate that will react with it are

$$\boxed{\text{mL } H_2SO_4} \xrightarrow{\frac{L}{10^3 \; mL}} \boxed{\text{L } H_2SO_4} \xrightarrow{\frac{mol \; H_2SO_4}{L}} \boxed{\text{mol } H_2SO_4} \xrightarrow{\frac{1 \; mol \; CaCO_3}{1 \; mol \; H_2SO_4}} \boxed{\text{mol } CaCO_3}$$

$$\xrightarrow{\text{molar mass}} \boxed{\text{g } CaCO_3} \xrightarrow{\frac{10^3 \; mg}{g}} \boxed{\text{mg g } CaCO_3}$$

**Solve** We calculate the mass of calcium carbonate that reacts starting with the volume of sulfuric acid available and its molarity:

$$16.75 \; \text{mL } H_2SO_4 \times \frac{1 \; \text{L}}{10^3 \; \text{mL}} \times \frac{1.00 \times 10^{-2} \; \text{mol } H_2SO_4}{1 \; \text{L } H_2SO_4} \times \frac{1 \; \text{mol } CaCO_3}{1 \; \text{mol } H_2SO_4}$$

$$\times \frac{100.09 \; \text{g } CaCO_3}{1 \; \text{mol } CaCO_3} \times \frac{1000 \; \text{mg } CaCO_3}{1 \; \text{g } CaCO_3} = 16.8 \; \text{mg } CaCO_3$$

Only 16.8 mg of the 25.00 mg sample will dissolve to form soluble calcium sulfate (Table 4.5).

**Think About It** The sulfuric acid is the limiting reactant, as only 67% of the $CaCO_3$ dissolves under the conditions in this exercise.

**Practice Exercise** Heavy metal ions like lead(II) can be precipitated from laboratory wastewater by adding sodium sulfide, $Na_2S$. Will all of the lead be removed from 15.00 mL of $7.52 \times 10^{-3} \; M \; Pb(NO_3)_2$ upon addition of 12.50 mL of 0.0115 M $Na_2S$?

*(Answers to Practice Exercises are in the back of the book.)*

## Saturated Solutions and Supersaturation

**CHEMTOUR** Saturated Solutions

A solution containing the maximum amount of solute that can dissolve is a **saturated solution** (Figure 4.21). The maximum amount of solute that can dissolve in a given quantity of solvent at a given temperature is called the **solubility** of that solute in the solvent. Precipitates form when the solubility of a solute is exceeded.

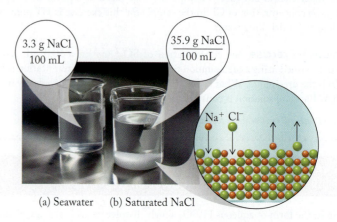

$$\frac{3.3 \text{ g NaCl}}{100 \text{ mL}}$$

$$\frac{35.9 \text{ g NaCl}}{100 \text{ mL}}$$

$Na^+$ $Cl^-$

(a) Seawater     (b) Saturated NaCl

**FIGURE 4.21** (a) Seawater, although a concentrated aqueous solution of NaCl, is far from saturated. (b) This solution is in equilibrium with solid NaCl (the white mass at the bottom of the beaker) at 20°C, so the liquid is a saturated solution of NaCl. Crystals in the solid are constantly dissolving, while ions in the solution are constantly precipitating. Because of this dynamic equilibrium, the concentration of solute [$Na^+(aq)$ and $Cl^-(aq)$ ions] in the solvent remains constant.

Solubility depends on temperature; often, the higher the temperature, the greater the solubility of solids in water. For example, more table sugar dissolves in hot water than in cold, a fact applied in the making of rock candy (Figure 4.22). After a large amount of sugar is dissolved in hot water, the solution is slowly cooled. An object with a slightly rough surface (like a string or a wooden stirrer), suspended in the solution serves as a site for crystallization. As the solution cools below the temperature at which the solubility of sugar is exceeded, crystals of sugar grow on the rough surface. The formation of crystals as the temperature of a saturated solution is lowered is a form of precipitation.

Sometimes, more solute dissolves in a volume of liquid than the amount predicted by the solute's solubility in that liquid, creating a **supersaturated solution**. This can happen when the temperature of a saturated solution drops slowly or when the volume of an unsaturated solution is reduced through slow evaporation. Slow evaporation of the solvent is partly responsible for the formation of stalactites and stalagmites, as well as of the salt flats in the American Southwest. Sooner or later, usually in response to a disruption such as a change in

(a) $T$ = close to boiling point

(b) Put in a wooden stirrer

(c) Cool to room temperature

(d) Rock candy

**FIGURE 4.22** Making rock candy. (a) A large quantity of table sugar is dissolved in hot water. The solution is not saturated at this high temperature, however. (b) A wooden stirrer is added and the solution is allowed to cool. (c) As the solution cools, it reaches the temperature at which the concentration of sugar exceeds the maximum that can remain in solution, and crystals ("rocks") of solid sugar ("candy") precipitate and attach to the stirrer. (d) Rock candy.

(a)                          (b)                          (c)

**FIGURE 4.23** (a) A seed crystal is added to a supersaturated solution of sodium acetate. (b) The seed crystal becomes a site for rapid growth of sodium acetate crystals. (c) Crystal growth continues until the solution is no longer supersaturated but merely saturated with sodium acetate.

temperature, a mechanical shock, or the addition of a *seed crystal* (a small crystal of the solute that provides a site for crystallization), the solute in a supersaturated solution rapidly comes out of solution (Figure 4.23).

## 4.8 Ion Exchange

We have seen how dissolved ions can exchange partners, forming insoluble precipitates. This idea of **ion exchange** is important in water purification. Water containing certain metal ions—principally $Ca^{2+}$ and $Mg^{2+}$—is called *hard water*, and it causes problems in industrial water supplies and in homes. Because hard water combines with soap to form a gray scum, clothes washed in hard water appear gray and dull. Hard water forms scale (an incrustation) in boilers, pipes, and kettles, diminishing their ability to conduct heat and carry water; and hard water sometimes has an unpleasant taste. In one common method of *water softening* (removing the ions responsible for hard water), hard water is passed through a system that exchanges calcium and magnesium ions for sodium ions (Figure 4.24).

The system consists of a cartridge packed with beads of a porous plastic resin (R) bonded to anions capable of binding with hard-water cations—mainly $Ca^{2+}$, $Mg^{2+}$, and $Fe^{2+}$. The carboxylate anion ($COO^-$) is often used, and water in the ion-exchange cartridge contains $Na^+$ to balance the negative charges on the anions. We can therefore think of the resin as beads containing $(R—COO^-)Na^+$ units. As hard water flows through the cartridge, 2+ ions in the water exchange places with sodium ions on the resin. This exchange takes place because cations with 2+ and 3+ charges bind more strongly to the $R—COO^-$ groups on the resin than sodium cations do. The ion-exchange reaction, with calcium ion as an example, is

$$2(R—COO^-)Na^+(s) + Ca^{2+}(aq) \rightarrow (R—COO^-)_2Ca^{2+}(s) + 2\,Na^+(aq)$$

Hard water that has been softened in this way contains increased concentrations of sodium ion. Although this may not be a problem for healthy children and adults, people suffering from high blood pressure often must limit their intake of $Na^+$ and so should not drink water softened by this kind of ion-exchange reaction.

**saturated solution** a solution that contains the maximum concentration of a solute possible at a given temperature.

**solubility** the maximum amount of a substance that dissolves in a given quantity of solvent at a given temperature.

**supersaturated solution** a solution that contains more than the maximum quantity of solute predicted to be soluble in a given volume of solution at a given temperature.

**ion exchange** a process by which one ion is displaced by another.

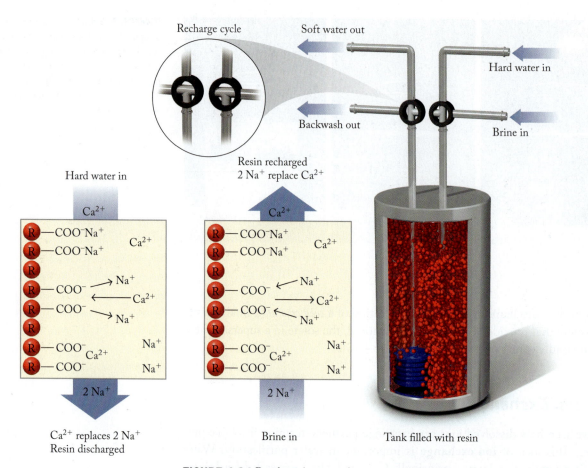

**FIGURE 4.24** Residential water softeners use ion exchange to remove 2+ ions (such as $Ca^{2+}$) that make water hard. The ion-exchange resin contains cation-exchange sites that are initially occupied by $Na^+$ ions. These ions are replaced by 2+ "hardness" ions as water flows through the resin. Eventually most of the ion-exchange sites are occupied by 2+ ions, and the system loses its water-softening ability. The resin is then backwashed with a saturated solution of NaCl (*brine*), displacing the hardness ions (which wash down the drain) and restoring the resin to its $Na^+$ form.

Not only synthetic materials perform ion exchange. Naturally occurring minerals called **zeolites** are extensively mined in many parts of the world and have huge commercial and technological utility as water softeners and purifiers, livestock feed additives, and odor suppressants. Zeolites, formed in nature as a result of the chemical reaction between molten lava from volcanic eruptions and salt water, are crystalline solids with tiny pores. Synthetic zeolites have been manufactured that have structures similar to the natural materials. All zeolites have a rigid three-dimensional structure like a honeycomb (Figure 4.25), consisting of a network of interconnecting tunnels and cages. They work just like the plastic resins: as water flows through tunnels (pores) lined with $Na^+$ ions, the sodium ions exchange with cations dissolved in the water.

Zeolite production reached 4.0 million tons in 2012 with numerous zeolite-containing products sold worldwide. Zeolites have replaced environmentally harmful phosphates that are added to detergents to bind $Ca^{2+}$ and $Mg^{2+}$ ions. They are used in municipal water filtration plants to treat drinking water, to remove heavy metals like $Pb^{2+}(aq)$ from contaminated waste streams, and in swimming pools to keep the water clean and clear. They are even used for odor control in barns and feedlots where animals are confined. Animal waste contains large quantities of

**zeolite** natural crystalline minerals or synthetic materials consisting of three-dimensional networks of channels that contain sodium or other 1+ cations.

**FIGURE 4.25** A sample of zeolite with a drawing showing the regular pattern of pores containing sodium ions that exchange for other cations dissolved in water that flows through the material.

ammonium ion [$NH_4^+(aq)$] that can be exchanged for the naturally occurring cations in zeolites. When $NH_4^+(aq)$ ions participate in acid–base reactions with basic materials (B = base), they liberate ammonia [$NH_3(g)$], which is partially responsible for the characteristic smell of cat boxes and barnyards:

$$NH_4^+(aq) + B \rightarrow NH_3(g) + HB^+(aq)$$

Zeolites reduce odors by selectively removing $NH_4^+(aq)$ through ion exchange.

Other ions can be exchanged for the sodium ion to prepare zeolites with special properties. Certain zeolites may be poured directly on wounds to stop bleeding by absorbing water from the blood, thereby concentrating clotting factors to promote coagulation. If the cations in these zeolites are exchanged for silver ion ($Ag^+$), the resulting product also has an antimicrobial effect and not only stems bleeding but reduces the risk of infection.

## 4.9 Oxidation–Reduction Reactions: Electron Transfer

In the 15th century, Leonardo da Vinci observed: "Where flame cannot live, no animal that draws breath can live." Although oxygen was not isolated and recognized as an element until the 18th century, Leonardo captured an important property of the substance as a requirement for combustion and metabolism. Oxygen makes up about 50% by mass of Earth's crust, it is almost 89% by mass of the water that covers 70% of Earth's surface, and it is about 20% of the volume of Earth's atmosphere. It is essential to the metabolism of most creatures, just as it is essential for combustion reactions. What type of reaction is involved in metabolism and in the combustion reactions in Figure 4.26? Methane and oxygen are neither acids nor bases. No precipitates are formed in the reactions, which produce water and carbon dioxide.

Combination reactions of oxygen with nonmetals such as carbon, sulfur, and nitrogen produce volatile oxides. Combination reactions of $O_2$ with metals and

**CONNECTION** In Chapter 3 we defined combustion as a heat-producing reaction between oxygen and another element or compound.

(a)

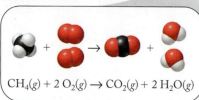

$$CH_4(g) + 2\,O_2(g) \rightarrow CO_2(g) + 2\,H_2O(g)$$

(b)

**FIGURE 4.26** (a) Forest fires and (b) the combustion of methane are oxidation–reduction reactions.

semimetals produce solid oxides. All of these are examples of an extremely important type of reaction: oxidation–reduction reactions. *Oxidation* was first defined as a reaction that increased the oxygen content of a substance. Hence two reactions involving oxygen—with methane to produce $CO_2(g)$ and $H_2O(\ell)$, and with elemental iron to produce $Fe_2O_3(s)$—are both oxidations from the point of view of the $CH_4$ and the Fe because the products contain more oxygen than the starting materials:

$$CH_4(g) + 2\,O_2(g) \rightarrow CO_2(g) + 2\,H_2O(\ell)$$
$$4\,Fe(s) + 3\,O_2(g) \rightarrow 2\,Fe_2O_3(s)$$

*Reduction* reactions were originally defined in a similar fashion—reactions in which the oxygen content of a substance is reduced. A classic example is the reduction of iron ore (mostly $Fe_2O_3$) to metallic iron:

$$Fe_2O_3(s) + 3\,CO(g) \rightarrow 2\,Fe(s) + 3\,CO_2(g)$$

The oxygen content of the iron ore is reduced—$Fe_2O_3$ is converted into Fe—as a result of its reaction with $CO(g)$. Note that the carbon in CO is oxidized because the oxygen content of CO increases as it is converted into $CO_2$.

**CONCEPT TEST**

The following equation describes the conversion of one iron mineral, called magnetite ($Fe_3O_4$), into another, called hematite ($Fe_2O_3$):

$$4\,Fe_3O_4(s) + O_2(g) \rightarrow 6\,Fe_2O_3(s)$$

Identify which element is oxidized and which is reduced and explain your choices.

*(Answers to Concept Tests are in the back of the book.)*

## Oxidation Numbers

Ultimately, the meanings of oxidation and reduction were expanded, so that **oxidation** now refers to any chemical reaction in which a substance *loses electrons* and **reduction** refers to any reaction in which a substance *gains electrons*.[3] The processes of oxidation and reduction always occur together. If one substance loses electrons, another substance must gain them; if one species is oxidized, another must be reduced. The defining event in oxidation–reduction is the *transfer* of electrons from one substance to another. The recognition that oxidation and reduction must occur together gives rise to the common way of referring to them as *redox reactions*.

To keep track of electron loss and gain and to distinguish between oxidation and reduction in redox reactions, chemists use a system of assigning oxidation numbers to atoms in reactants and products. If the oxidation number of an atom changes as a result of a reaction—if the oxidation number of an atom is different on the left-hand side of an equation than on the right-hand side—then the reaction is redox. The change in oxidation number reflects the change in the number of electrons associated with an atom or ion. This change in the number of electrons happens because electrons are transferred from one atom or ion to another during the course of the reaction.

---

[3]The mnemonic OIL RIG may be helpful for remembering these expanded definitions: "Oxidation Is Loss; Reduction Is Gain."

The **oxidation number (O.N.)**, or **oxidation state**, of an atom in a molecule or ion may be a positive or negative number or zero. The oxidation number represents the positive or negative character an atom has or appears to have as determined by the following rules:

1. The oxidation numbers of the atoms in a neutral molecule sum to zero; those of the atoms in an ion sum to the charge on the ion.
2. Each atom in a pure element has an oxidation number of zero:

| | | | |
|---|---|---|---|
| $F_2$ | O.N. = 0 for each F | $O_2$ | O.N. = 0 for each O |
| Fe | O.N. = 0 | Na | O.N. = 0 |

3. In monatomic ions, the oxidation number is the charge on the ion:

| | | | |
|---|---|---|---|
| $F^-$ | O.N. = −1 | $O^{2-}$ | O.N. = −2 |
| $Fe^{3+}$ | O.N. = +3 | $Na^+$ | O.N. = +1 |

Note that the number comes first when we write the symbol of an ion, $Fe^{3+}$, but the sign comes first when we write an oxidation number, +3.

4. In compounds containing fluorine and one or more other elements, the oxidation number of the fluorine is *always* −1:

| | |
|---|---|
| KF | O.N. = −1 for F, which means O.N. = +1 for K (rule 1) |
| $OF_2$ | O.N. = −1 for each F, which means O.N. = +2 for O (rule 1) |
| $CF_4$ | O.N. = −1 for each F, which means O.N. = +4 for C (rule 1) |

5. In most compounds, the oxidation number of hydrogen is +1 and that of oxygen is −2. Exceptions include hydrogen in metal hydrides (for example, LiH), where O.N. = −1 for H, and oxygen in the peroxide ion, $O_2^{2-}$ (for example, $H_2O_2$), where O.N. = −1 for each O.
6. Unless combined with oxygen or fluorine, chlorine and bromine have an oxidation number of −1:

| | |
|---|---|
| $CaCl_2$ | O.N. = −1 for Cl, which means O.N. = +2 for Ca (rule 1) |
| $AlBr_3$ | O.N. = −1 for Br, which means O.N. = +3 for Al (rule 1) |

*but*

| | |
|---|---|
| $ClO_4^-$ | O.N. = −2 for O (rule 5), which means O.N. = +7 for Cl (rule 1) |

**oxidation** a chemical change in which a species loses electrons; the oxidation number of the species increases.

**reduction** a chemical change in which a species gains electrons; the oxidation number of the species decreases.

**oxidation number (O.N.)** (also called **oxidation state**) a positive or negative number based on the number of electrons an atom gains or loses when it forms an ion, or that it shares when it forms a covalent bond with another element; pure elements have an oxidation number of zero.

---

**SAMPLE EXERCISE 4.16**   **Determining Oxidation Numbers**   **LO7**

What is the oxidation number of sulfur in (a) $SO_2$, (b) $Na_2S$, and (c) $CaSO_4$?

**Collect and Organize** We are to assign the oxidation number of sulfur in three of its compounds.

**Analyze** We apply the rules for determining the oxidation numbers of the elements in each compound. $SO_2$ is the molecule sulfur dioxide. $Na_2S$ is a binary ionic compound, and the oxidation number of an ion in an ionic compound equals its charge. Sulfur in $CaSO_4$ is part of the sulfate ion, which means its oxidation number added to those of the four O atoms must add up to the charge of the ion; we determine the oxidation number for calcium from rule 1.

**Solve**

a. From rule 5, O.N. = −2 for the O in $SO_2$. Rule 1 says that the sum of the oxidation numbers for S and the two O in the neutral molecule must be zero. Letting O.N. for S be $x$:

$$x + 2(-2) = 0$$
$$x = +4 \qquad \text{O.N. for S in } SO_2 = +4$$

b. Sodium forms only one ion, $Na^+$, which means that, according to rule 3, O.N. = +1. To balance the O.N. values in $Na_2S$ (rule 1), we let $y$ stand for the O.N. of sulfur:

$$2(+1) + y = 0$$
$$y = -2 \qquad \text{O.N. for S in } Na_2S = -2$$

c. The charge on the calcium ion is always 2+. This means that the charge on the sulfate ion must be 2−. Assigning O.N. = −2 for oxygen (rule 5) and $z$ for sulfur:

$$z + 4(-2) = -2$$
$$z = +6 \qquad \text{O.N. for S in } CaSO_4 = +6$$

**Think About It** We found oxidation numbers for sulfur ranging from −2 to +6. Many other elements have a range of oxidation numbers in their compounds. For example, the oxidation numbers of iodine range from −1 in KI to +7 in $KIO_4$.

⚙ **Practice Exercise** Determine the oxidation number of nitrogen in (a) $NO_2$, (b) $N_2O$, and (c) $HNO_3$.

*(Answers to Practice Exercises are in the back of the book.)*

# Considering Electron Transfer in Redox Reactions

Because redox reactions involve the transfer of electrons among atoms, we must learn how to track changes in oxidation numbers when we consider these reactions. To see how this works, let's look at a reaction that we balanced in Chapter 3: the combustion of methane ($CH_4$) to produce $CO_2$ and $H_2O$.

$$CH_4(g) + 2\,O_2(g) \rightarrow CO_2(g) + 2\,H_2O(g) \qquad (4.12)$$

First we assign oxidation numbers to all the atoms in the reactants and products:

| OXIDATION NUMBERS | | | | |
|---|---|---|---|---|
| **Atoms in Reactants** | | | **Atoms in Products** | |
| H in $CH_4$: | +1 | (rule 2) | H in $H_2O$: | +1 |
| O in $O_2$: | 0 | (rule 1) | O in $CO_2$: | −2 |
| | | | O in $H_2O$: | −2 |
| C in $CH_4$: | −4 | (rule 5) | C in $CO_2$: | +4 |

The oxidation numbers of the carbon and oxygen atoms change; such a change for any element defines a reaction as redox. One carbon atom on the left goes from O.N. = −4 in $CH_4$ to +4 in $CO_2$ for a net loss of 8 electrons ($e^-$). Loss of electrons is oxidation, so carbon is oxidized in this reaction. Oxygen atoms go from O.N. = 0 in $O_2$ to −2 in $CO_2$ and $H_2O$; each oxygen atom gains 2 $e^-$. In the balanced equation, there are 4 O atoms on the left and 4 O atoms on the right, so the total number of electrons gained by all the oxygen atoms is $(4 \times 2\ e^-) = 8\ e^-$. Gain of electrons is reduction, so oxygen is reduced in this reaction. The electrons gained by oxygen ($8\ e^-$) offset the electrons lost by carbon ($8\ e^-$). In redox reactions, we can keep track of the oxidation numbers and the electrons transferred by annotating the equation in the following way:

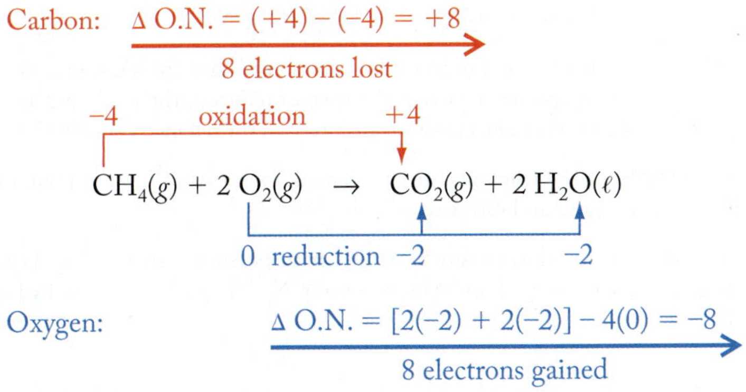

Carbon:  $\Delta$ O.N. $= (+4) - (-4) = +8$
8 electrons lost

$-4$   oxidation   $+4$

$$CH_4(g) + 2\ O_2(g) \rightarrow CO_2(g) + 2\ H_2O(\ell)$$

0 reduction $-2$   $-2$

Oxygen:  $\Delta$ O.N. $= [2(-2) + 2(-2)] - 4(0) = -8$
8 electrons gained

**oxidizing agent** a substance in a redox reaction that contains the element being reduced.

**reducing agent** a substance in a redox reaction that contains the element being oxidized.

Note that when we assigned oxidation numbers to the atoms in this reaction, the value for hydrogen did not change. This means that hydrogen is neither oxidized nor reduced in this reaction: it is not involved in our consideration of electron transfer, and we do not need to incorporate it into our annotation above.

Because oxygen in Equation 4.12 takes electrons from the methane carbon atom, oxygen is called the **oxidizing agent** in the combustion reaction. We say that "oxygen oxidizes methane." As an oxidizing agent, the oxygen is reduced. This is always the case: the *oxidizing* agent in any redox reaction is always *reduced* in the process. Any species that is reduced experiences a *reduction in oxidation number*, for instance, the oxidizing agent in Equation 4.12, where oxygen goes from O.N. $= 0$ to O.N. $= -2$.

The methane carbon atom is the species being oxidized in Equation 4.12, which means this C gives electrons to another substance and reduces that substance. Methane in this reaction is therefore called the **reducing agent**. The *reducing* agent is the source of the electrons that are transferred, and it is always *oxidized*. The material that is oxidized experiences an *increase in oxidation number*; in this reaction, the carbon atom in methane goes from O.N. $= -4$ to O.N. $= +4$.

Every redox reaction has an oxidizing agent and a reducing agent. The reducing agent is the electron donor and the oxidizing agent is the electron acceptor. Electrons are transferred from the donor to the acceptor.

**CONCEPT TEST**

Determine whether the following are redox reactions. For the redox reactions, identify the oxidizing agent and the reducing agent.

a. $Sn^{2+}(aq) + Br_2(aq) \rightarrow Sn^{4+}(aq) + 2\ Br^-(aq)$

b. $2\ F_2(g) + 2\ H_2O(\ell) \rightarrow 4\ HF(aq) + O_2(g)$

c. $NaHCO_3(aq) + HCl(aq) \rightarrow NaCl(aq) + CO_2(g) + H_2O(\ell)$

*(Answers to Concept Tests are in the back of the book.)*

**SAMPLE EXERCISE 4.17** **Identifying Oxidizing Agents and Reducing Agents and Number of Electrons Transferred** **LO7**

The reaction of oxygen with hydrazine is used to remove dissolved oxygen gas from aqueous solutions:

$$O_2(aq) + N_2H_4(aq) \rightarrow 2\,H_2O(\ell) + N_2(g)$$

and the combustion of hydrazine produces enough energy that the substance is used as a rocket fuel. Identify the species oxidized, the species reduced, the oxidizing agent, the reducing agent, and the number of electrons transferred in the balanced chemical equation.

**Collect and Organize** The reactants are the element oxygen and the compound hydrazine. The products are water and nitrogen.

**Analyze** The O.N. of the oxygen atoms in $O_2$ changes from zero to −2 in $H_2O$. The O.N. of nitrogen changes from −2 in $N_2H_4$ to zero in $N_2$. The O.N. = +1 of hydrogen is unchanged.

**Solve**

Nitrogen: Δ O.N. = 2(0) − 2(−2) = +4

4 electrons lost

−2    oxidation    0

$$O_2(aq) + N_2H_4(aq) \rightarrow 2\,H_2O(\ell) + N_2(g)$$

0    reduction    −2

Oxygen:    Δ O.N. = 2(−2) − 2(0) = −4

4 electrons gained

We see that oxygen is the oxidizing agent because the nitrogen in $N_2H_4$ loses electrons and is oxidized. That means that $O_2$ is reduced and $N_2H_4$ is the reducing agent. Four electrons are transferred in the process.

**Think About It** We tend to identify the specific *atoms* that are oxidized or reduced, whereas whole *molecules or ions* are identified as the oxidizing agents or reducing agents. In the equation in this exercise, nitrogen is oxidized but $N_2H_4$ is the reducing agent.

⚙ **Practice Exercise** In the reaction between $O_2(g)$ and $SO_2(g)$ to make $SO_3(g)$, identify the species oxidized, the species reduced, the oxidizing agent, the reducing agent, and the number of electrons transferred.

(*Answers to Practice Exercises are in the back of the book.*)

## Balancing Redox Reactions Using Half-Reactions

When a piece of copper wire [elemental copper, $Cu(s)$] is placed in a colorless solution of silver nitrate [$Ag^+(aq) + NO_3^-(aq)$], the solution gradually turns blue, the color of a solution of $Cu^{2+}(aq)$, and branchlike structures of $Ag(s)$ form in the medium (Figure 4.27). The process can be summarized as

Silver:    +1    reduction    0

$$Ag^+(aq) + Cu(s) \rightarrow Ag(s) + Cu^{2+}(aq)$$

Copper:    0    oxidation    +2

**half-reaction** one of the two halves of an oxidation–reduction reaction; one half-reaction is the oxidation component, and the other is the reduction component.

At first glance, this equation may seem balanced—and in a material sense it is: the number of silver and copper atoms or ions is the same on both sides. However, the charges are not balanced: the sum of the charges is 1+ for the reactants but 2+ for the products. The reason for this imbalance is that one electron is involved in the reduction reaction but two are involved in the oxidation reaction. Equal numbers of electrons must be lost and gained, so to balance equations describing redox reactions like this one, we must develop a method that accounts for electron transfer.

Think of this reaction as consisting of one oxidation and one reduction. Each of these reactions represents *half* of the overall reaction; hence, each is called a **half-reaction**. We balance the equation by following these five steps:

1. *Write one equation for the oxidation half-reaction and a separate equation for the reduction half-reaction*:

   Oxidation: $\qquad\qquad\qquad$ $Cu(s) \rightarrow Cu^{2+}(aq)$
   Reduction: $\qquad\qquad\qquad$ $Ag^+(aq) \rightarrow Ag(s)$

2. *Balance the number of particles in each half-reaction*. This example requires no changes at this point because both equations are balanced in terms of mass.

3. *Balance the charge in each half-reaction* by adding electrons to the appropriate side. Always *add* electrons; never subtract them. We add 2 electrons to the product side of the Cu half-reaction:

   Oxidation: $\qquad\qquad\qquad$ $Cu(s) \rightarrow Cu^{2+}(aq) + 2e^-$

   and 1 electron to the reactant side of the Ag half-reaction:

   Reduction: $\qquad\qquad\qquad$ $1\,e^- + Ag^+(aq) \rightarrow Ag(s)$

   This gives us a total charge of 0 on both sides of both half-reactions. Note that electrons are added on the reactant side of the reduction equation and on the product side of the oxidation equation. This is always the case.

4. *Multiply each half-reaction by the appropriate whole number* to make the number of electrons in the oxidation equation equal the number in the reduction equation:

   Oxidation: $\qquad\qquad\qquad$ $Cu(s) \rightarrow Cu^{2+}(aq) + 2\,e^-$
   Reduction: $\qquad$ $2 \times [1\,e^- + Ag^+(aq) \rightarrow Ag(s)]$

   As a result we have the following half-reactions:

   Oxidation: $\qquad\qquad\qquad$ $Cu(s) \rightarrow Cu^{2+}(aq) + 2\,e^-$
   Reduction: $\qquad$ $2\,e^- + 2\,Ag^+(aq) \rightarrow 2\,Ag(s)$

5. *Add the two half-reactions to generate the equation representing the redox reaction*:

   Oxidation: $\qquad\qquad\qquad$ $Cu(s) \rightarrow Cu^{2+}(aq) + \cancel{2\,e^-}$
   Reduction: $\qquad$ $\cancel{2\,e^-} + 2\,Ag^+(aq) \rightarrow 2\,Ag(s)$
   _____
   Overall equation: $2\,Ag^+(aq) + Cu(s) \rightarrow 2\,Ag(s) + Cu^{2+}(aq)$

If we have carried out step 4 correctly, the number of electrons in the oxidation half-reaction is the same as the number of electrons in the reduction half-reaction, and they cancel out when we write the net ionic equation. The reaction is now balanced in terms of mass and charge. The overall equation is a net ionic equation, which means that only species involved in the redox reaction are shown. The nitrate ions from the silver nitrate solution (Figure 4.27) are spectator ions

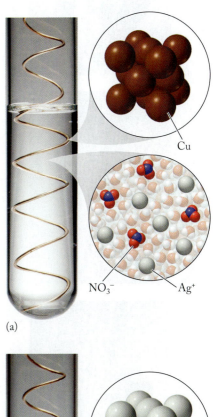

(a)

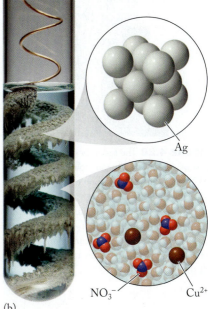

(b)

**FIGURE 4.27** (a) When a Cu wire is immersed in a solution of $AgNO_3$, Cu metal oxidizes to $Cu^{2+}$ ions as $Ag^+$ ions are reduced to Ag metal. (b) A day later, the solution has the blue color of a solution of $Cu(NO_3)_2$, and the wire is coated with Ag metal.

(a)

(b)

(c)

**FIGURE 4.28** The test tubes in these photos contain an aqueous solution of iodine. (a) When drops of a solution of $SnCl_2$ are added to the tube on the right, the yellow-brown color of the iodine solution begins to fade (b) as $Sn^{2+}$ ions reduce $I_2$ to colorless $I^-$ ions. (c) The color disappears completely when enough $Sn^{2+}$ has been added to reduce all the $I_2$.

and are not shown in the final net ionic equation or in any of the intermediate equations we used to balance it.

We will review the methods for balancing chemical equations as described here when we discuss applications of oxidation–reduction reactions in Chapter 19.

---

**SAMPLE EXERCISE 4.18** **Balancing Redox Reactions with Half-Reactions**   LO7

Iodine is slightly soluble in water and dissolves to make a yellow-brown solution of $I_2(aq)$. When colorless $Sn^{2+}(aq)$ is dissolved in it (Figure 4.28), the solution turns colorless as $I^-(aq)$ forms and $Sn^{2+}(aq)$ is converted into $Sn^{4+}(aq)$. (a) Is this a redox reaction? (b) Balance the equation that describes the reaction.

**Collect and Organize** The reactants in this process are $I_2(aq)$ and $Sn^{2+}(aq)$; the products are $I^-(aq)$ and $Sn^{4+}(aq)$.

**Analyze** The oxidation number of tin changes from +2 in $Sn^{2+}(aq)$ to +4 in $Sn^{4+}(aq)$. The oxidation number of iodine changes from 0 in $I_2(aq)$ to −1 in $I^-(aq)$. We write tin and iodine in separate equations when we write the respective half-reactions.

**Solve**
a. Because the oxidation numbers change, this is a redox reaction.
b. The unbalanced equation is

$$Sn^{2+}(aq) + I_2(aq) \rightarrow I^-(aq) + Sn^{4+}(aq)$$

We balance it via our five-step procedure.
1. Separate the half-reactions.

Oxidation: $\qquad\qquad\qquad Sn^{2+}(aq) \rightarrow Sn^{4+}(aq)$
Reduction: $\qquad\qquad\qquad I_2(aq) \rightarrow I^-(aq)$

2. Balance masses.

Oxidation: $\qquad\qquad\qquad Sn^{2+}(aq) \rightarrow Sn^{4+}(aq)$
Reduction: $\qquad\qquad\qquad I_2(aq) \rightarrow 2\,I^-(aq)$

3. Balance the charges by adding electrons.

Oxidation: $\qquad\qquad\qquad Sn^{2+}(aq) \rightarrow Sn^{4+}(aq) + 2\,e^-$
Reduction: $\qquad\qquad 2\,e^- + I_2(aq) \rightarrow 2\,I^-(aq)$

4. Balance the numbers of electrons. This is not needed here because the numbers of electrons are the same in the two half-reactions.
5. Add the two half-reactions.

Oxidation: $\qquad\qquad\qquad Sn^{2+}(aq) \rightarrow Sn^{4+}(aq) + \cancel{2\,e^-}$
Reduction: $\qquad\qquad \cancel{2\,e^-} + I_2(aq) \rightarrow 2\,I^-(aq)$
Overall equation: $\quad Sn^{2+}(aq) + I_2(aq) \rightarrow 2\,I^-(aq) + Sn^{4+}(aq)$

Check the overall equation for mass balance: 1 Sn + 2 I = 2 I + 1 Sn. Check the charge balance: reactant side 2+, product side 2(1−) + (4+) = 2+.

**Think About It** We can balance redox reactions using half-reactions. Adding the half-reactions gives us the net ionic reaction and tells us the number of electrons transferred from the reducing agent to the oxidizing agent. Two moles of electrons are transferred to iodine for each mole of $Sn^{2+}(aq)$ that is oxidized. A balanced redox equation must be balanced with regard to electrons transferred as well as numbers of atoms.

⚙ **Practice Exercise** A nail made of Fe(s) that is placed in an aqueous solution of a soluble palladium(II) salt [containing $Pd^{2+}(aq)$] gradually disappears as the iron enters the solution as $Fe^{3+}(aq)$ and palladium metal [Pd(s)] forms. (a) Is this a redox reaction? (b) Balance the equation that describes this reaction.

*(Answers to Practice Exercises are in the back of the book.)*

## The Activity Series for Metals

If we replace the solution in Figure 4.27a with a $Zn(NO_3)_2$ solution, no reaction is observed. Why do silver ions oxidize elemental copper but zinc ions do not? By combining different metal cations with copper wire, for example, we can establish that $Ag^+$ and $Au^{3+}$ ions both oxidize copper but $Zn^{2+}$, $Ni^{2+}$, and $Al^{3+}$ ions do not. Similarly, we could explore the oxidation of zinc metal with these same cations. We would find that all these ions except $Al^{3+}$ (and $Zn^{2+}$) will oxidize zinc. Although an explanation for these observations will have to wait until Chapter 19, for now we can create an **activity series** that allows us to predict whether a particular metal ion will oxidize a different metal. A metal cation on the list in Table 4.6 will oxidize any metal above it in the activity series. Conversely, a metal will be oxidized by any cation below it.

**activity series** a qualitative ordering of the oxidizing ability of metals and their cations.

### TABLE 4.6 An Activity Series for Metals in Aqueous Solution

| Metal | Oxidation Reaction |
|---|---|
| Lithium | $Li(s) \rightarrow Li^+(aq) + e^-$ |
| Potassium | $K(s) \rightarrow K^+(aq) + e^-$ |
| Barium | $Ba(s) \rightarrow Ba^{2+}(aq) + 2\,e^-$ |
| Calcium | $Ca(s) \rightarrow Ca^{2+}(aq) + 2\,e^-$ |
| Sodium | $Na(s) \rightarrow Na^+(aq) + e^-$ |
| Magnesium | $Mg(s) \rightarrow Mg^{2+}(aq) + 2\,e^-$ |
| Aluminum | $Al(s) \rightarrow Al^{3+}(aq) + 3\,e^-$ |
| Manganese | $Mn(s) \rightarrow Mn^{2+}(aq) + 2\,e^-$ |
| Zinc | $Zn(s) \rightarrow Zn^{2+}(aq) + 2\,e^-$ |
| Chromium | $Cr(s) \rightarrow Cr^{3+}(aq) + 3\,e^-$ |
| Iron | $Fe(s) \rightarrow Fe^{2+}(aq) + 2\,e^-$ |
| Cobalt | $Co(s) \rightarrow Co^{2+}(aq) + 2\,e^-$ |
| Nickel | $Ni(s) \rightarrow Ni^{2+}(aq) + 2\,e^-$ |
| Tin | $Sn(s) \rightarrow Sn^{2+}(aq) + 2\,e^-$ |
| Lead | $Pb(s) \rightarrow Pb^{2+}(aq) + 2\,e^-$ |
| Hydrogen | $H_2(g) \rightarrow 2\,H^+(aq) + 2\,e^-$ |
| Copper | $Cu(s) \rightarrow Cu^{2+}(aq) + 2\,e^-$ |
| Silver | $Ag(s) \rightarrow Ag^+(aq) + e^-$ |
| Mercury | $Hg(\ell) \rightarrow Hg^{2+}(aq) + 2\,e^-$ |
| Platinum | $Pt(s) \rightarrow Pt^{2+}(aq) + 2\,e^-$ |
| Gold | $Au(s) \rightarrow Au^{3+}(aq) + 3\,e^-$ |

### SAMPLE EXERCISE 4.19 Using the Activity Series LO7

An aluminum wire is dipped into a solution containing sodium, tin, and copper ions. Which ions will be present in the solution after the reaction is complete?

**Collect and Organize** We are asked to predict if the metal ions in solution will oxidize the aluminum wire. We use the activity series in Table 4.6 to guide our answer.

**Analyze** The activity series is arranged in terms of increasing oxidizing ability of cations as we proceed down Table 4.6. Alternatively, the ease of oxidation of the corresponding metal decreases down the list. Any metal in Table 4.6 is oxidized by a cation below it in the activity series.

**Solve** $Cu^{2+}$ and $Sn^{2+}$ both lie below Al metal so they will oxidize the aluminum wire. Sodium ion is above Al in the activity series so it will not react. Thus, both $Cu^{2+}$ and $Sn^{2+}$ ions are removed from the solution but $Na^+$ ions remain. Since electron transfer from aluminum to both $Cu^{2+}$ and $Sn^{2+}$ occurs, $Al^{3+}$ ions are also present in the solution at the end of the experiment.

**Think About It** The activity series allows us to predict which metals are oxidized by other members of the series. Table 4.6 does not explain *why* a particular metal is a better oxidizing agent than another.

⚙ **Practice Exercise** Of the metals listed in Table 4.6, which is oxidized by half of the metal cations in the table?

*(Answers to Practice Exercises are in the back of the book.)*

Why is hydrogen gas listed in Table 4.6? What does its position in the activity series tell us? Table 4.6 allows us to predict which metals will react with strong aqueous acids like HCl. Any metal below $H_2$ in Table 4.6 will not react with $H^+$ and hence will be resistant to strong acids. One reason coins and jewelry made from silver, gold, and platinum last is that they do not react with most acids.

---

**SAMPLE EXERCISE 4.20** **Writing Net Ionic Equations for Redox Reactions** LO7

Predict the products of the reaction between zinc metal and aqueous hydrochloric acid. Write balanced molecular and net ionic equations for the reaction. Identify the oxidation and reduction half reactions.

**Collect and Organize** The reactants in this process are Zn(*s*) and HCl(*aq*). The products are determined by the activity series.

**Analyze** Zinc metal lies above $H^+$ in Table 4.6, thus it is oxidized by $H^+$ from hydrochloric acid, producing $Zn^{2+}$ and hydrogen gas. After writing a molecular equation for the reaction, we will convert it to a net ionic equation, as in Sample Exercise 4.8. After assigning oxidation numbers, we can identify the half-reactions.

**Solve** The unbalanced molecular equation is

$$Zn(s) + HCl(aq) \rightarrow ZnCl_2(aq) + H_2(g)$$

Adding a coefficient of 2 in front of the HCl brings all the elements into balance:

$$Zn(s) + 2\,HCl(aq) \rightarrow ZnCl_2(aq) + H_2(g)$$

The ionic equation is:

$$Zn(s) + 2\,H^+(aq) + 2\,Cl^-(aq) \rightarrow Zn^{2+}(aq) + 2\,Cl^-(aq) + H_2(g)$$

The net ionic equation becomes:

$$Zn(s) + 2\,H^+(aq) + \cancel{2\,Cl^-(aq)} \rightarrow Zn^{2+}(aq) + \cancel{2\,Cl^-(aq)} + H_2(g)$$
$$Zn(s) + 2\,H^+(aq) \rightarrow Zn^{2+}(aq) + H_2(g)$$

The oxidation numbers of Zn, $Zn^{2+}$, and $H^+$ are 0, +2, and +1, respectively. The oxidation number of $H_2$ gas is zero because it is the elemental form of hydrogen. Zinc is oxidized and hydrogen ion is reduced. The two half-reactions are

| | |
|---|---|
| Oxidation: | $Zn(s) \rightarrow Zn^{2+}(aq) + 2\,e^-$ |
| Reduction: | $2\,e^- + 2\,H^+(aq) \rightarrow H_2(g)$ |

**Think About It** Including the spectator ions in a redox equation makes it more difficult to identify which atoms or ions are involved in the electron transfer. That is why most redox equations in this chapter and elsewhere in this book are written as net ionic equations.

⚙ **Practice Exercise** Predict the products of the reaction between aluminum metal and silver nitrate. Write balanced molecular and net ionic equations for the reaction. Identify the oxidation and reduction half-reactions.

*(Answers to Practice Exercises are in the back of the book.)*

## Redox in Nature

Redox processes play a major role in determining the character of Earth's rocks and soils. For example, the orange-red rocks of Bryce Canyon National Park in Utah contain an iron(III) oxide mineral called hematite (Figure 4.29). Rocks containing hematite are relatively weak and easily broken, causing the fascinating shapes of such deposits. Iron(III) oxide is also the form of iron known as rust—the crumbly, red-brown solid that forms on iron objects exposed to water and oxygen.

Soil color is influenced by mineral content. Wetlands, which are areas that are either saturated or flooded with water during much of the year, are protected environments in many areas of the United States. Defining a given area as a wetland relies in part on the characteristics of the soils in the area, and soil color is an important diagnostic feature (Figure 4.30).

The redox reactions of iron(II) and iron(III) compounds can be modeled in the laboratory. Figure 4.31(a) shows a solution of iron(II) ammonium sulfate, $(NH_4)_2Fe(SO_4)_2(aq)$, upon addition of $NaOH(aq)$. The pale, greenish color of the liquid is characteristic of $Fe^{2+}(aq)$. Addition of $NaOH(aq)$ causes $Fe(OH)_2(s)$ to precipitate as a blue-gray solid, similar in color to the wetland soils shown in Figure 4.30. When this mixture is filtered, the iron(II) hydroxide residue immediately starts to darken. After about 20 minutes of exposure to oxygen in the air (Figure 4.31c), the precipitate turns the orange-red color of iron(III) hydroxide. The iron(II) has been oxidized—exactly the way the iron(II) compounds in a wetland soil are oxidized when exposed to oxygen.

Many redox reactions take place either in acidic solutions or in basic solutions, and the $H^+(aq)$, $OH^-(aq)$, or even the water may play a role in the reaction. Let's look at what we must do to balance the equations for such reactions.

A method for determining the concentration of $Fe^{2+}(aq)$ in an acidic solution involves its oxidation to $Fe^{3+}(aq)$ by the intensely purple permanganate ion, $MnO_4^-(aq)$ (Figure 4.32). In the reaction, the permanganate ion is reduced to $Mn^{2+}(aq)$:

$$Fe^{2+}(aq) + MnO_4^-(aq) \rightarrow Fe^{3+}(aq) + Mn^{2+}(aq)$$

| Appears colorless | Deep purple | Yellow-orange | Pale pink |

This reaction occurs in an acidic solution, which suggests that it involves protons, and it is clearly redox because $Fe^{2+}(aq)$ is oxidized to $Fe^{3+}(aq)$. The permanganate ion must be the oxidizing agent, so it must be reduced. Indeed, the O.N. of the Mn atom in $MnO_4^-$ is +7, and it is reduced to +2 in $Mn^{2+}$.

A chemical equation describing the reaction will be balanced when (a) the number of atoms of each element on both sides of the reaction is the same and (b) the total charges on each side of the reaction arrow are the same. We proceed in

**FIGURE 4.29** The orange-red rock in Bryce Canyon contains oxides of the iron(III) ion that result from weathering of the rock.

(a)

(b)

**FIGURE 4.30** (a) The soil in this wetland has a blue-gray color because of the presence of iron(II) compounds. (b) Orange-red mottling indicates the presence of iron(III) oxides formed as a result of $O_2$ permeation through channels made by plant roots.

(a)

(b)

(c)

**FIGURE 4.31** (a) A solution of iron(II) ammonium sulfate upon addition of NaOH(*aq*). The precipitate is iron(II) hydroxide. (b) The iron(II) hydroxide precipitate immediately after filtration. (c) The same precipitate after about 20 minutes of exposure to air. The color change indicates oxidation of Fe(II) to Fe(III).

**FIGURE 4.32** The test tubes on the left and right initially contain the same solution of $Fe^{2+}$ ions, which appear colorless. When drops of the purple solution of $KMnO_4$ from the middle test tube are added to the solution on the right, the mixture turns yellow-orange, which is the color of $Fe^{3+}$ ions in solution. The purple color disappears as $MnO_4^-$ ions are reduced to $Mn^{2+}$ ions.

the same way as for the previous examples in neutral medium, but we must add a few new steps to manage the role of the protons in the reaction.

1. Separate iron and manganese in their respective half-reactions.

    Oxidation: $$Fe^{2+}(aq) \rightarrow Fe^{3+}(aq)$$
    Reduction: $$MnO_4^-(aq) \rightarrow Mn^{2+}(aq)$$

2. Balance the masses in three steps to account for any role played by the aqueous acid. First we balance all the elements *except hydrogen and oxygen:*

    2a.  Oxidation: $Fe^{2+}(aq) \rightarrow Fe^{3+}(aq)$    (mass balanced)
    
    Reduction: $MnO_4^-(aq) \rightarrow Mn^{2+}(aq)$    (O mass not balanced)

    Next we balance oxygen by adding water as a product.

    2b.  Oxidation: $Fe^{2+}(aq) \rightarrow Fe^{3+}(aq)$    (mass balanced)

    Reduction: $MnO_4^-(aq) \rightarrow Mn^{2+}(aq) + 4\,H_2O(\ell)$
    (O mass balanced, H mass not)

    In acidic solutions, the only source of H to make $H_2O(\ell)$ is $H^+$ from a strong acid, so we balance the hydrogen by adding $H^+(aq)$:

    2c.  Oxidation: $Fe^{2+}(aq) \rightarrow Fe^{3+}(aq)$    (mass balanced)

    Reduction: $8\,H^+(aq) + MnO_4^-(aq) \rightarrow Mn^{2+}(aq) + 4\,H_2O(\ell)$
    (mass balanced)

3. Balance charge by adding electrons.

    Oxidation: $$Fe^{2+}(aq) \rightarrow Fe^{3+}(aq) + 1\,e^-$$
    Reduction: $$5\,e^- + 8\,H^+(aq) + MnO_4^-(aq) \rightarrow Mn^{2+}(aq) + 4\,H_2O(\ell)$$

4. Balance numbers of electrons lost and gained.

    Oxidation: $$5 \times [Fe^{2+}(aq) \rightarrow Fe^{3+}(aq) + 1\,e^-]$$
    Reduction: $$1 \times [5\,e^- + 8\,H^+(aq) + MnO_4^-(aq) \rightarrow Mn^{2+}(aq) + 4\,H_2O(\ell)]$$

5. Add the two equations.

Oxidation: $\qquad\qquad 5\,Fe^{2+}(aq) \rightarrow 5\,Fe^{3+}(aq) + \cancel{5\,e^-}$

Reduction: $\cancel{5\,e^-} + 8\,H^+(aq) + MnO_4^-(aq) \rightarrow Mn^{2+}(aq) + 4\,H_2O(\ell)$

$\overline{8\,H^+(aq) + MnO_4^-(aq) + 5\,Fe^{2+}(aq) \rightarrow 5\,Fe^{3+}(aq) + Mn^{2+}(aq) + 4\,H_2O(\ell)}$

Always check the mass and charge balance of the final net ionic reaction. The reactant side contains 8 H, 1 Mn, 4 O, and 5 Fe atoms and the product side contains 8 H, 1 Mn, 4 O, and 5 Fe, so mass is balanced. On the reactant side, the charge is $(8+) + (1-) + (10+) = 17+$, and on the product side it is $(15+) + (2+) = 17+$, so charge is balanced.

This method works for balancing the equation for any redox reaction taking place in an aqueous acid. The oxidation in Figure 4.31, however, was carried out in a basic medium, so now let's look at how to balance the equation for that reaction. We balance these reactions in almost the same way we do for acidic media: in fact, we carry out exactly the same steps outlined in the example we just worked, as though the reaction were run in acidic medium, except we add a final step in which sufficient hydroxide ion is added to both sides of the equation to convert any $H^+(aq)$ into $H_2O(\ell)$. This step converts the medium into an aqueous basic solution, as specified by the conditions.

---

**SAMPLE EXERCISE 4.21**   **Balancing Redox Reactions that Involve Hydroxide Ions**   **LO7**

Wetland soil is blue-gray due to $Fe(OH)_2(s)$, whereas well-aerated soils are often orange-red due to the presence of $Fe(OH)_3(s)$ (Figure 4.30). Write the balanced equation for the reaction of $O_2(g)$ with $Fe(OH)_2(s)$ in soil that produces $Fe(OH)_3(s)$ in basic solution.

**Collect and Organize** We know the reactants and product and can write the unbalanced equation:

$$Fe(OH)_2(s) + O_2(g) \rightarrow Fe(OH)_3(s)$$

Our task is to balance the equation with respect to both mass and charge.

**Analyze** This reaction takes place with both iron compounds in the solid phase in the presence of hydroxide ions in water. The oxidation half-reaction is clear: iron(II) hydroxide is oxidized to iron(III) hydroxide. For the reduction half-reaction, $O_2$ must be converted into the additional $OH^-$ ion in the iron(III) hydroxide product.

**Solve**
1. Separate the equations.

Oxidation: $\qquad\qquad Fe(OH)_2(s) \rightarrow Fe(OH)_3(s)$

Reduction: $\qquad\qquad O_2(g) \rightarrow OH^-(aq)$

2a. Balance all masses except H and O. This step is not needed because the only mass not attributed to H and O is Fe, which is balanced.

2b. Balance O by adding water as needed.

Oxidation: $\quad H_2O(\ell) + Fe(OH)_2(s) \rightarrow Fe(OH)_3(s)$

Reduction: $\qquad\qquad O_2(g) \rightarrow OH^-(aq) + H_2O(\ell)$

2c. Balance H by adding $H^+(aq)$ as needed.

$$H_2O(\ell) + Fe(OH)_2(s) \rightarrow Fe(OH)_3(s) + H^+(aq)$$
$$3\,H^+(aq) + O_2(g) \rightarrow OH^-(aq) + H_2O(\ell)$$

3. Balance the charges.

$$H_2O(\ell) + Fe(OH)_2(s) \rightarrow Fe(OH)_3(s) + H^+(aq) + 1\,e^-$$
$$4\,e^- + 3\,H^+(aq) + O_2(g) \rightarrow OH^-(aq) + H_2O(\ell)$$

4. Balance the numbers of electrons lost and gained.

$$4 \times [H_2O(\ell) + Fe(OH)_2(s) \rightarrow Fe(OH)_3(s) + H^+(aq) + 1\,e^-]$$
$$1 \times [4\,e^- + 3\,H^+(aq) + O_2(g) \rightarrow OH^-(aq) + H_2O(\ell)]$$

5. Add the two equations.

$$3~4\,H_2O(\ell) + 4\,Fe(OH)_2(s) \rightarrow 4\,Fe(OH)_3(s) + 4\,H^+(aq) + \cancel{4\,e^-}$$
$$\cancel{4\,e^-} + 3\,\cancel{H^+(aq)} + O_2(g) \rightarrow OH^-(aq) + \cancel{H_2O(\ell)}$$

This gives us the balanced equation:

$$3\,H_2O(\ell) + 4\,Fe(OH)_2(s) + O_2(g) \rightarrow 4\,Fe(OH)_3(s) + OH^-(aq) + H^+(aq)$$

which can be simplified:

$$\overset{2}{\cancel{3}}\,H_2O(\ell) + 4\,Fe(OH)_2(s) + O_2(g) \rightarrow 4\,Fe(OH)_3(s) + \underbrace{H^+(aq) + OH^-(aq)}_{H_2O(\ell)}$$

to

$$2\,H_2O(\ell) + 4\,Fe(OH)_2(s) + O_2(g) \rightarrow 4\,Fe(OH)_3(s)$$

**Think About It** This reaction occurs in neutral and basic soils as $H_2O$ and $O_2$ combine to form the $OH^-$ ions needed in the conversion of $Fe(OH)_2(s)$ to $Fe(OH)_3(s)$.

**Practice Exercise** The hydroperoxide ion, $HO_2^-(aq)$, reacts with permanganate ion, $MnO_4^-(aq)$, to produce $MnO_2(s)$ and oxygen gas. Balance the equation for the oxidation of hydroperoxide ion to $O_2(g)$ by permanganate ion in a basic solution.

*(Answers to Practice Exercises are in the back of the book.)*

Reactions in aqueous solutions are an integral part of our daily lives. On a large scale, in oceans, rivers, and rain, they shape our physical world. Many reactions that produce the substances that are part of modern life—from paint pigments to drugs—are run in water, and many analytical procedures rely on reactions in water to determine the content of aqueous solutions that we drink, swim in, and use in countless consumer products like car batteries and shampoos. On a small scale, reactions in the water within the cells of our bodies and in all living organisms make the chemical processes that are essential to life possible. On Earth, there may be water without life, but there is no life without water.

## SAMPLE EXERCISE 4.22 Integrating Concepts: Shelf-Stability of Drugs

Commercial pharmaceutical agents undergo extensive analysis to establish how long they may be stored without degradation and loss of potency. A candidate drug has the following percent composition from combustion analysis: C, 62.50%; H, 4.20%. The drug is a diprotic acid. A standard tablet containing 325 mg of the drug is dissolved in 100.00 mL of water and titrated to the equivalence point with 16.45 mL of 0.2056 $M$ NaOH($aq$). After storage at 50°C under high humidity for one month, a second tablet, when dissolved in 50.00 mL of water, requires 10.10 mL of 0.1755 $M$ NaOH($aq$) to be completely neutralized.

a. What is the empirical formula of the drug?
b. What is its molar mass?
c. What is its molecular formula?
d. What is the percent of active drug substance remaining in the tablet after storage?

**Collect and Organize** We are given the percent composition of the drug and information from titration experiments. From these data we can find the empirical formula, the molar mass, and the amount of active drug in the stored tablet.

**Analyze** The percent composition data from the combustion analysis do not add to 100%. We may assume that the amount missing is due to oxygen in the molecule. The drug is a diprotic acid, so neutralizing 1 mole of the drug requires 2 moles of NaOH. The titration of the drug requires about 20 mL of 0.2 $M$ NaOH, which is about 0.004 mol $OH^-$. That means the pure sample is about 0.002 moles of the drug; if about 0.3 grams contains 0.002 moles, the drug must have a molar mass of about 150 g. If any of the drug decomposes upon storage, we should have less than 325 mg in the second sample.

**Solve**
a. The percent oxygen in the drug molecule is

$$100.00\% - (62.50\% + 4.20\%) = 33.30\%$$

Calculating the moles of C, H, and O in exactly 100 g of the drug:

$$\frac{62.50 \text{ g C}}{12.01 \text{ g/mol}} = 5.20 \text{ mol C}$$

$$\frac{4.20 \text{ g H}}{1.008 \text{ g/mol}} = 4.17 \text{ mol H}$$

$$\frac{33.30 \text{ g O}}{16.00 \text{ g/mol}} = 2.08 \text{ mol O}$$

Reducing these quantities to ratios of small whole numbers:

$$\frac{5.20 \text{ mol C}}{2.08} = 2.50 \text{ mol C}$$

$$\frac{4.17 \text{ mol H}}{2.08} = 2.00 \text{ mol H}$$

$$\frac{2.08 \text{ mol O}}{2.08} = 1.00 \text{ mol O}$$

requires multiplying by 2, which gives us

$$5 \text{ mol C} : 4 \text{ mol H} : 2 \text{ mol O}$$

and the empirical formula: $C_5H_4O_2$.

b. To find the molar mass of the drug from the titration data, we first find the number of moles of drug in the sample:

$$0.01645 \text{ L NaOH} \times \frac{0.2056 \text{ mol NaOH}}{1 \text{ L NaOH}} \times \frac{1 \text{ mol drug}}{2 \text{ mol NaOH}}$$

$$= 0.001691 \text{ mol drug}$$

If 0.325 g of drug is 0.001691 mol of drug, then the mass of 1 mole is

$$\frac{0.325 \text{ g}}{0.001691 \text{ mol}} = 192.2 \text{ g/mol} = \mathcal{M}_{drug}$$

c. The mass of one mole of empirical formula units ($C_5H_4O_2$) is

$$(5 \times 12.01 \text{ g/mol}) + (4 \times 1.008 \text{ g/mol}) + (2 \times 16.00 \text{ g/mol})$$

$$= 96.08 \text{ g/mol}$$

The number of moles of formula units per mole of molecules is

$$\frac{192.16 \frac{\text{g}}{\text{mol}}}{96.08 \frac{\text{g}}{\text{mol}}} = 2$$

So the molecular formula of the drug is

$$(C_5H_4O_2)_2 = C_{10}H_8O_4$$

d. The stored tablet contains

$$0.01010 \text{ L NaOH} \times \frac{0.1755 \text{ mol NaOH}}{1 \text{ L NaOH}} \times \frac{1 \text{ mol drug}}{2 \text{ mol NaOH}}$$

$$= 0.0008863 \text{ mol drug} = 8.863 \times 10^{-4} \text{ mol}$$

Comparing this value to the original amount:

$$\frac{8.863 \times 10^{-4} \text{ mol}}{1.691 \times 10^{-3} \text{ mol}} \times 100\% = 52.41\%$$

of the active drug is left in the tablet.

**Think About It** The molar mass we calculated is close to our estimated value, so it seems reasonable. If only half the active agent is present in the tablet after storage under these conditions, the manufacturer may have to reformulate the drug or package it differently to protect it from its surroundings.

# Calcium: In the Limelight

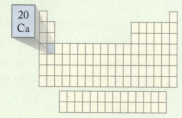

Calcium is the fifth most abundant element in Earth's crust and the fifth most abundant ion in seawater (after $Na^+$, $Cl^-$, $Mg^{2+}$, and $S^{2-}$). Vast deposits of calcium carbonate, $CaCO_3$, occur over large areas of the planet as limestone, marble, and chalk. These minerals are the fossilized remains of life in Earth's early oceans. Calcium carbonate is called a biomineral because it is an inorganic substance synthesized as an integral part of a living organism. Corals and seashells are mainly calcium carbonate, and islands like the Florida Keys and the Bahamas are made of large beds of $CaCO_3$.

Calcium sulfate, $CaSO_4$, commonly called gypsum, is another important calcium-containing mineral. If you are reading this in a room whose walls are finished with drywall, you are surrounded by gypsum because it is the primary component of drywall.

Although mostly known for their use as bulk chemicals in heavy industry and construction, special forms of calcium carbonate and calcium sulfate are valued for their beauty. Pearls (Figure 4.33a) are gemstones consisting mainly of calcium carbonate—specifically, a heterogeneous mixture of $CaCO_3(s)$ and a small amount of water suspended within the solid. Light travels one way through the solid material and a different way through the suspended water. The same phenomenon occurs when a thin layer of oil or gasoline floats on the surface of water: we see an array of colors. Diffraction of the light gives a high-quality pearl its special iridescence. The substance known as alabaster, prized for its translucence and subtle patterns, is a fine-grained form of calcium sulfate (Figure 4.33b).

Both calcium carbonate and calcium sulfate are used in *desiccants*, those small packets of granules included in boxes of electronic devices, shoes, pharmaceuticals, and foods. Desiccants (from the Latin *siccus*, "dry") remove water from the air, thereby keeping dry the items with which they are packed.

Another calcium compound used as a desiccant is calcium silicate, $CaSiO_3$. About 0.5% by mass is added to table salt to keep it flowing freely in damp weather. It is estimated that calcium silicate can absorb almost 600 times its mass in water and still remain a free-flowing powder.

Calcium hypochlorite, $Ca(OCl)_2$, is one of the sources of chlorine used to disinfect the water in swimming pools.

Calcium carbonate is an industrial chemical of great utility. High-quality paper contains $CaCO_3(s)$, which improves brightness as well as the paper's ability to bind ink. The pharmaceutical industry sells calcium carbonate as an antacid and in dietary supplements.

However, $CaCO_3$ has an even greater industrial use as the precursor of two other important materials: quicklime (CaO) and slaked lime [$Ca(OH)_2$]. To produce quicklime, calcium carbonate (derived usually from mining it or from oyster shells) is roasted:

$$CaCO_3(s) \xrightarrow{\text{heat}} CaO(s) + CO_2(g)$$

Slaked lime is produced from the reaction of quicklime with water:

$$CaO(s) + H_2O(\ell) \rightarrow Ca(OH)_2(s)$$

To be "in the limelight" is often used to describe anyone who is the center of attention. The expression derives from the calcium oxide lighting used in 19th-century theaters. When heated in an oxygen–hydrogen flame, calcium oxide emits a brilliant but soft white light. By the mid-1860s, limelights fitted with reflectors to direct and spread illumination were in wide use in theaters.

In modern times, lime has moved out of the entertainment business and into heavy industry. About 75 kg of lime is used to produce 1 ton of steel. Lime added to molten iron removes phosphorus, sulfur, and silicon, purifying the metal. It is used in large amounts in municipal water supplies to coagulate suspended solids. Lime is used

(a)

(b)

**FIGURE 4.33** Prized examples of calcium minerals are (a) pearls, which are heterogeneous mixtures of calcium carbonate and water, and (b) alabaster, fine-grained calcium sulfate.

in smokestack scrubbers to remove oxides of sulfur from the gases emitted by coal-burning power plants (Figure 4.34). During combustion, the sulfur in coal forms $SO_2$ gas. To "scrub" $SO_2$ from stack gases, the gases are passed through a suspension of calcium oxide in water. The $SO_2$ gas is trapped as solid calcium sulfite:

$$CaO(s) + SO_2(g) \rightarrow CaSO_3(s)$$

which is then oxidized to $CaSO_4$ (gypsum):

$$CaSO_3(s) + \tfrac{1}{2}O_2(g) \rightarrow CaSO_4(s)$$

Calcium is essential for life, and most animals—including humans—must absorb calcium every day from their diet. Over 95% of the calcium in the human system is in bones and teeth, and the average blood concentration of calcium is 100 mg per liter of blood. The body needs calcium to clot blood, transmit nerve impulses, and regulate the heartbeat. If the amount of calcium in the blood is insufficient for these uses, your body breaks down bone to extract it. Hence, blood calcium concentration is not an accurate indicator of your calcium status; bone density measurements are required to determine that status.

In addition, bones are not static: even in the absence of demand from other parts of your body, your bones are constantly being remodeled. This means that a small amount of old bone is removed each day and new bone formed in its place. After about age 35, however, bone is removed faster than it is deposited, resulting in net bone loss as we grow older.

**FIGURE 4.34** A scrubber capitalizes on the acid–base chemistry of aqueous solutions of calcium oxide and the oxides of sulfur to trap emission gases that would produce acid rain if released into the atmosphere.

## SUMMARY

**Learning Outcome 1** The concentration of **solute** in a solution can be expressed many different ways: as mass of solute per mass of solution (such as grams of solute per kilogram of solution), parts per million (1 ppm = 1 μg solute/g solution = 1 mg solute/kg solution), or parts per billion (1 ppb = 1 μg solute/kg solution). Solute concentration can also be expressed as mass of solute per volume of solvent and as moles of solute per liter of solution, or **molarity (M)**. One set of units can be converted into another using dimensional analysis. (Sections 4.1 and 4.2)

**Learning Outcome 2** Two common techniques to make solutions of desired concentration are determining the mass of solute to be dissolved in an appropriate amount of solvent to produce the quantity of solution needed and the **dilution** of a **stock solution**. (Sections 4.2 and 4.3)

**Learning Outcome 3** We can determine the concentration of a solution by **spectrophotometry** and the application of **Beer's Law**, which relates the **absorbance** of a solution ($A$) to concentration ($c$), path length ($b$) and the **molar absorptivity** ($\varepsilon$) by the equation $A = \varepsilon b c$. We can also use **titrations** based on acid–base **neutralizations** or **precipitation reactions**. (Sections 4.3, 4.6, and 4.7)

**Learning Outcome 4** Measuring the conductivity of a solution enables us to identify solutes as **strong electrolytes**, **weak electrolytes**, or **nonelectrolytes**. In aqueous solution, **Brønsted–Lowry acids** are proton donors and **Brønsted–Lowry bases** are proton acceptors; they may also be strong or weak electrolytes. (Sections 4.4 and 4.5)

**Learning Outcome 5** There are three different ways to write chemical equations.

**Molecular equations** may be the easiest to balance; **overall ionic equations** show all the species present in solution; **net ionic equations** eliminate **spectator ions** and show only the species that are involved in the reaction. (Sections 4.5, 4.6, and 4.7)

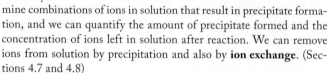

**Learning Outcome 6** Solubility rules define substances that dissolve readily in water and those that have very limited solubility. We can use them to determine combinations of ions in solution that result in precipitate formation, and we can quantify the amount of precipitate formed and the concentration of ions left in solution after reaction. We can remove ions from solution by precipitation and also by **ion exchange**. (Sections 4.7 and 4.8)

**Learning Outcome 7** In a redox reaction, substances either gain electrons (and thereby undergo **reduction**) or lose electrons (undergo **oxidation**). A reaction is a redox reaction if the **oxidation numbers (O.N.)**, or oxidation states, of the atoms in the reactants change during the reaction. An **activity series** allows us to predict whether a particular metal ion will oxidize a different metal. In balancing equations for redox reactions, we must consider the number of electrons transferred as well as the number of atoms involved. (Section 4.9)

## PROBLEM-SOLVING SUMMARY

| TYPE OF PROBLEM | CONCEPTS AND EQUATIONS | SAMPLE EXERCISES |
|---|---|---|
| **Comparing ion concentrations in aqueous solutions** | Use conversion factors to express concentrations in different units. | 4.1 |
| **Calculating molarity from mass and volume or from density** | Convert the solute mass into grams and then into moles. Convert the solution mass into volume by dividing by the density of the solvent. Divide moles of solute by liters of solution:<br><br>$$\text{Molarity} = \frac{\text{moles of solute}}{\text{liter of solution}}$$ | 4.2, 4.3 |
| **Calculating quantities of solute as a (a) mass of pure solid or (b) volume of stock solution** | a. Multiply the known concentration (in mol/L) by the target volume (in L) to obtain the moles of solute needed. Then multiply moles of solute by solute molar mass, $\mathcal{M}$, to get mass of solute needed:<br><br>$$\text{Mass of solute (g)} = \mathcal{M} \times V \times M \qquad (4.2)$$<br><br>b. Given three of the four variables, use<br><br>$$V_{\text{initial}} \times M_{\text{initial}} = V_{\text{diluted}} \times M_{\text{diluted}} \qquad (4.3)$$<br><br>to solve for the fourth. | 4.4, 4.5 |

| TYPE OF PROBLEM | CONCEPTS AND EQUATIONS | SAMPLE EXERCISES |
|---|---|---|
| **Applying Beer's law** | Substitute for $A$, $b$, and $c$ in Equation 4.4 and solve for $\varepsilon$ for a solution of known concentration:<br><br>$$\varepsilon = \frac{A}{bc}$$<br><br>Use $A$, $b$, and $\varepsilon$ to find $c$ for a solution of unknown concentration:<br><br>$$c = \frac{A}{\varepsilon b}$$ | 4.6 |
| **Comparing electrolytes, acids, and bases** | Use the definitions of a strong electrolyte, a weak electrolyte, a nonelectrolyte, an acid, and a base to classify compounds into one or more categories. | 4.7 |
| **Writing neutralization reaction equations** | Balance the molecular equation by balancing the moles of $H^+$ ions donated by the acid and accepted by the base. Next create the overall ionic equation by writing strong electrolytes in their ionic form. Finally, create the net ionic equation by eliminating spectator ions from the overall ionic equation. | 4.8, 4.11 |
| **Calculating molarity from titration data** | Use the volume and concentration of titrant used to neutralize a sample along with a balanced chemical equation to calculate the number of moles in a sample of known volume and determine its molarity. | 4.9, 4.10 |
| **Predicting precipitation reactions** | Write all the ions present in the solutions being mixed. If any cation/anion pair forms an insoluble compound, that compound will precipitate. | 4.12 |
| **Calculating the mass of a precipitate** | Find the limiting reactant. Use the stoichiometry of the net ionic reaction to calculate the moles of precipitate, then convert moles into mass of precipitate using the precipitate's molar mass. | 4.13, 4.15 |
| **Calculating a solute concentration from a precipitate mass** | Convert precipitate mass into moles by dividing by its molar mass. Convert moles of precipitate into moles of solute. Calculate molarity of the solute in the sample by dividing the moles of solute by the volume of sample in liters. | 4.14 |
| **Determining oxidation numbers (O.N.)** | O.N. of a monatomic ion is equal to the ion's charge. O.N. of a pure element is 0. To assign O.N. in a molecule containing more than one type of atom, assign O.N. +1 to H, −2 to O, and then calculate O.N. for any remaining atoms such that all the O.N. sum to 0. For polyatomic ions, the sum of the O.N. values must equal the charge on the ion. | 4.16 |
| **Identifying oxidizing and reducing agents and number of electrons transferred** | The oxidizing agent contains an atom whose O.N. decreases during the reaction; the reducing agent contains an atom whose O.N. increases; the change in O.N. determines the number of electrons transferred. | 4.17 |
| **Balancing redox reactions with half-reactions** | Multiply one or both half-reactions by the appropriate coefficient(s) to balance the loss and gain of electrons. Combine the two half-reactions and simplify. | 4.18 |
| **Using the activity series** | Any metal in Table 4.6 will be oxidized by a cation below it in the activity series. | 4.19 |
| **Balancing redox reactions that involve acidic or basic conditions** | Balance half-reactions for elements except H and O. In acid, balance H by adding $H^+$, O by adding $H_2O$. In base, add $OH^-$ to each side to neutralize $H^+$ and simplify. | 4.20, 4.21 |

## VISUAL PROBLEMS ● ● ● ● ● ● ● ● ● ● ● ● ● ● ● ● ● ● ● ● ● ● ● ● ● ● ● ● ● ●

*(Answers to boldface end-of-chapter questions and problems are in the back of the book.)*

**4.1.** In Figure P4.1, which shows a solution containing three binary acids; one of the three is a weak acid and the other two are strong acids. Which color sphere represents the dissociated weak acid?

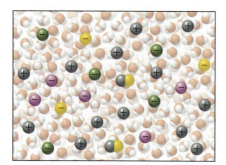

**FIGURE P4.1**

**4.2.** Solutions of sodium chloride and silver nitrate are mixed together and vigorously shaken. Which colored spheres in Figure P4.2 represent the following ions? (a) $Na^+$; (b) $Cl^-$; (c) $NO_3^-$ (*Note*: The silver spheres represent $Ag^+$ ions.)

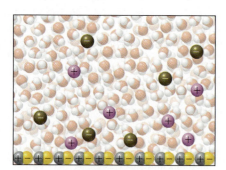

**FIGURE P4.2**

**4.3.** Which of the highlighted elements in Figure P4.3 forms an acid with the following generic formulas? (a) HX; (b) $H_2XO_4$; (c) $HXO_3$; (d) $H_3XO_4$

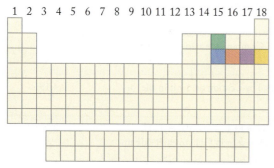

**FIGURE P4.3**

**4.4.** In which of the highlighted groups of elements in Figure P4.4 will you find an element that forms the following? (a) insoluble halides; (b) insoluble hydroxides; (c) hydroxides that are soluble; (d) binary compounds with hydrogen that are strong acids

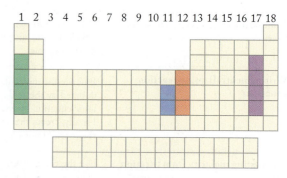

**FIGURE P4.4**

**4.5.** Which of the following drawings depicts a strong electrolyte? A weak electrolyte? A strong acid? A weak acid? A nonelectrolyte? Legend: individual white spheres represent hydronium ions; yellow spheres are ions with 2+ charges. Each drawing may fit more than one category.

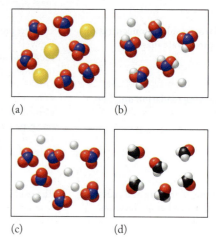

**FIGURE P4.5**

**4.6.** Which ions in Figure P4.6 are being oxidized and which are being reduced? Legend: (⊕) = hydronium ion

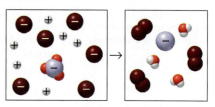

**FIGURE P4.6**

4.7. Which ions in Figure P4.7 will remain in solution?
Legend: (➕) = hydronium ions; (⚫) = bromide ions;
(🔴) = nitrate ions; (⚫) = lead(II) ions.

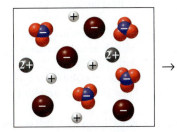

**FIGURE P4.7**

4.8. Which of the solutions in Figure P4.8 is the most dilute and which has the highest absorbance?

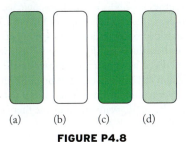

(a)    (b)    (c)    (d)

**FIGURE P4.8**

## QUESTIONS AND PROBLEMS

### Concentration Units

#### CONCEPT REVIEW

**4.9.** How do we decide which component in a solution is the solvent?

**4.10.** Can a solid ever be a solvent?

**4.11.** What is the molarity of a solution that contains 1.00 mmol of solute per milliliter of solution?

*__4.12.__ A beaker contains 100 g of 1.00 $M$ NaCl. If you transfer 50 g of the solution to another beaker, what is the molarity of the solution remaining in the first beaker?

#### PROBLEMS

**4.13.** Calculate the molarity of each of the following solutions:
a. 0.56 mol of $BaCl_2$ in 100.0 mL of solution
b. 0.200 mol of $Na_2CO_3$ in 200.0 mL of solution
c. 0.325 mol of $C_6H_{12}O_6$ in 250.0 mL of solution
d. 1.48 mol of $KNO_3$ in 250.0 mL of solution

**4.14.** Calculate the molarity of each of the following solutions:
a. 0.150 mol of urea ($CH_4N_2O$) in 250.0 mL of solution
b. 1.46 mol of $NaC_2H_3O_2$ in 1.000 L of solution
c. 1.94 mol of methanol ($CH_3OH$) in 5.000 L of solution
d. 0.045 mol of sucrose ($C_{12}H_{22}O_{11}$) in 50.0 mL of solution

**4.15.** Calculate the molarity of each of the following ions:
a. 0.33 g $Na^+$ in 100.0 mL of solution
b. 0.38 g $Cl^-$ in 100.0 mL of solution
c. 0.46 g $SO_4^{2-}$ in 50.0 mL of solution
d. 0.40 g $Ca^{2+}$ in 50.0 mL of solution

**4.16.** Calculate the molarity of each of the following solutions:
a. 64.7 g LiCl in 250.0 mL of solution
b. 29.3 g $NiSO_4$ in 200.0 mL of solution
c. 50.0 g KCN in 500.0 mL of solution
d. 0.155 g $AgNO_3$ in 100.0 mL of solution

**4.17.** How many grams of solute are needed to prepare each of the following solutions?
a. 1.000 L of 0.200 $M$ NaCl
b. 250.0 mL of 0.125 $M$ $CuSO_4$
c. 500.0 mL of 0.400 $M$ $CH_3OH$

**4.18.** How many grams of solute are needed to prepare each of the following solutions?
a. 500.0 mL of 0.250 $M$ KBr
b. 25.0 mL of 0.200 $M$ $NaNO_3$
c. 100.0 mL of 0.375 $M$ $CH_3OH$

**4.19.** **River Water** The Mackenzie River in northern Canada contains, on average, 0.820 m$M$ $Ca^{2+}$, 0.430 m$M$ $Mg^{2+}$, 0.300 m$M$ $Na^+$, 0.0200 $M$ $K^+$, 0.250 m$M$ $Cl^-$, 0.380 m$M$ $SO_4^{2-}$, and 1.82 m$M$ $HCO_3^-$. What, on average, is the total mass of these ions in 2.75 L of Mackenzie River water?

**4.20.** Zinc, copper, lead, and mercury ions are toxic to Atlantic salmon at concentrations of $6.42 \times 10^{-2}$ m$M$, $7.16 \times 10^{-3}$ m$M$, 0.965 m$M$, and $5.00 \times 10^{-2}$ m$M$, respectively. What are the corresponding concentrations in milligrams per liter?

**4.21.** Calculate the number of moles of solute contained in the following volumes of aqueous solutions of four pesticides:
a. 0.400 L of 0.024 $M$ lindane
b. 1.65 L of 0.473 m$M$ dieldrin
c. 25.8 L of 3.4 m$M$ DDT
d. 154 L of 27.4 m$M$ aldrin

**4.22.** **Hemoglobin in Blood** A typical adult body contains 6.0 L of blood. The hemoglobin content of blood is about 15.5 g/100.0 mL of blood. The approximate molar mass of hemoglobin is 64,500 g/mol. How many moles of hemoglobin are present in a typical adult?

**4.23.** **DDT Affects Neurons** The pesticide DDT ($C_{14}H_9Cl_5$) kills insects such as malaria-carrying mosquitoes by opening sodium ion channels in neurons, causing them to fire spontaneously, which leads to spasms and eventual death. However, its toxicity in wildlife and humans led to the banning of its use in the United States in 1972. Analysis of DDT concentrations in groundwater samples between 1969 and 1971 in Pennsylvania yielded the following results:

| Location | Sample Size | Mass of DDT |
|---|---|---|
| Orchard | 250.0 mL | 0.030 mg |
| Residential | 1.750 L | 0.035 mg |
| Residential after a storm | 50.0 mL | 0.57 mg |

Express these concentrations in ppm and in millimoles per liter.

**4.24.** Pesticide concentrations in the Rhine River between Germany and France between 1969 and 1975 averaged 0.55 mg/L of hexachlorobenzene ($C_6Cl_6$), 0.06 mg/L of dieldrin ($C_{12}H_8Cl_6O$), and 1.02 mg/L of hexachlorocyclohexane ($C_6H_6Cl_6$). Express these concentrations in ppb and in millimoles per liter.

**4.25.** Nitrogen trifluoride, $NF_3$, is used in the production of flat panel displays. It is also a potent greenhouse gas. The average concentration of $NF_3$ in the atmosphere increased from 0.051 parts per trillion by mass (ppt) in 1978 to 1.15 ppt in 2008. What is the concentration of $NF_3$ in mg per kg of air in 2008?

**4.26.** Sulfur hexafluoride, $SF_6$, is used in electrical transformers. Like $NF_3$, it has a potential impact on climate. Between 1978 and 2012, the concentration of $SF_6$ increased from 2.7 parts per trillion by mass (ppt) to 39.0 ppt. How many *more* molecules of $SF_6$ were found in one liter (1.29 g) of air in 2012 than in 1978?

**\*4.27.** The concentration of copper(II) sulfate in one brand of soluble plant fertilizer is 0.07% by mass. If a 20 g sample of this fertilizer is dissolved in 2.0 L of solution, what is the molarity of $Cu^{2+}$?

**\*4.28.** For which of the following compounds is it possible to make a 1.0 $M$ solution at 20°C?
   a. $CuSO_4$, solubility = 32.0 g/100 mL
   b. $Ba(OH)_2$, solubility = 3.9 g/100 mL
   c. $FeCl_2$, solubility = 68.5 g/100 mL
   d. $Ca(OH)_2$, solubility = 0.173 g/100 mL

## Dilutions
### PROBLEMS
**4.29.** Calculate the final concentrations of the following aqueous solutions after each has been diluted to a final volume of 25.0 mL:
   a. 1.00 mL of 0.452 $M$ $Na^+$
   b. 2.00 mL of 3.4 m$M$ LiCl
   c. 5.00 mL of $6.42 \times 10^{-2}$ m$M$ $Zn^{2+}$

**4.30. Dilution of Adult-Strength Cough Syrup** A standard dose of an over-the-counter cough suppressant for adults is 20.0 mL. A portion this size contains 35 mg of the active pharmaceutical ingredient (API). Your pediatrician says you may give this medication to your 6-year-old child, but the child may only take 10.0 mL at a time and receive a maximum of 4.00 mg of the API. What is the concentration in mg/mL of the adult-strength medication, and how many millimeters of it would you need to dilute to make 100.0 mL of child-strength cough syrup?

**4.31.** The concentration of $Na^+$ in seawater, 0.481 $M$, is higher than in the cytosol, the fluid inside human cells (12 m$M$). How much water must be added to 1.50 mL of seawater to make the $Na^+$ concentration equal to that found in cytosol?

**4.32.** The concentration of chloride ion in blood, 116 m$M$, is less than that in the ocean, 0.559 $M$. Describe how you would prepare 2.50 mL of a solution of 116 m$M$ chloride ion from seawater.

**\*4.33.** A puddle of coastal seawater, caught in a depression formed by some coastal rocks at high tide, begins to evaporate on a hot summer day as the tide goes out. If the volume of the puddle decreases to 23% of its initial volume, what is the concentration of $Na^+$ after evaporation if initially it was 0.449 $M$?

**\*4.34. Mixing Fertilizer** The label on a bottle of "organic" liquid fertilizer concentrate states that it contains 8 grams of phosphate per 100.0 mL and that 16 fluid ounces should

be diluted with water to make 32 gallons of fertilizer to be applied to growing plants. What is the phosphate concentration in grams per liter in the diluted fertilizer? (1 gallon = 128 fluid ounces.)

**4.35.** If the absorbance of a solution of copper ion decreases by 45% upon dilution, how much water was added to 15.0 mL of a 1.00 $M$ solution of $Cu^{2+}$?

**4.36.** By what percent does the absorbance decrease if 12.25 mL of water is added to a 16.75 mL sample of 0.500 $M$ $Cr^{3+}$?

**\*4.37.** The reaction of $SnCl_2(aq)$ with $Pt^{4+}(aq)$ in aqueous HCl yields a yellow-orange solution of a 1:1 Pt-Sn compound with a molar absorptivity ($\varepsilon$) of $1.3 \times 10^4$ $M^{-1}cm^{-1}$ at $\lambda$ = 400 nm. What is the absorbance in a cell with a path length of 1.00 cm of a solution prepared by adding 100 mL of an aqueous solution of 5.2 mg $(NH_4)_2PtCl_6$ to 100 mL of an aqueous solution of 2.2 mg $SnCl_2$?

**\*4.38.** The reaction of $SnCl_2(aq)$ with $RhCl_3(aq)$ in aqueous HCl yields a red solution of a 1:1 Rh-Sn compound. If a solution prepared by adding 150 mL of a 0.272 m$M$ aqueous solution of $SnCl_2$ to 50 mL of an aqueous solution of 8.5 mg $RhCl_3$ has an absorbance of 0.85 $M^{-1}cm^{-1}$, as measured in a 1.00 cm cell, what is the molar absorptivity of the red compound?

## Electrolytes and Nonelectrolytes
### CONCEPT REVIEW
**4.39.** A solution of table salt is a good conductor of electricity, but a solution containing an equal molar concentration of table sugar is not. Why?

**4.40.** Metallic fixtures on the bottom of a ship corrode more quickly in seawater than in freshwater. Why?

**4.41.** Explain why liquid methanol, $CH_3OH$, cannot conduct electricity, whereas molten NaOH can.

**4.42. Fuel Cells** The electrolyte in an electricity-generating device called a *fuel cell* consists of a mixture of $Li_2CO_3$ and $K_2CO_3$ heated to 650°C. At this temperature the ionic solids melt. Explain how this mixture of molten carbonates can conduct electricity.

**4.43.** Rank the following solutions on the basis of their ability to conduct electricity, starting with the most conductive: (a) 1.0 $M$ NaCl; (b) 1.2 $M$ KCl; (c) 1.0 $M$ $Na_2SO_4$; (d) 0.75 $M$ LiCl.

**4.44.** Rank the conductivities of 1 $M$ aqueous solutions of each of the following solutes, starting with the most conductive: (a) acetic acid; (b) methanol; (c) sucrose (table sugar); (d) hydrochloric acid.

### PROBLEMS
**4.45.** Calculate the molarity of $Na^+$ ions in a 0.025 $M$ aqueous solution of: (a) NaBr; (b) $Na_2SO_4$; (c) $Na_3PO_4$.

**4.46.** Calculate the molarity of each ion in a 0.025 $M$ aqueous solution of: (a) KCl; (b) $CuSO_4$; (c) $CaCl_2$.

**4.47.** Which of the following solutions has the greatest number of particles (atoms or ions) of solute per liter? (a) 1 $M$ NaCl; (b) 1 $M$ $CaCl_2$; (c) 1 $M$ ethanol; (d) 1 $M$ acetic acid

**4.48.** Which of the following solutions contains the most solute particles per liter? (a) 1 $M$ KBr; (b) 1 $M$ $Mg(NO_3)_2$; (c) 4 $M$ ethanol; (d) 4 $M$ acetic acid

## Acid–Base Reactions: Proton Transfer

### CONCEPT REVIEW

**4.49.** What name is given to a proton donor?

**4.50.** What is the difference between a strong acid and a weak acid?

**4.51.** Give the formulas of two strong acids and two weak acids.

**4.52.** Why is $HSO_4^-(aq)$ a weaker acid than $H_2SO_4(aq)$?

**4.53.** What name is given to a proton acceptor?

**4.54.** What is the difference between a strong base and a weak base?

**4.55.** Give the formulas of two strong bases and two weak bases.

**4.56.** Write the net ionic equation for the neutralization of a strong acid by a strong base.

### PROBLEMS

**4.57.** For each of the following acid–base reactions, identify the acid and the base, and then write the overall ionic and net ionic equations.
a. $H_2SO_4(aq) + Ca(OH)_2(aq) \rightarrow CaSO_4(s) + 2\,H_2O(\ell)$
b. $PbCO_3(s) + H_2SO_4(aq) \rightarrow$
$$PbSO_4(s) + CO_2(g) + H_2O(\ell)$$
c. $Ca(OH)_2(s) + 2\,CH_3COOH(aq) \rightarrow$
$$Ca(CH_3COO)_2(aq) + 2\,H_2O(aq)$$

**4.58.** Complete and balance each of the following neutralization reactions, name the products, and write the overall ionic and net ionic equations.
a. $HBr(aq) + KOH(aq) \rightarrow$
b. $H_3PO_4(aq) + Ba(OH)_2(aq) \rightarrow$
c. $Al(OH)_3(s) + HCl(aq) \rightarrow$
d. $CH_3COOH(aq) + Sr(OH)_2(aq) \rightarrow$

**4.59.** Write a balanced molecular equation and a net ionic equation for the following reactions:
a. Solid magnesium hydroxide reacts with a solution of sulfuric acid.
b. Solid magnesium carbonate reacts with a solution of hydrochloric acid.
c. Ammonia gas reacts with hydrogen chloride gas.
d. Gaseous sulfur trioxide is dissolved in water and reacts with a solution of sodium hydroxide.

**4.60.** Write a balanced molecular equation and a net ionic equation for the following reactions:
a. Solid aluminum hydroxide reacts with a solution of hydrobromic acid.
b. A solution of sulfuric acid reacts with solid sodium carbonate.
c. A solution of calcium hydroxide reacts with a solution of nitric acid.
d. Solid potassium oxide is dissolved in water and reacts with a solution of sulfuric acid.

**4.61. Toxicity of Lead Pigments** The use of lead(II) carbonate and lead(II) hydroxide as white pigments in paint was discontinued because children have been known to eat paint chips. The pigments dissolve in stomach acid, and lead ions enter the nervous system and interfere with neurotransmissions in the brain, causing neurological disorders. Using net ionic equations, show why lead(II) carbonate and lead(II) hydroxide dissolve in acidic solutions.

**4.62. Lawn Care** Many homeowners treat their lawns with $CaCO_3(s)$ to reduce the acidity of the soil. Write a net ionic equation for the reaction of $CaCO_3(s)$ with a strong acid.

## Titrations

### PROBLEMS

**4.63.** How many milliliters of $0.100\,M$ NaOH are required to neutralize the following solutions?
a. 10.0 mL of $0.0500\,M$ HCl
b. 25.0 mL of $0.126\,M$ $HNO_3$
c. 50.0 mL of $0.215\,M$ $H_2SO_4$

**4.64.** How many milliliters of $0.100\,M$ $HNO_3$ are needed to neutralize the following solutions?
a. 45.0 mL of $0.667\,M$ KOH
b. 58.5 mL of $0.0100\,M$ $Al(OH)_3$
c. 34.7 mL of $0.775\,M$ NaOH

**\*4.65.** The solubility of slaked lime, $Ca(OH)_2$, in water at 20°C is 0.185 g/100.0 mL. What volume of $0.00100\,M$ HCl is needed to neutralize 10.0 mL of a saturated $Ca(OH)_2$ solution?

**\*4.66.** The solubility of magnesium hydroxide, $Mg(OH)_2$, in water is $9.0 \times 10^{-4}$ g/100.0 mL. What volume of $0.00100\,M$ $HNO_3$ is required to neutralize 1.00 L of saturated $Mg(OH)_2$ solution?

**4.67.** A 10.0 mL dose of the antacid in Figure P4.67 contains 830 mg of magnesium hydroxide. What volume of $0.10\,M$ stomach acid (HCl) could one dose neutralize?

**FIGURE P4.67**

**\*4.68. Exercise Physiology** The ache, or "burn," you feel in your muscles during strenuous exercise is caused by the accumulation of lactic acid, which has the structure shown in Figure P4.68. Only the hydrogen atom in the −COOH group is acidic, that is, can be released as an $H^+$ ion in aqueous solutions. To determine the concentration of a solution of lactic acid, a chemist titrates a 20.00 mL sample of it with $0.1010\,M$ NaOH and finds that 12.77 mL of titrant is required to reach the equivalence point. What is the concentration of the lactic acid solution in moles per liter?

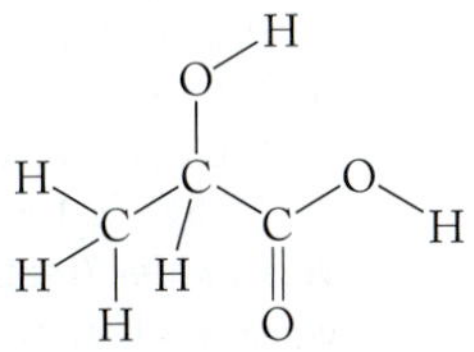

**FIGURE P4.68**

## Precipitation Reactions

### CONCEPT REVIEW

**4.69.** What is the difference between a saturated solution and a supersaturated solution?

**4.70.** What are common solubility units?

**4.71.** An aqueous solution containing $Ca^{2+}$, $Cl^-$, $CO_3^{2-}$, and $NO_3^-$ is allowed to evaporate. Which compound will precipitate first?

**4.72.** A precipitate may appear when two completely clear aqueous solutions are mixed. What circumstances are responsible for this event?

**4.73.** Is a saturated solution always a concentrated solution? Explain.

**4.74.** Honey is a concentrated solution of sugar molecules in water. Clear, viscous honey becomes cloudy after being stored for long periods. Explain how this transition illustrates supersaturation.

### PROBLEMS

**4.75.** According to the solubility rules in Table 4.5, which of the following compounds have limited solubility in water? (a) barium sulfate; (b) barium hydroxide; (c) lanthanum nitrate; (d) sodium acetate; (e) lead hydroxide; (f) calcium phosphate

**4.76.** **Ocean Vents** The black "smoke" that flows out of deep ocean hydrothermal vents (Figure P4.76) is made of insoluble metal sulfides suspended in seawater. Of the following cations that are present in the water flowing up through these vents, which ones could contribute to the formation of the black smoke?
$Na^+$, $Li^+$, $Mn^{2+}$, $Fe^{2+}$, $Ca^{2+}$, $Mg^{2+}$, $Zn^{2+}$, $Pb^{2+}$, $Cu^{2+}$

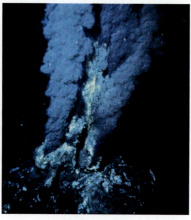

**FIGURE P4.76**

**4.77.** Complete and balance the molecular equations for the precipitation reactions, if any, between the following pairs of reactants, and write the overall and net ionic equations.
a. $Pb(NO_3)_2(aq) + Na_2SO_4(aq) \rightarrow$
b. $NiCl_2(aq) + NH_4NO_3(aq) \rightarrow$
c. $FeCl_2(aq) + Na_2S(aq) \rightarrow$
d. $MgSO_4(aq) + BaCl_2(aq) \rightarrow$

**\*4.78.** **Wastewater Treatment** Show with appropriate net ionic equations how $Cr^{3+}$ and $Cd^{2+}$ can be removed from wastewater by treatment with solutions of sodium hydroxide.

**4.79.** Calculate the mass of $MgCO_3$ precipitated by mixing 10.0 mL of a 0.200 $M$ $Na_2CO_3$ solution with 5.00 mL of 0.0500 $M$ $Mg(NO_3)_2$ solution.

**4.80.** Toxic chromate can be precipitated from an aqueous solution by bubbling $SO_2$ through the solution. How many grams of $SO_2$ are required to treat $3.0 \times 10^8$ L of 0.050 m$M$ Cr(VI)?

$$2\,CrO_4^{2-}(aq) + 3\,SO_2(g) + 4\,H^+(aq) \rightarrow Cr_2(SO_4)_3(s) + 2\,H_2O(\ell)$$

**4.81.** Fe(II) can be precipitated from a slightly basic aqueous solution by bubbling oxygen through the solution, which converts Fe(II) to insoluble Fe(III). How many grams of $O_2$ are consumed to precipitate all of the iron in 75 mL of 0.090 $M$ Fe(II)?

$$4\,Fe(OH)^+(aq) + 4\,OH^-(aq) + O_2(g) + 2\,H_2O(\ell) \rightarrow 4\,Fe(OH)_3(s)$$

**4.82.** Given the following equation, how many grams of $PbCO_3$ will dissolve when 1.00 L of 1.00 $M$ $H^+$ is added to 5.00 g of $PbCO_3$?

$$PbCO_3(s) + 2\,H^+(aq) \rightarrow Pb^{2+}(aq) + H_2O(\ell) + CO_2(g)$$

**\*4.83.** **Treating Drinking Water** Phosphate can be removed from drinking-water supplies by treating the water with $Ca(OH)_2$. How much $Ca(OH)_2$ is required to remove 90% of the $PO_4^{3-}$ from $4.5 \times 10^6$ L of drinking water containing 25 mg/L of $PO_4^{3-}$?

$$5\,Ca(OH)_2(aq) + 3\,PO_4^{3-}(aq) \rightarrow Ca_5OH(PO_4)_3(s) + 9\,OH^-(aq)$$

**4.84.** Toxic cyanide ions can be removed from wastewater by adding hypochlorite.

$$2\,CN^-(aq) + 5\,OCl^-(aq) + H_2O(\ell) \rightarrow$$
$$N_2(g) + 2\,HCO_3^-(aq) + 5\,Cl^-(aq)$$

a. If $1.50 \times 10^3$ liters of 0.125 $M$ $OCl^-$ are required to remove the $CN^-$ in $3.4 \times 10^6$ L of wastewater, what is the $CN^-$ concentration in the water in mg/L?
*b. How many mL of 0.575 $M$ $AgNO_3$ would you need to add to a 50.00 mL aliquot of the final solution (consider the volumes simply additive) to precipitate the chloride ions formed in the reaction?

**4.85.** For each of the following aqueous mixtures, determine which ionic concentrations decrease and which remain the same.
a. Sodium chloride and silver nitrate are dissolved in 100 mL water.
b. Equimolar amounts of sodium hydroxide and hydrochloric acid react.
b. Ammonium sulfate and potassium bromide are dissolved in 100 mL water.

**4.86.** For each of the following aqueous mixtures, determine which ionic concentrations decrease and which remain the same.
a. Sodium chloride and iron(II) chloride are dissolved in 100 mL water.
b. Equimolar amounts of sodium carbonate and sulfuric acid react.
c. Potassium sulfate and barium nitrate are dissolved in 100 mL water.

## Ion Exchange

### CONCEPT REVIEW

**4.87.** Explain how a mixture of anion and cation exchangers can be used to deionize water.

**4.88.** Describe the process by which the ion exchanger in a home water softener is regenerated for further use.

**4.89.** If an ion-exchange resin is to be used to deionize water, what ions must be at the cation and anion exchange sites on the resin?

*4.90. A piece of Zn metal is placed in a solution containing $Cu^{2+}$ ions. At the surface of the Zn metal, $Cu^{2+}$ ions react with Zn atoms, forming Cu atoms and $Zn^{2+}$ ions. Is this reaction an example of ion exchange? Explain why or why not.

## Oxidation–Reduction Reactions: Electron Transfer

### CONCEPT REVIEW

4.91. How are the gains or losses of electrons related to changes in oxidation numbers?

4.92. What is the sum of the oxidation numbers of the atoms in a molecule?

4.93. What is the sum of the oxidation numbers of all the atoms in each of the following polyatomic ions? (a) $OH^-$; (b) $NH_4^+$; (c) $SO_4^{2-}$; (d) $PO_4^{3-}$

4.94. Gold does not dissolve in concentrated $H_2SO_4$ but readily dissolves in $H_2SeO_4$ (selenic acid). Which acid is the stronger oxidizing agent?

4.95. Silver dissolves in sulfuric acid to form silver sulfate and $H_2$, but gold does not dissolve in sulfuric acid to form gold sulfate. Which of the two metals is the better reducing agent?

4.96. What is meant by a half-reaction?

4.97. What are the half-reactions that take place in the electrolysis of molten NaCl?

4.98. Electron gain is associated with _____ half-reactions, and electron loss is associated with _____ half-reactions.

### PROBLEMS

4.99. Give the oxidation number of chlorine in each of the following: (a) hypochlorous acid (HClO); (b) chloric acid ($HClO_3$); (c) perchlorate ion ($ClO_4^-$).

4.100. Give the oxidation number of nitrogen in each of the following: (a) elemental nitrogen ($N_2$); (b) hydrazine ($N_2H_4$); (c) ammonium ion ($NH_4^+$).

4.101. Balance the following half-reactions by adding the appropriate number of electrons. Identify the oxidation half-reactions and the reduction half-reactions.
a. $Br_2(\ell) \rightarrow 2\ Br^-(aq)$
b. $Pb(s) + 2\ Cl^-(aq) \rightarrow PbCl_2(s)$
c. $O_3(g) + 2\ H^+(aq) \rightarrow O_2(g) + H_2O(\ell)$
d. $2\ H_2SO_3(aq) + H^+(aq) \rightarrow HS_2O_4^-(aq) + 2\ H_2O(\ell)$

4.102. Balance the following half-reactions by adding the appropriate number of electrons. Which are oxidation half-reactions and which are reduction half-reactions?
a. $Fe^{2+}(aq) \rightarrow Fe^{3+}(aq)$
b. $AgI(s) \rightarrow Ag(s) + I^-(aq)$
c. $VO_2^+(aq) + 2\ H^+(aq) \rightarrow VO^{2+}(aq) + H_2O(\ell)$
d. $I_2(s) + 6\ H_2O(\ell) \rightarrow 2\ IO_3^-(aq) + 12\ H^+(aq)$

4.103. **Earth's Crust** The following chemical reactions have helped to shape Earth's crust. Determine the oxidation numbers of all the elements in the reactants and products, and identify which elements are oxidized and which are reduced.
a. $3\ SiO_2(s) + 2\ Fe_3O_4(s) \rightarrow 3\ Fe_2SiO_4(s) + O_2(g)$
b. $SiO_2(s) + 2\ Fe(s) + O_2(g) \rightarrow Fe_2SiO_4(s)$
c. $4\ FeO(s) + O_2(g) + 6\ H_2O(\ell) \rightarrow 4\ Fe(OH)_3(s)$

4.104. Determine the oxidation numbers of each of the elements in the following reactions, and identify which of them are oxidized or reduced, if any.
a. $SiO_2(s) + 2\ H_2O(\ell) \rightarrow H_4SiO_4(aq)$
b. $2\ MnCO_3(s) + O_2(g) \rightarrow 2\ MnO_2(s) + 2\ CO_2(g)$
c. $3\ NO_2(g) + H_2O(\ell) \rightarrow$
$$2\ NO_3^-(aq) + NO(g) + 2\ H^+(aq)$$

4.105. Combine the half-reaction for the reduction of $O_2$
$$O_2(aq) + 4\ H^+(aq) + 4\ e^- \rightarrow 2\ H_2O(\ell)$$
with the following oxidation half-reactions (which are based on common iron minerals) to develop complete redox reactions:
a. $2\ FeCO_3(s) + H_2O(\ell) \rightarrow$
$$Fe_2O_3(s) + 2\ CO_2(g) + 2\ H^+(aq) + 2\ e^-$$
b. $3\ FeCO_3(s) + H_2O(\ell) \rightarrow$
$$Fe_3O_4(s) + 3\ CO_2(g) + 2\ H^+(aq) + 2\ e^-$$
c. $2\ Fe_3O_4(s) + H_2O(\ell) \rightarrow 3\ Fe_2O_3(s) + 2\ H^+(aq) + 2\ e^-$

4.106. Uranium is found in Earth's crust as $UO_2$ and an assortment of compounds containing $UO_2^{n+}$ cations. Add the following pairs of reduction and oxidation equations to develop overall equations for converting soluble uranium polyatomic ions into insoluble $UO_2$.
a. $6\ H^+(aq) + UO_2(CO_3)_3^{4-}(aq) + 2\ e^- \rightarrow$
$$UO_2(s) + 3\ CO_2(g) + 3\ H_2O(\ell)$$
$Fe^{2+}(aq) + 3\ H_2O(\ell) \rightarrow Fe(OH)_3(s) + 3\ H^+(aq) + e^-$
b. $6\ H^+(aq) + UO_2(CO_3)_3^{4-}(aq) + 2\ e^- \rightarrow$
$$UO_2(s) + 3\ CO_2(g) + 3\ H_2O(\ell)$$
$HS^-(aq) + 4\ H_2O(\ell) \rightarrow SO_4^{2-}(aq) + 9\ H^+(aq) + 8\ e^-$
c. $2\ e^- + UO_2(HPO_4)_2^{2-}(aq) \rightarrow UO_2(s) + 2\ HPO_4^{2-}(aq)$
$3\ OH^-(aq) \rightarrow H_2O(\ell) + HO_2^-(aq) + 2\ e^-$

4.107. Nitrogen in the hydrosphere is found primarily as ammonium ions and nitrate ions. Complete and balance the following chemical equation describing the oxidation of ammonium ions to nitrate ions in acid solution:
$$NH_4^+(aq) + O_2(g) \rightarrow NO_3^-(aq)$$

4.108. In sediments and waterlogged soil, dissolved $O_2$ concentrations are so low that the microorganisms living there must rely on other sources of oxygen for respiration. Some bacteria can extract the oxygen from sulfate ions, reducing the sulfur in them to hydrogen sulfide gas and giving the sediments or soil a distinctive rotten-egg odor.
a. What is the change in oxidation state of sulfur as a result of this reaction?
b. Write the balanced net ionic equation for the reaction, under acidic conditions, that releases $O_2$ from sulfate and forms hydrogen sulfide gas.

4.109. Chromium is more toxic and more soluble in natural waters as $HCrO_4^-$ than as chromium(III) ion. In the presence of $H_2S$, the following reaction takes place in neutral solution:
$$HCrO_4^-(aq) + H_2S(aq) \rightarrow Cr_2O_3(s) + SO_4^{2-}$$
a. Assign oxidation numbers to the reactants and products.
b. Balance the equation.
c. How many electrons are transferred for each atom of chromium that reacts?

**4.110.** The water soluble uranyl cation, $UO_2^+$, can be removed by reaction with methane gas:

$$UO_2^+(aq) + CH_4(g) \rightarrow UO_2(s) + HCO_3^-(aq)$$

a. Assign oxidation numbers to the reactants and products.
b. Balance the equation in acidic solution.
c. How many electrons are transferred for each atom of uranium that reacts?

**4.111.** The solubilities of Fe and Mn in freshwater streams are affected by changes in their oxidation states. Complete and balance the following redox reaction in which soluble $Mn^{2+}$ becomes solid $MnO_2$:

$$Fe(OH)_2^+(aq) + Mn^{2+}(aq) \rightarrow MnO_2(s) + Fe^{2+}(aq)$$

**4.112.** A method for determining the quantity of dissolved oxygen in natural waters requires a series of redox reactions. Balance the following chemical equations in that series under the conditions indicated:
a. $Mn^{2+}(aq) + O_2(g) \rightarrow MnO_2(s)$ (basic solution)
b. $MnO_2(s) + I^-(aq) \rightarrow Mn^{2+}(aq) + I_2(s)$ (acidic solution)
c. $I_2(s) + S_2O_3^{2-}(aq) \rightarrow I^-(aq) + S_4O_6^{2-}(aq)$ (neutral solution)

**4.113.** Complete and balance the following reactions for the removal of sulfide, cyanide, and sulfite. Assume that reaction conditions are basic.
a. $MnO_4^-(aq) + S^{2-}(aq) \rightarrow MnS(s) + S_8(s)$
b. $MnO_4^-(aq) + CN^-(aq) \rightarrow CNO^-(aq) + MnO_2(s)$
c. $MnO_4^-(aq) + SO_3^{2-}(aq) \rightarrow MnO_2(s) + SO_4^{2-}(aq)$
d. $Ag(s) + CN^-(aq) + O_2(g) \rightarrow Ag(CN)_2^-(aq)$

**4.114. Bacteriocide and Viruscide** The water-soluble gas $ClO_2$ is known as an oxidative biocide. It destroys bacteria by oxidizing their cell walls and viruses by attacking their viral envelopes. $ClO_2$ may be prepared for use as a decontaminating agent from several different starting materials in slightly acidic solutions. Complete and balance the following chemical reactions for the synthesis of $ClO_2$.
a. $ClO_3^-(aq) + SO_2(g) \rightarrow ClO_2(g) + SO_4^{2-}(aq)$
b. $ClO_3^-(aq) + Cl^-(aq) \rightarrow ClO_2(g) + Cl_2(g)$
c. $ClO_3^-(aq) + Cl_2(g) \rightarrow ClO_2(g) + O_2(g)$

**4.115.** Refer to Table 4.6 to determine which of the following metals will reduce aqueous $Fe^{2+}$ to iron metal: lead, copper, zinc, or aluminum.

**4.116.** Which ions will oxidize aluminum? $Li^+$; $Ca^{2+}$; $Ag^+$; $Sn^{2+}$

**4.117.** Through appropriate experiments, we could expand the activity series in Table 4.6 to include additional metals. If aluminum is oxidized by $V^{3+}$ but aluminum does not reduce $Sc^{3+}$, where would you place vanadium and scandium in the activity series? Which metal would you test to firmly establish scandium's position?

**4.118.** If iron is oxidized by $Cd^{2+}$ but iron does not reduce $Ga^{3+}$, where would you place cadmium and gallium in the activity series? Which metal would you test to firmly establish gallium's position?

**4.119.** Dichromate ion oxidizes $Fe^{2+}$ ion in aqueous, acidic solution producing $Fe^{3+}$ and $Cr^{3+}$ by the unbalanced chemical equation:

$$Cr_2O_7^{2-}(aq) + Fe^{2+}(aq) \rightarrow Fe^{3+} + 2\,Cr^{3+}$$

a. Balance the equation.
b. If 15.2 mL of 0.135 $M$ $Cr_2O_7^{2-}$ is required to completely react with 100.0 mL of $Fe^{2+}$, what is the concentration of the $Fe^{2+}$ solution?

**\*4.120.** Ozone, $O_3$, reacts with iodide ion ($I^-$) in basic solution to form $O_2$ and $I_2$ by the unbalanced chemical equation:

$$O_3(aq) + I^-(aq) \rightarrow O_2(g) + I_2(aq)$$

a. Balance the equation.
b. A saturated solution of ozone in 125 mL of water at 0°C is treated with 10 mL 2.0 $M$ KI. After the reaction is complete, the solution is titrated with 0.100 $M$ $H^+$. If 54.7 mL of acid is needed, what is the concentration of $O_3$ in a saturated solution?

## Additional Problems

**4.121.** To determine the concentration of $SO_4^{2-}$ ion in a sample of groundwater, 100.0 mL of the sample is titrated with 0.0250 $M$ $Ba(NO_3)_2$, forming insoluble $BaSO_4$. If 3.19 mL of the $Ba(NO_3)_2$ solution is required to reach the end point of the titration, what is the molarity of the $SO_4^{2-}$?

**4.122. Antifreeze** Ethylene glycol is the common name for the liquid used to keep the coolant in automobile cooling systems from freezing. It is 38.7% carbon, 9.7% hydrogen, and 51.6% oxygen by mass. Its molar mass is 62.07 g/mol, and its density is 1.106 g/mL at 20°C.
a. What is the empirical formula of ethylene glycol?
b. What is the molecular formula of ethylene glycol?
c. In a solution prepared by mixing equal volumes of water and ethylene glycol, which ingredient is the solute and which is the solvent?

**4.123.** According to the label on a bottle of concentrated hydrochloric acid, the contents are 36.0% HCl by mass and have a density of 1.18 g/mL.
a. What is the molarity of concentrated HCl?
b. What volume of it would you need to prepare 0.250 L of 2.00 $M$ HCl?
c. What mass of sodium hydrogen carbonate would be needed to neutralize the spill if a bottle containing 1.75 L of concentrated HCl dropped on a lab floor and broke open?

**4.124. Synthesis and Toxicity of Chlorine** Chlorine was first prepared in 1774 by heating a mixture of NaCl and $MnO_2$ in sulfuric acid:

$$NaCl(aq) + H_2SO_4(aq) + MnO_2(s) \rightarrow$$
$$Na_2SO_4(aq) + MnCl_2(aq) + H_2O(\ell) + Cl_2(g)$$

a. Assign oxidation numbers to the elements in each compound, and balance the redox reaction in acid solution.
b. Write a net ionic equation describing the reaction for formation of chlorine.
c. If chlorine gas is inhaled, it causes pulmonary edema (fluid in the lungs) because it reacts with water in the alveolar sacs of the lungs to produce the strong acid HCl and the weaker acid HOCl. Balance the equation for the conversion of $Cl_2$ to HCl and HOCl.

**\*4.125.** When a solution of dithionate ions ($S_2O_4^{2-}$) is added to a solution of chromate ions ($CrO_4^{2-}$), the products of the ensuing chemical reaction that occurs under basic conditions include soluble sulfite ions and solid chromium(III) hydroxide. This reaction is used to remove Cr(VI) from wastewater generated by factories that make chrome-plated metals.

a. Write the net ionic equation for this redox reaction.

b. Which element is oxidized and which is reduced?

c. Identify the oxidizing and reducing agents in this reaction.

d. How many grams of sodium dithionate would be needed to remove the Cr(VI) in 100.0 L of wastewater that contains 0.00148 $M$ chromate ion?

4.126. A prototype battery based on iron compounds with large, positive oxidation numbers was developed in 1999. In the following reactions, assign oxidation numbers to the elements in each compound, and balance the redox reactions in basic solution.

a. $FeO_4^{2-}(aq) + H_2O(\ell) \rightarrow FeOOH(s) + O_2(g) + OH^-(aq)$

b. $FeO_4^{2-}(aq) + H_2O(\ell) \rightarrow Fe_2O_3(s) + O_2(g) + OH^-(aq)$

4.127. **Polishing Silver** Silver tarnish is the result of silver metal reacting with sulfur compounds, such as $H_2S$, in the air. The tarnish on silverware ($Ag_2S$) can be removed by soaking in a solution of $NaHCO_3$ (baking soda) in a basin lined with aluminum foil.

a. Write a balanced equation for the tarnishing of Ag to $Ag_2S$, and assign oxidation numbers to the reactants and products. How many electrons are transferred per mole of silver?

b. Write a balanced equation for the reaction of $Ag_2S$ with Al metal, $NaHCO_3$, and water to produce $Al(OH)_3$, $H_2S$, $H_2$, and Ag metal.

4.128. Many nonmetal oxides react with water to form acidic solutions. Give the formula and name for the acids produced from the following reactions:

a. $P_4O_{10} + 6\ H_2O \rightarrow$

b. $SeO_2 + H_2O \rightarrow$

c. $B_2O_3 + 3\ H_2O \rightarrow$

4.129. Write overall and net ionic equations for the reactions that occur when

a. a sample of acetic acid is titrated with a solution of KOH.

b. a solution of sodium carbonate is mixed with a solution of calcium chloride.

c. calcium oxide dissolves in water.

*4.130. **Fluoride Ion in Drinking Water** Sodium fluoride is added to drinking water in many municipalities to protect teeth against cavities. The target of the fluoridation is hydroxyapatite, $Ca_{10}(PO_4)_6(OH)_2$, a compound in tooth enamel. There is concern, however, that fluoride ions in water may contribute to skeletal fluorosis, an arthritis-like disease.

a. Write a net ionic equation for the reaction between hydroxyapatite and sodium fluoride that produces fluorapatite, $Ca_{10}(PO_4)_6F_2$.

b. The EPA currently restricts the concentration of $F^-$ in drinking water to 4 mg/L. Express this concentration of $F^-$ in molarity.

c. One study of skeletal fluorosis suggests that drinking water with a fluoride concentration of 4 mg/L for 20 years raises the fluoride content in bone to 6 mg/g, a level at which a patient may experience stiff joints and other symptoms. How much fluoride (in milligrams) is present in a 100 mg sample of bone with this fluoride concentration?

*4.131. **Rocket Fuel in Drinking Water** Near Las Vegas, NV, improper disposal of perchlorates used to manufacture rocket fuel has contaminated a stream that flows into Lake Mead, the largest artificial lake in the United States and a major supply of drinking and irrigation water for the American Southwest. The EPA has proposed an advisory range for perchlorate concentrations in drinking water of 4 to 18 µg/L. The perchlorate concentration in the stream averages 700.0 µg/L, and the stream flows at an average rate of 161 million gallons per day (1 gal = 3.785 L).

a. What are the formulas of sodium perchlorate and ammonium perchlorate?

b. How many kilograms of perchlorate flow from the Las Vegas stream into Lake Mead each day?

c. What volume of perchlorate-free lake water would have to mix with the stream water each day to dilute the stream's perchlorate concentration from 700.0 to 4 µg/L?

d. Since 2003, Maryland, Massachusetts, and New Mexico have limited perchlorate concentrations in drinking water to 0.1 µg/L. Five replicate samples were analyzed for perchlorates by laboratories in each state, and the following data (µg/L) were collected:

| MD | MA | NM |
|-----|------|-----|
| 1.1 | 0.90 | 1.2 |
| 1.1 | 0.95 | 1.2 |
| 1.4 | 0.92 | 1.3 |
| 1.3 | 0.90 | 1.4 |
| 0.9 | 0.93 | 1.1 |

Which of the labs produced the most precise analytical results?

*4.132. **Acidic Mine Drainage** Water draining from abandoned mines on Iron Mountain in California is extremely acidic and leaches iron, zinc, and other metals from the underlying rock (Figure P4.132). One liter of drainage contains as much as 80.0 g of dissolved iron and 6 g of zinc.

**FIGURE P4.132**

a. Calculate the molarity of iron and of zinc in the drainage.

b. One source of the dissolved iron is the reaction between water containing $H_2SO_4$ and solid $Fe(OH)_3$. Complete the following chemical equation, and write a net ionic equation for the process.

$$2\ Fe(OH)_3(s) + 3\ H_2SO_4(aq) \rightarrow$$

c. Sources of zinc include the mineral smithsonite, $ZnCO_3$. Write a balanced net ionic equation for the reaction between smithsonite and $H_2SO_4$ that produces $Zn^{2+}(aq)$.

d. One member of a class of minerals called ferrites is found to contain a mixture of Zn(II), Fe(II), and Fe(III) oxides. The generic formula for the mineral is $Zn_xFe_{1-x}O \cdot Fe_2O_3$. If acidic mine waste flowing through a deposit of this mineral contains 80 g of Fe and 6 g of Zn as a result of dissolution of the mineral, what is the value of $x$ in the formula of the mineral in the deposit?

**\*4.133. Making Apple Cider Vinegar** Some people who prefer natural foods make their own apple cider vinegar. They start with freshly squeezed apple juice that contains about 6% natural sugars. These sugars, which all have nearly the same empirical formula, $CH_2O$, are fermented with yeast in a chemical reaction that produces equal numbers of moles of ethanol (Figure P4.133a) and carbon dioxide. The product of this fermentation, called hard cider, undergoes an acid fermentation step in which ethanol and dissolved oxygen gas react together to form acetic acid (Figure P4.133b) and water. This acetic acid is the principal solute in vinegar.

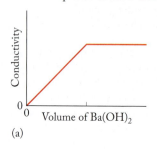

Ethanol
$CH_3-CH_2-OH$
(a)

Acetic acid
$CH_3-COOH$
(b)

**FIGURE P4.133**

a. Write a balanced chemical equation describing the fermentation of natural sugars to ethanol and carbon dioxide. You may use the empirical formula given in the above paragraph.

b. Write a balanced chemical equation describing the acid fermentation of ethanol to acetic acid.

c. What are the oxidation states of carbon in the reactants and products of the two fermentation reactions?

d. If a sample of apple juice contains $1.00 \times 10^2$ g of natural sugar, what is the maximum quantity of acetic acid that could be produced by the two fermentation reactions?

**\*4.134.** A food chemist determines the concentration of acetic acid in a sample of apple cider vinegar (see Problem 4.133) by acid–base titration. The density of the sample is 1.01 g/mL. The titrant is 1.002 $M$ NaOH. The average volume of titrant required to titrate 25.00 mL subsamples of the vinegar is 20.78 mL. What is the concentration of acetic acid in the vinegar? Express your answer the way a food chemist probably would: as percent by mass.

**\*4.135.** One way to follow the progress of a titration and detect its equivalence point is by monitoring the conductivity of the titration reaction mixture. For example, consider the way the conductivity of a sample of sulfuric acid changes as it is titrated with a standard solution of barium hydroxide before and then after the equivalence point.

a. Write the overall ionic equation for the titration reaction.

b. Which of the four graphs in Figure P4.135 comes closest to representing the changes in conductivity during the titration? (The zero point on the $y$-axis of these graphs represents the conductivity of pure water; the break points on the $x$-axis represent the equivalence point.)

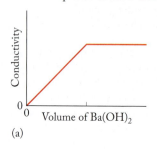

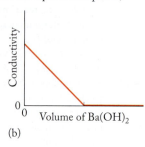
(a)

(b)

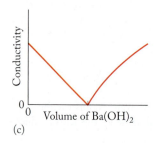
(c)

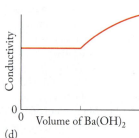
(d)

**FIGURE P4.135**

**\*4.136.** Which of the graphs in Figure P4.136 best represents the changes in conductivity that occur before and after the equivalence point in each of the following titrations:

a. sample, $AgNO_3(aq)$; titrant, $KCl(aq)$

b. sample, $HCl(aq)$; titrant, $LiOH(aq)$

c. sample, $CH_3COOH(aq)$; titrant, $NaOH(aq)$

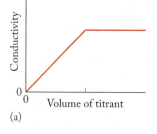
(a)

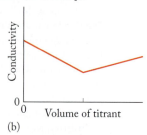
(b)

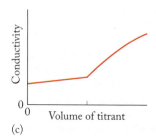
(c)

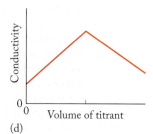
(d)

**FIGURE P4.136**

**4.137.** When electrodes connected to a lightbulb are inserted into an aqueous solution of acetic acid the bulb glows dimly. Will the bulb become brighter, remain the same, or turn off after one equivalent of aqueous NaOH is added to the solution? Write a balanced net ionic equation that supports your answer.

***4.138.** When electrodes connected to a lightbulb are inserted into a beaker containing silver carbonate and water, will the bulb not glow, glow dimly, or glow brightly? What do you think will happen after addition of one equivalent of aqueous HCl? Write a balanced net ionic equation that supports your answer.

**4.139. Superoxide Dismutases** Oxygen in the form of superoxide ions, $O_2^-$, is quite hazardous to human health. Superoxide dismutases represent a class of enzymes that convert superoxide ions to hydrogen peroxide and oxygen by the unbalanced chemical equation:

$$O_2^-(aq) + H^+(aq) \rightarrow H_2O_2(aq) + O_2(aq)$$

a. Identify the oxidation and reduction half-reactions.
b. Balance the equation.

**4.140. Nitrogen Fixing Bacteria** Bacteria found among the roots of legumes perform an important biological function, converting nitrogen to ammonia in a process known as nitrogen fixation. The electrons required for this redox reaction are supplied by transition-metal-containing enzymes called nitrogenases. The unbalanced chemical equation for this process is

$$N_2(g) + H^+(aq) + M^{2+}(aq) \rightarrow NH_3(aq) + H_2(g) + M^{3+}(aq)$$

where M represents a transition metal such as iron. Nitrification is a multistep process in which the nitrogen in organic and inorganic compounds is biochemically oxidized. Bacteria and fungi are responsible for a part of the *nitrification process* described by the reaction:

$$NH_4^+(aq) + M^{3+}(aq) \rightarrow NO_2^-(aq) + M^{2+}(aq)$$

a. What are the oxidation numbers of nitrogen in the reactants and products of each reaction?
b. Which compounds or ions are being reduced in each reaction?
c. Balance the equations in acidic solution.

## Calcium: In the Limelight

**4.141. Rocks in Caves** The stalactites and stalagmites in most caves are made of limestone (calcium carbonate; see Figure 4.14). However, in the Lower Kane Cave in Wyoming they are made of gypsum (calcium sulfate). The presence of $CaSO_4$ is explained by the following sequence of reactions:

$$H_2S(aq) + 2\ O_2(g) \rightarrow H_2SO_4(aq)$$

$$H_2SO_4(aq) + CaCO_3(s) \rightarrow CaSO_4(s) + H_2O(\ell) + CO_2(g)$$

a. Which (if either) of these reactions is a redox reaction? How many electrons are transferred?
b. Write a net ionic equation for the reaction of $H_2SO_4$ with $CaCO_3$.
c. How would the net ionic equation be different if the reaction were written as follows?

$$H_2SO_4(aq) + CaCO_3(s) \rightarrow CaSO_4(s) + H_2CO_3(aq)$$

**4.142.** The alkaline earth elements react with nitrogen to form nitrides (with the general formula $M_3N_2$) for M = Be, Mg, Ca, Sr, and Ba. Like the alkaline earth oxides (general formula MO), the nitrides react with water to form alkaline earth hydroxides, $M(OH)_2$. Predict the other product for the following reaction, and balance the equation.

$$M_3N_2(s) + H_2O(\ell) \rightarrow M(OH)_2(s) + \underline{\hspace{1cm}}$$

**4.143.** Which of the following reactions of calcium compounds is/are redox reactions?
a. $CaCO_3(s) \rightarrow CaO(s) + CO_2(g)$
b. $CaO(s) + SO_2(g) \rightarrow CaSO_3(s)$
c. $CaCl_2(s) \rightarrow Ca(s) + Cl_2(g)$
d. $3\ Ca(s) + N_2(g) \rightarrow Ca_3N_2(s)$

**4.144.** HF is prepared by reacting $CaF_2$ with $H_2SO_4$:

$$CaF_2(s) + H_2SO_4(\ell) \rightarrow 2\ HF(g) + CaSO_4(s)$$

HF can in turn be electrolyzed when dissolved in molten KF to produce fluorine gas:

$$2\ HF(\ell) \rightarrow F_2(g) + H_2(g)$$

Fluorine is extremely reactive, so it is typically sold as a 5% mixture by volume in an inert gas such as helium. How much $CaF_2$ is required to produce 500.0 L of 5% $F_2$ in helium? Assume the density of $F_2$ gas is 1.70 g/L.

If your instructor assigns problems in **smartwork**, log in at **smartwork.wwnorton.com**.

# 5

# Thermochemistry: Energy Changes in Reactions

5.1 Energy: Basic Concepts and Definitions

5.2 Systems, Surroundings, and Energy Transfer

5.3 Enthalpy and Enthalpy Changes

5.4 Heating Curves and Heat Capacity

5.5 Calorimetry: Measuring Heat Capacity and Enthalpies of Reaction

5.6 Hess's Law

5.7 Standard Enthalpies of Formation and Reaction

5.8 Fuel Values and Food Values

## Learning Outcomes

**LO1** Explain kinetic and potential energies at the molecular level
**Sample Exercise 5.1**

**LO2** Identify familiar endothermic and exothermic processes
**Sample Exercise 5.2**

**LO3** Calculate changes in the internal energy of a system
**Sample Exercises 5.3, 5.4**

**LO4** Calculate the amount of heat transferred in physical or chemical processes
**Sample Exercises 5.5, 5.6, 5.7, 5.8**

**LO5** Calculate thermochemical values using data from calorimetry experiments
**Sample Exercises 5.9, 5.10, 5.11**

**LO6** Calculate enthalpies of reaction
**Sample Exercises 5.12, 5.13, 5.15**

**LO7** Recognize and write equations for formation reactions
**Sample Exercises 5.14, 5.16**

**LO8** Calculate and compare fuel and food values and fuel densities
**Sample Exercises 5.17, 5.18**

## Sunlight Unwinding

All of the chemical reactions in Chapter 4 involve changes in energy. Energy—to power an automobile, heat a home, or support life—may seem an abstract idea. Energy has no mass and no volume. However, we see its effects very clearly, from changing matter from one state to another—sunlight melts snow, a gas flame boils water and converts it to steam—to the transformation of energy from one form to another, as when the chemical energy of gasoline is converted to mechanical energy to move a car. Part of the energy in the gasoline contributes nothing to moving the vehicle and is lost to the surroundings as heat. Adding the energy used to move the vehicle to the energy lost as heat equals the energy in the gasoline that was burned. In other words, energy is neither created nor destroyed during chemical reactions.

We can roast marshmallows using energy from a campfire, but where does that energy come from? R. Buckminster Fuller (1895–1983), a 20th-century architect, inventor, and futurist, described a burning log like this: Trees gather the energy in sunlight and combine it with water and carbon dioxide to make the molecules that compose wood. When the wood is burned, the chemical products are carbon dioxide and water, and the fire is, as Fuller said, "all that sunlight unwinding." The sunlight unwinding is the release of chemical energy stored in the molecules of the wood. Through the transforming power of green plants, sunlight is the source of the chemical energy stored in all the substances we consume as food and fuel.

Nearly every chemical reaction, from combustion of fuels to neutralization reactions to the dissolution of salts, involves energy as either a product or a reactant. All physical changes of matter, such as ponds freezing in winter and thawing in spring, involve changes in energy. How do we measure the amount of energy involved in physical and chemical processes? We cannot directly measure the amount of energy in a chemical reaction, but by measuring changes in the temperature, we can relate *the change* in energy to the identities and amounts of

**Sunlight and Life** The sun is the ultimate source of energy for most forms of life on Earth. ▶

**thermochemical equation** the chemical equation of a reaction that includes energy as a reactant or a product.

**work** a form of energy: the energy required to move an object through a given distance.

**thermodynamics** the study of energy and its transformations.

**thermochemistry** the study of the relation between chemical reactions and changes in energy.

**heat** the energy transferred between objects because of a difference in their temperatures.

**thermal equilibrium** a condition in which temperature is uniform throughout a material and no energy flows from one point to another.

**potential energy (PE)** the energy stored in an object because of its position.

**state function** a property of an entity based solely on its chemical or physical state or both, but not on how it achieved that state.

**CONNECTION** We introduced energy as the capacity to do work in Chapter 1.

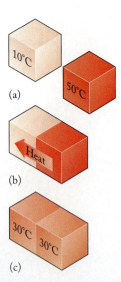

**FIGURE 5.1** (a) Two identical blocks at different temperatures are brought into contact (b). Heat is transferred from the block at higher temperature to the block at lower temperature until thermal equilibrium (same temperature) (c) is reached.

reactants and products involved in changes of state and in chemical reactions. Studying energy changes gives us important insights into the way nature works and how human activities impact our world. ■

# 5.1 Energy: Basic Concepts and Definitions

The reaction in which hydrogen combines with oxygen to form water releases energy: this energy may be converted into motion, as when spacecraft lift off, or into electrical energy. The equation representing a reaction in which energy is either a reactant or product is called a **thermochemical equation** because it describes whether energy is absorbed or released when the reaction occurs. A thermochemical equation can be written for any reaction. For instance, we represent the reaction of hydrogen and oxygen to form water vapor with the thermochemical equation

$$2\,H_2(g) + O_2(g) \rightarrow 2\,H_2O(g) + \text{energy}$$

Many chemical reactions produce energy. The energy from a thermochemical equation can be used to do **work**, can be transferred to an object to raise its temperature, or both. Energy that does work includes electrical, mechanical, light, and sound energy. Whatever the form, energy used to do work causes motion: the location or shape of an object changes when an energy source does work on the object. Changes in energy can cause changes in the state of a material, as when a solid melts or a liquid freezes. The study of energy and its transformations from one form to another is called **thermodynamics**. The part of thermodynamics that deals with changes in energy accompanying chemical reactions is known as **thermochemistry**.

When we put an ice cube, initially at −18°C—the typical temperature of a freezer—into room-temperature water (25°C), the ice cube melts and the water cools because energy moves from the room-temperature water into the colder ice cube. The process by which energy moves from a warmer object to a cooler object is called *heat transfer*, or more generally *energy transfer*. The difference in temperature defines the direction of energy flow when two objects come into contact: energy transfer in the form of **heat** always flows from the hotter object to the colder one (Figure 5.1). Energy transfer changes the temperature of matter; it can also change its physical state. The ice cube, for example, changes state from solid to liquid as energy is transferred to it from the water. The water remains in the liquid state, but its temperature drops as energy from it is transferred to the ice cube. Ultimately, these two portions of matter achieve the same temperature, higher than the initial temperature of the ice cube but lower than the initial temperature of the water. At this point, **thermal equilibrium** has been reached, which means the temperature is the same throughout the combined material and no further energy transfer occurs.

## Work, Potential Energy, and Kinetic Energy

In the physical sciences, work ($w$) is done whenever a force ($F$) moves an object through a distance ($d$). The amount of mechanical work done is

$$w = F \times d \tag{5.1}$$

**FIGURE 5.2** Work is done as skiers ascend to the top of a mountain. The amount of work may differ, depending on whether the skiers (1) ride a gondola on a direct route to the top or (2) hike to the top along a winding path.

Consider Equation 5.1 as it relates to skiers ascending a mountain (Figure 5.2). The work ($w$) done by the lift on a skier equals the length of the ride ($d$) times the force ($F$) needed to overcome gravity and transport the skier up the mountain. Some of the work done is stored in the skier as **potential energy (PE)**, which is the energy an object has because of its position. The farther up the mountain the skier is carried, the greater his potential energy. The mathematical expression for the skier's potential energy is

$$PE = m \times g \times h$$

where $m$ is the skier's mass, $g$ is the acceleration due to the force of gravity, and $h$ is the vertical distance between the skier's location on the mountain and his starting point. How he gets to that position is not important. Because the potential energy of any object does not depend on how the object gets to a particular point, potential energy is a **state function**, which means it is independent of the path followed to acquire the potential energy. Only position is important in considering potential energy (Figure 5.3). The term *state function* refers to a property of

▶❚❚ **CHEMTOUR** State Functions and Path Functions

**FIGURE 5.3** The potential energy of a skier depends only on the skier's mass and height above the base. If two skiers are at the same height ($h_1 = h_2$) and both skiers have the same mass, then they have the same potential energy, no matter how each skier got to that height.

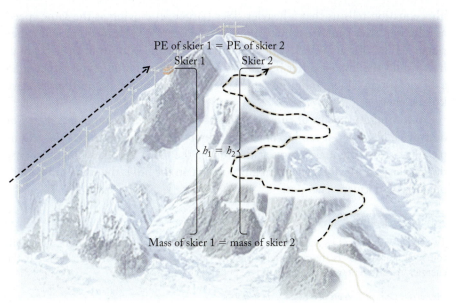

PE of skier 1 = PE of skier 2
Skier 1     Skier 2

$h_1 = h_2$

Mass of skier 1 = mass of skier 2

**FIGURE 5.4** (a) A skier at the starting gate of a ski jump has potential energy (PE) due to his position ($h_1$) above the bottom of the slope, his mass ($m$), and the acceleration due to gravity ($g$): PE = $mgh_1$. (b) During his run, the skier's potential energy is converted into kinetic energy: KE = $\frac{1}{2}mu^2$. While he moves down the slope, he has both KE and PE. (c) At the end of the run, the skier's potential energy is 0. His KE decreases from its maximum value to 0 as he slows to a stop at some point along the flat region.

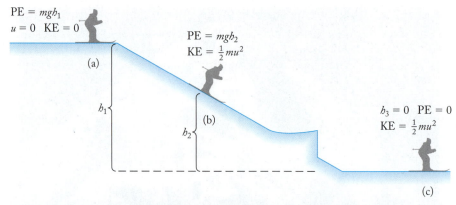

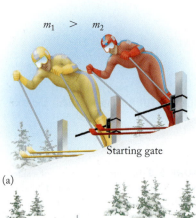

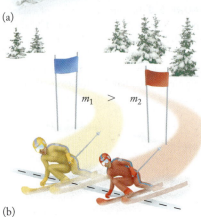

**FIGURE 5.5** (a) Two skiers of different mass are at the same position at the start of a race with respect to the bottom of the hill; (b) the same two skiers during the course of a race pass the same elevation at the same time.

a system that is determined by the position or condition of the system; do not confuse it with the term *change of physical state*, as in a phase change.

Now consider the potential energy of a skier standing still at the top of a ski jump (Figure 5.4a). At this position, his energy is all potential energy, but as he moves down the slope, his potential energy is converted into **kinetic energy (KE)**, the energy of motion (Figure 5.4b). At any moment between the start of the run and coming to a stop at the bottom of the hill, the jumper's kinetic energy is proportional to the product of his mass ($m$) times the square of his speed ($u$):

$$KE = \tfrac{1}{2}mu^2 \qquad (5.2)$$

This equation tells us that a heavier skier (larger $m$) moving at the same speed as a lighter skier (smaller $m$) has more kinetic energy. Our intuition and experience tell us this as well. If you were standing at the bottom of the ski jump, how would the impact differ if a 136 kg (300 lb) skier ran into you rather than a 45 kg (100 lb) skier going the same speed? The difference lies in their relative kinetic energies.

According to the **law of conservation of energy**, energy cannot be created or destroyed. However, it can be converted from one form to another, as this example illustrates. Potential energy at the top of the slope becomes kinetic energy during the run. The total energy at any position on the hill is the sum of the skier's potential and kinetic energies.

**CONCEPT TEST**

Two skiers with masses $m_1$ and $m_2$, where $m_1 > m_2$, are poised at the starting gate of a downhill course (Figure 5.5a). Is the potential energy of skier 1 the same as that of skier 2? If the energies are different, which skier has more potential energy?

*(Answers to Concept Tests are in the back of the book.)*

• • • • • • • • • • • • • • • • • • • • • • • • • • • • • • • • • • • • • • • • • • • •

Two skiers with masses $m_1$ and $m_2$, where $m_1 > m_2$, go past the same elevation on parallel race courses at the same time (Figure 5.5b). At that moment, is the potential energy of skier 2 more than, less than, or equal to the PE of skier 1? If the two are moving at the same speed, which has the greater kinetic energy?

*(Answers to Concept Tests are in the back of the book.)*

• • • • • • • • • • • • • • • • • • • • • • • • • • • • • • • • • • • • • • • • • • • • • • • • • • • • • • • • • • •

## Kinetic Energy and Potential Energy at the Molecular Level

The relation just described between kinetic and potential energies holds for atoms and molecules as well. There is no direct analogy with the kinetic and potential energies of a skier, however, because gravitational forces that control the skier play no role in the interactions of very small objects. At the molecular level, temperature and charge dominate the relation between kinetic and potential energies. Temperature governs motion at this level. Chemical bonds and differential electric charges cause interactions between particles that give rise to the potential energy stored in the arrangements of the atoms, ions, and molecules in matter.

The kinetic energy of a microscopic particle depends on its mass and speed, just as with macroscopic objects. However, because the particle's speed depends on temperature, its kinetic energy does, too. As the temperature of a population of particles increases, their average kinetic energy also increases. Consider, for example, how the molecules in the vapor phase above a liquid behave at different temperatures. We pick the gas phase because the molecules in a gas at normal pressures are widely separated and behave essentially independently of one another, and they all behave the same way regardless of their identities. If we have two samples of water vapor (molecular mass 18.02 amu) at room temperature, the two populations of $H_2O$ molecules have the same average kinetic energy. The average speeds of the molecules are the same because their masses are identical. If we increase the temperature of one sample, that population of $H_2O$ molecules acquires a higher average kinetic energy and the average speed of the molecules increases. An equivalent population of ethanol molecules (molecular mass 46.07 amu) in the gas phase at room temperature has the same average kinetic energy as the water molecules at room temperature, but the average speed of the ethanol molecules is lower because their mass is higher (Figure 5.6).

The kinetic energy associated with the total random motion of molecules is called **thermal energy**, and the thermal energy of a given sample of matter is proportional to the temperature of the sample. However, thermal energy also depends on the number of particles in a sample. The water in a swimming pool and in a cup of water taken from the pool have the same temperature, so their molecules have the same average kinetic energy. The water in the pool has much more thermal energy than the water in the cup, however, simply because there is a larger number of molecules in the pool. A large number of particles at a given temperature has a higher total thermal energy than a small number of particles at the same temperature.

• • • • • • • • • • • • • • • • • • • • • • • • • • • • • • • • •

If we heat a cup of water from a swimming pool almost to the boiling point, will its thermal energy be more than, less than, or the same as the thermal energy of all the water in the pool?

*(Answers to Concept Tests are in the back of the book.)*

• • • • • • • • • • • • • • • • • • • • • • • • • • • • • • • • • • • • • • • • • • • • • •

• • • • • • • • • • • • • • • • • • • • • • • • • • • • • • • • • •

**kinetic energy (KE)** the energy of an object in motion due to its mass ($m$) and its speed ($u$): $KE = \frac{1}{2}mu^2$.

**law of conservation of energy** energy cannot be created or destroyed but can be converted from one form into another.

**thermal energy** the kinetic energy of atoms, ions, and molecules.

**FIGURE 5.6** Two populations of gas-phase molecules have the same temperature and therefore the same average kinetic energy. Because ethanol molecules have a greater mass than water molecules, the average speed of the water molecules in the water vapor above the liquid water is greater than the average speed of the gas-phase ethanol molecules above the liquid ethanol.

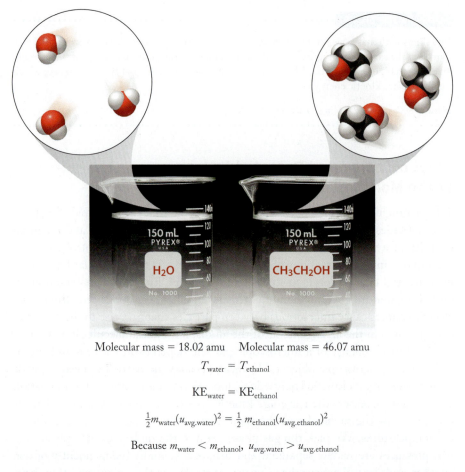

Molecular mass = 18.02 amu    Molecular mass = 46.07 amu

$$T_{water} = T_{ethanol}$$

$$KE_{water} = KE_{ethanol}$$

$$\tfrac{1}{2}m_{water}(u_{avg.water})^2 = \tfrac{1}{2}m_{ethanol}(u_{avg.ethanol})^2$$

Because $m_{water} < m_{ethanol}$, $u_{avg.water} > u_{avg.ethanol}$

An important form of potential energy at the atomic level arises from electrostatic interactions between charged particles. Just as the potential energy of skiers is determined by their positions above the bottom of the slope, the **electrostatic potential energy ($E_{el}$)** of charged particles is determined by the distance between them. The magnitude of this electrostatic potential energy, also known as *coulombic interaction*, is directly proportional to the product of the charges ($Q_1$ and $Q_2$) on the particles and is inversely proportional to the distance ($d$) between them:

$$E_{el} \propto \frac{Q_1 \times Q_2}{d} \tag{5.3}$$

where the symbol $\propto$ means "is proportional to." Coulombic interactions determine the potential energy of matter at the atomic level because they determine the relative positions of particles. Equation 5.3 is called Coulomb's Law because it traces back to the work by a French engineer, Charles Augustin de Coulomb (1736–1806), who first measured the interactions between charged particles.

If the two charges in Equation 5.3 are either both positive or both negative, their product is positive, $E_{el}$ is positive, and the particles repel each other. If one particle is positive and the other negative, $E_{el}$ is negative and the particles attract each other (Figure 5.7b–c). A lower electrostatic potential energy (a more negative value of $E_{el}$) corresponds to greater stability, so particles that attract each other because of their charges form an arrangement with a lower electrostatic potential energy than particles that repel each other. However, there is a limit to how close two particles can be. Recall that ions, whatever their charge, are surrounded by clouds of electrons that repel each other if they are pushed together too closely (Figure 5.7d).

**electrostatic potential energy ($E_{el}$)** the energy a particle has because of its electrostatic charge and its position relative to another particle; it is directly proportional to the product of the charges of the particles and inversely proportional to the distance between them.

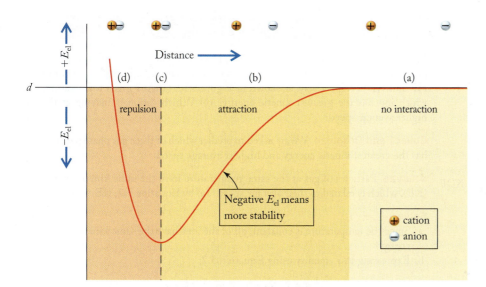

**FIGURE 5.7** Ionic interaction. (a) A positive
ion and a negative ion are so far apart ($d$ is
large) that they do not interact at all. (b) As
the ions move closer together ($d$ decreases),
the electrostatic potential energy between
them becomes more negative. (c) At this
distance, the attraction between them
produces an arrangement that is the most
favorable energetically because the ions have
the lowest electrostatic potential energy.
(d) If the ions are forced even closer together,
they repel each other.

Ions are not the only particles that experience coulombic interactions. Neutral species such as water molecules attract each other as well, due to distortions in the electron distribution about the nuclei of their atoms. The same ideas we use to describe the behavior of oppositely charged ions apply to molecules as well. Whether dealing with matter composed of atoms, molecules, or ions, the total energy at the microscopic level is the sum of the kinetic energy due to the random motion of particles and the potential energy due to their arrangement.

The energy given off or absorbed during a chemical reaction is equal to the difference in the energy of the reactants and products. For example, when hydrogen molecules burn in oxygen, the products are water and a considerable amount of energy (Figure 5.8). The energy given off by this reaction can be used to power rockets and is now being used to run some buses and automobiles. The fact that energy is given off in the reaction tells us that the product molecules must be at a lower potential energy than the reactant molecules. The difference between the energy of the products and the energy of the reactants is the energy released. In the case of hydrogen combustion (Figure 5.9a), this energy is sufficient to run vehicles once powered by fossil fuels (Figure 5.9b).

**FIGURE 5.8** Energy from the combustion of hydrogen can be used to launch rockets.

$$2\,H_2(g) \;+\; O_2(g) \;\longrightarrow\; 2\,H_2O(g) \;+\; \text{energy}$$

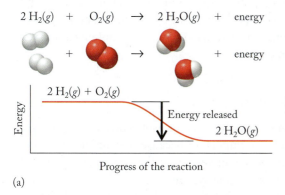

**FIGURE 5.9** (a) Hydrogen reacts with oxygen to produce water. Because this reaction releases energy as it runs, the molecules in the product are at a lower energy than those in the reactants. (b) Fueling a hydrogen-powered vehicle.

$+$ = H$^+$  $-$ = F$^-$  = HF  = NOF

$+$ = NO$^+$

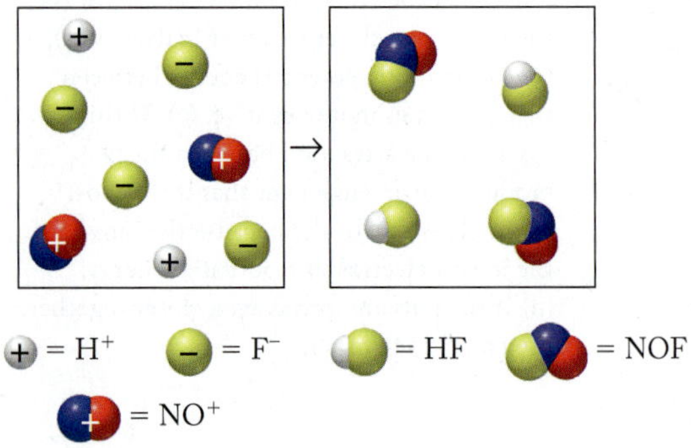

FIGURE 5.10 Protons and nitrosonium ions react with fluoride ions to form HF and NOF.

Protons (H$^+$) and nitrosonium ions (NO$^+$) react with fluoride ions (F$^-$) in the gas phase under laboratory conditions (constant temperature and pressure) to form HF and NOF, respectively, as shown in Figure 5.10. (a) At a given temperature, which ion has the greatest kinetic energy? (b) Which cation is moving at the higher average speed?

**Collect and Organize** We are asked to predict which of three gas phase particles has the greatest kinetic energy and highest average speed.

**Analyze** Particles of gas at the same temperature have the same kinetic energy (KE), which is related to their average speeds ($u$) by Equation 5.2, KE $= \frac{1}{2}mu^2$.

**Solve**
a. Since the temperature is constant, all of the ions have the same kinetic energy.
b. Expressing this equality using Equation 5.2:

$$\frac{1}{2}m_{H^+} \times u_{H^+}^2 = \frac{1}{2}m_{NO^+} \times u_{NO^+}^2$$

Rearranging the terms and simplifying:

$$\frac{m_{NO^+}}{m_{H^+}} = \left(\frac{u_{H^+}}{u_{NO^+}}\right)^2$$

The mass of a NO$^+$ ion is much greater than the mass of a proton: therefore, the protons in the reaction mixture have a higher average speed.

**Think About It** Heavier particles move slower than lighter particles with the same kinetic energy.

**Practice Exercise** Consider two positively charged particles, A$^+$ and B$^+$. If A$^+$ is 10% heavier than B$^+$, how much slower must it move to equal the kinetic energy of B$^+$?

*(Answers to Practice Exercises are in the back of the book.)*

# 5.2 Systems, Surroundings, and Energy Transfer

In both thermochemistry and thermodynamics, the specific part of the universe we are studying is called the **system** and everything else is called the **surroundings**. Although a system can be as large as a galaxy or as small as a living cell, most of the systems we examine in this chapter fit on a laboratory bench. Typically, we limit our concern about surroundings to that part of the universe that can exchange energy and matter with the system. In evaluating the energy gained or lost in a chemical reaction, the system may be just the particles involved in the reaction, or it may also include the contents along with the vessel in which the reaction occurs.

## Isolated, Closed, and Open Systems

There are three types of thermodynamic systems: isolated, closed, and open (Figure 5.11). These designations are important because they define the system we are dealing with and the part of the universe that the system interacts with. In the following discussion, hot soup is the system.

Consider hot soup in an ideal, closed thermos bottle. An ideal thermos bottle takes no energy away from the soup, thereby completely insulating the soup from the rest of the universe, and it loses no energy. Actually, this is

**system** the part of the universe that is the focus of a thermochemical study.

**surroundings** everything that is not part of the system.

**isolated system** a system that exchanges neither energy nor matter with the surroundings.

**closed system** a system that exchanges energy but not matter with the surroundings.

**open system** a system that exchanges both energy and matter with the surroundings.

**exothermic process** one in which energy flows from a system into its surroundings.

impossible, but sometimes in thermochemistry we must discuss systems that do not exist in the real world in order to define the total range of possibilities. The soup in the ideal thermos bottle is an example of an **isolated system**, which is a system that exchanges no energy or matter with its surroundings. The ideal thermos bottle prevents matter from being exchanged with the surroundings, and the thermal insulation provided by the ideal thermos also prevents heat from being transferred from the system to the surroundings, including the bottle itself. From the soup's point of view, the soup *is* the entire universe; it neither picks up nor donates any matter to the rest of the world and it loses no energy. In an isolated system, the system has no surroundings, which is why it is called *isolated*. The mass of the system does not change, and its energy content is constant; the soup stays at one constant temperature. Actually, even the best real thermos bottle cannot maintain the soup as an isolated system over time because heat will leak out, and the contents will cool. But for short time periods, the soup in a good thermos bottle approximates an isolated system.

Hot soup in a cup with a lid is an example of a **closed system**, which is a system that exchanges energy but not matter with its surroundings. Because the cup has a lid, no vapor (which is matter) escapes from the soup and no matter can be added. Only energy is exchanged between the soup and its surroundings. Heat from the soup is transferred—first into the cup walls, then into the air and the tabletop—and the soup gradually cools. Many real systems are closed systems.

Soup in an open cup is an **open system**, one that can exchange both energy and matter with its surroundings. Energy from the soup is transferred to the surroundings (cup, air, tabletop), and matter in the form of water vapor leaves the system and enters the air. We may also add matter to the system from the surroundings by sprinkling on a little grated cheese, some ground pepper, or a few crumbled crackers.

Most of the real systems we deal with are closed, and we may treat some of them as isolated, at which point we assume behavior that is more ideal than real. We do this because isolated and closed systems are easier to model quantitatively than open systems. However, many important systems are open—including cells, organisms, and Earth itself.

(a) **Isolated system:** A thermos bottle containing hot soup with the lid screwed on tightly

(b) **Closed system:** A cup of hot soup with a lid

(c) **Open system:** An open cup of hot soup

**FIGURE 5.11** Transfer of energy and matter in isolated, closed, and open systems. (a) Hot soup in a tightly sealed thermos bottle approximates an isolated system: no vapor escapes, no matter is added or removed, and no heat escapes to the surroundings. (b) Hot soup in a cup with a lid is a closed system; the soup transfers heat to the surroundings as it cools; however, no matter escapes and none is added. (c) Hot soup in a cup with no lid is an open system; it transfers both matter (steam) and energy (heat) to the surroundings as it cools. Matter in the form of pepper, grated cheese, crackers, or other matter from the surroundings may also be added to the soup.

**CONCEPT TEST** ...........................................

Identify the following systems as isolated, closed, or open: (a) the water in a pond; (b) a carbonated beverage in a sealed bottle; (c) a sandwich wrapped in thermally conducting plastic wrap; (d) a live chicken.

*(Answers to Concept Tests are in the back of the book.)*

## Exothermic and Endothermic Processes

Chemists classify thermochemical processes based on whether they give off or absorb energy. A chemical reaction or a change of state that results in the transfer of heat from a system to its surroundings is **exothermic** from the point of view of the system (Figure 5.12a). This flow of energy can be detected because it causes an increase in the temperature of the surroundings. Combustion reactions (the reactants are the system) release energy and are examples of exothermic reactions. In contrast, a chemical reaction or change of state that absorbs energy from the

**CONNECTION** In Chapter 3 we defined combustion as the reaction of oxygen with another element.

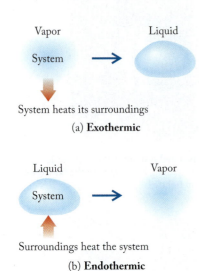

(a) **Exothermic**

Vapor    Liquid

System

System heats its surroundings

Liquid    Vapor

System

Surroundings heat the system

(b) **Endothermic**

**FIGURE 5.12** A process that is exothermic in one direction, such as (a) the condensation of a vapor, is endothermic in the reverse direction, such as (b) the evaporation of a liquid. Reversing a process changes the direction in which energy is transferred but not the quantity of energy transferred.

**FIGURE 5.13** Matter can be transformed from one physical state to another by adding or removing heat. Upward-pointing arrows represent endothermic processes (heat enters the system from the surroundings; $q > 0$). Downward-pointing arrows represent exothermic processes (heat leaves the system and enters the surroundings; $q < 0$).

surroundings is **endothermic** from the point of view of the system (Figure 5.12b). For example, ice cubes (the system) in a glass of warm water (the surroundings) absorb energy from the water, which causes the cubes to melt.

In another phase change, condensation of water vapor (the system) from humid air, forming droplets of liquid water on the outside of a glass containing an ice-cold drink (the surroundings), requires that energy flow from the system to the surroundings. From the point of view of the system, this process is exothermic. If we pour out the cold drink and pour hot coffee into the glass, the water droplets (the system) on the outside surface of the glass absorb energy from the coffee and vaporize in a process that is endothermic from the point of view of the system. This illustrates an important concept: a process that is exothermic in one direction (vapor → liquid; releases energy) is endothermic in the reverse direction (liquid → vapor; absorbs energy).

We use the symbol $q$ to represent the *quantity* of energy transferred during a chemical reaction or a change of state. For historical reasons, $q$ is called heat and represents energy transferred directly because of a difference in temperature. If the reaction or process is exothermic, $q$ is negative, meaning that the system *loses* energy to its surroundings. If the reaction or process is endothermic, $q$ is positive, indicating that energy is *gained* by the system. In Figure 5.13, endothermic changes of state are represented by arrows pointing upward: solid → liquid, liquid → gas, and solid → gas. The opposite changes: liquid → solid, gas → liquid, and gas → solid, represented by arrows pointing downward, are exothermic. To summarize:

Exothermic:   $q < 0$        Endothermic:   $q > 0$

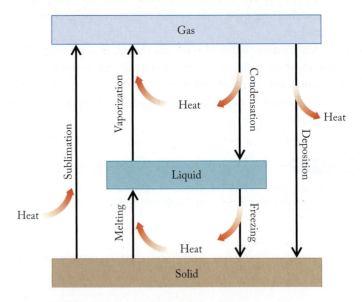

**SAMPLE EXERCISE 5.2** **Identifying Exothermic and Endothermic Processes**    LO2

Describe the flow of energy during the purification of water by distillation (Figure 5.14), identify the steps in the process as either endothermic or exothermic, and give the sign of $q$ associated with each step. Consider the water being purified to be the system.

**Collect and Organize** We are given that the water is the system. We must evaluate how the water gains or loses energy during distillation.

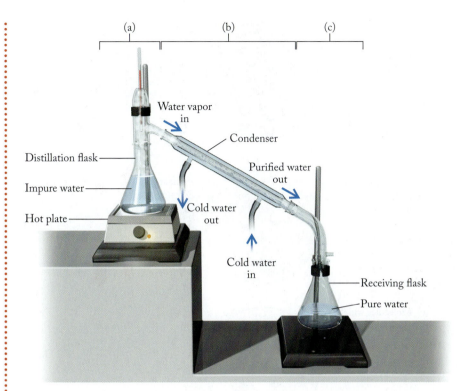

**FIGURE 5.14** A laboratory setup for distilling water. (a) Impure water is heated to the boiling point in the distillation flask. (b) Water vapor rises and enters the condenser, where it is liquefied. (c) The purified liquid is collected in the receiving flask.

**Analyze** In distillation, energy flows in three steps: (1) liquid water is heated to the boiling point and (2) vaporizes. (3) The vapors are cooled and condense as they pass through the condenser.

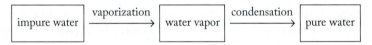

**Solve** Energy flows from the surroundings (hot plate) to heat the impure water (the system) to its boiling point and then to vaporize it. Therefore, processes 1 and 2 are endothermic. The sign of $q$ is positive for both. Because energy flows from the system (water vapor) into the surroundings (condenser walls), process 3 is exothermic. Therefore, the sign of $q$ is negative.

**Think About It** Endothermic means that energy is transferred from the surroundings into the system—the water in the distillation flask. When the water vapor is cooled in the condenser, energy flows from the vapor as it is converted from a gas to a liquid; the process is exothermic.

⚙ **Practice Exercise** What is the sign of $q$ as (a) a match burns, (b) drops of molten candle wax solidify, and (c) perspiration evaporates from skin? In each case, define the system and indicate whether the process is endothermic or exothermic.

*(Answers to Practice Exercises are in the back of the book.)*

Let's look at what happens at the molecular level as ice melts. These same types of changes occur when any substance goes through a phase change. Consider the flow of energy when an ice cube is left on a kitchen counter (Figure 5.15). Qualitatively, as the cube (the system) absorbs energy from the air and the counter (the surroundings) and starts to melt, the attractive forces that hold the water molecules in place in solid ice are overcome. The water molecules now have more positions available that they can occupy because they are not held in the rigid lattice, so they have more potential energy. After all the ice at 0°C has melted into liquid water at 0°C, the temperature of the water (the system) slowly rises to

◉◉ **CONNECTION** In Chapter 1 we discussed the arrangement of molecules in ice, water, and water vapor.

**endothermic process** one in which energy flows from the surroundings into the system.

**FIGURE 5.15** Changes of state. (a) The molecules in solid ice are held together in a rigid three-dimensional arrangement; they have the same nearest neighbors over time. Solid ice absorbs energy and is converted to liquid water. (b) As the solid melts, the molecules in the liquid state exchange nearest neighbors and occupy many more positions relative to one another than were possible in the solid. The liquid absorbs energy and is converted to a gas. (c) The molecules in a gas are widely separated from one another and move rapidly. The reverse of these processes occurs when water in the gas state loses energy and condenses to a liquid. The liquid also loses energy when it is converted to a solid.

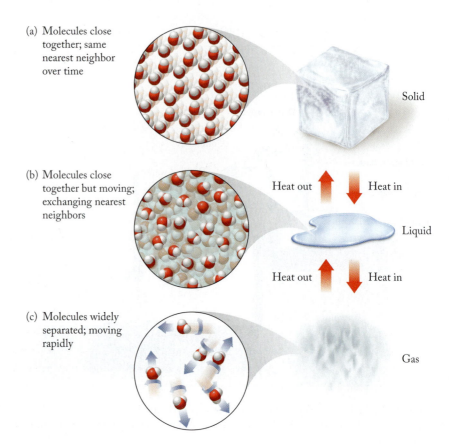

(a) Molecules close together; same nearest neighbor over time

Solid

(b) Molecules close together but moving; exchanging nearest neighbors

Heat out    Heat in

Liquid

Heat out    Heat in

(c) Molecules widely separated; moving rapidly

Gas

▶‖ **CHEMTOUR** Internal Energy

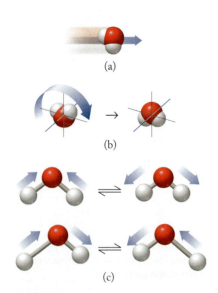

(a)

(b)

(c)

**FIGURE 5.16** Some of the types of molecular motion that contribute to the overall internal energy of a system: (a) translational motion, motion from place to place along a path; (b) rotational motion, motion about a fixed axis; and (c) vibrational motion, movement back and forth from some central position.

room temperature. As the temperature increases, the average kinetic energy of the molecules increases and they move more rapidly. (In Section 5.4 we present a quantitative view of this process.)

The kinetic energy of a system is part of its **internal energy ($E$)**, defined as the sum of the kinetic and potential energies of all the components of the system (Figure 5.16). It is not possible to determine the absolute values of kinetic and potential energies, but *changes* in internal energy ($\Delta E$) are fairly easy to measure because a change in a system's physical state or temperature is a measure of the change in its internal energy. (The capital Greek delta, $\Delta$, is the standard way scientists symbolize change in a quantity.) The change in internal energy is the difference between the final internal energy of the system and its initial internal energy:

$$\Delta E = E_{final} - E_{initial} \qquad (5.4)$$

Internal energy is a state function because $\Delta E$ depends only on the initial and final states. How the change occurs in the system does not matter.

The law of conservation of energy (Section 5.1) applies to the transfer of energy in materials. The total energy change experienced by a system must be balanced by the total energy change experienced by its surroundings.

## Energy Units and *P–V* Work

Energy changes that accompany chemical reactions and changes in physical state are sometimes expressed in calories. A **calorie (cal)** is the quantity of energy required to raise the temperature of 1 g of water from 14.5°C to 15.5°C. The SI unit of energy, used throughout this text, is the **joule (J)**; 1 cal = 4.184 J. The *Calorie* (Cal; note the capital C) in nutrition is actually one kilocalorie (kcal): 1 Cal = 1 kcal = 1000 cal.

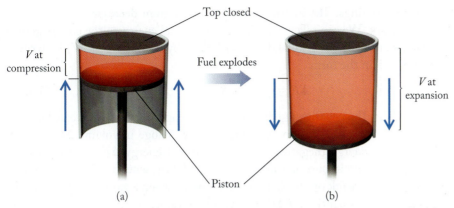

**FIGURE 5.17** Work performed by changing the volume of a gas. (a) Highest position of the piston after it compresses the gas in the cylinder, causing the fuel to ignite in a diesel engine. The piston does work on the gas in decreasing the gas volume. (b) The exploding fuel releases energy, causing the gas in the cylinder to expand and push the piston down. The gas has done work on the piston.

**internal energy (E)** the sum of all the kinetic and potential energies of all of the components of a system.

**calorie (cal)** the amount of energy necessary to raise the temperature of 1 g of water by 1°C, from 14.5°C to 15.5°C.

**joule (J)** the SI unit of energy; 4.184 J = 1 cal.

**first law of thermodynamics** the energy gained or lost by a system must equal the energy lost or gained by the surroundings.

**pressure–volume (P–V) work** the work associated with the expansion or compression of a gas.

Doing work on a system is a way to add to its internal energy. For example, compressing a quantity of gas (the system) into a smaller volume does work on the gas, and that work causes the temperature of the gas to rise, meaning its internal energy increases. The total increase in the internal energy of a closed system is the sum of the work done on it ($w$) and any other energy ($q$) gained:

$$\Delta E = q + w \qquad (5.5)$$

Equation 5.5 expresses the **first law of thermodynamics**: the energy changes ($\Delta E = q + w$) of a closed system and its surroundings are equal in magnitude but opposite in sign, so their sum is zero:

$$\Delta E_{sys} + \Delta E_{surr} = 0$$

When work is done *by* a system on its surroundings, the internal energy of the system decreases. For example, when fuel in the cylinder of a diesel engine ignites and produces hot gases, the gases (the system) expand and do work on the surroundings by pushing on the piston (Figure 5.17).

As another example of work being done by a system on its surroundings, consider a hot-air balloon (Figure 5.18), where we define the air in the balloon as the system. The air in the balloon is heated by a burner located at the balloon's bottom, which is open. Heating this air causes the balloon to expand. As the volume of the balloon increases, the balloon presses against the air outside the balloon, thus doing work on that outside air (the surroundings). This type of work, in which the pressure on a system remains constant but the volume of the system changes, is called **pressure–volume (P–V) work**.

The pressure referred to in the example of the hot-air balloon is atmospheric pressure. The pressure of the atmosphere on the balloon in this illustration, and indeed on all objects, is a result of Earth's gravity pulling the atmosphere toward the surface of the planet, where it exerts a force on all things because of its mass. The pressure at sea level on a dry day is about 1.00 atmosphere (atm). Atmospheric pressure varies slightly with the weather, but in an example involving P–V work like this one with the balloon, the important feature is that the pressure is constant.

Let's examine what happens to the balloon in terms of heat transferred and work done under constant pressure. First, adding hot air to the balloon increases its internal energy and causes the balloon to expand. Second, the expansion of the balloon against the pressure of its surroundings is P–V work done by the system

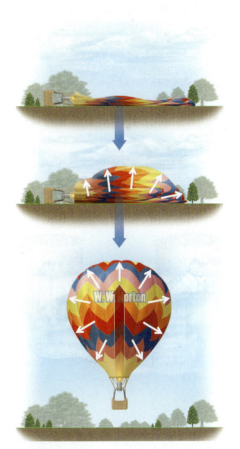

**FIGURE 5.18** In a deflated hot-air balloon, $V = 0$. Inflating the balloon causes it to do work on the atmosphere by pushing against the air. Because this work involves a change in volume, it is called P–V work.

▶❚❚ **CHEMTOUR** Pressure–Volume Work

on its surroundings. The internal energy of the system decreases as it performs this *P–V* work. We can relate this change in internal energy to the energy gained by the system by heating (*q*) and the work done by the system (*P*Δ*V*) by writing Equation 5.5 in the form

$$\Delta E = q + (-P\Delta V) = q - P\Delta V \qquad (5.6)$$

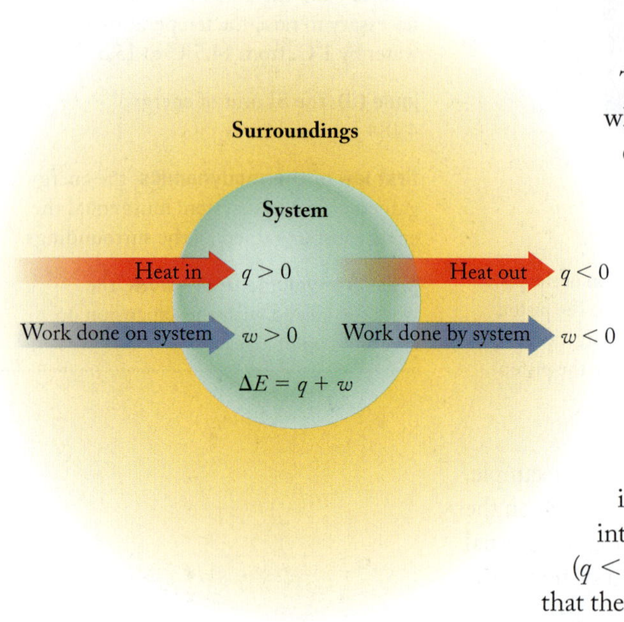

The negative sign in front of *P*Δ*V* is appropriate in this case because, when the system expands (positive change in volume Δ*V*), it loses energy (negative change in internal energy Δ*E*) as it does work on its surroundings. Correspondingly, when the surroundings do work on the system (for example, when a gas is compressed), the quantity (−*P*Δ*V*) has a positive value.

The sign of *q* may also be either positive or negative (Figure 5.19). If the system is heated by its surroundings, then *q* is positive (*q* > 0). Energy is added to the balloon, for instance, when it is being inflated, because energy flows from the surroundings (the burner) into the system (the air in the balloon). When the balloon expands, work is done by the system, *w* is negative (*w* < 0), and the internal energy decreases. If energy is transferred from the system into the surroundings due to a temperature difference, *q* is negative (*q* < 0), and the internal energy decreases. From Equation 5.6, we see that the change in internal energy of a system is positive when more energy enters the system than leaves and negative when more energy leaves the system than enters. Figure 5.19 also illustrates the first law of thermodynamics: the energy changes (Δ*E* = *q* + *w*) of the system and the surroundings are equal in magnitude but opposite in sign, so their sum is zero.

**FIGURE 5.19** Energy entering a system by heating or by work done on the system by the surroundings are both positive quantities because both increase the internal energy of the system. Energy released by a system to the surroundings and work done by a system on the surroundings are both negative quantities because both decrease the internal energy of a system.

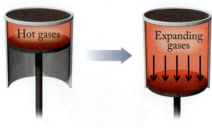

**FIGURE 5.20**

---

**SAMPLE EXERCISE 5.3** **Calculating Changes in Internal Energy** **LO3**

Figure 5.20 shows a simplified version of a piston and cylinder in an engine. Suppose combustion of fuel injected into the cylinder produces 155 J of energy. The hot gases in the cylinder expand, pushing the piston down and doing 93 J of *P–V* work on the piston. If the system is the gases in the cylinder, what is the change in internal energy of the system?

**Collect and Organize** We are given *q* and *w* and asked to calculate Δ*E*. We will need to decide whether *q* and *w* are positive or negative according to the sign convention in Figure 5.19.

**Analyze** The change in internal energy is related to the work done by or on a system and the energy gained or lost by the system (Equation 5.5). The system (gases in the cylinder) absorbs energy by heating, so *q* > 0, and the system does work on the surroundings (the piston), so *w* < 0.

**Solve** Substituting values into Equation 5.5, the change in internal energy for the system is

$$\Delta E = q + w = (155 \text{ J}) + (-93 \text{ J}) = 62 \text{ J}$$

**Think About It** More energy enters the system (155 J) than leaves it (93 J), so a positive value of Δ*E* is reasonable.

⚙ **Practice Exercise** In another event, the piston in Figure 5.20 compresses the air in the cylinder by doing 64 J of work on the gas. As a result, the air gives off 32 J of energy to the surroundings. If the system is the air in the cylinder, what is the change in its internal energy?

*(Answers to Practice Exercises are in the back of the book.)*

### SAMPLE EXERCISE 5.4 Calculating *P–V* Work    L03

A tank of compressed helium is used to inflate balloons for sale at a carnival on a day when the atmospheric pressure is 0.99 atm. If each balloon is inflated from an initial volume of 0.0 L to a final volume of 4.8 L, how much *P–V* work is done on the surrounding atmosphere by 100 balloons when they are inflated? The atmospheric pressure remains constant during the filling process.

**Collect and Organize** Each of the 100 balloons goes from empty ($V = 0.0$ L) to 4.8 L, which means $\Delta V = 4.8$ L, and the atmospheric pressure $P$ is constant at 0.99 atm. The identity of the gas used to fill the balloons does not matter because under these conditions all gases behave the same way, regardless of their identity. Our task is to determine how much *P–V* work is done by the 100 balloons on the air that surrounds them.

**Analyze** We focus on the work done on the atmosphere (the surroundings) by the system; the balloons and the helium they contain are the system.

**Solve** The volume change ($\Delta V$) as all the balloons are inflated is

$$100 \; \text{balloons} \times 4.8 \; \text{L/balloon} = 4.8 \times 10^2 \; \text{L}$$

The work ($w$) done by our system (the balloons) as they inflate against an external pressure of 0.99 atm is

$$w = -P\Delta V = -0.99 \; \text{atm} \times 4.8 \times 10^2 \; \text{L}$$

$$= -4.8 \times 10^2 \; \text{L} \cdot \text{atm}$$

**Think About It** Because work is done *by* the system on its surroundings, the work is negative from the point of view of the system.

⚙ **Practice Exercise** The balloon *Spirit of Freedom* (Figure 5.21), flown around the world by American Steve Fossett in 2002, contained 550,000 cubic feet of helium. How much *P–V* work was done by the balloon on the surrounding atmosphere while the balloon was being inflated, assuming atmospheric pressure was 1.00 atm? (1 m³ = 1000 L = 35.3 ft³.)

*(Answers to Practice Exercises are in the back of the book.)*

**FIGURE 5.21** The balloon *Spirit of Freedom* was flown around the world in 2002.

Note that units "L · atm" (liters times atmospheres) may seem strange for expressing work. However, a conversion factor, 101.32 J/(L · atm), connects joules, the SI unit for energy and work, to liters times atmospheres. Using this factor for the work done in filling the balloons in Sample Exercise 5.4, we get

$$w = -4.8 \times 10^2 \; \text{L} \cdot \text{atm} \times \frac{101.32 \; \text{J}}{\text{L} \cdot \text{atm}} = -4.9 \times 10^4 \; \text{J} = -49 \; \text{kJ}$$

## 5.3 Enthalpy and Enthalpy Changes

Many physical and chemical changes take place at constant atmospheric pressure ($P$). For example, boiling water for tea or any of the chemical reactions described in Chapter 4 take place in vessels open to the air. The thermodynamic parameter that relates the flow of energy into or out of a system during chemical reactions or physical changes at constant pressure is called the **enthalpy change ($\Delta H$)**. We symbolize this enthalpy change at constant pressure as $q_P$, where the subscript P specifies a process taking place at constant pressure. We then rearrange Equation 5.6 to define $\Delta H$ as

$$\Delta H = q_P = \Delta E + P\Delta V \qquad (5.7)$$

**enthalpy change ($\Delta H$)** the heat absorbed by an endothermic process or given off by an exothermic process occurring at constant pressure.

**enthalpy (H)** the sum of the internal energy and the pressure–volume product of a system; $H = E + PV$.

Thus the change in enthalpy is the change in the internal energy of the system at constant pressure plus the $P$–$V$ work done by the system on its surroundings.

The **enthalpy (H)** of a thermodynamic system is the sum of the internal energy and the pressure–volume product ($H = E + PV$). However, as we saw with internal energy, determining the absolute values of these parameters is difficult, whereas determining *changes* is fairly easy. We therefore concentrate on the *change* in enthalpy ($\Delta H$) of a system or the surroundings. Equation 5.7 tells us that for a reaction run at constant pressure, the enthalpy change is equal to $q_P$, which is the heat gained or lost by the system during the reaction. This statement also means that the units for $\Delta H$ are the same as those for $q$. Enthalpy has units of joules (J), and, if it is reported with respect to the quantity of a substance, J/g or J/mol.

Both $\Delta H$ and $\Delta E$ represent changes in a state function of a system. These two terms are similar, but the difference between them is important. $\Delta E$ includes *all* the energy (heat and work) exchanged by the system with the surroundings: $\Delta E = q + w$. However, $\Delta H$ is only $q$, the heat, exchanged at constant pressure: $\Delta H = q_P$. If a chemical reaction does not involve changes in volume, then $\Delta E$ and $\Delta H$ have very similar values. If a reaction consumes or produces gas and thereby the system experiences large changes in volume, $\Delta E$ and $\Delta H$ can be quite different.

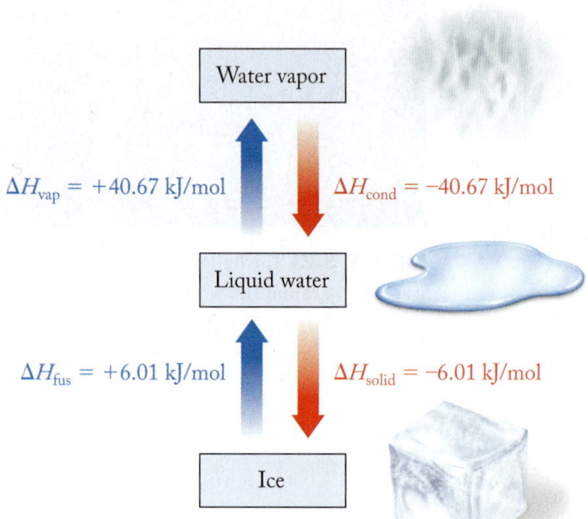

When energy flows out of a system, $q$ is negative according to our sign convention (Figure 5.19), so the enthalpy change is negative: $\Delta H < 0$. When heat flows into a system, $q$ is positive, so $\Delta H > 0$. As we saw in Section 5.2, positive $q$ values indicate endothermic processes, and negative $q$ values indicate exothermic processes. Knowing that, we can relate the terms *exothermic* and *endothermic* to enthalpy changes as well. For example, energy flows into a melting ice cube (the system; Figure 5.22) from its surroundings; this process is endothermic, meaning that $q$—and therefore $\Delta H$—is positive. However, to make ice cubes in a freezer, the water (the system) must lose energy, which means the process is exothermic and both $q$ and $\Delta H$ are negative. The enthalpy changes for the two processes—melting and freezing—have different signs but, for a given amount of water, they have the same absolute value. As Figure 5.22 illustrates, the energy required to melt a given quantity of a substance has the same magnitude but is opposite in sign to the energy given off when that same quantity of material freezes. For example, $\Delta H_{fus} = 6.01$ kJ if 1 mole of ice melts, but $\Delta H_{solid} = -6.01$ kJ if 1 mole of water freezes. We add the subscript *fus* or *solid* to $\Delta H$ to identify which process is occurring: melting (*fusion*) or freezing (*solidification*). The magnitudes and signs of enthalpies associated with other paired phase changes—vaporization and condensation, and sublimation and deposition—also work this way.

In terms of symbols, we put subscripts on $\Delta H$ to clarify not only the specific process but also the part of the universe to which the value applies. If we wish to indicate the enthalpy change associated with the system, we may write $\Delta H_{sys}$.

$\Delta H_{vap} = +40.67$ kJ/mol     $\Delta H_{cond} = -40.67$ kJ/mol

$\Delta H_{fus} = +6.01$ kJ/mol     $\Delta H_{solid} = -6.01$ kJ/mol

**FIGURE 5.22** The enthalpy changes for vaporization, $\Delta H_{vap} = 40.67$ kJ/mol, and condensation, $\Delta H_{cond} = -40.67$ kJ/mol of water at 100° are equal in magnitude but opposite in sign. So, too, are the enthalpy changes for melting one mole of ice, $\Delta H_{fus} = 6.01$ kJ/mol, and freezing one mole of water, $\Delta H_{solid} = -6.01$ kJ/mol at 0°C.

**SAMPLE EXERCISE 5.5  Determining the Value and Sign of $\Delta H$  LO4**

Between periods of a hockey game, an ice-resurfacing machine (Figure 5.23) spreads 855 L of water across the surface of a hockey rink. (a) If the system is the water, what is the sign of $\Delta H_{sys}$ as the water freezes? (b) To freeze this volume of water at 0°C, what is the value of $\Delta H_{sys}$? The density of water is 1.00 g/mL.

**Collect and Organize** We are given the volume and density of the water and are asked to determine the sign and value of $\Delta H_{sys}$.

**Analyze** The water from the ice-resurfacing machine is the system, so we can determine the sign of $\Delta H$ by determining whether energy transfer is into or out of the water. (a) The freezing takes place at constant pressure, so $\Delta H_{sys} = q_P$. (b) To calculate the amount of energy lost from the water as it freezes, we must convert 855 L of water into moles of water because the conversion factor between the quantity of energy removed and the quantity of water that freezes is the enthalpy of solidification of water, $\Delta H_{solid} = -6.01$ kJ/mol.

**Solve**

a. For the water (the system) to freeze, energy must be removed from it. Therefore, $\Delta H_{sys}$ must be a negative value.

b. We convert the volume of water into moles:

$$855 \text{ L} \times \frac{1000 \text{ mL}}{1 \text{ L}} \times \frac{1.00 \text{ g}}{1 \text{ mL}} \times \frac{1 \text{ mol}}{18.02 \text{ g}} = 4.745 \times 10^4 \text{ mol}$$

Then we calculate $\Delta H_{sys}$:

$$\Delta H_{sys} = 4.745 \times 10^4 \text{ mol} \times \frac{-6.01 \text{ kJ}}{1 \text{ mol}} = -2.85 \times 10^5 \text{ kJ}$$

**Think About It** The enthalpy change $\Delta H$ varies with pressure, but the magnitude of the change is negligible for pressures near normal atmospheric pressure. The magnitude of the answer to part b is reasonable given the large amount of water that is frozen to refinish the rink's surface.

⚙ **Practice Exercise** The flame in a torch used to cut metal is produced by burning acetylene ($C_2H_2$) in pure oxygen. Assuming the combustion of 1 mole of acetylene releases 1251 kJ of energy, what mass of acetylene is needed to cut through a piece of steel if the process requires $5.42 \times 10^4$ kJ of energy?

*(Answers to Practice Exercises are in the back of the book.)*

**FIGURE 5.23** An ice-resurfacing machine spreads a thin layer of liquid water on top of the existing ice. The liquid freezes, providing a fresh, smooth surface for skaters. If the ice/water represent the system, the energy change for the freezing of the water is described by $\Delta H_{sys}$.

---

**SAMPLE EXERCISE 5.6** **The Sign of $\Delta H$ in a Chemical Reaction** **LO4**

A chemical cold pack (Figure 5.24) contains a small pouch of water inside a bag of solid ammonium nitrate. To activate the pack, you press on it to rupture the pouch, allowing the ammonium nitrate to dissolve in the water. The result is a cold, aqueous solution of ammonium nitrate. Write a balanced chemical equation describing the dissolution of solid $NH_4NO_3$. If we define the system as the $NH_4NO_3$ and the surroundings as the water, what are the signs of $\Delta H_{sys}$ and $q_{surr}$?

**FIGURE 5.24** A chemical cold pack contains a bag of water inside a larger bag of ammonium nitrate. Breaking the inner bag and allowing the $NH_4NO_3$ to dissolve leads to a decrease in temperature.

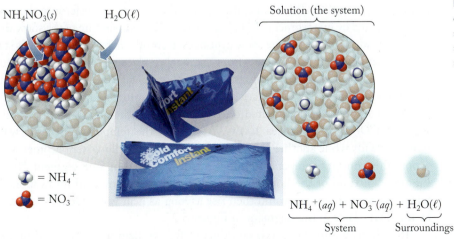

$NH_4NO_3(s)$    $H_2O(\ell)$      Solution (the system)

= $NH_4^+$

= $NO_3^-$

$NH_4^+(aq) + NO_3^-(aq) + H_2O(\ell)$

System      Surroundings

**Collect and Organize** The solution's temperature drops when we dissolve $NH_4NO_3$ in water. We want to write a balanced chemical equation for the dissolution of ammonium nitrate and determine the sign of $\Delta H$ for the solution (the system) and $q_{surr}$.

**Analyze** The first law of thermodynamics says energy must be conserved, so all the energy must be accounted for between system and surroundings.

**Solve** The solubility rules in Table 4.5 tell us that all ammonium salts are soluble. We can write a balanced chemical equation for the dissolution of $NH_4NO_3$ (the system) as

$$NH_4NO_3(s) \rightarrow NH_4^+(aq) + NO_3^-(aq)$$

The low temperature of the cold pack (the system) means that heat flows into it from its warmer surroundings. Therefore, the sign of $\Delta H_{sys}$ is positive ($\Delta H_{sys} > 0$), and the sign of $q_{surr}$ is negative ($q_{surr} < 0$).

**Think About It** The first law of thermodynamics tells us that the signs of $\Delta H_{sys}$ and $q_{surr}$ must be opposite.

⚙ **Practice Exercise** Potassium hydroxide can be used to unclog sink drains. The reaction between potassium hydroxide and water is quite exothermic. What are the signs of $\Delta H_{sys}$ and $q_{surr}$?

*(Answers to Practice Exercises are in the back of the book.)*

# 5.4 Heating Curves and Heat Capacity

Winter hikers and high-altitude mountain climbers use portable stoves fueled by propane or butane to prepare hot meals. Ice or snow may be their only source of water. In this section, we use this scenario to examine the transfer of energy into water that begins as snow and ends up as vapor.

## Hot Soup on a Cold Day

Let's consider the changes of temperature and state that water undergoes as some hikers melt snow as a first step in preparing soup from a dry soup mix. Suppose they start with a saucepan filled with snow at −18°C at constant pressure. They place the pan above the flame of a portable stove, and energy begins to flow into the snow. The temperature of the snow immediately begins to rise. If the flame is steady so that energy flow is constant, the temperature of the snow changes as indicated on the graph shown in Figure 5.25. First, heating increases the temperature of the snow to its melting point, 0°C. This temperature rise and the energy transfer into the snow are represented by line segment $\overline{AB}$ in Figure 5.25. The temperature of the snow then remains steady at 0°C while the snow continues to absorb energy and melt. This state change that produces liquid water from snow takes place at constant temperature and is represented by the constant-temperature (horizontal) line segment $\overline{BC}$. Phase changes of pure materials take place at constant temperature and pressure.

When all the snow has melted, the temperature of the liquid water rises (line segment $\overline{CD}$) until it reaches 100°C. At 100°C, the temperature of the water remains steady while another state change takes place: liquid water becomes water vapor. This constant-temperature process is represented by line segment $\overline{DE}$ in Figure 5.25. If all of the liquid water in the pan were converted into vapor, the temperature of the vapor would then rise as long as heat was added, as indicated by the final (slanted) line segment in Figure 5.25.

The difference in the $x$-axis coordinates for the beginning and end of each line segment in Figure 5.25 indicates how much energy is required in each step in

▶❙❙ **CHEMTOUR** Heating Curves

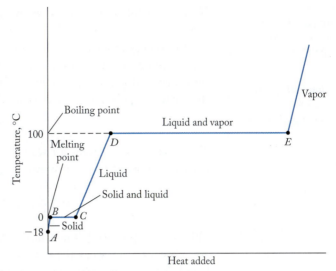

**FIGURE 5.25** The energy required to melt snow and boil the resultant water is illustrated by the four line segments on the heating curve of water: heating snow to its melting point $\overline{AB}$; melting the snow to form liquid water $\overline{BC}$; heating the water to its boiling point $\overline{CD}$; and boiling the water to convert it to vapor $\overline{DE}$.

this process. In the first step (line segment $\overline{AB}$), the energy required to raise the temperature of the snow from −18°C to 0°C can be calculated if we know how many moles of snow we have and how much energy is required to change the temperature of 1 mole of snow.

**Molar heat capacity ($c_P$)** is the quantity of energy required to raise the temperature of 1 mole of a substance by 1°C. The symbol we use for molar heat capacity is $c_P$, where the subscript P again indicates the value for a process taking place at *constant pressure*. Molar heat capacities for several substances are given in Table 5.1, from which we see that ice has a molar heat capacity of 37.1 J/(mol · °C) at constant pressure. From these units, we see that if we know the number of moles of snow we have and the temperature change it experiences as we bring it to its melting point, we can calculate $q$ for the process:

$$q = nc_P\Delta T \tag{5.8}$$

where $n$ is the number of moles of the substance absorbing or releasing energy and $\Delta T$ is the temperature change in degrees Celsius.

In addition to molar heat capacity, other measures exist that specify how much energy is required to increase the temperature of a substance, each referring to a specific amount of the substance. For example, some tables of thermodynamic data list values of **specific heat ($c_s$)**, which is the energy required to raise the temperature of *1 g of a substance* by 1°C; $c_s$ has units of J/(g · °C). Later in this chapter we use **heat capacity ($C_P$)**, which is the quantity of energy required to increase the temperature of *a particular object* (for example, a calorimeter, used to measure energy changes) by 1°C at constant pressure.

Let's assume the hikers decide to cook their meal with 270 g of snow. Dividing that mass by the molar mass of water (18.02 g/mol), we find that they melt 15.0 moles of snow. To calculate how much energy is needed to raise the temperature of 15.0 moles of $H_2O(s)$ from −18°C to 0°C, we use Equation 5.8:

$$q = nc_P\Delta T$$

$$= 15.0 \ \cancel{\text{mol}} \times \frac{37.1 \ \text{J}}{\cancel{\text{mol}} \cdot \cancel{°C}} \times (+18°\cancel{C}) = 1.0 \times 10^4 \ \text{J} = 10 \ \text{kJ}$$

| TABLE 5.1 | Molar Heat Capacities at 25°C of Selected Substances | |
|---|---|
| **Substance** | **$c_P$ [J/(mol · °C)]** |
| $H_2O(s)$ | 37.1 |
| $H_2O(\ell)$ | 75.3 |
| $H_2O(g)$ | 33.6 |
| $CH_3CH_2OH(\ell)$, ethanol | 113.1 |
| $C(s)$, graphite | 8.54 |
| $Al(s)$ | 24.4 |
| $Cu(s)$ | 24.5 |
| $Fe(s)$ | 25.1 |

**molar heat capacity ($c_P$)** the energy required at constant pressure to raise the temperature of 1 mole of a substance by 1°C.

**specific heat ($c_s$)** the energy required to raise the temperature of 1 g of a substance 1°C at constant pressure.

**heat capacity ($C_P$)** the quantity of energy needed to raise the temperature of an object 1°C at constant pressure.

**molar heat of fusion ($\Delta H_{fus}$)** the energy required to convert 1 mole of a solid substance at its melting point into the liquid state.

**molar heat of vaporization ($\Delta H_{vap}$)** the energy required to convert 1 mole of a liquid substance at its boiling point to the vapor state.

Notice that this value is positive, which means that the system (the snow) gains energy as it warms up.

The energy absorbed as the snow melts, or *fuses* (line segment $\overline{BC}$ in Figure 5.25), can be calculated using the enthalpy change that takes place as 1 mole of snow melts. This enthalpy change is called the **molar heat of fusion ($\Delta H_{fus}$)**, and the energy absorbed as *n* moles of a substance melts is given by

$$q = n\Delta H_{fus} \qquad (5.9)$$

The molar heat of fusion for water is 6.01 kJ/mol, and using this value and the known number of moles of snow, we get

$$q = 15.0 \ \cancel{mol} \times \frac{6.01 \ kJ}{\cancel{mol}} = 90.0 \ kJ$$

This value is positive because heat enters the system. This energy overcomes the attractive forces between the water molecules in the solid, and the snow becomes a liquid. Note that no factor for temperature appears in this calculation because state changes of pure substances take place at constant temperature, as the two horizontal line segments in Figure 5.25 indicate. Snow at 0°C becomes liquid water at 0°C.

While the water temperature increases from 0°C to 100°C, the relation between temperature and energy absorbed is again defined by Equation 5.8, but this time $c_P$ represents not the molar heat capacity of $H_2O(s)$ but the molar heat capacity of water($\ell$), 75.3 J/(mol · °C) (see Table 5.1):

$$q = nc_P\Delta T$$

$$= 15.0 \ \cancel{mol} \times \frac{75.3 \ J}{\cancel{mol} \cdot \cancel{°C}} \times 100\cancel{°C} = 1.13 \times 10^5 \ J = 113 \ kJ$$

This value is positive because the system takes in energy as its temperature rises.

At this point in our story, our hiker–chefs are ready to make their soup and enjoy a hot meal. However, if they accidentally leave the boiling water unattended, it will eventually vaporize completely. How much heat is needed to boil away all of the water in the pot? As line segment $\overline{DE}$ in Figure 5.25 shows, the temperature of the water remains at 100°C until all of it is vaporized. The enthalpy change associated with changing 1 mole of a liquid to a gas is the **molar heat of vaporization ($\Delta H_{vap}$)**, and the quantity of energy absorbed as *n* moles of a substance vaporizes is given by

$$q = n\Delta H_{vap} \qquad (5.10)$$

The molar heat of vaporization for water is 40.67 kJ/mol, so to completely vaporize the water we need an additional 610 kJ of energy:

$$q = (15.0 \ \cancel{mol})(40.67 \ kJ/\cancel{mol}) = +610 \ kJ$$

This value is positive because the system takes in energy as it converts liquid into vapor. Again, no factor for temperature appears because the phase change takes place at constant temperature. Only after all the water has vaporized does its temperature increase above 100°C, along the line above point *E* in Figure 5.25. The molar heat capacity of steam, 33.6 J/(mol · °C), is used to calculate the energy required to heat the vapor to any temperature above 100°C.

The $c_P$ values for $H_2O(s)$, $H_2O(\ell)$, and $H_2O(g)$ are different. Nearly all substances have different molar heat capacities in their different physical states.

Notice in Figure 5.25 that line segment $\overline{DE}$, the phase change from liquid water to water vapor, is much longer than the line segment $\overline{BC}$ representing the change from solid snow to liquid water. The relative lengths of these lines indicate that the molar heat of vaporization of water (40.67 kJ/mol) is much larger than the molar heat of fusion of snow (6.01 kJ/mol). Why does it take more energy to boil 1 mole of water than to melt 1 mole of snow? The answer is related to the extent to which attractive forces between molecules must be overcome in each process and to how the internal energy of the system changes as energy is added (Figure 5.26).

Attractive forces determine both the organization of the molecules in any substance and their relation to their nearest neighbors. The kinetic and potential energies of the molecules in snow are low, and attractive forces are strong enough to hold the molecules in place relative to one another. Melting the snow requires that these attractive forces be overcome. The energy added to the system at the melting point is sufficient to overcome the attractive forces and change the arrangement (and hence the potential energy) of the molecules.

Intermolecular attractive forces still exist in liquid water, but the energy added at the melting point increases the energy of the molecules and enables them to move with respect to their nearest neighbors. The molecules are still almost as close together in the liquid as they are in the solid, but their relative positions constantly change; they have different nearest neighbors over time.

Once the snow has completely melted, added energy causes the temperature of the liquid water to rise, increasing its internal energy. When water vaporizes at the boiling point, the attractive forces between water molecules must be overcome to separate the molecules from one another as they enter the gas state. Separating the molecules widely in space requires work (that is, energy), and the amount required is much larger than that for the solid-to-liquid state change. This difference is consistent with the relative lengths of $\overline{BC}$ and $\overline{DE}$ in Figure 5.25.

**FIGURE 5.26** Macroscopic and molecular-level views of (a) ice, (b) water, and (c) water vapor.

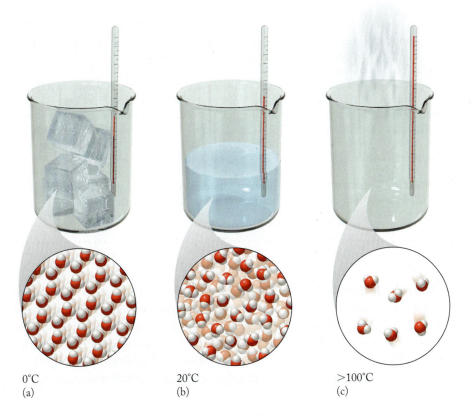

0°C
(a)

20°C
(b)

>100°C
(c)

**SAMPLE EXERCISE 5.7  Calculating the Energy Required   L04
to Raise the Temperature of Water**

Calculate the amount of energy required to convert 237 g of solid ice at 0.0°C to hot water at 80.0°C. The molar heat of fusion ($\Delta H_{fus}$) of ice is 6.01 kJ/mol. The molar heat capacity of liquid water is 75.3 J/(mol · °C).

**Collect and Organize** A given mass of ice (237 g) at 0.0°C is heated to 80.0°C. Using $\Delta H_{fus}$ and $c_{P,water}$, we can determine the energy required for this process.

**Analyze** This problem refers to the process symbolized by segments $\overline{BC}$ and $\overline{CD}$ in Figure 5.25. We can calculate the temperature change ($\Delta T$) of the water from the initial and final temperatures. Four mathematical steps are required: (1) calculate the number of moles of ice; (2) determine the amount of energy required to melt the ice using Equation 5.9; (3) calculate the amount of energy required to raise the liquid water temperature from 0.0°C to 80.0°C using Equation 5.8; (4) add the results of steps (2) and (3). We can calculate the number of moles in 237 g of water using the molar mass of water ($\mathcal{M}$ = 18.02 g/mol).

**Solve**
1. Calculate the number of moles of water:

$$n = 237 \; \cancel{\text{g H}_2\text{O}} \times \frac{1 \; \text{mol H}_2\text{O}}{18.02 \; \cancel{\text{g H}_2\text{O}}} = 13.2 \; \text{mol H}_2\text{O}$$

2. Determine the amount of energy needed to melt the ice (Equation 5.9):

$$q_1 = n\Delta H_{fus} = 13.2 \; \cancel{\text{mol}} \times 6.01 \; \text{kJ}/\cancel{\text{mol}} = 79.3 \; \text{kJ}$$

3. Determine the amount of energy needed to warm the water (Equation 5.8):

$$q_2 = nc_P\Delta T$$

$$= 13.2 \; \cancel{\text{mol}} \times \frac{75.3 \; \text{J}}{\cancel{\text{mol}} \cdot \cancel{°\text{C}}} \times (80.0 - 0.00)\cancel{°\text{C}} = 79,517 \; \text{J} = 79.5 \; \text{kJ}$$

4. Add the results of parts 2 and 3:

$$q_1 + q_2 = 79.3 \; \text{kJ} + 79.5 \; \text{kJ} = 158.8 \; \text{kJ}$$

**Think About It** Using the definitions of molar heat of fusion and molar heat capacity, we can solve this exercise without referring back to Equations 5.8 and 5.9. Molar heat of fusion defines the amount of energy needed to melt 1 mole of ice. Multiplying that value (6.01 kJ/mol) by the number of moles (13.2 mol) gives us the energy required to melt the given amount of ice. By the same token, the molar heat capacity of water [75.3 J/(mol · °C)] defines the amount of energy needed to raise the temperature of 1 mole of liquid water by 1°C. We know the number of moles (13.2 mol), and we know the number of degrees by which we want to raise the temperature (80.0°C − 0.00°C = 80.0°C); multiplying those factors together gives us the heat needed to raise the temperature of the water.

⚙ **Practice Exercise** Calculate the change in energy when 125 g of water vapor at 100.0°C condenses to liquid water and then cools to 25.0°C.

*(Answers to Practice Exercises are in the back of the book.)*

Water is an extraordinary substance for many reasons, but its high molar heat capacity is one of the more important. The ability of water to absorb large quantities of energy is one reason it is used as a *heat sink,* both in automobile radiators and in our bodies. The term "heat sink" is often used to identify matter that can absorb energy without changing phase or significantly changing its temperature. Weather and climate changes are largely driven and regulated by cycles involving retention of energy by our planet's oceans, which serve as giant heat sinks for solar energy.

## Cold Drinks on a Hot Day

Let's consider another familiar application involving heat transfer. Suppose we throw a party and plan to chill three cases (72 aluminum cans, each containing 355 mL) of beverages by placing the cans in an insulated cooler and covering them with ice cubes (Figure 5.27). If the temperature of the ice (sold in 10-pound bags) is −8.0°C and the temperature of the beverages is initially 25.0°C, how many bags of ice do we need to chill the cans and their contents to 0.0°C (as in "ice cold")?

You may already have an idea that more than 1 bag, but probably fewer than 10, will be needed. We can use the heat transfer relationships we have defined to predict more accurately how much ice is required. In doing so, we assume that whatever energy is absorbed by the ice is lost by the cans and the beverages in them. As the ice absorbs heat from the cans and the beverages, the temperature of the ice increases from −8.0°C to 0.0°C, and, as we saw in analyzing Figure 5.25, the resulting liquid water remains at 0.0°C until all the ice has melted. We need enough ice so that the last of it melts just as the temperature of the beverages and the cans reaches 0.0°C. Our analysis of this cooling process is a little simpler than our snow/liquid water/water vapor analysis because here there is no change of state. All of the heat transferred goes only into cooling the cans and beverages to 0.0°C. The cans stay in the solid state, and the beverages stay in the liquid state.

First let's consider the energy lost in the cooling process. Two materials are to be chilled: 72 aluminum cans and 72 × 355 mL = 25,600 mL of beverages. The beverages are mostly water. The other ingredients are present in such small concentrations that they will not affect our calculation, so we can assume that we need to reduce the temperature of 25,600 mL of water by 25.0°C. We can calculate the amount of energy lost with Equation 5.8 if we first calculate the number of moles of water in 25,600 mL, assuming a density of 1.000 g/mL:

$$25{,}600 \; \cancel{\text{mL H}_2\text{O}} \times 1.000 \; \frac{\cancel{\text{g}}}{\cancel{\text{mL}}} \times \frac{1 \; \text{mol H}_2\text{O}}{18.02 \; \cancel{\text{g H}_2\text{O}}} = 1420 \; \text{mol H}_2\text{O}$$

We can use the molar heat capacity of water and Equation 5.8 to calculate the energy lost by 1420 moles of water as its temperature decreases from 25.0°C to 0.0°C:

$$q = nc_\text{p}\Delta T$$

$$= 1420 \; \cancel{\text{mol}} \times \frac{75.3 \; \text{J}}{\cancel{\text{mol}} \cdot \cancel{°\text{C}}} \times (-25.0\cancel{°\text{C}})$$

$$= -2.67 \times 10^6 \; \text{J}$$

We must also consider the energy released in lowering the temperature of 72 aluminum cans by 25.0°C. The typical mass of a soda can is 12.5 g. The molar heat capacity of solid aluminum (Table 5.1) is 24.4 J/(mol · °C), and the molar mass of aluminum is 26.98 g/mol. Thus

$$q = nc_\text{p}\Delta T$$

$$= 72 \; \cancel{\text{cans}} \times \frac{12.5 \; \cancel{\text{g Al}}}{\cancel{\text{can}}} \times \frac{1 \; \cancel{\text{mol}}}{26.98 \; \cancel{\text{g Al}}} \times \frac{24.4 \; \text{J}}{\cancel{\text{mol}} \cdot \cancel{°\text{C}}} \times (-25.0\cancel{°\text{C}})$$

$$= -2.03 \times 10^4 \; \text{J}$$

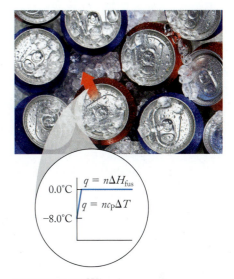

**FIGURE 5.27** When beverage cans are chilled in an ice-filled cooler, heat flows from the cans to the ice until the temperature of the cans equals that of the ice.

The total quantity of energy that must be removed from the cans and beverages is

$$q_{total} = q_{beverage} + q_{cans} = [(-2.67 \times 10^6) + (-2.03 \times 10^4)] \text{ J}$$

$$= (-2.67 - 0.0203) \times 10^6 \text{ J} = -2.69 \times 10^6 \text{ J}$$

$$= -2.69 \times 10^3 \text{ kJ}$$

This quantity of energy must be absorbed by the ice as it warms to its melting point and then melts. The calculation of the amount of ice needed is an algebra problem. Let $n$ be the number of moles of ice needed. The energy absorbed is the sum of (1) the energy needed to raise the temperature of $n$ moles of ice from $-8.0°$ to $0.0°C$ and (2) the energy needed to melt $n$ moles of ice. These quantities can be calculated with Equation 5.8 for step 1 and Equation 5.9 for step 2. Note that the $c_P$ values in Table 5.1 have units of joules, while $\Delta H_{fus}$ values given earlier have units of kilojoules. We need to have both terms in the same units, so we use 0.0371 kJ/mol · °C for $c_P$:

$$q_{total\ gained} = q_1 + q_2$$

$$= nc_P\Delta T + n\Delta H_{fus}$$

$$= n\left(\frac{0.0371 \text{ kJ}}{\text{mol} \cdot \cancel{°C}}\right)(8.0\cancel{°C}) + n(6.01 \text{ kJ/mol})$$

$$= n(6.31 \text{ kJ/mol})$$

The energy lost by the cans and the beverages balances the energy gained by the ice:

$$-q_{total\ lost} = +q_{total\ gained}$$

$$2.69 \times 10^3 \cancel{\text{ kJ}} = n(6.31 \cancel{\text{ kJ}}/\text{mol})$$

$$n = 4.26 \times 10^2 \text{ mol ice}$$

Converting $4.26 \times 10^2$ mol of ice into pounds and adjusting the significant figures gives us

$$(4.26 \times 10^2 \cancel{\text{ mol}}) \times \frac{18.02 \cancel{\text{ g}}}{\cancel{\text{mol}}} \times \frac{1 \text{ lb}}{453.6 \cancel{\text{ g}}} = 16.9 \text{ lb of ice}$$

Thus we need at least two 10-pound bags of ice to chill three cases of our favorite beverages.

**CONCEPT TEST**

The energy lost by the beverages inside the 72 cans in the preceding discussion was more than 100 times the heat lost by the cans. What factors contributed to this large difference between the energy lost by the cans and the energy lost by their contents?

*(Answers to Concept Tests are in the back of the book.)*

**SAMPLE EXERCISE 5.8** **Calculating the Temperature of Iced Tea** **LO4**

If you add 250.0 g of ice, initially at $-18.0°C$, to 237 g (1 cup) of freshly brewed tea, initially at $100.0°C$, and the ice melts, what is the final temperature of the tea? Assume that the mixture is an isolated system (in an ideal insulated container) and that tea has the same molar heat capacity, density, and molar mass as water.

**Collect and Organize** We know the mass of the tea, its initial temperature, and the molar heat capacity of tea, for which we use the $c_p$ value for water. We know the amount of ice, its initial temperature, and the heat of fusion of ice. Our task is to find the final temperature of the tea–ice–water mixture.

**Analyze** The amount of energy released when the tea is cooled will be the same as the amount of energy gained by the ice and, once it is melted, the amount gained by the water that is formed as it warms to the final temperature. Assuming the ice melts completely, three heat transfers account for the energy lost by the tea:

1. $q_1$: raising the temperature of the ice to 0.0°C
2. $q_2$: melting the ice
3. $q_3$: bringing the mixture to the final temperature; the temperature of the water from the melted ice rises and the temperature of the tea ($T_{initial} = 100.0$°C) falls to the final temperature of the mixture ($T_{final} = ?$)

The energy gained by the ice equals the energy lost by the tea:

$$q_{ice} = {}^-q_{tea}$$

Based on our analysis, we know that

$$q_{ice} = q_1 + q_2 + q_3$$

**Solve** The energy lost by the tea as it cools from 100.0°C to $T_{final}$ is, from Equation 5.8,

$$q_{tea} = nc_p\Delta T_{tea}$$

$$= 237 \text{ g} \times \frac{1 \text{ mol}}{18.02 \text{ g}} \times \frac{75.3 \text{ J}}{\text{mol} \cdot {}^\circ\text{C}} \times (T_{final} - 100.0{}^\circ\text{C})$$

$$= (990 \text{ J/}{}^\circ\text{C})(T_{final} - 100.0{}^\circ\text{C})$$

The transfer of this heat is responsible for the changes in the ice. We can treat the heat transfer from the hot tea to the ice in terms of the three processes we identified in Analyze. In step 1, the ice is warmed from −18.0°C ($T_{initial}$) to 0.0°C ($T_{final}$ for this step):

$$q_1 = n_{ice}c_{ice}\Delta T_{ice}$$

$$= 250.0 \text{ g} \times \frac{1 \text{ mol}}{18.02 \text{ g}} \times \frac{37.1 \text{ J}}{\text{mol} \cdot {}^\circ\text{C}} \times [0.0{}^\circ\text{C} - (-18.0{}^\circ\text{C})]$$

$$= 9.26 \times 10^3 \text{ J} = 9.26 \text{ kJ}$$

In step 2, the ice melts, requiring the absorption of energy:

$$q_2 = n_{ice}\Delta H_{fus,ice}$$

$$= 250.0 \text{ g} \times \frac{1 \text{ mol}}{18.02 \text{ g}} \times \frac{6.01 \text{ kJ}}{\text{mol}}$$

$$= 83.4 \text{ kJ}$$

In step 3, the water from the ice, initially at 0.0°C, warms to the final temperature (where $\Delta T = T_{final} - T_{initial} = T_{final} - 0.0$°C):

$$q_3 = n_{water}c_{water}\Delta T_{water}$$

$$= 250.0 \text{ g} \times \frac{1 \text{ mol}}{18.02 \text{ g}} \times \frac{75.3 \text{ J}}{\text{mol} \cdot {}^\circ\text{C}} \times (T_{final} - 0.0{}^\circ\text{C})$$

$$= (1045 \text{ J/}{}^\circ\text{C})(T_{final})$$

The sum of the quantities of energy absorbed by the ice and the water from it during steps 1 through 3 must balance the energy lost by the tea:

$$q_{ice} = q_1 + q_2 + q_3 = {}^-q_{tea}$$

$$9.26 \text{ kJ} + 83.4 \text{ kJ} + (1045 \text{ J/}{}^\circ\text{C})(T_{final}) = -[(990 \text{ J/}{}^\circ\text{C})(T_{final} - 100.0{}^\circ\text{C})]$$

100.0°C

23.5 g
Aluminum beads

Heat

(a)

23.0°C

Water
(130.0 g)

Insulation

(b)

26.0°C

Insulation

(c)

**FIGURE 5.28** Experimental setup to determine the molar heat capacity of a metal. (a) Pure aluminum beads having a combined mass of 23.5 g are heated to 100.0°C in boiling water; (b) 130.0 g of water at 23.0°C is placed in a Styrofoam box; (c) the hot Al beads are dropped into the water, and the temperature at thermal equilibrium is 26.0°C.

Expressing all values in kilojoules:

$$9.26 \text{ kJ} + 83.4 \text{ kJ} + (1.045 \text{ kJ/°C})(T_{final}) = -[(0.990 \text{ kJ/°C}) (T_{final} - 100.0°C)]$$

Rearranging the terms to solve for $T_{final}$, we have

$$(2.04 \text{ kJ/°C})T_{final} = -9.26 \text{ kJ} - 83.4 \text{ kJ} + 99.0 \text{ kJ} = 6.34 \text{ kJ}$$

$$T_{final} = 3.1 \text{ °C}$$

**Think About It** This calculation was carried out assuming the system (the tea plus the ice) is isolated and the vessel is a perfect insulator. Our answer, therefore, is an "ideal" answer and reflects the coldest temperature we can expect the tea to reach. In the real world, the ice would absorb some energy from the surroundings (the container and the air) and the tea would lose some energy to the surroundings, so the final temperature of the beverage could be different from the ideal value we calculated.

**Practice Exercise** Calculate the final temperature of a mixture of 350 g of ice, initially at −18°C, and 237 g of water, initially at 100.0°C.

*(Answers to Practice Exercises are in the back of the book.)*

# 5.5 Calorimetry: Measuring Heat Capacity and Enthalpies of Reaction

Up to this point we have discussed molar heat capacities ($c_P$) and the enthalpy changes $\Delta H_{fus}$ and $\Delta H_{vap}$ associated with phase changes, but we have not broached the issue of how we know the values of these parameters. The experimental method of measuring the quantities of energy associated with chemical reactions and physical changes is called **calorimetry**. The device used to measure the heat released or absorbed during a process is a **calorimeter**.

## Determining Molar Heat Capacity and Specific Heat

When we determined the amount of ice needed to cool 72 aluminum beverage cans in Section 5.4, we used the molar heat capacity of aluminum to determine the quantity of energy lost by the cans. How are heat capacities measured?

We can apply the first law of thermodynamics and design an experiment to determine the specific heat of aluminum, from which we can calculate its molar heat capacity. Recall from Section 5.4 that specific heat $c_s$ is defined as the quantity of energy required to raise the temperature of 1 g of a substance by 1°C. The units of specific heat, J/(g · °C), tell us exactly what we have to do. "Determine the specific heat of aluminum" means determine the quantity of energy (in joules) required to change the temperature of 1 g of aluminum by 1°C. If we determine how much energy is required to change the temperature of any known mass of aluminum by any measured number of degrees, we can calculate how much energy is required to change the temperature of 1 g of aluminum by 1°C—in other words, we will have the value of the specific heat of aluminum.

Suppose we have beads of pure aluminum with a total mass of 23.5 g. We want to transfer a known amount of energy to or from the aluminum (Figure 5.28).

First, we must heat the known quantity of aluminum to a known temperature. We can do this by boiling some water in a beaker containing a test tube holding the aluminum beads (the system). We wait a few minutes while the beads and the boiling water bath all come to thermal equilibrium at 100.0°C (Figure 5.28a). Because water (or any pure liquid) boils at a constant temperature, we can use this step to bring the beads to a known temperature. No matter how rapidly we boil the water, it will maintain a temperature of 100.0°C.

While the metal is heating, we place a measured mass of water (in this case, 130.0 g) in a beaker in a Styrofoam box (Figure 5.28b). We consider this box a perfect insulator that effectively isolates its contents from the rest of the universe. Through a small hole in the box lid (no energy escapes through the hole because the box is a perfect insulator), we insert a thermometer into the water, read the temperature (23.0°C), and leave the thermometer in place. When we have waited long enough for the beads to heat up to 100.0°C, we remove the test tube from the boiling water, remove the lid from the insulated box, pour the beads out of the test tube into the water in the beaker, and quickly close the lid (Figure 5.28c). The temperature of the water rises because it is now in contact with the hot aluminum. We assume that all the heat coming from the beads goes into the water in the beaker. Suppose the temperature of the aluminum–water mixture in the box rises to 26.0°C and stays at that temperature.

Energy flows from the beads to the water. From the first law of thermodynamics we know

$$-q_{aluminum} = q_{water}$$

where we show a minus sign on $q_{aluminum}$ because energy leaves the aluminum and a plus sign on $q_{water}$ because energy enters the water. To determine the amount of energy transferred from the aluminum, we need to know how much energy enters the water. We know the mass of water (130.0 g) and the temperature change the water experiences ($\Delta T_{water} = 26.0°C - 23.0°C = 3.0°C$); we need the specific heat of water to calculate $q_{water}$.

When we defined units of energy in Section 5.2, we defined 4.184 joules (J) as the amount of energy needed to raise the temperature of 1 g of water by 1°C; that value is the specific heat of water. We could also calculate $c_{s,water}$ by dividing the molar heat capacity of water [75.3 J/(mol · °C); Table 5.1] by its molar mass ($\mathcal{M} = 18.02$ g/mol). With that value we now have everything we need to calculate the specific heat $c_s$ of aluminum.

First, we determine the amount of energy gained by the water in the box. The units of specific heat tell us what to do:

$$c_{s,water} = \frac{4.184\ \text{J}}{\text{g} \cdot °\text{C}}$$

If we multiply the mass of the water in grams by the specific heat and the temperature change of the water, the result is the energy absorbed by the water:

$$q_{water} = (130.0\ \cancel{\text{g H}_2\text{O}})\left(\frac{4.184\ \text{J}}{\cancel{\text{g H}_2\text{O}} \cdot \cancel{°\text{C}}}\right)(3.0\ \cancel{°\text{C}})$$

$$= 1.6 \times 10^3\ \text{J}$$

The energy gained by the water has a positive value because the water absorbs energy from the beads. This expression can be generalized to calculate energy absorbed or released from any mass of matter over any temperature change:

$$q = mc_s\Delta T \qquad (5.11)$$

where $m$ is the mass, $c_s$ is specific heat, and $\Delta T = T_{final} - T_{initial}$.

▶❙❙ **CHEMTOUR** Calorimetry

To find the specific heat of aluminum, we recognize that the 1600 J of energy that increased the temperature of the water came from the aluminum beads. Therefore,

$$q_{aluminum} = -q_{water} = -1.6 \times 10^3 \, J = mc_s \Delta T_{aluminum}$$

where $m$ is the mass of aluminum, $c_s$ is the specific heat of aluminum, and $\Delta T_{aluminum} = T_{final} - T_{initial} = (26.0 - 100.0)°C = -74.0°C$. Note that $\Delta T$ for the aluminum is different from $\Delta T$ of the water. The aluminum is initially at 100°C and drops to a final temperature of 26.0°C. Thus

$$-1.6 \times 10^3 \, J = (23.5 \, g)(c_s)(-74.0°C)$$

where $c_s$ is the unknown value we are after.

Solve for $c_s$:

$$c_s = \frac{-1.6 \times 10^3 \, J}{(23.5 \, g)(-74.0°C)} = \frac{0.92 \, J}{g \cdot °C}$$

From this specific heat value, we can also calculate the molar heat capacity of aluminum by multiplying $c_s$ by the mass of 1 mole of aluminum:

$$c_P = c_s \times \mathcal{M} = \frac{0.92 \, J}{g \cdot °C} \times 26.98 \frac{g}{mol} = 25 \frac{J}{mol \cdot °C}$$

---

**SAMPLE EXERCISE 5.9** **Calculating the Final Temperature of a Hot Object Dropped Into Water** **LO5**

A blacksmith drops a 1.50 kg piece of iron heated to 525°C into 2.00 kg of water at 15.0°C. Given the molar heat capacities, $c_{P,Fe} = 25.1 \, J/(mol \cdot °C)$ and $c_{P,H_2O} = 75.3 \, J/(mol \cdot °C)$, calculate the final temperature of the water.

**Collect and Organize** We are given the masses and initial temperatures of a piece of iron and water. We need to calculate the final temperature of the water after the iron is dropped into it. We are given the molar heat capacities of both materials.

**Analyze** The temperature of the iron decreases and the temperature of the water increases until they reach thermal equilibrium. This means that the heat lost by the iron will equal the heat gained by the water, $q_{Fe} = -q_{water}$. We use Equation 5.8 to relate $q_{Fe}$ and $q_{water}$ to the final temperature. The mass of iron and the mass of water are similar, but the molar heat capacity of water is much greater than that of iron so we expect the final temperature to be closer to the initial temperature of the water.

**Solve** We start with the equation $q_{Fe} = -q_{H_2O}$ and substitute $q = nc_P \Delta T$:

$$n_{Fe}c_{P,Fe}\Delta T = -n_{H_2O}c_{P,H_2O}\Delta T$$

where $\Delta T = T_{final} - T_{initial}$ for each material. The final temperature of the water and iron must be equal,

$$T_{final,Fe} = T_{final,H_2O} = T_{final}$$

so

$$n_{Fe}c_{P,Fe}\left(T_{final,Fe} - T_{initial,Fe}\right) = -n_{H_2O}c_{P,H_2O}\left(T_{final,H_2O} - T_{initial,H_2O}\right)$$

We calculate $n_{Fe}$ and $n_{H_2O}$ using the appropriate molar masses:

$$n_{Fe} = 1.50 \, kg \, Fe \times \frac{10^3 \, g}{1 \, kg} \times \frac{1 \, mol \, Fe}{55.85 \, g \, Fe} = 26.9 \, mol \, Fe$$

$$n_{H_2O} = 2.00 \, kg \, H_2O \times \frac{10^3 \, g}{1 \, kg} \times \frac{1 \, mol \, H_2O}{18.02 \, g \, H_2O} = 1.11 \times 10^2 \, mol \, H_2O$$

Substitute the values for $n$, $c_P$, and $T_{\text{inital}}$:

$$26.9 \ \cancel{\text{mol Fe}} \times \frac{25.1 \ \text{J}}{\cancel{\text{mol Fe}} \cdot {}^{\circ}\text{C}} \times (T_{\text{final}} - 525{}^{\circ}\text{C})$$

$$= -1.11 \times 10^2 \ \cancel{\text{mol H}_2\text{O}} \times \frac{75.3 \ \text{J}}{\cancel{\text{mol H}_2\text{O}} \cdot {}^{\circ}\text{C}} (T_{\text{final}} - 15.0{}^{\circ}\text{C})$$

Solving, we get $T_{\text{final}} = 53.1{}^{\circ}\text{C}$.

**Think About It** The final temperature, 53.1°C, is much cooler than the initial temperature of the iron and less than 40°C warmer than the initial temperature of the water, reflecting the large difference in molar heat capacities of Fe and water. Notice that there is not enough heat in the hot piece of iron to boil away any of the water.

⚙ **Practice Exercise** A 255 g piece of granite, heated to 600°C in a campfire, is dropped into 1.00 L water ($d = 1.00$ g/mL) at 25°C. The molar heat capacity of water is $c_{P,\text{H}_2\text{O}} = 75.3$ J/(mol · °C) and the specific heat of the granite is $c_{s,\text{granite}} = 0.79$ J/(g · °C). Calculate the final temperature of the granite.

*(Answers to Practice Exercises are in the back of the book.)*

## Enthalpies of Reaction

The heat transfer accompanying any chemical reaction is defined by a quantity known as the **enthalpy of reaction ($\Delta H_{\text{rxn}}$)**, also called the *heat of reaction*. The subscript may be changed to reflect a specific type of reaction being studied: for example, $\Delta H_{\text{comb}}$ for the enthalpy of a combustion reaction.

Much of the energy we use every day can be traced to combustion of a fuel such as natural gas (mostly methane, $CH_4$), propane ($C_3H_8$), or gasoline (a mixture of hydrocarbons). Our bodies obtain energy from respiration, which is essentially the combustion of glucose ($C_6H_{12}O_6$). We have already written balanced chemical equations for the combustion of $CH_4$, $C_3H_8$, and $C_6H_{12}O_6$ in Chapter 3. Now let's add their $\Delta H_{\text{rxn}}$ values:

$$CH_4(g) + 2\,O_2(g) \rightarrow CO_2(g) + 2\,H_2O(g) \qquad \Delta H_{\text{rxn}} = -802 \ \text{kJ/mol CH}_4$$

$$C_3H_8(g) + 5\,O_2(g) \rightarrow 3\,CO_2(g) + 4\,H_2O(g) \qquad \Delta H_{\text{rxn}} = -2220 \ \text{kJ/mol C}_3\text{H}_8$$

$$C_6H_{12}O_6(s) + 6\,O_2(g) \rightarrow 6\,CO_2(g) + 6\,H_2O(g) \quad \Delta H_{\text{rxn}} = -2803 \ \text{kJ/mol C}_6\text{H}_{12}\text{O}_6$$

All three of these reactions are *exothermic*, so the enthalpy change for each reaction is less than zero: $\Delta H_{\text{rxn}} < 0$.

Knowledge of $\Delta H_{\text{comb}}$ values allows us to compare fuels and to calculate how much fuel is needed to supply a desired quantity of energy. For example, we can calculate how much propane we need to melt the 270 g (15.0 mol) of ice in Figure 5.25 and raise its temperature to 100°C. In the beginning of Section 5.4, we calculated that the heat required to melt this much ice is 113 kJ. If we get 2220 kJ/mol $C_3H_8$, then we need to burn at least 2.24 g $C_3H_8$:

$$113 \ \cancel{\text{kJ}} \times \frac{1 \ \cancel{\text{mol C}_3\text{H}_8}}{2220 \ \cancel{\text{kJ}}} \times \frac{44.09 \ \text{g C}_3\text{H}_8}{1 \ \cancel{\text{mol C}_3\text{H}_8}} = 2.24 \ \text{g C}_3\text{H}_8$$

How much energy would we get from burning 2.24 g of glucose? Converting the mass of glucose to moles and multiplying by $\Delta H_{\text{comb}}$:

$$2.24 \ \cancel{\text{g C}_6\text{H}_{12}\text{O}_6} \times \frac{1 \ \cancel{\text{mol C}_6\text{H}_{12}\text{O}_6}}{180.16 \ \cancel{\text{g C}_6\text{H}_{12}\text{O}_6}} \times \frac{2803 \ \text{kJ}}{1 \ \cancel{\text{mol C}_6\text{H}_{12}\text{O}_6}} = 34.9 \ \text{kJ}$$

**enthalpy of reaction ($\Delta H_{\text{rxn}}$)** the energy absorbed or given off by a chemical reaction; also called *heat of reaction*.

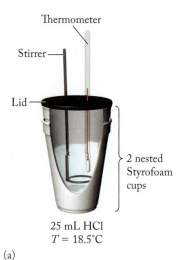

Thermometer

Stirrer

Lid

2 nested
Styrofoam
cups

25 mL HCl
$T = 18.5°C$

(a)

25 mL HCl + 25 mL NaOH
$T = 25.0°C$

(b)

**FIGURE 5.29** A cut-away image of a coffee-cup calorimeter made of two nested Styrofoam cups. A thermometer enables us to observe the temperature of the contents, which we can agitate using a simple stirrer to ensure mixing of solutions. (a) The inside cup contains 25 mL of 1.0 $M$ HCl. (b) The rapid addition of 25 mL of 1.0 $M$ NaOH causes the temperature to rise because of the heat given off in the ensuing neutralization reaction.

Calorimetry can be used to measure $\Delta H_{rxn}$ for chemical reactions, including the neutralization reactions discussed in Chapter 4. A simple laboratory apparatus for measuring the heat of reaction between a strong acid and a strong base is shown in Figure 5.29. A nested pair of Styrofoam coffee cups acts as our calorimeter, much like the apparatus in Figure 5.28. Note that pressure remains constant in a coffee-cup calorimeter, so $q_{cal}$ in this case will be the same as the enthalpy of reaction ($\Delta H_{rxn}$).

Let's say we place 25.0 mL of 1.0 $M$ aqueous HCl in the coffee cup and measure the temperature as 18.5°C. Rapid addition of 25.0 mL of 1.0 $M$ NaOH causes the temperature to rise to 25.0°C. What is $\Delta H_{rxn}$ for the reaction?

$$HCl(aq) + NaOH(aq) \rightarrow NaCl(aq) + H_2O(\ell)$$

The rise in temperature is the result of heat flowing from the reaction (the system) to the water (the surroundings). The amount of heat, $q$, gained by the surroundings is

$$q_{H_2O} = n_{H_2O}c_P\Delta T$$

First we need to calculate the moles of water, using the density of $H_2O$ as a reasonable approximation of the density of the solution:

$$n_{H_2O} = 50.0 \text{ mL } H_2O \times \frac{1.00 \text{ g } H_2O}{1 \text{ mL } H_2O} \times \frac{1.00 \text{ mol } H_2O}{18.02 \text{ g } H_2O} = 2.77 \text{ mol } H_2O$$

Notice that the total volume of the solution is 50.0 mL because we have mixed two 25.0 mL solutions in our calorimeter. Substituting $n_{H_2O}$ and the values of $c_P$ and $\Delta T$:

$$q_{H_2O} = n_{H_2O}c_P\Delta T = (2.77 \text{ mol } H_2O) \times \left(\frac{75.3 \text{ J}}{1 \text{ mol } H_2O \cdot °C}\right) \times (25.0°C - 18.5°C)$$

$$= 1356 \text{ J}$$

The heat gained by the surroundings must equal the heat lost by the system,

$$q_{H_2O} = -q_{rxn} = -1356 \text{ J}$$

Since 25.0 mL of 1.0 $M$ HCl contains 0.0250 mol of HCl, $\Delta H_{rxn}$ equals

$$\Delta H_{rxn} = \frac{-1356 \text{ kJ}}{0.0250 \text{ mol HCl}} = -54 \text{ kJ/mol}$$

In this experiment we made three important assumptions:

1. The coffee cup does not absorb any of the heat generated by the reaction.
2. The densities of the relatively dilute solutions are the same as that of pure water.
3. The molar heat capacity of the solution is the same as that of pure water.

These assumptions are reasonably valid for our example, but not in cases where the concentrations of the solutions are higher. Calorimeters suitable for accurate determination of $\Delta H$ are commercially available.

**SAMPLE EXERCISE 5.10** **Coffee-Cup Calorimetry: $\Delta H_{soln}$ of $NH_4NO_3$** **L05**

Dissolving 80.0 g $NH_4NO_3$ in 505 g of water in a coffee-cup calorimeter causes the temperature of the water to decrease by 13.3°C. What is the enthalpy change that accompanies the dissolution process, that is $\Delta H_{soln}$, expressed in kJ/mol?

**Collect and Organize** We are given the masses of solute and solvent used to prepare a solution of ammonium nitrate. The temperature decreases by 13.3°C. We are to determine the value of $\Delta H_{soln}$.

**Analyze** The ammonium nitrate is the system and the water represents the surroundings.

a. First, we use Equation 5.8 to find $q_{surr}$

$$\boxed{\text{mass } H_2O} \xrightarrow{\text{molar mass}} \boxed{\text{mol } H_2O} \xrightarrow{q = nc_p\Delta T} \boxed{\text{heat lost by } H_2O}$$

b. Then we use the relationship $q_{sys} = -q_{surr}$ to obtain $q_{sys}$.

c. Finally, we calculate $\Delta H_{soln}$ by dividing $q_{sys}$ by the number of moles of $NH_4NO_3$.

$$\boxed{\text{heat lost by } H_2O} \xrightarrow{q_{surr} = -q_{sys}} \boxed{\text{heat gained by system}} \xrightarrow{\dfrac{1}{\text{mol } NH_4NO_3}} \boxed{\Delta H_{soln}}$$

**Solve**

a. The heat lost by the water, $q_{surr}$, is

$$505 \text{ g } H_2O \times \frac{1 \text{ mol } H_2O}{18.02 \text{ g } H_2O} = 28.0 \text{ mol } H_2O$$

$$q_{lost,surr} = n_{H_2O}c_{P,H_2O}\Delta T = 28.0 \text{ mol } H_2O \times \frac{75.3 \text{ J}}{\text{mol } H_2O \cdot {}^\circ C} \times (-13.3{}^\circ C) = -2.81 \times 10^4 \text{ J}$$

b. The heat gained by the system, $q_{sys}$ is

$$q_{gained,sys} = -q_{lost,surr} = -(-2.81 \times 10^4 \text{ J}) = 2.81 \times 10^4 \text{ J}$$

c. The number of moles of $NH_4NO_3$ is

$$80.0 \text{ g } NH_4NO_3 \times \frac{1 \text{ mol } NH_4NO_3}{80.05 \text{ g } NH_4NO_3} = 0.999 \text{ mol } NH_4NO_3$$

and the enthalpy of solution of $NH_4NO_3$ is

$$\Delta H_{soln} = \frac{2.81 \times 10^4 \text{ J}}{0.999 \text{ mol } NH_4NO_3} = 2.81 \times 10^4 \text{ J/mol } NH_4NO_3 = 28.1 \text{ kJ/mol } NH_4NO_3$$

**Think About It** The value for $\Delta H_{soln}$, 28.1 kJ/mol, is positive, and consistent with the use of ammonium nitrate in chemical cold packs.

⚙ **Practice Exercise** Addition of 114 g potassium fluoride to 600 mL of water ($d = 1.00$ g/mL) causes the temperature to rise 3.6°C. What is $\Delta H_{soln}$ for KF?

*(Answers to Practice Exercises are in the back of the book.)*

**bomb calorimeter** a constant-volume device used to measure the energy released during a combustion reaction.

## Determining Calorimeter Constants

Although the apparatus in Figure 5.29 is useful for measuring the enthalpy change of a reaction in solution, heats of combustion are best measured with a device called a **bomb calorimeter** (Figure 5.30). To measure the heat of a combustion reaction, the sample is placed in a sealed vessel (called a *bomb*) capable of withstanding high pressures and submerged in a large volume of water in a heavily insulated container. Oxygen is introduced into the bomb, and the mixture is ignited with an electric spark. As combustion occurs, energy generated by the reaction flows into the walls of the bomb and then into the water surrounding the bomb. A good bomb calorimeter keeps the system contained within the bomb and ensures that all energy generated by the reaction stays in the calorimeter. The system consists of the bomb, the water, the insulated container, and minor components (stirrer, thermometer, and any other materials).

The energy produced by the reaction is determined by measuring the temperature of the water before and after the reaction. The water is at the same temperature as the parts of the calorimeter it contacts—the walls of the bomb, the

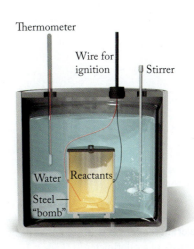

**FIGURE 5.30** A bomb calorimeter.

thermometer, and the stirrer—so the temperature change of the water takes the entire calorimeter into account.

Measuring the change in temperature of the water is not the whole story, however. We also need to know the heat capacity of the calorimeter. Recall that we have previously defined *specific heat* (J/g · °C) as the energy required to raise the temperature of *1 gram* of a substance 1°C, *molar heat capacity* (J/mol · °C) as the energy required to raise the temperature of 1 *mole* of a substance 1°C, and *heat capacity* (J/°C) as the energy required to raise the temperature of an *object* 1°C. Because specific heat and molar heat capacity refer to quantities of substances, units for these quantities include grams and moles. Because the heat capacity is a value unique to every calorimeter, it is frequently referred to as a **calorimeter constant ($C_{calorimeter}$)** and has units of J/°C, which is understood to mean "J/°C for this specific calorimeter." If we know the value of $C_{calorimeter}$ and if we can measure the change in water temperature, we can calculate the quantity of energy that flowed from the reactants into the calorimeter ($q_{calorimeter}$) to cause the temperature change of the water:

$$q_{calorimeter} = C_{calorimeter} \Delta T \qquad (5.12)$$

Rearranging terms:

$$C_{calorimeter} = \frac{q_{calorimeter}}{\Delta T} \qquad (5.13)$$

This equation indicates that heat capacity is expressed in units of energy divided by temperature, usually kilojoules per degree Celsius (kJ/°C).

Equation 5.13 can be used to determine $C_{calorimeter}$ for a bomb calorimeter. To do this, we burn a quantity of material in the calorimeter that produces a known quantity of energy when it burns—in other words, a material whose $\Delta H_{comb}$ value is known. Benzoic acid ($C_7H_6O_2$) is often used for this purpose because it can be obtained in a very pure form. Once $C_{calorimeter}$ has been determined, the calorimeter can be used to determine $\Delta H_{comb}$ for other substances. The observed increases in water temperature can be used to calculate the quantities of energy produced by combustion reactions on a per-gram or per-mole basis.

Because there is no change in the volume of the reaction mixture in a bomb calorimeter, this technique is referred to as *constant-volume calorimetry*. No *P–V* work is done, so according to Equation 5.12 the energy gained by the calorimeter during the combustion equals the internal energy lost by the reaction system during the combustion:

$$q_{calorimeter} = -\Delta E_{comb} \qquad (5.14)$$

The pressure inside a bomb calorimeter may change as a result of a combustion reaction, and in such cases $\Delta E_{comb}$ is not *exactly* the same as $\Delta H_{comb}$. However, the pressure effects are usually so small that $\Delta E_{comb}$ is *nearly* the same as $\Delta H_{comb}$, and we do not worry about the very small differences. Hence we discuss enthalpies of reactions ($\Delta H_{rxn}$) throughout and apply the approximate relation

$$q_{calorimeter} = -\Delta H_{comb}$$

or even more generally

$$q_{calorimeter} = -\Delta H_{rxn} \qquad (5.15)$$

**calorimeter constant ($C_{calorimeter}$)** the heat capacity of a calorimeter.

Other types of calorimeters (like the coffee-cup calorimeter in Figure 5.29) allow the volume to change while the pressure remains constant, so $q_{calorimeter}$ in those cases is exactly the same as the enthalpy of reaction ($\Delta H_{rxn}$).

**SAMPLE EXERCISE 5.11** **Determining a Calorimeter Constant** **LO5**

What is the calorimeter constant of a bomb calorimeter if burning 1.000 g of benzoic acid in it causes the temperature of the calorimeter to rise by 7.248°C? The heat of combustion of benzoic acid is $\Delta H_{comb} = -26.38$ kJ/g.

**Collect and Organize** We are asked to find the calorimeter constant of a calorimeter. We are given data describing how much the temperature of the calorimeter rises when a known amount of benzoic acid is burned in it, and we are given the heat of combustion of benzoic acid.

**Analyze** We need to determine the amount of energy required to raise the temperature of the calorimeter by 1°C. The heat capacity of a calorimeter can be calculated using Equation 5.13 and the knowledge that the combustion of 1.000 g of benzoic acid produces 26.38 kJ of heat.

**Solve**

$$C_{calorimeter} = \frac{q_{calorimeter}}{\Delta T}$$

$$= \frac{26.38 \text{ kJ}}{7.248°C}$$

$$= 3.640 \text{ kJ/°C}$$

This calorimeter can now be used to determine the heat of combustion of any combustible material.

**Think About It** The calorimeter constant is determined for a specific calorimeter. If anything changes—if the thermometer breaks and has to be replaced, or if the calorimeter loses any of the water it contains—a new constant must be determined.

⚙ **Practice Exercise** Assume that when 0.500 g of a mixture of hydrocarbons is burned in the bomb calorimeter from Sample Exercise 5.11, its temperature rises by 6.76°C. How much energy (in kilojoules) is released during combustion? How much energy is released with the combustion of 1.000 g of the same mixture?

*(Answers to Practice Exercises are in the back of the book.)*

# 5.6 Hess's Law

In Section 3.3 we described a reaction between methane and steam that produces hydrogen at high temperatures. The reaction can be carried out on an industrial scale in two steps. The first step is an endothermic reaction between methane and a limited supply of high-temperature steam, producing carbon monoxide and hydrogen gas in a reaction that has an enthalpy of reaction of +206 kJ:

▶❚❚ **CHEMTOUR** Hess's Law

(1)     $CH_4(g) + H_2O(g) \rightarrow CO(g) + 3\,H_2(g)$     $\Delta H_1 = +206$ kJ    (5.16)

In the second step, the carbon monoxide from the first reaction is allowed to react with more steam, producing carbon dioxide and more hydrogen gas in a reaction that has an enthalpy of reaction of −41 kJ:

(2)     $CO(g) + H_2O(g) \rightarrow CO_2(g) + H_2(g)$     $\Delta H_2 = -41$ kJ     (5.17)

We can write an overall reaction equation that results from adding reactions 1 and 2 and simplifying:

(1)      $CH_4(g) + H_2O(g) \rightarrow \cancel{CO(g)} + 3\,H_2(g)$

(2)      $\cancel{CO(g)} + H_2O(g) \rightarrow CO_2(g) + H_2(g)$

(3)      $CH_4(g) + 2\,H_2O(g) \rightarrow CO_2(g) + 4\,H_2(g)$     (5.18)

**Hess's law** the standard enthalpy of reaction $\Delta H_{rxn}$ for a reaction that is the sum of two or more reactions is equal to the sum of the $\Delta H_{rxn}$ values of the constituent reactions; also called *Hess's law of constant heat of summation*.

Just as we obtain the overall chemical equation 3 by adding equations 1 and 2, we obtain the enthalpy of reaction for the overall reaction by adding the $\Delta H$ values for reactions 1 and 2. The thermochemical equation for the overall reaction is

$$\Delta H_1 + \Delta H_2 = \Delta H_3$$

$$+206 \text{ kJ} + (-41 \text{ kJ}) = +165 \text{ kJ}$$

This calculation for $\Delta H_{rxn}$ is an application of **Hess's law**. Also known as *Hess's law of constant heat of summation*, it states that the enthalpy of reaction $\Delta H_{rxn}$ for a process that is the sum of two or more other reactions is equal to the sum of the $\Delta H_{rxn}$ values of the constituent reactions. This relation is illustrated for the reaction to synthesize hydrogen shown in Figure 5.31.

Hess's law is especially useful for calculating enthalpy changes that are difficult to measure directly. For example, $CO_2$ is the principal product of the combustion of carbon in the form of charcoal:

$$\text{Reaction A:} \quad C(s) + O_2(g) \rightarrow CO_2(g) \qquad \Delta H_{comb} = -393.5 \text{ kJ}$$

When the oxygen supply is limited, however, the products include carbon monoxide:

$$\text{Reaction B:} \quad C(s) + \tfrac{1}{2}O_2(g) \rightarrow CO(g)$$

It is difficult to measure the enthalpy of combustion of this reaction directly because, as long as any oxygen is present, some of the $CO(g)$ formed reacts with the $O_2$ to form $CO_2(g)$, yielding a mixture of CO and $CO_2$ as the product. However, we can use Hess's law to obtain this value *indirectly* by working with $\Delta H_{comb}$ values we can measure.

Because we can run reaction A with excess oxygen and thereby force it to completion, we can measure the enthalpy of combustion, which is −393.5 kJ. We can also react a sample of pure $CO(g)$ with oxygen and measure $\Delta H_{comb}$ for that reaction:

$$\text{Reaction C:} \quad CO(g) + \tfrac{1}{2}O_2(g) \rightarrow CO_2(g) \qquad \Delta H_{comb} = -283.0 \text{ kJ}$$

Hess's law gives us a way to calculate $\Delta H_{comb}$ for reaction B from the measured values for reactions A and C. To do this, we must find a way to combine the equations for reactions A and C so that the sum equals reaction B. Once we have that combination, because we know two of the $\Delta H_{comb}$ values, we can calculate the one we do not know.

One approach to this analysis is to focus on the reactants and products in the reaction whose $\Delta H_{comb}$ value is unknown—reaction B in this example. This reaction has carbon and oxygen as reactants and carbon monoxide as a product.

**FIGURE 5.31** Hess's law predicts that the enthalpy change for the production of 4 moles of $H_2(g)$ and 1 mole of $CO_2(g)$ from 1 mole of $CH_4(g)$ and 2 moles of $H_2O(g)$ ($\Delta H_3$) is the sum of the enthalpies of two reactions: $\Delta H_3 = \Delta H_1 + \Delta H_2$.

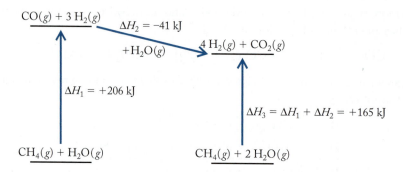

Note that reaction C has carbon monoxide as a reactant. If we add reaction C to reaction B, the carbon monoxide cancels out, and we end up with reaction A as the sum of C and B:

Reaction B:    $C(s) + \frac{1}{2} O_2(g) \rightarrow \cancel{CO(g)}$      $\Delta H_{comb} = ?$

Reaction C:    $\cancel{CO(g)} + \frac{1}{2} O_2(g) \rightarrow CO_2(g)$      $\Delta H_{comb} = -283.0 \text{ kJ}$

Reaction A:    $C(s) + O_2(g) \rightarrow CO_2(g)$      $\Delta H_{comb} = -393.5 \text{ kJ}$

We now use algebra to find $\Delta H_{comb}$ for reaction B:

$$\Delta H_B + \Delta H_C = \Delta H_A$$

$$\Delta H_B = \Delta H_A - \Delta H_C$$

$$= -393.5 \text{ kJ} - (-283.0 \text{ kJ}) = -110.5 \text{ kJ}$$

Recall that $\Delta H$ is a state function. This means we can manipulate equations in two important ways when applying Hess's law, should the need arise. (1) We can multiply the coefficients in a balanced equation and the $\Delta H$ for the reaction by a factor to change the quantity of material we are dealing with. (2) We can reverse a reaction (make the reactants the products and the products the reactants) if we also change the sign of $\Delta H$.

Hess's law is a direct consequence of the fact that enthalpy is a state function. In other words, for a particular set of reactants and products, the enthalpy change of the reaction is the same whether the reaction takes place in one step or in a series of steps. The concept of enthalpy and its expression in Hess's law are very useful because the heats associated with a large number of reactions can be calculated from a few that have been measured.

---

**SAMPLE EXERCISE 5.12    Applying Hess's Law in Biology    LO6**

One source of energy in our bodies is the conversion of sugars to $CO_2$ and $H_2O$ (called respiration; see Section 3.5). Maltose is one of the sugars found in foods. Show how we could combine reactions A and B below to get reaction C. How is $\Delta H_C$ related to the values of $\Delta H_A$ and $\Delta H_B$?

Reaction A:    maltose(s) + $H_2O(\ell)$ → 2 glucose(s)      $\Delta H_A$

Reaction B:    6 $CO_2(g)$ + 6 $H_2O(\ell)$ → glucose(s) + 6 $O_2(g)$      $\Delta H_B$

Reaction C:    maltose(s) + 12 $O_2(g)$ → 12 $CO_2(g)$ + 11 $H_2O(\ell)$      $\Delta H_C$

**Collect and Organize** We are asked to combine two chemical equations to write the equation of an overall reaction and to combine two $\Delta H_{rxn}$ values to obtain an overall $\Delta H_{rxn}$ value. Hess's law allows us to combine the enthalpy changes that accompany the steps in an overall chemical reaction to calculate the value of $\Delta H_{rxn}$ of the overall reaction.

**Analyze** Maltose is a reactant in equations A and C, so we can leave reaction A unchanged. The products of reaction C appear as reactants in equation B so we need to reverse B in order to get $CO_2$ and $H_2O$ on the product side. This delivers only 6 moles of $CO_2$ and $H_2O$ to the product side, so we need to multiply equation B by two before adding it to equation A.

Reversing equation B and multiplying it by 2 means changing the sign of $\Delta H_B$ and multiplying it by 2.

**Solve** We start with reaction A as written and reversing and doubling reaction B:

maltose(s) + $H_2O(\ell)$ → 2 glucose(s)      $\Delta H_A$

2 × [glucose(s) + 6 $O_2(g)$ → 6 $CO_2(g)$ + 6 $H_2O(\ell)$]      2 × [$-\Delta H_B$]

Adding A + 2(−B):

$$
\begin{array}{lll}
\text{A:} & \text{maltose}(s) + \cancel{\text{H}_2\text{O}(\ell)} \rightarrow \cancel{2\ \text{glucose}(s)} & \Delta H_A \\
-(2 \times \text{B}): & \cancel{2\ \text{glucose}(s)} + 12\ \text{O}_2(g) \rightarrow 12\ \text{CO}_2(g) + \overset{11}{\cancel{12}}\ \text{H}_2\text{O}(\ell) & -2\Delta H_B \\
\hline
\text{C:} & \text{maltose}(s) + 12\ \text{O}_2(g) \rightarrow 12\ \text{CO}_2(g) + 11\ \text{H}_2\text{O}(\ell) & \Delta H_C
\end{array}
$$

Hess's law allows us to add the respective enthalpies: $\Delta H_C = \Delta H_A - 2\Delta H_B$

**Think About It** We can add chemical equations just like algebraic equations, multiplying them by coefficients if necessary, to obtain an equation for a new chemical reaction. Then we can use Hess's Law to find the enthalpy change for the new reaction.

**Practice Exercise** How would you combine reactions A–C, shown below, to obtain reaction D?

A: $2\ \text{H}_2(g) + \text{O}_2(g) \rightarrow 2\ \text{H}_2\text{O}(g)$

B: $\text{H}_3\text{BNH}_3(s) \rightarrow \text{NH}_3(g) + \text{BH}_3(g)$

C: $\text{H}_3\text{BNH}_3(s) \rightarrow 2\ \text{H}_2(g) + \text{HBNH}(s)$

D: $\text{NH}_3(g) + \text{BH}_3(g) + \text{O}_2(g) \rightarrow 2\ \text{H}_2\text{O}(g) + \text{HBNH}(s)$

*(Answers to Practice Exercises are in the back of the book.)*

---

**SAMPLE EXERCISE 5.13** **Calculating Enthalpies of Reaction Using Hess's Law** **LO6**

Hydrocarbons burned in a limited supply of air may not burn completely, and $CO(g)$ may be generated. One reason furnaces and hot-water heaters fueled by natural gas need to be vented is that incomplete combustion can produce toxic carbon monoxide:

$$\text{Reaction A:} \quad 2\ \text{CH}_4(g) + 3\ \text{O}_2(g) \rightarrow 2\ \text{CO}(g) + 4\ \text{H}_2\text{O}(g) \quad \Delta H_A = ?$$

Use reactions B and C to calculate the $\Delta H_{\text{comb}}$ for reaction A.

$$\text{Reaction B:} \quad \text{CH}_4(g) + 2\ \text{O}_2(g) \rightarrow \text{CO}_2(g) + 2\ \text{H}_2\text{O}(g) \quad \Delta H_B = -802\ \text{kJ}$$

$$\text{Reaction C:} \quad 2\ \text{CO}(g) + \text{O}_2(g) \rightarrow 2\ \text{CO}_2(g) \quad \Delta H_C = -566\ \text{kJ}$$

**Collect and Organize** We are given two reactions (B and C) with thermochemical data and a third (A) for which we are asked to find $\Delta H_A$. All the reactants and products of reaction A are present in reactions B and/or C.

**Analyze** We can manipulate the equations for reactions B and C so that they sum to give the equation for which $\Delta H_{\text{comb}}$ is unknown. Then we can calculate this unknown value by applying Hess's law.

The reaction of interest (A) has methane on the reactant side. Because reaction B also has methane as a reactant, we can use B as written. Reaction A has CO as a product. Reaction C involves CO as a reactant, so we have to reverse C in order to get CO on the product side. Once we reverse C, we must change the sign of $\Delta H_C$. If the coefficients as given do not allow us to sum the two reactions to yield reaction A, we can multiply one or both reactions by other factors.

**Solve** We start with reaction B as written and add the reverse of reaction C, remembering to change the sign of its $\Delta H_{\text{comb}}$:

$$\text{B:} \quad \text{CH}_4(g) + 2\ \text{O}_2(g) \rightarrow \text{CO}_2(g) + 2\ \text{H}_2\text{O}(g) \quad \Delta H_B = -802\ \text{kJ}$$

$$\text{C (reversed):} \quad 2\ \text{CO}_2(g) \rightarrow 2\ \text{CO}(g) + \text{O}_2(g) \quad -\Delta H_C = 566\ \text{kJ}$$

Because methane has a coefficient of 2 in reaction A, we multiply all the terms in reaction B, including $\Delta H_{\text{comb}}$, by 2:

$$2 \times [\text{CH}_4(g) + 2\ \text{O}_2(g) \rightarrow \text{CO}_2(g) + 2\ \text{H}_2\text{O}(g)] \quad 2 \times [\Delta H_B = -802\ \text{kJ}]$$

---

**standard enthalpy of formation ($\Delta H_f^\circ$)** the enthalpy change of a formation reaction; also called *standard heat of formation* or *heat of formation*.

**formation reaction** a reaction in which 1 mole of a substance is formed from its component elements in their standard states.

**standard state** the most stable form of a substance under 1 bar pressure and some specified temperature (25°C unless otherwise stated).

**standard conditions** in thermodynamics: a pressure of 1 bar (~1 atm) and some specified temperature, assumed to be 25°C unless otherwise stated; for solutions, a concentration of 1 $M$ is specified.

**standard enthalpy of reaction ($\Delta H_{\text{rxn}}^\circ$)** the energy associated with a reaction that takes place under standard conditions; also called *standard heat of reaction*.

Because the carbon monoxide in reaction A has a coefficient of 2, we do not need to multiply reaction C by any factor. Now we add (2 × B) to the reverse of C and cancel out common terms:

$$2 \times \text{B}: \quad 2\,\text{CH}_4(g) + \overset{3}{4}\,\text{O}_2(g) \rightarrow \cancel{2\,\text{CO}_2(g)} + 4\,\text{H}_2\text{O}(g) \qquad \Delta H_\text{B} = -1604 \text{ kJ}$$

$$\text{C (reversed)}: \quad \cancel{2\,\text{CO}_2(g)} \rightarrow 2\,\text{CO}(g) + \cancel{\text{O}_2(g)} \qquad -\Delta H_\text{C} = 566 \text{ kJ}$$

$$\text{A}: \quad 2\,\text{CH}_4(g) + 3\,\text{O}_2(g) \rightarrow 2\,\text{CO}(g) + 4\,\text{H}_2\text{O}(g) \qquad \Delta H_\text{A} = -1038 \text{ kJ}$$

**Think About It** We used Hess's law to calculate the heat of combustion of methane to make CO, which is impossible to achieve in an experiment because any CO produced can react with $O_2$ to give $CO_2$, resulting in a product mixture of CO and $CO_2$. The answer of −1038 kJ is less negative than the enthalpy change accompanying the complete combustion of 2 mol $CH_4$ [2 × (−802 kJ)], so the answer is reasonable.

**Practice Exercise** It does not matter how you assemble the equations in a Hess's law problem. Show that reactions A and C can be summed to give reaction B and result in the same value for $\Delta H_\text{comb}$.

*(Answers to Practice Exercises are in the back of the book.)*

# 5.7 Standard Enthalpies of Formation and Reaction

As noted in Section 5.2, it is impossible to measure the *absolute* value of the internal energy of a substance. The same is true for the enthalpy of a substance. However, we can establish *relative* enthalpy values that are referenced to a convenient standard. This approach is similar to using the freezing point of water as the zero point on the Celsius temperature scale or sea level as the zero point for expressing altitude. Enthalpy values referenced to this zero point are called the **standard enthalpies of formation ($\Delta H_\text{f}^\circ$)**, defined as the enthalpy change that takes place at constant pressure when 1 mole of a substance is formed from its constituent elements in their standard states. The superscript degree in $\Delta H_\text{f}^\circ$ always signals that the symbol refers to *standard* pressure (1 bar). A reaction that fits this description is known as a **formation reaction**.

The **standard state** of an element is its most stable physical form under **standard conditions** of pressure (1 bar) and some specified temperature: in their standard states, oxygen is a gas, water is a liquid, and carbon is solid graphite. The value of 1 bar of pressure is very close to 1 atm; for the level of precision used in this book, a standard pressure of 1 bar will be considered equivalent to 1 atm.

By definition, for a pure element in its most stable form under standard conditions, $\Delta H_\text{f}^\circ = 0$. This is the zero point of enthalpy values. Table 5.2 lists standard heats of formation for several substances. The symbol for the enthalpy change associated with a reaction that takes place under standard conditions is $\Delta H_\text{rxn}^\circ$, and the value is called either a **standard enthalpy of reaction** or a *standard heat of reaction*. Implied in our notion of standard states and standard conditions is the assumption that parameters such as $\Delta H$ change with temperature and pressure. That assumption is correct, although the changes are so small that we ignore them in the calculations in this textbook.

Because the definition of a formation reaction specifies 1 mole of product, writing balanced equations for formation reactions may require the use of something we avoided in Chapter 3: fractional coefficients in the final form of our balanced equations. For example, the reaction for the production of ammonia from nitrogen and hydrogen is usually written

$$\text{N}_2(g) + 3\,\text{H}_2(g) \rightarrow 2\,\text{NH}_3(g) \qquad \Delta H_\text{rxn}^\circ = -92.2 \text{ kJ}$$

| TABLE 5.2 | Standard Enthalpies of Formation at 25°C for Selected Substances |
|---|---|
| **Substance** | **$\Delta H_\text{f}^\circ$ (kJ/mol)** |
| $O_2(g)$ | 0 |
| $H_2(g)$ | 0 |
| $H_2O(g)$ | −241.8 |
| $H_2O(\ell)$ | −285.8 |
| $C_\text{graphite}(s)$ | 0 |
| $CH_4(g)$, methane | −74.8 |
| $C_2H_2(g)$, acetylene | 226.7 |
| $C_2H_4(g)$, ethylene | 52.26 |
| $C_2H_6(g)$, ethane | −84.68 |
| $C_3H_8(g)$, propane | −103.8 |
| $C_4H_{10}(g)$, butane | −125.6 |
| $CO_2(g)$ | −393.5 |
| $CO(g)$ | −110.5 |
| $N_2(g)$ | 0 |
| $NH_3(g)$, ammonia | −46.1 |
| $N_2H_4(g)$, hydrazine | 95.4 |
| $N_2H_4(\ell)$ | 50.63 |
| $NO(g)$ | 90.3 |
| $Br_2(\ell)$ | 0 |
| $CH_3OH(\ell)$, methanol | −238.7 |
| $CH_3CH_2OH(\ell)$, ethanol | −277.7 |
| $CH_3COOH(\ell)$, acetic acid | −484.5 |

Although all reactants and products in this equation are in their standard states, it is not a formation reaction because 2 moles of product are formed, which is why we denote the enthalpy change of this reaction by $\Delta H^{\circ}_{rxn}$ rather than $\Delta H^{\circ}_{f}$. The formation reaction for ammonia must show 1 mole of product being formed, so we divide each coefficient in the above equation by 2. Because energy is a stoichiometric quantity, the heat of reaction is divided by 2 as well. Thus, the equation representing the formation reaction of ammonia is

$$\tfrac{1}{2}\,N_2(g) + \tfrac{3}{2}\,H_2(g) \rightarrow NH_3(g) \qquad \Delta H^{\circ}_{f} = -46.1 \text{ kJ}$$

### SAMPLE EXERCISE 5.14 Recognizing Formation Reactions LO7

Which of the following reactions are formation reactions at 25°C? For those that are not, explain why not.

a. $H_2(g) + \tfrac{1}{2}\,O_2(g) \rightarrow H_2O(g)$

b. $C_{graphite}(s) + 2\,H_2(g) + \tfrac{1}{2}\,O_2(g) \rightarrow CH_3OH(\ell)$
   ($CH_3OH$ is methanol, a liquid in its standard state.)

c. $CH_4(g) + 2\,O_2(g) \rightarrow CO_2(g) + 2\,H_2O(\ell)$

d. $P_4(s) + 2\,O_2(g) + 6\,Cl_2(g) \rightarrow 4\,POCl_3(\ell)$
   (In its standard state, $P_4$ is a solid; $Cl_2$ is a gas, and $POCl_3$ is a liquid.)

**Collect and Organize** We are given four balanced chemical equations and information about the standard states of specific reactants and products. We want to determine which of these are formation reactions.

**Analyze** For a reaction to be a formation reaction, it must meet the criteria that it produces 1 mole of a substance from its component elements in their standard states. Therefore we must evaluate each reaction for the quantity of product and for the state of each reactant and product.

**Solve**

a. The reaction shows 1 mole of water vapor formed from its constituent elements in their standard states. However, the product is $H_2O(g)$, which is not the standard state of water. Because all products and reactants must be in their standard state for the reaction to be considered a formation reaction, the reaction as written is not a formation reaction.

b. The reaction shows 1 mole of liquid methanol formed from its constituent elements in their standard states. This reaction is a formation reaction.

c. This is not a formation reaction because the reactants are not elements in their standard states and because more than one product is formed.

d. This is not a formation reaction because the product is 4 moles of $POCl_3$. Note, however, that all of the constituent elements are in their standard states and that only one compound is formed. We could easily convert this into a formation reaction by dividing all the coefficients by 4.

**Think About It** Just because we can write formation reactions for substances like methanol does not mean that anyone would ever use that reaction to make methanol. Remember that formation reactions are defined to provide a standard against which other reactions can be compared when evaluating their thermochemistry.

**Practice Exercise** Write formation reactions for (a) $CaCO_3(s)$, (b) $CH_3COOH(\ell)$ [acetic acid], and (c) $KMnO_4(s)$.

*(Answers to Practice Exercises are in the back of the book.)*

Table 5.2 lists standard enthalpies of formation for some substances, including hydrocarbons. (A more complete list can be found in Appendix 4.[1]) Note that the standard enthalpies of formation of acetylene and ethylene are positive, which means that the formation reactions for these compounds are endothermic. The

[1]The NIST Web site contains much additional data: http://webbook.nist.gov.

other hydrocarbon fuels in Table 5.2 have negative enthalpies of formation, but the values for all of them are less negative than the values for water and carbon dioxide. Recall that the products of the complete combustion of hydrocarbons are carbon dioxide and water. The more negative the heat of formation ($\Delta H_f^{\circ}$) of a substance, the more stable it is. The implication of this, stated earlier, is: the reaction of fuels with oxygen to produce $CO_2$ and $H_2O$ is exothermic. The reactions produce energy, which is why hydrocarbons are useful as fuels. Note also that the values for the $C_2$ to $C_4$ hydrocarbons become more negative with increasing molar mass.

Standard enthalpies of formation $\Delta H_f^{\circ}$ are used to predict standard heats of reaction $\Delta H_{rxn}^{\circ}$. We can calculate the standard heat of reaction for any reaction by determining the difference between the $\Delta H_f^{\circ}$ values of the products and the $\Delta H_f^{\circ}$ values of the reactants. To see how this approach works, consider the reaction we saw earlier (Equation 5.18):

$$CH_4(g) + 2\,O_2(g) \rightarrow CO_2(g) + 2\,H_2O(g)$$

This reaction represents the combustion of methane, an energy source of increasing importance because of its increased production from new geological as well as renewable sources (Figure 5.32). In this calculation, we use data from Table 5.2 in Equation 5.19:

$$\Delta H_{rxn}^{\circ} = \sum n_{products}\,\Delta H_{f,products}^{\circ} - \sum n_{reactants}\,\Delta H_{f,reactants}^{\circ} \quad (5.19)$$

where $n_{products}$ is the number of moles of each product in the balanced equation and $n_{reactants}$ is the number of moles of each reactant. Equation 5.19 states that the value of $\Delta H_{rxn}^{\circ}$ for any chemical reaction equals the sum ($\sum$) of the $\Delta H_f^{\circ}$ value for each product times the number of moles of that product in the balanced equation minus the sum of the $\Delta H_f^{\circ}$ value for each reactant times the number of moles of that reactant in the balanced chemical equation. Thus we multiply the value of $\Delta H_f^{\circ}$ for $O_2(g)$ in Equation 5.19 by 2 before summing with $\Delta H_f^{\circ}$ for $CH_4(g)$ because the $O_2$ coefficient is 2 in the balanced equation. We do the same with $\Delta H_f^{\circ}$ of water in the product.

Once we insert $\Delta H_f^{\circ}$ values from Table 5.2, remembering that for $O_2$ gas $\Delta H_f^{\circ} = 0$ because it is a pure element in its most stable form, Equation 5.19 becomes

$$\Delta H_{rxn}^{\circ} = [(1\ \text{mol}\ CO_2)(-393.5\ \text{kJ/mol}) + (2\ \text{mol}\ H_2O)(-241.8\ \text{kJ/mol})]$$
$$- [(1\ \text{mol}\ CH_4)(-74.8\ \text{kJ/mol}) + (2\ \text{mol}\ O_2)(0.0\ \text{kJ/mol})]$$
$$= [(-393.5\ \text{kJ}) + (-483.6\ \text{kJ})] - [(-74.8\ \text{kJ}) + (0.0\ \text{kJ})]$$
$$= -802.3\ \text{kJ}$$

Calculations of standard enthalpies of reaction $\Delta H_{rxn}^{\circ}$ from standard enthalpies of formation $\Delta H_f^{\circ}$ can be carried out for all kinds of chemical reactions, including reactions that occur in solution. Consider the reaction between methane and steam, which yields a mixture of hydrogen and carbon monoxide known as *water gas*:

$$CH_4(g) + H_2O(g) \rightarrow CO(g) + 3\,H_2(g)$$

This reaction is used to synthesize the hydrogen used in the steel industry to remove impurities from molten iron and in the chemical industry to make hundreds of compounds, including ammonia ($NH_3$) for use in fertilizers and as a refrigerant. The reaction is also important in the manufacture of hydrogen fuel for fuel cells, which are used to generate electricity directly from a chemical reaction.

**FIGURE 5.32** Methane ($CH_4$) is a renewable source of energy because it can be produced by degradation of organic matter by methanogenic bacteria. Methane is produced in swamps; hence it is commonly called swamp gas. Experiments in the bulk production of methane from natural sources, like the one shown here of a plastic tarp covering a lagoon of animal waste, are being carried out worldwide to augment the fuel supply.

Inserting the appropriate values for the compounds from Table 5.2 into Equation 5.19, along with the coefficient 3 for $H_2$, we calculate

$$\Delta H^{\circ}_{rxn} = [(1 \text{ mol } CO)(-110.5 \text{ kJ/mol}) + (3 \text{ mol } H_2)(0.0 \text{ kJ/mol})]$$

$$- [(1 \text{ mol } CH_4)(-74.8 \text{ kJ/mol}) + (1 \text{ mol } H_2O)(-241.8 \text{ kJ/mol})]$$

$$= +206.1 \text{ kJ}$$

The positive standard enthalpy of reaction tells us that this reaction between water vapor and methane (called steam–methane reforming, or simply *steam reforming*) is endothermic. Therefore, energy must be added to make the reaction take place. (It is typically conducted at temperatures near 1000°C.) Thus, although hydrogen is attractive as a fuel because it burns vigorously and produces only water as a product,

$$2 \, H_2(g) + O_2(g) \rightarrow 2 \, H_2O(\ell) + \text{energy}$$

its production from synthesis gas does not reduce the use of fossil fuels because that reaction consumes methane, most of which currently comes from fossil fuel. (This could change if production of methane from renewable sources becomes more feasible.) Fossil fuels are also burned to generate the heat that must be added to the reaction to make it run. Using hydrogen-powered vehicles merely shifts consumption of fossil fuels from the consumption of gasoline at the filling station to an earlier step in the process.

Hydrocarbons other than methane are also used as the starting materials in the production of $H_2(g)$. However, the reactions are still usually endothermic and require the input of energy, which means fossil fuels are consumed to generate that energy. Also, $CO_2$ is released into the environment as a result of the production of hydrogen in any of these processes.

Enthalpy is a state function, and, as noted in Section 5.1, the value of a state function is independent of the path taken to achieve that state. Thus values $\Delta H_{rxn}$ are independent of pathway. It does not matter what path we take to get from the reactants to the products; the enthalpy difference between them is always the same, as Figure 5.33 illustrates. Just as we saw in Section 5.2 with state changes, once we know the value of the energy associated with running a reaction in one direction, we also know the value of the energy associated with running the reaction in the reverse direction because it has the same magnitude but the opposite sign. This means that, having calculated a value of $\Delta H^{\circ}_{rxn} = +206.1$ kJ for the steam reforming reaction, we can write

$$CO(g) + 3 \, H_2(g) \rightarrow CH_4(g) + H_2O(g) \qquad \Delta H^{\circ}_{rxn} = -206.1 \text{ kJ}$$

for the exothermic reverse reaction.

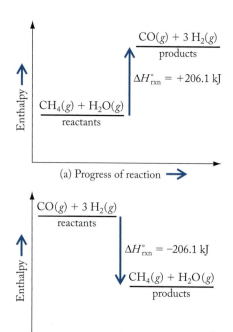

**FIGURE 5.33** (a) The reaction of methane with water to produce carbon monoxide and hydrogen is endothermic. (b) The reverse reaction, carbon monoxide plus hydrogen producing methane and water, is exothermic. The value of the enthalpy change of the two reactions has the same magnitude but is positive for the endothermic reaction and negative for the exothermic reaction.

SAMPLE EXERCISE 5.15  Calculating Enthalpies of Reaction  LO6

Using the appropriate values from Table 5.2, calculate $\Delta H^{\circ}_{rxn}$ for the combustion of the fuel propane ($C_3H_8$) in air.

**Collect and Organize** The reactants are propane and elemental oxygen, $O_2$, and the products are carbon dioxide and water. Even though combustion is an exothermic reaction, we assume that water is produced as liquid, $H_2O(\ell)$, so that all products and reactants are in their standard states. The heats of formation of propane, carbon dioxide, and $H_2O(\ell)$ are given in Table 5.2. The heat of formation of the element $O_2$ in its standard state is zero. Our task is to use the balanced equation and the data from Table 5.2 to calculate the heat of combustion.

**Analyze** Equation 5.19 defines the relation between heats of formation of reactants and products and the standard enthalpy of reaction. We also need the balanced equation for combustion of propane:

$$C_3H_8(g) + 5\,O_2(g) \rightarrow 3\,CO_2(g) + 4\,H_2O(\ell)$$

**Solve** Inserting $\Delta H_f^\circ$ values for the products [$CO_2(g)$ and $H_2O(\ell)$] and reactants [$C_3H_8(g)$ and $O_2(g)$] from Table 5.2 and the coefficients in the balanced chemical equation into Equation 5.17, we get

$$\Delta H_{rxn}^\circ = \left[(3\;\cancel{mol}\;CO_2)(-393.5\;kJ/\cancel{mol}) + (4\;\cancel{mol}\;H_2O)(-285.8\;kJ/\cancel{mol})\right]$$
$$- \left[(1\;\cancel{mol}\;C_3H_8)(-103.8\;kJ/\cancel{mol}) + (5\;\cancel{mol}\;O_2)(0.0\;kJ/\cancel{mol})\right]$$
$$= -2219.9\;kJ$$

**Think About It** The result of the calculation has a large negative value, which means that the combustion reaction is highly exothermic, as expected for a hydrocarbon fuel.

 **Practice Exercise** Calculate $\Delta H_{rxn}^\circ$ for the *water–gas shift reaction*:

$$CO(g) + H_2O(g) \rightarrow CO_2(g) + H_2(g)$$

*(Answers to Practice Exercises are in the back of the book.)*

Hydrocarbons such as methane, ethane, and propane are components of natural gas and are excellent fuels. However, they are currently classified as nonrenewable fuels, and their combustion produces carbon dioxide, contributing to climate change.

**CONCEPT TEST**

What is the enthalpy of reaction $\Delta H_{rxn}^\circ$ for the production of one mole of $C_3H_8$ and oxygen from $CO_2(g)$ and $H_2O(\ell)$? Explain the reasoning you used to arrive at your answer. (See information in Sample Exercise 5.15, and do not do any mathematical calculations in answering this question.)

*(Answers to Concept Tests are in the back of the book.)*

When we use standard enthalpies of formation ($\Delta H_f^\circ$) to determine standard enthalpies of reaction $\Delta H_{rxn}^\circ$, we are applying Hess's law. To see that this is the case, consider the reaction of ammonia with oxygen to make $NO(g)$ and liquid water (this is the first step of the industrial synthesis of nitric acid):

$$4\,NH_3(g) + 5\,O_2(g) \rightarrow 4\,NO(g) + 6\,H_2O(\ell)$$

First we write an equation for the formation reaction for each reactant and product that is not an element in its standard state (remember $\Delta H_f^\circ = 0$ for elements in their standard states), using the $\Delta H_f^\circ$ values in Table 5.2:

$$\tfrac{1}{2}N_2(g) + \tfrac{3}{2}H_2(g) \rightarrow NH_3(g) \qquad \Delta H_f^\circ = -46.1\;kJ/mol$$
$$\tfrac{1}{2}N_2(g) + \tfrac{1}{2}O_2(g) \rightarrow NO(g) \qquad \Delta H_f^\circ = +90.3\;kJ/mol$$
$$H_2(g) + \tfrac{1}{2}O_2(g) \rightarrow H_2O(\ell) \qquad \Delta H_f^\circ = -285.8\;kJ/mol$$

We reverse the $NH_3$ equation so that the $NH_3$ is a reactant and multiply each equation by a factor that matches the product's or reactant's coefficient in the balanced equation.

$$4\left[NH_3(g) \rightarrow \tfrac{1}{2}N_2(g) + \tfrac{3}{2}H_2(g)\right] \qquad 4 \times (\Delta H_f^\circ = +46.1\;kJ/mol)$$
$$4\left[\tfrac{1}{2}N_2(g) + \tfrac{1}{2}O_2(g) \rightarrow NO(g)\right] \qquad 4 \times (\Delta H_f^\circ = +90.3\;kJ/mol)$$
$$6\left[H_2(g) + \tfrac{1}{2}O_2(g) \rightarrow H_2O(\ell)\right] \qquad 6 \times (\Delta H_f^\circ = -285.8\;kJ/mol)$$

(a)

(b)

(c)

**FIGURE 5.34** (a) Charles M. Hall and (b) Paul Louis-Toussiant Héroult independently developed the same electrolytic process for producing aluminum metal from alumina. (c) Hall's sister Julia, also a chemistry major at Oberlin College, assisted her brother in the lab, and her business skills made their aluminum production company a financial success. The company became the Aluminum Company of America, shortened to Alcoa, Inc.

Their sum yields the reaction of interest:

$$4\ NH_3(g) \rightarrow \cancel{2\ N_2(g)} + \cancel{6\ H_2(g)} \qquad \Delta H^\circ_{rxn} = +184.4\ kJ$$

$$\cancel{2\ N_2(g)} + 2\ O_2(g) \rightarrow 4\ NO(g) \qquad \Delta H^\circ_{rxn} = +361.2\ kJ$$

$$\underline{\cancel{6\ H_2(g)} + 3\ O_2(g) \rightarrow 6\ H_2O(\ell) \qquad \Delta H^\circ_{rxn} = -1714.8\ kJ}$$

$$4\ NH_3(g) + 5\ O_2(g) \rightarrow 4\ NO(g) + 6\ H_2O(\ell) \qquad \Delta H^\circ_{rxn} = -1169.2\ kJ$$

This is exactly the same mathematical operation that results when we use Equation 5.19:

$$\Delta H^\circ_{rxn} = \sum n_{products}\ \Delta H^\circ_{f,products} - \sum n_{reactants}\ \Delta H^\circ_{f,reactants}$$

$$= \left[ 4\ \cancel{mol}\ NO(+90.3\ kJ/\cancel{mol}) + 6\ \cancel{mol}\ H_2O(-285.8\ kJ/\cancel{mol}) \right]$$

$$- \left[ 4\ \cancel{mol}\ NH_3(-46.1\ kJ/\cancel{mol}) + 5\ \cancel{mol}\ O_2(0\ kJ/\cancel{mol}) \right] = -1169.2\ kJ$$

## A Practical Application of Thermochemistry

Chemical engineers frequently analyze the energy required to carry out industrial procedures to assess costs of operations and support new approaches to producing materials. The following example of a simple evaluation of the energy requirements associated with the production and recycling of aluminum illustrates a practical application of thermochemical concepts.

Over the last century, aluminum—both alone and in combination with other metals—has replaced steel for use where high strength-to-weight ratios and corrosion resistance are paramount. Airplanes, motor vehicles, and the facades of buildings may now all be made from aluminum. The industrial process for converting $Al_2O_3$ (alumina), the form of aluminum in the ore bauxite, into aluminum was developed by two 23-year-old chemists, Charles Hall and Paul Louis-Toussaint Héroult (both 1863–1914), working independently in the United States (Hall) and France (Héroult; Figure 5.34). The process is based on passing an electric current through a solution of alumina dissolved in molten cryolite ($Na_3AlF_6$). As electricity passes through the solution, aluminum ions are reduced to aluminum metal, while the positively charged carbon electrode is oxidized to carbon dioxide. The process is described by the following reaction:

$$2\ Al_2O_3(\text{in molten } Na_3AlF_6) + 3\ C(s) \rightarrow 4\ Al(\ell) + 3\ CO_2(g)$$

The principal energy cost of the Hall–Héroult process is the electricity needed to reduce $Al_2O_3$; the major cost in recycling is the energy required to melt aluminum metal. We can use thermochemistry principles to estimate the energy requirements for the Hall–Héroult process and compare it to the energy needed for recycling.

Using Equation 5.19, we can calculate the standard energy change for the reduction reaction of alumina from the standard enthalpies of formation:

$$\Delta H^\circ_{rxn} = \left[ 3(\Delta H^\circ_{f,CO_2} + 4(\Delta H^\circ_{f,Al})] - [2(\Delta H^\circ_{f,Al_2O_3}) + 3(\Delta H^\circ_{f,C}) \right]$$

$$= \left[ (3\ \cancel{mol\ CO_2})\left( \frac{-393.5\ kJ}{1\ \cancel{mol\ CO_2}} \right) + (4\ \cancel{mol\ Al})\left( \frac{10.79\ kJ}{1\ \cancel{mol\ Al}} \right) \right]$$

$$- \left[ (2\ \cancel{mol\ Al_2O_3})\left( \frac{-1675.7\ kJ}{1\ \cancel{mol\ Al_2O_3}} \right) + (3\ \cancel{mol\ C})\left( \frac{0.0\ kJ}{1\ \cancel{mol\ C}} \right) \right]$$

$$= +2214.1\ kJ$$

Dividing this value by the 4 moles of aluminum produced in the reaction as written, we get

$$\frac{2214.1 \text{ kJ}}{4 \text{ mol Al}} = 553.52 \text{ kJ/mol}$$

The energy cost is 553.52 kJ for every mole of aluminum produced.

We can estimate the energy required to recycle 1.00 mole of aluminum by heating it from 25°C to its melting point (660°C) until all the aluminum melts. The energy needed to heat 1 mole of aluminum from 25°C to 660°C can be calculated from its molar heat capacity [24.4 J/(mol · °C)] and Equation 5.8:

$$q = nc_\text{P}\Delta T$$
$$= 1.00 \text{ mol} \times 24.4 \frac{\text{J}}{\text{mol} \cdot \text{°C}} \times (660 - 25)\text{°C} \times \frac{1 \text{ kJ}}{1000 \text{ J}}$$
$$= 15.5 \text{ kJ}$$

Once the aluminum is at its melting point, the energy required to melt 1 mole of Al is its heat of fusion ($\Delta H_\text{fus} = 10.79$ kJ/mol):

$$10.79 \text{ kJ/mol} \times 1.00 \text{ mol} = 10.8 \text{ kJ}$$

The estimated total energy to heat and melt 1.00 mole of aluminum is

$$15.5 \text{ kJ} + 10.8 \text{ kJ} = 26.3 \text{ kJ}$$

This value represents

$$\frac{26.3 \text{ kJ}}{553.52 \text{ kJ}} \times 100\% = 4.75\%$$

of the energy needed electrically to produce 1 mole of aluminum from its ore. The high cost of electricity makes recycling aluminum economically attractive as well as environmentally sound.

Other energy costs arise in both the production and recycling of aluminum, and the numbers calculated here should be viewed as estimates based on ideal situations; overall, however, recycling saves aluminum manufacturers about 95% of the energy required to produce the metal from the ore. This energy savings has inspired the rapid growth of a global aluminum recycling industry, and in the United States alone, aluminum recycling is a $1 billion per year business. Junkyards in the United States currently recycle 85% of the aluminum in cars and over 50% of the aluminum in food and beverage containers.

### SAMPLE EXERCISE 5.16 Recycling Metals LO6

Iron metal is obtained by reducing iron oxide with carbon. The balanced chemical equation for making iron from $Fe_2O_3$ is:

$$2 \text{ Fe}_2\text{O}_3(s) + 3 \text{ C}(s) \rightarrow 4 \text{ Fe}(s) + 3 \text{ CO}_2(g)$$

Iron melts at 1538°C with $\Delta H_\text{fus} = 19.4$ kJ/mol. The molar heat capacity of iron is $c_\text{P,Fe} = 25.1$ J/mol · °C. Is the energy required to melt down recycled iron less than that needed to reduce the iron in $Fe_2O_3$ to the free metal?

**Collect and Organize** We are given a balanced chemical equation, the melting point of iron, $\Delta H_\text{fus} = 19.4$ kJ/mol, and $c_\text{P,Fe}$. Additional thermodynamic data are available in Appendix 4 to calculate $\Delta H_\text{rxn}$. We are asked to compare the energy savings for recycling iron to reducing it from $Fe_2O_3$.

**Analyze** We will follow the example in the text for recycling Al in order to compare the energy cost of making 1.00 mole of Fe with the cost to melt 1.00 mole of Fe. (a) The energy needed to make one mole of iron is calculated using standard heats of formation and Equation 5.19. (b) The energy to melt a mole of iron is the sum of the energy to heat the iron to its melting point and the heat of fusion.

**Solve** Substituting for $\Delta H^\circ_{f,reactants}$ and $\Delta H^\circ_{f,products}$ in Equation 5.19 using the values in Appendix 4:

$$\Delta H^\circ_{rxn} = \sum n_{products} \, \Delta H^\circ_{f,products} - \sum n_{reactants} \, \Delta H^\circ_{f,reactants}$$

$$= \left( 4 \, \Delta H^\circ_{f,Fe} + 3 \, \Delta H^\circ_{f,CO_2} \right) - \left( 2 \, \Delta H^\circ_{f,Fe_2O_3} + 3 \, \Delta H^\circ_{f,C} \right)$$

$$= \left[ 4 \text{ mol Fe} \times \left( \frac{0 \text{ kJ}}{\text{mol Fe}} \right) + 3 \text{ mol CO}_2 \times \left( \frac{-393.5 \text{ kJ}}{\text{mol CO}_2} \right) \right]$$

$$- \left[ 2 \text{ mol Fe}_2\text{O}_3 \left( \frac{-824.2 \text{ kJ}}{\text{mol Fe}_2\text{O}_3} \right) + 3 \text{ mol C} \left( \frac{0 \text{ kJ}}{\text{mol C}} \right) \right]$$

$$= 467.9 \text{ kJ}$$

So, recovering 1.00 mole of Fe from $Fe_2O_3$ requires:

$$\frac{467.9 \text{ kJ}}{4 \text{ mol Fe}} = \frac{117.0 \text{ kJ}}{\text{mol Fe}}$$

The energy to melt 1.00 mole of Fe is:

$$q = n_{Fe} c_{P,Fe} \Delta T + n_{Fe} \Delta H_{fus}$$

$$= \left[ 1.00 \text{ mol Fe} \times \frac{25.1 \text{ J}}{\text{mol Fe} \cdot \text{°C}} \times \frac{1 \text{ kJ}}{10^3 \text{ J}} \times (1538\text{°C} - 25\text{°C}) \right] + \left[ 1.00 \text{ mol Fe} \times \frac{19.4 \text{ kJ}}{\text{mol Fe}} \right]$$

57.4 kJ for 1.00 mol Fe

Recycling iron is more energy efficient than refining it from $Fe_2O_3$.

**Think About It** The energy saved by recycling iron,

$$\frac{57.4 \text{ kJ}}{117.0 \text{ kJ}} \times 100\% = 49.1\%$$

is not nearly as great as for aluminum because the energy required to melt iron is higher and the energy to make iron is less than for aluminum. However, if you account for the costs of mining iron ore and the environmental impact of iron production, recycling is probably worthwhile.

**Practice Exercise** How does the energy required to recycle 1.00 mole of copper compare with that required to recover copper from CuO? The balanced chemical equation for the smelting of copper is: $CuO(s) + CO(g) \rightarrow Cu(s) + CO_2(g)$. Copper melts at 1084.5°C with $\Delta H^\circ_{fus} = 13.0$ kJ/mol and a molar heat capacity, $c_{P,Cu} = 24.5$ J/mol · °C. In addition, $\Delta H^\circ_{f,CuO} = -155$ kJ/mol.

*(Answers to Practice Exercises are in the back of the book.)*

# 5.8 Fuel Values and Food Values

In Section 5.5, we saw that the enthalpy of reaction for one mole of propane (−2220 kJ) is much greater than for one mole of methane (−802 kJ)—that is, much more energy is released in the combustion of propane. Does this make propane an inherently better (higher-energy) fuel? Not necessarily. Expressing $\Delta H_{rxn}$ values on a per-mole basis is the only way to ensure that we are talking about the same number of molecules. However, we do not purchase fuels, or anything else for that matter, in units of moles. Depending on the fuel, we buy it either by mass (coal by the ton) or by volume (gasoline by the gallon or the liter).

## Fuel Value

To calculate the enthalpy change that takes place when 1 g of methane or 1 g of propane burns in air, producing $CO_2$ and liquid water, we divide the absolute value of $\Delta H_{rxn}$ (or $\Delta H_{comb}$ in kilojoules per mole) for each reaction by the molar mass of the hydrocarbon. This gives us the number of kilojoules of energy released per gram of substance:

$$CH_4: \quad \frac{802.3 \text{ kJ}}{\text{mol}} \times \frac{1 \text{ mol}}{16.04 \text{ g}} = 50.02 \text{ kJ/g}$$

$$C_3H_8: \quad \frac{2219.9 \text{ kJ}}{\text{mol}} \times \frac{1 \text{ mol}}{44.10 \text{ g}} = 50.34 \text{ kJ/g}$$

These quantities of energy per gram of fuel are called **fuel values**. If we carry out similar calculations for the other hydrocarbons larger than methane in Table 5.2, we can determine their fuel values. When we do this, we find that the values decrease with increasing molar mass. Why is that the case?

The answer lies in the hydrogen-to-carbon ratios in these compounds. As the number of carbon atoms per molecule increases, the hydrogen-to-carbon ratio decreases. Consider that there are 12 times as many H atoms in 1 g of hydrogen as there are C atoms in 1 g of carbon; this means that, *gram for gram*, 1 g of hydrogen can form six times as many moles of $H_2O$ as 1 g of carbon can form moles of $CO_2$. More energy is released in a combustion reaction by the formation of 1 mole of $CO_2(g)$ (393.5 kJ) than by 1 mole of $H_2O(g)$ (241.8 kJ), but even if we take this difference into account, *gram for gram* hydrogen has many times the fuel value of carbon.

Another term frequently used to compare the energy content of liquid fuels is **fuel density**. Fuel density describes the amount of energy available per unit volume of a liquid fuel and is typically reported as energy released when 1 liter of liquid is completely burned. Both fuel value and fuel density are reported as positive numbers, and it is simply understood that these values refer to the energy released from the fuel when it is burned.

**fuel value** the energy released during the complete combustion of 1 g of a substance.

**fuel density** the amount of energy released during the complete combustion of 1 liter of a liquid fuel.

**CONCEPT TEST**

Without doing any calculations, predict which compound in each pair releases more energy during combustion in air: (a) 1 mole of $CH_4$ or 1 mole of $H_2$; (b) 1 g of $CH_4$ or 1 g of $H_2$.

*(Answers to Concept Tests are in the back of the book.)*

**SAMPLE EXERCISE 5.17    Comparing Fuel Values      L08
and Fuel Densities**

Most automobiles run on either gasoline or diesel fuel. Both fuels are mixtures, but the energy content in gasoline can be approximated by a hydrocarbon with the formula $C_9H_{20}$ ($d = 0.718$ g/mL). Diesel fuel may be considered to be $C_{14}H_{30}$ ($d = 0.763$ g/mL). Using

| TABLE 5.3 | Standard Enthalpies of Combustion |
|---|---|
| **Substance** | $\Delta H^{\circ}_{comb}$ **(kJ/mol)** |
| $CO(g)$ | −283.0 |
| $CH_4(g)$, methane | −802.3 |
| $C_3H_8(g)$, propane | −2219.9 |
| $C_5H_{12}(\ell)$, pentane | −3535 |
| $C_9H_{20}(\ell)$, avg. gasoline compound | −6160 |
| $C_{14}H_{30}(\ell)$, avg. diesel compound | −7940 |

these two formulas, compare (a) the fuel value per gram of each fuel and (b) the fuel density per liter of each fuel.

**Collect and Organize** We are given representative chemical formulas for gasoline ($C_9H_{20}$) and diesel fuel ($C_{14}H_{30}$) and can calculate the molar mass for each fuel. Enthalpy of combustion data are given in Table 5.3. We want to use these data and the respective molar masses to obtain the fuel values. We can then use the given density of each fuel to convert the fuel value to fuel density.

**Analyze** The enthalpies of combustion in Table 5.3 are given in kilojoules per mole, so to answer this question in terms of grams, we need molar masses to convert moles to grams in part a. For part b we need density ($d = m/V$, grams per milliliter) to convert grams to liters. Taking $C_9H_{20}$ and $C_{14}H_{30}$ as the average molecules in regular gasoline and diesel fuel, respectively, we can determine their molar masses. Then the fuel value can be calculated from the heats of combustion. Using the densities given in the problem, we can calculate fuel densities in kJ/L.

**Solve**

a. Fuel values

$$\text{Gasoline as } C_9H_{20}: \quad 6160 \frac{kJ}{mol} \times \frac{1 \: mol}{128.25 \: g} = 48.0 \text{ kJ/g}$$

$$\text{Diesel as } C_{14}H_{30}: \quad 7940 \frac{kJ}{mol} \times \frac{1 \: mol}{198.38 \: g} = 40.0 \text{ kJ/g}$$

b. Fuel densities

$$\text{Gasoline as } C_9H_{20}: \quad 48.0 \frac{kJ}{g} \times 0.718 \frac{g}{mL} \times \frac{10^3 \: mL}{L} = 34{,}500 \text{ kJ/L}$$

$$\text{Diesel as } C_{14}H_{30}: \quad 40.0 \frac{kJ}{g} \times 0.763 \frac{g}{mL} \times \frac{10^3 \: mL}{L} = 30{,}500 \text{ kJ/L}$$

**Think About It** We would not expect the fuel values of gasoline and diesel fuel to be very different, or there would be a strong preference for gasoline-fueled cars over diesel or vice versa. The values are similar to the fuel values of methane and propane calculated in the text, so these answers seem reasonable. In the United States, car and truck fuel is purchased by the gallon. Its consumption is rated in miles per gallon, whereas in most countries it is bought by the liter (1 U.S. gal = 3.785 L). Because of the way we buy automobile fuels, it makes sense to compare fuel densities rather than fuel values. On either basis, diesel is a slightly inferior fuel compared to gasoline. Diesel engines, however, tend to deliver greater fuel economy than gasoline engines.

⚙ **Practice Exercise** Kerosene, used as a fuel in high-performance aircraft and in space heaters, is a hydrocarbon intermediate in composition between gasoline and diesel fuel and may be approximated as $C_{12}H_{26}$ ($d = 0.750$ g/mL; $\Delta H^{\circ}_{comb} = -7050$ kJ/mol). Estimate the fuel value and the fuel density of kerosene.

*(Answers to Practice Exercises are in the back of the book.)*

■

## Food Value

Food serves the same purpose in living systems as fuel does in mechanical systems. The chemical reactions that convert food into energy resemble combustion but consist of many more steps that are much more highly controlled. Carbon dioxide and water are the ultimate end products, however, and in a fundamental way, metabolism of food by a living system and combustion of fuel in an engine are the same process. The **food value** of the material we eat—the amount of energy produced when food is burned completely—can be determined using the

**food value** the quantity of energy produced when a material consumed by an organism for sustenance is burned completely; it is typically reported in Calories (kilocalories) per gram of food.

same equipment and applying the same concepts of thermochemistry we have developed to evaluate fuels for vehicles. We can analyze the relative food value of the items we consume in the same way we analyzed fuel value: by burning material in a bomb calorimeter and measuring the quantity of energy released.

As an illustration, let's consider the food value of a serving of peanuts, about 28 grams or 40 peanuts. To determine its heat energy content per unit of mass, we burn the peanuts in a calorimeter. A portion of the mass of the peanuts (about 1.5 g) is water, so we first prepare the sample by drying it. This is a necessary step to get an accurate value for the enthalpy of reaction $\Delta H_{rxn}$ because we need to make sure that all of the sample mass is due to its carbon-containing components.

Once the peanuts are dry, suppose they have a mass of 26.5 g. We put them in a calorimeter for which $C_{calorimeter} = 41.8$ kJ/°C and burn them completely in excess oxygen. If the temperature of the calorimeter rises by 16.0°C, what is the food value of the peanuts?

To answer this question, we use Equation 5.12 to determine the quantity of energy that flowed from the peanuts to the calorimeter:

$$q_{calorimeter} = C_{calorimeter}\Delta T = (41.8 \text{ kJ/°C})(16.0°\text{C}) = +669 \text{ kJ}$$

where the plus sign reminds us that energy flowed into the calorimeter. The energy lost by the peanuts as they burned has the same value but opposite sign:

$$-q_{peanuts} = q_{calorimeter}$$

$$= -669 \text{ kJ}$$

Just as with fuel values, however, food value is reported as a positive number, and it is understood that the energy contained in the food is released when the food is metabolized. The peanuts' food value is therefore

$$\frac{669 \text{ kJ}}{1 \text{ serving of peanuts}} \times \frac{1 \text{ serving of peanuts}}{26.5 \text{ g}} = 25.2 \text{ kJ/g}$$

Most of us still think in terms of nutritional Calories (kilocalories), and we can use the definition of Calorie from Section 5.2 to convert energy in kilojoules into the familiar unit used by nutritionists:

$$\frac{669 \text{ kJ}}{\text{serving of peanuts}} \times \frac{1 \text{ Cal}}{4.184 \text{ kJ}} = \frac{160 \text{ Cal}}{\text{serving of peanuts}}$$

A single serving of peanuts represents 6% to 8% of the Calories needed by an adult with normal activity in one day. Figure 5.35 illustrates some other foods that have a similar food value.

**FIGURE 5.35** (a) A can of soda, (b) a protein bar, and (c) a stack of saltines all contain between 120 and 160 Calories, or about 6% to 8% of our daily energy needs.

---

● **SAMPLE EXERCISE 5.18** **Calculating Food Value** **LO8**

Glucose ($C_6H_{12}O_6$) is a simple sugar formed by photosynthesis in plants. The complete combustion of 0.5763 g of glucose in a calorimeter ($C_{calorimeter} = 6.20$ kJ/°C) raises the temperature of the calorimeter by 1.45°C. What is the food value of glucose in Calories per gram?

**Collect and Organize** We are asked to determine the food value of glucose, which means the energy given off when 1 g is burned. We have the mass of glucose burned and the calorimeter constant. We can convert kJ to Calories by using the conversion factor 1 Cal = 4.184 kJ.

**Analyze** We can relate the energy given off by the glucose to the energy gained by the calorimeter (Equation 5.15) and then to the temperature change and the calorimeter constant (Equation 5.12).

**Solve**

$$q_{\text{calorimeter}} = C_{\text{calorimeter}} \Delta T = (6.20 \text{ kJ/}°\text{C})(1.45°\text{C}) = 8.99 \text{ kJ}$$

To convert this quantity of energy to a food value, we divide by the sample mass:

$$\frac{8.99 \text{ kJ}}{0.5763 \text{ g}} = 15.6 \text{ kJ/g}$$

We can then convert this value into Calories:

$$(15.6 \text{ kJ/g})\left(\frac{1 \text{ Cal}}{4.184 \text{ kJ}}\right) = 3.73 \text{ Cal/g}$$

**Think About It** One gram of peanuts has about 50% more fuel value than 1 g of glucose (25.2 kJ/g and 15.6 kJ/g, respectively), so peanuts represent food with a greater energy density. This is important to hikers who carry their own food.

⚙ **Practice Exercise** Sucrose (table sugar) has the formula $C_{12}H_{22}O_{11}$ ($\mathcal{M} =$ 342.30 g/mol) and a food value of 16.4 kJ/g. Determine the calorimeter constant of the calorimeter in which the combustion of 1.337 g of sucrose raises the temperature by 1.96°C.

*(Answers to Practice Exercises are in the back of the book.)*

◉◉ **CONNECTION** In Chapter 3, we introduced the sugar glucose as a carbohydrate: a compound composed of carbon, hydrogen, and oxygen.

---

**SAMPLE EXERCISE 5.19** | **Integrating Concepts: Flameless Heat Sources**

Survival gear frequently includes flameless heat sources (FHSs) for warming food and water when electrical service is unavailable or when open flames are dangerous. A typical commercially available FHS consists of an insulated container with a cylindrical chamber for food surrounded by a jacket that contains a substance that releases heat when treated with water. The material to be heated goes into the chamber; water is poured into the outer jacket, which is then sealed. A chemical reaction ensues that heats the chamber's contents.

The package insert on an FHS states that it can heat 250 mL of water from 25°C to boiling (100°C). You dissemble the FHS and note that the jacket contains 70.0 g of a white powder. When the powder dissolves in water, the resultant solution, in addition to being very hot, is strongly basic. The insert states that the white powder "is derived from a natural mineral, readily available." You suspect that the powder could be calcium oxide. You make a solution by dissolving 10.0 g of the solid in 50 mL of water and titrate it with 94.30 mL of 3.785 $M$ HCl.

a. Write a balanced chemical equation for the reaction of calcium oxide with water.
b. Using thermochemical data from Appendix 4 and this chapter, determine if this reaction could liberate sufficient energy to boil 250 mL of water. The $\Delta H_f°$ of $Ca(OH)_2(aq)$ is −1002.82 kJ/mol.
c. Do the results of the titration confirm that the substance could be calcium oxide?

**Collect and Organize** We have the suggestion that the material in the FHS is calcium oxide. We can use the thermochemical information given about heating water to determine how much energy the reaction provides. We can calculate the $\Delta H_{\text{rxn}}$ of calcium oxide and water. We can confirm the identification using the titration data.

**Analyze** Calcium oxide is a reasonable guess for the identity of the material. It is the readily available commercial substance known as quicklime (see Chapter 4), derived from limestone by roasting. It produces a basic solution when treated with water.

**Solve**
a. $CaO(s) + H_2O(\ell) \rightarrow Ca(OH)_2(aq)$
b. We can calculate the amount of energy needed to take 250 mL of water from 25°C to 100°C using the molar heat capacity of water: 75.3 J/mol · °C (Table 5.1). Recall that 1 mL water = 1.0 g.

$$250 \text{ mL H}_2\text{O} = 250 \text{ g H}_2\text{O}$$

$$250 \text{ g H}_2\text{O} \times \frac{1 \text{ mol H}_2\text{O}}{18.02 \text{ g H}_2\text{O}} = 13.9 \text{ mol H}_2\text{O}$$

Using Equation 5.8

$$13.9 \text{ mol } H_2O \times \frac{75.3 \text{ J}}{\text{mol} \cdot {}^\circ C} \times (100{}^\circ C - 25{}^\circ C) = 79 \text{ kJ}$$

To heat the water, we need at least 79 kJ of heat. To determine if the hydrolysis of 70.0 g of CaO can provide that much heat, we can use heats of formation (Appendix 4 and given in the problem) and Equation 5.19 to determine $\Delta H_{rxn}$ and then calculate how much heat 70.0 g of CaO would provide.

$$\Delta H_f^\circ \, CaO(s) = -634.9 \text{ kJ/mol}$$

$$\Delta H_f^\circ \, Ca(OH)(aq) = -1002.82 \text{ kJ/mol}$$

$$\Delta H_f^\circ \, H_2O(\ell) = -285.8 \text{ kJ/mol}$$

$$\Delta H_{rxn}^\circ = [1 \text{ mol } (-1002.82 \text{ kJ/mol})]$$

$$- [1 \text{ mol}(-634.9 \text{ kJ/mol}) + 1 \text{ mol}(-285.8 \text{ kJ/mol})]$$

$$= -82.12 \text{ kJ}$$

The heat given off by the hydrolysis of 1 mole of CaO(s) is $-82.12$ kJ. The container has 70.0 g CaO in it. Converting that to moles and calculating the energy it would give off yields:

$$70.0 \text{ g CaO} \times \frac{1 \text{ mol CaO}}{56.08 \text{ g CaO}} \times \frac{-82.12 \text{ kJ}}{1 \text{ mol CaO}} = -102.5 \text{ kJ}$$

The amount of energy needed to boil the water is 79 kJ, so the reaction of CaO provides enough energy to do this.

c. We can use the titration data to calculate the molar mass of the product to see if it might be $Ca(OH)_2$. The reaction in the titration is

$$Ca(OH)_2(aq) + 2 \, HCl(aq) \rightarrow CaCl_2(aq) + 2 \, H_2O(\ell)$$

The number of moles of base in the solution is

$$0.09430 \text{ L HCl} \times \frac{3.785 \text{ mol HCl}}{1 \text{ L HCl}} \times \frac{1 \text{ mol base}}{2 \text{ mol HCl}} = 0.1785 \text{ mol base}$$

The amount of CaO dissolved was 10.0 g. If the material titrated was $Ca(OH)_2$, we need to consider the mass of $Ca(OH)_2$ produced by the hydrolysis of 10.0 g CaO:

$$10.0 \text{ g CaO} \times \frac{1 \text{ mol CaO}}{56.08 \text{ g CaO}} \times \frac{1 \text{ mol Ca(OH)}_2}{1 \text{ mol CaO}} \times \frac{74.09 \text{ g}}{1 \text{ mol Ca(OH)}_2}$$

$$= 13.21 \text{ g Ca(OH)}_2$$

That means the molar mass of the base is

$$\frac{13.21 \text{ g base}}{0.1785 \text{ mol base}} = 74.1 \text{ g/mol}$$

The molar mass of $Ca(OH)_2$ is 74.1 g/mol, and so the initial suggestion that the material in the FHS is CaO(s) is certainly justified by the calculations.

**Think About It** Simple chemical reactions can supply significant amounts of heat.

In this chapter we have considered the flow of energy and its role in defining the behavior of physical, chemical, and biological processes. The interaction of energy with matter—the transfer of energy into and out of materials—causes phase changes and alters the temperature of matter. The magnitude of these changes and the temperatures and temperature ranges over which they occur are characteristic of the quantity and identity of the matter involved. Energy is also a product or a reactant in virtually all chemical reactions and, as such, behaves stoichiometrically, which means the chemical changes that a specific quantity of material undergoes are characterized by the release or consumption of a specific amount of energy. Understanding the energy contained in fuels is particularly important because fuels provide the bulk of the power needed by the equipment and devices of modern life—from cars and airplanes to computers, air conditioners, and lightbulbs. Just as significant is the energy contained in foods that sustain life. How we deal with the needs for energy in the near future—in terms of both fuel and food—will determine the quality of all our lives and the health of our planet.

# Carbon: Diamonds, Graphite, and the Molecules of Life

The element carbon is central to our existence, both because we are a carbon-based life-form and because carbon is the central element in hydrocarbons—currently our most significant sources of energy and causes of environmental changes. We also value carbon in its crystalline forms (Figure 5.36). One of these is *diamond*; we prize diamonds for their beauty and take advantage of their hardness by using them in drill bits and as abrasives. Another crystalline form of carbon, *graphite*, is black and opaque and prized for its strength and flexibility (graphite fibers are used in golf clubs and tennis rackets) and for its softness and smoothness (flakes are used as lubricants and baked with clay to make the "lead" in pencils). A large variety of specialized noncrystalline materials known as *activated carbon* are synthesized for wastewater treatment, gas purification, and sugar refining (activated carbon decolorizes solutions of raw sugar by selectively binding colored impurities).

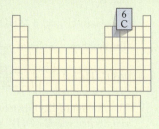

The 1996 Nobel Prize in Chemistry was awarded to Robert Curl, Harold Kroto, and Richard Smalley for their 1985 discovery of a third crystalline form of carbon, buckminsterfullerene. Each molecule consists of a cluster of carbon atoms with the formula $C_{60}$; this molecule is also known as a *buckyball* because it looks like a soccer ball and is reminiscent of the architectural designs of Buckminster Fuller. Buckminsterfullerene is the most abundant form of a class of carbon-clustering molecules called *fullerenes*. Their discovery has spawned research into applications ranging from rocket fuels to drug-delivery systems for treating cancer and AIDS.

Carbon is found in minor amounts in Earth's crust as the free element. Diamonds are found in ancient volcanic pipes (openings in the crust through which lava flowed); the exact way they were formed is still debated. Graphite deposits are found worldwide. Over half of the carbon on Earth is contained in carbonate minerals, such as limestone, dolomite, and chalk, in which carbon is in its highest oxidation state as the $CO_3^{2-}$ ion (O.N. = +4), and in fossil fuels (petroleum, coal, natural gas), in which carbon is in reduced states (O.N. as low as −4). These two carbon reservoirs are interconnected by a dynamic system described in Section 3.5, the *carbon cycle*.

Atmospheric $CO_2$ is the source of all carbon-containing compounds produced in the leaves of green plants by photosynthesis:

$$\text{Sunlight} + 6\ CO_2(g) + 6\ H_2O(\ell) \rightarrow C_6H_{12}O_6(aq) + 6\ O_2(g)$$

The most fundamental carbon-containing product of photosynthesis, the sugar glucose ($C_6H_{12}O_6$), is the main food for biological systems. Glucose is also the primary structural subunit of cellulose, which is the chief component of plant fibers such as wood and cotton. A single molecule of cellulose may contain over 1500 glucose units. Plants produce over 100 billion tons of cellulose each year. The oxidation number of carbon in glucose is 0, so green plants may be thought of as carbon-reducing machines. Other life-forms eat plants and derive large amounts of energy from metabolism through a series of oxidation reactions that are the reverse of photosynthesis, which takes carbon at O.N. = 0 (glucose) back to O.N. +4 (carbon dioxide). Life-forms are carbon-oxidizing machines.

When plants and animals die, their carbon-containing molecules are oxidized to $CO_2$ and water, if decomposition is complete. However,

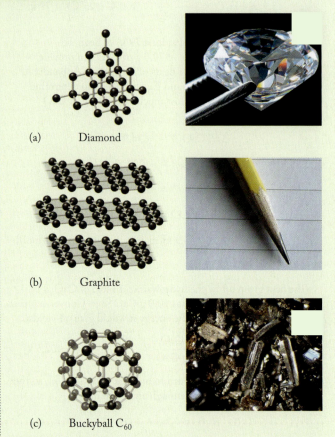

(a)    Diamond

(b)    Graphite

(c)    Buckyball $C_{60}$

**FIGURE 5.36** Three crystalline forms of carbon: (a) diamond, (b) graphite, and (c) buckminsterfullerene (a "buckyball").

plant material deprived of oxygen may be further reduced by bacteria and end up as petroleum and coal. When these fuels react with oxygen in a combustion reaction, energy is released because reduced carbon is oxidized when the hydrocarbons are converted into $CO_2$.

Some of the $CO_2$ removed from the atmosphere in photosynthesis is returned to the atmosphere as a result of respiration by animals and plants, but our burning of fossil fuels inserts additional carbon into the atmosphere as $CO_2$ that was formerly sequestered underground. Atmospheric $CO_2$ dissolves in seawater to form carbonic acid, which increases the acidity of the ocean. Changing the acidity affects the solubility of $CaCO_3$ in seawater, which in turn has potentially disastrous consequences for marine life, especially the many oceanic life-forms that build shells, reefs, and other structures from $CaCO_3$ and other carbonates.

Over 16 million compounds of carbon are known, and over 90% of the thousands of new materials synthesized each year contain carbon. The chemistry of carbon is the chemistry of pharmaceuticals, soft contact lenses, paints, synthetic fibers, plastics, and gasoline. It is the chemistry of the food we eat and the fuel that powers our cars and airplanes. Fossil fuels are the primary source of molecules that industry modifies to make the items we regard as part of modern life, so they can be regarded as fuel for the manufacturing industry as well as for transportation. Combined in precise ways with a small number of other elements—such as sulfur, phosphorus, oxygen, nitrogen, and hydrogen—carbon serves as the foundation of life.

## SUMMARY

**Learning Outcome 1** **Potential energy (PE)** is the energy of position or composition and is a **state function**. **Kinetic energy (KE)** is the energy of motion. Heating a sample increases the average kinetic energy of the atoms in the sample. Energy is stored in compounds, and energy is absorbed or released when they are transformed into different compounds or when a change of state occurs. (Section 5.1)

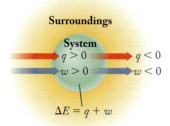

**Learning Outcome 2** In an **exothermic process** the system loses energy by heating its surroundings ($q < 0$); in an **endothermic process** the system absorbs energy ($q > 0$) from its surroundings. (Section 5.2)

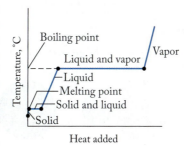

**Learning Outcome 3** The sum of the kinetic and potential energies of a system is called its **internal energy (E)**. The internal energy of a system is increased ($\Delta E = E_{final} - E_{initial}$ is positive) when it is heated ($q > 0$) or if work is done on it ($w > 0$). When the pressure on a system is constant but the volume of the system changes, the work done is called **pressure–volume (P–V) work**. (Section 5.2)

**Learning Outcome 4** The **enthalpy (H)** of a system is given by $H = E + PV$. The **enthalpy change ($\Delta H$)** of a system is equal to the energy in the form of heat ($q_P$) added to or removed from the system at constant pressure: $\Delta H > 0$ for endothermic reactions and $\Delta H < 0$ for exothermic reactions.

Thermodynamic values like **heat capacity**, **molar heat capacity**, and **specific heat** can be used to quantify changes in systems involved in physical and chemical changes. (Sections 5.3 and 5.4)

**Learning Outcome 5** A calorimeter, characterized by its **calorimeter constant** (heat capacity), is a device used to measure the amount of energy involved in physical and chemical processes. The enthalpy change associated with a reaction is defined by the **enthalpy of reaction ($\Delta H_{rxn}$)** or the heat of reaction. (Section 5.5)

**Learning Outcome 6** **Hess's law** states that the enthalpy of a reaction ($\Delta H_{rxn}$) that is the sum of two or more other reactions is equal to the sum of the $\Delta H_{rxn}$ values of the constituent reactions. It can be used to calculate enthalpy changes in reactions that are hard or impossible to measure directly. (Section 5.6)

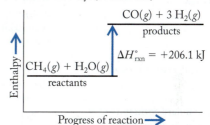

**Learning Outcome 7** The **standard enthalpy of formation** of a substance, $\Delta H_f^\circ$, is the amount of energy involved in a **formation reaction**, in which 1 mole of the substance is made from its constituent elements in their **standard states** (under **standard conditions**). Enthalpy changes for physical changes and chemical reactions can be calculated from the enthalpies of formation of the reactants and products. (Section 5.7)

**Learning Outcome 8** **Fuel value** is the amount of energy released on complete combustion of 1 g of a fuel. **Food value** is the amount of energy released when a material consumed by an organism for sustenance is burned completely; nutritionists often express food values in Calories (kilocalories) rather than in the SI unit kilojoules. (Section 5.8)

## PROBLEM-SOLVING SUMMARY

| TYPE OF PROBLEM | CONCEPTS AND EQUATIONS | | SAMPLE EXERCISES |
|---|---|---|---|
| **Kinetic and potential energy** | $KE = \frac{1}{2}mu^2 \qquad E_{el} \propto \dfrac{Q_1 \times Q_2}{d}$ | (5.3) | 5.1 |
| **Identifying endothermic and exothermic processes, and calculating internal energy change ($\Delta E$) and P–V work** | For the system:<br>$\Delta E = q + w$<br>where $w = -P\Delta V$. | (5.5) | 5.2, 5.3, 5.4 |
| **Predicting the sign of $\Delta H_{sys}$ for physical and chemical changes** | Exothermic, $\Delta H_{sys} < 0$<br>Endothermic, $\Delta H_{sys} > 0$ | | 5.5, 5.6 |

| TYPE OF PROBLEM | CONCEPTS AND EQUATIONS | SAMPLE EXERCISES |
|---|---|---|
| Determining the flow of energy ($q$) associated with a change of state or with changing the temperature of a substance | Melting a solid at its melting point:<br><br>$$q = n\Delta H_{fus} \qquad (5.9)$$<br><br>vaporizing a liquid at its boiling point:<br><br>$$q = n\Delta H_{vap} \qquad (5.10)$$<br><br>or heating a substance:<br><br>$$q = nc_P\Delta T \qquad (5.8)$$ | 5.7, 5.8, 5.9 |
| Measuring the heat of reaction | $$n_{sys}c_{P,sys}\,\Delta T = q_{gained,sys} = -q_{lost,surr} = -\left(n_{surr}c_{P,surr}\,\Delta T\right)$$ | 5.10 |
| Measuring the heat capacity (calorimeter constant) of a calorimeter | $$C_{calorimeter} = q/\Delta T \qquad (5.13)$$<br><br>where $C_{calorimeter}$ is the heat capacity of the calorimeter, $q$ is the heat released by a standard combustion reaction, and $\Delta T$ is the temperature change of the calorimeter. | 5.11 |
| Using Hess's law | Reorganize the information so that the reactions add together as desired. Reversing a reaction changes the sign of the reaction's $\Delta H_{rxn}$ value. Multiplying the coefficients in a reaction by a factor requires that the reaction's $\Delta H_{rxn}$ value be multiplied by the same factor. | 5.12, 5.13 |
| Recognizing and writing formation reactions | In a formation reaction, the reactants are elements in their standard states and the product is 1 mole of a single compound. | 5.14 |
| Calculating standard enthalpies of reaction from heats of formation | $$\Delta H^{\circ}_{rxn} = \sum n_{products}\,\Delta H^{\circ}_{f,products} - \sum n_{reactants}\,\Delta H^{\circ}_{f,reactants} \qquad (5.19)$$ | 5.15, 5.16 |
| Calculating fuel value and food value | The fuel value or food value of a substance is the energy released by the complete combustion of 1 g of the substance. | 5.17, 5.18 |

## VISUAL PROBLEMS

*(Answers to boldface end-of-chapter questions and problems are in the back of the book.)*

**5.1.** A brick lies perilously close to the edge of the flat roof of a building (Figure P5.1). The roof edge is 50 ft above street level, and the brick has 500 J of potential energy with respect to street level. Someone edges the brick off the roof, and it begins to fall. What is the brick's kinetic energy when it is 35 ft above street level? What is its kinetic energy the instant before it hits the street surface?

**FIGURE P5.1**

**5.2.** Figure P5.2 shows pairs of cations and anions in contact. The energy of each interaction is proportional to $Q_1Q_2/d$, where $Q_1$ and $Q_2$ are the charges on the cation and anion, and $d$ is the distance between their nuclei. Which pair has the greatest interaction energy and which has the smallest?

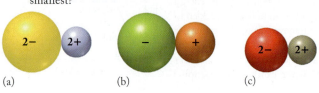

**FIGURE P5.2**

**5.3.** Figure P5.3 shows a sectional view of an assembly, consisting of a gas trapped inside a stainless steel cylinder with a stainless steel piston and a block of iron on top of the piston to keep it in place.

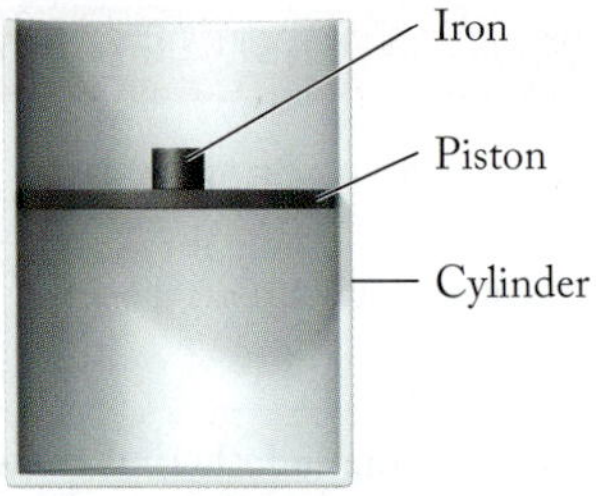

a. Sketch the situation after the assembly has been heated for a few seconds with a blowtorch.

b. Is the piston higher or lower in the cylinder?

c. Has heat ($q$) been added to the system?

d. Has the system done work ($w$) on its surroundings, or have the surroundings done work on the system?

**FIGURE P5.3**

5.4. Figure P5.4 represents the energy change in a chemical reaction. The reaction results in a very small decrease in volume.
   a. Has the internal energy of the system decreased or increased as a result of the reaction?
   b. What sign is given to $\Delta E$ for the reaction system?
   c. Is the reaction endothermic or exothermic?

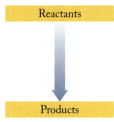

**FIGURE P5.4**

5.5. The closed, rigid, metal box lying on the wooden kitchen table in Figure P5.5 is about to be involved in an accident.
   a. Are the contents of the box an isolated system, a closed system, or an open system?
   b. What will happen to the internal energy of the system if the table catches fire and burns?
   c. Will the system do any work on the surroundings while the fire is burning or after the table has burned away and collapsed?

**FIGURE P5.5**

5.6. The diagram in Figure P5.6 shows how a chemical reaction in a cylinder with a piston affects the volume of the system.
   a. In this reaction, does the system do work on the surroundings?
   b. If the reaction is endothermic, does the internal energy of the system increase or decrease when the reaction is proceeding?

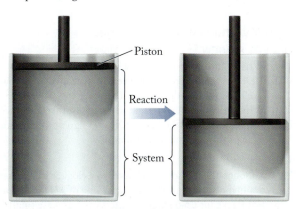

**FIGURE P5.6**

5.7. The enthalpy diagram in Figure P5.7 indicates the enthalpies of formation of four compounds made from the elements listed on the "zero" line of the vertical axis.
   a. Why are the elements all put on the same horizontal line?
   b. Why is $C_2H_2(g)$ sometimes called an "endothermic" compound?
   c. How could the data be used to calculate the heat of the reaction that converts a stoichiometric mixture of $C_2H_2(g)$ and $O_2(g)$ to $CO(g)$ and $H_2O(g)$?

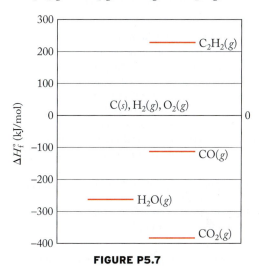

**FIGURE P5.7**

5.8. The process illustrated in Figure P5.8 takes place at constant pressure. (The figure is not drawn to scale; the molecules are actually much, much smaller relative to the container.)
   a. Write a balanced equation for the process.
   b. Is $w$ positive, negative, or zero for this reaction?
   c. Using data from Appendix 4, calculate $\Delta H^{\circ}_{rxn}$ for the formation of 1 mole of the product.

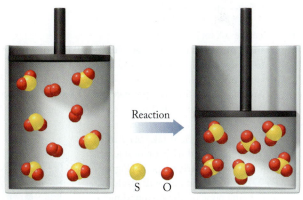

**FIGURE P5.8**

## QUESTIONS AND PROBLEMS

### Energy: Basic Concepts and Definitions

**CONCEPT REVIEW**

5.9. How are energy and work related?

5.10. Explain the difference between potential energy and kinetic energy.

5.11. Explain what is meant by a state function.

5.12. Are kinetic energy and potential energy both state functions?

5.13. Explain the nature of the potential energy in the following: (a) the new battery for your remote control; (b) a gallon of gasoline; (c) the crest of a wave before it crashes onto shore.

5.14. Explain the kinetic energy in a stationary ice cube.

## Systems, Surroundings, and Energy Transfer

### CONCEPT REVIEW

**5.15.** What is meant by the terms *system* and *surroundings*?

**5.16.** Do all exothermic processes that do work on the surroundings have the same sign for $\Delta E$?

**5.17.** Why don't all endothermic processes that do work on the surroundings have the same sign for $\Delta E$?

**5.18.** From the perspective of thermodynamics, why is the water in a pond warmer in August than in April?

### PROBLEMS

**5.19.** Which of the following processes are exothermic, and which are endothermic? (a) a candle burns; (b) rubbing alcohol feels cold on the skin; (c) a supersaturated solution crystallizes, causing the temperature of the solution to rise

**5.20.** Which of the following processes are exothermic, and which are endothermic? (a) frost forms on a car window in the winter; (b) water condenses on a glass of ice water on a humid summer afternoon; (c) adding ammonium nitrate to water causes the temperature of the solution to decrease

**5.21.** What happens to the internal energy of a liquid at its boiling point when it vaporizes?

**5.22.** What happens to the internal energy of a gas when it expands (with no heat flow)?

**5.23.** How much $P–V$ work does a gas system do on its surroundings at a constant pressure of 1.00 atm if the volume of gas triples from 250.0 mL to 750.0 mL? Express your answer in L · atm and joules (J).

**5.24.** An expanding gas does 150.0 J of work on its surroundings at a constant pressure of 1.01 atm. If the gas initially occupied 68 mL, what is the final volume of the gas?

**5.25.** Calculate $\Delta E$ for the following situations:
 a. $q = 120.0$ J; $w = -40.0$ J
 b. $q = 9.2$ kJ; $w = 0.70$ J
 c. $q = -625$ J; $w = -315$ J

**5.26.** Calculate $\Delta E$ for
 a. the combustion of a gas that releases 210.0 kJ of heat to its surroundings and does 65.5 kJ of work on its surroundings.
 b. a chemical reaction that produces 90.7 kJ of heat but does no work on its surroundings.

**\*5.27.** The following reactions take place in a cylinder equipped with a movable piston at atmospheric pressure (Figure P5.27). Which reactions will result in work being done on the surroundings? Assume the system returns to an initial temperature of 110°C. (*Hint*: The volume of a gas is proportional to number of moles ($n$) at constant temperature and pressure.)
 a. $CH_4(g) + 2\,O_2(g) \rightarrow CO_2(g) + 2\,H_2O(g)$
 b. $C_3H_8(g) + 5\,O_2(g) \rightarrow 3\,CO_2(g) + 4\,H_2O(g)$
 c. $N_2(g) + 2\,O_2(g) \rightarrow 2\,NO_2(g)$

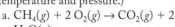

**FIGURE P5.27**

**\*5.28.** In which direction will the piston shown in Figure P5.27 move when the following reactions are carried out at atmospheric pressure inside the cylinder and after the system has returned to its initial temperature of 110°C? (*Hint*: The volume of a gas is proportional to number of moles ($n$) at constant temperature and pressure.)
 a. $N_2(g) + 3\,H_2(g) \rightarrow 2\,NH_3(g)$
 b. $C(s) + O_2(g) \rightarrow CO_2(g)$
 c. $CH_3CH_2OH(g) + 3\,O_2(g) \rightarrow 2\,CO_2(g) + 3\,H_2O(g)$

**\*5.29.** Automobile airbags produce nitrogen gas based on the reaction:

$$2\,NaN_3(s) \rightarrow 2\,Na(\ell) + 3\,N_2(g)$$

 a. If 2.25 g of $NaN_3$ react to fill an air bag, how much $P–V$ work will the $N_2$ do against an external pressure of 1.00 atm given that the density of nitrogen is 1.165 g/L at 20°C?
 b. If the process releases 2.34 kJ of heat, what is $\Delta E$ for the system?

**\*5.30.** Igniting gunpowder produces nitrogen and carbon dioxide gas that propels the bullet by the reaction:

$$2\,KNO_3(s) + \tfrac{1}{8}\,S_8(s) + 3\,C(s) \rightarrow K_2S(s) + N_2(g) + 3\,CO_2(g)$$

 a. If 1.00 g of $KNO_3$ reacts, how much $P–V$ work will the gases do against an external pressure of 1.00 atm given the densities of nitrogen and $CO_2$ are 1.165 g/L and 1.830 g/L, respectively, at 20°C?
 b. If the reaction produces 21.6 kJ of heat, what is $\Delta E$ for the system?

## Enthalpy and Enthalpy Changes

### CONCEPT REVIEW

**5.31.** What is meant by an *enthalpy change*?

**5.32.** Describe the difference between an internal energy change ($\Delta E$) and an enthalpy change ($\Delta H$).

**5.33.** Why is the sign of $\Delta H$ negative for an exothermic process?

**5.34.** What happens to the magnitude and sign of the enthalpy change when a process is reversed?

### PROBLEMS

**5.35.** **A Clogged Sink** Adding Drano® to a clogged sink causes the drainpipe to get warm. What is the sign of $\Delta H$ for this process?

**5.36.** **Hot Packs for Cold Fingers** Skiers can purchase small packets of finely divided iron that, after reaction with water, provide heat to frozen fingers and toes (Figure P5.36). Define the system and the surroundings in this example, and indicate the sign of $\Delta H_{sys}$.

(a)

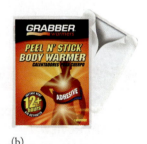

(b)

**FIGURE P5.36**

**5.37. Break a Bond** The stable form of oxygen at room temperature and pressure is the diatomic molecule $O_2$. What is the sign of $\Delta H$ for the following process?

$$O_2(g) \rightarrow 2\ O(g)$$

**5.38. Plaster of Paris** Gypsum is the common name of calcium sulfate dihydrate ($CaSO_4 \cdot 2\ H_2O$). When gypsum is heated to 150°C, it loses most of the water in its formula and forms plaster of Paris ($CaSO_4 \cdot 0.5\ H_2O$):

$$2\ CaSO_4 \cdot 2\ H_2O(s) \rightarrow 2\ CaSO_4 \cdot 0.5\ H_2O(s) + 3\ H_2O(g)$$

What is the sign of $\Delta H$ for making plaster of Paris from gypsum?

**5.39. Metallic Hydrogen** A solid with metallic properties is formed when hydrogen gas is compressed under extremely high pressures. Predict the sign of the enthalpy change for the following reaction:

$$H_2(g) \rightarrow H_2(s)$$

**5.40. Kitchen Chemistry** A simple "kitchen chemistry" experiment requires placing some vinegar in a soda bottle. A deflated balloon containing baking soda is stretched over the mouth of the bottle. Adding the baking soda to the vinegar starts the following reaction and inflates the balloon:

$$NaHCO_3(aq) + CH_3COOH(aq) \rightarrow$$
$$CH_3COONa(aq) + CO_2(g) + H_2O(\ell)$$

If the contents of the bottle are considered the system, is work being done on the surroundings or on the system?

## Heating Curves and Heat Capacity

### CONCEPT REVIEW

**5.41.** What is the difference between *specific heat* and *heat capacity*?

**5.42.** What happens to the heat capacity of a material if its mass is doubled? Is the same true for the specific heat?

**5.43.** Are the heats of fusion and vaporization of a given substance usually the same?

**5.44.** An equal amount of heat is added to pieces of metal A and metal B having the same mass. Does the metal with the larger heat capacity reach the higher temperature?

**\*5.45. Cooling an Automobile Engine** Most automobile engines are cooled by water circulating through them and a radiator. However, the original Volkswagen Beetle had an air-cooled engine. Why might car designers choose water cooling over air cooling?

**\*5.46. Nuclear Reactor Coolants** The reactor-core cooling systems in some nuclear power plants use liquid sodium as the coolant. Sodium has a thermal conductivity of 1.42 J/(cm · s · K), which is quite high compared with that of water [$6.1 \times 10^{-3}$ J/(cm · s · K)]. The respective molar heat capacities are 28.28 J/(mol · K) and 75.31 J/(mol · K). What is the advantage of using liquid sodium over water in this application?

### PROBLEMS

**5.47.** How much heat is needed to raise the temperature of 100.0 g of water from 30.0°C to 100.0°C?

**5.48.** At an elevation where the boiling point of water is 93°C, 100.0 g of water at 30°C absorbs 290.0 kJ of heat from a mountain climber's stove. Is this amount of energy sufficient to heat the water to its boiling point?

**5.49.** Use the following data to sketch a heating curve for 1 mole of methanol. Start the curve at −100°C and end it at 100°C.

| | |
|---|---|
| Boiling point | 65°C |
| Melting point | −94°C |
| $\Delta H_{vap}$ | 37 kJ/mol |
| $\Delta H_{fus}$ | 3.18 kJ/mol |
| Molar heat capacity ($\ell$) | 81.1 J/(mol · °C) |
| ($g$) | 43.9 J/(mol · °C) |
| ($s$) | 48.7 J/(mol · °C) |

**5.50.** Use the following data to sketch a heating curve for 1 mole of octane. Start the curve at −57°C and end it at 150°C.

| | |
|---|---|
| Boiling point | 125.7°C |
| Melting point | −56.8°C |
| $\Delta H_{vap}$ | 41.5 kJ/mol |
| $\Delta H_{fus}$ | 20.7 kJ/mol |
| Molar heat capacity ($\ell$) | 254.6 J/(mol · °C) |
| ($g$) | 316.9 J/(mol · °C) |

**5.51. Keeping an Athlete Cool** During a strenuous workout, an athlete generates 2000.0 kJ of heat energy. What mass of water would have to evaporate from the athlete's skin to dissipate this much heat?

**5.52.** Damp clothes can prove fatal when outdoor temperatures drop (hypothermia).
   a. If the clothes you are wearing absorb 1.00 kg of water and then dry in a cold wind on Mount Washington, how much heat would your body lose during this process? $c_{P,H_2O} = 75.3$ J/(mol · °C), and $\Delta H_{vap,H_2O} = 40.67$ kJ/mol.
   \*b. If the heat lost by your body is not replaced, what will the final temperature of your body be after the 1.00 kg of water has evaporated? Use your own body weight and assume that the specific heat of your body is 4.25 J/g. (Note that the specific heat is given in units of J/g.)

**5.53.** The same quantity of energy is added to 10.00 g pieces of gold, magnesium, and platinum, all initially at 25°C. The molar heat capacities of these three metals are 25.41 J/(mol · °C), 24.79 J/(mol · °C), and 25.95 J/(mol · °C), respectively. Which piece of metal has the highest final temperature?

**5.54.** Which of the following would reach the higher temperature after 10.00 g of iron [$c_P = 25.1$ J/(mol · °C)] at 150°C is added: 100 mL of water [$d = 1.00$ g/mL, $c_P = 75.3$ J/(mol · °C)] or 200 mL of ethanol [$CH_3CH_2OH$, $d = 0.789$ g/mL, $c_P = 113.1$ J/(mol · °C)]?

\*5.55. Exactly 10.0 mL of water at 25.0°C is added to a hot iron skillet. All of the water is converted into steam at 100.0°C. The mass of the pan is 1.20 kg and the molar heat capacity of iron is 25.19 J/(mol · °C). What is the temperature change of the skillet?

\*5.56. A 20.0 g piece of iron and a 20.0 g piece of gold at 100.0°C were dropped into the same 1.00 L of water at 20.0°C. The molar heat capacities of iron and gold are 25.19 J/(mol · °C) and 25.41 J/(mol · °C), respectively. What is the final temperature of the water and pieces of metal?

## Calorimetry: Measuring Heat Capacity and Enthalpies of Reaction

### CONCEPT REVIEW

5.57. Why is it necessary to know the heat capacity of a calorimeter?

5.58. Could an endothermic reaction be used to measure the heat capacity of a calorimeter?

5.59. If we replace the water in a bomb calorimeter with another liquid, do we need to redetermine the heat capacity of the calorimeter?

5.60. When measuring the heat of combustion of a very small amount of material, would you prefer to use a calorimeter having a heat capacity that is small or large?

### PROBLEMS

5.61. If you were designing a chemical cold pack, which of the following salts would you choose to provide the greatest drop in temperature per gram: $NH_4Cl$ ($\Delta H_{soln}$ = 14.6 kJ/mol), $NH_4NO_3$ ($\Delta H_{soln}$ = 25.7 kJ/mol), or $NaNO_3$ ($\Delta H_{soln}$ = 20.4 kJ/mol)?

5.62. Dissolving calcium hydroxide ($\Delta H_{soln}$ = −16.2 kJ/mol) in water is an exothermic process. How much lithium hydroxide ($\Delta H_{soln}$ = −23.6 kJ/mol) is needed to deliver the same amount of heat as 15.00 g of $Ca(OH)_2$?

5.63. Calculate the heat capacity of a calorimeter if the combustion of 5.000 g of benzoic acid led to a temperature increase of 16.397°C.

5.64. Calculate the heat capacity of a calorimeter if the combustion of 4.663 g of benzoic acid led to an increase in temperature of 7.149°C.

5.65. The complete combustion of 1.200 g of cinnamaldehyde ($C_9H_8O$, one of the compounds in cinnamon) in a bomb calorimeter ($C_{calorimeter}$ = 3.640 kJ/°C) produced an increase in temperature of 12.79°C. Calculate the molar enthalpy of combustion of cinnamaldehyde ($\Delta H_{comb}$) in kilojoules per mole of cinnamaldehyde.

5.66. **Aromatic Spice** The aromatic hydrocarbon cymene ($C_{10}H_{14}$) is found in nearly 100 spices and fragrances including coriander, anise, and thyme. The complete combustion of 1.608 g of cymene in a bomb calorimeter ($C_{calorimeter}$ = 3.640 kJ/°C) produced an increase in temperature of 19.35°C. Calculate the molar enthalpy of combustion of cymene ($\Delta H_{comb}$) in kilojoules per mole of cymene.

5.67. **Hormone Mimics** Phthalates, used to make plastics flexible, are among the most abundant industrial contaminants in the environment. Several have been shown to act as hormone mimics in humans by activating the receptors for estrogen, a female sex hormone. In characterizing the compounds completely, the value of $\Delta H_{comb}$ for dimethyl phthalate ($C_{10}H_{10}O_4$) was determined to be −4685 kJ/mol. Assume that 1.00 g of dimethyl phthalate is combusted in a calorimeter whose heat capacity ($C_{calorimeter}$) is 7.854 kJ/°C at 20.215°C. What is the final temperature of the calorimeter?

5.68. **Flavorings** The flavor of anise is due to anethole, a compound with the molecular formula $C_{10}H_{12}O$. The $\Delta H_{comb}$ value for anethole is −5541 kJ/mol. Assume 0.950 g of anethole is combusted in a calorimeter whose heat capacity ($C_{calorimeter}$) is 7.854 kJ/°C at 20.611°C. What is the final temperature of the calorimeter?

5.69. Adding 2.00 g Mg metal to 95.0 mL 1.00 *M* HCl in a coffee-cup calorimeter leads to an increase of 9.2°C.
   a. Write a balanced net ionic equation for the reaction.
   b. If the molar heat capacity of 1.00 *M* HCl is the same as that for water [$c_P$ = 75.3 J/(mol · °C)], what is $\Delta H_{rxn}$?

5.70. What is the $\Delta H_{rxn}$ for the precipitation of AgCl if adding 125 mL of 1.00 *M* $AgNO_3$ to 125 mL of 1.00 *M* NaCl at 18.6°C causes the temperature to increase to 26.4°C? (Assume $c_{P,soln}$ = $c_{P,water}$.)

5.71. How much glucose must be metabolized to completely evaporate 1.00 g of water at 37°C, given $\Delta H_{comb,glucose}$ = −2803 kJ/mol?

5.72. If all the energy obtained from burning of 275 g propane ($\Delta H_{comb,C_3H_8}$ = −2220 kJ/mol) is used to heat water, how many liters of water can be heated from 20.0°C to 100.0°C?

## Hess's Law

### CONCEPT REVIEW

5.73. How is Hess's law consistent with the law of conservation of energy?

5.74. Why is it important for Hess's law that enthalpy is a state function?

### PROBLEMS

5.75. How can the first two of the following reactions be combined to obtain the third reaction?
   a. $CO(g) + NH_3(g) \rightarrow HCN(g) + H_2O(g)$
   b. $CO(g) + 3 H_2(g) \rightarrow CH_4(g) + H_2O(g)$
   c. $CH_4(g) + NH_3(g) \rightarrow HCN(g) + 3 H_2(g)$

5.76. **Cleansing the Atmosphere** The atmosphere contains the highly reactive molecule, OH, which acts to remove selected pollutants. Use the values for $\Delta H_{rxn}$ given below to find the $\Delta H_{rxn}$ for the formation of OH and H from water.

$$\tfrac{1}{2} H_2(g) + \tfrac{1}{2} O_2(g) \rightarrow OH(g) \qquad \Delta H_{rxn} = 42.1 \text{ kJ}$$

$$H_2(g) \rightarrow 2 H(g) \qquad \Delta H_{rxn} = 435.9 \text{ kJ}$$

$$H_2(g) + \tfrac{1}{2} O_2(g) \rightarrow H_2O(g) \qquad \Delta H_{rxn} = -241.8 \text{ kJ}$$

$$H_2O(g) \rightarrow H(g) + OH(g) \qquad \Delta H_{rxn} = ?$$

**5.77. Ozone Layer** The destruction of the ozone layer by chlorofluorocarbons (CFCs) can be described by the following reactions:

$$ClO(g) + O_3(g) \rightarrow Cl(g) + 2\,O_2(g) \qquad \Delta H_{rxn} = -29.90 \text{ kJ}$$
$$2\,O_3(g) \rightarrow 3\,O_2(g) \qquad \Delta H_{rxn} = 24.18 \text{ kJ}$$

Determine the value of heat of reaction for the following:

$$Cl(g) + O_3(g) \rightarrow ClO(g) + O_2(g) \qquad \Delta H_{rxn} = ?$$

**5.78.** What is $\Delta H_{rxn}$ for the reaction between $H_2S$ and $O_2$ that yields $SO_2$ and water,

$$2\,H_2S(g) + 3\,O_2(g) \rightarrow 2\,SO_2(g) + 2\,H_2O(g) \qquad \Delta H_{rxn} = ?$$

given $\Delta H_{rxn}$ for the following reactions?

$$H_2(g) + \tfrac{1}{2}O_2(g) \rightarrow H_2O(g) \qquad \Delta H_{rxn} = -241.8 \text{ kJ}$$
$$SO_2(g) + 3\,H_2(g) \rightarrow 2\,H_2S(g) + H_2O(g) \qquad \Delta H_{rxn} = 34.8 \text{ kJ}$$

## Standard Enthalpies of Formation and Reaction

### CONCEPT REVIEW

**5.79.** Explain how the use of $\Delta H_f^\circ$ to calculate $\Delta H_{rxn}^\circ$ is an example of Hess's law.

**5.80.** Why is the standard enthalpy of formation of $CO(g)$ difficult to measure experimentally?

**5.81.** Oxygen and ozone are both forms of elemental oxygen. Are the standard enthalpies of formation of oxygen and ozone the same?

**5.82.** Why are the standard enthalpies of formation of elements in their standard states assigned a value of zero?

### PROBLEMS

**5.83.** For which of the following reactions does $\Delta H_{rxn}^\circ$ represent an enthalpy of formation?
a. $C(s) + O_2(g) \rightarrow CO_2(g)$
b. $CO_2(g) + C(s) \rightarrow 2\,CO(g)$
c. $CO_2(g) + H_2(g) \rightarrow H_2O(g) + CO(g)$
d. $2\,H_2(g) + C(s) \rightarrow CH_4(g)$

**5.84.** For which of the following reactions does $\Delta H_{rxn}^\circ$ also represent an enthalpy of formation?
a. $2\,N_2(g) + 3\,O_2(g) \rightarrow 2\,NO_2(g) + 2\,NO(g)$
b. $N_2(g) + O_2(g) \rightarrow 2\,NO(g)$
c. $2\,NO_2(g) \rightarrow N_2O_4(g)$
d. $N_2(g) + 2\,O_2(g) \rightarrow 2\,NO_2(g)$

**5.85. Methanogenesis** Use standard enthalpies of formation from Appendix 4 to calculate the standard enthalpy of reaction for the following methane-generating reaction of methanogenic bacteria:

$$4\,H_2(g) + CO_2(g) \rightarrow CH_4(g) + 2\,H_2O(\ell)$$

**5.86.** Use standard enthalpies of formation from Appendix 4 to calculate the standard enthalpy of reaction for the following methane-generating reaction of methanogenic bacteria, given $\Delta H_f^\circ$ of $CH_3NH_2(g) = -22.97$ kJ/mol:

$$4\,CH_3NH_2(g) + 2\,H_2O(\ell) \rightarrow 3\,CH_4(g) + CO_2(g) + 4\,NH_3(g)$$

**5.87.** Ammonium nitrate decomposes to $N_2O$ and water vapor at temperatures between 250°C and 300°C. Write a balanced chemical reaction describing the decomposition of ammonium nitrate, and calculate the heat of reaction using the appropriate enthalpies of formation from Appendix 4.

**5.88. Military Explosives** Explosives called amatols are mixtures of ammonium nitrate and TNT introduced during World War I when TNT was in short supply. The mixtures can provide 30% more explosive power than TNT alone. Above 300°C, ammonium nitrate decomposes to $N_2$, $O_2$, and $H_2O$. Write a balanced chemical reaction describing the decomposition of ammonium nitrate, and determine the standard heat of reaction by using the appropriate standard enthalpies of formation from Appendix 4.

**5.89. Explosives** Mixtures of fertilizer (ammonium nitrate) and fuel oil (a mixture of long-chain hydrocarbons similar to decane, $C_{10}H_{22}$) are the basis for a powerful explosion. Determine the enthalpy change of the following explosive reaction by using the appropriate enthalpies of formation $\Delta H_{f,C_{10}H_{22}}^\circ = 249.7$ kJ/mol):

$$3\,NH_4NO_3(s) + C_{10}H_{22}(\ell) + 14\,O_2(g) \rightarrow$$
$$3\,N_2(g) + 17\,H_2O(g) + 10\,CO_2(g)$$

**\*5.90. A Little TNT** Trinitrotoluene (TNT) is a highly explosive compound. The thermal decomposition of TNT is described by the following chemical equation:

$$2\,C_7H_5N_3O_6(s) \rightarrow 12\,CO(g) + 5\,H_2(g) + 3\,N_2(g) + 2\,C(s)$$

If $\Delta H_{rxn}$ for this reaction is $-10{,}153$ kJ/mol, how much TNT is needed to equal the explosive power of 1 mole of ammonium nitrate (Problem 5.89)?

**5.91.** How can the standard enthalpy of formation of $CO(g)$ be calculated from the standard enthalpy of formation $\Delta H_f^\circ$ of $CO_2(g)$ and the standard heat of combustion $\Delta H_{comb}^\circ$ of $CO(g)$?

**5.92.** Calculate the standard enthalpy of formation of $SO_2(g)$ from the standard enthalpy changes of the following reactions:

$$2\,SO_2(g) + O_2(g) \rightarrow 2\,SO_3(g) \qquad \Delta H_{rxn}^\circ = -196 \text{ kJ}$$
$$\tfrac{1}{4}S_8(s) + 3\,O_2(g) \rightarrow 2\,SO_3(g) \qquad \Delta H_{rxn}^\circ = -790 \text{ kJ}$$
$$\tfrac{1}{8}S_8(s) + O_2(g) \rightarrow SO_2(g) \qquad \Delta H_f^\circ = ?$$

**5.93.** Use the following data to calculate the heat of formation of $NO_2Cl$ from $N_2$, $O_2$, and $Cl_2$:

$$NO_2Cl(g) \rightarrow NO_2(g) + \tfrac{1}{2}Cl_2(g) \quad \Delta H_{rxn}^\circ = +20.6 \text{ kJ}$$
$$\tfrac{1}{2}N_2(g) + O_2(g) \rightarrow NO_2(g) \qquad \Delta H_f^\circ = +33.2 \text{ kJ}$$
$$\tfrac{1}{2}N_2(g) + O_2(g) + \tfrac{1}{2}Cl_2(g) \rightarrow NO_2Cl(g) \qquad \Delta H_f^\circ = ?$$

**5.94. Formation of CO₂** Baking soda decomposes on heating as follows, creating the holes in baked bread:

$$2\,NaHCO_3(s) \rightarrow Na_2CO_3(s) + CO_2(g) + H_2O(\ell)$$

Calculate the standard enthalpy of formation of $NaHCO_3(s)$ from the following information:

$$\Delta H_{rxn}^\circ = -129.3 \text{ kJ} \qquad \Delta H_f^\circ[Na_2CO_3(s)] = -1131 \text{ kJ/mol}$$
$$\Delta H_f^\circ[CO_2(g)] = -394 \text{ kJ/mol} \qquad \Delta H_f^\circ[H_2O(\ell)] = -286 \text{ kJ/mol}$$

## Fuel Values and Food Values

### CONCEPT REVIEW

**5.95.** What is meant by *fuel value*?

**5.96.** What are the units of fuel values?

**5.97.** How are fuel values calculated from molar heats of combustion?

**5.98.** Is the fuel value of liquid propane the same as that of propane gas?

### PROBLEMS

**5.99. Flex Fuel Vehicles** An increasing number of vehicles in the United States can run on either gasoline [$C_9H_{20}(\ell)$, $\Delta H_{comb,gasoline} = -6160$ kJ/mol] or ethanol [$CH_3CH_2OH(\ell)$]. Which fuel has the greater fuel value?

**5.100. Arctic Exploration** Food contains three main categories of compounds: carbohydrate, protein, and fat. Arctic explorers often eat a high-fat diet because fats have a high food value. The average food values for carbohydrate, protein, and fat are 4 Cal/g, 4 Cal/g, and 9 Cal/g, respectively. Consider a typical fat to have the chemical formula $C_{18}H_{36}O_2$, glucose ($C_6H_{12}O_6$) to be a representative carbohydrate, and the amino acid alanine ($C_3H_6NO_2$) a building block of proteins. Express the food values given above as $\Delta H_{comb}$ in kJ/mol.

---

**5.101. The Joys of Camping** Lightweight camping stoves typically use *white gas*, a mixture of $C_5$ and $C_6$ hydrocarbons.
  a. Calculate the fuel value of $C_5H_{12}$, given that $\Delta H°_{comb} = -3535$ kJ/mol.
  b. How much heat is released during the combustion of 1.00 kg of $C_5H_{12}$?
  c. How many grams of $C_5H_{12}$ must be burned to heat 1.00 kg of water from 20.0°C to 90.0°C? Assume that all the heat released during combustion is used to heat the water.

**5.102.** The heavier hydrocarbons in white gas are hexanes ($C_6H_{14}$).
  a. Calculate the fuel value of $C_6H_{14}$, given that $\Delta H°_{comb} = -4163$ kJ/mol.
  b. How much heat is released during the combustion of 1.00 kg of $C_6H_{14}$?
  c. How many grams of $C_6H_{14}$ are needed to heat 1.00 kg of water from 25.0°C to 85.0°C? Assume that all of the heat released during combustion is used to heat the water.
  d. Assume white gas is 25% $C_5$ hydrocarbons and 75% $C_6$ hydrocarbons; how many grams of white gas are needed to heat 1.00 kg of water from 25.0°C to 85.0°C?

## Additional Problems

**5.103. Diagnosing Car Trouble** Mechanics sometimes use diethyl ether ($CH_3CH_2OCH_2CH_3$) to diagnose starting problems in vehicles because it has a low ignition temperature. If sprayed into the air intake, it can help determine whether the spark and ignition system of the car is functioning and the fuel delivery system is not. If the normal fuel is not entering the engine, the engine will run only until the diethyl ether vapors are completely burned. Compare the fuel value and fuel density of diethyl ether ($\Delta H°_{comb} = -2726.3$ kJ/mol; density = 0.7134 g/mL) to that of diesel fuel (determined in Sample Exercise 5.17).

**5.104.** Polychlorinated biphenyls (PCBs) are efficient coolants whose use in transformers and other electrical devices has been banned because of their toxicity. What is the specific heat ($c_s$) of the PCB with the molecular formula $C_{12}Cl_{10}$ if its molar heat capacity ($c_P$) is 345.7 J/(mol · °C)?

**5.105.** Carbon tetrachloride ($CCl_4$) was at one time used as a fire-extinguishing agent. It has a molar heat capacity $c_P$ of 131.3 J/(mol · °C). How much heat is required to raise the temperature of 275 g of $CCl_4$ from room temperature (22°C) to its boiling point (77°C)?

**5.106.** Ethylene glycol ($HOCH_2CH_2OH$) is mixed with the water in radiators to cool car engines. How much heat will 725 g of pure ethylene glycol remove from an engine as it is warmed from 0°C to its boiling point of 196°C? The $c_P$ of ethylene glycol is 149.5 J/(mol · °C).

**5.107.** Sodium may be used as a heat-storage material in some devices. The specific heat ($c_s$) of sodium metal is 1.23 J/(g · °C). How many moles of sodium metal are required to absorb $1.00 \times 10^3$ kJ of heat?

**5.108.** The water in a calorimeter was replaced with the organic compound methylene chloride ($CH_2Cl_2$). Burning 2.23 g of glucose ($C_6H_{12}O_6$; $\Delta H_{comb} = -2801$ kJ/mol) in the calorimeter causes its temperature to rise 9.64°C. What is the heat capacity of the calorimeter?

**5.109.** The standard enthalpy of formation of $NH_3$ is −46.1 kJ/mol. What is $\Delta H°_{rxn}$ for the following reactions?
  a. $N_2(g) + 3\,H_2(g) \rightarrow 2\,NH_3(g)$
  b. $NH_3(g) \rightarrow \frac{1}{2}N_2(g) + \frac{3}{2}H_2(g)$

**\*5.110. Hung Out to Dry** Laundry forgotten and left outside to dry on a clothesline in the winter slowly dries by "ice vaporization" (sublimation). The increase in internal energy of water vapor produced by sublimation is less than the amount of heat absorbed. Explain.

**5.111.** Chlorofluorocarbons (CFCs) such as $CF_2Cl_2$ are refrigerants whose use has been phased out because of their destructive effect on Earth's ozone layer. The standard enthalpy of evaporation of $CF_2Cl_2$ is 17.4 kJ/mol, compared with $\Delta H_{vap} = 40.67$ kJ/mol for liquid water. How many grams of liquid $CF_2Cl_2$ are needed to cool 200.0 g of water from 50.0°C to 40.0°C? The specific heat of water is 4.184 J/(g · °C).

**5.112.** A 100.0 mL sample of 1.0 *M* NaOH is mixed with 50.0 mL of 1.0 *M* $H_2SO_4$ in a large Styrofoam coffee cup; the cup is fitted with a lid through which a calibrated thermometer passes. The temperature of each solution before mixing is 22.3°C. After adding the NaOH solution to the coffee cup and stirring the mixed solutions with the thermometer, the maximum temperature measured is 31.4°C. Assume that the density of the mixed solutions is 1.00 g/mL, the specific heat of the mixed solutions is 4.18 J/(g · °C), and no heat is lost to the surroundings.
  a. Write a balanced chemical equation for the reaction that takes place in the Styrofoam cup.
  b. Is any NaOH or $H_2SO_4$ left in the Styrofoam cup when the reaction is over?
  c. Calculate the enthalpy change per mole of $H_2SO_4$ in the reaction.

**5.113.** Varying the scenario in Problem 5.112, assume this time that 65.0 mL of 1.0 *M* $H_2SO_4$ is mixed with 100.0 mL of 1.0 *M* NaOH and that both solutions are initially at 25.0°C. Assume that the mixed solutions in the Styrofoam cup have the same density and specific heat as in Problem 5.112 and no heat is lost to the surroundings. What is the maximum measured temperature in the Styrofoam cup?

*5.114. An insulated container is used to hold 50.0 g of water at 25.0°C. A 7.25 g sample of copper is placed in a dry test tube and heated for 30 minutes in a boiling water bath at 100.1°C. The heated test tube is carefully removed from the water bath with laboratory tongs and inclined so that the copper slides into the water in the insulated container. Given that the specific heat of solid copper is 0.385 J/(g · °C), calculate the maximum temperature of the water in the insulated container after the copper metal is added.

5.115. The mineral magnetite ($Fe_3O_4$) is magnetic, whereas iron(II) oxide is not.
   a. Write and balance the chemical equation for the formation of magnetite from iron(II) oxide and oxygen.
   b. Given that 318 kJ of heat is released for each mole of $Fe_3O_4$ formed, what is the enthalpy change of the balanced reaction of formation of $Fe_3O_4$ from iron(II) oxide and oxygen?

5.116. Which of the following substances has a standard heat of formation $\Delta H_f^\circ$ of zero? (a) Pb at 1000°C; (b) $C_3H_8(g)$ at 25.0°C and 1 atm pressure; (c) solid glucose at room temperature; (d) $N_2(g)$ at 25.0°C and 1 atm pressure

*5.117. The standard heat of formation of liquid water is −285.8 kJ/mol.
   a. What is the significance of the negative sign associated with this value?
   b. Why is the magnitude of this value so much larger than the heat of vaporization of water ($\Delta H_{vap}^\circ$ = 40.67 kJ/mol)?
   c. Calculate the amount of heat produced in making 50.0 mL of water from its elements under standard conditions.

5.118. Endothermic compounds are unusual: they have positive heats of formation. An example is acetylene, $C_2H_2$ ($\Delta H_f^\circ$ = 226.7 kJ/mol). The combustion of acetylene is used to melt and weld steel. Use Appendix 4 to answer the following questions.
   a. Give the chemical formulas and names of three other endothermic compounds.
   b. Calculate the standard molar heat of combustion of acetylene.

*5.119. Balance the following chemical equation, name the reactants and products, and calculate the standard enthalpy change using the data in Appendix 4.

$$FeO(s) + O_2(g) \rightarrow Fe_2O_3(s)$$

*5.120. Add reactions 1, 2, and 3, and label the resulting reaction 4. Consult Appendix 4 to find the standard enthalpy change for the balanced reaction 4.

   (1)  $Zn(s) + \frac{1}{8} S_8(s) \rightarrow ZnS(s)$
   (2)  $ZnS(s) + 2 O_2(g) \rightarrow ZnSO_4(s)$
   (3)  $\frac{1}{8} S_8(s) + O_2(g) \rightarrow SO_2(g)$

5.121. Conversion of 0.90 g of liquid water to steam at 100.0°C requires 2.0 kJ of heat. Calculate the molar enthalpy of evaporation of water at 100.0°C.

5.122. The specific heat of solid copper is 0.385 J/(g · °C). What thermal energy change occurs when a 35.3 g sample of copper is cooled from 35.0°C to 15.0°C? Be sure to give your answer the proper sign. This amount of heat is used to melt solid ice at 0.0°C. The molar heat of fusion of ice is 6.01 kJ/mol. How many moles of ice are melted?

*5.123. **Metabolism of Methanol** Methanol is toxic because it is metabolized in a two-step process *in vivo* to formic acid (HCOOH). Consider the following overall reaction under standard conditions:

$$O_2(g) + 2 CH_3OH(\ell) \rightarrow 2 HCOOH(\ell) + 2 H_2O(\ell) + 1019.6 \text{ kJ}$$

   a. Is this reaction endothermic or exothermic?
   b. What is the value of $\Delta H_{rxn}^\circ$ for this reaction?
   c. How much heat would be absorbed or released if 60.0 g of methanol were metabolized in this reaction?
   d. In the first step of metabolism, methanol is converted into formaldehyde ($CH_2O$), which is then converted into formic acid. Would you expect $\Delta H_{rxn}^\circ$ for the metabolism of 1 mole of $CH_3OH(\ell)$ to give 1 mole of formaldehyde to be larger or smaller than 509.8 kJ?

5.124. Using the information given below, write an equation for the enthalpy change in reaction 3:

   (1)  $B(g) + A(s) \rightarrow C(g)$     $\Delta H_1$
   (2)  $C(g) \rightarrow C(s)$     $\Delta H_2$
   (3)  $C(s) \rightarrow A(s) + B(g)$     $\Delta H_3$

5.125. From the information given below, write an equation for the enthalpy change in reaction 3:

   (1)  $F(g) \rightarrow D(s) + E(g)$     $\Delta H_1$
   (2)  $F(s) \rightarrow F(g)$     $\Delta H_2$
   (3)  $D(s) + E(g) \rightarrow F(s)$     $\Delta H_3$

5.126. The reaction of $CH_3OH(g)$ with $N_2(g)$ to give $HCN(g)$ and $NH_3(g)$ requires 164 kJ/mol of heat.
   a. Write a balanced chemical equation for this reaction.
   b. Should the thermal energy involved be written as a reactant or as a product?
   c. What is the enthalpy change in the reaction of 60.0 g of $CH_3OH(g)$ with excess $N_2(g)$ to give $HCN(g)$ and $NH_3(g)$ in this reaction?

5.127. Calculate $\Delta H_{rxn}^\circ$ for the reaction

$$2 Ni(s) + \frac{1}{4} S_8(s) + 3 O_2(g) \rightarrow 2 NiSO_3(s)$$

from the following information:

   (1)  $NiSO_3(s) \rightarrow NiO(s) + SO_2(g)$     $\Delta H_{rxn}^\circ$ = 156 kJ
   (2)  $\frac{1}{8} S_8(s) + O_2(g) \rightarrow SO_2(g)$     $\Delta H_{rxn}^\circ$ = −297 kJ
   (3)  $Ni(s) + \frac{1}{2} O_2(g) \rightarrow NiO(s)$     $\Delta H_{rxn}^\circ$ = −241 kJ

5.128. Use the following information to calculate the enthalpy change involved in the complete reaction of 3.0 g of carbon to form $PbCO_3(s)$ in reaction 4. Be sure to give the proper sign (positive or negative) with your answer.

   (1)  $Pb(s) + \frac{1}{2} O_2(g) \rightarrow PbO(s)$     $\Delta H_{rxn}^\circ$ = −219 kJ
   (2)  $C(s) + O_2(g) \rightarrow CO_2(g)$     $\Delta H_{rxn}^\circ$ = −394 kJ
   (3)  $PbCO_3(s) \rightarrow PbO(s) + CO_2(g)$     $\Delta H_{rxn}^\circ$ = 86 kJ
   (4)  $Pb(s) + C(s) + \frac{3}{2} O_3(g) \rightarrow PbCO_2(s)$     $\Delta H_{rxn}^\circ$ = ?

*5.129. **Ethanol as Automobile Fuel** Brazilians are quite familiar with fueling their automobiles with ethanol, a fermentation product from sugarcane. Calculate the standard molar enthalpy for the complete combustion of liquid ethanol using the standard enthalpies of formation of the reactants and products as given in Appendix 4.

5.130. Adding 1.56 g $K_2SO_4$ to 6.00 mL of water at 16.2°C causes the temperature of the solution to drop by 7.70°C. How many grams of NaOH ($\Delta H_{soln} = -44.3$ kJ/mol) would you need to add to raise the temperature back to 16.2°C?

5.131. Two solids, 5.00 g NaOH and 4.20 g KOH, are added to 150 mL of water [$c_P = 75.3$ J/(mol · °C); $T = 23$°C] in a calorimeter. Given $\Delta H_{soln,NaOH} = -44.3$ kJ/mol and $\Delta H_{soln,KOH} = -56.0$ kJ/mol, what is the final temperature of the solution?

5.132. A 1995 article in *Discover* magazine on world-class sprinters contained the following statement: "In one race, a field of eight runners releases enough energy to boil a gallon jug of ice at 0.0°C in ten seconds!" How much "energy" do the runners release in 10 seconds? Assume that the ice has a mass of 128 ounces.

*5.133. **Specific Heats of Metals** In 1819, Pierre Dulong and Alexis Petit reported that the product of the atomic mass of a metal times its specific heat is approximately constant, an observation called the *law of Dulong and Petit*. Use the data in the table below to answer the following questions.

| Element | $\mathcal{M}$ (g/mol) | $c_s$ [J/(g · °C)] | $\mathcal{M} \times c_s$ |
| --- | --- | --- | --- |
| Bismuth | | 0.120 | |
| Lead | 207.2 | 0.123 | 25.5 |
| Gold | 197.0 | 0.125 | |
| Platinum | 195.1 | | |
| Tin | 118.7 | 0.215 | |
| | | 0.233 | |
| Zinc | 65.4 | 0.388 | |
| Copper | 63.5 | 0.397 | |
| | | 0.433 | |
| Iron | 55.8 | 0.460 | |
| Sulfur | 32.1 | | |
| | | Average value: | |

a. Complete each row in the table by multiplying each given molar mass and specific heat pair (one result has been entered in the table). What are the units of the resulting values in column 4?

b. Next, calculate the average of the values in column 4.

c. Use the mean value from part b to calculate the missing atomic masses in the table. Do you feel confident in identifying the elements from the calculated atomic masses?

d. Use the average value from part b to predict the missing specific heat values in the table.

*5.134. **Odor of Urine** Urine odor gets worse with time because urine contains the metabolic product urea [$CO(NH_2)_2$], a compound that is slowly converted to ammonia, which has a sharp, unpleasant odor, and carbon dioxide:

$$CO(NH_2)_2(aq) + H_2O(\ell) \rightarrow CO_2(aq) + 2\,NH_3(aq)$$

This reaction is much too slow for the enthalpy change to be measured directly using a temperature change. Instead, the enthalpy change for the reaction may be calculated from the following data:

| Compound | $\Delta H_f^\circ$ (kJ/mol) |
| --- | --- |
| Urea(*aq*) | −319.2 |
| $CO_2$(*aq*) | −412.9 |
| $H_2O(\ell)$ | −285.8 |
| $NH_3$(*aq*) | −80.3 |

Calculate the standard molar enthalpy change for the reaction.

5.135. **Experiment with a Metal** Explain how the specific heat of a metal sample could be measured in the lab. (*Hint*: You'll need a test tube, a boiling water bath, a Styrofoam cup calorimeter containing a known mass of water, a calibrated thermometer, and a known mass of the metal.)

*5.136. **Rocket Fuels** The payload of a rocket includes a fuel and oxygen for combustion of the fuel. Reactions 1 and 2 describe the combustion of dimethylhydrazine and hydrogen, respectively. Pound for pound, which is the better rocket fuel, dimethylhydrazine or hydrogen?

(1) $(CH_3)_2NNH_2(\ell) + 4\,O_2(g) \rightarrow$
$N_2(g) + 4\,H_2O(g) + 2\,CO_2(g)$    $\Delta H_{rxn}^\circ = -1694$ kJ

(2) $H_2(g) + \frac{1}{2}O_2(g) \rightarrow H_2O(g)$    $\Delta H_{rxn}^\circ = -242$ kJ

5.137. At high temperatures, such as those in the combustion chambers of automobile engines, nitrogen and oxygen form nitrogen monoxide:

$$N_2(g) + O_2(g) \rightarrow 2\,NO(g) \qquad \Delta H_{comb}^\circ = +180 \text{ kJ}$$

Any NO released into the environment is oxidized to $NO_2$:

$$2\,NO(g) + O_2(g) \rightarrow 2\,NO_2(g) \qquad \Delta H_{comb}^\circ = -112 \text{ kJ}$$

Is the overall reaction

$$N_2(g) + 2\,O_2(g) \rightarrow 2\,NO_2(g)$$

exothermic or endothermic? What is $\Delta H_{comb}^\circ$ for this reaction?

5.138. You are given the following data:

$\frac{1}{2}N_2(g) + \frac{1}{2}O_2(g) \rightarrow NO(g)$    $\Delta H_{rxn}^\circ = +90.3$ kJ
$NO(g) + \frac{1}{2}Cl_2(g) \rightarrow NOCl(g)$    $\Delta H_{rxn}^\circ = -38.6$ kJ
$2\,NOCl(g) \rightarrow N_2(g) + O_2(g) + Cl_2(g)$    $\Delta H_{rxn}^\circ = ?$

a. Which of the $\Delta H_{rxn}^\circ$ values represent enthalpies of formation?

b. Determine $\Delta H_{rxn}^\circ$ for the decomposition of NOCl.

5.139. **Hydrogen as Fuel** Hydrogen is attractive as a fuel because it has a high fuel value and produces no $CO_2$.

$$H_2(g) + \frac{1}{2}O_2(g) \rightarrow H_2O(g)$$

Unfortunately, the production and storage of hydrogen fuel remain problematic.

a. One problem with hydrogen is its low fuel density. What is the fuel density (kJ/L) for $H_2$ given the density of hydrogen gas is 0.0899 g/L? If we could liquefy hydrogen, what would the fuel density of hydrogen be, given the density of liquid hydrogen is 70.8 g/L?

b. One solution to the problem of hydrogen storage is to use solid, hydrogen-containing compounds that release hydrogen upon heating at low temperature. One such compound is ammonia-borane, $H_3NBH_3$. Calculate the enthalpy change ($\Delta H$) for the reactions shown below given $\Delta H^\circ_{f,H_3NBH_3} = -38.1$ kJ/mol, $\Delta H^\circ_{f,BH_3} = +110.2$ kJ/mol, $\Delta H^\circ_{f,NH_3} = -46.1$ kJ/mol, $\Delta H^\circ_{f,H_2NBH_2} = -66.5$ kJ/mol, and $\Delta H^\circ_{f,HNBH} = +56.9$ kJ/mol.

$$H_3NBH_3(g) \rightarrow NH_3(g) + BH_3(g)$$
$$H_3NBH_3(g) \rightarrow H_2(g) + H_2NBH_2(g)$$
$$H_2NBH_2(g) \rightarrow H_2(g) + HNBH(g)$$

c. How many kg of ammonia-borane are needed to supply 10.0 kg of hydrogen?

## Carbon: Diamonds, Graphite, and the Molecules of Life

*5.140. **Industrial Use of Cellulose** Research is being carried out on cellulose as a source of chemicals for the production of fibers, coatings, and plastics. Cellulose consists of long chains of glucose molecules ($C_6H_{12}O_6$), so for the purposes of modeling the reaction, we can consider the conversion of glucose to formaldehyde.

a. Is the reaction to convert glucose into formaldehyde an oxidation or a reduction?

b. Calculate the heat of reaction for the conversion of 1 mole of glucose into formaldehyde, given the following thermochemical data:

$\Delta H^\circ_{comb}$ of formaldehyde gas   $= -572.9$ kJ/mol

$\Delta H^\circ_f$ of solid glucose   $= -1274.4$ kJ/mol

$$C_6H_{12}O_6(s) \rightarrow 6\ CH_2O(g)$$
$$\text{Glucose} \qquad \text{Formaldehyde}$$

5.141. Use the following data to determine whether the conversion of diamond into graphite is exothermic or endothermic:

| | |
|---|---|
| $C_{diamond}(s) + O_2(g) \rightarrow CO_2(g)$ | $\Delta H^\circ = -395.4$ kJ |
| $2\ CO_2(g) \rightarrow 2\ CO(g) + O_2(g)$ | $\Delta H^\circ = 566.0$ kJ |
| $2\ CO(g) \rightarrow C_{graphite}(s) + CO_2(g)$ | $\Delta H^\circ = -172.5$ kJ |
| $C_{diamond}(s) \rightarrow C_{graphite}(s)$ | $\Delta H^\circ = ?$ |

5.142. **Converting Diamond to Graphite** The standard state of carbon is graphite. $\Delta H^\circ_{f,diamond}$ is $+1.896$ kJ/mol. Diamond masses are normally given in carats, where 1 carat $= 0.20$ g. Determine the standard enthalpy of the reaction for the conversion of a 4-carat diamond into an equivalent mass of graphite. Is this reaction endothermic or exothermic?

5.143. A 1-carat diamond ring (0.20 g diamond) in a 23 g gold setting is heated to 100°C and dropped into cold water at 15°C. What volume of water is needed to cool the ring to 40°C? The molar heat capacities of diamond and gold are 6.23 J/(mol · °C) and 25.41 J/(mol · °C), respectively.

If your instructor assigns problems in **smartwork**, log in at **smartwork.wwnorton.com**.

# 6

# Properties of Gases: The Air We Breathe

## An Invisible Necessity

How often do you think about air? You remember how important air is when you do not have enough, such as when you dive into a pool of water or hike to the top of a tall mountain. Otherwise you probably do not give much thought to breathing or to the mixture of gases that make up the air you breathe. You may not think much about the oxygen needed to stay alive because it is colorless, tasteless, odorless—and free. Despite not thinking about the process, the exchange of gases between your lungs and the surrounding atmosphere is crucial to life; air is an invisible necessity.

Having the right mixture of gases in our bodies is critical during surgery, which is why anesthesiologists in a hospital operating room constantly monitor blood levels of oxygen and carbon dioxide. The management of the delicate balance of gases entering and leaving a patient can mean the difference between a normal recovery and irreversible coma.

We have seen how dissolved compounds react in aqueous solution. Chemical reactions also take place in the gas phase, and gases are intimately involved in chemical reactions in living systems as well as in the material world. Most life in our biosphere requires oxygen. Insects, birds, mammals, plants, and even organisms that live under water need $O_2$ to metabolize nutrients. All the fuels we burn, to provide heat and light for our buildings and power for our vehicles, rely on atmospheric $O_2$ to support combustion.

In this chapter we examine the key properties of gases and how they relate to one another. Often both physical and chemical properties are involved when gases take part in reactions. Our ability to understand how the properties of gases fit into our lives—whether in living rooms, operating rooms, scuba tanks, welding torches, or automobiles—is based on our knowledge of how gases behave in response to changes in pressure, volume, and temperature. We will see once again that the observable properties and behaviors of matter are ultimately based on the properties and behavior of atoms and molecules. ■

**6.1**  The Gas Phase

**6.2**  Atmospheric Pressure

**6.3**  The Gas Laws

**6.4**  The Ideal Gas Law

**6.5**  Gases in Chemical Reactions

**6.6**  Gas Density

**6.7**  Dalton's Law and Mixtures of Gases

**6.8**  The Kinetic Molecular Theory of Gases

**6.9**  Real Gases

## Learning Outcomes

**LO1** Distinguish gases from liquids and solids

**LO2** Measure pressure and convert between the different units used to quantify it
Sample Exercises 6.1, 6.2

**LO3** Calculate changes in the volume, temperature, pressure, and number of moles of a gas using the individual, combined, and ideal gas laws
Sample Exercises 6.3, 6.4, 6.5, 6.6, 6.7

**LO4** Use balanced chemical equations to relate the volume of a gas-phase reactant to the amount of a product using the stoichiometry of the reaction and the ideal gas law
Sample Exercises 6.8, 6.9

**LO5** Calculate the density or molar mass of any gas
Sample Exercises 6.10, 6.11

**LO6** Determine the mole fraction and the partial pressure of a gas in a mixture
Sample Exercises 6.12, 6.13, 6.14

**LO7** Use kinetic molecular theory to explain the behavior of gases

**LO8** Calculate the root-mean-square speed of a gas and relative rates of effusion and diffusion
Sample Exercises 6.15, 6.16

**LO9** Use the van der Waals equation to correct for nonideal behavior
Sample Exercise 6.17

**Scuba Diver and Sea Goldies** The gas exhaled by the diver forms bubbles that increase in size as they rise to the surface. ▶

| TABLE 6.1 | Composition of Earth's Atmosphere |
|---|---|
| **Component** | **% (by volume)** |
| Nitrogen | 78.08 |
| Oxygen | 20.95 |
| Argon | 0.934 |
| Carbon dioxide | 0.0386 |
| Methane | $2 \times 10^{-4}$ |
| Hydrogen | $5 \times 10^{-5}$ |

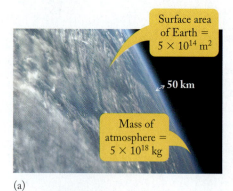

Surface area of Earth = $5 \times 10^{14}$ m²

50 km

Mass of atmosphere = $5 \times 10^{18}$ kg

(a)

(b)

**FIGURE 6.1** (a) Atmospheric pressure results from the force exerted by the atmosphere on Earth's surface. (b) If you stretch out your hand palm upward, the mass of the column of air above your palm is about 100 kg. This textbook has a mass of about 2.5 kg, so the mass of the atmosphere on your palm is equivalent to the mass of about 40 textbooks.

# 6.1 The Gas Phase

Gases have neither definite volumes nor definite shapes; they expand to occupy the entire volume of their container and assume the container's shape. Under everyday conditions, other properties also distinguish gases from liquids and solids.

1. Unlike the volume occupied by a liquid or solid, the volume occupied by a gas changes significantly with pressure. If we carry an inflated balloon from sea level (0 m) to the top of a 1600 m mountain, the balloon volume increases by about 20%. The volume of a liquid or solid is unchanged under these conditions.
2. The volume of a gas changes with temperature. For example, the volume of a balloon filled with room-temperature air decreases when the balloon is taken outside on a cold winter's day. A temperature decrease from 20°C to 0°C leads to a volume decrease of about 7%, whereas the volume of a liquid or solid remains practically unchanged by this modest temperature change.
3. Gases are **miscible**, which means they can be mixed together in any proportion (unless they chemically react with one another). A hospital patient experiencing respiratory difficulties may be given a mixture of nitrogen and oxygen in which the proportion of oxygen is much higher than its proportion in air, and a scuba diver may leave the ocean surface with a tank of air containing a homogeneous mixture of 17% oxygen, 34% nitrogen, and 49% helium. In contrast, many liquids are immiscible, such as oil and water.
4. Gases are typically much less dense than liquids or solids. One indicator of this large difference is that gas densities are expressed in grams per *liter* but liquid densities are expressed in grams per *milliliter*. The density of dry air at 20°C at typical atmospheric pressure is 1.20 g/L, for example, whereas the density of liquid water under the same conditions is 1.00 g/mL—over 800 times greater than the density of dry air.

These four observations about gases are consistent with the idea that the molecules or atoms in gases are farther apart than in solids and liquids. The larger spaces between the gas molecules in air, for example, account for the ready compressibility of air into scuba tanks. Greater distances between molecules account for both the lower densities and the miscibility of gases.

# 6.2 Atmospheric Pressure

Earth is surrounded by a layer of gases 50 kilometers thick. We call this mixture of gases either *air* or *the atmosphere*. By volume it is composed primarily of nitrogen (78%) and oxygen (21%), with lesser amounts of other gases (Table 6.1). To put the thickness of the atmosphere in perspective, if Earth were the size of an apple, the atmosphere would be about as thick as the apple's skin.

Earth's atmosphere is pulled toward Earth by gravity and exerts a force that is spread across the entire surface of the planet (Figure 6.1). The ratio of force ($F$) to surface area ($A$) is called **pressure ($P$)**.

$$P = \frac{F}{A} \tag{6.1}$$

The force exerted by the atmosphere on Earth's surface is called **atmospheric pressure ($P_{atm}$)**. Atmospheric pressure is measured with an instrument called a **barometer**. A simple but effective barometer design consists of a tube nearly

1 meter long, filled with mercury, and closed at one end (Figure 6.2). The tube is inverted with its open end immersed in a pool of mercury that is open to the atmosphere. Gravity pulls the mercury in the tube downward, creating a vacuum at the top of the tube, while atmospheric pressure pushes the mercury in the pool up into the tube. The net effect of these opposing forces is indicated by the height of the mercury in the tube, which provides a measure of atmospheric pressure.

Atmospheric pressure varies from place to place and with changing weather conditions. Several units are used to express pressure. An **atmosphere (1 atm)** of pressure is the pressure capable of supporting a column of mercury 760 mm high in a barometer. This column height is the average height of mercury in a barometer at sea level. Another pressure unit, **millimeters of mercury (mmHg)**, is more explicitly related to column height, where 1 atm = 760 mmHg. Pressure is also expressed in units called **torr** in honor of Evangelista Torricelli (1608–1647), the Italian mathematician and physicist who invented the barometer. There are exactly 760 torr in 1 atm. Thus

$$1 \text{ atm} = 760 \text{ torr} = 760 \text{ mmHg}$$

The SI unit of pressure is the *pascal* (Pa), named in honor of French mathematician and physicist Blaise Pascal (1623–1662), who was the first to propose that atmospheric pressure decreases with increasing altitude. We can explain this phenomenon by noting that the atmospheric pressure at any given location on Earth's surface is related to the mass of the column of air *above* that location (Figure 6.3).

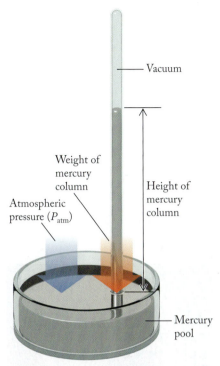

**FIGURE 6.2** The height of the mercury column in this simple barometer designed by Evangelista Torricelli is proportional to atmospheric pressure.

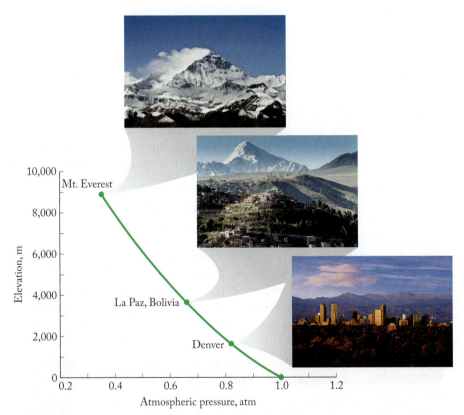

**FIGURE 6.3** Atmospheric pressure decreases with increasing altitude because the mass of the column of air above a given area decreases with increasing altitude. If the weather were the same nice, clear day in all three of these locations, identical barometers would read 1.00 atm for the atmospheric pressure at a city at sea level, 0.83 atm in Denver, 0.62 atm in La Paz, Bolivia, and a mere 0.35 atm on the summit of Mt. Everest.

**miscible** capable of being mixed in any proportion (without reacting chemically).

**pressure (P)** the ratio of force to surface area over which the force is applied.

**atmospheric pressure ($P_{atm}$)** the force exerted by the gases surrounding Earth on Earth's surface and on all surfaces of all objects.

**barometer** an instrument that measures atmospheric pressure.

**atmosphere (1 atm)** a unit of pressure based on Earth's average atmospheric pressure at sea level.

**millimeters of mercury (mmHg)** (also called **torr**) a unit of pressure where 1 atm = 760 torr = 760 mmHg.

As altitude increases, the height of the column of air above a location decreases, which means the mass of the gases in the column decreases. Less mass means a smaller force is exerted by the air, and, according to Equation 6.1, a smaller force means less pressure.

The pascal is a derived SI unit; it is defined using the SI base units kilogram, meter, and second:

$$1 \text{ Pa} = \frac{1 \text{ kg}}{\text{m} \cdot \text{s}^2}$$

To understand the logic of this combination of units, consider the relationship in physics that states that pushing an object of mass $m$ with a force $F$ causes the object to accelerate at the rate $a$:

$$F = ma \tag{6.2}$$

Combining Equations 6.1 and 6.2 we have

$$\frac{F}{A} = P = \frac{ma}{A} \tag{6.3}$$

If $m$ is in kilograms, $a$ is in meters per second-squared ($\text{m/s}^2$), and $A$ is in square meters ($\text{m}^2$), the units for $P$ are

$$\frac{\text{kg} \frac{\text{m}}{\text{s}^2}}{\text{m}^2} = \text{kg} \frac{\cancel{\text{m}}}{\text{s}^2} \times \frac{1}{\cancel{\text{m}^2}} = \frac{\text{kg}}{\text{m} \cdot \text{s}^2}$$

The relation between atmospheres and pascals is

$$1 \text{ atm} = 101{,}325 \text{ Pa}$$

Clearly, 1 Pa is a tiny quantity of pressure. In many applications it is more convenient to express pressure in kilopascals.

For many years, meteorologists have expressed atmospheric pressure in *millibars* (*mbar*). The weather maps prepared by the U.S. National Weather Service show changes in atmospheric pressure by constant-pressure contour lines, called *isobars*, spaced 4 mbar apart (Figure 6.4). There are exactly 10 mbar in 1 kPa; thus

$$1 \text{ atm} = (101.325 \cancel{\text{ kPa}})(10 \text{ mbar}/\cancel{\text{kPa}})$$
$$= 1013.25 \text{ mbar}$$

Other units of pressure are derived from masses and areas in the U.S. Customary System, such as pounds per square inch ($\text{lb/in}^2$, psi) for tire pressures and inches of mercury for atmospheric pressure in weather reports. The relationships between different units for pressure and 1 atm are summarized in Table 6.2.

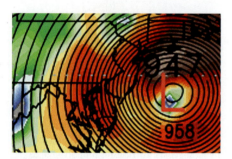

**FIGURE 6.4** Small differences in atmospheric pressure are associated with major changes in weather. Adjacent isobars on this map of Hurricane Sandy differ by 4 mbar of pressure.

| TABLE 6.2 | Units for Expressing Pressure |
|---|---|
| **Unit** | **Value** |
| Atmosphere (atm) | 1 atm |
| Pascal (Pa) | 1 atm = $1.01325 \times 10^5$ Pa |
| Kilopascal (kPa) | 1 atm = 101.325 kPa |
| Millimeter of mercury (mmHg) | 1 atm = 760 mmHg |
| Torr | 1 atm = 760 torr |
| Bar | 1 atm = 1.01325 bar |
| Millibar (mbar) | 1 atm = 1013.25 mbar |
| Pounds per square inch (psi) | 1 atm = 14.7 psi |
| Inches of mercury | 1 atm = 29.92 inches of Hg |

**SAMPLE EXERCISE 6.1** **Calculating Atmospheric Pressure** **LO2**

The atmosphere on Titan, one of Saturn's moons, is composed almost entirely of nitrogen (98%). However, its oceans contain molecules thought to be important in the evolution of life, making Titan interesting for planetary scientists. The mass of Titan's atmosphere is estimated to be $9.0 \times 10^{18}$ kg. The surface area of Titan is $8.3 \times 10^{13}$ $\text{m}^2$. The acceleration due to gravity on Titan's atmosphere is 1.35 $\text{m/s}^2$. From these values calculate an average atmospheric pressure in kilopascals.

**Collect and Organize** We are given the mass of the atmosphere in kilograms, the surface area of Titan in square meters, and the acceleration due to gravity in meters per second-squared. These units can be combined to calculate atmospheric pressure in pascals.

**Analyze** Equation 6.3 describes the relationship between pressure, force, mass, and the acceleration due to gravity. We can estimate the magnitude of the force by considering that the mass is about $10^{19}$ kg and $a$ is approximately 1.5, so the force is about $10^{19}$. If the force is $\sim 10^{19}$ and the area is $\sim 10^{14}$, then the pressure will be about $10^5$ Pa or $10^2$ kPa.

**Solve** Substituting the values into Equation 6.3 gives us

$$P = \frac{ma}{A}$$

$$= \frac{(9.0 \times 10^{18}\,\text{kg})(1.35\,\text{m/s}^2)}{8.3 \times 10^{13}\,\text{m}^2}$$

$$= \frac{1.46 \times 10^5\,\text{kg}}{\text{m} \cdot \text{s}^2} = 1.5 \times 10^5\,\text{Pa}$$

$$= 1.5 \times 10^2\,\text{kPa}$$

**Think About It** The calculated atmospheric pressure, $1.5 \times 10^2$ kPa, is close to our estimate. The atmospheric pressure on Titan is similar to that on Earth. The greater mass of Titan's atmosphere and its smaller surface area compensate for the lower acceleration due to gravity.

⚙ **Practice Exercise** Calculate the pressure in pascals exerted on a tabletop by a cube of iron that is 1.00 cm on each side and has a mass of 7.87 g.

*(Answers to Practice Exercises are in the back of the book.)*

Scientists conducting experiments with gases may monitor the pressures exerted by the gases using a **manometer**. Two types of manometers are illustrated in Figure 6.5. In each case, a U-shaped tube filled with mercury (or another dense liquid) is connected to a container holding the gas whose pressure is being measured. When the valve is opened, the gas exerts pressure on the mercury.

The difference between the two manometers is whether the end of the tube not connected to the evacuated flask is closed or open to the atmosphere. In a closed-end manometer, the difference in the heights of the mercury columns in the two arms of the U-shaped tube is a direct measure of the gas pressure (Figure 6.5b). In an open-end manometer, the difference in heights represents the difference between the pressure in the flask and atmospheric pressure. When the pressure in the flask is greater than atmospheric pressure (Figure 6.5d), the level in the arm open to the atmosphere is higher than the level in the other arm; when the pressure in the flask is lower than atmospheric pressure, the level in the arm attached to the flask is higher than the level in the other arm. Torricelli's barometer (Figure 6.2), which he used to measure atmospheric pressure, is an example of a closed-end manometer.

Manometers have been largely displaced by pressure sensors based on flexible metallic or ceramic diaphragms. As the pressure on one side of the diaphragm increases, it distorts away from that side. This is the mechanism used to sense changes in atmospheric pressure in most present-day barometers, including the recording barometer, or *barograph*, shown in Figure 6.6.

**manometer** an instrument for measuring the pressure exerted by a gas.

**FIGURE 6.5** (a) A closed-end manometer is designed to measure the pressure of a gas sample that is less than atmospheric pressure. (b) The pressure of the gas sample is the difference between the heights of the mercury columns. (c) An open-end manometer is designed to measure the pressure of a gas sample that is greater than atmospheric pressure. (d) The difference in the heights of the two mercury columns is the difference between the pressure of the sample and that of the atmosphere.

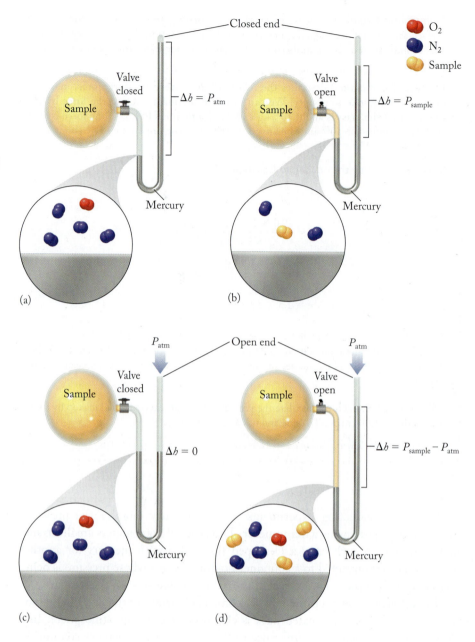

**FIGURE 6.6** (a) The pressure sensor in this barograph is a partially evacuated corrugated metal can. (b) As atmospheric pressure decreases, the lid of the can distorts outward. This motion is amplified by a series of levers and transmitted via the horizontal arm to a pen tip that records the pressure on the graph paper as the drum slowly turns. A week's worth of barometric data can be recorded in this way.

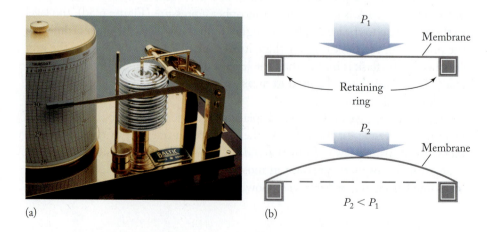

**Measuring Gas Pressure with a Manometer**     L02

Oyster shells are composed of calcium carbonate ($CaCO_3$). When the shells are roasted, they produce solid calcium oxide, CaO (quicklime), and $CO_2$. A chemist roasts some oyster shells in a flask from which the air has been removed and that is attached to a closed-end manometer (Figure 6.7a). When the roasting is complete and the system has cooled to room temperature, the difference in levels of mercury ($\Delta h$) in the arms of the manometer is 143.7 mm (Figure 6.7b). Calculate the pressure of the $CO_2(g)$ in (a) torr, (b) atmospheres, and (c) kilopascals.

**Collect and Organize** We are given the value of the height of the mercury column in millimeters, which we need to convert to the pressure of $CO_2$ produced in the reaction and express in three different units of pressure. This requires using the appropriate conversion factors in Table 6.2.

**Analyze** The difference in the mercury levels is a direct measure of the pressure in the flask, measured in millimeters of mercury. The appropriate conversion factors are 1 mmHg = 1 torr, 760 torr = 1 atm, and 1 atm = 101.325 kPa. We can estimate answers quite readily. Because units of mmHg and torr are equal, the numerical value of pressure is the same using either unit. Because one atmosphere contains hundreds of torr, the pressure in atm should be a much smaller number than the value of the pressure expressed in torr. On the other hand, 1 atmosphere is approximately 100 kPa, so the pressure in kPa should be about $10^2$ times the pressure in atm.

**Solve**

a. Converting torr to atmospheres:

$$143.7 \ \text{torr} \times \frac{1 \ \text{atm}}{760 \ \text{torr}} = 0.1891 \ \text{atm}$$

b. Converting atmospheres to kilopascals:

$$0.1891 \ \text{atm} \times \frac{101.325 \ \text{kPa}}{1 \ \text{atm}} = 19.16 \ \text{kPa}$$

**Think About It** The calculated pressure values make sense based on our estimates: a pressure less than 760 torr should also be less than 1 atm; by the same token, a pressure less than 1 atm should also be less than 101.325 kPa.

⚙ **Practice Exercise** The same quantity of $CO_2(g)$ produced in Sample Exercise 6.2 is produced in a flask connected to an open-end manometer (Figure 6.8). The atmospheric pressure and the pressure inside the flask containing the oyster shells are both 760 torr at the start of the experiment. Indicate on the drawing where the mercury levels in the two arms should be at the conclusion of the experiment. What is the value of $\Delta h$ in millimeters?

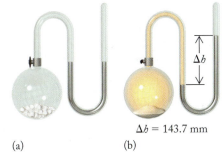

$\Delta h = 143.7 \ \text{mm}$

(a)     (b)

**FIGURE 6.7** Roasting $CaCO_3$ in a closed-end manometer. (a) An evacuated flask containing oyster shells. (b) The same setup after the shells have been roasted, a decomposition reaction that releases $CO_2$ in the flask.

**FIGURE 6.8** Before roasting oyster shells in an open-end manometer.

# 6.3 The Gas Laws

In Section 6.1, we summarized some of the properties of gases and described the effect of pressure and temperature on volume, mostly in qualitative terms. Our knowledge of the quantitative relationships among $P$, $T$, and $V$ goes back more than three centuries, to a time before the field of chemistry as we know it even existed. Some of the experiments that led to our understanding of how gases behave were driven by human interest in hot-air balloons, and today balloons are still used to study weather and atmospheric phenomena (Figure 6.9). Their successful and safe use requires an understanding of gas properties that was first gained in the 17th and 18th centuries.

## Boyle's Law: Relating Pressure and Volume

Did you notice that when the diver in the photograph on the first page of this chapter exhaled, the bubbles increased in size as they rose toward the surface? The

**FIGURE 6.9** Weather balloons are used to carry instruments aloft in the atmosphere.

reason is that as bubbles rise, the pressure on them decreases and as the pressure on a gas decreases, its volume increases. In other words, gases are compressible, a property that allows us to store large volumes of gases (at normal pressure) in relatively small metal cylinders at high pressure.

The relation between the pressure and volume of a fixed quantity of gas (constant value of $n$, where $n$ is the number of moles) at constant temperature was investigated by British chemist Robert Boyle (1627–1691) and is known as **Boyle's law**. Boyle observed that the volume of a given amount of a gas is inversely proportional to its pressure when kept at a constant temperature (Figure 6.10):

$$P \propto \frac{1}{V} \qquad (T \text{ and } n \text{ fixed}) \qquad (6.4)$$

In other words, as the pressure on a constant amount of gas at a constant temperature increases, the volume of the gas decreases, and conversely, as the pressure on the gas decreases, its volume increases.

How can we use Boyle's law to predict exactly how much a balloon expands when the pressure decreases? In his experiments, Boyle used a J-shaped tube open at one end and closed at the other end (Figure 6.11). When a small amount of mercury was poured into the tube, a column of air was trapped at the closed end. The pressure exerted on this trapped air depended on the difference in the height of the mercury in the two sides of the tube and the atmospheric pressure. The *amount* (number of moles) of trapped air remained constant, and the temperature was also held constant. The pressure on the trapped air was changed by varying the amount of mercury poured into the open arm. The more mercury added, the greater the force exerted by the mercury ($F = ma$), and hence the greater the force

**FIGURE 6.10** The change in the size of the balloon demonstrates the relationship between pressure and volume. (a) The balloon inside the bell jar is at a pressure of 760 mmHg (1 atm). (b) A slight vacuum applied to the bell jar drops the interior pressure to 266 mmHg and the balloon expands. (c) At constant temperature, the pressure of a given quantity of gas is inversely proportional to the volume occupied by the gas; the graph of an inverse proportion is a hyperbola. (d) The inverse proportion between $P$ and $V$ means that a plot of $P$ versus $1/V$ is a straight line.

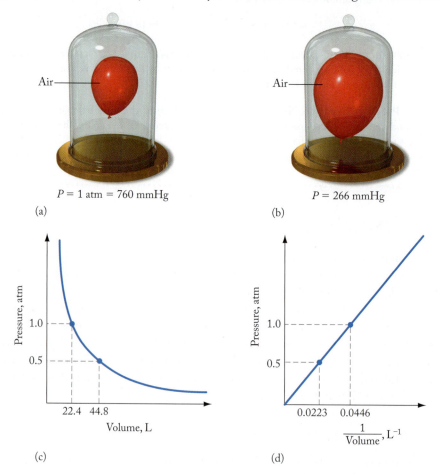

**Boyle's law** the volume of a given amount of gas at constant temperature is inversely proportional to its pressure.

per unit area ($P = F/A$) on the column of trapped air in the closed arm. As Boyle added mercury to the open arm, the volume occupied by the trapped air decreased in a way described by Equation 6.4. Mathematically, we can replace the proportionality symbol with an equals sign and a constant:

$$P = (\text{constant})\frac{1}{V}$$

$$PV = \text{constant} \qquad (6.5)$$

The value of the constant depends on the mass of trapped air and its temperature.

Because the value of the product $PV$ in Equation 6.5 does not change for a given mass of trapped air at constant temperature, any two combinations of pressure and volume are related as follows:

$$P_1V_1 = P_2V_2 \qquad (6.6)$$

This relationship, which applies to all gases, is illustrated by the dashed lines in Figure 6.10. When, for example, 44.8 L ($V_1$) of gas is held at a pressure of 0.500 atm ($P_1$), we have

$$P_1V_1 = (0.500 \text{ atm})(44.8 \text{ L}) = 22.4 \text{ L} \cdot \text{atm}$$

Equation 6.6 enables us to calculate the pressure required to compress this quantity of gas to any other volume. For example, if we want to hold this quantity of the gas in a 22.4-L container ($V_2$):

$$P_1V_1 = P_2V_2$$

$$P_2 = \frac{P_1V_1}{V_2} = \frac{(0.500 \text{ atm})(44.8 \text{ L})}{22.4 \text{ L}}$$

$$= 1.00 \text{ atm}$$

The products of $P$ and $V$ are the same for each point on either graph in Figure 6.10.

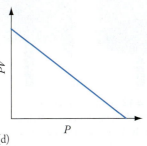

**FIGURE 6.11** Boyle used a J-shaped tube for his experiments on the relation between $P$ and $V$. The volume of air trapped in the closed end of the tube decreases as the difference in height of mercury in the tube increases. The total pressure on the gas is equal to the pressure exerted by the added mercury ($\Delta h$) plus atmospheric pressure.

**CONCEPT TEST** ..................................................

Which of the graphs in Figure 6.12 correctly describes the relationship between the product of pressure and volume ($PV$) as a function of pressure ($P$) for a given quantity of gas at constant temperature?

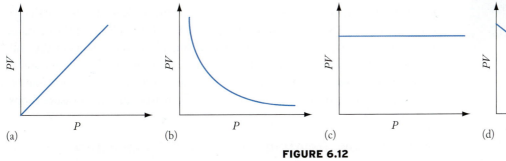

**FIGURE 6.12**

*(Answers to Concept Tests are in the back of the book.)*

.......................................................................

**SAMPLE EXERCISE 6.3**  **Using Boyle's Law**  **LO3**

A balloon is partly inflated with 5.00 L of helium at sea level, where the atmospheric pressure is 1.00 atm. The balloon ascends to an altitude of 1600 m, where the pressure is 0.83 atm. (a) What is the volume of the balloon at the higher altitude if the temperature of the helium does not change during the ascent? (b) What is the percent increase in volume?

**Charles's law** the volume of a fixed quantity of gas at constant pressure is directly proportional to its absolute temperature.

**absolute temperature** temperature expressed in kelvins on the absolute (Kelvin) temperature scale, on which 0 K is the lowest possible temperature.

**Collect and Organize** We are given the volume ($V_1$ = 5.00 L) of a gas and its pressure ($P_1$ = 1.00 atm) and asked to find the volume ($V_2$) of the gas when the pressure changes ($P_2$ = 0.83 atm). The problem states that the temperature of the gas does not change.

**Analyze** The balloon contains a fixed amount of gas and its temperature is constant, so pressure and volume are related by Boyle's law (Equation 6.6). We predict that the volume should increase because pressure decreases.

**Solve**

a. Rearranging Equation 6.6 to solve for $V_2$ and inserting the given values of $P_1$, $P_2$, and $V_1$ gives us

$$V_2 = \frac{P_1 V_1}{P_2}$$

$$V_2 = \frac{(1.00 \text{ atm})(5.00 \text{ L})}{0.83 \text{ atm}} = 6.0 \text{ L}$$

b. To calculate the percent increase in the volume, we must determine the difference between $V_1$ and $V_2$ and compare the result with $V_1$:

$$\% \text{ increase in volume} = \frac{V_2 - V_1}{V_1} \times 100\%$$

$$= \frac{6.0 \text{ L} - 5.0 \text{ L}}{5.0 \text{ L}} \times 100\% = 20\%$$

**Think About It** The prediction we made about the volume increasing is confirmed. Notice how Equation 6.6 can also be used to calculate the change in pressure that takes place at constant temperature when the volume of a quantity of gas changes.

⚙ **Practice Exercise** A scuba diver exhales 3.50 L of air while swimming at a depth of 20.0 m, where the sum of atmospheric pressure and water pressure is 3.00 atm. By the time the exhaled air rises to the surface, where the pressure is 1.00 atm, what is its volume?

■ • • • • • • • • • • • • • • • • • • • • • • • • • • • • • • • • • • •

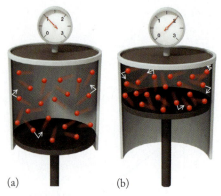

**FIGURE 6.13** (a) Gas molecules are in constant random motion, exerting pressure through collisions with the interior surface of their container. (b) When a quantity of gas is squeezed into half its original volume, the frequency of collisions increases by a factor of 2, and so does the pressure.

At the time Boyle discovered the relationship between $P$ and $V$, scientists lacked a clear understanding of atoms or molecules. Given what we now know about matter, how can Boyle's law be explained on a molecular basis? Consider the effect of compressing a collection of gas molecules into a smaller space (Figure 6.13). The molecules are in constant random motion, which means that they collide with one another and with the interior surface of their container. The collisions with the interior surface are responsible for the pressure exerted by the gas. If the molecules are squeezed into a smaller space—if the volume of the container is decreased—more collisions occur per unit time. The more frequent the collisions, the greater the force exerted by the molecules against the interior surface and thus the greater the pressure. The converse is also true: as the volume of a container is increased, the collision frequency decreases, and the pressure drops.

## Charles's Law: Relating Volume and Temperature

Nearly a century after Boyle's discovery of the inverse relation between the pressure exerted by a gas and the volume of the gas, French scientist Jacques Charles (1746–1823) documented the linear relation between the volume and temperature of a fixed quantity of gas at constant pressure. Now known as **Charles's law**, the relation states that, when the pressure exerted on a gas is held constant, the volume of a fixed quantity of gas is directly proportional to the **absolute temperature** (the temperature on the Kelvin scale) of the gas:

$$V \propto T \qquad (P \text{ and } n \text{ fixed}) \qquad (6.7)$$

Because the lowest temperature on the absolute temperature scale is 0 K, temperatures expressed in kelvins are always positive numbers.

The effects of Charles's law can be seen in Figure 6.14, in which a balloon has been attached to a flask, trapping a fixed amount of gas in the apparatus. Heating the flask causes the gas to expand, inflating the balloon. The higher the temperature, the greater the volume occupied by the gas, and the bigger the balloon.

As with Boyle's law, we can replace the proportionality symbol in Equation 6.7 by an equals sign if we include a proportionality constant:

$$V = (\text{constant})\,T$$

$$\frac{V}{T} = \text{constant} \tag{6.8}$$

The value of the constant depends on the number of moles of gas in the sample ($n$) and on the pressure of the gas ($P$). When those two parameters are held constant, the ratio $V/T$ does not change, and any two combinations of volume and temperature are related as follows:

$$\frac{V_1}{T_1} = \frac{V_2}{T_2} \tag{6.9}$$

Charles's law allows us to predict that a decrease in temperature from 20°C to 0°C reduces volume by about 7%. Let's see how we arrive at those numbers. Consider what happens when a balloon containing 2.00 L of air at 20°C is taken outside on a day when the temperature is 0°C. The amount of gas in the balloon is fixed, and the atmospheric pressure is constant. Experience tells us that the volume of the balloon decreases; Charles's law enables us to quantify that change. We can calculate the final volume with Equation 6.9, provided we express $T_1$ and $T_2$ in kelvins, not degrees Celsius. Solving Equation 6.9 for $V_2$:

$$V_2 = \frac{V_1 T_2}{T_1}$$

and substituting the known volume and temperature values:

$$V_2 = \frac{2.00\ \text{L} \times 273\ \cancel{\text{K}}}{293\ \cancel{\text{K}}} = 1.86\ \text{L}$$

As predicted, the volume of the balloon decreases when temperature decreases. The percent change in the volume of the balloon is

$$\% \text{ decrease} = \frac{V_1 - V_2}{V_1} \times 100\% = \frac{2.00\ \cancel{\text{L}} - 1.86\ \cancel{\text{L}}}{2.00\ \cancel{\text{L}}} \times 100\% = 7.0\%$$

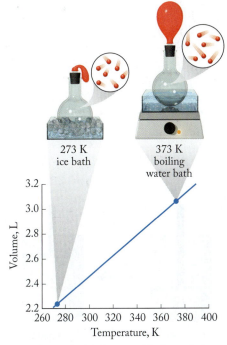

**FIGURE 6.14** A balloon attached to a flask inflates as the temperature of the gas inside the flask increases from 273 K to 373 K at constant atmospheric pressure. This behavior is described by Charles's law.

**CONNECTION** We learned in Chapter 1 that Kelvin and Celsius temperatures are related by the equation K = °C + 273.15.

**CONCEPT TEST** • • • • • • • • • • • • • • • • • • • • • • • • • • • • • • • • • •

Suppose graph "A" is a plot of volume ($V$) as a function of temperature ($T$) for 1 mole of a gas at a pressure of 1.00 atm, and graph "B" is a plot of volume ($V$) as a function of ($T$) for 1 mole of a gas at 2.00 atm pressure. How do the slopes of the two graphs differ?

• • • • • • • • • • • • • • • • • • • • • • • • • • • • • • • • • • • • • • • • • •

**SAMPLE EXERCISE 6.4    Using Charles's Law**        **LO3**

Charles was drawn to the study of gases in the late 18th century because of his interest in hot-air balloons. Scientists of his time might have designed an experiment to answer the question: What Celsius temperature is required to increase the volume of a sealed balloon from 2.00 L to 3.00 L if the initial temperature is 15°C and the atmospheric pressure is constant?

**Collect and Organize** We are given the volume ($V_1 = 2.00$ L) of a gas at an initial Celsius temperature and asked to calculate the Celsius temperature needed to make the gas reach a different volume ($V_2 = 3.00$ L). The container (a balloon) is sealed, so $n$ is constant, as is atmospheric pressure.

**Analyze** The relationship between volume and temperature is given by Charles's law in Equation 6.9. Charles's law tells us that for volume to increase, temperature must also increase. If the volume increases by 50%, we predict that the absolute temperature must also increase by 50%.

**Solve** First we rearrange Equation 6.9 to solve for $T_2$:

$$T_2 = \frac{V_2 T_1}{V_1}$$

We then substitute for $V_1$, $V_2$, and $T_1$, remembering to convert temperature to kelvins:

$$T_2 = \frac{V_2 T_1}{V_1} = \frac{(3.00 \text{ L})(288 \text{ K})}{2.00 \text{ L}} = 432 \text{ K}$$

The final step is to convert the final temperature to degrees Celsius:

$$T_2 = 432 \text{ K} - 273 = 159°C$$

**Think About It** To increase the volume by 1.5 times [(1.5)(2.00 L) = 3.00 L], *absolute* temperature must increase 1.5 times [(1.5)(288 K) = 432 K].

⚙ **Practice Exercise** Hot expanding gases can be used to perform useful work in a cylinder fitted with a movable piston, as in Figure 6.15. If the temperature of a gas confined to such a cylinder is raised from 245°C to 560°C, what is the ratio of the initial volume to the final volume if the pressure exerted on the gas remains constant?

$T_1 = 245°C \qquad T_2 = 560°C$

**FIGURE 6.15** The volume of a gas in a cylinder changes as the temperature of the gas increases from 245°C to 560°C.

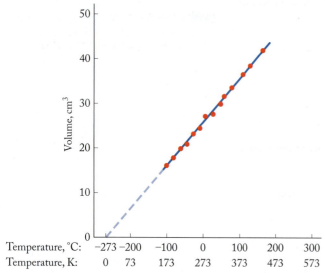

| Temperature, °C: | −273 | −200 | −100 | 0 | 100 | 200 | 300 |
|---|---|---|---|---|---|---|---|
| Temperature, K: | 0 | 73 | 173 | 273 | 373 | 473 | 573 |

**FIGURE 6.16** The volume of a gas for which $n$ and $P$ are held constant plotted against temperature on the Celsius and Kelvin scales. As predicted by Charles's law, volume decreases as the temperature decreases; the relationship is linear on both scales. The dashed line shows the extrapolation from the experimental data to a point corresponding to zero volume. This temperature is known as absolute zero: −273.15°C on the Celsius scale and 0 K on the Kelvin scale.

Charles's law also allowed the original determination that absolute zero (0 K) is equal to −273.15°C. Consider what happens when we plot the volume of a fixed quantity of gas at constant pressure as a function of temperature. Some typical results are graphed in Figure 6.16, showing that the decrease in volume with decreasing temperature is linear. We are limited by how low we can make the temperature of a gas because at some temperature the gas liquefies, at which point the gas laws no longer describe its behavior because it is no longer a gas. However, we can *extrapolate* from our measured data to the point where the volume of a gas would reach zero if liquefaction did not occur. The dashed line in Figure 6.16 crosses the temperature axis at −273.15°C, the temperature defined as 0 K, or absolute zero.

## Avogadro's Law: Relating Volume and Quantity of Gas

Boyle's and Charles's experiments involved a constant quantity of gas in a confined sample. Now consider what happens when the quantity of gas is not constant, as when a balloon is inflated. The larger the quantity of gas blown into the balloon, the larger the balloon gets until it finally bursts. Clearly, the volume depends on the quantity (number of grams or moles) of the gas. If you poke a hole in the balloon, the gas escapes, decreasing both the quantity of gas and the volume. The relationship between the pressure, volume, and the quantity of the gas can be illustrated in several ways.

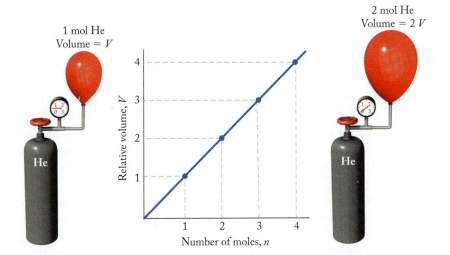

**FIGURE 6.17** Because it is filled with twice the quantity of helium, the balloon on the right has twice the volume of the balloon on the left. This direct relation between volume and number of moles of a gas at constant temperature and pressure is Avogadro's law.

Consider a bicycle tire that is inflated enough to hold its shape but is nevertheless too soft to ride on. We can increase the pressure inside the tire by pumping more air into it. In the process, the tire does not expand much because the heavy rubber is much more rigid than the material in a balloon. As another example, consider a carnival vendor filling balloons from a helium tank fitted with a pressure gauge (Figure 6.17). When sales are brisk, it is possible to observe the pressure decreasing as the vendor fills more and more balloons. The pressure in the tank decreases as the amount of gas in the cylinder decreases. Because the balloons are much more flexible than a tire, the size of the balloons the vendor inflates depends on the amount of gas he dispenses: the more gas he adds to a balloon, the larger it becomes.

A few decades after Charles discovered the relation between $V$ and $T$, Amedeo Avogadro (1776–1856), whom we know from Avogadro's number ($N_A$), recognized the relationship between $V$ and $n$. This relation is described by **Avogadro's law**, which states that the volume of a gas at a given temperature and pressure is proportional to the quantity of the gas in moles:

**CONNECTION** In Chapter 3 the number of particles in a mole was defined as Avogadro's number, in honor of his early work with gases that led to determining atomic masses.

$$V \propto n \quad \text{or} \quad \frac{V}{n} = \text{constant} \qquad (P \text{ and } T \text{ fixed}) \qquad (6.10)$$

**CONCEPT TEST** ················································

Which graph in Figure 6.18 correctly describes the relationship between the value of $V/n$ as $n$ is increased (at constant $P$ and $T$)?

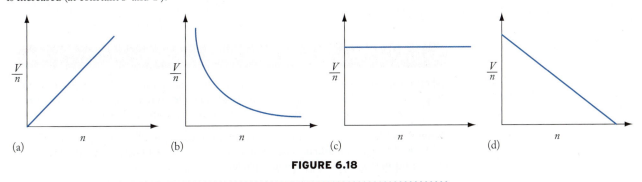

**FIGURE 6.18**

## Amontons's Law: Relating Pressure and Temperature

Both Boyle's law ($PV = $ constant) and Charles's law ($V/T = $ constant) describe effects on volume, but is there a relation between pressure and temperature?

**Avogadro's law** the volume of a gas at a given temperature and pressure is proportional to the quantity of the gas.

Experiments show that pressure is directly proportional to temperature when $n$ and $V$ are constant:

$$P \propto T \quad \text{or} \quad \frac{P}{T} = \text{constant} \qquad (V \text{ and } n \text{ fixed}) \qquad (6.11)$$

This relationship means that as the absolute temperature of a fixed amount of gas held at a constant volume increases, the pressure of the gas increases (Figure 6.19). This statement is referred to as **Amontons's law** in honor of French physicist Guillaume Amontons (1663–1705), a contemporary of Robert Boyle's, who constructed a thermometer based on the observation that the pressure of a gas is directly proportional to its temperature.

Where do we see evidence of Amontons's law? Consider a bicycle tire inflated to a prescribed pressure of 100 psi (6.8 atm) on an afternoon in late autumn when the temperature is 25°C. The next morning, following an early freeze in which the temperature dropped to 0°C, the pressure in the tire drops to about 91 psi (6.2 atm). Did the tire leak? Perhaps, but the decrease in pressure could also be explained by Amontons's law.

We can confirm that the decrease in tire pressure is consistent with the decrease in temperature by applying the logic we used to develop Equations 6.6 and 6.9. Relating initial pressure and temperature to final pressure and temperature, we can write

$$\frac{P_1}{T_1} = \frac{P_2}{T_2} \qquad (6.12)$$

Solving for $P_2$, we get

$$P_2 = \frac{P_1 T_2}{T_1} = \frac{(6.8 \text{ atm})(273 \text{ K})}{298 \text{ K}} = 6.2 \text{ atm}$$

Why is pressure directly proportional to temperature? Like pressure, temperature is directly related to molecular motion. The average speed at which a population of molecules moves increases with increasing temperature (see Figure 6.19). For a given number of gas molecules, increasing the temperature increases the average speed and therefore increases the frequency and force with which the molecules collide with the walls of their container. Because pressure is related to the frequency and force of the collisions, it follows that higher temperatures produce higher pressures, as long as other factors such as volume and quantity of gas are constant.

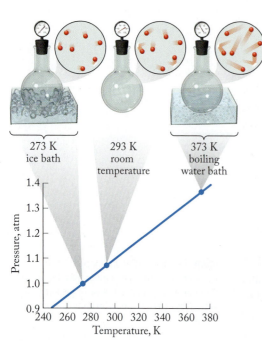

**FIGURE 6.19** The pressure of a given quantity of gas is directly proportional to its absolute temperature when the gas volume is held constant. Each flask has the same volume and contains the same number of molecules. The relative speeds of the molecules (represented by the lengths of their "tails") increase with increasing temperature, causing more frequent and more forceful collisions with the walls of the flasks and hence higher pressures.

---

Labels on aerosol cans caution against incineration because the cans may explode when the pressure inside them exceeds 3.00 atm. At what temperature in degrees Celsius will an aerosol can burst if the initial pressure inside the can is 2.20 atm at 25°C?

**Collect and Organize** We are given the temperature ($T_1 = 25°C$) and pressure ($P_1 = 2.20$ atm) of a gas and asked to determine the temperature ($T_2$) at which the pressure ($P_2 = 3.00$ atm) will cause the can to explode.

**Analyze** Because the gas is enclosed in an aerosol can, we know that the quantity of gas and volume of the gas are constant. We can use Amontons's law (Equation 6.12) because it describes what happens to pressure as the temperature of a gas is changed. We can estimate the answer by considering that the pressure in the can must increase by about 50% in order for the pressure to exceed 3 atm. Pressure is directly proportional to temperature, so the absolute temperature must also increase by about 50%. We should note that the initial temperature is given in degrees Celsius. Because Equation 6.12 works only with absolute temperatures, we must convert the initial temperature to kelvins. After solving for $T_2$, we need to convert kelvins back to degrees Celsius to report our answer.

**Solve** We rearrange Equation 6.12 to solve for $T_2$:

$$T_2 = \frac{T_1 P_2}{P_1}$$

After converting $T_1$ from degrees Celsius to kelvins, we have

$$T_2 = \frac{(25 + 273) \text{ K} \times 3.00 \text{ atm}}{2.20 \text{ atm}} = 406 \text{ K}$$

Converting $T_2$ to degrees Celsius gives us

$$T_2 = 406 \text{ K} - 273 = 133°\text{C}$$

**Think About It** This temperature is certainly higher than the original temperature. In fact 406 K is about 1.5 times (or 50% higher than) the initial temperature, 298 K, so the answer makes sense.

⚙ **Practice Exercise** The air pressure in the tires of an automobile is adjusted to 28 psi at a gas station in San Diego, where the air temperature is 68°F. The air in the tires is at the same temperature as the atmosphere. The automobile is then driven east along a hot desert highway, and the temperature inside the tires reaches 140°F. What is the pressure in the tires? ∎

Gases that behave in accordance with the linear relations discovered by Boyle, Charles, Avogadro, and Amontons are called **ideal gases**. Most gases exhibit ideal behavior at the pressures and temperatures typically encountered in the atmosphere. Under these conditions, the volumes occupied by the gas molecules or atoms are insignificant compared with the overall volume occupied by the gas. All the open space between molecules makes gases compressible. In an ideal gas, the atoms or molecules are assumed not to interact with one another; rather, they move independently with speeds that are related to their masses and to the temperature of the gas.

# 6.4 The Ideal Gas Law

What would happen if we launched a weather balloon from the surface of Earth and allowed it to drift to an elevation of 10,000 m? The volume of the balloon would expand as the atmospheric pressure decreased, but the air temperature would decrease as the balloon ascended, tending to make the volume smaller. How do we determine the final volume of the balloon when we are changing pressure and temperature while keeping the quantity of gas constant? Taken individually, none of the four gas laws and their accompanying mathematical relations fit this situation. Nevertheless, we can derive a relation, known as the **ideal gas equation**, that combines all four variables:

$$PV = nRT \tag{6.13}$$

where $R$ is the **universal gas constant**. This equation is also called the **ideal gas law**.

As Table 6.3 shows, $R$ has different values depending on the units used. For many calculations in chemistry, it is convenient to use $R = 0.08206$ L · atm/(mol · K). As the units tell us, however, we can use this value only when the quantity of gas is expressed in moles, the volume in liters, the pressure in atmospheres, and the temperature in kelvins.

---

**Amontons's law** the pressure of a quantity of gas at constant volume is directly proportional to its absolute temperature.

**ideal gas** a gas whose behavior is predicted by the linear relations defined by Boyle's, Charles's, Avogadro's, and Amontons's laws.

**ideal gas equation** (also called **ideal gas law**) relates the pressure, volume, number of moles, and temperature of an ideal gas; expressed as $PV = nRT$, where $R$ is the universal gas constant.

**universal gas constant** the constant $R$ in the ideal gas equation; its value and units depend on the units used for the variables in the equation.

---

| TABLE 6.3 | Values for the Universal Gas Constant (R) |
|---|---|
| **Value of R** | **Units** |
| 0.08206 | L · atm/(mol · K) |
| 8.314 | kg · m²/(s² · mol · K) |
| 8.314 | J/(mol · K) |
| 8.314 | m³ · Pa/(mol · K) |
| 62.37 | L · torr/(mol · K) |

**combined gas law** (also called **general gas equation**) relates the pressure, volume, and temperature of a quantity of an ideal gas:

$$\frac{P_1 V_1}{T_1} = \frac{P_2 V_2}{T_2}$$

**standard temperature and pressure (STP)** 0°C and 1 bar as defined by IUPAC; in the United States, 0°C and 1 atm.

**molar volume** volume occupied by 1 mole of an ideal gas at STP; 22.4 L.

▶❚❚ **CHEMTOUR** The Ideal Gas Law

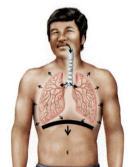

(a)                    (b)

**FIGURE 6.20** Breathing illustrates the relationship between *P*, *V*, and *n*. (a) When you inhale, your rib cage expands and your diaphragm moves down, (b) increasing the volume of your lungs. Increased volume decreases the pressure inside your lungs, in accord with Boyle's law: *PV* = constant. Decreased pressure allows more air to enter until the pressure inside your lungs matches atmospheric pressure.

Where does the ideal gas equation come from? Boyle's law states that volume and pressure are inversely proportional. Charles's law tells us that volume is directly proportional to temperature. Putting Boyle's and Charles's laws together (Equations 6.5 and 6.8) allows us to write an equation relating *P*, *V*, and *T* and combining the constants:

$$PV = \text{constant} \qquad \text{Boyle's Law}$$

$$\frac{V}{T} = \text{constant} \qquad \text{Charles's Law}$$

$$\frac{PV}{T} = \text{combined constant} \qquad (6.14)$$

Avogadro's law tells us that volume is directly proportional to the number of moles of gas when *T* and *P* are constant, so we can include *n* in Equation 6.14 and combine its constant with the "combined constant" in that equation:

$$\frac{PV}{nT} = \text{combined constant} \quad \text{or} \quad PV = (\text{combined constant}) \times nT \quad (6.15)$$

We turn Equation 6.15 into a meaningful equality by inserting the appropriate constant. By convention, this constant is the universal gas constant *R*, giving us the ideal gas law (Equation 6.13).

We can derive another relation involving these four variables that is useful when a system starts in an initial state ($P_1$, $V_1$, $T_1$, $n_1$) and moves to a final state ($P_2$, $V_2$, $T_2$, $n_2$):

$$P_1 V_1 = n_1 R T_1 \quad \text{so} \quad \frac{P_1 V_1}{n_1 T_1} = R \quad \text{and}$$

$$P_2 V_2 = n_2 R T_2 \quad \text{so} \quad \frac{P_2 V_2}{n_2 T_2} = R \quad (6.16)$$

Because both the initial-state and final-state expressions are equal to *R*, they are also equal to each other, allowing us to simplify Equation 6.16 to

$$\frac{P_1 V_1}{n_1 T_1} = \frac{P_2 V_2}{n_2 T_2} \quad (6.17)$$

Furthermore, because we frequently work with systems like weather balloons and gas canisters in which the amount of gas is constant ($n_1 = n_2$), but pressure, temperature, and volume vary, we define an especially useful form of this equation:

$$\frac{P_1 V_1}{T_1} = \frac{P_2 V_2}{T_2} \qquad \text{for constant } n \quad (6.18)$$

Equation 6.18 is known as the **combined gas law**, or the **general gas equation**. Sample Exercise 6.6 gives an example of how to use this equation for determining the effect of changes in *P* and *T* on the volume (*V*) of a weather balloon. Different reduced versions of Equation 6.17 are certainly possible when other variables besides *n* are fixed (Figure 6.20), but only Equation 6.18 is known as the combined gas law. Figure 6.21 summarizes the four laws expressed within the ideal gas law.

(a) **Boyle's law:** volume inversely proportional to pressure; n and T fixed

$PV = \text{constant}$

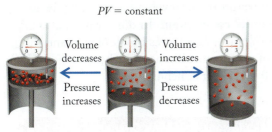

(b) **Charles's law:** volume directly proportional to temperature; n and P fixed

$\dfrac{V}{T} = \text{constant}$

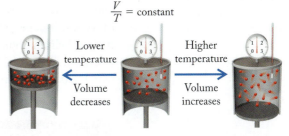

(c) **Avogadro's law:** volume directly proportional to number of moles; T and P fixed

$\dfrac{V}{n} = \text{constant}$

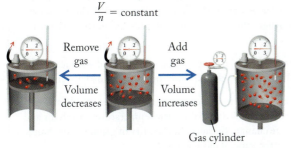

Gas cylinder

(d) **Amontons's law:** pressure directly proportional to temperature; n and V fixed

$\dfrac{P}{T} = \text{constant}$

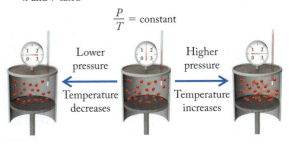

**FIGURE 6.21** Relationships between pressure, volume, temperature, and/or moles of gas. (a) Increasing or decreasing V at constant n and T: Boyle's law. (b) Increasing or decreasing T at constant n and P: Charles's law. (c) Increasing or decreasing n at constant T and P: Avogadro's law. (d) Increasing or decreasing P at constant n and V: Amontons's law.

**CONCEPT TEST**

Which of the following variations of Equation 6.17 are incorrect?

a. $\dfrac{n_2 T_2}{P_2} = \dfrac{n_1 T_1}{P_1}$ at constant $V$

b. $\dfrac{n_2 V_2}{P_2} = \dfrac{n_1 V_1}{P_1}$ at constant $T$

c. $\dfrac{n_2 T_2}{V_2} = \dfrac{n_1 T_1}{V_1}$ at constant $P$

d. $\dfrac{T_1}{n_1 V_1} = \dfrac{T_2}{n_2 V_2}$ at constant $P$

A useful reference point defined by the International Union of Pure and Applied Chemistry (IUPAC) and used in studying the properties of gases is **standard temperature and pressure (STP)**, defined as 0°C and 1 bar. The more familiar unit of 1 atm is very close to 1 bar, and at the level of accuracy of calculations in this text, this substitution makes little difference, so we consider STP to be 0°C and 1 atm.

Another useful reference point is **molar volume**, which is the volume that 1 mole of an ideal gas occupies at STP (Figure 6.22). We can calculate the molar volume from the ideal gas equation by solving the equation for V and inserting the values of n, P, and T at STP:

$$V = \dfrac{(1 \text{ mol})\left(0.08206 \dfrac{\text{L} \cdot \text{atm}}{\text{mol} \cdot \text{K}}\right)(273 \text{ K})}{1 \text{ atm}} = 22.4 \text{ L}$$

Many chemical and biochemical processes take place at pressures near 1 atm and at temperatures between 0°C and 40°C. Within this range, the volume that 1 mole of gaseous reactant or product occupies is no more than about 15%

**FIGURE 6.22** The box contains 1 mole of gas; the molar volume of an ideal gas is 22.4 L at 0° and 1 atm of pressure. A basketball fits loosely into a box having this volume.

greater than the molar volume. Therefore volumes can be estimated easily if molar amounts are known. An important feature of molar volume is that it applies to any ideal gas, independent of its chemical composition. In other words, at STP, 1 mole of helium occupies the same volume—22.4 L—as 1 mole of methane ($CH_4$), or carbon dioxide ($CO_2$), or even of a compound like uranium hexafluoride ($UF_6$), which has a molar mass almost 90 times that of helium.

---

**SAMPLE EXERCISE 6.6**    **Calculations Involving Changes in *P*, *V*, and *T***    **LO3**

A weather balloon filled with 100.0 L of He is launched from ground level ($T = 20°C$, $P = 755$ torr). No gas is added or removed from the balloon during its flight. Calculate its volume at an altitude of 10 km, where the temperature and pressure in the balloon are $-52°C$ and 195 torr, respectively.

**Collect and Organize** We are given the initial temperature, pressure, and volume of a gas and are asked to determine the final volume after the pressure and temperature have changed.

**Analyze** The decrease in temperature leads to a decrease in volume, but the decrease in pressure leads to an increase in volume. We need to express the temperatures in kelvins and estimate which variable dominates the change. Because the quantity of gas does not change ($n$ is constant), we can solve for $V_2$ in the general gas equation.

The decrease in pressure to 195 torr (about 25% of the starting value) is considerably larger than the decrease in temperature from 20°C to −52°C (about 75% of the starting value). Therefore, we conclude that pressure is the dominant variable and that the volume of the balloon increases.

**Solve** First we convert the given Celsius temperatures to kelvins:

$$T_1 = 20°C + 273 = 293 \text{ K}$$
$$T_2 = -52°C + 273 = 221 \text{ K}$$

Then we use the general gas equation to solve for $V_2$:

$$\frac{P_1 V_1}{T_1} = \frac{P_2 V_2}{T_2}$$

$$V_2 = V_1 \times \frac{P_1}{P_2} \times \frac{T_2}{T_1}$$

$$V_2 = (100.0 \text{ L}) \times \frac{755 \text{ torr}}{195 \text{ torr}} \times \frac{221 \text{ K}}{293 \text{ K}} = 292 \text{ L}$$

**Think About It** The volume increases nearly threefold as the balloon ascends to 10 km. This result makes sense because atmospheric pressure decreases to about one-quarter of its ground-level value during the ascent. The volume would have increased more if not for the countervailing effect of the lower temperature at 10 km.

⚙ **Practice Exercise** The balloon in Sample Exercise 6.6 is designed to continue its ascent to an altitude of 30 km, where it bursts, releasing a package of meteorological instruments that parachute back to Earth. If the atmospheric pressure at 30 km is 28.0 torr and the temperature is −45°C, what is the volume of the balloon when it bursts?

---

The ideal gas law describes the relations among number of moles, pressure, volume, and temperature for any gas, provided it behaves as an ideal gas. Because many gases behave ideally at typical atmospheric pressures, we can apply the ideal gas equation to many situations outside the laboratory. In Sample Exercise 6.7, we calculate the mass of oxygen in an alpine climber's compressed-oxygen cylinder.

**SAMPLE EXERCISE 6.7** **Applying the Ideal Gas Law** **LO3**

Bottles of compressed $O_2$ carried by climbers ascending Mt. Everest have an internal volume of 5.90 L. Assume that such a bottle has been filled with $O_2$ to a pressure of 2025 psi at 25°C. Also assume that $O_2$ behaves as an ideal gas. (a) How many moles of $O_2$ are in the bottle? (b) What is the mass in grams of $O_2$ in the bottle?

**Collect and Organize** We are given the pressure, volume, and temperature of $O_2$ and asked to determine its quantity in moles and its mass in grams.

**Analyze** The ideal gas equation enables us to use $P$, $V$, and $T$ to calculate $n$, the number of moles of $O_2$. Then we can use its molar mass (32.00 g/mol) to calculate the mass of $O_2$ in the bottle. We can estimate the answer by considering that 1 mole of gas at STP occupies 22.4 L. Our oxygen bottle has a volume of about 6 L, which is about one-quarter of the molar volume at STP, but the pressure (2025 psi) is more than 100 times greater than 1 atm (see Table 6.2). Thus, we predict that the bottle contains more than 25 moles of $O_2$.

**Solve**

a. Let's start with the ideal gas equation rearranged to solve for $n$:

$$PV = nRT$$

$$n = \frac{PV}{RT}$$

Before using this expression for $n$, we need to convert pressure into atmospheres and temperature into kelvins:

$$P = (2025 \text{ psi})\left(\frac{1 \text{ atm}}{14.7 \text{ psi}}\right) = 138 \text{ atm} \qquad T = 25°\text{C} + 273 = 298 \text{ K}$$

$$n = \frac{(138 \text{ atm})(5.90 \text{ L})}{\left(0.08206 \frac{\text{L} \cdot \text{atm}}{\text{mol} \cdot \text{K}}\right)(298 \text{ K})} = 33.3 \text{ mol}$$

b. Converting moles into grams is a matter of multiplying by the molar mass:

$$(33.3 \text{ mol})\left(\frac{32.00 \text{ g}}{1 \text{ mol}}\right) = 1.07 \times 10^3 \text{ g}$$

**Think About It** Our answer in part a is certainly reasonable based on our estimate. Most climbers require several bottles to climb Mt. Everest and return.

⚙ **Practice Exercise** Starting with the moles of $O_2$ calculated in Sample Exercise 6.7, calculate the volume of $O_2$ the bottle could deliver to a climber at an altitude where the temperature is −38°C and the atmospheric pressure is 0.35 atm.

# 6.5 Gases in Chemical Reactions

Gases are reactants or products in many important reactions. For example, if a commercial airliner flying at high altitudes loses cabin pressure, oxygen masks are deployed automatically for the passengers to breathe until the aircraft can descend or the cabin can be repressurized. The oxygen gas that flows in the masks is generated by the decomposition of sodium chlorate. Similarly, when an air bag deploys during an automobile accident, the nitrogen gas that rapidly inflates the protective air bag is the product of the decomposition of sodium azide. Gases are also reactants, as in the combustion of charcoal in a backyard grill. Solid carbon reacts with oxygen gas in the air to produce carbon dioxide and the heat used to cook our food:

$$C(s) + O_2(g) \rightarrow CO_2(g) + \text{heat} \qquad (6.19)$$

In any chemical reaction that involves a gas as either reactant or product, the volume of the gas indirectly defines the amount of it in the reaction. If $T$ and $P$ are known, we can use the ideal gas equation to relate volume to the number of moles of gas in the system. Once we know that, we can use stoichiometric calculations to relate quantities of gas to quantities of other reactants and products, including heat. For example, consider the volume of oxygen needed to completely burn 1.00 kg (about 2 lb) of charcoal at 1.00 atm of pressure on an average summer day (25°C). Before starting the calculation we must first write a balanced chemical equation for the reaction. Equation 6.19 is indeed balanced, and 1 mole of C reacting with 1 mole of oxygen should yield 1 mole of $CO_2$. If we start with 1.00 kg of C we have

$$1.00 \, \cancel{kg \, C} \times \frac{10^3 \, \cancel{g}}{1 \, \cancel{kg}} \times \frac{1 \, mol \, C}{12.01 \, \cancel{g \, C}} = 83.3 \, mol \, C$$

The stoichiometry of the reaction tells us that we need 83.3 moles of $O_2$ to completely react with the given amount of carbon. The volume of $O_2$ that corresponds to 83.3 moles of $O_2$ is calculated using the ideal gas equation by first rearranging the equation to solve for $V$ and then substituting the values of $n$, $R$, $T$ (in kelvins), and $P$:

$$V = \frac{nRT}{P} = \frac{(83.3 \, \cancel{mol})\left(0.08206 \frac{L \cdot \cancel{atm}}{\cancel{mol} \cdot \cancel{K}}\right)(298 \, \cancel{K})}{1.00 \, \cancel{atm}} = 2.04 \times 10^3 \, L$$

---

**SAMPLE EXERCISE 6.8    Combining Stoichiometry and    LO4**
**the Ideal Gas Law**

Oxygen generators in some airplanes are based on the chemical reaction between sodium chlorate and iron:

$$NaClO_3(s) + Fe(s) \rightarrow O_2(g) + NaCl(s) + FeO(s)$$

The resultant $O_2$ is blended with cabin air to provide 10–15 minutes of breathable air for passengers. How many grams of $NaClO_3$ are needed in a typical generator to produce 125 L of $O_2$ gas at 1.00 atm and 20.0°C?

**Collect and Organize**  We are given the volume of $O_2$ we need to prepare at a particular pressure and temperature. With this information we can determine the mass of $NaClO_3$ needed based on the stoichiometric relations in the balanced chemical equation.

**Analyze**  The solution requires two calculations. (1) We start by recognizing that the balanced chemical equation tells us that 1 mole of $NaClO_3$ is needed to produce 1 mole of oxygen. If we can determine how many moles of $O_2$ occupy a volume of 125 L at 1.00 atm pressure and 20.0°C, we can determine the number of moles of $NaClO_3$ we need. To determine the moles of oxygen ($n_{O_2}$), we can use the ideal gas law (Equation 6.13). (2) Then we use our calculated value of $n_{O_2}$ and the balanced chemical equation to determine the number of moles and number of grams of $NaClO_3(s)$ required.

We can estimate the answer by comparing the volume of gas we wish to make (125 L) with the molar volume of an ideal gas at STP (22.4 L). Considering that the difference between the temperature at STP (0°C = 273 K) and the temperature in this problem (20°C = 293 K) is relatively small, we estimate that we need about 5 moles of $O_2$, requiring 5 moles of $NaClO_3$.

**Solve**  We use the rearranged ideal gas equation to solve for moles of $O_2$:

$$n = \frac{PV}{RT}$$

$$n = \frac{(1.00 \, \cancel{atm})(125 \, \cancel{L})}{\left(0.08206 \frac{\cancel{L} \cdot \cancel{atm}}{mol \cdot \cancel{K}}\right)(273 + 20.0) \, \cancel{K}} = 5.20 \, mol \, O_2$$

To convert moles of $O_2$ into an equivalent mass of $NaClO_3$, we use the stoichiometry of the reaction to calculate the equivalent number of moles of $NaClO_3$, and use molar mass to determine the number of grams of $NaClO_3$ required:

$$5.20 \; \cancel{mol \; O_2} \times \frac{1 \; \cancel{mol \; NaClO_3}}{1 \; \cancel{mol \; O_2}} \times \frac{106.44 \; g \; NaClO_3}{1 \; \cancel{mol \; NaClO_3}} = 553 \; g \; NaClO_3$$

**Think About It** We predicted that about 5 moles of $NaClO_3$ would be needed to produce 125 L of oxygen, which is quite close to the calculated value of 5.20 moles of $NaClO_3$.

⚙ **Practice Exercise** Automobile air bags (Figure 6.23) inflate during a crash or sudden stop by the rapid generation of nitrogen gas from sodium azide, according to the reaction

$$2 \; NaN_3(s) \rightarrow 2 \; Na(s) + 3 \; N_2(g)$$

How many grams of sodium azide are needed to provide sufficient nitrogen gas to fill a 45.0 cm × 45.0 cm × 25.0 cm bag to a pressure of 1.20 atm at 15°C?

**FIGURE 6.23** An automobile air bag inflates when solid $NaN_3$ rapidly decomposes, producing $N_2$ gas.

---

| SAMPLE EXERCISE 6.9 | **Reactant Volumes and the Ideal Gas Law** | LO4 |

Reactions between methane and ammonia are of interest to scientists exploring non-biological reactions that lead to amino acids. One such reaction produces $H_2NCN$:

$$CH_4(g) + 2 \; NH_3(g) \rightarrow H_2NCN(g) + 4 \; H_2(g)$$

If a reaction mixture contains 1.25 L methane and 3.00 L ammonia, what volume of hydrogen is produced at constant temperature and pressure if the reaction proceeds to 100% yield?

**Collect and Organize** We are given a balanced chemical equation and the volumes of the starting materials. We are asked to calculate the volume of hydrogen produced. Temperature and pressure are constant but their values are not specified.

**Analyze** The balanced chemical equation relates the numbers of moles of reactants and products. In this case, 4 moles of $H_2$ are expected for every 1 mole of $CH_4$ and every 2 moles of $NH_3$. We could calculate the moles of $CH_4$ and $NH_3$ using the ideal gas law, Equation 6.13, but we lack values for $P$ and $T$. However, since $P$ and $T$ are constant, we can use Avogadro's Law instead, which states that $V \propto n$ (Equation 6.10), and use the volumes of the starting materials in place of moles.

**Solve** The balanced chemical equation tells us that complete reaction of 1.25 L methane requires 2.5 L of ammonia:

$$V \propto n$$

$$1.25 \; \cancel{L \; CH_4} \times \frac{2 \; L \; NH_3}{1 \; \cancel{L \; CH_4}} = 2.50 \; L \; NH_3$$

Since we have 3.00 L of ammonia available, $NH_3$ is in excess and $CH_4$ is the limiting reactant. The volume of hydrogen produced by the reaction is

$$1.25 \; \cancel{L \; CH_4} \times \frac{4 \; L \; H_2}{1 \; \cancel{L \; CH_4}} = 6.00 \; L \; H_2$$

**Think About It** At a constant temperature and pressure, the ideal gas law tells us that the volume of 1.00 mol of gas is the same, regardless of its identity. This means that we can compare the volumes of gases the same way we compare moles of gases. It is sometimes easier to measure the volumes of gases rather than their masses in reactions.

⚙ **Practice Exercise** If 316 mL nitrogen is combined with 178 mL oxygen, what volume of $N_2O$ is produced at constant temperature and pressure if the reaction proceeds to 82% yield?

$$2 \; N_2(g) + O_2(g) \rightarrow 2 \; N_2O(g)$$

**FIGURE 6.24** The release of $CO_2$ from volcanic centers on the Dieng Plateau in 1979 killed many people and animals as the dense gas flowed over the valley floor, effectively displacing the air (and oxygen) that the inhabitants and their livestock needed to survive.

# 6.6 Gas Density

Carbon dioxide is a relatively minor component of our atmosphere. Because $CO_2$ is produced by the combustion of fuels, however, the amount of it in the atmosphere has increased since the Industrial Revolution. This increase is of great concern because of its impact on climate change.

Another source of atmospheric carbon dioxide is volcanoes. Because $CO_2(g)$ is denser than air, sudden releases of large quantities from areas of volcanic activity are life-threatening events. On the Dieng Plateau in Indonesia in 1979, 149 people died from asphyxiation in a valley after a massive, sudden release of carbon dioxide from one of the volcanoes around the valley (Figure 6.24). Being denser than air, the carbon dioxide settled in the valley, displacing the air containing the oxygen necessary for life.

The density of a gas at STP can be calculated by dividing its molar mass by its molar volume. Carbon dioxide, for example, has a molar mass of 44.01 g/mol. Therefore the density of $CO_2$ at STP is

$$\frac{44.01 \text{ g/\cancel{mol}}}{22.4 \text{ L/\cancel{mol}}} = 1.96 \text{ g/L}$$

The density of air at STP is about 1.3 g/L, so it is not surprising that the $CO_2$ in the Dieng Plateau disaster concentrated at the bottom of the valley. That the density of $CO_2$ is greater than that of air also gives rise to its effectiveness in fighting fires. $CO_2$ blankets a fuel, separating it from the $O_2$ it needs to burn and thus extinguishing the fire.

> **CONCEPT TEST**
>
> Which gas has the highest density at STP: $CH_4$, $Cl_2$, Kr, or $C_3H_8$?

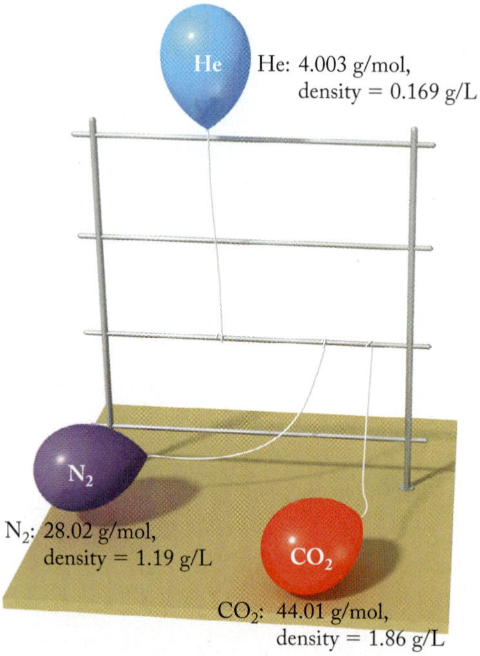

We can calculate the density of an ideal gas at any temperature and pressure using the ideal gas equation. Because density is the mass of a sample divided by its volume, $d = m/V$, we need to identify the variables in the ideal gas equation that represent the density. Sample mass can be determined from the number of moles, and of course the $V$ in the ideal gas equation is the volume. We can rearrange Equation 6.13 so that these two factors are isolated:

$$\frac{P}{RT} = \frac{n}{V} \qquad (6.20)$$

For a given sample of gas, $n$ equals the mass of the sample in grams divided by the molar mass of the gas: $m/\mathcal{M}$. We can substitute this ratio for $n$ in Equation 6.20 and solve the resulting expression for $m/V$ (the density):

$$\frac{P}{RT} = \frac{m/\mathcal{M}}{V} = \frac{m}{\mathcal{M}V}$$

$$d = \frac{m}{V} = \frac{P\mathcal{M}}{RT} \qquad (6.21)$$

If the term on the right in this last equation has $P$ in atmospheres, $\mathcal{M}$ in grams per mole, $R$ in L · atm/(mol · K), and $T$ in kelvins, then the units of density are g/L:

$$\frac{(\cancel{\text{atm}})\left(\dfrac{\text{g}}{\cancel{\text{mol}}}\right)}{\left(\dfrac{\text{L} \cdot \cancel{\text{atm}}}{\cancel{\text{mol}} \cdot \text{K}}\right)(\cancel{\text{K}})} = \text{g/L}$$

He: 4.003 g/mol, density = 0.169 g/L

$N_2$: 28.02 g/mol, density = 1.19 g/L

$CO_2$: 44.01 g/mol, density = 1.86 g/L

**FIGURE 6.25** At 15°C and 1 atm, the density of the gas inside each balloon determines whether the balloon floats (helium), hovers just slightly above the benchtop (nitrogen), or sinks (carbon dioxide). Note the correlation between molar mass and density: the higher the molar mass of a gas, the higher the density.

Figure 6.25 illustrates the relationship between density and molar mass of three pure gases and air. Note how the balloon containing $CO_2(g)$ does not float at all but sinks below the benchtop because the density of this pure gas is greater than the density of air. In the absence of mixing, carbon dioxide moves to the lowest level accessible to it, just as it filled the valley in the Dieng Plateau disaster, cutting the inhabitants off from their supply of oxygen, and just as it separates fuel from oxygen, extinguishing fires.

Equation 6.21 also explains why a balloon inflated with hot air rises (Figure 6.26). Like a bicycle tire, the volume of the balloon is essentially constant once inflated. As the temperature of the air inside the balloon increases, the balloon can no longer expand. However, the balloons in Figure 6.26 represent open systems—gas can escape from the bottom of the balloon and $n$ decreases. This can be seen in equation 6.20 where increasing $T$ at constant $P$ and $V$ means that $n$ decreases. If $n$ decreases, $d$ will also decrease. Put another way, the density of a given quantity of gas decreases with increasing temperature, giving the balloon buoyancy.

**FIGURE 6.26** Increasing the temperature of the air in the balloon leads to a decrease in density as the air expands. Hot air in a balloon is less dense than cooler air, making the balloon buoyant.

**CONCEPT TEST**

Which graph in Figure 6.27 best approximates (a) the relationship between density and pressure ($n$ and $T$ constant) and (b) the relationship between density and temperature ($n$ and $P$ constant) for an ideal gas?

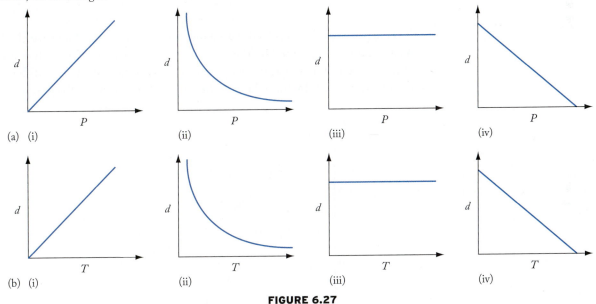

**FIGURE 6.27**

**CONCEPT TEST**

The density of methane is measured at four sets of conditions. Under which set of conditions is the density of $CH_4$ the greatest? (a) $T = 300$ K, $P = 1$ atm; (b) $T = 325$ K, $P = 1$ atm; (c) $T = 275$ K, $P = 2$ atm; (d) $T = 300$ K, $P = 2$ atm

**SAMPLE EXERCISE 6.10**   **Calculating the Density of a Gas**   **LO5**

Calculate the density of air at 1.00 atm and 302 K and compare your answer with the density of air at STP (1.29 g/L). Assume that air has an average molar mass of 28.8 g/mol. (This average molar mass is the weighted average of the molar masses of the gases in air.)

**Collect and Organize** We are provided with the average molar mass, temperature, and atmospheric pressure of air and asked to calculate its density.

**Analyze** The given quantities are related by Equation 6.21. The density of a gas is inversely proportional to temperature. The temperature given in this problem (302 K) is higher than the temperature at STP, so we predict that the density of the air will be less than 1.29 g/L.

**Solve** Inserting the values of $P$, $T$, and $\mathcal{M}$ into Equation 6.21, we have

$$d = \frac{P\mathcal{M}}{RT} = \frac{(1.00 \text{ atm})\left(28.8 \frac{g}{mol}\right)}{[0.08206 \text{ L} \cdot atm/(mol \cdot K)](302 \text{ K})} = 1.16 \text{ g/L}$$

**Think About It** The solution confirms that the density of air calculated at 302 K (1.16 g/L) is less than the value at 273 K (1.29 g/L).

**Practice Exercise** Air is a mixture of mostly nitrogen and oxygen. A balloon filled with only oxygen is released in a room full of air. Will it sink to the floor or float to the ceiling?

Equation 6.21 tells us that density decreases as temperature increases (Figure 6.28). We can rearrange this equation to calculate the molar mass of a gas from its density at any temperature and pressure:

$$\mathcal{M} = \frac{dRT}{P} \tag{6.22}$$

Gas density can be measured with a glass tube of known volume attached to a vacuum pump (Figure 6.29). The mass of the tube is determined when it has essentially no gas in it and again when it is filled with the test gas. The difference in the masses divided by the volume of the tube is the density of the gas.

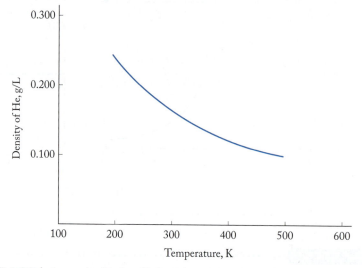

**FIGURE 6.28** At 1 atm, the density of helium decreases with increasing temperature. The density of any ideal gas at constant pressure is inversely proportional to its absolute temperature. This graph shows a portion of the hyperbola that describes the behavior of the gas from approximately 200 to 500 K.

**partial pressure** the contribution to the total pressure made by a component in a mixture of gases.

**Dalton's law of partial pressures** the total pressure of any mixture of gases equals the sum of the partial pressures of all the gases in the mixture.

**mole fraction ($X_x$)** the ratio of the number of moles of a component in a mixture to the total number of moles in the mixture.

**SAMPLE EXERCISE 6.11** **Calculating Molar Mass from Density** **L05**

Vent pipes at solid-waste landfills often emit foul-smelling gases that may be either relatively pure substances or mixtures of several gases. A sample of such an emission has a density of 0.650 g/L at 25.0°C and 757 mmHg. What is the molar mass of the gas emitted?

(Note that if the sample is a mixture, the answer will be the weighted average of molar masses of the individual gases.)

**Collect and Organize** We are given the density, temperature, and pressure of a gaseous sample and asked to calculate its molar mass.

**Analyze** Equation 6.22 relates molar mass to the density, pressure, and temperature of a gas, so we can use it to calculate the molar mass, remembering to convert the temperature to kelvins and the pressure to atmospheres before using these values in the equation.

$$T = 273 + 25.0 = 298 \text{ K}$$

$$P = 757 \text{ mmHg} \times \frac{1 \text{ atm}}{760 \text{ mmHg}} = 0.996 \text{ atm}$$

Many gases have relatively low molar masses, so we expect that the molar mass should be similar to that of typical gases we have discussed in this chapter.

**Solve**

$$\mathcal{M} = \frac{dRT}{P} = \left(\frac{0.650 \text{ g}}{L}\right)\left(\frac{0.08206 \text{ L} \cdot \text{atm}}{\text{mol} \cdot \text{K}}\right)\left(\frac{298 \text{ K}}{0.996 \text{ atm}}\right)$$

$$= 16.0 \text{ g/mol}$$

**Think About It** The molar mass of the gas is 16.0 g/mol, a fairly small molar mass consistent with methane, a principal component in the mixture of gases emitted by decomposing solid waste.

⚙ **Practice Exercise** When HCl($aq$) and NaHCO$_3$($aq$) are mixed together, a chemical reaction takes place in which a gas is one of the products. A sample of the gas has a density of 1.81 g/L at 1.00 atm and 23.0°C. What is the molar mass of the gas? Can you identify the gas?

(a)

(b)

(c)

**FIGURE 6.29** (a) A gas collection tube with an internal volume of 235 mL is evacuated by connecting it to a vacuum pump. (b) The mass of the evacuated tube is measured. (c) Then the tube is opened to the atmosphere and refills with air. The difference in mass between the filled and evacuated tubes (0.273 g or 273 mg) is the mass of the air inside it. The density of the air sample is 273 mg/235 mL or 1.16 mg/mL.

▶❚❚ **CHEMTOUR** Dalton's Law

# 6.7 Dalton's Law and Mixtures of Gases

As noted in Table 6.1, air is mostly N$_2$ and O$_2$ plus smaller quantities of other gases. Each gas in air, and in any other gas mixture, exerts its own pressure, called a **partial pressure**. Atmospheric pressure is the sum of the partial pressures of all of the gases in the air:

$$P_{\text{atm}} = P_{\text{N}_2} + P_{\text{O}_2} + P_{\text{Ar}} + P_{\text{CO}_2} + \ldots$$

A similar expression can be written for any mixture of gases:

$$P_{\text{total}} = P_1 + P_2 + P_3 + P_4 + \ldots \tag{6.23}$$

This is a mathematical expression of **Dalton's law of partial pressures**: the total pressure of any mixture of gases equals the sum of the partial pressures of the gases in the mixture.

You might guess that the most abundant gases in a mixture have the greatest partial pressures and contribute the most to the total pressure of the mixture. The mathematical term used to express the abundance of each component, $x$, is its **mole fraction ($X_x$)**,

$$X_x = \frac{n_x}{n_{\text{total}}} \tag{6.24}$$

where $n_x$ is the number of moles of the component and $n_{\text{total}}$ is the sum of the number of moles of all the components of the mixture.

Expressing concentration using mole fractions has three characteristics worth noting: (1) unlike molarity, mole fractions have no units; (2) unlike molarity, mole fractions are based on numbers of moles and can be used for any kind of mixture

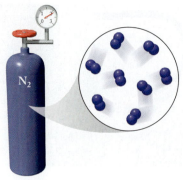

8 mol $N_2$

8 mol $O_2$

8 mol gas

**FIGURE 6.30** Pressure is directly proportional to number of moles of an ideal gas, independent of the identity of the gas, and independent of whether the sample is a pure gas or a mixture. In three containers that have the same volume and are at the same temperature, 8 moles of nitrogen, 8 moles of oxygen, and 8 moles of a 50:50 mixture of nitrogen and oxygen all exert the same pressure, as indicated by the gauges.

or solution (solid, liquid, or gas); and (3) the mole fractions of all the components in a mixture should add up to 1.

To see how mole fractions work, consider a sample of the atmosphere (air) that contains 100.0 mol of atmospheric gases. Making up this sample are 21.0 mol of $O_2$ and 78.1 mol of $N_2$. The mole fraction of $O_2$ in the air sample is thus

$$X_{O_2} = \frac{n_{O_2}}{n_{total}} = \frac{21.0}{100.0} = 0.210$$

Similarly, the mole fraction of $N_2$ is

$$X_{N_2} = \frac{n_{N_2}}{n_{total}} = \frac{78.1}{100.0} = 0.781$$

Note that mole fraction is an expression of concentration based on numbers of moles of material present. In the example, the numbers of moles indicate that there are more nitrogen molecules in air than there are molecules of oxygen:

$N_2$  78.1 ~~mol~~ $\times$ (6.022 $\times$ $10^{23}$ molecules/~~mol~~) = 4.70 $\times$ $10^{25}$ molecules

$O_2$  21.0 ~~mol~~ $\times$ (6.022 $\times$ $10^{23}$ molecules/~~mol~~) = 1.26 $\times$ $10^{25}$ molecules

**CONCEPT TEST** ................................................................

The partial pressure of one component of a gas mixture cannot be directly measured. Why?

................................................................

The partial pressure of any gas in air is the product of the mole fraction of the gas times total atmospheric pressure. If the total pressure is 1.00 atm, for instance, as it can be at sea level, the partial pressures of $O_2$ and $N_2$ are

$$P_{O_2} = X_{O_2}P_{total} = (0.210)(1.00 \text{ atm}) = 0.210 \text{ atm}$$
$$P_{N_2} = X_{N_2}P_{total} = (0.781)(1.00 \text{ atm}) = 0.781 \text{ atm}$$

These two equations are specific applications of the general equation for the partial pressure of a gas in a mixture of gases:

$$P_x = X_x P_{total} \tag{6.25}$$

The pressure of a gas is proportional only to the quantity of the gas in a given volume and does not depend on the identity of the gas or whether the gas is pure or a mixture (Figure 6.30).

**SAMPLE EXERCISE 6.12**  **Calculating Mole Fraction**  **LO6**

Scuba divers who descend more than 45 m below the surface may breathe a gas mixture called Trimix. Trimix is available in several different concentrations; one is 11.7% He, 56.2% $N_2$, and 32.1% $O_2$, by mass. This mixture of gases avoids high concentrations of dissolved nitrogen in the blood at great depth, which can lead to *nitrogen narcosis*, a potentially deadly condition for divers. Calculate the mole fraction of each gas in this mixture.

**Collect and Organize** We are given the composition of a gas mixture in terms of mass percentages and asked to determine mole fractions.

**Analyze** If we consider a 100.0 g sample of Trimix of the stated composition, we can calculate the mass of each gas from the mass percentages. Then we can convert each mass to moles, from which we can determine the total number of moles in the sample and the mole fraction of each gas. Mole fraction is defined in Equation 6.24, and to calculate it we need the total number of moles of gas in the mixture and the number of moles of each component.

We can estimate the answer by considering the relative molar masses of the three gases. The molar masses of nitrogen and oxygen are similar and seven to eight times larger, respectively, than the molar mass of helium. Because $N_2$ weighs less than $O_2$ and also is a higher percentage of the mass of the mixture than is $O_2$, we predict that the mole fraction of nitrogen will be larger than that of oxygen. Furthermore, we predict that the mole fractions of He, $N_2$, and $O_2$ will be quite different from the mass percentages of the three gases.

**Solve** In 100.0 g of this mixture we have 11.7 g of He, 56.2 g of $N_2$, and 32.1 g of $O_2$. Using molar masses to convert these masses into moles yields

$$(11.7 \text{ g He})\left(\frac{1 \text{ mol He}}{4.003 \text{ g He}}\right) = 2.92 \text{ mol He}$$

$$(56.2 \text{ g N}_2)\left(\frac{1 \text{ mol N}_2}{28.02 \text{ g N}_2}\right) = 2.01 \text{ mol N}_2$$

$$(32.1 \text{ g O}_2)\left(\frac{1 \text{ mol O}_2}{32.00 \text{ g O}_2}\right) = 1.00 \text{ mol O}_2$$

Mole fractions are calculated with Equation 6.24. The total number of moles in 100.0 g of Trimix is the sum of the number of moles of the constituent gases: He, $N_2$, and $O_2$.

$$n_{\text{total}} = 2.92 + 2.01 + 1.00 = 5.93 \text{ mol}$$

Thus, $X_{\text{He}}$, $X_{\text{N}_2}$, and $X_{\text{O}_2}$ are

$$X_{\text{He}} = \frac{2.92 \text{ mol He}}{5.93 \text{ mol}} = 0.492$$

$$X_{\text{N}_2} = \frac{2.01 \text{ mol N}_2}{5.93 \text{ mol}} = 0.339$$

$$X_{\text{O}_2} = \frac{1.00 \text{ mol O}_2}{5.93 \text{ mol}} = 0.169$$

Alternatively, we could use Equation 6.24 for all but one of the gases in the sample and then calculate the mole fraction of the final component by subtracting the sum of the calculated mole fractions from 1.00. For example, we could use the equation to calculate the mole fractions of He and $N_2$ and then determine the $O_2$ mole fraction by difference:

$$X_{\text{O}_2} = 1 - (X_{\text{He}} + X_{\text{N}_2}) = 1 - (0.492 + 0.339) = 0.169$$

**Think About It** We can check the answer by summing the mole fractions; they should add up to 1.

$$X_{\text{He}} + X_{\text{N}_2} + X_{\text{O}_2} = 1$$
$$0.492 + 0.339 + 0.169 = 1.00$$

Finally, as we predicted, the mole fractions of the three gases do not reflect their mass percentages. Although nitrogen is present in the greatest amount by mass in Trimix, helium has the largest mole fraction.

⚙ **Practice Exercise** A gas mixture called Heliox, 52.17% $O_2$ and 47.83% He by mass, is used in scuba tanks for descents more than 65 m below the surface. Calculate the mole fractions of He and $O_2$ in this mixture.

**CONCEPT TEST**

Does each of the gases in an equimolar mixture of four gases have the same mole fraction?

Let's consider the implications of Equation 6.25 in the context of the "thin" air at high altitudes. The mole fraction of oxygen in air is 0.210, which does not change significantly with increasing altitude. However, the total pressure of the atmosphere, and therefore the partial pressure of oxygen, decreases with increasing altitude, as illustrated in the following Sample Exercise.

**FIGURE 6.31** At high altitudes, atmospheric pressure is low and the partial pressure of oxygen falls below 0.21 atm, the optimum value for humans. Therefore mountaineers carry supplemental oxygen to compensate for the "thinner" air.

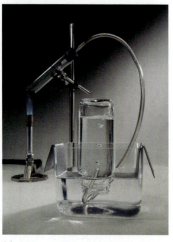

(a)

Equalize water levels
to equalize pressures

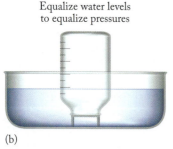

(b)

**FIGURE 6.32** (a) During the thermal decomposition of $KClO_3$, the product $O_2(g)$ is collected by displacing water from an inverted bottle initially filled with water. (b) When the reaction is over, the collection jar is lowered until the surface of the water in the jar is at the same level as the water in the large container. At that point the pressure in the jar equals atmospheric pressure. The temperature of the gas is the same as room temperature.

---

**SAMPLE EXERCISE 6.13   Calculating Partial Pressure        LO6**

Calculate the partial pressure in atmospheres of $O_2$ in the air outside an airplane cruising at an altitude of 10 km, where the atmospheric pressure is 190.0 mmHg. The mole fraction of $O_2$ in the air is 0.210.

**Collect and Organize** We are given the mole fraction of oxygen and atmospheric pressure and are asked to calculate the partial pressure of oxygen.

**Analyze** Equation 6.25 relates the partial pressure of a gas in a gas mixture to its mole fraction in the mixture and to the total gas pressure.

**Solve**

$$P_{O_2} = X_{O_2}P_{total} = (0.210)(190.0 \text{ mmHg})\left(\frac{1 \text{ atm}}{760 \text{ mmHg}}\right)$$

$$= 0.0525 \text{ atm}$$

**Think About It** As noted in the text, the partial pressure of oxygen in air at sea level is 0.210 atm. As we move upward, the atmosphere becomes thinner (less dense). Our answer of 0.0525 atm makes sense because, at an altitude of 10 km, the air is much less dense than at sea level.

**Practice Exercise** Assume a scuba diver is working at a depth where the total pressure is 5.0 atm (about 50 m below the surface). What mole fraction of oxygen is necessary for the partial pressure of oxygen in the gas mixture the diver breathes to be 0.21 atm?

---

Sample Exercises 6.12 and 6.13 illustrate how the properties of gases make our dependence on the atmosphere so great that if we venture very far from Earth's surface, we have to take the right blend of gases with us to sustain life. The scuba diver can't breathe the same mixture of gases as we do on the surface because $O_2$ can be toxic at the increased partial pressure of $O_2$ below 30 m. Similarly, the higher partial pressure of $N_2$ can lead to nitrogen narcosis, a dangerous condition caused by a high concentration of nitrogen in the blood that leads to hallucinations. Alpine climbers on the tallest peaks on Earth experience the opposite problem. For them, the lower atmospheric pressure leads to a lower partial pressure of oxygen, making it hard to get enough oxygen into the blood. Most mountaineers in these locations carry bottled oxygen (Figure 6.31). The lower partial pressure of oxygen at high elevations is also the reason why aircraft cabins are pressurized.

Dalton's law of partial pressures is useful in the laboratory when we need to know the pressure exerted by a gaseous product of a chemical reaction. For example, heating potassium chlorate ($KClO_3$) in the presence of $MnO_2$ causes it to decompose into $KCl(s)$ and $O_2(g)$. The oxygen gas produced by this reaction can be collected by bubbling the gas into an inverted bottle that is initially filled with water (Figure 6.32). As the reaction proceeds, $O_2(g)$ displaces the water in the bottle. When the reaction is complete, the volume of water displaced provides a measure of the volume of $O_2$ produced. If you know the temperature of the water and the barometric pressure, you can use the ideal gas law to determine the number of moles of $O_2$ produced.

The procedure works for any gas that neither reacts with nor dissolves appreciably in water. However, one additional step is needed to apply the ideal gas law in calculating the number of moles of gas produced. At room temperature (nominally 20°C), any enclosed space above a pool of liquid water contains some $H_2O(g)$, meaning some water vapor is in the collection flask in addition to the gas produced by the reaction. Thus in the $KClO_3$ reaction in Figure 6.32, the gas collected is a mixture of $O_2(g)$ and $H_2O(g)$. A mixture of a gas with water vapor is said to be *wet*, in contrast to a gas that contains no water vapor, which is said to be *dry*. Dalton's law of partial pressures gives us the total pressure of the mixture at the point where the water level inside the collection vessel matches the water level outside the vessel:

$$P_{\text{total}} = P_{\text{atm}} = P_{O_2} + P_{H_2O} \qquad (6.26)$$

To calculate the quantity of oxygen produced using the ideal gas law, we must know $P_{O_2}$, which we get by subtracting $P_{H_2O}$ (Table 6.4) from $P_{\text{atm}}$. If we know the values of $P_{O_2}$, $T$, and $V$, we can calculate the number of moles or number of grams of oxygen produced.

| TABLE 6.4 | Partial Pressure of $H_2O(g)$ at Selected Temperatures |
|---|---|
| Temperature (°C) | Pressure (mmHg) |
| 5 | 6.5 |
| 10 | 9.2 |
| 15 | 12.8 |
| 20 | 17.5 |
| 25 | 23.8 |
| 30 | 31.8 |
| 35 | 42.2 |
| 40 | 55.3 |
| 45 | 71.9 |
| 50 | 92.5 |

### SAMPLE EXERCISE 6.14 Determining the Partial Pressure of a Gas Collected over Water    LO6

During the decomposition of $KClO_3$, 92.0 mL of gas is collected by the displacement of water at 25.0°C. If atmospheric pressure is 756 mmHg, what mass of $O_2$ is collected?

**Collect and Organize** We are given the atmospheric pressure, the volume of oxygen, and the temperature, and asked to calculate the mass of oxygen collected by water displacement.

**Analyze** A quick estimation can give us an idea about the value of our final answer. The molar volume of oxygen is about 20 L at 25.0°C; if the oxygen were dry, about 100 mL would be collected, or (0.100 L)/(20 L/mol) = 0.005 mol of oxygen. Considering that oxygen has a molar mass of 32.00 g, 0.005 mol is about 0.16 g.

When a gas is collected over water, however, the collection vessel contains both water vapor and the gas of interest. According to Equation 6.26, we need the partial pressure of the $H_2O(g)$ at the temperature of the apparatus to determine the partial pressure due to $O_2$ in the collection flask. We are not given this pressure for the $H_2O(g)$ but can get it from Table 6.4. The presence of water vapor in the gas that we collect means that the partial pressure of the gas being collected is less than the total pressure in the vessel. This also means that our estimated value for the mass of oxygen, based on the entire volume of gas being $O_2$, is a high estimate.

**Solve** Table 6.4 indicates that $P_{H_2O}$ at 25°C is 23.8 mmHg. To calculate $P_{O_2}$ in the collected gas, we subtract this value from $P_{\text{total}}$:

$$P_{O_2} = P_{\text{total}} - P_{H_2O} = (756 - 23.8) \text{ mmHg}$$
$$= 732 \text{ mmHg}$$

We can use the ideal gas equation to calculate the moles of $O_2$ produced; recall that we need to convert $P_{O_2}$ to atmospheres, $V$ to liters, and $T$ to kelvins. The number of moles, $n$, is equal to the grams of $O_2$ collected (the unknown in this problem) divided by the molar mass of $O_2$ (32.00 g/mol).

$$732 \text{ mmHg} \times \frac{1 \text{ atm}}{760 \text{ mmHg}} = 0.963 \text{ atm}$$

$$92.0 \text{ mL} \times \frac{10^{-3} \text{ L}}{1 \text{ mL}} = 0.0920 \text{ L}$$

$$25.0°C = 298 \text{ K}$$

$$n = \frac{PV}{RT} = \frac{(0.963 \text{ atm})(0.0920 \text{ L})}{[0.08206 \text{ L} \cdot \text{atm}/(\text{mol} \cdot \text{K})](298 \text{ K})} = 0.00362 \text{ mol}$$

$$m_{O_2} = 0.00362 \text{ mol} \times \frac{32.00 \text{ g}}{1 \text{ mol}} = 0.116 \text{ g}$$

**Think About It** The pressure of the $O_2(g)$ is slightly less than $P_{\text{total}}$, as we predicted. Our answer (0.116 g) is close enough to our estimate that we can be confident our answer is reasonable.

**Practice Exercise** Electrical energy can be used to separate water into $O_2(g)$ and $H_2(g)$. In one demonstration of this reaction, 27 mL of $H_2$ is collected over water at 25°C. Atmospheric pressure is 761 mmHg. How many grams of $H_2$ are collected?

# 6.8 The Kinetic Molecular Theory of Gases

The fundamental discoveries of Boyle, Charles, and Amontons all took place before scientists recognized the existence of molecules. Only Avogadro, because he lived after Dalton, had the advantage of knowing Dalton's proposal that matter was composed of tiny particles called atoms. Thus Avogadro recognized that the volume of a gas was directly proportional to the number of particles (moles) of gas. Still, Avogadro's law, $V/n$ = constant (Equation 6.10), does not explain why 1 mole of relatively small He($g$) atoms at STP has the same volume as 1 mole of, for example, much larger $SF_6(g)$ molecules. Dalton's law of partial pressures (Equation 6.23) does not explain *why* each gas in a mixture contributes a partial pressure based on its mole fraction. Similarly, Boyle's, Charles's, and Amontons's laws demand explanations for why pressure and volume are inversely proportional to each other but both are directly proportional to temperature.

In this section we examine the **kinetic molecular theory** of gases, a unifying theory developed in the late 19th century that explains the relationships described by the ideal gas law and its predecessors, the laws of Boyle, Charles, Dalton, Amontons, and Avogadro.

The main assumptions of the kinetic molecular theory are:

1. Gas molecules have tiny volumes compared with the collective volume they occupy. Their individual volumes are so small as to be considered negligible, allowing particles in a gas to be treated as *point masses*—masses with essentially no volume. Gas molecules are separated by large distances; hence a gas is mostly empty space.
2. Gas molecules move constantly and randomly throughout the volume they collectively occupy.
3. The motion of these molecules is associated with an average kinetic energy that is proportional to the absolute temperature of the gas. All populations of gas molecules at the same temperature have the same average kinetic energy.
4. Gas molecules continually collide with one another and with their container walls. These collisions are *elastic;* that is, they result in no net transfer of energy to the walls. Therefore, the average kinetic energy of gas molecules is not affected by these collisions and remains constant as long as there is no change in temperature.
5. Each gas molecule acts independently of all other molecules in a sample. We assume there are no forces of attraction or repulsion between the molecules.

To illustrate and test the kinetic molecular theory, let's consider the examples provided by air and by the $N_2/O_2$/He mixture in a scuba tank.

## Explaining Boyle's, Dalton's, and Avogadro's Laws

We already have a picture in our minds that describes the origin of gas pressure. Every collision between a gas molecule and a wall of its container generates a force. The more frequent the collisions, the greater the force and thus the greater the pressure. Compressing molecules into a smaller space by reducing the volume of the container means more collisions take place per unit time, and the pressure increases (Boyle's law: $P \propto 1/V$). Because $P \propto n$, we can also increase the number of collisions by increasing the number of gas molecules (number of moles, $n$) in a container of fixed volume. We can increase $n$ either by adding more of the same gas to a container of fixed volume or by adding some quantity of a different gas

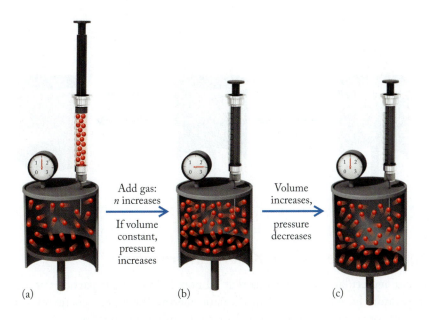

**FIGURE 6.33** (a) Gas contained in a cylinder fitted with a movable piston exerts a pressure equal to atmospheric pressure ($P_{gas} = P_{atm}$). (b) When more gas is added to the cylinder ($n$ increases), $P_{gas}$ increases because the number of collisions with the walls of the container increases. (c) In order for the pressure to remain the same as in part a, the volume must increase.

to the container. If we add two different gases, such as $N_2$ and $O_2$, to a container in a 4:1 mole ratio, then the number of collisions involving nitrogen molecules and a container wall will be greater than the number involving oxygen molecules. Therefore, the pressure due to nitrogen ($P_{N_2}$) will be greater than that due to oxygen ($P_{O_2}$), and the total pressure will be $P_{total} = P_{O_2} + P_{N_2}$. This statement is precisely what Dalton proposed in his law of partial pressures.

Avogadro's law ($V \propto n$) is also explained by the kinetic molecular theory. Avogadro observed that the volume of a gas *at constant pressure* is directly proportional to the number of moles of the gas. Because additional gas molecules in a given volume lead to increased numbers of collisions and an increase in pressure, the only way to reduce the pressure is to allow the gas to expand (Figure 6.33).

**CONCEPT TEST** ...........................................................................

If the collisions between molecules and the walls of the container were not elastic and energy was lost to the walls, would the pressure of the gas be higher or lower than that predicted by the ideal gas law?

..........................................................................................................

## Explaining Amontons's and Charles's Laws

Amontons's law ($P \propto T$) and Charles's law ($V \propto T$) both depend on temperature, and both can be explained in terms of kinetic molecular theory. Why is pressure directly proportional to temperature? Like pressure, temperature is directly related to molecular motion. The average speed at which a population of molecules moves increases with increasing temperature. For a given number of gas molecules, increasing the temperature increases velocity and therefore increases the frequency and average kinetic energy with which the molecules collide with the walls of their container. Because pressure is related to the frequency and force of these collisions, it follows that higher temperatures produce higher pressures, as long as volume and quantity of gas are constant. Figure 6.34(a,b) illustrates Amontons's law on a molecular level.

How do increased frequency of collisions and higher average kinetic energy explain Charles's law? Remember that Charles's law relates volume and temperature at *constant P* and *n*. As temperature increases, the system must expand—increase its volume—in order to maintain a constant pressure (Figure 6.34c).

**FIGURE 6.34** (a) As the temperature of a given amount of gas in a cylinder increases while the volume is held constant, the number and force of molecular collisions with the walls of the container increase, causing (b) the pressure to increase, as stated by Amontons's law. For the pressure to remain constant as the temperature of the gas increases, (c) the volume must increase until $P_{gas} = P_{atm}$. Hence, as temperature increases at constant pressure, volume increases, as stated by Charles's law.

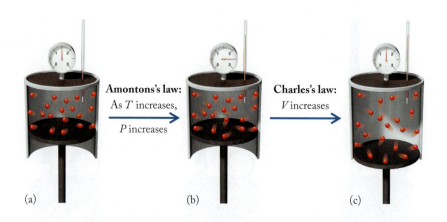

(a)   (b)   (c)

Amontons's law: As $T$ increases, $P$ increases

Charles's law: $V$ increases

## Molecular Speeds and Kinetic Energy

Kinetic molecular theory tells us that all populations of gas particles at a given temperature have the same *average* kinetic energy. The kinetic energy of a single molecule or atom of a gas can be calculated by using the equation

$$KE = \tfrac{1}{2}mu^2$$

where $m$ is the mass of a molecule of the gas and $u$ is its speed. At any given moment, however, not all gas molecules in a population are traveling at exactly the same speed. Even elastic collisions between two molecules may result in one of them moving with a greater speed than the other after the collision. One molecule might even stop completely. Thus collisions between gas molecules cause the molecules in any sample to have a range of speeds.

Figure 6.35(a) shows a typical distribution of speeds in a population of gas molecules. The peak in the curve represents the *most probable speed* ($u_m$) of molecules in the population. It is the speed that characterizes the largest number of molecules in the sample. Because the distribution of speeds is not symmetrical, the *average speed* ($u_{avg}$), which is simply the arithmetic average of

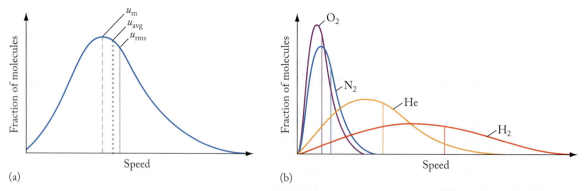

(a)   (b)

**FIGURE 6.35** At any given temperature, the speeds of gas molecules cover a range of values. (a) The most probable speed ($u_m$, dashed line), at the highest point on the curve, is the speed of the largest fraction of molecules in the population; in other words, more molecules have the most probable speed than any other speed. The average speed ($u_{avg}$, dotted line), a little faster than the most probable speed, is the arithmetic average of all the speeds. The root-mean-square speed ($u_{rms}$, solid line), a little faster than the average speed, is directly proportional to the square root of the absolute temperature of the gas and inversely proportional to the square root of its molar mass. (b) Molecular speed distributions for samples of oxygen, nitrogen, helium, and hydrogen gas at the same temperature. On each curve, the vertical solid line is the $u_{rms}$. The lower the molar mass of the molecules, the higher is their root-mean-square speed ($u_{rms}$) and the broader the distribution of speeds. Having the lowest molar mass of the four gases, the $H_2$ molecules have the highest $u_{rms}$ and the broadest curve. Having the highest molar mass in the group, the $O_2$ molecules have the lowest $u_{rms}$ and the narrowest curve.

all the speeds of all the molecules in the population, is a little higher than the most probable speed.

A very important value is the **root-mean-square speed ($u_{rms}$)**; this is the speed of a molecule possessing the average kinetic energy. The absolute temperature of a gas is a measure of the average kinetic energy of the population of gas molecules. At a given temperature, the population of molecules in a gas has the same *average* kinetic energy as every other population of gas molecules at that same temperature, and this *average* kinetic energy is defined as

$$KE_{avg} = \tfrac{1}{2}m(u_{rms})^2 \qquad (6.27)$$

The root-mean-square speed ($u_{rms}$) of a gas with molar mass $\mathcal{M}$ at temperature $T$ is defined as

$$u_{rms} = \sqrt{\frac{3RT}{\mathcal{M}}} \qquad (6.28)$$

Because different gases have different molar masses, this equation indicates that more massive molecules move more slowly at a given temperature than lighter molecules. Figure 6.35(b) shows the different distributions of speeds for several gases at a constant temperature.

Because $u_{rms}$ is typically expressed in meters per second, care must be taken in choosing the units for $R$ in Equation 6.28. One value of $R$ (Table 6.3) that has meters in its units is

$$R = 8.314 \text{ kg} \cdot \text{m}^2/(\text{s}^2 \cdot \text{mol} \cdot \text{K})$$

and this is the most convenient value to use when working with Equation 6.28. Using this value for $R$ requires that we express molar mass in kilograms per mole rather than the more common grams per mole.

Figure 6.36 shows how $u_m$ increases with temperature. How fast do gas molecules move? Very rapidly at or slightly above room temperature—many hundreds of meters per second, depending on molar mass.

▶II CHEMTOUR  Molecular Speed

▶II CHEMTOUR  Molecular Motion

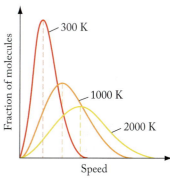

**FIGURE 6.36** The most probable speed ($u_m$, dashed lines) increases with increasing temperature. Notice that the distributions broaden as the temperature increases: a smaller fraction of the molecules moves at any given speed, and more speeds are represented by a significant fraction of the population.

---

**CONCEPT TEST** · · · · · · · · · · · · · · · · · · · · · · · · · · · · · · · · · · · · · · · · · · · · · · · ·

Rank these gases in order of increasing root-mean-square speed at 20°C, lowest speed first: $H_2$, $CO_2$, Ar, $SF_6$, $UF_6$, Kr.

· · · · · · · · · · · · · · · · · · · · · · · · · · · · · · · · · · · · · · · · · · · · · · · · · · · · · · · · · · · ·

---

**SAMPLE EXERCISE 6.15   Calculating Root-Mean-Square      LO8
                          Speeds**

Calculate the root-mean-square speed of nitrogen molecules at 300.0 K in meters per second and miles per hour.

**Collect and Organize** Given only the temperature of a sample of nitrogen gas, we are asked to calculate the root-mean-square speed of the molecules in the sample. According to Equation 6.28, we need the molar mass of the gas in addition to the temperature. We also need $R = 8.314 \text{ kg} \cdot \text{m}^2/(\text{s}^2 \cdot \text{mol} \cdot \text{K})$ because it gives us a speed in meters per second.

**Analyze** Nitrogen gas ($N_2$) has a molar mass of 28.02 g/mol. Having chosen the value of $R$, we need to express the molar mass in kilograms. We can estimate the magnitude of the answer by considering the relative magnitudes of $R$, $T$, and molar mass. When $T$ is expressed in kelvins, it is on the order of $10^2$, whereas the correct value of $R$ is approximately 10. The molar mass of nitrogen expressed as kilograms per mole is a small number, on the order of $10^{-2}$ kg/mol. When these values are substituted into Equation 6.28, the answer will be a large number, even after taking the square root: $(10 \times 10^2/10^{-2})^{1/2} \approx 10^2$ to $10^3$ m/s.

**root-mean-square speed ($u_{rms}$)** the square root of the average of the squared speeds of all the molecules in a population of gas molecules; a molecule possessing the average kinetic energy moves at this speed.

**Solve**

$$u_{rms,N_2} = \sqrt{\frac{3RT}{\mathcal{M}}} = \sqrt{\frac{3\left(8.314\,\frac{kg \cdot m^2}{s^2 \cdot mol \cdot K}\right)(300.0\,K)}{0.02802\,\frac{kg}{mol}}}$$

$$= 516.8\,m/s = 5.168 \times 10^2\,m/s$$

$$u_{rms,N_2} = \left(516.8\,\frac{m}{s}\right)\left(\frac{1\,mi}{1.6093 \times 10^3\,m}\right)\left(3600\,\frac{s}{hr}\right)$$

$$= 1156\,mi/hr = 1.156 \times 10^3\,mi/hr$$

**Think About It** The average speed of a nitrogen molecule does indeed fall between $10^2$ and $10^3$ m/s as predicted. The relatively small molar mass contributes to the large root-mean-square speed.

⚙ **Practice Exercise** Calculate the root-mean-square speed of helium at 300.0 K in meters per second, and compare your result with the root-mean-square speed of nitrogen calculated in Sample Exercise 6.15.

It is not immediately obvious why the pressure exerted by two different gases is the same at a given temperature when their root-mean-square speeds are quite different. The answer lies in the assumption in the kinetic molecular theory that all populations of gas molecules have the same average kinetic energy at a given temperature, independent of their molar mass. To examine this issue, let's calculate the average kinetic energy of 1 mole of $N_2$ and 1 mole of He at 300 K. The solution to Sample Exercise 6.15 reveals that the $u_{rms}$ of $N_2$ molecules at 300 K is 517 m/s. The root-mean-square speed of He atoms is 1370 m/s (the result for the Practice Exercise of Sample Exercise 6.15). Using these values of speed and the mass of 1 mole of each gas ($m = \mathcal{M}$) in Equation 6.27 gives

$$KE_{N_2} = \tfrac{1}{2}\mathcal{M}_{N_2}(u_{rms,N_2})^2 = \tfrac{1}{2}(2.802 \times 10^{-2}\,kg)(517\,m/s)^2$$

$$= 3.74 \times 10^3\,kg \cdot m^2/s^2$$

$$= \tfrac{1}{2}\mathcal{M}_{He}(u_{rms,He})^2 = \tfrac{1}{2}(4.003 \times 10^{-3}\,kg)(1370\,m/s)^2$$

$$= 3.76 \times 10^3\,kg \cdot m^2/s^2$$

This calculation demonstrates that the same quantities of two *different* gases have essentially the same average kinetic energy (when the correct number of significant figures is used). Therefore at the same temperature, they exert the same pressure. The slower $N_2$ molecules collide with the container walls less often, but because they are more massive than He atoms, the $N_2$ molecules exert a greater force during each collision.

Equation 6.28 enables us to compare the relative root-mean-square speeds of two gases at the same temperature. Consider an equimolar mixture of $N_2(g)$ and He($g$). We have just demonstrated that at any given temperature their average kinetic energies are the same:

$$KE_{N_2} = \tfrac{1}{2}m_{N_2}(u_{rms,N_2})^2 = \tfrac{1}{2}m_{He}(u_{rms,He})^2 = KE_{He}$$

or

$$m_{N_2}(u_{rms,N_2})^2 = m_{He}(u_{rms,He})^2$$

**effusion** the process by which a gas escapes from its container through a tiny hole into a region of lower pressure.

**Graham's law of effusion** the rate of effusion of a gas is inversely proportional to the square root of its molar mass.

Rearranging this equation to express the ratio of the root-mean-square speeds in terms of the ratio of the molar masses and then taking the square root of each side, we get

$$\frac{(u_{rms,He})^2}{(u_{rms,N_2})^2} = \frac{\mathcal{M}_{N_2}}{\mathcal{M}_{He}}$$

$$\frac{u_{rms,He}}{u_{rms,N_2}} = \sqrt{\frac{\mathcal{M}_{N_2}}{\mathcal{M}_{He}}} = \frac{28.02\ \text{g}}{4.003\ \text{g}} = 2.646$$

The root-mean-square speed of helium atoms at 300 K is 2.646 times higher than that of nitrogen molecules. We can test this using the values for $u_{rms}$ we used above to calculate $KE_{N_2}$ and $KE_{He}$:

$$\frac{u_{rms,He}}{u_{rms,N_2}} = \frac{1370\ \text{m/s}}{517\ \text{m/s}} = 2.65$$

The relation between root-mean-square speeds and molar masses applies to any pair of gases $x$ and $y$:

$$\frac{u_{rms,x}}{u_{rms,y}} = \sqrt{\frac{\mathcal{M}_y}{\mathcal{M}_x}} \qquad (6.29)$$

This equation leads to the conclusion that the root-mean-square speed of more massive particles is lower than the root-mean-square speed of lighter particles.

> **CONCEPT TEST** ................................................
>
> Since root-mean-square speed depends on temperature, why doesn't the ratio of the $u_{rms}$ of two gases change with temperature?

## Graham's Law: Effusion and Diffusion

Let's consider what happens to the two balloons in Figure 6.37, one filled with nitrogen and the other with helium. The volume, temperature, and pressure of the gases are identical in the two balloons, and the pressure inside each balloon is greater than atmospheric pressure. Over time, the volume of the helium balloon decreases significantly, but the volume of the nitrogen balloon does not.

The skin of any balloon is slightly permeable; that is, it has microscopic holes that allow gas to escape, reducing the pressure inside the balloon. Why does the helium leak out faster than the nitrogen? We now know the relation between root-mean-square speeds and molar masses of two gases, and we have calculated that helium atoms move about 2.65 times faster than nitrogen atoms at the same temperature. The gas with the greater root-mean-square speed (He) leaks out of the balloon at a higher rate. The process of moving through a small opening from a higher-pressure region to a lower-pressure region is called **effusion**.

In the 19th century, Scottish chemist Thomas Graham (1805–1869) recognized that the effusion rate of a gas is related to its molar mass. Today this relation is known as **Graham's law of effusion**, and it states that the effusion rate of any gas is inversely proportional to the square root of its molar mass. We can derive

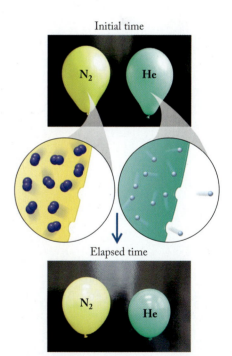

**FIGURE 6.37** Two balloons at the same volume, temperature and pressure—one filled with nitrogen gas and the other filled with helium gas. Over time, the volume of the helium balloon decreases much more than the nitrogen balloon. The helium atoms are lighter than the nitrogen molecules and therefore have higher root-mean-square speeds. As a consequence, the helium atoms escape from the balloon faster, causing the helium-containing balloon to shrink more quickly than the nitrogen balloon.

**diffusion** the spread of one substance (usually a gas or liquid) through another.

a mathematical representation of Graham's law starting from Equation 6.28 for two gases $x$ and $y$:

$$u_{\text{rms},x} = \sqrt{\frac{3RT}{\mathcal{M}_x}}$$

$$u_{\text{rms},y} = \sqrt{\frac{3RT}{\mathcal{M}_y}}$$

We take the ratio of the root-mean-square speeds, which we designate as the effusion rates $r_x$ and $r_y$:

$$\frac{r_x}{r_y} = \frac{u_{\text{rms},x}}{u_{\text{rms},y}} = \frac{\sqrt{\dfrac{3RT}{\mathcal{M}_x}}}{\sqrt{\dfrac{3RT}{\mathcal{M}_y}}} \tag{6.30}$$

This equation simplifies to the usual form of Graham's law:

$$\frac{r_x}{r_y} = \sqrt{\frac{\mathcal{M}_y}{\mathcal{M}_x}} \tag{6.31}$$

Using Equation 6.31 for the helium and nitrogen balloons, we see that the effusion rate of helium compared to nitrogen is:

$$\frac{r_{\text{He}}}{r_{\text{N}_2}} = \sqrt{\frac{\mathcal{M}_{\text{N}_2}}{\mathcal{M}_{\text{He}}}} = \sqrt{\frac{28.02 \ \text{g/mol}}{4.003 \ \text{g/mol}}} = 2.646$$

Notice that this expression has the same form—and hence the ratio has the same value—as the ratio of the root-mean-square speeds of the $N_2$ and He molecules.

Two things about Equation 6.31 are worth noting: (1) like the root-mean-square speed in Equation 6.29, Graham's law involves a square root. (2) With labels $x$ and $y$ for two gases in a mixture, the left side of Equation 6.31 has the $x$ term in the numerator and the $y$ term in the denominator, whereas the right side of the equation has the opposite: the $y$ term is in the numerator and $x$ term in the denominator. Keeping the labels straight is the key to solving problems involving Graham's law.

How does the kinetic molecular theory explain Graham's law of effusion? The escape of a gas molecule from a balloon requires that the molecules encounter one of the microscopic holes in the balloon. The faster a gas molecule moves, the more likely it is to find one of these holes.

Effusion of gas molecules is related to **diffusion**, which is the spread of one substance through another. If we focus our attention on gases, then diffusion contributes to the passage of odors such as a perfume throughout a room and the smell of baking bread throughout a house. The rates of diffusion of gases depend on the average speeds of the gas molecules and on their molar masses. Like effusion, diffusion of gases is described by Graham's law.

---

**SAMPLE EXERCISE 6.16** **Applying Graham's Law to Diffusion of a Gas**    **LO8**

An odorous gas emitted by a hot spring was found to diffuse 2.92 times slower than helium. What is the molar mass of the emitted gas?

**Collect and Organize** We are asked to determine the molar mass of an unknown gas based on its rate of diffusion relative to the rate for helium.

**Analyze** Graham's law (Equation 6.31) relates the relative rate of diffusion of two gases to their molar masses. We know that helium ($\mathcal{M}_{He}$ = 4.003 g/mol) diffuses 2.92 times faster than the unidentified gas (y). Because helium diffuses faster than the unknown gas, we predict that the unidentified gas has a larger molar mass. In fact, it should be about nine (or $3^2$) times the mass of He based on Graham's law, or about 36 g/mol, because the ratio of the molar masses will equal the square of the ratio of the diffusion rates.

**Solve** Rearranging Equation 6.31 we obtain an expression for $\mathcal{M}_y$:

$$\mathcal{M}_{He}\left(\frac{r_{He}}{r_y}\right)^2 = \mathcal{M}_y$$

Substituting for the ratio of the diffusion rates and the mass of 1 mole of helium gives

$$\mathcal{M}_y = \mathcal{M}_{He}\left(\frac{r_{He}}{r_y}\right)^2 = (4.003 \text{ g/mol})(2.92)^2 = 34.1 \text{ g/mol}$$

**Think About It** The molar mass of the unidentified gas is 34.1 g/mol, which is consistent with our prediction. One possibility for the identity of this gas is $H_2S(g)$, $\mathcal{M}$ = 34.08 g/mol, a foul-smelling and toxic gas frequently emitted from volcanoes and also responsible for the odor of rotten eggs.

 **Practice Exercise** Helium effuses 3.16 times as fast as which other noble gas?

# 6.9 Real Gases

Up to now we have treated all gases as ideal. This is acceptable because, under typical atmospheric pressures and temperatures, most gases *do* behave ideally. We have also assumed, according to kinetic molecular theory that the volume occupied by individual gas molecules is negligible compared with the total volume occupied by the gas. In addition, we have assumed that no interactions occur between gas molecules other than random elastic collisions. These assumptions are not valid, however, when we begin to compress gases into increasingly smaller volumes.

## Deviations from Ideality

Let's consider the behavior of 1.0 mole of a gas as we increase the pressure on it. From the ideal gas law, we know that $PV/RT = n$, so for 1 mole of gas, $PV/RT$ should remain equal to 1.0 regardless of how we change the pressure. The relationship between $PV/RT$ and $P$ for an ideal gas is shown by the purple line in Figure 6.38. However, the curves for $PV/RT$ versus $P$ for $CH_4$, $H_2$, and $CO_2$ at pressures above 10 atm are not horizontal, straight lines like this ideal curve. Not only do the curves diverge from the ideal, but the shapes of the curves also differ for each gas, indicating that when we are dealing with real gases rather than ideal ones, the identity of the gas does matter.

Why don't real gases behave like ideal gases at high pressure? One reason is that the ideal gas law considers gas molecules to have so little volume compared with the volume of their container that they are assumed to have no volume at all. However, at high pressures, more molecules are squeezed into a given volume (Figure 6.39). Under these conditions the volume occupied by the molecules can become significant. What is the impact of this on the graph of $PV/RT$ versus $P$?

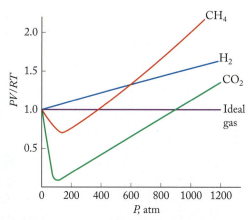

**FIGURE 6.38** The effect of pressure on the behavior of real and ideal gases. The curves diverge from ideal behavior in a manner unique to each gas. When gases deviate from ideal behavior, their identity matters.

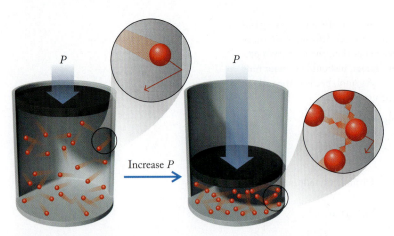

**FIGURE 6.39** At high pressures, a greater fraction of the volume of a gas is occupied by the gas molecules. The greater density of molecules also results in more interactions between them (broad red arrows in expanded view). These interactions reduce the frequency and force of collisions with the walls of the container (red arrow), thereby reducing the pressure.

In *PV/RT*, the *V* actually refers to the *free volume* ($V_{\text{free volume}}$), the empty space not occupied by gas molecules. Because $V_{\text{free volume}}$ is difficult to measure, we instead measure $V_{\text{total}}$, the total volume of the container holding the gas:

$$V_{\text{total}} = V_{\text{free volume}} + V_{\text{molecules}}$$

In an ideal gas and in a real gas at low pressure, $V_{\text{free volume}} \gg V_{\text{molecules}}$, so $V_{\text{molecules}}$ is essentially 0, and we can use the ideal gas equation with confidence. Consider, however, what happens to a real gas in a closed but flexible container as we increase the external pressure, causing the container—and hence the volume occupied by the gas—to shrink. As the external pressure increases, the assumption that $V_{\text{molecules}} = 0$ becomes less and less valid because the proportion of the container's volume taken up by the molecules increases. The relation $V_{\text{total}} = V_{\text{free volume}} + V_{\text{molecules}}$ still applies, but now $V_{\text{molecules}} > 0$, which means $V_{\text{free volume}}$ can no longer be approximated by $V_{\text{total}}$. Since $V_{\text{total}} > V_{\text{free volume}}$, the ratio *PV/RT* is also larger for a real gas than for an ideal gas:

$$\frac{PV_{\text{total}}}{RT} > \frac{PV_{\text{free volume}}}{RT}$$

As a consequence, the curve for a real gas in Figure 6.38 diverges upward from the line for an ideal gas.

A second factor also causes the ratio *PV/RT* to diverge from 1. Kinetic molecular theory assumes that molecules do not interact, but real molecules do attract one another. These attractive forces function over short distances, so the assumption that molecules behave independently is a good one as long as the molecules are far apart. However, as pressure increases on a population of gas molecules and they are pushed closer and closer together, intermolecular attractive forces can become significant. This causes the molecules to associate with one another, which decreases the force of their collisions with the walls of their container, thereby decreasing the pressure exerted by the gas. If the value of *P* in *PV/RT* is smaller in the real gas than the ideal value, the value of the ratio decreases

$$\frac{P_{\text{real}}V}{RT} < \frac{P_{\text{ideal}}V}{RT}$$

In this case, the curve for a real gas diverges from the ideal gas line by moving below it on the graph.

**CONCEPT TEST** •••••••••••••••••••••••••••••••••••••••

For which gases in Figure 6.38 does the effect of intermolecular attractive forces outweigh the effect of pressure on free volume at 200 atm?

•••••••••••••••••••••••••••••••••••••••••••••••••••••

## The van der Waals Equation for Real Gases

Because the ideal gas equation does not hold at high pressures, we need another equation that can be used under nonideal conditions—one that accounts for the following facts:

1. The free volume of a real gas is less than the total volume because its molecules occupy significant space.
2. The observed pressure is less than the pressure of an ideal gas because of intermolecular attractions.

The **van der Waals equation**

$$\left(P + \frac{n^2 a}{V^2}\right)(V - nb) = nRT \qquad (6.32)$$

includes terms to correct for pressure ($n^2 a/V^2$) and volume ($nb$). The values of $a$ and $b$, called *van der Waals constants*, have been determined experimentally for many gases (Table 6.5). Both $a$ and $b$ increase with increasing molar mass and with the number of atoms in each molecule of a gas.

| TABLE 6.5 | Van der Waals Constants of Selected Gases | |
|---|---|---|
| **Substance** | **$a$ ($L^2 \cdot atm/mol^2$)** | **$b$ (L/mol)** |
| He | 0.0341 | 0.02370 |
| Ar | 1.34 | 0.0322 |
| $H_2$ | 0.244 | 0.0266 |
| $N_2$ | 1.39 | 0.0391 |
| $O_2$ | 1.36 | 0.0318 |
| $CH_4$ | 2.25 | 0.0428 |
| $CO_2$ | 3.59 | 0.0427 |
| CO | 1.45 | 0.0395 |
| $H_2O$ | 5.46 | 0.0305 |
| NO | 1.34 | 0.02789 |
| $NO_2$ | 5.28 | 0.04424 |
| HCl | 3.67 | 0.04081 |
| $SO_2$ | 6.71 | 0.05636 |

**SAMPLE EXERCISE 6.17  Calculating Pressure with the van der Waals Equation  LO9**

Calculate the pressure of 1.00 mol of $N_2$ in a 1.00 L container at 300.0 K, using first the van der Waals equation and then the ideal gas equation.

**Collect and Organize** We are given the amount of nitrogen, its volume, and its temperature and are to calculate its pressure using the van der Waals equation and the ideal gas equation. Using the van der Waals equation means we need to know the values of the van der Waals constants for nitrogen.

**Analyze** Table 6.5 lists the van der Waals constants for nitrogen gas as $a = 1.39$ $L^2 \cdot atm/mol^2$ and $b = 0.0391$ L/mol. The values of $a$ and $b$ represent experimentally determined corrections for the interactions between molecules and the portion of the total volume occupied by the gas molecules. One mole of ideal gas at STP occupies 22.4 L. Compressing the gas to a volume of about 1/20 of the initial volume requires a pressure of about 20 atm. Heating it to 300 K at constant volume causes a relatively minor change in the pressure. We can estimate the effect of a pressure of about 20 atm on the ideality of the nitrogen from the curves in Figure 6.38. Although we do not know which curve best describes the behavior of $N_2$, the scale of the x-axis is rather large. At a pressure of 20 atm, none of the curves deviate substantially from ideality, so we expect a relatively small difference in the pressure calculated for $N_2$ as an ideal gas or by using the van der Waals equation.

**Solve** We solve Equation 6.32 for pressure:

$$P = \frac{nRT}{V - nb} - \frac{n^2 a}{V^2}$$

$$P_{N_2} = \frac{(1.00 \text{ mol})\left(0.08206 \frac{L \cdot atm}{mol \cdot K}\right)(300.0 \text{ K})}{1.00 \text{ L} - (1.00 \text{ mol})\left(0.0391 \frac{L}{mol}\right)} - \frac{(1.00 \text{ mol})^2\left(1.39 \frac{L^2 \cdot atm}{mol^2}\right)}{(1.00 \text{ L})^2}$$

$$= 24.2 \text{ atm}$$

**van der Waals equation** an equation that includes experimentally determined factors $a$ and $b$ that quantify the contributions of non-negligible molecular volume and non-negligible intermolecular interactions to the behavior of real gases with respect to changes in $P$, $V$, and $T$.

If nitrogen behaved as an ideal gas, we would have

$$P = \frac{nRT}{V}$$

$$P_{N_2} = \frac{(1.00 \;\cancel{\text{mol}}) \left(0.08206 \;\frac{\cancel{L} \cdot atm}{\cancel{\text{mol}} \cdot \cancel{K}}\right)(300.0 \;\cancel{K})}{1.00 \;\cancel{L}}$$

$$= 24.6 \; atm$$

**Think About It** This pressure is about 25 times greater than normal atmospheric pressure, so we expect some difference in the calculated pressures. However, the small deviation from ideality, $0.4/24.6 = 2\%$, supports our prediction. Furthermore, because of the likely deviation from ideal behavior, we expect the pressure calculated using the ideal gas equation to be too high. In this case, we observe that $P_{ideal}$ is greater than $P_{real}$.

⚙ **Practice Exercise** Assuming the conditions stated in Sample Exercise 6.17, use the van der Waals equation and the ideal gas equation to calculate the pressure for 1.00 mol of He gas. Which gas behaves more ideally at 300 K, He or $N_2$?

■ ·····················································

**SAMPLE EXERCISE 6.18** **Integrating Concepts: Scuba Diving**

The photograph on the opening page of this chapter shows a scuba diver whose exhaled air bubbles expand as they rise to the surface. The lungs of a diver behave similarly unless certain precautions are taken, and serious medical consequences may result if divers deviate from these precautions. An important rule in scuba diving is "never hold your breath." Why? In the following calculations, assume that the pressure at sea level is 1.0 atm and that the pressure increases by 1.0 atm for every 10 m you descend under the surface. Assume your lungs have a volume of 5.1 L and you are breathing dry air that is 78.1% $N_2$, 20.9% $O_2$, and 1.00% Ar by volume. Because you are breathing constantly as you descend, your lungs maintain this volume throughout the dive. (a) What is the partial pressure of each gas in your lungs before the dive? (b) What is the partial pressure of each gas in your lungs when you reach a depth of 30 m? (c) Suppose you are accompanied by a breath-hold diver on this descent. She fills her lungs (which also have a volume of 5.1 L) with air at the surface, dives to 30 m with you, and ascends to the surface, all without exhaling. What is the volume of her lungs at a depth of 30 m and upon the return to the surface? (d) As you return to the surface from this depth, you hold your breath. What is the volume of your lungs when you reach the surface?

Air may be used in scuba for shallow dives. However, a condition called nitrogen narcosis (dulling of the senses, inactivity, even unconsciousness) can result during deeper dives because of the anesthetic qualities of nitrogen under pressure. Breathing pure oxygen is not a solution because oxygen at high levels is quite toxic to the central nervous system and other parts of the body. Therefore mixtures of $N_2$, $O_2$, and He, generically called Trimix, are used to minimize nitrogen narcosis and oxygen toxicity. For most dives, the oxygen concentration in the tank is adjusted to a maximum partial pressure of 1.5 atm at the working depth; the "equivalent narcotic depth" (END) for $N_2$ pressure is normally set at 4.0 atm. Helium is

added as needed to achieve these values. (e) What is the best Trimix composition for a dive 225 ft below the surface?

**Collect and Organize** We are given initial and final conditions for mixtures of gases and desired partial pressures for gases at known total pressures. We can use Boyle's law and Dalton's law of partial pressures to answer these questions about gas mixtures in scuba tanks.

**Analyze** We can convert percent composition data into partial pressures at 1 atm and then into partial pressures under other conditions. The pressure at 30 m is about 4 atm, so the partial pressures of gases in the scuba diver's lungs will increase by a factor of 4. The scuba diver's lungs will have the same volume at 30 m; the breath-holding diver's lungs will shrink to 1/4 their volume on the surface. As the two ascend, both divers' lungs will expand as pressure drops, but the scuba diver's lungs start at a much higher volume than the breath-hold diver's. A dive 225 ft below the surface is around 70 m, where the pressure will be about 8 atm. Using that as the total pressure, we can estimate that the $P_{He}$ should be about 2.5 atm ($P_{total} \approx 8 \; atm = P_{N_2} + P_{O_2} + P_{He} = 4.0 + 1.5 + x$).

**Solve**

a. The percent composition by volume can be converted into mole fractions by recognizing, for example, that 78.1% nitrogen means 0.781 moles of nitrogen for every 1 mole of gas:

$$X_{N_2} = \frac{n_{N_2}}{n_{total}} = \frac{0.781}{1} = 0.781$$

The mole fraction of oxygen in dry air is 0.209, and that of argon is 0.010.

b. For every 10 m of descent below the surface, the pressure increases by 1.0 atm. At 30 m, the pressure would increase by

3.0 atm, so the total pressure is 3.0 atm + 1.0 atm = 4.0 atm. The partial pressure of each gas at 4.0 atm total pressure is:

$$P_{N_2} = X_{N_2} P_{total} = 0.781 \times 4.0 \text{ atm} = 3.1 \text{ atm}$$
$$P_{O_2} = X_{O_2} P_{total} = 0.209 \times 4.0 \text{ atm} = 0.84 \text{ atm}$$
$$P_{Ar} = X_{Ar} P_{total} = 0.010 \times 4.0 \text{ atm} = 0.04 \text{ atm}$$

c. The breath-hold diver does not breathe in more air, so the quantity of gas in her lungs remains constant. As the pressure increases, Boyle's law predicts her lungs will decrease in volume:

$$V_{final} = V_{initial} \times \frac{P_{initial}}{P_{final}} = 5.1 \text{ L} \times \frac{1.0 \text{ atm}}{4.0 \text{ atm}} = 1.3 \text{ L}$$

Upon return to the surface, her lungs will re-expand to their original volume of 5.1 L.

d. Because you have been breathing air supplied by your scuba tank, your lungs retain their volume of 5.1 L at a depth of 30 m. If you hold your breath as you ascend to the surface, your lungs will expand (or try to):

$$5.1 \text{ L} \times \frac{4.0 \text{ atm}}{1.0 \text{ atm}} = 20.4 \text{ L} = 20 \text{ L}$$

e. A dive 225 ft below the surface in terms of meters is

$$225 \text{ ft} \times \frac{12 \text{ in}}{1 \text{ ft}} \times \frac{2.54 \text{ cm}}{1 \text{ in}} \times \frac{1 \text{ m}}{100 \text{ cm}} = 68.6 \text{ m}$$

The pressure would rise:

$$68.6 \text{ m} \times \frac{1.0 \text{ atm}}{10 \text{ m}} = 6.9 \text{ atm}$$

The total pressure at 225 ft is 6.9 atm + 1.0 atm = 7.9 atm. The partial pressure of helium in the tank must be

$$7.9 \text{ atm} = 4.0 \text{ atm} + 1.5 \text{ atm} + x; \quad x = 2.4 \text{ atm}$$

To achieve these partial pressures at a depth of 225 ft, the percent composition of the gas mixture in the scuba tank must be:

$$N_2 \quad \frac{P_{N_2}}{P_{total}} \times 100\% = \frac{4.0 \text{ atm}}{7.9 \text{ atm}} \times 100\% = 50.6 = 51\%$$

$$O_2 \quad \frac{P_{O_2}}{P_{total}} \times 100\% = \frac{1.5 \text{ atm}}{7.9 \text{ atm}} \times 100\% = 19.0 = 19\%$$

$$He \quad \frac{P_{He}}{P_{total}} \times 100\% = \frac{2.4 \text{ atm}}{7.9 \text{ atm}} \times 100\% = 30.4 = 30\%$$

**Think About It** If a scuba diver holds his breath on an ascent, the resulting expansion of his lungs could rupture his lungs. In addition, air may also be pushed into the skin, the spaces between tissues, and even into blood vessels, causing embolisms. The lesson is clear: don't hold your breath when scuba diving. A breath-hold diver is not immune to damage from pressure changes. During deep dives, a condition called thoracic squeeze may develop, which can result in hemorrhage of lung tissues.

We began this chapter with a description of our atmosphere as an invisible necessity. Looking back, we can now answer the questions we raised about gases and our atmosphere. We now understand how elevation affects the quantity of gas available, making pressurized cabins on commercial airliners a necessity for safe, comfortable travel. We have seen the chemical reactions used to supply oxygen in aircraft under emergency conditions and applied principles of stoichiometry to calculate the amount of material required to produce sufficient breathable air. Undersea explorers carry a different mixture of gases in a scuba tank than we breathe under normal conditions because the high pressures under water increase the partial pressures of oxygen and nitrogen to unhealthy levels. Hot-air balloons rise because volume is directly proportional to temperature and because the density of air decreases as temperature rises. Understanding the physical properties of gases, especially their compressibility and their responses to changing conditions, makes their use in many everyday items, ranging from automobile tires to weather balloons and fire extinguishers, possible because their behavior is reliable and predictable. Our dependence on air for the maintenance of the chemical processes essential to life is so great that if we venture far from Earth's surface, we have to take the right blend of gases with us to sustain that life.

# Nitrogen: Feeding Plants and Inflating Air Bags

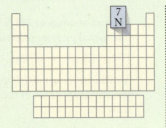

Nitrogen is the most abundant element in the atmosphere, 78% by volume. It is the most abundant free element on Earth and the sixth most abundant element in the universe. However, it is a relatively minor constituent of Earth's crust (only about 0.003% by mass), present mostly in deposits of potassium nitrate ($KNO_3$) or sodium nitrate ($NaNO_3$). Common names for $KNO_3$ include saltpeter and niter. The latter name is the source of the name of the element: in Greek *nitro-* and *-gen* mean "niter-forming."

Nitrogen in combined forms is essential to all life: it is the nutrient needed in greatest quantity by food crops, animals need it to make proteins, and all genetic material (DNA and RNA) contains nitrogen. The continuous exchange of nitrogen between the atmosphere and the biosphere (Figure 6.40) is mediated by bacteria called diazotrophs ("nitrogen eaters"), which live in the roots of leguminous plants such as peas, beans, and peanuts (Figure 6.41). These bacteria convert atmospheric nitrogen ($N_2$) into ammonia ($NH_3$) in a process called *nitrogen fixation*. Other microorganisms then oxidize $NH_3$ to oxides of nitrogen. Nitrogen fixation is necessary to sustain life on Earth because ammonia is required for the formation of biologically essential nitrogen-containing molecules, such as proteins and DNA.

A simple laboratory preparation of nitrogen is based on the thermal decomposition of ammonium nitrite:

$$NH_4NO_2(s) \rightarrow N_2(g) + 2\,H_2O(g)$$

On an industrial scale, nitrogen is produced by distilling liquid air. Nitrogen boils at −196°C, so it distills before oxygen, which boils at −183°C.

Liquid nitrogen is a cryogen, which means a very cold fluid. It is used in commercial refrigeration. Nitrogen's relative lack of reactivity at ordinary temperatures and pressures, coupled with its availability, makes it popular as a protective gas in the semiconductor industry to keep oxygen away from sensitive materials and for several industrial processes involving welding or soldering. Oil companies pump nitrogen under high pressure into oil deposits to force crude oil to the surface (Figure 6.42).

In the early 20th century, mined nitrates were widely used in the manufacture of gunpowder and other explosives. For example, a mixture

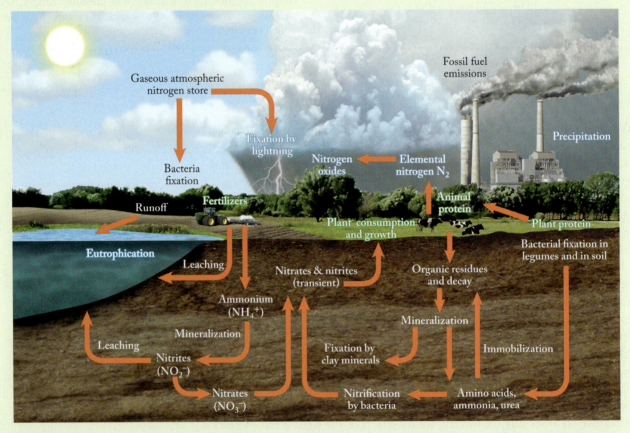

**FIGURE 6.40** The nitrogen cycle describes the relationship between atmospheric nitrogen and the compounds of nitrogen necessary to support life on Earth. At any time, a large portion of nitrogen is found in the biosphere in living organisms and their dead remains. When organic matter decomposes, ammonium ion and other simple nitrogen compounds are released. Nitrogen fixation provides ammonia to organisms. Microorganisms take up $NH_4^+$ and $NO_3^-$. $NH_4^+$ is released during decomposition of organisms upon their death. Microorganisms oxidize $NH_4^+$ to various oxides of nitrogen. Microorganisms use $NO_3^-$ as an oxidizing agent and convert nitrate ion back into $N_2$.

**FIGURE 6.41** The root system of a soybean plant contains nodules caused by the presence of nitrogen-fixing bacteria that live in a symbiotic relationship with the plants. This plant is a member of the legume family, and all other members of the family have the same type of root system.

of potassium nitrate, sulfur, and carbon known as *black powder* is still the best fuse material ever discovered. It reacts exothermically and explosively, producing nitrogen and carbon dioxide. Rapid expansion of these hot gases adds to the explosive character of the reaction:

$$16\,KNO_3(s) + S_8(s) + 24\,C(s) \rightarrow 8\,K_2S(s) + 8\,N_2(g) + 24\,CO_2(g)$$

When naval blockades cut off Germany's supplies of $KNO_3$ during World War I, German chemist Fritz Haber (1868–1934) developed a process for making nitrates that starts with the synthesis of ammonia from hydrogen gas and nitrogen from the atmosphere. The *Haber–Bosch process* relies on high temperature and pressure to promote the reaction:

$$N_2(g) + 3\,H_2(g) \rightarrow 2\,NH_3(g)$$

In this reaction, the very strong nitrogen–nitrogen bond in $N_2$ must be broken, which requires an energy investment. Today, the Haber–Bosch process is still the principal source of ammonia and the nitrogen compounds derived from ammonia that are widely used in industry and agriculture.

The first step in converting ammonia into other important nitrogen compounds entails its oxidation to $NO_2$ and $H_2O$:

$$4\,NH_3(g) + 5\,O_2(g) \rightarrow 4\,NO(g) + 6\,H_2O(g)$$
$$2\,NO(g) + O_2(g) \rightarrow 2\,NO_2(g)$$

Nitrogen dioxide dissolves in water, producing a mixture of nitrous and nitric acids:

$$2\,NO_2(g) + H_2O(\ell) \rightarrow HNO_2(aq) + HNO_3(aq)$$

Heating the mixture converts nitrous acid into nitric acid, as steam and NO are given off:

$$3\,HNO_2(aq) \rightarrow HNO_3(aq) + H_2O(g) + 2\,NO(g)$$

About 75% of nitric acid produced in this way is combined with more ammonia to produce ammonium nitrate:

$$NH_3(g) + HNO_3(aq) \rightarrow NH_4NO_3(s)$$

Ammonium nitrate is an important industrial chemical and the source of water-soluble nitrogen in many formulations of fertilizers.

All of this chemistry described for industrial use—some of which takes place at high temperature and pressure, requiring energy derived from fossil fuels—also takes place in bacteria but at normal temperatures and pressures. Studying the mechanism of these reactions in bacteria remains an active area of research. To date, efforts at finding alternative methods of producing the large quantities of nitrogen-containing substances needed by commerce and agriculture in ways that minimize the high costs associated with industrial production have not been successful.

Nitrogen has a wide and varied chemistry beyond the synthesis and uses of ammonia. Hydrazine ($N_2H_4$) is used as a rocket fuel. The reaction of metals with nitrogen produces metal nitrides containing the $N^{3-}$ anion. Iron(III) nitride (FeN) is used to harden the surface of steel. Sodium azide is a bactericide used in many biology laboratories to control bacterial growth. Metal salts containing the azide ion ($N_3^-$) are often explosive and are used as a ready source of nitrogen gas in some automobile air bags. The overall reaction in an air bag is the sum of several reactions, summarized as:

$$20\,NaN_3(s) + 6\,SiO_2(s) + 4\,KNO_3(s) \rightarrow$$
$$32\,N_2(g) + 5\,Na_4SiO_4(s) + K_4SiO_4(s)$$

Azides are contact explosives, which means mechanical contact causes them to detonate. The trigger mechanism in a steering column that causes the azide to explode and produce nitrogen to inflate the bag involves the transmission of the force of an accident to the azide contained in the air bags. Highly corrosive and reactive sodium metal is produced in the azide reaction along with the nitrogen, and other materials—the $SiO_2$ and $KNO_3$ included in the overall reaction shown—react with the sodium to convert it into relatively harmless sodium silicates.

**FIGURE 6.42** Nitrogen under high pressure is pumped into oil deposits to bring crude oil to the surface.

## SUMMARY

**Learning Outcome 1** Gases occupy the entire volume of their container. The volume occupied by a gas changes significantly with pressure and temperature. Gases are **miscible**, mixing in any proportion. They are much less dense than liquids or solids. (Section 6.1)

**Learning Outcome 2** **Pressure** is defined as the ratio of force to surface area, and gas pressure is measured with **barometers** and manometers. Pressure can be expressed in a variety of units, and we can convert between them using dimensional analysis. (Section 6.2)

**Learning Outcome 3** **Boyle's law, Charles's law, Amontons's law,** and **Avogadro's law** describe the behavior of gases under different conditions. They may be used individually or in a combined form called the **ideal gas law** **(PV = nRT)** to calculate the volume, temperature, pressure, or number of moles of a gas. (Sections 6.3 and 6.4)

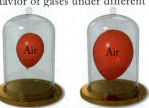

**Learning Outcome 4** The ideal gas equation and the stoichiometry of a chemical reaction can be used to calculate the volumes of gases required or produced in the reaction. (Section 6.5)

**Learning Outcome 5** The density of a gas can be calculated for a given set of conditions of volume, temperature, and pressure. The density of a gas can also be used to calculate its molar mass. (Section 6.6)

**Learning Outcome 6** In a gas mixture, the contribution each component gas makes to the total gas pressure is called the **partial pressure** of that gas. **Dalton's law of partial pressures** allows us to calculate the partial pressure ($P_x$) of any constituent gas $x$ in a gas mixture if we know its **mole fraction ($X_x$)** and the total pressure. (Section 6.7)

**Learning Outcome 7** **Kinetic molecular theory** describes gases as particles in constant random motion, moving at **root-mean-square speeds ($u_{rms}$)** that are inversely proportional to the square root of their molar masses and directly proportional to the square root of their temperature. Pressure arises from elastic collisions between gases and the walls of their container. (Section 6.8)

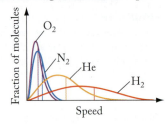

**Learning Outcome 8** **Graham's law of effusion** states that the rate of **effusion** (escape through a pinhole) or **diffusion** (spreading) of a gas at a fixed temperature is inversely proportional to the square root of its molar mass. (Section 6.8)

**Learning Outcome 9** At high pressures, the behavior of real gases deviates from the predictions of the ideal gas law. The **van der Waals equation**, a modified form of the ideal gas equation, accounts for real gas properties. (Section 6.9)

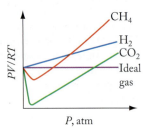

## PROBLEM-SOLVING SUMMARY

| TYPE OF PROBLEM | CONCEPTS AND EQUATIONS | SAMPLE EXERCISES |
|---|---|---|
| **Calculating pressure of any gas, calculating atmospheric pressure** | Divide the force by area over which the force is applied, using the equation $$P = \frac{F}{A} \qquad (6.1)$$ | 6.1, 6.2 |
| **Calculating changes in $P$, $V$, and/or $T$ in response to changing conditions** | Rearrange $$\frac{P_1 V_1}{T_1} = \frac{P_2 V_2}{T_2} \qquad (6.18)$$ for whichever variable is sought and then substitute given values. ($T$ must be in kelvins and $n$ is constant.) | 6.3, 6.4, 6.5, 6.6 |
| **Determining $n$ from $P$, $V$, and $T$** | Rearrange $$PV = nRT \qquad (6.13)$$ for $n$ and then substitute given values of $P$, $T$, and $V$. ($T$ must be in kelvins.) | 6.7, 6.8, 6.9 |

| TYPE OF PROBLEM | CONCEPTS AND EQUATIONS | SAMPLE EXERCISES |
|---|---|---|
| **Calculating the density of a gas and calculating molar mass from density** | Substitute values for pressure, absolute temperature, and molar mass into the equation $$d = \frac{P\mathcal{M}}{RT} \qquad (6.21)$$ Substitute values for pressure, absolute temperature, and density into the equation $$\mathcal{M} = \frac{dRT}{P} \qquad (6.22)$$ | 6.10, 6.11 |
| **Calculating mole fraction for one component gas in a mixture** | Divide the number of moles of the component gas by the total number of moles in the mixture: $$X_x = \frac{n_x}{n_{\text{total}}} \qquad (6.24)$$ | 6.12 |
| **Calculating partial pressure of one component gas in a mixture and total pressure in the mixture** | Substitute the mole fraction of the component gas and the total pressure in the equation $$P_x = X_x P_{\text{total}} \qquad (6.25)$$ Solve the equation $$P_{\text{total}} = P_1 + P_2 + P_3 + P_4 + \ldots \qquad (6.23)$$ for the partial pressure of the component gas and then substitute given values for other partial pressures and total pressure. | 6.13, 6.14 |
| **Calculating root-mean-square speeds** | Substitute absolute temperature, molar mass, and the value $8.314 \text{ kg} \cdot \text{m}^2/(\text{s}^2 \cdot \text{mol} \cdot \text{K})$ for $R$ in the equation $$u_{\text{rms}} = \sqrt{\frac{3RT}{\mathcal{M}}} \qquad (6.28)$$ | 6.15 |
| **Calculating relative rate of effusion or diffusion from molar masses** | Substitute values for molar masses into the equation $$\frac{r_x}{r_y} = \sqrt{\frac{\mathcal{M}_y}{\mathcal{M}_x}} \qquad (6.31)$$ | 6.16 |
| **Calculating $P$ for a real gas** | Solve the van der Waals equation $$\left(P + \frac{n^2 a}{V^2}\right)(V - nb) = nRT \qquad (6.32)$$ for $P$ and substitute given values of $n$, $V$, and $T$ (in kelvins), plus values of $a$ and $b$ from Table 6.5. | 6.17 |

## VISUAL PROBLEMS

*(Answers to boldface end-of-chapter questions and problems are in the back of the book.)*

**6.1.** Shown in Figure P6.1 are three barometers. The one in the center is located at sea level. Which barometer is most likely to reflect the atmospheric pressure in Denver, CO, where the elevation is approximately 1500 m? Explain your answer.

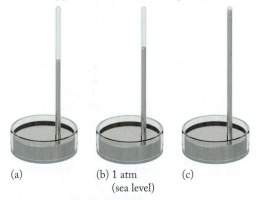

(a)    (b) 1 atm    (c)
      (sea level)

**FIGURE P6.1**

**6.2.** A rubber balloon is filled with helium gas. Which of the drawings in Figure P6.2 most accurately reflects the gas in the balloon on a molecular level? The blue spheres represent helium atoms. Explain your answer. Because atoms are in constant motion, the spheres represent their average position.

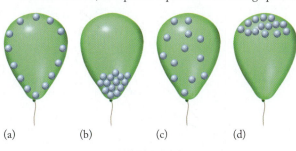

(a)    (b)    (c)    (d)

**FIGURE P6.2**

**6.3.** Which of the three changes shown in Figure P6.3 best illustrates what happens when the atmospheric pressure on a helium-filled rubber balloon is increased at constant temperature? Because atoms are in constant motion, the spheres represent their average position.

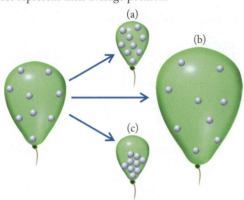

**FIGURE P6.3**

**6.4.** Which of the drawings in Figure P6.3 best illustrates what happens when the temperature of a helium-filled rubber balloon is increased at constant pressure?

**6.5.** Which of the three changes shown in Figure P6.5 best illustrates what happens when the amount of gas in a helium-filled rubber balloon is increased at constant temperature and pressure?

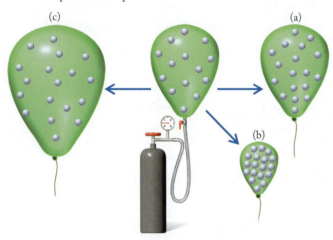

**FIGURE P6.5**

**6.6.** Which line plotting volume versus reciprocal pressure in Figure P6.6 corresponds to the higher temperature?

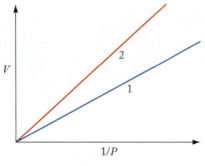

**FIGURE P6.6**

**6.7.** In Figure P6.7, which line of volume versus temperature represents a gas at higher pressure? Is the $x$-axis an absolute temperature scale?

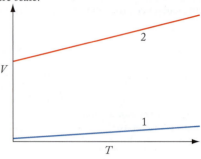

**FIGURE P6.7**

**6.8.** In Figure P6.8, which of the two plots of volume versus pressure at constant temperature is not consistent with the ideal gas law?

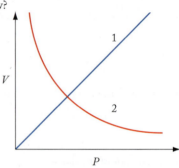

**FIGURE P6.8**

**6.9.** In Figure P6.9, which of the two plots of volume versus temperature at constant pressure is not consistent with the ideal gas law?

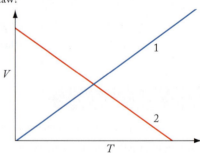

**FIGURE P6.9**

**6.10.** In Figure P6.10, which line in the graph of density versus pressure at constant temperature for methane ($CH_4$) and nitrogen ($N_2$) should be labeled *methane*?

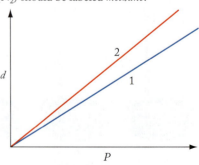

**FIGURE P6.10**

**6.11.** Add lines showing the densities of He and NO as a function of pressure to a copy of the graph in Figure P6.10.

**6.12.** Which of the graphs in Figure P6.12 best represents the relationship between number of collisions with the container and pressure?

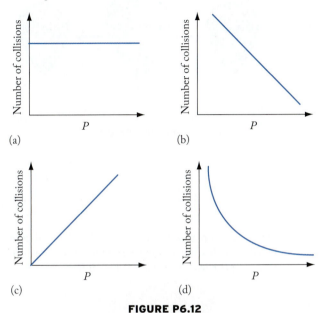

**FIGURE P6.12**

**6.13.** Which of the drawings in Figure P6.13 best depicts the arrangement of molecules in a mixture of gases? The red and blue spheres represent atoms of two different gases, such as helium and neon.

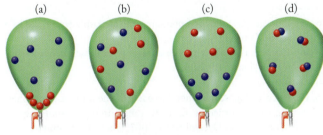

**FIGURE P6.13**

**6.14.** The drawings in Figure P6.14 illustrate four mixtures of two diatomic gases (such as $N_2$ and $O_2$) in flasks of identical volume and at the same temperature. Is the total pressure in each flask the same? Which flask has the highest partial pressure of nitrogen, depicted by the blue molecules?

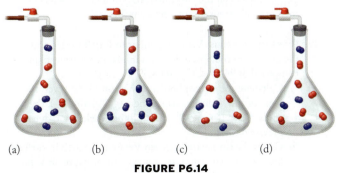

**FIGURE P6.14**

**6.15.** Figure P6.15 shows the distribution of molecular speeds of $CO_2$ and $SO_2$ molecules at 25°C. Which curve is the profile for $SO_2$? Which of these profiles should match that of propane ($C_3H_8$), a common fuel in portable grills?

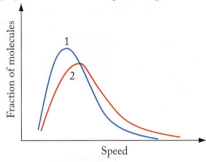

**FIGURE P6.15**

**6.16.** How would a graph showing the distribution of molecular speeds of $CO_2$ at −100°C differ from the curve for $CO_2$ shown in Figure P6.15?

**6.17.** A container with a pinhole leak contains a mixture of the elements highlighted in Figure P6.17. Which element leaks the slowest from the container?

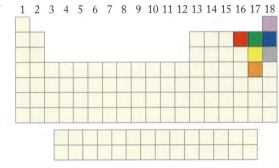

**FIGURE P6.17**

**6.18.** A container with a pinhole leak contains a mixture of the elements highlighted in Figure P6.17. Which element has the smallest root-mean-square speed?

**6.19.** The rate of effusion of a gas increases with temperature. Which graph in Figure P6.19 best describes the ratio of the rates of effusion of two gases ($x$ and $y$) as a function of temperature?

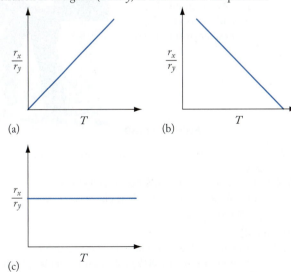

**FIGURE P6.19**

6.20. Which of the elements highlighted in Figure P6.17 has the smallest van der Waals *b* constant?

6.21. Which of the two outcomes diagrammed in Figure P6.21 more accurately illustrates the effusion of helium from a balloon at constant atmospheric pressure?

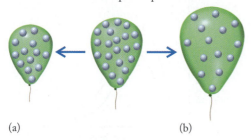

(a)                    (b)

**FIGURE P6.21**

6.22. Which of the two outcomes shown in Figure P6.22 more accurately illustrates the effusion of gases from a balloon at constant atmospheric pressure if the red spheres have a greater root-mean-square speed than the blue spheres?

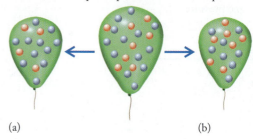

(a)                    (b)

**FIGURE P6.22**

## QUESTIONS AND PROBLEMS

### The Gas Phase; Atmospheric Pressure

**CONCEPT REVIEW**

6.23. Describe the difference between force and pressure.

6.24. How does Torricelli's barometer measure atmospheric pressure?

6.25. What is the relation between *torr* and *atmospheres* of pressure?

6.26. What is the relation between *millibars* and *pascals* of pressure?

6.27. Three barometers based on Torricelli's design are constructed using water ($d = 1.00$ g/mL), ethanol ($d = 0.789$ g/mL), and mercury ($d = 13.546$ g/mL). Which barometer contains the tallest column of liquid?

6.28. In constructing a barometer, what advantage is there in choosing a dense liquid?

6.29. Why does an ice skater exert more pressure on ice when wearing newly sharpened skates than when wearing skates with dull blades?

6.30. Why is it easier to travel over deep snow when wearing boots and snowshoes (Figure P6.30) rather than just boots?

**FIGURE P6.30**

6.31. Why does atmospheric pressure decrease with increasing elevation?

*6.32. Pieces of different metals have exactly the same mass but different densities. Could these objects ever exert the same pressure?

**PROBLEMS**

6.33. Calculate the downward pressure due to gravity exerted by the bottom face of a 1.00 kg cube of iron that is 5.00 cm on a side.

6.34. The gold block represented in Figure P6.34 has a mass of 38.6 g. Calculate the pressure exerted by the block when it is on (a) a square face and (b) a rectangular face.

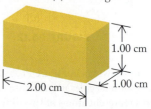

1.00 cm

2.00 cm    1.00 cm

**FIGURE P6.34**

6.35. Convert the following pressures into atmospheres: (a) 2.0 kPa; (b) 562 mmHg.

6.36. Convert the following pressures into millimeters of mercury: (a) 0.541 atm; (b) 2.8 kPa.

6.37. **Record High Atmospheric Pressure** The highest atmospheric pressure recorded on Earth was measured at Tosontsengel, Mongolia, on December 19, 2001, when the barometer read 108.6 kPa. Express this pressure in (a) millimeters of mercury, (b) atmospheres, and (c) millibars.

6.38. **Record Low Atmospheric Pressure** Hurricane Irene registered an atmospheric pressure of 982 mbar in August 2011. Hurricane Katrina registered an atmospheric pressure of 86.2 kPa in September 2005. What was the *difference* in pressure between the two hurricanes in (a) millimeters of mercury, (b) atmospheres, and (c) millibars?

*6.39. **Pressure on Venus** Venus is similar to Earth in size and distance from the Sun but has an inhospitable atmosphere composed of 96.5% $CO_2$ and 3.5% $N_2$ by mass.
   a. Calculate the atmospheric pressure on Venus given the mass of the Venusian atmosphere ($4.8 \times 10^{20}$ kg), the surface area ($4.6 \times 10^{14}$ m$^2$) and the acceleration due to gravity (8.87 m/s$^2$).
   *b. Given the similarity between Venus and Earth in surface area why do you suppose atmospheric pressure on Venus is so much higher than Earth?

*6.40. **Pressure on Uranus** Uranus is much larger than Earth, with a surface area of $8.1 \times 10^{15}$ m$^2$ with an atmosphere composed of low-density gases: 82.5% H$_2$, 15.2% He, and 2.3% CH$_4$. The acceleration due to gravity is similar to that on Earth, 8.7 m/s$^2$, yet the atmospheric pressure on Uranus is estimated to exceed $10^8$ Pa. What is the mass of the atmosphere on Uranus?

## The Gas Laws

### CONCEPT REVIEW

6.41. From the molecular perspective, why is pressure directly proportional to temperature at fixed volume (Amontons's law)?
6.42. How do we explain Boyle's law on a molecular basis?
6.43. A balloonist is rising too fast for her taste. Should she increase the temperature of the gas in the balloon or decrease it?
6.44. Could the pilot of the balloon in Problem 6.43 reduce her rate of ascent by allowing some gas to leak out of the balloon? Explain your answer.
6.45. The volume of a quantity of gas decreases by 50% when it cools from 20°C to 10°C. Does the pressure of the gas increase, decrease, or remain the same?
6.46. If the volume of gasoline vapor and air in an automobile engine cylinder is reduced to 1/10 of its original volume before ignition, by what factor does the pressure in the cylinder increase? (Assume there is no change in temperature.)

### PROBLEMS

6.47. What is the final pressure of 1.00 mol of ammonia gas, initially at 1.00 atm, if the volume is
  a. gradually decreased from 78.0 mL to 39.0 mL at constant temperature.
  b. increased from 43.5 mL to 65.5 mL at constant temperature.
  c. decreased by 40% at constant temperature.
6.48. The behavior of 485 mL of an ideal gas in response to pressure is studied in a vessel with a movable piston. What is the final volume of the gas if the pressure on the sample is
  a. increased from 715 mmHg to 3.55 atm at constant temperature.
  b. decreased from 1.15 atm to 520 mmHg at constant temperature.
  c. increased by 26% at constant temperature.

6.49. A scuba diver releases a balloon containing 153 L of helium attached to a tray of artifacts at an underwater archaeological site (Figure P6.49). When the balloon reaches the surface, it has expanded to a volume of 352 L. The pressure at the surface is 1.00 atm; what is the pressure at the underwater site? Pressure increases by 1.0 atm for every 10 m of depth; at what depth was the diver working? Assume the temperature remains constant.

**FIGURE P6.49**

6.50. **Breath-Hold Diving** The world record for diving without supplemental air tanks (breath-hold diving) is about 125 m, a depth at which the pressure is about 12.5 atm. If a diver's lungs have a volume of 6 L at the surface of the water, what is their volume at a depth of 125 m?

6.51. Use the following data to draw a graph of the volume of 1 mole of H$_2$ as a function of the reciprocal of pressure at 298 K:

| P (mmHg) | V (L) |
| --- | --- |
| 100 | 186 |
| 120 | 155 |
| 240 | 77.5 |
| 380 | 48.9 |
| 500 | 37.2 |

Would the graph be the same for the same number of moles of argon?

6.52. The following data for P and V were collected for 1 mole of argon at 300 K. Draw a graph of the volume of 1 mole of Ar as a function of the reciprocal of pressure. Does this graph look like the graph you drew for Problem 6.51?

| P (atm) | V (L) |
| --- | --- |
| 0.10 | 246.3 |
| 0.25 | 98.5 |
| 0.50 | 49.3 |
| 0.75 | 32.8 |
| 1.0 | 24.6 |

6.53. Use the following data to draw a graph of the volume of He as a function of temperature for 1.0 mol of He gas at a constant pressure of 1.00 atm:

| V (L) | T (K) |
| --- | --- |
| 7.88 | 96 |
| 3.94 | 48 |
| 1.97 | 24 |
| 0.79 | 9.6 |
| 0.39 | 4.8 |

How would the graph change if the amount of gas were halved?

6.54. Use the following data to draw a graph of the volume of He as a function of temperature for 0.50 mol of He gas at a constant pressure of 1.00 atm:

| V (L) | T (K) |
| --- | --- |
| 3.94 | 96 |
| 1.97 | 48 |
| 0.79 | 24 |
| 0.39 | 9.6 |
| 0.20 | 4.8 |

Does this graph match your prediction from Problem 6.53?

**6.55.** A cylinder with a piston (Figure P6.55) contains a sample of gas at 25°C. The piston moves in response to changing pressure inside the cylinder. At what gas temperature would the piston move so that the volume inside the cylinder doubled?

**FIGURE P6.55**

**6.56.** The temperature of the gas in Problem 6.55 is reduced to a temperature at which the volume inside the cylinder has decreased by 25% from its initial volume at 25.0°C. What is the new temperature?

**6.57.** Calculate the volume of a 2.68 L sample of gas in the cylinder shown in Figure P6.55 after it is subjected to the following changes in conditions:
   a. The gas is warmed from 250 K to a final temperature of 398 K at constant pressure.
   b. The pressure is increased by 33% at constant temperature.
   c. Ten percent of the gas leaks from the cylinder at constant temperature and pressure.

**6.58.** Calculate the temperature of a 5.6 L sample of gas in the cylinder in Figure P6.55 after it is subjected to the following changes in conditions:
   a. The gas, at constant pressure, is cooled from 78°C to a temperature at which its volume is 4.3 L.
   b. The external pressure is doubled at constant volume.
   c. The number of molecules in the cylinder is increased by 15% at constant pressure and volume.

**\*6.59.** An empty balloon is filled with 1.75 mol of He at 17°C and 1.00 atm pressure. How much work is done on the system if the temperature remains constant?

**\*6.60.** A student holding a 50.0 L balloon containing 800 mmHg of hydrogen at 19°C lets go of the balloon, allowing the gas to escape. How many moles of gas did the balloon contain and how much work was done on the surroundings if the temperature remains constant?

**6.61.** Which of the following actions would produce the greatest increase in the volume of a gas sample: (a) lowering the pressure from 760 mmHg to 720 mmHg at constant temperature or (b) raising the temperature from 10°C to 40°C at constant pressure?

**6.62.** Which of the following actions would produce the greatest increase in the volume of a gas sample: (a) doubling the amount of gas in the sample at constant temperature and pressure or (b) raising the temperature from 244°C to 1100°C at constant pressure?

**\*6.63.** What happens to the volume of gas in a cylinder with a movable piston under the following conditions?
   a. Both the absolute temperature and the external pressure on the piston double.
   b. The absolute temperature is halved, and the external pressure on the piston doubles.
   c. The absolute temperature increases by 75%, and the external pressure on the piston increases by 50%.

**\*6.64.** What happens to the pressure of a gas under the following conditions?
   a. The absolute temperature is halved and the volume doubles.
   b. Both the absolute temperature and the volume double.
   c. The absolute temperature increases by 75%, and the volume decreases by 50%.

**6.65.** A 150.0 L weather balloon filled with 6.1 mol of He has a small leak. If the helium leaks at a rate of 10 mmol/hr, what is the volume of the balloon after 24 hr?

**6.66.** Which causes the larger change in the volume of a gas at constant temperature, doubling the number of moles or doubling the pressure?

**6.67.** **Temperature Effects on Bicycle Tires** A bicycle racer inflates his tires to 7.1 atm on a warm autumn afternoon when temperatures reached 27°C. By morning the temperature has dropped to 5.0°C. What is the pressure in the tires if we assume that the volume of the tire does not change significantly?

**\*6.68.** A balloon vendor at a street fair is using a tank of helium to fill her balloons. The tank has a volume of 145 L and a pressure of 136 atm at 25°C. After a while she notices that the valve has not been closed properly, and the pressure has dropped to 94 atm. How many moles of gas have been lost?

## The Ideal Gas Law; Gases in Chemical Reactions
### CONCEPT REVIEW

**6.69.** What is meant by standard temperature and pressure (STP)? What is the volume of 1 mole of an ideal gas at STP?

**6.70.** Which of the following are not characteristics of an ideal gas?
   a. The molecules of gas have little volume compared with the volume that they occupy.
   b. Its volume is independent of temperature.
   c. The density of all ideal gases is the same.
   d. Gas atoms or molecules do not interact with one another.

**\*6.71.** What does the slope represent in a graph of pressure as a function of $1/V$ at constant temperature for an ideal gas?

**\*6.72.** How would the graph in Problem 6.71 change if we increased the temperature to a larger but constant value?

### PROBLEMS

**6.73.** How many moles of air must there be in a bicycle tire with a volume of 2.36 L if it has an internal pressure of 6.8 atm at 17.0°C?

**6.74.** At what temperature will 1.00 mole of an ideal gas in a 1.00 L container exert a pressure of
   a. 1.00 atm?
   b. 2.00 atm?
   c. 0.75 atm?

**6.75.** **Hyperbaric Oxygen Therapy** Hyperbaric oxygen chambers are used to treat divers suffering from decompression sickness (the "bends") with pure oxygen at greater than atmospheric pressure. Other clinical uses include treatment of patients with thermal burns, necrotizing fasciitis, and CO poisoning. What is the pressure in a chamber with a volume of $2.36 \times 10^3$ L that contains 4635 g of $O_2(g)$ at a temperature of 298 K?

**6.76.** Hydrogen holds promise as an "environment friendly" fuel. How many grams of $H_2$ gas are present in a 50.0 L fuel tank at a pressure of 2850 lb/in$^2$ (psi) at 20°C? Assume that 1 atm = 14.7 psi.

**6.77.** A weather balloon with a volume of 200.0 L is launched at 20°C at sea level, where the atmospheric pressure is 1.00 atm.
   a. What is the volume of the balloon at 20,000 m where atmospheric pressure is 63 mmHg and the temperature is 220 K?
   b. At even higher elevations (the stratosphere, up to 50,000 m), the temperature of the atmosphere increases to 270 K but the pressure decreases to 0.80 mmHg. What is the volume of the balloon under these conditions?
   c. If the balloon is designed to rupture when the volume exceeds 400.0 L, will the balloon break under either of these sets of conditions?

**6.78.** A sealed, flexible, foil bag of potato chips, containing 0.500 L of air, is carried from Boston ($P_{atm}$ = 1.0 atm) to Denver ($P_{atm}$ = 0.83 atm).
   a. What would the volume of the bag be upon arrival in Denver?
   b. If the structural limitations of the bag only allow for a 10% expansion in the volume of the bag, what is the pressure in the bag?
   c. If the bag is in our checked luggage, the pressure and temperature in the hold will decrease considerably during flight to $T$ = 210 K and $P$ = 126 torr. Calculate the pressure in the bag during flight.

**6.79.** **Miners' Lamps** Before the development of reliable batteries, miners' lamps burned acetylene produced by the reaction of calcium carbide with water:

$$CaC_2(s) + H_2O(\ell) \rightarrow C_2H_2(g) + CaO(s)$$

A lamp uses 1.00 L of acetylene per hour at 1.00 atm pressure and 18°C.
   a. How many moles of $C_2H_2$ are used per hour?
   b. How many grams of calcium carbide must be in the lamp for a 4 hr shift?

**6.80.** Acid precipitation dripping on limestone produces carbon dioxide by the following reaction:

$$CaCO_3(s) + 2 H^+(aq) \rightarrow Ca^{2+}(aq) + CO_2(g) + H_2O(\ell)$$

If 15.0 mL of $CO_2$ were produced at 25°C and 760 mmHg, then
   a. how many moles of $CO_2$ were produced?
   b. how many milligrams of $CaCO_3$ were consumed?

**\*6.81.** Air is about 78% nitrogen by volume and 21% oxygen. Pure nitrogen is produced by the decomposition of ammonium dichromate:

$$(NH_4)_2Cr_2O_7(s) \rightarrow N_2(g) + Cr_2O_3(s) + 4 H_2O(g)$$

Oxygen is generated by the thermal decomposition of potassium chlorate:

$$2 KClO_3(s) \rightarrow 2 KCl(s) + 3 O_2(g)$$

How many grams of ammonium dichromate and how many grams of potassium chlorate would be needed to make 200.0 L of "air" at 0.85 atm and 273 K?

**\*6.82.** Nitrogen can also be produced from sodium metal and potassium nitrate by the reaction:

$$10 Na(s) + 2 KNO_3(s) \rightarrow K_2O(s) + 5 Na_2O(s) + N_2(g)$$

If we generate oxygen by the thermal decomposition of potassium chlorate:

$$2 KClO_3(s) \rightarrow 2 KCl(s) + 3 O_2(g)$$

how many grams of potassium nitrate and how many grams of potassium chlorate would be needed to make 200.0 L of gas containing 160.0 L $N_2$ and 40.0 L $O_2$ at 1.00 atm and 290 K?

**6.83.** **Healthy Air for Sailors** The $CO_2$ that builds up in the air of a submerged submarine can be removed by reacting it with sodium peroxide:

$$2 Na_2O_2(s) + 2 CO_2(g) \rightarrow 2 Na_2CO_3(s) + O_2(g)$$

If a sailor exhales 150.0 mL of $CO_2$ per minute at 20°C and 1.02 atm, how much sodium peroxide is needed per sailor in a 24 hr period?

**6.84.** **Rescue Breathing Devices** Self-contained self-rescue breathing devices, like the one shown in Figure P6.84, convert $CO_2$ into $O_2$ according to the following reaction:

$$4 KO_2(s) + 2 CO_2(g) \rightarrow 2 K_2CO_3(s) + 3 O_2(g)$$

How many grams of $KO_2$ are needed to produce 100.0 L of $O_2$ at 20°C and 1.00 atm?

**FIGURE P6.84**

## Gas Density

### CONCEPT REVIEW

**6.85.** Do all gases at the same pressure and temperature have the same density? Explain your answer.

**6.86.** Birds and sailplanes take advantage of thermals (rising columns of warm air) to gain altitude with less effort than usual. Why does warm air rise?

**6.87.** How does the density of a gas sample change when (a) its pressure is increased and (b) its temperature is decreased?

**6.88.** How would you measure the density of a gas sample of known molar mass?

### PROBLEMS

**6.89. Biological Effects of Radon Exposure** Radon is a naturally occurring radioactive gas found in the ground and in building materials. It is easily inhaled and emits $\alpha$ particles when it decays. Cumulative radon exposure is a significant risk factor for lung cancer.
   a. Calculate the density of radon at 298 K and 1.00 atm of pressure.
   b. Are radon concentrations likely to be greater in the basement or on the top floor of a building?

*__6.90.__ Four empty balloons, each with a mass of 10.0 g, are inflated to a volume of 20.0 L. The first balloon contains He, the second Ne, the third $CO_2$, and the fourth CO. If the density of air at 25°C and 1.00 atm is 0.00117 g/mL, will any of the balloons float in this air?

_____

**6.91.** A 150.0 mL flask contains 0.391 g of a volatile oxide of sulfur. The pressure in the flask is 750 mmHg, and the temperature is 22°C. Is the gas $SO_2$ or $SO_3$?

**6.92.** A 100.0 mL flask contains 0.193 g of a volatile oxide of nitrogen. The pressure in the flask is 760 mmHg at 17°C. Is the gas NO, $NO_2$, or $N_2O_5$?

_____

**6.93.** The density of an unknown gas is 1.107 g/L at 300 K and 740 mmHg. Could this gas be CO or $CO_2$?

**6.94.** A gas containing chlorine and oxygen has a density of 2.875 g/L at 756 mmHg and 11°C. What is the most likely molecular formula of the gas?

## Dalton's Law and Mixtures of Gases

### CONCEPT REVIEW

**6.95.** What is meant by the *partial pressure* of a gas?

**6.96.** Can a barometer be used to measure just the partial pressure of oxygen in the atmosphere? Why or why not?

**6.97.** Which gas sample has the largest volume at 25°C and 1 atm pressure? (a) 0.500 mol dry $H_2$; (b) 0.500 mol dry $N_2$; (c) 0.500 mol wet $H_2$ ($H_2$ collected over water)

**6.98.** Two identical balloons are filled to the same volume at the same pressure and temperature. One balloon is filled with air and the other with helium. Which balloon contains more particles (atoms and molecules)?

### PROBLEMS

**6.99.** A gas mixture contains 0.70 mol of $N_2$, 0.20 mol of $H_2$, and 0.10 mol of $CH_4$. What is the mole fraction of $H_2$ in the mixture? Calculate the pressure of the gas mixture and the partial pressure of each constituent gas if the mixture is in a 10.0 L vessel at 27°C.

**6.100.** A gas mixture contains 7.0 g of $N_2$, 2.0 g of $H_2$, and 16.0 g of $CH_4$. What is the mole fraction of $H_2$ in the mixture? Calculate the pressure of the gas mixture and the partial pressure of each constituent gas if the mixture is in a 1.00 L vessel at 0°C.

_____

**6.101. The X-15 and Mach 6** On November 9, 1961, the Bell X-15 test aircraft exceeded Mach 6, or six times the speed of sound. A few weeks later it reached an elevation above 300,000 feet, effectively putting it in space. The combustion of ammonia and oxygen provided the fuel for the X-15.
   a. Write a balanced chemical equation for the combustion of ammonia given that nitrogen ends up as nitrogen dioxide.
   b. What ratio of partial pressures of ammonia and oxygen are needed for this reaction?

**6.102. Rocket Fuels** Before settling on hydrogen as the fuel of choice for space vehicles, a number of other fuels were explored, including hydrazine, $N_2H_4$, and pentaborane, $B_5H_{11}$. Both are gases under conditions in space.
   a. Write balanced chemical equations for the combustion of hydrazine and pentaborane given that the products in addition to water are $NO_2(g)$ and solid boron(III) oxide, respectively.
   b. What ratios of partial pressures of hydrazine to oxygen and pentaborane to oxygen are needed for these reactions?

_____

**6.103.** A sample of oxygen was collected over water at 25°C and 1.00 atm.
   a. If the total sample volume was 0.480 L, how many moles of $O_2$ were collected?
   b. If the same volume of oxygen is collected over ethanol instead of water, does it contain the same number of moles of $O_2$?

**6.104.** Water and ethanol were removed from the $O_2$ samples in Problem 6.103.
   a. What is the volume of the dry $O_2$ gas sample at 25°C and 1.00 atm?
   b. What is the volume of the dry $O_2$ gas sample at 25°C and 1.00 atm if $P_{ethanol} = 50$ mmHg at 25°C?

_____

**6.105.** The following reactions were carried out in sealed containers. Will the total pressure after each reaction is complete be greater than, less than, or equal to the total pressure before the reaction? Assume all reactants and products are gases at the same temperature.
   a. $N_2O_5(g) + NO_2(g) \rightarrow 3\,NO(g) + 2\,O_2(g)$
   b. $2\,SO_2(g) + O_2(g) \rightarrow 2\,SO_3(g)$
   c. $C_3H_8(g) + 5\,O_2(g) \rightarrow 3\,CO_2(g) + 4\,H_2O(g)$
   d. $4\,NH_3(g) + 5\,O_2(g) \rightarrow 4\,NO(g) + 6\,H_2O(g)$

**6.106.** The following reactions were carried out in a cylinder with a piston. If the external pressure is constant, in which reactions will the volume of the cylinder increase? Assume all reactants and products are gases at the same temperature.
   a. $CH_4(g) + NH_3(g) \rightarrow HCN(g) + 3\,H_2(g)$
   b. $H_2S(g) + 2\,O_2(g) \rightarrow H_2O(g) + SO_3(g)$
   c. $H_2(g) + Cl_2(g) \rightarrow 2\,HCl(g)$
   d. $2\,NO_2(g) \rightarrow 2\,NO(g) + O_2(g)$

_____

**6.107. High-Altitude Mountaineering** Alpine climbers use pure oxygen near the summits of 8000 m peaks, where $P_{atm} = 0.35$ atm (Figure P6.107). How much more $O_2$ is there in a lung full of pure $O_2$ at this elevation than in a lung full of air at sea level?

**FIGURE P6.107**

**6.108. Scuba Diving** A scuba diver is at a depth of 50 m, where the pressure is 5.0 atm. What should be the mole fraction of $O_2$ in the gas mixture the diver breathes to produce the same partial pressure of oxygen as the gas mixture at sea level?

**6.109.** Carbon monoxide at a pressure of 680 mmHg reacts completely with $O_2$ at a pressure of 340 mmHg in a sealed vessel to produce $CO_2$. What is the final pressure in the flask?

**6.110.** Ozone reacts completely with NO, producing $NO_2$ and $O_2$. A 10.0 L vessel is filled with 0.280 mol of NO and 0.280 mol of $O_3$ at 350 K. Find the partial pressure of each product and the total pressure in the flask at the end of the reaction.

**\*6.111.** Ammonia is produced industrially from the reaction of hydrogen with nitrogen under pressure in a sealed reactor. What is the percent decrease in pressure of a sealed reaction vessel during the reaction between $3.60 \times 10^3$ mol of $H_2$ and $1.20 \times 10^3$ mol of $N_2$ if half of the $N_2$ is consumed?

**\*6.112.** A mixture of 0.156 mol of C is reacted with 0.117 mol of $O_2$ in a sealed, 10.0 L vessel at 500 K, producing a mixture of CO and $CO_2$. The total pressure is 0.640 atm. What is the partial pressure of CO?

## The Kinetic Molecular Theory of Gases

### CONCEPT REVIEW

**6.113.** What is meant by the *root-mean-square speed* of gas molecules?

**6.114.** Why don't all molecules in a sample of air move at exactly the same speed?

**6.115.** How does the root-mean-square speed of the molecules in a gas vary with (a) molar mass and (b) temperature?

**6.116.** Does pressure affect the root-mean-square speed of the molecules in a gas? Explain your answer.

**6.117.** How can Graham's law of effusion be used to determine the molar mass of an unknown gas?

**6.118.** Is the ratio of the rates of effusion of two gases the same as the ratio of their root-mean-square speeds?

**6.119.** What is the difference between *diffusion* and *effusion*?

**6.120.** If gas X diffuses faster in air than gas Y, is gas X also likely to effuse faster than gas Y?

### PROBLEMS

**6.121.** Rank the gases $SO_2$, $CO_2$, and $NO_2$ in order of increasing root-mean-square speed at 0°C.

**6.122.** In a mixture of $CH_4$, $NH_3$, and $N_2$, which gas molecules are, on average, moving fastest?

**6.123.** At 286 K, three gases, A, B, and C, have root-mean-square speeds of 360 m/s, 441 m/s, and 472 m/s, respectively. Which gas is $O_2$?

**6.124.** Determine the root-mean-square speed of $CO_2$ molecules that have an average kinetic energy of $4.2 \times 10^{-21}$ J per molecule.

**\*6.125.** The root-mean-square speed of He gas at 300 K is $1.370 \times 10^3$ m/s. Sketch a graph of $u_{rms,He}$ versus $T$ for $T = 300, 450, 600, 750,$ and 900 K. Is the graph linear?

**\*6.126.** The root-mean-square speed of $N_2$ gas at 300 K is 516.8 m/s. Why doesn't doubling the temperature also double $u_{rms,N_2}$?

**6.127.** What is the ratio of the root-mean-square speed of $D_2$ to that of $H_2$ at constant temperature?

**6.128. Enriching Uranium** The two isotopes of uranium, $^{238}U$ and $^{235}U$, can be separated by diffusion of the corresponding $UF_6$ gases. What is the ratio of the root-mean-square speed of $^{238}UF_6$ to that of $^{235}UF_6$ at constant temperature?

**6.129.** A student measured the relative rate of effusion of carbon dioxide and propane, $C_3H_8$, and found that they effuse at exactly the same rate. Did the student make a mistake?

**\*6.130.** The density of Ne at 760 mmHg is exactly half its density at 2.00 atm at constant temperature. Is the root-mean-square speed of Ne at 2.00 atm half, twice, or the same as $u_{rms,Ne}$ at 760 mmHg?

**6.131.** If an unknown gas has one-third the root-mean-square speed of $H_2$ at 300 K, what is its molar mass?

**6.132.** A flask of ammonia is connected to a flask of an unknown acid, HX, by a 1.00 m glass tube. As the two gases diffuse down the tube, a white ring of $NH_4X$ forms 68.5 cm from the ammonia flask. Identify element X.

**6.133. Isotope Use by Plants** During photosynthesis, green plants preferentially use $^{12}CO_2$ over $^{13}CO_2$ in making sugars, and food scientists can frequently determine the source of sugars used in foods on the basis of the ratio of $^{12}C$ to $^{13}C$ in a sample.
   a. Calculate the relative rates of diffusion of $^{13}CO_2$ and $^{12}CO_2$.
   b. Specify which gas diffuses faster.

**6.134.** At fixed temperature, how much faster does NO effuse than $NO_2$?

**6.135.** Two balloons were filled with $H_2$ and He, respectively. The person responsible for filling them neglected to label them. After 24 hr the volumes of both balloons had decreased but by different amounts. Which balloon contained hydrogen?

**6.136.** Compounds sensitive to oxygen are often manipulated in *glove boxes* that contain a pure nitrogen or pure argon atmosphere. A balloon filled with carbon monoxide was placed in a glove box. After 24 hr, the volume of the balloon was unchanged. Did the glove box contain $N_2$ or Ar?

# Real Gases

## CONCEPT REVIEW

**6.137.** Explain why real gases behave nonideally at low temperatures and high pressures.

**6.138.** Under what conditions is the pressure exerted by a real gas *less* than that predicted for an ideal gas?

**6.139.** Explain why the values of the van der Waals constant *b* of the noble gas elements increase with atomic number.

**6.140.** Explain why the constant *a* in the van der Waals equation generally increases with the molar mass of the gas.

## PROBLEMS

**6.141.** The graphs of *PV/RT* versus *P* (see Figure 6.38) for 1 mole of $CH_4$ and 1 mole of $H_2$ differ in how they deviate from ideal behavior. For which gas is the effect of the volume occupied by the gas molecules more important than the attractive forces between molecules?

**6.142.** Which noble gas is expected to deviate the most from ideal behavior in a graph of *PV/RT* versus *P*?

---

**6.143.** At high pressures, real gases do not behave ideally. (a) Use the van der Waals equation and data in the text to calculate the pressure exerted by 40.0 g $H_2$ at 20°C in a 1.00 L container. (b) Repeat the calculation assuming that the gas behaves like an ideal gas.

**6.144.** (a) Calculate the pressure exerted by 1.00 mol of $CO_2$ in a 1.00 L vessel at 300 K, assuming that the gas behaves ideally. (b) Repeat the calculation using the van der Waals equation.

# Additional Problems

**6.145.** The volume of a sample of propane gas at 12.5 atm is 10.6 L. What volume does the gas occupy if the pressure is reduced to 1.05 atm and the temperature remains constant?

**6.146.** The pressure on a sample of carbon dioxide gas collected during a laboratory experiment is increased from 10.5 mmHg to 765 mmHg. If the initial volume of the gas is 185 mL, what volume does the gas occupy at the higher pressure if the temperature remains constant?

**6.147.** A 22.4 L sample of hydrogen chloride gas is heated from 15°C to 78°C. What volume does it occupy at the higher temperature if the pressure remains constant?

**6.148.** The reaction between 1.00 L sulfur dioxide and 0.500 L oxygen producing sulfur trioxide proceeds to 50% yield at constant temperature in a sealed vessel. Does the pressure in the vessel increase, decrease, or stay the same?

**6.149.** A gas cylinder in the back of an open truck experiences a temperature change from 12°F to 143°F as it is driven from high in the Rocky Mountains of Colorado to Death Valley, CA. If the pressure gauge on the tank reads 2200 psi in Colorado, what will the gauge read in Death Valley to the nearest psi?

**6.150.** The temperature of a quantity of methane gas at a pressure of 761 mmHg is 18.6°C. Predict the temperature of the gas if the pressure is reduced to 355 mmHg while the volume remains constant.

**6.151.** A souvenir soccer ball is partially deflated and then put into a suitcase. At a pressure of 0.947 atm and a temperature of 27°C, the ball has a volume of 1.034 L. What volume does it occupy during an airplane flight if the pressure in the baggage compartment is 0.235 atm and the temperature is −35°C?

**6.152.** **Asthma Therapy** A gas mixture used experimentally for asthma treatments contains 17.5 mol of helium for every 0.938 mol of oxygen. What is the mole fraction of oxygen in the mixture?

**6.153.** Scientists have used laser light to slow atoms to speeds corresponding to temperatures below 0.00010 K. At this temperature, what is the root-mean-square speed of argon atoms?

**6.154.** **Blood Pressure** A typical blood pressure in a resting adult is "120 over 80," meaning 120 mmHg with each beat of the heart and 80 mmHg of pressure between heartbeats. Express these pressures in the following units: (a) torr; (b) atm; (c) bar; (d) kPa.

**\*6.155.** A popular scuba tank is the "aluminum 80," so named because it can deliver 80 cubic feet of air at "normal" temperature (72°F) and pressure (1.00 atm, 14.7 psi) when filled with air at a pressure of 3000 psi. A particular aluminum 80 tank has a mass of 15 kg empty. What is its mass when filled with air at 3000 psi?

**6.156.** The flame produced by the burner of a gas (propane) grill is a pale blue color when enough air mixes with the propane ($C_3H_8$) to burn it completely. For every gram of propane that flows through the burner, what volume of air is needed to burn it completely? Assume that the temperature of the burner is 200°C, the pressure is 1.00 atm, and the mole fraction of $O_2$ in air is 0.21.

**6.157.** Which noble gas effuses at about half the effusion rate of $O_2$?

**6.158.** **Anesthesia** A common anesthesia gas is halothane, with the formula $C_2HBrClF_3$ and the structure shown in Figure P6.158. Liquid halothane boils at 50.2°C and 1.00 atm. If halothane behaved as an ideal gas, what volume would 10.0 mL of liquid halothane (*d* =1.87 g/mL) occupy at 60°C and 1.00 atm of pressure? What is the density of halothane vapor at 55°C and 1.00 atm of pressure?

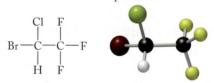

**FIGURE P6.158**

**6.159.** One cotton ball soaked in ammonia and another soaked in hydrochloric acid were placed at opposite ends of a 1.00 m glass tube (Figure P6.159). The vapors diffused toward the middle of the tube and formed a white ring of ammonium chloride where they met.

a. Write the chemical equation for this reaction.

b. Will the ammonium chloride ring form closer to the end of the tube with ammonia or the end with hydrochloric acid? Explain your answer.

c. Calculate the distance from the ammonia end to the position of the ammonium chloride ring.

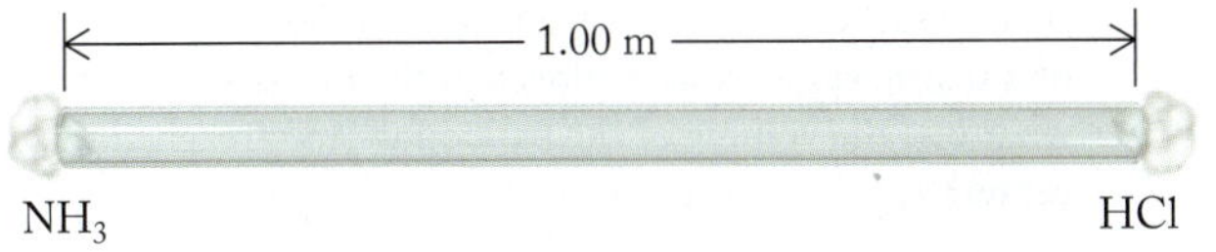

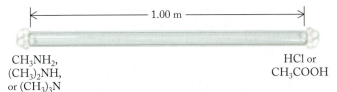

**FIGURE P6.159**

*6.160. The same apparatus described in Problem 6.159 was used in another series of experiments. A cotton ball soaked in either hydrochloric acid (HCl) or acetic acid ($CH_3COOH$) was placed at one end. Another cotton ball soaked in one of three amines (a class of organic compounds)—$CH_3NH_2$, $(CH_3)_2NH$, or $(CH_3)_3N$—was placed in the other end (Figure P6.160).
   a. In one combination of acid and amine, a white ring was observed almost exactly halfway between the two ends. Which acid and which amine were used?
   b. Which combination of acid and amine would produce a ring closest to the amine end of the tube?
   c. Do any two of the six combinations result in the formation of product at the same position in the ring? Assume measurements can be made to the nearest centimeter.

|← —————— 1.00 m —————— →|

$CH_3NH_2$,
$(CH_3)_2NH$,
or $(CH_3)_3N$                                           HCl or
                                                        $CH_3COOH$

**FIGURE P6.160**

**6.161.** **Liquid Nitrogen–Powered Car** Students at the University of North Texas and the University of Washington built a car propelled by compressed nitrogen gas. The gas was obtained by boiling liquid nitrogen stored in a 182 L tank. What volume of $N_2$ is released at 0.927 atm of pressure and 25°C from a tank full of liquid $N_2$ ($d = 0.808$ g/mL)?

**6.162.** The pressure in an aerosol can is 1.5 atm at 27°C. The can will withstand a pressure of about 2 atm. Will it burst if heated in a campfire to 450°C?

**6.163.** **Hydrogen-Powered Vehicles** Combustion of hydrogen supplies a great deal of energy per gram. One challenge in designing hydrogen-powered vehicles is storing hydrogen. The pressure limit for carbon-fiber reinforced gas cylinders is 10,000 psi (1 atm = 14.7 psi). A fuel tank capable of holding up to 10 kg of $H_2$ is required to meet a reasonable driving range for a vehicle. If we limit the volume of the fuel tank to 400 L, what pressure must the hydrogen be under at 300 K? Does your answer conform to the safety requirements?

**6.164.** A sample of 11.4 L of an ideal gas at 25.0°C and 735 torr is compressed and heated so that the volume is 7.9 L and the temperature is 72.0°C. What is the pressure in the container?

**6.165.** A sample of argon gas at STP occupies 15.0 L. What mass of argon is present in the container?

**6.166.** A sample of a gas has a mass of 2.889 g and a volume of 940 mL at 735 torr and 31°C. What is its molar mass?

**6.167.** What pressure is exerted by a mixture of 2.00 g of $H_2$ and 7.00 g of $N_2$ at 273°C in a 10.0 L container?

**6.168.** Uranus has a total atmospheric pressure of 130 kPa and consists of the following gases: 83% $H_2$, 15% He, and 2% $CH_4$ by volume. Calculate the partial pressure of each gas in Uranus's atmosphere.

**6.169.** A sample of $N_2(g)$ requires 240 s to diffuse through a porous plug. It takes 530 s for an equal number of moles of an unknown gas X to diffuse through the plug under the same conditions of temperature and pressure. What is the molar mass of gas X?

**6.170.** The rate of effusion of an unknown gas is 0.10 m/s and the rate of effusion of $SO_3(g)$ is 0.052 m/s under identical experimental conditions. What is the molar mass of the unknown gas?

**6.171.** Derive a equation that describes the relationship between root-mean-square speed, $u_{rms}$, of a gas and its density.

**6.172.** Derive an equation that expresses the ratio of the densities ($d_1$ and $d_2$) of a gas under two different combinations of temperature and pressure, ($T_1$, $P_1$) and ($T_2$, $P_2$).

**6.173.** **Denitrification in the Environment** In some aquatic ecosystems, nitrate ($NO_3^-$) is converted to nitrite ($NO_2^-$), which then decomposes to nitrogen and water. As an example of this second reaction, consider the decomposition of ammonium nitrite:

$$NH_4NO_2(aq) \rightarrow N_2(g) + 2\ H_2O(\ell)$$

What would be the change in pressure in a sealed 10.0 L vessel due to the formation of $N_2$ gas when the ammonium nitrite in 1.00 L of 1.0 $M$ $NH_4NO_2$ decomposes at 25°C?

*6.174. When sulfur dioxide bubbles through a solution containing nitrite, chemical reactions that produce gaseous $N_2O$ and NO may occur.
   a. How much faster, on average, would NO molecules be moving than $N_2O$ molecules in such a reaction mixture?
   b. If these two nitrogen oxides were to be separated based on differences in their rates of effusion, would unreacted $SO_2$ interfere with the separation? Explain your answer.

*6.175. **Using Wetlands to Treat Agricultural Waste** Wetlands can play a significant role in removing fertilizer residues from rain runoff and groundwater; one way they do this is through denitrification, which converts nitrate ions to nitrogen gas:

$$2\ NO_3^-(aq) + 5\ CO(g) + 2\ H^+(aq) \rightarrow N_2(g) + H_2O(\ell) + 5\ CO_2(g)$$

Suppose 200.0 g of $NO_3^-$ flows into a swamp each day. What volume of $N_2$ would be produced at 17°C and 1.00 atm if the denitrification process were complete? What volume of $CO_2$ would be produced? Suppose the gas mixture produced by the decomposition reaction is trapped in a container at 17°C; what is the density of the mixture assuming $P_{total} = 1.00$ atm?

**6.176.** Ammonium nitrate decomposes on heating. The products depend on the reaction temperature:

$$NH_4NO_3(s) \xrightarrow{\ >300°C\ } N_2(g) + \tfrac{1}{2} O_2(g) + 2\ H_2O(g)$$
$$\xrightarrow{\ 200–260°C\ } N_2O(g) + 2\ H_2O(g)$$

A sample of $NH_4NO_3$ decomposes at an unspecified temperature, and the resulting gases are collected over water at 20°C.
   a. Without completing a calculation, predict whether the volume of gases collected can be used to distinguish

between the two reaction pathways. Explain your answer.

b. The gas produced during the thermal decomposition of 0.256 g of $NH_4NO_3$ displaces 79 mL of water at 20°C and 760 mmHg of atmospheric pressure. Is the gas $N_2O$ or a mixture of $N_2$ and $O_2$?

**6.177. Hydrogen–Producing Enzymes** Generating hydrogen from water or methane is energy intensive. A non-natural enzymatic process has been developed that produces 12 moles of hydrogen per mole of glucose by the reaction:

$$C_6H_{12}O_6(aq) + 6\,H_2O(\ell) \rightarrow 12\,H_2(g) + 6\,CO_2(g)$$

What volume of hydrogen could be produced from 256 g glucose at STP?

**6.178.** A number of devices are available for generating oxygen. Breathing pure oxygen for any length of time, however, can be dangerous to the human body, so a better rebreathing apparatus would be one that also produces nitrogen. One possible reaction that produces both $N_2$ and $O_2$ is the thermal decomposition of ammonium nitrate:

$$2\,NH_4NO_3 \rightarrow 2\,N_2(g) + O_2(g) + 4\,H_2O(g)$$

a. The respiratory rate at rest for an average, healthy adult is 12 breaths per minute. If the average breath takes in 500 mL of gas into our lungs, what mass in grams of $NH_4NO_3$ is needed to satisfy these needs for 1 hour if $P = 1.00$ atm and $T = 37°C$?

b. What are the partial pressures of $N_2$ and $O_2$ in this mixture?

**6.179. Tropical Storms** The severity of a tropical storm is related to the depressed atmospheric pressure at its center. Figure P6.179 is a photograph of Typhoon Odessa taken from the space shuttle *Discovery* in August 1985, when the maximum winds of the storm were about 90 mi/hr and the pressure was 40 mbar lower at the center than normal atmospheric pressure. In contrast, the central pressure of Hurricane Andrew was 90 mbar lower than its surroundings when it hit southern Florida with winds as high as 165 mi/hr. If a

**FIGURE P6.179**

small weather balloon with a volume of 50.0 L at a pressure of 1.0 atmosphere was deployed at the center of Andrew, what was the volume of the balloon when it reached the center?

## Nitrogen: Feeding Plants and Inflating Air Bags

**6.180.** Write balanced chemical equations for the following reactions: (a) nitrogen reacts with hydrogen; (b) nitrogen dioxide dissolves in water.

**6.181.** Write balanced chemical equations for reactions of ammonia with (a) oxygen and (b) nitric acid.

**6.182. Military Signal Flares** Flares are used for signaling or as defensive countermeasures in both civilian and military applications. The U.S. Army has studied the reaction of black powder as well as materials with more consistent ignition properties for a possible replacement to use in handheld signal flares. The chemical reaction that occurs when black powder is ignited is

$$2\,KNO_3(s) + \tfrac{1}{8}\,S_8(s) + 3\,C(s) \rightarrow K_2S(s) + N_2(g) + 3\,CO_2(g)$$

a. Identify the oxidizing and reducing agents.

b. How many electrons are transferred per mole of $KNO_3$?

**6.183.** A simple laboratory-scale preparation of nitrogen involves the thermal decomposition of ammonium nitrite ($NH_4NO_2$).

a. Write a balanced chemical equation for the decomposition of $NH_4NO_2$ to give nitrogen and water.

b. Is the thermal decomposition of $NH_4NO_2$ a redox reaction?

**6.184. Oklahoma City Bombing** The 1995 explosion that destroyed the Murrah Federal Office Building in Oklahoma City was believed to be caused by the reaction of ammonium nitrate with fuel oil:

$$3\,NH_4NO_3(s) + C_{10}H_{22}(\ell) + 14\,O_2(g) \rightarrow$$
$$3\,N_2(g) + 17\,H_2O(g) + 10\,CO_2(g)$$

a. What common use for ammonium nitrate made it easy for the conspirators to acquire this compound in large quantities without suspicion?

b. Which elements are oxidized and which are reduced in the reaction?

*c. Write a chemical equation for the thermal decomposition of ammonium nitrate in the absence of fuel and oxygen.

**\*6.185. Automobile Air Bags** Sodium azide is used in automobile air bags as a source of nitrogen gas in the reaction

$$2\,NaN_3(s) \rightarrow 2\,Na(s) + 3\,N_2(g)$$

The reaction is rapid at 350°C but produces elemental sodium. Potassium nitrate and silica ($SiO_2$) are added to remove the liquid sodium metal formed in the reaction. Why is it necessary from a safety perspective to avoid molten alkali metals in an air bag?

**6.186. Manufacturing a Proper Air Bag** Here is the overall reaction in an automobile air bag:

$$20 \text{ NaN}_3(s) + 6 \text{ SiO}_2(s) + 4 \text{ KNO}_3(s) \rightarrow$$
$$32 \text{ N}_2(g) + 5 \text{ Na}_4\text{SiO}_4(s) + \text{K}_4\text{SiO}_4(s)$$

Calculate how many grams of sodium azide ($NaN_3$) are needed to inflate a $40 \times 40 \times 20$ cm bag to a pressure of 1.25 atm at a temperature of 20°C. How much more sodium azide is needed if the air bag must produce the same pressure at 10°C?

**6.187.** The first rocket-propelled aircraft used a mixture of hydrogen peroxide and hydrazine as the propellant. Using the thermochemical data in the appendices and $\Delta H_f^\circ$ of $N_2H_4(\ell)$ = 50.63 kJ/mol, calculate the enthalpy change for the following reaction:

$$2 \text{ H}_2\text{O}_2(\ell) + \text{N}_2\text{H}_4(\ell) \rightarrow \text{N}_2(g) + 4 \text{ H}_2\text{O}(g)$$

*\*6.188.* Nitrogen gas obtained by boiling liquid nitrogen ($d = 0.808$ g/mL) from a 182 L tank can be used to power a car. How much hydrazine (in grams) is needed to produce an equivalent amount of $N_2$ gas by reaction with hydrogen peroxide in Problem 6.187?

If your instructor assigns problems in **smartwork**, log in at **smartwork.wwnorton.com**.

# 7

# A Quantum Model of Atoms: Waves and Particles

## Can Nature Be as Absurd as It Seems?

On a typical visit to the grocery store, the doors open automatically, and your purchases are totaled with a scanner. How does this technology work? Science in the early 20th century focused on the interaction between matter and energy. Such inquiries led to the discovery of electrons, protons, and neutrons as the fundamental particles from which all matter is made. A century later we reap the benefits of these investigations in myriad electronic devices.

At the end of the 19th century, one of the lingering problems in science was the photoelectric effect—the observation that light shining on a metal surface led to the flow of electrons. Another involved the range of energies given off by hot objects. A third dealt with the light emitted from gases such as neon when an electric current was passed through them.

In this chapter we see that a new vision of matter at the atomic level and a new theory called quantum theory account for each of these observations. In addition to giving us powerful insights into the way the world works on the atomic level, these developments allow us to understand the organization of the periodic table, the sizes of atoms, and even biological processes like photosynthesis.

The world at the atomic level is governed by a set of rules entirely different from those describing behavior in the macroscopic world. Quantum theory explains these rules. Its development was a towering intellectual achievement, creating a surprising model of how electrons and light behave. In fact, the rules that apply at the atomic level often seem to contradict our understanding of how nature works at the macroscopic level.

Two of the giants of quantum theory, Niels Bohr and Werner Heisenberg, wrestled with these contradictions. They briefly worked together at the University of Copenhagen, and after a particularly heated discussion about a feature of quantum theory, Heisenberg

**Incandescent Steel** Hot sparks from the metal being machined by a metalworker emit a broad range of electromagnetic radiation, including visible light. Scientists' efforts to explain the spectra of hot objects led to the development of quantum theory and an entirely new way of viewing matter at the atomic level.

▶

7.1 Light Waves

7.2 Atomic Spectra

7.3 Particles of Light and Quantum Theory

7.4 The Hydrogen Spectrum and the Bohr Model

7.5 Electron Waves

7.6 Quantum Numbers and Electron Spin

7.7 The Sizes and Shapes of Atomic Orbitals

7.8 The Periodic Table and Filling the Orbitals of Multielectron Atoms

7.9 Electron Configurations of Ions

7.10 The Sizes of Atoms and Ions

7.11 Ionization Energies

7.12 Electron Affinities

## Learning Outcomes

**LO1** Explain the complementary nature of the absorption and emission lines of atomic spectra, calculate their wavelengths or frequencies, and describe the relationship of those lines to electron transitions between energy levels in atoms
**Sample Exercises 7.1, 7.2, 7.4, 7.5**

**LO2** Explain the photoelectric effect using quantum theory
**Sample Exercise 7.3**

**LO3** Describe the wavelike and particle-like properties of electromagnetic radiation and electrons
**Sample Exercise 7.6**

**LO4** Calculate the uncertainty in the velocity or position of an object
**Sample Exercise 7.7**

**LO5** Assign quantum numbers to orbitals and use their values to describe the size, energy, and orientation of orbitals
**Sample Exercises 7.8, 7.9**

**LO6** Use the aufbau principle and Hund's rule to write electron configurations and draw orbital diagrams of atoms and monatomic ions
**Sample Exercises 7.10, 7.11, 7.12, 7.13**

**LO7** Explain the energies of orbitals based on the concept of effective nuclear charge

**LO8** Relate the sizes of atoms and ions, ionization energies, and electron affinities of the elements to their positions in the periodic table
**Sample Exercises 7.14, 7.15**

**electromagnetic spectrum** a continuous range of radiant energy that includes radio waves, infrared radiation, visible light, ultraviolet radiation, X-rays, and gamma rays.

**electromagnetic radiation** any form of radiant energy in the electromagnetic spectrum.

**wavelength (λ)** the distance from crest to crest or trough to trough on a wave.

**frequency (ν)** the number of crests of a wave that pass a stationary point of reference per second.

**hertz (Hz)** the SI unit of frequency, measured in reciprocal seconds: $1 \text{ Hz} = 1 \text{ s}^{-1} = 1$ cycle per second (cps).

posed this question to himself: "Can nature possibly be as absurd as it seems?" The answer is probably yes.

In this chapter we explore the evolution of quantum theory. How do matter and energy interact at the atomic level? In addressing this question, we link the spectra of atoms to the structure of atoms, particularly to patterns in the distribution of electrons around the nuclei of atoms. These patterns help explain the properties of the elements, including the charges on the ions they form and how elements interact with other elements as they form compounds. ■

## 7.1 Light Waves

In Chapter 4 we discussed electron transfer reactions between atoms, ions, and molecules. When electrons were discovered in the 19th century, it seemed clear that electrons behaved as particles. As scientists explored the properties of electrons, however, they found that on the atomic scale the behavior of electrons

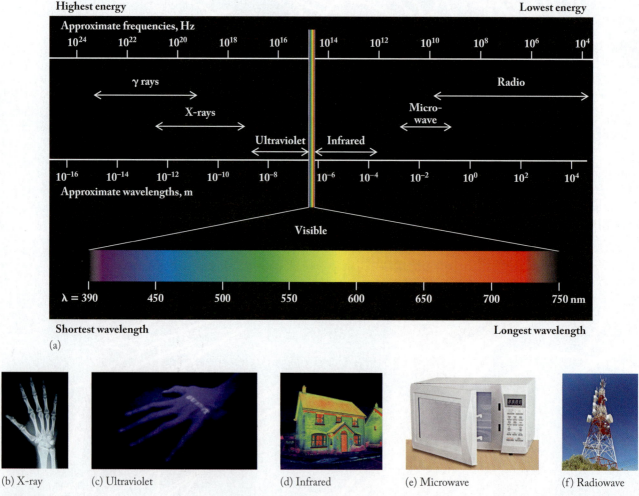

**FIGURE 7.1** (a) The visible light region between about 390 and 750 nm is a tiny fraction of the electromagnetic spectrum, which ranges from very-short-wavelength, high-frequency gamma (γ) rays to long-wavelength, low-frequency radio waves. Note the inverse relation between frequency and wavelength: frequencies increase from right to left here, and wavelengths increase from left to right. (b) X-rays used in medical imaging. (c) Black lights (ultraviolet radiation) to illuminate hand stamps. (d) Infrared radiation to monitor heat leakage from buildings. (e) Microwaves used to cook food. (f) Radiowaves carrying sound to listeners.

also exhibited properties associated with waves. The consequences of both wave-like and particle-like behavior are fundamental to our understanding of atoms, chemical bonds, and molecular structure.

## Properties of Waves

Visible light is a small part of the **electromagnetic spectrum** (Figure 7.1). This spectrum is a continuous range of radiant energy that extends from low-energy radio waves to ultrahigh-energy gamma rays. All forms of radiant energy are examples of **electromagnetic radiation**.

The term *electromagnetic* comes from the theory proposed by Scottish scientist James Clerk Maxwell (1831–1879) that radiant energy moves through space (or the air) in a way that resembles waves flowing across a body of water. Unlike water waves, which only oscillate up and down, Maxwell's waves of radiant energy have two components: an oscillating electric field and an oscillating magnetic field (Figure 7.2). These two fields are perpendicular to each other and travel together through space. Maxwell derived a set of equations based on his oscillating-wave model that accurately describes nearly all the observed properties of light.

A wave of electromagnetic radiation, like any wave traveling through any medium, has a characteristic **wavelength** ($\lambda$, the distance from crest to crest) and **frequency** ($\nu$, the number of crests that pass a stationary point of reference per second), as shown in Figure 7.3(a). Frequencies have units of **hertz (Hz)**, also called *cycles per second* (cps): $1 \text{ Hz} = 1 \text{ cps} = 1 \text{ s}^{-1}$. The product of the wavelength and frequency of any electromagnetic radiation is the universal constant $c$, which is the symbol for the speed of light in a vacuum:

$$\lambda\nu = c = 2.998 \times 10^8 \text{ m/s} \qquad (7.1)$$

Thus wavelength and frequency have a reciprocal relationship: as wavelength decreases, frequency increases. Another characteristic of a wave is its **amplitude**, the height of the crest or the depth of the trough with respect to the center line of the wave (Figure 7.3b).

---

**SAMPLE EXERCISE 7.1** **Calculating Frequency from Wavelength**   LO1

What is the frequency of the yellow-orange light ($\lambda = 589$ nm) produced by sodium-vapor streetlights?

**Collect and Organize** We are given light of a specific wavelength in nanometers and are asked to find the frequency. Frequency and wavelength are related by Equation 7.1 ($\lambda\nu = c$).

**Analyze** The value for $c$ given in Equation 7.1 is in meters per second, but the wavelength is given in nanometers. Therefore we need to convert nanometers to meters: $1 \text{ nm} = 10^{-9}$ m. The unit labels in Figure 7.1 indicate that frequencies of visible light are in the $10^{14}$ Hz range, so a correct answer should have a value in this range.

**Solve** We rearrange Equation 7.1 to solve for frequency

$$\nu = \frac{c}{\lambda}$$

Then we convert the wavelength units from nanometers to meters:

$$589 \text{ nm} \times \frac{10^{-9} \text{ m}}{1 \text{ nm}} = 589 \times 10^{-9} \text{ m} = 5.89 \times 10^{-7} \text{ m}$$

---

**CONNECTION** In Chapter 1 we discussed the background microwave radiation of the universe, discovered by Penzias and Wilson, that provided evidence for the Big Bang.

**CHEMTOUR** Electromagnetic Radiation

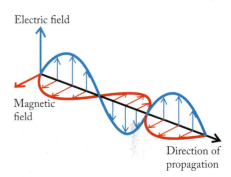

**FIGURE 7.2** Electromagnetic waves consist of electric and magnetic fields that oscillate in planes oriented at right angles to each other.

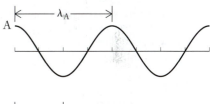
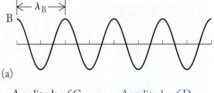
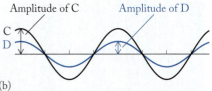

**FIGURE 7.3** Every wave has a characteristic wavelength ($\lambda$), frequency ($\nu$), and amplitude (intensity). (a) Wave A has a longer wavelength (and lower frequency) than wave B. (b) Waves C and D have the same wavelength and frequency but the amplitude of C is greater than that of D.

The frequency is

$$\nu = \frac{2.998 \times 10^8 \text{ m/s}}{5.89 \times 10^{-7} \text{ m}}$$

$$= 5.09 \times 10^{14} \text{ s}^{-1} = 5.09 \times 10^{14} \text{ Hz}$$

**Think About It** This large value is in the range we expected. Remember that $\lambda$ and $\nu$ are reciprocal: as one increases, the other decreases.

**Practice Exercise** If the radio waves transmitted by a radio station have a frequency of 90.9 MHz, what is the wavelength of the waves in meters?

*(Answers to Practice Exercises are in the back of the book.)*

**CONCEPT TEST**

The ultraviolet (UV) region of the electromagnetic spectrum contains waves with wavelengths from about $10^{-7}$ m to $10^{-9}$ m; the infrared (IR) region contains waves with wavelengths from about $10^{-4}$ m to $10^{-6}$ m. Are waves in the UV region higher in frequency or lower in frequency than waves in the IR region?

*(Answers to Concept Tests are in the back of the book.)*

## The Behavior of Waves

A rainbow shows us that sunlight is composed of an array of colors, ranging in wavelength from about 390 to 750 nm. A combination of refraction and reflection of sunlight inside raindrops produces rainbows. The path of a beam of light entering a raindrop is bent, or *refracted*, as it moves from air into the denser medium of water. Light of each different wavelength is bent at a slightly different angle, leading to the dispersion of colors in the rainbow.

Another property of electromagnetic radiation is that it undergoes **diffraction**. Diffraction happens when a beam of light passes through narrow slits, as shown in Figure 7.4(a). As the waves emerge from the slits, they spread out in circles. This pattern means that the waves passing through the slits bend around, or are *diffracted* by, the slits.

▶❙❙ **CHEMTOUR** Light Diffraction

◉◉ **CONNECTION** Diffraction of light caused by water suspended in layers of calcium carbonate was originally mentioned in Chapter 4 as the phenomenon responsible for the iridescence of pearls.

Parallel wave crests

Barrier with narrow slits          Screen

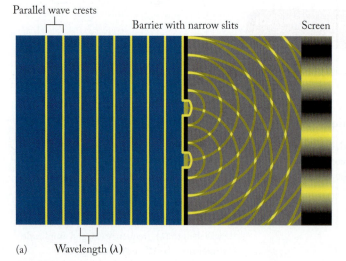

(a)          Wavelength ($\lambda$)

(b)

(c)

**FIGURE 7.4** Diffraction patterns and interference: (a) When waves approach a barrier that has gaps about the size of the wavelength, the waves bend around the edges of the gaps and radiate out in circular wave patterns. This is diffraction. When these circular diffracted waves overlap, they produce a pattern of bright and dark bands. (b) A similar pattern is produced when two pebbles are dropped into a pool at the same time. (c) Hold your hand in front of a light with your palm facing you; slowly bring your fingers together until they form a narrow slit. The dark lines in the gap between your fingers are an interference pattern generated by diffraction of the light.

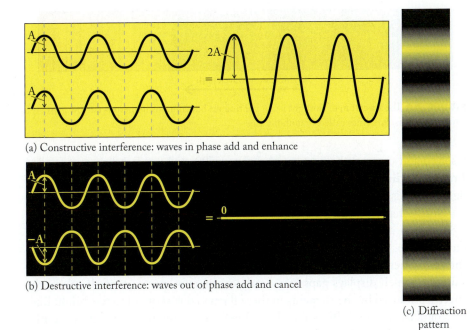

(a) Constructive interference: waves in phase add and enhance

(b) Destructive interference: waves out of phase add and cancel

(c) Diffraction pattern

**FIGURE 7.5** Constructive and destructive interference. (a) When the crests of two identical waves of amplitude A overlap, the waves are in phase and experience constructive interference; they enhance each other (amplitude = 2A). (b) When a crest from one pattern overlaps a trough of the other, the waves are out of phase and experience destructive interference; they cancel each other (amplitude = 0). (c) Constructive interference results in bright bands on the screen on the left, and destructive interference results in dark bands. The bright and dark bands form a diffraction pattern.

When two sets of these circular waves spread out from a pair of slits, as in Figure 7.4, they run into one another. In doing so they **interfere** with one another, much like the spreading waves produced when two pebbles are simultaneously dropped in a pool (Figure 7.4b). In both cases, the overlapping waves produce diffraction patterns. The waves in a pool produce series of concentric arcs, as shown in Figure 7.4(b). When interfering waves of light diffracted by the pair of slits in Figure 7.4(a) are projected onto a screen, we see a pattern of light and dark bands. Figure 7.4(c) shows how an interference pattern is generated by diffraction through a slit formed with your fingers.

To understand why interference patterns form, consider two waves with the same wavelength traveling in the same direction, as shown in Figure 7.5. When the crests of the two waves overlap, the waves are said to be *in phase*. They combine *constructively*, and the amplitude of the resulting wave is twice the amplitude of the two original waves. However, when the crest of one wave overlaps the trough of the other, the waves are said to be *out of phase*. They combine *destructively*, the amplitude of the resultant wave is zero, and the wave disappears. We can use this notion of constructive and destructive interference to explain the pattern on the screen in Figure 7.5(c): the light bands on the screen are the result of constructive interference, and the dark bands are caused by destructive interference.

**CONCEPT TEST** •••••••••••••••••••••••••••••••••••••••••••••••••••••••

When two identical waves of red light ($\lambda$ = 660 nm) interfere constructively, is there any change in the wavelength and the frequency of the light?

••••••••••••••••••••••••••••••••••••••••••••••••••••••••••••••••••••••••••

# 7.2 Atomic Spectra

In 1800, English scientist William Hyde Wollaston (1766–1828) made a startling discovery about sunlight. Using carefully ground prisms to separate a beam of sunlight into its component colors, Wollaston discovered that the spectrum was not continuous. Instead, the spectrum contained narrow, dark lines (Figure 7.6). Using even higher-quality prisms, German physicist Joseph von Fraunhofer (1787–1826)

**diffraction** bending of electromagnetic radiation as it passes around an edge of an object or through a narrow opening.

**interference** the interaction of waves that results in either reinforcing their amplitudes (constructive interference) or canceling them out (destructive interference).

**Fraunhofer lines** a set of dark lines in the otherwise continuous solar spectrum.

**atomic emission spectrum** (also called **bright-line spectrum**) a characteristic series of bright lines produced by high-temperature atoms.

**atomic absorption spectrum** (also called **dark-line spectrum**) a characteristic series of dark lines produced when free, gaseous atoms are illuminated by an external source of radiation.

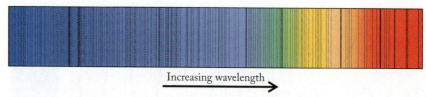

Increasing wavelength

**FIGURE 7.6** The spectrum of sunlight is not continuous but contains numerous narrow gaps, which appear as dark lines called Fraunhofer lines.

resolved and mapped the wavelengths of over 500 of these dark lines, now called **Fraunhofer lines**. Fraunhofer and his contemporaries knew that narrow regions of light were missing from the Sun's spectrum, but they did not know why.

Nearly half a century later, German chemist Robert Wilhelm Bunsen (1811–1899) and physicist Gustav Robert Kirchhoff (1824–1887) collaborated on extensive studies of the light emitted (given off) by elements vaporized in the transparent flame of a burner designed by Bunsen. Unlike the spectrum of sunlight, which displays gaps in a nearly continuous spectrum of all colors, the spectra produced by the elements in these flames consist of only a few bright lines on a dark background. Bunsen and Kirchhoff discovered that these **atomic emission spectra** (or **bright-line spectra**) consist of bright (colored) lines at *exactly the same wavelengths* as some of the dark (black) Fraunhofer lines in the spectrum of sunlight. For example, the Fraunhofer D line (actually a pair of closely spaced lines at 589.0 and 589.6 nm) exactly matched the two bright lines corresponding to the yellow-orange light produced by hot sodium vapor (Figure 7.7).

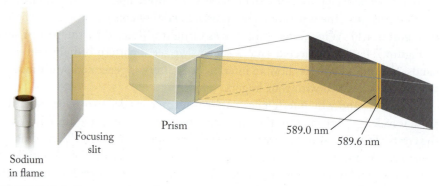

Sodium in flame

Focusing slit

Prism

589.0 nm

589.6 nm

**FIGURE 7.7** Sodium atoms heated in a flame produce a bright yellow-orange light due to two emission lines at 589.0 and 589.6 nm.

▶❚❚ **CHEMTOUR** Light Emission and Absorption

These experiments, and others employing light sources called gas discharge tubes, showed that each element *emits* characteristic electromagnetic radiation when its atoms are heated to a sufficiently high temperature. This emission is captured in the element's atomic emission spectrum (Figure 7.8). Also, elements *absorb* electromagnetic radiation when illuminated by an external source of radiation. This absorption of radiation by atoms produces an **atomic absorption spectrum** (or **dark-line spectrum**), which consists of dark lines in an otherwise colored

(a) Emission spectrum of hydrogen

(b) Emission spectrum of helium

(c) Emission spectrum of neon

**FIGURE 7.8** The colored light emitted from gas discharge tubes filled with various gaseous elements produces atomic emission spectra that are characteristic of the element: (a) hydrogen, (b) helium, (c) neon.

spectrum (Figure 7.9). For any given element, the dark lines in its absorption spectrum are at the same wavelengths as the bright lines in its emission spectrum. Thus the two processes we know as *emission* and *absorption* involve electromagnetic radiation of the same wavelength (and energy). The phenomenon of absorption explains the origins of the Fraunhofer lines: gaseous atoms in the outer regions of the Sun absorb characteristic wavelengths of the sunlight passing through them on its way to Earth.

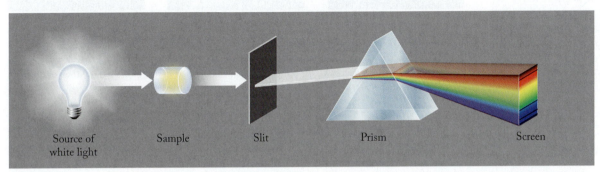

(a)

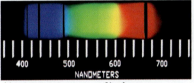
Absorption spectrum of hydrogen

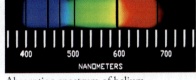
Absorption spectrum of helium

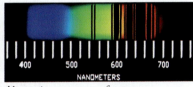
Absorption spectrum of neon

(b)

**FIGURE 7.9** (a) When gaseous atoms of hydrogen, helium, and neon are illuminated by an external source of white light (containing all colors of the visible spectrum), the resultant atomic absorption spectra contain dark lines that are characteristic of each element. (b) The dark lines in the atomic absorption spectra of these elements have the same wavelengths as the bright lines in their atomic emission spectra, shown in Figure 7.8.

**CONCEPT TEST**

The top spectrum in Figure 7.10 is the emission spectrum of mercury vapor. Select the absorption spectrum of mercury vapor from the other four spectra.

# 7.3 Particles of Light and Quantum Theory

As studies of electromagnetic radiation progressed in the 19th century, scientists discovered limits to the wave model for describing radiation. One limit was encountered when they tried to account for radiation given off by very hot objects. Consider, for example, what happens when a metal rod is heated in a flame. At first the rod gives off only heat in the form of infrared radiation, which we can feel but not see. With more heating, the metal begins to glow a dull red, shown in Figure 7.11(b). As the temperature of the metal is increased from 500 to 1000 K (Figure 7.11a), the most intense radiation is still in the infrared region of the spectrum, but a small fraction of the emitted light has $\lambda \approx 750$ nm, or red light, accounting for the observed color of the hot metal rod. With still more heating, the intensity of the red glow increases; the color begins to shift to red-orange, orange, yellow (Figure 7.11c), and eventually to white (Figure 7.11d) as the metal emits all wavelengths of visible light.

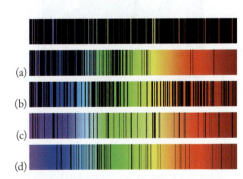

(a)

(b)

(c)

(d)

**FIGURE 7.10** Atomic emission spectrum of mercury vapor (top) and four atomic absorption spectra (a–d).

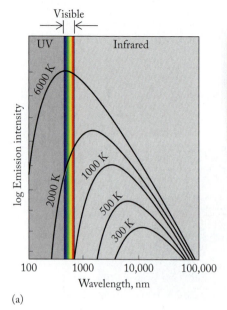

(a)

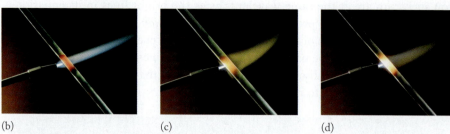

(b)   (c)   (d)

**FIGURE 7.11** (a) Plots of emission intensity versus wavelength show that as the temperature of a solid increases, the intensity of the radiation that it emits increases and the wavelength of maximum intensity decreases. As a metal rod is heated, (b) at first it glows red, (c) then orange, and (d) finally becomes white-hot as light of all colors is emitted.

Even white-hot metal emits little radiation in the ultraviolet region and none at even shorter wavelengths.

This phenomenon was well known at the end of the 19th century. However, none of Maxwell's equations for electromagnetic radiation could account for the emission spectrum of a heated metal filament. Another explanation—indeed, another model of radiation behavior—was required.

## Quantum Theory

In 1900, German scientist Max Planck (1858–1947) proposed such a model. It was based on the view that light and all other forms of electromagnetic radiation have both wavelike properties, as Maxwell had proposed, *and*, at the atomic level, particle-like properties. Planck (Figure 7.12) called a particle of radiant energy a **quantum** (plural *quanta*). In his model, which we call **quantum theory**, quanta are the smallest amounts of radiant energy in nature, the sort that a single atom or molecule might absorb or emit. An object made of a large but discrete number of atoms or molecules could therefore emit only a discrete number of quanta. In other words, light from such an object must be **quantized**.

To visualize the meaning of Planck's quanta, consider two ways you might get from the sidewalk to the entrance of a building (Figure 7.13). If you walked up the steps, you would be able to stand at only discrete heights above the sidewalk—each height equal to the rise of a single step. You could not stand at a height between two adjacent steps because there would be nothing to stand on at that height. If you walked up the ramp, however, you could stop at any height between the sidewalk and the entrance. The discrete height changes represented by the steps are also a model of Planck's hypothesis that energy is released (analogous to walking down the steps) or absorbed (walking up the steps) in discrete packets, or quanta, of energy.

**FIGURE 7.12** German scientist Max Planck is considered the father of quantum physics. He won the 1918 Nobel Prize in Physics for his pioneering work on the quantized nature of electromagnetic radiation. Planck was revered by his colleagues for his personal qualities as well as his scientific accomplishments.

**CONCEPT TEST** • • • • • • • • • • • • • • • • • • • • • • • • • • • • • • • • • •

Which of these quantities vary continuously (are not quantized) and which vary by discrete values (are quantized)?

a. the volume of water that evaporates from a lake each day during a summer heat wave
b. the number of eggs remaining in a carton
c. the time it takes you to get ready for class in the morning
d. the number of red lights encountered when driving the length of Fifth Avenue in New York City

• • • • • • • • • • • • • • • • • • • • • • • • • • • • • • • • • • • • • • • • • • • • • •

Quantum theory also connects the wave and particle natures of electromagnetic radiation. Planck proposed that the energy ($E$) of a quantum of light, which is called a **photon**, was directly proportional to the frequency of the wave of its corresponding radiation:

$$E = h\nu \qquad (7.2)$$

The value of the constant of proportionality ($h$) relating these two quantities is $6.626 \times 10^{-34}$ J · s. It is now called the **Planck constant**. When we solve Equation 7.1, $\lambda\nu = c$, for $\nu$ and substitute into Equation 7.2, we find that the energy of a photon is inversely proportional to its wavelength:

$$E = \frac{hc}{\lambda} \qquad (7.3)$$

**FIGURE 7.13** Quantized and unquantized heights. A flight of stairs exemplifies quantization: each step rises by a discrete height to the next step. In contrast, the heights on a ramp are not quantized.

SAMPLE EXERCISE 7.2    **Calculating the Energy of a Photon**    LO1

Photosynthesis relies on the absorption of photons of a particular wavelength by chlorophyll molecules, such as chlorophyll *b*. Chlorophyll *b* absorbs light at 462 nm and 647 nm. What are the energies of photons with these wavelengths? What colors of light does chlorophyll *b* absorb?

**Collect and Organize** We are given two wavelengths of light and are to calculate the energies of the photons of this electromagnetic radiation. The value of the Planck constant ($h$) is $6.626 \times 10^{-34}$ J · s and the speed of light ($c$) is $2.998 \times 10^8$ m/s.

**Analyze** Equation 7.3 relates the energy of a photon to its wavelength:

$$E = \frac{hc}{\lambda}$$

The wavelength is given in nanometers, but the value of the speed of light has units of meters per second. Therefore, we need to convert nanometers to meters so the distance units cancel. The value of $h$ is extremely small, so it is likely the results of our calculation, even factoring in the speed of light, will be very small, too.

**Solve**

$$E = \frac{hc}{\lambda} = \frac{(6.626 \times 10^{-34}\,\text{J}\cdot\text{s})\left(2.998 \times 10^8\,\frac{\text{m}}{\text{s}}\right)}{462\,\text{nm} \times \frac{10^{-9}\,\text{m}}{1\,\text{nm}}}$$

$$= 4.30 \times 10^{-19}\,\text{J}$$

$$E = \frac{hc}{\lambda} = \frac{(6.626 \times 10^{-34}\,\text{J}\cdot\text{s})\left(2.998 \times 10^8\,\frac{\text{m}}{\text{s}}\right)}{647\,\text{nm} \times \frac{10^{-9}\,\text{m}}{1\,\text{nm}}}$$

$$= 3.07 \times 10^{-19}\,\text{J}$$

We can use Figure 7.1 to determine which color light is absorbed. A wavelength of 462 nm is blue light and $\lambda = 647$ nm is orange light.

**Think About It** This quantity of energy is extremely small, as it should be, because a photon is an atomic-level particle of radiant energy. The results of the calculations confirm that photons of orange light have less energy than photons of blue light.

⚙ **Practice Exercise** Leaves contain an array of different light-absorbing molecules to harvest the full spectrum of visible light. Some of these include β-carotene (the compound that makes carrots orange), which absorbs at $\lambda = 450$ and 470 nm. How much energy do individual photons of 450 nm and 470 nm light have?

**quantum** (plural *quanta*) the smallest discrete quantity of a particular form of energy.

**quantum theory** a model based on the idea that energy is absorbed and emitted in discrete quantities of energy called quanta.

**quantized** having values restricted to whole-number multiples of a specific base value.

**photon** a quantum of electromagnetic radiation.

**Planck constant ($h$)** the proportionality constant between the energy and frequency of electromagnetic radiation expressed in $E = h\nu$; $h = 6.626 \times 10^{-34}$ J · s.

## The Photoelectric Effect

Although Planck's quantum theory fit the emission spectra of hot objects, Planck had no experimental evidence to support the existence of quanta. That evidence was supplied by Albert Einstein just five years after Planck published his quantum theory.

When light shines on a metal surface, the surface may give off electrons. Because light produces these electrons, this phenomenon is called the **photoelectric effect**, and the electrons emitted are called *photoelectrons*. (*Photo-* comes from the Greek word meaning "light.")

Photoelectrons flow only when the frequency of the light is above some minimum **threshold frequency** ($\nu_0$; Figure 7.14). Light of frequencies less than the threshold value produces no photoelectrons, no matter how intense the light. On the other hand, even a dim source of light produces at least a few photoelectrons when the frequencies it emits are equal to or greater than the threshold frequency. Einstein used Planck's quantum theory to explain this behavior. He proposed that the threshold frequency, $\nu_0$, is the frequency of the minimum quantum of absorbed energy needed to remove a single electron from the metal surface. This minimum quantity of energy is called the metal's **work function ($\Phi$)**:

$$\Phi = h\nu_0 \tag{7.4}$$

The value of $\Phi$ is related to the strength of the attraction between the nuclei of the metal atoms and the electrons surrounding those nuclei. Thus, it is different for different metals.

When a metal surface is illuminated by a beam of photons of sufficient energy, each photon is absorbed by an individual electron, giving that electron enough energy to break free of the surface. If the beam includes photons with frequencies above the threshold frequency ($\nu > \nu_0$), any energy in excess of $\Phi$ is imparted to each ejected electron as kinetic energy:

$$\mathrm{KE}_{\mathrm{electron}} = h\nu - h\nu_0 = h\nu - \Phi \tag{7.5}$$

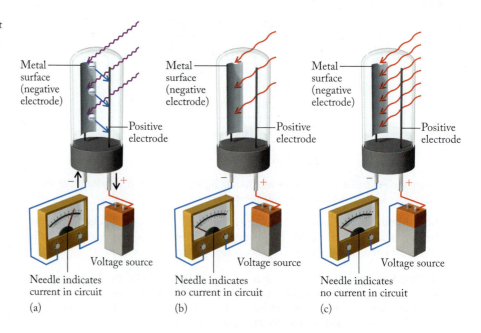

**FIGURE 7.14** A phototube includes a positive electrode and a negative metal electrode. (a) If radiation of high enough frequency and energy (violet) illuminates the negative electrode, electrons are dislodged from the surface and flow toward the positive electrode. This flow of electrons completes the circuit and produces an electric current. The size of the current is proportional to the intensity of the radiation—to the number of photons per unit time striking the negative electrode. (b) Photons of lower frequency (red), and hence lower energy, do not have sufficient energy to dislodge electrons and do not produce the photoelectric effect. (c) Even if many low-frequency photons bombard the surface of the metal, electrons are not emitted, the circuit is not complete, and there is no current.

The higher the frequency above the threshold frequency, the higher the kinetic energy and the higher the velocity of the ejected electrons.

## Wave–Particle Duality

Einstein's explanation of the photoelectric effect was of such profound significance that he was awarded the Nobel Prize for it in 1921. Why does the concept of photons have such far-reaching significance in science? Classical physics had identified the wave nature of light, but could not account for light behaving as a particle. Only quantum physics recognizes the **wave–particle duality** of light: light behaves as *both* a wave and a particle. Diffraction demonstrates the wave nature of light, but photons are needed to explain the photoelectric effect and the emission of light by hot objects.

Photosynthesis depends on the absorption of photons of a particular wavelength by chlorophyll molecules. These molecules use the energy in the light to power chemical reactions that convert carbon dioxide and water to glucose and oxygen. The photoelectric effect itself finds its way into our daily lives through technology such as TV remote control devices and some motion sensors. When we press a button on a TV remote, the remote emits a beam of infrared light that impinges on the "electric eye" in the television set. The resulting emission of electrons signals, for example, the television to turn on or off. The same principle applies to photosensors found in some motion detectors. In this case, a beam of light shines continually on a detector that emits electrons and causes a current to flow. If someone interrupts the light beam, the current drops. A door may open or an alarm may be triggered as a result of the drop in current. Other applications of the photoelectric effect include turning on an automatic faucet or hand dryer in a public restroom.

**photoelectric effect** the phenomenon of light striking a metal surface and producing an electric current (a flow of electrons).

**threshold frequency ($\nu_0$)** the minimum frequency of light required to produce the photoelectric effect.

**work function ($\Phi$)** the amount of energy needed to dislodge an electron from the surface of a metal.

**wave–particle duality** occurs when an object exhibits the properties of both a wave and a particle.

---

**SAMPLE EXERCISE 7.3** **Applying the Photoelectric Effect** **LO2**

If we want to use infrared radiation with a wavelength of 902 nm in a TV remote control, can we use germanium to generate a photocurrent? The work function of germanium is $7.61 \times 10^{-19}$ J.

**Collect and Organize** We need to convert the value of the work function ($\Phi$) of a metal into the threshold (minimum) frequency ($\nu_0$) and compare it to the frequency of 902 nm radiation.

**Analyze** The frequency of infrared radiation must exceed the threshold frequency of germanium to establish a photocurrent in the TV. Calculating the frequency of 902 nm radiation requires Equation 7.1. We need to rearrange the terms in Equation 7.4 to solve for $\nu_0$. The result of the calculation must be a $\nu_0$ value around $10^{14}$ Hz to use infrared light to eject an electron.

**Solve** The frequency of a 902 nm photon is calculated using Equation 7.1:

$$\lambda\nu = c = 2.998 \times 10^8 \text{ m/s}$$

$$\nu = \frac{c}{\lambda} = \frac{2.998 \times 10^8 \text{ m/s}}{902 \text{ nm} \times \frac{10^{-9} \text{ m}}{1 \text{ nm}}} = 3.33 \times 10^{14} \text{ s}^{-1}$$

Rearranging Equation 7.4 to solve for the threshold frequency for germanium:

$$\Phi = h\nu_0$$

$$\nu_0 = \frac{\Phi}{h} = \frac{7.61 \times 10^{-19} \text{ J}}{6.626 \times 10^{-34} \text{ J} \cdot \text{s}} = 1.14 \times 10^{15} \text{ s}^{-1}$$

Infrared radiation with a wavelength of 900 nm has a frequency that is less than the calculated threshold frequency and represents less energy than that required to liberate a photoelectron from germanium.

**Think About It** The calculated threshold frequency is close to, but still greater than, that of the highest-frequency (violet) visible light.

☀ **Practice Exercise** The work function of silver is $7.59 \times 10^{-19}$ J. What is the longest wavelength (nm) of electromagnetic radiation that can eject an electron from the surface of a piece of silver?

---

**CONCEPT TEST**

Consult Figure 7.1 to answer this question without making a calculation. If a photon of orange light has sufficient energy to eject a photoelectron from the surface of a metal, does a photon of green light have enough energy to do so?

## 7.4 The Hydrogen Spectrum and the Bohr Model

In formulating his quantum theory, Planck was influenced by the results of investigations of the emission spectra produced by free (gas-phase) atoms, which led him to question whether any spectrum, even that of an incandescent light bulb, was truly continuous. Among these earlier results was a discovery made in 1885 by a Swiss mathematician and schoolteacher named Johann Balmer (1825–1898).

### The Hydrogen Emission Spectrum

Balmer determined that the wavelengths of the four brightest lines in the visible region of the emission spectrum of hydrogen (Figure 7.8a) fit the simple equation

$$\lambda = 364.5 \text{ nm} \left( \frac{m^2}{m^2 - n^2} \right) \tag{7.6}$$

where $n$ is 2 and $m$ is a whole number greater than 2—specifically, 3 for the red line, 4 for the green, 5 for the blue, and 6 for the violet. Without having seen any other lines in hydrogen's visible emission spectrum, Balmer predicted there should be at least one more ($n = 7$) at the edge of the violet region, and indeed such a line was later discovered.

Balmer also predicted that a series of hydrogen emission lines should exist in regions outside the visible range, lines with wavelengths calculated by replacing $n = 2$ in Equation 7.6 with $n = 1, 3, 4$, and so forth. He was right. In 1908 German physicist Friedrich Paschen (1865–1947) discovered hydrogen emission lines in the infrared region, corresponding to $n = 3$ in Balmer's equation. A few years later, Theodore Lyman (1874–1954) at Harvard University discovered hydrogen emission lines in the UV region corresponding to $n = 1$. By the 1920s, the $n = 4$ and $n = 5$ series of emission lines had been discovered. Like the $n = 3$ lines, they are in the infrared region.

Later, Swedish physicist Johannes Robert Rydberg (1854–1919) revised Balmer's equation, changing wavelength to *wave number* ($1/\lambda$), which is the number of wavelengths per unit of distance. Rydberg's equation is

$$\frac{1}{\lambda} = \left[ 1.097 \times 10^{-2} \text{ nm}^{-1} \right] \left( \frac{1}{n_1^2} - \frac{1}{n_2^2} \right) \tag{7.7}$$

where $n_1$ is a whole number that remains fixed for a series of emission lines and $n_2$ is a whole number equal to $n_1 + 1$, $n_1 + 2$, . . . , for successive lines in the series.

When Balmer and Rydberg derived their equations describing the hydrogen spectrum, they did not know why the equations worked. The discrete frequencies of hydrogen's emission lines indicated that only certain levels of internal energy were available in hydrogen atoms. However, classical (macroscale) physics could not explain the existence of these internal energy levels. A new model that works at the atomic level was needed.

---

**SAMPLE EXERCISE 7.4** **Calculating the Wavelength of a Line** **LO1**
**in the Hydrogen Emission Spectrum**

In the visible portion of the hydrogen emission spectrum (Figure 7.8a), what is the wavelength of the bright line corresponding to $n_2 = 3$ in Equation 7.7?

**Collect and Organize** We are to calculate the wavelength of a line in the hydrogen emission spectrum given an $n_2$ value of 3. The $n_1$ value of a visible hydrogen emission line is 2.

**Analyze** Equation 7.7 relates $n$ values to wavelength of light. We know that the emission line is in the visible region, so we should obtain a wavelength between 400 and 750 nanometers.

**Solve**

$$\frac{1}{\lambda} = \left[1.097 \times 10^{-2}\ \text{nm}^{-1}\right]\left(\frac{1}{2^2} - \frac{1}{3^2}\right) = \left[1.097 \times 10^{-2}\ \text{nm}^{-1}\right]\left(\frac{1}{4} - \frac{1}{9}\right)$$

$$= \left[1.097 \times 10^{-2}\ \text{nm}^{-1}\right](0.1389) = 1.524 \times 10^{-3}\ \text{nm}^{-1}$$

$$\lambda = 656\ \text{nm}$$

**Think About It** The calculated wavelength is in the visible region of the electromagnetic spectrum, so the answer is reasonable.

⚙ **Practice Exercise** What is the wavelength, in nanometers, of the line in the hydrogen spectrum corresponding to $n_2 = 4$ in Equation 7.7?

---

## The Bohr Model of Hydrogen

Scientists in the early 20th century faced yet another dilemma. Ernest Rutherford had established a model of the atom that was mostly empty space, but occupied by negatively charged electrons with a tiny nucleus containing virtually all the mass and all the positive charge. What kept the electrons from falling into the nucleus? Rutherford had suggested that the electrons had to be in motion and hypothesized that they might orbit the nucleus the way planets orbit the sun. However, classical physics predicts that negative electrons orbiting a positive nucleus should emit energy in the form of electromagnetic radiation and eventually spiral into the nucleus; but this does not happen.

The problem of explaining why a hydrogen atom's electron is not pulled into its nucleus was addressed by Danish physicist Niels Bohr (1885–1962), who was well acquainted with the issue because he had studied with Rutherford. Bohr designed a theoretical model based on the electron in a hydrogen atom traveling around the nucleus in one of an array of concentric orbits. Each orbit represents an allowed energy level and is designated by the value of $n$ as shown in Equation 7.8:

$$E = -2.178 \times 10^{-18}\ \text{J}\left(\frac{1}{n^2}\right) \qquad (7.8)$$

◉◉ **CONNECTION** We discussed Rutherford's gold-foil experiment and the development of the idea of the nuclear atom in Section 2.1.

▶❚❚ **CHEMTOUR** Bohr Model of the Atom

where $n = 1, 2, 3, \ldots, \infty$. In the Bohr model, an electron in the orbit closest to the nucleus ($n = 1$) has the lowest energy:

$$E = -2.178 \times 10^{-18} \,\text{J}\left(\frac{1}{1^2}\right) = -2.178 \times 10^{-18} \,\text{J}$$

The next closest orbit has an $n$ value of 2 and an electron in it has an energy of

$$E = -2.178 \times 10^{-18} \,\text{J}\left(\frac{1}{2^2}\right) = -5.445 \times 10^{-19} \,\text{J}$$

Note that this value is less negative than the value for the electron in the $n = 1$ orbit. As the value of $n$ increases, the radius of the orbit increases and so, too, does the energy of an electron in the orbit; its value becomes less negative. As $n$ approaches $\infty$, $E$ approaches zero:

$$E = -2.178 \times 10^{-18} \,\text{J}\left(\frac{1}{\infty^2}\right) = 0$$

Zero energy means that the electron is no longer part of the atom. In other words, the H atom has become a $H^+$ ion and a free electron.

An important feature of the Bohr model is that it provides a theoretical framework for explaining the experimental observations of Balmer, Rydberg, and others. To see the connection, consider what happens when an electron moves between two allowed energy levels in Bohr's model. If we label the initial energy level (the level where the electron starts) $n_{\text{initial}}$, and we label the second level (the level where the electron ends up) $n_{\text{final}}$, then the change in energy of the electron is

$$\Delta E = -2.178 \times 10^{-18} \,\text{J}\left(\frac{1}{n_{\text{final}}^2} - \frac{1}{n_{\text{initial}}^2}\right) \qquad (7.9)$$

If the electron moves to an orbit farther from the nucleus, then $n_{\text{final}} > n_{\text{initial}}$, and the value of the terms inside the parentheses in Equation 7.9 is negative because

$$\frac{1}{n_{\text{final}}^2} < \frac{1}{n_{\text{initial}}^2}$$

This negative value multiplied by the negative coefficient gives us a positive $\Delta E$ and represents an increase in electron energy. On the other hand, if an electron moves from an outer orbit to one closer to the nucleus, then $n_{\text{final}} < n_{\text{initial}}$, and the sign of $\Delta E$ is negative. This means the electron loses energy. Equation 7.9 demonstrates that energy in the hydrogen atom is *quantized* because $\Delta E$ can have only certain values determined by $n_{\text{final}}$ and $n_{\text{initial}}$. The parameter $n$ is a called a **quantum number**—that is, a number that indicates an allowed energy of an electron in an atom.

When the electron in a hydrogen atom is in the lowest ($n = 1$) energy level, the atom is said to be in its **ground state**. If the electron in a hydrogen atom is in an energy level above $n = 1$, then the atom is said to be in an **excited state**. According to the Bohr model, the hydrogen electron can move from the $n = 1$ (ground state) energy level to a higher level (for example, $n = 3$) by absorbing a quantity of energy ($\Delta E$) that exactly matches the energy difference between the two states. Similarly, an electron in an excited state can move to an even higher energy level by absorbing a quantity of energy that exactly matches the energy difference between the two excited states. An electron in an excited state can also move to a lower-energy excited state, or to the ground state, by emitting a quantity of energy

**quantum number** a number that specifies the energy and location of an electron in an atom.

**ground state** the most stable, lowest-energy state available to an atom or ion.

**excited state** any energy state above the ground state in an atom or ion.

**electron transition** movement of an electron between energy levels.

that exactly matches the energy difference between those two states. This change in electron energy is called an **electron transition**.

*Energy-level diagrams* show the transitions from one energy level to another that electrons in atoms can make. Figure 7.15 is such a diagram for the hydrogen atom. The black arrow pointing upward represents absorption of sufficient energy to completely remove the electron from a hydrogen atom (ionization). The downward-pointing colored arrows represent decreases in the internal energy of the hydrogen atom that occur when photons are emitted as the electron moves from a higher energy level to a lower energy level. If the colored arrows pointed up, they would represent *absorption* of photons leading to *increases* in the internal energy of the atom. In every case the energy of the photon matches the absolute value of $\Delta E$.

If you compare Equation 7.9 with Equation 7.7, you will see that they are much alike. The coefficients differ only because of the different units used to express wave number and energy. In addition, Bohr was able to derive the coefficient as a combination of physical constants, including the Planck constant and the charge of an electron. It is not an arbitrary value that just happens to work, it is part of atomic structure. The key point is that the equation developed to fit the absorption and emission spectra of hydrogen has the same form as the theoretical equation developed by Bohr to explain the internal structure of the hydrogen atom. Thus, atomic emission and absorption spectra reveal the energies of electrons inside atoms.

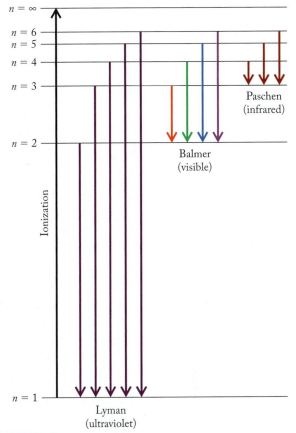

**FIGURE 7.15** An energy-level diagram showing electron transitions for the electron in the hydrogen atom. The arrow pointing up represents ionization. Arrows pointing down represent the electron emitting energy and falling to a lower energy level. This diagram shows several possible transitions for the single electron in hydrogen; each arrow *does not* represent a different electron in the atom.

**CONCEPT TEST**

Based on the lengths of the arrows in Figure 7.15, rank the following transitions in order of greatest change in electron energy to smallest change:

   a.   $n = 4 \rightarrow n = 2$
   b.   $n = 3 \rightarrow n = 2$
   c.   $n = 2 \rightarrow n = 1$
   d.   $n = 4 \rightarrow n = 3$

**SAMPLE EXERCISE 7.5**    **Calculating the Energy Needed for an Electron Transition**    **LO1**

How much energy is required to ionize a ground-state hydrogen atom?

**Collect and Organize** We are asked to determine the energy required to remove an electron from a hydrogen atom in its ground state.

**Analyze** Equation 7.9 enables us to calculate the energy change associated with any electron transition. To use Equation 7.9, we need to identify the initial ($n_{\text{initial}}$) and final ($n_{\text{final}}$) energy levels. The ground state of a H atom corresponds to the $n = 1$ energy level. If the atom is ionized, $n = \infty$ and the electron is no longer associated with the nucleus.

**Solve**

$$\Delta E = -2.178 \times 10^{-18} \text{ J}\left(\frac{1}{n_{\text{final}}{}^2} - \frac{1}{n_{\text{initial}}{}^2}\right)$$

$$= -2.178 \times 10^{-18} \text{ J}\left(\frac{1}{\infty^2} - \frac{1}{1^2}\right)$$

Dividing by $\infty^2$ yields zero, so the difference in parentheses simplifies to $-1$, which gives

$$\Delta E = 2.178 \times 10^{-18}\,\text{J}$$

**Think About It** This is a small amount of energy, but it is comparable to work function values (see Sample Exercise 7.3), which involve removing individual electrons from metal surfaces. The sign of $\Delta E$ is positive because energy must be added to remove an electron from the atom and away from its positive nucleus.

⚙ **Practice Exercise** Calculate the energy required to ionize a hydrogen atom when its electron is initially in the $n = 3$ energy level. Before doing the calculation, predict whether this energy is greater than or less than the $2.178 \times 10^{-18}$ J needed to ionize a ground-state hydrogen atom.

One of the strengths of the Bohr model of the hydrogen atom is that it accurately predicts the energy needed to remove the electron. This energy is called the *ionization energy* of the hydrogen atom. We examine the ionization energies of other elements in Section 7.11. However, the Bohr model applies only to hydrogen atoms and to ions that have only a single electron. The model does not account for the observed spectra of multielectron elements and ions because it does not account for the way electrons interact with each other. Thus, the picture of the atom provided by Bohr's model is limited. However, it enabled scientists to begin using quantum theory to explain the behavior of matter at the atomic level.

**CONCEPT TEST**

Can the Bohr model be used to explain the emission spectrum of He$^+$ ions? Why, or why not?

## 7.5 Electron Waves

A decade after Bohr published his model of the hydrogen atom, a French graduate student named Louis de Broglie (1892–1987) provided a theoretical basis for the stability of electron orbits. His approach incorporated yet another significant advance in the way early-20th-century scientists viewed atoms and subatomic particles—namely, to think of them not only as particles of matter but also as waves. His work provided an answer to the question of why an electron in the hydrogen atom does not spiral into the nucleus.

### De Broglie Wavelengths

De Broglie proposed that if light, which we normally think of as a wave, has particle properties, perhaps the electron, which we normally think of as a particle, has wave properties. If that were true, an electron moving around in an atom should have a characteristic wavelength. De Broglie calculated electron wavelengths from Einstein's equations relating energy and mass, $E = mc^2$, and the energy of a photon, $E = h\nu = hc/\lambda$:

$$\lambda = \frac{hc}{E} = \frac{hc}{mc^2} = \frac{h}{mc} \qquad (7.10)$$

To apply Equation 7.10 to electrons, de Broglie replaced $c$ (the speed of light) with $v$, the speed of an orbiting electron in an atom:

$$\lambda = \frac{h}{mv} \qquad (7.11)$$

where $h$ is the Planck constant, $m$ is the mass of the electron in kilograms, and $v$ is its velocity in meters per second. The wavelength of an electron calculated in this way is often called the *de Broglie wavelength* of the electron.

De Broglie's equation is not restricted to electrons. It tells us that *any moving particle* has wavelike properties. In other words, the particle behaves as a **matter wave**. De Broglie predicted that moving particles much bigger than electrons, such as atomic nuclei, molecules, and even macroscopic objects like tennis balls and airplanes, have characteristic wavelengths that can be calculated using Equation 7.11. The wavelengths of such large objects are extremely small, of course, given the tiny size of the Planck constant in the numerator of Equation 7.11, and for this reason we never notice the wave nature of large objects in motion.

▶❙❙ **CHEMTOUR** De Broglie Wavelength

---

**SAMPLE EXERCISE 7.6**   **Calculating the Wavelength of a Particle in Motion**   **LO3**

(a) Compare the wavelength of a 142 g baseball thrown at 44 m/s (98 mi/hr) with the size of the ball, which has a diameter of 7.5 cm. (b) Compare the wavelength of an electron ($m_e = 9.109 \times 10^{-31}$ kg) moving at one-tenth the speed of light in a hydrogen atom with the size of the atom (diameter $= 1.06 \times 10^{-10}$ m).

**Collect and Organize**  We know the masses and velocities of two moving objects and are to calculate the wavelengths of their matter waves.

**Analyze**  Equation 7.11 may be used to calculate these wavelengths. Given the small value of $h$, it is likely that the wavelength of a pitched baseball is only a tiny fraction of the size of the baseball. The wavelength of an electron moving at one-tenth the speed of light may be a much greater fraction of the size of the atom that it orbits. The fraction on the right side of Equation 7.11 has units of joule-seconds in the numerator and mass and velocity in the denominator. To combine these units in a way that gives us a unit of length, we need to use the following conversion factor:

$$1\,J = 1\,\text{kg} \cdot \text{m}^2/\text{s}^2$$

To use this equality, we must express the mass of the baseball in kilograms: 142 g = 0.142 kg.

**Solve**
a. For the baseball:

$$\lambda = \frac{h}{mv} = \frac{6.626 \times 10^{-34}\,\text{J} \cdot \text{s}}{(0.142\,\text{kg})(44\,\text{m/s})} \times \frac{1\,\text{kg} \cdot \text{m}^2/\text{s}^2}{1\,\text{J}}$$

$$= 1.06 \times 10^{-34}\,\text{m}$$

The wavelength of the baseball is

$$\frac{1.06 \times 10^{-34}\,\text{m}}{0.075\,\text{m}} \times 100\% = 1.4 \times 10^{-31}\%$$

of the ball's diameter.
b. The wavelength of an electron moving at one-tenth the speed of light is

$$\lambda = \frac{h}{mv} = \frac{6.626 \times 10^{-34}\,\text{J} \cdot \text{s}}{(9.109 \times 10^{-31}\,\text{kg})(2.998 \times 10^{7}\,\text{m/s})} \times \frac{1\,\text{kg} \cdot \text{m}^2/\text{s}^2}{1\,\text{J}}$$

$$= 2.42 \times 10^{-11}\,\text{m}$$

**matter wave**  the wave associated with any particle.

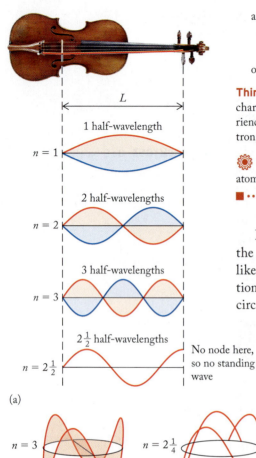

**FIGURE 7.16** Linear and circular standing waves. (a) The wavelength of a standing wave in a violin string fixed at both ends is related to the distance $L$ between the ends of the string by the equation $L = n(\lambda/2)$. In the standing waves shown, $n = 1, 2,$ and 3. An $n$ value of $2\frac{1}{2}$ does not produce a standing wave because a standing wave can have no string motion at either end. (b) The circular standing waves proposed by de Broglie account for the stability of the energy levels in Bohr's model of the hydrogen atom. Each stable wave must have a circumference equal to $n\lambda$, with $n$ restricted to being an integer, such as the $n = 3$ circular standing wave shown here. If the circumference is not an exact multiple of $\lambda$, as shown in the $n = 2\frac{1}{4}$ image, no standing wave occurs.

and

$$\frac{2.42 \times 10^{-11} \text{ m}}{1.06 \times 10^{-10} \text{ m}} \times 100\% = 23\%$$

of the diameter of a hydrogen atom.

**Think About It** The matter wave of the baseball is much too small to be observed, so its character contributes nothing to the behavior of the baseball. We expected that. Our experience with objects in the world is that they behave like matter, not like waves. For the electron, however, the wavelength is a significant percentage of the size of the hydrogen atom.

**Practice Exercise** The velocity of the electron in the ground state of a hydrogen atom is $2.2 \times 10^6$ m/s. What is the wavelength of this electron in meters?

De Broglie explained the stability of the electron levels in Bohr's model of the hydrogen atom by proposing that the electron in a hydrogen atom behaves like a circular wave oscillating around the nucleus. To understand the implications of this statement, we need to examine what is required to make a stable, circular wave. Consider the motion of a vibrating violin string of length $L$ (Figure 7.16a). Because the string is fixed at both ends, there is no vibration at the ends and the vibration is at maximum in the middle. The wave created by this combination of fixed ends and maximum vibration in the middle is a **standing wave**; that is, a wave that oscillates back and forth within a fixed space rather than moving through space the way waves of light travel through space. On a standing wave, any points that have zero displacement, such as the two ends of the string, are called **nodes**. The sound wave produced in this way on a violin string is called the *fundamental* of the string. The wavelength of the fundamental is $2L$. It has the lowest frequency and longest wavelength possible for that string. The length of the string is one-half the wavelength of the fundamental ($L = \lambda/2$).

If the string is held down in the middle (creating a third node) and plucked halfway between the middle and one end, a new, higher-frequency wave called the *first harmonic* is produced. The wavelength of this harmonic is equal to $L$ because $L = 2(\lambda/2) = \lambda$. We can continue generating higher frequencies with wavelengths that are related to the length of the string by the equation

$$L = \frac{n\lambda}{2}$$

where $n$ is a whole number and $L = 3(\lambda/2), 4(\lambda/2), 5(\lambda/2),$ and so on.

**CONCEPT TEST**

Why are the various sound waves possible from a violin string examples of quantization?

The standing-wave pattern for a circular wave generated by an electron differs slightly from that of a vibrating violin string in that the circular wave has no defined stationary ends. Instead, the electron vibrates in an endless series of waves. However, standing waves are produced if, as shown in Figure 7.16(b), the circumference of the circle equals a whole-number multiple of the electron's wavelength:

$$\text{Circumference} = n\lambda \qquad (7.12)$$

Equation 7.12 gives a new meaning to Bohr's quantum number $n$: it represents the number of matter-waves in a given energy level. De Broglie's work also solved the problem of a negatively charged electron spiraling into the positively charged nucleus. Since $n = 1$ represents the minimum circumference of the circular wave of a moving electron, the electron must remain a minimum distance from the nucleus at all times. Furthermore, the lowest energy possible for an electron in a hydrogen atom occurs for a standing wave with circumference $\lambda$, so the electron cannot get any closer to the nucleus than that.

De Broglie's research created a quandary for the graduate faculty at the University of Paris, where he studied. Bohr's model of electrons moving between allowed energy levels had been widely criticized as an arbitrary suspension of well-tested physical laws. De Broglie's rationalization of Bohr's model seemed even more outrageous to many scientists. Before the faculty would accept his thesis, they wanted another opinion, so they sent it to Albert Einstein for review. Einstein wrote back that he found the young man's work "quite interesting." That endorsement was good enough for the faculty: de Broglie's thesis was accepted in 1924 and immediately submitted for publication. Five years later, he was awarded the Nobel Prize.

## The Heisenberg Uncertainty Principle

After de Broglie proposed that electrons exhibited both particle-like and wave-like behavior, questions arose about the impact of wave behavior on our ability to locate the electron. A wave, by its very nature, is spread out in space. The question "Where is the electron?" has one answer if we treat the electron as a particle and a different answer if we treat it as a wave. This issue was addressed by German physicist Werner Heisenberg (1901–1976), who proposed the following thought experiment: watch an electron with a hypothetical gamma-ray microscope (a microscope that uses gamma rays instead of visible light) to "see" the electron's path around an atom. The microscope (if it existed) would need to use gamma rays for illumination because they are the only part of the electromagnetic spectrum with wavelengths short enough to match the diminutive size of electrons. However, Equations 7.2 and 7.3 tell us that the short wavelengths and high frequencies of gamma rays mean that they have enormous energies—so large that any gamma ray striking an electron would knock the electron off course. The only way not to affect the electron's motion would be to use a much lower-energy, longer-wavelength source of radiation to illuminate it—but then we would not be able to see the tiny electron clearly.

This situation presents a quantum mechanical dilemma. The only means for clearly observing an electron make it impossible to know the electron's motion or, more precisely, its momentum, which is defined as an object's velocity times its mass. Therefore we can never know exactly both the position and the momentum of the electron simultaneously. This conclusion is known as the **Heisenberg uncertainty principle**, and it is mathematically expressed by

$$\Delta x \cdot m\Delta v \geq \frac{h}{4\pi} \qquad (7.13)$$

where $\Delta x$ is the uncertainty in the position of the electron, $m$ is its mass, $\Delta v$ is the uncertainty in its velocity, and $h$ is the Planck constant. To Heisenberg, this uncertainty was the essence of quantum mechanics. Its message for us is that there are limits to what we can observe, measure, and therefore know.

**standing wave** a wave confined to a given space, with a wavelength ($\lambda$) related to the length $L$ of the space by $L = n(\lambda/2)$, where $n$ is a whole number.

**node** a location in a standing wave that experiences no displacement.

**Heisenberg uncertainty principle** we cannot determine both the position and the momentum of an electron in an atom at the same time.

**SAMPLE EXERCISE 7.7  Calculating Uncertainty**  **LO4**

Using the data in Sample Exercise 7.6, compare the uncertainty in the velocity of the baseball with the uncertainty in the velocity of the electron. Assume that the position of the baseball is known to within one wavelength of red light ($\Delta x_{baseball} = 680$ nm) and that the position of the electron is known to within the radius of the hydrogen atom ($\Delta x_{electron} = 5.3 \times 10^{-11}$ m).

**Collect and Organize**  We are asked to calculate the uncertainty in the velocities of two particles given the uncertainties in their positions. From Sample Exercise 7.6 we know that the mass of a baseball is 0.142 kg and the mass of an electron is $9.109 \times 10^{-31}$ kg.

**Analyze**  Equation 7.13 provides a mathematical connection between these variables. According to Equation 7.13, the uncertainty in the velocity and position of a particle is inversely proportional to its mass. Therefore, we can expect little uncertainty in the velocity of the baseball but much greater uncertainty in the velocity of the electron. We need to rearrange the terms in the equation to solve for the uncertainty in velocity ($\Delta v$):

$$\Delta v \geq \frac{h}{4\pi \Delta x m}$$

**Solve**  For the baseball:

$$\Delta v \geq \frac{6.626 \times 10^{-34} \text{ J} \cdot \text{s}}{4\pi (6.80 \times 10^{-7} \text{ m})(0.142 \text{ kg})} \times \frac{1 \text{ kg} \cdot \text{m}^2/\text{s}^2}{1 \text{ J}}$$

$$\geq 5.46 \times 10^{-28} \text{ m/s}$$

For the electron:

$$\Delta v \geq \frac{6.626 \times 10^{-34} \text{ J} \cdot \text{s}}{4\pi (5.3 \times 10^{-11} \text{ m})(9.109 \times 10^{-31} \text{ kg})} \times \frac{1 \text{ kg} \cdot \text{m}^2/\text{s}^2}{1 \text{ J}}$$

$$\geq 1.09 \times 10^{6} \text{ m/s}$$

Comparing the two values, we see that the uncertainty in the velocity of the baseball is extremely small (about $10^{-28}$ m/s), whereas that of the electron is huge (over 1 million m/s).

**Think About It**  The uncertainty in the measurement of the baseball's velocity is so minuscule it is insignificant. This result is expected for objects in the macroscopic world. The uncertainty in the measurement of the velocity of the electron is huge, which is what we must expect at the atomic level, where "particles" such as the electron also behave like waves.

⚙ **Practice Exercise**  What is the uncertainty, in meters, in the position of an electron moving near a nucleus at a speed of $8 \times 10^{7}$ m/s? Assume the uncertainty in the velocity of the electron is 1% of its value—that is, $\Delta v = (0.01)(8 \times 10^{7}$ m/s$)$.

**wave mechanics** (also called **quantum mechanics**) a mathematical description of the wavelike behavior of particles on the atomic level.

**Schrödinger wave equation** a description of how the electron matter wave varies with location and time around the nucleus of a hydrogen atom.

**wave function ($\psi$)** a solution to the Schrödinger wave equation.

**orbital** a region around the nucleus of an atom where the probability of finding an electron is high; each orbital is defined by the square of the wave function ($\psi^2$) and identified by a unique combination of three quantum numbers.

**principal quantum number (n)** a positive integer describing the relative size and energy of an atomic orbital or group of orbitals in an atom.

When Heisenberg proposed his uncertainty principle, he was working with Bohr at the University of Copenhagen. The two scientists had widely different views about the significance of the uncertainty principle and the idea that particles could behave like waves. To Heisenberg, uncertainty was a fundamental characteristic of nature. To Bohr, it was merely a mathematical consequence of the wave–particle duality of electrons; there was no physical meaning to an electron's position and path. The debate between these two gifted scientists was heated at times. Heisenberg later wrote about one particularly emotional debate:

> [A]t the end of the discussion I went alone for a walk in the neighboring park [and] repeated to myself again and again the question: "Can nature possibly be as absurd as it seems?"[1]

The Heisenberg uncertainty principle is fundamental to our present understanding of the atom. If we cannot know both the position and momentum of an electron in

[1]Heisenberg, W. *Physics and Philosophy: The Revolution in Modern Science* (Harper & Row, 1958), p. 42.

a hydrogen atom, then the electron cannot be moving in circular orbits as implied by Bohr's original model. As we will see in the next section, the Heisenberg uncertainty principle limits us to knowing only the probability of finding an electron at a particular location in an atom.

# 7.6 Quantum Numbers and Electron Spin

Many of the leading scientists of the 1920s were unwilling to accept the dual wave–particle nature of electrons proposed by de Broglie until the model could be used to predict the features of the hydrogen emission spectrum. Such application required the development of equations describing the behavior of electron waves. Over his Christmas vacation in 1925, Austrian physicist Erwin Schrödinger (1887–1961) did just that, developing in a few weeks the mathematical foundation for what came to be called **wave mechanics** or **quantum mechanics**.

Schrödinger's mathematical description of electron waves is called the **Schrödinger wave equation**. Although it is not discussed in detail in this book, you should know that solutions to the wave equation are called **wave functions** ($\psi$): mathematical expressions that describe how the matter wave of an electron in an atom varies both with time and with the location of the electron in the atom. Wave functions define the energy levels in the hydrogen atom. They can be simple trigonometric functions, such as sine or cosine waves, or they can be very complex.

What is the physical significance of a wave function? Actually there is none. However, the *square of a wave function* ($\psi^2$) does have physical meaning. Initially, Schrödinger believed that a wave function depicted the "smearing" of an electron through three-dimensional space. This notion of subdividing a discrete particle was later rejected in favor of the model developed by German physicist Max Born (1882–1970), who proposed that $\psi^2$ defines an **orbital**: the space around the nucleus of an atom where the probability of finding an electron is high. Born later showed that his interpretation could be used to calculate the probability of a transition between two orbitals, as happens when an atom absorbs or emits a photon.

To help visualize the probabilistic meaning of $\psi^2$, consider what happens when we spray ink onto a flat surface (Figure 7.17). If we then draw a circle encompassing most of the ink spots, we are identifying the region of maximum probability for finding the spots.

It is important to understand that quantum mechanical orbitals in an atom are not two-dimensional concentric orbits, as in Bohr's model of the hydrogen atom, or even two-dimensional circles, as in Figure 7.17. Instead, they are three-dimensional regions of space with distinctive shapes, orientations, and average distances from the nucleus. Each orbital is a solution to Schrödinger's wave equation and is identified by a unique combination of three integers, or quantum numbers, whose values flow directly from the mathematical solutions to the wave equation. The quantum numbers are as follows:

➤ The **principal quantum number** *n* is like Bohr's quantum number *n* for the hydrogen atom in that it is a positive integer that indicates the relative size and energy of an orbital or group of orbitals in an atom. Orbitals with the same value of *n* are in the same *shell*. Orbitals with larger values of *n* are farther from the nucleus and, in the hydrogen atom, represent higher energy levels, consistent with Bohr's model of the hydrogen atom. In multielectron atoms, the relationship between energy levels and orbitals is more complex, but increasing values of *n* generally represent higher energy levels.

**FIGURE 7.17** The probability of finding an ink spot in the pattern produced by a source of ink spray decreases with increasing distance from the center of the pattern.

▶❙❙ **CHEMTOUR** Quantum Numbers

**angular momentum quantum number ($\ell$)** an integer having any value from 0 to $n-1$ that defines the shape of an orbital.

**magnetic quantum number ($m_\ell$)** defines the orientation of an orbital in space; an integer that may have any value from $-\ell$ to $+\ell$, where $\ell$ is the angular momentum quantum number.

➤ The **angular momentum quantum number** $\ell$ is an integer with a value ranging from zero to $n-1$ that defines the shape of an orbital. Orbitals with the same values of $n$ and $\ell$ are in the same *subshell* and represent equal energy levels. Orbitals with a given value of $\ell$ are identified with a letter according to the following scheme:

| Value of $\ell$ | 0 | 1 | 2 | 3 |
|---|---|---|---|---|
| Letter Identifier | s | p | d | f |

➤ The choice of letters to designate the values of $\ell$ (that is, *s*, *p*, *d*, and *f*) may seem a bit odd. Before quantum mechanics was developed, scientists recording the line spectra of the elements described the lines they observed as **sharp**, **principal**, **diffuse**, and **fundamental**. The designations for the orbitals with the letters *s*, *p*, *d*, and *f* retained this convention.

➤ The **magnetic quantum number** $m_\ell$ is an integer with a value from $-\ell$ to $+\ell$. It defines the orientation of an orbital in the space around the nucleus of an atom.

Each subshell in an atom has a two-part designation containing the appropriate value of $n$ and a letter designation for $\ell$. For example, orbitals with $n=3$ and $\ell=1$ are called 3p orbitals, and electrons in 3p orbitals are called 3p electrons. How many 3p orbitals are there? We can answer this question by finding all possible values of $m_\ell$. Because *p* orbitals are those for which $\ell=1$, they have $m_\ell$ values of $-1$, 0, and $+1$. These three values mean that there are three 3p orbitals, each with a unique combination of $n$, $\ell$, and $m_\ell$ values. All the possible combinations of these three quantum numbers for the orbitals of the first four shells are given in Table 7.1.

| | | | | NUMBER OF ORBITALS IN: | |
|---|---|---|---|---|---|
| Value of $n$ | Allowed Value of $\ell$ | Subshell Label | Allowed Values of $m_\ell$ | Subshell | Shell |
| 1 | 0 | s | 0 | 1 | 1 |
| 2 | 0 | s | 0 | 1 | |
| | 1 | p | $-1, 0, +1$ | 3 | 4 |
| 3 | 0 | s | 0 | 1 | |
| | 1 | p | $-1, 0, +1$ | 3 | |
| | 2 | d | $-2, -1, 0, +1, +2$ | 5 | 9 |
| 4 | 0 | s | 0 | 1 | |
| | 1 | p | $-1, 0, +1$ | 3 | |
| | 2 | d | $-2, -1, 0, +1, +2$ | 5 | |
| | 3 | f | $-3, -2, -1, 0, +1, +2, +3$ | 7 | 16 |

**TABLE 7.1  Quantum Numbers of the Orbitals in the First Four Shells**

**SAMPLE EXERCISE 7.8   Identifying the Subshells and Orbitals in an Energy Level   LO5**

(a) What are the designations of all the subshells in the $n=4$ shell? (b) How many orbitals are in these subshells?

**Collect and Organize** We are asked to describe the subshells in the fourth shell and to determine how many orbitals are in all these subshells. Table 7.1 contains an inventory of all the subshells in the first four shells.

**Analyze** The designations of subshells are based on the possible values of quantum numbers $n$ and $\ell$. The allowed values of $\ell$ depend on the value of $n$, in that $\ell$ is an integer between 0 and $n - 1$. The number of orbitals in a subshell depends on the number of possible values of $m_\ell$, from $-\ell$ to $+\ell$.

**Solve**

a. The allowed values of $\ell$ for $n = 4$ range from 0 to 3 ($n - 1$), so they are 0, 1, 2, and 3. These $\ell$ values correspond to the subshell designations $s$, $p$, $d$, and $f$, respectively. The appropriate subshell names are thus $4s$, $4p$, $4d$, and $4f$.

b. The possible values of $m_\ell$ from $-\ell$ to $+\ell$ are as follows:

$\ell = 0$; $m_\ell = 0$: This combination of $\ell$ and $m_\ell$ values for the $n = 4$ shell represents a single $4s$ orbital.

$\ell = 1$; $m_\ell = -1, 0,$ or $+1$: These three combinations of $\ell$ and $m_\ell$ values for the $n = 4$ shell represent the three $4p$ orbitals.

$\ell = 2$; $m_\ell = -2, -1, 0, +1,$ or $+2$: These five combinations of $\ell$ and $m_\ell$ values represent the five $4d$ orbitals.

$\ell = 3$; $m_\ell = -3, -2, -1, 0, +1, +2,$ or $+3$: These seven combinations of $\ell$ and $m_\ell$ values represent the seven $4f$ orbitals.

Thus there are $1 + 3 + 5 + 7 = 16$ orbitals in the $n = 4$ shell.

**Think About It** We determined that there are 16 orbitals in the fourth shell. The number of orbitals in each shell is equal to the square of the principal quantum number of the shell.

 **Practice Exercise** How many orbitals are there in the $n = 2$ shell?

Several relationships are worth noting in the quantum numbering system (Figure 7.18):

➤ There are $n$ subshells in the $n$th shell: one subshell ($1s$) in the $n = 1$ shell, two subshells ($2s$ and $2p$) in the $n = 2$ shell, and so on.

➤ There are $n^2$ orbitals in the $n$th shell: $1^2 = 1$ in the $n = 1$ shell, $2^2 = 4$ in the $n = 2$ shell, and so on.

➤ There are ($2\ell + 1$) orbitals in each subshell: one $s$ orbital ($2 \times 0 + 1 = 1$) in each $s$ subshell, three $p$ orbitals ($2 \times 1 + 1 = 3$) in each $p$ subshell, five $d$ orbitals ($2 \times 2 + 1 = 5$) in each $d$ subshell, and so on.

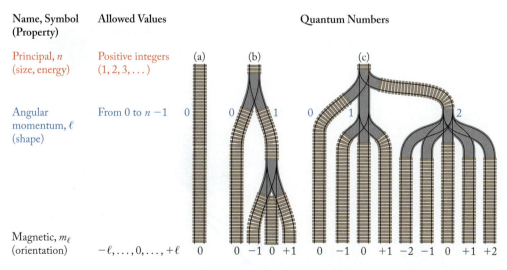

| Name, Symbol (Property) | Allowed Values | Quantum Numbers |
|---|---|---|
| Principal, $n$ (size, energy) | Positive integers $(1, 2, 3, \ldots)$ | |
| Angular momentum, $\ell$ (shape) | From 0 to $n - 1$ | |
| Magnetic, $m_\ell$ (orientation) | $-\ell, \ldots, 0, \ldots, +\ell$ | |

**FIGURE 7.18** These three sets of train tracks provide a visual metaphor of the orbitals in the first three energy shells in an atom. The single track labeled (a) represents the first shell and its $1s$ orbital, which has $\ell$ and $m_\ell$ values of 0. The set labeled (b) has two $\ell$ branches, labeled 0 and 1, representing the $2s$ and $2p$ orbitals. The 1 branch further divides into branches with $m_\ell$ values of $-1$, 0, and $+1$, representing the three $2p$ orbitals. The set labeled (c) divides into three main branches with $\ell$ values of 0, 1, 2, representing the $3p$, $3p$, and $3d$ subshells, respectively.

The Schrödinger wave equation accounts for most, but not all, aspects of atomic spectra. The emission spectrum of hydrogen, for example, has a pair of red lines at 656 nm where Balmer thought there was only one line (Figure 7.19).

**FIGURE 7.19** The Schrödinger equation does not account for the appearance of closely spaced pairs of bright lines in the emission spectra of atoms, such as the red lines at 656.272 and 656.285 nm in the spectrum of hydrogen.

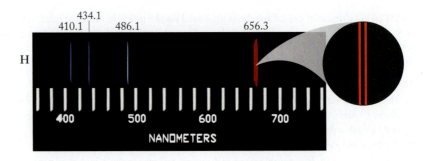

There are also pairs of lines in the spectra of multielectron atoms that have a single electron in their outermost shells. The Schrödinger equation cannot explain these pairs of lines.

In 1925 two students at the University of Leiden in the Netherlands, Samuel Goudsmit (1902–1978) and George Uhlenbeck (1900–1988), proposed that the pairs of lines, called *doublets*, were caused by a property they called electron spin. In their model, electrons spin in one of two directions, designated spin "up" and spin "down." A moving electron (or any charged particle) creates a magnetic field by virtue of its motion. The spinning motion produces a second magnetic field oriented up or down. To account for these two spin orientations, Goudsmit and Uhlenbeck proposed a fourth quantum number, the **spin magnetic quantum number ($m_s$)**. The values of $m_s$ are $+\frac{1}{2}$ for spin up and $-\frac{1}{2}$ for spin down.

Even before Goudsmit and Uhlenbeck proposed the electron-spin hypothesis, two other scientists, Otto Stern (1888–1969) and Walther Gerlach (1889–1979), observed the effect of electron spin when they shot a beam of silver ($Z = 47$) atoms through a magnetic field (Figure 7.20). Those atoms in which the net electron spin was "up" were deflected in one direction by the field; those in which the net electron spin was "down" were deflected in the opposite direction.

In 1925 Austrian physicist Wolfgang Pauli (1900–1958; Figure 7.21) proposed that no two electrons in a multielectron atom have the same set of four quantum numbers. This idea is known as the **Pauli exclusion principle**. The three quantum numbers from Schrödinger's wave equation define the orbitals where

**FIGURE 7.20** A narrow beam of silver atoms passed through a magnetic field is split into two beams because of the interactions between the field and the spinning electrons in the atoms. This observation led to proposing the fourth quantum number, $m_s$. (The blue arrows indicate the direction of the beam.)

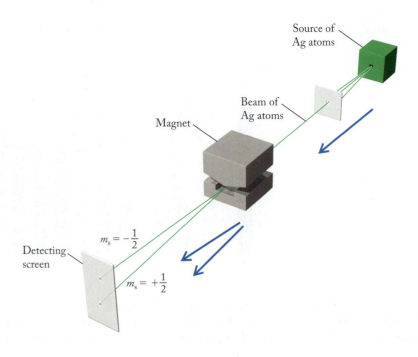

an atom's electrons are likely to be. The two allowed values of the spin magnetic quantum number indicate that each orbital can hold two electrons, one with $m_s = +\frac{1}{2}$ and the other with $m_s = -\frac{1}{2}$. Thus, each electron in an atom has a unique "quantum address" defined by a particular combination of $n$, $\ell$, $m_\ell$, and $m_s$ values.

**FIGURE 7.21** Wolfgang Pauli (left) and Niels Bohr are apparently amused by the behavior of a toy called a tippe top, which, when spun on its base, tips itself over and spins on its stem. Although the toy's behavior is caused by a combination of friction and the top's angular momentum and is not quantum mechanical in origin, it provides a visual metaphor of the two spin orientations of an electron in an orbital.

---

**SAMPLE EXERCISE 7.9**    **Identifying Valid Quantum Number Sets**    LO5

Which of these five combinations of quantum numbers are valid?

|     | $n$ | $\ell$ | $m_\ell$ | $m_s$ |
|-----|-----|--------|----------|-------|
| (a) | 1 | 0 | −1 | $+\frac{1}{2}$ |
| (b) | 3 | 2 | −2 | $+\frac{1}{2}$ |
| (c) | 2 | 2 | 0 | 0 |
| (d) | 2 | 0 | 0 | $-\frac{1}{2}$ |
| (e) | −3 | −2 | −1 | $-\frac{1}{2}$ |

**Collect and Organize** We are asked to validate five sets of possible quantum numbers. We can apply the definitions and ranges in value of each quantum number.

**Analyze** The principal quantum number ($n$) can be any positive integer. The valid values of $\ell$ in a given shell are integers from 0 to $(n − 1)$, and the values of $m_\ell$ in a given subshell include all integers from $-\ell$ to $+\ell$ including 0. The only two options for $m_s$ are $+\frac{1}{2}$ or $-\frac{1}{2}$.

**Solve**
a. Because $n$ is 1, the maximum (and only) value of $\ell$ is $(n − 1) = 1 − 1 = 0$. Therefore the values of $n$ and $\ell$ are valid. However, if $\ell = 0$, then $m_\ell$ must be 0; it cannot be −1. Therefore, this set is not valid. The spin quantum number is a possible value.
b. Because $n$ is 3, $\ell$ can be 2 and $m_\ell$ can be −2. Also, $m_s = +\frac{1}{2}$ is a valid choice for the spin magnetic quantum number. This set is valid.
c. Because $n$ is 2, $\ell$ cannot be 2, making this set invalid. In addition, $m_s$ has an invalid value (0).
d. Because $n$ is 2, $\ell$ can be 0, and for that value of $\ell$, $m_\ell$ must be 0. The value of $m_s$ is also valid, and so is the set.
e. This set contains two impossible values, $n = −3$ and $\ell = −2$, so it is invalid.

**Think About It** The values of $n$, $\ell$, and $m_\ell$ are related mathematically, and $m_s$ can be either $+\frac{1}{2}$ or $-\frac{1}{2}$. The quantum numbers are the address of the electron in an atom, and every electron has its own unique address—its own unique set of four quantum numbers.

⚙ **Practice Exercise** Write all the possible sets of quantum numbers for an electron in the $n = 3$ shell that has an angular momentum quantum number $\ell = 1$ and a spin quantum number $m_s = +\frac{1}{2}$.

---

Momentous advances in chemistry and physics were made in the first three decades of the 20th century due to the brilliant discoveries of Einstein, Planck, Rutherford, Bohr, de Broglie, Schrödinger, and others, which have forever changed scientists' view of the fundamental structure of matter and the universe. Figure 7.22 summarizes and connects some of these advances.

Consensus in the scientific community on the ideas of quantum theory did not come easily. We have seen how hot sodium atoms in the excited state emit photons of yellow-orange light as they fall to the ground state. Einstein puzzled over this phenomenon for several years before deciding that the moment when

**spin magnetic quantum number ($m_s$)** either $+\frac{1}{2}$ or $-\frac{1}{2}$, indicating that the spin orientation of an electron is either up or down.

**Pauli exclusion principle** no two electrons in an atom can have the same set of four quantum numbers.

**FIGURE 7.22** During the first three decades of the 20th century, quantum theory evolved from classical 19th-century theories of the nature of matter and energy. The arrows trace the development of modern quantum theory, which assumes that radiant energy has both wavelike and particle-like properties and that mass (matter) also has both wavelike and particle-like properties.

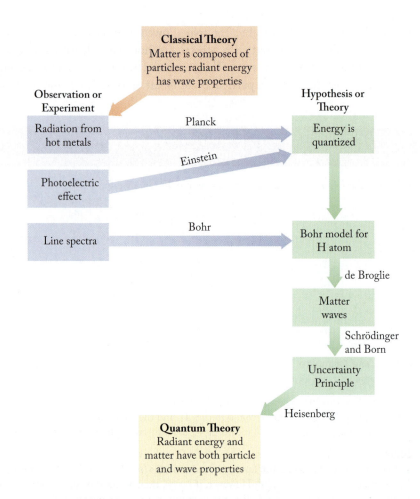

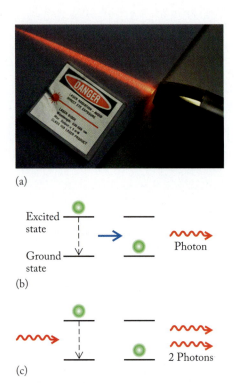

(a)

(b)

(c)

**FIGURE 7.23** (a) Laser pointers emitting different colors are commonplace. (b) When an excited-state electron spontaneously returns to the ground state, it emits a photon of characteristic wavelength. (c) If the emitted photon encounters another excited-state atom, then it causes a second photon of the same wavelength to be emitted. If a sufficient number of excited states are present, this 2-fold amplification is repeated many times and a bright pulse of light is produced.

emission occurs and the direction of the photon emitted could not be predicted exactly. He concluded that quantum theory allows us to calculate only the probability of a spontaneous electron transition; the details of the event are left to chance. In other words, no force of nature causes a hot sodium atom to fall to a lower energy level at a particular instant.

How different is the view of the interaction of matter and energy provided by quantum mechanics from the laws governing the behavior of large objects? A pebble picked up and dropped immediately falls to the ground. However, electrons remain in excited states for indeterminate (though usually short) times before falling to their ground states. This lack of determinacy bothered Einstein and many of his colleagues. Had they discovered an underlying theme of nature—that some processes cannot be described or known with certainty? Are there fundamental limits to how well we can know and understand our world and the events that change it?

The variable lifetimes of excited states is the basis for the lasers used in barcode scanners at grocery stores, laser pointers, DVD players, and in surgery (Figure 7.23a). The word laser is actually an acronym for **L**ight **A**mplification by **S**timulated **E**mission of **R**adiation. The phrase *stimulated emission* is linked to the fact that the atoms in lasers linger in particular excited states for unusually long times, which allows the number of these excited states to build up. Then, if a photon with just the right energy encounters one of these excited-state atoms, it *stimulates* the decay of the excited state, producing a second photon that matches the first: same wavelength, same phase, and going in the same direction (Figure 7.23c). When these two photons encounter two more excited state atoms, they could produce four in-phase photons, which could produce four more, and so on. In this way, one incident photon is rapidly amplified into an intense pulse of monochromatic light. To produce a steady beam of this light, electrical energy continuously *pumps up* the lasing material, repopulating the excited state. By manipulating the composition of lasers, scientists can adjust the electronic energy levels inside them and the color of the light they produce.

# 7.7 The Sizes and Shapes of Atomic Orbitals

As noted earlier, the orbitals in atoms have three-dimensional shapes that are graphical representations of $\psi^2$. In this section we examine the shapes of atomic orbitals and how those shapes impact the energies of the electrons in them. We will build upon the material in this section to connect quantum mechanics and atomic orbitals with the periodic table and the properties of elements.

## s Orbitals

Figure 7.24 provides several representations of the 1s orbital of the hydrogen atom. In Figure 7.24(a), electron density is plotted against distance from the nucleus and shows that density decreases with increasing distance. However, Figure 7.24(b) provides a more useful profile of electron distribution. To understand why, think of the hydrogen atom as being like an onion, made of many concentric spherical layers, all of the same thickness. A cross section of this image of the atom is shown in Figure 7.24(c). What is the probability of finding the electron in one of these spherical layers? A layer very close to the nucleus has a very small radius, so it accounts for only a small fraction of the total volume of the atom. A layer with a larger radius makes up a much larger fraction of the volume of the atom because the volume of the layers increases as a function of $r^2$. (Note that the volume of a sphere depends on $r^3$, but here we are discussing a spherical shell, the volume of which depends on $r^2$.) Even though electron densities are higher closer to the nucleus (as Figure 7.24a shows), the volumes of the spherical layers closest to the nucleus are so small that the chances of the electron being near the center of an atom are extremely low; this is shown in Figure 7.24(b), where the curve starts off at essentially zero for electron distribution values at distances very close to the nucleus. Farther from the nucleus, electron densities are lower but the volumes of the layers are much larger, so the probability of the electron being in one of these layers is relatively high, represented by the peak in the curve of Figure 7.24(b). At greater distances, volumes of the layers are very large but $\psi^2$ drops to nearly zero (see Figure 7.24a); therefore, the chances of finding an electron in layers far from the nucleus are very small.

Thus Figure 7.24(b) represents a combination of two competing factors: increasing layer volume and decreasing probability of finding an electron in a given layer. This

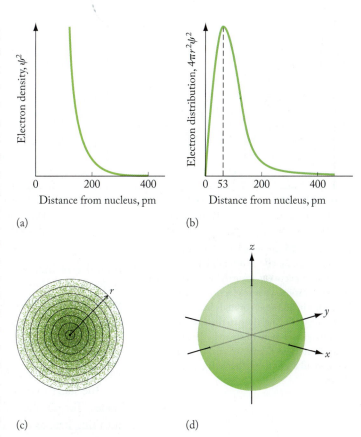

(a)      (b)

(c)      (d)

**FIGURE 7.24** (a) Probable electron density in the 1s orbital of the hydrogen atom represented by a plot of electron density ($\psi^2$) versus distance from the nucleus. (b) Electron distribution in the 1s orbital versus distance from the nucleus. The distribution is essentially zero both for very short distances from the nucleus and for very long distances from the nucleus. The maximum probability occurs at $r = 53$ pm. (c) Cross section through the hydrogen atom, with the space surrounding the nucleus divided into an arbitrary number of thin, concentric, hollow layers. Each layer has a unique value for radius $r$. The probability of finding an electron in a particular layer of radius $r$ depends on the volume of the layer and the density of electrons in the layer. (d) Boundary–surface representation of a sphere within which the probability of finding a 1s electron is 90%.

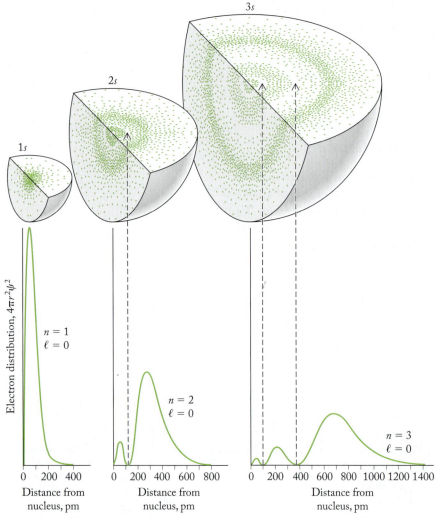

**FIGURE 7.25** These radial distribution profiles of 1s, 2s, and 3s orbitals have 0, 1, and 2 nodes, respectively, identifying (with dashed arrows) locations of zero electron density. Electrons in all these s orbitals have some probability of being close to the nucleus, but 3s electrons are more likely to be farther away from the nucleus than 2s electrons, which are more likely to be farther away than 1s electrons.

combination produces a *radial distribution profile* for the electron. Figure 7.24(b) is a plot, not of $\psi^2$ versus distance from the nucleus as in Figure 7.24(a), but rather of $4\pi r^2 \psi^2$ versus distance from the nucleus. In geometry, $4\pi r^2$ is the formula for the area of a sphere, but here it represents the volume of one of the thin spherical layers in Figure 7.24(c).

A significant feature of the curve in Figure 7.24(b) is that its maximum value corresponds to the most likely radial distance of the electron from the nucleus. The value of *r* corresponding to this maximum for the 1s orbital of hydrogen is 53 pm.

Figure 7.24(d) provides a view of the spherical shape of this (or any other) *s* orbital. The surface of the sphere encloses the volume within which the probability of finding a 1s electron is 90%. This type of depiction, called a *boundary–surface representation*, is one of the most useful ways to view the relative sizes, shapes, and orientations of orbitals. All *s* orbitals are spheres, which have only one orientation and in which electron density depends only on distance from the nucleus. Boundary surfaces are a useful way to depict the shape and relative size of an orbital.

The relative sizes of 1s, 2s, and 3s orbitals are shown in Figure 7.25. Note that orbital size increases with increasing values of the principal quantum number *n*. Note also that the sections of spheres above the profile curves show bands in which the density of dots is high. The dots represent the probability of an electron being in these regions of three-dimensional space, and each band is called a *local maximum* of electron density. In all three profiles, a local maximum occurs close to the nucleus. This means that electrons in *s* orbitals—even *s* orbitals with high values of *n*—have some probability of being close to the nucleus.

The local maxima in any *s* orbital are separated from other local maxima by nodes. The number of nodes in any *s* orbital is equal to *n* − 1. Nodes have the same meaning here as they do in one-dimensional standing waves: they are places where the wave has an amplitude of zero. In the context of electrons as three-dimensional matter waves, nodes are locations at which electron density goes to zero.

**CONCEPT TEST** ••••••••••••••••••••••••••••••••••••••••••

How many nodes are there in the electron distribution profile of the 6s orbital?

••••••••••••••••••••••••••••••••••••••••••••••••••••••••••••

## *p* and *d* Orbitals

We have noted that *s* orbitals are spherical, so in these orbitals electron density depends only on distance from the nucleus, not on angle of orientation. Let's now consider the shapes of *p* and *d* orbitals.

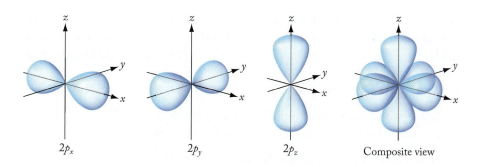

$2p_x$  $2p_y$  $2p_z$  Composite view

**FIGURE 7.26** Boundary–surface views of the three $p$ orbitals, showing their orientation along the $x$-, $y$-, and $z$-axes. The nucleus of the atom in which these orbitals appear is located at the origin. These shapes are an elongated version of the theoretical shapes of the orbitals. We use this version throughout the book to make it easier to see the orientation of the lobes.

All shells with $n \geq 2$ have a subshell containing three $p$ orbitals ($\ell = 1$; $m_\ell = -1, 0, +1$). Each of these orbitals has two teardrop-shaped lobes, oriented one on either side of the nucleus, along one of the three perpendicular Cartesian axes $x$, $y$, $z$ (Figure 7.26). These orbitals are designated $p_x$, $p_y$, and $p_z$, depending on the axis along which the lobes are aligned. The two lobes of a $p$ orbital are sometimes labeled with plus and minus signs, indicating that the sign of the wave function defining them is either $+\psi$ or $-\psi$. (These signs are not in any way connected with electric charges.) An electron in a $p$ orbital occupies *both* lobes. Because a node of zero probability separates the two lobes, you may wonder how an electron gets from one lobe to the other. One way to think about how this happens is to remember that an electron behaves as a three-dimensional standing wave, and waves have no difficulty passing through nodes. After all, Figure 7.16 shows a node right in the middle of the $n = 2$ standing wave separating the positive and negative displacement regions in the vibrating string.

The five $d$ orbitals ($\ell = 2$, $m_\ell = -2, -1, 0, +1, +2$) found in shells of $n \geq 3$ all have different orientations (Figure 7.27). Four of them consist of four teardrop-shaped lobes oriented like the leaves in a four-leaf clover. In three of these orbitals, the lobes lie between, not on, the $x$-, $y$-, and $z$-axes. These orbitals are designated $d_{xy}$, $d_{xz}$, and $d_{yz}$. In the fourth orbital in this set, designated $d_{x^2-y^2}$, the four lobes lie along the $x$- and $y$-axes. The fifth $d$ orbital, designated $d_{z^2}$, is mathematically equivalent to the other four but has a much different shape, with two teardrop-shaped lobes oriented along the $z$-axis and a doughnut shape called a *torus* in the $x$−$y$ plane that surrounds the middle of the two lobes.

We will not address the shapes and geometries of other types of orbitals in this text, although we will refer to them. The $s$, $p$, and $d$ orbitals are most crucial to the discussions of chemical bonding in the chapters to come.

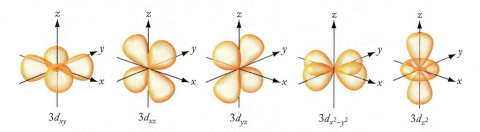

$3d_{xy}$  $3d_{xz}$  $3d_{yz}$  $3d_{x^2-y^2}$  $3d_{z^2}$

**FIGURE 7.27** Boundary–surface views of the five $d$ orbitals, showing their orientation relative to the $x$-, $y$-, and $z$-axes. The $d_{xy}$, $d_{xz}$, and $d_{yz}$ orbitals are not aligned along any axis; the $d_{x^2-y^2}$ orbital lies along the $x$ and $y$ axes; the $d_{z^2}$ orbital consists of two teardrop-shaped lobes along the $z$ axis with a donut-shaped torus ringing the point where the two lobes meet.

# 7.8 The Periodic Table and Filling the Orbitals of Multielectron Atoms

Now that we have developed a model describing the atomic orbitals, we can use that model to fill orbitals in atoms containing more than one electron. To explore the orbital-filling sequence, let's start at the beginning of the periodic table with hydrogen and put each successive electron into the lowest-energy orbital available

◉◯ **CONNECTION** We described in Section 2.4 how Mendeleev developed the first useful periodic table of the elements, not only decades before de Broglie gave us electron waves and Schrödinger pioneered quantum mechanics, but even before atoms were known to consist of electrons, protons, and neutrons.

▶❚❚ **CHEMTOUR** Electron Configuration

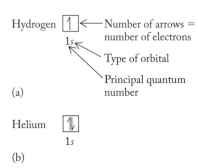

(a)

(b)

**FIGURE 7.28** Orbital diagrams for hydrogen and helium indicate the orbitals in which the electrons are found and the number of electrons in each orbital. The label on the orbital, 1*s*, indicates the value of the principal quantum number for the orbital (*n* = 1) and the type of orbital (*s*). (a) Hydrogen atoms have one electron in the 1*s* orbital, whereas (b) helium atoms have two.

Lithium ⬆⬇ ⬆ = [He] ⬆ ←Valence electron
      1*s*  2*s*        2*s* ←Valence shell
                    ↑
              Core electrons

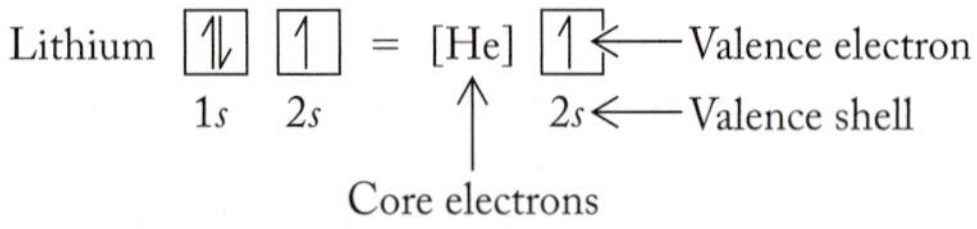

**FIGURE 7.29** An orbital diagram for lithium shows the two electrons in the 1*s* orbital with a single electron in the 2*s* orbital, corresponding to a ground-state electron configuration of $1s^2 2s^1$. Since the fully filled 1*s* orbital is equivalent to the ground-state electron configuration of He, we can write a condensed electron configuration for lithium as [He]2$s^1$ where the 2*s* orbital is called the valence orbital and the electron in the 2*s* orbital is the valence electron. The symbol [He] represents the Li atom's core electrons.

as we move through the table element by element. This method is based on the **aufbau principle** (German *aufbauen*, "to build up"), which states that the most stable atomic structures are those in which the electrons are in the lowest-energy orbitals available.

To decide which orbitals should contain electrons, we start with two rules:

1. Electrons always go into the lowest-energy orbital available.
2. Each orbital has a maximum occupancy of two electrons.

Using these rules, let's assign the single electron in a hydrogen (*Z* = 1) atom to the appropriate orbital. Actually, we have already discussed how the single electron in a ground-state atom of hydrogen is in the 1*s* orbital. We represent this arrangement with the **electron configuration** 1$s^1$, where the first "1" indicates the principal quantum number (*n*) of the orbital, "*s*" indicates the type of orbital, and the superscript "1" indicates that there is *one* electron in this 1*s* orbital. Because the hydrogen atom has only one electron, 1$s^1$ is the complete electron configuration for a hydrogen atom in its ground state (Figure 7.28).

To be unambiguous about the electron configuration, we use **orbital diagrams** to show how electrons, represented by single-headed arrows, are distributed among orbitals, represented by boxes. A single-headed arrow pointing upward represents an electron with spin up ($m_s = +\frac{1}{2}$), and a downward-pointing single-headed arrow represents an electron with spin down ($m_s = -\frac{1}{2}$).

The atomic number of helium is 2, which tells us there are two protons in the nucleus and two electrons in the neutral atom. Using the aufbau principle, we simply add another electron to the 1*s* orbital. This orbital contains only one electron, so it has space for one more. The spin quantum numbers for the two electrons cannot be the same. One must be $+\frac{1}{2}$ and the other must be $-\frac{1}{2}$. These two electrons are said to be *spin-paired*. Their presence gives helium a ground-state electron configuration of 1$s^2$. With two electrons, the 1*s* orbital is filled to capacity and so is the *n* = 1 shell (Figure 7.28b).

The concept of a *filled shell* is key to understanding chemical properties: elements composed of atoms that have filled *s* and *p* subshells in their outermost shells are chemically stable and generally unreactive. Helium is such an element, as are all the other elements in group 18.

The location of lithium (*Z* = 3) in the periodic table—the first element in the second period—is a signal that an atom of lithium has one electron in its *n* = 2 shell. The row numbers in the periodic table correspond to the *n* values of the outermost shells of the atoms in the rows. The second shell has four orbitals (one 2*s* and three 2*p*), so it can hold up to eight electrons. Which of the four orbitals in the second shell contains the third electron in a lithium atom? The answer is linked to the fact that an electron in a 2*s* orbital has a lower average energy than an electron in a 2*p* orbital. Therefore, the third electron in a lithium atom occupies the 2*s* orbital, making the electron configuration $1s^2 2s^1$ (Figure 7.29).

We can simplify the electron configuration of Li and all the elements that follow it in the periodic table. The simplified form is called a *condensed electron configuration*. In this form the symbols representing all of the electrons in orbitals that were filled in the rows above the element of interest are replaced by the symbol of the group 18 element at the end of the row above the element. For example, the condensed electron configuration of Li is [He]2$s^1$. Condensed electron configurations are useful because they eliminate the symbols of the **core electrons** in filled shells and subshells that are not involved in the chemistry of an element. They include only the symbols of the electrons in the outermost shells and subshells that are involved in bond formation. These electrons are called **valence electrons**.

In any atom, the shell containing the valence electrons is referred to as the **valence shell**. Notice that lithium has a single electron in its valence-shell *s* orbital, as does hydrogen, the element directly above it in the periodic table; therefore both atoms have the valence-shell configuration $ns^1$. Here *n* represents both the number of the row in which the atom is located on the table and the principal quantum number of the valence shell in the atom.

Beryllium ($Z = 4$) is the fourth element in the periodic table and the first in group 2. The configuration for its four electrons is $1s^2 2s^2$ or $[He]2s^2$. The other elements in group 2 also have two spin-paired electrons in the *s* orbital of their outermost occupied shell. The second shell is not full at this point because it also has three *p* orbitals, which are all empty but fill as we move to the next elements in the periodic table.

Boron ($Z = 5$) is the first element in group 13. Its fifth electron is in one of its three $2p$ orbitals, resulting in the condensed electron configuration $[He]2s^2 2p^1$. Which of the three $2p$ orbitals contains the fifth electron is not important because these three orbitals all have the same energy; we say they are **degenerate**.

The next element is carbon ($Z = 6$). It has four electrons in its valence shell (the $n = 2$ shell), so its condensed electron configuration is $[He]2s^2 2p^2$. Note that there are two electrons in the $2p$ orbitals. Are they both in the same orbital? Remember that all electrons have a negative charge and repel one another. Thus they tend to occupy orbitals that allow them to be as far away from each other as possible, which means the two $2p$ electrons in carbon occupy separate $2p$ orbitals. This distribution pattern is an application of **Hund's rule**, which states that, for degenerate orbitals, like the three $2p$ orbitals in the second shell, the lowest-energy electron configuration is the one with the maximum number of unpaired valence electrons, all of which have the same spin.

Hund's rule is the third rule we apply when determining electron configurations. By convention, the first electron placed in an orbital has a positive spin. When electrons fill a degenerate set of orbitals, Hund's rule requires that an electron with a positive spin occupies each valence orbital before any electron with a negative spin enters any orbital. Electrons do not pair in degenerate orbitals until they have to. To obey Hund's rule, the orbital diagram for carbon must be

Carbon:  ⇅   ⇅   ↑|↑|
         1s  2s    2p

which shows that the two $2p$ electrons are unpaired and have the same spin.

The next element is nitrogen ($Z = 7$), represented by either $1s^2 2s^2 2p^3$ or $[He]2s^2 2p^3$. According to Hund's rule, the third $2p$ electron resides alone in the third $2p$ orbital, so that the electron distribution is

Nitrogen:  ⇅   ⇅   ↑|↑|↑
           1s  2s    2p

As we proceed across the second row to neon ($Z = 10$), we fill the $2p$ orbitals as shown in Figure 7.30. The last three $2p$ electrons added (in oxygen, fluorine, and neon) pair with the first three, so that in neon, the three $2p$ orbitals are all filled to capacity. At this point the $n = 2$ shell is completely filled. Note that neon is directly below helium in group 18. Helium, neon, and all the other noble gases in group 18 have filled *s* and *p* orbitals in their valence shells. This same trend is seen throughout the periodic table: *elements in the same column of the periodic table have the same valence-shell configuration.* Argon, krypton, xenon, and radon also

**aufbau principle** the method of building electron configurations of atoms by adding one electron at a time as atomic number increases across the rows of the periodic table; each electron goes into the lowest-energy orbital available.

**electron configuration** the distribution of electrons among the orbitals of an atom or ion.

**orbital diagram** depiction of the arrangement of electrons in an atom or ion using boxes to represent orbitals.

**core electrons** electrons in the filled, inner shells in an atom or ion that are not involved in chemical reactions.

**valence electrons** electrons in the outermost occupied shell of an atom having the most influence on the atom's chemical behavior.

**valence shell** the shell in an atom containing the valence electrons.

**degenerate** describes orbitals of the same energy.

**Hund's rule** the lowest-energy electron configuration of an atom has the maximum number of unpaired electrons, all of which have the same spin, in degenerate orbitals.

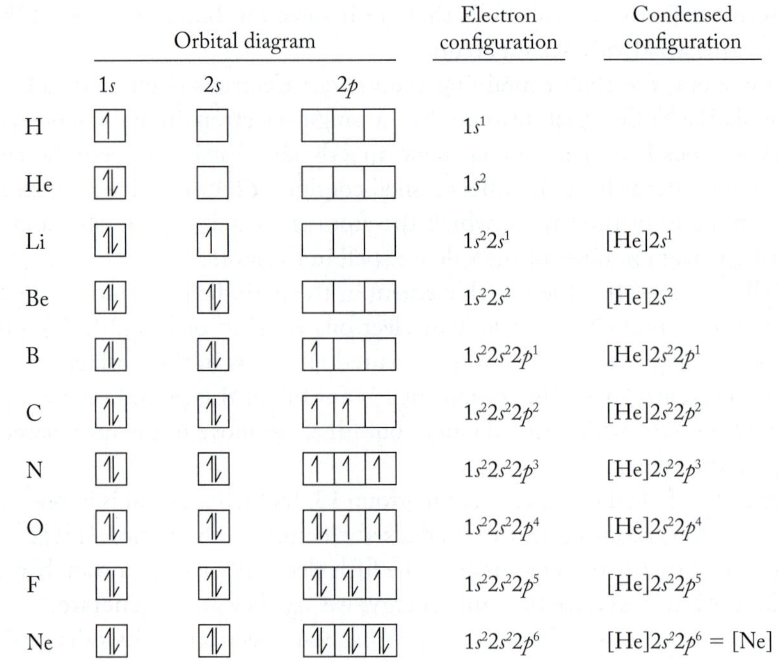

**FIGURE 7.30** Orbital diagrams and condensed electron configurations for the first ten elements show that each orbital (indicated by a square in the orbital diagrams) holds a maximum of two electrons and the two electrons must be of opposite spin. The orbitals are filled in order of increasing quantum numbers $n$ and $\ell$. Condensed electron configurations for all elements are given in Appendix 3.

| | Orbital diagram | | | Electron configuration | Condensed configuration |
|---|---|---|---|---|---|
| | $1s$ | $2s$ | $2p$ | | |
| H | ↑ | | | $1s^1$ | |
| He | ↑↓ | | | $1s^2$ | |
| Li | ↑↓ | ↑ | | $1s^2 2s^1$ | $[\text{He}]2s^1$ |
| Be | ↑↓ | ↑↓ | | $1s^2 2s^2$ | $[\text{He}]2s^2$ |
| B | ↑↓ | ↑↓ | ↑ | $1s^2 2s^2 2p^1$ | $[\text{He}]2s^2 2p^1$ |
| C | ↑↓ | ↑↓ | ↑ ↑ | $1s^2 2s^2 2p^2$ | $[\text{He}]2s^2 2p^2$ |
| N | ↑↓ | ↑↓ | ↑ ↑ ↑ | $1s^2 2s^2 2p^3$ | $[\text{He}]2s^2 2p^3$ |
| O | ↑↓ | ↑↓ | ↑↓ ↑ ↑ | $1s^2 2s^2 2p^4$ | $[\text{He}]2s^2 2p^4$ |
| F | ↑↓ | ↑↓ | ↑↓ ↑↓ ↑ | $1s^2 2s^2 2p^5$ | $[\text{He}]2s^2 2p^5$ |
| Ne | ↑↓ | ↑↓ | ↑↓ ↑↓ ↑↓ | $1s^2 2s^2 2p^6$ | $[\text{He}]2s^2 2p^6 = [\text{Ne}]$ |

have filled $s$ and $p$ orbitals in their outermost occupied shells, and they also are chemically inert gases at room temperature. Thus chemical inertness is associated with filled $s$ and $p$ orbitals in the valence shell.

Sodium ($Z = 11$) follows neon in the periodic table. It is the third element in group 1 and the first element in the third period. Ten of its electrons are distributed as in neon. The 11th electron is in the lowest-energy orbital available after $2p$ has been filled, which is $3s$. The condensed electron configuration of Na is $[\text{Ne}]3s^1$. Just as we write condensed electron configurations that provide detailed information for only the outermost occupied shell, we can condense orbital diagrams in this same way, so that the condensed orbital diagram for sodium is

$$\text{Sodium:} \quad [\text{Ne}] \underset{3s}{\boxed{\uparrow}}$$

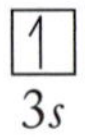

This diagram reinforces the message that the electron configuration of a sodium atom consists of a neon core plus a single electron in the $3s$ orbital of the outermost occupied shell. The sodium atom has the same generic valence-shell configuration as lithium and hydrogen, namely $ns^1$, where $n$ is the period number. This pattern for elements in the same group continues throughout the periodic table. The electron configuration of magnesium ($Z = 12$), for instance, is $[\text{Ne}]3s^2$, and the electron configuration of every other element in group 2 consists of the immediately preceding noble gas core followed by $ns^2$.

The next six elements in the periodic table—aluminum, $[\text{Ne}]3s^2 3p^1$, to argon, $[\text{Ne}]3s^2 3p^6$—show a pattern of increasing numbers of $3p$ electrons, a trend that continues until all three $3p$ orbitals are filled (six electrons), which means the $s$ and $p$ orbitals of the $n = 3$ shell are filled (eight electrons). Thus, argon is chemically inert, as predicted by its position in group 18.

After argon comes potassium ($Z = 19$) in group 1 of row 4 ($[\text{Ar}]4s^1$), followed by calcium ($Z = 20$) in group 2 ($[\text{Ar}]4s^2$). At this point, the $4s$ orbital is filled but the $3d$ orbitals are still empty. Why were the $3d$ orbitals not filled before 4s? (Review Table 7.1 if you need help in recalling that the $n = 3$ shell contains a *d* subshell.)

Applying the first aufbau rule—in building atoms, each electron goes in the lowest-energy orbital available—would be straightforward were it not for the fact that the differences in energy between shells get smaller as *n* gets larger (Figure 7.31). These smaller differences result in orbitals with large $\ell$ values in one shell having energies similar to orbitals with small $\ell$ values in the next higher shell. Note in Figure 7.31 that the energy of the 4s orbital is slightly lower than that of the 3d orbitals. The 4s orbitals in potassium and calcium are the lowest-energy orbitals available and are filled before any electrons go into a 3d orbital.

The element after calcium is scandium ($Z = 21$). It is the first element in the central region of the periodic table, the region populated by transition metals. Scandium has the condensed electron configuration [Ar]$3d^14s^2$. Note that the orbitals are arranged in order of increasing principal quantum number, not necessarily in the order in which they were filled. The 3d orbitals are filled in the transition metals from scandium to zinc ($Z = 30$). This pattern of filling the *d* orbitals of the shell whose principal quantum number is 1 less than the period number, $(n - 1)d$, is followed throughout the periodic table: the 4d orbitals are filled in the transition metals of the fifth period, and so on (Figure 7.31).

The element after scandium is titanium ($Z = 22$), which has one more *d* electron than scandium, so its condensed electron configuration is [Ar]$3d^24s^2$. At this point, you may feel you can accurately predict the electron configurations of the remaining transition metals in the fourth period. However, because the energies of the 3d and 4s orbitals are similar, the sequence of *d*-orbital filling deviates in two spots from the pattern you might expect. Vanadium ($Z = 23$) has the expected configuration [Ar]$3d^34s^2$, but the next element, chromium ($Z = 24$), has the configuration [Ar]$3d^54s^1$:

Chromium:     [Ar]
                   $3d^5$      $4s^1$

This half-filled set of *d* orbitals is an energetically favored configuration. Apparently, the stability of having five half-filled 3d orbitals compensates for the energy needed to raise a 4s electron to a 3d orbital. As a result, [Ar]$3d^54s^1$ is a more stable electron configuration than [Ar]$3d^44s^2$.

Another deviation from the expected filling pattern is observed near the end of a row of transition metals. Copper ($Z = 29$) has the electron configuration [Ar]$3d^{10}4s^1$ instead of the expected [Ar]$3d^94s^2$ because a completely filled set of *d* orbitals also represents a stable electron configuration.

The periodic table is a useful reference for writing the electron configurations of the ground states of atoms. The version of the periodic table in Figure 7.32 is especially helpful because the color patterns and labels indicate which type of orbital is being filled as we move across each row from left to right. For example, groups 1 and 2 are called *s* block elements because their outermost electrons are in *s* orbitals. Similarly, groups 13 through 18 (except for helium) are called the *p* block elements because their outermost electrons are in *p* orbitals. The principal quantum numbers (*n*) of the outermost electrons in the *s* and *p* blocks match their row numbers. For example, barium is the group 2 element in the sixth row. This location means that a ground-state Ba atom has 2 electrons in its 6s orbital, so its condensed electron configuration is

Ba:     [Xe]$6s^2$

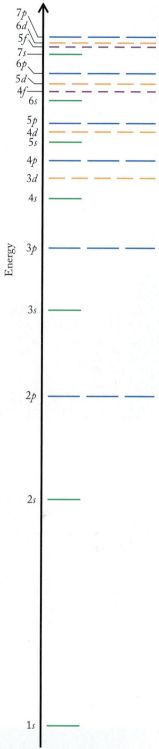

**FIGURE 7.31** The energy levels in multielectron atoms increase with increasing values of *n* and with increasing values of $\ell$ within a shell. The difference in energy between adjacent shells decreases with increasing values of *n*, which may cause the energies of subshells in two adjacent shells to overlap. For example, electrons in 3d orbitals have slightly higher energy than those in the 4s orbital, resulting in the order of subshell filling $4s \rightarrow 3d \rightarrow 4p$.

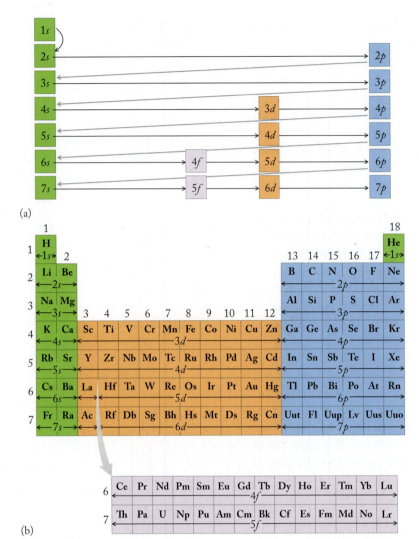

(a)

(b)

**FIGURE 7.32** (a) This diagram shows the sequence in which atomic orbitals fill. (b) The same color coding in this version of the periodic table highlights the four "blocks" of elements in which valence shell $s$ (green), $p$ (blue), $d$ (orange), $f$ (purple) orbitals are filled as atomic number increases across a row of the table.

Between the $s$ and $p$ blocks are the transition metals in groups 3 through 12, which make up the $d$ block, and the two rows of elements at the bottom of the periodic table, called the lanthanides and actinides, which make up the $f$ block. Both of the $f$ block series are 14 elements long.

The $n$ value of the $d$ orbitals being filled in a row is always one less than the row numbers and the $n$ value of the $f$ orbitals being filled is two less than the row number. With these rules in mind, let's write the condensed electron configuration of lead ($Z = 82$), which is the group 14 element in the sixth row. The nearest noble gas above it is Xe ($Z = 54$). The difference in atomic numbers means that we need to account for $82 - 54 = 28$ electrons in the electron configuration symbols. The location of Pb and the block labels in Figure 7.32 tell us that these 28 electrons are distributed as follows:

2 electrons in $6s$

14 electrons in $4f$

10 electrons in $5d$

2 electrons in $6p$

The electron configuration of ground-state lead atoms reflects this distribution:

Pb:     $[Xe]4f^{14}5d^{10}6s^26p^2$

---

**SAMPLE EXERCISE 7.10** **Writing Electron Configurations of Main Group Atoms** **LO6**

Strontium salts give off a bright red light at the high temperature of fireworks, explosions, and signal flares (Figure 7.33). What is the ground-state electron configuration of strontium atoms?

**Collect and Organize** We are asked to write the electron configuration of strontium, which is in the fifth row of group 2 of the periodic table.

**Analyze** The atomic number of strontium, $Z = 38$, tells us that a Sr atom has 38 electrons. Applying the first aufbau rule, each electron goes in the lowest-energy orbital available, starting with the $1s$ orbital.

**Solve** Figures 7.31 and 7.32 tell us that the orbitals fill in the order: $1s$, $2s$, $2p$, $3s$, $3p$, $4s$, $3d$, $4p$, and $5s$. The maximum occupancy of $s$ orbitals = 2 electrons, $p$ orbitals = 6 electrons and $d$ orbitals = 10 electrons. Therefore the distribution of the 38 electrons is: $1s^22s^22p^63s^23p^63d^{10}4s^24p^65s^2$.

**Think About It** The location of Sr in Figure 7.32 tells us that the condensed electron configuration of Sr is $[Kr]5s^2$.

⊙ **Practice Exercise** Gallium arsenide, GaAs, is used in the red lasers in bar-code readers. Write the electron configuration of a ground-state atom of gallium ($Z = 31$) and a ground-state arsenic atom ($Z = 33$).

**FIGURE 7.33** Strontium nitrate is a common ingredient in road flares that give off red light.

**SAMPLE EXERCISE 7.11** **Writing Condensed Electron Configurations of Transition Metal Atoms** **LO6**

Write the condensed electron configuration of an atom of silver ($Z = 47$).

**Collect and Organize** In a condensed electron configuration, the filled sets of orbitals in the inner shells of the atom are represented by the atomic symbol of the noble gas immediately preceding the element of interest in the periodic table.

**Analyze** Silver ($Z = 47$) is the group 11 element in the fifth row of the periodic table; krypton ($Z = 36$) is the immediately preceding noble gas at the end of the fourth row. The difference between these atomic numbers means that we need to account for $47 - 36 = 11$ electrons.

**Solve** We initially predict that the first two of the 11 electrons would be in the $5s$ orbital and the next nine would be in $4d$ orbitals, resulting in a condensed electron configuration of $[Kr]4d^9 5s^2$. However, a completely filled set of $d$ orbitals is more stable than a partially filled set, so silver, like copper just above it in the periodic table, has ten electrons in its occupied $d$ orbitals. Therefore it has only one electron in its outermost occupied shell, which is the $n = 5$ shell: $[Kr]4d^{10}5s^1$.

**Think About It** We can generate a tentative electron configuration by simply moving across a row in Figure 7.32 until we come to the element of interest. However, in the transition metals, we have to remember the special stability of half-filled and filled $d$ orbitals and make the appropriate adjustments in our configuration.

⚙ **Practice Exercise** Write the condensed electron configuration of a ground-state atom of cobalt ($Z = 27$).

# 7.9 Electron Configurations of Ions

To write the electron configuration of an ion, we begin with the electron configuration of the atom from which the ion was formed. If the ion has a positive charge, we remove the appropriate number of electrons from the orbital(s) with the highest principal quantum number. If the ion has a negative charge, we add the appropriate number of electrons to one or more partially filled outer-shell orbitals.

## Ions of the Main Group Elements

The $s$ block elements (see Figure 7.32) form monatomic cations by losing all their outer-shell electrons, leaving their ions with the electron configuration of the noble gas immediately preceding them in the periodic table. For example, an atom of sodium forms a $Na^+$ ion by losing its single $3s$ electron:

$$Na: \quad [Ne]3s^1$$
$$Na^+: \quad [Ne]$$

Nonmetallic elements of the $p$ block that form monatomic anions do so by gaining enough electrons to completely fill their outer-shell $p$ orbitals, forming ions with the electron configurations of the noble gases at the right ends of their rows in the periodic table. For example, an atom of fluorine forms a $F^-$ ion by gaining one electron, which completely fills its set of three $2p$ orbitals and gives it the electron configuration of neon:

$$F: \quad [He]2s^2 2p^5$$
$$F^-: \quad [He]2s^2 2p^6 = [Ne]$$

Thus, a $Na^+$ ion and a $F^-$ ion both have the same electron configuration as an atom of Ne. We say that these three species, $Na^+$, $F^-$, and Ne, are **isoelectronic**, meaning they have the same electron configuration.

⊙⊙ **CONNECTION** In previous chapters, we learned the charges on the ions of common elements. Electron configurations help us understand why these ions have the charges they do.

**isoelectronic** describes atoms or ions that have identical electron configurations.

**SAMPLE EXERCISE 7.12 Determining Isoelectronic Species in Main Group Ions** LO6

(a) Write the electron configurations of the ions in CsF, $MgCl_2$, CaO, and KBr. (b) Which ions in part a are isoelectronic with neon atoms?

**Collect and Organize** In part a, we must determine the electron configuration of each ion in four binary ionic compounds. In part b, our task is to compare the configurations from part a with the configuration for a neon atom and determine which have the same electron configuration.

**Analyze** The elements in the compounds include

- two from group 1, Cs and K, which are present as 1+ cations;
- two from group 2, Mg and Ca, which are present as 2+ cations;
- one from group 16, O, which is present as a 2− anion;
- three from group 17, F, Cl, and Br, which are present as 1− anions.

Let's arrange these atoms and ions in a table, remembering that an atom loses valence electrons in becoming a cation and gains valence electrons in becoming an anion:

| Element | Electron Configuration of Atom | Atomic Number (*Z*) | Formula of Ion | Electrons per Ion |
|---|---|---|---|---|
| Cs | $[Xe]6s^1$ | 55 | $Cs^+$ | 54 |
| K | $[Ar]4s^1$ | 19 | $K^+$ | 18 |
| Mg | $[Ne]3s^2$ | 12 | $Mg^{2+}$ | 10 |
| Ca | $[Ar]4s^2$ | 20 | $Ca^{2+}$ | 18 |
| O | $[He]2s^22p^4$ | 8 | $O^{2-}$ | 10 |
| F | $[He]2s^22p^5$ | 9 | $F^-$ | 10 |
| Cl | $[Ne]3s^23p^5$ | 17 | $Cl^-$ | 18 |
| Br | $[Ar]3d^{10}4s^24p^5$ | 35 | $Br^-$ | 36 |

**Solve**

a. The electron configurations for the eight ions are

$Cs^+$: [Xe]

$K^+$: [Ar]

$Mg^{2+}$: [Ne]

$Ca^{2+}$: [Ar]

$O^{2-}$: $[He]2s^22p^6 = [Ne]$

$F^-$: $[He]2s^22p^6 = [Ne]$

$Cl^-$: $[Ne]3s^23p^6 = [Ar]$

$Br^-$: $[Ar]3d^{10}4s^24p^6 = [Kr]$

b. Three of the ions formed—$Mg^{2+}$, $O^{2-}$, and $F^-$—have the same number of electrons as an atom of neon (10) and are isoelectronic with Ne and with one another.

**Think About It** Each of the ions has an electron configuration of a noble gas atom. This configuration is associated with stable ions of many main group elements.

**Practice Exercise** Write the electron configurations of the ions in KI, BaO, $Rb_2O$, and $AlCl_3$. Which of these ions are isoelectronic with Ar?

## Transition Metal Cations

As with the main group elements, writing the electron configurations of transition metal cations begins with the atoms from which the cations form. Nickel atoms, like those of many transition metals, form ions with 2+ charges by losing both electrons from the shell of highest $n$, in this case the outer-shell $s$ electrons:

Ni: [Ar]$3d^84s^2$

Ni$^{2+}$: [Ar]$3d^8$

A few transition metals, including silver, have only one outer-shell $s$ electron in their atoms and typically form singly charged ions:

Ag: [Kr]$4d^{10}5s^1$

Ag$^+$: [Kr]$4d^{10}$

We might have expected Ni and Ag atoms to lose their slightly higher-energy $3d$ and $4d$ electrons, respectively, reasoning that the last orbitals to be filled should be the first to be emptied when an atom forms a positive ion. However, the rule that the electrons in orbitals with the highest $n$ value ionize first applies to these and the other transition metals. Preferential loss of outer-shell $s$ electrons explains why the most frequently encountered charge on transition metal ions is 2+.

Many transition metal atoms also lose one or more $d$ electrons as they form ions with charges ≥2+. An atom of scandium, [Ar]$3d^14s^2$, for example, loses both $4s$ electrons and its $3d$ electron as it forms a Sc$^{3+}$ ion. The chemistry of titanium, [Ar]$3d^24s^2$, is dominated by its atoms losing their $4s$ and $3d$ electrons to form Ti$^{4+}$ ions. In general, when we are determining electron configurations for transition metal ions, we first remove electrons from the outermost occupied $s$ orbital, then from the outermost occupied $d$ subshell, until the charge on the ion is achieved.

---

**SAMPLE EXERCISE 7.13** **Writing Electron Configurations of Transition Metal Ions** **LO6**

What are the electron configurations of Fe$^{2+}$ and Fe$^{3+}$?

**Collect and Organize** We are asked to write the electron configurations of two ions formed by iron ($Z = 26$). Iron is the group 8 element of the fourth row of the periodic table. Figure 7.32 provides information on the order in which orbitals are filled. Transition metal atoms preferentially lose their outermost $s$ electrons when they form ions.

**Analyze** The location of iron in the periodic table tells us that the atom has two $4s$ and six $3d$ electrons built on an argon core: the electron configuration of Fe is

Fe: [Ar]$3d^64s^2$

**Solve** We remove the two $4s$ electrons to form Fe$^{2+}$ and the two $4s$ and one of the $3d$ electrons to form Fe$^{3+}$:

Fe$^{2+}$: [Ar]$3d^6$

Fe$^{3+}$: [Ar]$3d^5$

**Think About It** It makes sense that Fe atoms form ions with a charge of 3+ because the loss of both $4s$ electrons and one $3d$ electron produces an ion with a stable, half-filled set of $3d$ orbitals.

⚙ **Practice Exercise** Write the electron configurations for the manganese atom and the ions Mn$^{3+}$ and Mn$^{4+}$.

---

We have yet to consider the lanthanides (elements 58 through 71) and actinides (elements 90 through 103), represented by the two purple rows in Figure 7.32. The lanthanides have partly filled $4f$ orbitals, and the actinides have partly filled $5f$ orbitals. There are 14 elements in each group, reflecting the capacity of the

**atomic radius** (also called **covalent radius**) half the distance between identical nuclear centers in a molecule.

**metallic radius** half the distance between nuclear centers in the crystal of a metal.

Bond length
198 pm

99 pm

(a)          Radius of Cl

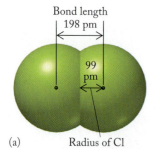

372 pm

186 pm

(b)          Metallic radius of Na

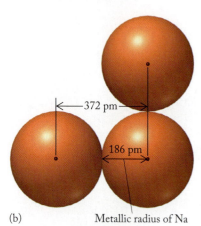

Na⁺     Cl⁻

$r_+$   $r_-$

102 pm   181 pm

(c) Ionic radii of $Na^+$ and $Cl^-$

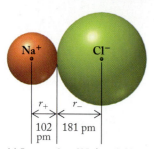

**FIGURE 7.34** A comparison of covalent, metallic, and ionic radii. (a) A covalent radius is half the length of the bond between identical atoms in a molecule, such as the bond in $Cl_2$. (b) A metallic radius is based on the distance of closest approach of adjacent atoms in a crystalline metal. (c) An ionic radius is determined by a series of comparisons among ionic compounds containing the ions of interest.

seven orbitals in each $f$ subshell ($\ell = 3$, $m_\ell = -3, -2, -1, 0, +1, +2$, and $+3$). As you can see in Figure 7.32, the $4f$ orbitals are not filled until after the $6s$ orbital has been filled. This order of filling is due to the similar energies of $6s$ and $4f$ orbitals (Figure 7.31). Similarly, the $5f$ orbitals are filled after the $7s$ orbital is filled.

The periodic table is a useful reference for predicting the physical and chemical properties of elements. Our quantum mechanical perspectives on atomic structure provide us with a theoretical basis for explaining why particular families of elements behave similarly. The original table was based on periodic trends in observable chemical properties. Now we know that the chemical properties of an element are closely linked to the electron configurations of the atoms of the element.

**CONCEPT TEST**

The electron configuration of Eu ($Z = 63$) is $[Xe]4f^76s^2$, but the electron configuration of Gd ($Z = 64$) is $[Xe]4f^75d^16s^2$. Suggest a reason why the additional electron in Gd is not in a $4f$ orbital, which would result in the electron configuration $[Xe]4f^86s^2$.

## 7.10 The Sizes of Atoms and Ions

The sizes of the atoms or ions making up any substance influence both its physical and chemical properties. As the Heisenberg uncertainty principle describes, we cannot know the exact location of electrons in ions and atoms; hence the boundary of an ion or atom cannot really be defined. Therefore we must define some convention for how atomic and ionic sizes are measured.

We calculate the sizes of atoms from the distances between nuclei bonded together in molecules. For elements such as $N_2$, $O_2$, and the halogens that exist as diatomic molecules, the **atomic radius**—also referred to as a **covalent radius**—is simply half the length of the covalent bond in these molecules (Figure 7.34a). For metals, the **metallic radius** is defined as half the distance between the nuclear centers in a crystal of the metal (Figure 7.34b). Based on these definitions, the relative sizes of atoms, along with their locations in the periodic table, are shown in Figure 7.35. The values of **ionic radii** are derived from the distances between nuclear centers in ionic crystals (Figure 7.34c).

Atomic size varies from top to bottom in a given group and from left to right across a given period. Two opposing factors contribute to these variations. The first is the principal quantum number $n$, which determines the most probable distance of the electrons from the nucleus. As $n$ increases, the probability that the electrons are farther from the nucleus increases. The second factor is the nuclear charge, the positive charge of the protons in the nucleus, which determines the attractive force holding the electrons in the atom. As nuclear charge increases, the positive charge felt by the electrons increases, and the electrons are pulled closer to the nucleus. In multielectron atoms, however, each electron is also repelled by all the other electrons. To determine the net force felt by any electron in an atom, we must consider both the attraction between the nucleus and the electron and the electron–electron repulsions.

Let's look at electron–nucleus attraction first. Consider an electron in a $2s$ orbital. The smaller peak on the $2s$ curve in Figure 7.36 shows that the $2s$ orbital has some electron distribution close to the nucleus. This is an example of **orbital penetration**, which occurs when an electron that resides mainly in an outer orbital has some probability of being close to the nucleus. Orbital penetration is important when outer-shell electrons are separated from the nucleus by one or more filled inner shells. Electrons in orbitals that penetrate closer to the nucleus experience more of the positive charge of the nucleus. Thus they are attracted more strongly to the nucleus than are electrons in orbitals that do not penetrate as much.

The absence of any secondary peak in the $2p$ curve of Figure 7.36 shows that electrons in a $2p$ orbital penetrate less effectively than those in a $2s$ orbital. Penetration by electrons in shells farther from the nucleus decreases in the orbitals in a given shell according to the order $s > p > d > f$. Since greater penetration means lower energy, the energies of the orbitals in a given shell are $s < p < d < f$. Because the electrons in the ground state of a multielectron atom occupy the lowest-energy orbitals available, the order of filling orbitals in a given shell, as we see in Figure 7.32, is $s$ first, then (if available) $p$, then $d$, then $f$.

We must also consider electron–electron repulsions in a multielectron atom. For example, an electron in a $3d$ orbital is repelled by every electron that lies between it and the nucleus. Because this repulsive effect cancels some of the electron–nuclear attraction, we can think of the electrons between a $3d$ electron and the nucleus as **shielding**, or **screening**, an electron from experiencing the full charge of the

**ionic radius** radius derived from the distance between nuclear centers in ionic crystals.

**orbital penetration** the probability that an electron in an outer orbital will be as close to the nucleus as an electron in an inner shell.

**shielding** (also called **screening**) the effect when inner-shell electrons prevent outer-shell electrons from experiencing the total nuclear charge.

**FIGURE 7.35** Atomic radii in picometers of the main group elements. Size generally increases from top to bottom in any group and generally decreases from left to right across any period.

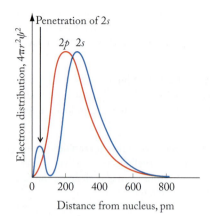

**FIGURE 7.36** Because the main peak of the 2s curve (blue) is farther from the nucleus than the 2p peak (red), the 2s orbital appears to be farther from the nucleus than the 2p orbital. However, the 2s orbital is of lower energy because electrons in it penetrate more closely to the nucleus, indicated by the smaller 2s peak at about 50 pm from the nucleus. The result is that 2s electrons experience a greater effective nuclear charge and have lower energy than 2p electrons.

nucleus. From the point of view of a 3d electron, the nuclear charge is reduced from its actual value to that of an **effective nuclear charge ($Z_{eff}$)**.

In general, the $Z_{eff}$ experienced by outer-shell electrons increases if the orbitals occupied by the electrons penetrate the lower-lying orbitals. For example, a 3s electron that penetrates close to the nucleus experiences a higher $Z_{eff}$ than does a 3d electron that has little penetration of inner-shell orbitals. Trends in properties that depend on the strength of attraction between electrons and nuclei are determined by how effectively electrons are shielded from nuclear charge.

**CONCEPT TEST**

Rank the following orbitals in an atom of silver in order of decreasing nuclear charge experienced by the electrons in them: (a) 1s; (b) 2s; (c) 3s; (d) 4s; (e) 2p; (f) 3p; (g) 4p.

## Trends in Atomic and Ionic Sizes

As you move down a group in the periodic table, the principal quantum number $n$ increases, and as $n$ increases, the probability that the outermost electrons are farther from the nucleus increases. As a result, atomic radii usually increase. However, as you move left to right across a row of main group elements or transition metals, the charge on the nucleus increases, but $n$ stays the same. As a result, $Z_{eff}$ increases. Increasing $Z_{eff}$ means that atomic size generally decreases with increasing atomic number across a row. The major deviation from this trend is seen in the transition metals, where size decreases for the first two elements but then decreases so gradually that it is virtually constant until the last element in the series. The $d$ electrons that fill in the transition series enter an inner shell ($n - 1$) and thus do not cause the size to change significantly.

The cations of the main group elements are much smaller than their parent atoms, but the anions are much larger (Figure 7.37). To understand these opposite trends, consider what happens when a Na atom forms a Na$^+$ ion: it loses its entire valence shell, leaving behind a much smaller neon core of electrons. On the other hand, when a Cl atom acquires an electron and forms a Cl$^-$ ion, it contains more electrons than protons; hence, the attractive force per electron decreases and the electron–electron repulsion increases. Anions are thus always larger than the atoms from which they form.

The trends in sizes for cations and anions are similar to the trends for atoms for much the same reasons. The sizes of cations and anions increase moving down a group and decrease moving across a row from left to right. Sizes of isoelectronic ions decrease with increasing atomic number or with increasing positive charge on the ion.

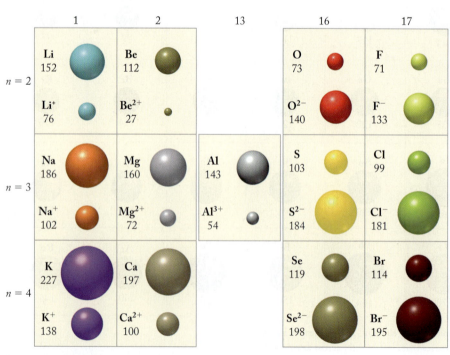

**FIGURE 7.37** Comparison of atomic and ionic radii, in picometers.

**effective nuclear charge ($Z_{eff}$)** the attractive force toward the nucleus experienced by an electron in an atom; the positive charge on the nucleus reduced by the extent to which other electrons in the atom shield the electron from the nucleus.

**ionization energy (IE)** the amount of energy needed to remove 1 mole of electrons from 1 mole of ground-state atoms or ions in the gas phase.

**SAMPLE EXERCISE 7.14**    **Ordering Atoms and Ions by Size**    **LO8**

Arrange each set by size, largest to smallest: (a) O, P, S; (b) $Na^+$, Na, K.

**Collect and Organize** We are to rank a set of three atoms based on their size and a set of two atoms and a cation of one of the atoms based on their size. The location of elements in the periodic table can be used to determine the relative sizes of their atoms.

**Analyze** Sizes decrease as we move left to right across a period and increase as we move down a column. In addition, a cation is always smaller than the atom from which it is made.

**Solve**

a. S is below O in group 16, so in terms of atomic size, S > O. S is to the right of P, so P > S. The size order is thus P > S > O.
b. Cations are smaller than their atoms, so Na > $Na^+$. Size increases down a group; K is below Na in the alkali metals group, so K > Na. Therefore, the size order is K > Na > $Na^+$.

**Think About It** The trend in the sizes of the atoms in set a reflects decreasing atomic size with increasing nuclear charge within a row of elements in the periodic table and increasing atomic size with increasing atomic number down a group. The relative sizes of the particles in set b are linked (1) to increasing atomic size as one goes down a column of elements in the periodic table and (2) to the smaller size of a cation relative to its parent atom.

**Practice Exercise** Arrange each set in order of increasing size (smallest to largest): (a) $Cl^-$, $F^-$, $Li^+$; (b) $P^{3-}$, $Al^{3+}$, $Mg^{2+}$.

# 7.11 Ionization Energies

In developing electron configurations, we followed a theoretical framework for the arrangement of electrons in orbitals that was developed in the early years of the 20th century. Is there actual experimental evidence for the existence of orbitals representing different energy levels inside atoms? Yes, there is. The evidence includes the measurement of the energies needed to remove electrons from atoms and ions.

**Ionization energy (IE)** is the energy needed to remove 1 mole of electrons from 1 mole of gas-phase atoms or ions in their ground state. Removing these electrons always costs energy because a negatively charged electron is attracted to a positively charged nucleus, and energy is required to overcome that attractive force. The amount of energy needed to remove 1 mole of electrons from 1 mole of atoms to make 1 mole of cations with a 1+ charge is called the *first ionization energy* ($IE_1$), the energy needed to remove 1 mole of electrons from 1 mole of 1+ cations to make 1 mole of cations with a 2+ charge is the *second ionization energy* ($IE_2$), and so forth. For example,

$$Mg(g) \rightarrow Mg^+(g) + 1\ e^- \qquad (IE_1 = 738 \text{ kJ/mol})$$
$$Mg^+(g) \rightarrow Mg^{2+}(g) + 1\ e^- \qquad (IE_2 = 1451 \text{ kJ/mol})$$

The total energy required to make 1 mole of $Mg^{2+}(g)$ cations from 1 mole of Mg(g) atoms is the sum of these two ionization energies:

$$Mg(g) \rightarrow Mg^{2+}(g) + 2e^-$$

$$\text{Total IE} = (738 + 1451) \text{ kJ/mol} = 2189 \text{ kJ/mol}$$

Figure 7.38 shows how the first ionization energies vary in the main group elements. The $IE_1$ of hydrogen is 1312 kJ/mol. The $IE_1$ of helium is nearly twice as big: 2372 kJ/mol. This difference seems reasonable because He atoms have two

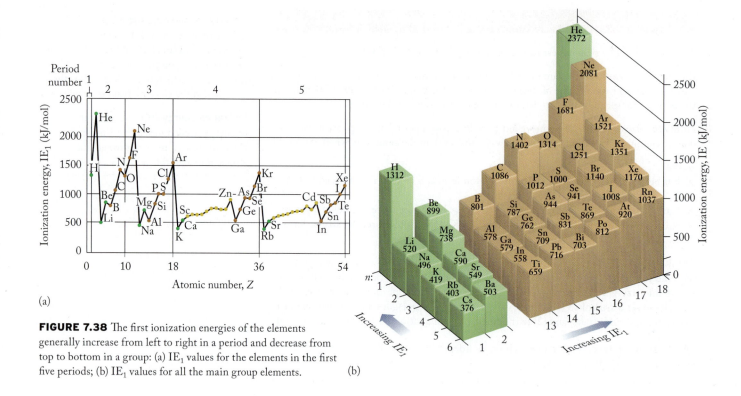

(a)

(b)

**FIGURE 7.38** The first ionization energies of the elements generally increase from left to right in a period and decrease from top to bottom in a group: (a) $IE_1$ values for the elements in the first five periods; (b) $IE_1$ values for all the main group elements.

protons per nucleus, whereas H atoms have only one. Twice the nuclear charge leads to twice the nuclear attraction, so twice the ionization energy is needed to pull an electron away from the nucleus. In general, first ionization energies increase from left to right across a period. Thus the easiest element to ionize in a period is the group 1 element and the hardest is the group 18 element. This pattern makes sense because the charge of the nucleus increases across a row and so does the attraction between the nucleus and the surrounding electrons.

Two anomalies occur in the general trend of increasing $IE_1$ with increasing $Z$ across a row. One shows up between the group 2 and group 13 elements that are next to each other in the second and third rows. The $IE_1$ decreases between Be and B and between Mg and Al. The reason is that B and Al lose a $p$ electron when they ionize, whereas Be and Mg must lose an $s$ electron. It is easier to remove a $p$ electron because it experiences less effective nuclear charge than an $s$ electron in the same shell.

Another anomaly occurs in the values of $IE_1$ between the group 15 and group 16 elements. The decrease here is associated with the enhanced stability that comes from having a half-filled set of $p$ orbitals. We observed in Section 7.9 a similar stability in a half-filled set of $d$ orbitals. For example, when an oxygen atom loses a $2p$ electron, it achieves the slightly more stable configuration of a half-filled set of $2p$ orbitals:

$$\boxed{\uparrow\downarrow}\ \boxed{\uparrow\downarrow}\ \boxed{\uparrow\downarrow\ \uparrow\ \uparrow} \quad \rightarrow \quad \boxed{\uparrow\downarrow}\ \boxed{\uparrow\downarrow}\ \boxed{\uparrow\ \uparrow\ \uparrow}\ +\ e^-$$

$\quad 1s\quad 2s\qquad 2p \qquad\qquad\quad 1s\quad 2s\qquad 2p$

$\qquad\qquad O \qquad\qquad\qquad\qquad\qquad O^+$

The enhanced stability of the $O^+$ ion means that less energy is required than would be predicted by applying the general trend of increasing $IE_1$ with increasing nuclear charge to produce the ion from a neutral O atom.

| TABLE 7.2 | | Successive Ionization Energies$^a$ of the First 10 Elements | | | | | | | |
|---|---|---|---|---|---|---|---|---|---|
| Element | Z | IE$_1$ | IE$_2$ | IE$_3$ | IE$_4$ | IE$_5$ | IE$_6$ | IE$_7$ | IE$_8$ | IE$_9$ |
| H | 1 | 1312 | | | | | | | | |
| He | 2 | 2372 | 5249 | | | | | | | |
| Li | 3 | 520 | 7296 | 12,040 | | | | | | |
| Be | 4 | 900 | 1758 | 15,050 | 21,070 | | | | | |
| B | 5 | 801 | 2426 | 3660 | 24,682 | 32,508 | | | | |
| C | 6 | 1086 | 2348 | 4617 | 6201 | 37,926 | 46,956 | | | |
| N | 7 | 1402 | 2860 | 4581 | 7465 | 9391 | 52,976 | 64,414 | | |
| O | 8 | 1314 | 3383 | 5298 | 7465 | 10,956 | 13,304 | 71,036 | 84,280 | |
| F | 9 | 1681 | 3371 | 6020 | 8428 | 11,017 | 15,170 | 17,879 | 92,106 | 106,554 |
| Ne | 10 | 2081 | 3949 | 6140 | 9391 | 12,160 | 15,231 | 19,986 | 23,057 | 115,584 |

$^a$All values in kilojoules per mole of atoms.

Now let's see how ionization energy changes as we go down a group of the periodic table. We begin with group 1 and the IE$_1$ values of hydrogen and lithium. A lithium atom has three times the nuclear charge of a hydrogen atom, so we might expect its ionization energy to be about three times larger. However, its IE$_1$ value is only 520 kJ/mol, or less than half that of hydrogen, 1312 kJ/mol. Why? The quantum mechanical model of atomic structure enables us to explain the much smaller IE$_1$ of lithium: the valence electron in a lithium atom is in a 2s orbital, which means the distance between that electron and the lithium nucleus is greater than the distance between the 1s valence electron and the nucleus in hydrogen. The lithium 2s electron is also shielded from the nucleus by the two electrons in the 1s orbital. Thus the effective nuclear charge felt by the lithium valence electron is much less than the 3+ due to the three protons in the nucleus. In general, the combination of larger atomic size and more shielding by inner-shell electrons leads to decreasing IE$_1$ values from top to bottom in every group of the periodic table.

Another perspective on energy levels in atoms is provided by looking at successive ionization energies. Consider the trend in successive ionization energies for the first 10 elements, shown in Table 7.2. For multielectron atoms, the second ionization energy (IE$_2$) is always greater than the IE$_1$ because the second electron is being removed from an ion that already has a positive charge. Because electrons carry a negative charge, they are held more strongly in a cation than in the atom from which the cation was formed.

The energy needed to remove a third electron is greater still because it is being removed from a 2+ ion. Superimposed on this trend is a much more dramatic increase in ionization energy (defined by the red line in Table 7.2) when all the valence electrons in an ionizing atom have been removed and the next electron must come from an inner shell. A core electron experiences much less shielding and much more effective nuclear charge and therefore requires much more energy to be removed from an atom than does an outer-shell electron.

**SAMPLE EXERCISE 7.15   Ranking Ionization Energies    LO8**

Arrange argon, magnesium, and phosphorus in order of increasing first ionization energy.

**Collect and Organize** We are to order three elements on the basis of their IE$_1$ values. All three elements are in the third row of the periodic table.

**electron affinity (EA)** the energy change that occurs when 1 mole of electrons combines with 1 mole of atoms or ions in the gas phase.

**Analyze** First ionization energies generally increase from left to right across a row.

**Solve** Assuming that increasing ionization energy with increasing atomic number is the dominant factor, the elements in order of increasing first ionization energies should be:

$$Mg < P < Ar$$

**Think About It** Magnesium forms stable 2+ cations, so we expect its ionization energy to be smaller than the values for phosphorus and argon, which do not form cations. Argon is a noble gas with a stable valence-shell electron configuration, so its first ionization energy would logically be the highest in the set. We can check our prediction against Figure 7.38(b): Mg 738 kJ/mol, P 1012 kJ/mol, Ar 1521 kJ/mol.

⚙ **Practice Exercise** Arrange cesium, calcium, and neon in order of decreasing first ionization energy.

---

**CONCEPT TEST**

Why does ionization energy decrease as you move down the noble gas family from helium to neon to argon?

---

# 7.12 Electron Affinities

In the preceding section we examined the periodic nature of the energy required to ionize atoms. Now we look at a complementary process and examine the change in energy when electrons are added to atoms to form monatomic anions. The energies involved are called **electron affinities (EA)**. They are the energy changes that occur when 1 mole of electrons is added to 1 mole of atoms or ions in the gas phase.

For example, the energy associated with adding 1 mole of electrons to 1 mole of chlorine atoms in the gas phase is

$$Cl(g) + e^- \rightarrow Cl^-(g) \qquad EA_1 = -349 \text{ kJ/mol}$$

The electron affinities of many elements are negative, meaning the formation of anions with a 1− charge is an exothermic process (Figure 7.39). However, an examination of the EA values in Figure 7.39 reveals that the trends in this property are not as regular as the trends in size and ionization energy. Electron affinity increases down a column only for the group 1 metals; other groups do not display a clear trend. In general, electron affinity becomes more negative with increasing atomic number across a row, but there are exceptions to that trend, too. The halogens of group 17 have the most negative EA values, which seems logical because each halogen atom needs one more electron to achieve a noble gas electron configuration.

On the other hand, adding an electron to an atom of a noble gas is an endothermic process, which makes sense because their atoms have stable electron configurations already. Beryllium and magnesium have very small but positive electron affinities because the added electrons have to occupy outer-shell *p* orbitals, which are significantly higher in energy than the outer-shell *s* orbitals. We can rationalize the positive electron affinity of nitrogen by noting that adding an

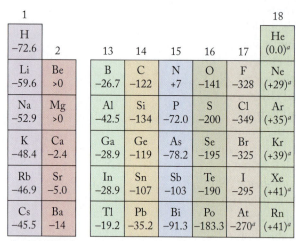

FIGURE 7.39 Electron affinity (EA) values of main group elements are expressed in kilojoules per mole. The more negative the value, the more energy is released when 1 mole of atoms combines with 1 mole of electrons to form 1 mole of anions with a 1− charge. A greater release of energy reflects more attraction between the atoms of the elements and free electrons.

| 1 | | | | | | | 18 |
|---|---|---|---|---|---|---|---|
| H −72.6 | 2 | 13 | 14 | 15 | 16 | 17 | He (0.0)^a |
| Li −59.6 | Be >0 | B −26.7 | C −122 | N +7 | O −141 | F −328 | Ne (+29)^a |
| Na −52.9 | Mg >0 | Al −42.5 | Si −134 | P −72.0 | S −200 | Cl −349 | Ar (+35)^a |
| K −48.4 | Ca −2.4 | Ga −28.9 | Ge −119 | As −78.2 | Se −195 | Br −325 | Kr (+39)^a |
| Rb −46.9 | Sr −5.0 | In −28.9 | Sn −107 | Sb −103 | Te −190 | I −295 | Xe (+41)^a |
| Cs −45.5 | Ba −14 | Tl −19.2 | Pb −35.2 | Bi −91.3 | Po −183.3 | At −270^a | Rn (+41)^a |

^aCalculated values.

electron to an N atom means that the atom loses the stability associated with a half-filled set of $2p$ orbitals:

N: ⇅ ⇅ ↑↑↑   + e⁻  →  N⁻: ⇅ ⇅ ⇅↑↑
$1s$ $2s$ $2p$              $1s$ $2s$ $2p$

Describe at least one similarity and one difference in the periodic trends in first ionization energies and electron affinities among the main group elements.

---

**SAMPLE EXERCISE 7.16** **Integrating Concepts: Red Fireworks**

The brilliant red color of some fireworks (Figure 7.40) is produced by the presence of lithium carbonate ($Li_2CO_3$) in the shells that are launched into the night sky and explode at just the right time. The explosions produce enough energy to vaporize the lithium compound and produce excited-state lithium atoms. These atoms quickly lose energy as they transition from their excited states to their ground states, emitting photons of red light that have a wavelength of 670.8 nm.

a. How much energy is in each photon of red light emitted by an excited-state lithium atom?
b. What is the frequency of the red light?
c. Lithium atoms in the lowest-energy state above the ground state emit these red photons. What is the difference in energy between the ground state of a Li atom and its lowest-energy excited state?
d. What is the electron configuration of a ground-state lithium atom?
e. What is the electron configuration of a lithium atom in its lowest-energy excited state?

**Collect and Organize** We are asked to calculate the energy of photons of red light from their wavelength. We are also asked to relate this energy to the difference in energies between a lithium atom's lowest-energy excited state and its ground state and to write the electron configurations of these states. The energy of a photon is related to its wavelength ($\lambda$) by Equation 7.3:

$$E = \frac{hc}{\lambda}$$

Lithium is the group 1 element in the second row of the periodic table.

**Analyze** The energy of a photon ($E$) emitted by an excited-state atom is the same as the difference in energies ($\Delta E$) between the excited state and the lower energy state (in this case, the ground state) of the transition that produced the photon. All the group 1 elements have one electron in their valence-shell $s$ orbital. In lithium this electron is in the $2s$ orbital, which is occupied after the $1s$ has filled. The next higher-energy orbitals are the $2p$ orbitals.

**Solve**
a. The energy of a photon with a wavelength of 670.8 nm is

$$E = \frac{hc}{\lambda} = \frac{(6.626 \times 10^{-34}\,\text{J} \cdot \text{s})(2.998 \times 10^8\,\frac{\text{m}}{\text{s}})}{670.8\,\text{nm}\left(\frac{10^{-9}\,\text{m}}{\text{nm}}\right)} = 2.961 \times 10^{-19}\,\text{J}$$

**FIGURE 7.40** Lithium carbonate is frequently used to obtain the red color in fireworks.

b. Frequency and wavelength are related:

$$c = \nu\lambda \qquad \text{so} \qquad \nu = \frac{c}{\lambda}$$

The frequency of red light with a wavelength of 670.8 nm is

$$\nu = \frac{2.998 \times 10^8\,\frac{\text{m}}{\text{s}}}{670.8\,\text{nm}\left(\frac{10^{-9}\,\text{m}}{\text{nm}}\right)} = 4.469 \times 10^{14}\,\text{s}^{-1}$$

c. The difference in energies between the ground state of a Li atom and its lowest-energy excited state must match exactly the energy of the photon emitted; therefore, $\Delta E = 2.961 \times 10^{-19}$ J.
d. The electron configuration of a ground-state Li atom is $1s^2 2s^1$.
e. In the lowest-energy excited state, the valence electron has moved to the next higher-energy orbital above the $2s$ orbital, which is one of the three $2p$ orbitals. Therefore, the electron configuration of the excited state is $1s^2 2p^1$.

**Think About It** You might expect that the colors that lithium and other elements make in fireworks explosions are the same colors they produce in Bunsen burner flames. You would be right. Many of the colors are produced by transitions from lowest-energy excited states to ground states because the lowest-energy excited state is the one most easily populated by the thermal energy available in a gas flame or exploding firework shell.

# A Noble Family: Special Status for Special Behavior

The descriptive chemistry boxes in previous chapters contain information about one or two elements. Now that we have developed electron configurations and highlighted their role in determining the periodic behavior of elements, we can discuss those features of the chemical and physical behavior of elements that vary periodically.

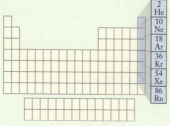

The six elements in the rightmost column of the periodic table (group 18) were originally called inert gases, but in the 1960s chemists discovered how to make compounds of several members of the group with very reactive substances like fluorine. After this discovery, the group name was changed to *noble gases*. Just as in human relations where the families of "nobility" tend not to interact with the "common people," the noble gases do not react with the common elements of other families on the periodic table. This distinct chemical behavior is the same for all elements in the noble gas family, even though each element is in a different period of the periodic table.

Helium is the second element in the periodic table and the second most abundant element in the universe (after hydrogen). Its presence in the absorption spectra of gases around the Sun was detected in 1868 as a set of dark lines that did not correspond with those of any known element on Earth. These dark lines were assigned to helium, whose name derives from *helios*, the Greek word for sun. The element was isolated on Earth by William Ramsay (1852–1916) in 1895.

After the hydrogen-filled dirigible *Hindenburg* exploded and burned in 1937, helium, which is chemically inert and therefore not a fire hazard, became the gas of choice for such craft (Figure 7.41).

Argon was isolated from the atmosphere the year before Ramsay discovered helium. The name of this element is from the Greek *argos*, meaning "lazy," in reference to the chemical inertness of this element and that of the others in group 18. Argon is the most abundant group 18 element, making up about 0.94% by volume of the atmosphere, and the 12th most abundant element in the universe. Most of the argon in the atmosphere is the product of radioactive decay, produced in rocks and minerals when nuclei of radioactive $^{40}K$ capture one of their own electrons and a proton is transformed into a neutron. The low decay rate allows scientists to determine the ages of rocks on the basis of their $^{40}Ar/^{40}K$ ratio. Eventually the argon leaks into the atmosphere, from which it is isolated by liquefying the air and separating Ar from $N_2$ and $O_2$ by distillation. Argon is used to provide an inert atmosphere for arc-welding certain metals, such as stainless steel. It is also used in light bulbs, to improve the lifetime of the filaments, and in photoelectric devices.

Neon, discovered by Ramsay and his coworkers as an impurity in argon in 1898, takes its name from the Greek *neos*, "new." Although neon is found in some minerals as a result of radioactive decay, its principal industrial source is the atmosphere. However, its tiny concentration in the atmosphere (0.0018% by volume) means that nearly 100 kg of air must be liquefied to obtain 1 g of neon. When an electric current is passed through a tube containing neon at low pressure, the gas emits the characteristic red-orange light we associate with neon signs. Electrical discharge tubes filled with the other noble gases emit

(a)                    (b)

**FIGURE 7.41** Both hydrogen gas and helium gas have been used to fill lighter-than-air craft, but (a) the hydrogen-filled *Hindenburg* ended its last trans-Atlantic flight in flames in Lakehurst, New Jersey, in 1937. (b) Modern blimps get their buoyancy from helium, which is chemically inert and hence not a fire or explosion hazard.

**FIGURE 7.42** Gas discharge tubes filled with different noble gases emit characteristic colors.

their own unique combinations of emission lines and characteristic colors (Figure 7.42). Neon was the gas analyzed by Francis Aston in his positive-ray analyzer that led to the discovery of stable isotopes and to the development of mass spectrometry.

Krypton (from the Greek *kryptos*, "hidden") and xenon (from the Greek *xenos*, "foreign") are even rarer components of the atmosphere. The concentration of krypton is only 1.1 ppm by volume, which means there are 1.1 atoms of krypton for every *million* atoms and molecules of the other components of air, on average. The atmospheric concentration of xenon is even smaller: only 0.086 ppm. Because the boiling points of krypton and xenon are higher than the boiling points of the principal components of the atmosphere, these two noble gases can be separated by fractional distillation of liquefied air, being recovered in the least-volatile fraction.

Radon is the heaviest known gas; it has no stable isotopes. It is produced by the radioactive decay of radium, a constituent in uranium ores. Its concentration varies widely from place to place, but it averages about $6 \times 10^{-18}$ % in air, which means 1 liter of air contains only about 160,000 atoms of radon. In 1985 heightened radon levels were identified in some homes, where its elevated presence caused concerns about health risks. In the United States, the highest concentrations have been found in Iowa and in southeastern Pennsylvania along a section of the Appalachian Mountains.

All the noble gases have a filled outermost occupied shell. The lack of reactivity of these elements is a direct result of this stable electron configuration. Helium has a filled $1s$ orbital, which corresponds to a full first shell, and only two valence electrons; the other noble gases have filled $s$ and $p$ orbitals and eight valence electrons.

The noble gases do not easily share electrons to form compounds with other atoms. However, xenon and krypton do react with fluorine to form, for example, $XeF_2$, $XeF_4$, $XeF_6$, and $KrF_2$. The three xenon compounds react with water, yielding compounds containing xenon, fluorine, and oxygen.

Each noble gas has the highest ionization energy of all the elements in its period. For example, the energy required to remove an electron from an atom of neon (period 2) is about 20% more than that required to remove an electron from an atom of the element immediately preceding it (fluorine) and 400% more than that required to remove an electron from an atom of the first element in period 2 (lithium). The ionization energies of the noble gases decrease as we move down the group (Figure 7.43) because the increasing numbers of core electrons shield the outermost electrons from the positive

charge on the nucleus. Atomic size increases from top to bottom in the group (Figure 7.44) because the valence electrons, which are a main determinant of atomic size, are farther and farther from the nucleus. Because covalent radii are determined by measuring internuclear distances in diatomic molecules, the radii of helium, neon, and argon—for which no stable compounds are known—are estimations.

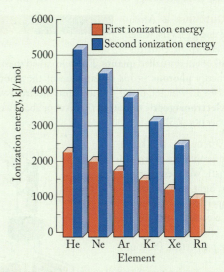

**FIGURE 7.43** The first ionization energies of the group 18 elements decrease down the group. The second ionization energies, which exhibit this same trend, are larger than the first ionization energies because of the greater amount of energy required to remove an electron from the cation formed in the first ionization.

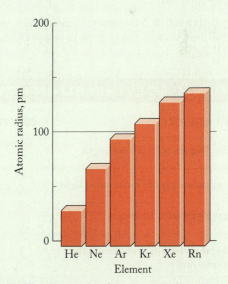

**FIGURE 7.44** The atomic radii of the group 18 elements increase down the group because the valence electrons occupy orbitals that are, on average, farther and farther from the nucleus.

## SUMMARY

**Learning Outcome 1** Free atoms in flames and in gas discharge tubes produce **atomic emission** (or bright-line) **spectra**. When radiation passes through an atomic gas, absorption produces a spectrum of dark lines, called an **atomic absorption** (or dark-line) **spectrum**. Electron transitions from higher to lower energy levels in an atom cause the atom to emit particular frequencies of radiation; absorption of the same frequencies accompanies transitions from corresponding lower to higher electron energy levels. (Sections 7.1, 7.2, 7.4)

**Learning Outcome 2** Atoms have discrete energy levels, which means they absorb or emit discrete amounts of energy called **quanta**. Einstein called these particles of radiant energy **photons** in his explanation of the **photoelectric effect**: a process in which a metal surface emits electrons when illuminated by **electromagnetic radiation** that is at or above a **threshold frequency**. (Section 7.3)

**Learning Outcome 3** Light has wavelike properties described by characteristic **wavelengths** and **frequencies**; light also has particle-like properties. Correspondingly, the electron, which we often think of as a particle, also has wavelike properties. (Sections 7.1, 7.3, 7.5)

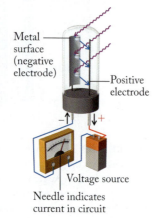

Metal surface (negative electrode)

Positive electrode

Voltage source

Needle indicates current in circuit

**Learning Outcome 4** The **Heisenberg uncertainty principle** states that the position and momentum of an electron cannot both be precisely known at the same time. (Section 7.5)

**Learning Outcome 5** Solutions to Schrödinger's wave equation define allowed electron energy levels in atoms. **Orbitals** are the three-dimensional regions inside an atom that describe the probability of finding an electron at a given distance from the nucleus. Orbitals have characteristic three-dimensional sizes, shapes, and orientations that are depicted by boundary–surface representations. Each orbital has a unique set of three **quantum numbers** that come directly from the solution to Schrödinger's wave equation. A fourth quantum number is necessary to explain certain characteristics of emission spectra. (Sections 7.6, 7.7)

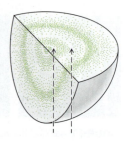

**Learning Outcome 6** An **electron configuration** is a set of numbers and letters expressing the number of electrons in each occupied orbital in an atom. All the orbitals of a given subshell are **degenerate**; they all have the same energy. In any set of degenerate orbitals, each orbital must contain one electron before any orbital in the set can accept a second electron. (Sections 7.6, 7.7, 7.8, 7.9)

**Learning Outcome 7** The **effective nuclear charge ($Z_{eff}$)** is the net nuclear charge felt by an electron when electrons closer to the nucleus **shield** (screen) the electron from experiencing the full nuclear charge. (Section 7.10)

**Learning Outcome 8** Trends in $Z_{eff}$ cause atomic size to decrease across a period and increase down a group. **Ionization energies** generally increase with increasing effective nuclear charge across a period. **Electron affinity** is most exothermic for the halogens, moderately exothermic for many main group elements, but endothermic for all the noble gases. (Sections 7.10, 7.11, 7.12)

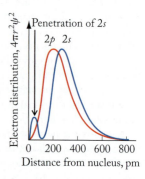

Penetration of 2s
2p 2s
Electron distribution, $4\pi r^2 \psi^2$
Distance from nucleus, pm

## PROBLEM-SOLVING SUMMARY

| TYPE OF PROBLEM | CONCEPTS AND EQUATIONS | | SAMPLE EXERCISES |
|---|---|---|---|
| **Calculating frequency from wavelength** | $\lambda\nu = c$   where $c = 2.998 \times 10^8$ m/s | (7.1) | 7.1 |
| **Calculating the energy of a quantum of light** | $E = \dfrac{hc}{\lambda}$ | (7.3) | 7.2 |
| **Applying the photoelectric effect** | $\Phi = h\nu_0$ | (7.4) | 7.3 |
| **Calculating the wavelength of a line in the hydrogen spectrum** | $\dfrac{1}{\lambda} = \left[1.097 \times 10^{-2} \text{ nm}^{-1}\right]\left(\dfrac{1}{n_1^2} - \dfrac{1}{n_2^2}\right)$ | (7.7) | 7.4 |
| **Calculating the energy needed for an electron transition** | $\Delta E = -2.178 \times 10^{-18} \text{ J}\left(\dfrac{1}{n_{final}^2} - \dfrac{1}{n_{initial}^2}\right)$ | (7.9) | 7.5 |
| **Calculating the wavelength of particles in motion** | $\lambda = \dfrac{h}{mv}$ | (7.11) | 7.6 |
| **Calculating uncertainty** | $\Delta x \cdot m\Delta v \geq \dfrac{h}{4\pi}$ | (7.13) | 7.7 |

| TYPE OF PROBLEM | CONCEPTS AND EQUATIONS | SAMPLE EXERCISES |
|---|---|---|
| **Identifying the subshells and orbitals in an energy level and valid quantum number sets** | $n$ is the shell number, $\ell$ defines both the subshell and the type of orbital; an orbital has a unique combination of allowed $n$, $\ell$, and $m_\ell$ values. $\ell$ is any integer from 0 to $n-1$; $m_\ell$ is any integer from $-\ell$ to $+\ell$, including zero. | 7.8, 7.9 |
| **Writing electron configurations of atoms and ions; determining isoelectronic species** | Orbitals fill up in the following sequence: $1s^2$, $2s^2$, $2p^6$, $3s^2$, $3p^6$, $4s^2$, $3d^{10}$, $4p^6$, $5s^2$, $4d^{10}$, $5p^6$, $6s^2$, $4f^{14}$, $5d^{10}$, $6p^6$ (superscripts represent maximum numbers of electrons). Arrange orbitals in electron configuration based on increasing value of $n$. In forming transition metal ions, electrons are removed to maximize the number of $d$ electrons; there is enhanced stability in half-filled and filled $d$ subshells. | 7.10, 7.11, 7.12, 7.13 |
| **Ordering atoms and ions by size** | In general, sizes decrease left to right across a row and increase down a column of the periodic table: cations are smaller, and anions are larger, than their parent atoms. | 7.14 |
| **Ranking ionization energies** | First ionization energies increase across a row and decrease down a column. | 7.15 |

## VISUAL PROBLEMS

*(Answers to boldface end-of-chapter questions and problems are in the back of the book.)*

**7.1.** Which of the elements highlighted in Figure P7.1 consists of ground state atoms with
  a. a single $s$ electron in their outermost shells? (More than one answer is possible.)
  b. filled sets of $s$ and $p$ orbitals in their outermost shells?
  c. filled sets of $d$ orbitals?
  d. half-filled sets of $d$ orbitals?
  e. two $s$ electrons in their outermost shells?

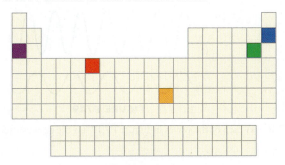

**FIGURE P7.1**

**7.2.** Which of the highlighted elements in Figure P7.1 has the greatest number of unpaired electrons per ground-state atom?

**7.3.** Which of the highlighted elements in Figure P7.1 form common monatomic ions that are larger than their parent atoms?

**7.4.** Which of the elements highlighted in Figure P7.4 forms monatomic ions by
  a. losing an $s$ electron?
  b. losing two $s$ electrons?
  c. losing two $s$ electrons and a $d$ electron?
  d. adding an electron to a $p$ orbital?
  e. adding electrons to two $p$ orbitals?

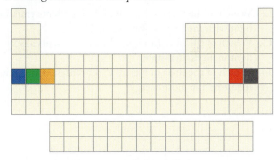

**FIGURE P7.4**

**7.5.** Which of the highlighted elements in Figure P7.4 form common monatomic ions that are smaller than their parent atoms?

**7.6.** Rank the elements highlighted in Figure P7.4 based on increasing atomic size.

**7.7.** Rank the highlighted elements in Figure P7.4 based on increasing size of their most common monatomic ions.

**7.8.** Which of the elements highlighted in Figure P7.4 has the largest first ionization energy, $IE_1$?

**7.9.** Which of the elements highlighted in Figure P7.4 has the largest second ionization energy, $IE_2$?

**7.10.** Consider the results of the experiment shown in Figure P7.10. The screen containing two narrow slits is illuminated by a distant source of light. Instead of just the images of the two slits appearing on the screen on the right, a whole series of slit images (more than the five shown here) appear. Explain how there could be many more than two slit images on the screen on the right.

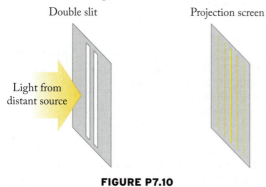

**FIGURE P7.10**

**7.11.** Which of the lines in Figure P7.11 represent absorption and which ones represent emission?

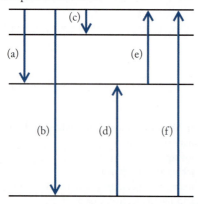

**FIGURE P7.11**

7.12. Which of the lines in Figure P7.11 represents emission of light with the longest wavelength?

**7.13.** Which of the lines in Figure P7.11 represents absorption of light in an excited state?

7.14. Which of the lines in Figure P7.11 represent transitions requiring the most energy?

**7.15.** In Figure P7.15, identify which sphere represents Na atoms, K atoms, and Na$^+$ ions.

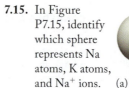

**FIGURE P7.15**

7.16. In Figure P7.16, identify which sphere represents O atoms, S atoms, and S$^{-2}$ ions.

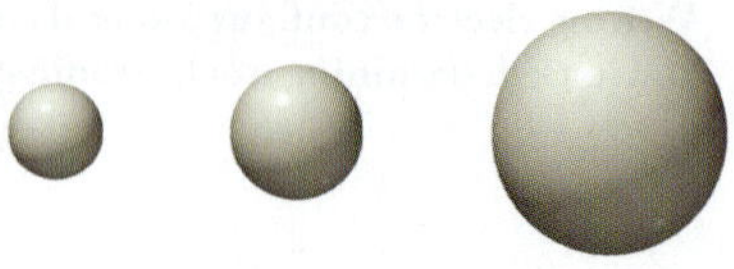

**FIGURE P7.16**

**7.17.** Which of the orbital diagrams in Figure P7.17 represents an excited state of a nitrogen atom?

7.18. Which of the orbital diagrams in Figure P7.17 represents the ground state of an oxygen cation, O$^+$?

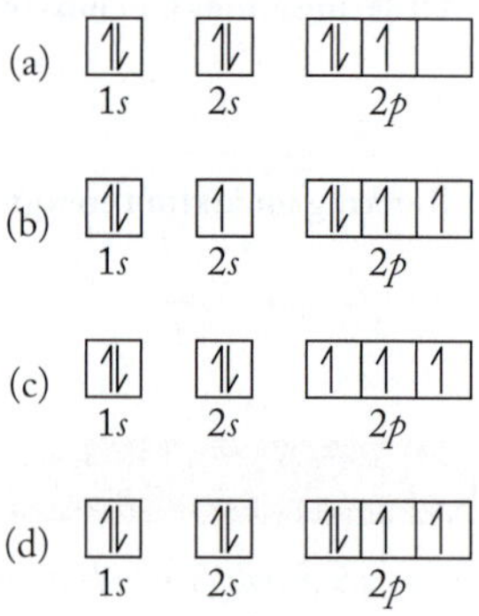

(a) $1s$ $2s$ $2p$
(b) $1s$ $2s$ $2p$
(c) $1s$ $2s$ $2p$
(d) $1s$ $2s$ $2p$

**FIGURE P7.17**

**7.19.** Which of the following waves has the lowest energy?

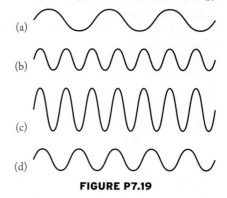

(a)
(b)
(c)
(d)

**FIGURE P7.19**

7.20. Which of the waves in Figure P7.19 has the greatest frequency?

## QUESTIONS AND PROBLEMS

### Light Waves

#### CONCEPT REVIEW

**7.21.** Why are the various forms of radiant energy called *electromagnetic* radiation?

**7.22.** Explain with a sketch why the frequencies of long-wavelength waves of electromagnetic radiation are lower than those of short-wavelength waves.

**7.23.** **Dental X-rays** When X-ray images are taken of your teeth and gums in the dentist's office, your body is covered with a lead shield. Explain the need for this precaution.

**7.24.** **UV Radiation and Skin Cancer** Ultraviolet radiation causes skin damage that may lead to cancer, but exposure to infrared radiation does not seem to cause skin cancer. Why do you think this is so?

**7.25.** Gamma rays are an example of "ionizing" radiation because they have the energy to break apart molecules into molecular ions and free electrons. What other forms of electromagnetic radiation could also be ionizing radiation?

*7.26. If light consists of waves, why don't things look "wavy" to us?

#### PROBLEMS

**7.27.** A neon light emits radiation of $\lambda = 616$ nm. What is the frequency of this radiation?

7.28. **Submarine Communications** In the 1990s the Russian and American navies developed extremely low frequency

communications networks to send messages to submerged submarines. The frequency of the carrier wave of the Russian network was 82 Hz, while the Americans used 76 Hz.
a. What was the ratio of the wavelengths of the Russian network to the American network?
b. To calculate the actual underwater wavelength of the transmissions in either network, what additional information would you need?

**7.29. Broadcast Frequencies** FM radio stations broadcast at different frequencies. Calculate the wavelengths corresponding to the broadcast frequencies of the following radio stations: (a) KKNB (Lincoln, NE), 104.1 MHz; (b) WFNX (Boston, MA), 101.7 MHz; (c) KRTX (Houston, TX), 100.7 MHz.

7.30. Which radiation has the longer wavelength? (a) radio waves from an AM radio station broadcasting at 680 kHz; (b) infrared radiation emitted by the surface of Earth ($\lambda = 15 \; \mu m$)

**7.31.** Which radiation has the lower frequency? (a) radio waves from an AM radio station broadcasting at 1090 kHz; (b) the green light ($\lambda = 550$ nm) from an LED (light-emitting diode) on a stereo system

7.32. Which radiation has the higher frequency? (a) the red light on a bar-code reader at a grocery store; (b) the green light on the battery charger for a laptop computer

**7.33. Speed of Light** How long does it take light to reach Earth from the Sun? (The distance from the Sun to Earth is 93 million miles.)

7.34. **Exploration of the Solar System** How long would it take an instruction to a Martian rover to travel from a NASA site on Earth to Mars? Assume the signal is sent when Earth and Mars are 75 million kilometers apart.

## Atomic Spectra
### CONCEPT REVIEW

**7.35.** Describe the similarities and differences in the atomic emission and absorption spectra of hydrogen.

7.36. Are the Fraunhofer lines in the spectra of stars the result of atomic emission or atomic absorption?

**7.37.** How did the study of the atomic emission spectra of elements lead to the identification of the Fraunhofer lines in sunlight?

*7.38. What would happen in terms of the appearance of the Fraunhofer lines in the solar spectrum if sunlight were passed through a flame containing high-temperature calcium atoms and then analyzed?

## Particles of Light and Quantum Theory
### CONCEPT REVIEW

**7.39.** What is the difference between a quantum and a photon?

7.40. A variable power supply is connected to an incandescent light bulb. At the lowest power setting, the bulb feels warm to the touch but produces no light. At medium power, the light bulb filament emits a red glow. At the highest power, the light bulb emits white light. Explain this emission pattern.

*7.41. What effect does the intensity (amplitude) of a wave have on the emission of electrons from a surface, assuming we are above the threshold frequency?

*7.42. Has a photon of radiation that is red-shifted 10% actually lost 10% of its energy, where $E = h\nu$?

7.43. Which of the following have quantized values? Explain your selections.
a. the elevation of the treads of a moving escalator
b. the elevations at which the doors of an elevator open
c. the speed of an automobile

7.44. Which of the following have quantized values? Explain your selections.
a. the pitch of a note played on a slide trombone
b. the pitch of a note played on a flute
c. the wavelengths of light produced by the heating elements in a toaster
d. the wind speed at the top of Mt. Everest

### PROBLEMS

*7.45. Thin layers of potassium ($\Phi = 3.68 \times 10^{-19}$ J) and sodium ($\Phi = 4.41 \times 10^{-19}$ J) are exposed to radiation of wavelength 300 nm. Which metal emits electrons with the greater velocity? What is the velocity of these electrons?

7.46. Titanium ($\Phi = 6.94 \times 10^{-19}$ J) and silicon ($\Phi = 7.24 \times 10^{-19}$ J) surfaces are irradiated with UV radiation with a wavelength of 250 nm. Which surface emits electrons with the longer wavelength? What is the wavelength of the electrons emitted by the titanium surface?

**7.47. Solar Power** Photovoltaic cells convert solar energy into electricity. Could tantalum ($\Phi = 6.81 \times 10^{-19}$ J) be used to convert visible light to electricity? Assume that most of the electromagnetic energy from the Sun is in the visible region near 500 nm.

7.48. With reference to Problem 7.47, could tungsten ($\Phi = 7.20 \times 10^{-19}$ J) be used to construct solar cells?

**7.49.** The power of a red laser ($\lambda = 630$ nm) is 1.00 watt (abbreviated W, where 1 W = 1 J/s). How many photons per second does the laser emit?

*7.50. The energy density of starlight in interstellar space is $10^{-15}$ J/m³. If the average wavelength of starlight is 500 nm, what is the corresponding density of photons per cubic meter of space?

## The Hydrogen Spectrum and the Bohr Model
### CONCEPT REVIEW

**7.51.** Why should hydrogen have the simplest atomic spectrum of all the elements?

7.52. For an electron in a hydrogen atom, how is the value of $n$ of its orbit related to its energy?

**7.53.** Does the electromagnetic energy emitted by an excited-state H atom depend on the individual values of $n_1$ and $n_2$, or only on the difference between them $(n_1 - n_2)$?

7.54. Explain the difference between a ground-state H atom and an excited-state H atom.

**7.55.** Without calculating any wavelength values, predict which of the following electron transitions in the hydrogen atom is associated with radiation having the shortest wavelength.
a. from $n = 1$ to $n = 2$      c. from $n = 3$ to $n = 4$
b. from $n = 2$ to $n = 3$      d. from $n = 4$ to $n = 5$

7.56. Without calculating any frequency values, rank the following transitions in the hydrogen atom in order of increasing frequency of the electromagnetic radiation that could produce them.
a. from $n = 4$ to $n = 6$      c. from $n = 9$ to $n = 11$
b. from $n = 6$ to $n = 8$      d. from $n = 11$ to $n = 13$

**7.57.** Electron transitions from $n = 2$ to $n = 3, 4, 5,$ or 6 in hydrogen atoms are responsible for some of the Fraunhofer lines in the Sun's spectrum. Are there any Fraunhofer lines due to transitions that start from $n = 3$ in hydrogen atoms?

**7.58.** In the visible portion of the atomic emission spectrum of hydrogen, are there any bright lines due to electron transitions to the $n = 1$ state?

**7.59.** Balmer observed a hydrogen emission line for the transition from $n = 6$ to $n = 2$, but not for the transition from $n = 7$ to $n = 2$. Why?

*__7.60.__ In what ways should the emission spectra of H and He$^+$ be alike, and in what ways should they be different?

## PROBLEMS

**7.61.** What is the wavelength of the photons emitted by hydrogen atoms when they undergo transitions from $n = 4$ to $n = 3$? In which region of the electromagnetic spectrum does this radiation occur?

**7.62.** What is the frequency of the photons emitted by hydrogen atoms when they undergo transitions from $n = 5$ to $n = 3$? In which region of the electromagnetic spectrum does this radiation occur?

_____

*__7.63.__ The energies of the photons emitted by one-electron atoms and ions fit the equation

$$E = (2.178 \times 10^{-18} \text{ J})Z^2\left(\frac{1}{n_1^2} - \frac{1}{n_2^2}\right)$$

where $Z$ is the atomic number, $n_2$ and $n_1$ are positive integers, and $n_2 > n_1$.
  a. As the value of $Z$ increases, does the wavelength of the photon associated with the transition from $n = 2$ to $n = 1$ increase or decrease?
  b. Can the wavelength associated with the transition from $n = 2$ to $n = 1$ ever be observed in the visible region of the spectrum?

*__7.64.__ Can transitions from higher energy states to the $n = 2$ state in He$^+$ ever produce visible light? If so, for what values of $n_2$? (*Hint:* The equation in Problem 7.63 may be useful.)

_____

**7.65.** The transition from $n = 3$ to $n = 2$ in a hydrogen atom produces a photon with a wavelength of 656 nm. What is the wavelength of the transition from $n = 3$ to $n = 2$ in a Li$^{2+}$ ion? (Use the equation in Problem 7.63.)

*__7.66.__ The hydrogen atomic emission spectrum includes a UV line with a wavelength of 92.3 nm.
  a. Is this line associated with a transition between different excited states or between an excited state and the ground state?
  b. What is the value of $n_1$ of this transition?
  c. What is the energy of the longest-wavelength photon that a ground-state hydrogen atom can absorb?

## Electron Waves

### CONCEPT REVIEW

**7.67.** Identify the symbols in the de Broglie relation $\lambda = h/mv$, and explain how the relation links the properties of a particle to those of a wave.

**7.68.** Explain how the observation of electron diffraction supports the description of electrons as waves.

**7.69.** Would the density or shape of an object have an effect on its de Broglie wavelength?

**7.70.** How does de Broglie's hypothesis that electrons behave like waves explain the stability of the electron orbits in a hydrogen atom?

## PROBLEMS

**7.71.** Calculate the wavelengths of the following objects:
  a. a muon (a subatomic particle with a mass of $1.884 \times 10^{-25}$ g) traveling at 325 m/s
  b. electrons ($m_e = 9.10938 \times 10^{-28}$ g) moving at $4.05 \times 10^6$ m/s in an electron microscope
  c. an 80 kg athlete running a 4-minute mile
  d. Earth (mass $= 6.0 \times 10^{27}$ g) moving through space at $3.0 \times 10^4$ m/s

**7.72.** Two objects are moving at the same velocity. Which (if any) of the following statements about them are true?
  a. The de Broglie wavelength of the heavier object is longer than that of the lighter one.
  b. If one object has twice as much mass as the other, its wavelength is one-half the wavelength of the other.
  c. Doubling the velocity of one of the objects will have the same effect on its wavelength as doubling its mass.

_____

**7.73.** Which (if any) of the following statements about the frequency of a particle is true?
  a. Heavy, fast-moving objects have lower frequencies than those of lighter, faster-moving objects.
  b. Only very light particles can have high frequencies.
  c. Doubling the mass of an object and halving its velocity results in no change in its frequency.

**7.74.** How rapidly would each of the following particles be moving if they all had the same wavelength as a photon of red light ($\lambda = 750$ nm)?
  a. an electron of mass $9.10938 \times 10^{-28}$ g
  b. a proton of mass $1.67262 \times 10^{-24}$ g
  c. a neutron of mass $1.67493 \times 10^{-24}$ g
  d. an $\alpha$ particle of mass $6.64 \times 10^{-24}$ g

_____

*__7.75.__ Kinetic molecular theory tells us that helium atoms at 500 K are in constant random motion.
  a. Calculate the root mean square speed of helium atoms at 500 K.
  b. What is the wavelength of a helium atom at 500 K?

*__7.76.__ Neon atoms are expected to move slower than helium atoms at the same temperature.
  a. Using Graham's Law of Effusion and your answer from Problem 7.75, calculate the root mean square speed of neon atoms at 500 K.
  b. What is the wavelength of a neon atom at 500 K?

_____

**7.77. Particles in a Cyclotron** The first cyclotron was built in 1930 at the University of California, Berkeley, and was used to accelerate molecular ions of hydrogen, H$_2^+$, to a velocity of $4 \times 10^6$ m/s. (Modern cyclotrons can accelerate particles to nearly the speed of light.) If the uncertainty in the velocity of the H$_2^+$ ion was 3%, what was the uncertainty of its position?

**7.78. Radiation Therapy** An effective treatment for some cancerous tumors involves irradiation with "fast" neutrons. The neutrons from one treatment source have an average velocity of $3.1 \times 10^7$ m/s. If the velocities of individual neutrons are known to within 2% of this value, what is the uncertainty in the position of one of them?

## Quantum Numbers and Electron Spin; The Sizes and Shapes of Atomic Orbitals

### CONCEPT REVIEW

**7.79.** How does the concept of an orbit in the Bohr model of the hydrogen atom differ from the concept of an orbital in quantum theory?

7.80. What properties of an orbital are defined by each of the three quantum numbers $n$, $\ell$, and $m_\ell$?

**7.81.** How many quantum numbers are needed to identify an orbital?

7.82. How many quantum numbers are needed to identify an electron in an atom?

### PROBLEMS

**7.83.** How many orbitals are there in an atom with each of the following principal quantum numbers? (a) 1; (b) 2; (c) 3; (d) 4; (e) 5

7.84. How many orbitals are there in an atom with the following combinations of quantum numbers?
   a. $n = 3$, $\ell = 2$     c. $n = 4$, $\ell = 2$, $m_\ell = 2$
   b. $n = 3$, $\ell = 1$

**7.85.** What are the possible values of quantum number $\ell$ when $n = 4$?

7.86. What are the possible values of $m_\ell$ when $\ell = 2$?

**7.87.** What set of orbitals corresponds to each of the following sets of quantum numbers? How many electrons could occupy these orbitals?
   a. $n = 2$, $\ell = 0$     c. $n = 4$, $\ell = 2$
   b. $n = 3$, $\ell = 1$     d. $n = 1$, $\ell = 0$

7.88. What set of orbitals corresponds to each of the following sets of quantum numbers? How many electrons could occupy these orbitals?
   a. $n = 2$, $\ell = 1$     c. $n = 3$, $\ell = 2$
   b. $n = 5$, $\ell = 3$     d. $n = 4$, $\ell = 3$

**7.89.** Which of the following combinations of quantum numbers are allowed?
   a. $n = 1$, $\ell = 1$, $m_\ell = 0$, $m_s = +\frac{1}{2}$
   b. $n = 3$, $\ell = 0$, $m_\ell = 0$, $m_s = -\frac{1}{2}$
   c. $n = 1$, $\ell = 0$, $m_\ell = 1$, $m_s = -\frac{1}{2}$
   d. $n = 2$, $\ell = 1$, $m_\ell = 2$, $m_s = +\frac{1}{2}$

7.90. Which of the following combinations of quantum numbers are allowed?
   a. $n = 3$, $\ell = 2$, $m_\ell = 0$, $m_s = -\frac{1}{2}$
   b. $n = 5$, $\ell = 4$, $m_\ell = 4$, $m_s = +\frac{1}{2}$
   c. $n = 3$, $\ell = 0$, $m_\ell = 1$, $m_s = +\frac{1}{2}$
   d. $n = 4$, $\ell = 4$, $m_\ell = 1$, $m_s = -\frac{1}{2}$

## The Periodic Table and Filling the Orbitals of Multielectron Atoms; Electron Configurations of Ions

### CONCEPT REVIEW

**7.91.** What is meant when two or more orbitals are said to be degenerate?

7.92. Explain how the electron configurations of the group 2 elements are linked to their location in the periodic table developed by Mendeleev (see Figure 2.9).

**7.93.** How do we know from examining the periodic table's structure that the $4s$ orbital is filled before the $3d$ orbital?

7.94. Explain why so many transition metals form ions with a 2+ charge.

**7.95.** Why is there only one ground-state electron configuration for an atom but many excited-state electron configurations?

7.96. Can the ground-state electron configuration for an atom ever be an excited-state electron configuration for a different atom?

### PROBLEMS

**7.97.** List the following orbitals in order of increasing energy in a multielectron atom:
   a. $n = 3$, $\ell = 2$     c. $n = 3$, $\ell = 0$
   b. $n = 5$, $\ell = 1$     d. $n = 4$, $\ell = 1$, $m_\ell = -1$

7.98. Place the following orbitals in order of increasing energy in a multielectron atom:
   a. $n = 2$, $\ell = 1$     c. $n = 3$, $\ell = 2$
   b. $n = 5$, $\ell = 3$     d. $n = 4$, $\ell = 3$

**7.99.** What are the electron configurations of Li, $Li^+$, Ca, $F^-$, $Na^+$, $Mg^{2+}$, and $Al^{3+}$?

7.100. Which species listed in Problem 7.99 are isoelectronic with Ne?

**7.101.** What are the condensed electron configurations of K, $K^+$, $S^{2-}$, N, Ba, $Ti^{4+}$, and Al?

7.102. In what way are the electron configurations of H, Li, Na, K, Rb, and Cs similar?

**7.103.** How many unpaired electrons are there in the following ground-state atoms and ions? (a) N; (b) O; (c) $P^{3-}$; (d) $Na^+$

7.104. How many unpaired electrons are there in the following ground-state atoms and ions? (a) Sc; (b) $Ag^+$; (c) $Cd^{2+}$; (d) $Zr^{4+}$

**7.105.** Identify the atom whose electron configuration is $[Ar]3d^2 4s^2$. How many unpaired electrons are there in the ground state of this atom?

7.106. Identify the atom whose electron configuration is $[Ne]3s^2 3p^3$. How many unpaired electrons are there in the ground state of this atom?

**7.107.** Which monatomic ion has a charge of 1– and the electron configuration $[Ne]3s^2 3p^6$? How many unpaired electrons are there in the ground state of this ion?

7.108. Which monatomic ion has a charge of 1+ and the electron configuration $[Kr]4d^{10} 5s^2$? How many unpaired electrons are there in the ground state of this ion?

**7.109.** Does the ground-state electron configuration of $Na^+$ represent an excited-state electron configuration of Ne?

7.110. Which excited state of the hydrogen atom is likely to be of higher energy, $1s^1 2s^1$ or $1s^1 3s^1$?

**7.111.** Predict the charge of the monatomic ions formed by Al, N, Mg, and Cs.

7.112. Predict the charge of the monatomic ions formed by S, P, Zn, and I.

**7.113.** Which of the following electron configurations represent an excited state?
   a. $[He]2s^1 2p^5$     c. $[Ar]3d^{10} 4s^2 4p^5$
   b. $[Kr]4d^{10} 5s^2 5p^1$     d. $[Ne]3s^2 3p^2 4s^1$

**7.114.** Which of the following electron configurations represent an excited state?
   a. $[Ne]3s^23p^1$
   b. $[Ar]3d^{10}4s^14p^2$
   c. $[Kr]4d^{10}5s^15p^1$
   d. $[Ne]3s^23p^64s^1$

**7.115.** Boric acid, $H_3BO_3$, gives off a green color (Figure P7.115) when heated in a flame.
   a. Write ground-state electron configurations for B and O.
   b. Assign oxidation numbers to each of the elements in boric acid. If the oxidation number reflects the ionic charge on the element, write ground-state electron configurations for each ion.

**FIGURE P7.115**

**7.116.** Introducing calcium chloride into a flame imparts an intense orange color (Figure P7.116).
   a. Write ground-state electron configurations for Ca and Cl.
   b. Calcium chloride contains calcium and chloride ions. Write ground-state electron configurations for each ion.

**FIGURE P7.116**

**7.117.** In which subshell are the highest-energy electrons in a ground-state atom of the isotope $^{131}I$? Are the electron configurations of $^{131}I$ and $^{127}I$ the same?

**7.118.** Although no currently known elements contain electrons in $g$ orbitals, such elements may be synthesized some day. What is the minimum atomic number of an element whose ground-state atoms have an electron in a $g$ orbital?

## The Sizes of Atoms and Ions

### CONCEPT REVIEW

**7.119.** Sodium atoms are much larger than chlorine atoms, but in NaCl sodium ions are much smaller than chloride ions. Why?

**7.120.** Why does atomic size tend to decrease with increasing atomic number across a row of the periodic table?

**7.121.** Which of the following group 1 elements has the largest atoms: Li, Na, K, Rb? Explain your selection.

**7.122.** Which of the following group 17 elements has the largest monatomic ions: F, Cl, Br, I? Explain your selection.

## Ionization Energies

### CONCEPT REVIEW

**7.123.** How do ionization energies change with increasing atomic number (a) down a group of elements in the periodic table and (b) from left to right across a period of elements?

**7.124.** The first ionization energies of the main group elements are given in Figure 7.38. Explain the differences in the first ionization energy between (a) He and Li; (b) Li and Be; (c) Be and B; (d) N and O

**7.125.** Explain why it is more difficult to ionize a fluorine atom than a boron atom.

**7.126.** Do you expect the ionization energies of anions of group 17 elements to be lower or higher than for neutral atoms of the same group?

**7.127.** Which of the following elements should have the smallest *second* ionization energy? Br, Kr, Rb, Sr, Y

**7.128.** Why is the first ionization energy ($IE_1$) of Al ($Z = 13$) less than the $IE_1$ of Mg ($Z = 12$) *and* less than the $IE_1$ of Si ($Z = 14$)?

## Additional Problems

**7.129. Interstellar Hydrogen** Astronomers have detected hydrogen atoms in interstellar space in the $n = 732$ excited state. Suppose an atom in this excited state undergoes a transition from $n = 732$ to $n = 731$.
   a. How much energy does the atom lose as a result of this transition?
   b. What is the wavelength of radiation corresponding to this transition?
   c. What kind of telescope would astronomers need in order to detect radiation of this wavelength? (*Hint:* It would not be one designed to capture visible light.)

**\*7.130.** When an atom absorbs an X-ray of sufficient energy, one of its $2s$ electrons may be emitted, creating a hole that can be spontaneously filled when an electron in a higher-energy orbital—a $2p$, for example—falls into it. A photon of electromagnetic radiation with an energy that matches the energy lost in the $2p \rightarrow 2s$ transition is emitted. Predict how the wavelengths of $2p \rightarrow 2s$ photons would differ between (a) different elements in the fourth row of the periodic table and (b) different elements in the same column (for example, between the noble gases from Ne to Rn).

**7.131.** A green flame is observed when copper(II) chloride is heated in a flame (Figure P7.131).
   a. Write the ground-state electron configurations for copper atoms.
   b. There are two common forms of copper chloride, copper(I) chloride and copper(II) chloride. What is the difference in the ground-state electron configurations of copper in these two compounds?

**FIGURE P7.131**

**\*7.132.** Two helium ions ($He^+$) in the $n = 3$ excited state emit photons of radiation as they return to the ground state. One ion does so in a single transition from $n = 3$ to $n = 1$. The other does so in two steps: $n = 3$ to $n = 2$ and then $n = 2$ to $n = 1$. Which of the following statements about these two pathways is true?
   a. The sum of the energies lost in the two-step process is the same as the energy lost in the single transition from $n = 3$ to $n = 1$.
   b. The sum of the wavelengths of the two photons emitted in the two-step process is equal to the wavelength of the single photon emitted in the transition from $n = 3$ to $n = 1$.
   c. The sum of the frequencies of the two photons emitted in the two-step process is equal to the frequency of the single photon emitted in the transition from $n = 3$ to $n = 1$.
   d. The wavelength of the photon emitted by the $He^+$ ion in the $n = 3$ to $n = 1$ transition is shorter than the wavelength of a photon emitted by an H atom in an $n = 3$ to $n = 1$ transition.

**\*7.133.** Use your knowledge of electron configurations to explain the following observations:
   a. Silver tends to form ions with a charge of 1+, but the elements to the left and right of silver in the periodic table tend to form ions with 2+ charges.

b. The heavier group 13 elements (Ga, In, Tl) tend to form ions with charges of 1+ or 3+ but not 2+.

c. The heavier elements of group 14 (Sn, Pb) and group 4 (Ti, Zr, Hf) tend to form ions with charges of 2+ or 4+.

**7.134.** Trends in ionization energies of the elements as a function of the position of the elements in the periodic table are a useful test of our understanding of electronic structure.

a. Should the same trend in the first ionization energies for elements with atomic numbers $Z = 31$ through $Z = 36$ be observed for the second ionization energies of the same elements? Explain why or why not.

b. Which element should have the greater second ionization energy: Rb ($Z = 37$) or Kr ($Z = 36$)? Why?

**7.135. Chemistry of Photo-Gray Glasses** "Photo-gray" lenses for eyeglasses darken in bright sunshine because the lenses contain tiny, transparent AgCl crystals. Exposure to light removes electrons from $Cl^-$ ions, forming a chlorine atom in an excited state (indicated below by the asterisk):

$$Cl^- + h\nu \rightarrow Cl^* + e^-$$

The electrons are transferred to $Ag^+$ ions, forming silver metal:

$$Ag^+ + e^- \rightarrow Ag$$

Silver metal is reflective, giving rise to the photo-gray color.

a. Write condensed electron configurations of $Cl^-$, Cl, Ag, and $Ag^+$.

b. What do we mean by the term *excited state*?

c. Would more energy be needed to remove an electron from a $Br^-$ ion or from a $Cl^-$ ion? Explain your answer.

*d. How might substitution of AgBr for AgCl affect the light sensitivity of photo-gray lenses?

**7.136.** The first ionization energy of a gas-phase atom of a particular element is $6.24 \times 10^{-19}$ J. What is the maximum wavelength of electromagnetic radiation that could ionize this atom?

**7.137.** Tin (in group 14) forms both $Sn^{2+}$ and $Sn^{4+}$ ions, but magnesium (in group 2) forms only $Mg^{2+}$ ions.

a. Write condensed ground-state electron configurations for the ions $Sn^{2+}$, $Sn^{4+}$, and $Mg^{2+}$.

b. Which neutral atoms have ground-state electron configurations identical to $Sn^{2+}$ and $Mg^{2+}$?

c. Which 2+ ion is isoelectronic with $Sn^{4+}$?

**7.138. Oxygen Ions in Space** Since 1999 the *Far Ultraviolet Spectroscopic Explorer (FUSE)* satellite has been analyzing the spectra of emission sources within the Milky Way. Among the satellite's findings are interplanetary clouds containing oxygen atoms that have lost five electrons.

a. Write an electron configuration for these highly ionized oxygen atoms.

b. Which electrons have been removed from the neutral atoms?

c. The ionization energies corresponding to removal of the third, fourth, and fifth electrons are 4581 kJ/mol, 7465 kJ/mol, and 9391 kJ/mol, respectively. Explain why removal of each additional electron requires more energy than removal of the previous one.

d. What is the maximum wavelength of radiation that will remove the fifth electron from an O atom?

*7.139. Effective nuclear charge ($Z_{eff}$) is related to atomic number ($Z$) by a parameter called the shielding parameter ($\sigma$) according to the equation $Z_{eff} = Z - \sigma$.

a. Calculate $Z_{eff}$ for the outermost $s$ electrons of Ne and Ar given $\sigma = 4.24$ (for Ne) and 11.24 (for Ar).

b. Explain why the shielding parameter is much greater for Ar than for Ne.

**7.140. Fog Lamp Technology** Sodium fog lamps and street lamps contain gas-phase sodium atoms and sodium ions. Sodium atoms emit yellow-orange light at 589 nm. Do sodium ions emit the same yellow-orange light? Explain why or why not.

**7.141.** How can an electron get from the (+) lobe of a $p$ orbital to the (−) lobe without going through the node between the lobes?

**7.142.** Einstein did not fully accept the uncertainty principle, remarking that "He [God] does not play dice." What do you think Einstein meant? Niels Bohr allegedly responded by saying, "Albert, stop telling God what to do." What do you think Bohr meant?

*7.143. The wavelengths of Fraunhofer lines in galactic spectra are not exactly the same as those in sunlight: they tend to be shifted to longer wavelengths (*redshift*), in part due to the Doppler effect. The Doppler effect is described by the equation

$$\frac{(\nu - \nu^1)}{\nu} = \frac{u}{c}$$

where $\nu$ is the unshifted frequency, $\nu^1$ is the perceived frequency, $c$ is the speed of light, and $u$ is the speed at which the object is moving. If hydrogen in a galaxy that is receding from Earth at half the speed of light emits radiation with a wavelength of 656 nm, will the radiation still be in the visible part of the electromagnetic spectrum when it reaches Earth?

**7.144.** The work function of mercury is $7.22 \times 10^{-19}$ J.

a. What is the minimum frequency of radiation required to eject photoelectrons from a mercury surface?

b. Could visible light produce the photoelectric effect in mercury?

## A Noble Family: Special Status for Special Behavior

*7.145. Compounds have been made from xenon and krypton with fluorine and oxygen, but not from the other noble gases. Suggest a reason, based on periodic properties, why xenon and krypton can be forced to form compounds but helium, neon, and argon cannot.

**7.146.** The successive ionization energies of the noble gases increase uniformly, in comparison to the successive ionization energies of the other elements, which at some place in the sequence show a rather large jump between values. Why is this the case?

*7.147. Helium is the only element that was discovered in space before it was found on Earth. Suggest a reason for this.

**7.148.** The first ionization energy of the noble gas neon is 2081 kJ/mol. The sodium ion ($Na^+$) is isoelectronic with neon, but the energy required to take one electron from the sodium ion is 4560 kJ/mol. Why is it so much harder to remove an electron from a sodium ion than from a neon atom if the two are isoelectronic?

If your instructor assigns problems in **smartwork**, log in at **smartwork.wwnorton.com**.

# 8

# Chemical Bonds: What Makes a Gas a Greenhouse Gas?

8.1 Types of Chemical Bonds

8.2 Lewis Structures

8.3 Polar Covalent Bonds

8.4 Vibrating Bonds and the Greenhouse Effect

8.5 Resonance

8.6 Formal Charge: Choosing among Lewis Structures

8.7 Exceptions to the Octet Rule

8.8 The Lengths and Strengths of Covalent Bonds

## Learning Outcomes

**LO1** Describe ways in which covalent, ionic, and metallic bonds are alike and ways in which they differ

**LO2** Draw Lewis structures of molecular compounds and polyatomic ions, including resonance structures when appropriate
**Sample Exercises 8.1, 8.2, 8.3, 8.4, 8.6, 8.7, 8.10, 8.11**

**LO3** Predict the polarity of covalent bonds based on differences in the electronegativities of the bonded elements
**Sample Exercise 8.5**

**LO4** Use formal charges to identify preferred resonance structures
**Sample Exercises 8.8, 8.9, 8.10**

**LO5** Describe how bond order, bond energy, and bond length are related

**LO6** Estimate enthalpy of reaction ($\Delta H_{rxn}$) from bond energies
**Sample Exercise 8.12**

## The Greenhouse Effect: Good News and Bad

**G**lobal warming is a hotly debated topic. However, most scientists agree that recent increases in the average temperature of Earth's surface (about half a Celsius degree in the last half century) are linked to increases in the concentration of carbon dioxide and other *greenhouse gases* in the atmosphere. These gases trap Earth's heat in its atmosphere much like panes of glass trap heat in a greenhouse. Increases in atmospheric concentrations of greenhouse gases have been linked to human activity, particularly to increasing rates of fossil fuel combustion and the destruction of forests that would otherwise consume $CO_2$ during photosynthesis.

Carbon dioxide allows solar radiation to pass through the atmosphere and warm Earth's surface, but it traps heat that would otherwise radiate from the warm surface back into space. Were it not for the presence of atmospheric $CO_2$ and other gases, Earth would be much colder than it is. Indeed, it would be too cold to be habitable. The presence of these gases has moderated Earth's average global temperature to within a fairly narrow range, 15°–17°C, for many thousands of years.

During the last 50 years, atmospheric concentrations of $CO_2$ have increased from about 315 to nearly 400 ppm. The atmosphere has not contained this much $CO_2$ in over half a million years, since long before our species evolved. We will not discuss the climatic and social consequences of global warming, but we will address the question of why $CO_2$ is so good at trapping heat. We noted in Chapter 7 that heat is associated with the infrared region of the electromagnetic spectrum. Carbon dioxide is good at trapping heat because it absorbs infrared radiation. To understand how $CO_2$ does this, whereas $O_2$ and $N_2$ do not, we need to consider the chemical bonds that hold atoms together in molecules.

**Greenhouse Effect** The glass windows of a greenhouse allow sunlight to enter ▶ but trap the warm air inside. Greenhouse gases, such as $CO_2$, behave in a similar fashion. Recent increases in atmospheric $CO_2$ concentrations have been linked to rising global temperatures and the threat of severe climate change.

**ionic bond** a chemical bond that results from the electrostatic attraction between cations and anions.

**bond length** the distance between the nuclear centers of two atoms joined together in a bond.

These bonds form when atoms share their outermost electrons. However "sharing" does not mean *equal* sharing. In this chapter we explore how unequal sharing of electron pairs, together with the motion of atoms held together by these bonds, allows some atmospheric gas, such as $CO_2$, to absorb some of the infrared radiation emitted by Earth's surface and perhaps contribute to global warming. ■

## 8.1 Types of Chemical Bonds

Fewer than 100 stable (nonradioactive) elements make up all the matter in our world. These elements combine to form over 60 million compounds, and the number grows daily as chemists synthesize new ones. This multitude of compounds exists because they are more stable than free elements. That is, atoms that are linked together by chemical bonds are in lower chemical energy states than the free atoms that form them.

Why are bonded atoms more stable? We introduced the principal reason in Chapter 5 when we discussed the concept of electrostatic potential energy ($E_{el}$) between charged particles. We noted that $E_{el}$ is directly proportional to the product of the charges ($Q_1$ and $Q_2$) on pairs of ions (or subatomic particles) and is inversely proportional to the distance ($d$) between them:

$$E_{el} \propto \frac{Q_1 \times Q_2}{d} \tag{5.3}$$

When we apply Equation 5.3 to particles of opposite charge, the product of $Q_1$ and $Q_2$ is negative and so is the value of $E_{el}$. Also, $E_{el}$ becomes more negative as the particles approach each other and the distance ($d$) between them decreases. This trend is illustrated in Figure 5.7 for a cation–anion pair. If such a pair of ions moved very close together they would experience electrostatic repulsion between their two clouds of negatively charged electrons and between their two positive nuclei. At the energy minimum in Figure 5.7, the forces of attraction and repulsion are in balance, and the cation–anion pair is held together by a stable **ionic bond**.

⊙⊙ **CONNECTION** Negative values of $E_{el}$ quantify the attraction between particles of opposite charge; positive values relate to the repulsion of particles with the same charge, as described in Chapter 5.

We can apply the concept of electrostatic potential energy to explain why two neutral atoms may come together to form a stable molecule. Let's consider the case of two hydrogen atoms (Figure 8.1a) that are sufficiently far apart that they do not interact with each other and have zero potential energy with respect to each other. As they approach each other, the proton in the nucleus of one H atom is attracted to the electron of the other, and vice versa. This mutual attraction produces a negative $E_{el}$ (Figure 8.1b). As the distance between the atoms decreases, the value of $E_{el}$ continues to decrease, eventually reaching a minimum when the nuclei are 74 pm apart (Figure 8.1c). If they come any closer together, the repulsion between their two positive nuclei more than offsets the mutual proton–electron attractions, and $E_{el}$ begins to rise rapidly with decreasing $d$ (Figure 8.1d).

▶️❚❚ **CHEMTOUR** Bonding

The distance 74 pm is the **bond length** of a H—H covalent bond. The value of $E_{el}$ at this distance, −432 kJ/mol, is the energy released when two moles of H atoms form one mole of $H_2$ molecules. You would have to add this much energy to break up a mole of $H_2$ molecules into free H atoms. Thus, 432 kJ/mol is the H—H bond energy or *bond strength*. We will explore the significance of bond length and bond strength in more detail in Section 8.8.

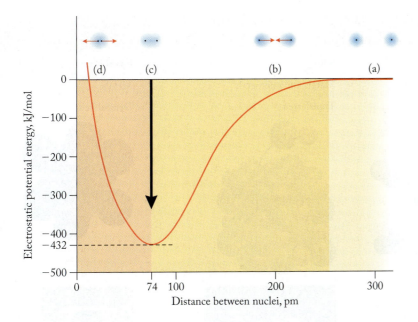

**FIGURE 8.1** The electrostatic potential energy ($E_{el}$) profile of a H—H bond: (a) Two H atoms that are far apart do not interact, so $E_{el} = 0$. (b) As the atoms approach each other, mutual attraction between their positive nuclei and negative electrons causes $E_{el}$ values to decrease. (c) At 74 pm $E_{el}$ reaches a minimum as the two atoms form a H—H bond. (d) If the atoms are even closer together, repulsion between their nuclei causes $E_{el}$ to increase and destabilizes the bond.

We can also use electrostatic potential energy to explain the interactions between the atoms in a solid piece of metal. As with atoms in molecules, the positive nucleus of each atom in a metallic solid is attracted to the electrons of the atoms that surround it. These attractions result in the formation of **metallic bonds**, which are unlike covalent bonds in that they are not pairs of electrons shared by pairs of atoms. Instead, the shared electrons in metallic bonds are highly delocalized, forming a "sea" of mobile electrons that move freely among all the atoms in a metallic solid. This electron mobility makes metals good conductors of heat and electricity.

Table 8.1 summarizes some of the differences between ionic, covalent, and metallic bonds. As you consider these properties, keep in mind that many bonds do not fall exclusively into just one category, but rather have some covalent, ionic, and even metallic character.

## 8.2 Lewis Structures

In 1916 American chemist G. N. Lewis (1875–1946) proposed that atoms form chemical bonds by sharing electrons. He further suggested that through this sharing, each atom acquires enough electrons to mimic the electron configuration of a noble gas. Today we associate such an electron configuration with completely filled $s$ and $p$ orbitals in the valence shells of atoms. Lewis's view of chemical bonding predated quantum mechanics and the notion of atomic orbitals, but it was consistent with what he called the **octet rule**: atoms of most main group elements tend to lose, gain, or share electrons so that each atom has eight valence electrons, or an *octet* of them. Hydrogen is an important exception to the octet rule in that each of its atoms shares only a single pair of bonding electrons, thereby acquiring a "duet" instead of an octet. In Table 8.1, we represent atoms involved in different types of bonding, but we make no attempt to show how electrons are distributed in bonds. Lewis devised a way of representing individual valence electrons in atoms so chemists could focus on how these electrons are distributed in various types of bonds.

**◉◉ CONNECTION** In Chapter 1 we described a chemical bond as the energy that holds two atoms in a molecule together. A covalent bond is defined in Chapter 2 as a bond between two atoms created by sharing pairs of electrons.

...................................................................................

**metallic bond** a chemical bond consisting of metal atoms surrounded by a "sea" of shared electrons.

**octet rule** atoms of main group elements make bonds by gaining, losing, or sharing electrons to achieve a valence shell containing eight electrons, or four electron pairs.

...................................................................................

| TABLE 8.1 | **Types of Chemical Bonds** | | |
|---|---|---|---|
| | **Covalent** | **Ionic** | **Metallic** |
| Elements involved: | Nonmetals and metalloids | Metals and nonmetals | Metals |
| Electron distribution: | Shared | Transferred | Delocalized |
| Microview: | | | |
| Macroview: | | | |

## Lewis Symbols

Lewis developed a system of symbols, called **Lewis symbols** or **Lewis dot symbols**, to explain the number of chemical bonds an atom typically forms to complete its octet. In other words, the Lewis symbol of an atom explains its **bonding capacity**. A Lewis symbol consists of the symbol of an element surrounded by dots representing the valence electrons in one of its atoms. The dots are placed on the four sides of the symbol (top, bottom, right, and left). The order in which they are placed does not matter *as long as one dot is placed on each side before any dots are paired*. The number of *unpaired* dots in a Lewis symbol indicates the typical bonding capacity of the atom.

Figure 8.2 shows the Lewis symbols of the main group elements. Because all elements in one of these groups have the same number of valence electrons, they all have the same arrangement of dots in their Lewis symbols. For example, the Lewis symbols of carbon and all other group 14 elements have four unpaired electrons. Four unpaired electrons means that a carbon atom tends to form four chemical bonds. In doing so, it completes its octet because these four bonds contain the carbon atom's original four valence electrons plus four more from the atoms with which it forms the bonds. Similarly, the Lewis symbols of nitrogen and all other group 15 elements have three unpaired electrons; a nitrogen atom *typically* forms three bonds to complete its octet.

Some important exceptions to Lewis's octet rule occur, several of which involve H, Be, and B. Compounds having fewer than eight valence electrons about an atom are referred to as *electron-deficient* compounds in Lewis theory. In another type of electron deficiency, atoms in some molecules have incomplete octets because there are odd numbers of valence electrons in those molecules. Examples include NO

**CONNECTION** The elements in groups 1, 2, and 13 through 18 in the periodic table are called main group elements (see Chapter 2). The monatomic ions of main group elements have filled *s* and *p* orbitals in their outermost shell and so are isoelectronic with a noble gas (Chapter 7).

**Lewis symbol** (also called **Lewis dot symbol**) the chemical symbol for an atom surrounded by one or more dots representing the valence electrons.

**bonding capacity** the number of covalent bonds an atom forms to have an octet of electrons in its valence shell.

and $NO_2$. We explore electron-deficient compounds and the chemical implications of the unpaired electrons in molecules in Section 8.7.

## Lewis Structures

The properties of molecular substances depend on how many atoms of which elements are in each of their molecules and how those atoms are bonded together. A **Lewis structure** is a two-dimensional representation of a molecule showing how the atoms in the molecule are connected. Because covalent bonds are *pairs* of shared valence electrons, Lewis structures focus on how these electron pairs, called **bonding pairs**, are distributed among the atoms in a molecule. In a Lewis structure, the pairs of electrons that are shared in chemical bonds are drawn with lines, as in the molecular structures we first saw in Chapter 1. A pair of electrons shared between two atoms is called a **single bond**. Electron pairs that are not involved in bond formation appear as pairs of dots on one atom. These unshared electron pairs are called **lone pairs**. Atoms with one or more pairs of dots in their Lewis symbols frequently have that same number of lone pairs of electrons in the Lewis structures of the molecules they form.

The following guidelines describe a five-step approach to drawing Lewis structures. The guidelines are particularly useful in drawing the structures of molecules and polyatomic ions that have a central atom bonded to atoms that have less bonding capacity. Many inorganic molecules, polyatomic ions, and many small organic molecules (fewer than four carbon atoms) fit this description. In addition, to build some larger structures, we may sometimes need to break the structure into subunits and consider more than one "central" atom, as in hydrogen peroxide:

$$H—\ddot{O}—\ddot{O}—H$$

In these cases, each atom that is bonded to two other atoms—each oxygen atom in $H_2O_2$—is the "central atom" within a three-atom subunit, as shown by the O atoms highlighted in red:

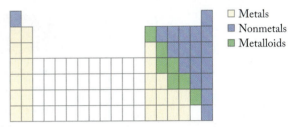

It is important to note that these initial guidelines are just a starting point in the discussion of molecular structure. We will continue to refine these guidelines as we draw structures for increasingly more challenging molecules and learn more about the distribution of electrons in bonds.

## Steps to Follow when Drawing Lewis Structures

1. *Determine the number of valence electrons.* For a neutral molecule, count the valence electrons in all the atoms in the molecule. For a polyatomic ion, count the valence electrons and then add or subtract the number of electrons needed to account for the charge on the ion. This total number of valence electrons is used to form bonds and lone pairs in the Lewis structure.

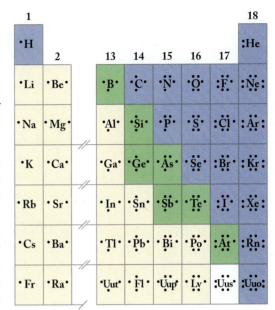

**FIGURE 8.2** Lewis symbols of the main group elements of the periodic table. Because elements in a family have similar outer-shell electron configurations, they have the same number of dots in their Lewis symbols, which represent valence electrons.

▶❚❚ **CHEMTOUR** Lewis Dot Structures

**Lewis structure** a two-dimensional representation of the bonds and lone pairs of valence electrons in a molecule or polyatomic ion.

**bonding pair** a pair of electrons shared between two atoms.

**single bond** a chemical bond that results when two atoms share one pair of electrons.

**lone pair** a pair of electrons that is not shared.

2. *Arrange the symbols of the elements in a pattern that shows how their atoms are bonded together and then connect them with single bonds (single pairs of bonding electrons).* Put the atom with the greatest bonding capacity in the center. Place the remaining atoms around the central atom and those that form the fewest bonds (hydrogen) around the periphery.

3. *Complete the octets or duets of the atoms bonded to the central atom by adding lone pairs of electrons.*

4. *Compare the number of valence electrons in the Lewis structure to the number determined in step 1.* If the numbers are the same, then all the electrons have been used and none need to be added or taken away. If there are not enough in the structure, add electrons to the central atom; if there are too many in the structure, delete lone pairs from bonded atoms.

5. *Complete the octet on the central atom.* If there is an octet on the central atom, the structure is done. If there is less than an octet on the central atom, create additional bonds to it by converting one or more lone pairs of electrons on atoms surrounding the central atom into bonding pairs.

---

**SAMPLE EXERCISE 8.1**    **Drawing the Lewis Structure of Chloroform**    **LO2**

Chloroform, $CHCl_3$, is a low-boiling liquid that was once used as an anesthetic in surgery. Draw its Lewis structure.

**Collect and Organize** We are given the molecular formula of chloroform, $CHCl_3$, and we can use the five-step approach described above to generate the Lewis structure.

**Analyze** The formula $CHCl_3$ tells us that a chloroform molecule contains one carbon atom, one hydrogen atom, and three chlorine atoms. Because carbon is a group 14 element, it has four valence electrons and needs four more to complete its octet. Hydrogen, in group 1, has one valence electron in the $1s$ orbital and needs one more to complete its duet. Chlorine, in group 17, has seven valence electrons (in the $3s$ and $3p$ atomic orbitals) and needs one more to complete its octet.

**Solve**

Step 1: The number of valence electrons in the $CHCl_3$ molecule is

| Element: | C | + | H | + | 3 Cl |
|---|---|---|---|---|---|
| Valence electrons: | 4 | + | 1 | + | $(3 \times 7) = 26$ |

Step 2: The carbon atom has the most (four) unpaired electrons in its Lewis symbol and hence has the greatest bonding capacity. Therefore C is the central atom in the molecule. Each H and each Cl needs one more electron to achieve the electron configuration of a noble gas, and so each forms one bond to the C atom:

$$
\begin{array}{c}
\text{H} \\
| \\
\text{Cl}\!-\!\overset{\displaystyle |}{\underset{\displaystyle |}{\text{C}}}\!-\!\text{Cl} \\
\text{Cl}
\end{array}
$$

Step 3: In this structure the hydrogen atom has a duet because it shares a bonding pair of electrons with the central carbon atom. We need to add lone pairs of electrons to complete the octets on the three chlorine atoms:

$$
\begin{array}{c}
\text{H} \\
| \\
:\!\ddot{\text{Cl}}\!-\!\overset{\displaystyle |}{\underset{\displaystyle |}{\text{C}}}\!-\!\ddot{\text{Cl}}\!: \\
:\!\ddot{\text{Cl}}\!: \\
\end{array}
$$

Step 4: This structure contains four pairs of bonding electrons and nine lone pairs, for a total of

$$(4 \times 2) + (9 \times 2) = 26 \text{ electrons}$$

which is the number of electrons determined in step 1.

Step 5: The carbon atom is surrounded by four bonds, which means eight electrons, so it has a full octet. The structure is done.

**Think About It** In this example, there was no need to change the structure during steps 4 and 5 because all the valence electrons are included and all the atoms have an octet (or duet in the case of the hydrogen atom).

 **Practice Exercise** Draw the Lewis structure of methane, $CH_4$.

*(Answers to Practice Exercises are in the back of the book.)*

---

**SAMPLE EXERCISE 8.2**    **Drawing the Lewis Structure of Ammonia**    **LO2**

Draw the Lewis structure of ammonia, $NH_3$.

**Collect and Organize** Ammonia has the molecular formula $NH_3$.

**Analyze** The formula $NH_3$ tells us that each molecule contains one atom of nitrogen and three atoms of hydrogen. Nitrogen is a group 15 element with five valence electrons, three of which are unpaired, for a bonding capacity of 3. Each hydrogen atom has one valence electron in the $1s$ orbital and a bonding capacity of 1.

**Solve**

Step 1: The number of valence electrons in the $NH_3$ molecule is

| Element: | N | + | 3 H |
|---|---|---|---|
| Valence electrons: | 5 | + | $(3 \times 1) = 8$ |

Step 2: The nitrogen atom has the greater bonding capacity and is the central atom. Connecting each H atom to the nitrogen atom with a covalent bond yields

$$H-N-H$$
$$\vert$$
$$H$$

Step 3: Each bonded H atom has a complete duet of electrons.

Step 4: The three bonds in the structure represent $3 \times 2 = 6$ valence electrons, but we need 8 to match the number determined in step 1. To add 2 electrons we add a lone pair to nitrogen:

$$H-\overset{..}{N}-H$$
$$\vert$$
$$H$$

Step 5: The N atom now has an octet of electrons, so the Lewis structure is complete.

**Think About It** This structure makes sense because the Lewis symbol of the nitrogen atom contains two electrons that are paired and three that are unpaired. Therefore, it is reasonable that the nitrogen atom in $NH_3$ has one lone pair and three bonding pairs of electrons.

 **Practice Exercise** Draw the Lewis structure of phosphorus trichloride.

---

## Lewis Structures of Molecules with Double and Triple Bonds

We can use Lewis structures to show the bonding in molecules in which two atoms share more than one pair of bonding electrons. A bond in which two atoms share two pairs of electrons is called a **double bond**. For example, when the two oxygen

**double bond** a chemical bond in which two atoms share two pairs of electrons.

**triple bond** a chemical bond in which two atoms share three pairs of electrons.

atoms in a molecule of $O_2$ share two pairs of electrons, they form an O=O double bond. When the two nitrogen atoms in a molecule of $N_2$ share three pairs of electrons, they form a N≡N **triple bond**. In Section 8.8 we discuss the characteristics of these multiple bonds and compare them with those of single bonds.

How do we know when a Lewis structure has a double or triple bond? Typically, we find out when we apply steps 4 and 5 in the guidelines. Suppose we fill the octets of all the atoms attached to the central atom (step 4), and in doing so we use all the valence electrons that we have. We then determine in step 5 that the central atom does not have an octet. We do not have any more electrons available, so the only way we can provide the central atom with an octet is to *convert a lone pair of electrons* from one of the other atoms into a bonding pair. If the central atom still does not have an octet, we may convert a second lone pair into a bonding pair. The following example illustrates this application of the guidelines.

Let's draw the Lewis structure for the organic molecule formaldehyde ($H_2CO$).

Step 1: The total number of valence electrons is

| Element: | C | + | 2 H | + | O | |
|----------|---|---|-----|---|---|---|
| Valence electrons: | 4 | + | (2 × 1) | + | 6 | = 12 |

Step 2: Of the three elements, carbon has the greatest bonding capacity (4). Therefore, C is the central atom. Connecting it with single bonds to the other three atoms we have

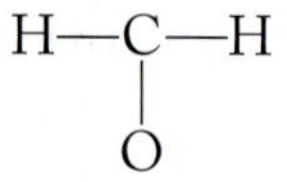

$$H—C—H$$
$$|$$
$$O$$

Step 3: Each H atom has a single covalent bond (2 electrons) and a complete valence shell. Oxygen needs three lone pairs of electrons to complete its octet:

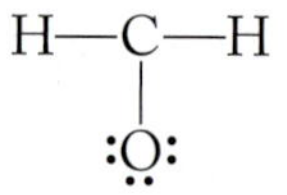

$$H—C—H$$
$$|$$
$$:\ddot{O}:$$

Step 4: There are 12 valence electrons in this structure, which matches the number determined in step 1.

Step 5: The central C atom has only 6 electrons. To provide the carbon atom with 2 more electrons so that it has an octet, but without removing any from oxygen, which already has an octet, we convert one of the lone pairs on the oxygen atom into a bonding pair between C and O:

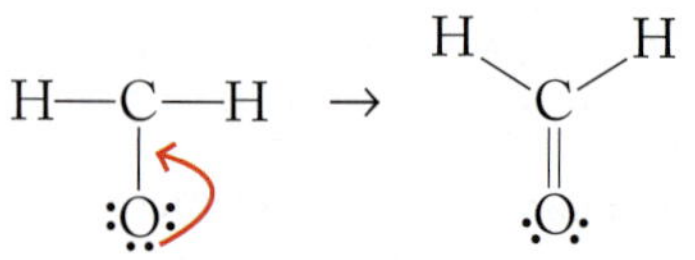

It does not matter which of the three lone pairs we move because they are equivalent. The central carbon atom now has a complete octet, and the oxygen atom still does. This structure makes sense because the four covalent bonds around carbon—two single bonds and one double bond—match its bonding capacity. The double bond to oxygen makes sense because oxygen is a group 16 element with a bonding capacity of 2, and it has two bonds in this structure.

Notice that we have drawn the double bond and the two single bonds on the central carbon atom at an angle of about 120° from each other in the final structure; we have done the same thing with the lone pairs of electrons on oxygen with respect to the double bond. Electrons are negatively charged and repel each other, so we draw them as far apart from each other as possible. This logic will be an important part of drawing three-dimensional structures of molecules in Chapter 9.

---

**SAMPLE EXERCISE 8.3**   **Drawing the Lewis Structure**          **LO2**
                          **of Acetylene**

Draw the Lewis structure of acetylene, $C_2H_2$, the hydrocarbon fuel used in oxyacetylene torches for welding and cutting metal.

**Collect and Organize**  We are to draw the Lewis structure of acetylene, which has the molecular formula $C_2H_2$. We follow the five steps in the guidelines.

**Analyze**  Each molecule contains two atoms of carbon and two atoms of hydrogen. Carbon is a group 14 element with a bonding capacity of 4.

**Solve**
Step 1:   The total number of valence electrons is

| Element: | 2 C | + | 2 H | |
|---|---|---|---|---|
| Valence electrons: | $(2 \times 4)$ | + | $(2 \times 1)$ | = 10 |

Step 2:   Of the two elements, carbon has the greater bonding capacity, so the two carbon atoms are placed in the center of the molecule. Connecting each carbon atom with single bonds to the other atoms, we have:

$$H—C—C—H$$

Step 3:   Each H atom has a single covalent bond and a complete valence shell.
Step 4:   The structure has 6 valence electrons, but should have 10. This means we have to add two pairs of electrons. One way to do that is to add two bonds between the carbon atoms. This gives the structure the right number of valence electrons, and it completes the octets of both carbon atoms (step 5):

$$H—C≡C—H$$

**Think About It**  This structure makes sense because carbon has a bonding capacity of 4, and there are four covalent bonds around each carbon atom—1 single bond and 1 triple bond—giving both atoms complete octets.

 **Practice Exercise**  Determine the Lewis structure for carbon dioxide.

---

## Lewis Structures of Ionic Compounds

We can use the octet rule and Lewis structures to explain the composition of many common ionic compounds, such as sodium chloride (NaCl). Crystals of NaCl are held together by attractive forces between oppositely charged ions. By giving away its $3s^1$ valence electron, a sodium atom becomes a positively charged $Na^+$ ion and achieves the same electron configuration as Ne: $1s^2 2s^2 2p^6$. In other words, $Na^+$ has a vacant valence shell. If a chlorine atom, which has the valence electron configuration $3s^2 3p^5$, gains an electron to form a $Cl^-$ ion, it achieves a

**polar covalent bond** unequal sharing of bonding pairs of electrons between atoms.

**bond polarity** a measure of the extent to which bonding electrons are unequally shared due to differences in electronegativity of the bonded atoms.

**CONNECTION** We defined ionization energy in Chapter 7 as the amount of energy required to remove 1 mole of electrons from 1 mole of particles in the gas phase.

$$H\!-\!\overset{..}{\underset{..}{Cl}}:$$

$$\overset{\delta+}{H}\!-\!\overset{\delta-}{\underset{..}{\overset{..}{Cl}}}:$$

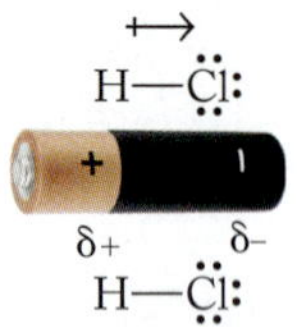

**FIGURE 8.3** Just as a battery has positive and negative poles, a polar molecule such as HCl has positive and negative ends, represented here by the arrow above the H—Cl bond. The tail of the arrow is on the hydrogen atom, which has a partial positive charge, and the arrowhead is pointed toward the Cl, which has a partial negative charge. The partial positive charge on the hydrogen atom may also be represented by a lowercase delta (δ) followed by a + sign. Correspondingly, the partial negative charge on the chlorine is represented by the δ− symbol.

filled outermost shell and becomes isoelectronic with the noble gas argon. We can illustrate this behavior using Lewis symbols as

$$Na^{\cdot} + \overset{..}{\underset{..}{\cdot Cl}}: \rightarrow Na^{+}\left[:\overset{..}{\underset{..}{Cl}}:\right]^{-}$$

The brackets around the chloride ion are used to emphasize that all eight valence electrons are associated with it; none of them are associated with the sodium ion.

All alkali metal elements and alkaline earth elements tend to lose, rather than share, their valence electrons, in part because they have low ionization energies. The cations they form by losing these electrons obey the octet rule because they have the electron configurations of the noble gases that precede the parent elements in the periodic table.

**SAMPLE EXERCISE 8.4** **Drawing Lewis Structures of Binary Ionic Compounds** LO2

Draw the Lewis structure of calcium oxide.

**Collect and Organize** We are to draw the Lewis structure of an ionic compound.

**Analyze** When a metal combines with a nonmetal, the metal forms a cation and the nonmetal forms an anion. The monatomic ions formed by calcium and oxygen are $Ca^{2+}$ and $O^{2-}$, respectively.

**Solve** To form a cation with a 2+ charge, a Ca atom loses both its valence electrons, leaving it with none:

$$\overset{\cdot}{Ca}\!\cdot \xrightarrow{-2e^-} Ca^{2+}$$

When an O atom accepts two electrons it has a complete octet:

$$:\overset{..}{O}\!\cdot \xrightarrow{+2e^-} \left[:\overset{..}{\underset{..}{O}}:\right]^{2-}$$

Combining the Lewis symbols of $Ca^{2+}$ and $O^{2-}$ ions, we have the Lewis structure of CaO:

$$Ca^{2+}\left[:\overset{..}{\underset{..}{O}}:\right]^{2-}$$

**Think About It** The lack of dots in the symbol of the $Ca^{2+}$ ion reinforces the fact that atoms of Ca, like all main group metal atoms, lose all the electrons in their valence shells when they form monatomic ions. In contrast, atoms of nonmetals form monatomic anions with complete octets.

 **Practice Exercise** Draw the Lewis structure of magnesium fluoride.

# 8.3 Polar Covalent Bonds

When Lewis proposed that atoms form chemical bonds by sharing electrons, he knew that electron sharing in covalent bonds did not necessarily mean *equal* sharing. For example, he knew, based on the chemical properties of HCl, that the H—Cl bond is a **polar covalent bond**. Lewis explained this **bond polarity** by assuming that the shared pair of electrons in HCl is closer to the chlorine end of the molecule than to the hydrogen end. In Figure 8.3 we show this unequal sharing using an arrow with a plus sign embedded in its tail (⟶) to indicate the *direction of polarity*: the arrow points toward the more negative, electron-rich atom in the bond, and the position of the plus sign indicates the more positive, electron-poor atom.

Figure 8.4 shows examples of equal and unequal sharing of bonding pairs of electrons. It includes (a) $Cl_2$, in which the two Cl atoms share a pair of electrons equally in a **nonpolar covalent bond**; (b) a polar covalent bond in HCl; and (c) an ionic bond in NaCl, which represents an extreme case of unequal sharing. In NaCl, the bonding pair has been completely transferred from the Na atom to the Cl atom, creating a $Na^+$ ion and a $Cl^-$ ion. The color bar at the top of Figure 8.4 indicates the values associated with the colors: yellow-green = no charge separation; violet and red = full 1+ and 1− charges, respectively; intermediate colors = partial charges ($\delta+$ and $\delta-$).

Another American chemist, Linus Pauling (1901–1994), developed the concept of **electronegativity** to explain bond polarity. Pauling assigned electronegativity values to the elements (Figure 8.5) based on the idea that the bonds between atoms of different elements are neither 100% covalent nor 100% ionic, but somewhere in between. The degree of **ionic character** of a bond depends on the differences in the abilities of the two atoms to attract the electrons they share: the greater the difference, the more ionic is the bond between them.

The data in Figure 8.5 show that electronegativity is a periodic property, with values generally increasing left to right across a row in the periodic table, and decreasing top to bottom down a group. The reasons for these trends are essentially the same ones that produce similar trends in first ionization energies, as we discussed in Chapter 7 (see Figure 7.38). Increasing attraction between the nuclei of atoms and their outer-shell electrons with increasing atomic number across a row produces both higher ionization energies and greater electronegativities (Figure 8.6a). Within a group of elements, the weaker attraction between nuclei and outer-shell electrons with increasing atomic number leads to lower ionization energies and smaller electronegativities (Figure 8.6b). For these two reasons, the most electronegative elements—fluorine, oxygen, and nitrogen—are in the upper

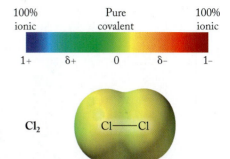

▶️II **CHEMTOUR** Partial Charges and Bond Dipoles

| 100% ionic | | Pure covalent | | 100% ionic |
|---|---|---|---|---|
| 1+ | $\delta+$ | 0 | $\delta-$ | 1− |

$Cl_2$

(a) Pure covalent: even charge distribution

HCl

(b) Polar covalent: uneven charge distribution

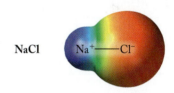

NaCl

(c) Ionic: complete transfer of electron

**FIGURE 8.4** Variations in valence electron density are represented using colored surfaces in these molecular models. (a) In the covalent bond in $Cl_2$, uniform electron density is represented by a uniformly colored surface, indicating that the two atoms share their bonding pair of electrons equally. (b) Unequal sharing of the bonding pair of electrons occurs in HCl, as shown by the orange color of the Cl end of the bond and the green color around the H end. (c) In ionic NaCl, the violet color on the surface of the sodium ion indicates that it has a full 1+ charge and the red of the chloride ion reflects its charge of 1−.

**nonpolar covalent bond** a bond characterized by an even distribution of charge; electrons in the bonds are shared equally by the two atoms.

**electronegativity** a relative measure of the ability of an atom to attract electrons in a bond to itself.

**ionic character** an estimate of the magnitude of charge separation in a covalent bond.

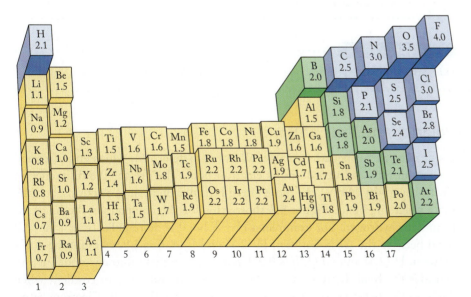

**FIGURE 8.5** The Pauling electronegativity values shown below the symbols of the elements are unitless values that indicate the relative ability of an atom in a bond to attract shared electrons to itself. Electronegativity generally increases from left to right across a period and decreases from top to bottom down a group.

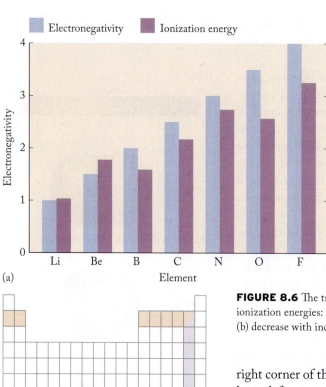

(a)

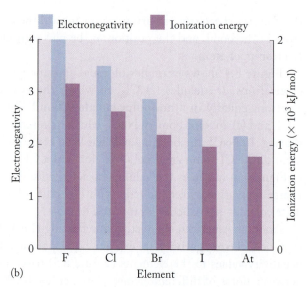

(b)

**FIGURE 8.6** The trends in the electronegativities of the main group elements follow those of ionization energies: both tend to (a) increase with increasing atomic number across a row and (b) decrease with increasing atomic number within a group.

right corner of the periodic table, and the least electronegative elements are in the lower left corner. Other electronegativity scales have been developed with slightly different values from those proposed by Pauling; however, whichever scale you use, the trends in electronegativity remain the same as shown in Figure 8.6.

Now that we have defined electronegativity, we can add to the second guideline for drawing Lewis structures: *The central atom in a Lewis structure is often the atom with the lowest electronegativity.*

**CONCEPT TEST** ........................................................................

Why does Figure 8.5 not include electronegativity values for the noble gases?

*(Answers to Concept Tests are in the back of the book.)*
.................................................................................................

**CONCEPT TEST** ........................................................................

Draw arrows on the Lewis structure of $CO_2$ to indicate the polarity of the bonds.
.................................................................................................

## Polarity and Type of Bond

Comparing electronegativity values allows us to determine which end of a bond is electron-rich and which end is electron-poor. The greater the difference in electronegativity ($\Delta EN$), the more uneven the distribution of electrons, and the more polar the bond. Figure 8.7 shows this for the compounds the halogens form with hydrogen. For example, a H—F bond is more polar than a H—Cl bond because the $\Delta EN$ between H (2.1) and F (4.0) is 1.9, whereas the $\Delta EN$ between H (2.1) and Cl (3.0) is 0.9. Under a somewhat arbitrary guideline, we consider the bond between two atoms to be ionic rather than covalent when $\Delta EN$ values between them are equal to or greater than 2.0. Electronegativity differences greater than 2.0 frequently exist in compounds formed between metals and nonmetals. For example, calcium oxide is considered an ionic compound because the difference in electronegativity between Ca (1.0) and O (3.5)

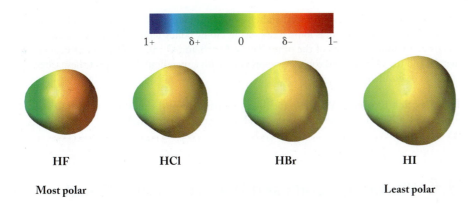

**FIGURE 8.7** The greater the electronegativity difference (ΔEN) between two atoms, the more polar is the bond they form. Among the hydrogen halides, HF has the largest ΔEN and HI has the smallest. Therefore the HF bond is the most polar in this group, and the HI bond is the least polar.

is 2.5. On the other hand, carbon–hydrogen bonds, while polar, are considered weakly polar because their electronegativity difference is only 0.4. The bromine–chlorine bond or boron–silicon bond (ΔEN = 0.2) is close to a nonpolar covalent bond.

Figure 8.8 provides an approximate scale for judging the degree of polarity in a chemical bond based on electronegativity differences. Keep in mind that the scale in Figure 8.8 should be used with caution—LiCl and HF both have an electronegativity difference of 1.9, but lithium chloride is an ionic compound whereas HF is a covalently bonded molecule.

Figure 8.9 illustrates both symbols used to indicate polarity in bonds: the arrow and δ+ and δ−. The lowercase Greek letter delta followed by a negative sign, δ−, is used to indicate which end of a bond has a partial negative charge, and δ+ indicates which end has a partial positive charge. The term *partial charge* indicates that there is unequal distribution of the shared electron pair in a bond, but some sharing still occurs. Complete transfer of the shared pair to the more electronegative atom would produce a full 1− charge on it and 1+ on the less electronegative atom.

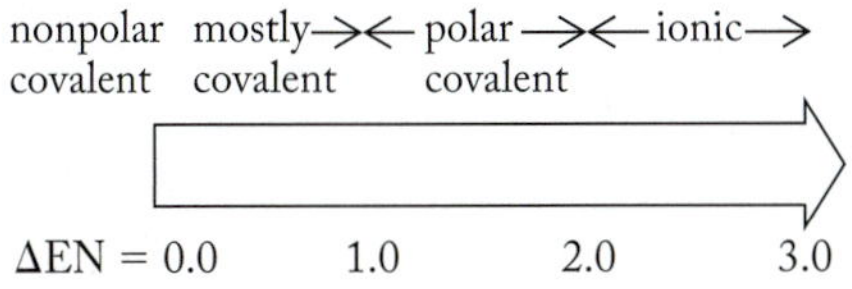

nonpolar   mostly→ ←polar→ ←ionic→
covalent   covalent        covalent

ΔEN = 0.0          1.0          2.0          3.0

**FIGURE 8.8** As the electronegativity difference (ΔEN) between two atoms increases, the bond between the two atoms becomes more ionic.

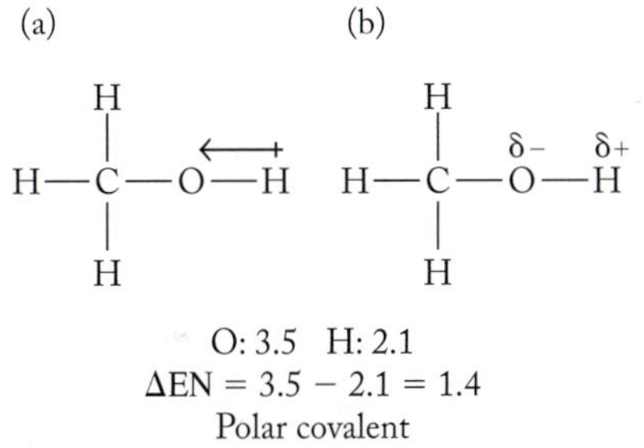

(a)                    (b)

$$\text{H} \qquad\qquad \text{H}$$
$$| \qquad\qquad\qquad |$$
$$\text{H—C—O—H} \quad \text{H—C—O—H}$$
$$| \qquad\qquad\qquad |$$
$$\text{H} \qquad\qquad \text{H}$$

O: 3.5   H: 2.1
ΔEN = 3.5 − 2.1 = 1.4
Polar covalent

**FIGURE 8.9** Two ways to indicate the polarity of the O—H bond in methanol include (a) the use of an arrow with a plus sign on its tail (see Figure 8.3) and (b) the use of δ+ and δ−. In structure a, the tail is placed next to the H atom with the partial positive charge and the head of the arrow points toward the more electronegative O atom. In structure b, the symbols δ+ and δ− indicate the presence of partial positive and negative charges on the H and O atoms, respectively.

---

**SAMPLE EXERCISE 8.5**  **Comparing the Polarity of Bonds**  **LO3**

Rank, in order of increasing polarity, the bonds formed between: O and C; Cl and Ca; N and S; O and Si. Are any of these bonds considered ionic?

**Collect and Organize**  We are given four pairs of atoms and are to rank them according to the polarity of the bond each pair forms and to identify any ionic bonds in the set.

**Analyze**  The polarity of a bond is related to the difference in electronegativities of the atoms in the bond. The guideline we apply is as follows: if the electronegativity difference is 2.0 or greater, the bond is considered ionic. We need to refer to the Pauling electronegativities (Figure 8.5) to judge the relative polarity.

**Solve**  Calculate the electronegativity difference between the atoms:

| O and C: | ΔEN = 3.5 − 2.5 = 1.0 |
| Cl and Ca: | ΔEN = 3.0 − 1.0 = 2.0 |
| N and S: | ΔEN = 3.0 − 2.5 = 0.5 |
| O and Si: | ΔEN = 3.5 − 1.8 = 1.7 |

These electronegativity differences are proportional to the polarity of the bonds formed between the pairs of atoms. Therefore, ranking them in order of increasing polarity we have

N—S < O—C < O—Si < Cl—Ca

The bond between Cl and Ca is considered ionic because ΔEN = 2.0.

**Think About It** Calcium is a metal and chlorine is a nonmetal, so the result indicating that the bond between them is ionic is reasonable. Ionic bonds tend to be formed between metals and nonmetals. Two of the other bonds, N—S and O—C, connect pairs of nonmetals, and the O—Si bond connects a nonmetal with a metalloid. We expect these three bonds to be covalent.

**Practice Exercise** Which of the following pairs forms the most polar bond: O and S; Be and Cl; N and H; C and Br? Is the bond between that pair ionic?

# 8.4 Vibrating Bonds and the Greenhouse Effect

As we noted in Chapter 5 in the discussion of thermal energy, chemical bonds are not rigid. They vibrate a little, stretching and bending like tiny atomic-sized springs (Figure 8.10). These vibrations have natural frequencies, which match the frequencies of infrared electromagnetic radiation. This match, coupled with the unequal sharing of bonding electrons, allows atmospheric molecules with polar bonds, such as $CO_2$, to absorb infrared radiation emitted by Earth's surface. As a result, heat that might have dissipated into space is trapped in the atmosphere by these molecules, contributing to what is called the greenhouse effect. The understanding of bond vibrations requires subtle and complex models that are beyond the scope of this book, but the basic ideas here give a useful picture of the interaction between bond vibrations and infrared radiation.

▶❚❚ **CHEMTOUR** Greenhouse Effect

Molecules with polar bonds may absorb and emit photons, much like electrons in atoms absorb radiation and form excited states (see Section 7.4). The radiation absorbed by polar bonds, leading to bond vibrations, is typically in the infrared region (see Figure 7.1). Vibrations result in tiny fluctuating electrical fields associated with the separation of partial charges in the molecule. These fluctuations can alter the strengths of the fields or even create new ones, depending on the nature of the vibration. When the frequencies of the bond vibrations match the frequencies of photons of infrared radiation, the molecules may absorb those photons. Scientists say these vibrations are *infrared active*. This is the molecular mechanism behind the greenhouse effect. The presence of fluctuating electric fields in molecules is further evidence that covalent bonds involve shared electrons.

Not all of the ways polar bonds vibrate absorb and emit infrared radiation. For example, two kinds of stretching vibrations can occur in a molecule of $CO_2$. One is a symmetric stretching vibration (Figure 8.10a) in which the two C=O bonds stretch and then compress in opposite directions. In this case the two fluctuating electrical fields produced by the two C=O bonds cancel each other out, and no infrared absorption or emission is possible. This vibration is said to be *infrared inactive*. However, when the bonds stretch in the same direction—that is, one gets shorter as the other gets longer (Figure 8.10b)—the changes in charge separation do not cancel. This asymmetric stretch produces a fluctuation in the electrical field surrounding the overall $CO_2$ molecule and enables it to absorb infrared radiation. Molecules can also bend (Figure 8.10c) in ways that produce fluctuating electrical fields. Because the frequencies of the asymmetric stretching and bending of the bonds in $CO_2$ are in the same range as the frequencies of infrared radiation emitted from Earth's surface, carbon dioxide is a potent greenhouse gas.

**FIGURE 8.10** Three modes of bond vibration in a molecule of $CO_2$ include (a) symmetric stretching of the C=O bonds, which produces no overall change in the polarity of the molecule; (b) asymmetric stretching, which does produce side-to-side fluctuations in polarity that may result in absorption of IR radiation; (c) bending, which produces up-and-down fluctuations that also may absorb IR radiation.

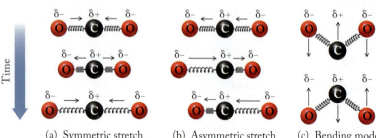

(a) Symmetric stretch (infrared inactive)

(b) Asymmetric stretch (infrared active)

(c) Bending mode (infrared active)

▶‖ **CHEMTOUR** Vibrational Modes

**CONCEPT TEST**

Nitrogen and oxygen make up about 99% of the gases in the atmosphere. Could the stretching of the N≡N and O=O bonds in these molecules result in the absorption of infrared radiation? Explain why or why not.

## 8.5 Resonance

The atmosphere contains two types of molecular oxygen. Most of it is $O_2$, but trace concentrations of ozone ($O_3$) are also present. Ozone in the lower atmosphere is sometimes referred to as "bad ozone" because high levels damage crops, harm trees, and lead to human health problems. Ozone is also present in the upper atmosphere, where it is considered "good ozone" because it shields life on Earth from potentially harmful UV radiation from the sun.

Ozone is a *triatomic* (three-atom) molecule produced naturally by lightning (Figure 8.11) and accounts for the pungent odor you may have smelled after a severe thunderstorm. Ozone ($O_3$) and diatomic oxygen ($O_2$) have the same empirical formula: O. Different molecular forms of the same element, such as $O_2$ and $O_3$, are called **allotropes** of the element and have different chemical and physical properties. Ozone, for example, is an acrid, pale blue gas that is toxic even at low concentrations, whereas $O_2$ is a colorless, odorless gas that is essential for most life forms. We saw another example of allotropes in Chapter 5—graphite, diamond, and buckyballs, which are all composed only of carbon atoms but differ in the arrangement of the atoms and hence in their physical and chemical properties.

The different molecular formulas of allotropes mean that they have different molecular structures. Let's draw the Lewis structure for ozone, $O_3$, following our five-step approach. Oxygen is a group 16 element and so has 6 valence electrons. Therefore the total number of valence electrons in an ozone molecule is $3 \times 6 = 18$ (step 1). Connecting the three O atoms with single bonds, we have (step 2)

$$O-O-O$$

Completing the octets of the noncentral atoms gives us (step 3)

$$:\ddot{O}-O-\ddot{O}:$$

This structure contains 16 electrons. We determined in step 1 that there are 18 valence electrons in the molecule, so we add 2 to the central oxygen atom (step 4):

$$:\ddot{O}-\underset{..}{O}-\ddot{O}:$$

This structure leaves the central atom two electrons short of an octet, so we convert one of the lone pairs on the O atom on the left end of the molecule into a bonding pair (step 5):

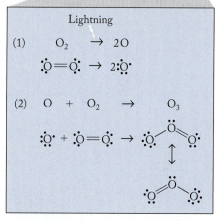

However, we could just as well have used a lone pair from the O atom on the right, which would have given us

**FIGURE 8.11** Lightning strikes contain sufficient energy to break O=O double bonds. The O atoms formed in this fashion collide with other $O_2$ molecules, forming ozone ($O_3$), an allotrope of oxygen.

**allotropes** different molecular forms of the same element, such as oxygen ($O_2$) and ozone ($O_3$).

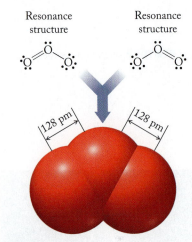

Resonance structure    Resonance structure

**FIGURE 8.12** The molecular structure of ozone is an average of the two resonance structures shown at the top of the figure. Both bonds in ozone are 128 pm long, a value between the average length of an O—O single bond (148 pm) and the average length of an O=O double bond (121 pm). The intermediate value for the ozone bond length indicates that in an ozone molecule, the bonds are neither single bonds nor double bonds, but something in between.

**resonance** a characteristic of electron distributions when two or more equivalent Lewis structures can be drawn for one compound.

**resonance structure** one of two or more Lewis structures with the same arrangement of atoms but different arrangements of bonding pairs of electrons.

Which structure is correct? Experimental evidence indicates that neither structure is accurate. Scientists have determined that both bonds in ozone have the same length. Since a double bond is always shorter than a single bond between the same two atoms (see Section 8.8), the structure of ozone cannot consist of one single bond and one double bond. Figure 8.12 shows that the length of the two bonds in $O_3$ (128 pm) is in between the length of an O—O single bond (148 pm) and an O=O double bond (121 pm). One way to explain this result is to assume that the bonding pattern of $O_3$ is an average of the two structures drawn above:

In other words, each pair of atoms has the equivalent of 1.5 bonds between each pair of atoms. It is very important to note that this average does *not* mean that the molecule spends half its time as the left-hand structure in Figure 8.12 and half as the right-hand structure. It always has three bonding pairs spread out evenly on the two sides of the central atom.

To better understand this bonding pattern, consider what happens when the bonding electrons and the lone pair electrons in structure a below are rearranged as shown by the red arrows:

This rearrangement produces structure b. The process is completely reversible, so that the electron pairs in structure b could just as easily be rearranged into the pattern in structure a:

The two structures are equivalent. We use a double-headed reaction arrow between the structures to indicate that they are equivalent and that the actual structure is a blend of the two:

The ability to draw two equivalent Lewis structures for the ozone molecule illustrates an important concept in Lewis's theory called **resonance**: the existence of multiple Lewis structures, called **resonance structures** (or sometimes *resonance hybrids*), for a given arrangement of atoms. To determine whether resonance occurs in a molecule, we need to determine whether there can be alternative bond arrangements inside the molecule: arrangements in which the positions of some bonding pairs of electrons change but the positions of the atoms stay the same. One indicator of the possibility of resonance is the presence of both single and double bonds from a central atom to two or more atoms of the same element, as we just saw in ozone.

Because all resonance structures are Lewis structures by definition, the five-step guidelines apply to drawing resonance structures. Because all the atoms in ozone are oxygen, an oxygen atom is clearly the central atom. Two additional issues, however, must be raised at this point, both of which relate to the concept of bonding capacity. First, note that the oxygen atoms in ozone have different numbers of bonds. Oxygen atoms *typically* form two bonds, but Lewis theory allows both more and fewer bonds than are implied by the Lewis symbol of an atom. Second, in previous examples we have used differences in typical bonding capacity as a criterion to select a central atom. As we deal with more molecules, the situation frequently arises that several atoms in one molecule may have the same bonding capacity. In that case, the selection of the central atom is based on electronegativity, with the least electronegative atom chosen as the central atom in the Lewis structure.

▶❚❚ **CHEMTOUR**  Resonance

---

**SAMPLE EXERCISE 8.6**   **Drawing Resonance Structures** **LO2**
                                       **of a Molecule**

Sulfur trioxide ($SO_3$) is a pollutant produced in the atmosphere when the $SO_2$ from natural and industrial sources combines with $O_2$. Draw all the possible resonance structures of $SO_3$ assuming sulfur obeys the octet rule.

**Collect and Organize**  Each $SO_3$ molecule contains one atom of sulfur and three atoms of oxygen. We are to draw all resonance structures, which means at least two different Lewis structures with different bonding patterns but the same atom placement.

**Analyze**  Both sulfur and oxygen are in group 16 and have 6 valence electrons per atom and a typical bonding capacity of 2. Sulfur is less electronegative than oxygen.

**Solve**

Step 1: The number of valence electrons in the $SO_3$ molecule is

| Element: | S | + | 3 O |
|---|---|---|---|
| Valence electrons: | 6 | + | $(3 \times 6) = 24$ |

Step 2: Because sulfur is less electronegative, we select it as the central atom. Connecting it with single bonds to the three O atoms:

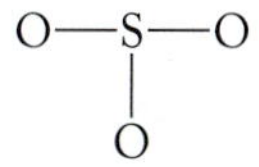

$$O\!-\!S\!-\!O$$
$$\overset{\displaystyle |}{O}$$

Step 3: Each O atom needs three lone pairs of electrons to complete its octet:

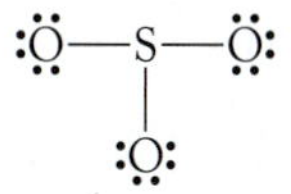

$$:\!\ddot{O}\!-\!S\!-\!\ddot{O}\!:$$
$$\overset{\displaystyle |}{\underset{..}{:\!\ddot{O}\!:}}$$

Step 4: There are 24 valence electrons in the structure in step 3, which matches the number determined in step 1.

Step 5: The central S atom has only 6 electrons in the structure of step 3. To complete its octet, we convert a lone pair on one of the oxygen atoms into a bonding pair:

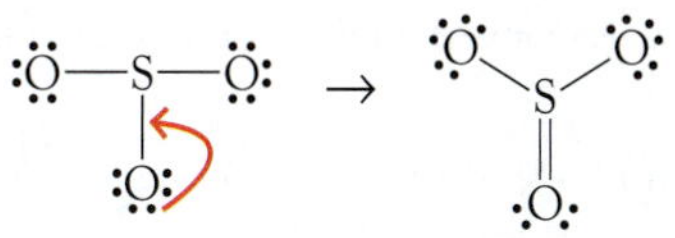

The sulfur atom now has a complete octet, and the Lewis structure is complete.

Because the three oxygen atoms are equivalent in the structure in step 3, we cannot arbitrarily choose one of them and ignore the others in forming a double bond. Therefore, we need three resonance structures to describe the bonding in $SO_3$:

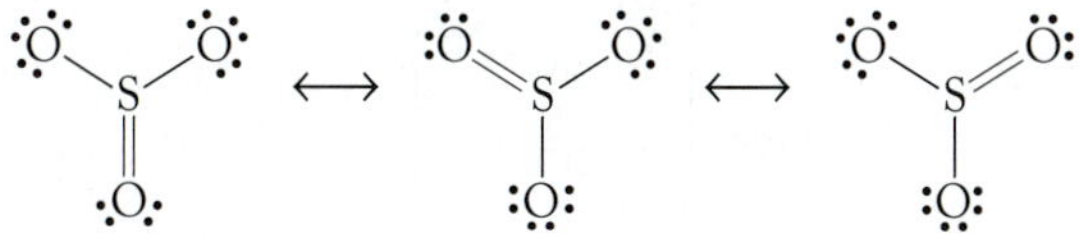

**Think About It** It makes sense that $SO_3$ has three resonance forms because three equivalent O atoms are bonded to the central S atom, and any one of the O atoms could be the one with the double bond.

⚙ **Practice Exercise** Draw all possible resonance structures for sulfur dioxide.

We can also draw resonance structures for polyatomic ions. In doing so, we need to account for the charge on each ion, which means adding the appropriate number of valence electrons to a polyatomic anion and subtracting the appropriate number from a polyatomic cation.

---

**SAMPLE EXERCISE 8.7** **Drawing Resonance Structures of a Polyatomic Ion** **LO2**

Draw all the resonance structures for the nitrate ion, $NO_3^-$.

**Collect and Organize** We are to draw the resonance structures for the $NO_3^-$ ion, which contains one nitrogen atom and three oxygen atoms. The charge of the ion is 1−.

**Analyze** Nitrogen is a group 15 element and has 5 valence electrons per atom and a bonding capacity of 3. Oxygen is in group 16 and has 6 valence electrons and a bonding capacity of 2. The 1− charge means the ion has an additional valence electron.

**Solve**

Step 1: The number of valence electrons is

| Element: | N | + | 3 O | |
|---|---|---|---|---|
| Valence electrons: | 5 | + | $(3 \times 6) = 23$ | |
| Additional electrons due to the 1− charge: | | | | + 1 |
| Total valence electrons: | | | | 24 |

Step 2: The nitrogen atom has the higher bonding capacity and the lower electronegativity, so N is the central atom in a $NO_3^-$ ion. Connecting N with single bonds to the three O atoms, we have

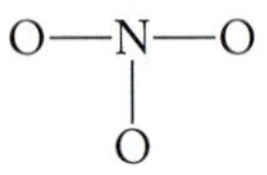

$$O—N—O$$
$$|$$
$$O$$

Step 3: Each O atom needs three lone pairs of electrons to complete its octet:

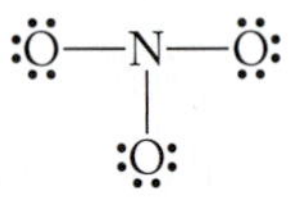

$$:\ddot{O}—N—\ddot{O}:$$
$$|$$
$$:\ddot{O}:$$

Step 4: There are 24 valence electrons in this structure, which matches the number determined in step 1.

Step 5: The central N atom has only 6 electrons. To provide it with the 2 more it needs to complete its octet, we convert a lone pair on one of the oxygen atoms into a bonding pair:

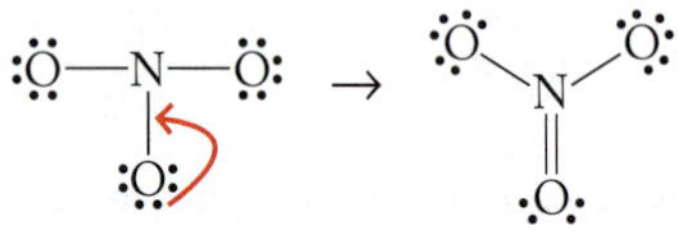

$$:\ddot{O}—N—\ddot{O}: \rightarrow \ddot{O} \diagdown N \diagup \ddot{O}$$
$$| \qquad \qquad \|$$
$$:\ddot{O}: \qquad \qquad \cdot\ddot{O}\cdot$$

The nitrogen atom now has a complete octet. Adding brackets and the ionic charge, we have a complete Lewis structure:

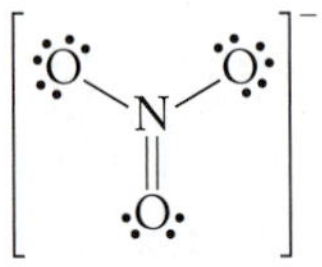

$$\left[ :\ddot{O} \diagdown N \diagup \ddot{O}: \atop \| \atop \cdot\ddot{O}\cdot \right]^-$$

Using the O atom to the left or right of the central N atom to form the double bond, instead of the O atom below N, creates two additional resonance forms, three in all:

$$\left[ \overset{..}{\underset{..}{:}O} \overset{}{\underset{N}{\diagdown}} \overset{..}{\underset{..}{O}:} \right]^{-} \longleftrightarrow \left[ :\overset{..}{O} \overset{}{\underset{N}{=}} \overset{..}{\underset{..}{O}:} \right]^{-} \longleftrightarrow \left[ :\overset{..}{\underset{..}{O}} \overset{}{\underset{N}{\diagdown}} \overset{..}{\underset{..}{O}:} \right]^{-}$$

**Think About It** It makes sense that the $NO_3^-$ ion has three resonance forms because three equivalent O atoms are bonded to the central N atom, and any one of the O atoms could be the one with the double bond. The additional electron from the negative charge on the ion means that it has the same number of valence electrons as $SO_3$ (Sample Exercise 8.6), which has the same number of resonance structures. Nitrogen in neutral molecules typically has a bonding capacity of 3, but in these resonance structures it makes four bonds.

**Practice Exercise** Draw all the resonance forms of the azide ion, $N_3^-$, and the nitronium ion, $NO_2^+$.

Resonance also occurs in organic (carbon-containing) molecules that have alternating single and double bonds. Molecules of benzene ($C_6H_6$), for example, contain six-membered rings of carbon atoms with alternating single and double bonds (Figure 8.13a). When we fix the atoms in a molecule of benzene in place, we have two equivalent ways to draw the single and double bonds. To depict their equivalency, chemists frequently draw benzene molecules with circles in the centers (Figure 8.13b). This symbol emphasizes that the six carbon–carbon bonds in the ring are all identical and intermediate in character between single and double bonds. (Note that experimental evidence confirms that all six C—C bonds in benzene are equivalent.)

# 8.6 Formal Charge: Choosing among Lewis Structures

Let's turn our attention to the molecular structure of another atmospheric pollutant, dinitrogen monoxide ($N_2O$), also known as nitrous oxide or, more commonly, laughing gas. Its common name is linked to the fact that people who inhale high concentrations of $N_2O$ usually laugh spontaneously. Gaseous nitrous oxide has long been used as an anesthetic in dentistry because it has a narcotic effect.

In the lower atmosphere, nitrous oxide concentrations range between 0.1 and 1.0 ppm. It is produced naturally by bacterial action in soil and through human activities such as agriculture and sewage treatment. It is occasionally in the news because its concentration in the troposphere (the atmosphere at ground level) has been increasing. Along with $CO_2$ and other gases, it may be contributing to global warming.

Let's draw the Lewis structure of nitrous oxide. First we count the number of valence electrons: 5 each from the two nitrogen atoms and 6 from the

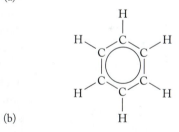

(a)

(b)

**FIGURE 8.13** (a) The molecular structure of benzene is an average of two equivalent structures. (b) The average is frequently represented by a circle inside the hexagonal ring, indicating completely uniform distribution of the electrons in the bonds around the ring.

oxygen atom for a total of $(2 \times 5) + 6 = 16$. The central atom is a nitrogen atom because N has a higher bonding capacity than O. Connecting the atoms with single bonds, we have

$$N—N—O$$

Completing the octets of the noncentral atoms gives us a structure with 16 valence electrons, which is the number determined in step 1:

$$:\ddot{N}—N—\ddot{O}:$$

However, there are only two bonds and thus only four valence electrons on the central N atom. To give it 4 more electrons, we need to convert lone pairs on the surrounding atoms to bonding pairs. Which lone pairs do we choose? We could use two lone pairs from the N atom on the left to form a N≡N triple bond:

$$:N≡N—\ddot{O}:$$

Alternatively, we could use two lone pairs on the O atom to make a N≡O triple bond:

$$:\ddot{N}—N≡O:$$

Or we could use one pair from each terminal atom to make two double bonds:

$$\ddot{N}=N=\ddot{O}$$

Which of these resonance structures is best? We have seen that in some sets of resonance structures, such as those of $O_3$, $SO_3$, and $NO_3^-$, all the structures are equivalent, so no one of them is any more important than another in giving us a sense of the actual bonding in the molecule. This is not the case with the resonance forms of $N_2O$. To decide which resonance form in a nonequivalent set is the most important and comes closest to representing the actual bonding pattern in a molecule, we use the concept of formal charge.

A **formal charge (FC)** is not a real charge but rather a measure of the number of electrons *formally assigned* to an atom in a molecular structure, compared to the number of electrons in the free atom. To determine a formal charge, we follow a series of steps to calculate the number of electrons formally assigned to each atom in each resonance structure. Once we have determined the formal charges, we then use these two criteria to select the preferred structure:

1. The preferred structure is the one with formal charges of zero; if no such structure can be drawn, the preferred structure is the one with the most formal charges equal to zero or closest to zero.
2. Any negative formal charges should be on the atom(s) of the most electronegative element(s).

**formal charge (FC)** a value calculated for an atom in a molecule or polyatomic ion by determining the difference between the number of valence electrons in the free atom and the sum of lone-pair electrons plus half of the electrons in the atom's bonding pairs.

## Calculating Formal Charge of an Atom in a Resonance Structure

First we select one atom in the molecule.

Step 1: Determine the number of valence electrons in the free atom.
Step 2: Count the number of lone-pair electrons on the atom in the molecular structure.

Step 3: Count the number of electrons in bonding pairs on the atom and divide that number by 2.

Step 4: Sum the results of steps 2 and 3 and subtract that sum from the number determined in step 1.

Summarizing these steps in the form of an equation, we have

$$\text{FC} = \left( \begin{array}{c} \text{number of} \\ \text{valence } e^- \end{array} \right) - \left[ \begin{array}{c} \text{number of} \\ \text{unshared } e^- \end{array} + \tfrac{1}{2} \left( \begin{array}{c} \text{number of } e^- \\ \text{in bonding pairs} \end{array} \right) \right] \qquad (8.1)$$

The calculation of formal charge assumes that each atom has exclusive "title" to the electrons in its lone pairs and shares its bonding electrons equally with the atoms at the other ends of the bonds. We can confirm that the three resonance forms of $N_2O$ are not equivalent by calculating the formal charges in each structure.

In Table 8.2 we have colored the lone pairs of electrons red and the shared pairs green to make it easier to track the quantities of these electrons in the formal charge calculation. The numbers of valence electrons in the free atoms are shown in blue.

| TABLE 8.2 | Formal Charge Calculations for the Resonance Structures of $N_2O$ | | | | | | | | |
|---|---|---|---|---|---|---|---|---|---|
| Step | | :N≡N—Ö: | | | :N̈=N=O: | | | :N̈—N≡O: | |
| 1 | Number of valence electrons | 5 | 5 | 6 | 5 | 5 | 6 | 5 | 5 | 6 |
| 2 | Number of electrons in lone pairs | 2 | 0 | 6 | 4 | 0 | 4 | 6 | 0 | 2 |
| 3 | Number of shared electrons | 6 | 8 | 2 | 4 | 8 | 4 | 2 | 8 | 6 |
| 4 | FC = [valence − (lone pair + $\tfrac{1}{2}$ shared)] | 0 | +1 | −1 | −1 | +1 | 0 | −2 | +1 | +1 |

To illustrate one of the formal charge calculations in the table, consider the N atom at the left end of the left resonance structure. The Lewis dot symbol of nitrogen reminds us that free nitrogen atoms have five valence electrons. In this structure, the N atom has 2 electrons in a lone pair and 6 electrons in three shared (bonding) pairs. Using Equation 8.1 to calculate the formal charge on this N atom,

$$\text{FC} = 5 - [2 + \tfrac{1}{2}(6)] = 0$$

which is the first value in the bottom row of the table. The results of similar formal charge calculations for all the other atoms in the three resonance structures complete the row. Note that the sum of the formal charges on the three atoms in each structure is zero, as it should be for a neutral molecule. When we do an analysis of the formal charges of atoms in a polyatomic ion, the formal charges on its atoms must add up to the charge on the ion.

Now we must apply the two criteria for selecting the preferred resonance structure of $N_2O$. The first thing to note about these sets of formal charges is that in none of them are all three formal charges zero, meaning that the first part of criterion 1 is not met. The next step is to identify which structure has the most formal charge values that are the closest to zero, such as −1 or +1. On this count we have a tie between the structure on the left (0, +1, −1) and the one in the middle (−1, +1, 0).

To break the tie, we invoke the second criterion and answer the question, "In which structure is the negative formal charge on the more electronegative atom?"

The answer is the structure on the left, which has an oxygen atom with a formal charge of −1. We conclude that this structure is the best of the three in representing the actual bonding in a molecule of $N_2O$.

From experimental measurements, we know that the N≡N—O structure contributes the most to the bonding in $N_2O$, the middle structure (N=N=O) also contributes to the bonding. We know this because the length of the bond between the two nitrogen atoms is between the length of a N=N bond and the length of a N≡N bond, and the nitrogen–oxygen bond is a little shorter than a typical N—O single bond.

**CONCEPT TEST**

What is the formal charge on a sulfur atom that has three lone pairs of electrons and one bonding pair?

---

**SAMPLE EXERCISE 8.8** **Selecting Resonance Structures Based on Formal Charges** **LO4**

Which of these resonance forms best describes the actual bonding in a molecule of $CO_2$?

$$:\ddot{O}—C≡O: \quad \longleftrightarrow \quad \ddot{O}=C=\ddot{O} \quad \longleftrightarrow \quad :O≡C—\ddot{O}:$$

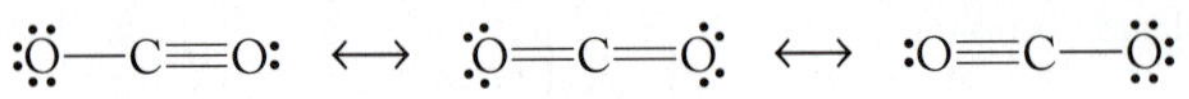

**Collect and Organize** We are given three resonance forms for $CO_2$. Formal charges can be used to select the most representative structure.

**Analyze** The preferred structure is the one in which the formal charges are closest to zero and any negative formal charges are on the more electronegative atom. In this case, oxygen is a group 16 element and is more electronegative than carbon, a group 14 element. Each free carbon atom has four valence electrons, and free oxygen atoms have six valence electrons each.

**Solve** We use Equation 8.1 to find the formal charge on each atom. We illustrate these results in a table:

| Formal Charge Calculations for the Resonance Structures of $CO_2$ | | | | | | | | | |
|---|---|---|---|---|---|---|---|---|---|
| **Step** | $:\ddot{O}—C≡O:$ | | | $\ddot{O}=C=\ddot{O}$ | | | $:O≡C—\ddot{O}:$ | | |
| 1 Number of valence electrons | 6 | 4 | 6 | 6 | 4 | 6 | 6 | 4 | 6 |
| 2 Number of electrons in lone pairs | 6 | 0 | 2 | 4 | 0 | 4 | 2 | 0 | 6 |
| 3 Number of shared electrons | 2 | 8 | 6 | 4 | 8 | 4 | 6 | 8 | 2 |
| 4 FC = [valence − (lone pair + $\frac{1}{2}$ shared)] | −1 | 0 | +1 | 0 | 0 | 0 | +1 | 0 | −1 |

The formal charges are zero on all the atoms in the middle resonance form with the two double bonds. Therefore this structure best represents the actual bonding in a molecule of $CO_2$.

**Think About It** Notice that the sum of the formal charges is zero in all three resonance forms. Valid Lewis structures of all neutral molecules have net formal charges of zero.

✦ **Practice Exercise** Which resonance forms of the azide ion, $N_3^-$, and the nitronium ion, $NO_2^+$, contribute the most to the actual bonding in each ion?

Let's take a final look at the resonance structures for $N_2O$. This time our purpose is to examine the link between the formal charge on an atom in a resonance structure and the bonding capacity of that atom. The Lewis symbol of nitrogen, which has three unpaired electrons, tells us that a nitrogen atom can complete its octet by forming three bonds. The two unpaired electrons in the Lewis symbol of oxygen tell us that the bonding capacity of an oxygen atom is 2. In the $N_2O$ resonance structures in Table 8.2, the nitrogen atom with three bonds has a formal charge of zero (the left N in the first structure), and the oxygen atom with two bonds has a formal charge of zero (the O in the middle structure). As a general rule—and assuming the octet rule is obeyed—atoms have zero formal charges in resonance structures in which the numbers of bonds they form match their bonding capacities. If an atom forms one more bond than its bonding capacity, such as an oxygen atom with three bonds (O in the third structure in the table), the formal charge is +1. If the atom forms one fewer bond than the bonding capacity, such as an oxygen atom with one bond (O in the first structure in the table), the formal charge is −1.

# 8.7 Exceptions to the Octet Rule

Earth's atmosphere contains trace concentrations of several compounds that illustrate the limitations of the octet rule. Two such compounds are the nitrogen oxides NO and $NO_2$, generated by human activities like welding (Figure 8.14) and driving gasoline-burning vehicles, which contribute to photochemical smog formation in urban areas. Each has an odd number of valence electrons per molecule, which means that at least one atom in each molecule cannot have a complete octet.

### Odd-Electron Molecules

Nitric oxide (NO) is produced when high temperatures in vehicle engines or other sources lead to the reaction

$$N_2(g) + O_2(g) \rightarrow 2\,NO(g) \qquad (8.2)$$
$$\text{Nitric oxide}$$

The nitric oxide then reacts with atmospheric oxygen to produce nitrogen dioxide:

$$2\,NO(g) + O_2(g) \rightarrow 2\,NO_2(g) \qquad (8.3)$$
$$\text{Nitric oxide} \qquad\qquad \text{Nitrogen}$$
$$\text{dioxide}$$

Nitric oxide is highly reactive because it is an odd-electron molecule. To understand the implications of this, let's draw its Lewis structure. Nitric oxide has 11 valence electrons: nitrogen contributes 5 and oxygen 6. There is no central atom so we start with a single bond between N and O and then complete the octet around O, which is the more electronegative element:

$$N—\overset{\cdot\cdot}{\underset{\cdot\cdot}{O}}:$$

We then place the remaining 3 electrons around the N atom:

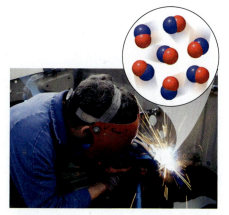

**FIGURE 8.14** The high temperatures of arc welding produce significant concentrations of nitric oxide (NO) as a result of the highly endothermic reaction $N_2 + O_2 \rightarrow 2\,NO$. The U.S. Environmental Protection Agency limit on NO concentrations in the air is 25 ppm, or 0.0025% by volume.

**free radical** an odd-electron molecule with an unpaired electron in its Lewis structure.

This leaves the N atom short three valence electrons. We can increase its number by converting a lone pair on the O atom into a bonding pair:

$$\cdot\ddot{N}{=}\ddot{O}{:}$$

This change has the added advantage of creating a double-bonded O atom, which gives it a formal charge of zero. The formal charge on the N atom is also zero. The only problem with the structure is that N does not have an octet: it has only seven electrons. Because nitrogen is less electronegative than oxygen, it is reasonable that we short-change it when there are not enough electrons to complete the octets of both atoms. The fact that NO exists indicates that there are exceptions to the complete octet requirement. When that happens, the most representative Lewis structures are those that come *as close as possible* to producing zero formal charges and complete octets.

Compounds that have odd numbers of valence electrons are called **free radicals**. They are typically very reactive species because it is often energetically favorable for them to acquire an electron from another molecule or ion. This characteristic makes them excellent oxidizing agents and is responsible for the damage they may cause to materials or living tissue with which they come in contact.

---

**SAMPLE EXERCISE 8.9** **Drawing Lewis Structures of Odd-Electron Molecules** **LO4**

Draw the resonance structures of nitrogen dioxide ($NO_2$), assign formal charges to the atoms, and suggest which structure may best represent the actual bonding in the molecule.

**Collect and Organize** We need to draw the resonance forms of $NO_2$ and assign formal charges to the atoms. Each molecule contains one atom of nitrogen (bonding capacity 3) and two atoms of oxygen (bonding capacity 2).

**Analyze** Nitrogen dioxide is an odd-electron molecule, so we anticipate that one of the atoms will not have a complete octet. We analyze the resonance structures by assigning formal charges to select the one most representative of the actual bonding in the molecule.

**Solve** The number of valence electrons is

| Element: | N | + | 2 O |
|---|---|---|---|
| Valence electrons: | 5 | + | $(2 \times 6) = 17$ |

Nitrogen is less electronegative than oxygen and has the greater bonding capacity, so it is the central atom:

$$O{-}N{-}O$$

Completing the octets on the O atoms gives

$$:\ddot{O}{-}N{-}\ddot{O}:$$

There are 16 valence electrons in this structure, but we need 17 to match the number available in the molecule. We add 1 more electron to the N atom:

$$:\ddot{O}{-}\dot{N}{-}\ddot{O}:$$

There are only 5 valence electrons around the N atom, fewer than the 8 we need. We can increase this number by converting a lone pair on one of the O atoms to a bonding pair, giving the formal charges shown in red:

$$:\underset{..}{\overset{0}{O}}{=}\overset{+1}{\dot{N}}{\diagdown}\underset{..}{\overset{-1}{\ddot{O}}}:$$

An equivalent resonance form can be drawn with the double bond on the right side:

$$\overset{-1}{:\!\ddot{O}\!\cdot}\!-\!\overset{+1}{\overset{\cdot}{N}}\!=\!\overset{0}{\ddot{O}\!:}$$

These structures are equivalent: each has a charge of 1+ on the nitrogen atom, one oxygen with a 1– charge, and one with a charge of 0 (zero).

**Think About It** The two Lewis structures are equivalent because the two O atoms in each structure are equivalent. The structures do not satisfy the octet rule, but the formal charges of the atoms are close to zero, and the negative formal charge is on the atom of the more electronegative element. Both O atoms have complete octets, leaving the less electronegative N atom one electron short in this odd-electron molecule.

⚙ **Practice Exercise** Nitrogen trioxide ($NO_3$) may form in polluted air when $NO_2$ reacts with $O_3$. Draw its Lewis structure(s).

## Atoms with More than an Octet

Another important atmospheric pollutant is sulfur hexafluoride ($SF_6$), which may be the most potent greenhouse gas (in terms of IR radiation absorbed per mole) present in the atmosphere (Figure 8.15). Each of its molecules contains six sulfur–fluorine covalent bonds, which means each S atom is surrounded by 12 valence electrons, not 8.

Let's consider the Lewis structure of $SF_6$. Its valence electron inventory (step 1) is

| Element: | S | + | 6 F |
|---|---|---|---|
| Valence electrons: | 6 | + | $(6 \times 7) = 48$ |

Sulfur has a greater bonding capacity (2) than fluorine (1), so S is the central atom (step 2). However, connecting six fluorine atoms to the sulfur atom means that sulfur's bonding capacity is significantly exceeded:

```
      F   F
       \ /
   F — S — F
       / \
      F   F
```

Completing the octets on the fluorine atoms (step 3)

```
    :F·  ·F:
       \ /
  :F — S — F:
       / \
    :F·  ·F:
```

gives us a structure with 48 valence electrons (step 4), a match with the number determined for the molecule. This structure is correct, even though it breaks the octet rule. It tells us that, in order to accommodate six fluorine atoms about a central sulfur atom, the sulfur atom is able to *expand its valence shell* to accommodate more than 8 electrons. Six S—F bonds and 12 valence electrons around S leave the $SF_6$ molecule with zero formal charge on each atom:

$$\text{Formal charge of S} = 6 - [0 + \tfrac{1}{2}(12)] = 0$$
$$\text{Formal charge of F} = 7 - [6 + \tfrac{1}{2}(2)] = 0$$

Based on our criteria for judging Lewis structures, this is a good one, but how can a sulfur atom have more than eight valence electrons?

**FIGURE 8.15** Electrical transformers use sulfur hexafluoride as an insulator because it is thermally stable and does not react with water. This compound, however, is classified as a possible contributor to global warming because it can leak from transformers and other electrical equipment and enter the atmosphere. Because it is so unreactive, it may remain in the atmosphere for thousands of years. It is also an exceptionally effective absorber of infrared radiation. The Intergovernmental Panel on Climate Change has identified $SF_6$ as the most potent greenhouse gas it has ever evaluated, with over $10^4$ times the global warming potential of $CO_2$.

The answer to this question is not completely clear. Some researchers believe the availability of *d* orbitals in large-Z atoms explains the larger valence shells, whereas others think it is simply due to the larger size of the atom. We discuss this further in Chapter 9. For the moment, we note that compounds with expanded octets are observed for elements with $Z > 12$. Nevertheless, the fact that some atoms have the ability to expand their valence shell does not mean these atoms *always* do so. Instead, they tend to do so in two situations:

▶❚❚ **CHEMTOUR** Expanded Valence Shells

1. When they form compounds with strongly electronegative elements, particularly F, O, and Cl.
2. When an expanded shell results in smaller formal charges on the atoms in a molecule.

To illustrate a situation in which expanding a valence shell results in smaller formal charges, let's consider the sulfate ion, $SO_4^{2-}$. When high-sulfur coal is burned, $SO_2$ is released into the atmosphere and reacts with atmospheric oxygen to form sulfur trioxide:

$$2\ SO_2(g) + O_2(g) \rightarrow 2\ SO_3(g)$$

Sulfur trioxide, in turn, reacts with water vapor to form sulfuric acid:

$$SO_3(g) + H_2O(g) \rightarrow H_2SO_4(\ell)$$

Sulfuric acid, a principal component of acidic precipitation ("acid rain") in eastern North America and in Europe, dissociates in dilute aqueous solutions:

$$H_2SO_4(aq) \rightarrow 2\ H^+(aq) + SO_4^{2-}(aq)$$

In the Lewis structure of the sulfate ion, a central S atom is bonded to four O atoms. Applying the method for drawing Lewis structures and assigning formal charges, we get

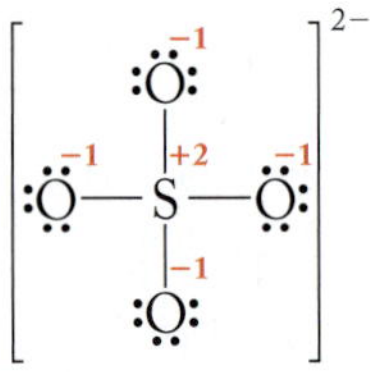

The sum of the formal charges on atoms in the ion is $1(+2) + 4(-1) = -2$. This calculation yields the correct ionic charge, but remember that the goal is to minimize formal charges in Lewis structures. We can do this by expanding the valence shell of sulfur:

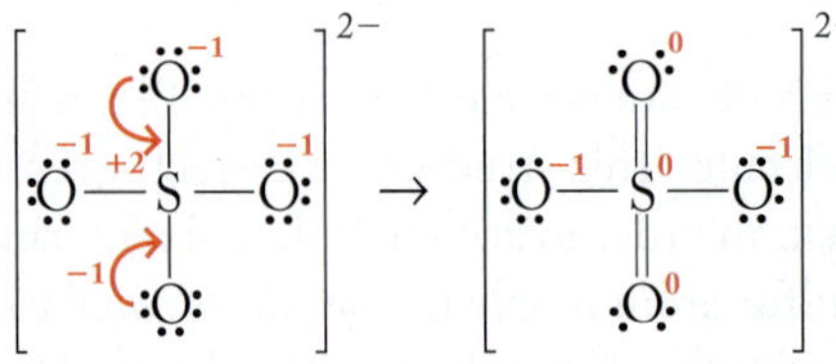

Each oxygen atom still has a complete octet, but now the sulfur has expanded its valence shell to accommodate 12 electrons. In this way, the formal charges change from +2 to 0 on sulfur and from −1 to 0 on two of the four oxygen atoms. The remaining two −1 values sum to −2, which is the value required to give the structure its overall 2− charge.

We can draw the two double bonds at any locations around the sulfur atom in the Lewis structure for the sulfate ion. Consequently, this structure has six equivalent resonance forms (not shown).

If we now wanted to draw the Lewis structure for sulfuric acid, we could bond two hydrogen ions to the two negative oxygen atoms:

$$H-\ddot{O}-S-\ddot{O}-H$$

Each hydrogen atom has achieved a duet of electrons, each oxygen atom has an octet, and every atom has a formal charge of zero.

---

**SAMPLE EXERCISE 8.10  Drawing Lewis Structures   LO2 and LO4
of Ions with an Expanded
Valence Shell**

Draw the Lewis structure for the phosphate ion ($PO_4^{3-}$) that minimizes the formal charges on its atoms.

**Collect and Organize** Each ion contains one atom of phosphorus and four atoms of oxygen and has an overall charge of 3−.

**Analyze** Phosphorus and oxygen are in groups 15 and 16 and have bonding capacities of 3 and 2, respectively. Phosphorus is in row 3 ($Z = 15$), so we can expand its octet if we need to.

**Solve** The number of valence electrons is

| Element: | P | + | 4 O |
|---|---|---|---|
| Valence electrons: | 5 | + | $(4 \times 6) = 29$ |
| Additional electrons due to the 3− charge: | | | $+ 3$ |
| Total valence electrons: | | | $32$ |

Phosphorus has the greater bonding capacity (3) and so is the central atom:

$$O-P-O \ (\text{with } O \text{ above and below})$$

Each O atom needs three lone pairs of electrons to complete its octet:

$$:\ddot{O}-P-\ddot{O}:$$

There are 32 valence electrons in this structure, which matches the number determined for the ion. Therefore, it is a complete Lewis structure of a polyatomic ion once we add the brackets and electrical charge:

$$\left[:\ddot{O}-P-\ddot{O}:\right]^{3-}$$

Each O has a single bond and a formal charge of −1; the four bonds around the P atom are one more than its bonding capacity, so its formal charge is +1. The sum of the formal charges, [+1 + 4(−1)] matches the charge on the ion, 3−.

We can reduce the formal charge on P by increasing the number of bonds to it, and we can do that by converting a lone pair on one of the O atoms into a bonding pair:

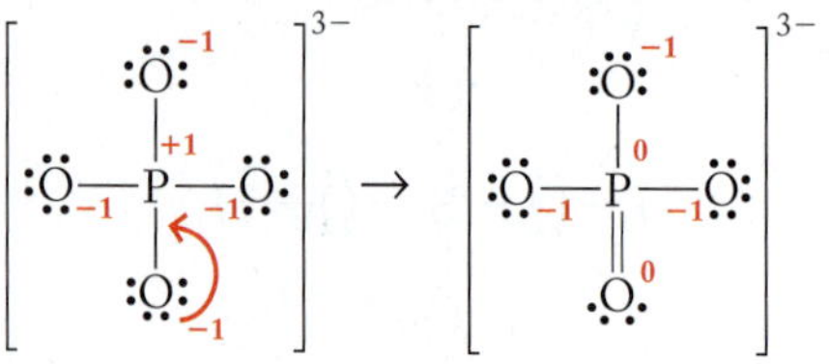

At the same time, we change a single-bonded O atom into a double-bonded O atom and thereby make its formal charge zero. Therefore, the structure on the right, in which the P atom has ten valence electrons, is the best Lewis structure we can draw for the phosphate ion.

**Think About It**  Note that three other equivalent resonance structures can be drawn with a double bond to one of the other O atoms.

**Practice Exercise**  Draw the resonance structures of the selenite ion ($SeO_3^{2-}$) that minimize the formal charges on the atoms.

## Atoms with Less than an Octet

We have seen that sulfur can combine with fluorine to form $SF_6$, a compound with an expanded octet. When fluorine combines with boron, a compound with the formula $BF_3$ is isolated. The electronegativity difference between boron and fluorine is 2.0, which puts $BF_3$ on the borderline between ionic and covalent compounds. If $BF_3$ contains three fluoride anions and a $B^{3+}$ cation, then the ions are isoelectronic with a noble gas. Boron trifluoride, however, does not have the physical properties associated with an ionic compound; it is a gas that boils at −101°C. What does the Lewis structure for $BF_3$ look like?

Boron trifluoride has a total of 24 valence electrons (step 1):

| Element: | B | + | 3 F |
|---|---|---|---|
| Valence electrons: | 3 | + | $(3 \times 7) = 24$ |

Boron has a greater bonding capacity (3) than fluorine (1), so B is the central atom (step 2). Connecting three fluorine atoms to the boron atom and completing the octets on fluorine (step 3) gives us a structure with the requisite number of valence electrons (step 4) on each F but not on the boron atom.

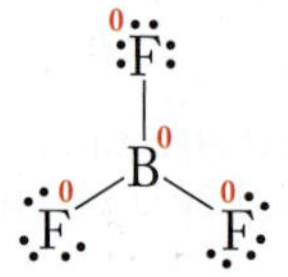

Formal charges are minimized in this structure; however, boron shares only six electrons and has no electrons in lone pairs. If we share an additional pair of electrons from one of the three F atoms, as shown in the three resonance forms on the next page, we can complete the octets of all the atoms. But forming a B═F double bond results in unfavorable formal charges, particularly for the F atom. We do not expect a positive formal charge on the most electronegative of all elements.

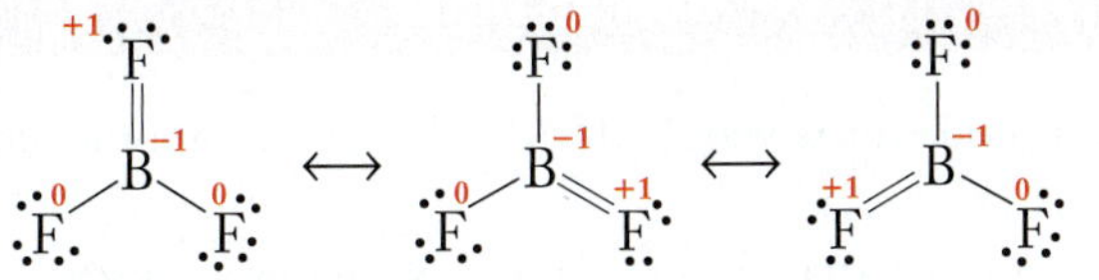

How do we resolve this conflict? The B—F bond lengths in $BF_3$ are all equal, but shorter than the B—F single bonds in the $BF_4^-$ ion in $NH_4BF_4$:

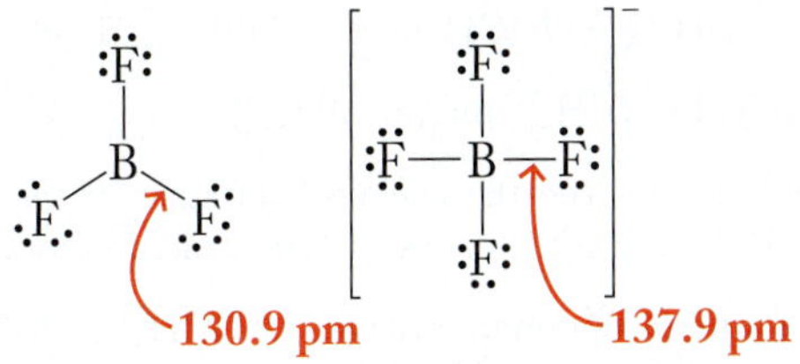

130.9 pm          137.9 pm

As we will see in the next section, multiple bonds are generally shorter than single bonds. This suggests that all four of the following resonance forms contribute to the molecular structure of $BF_3$:

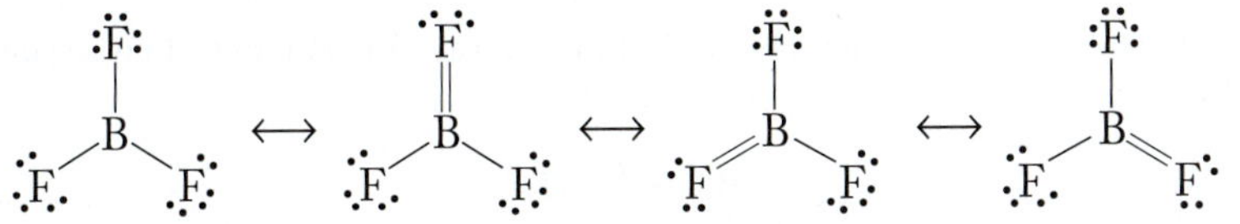

Dominant contributor

The atoms in the other group 13 halides complete their octets in different ways. For example, in aluminum(III) chloride, the aluminum atom shares a lone pair of electrons with a chlorine atom in a second molecule, so that all the atoms in $Al_2Cl_6$ have a complete octet (Figure 8.16). The structure for $Al_2Cl_6$ is observed only in the liquid (or molten) phase. In the gas phase, aluminum(III) chloride exists as discrete molecules of $AlCl_3$, like $BF_3$.

Do compounds containing elements other than group 13 ever have less than an octet around the central atom? The answer depends on which phase of the material we are considering. For example, as a solid, beryllium chloride consists of long chains of chlorine-bridged Be atoms (Figure 8.17) where all atoms have completed octets. Above 800 K, $BeCl_2$ exists in the gas phase as discrete molecules where the beryllium shares only two pairs of electrons with chlorine—much less than an octet.

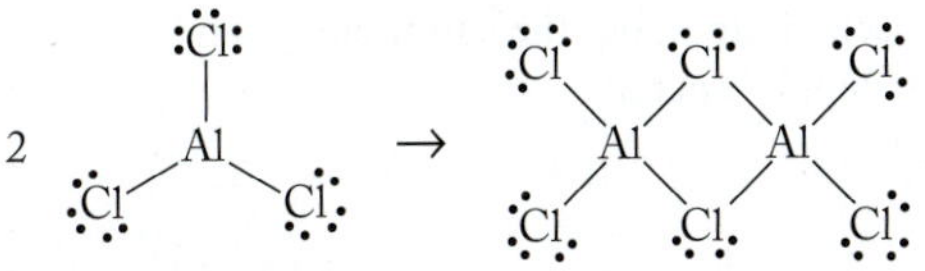

**FIGURE 8.16** Aluminum(III) chloride completes the octet on aluminum by sharing a pair of electrons with one Cl atom from a second molecule of $AlCl_3$.

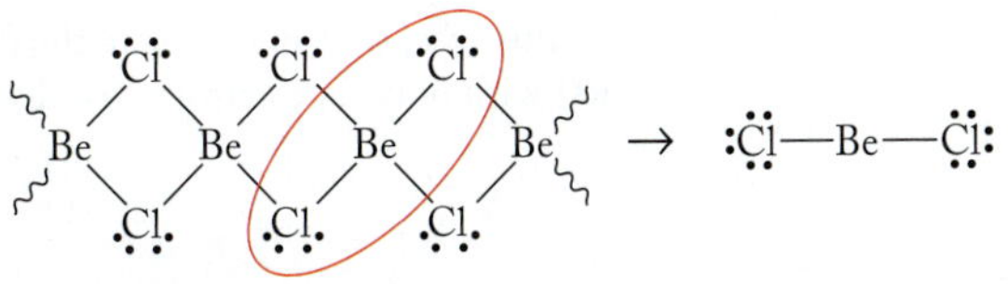

**FIGURE 8.17** The chain structure of solid beryllium chloride, with bridging chlorine atoms, changes to discrete $BeCl_2$ molecules in the gas phase. Each beryllium atom in the solid has a completed octet. In the gas phase, each Be atom is surrounded by only four electrons.

Lithium aluminum hydride reacts with $[(CH_3)_3NH]^+[Cl]^-$ to form the products in this chemical equation:

$$Li^+[AlH_4]^-(s) + [(CH_3)_3NH]^+Cl^-(s) \rightarrow AlH_3 \cdot N(CH_3)_3(s) + LiCl(s) + H_2(g)$$

The product $AlH_3 \cdot N(CH_3)_3$ is used in the electronics industry as a source of aluminum to make thin films of aluminum metal by the thermal decomposition of $AlH_3$:

$$AlH_3 \cdot N(CH_3)_3(g) \rightarrow AlH_3(g) + N(CH_3)_3(g) \rightarrow Al(s) + \tfrac{3}{2} H_2(g) + N(CH_3)_3(g)$$

Draw Lewis structures for (a) $Li^+[AlH_4]^-$ and (b) $AlH_3(g)$.

**Collect and Organize** We are given the chemical formulas of two aluminum compounds, $Li^+[AlH_4]^-$ and $AlH_3$, and asked to draw Lewis structures for each.

**Analyze** The formula of lithium aluminum hydride, $Li^+[AlH_4]^-$, tells us that it contains $Li^+$ cations and $AlH_4^-$ anions. Lithium and hydrogen each have one valence electron and aluminum has three valence electrons. We use the steps outlined in Section 8.2 to draw Lewis structures for both compounds. Aluminum is the central atom in $AlH_4^-$ and $AlH_3$ because each hydrogen atom can form only one bond.

**Solve**

a. The lithium atom has lost its valence electron to form a $Li^+$ ion, so its Lewis symbol is $Li^+$:

$$\cdot Li \xrightarrow{\ -e^-\ } Li^+$$

The total number of valence electrons in $AlH_4^-$ is:

| Element: | Al | + | 4 H |
|---|---|---|---|
| Valence electrons: | 3 | + | $(4 \times 1) = 7$ |
| Additional electron due to 1− charge: | | | + 1 |
| Total valence electrons: | | | 8 |

Connecting the four H atoms to the Al with single bonds results in a Lewis structure where aluminum has a complete octet and each hydrogen atom has a complete duet.

$$Li^+ \left[ \begin{array}{c} H \\ | \\ H-Al-H \\ | \\ H \end{array} \right]^-$$

b. The total number of valence electrons in $AlH_3$ is:

| Element: | Al | + | 3 H |
|---|---|---|---|
| Valence electrons: | 3 | + | $(3 \times 1) = 6$ |

In this case, we can complete the duets of each H atom, but the aluminum atom is left with only six electrons, two less than a complete octet.

**Think About It** The aluminum atom in $AlH_3$ must have less than an octet because there are only six total valence electrons. As a result, $AlH_3$ reacts readily with the lone pair on nitrogen in $N(CH_3)_3$, for example, to form $AlH_3 \cdot N(CH_3)_3$. In $AlH_4^-$, the additional H atom and the overall negative charge on the anion completes the octet on aluminum.

⚙ **Practice Exercise**  The boiling point of sodium chloride is approximately 1700 K ($\approx$ 1427°C). In the gas phase, NaCl exists as molecules of NaCl and as molecular species with a formula $Na_2Cl_2$. Draw Lewis structures for these two compounds. Which atoms do not have complete octets?

### The Limits of Bonding Models

After studying the structures of $SF_6$, $B(CH_3)_3$, and other exceptions to the octet rule, you may be wondering whether the octet rule really has any validity at all. The point to remember is that the octet rule is just a *model*. The octet rule works remarkably well in many cases to predict the number of covalent bonds between main group elements and to predict the composition of ionic compounds. The true nature of the chemical bond is much more subtle and shows great variability among the myriad chemical compounds that have been discovered.

What about the validity of formal charges? How can F, the most electronegative atom, carry a +1 formal charge in $BF_3$? Like Lewis structures, formal charges represent a *model* for identifying resonance forms that contribute most to the bonding in a molecule. Sometimes our model does not fit the observed data as well as we would like. In the case of $BF_3$, we find that the observed B—F distances suggest that resonance forms containing B=F bonds are important.

If aluminum(III) chloride is a covalently bonded molecule, why do aqueous solutions of $AlCl_3$ conduct electricity? Why doesn't $BF_3$, with an electronegativity difference of 2.0, behave similarly? Why does $AlCl_3$ display ionic properties when $\Delta$EN for Al—Cl bonds is only 1.5? In part, the answer lies in the partial ionic character of polar bonds. The greater difference in electronegativity between B and F suggests that B—F bonds have more ionic character, an estimate of the magnitude of charge separation ($\delta+$ and $\delta-$ in Figures 8.7 and 8.9) in a covalent bond. When $AlCl_3$ dissolves in water, it is not only the ionic character of the bond that matters, but its bond energy and the resultant chemical reactivity of each compound.

The significance of molecules with incomplete or expanded octets is that they challenge our models for bonding and push us toward better explanations for what we observe. We will explore questions about bonding in more detail in Chapter 9, but before we do, let's look at some experimental data about covalent bonds: bond lengths and bond energies. As measurable quantities, these parameters allow us to test our model of the chemical bond as proposed by Gilbert Lewis.

## 8.8 The Lengths and Strengths of Covalent Bonds

In Section 8.5 we discussed the equivalent resonance structures of ozone. We noted that the true nature of the two oxygen–oxygen bonds in $O_3$ is reflected in their equal bond length, 128 pm, which is between the length of a typical O=O double bond (121 pm) and an O—O single bond (148 pm), as shown in Figure 8.18. We used these results to conclude that there are effectively 1.5 bonds between the atoms in a molecule of $O_3$. In this section we explore this use of bond length, and also bond strength, to rationalize and validate molecular structures. We also use bond strengths to estimate the enthalpy changes that occur in chemical reactions.

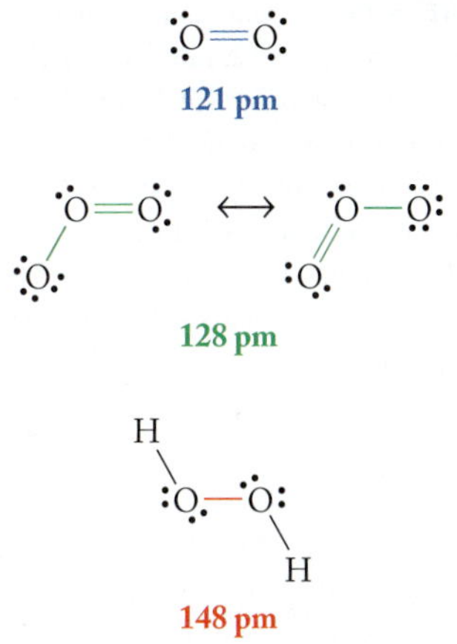

**FIGURE 8.18** Resonance influences bond length. Because of resonance, the lengths of the two bonds in ozone are identical and in between the lengths of the O=O double bond in $O_2$ and the O—O single bond in $H_2O_2$.

**bond order** the number of bonds between atoms: 1 for a single bond, 2 for a double bond, and 3 for a triple bond.

## Bond Length

The length of the bond between any two atoms depends on the identity of the atoms and on whether the bond is single, double, or triple (Table 8.3). The number of bonds between two atoms is called **bond order**. As bond order increases, bond length decreases, as we can see by comparing the lengths of C—C, C═C, and C≡C bonds in Table 8.3. Similarly, for carbon–oxygen bonds the C≡O triple bond in carbon monoxide is shorter than the C═O double bond in carbon dioxide (Figure 8.19a).

Measurements of the bond lengths in many molecules indicate that there are small differences in bond lengths for any given covalent bond. For example, the C—H and C═O bond lengths in formaldehyde (Figure 8.19b) are close to but not exactly the same as the C—H bond length in $CH_4$ and the C═O bond length in $CO_2$.

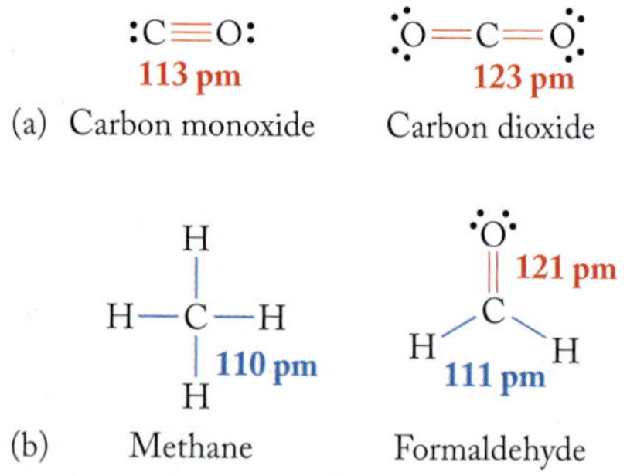

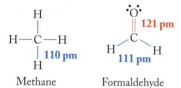

(a) Carbon monoxide    Carbon dioxide

(b)    Methane     Formaldehyde

**FIGURE 8.19** (a) Bond length depends on the identity of the two atoms forming the bond and on bond order. (b) A bond between the same two atoms can have different lengths in different molecules. Compare the C—H length in the formaldehyde molecule with the C—H length in $CH_4$. Also compare the C═O length in formaldehyde with the C═O length in $CO_2$.

| TABLE 8.3 | Selected Average Covalent Bond Lengths and Bond Energies | | | | | |
|---|---|---|---|---|---|---|
| Bond | Bond Length (pm) | Bond Energy (kJ/mol) | | Bond | Bond Length (pm) | Bond Energy (kJ/mol) |
| C—C | 154 | 348 | | N═O | 106 | 678 |
| C═C | 134 | 614 | | O—O | 148 | 146 |
| C≡C | 120 | 839 | | O═O | 121 | 495 |
| C—N | 147 | 293 | | O—H | 96 | 463 |
| C═N | 127 | 615 | | S—O | 151 | 265 |
| C≡N | 116 | 891 | | S═O | 143 | 523 |
| C—O | 143 | 358 | | S—S | 204 | 266 |
| C═O | 123 | 743[a] | | S—H | 134 | 347 |
| C≡O | 113 | 1072 | | H—H | 75 | 436 |
| C—H | 110 | 413 | | H—F | 92 | 567 |
| C—F | 133 | 485 | | H—Cl | 127 | 431 |
| C—Cl | 177 | 328 | | H—Br | 141 | 366 |
| N—H | 104 | 388 | | H—I | 161 | 299 |
| N—N | 147 | 163 | | F—F | 143 | 155 |
| N═N | 124 | 418 | | Cl—Cl | 200 | 243 |
| N≡N | 110 | 941 | | Br—Br | 228 | 193 |
| N—O | 136 | 201 | | I—I | 266 | 151 |
| N═O | 122 | 607 | | | | |

[a] The bond energy of the C═O bond in $CO_2$ is 799 kJ/mol.

> **CONCEPT TEST**
>
> Rank the following molecules in order of decreasing lengths of their nitrogen–oxygen bonds: NO, $NO_2$, $N_2O$.

## Bond Energies

The energy changes associated with chemical reactions depend on how much energy is required to break the bonds in the reactants and how much is released

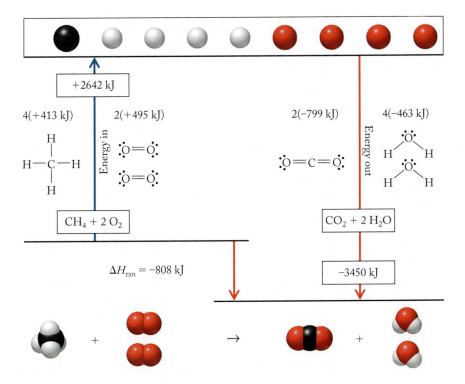

**FIGURE 8.20** The combustion of 1 mole of methane requires that 4 moles of C—H bonds and 2 moles of O=O bonds be broken. These processes require an enthalpy change of about +2642 kJ. In the formation of 4 moles of O—H bonds and 2 moles of C=O bonds, there is an enthalpy change of about −3450 kJ. The overall reaction is exothermic: 2642 kJ − 3450 kJ = −808 kJ.

as the atoms recombine to form products. For example, in the methane combustion reaction

$$CH_4(g) + 2\ O_2(g) \rightarrow CO_2(g) + 2\ H_2O(g)$$

the C—H bonds in $CH_4$ and the O=O bonds in $O_2$ must be broken before the C=O bonds in $CO_2$ and the O—H bonds in $H_2O$ can form. (In reality, some bond formation occurs simultaneously with bond breaking; it is not completely sequential.) Breaking bonds is endothermic (the blue "energy in" arrow in Figure 8.20), and forming bonds is exothermic (the red "energy out" arrow). If a chemical reaction is exothermic, as is methane combustion, more energy is released in forming the bonds in molecules of products than is consumed in breaking the bonds in molecules of reactants.

**Bond energy**, or *bond strength*, is usually expressed in terms of the enthalpy change ($\Delta H$) that occurs when 1 mole of a particular bond in the gas phase is broken. Bond energies for some common covalent bonds are given in Table 8.3. As we noted in Section 8.1, the quantity of energy needed to break a particular bond is equal in magnitude but opposite in sign to the quantity of energy released when that same bond forms.

Like bond lengths, the bond energies in Table 8.3 are average values because bond energies vary depending on the structure of the rest of the molecule. For example, the bond energy of a C=O bond in carbon dioxide is 799 kJ/mol, but C=O bond energy in formaldehyde is only 743 kJ/mol. Bond energies are always positive quantities because breaking bonds is endothermic.

Another view of the variability in bond energy comes from breaking the C—H bonds in $CH_4$ in a step-by-step fashion, shown in the table on the next page. These results tell us that the chemical environment of a bond affects the energy required to break it: breaking the first C—H bond in methane is easier (requires less energy) than breaking the second but more difficult than breaking the third or fourth. The total energy needed to break all four C—H bonds is 1652 kJ/mol, an average of 413 kJ/mol per bond.

**bond energy** the energy needed to break 1 mole of a particular covalent bond in a molecule or polyatomic ion in the gas phase.

| Decomposition Step | Energy Needed (kJ/mol) |
|---|---|
| $CH_4 \rightarrow CH_3 + H$ | 435 |
| $CH_3 \rightarrow CH_2 + H$ | 453 |
| $CH_2 \rightarrow CH + H$ | 425 |
| $CH \rightarrow C + H$ | 339 |
| Total: | 1652 |
| Average: | 413 |

The relationship between bond order and bond energy is also apparent in Table 8.3. At 495 kJ/mol, the bond energy of the $O{=}O$ double bond is more than three times that of the $O{-}O$ single bond. This correlation between bond order and bond energy is true for other pairs of atoms: the higher the bond order, the greater the bond energy. The bond energy of the $N{\equiv}N$ triple bond (941 kJ/mol) is more than twice the bond energy of the $N{=}N$ double bond and more than five times that of the $N{-}N$ single bond. The large quantity of energy required to break a mole of $N{\equiv}N$ triple bonds is one of the reasons $N_2$ participates in so few chemical reactions.

In the combustion of 1 mole of $CH_4$ (Figure 8.20), 4 moles of $C{-}H$ bonds and 2 moles of $O{=}O$ bonds must be broken. The formation of 1 mole of $CO_2$ and 2 moles of $H_2O$ requires the formation of 2 moles of $C{=}O$ bonds and 4 moles of $O{-}H$ bonds. The net change in energy resulting from breaking and forming these bonds can be estimated from average bond energies. We can take an inventory of the bond energies involved:

| **Bond Energies ($\Delta H$) in Methane Combustion** | | | |
|---|---|---|---|
| Bond | Number of Bonds (mol) | Bond Energy (kJ/mol) | Bond Breaking or Forming? |
| $C{-}H$ | 4 | 413 | Breaking |
| $O{=}O$ | 2 | 495 | Breaking |
| $O{-}H$ | 4 | 463 | Forming |
| $C{=}O$ | 2 | 799 | Forming |

▶❚❚ **CHEMTOUR** Estimating Enthalpy Changes

Next we estimate the enthalpy change of the reaction by calculating the difference between the sum of the bond energies of the reactants and the sum of the bond energies of the products. Equation 8.4 can be used to do this calculation:

$$\Delta H_{rxn} = \sum \Delta H_{bonds\ breaking} - \sum \Delta H_{bonds\ forming} \qquad (8.4)$$

$$= [(4\ mol \times 413\ kJ/mol) + (2\ mol \times 495\ kJ/mol)]$$

$$- [(4\ mol \times 463\ kJ/mol) + (2\ mol \times 799\ kJ/mol)]$$

$$= -808\ kJ$$

⊙⊙ **CONNECTION** In Chapter 5 we calculated the difference between the sums of heats of formation of products and of reactants to estimate the heat of reaction.

**SAMPLE EXERCISE 8.12** **Estimating Heats of Reaction from Average Bond Energies** L06

Use the average bond energies in Table 8.3 to estimate $\Delta H_{rxn}$ for the reaction in which $HCl(g)$ is formed from $H_2(g)$ and $Cl_2(g)$:

$$H{-}H \quad + \quad :\ddot{C}l{-}\ddot{C}l: \quad \rightarrow \quad 2\ H{-}\ddot{C}l:$$

**Collect and Organize** We are to estimate the value of $\Delta H_{rxn}$ of a gas-phase reaction from the bond energies of the reactants and products.

**Analyze** The reaction between $H_2$ and $Cl_2$ requires that 1 mole of H—H bonds and 1 mole of Cl—Cl bonds be broken. Two moles of H—Cl bonds are formed. Table 8.3 lists these average bond energies:

$$\begin{array}{ll} \text{H—H} & 436 \text{ kJ/mol} \\ \text{Cl—Cl} & 243 \text{ kJ/mol} \\ \text{H—Cl} & 431 \text{ kJ/mol} \end{array}$$

**Solve** Using the above information in Equation 8.4:

$$\begin{aligned} \Delta H_{rxn} &= \sum \Delta H_{\text{bond breaking}} - \sum \Delta H_{\text{bond forming}} \\ &= [(1 \text{ mol} \times 436 \text{ kJ/mol}) + (1 \text{ mol} \times 243 \text{ kJ/mol})] \\ &\quad - [(2 \text{ mol} \times 431 \text{ kJ/mol})] \\ &= -183 \text{ kJ} \end{aligned}$$

**Think About It** An enthalpy change of −183 kJ (note the minus sign) means that the amount of energy consumed as the bonds in the reactants are broken is 183 kJ less than the amount of energy released as the bonds in the products are formed. In other words, the reaction is exothermic. The reaction involves the formation of 2 moles of a gaseous compound from its component elements in their standard states. Therefore the result of this calculation should be close to two times the standard heat of formation ($\Delta H_f^\circ$) of HCl. That value (see Appendix 4) is −92.3 kJ/mol. Multiplying by 2 moles we get −184.6 kJ, which is quite close to the estimated value.

**Practice Exercise** Use average bond energies to calculate $\Delta H_{rxn}$ for the reaction of $H_2$ and $N_2$ to form ammonia:

$$:N\equiv N: \quad + \quad 3 \text{ H—H} \quad \rightarrow \quad 2 \text{ H—}\overset{\displaystyle ..}{\underset{\displaystyle |}{N}}\text{—H}$$
$$\phantom{2 \text{ H—N—H}}$$
$$\phantom{aaaaaaaaaaaaaaaaaaaaaaaaaaaaaaa} H$$

---

**CONCEPT TEST**

Suggest a reason, based on bond energies, why $O_2$ is much more reactive than $N_2$.

---

In this chapter we have explored the nature of the covalent bonds that hold together molecules and polyatomic ions, observing that these bonds owe their strength to the presence of pairs of electrons shared between nuclei of atoms. Sharing does not necessarily mean equal sharing, and unequal sharing coupled with bond vibration accounts for the ability of some atmospheric gases to absorb and emit infrared radiation. As a result, these gases function as greenhouse gases.

Early in the chapter we noted that moderate concentrations of greenhouse gases are required for climate stability and to make our planet habitable. The escalating concern of many is that Earth's climate is currently being destabilized by too much of a good thing. Policies made by the world's governments in the near future will have a significant impact on the problem of global warming, one way or the other. As an informed member of the world community, you will have the opportunity to influence how those policy decisions are made.

**SAMPLE EXERCISE 8.13    Integrating Concepts: The Search for Gallium(II) Chloride**

Gallium(III) chloride exists as $Ga_2Cl_6$ in both the solid and liquid phases, dissociating to $GaCl_3$ in the gas phase. Heating gallium metal and gallium(III) chloride yields $[Ga]^+[GaCl_4]^-$.

a. Draw Lewis structures for $GaCl_3$, $Ga_2Cl_6$, and $[Ga]^+[GaCl_4]^-$.
b. Assign formal charges to the atoms in each structure.
c. Write a balanced chemical equation for the redox reaction between Ga and $Ga_2Cl_6$. Identify which atoms are oxidized and which are reduced.
d. What is the empirical formula of $[Ga]^+[GaCl_4]^-$? Why did the first chemists to discover $[Ga]^+[GaCl_4]^-$ think they had made gallium(II) chloride?
e. An actual gallium(II) halide complex has been prepared: $Ga_2Cl_6{}^{2-}$, containing a Ga—Ga bond. Draw a Lewis structure for this anion.

**Collect and Organize** There are five parts to this sample exercise. We are given the formulas of three gallium compounds and asked to draw Lewis structures for each, as well as assigning formal charges. We are also asked to balance the chemical equation for the redox reaction between $Ga_2Cl_6$ and Ga metal.

**Analyze** We use the steps outlined in Section 8.2 to draw Lewis structures. Gallium is the central atom in all four compounds, as it is less electronegative than Cl. The formula for $Ga^+[GaCl_4]^-$ is written in a way to suggest that it contains a $Ga^+$ cation and a $GaCl_4^-$ anion. In $Ga_2Cl_6{}^{2-}$ there are two Ga atoms connected by a Ga—Ga bond. Formal charges are assigned using the procedure in Section 8.6.

**Solve**

a. To draw Lewis structures for $GaCl_3$ and $Ga^+[GaCl_4]^-$, we first determine the number of valence electrons in each:

$GaCl_3$:

| Element: | Ga | + | 3 Cl |
|---|---|---|---|
| Valence electrons: | 3 | + | $(3 \times 7) = 24$ |

$Ga^+$:    $\cdot\dot{G}a \xrightarrow{-e^-} \dot{G}a^+$

$GaCl_4^-$:

| Element: | Ga | + | 4 Cl |
|---|---|---|---|
| Valence electrons: | 3 | + | $(4 \times 7) = 31$ |
| Additional electron due to 1− charge: | | | + 1 |
| Total valence electrons: | | | 32 |

Arranging the chlorine atoms around Ga in each polyatomic species and completing the octets of the chlorine atoms, we arrive at the Lewis structures for $GaCl_3$ and $GaCl_4^-$:

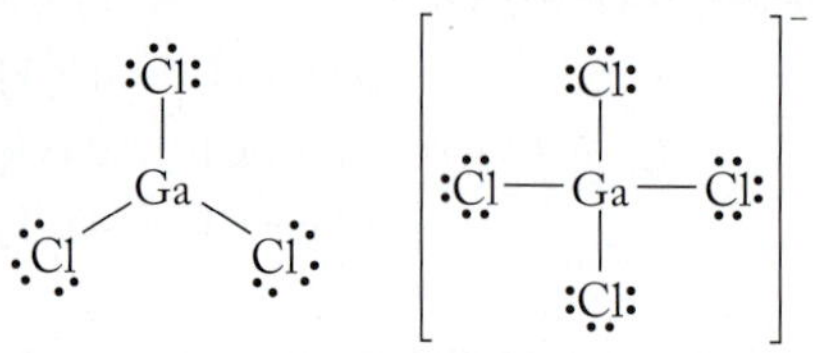

The gallium in $GaCl_3$ has an incomplete octet. We can draw three additional resonance forms for $GaCl_3$, each containing a Ga=Cl double bond:

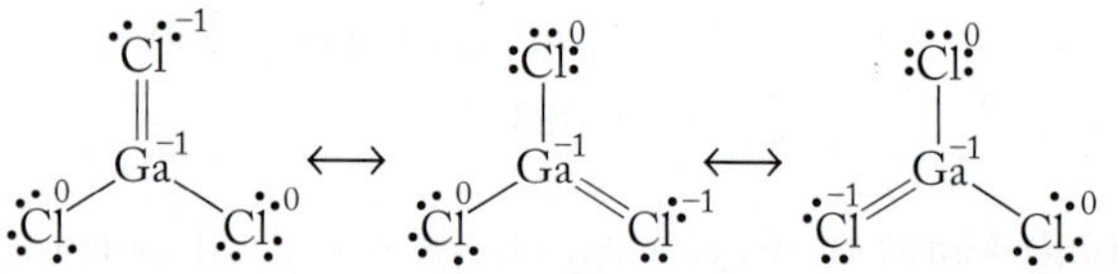

However, like $AlCl_3$, $GaCl_3$ forms $Ga_2Cl_6$, a structure in which two chlorine atoms are shared by two gallium atoms. This completes the octet on each gallium without requiring Ga=Cl bonds:

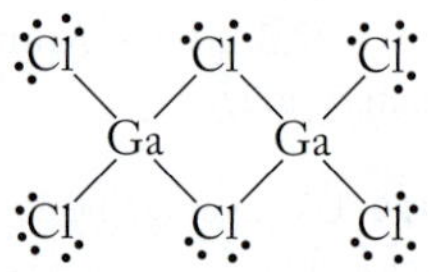

b. Formal charges are assigned using the procedure in Section 8.6 (see Sample Exercise 8.8).

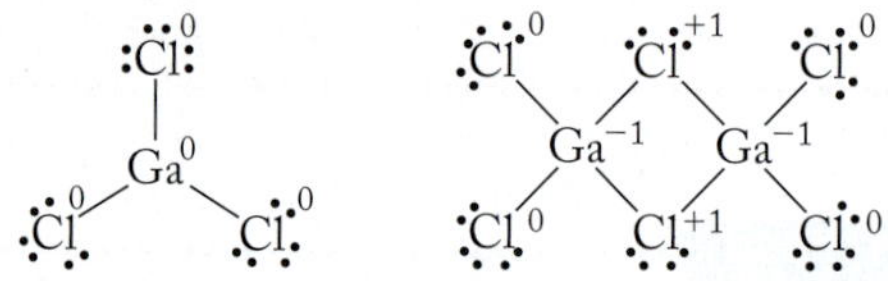

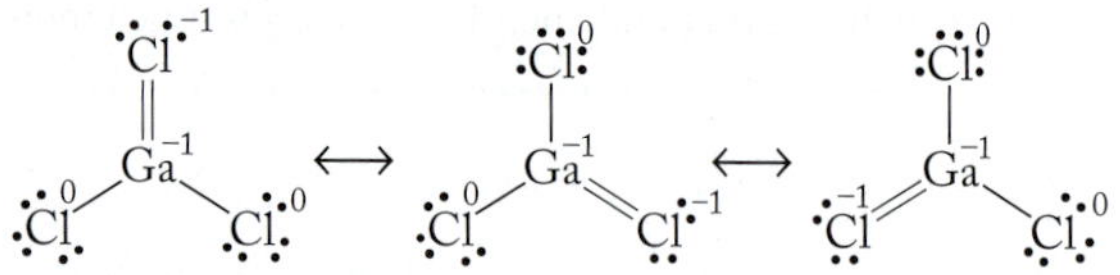

$GaCl_4^-$:

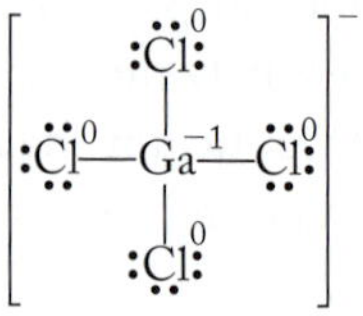

c. A preliminary expression describing the reaction of Ga with $Ga_2Cl_6$ is:

$$Ga(s) + Ga_2Cl_6(s) \rightarrow Ga^+[GaCl_4]^-(s)$$

To make it easier to write a balanced chemical equation, let's rewrite the above expression using the empirical formula of the product:

$$Ga(s) + Ga_2Cl_6(s) \rightarrow Ga_2Cl_4(s)$$

The oxidation states of Ga in the three species are 0 in Ga, +3 in $Ga_2Cl_6$, and +2 in $Ga_2Cl_4$. The last value is consistent with the formula and with the ionic composition of the compound which contains one $Ga^+$ cation and a $GaCl_4^-$ anion in which the charge of central Ga ion is 3+. The average of these charges is +2. Therefore, the oxidation state of elemental Ga increases from 0 to an average of +2 while the oxidation state of Ga in $Ga_2Cl_6$ decreases from +3 to +2. To balance these changes in oxidation state (and electron loss and gain) we need one atom of elemental Ga for every two $Ga^{3+}$ ions in $Ga_2Cl_6$. The expression above already has this ratio of reactants, but there are a total of three Ga atoms and ions on the reactant side and only two on the product side. To balance these total numbers of Ga atoms and ions, we multiply the reactant terms by 2 and the product by 3:

$$2\,Ga(s) + 2\,Ga_2Cl_6(s) \rightarrow 3\,Ga_2Cl_4(s)$$

This equation is now balanced.

d. The formula of $Ga^+[GaCl_4]^-$ is $Ga_2Cl_4$. The empirical formula (simplest whole number ratio) is $GaCl_2$. From Section 2.5 we know that the charge on a chloride ion is 1−, making the *apparent* charge on gallium $Ga^{2+}$. This calculation misled investigators to believe they had isolated "gallium(II) chloride" rather than the $Ga^+$ salt of a gallium(III) complex containing polyatomic ions.

e. A true gallium(II) compound is the anion $Ga_2Cl_6^{2-}$. The Lewis structure of $Ga_2Cl_6^{2-}$ can be drawn using the procedure in part a.

Determining the number of valence electrons in the molecular ion:

| Element: | 2 Ga | + | 6 Cl | + | charge | |
|---|---|---|---|---|---|---|
| Valence electrons: | $(2 \times 3)$ | + | $(6 \times 7)$ | + | 2 = 36 | |

We are told $Ga_2Cl_6^{2-}$ has a Ga—Ga bond. Distributing the six chlorine atoms evenly around two gallium atoms and completing the octets of the chlorine atoms, we arrive at the Lewis structure for $Ga_2Cl_6^{2-}$:

All eight atoms in $Ga_2Cl_6^{2-}$ have a complete octet. We can confirm the +2 oxidation state of gallium as follows:

$$2x + 6(-1) = -2$$
$$2x = -2 + 6 = 4$$

so $x = +2$.

**Think About It** Gallium chloride illustrates several important features from this chapter. Four resonance structures are needed to describe gas phase $GaCl_3$. A positive formal charge on the more electronegative Cl atoms in structures containing Ga=Cl double bonds suggests that these structures might contribute less to the bonding in $GaCl_3$ than a structure in which gallium has an incomplete octet. In the solid state, gallium chloride completes its octet by sharing chlorine atoms between two gallium atoms. This Lewis structure places an "unfavorable" +1 formal charge on two Cl atoms, yet this is the observed structure of "$GaCl_3$" at room temperature.

# Fluorine and Oxygen: Location, Location, Location

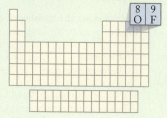

Fluorine does not occur as the free element ($F_2$) in nature because it reacts with nearly anything with which it comes into contact, including Kr and Xe. If fluorine gas is allowed to flow over the surface of water, the water actually burns (rapidly oxidizes) as a result of the reaction:

$$5\ F_2(g) + 5\ H_2O(\ell) \rightarrow 8\ HF(g) + O_2(g) + H_2O_2(\ell) + OF_2(g)$$

In an atmosphere of fluorine gas, wood, plastic, and even some metals burst into white-hot, intense flames. Even "fireproof" asbestos burns in fluorine. Fluorine reacts with hydrocarbons producing *fluorocarbons*, which are completely fluorinated molecules, meaning that every hydrogen in the hydrocarbon is replaced by fluorine.

Although fluorine is extremely reactive, fluorocarbons are among the most stable compounds known because of the stability of the carbon–fluorine bond. Fluorocarbons are very rare in nature; most are manufactured. They are generally unreactive toward most chemical reagents, inert to solvents, and nonflammable. The stability of the carbon–fluorine bond has been used to advantage in commercial products. For example, the addition of a single carbon-bound fluorine to certain pharmaceutical compounds improves their potency because it makes the molecules more difficult for the body to degrade.

The anesthetic halothane ($C_2HBrClF_3$) was introduced in 1956 as the first modern inhaled anesthetic. The low toxicity, low flammability, and high stability of halothane enabled it to replace much more hazardous anesthetics, such as ether ($CH_3CH_2OCH_2CH_3$) and chloroform ($CHCl_3$). One synthetic polymer (a very large molecule), made when molecules of tetrafluoroethylene bind together, is commonly known as Teflon, a substance so stable to moderate temperatures and so much less reactive than other nonmetallic materials that it is used on non-stick cookware (Figure 8.21), electrical insulation, and shipping containers for fluorine gas.

**FIGURE 8.21** A frying pan coated with Teflon.

In contrast to the free element, compounds containing fluoride ions ($F^-$) are relatively abundant in nature. For decades, small quantities of fluoride salts have been added to municipal water supplies in the United States to prevent cavities in adults and children. Fluorides are also added in small amounts (about 0.15% by mass) to toothpaste for the same reason.

Oxygen is one of the most reactive elements, though less reactive than fluorine. It is the second most abundant gas in the atmosphere and is essential to many life forms on Earth. Physical activity for human beings becomes difficult if the partial pressure of oxygen in the air drops from the normal value of 0.21 atm to around 0.12, and breathing air that contains only 0.06 atm oxygen can result in death in 6 to 8 minutes.

$$n \quad \begin{matrix} F & & F \\ & \diagdown \ \diagup & \\ & C = C & \\ & \diagup \ \diagdown & \\ F & & F \end{matrix} \quad \rightarrow \quad \begin{bmatrix} F & F \\ | & | \\ -C-C- \\ | & | \\ F & F \end{bmatrix}_n \quad \text{where } n \approx 1000$$

Tetrafluoroethylene          Teflon

Oxygen is also an important commercial bulk chemical, heavily used in the steel industry and in medicine. It is frequently transported and stored as a cryogenic (around −183°C) liquid referred to as LOX (*l*iquid *ox*ygen). LOX is extremely reactive and must be handled with great care. Metallic iron burns so vigorously in liquid oxygen that the iron melts from the heat of combustion, even though it is surrounded by frigid liquid. Many materials normally considered noncombustible burn at explosive rates in liquid oxygen, and oils, greases, paint, and paper burn immediately on contact. LOX penetrates into porous materials such as wood and asphalt; even when combustion does not occur immediately, the material remains hazardous. Asphalt has been known to burst into flames days after LOX had been spilled on it. Large tanks containing cryogenic oxygen at hospitals or manufacturing sites are always mounted on concrete slabs, not asphalt, as a safety precaution.

The chemistries of oxygen and fluorine intersect in Earth's upper atmosphere. The substances at the intersection are ozone ($O_3$) and chlorofluorocarbons (CFCs). CFCs were used for years as "clean" fire-extinguishing agents and as refrigerants and aerosol propellants. CFCs have now been banned because of their role in destroying ozone in the stratosphere.

The quantity of ozone in the upper atmosphere is not large—if compressed by atmospheric pressure at Earth's surface, it would be squashed to a band only a few millimeters thick—but in the upper atmosphere it exists in a layer more than 20 km thick. This ozone absorbs UV radiation from the sun and protects life on Earth from these energetic rays. Ozone is produced from $O_2$ in the upper atmosphere in a process driven by UV radiation:

$$3\ O_2(g) \xrightarrow{\text{UV radiation}} 2\ O_3(g)$$

The conversion of ozone back to $O_2$ occurs but is slow in the upper atmosphere, so a protective ozone layer remains in place unless disrupted by other agents.

Those disrupting agents are CFCs. When these gases are released into the atmosphere, they persist because of the lack of reactivity that makes them so useful. Wind and air currents carry them to the upper reaches of the atmosphere, where UV radiation provides sufficient energy to break chemical bonds and form chlorine atoms:

$$CF_2Cl_2(g) \xrightarrow{\text{UV radiation}} CF_2Cl(g) + Cl(g)$$

Chlorine atoms are free radicals that react with ozone, producing chlorine monoxide (another free radical) and oxygen:

$$Cl(g) + O_3(g) \rightarrow ClO(g) + O_2(g) \qquad (1)$$

Chlorine monoxide reacts with more ozone, generating more oxygen and another chlorine free radical:

$$ClO(g) + O_3(g) \rightarrow Cl(g) + 2\ O_2(g) \qquad (2)$$

If we add Equations 1 and 2, we see that the end result is the conversion of ozone to oxygen:

$$\begin{aligned}
\cancel{Cl(g)} + O_3(g) &\rightarrow \cancel{ClO(g)} + O_2(g) \\
\cancel{ClO(g)} + O_3(g) &\rightarrow \cancel{Cl(g)} + 2\ O_2(g) \\
\hline
2\ O_3(g) &\rightarrow 3\ O_2(g)
\end{aligned}$$

Not only does this process take place much faster than it would without Cl, but chlorine free radicals are not consumed in the reaction. Therefore one chlorine free radical can destroy thousands of ozone molecules before being consumed in a reaction with another free radical in the stratosphere.

The Montreal Protocol, an international agreement now signed by over 190 nations, mandated the phasing out of synthetic substances responsible for depletion of Earth's ozone layer. The search continues for new substances that will be as useful for fire extinguishers and refrigerants as CFCs without damaging the stratospheric ozone layer.

## SUMMARY

**Learning Outcome 1** A chemical bond results when two atoms share electrons (a covalent bond) or when two ions are attracted to each other (an **ionic bond**). The atoms in metallic solids pool their electrons in forming **metallic bonds**. (Section 8.1)

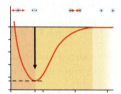

**Learning Outcome 2** **Lewis symbols** use dots to represent paired and unpaired electrons in the ground states of atoms. A **Lewis structure** shows the bonding pattern in molecules and polyatomic ions. Two or more equivalent Lewis structures—called **resonance structures**—can sometimes be drawn for one molecule or polyatomic ion. The actual bonding pattern in a molecule is an average of equivalent resonance structures. (Sections 8.2, 8.5, and 8.7)

**Learning Outcome 3** Unequal electron sharing between atoms of different elements results in **polar covalent bonds**. **Bond polarity** is a measure of how unequally the electrons in covalent bonds are shared. More polarity results from greater differences in the **electronegativi-** ties of the bonded atoms. Polarity of molecules of atmospheric gases may lead to the absorption of IR radiation which contributes to the greenhouse effect. (Sections 8.3 and 8.4)

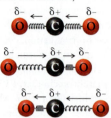

**Learning Outcome 4** The preferred resonance structure of a molecule is one in which the **formal charges** on its atoms are zero or closest to zero and any negative formal charges are on the more electronegative atoms. (Section 8.6)

**Learning Outcome 5** **Bond order** is the number of bonding pairs in a covalent bond, **bond energy** is the enthalpy change, $\Delta H$, required to break 1 mole of a particular covalent bond in the gas phase, and **bond length** is the distance between the nuclear centers of two atoms joined together in a bond. (Section 8.8)

**Learning Outcome 6** Energy changes in chemical reactions depend on the energy required to break bonds in reactants and the energy released when new bonds are formed in the products. (Section 8.8)

## PROBLEM-SOLVING SUMMARY

| TYPE OF PROBLEM | CONCEPTS AND EQUATIONS | SAMPLE EXERCISES |
|---|---|---|
| **Drawing Lewis structures for molecules, monatomic ions, and ionic compounds** | Connect the atoms with single covalent bonds, distribute the valence electrons to give each noncentral atom eight valence electrons (except two for H); use multiple bonds where necessary to complete the central atom's octet. | 8.1–8.4 |
| **Comparing bond polarities** | Calculate the difference in electronegativity ($\Delta EN$) between the two bonded atoms; if $\Delta EN \geq 2.0$, the bond is considered ionic. | 8.5 |
| **Drawing resonance structures of molecules and polyatomic ions** | Include all possible arrangements of covalent bonds in the molecule if more than one equivalent structure can be drawn. | 8.6, 8.7 |
| **Selecting resonance structures based on formal charges** | Calculate formal charge using $$FC = \left(\begin{array}{c}\text{number of}\\\text{valence e}^-\end{array}\right) - \left[\begin{array}{c}\text{number of}\\\text{unshared e}^-\end{array} + \frac{1}{2}\left(\begin{array}{c}\text{number of e}^-\\\text{in bonding pairs}\end{array}\right)\right] \quad (8.1)$$ Select structures with formal charges closest to zero and with negative formal charges on the most electronegative atoms. | 8.8 |
| **Drawing Lewis structures of odd-electron molecules** | Distribute the valence electrons in the Lewis structure to leave the most electronegative atom(s) with eight valence electrons and the least electronegative atom with the odd number of electrons. | 8.9 |

| TYPE OF PROBLEM | CONCEPTS AND EQUATIONS | SAMPLE EXERCISES |
|---|---|---|
| **Drawing Lewis structures with an expanded valence shell** | Distribute the valence electrons in the Lewis structure, allowing atoms of elements in period 3 and beyond to have more than eight valence electrons if more than four bonds are needed or if the structure with the expanded valence shell results in formal charges closer to zero. | 8.10 |
| **Drawing Lewis structures with an incomplete octet** | Distribute the valence electrons in the Lewis structure, allowing atoms of elements in period 3 and beyond to have less than eight valence electrons if needed. | 8.11 |
| **Estimating heats of reaction from average bond energies** | Multiply bond energies by the number of bonds and calculate using $$\Delta H_{rxn} = \sum \Delta H_{\text{bonds breaking}} - \sum \Delta H_{\text{bonds forming}} \qquad (8.4)$$ | 8.12 |

## VISUAL PROBLEMS

*(Answers to boldface end-of-chapter questions and problems are in the back of the book.)*

**8.1.** Which group highlighted in Figure P8.1 contains atoms that have the following? (a) 1 valence electron; (b) 4 valence electrons; (c) 6 valence electrons

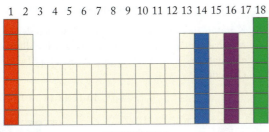

1 2 3 4 5 6 7 8 9 10 11 12 13 14 15 16 17 18

**FIGURE P8.1**

**8.2.** Which of the groups highlighted in Figure P8.2 contains atoms with the following? (a) 2 valence electrons; (b) 3 valence electrons; (c) 5 valence electrons

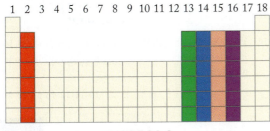

1 2 3 4 5 6 7 8 9 10 11 12 13 14 15 16 17 18

**FIGURE P8.2**

**8.3.** Which of the Lewis symbols in Figure P8.3 correctly portrays the most stable ion of magnesium?

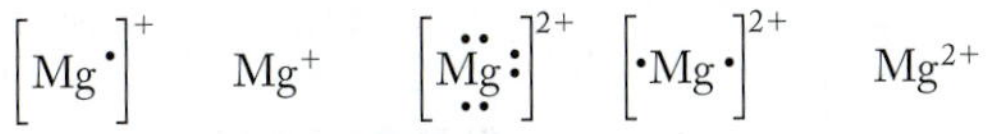

$$\left[ \text{Mg} \cdot \right]^{+} \qquad \text{Mg}^{+} \qquad \left[ \ddot{\text{Mg}}\!: \right]^{2+} \qquad \left[ \cdot\text{Mg}\cdot \right]^{2+} \qquad \text{Mg}^{2+}$$

**FIGURE P8.3**

**8.4.** What changes must be made to the Lewis symbols in Figure P8.4 to make them correct?

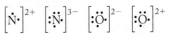

$$\left[ \text{N} \cdot \right]^{2+} \quad \left[ :\ddot{\text{N}}\cdot \right]^{3-} \quad \left[ :\ddot{\text{O}}\cdot \right]^{2-} \quad \left[ :\ddot{\text{O}}\cdot \right]^{2+}$$

**FIGURE P8.4**

**8.5.** Which of the highlighted elements in Figure P8.5 has the greatest bonding capacity?

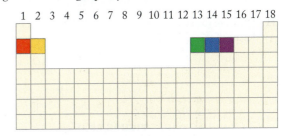

1 2 3 4 5 6 7 8 9 10 11 12 13 14 15 16 17 18

**FIGURE P8.5**

**8.6.** Which of the highlighted elements in Figure P8.6 has the greatest electronegativity?

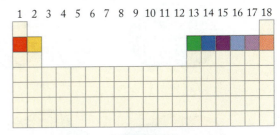

1 2 3 4 5 6 7 8 9 10 11 12 13 14 15 16 17 18

**FIGURE P8.6**

**8.7.** Which two of the highlighted elements in Figure P8.6 is the pair that forms the bond with the most ionic character?

NOTE: The color scale used in Problems 8.8, 8.9, 8.12, and 8.13 is the same as in Figure 8.4, where violet is a charge of 1+, red is a charge of 1−, and green is 0. The larger the size, the greater the electron density.

8.8. Which of the drawings in Figure P8.8 is the best description of the distribution of electrical charge in ClBr?

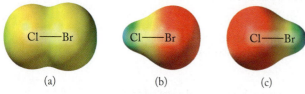

FIGURE P8.8

*8.9. Which of the drawings in Figure P8.9 best describes the distribution of electrical charge in LiF?

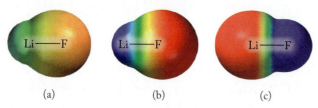

FIGURE P8.9

8.10. Are the three structures in Figure P8.10 resonance forms of the thiocyanate ion (SCN⁻)?

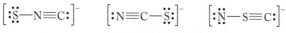

FIGURE P8.10

8.11. Why are the structures in Figure P8.11 not all resonance forms of the molecule $S_2O$?

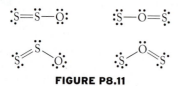

FIGURE P8.11

*8.12. Which of the drawings in Figure P8.12 most accurately describes the distribution of electrical charge in ozone? Explain your answer.

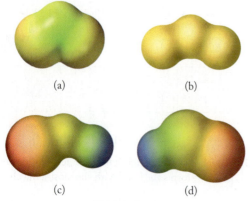

FIGURE P8.12

*8.13. Which of the drawings in Figure P8.13 most accurately describes the distribution of electrical charge in $SO_2$? Explain your answer.

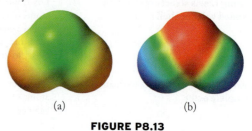

FIGURE P8.13

8.14. How many electron pairs are shared in each of the molecules and ions in Figure P8.14?

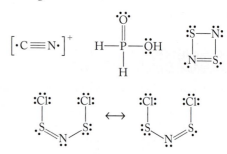

FIGURE P8.14

8.15. Krypton and xenon form compounds with only the most reactive of other elements. Which of the highlighted elements in Figure P8.15 is one of these highly reactive elements?

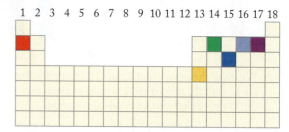

FIGURE P8.15

8.16. Which of the highlighted elements in Figure P8.16 expands its valence shell when bonding to a highly electronegative element?

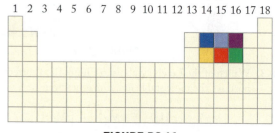

FIGURE P8.16

8.17. What changes must be made to the Lewis structures in Figure P8.17 to make them correct? Assume the skeletal structures shown are correct.

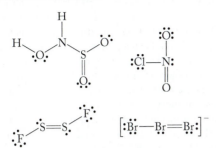

FIGURE P8.17

8.18. In each pair of resonance structures in Figure P8.18, which one contributes more to the bonding in the molecule or molecular ion?

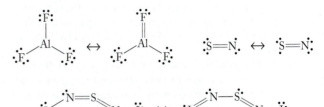

FIGURE P8.18

## QUESTIONS AND PROBLEMS

### Types of Chemical Bonds

#### CONCEPT REVIEW

8.19. Does the number of valence electrons in a neutral atom ever equal the atomic number?

8.20. Does the number of valence electrons in a neutral atom ever equal the group number?

8.21. Do all the elements in a group in the periodic table have the same number of valence electrons?

8.22. Describe the differences in bonding in *covalent* and *ionic* compounds.

### Lewis Structures

#### CONCEPT REVIEW

8.23. Some of his critics described G. N. Lewis's approach to explaining covalent bonding as an exercise in double counting and therefore invalid. Explain the basis for this criticism.

8.24. Does the octet rule mean that a diatomic molecule must have 16 valence electrons?

8.25. Why is the bonding pattern in water H—O—H and not H—H—O?

8.26. Does each atom in a pair that is covalently bonded always contribute the same number of valence electrons to form the bonds between them?

#### PROBLEMS

8.27. Draw Lewis symbols of atoms of lithium, magnesium, and aluminum.

8.28. Draw Lewis symbols of atoms of nitrogen, oxygen, fluorine, and chlorine.

8.29. Find the error in the following Lewis symbols:

$$\left[:\ddot{N}a:\right]^{+} \quad \left[Pb\right]^{2+} \quad \left[:\ddot{S}:\right]^{2-} \quad \left[\cdot Al:\right]^{3+}$$

FIGURE P8.29

8.30. Find the error in the following Lewis symbols:

Li:   $\left[Ga\right]^{+}$   $\left[:\ddot{M}g:\right]^{2+}$   $\left[:\ddot{F}:\right]^{-}$

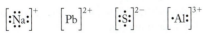
FIGURE P8.30

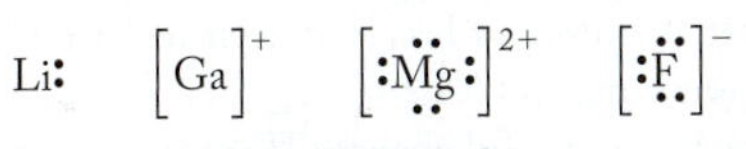

8.31. Draw Lewis symbols for $In^+$, $I^-$, $Ca^{2+}$, and $Sn^{2+}$. Which ions have a complete valence-shell octet?

8.32. Draw Lewis symbols of Xe, $Sr^{2+}$, Cl, and $Cl^-$. How many valence electrons are in each atom or ion?

8.33. Draw the Lewis symbol of an ion that has the following:
   a. 1+ charge and 1 valence electron
   b. 3+ charge and 0 valence electrons

8.34. Draw the Lewis symbol of an ion that has the following:
   a. 1− charge and 8 valence electrons
   b. 1+ charge and 5 valence electrons

8.35. How many valence electrons does each of the following species contain? (a) BN; (b) HF; (c) $OH^-$; (d) $CN^-$

8.36. How many valence electrons does each of the following species contain? (a) $N_2^+$; (b) $CS^+$; (c) CN; (d) CO

8.37. Draw Lewis structures for the following diatomic molecules and ions: (a) CO; (b) $O_2$; (c) $ClO^-$; (d) $CN^-$.

8.38. Draw Lewis structures for the following diatomic molecules and ions: (a) $F_2$; (b) $NO^+$; (c) SO; (d) HI.

8.39. Which groups among main group elements have an odd number of valence electrons?

8.40. Which of the groups in the periodic table will carry negative partial charges in diatomic compounds with hydrogen, HX and $H_2X$?

8.41. **Greenhouse Gases** Chlorofluorocarbons (CFCs) are linked to the depletion of stratospheric ozone. They are also greenhouse gases. Draw Lewis structures for the following CFCs:
   a. $CF_2Cl_2$ (Freon 12)
   b. $Cl_2FCCF_2Cl$ (Freon 113, containing a C—C bond)
   c. $C_2ClF_3$ (Freon 1113, containing a C=C bond)

8.42. Draw Lewis structures for the following silicon compounds:
   a. $HSiCl_3$
   b. $CH_3SiCl_3$ (contains a Si—C bond)
   c. $N(SiH_3)_3$

**8.43.** **Skunks and Rotten Eggs** Many sulfur-containing organic compounds have characteristically foul odors: butanethiol ($CH_3CH_2CH_2CH_2SH$) is responsible for the odor of skunks, and rotten eggs smell the way they do because they produce tiny amounts of pungent hydrogen sulfide, $H_2S$. Draw the Lewis structures for $CH_3CH_2CH_2CH_2SH$ and $H_2S$.

**8.44.** **Acid in Ants** Formic acid, HCOOH, is the smallest organic acid and was originally isolated by distilling red ants. Draw its Lewis structure given the connectivity of the atoms as shown in Figure P8.44.

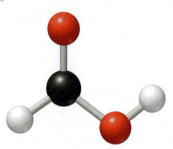

**FIGURE P8.44**

**8.45.** **Chlorine Bleach** Chlorine combines with oxygen in several proportions. Dichlorine monoxide ($Cl_2O$) is used in the manufacture of bleaching agents. Potassium chlorate ($KClO_3$) is used in oxygen generators aboard aircraft. Draw the Lewis structures for $Cl_2O$ and $ClO_3^-$. Cl is the central atom in each case.

**8.46.** **Dangers of Mixing Cleansers** Labels on household cleansers caution against mixing bleach with ammonia (Figure P8.46) because the reaction produces monochloramine ($NH_2Cl$) and hydrazine ($N_2H_4$), both of which are toxic:

$$NH_3(aq) + OCl^-(aq) \rightarrow NH_2Cl(aq) + OH^-(aq)$$

$$NH_2Cl(aq) + NH_3(aq) + OH^-(aq) \rightarrow$$
$$N_2H_4(aq) + Cl^-(aq) + H_2O(\ell)$$

Draw the Lewis structures for monochloramine and hydrazine.

**FIGURE P8.46**

## Polar Covalent Bonds
### CONCEPT REVIEW

**8.47.** How can we use electronegativity to predict whether a bond between two atoms is likely to be covalent or ionic?

**8.48.** How do the electronegativities of the elements change across a period and down a group?

**8.49.** Explain on the basis of atomic structure why trends in electronegativity are related to trends in atomic size.

**8.50.** Is the element with the most valence electrons in a period also the most electronegative?

**8.51.** What is meant by the term *polar covalent bond*?

**8.52.** Why are the electrons in bonds between different elements not shared equally?

### PROBLEMS

**8.53.** Which of the following bonds are polar? C—Se, C—O, Cl—Cl, O=O, N—H, C—H. In the bond or bonds that you selected, which atom has the greater electronegativity?

**8.54.** Which is the least polar bond: C—Se, C=O, Cl—Br, O=O, N—H, C—H?

**8.55.** In which of the following binary compounds is the bond expected to have the *least* ionic character? LiCl; CsI; KBr; NaF

**8.56.** In which of the following compounds is the bond between the atoms expected to have the most covalent character? $AlCl_3$; $AlBr_3$; $AlI_3$; $GaF_3$

## Vibrating Bonds and the Greenhouse Effect
### CONCEPT REVIEW

**8.57.** Describe how atmospheric greenhouse gases act like the panes of glass in a greenhouse.

**\*8.58.** Water vapor in the atmosphere contributes more to the greenhouse effect than carbon dioxide, yet water vapor is not considered an important factor in global warming. Propose a reason why.

**8.59.** Increasing concentrations of nitrous oxide in the atmosphere may be contributing to climate change. Is the ability of $N_2O$ to absorb IR radiation due to nitrogen–nitrogen bond stretching, nitrogen–oxygen bond stretching, or both? Explain your answer.

**8.60.** Is the ability of $H_2O$ molecules to absorb photons of IR radiation due to symmetrical stretching or asymmetrical stretching of its O—H bonds, or both? Explain your answer. (*Hint:* The angle between the two O—H bonds in $H_2O$ is 104.5°.)

**8.61.** Can molecules of carbon monoxide in the atmosphere absorb photons of IR radiation? Explain why or why not.

**8.62.** One of the activities that may be contributing to climate change is the production of cement. Why? (*Hint:* see Feature Box in Chapter 4.)

**8.63.** Why does IR radiation cause bonds to vibrate but not break (as UV radiation can)?

**8.64.** Argon is the third most abundant species in the atmosphere. Why isn't it a greenhouse gas?

**\*8.65.** Would the energy required to cause the bond in CO to vibrate be more or less than that required by the carbon–oxygen bond in $CO_2$?

**\*8.66.** Which compound absorbs IR radiation of a longer wavelength, NO or $NO_2$?

## Resonance

### CONCEPT REVIEW

**8.67.** Explain the concept of resonance.

**8.68.** How does resonance influence the stability of a molecule or an ion?

**8.69.** What factors determine whether or not a molecule or ion exhibits resonance?

**8.70.** What structural features do all the resonance forms of a molecule or ion have in common?

**8.71.** Explain why $NO_2$ is more likely to exhibit resonance than $CO_2$.

**8.72.** Are these two skeletal structures resonance forms: X—X—O and X—O—X?

### PROBLEMS

**8.73.** Draw Lewis structures for fulminic acid (HCNO) showing all resonance forms.

**8.74.** Draw Lewis structures for hydrazoic acid ($HN_3$) showing all resonance forms.

---

**\*8.75.** Oxygen and nitrogen combine to form a variety of nitrogen oxides, including the following two unstable compounds each with two nitrogen atoms per molecule: $N_2O_2$ and $N_2O_3$. Draw Lewis structures for these molecules showing all resonance forms.

**\*8.76.** Oxygen and sulfur combine to form a variety of different sulfur oxides. Some are stable molecules and some, including $S_2O_2$ and $S_2O_3$, decompose when they are heated. Draw Lewis structures for these two compounds showing all resonance forms.

---

**\*8.77.** The oxygen–oxygen distance in $F_2O_2$ is about 20% shorter than in hydrogen peroxide.
  a. Draw Lewis structures for both $F_2O_2$ and $H_2O_2$.
  b. It was proposed that $F_2O_2$ has a resonance form of $[FO_2]^+F^-$. Draw a Lewis structure for this "ionic" form of $F_2O_2$ and explain how the structure is consistent with the observed O–O distance.

**\*8.78.** The nitrogen–oxygen bond distance in $NOF_3$ is shorter than in $NO(CH_3)_3$.
  a. Draw Lewis structures for $NOF_3$ and $NO(CH_3)_3$.
  b. It was proposed that $NOF_3$ has a resonance form of $[NOF_2]^+F^-$. Draw a Lewis structure for this "ionic" form of $NOF_3$ and explain how the structure is consistent with the observed N–O distance.

---

**8.79.** Chemists can use the octet rule to predict the structures of new compounds to synthesize. Draw Lewis structures showing all resonance forms for the hypothetical compound ClSeNSO, where the atoms are connected in the order they are written.

**8.80.** Sulfur and nitrogen combine to make a number of cyclic compounds. Draw Lewis structures for the $[S_3N_2]^{2+}$ cation. Include all resonance forms. The atoms are connected as shown in Figure P8.80.

**FIGURE P8.80**

## Formal Charge: Choosing among Lewis Structures

### CONCEPT REVIEW

**8.81.** Describe how formal charges are used to choose between possible molecular structures.

**8.82.** How do the electronegativities of elements influence the selection of which Lewis structure is favored?

**8.83.** In a molecule containing S and O atoms, is a structure with a negative formal charge on sulfur more likely to contribute to bonding than an alternative structure with a negative formal charge on oxygen?

**8.84.** In a cation containing N and O, why do Lewis structures with a positive formal charge on nitrogen contribute more to the actual bonding in the molecule than do those structures with a positive formal charge on oxygen?

### PROBLEMS

**8.85.** Hydrogen isocyanide (HNC) has the same elemental composition as hydrogen cyanide (HCN) but the H in HNC is bonded to the nitrogen atom. Draw a Lewis structure for HNC, and assign formal charges to each atom. How do the formal charges on the atoms differ in the Lewis structures for HCN and HNC?

**8.86.** **Molecules in Interstellar Space** Hydrogen cyanide (HCN) and cyanoacetylene ($HC_3N$) have been detected in the interstellar regions of space and in comets close to Earth (Figure P8.86). Draw Lewis structures for these molecules, and assign formal charges to each atom. The hydrogen atom is bonded to carbon in both cases.

**FIGURE P8.86**

---

**8.87.** **Origins of Life** The discovery of polyatomic organic molecules such as cyanamide ($H_2NCN$) in interstellar space has led some scientists to believe that the molecules from which life began on Earth may have come from space. Draw Lewis structures for cyanamide, and select the preferred structure on the basis of formal charges.

**8.88.** Nitromethane ($CH_3NO_2$) reacts with hydrogen cyanide to produce $CNNO_2$ and $CH_4$:

$$HCN(g) + CH_3NO_2(g) \rightarrow CNNO_2(g) + CH_4(g)$$

a. Draw Lewis structures for $CH_3NO_2$, showing all resonance forms.
b. Draw Lewis structures for $CNNO_2$, showing all resonance forms, based on the two possible skeletal structures for it in Figure P8.88. Assign formal charges, and predict which structure is more likely to exist.
c. Are the two structures of $CNNO_2$ resonance forms of each other?

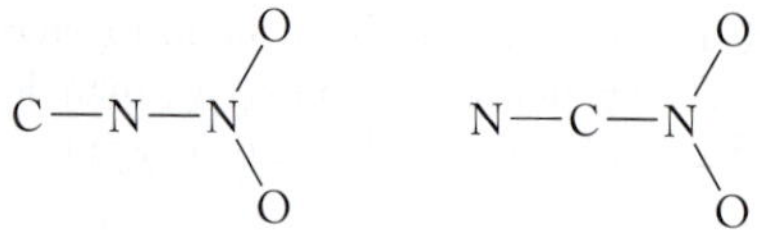

**FIGURE P8.88**

**\*8.89.** Sulfur chemistry is rich in molecules where resonance is needed to explain the bonding. Complete the Lewis structures for $AsN_2S_2$ as shown in Figure P8.89 and assign formal charges to the atoms in each structure.

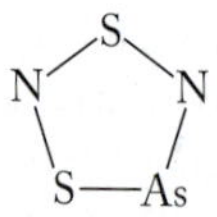

**FIGURE P8.89**

**8.90.** Draw all resonance forms of the sulfur–nitrogen anion, $S_4N^-$, and assign formal charges. The atoms are arranged as SSNSS.

**\*8.91.** Nitrogen is the central atom in molecules of nitrous oxide ($N_2O$). Draw Lewis structures for another possible arrangement: N—O—N. Assign formal charges and suggest a reason why this structure is not likely to be stable.

**8.92.** Use formal charges to determine which resonance form of each of the following ions is preferred: $CNO^-$; $NCO^-$; $CON^-$.

## Exceptions to the Octet Rule

### CONCEPT REVIEW

**8.93.** Are all odd-electron molecules exceptions to the octet rule?
**8.94.** Describe the factors that contribute to the stability of structures in which the central atoms have more than eight valence electrons.
**8.95.** Why do C, N, O, and F atoms in covalently bonded molecules and ions have no more than eight valence electrons?
**8.96.** Do atoms in rows 3 and below always expand their valence shell? Explain your answer.

### PROBLEMS

**8.97.** In which of the following molecules does the sulfur atom have an expanded valence shell? (a) $SF_6$; (b) $SF_5$; (c) $SF_4$; (d) $SF_2$

**8.98.** In which of the following molecules does the phosphorus atom have an expanded valence shell? (a) $POCl_3$; (b) $PF_5$; (c) $PF_3$; (d) $P_2F_4$ (which has a P—P bond)

**8.99.** How many electrons are there in the covalent bonds surrounding the central atom in the following species? (a) $(CH_3)_3Al$; (b) $B_2Cl_4$; (c) $SO_3$; (d) $SF_5^-$

**8.100.** How many electrons are there in the covalent bonds surrounding the central atom in the following species? (a) $POCl_3$; (b) $InCl_5^{2-}$; (c) $FBO$; (d) $PF_4^-$

**\*8.101.** Draw Lewis structures for $NOF_3$ and $POF_3$ in which the group 15 element is the central atom and the other atoms are bonded to it. What differences are there in the types of bonding in these molecules?

**\*8.102.** The phosphate anion is common in minerals. The corresponding nitrogen-containing anion, $NO_4^{3-}$, is unstable but can be prepared by reacting sodium nitrate with sodium oxide at 300°C. Draw Lewis structures for each anion. What are the differences in bonding between these ions?

**8.103.** Dissolving NaF in selenium tetrafluoride ($SeF_4$) produces $NaSeF_5$. Draw Lewis structures for $SeF_4$ and $SeF_5^-$. In which structure does Se have more than eight valence electrons?

**8.104.** Reaction between $NF_3$, $F_2$, and $SbF_3$ at 200°C and 100 atm pressure gives the ionic compound $NF_4SbF_6$:

$$NF_3(g) + 2\,F_2(g) + SbF_3(g) \rightarrow NF_4SbF_6(s)$$

Draw Lewis structures for the ions in this product.

**8.105. Ozone Depletion** The compound $Cl_2O_2$ may play a role in ozone depletion in the stratosphere. In the laboratory, reaction of $FClO_2$ with aluminum chloride produces $Cl_2O_2$ and $AlCl_2F$:

$$FClO_2(g) + AlCl_3(s) \rightarrow Cl_2O_2(g) + AlFCl_2(s)$$

Draw a Lewis structure for $Cl_2O_2$ based on the arrangement of atoms in Figure P8.105. Does either of the chlorine atoms in the structure have an expanded valence shell?

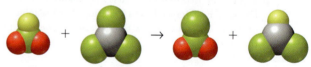

**FIGURE P8.105**

**\*8.106.** Trimethylaluminum reacts with dimethylamine, $(CH_3)_2NH$, forming methane and $Al_2(CH_3)_4[N(CH_3)_2]_2$ by the balanced chemical equation:

$$2\,(CH_3)_3Al + 2\,HN(CH_3)_2 \rightarrow$$
$$2\,CH_4 + Al_2(CH_3)_4[N(CH_3)_2]_2$$

Draw the Lewis structure for the reactants and products. Must the structure contain atoms with less than eight valence electrons?

**8.107.** Which of the following chlorine oxides are odd-electron molecules? (a) $Cl_2O_7$; (b) $Cl_2O_6$; (c) $ClO_4$; (d) $ClO_3$; (e) $ClO_2$

**8.108.** Which of the following nitrogen oxides are odd-electron molecules? (a) $NO$; (b) $NO_2$; (c) $NO_3$; (d) $N_2O_4$; (e) $N_2O_5$

**8.109.** Which of the Lewis structures in Figure P8.109 contributes most to the bonding in CNO?

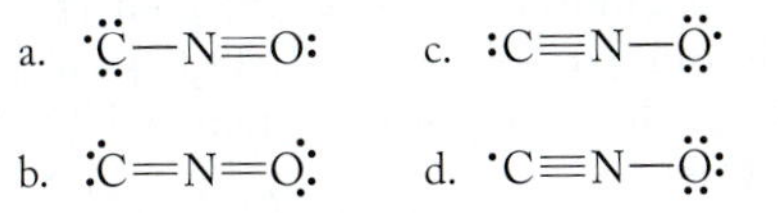

a. $\cdot\ddot{C}-N\equiv O\colon$     c. $\colon C\equiv N-\ddot{\underset{\cdot}{O}}\cdot$

b. $\colon\ddot{C}=N=\ddot{O}\colon$     d. $\cdot C\equiv N-\ddot{\underset{\cdot\cdot}{O}}\colon$

**FIGURE P8.109**

**8.110.** Why is the Lewis structure in Figure P8.110 unlikely to contribute much to the bonding in NCO?

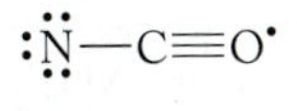

$$\colon\ddot{\underset{\cdot\cdot}{N}}-C\equiv O\cdot$$

**FIGURE P8.110**

**8.111.** Using Lewis structures explain why dimethylaluminum chloride is more likely to exist as $(CH_3)_4Al_2Cl_2$ than as $(CH_3)_2AlCl$.

**8.112.** When left at room temperature, mixtures of $BF_3$ and $BCl_3$ are found to contain significant amounts of $BF_2Cl$ and $BFCl_2$. Using Lewis structures, explain the origin of the latter two products.

**8.113.** Some have argued that $SF_6$ has ionic resonance forms that do not require an expanded octet for S. Draw a resonance structure consistent with this hypothesis and assign formal charges to each atom. Is this resonance form better than or the same as the one with an expanded octet?

**8.114.** The synthesis of an extremely unusual molecule, $[NO]_2{}^+[XeF_8]^{2-}$, was reported close to 50 years ago. The structure of $[XeF_8]^{2-}$ shows that all eight fluorine atoms are within bonding distance of the Xe. Draw a Lewis structure for the $[XeF_8]^{2-}$ anion.

## The Lengths and Strengths of Covalent Bonds

### CONCEPT REVIEW

**8.115.** Do you expect the nitrogen–oxygen bond length in the nitrate ion to be the same as in the nitrite ion?

**8.116.** Why is the oxygen–oxygen bond length in $O_3$ not the same as in $O_2$?

**8.117.** Explain why the nitrogen–oxygen bond lengths in $N_2O_4$ (which has a nitrogen–nitrogen bond) and $N_2O$ are nearly identical (118 and 119 pm, respectively).

**8.118.** Do you expect the sulfur–oxygen bond lengths in $SO_3{}^{2-}$ and $SO_4{}^{2-}$ ions to be about the same? Why?

**8.119.** Rank the following ions in order of (a) increasing nitrogen–oxygen bond lengths and (b) increasing bond energies: $NO_2{}^-$; $NO^+$; $NO_3{}^-$.

**8.120.** Rank the following compounds and ions in (a) order of increasing carbon–oxygen bond lengths and (b) increasing bond energies: $CO$; $CO_2$; $CO_3{}^{2-}$.

**\*8.121.** Do you expect the boron–fluorine bond energy to be the same in $BF_3$ and $F_3BNH_3$?

**\*8.122.** The boron–oxygen distances in the $BO_2{}^+$ cation are equal. Does this mean the bond order of the B–O bond is two?

**8.123.** Why must the stoichiometry of a reaction be known in order to estimate the enthalpy change from bond energies?

**8.124.** Why must the structures of the reactants and products be known in order to estimate the enthalpy change of a reaction from bond energies?

**\*8.125.** When calculating the enthalpy change for a chemical reaction using bond energies, why is it important to know the phase (solid, liquid, or gaseous) for every compound in the reaction?

**\*8.126.** If the energy needed to break 2 moles of $C=O$ bonds is greater than the sum of the energies needed to break the $O=O$ bonds in 1 mole of $O_2$ and vaporize 1 mole of carbon, why does the combustion of pure carbon release heat?

### PROBLEMS

NOTE: Use the average bond energies in Table 8.3 to answer Problems 8.127 to 8.136.

**8.127.** Use average bond energies to estimate the enthalpy changes of the following reactions:
a. $N_2(g) + 3\,H_2(g) \rightarrow 2\,NH_3(g)$
b. $N_2(g) + 2\,H_2(g) \rightarrow H_2NNH_2(g)$
c. $2\,N_2(g) + O_2(g) \rightarrow 2\,N_2O(g)$

**8.128.** Use average bond energies to estimate the enthalpy changes of the following reactions:
a. $CO_2(g) + H_2(g) \rightarrow H_2O(g) + CO(g)$
b. $N_2(g) + O_2(g) \rightarrow 2\,NO(g)$
\*c. $C(s) + CO_2(g) \rightarrow 2\,CO(g)$
(*Hint*: The enthalpy of sublimation of graphite, $C(s)$, is 719 kJ/mol.)

**8.129.** The combustion of CO to $CO_2$ releases 283 kJ/mol. What is the bond energy of the carbon–oxygen bond in carbon monoxide?

**8.130.** Use average bond energies to estimate the standard enthalpy of formation of HF gas.

**8.131.** Estimate how much less energy is released during the incomplete combustion of 1 mole of methane to carbon monoxide and water vapor than in the complete combustion to carbon dioxide and water vapor.

**8.132.** Estimate how much more energy is released by the reaction

$$C(s) + O_2(g) \rightarrow CO_2(g)$$

than by the reaction

$$C(s) + \tfrac{1}{2}O_2(g) \rightarrow CO(g)$$

**\*8.133.** Estimate $\Delta H_{rxn}$ for the following reaction:

$$4\,NH_3(g) + 7\,O_2(g) \rightarrow 4\,NO_2(g) + 6\,H_2O(g)$$

**\*8.134.** The value of $\Delta H_{rxn}$ for the reaction

$$2\,H_2S(g) + 3\,O_2(g) \rightarrow 2\,SO_2(g) + 2\,H_2O(g)$$

is $-1036$ kJ. Estimate the energy of the bonds in $SO_2$.

**8.135.** A molecular view of the combustion of $CS_2$ is shown in Figure P8.135. If the standard enthalpy of combustion of $CS_2$ is $-1102$ kJ/mol, what is the average bond energy for the carbon–sulfur bonds in $CS_2$?

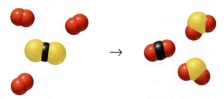

**FIGURE P8.135**

**\*8.136.** The standard enthalpy of reaction for the decomposition of carbon oxysulfide (COS) to $CO_2$ and $CS_2$, as shown in Figure P8.136, is $-1.9$ kJ/mol COS. Are the apparent bond energies of the carbon–sulfur and carbon–oxygen bonds in COS stronger or weaker than in $CS_2$ and $CO_2$, respectively?

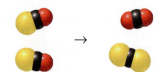

**FIGURE P8.136**

**\*8.137.** Carbon and oxygen form three oxides: CO, $CO_2$, and carbon suboxide ($C_3O_2$). Draw a Lewis structure for $C_3O_2$ in which the three carbon atoms are bonded to each other, and predict whether the carbon–oxygen bond lengths in the carbon suboxide molecule are equal.

**\*8.138.** Spectroscopic analysis of the linear molecule $N_4O$ reveals that the nitrogen–oxygen bond length is 135 pm and that there are three nitrogen–nitrogen bond lengths: 148, 127, and 115 pm. Draw the Lewis structure for $N_4O$ consistent with these observations.

## Additional Problems

**8.139.** The unpaired dots in Lewis symbols of the elements represent valence electrons available for covalent bond formation. In Figure P8.139, which of the options for placing dots around the symbol for each element is preferred?

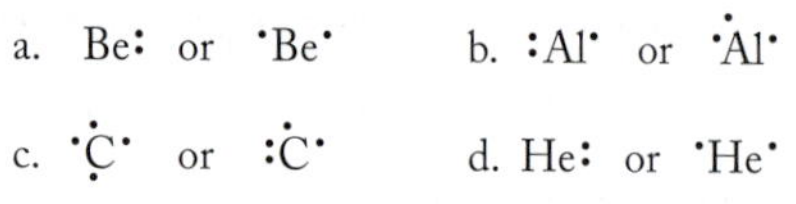

     a. Be: or ·Be·     b. :Al· or ·Al·

     c. ·Ċ· or :C·     d. He: or ·He·

**FIGURE P8.139**

**8.140.** Based on the Lewis symbols in Figure P8.140, predict to which group in the periodic table element X belongs.

     a. ·Ẋ·   b. ·Ẋ:   c. :Ẍ:   d. :Ẍ:

**FIGURE P8.140**

**8.141.** Use formal charges to predict whether the atoms in carbon disulfide are arranged CSS or SCS.

**8.142.** Use formal charges to predict whether the atoms in hypochlorous acid are arranged HOCl or HClO.

**\*8.143. Chemical Weapons** Phosgene is a poisonous gas first used in chemical warfare during World War I. It has the formula $COCl_2$ (C is the central atom).
     a. Draw its Lewis structure.
     b. Phosgene kills because it reacts with water in nasal passages, in the lungs, and on the skin to produce carbon dioxide and hydrogen chloride. Write a balanced chemical equation for this process showing the Lewis structures for reactants and products.

**8.144.** The dinitramide anion $[N(NO_2)_2^-]$ was first isolated in 1996. The arrangement of atoms in $N(NO_2)_2^-$ is in Figure P8.144.
     a. Complete the Lewis structure for $N(NO_2)_2^-$, including any resonance forms, and assign formal charges.
     b. Explain why the nitrogen–oxygen bond lengths in $N(NO_2)_2^-$ and $N_2O$ should (or should not) be similar.
     c. $N(NO_2)_2^-$ was isolated as $[NH_4^+][N(NO_2)_2^-]$. Draw the Lewis structure for $NH_4^+$.

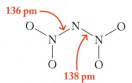

**FIGURE P8.144**

**\*8.145.** Silver cyanate (AgOCN) is a source of the cyanate ion ($OCN^-$), which reacts with a number of small molecules. Under certain conditions the species OCN is an anion with a charge of $1-$; under others it is a neutral, odd-electron molecule.
     a. Two molecules of OCN combine to form OCNNCO. Draw the Lewis structures for this molecule, including all resonance forms.
     b. The $OCN^-$ ion reacts with BrNO, forming the unstable molecule OCNNO. Draw the Lewis structures for BrNO and OCNNO, including all resonance forms.
     c. The $OCN^-$ ion reacts with $Br_2$ and $NO_2$ to produce $N_2O$, $CO_2$, BrNCO, and OCN(CO)NCO. Draw three of the resonance forms of OCN(CO)NCO, which has the arrangement of atoms shown in Figure P8.145.

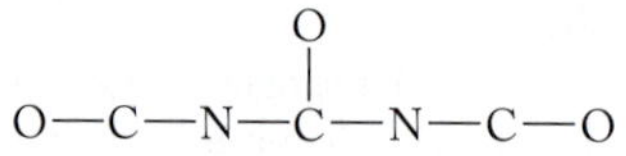

$$O-C-N-\overset{\overset{\displaystyle O}{|}}{C}-N-C-O$$

**FIGURE P8.145**

**\*8.146.** During the reaction of the cyanate ion ($OCN^-$) with $Br_2$ and $NO_2$, a very unstable substance called an *intermediate* forms and then quickly falls apart. Its formula is $O_2NNCO$.
     a. Draw three of the resonance forms for $O_2NNCO$, assign formal charges, and predict which of the three contributes the most to the bonding in $O_2NNCO$. The connectivity of the atoms is shown in Figure P8.146a.

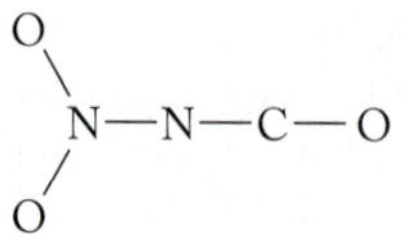

$$\begin{array}{c} O \\ \diagdown \\ N-N-C-O \\ \diagup \\ O \end{array}$$

**FIGURE P8.146a**

b. Which bond in $O_2NNCO$ must break in the reaction with $Br_2$ to form BrNCO? What other product forms?

c. Draw Lewis structures for a different arrangement of the N, C, and O atoms in $O_2NNCO$ as shown in Figure P8.146b.

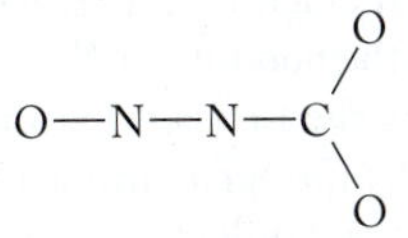

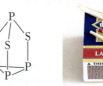

**FIGURE P8.146b**

**8.147.** A compound with the formula $Cl_2O_6$ decomposes to a mixture of $ClO_2$ and $ClO_4$. Draw two Lewis structures for $Cl_2O_6$: one with a chlorine–chlorine bond and one with a Cl—O—Cl arrangement of atoms. Draw a Lewis structure for $ClO_2$.

$$Cl_2O_6 \rightarrow ClO_2 + ClO_4$$

*8.148. A compound consisting of chlorine and oxygen, $Cl_2O_7$, decomposes by the following reaction:

$$Cl_2O_7 \rightarrow ClO_4 + ClO_3$$

a. Draw two Lewis structures for $Cl_2O_7$: one with a chlorine–chlorine bond and one with a Cl—O—Cl arrangement of atoms.

b. Draw a Lewis structure for $ClO_3$.

*8.149. The odd-electron molecule CN dimerizes to give cyanogen ($C_2N_2$).

a. Draw a Lewis structure for CN, and predict which arrangement for cyanogen is more likely: NCCN or CNNC.

b. Cyanogen reacts slowly with water to produce oxalic acid ($H_2C_2O_4$) and ammonia; the Lewis structure for oxalic acid is shown in Figure P8.149. Compare this structure to your answer in part a. When the actual structures of molecules have been defined experimentally, the structures have been used to refine Lewis structures. Does this structure increase your confidence that the structure you selected in part a may be the better one?

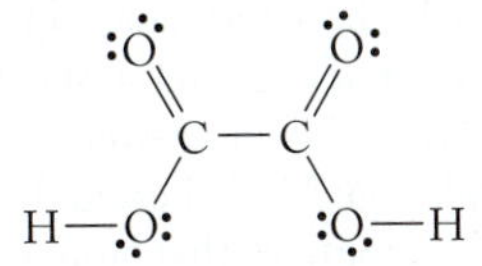

**FIGURE P8.149**

*8.150. Draw all resonance forms for the molecules NSF and HBS where S and B are the central atoms. Include possible ionic structures for NSF.

*8.151. In electron diffraction, an interference pattern is produced based on the location of atoms when a beam of electrons is fired at a sample. The electrons are diffracted by the arrangement of the atoms in a crystal just the way waves are diffracted by passing through gaps in a barrier, as we saw in Chapter 7. The positions of the atoms can be determined from the interference pattern. The molecular structure of sulfur cyanide trifluoride ($SF_3CN$) has been shown by electron diffraction studies to have the arrangement of atoms with the indicated bond lengths in Figure P8.151.

Using the observed bond lengths as a guide, complete the Lewis structure for $SF_3CN$, and assign formal charges.

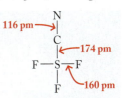

**FIGURE P8.151**

8.152. **Strike-Anywhere Matches** Heating phosphorus with sulfur gives $P_4S_3$, a solid used in the heads of strike-anywhere matches (Figure P8.152). $P_4S_3$ has the Lewis structure framework shown. Complete the Lewis structure so that each atom has the optimum formal charge.

**FIGURE P8.152**

*8.153. The heavier group 16 elements can expand their valence shell. The $TeOF_6^{2-}$ anion was first prepared in 1993. Draw the Lewis structure for $TeOF_6^{2-}$.

*8.154. **Sulfur in the Environment** Sulfur is cycled in the environment through compounds such as dimethyl sulfide ($CH_3SCH_3$), hydrogen sulfide ($H_2S$), and sulfite and sulfate ions. Draw Lewis structures for these four species. Are expanded valence shells needed to minimize the formal charges for any of these species?

8.155. How many pairs of electrons does xenon share in the following molecules and ions? (a) $XeF_2$; (b) $XeOF_2$; (c) $XeF^+$; (d) $XeF_5^+$; (e) $XeO_4$

8.156. Consider a hypothetical structure of ozone that is cyclic (the atoms form a ring) such that its three O atoms are at the corners of a triangle. Draw the Lewis structure for this molecule.

*8.157. How might electron diffraction (see Problem 8.151) help characterize the bonding in a compound with the formula $A_2X$ in terms of the following?

a. distinguishing between these two bonding patterns: X—A—A and A—X—A

b. distinguishing between these resonance forms: A—X≡A, A=X=A, and A≡X—A

8.158. The highly explosive $N_5^+$ cation was first isolated in 1999 by reaction of $N_2F^+$ with $HN_3$:

$$N_2F^+ + HN_3 \rightarrow N_5^+ + HF$$

Draw the Lewis structures for the reactants and products, including all resonance forms.

*8.159. **Jupiter's Atmosphere** The ionic compound $NH_4SH$ was detected in the atmosphere of Jupiter (Figure P8.159) by the Galileo space probe in 1995. Draw the Lewis structure for $NH_4SH$. Why couldn't there be a covalent bond between the nitrogen and sulfur atoms, making $NH_4SH$ a molecular compound?

**FIGURE P8.159**

8.160. **Antacid Tablets** Antacids commonly contain calcium carbonate and/or magnesium hydroxide. Draw the Lewis structures for calcium carbonate and magnesium hydroxide.

*8.161. An allotrope of nitrogen, $N_4$, was reported in 2002. The compound has a lifetime of 1 μs at 298 K and was prepared by adding an electron to $N_4^+$. Since the compound cannot be isolated, its structure is unconfirmed experimentally.
  a. Draw the Lewis structures for all the resonance forms of linear $N_4$. (Linear means that all four nitrogens are in a straight line.)
  b. Assign formal charges, and determine which structure is the best description of $N_4$.
  c. Draw a Lewis structure for a ring (cyclic) form of $N_4$, and assign formal charges.

*8.162. Scientists have predicted the existence of $O_4$ even though this compound has never been observed. However, $O_4^{2-}$ has been detected. Draw the Lewis structures for $O_4$ and $O_4^{2-}$.

8.163. Draw a Lewis structure for $AlFCl_2$, the second product in the synthesis of $Cl_2O_2$ in the following reaction:

$$FClO_2(g) + AlCl_3(s) \rightarrow Cl_2O_2(g) + AlFCl_2(s)$$

*8.164. Draw Lewis structures for $BF_3$ and $(CH_3)_2BF$. The B—F distance in both molecules is the same (130 pm). Does this observation support the argument that all the boron–fluorine bonds in $BF_3$ are single bonds?

8.165. Which of the following molecules and ions contains an atom with an expanded valence shell? (a) $Cl_2$; (b) $ClF_3$; (c) $ClI_3$; (d) $ClO^-$

8.166. Which of the following molecules contains an atom with an expanded valence shell? (a) $XeF_2$; (b) $GaCl_3$; (c) $ONF_3$; (d) $SeO_2F_2$

*8.167. A linear nitrogen anion, $N_5^-$, was isolated for the first time in 1999.
  a. Draw the Lewis structures for four resonance forms of linear $N_5^-$.
  b. Assign formal charges to the atoms in the structures in part a, and identify the structures that contribute the most to the bonding in $N_5^-$.
  c. Compare the Lewis structures for $N_5^-$ and $N_3^-$. In which ion do the nitrogen–nitrogen bonds have the higher average bond order?

*8.168. Carbon tetraoxide ($CO_4$) was discovered in 2003.
  a. The four atoms in $CO_4$ are predicted to be arranged as shown in Figure P8.168. Complete the Lewis structure for $CO_4$.
  b. Are there any resonance forms for the structure you drew that have zero formal charges on all atoms?
  c. Can you draw a structure in which all four oxygen atoms in $CO_4$ are bonded to carbon?

**FIGURE P8.168**

8.169. Plot the electronegativities of elements with $Z = 3$ to 9 ($y$-axis) versus their first ionization energy ($x$-axis). Is the plot linear? Use your graph to predict the electronegativity of neon, whose first ionization energy is 2080 kJ/mol.

8.170. Use the data in Figures 7.38 and 8.5 to plot electronegativity as a function of first ionization energy for the main group elements of the fifth row of the periodic table. From this plot estimate the electronegativity of xenon, whose first ionization energy is 1170 kJ/mol.

8.171. The cation $N_2F^+$ is isoelectronic with $N_2O$.
  a. What does it mean to be isoelectronic?
  b. Draw the Lewis structure for $N_2F^+$. (*Hint*: The molecule contains a nitrogen–nitrogen bond.)
  c. Which atom has the +1 formal charge in the structure you drew in part b?
  d. Does $N_2F^+$ have resonance forms?
  e. Could the middle atom in the $N_2F^+$ ion be a fluorine atom? Explain your answer.

8.172. **Ozone Depletion** Methyl bromide ($CH_3Br$) is produced naturally by fungi. Methyl bromide has also been used in agriculture as a fumigant, but this use is being phased out because the compound has been linked to ozone depletion in the upper atmosphere.
  a. Draw the Lewis structure for $CH_3Br$.
  b. Which bond in $CH_3Br$ is more polar, carbon–hydrogen or carbon–bromine?

## Fluorine and Oxygen: Location, Location, Location

**8.173.** Write a balanced equation for the reaction of fluorine gas with water.
  a. Identify the oxidizing and reducing agents.
  b. Draw Lewis structures for all the compounds involved in the reaction, and place arrows on each polar bond to indicate the direction of the polarity.

*8.174. Atoms of Xe have complete octets, yet the compound $XeO_2$ exists. How is this possible?

*8.175. All the carbon–fluorine bonds in tetrafluoroethylene (Figure P8.175) are polar, but experimental results show that the molecule as a whole is nonpolar. Draw arrows on each bond to show the direction of polarity, and suggest a reason why the molecule is nonpolar.

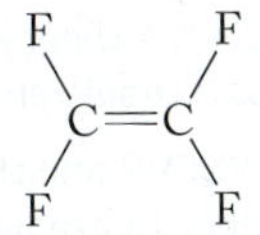

$$\begin{array}{c} F \qquad F \\ \diagdown \qquad \diagup \\ C {=\!=} C \\ \diagup \qquad \diagdown \\ F \qquad F \end{array}$$

Tetrafluoroethylene

**FIGURE P8.175**

*8.176. Free radicals increase in stability, and hence decrease in reactivity, if they have more than one atom in their structure that can carry the unpaired electron.
  a. Draw Lewis structures for the three free radicals formed from $CF_2Cl_2$ in the stratosphere—ClO, Cl, and $CF_2Cl$—and use this principle to rank them in order of reactivity.
  b. Free radicals are "neutralized" when they react with each other, forming an electron pair (a single covalent bond) between two atoms. This is called a *termination reaction* because it shuts down any reaction that was powered by the free radical. Predict the formula of the molecules that are produced when the chlorine free radical reacts with each of the free radicals in part a.

If your instructor assigns problems in **smartwork**, log in at **smartwork.wwnorton.com**.

# 9

# Molecular Geometry: Shape Determines Function

9.1    Molecular Shape

9.2    Valence-Shell Electron-Pair Repulsion Theory (VSEPR)

9.3    Polar Bonds and Polar Molecules

9.4    Valence Bond Theory

9.5    Shape and Interactions with Large Molecules

9.6    Chirality and Molecular Recognition

9.7    Molecular Orbital Theory

## Learning Outcomes

**LO1** Explain the theory of valence-shell electron-pair repulsion (VSEPR)

**LO2** Use VSEPR and the concept of steric number to predict the bond angles in molecules and the shapes of molecules with one central atom
**Sample Exercises 9.1, 9.2, 9.3**

**LO3** Predict whether a substance is polar or nonpolar based on its molecular structure
**Sample Exercise 9.4**

**LO4** Use valence bond theory to explain orbital overlap, bond angles, and molecular shapes
**Sample Exercises 9.5, 9.6, 9.7, 9.8**

**LO5** Recognize molecular shapes that are stabilized by delocalization of π electrons

**LO6** Recognize chiral molecules

**LO7** Draw molecular orbital (MO) diagrams of diatomic molecules and use MO diagrams to predict magnetic properties and explain spectra
**Sample Exercises 9.9, 9.10, 9.11**

## Biological Activity and Molecular Shape

Hold your hands out in front of you, palms up, fingers extended. Now rotate your wrists inward so that your thumbs point straight up. Your right hand looks the same as the image your left hand makes in a mirror. Does that mean your two hands have the same shape? If you ever tried to put your right hand in a glove made for your left, you know that they do not have the same shape. Many other objects in our world have a "handedness" about them, from headphones to scissors to golf clubs.

Handedness, or chirality, is a three-dimensional phenomenon. For example, the compound that produces the refreshing aroma of spearmint has the molecular formula $C_{10}H_{14}O$. The compound responsible for the musty aroma of caraway seeds has the same molecular formula *and* the same Lewis structure. To understand how two compounds can be so much alike and still have different properties, we have to consider their structures in three dimensions.

We perceive a difference in aromas in part because each molecule has a unique site where it attaches to our nasal membranes. Just as a left hand only fits a left glove, the spearmint molecule fits only the spearmint-shaped site, and the caraway molecule, the caraway site. This behavior, called molecular recognition, enables biomolecular structures, such as nasal membranes, to recognize and react when a particular molecule with a particular shape fits into a part of the membrane known as an *active site*. Many of the substances we ingest, from food to pharmaceuticals, exert physiological effects because their molecules are recognized by and become bound to active sites in biomolecules.

In Chapter 8 we studied a simple model for describing the arrangement of atoms in molecules without considering their three-dimensional shapes. In this chapter, we first discuss the parameters that determine a molecule's shape. Next we examine theories of bonding that explain the molecular shapes we observe. Finally, we begin to

**Molecules Accelerate Ripening** Ripening tomatoes give off the gas ethylene that speeds up the ripening process. Ethylene molecules have a unique shape that enables them to fit into a biomolecular site that controls ripening. ▶

explore the impact of molecular shape on the physical, chemical, and biological properties of compounds. We will see in later chapters that molecular geometry is an important factor in determining how molecules interact with one another. ■

# 9.1 Molecular Shape

The shape of a compound's molecules can affect many of its properties: its physical state at room temperature, its solubility in water and other solvents, its aroma, its biological activity, and its reactivity with other compounds. In Chapter 8 we drew Lewis structures to describe bonding in molecules, but Lewis structures are only two-dimensional representations of how atoms and the electron pairs that surround them are arranged in molecules. Lewis structures show how atoms are connected in molecules, but they do not show how the atoms are arranged in three dimensions, nor do they necessarily show the overall shape of the molecule.

To illustrate this point, let's consider the Lewis structures and ball-and-stick models (Figure 9.1) of two compounds: carbon dioxide and methane. The linear array of atoms and bonding electrons in the Lewis structure for $CO_2$ corresponds to the actual linear shape of the molecule, as represented by the ball-and-stick model. The angle between the two C=O bonds is 180°, just as in the Lewis structure. On the other hand, the Lewis structure of methane does not convey the true three-dimensional orientation of the four C—H bonds in each molecule. The 90° angles between bonds in the planar Lewis structure are not close to the three-dimensional H—C—H **bond angles** of 109.5°.

In this chapter, we explore several theories of covalent bonding that account for and predict the shapes of molecules. We focus on small molecules with single central atoms, though these theories can be applied to molecules of any size. We start with the shared pairs and lone pairs of electrons described by Lewis theory and then predict how those pairs should be oriented about a central atom to minimize their interactions and produce the most stable molecular structure. As we shall see, each theory accurately predicts electron-pair orientations, molecular shapes, and the properties of molecular compounds.

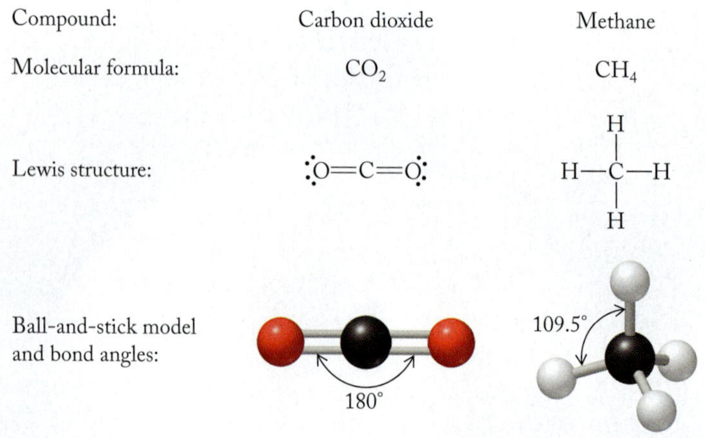

**FIGURE 9.1** The Lewis structure of $CO_2$ matches its true molecular structure, but the Lewis structure of $CH_4$ does not because the C—H bonds in $CH_4$ extend in all three dimensions.

# 9.2 Valence-Shell Electron-Pair Repulsion Theory (VSEPR)

▶❙❙ **CHEMTOUR** VSEPR Model

Let's start with a theory based on a fundamental chemical principle: electrons have negative charges and repel each other. **Valence-shell electron-pair repulsion theory** (**VSEPR**) applies this principle by assuming that pairs of valence electrons are arranged about central atoms in ways that minimize repulsions between the pairs. To predict molecular shape using VSEPR we must consider two things:

➤ **Electron-pair geometry**, which defines the relative positions in three-dimensional space of all the bonding pairs and lone pairs of valence electrons on the central atom
➤ **Molecular geometry**, which defines the relative positions in three-dimensional space of the atoms in a molecule

Electron-pair geometry may be the same as molecular geometry, but it does not have to be. The presence of lone pairs on the central atom makes the difference.

To accurately predict molecular geometry, we first need to know electron-pair geometry. If there are no lone pairs of electrons, then the process is simplified because the electron-pair geometry *is* the molecular geometry. Let's begin with this simple case and consider the shapes of molecules that have various numbers of bonds around a central atom and no lone pairs.

**CONCEPT TEST**

In your own words, explain how electron-pair geometry and molecular geometry are related and how they are different.

*(Answers to Concept Tests are in the back of the book.)*

## Central Atoms with No Lone Pairs

To determine the geometry of a molecule, we start by drawing its Lewis structure. From the Lewis structure, we determine the **steric number (SN)** of the central atom, which is the sum of the number of atoms bonded to that atom and the number of lone pairs on it:

$$\text{SN} = \left(\begin{array}{c}\text{number of atoms}\\\text{bonded to central atom}\end{array}\right) + \left(\begin{array}{c}\text{number of lone pairs}\\\text{on central atom}\end{array}\right) \quad (9.1)$$

Because we are focused on molecules in which the central atom has no lone pairs, the steric number equals the number of atoms bonded to the central atom. In evaluating the shapes of these molecules, we generate five common shapes that describe both the electron-pair geometries and the molecular geometries of many covalent compounds.

Let's start by thinking of the central atom as the center of a sphere, with all the other atoms in the molecule placed on the surface of the sphere and linked to the central atom by covalent bonds. If the central atom is bonded covalently to only two other atoms, then SN = 2. How do the electron pairs in the two bonds arrange themselves to minimize their mutual repulsion? The answer is that they are as far from each other as possible: on opposite sides of the sphere (Figure 9.2a). This gives a linear electron-pair geometry and a linear molecular geometry.

**bond angle** the angle (in degrees) defined by lines joining the centers of two atoms to the center of a third atom to which they are chemically bonded.

**valence-shell electron-pair repulsion theory (VSEPR)** a model predicting that the arrangement of valence electron pairs around a central atom minimizes their mutual repulsion to produce the lowest-energy orientations.

**electron-pair geometry** the three-dimensional arrangement of bonding pairs and lone pairs of electrons about a central atom.

**molecular geometry** the three-dimensional arrangement of the atoms in a molecule.

**steric number (SN)** the sum of the number of atoms bonded to a central atom and the number of lone pairs of electrons on the central atom.

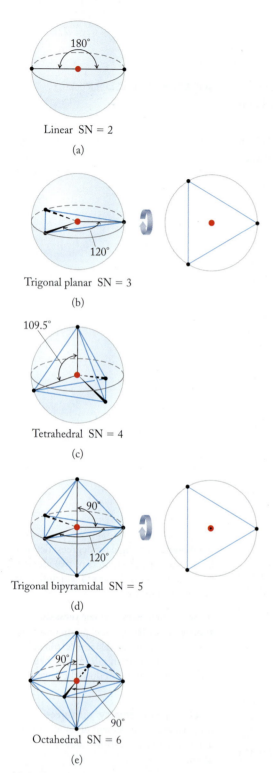

**FIGURE 9.2** Electron-pair geometries depend on the steric number (SN) of the central atom. In these images, there are no lone pairs of electrons on the central atoms (red dots), so the molecular geometries are the same as the electron-pair geometries. The images show bond angles for different numbers of atoms located on the surface of a sphere and bonded to an atom at the center of the sphere. The blue lines define the geometric forms that give the molecular shapes their names. Figures (b) and (d) show spheres rotated 90° so equilateral triangles in the equatorial plane are clearly visible.

The three atoms in the molecule are arranged in a straight line, and the bond angle is 180°.

If three atoms are bonded to a central atom with no lone pairs, then SN = 3. The three bonding pairs are as far apart as possible when they are located at the three corners of an equilateral triangle. The angle between each pair of bonds is the same: 120° (Figure 9.2b). The name of this electron-pair and molecular geometry is **trigonal planar**.

With four atoms around the central atom and no lone pairs (SN = 4), we have the first case in which the atoms are not all in the same plane. Instead, the atoms bonded to the central atom occupy the four vertices of a tetrahedron, which is a four-sided pyramid (*tetra* is Greek, meaning "four"). The bonding pairs form bond angles of 109.5° with each other, as shown in Figure 9.2(c). The electron-pair geometry and molecular geometry are both **tetrahedral**.

When five atoms are bonded to a central atom with no lone pairs, SN = 5 and the atoms occupy the five corners of two triangular pyramids that share the same base. The central atom of the molecule is in the center of the sphere at the common center of the two bases, as shown in Figure 9.2(d). One bonding pair points to the tip of the top pyramid, one points to the tip of the bottom pyramid, and the other three point to the three vertices of the shared triangular base. These three vertices lie along the equator of the sphere, so the atoms that occupy these sites and the bonds that connect them to the central atom are called *equatorial* atoms and bonds. The bond angles between the three equatorial bonds are 120° (just as in the triangle of the trigonal planar geometry in Figure 9.2b). The bond angle between an equatorial bond and either vertical, or *axial*, bond is 90°, and the angle between the two axial bonds is 180°. A molecule in which the atoms are arranged this way is said to have a **trigonal bipyramidal** electron-pair and molecular geometry.

For SN = 6, picture two pyramids that have a square base (Figure 9.2e). Put them together, base-to-base, and you form a shape in which all six positions around the sphere are equivalent. We can think of the six bonding pairs of electrons as three sets of two electron pairs each. The two pairs in each set are oriented at 180° to each other and at 90° to the other two pairs, just like the axes of an *xyz* coordinate system. Four equatorial atoms lie at the four vertices of the common square base and are 90° apart, another atom lies at the tip of the top pyramid, and a sixth lies at the tip of the bottom pyramid. This arrangement defines **octahedral** electron-pair and molecular geometries.

To demonstrate a real-world analogy to the above bond orientations, we start with a batch of fully inflated yellow balloons. We tie them together in clusters of two, three, four, five, and six balloons (Figure 9.3). Let us assume that all the balloons have acquired a static electrical charge, so they repel each other. If the tie points of the clusters represent the central atom in our balloon models, then the opposite ends of the balloons represent the atoms that are bonded to the central atom. Note how our clusters of two, three, four, five, and six balloons produce the same orientations that resulted from selecting points on a sphere that were as far apart as possible. The long axes of the balloons in Figure 9.3 provide an accurate representation of the bond directions in Figure 9.2.

| SN = 2 | 3 | 4 | 5 | 6 |

**FIGURE 9.3** Balloons charged with static electricity repel each other and align themselves as far apart as possible when tied together. In doing so, they mimic the locations of different numbers of electron pairs about a central atom. Each balloon represents one electron pair. Note the similarities between the patterns of the balloons and the diagrams in Figure 9.2.

**trigonal planar** molecular geometry about a central atom with a steric number of 3 and no lone pairs of electrons.

**tetrahedral** molecular geometry about a central atom with a steric number of 4 and no lone pairs of electrons.

**trigonal bipyramidal** molecular geometry about a central atom with a steric number of 5 and no lone pairs of electrons; three atoms occupy equatorial sites and two other atoms occupy axial sites above and below the equatorial plane.

**octahedral** molecular geometry about a central atom with a steric number of 6 and no lone pairs of electrons, in which all six sites are equivalent.

Now let's look at five simple molecules that have no lone pairs about the central atom and apply VSEPR and the concept of steric number to predict their electron-pair and molecular geometries. To do so, we follow these three steps:

1. Draw the Lewis structure for the molecule.
2. Determine the steric number of the central atom.
3. Use the steric number to predict the electron-pair and molecular geometries using the images in Figure 9.2.

## Example I: Carbon Dioxide, $CO_2$

1. Lewis structure:

$$:\overset{..}{O}=C=\overset{..}{O}:$$

2. The central carbon atom has two atoms bonded to it and no lone pairs, so the steric number is 2.
3. Because SN = 2, the O—C—O bond angle is 180° (see Figure 9.2a) and the electron-pair and molecular geometries are both linear.

## Example II: Boron Trifluoride, $BF_3$

1. Lewis structure:

$$:\overset{..}{\underset{..}{F}}\diagdown_{B}\diagup\overset{..}{\underset{..}{F}}:$$
$$:\overset{..}{\underset{..}{F}}:$$

2. Three fluorine atoms are bonded to the central boron atom, and there are no lone pairs on boron. Therefore, SN = 3.
3. Because SN = 3, all four atoms lie in the same plane, forming a triangle with F—B—F bond angles of 120°. The electron-pair and molecular geometries are trigonal planar (Figure 9.4).

## Example III: Carbon Tetrachloride, $CCl_4$

1. Lewis structure:

$$:\overset{..}{\underset{..}{Cl}}:$$
$$:\overset{..}{\underset{..}{Cl}}-C-\overset{..}{\underset{..}{Cl}}:$$
$$:\overset{..}{\underset{..}{Cl}}:$$

🔴🔵 **CONNECTION** We learned in Chapter 8 that some molecules have central atoms that do not have an octet of electrons. The boron atom in $BF_3$ shares only three pairs of electrons, but because this distribution results in formal charges of zero on all the atoms, it is the preferred structure.

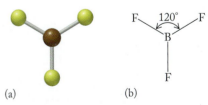

(a)  (b)

**FIGURE 9.4** (a) The ball-and-stick model of $BF_3$ shows the orientation of the atoms. (b) All F—B—F bond angles are 120° in this trigonal planar molecular geometry.

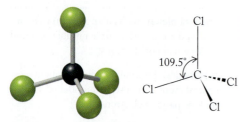

**FIGURE 9.5** The ball-and-stick model shows the actual orientation of the atoms in carbon tetrachloride in three dimensions. All Cl—C—Cl bond angles are 109.5° in this tetrahedral molecular geometry.

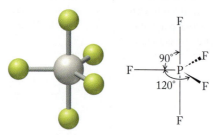

**FIGURE 9.6** The ball-and-stick model shows the actual orientation of the atoms in phosphorus pentafluoride. The P—F bonds located around the equator of the imaginary sphere are 120° apart. Each of them is 90° from the two bonds connecting F atoms in the axial positions (at the north and south poles). The resulting molecular geometry is trigonal bipyramidal.

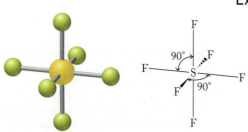

**FIGURE 9.7** The ball-and-stick model shows how the six fluorine atoms are oriented in three dimensions about the sulfur atom. All the F—S—F bond angles are 90° in this octahedral molecular geometry.

2. Four chlorine atoms are bonded to the central carbon atom, which gives carbon a full octet. Therefore, SN = 4.
3. Because SN = 4, the chlorine atoms are located at the vertices of a tetrahedron and all four Cl—C—Cl bond angles are 109.5°, producing tetrahedral electron-pair and molecular geometries (Figure 9.5).

Before we explore any more molecular geometries, we should explain the conventions used in drawing three-dimensional structures in two dimensions. To convey the 3D structure of a molecule, we use a solid wedge (➛) to indicate a bond that comes out of the paper toward the viewer. The solid wedge in the CCl₄ structure in Figure 9.5, for example, means that the chlorine in that position projects out of the plane of the paper and toward the viewer at a downward angle. A dashed wedge (⋯•) indicates a bond that goes behind the plane of the paper and away from the viewer at a downward angle. Solid lines are used for bonds that lie in the plane of the paper. We typically orient a structure so that the maximum number of bonds lie in the plane of the paper.

## Example IV: Phosphorus Pentafluoride, PF₅

1. Lewis structure:

2. Five fluorine atoms are bonded to the central phosphorus atom, which has no lone pairs. Therefore, SN = 5.
3. Because SN = 5, the fluorine atoms are located at the five vertices of a trigonal bipyramid and the electron-pair geometry and molecular geometry are trigonal bipyramidal (Figure 9.6).

## Example V: Sulfur Hexafluoride, SF₆

1. Lewis structure:

2. Six fluorine atoms are bonded to the central sulfur atom, which has no lone pairs. Therefore, SN = 6.
3. Because SN = 6, the fluorine atoms are located at the vertices of an octahedron and the electron-pair and molecular geometries are octahedral (Figure 9.7).

---

**SAMPLE EXERCISE 9.1   Using VSEPR to Predict Geometry I   LO2**

Formaldehyde, CH₂O, is a gas at room temperature (boiling point −21°C). Aqueous solutions of formaldehyde are used to preserve biological samples. It is also a product of the incomplete combustion of hydrocarbons. Use VSEPR to predict the molecular geometry of formaldehyde.

**Collect and Organize** We are given the molecular formula of formaldehyde. Assuming there are no lone pairs of valence electrons on the central atom, the solution requires (1) drawing the Lewis structure, (2) determining the steric number, and (3) identifying the molecular geometry using Figure 9.2.

**Analyze** Carbon is the likely central atom of the molecule because it has a bonding capacity of 4 and is less electronegative than oxygen, whose bonding capacity is 2. If the three atoms bonded to C are as far from each other as possible, the probable molecular geometry is trigonal planar.

**Solve** Following the procedure developed in Chapter 8 for drawing Lewis structures, we draw this one for formaldehyde:

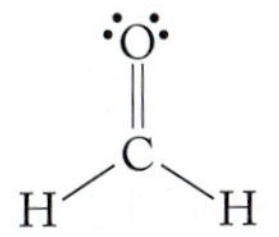

As predicted, SN = 3; and according to Figure 9.2, the molecular geometry is trigonal planar.

**Think About It** The key to predicting the correct molecular geometry of a molecule with no lone pairs on its central atom is to determine the steric number, which is simply a matter of counting the number of atoms bonded to the central atom.

⚙ **Practice Exercise** Use VSEPR to determine the molecular geometry of the chloroform molecule, $CHCl_3$, and draw the molecule using the solid-wedge, dashed-wedge convention.

*(Answers to Practice Exercises are in the back of the book.)*

■

Measurements of the bond angles in formaldehyde show that the H—C—H bond angle is slightly smaller than the 120° predicted for trigonal planar geometry (Figure 9.2b). The C=O double bond consists of two pairs of bonding electrons (four electrons) and exerts greater repulsion than a single bond would. This greater repulsion decreases the H—C—H bond angle. VSEPR does not enable us to predict the actual values of the bond angles in molecules containing double bonds to the central atom, but it does allow us to correctly predict how a bond angle deviates from the ideal value due to the presence of a double bond.

**CONCEPT TEST**

Rank the following bond angles from largest to smallest:
   a. The F—B—F bond angles in $BF_3$
   b. The O—C—H bond angles in $CH_2O$
   c. The H—C—H bond angle in $CH_2O$

## Central Atoms with Lone Pairs

We have established the electron-pair and molecular geometries of molecules whose central atoms have no lone pairs of electrons. Now let's explore what happens when a central atom has one or more lone pairs. For SN = 2, the only bonding pattern possible is two atoms bound to a central atom. If we replace one bonding pair with a lone pair, we have a molecule with only two atoms, which means there is no central atom and no bond angle; it takes three atoms to define a bond angle. Such diatomic molecules must have a linear geometry. Therefore, we begin the discussion of molecules containing lone pairs with SN = 3.

Consider sulfur dioxide, one of the gases produced when high-sulfur coal is burned. The resonance structures showing the distribution of electrons are

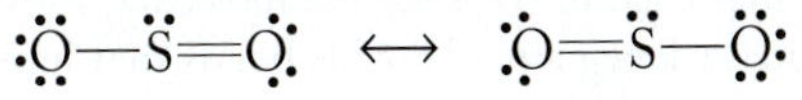

**FIGURE 9.8** (a) The electron-pair geometry of $SO_2$ is trigonal planar because the steric number of the S atom is 3; it is bonded to two atoms and has one lone pair of electrons. (b) The molecular geometry is bent because the central sulfur atom in the structure has no bonded atom in the third position, only a lone pair of electrons.

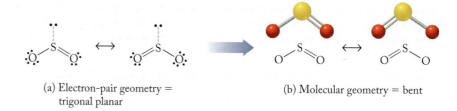

(a) Electron-pair geometry = trigonal planar

(b) Molecular geometry = bent

O—S—O bond angle < 120°

**FIGURE 9.9** The lone pair of electrons on the central sulfur atom in $SO_2$ occupies more space (larger purple region) than the electron pairs in the S—O bonds (smaller purple regions). The increased repulsion (represented by the double-headed arrows), resulting from the larger volume occupied by the lone pair, forces the oxygen atoms closer together, making the O—S—O bond angle slightly less than the ideal value of 120°.

To calculate the steric number of the central S atom we need to add up the number of atoms and lone pairs of electrons that surround it. In both resonance structures, S is bonded to two atoms and has one lone pair of electrons. Therefore its SN value in either resonance structure is $2 + 1 = 3$. When SN = 3, the arrangement of atoms and lone pairs about the sulfur atom is trigonal planar (Figure 9.2b). This means that the electron-pair geometry, which takes into account both the atoms and the nonbonding pairs of electrons around the central atom, is trigonal planar (as shown in Figure 9.8a). However, the dotted lines in the structures in Figure 9.8(a) do not indicate a bond, but rather point to a lone pair of electrons. The lone pair is located on the S atom, and the attachment of two oxygen atoms gives the molecular geometry shown in Figure 9.8(b). This molecular geometry is called *bent*, or *angular*.

Experimental measurements establish that $SO_2$ is indeed a bent molecule, but the O—S—O bond angle is actually a little smaller than 120°. We explain the smaller angle using VSEPR by comparing the space occupied by bonding electrons with the space occupied by electrons in a lone pair. Because the bonding electrons are attracted to two nuclei, they have a high probability of being located between the two atomic centers that share them. In contrast, the lone pair is not shared with a second atom and is spread out around the sulfur atom, as shown in Figure 9.9. This puts the lone pair of electrons closer to the bonding pairs and produces greater repulsion. As a result, the lone pair pushes the bonding pairs closer together, thereby reducing the bond angle. In general:

➤ Repulsion between lone pairs and bonding pairs is greater than the repulsion between bonding pairs.
➤ Repulsion caused by a lone pair is greater than the repulsion caused by a double bond.
➤ Repulsion caused by a multiple bond is greater than that caused by a single bond.
➤ Two lone pairs of electrons on a central atom exert a greater repulsive force on the atom's bonding pairs than does one lone pair.

**SAMPLE EXERCISE 9.2** **Predicting Relative Sizes of Bond Angles** **LO2**

Rank $NH_3$, $CH_4$, and $H_2O$ in order of decreasing bond angles in their molecular structures.

**Collect and Organize** We are asked to predict the relative sizes of the bond angles in three molecules, given their molecular formulas. The information in Figure 9.2 links steric numbers to electron-pair geometries and bond angles. Bond angles are also influenced by the presence of double bonds and lone pairs of electrons on the central atom.

**Analyze** We need to determine the steric numbers of the central atoms in these three molecules. To do that we first need to translate the molecular formulas into Lewis structures and determine how many lone pairs or double bonds they have.

**Solve** Using the method for drawing Lewis structures from Chapter 8 we obtain these results:

$$H-\overset{\cdot\cdot}{N}-H \qquad H-\overset{\displaystyle H}{\underset{\displaystyle H}{C}}-H \qquad \overset{\cdot\cdot}{\underset{H\quad H}{O}}$$

In each of these structures the central atom is surrounded by a total of four atoms or lone pairs of electrons, which means a steric number of 4. There are no double bonds.

According to Figure 9.2, the electron-pair geometry for SN = 4 is tetrahedral. In a molecule such as $CH_4$, in which all four tetrahedral electron pairs are equivalent bonding pairs, all bond angles are 109.5°. However, repulsion from the lone pairs of electrons in $NH_3$ and $H_2O$ squeezes their bonds together, reducing the bond angles. The two lone pairs on the O atom in $H_2O$ exert a greater repulsive force on its bonding pairs than the single lone pair on the N atom exerts on the bonding pairs in $NH_3$. Therefore, the bond angle in $H_2O$ should be less than the bond angles in $NH_3$. Ranking the three molecules in order of decreasing bond angle we have:

$$CH_4 > NH_3 > H_2O$$

**Think About It** The logic used in answering this problem is supported by experimental evidence: the bond angles in molecules of $CH_4$, $NH_3$, and $H_2O$ are 109.5°, 107.0°, and 104.5°, respectively.

**Practice Exercise** Are the O—S—O bond angles greater in $SO_2$ or $SO_3$?

As we saw in Sample Exercise 9.2, three combinations of atoms and lone pairs are possible about a central atom with a steric number of 4: four atoms and no lone pairs, three atoms and one lone pair, or two atoms and two lone pairs. The first case is illustrated by the molecular structure of methane, $CH_4$, in which a tetrahedral electron-pair geometry translates into a tetrahedral molecular geometry. A molecule of ammonia, $NH_3$, has only three bonding pairs and one lone pair. This means that the Lewis structure in Figure 9.10(a) translates into a tetrahedral electron-pair geometry in which one of the vertices is the lone pair on the N atom (Figure 9.10b). The resulting molecular geometry, which is essentially a flattened tetrahedron, is called *trigonal pyramidal* (Figure 9.10c). As we discussed in Sample Exercise 9.2, the strong repulsion produced by the diffuse lone pair of electrons on the N atom pushes the N—H bonds closer together in $NH_3$ and reduces the angles between them as compared to the H—C—H bond angles in $CH_4$. Thus, the H—N—H bond angles in ammonia are only 107.0°.

The central O atom in a molecule of $H_2O$ has a steric number of 4 because it is bonded to two atoms and has two lone pairs of electrons. The result is a tetrahedral electron-pair geometry (Figure 9.11). However, the presence of the two lone pairs means that two of the four tetrahedral vertices are not occupied by atoms, which leaves us with a bent (or angular) molecular geometry. As we also discussed in Sample Exercise 9.2, the H—O—H bond angle in water is smaller than 109.5° due to repulsion between the two lone pairs and each bonding pair. The actual bond angle is 104.5°.

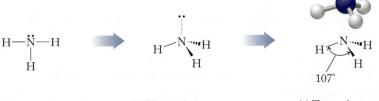

(a) Lewis structure   (b) Tetrahedral electron-pair geometry   (c) Trigonal pyramidal molecular geometry

**FIGURE 9.10** (a) The steric number of the N atom in $NH_3$ is 4, (b) so its electron-pair geometry is tetrahedral. However, one of the vertices of the tetrahedron is occupied by a lone pair of electrons, not an atom, (c) so the molecular geometry is trigonal pyramidal.

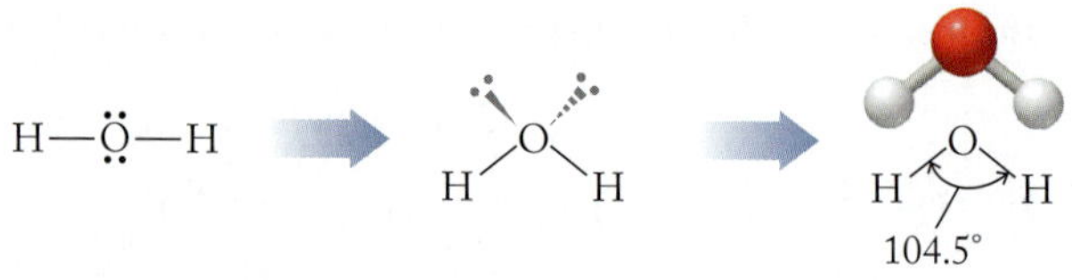

**FIGURE 9.11** (a) The steric number of the O atom in $H_2O$ is 4 (it is bonded to two atoms and has two lone pairs of electrons). Therefore, (b) its electron-pair geometry is tetrahedral. However, two of the vertices of the tetrahedron are occupied by lone pairs of electrons, meaning (c) only three atoms define the molecular geometry, which is bent.

(a) Lewis structure

(b) Tetrahedral electron-pair geometry

(c) Bent (angular) molecular geometry

104.5°

**CONCEPT TEST**

The bond angles in silane ($SiH_4$), phosphine ($PH_3$), and hydrogen sulfide ($H_2S$) are 109.5°, 93.6°, and 92.1°, respectively. Use VSEPR to explain this trend.

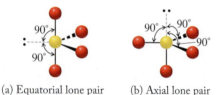

(a) Equatorial lone pair    (b) Axial lone pair

**FIGURE 9.12** (a) A lone pair of electrons in an equatorial position of a molecule with trigonal bipyramidal electron-pair geometry interacts through 90° with two other electron pairs. (b) A lone pair in an axial position interacts through 90° with *three* other electron pairs. Fewer 90° interactions reduce internal electron-pair repulsion and lead to greater stability.

Molecules with trigonal bipyramidal electron-pair geometry have four possible molecular geometries, depending on the number of lone pairs per molecule (Table 9.1). These options are possible because of the different axial and equatorial vertices in a trigonal bipyramid (Figure 9.2d). VSEPR enables us to predict which vertices are occupied by atoms and which are occupied by lone pairs. The key to these predictions is the fact that the repulsions between pairs of electrons decrease as the angle between them increases: two electron pairs at 90° experience a greater mutual repulsion than two at 120°, which have a greater repulsion than two at 180°. To minimize repulsions involving lone pairs, VSEPR predicts that they preferentially occupy equatorial rather than axial vertices. Why? Because an equatorial lone pair has *two* 90° repulsions with the two axial electron pairs as shown in Figure 9.12(a), but an axial lone pair has *three* 90° repulsions (with three equatorial electron pairs), as shown in Figure 9.12(b).

When we assign one, two, or three lone pairs of valence electrons to equatorial vertices, we get three of the molecular geometries for SN = 5 in Table 9.1. When a single lone pair occupies an equatorial site, we get a molecular geometry called **seesaw** (Figure 9.13) because its shape, when rotated 90° clockwise, resembles a playground seesaw. (The formal name for this shape is *disphenoidal*.)

**FIGURE 9.13** A single lone pair of electrons in an equatorial position of a trigonal bipyramidal electron-pair geometry produces a seesaw molecular geometry.

(a) Trigonal bipyramidal electron-pair geometry

(b) Rotated clockwise 90° about horizontal axis

(c) Seesaw molecular geometry

When two lone pairs occupy equatorial sites (Figure 9.14a), the molecular geometry that results is called **T-shaped**. This designation becomes more apparent when the structure in Figure 9.14(a) is rotated 90° counterclockwise. Lone-pair repulsions result in bond angles in T-shaped molecules that are slightly less than

**seesaw** molecular geometry about a central atom with a steric number of 5 and one lone pair of electrons in an equatorial position.

**T-shaped** molecular geometry about a central atom with a steric number of 5 and two lone pairs of electrons that occupy equatorial positions; the atoms occupy two axial sites and one equatorial site.

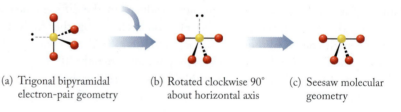

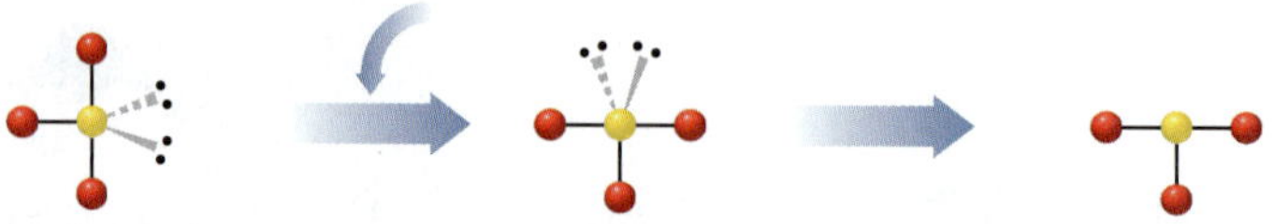

(a) Trigonal bipyramidal electron-pair geometry

(b) Rotated counterclockwise 90° about horizontal axis

(c) T-shaped molecular geometry

**FIGURE 9.14** Two lone pairs of electrons in equatorial positions of a trigonal bipyramidal electron-pair geometry produce a T-shaped molecular geometry.

**TABLE 9.1**   **Electron-Pair Geometries and Molecular Geometries**

| SN = 3 | Electron-Pair Geometry | No. of Bonded Atoms | No. of Lone Pairs | Molecular Geometry | Structure |
|---|---|---|---|---|---|
| | Trigonal planar | 3 | 0 | Trigonal planar | |
| | Trigonal planar | 2 | 1 | Bent (angular) | |
| **SN = 4** | | | | | |
| | Tetrahedral | 4 | 0 | Tetrahedral | |
| | Tetrahedral | 3 | 1 | Trigonal pyramidal | |
| | Tetrahedral | 2 | 2 | Bent (angular) | |
| **SN = 5** | | | | | |
| | Trigonal bipyramidal | 5 | 0 | Trigonal bipyramidal | |
| | Trigonal bipyramidal | 4 | 1 | Seesaw | |
| | Trigonal bipyramidal | 3 | 2 | T-shaped | |
| | Trigonal bipyramidal | 2 | 3 | Linear | |
| **SN = 6** | | | | | |
| | Octahedral | 6 | 0 | Octahedral | |
| | Octahedral | 5 | 1 | Square pyramidal | |
| | Octahedral | 4 | 2 | Square planar | |
| | Octahedral | 3 | 3 | Although these geometries are possible, we will not encounter any molecules with them | |
| | Octahedral | 2 | 4 | | |

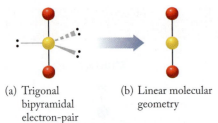

(a) Trigonal bipyramidal electron-pair geometry

(b) Linear molecular geometry

**FIGURE 9.15** Three lone pairs of electrons in equatorial positions of a trigonal bipyramidal electron-pair geometry produce a linear molecular geometry.

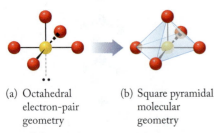

(a) Octahedral electron-pair geometry

(b) Square pyramidal molecular geometry

**FIGURE 9.16** A lone pair of electrons in an octahedral electron-pair geometry produces a square pyramidal molecular geometry.

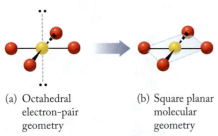

(a) Octahedral electron-pair geometry

(b) Square planar molecular geometry

**FIGURE 9.17** Two lone pairs of electrons on opposite sides of an octahedral electron-pair geometry produce a square planar molecular geometry.

the 90° and 180° angles we would expect from a perfectly shaped "T" geometry. Finally, a SN = 5 molecule with three lone pairs and two bonded atoms has a linear geometry because all three lone pairs occupy equatorial sites, and the bonding pairs are in the two axial positions (Figure 9.15).

**CONCEPT TEST**

The bond angles in a trigonal bipyramidal molecular structure are either 90°, 120°, or 180°. Are the corresponding bond angles in a seesaw structure likely to be larger than, the same as, or smaller than these values?

Molecules that have a central atom with a steric number of 6 have an octahedral electron-pair geometry (Figure 9.2e) and the molecular geometries listed in Table 9.1. With one lone pair of electrons, there is only one possible molecular geometry because all the sites in an octahedron are equivalent (Figure 9.16). That one molecular geometry is called **square pyramidal**: a pyramid with a square base, four triangular sides, and the central atom "embedded" in the base.

Because of stronger repulsion between lone pairs and bonding pairs, we predict the bond angles in a square pyramidal molecule to be slightly less than the ideal angle of 90°. The molecule $BrF_5$, for example, has a square pyramidal molecular geometry, and the angles between its equatorial and axial bonds are 85°.

When two lone pairs are present at the central atom in a molecule with octahedral electron-pair geometry, they occupy vertices on opposite sides of the octahedron to minimize the interactions between them. The resultant molecular geometry is called **square planar** because the molecule is shaped like a square and all five atoms reside in the same plane (Figure 9.17). Because the two lone pairs are on opposite sides of the bonding pairs, the presence of the lone pairs does not distort the bond angles: they are all 90°.

**SAMPLE EXERCISE 9.3**  Using VSEPR to Predict Geometry II  **LO2**

The Lewis structure of sulfur tetrafluoride ($SF_4$) is

What is its molecular geometry and what are the angles between the S—F bonds?

**Collect and Organize** We are given the Lewis structure of $SF_4$ and can determine from it the steric number of the central atom in the molecule and its electron-pair geometry.

**Analyze** Four atoms are bonded to the central S atom, which also has one lone pair of electrons. This means that SN = 5.

**Solve** With a steric number of 5 for its central atom, the electron-pair geometry of $SF_4$ is trigonal bipyramidal. The presence of one lone pair of electrons on the S atom means that its molecular geometry is not the same as its electron-pair geometry. Table 9.1 tells us that seesaw molecular geometry results when a lone pair occupies one of the three equatorial positions in a trigonal bipyramidal electron-pair geometry:

Electron-pair geometry = Molecular geometry = Frequently drawn from
trigonal bipyramidal             seesaw                  this perspective as well

The equatorial lone pair slightly reduces the bond angles from their normal values of 90° between the axial and equatorial bonds, 120° between the two equatorial bonds, and 180° between the two axial bonds.

**Think About It** Any molecule with SN = 5 and one lone pair should have the same geometry as $SF_4$.

✺ **Practice Exercise** Determine the molecular geometry and the bond angles of $SO_2Cl_2$.

■

**square pyramidal** molecular geometry about a central atom with a steric number of 6 and one lone pair of electrons; as typically drawn, the atoms occupy four equatorial and one axial site.

**square planar** molecular geometry about a central atom with a steric number of 6 and two lone pairs of electrons that occupy axial sites; the atoms occupy four equatorial positions.

**bond dipole** separation of electrical charge created when atoms with different electronegativities form a covalent bond.

# 9.3 Polar Bonds and Polar Molecules

Water is the only naturally occurring atmospheric gas that is a liquid at standard temperature and pressure. Why is $H_2O$ a liquid when all other compounds of comparable molar mass, such as $N_2$, $O_2$, and $CO_2$, and the much heavier compound $CF_4$, are gases? The answer, in part, is that water is a polar substance.

One of the reasons why $H_2O$ is a polar molecule is because its O—H bonds are polar bonds. However, some nonpolar molecules also contain polar bonds. For example, the C=O double bonds in $CO_2$ are polar because carbon and oxygen have different electronegativities. The carbon atom in the molecule has a partial positive charge and the oxygen atoms have a partial negative charge:

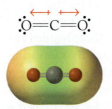

See Figures 8.3 and 8.4 to review the use of these symbols and electron-density surfaces to indicate bond polarity.

As we noted in Chapter 8, an unequal distribution of bonding electrons between two atoms produces a partial negative charge in one region of the bond and a partial positive charge in the other. This charge separation creates a **bond dipole**. We can determine the overall polarity of a molecule by summing the polarities of all the bond dipoles in the molecule. However, this summing must take into account both the strengths of the individual bond dipoles and their orientations with respect to one another. In $CO_2$, for example, the two dipoles are equivalent in strength because they involve the same atoms and the same kind of bond (C=O). In addition, the linear shape of the molecule means that the direction of one C=O bond dipole is opposite the direction of the other. Thus, the two dipoles exactly offset so that, overall, $CO_2$ is a nonpolar substance.

Similarly, $CF_4$ is a nonpolar substance even though its molecules contain polar C—F bonds. The four C—F bond dipoles are all the same because the same atoms are involved. In addition, the tetrahedral molecular geometry means that the bond dipoles offset one another so that, overall, the $CF_4$ molecule is nonpolar:

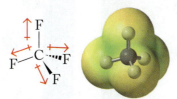

Water molecules are bent, not linear like molecules of $CO_2$. Therefore, the dipoles of the two O—H bonds in a molecule of $H_2O$ do not offset each other.

**CONNECTION** In Chapter 8, we introduced electronegativity and the unequal distribution of electrons in polar covalent bonds.

Instead, the molecule has an overall dipole with the negative end directed toward the O atom:

The presence of this overall dipole means that water is a polar molecule. This polarity leads to interactions between H and O atoms on adjacent $H_2O$ molecules in liquid water or solid ice. Other polar molecules exhibit similar interactions. We discuss these interactions in Chapter 10.

**CONNECTION** In Chapter 8 we discussed the impact of bond polarity on the physical properties of compounds like the hydrogen halides.

**CONCEPT TEST**

Each molecule of ethane contains six polar C—H bonds, yet ethane is a nonpolar substance. Explain how this is possible.

We have seen how the overall polarity of a molecule depends on the differences in the electronegativities of the bonded pairs of atoms in its molecular structure and on the arrangement of those bonded atoms. Experimentally, the polarity of a molecule can be determined by measuring its permanent **dipole moment** (**$\mu$**). The value of $\mu$ expresses the extent of the overall separation of positive and negative charge in the molecules of a substance and is determined by measuring the degree to which the molecules are aligned when they are placed in a strong electric field (Figure 9.18). Polar molecules align with the field so that their negative regions are oriented toward the positive plate and their positive regions are oriented toward the negative plate. The more polar the molecules are, the more strongly they align with the field. Dipole moments are usually expressed in units of *debyes* (D), where $1 D = 3.34 \times 10^{-30}$ coulomb-meter.

**FIGURE 9.18** Gaseous HF molecules are oriented randomly in the absence of an electric field but align themselves when an electric field is applied to two metal plates. The negative (fluorine) end of each molecule is directed toward the positively charged plate; the positive (hydrogen) end, toward the negative plate.

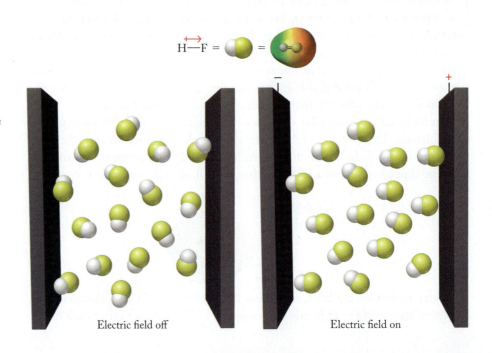

Electric field off          Electric field on

| TABLE 9.2 | Permanent Dipole Moments of Several Polar Molecules | | |
|---|---|---|---|
| **Formula** | **Bond Dipole(s)** | **Overall Dipole** | **Dipole Moment (debyes)** |
| HF | H—F | ⟶ | 1.91 |
| $H_2O$ | (H—O—H structure) | ↕ | 1.85 |
| $NH_3$ | (H—N—H structure) | ↕ | 1.47 |
| $CHCl_3$ | (H, C, Cl structure) | ↓ | 1.04 |
| $CCl_3F$ | (F, C, Cl structure) | ↕ | 0.45 |

**dipole moment ($\mu$)** a measure of the degree to which a molecule aligns itself in a strong electric field; a quantitative expression of the polarity of a molecule.

The dipole moments of several polar substances are shown in Table 9.2. Note the structures in the last two rows of the table. Chloroform ($CHCl_3$) has a relatively strong dipole moment because its bond dipoles differ both in the *degree* of polarity and in the *direction* of the polarity (Figure 9.19a). Chlorine is more electronegative than carbon, and the two electrons in each C—Cl bond are pulled away from the carbon atom and toward the chlorine atom. Because hydrogen is less electronegative than carbon, the two electrons in the H—C bond are pulled away from the hydrogen atom and toward the carbon atom. Consequently the electron distribution is away from the top part of the molecule as drawn and toward the bottom.

In trichlorofluoromethane ($CCl_3F$) all four bond dipoles point away from the central carbon atom because fluorine and chlorine are more electronegative than carbon (Figure 9.19b). However, a C—F bond dipole is stronger than a C—Cl bond dipole because fluorine is more electronegative than chlorine. As a result, bonding electrons are pulled more toward the C—F (top) side of the molecule than toward the C—Cl (bottom) side. The direction of the overall dipole is upward.

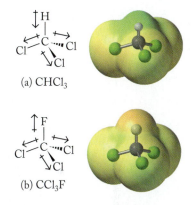

(a) $CHCl_3$

(b) $CCl_3F$

**FIGURE 9.19** (a) The four bonds in chloroform ($CHCl_3$) are not equivalent in terms of direction or degree of polarity. The molecule has a dipole moment. (b) The C—F bond dipole in $CCl_3F$ is stronger than the C—Cl bond dipoles and is not offset by them. This molecule also has a dipole moment, but not as large as chloroform.

**SAMPLE EXERCISE 9.4** **Predicting the Polarity of a Substance** **LO3**

Does formaldehyde gas ($CH_2O$) have a permanent dipole moment?

**Collect and Organize** To predict whether a molecule has a permanent dipole moment, we need to determine whether or not it contains polar bonds and whether the bond dipoles offset each other. A bond's polarity depends on the difference in the electronegativity of the bonded pair of atoms. The extent to which bond dipoles offset each other depends on the molecular geometry of the molecule. In Sample Exercise 9.1 we determined that formaldehyde has a trigonal planar structure:

(O=C with H and H structure)

**Analyze** The electronegativities of the elements in the compound are H = 2.1, C = 2.5, and O = 3.5. Given the differences in the electronegativities of the atoms bonded to the central atom, it is likely that formaldehyde has a permanent dipole moment.

**Solve** The hydrogen atoms are the least electronegative atoms in the molecule and the oxygen atom the most, so each of the bonds in the molecule has a bond dipole directed toward the oxygen atom:

Thus, the formaldehyde molecule has a permanent dipole moment.

**Think About It** The presence of different atoms bonded to a central atom makes it highly likely that the molecule has a permanent dipole moment.

⚙ **Practice Exercise** Does carbon disulfide ($CS_2$), a gas present in small amounts in crude petroleum, have a dipole moment?

---

**CONCEPT TEST**

Water and hydrogen sulfide both have a bent molecular geometry with dipole moments of 1.85 D and 0.98 D, respectively. Why is the dipole moment of $H_2S$ less than that of $H_2O$?

# 9.4 Valence Bond Theory

Drawing Lewis structures and applying VSEPR enable us to predict the geometry of many molecules reliably. However, we have yet to make a connection between molecular geometry and the arrangement of electrons in atoms in atomic orbitals as described in Chapter 7. The absence of a connection between the electronic structure of atoms and the electronic structure of molecules led to the development of bonding theories that use atomic orbitals to account for the molecular geometries predicted by VSEPR. We explore one of these theories next.

## Bonds from Orbital Overlap

**Valence bond theory** arose in the late 1920s, largely as a result of the genius and efforts of Linus Pauling, who developed this theory of molecular bonding based on quantum mechanics. Valence bond theory assumes that (a) a chemical bond between two atoms results from **overlap** of the atoms' atomic orbitals, and (b) the greater the overlap, the stronger and more stable the bond.

Shared electrons in a chemical bond are located between the nuclei of two atoms and are attracted to both. This attraction leads to lower potential energy and greater chemical stability than if the atoms were completely free (see Figure 8.1). This view of chemical bonding is especially useful for analyzing the physical and chemical properties of covalent substances because it provides a model of where the electrons are in a molecule. Just as the locations of electrons in atoms define atomic properties, the locations of electrons in molecules define the properties of those molecules.

Let's begin by applying Pauling's valence bond theory to the simplest of diatomic molecules: $H_2$. A free H atom has a single electron in a $1s$ atomic orbital. According to valence bond theory, the overlap between two of these half-filled $1s$ orbitals produces a single H—H bond (Figure 9.20). Overlapping two orbitals from two atoms in this way increases electron density along the axis connecting

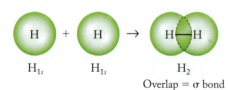

**FIGURE 9.20** The overlap of the $1s$ orbitals on two hydrogen atoms produces a single σ bond that holds the two hydrogen atoms together in a $H_2$ molecule.

the two nuclei. Whenever the region of high electron density lies along the bond axis, the resulting covalent bond is called a **sigma (σ) bond**.

---

**SAMPLE EXERCISE 9.5** **Identifying Overlapping** **LO4**
**Orbitals in a Molecule**

Identify the atomic orbitals responsible for the covalent bond in gaseous HCl.

**Collect and Organize** We are given the formula of HCl and asked to identify which atomic orbitals are involved in bonding.

**Analyze** First we must determine the number of valence electrons in each atom and then draw Lewis structures for the molecule. Next we identify which atomic orbitals are half-filled; these are the orbitals that form the covalent bond between H and Cl.

**Solve** Atoms from group 17 (halogens) all have seven valence electrons and hydrogen has one valence electron. Thus, the Lewis structure for HCl is:

$$H\!-\!\ddot{\underset{..}{Cl}}\!:$$

The electron configurations of H and Cl atoms are

$$H: 1s^1 \quad \text{and} \quad Cl: [Ne]3s^23p^5$$

The covalent bond between H and Cl in HCl results from overlap between the half-filled hydrogen 1s and chlorine 3p atomic orbitals.

**Think About It** The σ bond between hydrogen and chlorine in HCl that arises from overlap of two atomic orbitals is consistent with the prediction from its Lewis structure.

⚙ **Practice Exercise** The halogens (group 17 elements) form a series of interhalogen compounds such as IBr. Identify the atomic orbitals that overlap to form the σ bond in IBr.

---

**valence bond theory** a quantum mechanics-based theory of bonding that assumes covalent bonds form when half-filled orbitals on different atoms overlap or occupy the same region in space.

**overlap** a term in valence bond theory describing bonds arising from two orbitals on different atoms that occupy the same region of space.

**sigma (σ) bond** a covalent bond in which the highest electron density lies between the two atoms along the bond axis.

**hybridization** in valence bond theory, the mixing of atomic orbitals to generate new sets of orbitals that are then available to form covalent bonds with other atoms.

**hybrid atomic orbital** in valence bond theory, one of a set of equivalent orbitals about an atom created when specific atomic orbitals are mixed.

## Hybridization

In molecules other than simple ones like $H_2$, an inconsistency arises between the molecular geometries we predict based on VSEPR and the atomic orbital shapes and orientations described in Chapter 7. Methane, for example, has a carbon atom at its center. We established in Chapter 7 that carbon atoms have the electron configuration $[He]2s^22p^2$, that the 2s orbital is spherical, and the three 2p orbitals, $p_x$, $p_y$, and $p_z$, are oriented at 90° to one another (Figure 9.21). Also, we learned in Chapter 8 that carbon atoms form four covalent bonds to complete their octets. How can four equivalent bonds oriented at 109.5° to one another in a tetrahedral arrangement form from a filled 2s orbital and two partly filled 2p orbitals? Furthermore, how can two double bonds 180° apart form around the central carbon atom in $CO_2$, or one double bond and two single bonds 120° apart form around the central carbon in $CH_2O$?

To account for the geometry of methane, carbon dioxide, formaldehyde, and many other molecules, valence bond theory describes atomic orbitals of different shapes and energies as being mixed together in a process called **hybridization**, to form **hybrid atomic orbitals**. Covalent bonds then result either from overlap of a hybrid orbital on one atom with an unhybridized orbital on another atom, or from overlap of two hybrid orbitals on two atoms. Let's examine the types of hybrid orbitals that form on carbon and other elements, and how these orbitals account for observed molecular geometries.

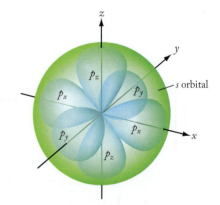

**FIGURE 9.21** Relative orientation in space of one 2s orbital (spherical) and three 2p orbitals oriented at 90° to one another along the x-, y-, and z-axes of a coordinate system.

**sp³ hybrid orbitals** a set of four hybrid orbitals with a tetrahedral orientation, formed by mixing one *s* and three *p* atomic orbitals.

**sp² hybrid orbitals** three hybrid orbitals in a trigonal planar orientation, formed by mixing one *s* and two *p* atomic orbitals.

## Tetrahedral Geometry: *sp*³ Hybrid Orbitals

We have noted that methane has a tetrahedral molecular geometry with H—C—H bond angles of 109.5°. To account for this geometry using orbital hybridization, we start with a free carbon atom with its $2s^2 2p^2$ valence-shell electron configuration (Figure 9.22a). Then we promote one electron from the $2s$ orbital to the empty $2p$ orbital. Promotion raises the energy of the $2s$ electron; however, this gain in energy is balanced by a slight decrease in the energies of the three $p$ orbitals. We now have four orbitals of equal energy on the carbon atom, each containing one electron.

These four orbitals have been *hybridized* (mixed and averaged) to make a set of four equivalent hybrid orbitals, each containing one electron. An atom's steric number always indicates how many of its orbitals must be mixed to generate the hybrid set, and the number of hybrid orbitals in a set is always equal to the number of atomic orbitals mixed. The carbon atom in methane has a steric number of 4, so four atomic orbitals on carbon are mixed to form four hybrid orbitals.

The four orbitals are called ***sp*³ hybrid orbitals** because they result from the mixing of one *s* and three *p* orbitals. Their energy level is the weighted average of the energies of the *s* and *p* orbitals that formed them. They are oriented 109.5° from one another and point toward the vertices of a tetrahedron (Figure 9.22b). Now four hydrogen atoms can form σ bonds with the *sp*³ hybridized orbitals on carbon and form a methane molecule that has the required tetrahedral geometry.

The *sp*³ hybrid orbitals in Figure 9.22(b) are shown as having one lobe each, but in fact every *sp*³ hybrid orbital consists of a major lobe and a minor lobe (Figure 9.23). Because the minor lobe is not involved in bonding, we ignore it in this chapter. It becomes important in the discussion of chemical reactions that are beyond the scope of this book. Also, to better show the orientation of bonds formed by hybrid orbitals, we elongate the major lobe in illustrations.

According to valence bond theory, any atom with a set of four equivalent *sp*³ hybrid orbitals has a tetrahedral orientation of its valence electrons. This includes atoms in which one or more hybrid orbitals are filled before any bonding takes place. For example, *sp*³ hybridization of the valence electrons on the nitrogen atom in ammonia produces three hybrid orbitals that are half-filled and one hybrid

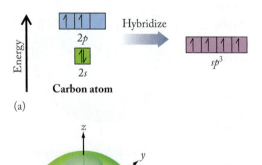

(a)

**FIGURE 9.22** (a) When the $2s$ and $2p$ atomic orbitals of carbon are hybridized, one electron is promoted from the filled $2s$ orbital to an unoccupied $2p$ orbital. The four orbitals are then mixed to create four *sp*³ hybrid orbitals. (b) The hybrid orbitals are oriented 109.5° from one another, exactly the orientation needed to form a tetrahedral methane molecule.

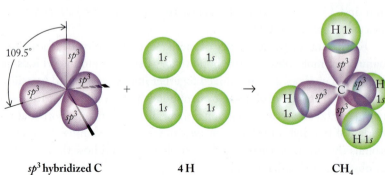

(b)

orbital that is completely filled (Figure 9.24a). The three half-filled orbitals form the three σ bonds in ammonia by overlapping with the 1s orbitals of three hydrogen atoms. The one filled hybrid orbital of N contains the lone pair of electrons. Similarly, the oxygen atom in water has four $sp^3$ hybrid orbitals in its valence shell, two filled before any bonding takes place and two half-filled and available for bond formation (Figure 9.24b). The two half-filled orbitals overlap with 1s orbitals from hydrogen atoms and form the two σ bonds in $H_2O$. Thus, a carbon atom forming four σ bonds, a nitrogen atom with three σ bonds and one lone pair of electrons, and an oxygen atom with two σ bonds and two lone pairs of electrons all have steric numbers equal to 4 and are all $sp^3$ hybridized atoms. As we shall see, the steric number of an atom and its hybridization are closely related in other electron pair geometries as well.

## Trigonal Planar Geometry: $sp^2$ Hybrid Orbitals

We saw in Sample Exercise 9.1 that formaldehyde has a trigonal planar molecular geometry (Figure 9.25). A hybridization scheme other than $sp^3$ must be used to generate an orbital array with this molecular geometry. The VSEPR model defines SN = 3 for the carbon atom in a molecule of formaldehyde, so three atomic orbitals must be mixed to make three hybrid orbitals. To produce a trigonal planar geometry, the 2s orbital on carbon is mixed with two of the carbon 2p orbitals, and the third 2p orbital is left unhybridized (Figure 9.26a).

Mixing and averaging one s and two p orbitals generates three hybrid orbitals called **$sp^2$ hybrid orbitals**. The energy level of $sp^2$ orbitals is slightly lower than that of $sp^3$ orbitals because only two p orbitals are mixed with one s orbital in a set of $sp^2$ orbitals. The orbitals in an $sp^2$ hybridized atom all lie in the same plane and

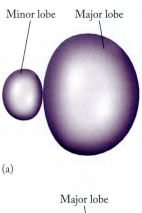

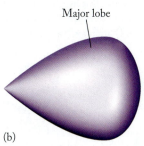

**FIGURE 9.23** (a) The $sp^3$ hybrid orbitals each consist of a major and minor lobe. (b) Because the minor lobes are not involved in orbital overlap leading to bond formation, they have been omitted from the orbital models in this book. The major lobes have also been elongated in illustrations to better show bond angles and orientations.

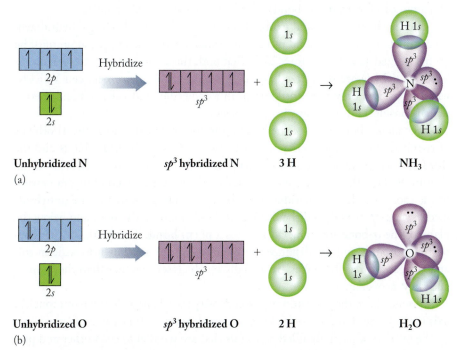

**FIGURE 9.24** A set of four $sp^3$ hybrid orbitals point toward the vertices of a tetrahedron. (a) The nonbonding electron pair of the N in $NH_3$ occupies one of the four hybrid orbitals, giving the molecule a trigonal pyramidal molecular geometry. (b) The O atom in $H_2O$ is also $sp^3$ hybridized. Only two of the four orbitals contain bonding pairs of electrons, giving $H_2O$ a bent molecular geometry.

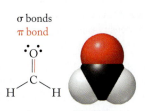

**FIGURE 9.25** Formaldehyde has three σ bonds and one π bond.

**FIGURE 9.26** (a) Mixing the 2*s* orbital of a carbon atom and *only two* of its three 2*p* orbitals produces three *sp²* hybrid orbitals, leaving one unhybridized *p* orbital available for π bond formation. (b) The 2*s* and 2*p* orbitals of oxygen can hybridize in the same way. (c) An *sp²* hybridized carbon forms three σ bonds in the plane with its hybrid orbitals and one π bond with the unhybridized *p* orbital above and below the plane defined by the three hybrid orbitals. An *sp²* hybridized oxygen atom forms one σ bond in the plane with one of its *sp²* hybrid orbitals, forms one π bond above and below the plane with an unhybridized *p* orbital, and has two lone pairs of electrons in the remaining two *sp²* hybrid orbitals. This hybridization scheme for the central C atom in a molecule of $CH_2O$ accounts for its double bond to an O atom and two single bonds to H atoms.

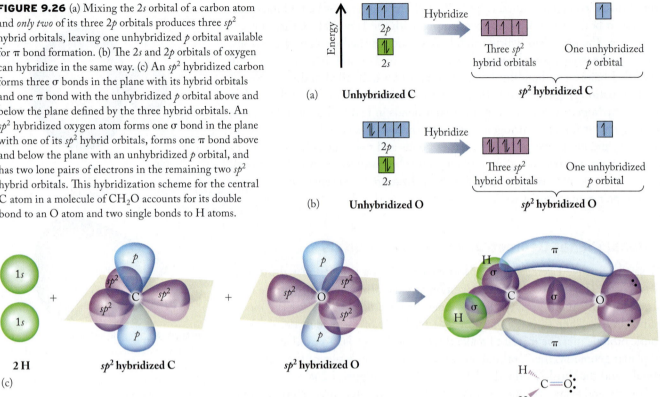

are 120° apart. The two lobes of the unhybridized *p* orbital lie above and below the plane of the triangle defined by the *sp²* hybrid orbitals. An *sp²* hybridized carbon atom can form three σ bonds with its three hybridized orbitals.

Valence electrons in *s* and *p* orbitals in other atoms can also be *sp²* hybridized. To complete the valence bond picture of formaldehyde, the oxygen atom must be *sp²* hybridized as well (Figure 9.26b). That both the carbon and the oxygen atoms in formaldehyde are hybridized points out another difference between VSEPR, which considers only the central atom in a molecule, and valence bond theory, which considers all the atoms in the molecule.

The valence bond view of the bonding in formaldehyde shows the 1*s* orbitals of two hydrogen atoms overlapping with two carbon *sp²* hybrid orbitals and the third carbon *sp²* hybrid orbital overlapping with one oxygen *sp²* hybrid orbital (Figure 9.26c). These overlapping orbitals constitute the σ-bonding framework of the molecule. The unhybridized carbon 2*p* orbital is parallel to the unhybridized oxygen 2*p* orbital, and the overlap of these two orbitals above and below the plane of the σ-bonding framework forms a **pi (π) bond**. The widths of the lobes on the *p* orbitals and the distances between atoms are not drawn to scale in this and similar figures. The lobes are actually much closer together than they appear and they *do* overlap.

Pi bonds have their greatest electron density above and below the internuclear axis (or in front of and in back of the internuclear axis). They can be formed only by the overlap of partially filled *p* orbitals that are parallel to each other and perpendicular to the σ bond joining the atoms.

The lone pairs of electrons on the oxygen atom are located in the two oxygen *sp²* hybrid orbitals not involved in the bonding with carbon. These two orbitals are

**pi (π) bond** a covalent bond in which electron density is greatest around—not along—the bonding axis.

***sp*** **hybrid orbitals** two hybrid orbitals on opposite sides of the hybridized atom, formed by mixing one *s* and one *p* orbital.

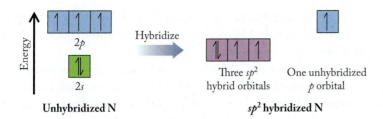

**FIGURE 9.27** A nitrogen atom with $sp^2$ hybrid orbitals has the capacity to form two σ bonds and one π bond.

oriented at an angle of about 120° in the plane of the molecule. A nitrogen atom can also form $sp^2$ hybrid orbitals, as shown in Figure 9.27. When it does, its lone pair occupies one of the three hybrid orbitals, the other two are half-filled and available to form two σ bonds, and the unhybridized half-filled $p$ orbital is available to form one π bond. This combination of one lone pair plus the capacity to form two σ bonds to two other atoms gives N a steric number of 3, the same SN value as $sp^2$ hybridized C and O atoms. This link to SN = 3 means that central atoms with trigonal planar electron-pair geometry are $sp^2$ hybridized.

## Linear Geometry: *sp* Hybrid Orbitals

We have seen hybrid orbitals generated by mixing one $s$ orbital with three $p$ orbitals and by mixing one $s$ orbital with two $p$ orbitals. The remaining mixture, one $s$ orbital with one $p$ orbital, forms the final set of important hybrid orbitals made from $s$ and $p$ atomic orbitals.

In the linear molecule acetylene, $C_2H_2$, both carbon atoms have SN = 2 according to VSEPR theory. Because steric number indicates the number of hybrid orbitals that must be created, we must mix two atomic orbitals to make two hybrid orbitals on each carbon atom.

The results of mixing one $2s$ and one $2p$ orbital on carbon and leaving the other two $2p$ orbitals unhybridized are shown in Figure 9.28(a). The two **$sp$ hybrid orbitals** have major lobes that are on opposite sides of the carbon

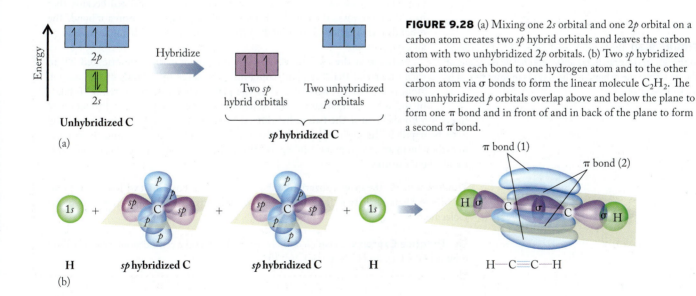

**FIGURE 9.28** (a) Mixing one $2s$ orbital and one $2p$ orbital on a carbon atom creates two $sp$ hybrid orbitals and leaves the carbon atom with two unhybridized $2p$ orbitals. (b) Two $sp$ hybridized carbon atoms each bond to one hydrogen atom and to the other carbon atom via σ bonds to form the linear molecule $C_2H_2$. The two unhybridized $p$ orbitals overlap above and below the plane to form one π bond and in front of and in back of the plane to form a second π bond.

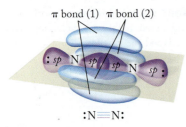

**FIGURE 9.29** Valence bond view of a molecule of $N_2$ and its *sp* hybrid N atoms.

atom. One set of the unhybridized *p* orbitals forms lobes above and below the axis of the *sp* hybrid orbitals; the second set is in front and in back of the plane of the hybrid orbitals.

The valence bond view of $C_2H_2$ shows a σ-bonding framework in which a hydrogen 1s orbital overlaps with one *sp* hybrid orbital on each carbon atom. The atom's other *sp* hybrid orbital overlaps with one *sp* hybrid orbital on the other carbon atom (Figure 9.28b). This arrangement brings the two sets of unhybridized *p* orbitals on the two carbon atoms into parallel alignment with each other. They then overlap to form two π bonds between the two carbon atoms, so that the one σ bond and the two π bonds form the triple bond of HC≡CH. The steric number of the *sp* hybridized carbon atom in acetylene is 2 (no lone pairs and bonds to two atoms). The link between SN = 2 and *sp* hybridization also applies to nitrogen atoms in $N_2$. A lone pair occupies one of the two *sp* orbitals on each nitrogen atom and the other *sp* hybrid orbital forms the σ bond (Figure 9.29).

---

**SAMPLE EXERCISE 9.6**  **Describing Bonding in a Molecule**  **LO4**

Use valence bond theory to account for the linear molecular geometry of $CO_2$, determine the hybridization of carbon and oxygen in this molecule, and describe the orbitals that overlap to form the bonds. Draw the molecule, showing the orbitals that overlap to form the bonds.

**Collect and Organize** We know that $CO_2$ is linear and that the Lewis structure of the molecule is

$$:\!\ddot{O}\!=\!C\!=\!\ddot{O}\!:$$

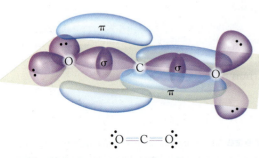

:Ö=C=Ö:

**FIGURE 9.30** The bonding pattern in $CO_2$: the π bond to the left in the drawing is above and below the plane of the molecule; the π bond to the right is in front of and in back of the plane.

**Analyze** Each double bond is composed of one σ bond and one π bond. Therefore, the carbon atom forms two σ bonds and two π bonds. It has no lone pairs and so SN = 2. Each oxygen atom forms one σ bond and one π bond.

**Solve** The carbon atom must be *sp* hybridized because half-filled *sp* hybrid orbitals have the capacity to form two σ bonds and the two half-filled unhybridized *p* orbitals are available to form the two π bonds (Figure 9.30). The lobes of *sp* hybrid orbitals are at an angle of 180°, which is consistent with the linear molecular geometry of $CO_2$. The oxygen atoms must be $sp^2$ hybridized because that hybridization leaves them with one unhybridized *p* orbital to form a π bond. The σ bonds form when each *sp* orbital on the carbon overlaps with one $sp^2$ orbital on an oxygen atom.

Notice that the two unhybridized *p* orbitals of carbon are oriented 90° to each other and that the plane containing the three $sp^2$ orbitals of one oxygen atom is rotated 90° with respect to the plane containing the three $sp^2$ orbitals of the other oxygen atom. This orientation is necessary for the formation of the two π bonds because the two unhybridized *p* orbitals that overlap to form a π bond must be parallel. The unshared pairs of electrons on each oxygen atom and the σ bond to carbon lie in a trigonal plane 120° apart. The $sp^2$ hybridization of the oxygen accounts for all these features.

**Think About It** The *sp* hybridization of the carbon atom is consistent with its steric number (SN = 2), its capacity to form two σ and two π bonds, and the linear geometry of the molecule.

⚙ **Practice Exercise** In which of these molecules does the central atom have $sp^3$ hybrid orbitals? (a) $CCl_4$; (b) HCN; (c) $SO_2$; (d) $PH_3$

Why can't a carbon atom with $sp^3$ hybrid orbitals form $\pi$ bonds?

## Octahedral and Trigonal Bipyramidal Geometries: $sp^3d^2$ and $sp^3d$ Hybrid Orbitals

Valence bond theory can also account for the molecular geometries of molecules with central atoms that have more than eight valence electrons. For example, the central sulfur atom in a molecule of $SF_6$ must expand its octet to bind six fluorine atoms. As we discussed in Chapter 8, atoms with unoccupied $d$ orbitals in their valence shells can expand their octets. Valence bond theory provides a way to describe the expansion of an octet by including these $d$ orbitals in hybridization schemes. To form six $\sigma$ bonds to six fluorine atoms, six atomic orbitals on sulfur—one $3s$ orbital, three $3p$ orbitals, and two $3d$ orbitals—hybridize, producing six equivalent **$sp^3d^2$ hybrid orbitals** and an octahedral molecular geometry (Figure 9.31a).

Other molecules, such as phosphorus pentafluoride ($PF_5$), have a trigonal bipyramidal geometry, and the central phosphorus atom shares 10 valence electrons. Five atomic orbitals—one $3s$ orbital, three $3p$ orbitals, and one $3d$ orbital—can be hybridized to produce five equivalent **$sp^3d$ hybrid orbitals** with lobes that point toward the vertices of a trigonal bipyramid (Figure 9.31b). A summary of the shapes associated with all of the hybridization schemes we have discussed so far is given in Figure 9.32.

**$sp^3d^2$ hybrid orbitals** six equivalent orbitals that point toward the vertices of an octahedron, formed by mixing one $s$ orbital, three $p$ orbitals, and two $d$ orbitals from the same shell.

**$sp^3d$ hybrid orbitals** five equivalent hybrid orbitals with lobes pointing toward the vertices of a trigonal bipyramid, formed by mixing one $s$ orbital, three $p$ orbitals, and one $d$ orbital from the same shell.

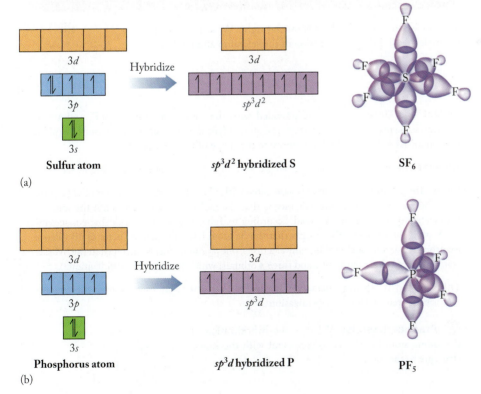

**FIGURE 9.31** Hybrid orbitals can be generated by combining $d$ orbitals with $s$ and $p$ orbitals in atoms that expand their octet. (a) One $3s$ orbital, three $3p$ orbitals, and two $3d$ orbitals mix on the central sulfur atom in $SF_6$ to form six $sp^3d^2$ hybrid orbitals. (b) One $3s$, three $3p$ orbitals, and one $3d$ orbital mix on the central phosphorus atom in $PF_5$ to form five $sp^3d$ hybrid orbitals. For simplicity, only the fluorine orbitals that overlap with the hybrid orbitals are shown.

**FIGURE 9.32** Summary of hybridization schemes and the orientations of orbitals derived from them. Purple orbitals are hybrids; blue are unhybridized atomic orbitals.

| Hybridization | Orientation of hybrid orbitals | Number of σ bonds | Molecular geometries | Theoretical angles between hybrid orbitals |
|---|---|---|---|---|
| $sp$ |  | 2 | Linear | 180° |
| $sp^2$ |  | 3 2 | Trigonal planar Bent | 120° |
| $sp^3$ | 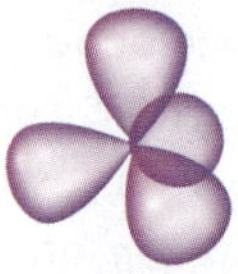 | 4 3 2 | Tetrahedral Trigonal pyramidal Bent | 109.5° |
| $sp^3d$ |  | 5 4 3 2 | Trigonal bipyramidal Seesaw T-shaped Linear | 90°, 120°, 180° |
| $sp^3d^2$ |  | 6 5 4 | Octahedral Square pyramidal Square planar | 90°, 180° |

▶❚❚ **CHEMTOUR** Hybridization

**SAMPLE EXERCISE 9.7** **Recognizing Hybridized Atoms in Molecules** **LO4**

The Lewis structure of $SeF_4$ is shown at right. What is the shape of the molecule and the hybridization of the selenium atom in $SeF_4$?

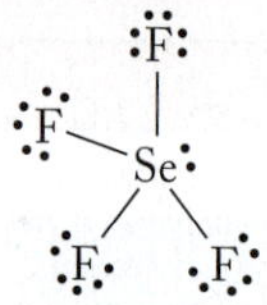

**Collect and Organize** We are provided with the Lewis structure of $SeF_4$ and are to determine the hybridization of the central atom. Hybridization schemes are used to explain molecular shapes, so we need to determine the shape of the molecule first.

**Analyze** The central atom has four single (σ) bonds and one lone pair, giving it SN = 5.

**Solve** The electron-pair geometry associated with SN = 5 is trigonal bipyramidal (Figure 9.2). The presence of the lone pair means that the molecular geometry is not the same as the electron-pair geometry. Instead, according to Table 9.1, the molecular shape is seesaw.

A steric number of 5 also means that the Se atom has an expanded octet of 10 valence electrons. To accommodate this many electrons, one $d$ orbital must be involved in the hybridization scheme along with one $s$ and three $p$ orbitals, which means $sp^3d$ hybridization.

**Think About It** The information in Figure 9.32 confirms that a seesaw molecular geometry is consistent with $sp^3d$ hybridization.

⚙ **Practice Exercise** What is the hybridization of the iodine atom in $IF_5$ that is consistent with the Lewis structure to the right?

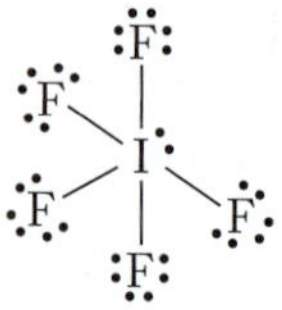

# 9.5 Shape and Interactions with Large Molecules

Up to now, most of the molecules we have considered have small molar masses and typically have a single central atom bonded to two or more other atoms. It is important to learn to apply the skills we have developed to larger molecules, however, because molecular shape is an important factor in determining the physical, chemical, and biological properties of all substances.

Living things respond to molecules that interact with molecular-scale regions in their tissues called *receptors* or *active sites*. The process by which these molecules and sites interact is known as **molecular recognition**. This recognition does not usually involve covalent bond formation. These noncovalent interactions require that the biologically active molecules and the receptors that respond to them fit tightly together, which means that they must have complementary three-dimensional shapes. An example of a biological effect caused by molecular recognition is the process by which produce such as green tomatoes ripens. Tomatoes ripen faster when stored in a paper or plastic bag instead of sitting on a kitchen counter. The reason why is that the tomatoes give off ethylene gas as they ripen, and this gas accelerates the ripening process when it is trapped in a bag with the tomatoes.

The molecular structure of ethylene is shown in Figure 9.33. Both carbon atoms are bonded to three atoms and have no lone pairs of electrons. Therefore SN = 3. This means that the geometry around each carbon atom is trigonal planar and that the carbon atoms are both $sp^2$ hybridized (see Figure 9.32). Two of these orbitals form σ bonds with hydrogen 1s orbitals. The third forms the C=C σ bond. The C=C π bond is formed by overlap of the unhybridized 2p orbitals on the carbon atoms. Taken together, the two trigonal planar carbon atoms produce an overall planar geometry for ethylene, which means that all six atoms lie in the same plane.

**molecular recognition** the process by which molecules interact with other molecules in living tissues to produce a biological effect.

FIGURE 9.33 Ethylene molecules contain two carbon atoms that are at the centers of two overlapping triangular planes that are also coplanar. This combination means that all of the atoms are in the same plane.

---

**SAMPLE EXERCISE 9.8** **Comparing Structures Using Valence Bond Theory** **LO4**

Use valence bond theory to describe the bonding and molecular geometry of ethane, $CH_3$—$CH_3$, and compare them to the bonding and molecular geometry of ethylene.

**Collect and Organize** We are given the molecular formula of ethane and asked to describe its geometry and bonding and then compare them with the shape and bonding of ethylene.

**Analyze** We can use the Lewis structure for ethane and VSEPR to determine the molecular geometry and then determine the hybridization of its carbon atoms. We can then compare atom locations and electron distributions in the two molecules.

**Solve** Each carbon atom in ethane has four single bonds and no lone pairs, which means SN = 4 for both carbon atoms, tetrahedral geometry around both, and $sp^3$ hybridization of both. Comparing the overall molecular structures of ethane and ethylene (Figure 9.34), we see the distinctive three-dimensionality of tetrahedral environments: only two of the six C—H bonds in ethane are in the plane of the page, whereas all the bonds in ethylene are coplanar.

**Think About It** Remember the analogy at the beginning of the chapter about how molecules fit receptors like hands fit gloves. It is clear from the structures that ethylene and ethane would fit into very different gloves.

**FIGURE 9.34** Molecular structures of ethane and ethylene.

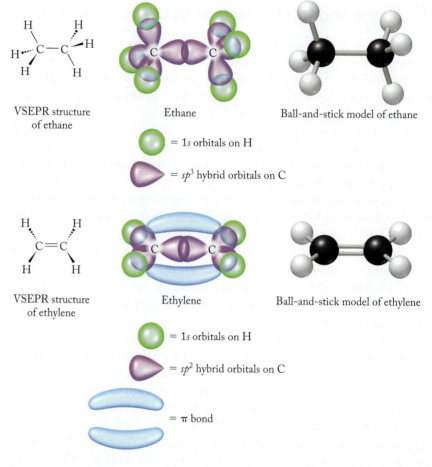

VSEPR structure of ethane

Ethane

Ball-and-stick model of ethane

● = 1s orbitals on H

● = sp³ hybrid orbitals on C

VSEPR structure of ethylene

Ethylene

Ball-and-stick model of ethylene

● = 1s orbitals on H

● = sp² hybrid orbitals on C

● = π bond

⚙ **Practice Exercise** Diazene ($N_2H_2$) and hydrazine ($NH_2NH_2$) are reactive nitrogen compounds. Use valence bond theory to compare the bonding in these two molecules, and describe the differences in their molecular structures.

As Sample Exercise 9.8 illustrates, ethane and ethylene have similar formulas and molar masses, but very different molecular geometries. Ethane does not trigger the ripening process because its shape does not allow it to fit the receptor in plant tissue that binds the planar molecule ethylene.

Now let's consider a biologically active molecule with three "central" atoms. Acrolein is one of the components of barbeque smoke that contributes to the distinctive odor of a cookout. It is also a possible cancer-causing compound. The Lewis structure of acrolein is

**delocalization** (*adjective*: delocalized) the sharing of bonding pairs of electrons in alternating single and double bonds by three or more atoms in a molecule.

**aromatic compound** a cyclic, planar compound with delocalized π (pi) electrons above and below the plane of the molecule.

Each molecule contains a C=O double bond and a C=C double bond. Both are formed by trigonal planar, $sp^2$ hybridized carbon atoms.

The pattern of alternating single and double bonds means that the electrons in the π bonds can be **delocalized** over the three carbon atoms and the oxygen atom. Delocalization can occur both when the atoms involved are all carbon atoms and when atoms of different elements are involved, as in acrolein.

Benzene, $C_6H_6$, is another molecule in barbeque smoke. As we saw in Chapter 8, each benzene molecule is a hexagon of six carbon atoms, each bonded to one hydrogen atom. Benzene also has three C=C bonds. Resonance structures for benzene are shown in Figure 9.35, along with a view of the π bonds located above and below the plane of the ring. Each carbon in benzene has a trigonal planar geometry and forms $sp^2$ hybrid orbitals. The carbon ring is made of σ bonds formed by overlapping $sp^2$ hybrid orbitals on adjacent carbon atoms. The C—H bonds are formed by the overlap of carbon $sp^2$ hybrid orbitals with hydrogen 1s orbitals.

The two resonance forms of benzene shown in Figure 9.35(a) correspond to shifts in the locations of the π bonds formed by the overlap of carbon 2p orbitals. However, because all the carbon 2p orbitals are identical, they are all equally likely to overlap with their neighbors, so the π bonds are actually delocalized over all six carbon atoms rather than fixed between alternating pairs of carbon atoms (Figure 9.35b). As noted in Section 8.5, the presence of these delocalized π bonds is often represented by a circle drawn in the middle of the hexagon of carbon atoms in the line-bond structure, corresponding to continuous rings of π electrons above and below the plane of the molecule (Figure 9.35b).

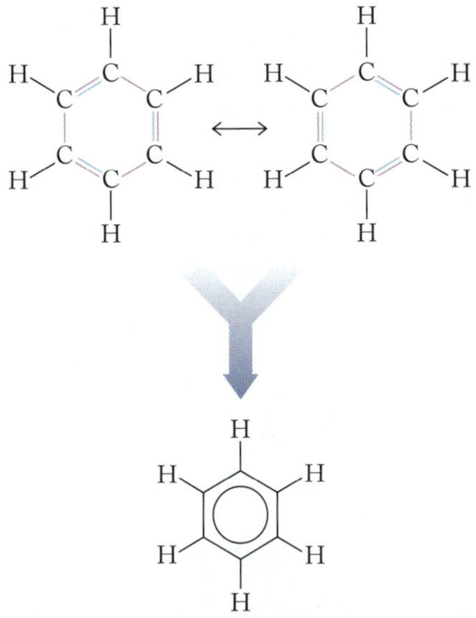

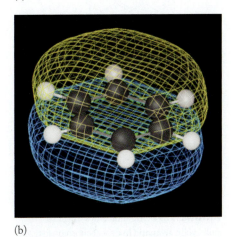

(a)

(b)

**FIGURE 9.35** (a) Resonance between the two Lewis structures of benzene leads to complete delocalization of the π bonds around the benzene ring. (b) A computer-generated view of benzene's σ bonds in the plane of the ring and its delocalized π bonds above and below the ring.

**CONCEPT TEST**

In which of the molecules and polyatomic ions in Figure 9.36 are the π bonds delocalized?

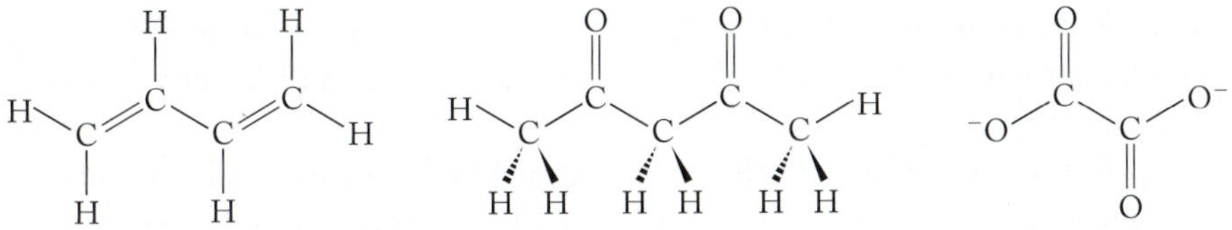

(a)                    (b)                    (c)

**FIGURE 9.36** Organic compounds with double bonds: (a) butadiene, used to make polymers; (b) 2,4-pentanedione, a reactive organic compound used in synthesis; and (c) the oxalate ion, present in spinach.

The molecular structures of many other compounds in addition to benzene contain carbon rings with delocalized π electrons above and below the plane of the ring. They are called **aromatic compounds**. An important class of these compounds, known as *polycyclic aromatic hydrocarbons* (PAHs), consists of molecules containing several benzene rings joined together (Figure 9.37). PAHs are formed any time coal, oil, gas, and most hydrocarbon fuels are burned, and they are also found in cigarette smoke. In 2004 they were discovered in interstellar space.

The shape of PAH molecules gives rise to a particular health hazard. After we inhale or ingest them, some PAHs associate with DNA in a process called *intercalation*. Because PAHs are flat, they can intercalate in DNA—they slide in between the two strands of DNA as shown in Figure 9.37. Once there, they sometimes form covalent bonds, altering or preventing DNA replication and thereby damaging or killing cells. Intercalation in DNA is one step in the process by which PAHs induce cancer.

The planarity of ethylene and aromatic compounds as described by VSEPR and valence bond theories plays a key role in determining the behavior of such molecules in biological systems. The shape of these molecules matters because it influences their interaction with other molecules, which in turn determines their biological activity.

Naphthalene

Anthracene

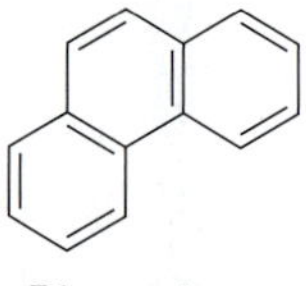

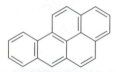

Phenanthrene

Benzo[a]pyrene

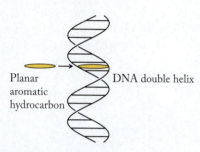

Planar aromatic hydrocarbon    DNA double helix

Intercalation of PAH in DNA

**FIGURE 9.37** The molecules of these polycyclic aromatic hydrocarbons consist of fused benzene rings whose π bonds are delocalized over all the rings in each molecule. Molecules with this shape can slip in between the strands of DNA and disrupt cell replication, which can lead to cell death or induce malignancy.

**FIGURE 9.38** The distinctive aromas of (a) caraway and (b) spearmint are primarily due to two compounds with nearly identical molecular structures. The only difference between the two is the orientation of the two groups attached to the carbon atoms highlighted with the red circles. Note that the hydrogen atom is in back of the plane of the paper and down in caraway (a) but is in front of the plane of the paper and down in spearmint (b).

# 9.6 Chirality and Molecular Recognition

Before ending our exploration of molecular shapes, we need to address the subject of handedness introduced at the beginning of this chapter. There we described how two molecules with the same molecular formula and Lewis structure can interact differently with receptors in our nasal membranes. As a result, one produces the smell of caraway seeds, and the other spearmint leaves. You may find these different odors surprising when you consider the molecular structures of these two compounds (Figure 9.38). At first glance they may seem identical, but look closely. Notice in particular the bonding pattern around the carbon atom at the bottom of the ring. The H atom bonded directly to the bottom carbon atom is on the front side of the ring in the spearmint compound, but it is on the back side in the caraway compound. This minor difference in bond orientation creates a difference in molecular shape that is easily recognized by receptors in our noses.

The structures in Figure 9.38 are called *optical isomers*. The term "optical" refers to the ways these compounds interact with a special kind of light called *plane-polarized* light. We explore this topic in more detail in Chapter 13, but it is sufficient now for you to know that when plane-polarized light passes through a solution of the caraway compound, the light twists in one direction. However, when the light passes through a solution of the spearmint compound, it twists in the opposite direction. The term "isomer" is derived from the Greek *iso*, meaning "same," and *mer*, meaning "unit" or "part." Isomers are compounds with the same formula, but within their molecules the atoms are arranged differently in three-dimensional space.

Optical isomerism has another name: **chirality**. Many molecules of biological importance, including the proteins and carbohydrates that we consume each day, are composed of chiral compounds. Chirality comes from the Greek word *chier*, meaning "hand," and is quite correctly called "handedness." Although several features within molecules can lead to chirality, the most common is the presence of a carbon atom that has four different atoms or groups of atoms attached to it.

To see how chirality works, let's look at a compound that contains a central carbon atom bonded to four different atoms. The compound is bromochlorofluoromethane (CHBrClF; Figure 9.39a), which is used in fire extinguishers on airplanes. We will compare its molecular structure to that of dibromochloromethane (CHBr$_2$Cl), a compound that may form during the purification of municipal water supplies with chlorine and that has only three different atoms bonded to its central carbon atom (Figure 9.39b). First we generate mirror images of both molecules, then we rotate the mirror images 180° in an attempt

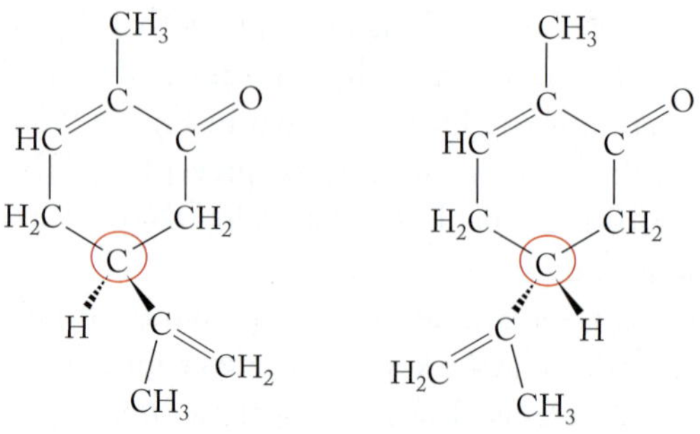

(a) (+)-Carvone (caraway)

(b) (−)-Carvone (spearmint)

to superimpose each mirror image on its original image. If the reflected, rotated image is superimposable on the original image, then the substance is not chiral. Scientists call it *achiral*. The molecular images of $CHBr_2Cl$ can be superimposed in this way, so it is not a chiral compound. However, the images of CHBrClF cannot be superimposed. For example, when we superimpose the F, C, and H atoms of the two images in Figure 9.39(a), the Br and Cl atoms are not aligned. This means that CHBrClF *is* a chiral compound.

Any structure like CHBrClF that has four different groups attached to an $sp^3$ hybridized carbon atom is chiral. It has two optical isomers that are mirror images of each other and are distinctly different compounds. You may be wondering where the chiral carbon atom is in the molecules of the caraway and spearmint compounds. If you guessed the circled carbon atoms, you were right. Two of the four different groups are easy to see: a hydrogen atom and the

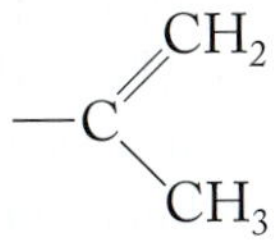

group. The other two "groups" on the bottom carbon atom are really the two sides of the ring. The left side contains a C=C double bond, and the right side contains a C=O double bond. These differences mean that the two sides are not equivalent, so the carbon atom is attached to four different groups. This makes it a chiral carbon atom, and the two compounds are optical isomers of each other.

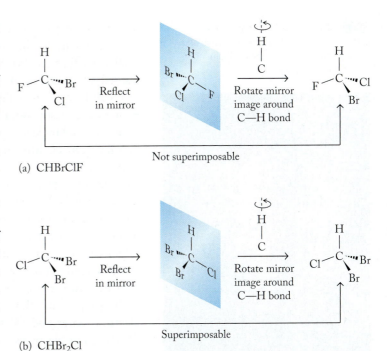

(a) CHBrClF

(b) $CHBr_2Cl$

**FIGURE 9.39** (a) A molecule of a chiral compound, such as CHBrClF, is not superimposable on its mirror image. (b) A molecule of a compound that is not chiral, such as $CHBr_2Cl$, is superimposable on its mirror image.

**CONCEPT TEST**

Identify the molecules in Figure 9.40 that are chiral.

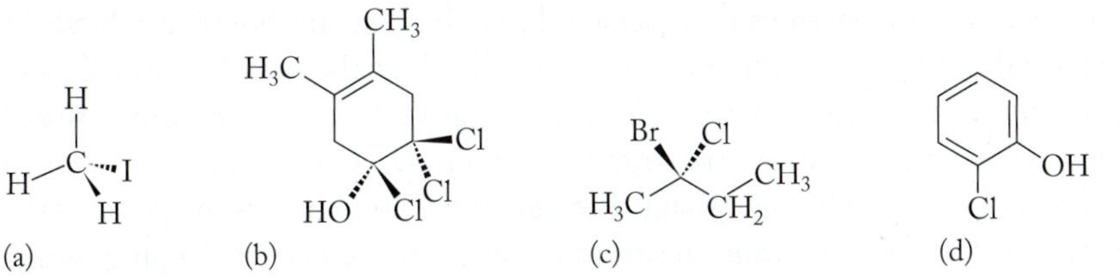

(a)     (b)     (c)     (d)

**FIGURE 9.40**

## 9.7 Molecular Orbital Theory

Lewis structures and valence bond theory help us understand the bonding patterns in molecules and molecular ions; VSEPR and valence bond theory also account for their molecular shapes. However, none of these models explains why $O_2$ is attracted to a magnetic field whereas $N_2$ is slightly repelled by one. For that explanation, we turn to **molecular orbital (MO) theory**.

Molecular orbital theory also explains the absorption and emission of light by molecules and molecular ions, including the colors produced in Earth's upper atmosphere in the shimmering displays known as *aurora* (Figure 9.41).

▶ **CHEMTOUR** Chemistry of the Upper Atmosphere

**chirality** property of a molecule that is not superimposable on its mirror image.

**molecular orbital (MO) theory** a bonding theory based on the mixing of atomic orbitals of similar shapes and energies to form molecular orbitals that extend to two or more atoms.

⊙⊙⊙ **CONNECTION** In Chapter 7 we discussed the role of atomic emission and absorption spectra in the development of quantum mechanics.

**FIGURE 9.41** Auroras are spectacular displays of color produced when the solar wind collides with Earth's upper atmosphere, producing excited-state atoms, ions, and molecules.

**molecular orbital** a region of characteristic shape and energy where electrons in a molecule are delocalized over two or more atoms in a molecule.

**bonding orbital** term in MO theory describing regions of increased electron density between nuclear centers that serve to hold atoms together in molecules.

**antibonding orbital** term in MO theory describing regions of electron density in a molecule that destabilize the molecule because they decrease the electron density between nuclear centers.

**sigma (σ) molecular orbital** in MO theory, a molecular orbital in which the greatest electron density is concentrated along an imaginary line drawn through the bonded atom centers.

**molecular orbital diagram** in MO theory, an energy-level diagram showing the relative energies and electron occupancy of the molecular orbitals for a molecule.

The colors of an aurora are caused by collisions between atoms, molecules, and molecular ions in the atmosphere with electrons and positive particles in the solar wind that are attracted to Earth's magnetic poles. These collisions produce excited-state species. As they return to their ground states, these species emit characteristic colors of light. We discussed in Chapter 7 how excited-state atoms emit characteristic atomic spectra. In this section we explore how excited-state molecules and molecular ions, such as $N_2$ and $N_2^+$, do the same thing. Our exploration requires a different view of the bonding in these species; MO theory provides us with that view.

Like valence bond theory, molecular orbital theory invokes the mixing of atomic orbitals. In valence bond theory, the mixing results in hybrid atomic orbitals, whereas molecular orbital theory is based on the formation of **molecular orbitals**. A key difference between the two types of orbitals is that hybrid atomic orbitals are associated with a particular atom in the molecule, but molecular orbitals are spread out over two or more of the atoms in a molecule.

Molecular orbitals represent discrete energy states in molecules, just as atomic orbitals represent allowed energy states in free atoms. As with atomic orbitals, electrons enter and fill the lowest-energy MOs first; higher-energy MOs are filled as more electrons are added. Electrons in molecules can be raised to higher-energy MOs by absorbing quanta of electromagnetic radiation. When they return to lower-energy MOs, distinctive wavelengths of UV and visible radiation are emitted, including some of the shimmering colors in an aurora.

According to MO theory, when two atomic orbitals combine, they form two molecular orbitals. One of the two molecular orbitals is a **bonding orbital**. When electrons occupy a bonding orbital, they hold the molecule together by increasing electron density between the atoms; in other words, they form a covalent bond. The second type of molecular orbital is an **antibonding orbital**. An antibonding orbital is higher in energy than its corresponding bonding orbital, and when electrons reside in an antibonding orbital, they destabilize the molecule and do not contribute to holding its atoms together.

Recall from Chapter 7 that the quantum-mechanical picture of atoms involves wave–particle duality: electrons behave like both matter and waves. We can invoke the wave behavior of electrons to understand bonding and antibonding orbitals in molecular orbital theory in the following way: Think of the bonding orbital arising from the constructive interference of two waves. Just as a one-dimensional standing wave has a positive and negative amplitude (Figure 9.42a), so does a three-dimensional wave like an orbital. The sign of the amplitude is the phase of the wave. Bonding orbitals arise from constructive interference when overlapping atomic orbitals have the same phase—the two orbitals add (Figure 9.42b). Antibonding orbitals arise out of destructive interference between two orbitals of opposite phase—the two orbitals subtract (Figure 9.42c). A bonding orbital has an increased electron density between the nuclei of the two atoms, which lowers its energy relative to the atomic orbitals from which it was formed. An antibonding orbital has a node (a region of no electron density) between the nuclear centers and has higher energy than the atomic orbitals from which it was formed.

To further explore the distinction between bonding and antibonding molecular orbitals, let's apply MO theory to the simplest molecular compounds: hydrogen and helium.

**CONCEPT TEST** ● ● ● ● ● ● ● ● ● ● ● ● ● ● ● ● ● ● ● ● ●

In your own words, describe one way in which hybrid orbitals and molecular orbitals are similar and one way in which they are different.

● ● ● ● ● ● ● ● ● ● ● ● ● ● ● ● ● ● ● ● ● ● ● ● ● ● ● ● ● ●

## Molecular Orbitals of Hydrogen and Helium

According to MO theory, a hydrogen molecule is formed when the $1s$ atomic orbitals on two hydrogen atoms combine to form two molecular orbitals. Molecular orbital theory stipulates that mixing two atomic orbitals creates two molecular orbitals and that these two orbitals represent two different energy states (Figure 9.43). As a general rule, *the number of molecular orbitals formed in a molecule equals the number of atomic orbitals that combine.*

When two $1s$ atomic orbitals combine, the lower-energy bonding molecular orbital formed is oval and spans the two atomic centers. Its shape corresponds to enhanced electron density between the two atoms that donated their atomic orbitals. This enhanced electron density is a covalent bond. When two electrons occupy a bonding MO, a single bond is formed. When the region of highest density lies along the bond axis, as it does in the bonding MO in $H_2$, the MO is designated a **sigma ($\sigma$) molecular orbital** and the covalent bond is a $\sigma$ bond. The bonding molecular orbital in $H_2$ is labeled $\sigma_{1s}$ in Figure 9.43(a) because it is formed by mixing two $1s$ atomic orbitals.

The higher-energy (less stable) antibonding molecular orbital formed from two hydrogen atomic orbitals is designated $\sigma_{1s}^*$ (pronounced "sigma star"). This antibonding orbital has two separate lobes of electron density and a region of zero electron density (a node) between the two hydrogen atoms, as shown in Figure 9.43(a).

Figure 9.43(b) is a **molecular orbital diagram**, analogous to an energy-level diagram for atomic orbitals. It shows that the $\sigma_{1s}$ MO is lower in energy and therefore more stable than the $1s$ atomic orbitals by nearly the same amount that the $\sigma_{1s}^*$ MO is higher in energy than the $1s$ atomic orbitals. Therefore, the formation of the two MOs does not significantly change the total energy of the system. A hydrogen molecule has two valence electrons, one from each H atom, both residing in the lower-energy $\sigma_{1s}$ orbital because that is the lowest-energy orbital available. As in atomic orbitals, these two $\sigma_{1s}$ electrons must have opposite spins. The electron configuration that corresponds to the molecular orbital diagram in Figure 9.43 is written $(\sigma_{1s})^2$ where the superscript indicates that there are two electrons in the $\sigma_{1s}$ molecular orbital. Because the energy of the electrons in a $(\sigma_{1s})^2$ configuration is lower than the energy of the electrons in two isolated hydrogen atoms, MO theory explains why hydrogen is a diatomic gas: $H_2$ molecules are lower in energy and so are more stable than H atoms.

Hydrogen is a diatomic gas, but helium exists as free atoms and not as molecular $He_2$. MO theory explains why. One helium atom has two valence electrons in a $1s$ atomic orbital. Mixing two He $1s$ orbitals yields the same set of molecular orbitals we generated for $H_2$, as shown in Figure 9.44. Unlike $H_2$, the two helium atoms have a total of four valence electrons. Each molecular orbital in Figure 9.44—the

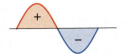

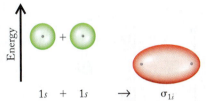

(a) One-dimensional standing wave

(b) Atomic orbitals add to make bonding orbital

$1s \quad + \quad 1s \quad \rightarrow \quad \sigma_{1s}$

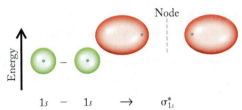

(c) Atomic orbitals subtract to make antibonding orbital

$1s \quad - \quad 1s \quad \rightarrow \quad \sigma_{1s}^*$

**FIGURE 9.42** (a) A standing wave has a positive and a negative amplitude. (b) Bonding orbitals, lower in energy than the atomic orbitals, arise from constructive interference when the atomic orbitals add. The molecular orbitals have increased electron density between the two nuclear centers. (c) Antibonding orbitals, higher in energy than the atomic orbitals, arise from the destructive interference when two atomic orbitals of opposite phase overlap; they have a node between the two nuclear centers.

**CONNECTION** We used energy-level diagrams in Chapter 7 to show the transitions that electrons in atoms can make from one energy level to another.

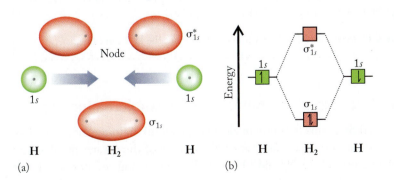

**FIGURE 9.43** Mixing the $1s$ orbitals of two hydrogen atoms creates two molecular orbitals: a filled bonding $\sigma_{1s}$ orbital containing two electrons and an empty antibonding $\sigma_{1s}^*$ orbital. (a) The lower red oval is the bonding orbital. The two red ovals at the top together make up the antibonding orbital. *Note*: The two top ovals represent only *one* molecular orbital with a node of zero electron density in between. Dots show the locations of hydrogen nuclei. (b) A molecular orbital diagram shows the relative energies of bonding and antibonding molecular orbitals and of the atomic orbitals that formed them.

**CONNECTION** We first defined bond order in Chapter 8 when we related the length and strength of bonds to the number of electron pairs shared by two atoms.

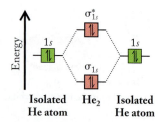

**FIGURE 9.44** The molecular orbital diagram for the hypothetical molecule He₂ indicates that the same number of electrons occupy the antibonding orbital and the bonding orbital. Therefore the bond order is 0; the molecule is not stable.

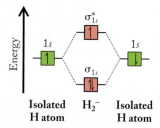

**FIGURE 9.45** The molecular orbital diagram of $H_2^-$.

bonding orbital and the antibonding orbital—has a maximum capacity of two electrons. Adding four valence electrons to the orbitals in Figure 9.44 means filling both orbitals. The presence of two electrons in the $\sigma_{1s}^*$ orbital cancels the stability gained from having two electrons in the $\sigma_{1s}$ orbital. Because there is no net gain in stability, He₂ does not form.

Another way of comparing the bonding in H₂ and He₂ is to look at the *bond order* in the two molecules. We have previously defined bond order as the number of bonds between two atoms: a bond order of 1 for X—X, 2 for X=X, and 3 for X≡X. In MO theory we define bond order as

$$\text{Bond order} = \frac{1}{2}\left(\begin{array}{c}\text{number of}\\\text{bonding electrons}\end{array} - \begin{array}{c}\text{number of}\\\text{antibonding electrons}\end{array}\right) \quad (9.2)$$

A molecule of H₂ has two electrons in the bonding MO and none in the antibonding MO, so

$$\text{Bond order in } H_2 = \tfrac{1}{2}(2-0) = 1$$

For He₂, the bond order is 0 because an equal number of electrons reside in bonding and antibonding orbitals:

$$\text{Bond order in } He_2 = \tfrac{1}{2}(2-2) = 0$$

A bond order of 0 means that He₂ is not a stable molecule. In general, the greater the bond order, the stronger the bond and the more stable the molecule.

---

**SAMPLE EXERCISE 9.9** **Using MO Diagrams to Predict Bond Order I** **LO7**

Draw the MO diagram for the molecular ion $H_2^-$, determine the bond order of the ion, and predict whether or not the ion is stable.

**Collect and Organize** We want to draw the MO diagram for the molecular ion $H_2^-$, and then determine bond order using Equation 9.2. If the value of the bond order is greater than zero, the ion may be stable.

**Analyze** We should be able to base the MO diagram for $H_2^-$ on the MO diagram for H₂ (Figure 9.43) because the $H_2^-$ ion has only one more electron than H₂ and the empty $\sigma_{1s}^*$ orbital in H₂ can accommodate up to two more electrons.

**Solve** The $\sigma_{1s}$ orbital is filled in H₂. The third electron goes into the $\sigma_{1s}^*$ orbital, so the MO diagram is as shown in Figure 9.45. The notation for this electron configuration is $(\sigma_{1s})^2(\sigma_{1s}^*)^1$ (listing the molecular orbitals in order of increasing energy). The bond order is

$$\text{Bond order} = \tfrac{1}{2}(2-1) = 0.5$$

We predict that the $H_2^-$ ion is less stable than H₂, but more stable than He₂.

**Think About It** We encountered the idea of fractional bonds in Chapter 8 in the discussion of resonance, and we encounter it again here in the MO treatment of $H_2^-$. A bond order of 0.5 in MO theory means that the bond between the two atoms in $H_2^-$ is weaker than the single bond in H₂, making $H_2^-$ a less stable species.

**Practice Exercise** Use MO theory to predict whether the $H_2^+$ ion can exist.

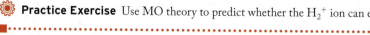

## Molecular Orbitals of Homonuclear Diatomic Molecules

Molecular orbital diagrams for homonuclear (same atom) diatomic molecules like N₂ and O₂ are more complex than that of H₂ because of the greater number and variety of atomic orbitals in N₂ and O₂. Not all combinations of atomic orbitals result in effective bonding, but there are some general guidelines for constructing the molecular orbital diagram for any molecule:

1. The number of molecular orbitals equals the number of atomic orbitals used to create them.
2. Atomic orbitals with similar energy and shape mix more effectively than do those that have different energies and shapes. For example, an *s* atomic orbital mixes more effectively with another *s* atomic orbital than with a *p* orbital.
3. Atomic orbitals of different principal quantum numbers (for example, 1*s* and 2*s*) have different sizes and energies, resulting in less effective mixing than two 1*s* or two 2*s* orbitals. Better mixing leads to a larger energy difference between bonding and antibonding orbitals and thus greater stabilization of the bonding MOs.
4. A molecular orbital can accommodate a maximum of two electrons; two electrons in the same MO have opposite spins.
5. Electrons occupy the lowest-energy molecular orbitals available, distributed among degenerate orbitals following Hund's rule.

In mixing atomic orbitals to create molecular orbitals, we consider *only the valence electrons* on the atoms because core electrons do not participate in bonding. Focusing on $N_2$ and $O_2$ as examples, we first mix their 2*s* orbitals. The mixing process is analogous to the one we used for $H_2$, except that the resulting MOs are designated $\sigma_{2s}$ and $\sigma_{2s}^*$.

Next we mix the three pairs of 2*p* orbitals, producing a total of six MOs. The different spatial orientations of the $2p_x$, $2p_y$, and $2p_z$ atomic orbitals result in different kinds of MOs (Figure 9.46). The $2p_z$ atomic orbitals point toward each other. When they mix, two molecular orbitals form, a $\sigma_{2p}$ bonding orbital and a $\sigma_{2p}^*$ antibonding orbital. The lobes of the $2p_y$ and $2p_x$ atomic orbitals are oriented at 90° to the bonding axis and also at 90° to each other. When the $2p_x$ orbitals mix together, and when the $2p_y$ orbitals mix together, they do so around the bonding axis instead of along it. This mixing produces two **pi (π) molecular orbitals** and two $\pi^*$ molecular orbitals. When electrons occupy a π orbital, they form a π bond.

The relative energies of σ and π molecular orbitals for $N_2$ and $O_2$ are shown in Figure 9.47. In each molecule, the energies of the MOs derived from two 2*s* atomic orbitals ($\sigma_{2s}$ and $\sigma_{2s}^*$) are lower than the energy of the $\sigma_{2p}$ MO for the same reason that a 2*s* atomic orbital is lower in energy than a 2*p* atomic orbital.

Now let's consider the relative energies of the MOs formed by mixing the 2*p* orbitals in $N_2$ and $O_2$. We begin with $O_2$ (Figure 9.47b) because it is representative of most homonuclear diatomic molecules, including all the halogens. In order of increasing energy, the MOs are $\sigma_{2p}$, $\pi_{2p}$, $\pi_{2p}^*$, and $\sigma_{2p}^*$. Keep in mind that there are groups of two $\pi_{2p}$ and two $\pi_{2p}^*$ orbitals. This means that each group of two can hold four electrons. Adding 12 valence electrons (six from each O atom) in the $O_2$ molecule into these MOs, starting with the lowest-energy MO first and working our way up, we get the following electron configuration:

$$O_2: (\sigma_{2s})^2(\sigma_{2s}^*)^2(\sigma_{2p})^2(\pi_{2p})^4(\pi_{2p}^*)^2$$

The distribution of electrons in the MO diagram follows this sequence. Note that the two $\pi_{2p}^*$ orbitals are degenerate (equivalent in energy), so each contains a single electron, in accordance with Hund's rule.

The MO diagram tells us that there are two unpaired electrons in a molecule of $O_2$. This is not the picture that we obtain from a Lewis structure, from VSEPR, or from valence bond theory, which all predict that all the valence electrons are paired. We will return to this point shortly, but for now let's consider the MO diagram for $N_2$.

**pi (π) molecular orbitals** in MO theory, molecular orbitals formed by the mixing of atomic orbitals oriented above and below, or in front of and behind, the bonding axis.

⊙⊙ **CONNECTION** In Chapter 7 we discussed Hund's rule, which states that the lowest-energy electron configuration of a set of degenerate orbitals is the one with the maximum number of unpaired electrons, all having the same spin.

▶❙❙ **CHEMTOUR** Molecular Orbitals

**FIGURE 9.46** Two atoms come together and their *p* atomic orbitals mix to form six molecular orbitals. (a) The $2p_z$ atomic orbitals create a $\sigma_{2p}$ bonding orbital and a $\sigma_{2p}^*$ antibonding orbital. (b) The $2p_x$ and $2p_y$ atomic orbitals mix to form two $\pi_{2p}$ bonding molecular orbitals and two $\pi_{2p}^*$ antibonding molecular orbitals.

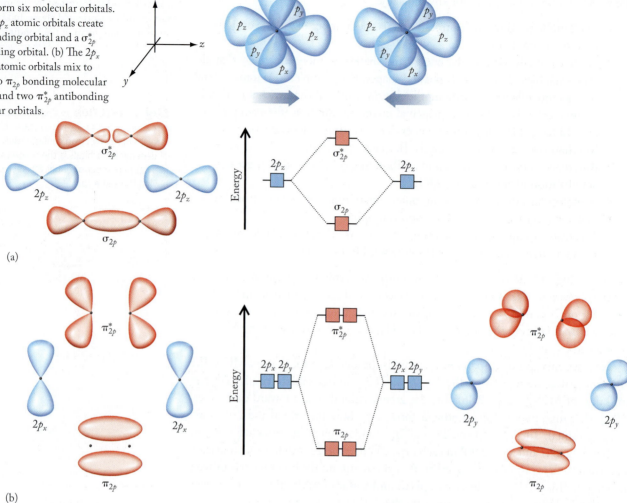

(a)

(b)

When we compare the MO diagrams of $N_2$ and $O_2$ in Figure 9.47, we see a difference in the relative energies of two of their MOs. The $\pi_{2p}$ molecular orbital is lower in energy than the $\sigma_{2p}$ orbital in $N_2$. This switch of energy levels from their relative positions in $O_2$ and many other diatomic molecules is thought to be due to the stability (and lower energy) of the three half-filled $2p$ orbitals in N atoms ($2s^2 2p^3$), which brings their energy closer to that of the $2s$ orbital. This proximity of orbitals has the effect of lowering the energy of the $\sigma_{2s}$ molecular orbital and raising the energy of the $\sigma_{2p}$ orbital—enough to put it above $\pi_{2p}$. Adding ten valence electrons to this stack of MOs in $N_2$, lowest energy first, produces the following electron configuration:

$$N_2: (\sigma_{2s})^2 (\sigma_{2s}^*)^2 (\pi_{2p})^4 (\sigma_{2p})^2$$

The distribution of electrons in the MO diagram in Figure 9.47(a) reflects this electron configuration. There are a total of eight electrons in bonding MOs and two in antibonding MOs. Using Equation 9.2 to calculate the bond order for $N_2$, we get

$$\text{Bond order} = \tfrac{1}{2}(8 - 2) = 3$$

In $O_2$, eight electrons occupy bonding orbitals and four electrons occupy antibonding orbitals, so

$$\text{Bond order} = \tfrac{1}{2}(8 - 4) = 2$$

**diamagnetic** describes a substance with no unpaired electrons that is weakly repelled by a magnetic field.

**paramagnetic** describes a substance with unpaired electrons that is attracted to a magnetic field.

On the basis of their Lewis structures, we predicted in Section 8.2 a triple bond in $N_2$ and a double bond in $O_2$, and molecular orbital theory leads us to the same predictions.

As we noted previously, one of the strengths of MO theory is that it enables us to explain properties of molecular compounds that cannot be explained by other bonding theories. The magnetic behavior of homonuclear diatomic molecules is a case in point. Electrons in atoms have two possible spin orientations, depending on the value of their spin magnetic quantum number $m_s$. Figure 9.48 shows the MO-based electron configurations in diatomic molecules of the row 2 elements. Note that in most of these molecules, all of the electrons are paired so that their spins cancel out. The molecules that make up most substances contain only paired electrons. This complete electron pairing means that these substances are repelled slightly by a magnetic field. These substances are said to be **diamagnetic**. If a substance's molecules contain unpaired electrons, as do those of $O_2$, then it is attracted by a magnetic field, as shown in Figure 9.49, and is **paramagnetic**. The more unpaired electrons in a molecule, the greater its paramagnetism. Only MO theory accounts for the magnetic behavior of oxygen and of many other substances as well.

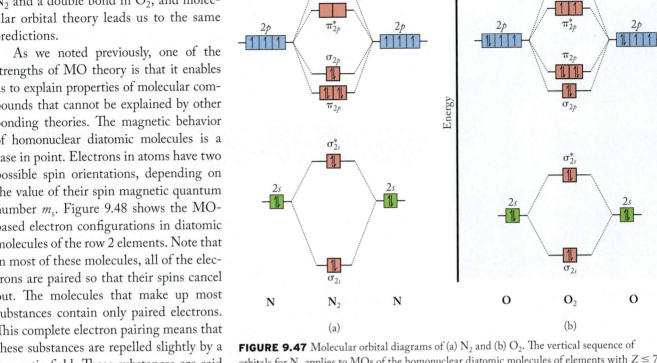

**FIGURE 9.47** Molecular orbital diagrams of (a) $N_2$ and (b) $O_2$. The vertical sequence of orbitals for $N_2$ applies to MOs of the homonuclear diatomic molecules of elements with $Z \leq 7$. The $O_2$ sequence applies to homonuclear diatomic molecules of all elements beyond oxygen ($Z \geq 8$) including the halogens.

**CONNECTION** In Chapter 7 we introduced the spin magnetic quantum number ($m_s$), which has a value of either $+\frac{1}{2}$ or $-\frac{1}{2}$ and defines the orientation of an electron in a magnetic field.

**CONCEPT TEST**

Can liquid $N_2$ and $O_2$ be separated from each other with a magnet?

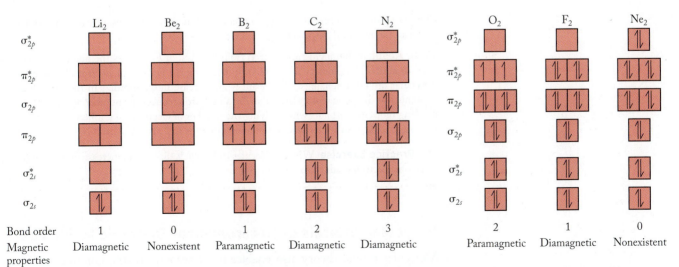

| | Li₂ | Be₂ | B₂ | C₂ | N₂ | | O₂ | F₂ | Ne₂ |
|---|---|---|---|---|---|---|---|---|---|
| Bond order | 1 | 0 | 1 | 2 | 3 | | 2 | 1 | 0 |
| Magnetic properties | Diamagnetic | Nonexistent | Paramagnetic | Diamagnetic | Diamagnetic | | Paramagnetic | Diamagnetic | Nonexistent |

**FIGURE 9.48** Valence-shell molecular orbital diagrams and magnetic properties of the homonuclear diatomic molecules of the second row elements.

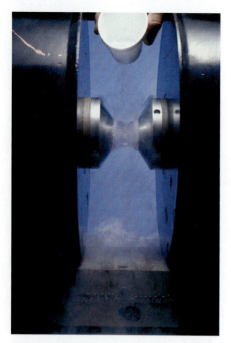

**FIGURE 9.49** Liquid $O_2$ poured from a Styrofoam cup is suspended in the space between the poles of this magnet because the unpaired electrons in its molecules make $O_2$ paramagnetic. Paramagnetic substances are attracted to magnetic fields.

Figure 9.48 enables us to make several predictions about diatomic molecules of the second row elements. First, we predict that $Be_2$ and $Ne_2$ do not exist for the same reason that $He_2$ does not exist: both $Be_2$ and $Ne_2$ have as many antibonding electrons as they have bonding electrons and therefore have a net bond order of 0. Second, $Li_2$, $B_2$, and $F_2$ have a bond order of 1, whereas $C_2$ has a bond order of 2. Like $O_2$, $B_2$ is paramagnetic, whereas $Li_2$, $C_2$, $N_2$, and $F_2$ are diamagnetic. (We have not seen $C_2$ molecules before, but they actually do exist in electric arcs and in comets, and they are responsible for the blue light of candle flames.)

---

**SAMPLE EXERCISE 9.10**   **Using MO Diagrams to Predict Bond Order II**   **LO7**

In which molecules in Figure 9.48 is there an increase in bond order when one electron is removed from the molecule?

**Collect and Organize** We are to determine which molecules in Figure 9.48 acquire a higher bond order when one electron is removed. Equation 9.2 relates bond order to the difference in the numbers of electrons in bonding and antibonding orbitals.

**Analyze** Removing an electron from $Li_2$, $Be_2$, $B_2$, $C_2$, $N_2$, $O_2$, and $F_2$ results in the molecular ions $Li_2^+$, $Be_2^+$, $B_2^+$, $C_2^+$, $N_2^+$, $O_2^+$, and $F_2^+$. We may assume that the molecular ions have MO diagrams with orbital energies in the same order as their parent molecules. Thus, the MO diagrams for the molecular ions are the same as those in Figure 9.48, but with one electron removed from the highest-energy orbital.

**Solve** Removing one electron from each MO diagram in Figure 9.48 gives us

| Ion | Electron Configuration | Bond Order |
|-----|------------------------|------------|
| $Li_2^+$ | $(\sigma_{2s})^1$ | $\frac{1}{2}(1-0) = 0.5$ |
| $Be_2^+$ | $(\sigma_{2s})^2(\sigma_{2s}^*)^1$ | $\frac{1}{2}(2-1) = 0.5$ |
| $B_2^+$ | $(\sigma_{2s})^2(\sigma_{2s}^*)^2(\pi_{2p})^1$ | $\frac{1}{2}(3-2) = 0.5$ |
| $C_2^+$ | $(\sigma_{2s})^2(\sigma_{2s}^*)^2(\pi_{2p})^3$ | $\frac{1}{2}(5-2) = 1.5$ |
| $N_2^+$ | $(\sigma_{2s})^2(\sigma_{2s}^*)^2(\pi_{2p})^4(\sigma_{2p})^1$ | $\frac{1}{2}(7-2) = 2.5$ |
| $O_2^+$ | $(\sigma_{2s})^2(\sigma_{2s}^*)^2(\sigma_{2p})^2(\pi_{2p})^4(\pi_{2p}^*)^1$ | $\frac{1}{2}(8-3) = 2.5$ |
| $F_2^+$ | $(\sigma_{2s})^2(\sigma_{2s}^*)^2(\sigma_{2p})^2(\pi_{2p})^4(\pi_{2p}^*)^3$ | $\frac{1}{2}(8-5) = 1.5$ |

Comparing these values with the bond orders listed in Figure 9.48, we see that bond order increases for only $Be_2^+$, $O_2^+$, and $F_2^+$.

**Think About It** Removing an electron from $Be_2$, $O_2$, or $F_2$ reduces the number of electrons in antibonding molecular orbitals while leaving the number of electrons in bonding molecular orbitals unchanged. The result is an increase in bond order. In the other four homonuclear diatomic molecules, removing an electron reduces the number of electrons in bonding molecular orbitals while leaving the number of electrons in antibonding orbitals unchanged. The result is a reduction in bond order for $Li_2^+$, $B_2^+$, $C_2^+$, and $N_2^+$.

⚙ **Practice Exercise** Which molecules in Figure 9.48 show an increase in bond order when one electron is added to the molecule?

---

## Molecular Orbitals of Heteronuclear Diatomic Molecules

Molecular orbital theory also enables us to account for the bonding in *heteronuclear* diatomic molecules, which are molecules containing two different atoms. The bonding in some of these molecules is difficult to explain using other bonding theories. For example, it is often difficult to draw a single Lewis structure for an odd-electron molecule like nitrogen monoxide (NO). In Chapter 8, we considered several arrangements of its valence electrons, such as

$$\ddot{\text{N}}=\ddot{\text{O}}\text{:} \qquad \text{:}\ddot{\text{N}}=\ddot{\text{O}}\text{:}$$

We predicted that oxygen was more likely to have a complete octet of valence electrons because it is the more electronegative element. In addition, experimental evidence allowed us to rule out structures with unpaired electrons on the oxygen atom. Our preferred structure was therefore the one shown in red. However, the bond length in NO (115 pm) is considerably shorter than the value in Table 8.2 for an average N=O double bond (122 pm). Molecular orbital theory is useful for explaining both the bonding in NO and the deviation from the expected bond length.

Let's look at the bonding first. The MO diagram for NO is different from the diagrams of homonuclear diatomic gases. Nitrogen and oxygen atoms have different numbers of protons and electrons, and the difference in effective nuclear charge in N and O atoms means that their atomic orbitals have different energies, as Figure 9.50 shows, for the 2s and 2p orbitals.

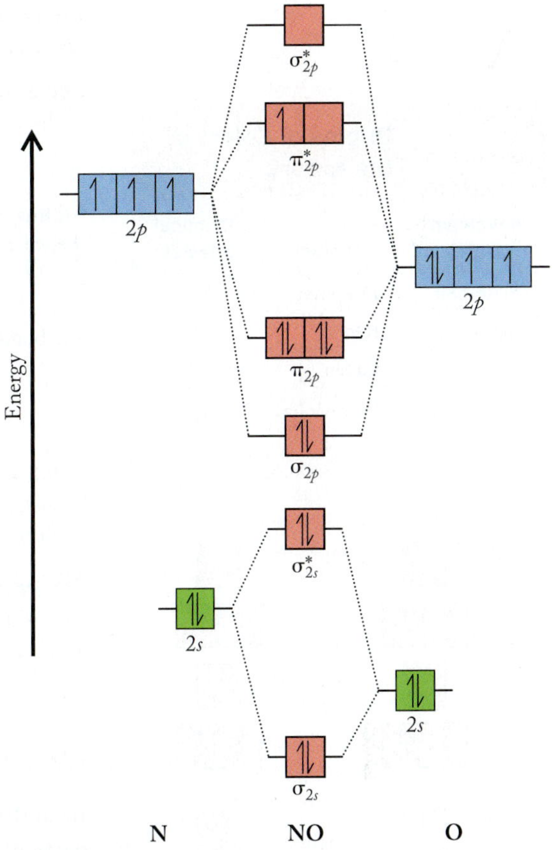

In constructing the MO diagram for NO, the guidelines described previously still apply. The number of MOs formed must equal the number of atomic orbitals combined, and the energy and orientation of the atomic orbitals being mixed must be considered. One additional factor influences the energies of the MOs in heteronuclear diatomic molecules: *bonding* MOs tend to be closer in energy to the atomic orbitals of the more electronegative atom and *antibonding* MOs tend to be closer in energy to the atomic orbitals of the less electronegative atom. The MO diagram for NO in Figure 9.50 illustrates this phenomenon. Note how the energy of the bonding $\sigma_{2s}$ orbital is closer to that of the 2s orbital of the O atom, and the energy of the antibonding $\sigma_{2s}^*$ orbital is closer to that of the 2s orbital of the N atom. Similarly, the $\pi_{2p}$ MOs in NO are closer in energy to the 2p orbitals of oxygen, and the $\pi_{2p}^*$ MOs are closer in energy to the 2p orbitals of nitrogen. The proximity of the nitrogen 2p atomic orbitals to the $\pi_{2p}^*$ MOs means that the single electron in the $\pi_{2p}^*$ MO is more likely to be on nitrogen than on oxygen. This prediction is consistent with our Lewis structure in which the odd electron in NO is on the nitrogen atom.

Molecular orbital theory also enables us to rationalize the relatively short bond length in NO. Equation 9.2 tells us that the bond order is $\frac{1}{2}(8-3)=2.5$, halfway between the bond orders for N=O and N≡O and consistent with a bond length of 115 pm, halfway between the lengths of the N=O bond (122 pm) and the N≡O bond (106 pm).

**FIGURE 9.50** The molecular orbital diagram for NO shows that the unpaired electron occupies a $\pi_{2p}^*$ antibonding orbital, which is closer in energy to the 2p atomic orbitals of nitrogen than to the 2p atomic orbitals of oxygen. As a result of this proximity, the electron has more nitrogen character and is more likely to be located on the nitrogen atom than on the oxygen atom.

---

**SAMPLE EXERCISE 9.11** **Using MO Diagrams for Heteronuclear Diatomic Molecules**  **LO7**

Nitrogen monoxide reacts with many transition metals, including the iron in our blood. In these compounds, NO is sometimes considered to be $NO^+$ and at other times $NO^-$. Use Figure 9.50 to predict the bond order of $NO^+$ and $NO^-$.

**Collect and Organize** We are to predict the bond order of two diatomic ions based on the MO diagram of their parent molecule (Figure 9.50). Equation 9.2 relates bond order to the numbers of electrons in bonding and antibonding orbitals.

| TABLE 9.3 | Origins of Colors in the Aurora | |
|---|---|---|
| Wavelength (nm) | Color | Chemical Species |
| 650–680 | Deep red | $N_2*$ |
| 630 | Red | $O*$ |
| 558 | Green | $O*$ |
| 391–470 | Blue-violet | $N_2^{+*}$ |

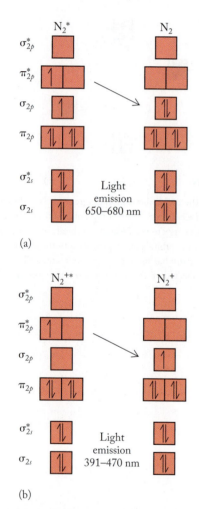

(a)

(b)

**FIGURE 9.51** Molecular orbital diagrams for (a) $N_2$ and (b) $N_2^+$ show electronic transitions that result in the emission of visible light. Collisions with ions in the solar wind result in the promotion of electrons from the $\sigma_{2p}$ orbitals in $N_2$ molecules to $\pi_{2p}^*$ orbitals. Higher-energy collisions create $N_2^{+*}$ molecular ions, which also have electrons in $\pi_{2p}^*$ orbitals. When these electrons return to the ground state, red and blue-violet light are emitted.

**Analyze** NO has 11 valence electrons. We can remove one electron from the MO diagram for NO to get the diagram for $NO^+$ and add one electron to get the diagram for $NO^-$.

**Solve** To generate $NO^+$, we remove the highest-energy electron in NO, which is the one in the $\pi_{2p}^*$ molecular orbital. This gives $NO^+$ the electron configuration

$$NO^+: \quad (\sigma_{2s})^2(\sigma_{2s}^*)^2(\sigma_{2p})^2(\pi_{2p})^4$$

Adding an electron to the lowest-energy MO available in NO (also $\pi_{2p}^*$) yields $NO^-$ with the valence electron configuration

$$NO^-: \quad (\sigma_{2s})^2(\sigma_{2s}^*)^2(\sigma_{2p})^2(\pi_{2p})^4(\pi_{2p}^*)^2$$

The bond orders of the two ions are

$$NO^+: \quad \text{bond order} = \tfrac{1}{2}(8-2) = 3$$
$$NO^-: \quad \text{bond order} = \tfrac{1}{2}(8-4) = 2$$

**Think About It** The bond orders in $N_2$ and $O_2$ are 3 and 2, respectively. The cation $NO^+$ is isoelectronic with $N_2$, so our calculated bond order for it makes sense. The anion $NO^-$ is isoelectronic with $O_2$, and so our calculated bond order for this ion is also reasonable.

**Practice Exercise** Using Figure 9.50 as a guide, draw the MO diagram for carbon monoxide, and determine the bond order for the carbon–oxygen bond.

## Molecular Orbitals of $N_2^+$ and Spectra of Auroras

In addition to predicting the magnetic properties of molecules, MO theory is particularly useful for predicting their spectroscopic properties—and the colors of auroras. In Section 7.3, we learned that the light emitted by excited free atoms is quantized and can be related to the movement of electrons between atomic orbitals. Broadly speaking, the same is true in molecules: electrons can move from one molecular orbital to another by absorbing or emitting light.

We can use this information to look again at the phenomenon described at the opening of this section: how the colors of the aurora are produced. The principal chemical species involved are listed in Table 9.3. An asterisk indicates a molecule or molecular ion in an excited state. Excited $N_2$ molecules produce deep crimson red (650–680 nm) light, and excited $N_2^+$ ions produce blue-violet (391–470 nm) light. The MO diagrams for these species are shown in Figure 9.51. Comparing the MO diagrams of $N_2*$ and $N_2$ in Figure 9.51(a), we find that one of the two electrons originally in the $\sigma_{2p}$ MO in $N_2$ has been raised to a $\pi_{2p}^*$ orbital in $N_2*$, leaving an unpaired $\sigma_{2p}$ electron behind. Figure 9.51(b) shows us that $N_2^{+*}$ also has one electron in a $\pi_{2p}^*$ orbital, but its $\sigma_{2p}$ orbital is empty because the other $\sigma_{2p}$ electron originally in the $N_2$ molecule was lost when the molecule was ionized. As $\pi_{2p}^*$ electrons return from their antibonding, excited-state orbitals to the bonding $\sigma_{2p}$ orbital in the ground state, the distinctive blue-violet and crimson emissions of $N_2^+$ and $N_2$ appear, as shown in the photograph in Figure 9.41.

**CONCEPT TEST**

Are the bond orders of the excited-state species in Figure 9.51 the same as the ground-state species?

In Chapter 8 and in this chapter we have presented several theories of chemical bonding. Each theory has its strengths and weaknesses. The best one to apply in a given situation depends on the question being asked and on the level of sophistication required in the answer. Molecular orbital theory may provide the most complete picture of covalent bonding, but it is also the most difficult to apply to large molecules.

Although we have focused on the small gas-phase molecules found in the atmosphere—nitrogen, oxygen, water, carbon dioxide, methane, ozone, and others—it is important to realize that the principles described in this chapter apply to larger and more complex molecules and ions. We return to the importance of molecular shape, particularly in defining the biological activity of both large and small molecules, in later chapters of this book.

---

**SAMPLE EXERCISE 9.12    Integrating Concepts: Molecules in Space**

Aminoacetonitrile, $NCCH_2NH_2$, was detected in space in 2008. It is a possible precursor for glycine, one of the twenty essential amino acids for life on Earth.
a. Draw a Lewis structure of aminoacetonitrile.
b. Determine the geometry at both carbon atoms and the hybridization scheme that best explains the geometry.
c. Aminoacetonitrile reacts with water to produce glycine and ammonia by the balanced chemical equation:

$$NCCH_2NH_2 + 2\,H_2O \rightarrow HOC(\!=\!O)CH_2NH_2 + NH_3$$

How many bonds of each type are broken and formed in the reaction?
d. Use the data for average bond energies in Table 8.3 to estimate the value of $\Delta H_{rxn}$ for the reaction. Assume the reaction occurs in the gas phase.

**Collect and Organize** We are given the formula of aminoacetonitrile and a balanced chemical equation describing its reaction with water. We are asked four questions about the bonding in the molecule and to use this information to estimate $\Delta H_{rxn}$.

**Analyze** The connectivity of the atoms in aminoacetonitrile is implied by the order in which they are written: N—C—C—N with hydrogen atoms bonded to the right-hand carbon and nitrogen atoms. Using the steps outlined in Section 8.2, we can draw the Lewis structure for the molecule and determine how many and which types of bonds are present in any molecule. Sections 9.2 and 9.4 provide guidelines for determining the molecular geometry about an atom and for assessing the hybrid orbitals that are consistent with those geometries. Estimating $\Delta H_{rxn}$ values from bond energies is described in Section 8.8.

**Solve**
a. A total of 22 valence electrons in $NCCH_2NH_2$ are distributed over eight atoms. In the following Lewis structure, each H has a complete duet and the remaining atoms have complete octets:

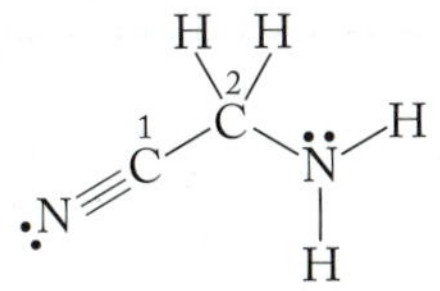

There are no resonance forms for this molecule.
b. The carbon labeled "1" (C1) has two bonded atoms and no lone pairs, while carbon "2" (C2) has four bonded atoms and no lone pairs. Based on Table 9.1, this means that the geometry about C1 is linear and the geometry about C2 is tetrahedral.

C1 forms two $sp$ hybrid orbitals that are used in two σ bonds: one to C2 and one to the nitrogen atom. The remaining $p$ orbitals on C1 are used to form the two π bonds to nitrogen.

C2 forms four $sp^3$ hybrid orbitals, which are used to form four σ bonds.
c. To determine which bonds are broken, we need to draw Lewis structures for the remaining reactants and products. The Lewis structures for water and ammonia were drawn in Chapter 8. Applying the steps in Section 8.2, we draw the Lewis structure for glycine as well. Taken together, we can write the balanced chemical equation as follows, using Lewis structures:

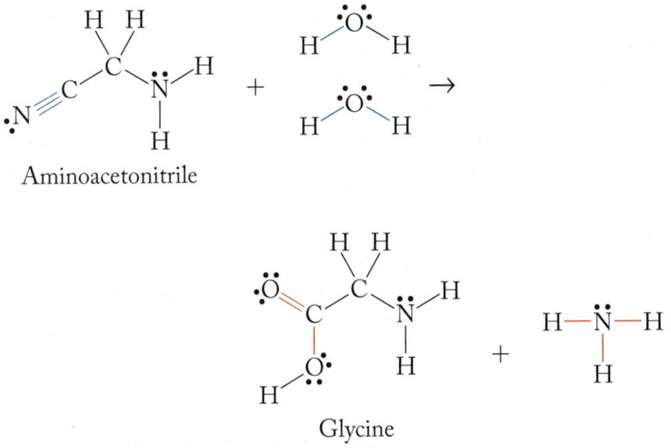

Aminoacetonitrile

Glycine

—— = bonds broken    —— = bonds formed

From the balanced equation we see that one C≡N bond and three O—H bonds (in blue) are broken, accompanied by the formation of three N—H bonds, one C=O, and one C—O bond (red).
d. We can estimate $\Delta H$ for the reaction by subtracting the sum of the energies of the bonds formed from the sum of the energies of the bonds broken:

$$\Delta H = \sum \Delta H_{broken} - \sum \Delta H_{formed}$$

$$= [3(\text{N—H}) + (\text{C}{=}\text{O}) + (\text{C—O})] - [(\text{C}{\equiv}\text{N}) + 3(\text{O—H})]$$

Substituting the values for average bond energies from Table 8.2:

$$\Delta H = [3(163\ \text{kJ/mol}) + (743\ \text{kJ/mol}) + (358\ \text{kJ/mol})]$$
$$- [(891\ \text{kJ/mol}) + 3(463\ \text{kJ/mol})]$$

$$= 1590\ \text{kJ/mol} - 2280\ \text{kJ/mol}$$

$$= -690\ \text{kJ/mol for aminoacetonitrile}$$

**Think About It** If we draw the Lewis structures for the reactants and products in a chemical reaction, we can easily see which bonds are broken and which new bonds are formed. This simplifies our calculation of $\Delta H$ because we do not need to account for bonds that stay intact throughout the reaction.

# Chalcogens: From Alcohol to Asparagus, the Nose Knows

Oxygen, sulfur, and the other group 16 elements of the periodic table are called *chalcogens*. Their atoms have valence-shell electron configurations $ns^2np^4$. Consequently, the formation of molecules that have two bonds on each atom determines the common chemical properties of these elements, since forming two bonds fills their valence octets.

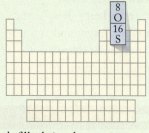

We have already discussed oxygen in comparison to its neighbor fluorine to the right in the periodic table. In this section, we compare oxygen to the element directly below it in the periodic table, sulfur. A significant difference between the two elements is illustrated in some of the compounds discussed in this chapter. In molecular compounds, oxygen tends to form two covalent bonds. Sulfur does too, but it also can expand its octet to accommodate five or even six electron pairs in its valence shell. Hence the bonding patterns and molecular shapes of sulfur compounds show greater variability. This, in turn, affects the properties of the compounds.

Compounds of hydrogen with oxygen, sulfur, and the other group 16 elements provide interesting contrasts with respect to molecular shape and properties. The data in the table below show how different water is from the other compounds. It is a liquid at ordinary temperatures and pressures; it has a large negative heat of formation; and it has a much larger bond angle than the other three hydrides. We also know that water is odorless and is absolutely essential for life. The hydrides of sulfur (S), selenium (Se), and tellurium (Te) are all gases under standard conditions; have bond angles close to 90°; and

are foul-smelling and poisonous. Hydrogen sulfide is responsible for the smell of rotten eggs. It is especially dangerous because it tends to very quickly fatigue the nasal sensory sites responsible for detecting it. This means that the intensity of the odor is a very poor indicator of the concentration of $H_2S$ in the air. Headache and nausea begin at air concentrations of $H_2S$ as low as 5 ppm, and at 100 ppm paralysis and death result.

Similarly, compounds of sulfur have properties that differ from those of their oxygen-containing counterparts, and many of them have characteristic odors. Methanol is an alcohol with the formula $CH_3—OH$. It is a liquid at room temperature and has an odor usually described as slightly alcoholic. Methanethiol, $CH_3—SH$, is a gas at room temperature and has the pungent odor of rotten cabbage. It is produced in the intestinal tract of animals by the action of bacteria on proteins and is one of the sulfur compounds responsible for the characteristic aroma of a feedlot or a barnyard.

If we go up one more carbon unit in size, the oxygen-containing compound is a liquid at room temperature called ethanol (beverage grade alcohol), $CH_3—CH_2—OH$. The corresponding sulfur compound is a very low-boiling liquid at room temperature called ethanethiol, which has a penetrating and unpleasant odor, like very powerful green onions. The human nose can detect the presence of ethanethiol at levels as low as 1 ppb (part per billion) in the air. This gives rise to its use as an odorant in natural gas. Natural gas has no odor, and natural gas leaks are such enormous fire hazards that ethanethiol is added to natural gas streams to make leaks immediately detectable.

If we rearrange the atoms in ethanol and ethanethiol, we produce two new compounds. In the case of ethanol we get dimethyl ether,

| Hydride[a] | Melting Point (°C) | Boiling Point (°C) | Bond Length (pm) | Bond Angle (degrees) | Heat of Formation (kJ/mol) |
|---|---|---|---|---|---|
| $H_2O$ | 0 | 100 | 96 | 104.5 | −285.8 |
| $H_2S$ | −86 | −60 | 134 | 92 | −20.17 |
| $H_2Se$ | −66 | −41 | 146 | 91 | 73.0 |
| $H_2Te$ | −51 | −4 | 169 | 90 | 99.6 |

[a]$H_2Po$ is excluded; too little is known of its chemistry. Polonium has no stable isotopes and is present on Earth only in very small quantities.

$CH_3$—O—$CH_3$, a colorless gas used in refrigeration systems. Its counterpart, dimethyl sulfide, $CH_3$—S—$CH_3$, is one of the compounds responsible for the "low-tide" smell of ocean shorelines.

Three of the sulfur compounds described—hydrogen sulfide, methanethiol, and dimethyl sulfide—are referred to as volatile sulfur compounds (VSCs) by dentists. They are produced by bacteria in the mouth and are the principal compounds responsible for bad breath. One of the reasons the odors of these compounds differ from those of their oxygen counterparts is that their molecular sizes and shapes are slightly different. Also, their polarities differ because of the electronegativity difference between oxygen and sulfur. At the beginning of this chapter, we discussed the importance of molecular shape in determining the extent of interaction of a compound with receptors in nasal membranes. In part, the vast differences in odor and sensory detectability of these compounds are due to their shapes and electron distributions.

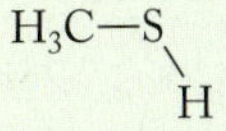

Not all sulfur compounds have an odor, but many odiferous compounds do contain sulfur. The characteristic and unpleasant smell of urine produced by some people after eating asparagus results from the inability of their bodies to convert odiferous sulfur compounds (Figure 9.52) into odor-free sulfate ions. Not all people are able to convert the sulfur compounds in asparagus to sulfate, and not all people are able to smell the odiferous sulfur compounds. Apparently, genetic differences determine how we metabolize these compounds and how well we can sense their odors.

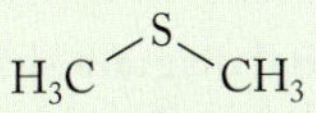

The odor of skunk is due mostly to butanethiol (Figure 9.53), and the odor of well-used athletic shoes is primarily due to the presence of sulfur compounds produced by bacteria. Not all sulfur compounds have aromas as unpleasant as these, however. A compound with the formula $C_{10}H_{18}S$ is responsible for the aroma of grapefruit. If the orientation of two atoms on one of the carbon atoms in the molecule is switched, the resulting molecule has the same Lewis structure but now has no aroma at all.

$H_3C$—S
        $\backslash$
         H

Methanethiol

$H_3C$—S—$CH_3$

Dimethyl sulfide

$H_3C$—S—S—$CH_3$

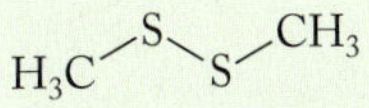

Dimethyl disulfide

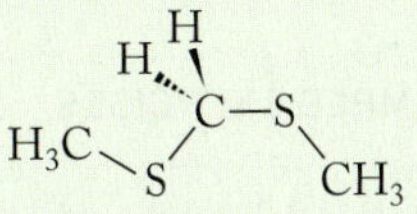

Bis(methylthio)methane

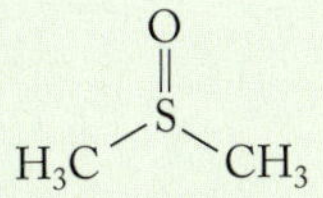

Dimethyl sulfoxide

**FIGURE 9.52** Structures of some of the volatile sulfur compounds responsible for the smell of "asparagus" urine. Compounds shown here toward the top have stronger (and more unpleasant) odors.

**FIGURE 9.53** The pungent smell of skunk spray is due to butanethiol, $CH_3(CH_2)_3SH$.

## SUMMARY

**Learning Outcome 1** Minimizing repulsion between pairs of valence electrons (the **VSEPR** model) results in the lowest-energy orientations of bonding and nonbonding electron pairs and accounts for the observed **molecular geometries** of molecules. (Section 9.2)

**Learning Outcome 2** The shape of a molecule reflects the arrangement of the atoms in three-dimensional space and is determined largely by characteristic **bond angles**. (Section 9.2)

**Learning Outcome 3** Two covalently bonded atoms with different electronegativities have partial electrical charges of opposite sign, creating a **bond dipole**. If the individual bond dipoles in a molecule do not offset each other, the molecule is polar. If they do offset each other, the molecule is nonpolar. (Section 9.3)

**Learning Outcome 4** In **valence bond theory**, the **overlap** of atomic orbitals results in covalent bonds between pairs of atoms in molecules. Molecular geometry is explained by the mixing, or **hybridizing**, of atomic orbitals to create **hybrid atomic orbitals**. (Section 9.4)

**Learning Outcome 5** Molecules with alternating single and double bonds are stabilized by electron delocalization over the system. (Section 9.5)

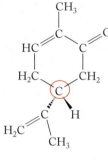

**Learning Outcome 6** **Chiral** molecules exist in left- and right-handed forms that have different properties. Many contain an $sp^3$ hybridized carbon atom with four different groups attached. (Section 9.6)

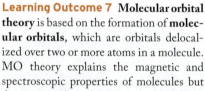

**Learning Outcome 7** **Molecular orbital theory** is based on the formation of **molecular orbitals**, which are orbitals delocalized over two or more atoms in a molecule. MO theory explains the magnetic and spectroscopic properties of molecules but does not explain their shapes. A **molecular orbital diagram** shows relative energies of the molecular orbitals in a molecule. (Section 9.7)

## PROBLEM-SOLVING SUMMARY

| TYPE OF PROBLEM | CONCEPTS AND EQUATIONS | SAMPLE EXERCISES |
|---|---|---|
| **Predicting molecular geometry** | Draw a Lewis structure for the molecule. Determine the steric number (SN) of the central atom, where $$SN = \left(\begin{array}{c}\text{number of atoms}\\\text{bonded to central atom}\end{array}\right) + \left(\begin{array}{c}\text{number of lone pairs}\\\text{on central atom}\end{array}\right) \quad (9.1)$$ Choose a geometry that minimizes repulsion between electron pairs. | 9.1, 9.3 |
| **Predicting relative sizes of bond angles** | Lone pairs on a central atom push bonded atoms closer together, decreasing bond angles. | 9.2 |
| **Predicting polarity of a substance** | Assign the direction of polarity to each bond dipole and use molecular geometry to determine whether the dipoles offset each other. | 9.4 |
| **Identifying overlapping orbitals in a molecule** | Identify the partially filled atomic orbitals on the atoms. | 9.5 |
| **Describing bonding in molecules and the shape of molecules using hybrid orbitals** | Identify the hybrid orbitals in molecules that result from mixing different numbers of $s$, $p$, and $d$ orbitals that result in the observed molecular geometry: $s + p$ = two $sp$ hybrid orbitals $s + $ two $p$ = three $sp^2$ hybrid orbitals $s + $ three $p$ = four $sp^3$ hybrid orbitals $s + $ three $p + d$ = five $sp^3d$ hybrid orbitals $s + $ three $p + $ two $d$ = six $sp^3d^2$ hybrid orbitals | 9.6, 9.7, 9.8 |
| **Using MO diagrams to predict bond order** | $$\text{Bond order} = \tfrac{1}{2}\left(\begin{array}{c}\text{number of}\\\text{bonding electrons}\end{array} - \begin{array}{c}\text{number of}\\\text{antibonding electrons}\end{array}\right) \quad (9.2)$$ | 9.9, 9.10, 9.11 |

## VISUAL PROBLEMS

*(Answers to boldface end-of-chapter questions and problems are in the back of the book.)*

**9.1.** Two compounds with the same formula, $S_2F_2$, have been isolated. The structures in Figure P9.1 show the arrangements of the atoms in these different compounds. Can these two compounds be distinguished by their dipole moments?

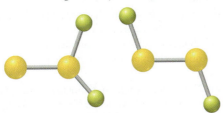

**FIGURE P9.1**

**9.2.** Could you distinguish between the two structures of $N_2H_2$ shown in Figure P9.2 by the magnitude of their dipole moments?

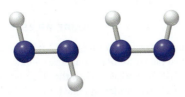

**FIGURE P9.2**

**9.3.** Which of the molecules shown in Figure P9.3 are planar, that is, all atoms are in a single plane? Are there delocalized $\pi$ electrons in any of these molecules?

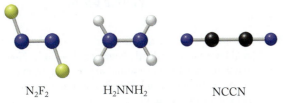

$N_2F_2$      $H_2NNH_2$      NCCN

**FIGURE P9.3**

**9.4.** Which of the molecules shown in Figure P9.4 is *not* planar? Are there delocalized $\pi$ electrons in any of these molecules?

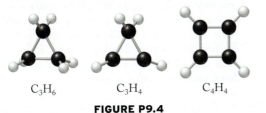

$C_3H_6$      $C_3H_4$      $C_4H_4$

**FIGURE P9.4**

**9.5.** Use the MO diagram in Figure P9.5 to predict whether $O_2^+$ has more or fewer electrons in antibonding molecular orbitals than $O_2^{2+}$.

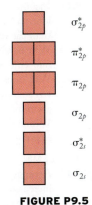

**FIGURE P9.5**

**9.6.** Under appropriate conditions, $I_2$ can be oxidized to $I_2^+$, which is bright blue. The corresponding anion, $I_2^-$, is not known. Use the molecular orbital diagram in Figure P9.6 to explain why $I_2^+$ is more stable than $I_2^-$.

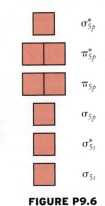

**FIGURE P9.6**

**9.7.** The molecular geometry of $ReF_7$ is an uncommon structure called a pentagonal bipyramid, which is shown in Figure P9.7. What are the bond angles in a pentagonal bipyramid?

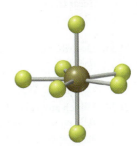

**FIGURE P9.7**

**\*9.8.** The molecular geometry of the transition metal-containing anions $MF_8^{2-}$ (M = Mo or W) is not known. What would be the F—M—F bond angles in $MF_8^{2-}$ if the anion has a cubic geometry, as shown in Figure P9.8?

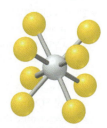

**FIGURE P9.8**

## QUESTIONS AND PROBLEMS

### Molecular Shape; Valence-Shell Electron-Pair Repulsion Theory (VSEPR)

#### CONCEPT REVIEW

**9.9.** Why is the shape of a molecule determined by repulsions between electron pairs and not by repulsions between nuclei?

**9.10.** Do all resonance forms of a molecule have the same molecular geometry? Explain your answer.

**9.11.** How can $SO_3$ and $BF_3$ have different numbers of bonds but the same trigonal planar geometry?

9.12. Account for the range of bond angles from less than 100° to 180° in triatomic molecules.

9.13. In a molecule of ammonia, why is the repulsion between the lone pair and a bonding pair of electrons on nitrogen greater than the repulsion between two N—H bonding pairs?

9.14. Why is it important to draw a correct Lewis structure for a molecule before predicting its geometry?

9.15. Why does the seesaw structure have lower energy than a trigonal pyramidal structure derived by removing an axial atom from a trigonal bipyramidal $AB_5$ molecule?

*9.16. Which geometry do you predict will have lower energy: a square pyramid or a trigonal bipyramid?

## PROBLEMS

9.17. Arrange the following molecular geometries in order of increasing bond angle: (a) trigonal planar; (b) octahedral; (c) tetrahedral.

9.18. Arrange the following molecular geometries in order of increasing bond angle: (a) seesaw; (b) tetrahedral; (c) square pyramidal.

9.19. Which of the molecular geometries discussed in this chapter have more than one characteristic bond angle?

*9.20. Which molecular geometries for molecules of the general formula $AB_x$ ($x = 2$ to 6) discussed in this chapter have the same bond angles when lone pairs replace one or more atoms?

9.21. Which of the following molecular geometries does not lead to linear triatomic molecules after removing one or more atoms? (a) tetrahedral; (b) octahedral; (c) T-shaped

9.22. Which of the following molecular geometries does not lead to linear triatomic molecules after removing one or more atoms? (a) trigonal bipyramidal; (b) seesaw; (c) trigonal planar

*9.23. Describe the molecular geometries that result from replacing one atom with a lone pair of electrons in an $AB_7$ molecule with a pentagonal bipyramidal geometry. (See Figure P9.7 for the shape of a pentagonal bipyramid.)

*9.24. Which atoms would you have to remove from the cubic $AB_8$ molecule to create a geometry that approximates an octahedron? (See Figure P9.8 for the shape of a cubic molecule.)

9.25. Determine the molecular geometries of the following molecules: (a) $GeH_4$; (b) $PH_3$; (c) $H_2S$; (d) $CHCl_3$.

9.26. Determine the molecular geometries of the following molecules and ions: (a) $NO_3^-$; (b) $NO_4^{3-}$; (c) $S_2O$; (d) $NF_3$.

9.27. Determine the bond angles in the following ions: (a) $NH_4^+$; (b) $SO_3^{2-}$; (c) $NO_2^-$; (d) $XeF_5^+$.

9.28. Determine the bond angles in the following ions: (a) $SCN^-$; (b) $BF_2^+$; (c) $ICl_2^-$; (d) $PO_3^{3-}$.

9.29. Determine the geometries of the following ions and molecules: (a) $S_2O_3^{2-}$; (b) $PO_4^{3-}$; (c) $NO_3$; (d) NCO.

9.30. Determine the geometries of the following molecules: (a) $ClO_2$; (b) $ClO_3$; (c) $IF_3$; (d) $SF_4$.

9.31. Which of the following triatomic molecules, $O_3$, $SO_2$, $N_2O$, $S_2O$, and $CO_2$, have the same molecular geometry?

9.32. Which of the following species, $N_3^-$, $O_3$, $CO_2$, $SCN^-$, $CNO^-$, and $NO_2^-$, have the same molecular geometry?

9.33. The anion $C(CN)_3^-$ is found to have a trigonal planar geometry about the central carbon atom. Draw Lewis structures for $C(CN)_3^-$, including resonance forms, and determine which structure contributes the most to the bonding picture.

9.34. The anion $C(NO_2)_3^-$ also has a trigonal planar geometry about the carbon atom. Draw Lewis structures for $C(NO_2)_3^-$, including resonance forms, and determine which structure contributes the most to the bonding picture.

9.35. The N—C bond angles in tri(methyl)amine, $N(CH_3)_3$, are approximately 109° while the N—Si bond angles in tri(silyl) amine, $N(SiH_3)_3$, are 120°. Explain the change in geometry when Si substitutes for C in this amine.

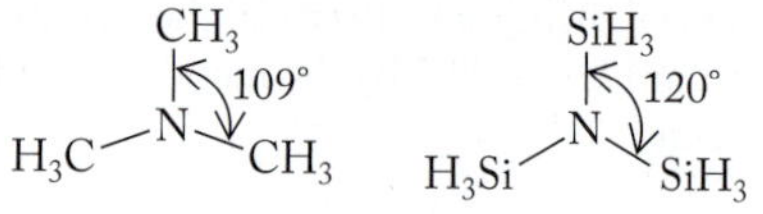

**FIGURE P9.35**

*9.36. The geometry about nitrogen in $N(CF_3)_3$ and $N(SCF_3)_3$ is trigonal planar for both complexes. Draw Lewis structures for each that are consistent with the observed geometry. (*Hint*: For $N(CF_3)_3$ consider an ionic form $[(CF_3)_2NCF_2]^+[F]^-$.)

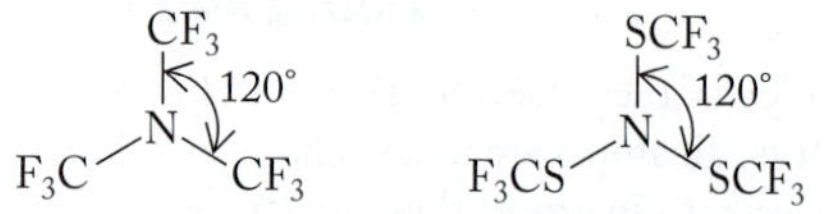

**FIGURE P9.36**

*9.37. For many years, it was believed that the noble gases could not form covalently bonded compounds. However, xenon reacts with fluorine and oxygen. Reaction between xenon tetrafluoride and fluoride ions produces the pentafluoroxenate anion:

$$XeF_4 + F^- \rightarrow XeF_5^-$$

Draw Lewis structures for $XeF_4$ and $XeF_5^-$, and predict the geometry around xenon in $XeF_4$. The crystal structure of $XeF_5^-$ compounds indicates a pentagonal bipyramidal orientation of valence pairs around Xe. Sketch the structure for $XeF_5^-$.

*9.38. The first compound containing a xenon–sulfur bond was isolated in 1998. Draw a Lewis structure for HXeSH and determine its molecular geometry.

*9.39. The Cl–O distances in $ClO_2^+$, $ClO_2$, and $ClO_2^-$ are found to be 131 pm, 147 pm, and 156 pm, respectively. The corresponding O—Cl—O bond angles are 122°, 118°, and 110°. Draw Lewis structures consistent with these data.

9.40. Complete the Lewis structures of $SCNCl_3$ in Figure P9.40. Is the geometry around nitrogen the same in both molecules?

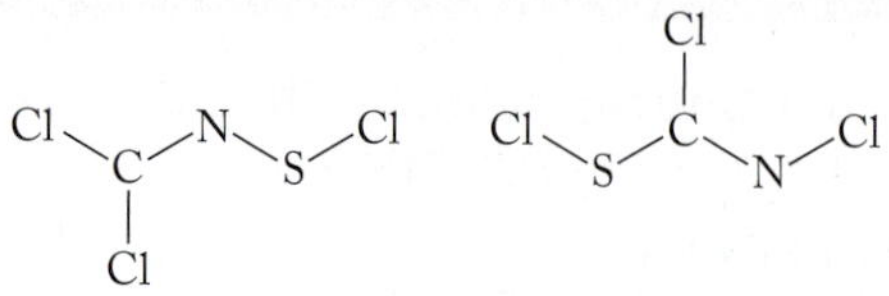

**FIGURE P9.40**

## Polar Bonds and Polar Molecules

### CONCEPT REVIEW

**9.41.** Explain the difference between a polar bond and a polar molecule.

**9.42.** Must a polar molecule contain polar covalent bonds? Why or why not?

**9.43.** Can a nonpolar molecule contain polar covalent bonds?

**9.44.** What does a dipole moment measure?

### PROBLEMS

**9.45.** Consider the following molecules: (a) $CCl_4$; (b) $CHCl_3$; (c) $CO_2$; (d) $H_2S$; (e) $SO_2$
  a. Which of them contain polar bonds?
  b. Which are polar molecules?
  c. Which are nonpolar molecules?

**9.46.** Simple diatomic molecules detected in interstellar space include CO, CS, SiO, SiS, SO, and NO. Arrange these molecules in order of increasing dipole moment based on the location of the constituent elements in the periodic table, and then calculate the electronegativity differences from the data in Figure 8.5.

---

**9.47.** **Freon Ban** Compounds containing carbon, chlorine, and fluorine are known as Freons or chlorofluorocarbons (CFCs). Widespread use of these substances was banned because of their effect on the ozone layer in the upper atmosphere. Which of the following CFCs are polar and which are nonpolar? (a) Freon 11 ($CFCl_3$); (b) Freon 12 ($CF_2Cl_2$); (c) Freon 113 ($Cl_2FCCF_2Cl$)

**9.48.** Which of the following chlorofluorocarbons (CFCs) are polar and which are nonpolar? (a) Freon C318 ($C_4F_8$, cyclic structure); (b) Freon 1113 ($C_2ClF_3$); (c) $Cl_2HCCClF_2$

---

**9.49.** Which molecule in each of the following pairs has the larger dipole moment?

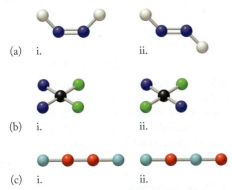

(a)  i.  ii.

(b)  i.  ii.

(c)  i.  ii.

**FIGURE P9.49**

**9.50.** Which molecule in each of the following pairs has the larger dipole moment? (a) $BF_3$ or $BCl_3$; (b) $BCl_2F$ or $BClF_2$

---

**9.51.** Aluminum chloride has the Lewis structure shown in Figure P9.51.
  a. What is the geometry about the aluminum?
  b. Is $Al_2Cl_6$ polar or nonpolar?

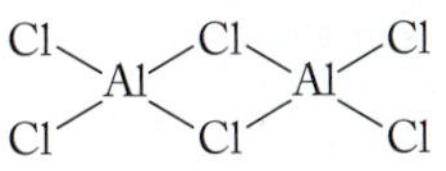

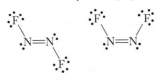

**FIGURE P9.51**

**9.52.** Nitrogen trifluoride, $NF_3$, is used in the electronics industry to clean surfaces. $NF_3$ is also a potent greenhouse gas.
  a. Draw the Lewis structure of $NF_3$ and determine its molecular geometry.
  b. $BF_3$ and $NF_3$ both have three covalently bonded fluorine atoms around a central atom. Do they have the same dipole moment?
  c. Could $BF_3$ also behave as a greenhouse gas?

## Valence Bond Theory

### CONCEPT REVIEW

**9.53.** Describe in your own words the differences between sigma and pi bonds.

**9.54.** Why aren't the orbitals on free atoms hybridized?

**9.55.** Do all resonance forms of $N_2O$ have the same hybridization at the central N atom?

*****9.56.** What combination of $s$, $p$, and $d$ orbitals would we need to form four $\sigma$ and two $\pi$ bonds to a sulfur atom?

### PROBLEMS

**9.57.** What is the hybridization of nitrogen in each of the following ions and molecules? (a) $NO_2^+$; (b) $NO_2^-$; (c) $N_2O$; (d) $N_2O_5$; (e) $N_2O_3$

**9.58.** What is the hybridization of sulfur in each of the following molecules? (a) SO; (b) $SO_2$; (c) $S_2O$; (d) $SO_3$

---

**9.59.** **Airbags** Azides such as sodium azide, $NaN_3$, are used in automobile airbags as a source of nitrogen gas. Another compound with three nitrogen atoms bonded together is $N_3F$. What differences are there in the arrangement of the electrons around the nitrogen atoms in the azide ion ($N_3^-$) and $N_3F$? Is there a difference in the hybridization of the central nitrogen atom?

**9.60.** $N_3F$ decomposes to nitrogen and $N_2F_2$ by the following reaction:

$$2\,N_3F \rightarrow 2\,N_2 + N_2F_2$$

$N_2F_2$ has two possible structures as shown in Figure P9.60. Are the differences between these structures related to differences in the hybridization of nitrogen in $N_2F_2$? Identify the hybrid orbitals that account for the bonding in $N_2F_2$. Are they the same as those in acetylene, $C_2H_2$?

**FIGURE P9.60**

---

**9.61.** How does the hybridization of the sulfur atom change in the series $SF_2$, $SF_4$, $SF_6$?

**9.62.** How does the hybridization of the central atom change in the series $CO_2$, $NO_2$, $O_3$, and $ClO_2$?

---

**9.63.** **Minoxidil** The drug minoxidil was originally developed for treating high blood pressure but is now used primarily for treating hair loss. The Lewis structure of minoxidil is shown in Figure P9.63. Complete the Lewis structure by adding lone pairs where needed. Assign formal charges to the nitrogen and oxygen highlighted in red. Describe the bonding around the nitrogen in the N–O group.

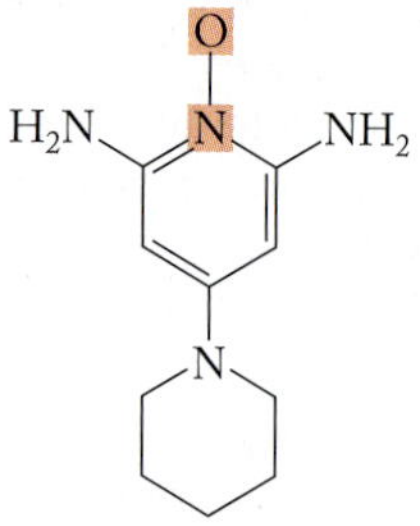

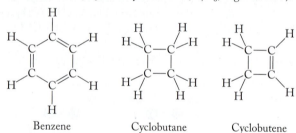

**FIGURE P9.63**

*9.64. Draw the Lewis structure of the chlorite ion, $ClO_2^-$, which is used as a bleaching agent. Include all resonance structures in which formal charges are closest to zero. What is the shape of the ion? Suggest a hybridization scheme for the central chlorine atom that accounts for the structures you have drawn.

*9.65. **Perchlorate Ion and Human Health** Perchlorate ion adversely affects human health by interfering with the uptake of iodine in the thyroid gland, but because of this behavior, it also provides a useful medical treatment for hyperthyroidism, or overactive thyroid. Draw the Lewis structure of the perchlorate ion, $ClO_4^-$. Include all resonance structures in which formal charges are closest to zero. What is the shape of the ion? Suggest a hybridization scheme for the central chlorine atom that accounts for this shape.

9.66. Draw a Lewis structure for $CF_3PCF_2$ where the fluorines are all bonded to C. Determine its molecular geometry and the hybridization of the phosphorus atom.

9.67. Synthesis of the first compound of argon was reported in 2000. HArF was made by reacting Ar with HF. Draw a Lewis structure for HArF, and determine the hybridization of Ar in this molecule.

9.68. The Lewis structure of $N_4O$, with the skeletal structure O—N—N—N—N, contains one N—N single bond, one N=N double bond, and a N≡N triple bond. Is the hybridization of all the nitrogen atoms the same?

*9.69. The trifluorosulfate anion was isolated in 1999 as the tetramethylammonium salt $[(CH_3)_4N]^+[SO_2F_3]^-$.
   a. Determine the geometry around the nitrogen atom in the cation and describe the C—N bonding according to valence bond theory.
   b. The S—O bond lengths in the anion are both 143 pm. Draw the Lewis structure that is consistent with this bond length.
   c. What is the molecular geometry of the anion?

9.70. **Treating Diabetes** The drug metformin (Figure P9.70) has been used to treat type 2 diabetes for a half-century by suppressing glucose production. Metformin contains five nitrogen atoms. Determine the geometry around each

nitrogen atom, and describe the bonding according to valence bond theory.

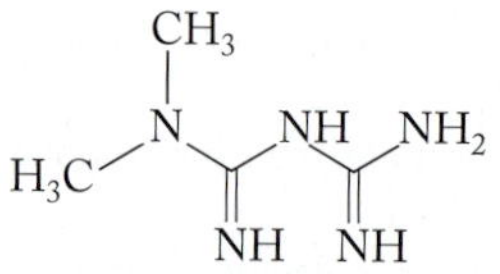

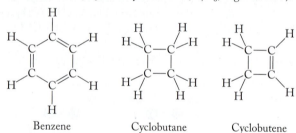

**FIGURE P9.70**

## Shape and Interactions with Large Molecules; Chirality and Molecular Recognition

### CONCEPT REVIEW

9.71. Can molecules with more than one central atom have resonance forms?

*9.72. Can hybrid orbitals be associated with more than one atom?

*9.73. Are resonance structures examples of electron delocalization? Explain your answer.

9.74. Can $sp^2$ and $sp$ hybridized carbon atoms be chiral centers?

9.75. Which of the following objects are chiral? (a) a baseball bat with no lettering on it; (b) a pair of scissors; (c) a boot; (d) a fork

9.76. Why is it difficult to assign a single geometry to a molecule with more than one central atom?

### PROBLEMS

9.77. Cyclic structures exist for many compounds of carbon and hydrogen. Describe the molecular geometry and hybridization around each carbon atom in benzene ($C_6H_6$), cyclobutane ($C_4H_8$), and cyclobutene ($C_4H_6$; Figure P9.77).

Benzene    Cyclobutane    Cyclobutene
**FIGURE P9.77**

9.78. What is the molecular geometry around sulfur and nitrogen in the sulfamate anion shown in Figure P9.78? Which atomic or hybrid orbitals overlap to form the S—O and S—N bonds in the sulfamate anion?

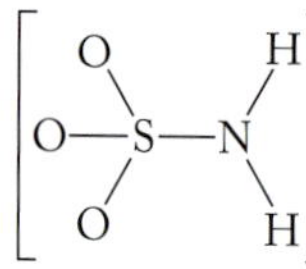

**FIGURE P9.78**

9.79. **Artificial Sweeteners** Acesulfame potassium is one of many artificial sweeteners used in food. It is 200 times sweeter than sugar and has the structure shown in Figure P9.79. What is the geometry at each of the atoms in the six-membered ring? Which atomic or hybrid orbitals overlap to form the C—O and C—N bonds? In which atomic or hybrid orbital is the extra electron on N located?

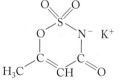

**FIGURE P9.79**

*9.80. Saccharine (Figure P9.80) was the first artificial sweetener, discovered in 1879. Like acesulfame potassium, it contains a sulfur atom adjacent to a nitrogen atom. Why don't all the atoms in the five-membered ring lie in the same plane? Why is it difficult to explain the bonding between S and O using the hybrid orbitals described in Section 9.4?

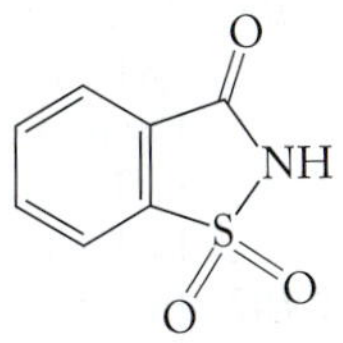

**FIGURE P9.80**

9.81. Which molecules in Figure P9.81 are chiral?

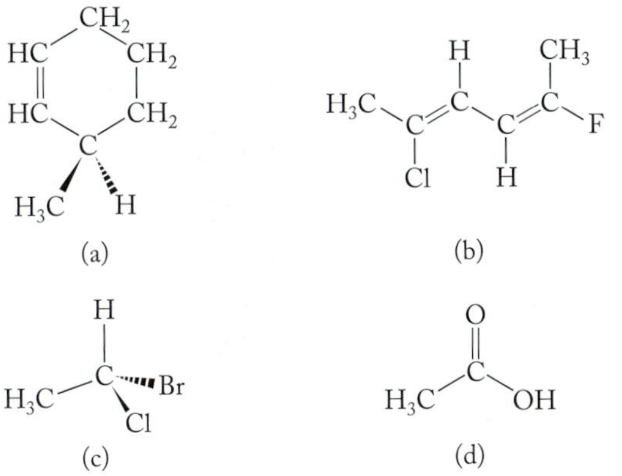

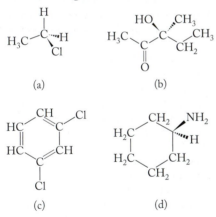

(a)

(b)

(c)

(d)

**FIGURE P9.81**

9.82. Which molecules in Figure P9.82 are chiral?

(a)

(b)

(c)

(d)

**FIGURE P9.82**

## Molecular Orbital Theory

### CONCEPT REVIEW

9.83. Do all $\sigma$ molecular orbitals result from the overlap of $s$ atomic orbitals?

9.84. Do all $\pi$ molecular orbitals result from the overlap of $p$ atomic orbitals?

9.85. Are $s$ atomic orbitals with different principal quantum numbers ($n$) as likely to overlap and form MOs as $s$ atomic orbitals with the same value of $n$?

9.86. Which atomic orbitals are more likely to mix to form a set of molecular orbitals—a $2s$ and a $3p$ orbital or a $4s$ and a $5p$ orbital?

9.87. Why might some molecules with even numbers of valence electrons be paramagnetic?

9.88. Which better explains the magnetic properties of a diatomic molecule: valence bond theory or molecular orbital theory?

### PROBLEMS

9.89. Make a sketch showing how two $1s$ orbitals overlap to form a $\sigma_{1s}$ bonding molecular orbital and a $\sigma_{1s}^*$ antibonding molecular orbital.

9.90. Make a sketch showing how two $2p_y$ orbitals overlap "sideways" to form a $\pi_{2p}$ bonding molecular orbital and a $\pi_{2p}^*$ antibonding molecular orbital.

9.91. Consider the following molecular ions: $N_2^+$, $O_2^+$, $C_2^+$, and $Br_2^{2-}$. Using MO theory,
   a. write their orbital electron configuration.
   b. predict their bond orders.
   c. do you expect any of these species to exist?

9.92. Diatomic noble gas molecules, such as $He_2$ and $Ne_2$, do not exist.
   a. Write their orbital electron configurations.
   b. Does removing one electron from each of these molecules create molecular ions ($He_2^+$ and $Ne_2^+$) that are more stable than $He_2$ and $Ne_2$?

9.93. Which of the following molecular ions is expected to have one or more unpaired electrons? (a) $N_2^+$; (b) $O_2^+$; (c) $C_2^{2+}$; (d) $Br_2^{2-}$; (e) $O_2^-$; (f) $O_2^{2-}$; (g) $N_2^{2-}$; (h) $F_2^+$

9.94. Which of the following molecular ions have electrons in $\pi$ antibonding orbitals? (a) $O_2^-$; (b) $O_2^{2-}$; (c) $N_2^{2-}$; (d) $F_2^+$; (e) $N_2^+$; (f) $O_2^+$; (g) $C_2^{2+}$; (h) $Br_2^{2+}$

9.95. The odd-electron molecule ClO is implicated in the atmospheric chemistry of chlorofluorocarbons as illustrated by the reaction (where the * indicates an excited-state oxygen atom):

$$CF_2Cl_2 + O^* \rightarrow ClO + CF_2Cl$$

Draw a molecular orbital diagram for ClO. Is the odd electron in a bonding or antibonding orbital?

9.96. The elusive molecule boron monoxide, BO, can be stabilized by bonding to platinum. Draw a molecular orbital diagram for BO. Is the odd electron in a bonding or antibonding orbital?

9.97. For which of the following diatomic molecules does the bond order increase with the gain of two electrons, forming the corresponding anion with a 2− charge?
   a. $B_2 + 2\,e^- \rightarrow B_2^{2-}$
   c. $N_2 + 2\,e^- \rightarrow N_2^{2-}$
   b. $C_2 + 2\,e^- \rightarrow C_2^{2-}$
   d. $O_2 + 2\,e^- \rightarrow O_2^{2-}$

9.98. For which of the following diatomic molecules does the bond order increase with the loss of two electrons, forming the corresponding cation with a 2+ charge?
   a. $B_2 \rightarrow B_2^{2+} + 2\,e^-$
   c. $N_2 \rightarrow N_2^{2+} + 2\,e^-$
   b. $C_2 \rightarrow C_2^{2+} + 2\,e^-$
   d. $O_2 \rightarrow O_2^{2+} + 2\,e^-$

9.99. Do the 1+ cations of homonuclear diatomic molecules of the second-row elements always have shorter bond lengths than the corresponding neutral molecules?

**9.100.** Do any of the anions of the homonuclear diatomic molecules formed by B, C, N, O, and F have shorter bond lengths than those of the corresponding neutral molecules? Consider only the anions with 1− or 2− charge.

## Additional Problems

**9.101.** Draw the Lewis structure for the two ions in ammonium perchlorate ($NH_4ClO_4$), which is used as a propellant in solid fuel rockets, and determine the molecular geometries of the two polyatomic ions.

**9.102.** **Arsenic-Based DNA?** The waters of Mono Lake in the eastern Sierra Mountains of California (Figure P9.102) are rich in arsenate ion ($AsO_4^{3-}$). Some biochemists have proposed that microorganisms in this environment actually incorporate arsenate into their DNA in place of the phosphate ion ($PO_4^{3-}$). Draw the Lewis structure of the arsenate ion that yields the most favorable formal charges. Predict the angles between the arsenic–oxygen bonds in the arsenate anion.

**FIGURE P9.102**

**9.103.** Consider the molecular structure of the amino acid glycine in Figure P9.103. What is the angle formed by the N—C—C bonds in this structure? What are the O—C—O and C—O—H bond angles?

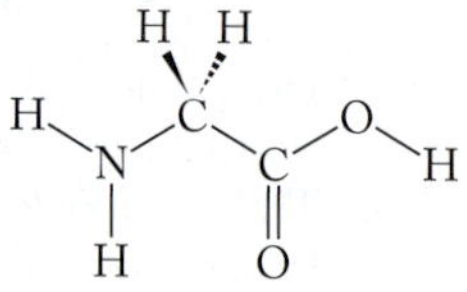

**FIGURE P9.103**

**9.104.** $Cl_2O_2$ may play a role in ozone depletion in the stratosphere. In the laboratory, a reaction between $ClO_2F$ and $AlCl_3$ produces $Cl_2O_2$ and $AlCl_2F$. Draw the Lewis structure for $Cl_2O_2$ based on the skeletal structure in Figure P9.104. What is the geometry about the central chlorine atom?

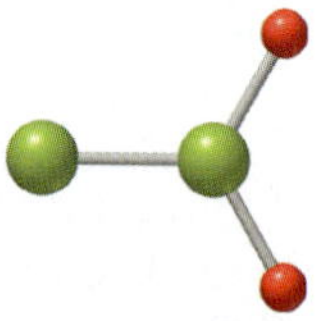

**FIGURE P9.104**

**9.105.** Bombardment of $Cl_2O_2$ molecules (Figure P9.104) with intense radiation is thought to produce the two compounds with the skeletal structures shown in Figure P9.105.
 a. Do both of these molecules have linear geometry?
 b. Do they have the same dipole moment?

**FIGURE P9.105**

*__9.106.__ Complete the Lewis structure for the cyclic structure of $Cl_2O_2$ shown in Figure P9.106.
 a. Is the cyclic $Cl_2O_2$ molecule planar?
 b. Is the molecule polar or nonpolar?

**FIGURE P9.106**

**9.107.** In 1999 the $ClO^+$ ion, a potential contributor to stratospheric ozone depletion, was isolated in the laboratory.
 a. Draw the Lewis structure for $ClO^+$.
 b. Using the molecular orbital diagram in Figure P9.107, determine the order of the Cl—O bond in $ClO^+$.

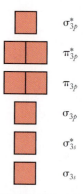

**FIGURE P9.107**

**9.108.** The molecule trinitramide, $N(NO_2)_3$, was first prepared in late 2010 by chemists in Sweden. Draw Lewis structures for trinitramide and predict the geometry about the central nitrogen atom. Do all resonance forms of $N(NO_2)_3$ have the same geometry?

**9.109.** **Cola Beverages** Phosphoric acid imparts a tart flavor to cola beverages. The skeletal structure of phosphoric acid is shown in Figure P9.109. Complete the Lewis structure for phosphoric acid in which formal charges are closest to zero. What is the molecular geometry around the phosphorus atom in your structure?

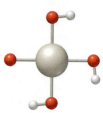

**FIGURE P9.109**

**9.110.** The fluoroaluminate anions $AlF_4^-$ and $AlF_6^{3-}$ have been known for over a century, but the structure of the pentafluoroaluminate ion, $AlF_5^{2-}$, was not determined until 2003. Draw the Lewis structures for $AlF_3$, $AlF_4^-$, $AlF_5^{2-}$, and $AlF_6^{3-}$. Determine the molecular geometry of each molecule or ion. Describe the bonding in $AlF_3$, $AlF_4^-$, $AlF_5^{2-}$, and $AlF_6^{3-}$ using valence bond theory.

*__9.111.__ Thermally unstable compounds can sometimes be synthesized using matrix isolation methods in which the compounds are isolated in a nonreactive medium such as frozen argon. The reaction of boron with carbon monoxide produces compounds with these skeletal structures: B—B—C—O and O—C—B—B—C—O. For each of these compounds, draw the Lewis structure that minimizes formal charges. Do any of your structures contain atoms with incomplete octets? Predict the molecular geometries of BBCO and OCBBCO.

*9.112. The products of the reaction between boron and NO can be trapped in solid argon matrices. Among the products is BNO. Draw the Lewis structure for BNO, including any resonance forms. Assign formal charges and predict which structure provides the best description of the bonding in this molecule. Do any of your structures contain atoms without complete octets? Predict the molecular geometry of BNO.

9.113. **Compounds May Help Prevent Cancer** Broccoli, cabbage, and kale contain compounds that break down in the human body to form isothiocyanates, whose presence may reduce the risk of certain types of cancer. The simplest isothiocyanate is methyl isothiocyanate, $CH_3NCS$. Draw the Lewis structure for $CH_3NCS$, including all resonance forms. Assign formal charges and determine which structure is likely to contribute the most to bonding. Predict the molecular geometry of the molecule at both carbon atoms.

9.114. **Toxic to Insects and People** Methyl thiocyanate ($CH_3SCN$) is used as an agricultural pesticide and fumigant. It is slightly water soluble and is readily absorbed through the skin; it is highly toxic if ingested. Its toxicity stems in part from its metabolism to cyanide ion. Draw three resonance structures for methyl thiocyanate. Assign formal charges and predict which structure would be the most stable. Predict the molecular geometry of the molecule at both carbon atoms.

9.115. Borazine, $B_3N_3H_6$ (a cyclic compound with alternating B and N atoms in the ring), is isoelectronic with benzene ($C_6H_6$). Are there delocalized $\pi$ electrons in borazine?

9.116. Unlike $O_2$, sulfur monoxide (SO) is highly unstable, decomposing to a mixture of $S_2O$ and $O_2$ in less than one second. Using the $O_{2s}$, $O_{2p}$, $S_{3s}$, and $S_{3p}$ atomic orbitals, construct an approximate molecular orbital diagram for SO. Is SO diamagnetic or paramagnetic?

*9.117. Some chemists think HArF consists of $H^+$ ions and $ArF^-$ ions. Using an appropriate MO diagram, determine the bond order of the Ar—F bond in $ArF^-$.

*9.118. Assuming HArF is a molecular compound:
   a. Draw its Lewis structure.
   b. What are the formal charges on Ar and F in the structure you drew?
   c. What is the shape of the molecule?
   d. Is HArF polar?

9.119. Which of the following unstable nitrogen oxides, $N_2O_2$, $N_2O_5$, and $N_2O_3$, are polar molecules? ($N_2O_2$ and $N_2O_3$ have N—N bonds; $N_2O_5$ does not.)

9.120. Explain why $O_2$ is paramagnetic.

9.121. Using an appropriate molecular orbital diagram, show that the bond order in the disulfide anion $S_2^{2-}$ is equal to 1. Is $S_2^{2-}$ diamagnetic or paramagnetic?

9.122. Use molecular orbital diagrams to determine the bond order of the peroxide ($O_2^{2-}$) and superoxide ions ($O_2^-$). Are these bond order values consistent with those predicted from Lewis structures?

9.123. Elemental sulfur has several allotropic forms including cyclic $S_8$ molecules. What is the orbital hybridization of sulfur atoms in this allotrope? The bond angles are about 108°.

*9.124. Which $3d$ atomic orbitals have the proper orientation to overlap with a $4p_z$ atomic orbital?

## Chalcogens: From Alcohol to Asparagus, the Nose Knows

*9.125. Ozone ($O_3$) has a dipole moment (0.54 D). How can a molecule with only one kind of atom have a dipole moment?

*9.126. The bond angle in $H_2O$ is 104.5°; the bond angles in $H_2S$, $H_2Se$, and $H_2Te$ are very close to 90°. Which theory would you apply to describe the geometry in $H_2S$, $H_2Se$, and $H_2Te$: VSEPR? Valence bond without invoking hybrid orbitals? Valence bond theory using hybrid orbitals? Why?

9.127. **Garlic** Garlic contains the molecule alliin (Figure P9.127). When garlic is crushed or chopped, a reaction occurs that converts alliin into the molecule allicin, which is primarily responsible for the aroma we associate with garlic.
   a. Describe the molecular geometry about the sulfur atoms in both compounds.
   b. Do any of the sulfur atoms in allicin have the same geometry as the sulfur atoms in the volatile sulfur compounds that cause bad breath ($H_2S$, $CH_3$—SH, and $CH_3$—S—$CH_3$)?

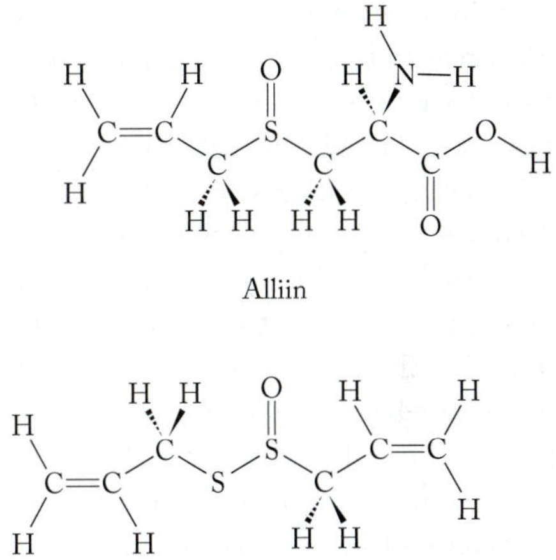

Alliin

Allicin

**FIGURE P9.127**

9.128. All of the group 16 elements form compounds with the generic formula $H_2E$ (E = O, S, Se, or Te). Which compound is the most polar? Which compound is the least polar?

If your instructor assigns problems in **smartwork**, log in at **smartwork.wwnorton.com**.

# 10

# Intermolecular Forces: The Uniqueness of Water

10.1 Interactions between Ions

10.2 Interactions Involving Polar Molecules

10.3 Dispersion Forces

10.4 Polarity and Solubility

10.5 Vapor Pressure of Pure Liquids

10.6 Phase Diagrams: Intermolecular Forces at Work

10.7 Some Remarkable Properties of Water

## Learning Outcomes

**LO1** Estimate the relative strengths of ion–ion interactions
**Sample Exercise 10.1**

**LO2** Explain the origins of ion–dipole forces, dipole–dipole forces, hydrogen bonds, and dispersion forces

**LO3** Explain the effect of intermolecular forces on the boiling points of compounds
**Sample Exercises 10.2, 10.3**

**LO4** Explain the effect of intermolecular forces on the solubilities of compounds in water and other solvents
**Sample Exercises 10.4, 10.5**

**LO5** Calculate the vapor pressure of a pure liquid
**Sample Exercise 10.6**

**LO6** Identify the regions of a phase diagram and explain the effect of temperature and pressure on phase changes
**Sample Exercise 10.7**

**LO7** Describe the role of hydrogen bonding in determining the unique properties of water

## Ubiquitous, Essential, and Remarkable

In the past few chapters we examined bonding in molecules and showed how the combination of attractive and repulsive electrical forces between pairs of electrons determines molecular geometry. We refer to these interactions *within* a molecule as "intramolecular forces." In this chapter we begin the study of the forces that act *between* molecules and *between* molecules and ions. These so-called intermolecular forces are also electrical in nature, but because they act over larger distances, they are weaker than the intramolecular forces.

Intermolecular forces have considerable influence on the physical properties of substances. Let's consider water as an example. Water molecules are polar, which makes water capable of dissolving many substances because of favorable interactions between water molecules and the particles of these substances. Seawater contains dissolved ionic compounds such as sodium chloride and dissolved molecular species such as oxygen gas. Aquatic life—indeed, all life on Earth—relies on the presence of these and many other substances in salt water, fresh water, and water-based biological fluids, such as blood in animals and sap in plants. The ability of water to dissolve and transport substances is governed by the strength and number of intermolecular interactions between water molecules and other ions and molecules.

The polarity of its molecules also influences other physical properties of water, such as its boiling point and melting point. We sometimes observe water in all three phases at the same time on an early spring day, as solid ice melts to liquid water in the sun while white clouds of condensed water vapor dot the sky. We know that ice cubes float in a glass of water and that ice floats on a pond or river during the spring thaw. These and other properties result from the strong interactions between water molecules.

In this chapter we discuss the interactions between particles of solvent and solute in solutions. We also look at how these interactions influence the phase of a substance and the conditions under which phase transitions occur. At the end of the chapter we return to

**Water and Life** All life on Earth depends on water—the substance itself, the substances dissolved in it, and its unique physical properties. ▶

water and show how its interactions with itself and other substances determine its unique properties. We often take water for granted because it is such a familiar substance, but water's physical properties are truly remarkable and so essential to life as we know it that astronomers looking for life on other planets look first for the presence of liquid water. ■

## 10.1 Interactions between Ions

Ocean waves crashing on a rocky shore create plumes of sea spray, carrying small drops of seawater into the atmosphere where they evaporate. As the water in these drops evaporates, the concentrations of dissolved ions such as $Cl^-$, $Na^+$, $Mg^{2+}$, $Br^-$, $Ca^{2+}$, and $SO_4^{2-}$ increase. Eventually, the decreased volume of the drops produces supersaturated solutions of the salts, which begin to precipitate. Among the first solids to form is $CaSO_4$. Among the last is NaCl, which does not precipitate until 90% of the seawater in a drop has evaporated. This sequence takes place even though the concentrations of $Ca^{2+}$ and $SO_4^{2-}$ ions are much lower than the concentrations of $Na^+$ and $Cl^-$ ions. Why does $CaSO_4$ precipitate before NaCl? Put another way: Why is NaCl more soluble in water than $CaSO_4$? The process of forming solutions is discussed in greater detail in the next chapter; here we address questions such as why are NaCl and $CaSO_4$ solids under ordinary conditions of temperature and pressure, whereas water is a liquid under the same conditions? To answer these questions, we need to examine the attractive forces between the particles that make up these substances.

Think about the three common states of matter (Figure 10.1) and how they differ based on the kinetic energy of the particles in them and their ability to overcome the attractive forces between particles. In a solid, the average kinetic energy of the particles is insufficient to overcome these forces of attraction. Consequently, the particles have the same nearest neighbors over time and do not move much. In a liquid, the average kinetic energy of the particles is sufficient to overcome some of the attractive forces; particles in a liquid experience more freedom of motion and can move past one another. In gases, the average kinetic energy of the particles is sufficient to overcome essentially all of the attractive forces between them, imparting nearly complete freedom of motion to the widely separated particles.

The stronger the attractive forces among the particles in a substance, the greater the amount of energy needed to overcome those forces and allow the substance to melt or vaporize. Thus, a substance made of particles that interact relatively strongly has high melting and boiling points, which means it is likely to be a solid at room temperature and normal pressure. Under the same conditions, a substance with somewhat weaker particle–particle interactions has a lower melting point and is more likely to be a liquid. A substance with very weak particle–particle interactions has even lower melting and boiling points and is more likely to be a gas.

**CONNECTION** In Chapter 5, we looked at the flow of energy accompanying phase changes of water (solid ice ⇌ liquid water ⇌ water vapor) in terms of the changes in kinetic and potential energies of the molecules.

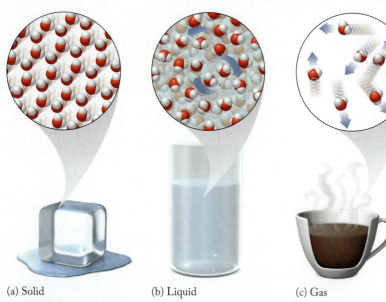

(a) Solid          (b) Liquid          (c) Gas

**FIGURE 10.1** (a) The molecules of $H_2O$ in solid ice are locked in place by the strength of intermolecular attraction. (b) In liquid water, molecules of $H_2O$ have more energy and are free to flow past one another. (c) In water vapor, the molecules have enough energy to overcome nearly all intermolecular attraction and move freely throughout the space they occupy.

## Ion–Ion Interactions

Ionic compounds are among the substances most likely to be solids at room temperature. They are solids because ion–ion interaction is the strongest kind of interactive force between particles; the interaction between ions of opposite charge results in the formation of ionic bonds. The strengths of ion–ion interactions are defined by *coulombic interaction* and the potential energy $E$ between charged particles separated by a distance, $d$, is

$$E \propto \frac{(Q_1 Q_2)}{d} \tag{5.3}$$

The value of $E$ in Equation 5.3 is negative when $Q_1$ and $Q_2$ have opposite signs, which means $E$ is negative for any salt because the dominant interaction is between cations ($+Q$) and anions ($-Q$). Oppositely charged ions attract each other to form an arrangement characterized by a lower potential energy than the potential energy of the separated ions. When the ions are far apart, the $d$ term in Equation 5.3 is large and $E$ is a small negative number. When the ions are close together in a salt, $d$ is small and $E$ is a large negative number, which corresponds to lower energy (see Figure 5.7).

Whenever cations and anions in the gas phase combine to make a solid ionic compound, energy is released and the process is exothermic. The same amount of energy must be absorbed to separate the compound into its cations and anions. We can use Equation 5.3 to compare the relative strengths of interactive forces between ions in salts if we know the charges on the ions and their radii; the attractive force between two ions increases as ionic charge increases and as ionic size decreases.

The inter-ion distance $d$ for ionic compounds is the sum of the ionic radii. Recall that the sizes of atoms and ions is a periodic property. In Chapter 7 we saw that cations are always smaller than the atoms from which they form and anions are always larger (Figure 10.2). Ions in a group of the periodic table, having the same charge, increase in size as you move down the group.

**CONNECTION** We introduced coulombic interactions in Chapter 5 when discussing the electrostatic interaction between particles in an ionic bond.

**CONNECTION** In Chapter 5 we learned that the energy associated with a process has the same absolute value but the opposite sign of the energy associated with the reverse process.

**CONNECTION** In Chapter 7 we discussed the use of ionic radii to calculate the distance between the nuclear centers in an ionic bond.

**CONCEPT TEST**

Rank the following sets of ions in order of increasing distance ($d$) between their nuclear centers: (a) KF, LiF, NaF; (b) CaBr$_2$, CaCl$_2$, CaF$_2$.

*(Answers to Concept Tests are in the back of the book.)*

To address the question of why $CaSO_4$ precipitates from sea spray before NaCl, we can use Equation 5.3 to compare the relative strengths of their ionic interactions. For NaCl the numerator is $(+1)(-1) = -1$, and for $CaSO_4$ the numerator is $(+2)(-2) = -4$, four times as large as for NaCl. For NaCl, the ionic radii are 102 pm for $Na^+$ and 181 pm for $Cl^-$, so $d$ is $102 + 181 = 283$ pm. For $CaSO_4$, the radii are 100 pm for $Ca^{2+}$ and 230 pm for $SO_4^{2-}$ (Table 10.1), making $d = 330$ pm. The denominator for $CaSO_4$ is about 1.2 times the size of the NaCl denominator, so the factor-of-four difference in the numerators dominates this comparison. Thus, the attraction between $Ca^{2+}$ ions and $SO_4^{2-}$ ions is stronger than the attraction between $Na^+$ ions and $Cl^-$ ions. The stronger attraction between its ions is not the only reason $CaSO_4$ precipitates first from sea spray, but it is a major factor.

| TABLE 10.1 | **Estimated Radii of Polyatomic Ions** |
| --- | --- |
| **Polyatomic Ion** | **Radius (pm)** |
| $CO_3^{2-}$ | 185 |
| $NO_3^-$ | 189 |
| $SO_4^{2-}$ | 230 |
| $PO_4^{3-}$ | 238 |

Values from Heslop, R. B. and K. Jones. *Inorganic Chemistry: A Guide to Advanced Study* (Elsevier, 1976), p. 123.

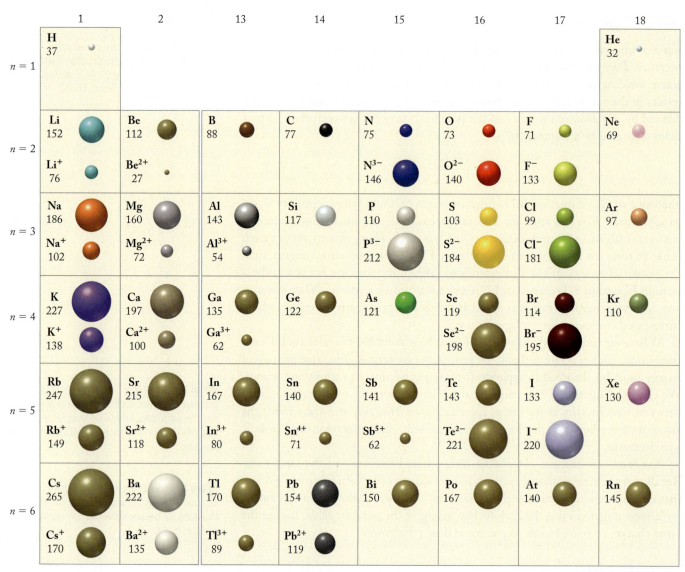

**FIGURE 10.2** The radii of anions formed by the main group elements are larger than the radii of their parent atoms. The radii of cations are smaller than the radii of their parent atoms. All values are in picometers. Values from Greenwood, N. N. and A. Earnshaw, *Chemistry of the Elements*, 2nd ed. (Boston: Butterworth-Heinemann, 1997); Housecroft, C. E. and Sharpe, A. G. *Inorganic Chemistry*, 3rd ed. (Upper Saddle River, NJ: Pearson Prentice Hall, 2008).

---

**SAMPLE EXERCISE 10.1** **Predicting Relative Strengths of Ion–Ion Interactions** **LO1**

List the ionic compounds $CaO$, $NaF$, and $CaF_2$ in order of decreasing strength of the attraction between their ions.

**Collect and Organize** We are given three ionic compounds and want to find the relative strength of the ion–ion interaction in each one. Equation 5.3 shows that the strength of an ion–ion attractive interaction depends on the charges on the ions and the distance between them. We know the charges on the ions in binary ionic compounds from Chapter 2. Interionic distances are determined from ionic radii in Figure 10.2.

**Analyze** We need to compare the charges on the calcium, sodium, fluoride, and oxide ions, which are 2+, 1+, 1−, and 2−, and the distances between the ions in $CaO$, $NaF$, and $CaF_2$, which are equal to the sum of their ionic radii. The strength of an ion–ion interaction is directly proportional to the product of the charges on the ions and inversely proportional to the distance between them.

**Solve** Each of these compounds has a different $Q_1Q_2$ value:

$$
\begin{array}{llll}
Ca^{2+} & O^{2-} & Q_1Q_2 = (+2)(-2) = -4 \\
Na^+ & F^- & Q_1Q_2 = (+1)(-1) = -1 \\
Ca^{2+} & F^- & Q_1Q_2 = (+2)(-1) = -2
\end{array}
$$

The values of $d$ in Equation 5.3 are

$$d_{CaO} = Ca^{2+}\ 100\ pm + O^{2-}\ 140\ pm = 240\ pm$$
$$d_{NaF} = Na^+\ 102\ pm + F^-\ 133\ pm = 235\ pm$$
$$d_{CaF_2} = Ca^{2+}\ 100\ pm + F^-\ 133\ pm = 233\ pm$$

Substituting the values of $d$ and $Q_1Q_2$ into Equation 10.1 for each compound:

$$E_{CaO} \propto (-4)/240 = -0.0167$$
$$E_{NaF} \propto (-1)/235 = -0.00426$$
$$E_{CaF_2} \propto (-2)/233 = -0.00858$$

Thus, the predicted order of decreasing strength of ionic interactions is $CaO > CaF_2 > NaF$.

**Think About It** The ion–ion distances are nearly the same for all three compounds, which means that the product of the ionic charges determines the relative interaction strengths.

⚙ **Practice Exercise** Arrange the ionic compounds $CaCl_2$, BaO, and KCl in order of decreasing strength of the interaction between their ions.

*(Answers to Practice Exercises are in the back of the book.)*

■

We have seen two instances where the product of the charges $Q_1Q_2$ was the dominant factor in determining the relative strengths of the ion–ion interactions. This result is generally true. The ionic radii of cations in Figure 10.2 range from 27 pm ($Be^{2+}$) to 170 pm ($Cs^+$), and those of monatomic anions and polyatomic anions (Figure 10.2 and Table 10.1) range from 133 pm ($F^-$) to 238 pm ($PO_4^{3-}$). If we put together the smallest cation and anion ($BeF_2$) and the largest cation and anion ($Cs_3PO_4$), the range of nucleus-to-nucleus distances is 160 to 408 pm, which is a factor of 2.6. Moreover, the radii of common ions cover an even smaller size range. Therefore, for common ionic compounds, the charge product is nearly always responsible for large differences in the strengths of ionic interactions. Solid ionic compounds have more than one ion–ion interaction in a three-dimensional lattice. We will learn more about the strengths of multiple ion–ion interactions in Chapter 11, and we will examine the structures of the resultant solids in Chapter 12.

# 10.2 Interactions Involving Polar Molecules

Ionic and covalent bonds involve the strong interactions that hold ions together in an ionic solid and atoms together in a molecule. Typical bond energies for covalent and ionic compounds range from several hundred to several thousand kilojoules per mole. In contrast, interactions between molecules and ions, such as those responsible for salts dissolving in water, are 10- to 100-fold weaker than bonding interactions.

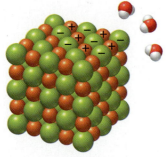

Hydrated Na⁺ ion    Hydrated Cl⁻ ion

Solid NaCl

**FIGURE 10.3** The hydrogen atoms (positive poles) of $H_2O$ molecules are attracted to the Cl⁻ ions of NaCl and the O atoms (negative poles) are attracted to the Na⁺ ions. Multiple ion–dipole interactions overcome the attractive forces holding ions at the surface of the solid NaCl, causing the NaCl to dissolve.

**⊙⊙ CONNECTION** In previous chapters, we have indicated the presence of a hydration sphere around an ion in aqueous solution by placing (*aq*) after its symbol or formula.

Even though molecule–ion interactions are much weaker than bonds, they do influence the properties of substances. For example, polar covalent molecules interact with ionic substances. These interactions are responsible for the solubility of salts in water. Polar molecules also attract one another, and these interactions are responsible for a range of physical properties. For example, boiling a liquid requires that nearly all the intermolecular interactions between molecules in the liquid be overcome so that these molecules can be separate from one another in the gas phase. Thus, polar substances with stronger interactions tend to have higher boiling points than substances with similar molar mass but weaker interactions.

## Ion–Dipole Interactions

One factor influencing the solubility of ionic compounds in water is the attraction between the ions and the polar water molecules, which carry both positive and negative partial charges. These attractions are **ion–dipole interactions**, and they occur between ions and water molecules in all aqueous solutions.

When an ionic solid dissolves in water, ion–dipole interactions between the ions and water molecules pull ions from the solid into solution (Figure 10.3). As an ion is pulled away from its solid-state neighbors, it is surrounded by water molecules, forming a **sphere of hydration**. If the solvent were something other than water, the cluster would be called a *sphere of solvation*. These dissolved ions are said to be *hydrated* or, for other solvents, *solvated*. The strengths of many ion–dipole interactions help to overcome the ion–ion interactions in an ionic solid, allowing these spheres of hydration (or solvation) to form and the compound to dissolve.

Within a sphere of hydration, the water molecules closest to the ion are oriented so that their oxygen atoms (negative poles) are directed toward a cation or their hydrogen atoms (positive poles) are directed toward an anion (Figure 10.4). The number of water molecules oriented in this way depends on the size of the ion. Typically, six water molecules hydrate an ion, but the number can range from four to nine. As Figure 10.4 shows, six water molecules surround the Na⁺ and Cl⁻ ions in an aqueous solution of NaCl.

**FIGURE 10.4** Each hydrated Na⁺ ion and Cl⁻ ion is surrounded by six water molecules, which create an inner hydration sphere. Water molecules in an outer hydration sphere surround the inner sphere. The outer sphere is the result of dipole–dipole interactions between the rest of the water (known as bulk water) and the water molecules of the inner sphere. Beyond the outer hydration sphere, dipole–dipole interactions also occur between outer-sphere water molecules and molecules in bulk water.

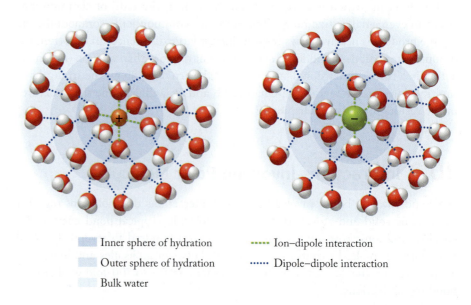

|  | Inner sphere of hydration | | ----- | Ion–dipole interaction |
|  | Outer sphere of hydration | | ⋯⋯ | Dipole–dipole interaction |
|  | Bulk water | | | |

## Dipole–Dipole Interactions

The water molecules closest to the ions in Figure 10.4 are surrounded by other water molecules that form an outer hydration sphere. The molecules in this outer sphere are more randomly oriented than those in the inner sphere, but not completely so. What ordering there is among outer-sphere water molecules is caused by another intermolecular force, **dipole–dipole interactions**, which operate between molecules that have permanent dipole moments—in other words, between polar molecules. In water molecules, the partial charges on oxygen and hydrogen atoms result in attractions between a hydrogen atom of one molecule and an oxygen atom of another. Dipole–dipole interactions are not as strong as ion–dipole interactions because dipole–dipole interactions involve only partial charges, caused by unequal sharing of electrons within the molecule. In contrast, the ion involved in an ion–dipole interaction has completely lost or gained electrons and has a full positive or negative charge. The magnitude of the dipole moment is reflected in the strength of the dipole–dipole interactions between molecules.

▶❚❚ **CHEMTOUR** Intermolecular Forces

◉◉ **CONNECTION** In Chapter 9 we learned that permanent dipole moments are experimentally measured values, expressed in units of debyes, that define the polarity of molecules.

**CONCEPT TEST** • • • • • • • • • • • • • • • • • • • • • • • • • • • • • • • • • • •

Dimethyl ether, $CH_3OCH_3$, and acetone, $CH_3C(O)CH_3$ (Figure 10.5), have similar formulas and molar masses. However, their dipole moments are quite different: 1.30 D for dimethyl ether and 2.88 D for acetone. Predict which compound has the higher boiling point.

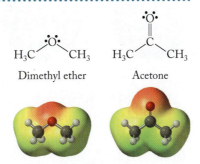

Dimethyl ether          Acetone

**FIGURE 10.5**

• • • • • • • • • • • • • • • • • • • • • • • • • • • • • • • • • • • • • • • • • • • • • • • • • • • • • • • • • • •

## Hydrogen Bonds

In polar molecules containing O—H, N—H, or F—H bonds, hydrogen atoms are bonded to small, highly electronegative atoms. This leads to relatively large separations of charge, large dipole moments, and stronger-than-average dipole–dipole interactions. Because of its strength, this particular dipole–dipole interaction—involving one of these H atoms on one molecule and an O, N, or F atom on an adjacent molecule—merits special distinction: it is called a **hydrogen bond** (Figure 10.6). Hydrogen bonds are the strongest dipole–dipole interactions and are about one-tenth the strength of covalent bonds; hydrogen bonds range in strength from about 5 to about 30 kJ/mol. Hydrogen bonds in water play a key role in defining the remarkable behavior of $H_2O$. The dipole–dipole interactions between water molecules in Figure 10.4 are hydrogen bonds.

One physical property strongly influenced by hydrogen bonds is boiling point. Water, HF, and ammonia all have boiling points that are much higher than the boiling points of other compounds of similar mass. One way to look at the relationship between hydrogen bonding and boiling point is shown in Figure 10.7. Note how boiling points increase with increasing period number for the period

**ion–dipole interaction** an attractive force between an ion and a molecule that has a permanent dipole moment.

**sphere of hydration** the cluster of water molecules surrounding an ion in aqueous medium; the general term applied to such a cluster forming in any solvent is *sphere of solvation.*

**dipole–dipole interaction** an attractive force between polar molecules.

**hydrogen bond** the strongest dipole–dipole interaction. It occurs between a hydrogen atom bonded to a small, highly electronegative element (O, N, F) and an atom of oxygen or nitrogen in another molecule. Molecules of HF also form hydrogen bonds.

**FIGURE 10.6** (a) Hydrogen bonds (blue dotted lines) occur between hydrogen atoms bonded to O, N, or F atoms and O, N, or F atoms in adjacent molecules. (b) Hydrogen bonding interactions are so strong in a carboxylic acid like acetic acid that two molecules stay together as a unit called a dimer in the liquid phase. (c) Hydrogen bonds between nitrogen and hydrogen in ammonia form extensive three-dimensional networks.

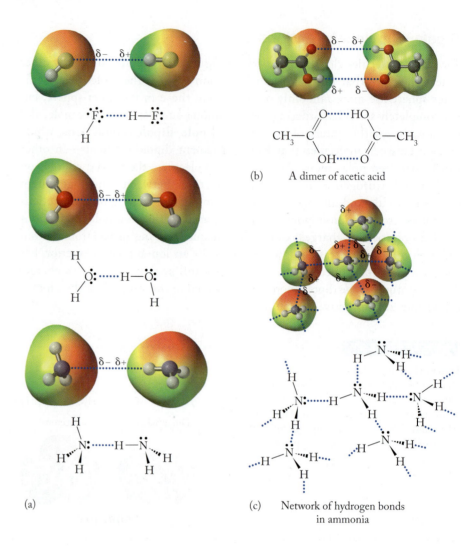

(b)    A dimer of acetic acid

(a)

(c)    Network of hydrogen bonds in ammonia

3, 4, and 5 hydrides. Because methane, $CH_4$, is a nonpolar molecule, it cannot participate in hydrogen bonding. As a result, the boiling points of the group 14 hydrogen compounds increase with increasing period number (and molar mass), as shown by the black line in Figure 10.7. Each of the three hydrogen-bonding substances, however, has a boiling point much higher than the boiling points of other hydrogen-containing compounds in the same group (blue, green, and red lines in Figure 10.7).

**CONCEPT TEST**

Predict which compound has the higher boiling point in each of the following pairs: (a) $CH_3Cl$, $CH_3Br$; (b) $CH_3CH_2OH$, $CH_3OH$; (c) $CH_3NH_2$, $(CH_3)_3N$.

Hydrogen bonds have a defining impact on the three-dimensional shape of many polymers and large biological molecules, such as proteins and DNA. As we have seen with small molecules, shape often determines behavior. The ability of the two strands of DNA to stay together in a double helix is due to many hydrogen bonds that hold the molecule together in aqueous solution (Figure 10.8). Because the hydrogen bonds are weaker than covalent bonds, however, the DNA strands can be pulled apart during DNA replication.

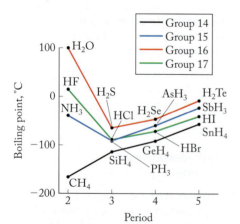

**FIGURE 10.7** The boiling points of binary hydrides plotted against the period in the periodic table of the atom bonded to hydrogen. The boiling points of $H_2O$, $NH_3$, and HF are unusually high because their molecules form hydrogen bonds.

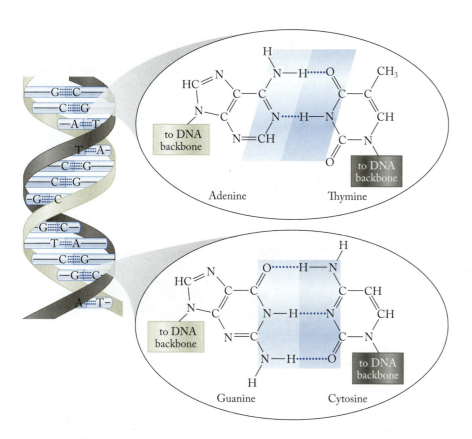

**FIGURE 10.8** Hydrogen bonds (blue dotted lines) occur between hydrogen and nitrogen or oxygen in adjacent strands of DNA, contributing to the double-helix structure. The two detailed views of the DNA double helix contain the names of four of the building blocks of DNA: guanine, cytosine, adenine, and thymine. Note that there are three hydrogen bonding sites between pairs of guanine and cytosine, but only two between adenine and thymine.

---

**SAMPLE EXERCISE 10.2** | **Explaining Differences in Boiling Points** | **LO3**

Dimethyl ether ($C_2H_6O$) has a molar mass of 46.07 g/mol and a boiling point of −24.9°C. Ethanol ($C_2H_6O$) has the same formula and therefore the same molar mass, but a boiling point of 78.5°C. Explain this difference in boiling points. The structures are shown in Figure 10.9.

**Collect and Organize** We are asked to explain the difference in boiling points of two compounds that have the same molar mass. Because the molar masses are the same, intermolecular interactions must account for the difference in the boiling points.

**Analyze** We are given the three-dimensional structures of both molecules. We need to consider:

| Polarity of individual bonds | → | Polarity of molecule | → | Type of intermolecular interactions | → | Influence on boiling point |

Molecules of both compounds contain O atoms bonded to atoms of less-electronegative elements. This means the molecules contain bond dipoles. The geometrical shapes of the molecules show that the bond dipoles do not cancel, so overall, both are polar molecules, as shown in Figure 10.10. Their polarities mean that both experience dipole–dipole interactions. Ethanol molecules contain −OH groups, which means their dipole–dipole interactions are hydrogen bonds.

**Solve** Hydrogen bonds are stronger than other types of dipole–dipole interactions. Therefore, the intermolecular interactions in ethanol are stronger than those in dimethyl ether. More energy is required to overcome these stronger interactions, and that is why ethanol has a higher boiling point.

**Think About It** Dimethyl ether and ethanol have the same molecular formula and similar molecular structures, yet their boiling points differ by more than 100°C: a difference directly linked to the presence of −OH groups capable of hydrogen bonding in ethanol, compared to weaker dipole–dipole interactions in dimethyl ether.

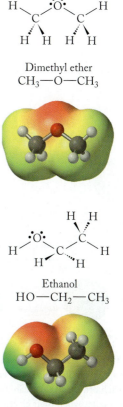

Dimethyl ether
$CH_3—O—CH_3$

Ethanol
$HO—CH_2—CH_3$

**FIGURE 10.9** Dimethyl ether and ethanol have the same molecular formula but different molecular structures and hence different boiling points.

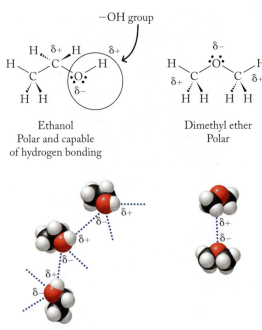

Ethanol
Polar and capable
of hydrogen bonding

Dimethyl ether
Polar

**FIGURE 10.10** The polar —OH group in ethanol leads to hydrogen bonding (blue dotted lines) between molecules of ethanol. Dimethyl ether does not form hydrogen bonds but does experience weaker dipole–dipole interactions (dashed black lines) between the δ+ and δ− regions of its molecules.

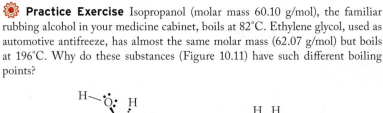

**Practice Exercise** Isopropanol (molar mass 60.10 g/mol), the familiar rubbing alcohol in your medicine cabinet, boils at 82°C. Ethylene glycol, used as automotive antifreeze, has almost the same molar mass (62.07 g/mol) but boils at 196°C. Why do these substances (Figure 10.11) have such different boiling points?

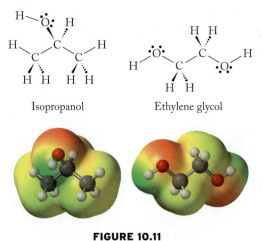

Isopropanol

Ethylene glycol

**FIGURE 10.11**

## 10.3 Dispersion Forces

If we can use dipole–dipole interactions and hydrogen bonding to explain differences in the boiling points of polar molecules, what accounts for the different boiling points of nonpolar molecules? One type of intermolecular interaction occurs between all molecules: **dispersion forces**, also called **London forces**, in honor of German-American physicist Fritz London (1900–1954), whose work explained attractions between atoms of the noble gases and between nonpolar molecules. Dispersion forces are the only intermolecular force between nonpolar molecules.

Higher dispersion forces result in increased intermolecular interaction and higher boiling points. Among small molecules, dispersion forces may be weaker than other intermolecular forces, but in large molecules they can be stronger. The relative strengths of intermolecular forces among small molecules are shown in Table 10.2.

London determined that the intermolecular attractive forces between atoms of the group 18 elements and between nonpolar molecules are caused by **temporary dipoles** (or **induced dipoles**), produced by momentary changes in the electron distribution in the atoms or molecules. The mutual repulsion between the electrons of the two atoms (or two molecules) causes the electrons to be distributed unevenly, creating two temporary dipoles (Figure 10.12). The presence of temporary dipoles in a noble gas or nonpolar molecular substance causes the atoms or molecules to interact more than they would in the absence of the temporary dipoles.

A molecule with a permanent dipole can induce a temporary dipole in a nonpolar molecule by perturbing the electron distribution in the nonpolar molecule (Figure 10.13). The intermolecular interaction in this case is weaker than the dipole–dipole force between two polar molecules and is of the same order of magnitude as the dispersion forces between temporary dipoles in nonpolar molecules (see Table 10.2).

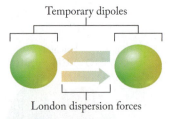

(a) Nonpolar atom    Nonpolar atom

Temporary dipoles

(b)    London dispersion forces

**FIGURE 10.12** (a) Two atoms of a noble gas with a spherical distribution of electrons approach each other and (b) create two temporary dipoles when their electron clouds repel one another. As soon as the atoms move away from each other, the electrons in each atom return to their original uniform distribution.

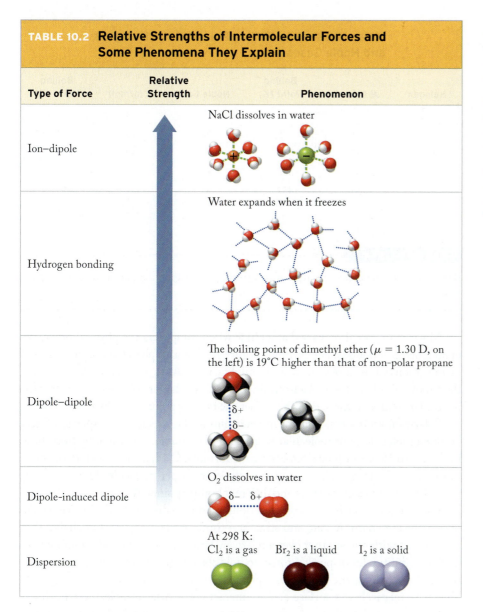

**TABLE 10.2** **Relative Strengths of Intermolecular Forces and Some Phenomena They Explain**

| Type of Force | Relative Strength | Phenomenon |
|---|---|---|
| Ion–dipole | | NaCl dissolves in water |
| Hydrogen bonding | | Water expands when it freezes |
| Dipole–dipole | | The boiling point of dimethyl ether ($\mu = 1.30$ D, on the left) is 19°C higher than that of non-polar propane<br>$\delta+$<br>$\delta-$ |
| Dipole-induced dipole | | $O_2$ dissolves in water<br>$\delta-$  $\delta+$ |
| Dispersion | | At 298 K:<br>$Cl_2$ is a gas   $Br_2$ is a liquid   $I_2$ is a solid |

**dispersion force** (also called **London force**) an intermolecular force between nonpolar molecules caused by the presence of temporary dipoles in the molecules.

**temporary dipole** (also called **induced dipole**) the separation of charge produced in an atom or molecule by a momentary uneven distribution of electrons.

**polarizability** the relative ease with which the electron cloud in a molecule, ion, or atom can be distorted, inducing a temporary dipole.

The magnitude of an induced dipole depends on the ease with which the electrons in a molecule, ion, or atom can be redistributed. The relative tendency of the electron density to be distorted by another charged particle is a measure of the **polarizability** of the electron cloud of the atom, ion, or molecule. What factors determine how easily an electron cloud can be polarized?

Atoms having larger molar masses also have larger numbers of electrons and electron density in orbitals farther from the nucleus than atoms with smaller molar masses and fewer electrons. Electrons in these larger atoms are held less tightly by the nucleus because of both their greater average distance from the nucleus and the screening of the nuclear charge by electrons in lower orbitals. Consequently, they are more easily polarized than electrons in smaller atoms or molecules. Greater polarizability leads to stronger temporary dipoles and stronger intermolecular interactions, so dispersion forces become stronger as atoms or ions become larger. Table 10.3 shows that for both monatomic noble gases and diatomic halogens (which are nonpolar), boiling point increases with increasing numbers of electrons.

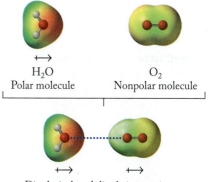

$H_2O$
Polar molecule

$O_2$
Nonpolar molecule

Dipole-induced dipole interaction

**FIGURE 10.13** The approach of a polar water molecule induces a temporary dipole in an initially nonpolar oxygen molecule by distorting the distribution of the electrons.

**CONNECTION** The concept of screening of outer electrons by inner electrons was presented in Chapter 7 when we discussed trends in the sizes of atoms and ions.

**TABLE 10.3** **Molar Masses and Boiling Points of the Halogens and Noble Gases**

| Halogen | $\mathcal{M}$ (g/mol) | Boiling Point (K) | Noble Gas | $\mathcal{M}$ (g/mol) | Boiling Point (K) |
|---------|-----------|----------|-----------|-----------|----------|
|         |           |          | He        | 4         | 4        |
| $F_2$   | 38        | 85       | Ne        | 20        | 27       |
| $Cl_2$  | 71        | 239      | Ar        | 40        | 87       |
| $Br_2$  | 160       | 332      | Kr        | 84        | 120      |
| $I_2$   | 254       | 457      | Xe        | 131       | 165      |
|         |           |          | Rn        | 222       | 211      |

**CONCEPT TEST**

Rank the following atoms in order of increasing polarizability: argon, hydrogen, krypton, neon.

Molar mass can also be used as a guide when comparing the physical properties of nonpolar compounds. For example, Figure 10.14 is a graph of the boiling points of hydrocarbons whose molecules are all straight chains. As their molar masses increase, their molecules have more electrons distributed over larger volumes. The larger the molecular volume in the series, the higher the boiling point of the hydrocarbon.

Polarizability is a major factor in determining the strength of dispersion forces between molecules, but molecular shape also plays a role. The three hydrocarbons in Figure 10.15, for instance, have the same formula, $C_5H_{12}$, and therefore the same molar mass, 72.15 g/mol, and number of electrons per molecule. However, they have different molecular shapes and different boiling points. All these molecules are nonpolar, so any intermolecular interactions are due to dispersion forces. Molecules of pentane are relatively long and straight—think of them as pieces of chalk. They can interact with one another over a relatively large surface area and therefore have more possibilities for dispersion forces to hold them together. In contrast, molecules of 2,2-dimethylpropane are almost spherical—think of them as Ping-Pong balls. They have relatively less surface area to interact with adjacent molecules; less interaction means weaker dispersion forces, resulting in the lowest boiling point in this set of compounds. The remaining molecule, 2-methylbutane, is neither as straight as pentane nor as spherical as 2,2-dimethylpropane—think of it as a short, forked stick. It should come as no surprise that this compound boils at a temperature lower than pentane but higher than 2,2-dimethylpropane.

**FIGURE 10.14** The boiling points of straight-chain hydrocarbons increase as masses increase and the magnitude of the dispersion forces between molecules increases.

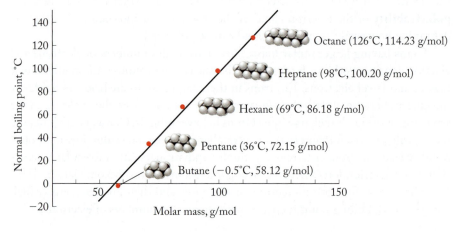

Octane (126°C, 114.23 g/mol)
Heptane (98°C, 100.20 g/mol)
Hexane (69°C, 86.18 g/mol)
Pentane (36°C, 72.15 g/mol)
Butane (−0.5°C, 58.12 g/mol)

**FIGURE 10.15** Three molecular shapes are possible for $C_5H_{12}$. The more spread out the molecule, the greater its polarizability, the stronger the opportunity for dispersion forces between molecules, and the higher the boiling point.

|  |  |  |
|---|---|---|
| $CH_3-CH_2-CH_2-CH_2-CH_3$ | $CH_3-CH_2-CH-CH_3$<br>$\quad\quad\quad\quad\quad \mid$<br>$\quad\quad\quad\quad\quad CH_3$ | $CH_3-\overset{\displaystyle CH_3}{\underset{\displaystyle CH_3}{\overset{\mid}{\underset{\mid}{C}}}}-CH_3$ |
| Pentane<br>Boiling point 36°C | 2-Methylbutane<br>Boiling point 28°C | 2,2-Dimethylpropane<br>Boiling point 9°C |

Dispersion forces between small atoms are weak, but they add up, and for large molecules they are particularly strong. A large number of weak interactions can sometimes dominate a much smaller number of strong interactions in a system, as we will see in the discussion of polarity and solubility in Section 10.4.

**CONCEPT TEST**

Do you think an attractive force exists between ions and nonpolar molecules? How would you describe such an interaction? Where would you place such an intermolecular force in Table 10.2?

**SAMPLE EXERCISE 10.3  Explaining Trends in Boiling Points  LO3**

Table 10.4 gives the formulas, molar masses, and boiling points of several hydrocarbons and alcohols that have comparable molar masses. Figure 10.16 shows the molecular structures of the compounds in rows 4 and 5 of the table. (a) Explain the trend in boiling points of the five hydrocarbons and the four alcohols. (b) Explain the difference in boiling point for each hydrocarbon and the alcohol of comparable molar mass.

**TABLE 10.4  Boiling Point Data for Sample Exercise 10.3**

| HYDROCARBON | | | ALCOHOL | | |
|---|---|---|---|---|---|
| Molecular Formula | $\mathcal{M}$ (g/mol) | Boiling Point (°C) | Molecular Formula | $\mathcal{M}$ (g/mol) | Boiling Point (°C) |
| 1. $CH_4$ | 16.04 | −161.5 | | | |
| 2. $CH_3CH_3$ | 30.07 | −88 | $CH_3OH$ | 32.04 | 64.5 |
| 3. $CH_3CH_2CH_3$ | 44.10 | −42 | $CH_3CH_2OH$ | 46.07 | 78.5 |
| 4. $CH_3CH(CH_3)CH_3$ | 58.12 | −11.7 | $CH_3CH(OH)CH_3$ | 60.10 | 82 |
| 5. $CH_3CH_2CH_2CH_3$ | 58.12 | −0.5 | $CH_3CH_2CH_2OH$ | 60.10 | 97 |

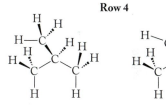

**Row 4**

$CH_3CH(CH_3)CH_3$
Boiling point −11.7°C
Isobutane

$CH_3CH(OH)CH_3$
Boiling point 82°C
Isopropanol

**Row 5**

$CH_3CH_2CH_2CH_3$
Boiling point −0.5°C
Butane

$CH_3CH_2CH_2OH$
Boiling point 97°C
Propanol

**FIGURE 10.16** Structures of isobutane (a hydrocarbon) and isopropanol (an alcohol) from row 4 and butane (a hydrocarbon) and propanol (an alcohol) from row 5 in Table 10.4.

**Collect and Organize** Interpretation of trends requires evaluation of the types and relative strengths of intermolecular forces between molecules. The stronger the interactions between the molecules of a compound, the greater its boiling point. Several types of intermolecular forces have been presented in this section and are summarized in Table 10.2.

**Analyze** All of the compounds are molecular, so we do not need to consider ion–ion or ion–dipole forces. Alcohols contain –OH groups, whereas hydrocarbons do not. The presence of the –OH group means that alcohols can form hydrogen bonds, which hydrocarbons cannot. For both sets of compounds we need to consider:

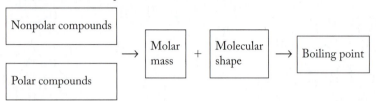

**Solve**

a. The trend in boiling point for both series of compounds in Table 10.4 reflects an increase in boiling point as molar mass increases. Larger molecules have more electrons distributed over larger volumes, leading to larger induced dipoles and greater dispersion forces between molecules. Based on the structures shown, the molecules in row 4 (Figure 10.16) are more spherical and have smaller surface areas than the molecules in row 5. The molecules in row 4 experience weaker dispersion forces. Consequently, both compounds in row 4 boil at lower temperatures than their counterparts in row 5.

b. The molar masses of ethane and methanol (row 2) are very similar, as are the molar masses of each hydrocarbon–alcohol pair; however, the alcohols all have much higher boiling points than their hydrocarbon counterparts. The presence of –OH groups means that the alcohols are capable of hydrogen bonding. Hydrogen bonds are particularly strong interactions, and the occurrence of hydrogen bonding among the alcohol molecules accounts for their higher boiling points. All of the hydrocarbons interact only via dispersion forces.

**Think About It** Identifying the types of intermolecular interactions between molecules of different substances helps us explain trends in their boiling points. Molecular size and shape is especially important in evaluating the contribution of dispersion forces to molecular interactions.

⚙ **Practice Exercise** Molecules of which of the following substances experience the strongest dipole–dipole interactions? The largest dispersion forces? The lowest boiling point? (a) $H_2NNH_2$; (b) $H_2C{=}CH_2$; (c) Ne; (d) $CH_3CH_2CH_2CH_2CH_3$

---

**CONCEPT TEST**

Explain why $CF_4$ is a gas at room temperature but $CCl_4$ is a liquid.

---

# 10.4 Polarity and Solubility

In Section 10.2, we explained the process by which an ionic salt dissolves in water in terms of ion–dipole interactions. The dissolving of one liquid in another or the dissolving of a gas in a liquid can also be explained in terms of intermolecular forces. We examine these processes by considering the situation where the solvent is water or some other liquid and the solute is molecular. To predict whether a given solute is soluble in a given solvent, we look at the balance between solute–solute interactions (that is, interactions between solute molecules), solvent–solvent interactions, and solvent–solute interactions.

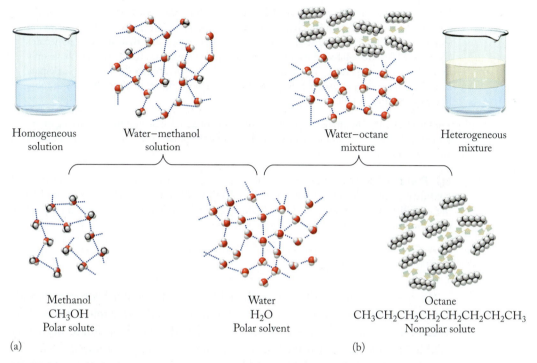

Homogeneous solution    Water–methanol solution

Water–octane mixture    Heterogeneous mixture

Methanol
CH₃OH
Polar solute

Water
H₂O
Polar solvent

Octane
CH₃CH₂CH₂CH₂CH₂CH₂CH₂CH₃
Nonpolar solute

(a)

(b)

**FIGURE 10.17** (a) A polar solvent like water dissolves polar materials like methanol because of favorable dipole–dipole interactions. In this specific case, both molecules are capable of hydrogen bonding, and solute–solvent interactions are of the same order of magnitude as solvent–solvent or solute–solute interactions. Methanol dissolves in all proportions in water. (b) Dispersion forces between long hydrocarbon chains provide the attractive force that holds octane molecules together in the liquid phase. The highly polar water molecules interact preferentially with each other, and any solvent–solute interactions are too weak to compete with the hydrogen-bonding network within the solvent. As a result, octane is virtually insoluble in water.

Just as ionic compounds dissolve in polar solvents because of strong ion–dipole interactions, polar solutes tend to dissolve in polar solvents because of dipole–dipole interactions between solute and solvent molecules (Figure 10.17a). Nonpolar solutes tend not to dissolve in polar solvents because the solvent–solute interactions that promote dissolution are weaker than the solute–solute interactions that keep solute molecules together and the solvent–solvent interactions that keep solvent molecules together (Figure 10.17b). This observation is the source of a common phrase used to describe solubility: like dissolves like.

Factors other than polarity influence solubility: temperature and pressure are very important. Nevertheless, we can use the attractive forces present in any mixture of molecules as a guide to predict the relative solubilities of solutes in a given solvent.

**SAMPLE EXERCISE 10.4**    **Predicting Solubility in Water**    **LO4**

Considering the compounds carbon tetrachloride ($CCl_4$), ammonia ($NH_3$), hydrogen fluoride (HF), and oxygen ($O_2$), which should be very soluble in water and which should have limited solubility in water?

**Collect and Organize** We are to predict the solubilities of four substances in water. We can make our predictions based on polarity. Because water is polar, it is likely that compounds soluble in water are also polar.

**Analyze** To judge which of the four substances is polar we need to draw all of their molecular structures and determine (1) which of them have polar bonds and (2) if there is symmetry in the way these bonds are oriented in these molecules that gives them permanent dipoles.

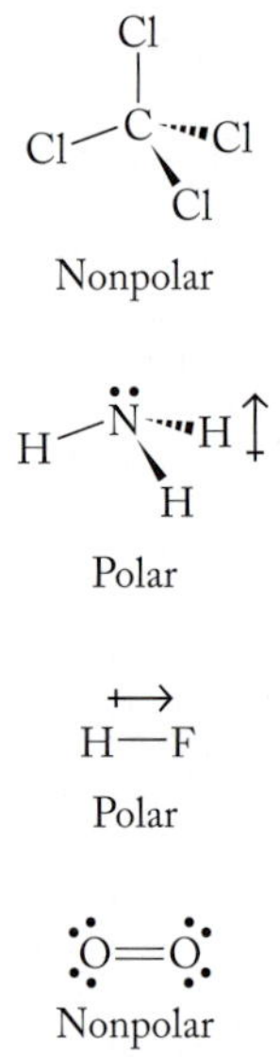

Cl

Cl—C····Cl

Cl

Nonpolar

H—N····H ↑

H

Polar

⟷

H—F

Polar

Ö=Ö

Nonpolar

**FIGURE 10.18**

**Solve** Applying the concepts from Chapter 9 we draw the molecular structures of the four substances and determine that $NH_3$ and HF are polar, but that $CCl_4$ and $O_2$ are not (Figure 10.18). On the premise that like dissolves like, we predict that the polar molecules ($NH_3$ and HF) are soluble in water, and the nonpolar molecules ($CCl_4$ and $O_2$) have only limited solubility in water. The ability of ammonia and hydrogen fluoride to participate in hydrogen bonding is another property they share with water, making them even more likely to be water soluble.

**Think About It** The three-dimensional structure of a molecule and the electronegativity difference between the atoms in its bonds determine the polarity of a molecule. The polarity serves as a guide to solubility.

⚙ **Practice Exercise** In Chapter 6, we mentioned that some deep-sea divers breathe a mixture of gases rich in helium because the solubility of helium gas in blood is lower than the solubility of nitrogen gas in blood. Assuming that blood behaves like water in terms of dissolving substances, why should helium be less soluble in water than nitrogen?

In Sample Exercise 10.4 we predicted that ammonia and hydrogen fluoride are water soluble, whereas carbon tetrachloride and oxygen are much less soluble. These predictions are correct. Ammonia and hydrogen fluoride are both very soluble in water. In fact, they not only dissolve in water, *they react with it.* The hydrogen bonds between $NH_3$ and $H_2O$ molecules are strong enough to ionize water molecules, producing $OH^-$ ions and $H^+$ ions that attach to $NH_3$ molecules, forming ammonium ($NH_4^+$) ions. The hydrogen bonds between molecules of HF and $H_2O$ are so strong they ionize some HF molecules, producing $F^-$ ions and $H^+$ ions that attach to $H_2O$ molecules, forming hydronium ($H_3O^+$) ions. We will return to these reactions in our discussion of acids and bases in Chapter 16.

In contrast, the nonpolar molecules carbon tetrachloride and oxygen have very limited solubility in water because the solvent–solute interactions are very weak. The large permanent dipole of a water molecule interacts more favorably with other water dipoles than it does with the weaker, induced dipoles in the nonpolar materials. Solvent–solute interactions between water and nonpolar molecules cannot compete with the much stronger solvent–solvent interactions in this case. However, the induced dipoles in molecules such as $CCl_4$ and $O_2$ are of the same order of magnitude as the dispersion forces between molecules of a nonpolar solvent; hence the solvent–solute interactions in that case compete favorably with the solvent–solvent interactions in the nonpolar solvent and the solute dissolves.

Nonpolar molecules are very sparingly soluble in polar solvents but can be quite soluble in nonpolar solvents. We explore these effects further in Chapter 11. Even though oxygen is only slightly soluble in water, its limited solubility of about 10 mg/L at normal atmospheric pressure is sufficient to sustain aquatic life.

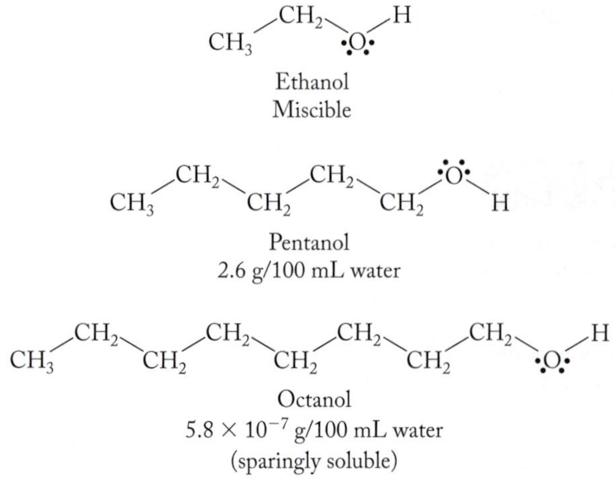

## Combinations of Intermolecular Forces

When larger molecules dissolve in a liquid solvent, we may need to take account of more than one type of intermolecular force. Consider the solubilities in water of the three organic alcohols with the structures shown in Figure 10.19. All three alcohols are liquids at room temperature, so we are looking at the solubility of one liquid in another. All three compounds contain a –OH group, which means they are like water in that they form hydrogen

CH₃—CH₂—Ö—H

Ethanol
Miscible

CH₃—CH₂—CH₂—CH₂—Ö—H

Pentanol
2.6 g/100 mL water

CH₃—CH₂—CH₂—CH₂—CH₂—CH₂—Ö—H

Octanol
$5.8 \times 10^{-7}$ g/100 mL water
(sparingly soluble)

**FIGURE 10.19** Ethanol, pentanol, and octanol have different solubilities in water.

bonds. Considering that they are all polar molecules, why do their solubilities differ so greatly in the polar solvent water?

Ethanol is miscible with water, which means it is soluble in all proportions. Pentanol has a finite solubility, which means that once a solution of pentanol in water is saturated, adding more pentanol results in a heterogeneous mixture in which two layers of liquid are visible. Octanol is sparingly soluble in water, and almost any measurable amount of octanol added to water results in a heterogeneous mixture.

The feature that differentiates these molecules from one another is the number of $-CH_2-$ groups. The long series of $-CH_2-$ groups in octanol makes the non-OH part of the molecule very nonpolar and thus unlike water. The dispersion forces between the long $-CH_2-$ chains of adjacent octanol molecules are quite strong (Figure 10.20), and these attractive forces keep the octanol molecules together, even though the polar ends of the molecules are attracted to water molecules. The minuscule solubility of octanol in water illustrates a situation where dispersion forces contribute more to intermolecular interactions than do dipole–dipole forces and hydrogen bonding, localized in one small region of a molecule.

In ethanol, the polarity and hydrogen bonding ability of the $-OH$ group dominate the interaction of ethanol with water because the hydrocarbon portion of the molecule is too short (only two carbon atoms long) to have significant dispersion forces. The solubility of pentanol is less than that of ethanol but much greater than that of octanol because of the differences in lengths of the $-CH_2-$ chains. The competing dispersion forces that limit pentanol solubility and the hydrogen-bonding interactions that promote it combine to produce moderate solubility for this alcohol.

Nonpolar interactions like the dispersion forces between hydrocarbon chains are called **hydrophobic** (literally, "water fearing") interactions, whereas interactions that promote solubility in water are called **hydrophilic** ("water loving") interactions. For molecules that contain both polar and nonpolar groups, as these alcohols do, solubility in water is due to the balance between hydrophilic and hydrophobic interactions. As the hydrophobic portion of the molecule increases in size, the entire molecule becomes more hydrophobic and solubility in water decreases.

**hydrophobic** a "water-fearing" or repulsive interaction between a solute and water that diminishes water solubility.

**hydrophilic** a "water-loving" or attractive interaction between a solute and water that promotes water solubility.

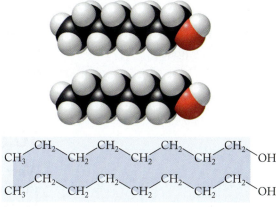

Dispersion forces

**FIGURE 10.20** Even though the interaction between two $-CH_2-$ units is weak, the interactions add up along the long hydrocarbon chains of octanol molecules and give rise to an attractive force that keeps the octanol molecules together in water despite the presence of the polar $-OH$ group in the alcohol.

---

**SAMPLE EXERCISE 10.5  Predicting Miscibility of Liquids   LO4**

Predict which of the following pairs of liquids are miscible. Explain your predictions. (a) Ethylene glycol ($HOCH_2CH_2OH$) and water; (b) benzene ($C_6H_6$) and pentane ($CH_3CH_2CH_2CH_2CH_3$); (c) octanol ($CH_3CH_2CH_2CH_2CH_2CH_2CH_2CH_2OH$) and octane ($CH_3CH_2CH_2CH_2CH_2CH_2CH_2CH_3$)

**Collect and Organize** We are given formulas of liquids and want to determine their miscibility. We can apply a molecular-level understanding of the common phrase "like dissolves like" to predict miscibility.

**Analyze** To predict mutual solubilities and explain our predictions, we must consider:

Polarity and size of each molecule → Number, type, and strength of each possible interaction → How alike are the two molecules?

Benzene

**Solve**

a. Ethylene glycol is an organic liquid with two $-OH$ groups per molecule that can form hydrogen bonds to water molecules. Prediction: the two liquids are miscible.

b. Benzene and pentane are both nonpolar hydrocarbons. Both molecules would interact via dispersion forces. They are both hydrophobic, and hence very alike. Prediction: the two are miscible.

c. Octanol has a polar –OH group capable of forming hydrogen bonds, whereas octane is nonpolar. However, octanol and octane both have long hydrocarbon chains of the same length. The dispersion forces associated with these chains interacting might lead to miscibility of the two substances. Prediction: the two are miscible.

**Think About It** Ethylene glycol and water are miscible in all proportions, as are benzene and pentane. Octanol and octane are also miscible and are actually used together as a mixed solvent system to separate mixtures of compounds containing metal ions.

⚙ **Practice Exercise** Predict whether methanol ($CH_3OH$) or hexanol ($CH_3CH_2CH_2$ $CH_2CH_2CH_2OH$) would be more soluble in acetone. Explain your answer.

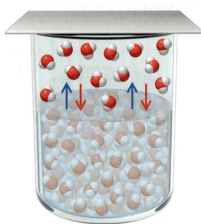

Acetone

# 10.5 Vapor Pressure of Pure Liquids

Water in a glass left on a countertop slowly disappears. Of course, it does not actually disappear, it just enters the gas phase and we can no longer see it. Molecules on the surface of the liquid vaporize, or evaporate, over time; they escape from the surface of the liquid and enter the gas phase. The rate at which the molecules make this transformation depends on the following factors:

1. The higher the temperature, the greater the number of molecules with sufficient kinetic energy to break the attractive forces that hold them together in the liquid and enter the gas phase.
2. The larger the surface area of the liquid, the greater the number of molecules on the surface in a position to enter the gas phase.
3. The stronger the intermolecular forces, the greater the kinetic energy needed for a molecule to escape the surface, and the smaller the number of molecules in the population that have this energy.

If a glass of water is covered (Figure 10.21), a different situation arises. Molecules at the surface still evaporate, but now they are confined to the space above the water. Some of them *condense* at the liquid surface and return to the liquid phase. In a short time, the rates of the two processes equalize, and as many molecules leave the surface as reenter it. At this point of dynamic equilibrium, no further change takes place in the level of the liquid in the glass, although molecules continue to evaporate and condense constantly. In such a situation at constant temperature, the pressure exerted by the gas in equilibrium with its liquid is called the **vapor pressure** of the liquid.

↑ Evaporation   ↓ Condensation

**FIGURE 10.21** A covered glass of water achieves a dynamic equilibrium when the rate at which the liquid water is lost to evaporation equals the rate at which liquid water is gained by condensation.

**vapor pressure** the pressure exerted by a gas at a given temperature in equilibrium with its liquid phase.

**normal boiling point** the temperature at which the vapor pressure of a liquid equals 1 atm (760 torr).

## Vapor Pressure and Temperature

Figure 10.22 is a graph of the vapor pressures of several liquids as a function of temperature, showing that vapor pressure increases with increasing temperature. At some point as temperature increases, the vapor pressure of any substance reaches 1 atm; the temperature at which this occurs is the **normal boiling point** of the substance. For water, to take one example from Figure 10.22, the vapor pressure reaches 1 atm at 100°C, which we recognize as the normal boiling point of water. The reason for calling it the *normal* boiling point arises from the observation that

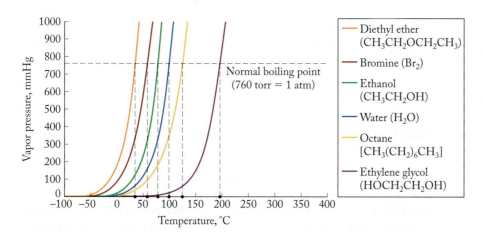

**FIGURE 10.22** A graph of vapor pressure versus temperature for six liquids shows that vapor pressure increases with increasing temperature. The temperature at which the vapor pressure equals 1 atm is the normal boiling point of the liquid.

most chemical reactions in nature, in our bodies, and in the laboratory take place at or near an atmospheric pressure of 1 atm.

Notice the order of the normal boiling points of the liquids in Figure 10.22 from lowest to highest: diethyl ether (74.12 g/mol) < bromine (159.81 g/mol) < ethanol (46.07 g/mol) < water (18.02 g/mol) < octane (114.23 g/mol) < ethylene glycol (62.07 g/mol). It is challenging to predict this order correctly because it is difficult to predict the relative strengths of the different types of intermolecular interactions. Taking only the organic molecules as an example, the molar mass of diethyl ether is over 1.5 times that of ethanol, but ethanol molecules are capable of hydrogen bonding, whereas diethyl ether molecules are not. Consequently, ethanol's boiling point (78.5°C) is over 40 degrees higher than diethyl ether's (34.6°C). Octane (126°C) has only dispersion forces between its molecules, but it has a long chain of eight carbon atoms over which the molecules interact. Its molar mass is almost 2.5 times that of ethanol and its boiling point is 48 degrees higher. Finally, ethylene glycol has a smaller molar mass than diethyl ether and just a little more than that of ethanol, but it has two groups capable of hydrogen bonding; its boiling point (197°C) is almost 120 degrees higher than that of ethanol.

**CONCEPT TEST** ••••••••••••••••••••••••••••••••••••••••••••••••••••

Diesel fuel is made of hydrocarbons with an average of 13 carbon atoms per molecule, and gasoline is made of hydrocarbons with an average of 7 carbon atoms per molecule. Which fuel has the higher vapor pressure at room temperature?

••••••••••••••••••••••••••••••••••••••••••••••••••••••••••••••••••••••

## Volatility and the Clausius–Clapeyron Equation

We describe substances as *volatile* if they evaporate readily at normal temperatures and pressures. At a given temperature, a more volatile substance has a higher vapor pressure than a less volatile substance, and as temperature increases, vapor pressure increases. The relationship between the vapor pressure of a pure substance and absolute temperature is not linear as can be seen in the curves in Figure 10.22. However, if we graph the natural logarithm of the vapor pressure versus $1/T$, where $T$ is the absolute temperature in kelvins, we get a straight line (Figure 10.23) described by the equation

$$\ln(P_{vap}) = -\frac{\Delta H_{vap}}{R}\left(\frac{1}{T}\right) + C$$

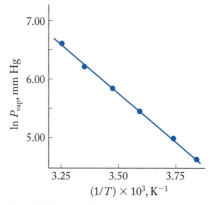

**FIGURE 10.23** Plotting the natural logarithm of the vapor pressure versus the reciprocal of the absolute temperature gives a straight line described by the Clausius–Clapeyron equation. This graph shows the plot for pentane.

where $\Delta H_{vap}$ is the enthalpy of vaporization, $R$ is the gas constant, $T$ is in kelvins, and $C$ is a constant that depends on the identity of the liquid. We can solve this equation for $C$ and write it in terms of two temperatures:

$$\ln(P_{vap,T_1}) + \left(\frac{\Delta H_{vap}}{RT_1}\right) = C = \ln(P_{vap,T_2}) + \frac{\Delta H_{vap}}{RT_2}$$

Rearranging the right and left sides of this equation, we get

$$\ln\left(\frac{P_{vap,T_1}}{P_{vap,T_2}}\right) = \frac{\Delta H_{vap}}{R}\left(\frac{1}{T_2} - \frac{1}{T_1}\right) \qquad (10.1)$$

We can use this expression, called the **Clausius–Clapeyron equation**, to calculate $\Delta H_{vap}$ if the vapor pressures at two temperatures are known. We can also calculate either the vapor pressure $P_{vap,T_2}$ at any given temperature $T_2$, or the temperature $T_2$ at any given vapor pressure $P_{vap,T_2}$, if $\Delta H_{vap}$, $P_{vap,T_1}$ and $T_1$ are known. Because the units of $\Delta H_{vap}$ are typically joules (J) or kilojoules (kJ) per mole, the value used for $R$ in the Clausius–Clapeyron equation is 8.314 J/(mol · K).

---

**SAMPLE EXERCISE 10.6** **Calculating Vapor Pressure** **LO5**

At its normal boiling point of 126°C, octane, $C_8H_{18}$, has a vapor pressure of 760 torr. What is its vapor pressure at 25°C? The enthalpy of vaporization of octane is 39.07 kJ/mol.

**Collect and Organize** The vapor pressure of octane at any temperature can be calculated using Equation 10.1, the Clausius–Clapeyron equation, given its enthalpy of vaporization and its normal boiling point.

**Analyze** To use the Clausius–Clapeyron equation, we must convert the given temperature to kelvins. Because we will use 8.314 J/(mol · K) for $R$, we must also convert $\Delta H_{vap}$ to joules per mole. Vapor pressure decreases as temperature decreases, so we expect the vapor pressure of octane to be significantly below 760 torr at 25°C.

**Solve** The two temperatures are

$$T_1 = 126°C + 273 = 399\ K \qquad \text{and} \qquad T_2 = 25°C + 273 = 298\ K$$

and $\Delta H_{vap}$ in joules is

$$39.07\frac{\cancel{kJ}}{mol} \times \frac{1000\ J}{1\ \cancel{kJ}} = 39,070\frac{J}{mol}$$

Inserting these values in Equation 10.2:

$$\ln\left(\frac{P_{vap,T_1}}{P_{vap,T_2}}\right) = \frac{\Delta H_{vap}}{R}\left(\frac{1}{T_2} - \frac{1}{T_1}\right)$$

$$\ln\left(\frac{760\ torr}{P_{vap,T_2}}\right) = \frac{39,070\ \dfrac{J}{\cancel{mol}}}{8.314\ \dfrac{J}{\cancel{mol} \cdot K}}\left(\frac{1}{298\ K} - \frac{1}{399\ K}\right)$$

$$P_{vap,T_2} = 14.0\ torr$$

**Think About It** We expect octane to have a low vapor pressure at a temperature much lower than its boiling point, so this number seems reasonable.

⚙ **Practice Exercise** Pentane, $C_5H_{12}$, boils at 36°C at $P$ = 1 atm. What is its molar heat of vaporization in kilojoules per mole if its vapor pressure at 25°C is 505 torr?

# 10.6 Phase Diagrams: Intermolecular Forces at Work

Whether a substance exists as a solid, liquid, or gas at a given temperature depends on the strength of forces between the particles making up the substance. However, temperature is not the only factor that influences the state of a substance. Pressure also plays a role.

To understand why temperature and pressure have these effects, think for a moment about gases. Our model for understanding the behavior of gases is the ideal gas law and kinetic molecular theory, which we discussed in Chapter 6. Recall that in the ideal gas law, we assume that particles in the gas phase do not interact, but that this idealization breaks down at high pressures and low temperatures. High pressures push molecules closer together; low temperatures reduce their average speed. Both of these conditions favor intermolecular interactions because molecules that are closer together collide and interact with each other more frequently, and low temperatures slow molecules down so they have more time to interact. Hence both temperature and pressure influence the degree to which molecules interact. Scientists use **phase diagrams** to show how the phases of substances depend on temperature and pressure.

## Phases and Phase Transformations

The phase diagram for water, like that for many other pure substances, has three regions corresponding to the three phases of matter, plus a fourth region called a *supercritical region* (Figure 10.24). The lines separating the regions are called *equilibrium lines* because the two states bordering them are at equilibrium. The blue equilibrium line separating the solid and liquid regions represents a series of freezing (or melting) points; the points on this line are combinations of temperature and pressure at which the solid and liquid states coexist. The red line separating the liquid and gas regions of Figure 10.24 represents a series of boiling points or condensation points; the points on this line are combinations of temperature and pressure at which the liquid and gaseous states coexist. The green line separating the solid and gaseous states represents a series of sublimation points (solid turning to gas) or deposition points (gas turning to solid); the points on this line are combinations of temperature and pressure at which the solid and gaseous states coexist.

Notice that the red line curves from the lower left to the upper right, separating the liquid and gas regions of the phase diagram. This line represents the changing boiling point of the liquid as a function of pressure. Its shape makes sense because when the pressure above a liquid increases, the temperature required for liquid molecules to overcome intermolecular attractive forces and enter the gas phase (i.e., the boiling point) increases. The shape of the solid–gas curve also makes sense: higher pressures make it more difficult for molecules of ice to overcome intermolecular forces and sublime to water vapor.

The trends in the boiling/condensation equilibrium line and the sublimation/deposition equilibrium line in Figure 10.24 show that both the boiling point (liquid ⇌ gas) and the sublimation point (solid ⇌ gas) of water decrease as pressure decreases. In both cases, a phase transition from a more dense phase to the gas phase occurs at a lower temperature when the pressure is lower. Lower

**Clausius–Clapeyron equation** relates the vapor pressures of a substance at different temperatures to its heat of vaporization.

**phase diagram** a graphical representation of how the stabilities of the physical states of a substance depend on temperature and pressure.

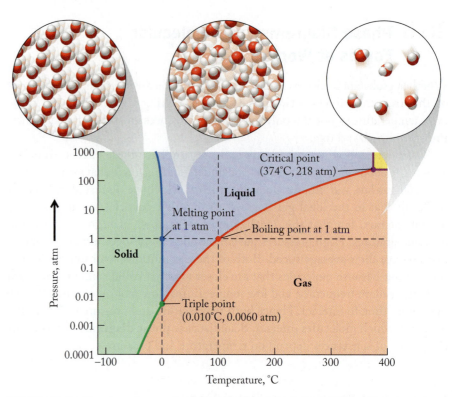

**FIGURE 10.24** The phase diagram for water indicates in which phase water exists at various combinations of pressure and temperature.

pressure allows molecules to be farther apart, weakening the effect of intermolecular interactions and offsetting the effects of a lower temperature. Typical solid and liquid phases of a compound, like ice and water, are similar in density and many times more dense than the vapor phase of the material. A decrease in pressure favors the phase change to the much less dense gas phase. Conversely, applying pressure to a gas forces its molecules closer together, which in turn increases the effect of their intermolecular interactions and favors the more dense liquid and solid phases. At some point, the pressure may change the gas into a liquid or solid, which takes up much less volume.

The trend in the melting/freezing equilibrium line for water, however, shows a decrease in the melting point of ice as pressure increases. This trend in the melting/freezing points for water is opposite the trend observed for almost all other substances (see, for example, the slope direction of the blue line for $CO_2$ in Figure 10.25). The reason for water's unusual melting/freezing behavior is that water expands when it freezes. Most other substances are denser in the solid state than in the liquid state because they contract when they freeze. Water, however, expands as it freezes because hydrogen bonds between molecules of water in the solid phase create a structure that is more open than the structure in liquid water, making ice less dense than liquid water. Applying enough pressure to ice forces it into a physical state (liquid water) in which it takes up less volume.

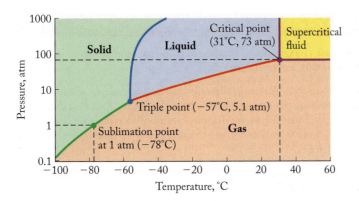

**FIGURE 10.25** Phase diagram for carbon dioxide.

---

**CONCEPT TEST**

A truck with a mass of 2000 kg is parked on an icy driveway where the ice is at −4°C. Is it possible that the ice under the tires of this vehicle will melt, even though the temperature remains constant?

---

A point of special interest on a phase diagram is the point where all three lines describing the phase transitions meet. Known as the **triple point**, it identifies the temperature and pressure at which all three states (liquid, solid, and gas) coexist. For water, the triple point is just above the normal melting temperature, at 0.010°C, but at a very low pressure of 0.0060 atm.

Another point of interest is the place where the boiling/liquefaction equilibrium line ends. At this **critical point**, the liquid and gaseous states are indistinguishable from each other. This point is reached because thermal expansion at this high temperature causes the liquid to become less dense, while the high pressure compresses the gas into a small volume, increasing its density. At the critical point, the densities of the liquid and gaseous states are equal, so one cannot be distinguished from the other.

At temperature–pressure combinations above its critical point, a substance exists as a **supercritical fluid**. A supercritical fluid has a density similar to that of a liquid, but it can penetrate materials like a gas and dissolve substances in them like a liquid. Supercritical carbon dioxide is used in the food-processing industry to decaffeinate coffee and remove fat from potato chips. Supercritical carbon dioxide and water are sometimes mixed to generate a range of fluids that are able to selectively dissolve specified materials from mixtures while leaving other components untouched.

The phase diagram of $CO_2$ is shown in Figure 10.25. The dashed line at $P = 1$ atm defines the phases that exist at 1 atm pressure, but note that the blue region representing liquid $CO_2$ does not extend below 5.1 atm. This means that solid $CO_2$ does not melt into a liquid at normal temperatures and pressures. Rather, it sublimes directly to $CO_2$ gas. This behavior gives rise to the common name for solid $CO_2$, *dry ice*; it is a solid that keeps things cool, as ice does, but it forms no "wet" liquid. The critical point of $CO_2$ is at 31°C and 73 atm, a pressure easily achieved with compressors in laboratories, factories, and food-processing plants, which means $CO_2$ is readily available for use as a supercritical fluid.

▶ II **CHEMTOUR** Phase Diagrams

---

**triple point** the temperature and pressure at which all three phases of a substance coexist. Under these conditions, freezing and melting, boiling and liquefaction, and sublimation and deposition all proceed at the same rate.

**critical point** a specific temperature and pressure at which the liquid and gas phases of a substance have the same density and are indistinguishable from each other.

**supercritical fluid** a substance at conditions above its critical temperature and pressure, where the liquid and vapor phases are indistinguishable and have some characteristics of both a liquid and a gas.

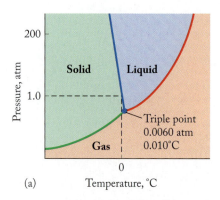

(a)

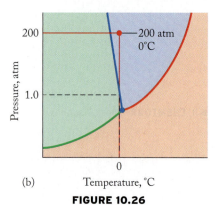

(b)

**FIGURE 10.26**

Describe the phase changes that take place as the pressure on a sample of water at 0°C is increased from 0.0001 atm to 200 atm.

**Collect and Organize** We are to describe the phase changes water undergoes at a constant temperature as the pressure is increased. We can use the phase diagram for water in Figure 10.26(a). To read a phase diagram, we need to remember that every point is characterized by a temperature and a pressure and that every time we cross an equilibrium line, the phase changes.

**Analyze** The changes we must describe take place along the vertical line that intersects the temperature axis at 0°C.

**Solve** Starting at the bottom of the phase diagram with the red line (Figure 10.26b) and following the dashed line at 0°C upward as pressure increases, we approximate the location of the starting pressure, 0.0001 atm, a little below and to the left of the triple point ($T = 0.010$°C and $P = 0.0060$ atm). Water at 0°C and 0.0001 atm is a gas, as indicated by the phase diagram. As we follow the 0°C dashed line up from the temperature axis, the point where this line intersects the green equilibrium line indicates that the gas solidifies to ice at a pressure below 1 atm. Increasing the pressure toward 1 atm, we see the dashed line intersect the blue equilibrium line at that pressure. This intersection means that at this pressure the ice melts to liquid water. Extending the 0°C dashed line farther upward tells us that the water remains a liquid at pressures up to and beyond 200 atm.

**Think About It** The phase changes in water, from gas to solid to liquid with increasing pressure at 0°C, make sense from a molecular point of view because higher pressures favor the densest phase. Water is denser than ice, so the transitions from gas to solid to liquid are transitions from the least dense to the most dense phase.

**Practice Exercise** Describe the phase changes that occur when the temperature of $CO_2$ (Figure 10.25) is increased from −100°C to 200°C at a pressure of 25 atm.

**CONCEPT TEST**

Look at the phase diagram in Figure 10.25. If a phase diagram favors the denser phase as pressure increases, does solid $CO_2$ float on liquid $CO_2$ at any point along the blue line?

# 10.7 Some Remarkable Properties of Water

Water has many remarkable properties. Its melting and boiling points are much higher than those of all other molecular substances with similar molar masses (Table 10.5), and its solid form (ice) is less dense than liquid water. These phenomena and many others are related to the strength of the hydrogen bonds that attract water molecules to one another. Let's look now at some other unique properties resulting from the hydrogen bonds in water.

## Surface Tension and Viscosity

▶II **CHEMTOUR** Hydrogen Bonding in Water

**Surface tension** is the resistance of a liquid to any increase in its surface area. Surface tension represents the energy required to move molecules apart so that an object can break through the surface.

| TABLE 10.5 | Melting Points and Boiling Points of Four Compounds of Similar Molar Mass | | |
|---|---|---|---|
| Substance | $\mathcal{M}$ (g/mol) | Melting Point (°C) | Boiling Point (°C) |
| $H_2O$ | 18.02 | 0 | 100 |
| HF | 20.01 | −83 | 19.5 |
| $NH_3$ | 17.03 | −78 | −33 |
| $CH_4$ | 16.04 | −182 | −164 |

Hydrogen bonding in water creates such a high surface tension, $7.29 \times 10^{22}$ J/m$^2$ at 25°C, that a carefully placed steel needle floats on water and insects called water striders can walk on it (Figure 10.27). The same needle and insect would sink in oil or gasoline.

Another illustration of the intermolecular forces acting in liquids is seen in the shape of the liquid surface when it is in a graduated cylinder or other small-diameter tube (Figure 10.28). The surface of water is concave in such a tube, but the surface of liquid mercury is convex. Either curved surface, concave or convex, is called a **meniscus**. In both liquids, the meniscus is the result of two competing forces: *cohesive forces*, which are interactions between like particles, and *adhesive forces*, which are interactions between unlike particles. In the water sample in Figure 10.28a, the cohesive forces are hydrogen bonds between water molecules and the adhesive forces are dipole–dipole interactions between water molecules and polar Si—O—Si groups on the surface of the glass. The adhesive

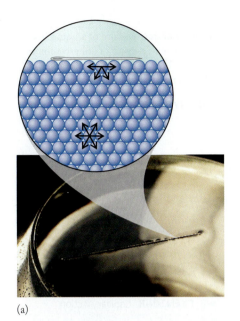

(a)

(b)

**FIGURE 10.27** (a) Intermolecular forces, including hydrogen bonding, are exerted equally in all directions in the interior of a liquid. Because forces in opposite directions cancel, a molecule in the interior does not experience surface tension. However, there is no liquid water above a surface to exert attractive intermolecular forces. The resulting imbalance causes the surface water molecules to adhere tightly to one another, creating surface tension. When the magnitude of the surface tension exceeds the downward force exerted on the surface water molecules by an object, such as a needle, the object cannot break through and floats on the surface. (b) Surface tension allows a water strider to rest on top of water without sinking.

**surface tension** the energy needed to separate the molecules at the surface of a liquid.

**meniscus** the concave or convex surface of a liquid in a small-diameter tube.

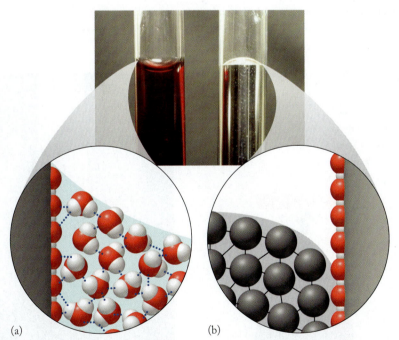

(a)  (b)

**FIGURE 10.28** (a) Hydrogen bonds between the H atoms of water molecules and the O atoms of silicon dioxide that makes up the glass are an adhesive force that causes water molecules to adhere to the glass surface and form a concave meniscus. (b) Because mercury atoms have no such attraction to the silicon dioxide in the glass, mercury forms a convex meniscus.

**capillary action** the rise of a liquid in a narrow tube as a result of adhesive forces between the liquid and the tube and cohesive forces within the liquid.

**viscosity** the resistance to flow of a liquid.

▶❚❚ **CHEMTOUR** Capillary Action

forces are strong enough to cause the water to climb upward on the glass, creating the concave meniscus. The adhesive forces are greater than the cohesive forces in this case because surface water molecules next to the glass have less contact with other water molecules and therefore experience smaller cohesive forces.

In the mercury sample (Figure 10.28b), the cohesive forces are metallic bonds between mercury atoms and the adhesive forces are interactions between induced dipoles in the mercury atoms and the polar Si—O—Si groups on the glass surface. In this case, the adhesive forces are much weaker than the cohesive forces, and mercury atoms are more attracted to one another than to the glass. As a result, they do not adhere to the wall, forming a convex meniscus (Figure 10.28b).

The observation that adhesive forces outweigh cohesive forces on the surface of water has consequences in other situations. In a narrow tube adhesion to the tube wall draws the outer ring of water molecules upward. At the same time, cohesive forces between the outer ring molecules and those adjacent to them draw the molecules upward (Figure 10.29). If the tube is narrow enough (a capillary tube), this combination of adhesion and cohesion draws a column of water up the tube in a phenomenon known as **capillary action**. The water column reaches its maximum height when the downward force of gravity balances the upward adhesive and cohesive forces.

In a test tube or pipette, water molecules in contact with the glass move only a very small distance up the glass before the force of gravity balances the adhesive and cohesive forces. However, when a nurse takes a blood sample after a pin prick in your finger, the blood (essentially an aqueous solution) spontaneously moves into a capillary tube because of capillary action. Capillary action also contributes to the process by which water rises 100 meters or more up the trunks of tall trees.

**CONCEPT TEST**

Would mercury be spontaneously drawn into a glass capillary tube?

**FIGURE 10.29** Because of capillary action, colored water rises (a) in a capillary tube and (b) in a stick of celery.

(a)

(b)

**Viscosity**, or resistance to flow, is another property of liquids related to the strength of intermolecular forces (Figure 10.30). In nonpolar compounds—for example, petroleum products like the fuels pentane and hexane—viscosity increases with increasing molar mass. Lubricating oil, for example, is much more viscous than gasoline because the hydrocarbons in a typical lubricating oil have molar masses that are two to three times those of the hydrocarbons in gasoline. Larger molar masses mean stronger dispersion forces between molecules. Because of the stronger interactions, molecules of lubricating oil do not slide past one another as easily as the shorter-chain hydrocarbons in gasoline, and bulk quantities of the larger molecules do not flow as easily when poured.

The viscosities of polar compounds are influenced by both London dispersion and dipole–dipole interactions. Considering two of the polar alcohols in Figure 10.19, we predict that octanol has a higher viscosity than ethanol, reflecting the greater dispersion forces in octanol. Water is more viscous than gasoline even though water molecules are much smaller than the nonpolar molecules in gasoline. The remarkable viscosity of water is another property directly related to the hydrogen bonds between water molecules.

Another property unique to water is how its density changes with temperature. The density of water increases as it is cooled to 4°C (Figure 10.31), a pattern observed for most liquids and solids and for all gases. However, as water is cooled from 4°C to 0°C, it expands, and its density decreases as the pattern of its hydrogen bonds changes with temperature. As water freezes at 0°C, its density drops even more, to about 0.92 g/mL for ice, causing ice to float on liquid water (Figure 10.32). This unusual behavior is caused by the formation of a network of hydrogen bonds in ice. With each oxygen atom covalently bonded to two hydrogen atoms and hydrogen-bonded to two other hydrogen atoms, the molecules form an extensive and open hexagonal network in the solid phase. Because of the space between the molecules in the network, the same number of molecules occupies more volume in ice than in liquid water. When ice melts, some of the hydrogen bonds in the rigid array break, allowing the molecules in the liquid to be arranged more compactly, up to 4°C, whereupon the density begins to decrease once more.

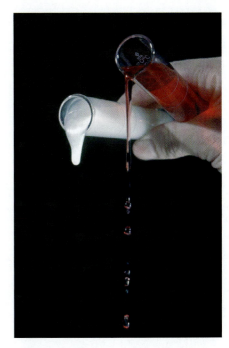

**FIGURE 10.30** High viscosity results from strong intermolecular forces between large polar molecules. The white liquid is more viscous than the red, which pours freely.

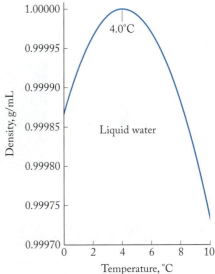

**FIGURE 10.31** As water is cooled, its density increases until its temperature reaches 4°C. At this temperature the density has its maximum value. As the water cools from 4°C to its freezing point at 0°C, its density decreases.

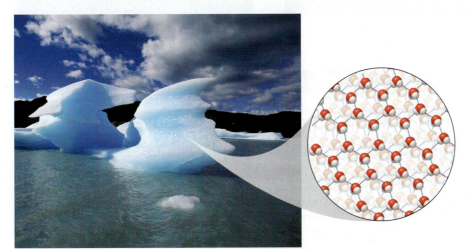

**FIGURE 10.32** Because of the changes in the density of water near the freezing point and because ice has a lower density (0.92 g/mL) than water, ice and ice-cold water float on top of warmer water (4°C) in lakes and rivers. In ice, each oxygen atom is linked to four hydrogen atoms: two by covalent bonds and two by hydrogen bonds.

## Water and Aquatic Life

The expanded structure of ice plays a crucial ecological role in temperate and polar climates. The lower density of ice means that lakes, rivers, and polar oceans freeze from the top down, allowing fish and other aquatic life to survive in the liquid water below. Each fall, surface water cools first, and its density increases until its temperature reaches 4°C, the peak in the graph of Figure 10.31. This maximum-density water at 4°C sinks to the bottom, bringing warmer water to the surface, which in turn cools to 4°C and sinks, pushing the next layer of warm water to the surface in a continuing cycle (Figure 10.33). This autumnal turnover stirs up dissolved nutrients, making them available for life in the sunlit surface during the next growing season. When all of the water has reached 4°C and the surface water cools further, it becomes less dense and ice may eventually form. The layer of ice insulates the 4°C water beneath it, allowing aquatic life to survive.

In spring, the ice melts and the surface water warms to 4°C. At this temperature, the entire column of water has nearly the same temperature and density; dissolved nutrients for plant growth are evenly distributed, and the stage is set for a burst of photosynthesis and biological activity called the spring bloom.

Further warming of the surface water creates a warm upper layer separated from colder, denser water by a *thermocline*, which is a sharp change in temperature between the two layers. Biological activity depletes the pool of nutrients above the thermocline as decaying biomass settles to the bottom. Consequently, photosynthetic activity drops from its spring maximum during the summer, even though there is much more energy available from the sun. The thermocline persists until the autumn turnover mixes the water column and the cycle begins anew.

**FIGURE 10.33** (a) As the surface water of a pond cools to 4°C, its density increases and it sinks to the bottom, bringing the warmer, less dense water to the surface. (b) Continued cooling of the surface water below 4°C produces a less dense layer that may eventually freeze while the more dense water deeper in the pond remains at 4°C.

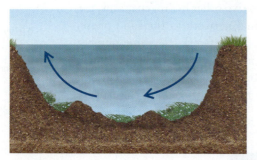

(a) Autumn

(b) Winter

---

**SAMPLE EXERCISE 10.8   Integrating Concepts: Testing Possible Drugs**

Candidate pharmaceutical agents may initially be evaluated for their partition coefficient, which is the ratio between their solubility in octanol (a mimic for cell membranes) and their solubility in water (a mimic for body fluids). A drug must be sufficiently hydrophilic to be carried in the blood and sufficiently hydrophobic to move across cell membranes. One way to determine this ratio is to put a weighed sample of the candidate agent in a vessel containing octanol and water. The container is vigorously shaken, and the amount of sub-stance that ends up in the water and the amount in the octanol are measured. The ratio is calculated from these two numbers:

$$\text{Partition coefficient} = \frac{\text{g substance in octanol}}{\text{g substance in water}}$$

a. Draw a rough sketch of a tube containing 10 mL of water (density 1.00 g/mL) and 10 mL of octanol (density 0.83 g/mL). Indicate each fluid clearly and suggest what the meniscus would look like.

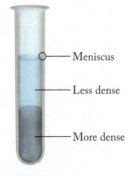

5-Fluorouracil

130.08 g/mol

BCNU

214.05 g/mol

**FIGURE 10.34**

b. Two anticancer drugs, 5-fluorouracil (5-FU) and BCNU, are shown in Figure 10.34. Predict which has the higher solubility in octanol and which has the higher solubility in water. Explain your prediction.

c. 5-FU has an octanol–water partition coefficient of 0.112. When 0.100 g of 5-FU was tested by this method, how much drug was found in the water and how much in the octanol?

**Collect and Organize** To run the experiment described, we have octanol and water together in a test tube. We are given the structures of two anticancer drugs and the value of one of their octanol–water partition coefficients.

**Analyze** We are first asked to show what the system looks like, so we need to think about the miscibility of water and octanol, and if they are immiscible, we need their densities to determine which would float on the other. We also need to think about how the top fluid would interact with the glass in forming a meniscus.

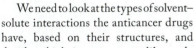

We need to look at the types of solvent–solute interactions the anticancer drugs have, based on their structures, and decide which is more water-like and which more octanol-like. Based on molecular masses alone, we predict that 5-fluorouracil is more soluble in water than is BCNU.

**Solve**

a. As we discussed in Section 10.4, both water and octanol can form hydrogen bonds, but the strong dispersion forces between the hydrocarbon chains in octanol make it only slightly soluble in water. Octanol is less dense than water, so the octanol layer floats on the water layer.

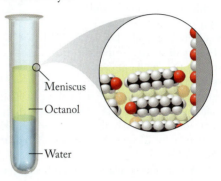

Regarding the meniscus, octanol is very oil-like, despite the hydrogen-bonding –OH groups. Its meniscus would be much less concave than the meniscus of water.

b. 5-Fluorouracil has two –NH groups and two polar C=O groups capable of hydrogen bonding with water. BCNU has only one –NH group and one C=O group; it also has two other nitrogen atoms, but one is $sp^2$ hybridized and attached to an electronegative oxygen atom, so it would not have electrons readily available to bond with a hydrogen atom in water. In addition, the molar mass of BCNU is almost twice that of 5-fluorouracil, which also indicates a lower water solubility. 5-Fluorouracil is most likely the more water-soluble of the two drugs and would therefore have the smaller octanol–water partition coefficient.

c. For 5-fluorouracil:

$$0.112 = \frac{\text{g drug in octanol}}{\text{g drug in water}}$$

If the total amount of drug added to the test tube is 0.100 g, we can call the amount dissolved in water $x$ and the amount dissolved in octanol $0.100 - x$. Substituting these expressions into the ratio, we can calculate $x$:

$$0.112 = \frac{0.100 - x}{x} \qquad 0.112x = 0.100 - x$$

$$x + 0.112x = 0.100 \qquad x = 0.0899 \text{ g}$$

Therefore the amount dissolved in water is 0.0899 g and the amount in octanol is 0.0101 g.

**Think About It** The known partition coefficient of BCNU is 34.7, so it is indeed less soluble in water and more soluble in octanol than 5-fluorouracil. This evaluation is usually carried out with an aqueous solution of sodium chloride rather than pure water to even more closely mimic the behavior of drugs in blood.

# The Halogens: The Salt of the Earth

The halogen family, group 17 of the periodic table, consists of four common elements—fluorine, chlorine, bromine, and iodine—and one, astatine, that is rarely encountered because all of its 24 isotopes are radioactive. Nothing is known of the bulk physical properties of astatine, and it may be the rarest naturally occurring terrestrial element. Estimates suggest that the outermost kilometer of Earth's crust contains less than 45 milligrams of astatine. We discussed fluorine in the descriptive chemistry section in Chapter 8, and we address the remaining halogens here.

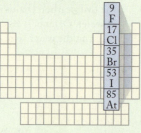

The word *halogen* means salt-former. Although the chemistry of the halogens as components of salts is important, these elements have a rich molecular chemistry as well.

Physical properties of the common halogens follow the expected trends:

|  | Fluorine (F₂) | Chlorine (Cl₂) | Bromine (Br₂) | Iodine (I₂) |
|---|---|---|---|---|
| Color: | Pale yellow | Yellow-green | Red-brown | Violet-black |
| State under standard conditions: | Gas | Gas | Liquid | Solid |
| Melting point (°C): | −219 | −101 | −7 | 114 |
| Boiling point (°C): | −188 | −34 | 59 | 185 |
| Atomic radius (pm): | 71 | 99 | 114 | 133 |
| Ionic radius (pm): | 133 | 181 | 195 | 220 |
| Energy of H—X bond (kJ/mol): | 567 | 431 | 366 | 299 |

Under standard conditions, fluorine and chlorine are both gases, bromine is a liquid, and iodine is a solid (Figure 10.35). Bromine and iodine have low boiling and sublimation points, respectively, so that a dark-red vapor always accompanies liquid bromine at room temperature, and violet vapor is frequently visible above solid iodine. Bromine is the only liquid nonmetallic element under standard conditions.

**FIGURE 10.35** Liquid bromine and solid iodine both vaporize considerably at room temperature.

All the halogens have strong, penetrating odors and are hazardous to humans. A concentration of chlorine in the air greater than 1 ppm is damaging to health, and a few breaths of air containing 1000 ppm of chlorine are fatal. Bromine is extremely corrosive to human tissue, and exposure of the skin causes burns that are painful and slow to heal. Iodine, as iodide ion, is an essential trace element, but ingestion of large amounts of iodine is dangerous, and consumption of 2 to 3 g is fatal to humans.

The halogens may be the most important group in the periodic table in terms of general industrial use. By far the most important use of chlorine is in the manufacture of chemicals used to sterilize water for drinking and for filling swimming pools, bleach for the paper industry, explosives, dyes, insecticides, cleaning agents, and plastics. The synthetic sweetener Splenda is made from sucrose by replacing three of its −OH groups with chlorine atoms. The molecule retains its sweetness but is not metabolized in our bodies and hence is considered calorie-free. Chloride ions are ubiquitous in living systems and play a vital role in biology. They balance the positive charge on the sodium and potassium ions in body fluids and thereby maintain electrical neutrality.

Bromine is used in making fumigants and insecticides, dyes, compounds for purifying water, and flame-proofing agents.

Tincture of iodine, typically part of emergency survival and first aid kits, is a solution of up to 10% iodine in ethanol that is used to disinfect wounds and purify water. The radioactive isotope [131]I is used to treat thyroid cancer and other diseases of the thyroid. Two other radioactive isotopes, [123]I and [125]I, are used as imaging agents to evaluate thyroid function. Table salt is often enriched with iodide ion (iodized salt) to ensure proper production of the thyroid hormones that regulate metabolism. Iodine is necessary for human health at the dietary level of 150 µg per day. Iodine deficiency in infants is a leading cause of preventable mental retardation and is a serious health problem in developing nations.

None of the halogens exist in nature as free diatomic elements. One source of chloride, bromide, and iodide ions is the ocean, from which these ions are extracted and then oxidized to produce the pure elements. It is estimated that $10^{16}$ tons of chloride ion are available in Earth's oceans. However, only about one-third of the NaCl used commercially is claimed from the currently existing oceans; the bulk of it is mined from rock salt deposits, which are the residues from the evaporation of ancient seas.

Chlorine is produced industrially by electrolysis of molten NaCl or aqueous solutions of NaCl. In the first reaction, sodium metal is produced; in the second, hydrogen gas and sodium hydroxide are produced, giving rise to the name *chlor-alkali process* for the reaction.

$$2\,NaCl(\ell) \rightarrow 2\,Na(s) + Cl_2(g)$$
$$2\,NaCl(aq) + 2\,H_2O(\ell) \rightarrow H_2(g) + Cl_2(g) + 2\,NaOH(aq)$$

Bromine is obtained from ocean water and from highly concentrated brine sources like the Dead Sea (Figure 10.36). The bromide ion is oxidized by treatment with $Cl_2$:

$$2\,Br^-(aq) + Cl_2(g) \rightarrow Br_2(\ell) + 2\,Cl^-(aq)$$

Elemental iodine is obtained from natural brines as well, and its collection involves the oxidation of $I^-$ to $I_2$. Another industrial source is Chilean saltpeter, in which iodine is in the form of iodate salts ($IO_3^-$). The iodate is reduced to $I^-$, which is then oxidized to $I_2$ by treatment with more $IO_3^-$:

$$IO_3^-(aq) + 3\,HSO_3^-(aq) \rightarrow I^-(aq) + 3\,SO_4^{2-}(aq) + 3\,H^+(aq)$$
$$5\,I^-(aq) + IO_3^-(aq) + 6\,H^+(aq) \rightarrow 3\,I_2(s) + 3\,H_2O(\ell)$$

Elemental chlorine reacts with water to form a mixture of hydrochloric and hypochlorous acids:

$$Cl_2(g) + H_2O(\ell) \rightarrow HCl(aq) + HOCl(aq)$$

All the other halogens react with water in this fashion to make the corresponding acids. Addition of NaOH to solutions of HOCl (a neutralization reaction) produces aqueous NaOCl, the active ingredient in household bleach:

$$HOCl(aq) + NaOH(aq) \rightarrow NaOCl(aq) + H_2O(\ell)$$

Hypochlorites are also used in the paper industry, to bleach wood pulp to produce white paper, and to sterilize water in swimming

**FIGURE 10.36** Salt formations at the shore of the Dead Sea indicate the high salt content of the water.

pools. The chlorine odor of pools is due to compounds called chloramines, produced when hypochlorous acid reacts with ammonia and nitrogen-containing compounds in bacteria:

$$HOCl(aq) + NH_3(aq) \rightarrow NH_2Cl(aq) + H_2O(\ell)$$

Chloramine is used as an alternative to direct chlorination to disinfect municipal drinking water. The reason some municipalities prefer chloramine to $Cl_2$ is related to the hazards associated with chlorine gas. Salts containing chlorate ($ClO_3^-$) and chlorite ($ClO_2^-$) ions are also used as bleaching agents. The perchlorate ion ($ClO_4^-$) reacts with organic matter rapidly, and some perchlorate salts are extremely reactive as contact explosives or as oxidizing agents when mixed with other substances that are easily oxidized. The oxidizers in many fireworks are perchlorate salts. When potassium perchlorate is mixed with charcoal, the material ignites spontaneously:

$$KClO_4(s) + 2\,C(s) \rightarrow KCl(s) + 2\,CO_2(g)$$

Mixtures of $KClO_4$, S, and Al provide the white flash and noise in fireworks. The flash powder used in stage shows is a mixture of Mg metal and $KClO_4$. Ammonium perchlorate decomposes explosively with heat and is also shock sensitive:

$$2\,NH_4ClO_4(s) \rightarrow N_2(g) + Cl_2(g) + 4\,H_2O(g) + 2\,O_2(g)$$

This behavior gives rise to its use as a solid propellant in space-shuttle booster rockets. Oxoanions of the other halogens, including bromate, iodate, perbromate, and periodate, have chemistries that are similar to those of the corresponding chlorine oxoanions.

## SUMMARY

**Learning Outcome 1** Ion–ion interactions hold ionic solids together. Their magnitude depends upon the charge on the ions and the distance between them. (Section 10.1)

**Learning Outcome 2** Ions interact with water through **ion–dipole inter-actions**. **Dipole–dipole interactions** take place between other polar molecules. The strongest dipole–dipole interactions are **hydrogen bonds**. **Dispersion (London) forces** are due to the **polarizability** of atoms and molecules and the existence of **temporary (induced) dipoles**. (Sections 10.2 and 10.3)

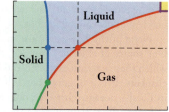

**Learning Outcome 3** Boiling points of compounds are strongly influenced by the strengths of the intermolecular interactions between particles. (Sections 10.2 and 10.3)

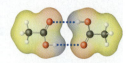

**Learning Outcome 4** Recognizing the types of intermolecular forces between particles helps explain solubilities. The relationship is summarized in the commonly applied phrase "like dissolves like." **Hydrophilic** substances are more soluble in water than are **hydrophobic** substances. (Section 10.4)

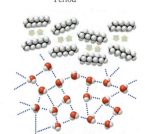

**Learning Outcome 5** Relative volatility of substances depends on the strength of intermolecular interactions between particles. The **Clausius–Clapeyron equation** relates vapor pressure to absolute temperature. (Section 10.5)

**Learning Outcome 6** The **phase diagram** of a substance indicates whether it exists as a solid, liquid, gas, or **supercritical fluid** at a particular pressure and temperature. (Section 10.6)

**Learning Outcome 7** The remarkable behavior of water, including its high melting and boiling points, its **surface tension**, its **capillary action**, and its **viscosity**, results from the strength of intermolecular hydrogen bonds. (Section 10.7)

## PROBLEM-SOLVING SUMMARY

| TYPE OF PROBLEM | CONCEPTS AND EQUATIONS | SAMPLE EXERCISES |
|---|---|---|
| **Predicting relative strengths of ion–ion interactions** | To predict relative interaction strengths, use $$E \propto \frac{(Q_1 Q_2)}{d} \qquad (5.3)$$ More negative values of $E$ correspond to stronger ion–ion attractions. | 10.1 |
| **Explaining differences in boiling points of liquids and trends in boiling points of pure substances** | Large molecules usually have higher boiling points than smaller molecules, with notable exceptions. The presence of polar –OH and –NH groups in molecules of a liquid leads to intermolecular hydrogen bonding that markedly increases the boiling point of the liquid. | 10.2, 10.3 |
| **Predicting solubility in water and miscibility of liquids** | Polar molecules are more soluble in water than nonpolar molecules. Molecules that form hydrogen bonds are more soluble in water than molecules that cannot form these bonds. Like dissolves like. | 10.4, 10.5 |
| **Calculating the vapor pressure, enthalpy of vaporization, or temperature of a pure liquid** | Use the Clausius–Clapeyron equation: $$\ln\left(\frac{P_{vap,T_1}}{P_{vap,T_2}}\right) = \frac{\Delta H_{vap}}{R}\left(\frac{1}{T_2} - \frac{1}{T_1}\right) \qquad (10.1)$$ and solve for any unknown value. | 10.6 |

| TYPE OF PROBLEM | CONCEPTS AND EQUATIONS | SAMPLE EXERCISES |
|---|---|---|
| **Reading a phase diagram** | Locate the point ($T$ and/or $P$) specified by the question as the starting point for the exercise. Temperature is on the horizontal axis; pressure is on the vertical axis. When you draw a line either horizontally rightward from the pressure axis or vertically up from the temperature axis, a phase change occurs wherever the line crosses an equilibrium line. | 10.7 |

## VISUAL PROBLEMS

*(Answers to boldface end-of-chapter questions and problems are in the back of the book.)*

**10.1.** Look at the pairs of ions in the structures of KF and KI represented in Figure P10.1. Which substance has the stronger cation–anion attractive forces and the higher melting point?

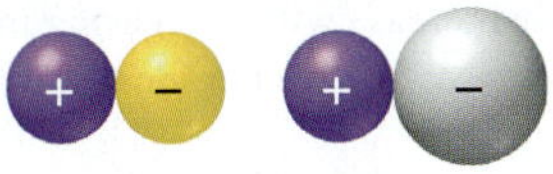

**FIGURE P10.1**

**10.2.** In Figure P10.2, identify the physical state (solid, liquid, or gas) of xenon and classify the attractive forces between the xenon atoms.

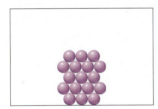

**FIGURE P10.2**

**10.3.** Figure P10.3 depicts molecules of $XH_3$ and $YH_3$ (not shown to scale) and the boiling points of $XH_3$ and $YH_3$ at 1 atm pressure, respectively. One substance is phosphine ($PH_3$) and the other substance is ammonia ($NH_3$). Which molecule is phosphine? Explain your answer.

XH₃
Boiling point −88°C

YH₃
Boiling point −33°C

**FIGURE P10.3**

**10.4.** Figure P10.4 shows representations of the molecules pentane, $C_5H_{12}$, and decane, $C_{10}H_{22}$. Which substance has the lower freezing point? Explain your answer.

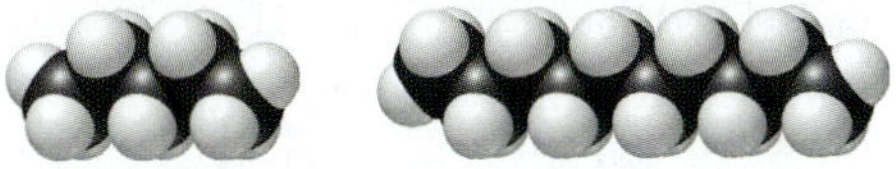

Pentane     Decane

**FIGURE P10.4**

**10.5.** The graphs in Figure P10.5 have the same scales and describe the change in $\ln P_{vap}$ of two pure liquids as a function of temperature. Which liquid has the stronger intermolecular attractive forces?

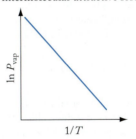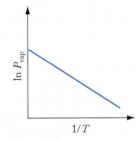

**FIGURE P10.5**

**10.6.** Which of the drawings in Figure P10.6, both of which are at constant temperature, most likely illustrates the pure liquid with the lower normal boiling point? Explain your choice.

A                                    B

**FIGURE P10.6**

**10.7.** Examine the phase diagram of substance Z in Figure P10.7. Does the freezing point of the substance increase or decrease with increasing pressure?

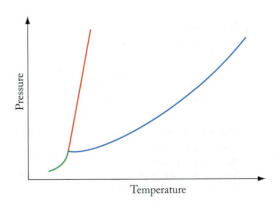

**FIGURE P10.7**

**10.8** Refer to Figure P10.7. Do you predict the solid phase of substance Z would float on the liquid phase? Why?

## QUESTIONS AND PROBLEMS

### Interactions between Ions

#### CONCEPT REVIEW

**10.9.** Indicate the substance that contains the largest anion. (a) $BaCl_2$; (b) $AlF_3$; (c) KI; (d) $SrBr_2$

10.10. Indicate the substance that contains the smallest cation. (a) $BaCl_2$; (b) $Al_2O_3$; (c) $Mg_3N_2$; (d) SrS

**10.11.** Why is $CaSO_4$ less soluble in water than NaCl?

10.12. Does the strength of an ion–ion attraction depend on the number of ions in the compound?

#### PROBLEMS

**10.13.** Rank the following ionic compounds in order of increasing attraction between their ions: KBr, $SrBr_2$, CsBr.

10.14. Rank the following ionic compounds in order of increasing attraction between their ions: BaO, $BaCl_2$, CaO.

### Interactions Involving Polar Molecules

#### CONCEPT REVIEW

**10.15.** How are the water molecules preferentially oriented around the anion in an aqueous solution of sodium chloride?

10.16. How are the water molecules preferentially oriented around the cation in an aqueous solution of potassium bromide?

**10.17.** Why are dipole–dipole interactions generally weaker than ion–dipole interactions?

10.18. Two liquids—one polar, one nonpolar—have the same molar mass. Which one is likely to have the higher boiling point?

**10.19.** Why are hydrogen bonds considered a special class of dipole–dipole interactions?

10.20. Can all polar hydrogen-containing molecules form hydrogen bonds?

#### PROBLEMS

**10.21.** In an aqueous solution containing chloride and iodide salts, which anion would you expect to be the more strongly hydrated?

10.22. In an aqueous solution containing Fe(II) and Fe(III) salts, which cation would you expect to be the more strongly hydrated?

**10.23.** Explain why the melting point of methyl fluoride, $CH_3F$ (−142°C), is higher than the melting point of methane, $CH_4$ (−182°C).

10.24. Explain why the boiling point of $Br_2$ (59°C) is lower than that of iodine monochloride, ICl (97°C), even though they have nearly the same molar mass.

**10.25.** Why doesn't fluoromethane ($CH_3F$) exhibit hydrogen bonding, whereas hydrogen fluoride, HF, does?

10.26. The boiling point of phosphine, $PH_3$ (−88°C), is lower than that of ammonia, $NH_3$ (−33°C), even though $PH_3$ has twice the molar mass of $NH_3$. Why?

**10.27.** In which of the following compounds do the molecules experience the strongest dipole–dipole attractions? (a) $CF_4$; (b) $CF_2Cl_2$; (c) $CCl_4$

10.28. Which of the following compounds, $CO_2$, $NO_2$, $SO_2$, or $H_2S$, is expected to have the weakest interactions between its molecules?

10.29. Which of the following molecules can hydrogen-bond among themselves in pure samples of bulk material? (a) methanol ($CH_3OH$); (b) ethane ($CH_3CH_3$); (c) dimethyl ether ($CH_3OCH_3$); (d) acetic acid ($CH_3COOH$)

10.30. Which of the following molecules can hydrogen-bond with molecules of water? (a) methanol ($CH_3OH$); (b) ethane ($CH_3CH_3$); (c) dimethyl ether ($CH_3OCH_3$); (d) acetic acid ($CH_3COOH$)

### Dispersion Forces

#### CONCEPT REVIEW

**10.31.** Which type of intermolecular force exists in all substances?

10.32. Why do the strengths of London (dispersion) forces generally increase with increasing molecular size?

**10.33.** Why do gases behave nonideally at high pressures and low temperatures?

10.34. Why are normal boiling points generally lower for branched hydrocarbons than for straight-chain hydrocarbons of the same molecular mass?

#### PROBLEMS

**10.35.** The permanent dipole moment of $CH_2F_2$ (1.93 D) is larger than that of $CH_2Cl_2$ (1.60 D), yet the boiling point of $CH_2Cl_2$ (40°C) is much higher than that of $CH_2F_2$ (−52°C). Why?

10.36. How is it that the permanent dipole moment of HCl (1.08 D) is larger than the permanent dipole moment of HBr (0.82 D), yet HBr boils at a higher temperature?

10.37. In each of the following pairs of molecules, which one experiences the stronger London (dispersion) forces? (a) $CCl_4$ or $CF_4$; (b) $CH_4$ or $C_3H_8$

10.38. What kinds of intermolecular forces must be overcome as (a) solid $CO_2$ sublimes, (b) $CHCl_3$ boils, and (c) ice melts?

*10.39. Consider the two molecules shown in Figure P10.39. One of these compounds is a liquid at room temperature; the other is a solid. Which is which? Explain why.

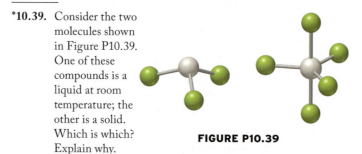

**FIGURE P10.39**

*10.40. Consider the two molecules shown in Figure P10.40. One has a boiling point of 57°C; the other, 84°C. Predict which has the higher boiling point, and explain your choice.

**FIGURE P10.40**

# Polarity and Solubility

## CONCEPT REVIEW

**10.41.** What is the difference between the terms *miscible* and *insoluble*?

**10.42.** Which of the following substances are essentially insoluble in water? (a) benzene, $C_6H_6(\ell)$; (b) KBr($s$); (c) $Br_2(\ell)$

**10.43.** One of the compounds in Figure P10.40 is insoluble in water; the other has a water solubility of 0.87 g/100 mL at 20°C. Identify which is which, and explain your reasoning.

**10.44.** Do you predict the two compounds in Figure P10.40 would be miscible with each other? Why?

**10.45.** In what context do the terms *hydrophobic* and *hydrophilic* relate to the solubilities of substances in water?

**10.46.** How does the presence of increasingly longer hydrocarbon chains in the structure affect the solubility of a series of structurally related molecules in water?

## PROBLEMS

**10.47.** In each of the following pairs of compounds, which compound is likely to be more soluble in water?
a. $CCl_4$ or $CHCl_3$     c. NaF or MgO
b. $CH_3OH$ or $C_6H_{11}OH$     d. $CaF_2$ or $BaF_2$

**10.48.** In each of the following pairs of compounds, which compound is likely to be more soluble in $CCl_4$?
a. $Br_2$ or NaBr     c. $CS_2$ or KOH
b. $CH_3CH_2OH$ or $CH_3OCH_3$     d. $I_2$ or $CaF_2$

---

**10.49.** Which of these pairs of substances is likely to be miscible?
a. $Br_2$ and benzene ($C_6H_6$)
b. $CH_3CH_2OCH_2CH_3$ (diethyl ether) and $CH_3COOH$ (acetic acid)
c. $C_6H_{12}$ (cyclohexane) and hexane ($CH_3CH_2CH_2CH_2CH_2CH_3$)
d. $CS_2$ (carbon disulfide) and $CCl_4$ (carbon tetrachloride)

**10.50.** Which of these pairs of substances is likely to be miscible?
a. $CH_3CH_2OH$ (ethanol) and $CH_3CH_2OCH_2CH_3$ (diethyl ether)
b. $CH_3OH$ (methanol) and methyl amine ($CH_3NH_2$)
c. $CH_3CN$ (acetonitrile) and acetone ($CH_3COCH_3$)
d. $CF_3CHF_2$ (a Freon replacement) and $CH_3CH_2CH_2CH_2CH_3$ (pentane)

---

**10.51.** Which of the following compounds is likely to be the most soluble in water? (a) NaCl; (b) KI; (c) $Ca(OH)_2$; (d) CaO

**10.52.** Based on the data in Figure P10.52, which has a greater effect on the solubility of oxygen in water: (a) decreasing the temperature from 20°C to 10°C or (b) raising the pressure from 1.00 atm to 1.25 atm?

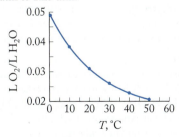

**FIGURE P10.52**

---

**10.53.** Which of these substances is the least soluble in water?
a. $CH_3(CH_2)_2CH_2OH$     c. $CH_3(CH_2)_6CH_2OH$
b. $CH_3(CH_2)_4CH_2OH$     d. $CH_3(CH_2)_8CH_2OH$

**10.54.** Which of these substances is the most soluble in water?
a. $CH_3(CH_2)_2CH_2NH_2$     c. $CH_3(CH_2)_6CH_2Br$
b. $CH_3(CH_2)_4CH_2Cl$     d. $CH_3(CH_2)_8CH_2I$

---

**10.55.** Which sulfur oxide would you predict to be more soluble in nonpolar solvents, $SO_2$ or $SO_3$?

# Vapor Pressure of Pure Liquids

## CONCEPT REVIEW

**10.56.** Why does the vapor pressure of a liquid increase as temperature increases?

**10.57.** Which of the following factors influences the vapor pressure of a pure liquid?
a. the volume of liquid present in a container
b. the temperature of the liquid
c. the surface area of the liquid

**10.58.** Is vapor pressure an intensive or extensive property of a liquid?

**10.59.** Do the molecules in the vapor phase above a liquid have the same formula as the molecules in the liquid phase?

**10.60.** A chef observes bubbles while heating a pot of water to 60°C to poach vegetables. What are the gases in the bubbles and where did they come from?

## PROBLEMS

**10.61.** Rank the following compounds in order of increasing vapor pressure at 298 K: (a) $CH_3CH_2OH$; (b) $CH_3OCH_3$; (c) $CH_3CH_2CH_3$

**10.62.** Rank the compounds in Figure P10.62 in order of increasing vapor pressure at 298 K.

Cyclopropane     Cyclobutane     Cyclopentane

**FIGURE P10.62**

---

**10.63. Pine Oil** The smell of fresh cut pine is due in part to the cyclic alkene pinene, whose carbon-skeleton structure is shown in Figure P10.63. (a) Use the data in the table to calculate the heat of vaporization, $\Delta H_{vap}$, of pinene. (b) Use the value of $\Delta H_{vap}$ to calculate the vapor pressure of pinene at room temperature (23°C).

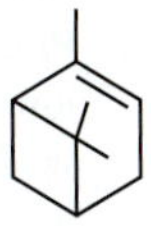

Pinene

**FIGURE P10.63**

| Vapor Pressure (torr) | Temperature (K) |
|---|---|
| 760 | 429 |
| 515 | 415 |
| 340 | 401 |
| 218 | 387 |
| 135 | 373 |

**10.64. Almonds and Cherries** Almonds and almond extracts are common ingredients in baked goods. Almonds contain the compound benzaldehyde (shown in Figure P10.64), which accounts for the odor of the nut. Benzaldehyde is also responsible for the aroma of cherries. (a) Use the data in the table to calculate the heat of vaporization, $\Delta H_{vap}$, of benzaldehyde. (b) Use the value of $\Delta H_{vap}$ to calculate the vapor pressure of benzaldehyde at room temperature (23°C).

Benzaldehyde

**FIGURE P10.64**

| Vapor Pressure (torr) | Temperature (K) |
| --- | --- |
| 50 | 373 |
| 111 | 393 |
| 230 | 413 |
| 442 | 433 |
| 805 | 453 |

## Phase Diagrams: Intermolecular Forces at Work

### CONCEPT REVIEW

**10.65.** Explain the difference between sublimation and evaporation.

**10.66.** Can ice be melted merely by applying pressure? Explain your answer.

**10.67.** Explain the effect of temperature and pressure on phase changes.

**10.68.** Explain how the solid–liquid line in the phase diagram of water differs in character from the solid–liquid line in the phase diagrams of most other substances, such as $CO_2$.

**10.69.** Which phase of a substance (gas, liquid, or solid) is more likely to be the stable phase: (a) at low temperatures and high pressures; (b) at high temperatures and low pressures?

**10.70.** At what temperatures and pressures does a substance behave as a supercritical fluid?

**10.71.** **Preserving Food** Freeze-drying is used to preserve food at low temperature with minimal loss of flavor. Freeze-drying works by freezing the food and then lowering the pressure with a vacuum pump to sublime the ice. Must the pressure be lower than the pressure at the triple point of $H_2O$?

**10.72.** Solid helium cannot be converted directly into the vapor phase. Does the phase diagram of helium have a triple point?

### PROBLEMS

For help in answering Problems 10.73 through 10.84, consult Figures 10.22, 10.24, and 10.25.

**10.73.** What is the normal boiling point of bromine?

**10.74.** If water boils at 50°C, what is the pressure?

**10.75.** Which molecules have stronger intermolecular attractive forces: ethylene glycol or ethanol? Explain your choice.

**10.76.** What is the boiling point of diethyl ether at 300 mmHg pressure?

**10.77.** List the steps you would take to convert a 10.0 g sample of water at 25°C and 1 atm pressure to water at its triple point.

**10.78.** List the steps you would take to convert a 10.0 g sample of water at 25°C and 2 atm pressure to ice at 1 atm pressure. At what temperature would the water freeze?

**10.79.** What phase changes, if any, does liquid water at 100°C undergo if the initial pressure of 5.0 atm is reduced to 0.5 atm at constant temperature?

**10.80.** What phase changes, if any, occur if $CO_2$ initially at −80°C and 8.0 atm is allowed to warm to −25°C at 5.0 atm?

**10.81.** Below what temperature can solid $CO_2$ (dry ice) be converted into $CO_2$ gas simply by lowering the pressure?

**10.82.** What is the maximum pressure at which solid $CO_2$ (dry ice) can be converted into $CO_2$ gas without melting?

**10.83.** Predict the phase of water that exists under the following conditions:
a. 2 atm of pressure and 110°C
b. 200 atm of pressure and 380°C
c. $6.0 \times 10^{-3}$ atm of pressure and 0°C

**10.84.** Which phase or phases of water exist under the following conditions?
a. 2.0 atm and 50°C       c. 1 atm and 0°C
b. 0.10 atm and 300°C

## Some Remarkable Properties of Water

### CONCEPT REVIEW

**10.85.** Explain why a needle floats on the surface of water but sinks in a container of methanol ($CH_3OH$).

**10.86.** Explain why different liquids do not reach the same height in capillary tubes of the same diameter.

**10.87.** Explain why pipes filled with water are in danger of bursting when the temperature drops below 0°C.

**10.88.** A hot needle sinks when put on the surface of cold water. Will a cold needle float in hot water?

**10.89.** The meniscus of mercury in a thermometer (Figure P10.89) is convex, rather than concave. Explain why.

*10.90. The mercury level in a capillary tube placed in a dish of mercury is below the surface of the mercury in the dish. Explain why.

**10.91.** Describe the origin of surface tension at the molecular level.

**FIGURE P10.89**

**10.92.** What is the cause of the high viscosity of molasses?

**10.93.** Describe how the surface tension and viscosity of a liquid are affected by increasing temperature.

**10.94.** Explain how strong intermolecular forces are expected to result in a relatively high surface tension and viscosity of a liquid.

### PROBLEMS

**10.95.** One of two glass capillary tubes of the same diameter is placed in a dish of water and the other in a dish of ethanol ($CH_3CH_2OH$). Which liquid will rise higher in its tube?

**10.96.** Would you expect water to rise to the same height in a tube made of a polyethylene plastic as it does in a glass capillary tube of the same diameter? The molecular structure of polyethylene is shown in Figure P10.96.

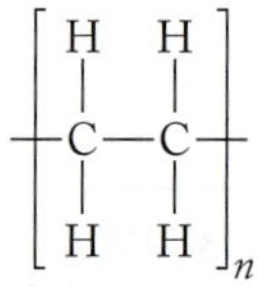

**FIGURE P10.96**

**10.97.** The normal boiling points of liquids A and B are 75.0°C and 151°C, respectively. Which of these liquids would you expect to have the higher surface tension and viscosity at 25°C? Explain your answer.

**10.98.** One beaker contains pure water and the other beaker contains pure methanol at the same temperature. Which liquid has the higher surface tension and viscosity? Explain your answer.

## Additional Problems

**10.99.** Does the sublimation point of ice increase or decrease with increasing pressure? Explain why.

**10.100.** Which substance contains the most positively charged cation? (a) $MgCl_2$; (b) $AlF_3$; (c) KI; (d) SrO

**10.101.** Why does methanol ($CH_3OH$) boil at a lower temperature than water, even though $CH_3OH$ has the greater molar mass?

**10.102.** Why is methanol ($CH_3OH$) miscible with water, whereas $CH_4$ is almost completely insoluble in water?

**10.103.** Does the sublimation point of ice increase or decrease with increasing pressure?

**10.104.** Sketch a phase diagram for element X, which has a triple point at 152 K and a pressure of 0.371 atm, a boiling point of 166 K at a pressure of 1.00 atm, and a normal melting point of 161 K.

**\*10.105.** The melting point of hydrogen is 14.96 K at 1.00 atm pressure. The temperature of its triple point is 13.81 K. Does $H_2$ expand or contract when it freezes?

**10.106.** Explain why water climbs higher in a capillary tube than in a test tube.

**10.107.** Explain why ice floats on water.

**10.108.** **Fish Dying in Summer Heat** Explain why fish in a pond die if water becomes too warm.

**\*10.109.** **Evaluation of Pharmaceuticals** A test done on new pharmaceutical agents early in the development process required the observation of their relative solubilities in octanol and water. A drug had to be sufficiently soluble in water (hydrophilic) to be carried in the bloodstream but also sufficiently hydrophobic (octanol soluble) to move across cell membranes. Pick the molecule from Figure P10.109 that you predict might have comparable solubility in both solvents.

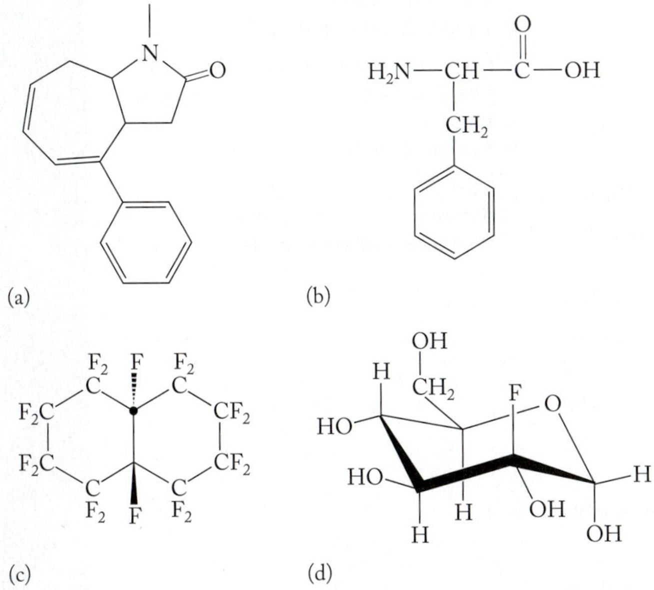

**FIGURE P10.109**

**10.110.** **First-Aid for Bruises** Compounds with low boiling points may be sprayed on skin as a topical anesthetic—they chill it as they evaporate, providing short-term relief from injuries. Predict which compound among those in Figure P10.110 has the lowest boiling point.

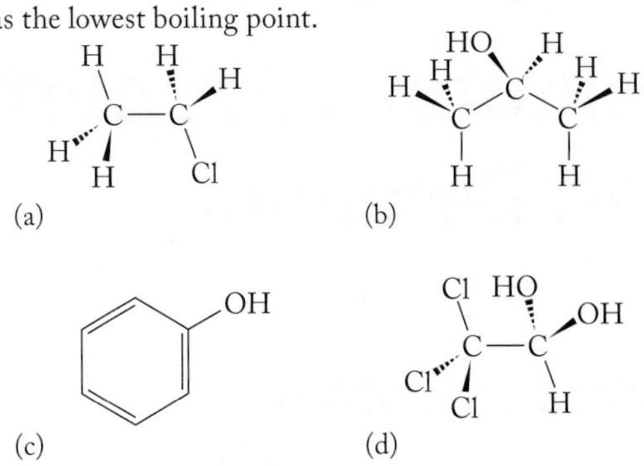

**FIGURE P10.110**

**10.111.** **Refrigerators** Refrigerators have a unit called a compressor that liquefies a gas. The refrigerator is cooled by a continuous cycle of compression of the gas to produce the liquid, followed by evaporation of the liquid to provide the cooling. Ammonia ($NH_3$) and sulfur dioxide ($SO_2$) were the gases used originally, and hexafluoroethane ($C_2F_6$) has been used since the 1990s. What are the intermolecular interactions that characterize these substances?

## The Halogens: The Salt of the Earth

**10.112.** What are the major natural sources of chlorine and bromine?

**10.113.** Under normal conditions, bromine is a volatile red liquid and iodine is a violet solid. Why are $Br_2$ and $I_2$ so volatile?

**10.114.** How do the boiling points of the halogens change in the progression $F_2$, $Cl_2$, $Br_2$, $I_2$?

**10.115.** Which is by far the least abundant of the halogens?

**10.116.** Give two industrial uses each of (a) chlorine and (b) bromine.

**10.117.** Give the chemical formula of chloramine. What shape is the chloramine molecule? Would you expect chloramine to be soluble in water? Why?

**10.118.** Which halide in the following pairs is the more easily oxidized? (a) fluoride or iodide ion; (b) bromide or chloride ion; (c) chloride or fluoride ion; (d) iodide or bromide ion

**10.119.** Give the formula of each of the following halogen species: (a) hypochlorite ion; (b) chlorite ion; (c) chlorate ion; (d) perchlorate ion

**10.120.** Which halogen species in each of the following pairs has the higher oxidation number? (a) hypochlorous acid, $Cl^-$; (b) chlorous acid, $HClO_3$; (c) chloric acid, $BrO_2^-$; (d) perchloric acid, $IO^-$

**10.121.** Balance the equation for the reduction of iodate by hydrogen sulfite ions to give iodide and sulfate in basic aqueous solution.

If your instructor assigns problems in **smartwork**, log in at **smartwork.wwnorton.com**.

# 11

# Solutions: Properties and Behavior

11.1 Vapor Pressure of Solutions

11.2 Solubility of Gases in Water

11.3 Energy Changes during Formation and Dissolution of Ionic Compounds

11.4 Mixtures of Volatile Solutes

11.5 Colligative Properties of Solutions

11.6 Measuring the Molar Mass of a Solute Using Colligative Properties

## Learning Outcomes

**LO1** Calculate the vapor pressure of a solution containing a nonvolatile solute using Raoult's law
**Sample Exercise 11.1**

**LO2** Calculate the solubility of gases in water using Henry's Law
**Sample Exercise 11.2**

**LO3** Explain the energy changes that accompany the formation and dissolution of an ionic compound
**Sample Exercises 11.3, 11.4, 11.5**

**LO4** Describe the process of fractional distillation and interpret graphs of temperature versus volume of distillate
**Sample Exercises 11.6, 11.7**

**LO5** Express the concentration of a solution in molality
**Sample Exercise 11.8**

**LO6** Calculate the freezing point and boiling point of a solution of a nonvolatile solute
**Sample Exercises 11.9, 11.10**

**LO7** Explain the significance of the van 't Hoff factor
**Sample Exercises 11.11, 11.12**

**LO8** Predict the direction of solvent flow in osmosis and calculate osmotic pressure
**Sample Exercises 11.13, 11.14, 11.15**

**LO9** Apply colligative properties to the determination of molar mass
**Sample Exercises 11.16, 11.17**

## A World of Solutions

In Chapter 10 we discussed the unique properties of water and other pure liquids in terms of the intermolecular forces that determine how molecules associate with like molecules or other particles. In this chapter we focus more closely on the ability of water and other liquids to dissolve solids, liquids, and gases and the properties exhibited by the resultant solutions. These properties are of prime importance to all living things, which depend on aqueous solutions to contain and transport molecules and ions within cells and between cells. In addition, the solubility of oxygen gas in water is crucial to the survival of myriad aquatic life forms. Solubility is an important feature of many substances in the material world as well, where the behavior and interactions of substances often depend on the identity and amount of solute dissolved in a solvent. We live in a world of solutions.

The concentration of solute in a solution can be of critical importance. Water dispensed intravenously in hospitals, for delivering medications to patients or restoring fluids to trauma victims, must contain specific concentration of solutes. Too much solute in the intravenous fluid causes dehydration, whereas too little causes retention of excess water by the body. The concentration of oxygen dissolved in lakes, oceans, ponds, and streams is of critical importance to aquatic life—it is much lower than in air, and even small reductions in the level of dissolved $O_2$ can make it difficult for fish and other creatures to extract enough $O_2$ from the water to sustain life. Our own bodies depend on an adequate supply of $O_2$ in the air we breathe and the solubility of $O_2$ in the water present in the tiny, moist alveolar sacs in our lungs. Oxygen must move from the water to plasma in the alveolar capillaries before it can be transported throughout the body by hemoglobin in the blood. Life in air and water depends on relatively narrow ranges of amounts of solutes ($O_2$, salts) dissolved in solutions (lakes, blood).

Solutions differ from pure liquids in several important respects. For example, solutions of 50–80 percent ethylene glycol in water, known as antifreeze, freeze at much lower temperatures than pure water or pure ethylene glycol, which is why these solutions are used in the radiators of vehicles. Antifreeze solutions also boil at higher temperatures than pure water, thereby increasing the temperature range over which fluid can transfer heat away from the engine. The temperatures

**Healthy Red Blood Cells** Red blood cells contain hemoglobin, which binds ▶ oxygen and transports it from the lungs to tissues.

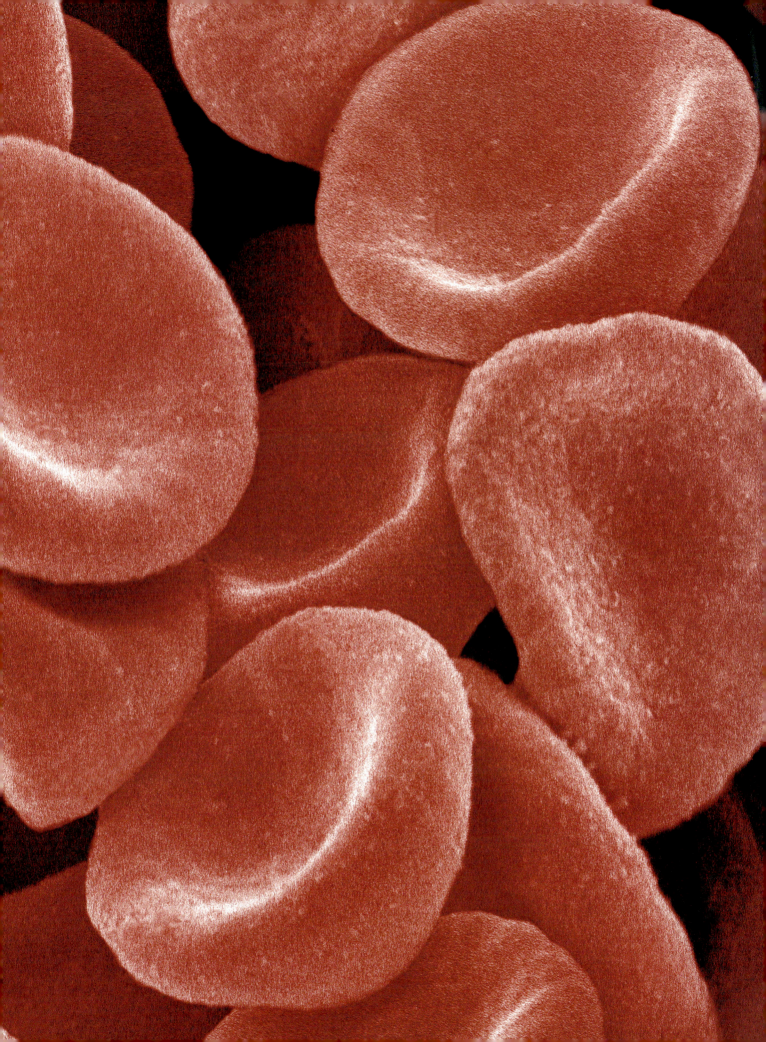

at which an antifreeze solution boils or freezes depend on the concentration of the solute (ethylene glycol) in the solvent (water).

The reasons solutions and pure liquids behave differently arise from the intermolecular forces between particles of solvent and solute. In this chapter we extend many of the ideas about pure liquids presented in Chapter 10. We examine how these ideas apply to the more common situations we face in dealing with solutions, such as how different kinds of particles interact and how the properties of a mixture change with variations in composition. ■

# 11.1 Vapor Pressure of Solutions

In Chapter 10 we discussed the influence of intermolecular interactions on the vapor pressure and the normal boiling point of pure liquids. Let's see what happens to vapor pressure when nonvolatile solutes such as salts are dissolved in water.

When adjoining compartments of seawater (water that contains numerous dissolved nonvolatile solutes) and pure water are sealed in a chamber (Figure 11.1), the volume of fluid on the seawater side increases over time, while the volume on the side of pure water decreases at the same rate. Eventually, nearly all the water ends up in the seawater compartment. The transfer of the pure water is due to the dissolved solutes in the seawater.

As the water in both compartments evaporates, the concentration of water vapor in the air space of the sealed chamber increases. As the concentration of water vapor increases, the pressure the water vapor exerts on the two liquid surfaces increases. At constant temperature, this pressure eventually stabilizes at a value equal to the vapor pressure of water at that temperature. At this point, the rate of evaporation from the compartments is equal to the rate of condensation. If the evaporation and condensation rates for the pure water and the seawater were the same, the liquid levels in the compartments would not change over time.

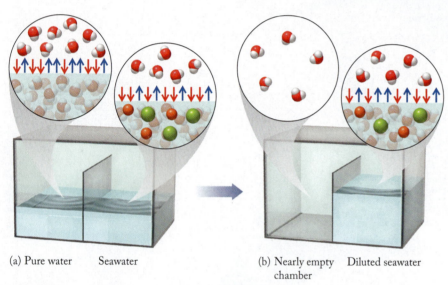

(a) Pure water    Seawater

(b) Nearly empty    Diluted seawater
    chamber

**FIGURE 11.1** (a) Adjoining compartments are partially filled with pure water and seawater. (b) The slightly higher vapor pressure of the pure water leads to a net transfer of water from the pure-water compartment to the seawater compartment.

However the liquid levels do change. Because the $H_2O(g)$ molecules in the sealed chamber are free to condense into either compartment, the rate of condensation into both compartments is the same. The presence of dissolved solute in the seawater affects its rate of evaporation. Since most of the pure water ends up in the seawater compartment, the pure water must have a higher rate of evaporation than the seawater. This conclusion leads to another: if the seawater evaporates at a lower rate, it must have a lower vapor pressure than the pure water at the same temperature.

More water vapor enters the air in the chamber from the pure water compartment than from the seawater compartment because the vapor pressure of the pure water is greater than the vapor pressure of the seawater. Because the condensation rates are the same, the pure water loses more water over time than is restored to it by condensation, while the seawater gains more water than it loses by evaporation. The process depicted in Figure 11.1 is an illustration of a more general observation: at a given temperature, the vapor pressure of the sol-

**Raoult's law** the vapor pressure of the solvent in a solution is equal to the vapor pressure of the pure solvent multiplied by the mole fraction of the solvent in the solution.

**colligative properties** characteristics of solutions that depend on the concentration and not the identity of particles dissolved in the solvent.

**ideal solution** one that obeys Raoult's law.

vent in a solution containing nonvolatile solutes is less than the vapor pressure of the pure solvent.

## Vapor Pressure of Solutions: Raoult's Law

The connection between the vapor pressure of a solution and the concentration of nonvolatile solutes dissolved in the solvent was studied extensively by French chemist François Marie Raoult (1830–1901). He discovered that the relation between the vapor pressure of a solution, $P_{solution}$, and that of the pure solvent, $P^{\circ}_{solvent}$ is

$$P_{solution} = X_{solvent}\, P^{\circ}_{solvent} \qquad (11.1)$$

where $X_{solvent}$ is the mole fraction of solvent. This relationship is now known as **Raoult's law**.

Let's take another look at the two compartments in Figure 11.1 and assume that the temperature in the chamber is 20°C. At 20°C, the vapor pressure of pure water is 0.0231 atm. What is the vapor pressure produced by evaporation of water from seawater if the mole fraction of water in the sample is 0.980? We can use Raoult's law as expressed by Equation 11.1 to answer this question:

$$P_{solution} = (0.980)(0.0231 \text{ atm}) = 0.0226 \text{ atm}$$

Because the vapor pressure of seawater at 20°C is slightly lower than the vapor pressure of pure water at 20°C, seawater evaporates more slowly than pure water.

The lower vapor pressure of a solution relative to the vapor pressure of pure solvent depends only on the concentration of solute particles, not on their identity. Properties of solutions that depend only on the concentration of particles and not on their identity are called **colligative properties**. We examine them in detail in Section 11.5.

Solutions that obey Raoult's law are called **ideal solutions**. In ideal solutions, solute and solvent experience similar intermolecular forces. For the most part, we treat solutions as ideal systems, but you should be aware that deviations from ideal behavior exist in liquids just as they do in gases. Deviations from ideal behavior typically occur when solute–solvent interactions are much stronger than solvent–solvent interactions. The definition of colligative properties states that the identity of the solute particles does not matter, but that is not always the case. We address such situations in Section 11.4 in the discussion of the distillation of crude oil to produce gasoline.

**CONNECTION** We defined mole fraction in Chapter 6 in the discussion of partial pressures of gases.

▶❙❙ **CHEMTOUR** Raoult's Law

**CONNECTION** As we discussed in Chapter 1, intensive properties of matter are independent of the amount of material present, while extensive properties vary with the quantity of substance present.

**CONCEPT TEST** ..........................................................................

Is the vapor pressure of a pure solvent an intensive or an extensive property? Is the vapor pressure of a solution an intensive or an extensive property?

*(Answers to Concept Tests are in the back of the book.)*
..........................................................................

**SAMPLE EXERCISE 11.1** **Calculating the Vapor Pressure of a Solution** LO1

The liquid used in automobile cooling systems is prepared by dissolving ethylene glycol ($HOCH_2CH_2OH$, molar mass 62.07 g/mol) in water. What is the vapor pressure of a solution prepared by mixing 1.000 L of ethylene glycol (density 1.114 g/mL) with 1.000 L of water (density 1.000 g/mL) at 100.0°C? Assume that the mixture obeys Raoult's law.

**Collect and Organize** The vapor pressure of a solution is a colligative property that depends on the number of solute particles, and hence the concentration. We have the volume and density of the components and can use them to determine the concentration of ethylene glycol. We may treat ethylene glycol as a nonvolatile solute whose solutions obey Raoult's law. The vapor pressure curves in Figure 10.22 show that the vapor pressure of pure water at 100°C (its normal boiling point) is 1.00 atm, whereas that of ethylene glycol is less than 0.05 atm.

**Analyze** We have a mixture of equal volumes of two liquids. The solvent is the one present in the greater number of moles:

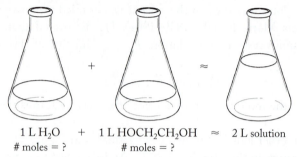

$$1\ \text{L}\ H_2O\ +\ 1\ \text{L}\ HOCH_2CH_2OH\ \approx\ 2\ \text{L solution}$$
$$\#\ \text{moles} = ?\qquad\qquad \#\ \text{moles} = ?$$

We can determine the numbers of moles of each by the following calculation:

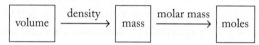

From these values we can decide which liquid is the solvent and calculate its mole fraction in the mixture. Water and ethylene glycol are both capable of hydrogen bonding, so their intermolecular interactions are similar, and we may treat the solution as ideal.

**Solve** For ethylene glycol:

$$\frac{1.114\ \text{g}}{1\ \text{mL}} \times 1000\ \text{mL} \times \frac{1\ \text{mol}}{62.07\ \text{g}} = 17.95\ \text{mol}$$

For water:

$$\frac{1.000\ \text{g}}{1\ \text{mL}} \times 1000\ \text{mL} \times \frac{1\ \text{mol}}{18.02\ \text{g}} = 55.49\ \text{mol}$$

The mole fraction of water is

$$X_{\text{water}} = \frac{55.49\ \text{mol}}{55.49\ \text{mol} + 17.95\ \text{mol}} = 0.7556$$

The number of moles of water is greater than the number of moles of ethylene glycol, so water is the solvent. Ethylene glycol is essentially nonvolatile, so the vapor pressure of the solution is due only to the solvent, and $P_{\text{solvent}} = P_{H_2O} = 1.00$ atm. Using this value and the calculated mole fraction of water in the mixture yields

$$P_{\text{solution}} = X_{H_2O} \times P^{\circ}_{H_2O} = (0.756)(1.00\ \text{atm}) = 0.756\ \text{atm}$$

**Think About It** The presence of a nonvolatile solute causes the vapor pressure of the solution to be less than the vapor pressure of a pure solvent (1 atm, in this case), giving us confidence in our result.

**Practice Exercise** Glycerol [$HOCH_2CH(OH)CH_2OH$] is considered to be a non-volatile, water-soluble liquid. Its density is 1.25 g/mL. Predict the vapor pressure of a solution of 275 mL of glycerol in 375 mL of water at the normal boiling point of water.

*(Answers to Practice Exercises are in the back of the book.)*

To boil the solution described in Sample Exercise 11.1, we must heat it to above 100°C—to a temperature at which the vapor pressure of the solution is 1 atm. This is why antifreeze works in an automobile engine: it not only lowers the freezing point of water, protecting the radiator, but it also raises the temperature of the coolant above the boiling point of $H_2O$, making it possible for the coolant to remove more heat from the engine while staying in the liquid state.

# 11.2 Solubility of Gases in Water

In Section 10.7 we discussed the turnover in natural water systems such as lakes and ponds, which distributes nutrients and contributes to the health of aquatic life forms. The ability of water to dissolve gases as well as nutrients is crucial to both terrestrial and aquatic life. The $O_2$ level in natural waters is normally sufficient to support life, but you may have noticed fish in lakes and rivers rising to the surface and gasping for air in very warm weather. The warm water lacks sufficient dissolved oxygen because the solubility of $O_2$, and of most other gases, in water decreases with increasing temperature. The solubility of gases in water also decreases with decreasing pressure, as is dramatically illustrated by popping the top off a bottle containing a carbonated beverage. The release of pressure causes the $CO_2$ gas to escape from the fluid, causing bubbles and fizz. When humans move to regions of low atmospheric pressure, like at the tops of mountains, they may need supplemental oxygen to function normally because of the decreased solubility of oxygen in bodily fluids at the lower pressures.

Relatively weak dipole-induced dipole interactions between water molecules and the nonpolar molecules of oxygen gas account for the solubility of $O_2$, or indeed any sparingly soluble gas in water. We can visualize the influence of temperature on the solubility of gases in terms of kinetic molecular theory. Think of a population of gas molecules dissolved in a liquid (Figure 11.2a). As temperature increases, the kinetic energy of the molecules increases, and more energy is available to disrupt intermolecular attractions, reducing solubility. At higher temperatures (Figure 11.2b), more gas molecules have the necessary kinetic energy to overcome the

**◉◉ CONNECTION** Kinetic molecular theory was introduced in Chapter 6 as a model to explain the behavior of gases.

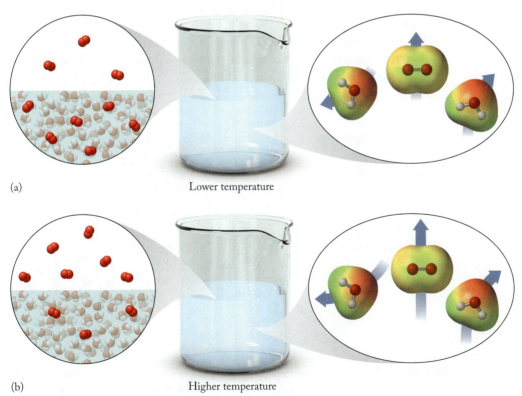

(a)   Lower temperature

(b)   Higher temperature

**FIGURE 11.2** (a) At lower temperature, molecules of gas dissolved in water have less kinetic energy, and dipole–induced dipole interactions between solute and solvent molecules keep more gas dissolved. (b) At higher temperatures, molecules have more kinetic energy, which overcomes solute–solvent interactions and reduces solubility.

**▶II CHEMTOUR** Henry's Law

dipole–induced dipole interactions between $O_2$ and $H_2O$ molecules, allowing dissolved gas to escape from the solution.

The solubility of a gas in a liquid such as water also depends on the partial pressure of the gas in the air above the surface of the liquid. For example, the partial pressure of $O_2$ at sea level remains fairly constant at about 0.21 atm, but the low partial pressure of oxygen at high altitudes can result in lower than normal concentrations of oxygen in liquids, including the blood and tissues of humans and animals. Climbers in the Himalayas may become weak and unable to think clearly because of lack of oxygen to the brain, a condition known as *hypoxia*.

We can also visualize the relationship between pressure and solubility of a gas in a liquid in terms of kinetic molecular theory. Think of a population of gas molecules above the surface of a liquid (Figure 11.3). The amount of gas that dissolves in a liquid depends on the frequency and number of collisions gas molecules have with the surface of the liquid: the more collisions, the more gas molecules may be entrapped by the solvent.

The relationship between gas solubility in a liquid and the partial pressure of the gas in the environment surrounding the liquid applies to all sparingly soluble gases. The quantitative statement based on this observation is known as **Henry's law**, in honor of William Henry (1775–1836), a British physician who first proposed the relationship:

$$C_{\text{gas}} = k_{\text{H}} P_{\text{gas}} \tag{11.2}$$

where $C_{\text{gas}}$ represents the concentration (solubility) of a gas in a particular solvent, $k_{\text{H}}$ is the Henry's law constant for the gas in that solvent, and $P_{\text{gas}}$ is the partial pressure of the gas in the environment surrounding the solvent. When $C_{\text{gas}}$ is expressed in molarity and $P_{\text{gas}}$ in atm, the units of the Henry's law constant are moles per liter-atmosphere, mol/(L · atm). Table 11.1 lists $k_{\text{H}}$ values for several common gases in water.

Henry's law states that the concentration of dissolved oxygen in blood is proportional to the partial pressure of oxygen in the air we inhale and thus is proportional to atmospheric pressure. This is an accurate statement of Henry's law, but resi-

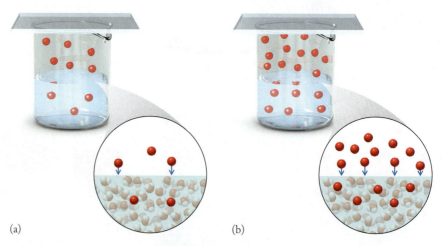

(a)                                        (b)

**FIGURE 11.3** The partial pressure of a gas (red particles) in a mixture of gases in the space above a liquid is twice as large in (b) as in (a). Consequently, the number of collisions the gas molecules have with the surface is greater in (b) than in (a), and the solubility of the gas in the liquid increases.

dents of Denver, Colorado, or Kimberley, Canada (average atmospheric pressure 0.85 atm), do not live with less blood oxygen than residents of New York City or Rome, Italy (average atmospheric pressure 1.00 atm). The reason is that the oxygen transport system in our bodies responds to local conditions.

The amount of oxygen in the blood is related to the concentration of hemoglobin and to the fraction of the hemoglobin sites that contain oxygen as the blood leaves the lungs. This saturation of binding sites depends on the partial pressure of oxygen as well as on proper lung function. For most people, breathing air with $P_{O_2} > 0.11$ atm results in nearly 100% saturation of hemoglobin binding sites. If $P_{O_2}$ decreases to about 0.066 atm (as it does on high mountains), the percent saturation decreases to about 80%. Over several weeks, the body responds to lower oxygen partial pressures by producing more red blood cells and more hemoglobin. This increase in the number of $O_2$ carriers compensates for the lower partial pressure of $O_2$. Even though the level of saturation decreases, the actual number of carriers of $O_2$ increases, and the same amount of $O_2$ is delivered to tissues. Some endurance athletes try to capitalize on these physiological effects by a technique known as "live high, train low." They acclimate to higher altitudes, typically defined as any elevation above 1500 meters (5000 ft), to effect the physiological changes, but they continue to train at lower elevations.

| TABLE 11.1 | Henry's Law Constants for Gas Solubility in Water at 20°C | |
|---|---|---|
| **Gas** | **$k_H$ [mol/(L · atm)]** | |
| He | $3.5 \times 10^{-4}$ | |
| $O_2$ | $1.3 \times 10^{-3}$ | |
| $N_2$ | $6.7 \times 10^{-4}$ | |
| $CO_2$ | $3.5 \times 10^{-2}$ | |

---

**SAMPLE EXERCISE 11.2**   **Calculating Gas Solubility Using Henry's Law**   **LO2**

Calculate the solubility of oxygen in water in moles per liter at 1.00 atm pressure and 20°C. The mole fraction of $O_2$ in air is 0.209. (Remember that the sum of the mole fractions of all the gases in a mixture equals 1.)

**Collect and Organize** We are given the mole fraction of oxygen in air and the total (atmospheric) pressure. Henry's law (Equation 11.2) relates solubility to partial pressure. Table 11.1 gives the Henry's law constant for oxygen at 20°C as $1.3 \times 10^{-3}$ mol/(L · atm).

**Analyze** We need the partial pressure of oxygen for Henry's law:

$$\boxed{\text{Mole fraction}} \xrightarrow{\times\ P_{total}} \boxed{\text{Partial pressure}} \xrightarrow{\times\ k_H} \boxed{\text{Solubility}}$$

The product of the mole fraction ($X_{O_2}$) of $O_2$ times its total pressure gives us its partial pressure, which we use in Equation 11.2 to calculate the solubility of oxygen. We predict that oxygen is not very soluble in water because the interactions involved are weak dipole-induced dipole forces.

**Solve** We calculate the partial pressure of oxygen using Equation 6.25:

$$P_{O_2} = X_{O_2} P_{total} = (0.209)(1.00 \text{ atm}) = 0.209 \text{ atm}$$

Substituting this value for $P_{O_2}$ and $k_H$ for $O_2$ in water in Equation 11.2 gives

$$C_{O_2} = k_H P_{O_2} = \left( \frac{1.3 \times 10^{-3} \text{ mol}}{\text{L} \cdot \text{atm}} \right)(0.209 \text{ atm}) = 2.7 \times 10^{-4} \text{ mol/L}$$

**Think About It** We predicted that oxygen is not very soluble in water because $O_2$ is a nonpolar solute and water is a polar solvent, and the answer agrees with that prediction.

**Practice Exercise** Calculate the solubility of oxygen in water at the top of Mt. Everest, where atmospheric pressure is 0.35 atm.

## 11.3 Energy Changes during Formation and Dissolution of Ionic Compounds

In Chapter 10 we discussed the role of ion–dipole interactions in making salts soluble in water and described qualitatively how the combined effect of many ion–dipole interactions could overcome ion–ion interactions in a crystal of solute to produce hydrated ions in solution. In Section 11.1 we described the influence of nonvolatile solutes on the vapor pressure of a solvent by examining the behavior of seawater and pure water in separate chambers within a closed container. Now that we have visual models to describe the dissolution of ionic compounds like NaCl at the microscopic level, and we see how the presence of a solute changes the behavior of a solvent, we can take the next step and quantify the actual changes in energy associated with these interactions. We can then use these quantitative observations to understand further what happens at the particle level when solutes dissolve in solvents.

We can apply the qualitative picture of the dissolution process we developed in Chapter 10 to describe the factors that contribute to the overall enthalpy change that accompanies dissolution of an ionic compound in a polar solvent. The ions must be separated from one another (Figure 11.4a), a process that requires energy to break the ion–ion interactions that hold the particles in the crystal lattice ($\Delta H_{\text{ion–ion}}$). The solvent molecules must also be separated from one another so they can bind to the ions (Figure 11.4b); this process requires energy to break the dipole–dipole interactions (in water, the hydrogen bonds) between solvent molecules ($\Delta H_{\text{dipole–dipole}}$). Last, energy is released when the solvent molecules associate with the solute ions via ion–dipole interactions ($\Delta H_{\text{ion–dipole}}$) and form hydrated ions (Figure 11.4c). The sum of the enthalpies of these interactions is the **enthalpy of solution ($\Delta H_{\text{solution}}$)**, which defines the overall change in enthalpy when an ionic solute is dissolved in a polar solvent.

$$\Delta H_{\text{solution}} = \Delta H_{\text{ion–ion}} + \Delta H_{\text{dipole–dipole}} + \Delta H_{\text{ion–dipole}} \qquad (11.3)$$

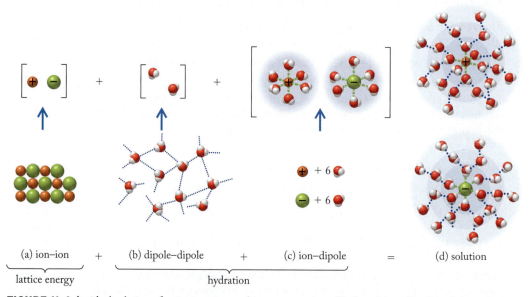

(a) ion–ion + (b) dipole–dipole + (c) ion–dipole = (d) solution

lattice energy        hydration

**FIGURE 11.4** An ideal solution of an ionic compound in water consists of hydrated ions distributed throughout the solvent. We can account for the energy associated with establishing a solution by considering (a) the energy required to separate the ions from their lattice ($\Delta H_{\text{ion}} = U$), (b) the energy required to overcome dipole–dipole interactions in the solvent ($\Delta H_{\text{dipole–dipole}}$), and (c) the energy released when ion–dipole bonds form between solvent and solute particles ($\Delta H_{\text{ion–dipole}}$). The sum of (b) and (c) is $\Delta H_{\text{hydration}}$, the enthalpy of hydration. (d) The sum of all three processes is $\Delta H_{\text{solution}}$, the enthalpy of solution.

Of the quantities in Equation 11.3, we can readily measure $\Delta H_{\text{solution}}$ using calorimetric methods. Measuring the remaining three terms independently is more difficult. The energy associated with ion–ion attraction was introduced in Equation 5.3 and is proportional to the product of the charges on the ions and inversely proportional to the distance between them:

$$E \propto \frac{(Q_1Q_2)}{d} \tag{5.3}$$

We can combine the strengths of dipole–dipole and ion–dipole interactions into a single term called the **enthalpy of hydration ($\Delta H_{\text{hydration}}$)** for the compound:

$$\Delta H_{\text{hydration}} = \Delta H_{\text{dipole–dipole}} + \Delta H_{\text{ion–dipole}} \tag{11.4}$$

Here *hydration* refers to the formation of solvated ions. Combining Equations 11.3 and 11.4, we have

$$\Delta H_{\text{solution}} = \Delta H_{\text{ion–ion}} + \Delta H_{\text{hydration}} \tag{11.5}$$

Let's examine each of these terms in more detail.

The strength of ion–ion interactions is described by the **lattice energy ($U$)** of an ionic compound, which is the change in energy when free ions in the gas phase combine to form 1 mole of a solid ionic compound. The lattice energies of some common binary ionic compounds are given in Table 11.2. The formula for lattice energy is

$$U = \frac{k(Q_1Q_2)}{d} \tag{11.6}$$

This formula resembles Equation 5.3 except that it includes a proportionality constant, $k$, the value of which depends on the structure of the ionic solid. We examine the structures of solids in Chapter 12, at which point we will learn about the different arrangements possible for ions in an ionic solid. The key point here is that the same value of $k$ is used for all compounds that have the same or nearly the same arrangement of ions. For now, we need only consider the charges on ions and the distances between them and use Equation 10.1 to predict relative values of lattice energies. We can then substitute the lattice energy, $U$, of an ionic compound in Equation 11.3 for $\Delta H_{\text{ion–ion}}$:

$$\Delta H_{\text{solution}} = -U + \Delta H_{\text{hydration}}$$

or

$$\Delta H_{\text{solution}} = \Delta H_{\text{hydration}} - U \tag{11.7}$$

There is a minus sign in front of the lattice energy term in Equation 11.7 because $U$ is defined as the enthalpy change when gas-phase ions *combine* to form an ionic solid. In Equation 11.7, we are calculating the enthalpy change associated with *separating* the ions in an ionic compound. Recall from Chapter 5 that the enthalpy change for a process in one direction has the same magnitude but opposite sign of the process in the reverse direction.

The lattice energy of an ionic solid not only affects its solubility in water, but also determines the temperature at which the solid melts. Melting an ionic structure in which the ions are held together tightly should require more thermal energy (a higher temperature) than melting a structure in which the ions are held together less tightly—a trend we observe experimentally. Consider two ionic compounds: LiF ($U = -1047$ kJ/mol) and MgO ($U = -3791$ kJ/mol). The greater lattice energy of MgO, nearly four times that of LiF, is reflected in its higher melting point, 2825°C, versus 848°C for LiF, and its higher boiling point, 3600°C for MgO, versus 1673°C for LiF.

**CONNECTION** Calorimetry as a method for measuring enthalpy changes was described in Chapter 5.

▶❚❚ **CHEMTOUR** Lattice Energy

| TABLE 11.2 | Lattice Energies ($U$) of Common Binary Ionic Compounds |
|---|---|
| **Compound** | **$U$ (kJ/mol)** |
| LiF | −1047 |
| LiCl | −864 |
| NaCl | −786 |
| KCl | −720 |
| KBr | −691 |
| MgCl$_2$ | −2540 |
| MgO | −3791 |

**enthalpy of solution ($\Delta H_{\text{solution}}$)** the overall energy change when a solute is dissolved in a solvent.

**enthalpy of hydration ($\Delta H_{\text{hydration}}$)** the energy change when gas-phase ions dissolve in a solvent.

**lattice energy ($U$)** the enthalpy change that occurs when 1 mole of an ionic compound forms from its free ions in the gas phase.

Rank these three ionic compounds in order of (a) increasing lattice energy, and (b) increasing melting point: NaF, KF, and RbF. Assume that these compounds have the same solid structure, which means they have the same value of $k$ in Equation 11.6.

**Collect and Organize**  We can determine relative lattice energies from Equation 11.6. To do so, we need to establish the charges and radii of the ions in each compound, using Figure 10.2.

**Analyze**  All the cations are alkali metal cations and have a charge of 1+; all the anions are fluoride ions, with a charge of 1−. We are told that $k$ is the same for all three solids. Therefore, any differences in lattice energy must be related to differences in the nucleus-to-nucleus distance, $d$, between ions. Because the fluoride ion is common to all the salts, variations in $d$ depend only on the size of the cation.

**Solve**  Periodic trends in size predict, and Figure 11.5 confirms, that the cation sizes are $Na^+ < K^+ < Rb^+$. Therefore, the compounds in order of increasing value of $d$ are NaF < KF < RbF.
a. As $d$ increases, lattice energy decreases, so we predict RbF has the lowest lattice energy, followed by KF, followed by NaF with the highest.
b. The same trend occurs in melting points: RbF < KF < NaF.

**Think About It**  We predicted the order of lattice energies and melting points for three alkali metal fluorides based on the radii of the ions. The predicted orders are confirmed by the experimentally measured melting points: 775°C for RbF; 846°C for KF; and 988°C for NaF.

⚙ **Practice Exercise**  Predict which compound has the highest melting point: $CaCl_2$, $PbBr_2$, or $TiO_2$. All three compounds have nearly the same structure and therefore the same value of $k$ in Equation 11.6. The radius of $Ti^{4+}$ is 60.5 pm.

$Na^+ = 102$ pm
$F^- = 133$ pm

$K^+ = 138$ pm
$F^- = 133$ pm

$Rb^+ = 149$ pm
$F^- = 133$ pm

**FIGURE 11.5** The distance, $d$, between the ions in NaF, KF, and RbF is in the order $d_{RbF} > d_{KF} > d_{NaF}$.

## Calculating Lattice Energies Using the Born–Haber Cycle

We predicted the relative order of lattice energies by considering atomic radii and bond distances, but we can also calculate values for lattice energies. Such calculations are necessary because measuring lattice energies directly is difficult. We can calculate the lattice energy of a binary ionic compound from its standard enthalpy of formation ($\Delta H_f^\circ$). We use Hess's law to determine the enthalpies of reaction, $\Delta H_{rxn}$, associated with a series of reactions that take the constituent elements from their standard states to ions in the gas phase and then to ions in the ionic solid.

Consider the formation of NaCl:

$$Na(s) + \tfrac{1}{2}Cl_2(g) \rightarrow NaCl(s) \qquad \Delta H_f^\circ = -411.2 \text{ kJ}$$

The reaction is exothermic, and the heat produced, 411.2 kJ per mole of NaCl formed, has been measured with a calorimeter and is among those tabulated in Table A4.3 in the Appendix. In terms of Hess's law, this enthalpy of formation is the algebraic sum of all the enthalpy changes associated with five reactions that together form a **Born–Haber cycle** (Figure 11.6):

1. Sublimation of 1 mole of Na(s) atoms into 1 mole of Na(g) atoms: $\Delta H_{sub}$, the molar heat of sublimation of sodium.
2. Breaking covalent bonds in $\tfrac{1}{2}$ mole of $Cl_2(g)$ molecules to make 1 mole of Cl(g) atoms: $\tfrac{1}{2}\Delta H_{BE}$, where $\Delta H_{BE}$ is the enthalpy change needed to break 1 mole of $Cl_2(g)$ bonds.

👁👁 **CONNECTION** We used Hess's law in Chapter 5 to determine heats of reactions that are difficult to measure experimentally. We also defined a formation reaction as a reaction in which 1 mole of a substance is produced from its constituent elements in their standard states.

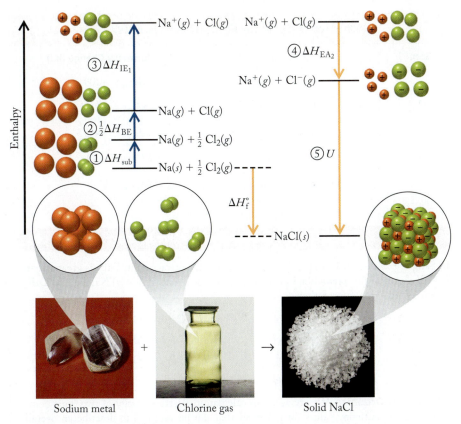

**FIGURE 11.6** The reaction between sodium metal and chlorine gas releases more than 400 kJ of energy per mole of NaCl produced. The Born–Haber cycle shows that the most exothermic step is the combination of free sodium ions $Na^+(g)$ and free chloride ions $Cl^-(g)$ to form NaCl($s$).

3. Ionization of 1 mole of Na($g$) atoms to 1 mole of $Na^+(g)$ ions and 1 mole of electrons: $IE_1$, the first ionization energy of sodium.
4. Combination of 1 mole of Cl($g$) atoms with 1 mole of electrons to form 1 mole of $Cl^-(g)$ ions: $EA_1$, the first electron affinity of chlorine.
5. Formation of 1 mole of NaCl($s$) from 1 mole of $Na^+(g)$ ions and 1 mole of $Cl^-(g)$ ions: $U = \Delta H_{lattice}$, the lattice energy of NaCl.

This reaction sequence is summarized in Table 11.3. To use the Born–Haber cycle to calculate the lattice energy of NaCl($s$), we start with an equation relating the value of $\Delta H_f^\circ$ for NaCl to the sum of the enthalpy changes of the five reactions in Table 11.3:

$$\Delta H_f^\circ = \Delta H_{\text{step 1}} + \Delta H_{\text{step 2}} + \Delta H_{\text{step 3}} + \Delta H_{\text{step 4}} + \Delta H_{\text{step 5}}$$
$$= \Delta H_{sub} + \tfrac{1}{2}\Delta H_{BE} + \Delta H_{IE_1} + \Delta H_{EA_1} + U$$

Inserting the values from Table 11.3 and solving for $U$:

$$-411.2 \text{ kJ} = (+109 \text{ kJ}) + \tfrac{1}{2}(+240 \text{ kJ}) + (+495 \text{ kJ}) + (-349 \text{ kJ}) + U$$
$$U = (-411.2 \text{ kJ}) - (+109 \text{ kJ}) - (+120 \text{ kJ}) - (+495 \text{ kJ}) - (-349 \text{ kJ}) = -786 \text{ kJ}$$

In addition to its usefulness for calculating lattice energies, the Born–Haber cycle can also be used to calculate other values. For example, electron affinities can be very difficult to measure, and if the thermochemical values are known for all other steps in the cycle, the Born–Haber cycle can be used to calculate electron affinities.

**Born–Haber cycle** a series of steps with corresponding enthalpy changes that describes the formation of an ionic solid from its constituent elements.

## TABLE 11.3 Born–Haber Cycle for Formation of NaCl(s)

| Step | Process | Enthalpy Change (kJ) |
|------|---------|----------------------|
| 1 | $Na(s) \rightarrow Na(g)$ | $\Delta H_{sub} = +109$ |
| 2 | $\frac{1}{2} Cl_2(g) \rightarrow Cl(g)$ | $\frac{1}{2}\Delta H_{BE} = \frac{1}{2}(240) = +120$ |
| 3 | $Na(g) \rightarrow Na^+(g) + e^-$ | $\Delta H_{IE_1} = +495$ |
| 4 | $Cl(g) + e^- \rightarrow Cl^-(g)$ | $\Delta H_{EA_1} = -349$ |
| 5 | $Na^+(g) + Cl^-(g) \rightarrow NaCl(s)$ | $\Delta H_{lattice} = U$ |

## SAMPLE EXERCISE 11.4 Calculating Lattice Energy  LO3

In Sample Exercise 10.1, we predicted that the ion–ion attraction in $CaF_2$ is greater than in NaF. Confirm this prediction by calculating the lattice energies of (a) NaF and (b) $CaF_2$ given the following information:

$$\Delta H_{sub}\ Na(s) = +109\ kJ/mol$$
$$\Delta H_{BE}\ F_2(g) = +154\ kJ/mol$$
$$\Delta H_{EA_1}\ F(g) = -328\ kJ/mol$$
$$\Delta H_{IE_1}\ Na(g) = +495\ kJ/mol$$

$$\Delta H_{sub}\ Ca(s) = +154\ kJ/mol$$
$$\Delta H_{IE_1}\ Ca(g) = +590\ kJ/mol$$
$$\Delta H_{IE_2}\ Ca(g) = +1145\ kJ/mol$$

**Collect and Organize** We are asked to calculate the lattice energy of two ionic compounds, using enthalpy values for processes that can be summed to describe an overall process that produces a salt from its constituent elements. We can use a Born–Haber cycle to calculate the unknown lattice energy. Table A4.3 in the Appendix contains standard enthalpy of formation values for NaF and $CaF_2$: −569.0 kJ/mol and −1228.0 kJ/mol, respectively. These enthalpy changes apply to the formation of one mole of the two compounds by combining their component elements in their standard states.

**Analyze** Figure 11.7 summarizes the Born–Haber cycles for calculating the lattice energies of NaF and $CaF_2$. The lattice energy $U$ is the only unknown value in both cycles. For calcium, we must include both the first and second ionization energies because Ca loses two electrons when it forms $Ca^{2+}$ ions. Because two fluorine atoms are needed to react with

**FIGURE 11.7** Born–Haber cycles for the formation of (a) NaF and (b) $CaF_2$.

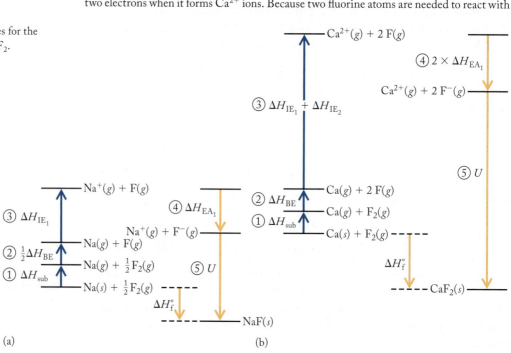

a single calcium atom, we do not need the factor of $\frac{1}{2}$ in front of the term for energy to break a mole of F—F bonds. However, we need to multiply the electron affinity of F by 2 because two moles of fluorine atoms gain two moles of electrons to form two moles of fluoride ions. Based on the charges on the ions, their ionic radii, and Equation 11.6, we predict the lattice energy of $CaF_2$ should be greater than the lattice energy of NaF.

**Solve**

a. The Born–Haber cycle for the formation of NaF(s) from Na(s) and $F_2(g)$ is illustrated in Figure 11.7(a). The overall enthalpy change in the reaction producing NaF is

$$\Delta H_f^\circ = \Delta H_{sub,Na(s)} + \tfrac{1}{2}\Delta H_{BE,F_2(g)} + \Delta H_{IE_1,Na(g)} + \Delta H_{EA_1,F(g)} + U_{NaF(s)}$$

Substituting the values given:

$$-569.0 \text{ kJ} = (+109 \text{ kJ}) + \tfrac{1}{2}(+154 \text{ kJ}) + (+495 \text{ kJ}) + (-328 \text{ kJ}) + U$$

Solving for $U$:

$$U = (-569 \text{ kJ}) - (+109 \text{ kJ} + 77 \text{ kJ} + 495 \text{ kJ} - 328 \text{ kJ})$$
$$= -922 \text{ kJ/mol NaF}(s) \text{ formed}$$

b. The Born–Haber cycle for the formation of $CaF_2(s)$ from Ca(s) and $F_2(g)$ is illustrated in Figure 11.7(b). The overall enthalpy change in the reaction producing $CaF_2$ is

$$\Delta H_f^\circ = \Delta H_{sub,Ca(s)} + \Delta H_{BE,F_2(g)} + [\Delta H_{IE_1,Ca(g)} + \Delta H_{IE_2,Ca(g)}] + 2\Delta H_{EA_1,F(g)} + U_{CaF_2(s)}$$

Substituting the values given:

$$-1228.0 \text{ kJ} = (+154 \text{ kJ}) + (+154 \text{ kJ}) + [(+590 \text{ kJ}) + (+1145 \text{ kJ})] + 2(-328 \text{ kJ}) + U$$

Solving for $U$:

$$U = (-1228.0 \text{ kJ}) - [+154 \text{ kJ} + 154 \text{ kJ} + (590 \text{ kJ} + 1145 \text{ kJ}) - 656 \text{ kJ}]$$
$$= -2615 \text{ kJ/mol CaF}_2(s) \text{ formed}$$

**Think About It** Based on Equation 11.6, we predicted that the lattice energy of $CaF_2$ would be larger than that of NaF. The calculated values of $U$ confirm that $U_{CaF2} > U_{NaF}$.

**Practice Exercise** Burning magnesium metal in air produces MgO and a very bright white light, making the reaction popular in fireworks and signaling devices:

$$Mg(s) + \tfrac{1}{2}O_2(g) \rightarrow MgO(s) + \text{light}$$

The energy change that accompanies this reaction is −602 kJ/mol MgO. Calculate the lattice energy of MgO from the following energy changes:

| Process | Enthalpy Change (kJ/mol) | Process | Enthalpy Change (kJ/mol) |
|---|---|---|---|
| $Mg(s) \rightarrow Mg(g)$ | 150 | $Mg(g) \rightarrow Mg^{2+}(g) + 2\,e^-$ | 2188 |
| $O_2(g) \rightarrow 2\,O(g)$ | 499 | $O(g) + 2\,e^- \rightarrow O^{2-}(g)$ | 603 |

The lattice energy of an ionic compound is one of many factors that affects its solubility in water. We might predict that NaF is more soluble in water than $CaF_2$ on the basis of the greater lattice energy of $CaF_2$, and that prediction is correct: at 20°C, the solubility of NaF in water is 4.0 g/100 mL and the solubility of $CaF_2$ is only 0.0015 g/100 mL.

**CONNECTION** The solubility trends for ionic compounds given in Chapter 4 are related to the strengths of intermolecular attractive forces.

## Enthalpies of Hydration

Once we have calculated the lattice energy of an ionic compound, we can use measured values of $\Delta H_{solution}$ and Equation 11.7 to calculate its enthalpy of hydration, $\Delta H_{hydration}$, as shown in Figure 11.8. Figure 11.8 is another example of a

**FIGURE 11.8** Born–Haber cycle for the dissolution of an ionic compound in water. The enthalpy of solution ($\Delta H_{\text{solution}}$) is the enthalpy of hydration ($\Delta H_{\text{hydration}}$) minus the lattice energy ($U$). The size of $\Delta H_{\text{solution}}$ is not drawn to scale; it is very small relative to the other values.

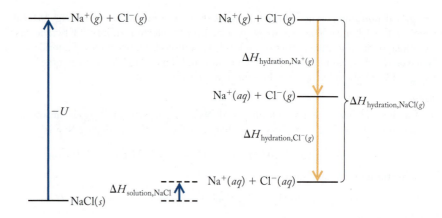

Born–Haber cycle. If the measured enthalpy of solution of sodium chloride is 4 kJ/mol and the lattice energy of NaCl is −786 kJ/mol, then:

$$\Delta H_{\text{solution,NaCl}} = \Delta H_{\text{hydration,NaCl}(aq)} - U_{\text{NaCl}}$$
$$\Delta H_{\text{hydration,NaCl}(aq)} = \Delta H_{\text{solution,NaCl}} + U_{\text{NaCl}}$$
$$= 4 \text{ kJ/mol} + (-786 \text{ kJ/mol}) = -782 \text{ kJ/mol}$$

The value for $\Delta H_{\text{hydration,NaCl}(aq)}$ is negative, indicating that when $Na^+$ and $Cl^-$ ions in the gas phase dissolve in water, the reaction is exothermic. The sign of $\Delta H_{\text{hydration,NaCl}(aq)}$ also tells us something about the relative strengths of the hydrogen bonding interactions in water and the ion–dipole interactions in aqueous NaCl. If $\Delta H_{\text{hydration,NaCl}(aq)}$ is less than zero, then according to Equation 11.7 the formation of ion–dipole interactions must release more energy than is needed to disrupt the dipole–dipole interactions (hydrogen bonds) in water.

With careful measurements and calculations, we can even separate $\Delta H_{\text{hydration,NaCl}(aq)}$ into enthalpies of hydration for the individual ions:

$$\Delta H_{\text{hydration,NaCl}(aq)} = \Delta H_{\text{hydration,Na}^+(g)} + \Delta H_{\text{hydration,Cl}^-(g)}$$

The values for $\Delta H_{\text{hydration}}$ for selected cations and anions are listed in Table 11.4 and can be used to calculate $\Delta H_{\text{solution}}$ for an ionic compound if we know the lattice energy, or to calculate the lattice energy if we measure the enthalpy of solution. The values in Table 11.4 are the results of several different experiments and include an inherent uncertainty. The sum of the values for $Na^+$ and $Cl^-$ is −786 kJ/mol; that is the same as the value calculated from the measured enthalpy of solution (Table 11.2).

| TABLE 11.4 | Enthalpies of Hydration for Selected Cations and Anions[a] | | |
|---|---|---|---|
| **Cation** | **$\Delta H_{\text{hydration}}$ (kJ/mol)** | **Anion** | **$\Delta H_{\text{hydration}}$ (kJ/mol)** |
| $Li^+$ | −536 | $F^-$ | −502 |
| $Na^+$ | −418 | $Cl^-$ | −368 |
| $K^+$ | −335 | $Br^-$ | −335 |
| $Rb^+$ | −305 | $I^-$ | −293 |
| $Cs^+$ | −289 | $ClO_4^-$ | −238 |
| $Mg^{2+}$ | −1903 | $NO_3^-$ | −301 |
| $Ca^{2+}$ | −1591 | $SO_4^{2-}$ | −1017 |

[a]Based on enthalpy of hydration of $H^+$ as −1105 kJ/mol.

**SAMPLE EXERCISE 11.5** **Calculating Lattice Energy** **LO3**
**Using Enthalpies of Hydration**

Use the appropriate enthalpy of hydration values in Table 11.4 and $\Delta H_{solution} = 0.914$ kJ/mol for NaF to calculate the lattice energy of NaF.

**Collect and Organize** The enthalpies of hydration for $Na^+$ and $F^-$ are $-418$ kJ/mol and $-502$ kJ/mol, respectively. We can use the enthalpy of solution and Equation 11.7 to calculate the lattice energy.

**Analyze** To calculate $U_{NaF}$, we must obtain a value for $\Delta H_{hydration}$ for NaF:

$$\boxed{\Delta H_{hydration} \text{ of cation}} + \boxed{\Delta H_{hydration} \text{ of anion}} \longrightarrow \boxed{\Delta H_{hydration} \text{ of ionic compound}}$$

From our calculations in Sample Exercise 11.4, we expect the lattice energy of NaF to be close to $-922$ kJ/mol.

**Solve** Solving for the enthalpy of hydration of NaF:

$$\Delta H_{hydration,NaF} = \Delta H_{hydration,Na^+(g)} + \Delta H_{hydration,F^-(g)}$$
$$= -418 \text{ kJ/mol} + (-502 \text{ kJ/mol}) = -920 \text{ kJ/mol}$$

Substituting $\Delta H_{hydration,NaF(aq)}$ and $\Delta H_{solution,NaF}$ into Equation 11.7:

$$\Delta H_{solution,NaF} = \Delta H_{hydration,NaF(aq)} - U_{NaF}$$
$$+0.91 \text{ kJ/mol} = (-920 \text{ kJ/mol}) - U_{NaF}$$

Solving for $U_{NaF}$:

$$U_{NaF} = -920 \text{ kJ/mol} - (0.91 \text{ kJ/mol}) = -921 \text{ kJ/mol}$$

**Think About It** The value of $U_{NaF}$ calculated from $\Delta H_{solution}$ and $\Delta H_{hydration}$ is within 1% of the value calculated in Sample Exercise 11.4 using a Born–Haber cycle, suggesting that the value calculated this way accurately reflects the lattice energy of NaF.

⚙ **Practice Exercise** Calculate the lattice energy for $NaClO_4$ using the data in Table 11.4 and $\Delta H_{solution,NaClO_4} = 14$ kJ/mol. ∎

# 11.4 Mixtures of Volatile Solutes

Thus far, we have considered only the effect of essentially nonvolatile solutes on the vapor pressure of a solvent. However, in our everyday lives we encounter many solutions containing volatile solutes. For example, the natural gas used for heating and the gasoline we use to power vehicles are mixtures of hydrocarbons. Gasoline comes from **crude oil**, a complex mixture of compounds composed mostly of carbon and hydrogen. Depending on its source, crude oil contains varying amounts of hydrocarbons with five or more carbon atoms in their molecular structures. The hydrocarbons with one to four carbon atoms are usually found in deposits of natural gas, although they are also dissolved in crude oil. Most hydrocarbons in gasoline have between five and nine carbon atoms per molecule. Each of these compounds is volatile and has a measurable vapor pressure at 25°C. Because of this volatility, we can separate gasoline from crude oil by distillation. In this section we explore distillation on a molecular level and consider the effects of volatile solutes on the vapor pressure of a solution.

**CONNECTION** We discussed simple distillation in Chapter 1 as a way to make drinking water from seawater.

## Vapor Pressures of Mixtures of Volatile Solutes

In Chapter 5 we discussed the process of distillation as a way to purify liquids, and in Chapter 10 we discussed the vaporization of pure liquids and the role of intermolecular forces in determining normal boiling point. In addition, we saw that the temperature at which a pure liquid boils remains constant throughout the

**crude oil** a combustible liquid mixture of hydrocarbons and other organic molecules formed under Earth's surface.

**FIGURE 11.9** Fractional distillation separates mixtures. Vapors rise through a fractionating column, where they repeatedly condense and vaporize. The most volatile component distills first and is first to pass through the condenser and into the collecting flask. Increasingly less volatile, higher-boiling components are distilled in turn. The progress of the distillation process is monitored using the thermometer at the top of the fractionating column.

process. Another type of distillation, called **fractional distillation** (Figure 11.9), is used to separate the volatile components of a mixture. This method of purifying the components of a liquid mixture is based on the observation that the boiling point of a mixture changes as the mixture is distilled.

As a mixture of volatile liquids is heated, the vapor that rises and fills the space above the liquid has a different composition from the composition of the mixture: the concentration of the component with the lowest boiling point is higher in the vapor than in the liquid. If this enriched vapor is collected, condensed, and redistilled, the vapor this time is even richer in the component with the lowest boiling point. In a fractional distillation apparatus, repeated distillation steps allow components with only slightly different boiling points to be separated from one another.

To see how a mixture behaves when boiled and how fractional distillation uses that behavior to separate the components, let's examine the heating curves of a pure substance and a solution of two volatile liquids. Figure 11.10(a) is the heating curve for pure octane ($C_8H_{18}$, bp 126°C). As in the heating curves of Section 5.4, the phase change from liquid to vapor takes place at a constant temperature of 126°C. Figure 11.10(b) shows the heating curve for a mixture of heptane ($C_7H_{16}$, bp 98°C) and octane. Here the portion of the curve from point 1 to point 2 again represents a vapor–liquid phase change, but in this case the temperature is not constant during the phase change. Instead, the two components co-distill over a range of temperatures.

Now let's analyze the graphs describing this process to understand how fractional distillation works. Figure 11.10(c) shows the composition of the vapor above a series of boiling solutions, and herein lies the key to how distillation separates

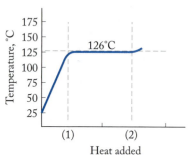

(a) Distillation of octane
(1) Octane begins to distill
(2) Octane finishes distilling

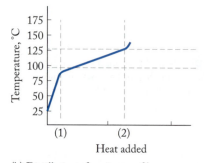

(b) Distillation of a mixture of heptane and octane
(1) Solution begins to distill
(2) Solution finishes distilling

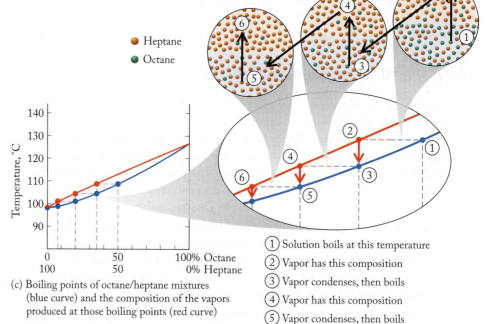

(c) Boiling points of octane/heptane mixtures (blue curve) and the composition of the vapors produced at those boiling points (red curve)

① Solution boils at this temperature
② Vapor has this composition
③ Vapor condenses, then boils
④ Vapor has this composition
⑤ Vapor condenses, then boils

**FIGURE 11.10** Heating curves describe the behavior of boiling liquids. (a) When pure octane is distilled, the distillation takes place at a constant temperature, indicated by the horizontal line at $T = 126$°C. The liquid in the collecting flask is pure octane. (b) When a mixture of heptane (bp 98°C) and octane is distilled, the distillation takes place over a range of temperatures. (c) The blue line shows the temperatures where solutions of a given composition boil. The red line shows the composition of the vapor arising from those solutions. The stair-step line illustrates what happens in a fractionating column. The balloons illustrate the relative composition of the two phases as the fractional distillation proceeds.

mixtures. In the vapor in equilibrium with a solution of two volatile substances, the concentration of the lower-boiling (more volatile) component is always greater in the vapor than in the liquid. Suppose we use the distillation apparatus from Figure 11.9 to heat 100 mL of a solution that is 50% by volume heptane and 50% octane. The blue curve in Figure 11.10(c) gives the temperature of the boiling solution, and point 1 on the curve tells us that the boiling point of the solution (50:50) is initially about 108°C. The red line on the graph gives the composition of the vapor above the solution at a given temperature. To find the composition of the vapor at 108°C, we move horizontally to the left along the dashed line between points 1 and 2. This horizontal line corresponds to a temperature of 108°C on the vertical (temperature) axis. Point 2 corresponds to a different composition (on the *x*-axis) for the vapor than for the solution at point 1. The vapor at point 2 is enriched in the lower-boiling component: it is about 65% heptane and only 35% octane.

Suppose this 65:35 vapor rises up in the distillation column, cools, and condenses in the next region of the column, a process represented by the red arrow from point 2 to point 3. The condensed liquid in this flask is 65% heptane and 35% octane. Continued heating of the column warms this liquid, and it begins boiling at about 104°C, which is the temperature at point 3. To find the concentration of the vapor above this boiling liquid, we move left from point 3 until we intersect the red curve (point 4). Reading down from point 4 to the concentration axis, we see that the vapor concentration is now about 80% heptane and only 20% octane. This vapor with 80:20 composition rises up, where it cools and condenses, and the distillation is repeated.

If we continue this process of redistilling mixtures with increasing concentrations of heptane and then cooling and condensing the vapors, we eventually obtain a condensate that is pure heptane. If we monitor the temperature at which vapors condense at the very top of our distillation column, we will see a profile of temperature versus volume of distillate produced, like that shown in Figure 11.11. The first liquid to be produced is pure heptane, which has a boiling point of 98°C. Ideally over time, all 50 mL of heptane in the original sample is recovered. As we continue to add heat at the bottom of the column, the temperature at the top rises to 126°C, signaling that the second component (pure octane, bp 126°C) is being collected. In the real world, the separation may not be as complete as described in this idealized presentation.

**CONCEPT TEST** ................................................

Dimethyl ether, $CH_3OCH_3$, and acetone, $CH_3C(O)CH_3$, shown in Figure 11.12, have similar molar masses but different dipole moments: 1.30 D and 2.88 D, respectively. Which compound would you expect to distill first from a mixture of acetone and dimethyl ether?

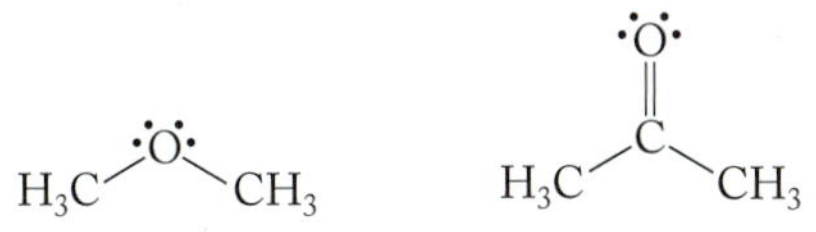

................................................

**SAMPLE EXERCISE 11.6**    **Predicting the Results of**    **LO4**
                            **Fractional Distillation**

How would the graph in Figure 11.11 be different if (a) we started with a mixture of 75 mL of octane ($C_8H_{18}$) and 25 mL of heptane ($C_7H_{16}$) and (b) we started with a mixture of 75 mL of octane ($C_8H_{18}$) and 25 mL of nonane ($C_9H_{20}$)? The normal boiling points of heptane, octane, and nonane are 98°C, 126°C, and 151°C, respectively.

**Collect and Organize** The graph in Figure 11.11 is an idealized plot of the temperature of the solution as a function of the volume of distillate for a 50:50 mixture of $C_7H_{16}$:$C_8H_{18}$. We want to describe the changes in the graph if we change the ratio of the components and if we change the identity of one of the components. We are given the normal boiling points of all compounds.

▶Ⅱ **CHEMTOUR** Fractional Distillation

**FIGURE 11.11** Fractional distillation of a mixture of 50 mL of heptane and 50 mL of octane produces two plateaus at the boiling points of the two components. If fractionation is perfect, the first 50 mL of distillate is pure heptane, and the second 50 mL is pure octane.

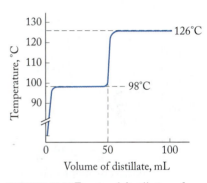

Dimethyl ether
$\mu = 1.30$ D

Acetone
$\mu = 2.88$ D

**FIGURE 11.12** Lewis structures for dimethyl ether and acetone.

........................................................

**fractional distillation** a method of separating a mixture of compounds on the basis of their different boiling points.

........................................................

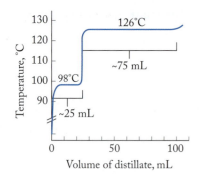

**FIGURE 11.13** Temperature as a function of distillate volume for a mixture of 25 mL of $C_7H_{16}$ and 75 mL of $C_8H_{18}$.

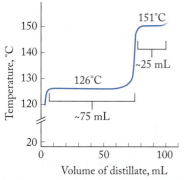

**FIGURE 11.14** Temperature as a function of distillate volume for a mixture of 75 mL of $C_8H_{18}$ and 25 mL of $C_9H_{20}$.

**Analyze** Fractional distillation separates solutions of volatile liquids into pure substances based on their different boiling points. The components distill in the order of their boiling points, with the lowest-boiling component distilling first. Ideally the total volume of distillate equals the total volume of the solution, and the volume of each fraction reflects the volume of the component in the original solution.

**Solve**

a. The first component that distills is the lower-boiling heptane. If our system were perfect, 25 mL of heptane would distill at 98°C. When the heptane was completely removed from the solution, the only remaining component (75 mL of octane) would distill at 126°C. The idealized graph (Figure 11.13) shows the boiling points of the two substances along the *y*-axis and the volume of distillate along the *x*-axis. The lengths of the horizontal lines at 98°C and 126°C are in a 1:3 ratio, reflecting the composition of the mixture, 25 mL of $C_7H_{16}$ and 75 mL of $C_8H_{18}$.

b. The first component that distills in the second mixture is the lower-boiling octane. If our system were perfect, 75 mL of octane would distill at 126°C. Once the octane is completely removed from the solution, the only remaining component (25 mL of nonane) would distill at 151°C. The idealized graph (Figure 11.14) shows the boiling points of the two substances along the *y*-axis and the volume of distillate along the *x*-axis.

**Think About It** Our graphs show the best results we could achieve. Actually, a small fraction that is a mixture of two hydrocarbons distills at temperatures between the boiling points of the two pure components.

⚙ **Practice Exercise** Draw an idealized graph of the temperature versus volume of distillate collected when a solution consisting of 30 mL of hexane (boiling point 69°C), 50 mL of heptane (boiling point 98°C), and 20 mL of nonane (boiling point 151°C) is fractionally distilled.

Now that we know how fractional distillation works, the next step is to understand *why* it works. We return to Raoult's law, which we used in Section 11.1 to describe the influence of nonvolatile solutes on the boiling point of a pure solvent. Raoult's law also applies to homogeneous mixtures of volatile compounds, such as crude oil. Because the solutes in a solution of crude oil are volatile, they contribute to the solution's overall vapor pressure. The total vapor pressure equals the sum of the vapor pressures of each component ($P_x^\circ$ is the equilibrium vapor pressure of the pure component at the temperature of interest) multiplied by the mole fraction of that component in the solution ($X_x$):

$$P_{total} = X_1 P_1^\circ + X_2 P_2^\circ + X_3 P_3^\circ + \cdots \qquad (11.8)$$

**SAMPLE EXERCISE 11.7** **Calculating the Vapor Pressure** **LO4**
**of a Solution of Volatile Substances**

Calculate the vapor pressure of a solution prepared by dissolving 13 g of heptane ($C_7H_{16}$) in 87 g of octane ($C_8H_{18}$) at 25°C. By what factor does the concentration of the more volatile component in the vapor exceed the concentration of this component in the liquid? The vapor pressures of octane and heptane at 25°C are 11 torr and 31 torr, respectively.

**Collect and Organize** We can use Equation 11.8 to determine the total vapor pressure of the solution from the vapor pressures of the components after we determine the composition of the solution in terms of mole fractions.

**Analyze** To calculate mole fractions, we need the molar masses of heptane and octane. The mole fraction is then equal to the number of moles of each component divided by the total number of moles of material in the solution.

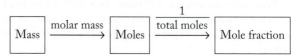

The vapor pressure of the solution must lie between the vapor pressures of the pure compounds.

**Solve** The number of moles of each component is

$$87 \text{ g } C_8H_{18} \times \frac{1 \text{ mol } C_8H_{18}}{114.23 \text{ g } C_8H_{18}} = 0.762 \text{ mol } C_8H_{18}$$

$$13 \text{ g } C_7H_{16} \times \frac{1 \text{ mol } C_7H_{16}}{100.20 \text{ g } C_7H_{16}} = 0.130 \text{ mol } C_7H_{16}$$

The total number of moles of material is $(0.762 \text{ mol} + 0.130 \text{ mol}) = 0.892 \text{ mol}$. Thus the mole fraction of each component in the mixture is

$$X_{\text{octane}} = \frac{0.762 \text{ mol}}{0.892 \text{ mol}} = 0.854$$

$$X_{\text{heptane}} = 1 - X_{\text{octane}} = 0.146$$

Using these values and the vapor pressures of the two hydrocarbons in Equation 11.8, we have

$$P_{\text{total}} = X_{\text{heptane}}P^\circ_{\text{heptane}} + X_{\text{octane}}P^\circ_{\text{octane}}$$

$$= 0.146(31 \text{ torr}) + 0.854(11 \text{ torr})$$

$$= 4.5 \text{ torr} + 9.4 \text{ torr} = 13.9 \text{ torr}$$

To calculate how enriched the vapor phase is in the more volatile component (the component with the greater vapor pressure, heptane), we need to recall Dalton's law of partial pressures (Section 6.7) and the concept that the partial pressure of a gas in a mixture of gases is proportional to its mole fraction in the mixture. Therefore, the ratio of the mole fraction of heptane to that of octane is the ratio of their two vapor pressures:

$$\frac{4.5 \text{ torr}}{9.4 \text{ torr}} = 0.48$$

The mole ratio of heptane to octane in the liquid mixture is

$$\frac{0.13 \text{ mol}}{0.76 \text{ mol}} = 0.17$$

Therefore, the vapor phase is enriched in heptane by a factor of

$$\frac{0.48}{0.17} = 2.8$$

**Think About It** As expected, the vapor pressure of the mixture (13.9 torr) is between the vapor pressures of the separate components. This result illustrates how fractional distillation works. The vapor is enriched in the lower-boiling component.

⚙ **Practice Exercise** Benzene ($C_6H_6$) is a trace component of gasoline. What is the mole ratio of benzene to octane in the vapor above a solution of 10% benzene and 90% octane by mass at 25°C? The vapor pressures of octane and benzene at 25°C are 11 torr and 95 torr, respectively.

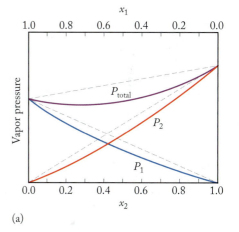

(a)

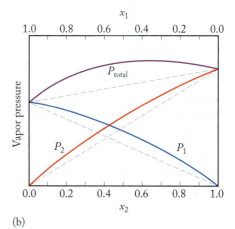

(b)

**FIGURE 11.15** In a mixture of two volatile substances $x_1$ and $x_2$, the vapor pressures $P_1$ and $P_2$ may deviate from the ideal behavior predicted by Raoult's law and described by the dashed lines. (a) If solute–solvent interactions are stronger than solvent–solvent or solute–solute interactions, the deviations from Raoult's law are negative. (b) If solute–solvent interactions are weaker than solvent–solvent or solute–solute interactions, the deviations are positive.

Solutions such as the hydrocarbons in crude oil obey Raoult's law when the strengths of solvent–solvent, solute–solute, and solute–solvent interactions are similar. Under these conditions, a solution behaves ideally. A mixture of hydrocarbons is expected to behave like an ideal solution because intermolecular interactions between the components are all London forces acting on molecules of similar structure and size.

If the solute–solvent interactions are stronger than solvent–solvent or solute–solute interactions, the solute inhibits the solvent from vaporizing and the solvent inhibits the solute from vaporizing. This situation produces negative deviations from the vapor pressures predicted by Raoult's law, as shown in Figure 11.15(a). In such a solution, the rate of evaporation is slower because the vapor pressure of the mixture is lower than predicted. Because solvent and solute molecules are held at the surface by solute–solvent attractive interactions, more energy is required to separate them from the surface, and fewer vaporize at a given temperature.

If solute–solvent interactions are much weaker than solvent–solvent interactions, less energy is required to separate solute molecules from the surface, and more solute molecules vaporize. In this case the vapor pressure is greater than the value predicted by Raoult's law, as shown in Figure 11.15(b).

**CONCEPT TEST** ● ● ● ● ● ● ● ● ● ● ● ● ● ● ● ● ● ● ● ● ● ● ● ● ● ● ● ● ● ● ● ● ● ● ● ● ● ● ● ●

Which of the following solutions is least likely to follow Raoult's law: (a) acetone/ethanol; (b) pentane/hexane; (c) pentanol/water? The structures of the compounds are given in Figure 11.16.

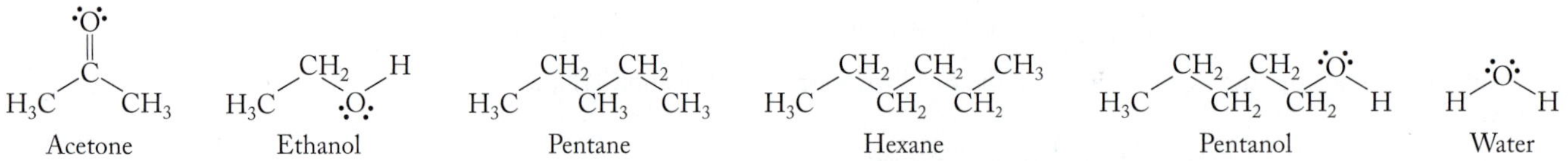

**FIGURE 11.16** Lewis structures for acetone, ethanol, pentane, hexane, pentanol, and water.

● ● ● ● ● ● ● ● ● ● ● ● ● ● ● ● ● ● ● ● ● ● ● ● ● ● ● ● ● ● ● ● ● ● ● ● ● ● ● ● ● ● ● ● ● ● ● ● ● ● ● ● ● ● ● ● ● ●

# 11.5 Colligative Properties of Solutions

We saw in Section 11.1 that the vapor pressure of a solution containing a nonvolatile solute is lower than the vapor pressure of the pure solvent. In this section we will see that many other physical properties of solvents are changed when a nonvolatile solute is added to it. Solutions generally have greater densities than the solvent alone, and an aqueous solution containing a nonvolatile solute has a higher boiling point and a lower freezing point than pure water. As we saw earlier, antifreeze, an aqueous solution of ethylene glycol used to cool automobile engines, boils at a temperature above 100°C. It also has a lower freezing point than pure water.

Figure 11.17 shows the combined phase diagram of water and an aqueous solution of a nonvolatile solute. Note that the red line representing boiling/condensation points of the solution lies below the orange line for pure water.

**FIGURE 11.17** Combined phase diagram for pure water and a solution of a nonvolatile solute in water. Notice that the solution's boiling point is higher than the boiling point of the water and that the solution's freezing point is lower than the freezing point of the water.

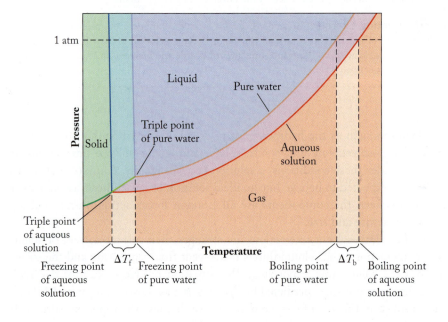

The blue melting/freezing line for the solution lies to the left of the light blue line for pure water. Following the dashed line from left to right at $P = 1$ atm, we see that the solution freezes at a lower temperature than pure water and boils at a higher temperature than pure water. Furthermore, the higher boiling point for the solution indicates that the solution has a lower vapor pressure than pure water.

Boiling point elevation and freezing point depression are both colligative properties. As noted in Section 11.1, only the number of particles in solution determines the impact of the solute on the colligative properties of the solvent. However, before we can quantify these properties, we must introduce a new concentration unit.

**molality (m)** concentration expressed as the number of moles of solute per kilogram of solvent.

## Molality

In Chapter 4, we introduced the concentration unit of molarity. Molarity is the concentration unit of choice when we run reactions in solutions, usually at constant temperature, so that we can measure volumes of solutions and know the number of moles of reactants we are dealing with. When we talk about colligative properties, however, we frequently deal with the properties of systems as a function of temperature. Because volumes change with temperature, molarity changes with temperature. If we want to quantify colligative properties, therefore, the approach is to use a different concentration unit, called **molality (m)**, defined as the number of moles of solute ($n_{solute}$) per kilogram of solvent:

$$m = \frac{n_{solute}}{\text{kg solvent}} \qquad (11.9)$$

The difference between these two similar-sounding concentration units is that molarity is the number of moles of solute *per liter of solution*, whereas molality is the number of moles of solute *per kilogram of solvent*. Molality is used when discussing colligative properties when temperature changes are involved. Because solvent volume changes with temperature but solvent mass does not change, a concentration expressed in molality does not change with temperature.

To illustrate the difference between the molarity and molality of a solution, let's use the procedure outlined in Figure 11.18 to calculate the molality of 1 L of an aqueous solution of sodium chloride that is 0.558 $M$ in NaCl (approximately the NaCl concentration in seawater) at 25°C. To calculate molality, we need the density of the solution so that we can convert liters of solution into kilograms of solvent. The density of the solution is 1.022 g/mL at 25°C, so 1 L of solution has a mass of

$$\frac{1.022 \text{ g}}{1 \text{ mL}} \times 1000 \text{ mL} = 1022 \text{ g}$$

Of this 1022 g of solution, the mass of dissolved NaCl is

$$\frac{0.558 \text{ mol NaCl}}{1 \text{ L solution}} \times \frac{58.44 \text{ g NaCl}}{1 \text{ mol NaCl}} = 32.6 \text{ g NaCl in 1 L of solution}$$

If the mass of 1 L of solution is 1022 g, and 32.6 g is due to the dissolved NaCl, then the mass of the solvent is (1022 g − 32.6 g) = 989 g = 0.989 kg. The molality of the solution (number of moles of solute per kilogram of solvent) is

$$\frac{0.558 \text{ mol NaCl}}{0.989 \text{ kg solvent}} = 0.564 \ m$$

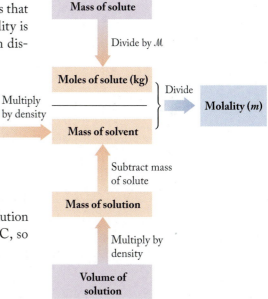

**FIGURE 11.18** Flow diagram for calculating molality.

The two concentrations are numerically close, but molality is always slightly higher than molarity.

**CONCEPT TEST** ······················································

The difference between molarity and molality of any dilute aqueous solution is small. Why?

······················································

**SAMPLE EXERCISE 11.8** | **Preparing a Solution of Known Molality** | **LO5**

How many grams of $Na_2SO_4$ should be added to 275 mL of water to prepare a 0.750 $m$ solution of $Na_2SO_4$? Assume the density of water is 1.000 g/mL.

**Collect and Organize** We want to determine the mass of sodium sulfate (the solute) needed to prepare a 0.750 $m$ solution. The volume of water (the solvent) is 275 mL.

**Analyze** Following the procedure shown in Figure 11.18,

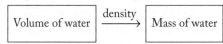

We can then work backward from molality to mass of solute:

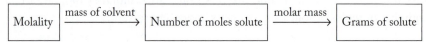

Our goal is to prepare a relatively small volume ($\sim\frac{1}{4}$ L) of a dilute solution ($< 1\ m$), so we predict that the mass of solute needed is probably less than $\frac{1}{4}$ mole of solute, which has a molar mass of about 142 g/mol, or about 35 g.

**Solve** Multiplying 275 mL of water by the density of water, we find that the mass of water we start with is

$$275\ \text{mL water} \times \frac{1.000\ \text{kg water}}{1000\ \text{mL water}} = 0.275\ \text{kg water}$$

We can rearrange Equation 11.9 to determine the number of moles of solute needed, using the fact that a molality of 0.750 is equivalent to 0.750 mol of solute in 1 kg of solvent:

$$n_{\text{solute}} = (m)(\text{kg solvent})$$

$$n_{Na_2SO_4} = \frac{0.750\ \text{mol}\ Na_2SO_4}{1\ \text{kg water}} \times 0.275\ \text{kg water} = 0.206\ \text{mol}\ Na_2SO_4$$

The molar mass of $Na_2SO_4$ is 142.04 g/mol, so the number of grams of $Na_2SO_4$ needed is

$$0.206\ \text{mol}\ Na_2SO_4 \times \frac{142.04\ \text{g}}{1\ \text{mol}} = 29.3\ \text{g}\ Na_2SO_4$$

Dissolving 29.3 g of $Na_2SO_4$ in 275 mL of water produces a 0.750 $m$ solution.

**Think About It** The calculated value of 29 g $Na_2SO_4$ is consistent with our prediction that about 35 g of solute would be required.

⚙ **Practice Exercise** What is the molality of a solution prepared by dissolving 78.2 g of ethylene glycol, $HOCH_2CH_2OH$, in 1.50 L of water? Assume the density of water is 1.00 g/mL.

······················································

## Boiling Point Elevation

In Section 11.1 we discussed how the vapor pressure of a solution is reduced and the boiling point is elevated with respect to pure solvent, and at the beginning of this section we examined the phase diagrams of solutions. Now we can look quantitatively at how much the boiling points and freezing points of liquids change when solutes are present.

*Boiling point elevation* is a colligative property of the solvent. It is described in equation form as

$$\Delta T_b = K_b m \qquad (11.10)$$

where $\Delta T_b$ is the increase in temperature above the boiling point of the pure solvent, $K_b$ is the *boiling-point-elevation constant* of the solvent, and $m$ is the molality of the solution. The units of $K_b$ are $°C/m$, so the concentration of particles in solution must also have units of molality ($m$):

$$\Delta T_b = K_b m = \frac{°C}{m} \times m$$

The $K_b$ of water is $0.52°C/m$, which means that for every mole of particles that dissolves in 1 kg of water (1 $m$), the boiling point of the solution rises by $0.52°C$:

$$\Delta T_b = K_b m = \frac{0.52°C}{m} \times 1\ m = 0.52°C$$

Therefore the boiling point ($T_b$) of the solution is

$$T_b = T_{bp} + \Delta T_b = 100.00°C + 0.52°C = 100.52°C$$

---

**SAMPLE EXERCISE 11.9** **Calculating the Boiling Point Elevation of an Aqueous Solution** **LO6**

What is the boiling point of seawater if the concentration of ions in seawater is 1.15 $m$?

**Collect and Organize** We are asked to calculate the boiling point of an aqueous solution given that the ion concentration is 1.15 $m$. The boiling-point-elevation constant of water is $K_b = 0.52°C/m$, and the normal boiling point of water is $100.0°C$.

**Analyze** We can calculate the boiling point elevation using Equation 11.10; the boiling-point-elevation constant of water, $0.52°C/m$; and the solute concentration. Because the concentration is close to 1 $m$, we predict that the boiling point elevation will be close to $0.5°C$.

**Solve**

$$\Delta T_b = K_b m = \frac{0.52°C}{m} \times 1.15\ m = 0.60°C$$

The temperature at which this seawater boils is $0.60°C$ higher than the normal boiling point of pure water: $100.0°C + 0.60°C = 100.6°C$.

**Think About It** As predicted, the boiling point of the solution is only slightly higher than the boiling point of the pure solvent.

⚙ **Practice Exercise** Crude oil pumped out of the ground may be accompanied by *formation water*, a solution that contains high concentrations of NaCl and other salts. If the boiling point of a sample of formation water is $2.3°C$ above the boiling point of pure water, what is the molality of dissolved particles in the sample?

---

## Freezing Point Depression

As we mentioned when discussing Figure 11.17, *freezing point depression* is a colligative property that is put to good use in car radiators to ensure that their fluid does not freeze in cold weather. The magnitude of the freezing point depression is directly proportional to the molal concentration of dissolved solute:

$$\Delta T_f = K_f m \qquad (11.11)$$

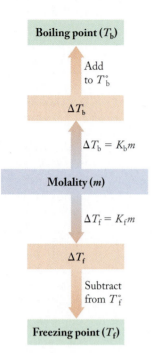

**FIGURE 11.19** Flow diagram for calculating freezing points and boiling points of solutions of known molality ($T_b^\circ$ = normal boiling point of pure solvent; $T_f^\circ$ = normal freezing point of pure solvent).

where $\Delta T_f$ is the change in the freezing temperature of the solvent; $K_f$ is the *freezing-point-depression constant* of the solvent; and $m$ is the molality of the solution. The freezing point ($T_f$) of a solution is

$$T_f = T_{fp} - \Delta T_f$$

Figure 11.19 summarizes calculations of freezing point depression and boiling point elevation.

---

**SAMPLE EXERCISE 11.10    Calculating the Freezing     LO6
                            Point of a Solution**

What is the freezing point of radiator fluid prepared by mixing 1.00 L of ethylene glycol ($HOCH_2CH_2OH$, density 1.114 g/mL) with 1.00 L of water (density 1.000 g/mL)? The freezing-point-depression constant of water, $K_f$, is 1.86°C/$m$.

**Collect and Organize** We are asked to determine the freezing point of a solution of ethylene glycol in water from the volumes of the two liquids, their densities, and $K_f$ for water.

**Analyze** To determine the freezing point, we need to calculate $\Delta T_f$ for the solution and then subtract that value from water's normal freezing point. Because using Equation 11.11 requires us to know $m$, the molality of the solution, we need to convert volumes of solute and solvent into moles of solute and kilograms of solvent.

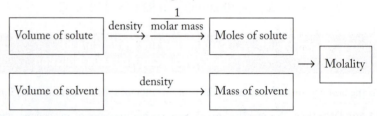

Knowing their densities allows us to make both conversions, and we use ethylene glycol's formula to calculate its molar mass. Based on Sample Exercise 11.8, we choose water as the solvent. It is a logical choice because ethylene glycol has a much higher molar mass and a density very similar to water, so we can assume that 1 L of ethylene glycol contains fewer moles of material than 1 L of water. The relatively large volume ($10^3$ mL) and a density close to 1 g/mL for the solute allow us to predict that the mass of solute will be on the order of $10^3$ g. With a solute molar mass of 62 g/mol, the molality of the solution should be between 10 and 20 $m$. Considering $K_f$ = 1.86°C/m, we predict $\Delta T_f$ to be in the range of 20° to 40°C and therefore the freezing point between −20° and −40°C.

**Solve** The solvent mass is

$$1.00 \text{ L} \times \frac{1000 \text{ mL}}{1 \text{ L}} \times \frac{1.000 \text{ g}}{1 \text{ mL}} \times \frac{0.001 \text{ kg}}{1 \text{ g}} = 1.00 \text{ kg}$$

After calculating the ethylene glycol molar mass to be 62.07 g/mol, we have for the solute

$$1.00 \text{ L solute} \times \frac{1000 \text{ mL}}{1 \text{ L}} \times \frac{1.114 \text{ g}}{1 \text{ mL}} \times \frac{1 \text{ mol}}{62.07 \text{ g}} = 17.9 \text{ mol solute}$$

Therefore the molal concentration is

$$m = \frac{17.9 \text{ mol solute}}{1.00 \text{ kg solvent}} = 17.9 \, m$$

Using Equation 11.11 gives

$$\Delta T_f = K_f m$$

$$= \frac{1.86°C}{m} \times 17.9 \, m = 33.3°C$$

Subtracting this temperature change from the normal freezing point of water, 0.0°C, we have

$$\text{Freezing point of radiator fluid} = 0.0°C - 33.3°C = -33.3°C$$

**Think About It** The answer makes sense because the reason we add antifreeze to radiators is to depress the freezing point of water, and the solution we evaluated here certainly has a freezing point lower than that of pure water. To be an effective radiator fluid, the solution must have a freezing point sufficiently below the coldest expected temperatures, so our answer of −33°C is reasonable and matches our prediction.

⚙ **Practice Exercise** What is the boiling point of the automobile radiator fluid in the preceding Sample Exercise? The $K_b$ of water is 0.52°C/$m$.

Figure 11.20 provides a molecular view of the influence of dissolved seasalt on the freezing point of seawater. At the freezing point of pure water (0°C, Figure 11.20a), ice and liquid water coexist: molecules of water that collide with the ice surface are captured (depicted by the blue arrows), while solid particles become detached from the ice layer and enter the liquid phase (red arrows). When these two processes occur at the same rate there is no change in the masses of ice and liquid water.

Now, consider the same equilibrium in seawater (Figure 11.20b). The presence of seasalt ions appears to block some water molecules from reaching the ice surface, which reduces the number of blue-arrow freezing events. With no reduction in the number of red arrow melting events, the ice melts. If the mixture is cooled, the rate of freezing increases and balance is restored (Figure 11.20c). This lower temperature is now the freezing point of seawater (−2°C).

The notion that solute particles can block solvent particles from freezing at their normal freezing point is an appealing one for explaining freezing point depression. It is also tempting to use it to explain the lower vapor pressure (see Figure 11.1) and higher boiling points of solutions compared to pure solvents. However, there is a problem with this model: common sense tells us that bigger solute particles would be better blockers than little ones. However, colligative properties such as freezing point depression depend only on *the number* of particles, *not on their size*. Thus, the particle-interference model is not supported by observation and experimental results. A model that is supported by the data is based on greater dispersion of the internal energy of a solution due to the presence of solute particles. We explore this concept in detail in Chapter 18.

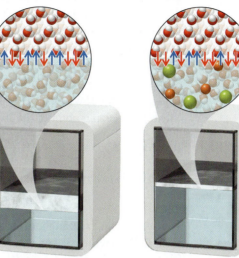
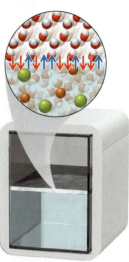
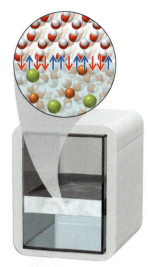

(a) Pure water at 0.0°C (water and ice coexist)

(b) Seawater at 0.0°C (ice melts)

(c) Seawater at −1.9°C (seawater and ice coexist)

**FIGURE 11.20** The effect of solute particles on the freezing point of water. (a) A layer of ice floats on pure water in an insulated container at 0.0°C. The two phases are in equilibrium as indicated by the equal numbers of red and blue arrows indicating that water is freezing (blue arrows) just as rapidly as ice is melting (red arrows). (b) The thickness of a layer of ice floating on seawater gets smaller at 0.0°C because the rate at which the water in seawater freezes is slower than the rate at which ice melts. (c) Ice and seawater coexist at −1.9°C because the rate at which ice melts and seawater freezes are the same at this temperature.

## The van 't Hoff Factor

Recall from Section 4.4 that compounds called *electrolytes* dissociate into ions when in solution, whereas *nonelectrolytes* remain intact. Think about what this means in terms of how we count the number of dissolved particles that various solutes produce upon dissolution. If a solute is a nonelectrolyte, then every mole of solute yields one mole of particles. If, however, the solute is an electrolyte, then the number of moles of particles depends on the chemical formula of the compound. For example, if we approximate seawater as 0.574 $m$ sodium chloride, then each kilogram of water contains 1.148 moles of particles because NaCl forms $Na^+$ and $Cl^-$ ions when it dissolves.

Because freezing point depression and boiling point elevation are colligative properties, the dissolution of 1 mole of a strong electrolyte such as NaCl in a given quantity of water produces the same changes in freezing point and boiling point as

∞ **CONNECTION** In Chapter 4, we defined an electrolyte as a substance that dissociates into ions when it dissolves.

▶❙❙ **CHEMTOUR** Boiling and Freezing Points

1 mole of the strong electrolyte $KNO_3$, even though the latter has a much higher molar mass. Each of these salts adds 2 moles of particles (1 mole of cations and 1 mole of anions) to the water for every 1 mole of salt that dissolves. However, when a nonelectrolyte such as ethylene glycol is dissolved in water, 1 mole of the solute produces only 1 mole of particles (1 mole of molecules) in the solution. Therefore, we would need to have a 2 $m$ ethylene glycol solution to achieve the same boiling point and freezing point as solutions of 1 $m$ NaCl or 1 $m$ $KNO_3$.

Dutch chemist Jacobus van 't Hoff (1852–1911) studied colligative properties and defined a term $i$, now called the **van 't Hoff factor** (or ***i* factor**), which is the ratio of the experimentally measured value of a colligative property to the value expected if the solute were a nonelectrolyte (that is, no dissociation into ions). For example, suppose we make a solution that is 0.010 $m$ in NaCl by dissolving 0.5844 g of NaCl in 1 kg of water. If we treat the NaCl as a nonelectrolyte, then according to Equation 11.11, the freezing point of the solution should be lower than that of pure water by

$$\Delta T_f = K_f m = \frac{1.86°C}{1 \text{ mol NaCl}/1 \text{ kg water}} \times \frac{0.0100 \text{ mol NaCl}}{1 \text{ kg water}} = 0.0186°C$$

which leads to a new freezing point of

$$T_f = 0.0000° - 0.0186° = -0.0186°C$$

However, the freezing point of the solution measured in the laboratory is $-0.0372°C$, which is twice as low as the value predicted for a nonelectrolyte. The ratio of the experimentally measured value to the theoretical value, which is the $K_f m$ term in Equation 11.11, is

$$\frac{\Delta T_{f,\text{measured}}}{K_f m} = \frac{0.0372°C}{0.0186°C} = 2 = i$$

This is the van 't Hoff factor for sodium chloride. This makes sense because sodium chloride is a strong electrolyte and forms two moles of ions from each mole of NaCl.

The significance of the $i$ factor is based on the definition of colligative properties: changes due to the *total concentration of dissolved particles* present in solution. When 1 mole of a strong electrolyte such as solid sodium chloride dissolves, it produces 2 moles of particles. Thus, we should expect that the freezing point depression (or boiling point elevation) that results from the dissolution of a given number of moles of NaCl should be twice as great as that produced when the same number of moles of a nonelectrolyte dissolves.

Equations 11.10 and 11.11 can be modified to include the $i$ factor:

$$\Delta T_b = iK_b m \qquad (11.12)$$

$$\Delta T_f = iK_f m \qquad (11.13)$$

If the solute is molecular (such as ethylene glycol) and therefore a nonelectrolyte, then $i = 1$ because each mole of solute produces 1 mole of dissolved particles. If the solute is a strong electrolyte, then $i =$ the number of ions in one formula unit. For NaCl, therefore, $i = 2$; for $Na_2SO_4$, $i = 3$ (two $Na^+$ ions and one $SO_4^{2-}$ ion). Note that we do *not* break a polyatomic ion such as $SO_4^{2-}$ into its atoms when determining an $i$ factor; polyatomic ions stay intact.

**van 't Hoff factor** (also called ***i* factor**) the ratio of the experimentally measured value of a colligative property to the theoretical value expected for that property if the solute were a nonelectrolyte.

**ion pair** a cluster formed when a cation and an anion associate with each other in solution.

---

**SAMPLE EXERCISE 11.11    Using the van 't Hoff Factor    LO7**

The salt lithium perchlorate ($LiClO_4$) is one of the most water-soluble salts known. At what temperature does a 0.130 $m$ solution of $LiClO_4$ freeze? The $K_f$ of water is 1.86°C/$m$; assume $i = 2$ for $LiClO_4$, and the freezing point of pure water is 0.00°C.

**Collect and Organize** We are asked to determine the freezing point of a salt solution. We know the formula of the solute, its molal concentration, and $K_f$ of the solvent. We know that the freezing point of the pure solvent is 0.00°C. We are given the value of the van 't Hoff factor as $i = 2$.

**Analyze** We use Equation 11.13 to solve for the freezing point depression. The value of $K_f$ is close to 2, the value of $i$ is 2, and the concentration of solute is close to 0.1 $m$, so we predict that the freezing point of the solution will be about 0.4°C lower than that of pure water.

**Solve**

$$\Delta T_f = iK_f m = (2)(1.86°C/m)(0.130\ m) = 0.48°C$$

The freezing point of the solution is $(0.00 - 0.48)°C = -0.48°C$.

**Think About It** Lithium perchlorate dissolves in aqueous solution to form 2 moles of ions for every mole of solute that dissolves: 1 mole of $Li^+$ cations and 1 mole of $ClO_4^-$ anions, consistent with $i = 2$. The calculated value is consistent with our prediction.

**◉ Practice Exercise** Determine the value of the van 't Hoff factor and calculate the boiling point of a 1.75 $m$ aqueous solution of barium nitrate, $Ba(NO_3)_2$. The $K_b$ of water is 0.52°C/$m$.

---

**CONCEPT TEST**

Which aqueous solution has the lowest freezing point: (a) 3 $m$ glucose ($C_6H_{12}O_6$), (b) 2 $m$ potassium iodide (KI), or (c) 1 $m$ sodium sulfate ($Na_2SO_4$)?

---

Using Equations 11.12 and 11.13 to calculate freezing point depressions and boiling point elevations for concentrated solutions of strong electrolytes often gives larger values than the experimentally measured values. The reason is that the cations and anions produced when strong electrolytes dissolve may not be totally independent of one another. As concentration increases, cations and anions may form ionic clusters. The simplest cluster, an **ion pair**, consists of a cation and an anion that associate in solution, acting as a single particle. Thus, the overall concentration of particles is reduced when ion pairs form, and experimentally measured freezing point depressions and boiling point elevations are smaller than the theoretical values obtained with Equations 11.12 and 11.13 (Figure 11.21).

**FIGURE 11.21** (a) The experimentally measured freezing point of a 0.010 $m$ solution of NaCl is the same as the theoretical value obtained with Equation 11.13. This agreement means that the solution behaves ideally and little or no ion pairing takes place. (b) The experimentally measured freezing point of a 0.10 $m$ solution is about 0.04°C higher than the theoretical value because some of the $Na^+$ and $Cl^-$ ions form ion pairs, as shown inside the red ovals. The formation of ion pairs causes the concentration of solute particles to be less than the theoretical number, and as a result the van 't Hoff factor for the solution is less than the theoretical value of 2, and the decrease in the freezing point is less than expected.

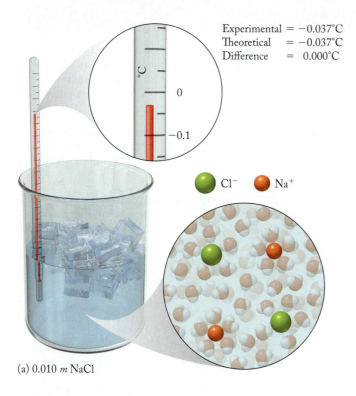

Experimental = −0.037°C
Theoretical = −0.037°C
Difference = 0.000°C

Cl⁻   Na⁺

(a) 0.010 $m$ NaCl

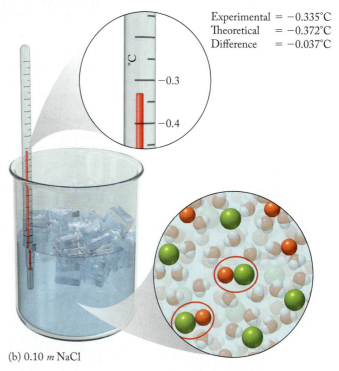

Experimental = −0.335°C
Theoretical = −0.372°C
Difference = −0.037°C

(b) 0.10 $m$ NaCl

**FIGURE 11.22** Theoretical and experimentally measured values for the van 't Hoff factors for 0.1 $m$ solutions of several electrolytes and the nonelectrolyte ethanol. The higher the charge on the ions, the greater the difference between theoretical and experimentally measured values.

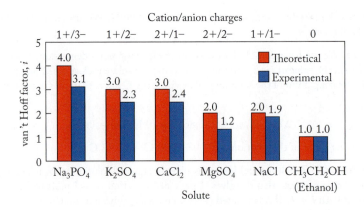

The extent to which free ions form when a strong electrolyte dissolves is expressed by the van 't Hoff factor. The van 't Hoff factor for NaCl in water is 2 if the solution behaves ideally, because ideally 2 moles of ions are produced for each mole of NaCl that dissolves. The value of $i$ is 2 for 0.010 $m$ NaCl, but a little less than 2 for 0.10 $m$ NaCl. Whenever a calculation for $i$ gives a noninteger value, solute particles are associating in solution, and the behavior is nonideal. Figure 11.22 gives some theoretical and experimentally measured values of the van 't Hoff factor for several substances.

**CONCEPT TEST**

The van 't Hoff factor for an aqueous solution of an ionic compound is 2. Which of the following is (are) not a possible explanation?
  a. The solute is a 1:1 salt behaving as an ideal solution.
  b. The solute is a nonelectrolyte.
  c. The solute is a 2:1 electrolyte behaving in a nonideal fashion.
  d. The solute is a 1:1 salt of a weak electrolyte.

**SAMPLE EXERCISE 11.12** **Assessing Particle Interactions in Solution** LO7

The experimentally measured freezing point of a 1.90 $m$ aqueous solution of NaCl is −6.57°C. What is the value of the van 't Hoff factor for this solution? Is the solution behaving ideally, or is there evidence that solute particles are interacting with one another? The freezing-point-depression constant of water is $K_f = 1.86°C/m$, and the freezing point of pure water is 0.00°C.

**Collect and Organize** We are asked to calculate the $i$ factor for a solution of known molality. We are given the measured freezing point and the $K_f$ value. Equation 11.13 relates $\Delta T_f$, $K_f$, the molality of the solution, and the value of $i$. We are also asked whether the solution is behaving ideally.

**Analyze** Equation 11.13 tells us that $\Delta T_f = iK_f m$. We can rearrange the equation to solve for $i$ before substituting the values for $\Delta T_f$, $K_f$, and $m$. If the solution behaves ideally, we would expect $i = 2$ for the strong electrolyte NaCl; however, given the concentration of NaCl (1.90 $m$), we predict a value less than 2.

**Solve** Rearranging Equation 11.13 gives us

$$i = \frac{\Delta T_{f,measured}}{K_f m}$$

$$i = \frac{6.57°C}{\left(\dfrac{1.86°C}{m}\right) 1.90\ m} = 1.86$$

That $i$ is not an integer tells us that the solution is not behaving ideally. Ion pairs must be forming in solution.

**Think About It** As predicted, the value of $i$ for this solution is less than the theoretical value of 2.

⚙ **Practice Exercise** The van 't Hoff factor for a 0.050 $m$ aqueous solution of magnesium sulfate is 1.3. What is the freezing point of the solution?

---

**osmosis** the flow of a fluid through a semipermeable membrane to balance the concentration of solutes in solutions on the two sides of the membrane. The solvent molecules' flow proceeds from the more dilute solution into the more concentrated one.

---

The extent of ion pairing in a solution of a strong electrolyte generally increases with solute concentration as the solution "runs out of water" needed to form spheres of hydration around the ions. For any salt, the theoretical value of $i$ obtained with Equation 11.12 or 11.13 is an upper limit of possible values. If ion pairing occurs, the experimentally measured value must be smaller than the theoretical value.

## Osmosis and Osmotic Pressure

The final colligative property we look at in this chapter is *osmotic pressure*, the result of the process called **osmosis**—the movement of a solvent through a semipermeable membrane from a region of lower solute concentration to a region of higher solute concentration. A *semipermeable membrane* is one in which the pores are so small that water molecules can pass through them but most solute particles are too large to pass through.

As an example of the importance of osmosis, we start with the observation that we all need water to survive. More than 97% of the water on Earth is seawater, and none of it is fit to drink. Let's look at the reason on a cellular level. The liquid inside each cell in the body is a complex solution of numerous solutes, with the average concentration of these solutes being about one-third the concentration of solutes in seawater. When cells are exposed to seawater, this substantial difference in solute concentration is the driving force for osmosis, with the cell membrane acting as the necessary semipermeable membrane (Figure 11.23). Because the solute concentration is higher outside the cell, water from inside the

**FIGURE 11.23** The membrane of a red blood cell is semipermeable, which means that water easily flows by osmosis into and out of the cell to equalize the solute concentrations on the two sides of the membrane. (a) When a cell is immersed in a solution in which the solute concentration equals the solute concentration inside the cell (isotonic conditions), the flow of water into the cell is exactly balanced by the flow of water out of the cell, and the cell size does not change. (b) When the cell is immersed in a solution in which the solute concentration is higher than the solute concentration inside the cell (hypertonic conditions), water flows by osmosis from the region of lower solute concentration to the region of higher solute concentration—in other words, out of the cell—and the cell shrinks. (c) When the cell is immersed in pure water, the solute concentration is higher inside the cell than outside (hypotonic conditions), and water flows by osmosis from the region of zero solute concentration to the region of high solute concentration—into the cell—and the cell expands.

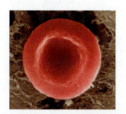
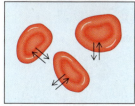

(a) Isotonic solution: total solute concentration in the solution matches that inside the cell

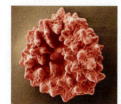

(b) Hypertonic solution: total solute concentration in the solution is greater than that inside the cell; water leaves cells, cells shrink

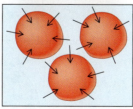

(c) Hypotonic solution: total solute concentration in the solution is less than that inside the cell; water enters cells, cells expand

▶‖ CHEMTOUR Osmotic Pressure

cell crosses the cell membrane and enters the seawater, moving from the low-solute-concentration side of the membrane to the high-solute-concentration side. As water leaves the cell, the cell shrivels and ultimately ceases to function.

Water molecules migrate through a cell membrane, or through any other semipermeable membrane, because a force makes it happen—a force caused by the different concentrations of solutes on the two sides of the membrane. When we divide the magnitude of this force, $F$, by the surface area, $A$, of the membrane, we get pressure: $P = F/A$. **Osmotic pressure ($\Pi$)** is the pressure required to halt the flow of solvent from the more dilute solution across the membrane (Figure 11.24). Osmotic pressure exactly balances the pressure ($F/A$) driving solvent through the membrane so that no net flow of solvent takes place.

The Greek letter pi ($\Pi$) is used as the symbol for osmotic pressure to distinguish it from the pressure exerted by gases. In Figure 11.24(a), the solution on the right has a lower concentration and lower osmotic pressure ($\Pi_{NaCl}$) than the solution on the left ($\Pi_{seawater}$): $\Pi_{NaCl} < \Pi_{seawater}$. The result is a net flow of solvent (water) from right to left until the pressure on both sides is equal. Figure 11.24(b) shows how the volumes of both solutions change: the difference in volume reflects the difference in osmotic pressures, $\Delta\Pi = \Pi_{seawater} - \Pi_{NaCl}$, in the original solutions. This process is profoundly important in living systems, where the solution inside each cell exerts an osmotic pressure on the cell membrane, a pressure pushing toward the outside of the cell. At the same time, blood or other liquid outside the cell exerts an osmotic pressure on the cell membrane, and this pressure pushes toward the cell's interior.

The magnitude of the osmotic pressure $\Pi$ required to stop the net flow of solvent across a semipermeable membrane separating a pure solvent from a solution depends on the solute concentration, the absolute temperature, and the constant $R$, 0.0821 L · atm/(mol · K):

$$\Pi = MRT$$

where $M$ is the molarity of the solute. Molarity is used to express concentration in calculating $\Pi$ because the expression for $\Pi$ can be derived from an equation similar to the ideal gas equation:

$$\Pi = P = \left(\frac{n}{V}\right)RT = MRT$$

**FIGURE 11.24** (a) The solute concentration of seawater is approximately 1.15 $M$. When equal volumes of seawater and a 0.10 $M$ solution of NaCl are separated by a semipermeable membrane, water moves by osmosis from 0.10 $M$ NaCl (low solute concentration) to the seawater (high solute concentration) side. (b) The volume on the NaCl side decreases until the osmotic pressure of the solution equals the osmotic pressure (and concentration) of the diluted seawater solution, resulting in a difference in the heights of the liquid levels. This difference in volume is proportional to the original difference in osmotic pressures ($\Delta\Pi$) between the two solutions.

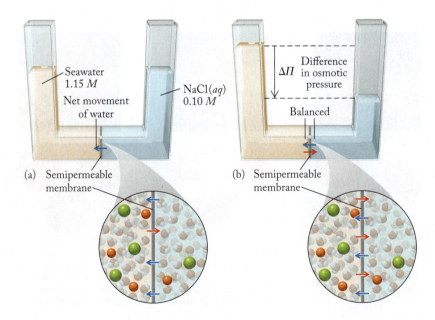

(a) Seawater 1.15 $M$ — Net movement of water — NaCl($aq$) 0.10 $M$ — Semipermeable membrane

(b) $\Delta\Pi$ Difference in osmotic pressure — Balanced — Semipermeable membrane

Note that the term $n/V$ has units of moles per liter, which matches the definition of molarity. Osmotic pressure is a colligative property because $\Pi$ is proportional to the concentration of solute and does not depend on the identity of the solute. Therefore the molarity must be multiplied by the van 't Hoff factor $i$ for the solute:

$$\Pi = iMRT \qquad (11.14)$$

Because the units of $R$ are $L \cdot atm/(mol \cdot K)$ and $i$ has no units, the product $iMRT$ gives $\Pi$ in units of atmospheres.

Osmotic pressure is a colligative property, so a 1.0 $M$ solution of NaCl produces the same osmotic pressure as 1.0 $M$ KCl or 1.0 $M$ NaNO$_3$ because all three solutions are 2.0 $M$ in total ions ($i = 2$). These solutions have twice the osmotic pressure of a 1.0 $M$ solution of glucose at a given temperature because glucose is a molecular substance ($i = 1$) and produces a solution that is only 1.0 $M$ in dissolved particles (glucose molecules).

> **osmotic pressure ($\Pi$)** the pressure applied across a semipermeable membrane to stop the flow of solvent from the compartment containing pure solvent or a less concentrated solution to the compartment containing a more concentrated solution. The osmotic pressure of a solution increases with solute concentration, $M$, and with solution temperature, $T$.

**CONCEPT TEST**

If a 1.0 $M$ glucose solution is on one side of a semipermeable membrane and a 1.0 $M$ KCl solution is on the other side, in which direction does the water flow?

---

**SAMPLE EXERCISE 11.13    Calculating Osmotic Pressure I    LO8**

At the beginning of this section, we mentioned that the concentration of solutes in a red blood cell is about a third of that of seawater—more precisely, about 0.30 $M$. If red blood cells are immersed in pure water, they swell, as shown in Figure 11.23(c). Calculate the osmotic pressure at 25°C of red blood cells across the cell membrane from pure water.

**Collect and Organize** We are asked to calculate an osmotic pressure. We are given the total particle concentration (0.30 $M$) and temperature (25°C) of the solution. We must convert the temperature to kelvins.

**Analyze** Osmotic pressure is related to the total concentration of all particles in solution and the absolute temperature. The total particle concentration, 0.30 $M$, is the value of the $iM$ term in Equation 11.14. Estimating the answer by taking $iM = 0.3$, $R$ at about 0.1, and $T$ at about 300, we get about 10 atm.

**Solve** First we need to convert the temperature from degrees Celsius to kelvins: $T(°C) + 273 = T(K)$. Inserting the values of $iM$, $R$, and $T$ in Equation 11.14 we have

$$\Pi = iMRT = \frac{0.30 \text{ mol}}{L} \times \frac{0.0821 \text{ L} \cdot atm}{mol \cdot K} \times (25 + 273) \text{ K} = 7.3 \text{ atm}$$

**Think About It** Our answer is very much in line with our estimation. The calculated pressure across the membranes of red blood cells in pure water is over 7 atmospheres: this is enough to rupture the membranes.

⚙ **Practice Exercise** Calculate the osmotic pressure across a semipermeable membrane separating pure water from seawater at 25°C. The total concentration of all the ions in seawater is 1.15 $M$.

---

**CONCEPT TEST**

Which has the greater effect on the osmotic pressure of a 1.0 $M$ solution: increasing the temperature from 10°C to 20°C or adding enough solute to raise the concentration to 2.0 $M$?

Calculating the osmotic pressure of a solution relative to pure solvent is straightforward. In Figure 11.24(a), however, the two solutions have different osmotic pressures, $\Pi_{NaCl}$ and $\Pi_{seawater}$. Let's derive an equation for the difference between the osmotic pressures, $\Delta\Pi$, in terms of $M$, $R$, and $T$ for the situation depicted in Figure 11.24.

Earlier in this section we said that $\Delta\Pi = \Pi_{seawater} - \Pi_{NaCl}$ and that the osmotic pressures of both solutions are described by Equation 11.14:

$$\Pi_{seawater} = iM_{seawater}RT \quad \text{and} \quad \Pi = iM_{NaCl}RT$$

Taking the difference in $\Pi$ expressions:

$$\Delta\Pi = \Pi_{seawater} - \Pi_{NaCl} = iM_{seawater}RT - iM_{NaCl}RT$$

$$\Delta\Pi = (iM_{seawater} - iM_{NaCl})RT \quad\quad\quad (11.15)$$

The total ion concentration in seawater is about 1.15 $M$; therefore, $iM_{seawater} = 1.15\ M$. If a solution of that concentration is put on one side of a semipermeable membrane and a solution of 0.10 $M$ NaCl on the other side, then $iM_{NaCl} = (2 \times 0.10\ M)$, and the osmotic pressure difference between the two solutions at 25°C is

$$\Delta\Pi = (iM_{seawater} - iM_{NaCl})RT$$

$$= \left(1.15\ \frac{mol}{L} - 2 \times 0.10\ \frac{mol}{L}\right)\left(0.0821\ \frac{L \cdot atm}{mol \cdot K}\right)(298\ K)$$

$$= 23\ atm$$

---

**SAMPLE EXERCISE 11.14    Calculating Osmotic Pressure II    LO8**

Red blood cells placed in seawater shrivel, as shown in Figure 11.23(b). Calculate the osmotic pressure across the semipermeable cell membrane separating the solution inside a red blood cell from seawater at 25°C if the total concentration of all the particles inside the cell is 0.30 $M$ and the total concentration of ions in seawater is 1.15 $M$. Compare the result with the answer from Sample Exercise 11.13.

**Collect and Organize** We are asked to calculate an osmotic pressure across a membrane. We can use Equation 11.15 to calculate the difference in osmotic pressure between the two solutions. As in Sample Exercise 11.13, we must convert the temperature to units of kelvins.

**Analyze** We know that the concentration of particles in seawater is greater than inside a red blood cell, so solvent will flow from the cell to the seawater. Thus, we predict the osmotic pressure in the cell ($\Pi_{cell}$) is less than the osmotic pressure of the seawater ($\Pi_{seawater}$). In addition, the osmotic pressure across the membrane is the difference between $\Pi_{seawater}$ and $\Pi_{cell}$, or $\Delta\Pi$. The value of $\Delta\Pi$ should be greater than the $\Pi$ calculated in Sample Exercise 11.13, where red blood cells were immersed in pure water, because the difference in solute concentrations is greater.

**Solve** First we need to convert the temperature from degrees Celsius to kelvins: $T(°C) + 273 = T(K)$. Inserting the values of $iM$, $R$, and $T$ in Equation 11.15 for seawater and cells we have

$$\Delta\Pi = (iM_{seawater} - iM_{cell})RT$$

$$= \left(1.15\ \frac{mol}{L} - 0.30\ \frac{mol}{L}\right)\left(0.0821\ \frac{L \cdot atm}{mol \cdot K}\right)(298\ K)$$

$$= 21\ atm$$

The pressure across a semipermeable membrane separating seawater from red blood cells, 21 atm, is greater than the pressure across the membrane separating red blood cells from pure water by almost a factor of three.

**Think About It** The values compare as we predicted. The effect of shrinking cells in the presence of hypertonic solutions is one of the reasons salt and sugar can be used to preserve foods. Salt rubbed on meat, for example, makes a salty solution; any bacteria exposed to this solution shrivel and die because of osmosis.

**Practice Exercise** Calculate the osmotic pressure at 25°C across a semipermeable membrane separating seawater (1.15 *M* total particles) from a 0.50 *M* solution of aqueous NaCl.

The osmotic pressures across cell membranes in Sample Exercises 11.13 and 11.14 are very large. The pressure of over 7 atm calculated in Sample Exercise 11.13 for red blood cells immersed in pure water is more than three times the air pressure in a typical automobile tire and about the same as the water pressure experienced by a diver at a depth of 80 m. A pressure of 21 atm across the walls of a steel-reinforced concrete building is sufficient to cause major structural damage or even collapse the building. The possibility of serious structural damage to cells as a result of such huge pressure differentials across membranes is one reason why solutions dispensed intravenously (IV) or intramuscularly (IM) must be carefully constituted.

During a medical emergency, medication may need to be administered to a patient intravenously (Figure 11.25), and it is crucial that the osmotic pressure exerted by the intravenous solution on the body's cells be identical to the osmotic pressure exerted by the solution inside the cells. The solute concentrations in such solutions are said to be *isotonic* because they exert the same osmotic pressure as the blood exerts. As we saw in Figure 11.23, solutions with higher (*hypertonic*) or lower (*hypotonic*) solute concentrations cause the body's cells to either shrink as water leaves or swell as water enters the cell.

Two solutions are widely used to administer intravenous medications, depending on the clinical situation. One is physiological saline, which contains 0.92% NaCl by mass: 0.92 g NaCl for every 100 g of solution. The density of dilute aqueous solutions is close to 1.00 g/mL, so 100 g of this solution has a volume of 100 mL, which means a concentration of 0.92 g/100 g is nearly the same as 0.92 g/100 mL = 9.2 g/L.

Another common IV solution is called D5W. The acronym stands for a 5.5% solution by mass of dextrose (another name for glucose; molar mass 180.16 g/mol) in water. Because this solution must be isotonic with blood and therefore isotonic with physiological saline solution, it must contain about the same concentration of solute particles as saline solution. A 5.5% by mass concentration means 5.5 g dextrose/100 g water, or 5.5 g dextrose/100 mL water, or 55 g dextrose/L.

To compare the solute levels of these two solutions, we compare molarities:

Physiological saline: $\dfrac{9.2 \text{ g}}{1 \text{ L}} \times \dfrac{1 \text{ mol NaCl}}{58.44 \text{ g NaCl}} = 0.16 \ M$

D5W: $\dfrac{55 \text{ g}}{1 \text{ L}} \times \dfrac{1 \text{ mol D}}{180.16 \text{ g D}} = 0.31 \ M$

With the molar concentration of dextrose about twice that of the saline solution, how can both solutions match the solute level in blood? The answer comes again from the definition of a colligative property and the role of the *i* factor. We must look at the number of particles in solution in both cases. Sodium chloride is a salt with $i = 2$. Dextrose is a sugar, a molecular material, and its molar concentration directly reflects the number of sugar molecules in solution; $i = 1$. The total particle

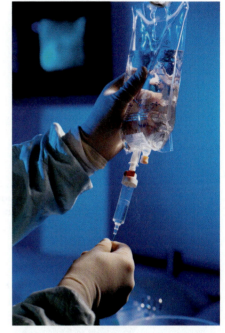

**FIGURE 11.25** A solution of physiological saline has a concentration of 0.92 g NaCl per 100 g of solution. The concentration of ions in this solution is equal to the concentration of ions in blood plasma. This solution is *isotonic* with blood plasma.

**reverse osmosis** a process in which solvent is forced through semipermeable membranes, leaving a more concentrated solution behind.

concentration in the NaCl solution is two times the molarity, or about 0.32 $M$, which is very close to the molar concentration of particles (molecules) in the dextrose solution. Therefore, the solute concentration in 0.92% NaCl is the same as the solute concentration in 5.5% D5W, which means that the two solutions exert the same osmotic pressure.

The values used in these examples are for normal conditions with a healthy patient. Depending on the clinical condition presented, doctors may need to use solutions with different concentrations of solutes to respond most effectively to the patient's needs.

## Reverse Osmosis

Many people in the world suffer from a lack of fresh water. A recent report from the United Nations (UN-Water Policy Brief on Water Quality, UN-Water 2011) suggested that by 2025, two out of three people in the world could be living in water-stressed areas—places without enough fresh water to drink or grow crops. The sea is already the source of drinking water in several desert countries bordering the Persian Gulf; to make the seawater drinkable, it must be *desalinated*, which means the salts must be removed. One way to desalinate saltwater is to distill it, but distillation requires a great deal of energy to heat seawater to its boiling point and convert it to steam.

The process of osmosis can also be applied to the desalination of seawater. As we have seen, osmosis is the movement of water from a region of low solute concentration to a region of high solute concentration. However, if a sufficiently high pressure is applied to the region of high solute concentration, the water can be forced to move from the region of high solute concentration to the region of low solute concentration. This technique, called **reverse osmosis**, is another way of desalinating seawater to make it drinkable.

The desalination apparatus shown in Figure 11.26 consists of an outer metal tube containing a large number of inner tubes made of a semipermeable membrane. Seawater is forced through the outer tube so that it washes over the exterior of all the inner tubes, which are initially filled with flowing pure water. Ordinarily, the direction of osmosis would be from the inner tubes (zero solute concentration) into the seawater (very high solute concentration), and indeed, the pure water does exert an osmotic pressure on the interior of the tube membranes. However, when an external pressure—a *reverse osmotic pressure*—greater than the osmotic pressure is exerted on the seawater side of the membranes, water molecules in the seawater move across the membranes into the inner tubes. The desalinated water entering the inner tubes flows into a collector and is ready for use.

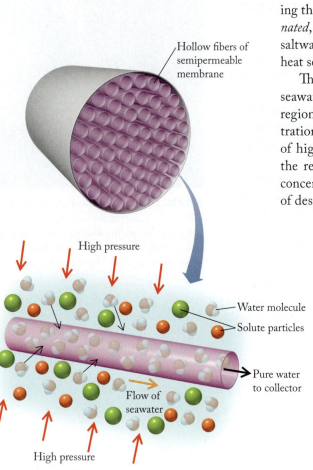

Hollow fibers of semipermeable membrane

High pressure

Water molecule

Solute particles

Pure water to collector

Flow of seawater

High pressure

**FIGURE 11.26** Seawater being desalinated by reverse osmosis flows at a pressure greater than its osmotic pressure around bundles of tubes with semipermeable walls. Water molecules pass from the seawater into the tubes and flow through the tubes to a collection vessel.

Some municipal water-supply systems use reverse osmosis to make saline (brackish) water fit to drink. Some industries use this method to purify conventional tap water, and it is a common way for ships at sea to desalinate ocean water. However, very tough semipermeable membranes are necessary because reverse osmosis systems operate at very high pressures. The continued development of technologies using reverse osmosis has resulted in millions of people being supplied with sanitary drinking water. Figure 11.27 shows the range of reverse osmosis facilities available, from the world's largest desalination plant to a portable truck-mounted unit to a small device for a home water supply.

(a)

| SAMPLE EXERCISE 11.15 | Calculating Pressure for Reverse Osmosis | LO8 |

What is the reverse osmotic pressure required at 20°C to purify brackish well water containing 0.355 $M$ dissolved particles if the purified water is to contain no more than 87 mg of dissolved solids (measured as NaCl equivalents) per liter?

**Collect and Organize** We are asked to calculate the reverse osmotic pressure needed to purify water to a stated solute concentration. We are given the temperature, the solute concentration in the water to be purified, and the amount of solute (expressed as NaCl) tolerable in the product water. We can adapt Equation 11.15 to calculate the difference in osmotic pressure between the two solutions.

**Analyze** We need the molarity of the less concentrated solution (the drinkable water) to calculate the osmotic pressure exerted by the drinkable water, and we can calculate that from the given information. We can also calculate the osmotic pressure once we know the difference in the molarities of the solutions on the two sides of the semipermeable membrane.

(b)

**Solve** First we convert 87 mg NaCl/L to molarity:

$$\frac{87 \text{ mg NaCl}}{1 \text{ L}} \times \frac{1 \text{ g}}{1000 \text{ mg}} \times \frac{1 \text{ mol NaCl}}{58.44 \text{ g NaCl}} = \frac{1.5 \times 10^{-3} \text{ mol NaCl}}{1 \text{ L}}$$

$$= 1.5 \times 10^{-3} \, M \text{ NaCl}$$

The total ion concentration for a $1.5 \times 10^{-3} \, M$ NaCl solution is $2(1.5 \times 10^{-3} \, M) = 0.0030 \, M$. Using Equation 11.15:

$$\Delta \Pi = \Pi_{\text{brackish water}} - \Pi_{\text{drinkable water}}$$

$$= (iM_{\text{brackish water}} - iM_{\text{drinkable water}})RT$$

$$= (0.355 \, M - 0.0030 \, M)\left(0.0821 \, \frac{\text{L} \cdot \text{atm}}{\text{mol} \cdot \text{K}}\right)(293 \, \text{K})$$

$$= 8.47 \text{ atm}$$

If we maintain an osmotic pressure of exactly 8.47 atm on the side of the membrane with the brackish well water, no net flow of solvent will occur between the two solutions. Pressures greater than 8.47 atm will force water molecules from the well water through the membrane, producing water containing less than 87 mg NaCl per liter. (We can never obtain absolutely pure water in this process; in practice, some $Na^+$ and $Cl^-$ ions inevitably pass through the membrane from the well water to the drinkable water.)

(c)

**Think About It** In the absence of any external pressure, solvent flows from the product water to the well water, driven by the concentration difference between the two solutions as shown in Figure 11.24. However, if we apply sufficient external pressure we can reverse the flow of water. The value of the external pressure, about 8.5 atm, represents the minimum external pressure that must be applied to make this device function.

**FIGURE 11.27** Reverse osmosis can supply sanitary water for personal, industrial, and commercial use. (a) The world's largest reverse osmosis desalination plant at Hadera, Israel, produces up to 120 million gallons of water daily. (b) The interior of a truck carrying a portable reverse osmosis unit that can purify 500 gallons per day of brackish water. (c) A small unit attached to a kitchen sink purifies a 16 oz glass of water in about 5 minutes.

⚙ **Practice Exercise** Calculate the minimum external pressure that must be applied in a reverse osmosis system to seawater with a total ion concentration of 1.15 $M$ at 20°C if the maximum concentration allowed in the product water is 174 mg NaCl per liter.

**CONCEPT TEST**

If you were trying to purify water by reverse osmosis using the apparatus in Figure 11.26, what advantage might there be in running the system at 50°C rather than 20°C?

## 11.6 Measuring the Molar Mass of a Solute Using Colligative Properties

**◐◉ CONNECTION** In Chapter 3 we used molar masses determined by mass spectrometry to convert empirical formulas derived from elemental analyses to molecular formulas. In Chapter 6, we used measurements of density and the ideal gas law to calculate molar masses of gases.

In principle, we can determine the molar mass of any solute by dissolving a known quantity of the solute in a known quantity of solvent and then measuring the effect the dissolved solute has on any colligative property of the solvent. In practice, this method works only for nonelectrolytes, which have a van 't Hoff factor of 1. Freezing-point-depression measurements, for example, can be used to find the molar mass of a molecular compound if it is sufficiently soluble in a solvent whose $K_f$ value is known, as illustrated in Sample Exercise 11.16.

---

**SAMPLE EXERCISE 11.16** **Using Freezing Point Depression to Determine Molar Mass** **LO9**

Eicosene is a molecular compound and nonelectrolyte with the empirical formula $CH_2$. The freezing point of a solution prepared by dissolving 100 mg of eicosene in 1.00 g of benzene was 1.75°C lower than the freezing point of pure benzene. What is the molar mass of eicosene? ($K_f$ for benzene is 4.90°C/$m$.)

**Collect and Organize** We are asked to determine the molar mass of a compound. We are given the mass of the compound that lowers the freezing point of a solvent by a known amount, and we have the $K_f$ of the solvent. Because the solute is a nonelectrolyte, $i = 1$. Equation 11.13 relates concentration to the change in freezing point.

**Analyze** The molar mass of a compound is expressed in units of grams per mole. Our sample consists of 100 mg, or 0.100 g. To find the molar mass we need to determine how many moles of eicosene are contained in the sample. The concentration term in Equation 11.13 is molality, or moles of solute per kilogram of solvent, so we can use the experimental data to first calculate molality. Once we know the molality of the eicosene, we know the number of moles of eicosene in 1 kg of benzene; we are given the mass of benzene used, so we can calculate the number of moles of eicosene in the sample, from which we may calculate the molar mass of eicosene. The empirical formula of eicosene enables us to check the validity of our answer, since the molar mass of eicosene must be a whole-number multiple of the mass of $CH_2$ (14 g/mol).

**Solve** The molality of the eicosene solution is

$$\Delta T_f = iK_f m = K_f m$$

$$m = \frac{\Delta T_f}{K_f} = \frac{1.75°\cancel{C}}{4.90°\cancel{C}/m} = 0.357\ m \text{ eicosene}$$

This means that 0.357 mol of eicosene is dissolved per kilogram of solvent. Only 1.00 g $(1.00 \times 10^{-3}$ kg) was used in this sample. Calculating the moles of eicosene in the sample:

$$m = \frac{\text{moles of solute}}{\text{kilograms of solvent}}$$

$$0.357\ m = \frac{\text{moles of eicosene}}{1.00 \times 10^{-3}\ \text{kg benzene}}$$

$$\text{Moles of eicosene} = \frac{0.357\ \text{mol eicosene}}{1\ \cancel{\text{kg benzene}}} \times 1.00 \times 10^{-3}\ \cancel{\text{kg benzene}}$$

$$= 3.57 \times 10^{-4}\ \text{mol eicosene}$$

Because the molar mass is the mass of one mole of eicosene, and 100 mg of eicosene was used to prepare the solution,

$$\text{Molar mass} = \frac{\text{mass of eicosene}}{\text{moles of eicosene}} = \frac{0.100\ \text{g eicosene}}{3.57 \times 10^{-4}\ \text{mol}} = 280\ \text{g/mol}$$

**Think About It** The molar mass of eicosene, 280 g/mol, is reasonable since it corresponds to $(CH_2)_n$ where $n = 20$. The molecular formula of eicosene is therefore $C_{20}H_{40}$.

⚙ **Practice Exercise** A solution prepared by dissolving 360 mg of a sugar (a molecular compound and a nonelectrolyte) in 1.00 g of water froze at −3.72°C. What is the molar mass of this sugar? The value of $K_f$ of water is 1.86°C/$m$.

■

For determining the molar mass of water-soluble substances, measuring osmotic pressure is a better choice than measuring either boiling point elevation or freezing point depression for several reasons. First, the $K_f$ and $K_b$ values for water are much smaller than those of other solvents (Table 11.5). Thus, in order to have a solution that gives a measurable boiling point or freezing point change for an aqueous solution, the concentration of the solution has to be much higher than is readily achievable. Second, biomaterials such as proteins and carbohydrates are nearly always available only in small quantities, and often they are not very soluble in nonaqueous solvents that have larger $K_f$ or $K_b$ values. Furthermore, these biomaterials often have high molar masses, which means that large quantities are needed to give high enough molal concentrations for reliable $\Delta T_f$ or $\Delta T_b$ measurements. Third, a solute might need to be recovered unchanged for other uses, which rules out boiling-point-elevation measurements for heat-sensitive solutes.

| TABLE 11.5 | Molal Freezing–Point–Depression and Boiling–Point–Elevation Constants for Selected Solvents | | | |
|---|---|---|---|---|
| Solvent | Freezing Point (°C) | $K_f$ (°C/m) | Boiling Point (°C) | $K_b$ (°C/m) |
| Water ($H_2O$) | 0.0 | 1.86 | 100.0 | 0.52 |
| Benzene ($C_6H_6$) | 5.5 | 4.90 | 80.1 | 2.53 |
| Ethanol ($CH_3CH_2OH$) | −114.6 | 1.99 | 78.4 | 1.22 |
| Carbon tetrachloride ($CCl_4$) | −22.3 | 29.8 | 76.8 | 5.02 |

Perhaps the most compelling reasons for using osmotic pressure are that very small osmotic pressures can be measured precisely, the measurement equipment can be miniaturized so that only minute quantities of solute are needed, and the measurements can be made at room temperature.

**SAMPLE EXERCISE 11.17** **Using Osmotic Pressure to Determine Molar Mass** **LO9**

A molecular compound that is a nonelectrolyte was isolated from a South African tree. A 47 mg sample was dissolved in water to make 2.50 mL of solution at 25°C, and the osmotic pressure of the solution was 0.489 atm. Calculate the molar mass of the compound.

**Collect and Organize** We are given the mass of a substance, the volume of its aqueous solution, the temperature, and its osmotic pressure. We can relate these parameters with Equation 11.14, using the value $i = 1$ because this is a nonelectrolyte.

**Analyze** We can calculate the molar concentration of the solution from the information given and Equation 11.14. Since we know the solution volume, we can calculate the number of moles of the solute from the molarity. Because we know the mass of this number of moles, we can calculate the molar mass of the solute.

**Solve** Rearranging Equation 11.14 to isolate $M$ and substituting the given values:

$$M = \Pi/iRT = \frac{0.489 \text{ atm}}{(1)(0.0821 \text{ L} \cdot \text{atm}/\text{mol} \cdot \text{K})(298 \text{ K})}$$

$$= \frac{2.00 \times 10^{-2} \text{ mol}}{\text{L}} = 2.00 \times 10^{-2} \text{ M}$$

Next we solve the defining equation for molarity, $M = n/V$, for $n$, the number of moles of solute:

$$n = MV = \frac{2.00 \times 10^{-2} \text{ mol}}{1 \text{ L}} \times 2.50 \times 10^{-3} \text{ L} = 5.00 \times 10^{-5} \text{ mol}$$

We know that this number of moles of solute has a mass of 47 mg. The molar mass of the solute is therefore

$$\text{Molar mass} = \frac{\text{g solute}}{\text{moles of solute}}$$

$$= \frac{47 \times 10^{-3} \text{ g}}{5.00 \times 10^{-5} \text{ mol}} = 9.4 \times 10^2 \text{ g/mol}$$

**Think About It** One of the advantages of determining molar mass by osmotic pressure is that only a small amount of material is required. With only a 47 mg sample of a rather large molecule (molar mass 940 g/mol), the osmotic pressure is sufficiently large (0.489 atm) to enable us to calculate the molar mass accurately.

**Practice Exercise** A solution was made by dissolving 5.00 mg of a polysaccharide (a polymer made of sugar molecules) in water to give a final volume of 1.00 mL. The osmotic pressure of this solution was $1.91 \times 10^{-3}$ atm at 25°C. Calculate the molar mass of the polysaccharide, which is a nonelectrolyte.

We very rarely deal with pure liquids in the real world. Milk, gasoline, tap water, shampoo, olive oil, cough syrup, and countless other fluids that we use in our daily lives are all solutions whose physical properties depend on the amounts of solutes dissolved in them. By understanding the role solutes play in determining the properties of solutions, we can make many extraordinarily useful materials that display exactly the behaviors we desire. The boiling and freezing points of water can be extended above 100°C and below 0°C to make antifreeze that takes heat away from an operating engine and also protects the fluid from freezing in the winter. Adding the right amount of dextrose to water for an IV drip keeps trauma patients hydrated. Having too much or too little solute in a solution can have catastrophic effects: engine blocks can crack if insufficient antifreeze is dissolved in the water in a radiator to protect the fluid from freezing; a patient can die if the concentration of sodium chloride in an IV drip is too low or too high. It is important to remember that changes in boiling point elevation, freezing point depression, and osmotic pressure—as well as other properties we have discussed in this chapter—depend solely on the concentration of solute particles present in the solutions and not on their identity.

**SAMPLE EXERCISE 11.18    Integrating Concepts: Fun with Eggs**

The shell of a chicken's egg is mostly calcium carbonate, and it is lined with a semipermeable membrane. Explain what happens during the following series of steps, all carried out at room temperature (21°C), and answer any additional questions.

1. Put a chicken egg in a beaker and pour enough pickling vinegar (aqueous acetic acid [$CH_3COOH(aq)$]) over it to submerge it completely. Gas bubbles out from the shell, and after several days the hard shell completely dissolves, leaving the sac-like semipermeable membrane containing the white and the yolk intact.

(a) Identify the gas given off and write a balanced chemical equation for the reaction. (b) The beaker is open to the air during this period. Does most of the gas dissolve in the fluid, or does most of it escape? Explain your answer.

2. At this point, you weigh the egg; its mass is 100 g.

3. You submerge the egg in a solution of corn syrup (75% by weight sucrose [$C_{12}H_{22}O_{11}$]; 25% water; density = 1.38 g/mL) and leave it there for 8 hours. The egg visibly shrivels. You remove it from the solution and weigh it: its mass is now 58 g. (c) Why does the egg shrivel?

4. You now submerge the egg in distilled water for 8 hours. The egg expands and its mass is now 103 g. (d) Why does the egg expand?

5. As a final step, you submerge the egg in a 0.75 $M$ solution of table salt. (e) Predict what happens. (f) If you were to start with two fresh eggs and put one in the syrup and the other in the salt solution, which would change more in volume?

**Collect and Organize** We are given a series of changes an egg goes through when exposed to several solutions. We have the starting materials (calcium carbonate and acetic acid) for a chemical reaction involving the egg's shell and are asked to identify the products. We can determine the concentration of the sugar and salt solutions that cause the egg to shrink or expand.

**Analyze** From Chapter 4, we recall that the reaction between a carbonate and an acid produces carbon dioxide gas. We can use Henry's law (Section 11.2) to calculate the solubility of $CO_2$ in water. The other changes must involve osmosis across a semipermeable membrane (Section 11.5). Equation 11.14 relates solute concentration to osmotic pressure, and we can use it to compare the effects of the sugar and salt solutions.

**Solve**

a. $CaCO_3(s) + 2\ CH_3COOH(aq) \rightarrow$
$\qquad\qquad Ca(CH_3COO)_2(aq) + H_2O(\ell) + CO_2(g)$
The gas given off as the eggshell dissolves is carbon dioxide ($CO_2$).

b. $CO_2$ is nonpolar, so it would be held in the aqueous phase by relatively weak dipole–induced dipole interactions. Its Henry's law constant $k_H$, listed in Table 11.1, is about ten times that of oxygen, but that is still small, so we would not expect $CO_2$ to be very water soluble. In addition, the description states that the beaker is open to the atmosphere, so we can expect most of the $CO_2$ to escape into the air, just as it does if we leave a carbonated beverage open.

c. The egg shrivels because the concentration of the syrup solution is higher than the concentration of the solution within the egg sac. Osmosis causes water to leave the egg sac and enter the surrounding solution to equalize the osmotic pressure.

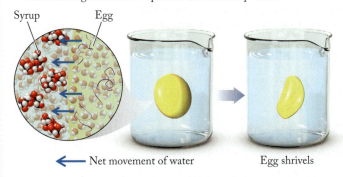

Syrup    Egg

← Net movement of water          Egg shrivels

d. The egg expands when put into distilled water (containing no solute). Osmosis causes water to flow through the sac and enter the egg.

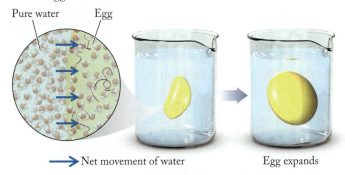

Pure water    Egg

→ Net movement of water          Egg expands

e. If the egg is then placed in a salt solution, as long as the concentration of salt is higher than the concentration of the solution inside the egg, the egg will shrivel again as water flows from inside the sac to the outside solution.

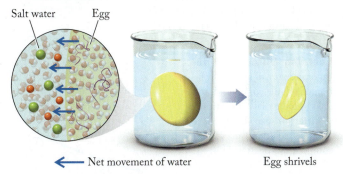

Salt water    Egg

← Net movement of water          Egg shrivels

f. The solution that causes the greatest osmotic pressure difference across the semipermeable membrane will cause the greatest change in the volume of the egg. The sugar solution is 75% by weight sucrose. That means 75 g of sucrose is dissolved in 100 g of total solution. We have the density of the solution, so we can convert this concentration into molarity:

$$\frac{75\ \text{g sucrose}}{100\ \text{g solution}} \times \frac{1\ \text{mol}}{342.30\ \text{g}} \times \frac{1.380\ \text{g solution}}{1\ \text{mL solution}} \times \frac{1000\ \text{mL}}{1\ \text{L}} = 3.0\ M$$

The concentration of the salt solution is 0.75 $M$ NaCl. We can compare the two solutions by comparing their osmotic pressures:

$$\begin{aligned}
\Pi_{\text{syrup}} &= iM_{\text{syrup}}RT \\
&= (1)(3.0\ \text{mol/L})(0.08204\ \text{L atm/K} \cdot \text{mol})(294\ \text{K}) \\
&= 72\ \text{atm}
\end{aligned}$$

$$\begin{aligned}
\Pi_{\text{salt water}} &= iM_{\text{salt water}}RT \\
&= (2)(0.75\ \text{mol/L})(0.08204\ \text{L atm/K} \cdot \text{mol})(294\ \text{K}) \\
&= 36\ \text{atm}
\end{aligned}$$

The syrup exerts a much higher osmotic pressure than the salt solution, so we would predict that an egg placed in the syrup would shrivel more than an egg placed in the salt water.

**Think About It** We have to treat the solutions ideally because we have no indication about the actual value of the $i$ factor. Because $i_{\text{salt water}} \leq 2$, $\Pi_{\text{syrup}}$ has to be greater than $\Pi_{\text{salt water}}$ here.

## SUMMARY

**Learning Outcome 1** The vapor pressure of a liquid is proportional to the fraction of its molecules that enter the gas phase. **Raoult's law** relates the vapor pressure of a solution to its composition and to the vapor pressure of the solvent. (Section 11.1)

**Learning Outcome 2** The solubility of a gas in water depends on the temperature and pressure. **Henry's law** gives the maximum concentration (the solubility) of a sparingly soluble gas in a liquid solvent. (Section 11.2)

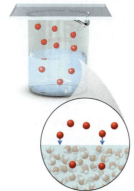

**Learning Outcome 3** The **enthalpy of solution** ($\Delta H_{solution}$) for an ionic compound is the sum of the **lattice energy** ($U$) and the **enthalpy of hydration** ($\Delta H_{hydration}$). Lattice energies can be calculated with a **Born–Haber cycle**, an application of Hess's law. (Section 11.3)

**Learning Outcome 4** The vapor pressure of an ideal solution of volatile compounds follows Raoult's law. **Fractional distillation** can be used to separate solutions of volatile compounds. (Section 11.4)

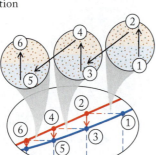

**Learning Outcome 5** The concentration units used for **colligative property** measurements include molarity and **molality**. (Section 11.5)

**Learning Outcome 6** Solutes in solution elevate the solvent's boiling point and depress its freezing point. (Section 11.5)

**Learning Outcome 7** The **van 't Hoff factor** accounts for the colligative properties of electrolytes and the formation of solute **ion pairs** in concentrated solutions. (Section 11.5)

**Learning Outcome 8** In **osmosis**, solvent flows through a semipermeable membrane. **Osmotic pressure** is the pressure required to halt the flow of solvent from the more dilute solution across the membrane. (Section 11.5)

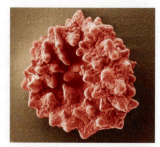

**Learning Outcome 9** The molar mass of a compound can be determined by measuring the freezing point depression, boiling point elevation, or osmotic pressure of a solution of the compound. (Section 11.6)

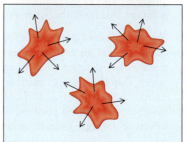

## PROBLEM-SOLVING SUMMARY

| TYPE OF PROBLEM | CONCEPTS AND EQUATIONS | SAMPLE EXERCISES |
|---|---|---|
| **Calculating vapor pressure of a solution** | Use Raoult's law,<br><br>$$P_{solution} = X_{solvent} P^\circ_{solvent} \qquad (11.1)$$<br><br>where $P_{solution}$ is the vapor pressure of the solution at a given temperature, $X_{solvent}$ is the mole fraction of the solvent in the solution, and $P^\circ_{solvent}$ is the vapor pressure of the pure solvent at the same temperature. | 11.1 |
| **Calculating the solubility of a gas using Henry's law** | Use Henry's law,<br><br>$$C_{gas} = k_H P_{gas} \qquad (11.2)$$<br><br>where $C_{gas}$ is the solubility of the gas, $k_H$ depends on the gas, the solvent, and the temperature, and $P_{gas}$ is the pressure of the gas (or partial pressure if the gas is part of a mixture of gases). | 11.2 |
| **Ranking lattice energies, solubility, and melting points** | Predict relative lattice energies, solubilities, and melting points using<br><br>$$U = \frac{k(Q_1 Q_2)}{d} \qquad (11.6)$$ | 11.3 |
| **Calculating lattice energy with a Born–Haber cycle** | The enthalpy of reaction, $\Delta H_{rxn}$, when an ionic solid forms from its constituent elements is equal to the sum of the enthalpies of reaction for every step in the process. Solve for lattice energy $U$ or any one unknown enthalpy change term. | 11.4 |

| TYPE OF PROBLEM | CONCEPTS AND EQUATIONS | SAMPLE EXERCISES |
|---|---|---|
| **Calculating lattice energy using enthalpies of hydration** | Use enthalpies of hydration ($\Delta H_{\text{hydration}}$) and enthalpies of solution ($\Delta H_{\text{solution}}$) to calculate lattice energy:$$\Delta H_{\text{solution}} = \Delta H_{\text{hydration}} - U$$where$$\Delta H_{\text{hydration}} = \Delta H_{\text{hydration,cation}} + \Delta H_{\text{hydration,anion}}$$ | 11.5 |
| **Interpreting fractional distillation data** | Draw an idealized graph showing distillation temperature versus volume of distillate. | 11.6 |
| **Calculating the vapor pressure of a solution of volatile substances** | Determine the mole fractions and vapor pressures of each component of the solution. Use Raoult's law to calculate the vapor pressure of the solution:$$P_{\text{total}} = X_1 P_1^{\circ} + X_2 P_2^{\circ} + X_3 P_3^{\circ} + \cdots \quad (11.8)$$ | 11.7 |
| **Calculating molal concentrations** | Molality is defined as$$m = \frac{n_{\text{solute}}}{\text{kg solvent}} \quad (11.9)$$where $n$ is the number of moles of solute. | 11.8 |
| **Calculating boiling point or freezing point of a solution** | For nonelectrolyte solutes, use$$\Delta T_b = K_b m \quad (11.10)$$where $\Delta T_b$ is the elevation in the boiling point of the solvent, $K_b$ is a constant that depends only on the solvent, and $m$ is the molality of the solution. For the freezing point, use$$\Delta T_f = K_f m \quad (11.11)$$where $\Delta T_f$ is the depression in the freezing point of the solvent, $K_f$ is a constant that depends only on the solvent, and $m$ is the molality of the solution. | 11.9, 11.10 |
| **Assessing interactions among particles in solution by comparing the theoretical value of the van 't Hoff factor $i$ with the experimentally measured value** | For electrolytes, use$$\Delta T_b = i K_b m \quad (11.12)$$and$$\Delta T_f = i K_f m \quad (11.13)$$where the theoretical value of $i$ is the number of particles created when an electrolyte dissociates completely: $i = 2$ for NaCl, 3 for $CaCl_2$, 4 for $Na_3PO_4$. For real solutions of electrolytes where $\Delta T_{b/f}$, $K_{b/f}$, and $m$ are known, rearrange $\Delta T_b = i K_b m$ and $\Delta T_f = i K_f m$ to calculate the value of $i$ and compare the result with the theoretical value. | 11.11, 11.12 |
| **Calculating osmotic pressure and reverse osmotic pressure** | Use$$\Pi = iMRT \quad (11.14)$$where $\Pi$ is the osmotic pressure, $i$ is the van 't Hoff factor, $M$ is the molar concentration of the solution, $R$ is the ideal gas constant, and $T$ is the absolute temperature of the solution. | 11.13, 11.14, 11.15 |
| **Determining molar mass from boiling point elevation, freezing point depression, or osmotic pressure** | Use $\Delta T_b = K_b m$, $\Delta T_f = K_f m$, or $\Pi = MRT$ to calculate the molality $m$ or molarity $M$ of the solution. Then use $m$ or $M$ to determine the moles of solute and molar mass of the solute. | 11.16, 11.17 |

## VISUAL PROBLEMS

*(Answers to boldface end-of-chapter questions and problems are in the back of the book.)*

**11.1.** Which of the drawings in Figure P11.1 best describes the effect of pressure on the solubility of a gas in a liquid?

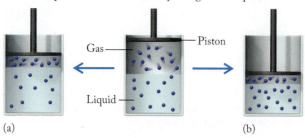

(a)        Gas — Piston        (b)

Liquid —

**FIGURE P11.1**

**11.2.** Which line drawn on the graph in Figure P11.2 represents the gas with the greatest value for the Henry's law constant, $k_H$?

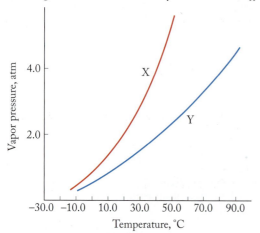

**FIGURE P11.2**

**\*11.3.** At constant temperature, a volume of gas (Figure P11.3a) is in dynamic equilibrium with dissolved gas in a liquid. The pressure is increased (Figure P11.3b). Which drawing in Figure P11.3c best describes the system after dynamic equilibrium is restored?

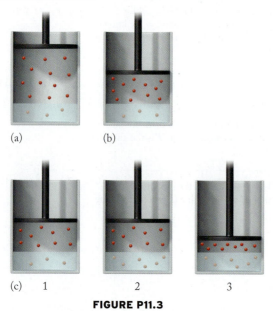

(a)        (b)

(c)    1      2      3

**FIGURE P11.3**

**11.4.** The graph in Figure P11.4 describes the variation in solubility of nitrogen, oxygen, and helium gases with pressure. Use the data in Table 11.1 to identify which gas gives rise to each line.

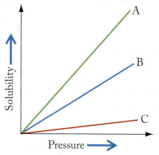

**FIGURE P11.4**

**11.5.** The graph in Figure P11.5 describes the volume of distillate collected during the fractional distillation of a liquid. Answer the following questions about the process: (a) Is the sample a pure liquid or a mixture? (b) If it is a mixture: (i) how many components are in the mixture? (ii) What are the relative ratios of the volumes in the mixture? (iii) What are their approximate boiling points?

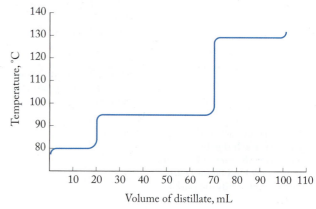

**FIGURE P11.5**

**11.6.** The graph in Figure P11.6 shows the decrease in the freezing point of water $\Delta T_f$ for solutions of two different substances, A (triangles) and B (circles), in water. Explain how you can reasonably conclude that (a) A and B are nonelectrolytes and (b) the freezing point depression constant $K_f$ of water is independent of the solute's identity.

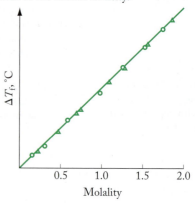

**FIGURE P11.6**

**11.7.** The arrow in Figure P11.7 indicates the direction of solvent flow through a semipermeable membrane in equipment designed to measure osmotic pressure. Which solution, A or B, is more concentrated? Explain your answer.

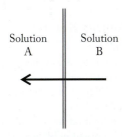

Solution A | Solution B

**FIGURE P11.7**

**11.8. Kidney Dialysis** Semipermeable membranes of the sort used in kidney dialysis do not allow large molecules and cells to pass but do allow small ions and water to pass. Figure

P11.8 shows such a membrane separating fluids of various compositions.

   a. In which direction does the water flow in each apparatus?

   b. In which direction do sodium ions flow in each apparatus?

   c. In which direction do the potassium ions flow in each apparatus?

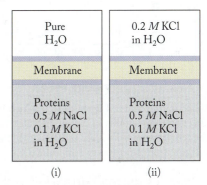

**FIGURE P11.8**

---

## QUESTIONS AND PROBLEMS

## Vapor Pressure of Solutions

### CONCEPT REVIEW

**11.9.** Explain the term *nonvolatile solute.*

**11.10.** Which has the higher vapor pressure at constant temperature, pure water or seawater? Explain your answer.

**11.11.** Why does the vapor pressure of a liquid increase with increasing temperature?

**11.12.** In the experiment shown in Figure 11.1, the vapor pressure of one of the solutions remains constant throughout the experiment, while the vapor pressure of the other solution changes. Which solution is which?

**11.13.** An experiment like that shown in Figure 11.1 is set up with the beaker containing pure ethanol full to the brim and the beaker containing a solution of sugar in ethanol half-full. Explain why the beaker that contained the ethanol–sugar solution will eventually overflow.

**11.14.** Explain to a nonscientist how the water gets from one beaker to the other in the experiment depicted in Figure 11.1.

### PROBLEMS

**11.15.** A solution contains 3.5 mol of water and 1.5 mol of nonvolatile glucose ($C_6H_{12}O_6$). What is the mole fraction of water in this solution? What is the vapor pressure of the solution at 25°C, given that the vapor pressure of pure water at 25°C is 23.8 torr?

**11.16.** A solution contains 4.5 mol of water, 0.3 mol of sucrose ($C_{12}H_{22}O_{11}$), and 0.2 mol of glucose. Sucrose and glucose are nonvolatile. What is the mole fraction of water in this solution? What is the vapor pressure of the solution at 35°C, given that the vapor pressure of pure water at 35°C is 42.2 torr?

**11.17.** Another way of stating Raoult's law is that the fractional lowering of the vapor pressure of a solvent $(P°_{solvent} - P_{solvent})/P°_{solvent}$ is equal to the mole fraction of the solute, $X_{solute}$. Use Equation 11.1 to show that this is true.

**11.18.** Use the statement of Raoult's law in Problem 11.17 to determine the mole fraction of glucose in Problem 11.15.

## Solubility of Gases in Water

### CONCEPT REVIEW

**11.19.** Why does the solubility of most gases in most liquids increase with decreasing temperature?

**11.20.** Which term, $k_H$ or $P$, in Henry's law is affected by temperature?

**11.21.** Air is primarily a mixture of nitrogen and oxygen. Is the Henry's law constant for the solubility of air in water the sum of $k_H$ for $N_2$ and $k_H$ for $O_2$? Explain why or why not.

**11.22.** Why is the Henry's law constant for $CO_2$ so much larger than those for $N_2$ and $O_2$ at the same temperature?

### PROBLEMS

**11.23. Arterial Blood** Arterial blood contains about 0.25 g of oxygen per liter at 37°C and standard atmospheric pressure. What is the Henry's law constant [mol/(L · atm)] for $O_2$ dissolution in blood? The mole fraction of $O_2$ in air is 0.209.

**11.24.** The solubility of $O_2$ in water is 6.5 mg/L at an atmospheric pressure of 1 atm and temperature of 40°C. Calculate the Henry's law constant of $O_2$ at 40°C. The mole fraction of $O_2$ in air is 0.209.

***11.25. Oxygen for Climbers and Divers** Use the Henry's law constant for $O_2$ dissolved in arterial blood from Problem 11.23 to calculate the solubility of $O_2$ in the blood of (a) a climber on Mt. Everest ($P_{atm}$ = 0.35 atm) and (b) a scuba diver at 100 feet ($P ≈ 3$ atm).

***11.26.** The solubility of air in water is approximately $7.9 × 10^{-4}$ $M$ at 20°C and 1.0 atm. Calculate the Henry's law constant for air. Is the $k_H$ value of air approximately equal to the sum of the $k_H$ values for $N_2$ and $O_2$ because these two gases make up 99% of the gases in air?

**\*11.27.** Use the graph of solubility of $O_2$ versus temperature in Figure P11.27 to calculate the value of the Henry's law constant $k_H$ for $O_2$ at 10°C, 20°C, and 30°C. Assume $P_{O_2} = 1$ atm.

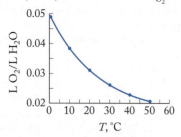

**FIGURE P11.27**

**11.28.** Based on the data in Figure P11.27, which has a greater effect on the solubility of oxygen in water: (a) decreasing the temperature from 20°C to 10°C or (b) raising the pressure from 1.00 atm to 1.25 atm?

## Energy Changes during Formation and Dissolution of Ionic Compounds

### CONCEPT REVIEW

**11.29.** Explain why the term for enthalpy of hydration ($\Delta H_{hydration}$) contains two terms: one for ion–dipole interactions and one for dipole–dipole interactions.

**11.30.** Explain why trends in lattice energies for ionic compounds parallel trends in melting points and are opposite to the trends in water solubility.

**\*11.31.** In Chapter 4, the solubility rules stated that all metal nitrates are water soluble, but many metal sulfides are not. Explain this observation.

**11.32.** Explain why it might be difficult to measure the enthalpy of hydration of a single ion.

### PROBLEMS

**11.33.** How do the melting points of the series of sodium halides NaX (X = F, Cl, Br, I) relate to the atomic number of X?

**11.34.** Rank the following ionic compounds in order of (a) increasing melting point and (b) increasing water solubility: $BaF_2$, $CaCl_2$, $MgBr_2$, and $SrI_2$.

**11.35.** Which substance has the least negative lattice energy? (a) $MgI_2$; (b) $MgBr_2$; (c) $MgCl_2$; (d) $MgF_2$

**11.36.** Rank the following from lowest to highest lattice energy: NaBr, $MgBr_2$, $CaBr_2$, KBr.

**11.37.** Use a Born–Haber cycle to calculate the lattice energy of potassium chloride (KCl) from the following data:
Ionization energy of K(g) = 425 kJ/mol
Electron affinity of Cl(g) = −349 kJ/mol
Energy to sublime K(s) = 89 kJ/mol
Bond energy of $Cl_2(g)$ = 240 kJ/mol

$$\Delta H_{rxn} \text{ for K}(s) + \tfrac{1}{2} Cl_2(g) \rightarrow KCl(s) = -438 \text{ kJ/mol}$$

**11.38.** Calculate the lattice energy of sodium oxide ($Na_2O$) from the following data:
Ionization energy of Na(g) = 495 kJ/mol
Electron affinity of O(g) for 2 electrons = 603 kJ/mol
Energy to sublime Na(s) = 109 kJ/mol
Bond energy of $O_2(g)$ = 499 kJ/mol

$$\Delta H_{rxn} \text{ for 2 Na}(s) + \tfrac{1}{2} O_2(g) \rightarrow Na_2O(s) = -416 \text{ kJ/mol}$$

**11.39.** Using the values in Table 11.4 and $\Delta H_{solution} = 19.9$ kJ/mol for KBr, calculate the lattice energy of KBr.

**11.40.** Using the values in Table 11.4 and $\Delta H_{solution} = -17.7$ kJ/mol for KF, calculate the lattice energy of KF.

## Mixtures of Volatile Solutes

### CONCEPT REVIEW

**11.41.** What physical property of the components of crude oil is used to separate them?

**11.42.** What is the difference between simple distillation and fractional distillation?

**11.43.** In an equimolar mixture of $C_5H_{12}$ and $C_7H_{16}$, which compound is present in higher concentration in the vapor above the solution?

**11.44.** Why does the boiling point of a mixture of volatile hydrocarbons increase over time during a simple distillation?

### PROBLEMS

**11.45.** At 20°C, the vapor pressure of ethanol is 45 torr and the vapor pressure of methanol is 92 torr. What is the vapor pressure at 20°C of a solution prepared by mixing 25 g of methanol and 75 g of ethanol?

**\*11.46.** At 90°C, the vapor pressure of styrene ($C_8H_8$) is 134 torr and that of ethylbenzene ($C_8H_{10}$) is 183 torr. What is the vapor pressure of a solution of 38% by weight styrene and 62% by weight ethylbenzene at 90°C?

**\*11.47.** The mixture described in Problem 11.46 is separated by fractional distillation at reduced pressure so that it begins to boil when the solution in the distillation flask reaches 90°C.
  a. What is the ratio of ethylbenzene to styrene in the vapor phase as the mixture first begins to boil?
  b. What will be the temperature of the first distillate that comes off the top of the column: (i) lower than 90°C; (ii) 90°C; (iii) higher than 90°C

**11.48.** A bottle is half-filled with a 50:50 (mole-to-mole) mixture of heptane ($C_7H_{16}$) and octane ($C_8H_{18}$) at 25°C. What is the mole ratio of heptane vapor to octane vapor in the air space above the liquid in the bottle? The vapor pressures of heptane and octane at 25°C are 31 torr and 11 torr, respectively.

## Colligative Properties of Solutions

### CONCEPT REVIEW

**11.49.** What is the difference between molarity and molality?

**11.50.** As a solution of NaCl becomes more concentrated, does the difference between its molarity and its molality increase or decrease?

**11.51.** As a solution of NaCl is heated from 5°C to 90°C, does the difference between its molarity and its molality increase or decrease?

**\*11.52.** The thermostat in a refrigerator filled with cans of soft drinks malfunctions and the temperature of the refrigerator drops below 0°C. The contents of the cans of diet soft drinks freeze, rupturing many of the cans and causing an awful mess. However, none of the cans containing regular, nondiet soft drinks rupture. Why?

**11.53.** Why is it important to know if a substance is a molecular compound or an ionic compound before predicting its effect on the boiling and freezing points of a solvent?

11.54. Refer to the phase diagram in Figure 11.17 and explain in your own words why the change in the vapor pressure of a solution caused by a nonvolatile solute results in a higher boiling point and a lower melting point.

**11.55.** Explain how the theoretical value of the van 't Hoff factor $i$ for substances such as $CH_3OH$, $NaBr$, and $K_2SO_4$ can be predicted from their formulas.

11.56. Is it possible for an experimentally measured value of a van 't Hoff factor to be greater than the theoretical value? Explain your answer.

**11.57.** What is a semipermeable membrane?

11.58. A pure solvent is separated from a solution containing the same solvent by a semipermeable membrane. In which direction does the solvent flow across the membrane, and why?

**11.59.** A dilute solution is separated from a more concentrated solution containing the same solvent by a semipermeable membrane. In which direction does the solvent tend to flow across the membrane, and why?

11.60. How is the osmotic pressure of a solution related to its molar concentration and its temperature?

**11.61.** What is reverse osmosis? List the basic components of equipment used to purify seawater by reverse osmosis.

11.62. Explain how the minimum pressure for purification of seawater by reverse osmosis can be estimated from its composition.

**11.63.** Instructions for serving salads made of fresh greens specify that salad dressings, especially vinaigrettes made of oil, vinegar, salt, and other seasonings, should be put on the salad immediately before serving to keep the greens from wilting. Why?

11.64. Why do cucumbers shrivel when put in pickling liquids?

## PROBLEMS

**11.65.** Calculate the molality of each of the following solutions:
a. 0.433 mol of sucrose ($C_{12}H_{22}O_{11}$) in 2.1 kg of water
b. 71.5 mmol of acetic acid ($CH_3COOH$) in 125 g of water
c. 0.165 mol of baking soda ($NaHCO_3$) in 375.0 g of water

*11.66. Table 4.1 lists molarities of major ions in seawater. Using a density of 1.022 g/mL for seawater, convert the concentrations into molalities.

**11.67.** What mass of the following solutions contains 0.100 mol of solute? (a) 0.135 $m$ $NH_4NO_3$; (b) 3.92 $m$ ethylene glycol, $HOCH_2CH_2OH$; (c) 1.07 $m$ $CaCl_2$

11.68. How many moles of solute are there in the following solutions?
a. 0.750 $m$ glucose solution made by dissolving the glucose in 10.0 kg of water
b. 0.183 $m$ $Na_2CrO_4$ solution made by dissolving the $Na_2CrO_4$ in 900.0 g of water
c. 1.425 $m$ urea solution made by dissolving the urea in 750.0 g of water

**11.69. Fish Kills** High concentrations of ammonia ($NH_3$), nitrite ion, and nitrate ion in water can kill fish. Lethal concentrations of these species for rainbow trout are 1.1 mg/L, 0.40 mg/L, and 1361 mg/L, respectively. Express these concentrations in molality units, assuming a solution density of 1.00 g/mL.

11.70. The concentrations of six important elements in a sample of river water are 0.050 mg/kg of $Al^{3+}$, 0.040 mg/kg of $Fe^{3+}$, 13.4 mg/kg of $Ca^{2+}$, 5.2 mg/kg of $Na^+$, 1.3 mg/kg of $K^+$, and 3.4 mg/kg of $Mg^{2+}$. Express each of these concentrations in molality units.

**11.71. Cinnamon** Cinnamon owes its flavor and odor to cinnamaldehyde ($C_9H_8O$). Determine the boiling point elevation of a solution of 100 mg of cinnamaldehyde dissolved in 1.00 g of carbon tetrachloride ($K_b = 5.02°C/m$).

11.72. **Spearmint** Determine the boiling point elevation of a solution of 125 mg of carvone ($C_{10}H_{14}O$, oil of spearmint) dissolved in 1.50 g of carbon disulfide ($K_b = 2.34°C/m$).

**11.73.** What molality of a nonvolatile, nonelectrolyte solute is needed to lower the melting point of camphor by 1.000°C ($K_f = 39.7°C/m$)?

11.74. What molality of a nonvolatile, nonelectrolyte solute is needed to raise the boiling point of water by 7.60°C ($K_b = 0.52°C/m$)?

**11.75. Saccharin** Determine the melting point of an aqueous solution made by adding 186 mg of saccharin ($C_7H_5O_3NS$) to 1.00 mL of water (density = 1.00 g/mL, $K_f = 1.86°C/m$).

11.76. Determine the boiling point of an aqueous solution that is 2.50 $m$ ethylene glycol ($HOCH_2CH_2OH$); $K_b$ for water is 0.52°C/m. Assume that the boiling point of pure water is 100.00°C.

**11.77.** Which aqueous solution has the lowest freezing point: 0.5 $m$ glucose, 0.5 $m$ NaCl, or 0.5 $m$ $CaCl_2$?

11.78. Which aqueous solution has the highest boiling point: 0.5 $m$ methanol ($CH_3OH$), 0.5 $m$ KI, or 0.5 $m$ $Na_2SO_4$?

**11.79.** Which of the following aqueous solutions should have the highest boiling point: 0.0200 $m$ ethanol ($CH_3CH_2OH$), 0.0125 $m$ $LiClO_4$, or 0.0100 $m$ $Mg(NO_3)_2$?

11.80. Which of the following aqueous solutions should have the lowest freezing point: 0.0500 $m$ $C_6H_{12}O_6$, 0.0300 $m$ KBr, or 0.0150 $m$ $Na_2SO_4$?

**11.81.** Arrange the following aqueous solutions in order of increasing boiling point:
a. 0.06 $m$ $FeCl_3$ ($i = 3.4$)
b. 0.10 $m$ $MgCl_2$ ($i = 2.7$)
c. 0.20 $m$ KCl ($i = 1.9$)

11.82. Arrange the following solutions in order of increasing freezing point depression:
a. 0.10 $m$ $MgCl_2$ in water, $i = 2.7$, $K_f = 1.86°C/m$
b. 0.20 $m$ toluene in diethyl ether, $i = 1.00$, $K_f = 1.79°C/m$
c. 0.20 $m$ ethylene glycol in ethanol, $i = 1.00$, $K_f = 1.99°C/m$

**11.83.** The following pairs of aqueous solutions are separated by a semipermeable membrane. In which direction will the solvent flow?
a. A = 1.25 $M$ NaCl; B = 1.50 $M$ KCl
b. A = 3.45 $M$ $CaCl_2$; B = 3.45 $M$ NaBr
c. A = 4.68 $M$ glucose; B = 3.00 $M$ NaCl

**11.84.** The following pairs of aqueous solutions are separated by a semipermeable membrane. In which direction will the solvent flow?
   a. A = 0.48 $M$ NaCl; B = 55.85 g of NaCl dissolved in 1.00 L of solution
   b. A = 100 mL of 0.982 $M$ $CaCl_2$; B = 16 g of NaCl in 100 mL of solution
   c. A = 100 mL of 6.56 m$M$ $MgSO_4$; B = 5.24 g of $MgCl_2$ in 250 mL of solution

**11.85.** Calculate the osmotic pressure of each of the following aqueous solutions at 20°C:
   a. 2.39 $M$ methanol ($CH_3OH$)
   b. 9.45 m$M$ $MgCl_2$
   c. 40.0 mL of glycerol ($C_3H_8O_3$) in 250.0 mL of aqueous solution (density of glycerol = 1.265 g/mL)
   d. 25 g of $CaCl_2$ in 350 mL of solution

**11.86.** Calculate the osmotic pressure of each of the following aqueous solutions at 27°C:
   a. 10.0 g of NaCl in 1.50 L of solution
   b. 10.0 mg/L of $LiNO_3$
   c. 0.222 $M$ glucose
   d. 0.00764 $M$ $K_2SO_4$

**11.87.** Determine the molarity of each of the following solutions from its osmotic pressure at 25°C. Include the van 't Hoff factor for the solution when the factor is given.
   a. $\Pi$ = 0.674 atm for a solution of ethanol ($CH_3CH_2OH$)
   b. $\Pi$ = 0.0271 atm for a solution of aspirin ($C_9H_8O_4$)
   c. $\Pi$ = 0.605 atm for a solution of $CaCl_2$, $i$ = 2.47

**11.88.** Determine the molarity of each of the following solutions from its osmotic pressure at 25°C. Include the van 't Hoff factor for the solution when the factor is given.
   a. $\Pi$ = 0.0259 atm for a solution of urea ($CH_4N_2O$)
   b. $\Pi$ = 1.56 atm for a solution of sucrose ($C_{12}H_{22}O_{11}$)
   c. $\Pi$ = 0.697 atm for a solution of KI, $i$ = 1.90

**11.89.** Is the following statement true or false? For solutions of the same reverse osmotic pressure at the same temperature, the molarity of a solution of NaCl will always be less than the molarity of a solution of $CaCl_2$. Explain your answer.

**11.90.** Suppose you have 1.00 $M$ aqueous solutions of each of the following solutes: glucose ($C_6H_{12}O_6$), NaCl, and acetic acid ($CH_3COOH$). Which solution has the highest pressure requirement for reverse osmosis?

## Measuring the Molar Mass of a Solute Using Colligative Properties

### CONCEPT REVIEW

**11.91.** What effect does dissolving a solute have on the following properties of a solvent? (a) its osmotic pressure; (b) its freezing point; (c) its boiling point

**11.92.** How can measurements of osmotic pressure, freezing point depression, and boiling point elevation be used to find the molar mass of a solute? Why are such determinations usually carried out on molecular substances as opposed to ionic ones?

### PROBLEMS

**11.93. Throat Lozenges** A 188 mg sample of a nonelectrolyte isolated from throat lozenges was dissolved in enough water to make 10.0 mL of solution at 25°C. The osmotic pressure of the resulting solution was 4.89 atm. Calculate the molar mass of the compound.

**\*11.94. Antibiotic** An unknown compound (27.40 mg) with antibiotic properties was dissolved in water to make 100.0 mL of solution. The solution did not conduct electricity and had an osmotic pressure of 9.94 torr at 23.6°C. Elemental analysis revealed the substance to be 42.34% C, 5.92% H, and 32.93% N. Determine the molecular formula of this compound.

**\*11.95. Cloves** Eugenol is one of the compounds responsible for the flavor of cloves. A 111 mg sample of eugenol was dissolved in 1.00 g of chloroform ($K_b$ 53.63°C/$m$), increasing the boiling point of the chloroform by 2.45°C. Calculate eugenol's molar mass. Eugenol is 73.17% C, 7.32% H, and 19.51% O by mass. What is the molecular formula of eugenol?

**\*11.96. Caffeine** The freezing point of a solution prepared by dissolving 150 mg of caffeine in 10.0 g of camphor is lower than that of pure camphor ($K_f$ = 39.7°C/$m$) by 3.07°C. What is the molar mass of caffeine? Elemental analysis of caffeine yields the following results: 49.49% C, 5.15% H, 28.87% N, and the remainder O. What is the molecular formula of caffeine?

## Additional Problems

**11.97.** Which substance has the least negative lattice energy? (a) $SrI_2$; (b) $CaBr_2$; (c) $CaCl_2$; (d) $MgF_2$

**11.98.** Explain why the boiling point of pure sodium chloride is much higher than the boiling point of an aqueous solution of sodium chloride.

**\*11.99. Melting Ice** $CaCl_2$ is often used to melt ice on sidewalks. Could $CaCl_2$ melt ice at −20°C? Assume that the solubility of $CaCl_2$ at this temperature is 70.1 g $CaCl_2$/100.0 g of $H_2O$ and that the van 't Hoff factor for a saturated solution of $CaCl_2$ is 2.5.

**\*11.100. Making Ice Cream** A mixture of table salt and ice is used to chill the contents of hand-operated ice-cream makers. What is the melting point of a mixture of 2.00 lb of NaCl and 12.00 lb of ice if exactly half of the ice melts? Assume that all the NaCl dissolves in the melted ice and that the van 't Hoff factor for the resulting solution is 1.44.

**11.101.** The freezing points of 0.0935 $m$ ammonium chloride and 0.0378 $m$ ammonium sulfate in water were found to be −0.322°C and −0.173°C, respectively. What are the values of the van 't Hoff factors for these salts?

**11.102.** The following data were collected for three compounds in aqueous solution. Determine the value of the van 't Hoff factor for each salt ($K_f$ for water = 1.86°C/$m$).

| Compound | Concentration | Experimentally Measured $\Delta T_f$ |
|---|---|---|
| LiCl | 5.0 g/kg | 0.410°C |
| HCl | 5.0 g/kg | 0.486°C |
| NaCl | 5.0 g/kg | 0.299°C |

**11.103. Physiological Saline** 100.0 mL of a solution of physiological saline (0.92% NaCl by mass) is diluted by the addition of 250.0 mL of water. What is the osmotic pressure of the final solution at 37°C? Assume that NaCl dissociates completely into $Na^+(aq)$ and $Cl^-(aq)$.

**11.104.** 100.0 mL of 2.50 m$M$ NaCl is mixed with 80.0 mL of 3.60 m$M$ $MgCl_2$ at 20°C. Calculate the osmotic pressure of each starting solution and that of the mixture, assuming that the volumes are additive and that both salts dissociate completely into their component ions.

**11.105.** A solution of 7.50 mg of a small protein in 5.00 mL aqueous solution has an osmotic pressure of 6.50 torr at 23.1°C. What is the molar mass of the protein?

**11.106. Kidney Dialysis** Hemodialysis, a method of removing waste products from the blood if the kidneys have failed, uses a tube made of a cellulose membrane that is immersed in a large volume of aqueous solution. Blood is pumped through the tube and is then returned to the patient's vein. The membrane does not allow passage of large protein molecules and cells but does allow small ions, urea, and water to pass through it. Assume that a physician wants to decrease the concentration of sodium ion and urea in a patient's blood, while maintaining the concentration of potassium ion and chloride ion in the blood. What materials must be dissolved in the aqueous solution in which the dialysis tube is immersed? How must the concentrations of ions in the immersion fluid compare to those in blood?

**11.107. IV Solution** Another solution used clinically in the hospital setting for IV administration is Ringer's lactate, a solution of sodium, potassium, and calcium cations and chloride and lactate anions. This solution is isotonic with 0.9% saline and D5W described in Section 11.5. Write a mathematical statement that indicates the relationship between the concentrations of cations and anions in this solution compared to 0.9% saline.

**11.108. Injections** The injection of pharmaceutical solutions that are hypertonic compared to human plasma can cause considerable pain at the site of injection. Why?

If your instructor assigns problems in **smartwork**, log in at **smartwork.wwnorton.com**.

# 12

# Solids: Structures and Applications

12.1   The Solid State

12.2   Structures of Metals

12.3   Alloys

12.4   Metallic Bonds and Conduction Bands

12.5   Semiconductors

12.6   Salt Crystals: Ionic Solids

12.7   Structures of Nonmetals

12.8   Ceramics: Insulators to Superconductors

12.9   X-ray Diffraction: How We Know Crystal Structures

## Stronger, Tougher, Harder

**M**etals play a major role in our lives, from copper electrical wires to titanium surgical screws to gold jewelry. In earlier chapters we explored the properties of liquids and gases; in this chapter we focus on solids. Metals are solids, but so are a host of other materials. Today's research into solid materials is yielding applications that would be unimaginable as recently as 50 years ago—small particles of gold (*nanoparticles*) used in medicine, materials for miniaturized electronic devices, and heat-resistant ceramic tiles for spacecraft.

Did you know that most gold rings and other gold jewelry are not made of pure gold? Gold (measured in *karats*, with 24-karat gold being pure) is one of the softer metals, and easy to bend. Blending gold with 8% of another metal yields 22-karat gold, a material more resistant to damage. Unlike pure gold, pure iron is strong, but it is susceptible to corrosion. Adding as little as 0.4% by mass of carbon results in steel, a much harder and more durable material. Stainless steel containing 18% chromium resists corrosion.

A mixture of two or more metals in the solid state is called an *alloy*, and over 500,000 alloys have been made and characterized. Some are solid solutions, which means they are homogeneous at the atomic level. Other alloys are heterogeneous mixtures. The purpose of making alloys is to control their properties by varying the proportions of their constituent metals. The construction industry uses steel I-beams and steel-reinforced concrete for most large projects. Interior fixtures are often made of alloys, such as brass, which is an alloy of copper and zinc. The transportation industry uses aluminum alloys that are ideal for making aircraft because they are strong and light in weight. Alloys of nickel and titanium can be shaped into stents used to prop open arteries in heart patients.

Other mixtures of solids play key roles in the operation of electronic devices. A cell phone, computer, or any device with a microchip inside depends on the selective conductivity of a mixture of silicon

### Learning Outcomes

**LO1** Recognize the differences in packing schemes for atoms in the solid state and identify face-centered, body-centered, and simple cubic unit cells
**Sample Exercise 12.1**

**LO2** Describe the differences between substitutional and interstitial alloys
**Sample Exercise 12.2**

**LO3** Describe the differences between n- and p-type semiconductors
**Sample Exercise 12.3**

**LO4** Describe the crystalline structures of metalloids and molecular and ionic compounds
**Sample Exercise 12.4**

**LO5** Calculate the interlayer spacing in crystalline solids using the Bragg equation
**Sample Exercise 12.5**

**Alloys for Strength** The metal cables in the Leonard P. Zakim Bunker Hill Bridge over the Charles River in Boston, MA, are steel, an alloy of iron and carbon with small amounts of other elements, sheathed in high-density polyethylene. Each cable can support over 5 million kilograms. The frame of the bridge is a hybrid construction using concrete and another type of steel. ▶

**crystalline solid** a solid made of an ordered array of atoms, ions, or molecules.

**amorphous solid** a solid that lacks long-range order for the atoms, ions, or molecules in its structure.

**molecular solid** a solid formed by neutral, covalently bonded molecules held together by intermolecular attractive forces.

**ionic solid** a solid consisting of monatomic or polyatomic ions held together by ionic bonds.

**CONNECTION** Some of the physical properties of metals were described in Chapter 2 and intermolecular forces were described in Chapter 10.

with another metalloid. Mixing gallium with arsenic yields materials used in the lasers found in CD players and price scanners.

In this chapter we examine the major classes of solids and explore the links between their physical properties at the macroscopic level and their structures at the atomic level. Materials science is a fast-changing field, and we are able to discuss only a few of its many applications. The search for new applications involving other elements is constant and constantly surprising. ■

## 12.1 The Solid State

In earlier chapters we studied the kinetic molecular behavior of gases (Chapter 6), and we examined the intermolecular forces between the molecules (and ions) in liquids (Chapter 10). In this chapter we turn our attention to the third state of matter: solids. Solid objects may be composed of pure substances or mixtures, compounds or elements. Solids may exist as **crystalline solids**—that is, ordered arrays of atoms, ions, or molecules—or as **amorphous solids**, which have random or disordered arrangements of particles (Figure 12.1).

Metallic elements generally form crystalline solids consisting of ordered arrays of atoms, whereas nonmetals crystallize as **molecular solids** consisting of neutral, covalently bonded molecules held together by intermolecular forces. Ionic compounds form crystalline **ionic solids** in which ions, either monatomic or polyatomic, are held together by ionic bonds. Molecular solids often have lower melting points than ionic solids because they are held together by relatively weak intermolecular forces compared to the strength of ionic bonds. The melting points

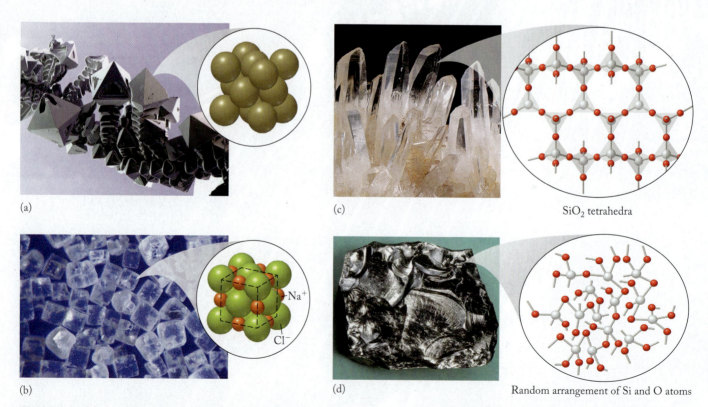

(a)

(b)

(c)

SiO$_2$ tetrahedra

(d)

Random arrangement of Si and O atoms

Na$^+$

Cl$^-$

**FIGURE 12.1** Many solids exist as (a) crystalline metallic solids, such as palladium; (b) ionic solids like sodium chloride; or (c) molecular solids like quartz, with an ordered array of atoms, ions, or molecules. Amorphous solids like (d) obsidian, a form of SiO$_2$, lack long-range order in their constituent particles.

of metals cover a broad range, from mercury, a liquid at room temperature, to tungsten, which melts at 3422°C. Metallic solids are malleable, and can be bent into a variety of shapes. Ionic solids tend to be hard but brittle. We start our exploration of the solid state by looking at the structures of some familiar metals.

## 12.2 Structures of Metals

People have placed a high value on gold for thousands of years, and for many different reasons: from jewelry, to coinage, to medicine. Gold is one of the few elements found naturally in metallic form (Figure 12.2). Metals, like gold, are typically shiny, malleable (easily shaped), ductile (easily drawn out), and able to conduct electricity.

For centuries, suspensions of gold have been used in medicine for treatment of diseases as varied as cancer and ulcers. In the 21st century, the medical application of gold **nanoparticles** (small particles with diameters less than $10^{-7}$ m) have been used to treat rheumatoid arthritis and to bind drugs (using intermolecular forces) and deliver them to specific target cells.

### Stacking Patterns

When a metal is heated above its melting point and then slowly allowed to cool, it solidifies into a crystalline solid; that is, a solid in which the atoms are arranged in an ordered three-dimensional array called a **crystal lattice**. Think of a crystal lattice as stacked layers (designated *a*, *b*, *c*, . . .) of particles packed together as tightly as possible. Each atom in layer *a* touches six others in that layer, as shown in Figure 12.3(a). The atoms in layer *b* nestle into some of the spaces created by the atoms of layer *a* (Figure 12.3b), just like oranges in a fruit-stand display or cannonballs at a 16th-century fort (Figure 12.4). Similarly, the atoms in a third layer, *c*, nestle among those in layer *b*. However, two different alignments are possible for the atoms in layer *c*. They can align

(a)

(b)

(c)

**FIGURE 12.2** Gold is familiar to us as (a) nuggets found naturally, (b) gold coins, or (c) tiny particles suspended in water touted as elixirs for good health. The gold in each of these forms is the same—an ordered arrangement of gold atoms—even if the colors of gold coins and small gold particles are different.

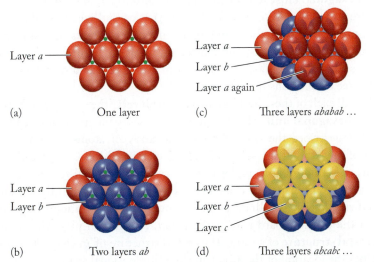

(a) One layer

(c) Three layers *ababab* . . .

(b) Two layers *ab*

(d) Three layers *abcabc* . . .

**FIGURE 12.3** Two equally efficient ways to stack layers of atoms (or any particles of equal size). In both stacking patterns the atoms in all layers (shown in different colors to distinguish one layer from another) are packed as closely together as possible. Layers (a) and (b) are the same in both patterns. (c) In the *abababab* . . . pattern, atoms in the third layer are directly above the atoms in the first layer. (d) In the *abcabcabc* . . . pattern, atoms in the third layer are directly above spaces (marked by red dots) between the atoms in the first layer.

**nanoparticle** approximately spherical sample of matter with dimensions less than 100 nanometers ($1 \times 10^{-7}$ m).

**crystal lattice** a three-dimensional array of particles (atoms, ions, or molecules) in a crystalline solid.

**hexagonal closest-packed (hcp)** a crystal lattice in which the layers of atoms or ions in hexagonal unit cells have an *ababab* . . . stacking pattern.

**unit cell** the basic repeating unit of the arrangement of atoms, ions, or molecules in a crystalline solid.

**hexagonal unit cell** an array of closest-packed particles that includes parts of four particles on the top and four on the bottom faces of a hexagonal prism and one particle in a middle layer.

**crystal structure** an ordered arrangement in three-dimensional space of the particles (atoms, ions, or molecules) that make up a crystalline solid.

(a)                                             (b)

**FIGURE 12.4** (a) Stacks of oranges in a grocery store and (b) cannonballs at a 16th-century fort illustrate closest-packed arrays of spherical objects. Both (a) and (b) exhibit an *abcabc* stacking pattern.

directly above the atoms in layer *a* (Figure 12.3c), or they can nestle into the atoms of layer *b* in such a way that they are not aligned directly above the layer *a* atoms (Figure 12.3d). When a fourth layer is nestled into the spaces of layer *c*, the fourth layer atoms lie directly above the layer *a* atoms. In Figure 12.3(c), we have a stacking pattern *ababab* . . . throughout the crystal, and in Figure 12.3(d) we show the stacking pattern *abcabc* . . . .

## Stacking Spheres and Unit Cells

Which of these two patterns do gold atoms adopt? Whether the atoms in a metal like gold are stacked in an *ababab* or an *abcabc* pattern determines the shape of the crystals the metal forms when it slowly cools and solidifies from the molten state. To see how crystal structures are linked to stacking patterns, let's take a closer look at a cluster of atoms in the *ababab* . . . stacking pattern (Figure 12.5a). This cluster forms a *hexagonal* (six-sided) prism of closely packed atoms. In fact, they are as tightly packed as they can be, so the crystal structure is called **hexagonal closest-packed (hcp)**. Titanium metal used in surgical repair of fractures (Figure 12.6) is one element that crystallizes in an hcp structure. As Figure 12.7 shows, the atoms in 16 metallic elements have an hcp crystal structure. In these metals, the cluster of atoms in Figure 12.5a serves as an atomic-scale building block—a pattern of atoms repeated over and over again in all three dimensions in the metal.

We call each of these building blocks a **unit cell**. The example in Figure 12.5a shows a **hexagonal unit cell**. A unit cell represents the minimum repeating pattern that describes the three-dimensional array of atoms forming the crystal lattice of any crystalline solid, including metals. Think of unit cells as three-dimensional microscopic analogs of the two-dimensional repeating pattern in fabrics, wrapping paper, or even a checkerboard. Look carefully at Figure 12.8 to confirm that the outlined portion represents the minimum repeating pattern in the checkerboard and in the paper. A unit cell plays the same role in the **crystal structure** of a solid. The crystal structure of an element or compound describes the location of the atoms in three-dimensional space.

Atoms in solid gold and 11 other metals adopt the *abcabc* . . . stacking pattern when they solidify (Figures 12.2 and 12.7). What is the unit cell in this stacking pattern? Consider what happens when we take the cluster of atoms on the left in Figure 12.5b, compact it, rotate the cluster, and tip it 45° to get the orientation shown in the center of Figure 12.5b. The black outline shows that the atoms form a

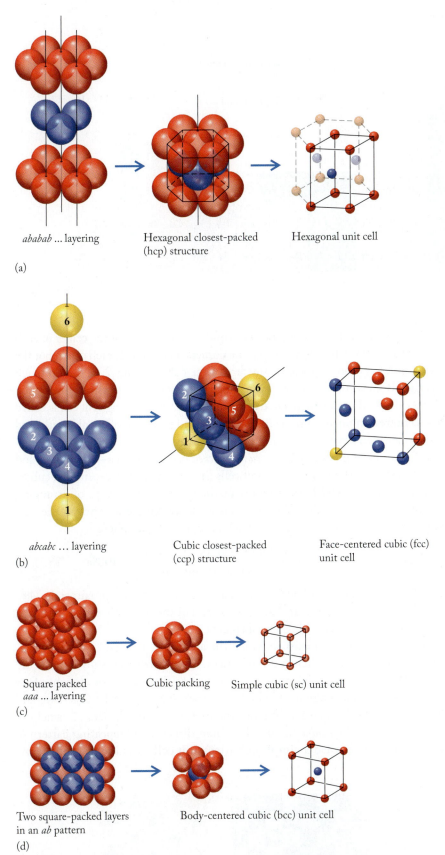

*ababab* ... layering

Hexagonal closest-packed (hcp) structure

Hexagonal unit cell

(a)

*abcabc* ... layering

Cubic closest-packed (ccp) structure

Face-centered cubic (fcc) unit cell

(b)

Square packed *aaa* ... layering

Cubic packing

Simple cubic (sc) unit cell

(c)

Two square-packed layers in an *ab* pattern

Body-centered cubic (bcc) unit cell

(d)

**FIGURE 12.5** (a) A hexagonal closest-packed (hcp) crystal structure and its hexagonal unit cell. (b) The stacking pattern *abcabc* . . . produces a face-centered cubic (fcc) unit cell. (c) In *cubic packing*, the atoms in all layers are directly above those in the *a* layer. The repeating unit of this pattern is called a *simple cubic* (sc) unit cell. (d) The atoms (blue spheres) in the second *b* layer nestle into the spaces between the square-packed atoms (red spheres) in the *a* layer. Atoms in the third layer are directly above those in the first, producing an *ababab* . . . stacking pattern and a *body-centered cubic* (bcc) unit cell.

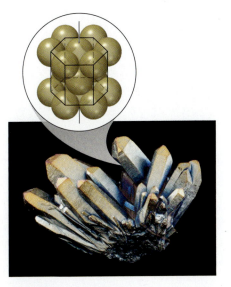

**FIGURE 12.6** Titanium crystallizes in an hcp pattern with a hexagonal unit cell. This photo shows quartz crystals (see Figure 12.30) coated with titanium metal.

**FIGURE 12.7** Unit cells of metals and metalloids in the periodic table. The five metals designated "Other" have unit cells more complicated than can easily be described in this book.

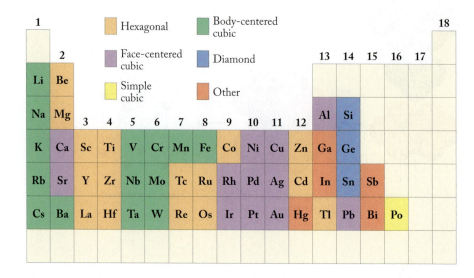

cube: one atom at each of the eight corners of the cube and one at the center of each of the six faces. Note that atoms at adjacent corners do not touch each other, but the three atoms along the diagonal of any face of the cube—atoms 2, 3, 4—do touch each other. Because the atoms are stacked together as closely as possible, this crystal lattice is called **cubic closest-packed (ccp)**, and the corresponding unit cell is called a **face-centered cubic (fcc) unit cell**. The unit cell edges are all of equal length, and the angle between any two edges is 90°.

So far we have introduced two *closest-packed* crystal lattices—hexagonal and cubic—along with their associated unit cells (hexagonal and face-centered cubic). The hcp and ccp crystal lattices represent the most efficient ways of arranging solid spheres of equal radius. We can express the **packing efficiency** as the percentage of the total volume of the unit cell occupied by the spheres:

$$\text{Packing efficiency (\%)} = \frac{\text{volume occupied by spheres}}{\text{volume of unit cell}} \times 100\% \qquad (12.1)$$

For both hcp and ccp crystal lattices, the packing efficiency is approximately 74%. We will come back to this calculation at the end of the next subsection, but first, let's look at some other packing arrangements.

Stacking patterns also exist in which the atoms are arranged close together, but not as efficiently as in hcp and ccp lattices. Two of these are shown in Figure 12.5c,d. We can arrange the atoms in an *a* layer so that each atom touches four adjacent atoms, an arrangement called *square packing* (Figure 12.5c). If we add a second layer of spheres directly above the first, we create the *aaa . . .* stacking pattern, which is called *cubic packing*. The three-dimensional repeating pattern of this arrangement is called a **simple cubic (sc) unit cell**. It is the least efficiently

**FIGURE 12.8** (a) The highlighted "unit cell" of a checkerboard is the smallest set of squares that defines the pattern repeated over the entire board. (b) This wrapping paper has a more complex pattern. One unit cell is highlighted. Can you outline another?

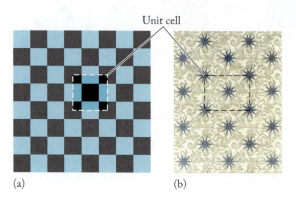

Unit cell

(a)          (b)

packed of the cubic unit cells and is quite rare among metals: only radioactive polonium (Po) forms a simple cubic unit cell.

If each atom in a second layer is nestled in the space created by four atoms in a square-packed *a* layer (Figure 12.5d), we have two layers in an *ab* stacking pattern. If the atoms in the third layer are directly above those in the first, then we have an *ababab* . . . stacking pattern based on layers of square-packed atoms. The simplest three-dimensional repeating unit of this pattern is called a **body-centered cubic (bcc) unit cell**. It consists of portions of nine atoms, one at each of the eight corners of a cube and one in the middle of the cube. All the group 1 metals and many transition metals have bcc unit cells (Figure 12.7). Included among the metals with a bcc unit cell is tantalum, a metal used to coat artificial joints (Figure 12.9). Table 12.1 summarizes the different stacking patterns, packing efficiencies, and unit cells described in this section.

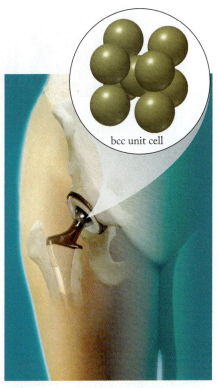

**FIGURE 12.9** Tantalum metal is used to coat artificial joints such as this hip joint because tantalum metal is wear-resistant and does not react with body fluids. The crystal structure of tantalum is body-centered cubic.

| TABLE 12.1 | **Summary of Unit Cells, Stacking Patterns, and Packing Efficiencies for Solid Spheres** | | | | |
|---|---|---|---|---|---|
| **Lattice Name** | **Unit Cell** | **Type of Packing** | **Stacking Pattern** | **Number of Nearest Neighbors** | **Packing Efficiency** |
| Hexagonal closest-packed (hcp) | Hexagonal | Close packing | *ababab* . . . | 12 | 74% |
| Cubic closest-packed (ccp) | Face-centered cubic (fcc) | Close packing | *abcabc* . . . | 12 | 74% |
| Body-centered cubic packing | Body-centered cubic (bcc) | Square packing | *ababab* . . . | 8 | 68% |
| Cubic packing | Simple cubic (sc) | Square packing | *aaa* . . . | 6 | 52% |

Crystalline solids with cubic unit cells form crystals that are cubic in appearance on a macroscopic scale (palladium in Figure 12.1a), whereas crystalline solids with hexagonal unit cells tend to form hexagonal crystals (titanium in Figure 12.6).

**CONCEPT TEST**

What is the difference between a crystal lattice and a unit cell?

*(Answers to Concept Tests are in the back of the book.)*

## Unit Cell Dimensions

Of all the metallic elements, only Li, Na, and K have densities less than 1 g/cm³. The density of the transition metals varies between 3 g/cm³ (Sc) and 22.5 g/cm³ (Os). The number of atoms in the unit cell of a metallic element, its atomic radius, and its molar mass all contribute to its density. Figure 12.10 shows whole-atom and cutaway views of sc, fcc, and bcc unit cells. These views also provide us with a way to determine how many atoms are in each type of cubic unit cell.

Let's start with the simple cubic unit cell (Figure 12.10a). Note how only a fraction of each corner atom is inside the unit cell boundary. In a crystal lattice with this unit cell, each atom is a corner atom in eight unit cells (Figure 12.11a). Thus each atom contributes the equivalent of one-eighth of an atom to the unit cell. A cube has eight corners, so there is a total of

$$\tfrac{1}{8} \text{ corner atom/corner} \times 8 \text{ corners/unit cell } = 1 \text{ atom/unit cell}$$

**cubic closest-packed (ccp)** a crystal structure composed of face-centered cubic unit cells and layers of particles having an *abcabc* . . . stacking pattern.

**face-centered cubic (fcc) unit cell** an array of closest-packed particles that has one particle at each of the eight corners of a cube and one more at the center of each face of the cube.

**packing efficiency** percentage of the total volume of a unit cell occupied by the spheres.

**simple cubic (sc) unit cell** a cell with atoms or molecules only at the eight corners of a cube.

**body-centered cubic (bcc) unit cell** a cell with one atom at each of the eight corners of a cube and one at the center of the cell.

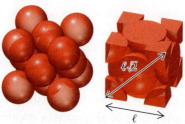

(a) Simple cubic:
Atoms touch along edge

(b) Face-centered cubic:
Atoms touch along face diagonal

(c) Body-centered cubic:
Atoms touch along body diagonal

**FIGURE 12.10** Whole-atom and cutaway views of cubic unit cells. (a) In a simple cubic unit cell, each corner atom of the unit cell is part of eight unit cells. Atoms along each edge touch with an edge length $\ell$. (b) In a face-centered cubic unit cell, the face atoms are part of two unit cells. Atoms along the face diagonal touch. (c) In a body-centered cubic unit cell, one atom in the center lies entirely in one unit cell. The atoms along the body diagonal touch.

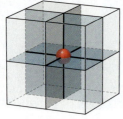

Corner atom
in 1 unit cell

Corner atom shared
by 8 unit cells

(a)

Face-centered
atom in 1 unit cell

Face-centered atom
shared by 2 unit cells

(b)

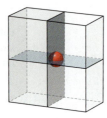

Body-centered
atom in 1 unit cell

Edge atom in
1 unit cell

Edge atom shared
by 4 unit cells

(c)

(d)

**FIGURE 12.11** Crystal lattices illustrating (a) corner atoms shared by eight unit cells, (b) face atoms shared by two unit cells, (c) center atoms entirely in one unit cell, and (d) edge atoms shared by four unit cells.

This calculation applies to the corner atoms in any type of cubic unit cell. Note that the two corner atoms along each edge in Figure 12.10(a) touch each other. Therefore the edge length $\ell$ in the simple cubic unit cell is equal to twice the atomic radius:

$$\ell = 2r$$

So, if we can measure the length of the unit cell ($\ell$), we can easily calculate the atomic radius of an element with this unit cell. We will describe one way of measuring the length of a unit cell in Section 12.9.

An fcc unit cell (Figure 12.10b) has eight corner atoms and one atom in the center of each of the six faces (Figure 12.11b). Each face atom is shared by the two unit cells that abut each other at that face. Therefore each unit cell "owns" half of each face atom, making a total of

$$\tfrac{1}{2} \text{ face atom/}\cancel{\text{face}} \times 6 \text{ }\cancel{\text{faces}}\text{/fcc unit cell} = 3 \text{ face atoms/fcc unit cell}$$

As just noted, every cubic unit cell owns the equivalent of one corner atom. Therefore, an fcc unit cell consists of

$$1 \text{ corner atom} + 3 \text{ face atoms} = 4 \text{ atoms per fcc unit cell}$$

To relate the size of these atoms to the dimensions of the fcc unit cell, note in the cutaway view of Figure 12.10(b) that the corner atoms do not touch one another, but adjacent atoms along the face diagonal do touch each other. Therefore, a face diagonal spans the radius $r$ of two corner atoms and the diameter (2 radii = $2r$) of a face atom. Therefore the length of a face diagonal is $1 + 2 + 1 = 4$ atomic radii = $4r$. A face diagonal connects the ends of two edges and forms a right triangle with those two edges, each of length $\ell$ (Figure 12.10b). Therefore, according to the Pythagorean theorem, the length of a face diagonal is

$$\text{Face diagonal} = 4r = \sqrt{\ell^2 + \ell^2} = \sqrt{2\ell^2} = \ell\sqrt{2}$$

$$r = \frac{\ell\sqrt{2}}{4} = 0.3536\ell \qquad (12.2)$$

As in the case of the simple cubic unit cell, if we know the value of $\ell$, we can calculate $r$ using Equation 12.2.

Now let's focus on the bcc unit cell in Figure 12.10(c). In addition to the one atom from one-eighth of an atom at each of the eight corners, a bcc cell also has one atom in the center of the cell that is entirely within the cell (Figure 12.11c). This means a bcc unit cell consists of

1 corner atom + 1 center atom = 2 atoms per bcc unit cell

Relating unit cell edge length to atomic radius in a bcc cell is complicated by the fact that, in addition to not touching along the edges, adjacent atoms along any face diagonal do not touch each other. However, each corner atom does touch the atom in the center of the cell, which means that the atoms touch along a *body diagonal*, which runs between opposite corners through the center of the cube. In the cutaway view in Figure 12.10(c), the body diagonal runs from the bottom left corner of the front face to the top right corner of the rear face. It spans (1) the radius of the front-face bottom left corner atom, (2) the diameter (2 radii) of the central atom, and (3) the radius of the rear-face top right atom, making the length of the body diagonal equivalent to $4r$.

We can again use the Pythagorean theorem to determine the relationship between $\ell$ and $r$. A right triangle is formed by an edge, a face diagonal, and a body diagonal serving as the hypotenuse of the triangle (Figure 12.10c). Using the face-diagonal value $\ell\sqrt{2}$, we get

$$\text{Body diagonal} = 4r = \sqrt{(\text{edge length})^2 + (\text{face diagonal})^2}$$
$$= \sqrt{\ell^2 + (\ell\sqrt{2})^2} = \sqrt{\ell^2 + \ell^2(2)} = \sqrt{3\ell^2} = \ell\sqrt{3}$$

so

$$r = \frac{\ell\sqrt{3}}{4} = 0.4330\ell \qquad (12.3)$$

Table 12.2 summarizes how atoms in different locations in sc, bcc, and fcc unit cells contribute to the total number of atoms in each unit cell. Table 12.3 summarizes the number of equivalent atoms and the relationship between $r$ and $\ell$ for the three cubic unit cells.

**TABLE 12.2 Contributions of Atoms to Cubic Unit Cells**

| Atom Position | Contribution to Unit Cell | Unit Cell Type |
|---|---|---|
| Center | 1 atom | bcc |
| Face | $\frac{1}{2}$ atom | fcc |
| Corner | $\frac{1}{8}$ atom | bcc, fcc, sc |

**TABLE 12.3 Summary of Unit Cells, Equivalent Atoms, and the Relationship between Radius of Atoms and Edge Length of Cubic Unit Cells**

| Unit Cell | Number of Equivalent Atoms per Unit Cell | Relationship between $r$ and $\ell$ |
|---|---|---|
| Simple cubic | 1 | $r = \dfrac{\ell}{2} = 0.5\ell$ |
| Body-centered cubic | 2 | $r = \dfrac{\ell\sqrt{3}}{4} = 0.4330\ell$ |
| Face-centered cubic | 4 | $r = \dfrac{\ell\sqrt{2}}{4} = 0.3536\ell$ |

**SAMPLE EXERCISE 12.1** **Calculating Atomic Radius and Density from Unit Cell Dimensions** **LO1**

Tantalum crystallizes with a bcc unit cell as shown in Figure 12.9. Its unit cell has an edge length of 330.3 pm. (a) Calculate the radius in picometers of the tantalum atoms. Check your answer against the data in Appendix 3. (b) Calculate the density of tantalum in grams per cubic centimeter at 25°C.

**Collect and Organize** We are given the unit cell (bcc) and edge length ($\ell = 330.3$ pm) of tantalum. Together with the atomic mass of tantalum (180.95 g/mol), we can calculate both the atomic radius of a Ta atom and the density of Ta metal.

**Analyze** The tantalum atoms do not touch along the unit cell edges or along any face diagonal, but they do touch along the body diagonals. Table 12.3 gives the relationship between $r$ and $\ell$ for a bcc unit cell. We know that the radius is a little less than half the edge length ($\ell/2$), so we expect a value less than 165 pm for $r$.

We assume that the density of the unit cell is the same as the density of solid Ta. The density of the Ta bcc unit cell is the mass of two Ta atoms divided by the volume of the cell. We can calculate the mass of two Ta atoms from the molar mass, which is 180.95 g/mol. The conversion includes dividing by Avogadro's number to calculate the mass of each Ta atom in grams. The formula for the volume of a cube of edge length $\ell$ is $V = \ell^3$. A piece of tantalum sinks when immersed in water, so we predict that the density of tantalum should be greater than 1.0 g/cm³.

**Solve**
a. Substituting the edge length into the formula for $r$ from Table 12.3:

$$r = 0.4330 \times 330.3 \text{ pm} = 143.0 \text{ pm}$$

This is close to the reference value of 146 pm for Ta (Table A3.1).
b. We first calculate the mass $m$ of two Ta atoms:

$$m = \frac{180.95 \text{ g Ta}}{1 \text{ mol Ta}} \times \frac{1 \text{ mol Ta}}{6.022 \times 10^{23} \text{ atoms Ta}} \times 2 \text{ atoms Ta} = 6.010 \times 10^{-22} \text{ g Ta}$$

The volume of the cell in cubic centimeters is

$$V = \ell^3 = (330.3 \text{ pm})^3 \times \frac{(10^{-10} \text{ cm})^3}{1 \text{ pm}^3} = 3.604 \times 10^{-23} \text{ cm}^3$$

The density is

$$d = \frac{m}{V} = \frac{6.010 \times 10^{-22} \text{ g}}{3.604 \times 10^{-23} \text{ cm}^3} = 16.68 \text{ g/cm}^3$$

**Think About It** The value of $r$, 143 pm, is indeed less than half the value of the edge length ($\ell/2 = 165$ pm), so the result of this calculation is reasonable. According to the data in Table A3.2, the density of tantalum is 16.65 g/mL, so the result of that calculation is reasonable and reflects the prediction we made. The density of the unit cell should equal the density of a bulk sample because the sample is composed of a crystal lattice of bcc unit cells.

⚙ **Practice Exercise** Silver and gold both crystallize in face-centered cubic unit cells with edge lengths of 407.7 and 407.0 pm, respectively. Calculate the radii and density of each metal and compare your answers with the data listed in Appendix 3.

*(Answers to Practice Exercises are in the back of the book.)*

Earlier in this section we mentioned that ccp and hcp are the most efficient packing schemes for spheres. Now that we know more about the fcc unit cell, let's take another look at the calculation of packing efficiency using Equation 12.1. The unit cell for a ccp arrangement of solid spheres of radius $r$ is an fcc unit cell that contains four equivalent spheres. The volume occupied by these spheres is

$$V_{\text{spheres}} = 4\left(\tfrac{4}{3}\pi r^3\right) = \tfrac{16}{3}\pi r^3$$

From Table 12.3, $r = (\ell\sqrt{2})/4$, and the unit cell edge ($\ell$) is

$$\ell = \frac{4r}{\sqrt{2}} = 2.828r$$

This equation enables us to express the volume of the unit cell in terms of $r$:

$$V_{\text{unit cell}} = \ell^3 = (2.828r)^3 = 22.62r^3$$

Substituting $V_{\text{spheres}}$ and $V_{\text{unit cell}}$ into Equation 12.1:

$$\text{Packing efficiency (\%)} = \frac{V_{\text{spheres}}}{V_{\text{unit cell}}} \times 100\% = \frac{\tfrac{16}{3}\pi r^3}{22.62r^3} \times 100\% = 74.1\%$$

**alloy** a blend of a host metal and one or more other elements, which may or may not be metals, that are added to change the properties of the host metal.

**substitutional alloy** an alloy in which atoms of the nonhost metal replace host atoms in the crystal lattice.

This means that the most efficient packing of spheres results in about 26% of the total volume being empty.

## 12.3 Alloys

Valuable as it is, pure gold suffers from one disadvantage: gold is relatively soft and very malleable, which means that pure gold objects are easily bent and damaged. We can explain the malleability of Au, Cu, and other metals in terms of the relatively weak bonds between the atoms in their cubic closest-packed crystal structure. This arrangement gives the atoms in one layer the ability, under stress, to slip past atoms in an adjacent layer (Figure 12.12), but the overall crystal structure is still cubic closest-packed. The ease with which copper atoms slip past each other makes it easy to bend copper pipes used in plumbing, but it also makes them susceptible to damage.

Both copper and gold can be strengthened by the addition of other metals. For example, adding tin to copper yields bronze, whereas mixing gold and copper generates materials known as rose gold (20–55% copper, Figure 12.13a) or the dark purplish-blue shakudo gold (96% Cu, 4% Au, Figure 12.13b). A mixture of two metals represents an **alloy**: a metallic material made when a host metal is blended with one or more other elements, which may or may not be metals, thereby changing the properties of the host metal. Like the mixtures discussed in Chapter 1, alloys can be classified according to their composition as homogeneous or heterogeneous mixtures. Bronze, rose gold, and shakudo gold are all *homogeneous alloys*, solid solutions in which the atoms of the added element(s) (in this case tin or gold) are randomly but uniformly distributed among the atoms of the host (copper). *Heterogeneous alloys* consist of matrices of atoms of host metals interspersed with small "islands" made up of individual atoms of other elements. In both cases, the compositions may vary over a limited range.

Every year thousands of patients suffering from cardiovascular disease undergo a procedure called balloon angioplasty to open clogged arteries. This surgery involves the insertion of small metal supports, or stents, to prop open the artery. Increasingly, stents are manufactured from an alloy containing nickel and titanium. The alloy NiTi (nitinol) is an example of an *intermetallic compound*, a substance that has a reproducible stoichiometry and constant composition (just like chemical compounds). Intermetallic compounds are still commonly referred to as alloys and are considered to be a subgroup within homogeneous alloys. What makes nickel–titanium alloys particularly useful is that they demonstrate shape memory. Figure 12.14 illustrates how memory alloys work. A piece of NiTi wire is heated and formed into the desired shape for the stent. On cooling, the NiTi undergoes a phase change to a different structure. The structural change allows the stent to be easily bent into a small coil for insertion into an artery. After insertion, the coiled stent is heated, springing back into its original shape as it rapidly reverts to the high-temperature form.

### Substitutional Alloys

If alloys are mixtures of metals, and pure metals have crystal lattices as described in Section 12.2, how are the metal atoms arranged in the crystal lattices and unit cells of alloys? How does the structure of NiTi memory metal change as a function of temperature? The answer to these questions gives rise to another classification system: a **substitutional alloy** is one in which atoms of the nonhost metal

External force

Metal is deformed

**FIGURE 12.12** Copper and other metals are malleable because their atoms are stacked in layers that can slip past each other under stress. Slippage is possible because of the diffuse nature of metallic bonds and the relatively weak interactions between pairs of atoms in adjoining layers.

⬤⬤ **CONNECTION** The classification of homogeneous and heterogeneous mixtures and the difference between compounds and mixtures were described in Chapter 1.

(a)

(b)

**FIGURE 12.13** (a) Rose gold is an alloy containing gold alloyed with 20 to 55% copper. (b) This shakudo gold ring is made of a copper alloy containing 4% gold and 96% copper. Compare the colors of these gold alloys with that of pure gold in Figure 12.2.

FIGURE 12.14 Shape memory alloys are used in stents for heart patients. This figure illustrates how shape memory alloys work. (a) The S shape on the right is the desired final shape of the wire. The left and middle wires have been coiled to take up less space. The white circle in the background is a hair dryer heating the wires. (b) When the wire in the middle is heated, it returns to its original shape, while the wire on the left is in the process of unraveling.

(a)

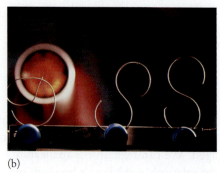
(b)

(a)

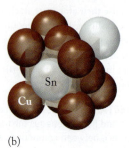
(b)

FIGURE 12.15 Two atomic-scale views of one type of bronze, a substitutional alloy. (a) A layer of close-packed copper (Cu) atoms interspersed with a few atoms of tin (Sn). (b) One possible unit cell for bronze. In this case tin atoms have replaced one corner Cu atom and one face Cu atom.

FIGURE 12.16 The larger Sn atoms in bronze disturb the Cu crystal lattice, producing atomic-scale bumps in the slip plane (wavy line) between layers of Cu atoms. These bumps make it more difficult for Cu atoms to slide past each other when an external force is applied.

replace host atoms in the crystal lattice. Bronze and nitinol are both examples of *homogeneous, substitutional alloys*. Substitutional alloys may form between metals that have the same crystal lattice and atomic radii that are within about 15% of each other.

CONCEPT TEST ............................................................

Is it accurate to call bronze a *solution* of tin *dissolved* in copper? Explain why or why not.

..............................................................................

Why is an alloy more difficult to bend than a pure metal? Figure 12.15 illustrates one layer of the crystal lattice of bronze. The radii of copper and tin atoms are similar—128 pm and 140 pm, respectively. Inserting the slightly larger Sn atoms in the cubic closest-packed Cu crystal lattice disturbs the structure a little, making the planes of copper atoms "bumpy" instead of uniform (Figure 12.16). This atomic-scale roughness makes it more difficult for the copper atoms to slip past one another. Less slippage makes bronze less malleable than copper, but being less malleable also means that bronze is harder and stronger.

## Interstitial Alloys

Steel is an alloy of iron. Molten iron produced in a blast furnace may contain up to 5% carbon. When the molten iron cools to its melting point of 1538°C, it crystallizes in a body-centered cubic structure before undergoing a phase transition at around 1390°C to austenite, a form of solid iron made up of face-centered cubic unit cells. The spaces, or *holes*, between iron atoms in austenite can accommodate carbon atoms, forming an **interstitial alloy**, so named because the carbon atoms occupy spaces, or *interstices*, between the iron atoms (Figure 12.17).

CONCEPT TEST ............................................................

Is a substitutional alloy a homogeneous or a heterogeneous alloy, or could it be both? Is an interstitial alloy a homogeneous or a heterogeneous alloy, or could it be both?

..............................................................................

Not all interstices in a crystal lattice are equivalent. Indeed, holes of two different sizes occur between the atoms in any closest-packed crystal lattice (Figure 12.18). The larger holes are surrounded by clusters of six host atoms in the shape of an octahedron and are called *octahedral holes*. The smaller holes are located between clusters of four host atoms and are called *tetrahedral holes*. The data in Table 12.4 show which holes are more likely to be occupied based on the relative

| TABLE 12.4 | Atomic Radius Ratios and Location of Nonhost Atoms in Unit Cells of Interstitial Alloys | |
|---|---|---|
| Lattice Name | Hole Type | Atomic Radius Ratio $r_{\text{nonhost}}/r_{\text{host}}{}^a$ |
| hcp or ccp | Tetrahedral | 0.22–0.41 |
| hcp or ccp | Octahedral | 0.41–0.73 |
| Cubic packing | Cubic | 0.73–1.00 |

$^a$Radius ratios as predictors of the crystal lattice in crystalline solids are of limited value because atoms are not truly solid spheres with a constant radius; because the radius of an atom may differ in different compounds, these ranges are approximate.

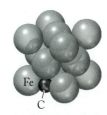

**FIGURE 12.17** Carbon steel is an interstitial alloy of carbon in iron. The fcc form of iron (austenite) that forms at high temperatures can accommodate carbon atoms in its octahedral holes.

sizes of nonhost and host atoms. According to Appendix 3, the atomic radii of C and Fe are 77 and 126 pm, respectively. According to Table 12.4, the ratio 77/126 = 0.61 means that C atoms should fit in the octahedral holes of austenite, as shown in Figure 12.17, but not in the tetrahedral holes.

As austenite with its fcc unit cell cools to room temperature, it converts into the crystalline solid form of iron called ferrite, which is body-centered cubic. The octahedral holes in ferrite are smaller than those in austenite, and they are too small to accommodate carbon atoms. As a result, much of the carbon precipitates as clusters of carbon atoms, or it may react with iron to form iron carbide, $Fe_3C$. The clusters of carbon and $Fe_3C$ disrupt ferrite's body-centered cubic lattice and inhibit the host iron atoms from slipping past each other when a stress is applied. This resistance to slippage, which is much like that experienced by the copper atoms in bronze (Figure 12.16), makes iron–carbon alloys, known as *carbon steel*, much harder and stronger than pure iron. In general, the higher the carbon concentration, the stronger the steel. However, there is a trade-off in this relationship; increased strength and hardness come at the cost of increased brittleness.

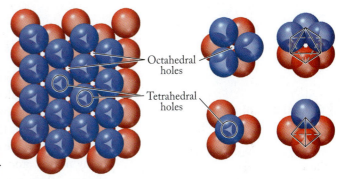

Octahedral holes

Tetrahedral holes

**FIGURE 12.18** Close-packed atoms in adjacent layers of a crystal lattice produce octahedral holes surrounded by six host atoms and tetrahedral holes surrounded by four host atoms. Octahedral holes are larger than tetrahedral holes and so can accommodate larger nonhost atoms in interstitial alloys. Note that all the atoms are identical here; the colors are only to distinguish one layer of atoms from another.

**SAMPLE EXERCISE 12.2** **Predicting the Crystal Structure of a Two-Element Alloy** LO2

Sterling silver, which is 93% Ag and 7% Cu by mass, is widely used in jewelry. The presence of Cu inhibits tarnishing and strengthens the alloy. Is this copper–silver alloy a substitutional or an interstitial alloy? Silver has a cubic closest-packed crystal lattice with face-centered cubic unit cells.

**Collect and Organize** We are asked whether in sterling silver the Cu–Ag alloy is substitutional or interstitial, given that the atoms in solid Ag form fcc unit cells and have a cubic closest-packed crystal lattice.

**Analyze** Atoms of two or more elements form a substitutional alloy when all the atoms are of similar size. The highest $r_{\text{nonhost}}/r_{\text{host}}$ ratio in Table 12.4 for a ccp lattice, 0.73, means that an interstitial alloy can form only when the radius of the nonhost atoms is less than 73% of the radius of the host atoms. The atomic radii found in Appendix 3 are 144 pm for Ag and 128 pm for Cu.

**Solve** The ratio of the atomic radii of Cu to Ag is

$$r_{\text{nonhost}}/r_{\text{host}} = \frac{128 \text{ pm}}{144 \text{ pm}} = 0.89$$

**interstitial alloy** atoms of the added element occupy the spaces between atoms of the host.

**band theory** an extension of molecular orbital theory that describes bonding in solids.

**valence band** a band of orbitals that are filled or partially filled by valence electrons.

**conduction band** in metals, an unoccupied band higher in energy than a valence band, in which electrons are free to migrate.

**band gap ($E_g$)** the energy gap between the valence and conduction bands.

The $r_{nonhost}/r_{host}$ ratio indicates that the copper atoms are too big to fit into interstices in the lattice of silver atoms, regardless of whether the holes are tetrahedral or octahedral. Thus, an interstitial alloy is impossible. Atoms of Cu can substitute for atoms of Ag in the Ag ccp lattice, however, with some room to spare. Figure 12.7 tells us that both elements form fcc unit cells, so little disruption to the Ag lattice should result from incorporating Cu atoms. Thus, sterling silver is a substitutional alloy.

**Think About It** The guidelines for two metals forming a substitutional alloy are that they have the same type of crystal lattice and that their atomic radii are within 15% of each other. The radii of Ag and Cu differ by only 11%, so we would expect copper and silver to form a substitutional alloy.

☀ **Practice Exercise** Would you expect gold (atomic radius 144 pm) to form a substitutional alloy with silver (atomic radius 144 pm)? With copper (atomic radius 128 pm)?

◐◑ **CONNECTION** In Chapter 9 we discussed the valence bond theory of chemical bond formation, as well as combining atomic orbitals into molecular orbitals.

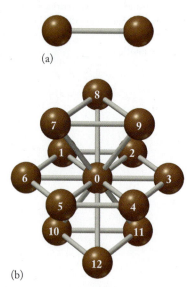

**FIGURE 12.19** Covalent bonds differ from metallic bonds. (a) A ball-and-stick model of the molecule $Cu_2$ is based on the assumption that the two atoms share their $4s$ electrons to form a covalent bond. (b) In solid copper, the atom labeled 0 shares its $4s$ electron with 12 other atoms. As a result, the bonds in copper and other metals are much more diffuse than the covalent bonds in molecules.

## 12.4 Metallic Bonds and Conduction Bands

According to valence bond theory, a covalent bond forms between two atoms when partially filled atomic orbitals—one from each atom—overlap. Our focus in Chapter 9 was on covalent bonding in gas-phase molecules. In this section we explore the bonds that form between the densely packed atoms in metallic solids. Dense packing means that the valence orbitals of atoms overlap with orbitals of many nearby atoms. This large number of interactions makes metals strong. At the same time, sharing a limited number of valence electrons with many bonding partners makes the bond linking any two metal atoms relatively weak.

To understand this point, consider the bond that would result if two Cu atoms came together and formed a molecule of $Cu_2$. Copper atoms have the electron configuration $[Ar]3d^{10}4s^1$. When the partially filled $4s$ orbitals of the two atoms overlap, they form a diatomic molecule held together by a single covalent bond (Figure 12.19a). However, the Cu atoms in solid copper are each surrounded by a total of 12 other Cu atoms and each is bonded to all 12 of its neighbors (Figure 12.19b). This means that each Cu atom must share its $4s$ electron with 12 other atoms, not just one other atom. Inevitably, this dispersion of bonding electrons weakens the Cu—Cu bond between each pair of Cu atoms.

Adding to the diffuse nature of this bonding is the fact that metallic elements have lower electronegativities than nonmetals. Recall from Chapter 8 that differences in electronegativity are important for defining how bonding electrons are distributed between pairs of atoms. Atoms in a pure metal have no such differences in electronegativity. However, the low electronegativities of metals mean that the bonding electrons are not held tightly to the nuclei of individual atoms; they have higher energy and hence are more readily removed.

In Chapter 8 we described metal atoms "floating" in seas of mobile bonding electrons where the electrons are shared by all of the nuclei in the sample. The diffuse nature of metallic bonding described in the preceding paragraphs certainly fits the sea-of-electrons model; however, a more sophisticated approach, called **band theory**, better explains the bonding in metals and other solids. Let's apply band theory, which is an extension of molecular orbital theory, to explain the bonding between the atoms in solid copper.

When the $4s$ atomic orbitals on two Cu atoms overlap to form $Cu_2$, the atomic orbitals combine to form two molecular orbitals with different energies, equally spaced above and below the initial energy value (Figure 12.20a). This is analogous to the formation of low-energy bonding and high-energy antibonding molecular orbitals (Section 9.7). If another two Cu atoms join the first two to form a molecule of $Cu_4$, the $4s$ atomic orbitals of four Cu atoms combine to form four molecular orbitals. If we add another four atoms to make $Cu_8$, a total of eight copper $4s$ atomic orbitals combine to form eight molecular orbitals. In all these molecules the lower energy orbitals are filled with the available $4s$ electrons and the upper orbitals are empty (Figure 12.20b). If we apply this model to the enormous number of atoms in a piece of copper wire, an equally enormous number of molecular orbitals are created. The lower energy half of them are occupied by electrons; the higher energy half are empty. There are so many of these orbitals that they form a continuous *band* of energies with no gap between the occupied lower half and the empty upper half. Because this band of MOs was formed by combining valence-shell orbitals, it is called a **valence band**.

Band theory explains the conductivity of copper and many other metals by assuming that essentially no gap exists between the energy of the occupied lower portion of the valence band and the empty upper portion. Therefore, valence electrons can move easily from the filled lower portion to the empty upper portion, where they are free to move from one empty orbital to the next and thus flow throughout the solid.

The model of a partially filled valence shell explains the conductivity of many metals, but not all. Consider, for example, zinc, copper's neighbor in the periodic table. Its electron configuration, $[Ar]3d^{10}4s^2$, tells us that all its valence-shell electrons reside in filled orbitals, which means the valence band in solid zinc is filled (Figure 12.21). With no empty space in the valence band to accommodate additional electrons, it might seem that the valence-shell electrons in Zn would be immobile, making Zn a poor electrical conductor. However, electrons in the valence band of Zn *do* migrate through the solid; zinc is a good conductor of electricity, and band theory explains why. The theory assumes that *all* atomic orbitals of comparable shape and energy, including the empty $4p$ orbitals on zinc, can combine to form additional energy bands. The energy band produced by combining empty $4p$ orbitals, called a **conduction band**, is also empty and is broad enough to overlap the valence band. This overlap means that electrons from the valence band can move to the conduction band, where they are free to migrate from atom to atom in solid zinc, thereby conducting electricity.

**CONCEPT TEST** ..........................................

Is the electrical conductivity of magnesium metal best explained in terms of overlapping conduction and valence bands or in terms of a partially filled valence band? Explain your answer.

. . . . . . . . . . . . . . . . . . . . . . . . . . . . . . . . . . . . . . . . . .

## 12.5 Semiconductors

To the right of the metals in the periodic table is a "staircase" of elements that tend to have the physical properties of metals and the chemical properties of nonmetals. These semimetals (or metalloids) are not so good at conducting electricity as metals, but they are much better at it than nonmetals. We can use band theory to explain this intermediate behavior. In semimetals, conduction and valence bands do not overlap, but instead are separated by an energy gap. In silicon, the most abundant semimetal, band theory predicts an energy gap, or **band gap ($E_g$)**, of 107 kJ/mol (Figure 12.22a).

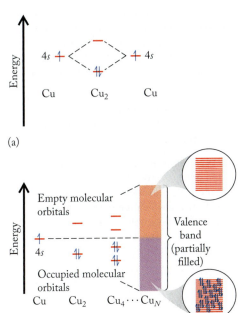

(a)

(b)

**FIGURE 12.20** (a) Two $4s$ atomic orbitals combine to form two molecular orbitals in the molecule $Cu_2$. (b) As the half-filled $4s$ atomic orbitals of an increasing number of Cu atoms overlap, more and more molecular orbitals are formed; half of them are occupied, the other half are empty. As more MOs form, their energies get closer together until a continuous energy band forms—a valence band that is only half-filled with electrons. Electrons can move from the filled half (purple) to the slightly higher energy upper half (orange), where they are free to migrate through delocalized empty orbitals throughout the entire solid.

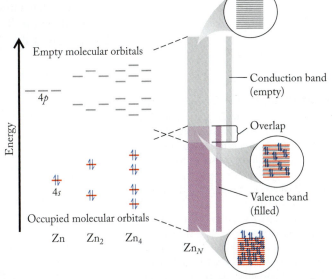

**FIGURE 12.21** As the filled $4s$ atomic orbitals of an increasing number of Zn atoms overlap, they form a filled valence band (purple). An empty conduction band (gray) is produced by combining the empty $4p$ orbitals. The valence and conduction bands overlap each other and electrons move easily from the filled valence band to the empty conduction band.

**semiconductor** a semimetal (metalloid) with electrical conductivity between that of metals and insulators that can be chemically altered to increase its electrical conductivity.

**n-type semiconductor** semiconductor containing electron-rich dopant atoms that contribute excess electrons.

**p-type semiconductor** semiconductor containing electron-poor dopant atoms that cause a reduction in the number of electrons, which is equivalent to the presence of positively charged holes.

Generally, only a few valence-band electrons in Si have sufficient energy to move to the conduction band, which limits silicon's ability to conduct electricity and makes it a **semiconductor**. However, we can enhance the conductivity of solid Si, or of any other elemental semimetal, by replacing some of the Si atoms with atoms of an element of similar atomic radius but with a different number of valence electrons. The replacement process is called *doping*, and the added element is called a *dopant*. Suppose the dopant is a group 15 element such as phosphorus. Each P atom has one more electron than the atom of Si that it replaced. The energy of these additional electrons is different from the energy of the silicon electrons. They populate a narrow band located in the silicon band gap (Figure 12.22b). This arrangement effectively reduces the size of the energy gap and increases electrical conductivity because electrons can move more easily across the remaining gaps between the valence and conduction bands. Phosphorus-doped silicon is an example of an **n-type semiconductor** because the dopant contributes extra negative charges (electrons) to the structure of the host element.

The conductivity of solid silicon can also be enhanced by replacing some Si atoms with atoms of a group 13 element, such as gallium (Figure 12.22c). Because Ga atoms have one fewer valence electron than Si atoms, substituting them in the Si structure means fewer valence electrons in the solid. The result is the creation of a narrow Ga conduction band (*acceptor band*, Figure 12.23a) in the Si band gap. Because the gap between the Si valence band and the acceptor band is smaller than the band gap in pure Si, electrons from the valence band move more easily to the acceptor band, increasing electrical conductivity. This array of bands makes Si doped with Ga a **p-type semiconductor** because a reduction in the number of negatively charged electrons in the valence band is equivalent to the presence of positively charged "holes" (Figure 12.23b). The semiconductors used in solid-state electronics are combinations of n-type and p-type.

Doping is not the only way to change the conductivity of semimetals. Compounds prepared from combinations of group 13 and group 15 elements may also behave as semiconductors. For example, gallium arsenide (GaAs) is a semiconductor that emits infrared radiation ($\lambda = 874$ nm) when connected to an electrical circuit. This emission is used in devices such as bar-code readers and CD players. Like silicon, solid gallium arsenide has both a valence band and

**◉◉ CONNECTION** We introduced the semimetals (metalloids) in Section 2.4 when we described the structure of the periodic table.

**FIGURE 12.22** The electrical conductivity of semiconductors can be greatly enhanced by doping. (a) Pure Si has a band gap $E_g$ of 107 kJ/mol. (b) Adding a few atoms of P, which has more valence electrons than Si, creates an n-type semiconductor with a narrow, filled valence band, called the donor level, $E_d$, 4 kJ/mol below the conduction band of Si. (c) Adding a few atoms of Ga, which has fewer valence electrons than Si, creates an empty conduction band in the band gap of a p-type semiconductor called the acceptor level, $E_a$, 7 kJ/mol above the valence band of Si.

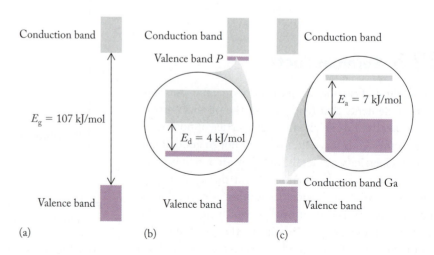

a conduction band separated by a characteristic band gap. The energy of each photon of light corresponds to the energy gap between the valence band and the conduction band. When electrical energy is applied to the material, electrons are raised to the conduction band. When they fall back to the valence band, they emit radiation. If some aluminum is substituted for gallium in GaAs, the band gap increases, and predictably the wavelength of emitted light decreases. For example, a material with the empirical formula $AlGaAs_2$ emits orange-red light ($\lambda = 620$ nm). Many of the multi-colored indicator lights in electronic devices use $AlGaAs_2$ semiconductors.

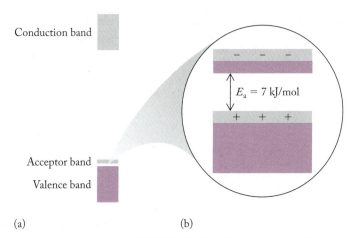

(a)

(b)

**FIGURE 12.23** (a) Band structure of a p-type semiconductor with no electrons in the conduction band of the dopant. The valence band is filled. (b) In a p-type semiconductor, some electrons have enough energy to move from the valence band to the empty conduction band of the dopant, leaving behind positively charged vacancies or "holes." The presence of holes makes the valence band partially filled, increasing electrical conductivity.

**CONNECTION** Emission spectra obtained from gas discharge tubes were discussed in Chapter 7.

---

**SAMPLE EXERCISE 12.3** **Distinguishing p- and n-Type Semiconductors** **LO3**

Which kind of semiconductor—n-type or p-type—does doping germanium with arsenic create?

**Collect and Organize** We are asked to determine the type of semiconductor formed when germanium (Ge, group 14) is doped with arsenic (As, group 15). Figure 12.22 helps us distinguish between n- and p-type semiconductors.

**Analyze** When the dopant has more valence electrons than the host semimetal, the two form an n-type semiconductor, as in Figure 12.22(b). If the dopant has fewer valence electrons than the host, the result is a p-type semiconductor (Figure 12.22c).

**Solve** Arsenic is in group 15, which means that its atoms have one more valence electron than the atoms of Ge, a group 14 element. This makes As-doped Ge an n-type semiconductor.

**Think About It** Arsenic is a good candidate for a dopant to make an n-type semiconductor with germanium because As atoms are nearly the same size as Ge atoms and fit easily into the structure of solid Ge.

**Practice Exercise** Gallium arsenide (GaAs) is a semiconductor used in optical scanners in retail stores. GaAs can be made an n-type or a p-type semiconductor by replacing some of the As atoms with another element. Which element—Se or Sn—would form an n-type semiconductor with GaAs?

---

**FIGURE 12.24** (a) Sodium chloride forms cubic crystals that reach various sizes. (b) A NaCl crystal lattice is based on cubic closest-packing: an fcc unit cell made of $Cl^-$ ions, showing the $Na^+$ ions in the octahedral holes. (c) Cutaway view of the unit cell in part b.

# 12.6 Salt Crystals: Ionic Solids

Most of Earth's crust is composed of ionic solids, consisting of monatomic or polyatomic ions held together by ionic bonds. Most of these solids are crystalline. The simplest crystal structures are those of binary salts, such as NaCl (Figure 12.24). The cubic shape of large NaCl crystals is due to the cubic shape of the NaCl unit cell. We can describe the unit cell of NaCl (Figure 12.24b,c) as a face-centered cubic arrangement of $Cl^-$ ions at the corners and in the center of each face, with the smaller $Na^+$ ions occupying the twelve octahedral holes along the edges of the unit cell and the single octahedral hole in the middle of the cell. The ordered arrangement in Figure 12.24 maximizes the coulombic forces of attraction

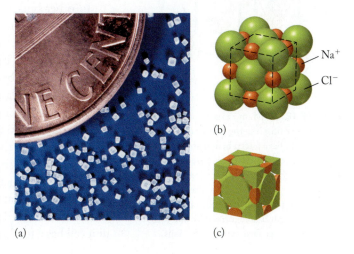

(a)

(b)

(c)

**CONNECTION** Coulombic forces and potential energy in ionic compounds were discussed in Section 8.1.

between oppositely charged particles and minimizes the repulsions between similarly charged ions. As a result, the potential energy of the crystal is minimized.

In Section 12.2 we used atomic radius ratios to determine the location of atoms in the interstices of a crystal lattice. The same ranges for radius ratios that applied to interstitial alloys guide us in understanding the structures of binary ionic compounds. Since all of the radius ratios in Table 12.4 are ≤1, it makes sense to focus on the unit cell of NaCl in Figure 12.24(b), where the smaller $Na^+$ ions occupy the interstices in the cubic closest-packed lattice of $Cl^-$ ions. The $Na^+$ ions fit better into the octahedral holes because the radius ratio of $Na^+$ to $Cl^-$ is

$$\frac{r_+}{r_-} = \frac{102 \text{ pm}}{181 \text{ pm}} = 0.564$$

The radius ratio 0.564 is too large for $Na^+$ to occupy a tetrahedral hole but well within the range for occupying an octahedral hole.

Let's take an inventory of the ions in a unit cell of NaCl. Like the metal atoms in the fcc unit cell in Figure 12.10(b), an isolated unit cell contains portions of 14 $Cl^-$ ions: one at each corner and one in each of the six faces of the cube. As in the analysis of Figure 12.10, accounting for partial ions gives us a total of four $Cl^-$ ions in the unit cell in Figure 12.24(c). To count the $Na^+$ ions, note that one $Na^+$ ion occupies the central octahedral hole, and one $Na^+$ ion fits into each of the 12 octahedral holes along the edges of the cell. Because each $Na^+$ ion along an edge is shared by four unit cells, only one-fourth of each $Na^+$ ion on an edge is in each cell (see Figure 12.11d). Only the $Na^+$ in the center belongs completely to the unit cell. Therefore, the total number of $Na^+$ ions in the unit cell is

$$\left(12 \times \tfrac{1}{4}\right) + 1 = 4 \text{ Na}^+ \text{ ions}$$

The ratio of $Na^+$ to $Cl^-$ ions in the unit cell is therefore 4:4, consistent with the chemical formula NaCl. Because the four $Na^+$ ions occupy all the octahedral holes in the unit cell, the result of this calculation also means that each fcc unit cell contains the equivalent of four octahedral holes.

Note in Figure 12.24 that adjacent $Cl^-$ ions along any face diagonal do not touch each other the way they do in Figure 12.10(b), because the $Cl^-$ ions have to spread out a little to accommodate the $Na^+$ ions in the octahedral holes. Sodium ions and chloride ions touch along each edge of the unit cell, however, which means that each $Na^+$ ion touches six $Cl^-$ ions and each $Cl^-$ ion touches six $Na^+$ ions. This arrangement of positive and negative ions is common enough among binary ionic compounds to be assigned its own name: the *rock salt structure*.

In other binary ionic solids, the smaller ion is small enough to fit into the tetrahedral holes formed by the larger ions. For example, in the unit cell of the mineral sphalerite (zinc sulfide), the $S^{2-}$ anions (ionic radius 184 pm) are arranged in an fcc unit cell (Figure 12.25), and half of the eight tetrahedral holes inside the cell are occupied by $Zn^{2+}$ cations (74 pm). Therefore the unit cell contains four $Zn^{2+}$ ions that balance the charges on the four $S^{2-}$ ions. This pattern of half-filled tetrahedral holes in an fcc unit cell is sometimes called the *sphalerite structure*. Compounds containing zinc, cadmium, or mercury cations with a 2+ charge and heavier anions of the group 16 elements with a 2− charge, ($S^{2-}$, $Se^{2-}$, and $Te^{2-}$) also have a sphalerite structure.

The crystal structure of the mineral fluorite ($CaF_2$) is based on an fcc unit cell of smaller $Ca^{2+}$ ions at the eight cube corners and six face centers, with all eight tetrahedral holes formed by neighboring $Ca^{2+}$ ions filled by larger $F^-$ ions. Because the unit cell has a total of four $Ca^{2+}$ ions and the eight $F^-$ ions are all completely inside

**FIGURE 12.25** Many crystals of the mineral sphalerite (ZnS), like the largest ones in this photograph, have a tetrahedral shape. The crystal lattice of sphalerite is based on an fcc unit cell of $S^{2-}$ ions with $Zn^{2+}$ ions in four of the eight tetrahedral holes. In the expanded view of the sphalerite unit cell, each $Zn^{2+}$ ion is in a tetrahedral hole formed by one corner $S^{2-}$ ion and three face-centered $S^{2-}$ ions.

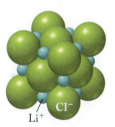

**FIGURE 12.26** The mineral fluorite ($CaF_2$) forms cubic crystals. The crystal lattice of $CaF_2$ is based on an fcc array of $Ca^{2+}$ ions, with $F^-$ ions occupying all eight tetrahedral holes. Because they are bigger than $Ca^{2+}$ ions, the $F^-$ ions do not fit in the tetrahedral holes of a cubic closest-packed array of $Ca^{2+}$ ions. Instead, the $Ca^{2+}$ ions, while maintaining the fcc unit cell arrangement, spread out to accommodate the larger $F^-$ ions. Note how adjacent $Ca^{2+}$ ions along any face diagonal do not touch each other the way they do in the ideal fcc unit cell in Figure 12.10(b).

the cell, this arrangement satisfies the 1:2 mole ratio of $Ca^{2+}$ ions to $F^-$ ions. This structure is so common that it too has its own name: the *fluorite structure* (Figure 12.26). Other compounds having this structure are $SrF_2$, $BaCl_2$, and $PbF_2$.

Some compounds in which the cation-to-anion mole ratio is 2:1 have an *antifluorite structure*. In the crystal lattices of these compounds, which include $Li_2O$ and $K_2S$, the smaller cations occupy the tetrahedral holes in an fcc unit cell formed by cubic closest-packing of the larger anions.

**CONNECTION** Periodic trends in atomic radii were discussed in Chapter 7.

---

**SAMPLE EXERCISE 12.4**    **Calculating an Ionic Radius and Density from a Unit Cell Dimension**    **LO4**

The unit cell of lithium chloride (LiCl) contains an fcc arrangement of $Cl^-$ ions (Figure 12.27). In LiCl, the $Li^+$ cations (radius 76 pm) are small enough to allow adjacent $Cl^-$ ions to touch along any face diagonal.

a. If the edge length of the LiCl fcc cell is 513 pm, what is the radius of the $Cl^-$ ion?
b. Use that value to predict the type of hole the $Li^+$ ion occupies.
c. What is the density of LiCl at 25°C?

**Collect and Organize** We are given the edge length (513 pm), the radius of the $Li^+$ cation (76 pm), and the type of unit cell for LiCl (fcc). We are asked to calculate the ionic radius of the $Cl^-$ ion, predict in which type of hole the $Li^+$ is found, and calculate the density of LiCl.

**Analyze** (a) The unit cell is an fcc array of $Cl^-$ ions that touch along the face diagonal. Equation 12.2 relates the edge length ($\ell$) of an fcc cell to the radius ($r$) of the atoms or ions that touch along its diagonals: $r = 0.3536\ell$. (b) Once we know the radius of the $Cl^-$ ion, the radius ratio of $Li^+$ to $Cl^-$ ions and Table 12.4 allow us to predict which type of hole $Li^+$ occupies. (c) The fact that LiCl has an fcc unit cell means that there are four $Cl^-$ ions and four $Li^+$ ions in the cell. The molar masses of these ions are 35.45 and 6.941 g/mol, respectively. As in Sample Exercise 12.1, we need to convert from molar masses to the masses of individual particles by dividing by Avogadro's number. Density is the ratio of mass to volume, and the volume of a cubic cell is the cube of its edge length: $V = \ell^3$.

**Solve**
a. Substituting the edge length into Equation 12.2, we calculate the radius of $Cl^-$:

$$r = 0.3536 \times 513 \text{ pm} = 181 \text{ pm}$$

b. The ratio of the radius of a $Li^+$ ion to the radius of a $Cl^-$ ion is

$$76 \text{ pm}/181 \text{ pm} = 0.42$$

According to Table 12.4, the $Li^+$ cations occupy octahedral holes in the lattice formed by the larger $Cl^-$ anions.

**FIGURE 12.27**

c. Calculating the mass of four $Cl^-$ ions:

$$m = \frac{35.45 \text{ g } Cl^-}{1 \text{ mol } Cl^-} \times \frac{1 \text{ mol } Cl^-}{6.022 \times 10^{23} \text{ ions } Cl^-} \times 4 \text{ ions } Cl^-$$

$$= 2.355 \times 10^{-22} \text{ g } Cl^-$$

The mass of four $Li^+$ ions is

$$m = \frac{6.941 \text{ g } Li^+}{1 \text{ mol } Li^+} \times \frac{1 \text{ mol } Li^+}{6.022 \times 10^{23} \text{ ions } Li^+} \times 4 \text{ ions } Li^+$$

$$= 0.4610 \times 10^{-22} \text{ g } Li^+$$

Combining the masses of the two kinds of ions in the unit cell:

$$2.355 \times 10^{-22} \text{ g} + 0.4610 \times 10^{-22} \text{ g} = 2.816 \times 10^{-22} \text{ g}$$

The volume of the cell in cubic centimeters is

$$V = \ell^3 = (513 \text{ pm})^3 \times \frac{(10^{-10} \text{ cm})^3}{(1 \text{ pm})^3} = 1.35 \times 10^{-22} \text{ cm}^3$$

Taking the ratio of mass to volume, we have

$$d = \frac{m}{V} = \frac{2.816 \times 10^{-22} \text{ g}}{1.35 \times 10^{-22} \text{ cm}^3} = 2.09 \text{ g/cm}^3$$

**Think About It** The average ionic radius value in Figure 10.2 for $Cl^-$ ions is 181 pm, so our calculation is correct. Figure 12.27 shows the structure of LiCl to be similar to the rock salt structure of NaCl in Figure 12.24(b). The calculated density for LiCl is consistent with the observation that most minerals are more dense than water but less dense than common metals.

**Practice Exercise** Assuming the $Cl^-$ radius in NaCl is also 181 pm, what is the radius of the $Na^+$ ion in NaCl if the edge length of the NaCl unit cell is 564 pm but the anions do not touch along any face diagonal? What is the density of NaCl?

## 12.7 Structures of Nonmetals

▶❙❙ CHEMTOUR Allotropes of Carbon

In Figure 12.7, the color key identifies the unit cells of silicon, germanium, and tin as "diamond." We introduced the structure of diamond in the descriptive chemistry feature at the end of Chapter 5, where we described the allotropes of carbon. Carbon, of course, is a nonmetal, but the specific arrangement of carbon atoms in diamond (Figure 12.28a) is also found in some metallic materials. In addition, some diamonds may include metal impurities in the crystal lattice that give rise to distinctive colors.

Diamond is one of three allotropes of carbon, the other two being graphite and fullerenes (Figures 12.28b and 12.29). Diamond is classified as a crystalline **covalent network solid** because it consists of atoms held together in an extended three-dimensional network of covalent bonds. Each carbon atom in diamond bonds by overlapping one of its $sp^3$ hybrid orbitals with an $sp^3$ hybrid orbital in each of four neighboring carbon atoms, creating a network of carbon tetrahedra. The atoms in these tetrahedra are connected by localized σ bonds, making diamond a poor electrical conductor. The sigma-bond network is extremely rigid, making diamond the hardest natural material known. The atoms of other group 14 elements, particularly silicon, germanium, and tin, also form covalent network solids based on the diamond crystal lattice.

Natural diamond forms from graphite under intense heat (>1700 K) and pressure (>50,000 atm) deep in Earth. Industrial diamonds are synthesized at high

**covalent network solid** a solid consisting of atoms held together by extended arrays of covalent bonds.

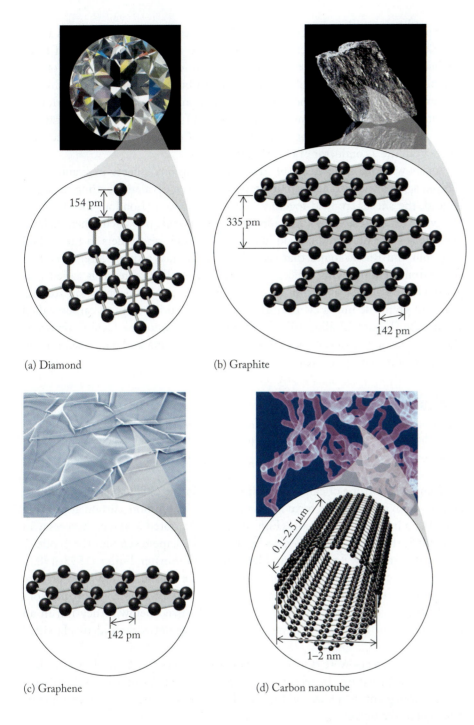

(a) Diamond

(b) Graphite

(c) Graphene

(d) Carbon nanotube

**FIGURE 12.28** Two of carbon's three allotropes. (a) Diamond is a three-dimensional covalent network solid made of carbon atoms, each connected by σ bonds to four adjacent carbon atoms. (b) Graphite is a collection of layers of carbon atoms connected by σ bonds and delocalized π bonds. (c) Graphene is a single layer of C atoms from graphite. (d) Carbon nanotube.

temperatures and pressures from graphite or any other source rich in carbon. Synthetic diamonds are used as abrasives and for coating the tips and edges of cutting tools. Diamond has the highest thermal conductivity of any natural substance (five times higher than copper and silver, the most thermally conductive metals), so tools made from diamond do not become overheated. The conditions used to manufacture industrial diamonds mostly yield stones that lack the size and optical clarity of gemstones. The inclusion of impurities in natural diamonds leads to rare and valuable colored diamonds, a process that is difficult to duplicate under laboratory conditions.

By far the most abundant allotrope of carbon is graphite, another covalent network solid, frequently the principal ingredient in soot and smoke and used to

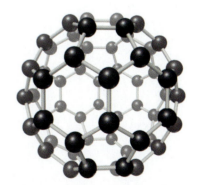

**FIGURE 12.29** Some solids are described as clusters, a category between covalent network solids and molecular solids. Fullerenes, the third allotrope of carbon, are clusters. One of them, buckminsterfullerene ($C_{60}$), is made up of 60 $sp^2$-hybridized carbon atoms. Both five- and six-membered rings are required to construct the nearly spherical molecule.

**CONNECTION** Allotropes are structurally different forms of the same physical state of an element, as explained in Chapter 8.

**CONNECTION** We introduced dispersion forces between molecules in Chapter 10.

make pencils, lubricants, and gunpowder. Graphite contains sheets of carbon atoms in which each atom is connected by $sp^2$ orbitals to a like orbital in each of three neighboring carbon atoms, forming a two-dimensional covalent network of six-membered rings (Figure 12.28b). Each carbon–carbon σ bond is 142 pm, which is shorter than the C—C σ bond in diamond (154 pm). Overlapping unhybridized $p$ orbitals on the carbon atoms form a network of π bonds that are delocalized across the plane defined by the rings. The mobility of these delocalized electrons makes graphite a conductor of electricity.

As shown in Figure 12.28(b), the two-dimensional sheets in graphite are 335 pm apart. This distance is much too long to be a covalent bonding distance. Instead, the sheets are held together only by dispersion (London) forces. These relatively weak interactions allow adjacent sheets to slide past each other, making graphite soft, flexible, and a good lubricant. In 2004, researchers in the United Kingdom and Russia found that they could isolate just a single layer of carbon atoms from graphite with a piece of adhesive tape. This material is called graphene (Figure 12.28c and front cover) and behaves as a semiconductor. Rolling a layer of graphene into a cylindrical shape yields another form of carbon called a nanotube (Figure 12.28d). Like graphene, carbon nanotubes exhibit semiconductor properties; however, they are also extremely strong and have been used to strengthen fibers in sports gear.

**CONCEPT TEST**

The diamond form of carbon is a semiconductor with a much larger band gap than silicon. With what elements might you choose to dope diamonds to form an n-type semiconductor?

Another allotrope of carbon was discovered in the 1980s. Networks of five- and six-atom carbon rings form molecules of 60, 70, or more carbon atoms that look like miniature soccer balls (Figure 12.29). As noted in the carbon essay in Chapter 5, they are called *fullerenes* because their shape resembles the geodesic domes designed by American architect R. Buckminster Fuller (1895–1983). Many chemists call them *buckyballs* for the same reason. Based both on size and properties, buckyballs are too small to be classified as covalent network solids but too large to be molecular solids (discussed below). They fall in an ambiguous zone between small molecules and large networks and are classified as *clusters*.

When fullerenes were discovered, they were believed to be a form of carbon rarely found in nature. In recent years, however, analyses of soot and emission spectra from giant stars have disclosed that fullerenes are present in trace amounts throughout the universe.

# 12.8 Ceramics: Insulators to Superconductors

**ceramic** a solid inorganic compound or mixture that has been transformed into a harder, more heat-resistant material by heating.

A **ceramic** is a solid inorganic compound or mixture of compounds that has been heated to transform it into a harder and more heat-resistant material. The use of ceramic materials preceded metal technology by many thousands of years. The first ceramics were probably made of *clay*, the fine-grained soil produced by the physical and chemical weathering of igneous rocks (rocks of volcanic origin). Moist clay is easily molded into a desired shape and then hardened over fires or in wood-burning kilns, an ancient process still in use today.

In this section we examine some of the physical properties of both primitive earthenware (ceramics fired at low temperatures) and modern ceramic materials (typically fired at high temperatures). We also relate those properties to the chemical compositions of the materials and to the chemical changes that occur when they are heated.

## Polymorphs of Silica

One of the most abundant families of minerals found in igneous rocks has the chemical composition $SiO_2$. The correct chemical name is silicon dioxide, but the more common name is *silica*. Silica is a covalent network solid in which each silicon atom is covalently bonded to four oxygen atoms, forming a tetrahedron with an oxygen atom at each corner and the silicon atom at the center (Figure 12.30). Each oxygen atom is covalently bonded to two silicon atoms, thereby linking the tetrahedra into an extended three-dimensional network. Because each corner oxygen atom is bonded to two silicon atoms, each silicon atom gets only half "ownership" of the four oxygen atoms to which it is covalently bonded—hence the formula $SiO_2$.

At least eight different minerals have the empirical formula $SiO_2$. The members of a family of substances with the same empirical formula but different crystal structures and properties are called *polymorphs*. The most abundant silica polymorph is quartz, a type of $SiO_2$ that can form impressively large, nearly transparent crystals (Figure 12.30). Note how the hexagonal ordering of the $SiO_2$ tetrahedra translates into hexagonal crystals.

Most, but not all, silica polymorphs are crystalline. When lava containing molten $SiO_2$ flows from a volcano into the sea or a lake, it cools so quickly that the Si and O atoms may not have enough time to achieve an ordered crystal lattice as the lava solidifies. The solid formed in this way is an *amorphous* (disordered, noncrystalline) polymorph of silica known as either volcanic glass or obsidian (Figure 12.31). *Glass* is a term scientists and engineers use to describe any solid that has either no crystalline structure or only very tiny crystals surrounded by disordered arrays of atoms. This definition applies to laboratory glassware and the drinking glasses we use at home, as well as to more exotic solids such as obsidian.

## Ionic Silicates

In addition to covalent silica, igneous rocks also contain ionic minerals made of silicon and oxygen. These minerals have some of the tetrahedral crystal structure of silica, but not all the oxygen corner atoms are bonded to two Si atoms. Instead, some of the O atoms have an extra electron. The result is a *silicate* anion. One of the common ionic silicates is chrysotile, a type of asbestos that consists of sheets of linked silicon–oxygen tetrahedra that form hexagonal clusters of six tetrahedra each (Figure 12.32). Each tetrahedron has three O atoms that it shares with other tetrahedra and one O atom—the one with the extra electron—that it does not share. Thus the basic tetrahedral unit consists of one Si atom and $[1 + 3(\frac{1}{2})] = 2.5$ oxygen atoms, as well as a negative charge. This gives the sheet the empirical formula $SiO_{2.5}^-$. We generally use whole-number subscripts in chemical formulas when possible. In this case, multiplying $SiO_{2.5}^-$ by 2 gives us the empirical formula $Si_2O_5^{2-}$ for this silicate anion. The subscript $n$ in the formula in Figure 12.32 indicates the presence of many empirical formula units in a single crystal.

Silicate minerals are neutral materials, so they must contain cations to balance the negative charges on the silicate layers. When the cation is $Al^{3+}$, the minerals

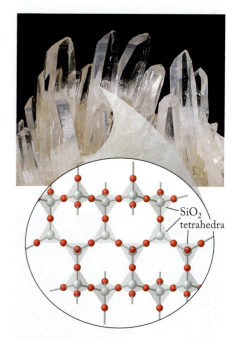

**FIGURE 12.30** Quartz crystals are hexagonal. Their crystal structures feature hexagonal arrays of silicon–oxygen tetrahedra that form an extended three-dimensional network. (All atoms are drawn undersized relative to the volume they actually occupy to make the structure easier to see.)

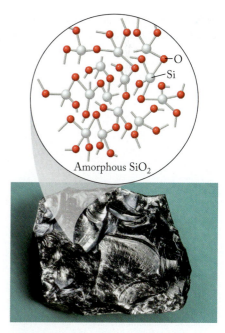

**FIGURE 12.31** Obsidian (volcanic glass) is an unusual form of silica in that it is not crystalline. Obsidian contains mostly amorphous silica with random arrangements of silicon and oxygen atoms.

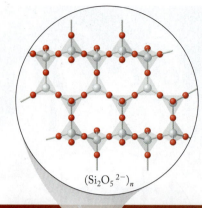

$(Si_2O_5{}^{2-})_n$

**FIGURE 12.32** Chrysotile, one of the two principal forms of asbestos formerly used in building construction as thermal insulation, is an ionic silicate compound. The ease with which thin fibers of chrysotile can flake off is related to its layered crystal lattice and the relatively weak intermolecular interactions between layers. The O atoms that are bonded to only one Si atom have an extra electron and a negative charge.

**CONNECTION** Ion-exchange reactions were discussed in Chapter 4.

are called *aluminosilicates*. One of the most common aluminosilicates is the clay mineral kaolinite (Figure 12.33). At least a little kaolinite is found in practically every soil, but rich deposits of nearly pure, brilliantly white kaolinite are found in highly weathered soils. For centuries these deposits have been mined to make fine china and white porcelain. Today the greatest demand for kaolinite is in the production of glossy white paper, which is used in most magazines and books (including this one).

Common metal ions found in igneous rocks—including $Na^+$, $K^+$, $Ca^{2+}$, $Mg^{2+}$, and $Fe^{3+}$—are largely absent in kaolinite. Their absence indicates that kaolinite deposits form under acidic weathering conditions. Under these conditions, $H^+$ ions displace other cations from ion-exchange sites. For example, $-O^-Na^+$ sites exchange $H^+$ for $Na^+$, leaving behind $-OH$ groups such as those shown in Figure 12.33.

The strong ionic interactions and hydrogen bonds in kaolinite make it hard to separate its layers. This means that water molecules cannot penetrate between kaolinite layers. For thousands of years this property has made kaolinite pots handy vessels for carrying water. Because water cannot penetrate between its layers, kaolinite does not expand when water is added. For the same reason, it does not shrink as much as most other clays when dehydrated at high temperatures. This is another property that has made kaolinite a desirable starting material for ceramics. Finally, moist kaolinite is *plastic*, meaning that it can be molded into a shape, and it keeps that shape during heating and cooling.

**CONCEPT TEST** ••••••••••••••••••••••••••••••••••••••••••

Magnesium ion, $Mg^{2+}$, can substitute for $Al^{3+}$ in kaolinite. What is the formula of the mineral obtained—that is, the values of $x$ and $y$ in $Mg_xAl_y(Si_2O_5)(OH)_4$ if $x = y$.

••••••••••••••••••••••••••••••••••••••••••••••••••••••••••••

## From Clay to Ceramic

Creating ceramic objects from kaolinite and other clays takes several steps. First, moist clay is formed into pots, bricks, and other objects on a potter's wheel or in molds or presses. Heating up to 1000°C leads to loss of water and structural changes in the clay (Equation 12.4). The result of the heat treatment is the formation of a heterogeneous mixture of $Al_6Si_2O_{13}$ (called mullite) and silica.

$$3\ Al_2Si_2O_5(OH)_4(s) \xrightarrow{100-1000°C} 6\ H_2O(g) + Al_6Si_2O_{13}(s) + 4\ SiO_2(s) \quad (12.4)$$

The formula of mullite is sometimes written $3\ Al_2O_3 \cdot 2\ SiO_2$ indicating that it is a blend of an $Al_2O_3$ (alumina) crystal structure and a $SiO_2$ (silica) crystal structure. Mullite is a ceramic material widely used in the manufacture of products that must tolerate temperatures as high as 1700°C: furnaces, boilers, ladles, and kilns. These products are used as containers of molten metals and in the glass, chemical, and cement industries. Mullite is also very hard and is widely used as an abrasive.

**CHEMTOUR** Superconductors

**critical temperature ($T_c$)** the temperature below which a material becomes a superconductor.

**superconductor** a material that has zero resistance to the flow of electric current.

## Superconductors

Band theory can be used to describe the properties of, and bonding in, all solids, including the compounds in ceramic materials. Earlier in the chapter we noted that metals are good conductors because of the ease with which their valence electrons can move to conduction bands, whereas semimetals are semiconductors because they have significant energy gaps between their valence and conduction bands (though doping can shrink the gaps). Similarly, the insulating properties

$Al_2(Si_2O_5)(OH)_4$

**FIGURE 12.33** This enormous kaolinite mine in Bulgaria contributes to a worldwide production of about 34 million metric tons of the mineral per year. Kaolinite is used in many products, from ceramics to glossy white paper. The edge-on view of the structure of kaolinite shows a top layer of $OH^-$ ions bonded to a middle layer of $Al^{3+}$ ions followed by a bottom layer of silicate ions ($Si_2O_5^{2-}$). The empirical formula of this crystal lattice is $Al_2(Si_2O_5)(OH)_4$.

of ceramics can be explained by the large energy gaps between their valence and conduction bands. As a result, their valence electrons have very limited mobility.

In conventional metallic conductors the vibration of the metal atoms in the crystal lattice can interfere with the flow of free electrons: the higher the temperature, the greater the atomic vibration and the greater the resistance to electron flow. However, at temperatures approaching absolute zero these vibrations become very small. In the early 20th century, scientists discovered that the lattice vibrations in mercury and some metal alloys are so small at temperatures below about 20 K, called the **critical temperature ($T_c$)** for each substance, that electrons can pass freely through these materials which become **superconductors**. Today superconducting alloys, such as $Nb_3Sn$, are widely used in devices that require very high electrical currents, such as the electromagnets of MRI (magnetic resonance imaging) instruments.

Unfortunately it is difficult and expensive to chill materials to 20 K. In 1986 it was discovered that $YBa_2Cu_3O_7$ and several other ceramic materials become superconductors when cooled to temperatures just above the boiling point of liquid nitrogen (77 K). This represented an important economical advance because it is much easier and cheaper to cool a material with liquid nitrogen than with liquid helium (boiling point = 4 K), which must be used to achieve superconductivity in Hg and metal alloys. Since the 1986 discovery, scientists have produced superconducting ceramic materials with critical temperatures as high as 133 K.

Let's take a closer look at the crystal structure of $YBa_2Cu_3O_7$, which has the unit cell shown in Figure 12.34. The unit cell is a stack of three cubes with a $Y^{3+}$ ion in the center of the middle cube, $Ba^{2+}$ ions in the center of the top and bottom cubes, 16 copper ions at the corners of each cube, and 20 $O^{2-}$ ions along the edges of the three stacked cubes.

Applying our usual practice of assigning fractions of atoms and ions to unit cells, we find that the stack in Figure 12.34 has 7 oxide ions ($\frac{1}{4}$ of the 12 that occupy edges of the top and bottom cubes in the stacked array plus $\frac{1}{2}$ of the remaining 8 ions that occupy faces—7 in all) and 3 copper ions ($\frac{1}{8}$ of the 8 ions on the top and bottom surfaces plus $\frac{1}{4}$ of the 8 ions around the

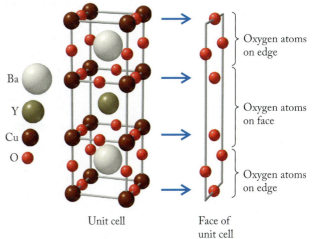

Ba
Y
Cu
O

Unit cell        Face of unit cell

Oxygen atoms on edge

Oxygen atoms on face

Oxygen atoms on edge

**FIGURE 12.34** The unit cell of $YBa_2Cu_3O_7$, a high-temperature superconductor, contains a central barium ion in the top and bottom cubes, a central yttrium ion in the middle cube, and oxide ions on the edges of the unit cell.

**FIGURE 12.35** The Meissner effect. A small rare earth magnet floats above a superconductive material that has been chilled below its critical temperature. The magnetic field produced by the magnet (foreground) cannot penetrate the magnetic field of the superconductor. As a result, the superconductor repels the magnet and the magnet floats above the superconductor.

▶‖ **CHEMTOUR** X-ray Diffraction

◉⟶ **CONNECTION** We discussed the constructive and destructive interference of waves in Chapter 7.

middle—3 in all). The two $Ba^{2+}$ ions and the $Y^{3+}$ ion lie entirely within the unit cell. Thus, the material with the crystal structure shown in Figure 12.34 has the chemical formula $YBa_2Cu_3O_7$.

Superconductors are of technological interest because, in principle, large amounts of electric charge can flow through them with no resistance. This property is attractive for the design of extremely fast computers. Superconductors also have the ability to exclude magnetic fields—a property known as the *Meissner effect* after German physicist Walther Meissner (1882–1974). Figure 12.35 shows a small magnet floating above a superconducting material cooled below its critical temperature because its magnetic lines of force are excluded from the superconductor, suspending it in air. This phenomenon, also called magnetic levitation, or maglev, has been used to build trains with superconducting undercarriages that float above electromagnetic tracks and guidance systems. The first high-speed maglev train went into commercial service near Shanghai, China in 2004. Its operational top speed is 431 km/h (268 mph), making it the fastest passenger train in the world.

## 12.9 X-ray Diffraction: How We Know Crystal Structures

Unit cell dimensions are determined using **X-ray diffraction (XRD)**. X-rays are well suited for the task of crystal structure determination because the wavelengths of X-rays ($10^2$ to $10^3$ pm) are similar to the radii of atoms and ions and, therefore, the distances between them in crystals. To see how XRD works we will look at atoms in a crystalline metal, but keep in mind that our description also applies to the particles in any crystalline solid.

A narrow beam of X-rays is directed at some angle of incidence $\theta$ at the surface of the metal, as shown in Figure 12.36(a), and is absorbed and reemitted by the layers of particles in the solid. Some of the X-rays collide with atoms in the surface layer, and some of those that do so are reradiated by atoms at an angle equal to the angle of incidence $\theta$. For these X-rays the total change in their direction is the sum of the two angles, or $2\theta$, which is called their *angle of diffraction*.

Some of the incident X-rays pass through the surface layer and hit atoms in a second layer a distance $d$ below the surface. For any X-ray reaching this layer, the angle of incidence and angle of reflection are again both $\theta$, and this ray is also diffracted through an angle $2\theta$. The two reflected rays can interfere with each other either constructively or destructively. They interfere constructively when they are in phase, as in Figure 12.36(b). To be in phase, the extra distance traveled by the X-ray reflecting off the second layer must be some whole-number multiple of the wavelength of the X-rays. This extra distance traveled is the sum of the lengths of line segments $\overline{XY}$ and $\overline{YZ}$ in Figure 12.36(b). The two right triangles incorporating these line segments share a hypotenuse $d$. Geometry tells us that the angles opposite $\overline{XY}$ and $\overline{YZ}$ are both equal to $\theta$. According to trigonometry, the ratio of either $\overline{XY}$ or $\overline{YZ}$ to $d$ is

$$\frac{\overline{XY}}{d} = \sin\theta \qquad \frac{\overline{YZ}}{d} = \sin\theta$$

which means

$$d\sin\theta = \overline{XY} \qquad d\sin\theta = \overline{YZ} \qquad (12.5)$$

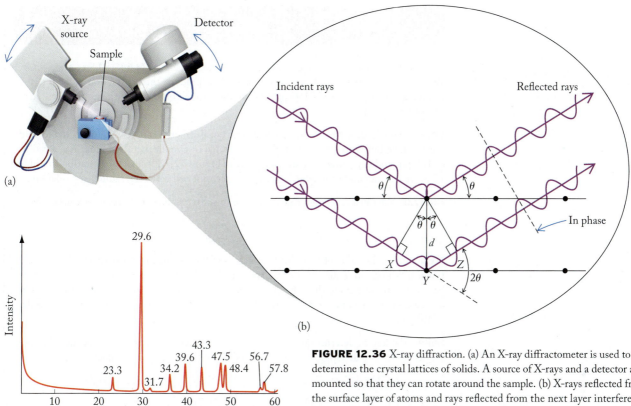

(a)

(c)

**FIGURE 12.36** X-ray diffraction. (a) An X-ray diffractometer is used to determine the crystal lattices of solids. A source of X-rays and a detector are mounted so that they can rotate around the sample. (b) X-rays reflected from the surface layer of atoms and rays reflected from the next layer interfere constructively when they are in phase, as they are at the distance *d* shown here. The angle between the incident and reflected X-rays is 2θ. (c) Moving the source and detector around the sample produces a scan such as this one for quartz. The peaks at different values of 2θ represent reflections from different planes of atoms in the solid.

As noted above, we are interested in $\overline{XY} + \overline{YZ}$, the extra distance traveled by the second ray. We therefore add Equations 12.5 to get

$$\overline{XY} + \overline{YZ} = d \sin \theta + d \sin \theta = 2d \sin \theta$$

When this extra distance $\overline{XY} + \overline{YZ}$ equals a whole-number multiple (*n*) of the wavelength (λ) of the X-rays, we have

$$\overline{XY} + \overline{YZ} = n\lambda \qquad \text{or}$$

$$n\lambda = 2d \sin \theta \qquad (12.6)$$

Equation 12.6 is called the **Bragg equation** after William Henry Bragg (1862–1942) and his son William Lawrence Bragg (1890–1971), who discovered the relationship between wavelength and the spacings between layers in crystalline solids. Whenever $2d \sin \theta$ equals $n\lambda$, the crests and troughs of the two X-rays are in phase as they emerge from the metal and so interfere constructively.

To measure this interference, the X-ray source is rotated through a range of incident angles θ, and the intensity of the reflected rays is plotted as a function of 2θ (2θ rather than θ because the angle between the diffracted and undiffracted beam is 2θ, as shown in Figure 12.36). Peaks in the intensity of scattered X-rays (Figure 12.36c) occur at angles of diffraction that satisfy Equation 12.6. From these angles, and knowing the wavelength (λ) of the X-rays, scientists can calculate the distance (*d*) between the layers of particles in a crystalline sample, and its type of lattice structure.

**X-ray diffraction (XRD)** a technique for determining the arrangement of atoms or ions in a crystal by analyzing the pattern that results when X-rays are scattered after bombarding the crystal.

**Bragg equation** relates the angle of diffraction (2θ) of X-rays to the spacing (*d*) between the layers of ions or atoms in a crystal: $n\lambda = 2d \sin \theta$.

An XRD analysis of a sample of copper has peaks at $2\theta = 24.64°$, $50.54°$, and $79.62°$. What distance $d$ between layers of Cu atoms can produce this diffraction pattern if the wavelength of the X-rays is 154 pm? Is the distance $d$ consistent with the unit cell edge length $\ell$ of Cu metal?

**Collect and Organize** We are given a series of diffraction angles $2\theta$ and asked to find the distance between layers of atoms in a sample of copper metal. Copper has a radius of 128 pm and crystallizes in an fcc unit cell (Figure 12.7). Using the equation in Table 12.3, we can calculate the edge length of an fcc unit cell from the radius: $r = 0.3536\ell$.

**Analyze** We must rearrange Equation 12.6 to solve for the parameter we are after, $d$:

$$d = \frac{n\lambda}{2 \sin \theta}$$

We are given the value of $\lambda$ and can get values of $\theta$ from the given values of $2\theta$. The additional unknown in Equation 12.6 is the wavelength multiplier $n$, which we must know before we can determine the value of $d$. Since we know that the radius of a copper atom is on the order of $10^2$ pm, we expect the distance between the layers to also be on the order of $10^2$ pm.

**Solve** The key to determining $n$ is to look for a pattern in the $\theta$ values. Notice that the higher values of $\theta$ are approximately two and three times the lowest value:

$$24.64°/2 = 12.32° \qquad 50.54°/2 = 25.27° \qquad 79.62°/2 = 39.81°$$

$$\frac{25.27°}{12.32°} = 2.051 \qquad \frac{39.81°}{12.32°} = 3.231$$

This pattern suggests that the values of $n$ for this set of data are 1, 2, and 3, so let's use these combinations of $n$ and $\theta$ to see whether they all give the same value of $d$:

$$d = \frac{(1)(154 \text{ pm})}{2 \sin 12.32°} = \frac{154 \text{ pm}}{(2)(0.2134)} = 361 \text{ pm}$$

$$d = \frac{(2)(154 \text{ pm})}{2 \sin 25.27°} = \frac{308 \text{ pm}}{(2)(0.4269)} = 361 \text{ pm}$$

$$d = \frac{(3)(154 \text{ pm})}{2 \sin 39.81°} = \frac{462 \text{ pm}}{(2)(0.6402)} = 361 \text{ pm}$$

We do indeed get the same value of $d$, so $d = 361$ pm is the distance between Cu atoms that produced the three peaks.

Rearranging the equation $r = 0.3536\ell$ and substituting for $r$:

$$\ell = \frac{r}{0.3536} = \frac{128 \text{ pm}}{0.3536} = 361 \text{ pm}$$

We find that the distance between the layers in copper metal determined by X-ray diffraction matches the edge length of the unit cell for Cu.

**Think About It** These consistent results mean that our assumption about the values of $n$ for the three values of $\theta$ was correct. Also, the value of $d$ is in the range of the edge lengths of unit cells we used in several Sample Exercises earlier in the chapter and so is reasonable. In practice, the reflections from $n > 1$ have weak intensities and are less useful for determining the value of the interlayer spacing, $d$.

⚙ **Practice Exercise** An X-ray diffraction analysis of crystalline CsCl using X-rays of wavelength 71.2 pm has a prominent peak at $2\theta = 19.9°$. If this peak corresponds to $n = 2$, what is the spacing between the ion layers? What are the values of $2\theta$ for $n = 3$ and $n = 4$?

The mantles in camping lanterns contain the mineral thorite, an oxide of thorium. Thorite crystallizes with the fluorite structure and contains 87.88% thorium and 12.12% oxygen.
a. Determine the empirical formula of thorite.
b. X-ray diffraction analysis of a thorite crystal using X-rays of 154 nm reveals reflections at values of $2\theta = 15.8°$ and $31.9°$. Calculate the interlayer spacing between the layers of Th ions in thorite.
c. Calculate the density of thorite. Does the calculated density match the measured density of 9.86 g/mL?
d. The ionic radii of $Th^{4+}$ and $O^{2-}$ ions are 102 pm and 140 pm, respectively. If thorite were to crystallize with the rock salt structure, what type of hole would the $Th^{4+}$ ions most likely occupy?
e. Calculate the density of a rock salt polymorph of thorite using the same unit cell edge length as in part c. Compare the value with the observed and calculated values in part c.

**Collect and Organize** We are given the percent composition, X-ray diffraction data, and density of a thorium mineral. We are asked to determine the empirical formula of thorite and to calculate the density of thorite in two possible crystal structures—the fluorite (observed) and rock salt (hypothetical) structures. Methods for determining empirical formulas were described in Chapter 3. We will also make use of the information in Sections 12.6 and 12.9.

**Analyze** We recognize that thorite contains only Th and O with the general formula $Th_xO_y$ because the sum of the %Th and %O equals 100%, and thus we know the masses of each element in thorite. We can convert the masses to the equivalent number of moles and determine the simplest whole-number ratio of the moles of the elements. Substituting the observed values for $\theta$ in Equation 12.6 allows us to determine the length of the unit cell. In the fluorite structure of thorite, the unit cell consists of a face-centered arrangement of Th ions with oxide ions in all of the tetrahedral holes. Following the procedure in Sample Exercise 12.4, we can calculate the density of thorite. If thorite were to crystallize in a ccp rock salt structure, then Th ions would be found in the octahedral holes of an fcc unit cell of oxide ions.

**Solve**
a. A 100.00 g sample of thorite contains 87.88 g Th and 12.12 g of O. Converting these masses to moles:

$$87.88 \text{ g Th} \times \frac{1 \text{ mol Th}}{232.04 \text{ g Th}} = 0.3787 \text{ mol Th}$$

$$12.12 \text{ g O} \times \frac{1 \text{ mol O}}{16.00 \text{ g O}} = 0.7575 \text{ mol O}$$

Dividing by the smallest number of moles gives us the empirical formula: $ThO_2$.
b. We can assume that the values for $n$ in Equation 12.6 are likely to be 1 and 2 and that $\theta = 7.9°$ and $16.0°$. Substituting into the form of Equation 12.6 in Sample Exercise 12.5:

$$d = \frac{n\lambda}{2\sin\theta} = \frac{1(154 \text{ nm})}{2\sin(7.9°)} = 560 \text{ nm}$$

$$d = \frac{n\lambda}{2\sin\theta} = \frac{2(154 \text{ nm})}{2\sin(16.0°)} = 559 \text{ nm}$$

These values for $d$ are very close to each other and represent the unit cell edge length $\ell$.
c. In the fluorite structure of $ThO_2$, the $Th^{4+}$ ions adopt an fcc unit cell with $O^{2-}$ ions in all the tetrahedral holes. The equivalent of

four $Th^{4+}$ and eight $O^{2-}$ are in the unit cell. The mass of the ions in the unit cell is

$$m = \frac{232.04 \text{ g Th}^{4+}}{1 \text{ mol Th}^{4+}} \times \frac{1 \text{ mol Th}^{4+}}{6.022 \times 10^{23} \text{ ions Th}^{4+}} \times 4 \text{ ions Th}^{4+}$$
$$= 1.541 \times 10^{-21} \text{ g Th}^{4+}$$

$$m = \frac{16.00 \text{ g O}^{2-}}{1 \text{ mol O}^{2-}} \times \frac{1 \text{ mol O}^{2-}}{6.022 \times 10^{23} \text{ ions O}^{2-}} \times 8 \text{ ions O}^{2-}$$
$$= 0.2126 \times 10^{-21} \text{ g O}^{2-}$$

Total mass of ions = $1.541 \times 10^{-21}$ g $Th^{4+}$ + $0.2126 \times 10^{-21}$ g $O^{2-}$
$$= 1.754 \times 10^{-21} \text{ g}$$

The volume of the unit cell is

$$V = \ell^3 = (560 \text{ pm})^3 \times \frac{(10^{-10} \text{ cm})^3}{(1 \text{ pm})^3} = 1.756 \times 10^{-22} \text{ cm}^3$$

and the calculated density of $ThO_2$ is

$$d = \frac{m}{V} = \frac{1.754 \times 10^{-21} \text{ g}}{1.756 \times 10^{-22} \text{ cm}^3} = 10.0 \text{ g/cm}^3$$

The calculated density differs from the observed density by less than 2%.
d. In the hypothetical rock salt structure of $ThO_2$, the lattice is composed of fcc $O^{2-}$ ions with $Th^{4+}$ in holes. The radius ratio of $Th^{4+}$ to $O^{2-}$ is

$$\frac{r_+}{r_-} = \frac{102 \text{ pm}}{140 \text{ pm}} = 0.728$$

Based on Table 12.4, the $Th^{4+}$ ions would occupy one-half of the octahedral holes.
e. By analogy to part c, the density of $ThO_2$ in the hypothetical rock salt structure would be

$$m = \frac{232.04 \text{ g Th}^{4+}}{1 \text{ mol Th}^{4+}} \times \frac{1 \text{ mol Th}^{4+}}{6.022 \times 10^{23} \text{ ions Th}^{4+}} \times 2 \text{ ions Th}^{4+}$$
$$= 7.706 \times 10^{-22} \text{ g Th}^{4+}$$

$$m = \frac{16.00 \text{ g O}^{2-}}{1 \text{ mol O}^{2-}} \times \frac{1 \text{ mol O}^{2-}}{6.022 \times 10^{23} \text{ ions O}^{2-}} \times 4 \text{ ions O}^{2-}$$
$$= 1.063 \times 10^{-22} \text{ g O}^{2-}$$

$$V = \ell^3 = (560 \text{ pm})^3 \times \frac{(10^{-10} \text{ cm})^3}{(1 \text{ pm})^3} = 1.756 \times 10^{-22} \text{ cm}^3$$

$$d = \frac{m}{V} = \frac{8.769 \times 10^{-22} \text{ g}}{1.756 \times 10^{-22} \text{ cm}^3} = 5.00 \text{ g/cm}^3$$

The calculated density for $ThO_2$ in a rock salt structure is about half of the observed value, and it is about half of the calculated value based on the fluorite structure.

**Think About It** By analogy to the fluorite structure shown in Figure 12.26, the larger oxide ions occupy the tetrahedral holes. The calculated value for the density of $ThO_2$ is very close to the observed value, supporting the assignment of the fluorite structure for $ThO_2$. The rock salt structure, where the smaller $Th^{4+}$ ions occupy octahedral holes in the fcc lattice of the larger $O^{2-}$ ions, can be ruled out because the calculated density is very different from the observed density.

## Silicon, Silica, Silicates, Silicone: What's in a Name?

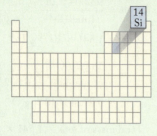

Group 14 of the periodic table is headed by carbon, an element so important to life, energy, and commerce that it merited its own descriptive chemistry section in Chapter 5. The element below carbon, silicon, is central both to ancient and modern technology and to the entire solid-state revolution that brought us computers and many of our electronic devices; it too deserves to be highlighted.

Silicon is the second most abundant element in Earth's crust, after oxygen. Elemental silicon never occurs free in nature. Mostly it is present as *silica* ($SiO_2$, a covalent network solid) and *silicate minerals*, also called just *silicates* (solids containing silicon–oxygen groups and metals). Silicates almost invariably consist of silicon–oxygen tetrahedral units bonded together in chains, rings, sheets, and three-dimensional arrays similar to those found in carbon (Section 12.7). Silica constitutes about 60% by mass of Earth's crust and is of great geological and commercial importance. It is most familiar to us as quartz crystals and beach sand.

People have used silica and silicates since prehistoric times in simple tools, arrowheads, and flint knives. The glass most familiar to us as window panes and bottles, called soda-lime glass, is principally a mixture of $SiO_2$ (73%), CaO (11%), and $Na_2O$ (13%), with the balance composed of other metal oxides. Leaded crystal stemware is made by adding PbO (24%) and $K_2O$ (15%) to silica (60%). Laboratory glassware is composed primarily of silica (81%), $B_2O_3$ (11%), and $Na_2O$ (5%).

Natural silicates are the raw materials of the ceramic and concrete industries. Silica is also used in detergents; in the pulp and paper industry; as anticaking agents in powdered foods such as cocoa, sugar, and spices; and as an ingredient in paints to provide a matte finish. Small packs of material labeled "desiccant" in packaged electronics, pharmaceuticals, and some foods may contain porous silica because it can absorb about 40% of its mass in water and thereby keep moisture-sensitive materials dry. Silica is a major component in toothpaste, functioning as both a mild abrasive and a thickening agent.

Most silica is now used in the production of highly purified elemental silicon, the most basic material of the microelectronics industry. The name "Silicon Valley" to describe the semiconductor industry around Palo Alto, CA, is an acknowledgment of the importance of the element to the industry. Computers, calculators, liquid-crystal displays, cell phones, video games, solar cells, and space vehicles are all products of the solid-state revolution that began in 1947 with the discovery of the transistor effect at Bell Laboratories in Murray Hill, NJ. The theoretical work and practical development of transistors and semiconductors by Walter H. Brattain, John Bardeen, and William B. Shockley resulted in their sharing the Nobel Prize in Physics in 1956.

Silicon is the element most widely used as a starting material in the production of semiconductors. Germanium, the element below silicon in the periodic table, is second. To function correctly in electronic devices, silicon must contain less than 1 ppb of impurities.

Such ultrapure silicon is produced in a series of reactions starting with

$$SiO_2(s) + C(s) \rightarrow Si(s) + CO_2(g)$$

Silicon prepared in this way is about 98% pure. Before it can be used to make microchips, it must be refined further, through the reaction

$$Si(s) + 2\,Cl_2(g) \rightarrow SiCl_4(g)$$

Impurities that do not form volatile chlorides are removed, and the $SiCl_4$ is heated above 700°C, where it decomposes to silicon and chlorine gas:

$$SiCl_4(g) \rightarrow Si(s) + 2\,Cl_2(g)$$

The final step in processing silicon results in both an ultrapure material (one that contains only silicon atoms) and a solid that is free of crystalline imperfections (every position in the lattice is occupied by a silicon atom, and no gaps or vacancies exist). A semiconductor is usually made from a single crystal of silicon grown by a process called "pulling" (Figure 12.37). A small seed crystal is allowed to touch the surface of a pool of melted silicon. The seed crystal is withdrawn from the melt at a rate that allows material from the melt to solidify

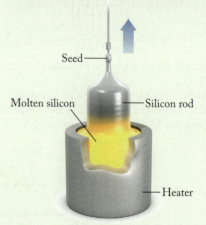

**FIGURE 12.37** "Pulling" a silicon seed crystal from a melt yields ultrapure and perfectly crystalline silicon.

on the seed crystal. Single crystals of silicon 25 cm long or longer are made in this fashion, and semiconductor devices are made from pieces of these single crystals. Crystalline silicon has the same crystal structure as diamond.

Silicones are another commercially important family of silicon compounds; they are made of carbon, hydrogen, oxygen, and silicon and hence are called organosilicon compounds. No natural organosilicon compounds exist, but they are the basis of a major industry. A silicone polymer is applied to the windshields of automobiles and aircraft to prevent bugs and grime from clouding vision, to reduce glare, and to disperse rain, sleet, and snow. The most common silicone used in this application, poly(dimethylsiloxane) or PDMS, has the monomer repeating unit shown in Figure 12.38.

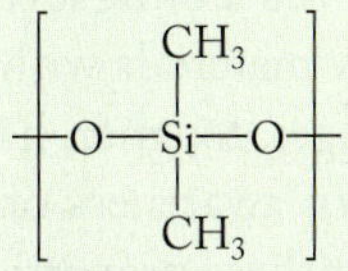

**FIGURE 12.38**

Glass contains silicates, which interact with water, as we discussed in Chapter 10 in terms of meniscus formation. PDMS molecules adhere readily to glass because their structure is similar to that of the silicates, but the $-CH_3$ groups make PDMS much more hydrophobic. When water contacts the PDMS coating, it does not adhere to the surface; raindrops are repelled (Figure 12.39).

Silicones are available as liquids, gels, and flexible solids. They are stable over a range of temperatures from below $-40°C$ to greater than $150°C$. Thousands of applications for silicones, beyond coatings for glass surfaces, include their use as moisture-proof sealants for ignition cables, spark plugs, medical appliances, and catheters, and as protective surfaces for everything from integrated circuits to textiles and skyscrapers. PDMS is also a key ingredient in Silly Putty.

The chemical properties of the other elements in group 14—germanium, tin, and lead—vary with increasing atomic number down the family (Figure 12.40). Germanium (like silicon) is a

**FIGURE 12.39** Rain beads on metal treated with a silicone polymer.

metalloid, whereas tin and lead are two of the oldest known metals. All three form oxides with the formula $MO_2$ (where M is the group 14 metal or metalloid) and react with halogens to form tetrahalides ($MX_4$, where X = F, Cl, Br, or I). The dichlorides of tin and lead, $SnCl_2$ and $PbCl_2$, also are stable. Both germanium and one of the allotropes of tin, α-tin, have the diamond structure, but β-tin has a more complicated structure. Elemental lead crystallizes in an fcc unit cell. Germanium and α-tin are semiconductors, but β-tin and lead are conductors. Tin is used extensively in metallurgy in combination with other metals. Tin oxides are used to make ceramics. Lead and lead(IV) oxide are the electrode materials in most automobile batteries.

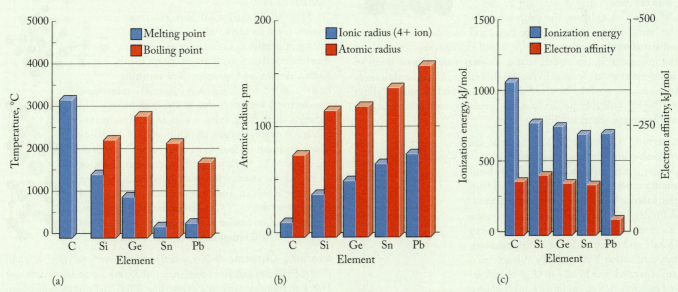

**FIGURE 12.40** (a) Melting points and boiling points of the group 14 elements; the boiling point of carbon has not yet been determined. (b) Atomic and ionic radii of the group 14 elements. (c) The ionization energies and electron affinities of the group 14 elements.

As we have seen throughout this text, the arrangement of atoms, ions, and molecules in a sample of matter is responsible for many of its macroscopic properties. XRD is a powerful tool for analyzing structure at the atomic level of a wide range of materials, including metals, minerals, polymers, plastics, ceramics, pharmaceuticals, and semiconductors. The technique is indispensable for scientific research and industrial production. The link between structure and properties is robust, and understanding the ordered arrangements of atoms and ions that characterize pure metals, alloys, semiconductors, ionic compounds, and ceramics provides tremendous insight into the behavior of bulk materials. Knowing the structure of solids at the atomic level enables us to explain properties of materials and to design new materials with improved properties. The increasingly rapid development of stronger, tougher, harder, and more heat-resistant solid materials is made possible by an awareness of the role of size, shape, and arrangement of atoms in determining these properties.

## SUMMARY

**Learning Outcome 1** Many metallic crystals are based on **crystal lattices** of the **cubic closest-packed (ccp)** and **hexagonal closest-packed (hcp)** types, which are the two most efficient ways of packing atoms in a solid. **Crystalline solids** contain repeating **unit cells**, which can be **simple cubic (sc)**, **body-centered cubic (bcc)**, or **face-centered cubic (fcc)**. The dimensions of the unit cell in a crystalline solid can be used to determine the radius of the atoms or ions and to predict density. (Sections 12.1 and 12.2)

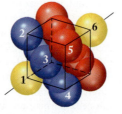

**Learning Outcome 2** **Alloys** are blends of a host metal and one or more other elements (which may or may not be metals) added to enhance the properties of the host, including strength, hardness, and corrosion resistance. In **substitutional alloys**, atoms of the added elements replace atoms of the host metal in the crystal lattice. In **interstitial alloys**, atoms of added elements are located in the tetrahedral and/or octahedral holes between atoms of the host metal. Aluminum and aluminum alloys are highly desirable for applications requiring corrosion resistance and low mass. (Section 12.3)

**Learning Outcome 3** Most metals are malleable and ductile. The electrical conductivity of metals can be explained by **band theory** as the easy movement of electrons from the **valence band** to the **conduction band**. Semimetals are **semiconductors**, intermediate in electrical conducting ability between metals and nonmetals. In semiconductors, the conduction band and the valence band are separated by a **band gap ($E_g$)**. Substituting electron-rich atoms into a semiconductor results in **n-type semiconductors**. Substituting electron-poor atoms results in **p-type semiconductors**. Both types of substitution

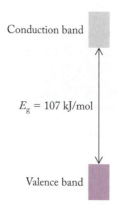

increase the conductivity of the semiconductor by decreasing its band gap. (Sections 12.4 and 12.5)

**Learning Outcome 4** Many **ionic solids** consist of crystals with some number of either cations or anions forming the unit cell and the opposite ion occupying octahedral and tetrahedral holes in the unit cell. The unit cell edge lengths of ionic solids can be used to calculate ionic radii and to predict densities. (Section 12.6)

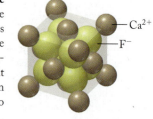

Two allotropes of carbon are the **covalent network solids**, graphite and diamond. (Section 12.7)

Heating selected solid inorganic compounds (such as clays, which are aluminosilicate minerals) to high temperature alters their chemical composition and makes them harder, denser, and stronger. The resulting heat- and chemical-resistant materials (**ceramics**) are electrical insulators due to the large energy gap between their filled valence and empty conduction bands. The polymorphs of silica consist of tetrahedra made up of four O at the corners surrounding a central Si. Each tetrahedron can share some or all its oxygen atoms with other tetrahedra, forming Si—O—Si bridges and two- and three-dimensional covalent networks. In **superconducting** ceramics, the electrical resistance of the material drops to zero below its **critical temperature ($T_c$)**. (Section 12.8)

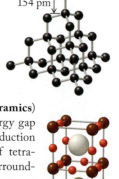

**Learning Outcome 5** X-ray diffraction is an analytical method that records constructive interference of X-rays scattered off different layers of atoms or ions in a crystalline solid. The distances between layers are calculated using the **Bragg equation**. X-ray diffraction makes it possible to determine the crystal structure of crystalline solids. (Section 12.9)

## PROBLEM-SOLVING SUMMARY

| TYPE OF PROBLEM | CONCEPTS AND EQUATIONS | SAMPLE EXERCISES |
|---|---|---|
| **Calculating atomic or ionic radii and density from unit cell dimensions** | Determine the length of a unit cell edge, face diagonal, or body diagonal along which adjacent atoms touch. Use the relationship between edge length $\ell$ and the atomic radius $r$ to calculate the value of $r$: $r = 0.5\,\ell$ for sc unit cells, $r = 0.3536\ell$ for fcc unit cells, and $r = 0.4330\ell$ for bcc unit cells. Determine the mass of the atoms in the unit cell from the molar mass and the volume of the unit cell from the unit cell edge length; then calculate the density: $$d = \frac{m}{V}$$ | 12.1, 12.4 |
| **Predicting the crystal structure of two-element alloys** | Compare the radii of the alloying elements. Similarly sized radii (within 15%) predict a substitutional alloy. When the radius of the smaller atom in an hcp or ccp lattice is <73% of the radius of the larger atom, an interstitial alloy forms. | 12.2 |
| **Distinguishing p- and n-type semiconductors** | Determine whether the dopant has more (n-type) or fewer (p-type) valence electrons than the host semiconductor. | 12.3 |
| **Determining interlayer distances by X-ray diffraction** | Apply the Bragg equation: $$n\lambda = 2d \sin \theta \qquad (12.6)$$ | 12.5 |

## VISUAL PROBLEMS

*(Answers to boldface end-of-chapter questions and problems are in the back of the book.)*

**12.1.** In Figure P12.1, which drawings are analogous to crystalline solids, and which are analogous to amorphous solids?

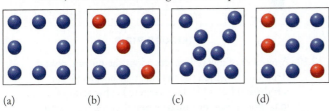

(a)  (b)  (c)  (d)

**FIGURE P12.1**

**12.2.** The unit cells in Figure P12.2 continue infinitely in two dimensions. Draw a box around the unit cell in each pattern. How many light squares and how many dark squares are in each unit cell?

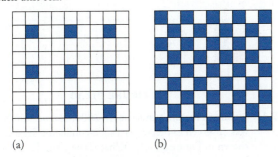

(a)  (b)

**FIGURE P12.2**

**12.3.** The pattern in Figure P12.3 continues indefinitely in three dimensions. Draw a box around the unit cell. If the red circles represent element A and the blue circles element B, what is the chemical formula of the compound?

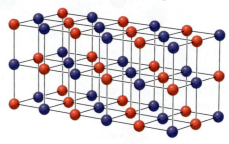

**FIGURE P12.3**

**12.4.** What is the chemical formula of the ionic compound a portion of whose unit cell is shown in Figure P12.4? (A and B are cations, X is an anion.)

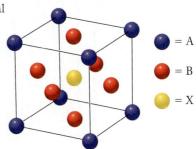

= A
= B
= X

**FIGURE P12.4**

**12.5.** What is the formula of the compound that crystallizes in a cubic closest-packed arrangement of A atoms (silver) with B atoms (red) occupying half of the octahedral holes and C atoms (blue) occupying one-eighth of the tetrahedral holes, as shown in Figure P12.5?

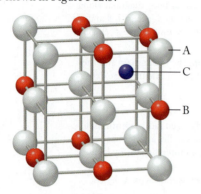

**FIGURE P12.5**

**12.6.** What is the formula of the compound that crystallizes with copper ions (the small spheres in Figure P12.6) occupying half of the tetrahedral holes in a cubic closest-packed arrangement of chloride ions?

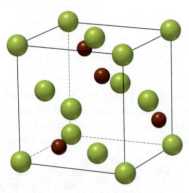

**FIGURE P12.6**

**12.7.** What is the formula of the compound that crystallizes with lithium ions occupying all of the tetrahedral holes in a cubic closest-packed arrangement of sulfide ions? See Figure P12.7.

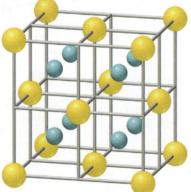

**FIGURE P12.7**

**12.8.** Figure P12.8 shows the unit cell of CsCl. From the information given and the radius of the chloride (corner) ions of 181 pm, calculate the radius of $Cs^+$ ions.

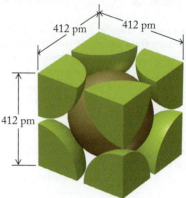

**FIGURE P12.8**

**12.9.** A number of metal chlorides adopt the rock salt crystal structure, in which the metal ions occupy all the octahedral holes in a face-centered cubic array of chloride ions. For at least one (maybe more) of the metallic elements highlighted in Figure P12.9, this crystal structure is not possible. Which one(s) cannot form a rock salt structure with $Cl^-$ ions?

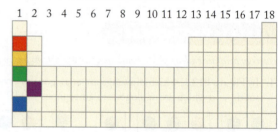

**FIGURE P12.9**

**12.10.** A number of metal fluorides adopt the fluorite crystal structure, in which the fluoride ions occupy all the tetrahedral holes in a face-centered cubic array of metal ions. For at least one (maybe more) of the highlighted metallic elements in Figure P12.10, this crystal structure is not possible. Which one(s)?

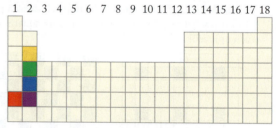

**FIGURE P12.10**

**\*12.11. Superconducting Materials** In 2000, magnesium boride was observed to behave as a superconductor. Its unit cell is shown in Figure P12.11. What is the formula of magnesium boride? A boron atom is in the center of the unit cell (on the left), which is part of the hexagonal closest-packed crystal structure (right).

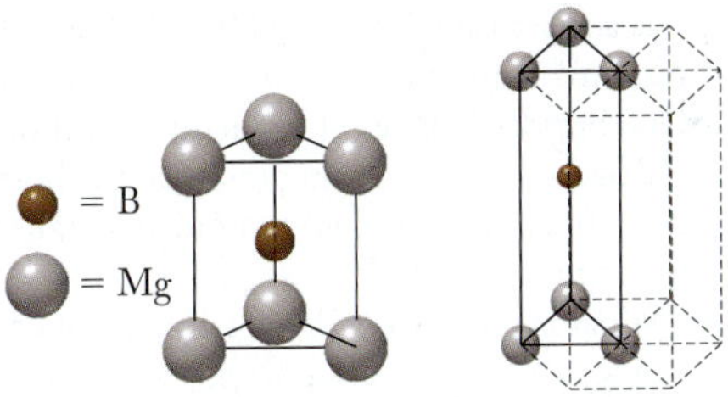

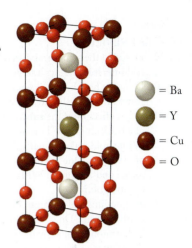

FIGURE P12.11

a. What is the chemical formula of this compound?

b. Eight oxygen atoms must be removed from the unit cell shown here to produce the unit cell of $YBa_2Cu_3O_7$. Does it make a difference which oxygen atoms are removed?

*12.12. **Superconducting Materials** The 1987 Nobel Prize in Physics was awarded to J. G. Bednorz and K. A. Müller for their discovery of superconducting ceramic materials such as $YBa_2Cu_3O_7$. Figure P12.12 shows the unit cell of another yttrium–barium–copper oxide.

FIGURE P12.12

## QUESTIONS AND PROBLEMS

### Structures of Metals

#### CONCEPT REVIEW

**12.13.** Explain the difference between cubic closest-packed and hexagonal closest-packed arrangements of identical spheres.

*12.14. Describe the features of simple cubic, body-centered cubic, and face-centered cubic crystal structures and rank these structures based on decreasing packing efficiency.

**12.15.** Which unit cell has the greater packing efficiency, simple cubic or body-centered cubic?

**12.16.** Consult Figure 12.10 to predict which unit cell has the greater packing efficiency: body-centered cubic or face-centered cubic.

**12.17.** The unit cell in iron metal is either fcc or bcc, depending on temperature. Are the fcc form of iron and the bcc form allotropes? Explain your answer.

*12.18. At low temperatures, the unit cell of calcium metal is found to be fcc, a closest-packed crystal lattice. At higher temperatures, the unit cell of calcium metal is found to be bcc, a crystal lattice that is not a closest-packed structure. What might be a reason for this difference?

#### PROBLEMS

**12.19.** Europium, one of the lanthanide elements used in television screens, crystallizes in a crystal lattice built on bcc unit cells, with a unit cell edge of 240.6 pm. Calculate the radius of a europium atom.

**12.20.** Nickel has an fcc unit cell with an edge length of 350.7 pm. Calculate the radius of a nickel atom.

**12.21.** What is the length of an edge of the unit cell when barium (atomic radius 222 pm) crystallizes in a crystal lattice of bcc unit cells?

**12.22.** What is the length of an edge of the unit cell when aluminum (atomic radius 143 pm) crystallizes in a crystal lattice of fcc unit cells?

**12.23.** A crystalline form of copper has a density of 8.95 $g/cm^3$. If the radius of copper atoms is 127.8 pm, is the copper unit cell (a) simple cubic; (b) body-centered cubic; or (c) face-centered cubic?

**12.24.** A crystalline form of molybdenum has a density of 10.28 $g/cm^3$ at a temperature at which the radius of a molybdenum atom is 139 pm. Which unit cell is consistent with these data? (a) simple cubic; (b) body-centered cubic; (c) face-centered cubic

**12.25.** Sodium metal crystallizes with a body-centered cubic structure at normal atmospheric pressure. The atomic radius of a sodium atom is 186 pm and the density of Na is 0.971 $g/cm^3$. At 613,000 atm (9 million psi) the structure of Na metal becomes fcc. Assuming the radius of Na is unchanged, what is the density of the fcc form of Na?

**12.26.** Calcium metal undergoes two phase changes under pressure. At atmospheric pressure, Ca crystallizes in an fcc unit cell which changes to a bcc unit cell above 20 GPa and to a simple cubic structure above 32 GPa.

a. If the radius of Ca remains unchanged, calculate the density of the simple cubic phase.

*b. Calculate the radius of a Ca atom in the simple cubic structure if the density of the sc and fcc phases is the same.

*12.27. The unit cell for hcp Ti is shown in Figure P12.27. Given the unit cell dimensions, calculate the density of titanium (atomic radius 147 pm).

*12.28. Cobalt crystallizes in the same unit cell as titanium (Figure P12.27). The edge length (251 pm) and height (407 pm) are both shorter than for titanium. Is cobalt denser than titanium?

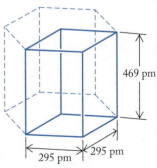

469 pm

295 pm   295 pm

FIGURE P12.27

## Alloys

### CONCEPT REVIEW

**12.29.** Describe the structural differences between substitutional and interstitial alloys and give an example of each alloy.

**12.30.** Describe how homogeneous alloys and intermetallic compounds are similar and how they are different.

**12.31.** What effect does the substitution of Ni for Ti in the center of a bcc unit cell of Ti have on the unit cell edge length?

**12.32.** White gold was originally developed to give the appearance of platinum. One formulation of white gold contains 25% nickel and 75% gold. Which is more malleable, white gold or pure gold?

**12.33.** Magnesium and hafnium have nearly the same atomic radii (within 1 pm). Does substitution of 25% of the magnesium by hafnium in an alloy increase or decrease the density of the alloy relative to pure magnesium?

**12.34.** Why are the alloys that second-row nonmetals—such as B, C, and N—form with transition metals more likely to be interstitial than substitutional?

### PROBLEMS

**12.35.** The unit cell of NiTi in Figure P12.35(a) shows a bcc arrangement of Ti atoms with Ni in the middle. Could we also draw the unit cell of NiTi as a bcc arrangement of Ni atoms with Ti in the middle as shown in Figure P12.35(b)?

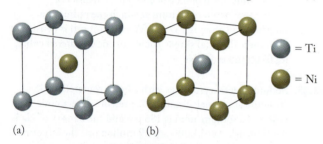

(a)                    (b)

= Ti

= Ni

**FIGURE P12.35**

**12.36.** Can both of the unit cells in Figure P12.36 represent the same substitutional alloy?

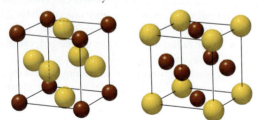

**FIGURE P12.36**

**12.37.** **Hydrogen Storage** Hydrogen is an attractive alternative fuel to hydrocarbons because it has a high fuel value and does not produce carbon dioxide when burned. One challenge to a hydrogen economy is storing hydrogen. Many transition metals absorb hydrogen by breaking the H—H bond and storing H atoms in the interstices between the metal atoms.
   a. What is the minimum radius for the host metal for a H atom (radius 37 pm) to fit in a tetrahedral hole of an fcc unit cell?
   b. If the H is present as a hydride ion, $H^-$ (ionic radius = 146 pm), is the hydrogen more likely to be found as an interstitial alloy in a tetrahedral or octahedral hole or

as a substitutional alloy? Hint: Consult Appendix 3 for atomic radii of metals.

**12.38.** Metal borides $MB_x$ exist for many transition metals.
   a. What is the minimum radius for the host metal for a B atom (radius 88 pm) to fit in an octahedral hole of an fcc unit cell?
   b. What is the value of $x$ if an average of one octahedral hole is occupied per unit cell?

**12.39.** **Dental Fillings** Dental fillings are mixtures of several alloys including one with the formula $Ag_3Sn$. Silver (radius 144 pm) and tin (140 pm) both crystallize in an fcc unit cell. Is this alloy likely to be a substitutional alloy or an interstitial alloy?

**12.40.** An alloy used in dental fillings has the formula $Sn_3Hg$. The radii of tin and mercury atoms are 140 pm and 151 pm, respectively. Which alloy has a smaller mismatch (percent difference in atomic radii), $Sn_3Hg$ or bronze (Cu/Sn alloys)?

**12.41.** What is the formula of vanadium carbide if the vanadium atoms have a cubic closest-packed structure and two of the octahedral holes are occupied by C atoms?

**12.42.** What is the formula of manganese nitride if the manganese atoms have an fcc unit cell and one of the tetrahedral holes is occupied by a N atom?

**12.43.** Calculate the density of nitinol, the memory alloy with the composition NiTi and the unit cell shown in Figure P12.43. The atomic radii of Ni and Ti are 124 and 147 pm, respectively.

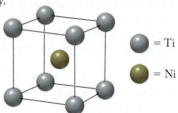

= Ti

= Ni

**FIGURE P12.43**

**12.44.** Calculate the density of a rose gold alloy composed of 25% Cu and 75% Au given the unit cell in Figure P12.44. The atomic radii of Cu and Au are 128 and 144 pm, respectively.

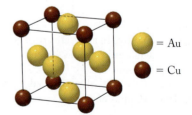

= Au

= Cu

**FIGURE P12.44**

**12.45.** One of the many alloys of copper and zinc has the unit cell shown in Figure P12.45.

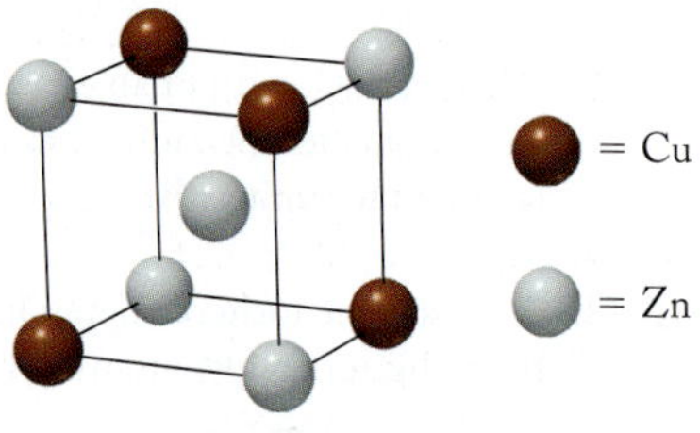

= Cu

= Zn

**FIGURE P12.45**

a. What type of crystal structure does this alloy have?
b. What are the proportions of copper and zinc in the alloy?

*12.46. The unit cell of an iron aluminum alloy is shown in Figure P12.46. (*Hint*: Consult Figure 12.7 for help in identifying the atoms.)
   a. Is this alloy a substitutional or interstitial alloy?
   b. What is the composition of this alloy?

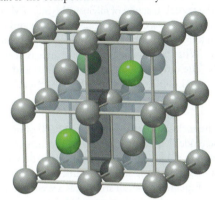

**FIGURE P12.46**

## Metallic Bonds and Conduction Bands
### CONCEPT REVIEW
**12.47.** How does the sea-of-electrons model (Chapter 8) explain the high electrical conductivity of gold?
*12.48. How does band theory explain the high electrical conductivity of mercury?
**12.49.** The melting and boiling points of sodium metal are much lower than those of sodium chloride. What does this difference reveal about the relative strengths of metallic bonds and ionic bonds?
*12.50. Which metal do you expect to have the higher melting point—Al or Na? Explain your answer.
**12.51.** Some scientists believe that the solid hydrogen that forms at very low temperatures and high pressures may conduct electricity. Is this hypothesis supported by band theory?
**12.52.** Would you expect solid helium to conduct electricity?

## Semiconductors
### CONCEPT REVIEW
**12.53.** Which groups in the periodic table contain metals with filled valence bands?
**12.54.** Insulators are materials that do not conduct electricity; conductors are substances that allow electricity to flow through them easily. Rank the following in order of increasing band gap: semiconductor, insulator, conductor.
**12.55.** Describe in general terms the differences in composition and conduction between n-type and p-type semiconductors.
**12.56.** How might doping of silicon with germanium affect the conductivity of silicon?
*12.57. Antimony (Sb) combines with sulfur to form the semiconductor compound $Sb_2S_3$. In which group of the periodic table might you find elements for doping $Sb_2S_3$ to form a p-type semiconductor?
*12.58. In which group of the periodic table might you find elements for doping $Sb_2S_3$ to form an n-type semiconductor?

### PROBLEMS
**12.59.** Thin films of doped diamond hold promise as semiconductor materials. Trace amounts of nitrogen impart a yellow color to otherwise colorless pure diamonds.
   a. Are nitrogen-doped diamonds examples of semiconductors that are p-type or n-type?
   b. Draw a picture of the band structure of diamond to indicate the difference between pure diamond and N-doped (nitrogen-doped) diamond.
   *c. Nitrogen-doped diamonds absorb violet light at about 425 nm. What is the magnitude of $E_g$ that corresponds to this wavelength?
**12.60. Hope Diamond** Trace amounts of boron give diamonds (including the Smithsonian's Hope Diamond) a blue color (Figure P12.60).
   a. Are boron-doped diamonds examples of semiconductors that are p-type or n-type?
   b. Draw a picture of the band structure of diamond to indicate the difference between pure diamond and B-doped diamond.

**FIGURE P12.60**

   *c. What is the band gap in energy if blue diamonds absorb red-orange light with a wavelength of 675 nm?

*12.61. Zinc sulfide is a semiconductor. The band gap of zinc sulfide increases as the pressure on a crystal increases. What effect does this have on the wavelength of light emitted by a ZnS crystal?
*12.62. Calculate the wavelengths of light emitted by the semiconducting phosphides AlP, GaP, and InP, which have band gaps of 241.1, 216.0, and 122.5 kJ/mol, respectively, and are used in the type of light source shown in Figure P12.62.

**FIGURE P12.62**

## Salt Crystals: Ionic Solids
### CONCEPT REVIEW
**12.63.** Crystals of both LiCl and KCl have the rock salt structure. In the unit cell of LiCl adjacent $Cl^-$ ions touch each other. In KCl they don't. Why?
**12.64.** Does the absolute size of an octahedral hole in an fcc lattice of halide ions change as we move down group 17 from fluoride to iodide?

**\*12.65.** In some books the unit cell of CsCl is described as being body-centered cubic (Figure P12.65); in others, as simple cubic (see Figure 12.10a). Explain how CsCl crystals might be described by either unit cell type.

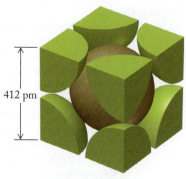

412 pm

**FIGURE P12.65**

**12.66.** If some of the sulfide ions in zinc sulfide are replaced by selenide ions, will the selenide ions occupy the same sites as the sulfide ions?

**12.67.** Why can't calcium chloride have a rock salt structure?

**12.68.** Why can't sodium fluoride have exactly the same structure as calcium fluoride?

**12.69.** If the unit cell of a substitutional alloy of copper and tin has the same unit cell edge as the unit cell of copper, will the alloy have a greater density than copper?

**12.70.** If the unit cell of an interstitial alloy of vanadium and carbon has the same unit cell edge as the unit cell of vanadium, will the alloy have a greater density than vanadium?

**\*12.71.** As the cation–anion radius ratio increases for an ionic compound with the rock salt crystal structure, is the calculated density more likely to be greater than, or less than, the measured value?

**\*12.72.** As the cation–anion radius ratio increases for an ionic compound with the rock salt crystal structure, is the length of the unit cell edge calculated from ionic radii likely to be greater than, or less than, the observed unit cell edge length?

## PROBLEMS

**12.73.** What is the formula of the oxide that crystallizes with $Fe^{3+}$ ions in one-fourth of the octahedral holes, $Fe^{3+}$ ions in one-eighth of the tetrahedral holes, and $Mg^{2+}$ in one-fourth of the octahedral holes of a cubic closest-packed arrangement of oxide ions ($O^{2-}$)?

**12.74.** What is the chemical formula of the compound that crystallizes a simple cubic arrangement of fluoride ions with $Ba^{2+}$ ions occupying half of the cubic holes?

**12.75. The Vinland Map** At Yale University there is a map, believed to date from the 1400s, of a landmass labeled "Vinland" (Figure P12.75). The map is thought to be evidence of early Viking exploration of North America. Debate over the map's authenticity centers on yellow stains on the map

**FIGURE P12.75**

paralleling the black ink lines. One analysis suggests the yellow color is from the mineral anatase, a form of $TiO_2$ that was not used in 15th-century inks.

　a. The crystal structure of anatase is approximated by a ccp arrangement of oxide ions with titanium(IV) ions in holes. Which type of hole are $Ti^{4+}$ ions likely to occupy?

　b. What fraction of these holes are likely to be occupied? (The radius of $Ti^{4+}$ is 60.5 pm.)

**\*12.76.** The crystal structure of olivine—$M_2SiO_4$ (M = Mg, Fe)—can be viewed as a ccp arrangement of oxide ions with silicon(IV) in tetrahedral holes and the metal ions in octahedral holes.

　a. What fraction of each type of hole is occupied?

　b. The unit cell volumes of $Mg_2SiO_4$ and $Fe_2SiO_4$ are $2.91 \times 10^{-26}$ cm³ and $3.08 \times 10^{-26}$ cm³. Why is the unit cell volume of $Fe_2SiO_4$ larger?

**12.77.** The rock salt structure (Figure 12.34) of magnesium selenide (MgSe) at atmospheric pressure changes to a cesium chloride (Figure P12.65) structure under pressure.

　a. How do these structures differ?

　b. Explain how both structures are consistent with the observed stoichiometry of MgSe.

**12.78.** The sphalerite structure of ZnS changes to a rock salt structure above 15 GPa.

　a. Describe the differences between these two structures.

　b. Explain how both structures are consistent with the observed stoichiometry of ZnS.

**12.79.** The unit cell of rhenium trioxide ($ReO_3$) consists of a cube with rhenium atoms at the corners and an oxygen atom on each of the 12 edges. The atoms touch along the edge of the unit cell. The radii of Re and O atoms in $ReO_3$ are 137 and 73 pm. Calculate the density of $ReO_3$.

**12.80.** With reference to Figure P12.65, calculate the density of simple cubic CsCl.

**12.81.** Magnesium oxide crystallizes in the rock salt structure. Its density is 3.60 g/cm³. What is the edge length of the fcc unit cell of MgO?

**12.82.** Crystalline potassium bromide (KBr) has a rock salt structure and a density of 2.75 g/cm³. Calculate its unit cell edge length.

## Structures of Nonmetals

### CONCEPT REVIEW

**12.83.** When amorphous red phosphorus is heated at high pressure, it is transformed into the allotrope black phosphorus, which can exist in one of several forms. One form consists of six-membered rings of phosphorus atoms (Figure P12.83a). Why are the six-atom rings in black phosphorus puckered, whereas the six-atom rings in graphite (Figure P12.83b) are planar?

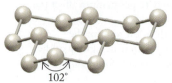

102°

(a) Black phosphorus　　　　　　(b) Graphite

**FIGURE P12.83**

*12.84. If the carbon atoms in graphite are replaced by alternating B and N atoms, would the resulting structure contain puckered rings like black phosphorus or flat ones like graphite (see Figure P12.83)?

**PROBLEMS**

**12.85.** In the fullerene known as buckminsterfullerene, $C_{60}$, molecules of $C_{60}$ form a cubic closest-packed array of spheres with a unit cell edge length of 1410 pm.
   a. What is the density of crystalline $C_{60}$?
   b. If we treat each $C_{60}$ molecule as a sphere of 60 carbon atoms, what is the radius of the $C_{60}$ molecule?
**12.86.** $C_{60}$ reacts with alkali metals to form $M_3C_{60}$ (where M = Na or K). The crystal structure of $M_3C_{60}$ contains cubic closest-packed spheres of $C_{60}$ with metal ions in holes. (Use the radius for $C_{60}$ that you calculated in 12.85.)
   a. If the radius of a $K^+$ ion is 138 pm, which type of hole is a $K^+$ ion likely to occupy? What fraction of the holes will be occupied?
   b. Under certain conditions, a different substance, $K_6C_{60}$, can be formed in which the $C_{60}$ molecules have a bcc unit cell. Calculate the density of a crystal of $K_6C_{60}$.

**12.87.** The distance between atoms in a cubic form of phosphorus is 238 pm (Figure P12.87). Calculate the density of this form of phosphorus.

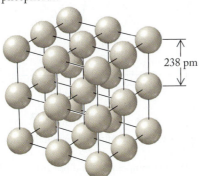

238 pm

**FIGURE P12.87**

12.88. **Ice under Pressure** Kurt Vonnegut's novel *Cat's Cradle* describes an imaginary, high-pressure form of ice called "ice nine." With the assumption that ice nine has a cubic closest-packed arrangement of oxygen atoms with hydrogen atoms in the appropriate holes, what type of hole will accommodate the H atoms?

## Ceramics: Insulators to Superconductors

**CONCEPT REVIEW**
**12.89.** Which of the following properties are associated with ceramics and which are associated with metals? Ductile; thermal insulator; electrically conductive; malleable
12.90. Many ceramics such as $TiO_2$ are electrical insulators. What differences are there in the band structure of $TiO_2$ compared with Ti metal that account for the different electrical properties?
**12.91.** Replacement of $Al^{3+}$ ions in kaolinite $[Al_2(Si_2O_5)(OH)_4]$ with $Mg^{2+}$ ions yields the mineral antigorite. What is its formula?
12.92. What is the formula of the silicate mineral talc, obtained by the replacement of $Al^{3+}$ ions in pyrophyllite $[Al_4Si_8O_{20}(OH)_4]$ with $Mg^{2+}$ ions?

**PROBLEMS**
**12.93.** Kaolinite $[Al_2(Si_2O_5)(OH)_4]$ is formed by weathering of the mineral $KAlSi_3O_8$ in the presence of carbon dioxide and water, as described by the following unbalanced equation:

$$KAlSi_3O_8(s) + H_2O(\ell) + CO_2(aq) \rightarrow$$
$$Al_2(Si_2O_5)(OH)_4(s) + SiO_2(s) + K_2CO_3(aq)$$

   Balance the equation and determine whether this is a redox reaction.
12.94. Albite, a feldspar mineral with an ideal composition of $NaAlSi_3O_8$, can be converted to jadeite ($NaAlSi_2O_6$) and quartz. Write a balanced chemical equation describing this transformation.

**12.95.** Under the high pressures in Earth's crust, the mineral anorthite ($CaAl_2Si_2O_8$) is converted to a mixture of three minerals: grossular $[Ca_3Al_2(SiO_4)_3]$, kyanite ($Al_2SiO_5$), and quartz ($SiO_2$).
   a. Write a balanced chemical equation describing this transformation.
   b. Determine the charges and formulas of the silicate anions in anorthite, grossular, and kyanite.
*12.96. The calcium silicate mineral grossular is also formed under pressure in a reaction between anorthite ($CaAl_2Si_2O_8$), gehlenite ($Ca_2Al_2SiO_7$), and wollastonite ($CaSiO_3$):

$$\underset{\text{Anorthite}}{CaAl_2Si_2O_8} + \underset{\text{Gehlenite}}{Ca_2Al_2SiO_7} + \underset{\text{Wollastonite}}{CaSiO_3} \rightarrow \underset{\text{Grossular}}{Ca_3Al_2(SiO_4)_3}$$

   a. Balance this chemical equation.
   b. Express the composition of gehlenite the way mineralogists often do: as the percentage of the metal and semimetal oxides in it, that is, %CaO, %$Al_2O_3$, and %$SiO_2$.

**12.97.** The ceramic material barium titanate ($BaTiO_3$) is used in devices that measure pressure. The radii of $Ba^{2+}$, $Ti^{4+}$, and $O^{2-}$ are 135, 60.5, and 140 pm, respectively. If the $O^{2-}$ ions are in a closest-packed structure, which hole(s) can accommodate the metal cations?
12.98. The mixed metal oxide $LiMnTiO_4$ has a structure with cubic closest-packed oxide ions and metal ions in both octahedral and tetrahedral holes. Which metal ion is most likely to be found in the tetrahedral holes? The ionic radii of $Li^+$, $Mn^{3+}$, $Ti^{4+}$, and $O^{2-}$ are 76, 67, 60.5, and 140 pm, respectively.

## X-ray Diffraction: How We Know Crystal Structures

**CONCEPT REVIEW**
**12.99.** Why does an amorphous solid not produce an XRD scan with sharp peaks?
12.100. X-ray diffraction cannot be used to determine the structures of compounds in solution—why?
**12.101.** Why are X-rays rather than microwaves chosen for diffraction studies of crystalline solids?
12.102. The radiation sources used in X-ray diffraction can be changed. Figure P12.102 shows a diffraction pattern made by a short-wavelength source. How would changing to a longer-wavelength source affect the pattern?

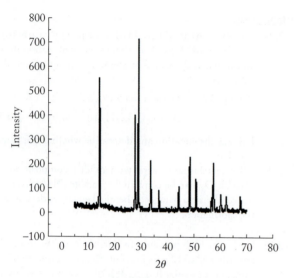

**FIGURE P12.102**

*__12.103.__ Why might a crystallographer (a scientist who studies crystal structures) use different X-ray wavelengths to determine a crystal structure? (*Hint*: Consider what mechanical limits are inherent in the design of the instrument depicted in Figure 12.36, and how those limits impact the 2θ scanning range.)

*__12.104.__ Where in earlier chapters have we seen diffraction used to acquire structural information?

**PROBLEMS**

__12.105.__ The spacing between the layers of ions in sylvite (the mineral form of KCl) is larger than in halite (NaCl). Which crystal will diffract X-rays of a given wavelength through larger 2θ values?

__12.106.__ Silver halides are used in black-and-white photography. In which compound would you expect to see a larger distance between ion layers, AgCl or AgBr? Which compound would you expect to diffract X-rays through larger values of 2θ if the same wavelength of X-ray were used?

__12.107.__ Galena, Illinois, is named for the rich deposits of lead(II) sulfide (PbS) found nearby. When PbS is exposed to X-rays with λ = 71.2 pm, strong reflections from a single crystal of PbS are observed at 13.98° and 21.25°. Determine the values of *n* to which these reflections correspond, and calculate the spacing between the crystal layers.

__12.108.__ **Pigments in Ceramics** Cobalt(II) oxide is used as a pigment in ceramics. It has the same type of crystal structure as NaCl. When cobalt(II) oxide is exposed to X-rays with λ = 154 pm, reflections are observed at 42.38°, 65.68°, and 92.60°. Determine the values of *n* to which these reflections correspond, and calculate the spacing between the crystal layers.

__12.109.__ Pyrophyllite [$Al_2Si_4O_{10}(OH)_2$] is a silicate mineral with a layered structure. The distances between the layers is 1855 pm. What is the smallest angle of diffraction of X-rays with λ = 154 pm from this solid?

__12.110.__ Minnesotaite [$Fe_3Si_4O_{10}(OH)_2$] is a silicate mineral with a layered structure similar to that of kaolinite. The distance between the layers in minnesotaite is 1940 ± 10 pm. What is the smallest angle of diffraction of X-rays with λ = 154 pm from this solid?

**Additional Problems**

__12.111.__ A unit cell consists of a cube that has an ion of element X at each corner, an ion of element Y at the center of the cube, and an ion of element Z at the center of each face. What is the formula of the compound?

__12.112.__ The unit cell of an oxide of uranium consists of cubic closest-packed uranium ions with oxide ions in all the tetrahedral holes. What is the formula of the oxide?

__12.113.__ The phase diagram for titanium is shown in Figure P12.113.
  a. Which structure does Ti metal have at 1500 K and 6 GPa of pressure?
  b. How many phase changes does Ti metal undergo as pressure is increased at 725°C?

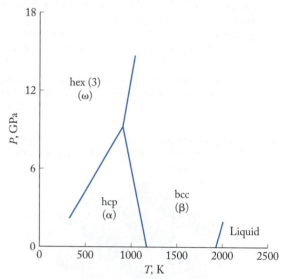

**FIGURE P12.113**

__12.114.__ The phase diagram of thallium metal is shown in Figure P12.114.
  a. At what temperature and pressure are all three solid phases of thallium metal present?
  b. Does the melting point of thallium increase or decrease with increasing pressure?

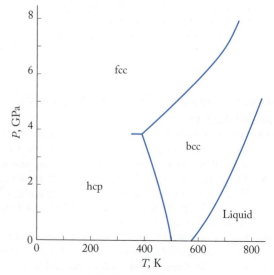

**FIGURE P12.114**

**12.115.** Silver nanoparticles are embedded in clothing fabric to kill odor-causing bacteria. Silver crystallizes in an fcc unit cell.
   a. How many unit cells are present in a cubic silver particle with an edge length of 25 nm?
   *b. How many silver atoms are there in this particle?
   c. If there is 1360 μg of silver per sock, how many of these particles does this correspond to?

**12.116.** In 2011, researchers at Duke University reported on the use of gold "nanorods," cylindrical gold particles to image brain tumors (Figure P12.116).
   a. If the nanorods have a diameter of 32 nm and a height of 67 nm, approximately how many gold atoms are in each nanorod?
   b. How many unit cells does this correspond to?
   c. What is the mass of each nanorod?

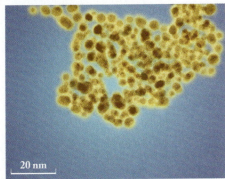

**FIGURE P12.116**

**12.117.** The center of Earth is composed of a solid iron core within a molten iron outer core. When molten iron cools, it crystallizes in different ways depending on pressure—in a bcc unit cell at low pressure and in a hexagonal unit cell at high pressures like those at Earth's center.
   a. Calculate the density of bcc iron given that the radius of an iron atom is 126 pm.
   b. Calculate the density of hexagonal iron given a unit cell volume of $5.414 \times 10^{-23}$ cm³.
   *c. Seismic studies suggest that the density of Earth's solid core is only about 90% of that of hexagonal Fe. Laboratory studies have shown that up to 4% by mass of Si can be substituted for Fe without changing the hcp crystal structure built on hexagonal unit cells. Calculate the density of such a crystal.

**12.118.** The unit cell of an alloy with a 1:1 ratio of magnesium and strontium is identical to the unit cell of CsCl. The unit cell edge of MgSr is 390 pm. What is the density of MgSr?

**12.119.** Gold and silver can be separately alloyed with zinc to form AuZn (unit cell edge 319 pm) and AgZn (unit cell edge 316 pm). The two alloys have the same unit cell. Which alloy is more dense?

**12.120.** Figure P12.120 shows four identical layers of atoms in an *aaaa* ... pattern. Using the yellow triangle as a guide, what is the coordination number of the holes between the layers? Are there any other kinds of holes in this structure?

**FIGURE P12.120**

**12.121.** Manganese steels are a mixture of iron, manganese, and carbon. Is the manganese likely to occupy holes in the austenite fcc unit cell, or are manganese steels substitutional alloys?

**12.122.** Aluminum forms alloys with lithium (LiAl), gold ($AuAl_2$), and titanium ($Al_3Ti$). Based on their crystal lattices, each of these alloys is considered to be a substitutional alloy.
   a. Do these alloys fit the general size requirements for substitutional alloys? The atomic radii for Li, Al, Au, and Ti are 152, 143, 144, and 147 pm, respectively.
   b. If the unit cell of LiAl is bcc, what is the density of LiAl?

**\*12.123.** The aluminum alloy $Cu_3Al$ crystallizes in a bcc unit cell. Propose a way that the Cu and Al atoms could be allocated between bcc unit cells that is consistent with the formula of the alloy.

**12.124.** **Light-Emitting Diodes** The colored lights on many electronic devices are light-emitting diodes (LEDs). One of the compounds used to make them is aluminum phosphide (AlP), which crystallizes in a sphalerite crystal structure.
   a. If AlP were an ionic compound, would the ionic radii of $Al^{3+}$ and $P^{3-}$ be consistent with the size requirements of the ions in a sphalerite crystal structure?
   b. If AlP were a covalent compound, would the atomic radii of Al and P be consistent with the size requirements of atoms in a sphalerite crystal structure?

## Silicon, Silica, Silicates, Silicone: What's in a Name?

**12.125.** Write a balanced chemical equation for each of the following reactions: (a) silicon dioxide reacts with carbon; (b) silicon reacts with chlorine; (c) germanium reacts with bromine.

**12.126.** Write balanced chemical equations for each of the following reactions: (a) tin reacts with chlorine to make a tetrahalide; (b) lead reacts with chlorine to make a dihalide; (c) silicon(IV) chloride is heated above 700°C.

**12.127.** Calcium silicide ($CaSi_2$) has a structure consisting of graphitelike layers of Si atoms with Ca atoms between the layers. In an X-ray diffraction analysis of a sample of $CaSi_2$, X-rays with a wavelength of 154 pm produced signals at $2\theta = 29.86°, 45.46°,$ and 62.00° for $n = 2, 3,$ and 4.
   a. What is the distance between the Si layers?
   b. If the $Ca^{2+}$ ion lies exactly halfway between the layers, what is the Ca–Si distance?
   c. $CaSi_2$ is sometimes used as a "deoxidizer" in converting iron ore to steel. Explain how $CaSi_2$ might fill this role. (*Hint*: What is the most common oxidation state of Si?)

**12.128.** The ionization energies and electron affinities of the group 14 elements have values close to the corresponding average values for all the elements in their rows in the periodic table. How does this "average" behavior explain the tendency for the group 14 elements to form covalent bonds rather than ionic bonds?

If your instructor assigns problems in smartwork, log in at **smartwork.wwnorton.com**.

# 13

# Organic Chemistry: Fuels, Pharmaceuticals, Materials, and Life

**13.1** Carbon: The Scope of Organic Chemistry

**13.2** Alkanes

**13.3** Alkenes and Alkynes

**13.4** Aromatic Compounds

**13.5** Amines

**13.6** Alcohols, Ethers, and Reformulated Gasoline

**13.7** Carbonyl-Containing Compounds

**13.8** Chirality

## Learning Outcomes

**LO1** Describe the differences among alkanes, alkenes, and alkynes
**Sample Exercise 13.1**

**LO2** Draw, name, and identify constitutional isomers and stereoisomers
**Sample Exercises 13.2, 13.3, 13.4, 13.5**

**LO3** Identify monomers given the structure of a polymer, and the structure of a polymer given the structure of a monomer
**Sample Exercises 13.6, 13.9, 13.11**

**LO4** Identify organic compounds based on their functional groups

**LO5** Compare the energy content of fuels
**Sample Exercise 13.7**

**LO6** Assess the properties of polymers
**Sample Exercises 13.8, 13.10**

**LO7** Identify chiral molecules
**Sample Exercises 13.12, 13.13**

## Carbon Everywhere

Today's world is a complex place. People have discovered disease-fighting drugs that save the lives of millions of patients each year who would otherwise suffer slow and painful deaths. But people are also responsible for pollutants in our air and water that present other health hazards and cost the lives of many. We can travel almost anywhere around the globe within a single day for business or pleasure, but getting there requires fossil fuels whose combustion contributes to climate change. Modern electronic devices enable us to be in touch with people on a scale never before seen, but today's sources of electrical energy are limited and new sources are difficult to develop.

All these advantages and problems have one thing in common—they deal with the chemistry of the compounds of carbon, an area called organic chemistry. Carbon is by far the most versatile element, able to form millions of different compounds with correspondingly millions of uses. Fossil fuels are carbon compounds, as are most molecules in living systems. The food we eat, the clothes we wear, the medicines we take are all compounds of carbon. Organic chemists are at the forefront of research in everything from alternative fuels to medicine to the search for extraterrestrial life. Knowledge of organic chemistry is essential if we are to extract energy from fossil fuels more efficiently and with less environmental impact, and to use these resources as feedstocks instead of fuels and turn them into more durable plastics, more resilient fabrics, and the next generation of life-saving drugs.

In this chapter we can give only an introduction to the vast area of organic chemistry. We concentrate on a few areas of major importance in daily life: fuels, drugs, and materials. We also lay the foundation for Chapter 20 on biochemistry, which addresses the molecules and reactions that sustain life. Although there are more than 50 million organic compounds, the types of chemical reactions they undergo are relatively

**Bionic Handshake** Scientists have used organic chemistry to create a sensitive ▶ skin for a robotic hand that is responsive to its environment, returning the correct pressure for a comfortable handshake.

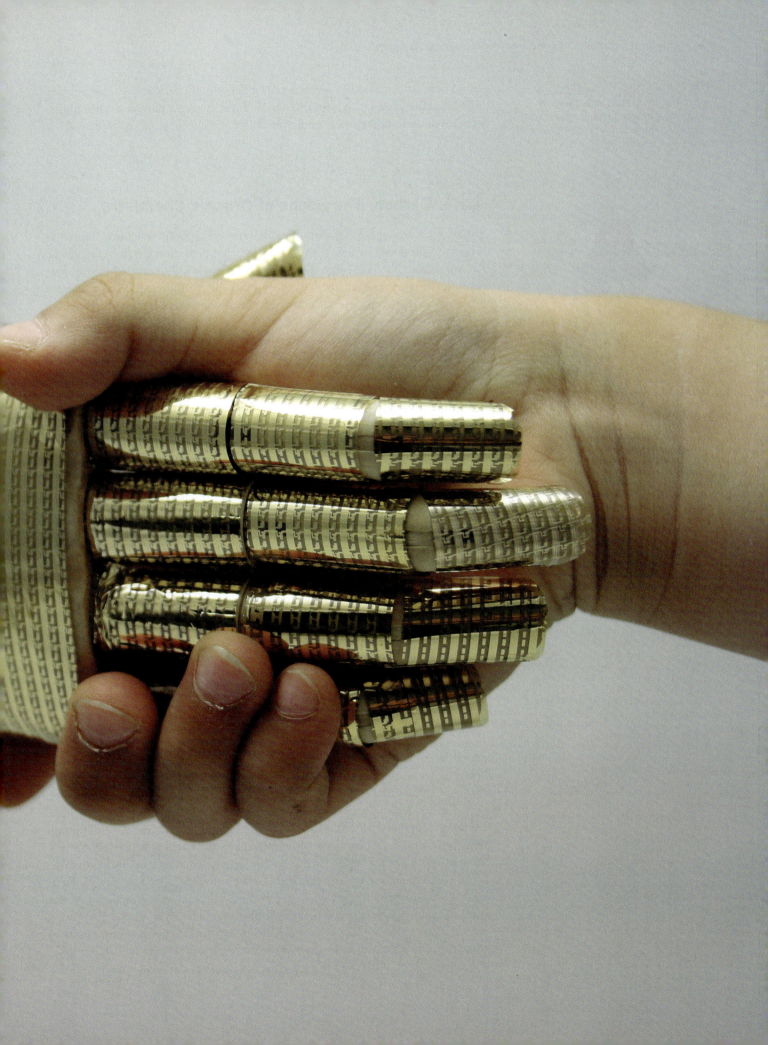

●○ **CONNECTION** In Chapter 10, we learned that compounds with the same molecular formula but different arrangements of atoms differ in physical properties such as melting point and boiling point.

few in number because they occur at only a handful of characteristic reaction sites called functional groups within molecules. These concepts provide a basic framework for understanding the amazing versatility of the compounds of carbon. ∎

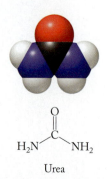

$$\begin{matrix} & O \\ & \| \\ H_2N & \overset{\displaystyle C}{\phantom{C}} & NH_2 \end{matrix}$$

Urea

**FIGURE 13.1** Urea was the first naturally occurring organic compound to be synthesized in the laboratory.

## 13.1 Carbon: The Scope of Organic Chemistry

The designation *organic* for carbon-containing compounds was once limited to substances produced by living organisms, but that definition has been broadened for two reasons. In 1828, Friedrich Wöhler (1800–1882) discovered how to prepare urea (Figure 13.1) in the laboratory "without the intervention of a kidney." Since then scientists have learned to synthesize many materials previously thought to be the products only of living systems. Second, chemists have learned how to synthesize many carbon-based materials that have never been produced by living systems. Today the study of **organic chemistry** encompasses the chemistry of all compounds containing carbon–carbon and carbon–hydrogen bonds, regardless of their origin.

So many compounds of carbon exist that we need some organizing principles to simplify learning about them. One of the most useful methods of characterization involves functional groups; this method groups molecules based on their reactivities. A second common method is based upon the size of molecules; this method divides molecules into categories of small molecules and polymers based primarily on the different physical behavior that characterizes the two groups.

### Families Based on Functional Groups

Much of the variety in the chemistry of carbon arises from a carbon atom's ability to form covalent bonds to other carbon atoms. These carbon atoms, in turn, can bond to more carbon atoms or atoms of other elements, making compounds containing a few atoms or many thousands of atoms. Additional variety is introduced because, in addition to long chains, these molecules may contain a multitude of branched structures as well as rings. Organic compounds can also contain nonmetallic elements in addition to carbon and hydrogen. If they contain metal atoms directly bound to carbon atoms, they are usually called organometallic compounds. We will discuss those substances in Chapter 17.

Managing the wealth of information about the millions of organic compounds requires some organizing concepts. Chemists group organic compounds into families on the basis of **functional groups**, which are subunits in a molecule that confer particular chemical and physical properties.

In earlier chapters (especially Chapter 10, Sections 10.2 and 10.4) we encountered two functional groups: the –OH group in alcohols and the –COOH group in carboxylic acids. Although we have yet to identify them as functional groups, we have seen carbon atoms make double bonds to oxygen in formaldehyde, $CH_2O$, and to carbon in ethylene, $C_2H_4$, and form delocalized π bonds in benzene, $C_6H_6$. In this chapter, we examine these and the other functional groups summarized in Table 13.1. When discussing functional groups, the convention is to use **R** to represent all of the molecule except the functional group. Think of R as "the rest of the molecule" (and don't confuse it with the italic *R* standing for the universal gas constant in the ideal gas equation).

**TABLE 13.1  Functional Groups of Organic Compounds**

| Name | Structural Formula of Group[a] | Example and Name |
|---|---|---|
| Alkane | R—H | $CH_3CH_2CH_3$  Propane |
| Alkene | C=C | C=C  Ethylene (ethene) |
| Alkyne | —C≡C— | H—C≡C—H  Acetylene (ethyne) |
| Aromatic | e.g, ⬡ | ⬡  Benzene |
| Amine | R—$NH_2$ <br> R—NHR′ <br> R—$NR_2$ | $H_3C$—$NH_2$  Methylamine |
| Alcohol | R—OH | $CH_3CH_2OH$  Ethanol |
| Ether | R—O—R′ | $CH_3CH_2OCH_2CH_3$  Diethyl ether |
| Aldehyde | R–C(=O)–H | $H_3C$–C(=O)–H  Acetaldehyde |
| Ketone | R–C(=O)–R′ | $H_3C$–C(=O)–$CH_3$  Acetone |
| Carboxylic acid | R–C(=O)–OH | $H_3C$–C(=O)–OH  Acetic acid |
| Ester | R–C(=O)–OR′ | $H_3C$–C(=O)–$OCH_3$  Methyl acetate |
| Amide | R–C(=O)–$NH_2$ | $H_3C$–C(=O)–$NH_2$  Acetamide |

[a]In compounds with two R groups, the two may be the same (R = R′) or they may be different (R ≠ R′).
*Note*: Functional groups are highlighted in yellow.

## Monomers and Polymers

A second organizing principle for organic compounds is based on molecular size. We have two categories, although the boundaries are somewhat arbitrary. All organic compounds having a molar mass typically less than 1000 g/mol are one category, and all larger ones, called **polymers**, are the second category, with molar masses up to and exceeding 1,000,000 g/mol. In between small molecules and polymers are midsize molecules called *oligomers*. The root word *meros* is Greek for "part" or "unit," so *polymer* literally means "many units" and *oligomer* means "a few units."

**organic chemistry**  the study of carbon-containing compounds, typically those with C—C and C—H bonds.

**functional group**  a structural subunit in organic molecules that imparts characteristic chemical and physical properties.

**R**  symbol in a general formula representing an organic group that has one available bond; it is used to indicate the variable part of a molecule so that the focus is placed on the functional group.

**polymer**  a very large molecule with high molar mass formed by bonding together a large number of small molecules of low molecular mass.

∞ **CONNECTION** We have already introduced many organic molecules in Chapters 4 through 11: acetic acid; methane, ethane, and other hydrocarbon fuels; acetylene used in welding; and ethanol and other alcohols.

**FIGURE 13.2** Items made of synthetic polymers are ubiquitous in modern society.

**monomer** a small molecule that bonds with others like it to form polymers.

**alkane** hydrocarbon in which all the bonds are single bonds with the general formula $C_nH_{2n+2}$.

**alkene** hydrocarbon containing one or more carbon–carbon double bonds.

**alkyne** hydrocarbon containing one or more carbon–carbon triple bonds.

**saturated hydrocarbon** an alkane.

**unsaturated hydrocarbon** an alkene or alkyne.

**hydrogenation** the reaction of an unsaturated hydrocarbon with hydrogen.

We have already seen examples of polymers in previous chapters, including polyvinyl chloride, used to make drain pipes and other building materials; and Teflon, used in nonstick cookware. Polymers are composed of small structural units called **monomers**. Polymers may be formed from more than one type of monomer unit; they may have more than one functional group; and they may have shapes other than long chains. However, the single feature that distinguishes them as a class is their large size.

## Small Molecules versus Polymers: Physical Properties

Before looking at how the composition and structure of polymers influence their behavior, let's first discuss the differences between the physical properties of small organic molecules and the physical properties of polymers. For now, we restrict our discussion to synthetic polymers made in the laboratory from small organic molecules (Figure 13.2).

All small organic compounds have well-defined properties. They have constant composition, for example, and their phase transitions take place at well-defined temperatures. In contrast, many synthetic polymers do not have constant composition and well-defined properties because they are mixtures of large molecules that are similar but not identical, because they differ in the number of monomers and the arrangement of monomers in the polymer chain. Their physical properties depend on the range and distribution of molecular masses in the sample.

Polymers typically do not have well-defined melting points and boiling points. Indeed, polymer molecules are so large that they cannot acquire sufficient kinetic energy to enter the gas phase. As temperature increases, some polymers gradually soften and become more malleable. Some may eventually melt and become very viscous liquids, whereas others remain solids up to temperatures at which their covalent bonds break, causing the material to decompose. For example, plastic grocery bags are made of the polymer polyethylene, the monomer unit of which is ethylene ($CH_2{=}CH_2$). The melting point of ethylene is −169°C; its boiling point is −104°C. Polyethylene is a tough and flexible solid at room temperature. It softens gradually over a range of temperatures from 85°C to 110°C, and at higher temperatures tends to decompose into molecules of low-molar-mass organic gases.

In addition, intermolecular forces between long chains in polymers can lead to both highly ordered, crystalline regions as well as less ordered, amorphous regions. When a polymer is heated, rearrangements of the molecules contribute to a broad range of melting temperatures. For now, we conclude that matter composed of polymers behaves very differently from matter composed of small organic molecules.

**CONCEPT TEST**

Which types of intermolecular forces are most likely to dominate between large molecules (polymers) containing long chains of carbon atoms bound to hydrogen atoms?

*(Answers to Concept Tests are in the back of the book.)*

# 13.2 Alkanes

The first three functional groups in Table 13.1 define the families of organic compounds called alkanes, alkenes, and alkynes. We introduced these compounds in Chapter 3 as a group called *hydrocarbons*—compounds composed only of hydrogen and carbon. We discussed some members of these families in Chapter 5 as compounds found in fuels. The three families within hydrocarbons are distinguished by the types of carbon–carbon bonds in their molecules: **alkanes** have only carbon–carbon single bonds; **alkenes** have one or more carbon–carbon double bonds; **alkynes** have one or more carbon–carbon triple bonds.

Alkanes are classified as **saturated hydrocarbons** because they contain the maximum ratio of hydrogen atoms to carbon atoms. The general molecular formula of linear alkanes is $C_nH_{2n+2}$, where $n$ is the number of carbon atoms per molecule. Alkenes and alkynes are **unsaturated hydrocarbons** that can combine with $H_2$ to form alkanes in a process called **hydrogenation**. A molecule with one C=C bond is described as having one *degree of unsaturation*. It combines with one molecule of $H_2$ to form one molecule of alkane (Figure 13.3). A molecule with one C≡C bond has two degrees of unsaturation and combines with two molecules of $H_2$; and a molecule with one double bond and one triple bond has three degrees of unsaturation. Determining the degree of unsaturation of an unknown compound using a hydrogenation reaction can help identify its structure.

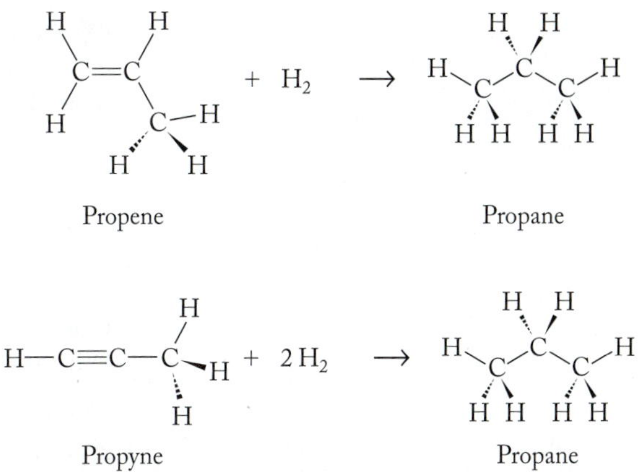

**FIGURE 13.3** Hydrogenation of propene requires 1 mole $H_2$ per mole propene; hydrogenation of propyne requires 2 moles $H_2$ per mole propyne.

●●● **CONNECTION** In Chapter 3, we defined hydrocarbons as compounds composed of only carbon and hydrogen atoms.

---

**SAMPLE EXERCISE 13.1   Distinguishing among         LO1
                         Alkanes, Alkenes, and Alkynes**

Three hydrocarbons, each containing four carbon atoms, are stored in separate, unlabeled flasks. If one is an alkane, one an alkene with one double bond, and one an alkyne with one triple bond, design an experiment based on their reactivity with $H_2$ gas to determine which is the alkane, which is the alkene, and which is the alkyne. Each flask contains one mole of compound.

**Collect and Organize** We react three compounds with hydrogen. The compounds contain the same number of carbon atoms, but one compound contains a carbon–carbon double bond and another contains a carbon–carbon triple bond.

**Analyze** The alkane is a saturated hydrocarbon, which means it already contains the maximum number of hydrogen atoms possible and so does not react with hydrogen. The alkene and alkyne are unsaturated, so they can both react with $H_2$ gas. One mole of the alkene with one double bond (or one degree of unsaturation) reacts with 1 mole of $H_2$. One mole of the alkyne with one triple bond (or two degrees of unsaturation) reacts with 2 moles of $H_2$. The flask containing the alkyne will require the most $H_2$ to react completely, while the flask containing the alkane will not react at all with $H_2$.

**Solve** If we allow 1 mole of each hydrocarbon to react with 2 moles of hydrogen and measure how much hydrogen is consumed, we can identify the compounds. We pick 2 moles of $H_2$ because that is the maximum amount that any of the three samples requires for complete reaction:

$$1 \text{ mol alkane} + 2 \text{ mol } H_2 \xrightarrow{\text{no reaction}} 2 \text{ mol } H_2 \text{ left}$$

$$1 \text{ mol alkene} + 2 \text{ mol } H_2 \rightarrow 1 \text{ mol alkane} + 1 \text{ mol } H_2 \text{ left}$$

$$1 \text{ mol alkyne} + 2 \text{ mol } H_2 \rightarrow 1 \text{ mol alkane} + 0 \text{ mol } H_2 \text{ left}$$

**Think About It** Measuring the amount of hydrogen that reacts with a hydrocarbon is a way to distinguish saturated from unsaturated hydrocarbons. As predicted, the alkyne, with two degrees of unsaturation, reacts with the most hydrogen. Unsaturated hydrocarbons may be distinguished if their degree of unsaturation differs.

⚙ **Practice Exercise** The labels have fallen off two containers in the lab. One of the fallen labels has structure A printed on it, and the other has structure B printed on it, as shown below:

$$CH_3—CH_2—CH\!\!=\!\!CH—CH\!\!=\!\!CH—CH_3$$
Compound A

$$CH_3—CH\!\!=\!\!CH—CH\!\!=\!\!CH—CH\!\!=\!\!CH_2$$
Compound B

Can you use hydrogenation reactions to decide which label belongs on which container? What is the structure of the product of hydrogenation in each case?

*(Answers to Practice Exercises are in the back of the book.)*

■ ···············································

**CONCEPT TEST** ·········································

Could we use information from hydrogenation reactions to distinguish a hydrocarbon with one triple bond from a hydrocarbon with two double bonds?

···················································

Now that we see how the three classes of hydrocarbons are related, let's look first in detail at the properties of the alkanes. We will develop an understanding of their behavior and a few guidelines regarding their names. In the next section, we will build on these concepts and do the same with alkenes and alkynes.

## Physical Properties and Structures of Alkanes

Alkanes are also known as *paraffins*, a name derived from Latin meaning "little affinity." This is a perfect description of the alkanes, which tend to be much less reactive than the other hydrocarbon families. Despite their lack of reactivity, alkanes are compounds of great importance because of their use as fuels, oils, and lubricants. *Unreactive* may not seem like the correct term to apply to compounds that are fuels, but alkanes do not react readily, even with oxygen. As evidence of this, consider that most fuels need some source of energy, such as a spark from a spark plug in an engine, to initiate combustion.

A brief word about alkane nomenclature will aid us in this discussion. The first four members of the alkane family are methane (1 carbon atom: a $C_1$ alkane), ethane (2 carbon atoms: a $C_2$ alkane), propane (a $C_3$ alkane), and butane (a $C_4$ alkane). The names of alkanes with more than 4 carbon atoms are derived from the Latin or Greek prefix for the number of carbon atoms per molecule, followed by -*ane*. Table 13.2 lists the prefixes for $C_1$ through $C_{10}$ alkanes.

In the language of organic chemistry, a **homologous series** is defined as a series of compounds in which members can be described by a general formula and have similar chemical properties. Table 13.2 is a homologous series of alkanes ranging from one to ten carbon atoms in length. They all have the general formula $C_nH_{2n+2}$ and each member of the series differs from the next by one $-CH_2-$ unit, called a **methylene group**. The terminal $-CH_3$ groups are called **methyl groups**. All of the alkanes in the table are **straight-chain alkanes**, which means a continuous sequence of carbon atoms with no branching. Straight-chain alkanes are identified by the chemical name: propane, butane, and so forth.

Table 13.3 shows a larger homologous series of alkanes with some of their physical properties. The data in Table 13.3 show similar trends in melting and

| TABLE 13.2 | Prefixes for Naming Alkanes | |
|---|---|---|
| **Prefix** | **Condensed Structure** | **Name** |
| Meth- | $CH_4$ | Methane |
| Eth- | $CH_3CH_3$ | Ethane |
| Prop- | $CH_3CH_2CH_3$ | Propane |
| But- | $CH_3(CH_2)_2CH_3$ | Butane |
| Pent- | $CH_3(CH_2)_3CH_3$ | Pentane |
| Hex- | $CH_3(CH_2)_4CH_3$ | Hexane |
| Hept- | $CH_3(CH_2)_5CH_3$ | Heptane |
| Oct- | $CH_3(CH_2)_6CH_3$ | Octane |
| Non- | $CH_3(CH_2)_7CH_3$ | Nonane |
| Dec- | $CH_3(CH_2)_8CH_3$ | Decane |

**TABLE 13.3  Melting Points and Boiling Points for Selected Straight-Chain Alkanes**

| Condensed Structure[a] | Use | Melting Point (°C) | Normal Boiling Point (°C) |
|---|---|---|---|
| $CH_3CH_2CH_3$ | Gaseous fuels | −190 | −42 |
| $CH_3(CH_2)_2CH_3$ | | −138 | −0.5 |
| $CH_3(CH_2)_3CH_3$ | | −130 | 36 |
| $CH_3(CH_2)_4CH_3$ | Gasoline | −95 | 69 |
| $CH_3(CH_2)_5CH_3$ | | −91 | 98 |
| $CH_3(CH_2)_6CH_3$ | | −57 | 126 |
| $CH_3(CH_2)_7CH_3$ | | −54 | 151 |
| $CH_3(CH_2)_{10}CH_3$ through $CH_3(CH_2)_{16}CH_3$ | Diesel fuel and heating oil | −10  28 | 216  316 |
| $CH_3(CH_2)_{18}CH_3$ through $CH_3(CH_2)_{32}CH_3$ | Paraffin candle wax | 37  72–75 | 343  na[b] |
| $CH_3(CH_2)_{34}CH_3$ and higher homologs | Asphalt | 72–76 | na |

[a]See Figure 13.4.
[b]na = not available; compound decomposes before boiling at 1 atm pressure.

---

**homologous series** a set of related organic compounds that differ from one another by the number of common subgroups, such as $-CH_2-$, in their molecular structures.

**methylene group** ($-CH_2-$) a structural unit that can make two bonds.

**methyl group** ($-CH_3$) a structural unit that can make only one bond.

**straight-chain alkane** a hydrocarbon in which the carbon atoms are bonded together in one continuous line. Linear alkane chains have a methyl group at each end with methylene groups connecting them.

**Kekulé structure** a structure showing all of the bonds in a covalently bonded molecule using lines but not showing lone pairs on the atoms.

---

boiling points: as straight-chain alkanes increase in molar mass, their melting and boiling points increase. This trend in physical properties is typical of all homologous series.

**CONCEPT TEST**

Alkanes have the general formula $C_nH_{2n+2}$. If an alkane has a molar mass of 114 g/mol, what is the value of $n$?

## Drawing Organic Molecules

Figure 13.4(a) shows the Lewis structure for the five-carbon alkane pentane. A Lewis structure shows all of the bonds in the molecule, as well as any lone pairs on the atoms. Alkanes do not have lone pairs on any of the atoms, but lone pairs are common in other organic molecules. When we draw structures of organic molecules showing all the bonds with lines but leaving off any lone pairs, the structural formulas are called **Kekulé structures** after August Kekulé (1829–1896), who first used this method for illustrating molecules. As you might imagine, it soon gets tedious to write Lewis or Kekulé structures for alkanes or any other organic molecule. For this reason, chemists use various shorter notations to convey structures in organic chemistry, such as those shown in Figure 13.4. The notations in Figure 13.4(b) are called *condensed structures*, a name that makes perfect sense when you compare the structure with the Lewis structure in Figure 13.4(a): a condensed structure does not show the individual bonds between atoms the way a Lewis structure does.

Structures that use subscripts to indicate the number of times a particular subgroup is repeated are also considered condensed structures. For example, the condensed structure of pentane can also be written as $CH_3(CH_2)_3CH_3$ (Figure 13.4b). The numerical subscript after the parenthetical methylene group

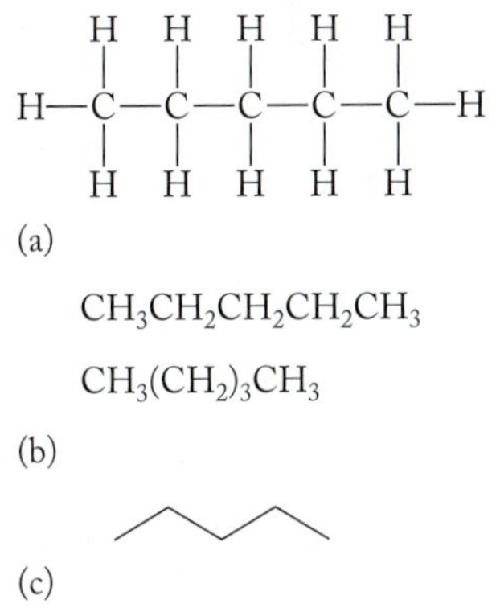

(a)

$CH_3CH_2CH_2CH_2CH_3$

$CH_3(CH_2)_3CH_3$

(b)

(c)

**FIGURE 13.4** Several representations of the molecular structure of pentane: (a) Lewis structure; (b) two condensed structures; (c) carbon-skeleton structure.

**CONNECTION** Kekulé structures are the same as the structural formulas we have been drawing throughout the text since Chapter 1.

or $CH_3(CH_2)_6CH_3$

(a) Condensed structural formula

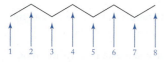

(b) Carbon-skeleton structure

**FIGURE 13.5** When converting from (a) a condensed structure to (b) a carbon-skeleton structure, the carbon atoms are represented by junctions between lines and each junction is assumed to have a sufficient number of H atoms to give that carbon atom four bonds.

means that three of these groups connect the two terminal methyl groups in this compound. (Note that the subscript indicating the number of methylene groups comes *after* the closing parenthesis. The subscript 2 inside the parentheses is for the two H atoms on the C atom of each methylene group.)

The most minimal notation, shown in Figure 13.4(c), is the *carbon-skeleton structure*, which has no alphabetic symbols for carbon and hydrogen atoms. (Atoms other than C and H are shown in carbon-skeleton structures, as we see in Section 13.5.) Figure 13.5 shows how a carbon-skeleton structure is created. Short line segments are drawn at angles to one another, and each line segment symbolizes one carbon–carbon bond in the molecule. The angles represent the bond angle between the two carbon atoms (109.5° in the case of $sp^3$ hybridized carbon atoms in alkanes). Each end of the zigzag line is a $-CH_3$ group, and every intersection of two line segments is a $-CH_2-$ group. In other words, it is understood that each carbon atom has the appropriate number of hydrogen atoms to give it a steric number of four. Hydrogen atoms are not shown in a carbon-skeleton structure because all carbon atoms are known to make four bonds, and any bonds not shown are understood to be C—H bonds.

---

**SAMPLE EXERCISE 13.2    Drawing Alkane Structures    LO2**

(a) Write the condensed structure and draw the carbon-skeleton structure for the 3-carbon straight-chain hydrocarbon propane. (b) Write the condensed structure and draw the carbon-skeleton structure for the 12-carbon dodecane.

**Collect and Organize** Figure 13.4 summarizes the differences between condensed and carbon-skeleton structures.

**Analyze** All the carbon atoms in each molecule are in one straight chain. Condensed structures show the symbol for each element in a molecule and subscripts indicating the numbers of atoms of each element, but they do not show C—H or single C—C bonds. We can group all methylene groups into one term by using parentheses, followed by a subscript showing the number of $-CH_2-$ groups. Carbon-skeleton structures use short line segments to represent the carbon–carbon bonds in molecules. In the carbon-skeleton structures, we use zigzag line segments because the intersections of these segments represent the carbon atoms in the skeleton.

**Solve**

a. The condensed structure for propane comes from its Lewis structure (on the left), and then we write each carbon followed by H plus a subscript showing the number of H atoms bonded to that carbon atom (structure in the center):

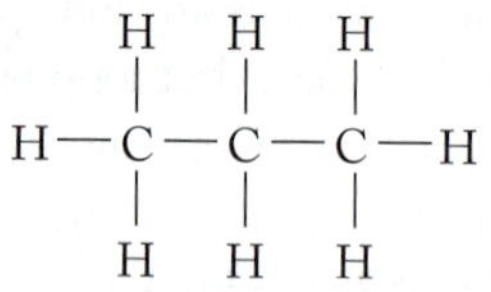

$$H-\overset{\displaystyle H}{\underset{\displaystyle H}{C}}-\overset{\displaystyle H}{\underset{\displaystyle H}{C}}-\overset{\displaystyle H}{\underset{\displaystyle H}{C}}-H \qquad CH_3CH_2CH_3$$

Lewis structure                Condensed structure            Carbon-skeleton structure

To draw the propane carbon-skeleton structure (on the right), first draw a short line slanted upward to the right; this represents the bond between the first and second carbons. Then without removing your pencil from the paper, draw a short line slanted downward to the right to represent the bond between the second and third carbons. This simple inverted V represents propane because there is a $CH_3-$ group on each end and a $-CH_2-$ group at the peak.

b. Dodecane is a continuous chain of 12 carbon atoms. There must be two methyl groups at the ends with ten methylene groups between them. The condensed structure is therefore

$$CH_3CH_2CH_2CH_2CH_2CH_2CH_2CH_2CH_2CH_2CH_2CH_3$$

or, gathering the ten $-CH_2-$ groups: $CH_3(CH_2)_{10}CH_3$. The carbon-skeleton structure must contain as many ends and intersections as there are carbon atoms in the molecule:

Here we have 2 ends and 10 intersections for a total of 12 carbon atoms.

**Think About It** Condensed and carbon-skeleton structures must both contain the same number of atoms and the same number of bonds, even though the bonds are not shown in the condensed structure and not all of the atoms are shown in the carbon-skeleton structure.

⚙ **Practice Exercise** Draw the carbon-skeleton structure of hexane, $CH_3(CH_2)_4CH_3$, and a condensed structure of heptane:

Heptane

## Constitutional (Structural) Isomers

The alkane family would be huge even if it consisted of only straight-chain molecules, but another structural possibility makes the family even larger. We first saw this possibility in Chapter 10 when we compared the boiling points of compounds having the same molecular formula, and hence the same molar mass, but different shapes. For example, alkanes with four or more carbon atoms can have straight-chain structures, such as the one on the left in Table 13.4, or *branched* structures, as shown on the right in the table. A *branch* is a side chain attached to the main carbon chain.

The molecular formula of both structures in Table 13.4 is $C_4H_{10}$, but they represent different compounds with different properties. In recognition of these differences, the two compounds have different formal names (butane and 2-methylpropane). At this point, however, it is not important to be able to name them but to recognize them as **constitutional isomers** (also called *structural isomers*): compounds with the same molecular formula but with their atoms connected in different ways. This difference makes structural isomers chemically distinct from one another. 2-Methylpropane is a **branched-chain hydrocarbon**, which means

| TABLE 13.4 | Comparing Structural Isomers | |
|---|---|---|
| | **Butane** | **2-Methylpropane** |
| **Condensed Structure** | $CH_3CH_2CH_2CH_3$ | $CH_3CH(CH_3)CH_3$ <br> or <br> $CH_3CHCH_3$ <br> $\vert$ <br> $CH_3$ |
| **Carbon-Skeleton Structure** |  |  |
| **Melting Point (°C)** | −138 | −160 |
| **Normal Boiling Point (°C)** | 0 | −12 |
| **Density of Gas (g/L)** | 0.5788 | 0.5934 |

**constitutional isomers** molecules having the same molecular formula but different arrangements of atoms; they are different compounds and have different chemical and physical properties.

**branched-chain hydrocarbon** an organic molecule in which the chain of carbon atoms is not linear.

that its molecular structure contains a main chain (the longest carbon chain in the molecule) and at least one side chain.

We can illustrate constitutional isomerism with either condensed or carbon-skeleton structures, as shown in Table 13.4. There are two ways to show the location of the methyl group. One way is to use condensed structures and to put the $CH_3$ group in parentheses next to the carbon atom to which it is bonded: $CH_3CH(CH_3)CH_3$. We can also show the carbon–carbon bond between the $CH_3$ group and the center carbon atom of the three-carbon main chain. In this instance, it is easier to see the constitutional isomers of $C_4H_{10}$ using the carbon-skeleton structures.

To determine whether two condensed structures or carbon-skeleton structures represent two different compounds, a single compound, or two constitutional isomers, we first translate each structure into a molecular formula. If the molecular formulas are different, the structures represent two different compounds. If the molecular formulas are the same, we must compare the way the atoms are connected in the two structures. In doing so, we may have to reverse or rotate one of the structures. For example, structures 1 and 2 below may seem to be constitutional isomers, but rotating structure 2 by 180° shows that it is the same as structure 1. The two represent the same compound: a four-carbon main chain with a one-carbon side chain connected to the carbon atom next to a terminal carbon.

(1)                  (2)              180° rotation          After rotation

To determine whether two structures are identical, draw the structures with the longest chain horizontal, and then check to see whether the same side chains are attached at the same positions along the longest chain.

**SAMPLE EXERCISE 13.3    Recognizing Constitutional Isomers  LO2**

Do the two structures in each set describe the same compound, constitutional isomers, or compounds with different molecular formulas?

a. $(CH_3)_2CHCH_2CH(CH_3)_2$

b.

c.

**Collect and Organize** We have structures showing connectivity of the atoms and from which we can determine molecular formulas. Identical compounds have the same molecular formula and the same connectivity of the atoms. Constitutional isomers have the same molecular formula but different connectivity.

**Analyze** In each pair, we check first to see if the molecular formulas are the same. That means we count the number of carbon atoms and hydrogen atoms. If the molecular formulas are the same, we may have either one compound drawn two ways or a pair of constitutional isomers. If the carbon skeletons are the same in any pair, the drawings represent the same hydrocarbon.

**Solve**

a. The molecular formulas are the same: $C_7H_{16}$. Converting the condensed structure to a carbon-skeleton structure gives us

$$(CH_3)_2CHCH_2CH(CH_3)_2 \quad \rightarrow$$

This structure is identical to the carbon-skeleton structure in this set. Therefore the condensed structure and carbon-skeleton structure represent the same molecule.

b. Both molecules in this set have the molecular formula $C_8H_{18}$. If we rotate the second structure 180° about its vertical axis, we get structure (1), so the two compounds are identical.

(1)  (2) 180° rotation  after rotation

c. The left structure contains nine carbon atoms, but the right structure contains only eight. Therefore the two structures represent different compounds.

**Think About It** Just because two compounds have the same molecular formula, they are not necessarily identical. We must determine whether they are constitutional isomers.

**Practice Exercise** Do the two structures in each set describe the same compound, constitutional isomers, or compounds with different molecular formulas?

a.

b.

c.

# Naming Alkanes

Now that we know how to draw alkanes, let's look at a few simple rules for naming them. The nomenclature system follows the same pattern for all families of organic compounds, so we begin with rules for naming alkanes and will add a few more for alkenes and alkynes. These rules are called IUPAC rules after the International Union of Pure and Applied Chemistry, the organization that defined them. Appendix 7 contains a more extensive treatment of nomenclature for organic compounds.

We already presented in Table 13.2 the convention for naming the first ten straight-chain alkanes. Because every alkane larger than propane has constitutional isomers, we need an unambiguous way to name branched-chain molecules. To name the branched-chain members of the set, follow these steps:

1. Select the longest chain of carbon atoms and use the prefixes in Table 13.2 to name this as the *parent structure*. For the structure

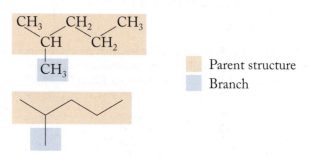

Parent structure
Branch

the parent name is pentane because the longest chain is 5 carbon atoms long.

2. Identify each branch and name it with the prefix from Table 13.2 that defines the number of carbons in the branch; append the suffix -*yl* to the prefix. A –$CH_3$ group is methyl, a –$CH_2CH_3$ group is ethyl, and so forth. The structure in step 1 has a methyl group branch. The name of the branch comes before the name of the parent structure, and the two are written together as one word: methylpentane.

3. To indicate the point where the branch is attached, number the carbon atoms in the parent chain so that the branch (or branches, if there is more than one) has the lowest possible number. In the molecule in step 1, start numbering from the left so that the methyl group is on carbon atom 2 in the parent chain.

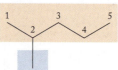

Indicate the position of the branch by a 2 followed by a hyphen in front of the name: 2-methylpentane.

4. If the same group is attached more than once to the parent structure, use the prefixes di-, tri-, tetra-, and so forth to indicate the number of groups present. The position of each group is indicated by the appropriate number before the group name, with the numbers separated by commas. Thus the name 2,4-dimethylpentane is the structure

$$CH_3CH(CH_3)CH_2CH(CH_3)CH_3$$
1 2 3 4 5

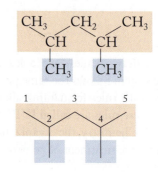

5. If different groups are attached to a parent chain, they are named in alphabetical order, as in 3-ethyl-4-methylheptane for the following structure (achieving the lowest numbering, however, takes precedence over alphabetization of the substituents):

$$\underset{1}{CH_3}\underset{2}{CH_2}\underset{3}{CH}(\underset{}{CH_2CH_3})\underset{4}{CH}(\underset{}{CH_3})\underset{5}{CH_2}\underset{6}{CH_2}\underset{7}{CH_3}$$

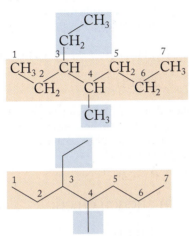

Note that numbering the parent chain in this structure begins at the carbon on the far left, in accord with the instruction in step 3, so that the branch locations have the lowest possible numbers. For example, the compound shown here is 5-ethyl-2-methyloctane, not 4-ethyl-7-methyloctane:

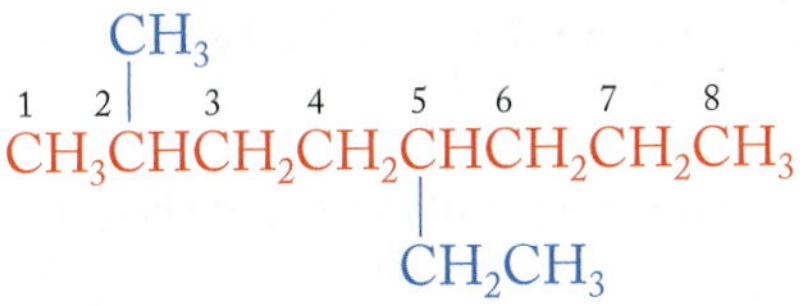

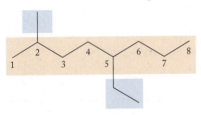

Correct numbering

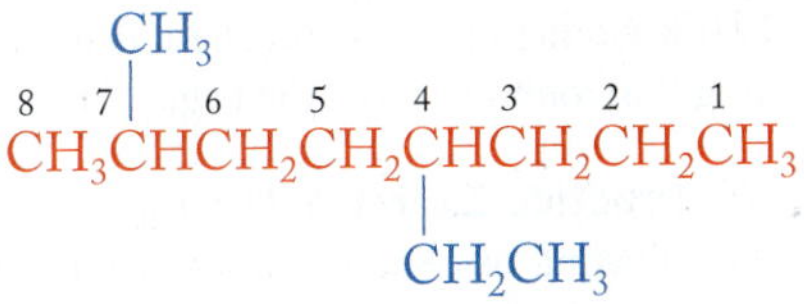

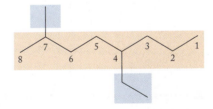

Incorrect numbering

---

**SAMPLE EXERCISE 13.4**    **Identifying the Longest Chain in Organic Molecules**      **LO2**

Determine the number of carbon atoms in the longest chain in the following two branched, saturated hydrocarbons:

a. 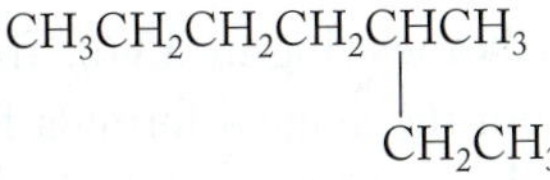 $CH_3CH_2CH_2CH_2CHCH_3$
        $|$
     $CH_2CH_3$

b. 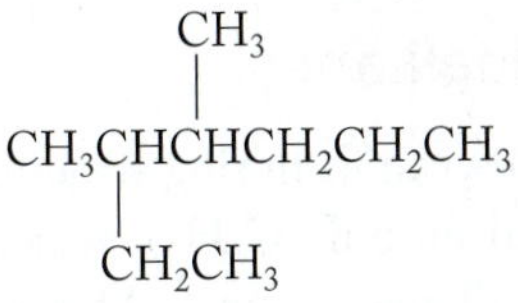 $CH_3CHCHCH_2CH_2CH_3$
      $|$   $CH_3$ (above)
   $CH_2CH_3$

**Collect and Organize** We are asked to identify the longest carbon chain or *main chain* in two compounds, given their condensed structures.

**Analyze** There are bonds between each two carbon atoms in the condensed structures. Carbon–carbon bonds to additional chains are indicated by a single line. We need to determine which of these branches belong to the *main chain* and which belong to the *side chains*.

**Solve** We start at one end of any branch and assign numbers to each carbon atom. If the chain branches, we must choose one branch to follow. Later, we may wish to return to the compound and number the carbons starting from a different end of the molecule or taking a different branch.

a. Starting with the carbon atom furthest to the left and numbering consecutively from left to right, we reach a branch at carbon atom 5. If we continue to the right, we find that the chain contains six carbons. If we take the branch at carbon atom 5, however, we find that the chain contains seven carbon atoms. This makes the longest chain in the compound seven carbon atoms long.

$$\underset{1}{CH_3}\underset{2}{CH_2}\underset{3}{CH_2}\underset{4}{CH_2}\underset{5}{CH}\underset{6}{CH_3}$$
$$|$$
$$CH_2CH_3$$

$$\underset{1}{CH_3}\underset{2}{CH_2}\underset{3}{CH_2}\underset{4}{CH_2}\underset{5}{CH}CH_3$$
$$|$$
$$\underset{6\quad7}{CH_2CH_3}$$

If we start at the right end of the molecule, we do not find a longer chain; the only other chain in the molecule is four carbon atoms long.

b. There are four places to start counting carbon atoms in structure b. Some possible ways to count them are as follows:

$$\overset{CH_3}{\underset{1\ \ 2\ \ |3\ \ 4\ \ 5\ \ 6}{CH_3CHCHCH_2CH_2CH_3}}$$
$$|$$
$$CH_2CH_3$$

$$\overset{CH_3}{\underset{5\ \ |4\ 3\ \ 2\ \ 1}{CH_3CHCHCH_2CH_2CH_3}}$$
$$|$$
$$\underset{6\quad7}{CH_2CH_3}$$

$$\overset{5}{\overset{CH_3}{\underset{|4\ 3\ \ 2\ \ 1}{CH_3CHCHCH_2CH_2CH_3}}}$$
$$|$$
$$CH_2CH_3$$

The longest chain in the molecule (the main chain) also has seven carbon atoms. (Note that we are just counting the numbers of carbon atoms in the chains to determine the longest; this does not reflect the way we number the carbon atoms to name the compounds.)

**Think About It** The main chain in an organic compound may not always be the one running horizontally across the page.

⚙ **Practice Exercise** Determine the number of carbon atoms in the longest chain in the following two branched, saturated hydrocarbons:

a.  b.

**CONCEPT TEST**

A useful way of determining whether two similar structures are the same is to name each according to the IUPAC rules. Why is this the case?

## Cycloalkanes

Alkanes can form ring structures called **cycloalkanes** (Figure 13.6). They have the general formula $C_nH_{2n}$, which is different from the general formula for straight-chain alkanes ($C_nH_{2n+2}$) because cycloalkanes have one more carbon–carbon bond and two fewer hydrogen atoms per molecule than the alkanes with the same number of carbon atoms. Although cycloalkanes have no C=C or C≡C bonds, 1 mole of a cycloalkane can, at least in theory, react with 1 mole of hydrogen to make 1 mole of a straight-chain alkane. In practice, however, only cycloalkanes with three-carbon and four-carbon rings react with hydrogen.

▶❚❚ **CHEMTOUR** Structure of Cyclohexane

**cycloalkane** ring-containing alkane with the general formula $C_nH_{2n}$.

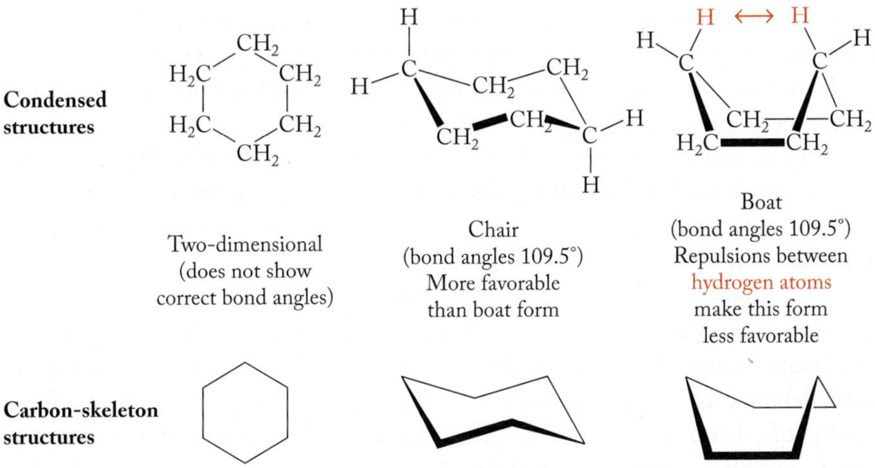

Condensed structures

Two-dimensional (does not show correct bond angles)

Chair (bond angles 109.5°) More favorable than boat form

Boat (bond angles 109.5°) Repulsions between hydrogen atoms make this form less favorable

Carbon-skeleton structures

**FIGURE 13.6** The six-membered ring of cyclohexane is drawn either flat or in styles that show its three-dimensional puckered forms.

**CONNECTION** We continue to use the drawing conventions introduced in Chapter 9, where a solid wedge indicates a bond that comes out of the paper toward the viewer. The thicker line then means a bond is closer to the viewer than a thinner line.

The left column of Figure 13.6 shows the condensed structure and carbon-skeleton structure of cyclohexane drawn in two dimensions, where the C—C—C bond angles appear to be 120°. Because the carbon atoms have $sp^3$ hybrid orbitals, however, we expect the angles to be 109.5°, and indeed they are in this molecule. A more accurate representation of the ring is shown in the adjacent structures, in which the ring is puckered instead of flat and the bond angles are 109.5°.

Ring structures are possible for other alkanes, but smaller rings are less stable than the six-membered ring. One important feature that determines the relative stability of rings is the size of the C—C—C bond angle. For example, cyclopropane (a $C_3$ ring) exists but is a more reactive species because the interior ring angles are 60°, far from the ideal bond angle of 109.5° for an $sp^3$ hybridized carbon atom (Figure 13.7). No puckering is possible in a three-membered ring to relieve the strain in this system, and cyclopropane tends to react in a fashion that opens up the ring and relieves the strain. The C—C—C bond angles in cyclobutane and cyclopentane are larger, so there is less ring strain. Rings of six $sp^3$ hybridized carbons and beyond are essentially the same as straight-chain alkanes in terms of bond angle and have no ring strain. Six-membered rings are the most favored, because other thermodynamic factors have an impact on the formation of rings with seven or more carbons.

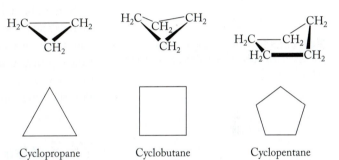

Cyclopropane          Cyclobutane          Cyclopentane

**FIGURE 13.7** Cyclic alkanes have the general formula $C_nH_{2n}$ and are considered to be unsaturated hydrocarbons because they have two fewer hydrogen atoms in their structures than the straight-chain hydrocarbon with the same number of carbon atoms. Shown here are cyclopropane ($n = 3$), cyclobutane ($n = 4$), and cyclopentane ($n = 5$).

▶❙❙ **CHEMTOUR** Cyclohexane in 3-D

## Sources and Uses of Alkanes

The principal source of liquid alkanes on Earth is crude oil. Natural gas is the major source for the simplest alkane, methane, as well as smaller quantities of low-molar-mass alkanes including ethane, propane, and perhaps some butanes. Methane is often associated with oil deposits, but it is also produced during bacterial decomposition of vegetable matter in the absence of air, a condition that frequently arises in swamps. Hence, methane's common name is swamp gas or marsh gas (Figure 13.8). Frequently in the reducing environment of a marsh, a compound known as phosphine ($PH_3$) is also formed. Phosphine and methane together spontaneously ignite; this produces a ghostly flame known as will-o'-the-wisp that features prominently in some legends and gothic mysteries. Methane can also be produced in coal mines, where it can be especially dangerous because it forms explosive mixtures with humid air, giving rise to another common name for the gas: firedamp.

**FIGURE 13.8** Methane bubbles trapped in a frozen pond. Methane is produced by rotting organic matter at the bottom of the pond.

By far the most common use of alkanes in our lives is as fuels. Combustion reactions between alkanes and oxygen provide energy to power vehicles, generate electricity, warm our homes, and prepare our meals. Gasoline, kerosene, and diesel fuels are mostly mixtures of alkanes containing up to 20 carbon atoms per molecule, as summarized in Table 13.3. Alkanes with higher boiling points are viscous liquids used as lubricating oils. Low-melting solid alkanes ($C_{20}$–$C_{40}$) are used in candles and in manufacturing matches. Very heavy hydrocarbon gums and solid residues ($C_{36}$ and up) are used for paving roads.

The combustion of fossil fuels is an important topic in the 21st century for several reasons. The production of $CO_2$, a greenhouse gas, leads to the potential for climate change. The supplies of crude oil are also limited and with increasing world demand for fossil fuels, society faces choices in how best to allocate finite supplies of oil and natural gas. The alkanes distilled from crude oil serve not only as fuels but also as the starting materials for building more complex molecules. They are the major source of carbon for the chemical manufacturing industry.

Some alkanes have therapeutic value. Mineral oil is a mixture of $C_{15}$–$C_{24}$ alkanes and is used as a skin ointment ("baby oil") to treat diaper rash and to alleviate some forms of eczema. It is used in many cosmetics, creams, and ointments. Taken orally, mineral oil acts as a laxative. As we continue our study of organic compounds, we will return to these dual themes of fuels and pharmaceuticals with the functional groups that we encounter.

**CONNECTION** In Chapter 11 we discussed the distillation of crude oil to produce gasoline, kerosene, and other hydrocarbon mixtures useful as fuels and as feedstocks for the chemical industry.

**CONNECTION** In Chapter 8 we discussed methane as one of the greenhouse gases.

## 13.3 Alkenes and Alkynes

In Section 13.2 we introduced alkenes, which are compounds having one or more carbon–carbon double bonds (two or more $sp^2$ hybridized carbon atoms), and alkynes, which are compounds having one or more carbon–carbon triple bonds (two or more $sp$ hybridized carbon atoms). Both of these families represent unsaturated hydrocarbons. Alkenes are minor components of crude oil and are prevalent in many plants, including pine needles, celery, and ginger (Figure 13.9).

Alkynes are found in crude oil as well, but they are generally more difficult to find in nature because the carbon–carbon triple bond is quite reactive. Nevertheless, carbon–carbon triple bonds are found in some drugs, along with other functional groups. Some examples are shown in Figure 13.10. Capillin is an antifungal drug; pargyline is used to treat hypertension; and panaxytriol is a potent antitumor compound isolated from ginseng. The simplest alkyne, $C_2H_2$, also known as acetylene, is used in welding.

**FIGURE 13.9** Some naturally occurring alkenes include (a) pinene in pine resin, (b) selinene in celery, and (c) zingiberene in oil of ginger.

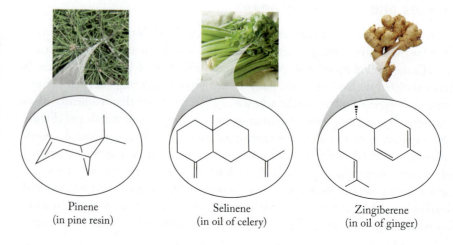

Pinene (in pine resin)

Selinene (in oil of celery)

Zingiberene (in oil of ginger)

Acetylene and some other alkynes are manufactured by the controlled oxidation of alkanes. When the simplest alkane, methane, is oxidized completely, the products are $CO_2$ and $H_2O$. But if this oxidation is carried out in a highly controlled process, acetylene can be produced:

$$6\ CH_4(g) + O_2(g) \rightarrow 2\ CH\equiv CH(g) + 2\ CO(g) + 10\ H_2(g) \quad (13.1)$$

This reaction illustrates an important difference between alkanes and the unsaturated hydrocarbons: alkanes are the most reduced form of carbon. The $sp^2$ hybridized carbons in alkenes are in a higher oxidation state than the $sp^3$ hybridized carbon atoms of alkanes, and the $sp$ hybridized carbons in alkynes are in an even higher oxidation state. Their capacity to be reduced makes alkenes and alkynes more reactive than alkanes, and the carbon–carbon double bond in alkenes is one of the most versatile functional groups in organic chemistry. Structures of a $C_5$ alkene and a $C_5$ alkyne are shown in Figure 13.11.

Alkenes and alkynes share many properties with alkanes. The melting and boiling points of homologous series of these compounds (Table 13.5) vary with molar mass and size, just as with the alkanes.

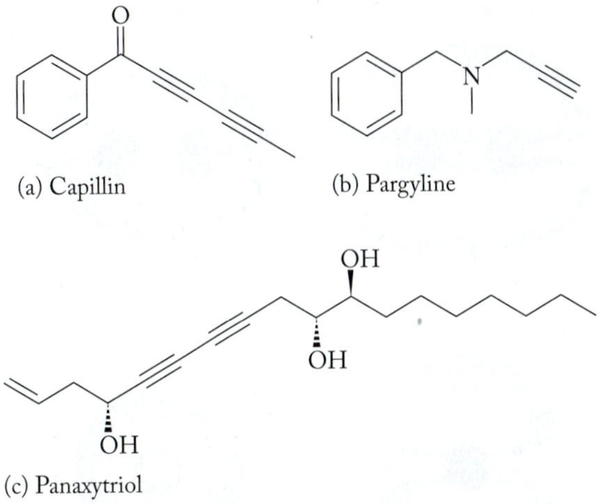

Molecules may contain more than one alkene or alkyne group, as shown in Figure 13.10. In particular, many molecules found in crude oil or produced by living systems have several double bonds. We discuss molecules with multiple double bonds in Chapter 20, but for now we concentrate on the properties associated with small molecules containing only one or a small number of double or triple bonds.

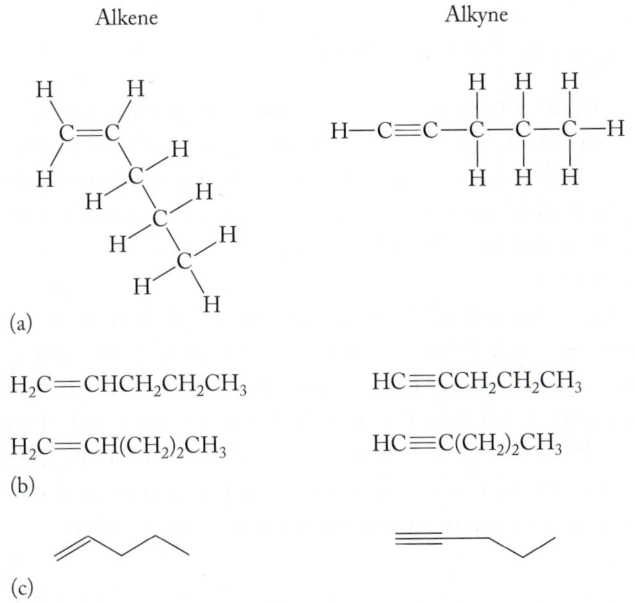

(a) Capillin    (b) Pargyline

(c) Panaxytriol

**FIGURE 13.10** The alkyne functional group is present in (a) capillin, an antifungal drug; (b) pargyline, used to treat hypertension; and (c) panaxytriol, a potent antitumor drug.

**FIGURE 13.11** Structures of a $C_5$ alkene and a $C_5$ alkyne. (a) Lewis structures. (b) Condensed structures. (c) Carbon-skeleton structures.

**TABLE 13.5  Melting Points and Normal Boiling Points of Homologous Series of Alkenes and Alkynes**

| Condensed Structure: Alkene | Melting Point (°C)[a] | Normal Boiling Point (°C) |
|---|---|---|
| $H_2C{=}CHCH_3$ | −185 | −47 |
| $H_2C{=}CHCH_2CH_3$ | −185 | −6 |
| $H_2C{=}CH(CH_2)_2CH_3$ | −138 | 30 |
| $H_2C{=}CH(CH_2)_3CH_3$ | −140 | 63 |
| $H_2C{=}CH(CH_2)_4CH_3$ | −119 | 94 |
| $H_2C{=}CH(CH_2)_5CH_3$ | −104 | 123 |
| $H_2C{=}CH(CH_2)_6CH_3$ | −81 | 146 |
| $H_2C{=}CH(CH_2)_7CH_3$ | −87 | 171 |
| **Condensed Structure: Alkyne** | | |
| $HC{\equiv}CCH_3$ | −102 | −23 |
| $HC{\equiv}CCH_2CH_3$ | −126 | 8 |
| $HC{\equiv}C(CH_2)_2CH_3$ | −90 | 40 |
| $HC{\equiv}C(CH_2)_3CH_3$ | −132 | 71 |
| $HC{\equiv}C(CH_2)_4CH_3$ | −81 | 100 |

[a]Melting points increase with molar mass but also depend on how molecules fit into crystal lattices. Melting points of alkenes with even numbers of carbon atoms form one series that follows this trend; alkenes with odd numbers of carbon atoms (gray shading) form another series.

## Chemical Reactivities of Alkenes and Alkynes

Figure 13.12 shows the electron distributions in the π bonding orbitals in alkenes and alkynes. The electrons in these orbitals, where electron density is greatest above and below the plane of the carbon skeleton, are more accessible to reactants than the electrons in the σ bonds of

Alkene                    Alkyne

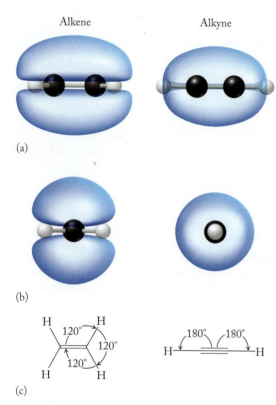

(a)

(b)

(c)

**FIGURE 13.12** The characteristic structural feature of alkenes and alkynes is the presence of electrons in π orbitals. (a) The electron distribution in a C=C double bond and a C≡C triple bond. (b) View of electron distribution looking down the carbon–carbon bond axis. (c) The idealized H–C–C bond angles in double and triple bonds.

alkanes, where electron density is greatest between the atoms. This is one reason unsaturated hydrocarbons are more reactive than saturated ones.

As an illustration of this difference in reactivity, consider how alkanes and alkenes react with the hydrogen halides HX = HCl, HBr, and HI. With alkanes, there is no reaction:

$$HX(g) + H_3C—CH_3(g) \rightarrow \text{no reaction} \qquad (13.2)$$

With an alkene, however, the hydrogen halides react with the double bond to make alkyl halides (alkanes in which a halogen has been substituted for one of the hydrogen atoms):

$$\underset{\substack{\text{Hydrogen} \\ \text{halide}}}{HX(g)} + \underset{\text{Alkene}}{H_2C=CH_2(g)} \rightarrow \underset{\text{Alkyl halide}}{CH_3CH_2X(\ell)} \qquad (13.3)$$

This reaction is called an **addition reaction** because one reactant adds across a multiple bond in another reactant to form one product.

Alkynes also react with hydrogen halides. They differ, however, in that two molecules of a hydrogen halide react with one triple bond and yield alkanes that bear two halogen atoms as products:

$$2\,HX(g) + HC≡CH(g) \rightarrow CH_3CHX_2(\ell) \qquad (13.4)$$

Because both double and triple bonds react with many reagents in addition to hydrogen halides, alkenes and alkynes are useful substances in the industrial production of other compounds.

## Isomers of Alkenes and Alkynes

Molecules that contain alkene and alkyne functional groups can have straight or branched chains, with the same types of constitutional isomers we saw with alkanes. One additional facet of constitutional isomerization involves the location of the double or triple bond in a molecule. For example, consider the straight-chain isomers of the alkene that contains five carbon atoms and one double bond (Figure 13.13).

Applying the test used in Section 13.2 for alkanes, we see that structures a and d in Figure 13.13 are the same. This is easier to see if we look at the carbon-skeleton structures in the figure. In both cases the double bond is between carbon atoms 1 and 2, that is, between C1 and C2. Recall that with branched-chain alkanes, we number the carbons from whichever end gives the carbon attached to the branch the lowest number. The same holds for functional groups, as shown here, where the carbons in structure d must be numbered from right to left.

**addition reaction** a reaction in which one reactant adds across a multiple bond in another reactant to form one product.

**cis isomer** (also called **Z isomer**) molecule with two like groups (such as two R groups or two hydrogen atoms) on the same side of the molecule.

**trans isomer** (also called **E isomer**) molecule with two like groups (such as two R groups or two hydrogen atoms) on opposite sides of the molecule.

**stereoisomers** molecules with the same formulas and the same connectivities between their atoms, but with different spatial arrangements of their atoms.

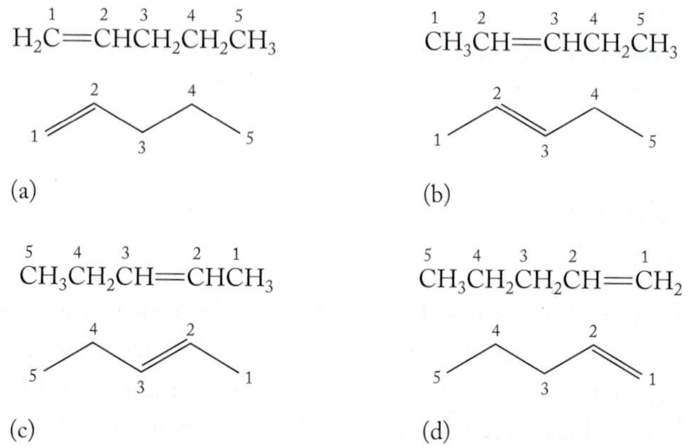

**FIGURE 13.13** Four possible constitutional isomers of pentene, $C_5H_{10}$. Notice that structures a and d are identical, as are b and c.

$$\underset{1 \quad 2 \quad 3 \quad 4 \quad 5}{H_2C=CHCH_2CH_2CH_3}$$

(a)

$$\underset{1 \quad 2 \quad 3 \quad 4 \quad 5}{CH_3CH=CHCH_2CH_3}$$

(b)

$$\underset{5 \quad 4 \quad 3 \quad 2 \quad 1}{CH_3CH_2CH=CHCH_3}$$

(c)

$$\underset{5 \quad 4 \quad 3 \quad 2 \quad 1}{CH_3CH_2CH_2CH=CH_2}$$

(d)

Structures b and c are equivalent and are constitutional isomers of a and d because they have the same chemical formula but their double bond is in a different location. Drawing the carbon-skeleton structures of b and c, however, presents us with a new situation. After we draw the first three atoms of structure b, as in Figure 13.14(a), and draw a straight dashed line through the double bond, we see that we have two options for how to orient the rest of the molecule relative to the double bond. We can place the bond between C3 and C4 on the same side of the dashed line as the methyl group at C1 (Figure 13.14b) or on the opposite side (Figure 13.14c).

These two molecules are isomers of each other because they have the same molecular formula but different structures and therefore different properties. The isomer in Figure 13.14(b) is called either the **Z isomer** (Z stands for the German word *zusammen* or "together") or the **cis isomer** (*cis* is Latin for "on this side"), which in this case translates to "the methyl group and the chain after the double bond are both *together* or *on this side* of the structure." The isomer in Figure 13.14(c) is called either the **E isomer** (E for *entgegen* or "opposite") or **trans isomer** (*trans* is Latin for "across"). These molecules are called **stereoisomers**, and they exist because there is no free rotation about the double bond. The system of naming using cis and trans is in wide use and is sufficient for the simple molecules discussed in this text. The *E/Z* system is routinely used for more complex molecules in which more than two different substituents are attached to a double bond.

Recall from Chapter 9 that a double bond forms from the overlap of two unhybridized *p* orbitals on adjacent carbon atoms. As Figure 13.15 shows, for the carbon atoms joined in a double bond to rotate freely, they have to twist about the bond axis, eliminating the orbital overlap and breaking the bond. Breaking a π bond in 1 mole of an alkene costs about 290 kJ of energy, and that much energy is not available to the molecules at room temperature. This situation gives rise to restricted rotation about a carbon–carbon double bond and to the existence of stereoisomers.

Now let's examine the stereoisomers of structure c from Figure 13.13. In Figure 13.16(a), the two hydrogen atoms are on the same side of the double bond; this is the cis isomer. In Figure 13.16(b), the two hydrogen atoms are on opposite sides of the double bond; this is the trans isomer. The two molecules are stereoisomers.

A comparison of the stereoisomers in Figures 13.14 and 13.16 shows that the two cis structures are identical and the two trans structures are identical. Therefore, the straight-chain alkenes with five carbon atoms exist as three isomers: structure a from Figure 13.13, plus the cis and trans isomers of structure b. All three isomers are chemically distinct compounds.

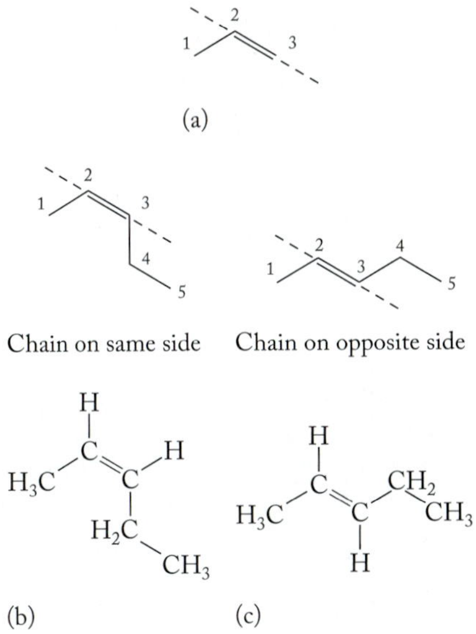

**FIGURE 13.14** (a) The first three atoms of a carbon chain with a double bond between C2 and C3. (b) The chain continues on the same side of the double bond as C1 in the cis isomer. (c) The chain continues on the opposite side of the double bond as C1 in the trans isomer.

**CONNECTION** In Chapter 5 we introduced rotations about bonds as one of the types of motion molecules experience as part of their overall kinetic energy.

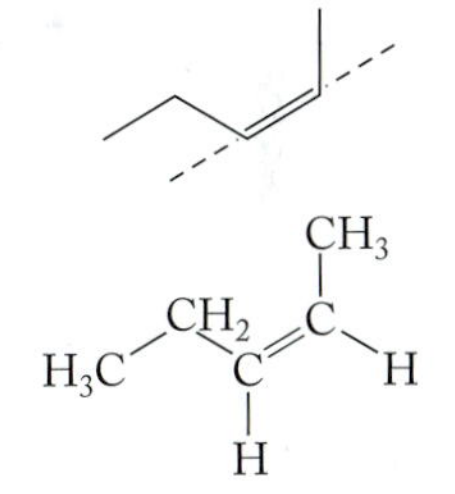

(a) The hydrogen atoms on the double bond are cis

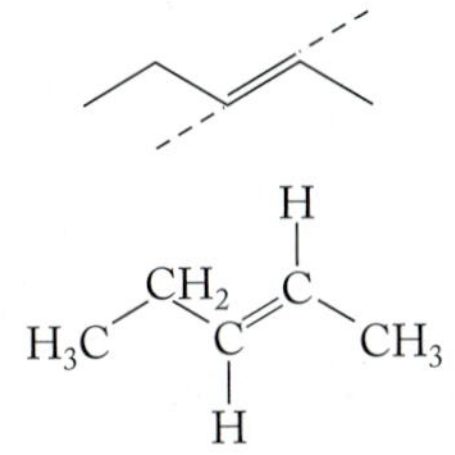

(b) The hydrogen atoms on the double bond are trans

**FIGURE 13.16** The positions of the H atoms in alkenes can be used to distinguish between (a) cis and (b) trans isomers.

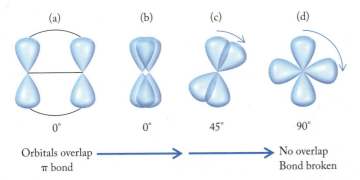

**FIGURE 13.15** (a) To form a π bond, $p_z$ orbitals overlap to establish a region of shared electron density above and below the plane of the C—C bond. (b) If you look down the carbon–carbon bond axis, the orbitals line up. (c) If you rotate one carbon atom while keeping the other fixed, the orbitals are no longer parallel and do not overlap. (d) If you rotate one of the two bonded C atoms far enough, the π bond breaks.

---

**CONCEPT TEST** • • • • • • • • • • • • • • • • • • • • • • • • • • • • • • • • • • • • • • • • • •

Which of the following alkenes has cis and trans isomers?

$$CH_2$$

a. $CH_2{=}CHCH_2CH_3$    b. $CH{=}CH$    c. $(CH_3)_2C{=}C(CH_3)_2$

d. $(CH_3)_2C{=}CH_2$    e. $CH_3CH{=}CHCH_3$

• • • • • • • • • • • • • • • • • • • • • • • • • • • • • • • • • • • • • • • • • • • • • • • • • • •

## Naming Alkenes and Alkynes

To name straight-chain alkenes and alkynes, the prefixes in Table 13.2 are used to identify the length of the chain. The suffix *-ene* is appended if the compound is an alkene and *-yne* if the substance is an alkyne. The carbon atoms in the chain are numbered so that the first carbon atom in the double or triple bond has the lowest number possible, and that number precedes the name, followed by a hyphen. Stereoisomers are identified by writing *cis-* or *trans-* before the number. Thus the compounds in Figure 13.16(a) and (b) are *cis*-2-pentene and *trans*-2-pentene, respectively.

---

**SAMPLE EXERCISE 13.5**    **Identifying and Naming**    **LO2**
                             **Stereoisomers and Constitutional**
                             **Isomers**

Write the condensed structure and draw the carbon-skeleton structure of the five isomers of the six-carbon straight-chain alkene containing one double bond, and name each isomer.

**Collect and Organize** We are to draw and name five isomers that each have six carbon atoms in a single chain and one C=C double bond. Figures 13.11, 13.13, 13.14, and 13.16 show examples of condensed and carbon-skeleton structures of alkenes.

**Analyze** We must consider two kinds of isomers: constitutional isomers, which depend on where the double bond is located in the chain, and stereoisomers, which depend on the orientation of groups about the double bond (cis/trans isomers). A straight-chain six-carbon alkene has a maximum of five places where a C=C can be placed; however, some locations may result in identical molecules. Not all alkenes have cis and trans isomers.

**Solve** Let's start with the constitutional isomers, which have the double bond at different locations, and then draw the stereoisomers (cis/trans) where possible. If all six carbons are

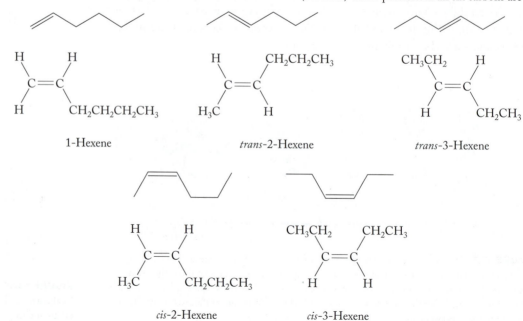

1-Hexene                  *trans*-2-Hexene                  *trans*-3-Hexene

*cis*-2-Hexene                  *cis*-3-Hexene

in one straight chain, then a double bond can be between C1 and C2, C2 and C3, or C3 and C4. These isomers are named 1-hexene, 2-hexene, and 3-hexene, respectively, where the number indicates the location of the double bond along the chain. Chains with a double bond between C4 and C5 or C5 and C6 are identical to the isomers with a double bond between C2 and C3 or C1 and C2, respectively.

1-Hexene does not have stereoisomers because the carbon atoms that form the double bond have three H atoms and only one nonhydrogen atom attached. Only 2-hexene and 3-hexene can have stereoisomers. *cis*-2-Hexene and *cis*-3-hexene have the two R groups on the same side of the double bond. *trans*-2-Hexene and *trans*-3-hexene have R groups on opposite sides of the double bond.

**Think About It** The number of isomers depends on the number of independent locations for a double bond in an alkene. Not all alkenes have stereoisomers.

**Practice Exercise** Draw and name the five isomers of the molecule with this carbon skeleton and one carbon–carbon double bond:

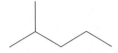

**CONCEPT TEST**

(a) Does a straight-chain hydrocarbon with a terminal double bond, such as 1-pentene, $CH_2{=}CHCH_2CH_2CH_3$, have stereoisomers? (b) Do alkynes have stereoisomers?

## Polymers of Alkenes

Some widely used polymeric alkanes are produced industrially from small alkenes. The alkane polymer with the simplest structure is linear polyethylene (PE), produced from ethylene ($CH_2{=}CH_2$) at high temperature and pressure:

$$n\,CH_2{=}CH_2 \rightarrow {\left[\!\!\left[ CH_2{-}CH_2 \right]\!\!\right]}_n \qquad (13.5)$$

Polyethylene is a **homopolymer**, which means it is composed of only one type of monomer. Its condensed structure is $CH_3(CH_2)_nCH_3$, but there are so many more methylene groups than methyl groups that the structure is frequently written ${\left[\!\!\left[ CH_2CH_2 \right]\!\!\right]}_n$ to highlight the structure and composition of the monomer. Polyethylene is also an example of an **addition polymer**, which is a polymer constructed by adding many molecules together to form the polymer chain without the loss of any atoms or molecules.

In most products made of polyethylene, *n* is a very large number, ranging from 1000 to almost 1 million. Polyethylene has a wide range of properties that depend on the value of *n* and on whether the polymer chains are straight or branched. In low-density PE (LDPE)—a stretchable, soft plastic used in films and wrappers—*n* ranges from 350 to 3500 and the chains are branched. When the bagger at the grocery store asks, "Paper or plastic?" the plastic in question is LDPE.

When the molar mass of a PE polymer is between 100,000 and 500,000 and the chains are straight, the polymer has physical properties different from those of grocery bags. This straight-chain polymer—a rigid, translucent solid called high-density polyethylene (HDPE)—is used in milk containers, electrical insulation, and toys.

Why are the properties of straight-chain HDPE (rigid, tough) so different from those of branched-chain LDPE (stretchable, soft)? Think of the branched polymer (Figure 13.17) as a tree branch with lots of smaller branches attached to it. The polymer has three dimensions, so it can have branches that come out of the plane of the paper. In contrast, the straight-chain polymer is like a long, straight pole. Suppose you had a pile of 100 tree branches and a pile of 100 poles, and your task was to stack each pile into the smallest possible volume to fit into a truck. You can certainly pile the branches on top of one another, but they will not fit together

**homopolymer** a polymer composed of only one kind of monomer unit.

**addition polymer** macromolecule prepared by adding monomers to a growing polymer chain without the loss of any atoms or molecules.

**FIGURE 13.17** Low-density polyethylene consists of branched chains, but high-density polyethylene consists of straight chains. Efficient stacking—and thus high density—is possible in large polymer molecules of HDPE but not in large polymer molecules of LDPE.

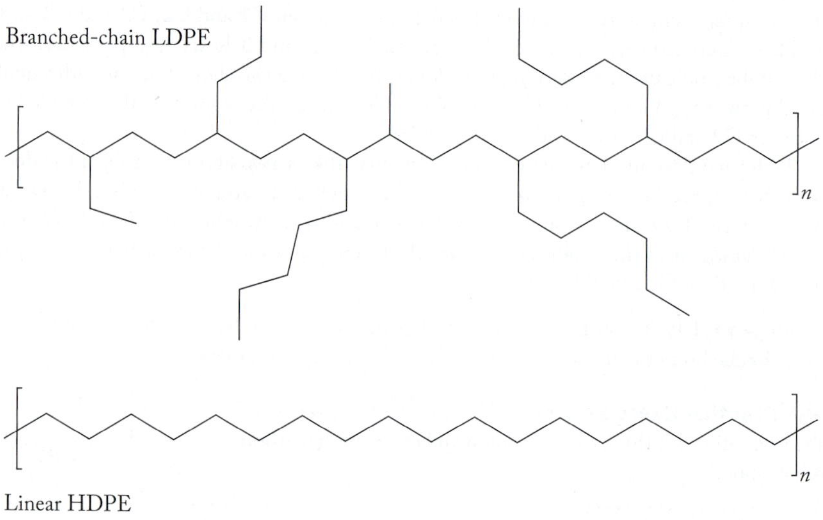

Branched-chain LDPE

Linear HDPE

**FIGURE 13.18** This prosthetic hip joint consists of a metal shaft (black) that fits into the thigh bone and ends in a silver ball embedded in a (white) plastic socket that is cemented into the pelvis. To make the joint last longer, the inner surface of the socket may be coated with a relatively new polymer called ultrahigh molecular weight PE (UHMWPE), which is a very tough material and highly resistant to wear. UHMWPE consists of linear molecules with an average molar mass over 3 million.

neatly, and probably the best you can do is to make the pile a bit more compact. In contrast, you can stack the poles into a very compact pile.

The same situation arises with the branched and linear molecules of polyethylene. Branched PE is low density because the molecules do not line up neatly. Their density is low compared to HDPE, primarily because the branched molecules stack less efficiently. This makes LDPE more deformable and softer; HDPE is more rigid and even has some regions that are crystalline because the packing is so uniform. Ultrahigh molecular weight PE (UHMWPE; $n > 100,000$) is an even tougher material, because not only are the molecules straight, they are significantly larger than the molecules in HDPE. UHMWPE is used as a coating on some artificial ball-and-socket joints (Figure 13.18) because it is extremely resistant to abrasion and makes the joints last longer. The different forms of polyethylene illustrate how the size and shape of its molecules affect the physical properties of a material.

**CONCEPT TEST**

During recycling, articles made from LDPE are separated from those made from HDPE (Figure 13.19). Why?

$$-\left[CH_2-CH_2\right]_n$$

Polyethylene

**FIGURE 13.19** Products made of polyethylene bear a recycle symbol that identifies them as straight-chain molecules (high-density) or branched-chain molecules (low-density).

Of course, chemical composition also plays a role in determining physical properties. If the hydrogen atoms in PE are all replaced with fluorine atoms, the resultant polymer is chemically very unreactive, capable of withstanding high temperatures, and has a very low coefficient of friction, which means other things do not stick to it. This polymer is Teflon, $+CF_2CF_2+_n$ (Figure 13.20), most familiar for its use as a nonstick surface in cookware. However, Teflon tubing is also used in the grafts inserted into small-diameter blood vessels during vascular surgery on limbs. The analogous material cannot be formed with chlorine, but a polymer does exist in which every other $-CH_2-$ in the polymer chain is $-CCl_2-$. Its repeating monomer unit is $+CH_2CCl_2+_n$ (Figure 13.21), and the polymer is the familiar thin, flexible plastic used as Saran wrap.

Hydrocarbons containing two or more double bonds are frequently used to manufacture polymers. For example, polymerization of butadiene yields a stretchy, synthetic rubber useful in rubber bands (Figure 13.22). Polyisoprene, prepared from 2-methyl-1,3-butadiene, is used in surgical gloves.

$+CF_2-CF_2+_n$

Teflon

**FIGURE 13.20** Repeating unit of polytetrafluoroethylene (Teflon).

$+CH_2-CCl_2+_n$

**FIGURE 13.21** Repeating unit of poly(1,1-dichloroethylene).

(a)

(b)

**FIGURE 13.22** Monomer units and polymer structures for (a) butadiene and (b) a methylated butadiene known as isoprene.

**FIGURE 13.23** Polypropylene is a common material for furniture, containers, clothing, lighting fixtures, and even objects of art. In addition to being moldable into many shapes, polypropylene is a good thermal insulator and does not absorb water easily.

---

**SAMPLE EXERCISE 13.6   Identifying Monomers       L03**

Polypropylene, $+CH_2CH(CH_3)+_n$, is an addition polymer used in the manufacture of fabrics, ropes, and other materials (Figure 13.23). Draw condensed and carbon-skeleton structures of the monomer used to prepare polypropylene and name it.

**Collect and Organize** We are given a condensed structure of a polymer and asked to identify the monomer used in its preparation. We know that polypropylene is an addition polymer, so the monomer must be an alkene.

**Analyze** To understand the relationship between the polymer and the monomer from which it is made, let's look at Equation 13.5 in the reverse direction (Equation 13.6):

$$+CH_2-CH_2+_n \rightarrow n\ CH_2{=}CH_2 \qquad (13.6)$$

Breaking the blue bonds in Equation 13.6 and making the red carbon–carbon single bond a double bond illustrates the relationship between polyethylene and ethylene, the alkene monomer. We need to apply a similar analysis to polypropylene.

**Solve** The relationship between polypropylene and its monomer is illustrated by Equation 13.7:

$$+CH_2-CH+_n \rightarrow n\ CH_2{=}CH(CH_3) \qquad (13.7)$$
$$\quad\quad\ |$$
$$\quad\ CH_3$$

$CH_3CH{=}CH_2$

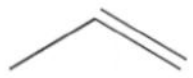

**FIGURE 13.24** The monomer propene is polymerized to make polypropylene.

Breaking the two blue bonds and making the red C—C bond a C=C double bond yields the alkene shown in Figure 13.24. This three-carbon alkene has the name propene. There is no need to precede the name of the monomer with a number to indicate the position of the C=C bond, because it has to be between the C1 and C2 carbon atoms.

**Think About It** The structural difference between propene and ethylene is the presence of a $CH_3-$ group bonded to one of the two $sp^2$ carbon atoms in propene instead of an H atom in ethylene.

**Practice Exercise** Draw the condensed and carbon-skeleton structures of the monomer used to make poly(methyl methacrylate), PMMA, a polymer used in shatterproof transparent plastic that can be used in place of glass:

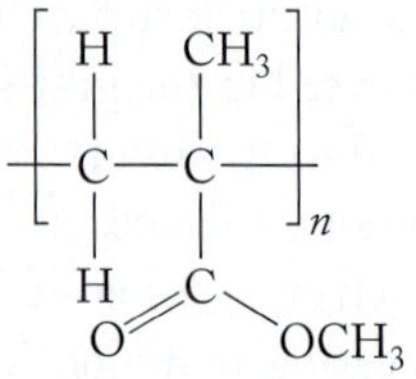

All addition polymers based on addition reactions of monosubstituted ethylene are called **vinyl polymers** because the $CH_2{=}CH-$ subunit is called the **vinyl group**, a name derived from *vinum* (Latin: wine). The name was given to the group by 18th-century chemists who prepared ethylene ($CH_2{=}CH_2$) from ethanol ($CH_3CH_2OH$), the alcohol in wine and other liquors.

Polyvinyl chloride, PVC, is widely used in commercial articles ranging from plastic pipes for plumbing to computer cases. Classic vinyl phonograph records are made from PVC. The condensed structures for the monomer and the polymer are shown in Equation 13.8:

$$n\,CH_2{=}CHCl \rightarrow -\!\!\left[CH_2\!-\!CHCl\right]\!\!\overline{\phantom{x}}_n \qquad (13.8)$$

**CONCEPT TEST**

Suggest a structural reason why Saran wrap (Figure 13.21), with two chlorine atoms on every other carbon atom, is a soft, flexible polymer, while PVC, with one chlorine atom on every other carbon atom, is more rigid.

Alkenes derived from crude oil are the primary source of monomers used to make most of the polymers used in construction, in fabrics, as wrapping and packaging material, and in medical devices. Vinyl polymers are the world's second largest selling plastics materials, and polymers in this category are extraordinarily versatile. You probably encounter five to ten vinyl polymers before you leave your room in the morning: vinyl shower curtains, vinyl drain pipes, vinyl flooring, vinyl insulation around electrical conduits. As we explore more organic functional groups, the basic concepts developed for the vinyl polymers will apply to polymers in other categories: the features of functional group, size, and shape determine the chemical and physical properties of these extraordinarily useful materials.

## 13.4 Aromatic Compounds

Among the components of gasoline that play an important role in increasing octane ratings (a measure of the ignition temperature of the fuel and its ability to resist engine "knock") is the class of compounds called *aromatic* hydrocarbons. Benzene is an example of an *aromatic compound*, a cyclic, planar molecule with delocalized π electrons above and below the plane of its six carbon and six hydrogen atoms.

As their class name implies, aromatic hydrocarbons have distinctive odors. However, *aromaticity* from a chemist's perspective is associated with the molecular and electronic structure of cyclic, planar molecules with $sp^2$ hybridized carbon atoms joined by a combination of alternating σ and π bonds (Figure 13.25). Aromatic compounds are relatives of alkenes because they contain carbon–carbon double bonds. However, because their chemical and physical properties are distinct from those of alkenes, they merit designation as a separate family.

The most common aromatic compound is benzene, $C_6H_6$. The different ways we view the bonding in benzene using Lewis theory and valence bond theory are summarized in Figure 13.25. As noted in Sections 8.5 and 9.5, the delocalized electrons in benzene lead to considerable resonance stability in this molecule, and this is true of all other aromatic molecules as well.

The stability of aromatic systems has an impact on their chemical reactivity. In Section 13.3 we saw that alkenes react rapidly with hydrogen halides to make halogenated alkanes. In contrast, benzene does not react at all if HBr gas is bubbled through it. Alkenes and aromatic compounds differ with respect to many other reactions, so the classification of aromatic systems as a unique family is justified.

## Isomers of Aromatic Compounds

Many compounds can be formed by replacing the hydrogen atoms in an aromatic ring with other substituents. For example, when one methyl group replaces a hydrogen atom in benzene, we get methylbenzene, also known by its common name, toluene:

Toluene

All the positions around the benzene ring are equivalent, so it does not matter which of its six carbon atoms is bonded to the methyl group. This is why we do not have to write a numerical prefix in the name methylbenzene.

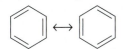 =

(a) Carbon-skeleton structures showing resonance forms of benzene

Skeletal symbol of benzene ring

(b) Sigma bonds in benzene

(c) Unhybridized *p* orbitals of carbon atoms

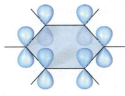

(d) Delocalized π cloud of electrons above and below plane of ring

**vinyl polymer** one of the family of polymers formed from monomers containing the subgroup $CH_2=CH-$.

**vinyl group** the subgroup $CH_2=CH-$.

▶❚❚ **CHEMTOUR** Structure of Benzene

**CONNECTION** In Chapters 8 and 9 we introduced benzene as an aromatic compound and described bonding in the benzene molecule using Lewis theory and valence bond theory.

**FIGURE 13.25** Different views of the bonding in benzene. (a) Carbon-skeleton structures showing resonance and double-bond delocalization. (b) Hexagonal array of σ bonds. (c) The unhybridized $p_z$ orbitals on the $sp^2$ hybridized carbons. (d) Delocalized π electrons above and below the ring.

We have, however, three options for attaching two methyl groups to a benzene ring, so there are three constitutional isomers of dimethylbenzene, also known as xylene. We distinguish between the three constitutional isomers by numbering the carbon atoms to give the substituents the lowest possible numbers. From left to right below, the three dimethylbenzenes are 1,2-dimethylbenzene; 1,3-dimethylbenzene; and 1,4-dimethylbenzene. Toluene and xylenes are used in inks, glues, and disinfectants.

1,2-Dimethylbenzene     1,3-Dimethylbenzene     1,4-Dimethylbenzene

The existence of isomers and the variety of substituents that can be attached to an aromatic system result in the occurrence of a large number of aromatic compounds.

Benzene rings can share one or more of their hexagonal sides and thereby form polycyclic aromatic molecules, often referred to as polycyclic aromatic hydrocarbons, or PAHs. Three such compounds are naphthalene, anthracene, and phenanthrene:

Naphthalene          Anthracene          Phenanthrene

Extensive delocalization of the $\pi$ electrons over all the rings makes these structures particularly stable. In addition to being found in fossil fuels, they may be formed during the incomplete combustion of hydrocarbons and are present in particularly high concentrations in the soot from incinerators and diesel engines. They have also been identified in interstellar dust clouds and in blackened portions of grilled meat. When introduced into the environment, these compounds persist and are among the most long-lived of hydrocarbons.

**CONCEPT TEST** ........................................................

Why isn't hexatriene considered an aromatic compound?

1,3,5-Hexatriene

........................................................

## Polymers Containing Aromatic Rings

Individual aromatic rings, as well as fused rings, are flat molecules. They tend to stack neatly (Figure 13.26), and this tendency gives rise to useful properties in materials that incorporate aromatic systems. Replacing one hydrogen atom in

**FIGURE 13.26** The aromatic rings on neighboring chains in polystyrene stack together and provide strength to the material.

benzene with a vinyl group gives the monomer styrene, and the polymer made from this monomer is polystyrene (PS):

Styrene          Polystyrene

Solid PS is a transparent, colorless, hard, inflexible plastic. In this form it is used for compact disc cases and plastic cutlery. A more common form of PS, however, is the *expanded solid* made by blowing $CO_2$ or pentane gas into molten polystyrene, which then expands and retains voids in its structure when it solidifies. One form of this expanded PS is Styrofoam, the familiar material of coffee cups and take-out food containers (Figure 13.27).

The difference in properties between transparent, colorless, inflexible nonexpanded PS and opaque, white, pliable Styrofoam can be explained by considering the role of the aromatic ring in aligning the polymer chains. Branches in the chains have the same effect that we saw with polyethylene, but the aromatic rings and their tendency to stack (Figure 13.26) provide additional interactions that make chain alignment more favorable energetically. The aromatic rings along two neighboring chains can stack, and this stacking makes nonexpanded PS rigid. When the chains are blown apart by a gas, the stacking is disrupted and the chains open to form cavities that fill with air, making expanded PS a good thermal insulator and packing material. The presence of air in Styrofoam is illustrated in Figure 13.28.

**FIGURE 13.27** Polystyrene can be made into either rigid or foamed products. In its nonexpanded form, it is rigid and strong, suitable for making such products as plastic knives, forks, and spoons. In its expanded form, it is Styrofoam, used in carry-out food containers, packing materials, and thermal insulation in buildings.

**FIGURE 13.28** When the Styrofoam coffee cup on the left is placed under pressure, some of the air between the polystyrene chains is forced out. The cup shrinks to the size on the right but retains its overall shape.

## 13.5 Amines

Nitrogen atoms are the defining components of functional groups in another important family of organic molecules called **amines**. In organic compounds, nitrogen atoms—and any atoms other than carbon, hydrogen, or metals—are called **heteroatoms** and are shown explicitly in Kekulé structures (without showing the lone pairs), in condensed structures, and in carbon-skeleton structures of organic compounds. Notice the N at the junction of three lines in Benadryl and pargyline in Figure 13.29.

Amines containing an alkyl or aromatic group are thought of as being derived from ammonia, $NH_3$. If one hydrogen atom in ammonia is replaced by an R group, the compound is called a *primary amine*. If two hydrogen atoms are replaced by R groups, the compound is a *secondary amine*; if all three hydrogen atoms are replaced by R groups, a *tertiary amine*:

| $RNH_2$ | $R_2NH$ | $R_3N$ |
|---|---|---|
| Primary amine | Secondary amine | Tertiary amine |

**amine** organic compound that contains a group with the general formula $RNH_2$, $R_2NH$, or $R_3N$, where R is any organic subgroup.

**heteroatom** any atom other than carbon, hydrogen, or metals in an organic compound.

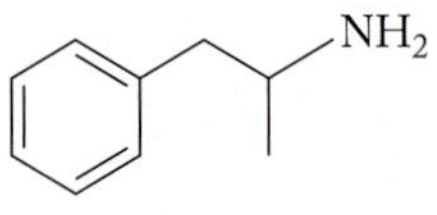

Amphetamine

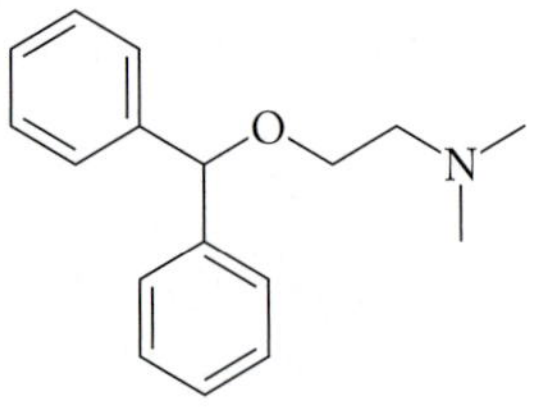

Benadryl

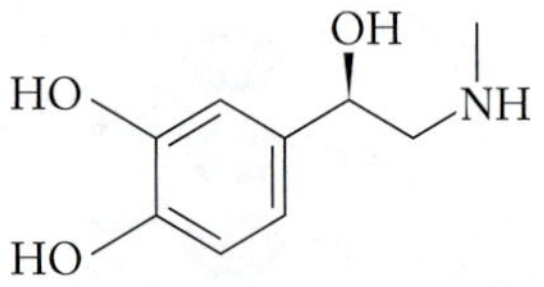

Adrenaline

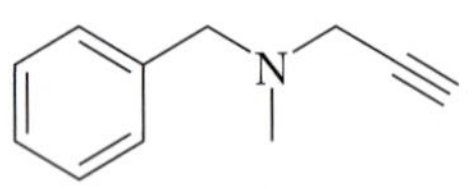

Pargyline

**FIGURE 13.29** Examples of physiologically active compounds that contain the amine functional group are amphetamine, a drug known to produce increased wakefulness and focus; Benadryl, an antihistamine; adrenaline, a hormone produced in our bodies and involved in the fight-or-flight response; and pargyline, a drug used to treat hypertension.

The R groups in a secondary or tertiary amine may be the same or different organic subunits. You may be familiar with the odor of trimethylamine, $(CH_3)_3N$, the compound responsible for the smell of decaying fish. When it comes to foul-smelling compounds, the names of other naturally occurring amines like putrescine and cadaverine speak for themselves.

The amine functional group is found in many natural products and drugs. Some relatively simple examples include amphetamine (Figure 13.29), a stimulant that also contains an aromatic group, and Benadryl, an antihistamine used to treat the symptoms associated with allergies. Another amine, adrenaline, is produced by our bodies in glands near the kidneys and plays an important role in our nervous system.

Amines are organic bases, and their basicity is their defining chemical characteristic. They all react with water to some extent to produce hydroxide ions and protonated cations, just like ammonia:

$$NH_3(aq) + H_2O(\ell) \rightleftharpoons NH_4^+(aq) + OH^-(aq) \qquad (13.9)$$
$$RNH_2(aq) + H_2O(\ell) \rightleftharpoons RNH_3^+(aq) + OH^-(aq) \qquad (13.10)$$

Amines also react readily with acids such as hydrochloric acid to form salts:

$$RNH_2(\ell) + HCl(aq) \rightarrow RNH_3^+(aq) + Cl^-(aq) \qquad (13.11)$$

In fact, many pharmaceuticals (like Benadryl) containing the amine functional group are sold as hydrochloride salts to improve their solubility in water.

Amines are also polar and form hydrogen bonds with water. Hydrogen bonding is possible with pure primary and secondary amines but not with pure tertiary amines, because the nitrogen atom in a tertiary amine bears no hydrogen atom.

A particularly important type of amine is one in which the nitrogen atom is part of a ring. We will explore the properties of these cyclic amines in Chapter 20.

Bacteria of the genus *Methanosarcina* convert primary, secondary, and tertiary methylamines to methane, carbon dioxide, and ammonia:

$$4\,CH_3NH_2(aq) + 2\,H_2O(\ell) \rightarrow 3\,CH_4(g) + CO_2(g) + 4\,NH_3(aq) \qquad (13.12)$$
$$2\,(CH_3)_2NH(aq) + 2\,H_2O(\ell) \rightarrow 3\,CH_4(g) + CO_2(g) + 2\,NH_3(aq) \qquad (13.13)$$
$$4\,(CH_3)_3N(aq) + 6\,H_2O(\ell) \rightarrow 9\,CH_4(g) + 3\,CO_2(g) + 4\,NH_3(aq) \qquad (13.14)$$

These reactions describe a pathway by which methane can be produced from the decay of biomass, serving as potential sources of methane for heating and as fuel for transportation. Amines represent only a small fraction of the biomass of plants, however, and the industrial development of fuel production from amines has been slow, mainly because fossil fuels are still plentiful enough and cheap enough to make processing amines not cost-effective. This economic imbalance may change as fossil fuels are depleted and become more expensive.

**CONCEPT TEST**

Are pargyline, amphetamine, adrenaline, and Benadryl (Figure 13.29) primary, secondary, or tertiary amines?

# 13.6 Alcohols, Ethers, and Reformulated Gasoline

In January 1995, air-quality regulations went into effect in many U.S. cities mandating reductions in atmospheric pollutants from gasoline-fueled engines. The regulations led to the widespread use of "reformulated" gasoline containing additives to promote complete combustion and boost octane ratings. These additives are often organic compounds that contain oxygen in addition to hydrogen and carbon. Oxygen atoms in an organic compound are also heteroatoms and are components of functional groups in two important families of organic molecules: alcohols and ethers.

**alcohol** organic compound containing the –OH functional group.

**CONCEPT TEST** ● ● ● ● ● ● ● ● ● ● ● ● ● ● ● ● ● ● ● ● ● ● ● ● ● ● ● ● ● ● ● ●

Explain why the solubility of low-molar-mass alcohols and ethers in water is greater than the solubility of hydrocarbons in water.

## Alcohols: Methanol and Ethanol

**Alcohols** have the general formula R—OH, where R is any alkyl group. The R group can be a straight chain, a branched chain, or a ring. Like the N atoms in amines, the O atoms in alcohols are shown explicitly in carbon-skeleton structures. The chemical and physical properties of alcohols can be understood if we recognize that an alcohol looks like a combination of an alkane and water:

R—H         H—OH         R—OH
Alkane       Water        Alcohol

As we saw in Section 10.4, if the R group in the molecule is small, the alcohol behaves like water; as the R group gets larger, the alcohol behaves more like a hydrocarbon. As an illustration of this behavior, Table 13.6 shows the water solubilities of a homologous series of alcohols. The polar –OH group makes one end of these molecules "water-like." However, as the number of carbon atoms increases, a greater portion of these alcohols is "oil-like." Therefore, their solubilities in water decrease until about $C_8$, beyond which their solubilities are comparable to those of the corresponding hydrocarbons.

As the names of the two simplest alcohols—methanol and ethanol—indicate, the chemical names of alcohols end in *-ol*; this suffix identifies the compound as an alcohol. Methanol ($CH_3OH$) is also known as methyl alcohol or wood alcohol. The latter name comes from one former source of this alcohol: it was made by collecting the vapors given off when wood is heated to the point of decomposition in the absence of oxygen. Methanol is a widely used industrial organic chemical, serving as the starting material in the preparation of several organic compounds used to make polymers. Its industrial synthesis is based on reducing carbon monoxide with hydrogen:

$$CO(g) + 2\,H_2(g) \rightarrow CH_3OH(\ell)$$

The CO and $H_2$ used to make methanol come from the reaction of methane and water in the steam reforming process we discussed in Chapter 5:

$$CH_4(g) + H_2O(g) \rightarrow CO(g) + 3\,H_2(g)$$

Methanol burns according to the thermochemical equation:

$$2\,CH_3OH(\ell) + 3\,O_2(g) \rightarrow 2\,CO_2(g) + 4\,H_2O(\ell) \qquad \Delta H^\circ_{comb} = -1454 \text{ kJ}$$

If we divide the absolute value of $\Delta H^\circ_{comb}$ by twice the molar mass of methanol (because the reaction consumes 2 moles of methanol), we get a fuel value for methanol of

$$\frac{1454 \text{ kJ}}{\left(\dfrac{32.04 \text{ g}}{\text{mol}}\right)(2 \text{ mol})} = 22.69 \text{ kJ/g}$$

**TABLE 13.6 Solubilities of a Homologous Series of Alcohols in Water at 20°C**

| Condensed Structure | Water Solubility (g/100 mL) |
|---|---|
| $CH_3OH$ | Miscible |
| $CH_3CH_2OH$ | Miscible |
| $CH_3(CH_2)_2OH$ | Miscible |
| $CH_3(CH_2)_3OH$ | 7.9 |
| $CH_3(CH_2)_4OH$ | 2.3 |
| $CH_3(CH_2)_5OH$ | 0.6 |
| $CH_3(CH_2)_6OH$ | 0.2 |
| $CH_3(CH_2)_7OH$ | 0.05 |

Let's compare the fuel value of methanol with that of octane:

$$2\,C_8H_{18}(\ell) + 25\,O_2(g) \rightarrow 16\,CO_2(g) + 18\,H_2O(\ell) \qquad \Delta H^\circ_{comb} = -1.091 \times 10^4\ \text{kJ}$$

$$\frac{1.091 \times 10^4\ \text{kJ}}{\left(\dfrac{114.22\ \text{g}}{1\ \cancel{\text{mol}}}\right)(2\ \cancel{\text{mol}})} = 47.76\ \text{kJ/g}$$

The fuel value of methanol is less than half that of octane (and most of the other hydrocarbons in gasoline). Why is this the case? The answer involves the composition of methanol. The amount of energy released during combustion depends on the number of carbon atoms available for forming $C{=}O$ bonds in $CO_2$ and the number of hydrogen atoms available for forming $O{-}H$ bonds in $H_2O$. The presence of oxygen in $CH_3OH$ adds significantly to its mass (methanol is 50% oxygen by mass) but adds nothing to its fuel value. The oxygen content of a combustible substance essentially dilutes its energy value. The higher the oxygen content of a fuel, the lower is its fuel value.

As noted earlier, ethanol ($CH_3CH_2OH$), also known as ethyl alcohol, is the alcohol in alcoholic beverages. It is formed by the fermentation of sugar from an amazing variety of vegetable sources. Indeed, any plant matter containing sufficient sugar may be used to produce ethanol. Grains are commonly used, from which ethanol derives its trivial name *grain alcohol*. Ethanol may be the earliest organic chemical used by humans, and it is still one of the most important. For industrial purposes, ethanol is prepared by the reaction of water and ethylene.

Most gasoline in the United States currently contains 10% ethanol, with efforts under way to increase this value to 15%. Most of the ethanol used in gasoline is produced by fermentation of sugar derived from corn. Ethanol burns readily in air:

$$CH_3CH_2OH(\ell) + 3\,O_2(g) \rightarrow 2\,CO_2(g) + 3\,H_2O(\ell) \qquad \Delta H^\circ_{comb} = -1367\ \text{kJ}$$

**○○ CONNECTION** We introduced fuel values in Chapter 5 as the amount of heat given off when one gram of fuel is burned.

---

**SAMPLE EXERCISE 13.7**  **Comparing the Energy Content of Fuel Mixtures**  **LO5**

Suppose the hydrocarbons in gasoline can be represented by nonane, $CH_3(CH_2)_7CH_3$ ($\Delta H^\circ_{comb} - 6160$ kJ/mol). The $\Delta H^\circ_{comb}$ of ethanol is $-1367$ kJ/mol. How much energy in the form of heat is available from a mixture of 10.0% ethanol / 90.0% gasoline by mass compared with the amount available from pure gasoline? Carry out the calculation based on $2.70 \times 10^3$ g of mixture, which is about the mass of 1.00 gallon of gasoline.

**Collect and Organize**  We are given the enthalpies of combustion of nonane and ethanol, the total mass of the fuels, and the composition of the fuel as a percentage by mass. Calculations of the heat ($q$) produced by a reaction given the enthalpy change for the process and the amount of reactant were illustrated in Section 5.6 for the recycling of aluminum.

**Analyze**  To use the thermochemical values, we need to calculate the number of moles of each component and then calculate the energy released:

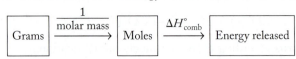

We can determine molar masses from the molecular formulas: $CH_3CH_2OH$, 46.07 g/mol; $CH_3(CH_2)_7CH_3$, 128.25 g/mol. The heat produced by each component is the product of the number of moles of the component and $\Delta H^\circ_{comb}$. The molar mass of ethanol is about one-third the molar mass of nonane, but the mass of nonane present in the mixture is 9 times the mass of the ethanol. This means that we have many more moles of nonane than ethanol in the mixture. The $\Delta H^\circ_{comb}$ of nonane is about 4–5 times greater than the $\Delta H^\circ_{comb}$ for ethanol. We predict that the heat produced from burning the mixture will be lower but not much lower than the heat produced by burning pure nonane.

**Solve** Let's calculate how many grams of each component are in the mixture and then convert that to moles:

10.0% ethanol:  $0.100(2.70 \times 10^3 \text{ g}) = 2.70 \times 10^2$ g ethanol

90.0% nonane:  $0.900(2.70 \times 10^3 \text{ g}) = 2.43 \times 10^3$ g nonane

Moles ethanol:  $2.70 \times 10^2 \text{ g} \times \dfrac{1 \text{ mol}}{46.07 \text{ g}} = 5.861$ mol

Moles nonane:  $2.43 \times 10^3 \text{ g} \times \dfrac{1 \text{ mol}}{128.25 \text{ g}} = 18.95$ mol

The amount of heat given off by the mixture is the sum of the heat given off by the two components:

$$5.861 \text{ mol} \left( \frac{-1367 \text{ kJ}}{\text{mol}} \right) + 18.95 \text{ mol} \left( \frac{-6160 \text{ kJ}}{\text{mol}} \right) = -124{,}750 \text{ kJ} = -1.25 \times 10^5 \text{ kJ}$$

When $2.70 \times 10^3$ g of nonane is burned, the heat given off is

$$2.70 \times 10^3 \text{ g} \times \frac{1 \text{ mol}}{128.25 \text{ g}} \times \frac{-6160 \text{ kJ}}{\text{mol}} = -129{,}700 \text{ kJ} = -1.30 \times 10^5 \text{ kJ}$$

Expressing the difference as a percentage gives us

$$\frac{-(129{,}700 - 124{,}750) \text{ kJ}}{-129{,}700 \text{ kJ}} \times 100\% = 3.82\%$$

The ethanol/gasoline mixture produces about 4% less energy than gasoline by itself.

**Think About It** The presence of oxygen in ethanol adds to its mass but does not add to its fuel value. It is logical that a blend of a hydrocarbon and an alcohol has slightly lower energy content than the hydrocarbon itself. Among the reasons we use ethanol anyway are that it produces less air pollution and, in principle, is a renewable resource.

⚙ **Practice Exercise** A fuel called E-85 is a mixture of 85% ethanol and 15% gasoline by volume. How much energy in the form of heat is available from this mixture compared with the amount from pure gasoline? Base your calculations on 100.00 mL of E-85 and express the difference as a percentage. The densities of ethanol and gasoline are 0.789 and 0.737 g/mL, respectively, and nonane may be used as a model hydrocarbon for gasoline.

The use of ethanol as a gasoline additive resulted in a sharp increase in ethanol production at the beginning of the 21st century, with annual consumption in the United States estimated as 13 billion gallons ($5 \times 10^{10}$ L) in 2012. However, several challenges limit the wide use of ethanol as an automobile fuel. Like methanol, ethanol has a fuel value that is less than that of a comparable mass or volume of gasoline. Furthermore, considerable energy, irrigation water, and valuable farmland are required for its production: growing and harvesting corn, converting corn starch into sugar, converting the sugar into alcohol, and finally distilling the alcohol from the fermentation mixture. It is estimated that more than two-thirds of the energy released in the combustion of ethanol derived from corn is consumed in its production. Ethanol produced in this way is more expensive than gasoline, even at today's prices for fossil fuels. Fuels are extraordinarily complex in terms of their composition, their combustion, the emissions they produce, and the issue of renewability, and this brief discussion addresses only a very small part of a challenging problem.

Alcohols are also prevalent in natural products and in pharmaceuticals. The distinctive aroma of mint leaves comes from menthol, which is an alcohol, as is terpineol (oil of turpentine), an oil distilled from the resin of pine trees (Figure 13.30). Notice the similarity in the carbon-skeleton structures of these two compounds— both contain a six-carbon ring and an −OH group. Only the location of the −OH group and the presence of a C=C bond distinguish these two compounds.

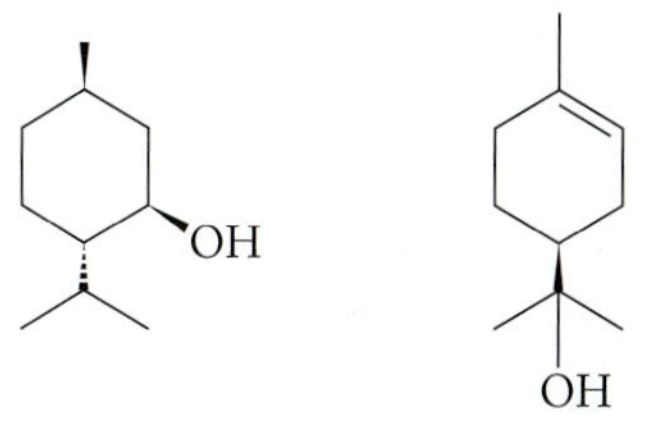

(a) Menthol  (b) Terpineol
(oil of mint)  (oil of turpentine)

**FIGURE 13.30** (a) Menthol and (b) terpineol are two naturally occurring alcohols present in mint leaves and pine needles, respectively.

**ether** organic compound with the general formula R—O—R′, where R is any alkyl group or aromatic ring; the two R groups may be different.

## Ethers: Diethyl Ether

**Ethers** have the general formula R—O—R′, where R is any alkyl group or an aromatic ring. Similarly to alcohols, we can think of an ether structurally as a water molecule in which the two H atoms have been replaced by two organic groups (R and R′, which may be the same or different):

| R—H | H—O—H | H—R′ | R—O—R′ |
|:---:|:---:|:---:|:---:|
| Alkane | Water | Alkane | Ether |

Because the C—O—C bond angle is close to the tetrahedral bond angle of 109.5°, the dipole moments of the two C—O bonds in an ether do not cancel, which means that ethers are polar molecules. This structural feature gives rise to the properties of typical ethers: their water solubility is comparable to that of alcohols of similar molar mass, but their boiling points are about the same as alkanes of comparable molar mass (Table 13.7). Note that Benadryl in Figure 13.29 also contains an ether functional group.

The most important ether industrially is diethyl ether, $CH_3CH_2OCH_2CH_3$. You may have first heard of this as the material simply called "ether" that has had wide use in medicine as an anesthetic since 1842. Although exactly how an anesthetic dulls nerves and puts patients to sleep is still unknown, certain properties of diethyl ether play a role in determining its behavior as a medicinal agent. Because diethyl ether has a low boiling point, 35°C, it vaporizes easily, and a patient can inhale it. Because diethyl ether has a significant solubility in water, it is soluble in blood, which means that once inhaled, it can be easily transported throughout the body. Its low polarity and short saturated hydrocarbon chains combine to make it soluble in cell membranes, where it blocks stimuli coming into nerves. Ether has the unfortunate side effect of inducing nausea and headaches and has been replaced by new anesthetics in modern hospitals, but for many years ether was the anesthetic of choice for surgical procedures.

A second common use of diethyl ether stems from another property that caused difficulty in the clinical setting—it is extremely flammable. Flammability is a liability in an operating room, but this property is used to our advantage when we spray ether in diesel engines to start them in cold weather when it is too cold for diesel fuel to ignite.

| TABLE 13.7 | Functional Groups Affect Physical Properties | | |
|:---|:---:|:---:|:---:|
| | **Molar Mass (g/mol)** | **Normal Boiling Point (°C)** | **Solubility in Water (g/100 mL at 20°C)** |
| $CH_3CH_2$—O—$CH_2CH_3$ <br><br> Diethyl ether | 74 | 35 | 6.9 |
| $CH_3CH_2CH_2CH_2CH_3$ <br><br> Pentane | 72 | 36 | 0.0038 |
| $CH_3CH_2CH_2CH_2OH$ <br><br> Butanol | 74 | 117 | 7.9 |

One ether widely used as a gasoline additive in the 1990s was methyl *tert*-butyl ether (MTBE, Figure 13.31), added as a replacement for tetraethyl lead to promote complete combustion (*tert-* is an abbreviation for *tertiary*, referring to a carbon atom bonded to three other carbon atoms). However, unlike the nonpolar hydrocarbons in gasoline, MTBE is soluble in water. Consequently, gasoline spills, leakage from storage tanks, and release from watercraft produce extensive MTBE contamination of groundwater and drinking water. After toxicity tests initially indicated MTBE to be a possible carcinogen (cancer-causing agent), several states—including California, where more than 25% of the world's production of MTBE was used in gasoline—banned the use of MTBE as a gasoline additive. Most oil companies stopped adding MTBE to their gasolines in 2006. As of 2011 MTBE has been dropped from the list of possible carcinogens by both the National Toxicology Program and the International Agency for Research on Cancer. These concerns, however, have led to the increased use of ethanol rather than MTBE as a fuel additive.

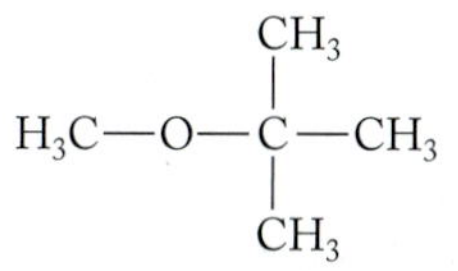

**FIGURE 13.31** MTBE, methyl *tert*-butyl ether, was used in the early 1990s as a fuel additive but its use was curtailed when it was found to be an environmental pollutant.

**CONCEPT TEST**

Rank the following compounds in order of decreasing fuel value: diethyl ether, MTBE, methanol, and ethanol.

## Polymers of Alcohols and Ethers

Over 400,000 tons of the addition polymer poly(vinyl alcohol) (PVAL) are produced annually in the United States. It is used in fibers, in adhesives, and in materials known as sizing, which change the surface properties of textiles and paper to make them less porous, less able to absorb liquids, and smooth. Because the polymer chains in PVAL are studded with –OH groups (Figure 13.32a), its surface is very polar and very water-like, and hydrocarbon solvents that are not soluble in water do not penetrate PVAL barriers. PVAL is the material of choice for laboratory gloves that are resistant to organic solvents.

PVAL is also impenetrable to carbon dioxide, and this property has led to its use in soda bottles, in which it is blended with the polymer poly(ethylene terephthalate) (PETE, Figure 13.32b). The two polymers do not mix but separate into layers (Figure 13.32c). The PETE makes the bottle strong enough to bear pressure changes due to temperature changes and survive the impact of falling off tables. Also, $CO_2$, the dissolved gas that makes soda fizz, passes readily through PETE but not through the PVAL layers, with the result that the soda does not go flat. Polymers with different properties are frequently combined like this to create new materials with desired properties.

The monomer from which poly(vinyl alcohol) is made is not what you might expect based on our discussion of how addition polymers are synthesized. The monomer "vinyl alcohol" does exist.

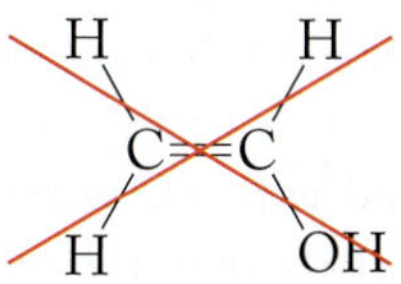

Vinyl acetate (a)   PVAC   PVAL

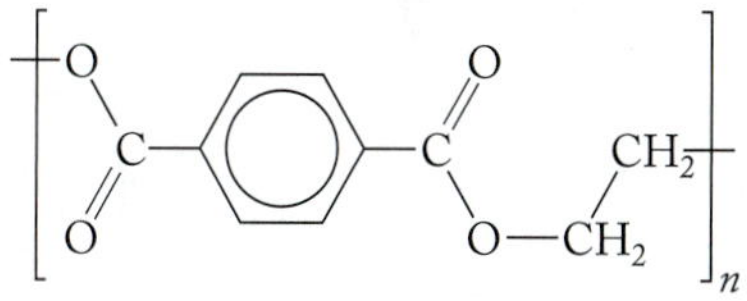

Repeating unit in PETE
(b)

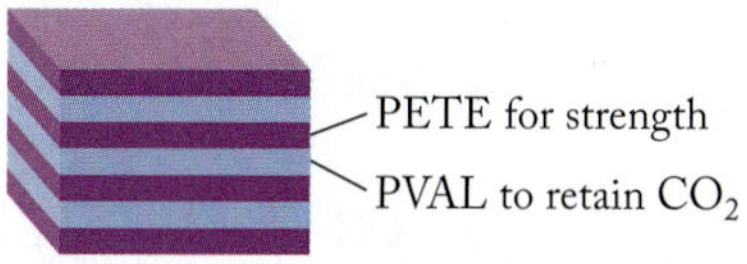

PETE for strength
PVAL to retain $CO_2$
(c)

**FIGURE 13.32** (a) Poly(vinyl alcohol), or PVAL, is synthesized from vinyl acetate. The intermediate polymer, poly(vinyl acetate), or PVAC, is reacted with water to produce PVAL. (b) The repeating unit in PETE. (c) Layers of the polymers PVAL and PETE are used to make soda bottles. PETE makes the bottle strong, and PVAL keeps the carbon dioxide from leaking out.

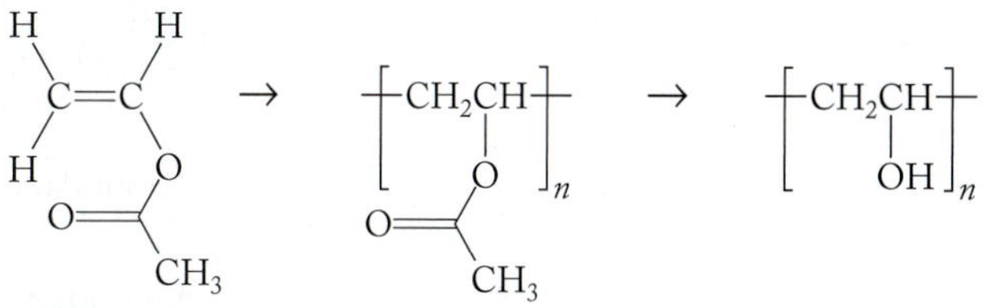

If you try to make vinyl alcohol, you get acetaldehyde instead. It is almost always true that an –OH group on a C=C double bond (called an *enol*) rearranges to the carbonyl form. In most cases the two forms are in equilibrium that favors the carbonyl:

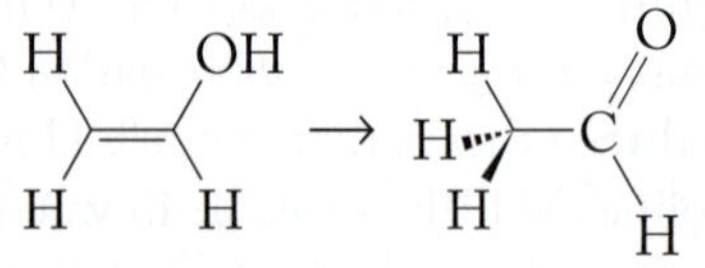

Vinyl alcohol rearranges to acetaldehyde

The monomer vinyl acetate is used to make poly(vinyl acetate) (PVAC), which is then reacted with water to replace the acetate group with an –OH group. This replacement reaction turns PVAC into PVAL, as shown in Figure 13.32(a).

The blend of PVAL and PETE in early soda bottles was a physical mixture of the two polymers. New materials can also be made by combining different monomer units in one polymer molecule. This type of molecule is called a **copolymer** when two different monomers are combined and a **heteropolymer** when three or more different monomers are combined. One example of an addition copolymer is a material called EVAL, made from ethylene and vinyl acetate (Figure 13.33). EVAL is used in food wrappings when preservation of aroma and flavor are required. Food usually deteriorates in the presence of oxygen, and packages made of EVAL provide an excellent barrier to the entry of oxygen while retaining the flavor and fragrance of the packaged food.

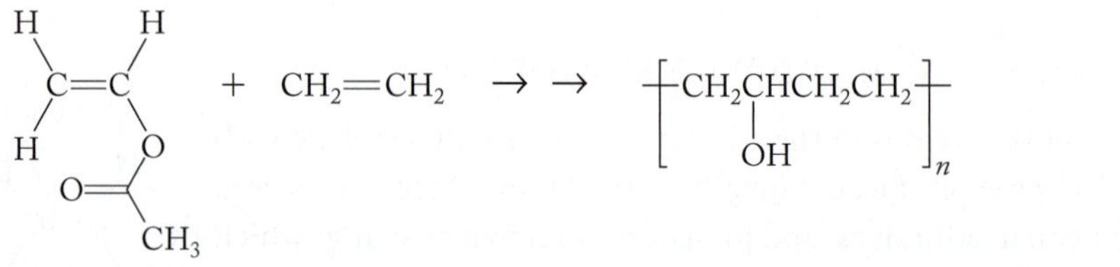

**FIGURE 13.33** EVAL is a copolymer of ethylene and vinyl acetate.

Vinyl acetate          Ethylene          Poly(ethylene-co-vinyl alcohol) = EVAL

Monomers forming hetero- or copolymers can combine in different ways. If we represent the monomer units making up a copolymer with the letters A and B, one possible way they can combine is

$$+A—B—A—B—A—B—A—B+$$

an arrangement called an *alternating copolymer*. Another possibility is

$$+A—A—A—A—B—B—B—B—A—A—A—A—B—B—B—B+$$

called a *block copolymer*. A third possibility is a *random copolymer*:

$$+A—A—B—A—B—B—A—B—A—B—B—A—A—A—A—B+$$

and this is the arrangement we see in the copolymer EVAL: it is a random copolymer of the monomers A = ethylene, B = vinyl acetate.

Commercially important polymers made from ethers include poly(ethylene glycol) (PEG) and poly(ethylene oxide) (PEO), which contain the same subunit (Figure 13.34). PEG is a low-molar-mass liquid made from ethylene glycol, and PEO is a higher-molar-mass solid made from ethylene oxide. As a polyether, PEG has properties closely related to those of diethyl ether, in that it is soluble in both polar and nonpolar liquids. It is a common component in toothpaste because it interacts both with water and with the water-insoluble materials in the paste and keeps the toothpaste uniform both in the tube and during use. PEGs of many lengths are finding increasing use as attachments to pharmaceutical agents to improve their solubility and biodistribution.

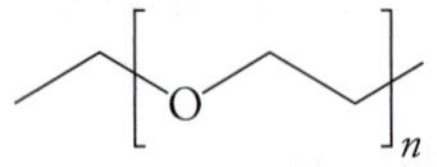

Repeating unit in PEG and PEO

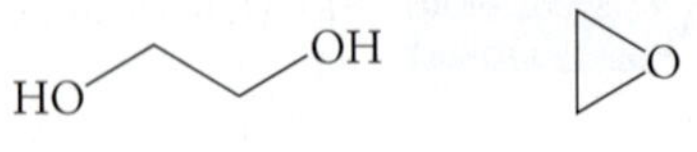

Ethylene glycol          Ethylene oxide

**Monomers**

**FIGURE 13.34** Poly(ethylene glycol) (PEG) and poly(ethylene oxide) (PEO) have the same repeating unit. The two polymers differ only in their molar masses. PEG is typically made from ethylene glycol, and ethylene oxide is the monomer of choice for making PEO.

**SAMPLE EXERCISE 13.8** **Assessing Properties of Polymers** **LO6**

The polymer PEG (Figure 13.34) is used to blend materials that are not soluble in each other. It is soluble both in water and in benzene, a nonpolar solvent. Describe the structural features of PEG that make it soluble in these two liquids of very different polarities.

**Collect and Organize** The relationship between structure and solubility of compounds was discussed in Chapter 10, where we learned that "like dissolves like."

**Analyze** The statement "like dissolves like" refers to the polarity of the molecules and to the attractive and repulsive forces between molecules. We need to describe the intermolecular forces in PEG and see if one part is water-like and another benzene-like. Water is a polar molecule and benzene is a nonpolar molecule, so we predict that PEG contains both polar and nonpolar regions.

**Solve** The structure of PEG consists of $-CH_2CH_2-$ groups connected by oxygen atoms. The oxygen atoms are capable of hydrogen bonding with water molecules, so the attractive force between PEG and water is due to hydrogen bonding. Benzene is nonpolar and is attracted to the $-CH_2CH_2-$ groups in the polymer. Nonpolar groups interact via dispersion forces, so those forces must be responsible for the solubility of PEG in nonpolar benzene.

**Think About It** As predicted, PEG contains both polar and nonpolar regions, allowing it to be solvated by both polar solvents like water and nonpolar solvents like benzene.

⚙ **Practice Exercise** When PEG is added to soft drinks it keeps $CO_2$, responsible for the fizz in soda, in solution longer when the soda is poured. What intermolecular attractive forces between PEG and $CO_2$ might make this application possible?

**CONCEPT TEST**

A polymer chemist decides to make a series of derivatives of PEG using the following alcohols in place of ethylene glycol:

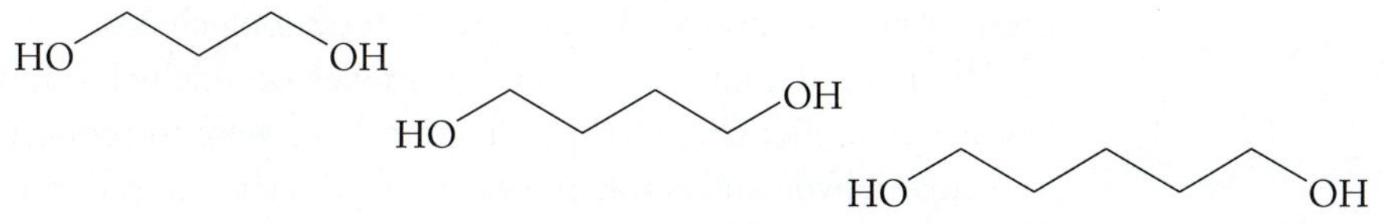

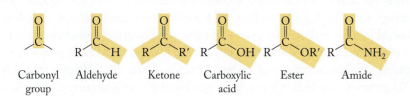

What effect will this have on the solubility of the resulting polymer in water?

# 13.7 Carbonyl-Containing Compounds

Five functional groups—aldehydes, ketones, carboxylic acids, esters, and amides—all contain a subunit called the **carbonyl group**: a carbon atom double-bonded to an oxygen atom (Figure 13.35). The R groups in the figure may be any organic group. Aldehydes and ketones are collectively referred to as *carbonyl compounds* because the carbonyl group determines their chemistry. Carboxylic acids, as their name implies, are acidic, and their chemistry is determined by the –COOH subunit, referred to as a *carboxylic acid group*. Esters and amides can be made from

**copolymer** a macromolecule formed from the chemical combination of two different monomers.

**heteropolymer** a polymer made of three or more different monomer units.

**carbonyl group** a carbon atom with a double bond to an oxygen atom.

👀 **CONNECTION** We first introduced carboxylic acids in our discussion of acids in Chapter 2.

**FIGURE 13.35** The carbonyl group is found in five important functional groups: aldehydes, ketones, carboxylic acids, esters, and amides.

| Carbonyl group | Aldehyde | Ketone | Carboxylic acid | Ester | Amide |

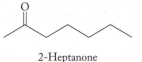

2-Heptanone

**FIGURE 13.36** The ketone 2-heptanone is found in many plants and dairy products.

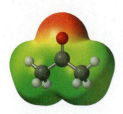

**FIGURE 13.37** The electron distribution in a carbonyl group is skewed toward the oxygen end of the bond because oxygen is more electronegative than carbon.

(a) Acrolein

(b) Acetone

(c) Formaldehyde

(d) Zingerone

(e) Carvone

(f) Cinnamaldehyde

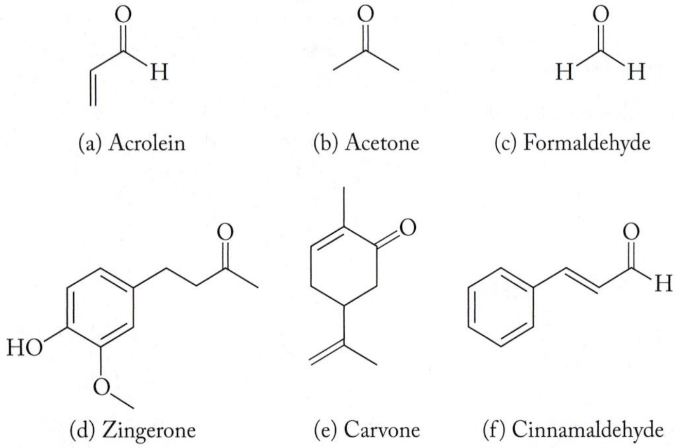

**FIGURE 13.38** The ketone and aldehyde functional groups are common among organic compounds: (a) acrolein is found in barbeque smoke, (b) acetone is used in nail polish remover, (c) aqueous solutions of formaldehyde are used to preserve biological specimens, (d) zingerone is found in the spice ginger, (e) carvone is found in the leaves of spearmint, and (f) cinnamon owes its flavor and odor to cinnamaldehyde.

carboxylic acids by reacting them with alcohols and amines. Other families of compounds also contain the carbonyl group, but we will focus on just these five groups.

## Aldehydes and Ketones

An **aldehyde** contains a carbonyl group bound to one R group and one hydrogen atom; its general formula is RCHO or RC(O)H. A **ketone** contains a carbonyl group bound to two R groups; its general formula is RCOR′ or RC(O)R′. The R groups may be the same, as in acetone, $CH_3C(O)CH_3$, or different, as in 2-heptanone (Figure 13.36), which is found in cloves, blue cheese, and many fruits and dairy products.

The double bond in the carbonyl group accounts for the reactivity of aldehydes and ketones. It is different from the double bond in an alkene, however, because it is polar (Figure 13.37). The electronegative oxygen pulls electron density toward itself, and the chemistry of aldehydes and ketones is linked to the polarity of the $C=O$ bond. Other polar species tend to react with carbonyls when electron-rich regions of their molecules approach the δ+ carbon atoms of the carbonyl groups.

Because aldehydes and ketones are polar, they tend to parallel the ethers with respect to water solubility. They cannot hydrogen bond with other aldehyde and ketone molecules because they contain only carbon-bonded hydrogen atoms, so they have lower boiling points than alcohols of comparable molar mass.

Because of the $C=O$ bond, aldehydes and ketones are in a higher oxidation state than alcohols, and indeed many of the smaller aldehydes and ketones are made by oxidizing alcohols with the same number of carbons. Aldehydes and ketones do not polymerize through their carbonyl groups. Many polymers have carbonyl functional groups as part of their structure, but these groups themselves do not react to form long chains.

We have already seen several examples of aldehydes and ketones in earlier chapters. In Chapter 9, we were introduced to formaldehyde and acrolein, two aldehydes shown in Figure 13.38. Acetone, $CH_3C(O)CH_3$, is used in nail polish remover and is a widely used solvent. The flavors of ginger (zingerone), spearmint (carvone), and cinnamon (cinnamaldehyde) all come from compounds that contain carbonyl groups.

**CONCEPT TEST** • • • • • • • • • • • • • • • • • • • • • • • • • • • • • • • • • • • •

What other functional groups are present in zingerone, carvone, and cinnamaldehyde besides the carbonyl group?

• • • • • • • • • • • • • • • • • • • • • • • • • • • • • • • • • • • • • • • • • • • • • • •

## Carboxylic Acids

**Carboxylic acids** are proton donors, which means they are Brønsted–Lowry acids (Section 4.5). The R group in RCOOH may be any organic subunit. Because it is attached to a δ+ carbon atom, the –OH group is polarized more than what is due to the electronegativity difference between the hydrogen and oxygen atoms, which explains two characteristics of carboxylic acids. First, the hydrogen atom on the –OH group of one molecule can hydrogen-bond to a neighboring carboxylic acid, either at the oxygen atom in the –OH group or at the C=O atom (Figure 13.39). This interaction results in high boiling points relative to those of other organic compounds of comparable molar mass.

Second, donating a proton leaves a negatively charged oxygen on the carboxylic acid that is delocalized over the whole carboxylate group. This delocalization contributes to the stability of the carboxylate anion. The common carboxylic acids are weak acids, which means that they are present in aqueous solutions as mostly neutral molecules. A small fraction of these molecules are ionized, donating $H^+$ ions to molecules of water, as shown in Figure 13.40 for acetic acid.

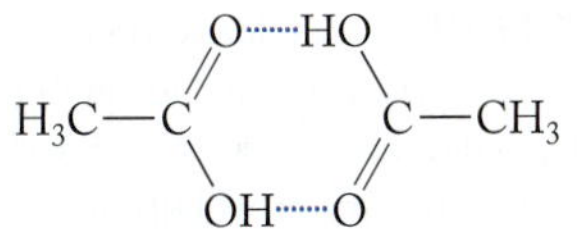

**FIGURE 13.39** The high boiling points of carboxylic acids are the result of strong hydrogen bonds between neighboring molecules, resulting in the formation of persistent dimers in condensed phases.

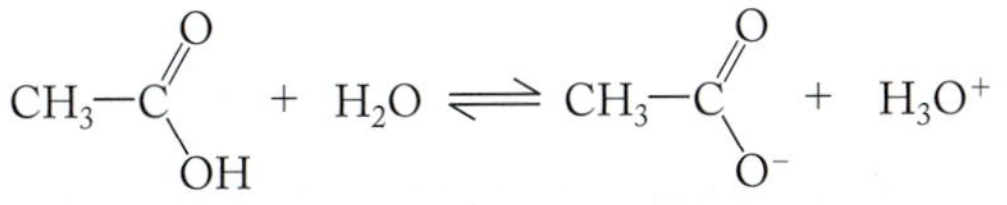

**FIGURE 13.40** Carboxylic acids such as acetic acid (vinegar) are weak acids in water.

Vinegar is a dilute aqueous solution of the carboxylic acid acetic acid. Large quantities of vinegar are produced commercially by the air oxidation of ethanol in the presence of enzymes from *Acetobacter* bacteria. Bacteria can also convert acetic acid and other constituents in biomass to methane. For example, the digestive systems of cows introduce significant amounts of methane to the atmosphere, about 100–200 liters per day per animal. Translating this process to an industrial scale is an attractive future source of hydrocarbons, provided the complexities of bacterial action can be adapted for large-scale production. Recall that we described the timescale for the formation of fossil fuels in Chapter 3. Converting organic matter into hydrocarbon fuel may be possible without waiting millennia for the anaerobic processes deep within Earth to do so.

The production of methane from plant residues that are mostly cellulose (a carbohydrate) requires the sequential action of several types of bacteria. In the first stages, selected bacteria break up cellulose into mixtures of small molecules. Depending on the bacterial strain, these small-molecule products include $H_2$ and $CO_2$, acetic acid, formic acid, or methanol or other small alcohols. All these products then undergo reactions promoted by the metabolism of **methanogenic** (methane-producing) **bacteria**, which consume hydrogen and simple organic compounds for energy and produce methane gas in the process:

$$4\,H_2(g) + CO_2(g) \rightarrow CH_4(g) + 2\,H_2O(\ell)$$

$$CH_3COOH(aq) \rightarrow CH_4(aq) + CO_2(g)$$
Acetic acid

$$4\,HCOOH(aq) \rightarrow CH_4(g) + 3\,CO_2(g) + 2\,H_2O(\ell)$$
Formic acid

$$4\,CH_3OH(aq) \rightarrow 3\,CH_4(g) + CO_2(g) + 2\,H_2O(\ell)$$
Methanol

The actions of methanogenic bacteria have a measurable effect on Earth's atmosphere and climate. Methane, like $CO_2$, is a greenhouse gas but is much more potent, trapping about 20 times more heat per molecule than carbon dioxide.

## Esters and Amides

Several chemical families are closely related to the carboxylic acids. We consider only two of them here: esters and amides. As Figure 13.41(a) shows, in **esters**, the –OH of the carboxylic acid is replaced by –OR, where R can be any organic

**aldehyde** organic compound containing a carbonyl group bonded to one R group and one hydrogen; its general formula is RCHO.

**ketone** organic molecule containing a carbonyl group bonded to two R groups; its general formula is RC(O)R′.

**carboxylic acid** an organic compound containing the –COOH functional group.

**methanogenic bacteria** bacteria using simple organic compounds and hydrogen for energy; their respiration produces methane, carbon dioxide, and water, depending on the compounds they consume.

**ester** organic compound in which the –OH of a carboxylic acid group is replaced by –OR, where R can be any organic group.

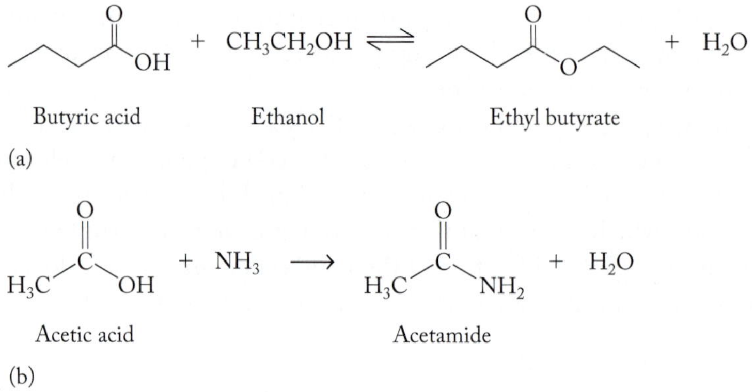

**FIGURE 13.41** (a) Condensation reactions between carboxylic acids and alcohols produce esters and water. Here butyric acid reacts with ethanol, forming ethyl butyrate. (b) Condensation reactions between carboxylic acids and ammonia (or amines) produce amides. Here acetic acid reacts with ammonia, forming acetamide.

Butyric acid    Ethanol    Ethyl butyrate

(a)

Acetic acid    Acetamide

(b)

group. The presence of the carbonyl group makes esters polar, and their boiling points are comparable to those of aldehydes and ketones of similar size.

In **amides**, the –OH is replaced by an amine group—either –NH$_2$, –NHR, or –NR$_2$ (Figure 13.41b). Amides are also polar and capable of intermolecular hydrogen bonding. The hydrogen atoms on the –NH$_2$ group can hydrogen-bond with the oxygen in the carbonyl group of an adjacent molecule. This causes their boiling points to be considerably higher than those of esters of comparable size. Although amines are bases, amides are actually weakly acidic. The carbonyl group pulls electron density away from the nitrogen, making it less favorable for the nitrogen to behave like a Brønsted–Lowry base and pick up a proton. It is actually more favorable for an amide to donate a proton and function as an acid.

Esters frequently have very pleasant fragrances, much different from the acids from which they are derived. For example, the carboxylic acid butyric acid, with a straight chain of 4 carbon atoms, is responsible for the odor of rancid butter. However, the ethyl ester of this carboxylic acid (ethyl butyrate) is the ester responsible for the aroma of ripe pineapples.

An ester is prepared by the *esterification* of an acid with an alcohol (Figure 13.41a). Esterification is a **condensation reaction**: two molecules combine ("condense") to create a larger molecule while a small molecule (typically water) is also formed. Esters are widely used in the personal products industry to provide pleasant scents for products like shampoos and soaps. Esters and carboxylic acids are also common in over-the-counter medications. Figure 13.42 illustrates the carbon-skeleton structures of three common pain relievers—aspirin, ibuprofen, and naproxen—each of which contains a carboxylic acid. Aspirin also contains an ester functional group.

Esters can be made from acids other than carboxylic acids. Important examples of this chemistry are phosphate esters, made from phosphoric acid and alcohols.

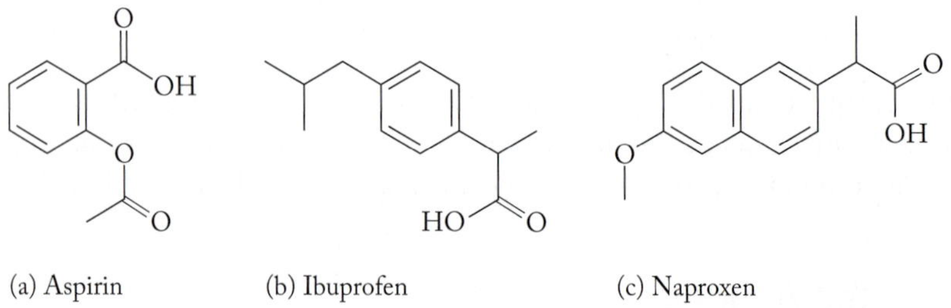

**amide** organic compound in which the same carbon atom is single bonded to a nitrogen atom and double bonded to an oxygen atom.

**condensation reaction** two molecules combining to form a larger molecule and a small molecule (typically water).

(a) Aspirin    (b) Ibuprofen    (c) Naproxen

**FIGURE 13.42** Three pain medications, all of which contain carboxylic acid functional groups and belong to a class of compounds called nonsteroidal anti-inflammatory drugs (NSAIDs). (a) Aspirin was the first medication to be available in tablet form. (b) Ibuprofen has fewer side effects than aspirin. (c) Naproxen is used for the management of mild to moderate pain.

In analogy to the carbonyl-containing esters, these compounds have a phosphorus–oxygen double bond (P=O) called the phosphoryl group. Because it is a triprotic acid, phosphoric acid can form mono-, di-, or triesters via condensation reactions analogous to those of carboxylic acids:

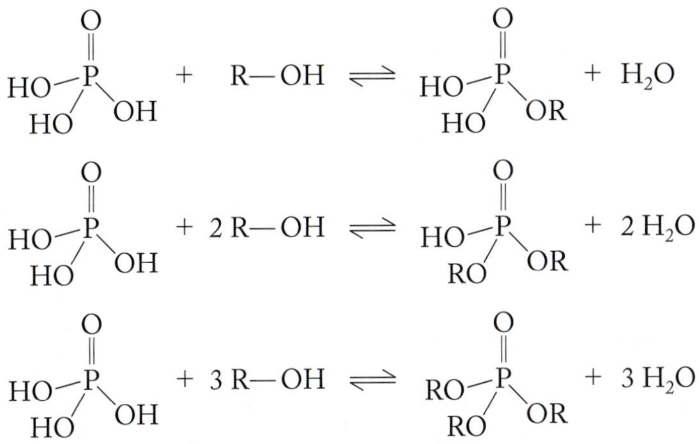

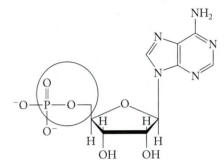

FIGURE 13.43 The molecule adenosine monophosphate (AMP). The phosphate ester linkage is circled.

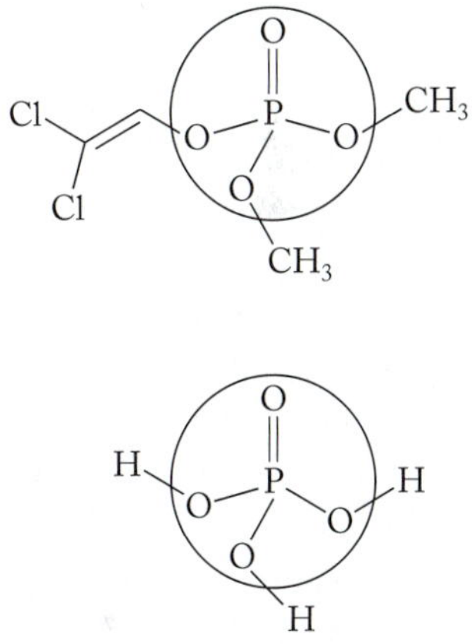

The R groups may be the same or different in these reactions. Many important biomolecules, including adenosine triphosphate (ATP) and the biopolymers DNA and RNA, are phosphate esters. Figure 13.43 shows the compound adenosine monophosphate (AMP), highlighting it as a monoester of phosphoric acid. We discuss molecules containing phosphate esters in greater detail in Chapter 20.

The insecticide dichlorvos (Figure 13.44a) is a triester of phosphoric acid. Other esters of phosphorus-containing acids, including sarin (Figure 13.44b) that have phosphorus–carbon bonds in addition to the phosphoryl linkage, are active insecticides, herbicides, and nerve gases. In addition to the phosphoryl group, these organophosphates may have a hydrophobic group (such as the methyl group in sarin) and often a halide (like the fluorine in sarin) bonded to the phosphorus.

Amides are made from carboxylic acids by several methods, but the net result is a condensation reaction in which the −OH of the carboxylic acid is replaced by the $-NH_2$ group of ammonia or by −NHR or $-NR_2$ groups of amines. Water is also formed in the process, as shown in Figure 13.41(b).

**CONCEPT TEST** ...........................................................

What is the principal structural difference between an amine and an amide?

...............................................................................

(a) Dichlorvos and phosphoric acid

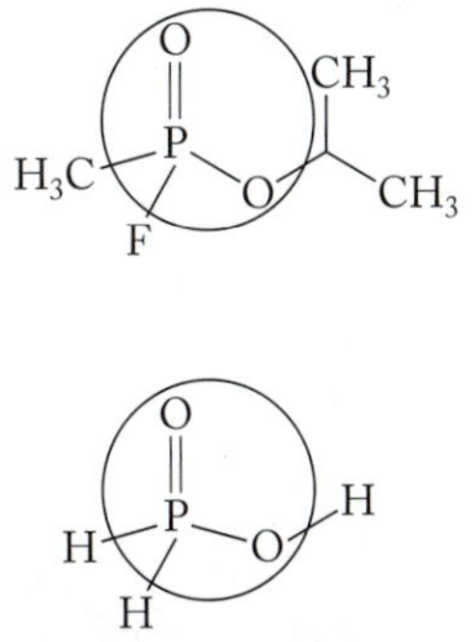

(b) Sarin and hypophosphorous acid

FIGURE 13.44 (a) The insecticide dichlorvos is a derivative of phosphoric acid in which all three R groups are different. (b) The nerve gas sarin may be considered a derivative of hypophosphorous acid, $H_3PO_2$, in which the one oxygen-bound hydrogen atom is replaced to form an ester and the two other hydrogen atoms, bound directly to the phosphorus, are replaced by other groups.

## Polyesters and Polyamides

Prior to this point, many of the organic compounds we have examined have been monofunctional, which means they have only one functional group that identifies their family. With the polymers of carboxylic acids and their derivatives, we enter the world of *difunctional* molecules, which are molecules with two functional groups. The key point to remember is that for the most part, the functional groups still retain their individual chemical reactivity even if they are in a molecule with another functional group. Also remember that the same features we enumerated for all other polymers still apply here: for polymers, function is determined by composition, structure, and size.

Consider the esterification reaction in Figure 13.41(a). The −COOH group of the acid reacts with the −OH group of the alcohol to form a carbon–oxygen

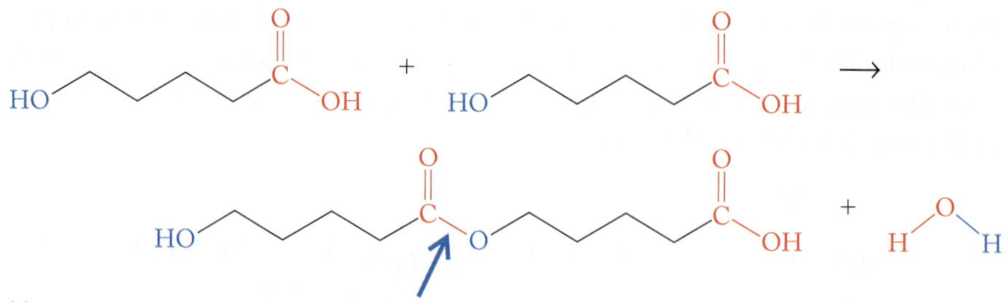

**FIGURE 13.45** (a) Synthesis of an ester from a condensation reaction between two identical difunctional molecules, each one containing an alcohol group and a carboxylic acid group. The diester can then react with additional difunctional molecules at its –OH and –COOH ends, forming a triester, and the reaction repeats over and over, forming (b) the polyester made up of the repeating monomer unit shown.

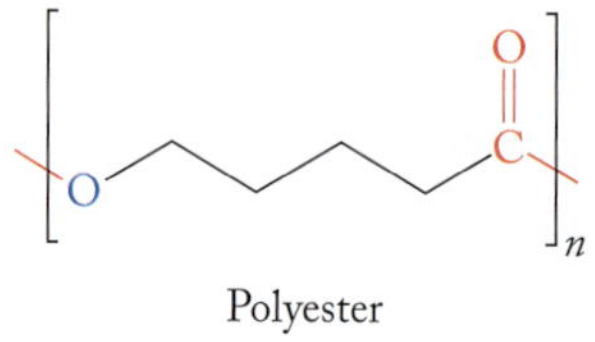

(a)

Ester linkage

(b)

Polyester

▶❙❙ **CHEMTOUR** Polymers

single bond and release a molecule of water. Think about what could happen at the molecular level if we had a single compound that contained a carboxylic acid functional group at one end and an alcohol functional group at the other (Figure 13.45). The carboxylic acid group of one molecule could react with the alcohol group of another molecule in a condensation reaction to generate a molecule that has a carboxylic acid group at one end, an alcohol at the other end, and an ester linkage in between. If this reaction happens repeatedly, a monomer containing one carboxylic acid and one hydroxyl group (a hydroxy acid) polymerizes, as shown in Figure 13.45, to form a polyester, a **condensation polymer**. We have already encountered one condensation polymer, PETE, in Section 13.6. In general, condensation polymers are formed by the reaction of monomers yielding a polymer and water as a by-product of the reaction. In addition to its use in plastic soda bottles, PETE is used extensively in medicine. Artificial heart valves and grafts for arteries are made from PETE.

A copolymer formed by reaction of glycolic acid and lactic acid (Figure 13.46) is used to support the growth of skin cells for burn victims. The polymer in Figure 13.46 is also used in making dissolving sutures. Esterification reactions used to make polyesters can be reversed by the addition of water, breaking their ester linkages and forming alcohol and acid functional groups. We examine this process in greater detail in Chapter 20.

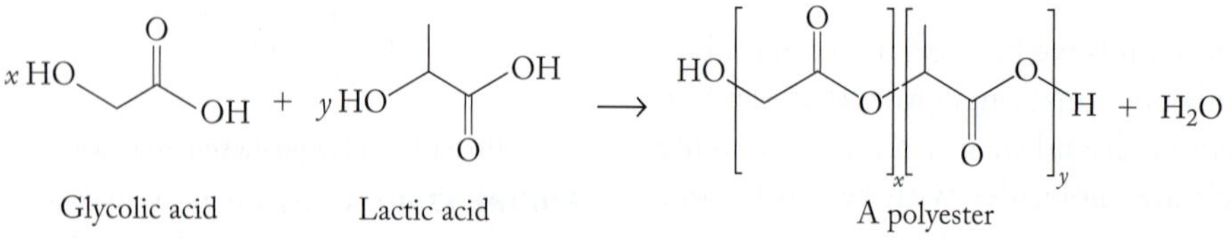

Glycolic acid      Lactic acid                A polyester

(a)

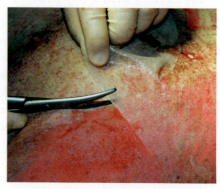

**FIGURE 13.46** (a) The condensation polymer prepared from glycolic acid and lactic acid is used to make sutures that dissolve and as artificial skin that protects against infection while promoting the regrowth of skin cells. (b) Synthetic skin being applied to a burn patient.

(b)

Show how a polyester can be synthesized from the difunctional alcohol $HO(CH_2)_3OH$ and the difunctional carboxylic acid $HOOC(CH_2)_3COOH$.

**Collect and Organize** An ester is the product of a reaction between a carboxylic acid and an alcohol. A polyester is a polymer with a repeating unit containing an ester functional group. We are given an alcohol and a carboxylic acid to react to make the ester repeating monomer unit.

**Analyze** Because the starting materials are difunctional, the alcohol can react with two molecules of carboxylic acid, and the acid can react with two molecules of alcohol.

**Solve** The reaction is

$$HOCH_2CH_2CH_2OH \quad + \quad \underset{HO}{\overset{O}{\|}}CCH_2CH_2CH_2\overset{O}{\overset{\|}{C}}OH \quad \longrightarrow$$

$$\underset{HOCH_2CH_2CH_2O}{\overset{O}{\|}}CCH_2CH_2CH_2\overset{O}{\overset{\|}{C}}OH \quad + \quad H_2O$$

The two –OH groups shown in blue react to form an ester at one end of the carboxylic acid. The product molecule has an alcohol group on one end (shown in red) that can react with another molecule of carboxylic acid and a carboxylic acid group (green) on the other end that can react with another molecule of alcohol. Continuing these condensation reactions results in the formation of a polymer whose repeating unit is

$$\left[ -CH_2CH_2CH_2O\underset{}{\overset{O}{\|}}CCH_2CH_2CH_2\overset{O}{\overset{\|}{C}}O- \right]_n$$

**Think About It** The repeating unit in the polyester contains one section that came from the alcohol and a second section that came from the carboxylic acid because esters are formed from alcohols and acids.

⚙ **Practice Exercise** A difunctional molecule may contain two different functional groups, such as this one with an alcohol group and a carboxylic acid group:

$$HO-CH_2CH_2CH_2\overset{O}{\overset{\|}{C}}OH$$

Draw the repeating unit of the polyester made from this molecule.

Clothes made from the polyester fabric known as Dacron can be less comfortable in hot weather than clothes made of cotton (also a polymer) because Dacron does not absorb perspiration as effectively as cotton. Based on the repeating units of these two polymers (Figure 13.47), suggest a structural reason why cotton absorbs perspiration (water) better than Dacron.

**Collect and Organize** Cotton and Dacron are polymers that differ in the functional groups in their repeating units. We need to identify the different functional groups and the interactions between these functional groups and water, a polar molecule.

**Analyze** The absorption of water by a polymer depends on the intermolecular forces present. Water is a polar molecule and is attracted to polar groups. We need to compare the

**condensation polymer** macromolecule formed by the reaction of monomers yielding a polymer and water or other small molecule as a by-product of the reaction.

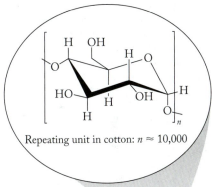

Repeating unit in Dacron

(a)

Repeating unit in cotton: $n \approx 10,000$

(b)

**FIGURE 13.47** Based on the molecular structures of (a) Dacron and (b) cotton, why is cotton the better material for making tee shirts worn during strenuous exercise?

groups in each monomer to see which has the greatest number of polar functional groups or atoms that can form hydrogen bonds. This will be the polymer more likely to absorb water.

**Solve** Each monomer unit in cotton has three –OH groups attached to it, all polar and capable of hydrogen bond formation, which means they are likely to interact with the water in perspiration and thereby draw it away from the body. The monomer unit in Dacron has oxygen atoms in it, but no –OH groups. Although regions in the Dacron monomer are polar, they are not nearly as polar as the –OH groups in the cotton monomer unit.

**Think About It** The principle of like interacting with like works for polymers just as it does for small molecules.

⚙ **Practice Exercise** Gloves made of a woven blend of cotton and polyester fibers protect the hands from exposure to oil and grease and are comfortable to wear because they "breathe"—they allow perspiration to evaporate and pass through them, thereby cooling the skin. Suggest how these gloves work at the molecular level.

Combining difunctional molecules in a condensation reaction can be used to make *polyamides*, another class of very useful synthetic polymers. The functional groups are a carboxylic acid and an amine. The difunctional monomer units can be identical (Figure 13.48a), each containing one carboxylic acid group and one amine group, or they can be different (Figure 13.48b), with one monomer containing two

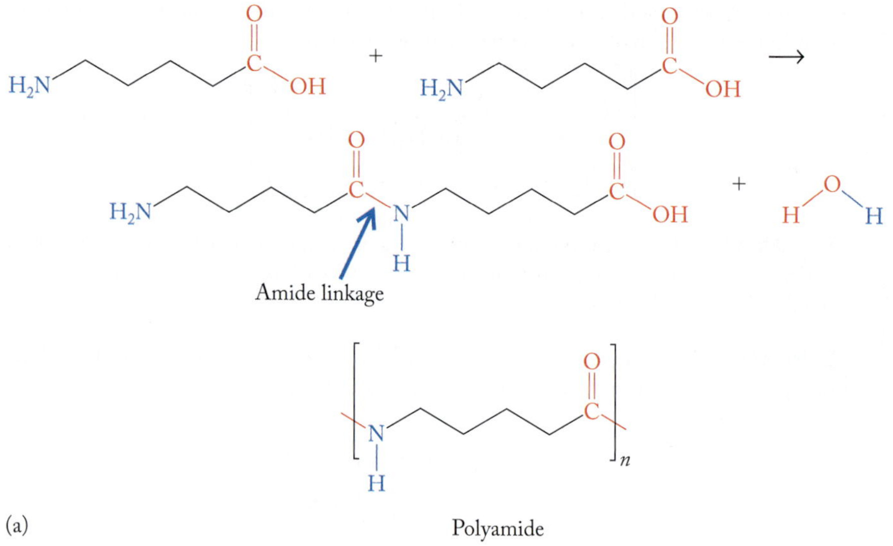

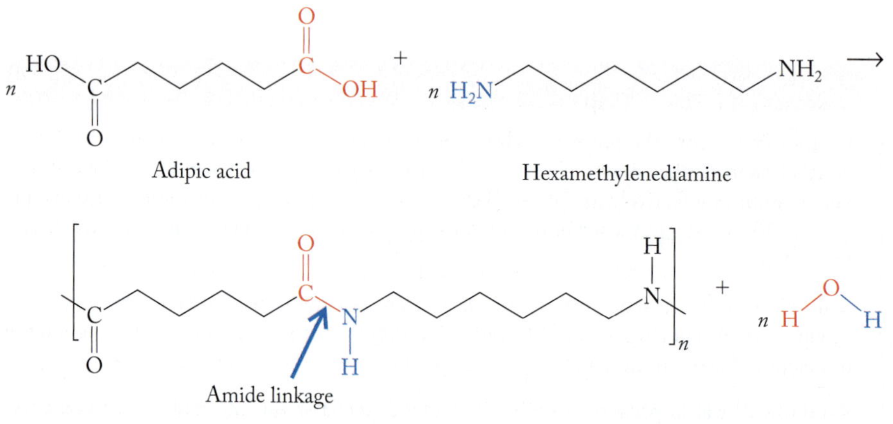

**FIGURE 13.48** (a) Synthesis of a polyamide from two identical monomers, each containing a carboxylic acid functional group and an amine functional group. (b) Synthesis of the polyamide nylon-6,6 from nonidentical monomers: adipic acid (a dicarboxylic acid) and hexamethylenediamine (a diamine).

carboxylic acid groups (a dicarboxylic acid) and the other containing two amine groups (a diamine).

Probably the most familiar polyamide is nylon-6,6, made from the monomers shown in Figure 13.48(b). Each monomer contains six carbon atoms, which is what the digits in the name represent.

---

**SAMPLE EXERCISE 13.11    Identifying Monomers    LO3**

Another form of nylon is nylon-6, with the single 6 indicating that the polymer is made from the reaction of a series of identical six-carbon monomers. By analogy with the polyester in Sample Exercise 13.9 and the accompanying Practice Exercise, draw the condensed structure of a compound that could polymerize to make nylon-6.

**Collect and Organize** We are asked to draw the condensed structure for a monomer that contains both functional groups that react to form an amide and that could react with identical monomers to form a polyamide with a repeating unit six carbon atoms long. The exercise refers us to an example with polyesters to use as an analogy.

**Analyze** By analogy to Sample Exercise 13.9 and its accompanying Practice Exercise, we should suggest a difunctional molecule that has a carboxylic acid on one end and an amine on the other because these are the two functional groups that react to form the amide linkage.

**Solve** We can build the required monomer by starting with one of the functional groups. It doesn't matter which one, so let's begin with the amine:

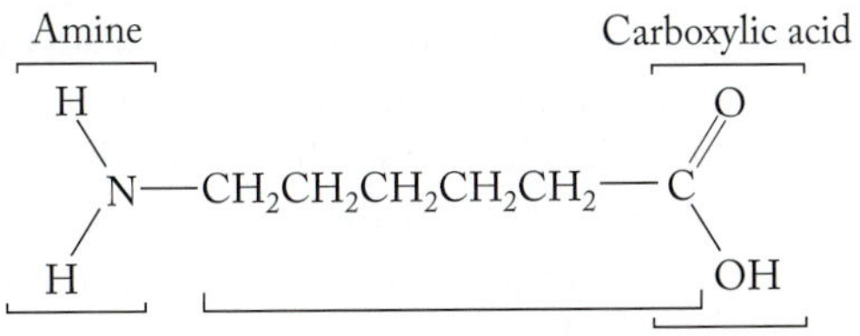

Five –$CH_2$– groups plus one C
from the –COOH = six C

We then add a chain of five –$CH_2$– units because the name nylon-6 indicates that six carbon atoms separate the ends of the repeat unit. Finally, we add the carboxylic acid functional group as the second functional group and the sixth carbon atom in the chain.

**Think About It** Many nylons with different properties can be made by varying the length of the carbon chain in a monomer like the one in this exercise or by varying the lengths of the chains in both the difunctional acid and difunctional amine in Figure 13.48(b).

⚙ **Practice Exercise** Draw the carbon-skeleton structures of two monomers that could react with each other to make nylon-5,4. Draw the carbon-skeleton structure of the repeating unit in the polymer. (Note: The first number refers to the carboxylic acid monomer; the second refers to the amine monomer.)

---

Polymers of nylon make long, straight fibers that are quite strong and excellent for weaving into fabrics. A special variety of nylon called Kevlar, invented by Stephanie Kwolek of DuPont in 1965, is so strong that it is used in body armor, puncture-resistant tires, and face masks for hockey goaltenders. The extraordinary combination of strength and flexibility in Kevlar arises directly from its molecular structure (Figure 13.49).

Nylon is flexible and stretchable because the hydrocarbon chains can bend and curl, much like a telephone cord or a Slinky spring toy. To produce a stronger nylon, researchers recognized they had to find some way to reduce the ability of the chains to form coils. They discovered this could be done by using monomer

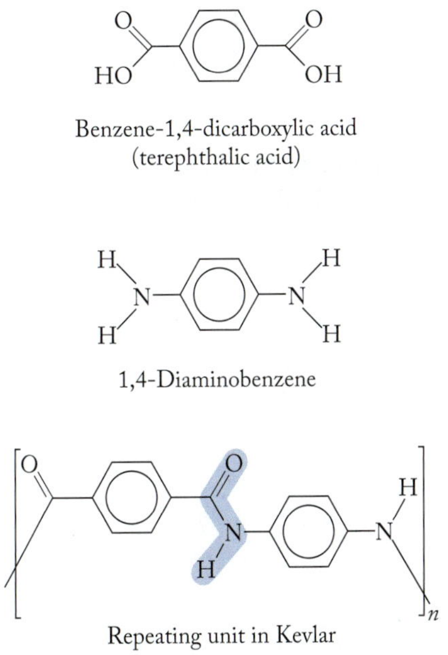

Benzene-1,4-dicarboxylic acid
(terephthalic acid)

1,4-Diaminobenzene

Repeating unit in Kevlar

**FIGURE 13.49** The monomers used to make Kevlar are a dicarboxylic acid and a diamine. The amide bond in the repeating unit is highlighted.

**FIGURE 13.50** Interactions between groups in Kevlar.

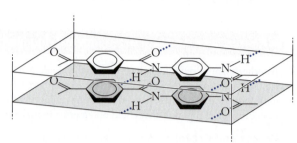

Hydrogen bonding between chains in the same plane

Stacking of benzene rings between chains in layered planes

units containing functional groups that made it difficult for the chains to bend. The result of this work was Kevlar, a polyamide formed from a dicarboxylic acid of benzene and a diamine of benzene. When these two monomers polymerize, the flat, rigid aromatic rings keep the chains straight.

Two additional intermolecular interactions orient the chains and hold them together very tightly (Figure 13.50). First, the –NH hydrogen atoms form hydrogen bonds with the oxygen atoms of carbonyl groups on adjacent chains. Second, the rings stack on top of one another (just as in polystyrene) and provide additional interactions that hold the chains together in parallel arrays. The result is a fiber that is very strong but still flexible. Fabrics and helmets made of Kevlar resist puncture, even by bullets fired at them, and they are also resistant to flames and reactive chemicals (Figure 13.51).

In this chapter we have seen many polymers and mentioned their uses in common products. The classification of these polymeric materials as addition or condensation polymers is summarized in Table 13.8. Table 13.8 also identifies the polymers by their distinctive functional groups and mentions some common uses in our lives.

**FIGURE 13.51** A bullet fired point-blank at a sheet of Kevlar does not puncture the fabric.

| TABLE 13.8 | **Summary of Common Polymers and Their Uses** | | |
|---|---|---|---|
| **NAME** | **ABBREVIATION** | **FUNCTIONAL GROUP** | **USE** |
| **Addition Polymers** | | | |
| Polyethylene | PE | Alkane | Plastic bags and films |
| Polytetrafluoroethylene | Teflon | Fluoroalkane | Nonstick coatings |
| Poly(1,1-dichloroethylene) | Saran | Chloroalkane | Plastic wrap |
| Poly(vinyl chloride) | PVC | Chloroalkane | Drain pipes |
| Poly(methyl methacrylate) | PMMA | Alkane and ester | Shatter-resistant glass, e.g., Plexiglas, Lucite |
| Polystyrene | PS | Aromatic hydrocarbon | Cups, dishes, insulation |
| Poly(vinyl alcohol) | PVAL | Alcohol | Gloves, bottles |
| **Condensation Polymers** | | | |
| Poly(ethylene glycol) | PEG | Ether | Pharmaceuticals, consumer products |
| Poly(ethylene oxide) | PEO | Ether | As for PEG |
| Poly(ethylene terephthalate) | PETE | Ester | Plastic bottles |
| Nylon | | Amide | Clothing |
| Kevlar | | Amide | Protective equipment |
| Dacron | | Ester | Clothing |

**optical isomers** molecules that are not superimposable on their mirror images.

**chiral** compounds having nonsuperimposable mirror images.

**achiral** not chiral; describes compounds that can be superimposed on their mirror images.

# 13.8 Chirality

In this chapter we have encountered two types of isomerism: constitutional isomers and stereoisomers. Cis and trans isomers of alkenes are stereoisomers: molecules having the same formula and the same bonds, but differing in the orientation in space of groups within the molecules. Another type of stereoisomerism, called *optical isomerism*, is especially important in organic molecules involved in biological processes.

## Optical Isomerism

**Optical isomers** are molecules that are not superimposable on their mirror images, just as your left hand is not superimposable on your right hand. As we explored in Section 9.6, such molecules are **chiral**, and their optical isomerism comes from their molecular structures containing *chiral centers*—most commonly a carbon atom that has four different groups attached to it. Figure 13.52 illustrates what is meant by mirror images and superimposition. The reflection in a mirror of an object like a plain coffee mug that is **achiral** (not chiral) is an image that can be superimposed on the original object. Your two hands, however, are mirror images that cannot be superimposed. Molecules that have this same property— molecules that cannot be superimposed on their mirror images—exist as optical isomers and are chiral. Review Section 9.6 and the discussion surrounding Figure 9.39 for a simple molecular example.

Let's now look at a slightly more complicated molecule, limonene (Figure 13.53), an alkene found in oranges and turpentine. Its chiral center is identified by the dotted circle. This chiral carbon atom is part of a six-membered ring, and

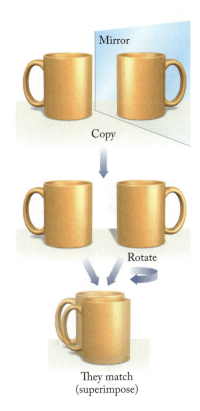

They do not match (they do not superimpose)

**FIGURE 13.52** A plain coffee mug is superimposable on its mirror image and is achiral. The mirror image of your left hand is your right hand, and the two are not superimposable.

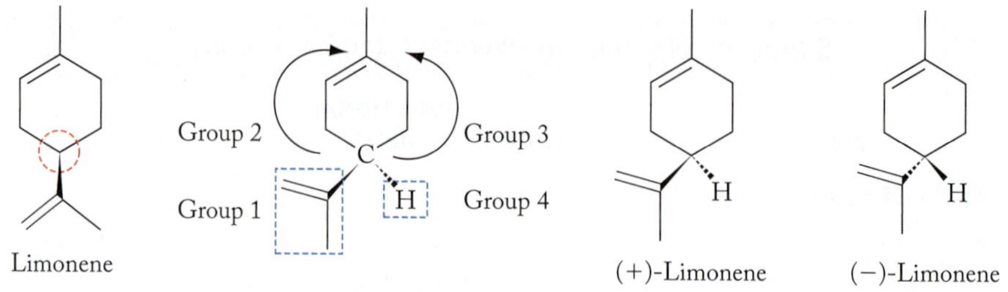

**FIGURE 13.53** Limonene is a chiral molecule with two enantiomeric forms. To show why it is chiral, we have highlighted its chiral carbon with a dashed circle and numbered the four groups bonded to it. Groups 1 and 4 are clearly different. Groups 2 and 3 are two halves of the same ring. They are different because group 2 has a C=C bond and group 3 does not. Therefore, the circled carbon atom is bonded to four different groups and is chiral.

⊙◑⊙ **CONNECTION** Chirality and optical isomerism were introduced in Section 9.6.

**CHEMTOUR** Chirality

if we examine the structure of limonene closely, we see that this carbon atom has four different groups bound to it. The three-carbon alkene group and the hydrogen atom are easy to see, but if we follow the carbon atoms around the ring, we find a C=C bond three bonds from the circled carbon on the left side, but only C—C single bonds on the right side. The presence of the C=C bond makes the groups attached to the circled carbon different, making limonene chiral. Therefore optical isomers exist. The two optical isomers are called **enantiomers**, nonsuperimposable mirror images. In no orientation do all of the groups on the chiral center of superimposed enantiomers coincide, because the molecules have different shapes in three dimensions.

Chirality is especially important in living systems, where molecules interact with other molecules in a process called **recognition**. Recognition in biochemistry means identifying a molecule based on its interaction with another molecule because of its shape. Just as your right hand fits into the correct glove, a molecule may fit into a three-dimensional cavity in a protein and cause an event based on that recognition. The protein that forms such a cavity is called a **receptor**.

Constitutional isomers and stereoisomers differ in their physical and chemical properties, but optical isomers have the same physical and chemical properties except for those that relate to a few specialized types of behavior. Recognition is one of those special behaviors. Limonene has two enantiomers, distinguished by the (+) and (−) signs in front of their names. One enantiomer smells like oranges; the other smells like turpentine. Part of the process of sensing different aromas involves recognition by receptors in your nasal passages. One receptor is shaped like (+)-limonene, the other like (−)-limonene, and the enantiomers bind to their own receptor just like your right and left hands fit into right and left gloves.

Enantiomers are also called **optically active molecules**. The (+) and (−) signs that distinguish the two isomers of limonene refer to the specific effect each isomer has on polarized light. In plane-polarized light, the electric fields that compose the beam oscillate in only one plane (Figure 13.54). When a beam of plane-polarized light passes through a solution containing one member of an enantiomeric pair, the beam rotates. If the beam rotates to the left (counterclockwise), the enantiomer is the *levorotary* form of the molecule and a (−) sign

**FIGURE 13.54** A beam of plane-polarized light contains an electric field that oscillates in only one direction. The plane of oscillation rotates if the beam passes through a solution of one enantiomer of an optically active compound. The (+) enantiomer causes the beam to rotate clockwise; the (−) enantiomer causes the beam to rotate counterclockwise.

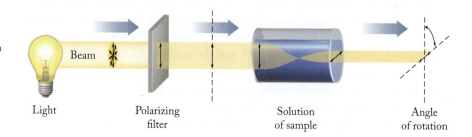

Light · Beam · Polarizing filter · Solution of sample · Angle of rotation

precedes its name. If the beam rotates to the right (clockwise), the enantiomer is the *dextrorotary* form and a (+) sign precedes its name. These molecules' effect on polarized light is why we call them *optical* isomers.

---

**SAMPLE EXERCISE 13.12**  **Recognizing Chiral Molecules**  **LO7**

Identify which of the molecules shown are chiral, and circle the chiral centers. Structures may have more than one chiral center.

a.

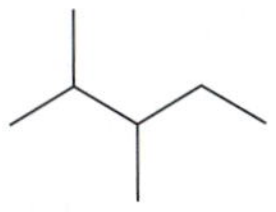

b.

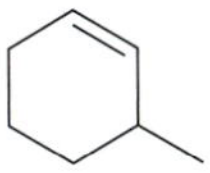

c.

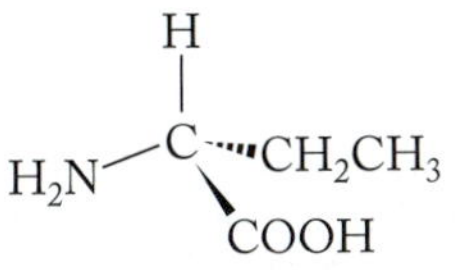

H
H₂N—C····CH₂CH₃
COOH

d.

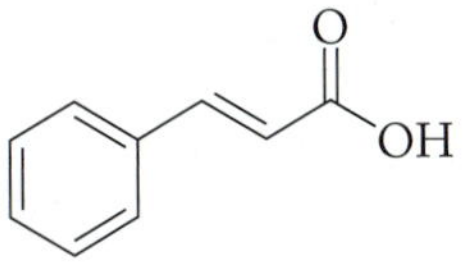

e.

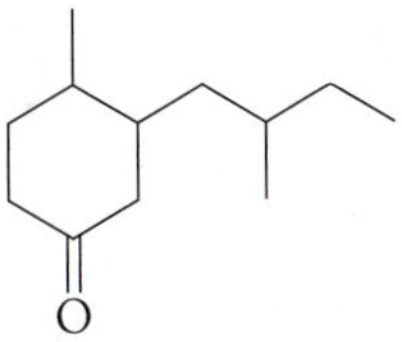

**Collect and Organize** Chiral molecules are nonsuperimposable on their mirror images. A chiral molecule has one or more chiral centers.

**Analyze** If molecules have carbon atoms with four different groups attached, they are chiral.

**Solve** The chiral carbon atoms in each structure are circled.

a.

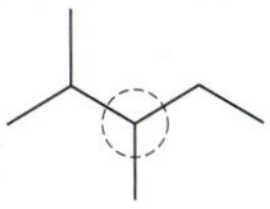

b.

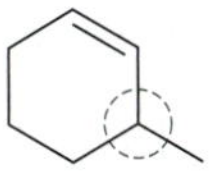

c.

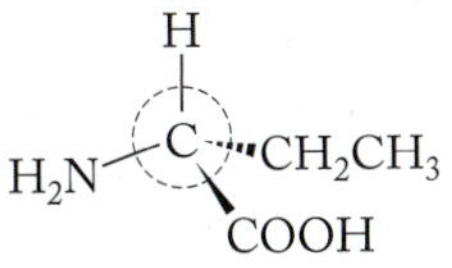

H
H₂N—C····CH₂CH₃
COOH

d.

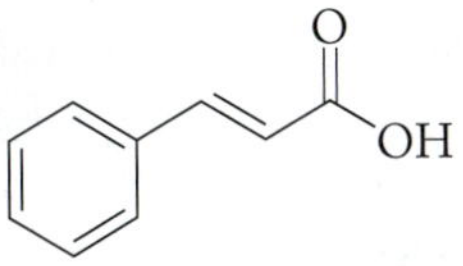

e.

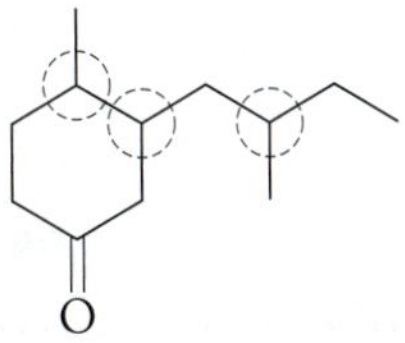

The C3 carbon of the pentane chain in compound a is bonded to a –CH₃ group, a –C₂H₅ group, a –C₃H₇ group, and an H atom, so the compound, 2,3-dimethylpentane, is chiral.

The chiral center in compound b is similar to the chiral center in limonene. The chiral carbon is bonded to a methyl group and an H. Tracing your finger around the ring in both directions from the circled carbon reveals the differences in the remaining two groups bonded to the chiral carbon.

The circled carbon atom in c is bonded to four different groups: –H, –NH₂, –COOH, and –CH₂CH₃, so this compound is chiral.

All of the carbon atoms in compound d are $sp^2$ hybridized, giving d a planar molecular geometry. Planar molecules cannot be chiral because they have a superimposable mirror image. Think of a mirror plane that contains all nine carbons in compound d: the mirror image is exactly the same.

Compound e has three chiral centers. Working from right to left, the first chiral center is similar to the chiral center in compound a. The other two chiral centers are in the cyclohexane ring. In each case, the carbon atom is bonded to four different groups, so compound e is chiral.

▶❚❚ **CHEMTOUR** Chiral Centers

**enantiomer** one of a pair of optical isomers of a compound.

**recognition** process by which molecules in living systems interact with one another.

**receptor** a cavity in a protein molecule that fits a particular molecule.

**optically active molecule** a chiral compound that causes rotation of a beam of plane-polarized light when it passes through a solution of the compound.

**Think About It** The presence of a chiral center in a molecule means that the molecule has two enantiomeric forms. The molecule is not superimposable on its mirror image. We should also note that although structure d has no chiral center, it has cis–trans isomers about the double bond in the chain. The isomer shown has the trans configuration.

**Practice Exercise** Identify which of the molecules below are chiral. Circle the chiral centers in each structure.

a.

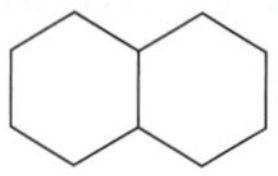

b.

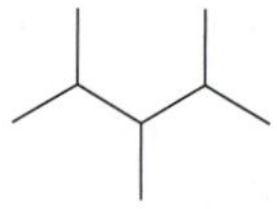

c.

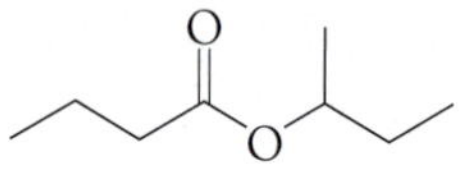

d.

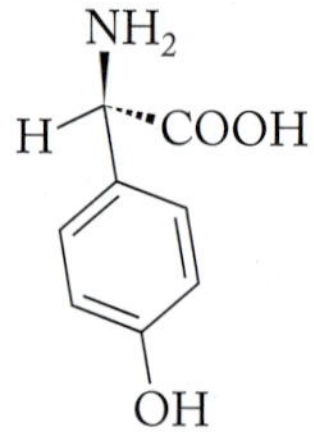

NH$_2$

H$\cdots$COOH

OH

e.

O

OH

OH

---

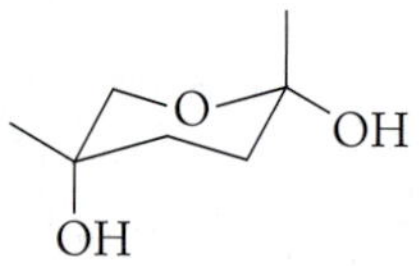

**CONCEPT TEST**

Quinine is a bitter-tasting substance extracted from the bark of the cinchona tree. It was the first effective substance used to treat malaria. Its enantiomer quinidine is a synthetic compound that suppresses cardiac arrhythmias (fast-beating heart). Describe the recognition between these molecules and their receptors in terms of the illustration on the right in Figure 13.55.

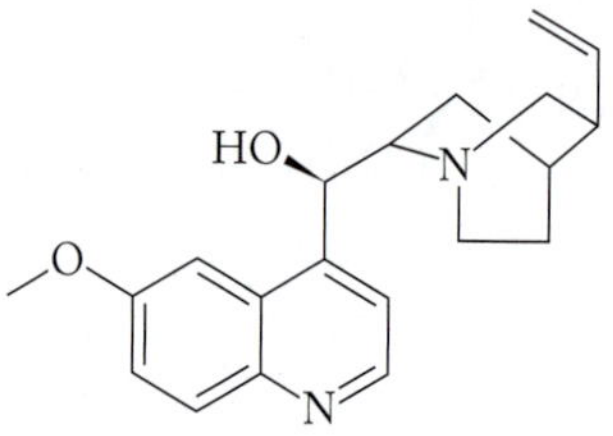

Quinine 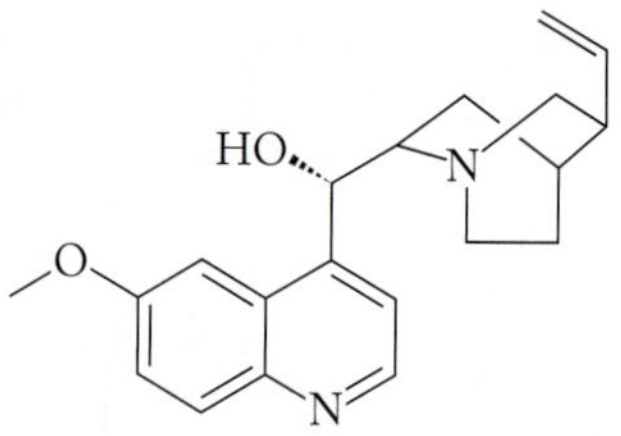                Quinidine

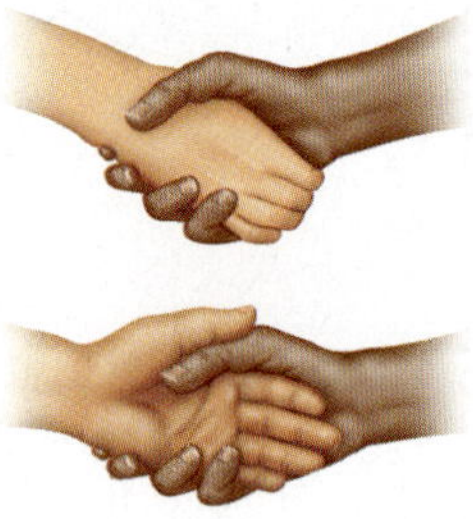

**FIGURE 13.55**

---

## Chirality in Nature

Chirality is ubiquitous in the organic compounds formed by living systems, and in most cases only one optical isomer occurs naturally in a particular organism. A molecule might even have more than one chiral center.

The origin of the fundamental preference of life for one enantiomer over another is unknown. Ongoing studies on the origin of chiral preference in living systems have produced no definitive answers but are providing increasingly interesting suggestions. Perhaps the origin of the preference involves the effect of slightly polarized sunlight.

Whatever the origin of these preferences, processes requiring molecular recognition in living systems often depend on the selectivity conveyed by chirality.

The human body is a chiral environment, so handedness of molecules matters. The perception of an aroma as either turpentine or orange, or as caraway or spearmint, depends on chirality, but other consequences of chiral recognition are much more dramatic. As many as half of the drugs made by large pharmaceutical companies are chiral and owe their function to recognition by a receptor that favors one enantiomer over the other. In 2012, six of the ten top-selling drugs globally were chiral. Typically, only one of the enantiomers of a chiral drug is active. A classic example of two enantiomers with different activity is the drug albuterol (Figure 13.56), used to treat wheezing and shortness of breath in people suffering from asthma and other lung disorders. One isomer causes bronchodilation (widening the air passages of the lungs and easing breathing), while the other causes bronchial constriction and may actually be detrimental to the patient.

When chiral compounds are produced by living systems, typically only one stereoisomer is produced. For example, the aroma of spearmint leaves is due to (+)-carvone (Figure 13.38). However, when a molecule such as albuterol is made in a laboratory, both isomers are produced unless special methods are used. When both optical isomers are present in equal amounts in a sample, the material is known as a **racemic mixture**. Because one isomer rotates the plane of polarized light in one direction, and the other to the same extent in the opposite direction, a racemic mixture does not rotate the plane of polarized light at all. Because the two isomers interact differently with receptors, the pharmaceutical industry routinely faces two choices: devise a special synthetic procedure that yields only the isomer of interest, or separate the two isomers at the end of the manufacturing process. Both approaches are widely used.

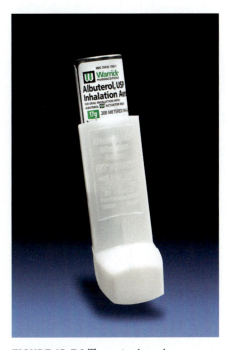

**FIGURE 13.56** The antiasthma drug albuterol; the chiral center is circled. Albuterol is frequently administered through an inhaler like the one shown.

---

**SAMPLE EXERCISE 13.13** **Recognizing Optical Properties of Chiral Molecules** **LO7**

The structure of the antidepressant drug bupropion is shown in the margin. When tested, a solution of a sample of the drug that comes directly from the laboratory does not rotate the plane of polarized light, so the analyst rejects the sample, saying it is not the pure drug, which is known to cause a beam of polarized light to rotate when it passes through a solution of the compound. The chemist who made the sample tells you that other analytical data (percent composition and mass spectrometry) prove the material is 100% bupropion. Both people are correct in what they say. Explain why.

**Collect and Organize** We are asked to explain how a sample that is 100% chemically pure is not 100% pure active drug. We are given the structure of the compound, and the information that a solution of the material does not rotate the plane of polarized light.

**Analyze** The sample was rejected by one analyst because a solution did not rotate the plane of polarized light. We must examine the structure to see if the molecule has a chiral center—that is, a carbon atom bonded to four different groups.

**Solve** Bupropion is a chiral molecule. The chiral center is circled in the structure shown. The sample must contain both enantiomers of bupropion in equal amounts. They have exactly the same chemical formula, so the sample is chemically pure, but one enantiomer rotates polarized light to the left and the other to the right. No net rotation is observed when polarized light passes through the sample. Only one enantiomer is the active drug. Therefore, the sample is chemically pure but not optically pure.

**Think About It** Enantiomers have the same chemical composition and the same molar mass. Their physical properties are identical except for the direction in which their solutions cause plane-polarized light to rotate when it passes through them.

Chiral center

Bupropion

**racemic mixture** a sample containing equal amounts of both optical isomers of a compound.

⚙ **Practice Exercise** Identify the chiral carbon atoms in the cationic natural product muscarine, found in some poisonous mushrooms (Figure 13.57).

Muscarine

**FIGURE 13.57** Several varieties of mushrooms, including *Amanita muscaria*, contain the toxic substance muscarine.

**CONCEPT TEST**

*The Documents in the Case,* a mystery written by Dorothy L. Sayers in 1930, involves the suspicious death of an authority on wild, edible mushrooms. Allegedly, the victim ate a stew made of poisonous mushrooms that contained muscarine, a toxic natural product. A forensic specialist evaluated the contents of the victim's stomach and observed that the fluid contained muscarine but did not rotate the plane of polarized light. Because of this fact, the coroner concluded that the man was murdered. Why did the coroner in the story reach this conclusion?

---

**SAMPLE EXERCISE 13.14    Integrating Concepts: Lotions and Creams**

Two materials used in personal care preparations (skin lotions, lubricants, cosmetics) are sometimes confused because they feel very similar on the skin. They are, however, very different. Petroleum jelly (Vaseline or cold cream) is a mixture of hydrocarbons with carbon numbers above 25, whereas glycerin (glycerol) is a pure substance and an alcohol.

$$CH_3(CH_2)_{26}CH_3 \qquad HOCH_2CH(OH)CH_2OH$$
Average molecule in              Glycerol
petroleum jelly

Glycerol (mp = 17.8°C; bp = 290°C) is a liquid at body temperature (37°C), while petroleum jelly is a semisolid, meaning it melts over a range of temperatures around body temperature.

a. Both materials coat the skin. Petroleum jelly smoothes and protects the skin from moisture loss but feels greasy; glycerol improves smoothness and produces a feeling of wetness on the skin. Refer to their structures and explain these observations.

b. Petroleum jelly was originally isolated from a waxy material that formed on oil rigs and is collected during the fractional distillation of crude oil. Why does it melt over a range of temperatures, rather than having a sharply defined melting point?

c. Glycerol may be produced from fats by a reaction called saponification; it is also a by-product of biodiesel production. Both of those reactions are shown in **Analyze**. Identify the functional groups in the numbered molecules and the reaction discussed in this chapter that is related to these reactions.

d. Items made of natural latex (structure below) are frequently used in the presence of skin softeners and lubricants. However, using latex items is not advised in the presence of materials with petroleum jelly as a base; only glycerin-containing creams are considered suitable. Refer to the structures and likely properties of these materials to explain this advice.

**Collect and Organize** We are given the structures and physical properties of two materials used in personal care products.

**Analyze** Petroleum jelly is a mixture of nonpolar alkanes, so van der Waals forces are the main interactive forces between its molecules. Glycerol is a polar alcohol and a pure liquid. Glycerol has three –OH groups per molecule and is capable of forming an extensive network of hydrogen bonds.

(1)                                        (2)

(3)              (4)                (5)

### Solve

a. Petroleum jelly protects the skin by coating it with a barrier that won't dissolve in water because it is a nonpolar hydrocarbon. The water-repelling property imparts the greasy feel. Glycerol remains a liquid on the skin, unlike other low-molecular-mass alcohols that evaporate and cool, because it has such a high boiling point. With its three –OH groups, it is very waterlike and would absorb water from the air, which contributes to the wet feeling.

b. Petroleum jelly is a mixture, not a pure compound. Table 13.3 indicates that individual alkanes in the mass range of those in petroleum jelly melt between 37°C and 75°C. Mixtures have lower melting points and broader melting ranges than their pure components, so a mixture of alkanes in this range could be expected to melt around body temperature.

c. The following functional groups are present: (1) ester, (2) carboxylic acid (shown is the carboxylate ion because the reaction takes place in basic medium), (3) ester, (4) alcohol, (5) ester.

Saponification results in the cleavage of an ester linkage in base to yield an alcohol and a carboxylic acid anion. That is the reverse of the esterification reaction, a condensation reaction discussed in Section 13.7. The reaction to produce biodiesel is related to esterification. In the reaction shown, the alcohol group in the starting material [an ester, (3)] is exchanged for another alcohol [in this case, ethanol, (4)], forming a new ester (5) and liberating the alcohol in the starting material.

d. Natural latex is a nonpolar hydrocarbon polymer that would likely be soluble in petroleum jelly. Items made of latex would lose their physical integrity on exposure to petroleum jelly. Glycerol, however, as a highly polar, waterlike material, would have no effect on latex.

**Think About It** A knowledge of functional groups and interactive forces enables us to predict the behavior of materials whose composition and structure we know.

In this chapter, we introduced the major functional groups in organic chemistry and discussed how these functional groups and the structure of the molecules determine their chemical and physical properties. We examined a few reactions of those groups that give rise to several types of polymeric materials common in the modern world, ranging from carpets and insulation to bulletproof vests. Other organic compounds are used as fuels, as pharmaceuticals, and as building materials. Despite the range of materials we have discussed, this treatment can give only a brief glimpse of the importance of organic compounds. Chemical Abstracts Service (CAS), an organization that tracks and collects all chemical information published worldwide, announced in 2009 that it had recorded the 50 millionth unique chemical substance (a new drug with pain-relieving properties) in its registry. CAS registered the 40 millionth substance just nine months earlier. There are now more than 70 million. In contrast, it took 33 years for CAS to register the 10 millionth compound in 1990. Typically, more than 95% of the new compounds registered in a given year contain carbon, which gives you some idea of why the study of organic chemistry occupies a special place within the discipline.

## SUMMARY

**Learning Outcome 1** The carbon atoms in **alkanes** (or **saturated hydrocarbons**) bear the maximum number of hydrogen atoms possible and contain carbon–carbon single bonds. **Cycloalkanes** are alkanes containing rings of carbon atoms. **Alkenes** and **alkynes** are **unsaturated hydrocarbons.** (Sections 13.1, 13.2, and 13.3)

**Learning Outcome 2** Alkanes and alkenes may have **constitutional isomers**, and alkenes may in addition have **stereoisomers**. (Sections 13.2 and 13.3)

**Learning Outcome 3** Alkenes undergo **addition reactions** with hydrogen and hydrogen halides. Alkenes can also be polymerized to **homopolymers**. Most of these polymers are **addition polymers**. (Sections 13.3 and 13.4)

**Learning Outcome 4** Organic compounds are categorized based on functional groups present in their molecular structures: in addition to alkanes, alkenes, and alkynes, other categories include aromatic hydrocarbons, **amines**, **alcohols**, **ethers**, and a number of **carbonyl**-containing compounds. (Sections 13.4, 13.5, 13.6, and 13.7)

**Learning Outcome 5** Ethers are frequently blended with gasoline to improve combustion, and most gasoline in the United States now contains around 10% ethanol. (Section 13.6)

**Learning Outcome 6** Monomer units derived from alcohols or ethers can be chemically combined to make **heteropolymers** or **copolymers**. Condensation reactions of carboxylic acids and amines are used to prepare **condensation polymers** such as polyesters and polyamides. Properties of the polymers depend on the identity and arrangement of the monomer units in the polymer. (Sections 13.6 and 13.7)

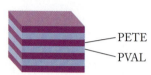

**Learning Outcome 7** Many biologically important molecules exhibit chirality, or optical isomerism, a type of stereoisomerism. Chirality is important in biological systems because **recognition** of molecules by **receptors** is often based on shape. Chiral molecules can rotate a beam of plane-polarized light. If two optical isomers of a compound are present in equal ratios in a sample, the material is known as a **racemic mixture** and there is no rotation of polarized light. (Section 13.8)

## PROBLEM-SOLVING SUMMARY

| TYPE OF PROBLEM | CONCEPTS AND EQUATIONS | SAMPLE EXERCISES |
|---|---|---|
| **Distinguishing among alkanes, alkenes, and alkynes** | Alkanes contain only C—C single bonds; they are saturated hydrocarbons and do not react with $H_2$. Alkenes contain at least one C=C double bond that can combine with a molecule of $H_2$. Alkynes contain at least one C≡C triple bond that can combine with two molecules of $H_2$. | 13.1 |
| **Drawing alkane structures** | Use the Lewis structure to create a condensed structure, then gather all methylene groups inside parentheses to create the final condensed structure. To create a carbon-skeleton structure, use lines to depict single covalent bonds between carbon atoms; a sufficient number of H atoms to complete the valency of the carbon atoms is assumed. | 13.2 |
| **Recognizing constitutional isomers** | Establish that the compounds have the same molecular formula, and if they do, look for different arrangements of C—C bonds. | 13.3 |
| **Identifying the longest chain in organic molecules** | Start at one end of any branch and assign numbers to each carbon atom. If the chain branches, choose one branch to follow. Repeat the process to explore other side chains and identify the longest chain. | 13.4 |
| **Identifying and naming stereoisomers and constitutional isomers** | After establishing that the compounds have the same molecular formula, look for different arrangements of C—C bonds. Molecules with two like groups on the same side of a C=C bond are cis; those with two like groups on opposite sides are trans isomers. | 13.5 |
| **Identifying monomers in polymers** | Find the smallest portion of the polymer that is repeated. | 13.6, 13.11 |

| TYPE OF PROBLEM | CONCEPTS AND EQUATIONS | SAMPLE EXERCISES |
|---|---|---|
| **Comparing the energy content of fuel mixtures** | Determine the number of moles of each component, then multiply that number by the corresponding value of $\Delta H^{\circ}_{comb}$. The total amount of energy available equals the sum of the heats available from each component. | 13.7 |
| **Assessing properties of polymers** | Evaluate the polarity of the functional groups in the polymer, and assess the relative importance of all types of intermolecular forces possible in the molecules. | 13.8, 13.10 |
| **Making a polyester** | Combine monomers with alcohol functional groups (ROH) and carboxylic acid functional groups (RCOOH) to form water ($H_2O$) and ester groups (RCOOR). | 13.9 |
| **Recognizing chiral molecules** | Identify carbon atoms with four different groups attached. | 13.12 |
| **Recognizing optical properties of chiral molecules** | Racemic mixtures contain equal amounts of two enantiomers and do not rotate the plane of polarized light. | 13.13 |

## VISUAL PROBLEMS

*(Answers to boldface end-of-chapter questions and problems are in the back of the book.)*

**13.1.** How many degrees of unsaturation are in each of the hydrocarbons shown in Figure P13.1?

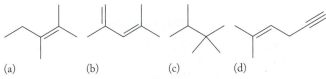

(a)          (b)          (c)          (d)

**FIGURE P13.1**

**13.2.** Which of the hydrocarbons in Figure P13.2 are constitutional isomers of each other?

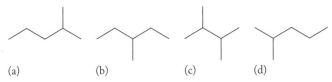

(a)          (b)          (c)          (d)

**FIGURE P13.2**

**13.3.** In Figure P13.3 are the carbon-skeleton structures of four organic compounds found in nature as fragrant oils. Which are alkenes?

Pine oil     Oil of peppermint     Oil of celery     Camphor

**FIGURE P13.3**

**13.4.** Figure P13.4 shows three molecules: acrolein (found in barbeque smoke), capillin (an antifungal drug), and pargyline (an antihypertensive drug). Which of these molecules does not contain the alkyne functional group?

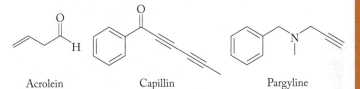

Acrolein          Capillin          Pargyline

**FIGURE P13.4**

**13.5.** Condensed structures may be ambiguous about constitutional isomers. Consider these formulas, identify possible cis and trans isomers, and draw them.
a. $CH_3CH_2CH\!\!=\!\!CHCH(CH_3)_2$
b. $CH_3CH_2CH(CH_3)CH\!\!=\!\!C(CH_3)_2$
c. $CH_2\!\!=\!\!C(CH_3)CH_2CH\!\!=\!\!CHCH_3$

\*13.6. Condensed structures may be ambiguous about constitutional isomers. Consider these formulas, identify possible constitutional isomers, and draw them.
a. $(C_6H_5)CH\!\!=\!\!C(CH_3)CH_2CH_3$

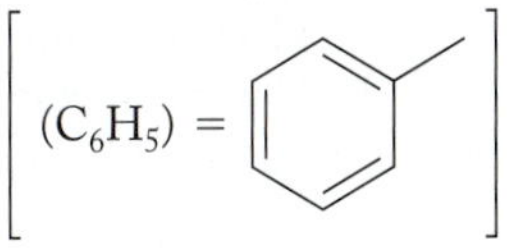

$$\left[ (C_6H_5) = \phantom{xxx} \right]$$

b. $CH_3C(Cl)\!\!=\!\!CHBr$
c. $CH_3CH\!\!=\!\!CHCH_2CH\!\!=\!\!CHCH_3$

**13.7.** Which molecules in Figure P13.7 are considered to be aromatic compounds?

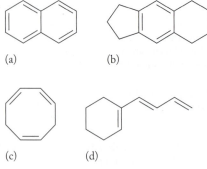

(a)        (b)

(c)        (d)

**FIGURE P13.7**

**13.8.** Benzyl acetate, carvone, and cinnamaldehyde are all naturally occurring oils. Their carbon-skeleton structures are shown in Figure P13.8. Which ones contain an aromatic ring?

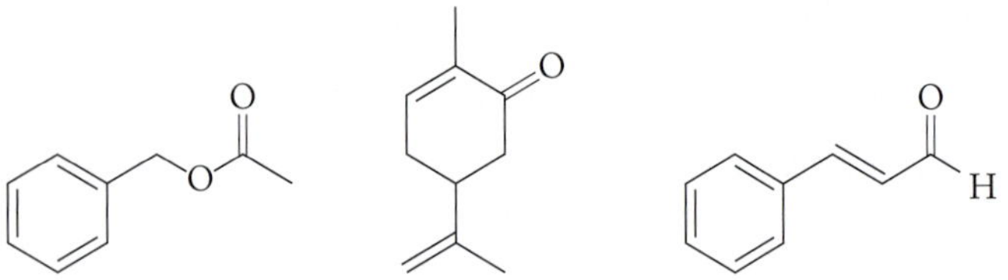

Benzyl acetate        Carvone        Cinnamaldehyde
(oil of jasmine)      (oil of spearmint)    (oil of cinnamon)

**FIGURE P13.8**

**13.9.** In addition to the aromatic ring, what other functional groups can you identify in the molecules in Problem 13.8?

*13.10.* The three polymers shown in Figure P13.10 are widely used in the plastics industry. In which of them are the intermolecular forces per mole of monomer the strongest?

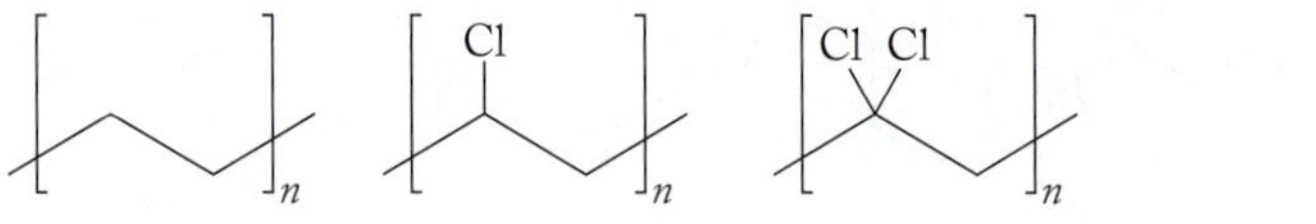

(a) Polyethylene   (b) Poly(vinyl chloride)   (c) Poly(1,1-dichloroethylene)

**FIGURE P13.10**

*13.11.* **Silly Putty** Silly Putty is a condensation polymer of dihydroxydimethylsilane (Figure P13.11). Draw the condensed structure of the repeating monomer unit in Silly Putty.

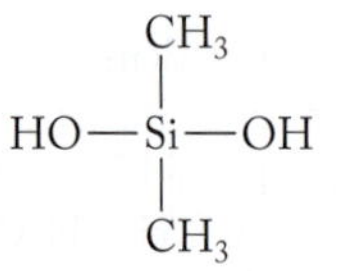

$$CH_3$$
$$HO-Si-OH$$
$$CH_3$$

Dihydroxydimethylsilane

**FIGURE P13.11**

**13.12.** The nectar that attracts hummingbirds to flowers is rich in sucrose (sugar). Sucrose is composed of one molecule each of glucose (6-membered ring) and fructose (5-membered ring).
 a. Identify the functional groups in sucrose.
 b. How many chiral centers does the molecule have?

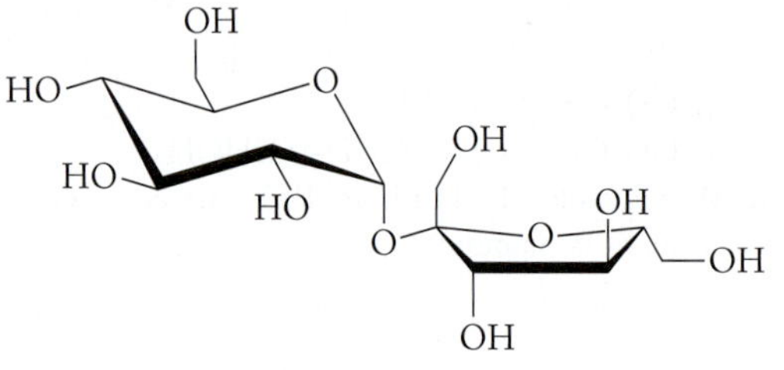

**FIGURE P13.12**

**13.13.** Rubber is a polymer of isoprene. It is sometimes called polyisoprene. There are two forms of polyisoprene (Figure P13.13): *cis*-polyisoprene is the soft, flexible material we associate with the term "rubber"; gutta percha, or *trans*-polyisoprene, is a much harder material. Draw the monomeric units of *cis*- and *trans*-polyisoprene.

*cis*-Polyisoprene        *trans*-Polyisoprene

**FIGURE P13.13**

**13.14.** The cholesterol-lowering drug Lipitor (Figure P13.14) is the best-selling drug in the history of pharmaceuticals, making over $125 billion in sales from 1997 to 2011. Circle the chiral carbon atom(s) in Lipitor, and identify the functional groups in the molecule.

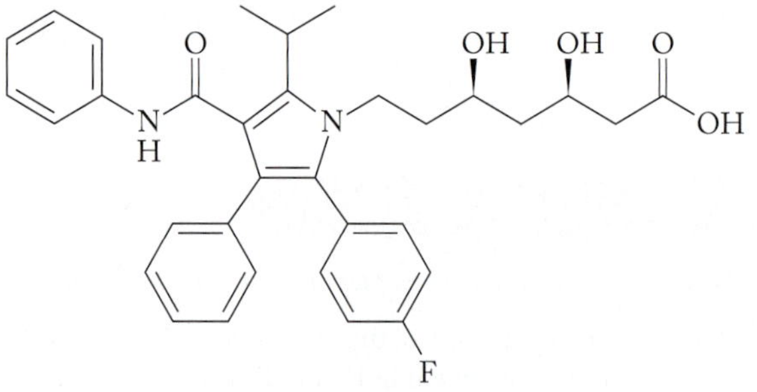

Lipitor

**FIGURE P13.14**

**13.15. Cholesterol-Lowering Drugs** High serum cholesterol levels often correlate with increased risk of heart attacks. The drug sold under the trade name Mevacor has proven to be effective in lowering serum cholesterol. How many chiral carbon atoms are there in Mevacor (Figure P13.15)? Identify the functional groups in the molecule.

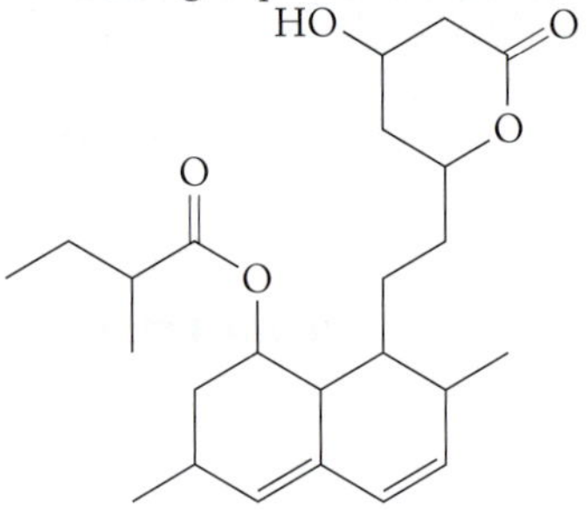

Mevacor

**FIGURE P13.15**

*13.16.* The two compounds shown in Figure P13.16 are both amino acids. Identify the structural difference between them and explain why they are both amino acids.

$H_2N$     $COOH$        $H_2N$      $COOH$

**FIGURE P13.16**

## QUESTIONS AND PROBLEMS

### Carbon: The Scope of Organic Chemistry

**CONCEPT REVIEW**

**13.17.** Describe three ways in which carbon atoms can form bonds to other carbon atoms using hybrid orbitals (valence bond theory).

**13.18.** Can a macromolecule be composed of more than one type of monomer?

**\*13.19.** Is the interstitial alloy tungsten carbide (WC) considered to be an organic compound?

**\*13.20.** Calcium carbide, $CaC_2$, was used in miner's lamps. Reaction of $CaC_2$ with water yields acetylene, which, when ignited, gives light. Is calcium carbide considered an organic compound?

**PROBLEMS**

**13.21.** Find an example of a small molecule with more than one functional group in Chapter 9.

13.22. Explain in your own words why polymers typically do not have well-defined melting points and boiling points.

**13.23.** Polyethylene is prepared from the monomer ethylene, $C_2H_4$. About how many monomers are needed to make a polymer with a molar mass of 100,000 g/mol?

13.24. Synthetic rubber is prepared from butadiene, $C_4H_6$. About how many monomers are needed to make a polymer with a molar mass of 100,000 g/mol?

### Alkanes

**CONCEPT REVIEW**

**13.25.** Do linear and branched alkanes with the same number of carbon atoms all have the same empirical formula?

13.26. If an alkane and a cycloalkane have equal numbers of carbon atoms per molecule, do they have the same number of hydrogen atoms?

**13.27.** What is the hybridization of carbon in alkanes?

13.28. Figure P13.28 shows the carbon-skeleton structures of hexane and cyclohexane. Are hexane and cyclohexane constitutional isomers?

Hexane      Cyclohexane

**FIGURE P13.28**

**13.29.** Why isn't cyclohexane a planar molecule?

13.30. Which of the simple cycloalkanes ($C_nH_{2n}$, $n = 3–8$) has a nearly planar geometry?

**13.31.** Are cycloalkanes saturated hydrocarbons?

13.32. Do constitutional isomers always have the same molecular formula?

**13.33.** Do constitutional isomers always have the same chemical properties?

13.34. Are constitutional isomers members of a homologous series?

**PROBLEMS**

**13.35.** Draw and name all the constitutional isomers of $C_5H_{12}$.

13.36. Draw and name all the constitutional isomers of $C_6H_{14}$.

**13.37.** Which of the molecules in Figure P13.37 are constitutional isomers of octane ($C_8H_{18}$)? Name these molecules.

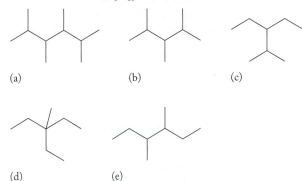

**FIGURE P13.37**

**13.38.** Which of the molecules in Figure P13.38 are constitutional isomers of heptane ($C_7H_{16}$)? Name these molecules.

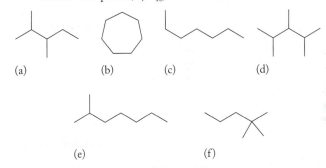

**FIGURE P13.38**

**13.39.** Convert the carbon-skeleton structures in Problem 13.37 to molecular formulas.

13.40. Convert the carbon-skeleton structures in Problem 13.38 to molecular formulas.

**13.41.** Using the average bond strengths given in Appendix 4, estimate the molar heat of hydrogenation, $\Delta H_{hydrogenation}$, for the conversion of $C_2H_4$ to $C_2H_6$.

$$CH_2{=}CH_2(g) + H_2(g) \rightarrow CH_3CH_3(g)$$

13.42. Using the average bond strengths given in Appendix 4, estimate the molar heat of hydrogenation, $\Delta H_{hydrogenation}$, for the conversion of $C_2H_2$ to $C_2H_6$.

$$CH{\equiv}CH(g) + 2\,H_2(g) \rightarrow CH_3CH_3(g)$$

**13.43.** Place the following molecules in order of increasing boiling point: $C_3H_8$, $C_{14}H_{30}$, cyclooctane ($C_8H_{16}$).

**\*13.44.** Rank the molecules in Figure P13.44 in order of decreasing van der Waals forces.

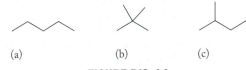

**FIGURE P13.44**

## Alkenes and Alkynes

### CONCEPT REVIEW

**13.45.** How do structural isomers differ from stereoisomers?

**13.46.** Explain why alkanes don't have stereoisomers.

**13.47.** Can combustion analysis distinguish between an alkene and a cycloalkane containing the same number of carbon atoms?

**13.48.** Can combustion analysis data be used to distinguish between an alkyne and a cycloalkene containing the same number of carbon atoms?

**13.49.** Why don't the alkenes in Figure P13.49 have cis and trans isomers?

**FIGURE P13.49**

**13.50.** Why don't alkynes have cis and trans isomers?

**\*13.51.** Figure P13.51 shows the carbon-skeleton structure of carvone, which is found in oil of spearmint. Why doesn't the molecule carvone have cis and trans isomers?

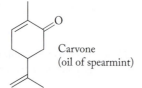

Carvone
(oil of spearmint)

**FIGURE P13.51**

**\*13.52.** Figure P13.52 shows the carbon-skeleton structure of the antifungal compound capillin. Are the π electrons in capillin delocalized?

Capillin

**FIGURE P13.52**

**13.53.** Ethylene reacts quickly with HBr at room temperature, but polyethylene is chemically unreactive toward HBr. Explain why these related substances have such different properties.

**\*13.54.** Identify the oxidation state of the carbon atoms indicated by the arrows in these structures:

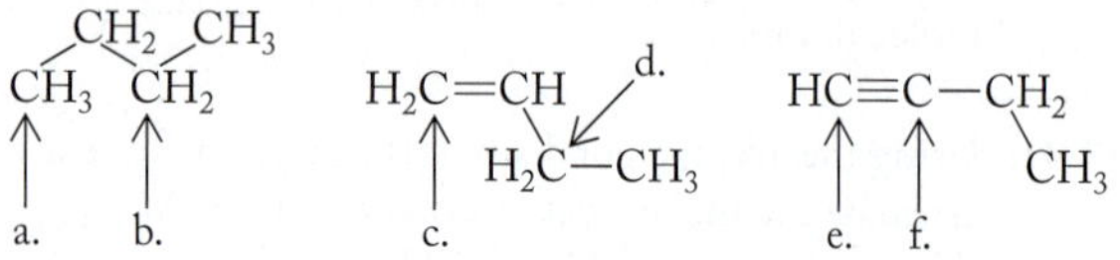

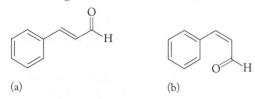

### PROBLEMS

**13.55. Cinnamon** Label the isomers of cinnamaldehyde (oil of cinnamon) in Figure P13.55 as cis or trans and *E* or *Z*.

**FIGURE P13.55**

**13.56.** Prostaglandins, naturally occurring compounds in our bodies that cause inflammation and other physiological responses, are formed from arachidonic acid, an unsaturated hydrocarbon containing four C=C double bonds and a carboxylic acid functional group. The stereoisomer containing all cis double bonds is shown in Figure P13.56. How many stereoisomers other than this one are possible? Draw the isomer containing all trans double bonds.

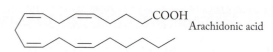

Arachidonic acid

**FIGURE P13.56**

**13.57.** Using data in Appendix 4, calculate $\Delta H_{rxn}$ for the production of acetylene from the controlled combustion of methane:

$$6\ CH_4(g) + O_2(g) \rightarrow 2\ C_2H_2(g) + 2\ CO(g) + 10\ H_2(g)$$

a. Is this an endothermic or an exothermic reaction?

b. Is the carbon atom in $CH_4$ oxidized or reduced in this reaction?

**13.58.** Using data in Appendix 4, calculate $\Delta H_{rxn}$ for the production of acetylene from the reaction between calcium carbide and water given $\Delta H_f^\circ$ of $CaC_2$ is −59.8 kJ/mol:

$$CaC_2(s) + 2\ H_2O(\ell) \rightarrow C_2H_2(g) + Ca(OH)_2(s)$$

a. Is this an endothermic or an exothermic reaction?

b. Are the carbon atoms in calcium carbide oxidized or reduced in this reaction?

**13.59. Making Glue** Wood glue or "carpenter's glue" is made of poly(vinyl acetate). Draw the carbon-skeleton structure of this polymer. The monomer is shown in Figure P13.59.

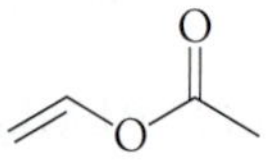

Vinyl acetate

**FIGURE P13.59**

**13.60.** The 2000 Nobel Prize in Chemistry was awarded for research on the electrically conductive polymer polyacetylene.

a. Draw the carbon-skeleton structure of three monomeric units of the addition polymer that results from polymerization of acetylene, HC≡CH.

\*b. There are two possible stereoisomers of polyacetylene. Describe the two isomeric forms.

## Aromatic Compounds

### CONCEPT REVIEW

**13.61.** Why is benzene a planar molecule?

**13.62.** Is aromaticity different from resonance?

**13.63.** Do tetramethylbenzene and pentamethylbenzene have constitutional isomers?

**13.64.** Why aren't butadiene, $C_4H_6$, and 1,3-cyclohexadiene, $C_6H_8$ (Figure P13.64), considered aromatic molecules?

Butadiene       1,3-Cyclohexadiene

**FIGURE P13.64**

**\*13.65.** Pyridine (Figure P13.65) has the molecular formula $C_6H_5N$. Is pyridine an aromatic molecule?

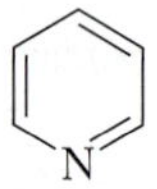

Pyridine

**FIGURE P13.65**

**\*13.66.** Is graphite (see Chapter 12) an aromatic compound?

## PROBLEMS

**13.67.** Draw all the constitutional isomers of trimethylbenzene.

**13.68.** Draw all the constitutional isomers of dimethylnaphthalene.

**13.69.** Calculate the fuel values of gaseous benzene ($C_6H_6$) and ethylene gas ($C_2H_4$). Does 1 mole of benzene have a higher or lower fuel value than 3 moles of ethylene?

**13.70.** Does 1 mole of gaseous benzene ($C_6H_6$) have a higher or lower fuel value than 3 moles of acetylene gas ($C_2H_2$)?

## Amines

### CONCEPT REVIEW

**13.71.** Explain why methylamine ($CH_3NH_2$) is more soluble in water than butylamine [$CH_3(CH_2)_3NH_2$].

*\*13.72.** Dimethylamine [$(CH_3)_2NH$] is considerably more basic than diphenylamine [$(C_6H_5)_2NH$]. Explain why.

### PROBLEMS

**13.73.** Serotonin and amphetamine both contain the amine functional group (Figure P13.73). Serotonin is responsible, in part, for signaling that we have had enough to eat. Amphetamine, an addictive drug, can be used as an appetite suppressant. Identify the primary and secondary amine functional groups in these molecules.

Serotonin        Amphetamine

**FIGURE P13.73**

**13.74.** Epinephrine (adrenaline) and dopamine are two naturally occurring amine-containing neural transmitters in the human body. Amphetamine and MDMA (Ecstasy) are two drugs of abuse that increase wakefulness and induce feelings of euphoria. Their structures are given in Figure P.13.74. Which are primary and which secondary amines?

Epinephrine        Dopamine

Amphetamine        MDMA

**FIGURE P13.74**

**13.75.** Bacteria of the genus *Methanosarcina* convert amines to methane. Their action helps make methane a renewable energy source. Determine the standard enthalpy of the

following reaction from the appropriate standard enthalpies of formation ($\Delta H^\circ_{f,CH_3NH_2} = -23.0$ kJ/mol):

$$4\ CH_3NH_2(g) + 2\ H_2O(\ell) \rightarrow 3\ CH_4(g) + CO_2(g) + 4\ NH_3(g)$$

**13.76.** Determine the $\Delta H^\circ_{rxn}$ values of these combustion reactions of methylamine ($\Delta H^\circ_{f,CH_3NH_2} = -23.0$ kJ/mol):

$$4\ CH_3NH_2(g) + 13\ O_2(g) \rightarrow 4\ CO_2(g) + 4\ NO_2(g) + 10\ H_2O(\ell)$$
$$4\ CH_3NH_2(g) + 6\ O_2(g) \rightarrow 4\ CO_2(g) + 4\ NH_3(g) + 4\ H_2O(\ell)$$

## Alcohols, Ethers, and Reformulated Gasoline

### CONCEPT REVIEW

**13.77.** Why are the fuel values of dimethyl ether and ethanol (Figure P13.77) lower than that of ethane?

Dimethyl ether        Ethanol

**FIGURE P13.77**

**13.78.** Would you expect the fuel value of alcohols to increase or decrease as the number of carbon atoms in the alcohol increases?

**13.79.** Why do ethers typically boil at lower temperatures than alcohols with the same molecular formula?

**13.80.** Which do you expect to be more soluble in water, MTBE or 2,2-dimethylbutane (Figure P13.80)? Explain your answer.

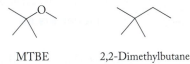

MTBE        2,2-Dimethylbutane

**FIGURE P13.80**

*\*13.81.** Disposable wipes used to clean the skin prior to receiving an immunization shot contain ethanol. After wiping your arm, your skin feels cold. Why?

*\*13.82.** During the winter months in cold climates, water condensing in a vehicle's gas tank reduces engine performance. An auto mechanic recommends adding "dry gas" to the tank during your next fill-up. Dry gas is typically an alcohol that dissolves in gasoline and absorbs water. Based on the structures shown in Figure P13.82, which product would you predict would do a better job—methanol or 2-propanol?

$CH_3OH$

Methanol        2-Propanol

**FIGURE P13.82**

### PROBLEMS

**13.83.** Which of the compounds in Figure P13.83 are alcohols and which ones are ethers? Place them in order of increasing boiling point.

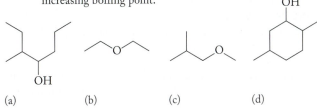

(a)        (b)        (c)        (d)

**FIGURE P13.83**

**13.84.** Which of the compounds in Figure P13.84 are alcohols and which ones are ethers? Place them in order of increasing vapor pressure at 25°C.

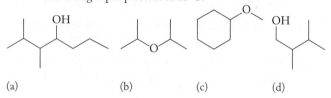

(a)         (b)         (c)         (d)

**FIGURE P13.84**

———

Consult tables of thermochemical data in Appendix 4 for any values you may need to solve Problems 13.85 through 13.88.

**13.85.** Calculate the fuel value of diethyl ether and butanol (Figure P13.85). Which has the higher fuel value?

Diethyl ether        Butanol

**FIGURE P13.85**

**13.86.** Calculate the fuel value of liquid diethyl ether and methyl propyl ether (Figure P13.86). Which has the higher fuel value? ($\Delta H^\circ_{f,\text{methyl propyl ether}} = -266.0$ kJ/mol.)

Diethyl ether        Methyl propyl ether

**FIGURE P13.86**

———

**13.87.** Problem 13.78 asked you to predict whether the fuel value of alcohols increased or decreased with the number of carbon atoms in the alcohol. Calculate the fuel values of liquid methanol and ethanol (Figure P13.87). Does your answer support the prediction you made in Problem 13.78?

$CH_3OH$            OH

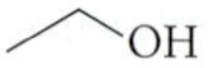

Methanol            Ethanol

**FIGURE P13.87**

**13.88.** Calculate the fuel values of liquid propanol and isopropanol (Figure P13.88). Which has the higher fuel value? ($\Delta H^\circ_{f,\text{propanol},\ell} = -302.5$ kJ/mol and $\Delta H^\circ_{f,\text{isopropanol},\ell} = -317.5$ kJ/mol.)

Propanol            Isopropanol

**FIGURE P13.88**

## Carbonyl-Containing Compounds

### CONCEPT REVIEW

**13.89.** Explain why carboxylic acids tend to be more soluble in water than aldehydes with the same number of carbon atoms.

**13.90.** In reference books, diethyl ether (Figure P13.90) is usually listed as "slightly soluble" in water, but 2-butanone is listed as "very soluble." Why do you suppose 2-butanone is more soluble?

Diethyl ether        2-Butanone

**FIGURE P13.90**

**13.91.** Are butanal and 2-butanone (Figure P13.91) constitutional isomers?

Butanal            2-Butanone

**FIGURE P13.91**

**13.92.** **Apples** The two esters shown in Figure P13.92 are both found in apples and contribute to the flavor and aroma of the fruit. Are the two compounds identical, constitutional isomers, or stereoisomers, or do they have different molecular formulas?

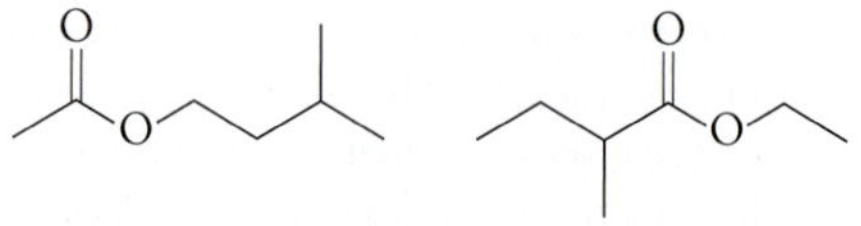

**FIGURE P13.92**

**13.93.** Can we distinguish between ketones and aldehydes with the same number of carbon atoms by combustion analysis?

**13.94.** Can we distinguish between ethers and ketones with the same number of carbon atoms by combustion analysis?

**13.95.** Resonance forms for acetic acid are shown in Figure P13.95. Which one contributes more to the bonding picture? Explain your choice.

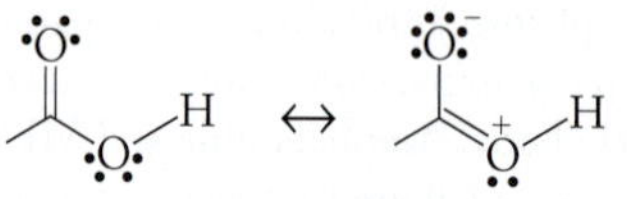

(a)         (b)

**FIGURE P13.95**

**13.96.** Figure P13.96 shows resonance forms for acetamide and acetic acid. Does resonance form a containing the C=N double bond contribute more to the bonding picture of acetamide than does resonance form b to the bonding picture in acetic acid? Explain your answer.

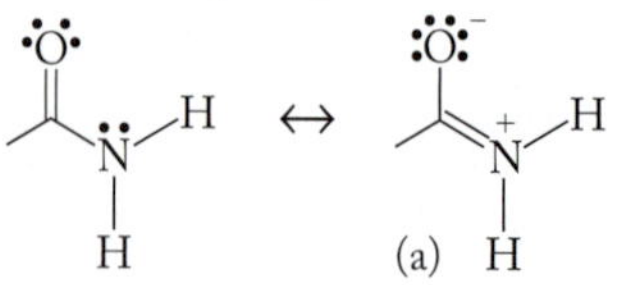

Acetamide

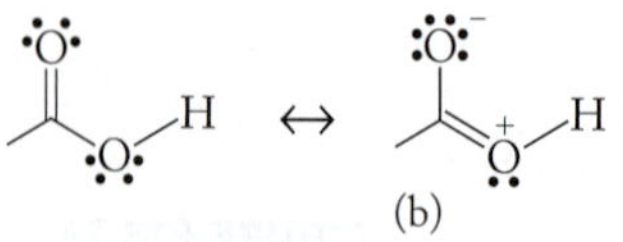

Acetic acid

**FIGURE P13.96**

**13.97.** What distinguishes an amine from an amide?

*13.98. Use the resonance forms of acetamide in Figure P13.96 to explain why an amide is acidic, not basic.

## PROBLEMS

13.99. Which of the compounds in Figure P13.99 are constitutional isomers of the aldehyde $C_5H_{10}O$?

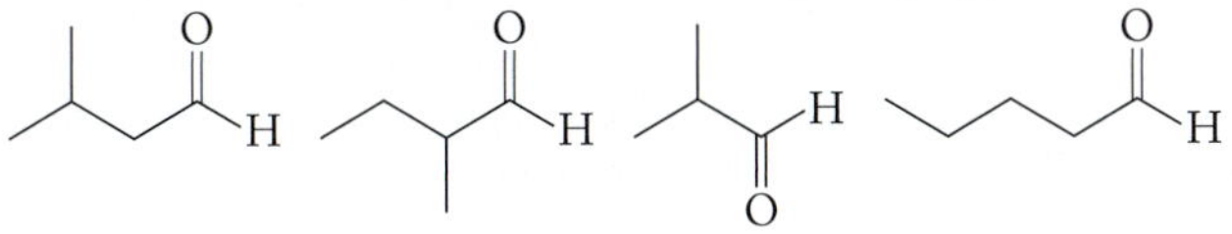

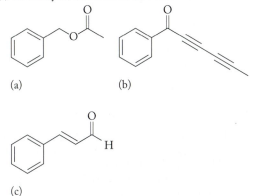

(a)  (b)  (c)  (d)

**FIGURE P13.99**

13.100. Each of the natural products in Figure P13.100 contains more than one functional group. Which of the compounds is (a) an aldehyde; (b) an ester; (c) a ketone?

(a)

(b)

(c)

**FIGURE P13.100**

13.101. Which of the compounds in Figure P13.101 is a ketone?

(a)  (b)  (c)  (d)

**FIGURE P13.101**

13.102. Propanal and acetone (2-propanone) have the same molecular formula, $C_3H_6O$, but different structures (Figure P13.102). Which compound is a ketone?

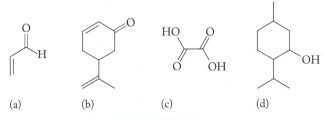

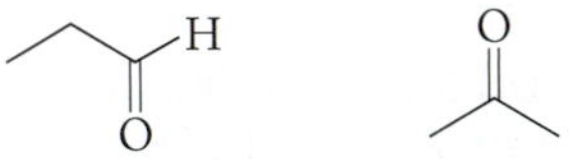

Propanal    Acetone

**FIGURE P13.102**

13.103. Plot the carbon-to-hydrogen ratio in aldehydes with one to six carbons as a function of the number of carbon atoms. Does this graph correlate better with the plot of C:H ratios for alkanes or for alkenes?

13.104. Plot the carbon-to-hydrogen ratio in carboxylic acids with one to six carbons as a function of the number of carbon atoms. Does this graph correlate better with the plot of C:H ratios for alkanes or for alkenes?

13.105. Esters are responsible for the odors of fruits, including apples, bananas, and pineapples. Figure P13.105 shows three esters from these fruits. Identify the alcohol and carboxylic acid that react to form each of these compounds.

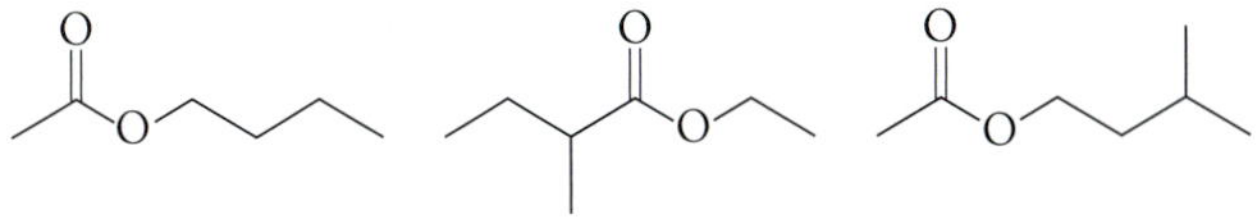

(a) Pineapples    (b) Bananas    (c) Apples

**FIGURE P13.105**

13.106. **Soap** One of the first chemical syntheses ever carried out was the making of soap. This same reaction is still carried out by modern-day manufacturers. The reaction is called saponification, and it is the reverse of esterification: an ester is taken apart to produce its component acid and alcohol. A compound called a glyceride, which is a fat, is reacted with aqueous base (usually NaOH). The ester bonds are cleaved to yield glycerol (an alcohol) and the sodium salt of a fatty acid, which is then used as soap. Given the glyceride shown in Figure P13.106, draw the structures of the alcohol and the fatty acids that result from its saponification.

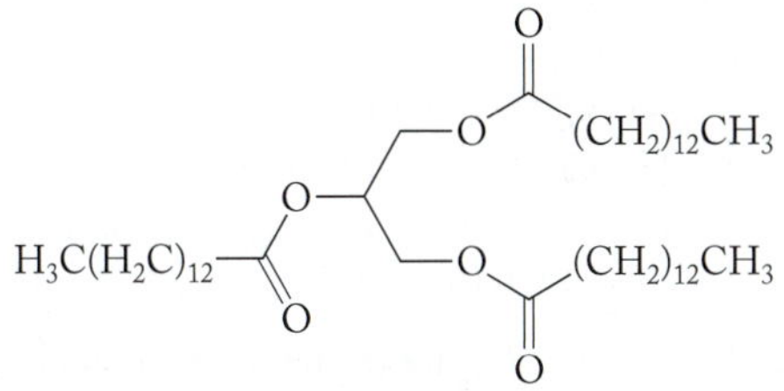

**FIGURE P13.106**

*13.107. **HIV Drug** In July 2012 a new drug named Truvada that actually kills the HIV virus was approved by the FDA. The drug is a mixture of several substances including the phosphorus-based ester shown in Figure P13.107. What is the formula of the acid of phosphorus from which this substance could be derived?

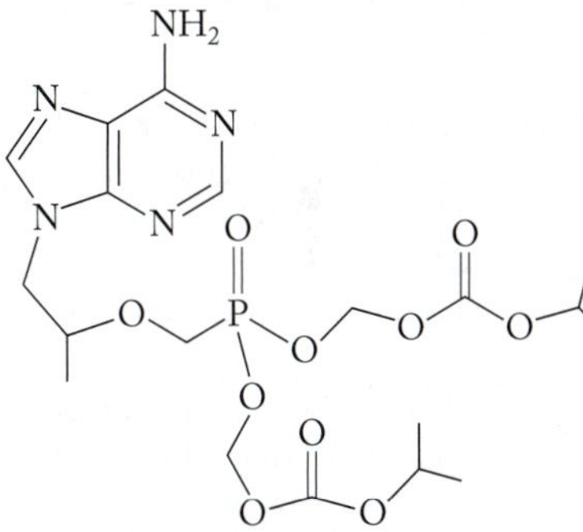

**FIGURE P13.107**

13.108. Beeswax (Figure P13.108) is an ester composed of an alcohol and a carboxylic acid with a long hydrocarbon chain. Identify the alcohol and acid in beeswax.

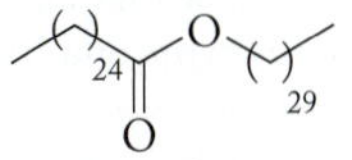

**FIGURE P13.108**

Consult tables of thermochemical data in Appendix 4 for any values you may need to solve Problems 13.109 through 13.112.

13.109. Calculate the fuel values of formaldehyde and formic acid (Figure P13.109). The $\Delta H_f^\circ$ of formaldehyde is −108.6 kJ/mol and the $\Delta H_f^\circ$ of formic acid is −378.7 kJ/mol. Which has the higher fuel value?

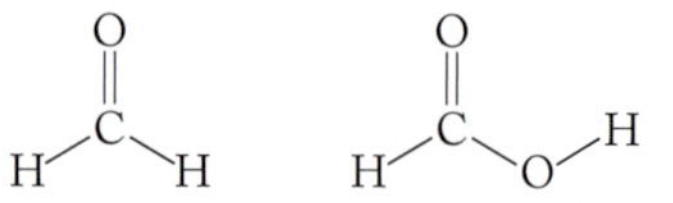

Formaldehyde          Formic acid

**FIGURE P13.109**

13.110. Calculate the fuel values of formamide and methyl formate (Figure P13.110), which have $\Delta H_f^\circ$ values of −251 and −391 kJ/mol, respectively. Assume $NO_2(g)$ is a product of formamide combustion.

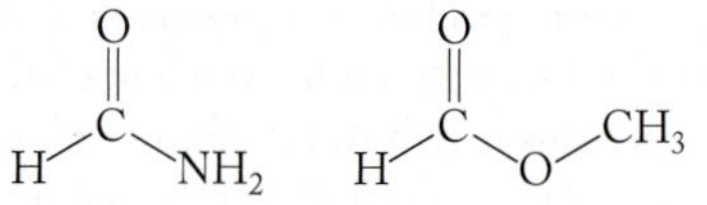

Formamide          Methyl formate

**FIGURE P13.110**

13.111. Calculate $\Delta H_{rxn}^\circ$ for the following reactions of methanogenic bacteria given $\Delta H_{f,HCOOH,g}^\circ = -378.7$ kJ/mol:

(1) $CH_3COOH(\ell) \rightarrow CH_4(g) + CO_2(g)$
(2) $4\,HCOOH(g) \rightarrow CH_4(g) + 3\,CO_2(g) + 2\,H_2O(\ell)$

13.112. Calculate $\Delta H_{rxn}^\circ$ for the following reactions of methanogenic bacteria:

(1) $4\,H_2(g) + CO_2(g) \rightarrow CH_4(g) + 2\,H_2O(\ell)$
(2) $4\,CH_3OH(\ell) \rightarrow 3\,CH_4(g) + CO_2(g) + 2\,H_2O(\ell)$

13.113. Reactions between 1,6-diaminohexane, $H_2N(CH_2)_6NH_2$, and different dicarboxylic acids, $HOOC(CH_2)_2COOH$, are used to prepare polymers that have a structure similar to that of nylon. How many carbon atoms ($n$) were in the dicarboxylic acids used to prepare the polymers with the repeating units shown in Figure P13.113?

(a)

(b)

(c)

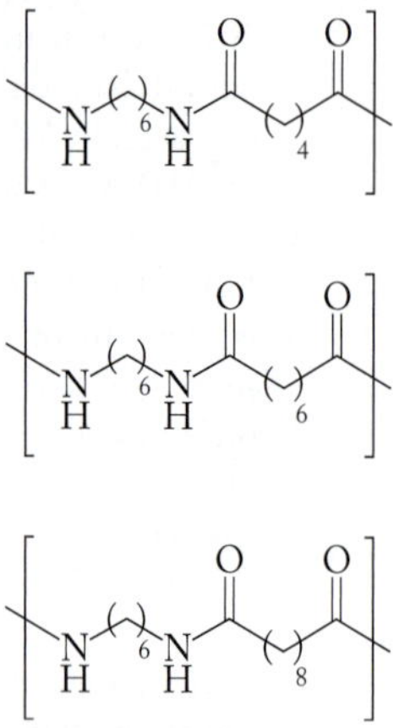

**FIGURE P13.113**

13.114. The two polymers in Figure P13.114 have the same empirical formula.
  a. What pairs of monomers could be used to make each of them?
  b. How might the physical properties of these two polymers differ?

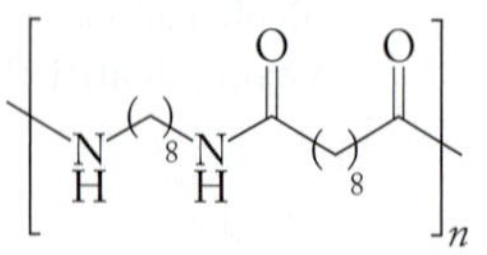

Polymer I

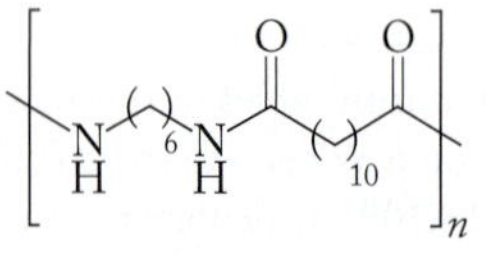

Polymer II

**FIGURE P13.114**

13.115. The polyester called Kodel is made with polymeric strands prepared by the reaction of dimethyl terephthalate with 1,4-di(hydroxymethyl)cyclohexane (Figure P13.115).
  a. Is Kodel a condensation polymer or an addition polymer? What is the other product of the reaction?
  *b. Dacron (see Sample Exercise 13.10) is made from dimethyl terephthalate and ethylene glycol. What properties of Kodel fibers might make them better than Dacron as a clothing material?

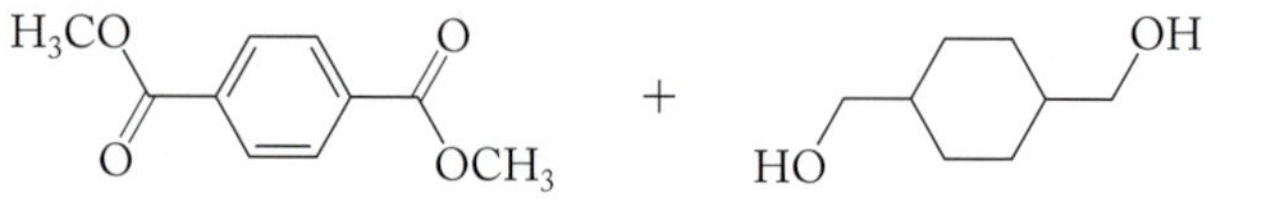

Dimethyl terephthalate          1,4-Di(hydroxymethyl)cyclohexane
(dimethyl benzene-1,4-dicarboxylate)

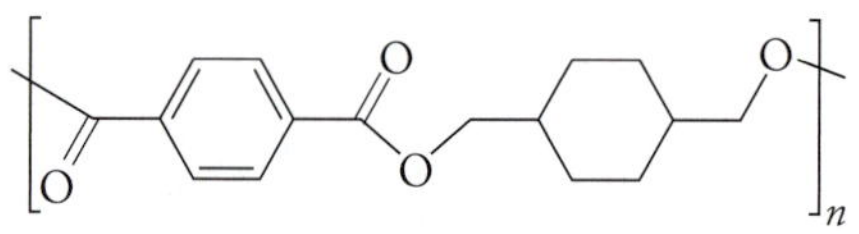

Kodel

**FIGURE P13.115**

13.116. Lexan is a polymer belonging to the class of materials called polycarbonates. Figure P13.116 shows the polymerization reaction for Lexan.
  a. What other compound is formed in the polymerization reaction?
  *b. Why is Lexan called a "polycarbonate"?

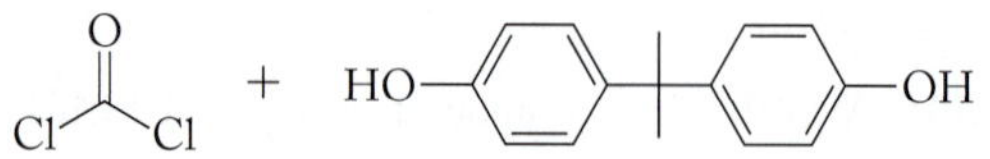

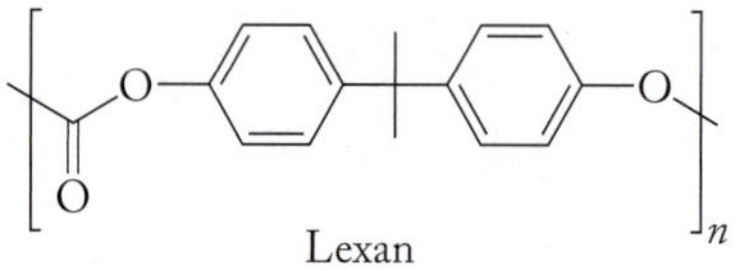

Lexan

**FIGURE P13.116**

## Chirality

### CONCEPT REVIEW

13.117. Can all of the terms *enantiomer*, *achiral*, and *optically active* be used to describe a single compound? Explain.

13.118. Two compounds have the same structure and the same physical properties but also have the same optical activity. Are they enantiomers or the same molecule?

13.119. Are racemic mixtures considered homogeneous or heterogeneous mixtures?

*13.120. Could a racemic mixture be distinguished from an achiral compound based on optical activity? Explain your answer.

13.121. Why is the amino acid glycine (Figure P13.121) achiral?

13.122. Can stereoisomers of molecules such as cis and trans RCH=CHR also have optical isomers? (R may be any of the functional groups we have encountered in this textbook.) Explain your answer.

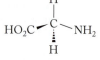

**FIGURE P13.121**

**13.123.** Which type of hybrid orbitals on a carbon atom, *sp*, *sp²*, or *sp³*, can give rise to enantiomers?

*__*13.124.** Could an oxygen atom in an alcohol, ketone, or ether ever be a chiral center in the molecule?

**13.125.** Which of the following objects are chiral? (a) a golf club; (b) a tennis racket; (c) a glove; (d) a shoe

**13.126.** Which of the following objects are chiral? (a) a key; (b) a screwdriver; (c) a lightbulb; (d) a baseball

## PROBLEMS

**13.127.** Which of the molecules in Figure P13.127 are chiral?

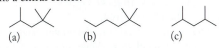

(a)  (b)  (c)

**FIGURE P13.127**

**13.128.** Which, if any, of the molecules shown in Figure P13.128 contains a chiral center?

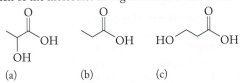

(a)  (b)  (c)

**FIGURE P13.128**

**13.129. Artificial Sweeteners** Artificial sweeteners are fundamental to the diet food industry. Figure P13.129 shows three artificial sweeteners that have been used in food. Saccharin is the oldest, dating to 1879. Cyclamates were banned in 1969 following research suggesting they led to tumors. Aspartame may be more familiar to you under the name NutraSweet. Each of these sweeteners contains between zero and two chiral carbon atoms. Circle the chiral center in each compound.

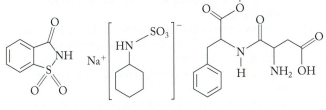

Saccharin  Sodium cyclamate  Aspartame

**FIGURE P13.129**

**13.130.** Identify the chiral centers in each of the molecules in Figure P13.130.

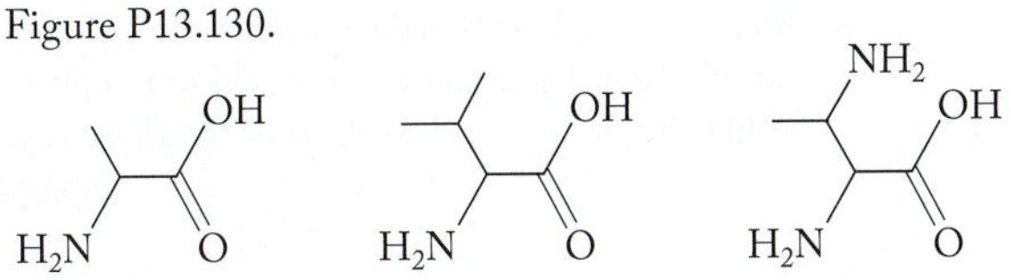

(a)  (b)  (c)

**FIGURE P13.130**

**13.131. Lowering Cholesterol** Crestor (shown as the carboxylic acid anion in Figure P13.131) is one of the top-10 best-selling pharmaceutical agents in the world.
   a. Is the anion chiral? If so, how many chiral centers does it have?

b. Are any other kinds of constitutional isomers possible for the anion? If so, draw them.

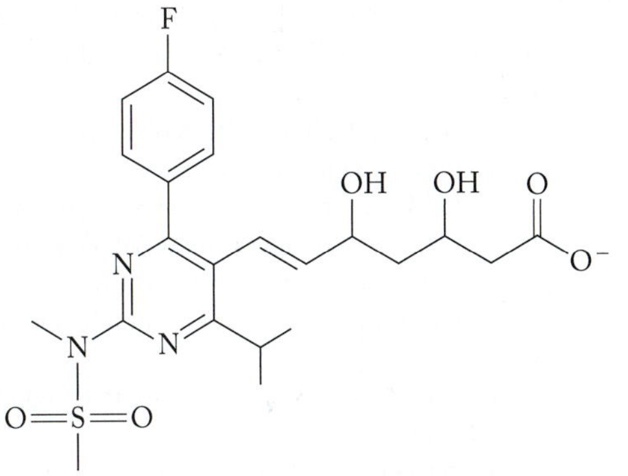

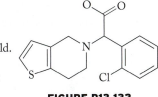

**FIGURE P13.131**

*__*13.132. Preventing Blood Clots**
   In 2009 Plavix (Figure P13.132) was the second highest selling pharmaceutical in the world. It is prescribed to prevent blood clots in heart attack and stroke patients.

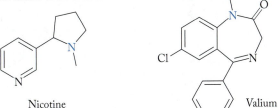

**FIGURE P13.132**

   a. Is Plavix optically active? If so, identify the chiral center(s) in the structure.
   b. Plavix contains an ester group. Identify it. Draw the carboxylic acid and the alcohol you could react together to make Plavix.

**13.133.** Nicotine is a stimulant found in tobacco. Valium is a tranquilizer. Both molecules contain two nitrogen atoms in addition to other functional groups. In Figure P13.133 identify the nitrogen atoms shown in blue as belonging to an amine or an amide.

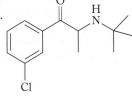

Nicotine  Valium

**FIGURE P13.133**

**13.134.** Bupropion (Figure P13.134) is a drug used to treat depression. Is it an amide or a secondary amine? Explain your answer.

Bupropion

**FIGURE P13.134**

## Additional Problems

**13.135.** How many grams of liquid methanol must be combusted to raise the temperature of 454 g of water from 20.0°C to 50.0°C? Assume that the transfer of heat to the water is 100% efficient. How many grams of carbon dioxide are produced in this combustion reaction?

**13.136.** How many grams of methylamine ($\Delta H_f^\circ = -23.0$ kJ/mol) must be combusted to raise the temperature of 454 g of water from 20°C to 50°C? Assume that the transfer of heat to the water is 100% efficient. Also assume $NO_2(g)$ is a product of the combustion of methylamine. How many grams of carbon dioxide are produced in this combustion reaction?

*13.137. Two compounds, both with molar masses of 74.12 g/mol, were combusted in a bomb calorimeter with $C_{calorimeter}$ = 3.640 kJ/°C. Combustion of 0.9842 g of compound A led to an increase in temperature of 10.33°C, while combustion of 1.110 g of compound B caused the temperature to rise 11.03°C. Which compound is butanol and which is diethyl ether?

13.138. Why should methane be more soluble in decane ($C_{10}H_{22}$) than in water?

13.139. **Salsa** Salsa has antibacterial properties because it contains dodecenal (Figure P13.139), a compound found in the cilantro used to make salsa.
   a. How many carbon atoms are in dodecenal?
   b. What functional groups are present in dodecenal?
   c. What types of isomerism are possible in dodecenal?

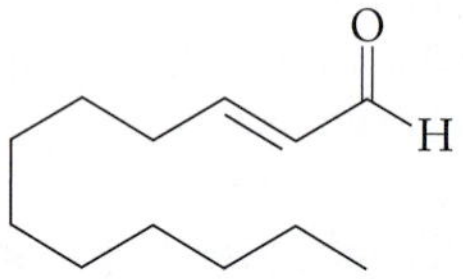

**FIGURE P13.139**

13.140. **Turmeric** Turmeric is commonly used as a spice in Indian and Southeast Asian dishes. Turmeric contains a high concentration of curcumin (Figure P13.140), a potential anticancer drug and a possible treatment for cystic fibrosis.
   a. Are the substituents on the C=C double bonds in cis or trans configurations?
   b. Draw two other stereoisomers of this compound.
   c. List all the types of valence shell hybridization of the carbon atoms in curcumin.

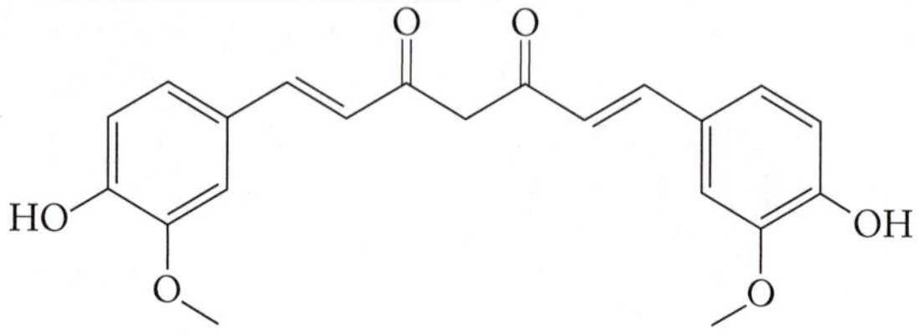

Curcumin
**FIGURE P13.140**

13.141. Polycyclic aromatic hydrocarbons are potent carcinogens. They are produced during combustion of fossil fuels and have also been found in meteorites. Can we use combustion analysis to distinguish between naphthalene and anthracene (Figure P13.141)?

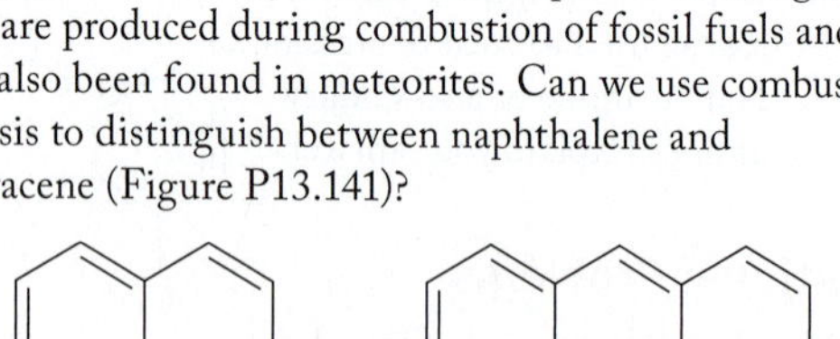

Naphthalene    Anthracene
**FIGURE P13.141**

13.142. **Flu Treatment** Tamiflu (Figure P13.142) is used to treat the flu. It is called a prodrug, because the actual active form is the carboxylic acid, not the ester that is administered.

The liver metabolizes the drug by breaking the ester link to produce the acid form and an alcohol.
   a. Draw the structures of the acid and alcohol formed when Tamiflu is metabolized.
   b. Identify the other functional groups and any chiral centers in the active form of the drug.

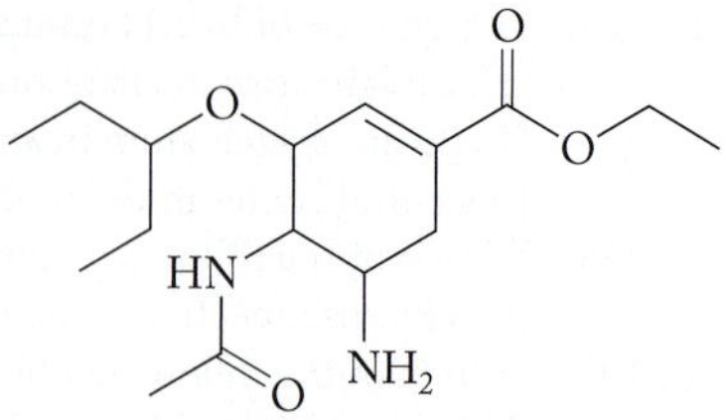

**FIGURE P13.142**

13.143. **WMD** Soman (Figure P13.143) is an extremely toxic nerve agent identified as a weapon of mass destruction. Circle the phosphoryl group in the structure and give the formula of the parent phosphorus-containing acid from which this agent could be derived.

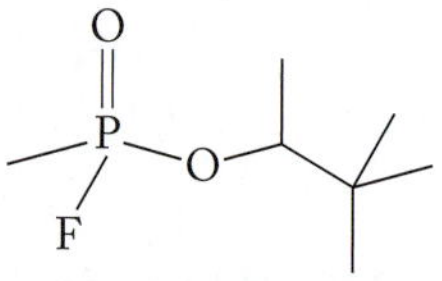

**FIGURE P13.143**

13.144. Identify the reactants in the polymerization reactions that produce the polymers shown in Figure P13.144.

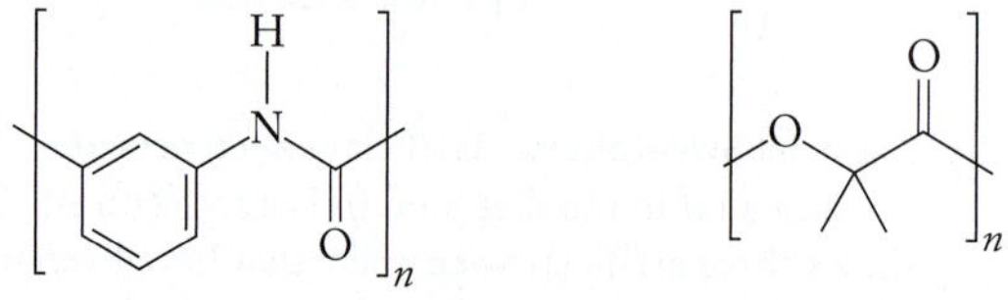

(a)          (b)
**FIGURE P13.144**

*13.145. **Raincoats** "Waterproof" nylon garments have a coating to prevent water from penetrating the hydrophilic fibers. Which functional groups in the nylon molecule make it hydrophilic?

13.146. Draw the carbon-skeleton structure of the condensation polymer of $H_2N(CH_2)_6COOH$. How does this polymer compare with nylon-6?

*13.147. Putrescine, $H_2N(CH_2)_4NH_2$, is one of the compounds that form in rotting meat.
   a. Draw the carbon-skeleton structures of all the trimers (a molecule formed from three monomers) that can be formed from putrescine, adipic acid, and terephthalic acid (Figure P13.147). The three monomers forming the trimer do not have to be different from one another.
   b. A chemist wishes to make a putrescine polymer containing a 1:1 ratio of adipic acid to terephthalic acid. What should be the mole ratio of the three reactants?

HOOCCH₂CH₂CH₂CH₂COOH

Adipic acid          Terephthalic acid
                     (benzene-1,4-dicarboxylic acid)
**FIGURE P13.147**

*13.148. Polymer chemists can modify the physical properties of polystyrene by copolymerizing divinylbenzene with styrene

(Figure P13.148). The resulting polymer has strands of polystyrene cross-linked with divinylbenzene. Predict how the physical properties of the copolymer might differ from those of 100% polystyrene.

Divinylbenzene (DVB)

Styrene (S)

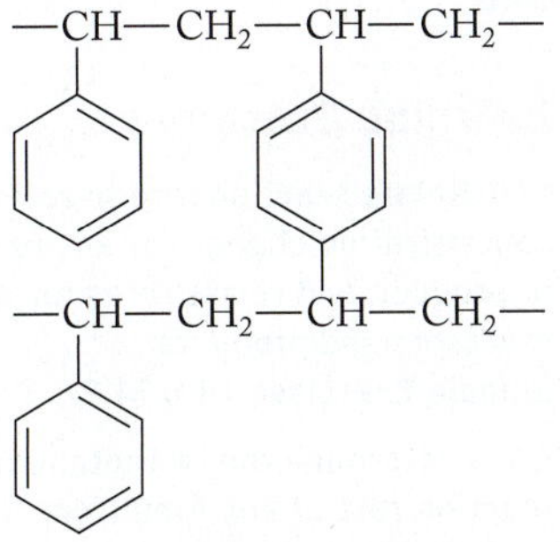

S cross-linked with DVB

**FIGURE P13.148**

**13.149.** Maleic anhydride and styrene (Figure P13.149) form a polymer with alternating units of each monomer.
  a. Draw two repeating monomer units of the polymer.
  b. Based on the structure of the copolymer, predict how its physical properties might differ from those of polystyrene.

Maleic anhydride

Styrene

**FIGURE P13.149**

**13.150.** **Superglue** The active ingredient in superglue is methyl 2-cyanoacrylate (Figure P13.150). The liquid glue hardens rapidly when methyl 2-cyanoacrylate polymerizes. This happens when it contacts a surface containing traces of water or other compounds containing –OH or –NH– groups. Draw the carbon-skeleton structure of two repeating monomer units of poly(methyl 2-cyanoacrylate).

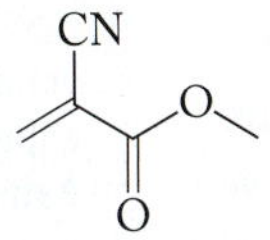

Methyl 2-cyanoacrylate

**FIGURE P13.150**

**\*13.151.** Silicones are polymeric materials with the formula $[R_2SiO]_n$ (Figure P13.151). They are prepared by reaction of $R_2SiCl_2$ with water yielding the polymer and aqueous HCl. Consider this reaction as taking place in two steps: (1) water reacts with 1 mole of $R_2SiCl_2$ to produce a new monomer and 1 mole of HCl(*aq*); (2) one new monomer molecule reacts with another new monomer molecule to eliminate one molecule of $H_2O$ and make a dimer with a Si—O—Si bond.
  a. Suggest two balanced equations describing these reactions that occur over and over again to produce a silicone polymer.
  b. Why are silicones water repellent?

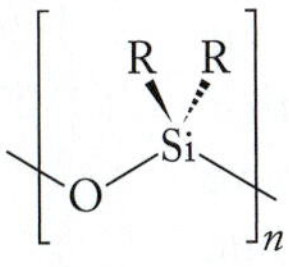

Silicone

**FIGURE P13.151**

**13.152.** Methadone is used in the treatment of heroin addiction; cyclizine is used for the control of nausea and dizziness arising from motion sickness. Their structures are shown in Figure P13.152.
  a. Are the nitrogen atoms present primary, secondary, or tertiary amines or are they amides?
  b. If the compound has a chiral carbon atom, circle that atom.

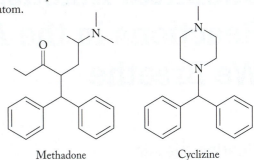

Methadone

Cyclizine

**FIGURE P13.152**

**13.153.** The heats of combustion of the two constitutional isomers in Table 13.4 calculated from bond energies have the same value, but the two experimentally determined values for $\Delta H°_{comb}$ are not the same. Why?

**13.154.** We studied the energy changes associated with the combustion of hydrocarbon fuels in Chapter 5. In principle, we could generate tremendous quantities of fuel by combining $CO_2$ and water vapor and chemically converting them back into $CH_4$ or other hydrocarbon fuels. Why is this not possible in practice?

**13.155.** One of the great intellectual tasks in the area of organic synthesis is natural product synthesis: the design of reactions that enable the production in the laboratory of the complex molecules found in nature. To aid in this process, small molecules containing reactive functional groups and chiral centers are made available to researchers to help them build more complex systems. Figure P13.155 shows several molecules prepared for use by natural product chemists. Identify the functional groups and circle the chiral centers in the molecules.

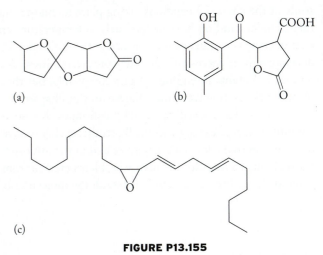

(a)

(b)

(c)

**FIGURE P13.155**

If your instructor assigns problems in **smartwork**, log in at **smartwork.wwnorton.com**.

# 14

# Chemical Kinetics: Reactions in the Air We Breathe

14.1 Cars, Trucks, and Air Quality

14.2 Reaction Rates

14.3 Effect of Concentration on Reaction Rate

14.4 Reaction Rates, Temperature, and the Arrhenius Equation

14.5 Reaction Mechanisms

14.6 Catalysis

## Learning Outcomes

**LO1** Relate reaction rate to rate of concentration change for any reactant or product, and relate rates for any two reactants and products
**Sample Exercises 14.1, 14.2**

**LO2** Determine the instantaneous reaction rate at any time from experimental data
**Sample Exercise 14.3**

**LO3** Derive a rate law from initial reaction rate data
**Sample Exercise 14.4**

**LO4** Use integrated rate laws to identify first- and second-order reactions
**Sample Exercises 14.5, 14.8**

**LO5** Calculate concentrations at given times from rate equations
**Sample Exercise 14.6**

**LO6** Calculate half-lives of first-order reactions
**Sample Exercise 14.7**

**LO7** Recognize pseudo-first-order conditions and calculate rate laws from experimental data
**Sample Exercises 14.9, 14.10**

**LO8** Calculate the activation energy of a reaction
**Sample Exercise 14.11**

**LO9** Link mechanisms to rate laws and identify catalysts and catalysis
**Sample Exercises 14.12, 14.13, 14.14**

## Clearing the Air

Look at the photographs to the right. That yellow-brown haze that colors the left side of the picture and obscures the mountains is the result of *smog*, so called because it combines the effects of smoke and fog, reducing visibility and causing respiratory difficulties. *Photochemical smog* forms when sunny weather and light winds allow vehicle emissions from burning fossil fuels to collect over urban centers. The interaction of sunlight with nitrogen oxides in the air initiates a host of chemical reactions that produce the brown haze.

Some approaches to minimizing this smog involve replacing our current vehicles with electric cars powered by high-performance batteries and/or fuel cells. Other approaches, based on an understanding of the rates of chemical reactions occurring in car engines, have enabled scientists and engineers to design internal combustion engines that burn fuel more efficiently than ever before. However, one device has removed many billions of tons of smog-forming pollutants from the air above U.S. cities since its development in the 1970s: the catalytic converter. The development of catalytic converters, which are now mandatory on vehicles in the United States, required extensive study of the chemical reactions taking place in engines burning fossil fuels. Understanding how factors such as temperature and the presence of catalysts influence *reaction rates* at the molecular level led to the development of successful catalytic converters. These devices do not remove pollutants in engine exhaust completely, but they almost do. We discuss reaction rates and catalysts in this chapter.

Air pollution caused by vehicles continues, but so too does our commitment to cleaning the air. By studying chemical reactions that occur in internal combustion engines and in the atmosphere, we have learned much about the air-quality problems created from automobile exhaust. In addition, we now understand the molecular basis of other

**Clearing the Air** A light snowfall and change in wind direction dramatically affect visibility and air quality over Salt Lake City, Utah, clearing the photochemical smog from the air. These photos were taken on January 28, 2007, at 4:15 PM (left) and January 31, 2007, at 3:53 PM (right).

**photochemical smog** a mixture of gases formed in the lower atmosphere when sunlight interacts with compounds produced in internal combustion engines and other pollutants.

atmospheric problems that impact life on Earth, such as the ozone ($O_3$) hole in the stratosphere (upper atmosphere).

The study of reaction rates, known as chemical kinetics, has broad application to many areas of chemistry. Studying the rate of a reaction can lead to understanding how the reaction occurs at the molecular level—the mechanism of the reaction. We can use this knowledge to optimize the yield, in the case of making a desired product like a medicine, or minimizing the production of an unwanted by-product, like the substances found in smog. The study of kinetics is important in industrial chemistry as well as in pharmaceutical and medical research. ■

## 14.1 Cars, Trucks, and Air Quality

**Photochemical smog** consists of various nitrogen oxides ($NO_x$), ozone ($O_3$), and organic molecules such as peroxyacetyl nitrate, a strong respiratory and eye irritant (Figure 14.1). This smog is created by the interaction of sunlight with nitrogen oxides from vehicle exhaust and volatile organic compounds (VOCs), from natural and man-made sources.

One of the first reactions leading to photochemical smog takes place inside an automobile engine—the formation of nitrogen monoxide from $N_2$ and $O_2$:

$$N_2(g) + O_2(g) \rightarrow 2\,NO(g) \qquad \Delta H° = 180.6 \text{ kJ} \qquad (14.1)$$

Once NO enters the atmosphere, it reacts with more oxygen, producing brown nitrogen dioxide gas:

$$2\,NO(g) + O_2(g) \rightarrow 2\,NO_2(g) \qquad \Delta H° = -114.2 \text{ kJ} \qquad (14.2)$$

Radiant energy ($hv$) from the Sun provides sufficient energy to break the bonds in $NO_2$, forming NO and very reactive oxygen atoms:

$$NO_2(g) \xrightarrow{hv} NO(g) + O(g) \qquad (14.3)$$

This photochemically generated atomic oxygen combines with molecular oxygen, producing ozone,

$$O_2(g) + O(g) \rightarrow O_3(g) \qquad (14.4)$$

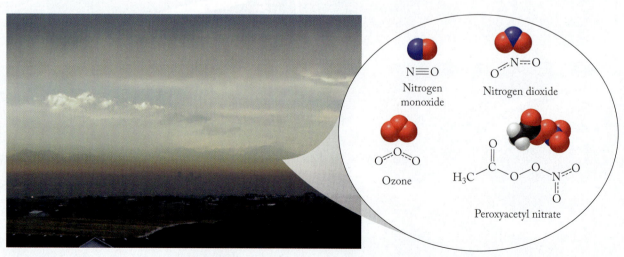

**FIGURE 14.1** Photochemical smog, like this orange-brown layer over Denver, contains a mixture of compounds including NO, $NO_2$, $O_3$, and peroxyacetyl nitrate ($CH_3CO_3NO_2$).

It also reacts with water vapor to produce hydroxyl radicals:

$$O(g) + H_2O(g) \rightarrow 2\, OH(g) \qquad (14.5)$$

If we examine just these few reactions, we see immediately how challenging is the task of understanding smog formation, hinging on the fact that most of the substances in photochemical smog are both reactants and products. The relationships among their concentrations, even in this simplified view, are more complex than others we have seen in previous chapters.

In the atmosphere, VOCs react with ozone, hydroxyl radicals, and atomic oxygen to form a series of oxygen compounds that subsequently react with $NO_2$ to form peroxyacetyl nitrate. One typical reaction involves the VOC acetaldehyde:

$$N_2(g) + 3\, O_2(g) + OH(g) + CH_3CHO(g) \rightarrow$$

Acetaldehyde

$(14.6)$

$$CH_3C(O)O_2NO_2(g) + H_2O(g) + NO_2(g)$$

Peroxyacetyl nitrate

To understand the production of $NO_x$ compounds in engines and the environment, we need to understand how the reactions producing $NO_x$ are linked. In particular, we need to know their **reaction rates**, which are the rates at which reactants are consumed and products are formed.

In the reactions involved in smog formation, the products of some reactions are the reactants in others. Therefore, the relative rates of these reactions influence at what time during the day pollutants appear, how long they persist, and what their concentrations are. Note in Figure 14.2, for instance, that the maximum NO concentration occurs during the morning rush hour. Later in the morning, the concentration of $NO_2$ reaches a maximum. This sequence makes sense because NO is a precursor of $NO_2$. The highest ozone concentrations are reached in the middle of the afternoon, when the reactions shown in Equations 14.2 and 14.3 are in full swing, producing a supply of free O atoms for the formation of ozone (Equation 14.4), hydroxyl radicals (Equation 14.5), and, indirectly, strong respiratory and eye irritants such as peroxyacetyl nitrate (PAN) (Equation 14.6).

The ozone formed in the reaction in Equation 14.4 also reacts with NO to form $NO_2$ and $O_2$:

$$O_3(g) + NO(g) \rightarrow O_2(g) + NO_2(g) \qquad (14.7)$$

The $NO_2$ concentration drops in the afternoon because $NO_2$, ozone, and hydrocarbons react to form an array of other compounds.

The catalytic converter, now a standard feature on automobiles in many countries, helps combat the production of photochemical smog by removing NO and unburned or partially oxidized hydrocarbons from exhaust gases. Knowledge of **chemical kinetics**, the study of the rates at which reactant and product concentrations change during a chemical reaction, has provided the understanding required to develop catalytic converters and other devices aimed at improving our air quality. At least 75% of the NO produced in the engines of vehicles equipped with catalytic converters is converted back to $N_2$ and $O_2$ before being emitted into the air.

## 14.2 Reaction Rates

In this section we begin the discussion of how scientists determine rates of reactions and the factors that influence rates. We use many examples derived from work on the reactions that determine the quality of the air we breathe. The principles we

⊙⊙ **CONNECTION** We introduced species like OH and O in Chapter 8 in the discussion of odd-electron molecules and free radicals.

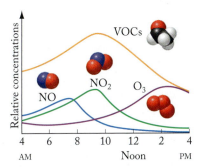

**FIGURE 14.2** In photochemical smog, NO from engine exhaust builds up in the early morning, then decreases as the NO reacts with atmospheric $O_2$, forming $NO_2$, the concentration of which is highest in late morning. Photodecomposition of $NO_2$ leads to the formation of high levels of $O_3$ in the afternoon.

**reaction rate** how rapidly a reaction occurs; it is related to rates of change in the concentrations of reactants and products over time.

**chemical kinetics** the study of the rates of change of concentrations of substances involved in chemical reactions.

develop here apply in many other situations, ranging from biochemical reactions within living systems to large-scale industrial processes that create the materials of our modern world. At the end of the chapter we present the molecular view of reactions provided by reaction mechanisms—stepwise views of what may actually happen at the level of the particles themselves. The development of a valid reaction mechanism is the real payoff both intellectually and practically for studying reaction rates.

We have already discussed the factors that influence reaction rates in earlier chapters. In Chapter 4, for example, we mentioned that water provides a medium in which reactions between molecules and ions can occur because these particles move freely in the liquid phase. In Chapter 5 we talked about the phases of matter and the types and extents of motions of particles in each of those phases. In Chapter 6 we presented kinetic molecular theory and the relationship between temperature and average speed of molecules in the gas phase. In addition, we need to keep in mind that the speed at which reactions proceed varies over a wide range (Figure 14.3), from so slow that we barely perceive they are occurring to as fast as explosions. With these concepts in mind, let's look quickly at four factors that influence the rates of reaction, remembering always the fundamental idea that, for a reaction to take place, the particles must interact. We then examine each of these factors in greater detail.

1. **Rates are affected by the physical state of the reactants.** The more particles interact—that is, the more they collide with one another—the faster a reaction proceeds. Particles in solids have the same nearest neighbors over time, whereas in liquids the particles exchange nearest neighbors and occupy more positions relative to each other. In gases, the particles are widely separated and move very rapidly. Reactions in the solid phase tend to be very slow, while those in liquid and gas phases tend to be much faster.

2. **Rates depend on the concentration of reactants.** Reactions depend on particles interacting. If the concentration of particles is increased in a given volume, the rates usually increase because the frequency with which the particles encounter each other increases.

(a)                                   (b)                                   (c)

**FIGURE 14.3** Hydrogen peroxide ($H_2O_2$) decomposes spontaneously to water and oxygen gas.
(a) A 30% aqueous solution of $H_2O_2$ decomposes so slowly that no change is observable.
(b) Pouring the solution on a wedge of potato, which contains the enzyme catalase, causes the reaction to proceed at a faster rate, clearly visible by the bubbles of $O_2$ gas given off.
(c) The addition of a small amount of $MnO_2$ causes the reaction to proceed so rapidly that the heat liberated causes the solution to boil.

3. **Rates depend on temperature.** As temperature increases, the rates of chemical reactions tend to increase. The average kinetic energy (KE) of particles increases as temperature rises. As KE increases, the likelihood of collisions increases, and the likelihood of reaction also increases.

4. **Rates are affected by the presence of catalysts.** Catalytic converters contain materials that increase the rate of desired reactions in automotive exhaust. In general, catalysts are materials that accelerate reactions without themselves being consumed in the process. Catalysts are also of tremendous importance in biochemistry, where proteins called enzymes catalyze most reactions in living systems.

With these four ideas in mind, let's begin our exploration of rates by returning to the reactions occurring in internal combustion engines.

The amount of NO that forms inside an automobile engine in a given time interval depends on how rapidly the reaction in Equation 14.1 proceeds:

$$N_2(g) + O_2(g) \rightarrow 2\,NO(g) \qquad (14.1)$$

We can express the reaction rate as the change in product (or reactant) concentration that occurs over some interval of time. For example, the change in concentration of NO is $\Delta[NO] = [NO]_{final} - [NO]_{initial}$ over the time interval $\Delta t = t_{final} - t_{initial}$. The reaction rate, expressed as the rate of formation of NO, is

▶❙❙ **CHEMTOUR** Reaction Rate

$$\text{Rate of NO formation} = \frac{\Delta[NO]}{\Delta t} = \frac{[NO]_{final} - [NO]_{initial}}{t_{final} - t_{initial}} \qquad (14.8)$$

We use square brackets in Equation 14.8 and throughout this chapter and beyond to indicate concentration, typically in units of moles per liter.

The rate of the reaction between $N_2$ and $O_2$ to form NO can also be expressed as the rate of consumption of either reactant. However, the rate of change in the concentration of a reactant ($\Delta[\text{reactant}]/\Delta t$) has a negative value because reactant concentrations decrease as a reaction proceeds. The measured rate of any reaction is defined as a positive quantity because it describes the rate at which reactants form products, so a minus sign is used with $\Delta[\text{reactant}]/\Delta t$ values to obtain an overall positive value for the reaction rate. Thus the rate at which $N_2$ is consumed is

$$\text{Rate of } N_2 \text{ consumption} = -\frac{\Delta[N_2]}{\Delta t} = -\frac{[N_2]_{final} - [N_2]_{initial}}{t_{final} - t_{initial}} \qquad (14.9)$$

Similarly for $O_2$:

$$\text{Rate of } O_2 \text{ consumption} = -\frac{\Delta[O_2]}{\Delta t} = -\frac{[O_2]_{final} - [O_2]_{initial}}{t_{final} - t_{initial}} \qquad (14.10)$$

In all three cases, the rate of formation or consumption is a change in concentration divided by a change in time, so the units are concentration per unit time, such as molarity per second (*M*/s).

**CONCEPT TEST** ••••••••••••••••••••••••••••••••••••••••••••••••••••••••••••••

A reaction rate is analogous to the speed at which you drive your car. If reaction rates reflect the "speed" of a reaction, why can't a reaction have a negative rate?

*(Answers to Concept Tests are in the back of the book.)*

••••••••••••••••••••••••••••••••••••••••••••••••••••••••••••••••••••••••••••••

Are the rates of $N_2$ consumption, $O_2$ consumption, and NO formation the same? The answer to this question lies in the balanced equation for the reaction. The fact that 2 moles of NO are formed from 1 mole of $N_2$ and 1 mole of $O_2$ means that the rate of consumption of $N_2$ ($-\Delta[N_2]/\Delta t$) is the same as the rate of consumption of $O_2$ ($-\Delta[O_2]/\Delta t$). However, the rate of formation of NO ($\Delta[NO]/\Delta t$) is twice the rate of consumption of either $N_2$ or $O_2$. These relations are expressed in equation form as

$$-2\frac{\Delta[N_2]}{\Delta t} = -2\frac{\Delta[O_2]}{\Delta t} = \frac{\Delta[NO]}{\Delta t} \qquad (14.11)$$

or, if we divide through by 2,

$$-\frac{\Delta[N_2]}{\Delta t} = -\frac{\Delta[O_2]}{\Delta t} = \frac{1}{2}\frac{\Delta[NO]}{\Delta t} \qquad (14.12)$$

Note that in Equation 14.12 the number 2 in the denominator of the $\Delta[NO]/\Delta t$ term matches the coefficient of NO in the balanced chemical equation. This pattern applies to all chemical reactions: the coefficient of each species in the balanced chemical equation appears in the denominator of its term in a rate expression such as equation 14.12. Notice that the numerators in Equation 14.13 all have a value of 1:

$$1\,N_2(g) + 1\,O_2(g) \rightarrow 2\,NO(g)$$

$$\text{Rate} = -\frac{1}{1}\frac{\Delta[N_2]}{\Delta t} = -\frac{1}{1}\frac{\Delta[O_2]}{\Delta t} = \frac{1}{2}\frac{\Delta[NO]}{\Delta t} \qquad (14.13)$$

(We have added "1" coefficients to the $N_2$ and $O_2$ terms in the chemical equation only to clarify the origin of the denominators in the rate equation.)

A balanced chemical equation enables us to predict the *relative* rates at which reactants are consumed and products are formed: the rate of formation of NO is twice the rate of consumption of $N_2$. However, Equations 14.11 to 14.13 provide no information on the numerical values of the rates, which are typically obtained experimentally. Although computational tools at our disposal today allow us to predict the rates of very simple reactions, we limit our discussion of reaction rates in this chapter to rates based on experimental data.

---

**SAMPLE EXERCISE 14.1** **Predicting a Relative Reaction Rate** **LO1**

The synthesis of ammonia via the reaction

$$N_2(g) + 3\,H_2(g) \rightarrow 2\,NH_3(g)$$

is an important reaction in the production of agricultural fertilizers. How is the rate of formation of $NH_3$ related to the rates of consumption of $N_2$ and $H_2$?

**Collect and Organize** The rate of formation of product is related to the rates of consumption of reactants by the balanced chemical equation. Rates are expressed as $-\Delta[\text{reactant}]/\Delta t$ or $\Delta[\text{product}]/\Delta t$.

**Analyze** Our task is to use the coefficients of the balanced chemical equation to determine the relative rates at which the concentrations of $N_2$, $H_2$, and $NH_3$ change during the reaction. Because nitrogen and hydrogen are consumed in the reaction, we include minus signs in their rate terms. Because ammonia is generated, its rate term is positive. Since the coefficients for the reactants and products in the balanced chemical equation are all different, the rate of formation of product and the rates of consumption of reactants are all different.

**Solve** We write the expressions for reactants and product with the correct signs and insert the coefficients from the balanced equation in the denominators:

$$-\frac{\Delta[N_2]}{\Delta t} = -\frac{1}{3}\frac{\Delta[H_2]}{\Delta t} = \frac{1}{2}\frac{\Delta[NH_3]}{\Delta t}$$

**Think About It** The balanced equation indicates that 2 moles of $NH_3$ are formed for every 1 mole of $N_2$ consumed. That means the rate of consumption of $N_2$ is half the rate of formation of ammonia. Similarly, the balanced equation tells us that 3 moles of $H_2$ are consumed for every 1 mole of $N_2$ consumed. Therefore, the rate at which $N_2$ is consumed is one-third the rate at which $H_2$ is consumed.

 **Practice Exercise** In the oxidation of carbon monoxide to carbon dioxide,

$$2\,CO(g) + O_2(g) \rightarrow 2\,CO_2(g)$$

which reactant is consumed at the higher rate? How is the rate of change in the concentration of $CO_2$ related to the rate of change in the concentration of $O_2$?

*(Answers to Practice Exercises are in the back of the book.)*

## Experimentally Determined Rates: Actual Values

Up to this point we have worked only with expressions for *relative* rates of reactions. Actual values for reaction rates are determined experimentally, and once we know the value for one reactant or product, we can use that value and information from the balanced chemical equation to express the reaction rate in terms of any other reactant or product.

The reaction in Sample Exercise 14.1 can serve as an example. If the rate at which ammonia forms ($\Delta[NH_3]/\Delta t$) under a given set of conditions is determined experimentally to be 0.472 *M*/s, we can calculate a numerical value for the rate of consumption of $N_2$ by using the relationships we derived in Sample Exercise 14.1:

$$-\frac{\Delta[N_2]}{\Delta t} = \frac{1}{2}\frac{\Delta[NH_3]}{\Delta t} = \frac{1}{2}(0.472\ M/s) = 0.236\ M/s$$

Because the coefficient of $N_2$ in the balanced equation is 1, the rate of $N_2$ consumption is equal to one-half the rate of the formation of ammonia, and thus we can state that the rate of consumption of $N_2$ is 0.236 *M*/s.

Which value for the rate of the reaction is correct? Both values are correct as long as we reference the compound in the chemical equation whose concentration was monitored as a function of time. However, it is conventional to use the rate of change of the reactant or product with a coefficient of "1" in the reaction equation as the basis for expressing the rate of the overall reaction.

---

**SAMPLE EXERCISE 14.2** **Converting Reaction Rates** **LO1**

Suppose that during the reaction between NO and $O_2$ to form $NO_2$,

$$2\,NO(g) + O_2(g) \rightarrow 2\,NO_2(g)$$

the rate of consumption of $O_2$ ($-\Delta[O_2]/\Delta t$) is measured as 0.033 *M*/s. What is the rate of formation of $NO_2$?

**Collect and Organize** We are to determine the rate of formation of a product in a reaction from the balanced chemical equation and the rate of consumption of one of the reactants. The rate of formation of $NO_2$ is related to the rates of consumption of NO and $O_2$ by the balanced chemical equation. The coefficients of $NO_2$ and $O_2$ are 2 and 1, respectively. Rates are expressed as $-\Delta[\text{reactant}]/\Delta t$ or $\Delta[\text{product}]/\Delta t$.

**Analyze** We can write an equation that expresses the relative rates of change in $[NO_2]$ and $[O_2]$ from the coefficients in the balanced chemical equation:

$$\frac{1}{2}\frac{\Delta[NO_2]}{\Delta t} = -\frac{\Delta[O_2]}{\Delta t}$$

The negative sign is needed because the concentration of $O_2$ decreases as the concentration of $NO_2$ increases. This stoichiometric relation means that the rate of consumption of $O_2$ is half the rate of formation of $NO_2$ because 2 moles of $NO_2$ are formed from every 1 mole of $O_2$ consumed.

**Solve** Solving for the rate of change of $[NO_2]$, we get

$$\frac{\Delta[NO_2]}{\Delta t} = -2\frac{\Delta[O_2]}{\Delta t} = 2(0.033 \; M/s) = 0.066 \; M/s$$

**Think About It** This result is twice the magnitude of $\Delta[O_2]/\Delta t$, which is consistent with the stoichiometry of the reaction: 2 moles of $NO_2$ produced for every 1 mole of $O_2$ consumed.

**Practice Exercise** The gas NO reacts with $H_2$, forming $N_2$ and $H_2O$:

$$2 \; NO(g) + 2 \; H_2(g) \rightarrow 2 \; H_2O(g) + N_2(g)$$

If $-\Delta[NO]/\Delta t = 21.5 \; M/s$ under a given set of conditions, what are the rates of change of $[N_2]$ and $[H_2O]$?

| TABLE 14.1 Changing Concentrations of Reactants and Products for the Reaction $N_2(g) + O_2(g) \rightarrow 2 \; NO(g)$ | | |
|---|---|---|
| Time (μs) | $[N_2], [O_2]$ (μM) | $[NO]$ (μM) |
| 0 | 17.0 | 0.0 |
| 5.0 | 13.1 | 7.8 |
| 10.0 | 9.6 | 14.8 |
| 15.0 | 7.6 | 18.6 |
| 20.0 | 5.8 | 22.2 |
| 25.0 | 4.5 | 24.8 |
| 30.0 | 3.6 | 26.7 |

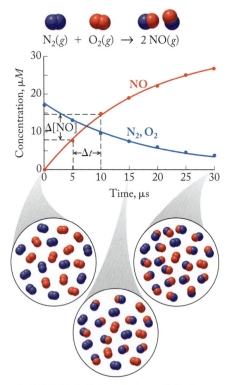

**FIGURE 14.4** Concentrations of $N_2$, $O_2$, and NO over 30.0 μs for the reaction $N_2(g) + O_2(g) \rightarrow 2 \; NO(g)$, plotted from data in Table 14.1.

## Average and Instantaneous Rates of Formation of NO

Suppose we run an experiment to determine the rate of formation of NO in an automobile engine. In the laboratory, we use a reaction vessel as hot as the combustion chambers in the engine and obtain the data in Table 14.1, which are plotted in Figure 14.4. We can use the data to calculate the reaction rate based on the change in the concentration of any participant in the reaction over a particular time interval. Calculations of reaction rates based on $-\Delta[\text{reactant}]/\Delta t$ or $\Delta[\text{product}]/\Delta t$ are *average* reaction rates. An average reaction rate could be the difference in NO concentration, $\Delta[NO] = [NO]_{\text{final}} - [NO]_{\text{initial}}$, at the beginning and end of the selected time interval $\Delta t = t_{\text{final}} - t_{\text{initial}}$. For example, the average rate of change in $[NO]$ between 5.0 and 10.0 μs is

$$\frac{\Delta[NO]}{\Delta t} = \frac{[NO]_{10.0 \; \mu s} - [NO]_{5.0 \; \mu s}}{(10.0 - 5.0) \; \mu s} = \frac{(14.8 - 7.8) \; \mu M}{5.0 \; \mu s}$$

$$= 1.4 \; M/s$$

During the same time interval, the average rate of consumption of $N_2$ is

$$-\frac{\Delta[N_2]}{\Delta t} = -\frac{[N_2]_{10.0 \; \mu s} - [N_2]_{5.0 \; \mu s}}{(10.0 - 5.0) \; \mu s} = -\frac{(9.6 - 13.1) \; \mu M}{5.0 \; \mu s}$$

$$= 0.70 \; M/s$$

These results give us two values for expressing the rate of the reaction. Recall that we use the rate of change of the reactant or product with a coefficient of 1 in the reaction equation as the basis for expressing the rate of the reaction; in this example, we can choose the rate of consumption of either $N_2$ or $O_2$, but not the rate of formation of NO. The average rate of this reaction is

$$\text{Rate} = -\frac{\Delta[N_2]}{\Delta t} = 0.70 \; M/s$$

The curvature of the lines in Figure 14.4 tells us that this value applies only to the interval from $t = 5.0$ μs to $t = 10.0$ μs and not over the entire 30.0-μs interval.

Any other 5.0-μs interval has a different average rate. For instance, from $t = 25.0$ μs to $t = 30.0$ μs, the rate of the reaction is

$$-\frac{\Delta[N_2]}{\Delta t} = -\frac{[N_2]_{30.0 \, \mu s} - [N_2]_{25.0 \, \mu s}}{(30.0 - 25.0) \, \mu s}$$

$$= -\frac{(3.6 - 4.5) \, \mu M}{5.0 \, \mu s} = 0.18 \, M/s$$

**instantaneous rate** the rate of a reaction at a specific instant during the course of the reaction.

Clearly, different reaction rates for the same reaction can be confusing, so why would we ever calculate average rates for a reaction? The usefulness of average rates is limited, but if we are comparing the rates of two different reactions over the same time period, an average rate can be sufficient to establish a relationship between the two rates. For example, if the average rate of reaction of nitrogen is 0.70 $M$/s between 5.0 and 10.0 μs, and the average rate of reaction for a different reaction at the same temperature is 2.10 $M$/s between 5.0 and 10.0 μs, then we know that the second reaction is three times faster than the first.

We can also determine an **instantaneous rate** of a reaction. An instantaneous rate is the reaction rate at a particular instant, which means the rate at a particular point on a curve of concentration as a function of time. The difference between average and instantaneous reaction rates is analogous to the difference between the average and instantaneous speeds of a runner. If a competitor in a marathon runs from mile 10 to mile 20 in 1 hr, her average speed over that distance is 10 mi/hr. At a given instant during the run, however, her instantaneous speed could be 12 mi/hr while going downhill and 8 mi/hr while going uphill.

Let's again consider the conversion of NO in engine exhaust into $NO_2$ (Equation 14.2), the gas responsible for much of the brown color in the photographs of photochemical smog at the beginning of this chapter. The conversion takes place when NO reacts with oxygen in the air:

$$2 \, NO(g) + O_2(g) \rightarrow 2 \, NO_2(g)$$

The rate of this reaction is described by the data in Table 14.2, specifically in terms of the rate of consumption of $O_2$, the reactant with a coefficient of 1. The data are plotted in Figure 14.5, where we calculate the instantaneous rate by drawing the tangent to a particular point on the curve.

**TABLE 14.2  Changing Concentrations of Reactants and Products for the Reaction 2 NO($g$) + O$_2$($g$) → 2 NO$_2$($g$) at 25°C**

| Time (s) | [NO] (M) | [O$_2$] (M) | [NO$_2$] (M) |
|---|---|---|---|
| 0 | 0.0100 | 0.0100 | 0.0000 |
| 285 | 0.0090 | 0.0095 | 0.0010 |
| 660 | 0.0080 | 0.0090 | 0.0020 |
| 1175 | 0.0070 | 0.0085 | 0.0030 |
| 1895 | 0.0060 | 0.0080 | 0.0040 |
| 2975 | 0.0050 | 0.0075 | 0.0050 |
| 4700 | 0.0040 | 0.0070 | 0.0060 |
| 7800 | 0.0030 | 0.0065 | 0.0070 |

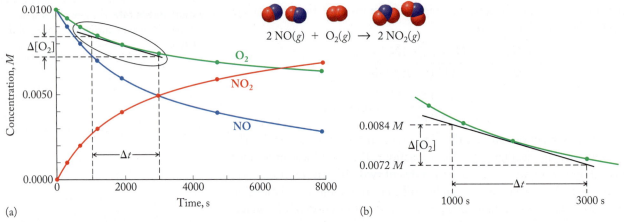

**FIGURE 14.5** (a) The instantaneous rate of change in $[O_2]$ in the reaction $2\,NO(g) + O_2(g) \rightarrow 2\,NO_2(g)$ is equal to the slope of a tangent to the curve of $[O_2]$ versus time. (b) An expanded view of the instantaneous rate of change in $[O_2]$ at $t = 2000$ s.

**CONCEPT TEST** ·································································

Which of the following statements is/are true about the instantaneous rate for the chemical reaction A → B as the reaction progresses?
a. $-\Delta[A]/\Delta t$ increases, $\Delta[B]/\Delta t$ decreases
b. $-\Delta[A]/\Delta t$ decreases, $\Delta[B]/\Delta t$ increases
c. $-\Delta[A]/\Delta t$ and $\Delta[B]/\Delta t$ both increase
d. $-\Delta[A]/\Delta t$ and $\Delta[B]/\Delta t$ both decrease

·····················································································

Any instantaneous reaction rate of a reaction can be determined from a graph of the concentration-versus-time data for the participants. For the $NO_2$ reaction, if we wish to determine the instantaneous rate of change of $[O_2]$ at $t = 2000$ s, we draw a tangent to the point on the green curve in Figure 14.4 corresponding to $t = 2000$ s and then select two convenient points, for example, $t = 1000$ s and $t = 3000$ s, along that tangent. From the differences in the coordinates on the $y$-axis for $[O_2]_{3000\,s}$ and $[O_2]_{1000\,s}$ at these two points and the difference in the $x$-axis coordinates (3000 s and 1000 s), we calculate the slope of the line. This is a measure of the instantaneous rate of change in $[O_2]$, the negative value of which is the instantaneous rate of the reaction at $t = 2000$ s:

$$\text{Slope} = \frac{\Delta[O_2]}{\Delta t} = \frac{(0.0072 - 0.0084)\,M}{(3000 - 1000)\,s} = -6.0 \times 10^{-7}\,M\,s^{-1}$$

$$\text{Rate} = -\frac{\Delta[O_2]}{\Delta t} = -(-6.0 \times 10^{-7}\,M\,s^{-1}) = 6.0 \times 10^{-7}\,M\,s^{-1}$$

Note that we substitute the concentration values corresponding to time points along the tangent line, not along the curve drawn from the data points, and that for the purpose of significant figures we consider the times exact.[1]

**SAMPLE EXERCISE 14.3    Determining an Instantaneous Rate    LO2**

(a) What is the instantaneous rate of change of [NO] at $t = 2000$ s in the experiment that produced the data in Table 14.2? (b) What is the rate of the reaction based on your result in part a?

[1]If you have studied calculus, you may realize that the average rate of a reaction approaches the instantaneous rate as $\Delta t$ approaches zero. Using calculus, we would say that the slope of the tangent to a curve at a given point is the derivative of the curve at that point and can be calculated using a scientific calculator or a graphing program. Appendix 1 discusses how to determine the slope and intercept of a line.

**initial rate** the rate of a reaction at $t = 0$, immediately after the reactants are mixed.

**Collect and Organize** We are asked to determine the instantaneous rate of change of [NO] and the corresponding reaction rate at $t = 2000$ s. The coefficient of NO is 2 in the balanced chemical equation:

$$2\ NO(g) + O_2(g) \rightarrow 2\ NO_2(g)$$

**Analyze** The instantaneous rate of change in [NO] at the stated time can be determined from a graph of the data. The corresponding reaction rate will be $(-\frac{1}{2})$ this value.

**Solve**

a. First we plot [NO] versus time and draw a tangent to the curve at the point $t = 2000$ s (Figure 14.6). We then choose two points along the tangent, $t = 1000$ and 3000 s, and determine the concentrations corresponding to those times along the vertical axis. By using those values, we calculate the slope of the line:

$$\frac{\Delta[NO]}{\Delta t} = \frac{(0.0046 - 0.0070)\ M}{(3000 - 1000)\ s} = -1.2 \times 10^{-6}\ M\ s^{-1}$$

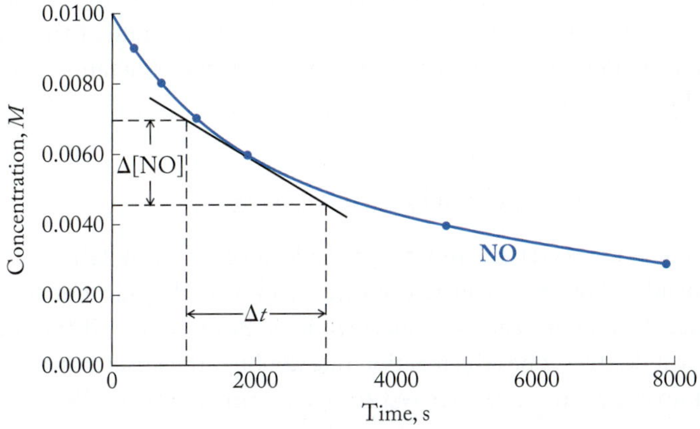

**FIGURE 14.6**

Note that we could also use a scientific calculator or a graphing program to calculate the slope of the tangent to a point on a curve.

b. The corresponding reaction rate is

$$\text{Rate} = -\frac{1}{2}\frac{\Delta[NO]}{\Delta t} = -\frac{1}{2}(-1.2 \times 10^{-6}\ M\ s^{-1}) = 6.0 \times 10^{-7}\ M\ s^{-1}$$

**Think About It** The sign of $\Delta[NO]/\Delta t$ is negative because NO is a reactant whose concentration decreases with time. However, the rates of chemical reactions have positive values. Therefore, we needed a minus sign in front of the $\Delta[NO]/\Delta t$ term in the solution to part b. Note that the instantaneous reaction rate calculated in this exercise is the same as the one we calculated earlier based on the rate of change of $[O_2]$.

**Practice Exercise** What is the instantaneous rate of change in $[NO_2]$ at $t = 2000$ s in the experiment that produced the data in Table 14.2?

## 14.3 Effect of Concentration on Reaction Rate

Figure 14.7 shows a typical result for a reactant concentration plotted as a function of time. Tangents have been drawn to the line at three points: (a) at the instant the reaction begins, (b) when the reaction is about halfway to completion, and (c) when the reaction is nearly over. Point (a) defines the **initial rate** of the reaction, which is the rate that occurs at the instant the reactants are mixed at $t = 0$.

The key observation from Figure 14.7 is that the slopes of the tangents approach zero as the reaction proceeds. When the reaction is over, no more change occurs in the concentration of any remaining reactant or in the concentration of any

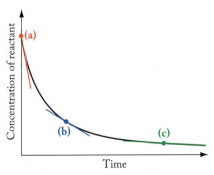

**FIGURE 14.7** Typical plot of reactant concentration as a function of time: (a) tangent at $t = 0$; (b) tangent at the midpoint of the reaction; (c) tangent close to the end of the reaction.

⊂⊃ **CONNECTION** We discussed kinetic molecular theory, a model that describes the behavior of gases, in Chapter 6.

▶️ **CHEMTOUR** Reaction Order

product, and in this region of the curve the slope of any tangent is zero. This behavior is typical: the most rapid changes in reactant and product concentrations take place early in most reactions.

Kinetic molecular theory and our picture of molecules in the gas phase provide us with a way to explain this trend. If we assume that most reactions take place as a result of collisions between reactant molecules, then the more reactant molecules there are in a given space, such as a flask, the more collisions per unit time and the more opportunities for reactants to turn into products. As reactant concentrations decrease, fewer reactant molecules occupy the space in the flask, so the frequency of collisions decreases and the rate of conversion of reactants to products slows down.

What happens to the rate of some of the reactions we have discussed when we change the initial concentrations of the reactants? Doubling the concentration of $O_2$ in Equation 14.2 doubles the rate of the reaction. However, doubling the concentration of NO *quadruples* the rate. Why does the rate of the reaction have different dependencies on the concentrations of the two reactants? We explore the reason why in this section.

## Reaction Order and Rate Constants

Experimental observations and theoretical considerations tell us that reaction rates depend on reactant concentrations. However, they do not tell us *to what extent* rates depend on reactant concentrations. For example, if the concentration of a reactant doubles, does the reaction rate double? The answer is expressed in the **reaction order**, a parameter derived from experiments that tells us how reaction rate depends on reactant concentrations. Knowing the order of a reaction provides insights into *how* the reaction takes place—which molecules collide with which other molecules as bonds break, new bonds form, and reactants are converted into products.

Let's look at one way in which reaction order is determined by revisiting the reaction between oxygen and nitrogen monoxide:

$$2\,NO(g) + O_2(g) \rightarrow 2\,NO_2(g)$$

In experiments to evaluate the kinetics of this reaction, different initial concentrations of NO and $O_2$ are introduced into a reaction vessel at 25°C, and the initial reaction rate is determined in each case. Table 14.3 shows three of these initial reaction rates. Recall that the initial rate is determined from the slope of a line tangent to the curve of concentration versus time. Why do we choose to calculate the initial rate (instantaneous rate at $t = 0$)? In the preceding section we observed that reaction rates depend on collisions between reactants. Choosing

**reaction order** an experimentally determined number defining the dependence of the reaction rate on the concentration of a reactant.

| TABLE 14.3 | **Effect of Reactant Concentrations on Initial Reaction Rates at 25°C for the Reaction $2\,NO(g) + O_2(g) \rightarrow 2\,NO_2(g)$** | | |
|---|---|---|---|
| **Experiment** | **$[NO]_0$ (M)** | **$[O_2]_0$ (M)** | **Initial Reaction Rate, $-\dfrac{1}{2}\dfrac{\Delta[NO]}{\Delta t}$ (M s$^{-1}$)** |
| 1 | 0.0100 | 0.0100 | $1.0 \times 10^{-6}$ |
| 2 | 0.0100 | 0.0050 | $0.5 \times 10^{-6}$ |
| 3 | 0.0050 | 0.0100 | $2.5 \times 10^{-7}$ |

$t = 0$ as our point on the concentration-versus-time curve means that the concentration of products is essentially zero. This choice is important because reactions can also run in reverse. If the product concentration is essentially zero, then no collisions occur between product molecules and the rate of the reverse reaction can be largely ignored. If we choose to work with rates at a point where $t \neq 0$, then we must account for the different rates of the forward and reverse reactions.

To interpret the data in Table 14.3, we select pairs of experiments in which the concentrations of one reactant differ but the concentrations of the other are the same. For example, $[NO]_0$ is the same in experiments 1 and 2, but $[O_2]_0$ in experiment 1 is twice $[O_2]_0$ in experiment 2. (The zero subscripts indicate that the concentrations of NO and $O_2$ are the values at the start of the experiments, when $t = 0$.) The initial reaction rate in experiment 1 is twice that in experiment 2, allowing us to state that doubling the concentration of $O_2$ while holding the NO concentration constant doubles the initial reaction rate. We conclude that the initial reaction rate is proportional to $O_2$ concentration:

$$\text{Rate} \propto [O_2]$$

In experiments 1 and 3, $[O_2]_0$ is the same but $[NO]_0$ in experiment 1 is twice $[NO]_0$ in experiment 3. Comparing the initial reaction rates for these two experiments, we find that the rate in experiment 1 is four times the rate in experiment 3. Thus, the reaction rate is proportional to [NO] squared:

$$\text{Rate} \propto [NO]^2$$

We combine these two rate expressions to get an overall rate expression for the reaction by multiplying the right sides of the rate expressions for the two reactants:

$$\text{Rate} \propto [NO]^2[O_2]$$

To understand why we multiply the concentration terms together, let's consider another reaction between oxygen and nitrogen monoxide. This one involves ozone ($O_3$) and NO and produces $NO_2$ and $O_2$:

$$NO(g) + O_3(g) \rightarrow NO_2(g) + O_2(g)$$

Figure 14.8 shows how different numbers of NO and $O_3$ molecules in a reaction vessel might collide together and react with each other. Note how increasing the numbers of molecules in the containers in Figure 14.8 produces increasing numbers of collisions that are proportional to the *product* of the number of molecules of each reactant. Increasing the numbers of molecules of each type in the vessels is equivalent to increasing the concentrations of the two gases. Therefore, the rate of the reaction should (and does) depend on the product of the concentrations of NO and $O_3$:

$$\text{Rate} \propto [NO][O_3] \tag{14.14}$$

**FIGURE 14.8** Increasing the concentration increases the number of possible collisions (double-headed arrows) and therefore the number of potential reaction events. Reaction rate depends on the number of collisions, which are shown for the reaction between NO and ozone ($O_3$) that produces $NO_2$ and $O_2$. With only one NO molecule and one $O_3$ molecule, as in a, each molecule can collide only with the other, giving a relative reaction rate of $1 \times 1 = 1$. In e, three molecules of NO can collide with three molecules of $O_3$ for a relative reaction rate of $3 \times 3 = 9$ times the rate in a.

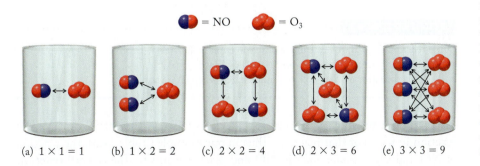

= NO   = $O_3$

(a) $1 \times 1 = 1$   (b) $1 \times 2 = 2$   (c) $2 \times 2 = 4$   (d) $2 \times 3 = 6$   (e) $3 \times 3 = 9$

Similar patterns occur with all chemical reactions whose rate depends on the concentration of more than one reactant. We can modify Equation 14.14 to obtain a **rate law**, an equation that defines the relation between reactant concentrations and reaction rate. We convert the proportionality to an equation by inserting a proportionality constant $k$, called the reaction **rate constant**:

$$\text{Rate} = k[\text{NO}][\text{O}_3] \tag{14.15}$$

Now let's consider a generic chemical reaction with two reactants, A and B:

$$A + B \rightarrow C$$

The rate law expression for this reaction may be written

$$\text{Rate} = k[\text{A}]^m[\text{B}]^n \tag{14.16}$$

where $m$ is the **partial reaction order** with respect to A and $n$ is the partial reaction order with respect to B. We determine partial reaction order values by comparing differences in reaction rates to differences in reactant concentrations. Sometimes these comparisons are easy to make, as with the data for the reaction between NO and $O_2$ in Table 14.3. Other times the comparisons are not so easy. On those occasions we can use a more mathematical approach. For example, suppose we run three experiments to solve for the values of $m$ and $n$ in Equation 14.16. In experiments 1 and 2 the value of [A] is the same, but the values of [B] are different. In experiments 2 and 3 we keep [B] constant and vary the concentrations of [A]. The ratio of the reaction rates in experiments 1 ($\text{Rate}_1$) and 2 ($\text{Rate}_2$) is related to the ratio of the concentrations of B: $[\text{B}]_1/[\text{B}]_2$, and to the dependence of reaction rate on [B]; that is, on the value of $n$. Expressing this relationship in equation form and then solving for $n$ by taking the logarithm of both sides:

$$\frac{\text{Rate}_1}{\text{Rate}_2} = \left(\frac{[\text{B}]_1}{[\text{B}]_2}\right)^n \qquad \log\left(\frac{\text{Rate}_1}{\text{Rate}_2}\right) = n\log\left(\frac{[\text{B}]_1}{[\text{B}]_2}\right)$$

Now rearrange the terms to solve for $n$:

$$n = \frac{\log\left(\dfrac{\text{Rate}_1}{\text{Rate}_2}\right)}{\log\left(\dfrac{[\text{B}]_1}{[\text{B}]_2}\right)}$$

An equation with this format could be used to calculate the value of $m$ from the results of experiments 2 and 3, or to calculate the partial order of any reaction with respect to a reactant (X) whose concentration differs in a pair of reaction rate experiments:

$$n = \frac{\log\left(\dfrac{\text{Rate}_1}{\text{Rate}_2}\right)}{\log\left(\dfrac{[\text{X}]_1}{[\text{X}]_2}\right)} \tag{14.17}$$

**rate law** an equation that defines the experimentally determined relation between the concentrations of reactants in a chemical reaction and the rate of that reaction.

**rate constant** the proportionality constant that relates the rate of a reaction to the concentrations of reactants.

**partial reaction order** the order of a reaction with respect to a single reactant.

**overall reaction order** the sum of the exponents of the concentration terms in the rate law.

**CONCEPT TEST**

In the reaction A → B, the rate of the reaction triples when [A] is tripled.
a. What is the correct value of $m$ in the rate law: $\text{Rate} = k[\text{A}]^m$?
b. What is the value of $m$ if the rate is unchanged when [A] is tripled?

The power to which a concentration term is raised is the partial order of the reaction in terms of that reactant. Thus in our example (Table 14.3), the reaction of NO and $O_2$ is *second order* in NO and *first order* in $O_2$. The **overall reaction**

**order** for a reaction is the sum of the powers in the rate equation, so the reaction of NO and $O_2$ is *third order* overall.

An exponent in a rate law may be a fraction, zero, or, in rare cases, negative. It is important to remember that rate laws and reaction orders are different from relative rates and must be determined experimentally. They cannot be predicted from the coefficients in a balanced chemical equation. The significance of reaction order in describing how a reaction takes place is addressed in Section 14.5.

The value of the rate constant $k$ is unique to each particular reaction at a given temperature, and right now we may consider $k$ simply as a proportionality constant. It does not change with concentration; in other words, *reaction rate depends on the concentration of the reactants but the rate constant does not*. The rate constant changes only with changing temperature or in the presence of a catalyst.

We can calculate $k$ for a reaction run at some specified temperature from initial reaction rate data. Let's do so for the reaction of $O_2$ and NO by selecting the results of one experiment in Table 14.3. Which experiment we use doesn't matter. As long as the temperature is the same, the value of $k$ is the same. To determine $k$, let's use Equation 14.16 and insert the data from experiment 1:

$$1.0 \times 10^{-6}\ M/s = k(0.0100\ M)^2(0.0100\ M)$$

$$k = \frac{1.0 \times 10^{-6}\ \cancel{M}/s}{(0.0100\ M)^2(0.0100\ \cancel{M})} = 1.0\ M^{-2}\ s^{-1} \qquad (14.18)$$

Remember that the value calculated from experimental data for a rate constant is valid only at the temperature at which the experiments were carried out.

The units of $k$ differ from one reaction to another. Remember that the units of reaction rates always change in concentration per unit time. Therefore, in a first-order reaction in which concentration is expressed in molarity and time in seconds, the units of $k$ must be per second ($s^{-1}$):

$$\text{Rate} = k[X]^1$$

$$k = \frac{\text{rate}}{[X]} = \frac{\cancel{M}/s}{\cancel{M}} = \frac{1}{s} = s^{-1}$$

The units of the rate constant for a reaction that is second order overall can be derived in a similar way. If the reaction is first order in reactants X and Y, we have

$$\text{Rate} = k[X][Y]$$

$$k = \frac{\text{rate}}{[X][Y]} = \frac{\cancel{M}/s}{M\ \cancel{M}} = M^{-1}\ s^{-1}$$

The units of $k$ for a reaction that is third order overall are $M^{-2}\ s^{-1}$, as we determined in Equation 14.18.

**CONCEPT TEST** ..................................................................

The reaction A + 2 B → C is found to be third order overall. How many possible rate laws, Rate = $k[A]^m[B]^n$, could we write assuming that $m$ and $n$ are positive integers?

..................................................................

**SAMPLE EXERCISE 14.4** **Deriving a Rate Law from** **LO3**
**Initial Reaction Rate Data**

Write the rate law for the reaction of $N_2$ with $O_2$

$$N_2(g) + O_2(g) \rightarrow 2\ NO(g)$$

using the data in Table 14.4. Determine the overall reaction order and the value of the rate constant.

**TABLE 14.4 Initial Reaction Rates for the Formation of NO from the Reaction of $N_2$ with $O_2$ at Constant Temperature**

| Experiment | $[N_2]_0$ (M) | $[O_2]_0$ (M) | Initial Reaction Rate (M/s) |
|---|---|---|---|
| 1 | 0.040 | 0.020 | 707 |
| 2 | 0.040 | 0.010 | 500 |
| 3 | 0.010 | 0.010 | 125 |

**Collect and Organize** We are to determine the rate law and rate constant for a reaction, given the initial reaction rate (note the subscript zero on the concentration terms in Table 14.4) for each of three sets of initial concentrations.

**Analyze** The general form of the rate law for any reaction between reactants A and B is

$$\text{Rate} = k[A]^m[B]^n$$

We can use the experimental data given to find the values of $k$, $m$, and $n$ for the reaction in which A = $N_2$ and B = $O_2$. The overall order of the reaction is the sum of the partial orders of the individual reactants. Once we have established the rate law, we can calculate the rate constant using concentrations of reactants from any row in Table 14.4. The rate constant must have units that express the reaction rate in $M\,s^{-1}$.

**Solve** To determine the partial order of the reaction ($m$) with respect to $N_2$, we use the data from experiments 2 and 3 because in these two experiments the values of $[N_2]$ are different but the $[O_2]$ values are the same. When the concentration $N_2$ is increased by a factor of 4, the rate increases by a factor of 4. Thus, the reaction rate is proportional to $N_2$ concentration, which means that $m = 1$. There are different values of $[O_2]$ in Experiments 1 and 2, but $[N_2]$ is the same. Therefore we can use these data to calculate the value of $n$. The ratio of the reaction rates (707/500) is not a whole number, so let's apply Equation 14.17:

$$n = \frac{\log\left(\dfrac{\text{Rate}_1}{\text{Rate}_2}\right)}{\log\left(\dfrac{[O_2]_1}{[O_2]_2}\right)} = \frac{\log\left(\dfrac{707\ \cancel{M/s}}{500\ \cancel{M/s}}\right)}{\log\left(\dfrac{0.020\ \cancel{M}}{0.010\ \cancel{M}}\right)} = 0.50$$

Thus, the reaction is $\frac{1}{2}$ order with respect to $O_2$, first order with respect to $N_2$, and $(\frac{1}{2} + 1) = \frac{3}{2}$ order overall:

$$\text{Rate} = k[N_2][O_2]^{1/2}$$

We can use the data from any experiment to obtain the value of $k$. Let's use experiment 1:

$$707\ M/s = k(0.040\ M)(0.020\ M)^{1/2}$$

$$k = 1.2 \times 10^5\ M^{-1/2}\ s^{-1}$$

**Think About It** A partial order of $\frac{1}{2}$ for oxygen in this reaction is determined from experimental data, not from the balanced chemical equation. The units of the rate constant, $M^{-1/2}\,s^{-1}$, are appropriate because when we substitute the units for each term into the rate law, we end up with the correct units for reaction rate: $(M^{-1/2}\,s^{-1})(M)(M^{1/2}) = M\,s^{-1}$.

**Practice Exercise** Nitrogen monoxide reacts rapidly with unstable nitrogen trioxide ($NO_3$) to form $NO_2$:

$$NO(g) + NO_3(g) \rightarrow 2\,NO_2(g)$$

Determine the rate law for the reaction and calculate the rate constant from the data in Table 14.5.

**TABLE 14.5 Initial Reaction Rates for the Formation of $NO_2$ from the Reaction of NO with $NO_3$ at 25°C**

| Experiment | $[NO]_0$ (M) | $[NO_3]_0$ (M) | Initial Reaction Rate (M/s) |
|---|---|---|---|
| 1 | $1.25 \times 10^{-3}$ | $1.25 \times 10^{-3}$ | $2.45 \times 10^4$ |
| 2 | $2.50 \times 10^{-3}$ | $1.25 \times 10^{-3}$ | $4.90 \times 10^4$ |
| 3 | $2.50 \times 10^{-3}$ | $2.50 \times 10^{-3}$ | $9.80 \times 10^4$ |

## Integrated Rate Laws: First-Order Reactions

Determining a rate law using initial reaction rate data has two distinct disadvantages. The method of initial rates requires several experiments with different concentrations of reactants, varied in a systematic fashion. We also must accurately determine the reaction rate at the instant the reaction begins. It would be much easier if we could determine the rate law and calculate the rate constant for a reaction from the plot of concentration versus time alone. In fact, we can do this for reactions in which the reaction rate depends on the concentration of only one substance. One such reaction is the photochemical decomposition of ozone in the stratosphere:

$$O_3(g) \xrightarrow{hv} O_2(g) + O(g)$$

This decomposition reaction can be studied in the laboratory using high-intensity ultraviolet lamps to simulate solar radiation. One such study yielded the results listed in Table 14.6 and plotted in Figure 14.9(a). Because ozone is the only reactant, the rate law for the reaction should depend only on the ozone concentration. In our interpretation of the data in Table 14.6, we can start with the assumption that the reaction is first order in $O_3$, which means that the rate law can be written

$$Rate = k[O_3]$$

**TABLE 14.6  Rate of Photochemical Decomposition of Ozone**

| Time (s) | $[O_3]$ (M) | $\ln[O_3]$ |
|---|---|---|
| 0 | $1.000 \times 10^{-4}$ | −9.2103 |
| 100 | $0.896 \times 10^{-4}$ | −9.320 |
| 200 | $0.803 \times 10^{-4}$ | −9.430 |
| 300 | $0.719 \times 10^{-4}$ | −9.540 |
| 400 | $0.644 \times 10^{-4}$ | −9.650 |
| 500 | $0.577 \times 10^{-4}$ | −9.760 |
| 600 | $0.517 \times 10^{-4}$ | −9.870 |

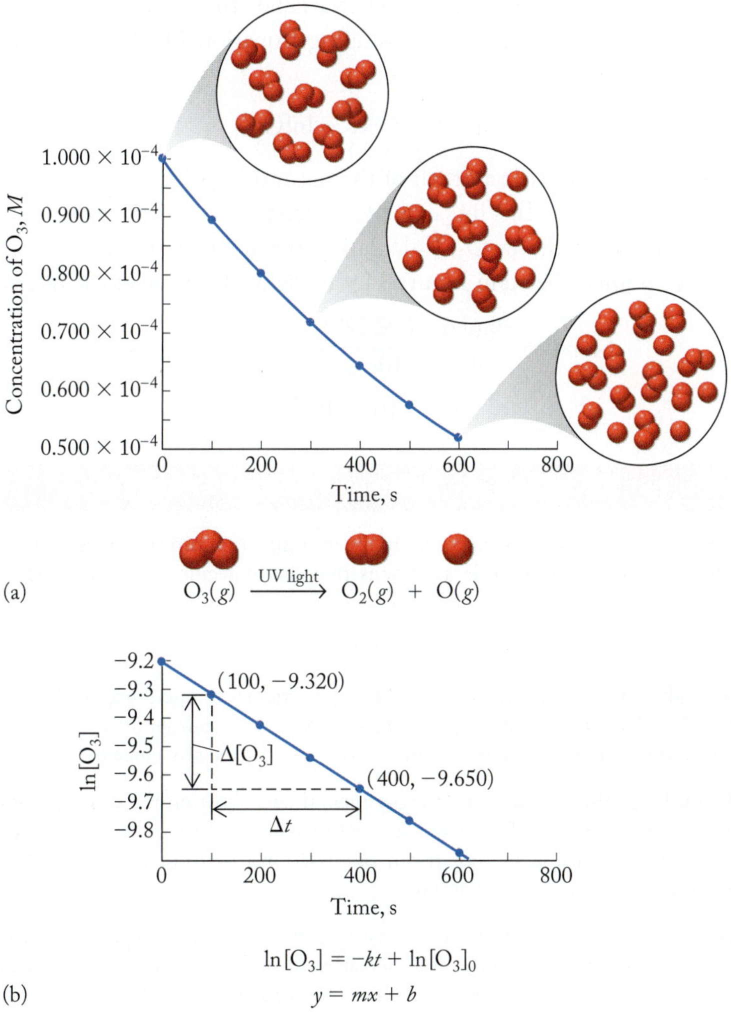

FIGURE 14.9 Data for the decomposition of $O_3$: (a) plot of $[O_3]$ versus time; (b) plot of $\ln[O_3]$ versus time. The line in (b) is straight, indicating that the decomposition reaction is first order in $O_3$.

$$\ln[O_3] = -kt + \ln[O_3]_0$$
$$y = mx + b$$

**integrated rate law** a mathematical expression that describes the change in concentration of a reactant in a chemical reaction with time.

Because the coefficient of $O_3$ in the balanced equation is 1, the $O_3$ consumption rate $-\Delta[O_3]/\Delta t$ is equal to the reaction rate, and we can write

$$\text{Rate} = -\frac{\Delta[O_3]}{\Delta t} = k[O_3]$$

This rate law can be transformed into an expression that relates the concentration of ozone $[O_3]$ at any instant during the reaction to the initial concentration $[O_3]_0$:

$$\ln\frac{[O_3]}{[O_3]_0} = -kt \tag{14.19}$$

This version of the rate law is called an **integrated rate law** because integral calculus is used to derive it, and it describes the change in reactant concentration with time. The general integrated rate law for any reaction that is first order in reactant X is

$$\ln\frac{[X]}{[X]_0} = -kt \tag{14.20}$$

Using the identity $\ln(a/b) = \ln a - \ln b$, we can rearrange this equation to

$$\ln[X] = -kt + \ln[X]_0 \tag{14.21}$$

This is the equation of a straight line of the form

$$y = mx + b$$

where $\ln[X]$ is the $y$ variable and $t$ is the $x$ variable. The slope of the line ($m$) is $-k$, and the $y$ intercept ($b$) is $\ln[X]_0$. Rearranging Equation 14.19 to fit the format of Equation 14.21 gives

$$\ln[O_3] = -kt + \ln[O_3]_0 \tag{14.22}$$

Note in Figure 14.9(b) that a graph of the natural logarithm of $[O_3]$ versus time is indeed a straight line. This linearity means that our assumption was correct and the reaction is first order in $O_3$. Calculating the slope of the line, $\Delta(\ln[O_3])/\Delta t$, from the two points labeled in Figure 14.9(b) yields the value of the rate constant:

$$\text{Slope} = \frac{-9.650 - (-9.320)}{400\text{ s} - 100\text{ s}} = -1.10 \times 10^{-3}\text{ s}^{-1}$$

$$k = 1.10 \times 10^{-3}\text{ s}^{-1}$$

**SAMPLE EXERCISE 14.5** **Using an Integrated Rate Law** **LO4**

One of the least abundant nitrogen oxides in the atmosphere is dinitrogen pentoxide. One reason concentrations of this oxide are low is that the molecule is unstable and rapidly decomposes to $N_2O_4$ and $O_2$:

$$2\,N_2O_5(g) \rightarrow 2\,N_2O_4(g) + O_2(g)$$

A kinetic study of the decomposition of $N_2O_5$ at a particular temperature yielded the data in Table 14.7(a). Assume that the decomposition of $N_2O_5$ is first order in $N_2O_5$. (a) Test the validity of your assumption, and (b) determine the value of the rate constant.

**Collect and Organize** We are given experimental data showing the concentration of a single reactant as a function of time and are told to assume a first-order reaction. We can verify the assumption using the integrated rate law for a first-order reaction (Equation 14.21) and then calculate the rate constant.

**Analyze** Equation 14.21 has the form $y = mx + b$, which means that a plot of $\ln[N_2O_5]$ ($y$) versus time ($x$) should be linear if the decomposition of $N_2O_5$ is first order. The slope ($m$) of the graph corresponds to $-k$, the negative of the rate constant.

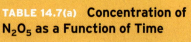

**TABLE 14.7(a)** **Concentration of $N_2O_5$ as a Function of Time**

| Time (s) | $[N_2O_5]$ (M) |
|----------|----------------|
| 0 | 0.1000 |
| 50 | 0.0707 |
| 100 | 0.0500 |
| 200 | 0.0250 |
| 300 | 0.0125 |
| 400 | 0.00625 |

**Solve** Our first step is to determine $\ln[N_2O_5]$ values (Table 14.7b).

a. The plot of $\ln[N_2O_5]$ versus $t$ is shown in Figure 14.10. The fact that the curve in Figure 14.10 is a straight line indicates that the reaction is first order in $N_2O_5$.

b. Arbitrarily choosing $t = 100$ s and $t = 300$ s as two points for calculating the slope:

$$\text{Slope} = \frac{\Delta y}{\Delta x} = \frac{-4.382 - (-2.996)}{300 \text{ s} - 100 \text{ s}} = \frac{-1.386}{200 \text{ s}}$$

$$= -0.00693 \text{ s}^{-1}$$

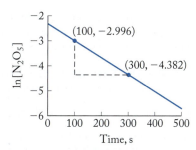

**FIGURE 14.10**

The slope of the line equals $-k$. Therefore, the rate constant $k = 0.00693$ s$^{-1}$. Note that we could also use a graphing calculator to plot $\ln[N_2O_5]$ versus $t$ and have the calculator determine the best fit of the data to a straight line. The equation of the line gives us the slope, which we then convert into the rate constant as described above.

**Think About It** We can test whether any reaction with a single reactant (X) is first order in X by determining whether a plot of $\ln[X]$ versus $t$ is linear.

 **Practice Exercise** Hydrogen peroxide ($H_2O_2$) decomposes into water and oxygen:

$$H_2O_2(\ell) \rightarrow H_2O(\ell) + \tfrac{1}{2} O_2(g)$$

Use the data in Table 14.8 to determine whether the decomposition of $H_2O_2$ is first order in $H_2O_2$, and calculate the value of the rate constant at the temperature of the experiment that produced the data.

---

**SAMPLE EXERCISE 14.6** **Calculating Concentration of a Reactant from an Integrated Rate Law** **LO5**

In Sample Exercise 14.5 we confirmed that the decomposition of dinitrogen pentoxide

$$2 N_2O_5(g) \rightarrow 2 N_2O_4(g) + O_2(g)$$

is first order in $N_2O_5$ and has a rate constant $k = 0.00693$ s$^{-1}$. If this reaction is run in the laboratory under the same conditions as in Sample Exercise 14.5 and the initial concentration of $N_2O_5$ in the reaction vessel is 0.375 $M$, what is the concentration of $N_2O_5$ after 3 minutes?

**Collect and Organize** We are given the rate constant for a first-order reaction, the initial concentration of reactant, and the reaction time. We are asked to calculate the concentration of reactant remaining at that time. We also have the general rate equation for a first-order reaction (Equation 14.21).

**Analyze** We know all the terms in the rate equation

$$\ln[X] = -kt + \ln[X]_0$$

except $[X]$, which is what we want to calculate. The rate constant $k$ is given in terms of seconds, so we must convert $t = 3$ min into $t = 180$ sec. In the data set for Sample Exercise 14.5, $[N_2O_5]$ dropped to about $\frac{1}{4}[N_2O_5]_0$ in 200 sec, so we estimate that the concentration of $N_2O_5$ in this example will drop to about one-quarter of its initial value or around 0.09 $M$.

**Solve** Using the given values in Equation 14.21:

$$\ln[X] = -kt + \ln[X]_0$$

$$\ln[N_2O_5] = -(0.00693 \text{ s}^{-1})(180 \text{ s}) + \ln(0.375)$$

$$\ln[N_2O_5] = -1.247 + (-0.981) = -2.228$$

$$[N_2O_5] = e^{-2.228} = 0.108 \ M$$

---

**TABLE 14.7(b)** **Concentration and Natural Logarithm of Concentration of $N_2O_5$ as a Function of Time**

| Time (s) | $[N_2O_5]$ (M) | $\ln[N_2O_5]$ |
| --- | --- | --- |
| 0 | 0.1000 | −2.303 |
| 50 | 0.0707 | −2.649 |
| 100 | 0.0500 | −2.996 |
| 200 | 0.0250 | −3.689 |
| 300 | 0.0125 | −4.382 |
| 400 | 0.00625 | −5.075 |

**TABLE 14.8** **Concentration of Hydrogen Peroxide as a Function of Time**

| Time (s) | $[H_2O_2]$ (M) |
| --- | --- |
| 0 | 0.500 |
| 100 | 0.460 |
| 200 | 0.424 |
| 500 | 0.330 |
| 1000 | 0.218 |
| 1500 | 0.144 |

**half-life (t$_{1/2}$)** the time in the course of a chemical reaction during which the concentration of a reactant decreases by half.

**Think About It** The answer is a lower number than the starting concentration, which we expect because some of the reactant has been transformed into product. Furthermore, our answer is close to our estimate.

**Practice Exercise** Under a different set of conditions from those in Sample Exercise 14.5, dinitrogen pentoxide decomposes into nitrogen dioxide and $O_2$:

$$2\,N_2O_5(g) \rightarrow 4\,NO_2(g) + O_2(g)$$

This reaction is also first order in $N_2O_5$ and has a rate constant $k = 9.55 \times 10^{-4}\,s^{-1}$. What is the concentration of $N_2O_5$ in this system after 10 min if $[N_2O_5]_0 = 0.763\,M$?

## Reaction Half-Lives

A parameter frequently cited in kinetic studies is the **half-life ($t_{1/2}$)** of a reaction, which is the interval during which the concentration of a reactant decreases by half. Half-life is inversely related to the rate constant of a reaction: the higher the reaction rate, the shorter the half-life.

Let's consider reaction half-life in the context of another nitrogen oxide found in the atmosphere: dinitrogen monoxide, also called nitrous oxide or laughing gas, an anesthetic sometimes used by dentists. Atmospheric concentrations of this potent greenhouse gas have been increasing in recent years, although the principal source is not automotive emissions but rather bacterial degradation of nitrogen compounds in soil. Dinitrogen monoxide is not a product in internal combustion engines because at typical engine temperatures, any $N_2O$ formed rapidly decomposes into nitrogen and oxygen:

$$N_2O(g) \rightarrow N_2(g) + \tfrac{1}{2}\,O_2(g)$$

The reaction rate for this first-order reaction is very high, which means that the reaction has a short half-life (Figure 14.11).

**FIGURE 14.11** The decomposition of $N_2O(g)$ is first order in $N_2O$. At a particular temperature the half-life of the reaction is 1.0 s, which means that, on average, half of a population of 16 $N_2O$ molecules decomposes in 1.0 s, half of the remaining 8 molecules decompose in the next 1.0 s, and so on.

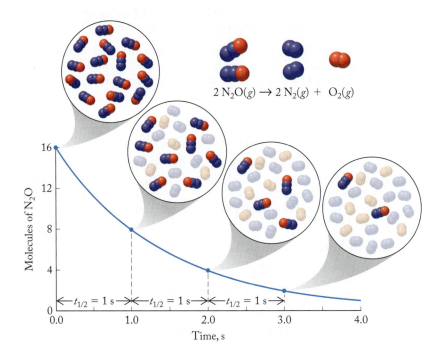

We can derive a mathematical relation between half-life $t_{1/2}$ and rate constant $k$ for this or any other first-order reaction by starting with Equation 14.20:

$$\ln\frac{[X]}{[X]_0} = -kt$$

After one half-life has passed, $t = t_{1/2}$, the concentration of X is half its original value: $[X] = \frac{1}{2}[X]_0$. Inserting these values for $[X]$ and $t$ into the equation yields

$$\ln\frac{\frac{1}{2}\cancel{[X]_0}}{\cancel{[X]_0}} = -kt_{1/2}$$

$$\ln\left(\tfrac{1}{2}\right) = -kt_{1/2}$$

The natural log of $\frac{1}{2}$ is $-0.693$, so

$$-0.693 = -kt_{1/2}$$

$$t_{1/2} = \frac{0.693}{k} \qquad\qquad (14.23)$$

Thus, the half-life of a first-order reaction is inversely proportional to the rate constant, as noted at the beginning of this discussion. The absence of any concentration term in Equation 14.23 means that no matter the initial concentration of reactant, half of it is consumed in one half-life.

---

**SAMPLE EXERCISE 14.7** **Calculating the Half-Life of a First-Order Reaction** **LO6**

The rate constant for the decomposition of $N_2O_5$ at a particular temperature is $7.8 \times 10^{-3}\ s^{-1}$. What is the half-life of $N_2O_5$ at that temperature?

**Collect and Organize** The half-life is the time required for the concentration of a reactant to decrease by half. We are asked to determine the half-life of $N_2O_5$ for its decomposition reaction. We know from Sample Exercise 14.5 that the decomposition of $N_2O_5$ is a first-order process.

**Analyze** Equation 14.23 relates $k$ for a first-order reaction to the reaction half-life. We are given no information about concentrations of reactants or products, but we know from Equation 14.23 that the half-life of a first-order reaction is independent of concentration. Because half-life is inversely proportional to $k$, and the value of $k$ is on the order of $10^{-2}\ s^{-1}$, we predict that the half-life will be on the order of $10^2$ s.

**Solve**

$$t_{1/2} = \frac{0.693}{k} = \frac{0.693}{7.8 \times 10^{-3}\ s^{-1}} = 89\ s$$

**Think About It** The calculated value of $t_{1/2}$ is a little less than 100 s, the predicted value. Remember that Equation 14.23 is valid only for first-order reactions.

⚙ **Practice Exercise** Environmental scientists calculating half-lives of pollutants often define a *transport rate constant* that is analogous to a reaction rate constant and describes how a pollutant moves out of an ecosystem. In a study of the gasoline additive MTBE in Donner Lake, California, scientists from the University of California, Davis, found that in the summer the half-life of MTBE in the lake was 28 days. Assume that the transport process is first order. What was the transport rate constant of MTBE out of Donner Lake during the study? Express your answer in reciprocal days.

---

**CONCEPT TEST**

Which has a shorter half-life, a fast reaction or a slow reaction?

## Integrated Rate Laws: Second-Order Reactions

In Section 14.1 we described how $NO_2$ exposed to UV rays from the Sun decomposes to NO and atomic oxygen, O (Equation 14.3). Nitrogen dioxide may also undergo thermal decomposition, producing NO and molecular oxygen, $O_2$:

$$2\,NO_2(g) \rightarrow 2\,NO(g) + O_2(g) \qquad (14.24)$$

Because it has only one reactant, like the reactions of $N_2O_5$ and $H_2O_2$ in Sample Exercise 14.5 and its Practice Exercise, we might expect this reaction to be first order. However, when we use the data in Table 14.9 to evaluate the reaction order and rate constant, we find that the plot of $\ln[NO_2]$ versus time (Figure 14.12a) is clearly not linear, which tells us that the thermal decomposition of $NO_2$ is *not* first order.

What is the reaction order? The answer to this question is hidden in how the reaction takes place. If each $NO_2$ molecule simply fell apart, the reaction would be first order, much like the decomposition of $N_2O_5$. However, if the reaction happened as a result of collisions between pairs of $NO_2$ molecules, the reaction would be first order in each and second order overall. In other words, if the reaction were second order, the decomposition described by Equation 14.24 would depend on collisions between pairs of molecules that just happen to be molecules of the same substance, $NO_2$. The rate law expression is

$$\text{Rate} = k[NO_2]^2 \qquad (14.25)$$

How can we determine whether this decomposition is really second order? One way is to assume that it is and then test that assumption. The test entails transforming the rate law in Equation 14.25 into the integrated rate law for a second-order reaction, again using calculus. The result of the transformation is

$$\frac{1}{[NO_2]} = kt + \frac{1}{[NO_2]_0} \qquad (14.26)$$

Like Equation 14.22, this has the form $y = mx + b$ and is the equation of a straight line, this time with $1/[NO_2]$ as the $y$ variable and $t$ as the $x$ variable.

The graph obtained using data from columns 1 and 4 of Table 14.9 is shown in Figure 14.12(b). The curve is linear, which tells us that decomposition of $NO_2$ is second order. The slope of the line provides a direct measure of $k$, which is 0.544 $M^{-1}\,s^{-1}$.

### TABLE 14.9 Rate of Decomposition of $NO_2$ to NO and $O_2$

| Time (s) | $[NO_2]$ (M) | $\ln[NO_2]$ | $1/[NO_2]$ (1/M) |
|---|---|---|---|
| 0 | $1.00 \times 10^{-2}$ | −4.605 | 100 |
| 100 | $6.48 \times 10^{-3}$ | −5.039 | 154 |
| 200 | $4.79 \times 10^{-3}$ | −5.341 | 209 |
| 300 | $3.80 \times 10^{-3}$ | −5.573 | 263 |
| 400 | $3.15 \times 10^{-3}$ | −5.760 | 317 |
| 500 | $2.69 \times 10^{-3}$ | −5.918 | 372 |
| 600 | $2.35 \times 10^{-3}$ | −6.057 | 426 |

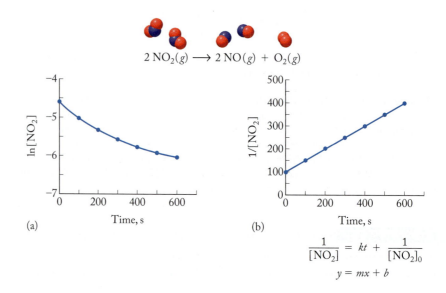

**FIGURE 14.12** At high temperatures, $NO_2$ slowly decomposes into NO and $O_2$. (a) The plot of $\ln[NO_2]$ versus time is not linear, indicating the reaction is not first order. (b) The plot of $1/[NO_2]$ versus time is linear, indicating that the reaction is second order in $NO_2$. The slope of the line in this graph equals the rate constant.

A general form of Equation 14.26 that applies to any reaction that is second order in a single reactant (X) is

$$\frac{1}{[X]} = kt + \frac{1}{[X]_0} \qquad (14.27)$$

---

**SAMPLE EXERCISE 14.8** **Distinguishing between First- and Second-Order Reactions** **LO4**

Chlorine monoxide accumulates in the stratosphere above Antarctica each winter and plays a key role in the formation of the ozone hole above the South Pole each spring. Eventually, ClO decomposes according to the equation

$$2\,ClO(g) \rightarrow Cl_2(g) + O_2(g)$$

The kinetics of this reaction were studied in a laboratory experiment at 298 K, and the data are shown in Table 14.10(a). Determine the order of the reaction, the rate law, and the value of $k$ at 298 K.

**Collect and Organize** We are given experimental data describing the variation in concentration of ClO with time at 298 K and we are to determine the order of the decomposition reaction of ClO. Two of the choices we have are first and second order.

**Analyze** To distinguish between first and second order for a reaction in which ClO is the single reactant, we plot ln[ClO] versus time and 1/[ClO] versus time. If the ln[ClO] plot is linear, the reaction is first order; if the 1/[ClO] plot is linear, the reaction is second order. The rate law has the form

$$Rate = k[ClO]^m$$

where $m = 1$ or 2. The overall order of the reaction equals the exponent for [ClO] in the rate law. We determine the rate constant from the slope of whichever plot is linear.

**Solve** To evaluate the two possibilities, we need to calculate ln[ClO] and 1/[ClO] values; sets of both these values are given in Table 14.10(b). The graphs for ln[ClO] and 1/[ClO] versus time are shown in Figure 14.13. The ln[ClO] plot is not linear, but the 1/[ClO] plot is, meaning that the reaction is second order in ClO and second order overall. Thus, $m = 2$ and the rate law is

$$Rate = k[ClO]^2$$

**TABLE 14.10(a)** **Concentration of Chlorine Monoxide as Function of Time**

| Time (ms) | [ClO] (M) |
|---|---|
| 0 | $1.50 \times 10^{-8}$ |
| 10 | $7.19 \times 10^{-9}$ |
| 20 | $4.74 \times 10^{-9}$ |
| 30 | $3.52 \times 10^{-9}$ |
| 40 | $2.81 \times 10^{-9}$ |
| 100 | $1.27 \times 10^{-9}$ |
| 200 | $0.66 \times 10^{-9}$ |

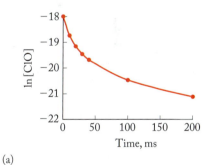

(a)

**TABLE 14.10(b)** **Concentration of Chlorine Monoxide, ln[ClO], and 1/[ClO] as Function of Time**

| Time (ms) | [ClO] (M) | ln[ClO] | 1/[ClO] (1/M) |
|---|---|---|---|
| 0 | $1.50 \times 10^{-8}$ | −18.015 | $6.67 \times 10^{7}$ |
| 10 | $7.19 \times 10^{-9}$ | −18.751 | $1.39 \times 10^{8}$ |
| 20 | $4.74 \times 10^{-9}$ | −19.167 | $2.11 \times 10^{8}$ |
| 30 | $3.52 \times 10^{-9}$ | −19.465 | $2.84 \times 10^{8}$ |
| 40 | $2.81 \times 10^{-9}$ | −19.690 | $3.56 \times 10^{8}$ |
| 100 | $1.27 \times 10^{-9}$ | −20.484 | $7.89 \times 10^{8}$ |
| 200 | $0.66 \times 10^{-9}$ | −21.139 | $1.51 \times 10^{9}$ |

Substituting [ClO] into the generic integrated rate law (Equation 14.27) gives

$$\frac{1}{[ClO]} = kt + \frac{1}{[ClO]_0}$$

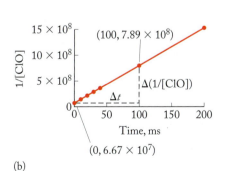

(b)

**FIGURE 14.13**

From the general equation $y = mx + b$, we know that $k$ is the slope of the graph of $1/[ClO]$ versus time. Arbitrarily choosing two convenient data points (at 0 and 100 ms), we can calculate $k$:

$$k = \text{slope} = \frac{\Delta y}{\Delta x} = \frac{\Delta\left(\frac{1}{[ClO]}\right)}{\Delta t}$$

$$= \frac{(7.89 - 0.667) \times 10^8 \ M^{-1}}{(100 - 0) \times 10^{-3} \ s}$$

$$= 7.22 \times 10^9 \ M^{-1} \ s^{-1}$$

As in Sample Exercise 14.5, we could also plot the data on a graphing calculator. By fitting the data to a straight line, we would discover that the graph of $1/[ClO]$ versus $t$ is a better fit and then find the slope of the line. This result allows us to conclude that the reaction is second order and calculate the rate constant $k$.

**Think About It** Sample Exercise 14.5 and this one show we can use integrated rate laws to distinguish between first- and second-order reactions in a single reactant.

**Practice Exercise** Experimental evidence shows that in the reaction

$$NO_2(g) + CO(g) \rightarrow NO(g) + CO_2(g)$$

the reaction rate depends only on the concentration of $NO_2$. Determine whether the reaction is first or second order in $NO_2$, and calculate the rate constant from the data in Table 14.11, which were obtained at 488 K.

---

**TABLE 14.11 Concentration of $NO_2$ as a Function of Time**

| Time (hr) | [$NO_2$] (M) |
|-----------|--------------|
| 0.00 | 0.250 |
| 1.39 | 0.198 |
| 3.06 | 0.159 |
| 4.72 | 0.132 |
| 6.39 | 0.114 |
| 8.06 | 0.099 |
| 9.72 | 0.088 |
| 11.39 | 0.080 |

---

The concept of half-life can also be applied to second-order reactions. The relation between rate constant and half-life for the decomposition of $NO_2$ can be derived from Equation 14.27 if we first rearrange the terms to solve for $kt$:

$$kt = \frac{1}{[X]} - \frac{1}{[X]_0}$$

After one half-life has elapsed ($t = t_{1/2}$), $[X]$ has decreased to half its initial concentration. Substituting this information into the preceding equation, we have

$$kt_{1/2} = \frac{1}{\frac{1}{2}[X]_0} - \frac{1}{[X]_0}$$

$$= \frac{2}{[X]_0} - \frac{1}{[X]_0} = \frac{1}{[X]_0}$$

or

$$t_{1/2} = \frac{1}{k[X]_0} \tag{14.28}$$

Note that this value of $t_{1/2}$ is inversely proportional to the initial concentration of X. This dependence on concentration is unlike the $t_{1/2}$ values of first-order reactions, which are independent of concentration.

## Pseudo-First-Order Reactions

The integrated rate law in Equation 14.27,

$$\frac{1}{[X]} = kt + \frac{1}{[X]_0}$$

applies only to reactions that are second order in a single reactant. It does not apply to reactions that are second order overall but are first order in two reactants, such as

$$NO(g) + O_3(g) \rightarrow NO_2(g) + O_2(g)$$

or

$$NO_2(g) + O_3(g) \rightarrow NO_3(g) + O_2(g)$$

Because the integrated rate law for a reaction that is first order in two reactants is complicated, kineticists frequently adjust reaction conditions so that a simpler rate law can be used. One approach is to have one of the reactants present at a much higher concentration than the other. This condition is common for many components in the urban atmosphere where, for example, ozone concentrations are often hundreds to thousands of times greater than NO concentrations. With such a large excess, the ozone concentration remains virtually constant over the course of the reaction

$$NO(g) + O_3(g) \rightarrow NO_2(g) + O_2(g)$$

Thus, the rate law for the reaction

$$\text{Rate} = k[NO][O_3]$$

may be simplified to

$$\text{Rate} = k'[NO] \tag{14.29}$$

where

$$k' = k[O_3]_0 \tag{14.30}$$

Here $[O_3]_0$ is the initial concentration of ozone, which remains virtually constant throughout the reaction.

Equation 14.29 looks like the rate law for a first-order reaction (Rate = $k[X]$). It is considered a **pseudo-first-order** rate law because the reaction *appears* to obey first-order kinetics. A pseudo-first-order reaction has the same integrated rate law as a first-order reaction, but the rate of the pseudo-first-order reaction depends on the concentration of more than one reactant.

This technique is frequently applied in biochemistry and medicine, especially when reactions of drugs with water are of interest. Active drug substances are usually present in a relatively small amount, whereas water is frequently present in great abundance. There is no hard and fast rule about when to apply pseudo-first-order methods, but any excess of around 100-fold should be sufficient to justify simplifying the treatment.

**pseudo-first-order** a reaction in which all the reactants but one are present at such high concentrations that they do not decrease significantly during the course of the reaction, so that reaction rate is controlled by the concentration of the limiting reactant.

---

**SAMPLE EXERCISE 14.9** **Deriving a Pseudo-First-Order Rate Law** **LO7**

The data in Table 14.12(a) were obtained in a study of the oxidation of trace levels of NO in the presence of a large excess of ozone at 298 K:

$$NO(g) + O_3(g) \rightarrow NO_2(g) + O_2(g)$$

(Note the NO concentration units in Table 14.12a. When the molar concentrations of reactants are very low, as in the case of many atmospheric pollutants, it is easier to express them in units of molecules/cm³.)

a. Verify that the reaction is pseudo-first-order and determine the pseudo-first-order rate constant $k'$.

b. If the ozone concentration is 100 times the NO concentration at $t = 0$, what is the second-order rate constant $k$? Express this $k$ in $M^{-1}\,s^{-1}$.

**Collect and Organize** Under pseudo-first-order conditions, a large excess of one reactant is maintained so that we can use the integrated rate law for a first-order reaction. We are to verify that the reaction is pseudo-first-order and to determine the pseudo-first-order

| TABLE 14.12(a) Concentration of NO as a Function of Time | |
|---|---|
| Time (μs) | Concentration of NO (molecules/cm³) |
| 0 | $1.00 \times 10^9$ |
| 100 | $8.36 \times 10^8$ |
| 200 | $6.98 \times 10^8$ |
| 300 | $5.83 \times 10^8$ |
| 400 | $4.87 \times 10^8$ |
| 500 | $4.07 \times 10^8$ |
| 1000 | $1.65 \times 10^8$ |

**TABLE 14.12(b) Concentration of NO and ln[NO] as a Function of Time**

| Time (μs) | [NO] (molecules/cm³) | ln[NO] |
|---|---|---|
| 0 | $1.00 \times 10^9$ | 20.723 |
| 100 | $8.36 \times 10^8$ | 20.544 |
| 200 | $6.98 \times 10^8$ | 20.364 |
| 300 | $5.83 \times 10^8$ | 20.184 |
| 400 | $4.87 \times 10^8$ | 20.004 |
| 500 | $4.07 \times 10^8$ | 19.824 |
| 1000 | $1.65 \times 10^8$ | 18.915 |

rate constant. We are given experimental data showing the change of concentration of NO with time. We are also given the initial concentration of $O_3$ and asked to calculate the second-order rate constant for the reaction.

**Analyze** We verify our assumption by treating the data as we would for a first-order reaction and plotting ln[NO] versus time. The plot should be a straight line with a slope equal to $-k'$. We can solve for the second-order rate constant $k$ using Equation 14.30. We expect the second-order rate constant to be rather large, or at least greater than 1, since the data (concentration changes) were collected on a microsecond scale.

**Solve**

a. The plot of ln[NO] values (Table 14.12b) versus time is linear as shown in Figure 14.14, which indicates that the reaction is indeed pseudo-first-order. We get the pseudo-first-order rate constant $k'$ by calculating the slope from the two points labeled in Figure 14.14:

$$\text{Slope} = -k' = \frac{\Delta y}{\Delta x} = \frac{(18.915 - 20.184)}{(1000 - 300)\ \mu s} = -1.81 \times 10^{-3}\ \mu s^{-1}$$

$$k' = 1.81 \times 10^{-3}\ \mu s^{-1} = 1.81 \times 10^3\ s^{-1}$$

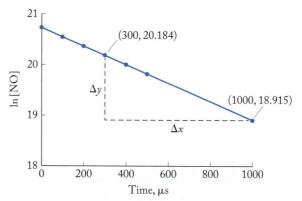

**FIGURE 14.14**

b. We calculate the second-order rate constant $k$ using Equation 14.30, $k' = k[O_3]_0$, and

$$[O_3]_0 = 100[NO]_0 = 100(1.00 \times 10^9\ \text{molecules/cm}^3)$$

Solving for $k$:

$$k = \frac{k'}{[O_3]_0} = \frac{1.81 \times 10^3\ s^{-1}}{1.00 \times 10^{11}\ \text{molecules/cm}^3} = \frac{1.81 \times 10^{-8}\ \text{cm}^3}{\text{molecules} \cdot s}$$

To convert the units to $M^{-1}\ s^{-1}$, we need to convert the reciprocal concentration units of $cm^3$/molecule into $M^{-1}$:

$$k = \frac{1.81 \times 10^{-8}\ \cancel{cm^3} \cdot s^{-1}}{\cancel{\text{molecules}}} \times \frac{6.022 \times 10^{23}\ \cancel{\text{molecules}}}{1\ \text{mol}} \times \frac{1\ L}{1000\ \cancel{cm^3}}$$

$$= 1.09 \times 10^{13}\ \frac{L \cdot s^{-1}}{\text{mol}} = 1.09 \times 10^{13}\ M^{-1}\ s^{-1}$$

**TABLE 14.13 Concentration of Chlorine Atoms as a Function of Time**

| Time (μs) | [Cl] (M) |
|---|---|
| 0 | $5.60 \times 10^{-14}$ |
| 100 | $5.27 \times 10^{-14}$ |
| 600 | $3.89 \times 10^{-14}$ |
| 1200 | $2.69 \times 10^{-14}$ |
| 1850 | $1.81 \times 10^{-14}$ |

**Think About It** As we predicted, the second-order rate constant for the reaction is greater than 1. The pseudo-first-order ($k'$) and second-order ($k$) rate constants are very different because $k'$ contains a term for the initial concentration of ozone, which is very small when expressed in moles per liter. The value of $k'$ will be different for every initial concentration of $O_3$, but the value of $k$ will be independent of $[O_3]_0$.

⚙ **Practice Exercise** The reaction

$$Cl(g) + O_3(g) \rightarrow ClO(g) + O_2(g)$$

is first order in both reactants. Determine the pseudo-first-order and second-order rate constants for the reaction from the data in Table 14.13 if the initial ozone concentration is $8.5 \times 10^{-11}\ M$.

**SAMPLE EXERCISE 14.10** **Identifying Pseudo-First-Order Conditions** **LO7**

A standard aspirin tablet in the United States contains 325 mg of acetylsalicylic acid ($\mathcal{M} = 180.15$ g/mol). Assume your stomach contains one aspirin tablet and exactly 55 mL of water. One molecule of acetylsalicylic acid reacts with one molecule of water to produce the actual drug substance (salicylic acid). If the reaction is first order in both reactants, do the conditions justify using pseudo-first-order kinetics to determine the value of the rate constant?

**Collect and Organize** We have the mass of a drug in a tablet and the volume of water in which it dissolves. We can calculate the numbers of moles of both reactants (drug and water) to determine if water is in 100-fold excess or more.

**Analyze** We have the volume of water but not the density of the solution. We can assume that it is close to 1.00 g/mL to calculate the mass of water and from that the number of moles. We estimate that we have more than 2 moles of water [55 g/(18 g/mol) $\sim$ 3] and less than about 0.01 mol of drug [0.325 g/(180 g/mol), or about 0.3/200 = 0.001] so we estimate that pseudo-first-order kinetics are suitable.

**Solve** The number of moles of water is

$$55 \text{ mL} \times \frac{1.00 \text{ g water}}{1 \text{ mL water}} \times \frac{1 \text{ mol}}{18.02 \text{ g}} = 3.05 \text{ mol water}$$

and the number of moles of aspirin is

$$0.325 \text{ g} \times \frac{1 \text{ mol}}{180.15 \text{ g}} = 1.8 \times 10^{-3} \text{ mol aspirin}$$

The ratio of the number of moles of water to aspirin is

$$\frac{3.05 \text{ mol}}{1.8 \times 10^{-3} \text{ mol}} = 1700$$

Water is 1700-fold in molar excess of aspirin, so we can definitely treat any data in terms of pseudo-first-order kinetics.

**Think About It** Our answer seems reasonable in light of our estimation. The kinetics of hydrolysis for most drugs taken orally probably follow pseudo-first-order kinetics.

⚙ **Practice Exercise** Another important issue involving pharmaceuticals is their shelf life. If aspirin tablets are left open to the humid air, they absorb water from the atmosphere. Suppose under extremely humid conditions, 1 aspirin tablet absorbs 1.0 g of water. Can we justify analyzing the hydrolysis reaction of 1 aspirin tablet and this quantity of absorbed water using pseudo-first-order kinetics?

**CONCEPT TEST**

A student measured the pseudo-first-order rate constant for the reaction of NO with $O_3$ in Sample Exercise 14.9 at four initial concentrations of ozone. Assuming that all four $[O_3]_0$ values were much greater than [NO], how could the student determine the second-order rate constant graphically?

## Zero-Order Reactions

In the Practice Exercise accompanying Sample Exercise 14.8, we introduced the reaction

$$NO_2(g) + CO(g) \rightarrow NO(g) + CO_2(g)$$

The solution revealed that the rate law for the reaction is

$$\text{Rate} = k[NO_2]^2 \qquad (14.31)$$

The rate of the reaction does not change with changing [CO], even when the concentrations of CO and $NO_2$ are comparable. This situation is not the same as in pseudo-first-order reactions, where the rate does not depend on the concentration of a reactant because that reactant is present in large excess.

One interpretation of Equation 14.31 is that it contains a [CO] term to the zeroth power, making the reaction *zero order* in that reactant. Because any value raised to the zeroth power is 1, we have

$$\text{Rate} = k[NO_2]^2[CO]^0 = k[NO_2]^2(1) = k[NO_2]^2$$

Reactions with a true zero-order rate law are rare, but let's consider a generic reaction involving a single reactant A, with the reaction zero order in reactant A and zero order overall. For the reaction

$$A \rightarrow B$$

the rate law is

$$\text{Rate} = -\Delta[A]/\Delta t = k[A]^0 = k$$

and the integrated rate law is

$$[A] = -kt + [A]_0$$

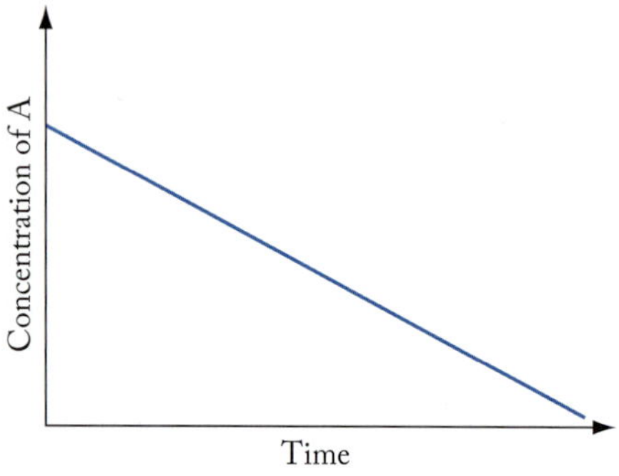

**FIGURE 14.15** The change in concentration of the reactant A in the zero-order reaction $A \rightarrow B$ is constant over time.

The slope of a plot of reactant concentration versus time (Figure 14.15) equals the negative of the zero-order rate constant $k$.

We can calculate the half-life of a zero-order reaction by substituting $t = t_{1/2}$ and $[A] = [A]_0/2$ into the integrated rate law:

$$[A]_0/2 = -kt_{1/2} + [A]_0$$
$$kt_{1/2} = [A]_0 - [A]_0/2 = [A]_0/2$$
$$t_{1/2} = [A]_0/2k$$

For now we leave the discussion of zero-order reactions with this purely mathematical treatment. We return to these reactions and examine their meaning at the molecular level in Section 14.5.

## 14.4 Reaction Rates, Temperature, and the Arrhenius Equation

**activation energy ($E_a$)** the minimum energy molecules need to react when they collide.

**Arrhenius equation** relates the rate constant of a reaction to absolute temperature ($T$), the activation energy of the reaction ($E_a$), and the frequency factor ($A$).

**frequency factor ($A$)** the product of the frequency of molecular collisions and a factor that expresses the probability that the orientation of the molecules is appropriate for a reaction to occur.

Why do rate constants for different reactions have different values? In Section 14.3, we introduced the idea that chemical reactions take place in the gas phase when molecules collide with sufficient energy to break bonds in reactants and allow bonds to form in products. The minimum amount of energy that enables this to happen is called the **activation energy ($E_a$)**. Every chemical reaction has a characteristic activation energy, usually expressed in kilojoules per mole. Activation energy is an energy barrier that must be overcome if a reaction is to proceed—like the mountain passes that must be climbed to hike the trails connecting two valleys shown in Figure 14.16. A hiker may choose to climb over a higher pass (a higher activation energy barrier), but reacting molecules go over the lowest barriers available. Generally, the greater the activation energy for a reaction, the slower the reaction.

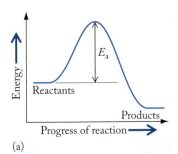

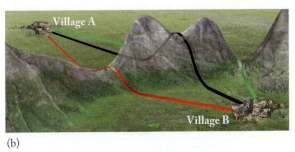

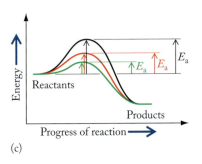

(a)

(b)

(c)

**FIGURE 14.16** (a) The energy profile of a reaction includes an activation energy barrier $E_a$ that must be overcome before the reaction can proceed. (b) A real-world analogy confronts a hiker climbing over mountain passes to get to a village in the next valley. (c) Although the hiker may choose one of the steeper routes (black or red), a molecule always goes over the lowest barrier to products (green). Note that the lowest barrier in going from Village A to Village B is also the lowest barrier for the return trip from B to A.

According to kinetic molecular theory, the fraction of molecules with kinetic energies greater than a given activation energy increases with increasing temperature, as shown in Figure 14.17(a). If the activation energy $E_a$ for a particular process is defined by the kinetic energy (KE), indicated by the dashed line, then the fraction of molecules in a given sample that have that amount of KE or more is defined by the area under the distribution curve at or above that value. Notice that at the higher temperature $T_2$, the total area under the curve (the purple area plus the green area in Figure 14.17a) is larger than the area under the curve at the lower temperature $T_1$ (the purple area alone). Because of this change, the rates of chemical reactions should increase with increasing temperature, and indeed they do (Figure 14.17b).

> **CONCEPT TEST** ···········································································
>
> Why does increasing temperature increase the frequency of collisions between molecules in the gas phase?
>
> ···········································································

In the late 19th century, experiments carried out in the laboratories of Jacobus van 't Hoff and the Swedish chemist Svante Arrhenius (1859–1927) led to a fundamental advance in understanding how temperature affects the rates of chemical reactions. The mathematical connection between temperature, the rate constant $k$ for a reaction, and its activation energy is given by the **Arrhenius equation**:

$$k = Ae^{-E_a/RT} \qquad (14.32)$$

where $R$ is the gas constant in J/(mol · K) and $T$ is the reaction temperature in kelvin. The factor $A$, called the **frequency factor**, is the product of collision frequency and a term that accounts for the fact that not every collision results in a chemical reaction.

Some collisions do not lead to products because the colliding molecules are not oriented relative to each other in the right way. To examine the importance of molecular orientation during collisions, let's revisit the reaction (Equation 14.7) between $O_3$ and NO:

$$O_3(g) + NO(g) \rightarrow O_2(g) + NO_2(g)$$

Two ways in which ozone and nitric oxide molecules might approach each other are shown in Figure 14.18. Only one of these orientations, the one in which an

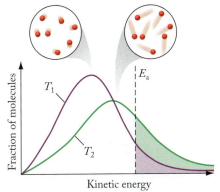

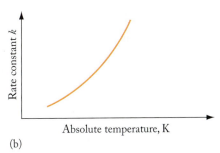

(a)

■ Fraction of molecules in sample with sufficient energy to react at $T_1$

■ Increase in number of molecules in sample with sufficient energy to react at $T_2$; $T_2 > T_1$

(b)

**FIGURE 14.17** (a) According to kinetic molecular theory, a fraction of reactant molecules have kinetic energies equal to or greater than the activation energy ($E_a$) of the reaction. As temperature increases from $T_1$ to $T_2$, the number of molecules with energies exceeding $E_a$ increases, leading to an increase in reaction rate. (b) The rate constant for any reaction increases with increasing temperature.

**CONNECTION** The van 't Hoff factor in Chapter 11 is named after the same Jacobus van 't Hoff who worked on the temperature dependence of reaction rates.

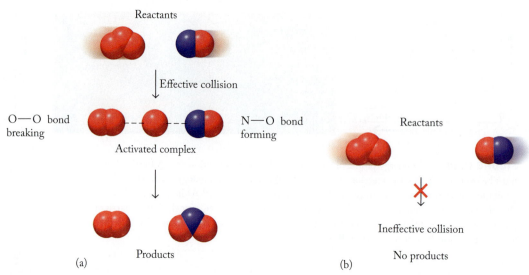

**FIGURE 14.18** The effect of molecular orientation on reaction rate. (a) When an $O_3$ and a NO molecule are oriented such that the collision is between an $O_3$ oxygen and the NO nitrogen, the collision is effective and an activated complex forms, which then yields the two product molecules $NO_2$ and $O_2$. (b) When the reactant molecules are oriented such that the collision is between an $O_3$ oxygen and the NO oxygen, no activated complex forms and no reaction occurs.

▶❚❚ **CHEMTOUR** Arrhenius Equation

▶❚❚ **CHEMTOUR** Collision Theory

**activated complex** a species formed in a chemical reaction when molecules have enough energy to react with each other.

**transition state** a high-energy state between reactants and products in a chemical reaction.

**energy profile** a graph showing the changes in potential energy for a reaction as a function of the progress of the reaction from reactants to products.

$O_3$ molecule collides with the nitrogen atom of NO, leads to a chemical reaction between the two molecules.

A collision between $O_3$ and NO molecules with the correct orientation and enough kinetic energy may result in the formation of the **activated complex** shown in Figure 14.18(a). In this species, one of the O—O bonds in the $O_3$ molecule has started to break and the new N—O bond is beginning to form. Activated complexes represent midway points in chemical reactions. They have extremely brief lifetimes and fall apart rapidly, either forming products or re-forming reactants. Activated complexes are formed by reacting species that have acquired enough energy to react with each other. The internal energy of an activated complex represents a high-energy **transition state** of the reaction. In fact, the energy of an activated complex for a reaction defines the height of the activation energy barrier for the reaction. The magnitudes of activation energies can vary from a few kilojoules to hundreds of kilojoules per mole.

We can draw an **energy profile** for a chemical reaction that shows the changes in energy for the reaction as a function of the *progress of the reaction* from reactants to products. We used similar energy profiles in Chapter 5 (Figure 5.33) when we discussed enthalpy changes in reactions. We have already seen one energy profile here in Figure 14.16(a). Now consider the energy profile for the reaction between nitric oxide and ozone, shown in Figure 14.19(a). The *x*-axis represents the progress of the reaction and the *y*-axis represents chemical energy. The activation energy is equivalent to the difference in energy between the transition state and the reactants. The size of the activation energy barrier depends on the direction from which it is approached. In the forward direction (NO + $O_3$ → $NO_2$ + $O_2$, Figure 14.19a), $E_a$ is smaller than in the reverse direction ($NO_2$ + $O_2$ → NO + $O_3$, Figure 14.19b). A smaller activation energy barrier means that the forward reaction proceeds at a higher rate than the reverse reaction if we have equal concentrations of reactants and products.

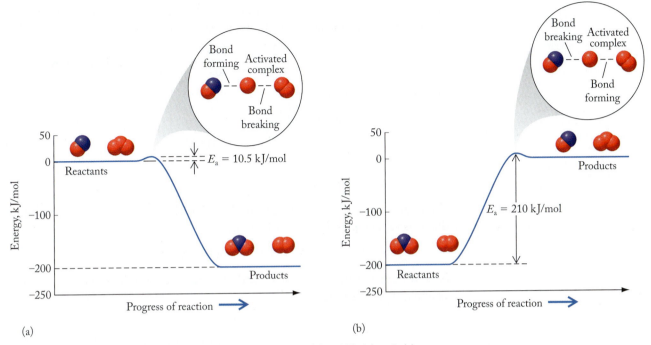

(a)

(b)

**FIGURE 14.19** (a) The energy profile for the reaction $NO(g) + O_3(g) \rightarrow NO_2(g) + O_2(g)$ includes an activation energy barrier of 10.5 kJ/mol. (b) The reverse reaction has a much larger activation energy of 210 kJ/mol.

One of the many uses of the Arrhenius equation is to calculate the value of $E_a$ for a chemical reaction. When we take the natural logarithm of both sides of Equation 14.32,

$$\ln k = -\frac{E_a}{R}\left(\frac{1}{T}\right) + \ln A \qquad (14.33)$$

the result fits the general equation of a straight line ($y = mx + b$) if we make ($\ln k$) the $y$ variable and ($1/T$) the $x$ variable. We can calculate $E_a$ by determining the rate constant $k$ for the reaction at several temperatures. Plotting $\ln k$ versus $1/T$ should give a straight line, the slope of which is $-E_a/R$. Table 14.14 and Figure 14.20 show

| TABLE 14.14 Temperature Dependence of the Rate of Reaction for the Reaction $NO(g) + O_3(g) \rightarrow NO_2(g) + O_2(g)$ | | | |
|---|---|---|---|
| **T (K)** | **k (M$^{-1}$ s$^{-1}$)** | **ln k** | **1/T (K$^{-1}$)** |
| 300 | $1.21 \times 10^{10}$ | 23.216 | $3.33 \times 10^{-3}$ |
| 325 | $1.67 \times 10^{10}$ | 23.539 | $3.08 \times 10^{-3}$ |
| 350 | $2.20 \times 10^{10}$ | 23.814 | $2.86 \times 10^{-3}$ |
| 375 | $2.79 \times 10^{10}$ | 24.052 | $2.67 \times 10^{-3}$ |
| 400 | $3.45 \times 10^{10}$ | 24.264 | $2.50 \times 10^{-3}$ |
| 425 | $4.15 \times 10^{10}$ | 24.449 | $2.35 \times 10^{-3}$ |

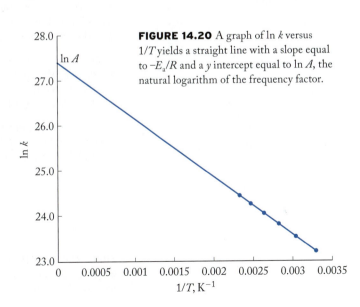

**FIGURE 14.20** A graph of ln $k$ versus $1/T$ yields a straight line with a slope equal to $-E_a/R$ and a $y$ intercept equal to ln $A$, the natural logarithm of the frequency factor.

data for the reaction between NO and $O_3$ at six temperatures. Arbitrarily picking two points on the line in Figure 14.20, we calculate its slope:

$$\text{Slope} = \frac{\Delta y}{\Delta x} = \frac{(23.814 - 24.264)}{(2.86 \times 10^{-3} - 2.50 \times 10^{-3})\ \text{K}^{-1}} = -1.25 \times 10^3\ \text{K}$$

The slope equals $-E_a/R$, so

$$E_a = -\text{slope} \times R$$

$$= -(-1.25 \times 10^3\ \text{K}) \times \left(\frac{8.314\ \text{J}}{\text{mol} \cdot \text{K}}\right) = 1.04 \times 10^4\ \text{J/mol}$$

$$= 10.4\ \text{kJ/mol}$$

The $y$ intercept $(1/T = 0)$ in Figure 14.20 is 27.41. From Equation 14.33, we know that this value represents $\ln A$, which means

$$A = e^{27.41} = 8.0 \times 10^{11}$$

Now we can use the values of $E_a$ and $A$ to calculate $k$ at any temperature. For example, at $T = 250$ K:

$$k = Ae^{-E_a/RT}$$

$$= (8.0 \times 10^{11})\ e^{-\left|\frac{1.04 \times 10^4\ \text{J/mol}}{8.314\ \frac{\text{J}}{\text{mol} \cdot \text{K}} \cdot 250\ \text{K}}\right|}$$

$$= 5.4 \times 10^9\ M^{-1}\ \text{s}^{-1}$$

---

**SAMPLE EXERCISE 14.11** **Calculating an Activation Energy from Rate Constants** **LO8**

The data in Table 14.15(a) were collected in a study of the effect of temperature on the rate of the decomposition reaction

$$2\ ClO(g) \rightarrow Cl_2(g) + O_2(g)$$

Determine the activation energy for the reaction.

**Collect and Organize** We can calculate activation energy by using the Arrhenius equation. We are given values of the rate constant $k$ as a function of absolute temperature. According to Equation 14.33, the slope of a plot of $\ln k$ against $1/T$ is equal to $-E_a/R$, where $E_a$ is the activation energy and $R$ is the ideal gas constant with units of J/(mol · K).

**Analyze** First, we need to convert the $T$ and $k$ values from Table 14.15(a) to $1/T$ and $\ln k$, respectively. We predict a positive value for $E_a$ because activation energy represents a barrier that costs energy to overcome. The rate constant for the reaction is fairly large, about $10^9\ M^{-1}\ \text{s}^{-1}$, so we predict that the activation energy barrier will be relatively low.

**Solve** Expanding the data table to include columns for $1/T$ and $\ln k$ yields Table 14.15(b). A plot of $\ln k$ versus $1/T$ gives us a straight line, the slope of which is $-1590$ K (Figure 14.21). We use the values of the slope and $R$ [8.314 J/(mol · K)] to calculate the value of $E_a$:

$$E_a = -\text{slope} \times R$$

$$= -(-1590\ \text{K}) \times \left(\frac{8.314\ \text{J}}{\text{mol} \cdot \text{K}}\right) = 1.3 \times 10^4\ \text{J/mol}$$

$$= 13\ \text{kJ/mol}$$

**Think About It** The activation energy is the height of a barrier that has to be overcome before a reaction can proceed. As predicted, the activation energy for ClO decomposition has a small positive value.

**TABLE 14.15(a)** **Rate Constant as a Function of Temperature for the Decomposition of ClO**

| $k$ ($M^{-1}$ s$^{-1}$) | $T$ (K) |
|---|---|
| $1.9 \times 10^9$ | 238 |
| $3.1 \times 10^9$ | 258 |
| $4.9 \times 10^9$ | 278 |
| $7.2 \times 10^9$ | 298 |

| TABLE 14.15(b) | Summary of *T*, 1/*T*, *k*, and ln *k* for the Decomposition of ClO | | |
|---|---|---|---|
| *T* (K) | 1/*T* (K$^{-1}$) | *k* (M$^{-1}$ s$^{-1}$) | ln *k* |
| 238 | $4.20 \times 10^{-3}$ | $1.9 \times 10^{9}$ | 21.365 |
| 258 | $3.88 \times 10^{-3}$ | $3.1 \times 10^{9}$ | 21.855 |
| 278 | $3.60 \times 10^{-3}$ | $4.9 \times 10^{9}$ | 22.313 |
| 298 | $3.36 \times 10^{-3}$ | $7.2 \times 10^{9}$ | 22.697 |

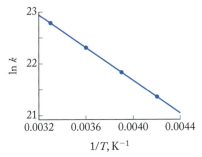

**FIGURE 14.21**

**Practice Exercise** The rate constant for the reaction

$$Br(g) + O_3(g) \rightarrow BrO(g) + O_2(g)$$

was determined at the four temperatures shown in Table 14.16. Calculate the activation energy for this reaction.

| TABLE 14.16 | Rate Constant as a Function of Temperature for the Reaction of Br with $O_3$ | |
|---|---|
| *T* (K) | *k* [cm$^3$/(molecule · s)] |
| 238 | $5.9 \times 10^{-13}$ |
| 258 | $7.7 \times 10^{-13}$ |
| 278 | $9.6 \times 10^{-13}$ |
| 298 | $1.2 \times 10^{-12}$ |

Calculation of activation energies using the graphical method generally requires measurement of the rate constant at a minimum of three different temperatures to verify that the plot of ln *k* versus 1/*T* is a straight line. Once we know the value of $E_a$, we can use it and the value of the rate constant ($k_1$) of a reaction at one temperature ($T_1$) to calculate the value of the rate constant ($k_2$) at another temperature ($T_2$). We start by substituting $k_1$, $k_2$, $T_1$, and $T_2$ into Equation 14.33:

$$\ln k_1 = -\frac{E_a}{R}\left(\frac{1}{T_1}\right) + \ln A \qquad \ln k_2 = -\frac{E_a}{R}\left(\frac{1}{T_2}\right) + \ln A$$

Subtracting these two expressions, ln $k_1$ − ln $k_2$, gives

$$\ln k_1 - \ln k_2 = \left[-\frac{E_a}{R}\left(\frac{1}{T_1}\right) + \ln A\right] - \left[-\frac{E_a}{R}\left(\frac{1}{T_2}\right) + \ln A\right]$$

Using the mathematical properties of logarithms, we can rearrange the terms to obtain this equation:

$$\ln\frac{k_1}{k_2} = \frac{E_a}{R}\left(\frac{1}{T_2}\right) - \frac{E_a}{R}\left(\frac{1}{T_1}\right)$$

$$= \frac{E_a}{R}\left(\frac{1}{T_2} - \frac{1}{T_1}\right) \qquad (14.34)$$

**CONCEPT TEST**

Which of the following statements is/are true about activation energies?
a. Exothermic reactions have negative activation energies.
b. Fast reactions have large rate constants *and* large activation energies.
c. The forward reaction sometimes has a lower activation energy than the reverse reaction.
d. Endothermic reactions always have large activation energies.

# 14.5 Reaction Mechanisms

Up to this point we have described reactions in terms of macroscopically observable quantities such as pressure, temperature, volume, and numbers of moles of substances. In this section we explore how reactions proceed at the molecular level. Chemists think about the behavior of matter at the molecular level, and the attempt to explain observations of reaction rates by considering the structure of particles at the molecular level is another way of relating bulk properties of matter to molecular structure. Thus, the microscopic interpretation of a kinetic process also enables us to refine our view of molecular structure. Being able to suggest what goes on at the molecular level by observing macroscopic properties is a remarkable achievement. The evaluation of the stepwise behavior of molecules—the mechanism of a reaction—provides valuable insights into reactions, ranging from the subtlest biochemical processes taking place in cells to the chemistry of industrial processes that yield tons of product.

▶‖ **CHEMTOUR** Reaction Mechanisms

Let's start by revisiting the thermal decomposition of $NO_2$:

$$2\ NO_2(g) \rightarrow 2\ NO(g) + O_2(g)$$

We noted in Section 14.3 that this reaction is second order in $NO_2$ because it takes place as a result of the collisions of *pairs* of $NO_2$ molecules. How do the atoms in two colliding $NO_2$ molecules rearrange themselves to form two NO molecules and one $O_2$ molecule? The answer to this question is contained in the mechanism of the reaction. A **reaction mechanism** describes the stepwise manner in which the bonds in reactant molecules break and the bonds in product molecules form. Chemists use the results of reaction rate measurements to develop explanations of how reactions actually happen. Sometimes they can test for the presence of the products formed in preliminary steps, but, until recently, this was not possible for most reactions. Today, technological advances allow chemists to follow the transformation of reactants to products in the time that it takes for individual covalent bonds to break and new ones to form. These processes occur as rapidly as the bonds vibrate, that is, in femtoseconds ($10^{-15}$ s), so the area of research based on monitoring these ultrafast processes is called *femtochemistry*.[2]

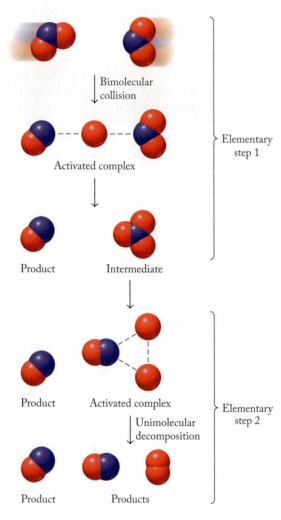

## Elementary Steps

A reaction mechanism proposed for the decomposition of $NO_2$ is shown in Figure 14.22. In the first step of the mechanism, a collision between two $NO_2$ molecules produces a molecule of NO and a molecule of $NO_3$. In a second step, the $NO_3$ decomposes to NO and $O_2$. Both steps in the mechanism involve very short-lived activated complexes. In the activated complex of the first step, two molecules share an oxygen atom. The bonds in the activated complex of the second step rearrange so that two oxygen atoms bond together, forming a molecule of $O_2$ and leaving behind a molecule of NO. The molecule $NO_3$ is an **intermediate** in this mechanism because it is produced in one step and consumed in the next. Intermediates are not considered reactants or products and do not appear in the equation describing a reaction. In some cases, intermediates in chemical reactions are sufficiently long-lived to be isolated. In

**Overall reaction:**

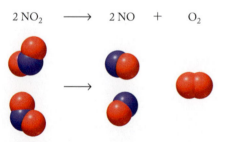

**FIGURE 14.22** The decomposition of $NO_2$ begins when two $NO_2$ molecules collide, producing NO and $NO_3$ (elementary step 1). The $NO_3$ then rapidly decomposes into NO and $O_2$ (elementary step 2).

[2]In 1999 Ahmed H. Zewail received the Nobel Prize in Chemistry for his pioneering research in the development of femtochemistry.

contrast, activated complexes have never been isolated, although they have been detected.

This reaction mechanism is a combination of two **elementary steps**. An elementary step that involves a single molecule is called **unimolecular**, and one that involves a collision between two molecules is **bimolecular**. Bimolecular elementary steps are much more common than **termolecular** (three-molecule) elementary steps because the chance of three molecules colliding at exactly the same time is much smaller. The terms *uni-*, *bi-*, and *termolecular* are used by chemists to describe the **molecularity** of an elementary step, which refers to the number of atoms, ions, or molecules involved in that step.

A valid reaction mechanism must be consistent with the stoichiometry of the reaction. In other words, the sum of the elementary steps in Figure 14.22 must be consistent with the observed proportions of reactants and products as defined in the balanced chemical equation. In this case, the sum matches the overall stoichiometry:

Elementary step 1 $\quad$ $2\,NO_2(g) \rightarrow NO(g) + NO_3(g)$

Elementary step 2 $\quad\quad$ $NO_3(g) \rightarrow NO(g) + O_2(g)$

Summing the two elementary steps and simplifying by cancelling out the intermediate ($NO_3$) terms:

$$2\,NO_2(g) + \cancel{NO_3(g)} \rightarrow 2\,NO(g) + \cancel{NO_3(g)} + O_2(g)$$

we get the overall reaction

$$2\,NO_2(g) \rightarrow 2\,NO(g) + O_2(g)$$

Before ending this discussion, let's consider how activation energy applies to a two-step reaction such as this one. The two elementary steps produce an energy profile with two maxima. In elementary step 1, collisions between pairs of $NO_2$ molecules result in the formation of an activated complex associated with the first transition state in Figure 14.23. As this activated complex transforms into NO and $NO_3$, the energy of the system drops to the bottom of the trough between the two maxima. In elementary step 2, $NO_3$ forms the activated complex associated with the second transition state. As this complex transforms into the final products NO and $O_2$, the energy of the system drops to its final level.

Figure 14.23 shows that the energy barrier for elementary step 1 is much greater than that for elementary step 2. This difference is consistent with the relative rates of the two steps: step 1 is slower than step 2. If the reaction were to proceed in the reverse direction (as NO and $O_2$ react, forming $NO_2$), the first energy barrier would be the smaller of the two, and the first elementary step would be the more rapid one. Experimental evidence supports these expectations.

**CONCEPT TEST** ··········

For a reaction mechanism with $n$ elementary steps, how many activation energies does the overall reaction have?

**reaction mechanism** a set of steps that describe how a reaction occurs at the molecular level; the mechanism must be consistent with the rate law for the reaction.

**intermediate** a species produced in one step of a reaction and consumed in a subsequent step.

**elementary step** a molecular view of a single process taking place in a chemical reaction.

**unimolecular step** a step in a reaction mechanism involving only one molecule on the reactant side.

**bimolecular step** a step in a reaction mechanism involving a collision between two molecules.

**termolecular step** a step in a reaction mechanism involving a collision among three molecules.

**molecularity** the number of ions, atoms, or molecules involved in an elementary step in a reaction.

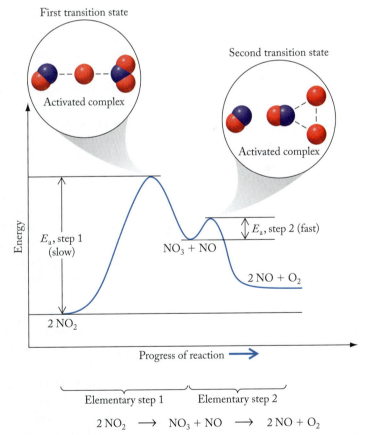

**FIGURE 14.23** The energy profile for the decomposition of $NO_2$ to NO and $O_2$ shows activation energy barriers for both elementary steps. The activation energy of the first step is larger than that of the second step, so the first step is the slower of the two.

## Rate Laws and Reaction Mechanisms

Any mechanism proposed for a reaction must be consistent with the rate law derived from experimental data. Is the mechanism proposed in Figure 14.22 consistent with the reaction's second-order rate law? For a reaction mechanism to be consistent with a rate law, the molecularity of one of the elementary steps in the mechanism must be the same as the reaction order expressed in the rate law.

As noted in Section 14.3, the experimentally determined rate law for $NO_2$ decomposition is second order in $NO_2$:

$$Rate = k[NO_2]^2 \tag{14.35}$$

Recall that the balanced chemical equation for the overall reaction does not give us enough information to write the rate law for the reaction. However, for any *elementary step* in a reaction mechanism, we can use the balanced chemical equation to write a rate law for that step. For the decomposition of $NO_2$, the rate law for the first elementary step, $2\ NO_2(g) \rightarrow NO_3(g) + NO(g)$, is

$$Rate_1 = k_1[NO_2]^2 \tag{14.36}$$

We obtain this rate law by writing the concentration for each reactant raised to a power equal to the coefficient of that reactant in the balanced equation. For the second elementary step in the $NO_2$ decomposition, $NO_3(g) \rightarrow NO(g) + O_2(g)$, the sole reactant, $NO_3$, has a coefficient of 1, so the rate law is

$$Rate_2 = k_2[NO_3] \tag{14.37}$$

How do we use the rate laws in Equations 14.36 and 14.37 to determine the validity of the mechanism, that is, to show that they conform to the observed rate law for the decomposition of $NO_2$ in Equation 14.35?

The two steps in the mechanism proceed at different rates, with different activation energies. One of the steps is slower than the other, and this is the **rate-determining step** in the reaction. The rate-determining step is the slowest elementary step in a chemical reaction, and the rate of this step controls the overall reaction rate. We can use the rate laws for the two steps to identify the rate-determining step. Because the rate law for step 1 matches the experimentally determined rate law, we may assume that step 1 defines how rapidly the reaction proceeds.

One way of visualizing the concept of a rate-determining step is to analyze the flow of people through a busy airport. Many travelers arrive at the airport with their boarding passes in hand or print them from convenient kiosks that allow them to avoid lines at ticket counters and move quickly. The next step in the process is passing through security. Typically, the number of people in the line outside the security point greatly exceeds the number of available security gates, so that the time required to make it to a flight depends mostly on the time needed to pass through security. Security screening is the rate-determining step on the way through the airport.

If the first step in the mechanism in Figure 14.22 is the rate-determining step, then the value of $k$ in Equation 14.35 is equal to $k_1$ from Equation 14.36. In addition, the value of $k_1$ must be smaller than the value of $k_2$. $NO_3$ is sufficiently stable that we can make small amounts of it and test whether $k_1 < k_2$. Experiments run at 300 K starting with $NO_2$ or $NO_3$ yield these values: $k_1 \approx 1 \times 10^{-10}\ M^{-1}\ s^{-1}$ and $k_2 \approx 6.3 \times 10^4\ s^{-1}$. Therefore, as soon as any $NO_3$ forms in step 1, it rapidly falls apart to NO and $O_2$ in step 2.

**rate-determining step** the slowest step in a multistep chemical reaction.

Now let's consider the reverse reaction of $NO_2$ decomposition, namely, the formation of $NO_2$ from NO and $O_2$:

$$\text{Overall reaction} \qquad 2\,NO(g) + O_2(g) \rightarrow 2\,NO_2(g)$$

We determined in Section 14.3 that this reaction is second order in NO and first order in $O_2$:

$$\text{Rate} = k[NO]^2[O_2] \qquad (14.38)$$

One proposed mechanism is shown in Figure 14.24. It has two elementary steps:

Step 1 $\quad NO(g) + O_2(g) \rightarrow NO_3(g) \qquad$ Rate $= k_1[NO][O_2] \qquad$ (14.39)

Step 2 $\quad NO_3(g) + NO(g) \rightarrow 2\,NO_2(g) \qquad$ Rate $= k_2[NO_3][NO] \qquad$ (14.40)

We obtained the rate laws in Equations 14.39 and 14.40 by writing the concentration for each reactant raised to a power equal to the reactant's coefficient in the balanced equation. If step 1 were the rate-determining step, the reaction would be first order in NO and $O_2$, but that is not what the experimentally determined rate law indicates. If step 2 were the rate-determining step, the reaction would be first order in NO and $NO_3$, but that is not consistent with the rate law either. So, how can we account for the experimental rate law?

Consider what happens if step 2 is slow while step 1 is fast *and reversible*, which means $NO_3$ forms rapidly from NO and $O_2$ but decomposes just as rapidly back into NO and $O_2$. Expressing these rates in equation form:

$$\text{Rate of forward reaction} = k_f[NO][O_2] = \text{fast}$$

$$\text{Rate of reverse reaction} = k_r[NO_3] = \text{equally fast}$$

The subscripts "f" and "r" refer to the forward and reverse reactions, respectively.

Combining these two expressions, we have

$$k_f[NO][O_2] = k_r[NO_3]$$

$$[NO_3] = \frac{k_f}{k_r}[NO][O_2] \qquad (14.41)$$

However, if we replace the $[NO_3]$ term in the rate law of Equation 14.40 (which we select because step 2 is the rate-determining step) with the right side of Equation 14.41, we get

$$\text{Rate} = k_2\frac{k_f}{k_r}[NO]^2[O_2]$$

The three rate constants can be combined,

$$k_{overall} = k_2\frac{k_f}{k_r}$$

and the rate law for the overall reaction becomes

$$\text{Rate} = k_{overall}[NO]^2[O_2]$$

This expression matches the overall rate law in Equation 14.38, so the proposed mechanism—a fast and reversible step 1 followed by a slow step 2—may be valid.

As a final point regarding reaction mechanisms, even though the proposed reaction mechanism is consistent with the overall stoichiometry of the reaction and with the experimentally derived rate law, these consistencies do not *prove* that the proposed mechanism is correct. On the other hand, *not* finding a reactive (and transient) intermediate would not necessarily disprove a reaction mechanism.

(1) Formation of intermediate: elementary step 1

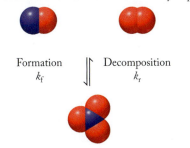

Formation
$k_f$

Decomposition
$k_r$

(2) Reaction of intermediate: elementary step 2

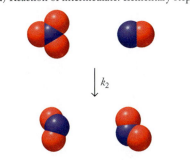

$k_2$

**Overall reaction:**

$$2\,NO \quad + \quad O_2 \quad \longrightarrow \quad 2\,NO_2$$

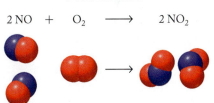

**FIGURE 14.24** A mechanism for the formation of $NO_2$ from NO and $O_2$ has two elementary steps: (1) a fast, reversible bimolecular reaction in which NO and $O_2$ form $NO_3$ and (2) a slower, rate-determining bimolecular reaction in which $NO_3$ reacts with a molecule of NO to form two molecules of $NO_2$.

**CONCEPT TEST** ...............................................................

Could the products of an elementary step in a chemical reaction include activated complexes?

...................................................................................

**SAMPLE EXERCISE 14.12** **Linking Reaction Mechanisms** **LO9**
**to Experimental Rate Laws**

The experimentally determined rate law for the reduction reaction

$$2\,NO(g) + 2\,H_2(g) \rightarrow N_2(g) + 2\,H_2O(g)$$

which occurs at high temperatures, is

$$\text{Rate} = k[NO]^2[H_2]$$

A proposed mechanism for the reaction is

Elementary step 1    $2\,NO(g) + H_2(g) \rightarrow N_2O(g) + H_2O(g)$
Elementary step 2    $N_2O(g) + H_2(g) \rightarrow N_2(g) + H_2O(g)$

Is this reaction mechanism consistent with the stoichiometry of the reaction and with the rate law? If so, which is the rate-determining step?

**Collect and Organize** We can use the two elementary steps to determine if the proposed mechanism is consistent with the reaction stoichiometry. To see if the mechanism is consistent with the experimentally determined rate law, we need to find the rate-determining step.

**Analyze** To answer the questions posed in this exercise, we need to
1. Determine whether the chemical equations of the elementary steps add up to the overall reaction equation.
2. Write rate laws for each elementary step.
3. Compare the rate laws of the elementary steps to the experimental rate law of the overall reaction to assess the validity of the mechanism.
4. Decide which step is rate determining by matching its rate law to the observed rate law.

**Solve** Let's test whether the elementary steps add up to the overall reaction:

(1)     $2\,NO(g) + H_2(g) \rightarrow \cancel{N_2O(g)} + H_2O(g)$
(2)     $\cancel{N_2O(g)} + H_2(g) \rightarrow N_2(g) + H_2O(g)$
-----------------------------------------------------------------
         $2\,NO(g) + 2\,H_2(g) \rightarrow N_2(g) + 2\,H_2O(g)$

This is indeed the equation of the overall reaction. The elementary steps are consistent with the stoichiometry of the overall reaction.

Next we need to focus on the reaction mechanism. Elementary step 1 involves two molecules of NO colliding with one molecule of $H_2$ in a termolecular reaction. Elementary step 2 is a bimolecular reaction between the $N_2O$ produced in step 1 and another molecule of $H_2$. We apply the fact that the rate law for any elementary step can be written directly from the balanced equations, using the equation coefficients as exponents in the rate law:

Step 1    $2\,NO(g) + H_2(g) \rightarrow N_2O(g) + H_2O(g)$    $\text{Rate} = k_1[H_2][NO]^2$
Step 2    $N_2O(g) + H_2(g) \rightarrow N_2(g) + H_2O(g)$    $\text{Rate} = k_2[H_2][N_2O]$

The rate law of step 1 matches the observed rate law of the overall reaction. Therefore, the proposed two-step mechanism is consistent with the experimental rate law, and step 1 is the rate-determining step.

**Think About It** We do not have direct proof that this is the correct mechanism, though it is consistent with the available data. A plausible mechanism for a reaction must yield a balanced equation whose stoichiometry matches the overall reaction and must be consistent with the rate law for the overall process. Both of these conditions are met in the solution to this exercise.

**Practice Exercise** The following mechanism is proposed for a reaction between compounds A and B:

| | | |
|---|---|---|
| Step 1 | $2\,A(g) + B(g) \rightleftharpoons C(g)$ | fast and reversible |
| Step 2 | $B(g) + C(g) \rightarrow D(g)$ | slow |
| Overall | $2\,A(g) + 2\,B(g) \rightarrow D(g)$ | |

What is the rate law for the overall reaction based on the proposed mechanism?

---

**SAMPLE EXERCISE 14.13** **Testing a Proposed Reaction Mechanism** **LO9**

One proposed mechanism for the decomposition of $N_2O_5$ to $NO_2$, shown below, involves three elementary steps:

| | | |
|---|---|---|
| Step 1 | $2\,N_2O_5(g) \rightleftharpoons N_4O_{10}(g)$ | fast and reversible |
| Step 2 | $N_4O_{10}(g) \rightarrow N_2O_3(g) + 2\,NO_2(g) + O_3(g)$ | slow |
| Step 3 | $N_2O_3(g) + O_3(g) \rightarrow 2\,NO_2(g) + O_2(g)$ | fast |
| Overall | $2\,N_2O_5(g) \rightarrow 4\,NO_2(g) + O_2(g)$ | |

What is the rate law for the overall reaction based on the proposed mechanism?

**Collect and Organize** The rate law for the overall reaction reflects the rate laws for the elementary reactions preceding and including the rate-determining (slowest) step in the mechanism. We are given three elementary steps and their relative rates.

**Analyze** The rate law for an elementary step can be written directly from the balanced chemical equation for that step. We are told that step 1 is fast *and reversible*, which means that as the reaction proceeds and $[N_4O_{10}]$ increases, eventually the rate of step 1 in the forward direction is matched by the rate of step 1 in the reverse direction. Only elementary steps 1 and 2 contribute to the rate law for the overall reaction since step 2 is the rate-determining step.

**Solve** The rate laws for step 1 in the forward and reverse directions are

$$\text{Rate of forward step 1} = k_f[N_2O_5]^2$$
$$\text{Rate of reverse step 1} = k_r[N_4O_{10}]$$

The rate law for step 2 is

$$\text{Rate of step 2} = k_2[N_4O_{10}]$$

To find the overall rate law, we need to express $[N_4O_{10}]$ in terms of $[N_2O_5]$ by making the step 1 rates equal to each other and rearranging the terms:

$$[N_4O_{10}] = \frac{k_f}{k_r}[N_2O_5]^2$$

Substituting this expression for $[N_4O_{10}]$ into the rate law for step 2:

$$\text{Rate} = \frac{k_f k_2}{k_r}[N_2O_5]^2$$

**Think About It** The rate law for the overall reaction must depend on the concentration of $N_2O_5$. The order of the reaction in $N_2O_5$ depends on the proposed mechanism for the reaction. Notice that the rate of step 3 has no impact on the rate law for this mechanism because it follows the slowest step. In addition, note that the rate law for the mechanism in this Sample Exercise does not match the experimentally determined rate law in Sample Exercise 14.5. Therefore, the mechanism that starts with the dimerization of $N_2O_5$ cannot be correct.

☀ **Practice Exercise** Here is another proposed mechanism for the reaction of NO with $H_2$ (Sample Exercise 14.12):

| Elementary step 1 | $H_2(g) + NO(g) \rightarrow N(g) + H_2O(g)$ |
| Elementary step 2 | $N(g) + NO(g) \rightarrow N_2(g) + O(g)$ |
| Elementary step 3 | $H_2(g) + O(g) \rightarrow H_2O(g)$ |

Is this a valid mechanism?

## Mechanisms and Zero-Order Reactions

Before we leave reaction mechanisms, let's revisit the reaction between $NO_2$ and CO, which has an experimentally determined rate law that is zero order in CO, second order in $NO_2$, and second order overall:

$$NO_2(g) + CO(g) \rightarrow NO(g) + CO_2(g) \qquad \text{Rate} = k[NO_2]^2$$

What does this overall rate law tell us about the reaction? Remember that the overall rate depends on the concentrations of the reactants in the rate-determining step, which means that CO is not a reactant in the rate-determining step. Carbon monoxide is clearly involved in the reaction—it is converted into $CO_2$—but whatever step involves CO must not be the rate-determining step. This leads us to conclude that the reaction must have at least two elementary steps, one rate-determining and one not.

It has been proposed that the reaction mechanism is

| (1) | $2\,NO_2(g) \rightarrow NO_3(g) + NO(g)$ | $\text{Rate} = k_1[NO_2]^2$ |
| (2) | $NO_3(g) + CO(g) \rightarrow NO_2(g) + CO_2(g)$ | $\text{Rate} = k_2[NO_3][CO]$ |

The experimentally determined overall rate law matches the rate law for the first step, which must be the slower, rate-determining step. The overall reaction is zero order in CO because CO is not a reactant in that step.

**CONCEPT TEST**

The rate law for the reaction $XO_2(g) + M(g) \rightarrow XO(g) + MO(g)$, where X and M represent metallic elements, is second order in $[XO_2]$ and independent of $[M]$ for a wide variety of compounds $XO_2$ and M. Why can we conclude that these reactions likely proceed by the same mechanism?

# 14.6 Catalysis

We noted in Section 14.4 that spontaneous reactions may be slow if they have a high activation energy. Suppose we wanted to increase the rate of such a reaction. How could we do it? One way is to increase the temperature of the reaction mixture. However, in some chemical reactions, elevated temperatures can lead to undesired products or to lower yields. Another way is to add a **catalyst**, a substance that increases the rate of a reaction but is not consumed in the process.

## Catalysts and the Ozone Layer

As we discussed in Chapter 8, the way we think about ozone depends on where the ozone is located. Ozone in the stratosphere between 10 and 40 km above Earth's surface is necessary to protect us from UV radiation, but ozone at ground level is hazardous to our health. In this section, we discuss the role of catalysis in the loss of stratospheric ozone that has led to the annual formation of ozone holes over Antarctica.

**catalyst** a substance added to a reaction that increases the rate of the reaction but is not consumed in the process.

The natural photodecomposition of ozone in the stratosphere occurs through the reaction

$$2\ O_3(g) \rightarrow 3\ O_2(g) \qquad\qquad (14.42)$$

It begins with the absorption of UV radiation from the sun and the generation of atomic oxygen:

$$(1) \qquad O_3(g) \rightarrow O_2(g) + O(g)$$

The oxygen atom may react with another ozone molecule to form two more molecules of oxygen:

$$(2) \qquad O_3(g) + O(g) \rightarrow 2\ O_2(g)$$

The rate of the second elementary step is low because its activation energy is relatively high: 17.7 kJ/mol.

In 1974, two American scientists, Sherwood Rowland and Mario Molina, predicted significant depletion of stratospheric ozone because of the release of a class of volatile compounds called chlorofluorocarbons (CFCs) into the atmosphere at ground level, which ultimately enter the stratosphere. This prediction was later supported by experimental evidence of a thinning of the ozone layer—and formation of ozone holes—over Antarctica. By 2000, stratospheric ozone concentrations over Antarctica were less than half of what they were in 1980 (Figure 14.25), and the ozone hole covered nearly all of Antarctica and the tip of South America. Less severe thinning of stratospheric ozone was observed in the Northern Hemisphere.

An international agreement known as the Montreal Protocol called for an end to the production of ozone-depleting CFCs. The Montreal Protocol has had a dramatic effect on CFC production and emission into the atmosphere. However,

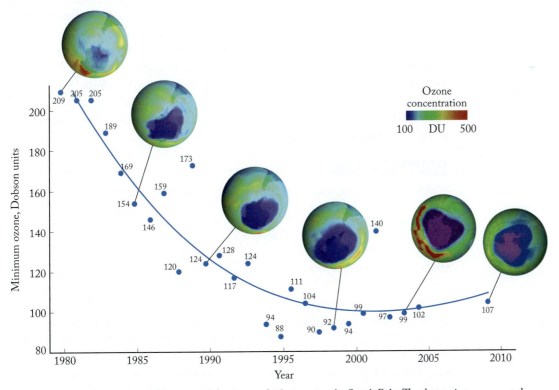

**FIGURE 14.25** A 30-year trend in stratospheric ozone depletion over the South Pole. The data points represent the minimum ozone concentration observed each year, expressed in Dobson units (DU). The colors of the six satellite images show the size of the holes in the ozone layer on a scale in which purple represents the most ozone depletion (see color chart). Values below 220 are considered dangerously low.

these compounds last for many years in the atmosphere, and recovery of the ozone layer may take most of the 21st century.

How do CFCs contribute to the destruction of ozone? Three of the more widely used CFCs were $CCl_2F_2$, $CCl_3F$, and $CClF_3$. In the stratosphere, they encounter UV radiation with enough energy to break C—Cl bonds, releasing chlorine atoms. For example,

$$CCl_3F(g) \xrightarrow{h\nu} CCl_2F(g) + Cl(g) \qquad (14.43)$$

Free chlorine atoms react with ozone, forming chlorine monoxide:

$$Cl(g) + O_3(g) \rightarrow ClO(g) + O_2(g) \qquad (14.44)$$

Chlorine monoxide then reacts with more ozone, producing oxygen and regenerating atomic chlorine:

$$ClO(g) + O_3(g) \rightarrow Cl(g) + 2\,O_2(g) \qquad (14.45)$$

If we add Equations 14.44 and 14.45 and cancel species as needed, we get

$$2\,O_3(g) \rightarrow 3\,O_2(g)$$

The overall reaction is exactly the same as the natural photodecomposition of ozone. The difference is the presence of chlorine atoms in Equations 14.44 and 14.45, which dramatically increase the rate of the overall reaction. Chlorine atoms act as a catalyst for the destruction of ozone because they speed up the reaction but are not consumed by it. Rather, they are consumed in one elementary step but then regenerated in a later elementary step. Chlorine is a catalyst, not an intermediate, because in a reaction mechanism, a catalyst is consumed in an early step and then regenerated in a later one, whereas an intermediate is produced before it is consumed. A single chlorine atom can catalyze the destruction of hundreds to thousands of stratospheric $O_3$ molecules before it combines with other atoms and forms a less reactive molecule.

The activation energy for the Cl-catalyzed reaction is only 2.2 kJ/mol, whereas the activation energy for the uncatalyzed reaction is 17.7 kJ/mol (Figure 14.26). Its smaller $E_a$ value means that the catalyzed destruction of ozone is faster than the natural photodecomposition process. In the reaction describing the destruction of ozone, the catalyst, Cl(g), and reactant, $O_3(g)$, exist in the same physical phase. When a catalyst and the reacting species are in the same phase, we call the catalyst a **homogeneous catalyst**.

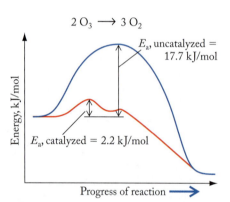

**FIGURE 14.26** The decomposition of $O_3$ in the presence of chlorine atoms has a smaller activation energy (2.2 kJ/mol) than the naturally occurring photodecomposition of $O_3$ to $O_2$ (17.7 kJ/mol). The catalytic effect of chlorine is a key factor in the depletion of stratospheric ozone and the formation of an ozone hole over the South Pole.

**CONCEPT TEST**

Which of the following statements is/are true about the elementary step shown in Equation 14.44?
a. Its rate law is first order in [Cl].
b. Its rate law does not include [Cl] because Cl is a catalyst.
c. Its rate is independent of [Cl].

The rates of the above reactions increase in the presence of drops of liquid, such as those present in the clouds that cover much of Antarctica each spring. These clouds form from crystals of ice and tiny drops of nitric acid in the winter, when temperatures in the lower stratosphere dip to −80°C. The clouds become collection sites for HCl, ClO, and other compounds containing chlorine. In August (the end of the Antarctic winter), the ice in the clouds melts and ClO is free to react on the surface of the drops of liquid water. These drops catalyze reactions by *adsorbing* (binding to the surface) the reactants. The drops are not reactants, but they provide a surface on which the reactants collect. The resulting proximity of the reactants increases the likelihood of reaction and, coupled with a decrease in activation energy, increases the reaction rate. When the catalyst is a liquid drop on which gas molecules adsorb, the drop is in a different phase than the reacting species and is called a **heterogeneous catalyst**. Both homogeneous and heterogeneous catalysts play a role in the reactions that diminish the amount of ozone in the stratosphere.

**homogeneous catalyst** a catalyst in the same phase as the reactants.

**heterogeneous catalyst** a catalyst in a different phase from the reactants.

**CONCEPT TEST** ....................................................

Is the ClO produced in Equation 14.44 a catalyst or an intermediate?

.......................................................................................

Evidence supporting the occurrence of these reactions in the stratosphere is presented in Figure 14.27. The concentration of ClO decreases sharply over the South Pole at the end of the Antarctic winter, as shown in satellite images taken one month apart, consistent with the reaction between ClO and $O_3$ shown in Equation 14.45. Although it is difficult to track the concentration of Cl atoms directly, some of them end up in molecules of HCl present in the icy polar clouds that form during the winter and melt in August. The images in Figure 14.27 show that a decrease in ClO concentration between late August and the Antarctic spring (late September) is accompanied by an increase in HCl concentration.

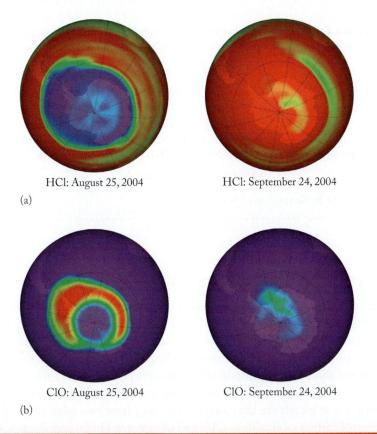

HCl: August 25, 2004        HCl: September 24, 2004

(a)

ClO: August 25, 2004        ClO: September 24, 2004

(b)

**FIGURE 14.27** (a) NASA satellite images of stratospheric concentrations of hydrogen chloride in August and September 2004 over the South Pole. Blue color at the pole indicates low levels in August; red color at the pole indicates high levels in September. (b) Stratospheric concentrations of chlorine monoxide in August and September 2004. Red indicates high levels in August and blue indicates low levels in September. The shifts in the concentrations of the two compounds are due in part to the reaction $OH + ClO \rightarrow HCl + O_2$.

**SAMPLE EXERCISE 14.14    Identifying Catalysts         LO10
                            in Reaction Mechanisms**

A reaction mechanism proposed for the decomposition of ozone in the presence of NO at high temperatures consists of three elementary steps:

(1)    $O_3(g) + NO(g) \rightarrow O_2(g) + NO_2(g)$

(2)    $NO_2(g) \rightarrow NO(g) + O(g)$

(3)    $O(g) + O_3(g) \rightarrow 2\,O_2(g)$

If the rate of the overall reaction is higher than the rate of the uncatalyzed decomposition of ozone to oxygen (Equation 14.42), is NO a catalyst in the reaction?

**Collect and Organize** We are asked to determine whether NO is a catalyst in a reaction. A catalyst increases the rate of a reaction and is not consumed by the overall reaction. We are given the elementary steps of the reaction and are told that the reaction is more rapid in the presence of NO.

**Analyze** We can sum the reactions to determine the overall reaction. If NO is consumed in an early step before it is regenerated in a later one, and is not consumed in the overall process, it is a catalyst.

**Solve** Summing the three elementary steps:

$$O_3(g) + \cancel{NO(g)} + \cancel{NO_2(g)} + \cancel{O(g)} + O_3(g) \rightarrow$$
$$O_2(g) + \cancel{NO_2(g)} + \cancel{NO(g)} + \cancel{O(g)} + 2\,O_2(g)$$

gives the overall reaction:

$$2\,O_3(g) \rightarrow 3\,O_2(g)$$

This equation does not include NO, and the rate of the reaction is higher when NO is present. Thus NO fulfills both requirements for being a catalyst.

**Think About It** NO behaves much like the Cl atoms in Equations 14.44–14.45. NO is not an intermediate because it is used in the reaction and then regenerated in a subsequent step.

⚙ **Practice Exercise** The combustion of fossil fuels results in the release of $SO_2$ into the atmosphere, where it reacts with oxygen to form $SO_3$:

$$2\,SO_2(g) + O_2(g) \rightarrow 2\,SO_3(g)$$

In the atmosphere, $SO_2$ may react with $NO_2$, forming $SO_3$ and NO:

$$NO_2(g) + SO_2(g) \rightarrow NO(g) + SO_3(g)$$

The rate of reaction of $SO_2$ with $NO_2$ is faster than the rate of reaction of $SO_2$ with oxygen. If the NO produced in the reaction of $NO_2$ and $SO_2$ is then oxidized to $NO_2$,

$$2\,NO(g) + O_2(g) \rightarrow 2\,NO_2(g)$$

is $NO_2$ a catalyst in the reaction of $SO_2$ with $O_2$?

## Catalysts and Catalytic Converters

We started this chapter with a discussion of air pollution caused by vehicles and of the technology that has been developed to clean the air. Figure 14.28 shows a catalytic converter in a car's exhaust system and the reactions that take place in the converter to remove one representative pollutant, NO, from the engine exhaust. Hot exhaust gases flowing through the converter pass through a fine honeycomb mesh coated with one or more of the transition metals palladium, platinum, and rhodium. These metals are the catalysts, and they have two roles: (1) to speed up oxidation of carbon monoxide to $CO_2$ and of unburned hydrocarbons to $CO_2$ and water vapor; and (2) to convert NO and $NO_2$ into $N_2$ and $O_2$.

The metals used in making catalytic converters are dissolved as metal salts and dispersed on the mesh, and they are then reduced to clusters 2 to 10 nm in diameter. The large surface area of the metal clusters provides sites where the oxidation and reduction of the gases take place.

Catalysts not only speed up reactions, but also allow them to take place at lower temperatures. For example, CO reacts rapidly with $O_2$ above 700°C,

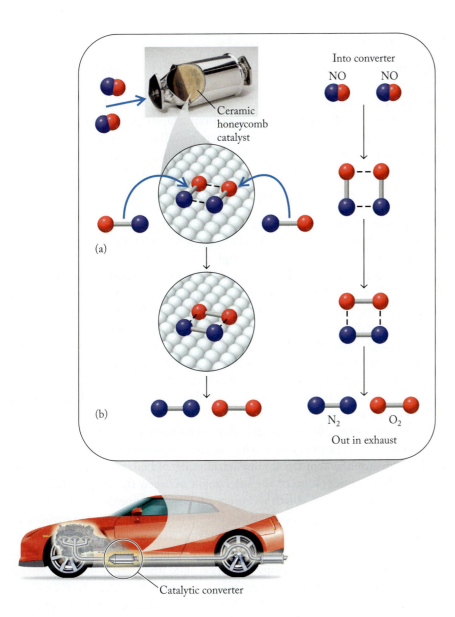

**FIGURE 14.28** Catalytic converters in automobiles reduce emissions of NO by lowering the activation energy of its decomposition into $N_2$ and $O_2$. Metal catalysts are supported on a porous ceramic honeycomb. (a) NO molecules are adsorbed onto the surface of metal clusters where their NO bonds are broken, and (b) pairs of O atoms and N atoms form $O_2$ and $N_2$. The $O_2$ and $N_2$ desorb from the surface and are released to the atmosphere.

but the presence of a Pd or Pt/Rh catalyst enables this reaction to take place rapidly at the much lower temperature of automobile exhaust, about 250°C. The catalysts have similar effects on the reduction reactions taking place in the converters.

The catalysts are selective in terms of the molecules they interact with and specific in the reactions they promote. What makes the catalysts selective is that several reactions are possible for each pollutant, but one reaction proceeds more rapidly than the others. For example, the preferred reduction of the nitrogen in NO is to $N_2$ rather than to $N_2O$ or to $NH_3$, and carbon monoxide and hydrocarbons are potential reducing agents for this reaction. Recall, however, that oxidizing CO and hydrocarbons is one of the two primary goals of a catalytic converter. If these compounds are oxidized, no NO reduction can take place, and the converter does only half its job. Fortunately, the reduction of NO by CO and hydrocarbons is much faster than the reactions of CO and hydrocarbons with $O_2$. As a result, NO reduction is essentially complete before any of the necessary reducing agents are consumed by reaction with oxygen.

**enzyme** a protein that catalyzes a reaction.

**biocatalysis** the use of enzymes to catalyze reactions on a large scale; it is becoming especially important in processes that involve chiral materials.

## Enzymes: Biological Catalysts

Large biomolecules called **enzymes** are highly selective catalysts; they catalyze very specific reactions in biological systems. For example, the chemical reactions involved in metabolism, which includes both the breaking down of molecules and the synthesis of complex substances from simpler precursors, are mediated in large part by enzymes. Sequences of reactions called *metabolic pathways* consist of steps, each of which is catalyzed by a specific enzyme. For example, carbonic anhydrase is an enzyme that speeds up the hydrolysis of $CO_2$:

$$CO_2(aq) + H_2O(\ell) \rightleftharpoons HCO_3^-(aq) + H^+(aq) \qquad (14.46)$$

In the presence of carbonic anhydrase, this reaction proceeds about 10 million times faster than in its absence. Without carbonic anhydrase, we would not be able to expel carbon dioxide fast enough to survive (as the reaction in Equation 14.46 runs in reverse). One molecule of carbonic anhydrase can hydrolyze from $1 \times 10^4$ to $1 \times 10^6$ molecules of $CO_2$ in 1 s. This value is called the *turnover number* for the enzyme; in general, the higher the turnover number, the faster the enzyme-catalyzed reaction proceeds. Turnover numbers for enzymes typically range from $10^3$ to $10^7$. The higher the turnover number, the lower the activation energy of the catalyzed reaction. As we learned earlier in this chapter, for reactions to proceed, molecules must collide with the proper orientation. Carbonic anhydrase is essentially a perfect enzyme because it catalyzes the hydrolysis reaction nearly every time it collides with $CO_2$.

Enzymes are specific because the idea of an effective collision has a different connotation in biochemistry than what we have depicted for other reactions and catalysts. A simplified approach to understanding and quantifying how enzymes work involves consideration of the interaction between the enzyme (E) and its *substrate* (S), the reactant molecule (Figure 14.29). We refine these ideas in Chapter 20 when we discuss specific enzymes, but for now it is sufficient to envision the substrate fitting into an *active site* in the enzyme, very much like a hand fits into a glove. Once in the active site, the substrate (S) is converted into product (P) via a pathway that has a lower energy transition state (the *enzyme–substrate complex* ES), which is absent without the enzyme.

Even in the simplest organism, hundreds of enzyme-catalyzed chemical reactions are constantly taking place. Most of these enzymes are effective only under limited reaction conditions: in aqueous media, at temperatures between 4°C and 37°C, and over a narrow pH range. These hundreds of catalyzed reactions require hundreds of different enzymes, many operating with exquisite efficiency and producing one pure product.

**FIGURE 14.29** In this simplified view, the substrate fits into the active site of the enzyme that catalyzes the conversion of substrate into product.

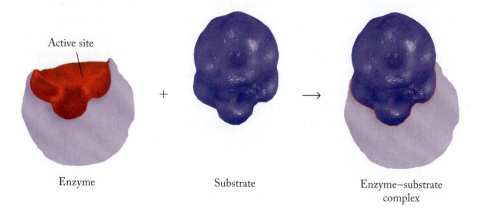

Enzyme + Substrate → Enzyme–substrate complex

In the pharmaceutical industry, research is focused on using enzymes outside living systems to produce high yields of specific products. This research area, called **biocatalysis**, uses enzymes to catalyze chemical reactions run in industrial-sized reactors. Because it deals with both isolated enzymes and microorganisms, it is considered a special type of heterogeneous catalysis.

The mathematical treatment of enzyme kinetics can be quite complex, but determining the rate of enzyme-catalyzed reactions is an important part of biochemical and medical research. We can think of the following elementary steps in the process:

$$\text{Step 1} \quad \text{E} + \text{S} \underset{k_{-1}}{\overset{k_1}{\rightleftharpoons}} \text{ES}$$

$$\text{Step 2} \quad \text{ES} \xrightarrow{k_2} \text{E} + \text{P}$$

Biochemists commonly assume that the formation of ES and its decomposition are both rapid and that step 2 is rate determining. A typical reaction profile for a system that meets these criteria is shown in Figure 14.30, and the rate of such a reaction is given by

$$\text{Rate} = \frac{\Delta[\text{P}]}{\Delta t} = k[\text{ES}]$$

Initially, the rate of reaction increases rapidly as the concentration of S increases (Figure 14.31). The rate of reaction is proportional to [S], so the reaction is first order and Rate = $k$[S]. At some specific concentration of S, all of the active sites in available enzymes are occupied, and the rate of the reaction becomes constant. At this stage, the rate is zeroth order in S (rate = $k$); adding more S has no effect because all the active sites are full.

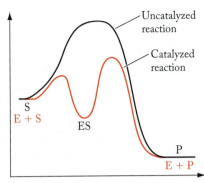

**FIGURE 14.30** These superimposed reaction profiles show (black) the uncatalyzed reaction of S → P and (red) the enzyme-catalyzed reaction (E + S → E + P). The enzyme-catalyzed reaction takes place in two steps, in which the second step (ES → E + P) is rate determining.

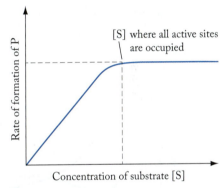

**FIGURE 14.31** Plot of the rate of formation of product P versus concentration of substrate S in an enzyme-catalyzed reaction. The dotted line identifies the concentration of substrate at which all active sites are occupied.

---

**SAMPLE EXERCISE 14.15    Integrating Concepts: Simulations of Smog**

As we have seen throughout this chapter, reactions in the atmosphere are complex. Figure 14.32 shows three graphs of data from programs that simulated atmospheric conditions arising from known reactions when the hydrocarbon propylene was introduced into air.[3] Propylene enters the atmosphere from the evaporation of unburned petroleum fuels. The three graphs show simulated concentration versus time profiles on (a) a sunny day, (b) a cloudy day, and (c) a sunny day when the concentration of propylene is twice the amount in (a).

a. Both alkanes and alkenes are present as dissolved gases in unburned fuels. In studies of both propylene ($CH_2\!\!=\!\!CHCH_3$)

and butane ($CH_3CH_2CH_2CH_3$), the alkene was shown to react more rapidly than the alkane. Suggest a reason for this observation.

b. In Figure 14.32(a), which substances are reactants and which are products? Are there any intermediates?

c. Compare Figures 14.32(a) and 14.32(c). How does doubling the concentration of propylene affect the level of irritants like ozone and peroxyacetyl nitrate (PAN)?

d. What are the major differences between the reactions on a sunny day (Figure 14.32a) and a cloudy day (Figure 14.32b)?

**Collect and Organize** We have three graphs for different conditions and we can use them to get relative quantitative information.

**Analyze** The graphs show differences in the appearance and disappearance of reactants, intermediates, and products of reactions involving an alkene, $NO_x$, ozone, and PAN.

[3]Graphs based on B. J. Hubert, *J. Chem. Educ.* 1974, *51*, 644–645; additional information from A. C. Baldwin, J. R. Barker, D. M. Golden, and D. G. Hendry, *J. Phys. Chem.* 1977, *81*(25), 2483–2492.

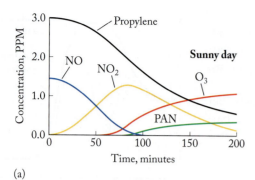

(a)

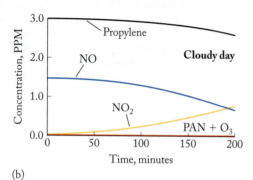

(b)

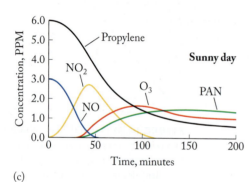

(c)

**FIGURE 14.32**

**Solve**

a. Alkenes react more rapidly than alkanes because of the presence of a π bond. The electrons in a π bond are much more accessible to reactants than are those in a σ bond, and the π bond is also weaker than the σ bond.

b. Propylene and NO are reactants. Their concentrations start out high and drop continuously throughout the time of observation. PAN and $O_3$ are products. Their concentration starts out at zero and gradually builds. $NO_2$ is an intermediate. Its concentration begins to increase at $t = 0$, reaches a peak around 75 min, and then decreases as PAN and $O_3$ form and accumulate. The decrease in propylene concentration begins immediately and accelerates as $O_3$ concentration builds. Some organic substances other than PAN must be forming to account for the drop; their concentrations are not shown on the graph.

c. Doubling the initial concentration of propylene causes it to disappear at a faster rate; its drop-off occurs at 25 min in (c) whereas it is at 50 min in (a). NO concentration also decreases more rapidly in (c). $NO_2$ level peaks in (c) at 3.0 ppm, which is twice the concentration it attains at its peak in (a), and then drops off almost twice as fast; it peaks at 75 min in (a) and 40 min in (c). At higher propylene concentrations in (c), $O_3$ and PAN appear earlier, at 35 and 45 min versus 50 and 80 min in (a). Also, $O_3$ appears at higher concentrations in (c) and gradually drops at later times, after about 100 min, compared to (a), where it continues to rise at the 200 min mark. PAN is almost 4 times as abundant at 200 min in (c) compared to its value at that same time in (a). Doubling the hydrocarbon nearly quadruples the level of PAN and also increases the level of ozone.

d. PAN and $O_3$ are not evident at all in (b) on the cloudy day, and propylene and NO disappear more slowly compared to a sunny day.

**Think About It** Photochemical smog is aptly named. In the absence of direct sunlight, it does not seem to form.

As we have seen with the reactions responsible for the production of photochemical smog, determinations of the rates of chemical reactions are crucial for us to understand processes on a molecular level. The mechanisms of the reactions of volatile oxides produced during combustion and by other natural events must be thoroughly understood so that problems arising because of the presence of these substances in the environment can be effectively managed. The continued development of catalytic converters for vehicles to diminish the problems caused by burning fossil fuels and to clean our air requires a thorough knowledge of the kinetics and mechanisms of many of the reactions discussed in this chapter. A large number of biological catalysts are responsible for the myriad reactions in living systems, and their behavior may be understood by applying the same methods of study we used for inorganic catalysts in reactions of small molecules.

# The Platinum Group: Catalysts, Jewelry, and Investment

The platinum group metals (PGMs) are a block of six elements in rows 5 and 6 and columns 8, 9, and 10 of the periodic table: platinum, palladium, rhodium, ruthenium, osmium, and iridium. In addition to having exceptional catalytic properties, all the PGMs have high melting points and are corrosion resistant. Besides their use in catalytic converters, they function as catalysts in fuel cells and in the industrial production of nitric acid and chlorine gas; they are alloyed with other metals in spark plugs and electronic devices (Figure 14.33); and several of them are key components of cancer chemotherapy drugs.

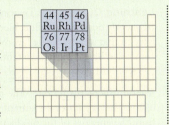

Platinum and palladium are used to catalyze a variety of chemical reactions. In this chapter, we learned about the use of Pd, Pt, and Rh in catalytic converters, where a key aspect of their catalytic activity is the ability to bind $NO_x$, CO, and hydrocarbons to the surfaces of small metallic clusters. Since 1996, Pd has become the most widely used PGM in this application. A fundamental measure of the binding ability of a heterogeneous catalyst is a parameter called the *heat of adsorption*, a measure of the strength with which a substance bonds to the surface of a material. If a substance is adsorbed (bound to the surface) too tightly, it cannot react with other adsorbed species. If it is bonded too weakly, it will desorb (move away from the surface) before it has a chance to react. Platinum and palladium operate between these two extremes, binding a wide range of substances with moderate strength. Both metals bind many gases with a moderate heat of adsorption, and this is the key to their catalytic activity.

The technological importance of PGMs extends far beyond their roles as heterogeneous catalysts of gas-phase reactions. They are very resistant to corrosion by acids, alkalis, and salts; they are stable at high temperatures; and they do not oxidize readily. As a result, they are frequently used as linings or coatings to protect other metals. The rigidity and hardness of individual PGMs can be improved by alloying them with other metals in the PGM family. Platinum is frequently alloyed with Rh, Ir, and Ru to produce hard, highly corrosion-resistant solids. Osmium is rarely used as a pure material and is usually alloyed with other metals in applications requiring objects that are resistant to wear.

The PGMs are rare, so their heavy use in technologically important applications like automotive catalytic converters has led to recycling efforts to recover the metals. In the United States, the most important source of Pt, Pd, and Rh is their collection from scrapped catalytic converters. It can be challenging to recover PGMs from ceramic support structures. One technique involves the selective dissolution of the metals by treatment with *aqua regia* ("water of kings"), a mixture of hydrochloric and nitric acids, so named because it is capable of dissolving gold.

All PGMs are used in medical devices and implants. Because of their lack of reactivity and tolerance for solutions that corrode most metals, they are highly biocompatible. Platinum alloyed with iridium is used in the electrodes of pacemakers and defibrillators and also in the tips of the guide wires surgeons use to direct the placement of catheters. One mode of cancer therapy uses radioactive [192]Ir wire wrapped in platinum as an implant to deliver radiation doses in the body.

Platinum is the PGM most commonly used for jewelry. In Japan it is actually preferred over gold in decorative items (Figure 14.34). As we saw with gold, pure platinum is too soft for jewelry, so it is usually alloyed with iridium or ruthenium to increase its wear resistance. Palladium is used in small quantities as a whitener in gold jewelry. Pure platinum is also made into coins and ingots and has considerable value as an investment.

**FIGURE 14.33** Discs for computer drives are coated with platinum.

**FIGURE 14.34** This platinum model of the Japanese robot Gundam is 12.5 cm tall and is valued at $250,000.

## SUMMARY

**Learning Outcome 1** The relative rates of disappearance of reactants and appearance of products are related by the stoichiometry of the reaction. (Sections 14.1 and 14.2)

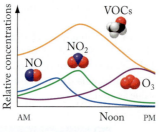

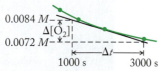

**Learning Outcome 2** The overall rate of a reaction is typically determined from experimental measurements and can be expressed as an **instantaneous rate**. (Section 14.2)

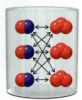

**Learning Outcome 3** The dependence of the rate of a reaction A + B → C on reactant concentrations is expressed in the **rate law** for the reaction: Rate = $k[A]^m[B]^n$, where $m$ and $n$ are the **reaction order** with respect to reactants A and B, respectively, and $k$ is the **rate constant**. The order of a reaction and the rate law for the reaction can be determined from differences in the **initial rates** of reaction. (Section 14.3)

**Learning Outcome 4** The units of a rate constant depend on the **overall reaction order**, which is the sum of the reaction orders with respect to individual reactants, observed with different concentrations of reactants or derived from the results of single kinetics experiments using **integrated rate laws**. (Section 14.3)

**Learning Outcome 5** An integrated rate law describes the change in concentration of a reactant over time. (Section 14.3)

**Learning Outcome 6** The **half-life** of a reaction is the time required for the concentration of a reactant to decrease to one-half its starting concentration: the higher the reaction rate, the shorter the half-life. (Section 14.3)

**Learning Outcome 7** Treating a reaction as pseudo-first order is frequently done in medicine, biochemistry, and environmental chemistry so that a simpler rate law may be used to deal with rates that are controlled by a limiting reactant. (Section 14.3)

**Learning Outcome 8** Increasing the temperature of a chemical reaction increases its rate. The **activation energy** of a reaction is a barrier that separates the sum of the internal energies of the reactants from the energies of the products. (Section 14.4)

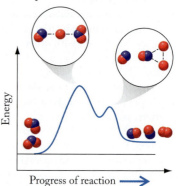

**Learning Outcome 9** Rate studies give insight into **reaction mechanisms**, which describe what is happening at a molecular level. A reaction mechanism consists of one or more **elementary steps** that describe how the reaction takes place on a molecular level. The proposed mechanism for any reaction must be consistent with the observed rate law and with the stoichiometry of the overall reaction. **Catalysts** increase reaction rates by changing the mechanism of a reaction and decreasing activation energies. **Enzymes** are catalysts in living systems. (Sections 14.5 and 14.6)

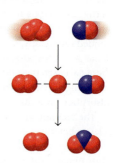

## PROBLEM-SOLVING SUMMARY

| TYPE OF PROBLEM | CONCEPTS AND EQUATIONS | SAMPLE EXERCISES |
|---|---|---|
| **Predicting a relative reaction rate** | A relative rate is determined from the stoichiometry of the balanced chemical equation. | 14.1 |
| **Converting reaction rates** | Use the balanced chemical equation and the known consumption or formation rate to determine the consumption or formation rate of another reaction participant. | 14.2 |
| **Determining an instantaneous rate** | Determine the slope of a line tangent to a point on the plot of concentration versus time. | 14.3 |
| **Deriving a rate law from initial reaction rate data** | Compare the change in rate when the concentration of one reactant is changed (while the concentrations of other reactants are kept constant) to determine the reaction order (usually whole numbers) with respect to that reactant: $$n = \frac{\log\left(\dfrac{\text{Rate}_1}{\text{Rate}_2}\right)}{\log\left(\dfrac{[X]_1}{[X]_2}\right)}$$ | 14.4 |

(14.17)

| TYPE OF PROBLEM | CONCEPTS AND EQUATIONS | SAMPLE EXERCISES |
|---|---|---|
| **Using an integrated rate law and distinguishing between first- and second-order reactions** | A linear plot of the natural logarithm of concentration versus time indicates a first-order reaction with a slope of $-k$, whereas a linear plot of the reciprocal of reactant concentration $1/[X]$ versus time indicates a second-order reaction. | 14.5, 14.8 |
| **Calculating concentration of a reactant from an integrated rate law** | Use the integrated rate law $$\ln[X] = -kt + \ln[X]_0 \qquad (14.21)$$ to calculate concentrations of reactant X at any given time $t$. | 14.6 |
| **Calculating the half-life of a first-order reaction** | In a first-order reaction, $$t_{1/2} = \frac{0.693}{k} \qquad (14.23)$$ | 14.7 |
| **Deriving a pseudo-first-order rate law** | A plot of the natural logarithm of concentration of the limiting reactant versus time is linear. The slope of the plot gives $k[\text{excess reactant}] = k'$. | 14.9, 14.10 |
| **Calculating an activation energy from rate constants** | Using the logarithmic form of the Arrhenius equation, $$\ln k = -\frac{E_a}{R}\left(\frac{1}{T}\right) + \ln A \qquad (14.33)$$ plot $\ln k$ versus $1/T$. The slope is $-E_a/R$. | 14.11 |
| **Linking reaction mechanisms to experimental rate laws and testing a proposed reaction mechanism** | The order of each reactant in an elementary step equals its coefficient in that step. The rate law for the mechanism must be the same as the observed rate law and does not include intermediates. | 14.12, 14.13 |
| **Identifying catalysts in reaction mechanisms** | Determine whether or not a potential catalyst is present by summing the elementary-step reactions to get the overall reaction. If that procedure reveals a potential catalyst, determine whether it increases the rate of reaction and whether it is initially consumed and then regenerated in the process. | 14.14 |

## VISUAL PROBLEMS ...........................................

*(Answers to boldface end-of-chapter questions and problems are in the back of the book.)*

**14.1.** Nitrous oxide decomposes to nitrogen and oxygen in the following reaction:

$$2\,N_2O(g) \rightarrow 2\,N_2(g) + O_2(g)$$

In Figure P14.1, which curve represents [$N_2O$] and which curve represents [$O_2$]?

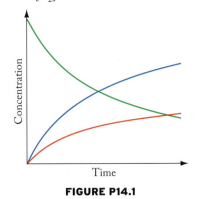

**FIGURE P14.1**

**14.2.** Sulfur trioxide is formed in the reaction

$$SO_2(g) + \tfrac{1}{2}\,O_2(g) \rightarrow SO_3(g)$$

In Figure P14.2, which curve represents [$SO_2$] and which curve represents [$O_2$]? All three gases are present initially.

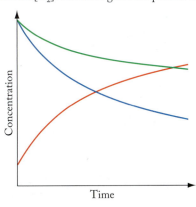

**FIGURE P14.2**

**14.3.** The rate law for the reaction 2 A → B is second order in A. Figure P14.3 represents samples with different concentrations of A; the red spheres represent molecules of A. In which sample will the reaction A → B proceed most rapidly?

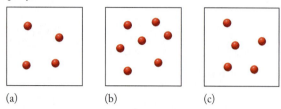

(a)          (b)          (c)

**FIGURE P14.3**

**14.4.** The rate law for the reaction A + B → C is first order in both A and B. Figure P14.4 represents samples with different concentrations of A (red spheres) and B (blue spheres). In which sample will the reaction A + B → C proceed most rapidly?

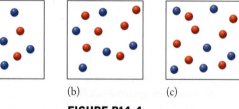

(a)          (b)          (c)

**FIGURE P14.4**

**14.5.** Which of the reaction profiles in Figure P14.5 represents (a) the slowest reaction? (b) the fastest reaction?

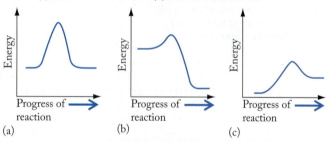

(a)          (b)          (c)

**FIGURE P14.5**

**14.6.** Figure P14.6 shows plots of reactant concentration versus time for four first-order reactions. Which has the greatest initial rate of reaction?

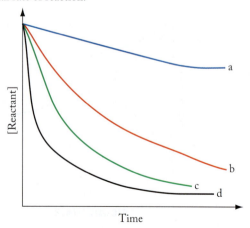

**FIGURE P14.6**

**14.7.** Which of the following mechanisms is consistent with the reaction profile shown in Figure P14.7?

a. $2 A \xrightarrow{\text{slow}} B$
   $B \xrightarrow{\text{fast}} C$
b. $A + B \rightarrow C$
c. $2 A \underset{}{\overset{\text{fast}}{\rightleftharpoons}} B$
   $B \xrightarrow{\text{slow}} C$

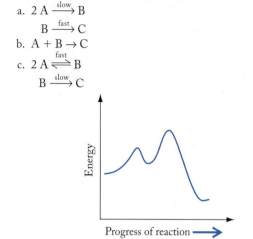

**FIGURE P14.7**

**14.8.** Which of the following mechanisms is consistent with the reaction profile shown in Figure P14.8?

a. $A + B \xrightarrow{\text{slow}} C$
   $C \xrightarrow{\text{fast}} D$
b. $A + B \rightarrow C$
c. $2 A \xrightarrow{\text{fast}} B$
   $B + C \xrightarrow{\text{slow}} D$

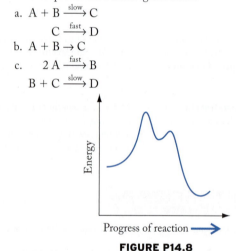

**FIGURE P14.8**

**14.9.** Which of the reaction profiles in Figure P14.9 represents the effect of a catalyst on the rate of a reaction?

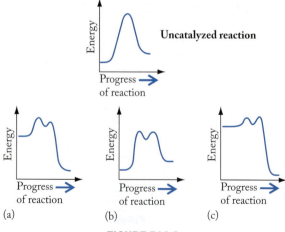

(a)          (b)          (c)

**FIGURE P14.9**

*14.10. The curves in Figure P14.10 show how the concentrations of four components of photochemical smog change during a sunny day. Describe how these curves would differ during a cloudy day.

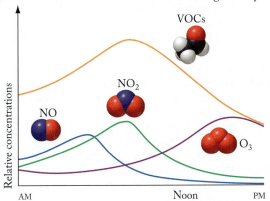

**FIGURE P14.10**

14.11. Refer to Figure P14.11 to answer the following questions:
   a. Which asterisk on Figure P14.11 identifies the transition state?
   b. Which arrow identifies the activation energy of the reaction in the forward direction?
   c. Which arrow identifies the activation energy of the reaction in the reverse direction?

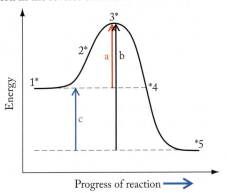

**FIGURE P14.11**

14.12. Refer to Figure P14.12 to answer the following questions:
   a. Which arrow identifies the activation energy of the reaction in the forward direction?
   b. Which arrow identifies the activation energy of the reaction in the reverse direction?
   c. Which arrow identifies the change in energy that accompanies the reaction, that is, the energy of the products less the energy of the reactants?

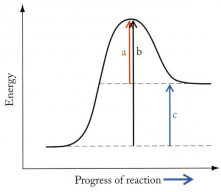

**FIGURE P14.12**

14.13. Which of the highlighted elements in Figure P14.13 forms volatile oxides associated with photochemical smog formation?

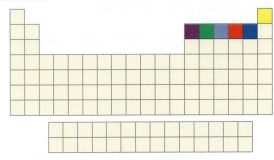

**FIGURE P14.13**

14.14. Which of the highlighted elements in Figure P14.13 forms noxious oxides that are removed from automobile exhaust as it passes through a catalytic converter?

14.15. Which of the highlighted elements in Figure P14.15 are widely used as heterogeneous catalysts?

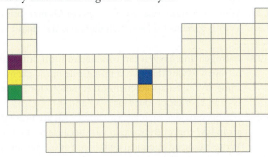

**FIGURE P14.15**

14.16. Which of the highlighted elements in Figure P14.16 forms volatile, odd-electron oxides that catalyze the destruction of stratospheric ozone?

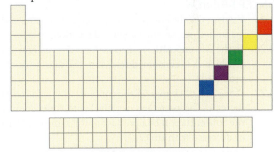

**FIGURE P14.16**

## Cars, Trucks, and Air Quality

### CONCEPT REVIEW

**14.17.** Why does the maximum concentration of ozone in Figure 14.2 occur much later in the day than the maximum concentration of NO and $NO_2$?

**14.18.** If we plot the concentration of reactants and products as a function of time for any sequence of two spontaneous chemical reactions, such as

$$A \rightarrow B \rightarrow C$$

will the maximum concentration of final product C always appear after the maximum concentration of B?

**14.19.** Why isn't there an increase in NO concentration after the evening rush hour?

**14.20.** If ozone can react with NO to form $NO_2$, why does the ozone concentration reach a maximum in the early afternoon?

### PROBLEMS

**14.21.** Which are more reactive: O atoms or $O_2$ molecules? Why?

**14.22.** Why is gaseous OH so much more reactive than $H_2O$ vapor?

**14.23.** Nitrogen and oxygen can combine to form different nitrogen oxides that play a minor role in the chemistry of smog. Write balanced chemical equations for the reaction of $N_2$ and $O_2$ that produce (a) $N_2O$ and (b) $N_2O_5$.

**14.24.** Nitrogen oxides such as $N_2O$ and $N_2O_5$ are present in the air in low concentrations, in part because of their reactivity. Write balanced chemical equations for (a) the conversion of $N_2O$ to $NO_2$ in the presence of oxygen and (b) the decomposition of $N_2O_5$ to $NO_2$ and $O_2$.

## Reaction Rates

### CONCEPT REVIEW

**14.25.** Explain the difference between the average rate and the instantaneous rate of a chemical reaction.

**14.26.** In the decomposition reaction $A \rightarrow B + C$, how is the rate at which A is consumed related to the rate at which B is produced?

**14.27.** If the rate of change in the concentration of a reactant increases (becomes less negative) with time, does the rate in the concentration of a product of the same reaction increase or decrease with time?

**14.28.** During the course of a reaction can there be a time when the instantaneous rate of the reaction does not change? If you think so, describe such a time.

### PROBLEMS

**14.29. Catalytic Converters in Automobiles (I)** Catalytic converters combat air pollution by converting NO into $N_2$ and $O_2$.
  a. How is the rate of formation of $O_2$ related to the rate of formation of $N_2$?
  b. How is the rate of formation of $N_2$ related to the rate of consumption of NO?

**14.30. Catalytic Converters in Automobiles (II)** Catalytic converters also combat air pollution by promoting the reaction between CO and $O_2$ that produces $CO_2$.

  a. How is the rate of formation of $CO_2$ related to the rate of consumption of $O_2$?
  b. How is the rate of formation of $CO_2$ related to the rate of consumption of CO?

**14.31.** Write expressions for the rate of formation of products and the rate of consumption of reactants in each of the following reactions:
  a. $F_2(g) + H_2O(\ell) \rightarrow HOF(g) + HF(g)$
  b. $Si(s) + 3\ HCl(g) \rightarrow SiHCl_3(\ell) + H_2(g)$
  c. $4\ NH_3(g) + 3\ O_2(g) \rightarrow 2\ N_2(g) + 6\ H_2O(g)$

**14.32.** Write expressions for the rate of formation of products and the rate of consumption of reactants in each of the following reactions:
  a. $SOF_2(g) + 2\ F_2(g) \rightarrow F_5SOF(g)$
  b. $B_2H_6(g) + 3\ Cl_2(g) \rightarrow 2\ BCl_3(g) + 6\ HCl(g)$
  c. $N_2H_4(g) + 2\ NH_2Cl(g) \rightarrow 2\ NH_4Cl(s) + N_2(g)$

**14.33.** In a study to determine the rate of formation of $NO(g)$ in the following reaction:

$$N_2(g) + O_2(g) \rightarrow 2\ NO(g)$$

the concentration of NO was 0.375 $M$ at $t = 50.0$ s and 0.407 $M$ at $t = 65.7$ s. What is the average rate of the reaction during this time period?

**14.34.** In the production of ammonia from nitrogen and hydrogen under a given set of conditions:

$$N_2(g) + 3\ H_2(g) \rightarrow 2\ NH_3(g)$$

the rate of ammonia formation is 7.37 $M$/s. What is the rate of disappearance of $H_2$?

**14.35. Power Plant Emissions** Sulfur dioxide emissions in stack gases at power plants may react with carbon monoxide as follows:

$$SO_2(g) + 3\ CO(g) \rightarrow 2\ CO_2(g) + COS(g)$$

Write an equation relating the rates for each of the following:
  a. The rate of formation of $CO_2$ to the rate of consumption of CO
  b. The rate of formation of COS to the rate of consumption of $SO_2$
  c. The rate of consumption of CO to the rate of consumption of $SO_2$

**14.36. Reducing Nitric Oxide Emissions from Power Plants** Nitric oxide (NO) can be removed from gas-fired power plant emissions by reaction with methane as follows:

$$CH_4(g) + 4\ NO(g) \rightarrow 2\ N_2(g) + CO_2(g) + 2\ H_2O(g)$$

Write an equation relating the rates for each of the following:
  a. The rate of formation of $N_2$ to the rate of formation of $CO_2$
  b. The rate of formation of $CO_2$ to the rate of consumption of NO
  c. The rate of consumption of $CH_4$ to the rate of formation of $H_2O$

**14.37. Stratospheric Ozone Depletion** Chlorine monoxide (ClO) plays a major role in the creation of the ozone holes in the stratosphere over Earth's polar regions.
  a. If $\Delta[\text{ClO}]/\Delta t$ at 298 K is $-2.3 \times 10^7$ $M$/s, what is the rate of change in $[\text{Cl}_2]$ and $[\text{O}_2]$ in the following reaction?

$$2\,\text{ClO}(g) \rightarrow \text{Cl}_2(g) + \text{O}_2(g)$$

  b. If $\Delta[\text{ClO}]/\Delta t$ is $-2.9 \times 10^4$ $M$/s, what is the rate of formation of oxygen and $\text{ClO}_2$ in the following reaction?

$$\text{ClO}(g) + \text{O}_3(g) \rightarrow \text{O}_2(g) + \text{ClO}_2(g)$$

**14.38.** The chemistry of smog formation includes $\text{NO}_3$ as an intermediate in several reactions.
  a. If $\Delta[\text{NO}_3]/\Delta t$ is $-2.2 \times 10^5$ m$M$/min in the following reaction, what is the rate of formation of $\text{NO}_2$?

$$\text{NO}_3(g) + \text{NO}(g) \rightarrow 2\,\text{NO}_2(g)$$

  b. What is the rate of change of $[\text{NO}_2]$ in the following reaction if $\Delta[\text{NO}_3]/\Delta t$ is $-2.3$ m$M$/min?

$$2\,\text{NO}_3(g) \rightarrow 2\,\text{NO}_2(g) + \text{O}_2(g)$$

**14.39.** Nitrite ion reacts with ozone in aqueous solution, producing nitrate ion and oxygen:

$$\text{NO}_2^-(aq) + \text{O}_3(g) \rightarrow \text{NO}_3^-(aq) + \text{O}_2(g)$$

The following data were collected for this reaction at 298 K. Calculate the average reaction rate between 0 and 100 μs (microseconds) and between 200 and 300 μs.

| Time (μs) | $[\text{O}_3]$ (M) |
|---|---|
| 0 | $1.13 \times 10^{-2}$ |
| 100 | $9.93 \times 10^{-3}$ |
| 200 | $8.70 \times 10^{-3}$ |
| 300 | $8.15 \times 10^{-3}$ |

**14.40.** Dinitrogen pentoxide ($\text{N}_2\text{O}_5$) decomposes as follows to nitrogen dioxide and nitrogen trioxide:

$$\text{N}_2\text{O}_5(g) \rightarrow \text{NO}_2(g) + \text{NO}_3(g)$$

Calculate the average rate of this reaction between consecutive measurement times in the following table.

| Time (s) | $[\text{N}_2\text{O}_5]$ (molecules/cm³) |
|---|---|
| 0 | $1.500 \times 10^{12}$ |
| 1.45 | $1.357 \times 10^{12}$ |
| 2.90 | $1.228 \times 10^{12}$ |
| 4.35 | $1.111 \times 10^{12}$ |
| 5.80 | $1.005 \times 10^{12}$ |

**14.41.** The following data were collected for the dimerization of ClO to $\text{Cl}_2\text{O}_2$ at 298 K.

| Time (s) | [ClO] (molecules/cm³) |
|---|---|
| 0 | $2.60 \times 10^{11}$ |
| 1 | $1.08 \times 10^{11}$ |
| 2 | $6.83 \times 10^{10}$ |
| 3 | $4.99 \times 10^{10}$ |
| 4 | $3.93 \times 10^{10}$ |
| 5 | $3.24 \times 10^{10}$ |
| 6 | $2.76 \times 10^{10}$ |

Plot [ClO] and $[\text{Cl}_2\text{O}_2]$ as a function of time and determine the instantaneous rates of change in both at 1 s.

**14.42. Tropospheric Ozone** Tropospheric (lower atmosphere) ozone is rapidly consumed in many reactions, including

$$\text{O}_3(g) + \text{NO}(g) \rightarrow \text{NO}_2(g) + \text{O}_2(g)$$

Use the following data to calculate the instantaneous rate of the preceding reaction at $t = 0.000$ s and $t = 0.052$ s.

| Time (s) | [NO] (M) |
|---|---|
| 0.000 | $2.0 \times 10^{-8}$ |
| 0.011 | $1.8 \times 10^{-8}$ |
| 0.027 | $1.6 \times 10^{-8}$ |
| 0.052 | $1.4 \times 10^{-8}$ |
| 0.102 | $1.2 \times 10^{-8}$ |

## Effect of Concentration on Reaction Rate

### CONCEPT REVIEW

**14.43.** Why do the rates of nearly all reactions decrease as reactants form products?

**14.44.** Why are the units of the rate constants different for reactions of different order?

**14.45.** Does the half-life of a second-order reaction have the same units as the half-life for a first-order reaction?

**14.46.** Does the half-life of a first-order reaction depend on the concentration of the reactants?

**14.47.** What effect does doubling the initial concentration of a reactant have on the half-life of a reaction that is second order in the reactant?

**14.48.** Suppose the decomposition reactions $A \rightarrow B + C$ and $X \rightarrow Y + Z$ are second order in A and X, respectively, and both have the same rate constant. Under what conditions do the two reactions also have the same half-life?

### PROBLEMS

**14.49.** For each of the following rate laws, determine the order with respect to each reactant and the overall reaction order.
  a. Rate $= k[\text{A}][\text{B}]$
  b. Rate $= k[\text{A}]^2[\text{B}]$
  c. Rate $= k[\text{A}][\text{B}]^3$

**14.50.** Determine the overall order of the following rate laws and the order with respect to each reactant.
  a. Rate $= k[\text{A}]^2[\text{B}]^{1/2}$
  b. Rate $= k[\text{A}]^2[\text{B}][\text{C}]$
  c. Rate $= k[\text{A}][\text{B}]^3[\text{C}]^{1/2}$

**14.51.** Write rate laws and determine the units of the rate constant (by using the units $M$ for concentration and s for time) for the following reactions:
   a. The reaction of oxygen atoms with $NO_2$ is first order in both reactants.
   b. The reaction between NO and $Cl_2$ is second order in NO and first order in $Cl_2$.
   c. The reaction between $Cl_2$ and chloroform ($CHCl_3$) is first order in $CHCl_3$ and one-half order in $Cl_2$.
   *d. The decomposition of ozone ($O_3$) to $O_2$ is second order in $O_3$ and an order of $-1$ in O atoms.

**14.52.** Compounds A and B react to give a single product, C. Write the rate law for each of the following cases and determine the units of the rate constant by using the units $M$ for concentration and s for time:
   a. The reaction is first order in A and second order in B.
   b. The reaction is first order in A and second order overall.
   c. The reaction is independent of the concentration of A and second order overall.
   d. The reaction is second order in both A and B.

---

**14.53.** Predict the rate law for the reaction $2\,BrO(g) \rightarrow Br_2(g) + O_2(g)$ if the following conditions hold true:
   a. The rate doubles when [BrO] doubles.
   b. The rate quadruples when [BrO] doubles.
   c. The rate is halved when [BrO] is halved.
   d. The rate is unchanged when [BrO] is doubled.

**14.54.** Predict the rate law for the reaction $NO(g) + Br_2(g) \rightarrow NOBr_2(g)$ if the following conditions apply:
   a. The rate doubles when [NO] is doubled and [$Br_2$] remains constant.
   b. The rate doubles when [$Br_2$] is doubled and [NO] remains constant.
   c. The rate increases by 1.56 times when [NO] is increased 1.25 times and [$Br_2$] remains constant.
   d. The rate is halved when [NO] is doubled and [$Br_2$] remains constant.

---

**14.55.** The rate of the reaction:

$$NO(g) + O_3(g) \rightarrow NO_2(g) + O_2(g)$$

quadruples when the concentrations of NO and $O_3$ are doubled. Does this prove that the reaction is first order in both reactants? Why or why not?

**14.56.** The reaction between chlorine monoxide and nitrogen dioxide,

$$ClO(g) + NO_2(g) + M(g) \rightarrow ClONO_2(g) + M(g)$$

produces chlorine nitrate ($ClONO_2$). A third molecule (M) takes part in the reaction but is unchanged by it. The reaction is first order in $NO_2$ and in ClO.
   a. Write the rate law for this reaction.
   b. What is the reaction order with respect to M?

---

**14.57. Rate Laws for Destruction of Tropospheric Ozone** The reaction of $NO_2$ with ozone produces $NO_3$ in a second-order reaction overall:

$$NO_2(g) + O_3(g) \rightarrow NO_3(g) + O_2(g)$$

   a. Write the rate law for the reaction if the reaction is first order in each reactant.
   b. The rate constant for the reaction is $1.93 \times 10^4\ M^{-1}\,s^{-1}$ at 298 K. What is the rate of the reaction when [$NO_2$] $= 1.8 \times 10^{-8}\ M$ and [$O_3$] $= 1.4 \times 10^{-7}\ M$?
   c. What is the rate of formation of $NO_3$ under these conditions?
   d. What happens to the rate of the reaction if the concentration of $O_3(g)$ is doubled?

**14.58. Sources of Nitric Acid in the Atmosphere** The reaction between $N_2O_5$ and water,

$$N_2O_5(g) + H_2O(g) \rightarrow 2\,HNO_3(g)$$

is a source of nitric acid in the atmosphere.
   a. The reaction is first order in each reactant. Write the rate law for the reaction.
   b. When [$N_2O_5$] is 0.132 m$M$ and [$H_2O$] is 230 m$M$, the rate of the reaction is $4.55 \times 10^{-4}$ m$M^{-1}$ min$^{-1}$. What is the rate constant for the reaction?

---

**14.59.** Each of the following reactions is first order in the reactants and second order overall. Which reaction is fastest if the initial concentrations of the reactants are the same? All reactions are at 298 K.
   a. $ClO_2(g) + O_3(g) \rightarrow ClO_3(g) + O_2(g)$
      $k = 3.0 \times 10^{-19}$ cm$^3$/(molecule $\cdot$ s)
   b. $ClO_2(g) + NO(g) \rightarrow NO_2(g) + ClO(g)$
      $k = 3.4 \times 10^{-13}$ cm$^3$/(molecule $\cdot$ s)
   c. $ClO(g) + NO(g) \rightarrow Cl(g) + NO_2(g)$
      $k = 1.7 \times 10^{-11}$ cm$^3$/(molecule $\cdot$ s)
   d. $ClO(g) + O_3(g) \rightarrow ClO_2(g) + O_2(g)$
      $k = 1.5 \times 10^{-17}$ cm$^3$/(molecule $\cdot$ s)

**14.60.** Two reactions in which there is a single reactant have nearly the same magnitude rate constant. One is first order; the other is second order.
   a. If the initial concentrations of the reactants are both 1.0 m$M$, which reaction will proceed at the higher rate?
   b. If the initial concentrations of the reactants are both 2.0 $M$, which reaction will proceed at the higher rate?

---

**14.61.** The rate constant for the decomposition of $N_2O_5$ to $NO_2$ and $O_2$,

$$2\,N_2O_5(g) \rightarrow 4\,NO_2(g) + O_2(g)$$

is $3.4 \times 10^{-5}$ s$^{-1}$ at 298 K. What is the rate law expression for the reaction at 298 K?

**14.62. Hydroperoxyl Radicals in the Atmosphere** During a smog event, trace amounts of many highly reactive substances are present in the atmosphere. One of these is the hydroperoxyl radical, $HO_2$, which reacts with sulfur trioxide, $SO_3$. The rate constant for the reaction,

$$2\,HO_2(g) + SO_3(g) \rightarrow H_2SO_3(g) + 2\,O_2(g)$$

at 298 K is $2.6 \times 10^{11}\ M^{-1}\,s^{-1}$. The initial rate of the reaction doubles when the concentration of $SO_3$ or $HO_2$ is doubled. What is the rate law for the reaction?

**14.63. Disinfecting Municipal Water Supplies** Chlorine dioxide ($ClO_2$) is a disinfectant used in municipal water treatment plants (Figure P14.63). It dissolves in basic solution, producing $ClO_3^-$ and $ClO_2^-$:

$$2\,ClO_2(g) + 2\,OH^-(aq) \rightarrow ClO_3^-(aq) + ClO_2^-(aq) + H_2O(\ell)$$

**FIGURE P14.63**

The following kinetic data were obtained at 298 K for the reaction:

| Experiment | $[ClO_2]_0$ (M) | $[OH^-]_0$ (M) | Initial Rate (M/s) |
|---|---|---|---|
| 1 | 0.060 | 0.030 | 0.0248 |
| 2 | 0.020 | 0.030 | 0.00827 |
| 3 | 0.020 | 0.090 | 0.0247 |

Determine the rate law and the rate constant for this reaction at 298 K.

**14.64.** The following kinetic data were collected at 298 K for the reaction of ozone with nitrite ion, producing nitrate and oxygen:

$$NO_2^-(aq) + O_3(g) \rightarrow NO_3^-(aq) + O_2(g)$$

| Experiment | $[NO_2^-]_0$ (M) | $[O_3]_0$ (M) | Initial Rate (M/s) |
|---|---|---|---|
| 1 | 0.0100 | 0.0050 | 25 |
| 2 | 0.0150 | 0.0050 | 37.5 |
| 3 | 0.0200 | 0.0050 | 50.0 |
| 4 | 0.0200 | 0.0200 | 200.0 |

Determine the rate law for the reaction and the value of the rate constant.

**14.65.** Hydrogen gas reduces NO to $N_2$ in the following reaction:

$$2\,H_2(g) + 2\,NO(g) \rightarrow 2\,H_2O(g) + N_2(g)$$

The initial reaction rates of four mixtures of $H_2$ and NO were measured at 900°C with the following results:

| Experiment | $[H_2]_0$ (M) | $[NO]_0$ (M) | Initial Rate (M/s) |
|---|---|---|---|
| 1 | 0.212 | 0.136 | 0.0248 |
| 2 | 0.212 | 0.272 | 0.0991 |
| 3 | 0.424 | 0.544 | 0.793 |
| 4 | 0.848 | 0.544 | 1.59 |

Determine the rate law and the rate constant for the reaction at 900°C.

**14.66.** The rate of the reaction

$$NO_2(g) + CO(g) \rightarrow NO(g) + CO_2(g)$$

was determined in three experiments at 225°C. The results are given in the following table.

| Experiment | $[NO_2]_0$ (M) | $[CO]_0$ (M) | Initial Rate, $-\Delta[NO_2]/\Delta t$ (M/s) |
|---|---|---|---|
| 1 | 0.263 | 0.826 | $1.44 \times 10^{-5}$ |
| 2 | 0.263 | 0.413 | $1.44 \times 10^{-5}$ |
| 3 | 0.526 | 0.413 | $5.76 \times 10^{-5}$ |

a. Determine the rate law for the reaction.
b. Calculate the value of the rate constant at 225°C.
c. Calculate the rate of formation of $CO_2$ when $[NO_2] = [CO] = 0.500\ M$.

**14.67.** The reaction between acetaldehyde ($CH_3CH_2CHO$) and hydrocyanic acid (HCN) has been studied in aqueous solution at 25°C. Concentrations of reactants as a function of time are shown in the following table.
a. What is the average rate of consumption of HCN from 11.12 min to 40.35 min?
b. What is the average rate of consumption of acetaldehyde over that same period?

| Time (min) | $[CH_3CH_2CHO]$ (M) | $[HCN]$ (M) |
|---|---|---|
| 3.28 | 0.0384 | 0.0657 |
| 11.12 | 0.0346 | 0.0619 |
| 24.43 | 0.0296 | 0.0569 |
| 40.35 | 0.0242 | 0.0515 |
| 67.22 | 0.0190 | 0.0463 |

**14.68.** Two structural isomers of $ClO_2$ are shown in Figure P14.68. The isomer with the Cl–O–O skeletal arrangement is unstable and rapidly decomposes according to the reaction

$2\,ClOO(g) \rightarrow Cl_2(g) + 2\,O_2(g)$. The following data were collected for the decomposition of ClOO at 298 K:

| Time (μs) | [ClOO] (M) |
|-----------|------------|
| 0.0 | $1.76 \times 10^{-6}$ |
| 0.7 | $2.36 \times 10^{-7}$ |
| 1.3 | $3.56 \times 10^{-8}$ |
| 2.1 | $3.23 \times 10^{-9}$ |
| 2.8 | $3.96 \times 10^{-10}$ |

**FIGURE P14.68**

Determine the rate law for the reaction and the value of the rate constant at 298 K.

**14.69.** Hydrogen peroxide decomposes spontaneously into water and oxygen gas via a first-order reaction:

$$2\,H_2O_2(\ell) \rightarrow 2\,H_2O(g) + O_2(g)$$

but in the absence of catalysts this reaction proceeds very slowly. If a small amount of a salt containing the $Fe^{3+}$ ion is added to a 0.437 $M$ solution of $H_2O_2$ in water, the reaction proceeds with a half-life of 17.3 minutes. What is the concentration of the solution after 10.0 minutes under these conditions?

**14.70.** Labels of many food products have expiration dates, at which point they are typically removed from supermarket shelves. A particular natural yogurt degrades with a half-life of 45 days. The manufacturer of the yogurt wants unsold product pulled from the shelves when it degrades to no more than 80% of its original quality. Assume the degradation process is first order. What should be the "best if used before" date on the container with respect to the date the yogurt was packaged?

**14.71.** Acetoacetic acid, $CH_3COCH_2COOH$, decomposes in aqueous acidic solution to form acetone and carbon dioxide:

$$CH_3COCH_2COOH(aq) \rightarrow CH_3COCH_3(aq) + CO_2(g)$$

The reaction is first order. At room temperature, the half-life of the reactant is 139 min.
a. What is the rate constant of the decomposition reaction?
b. If the initial concentration of acetoacetic acid is 2.75 $M$, what is its concentration after 5 hours?

**14.72.** $p$-Toluenesulfinic acid undergoes a second-order redox reaction at room temperature (Figure P14.72).

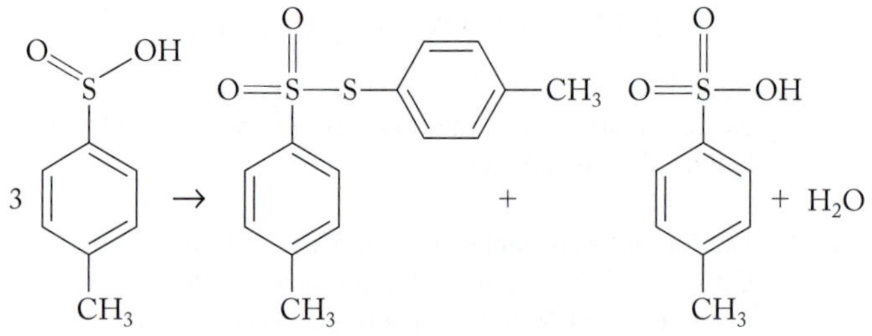

**FIGURE P14.72**

a. The value of the rate constant $k$ for the reaction is 0.141 L mol$^{-1}$ min$^{-1}$. What is the half-life of $p$-toluenesulfinic acid?
b. If the initial concentration of $p$-toluenesulfinic acid is 0.355 $M$, at what time will its concentration be 0.0355 $M$?

**14.73.** **Laughing Gas** Nitrous oxide ($N_2O$) is used as an anesthetic (laughing gas) and in aerosol cans to produce whipped cream. It is a potent greenhouse gas and decomposes slowly to $N_2$ and $O_2$:

$$2\,N_2O(g) \rightarrow 2\,N_2(g) + O_2(g)$$

a. If the plot of $\ln[N_2O]$ as a function of time is linear, what is the rate law for the reaction?
b. How many half-lives will it take for the concentration of the $N_2O$ to reach 6.25% of its original concentration? [*Hint*: The amount of reactant remaining after time $t$ ($A_t$) is related to the amount initially present ($A_0$) by the equation $A_t/A_0 = (0.5)^n$, where $n$ is the number of half-lives in time $t$.]

**14.74.** The unsaturated hydrocarbon butadiene ($C_4H_6$) dimerizes to 4-vinylcyclohexene ($C_8H_{12}$). When data collected in studies of the kinetics of this reaction were plotted against reaction time, plots of $[C_4H_6]$ or $\ln[C_4H_6]$ produced curved lines, but the plot of $1/[C_4H_6]$ was linear.
a. What is the rate law for the reaction?
b. How many half-lives will it take for the $[C_4H_6]$ to decrease to 3.1% of its original concentration?

**14.75. Tracing Phosphorus in Organisms** Radioactive isotopes such as $^{32}P$ are used to follow biological processes. The following radioactivity data (in relative radioactivity values) were collected for a sample containing $^{32}P$:

| Time (days) | Radioactivity (relative radioactivity values) |
|---|---|
| 0 | 10.0 |
| 1 | 9.53 |
| 2 | 9.08 |
| 5 | 7.85 |
| 10 | 6.16 |
| 20 | 3.79 |

a. Write the rate law for the decay of $^{32}P$.
b. Determine the value of the first-order rate constant.
c. Determine the half-life of $^{32}P$.

**14.76.** Nitrous acid slowly decomposes to NO, $NO_2$, and water in the following second-order reaction:

$$2\ HNO_2(aq) \rightarrow NO(g) + NO_2(g) + H_2O(\ell)$$

a. Use the data below to determine the rate constant for this reaction at 298 K.

| Time (min) | $[HNO_2]$ ($\mu M$) |
|---|---|
| 0 | 0.1560 |
| 1000 | 0.1466 |
| 1500 | 0.1424 |
| 2000 | 0.1383 |
| 2500 | 0.1345 |
| 3000 | 0.1309 |

b. Determine the half-life for the decomposition of $HNO_2$.

**14.77.** The dimerization of ClO,

$$2\ ClO(g) \rightarrow Cl_2O_2(g)$$

is second order in ClO. Use the following data to determine the value of $k$ at 298 K:

| Time (s) | [ClO] (molecules/cm³) |
|---|---|
| 0 | $2.60 \times 10^{11}$ |
| 1 | $1.08 \times 10^{11}$ |
| 2 | $6.83 \times 10^{10}$ |
| 3 | $4.99 \times 10^{10}$ |
| 4 | $3.93 \times 10^{10}$ |

Determine the half-life for the dimerization of ClO.

**14.78.** Kinetic data for the reaction $Cl_2O_2(g) \rightarrow 2\ ClO(g)$ are summarized in the following table. Determine the value of the first-order rate constant.

| Time ($\mu s$) | $[Cl_2O_2]$ (M) |
|---|---|
| 0 | $6.60 \times 10^{-8}$ |
| 172 | $5.68 \times 10^{-8}$ |
| 345 | $4.89 \times 10^{-8}$ |
| 517 | $4.21 \times 10^{-8}$ |
| 690 | $3.62 \times 10^{-8}$ |
| 862 | $3.12 \times 10^{-8}$ |

Determine the half-life for the decomposition of $Cl_2O_2$.

**14.79. Kinetics of Sucrose Hydrolysis** The metabolism of table sugar (sucrose, $C_{12}H_{22}O_{11}$) begins with the hydrolysis of the disaccharide to glucose and fructose (both $C_6H_{12}O_6$):

$$C_{12}H_{22}O_{11}(aq) + H_2O(\ell) \rightarrow 2\ C_6H_{12}O_6(aq)$$

The kinetics of the reaction were studied at 24°C in a reaction system with a large excess of water, so the reaction was pseudo-first-order in sucrose. Determine the rate law and the pseudo-first-order rate constant for the reaction from the following data:

| Time (s) | $[C_{12}H_{22}O_{11}]$ (M) |
|---|---|
| 0 | 0.562 |
| 612 | 0.541 |
| 1600 | 0.509 |
| 2420 | 0.484 |
| 3160 | 0.462 |
| 4800 | 0.4417 |

**14.80.** Hydroperoxyl radicals react rapidly with ozone to produce oxygen and OH radicals:

$$HO_2(g) + O_3(g) \rightarrow OH(g) + 2\ O_2(g)$$

The rate of this reaction was studied in the presence of a large excess of ozone. Determine the pseudo-first-order rate constant and the second-order rate constant for the reaction from the following data:

| Time (ms) | $[HO_2]$ (M) | $[O_3]$ (M) |
|---|---|---|
| 0 | $3.2 \times 10^{-6}$ | $1.0 \times 10^{-3}$ |
| 10 | $2.9 \times 10^{-6}$ | $1.0 \times 10^{-3}$ |
| 20 | $2.6 \times 10^{-6}$ | $1.0 \times 10^{-3}$ |
| 30 | $2.4 \times 10^{-6}$ | $1.0 \times 10^{-3}$ |
| 80 | $1.4 \times 10^{-6}$ | $1.0 \times 10^{-3}$ |

# Reaction Rates, Temperature, and the Arrhenius Equation

## CONCEPT REVIEW

**14.81.** Why are some reactions that occur in nature slower than others?

**14.82.** In many familiar reactions, high-energy reactants form lower-energy products. In such a reaction, is that activation energy barrier higher in the forward or in the reverse direction?

***14.83.** The order of a reaction is independent of temperature, but the value of the rate constant varies with temperature. Why?

**14.84.** Why do so many gas-phase reactions proceed more rapidly at higher temperatures?

***14.85.** Two first-order reactions have activation energies of 15 and 150 kJ/mol. Which reaction will show the larger increase in rate as temperature is increased?

**14.86.** According to the Arrhenius equation, does the activation energy of a chemical reaction depend on temperature? Explain your answer.

## PROBLEMS

**14.87.** The rate constant for the reaction of ozone with oxygen atoms was determined at four temperatures. Calculate the activation energy and frequency factor $A$ for the reaction

$$O(g) + O_3(g) \rightarrow 2\,O_2(g)$$

given the following data:

| $T$ (K) | $k$ [cm³/(molecule · s)] |
|---|---|
| 250 | $2.64 \times 10^{-4}$ |
| 275 | $5.58 \times 10^{-4}$ |
| 300 | $1.04 \times 10^{-3}$ |
| 325 | $1.77 \times 10^{-3}$ |

**14.88.** The rate constant for the reaction

$$NO_2(g) + O_3(g) \rightarrow NO_3(g) + O_2(g)$$

was determined over a temperature range of 40 K, with the following results:

| $T$ (K) | $k$ ($M^{-1}\,s^{-1}$) |
|---|---|
| 203 | $4.14 \times 10^5$ |
| 213 | $7.30 \times 10^5$ |
| 223 | $1.22 \times 10^6$ |
| 233 | $1.96 \times 10^6$ |
| 243 | $3.02 \times 10^6$ |

a. Determine the activation energy for the reaction.
b. Calculate the rate constant of the reaction at 300 K.

**14.89. Activation Energy for Smog-Forming Reactions** The initial step in the formation of smog is the reaction between nitrogen and oxygen. The activation energy of the reaction can be determined from the temperature dependence of the rate constants. At the temperatures indicated, values of the rate constant of the reaction

$$N_2(g) + O_2(g) \rightarrow 2\,NO(g)$$

are as follows:

| $T$ (K) | $k$ ($M^{-1/2}\,s^{-1}$) |
|---|---|
| 2000 | 318 |
| 2100 | 782 |
| 2200 | 1770 |
| 2300 | 3733 |
| 2400 | 7396 |

a. Calculate the activation energy of the reaction.
b. Calculate the frequency factor for the reaction.
c. Calculate the value of the rate constant at ambient temperature, $T = 300$ K.

**14.90.** Values of the rate constant for the decomposition of $N_2O_5$ gas at four different temperatures are as follows:

| $T$ (K) | $k$ ($s^{-1}$) |
|---|---|
| 658 | $2.14 \times 10^5$ |
| 673 | $3.23 \times 10^5$ |
| 688 | $4.81 \times 10^5$ |
| 703 | $7.03 \times 10^5$ |

a. Determine the activation energy of the decomposition reaction.
b. Calculate the value of the rate constant at 300 K.

**14.91. Activation Energy of Stratospheric Ozone Destruction Reactions** The kinetics of the reaction between chlorine dioxide and ozone is relevant to the study of atmospheric ozone destruction. The activation energy of the reaction can be determined from the temperature dependence of the rate constant. The value of the rate constant for the reaction between chlorine dioxide and ozone was measured at four temperatures between 193 and 208 K. The results are as follows:

| $T$ (K) | $k$ ($M^{-1}\,s^{-1}$) |
|---|---|
| 193 | 34.0 |
| 198 | 62.8 |
| 203 | 112.8 |
| 208 | 196.7 |

Calculate the values of the activation energy and the frequency factor for the reaction.

**14.92.** Chlorine atoms react with methane, forming HCl and $CH_3$. The rate constant for the reaction is $6.0 \times 10^7\ M^{-1}\,s^{-1}$ at

298 K. When the experiment was repeated at three other temperatures, the following data were collected:

| T (K) | k (M⁻¹ s⁻¹) |
|-------|-------------|
| 303 | $6.5 \times 10^7$ |
| 308 | $7.0 \times 10^7$ |
| 313 | $7.5 \times 10^7$ |

Calculate the values of the activation energy and the frequency factor for the reaction.

## Reaction Mechanisms

### CONCEPT REVIEW

**14.93.** The reaction between NO and $Cl_2$ is first order in each reactant. Does this mean that the reaction could occur in just one step?

**14.94.** The reaction between NO and $H_2$ is second order in NO. Does this mean that the reaction could occur in just one step?

***14.95.** Under what reaction conditions does a bimolecular reaction obey pseudo-first-order reaction kinetics?

***14.96.** If a reaction is zero order in a reactant, does that mean the reactant is never involved in collisions with other reactants? Explain your answer.

### PROBLEMS

**14.97.** If the reaction A → B is first order in A and first order overall, does it occur in just one step?

**14.98.** Substance A decomposes slowly into substance B, which then rapidly decomposes into substances C and D. Sketch a reaction profile for the reaction A→ C + D adding labels in the appropriate locations for the four substances involved.

**14.99.** Write the rate laws for the following elementary steps and identify them as uni-, bi-, or termolecular steps:
a. $SO_2Cl_2(g) \rightarrow SO_2(g) + Cl_2(g)$
b. $NO_2(g) + CO(g) \rightarrow NO(g) + CO_2(g)$
c. $2\,NO_2(g) \rightarrow NO_3(g) + NO(g)$

**14.100.** Write the rate laws for the following elementary steps and identify them as uni-, bi-, or termolecular steps:
a. $Cl(g) + O_3(g) \rightarrow ClO(g) + O_2(g)$
b. $2\,NO_2(g) \rightarrow N_2O_4(g)$
*c. $^{14}_{6}C \rightarrow {}^{14}_{7}N + {}^{0}_{-1}\beta$

**14.101.** Write the overall reaction that consists of the following elementary steps:

$$N_2O_5(g) \rightarrow NO_3(g) + NO_2(g)$$
$$NO_3(g) \rightarrow NO_2(g) + O(g)$$
$$2\,O(g) \rightarrow O_2(g)$$

**14.102.** What overall reaction consists of the following elementary steps?

$$ClO^-(aq) + H_2O(\ell) \rightarrow HClO(aq) + OH^-(aq)$$
$$I^-(aq) + HClO(aq) \rightarrow HIO(aq) + Cl^-(aq)$$
$$OH^-(aq) + HIO(aq) \rightarrow H_2O(\ell) + IO^-(aq)$$

***14.103.** In the following mechanism for NO formation, oxygen atoms are produced by breaking O=O bonds at high temperature in a fast, reversible reaction. If $\Delta[NO]/\Delta t = k[N_2][O_2]^{1/2}$, which step in the mechanism is the rate-determining step?

(1)       $O_2(g) \rightleftharpoons 2\,O(g)$
(2)       $O(g) + N_2(g) \rightarrow NO(g) + N(g)$
(3)       $N(g) + O(g) \rightarrow NO(g)$
Overall    $N_2(g) + O_2(g) \rightarrow 2\,NO(g)$

**14.104.** A proposed mechanism for the decomposition of hydrogen peroxide consists of three elementary steps:

$$H_2O_2(g) \rightarrow 2\,OH(g)$$
$$H_2O_2(g) + OH(g) \rightarrow H_2O(g) + HO_2(g)$$
$$HO_2(g) + OH(g) \rightarrow H_2O(g) + O_2(g)$$

If the rate law for the reaction is first order in $H_2O_2$, which step in the mechanism is the rate-determining step?

**14.105.** At a given temperature, the rate of the reaction between NO and $Cl_2$ is proportional to the product of the concentrations of the two gases: $[NO][Cl_2]$. The following two-step mechanism was proposed for the reaction:

(1)       $NO(g) + Cl_2(g) \rightarrow NOCl_2(g)$
(2)     $NOCl_2(g) + NO(g) \rightarrow 2\,NOCl(g)$
Overall    $2\,NO(g) + Cl_2(g) \rightarrow 2\,NOCl(g)$

Which step must be the rate-determining step if this mechanism is correct?

**14.106. Mechanism of Ozone Destruction** Ozone decomposes thermally to oxygen in the following reaction:

$$2\,O_3(g) \rightarrow 3\,O_2(g)$$

The following mechanism has been proposed:

$$O_3(g) \rightarrow O(g) + O_2(g)$$
$$O(g) + O_3(g) \rightarrow 2\,O_2(g)$$

The reaction is second order in ozone. What properties of the two elementary steps (specifically, relative rate and reversibility) are consistent with this mechanism?

**14.107. Mechanism of NO$_2$ Destruction** The rate laws for the thermal and photochemical decomposition of NO$_2$ are different. Which of the following mechanisms are possible for the thermal decomposition of NO$_2$, and which are possible for the photochemical decomposition of NO$_2$? For thermal decomposition, Rate $= k[NO_2]^2$, and for photochemical decomposition, Rate $= k[NO_2]$.

a. $$NO_2(g) \xrightarrow{\text{slow}} NO(g) + O(g)$$
$$O(g) + NO_2(g) \xrightarrow{\text{fast}} NO(g) + O_2(g)$$

b. $$NO_2(g) + NO_2(g) \xrightarrow{\text{fast}} N_2O_4(g)$$
$$N_2O_4(g) \xrightarrow{\text{slow}} NO(g) + NO_3(g)$$
$$NO_3(g) \xrightarrow{\text{fast}} NO(g) + O_2(g)$$

c. $$NO_2(g) + NO_2(g) \xrightarrow{\text{slow}} NO(g) + NO_3(g)$$
$$NO_3(g) \xrightarrow{\text{fast}} NO(g) + O_2(g)$$

**14.108.** The rate laws for the thermal and photochemical decomposition of NO$_2$ are different. Which of the following mechanisms are possible for the thermal decomposition of NO$_2$, and which are possible for the photochemical decomposition of NO$_2$? For thermal decomposition, Rate $= k[NO_2]^2$, and for photochemical decomposition, Rate $= k[NO_2]$.

a. $$NO_2(g) + NO_2(g) \xrightarrow{\text{slow}} N_2O_4(g)$$
$$N_2O_4(g) \xrightarrow{\text{fast}} N_2O_3(g) + O(g)$$
$$N_2O_3(g) + O(g) \xrightarrow{\text{fast}} N_2O_2(g) + O_2(g)$$
$$N_2O_2(g) \xrightarrow{\text{fast}} 2 NO(g)$$

b. $$NO_2(g) + NO_2(g) \xrightarrow{\text{slow}} NO(g) + NO_3(g)$$
$$NO_3(g) \xrightarrow{\text{fast}} NO(g) + O_2(g)$$

c. $$NO_2(g) \xrightarrow{\text{slow}} N(g) + O_2(g)$$
$$N(g) + NO_2(g) \xrightarrow{\text{fast}} N_2O_2(g)$$
$$N_2O_2(g) \xrightarrow{\text{slow}} 2 NO(g)$$

## Catalysis

### CONCEPT REVIEW

**14.109.** Does a catalyst affect both the rate and the rate constant of a reaction?

**14.110.** Is the rate law for a catalyzed reaction the same as that for the uncatalyzed reaction?

**14.111.** Does a substance that increases the rate of a reaction also increase the rate of the reverse reaction?

**14.112.** The rate of the reaction between NO$_2$ and CO is independent of [CO]. Does this mean that CO is a catalyst for the reaction?

**14.113.** Why doesn't the concentration of a homogeneous catalyst appear in the rate law for the reaction it catalyzes?

*14.114. The rate of a chemical reaction is too slow to measure at room temperature. We could either raise the temperature or add a catalyst. Which would be a better solution for making an accurate determination of the rate constant?

### PROBLEMS

**14.115.** Is NO a catalyst for the decomposition of N$_2$O in the following two-step reaction mechanism, or is N$_2$O a catalyst for the conversion of NO to NO$_2$?

(1) $$NO(g) + N_2O(g) \rightarrow N_2(g) + NO_2(g)$$
(2) $$2 NO_2(g) \rightarrow 2 NO(g) + O_2(g)$$

**14.116. NO as a Catalyst for Ozone Destruction** Explain why NO is a catalyst in the following two-step process that results in the depletion of ozone in the stratosphere:

(1) $$NO(g) + O_3(g) \rightarrow NO_2(g) + O_2(g)$$
(2) $$O(g) + NO_2(g) \rightarrow NO(g) + O_2(g)$$
Overall $$O(g) + O_3(g) \rightarrow 2 O_2(g)$$

---

**14.117.** On the basis of the frequency factors and activation energy values of the following two reactions, determine which one will have the larger rate constant at room temperature (298 K).

$$O_3(g) + O(g) \rightarrow O_2(g) + O_2(g)$$
$A = 8.0 \times 10^{-12}$ cm$^3$/(molecules $\cdot$ s)     $E_a = 17.1$ kJ/mol

$$O_3(g) + Cl(g) \rightarrow ClO(g) + O_2(g)$$
$A = 2.9 \times 10^{-11}$ cm$^3$/(molecules $\cdot$ s)     $E_a = 2.16$ kJ/mol

**14.118.** On the basis of the frequency factors and activation energy values of the following two reactions, determine which one will have the larger rate constant at room temperature (298 K).

$$O_3(g) + Cl(g) \rightarrow ClO(g) + O_2(g)$$
$A = 2.9 \times 10^{-11}$ cm$^3$/(molecules $\cdot$ s)     $E_a = 2.16$ kJ/mol

$$O_3(g) + NO(g) \rightarrow NO_2(g) + O_2(g)$$
$A = 2.0 \times 10^{-12}$ cm$^3$/(molecules $\cdot$ s)     $E_a = 11.6$ kJ/mol

## Additional Problems

**14.119.** A student inserts a glowing wood splint into a test tube filled with O$_2$. The splint quickly catches fire (Figure P14.119). Why does the splint burn so much faster in pure O$_2$ than in air?

**FIGURE P14.119**

*14.120. A backyard chef turns on the propane gas to a barbecue grill. Even though the reaction between propane and oxygen is spontaneous, the gas does not begin to burn until the chef pushes an igniter button to produce a spark. Why is the spark needed?

14.121. On average, someone who falls through the ice covering a frozen lake is less likely to experience anoxia (lack of oxygen) than someone who falls into a warm pool and is underwater for the same length of time. Why?

*14.122. Why doesn't a quadrupling of the rate correspond to a reaction order of 4, for example, Rate $\propto [NO]^4$?

14.123. If the rate of the reverse reaction is much slower than the rate of the forward reaction, does the method used to determine a rate law from initial concentrations and initial rates also work at some other time $t$? What concentrations would we use in the case where we use the rate when $t \neq 0$?

14.124. What is wrong with the following statement: The reaction rate and the rate constant for a reaction both depend on the number of collisions and on the concentrations of the reactants.

14.125. How do we find $k$ if we plot $1/[X] - 1/[X]_0$ as a function of $t$?

14.126. Many reactions are first order, fewer are second order in a single reactant, and third-order reactions in a single reactant are practically nonexistent. Can you suggest why?

14.127. Why can't an elementary step in a mechanism have a rate law that is zero order in a reactant?

14.128. During the decomposition of dinitrogen pentoxide,

$$2 N_2O_5(g) \rightarrow 4 NO_2(g) + O_2(g)$$

how is the rate of consumption of $N_2O_5$ related to the rate of formation of $NO_2$ and $O_2$?

14.129. In the reaction between nitrogen dioxide and ozone,

$$2 NO_2(g) + O_3(g) \rightarrow N_2O_5(g) + O_2(g)$$

how are the rates of change in the concentrations of the reactants and products related?

14.130. **Test for Herbicide** Sodium chlorate was used in weed-control preparations, but its sale has been banned in EU countries since 2009. A simple colorimetric test for the presence of the chlorate ion in solution of herbicide relies on the following reaction:

$$2 MnO_4^-(aq) + 5 ClO_3^-(aq) + 6 H^+(aq) \rightarrow$$
$$2 Mn^+(aq) + 5 ClO_4^-(aq) + 3 H_2O(\ell)$$

The table below contains rate data for this reaction.

| Experiment | $[MnO_4^-]$ (M) | $[ClO_3^-]$ (M) | $[H^+]$ (M) | Initial rate (M/s) |
|---|---|---|---|---|
| 1 | 0.10 | 0.10 | 0.10 | $5.2 \times 10^{-3}$ |
| 2 | 0.25 | 0.10 | 0.10 | $3.3 \times 10^{-2}$ |
| 3 | 0.10 | 0.30 | 0.10 | $1.6 \times 10^{-2}$ |
| 4 | 0.10 | 0.10 | 0.20 | $7.4 \times 10^{-3}$ |

Determine the rate law and the rate constant for this reaction.

14.131. At the temperature at which the experiments were carried out in Problem 14.129, what is the rate constant for the decomposition of $N_2O_5$? Write the complete rate law for the decomposition reaction.

14.132. The table below contains reaction rate data for the reaction

$$2 NO(g) + Cl_2(g) \rightarrow 2 NOCl(g)$$

| Experiment | $[NO]_0$ (M) | $[Cl_2]_0$ (M) | Initial Rate (M/s) |
|---|---|---|---|
| 1 | 0.20 | 0.10 | 0.63 |
| 2 | 0.20 | 0.30 | 5.70 |
| 3 | 0.80 | 0.10 | 2.58 |
| 4 | 0.40 | 0.20 | ? |

Predict the initial rate of reaction in experiment 4.

14.133. An important reaction in the formation of photochemical smog is the reaction between ozone and NO:

$$NO(g) + O_3(g) \rightarrow NO_2(g) + O_2(g)$$

The reaction is first order in NO and $O_3$. The rate constant of the reaction is 80 $M^{-1}$ $s^{-1}$ at 25°C and 3000 $M^{-1}$ $s^{-1}$ at 75°C.

a. If this reaction were to occur in a single step, would the rate law be consistent with the observed order of the reaction for NO and $O_3$?

b. What is the value of the activation energy of the reaction?

c. What is the rate of the reaction at 25°C when $[NO] = 3 \times 10^{-6}$ $M$ and $[O_3] = 5 \times 10^{-9}$ $M$?

d. Predict the values of the rate constant at 10°C and 35°C.

14.134. Ammonia reacts with nitrous acid to form an intermediate, ammonium nitrite ($NH_4NO_2$), which decomposes to $N_2$ and $H_2O$:

$$NH_3(g) + HNO_2(aq) \rightarrow NH_4NO_2(aq) \rightarrow N_2(g) + 2 H_2O(\ell)$$

a. The reaction is first order in ammonia and second order in nitrous acid. What is the rate law for the reaction? What are the units of the rate constant if concentrations are expressed in molarity and time in seconds?

b. The rate law for the reaction has also been written as Rate $= k[NH_4^+][NO_2^-][HNO_2]$. Is this expression equivalent to the one you wrote in part a?

c. With the data in Appendix 4, calculate the value of $\Delta H°_{rxn}$ of the overall reaction ($\Delta H°_f$, $HNO_2 = -43.1$ kJ/mol).

d. Draw an energy profile for the process with the assumption that $E_a$ of the first step is lower than $E_a$ of the second step.

\*14.135. When ionic compounds such as NaCl dissolve in water, the sodium ions are surrounded by six water molecules. The bound water molecules exchange with those in bulk solution as described by the reaction involving $^{18}O$-enriched water:

$$Na(H_2O)_6^+(aq) + H_2^{18}O(\ell) \rightarrow Na(H_2O)_5(H_2^{18}O)^+(aq) + H_2O(\ell)$$

a. The following reaction mechanism has been proposed:

(1)  $$Na(H_2O)_6^+(aq) \rightarrow Na(H_2O)_5^+(aq) + H_2O(\ell)$$

(2)  $$Na(H_2O)_5^+(aq) + H_2^{18}O(\ell) \rightarrow Na(H_2O)_5(H_2^{18}O)^+(aq)$$

What is the rate law if the first step is the rate-determining step?

b. If you were to sketch an energy profile, which would you draw with the higher energy, the reactants or the products?

14.136. **Lachrymators in Smog** The combination of ozone, volatile hydrocarbons, nitrogen oxide, and sunlight in urban environments produces peroxyacetyl nitrate (PAN), a potent lachrymator. PAN decomposes to acetyl radicals and nitrogen dioxide in a process that is second order in PAN, as shown in Figure P14.136:

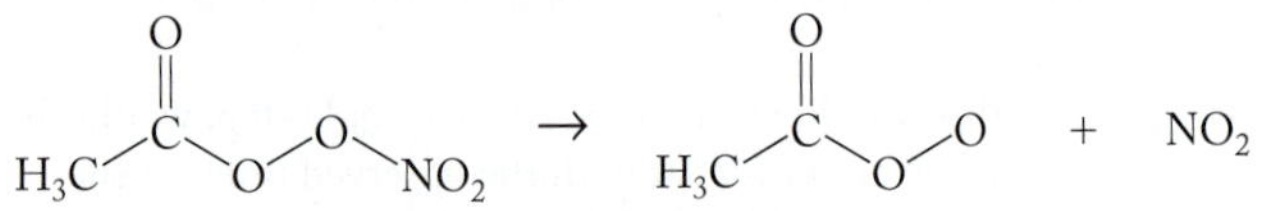

**FIGURE P14.136**

a. The half-life of the reaction, at 23°C and $P_{CH_3CO_3NO_2} = 10.5$ torr, is 100 hr. Calculate the rate constant for the reaction.

b. Determine the rate of the reaction at 23°C and $P_{CH_3CO_3NO_2} = 10.5$ torr.

c. Draw a graph showing $P_{PAN}$ as a function of time from 0 to 200 hr starting with $P_{CH_3CO_3NO_2} = 10.5$ torr.

14.137. **Nitric Oxide in the Human Body** Nitric oxide (NO) is a gaseous free radical that plays many biological roles including regulating neurotransmission and the human immune system. One of its many reactions involves the peroxynitrite ion (ONOO⁻):

$$NO(g) + ONOO^-(aq) \rightarrow NO_2(g) + NO_2^-(aq)$$

a. Use the following data to determine the rate law and rate constant of the reaction at the experimental temperature at which these data were generated.

| Experiment | [NO]$_0$ (M) | [ONOO⁻]$_0$ (M) | Rate (M/s) |
|---|---|---|---|
| 1 | $1.25 \times 10^{-4}$ | $1.25 \times 10^{-4}$ | $2.03 \times 10^{-11}$ |
| 2 | $1.25 \times 10^{-4}$ | $0.625 \times 10^{-4}$ | $1.02 \times 10^{-11}$ |
| 3 | $0.625 \times 10^{-4}$ | $2.50 \times 10^{-4}$ | $2.03 \times 10^{-11}$ |
| 4 | $0.625 \times 10^{-4}$ | $3.75 \times 10^{-4}$ | $3.05 \times 10^{-11}$ |

b. Draw the Lewis structure of peroxynitrite ion (including all resonance forms) and assign formal charges. Note which form is preferred.

c. Use the average bond energies in Table A4.1 to estimate the value of $\Delta H^\circ_{rxn}$ using the preferred structure from part b.

14.138. **Kinetics of Protein Chemistry** In the presence of $O_2$, NO reacts with sulfur-containing proteins to form S-nitrosothiols, such as $C_6H_{13}SNO$. This compound decomposes to form a disulfide and NO:

$$2\ C_6H_{13}SNO(aq) \rightarrow 2\ NO(g) + C_{12}H_{26}S_2(aq)$$

The following data were collected for the decomposition reaction at 69°C.

| Time (min) | [C$_6$H$_{13}$SNO] (M) |
|---|---|
| 0 | $1.05 \times 10^{-3}$ |
| 10 | $9.84 \times 10^{-4}$ |
| 20 | $9.22 \times 10^{-4}$ |
| 30 | $8.64 \times 10^{-4}$ |
| 60 | $7.11 \times 10^{-4}$ |

a. Calculate the value of the first-order rate constant for the reaction.

b. Which amino acids might act as sources of S-nitrosothiols?

14.139. Solutions of nitrous acid, $HNO_2$, in $^{18}O$-labeled water undergo isotope exchange:

$$HNO_2(aq) + H_2^{18}O(\ell) \rightarrow HN^{18}O_2(aq) + H_2O(\ell)$$

a. Use the following data at 24°C to determine the dependence of the reaction rate on the concentration of $HNO_2$.

| Time (min) | [HN$^{18}$O$_2$] (M) |
|---|---|
| 0 | $5.4 \times 10^{-2}$ |
| 20 | $1.5 \times 10^{-3}$ |
| 40 | $7.7 \times 10^{-4}$ |
| 60 | $5.2 \times 10^{-4}$ |

b. Does the reaction rate depend on the concentration of $H_2^{18}O$?

14.140. Ethylene ($C_2H_4$) reacts with ozone to form 2 mol of formaldehyde (a probable human carcinogen) per mole of ethylene, as shown in Figure P14.140. The following kinetic data were collected at 298 K.

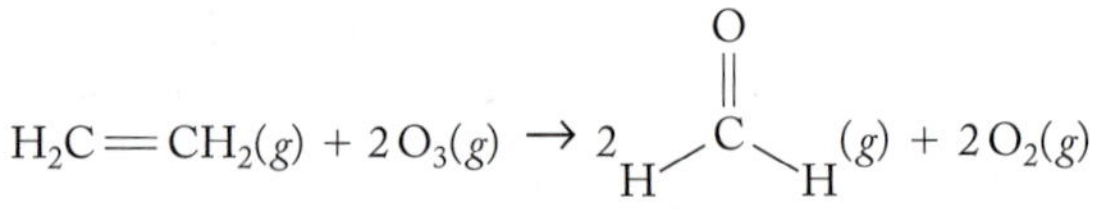

**FIGURE P14.140**

| Experiment | [O$_3$]$_0$ (M) | [C$_2$H$_4$]$_0$ (M) | Rate (M/s) |
|---|---|---|---|
| 1 | $0.86 \times 10^{-2}$ | $1.00 \times 10^{-2}$ | 0.0877 |
| 2 | $0.43 \times 10^{-2}$ | $1.00 \times 10^{-2}$ | 0.0439 |
| 3 | $0.22 \times 10^{-2}$ | $0.50 \times 10^{-2}$ | 0.0110 |

a. Determine the rate law and the value of the rate constant of the reaction at 298 K.

b. The rate constant was determined at several additional temperatures. Calculate the activation energy of the reaction from the following data.

| T (K) | k (M⁻¹ s⁻¹) |
|---|---|
| 263 | $3.28 \times 10^2$ |
| 273 | $4.73 \times 10^2$ |
| 283 | $6.65 \times 10^2$ |
| 293 | $9.13 \times 10^2$ |

**14.141.** **Reducing NO Emissions** Adding $NH_3$ to the stack gases at an electric-power-generating plant can reduce $NO_x$ emissions. This selective noncatalytic reduction (SNR) process depends on the reaction between $NH_2$ (an odd-electron molecule) and NO:

$$NH_2(g) + NO(g) \rightarrow N_2(g) + H_2O(g)$$

The following kinetic data were collected at 1200 K.

| Experiment | [NH₂]₀ (M) | [NO]₀ (M) | Rate (M/s) |
|---|---|---|---|
| 1 | $1.00 \times 10^{-5}$ | $1.00 \times 10^{-5}$ | 0.12 |
| 2 | $2.00 \times 10^{-5}$ | $1.00 \times 10^{-5}$ | 0.24 |
| 3 | $2.00 \times 10^{-5}$ | $1.50 \times 10^{-5}$ | 0.36 |
| 4 | $2.50 \times 10^{-5}$ | $1.50 \times 10^{-5}$ | 0.45 |

a. What is the rate law for the reaction?
b. What is the value of the rate constant at 1200 K?

If your instructor assigns problems in **smartwork**, log in at **smartwork.wwnorton.com**.

# 15

# Chemical Equilibrium: How Much Product Does a Reaction Really Make?

**15.1** The Dynamics of Chemical Equilibrium

**15.2** Writing Equilibrium Constant Expressions

**15.3** Relationships between $K_c$ and $K_p$ Values

**15.4** Manipulating Equilibrium Constant Expressions

**15.5** Equilibrium Constants and Reaction Quotients

**15.6** Heterogeneous Equilibria

**15.7** Le Châtelier's Principle

**15.8** Calculations Based on $K$

## Learning Outcomes

**LO1** Write mass action or equilibrium constant expressions for reversible reactions including those involving heterogeneous equilibria
**Sample Exercises 15.1, 15.9**

**LO2** Calculate the value of an equilibrium constant or a reaction quotient and use it to predict the direction of a reversible chemical reaction
**Sample Exercises 15.2, 15.3, 15.8**

**LO3** Interconvert the $K_c$ and $K_p$ values of gas-phase reactions
**Sample Exercise 15.4**

**LO4** Calculate the value of $K$ for a reverse reaction, for a reaction with different coefficients, and for combined reactions
**Sample Exercises 15.5, 15.6, 15.7**

**LO5** Predict how a reaction at equilibrium responds to changes in conditions
**Sample Exercises 15.10, 15.11, 15.12**

**LO6** Calculate the concentrations or partial pressures of reactants and products in a reaction mixture at equilibrium from their starting values and the value of $K$
**Sample Exercises 15.13, 15.14, 15.15**

## Making Commercial Goods Economically

Consider a simple aspirin tablet—the kind you take for relief from a headache or the symptoms of a cold. If you trace this common medication back to its ultimate sources, the precursors of aspirin are air, water, sulfur, petroleum, and natural gas. The chemical industry that produces aspirin and other drugs—along with polymers, fertilizers, fuels, fabrics, and virtually all the items we associate with modern life—from these raw materials is a $3 trillion enterprise that has an impact on almost every sector of the world's economy. Many industrial processes involve the manipulation of physical and chemical equilibria, the subject of the next two chapters.

Sulfur does not appear in molecules of aspirin but is involved in making the drug because it is needed to produce sulfuric acid, perhaps the most important industrial chemical worldwide; indeed, the amount of sulfuric acid produced by a nation can be viewed as a good indication of that nation's economic health and prosperity. The acid has widely varied uses and plays some part in the production of nearly all manufactured goods. The industrial production of sulfuric acid involves the reversible reaction between sulfur dioxide gas (formed by the combustion of elemental sulfur) and oxygen to form sulfur trioxide:

$$2\,SO_2(g) + O_2(g) \rightleftharpoons 2\,SO_3(g)$$

The $SO_3$ gas is dissolved in water to produce sulfuric acid. The double arrows between reactants and products mean that the reaction proceeds in both directions simultaneously: $SO_2$ and $O_2$ react to form $SO_3$, and $SO_3$ decomposes to reform $SO_2$ and $O_2$. When no further change in concentration of any component occurs, the system is at *equilibrium*, but that equilibrium is *dynamic*: the reactions continue to take place.

This reaction is exothermic in the forward direction (from left to right), and the equilibrium favors product under standard conditions. However, for various reasons the reaction is routinely run at high temperature, which actually decreases the yield. Because high pres-

**Manufacturing Sulfuric Acid** More sulfuric acid is manufactured each year than any other chemical. Large industrial plants like the one pictured here produce up to 4500 tons of acid per day. ▶

**chemical equilibrium** a dynamic process in which the concentrations of reactants and products remain constant over time and the rate of a reaction in the forward direction matches its rate in the reverse direction.

sure favors production of $SO_3$, typical industrial conditions are a compromise between temperatures high enough to ensure a suitable rate of reaction and pressures high enough to produce yields close to 99%.

A chemical reaction like the conversion of $SO_2$ to $SO_3$ "goes to completion," meaning reactants produce very close to 100% of the products you would predict based on the stochiometry of the reaction. Many other reactions do not seem to proceed at all, and little or no product is formed. Most processes are somewhere in between these two extremes, including many reactions of great commercial importance.

In Chapter 14 we studied kinetics, which tell us how fast a reaction occurs. The study of chemical equilibria is just as important, for it tells us the extent to which a reaction proceeds. Just as we saw how altering conditions such as temperature, pressure, and quantity of materials changes the rate at which a process occurs, these same variables influence the extent to which a reaction proceeds. Chemists use the principles of kinetics and equilibrium routinely to select the reaction conditions that give the highest possible yield of a product in a reasonable length of time. ■

## 15.1 The Dynamics of Chemical Equilibrium

**Chemical equilibrium** is a dynamic process in which the reactants in a reversible chemical reaction are constantly converted into products, while at the same time products are constantly converted back into reactants. This reversibility is represented by a pair of arrows pointed in opposite directions:

$$\text{Reactants} \rightleftharpoons \text{products}$$

At equilibrium, the rate at which reactants become products equals the rate at which products turn back into reactants; no net change in the concentrations of the reactants and products occurs over time. Said another way, the rate of the forward reaction is the same as the rate of the reverse reaction:

$$\text{Rate}_{forward} = \text{rate}_{reverse}$$

▶❚❚ **CHEMTOUR** Equilibrium

Some reactions reach equilibrium only after nearly all the reactants have formed products. We say that these equilibria *lie far to the right* (the direction of the forward reaction arrow). In other reactions, little product has formed at the point when equilibrium is reached. We say that these equilibria favor reactants and that they *lie far to the left* (the direction of the reverse reaction arrow). Keep in mind that we cannot infer anything from the position of the equilibrium regarding how much time the reaction takes to reach equilibrium. Studies of equilibrium reveal only the extent to which a reaction proceeds, not how rapidly it proceeds.

To explore the dynamics of chemical equilibrium, let's look at a two-step industrial process for making $H_2$ gas. Approximately 9 million tons of hydrogen per year is made via this process in the United States. The initial reactants are methane and steam. In the first step, methane reacts with steam in the presence of a nickel or iron oxide catalyst at temperatures near 1000°C:

$$CH_4(g) + H_2O(g) \rightleftharpoons CO(g) + 3\,H_2(g) \qquad (15.1)$$

◑◐ **CONNECTION** We introduced the steam–methane reforming reaction and the water–gas shift reaction in the descriptive chemistry box at the end of Chapter 3.

This reaction, called the steam–methane reforming reaction, is followed by another, called the water–gas shift reaction. In the second step, carbon monoxide formed in the first step reacts with more steam in the presence of a Cu/ZnO catalyst at around 200°C:

$$H_2O(g) + CO(g) \rightleftharpoons H_2(g) + CO_2(g) \qquad (15.2)$$

Let's explore the dynamics of the second reaction. If we put an equal number of moles of water vapor and carbon monoxide in a closed chamber and allow them to react, the concentrations of CO and $H_2O$ initially fall as the concentrations of $H_2$ and $CO_2$ increase, as shown in Figure 15.1. Because $H_2O$ and CO react in a 1:1 stoichiometric ratio, their concentrations decrease at the same rate and are always the same in the reaction chamber. Similarly, $H_2$ and $CO_2$ are formed in a 1:1 ratio, so their concentrations increase at the same rate and are always equal.

The water–gas shift reaction is reversible. Reversibility means that if we intervene in the course of the reaction by, for example, changing reaction temperature or removing or adding a reactant or product, we might be able to force the reaction to run in reverse, reforming reactants from products. One requirement for reversibility is that the products must remain in contact with each other. If one or more of the products is a gas, then we need to run the reaction in a sealed chamber.

The graph in Figure 15.1 shows that eventually the concentrations of reactants and products in the water–gas shift reaction no longer change with time, at which point the reaction has reached chemical equilibrium. Note, however, that the concentrations of the reactants (CO and $H_2O$) do not go to zero. The presence of reactants in the mixture when the reaction has reached equilibrium means that the yield of the reaction never reaches 100%.

Now let's think about the changes in the *rates* of the forward and reverse reactions during the course of the water–gas shift reaction. When the CO and $H_2O$ are initially mixed, their concentrations are at their maximum and the rate of the forward reaction, as indicated by the slope of the [CO] and [$H_2O$] curve at $t = 0$ in Figure 15.1, is also at its maximum. As the reaction proceeds, reactant concentrations decrease, and the rate of the forward reaction also decreases because the likelihood of collisions between reactant molecules decreases.

The concentrations of products are initially zero and so is the rate of the reverse reaction. As product molecules form, the likelihood of their colliding with one

**CONNECTION** The method for calculating the percent yield of a reaction was described in Section 3.9.

**CONNECTION** In Chapter 14 we discussed how reactions occur when molecules collide; the higher the concentrations of reactants, the more frequently molecules collide, and the faster a reaction proceeds.

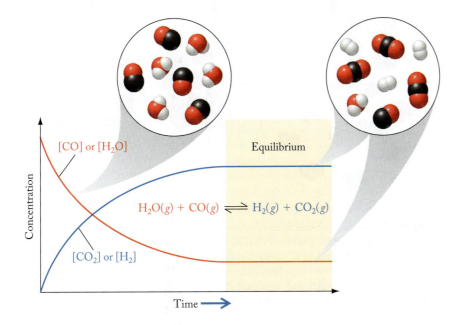

**FIGURE 15.1** Concentrations of reactants and products in the water–gas shift reaction change over time until equilibrium is reached. At equilibrium in the water–gas shift reaction, $H_2$ and $CO_2$ are being formed at the same rate at which they are reacting to reform $H_2O$ and CO.

[CO] or [$H_2O$]

Equilibrium

$H_2O(g) + CO(g) \rightleftharpoons H_2(g) + CO_2(g)$

[$CO_2$] or [$H_2$]

Concentration

Time

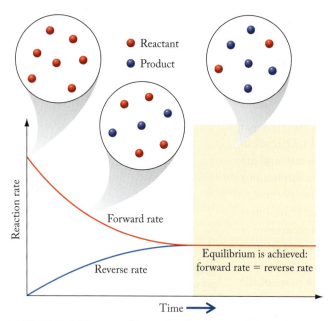

**FIGURE 15.2** The rates of forward and reverse reactions are the same when equilibrium is achieved.

another to reform reactant molecules increases and so does the rate of the reverse reaction. Ultimately, the rate of the reverse reaction equals the rate of the forward reaction, as shown in Figure 15.2. At this point equilibrium has been achieved.

We can explore the concept of chemical equilibrium quantitatively by first examining the kinetics of a reversible chemical reaction. We consider a reaction with a single reactant: $NO_2$, the gas that gives photochemical smog its distinctive brown color (Figure 15.3). Suppose we fill a large transparent syringe with an equilibrium mixture of $NO_2$ and $N_2O_4$, as shown in Figure 15.4(a). Then we push on the syringe plunger so that the gas is rapidly squeezed into half its original volume (Figure 15.4b). The color of the gas is instantly darker brown because we doubled its concentration by halving its volume. However, the darker brown color rapidly fades, as shown in Figure 15.4(c). The loss of color is due to a reaction in which pairs of molecules of $NO_2$ combine to form $N_2O_4$, which is a colorless gas:

$$2\ NO_2(g) \rightleftharpoons N_2O_4(g) \qquad (15.3)$$
$$\text{(brown)} \qquad \text{(colorless)}$$

The color never fades completely, however, because the reaction reaches an equilibrium in which some $NO_2$ is still present.

**FIGURE 15.3** Air quality in Los Angeles has improved since this photograph of "brown LA haze" was taken. The color was caused by high concentrations of $NO_2$.

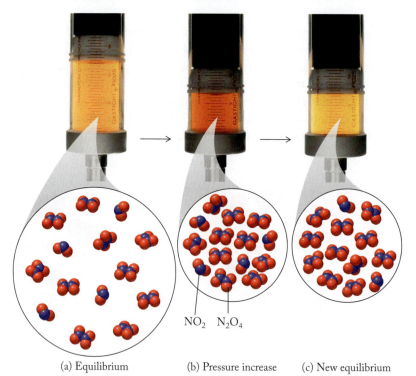

(a) Equilibrium      (b) Pressure increase      (c) New equilibrium

**FIGURE 15.4** Changing pressure affects equilibrium in a gas-phase reaction. (a) A gastight syringe contains an equilibrium reaction mixture of brown $NO_2(g)$ and colorless $N_2O_4(g)$. (b) The plunger is rapidly pushed in, increasing the pressure inside the syringe. The color of the mixture is instantly darker as the $NO_2$ molecules are compressed into a smaller volume. (c) The color fades rapidly as brown $NO_2$ forms colorless $N_2O_4$. Two moles of reactant ($NO_2$) are consumed for every one mole of $N_2O_4$ formed. The total number of moles of gas in the syringe is reduced, partly relieving the increase in pressure.

Chemists know from experimental data that the rate laws of the forward and reverse reactions are

$$\text{Rate}_f = k_f[NO_2]^2 \qquad (15.4)$$

$$\text{Rate}_r = k_r[N_2O_4] \qquad (15.5)$$

where "f" represents the forward reaction and "r" represents the reverse reaction. When the reaction achieves chemical equilibrium,

$$\text{Rate}_f = \text{rate}_r \qquad (15.6)$$

Replacing the terms in Equation 15.6 with the right sides of Equations 15.4 and 15.5:

$$k_f[NO_2]^2 = k_r[N_2O_4]$$

We can rearrange this equation to write the ratio:

$$\frac{k_f}{k_r} = \frac{[N_2O_4]}{[NO_2]^2} \qquad (15.7)$$

The ratio $k_f/k_r$ is the ratio of two constants, which is simply another constant. This constant has a special name: it is called an **equilibrium constant (K)**. Equating it to the ratio of product to reactant concentrations on the right side of Equation 15.7 gives us the **equilibrium constant expression** for this reaction:

$$K = \frac{[N_2O_4]}{[NO_2]^2} \qquad (15.8)$$

We will shortly explore other ways to formulate equilibrium constant expressions.

> **equilibrium constant (K)** the value of the ratio of concentration (or partial pressure) terms in the equilibrium constant expression at a specific temperature.
>
> **equilibrium constant expression** the ratio of the equilibrium concentrations or partial pressures of products to reactants, each term raised to a power equal to the coefficient of that substance in the balanced chemical equation for the reaction.

> **⊙⊙ CONNECTION** We discussed the rates of reactions of nitrogen oxides in the atmosphere in Chapter 14 in our discussion of the chemistry of photochemical smog.

**CONCEPT TEST** · · · · · · · · · · · · · · · · · · · · · · · · · · · · · · · ·

In the reversible reaction A ⇌ B, the rate constant of the forward reaction at a particular temperature is 3.0 times the value of the rate constant of the reverse reaction. What is the value of the equilibrium constant at that temperature?

*(Answers to Concept Tests are in the back of the book.)*

· · · · · · · · · · · · · · · · · · · · · · · · · · · · · · · · · · · · · · · · · · · · · · · · · · ·

Equilibrium constants can have values ranging from those that are almost too large to determine to those that are almost too small to determine. If $K$ is very large ($K \gg 1$), the concentrations of products in a reaction mixture at equilibrium are much larger than the concentrations of reactants. If $K$ is very small ($K \ll 1$), the concentrations of products at equilibrium are much smaller than the concentrations of reactants. Note that $K$ is a ratio of concentrations, so it is always positive. Also keep in mind that the equilibrium constant says nothing about how fast a reaction reaches equilibrium; it only tells us the extent of the reaction once equilibrium is reached.

## 15.2 Writing Equilibrium Constant Expressions

Let's revisit the water–gas shift reaction:

$$H_2O(g) + CO(g) \rightleftharpoons H_2(g) + CO_2(g) \qquad (15.2)$$

The data in Table 15.1 describe the results of four experiments in which different quantities of $H_2O$ gas, CO, $H_2$, and $CO_2$ are injected into a sealed reaction vessel heated to 500 K. The four gases are allowed to react and the final concentrations of all four are determined when the reaction has reached equilibrium.

**TABLE 15.1 Initial and Equilibrium Concentrations of the Reactants and Products in the Water–Gas Shift Reaction $H_2O(g) + CO(g) \rightleftharpoons H_2(g) + CO_2(g)$ at 500 K**

| Experiment | INITIAL CONCENTRATION (M) | | | | EQUILIBRIUM CONCENTRATION (M) | | | |
|---|---|---|---|---|---|---|---|---|
| | $[H_2O]$ | $[CO]$ | $[H_2]$ | $[CO_2]$ | $[H_2O]$ | $[CO]$ | $[H_2]$ | $[CO_2]$ |
| 1 | 0.0200 | 0.0200 | 0 | 0 | 0.0034 | 0.0034 | 0.0166 | 0.0166 |
| 2 | 0 | 0 | 0.0200 | 0.0200 | 0.0034 | 0.0034 | 0.0166 | 0.0166 |
| 3 | 0.0100 | 0.0200 | 0.0300 | 0.0400 | 0.0046 | 0.0146 | 0.0354 | 0.0454 |
| 4 | 0.0200 | 0.0100 | 0.0200 | 0.0100 | 0.0118 | 0.0018 | 0.0282 | 0.0182 |

In Experiment 1 the reaction vessel initially contains equimolar concentrations of $H_2O$ and CO but no $H_2$ or $CO_2$. In Experiment 2 the vessel initially contains equimolar concentrations of $H_2$ and $CO_2$ but no $H_2O$ or CO. The data in Table 15.1 indicate that when the reaction mixtures in Experiments 1 and 2 achieve chemical equilibrium, the concentrations of $H_2O$ and CO are the same (0.0034 $M$) in both experiments and so too are the concentrations of $H_2$ and $CO_2$ (0.0166 $M$). (Note that these statements about equilibrium concentrations are true in this case because the stoichiometric coefficients are all "1.") These results indicate that the composition of a reaction mixture at equilibrium is independent of the direction in which a particular reaction proceeds to achieve equilibrium.

Additional data in Table 15.1 show that the forward reaction takes place when the concentrations of the products are initially the same as the concentrations of the reactants (Experiment 4), or even higher (Experiment 3). The significance of these observations, taken with those from Experiments 1 and 2, can be appreciated if we do the following math: multiply the equilibrium concentrations of the products ($H_2$ and $CO_2$) together and divide that product by the product of the equilibrium concentrations of the reactants ($H_2O$ and CO):

$$\text{Experiments 1 and 2} \quad \frac{[H_2][CO_2]}{[H_2O][CO]} = \frac{(0.0166)(0.0166)}{(0.0034)(0.0034)} = 24$$

$$\text{Experiment 3} \quad \frac{[H_2][CO_2]}{[H_2O][CO]} = \frac{(0.0354)(0.0454)}{(0.0046)(0.0146)} = 24$$

$$\text{Experiment 4} \quad \frac{[H_2][CO_2]}{[H_2O][CO]} = \frac{(0.0282)(0.0182)}{(0.0118)(0.0018)} = 24$$

This calculation yields the same result in every experiment.

As you might guess, we would get the same ratio of product to reactant concentrations at equilibrium from *any* combination of initial concentrations of these four gases at 500 K. This constancy applies to other reaction mixtures, and has been known since the mid-19th century, from the work of Norwegian chemists Cato Guldberg (1836–1902) and Peter Waage (1833–1900). They discovered that any reversible reaction eventually reaches a state in which the ratio of the concentrations of products to reactants, with each value raised to a power corresponding to the coefficient for that substance in the balanced chemical equation for the reaction, has a characteristic value at a given temperature. They called this phenomenon the **law of mass action**. This ratio of concentration terms is the equilibrium constant expression for the reaction. It is also called the **mass action expression**, but we will limit the use of that term to those situations when the

**law of mass action** the ratio of the concentrations or partial pressures of products to reactants at equilibrium has a characteristic value at a given temperature when each term is raised to a power equal to the coefficient of that substance in the balanced chemical equation for the reaction.

**mass action expression** equivalent to the equilibrium constant expression, but applied to reaction mixtures that may, or may not, be at equilibrium.

system has not yet achieved equilibrium and in which the ratio of products to reactants does not match the value of the equilibrium constant. For the water–gas shift reaction at 500 K, the value of $K$ and the equilibrium constant expression are

$$K = \frac{[CO_2][H_2]}{[CO][H_2O]} = 24$$

Remember that the exponents in the equilibrium constant expression must be the same as the coefficients in the balanced equation. This is in contrast to rate law expressions, where the exponents are frequently not the same as the coefficients in the balanced equation because they reflect the stoichiometry of the rate-determining step, not necessarily the overall reaction.

**CONCEPT TEST** ....................................................................

Refer to the water–gas shift reaction on p. 735. Equal numbers of moles of water vapor, carbon dioxide, carbon monoxide, and hydrogen gas are injected into a rigid, sealed reaction vessel and heated to 500 K. Which expression about the composition of the reaction mixture at equilibrium is true?

a. $[CO] = [H_2] = [CO_2] = [H_2O]$
b. $[CO_2] = [H_2] > [CO] = [H_2O]$
c. $[CO_2] = [H_2] < [CO] = [H_2O]$

d. $[H_2O] = [H_2] > [CO_2] = [CO]$
e. $[H_2O] = [H_2] < [CO_2] = [CO]$

....................................................................................

In the preceding paragraphs we used equilibrium constant expressions based on the molar concentrations of products and reactants. For the generic reaction in which $a$ moles of reactant A react with $b$ moles of B to form $c$ moles of substance C and $d$ moles of D:

$$aA + bB \rightleftharpoons cC + dD$$

the equilibrium constant expression is

$$K_c = \frac{[C]^c[D]^d}{[A]^a[B]^b} \qquad (15.9)$$

Here the subscript "c" of the equilibrium constant represents *concentration*. If substances A, B, C, and D are gases, then the equilibrium constant may also be expressed in terms of their partial pressures:

$$K_p = \frac{(P_C)^c(P_D)^d}{(P_A)^a(P_B)^b} \qquad (15.10)$$

As we see later in this chapter, the values of $K_c$ and $K_p$ for a given reaction and temperature may or may not be the same. It depends on whether the number of moles of gaseous reactants is the same as the number of moles of gaseous products. Also, as a general practice, *equilibrium constants do not include units*. This is true even when there are different numbers of moles of reactants and products. Values of $K$ have no units because, in thermodynamics, $K$ is defined in terms of *activities*. The activity of a substance is the ratio of its concentration or partial pressure to a conventional standard concentration (1.000 $M$) or partial pressure (1.000 atm), multiplied by an activity coefficient that accounts for nonideal behavior. Because the terms are ratios of quantities with the same units, their units cancel out, leaving unitless equilibrium constants. When we use concentration and partial pressure values directly in equilibrium calculations in this text, we are making the assumption that all the reactants and products are behaving ideally.

▶❚❚ **CHEMTOUR** Equilibrium in the Gas Phase

◐◑ **CONNECTION** We introduced the concept of nonideal behavior in Section 6.9 when we described the reasons why *real* gases may not behave exactly like *ideal* gases.

SAMPLE EXERCISE 15.1    Writing Equilibrium    LO1
                        Constant Expressions

A key reaction in the formation of acid rain involves the reversible combination of $SO_2$ and $O_2$ in the atmosphere, producing $SO_3$:

$$2\,SO_2(g) + O_2(g) \rightleftharpoons 2\,SO_3(g)$$

Write the $K_c$ and $K_p$ expressions for this reaction.

**Collect and Organize** We are given the balanced chemical equation for a reaction and asked to write $K_c$ and $K_p$ expressions for it. Equilibrium constant expressions are ratios of the concentrations ($K_c$) or partial pressures ($K_p$) of products to reactants, with each term raised to the power equal to its coefficient in the balanced chemical equation of the reaction.

**Analyze** In the given reaction, the coefficients of $SO_2$ and $SO_3$ are both 2, so the $SO_2$ and $SO_3$ terms in the $K_c$ and $K_p$ expressions must be squared.

**Solve**

$$K_c = \frac{[SO_3]^2}{[SO_2]^2[O_2]}$$

$$K_p = \frac{(P_{SO_3})^2}{(P_{SO_2})^2(P_{O_2})}$$

**Think About It** The $K_c$ and $K_p$ expressions have the same format: their numerators and denominators contain terms for the same products and reactants, each raised to the same power. The difference between them is the nature of the terms: molar concentrations in the $K_c$ expression and partial pressures in the $K_p$ expression.

⚙ **Practice Exercise** Write the equilibrium constant expressions $K_c$ and $K_p$ for this reaction:

$$CH_4(g) + H_2O(g) \rightleftharpoons CO(g) + 3\,H_2(g)$$

*(Answers to Practice Exercises are in the back of the book.)*

SAMPLE EXERCISE 15.2    Calculating the Value of $K_c$    LO2

Table 15.2 contains data from four experiments on the dimerization of $NO_2$:

$$2\,NO_2(g) \rightleftharpoons N_2O_4(g)$$

The experiments were run at 100°C in a rigid, closed container. Use the data from each experiment to calculate a value of the equilibrium constant $K_c$ for the reaction.

**TABLE 15.2  Data for the Reaction $2\,NO_2(g) \rightleftharpoons N_2O_4(g)$ at 100°C**

| Experiment | INITIAL CONCENTRATION (M) | | EQUILIBRIUM CONCENTRATION (M) | |
|:---:|:---:|:---:|:---:|:---:|
| | $[NO_2]$ | $[N_2O_4]$ | $[NO_2]$ | $[N_2O_4]$ |
| 1 | 0.0200 | 0.0000 | 0.0172 | 0.00139 |
| 2 | 0.0300 | 0.0000 | 0.0244 | 0.00280 |
| 3 | 0.0400 | 0.0000 | 0.0310 | 0.00452 |
| 4 | 0.0000 | 0.0200 | 0.0310 | 0.00452 |

**Collect and Organize** We are given four sets of data that contain initial and equilibrium concentrations of a reactant and product. We are asked to determine the value of the equilibrium constant $K_c$ in each experiment. The equilibrium constant expression (Equation 15.8) for this reaction is

$$K_c = \frac{[N_2O_4]}{[NO_2]^2}$$

**Analyze** In each experiment, the concentration of $NO_2$ at equilibrium is nearly 10 times the concentration of $N_2O_4$. However, both values in each experiment are much less than unity, and the $[NO_2]$ term is squared. These two factors taken together mean that the values of the numerators in the equilibrium constant expressions will probably be greater than the denominators, so the calculated values of $K_c$ will be greater than 1.

**Solve**

$$\text{Experiment 1:} \quad K_c = \frac{[N_2O_4]}{[NO_2]^2} = \frac{0.00139}{(0.0172)^2} = 4.70$$

$$\text{Experiment 2:} \quad K_c = \frac{0.00280}{(0.0244)^2} = 4.70$$

$$\text{Experiment 3:} \quad K_c = \frac{0.00452}{(0.0310)^2} = 4.70$$

$$\text{Experiment 4:} \quad K_c = \frac{0.00452}{(0.0310)^2} = 4.70$$

**Think About It** The values calculated for $K_c$ are the same, as they should be for the same reaction at the same temperature, and, as predicted, are greater than 1.

**Practice Exercise** A mixture of gaseous CO and $H_2$, called *synthesis gas,* is used commercially to prepare methanol ($CH_3OH$), a compound considered an alternative fuel to gasoline. Under equilibrium conditions at 700 K, $[H_2] = 0.074$ mol/L, $[CO] = 0.025$ mol/L, and $[CH_3OH] = 0.040$ mol/L. What is the value of $K_c$ for this reaction at 700 K?

---

**SAMPLE EXERCISE 15.3    Calculating the Value of $K_p$    LO2**

A sealed chamber contains an equilibrium mixture of $NO_2$ and $N_2O_4$ at 300°C and partial pressures $P_{NO_2} = 0.101$ atm and $P_{N_2O_4} = 0.074$ atm. What is the value of $K_p$ for the following reaction under these conditions?

$$2\,NO_2(g) \rightleftharpoons N_2O_4(g)$$

**Collect and Organize** We are asked to determine the value of the $K_p$ equilibrium constant for the dimerization reaction of $NO_2$ to form $N_2O_4$. We are given the partial pressure values of both gases at equilibrium. The equilibrium constant expression for the reaction in terms of partial pressures (Equation 15.10) is

$$K_p = \frac{P_{N_2O_4}}{(P_{NO_2})^2}$$

**Analyze** The partial pressure of $NO_2$ is slightly greater than the partial pressure of $N_2O_4$ at equilibrium, but both values are less than one. Moreover, the $P_{NO_2}$ term is squared. These two factors taken together mean that the value of the numerator in the $K_p$ expression will probably be greater than that of the denominator, so the calculated value of $K_p$ will be greater than 1.

**Solve**

$$K_p = \frac{0.074}{(0.101)^2} = 7.3$$

**Think About It** As we predicted, the value of $K_p$ is greater than 1, even though the equilibrium partial pressure of the product ($P_{N_2O_4} = 0.074$ atm) is actually less than that of the reactant ($P_{NO_2} = 0.101$ atm). Relatively little $N_2O_4$ forms in this case because $P_{NO_2}$ is so low, and 2 moles of $NO_2$ are required to make 1 mole of $N_2O_4$. As we see in Section

15.7, low pressures favor the side of a gas-phase reaction that has more moles of gas. To illustrate this point, let's calculate the value of $P_{N_2O_4}$ that would be in equilibrium with 1 atm of $NO_2$ at 300°C:

$$K_p = \frac{(P_{N_2O_4})}{(P_{NO_2})^2} = 7.3 = \frac{(P_{N_2O_4})}{1^2}$$

$$(P_{N_2O_4}) = 7.3 \text{ atm}$$

The value of $P_{N_2O_4}$ is 7.3 times the $P_{NO_2}$ value at these higher overall pressures, whereas $P_{N_2O_4}$ was lower than $P_{NO_2}$ at the lower pressures of this exercise.

⚙ **Practice Exercise** A reaction vessel contains an equilibrium mixture of $SO_2$, $O_2$, and $SO_3$. Given the partial pressures $P_{SO_2} = 0.0018$ atm, $P_{O_2} = 0.0032$ atm, and $P_{SO_3} = 0.0166$ atm, calculate the value of $K_p$ for the reaction:

$$2 SO_2(g) + O_2(g) \rightleftharpoons 2 SO_3(g)$$

The value of $K$ indicates how far a reaction proceeds at a given temperature. The values we have seen thus far—$K_c = 24$ for the water–gas shift reaction at 500 K, and $K_p = 7.3$, $K_c = 4.7$ for the dimerization of $NO_2$ at 300°C and 100°C, respectively—are considered intermediate values. Because the range of $K$ values is so large ($0 < K < \infty$), all three of these values are considered *close* to 1, which means that comparable concentrations of reactants and products are likely to be present at equilibrium.

At equilibrium, some product and some reactant are always present, because equilibrium constants are never zero or infinity. An extremely large equilibrium constant means that the amount of reactant remaining is negligibly small, and an extremely small equilibrium constant means that the amount of product formed is negligibly small.

For example, the reaction between $H_2$ and $O_2$ to form water proceeds until one of the reactants is, for all practical purposes, completely consumed. This observation is consistent with a large value of $K$ at 25°C:

$$2 H_2(g) + O_2(g) \rightleftharpoons 2 H_2O(g) \qquad K_c = 3 \times 10^{81}$$

Also, the decomposition of $CO_2$ to CO and $O_2$ at 25°C proceeds hardly at all and is consistent with a very small value of $K$ and virtually no product being formed:

$$2 CO_2(g) \rightleftharpoons 2 CO(g) + O_2(g) \qquad K_c = 3 \times 10^{-92}$$

# 15.3 Relationships between $K_c$ and $K_p$ Values

As we noted in Section 15.2, the values of $K_c$ and $K_p$ for a given reaction and temperature may or may not be the same, depending on the numbers of moles of gaseous reactants and products. To better understand this relationship, we begin with the ideal gas law:

$$PV = nRT$$

If we solve for $P$ and express volume in liters, then $n/V$ has units of moles per liter, which is the same as molarity ($M$):

$$P = \frac{n}{V}RT$$

$$P = MRT \qquad\qquad (15.11)$$

Let's apply Equation 15.11 to the gases in the $NO_2/N_2O_4$ equilibrium from Sample Exercise 15.3:

$$P_{NO_2} = \frac{n_{NO_2}}{V}RT = [NO_2]RT$$

$$P_{N_2O_4} = \frac{n_{N_2O_4}}{V}RT = [N_2O_4]RT$$

Substituting these values into the expression for $K_p$ from Sample Exercise 15.3, we get

$$K_p = \frac{(P_{N_2O_4})}{(P_{NO_2})^2} = \frac{[N_2O_4]RT}{([NO_2]RT)^2} = \frac{[N_2O_4]RT}{[NO_2]^2(RT)^2}$$

The ratio of concentration terms in the expression on the right, $[N_2O_4]/[NO_2]^2$, is the same as the $K_c$ expression for this reaction. Substituting $K_c$ for those terms and simplifying the $RT$ terms:

$$K_p = K_c \frac{1}{RT}$$

This last equation defines the relationship between $K_c$ and $K_p$ for this specific reaction. A more general expression can be derived for the generic reaction of gases A and B forming gases C and D:

$$aA + bB \rightleftharpoons cC + dD$$

As noted on p. 739, the $K_p$ expression for this reaction is

$$K_p = \frac{(P_C)^c(P_D)^d}{(P_A)^a(P_B)^b} \qquad (15.10)$$

Replacing each partial pressure term in Equation 15.10 with the corresponding molar concentration term $\times RT$ (from Equation 15.11) gives us the general expression

$$K_p = \frac{([C]RT)^c([D]RT)^d}{([A]RT)^a([B]RT)^b} \qquad (15.12)$$

Combining the $RT$ terms, we get

$$K_p = \frac{[C]^c[D]^d}{[A]^a[B]^b} \times (RT)^{[(c+d)-(a+b)]} \qquad (15.13)$$

The concentration ratio in this equation matches the $K_c$ expression for this generic reaction (see Equation 15.9). Substituting this equality into Equation 15.13 gives us

$$K_p = K_c(RT)^{[(c+d)-(a+b)]} \qquad (15.14)$$

To simplify Equation 15.14, consider that $(c + d)$ represents the sum of the coefficients of the gaseous products in the reaction—the sum of the number of moles of gases produced. Similarly, $(a + b)$ represents the sum of the number of moles of gaseous reactants consumed. The difference between the two sums, $(c + d) - (a + b)$, represents the *change in the number of moles of gases* ($\Delta n$) between the product and reactant sides of the balanced chemical equation. We use the symbol $\Delta n$ to represent this change. Substituting $\Delta n$ for $(c + d) - (a + b)$ in Equation 15.14 gives us

$$K_p = K_c(RT)^{\Delta n} \qquad (15.15)$$

Equation 15.15 provides a quantitative interpretation of the opening statement of this section: the relationship between the $K_p$ and $K_c$ values of a chemical reaction involving gases depends on the number of moles of gaseous reactants and products. In reactions such as the water–gas shift reaction:

$$H_2O(g) + CO(g) \rightleftharpoons H_2(g) + CO_2(g) \qquad (15.2)$$

in which the number of moles of gas on both sides of the reaction arrow is the same, $\Delta n = 0$ and $K_p = K_c$. However, in the steam–methane reforming reaction:

$$CH_4(g) + H_2O(g) \rightleftharpoons CO(g) + 3\,H_2(g) \qquad (15.1)$$

2 moles of gaseous reactants form 4 moles of gaseous products. Therefore,

$$\Delta n = 4\text{ mol} - 2\text{ mol} = 2\text{ mol}$$

Inserting this value for $\Delta n$ in Equation 15.15 gives us this reaction's relationship between $K_p$ and $K_c$:

$$K_p = K_c(RT)^2$$

A final point about the relative sizes of $K_p$ and $K_c$ values: One mole of an ideal gas at STP occupies a volume of 22.4 liters. Therefore, its molar concentration is 1 mol/22.4 L = 0.0446 $M$. Thus the value of the pressure of a pure gas at STP (1.00 atm) is 22.4 times its molar concentration. The $RT$ term in Equation 15.15 essentially corrects for this difference in how we express how much of a gas is present in a reaction mixture. It is a *conversion factor* for changing molar concentrations into partial pressures. The value of $RT$ at STP is

$$0.08206\,\frac{\text{L} \cdot \text{atm}}{\text{mol} \cdot \cancel{\text{K}}} \times 273\,\cancel{\text{K}} = 22.4\,\frac{\text{L} \cdot \text{atm}}{\text{mol}}$$

---

**SAMPLE EXERCISE 15.4**    **Calculating $K_c$ from $K_p$**    **LO3**

In Sample Exercise 15.3 we calculated the value of $K_p$ (7.3) for the dimerization of $NO_2$ to $N_2O_4$ at 300°C. What is the value of $K_c$ for this reaction at 300°C?

**Collect and Organize** We are given a $K_p$ value and asked to calculate the corresponding $K_c$ value for the same reaction at the same temperature. Equation 15.15 relates $K_p$ and $K_c$ values:

$$K_p = K_c(RT)^{\Delta n}$$

where $\Delta n$ represents the change in the number of moles of gas when going from reactant to product.

**Analyze** We need a balanced chemical equation to determine the value of $\Delta n$. From Sample Exercise 15.2 we know that the equation is

$$2\,NO_2(g) \rightleftharpoons N_2O_4(g)$$

Because the number of moles of gas is not the same on both sides of the reaction arrow, we predict that the values of $K_c$ and $K_p$ will not be the same.

**Solve** According to the balanced equation, 2 moles of gaseous reactants yield 1 mole of gaseous product. Therefore,

$$\Delta n = 1\text{ mol} - 2\text{ mol} = -1\text{ mol}$$

Inserting this value, the given value of $K_p$, and temperature into Equation 15.15:

$$K_p = K_c(RT)^{\Delta n}$$

$$7.3 = K_c[0.08206 \times (273 + 300)]^{-1}$$

$$K_c = (7.3)(0.08206)(573) = 3.4 \times 10^2$$

**Think About It** The value of $K_c$ differs from the value of $K_p$ because two moles of gaseous reactants produce only one mole of gaseous product. Actually, the value of $K_c$ is nearly 50 times larger than that of $K_p$. The $K_c/K_p$ ratio is more than twice the STP molar volume factor (22.4 L · atm/mol) because the absolute temperature of the reaction is more than twice the temperature at STP.

⦿ **Practice Exercise** An important industrial process for synthesizing the ammonia used in agricultural fertilizers involves the combination of $N_2$ and $H_2$:

$$N_2(g) + 3\, H_2(g) \rightleftharpoons 2\, NH_3(g) \qquad K_c = 2.8 \times 10^{-9} \text{ at } 30°C$$

What is the value of $K_p$ of this reaction at 30°C?

## 15.4 Manipulating Equilibrium Constant Expressions

We can write equilibrium constant expressions for reactions running in reverse, for chemical equations that have been multiplied or divided by a value that gives a key component a coefficient of 1, and for overall reactions that are combinations of other reactions.

### K for Reverse Reactions

The $K$ values for the forward and reverse directions of a reversible reaction are related. For example, in Sample Exercise 15.2, we wrote the $K_c$ expression for the dimerization reaction

$$2\, NO_2(g) \rightleftharpoons N_2O_4(g)$$

this way:

$$K_c = \frac{[N_2O_4]}{[NO_2]^2}$$

In the reverse of this process, the decomposition of $N_2O_4$:

$$N_2O_4(g) \rightleftharpoons 2\, NO_2\,(g)$$

the reactant becomes the product and the product becomes the reactant. This reversal is reflected in the equilibrium constant expression for the decomposition reaction:

$$K_c = \frac{[NO_2]^2}{[N_2O_4]}$$

The numerator has become the denominator, and the denominator, the numerator. Expressing the relation between the forward ($K_f$) and reverse ($K_r$) equilibrium constants mathematically, we have

$$K_f = \frac{1}{K_r} \qquad (15.16)$$

To explore the meaning of Equation 15.16, consider a generic reversible reaction in which the equilibrium favors the formation of product, which means the equilibrium lies far to the right and is reflected in a large $K$ value:

$$A \rightleftharpoons B \qquad K \gg 1 \qquad (15.17)$$

Because the reciprocal of a large value is a small one, Equation 15.16 tells us that the equilibrium constant for the reverse reaction must be small:

$$B \rightleftharpoons A \qquad K \ll 1 \qquad (15.18)$$

If we apply Equation 15.16 to the dimerization of $NO_2$ at 300°C (Sample Exercise 15.3):

$$2\,NO_2(g) \rightleftharpoons N_2O_4(g) \qquad K_p = 7.3$$

we can calculate the $K_p$ value for the decomposition of $N_2O_4$:

$$N_2O_4(g) \rightleftharpoons 2\,NO_2(g) \qquad K_p = 1/7.3 = 0.14$$

---

**SAMPLE EXERCISE 15.5    Calculating the Value of $K$ for a Reverse Reaction    LO4**

Atmospheric NO combines with $O_2$ to form $NO_2$. The reverse reaction is the decomposition of $NO_2$ to NO and $O_2$. At 184°C, the value of $K_c$ for the forward reaction is $1.48 \times 10^4$. Write the equilibrium constant expressions for both reactions and calculate the value of $K_c$ for the decomposition of $NO_2$ at 184°C.

**Collect and Organize** We are to write equilibrium constant expressions for the forward and reverse reactions for a process at equilibrium and calculate the equilibrium constant for the reverse reaction. We know the identities of the reactants and products and the $K_c$ value of the forward reaction. We need to write balanced chemical equations for the forward and reverse reactions and then use the coefficients from those equations to write equilibrium constant expressions.

**Analyze** We start by writing balanced chemical equations for the forward and reverse reactions:

Forward reaction     $2\,NO(g) + O_2(g) \rightleftharpoons 2\,NO_2(g)$
Reverse reaction     $2\,NO_2(g) \rightleftharpoons 2\,NO(g) + O_2(g)$

The coefficient of 2 in front of NO and $NO_2$ means that the concentration terms for these compounds are squared in the $K_c$ expressions. The large ($>10^4$) value of $K_c$ for the forward reaction means that the $K_c$ value of the reverse reaction should be less than $10^{-4}$.

**Solve**

$$K_f = \frac{[NO_2]^2}{[NO]^2[O_2]}$$

$$K_r = \frac{[NO]^2[O_2]}{[NO_2]^2}$$

$$K_r = \frac{1}{K_f} = \frac{1}{1.48 \times 10^4} = 6.76 \times 10^{-5}$$

**Think About It** The value of $K_f$ is large, so it makes sense that its reciprocal, $K_r$, is small. Our calculated value is indeed close to our estimate of $10^{-4}$.

**Practice Exercise** At 300°C, the value of $K_p$ for the combination reaction of $N_2$ and $H_2$ that produces $NH_3$ gas is $4.3 \times 10^{-3}$. What is the equilibrium constant at the same temperature for the decomposition reaction of $NH_3$ that produces $N_2$ and $H_2$?

---

## K for an Equation Multiplied by a Number

As we saw in Chapter 5, sometimes we must represent a reaction by an equation containing fractional coefficients to have exactly 1 mole of a particular reactant or product. For example, we can rewrite the forward reaction from Sample Exercise 15.5:

$$2\,NO(g) + O_2(g) \rightleftharpoons 2\,NO_2(g)$$

To describe the preparation of only 1 mole of $NO_2$, we multiply each coefficient by $\frac{1}{2}$:

$$NO(g) + \tfrac{1}{2}O_2(g) \rightleftharpoons NO_2(g) \qquad (15.19)$$

Because the equilibrium constant expression for any reaction must be written to match the balanced chemical equation, the appropriate $K_c$ expression for Equation 15.19 is

$$K_c = \frac{[NO_2]}{[NO][O_2]^{1/2}} \qquad (15.20)$$

Compare this $K_c$ expression with the $K_c$ expression for the forward reaction from Sample Exercise 15.5:

$$K_c = \frac{[NO_2]^2}{[NO]^2[O_2]} \qquad (15.21)$$

Note that the right side of Equation 15.20 is equal to the square root of the right side of Equation 15.21: the NO and $NO_2$ terms are squared in Equation 15.21 but are raised to only the first power in Equation 15.20, and the $O_2$ term is raised to the first power in Equation 15.21 but to the $\frac{1}{2}$ power in Equation 15.20. Therefore, we may conclude that the value of the equilibrium constant $K_c$ in Equation 15.20 is the square root of the value of $K_c$ in Equation 15.21. We can extend this pattern to all chemical equilibria with the following rule: If the balanced chemical equation of a reaction is multiplied by some factor $n$, then the value of $K$ is raised to the $n$th power.

---

**SAMPLE EXERCISE 15.6**    **Calculating $K$ for Different Coefficients**     **LO4**

An important reaction in the industrial production of sulfuric acid is also involved in the formation of atmospheric aerosols of sulfuric acid (acid rain): the oxidation of $SO_2$ to $SO_3$. One way to write a chemical equation for the oxidation reaction is

$$\text{Equation A} \qquad SO_2(g) + \tfrac{1}{2}O_2(g) \rightleftharpoons SO_3(g)$$

If the value of $K_c$ for this reaction at 298 K is $2.8 \times 10^{12}$, what is the value of $K_c$ at 298 K for the following reaction?

$$\text{Equation B} \qquad 2\,SO_2(g) + O_2(g) \rightleftharpoons 2\,SO_3(g)$$

**Collect and Organize** We are given the value of $K_c$ for a reaction that describes the production of 1 mole of $SO_3$ and are asked to recalculate the value of $K_c$ for the same reaction but based on producing 2 moles of $SO_3$.

**Analyze** Equation A is multiplied by 2 to generate Equation B. This means that the $K_c$ for Equation A must be raised to the second power to generate $K_c$ for Equation B. The value of $K_c$ for Equation A is large; the value of $K_c$ for Equation B should be even larger.

**Solve**

$$K_B = (K_A)^2$$
$$= (2.8 \times 10^{12})^2 = 7.8 \times 10^{24}$$

**Think About It** Doubling the coefficients means squaring the value of $K$. Our prediction that the second $K$ would have a very large value is correct.

⊛ **Practice Exercise** The industrial production of ammonia for fertilizers involves the reaction of nitrogen and hydrogen. If $K_c = 2.4 \times 10^{-3}$ for the reaction

$$N_2(g) + 3\,H_2(g) \rightleftharpoons 2\,NH_3(g)$$

at 1000 K, what is $K_c$ at 1000 K for this reaction?

$$\tfrac{1}{3}N_2(g) + H_2(g) \rightleftharpoons \tfrac{2}{3}NH_3(g)$$

Given two equally legitimate equilibrium constant expressions for the same reaction, you may wonder how the same reaction having the same reactants and products can have two or more equilibrium constant values. Surely the same ingredients should be present in the same proportions at equilibrium, no matter how we choose to write a balanced equation describing their reaction. In fact, they are. The difference in $K$ values is not chemical; it is mathematical. It is related to how we use the equilibrium concentrations to calculate $K$ values. It is, for example, affected by our choice to use the value of [NO] in Equation 15.20 and not the value of $[NO]^2$ as in Equation 15.21. However, that choice does not affect the concentration of NO at equilibrium. Always remember that whenever you write an expression for either $K_c$ or $K_p$, you must identify the specific balanced equation you used to obtain the expression.

## Combining $K$ Values

In Chapter 5, we applied Hess's law to calculate the enthalpies of combined reactions. We carry out a similar process here to determine the overall $K$ for a reaction that is the sum of two or more other reactions.

Consider two reactions from Chapter 14 involved in the formation of photochemical smog, wherein the NO produced in a car's engine at high temperature is oxidized to $NO_2$ in the atmosphere:

$$
\begin{aligned}
(1) \qquad & N_2(g) + O_2(g) \rightleftharpoons 2\,\cancel{NO(g)} \\
(2) \qquad & 2\,\cancel{NO(g)} + O_2(g) \rightleftharpoons 2\,NO_2(g) \\
\hline
\text{Overall:} \qquad & N_2(g) + 2\,O_2(g) \rightleftharpoons 2\,NO_2(g)
\end{aligned}
$$

The equilibrium constant expression for the overall reaction is

$$
K_c = \frac{[NO_2]^2}{[N_2][O_2]^2}
$$

We can derive this expression from the equilibrium constant expressions for reactions 1 and 2,

$$
K_1 = \frac{[NO]^2}{[N_2][O_2]} \qquad \text{and} \qquad K_2 = \frac{[NO_2]^2}{[NO]^2[O_2]}
$$

if we multiply $K_1$ by $K_2$:

$$
K_1 \times K_2 = \frac{\cancel{[NO]^2}}{[N_2][O_2]} \times \frac{[NO_2]^2}{\cancel{[NO]^2}[O_2]} = \frac{[NO_2]^2}{[N_2][O_2]^2} = K_{overall}
$$

This approach works for all series of reactions, and as a general rule

$$
K_{overall} = K_1 \times K_2 \times K_3 \times K_4 \times \cdots \times K_n \tag{15.22}
$$

The overall equilibrium constant for a sum of two or more reactions is the product of the equilibrium constants of the individual reactions. Thus, the value of $K_c$ for the overall reaction for the formation of $NO_2$ from $N_2$ and $O_2$ at 1000 K is the product of the equilibrium constants for reaction 1:

$$
K_1 = \frac{[NO]^2}{[N_2][O_2]} = 7.2 \times 10^{-9}
$$

and reaction 2:

$$
K_2 = \frac{[NO_2]^2}{[NO]^2[O_2]} = 0.020
$$

Using Equation 15.22:

$$K_{overall} = K_1 \times K_2 = 7.2 \times 10^{-9} \times 0.020 = 1.4 \times 10^{-10}$$

The equilibrium constant expression for the overall reaction must contain the appropriate terms for the products and reactants of that reaction. Just as with Hess's law in thermochemical calculations, we may need to reverse an equation or multiply an equation by a factor when we combine it with another to create the equation of interest. If we reverse a reaction, we must take the reciprocal of its $K$. If we multiply a reaction by a constant, we must raise its $K$ to that power.

---

**SAMPLE EXERCISE 15.7    Calculating Overall $K$ of    LO4
Combined Reactions**

At 1000 K, the $K_c$ value of the following reaction is $1.5 \times 10^6$:

$$(1) \; N_2O_4(g) \rightleftharpoons 2\,NO_2(g)$$

At 1000 K, the $K_c$ value of the following reaction is $1.4 \times 10^{-10}$:

$$(2) \; N_2(g) + 2\,O_2(g) \rightleftharpoons 2\,NO_2(g)$$

Calculate the $K_c$ value at 1000 K of the reaction

$$N_2(g) + 2\,O_2(g) \rightleftharpoons N_2O_4(g)$$

**Collect and Organize** We are given two reactions and their $K_c$ values. We need to combine the two reactions in such a way that $N_2$ and $O_2$ are on the reactant side of the overall equation and $N_2O_4$ is on the product side. We can then calculate the $K_c$ value of the overall reaction. When two reactions are added, the value of $K_c$ of the overall reaction is the product of the $K_c$ values of the two reactions.

**Analyze** The overall reaction is the sum of the reverse of reaction 1 and reaction 2 as written:

| | |
|---|---|
| Reaction 1 reversed | $\cancel{2\,NO_2(g)} \rightleftharpoons N_2O_4(g)$ |
| + Reaction 2 | $N_2(g) + 2\,O_2(g) \rightleftharpoons \cancel{2\,NO_2(g)}$ |
| Overall | $N_2(g) + 2\,O_2(g) \rightleftharpoons N_2O_4(g)$ |

Reversing a chemical reaction requires taking the reciprocal of its $K_c$ value. The $K_c$ value of reaction 1 is large ($>10^6$), which means its reciprocal is small ($<10^{-6}$). The product of this small value times the even smaller $K_c$ value of reaction 2 ($\sim 10^{-10}$) should be a very small value—about $10^{-16}$.

**Solve**

$$K_{overall} = \frac{1}{K_1} \times K_2 = \frac{1}{1.5 \times 10^6} \times 1.4 \times 10^{-10} = 9.3 \times 10^{-17}$$

**Think About It** The result of the calculation is close to the ballpark value we predicted. The overall equilibrium of the combined reactions lies far to the left; little $N_2O_4$ forms from $N_2$ and $O_2$ at 1000 K.

 **Practice Exercise** Calculate the value of $K_c$ for the hypothetical reaction

$$Q(g) + X(g) \rightleftharpoons M(g)$$

from the following information:

$$2\,M(g) \rightleftharpoons Z(g) \qquad\qquad K_c = 6.2 \times 10^{-4}$$

$$Z(g) \rightleftharpoons 2\,Q(g) + 2\,X(g) \qquad K_c = 5.6 \times 10^{-2}$$

To summarize the key points for manipulating equilibrium constants:

➤ The $K$ value of a reaction running in reverse is the reciprocal of the $K$ of the forward reaction.
➤ If the original chemical equation describing an equilibrium is multiplied by a factor $n$, the value of $K$ of the new equilibrium constant expression is the value of the original $K$ raised to the $n$th power.
➤ If an overall chemical reaction is the sum of two or more other reactions, the overall value of $K$ is the product of the $K$ values of the other reactions.

## 15.5 Equilibrium Constants and Reaction Quotients

In Sections 15.1 and 15.2 we introduced two key terms: *equilibrium constant expression* and *mass action expression*. Until now we have used the first term almost exclusively because we have been dealing with chemical reactions that have achieved equilibrium. Now we want to reintroduce the concept of a mass action expression because we can apply it not only to concentrations (or partial pressures) of products and reactants in reaction mixtures that have reached equilibrium, but also to reaction mixtures that are on their way to equilibrium but are not yet there.

Even if a reversible chemical reaction has not reached equilibrium, we can still insert reactant and product concentrations (or partial pressures) into its mass action expression. The mathematical result is not a $K$ value because the reaction is not yet at equilibrium. Instead it is a $Q$ value, where $Q$ stands for **reaction quotient**. The value of $Q$ provides us with a status report on how a reaction is proceeding.

To see how this works, let's revisit the water–gas shift reaction

$$H_2O(g) + CO(g) \rightleftharpoons H_2(g) + CO_2(g) \qquad K_c = 24 \text{ at } 500 \text{ K}$$

and the data from Experiment 3 in Table 15.1. In Experiment 3, the initial concentrations of reactants and products are

| [H$_2$O] (M) | [CO] (M) | [H$_2$] (M) | [CO$_2$] (M) |
|---|---|---|---|
| 0.0100 | 0.0200 | 0.0300 | 0.0400 |

Inserting these values into the mass action expression for the reaction yields a value for $Q$ based on concentration; that is, $Q_c$:

$$Q_c = \frac{[H_2][CO_2]}{[H_2O][CO]} = \frac{(0.0300)(0.0400)}{(0.0100)(0.0200)} = 6.00$$

Now compare this value of $Q_c$ to the reaction's $K_c$ value, which is 24. Clearly $Q_c$ is less than $K_c$, which means there are proportionally smaller concentrations of products and larger concentrations of reactants in the initial reaction mixture than there will be at equilibrium. To achieve equilibrium, some of the reactants must react and form products, increasing the value of $Q_c$ until it matches the value of $K_c$. At that point, the following equilibrium concentrations of reactants and products are present in the reaction vessel:

| [H$_2$O] (M) | [CO] (M) | [H$_2$] (M) | [CO$_2$] (M) |
|---|---|---|---|
| 0.0046 | 0.0146 | 0.0354 | 0.0454 |

**reaction quotient (Q)** the numerical value of the mass action expression for *any values* of the concentrations (or partial pressures) of reactants and products; at equilibrium, $Q = K$.

To put the results from the data in Table 15.1 in context, let's consider the curves in Figure 15.5. Starting at the left side (zone a) of the graph, we have the ini-

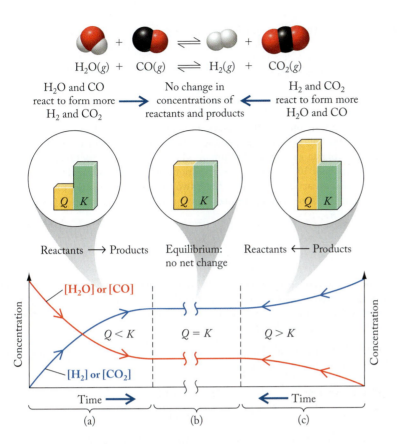

$$H_2O(g) + CO(g) \rightleftharpoons H_2(g) + CO_2(g)$$

H$_2$O and CO react to form more H$_2$ and CO$_2$ ⟶ No change in concentrations of reactants and products ⟵ H$_2$ and CO$_2$ react to form more H$_2$O and CO

Reactants ⟶ Products     Equilibrium: no net change     Reactants ⟵ Products

$Q < K$    $Q = K$    $Q > K$

[H$_2$O] or [CO]

[H$_2$] or [CO$_2$]

Concentration

Time ⟶     ⟵ Time

(a)     (b)     (c)

**FIGURE 15.5** The value of the reaction quotient $Q$ relative to the equilibrium constant $K$ for the water–gas shift reaction. (a) Reactant concentrations (red) are higher than they are once equilibrium is reached, and product concentrations (blue) are lower than at equilibrium; $Q < K$, and the reactants form more products. (b) Equilibrium concentrations are achieved; $Q = K$, and no net change in concentrations takes place. (c) Product concentrations are higher than what they are at equilibrium, and reactant concentrations are lower than at equilibrium; $Q > K$, and products form more reactants as the reaction runs in reverse.

tial conditions of Experiment 1 from Table 15.1: equal concentrations of reactants are present, but no products. Over time, reactant concentrations (the red curve) decrease as product concentrations (the blue curve) increase. At the right side of the graph in zone (c), we have the initial conditions of Experiment 2: products are present, but no reactants. Over time, reactant concentrations increase as product concentrations decrease. In zone (b) in the middle of the graph, no net change occurs in the composition because the reaction is at equilibrium.

We can also characterize the three zones on the basis of the value of $Q$ compared to $K$. In zone (a), $Q$ values are less than $K$ and there is a net conversion of reactants into products as the forward reaction dominates. In zone (c), $Q$ values are greater than $K$ and a net conversion of products into reactants takes place as the reverse reaction dominates. In the middle zone (b), $Q = K$ and no change in the composition of the reaction mixture occurs over time. The relative values of $Q$ and $K$ and their consequences are summarized in Table 15.3.

**TABLE 15.3  Comparison of $Q$ and $K$ Values**

| Value of Q | What It Means |
|---|---|
| $Q < K$ | Reaction as written proceeds in forward direction ($\rightarrow$) |
| $Q = K$ | Reaction is at equilibrium ($\rightleftharpoons$) |
| $Q > K$ | Reaction as written proceeds in reverse direction ($\leftarrow$) |

**CONCEPT TEST**

In which of the three zones do the initial reaction conditions of Experiments 3 and 4 from Table 15.1 (p. 738) fall?

**SAMPLE EXERCISE 15.8**   **Using Q and K Values to Predict the Direction of a Reaction**   **LO2**

At 2300 K the value of $K$ of the following reaction is $1.5 \times 10^{-3}$:

$$N_2(g) + O_2(g) \rightleftharpoons 2\,NO(g)$$

At the instant when a reaction vessel at 2300 K contains 0.50 $M$ N$_2$, 0.25 $M$ O$_2$, and 0.0042 $M$ NO, is the reaction mixture at equilibrium? If not, in which direction will the reaction proceed to reach equilibrium?

**homogeneous equilibria** involve reactants and products in the same phase.

**heterogeneous equilibria** involve reactants and products in more than one phase.

**Collect and Organize** We are asked whether or not a reaction mixture is at equilibrium. We are given the value of $K$ and the concentrations of reactants and product.

**Analyze** The mass action expression for this reaction based on concentrations is

$$Q = \frac{[NO]^2}{[N_2][O_2]}$$

If we insert the given concentration values into this expression, we can determine the value of $Q$. Comparing $Q$ with $K$ enables us to determine (a) whether the reaction is at equilibrium and (b) if it is not at equilibrium, in which direction the reaction proceeds. The value of $[NO]$ is about $10^{-2}$ times the concentrations of the reactants, and the $[NO]$ term is squared in the mass action expression. These two factors together make the value of the numerator about $10^{-4}$ times that of the denominator. Therefore, the value of $Q$ may be less than the value of $K$.

**Solve**

$$Q = \frac{(0.0042)^2}{(0.50)(0.25)} = 1.4 \times 10^{-4}$$

The value of $Q$ is indeed less than that of $K$; so, the reaction mixture is not at equilibrium. To achieve equilibrium, more reactant must form product so that the numerator of the mass action expression increases and the denominator decreases. This happens if the reaction proceeds in the forward direction.

**Think About It** Our estimate that $Q$ would be less than $K$ was correct. The value of $K$ is small, but the value of $Q$ is even smaller.

**Practice Exercise** The value of $K_c$ for the reaction

$$2\,NO_2(g) \rightleftharpoons N_2O_4(g)$$

is 4.7 at 373 K. Is a mixture of the two gases in which $[NO_2] = 0.025\ M$ and $[N_2O_4] = 0.0014\ M$ in chemical equilibrium? If not, in which direction does the reaction proceed to achieve equilibrium?

(a)

(b)

**FIGURE 15.6** (a) For centuries, primitive kilns were used to decompose limestone ($CaCO_3$) into lime ($CaO$) and $CO_2$. These kilns were charged with layers of fuel (originally wood, later coal) and crushed limestone, and the fuel was ignited. Farmers use lime to "sweeten" (neutralize the acidity of) soil. It is also used to make mortar and cement. (b) A modern limestone kiln.

## 15.6 Heterogeneous Equilibria

Thus far in this chapter we have focused on reactions in the gas phase. However, the principles of chemical equilibrium also apply to reactions in the liquid phase, particularly reactions in solution. Equilibria in which products and reactants are all in the same phase are called **homogeneous equilibria**. Equilibria in which reactants and products are in different phases are **heterogeneous equilibria**.

In Sample Exercise 15.6 we considered the equilibrium associated with the oxidation of $SO_2$ to $SO_3$, a key step in the industrial synthesis of sulfuric acid but also significant in the formation of aerosols of $H_2SO_4$ in the atmosphere when oxides of sulfur are produced by combustion of sulfur-containing fuels. One way to prevent this reaction from happening in the atmosphere is to "scrub" $SO_2$ from the exhaust gases emitted at factories that burn sulfur-containing fuels. Solid lime, $CaO$, is a widely used scrubbing agent. Suspended in water and sprayed into the exhaust gases, it combines with $SO_2$ to form calcium sulfite:

$$CaO(s) + SO_2(g) \rightleftharpoons CaSO_3(s)$$

The large quantities of lime needed for this reaction, and for many other industrial and agricultural uses, come from heating pulverized limestone, which is mostly $CaCO_3$, in kilns (Figure 15.6) operated at temperatures near 900°C. At these temperatures, $CaCO_3$ decomposes into lime ($CaO$) and $CO_2$ gas:

$$CaCO_3(s) \rightleftharpoons CaO(s) + CO_2(g) \qquad \Delta H° = 178.1\ \text{kJ/mol}$$

We might write the concentration-based equilibrium constant expression for this reaction as follows:

$$K_c = \frac{[CaO][CO_2]}{[CaCO_3]}$$

This expression contains concentration terms for two solids: $CaO$ and $CaCO_3$. But what do we mean by the concentration of a solid? Any pure solid has a constant concentration because its mass (and number of moles) per unit volume is always the same. As long as there is *any* $CaO$ or $CaCO_3$ present, there is no change in the "concentration" of either substance. Consequently, we remove them from the equilibrium constant expression, leaving us with

$$K_c = [CO_2]$$

This expression means that, as long as some $CaO$ and $CaCO_3$ are present, the concentration of $CO_2$ gas does not vary at a given temperature, as shown in Figure 15.7. Instead, the concentration of $CO_2$ is the same as the value of $K_c$ at that temperature.

The same concept of constant concentration applies to pure liquids involved in reversible chemical reactions. As long as the liquid is present, its "concentration" is considered constant during the course of the reaction and does not appear in the equilibrium constant expression. Similarly, $K_c$ expressions for most reactions in aqueous solutions do not include a term for $[H_2O]$, even when water is a reactant or product, because its concentration does not change significantly. In writing equilibrium constant expressions for heterogeneous equilibria, we follow the rules we learned earlier, with the additional rule that pure liquids and solids do not appear in the expression.

**CONNECTION** Scrubbing exhaust gases with CaO was discussed in the descriptive chemistry box in Chapter 4.

(a)

(b)

**FIGURE 15.7** The position of a heterogeneous equilibrium between $CaCO_3$, $CaO$, and $CO_2$ at constant temperature depends only on the concentration of $CO_2$ gas present. As long as some of each solid is in the system, the equilibrium concentration of $CO_2$ remains the same. (a) Muffle furnace for heating crucibles containing $CaCO_3$ and $CaO$. (b) Two crucibles containing different amounts of $CaCO_3$ and $CaO$ have the same concentration of $CO_2$ gas.

**Le Châtelier's principle** a system at equilibrium responds to a stress in such a way that it relieves that stress.

**Writing Equilibrium Constant Expressions for Heterogeneous Equilibria** LO1

Write $K_c$ expressions for
a. $CaO(s) + SO_2(g) \rightleftharpoons CaSO_3(s)$
b. $CO_2(g) + H_2O(\ell) \rightleftharpoons H_2CO_3(aq)$

**Collect and Organize** We are to write equilibrium constant expressions for equilibria involving reactants and products in more than one phase. We need to identify the pure liquids and solids involved in the equilibrium so we can exclude terms for them from the equilibrium constant expressions.

**Analyze** The first equilibrium involves two solids: $CaO$ and $CaSO_3$. The second involves liquid $H_2O$.

**Solve**
a. An expression with terms for all reactants and products is

$$K_c = \frac{[CaSO_3]}{[CaO][SO_2]}$$

Once we remove the terms representing pure solids, we have

$$K_c = \frac{1}{[SO_2]}$$

b. In the equilibrium constant expression for reaction b, $[H_2O]$ is a constant and is not included, leaving

$$K_c = \frac{[H_2CO_3]}{[CO_2]}$$

**Think About It** The equilibrium constant expressions we have written in this and previous exercises included terms for reactants and products whose concentrations or partial pressures were likely to change significantly during the course of the reaction. We exclude terms for pure solids and liquids because their concentrations do not change.

⚙ **Practice Exercise** Write $K_p$ expressions for the reactions
a. $C(s) + CO_2(g) \rightleftharpoons 2\,CO(g)$
b. $CO_2(g) + H_2(g) \rightleftharpoons CO(g) + H_2O(\ell)$

**CONCEPT TEST**

Explain why $K_c = [H_2O(g)]$ is the equilibrium constant expression for the equilibrium $H_2O(\ell) \rightleftharpoons H_2O(g)$.

# 15.7 Le Châtelier's Principle

We can perturb chemical reactions at equilibrium in several ways—for example, by changing the concentration or partial pressure of a reactant or product, or by changing the temperature of the system. Adding or removing an ingredient in the system alters the value of the reaction quotient $Q$ so that it is no longer equal to the value of $K$. On the other hand, changing the temperature of a system changes the value of $K$. Either way, the perturbed system is not at equilibrium and the composition of the system must change to restore equilibrium.

▶❚❚ **CHEMTOUR** Le Châtelier's Principle

One of the first scientists to study and then successfully predict how chemical equilibria respond to such perturbations was French chemist Henri Louis

Le Châtelier (1850–1936). He articulated **Le Châtelier's principle**, which states that, if a system at equilibrium is perturbed (or *stressed*), the position of the equilibrium shifts in the direction that relieves that stress. Through the years, chemists have used Le Châtelier's principle to increase the yields of chemical reactions that would otherwise have yielded very little of a desired product.

## Effects of Adding or Removing Reactants or Products

When a reactant or product is added or removed, a system at chemical equilibrium is perturbed. Following Le Châtelier's principle, the system responds in such a way as to restore equilibrium (Figure 15.8).

To explore how industrial chemists exploit Le Châtelier's principle, let's look again at the water–gas shift reaction for making hydrogen:

$$H_2O(g) + CO(g) \rightleftharpoons H_2(g) + CO_2(g) \qquad (15.2)$$

To shift the equilibrium toward the production of more $H_2$, chemists pass the reaction mixture through a scrubber containing a concentrated aqueous solution of $K_2CO_3$. Doing this removes $CO_2$ from the gaseous mixture because of the following reaction:

$$CO_2(g) + H_2O(\ell) + K_2CO_3(aq) \rightleftharpoons 2\,KHCO_3(s)$$

Removing $CO_2$ means fewer molecules of it are available to collide with molecules of $H_2$ and drive the reverse reaction in Equation 15.2. As a result, the rate of the reverse reaction becomes slower than the rate of the forward reaction. This means the system is no longer in equilibrium. To return to equilibrium, the reaction proceeds in the forward direction (chemists say the reaction *shifts to the right*), making more product to restore some of what was removed until a new equilibrium is achieved. The new equilibrium is like the old one in that the value of the mass action expression:

$$\frac{[H_2][CO_2]}{[H_2O][CO]}$$

is equal to the value of $K$. This is true even though the concentrations of the individual reactants and products have changed; the *overall* ratio in the $K$ expression is restored.

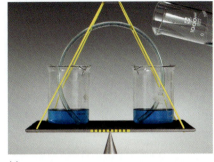

(a)

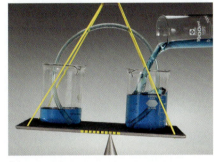

(b)

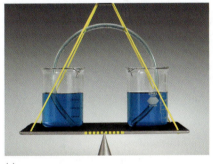

(c)

**FIGURE 15.8** Chemical equilibrium is a bit like two beakers of water connected by a siphon. (a) The system is at equilibrium when the levels of water in both beakers are the same. (b) If we add water to one beaker, the added water drives a flow of water through the siphon into the opposite beaker until (c) equilibrium is restored. The opposite response would occur if we removed some water from one of the beakers; the flow through the siphon would be toward that beaker.

**CONCEPT TEST**

The reaction mixture in the water–gas shift reaction is at equilibrium and some $CO_2$ is rapidly removed.
a. How does this removal affect the value of the reaction quotient $Q$?
b. When equilibrium is restored, which of the four compounds in the system is present at a higher concentration, which is present at a lower concentration, and which, if any, has the same concentration as in the original equilibrium mixture?

**SAMPLE EXERCISE 15.10   Adding or Removing Reactants       L05
or Products to Stress an Equilibrium**

Suggest three ways the production of ammonia via the reaction

$$N_2(g) + 3\,H_2(g) \rightleftharpoons 2\,NH_3(g)$$

could be increased without changing the reaction temperature.

**Collect and Organize** We are given the balanced chemical equation of a reversible reaction and are to suggest three ways to increase its yield; that is, to shift its equilibrium to the right.

**Analyze** The mass action expression for this reaction is

$$Q_p = \frac{(P_{NH_3})^2}{(P_{H_2})^3(P_{N_2})}$$

We want to make changes in the system that will have the effect of producing a smaller reaction quotient $Q_p$. If $Q_p < K_p$, the reaction system will respond by consuming $N_2$ and $H_2$ and forming more $NH_3$. Therefore we want to decrease the numerator or increase the denominator of the mass action expression.

**Solve** If we (1) increase the partial pressure of $N_2$, (2) increase the partial pressure of $H_2$, or (3) remove $NH_3$ from the system, the equilibrium will shift to the right.

**Think About It** The industrial synthesis of ammonia relies on shifting the equilibrium to the right by (1) running the reaction at high partial pressures of the reactants, and (2) removing the product $NH_3$ by passing the reaction mixture through chilled condensers. Chilling the mixture removes ammonia because it condenses at a higher temperature than $N_2$ or $H_2$.

⚙ **Practice Exercise** Describe the changes that occur in a gas-phase equilibrium based on the reaction

$$2\,H_2S(g) + 3\,O_2(g) \rightleftharpoons 2\,SO_2(g) + 2\,H_2O(g)$$

if (a) the mixture is cooled and water vapor condenses; (b) $SO_2$ gas dissolves in liquid water as it condenses; (c) more $O_2$ is added.

---

We can express the effects of stressing equilibria in general terms. First, increasing the partial pressure (or concentration) of a reactant or product shifts the equilibrium so that more of that substance is consumed in the reaction. Second, decreasing the partial pressure (or concentration) of a reactant or product shifts the equilibrium toward the production of more of that substance.

## Effects of Pressure and Volume Changes

A reaction involving gaseous reactants or products may be perturbed by altering the volume of the system, thereby changing the partial pressures of the reactants and products. To see how, let's revisit the equilibrium between $NO_2$ and its dimer, $N_2O_4$,

$$2\,NO_2(g) \rightleftharpoons N_2O_4(g) \tag{15.3}$$

which is shown in Figure 15.4. Suppose we fill a syringe with brown $NO_2$ gas. A fraction of the $NO_2$ combines to form colorless $N_2O_4$ according to the reaction in Equation 15.3, and as shown in the molecular view of Figure 15.4(a). When the mixture is compressed at constant $T$ (Figure 15.4b), the volume of the system decreases. Therefore, the partial pressures of both gases increase. These increases in partial pressure stress the equilibrium and change the value of the reaction quotient.

To understand this behavior, let's suppose that we have an equilibrium mixture of these gases in which $P_{NO_2} = X$ and $P_{N_2O_4} = Y$. Then the value of $K_p$ is

$$K_p = \frac{(P_{N_2O_4})}{(P_{NO_2})^2} = \frac{Y}{X^2}$$

When the volume of this equilibrium mixture is decreased to half its initial volume, the partial pressure of each gas is doubled. Inserting these new values into the mass action expression gives us a reaction quotient that is half the value of $K_p$:

$$Q_p = \frac{2Y}{(2X)^2} = \frac{2}{4}\frac{Y}{X^2} = \frac{1}{2}K_p$$

When the value of $Q_p$ for any reaction is less than the value of $K_p$, the reaction proceeds in the forward direction, consuming reactants and forming products. In this case, brown $NO_2$ is consumed, colorless $N_2O_4$ forms, and the color of the mixture of gases fades as shown in Figure 15.4(c).

Another way to explain this response to changing the volume of a mixture of reacting gases involves the ideal gas law, $PV = nRT$. If we rearrange the terms to solve for $V$, we obtain

$$V = \left(\frac{RT}{P}\right)n \tag{15.23}$$

This equation shows the direct proportionality between the volume of a gas mixture at constant temperature and pressure and the number of moles of gas in the mixture. Equation 15.23 tells us that one way to relieve the stress of a decrease in the volume of a mixture of reacting gases is for the reaction to proceed in a direction that reduces the number of moles of gas. Because 2 moles of $NO_2$ form 1 mole of $N_2O_4$, producing more $N_2O_4$ reduces the overall volume of the system.

In any reaction between gases in which the number of moles of gas changes as the reaction proceeds at constant temperature and pressure, changing the overall volume occupied by the reactant and product gases shifts the equilibrium of the mixture (Figure 15.9). *Increasing* the volume shifts the equilibrium toward the side of the reaction with *more* moles of gas. *Decreasing* the volume shifts the equilibrium toward the side of the reaction with *fewer* moles of gas.

**FIGURE 15.9** The effect of changing the volume (and pressure) on an equilibrium involving gases. An increase in volume (decrease in pressure) causes the reaction to shift to the side with the greater number of particles. A decrease in volume (increase in pressure) causes the reaction to shift to the side with the smaller number of particles.

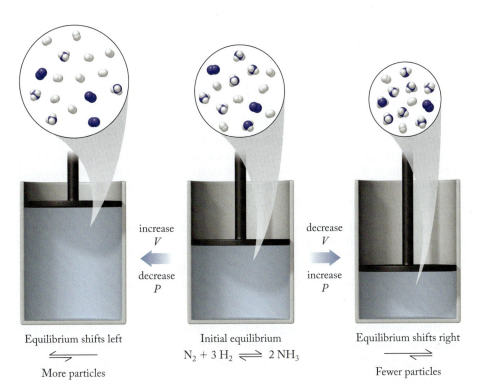

increase
V

decrease
P

decrease
V

increase
P

Equilibrium shifts left

More particles

Initial equilibrium

$N_2 + 3\,H_2 \rightleftharpoons 2\,NH_3$

Equilibrium shifts right

Fewer particles

CONCEPT TEST

Changing the overall volume of the reaction of the water–gas shift reaction at equilibrium:

$$H_2O(g) + CO(g) \rightleftharpoons H_2(g) + CO_2(g)$$

does not shift the equilibrium. Why?

---

SAMPLE EXERCISE 15.11 **Assessing Volume Effects on Gas-Phase Equilibria** LO5

In which of the following equilibria would a decrease in volume promote the formation of more product(s)?
a. $N_2(g) + O_2(g) \rightleftharpoons 2 NO(g)$
b. $2 NO(g) + O_2(g) \rightleftharpoons 2 NO_2(g)$
c. $N_2O_4(g) \rightleftharpoons 2 NO_2(g)$
d. $H_2O(\ell) + CO_2(g) \rightleftharpoons H_2CO_3(aq)$
e. $CaCO_3(s) \rightleftharpoons CaO(s) + CO_2(g)$

**Collect and Organize** We are asked to identify the reactions for which a decrease in volume causes an increase in product formation. We know that decreasing volume shifts a chemical equilibrium involving gases toward the side of the reaction with fewer moles of gas.

**Analyze** We need to identify those reactions in which there are fewer moles of gaseous products than there are gaseous reactants.

**Solve** Summing the number of moles of gas on the reactant side and product side in each reaction, we have

| Reaction | Moles of Gaseous Reactants | Moles of Gaseous Products |
|---|---|---|
| a | 2 | 2 |
| b | 3 | 2 |
| c | 1 | 2 |
| d | 1 | 0 |
| e | 0 | 1 |

The only two reactions with fewer moles of gaseous products than reactants are reactions b and d. Therefore, they are the only two in which a decrease in the total volume of the reacting gases increases product formation.

**Think About It** In the case of reaction a, the number of moles of gas on the reactant side is the same as the number of moles on the product side, so changing volume does not cause the equilibrium to shift. In reactions c and e, the number of moles on the product side is greater than on the reactant side, so decreasing volume favors the reverse of the reaction as written and decreases product formation.

⚙ **Practice Exercise** How does decreasing the volume of a reaction mixture of CO, $Cl_2$, and $COCl_2$ affect the following equilibrium?

$$CO(g) + Cl_2(g) \rightleftharpoons COCl_2(g)$$

---

Reaction d in Sample Exercise 15.11 contributes to the solubility of $CO_2$ in water and is one reason why carbonated beverages under pressure fizz when their tops are opened. Because 1 mole of gas appears on the reactant side and no moles of gas appear on the product side, the small volume maintained by the pressure in the space above the liquid in the can results in an increase in the solubility of $CO_2$

in the fluid. When the container is opened, the pressure decreases and the volume occupied by the gas increases rapidly, decreasing the partial pressure of the $CO_2$, favoring the reactants and causing gaseous $CO_2$ to fizz out of the beverage.

It is possible to maintain a constant volume of a system in which a reaction is run and change the pressure by adding or removing a reactant or product. We saw in the previous section how such changes influence the position of equilibrium. Pressure can also be changed by adding an inert gas (a gas not involved in the reaction) to the system, but this action does not affect the equilibrium because it does not change the partial pressure of any of the components of the reaction. Therefore it does not cause a shift in the equilibrium.

## Effect of Temperature Changes

In Chapter 5 we examined the energy changes that accompany many chemical reactions. Now we explore the stresses produced on chemical equilibria by adding or removing energy by increasing or decreasing temperature. Let's start with an exothermic reaction, the synthesis of ammonia:

$$N_2(g) + 3\,H_2(g) \rightleftharpoons 2\,NH_3(g) + \text{heat}$$

If we think of energy as a product in the forward reaction, then raising the temperature of a reaction mixture favors the reverse reaction; lowering the temperature favors the forward reaction.

One major difference arises, however, between applying Le Châtelier's principle to concentration or pressure changes and applying it to temperature changes: *changes in temperature change the value of K.* Increasing temperature reduces the yield of ammonia synthesis because increasing temperature reduces the value of $K$.

In general, the value of $K$ decreases as temperature increases for exothermic reactions and increases as temperature increases for endothermic reactions (Figure 15.10). We will look more closely at the influence of temperature on $K$ values in Chapter 18, but for now this general analysis enables us to predict the direction of a shift in equilibrium with changing temperature.

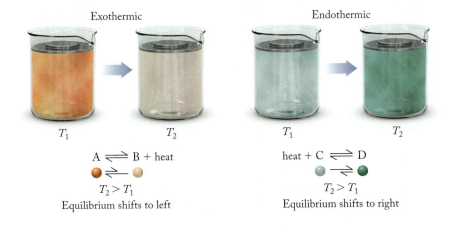

Exothermic

$T_1$   $T_2$

A $\rightleftharpoons$ B + heat

$T_2 > T_1$
Equilibrium shifts to left

Endothermic

$T_1$   $T_2$

heat + C $\rightleftharpoons$ D

$T_2 > T_1$
Equilibrium shifts to right

**FIGURE 15.10** The direction in which a temperature change shifts an equilibrium depends on whether the reaction is exothermic (heat is a product) or endothermic (heat is a reactant).

**SAMPLE EXERCISE 15.12**   **Predicting Changes in Equilibrium with Temperature**   **LO5**

The color of an aqueous acidic solution of cobalt(II) chloride depends on its temperature (Figure 15.11). In aqueous HCl, the solution is pink at 0°C, magenta at 25°C, and dark blue at 75°C. Is the reaction producing the pink-to-blue color change exothermic or endothermic?

Temperature = 5°C

Temperature = 75°C

$$Co(H_2O)_6^{2+}(aq) + 4\,Cl^-(aq) \rightleftharpoons CoCl_4^{2-}(aq) + 6\,H_2O(\ell)$$

Pink  Royal blue

**FIGURE 15.11** Two compounds of cobalt, one pink and one blue, are in equilibrium in aqueous hydrochloric acid solution. The position of equilibrium shifts to the right as the temperature changes, causing the color of the solution to change as more and more of the blue $CoCl_4^{2-}$ ion forms.

**Collect and Organize** We are asked to determine whether a reversible reaction is exothermic or endothermic. Asked another way, is heat a reactant or product?

**Analyze** If the reaction is exothermic, then increasing the temperature is the equivalent of adding a product, the impact of which will be to shift the reaction toward the pink reactant side. If the reaction is endothermic, then increasing the temperature is the equivalent of adding a reactant, the impact of which will be to shift the reaction toward the side of the blue product.

**Solve** When heat is added to the system, its color changes from pink to blue. This means heat is a reactant and the reaction as written must be endothermic:

$$\text{Heat} + \text{pink} \rightleftharpoons \text{blue}$$

**Think About It** This exercise illustrates how determining the effect of changing the temperature of an equilibrium reaction mixture can tell us whether the reaction is exothermic or endothermic. The fact that the reaction mixture in this exercise is magenta at room temperature tells us that both the pink and blue forms are present. This suggests that the value of $K$ at room temperature is close to 1, but the concentration of $Cl^-(aq)$ is also an important factor, because it is raised to the fourth power in the equilibrium expression.

⚙ **Practice Exercise** Predict how the value of the equilibrium constant of the reaction

$$N_2(g) + O_2(g) \rightleftharpoons 2\,NO(g) \qquad \Delta H^\circ_{rxn} = 181\text{ kJ}$$

changes with increasing temperature.

Table 15.4 summarizes how an exothermic system at equilibrium responds to various stresses.

| TABLE 15.4 | Responses of an Exothermic Reaction $2\,A(g) \rightleftharpoons B(g)$ at Equilibrium to Different Kinds of Stress | |
|---|---|---|
| **Kind of Stress** | **How Stress Is Relieved** | **Direction of Shift** |
| Add A | Consume A | To the right |
| Remove A | Produce A | To the left |
| Remove B | Produce B | To the right |
| Add B | Consume B | To the left |
| Increase temperature by adding heat | Consume some of the heat | To the left |
| Decrease temperature by removing heat | Generate heat | To the right |
| Decrease volume by compressing the reaction mixture | Reduce moles of gas to relieve pressure increase | To the right |
| Increase volume | Increase moles of gas to maintain equilibrium pressure | To the left |
| Increase pressure by adding inert gas | No change in partial pressures of reactants or products | No shift |

 **CONNECTION** In Chapter 14 we discovered that a catalyst simultaneously increases the rate of a reaction in both the forward and reverse directions, when we discussed the effect of temperature on the rates of reactions occurring in a catalytic converter.

## Catalysts and Equilibrium

The industrial production of ammonia (see Sample Exercise 15.10 and the descriptive chemistry box in Chapter 6) was developed by the German chemists Fritz Haber (1868–1934) and Carl Bosch (1874–1940) in the early 20th century and is still widely referred to as the Haber–Bosch process. What makes the process commercially feasible is the use of catalysts. As discussed in Chapter 14, a catalyst increases the rate of a chemical reaction by lowering its activation energy. The question is this: If a catalyst increases the rate of a reaction, does that catalyst affect the equilibrium constant of the reaction?

To answer this question, consider the energy profiles of the catalyzed and uncatalyzed reaction in Figure 15.12. The catalyst increases the rate of the reaction by decreasing the height of the energy barrier. However, the barrier height is reduced by the same amount whether the reaction proceeds in the forward direction or in reverse. As a result, the increase in reaction rate produced by the catalyst is the same in both directions. Therefore a catalyst has no effect on the equilibrium constant of a reaction or on the composition of an equilibrium reaction mixture. A catalyst does, however, decrease the amount of time needed for a reaction to reach equilibrium.

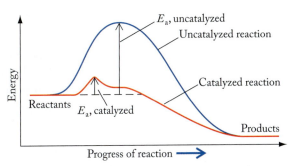

**FIGURE 15.12** The effect of a catalyst on a reaction. A catalyst lowers the activation energy barrier, and as a result the rate of the reaction increases. However, because both the forward reaction and the reverse reaction occur more rapidly, the position of equilibrium (that is, the value of *K*) does not change. The system comes to equilibrium more rapidly, but the relative amounts of product and reactant present at equilibrium do not change.

# 15.8 Calculations Based on *K*

Reference books and the tables in Appendix 5 of this book contain lists of equilibrium constants for chemical reactions. We can use these values in several kinds of calculations, including those in which:

1. We want to determine whether a reaction mixture has reached equilibrium (Sample Exercise 15.8).
2. We know the value of *K* and the starting concentrations or partial pressures of reactants and/or products, and we want to calculate their equilibrium concentrations or pressures.

▶ⅠⅠ **CHEMTOUR** Solving Equilibrium Problems

In this section we focus on the second type of calculation and introduce a useful way of handling such problems: a table of reactant and product concentrations (or partial pressures) called a *RICE table*. The acronym RICE means that the table starts with the balanced chemical equation describing the **R**eaction followed by rows that contain **I**nitial concentration values, **C**hanges in those initial values as the reaction proceeds toward equilibrium, and **E**quilibrium values.

In our first example, we calculate how much hydrogen iodide forms from the reaction of hydrogen gas and iodine gas at 445°C. At this temperature, the $K_p$ of the reaction is 50.2:

$$H_2(g) + I_2(g) \rightleftharpoons 2\,HI(g)$$

The initial partial pressures are $P_{H_2} = 1.00$ atm and $P_{I_2} = 1.03$ atm, and no HI is present.

We start by writing the given (initial) information in our RICE table:

| Reaction (R) | $H_2(g)$ + | $I_2(g)$ ⇌ | 2 HI(g) |
|---|---|---|---|
| | $P_{H_2}$ (atm) | $P_{I_2}$ (atm) | $P_{HI}$ (atm) |
| Initial (I) | 1.00 | 1.03 | 0 |
| Change (C) | | | |
| Equilibrium (E) | | | |

We know the reaction will proceed in the forward direction because there is no product initially present, so $Q_p = 0 < K_p$.

We need to use algebra to fill in rows C and E. We don't know how much $H_2$ or $I_2$ will be consumed or how much HI will be made. We can define the change in partial pressure of $H_2$ as $-x$ because $H_2$ is consumed during the reaction. Because the mole ratio of $H_2$ to $I_2$ in the reaction is 1:1, the change in $P_{I_2}$ is also $-x$. Two moles of HI is produced from each mole of $H_2$ and $I_2$, so the change in $P_{HI}$ is $+2x$. Inserting these values in the C row, we have

| Reaction (R) | $H_2(g)$ | + | $I_2(g)$ | ⇌ | 2 HI(g) |
|---|---|---|---|---|---|
| | $P_{H_2}$ (atm) | | $P_{I_2}$ (atm) | | $P_{HI}$ (atm) |
| Initial (I) | 1.00 | | 1.03 | | 0 |
| Change (C) | $-x$ | | $-x$ | | $+2x$ |
| Equilibrium (E) | | | | | |

Combining the I and C rows, we obtain the three partial pressures at equilibrium:

| Reaction (R) | $H_2(g)$ | + | $I_2(g)$ | ⇌ | 2 HI(g) |
|---|---|---|---|---|---|
| | $P_{H_2}$ (atm) | | $P_{I_2}$ (atm) | | $P_{HI}$ (atm) |
| Initial (I) | 1.00 | | 1.03 | | 0 |
| Change (C) | $-x$ | | $-x$ | | $+2x$ |
| Equilibrium (E) | $1.00 - x$ | | $1.03 - x$ | | $2x$ |

The next step is to substitute the terms from the E row into the $K_p$ expression for the reaction:

$$K_p = \frac{(P_{HI})^2}{(P_{H_2})(P_{I_2})}$$

$$= \frac{(2x)^2}{(1.00 - x)(1.03 - x)} \tag{15.24}$$

Expanding the terms in the numerator and denominator of Equation 15.24 gives

$$K_p = \frac{4x^2}{1.03 - 2.03\,x + x^2} = 50.2$$

Cross-multiplying, we get

$$1.03 \times 50.2 - (2.03 \times 50.2)x + 50.2x^2 = 4x^2$$

Combining the $x^2$ terms and rearranging, we have

$$46.2x^2 - 101.9x + 51.7 = 0$$

You may recognize this equation as one that fits the general form of a quadratic equation:

$$ax^2 + bx + c = 0$$

We can solve for $x$ either with a scientific calculator or using the quadratic formula

$$x = \frac{-b \pm \sqrt{b^2 - 4ac}}{2a}$$

Two values are possible for $x$: 0.791 and 1.415. We focus on 0.791 atm because it is the only one of the two values that is physically possible. (Note that we start with a partial pressure of iodine gas of 1.03 atm. If we were to subtract 1.415 atm from that value, the answer would be a negative number. Because we cannot have a negative pressure, that number is a physically impossible value for $x$.) Using $x = 0.791$ atm and calculating equilibrium partial pressures to two significant figures:

$$P_{H_2} = 1.00 - x$$
$$= 1.00 - 0.791 = 0.21 \text{ atm}$$
$$P_{I_2} = 1.03 - x$$
$$= 1.03 - 0.791 = 0.24 \text{ atm}$$
$$P_{HI} = 2x = 2(0.791) = 1.582 = 1.58 \text{ atm}$$

The value of $x$ makes sense because it produces pressures of $H_2$ and $I_2$ that are less than their initial values but positive. The sum of the partial pressures of the components in the system is $(0.21 + 0.24 + 1.58) = 2.03$ atm, which is the same as the starting pressure. This is expected, because we have 2 moles of gas on the reactant side and 2 moles of gas on the product side, so we do not expect the pressure to change as the reaction proceeds. As a final check, we can use the calculated partial pressures in the expression for $K_p$ and calculate its value. When we do so, we calculate $K_p = (1.58)^2/[(0.21)(0.24)] = 50$, which is not significantly different from the value given of 50.2.

**SAMPLE EXERCISE 15.13** **Calculating an Equilibrium Partial Pressure I** **LO6**

The gas-phase reaction for the production of $PCl_5$ from $PCl_3$ and chlorine gas at 250°C has an equilibrium constant of 24.2. If the initial partial pressures of the reactants in a vessel are $P_{PCl_3} = 0.43$ atm and $P_{Cl_2} = 0.87$ atm, and no $PCl_5$ is present, what are the partial pressures of the gases when equilibrium is achieved?

**Collect and Organize** We want to determine the equilibrium partial pressures of the three gases given a description of the reaction and the equilibrium constant. The equation is

$$PCl_3(g) + Cl_2(g) \rightleftharpoons PCl_5(g) \qquad K_p = 24.2$$

**Analyze** The system initially contains no product, so $Q < K_p$ and the reaction will proceed in the forward direction as written. We can construct a RICE table from the information given in the statement, write the partial pressures in terms of amount of product formed $x$, and then solve for $x$ using the mass-action expression and the quadratic equation, if we need it.

**Solve**

| Reaction (R) | $PCl_3(g)$ | + | $Cl_2(g)$ | $\rightleftharpoons$ | $PCl_5(g)$ |
|---|---|---|---|---|---|
| | $P_{PCl_3}$ (atm) | | $P_{Cl_2}$ (atm) | | $P_{PCl_5}$ (atm) |
| Initial (I) | 0.43 | | 0.87 | | 0 |
| Change (C) | $-x$ | | $-x$ | | $+x$ |
| Equilibrium (E) | $0.43 - x$ | | $0.87 - x$ | | $x$ |

We substitute the terms from row E into the $K_p$ expression for the reaction:

$$K_p = \frac{(P_{PCl_5})}{(P_{PCl_3})(P_{Cl_2})} = \frac{x}{(0.43 - x)(0.87 - x)}$$

Then we expand the denominator and cross-multiply:

$$K_p = \frac{x}{0.3741 - 1.30x + x^2} = 24.2$$

$$0.3741 \times 24.2 - (1.30 \times 24.2)x + 24.2x^2 = x$$

Combining the $x$ terms and rearranging, we have

$$24.2x^2 - 32.46x + 9.05322 = 0$$

Using the quadratic equation or a scientific calculator and solving for $x$, we get two roots: 0.3955 and 0.9458. Looking at the initial values of $P$, only 0.3955 gives physically meaningful values for partial pressures, so we continue with $x = 0.3955$ and calculate equilibrium partial pressures to two significant figures:

$$P_{PCl_3} = 0.43 - 0.3955 = 0.03 \text{ atm}$$

$$P_{Cl_2} = 0.87 - 0.3955 = 0.47 \text{ atm}$$

$$P_{PCl_5} = 0.3955 = 0.40 \text{ atm}$$

**Think About It** First, the values make sense because the partial pressures of both reactants have decreased and that of the product has increased. When we check our answers by substituting them into the equation for $K_p$:

$$K_p = \frac{(P_{PCl_5})}{(P_{PCl_3})(P_{Cl_2})} = \frac{0.40}{(0.03)(0.47)} = 28$$

we get 28, which is close to the given value of 24.2.

⚙ **Practice Exercise** Consider the reverse reaction in Sample Exercise 15.13 at the same temperature. If the vessel initially contains 1.35 atm $PCl_5$, what are the partial pressures when equilibrium is achieved?

We have to be alert to mathematical opportunities that simplify our calculations when we solve problems like these. Sometimes we can simplify an expression and avoid having to solve the quadratic equation. One such opportunity occurs when two moles of reactants form two moles of products and the initial concentrations of both reactants are the same.

Suppose we have a vessel containing 0.100 $M$ $N_2$ and 0.100 $M$ $O_2$ at a temperature where $K_c$ for the NO formation reaction is 0.100:

$$N_2(g) + O_2(g) \rightleftharpoons 2\,NO(g) \qquad K_p = 0.100$$

What is the equilibrium concentration of NO? The RICE table in this case is

| Reaction (R) | $N_2(g)$ | + | $O_2(g)$ | $\rightleftharpoons$ | 2 NO(g) |
|---|---|---|---|---|---|
| | $[N_2]$ (M) | | $[O_2]$ (M) | | [NO] (M) |
| Initial (I) | 0.100 | | 0.100 | | 0 |
| Change (C) | $-x$ | | $-x$ | | $+2x$ |
| Equilibrium (E) | $0.100 - x$ | | $0.100 - x$ | | $2x$ |

Inserting the values from the E row into the expression for $K_c$ gives

$$\frac{(2x)^2}{(0.100 - x)(0.100 - x)} = 0.100$$

Taking the square root of each side:

$$\frac{2x}{0.100 - x} = 0.316$$

Solving for $x$, we get $x = 0.0136$ $M$, and the equilibrium concentrations are $[N_2] = [O_2] = 0.100\ M - 0.0136\ M = 0.086\ M$ and $[NO] = 2x = 0.0272\ M$.

It's a good idea to check that these concentrations are consistent with the known value of $K_c$. Substituting into the $K_c$ expression, we get

$$K_c = \frac{(0.0272)^2}{(0.086)^2} = 0.10$$

which is the value of $K$ given for the reaction in question.

---

**SAMPLE EXERCISE 15.14** **Calculating an Equilibrium Partial Pressure II** **LO6**

Some of the $H_2$ used in the Haber–Bosch process is produced by the water–gas shift reaction:

$$CO(g) + H_2O(g) \rightleftharpoons CO_2(g) + H_2(g) \qquad (15.2)$$

If a reaction vessel at 400°C is filled with an equimolar mixture of CO and steam, such that $P_{CO} = P_{H_2O} = 2.00$ atm, what is the partial pressure of $H_2$ at equilibrium? The equilibrium constant $K_p = 10$ at 400°C.

**Collect and Organize** We are asked to find the partial pressure of a gaseous product in an equilibrium mixture, given the initial partial pressures of reactants and the value of $K_p$. We can (1) set up a RICE table, (2) use the partial pressures from the E row in the equilibrium constant expression, and (3) solve for $P_{H_2}$.

**Analyze** The system initially contains no product. This means that the reaction quotient $Q_p$ is equal to zero and thus less than $K$. Therefore, the reaction proceeds in the forward direction, decreasing the partial pressures of the reactants while increasing those of the products. A $K_p$ value of 10 means that reactants should be converted into products, but there should be significant partial pressures of both reactants and products at equilibrium.

**Solve** Let $x$ be the increase in partial pressure of $H_2$ as a result of the reaction. The stoichiometry of the reaction tells us that the change in $P_{CO_2}$ is also $x$ and that the changes in both $P_{CO}$ and $P_{H_2O}$ are $-x$:

| Reaction (R) | CO(g) | + | H₂O(g) | ⇌ | CO₂(g) | + | H₂(g) |
|---|---|---|---|---|---|---|---|
| | $P_{CO}$ (atm) | | $P_{H_2O}$ (atm) | | $P_{CO_2}$ (atm) | | $P_{H_2}$ (atm) |
| Initial (I) | 2.00 | | 2.00 | | 0.00 | | 0.00 |
| Change (C) | $-x$ | | $-x$ | | $+x$ | | $+x$ |
| Equilibrium (E) | $2.00 - x$ | | $2.00 - x$ | | $x$ | | $x$ |

Inserting these equilibrium terms into the equilibrium constant expression for the reaction gives

$$K_p = \frac{(P_{CO_2})(P_{H_2})}{(P_{CO})(P_{H_2O})}$$

$$= \frac{(x)(x)}{(2.00 - x)(2.00 - x)} = 10$$

Note that we can simplify this equation by taking the square root of both sides:

$$\frac{x}{2.00 - x} = \sqrt{10} = 3.16$$

Solving for $x$ gives $x = 1.52$ atm, which is the equilibrium partial pressure of $H_2$ (and $CO_2$).

**Think About It** Our prediction that both reactants and products would be present at equilibrium is justified. It is a good idea to substitute the results into the equilibrium constant expression as a check on the validity of the solution. The calculated partial pressures are $P_{H_2} = P_{CO_2} = 1.52$ atm and $P_{H_2O} = P_{CO} = 2.00 - 1.52 = 0.48$ atm. Inserting these values in the equilibrium constant expression gives $K_p = (1.52)^2/(0.48)^2 = 10$, which is the same as the given value. This confirms that our calculation is correct. We did not have to use the quadratic equation because we could simplify the expression algebraically.

⚙️ **Practice Exercise** The chemical equation for the reaction of chlorine and bromine to produce BrCl is

$$Cl_2(g) + Br_2(g) \rightleftharpoons 2\, BrCl(g)$$

The value for $K_p$ for the reaction under a given set of conditions is $4.7 \times 10^{-2}$. What is the partial pressure of each reactant and product in a sealed reaction vessel if the initial partial pressures of $Cl_2$ and $Br_2$ are both 0.100 atm and no BrCl is present?

■ ● ● ● ● ● ● ● ● ● ● ● ● ● ● ● ● ● ● ● ● ● ● ● ● ● ● ● ● ● ● ● ● ● ● ● ● ● ● ● ● ● ● ●

All of the problems we have worked thus far have had equilibrium constants between 0.1 and 100 and initial partial pressures of reactants around 1 atm. This situation means that we could reasonably expect that significant amounts of both reactants and products would be present at equilibrium. If equilibrium constants are very small, however, we can frequently make an approximation at the outset that enables us to simplify our mathematical operations by considering the impact that the size of an equilibrium constant has on the values in our equation. For example, the $K_p$ for the following reaction at 1500 $K$ is small:

$$N_2(g) + O_2(g) \rightleftharpoons 2\, NO(g) \qquad K_p = 1.0 \times 10^{-5}$$

If we are given that the initial partial pressures are 0.79 atm of $N_2$, 0.21 atm of $O_2$, and no NO is present, we can use the information in a RICE table and the mass action expression for the reaction to calculate the partial pressures of reactants and products at equilibrium as we have always done.

| Reaction (R) | $N_2(g)$ | + | $O_2(g)$ | $\rightleftharpoons$ | 2 NO(g) |
|---|---|---|---|---|---|
| | $P_{N_2}$ (atm) | | $P_{O_2}$ (atm) | | $P_{NO}$ (atm) |
| Initial (I) | 0.79 | | 0.21 | | 0 |
| Change (C) | $-x$ | | $-x$ | | $+2x$ |
| Equilibrium (E) | $0.79 - x$ | | $0.21 - x$ | | $2x$ |

We can write the expression for $K_P$ using these terms:

$$K_p = 1.0 \times 10^{-5} = \frac{(P_{NO})^2}{(P_{N_2})(P_{O_2})}$$

$$= \frac{(2x)^2}{(0.79 - x)(0.21 - x)} = \frac{4x^2}{(0.79 - x)(0.21 - x)}$$

Before we take the next usual step of expanding the equation, let's think about the size of $x$. The equilibrium constant is very small, which means very little product will form. The value of $x$ will be insignificant compared to the values of the initial partial pressures, so we can simplify the expression by ignoring the $x$ terms in the denominator and using the initial values for $P_{N_2}$ and $P_{O_2}$ instead:

$$\frac{4x^2}{(0.79 - x)(0.21 - x)} \approx \frac{4x^2}{(0.79)(0.21)}$$

Note that we cannot ignore the $x$ in the numerator, only in the denominator where it would be subtracted from a term much larger than itself. We then calculate $x$:

$$\frac{4x^2}{(0.79)(0.21)} = 1.0 \times 10^{-5}$$

$$4x^2 = (0.79)(0.21)(1.0 \times 10^{-5}) = 1.659 \times 10^{-6}$$

$$x^2 = 4.148 \times 10^{-7}$$

$$x = 6.44 \times 10^{-4} \text{ atm}$$

This value does not differ significantly from that obtained by solving the quadratic equation ($6.43 \times 10^{-4}$), given that we know the initial partial pressures to only two significant figures. Generally speaking, we can ignore the $x$ component of an equilibrium concentration or partial pressure if the value of $-x$ or $+x$ is less than 5% of the initial value. This frequently happens when $K$ values are small ($< \sim 3 \times 10^{-5}$) and initial concentration or partial pressure values are greater than about 0.03.

---

**SAMPLE EXERCISE 15.15** **Calculating an Equilibrium Partial Pressure III** **L06**

The decomposition of phosgene ($COCl_2$) into carbon monoxide and chlorine at 373 K in the absence of water has an equilibrium constant $K_p = 2.2 \times 10^{-10}$. If a sealed vessel contains only phosgene at a partial pressure of 2.75 atm, what are the partial pressures of reactants and products when the system comes to equilibrium?

**Collect and Organize** We can write a balanced chemical equation for the decomposition of phosgene and the mass action expression for $K_p$:

$$COCl_2(g) \rightleftharpoons CO(g) + Cl_2(g) \qquad K_p = \frac{(P_{CO})(P_{Cl_2})}{(P_{COCl_2})} = 2.2 \times 10^{-10}$$

**Analyze** The system initially contains no product. This means that the reaction quotient $Q_p$ is equal to zero and thus less than $K$. Therefore, the reaction proceeds in the forward direction, decreasing the partial pressures of the reactants while increasing those of the products. The $K_p$ value of $2.2 \times 10^{-10}$ is very small, which means that only a very small amount of products should be formed. Hence $x$ in the RICE table may be insignificant with respect to the partial pressure of reactant and we can simplify the calculation accordingly.

**Solve** Making a RICE table for this system:

| Reaction (R) | $COCl_2(g)$ $\rightleftharpoons$ | $CO(g)$ + | $Cl_2(g)$ |
|---|---|---|---|
| | $P_{COCl_2}$ (atm) | $P_{CO}$ (atm) | $P_{Cl_2}$ (atm) |
| Initial (I) | 2.75 | 0 | 0 |
| Change (C) | $-x$ | $+x$ | $+x$ |
| Equilibrium (E) | $2.75 - x$ | $x$ | $x$ |

Using the values in row E in the mass action expression, we can simplify and solve for $x$:

$$K_p = \frac{(x)(x)}{(2.75 - x)} \approx \frac{x^2}{(2.75)} = 2.2 \times 10^{-10}$$

$$x^2 = 2.75 \times (2.2 \times 10^{-10}) = 6.05 \times 10^{-10}$$

$$x = 2.5 \times 10^{-5}$$

The value of $x$ is indeed insignificant with respect to the initial concentration of reactant present, so at equilibrium $P_{COCl_2} = 2.75$ atm and $P_{CO} = P_{Cl_2} = 2.5 \times 10^{-5}$ atm.

**Think About It** The partial pressure of the reactant does not drop by a significant amount and very small quantities of product are formed. When we check our answers by substituting them into the mass action expression, we get $K_p = (2.5 \times 10^{-5})^2/2.75 = 2.3 \times 10^{-10}$, which is very close to the given value of $2.2 \times 10^{-10}$.

⚙ **Practice Exercise** At a temperature below room temperature but with all substances still in the gas phase, the equilibrium constant for the reaction for the decomposition of iodine monochloride (ICl) into $I_2$ and $Cl_2$ has an equilibrium constant of $1.4 \times 10^{-5}$. If a sealed vessel has ICl at a partial pressure of 1.37 atm and no $I_2$ or $Cl_2$ is present initially, what are the partial pressures of all substances involved in the reaction when it comes to equilibrium?

---

## SAMPLE EXERCISE 15.16 Integrating Concepts: Making Nitric Acid

We have mentioned the Haber process for the production of ammonia, much of which is used for fertilizer. Around 15% of the ammonia produced industrially is later converted into nitric acid by the Ostwald process, named after the German chemist who developed it. In the Ostwald process, ammonia is burned in air at 900°C in the presence of a platinum–rhodium catalyst:

Step 1:     $4\,NH_3(g) + 5\,O_2(g) \rightleftharpoons 4\,NO(g) + 6\,H_2O(g)$

Under these conditions, 90% of the starting ammonia is converted into NO.

a. Air is 21% $O_2$. Typical instructions for the Ostwald process specify that the volume of air used is 10 times the volume of ammonia reacted. (i) Is oxygen in excess under these conditions? (ii) Write the mass action expression for step 1 and explain how the quantity of oxygen present influences the position of equilibrium.

b. In the second step of the process, the temperature is lowered and more air is mixed with the products of step 1.

Step 2:     $2\,NO(g) + O_2(g) \rightleftharpoons 2\,NO_2(g)$     $\Delta H° = -114.0\ kJ$

Predict what would happen to the yield in step 2 if the temperature were increased instead of lowered.

c. In step 3, the final step of the Ostwald process, $NO_2$ is passed through liquid water to form a solution of nitric acid ($HNO_3$); nitrogen monoxide is an additional product. Write a balanced net ionic equation for the reaction.

**Collect and Organize** We have balanced chemical equations and we can write mass action expressions for each step. For step 3, we are given reactants and products and asked for a balanced chemical equation.

**Analyze** Part a involves a problem in limiting reagents. Part b requires us to apply Le Châtelier's principle once we have the mass action equation; we also know step 2 is exothermic, based on the sign of $\Delta H°$. Step 3 is a redox reaction; nitrogen is in the +5 oxidation state in nitric acid and +4 in $NO_2$.

**Solve**

a. (i)  If air is present at a level 10 times the volume of $NH_3$, using the coefficients in the balanced equation, for every 4 volumes of $NH_3$ we must have 40 volumes of air. Air is 21% $O_2$:

40 volumes air $\times$ 0.21 = 8.4 volumes of $O_2$

We need 5 volumes of $O_2$ for every 4 volumes of $NH_3$; we have 8.4 volumes of $O_2$, so $O_2$ is in excess by (8.4 − 5 = 3.4) volumes.

(ii)  $K_p = \dfrac{(P_{NO})^4(P_{H_2O})^6}{(P_{NH_3})^4(P_{O_2})^5}$

Oxygen is in excess, and its concentration appears in the denominator, so it will shift the equilibrium toward the production of more product.

b. The reaction is exothermic, meaning that heat is a product of the reaction. If the temperature were increased, the reaction would shift in the direction of more reactants, and less product would be present at equilibrium.

c. The unbalanced equation is

$$NO_2(g) + H_2O(\ell) \rightleftharpoons NO(g) + HNO_3(aq)$$

Nitrogen in the +4 oxidation state in $NO_2$ is both reduced to NO (ON = +2) and oxidized to $HNO_3$ (ON = +5). Remember also that nitric acid is a strong acid, so it exists in water as dissociated ions. Using the half-reaction method:

$$2\,e^- + 2\,H^+ + NO_2 \rightarrow NO + H_2O$$
$$2(H_2O + NO_2 \rightarrow NO_3^- + 2\,H^+ + e^-)$$

$$\overline{\cancel{2\,e^-} + \cancel{2\,H^+} + NO_2 + 2\,H_2O + 2\,NO_2 \rightarrow}$$
$$NO + \cancel{H_2O} + 2\,NO_3^- + \overset{2}{4}\,H^+ + \cancel{2\,e^-}$$
$$NO_2 + H_2O + 2\,NO_2 \rightarrow NO + 2\,NO_3^- + 2\,H^+$$
$$3\,NO_2 + H_2O \rightarrow NO + 2\,NO_3^- + 2\,H^+$$

**Think About It** The industrial production of nitric acid, like the industrial production of ammonia, requires balancing several competing effects. Temperatures and pressures must be adjusted so that materials can combine rapidly enough to produce product over a reasonable time frame but also to ensure that unfavorable equilibria are stressed in the direction of product formation.

We have seen throughout this chapter that every reversible chemical reaction has a unique equilibrium constant at a specific temperature. If we know the value of that equilibrium constant, we can calculate the concentrations of reactants and products at equilibrium from any combination of initial concentrations. We can carry out other calculations that enable us to optimize the yields of a chemical process by adjusting factors such as temperature, and to predict the response of the process to changing reaction conditions.

## SUMMARY

**Learning Outcome 1** According to the **law of mass action**, the **equilibrium constant expression** for $K_c$ is the ratio of the equilibrium molar concentrations of the products divided by the equilibrium molar concentrations of the reactants, each raised to the

respective stoichiometric coefficient in the balanced equation. If the substances involved in an equilibrium are gases, then an equilibrium constant ($K_p$) may also be written based on the partial pressures of the gases. The concentrations of pure liquids and solids do not change during a reaction and so are omitted from equilibrium constant expressions. (Sections 15.1, 15.2, 15.6)

**Learning Outcome 2** The **reaction quotient ($Q$)** is the value of the mass action expression at any instant during a reaction. At equilibrium $Q = K$, but for nonequilibrium conditions, $Q$ indicates how a reaction is proceeding. (Sections 15.2, 15.5)

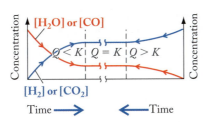

**Learning Outcome 3** The relationship between $K_c$ and $K_p$ depends on the number of moles of gaseous reactants and products in the balanced chemical equation. (Section 15.3)

**Learning Outcome 4** The reverse of a reaction has an equilibrium constant that is the reciprocal of $K$ for the forward reaction. If the balanced equation for a reaction is multiplied by some factor $n$, the value of $K$ for that reaction is raised to the $n$th power. If reactions are summed to give an overall reaction, their equilibrium constants are multiplied together to obtain an overall equilibrium constant. (Section 15.4)

**Learning Outcome 5** According to **Le Châtelier's principle**, chemical reactions at equilibrium respond to stress by shifting position to relieve the stress. A catalyst decreases the time it takes a system to achieve equilibrium but does not change the value of the equilibrium constant. (Section 15.7)

**Learning Outcome 6** Equilibrium concentrations or partial pressures of reactants and products can be calculated from initial concentrations or pressures, the reaction stoichiometry, and the value of the equilibrium constant. (Section 15.8)

## PROBLEM-SOLVING SUMMARY

| TYPE OF PROBLEM | CONCEPTS AND EQUATIONS | | SAMPLE EXERCISES |
|---|---|---|---|
| **Writing equilibrium constant expressions** | For the reaction $$aA + bB \rightleftharpoons cC + dD$$ $$K_c = \frac{[C]^c[D]^d}{[A]^a[B]^b}$$ and $$K_p = \frac{(P_C)^c(P_D)^d}{(P_A)^a(P_B)^b}$$ | (15.9) (15.10) | 15.1 |
| **Calculating and interconverting $K_c$ and $K_p$** | Insert equilibrium molar concentrations or partial pressures into the equilibrium constant expression and use the relation $$K_p = K_c(RT)^{\Delta n}$$ where $R = 0.08206$ L · atm/(mol · K), $T$ is the absolute temperature, and $\Delta n$ is the number of moles of product gas minus the number of moles of reactant gas in the balanced chemical equation. | (15.15) | 15.2, 15.3, 15.4 |

| TYPE OF PROBLEM | CONCEPTS AND EQUATIONS | SAMPLE EXERCISES |
|---|---|---|
| Recalculating $K$ | To calculate $K$ for the reverse of a reaction, take the reciprocal of $K$ of the forward reaction. If all the coefficients in a chemical equation are multiplied by $n$, the value of $K$ increases by the power of $n$. If reactions are summed to give an overall reaction, their equilibrium constants are multiplied together to obtain an overall $K$. | 15.5, 15.6, 15.7 |
| Using $Q$ and $K$ values to predict the direction of a reaction | If $Q < K$, the reaction proceeds in the forward direction to make more products; if $Q = K$, the reaction is at equilibrium; if $Q > K$, the reaction proceeds in the reverse direction to make more reactants. | 15.8 |
| Writing equilibrium constant expressions for heterogeneous equilibria | Molar concentrations of pure liquids and pure solids are omitted from equilibrium constant expressions because such concentrations are constant. | 15.9 |
| Adding or removing reactants or products to stress an equilibrium | Decreasing the concentration of a substance involved in an equilibrium shifts the equilibrium toward the production of more of that substance. Increasing the concentration of a substance shifts the equilibrium so that some of that substance is consumed in the reaction. | 15.10 |
| Predicting the effect of changing volume on gas-phase equilibria | Equilibria involving different numbers of moles of gaseous reactants and products shift in response to an increase (or decrease) in volume caused by a decrease (or increase) in pressure toward the side with fewer (or more) moles of gases. | 15.11 |
| Predicting changes in equilibrium with temperature | The value of $K$ for an endothermic reaction increases with increasing temperature; the value of $K$ for an exothermic reaction decreases with increasing temperature. | 15.12 |
| Calculating concentrations or partial pressures of reactants and products at equilibrium | Use a RICE table to develop algebraic terms for each reactant's and product's partial pressure or concentration at equilibrium. Let $x$ be the change in concentration or partial pressure of one component of the reaction. Express the changes in the other components in terms of $x$. Substitute these terms into the expression for $K$ and solve for $x$. | 15.13, 15.14, 15.15 |

## VISUAL PROBLEMS

*(Answers to boldface end-of-chapter questions and problems are in the back of the book.)*

**15.1.** Consider the graph of concentration versus time in Figure P15.1.
   a. What is the mass action expression for the reaction?
   b. What is the value of $K_c$?

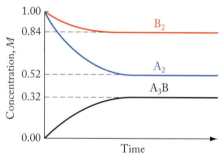

**FIGURE P15.1**

**15.2.** The progress with time of a reaction system is depicted in Figure P15.2. Red spheres represent the molar concentration of substance A and blue spheres represent the molar concentration of substance B.
   a. Does the system reach equilibrium?
   b. In which direction (A → B or B → A) is equilibrium attained?
   c. What is the value of the equilibrium constant $K_c$?

Time

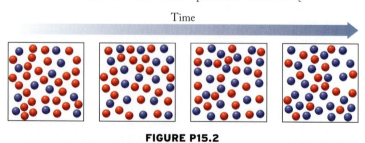

**FIGURE P15.2**

**15.3.** In Figure P15.3 the red spheres represent reactant A and the blue spheres represent product B in equilibrium with A.

a. Write a chemical equation that describes the equilibrium.
b. What is the value of the equilibrium constant $K_c$?

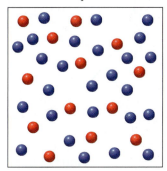

**FIGURE P15.3**

**15.4.** The equilibrium constant $K_c$ for the reaction

A (red spheres) + B (blue spheres) $\rightleftharpoons$ AB

is 3.0 at 300.0 K. Does the situation depicted in Figure P15.4 correspond to equilibrium? If not, in what direction (to the left or to the right) will the system shift to attain equilibrium?

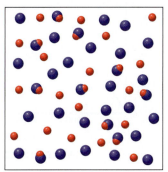

**FIGURE P15.4**

**15.5.** The diagrams in Figure P15.5 represent equilibrium states of the reaction

A (red spheres) + B (blue spheres) $\rightleftharpoons$ AB

at 300 K and 400 K, respectively. Is this reaction endothermic or exothermic? Explain.

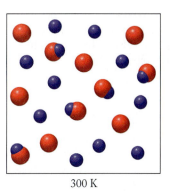

300 K                    400 K

**FIGURE P15.5**

*__15.6.__ At a temperature of 1000 K, when equilibrium is established in chamber 1 (Figure P15.6) for this reaction:

$$2\ SO_2(g) + O_2(g) \rightleftharpoons 2\ SO_3(g)$$

the quantity of each gas in the system is $SO_3 = 1.00 \times 10^2$ mol, $SO_2 = 4.15$ mol, $O_2 = 2.07$ mol. The pressure on the contents is increased so that the volume in chamber 2 is 1/10 the volume in chamber 1. The temperature is maintained at 1000 K.

a. What is the value of the equilibrium constant $K_c$ in chamber 1?
b. What is the value of the equilibrium constant $K_c$ in chamber 2?
c. Describe how the number of moles of each gas in the chamber changes in response to the change in conditions.

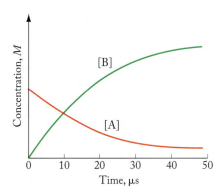

$1.00 \times 10^2$ mol $SO_3$
4.15 mol $SO_2$
2.07 mol $O_2$   10.0 L

1.0 L

1                    2

**FIGURE P15.6**

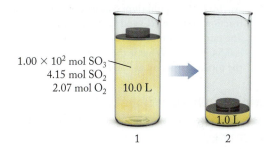

## QUESTIONS AND PROBLEMS

### The Dynamics of Chemical Equilibrium

**CONCEPT REVIEW**

**15.7.** How are forward and reverse reaction rates related in a system at chemical equilibrium?

**15.8.** Are ice cubes floating in water in an insulated container an example of a system in dynamic equilibrium? Explain why, or why not.

**15.9.** Does the reaction A → 2 B represented in Figure P15.9 reach equilibrium in 20 μs? Explain your answer.

**15.10.** At equilibrium, is the sum of the concentrations of all the reactants always equal to the sum of the concentrations of the products? Explain.

**15.11.** How is the value of the equilibrium constant of the reaction A(g) $\rightleftharpoons$ B(g) related to the values of the forward and reverse rate constants of the reaction?

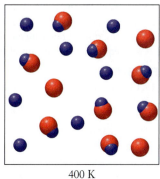

**FIGURE P15.9**

**15.12.** Suppose the value of the forward rate constant of the reaction $C(g) \rightleftharpoons D(g)$ is 5/s at 298 K, and the value of the reverse rate constant is 10/s. What is the value of the equilibrium constant of the reaction at 298 K?

## PROBLEMS

**15.13.** In a study of the reaction

$$2 N_2O(g) \rightleftharpoons 2 N_2(g) + O_2(g)$$

quantities of all three gases were injected into a reaction vessel. The $N_2O$ consisted entirely of isotopically labeled $^{15}N_2O$. Analysis of the reaction mixture after 1 day revealed the presence of compounds with molar masses 28, 29, 30, 32, 44, 45, and 46 g/mol. Identify the compounds and account for their appearance.

**15.14.** A mixture of $^{13}CO$, $^{12}CO_2$, and $O_2$ in a sealed reaction vessel was used to follow the reaction

$$2 CO(g) + O_2(g) \rightleftharpoons 2 CO_2(g)$$

Analysis of the reaction mixture after 1 day revealed the presence of compounds with molar masses 28, 29, 32, 44, and 45 g/mol. Identify the compounds and account for their appearance.

**15.15.** Suppose the reaction $A \rightleftharpoons B$ in the forward direction is first order in A and the rate constant is $1.50 \times 10^{-2}$ s$^{-1}$. The reverse reaction is first order in B and the rate constant is $4.50 \times 10^{-2}$ s$^{-1}$ at the same temperature. What is the value of the equilibrium constant for the reaction $A \rightleftharpoons B$ at this temperature?

**15.16.** At 700 K the equilibrium constant $K_c$ for the gas-phase reaction between NO and $O_2$ forming $NO_2$ is $8.7 \times 10^6$. The rate constant for the reverse reaction at this temperature is $0.54$ $M^{-1}$ s$^{-1}$. What is the value of the rate constant for the forward reaction at 700 K?

## Writing Equilibrium Constant Expressions; Relationships between $K_c$ and $K_p$ Values

### CONCEPT REVIEW

**15.17.** Under what conditions are the numerical values of $K_c$ and $K_p$ equal?

**15.18.** At 298 K, is $K_p$ greater than or less than $K_c$ if there is a net increase in the number of moles of gas in the reaction and if $K_c > 1$? Explain your answer.

**15.19.** Write $K_c$ and $K_p$ expressions for the following gas phase reactions
a. $C_2H_4(g) + H_2(g) \rightleftharpoons C_2H_6(g)$
b. $2 SO_2(g) + O_2(g) \rightleftharpoons 2 SO_3(g)$

**15.20.** Write $K_c$ and $K_p$ expressions for the following reactions at 500 K.
a. $NH_2Cl(g) + NH_3(g) \rightleftharpoons N_2H_4(g) + HCl(g)$
b. $CO(g) + 2 H_2(g) \rightleftharpoons CH_3OH(g)$

### PROBLEMS

**15.21.** Use the graph in Figure P15.21 to estimate the value of the equilibrium constant $K_c$ for the reaction

$$N_2O(g) \rightleftharpoons N_2(g) + \tfrac{1}{2} O_2(g)$$

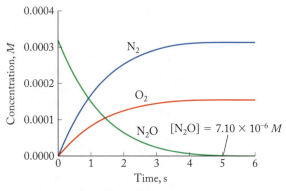

**FIGURE P15.21**

**15.22.** Estimate the value of the equilibrium constant $K_c$ for the reaction

$$2 NO(g) + O_2(g) \rightleftharpoons 2 NO_2(g)$$

from the data in Figure P15.22.

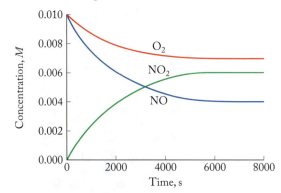

**FIGURE P15.22**

**15.23.** At 1200 K the partial pressures of an equilibrium mixture of $H_2S$, $H_2$, and S are 0.020, 0.045, and 0.030 atm, respectively. Calculate the value of the following equilibrium constant at 1200 K.

$$H_2S(g) \rightleftharpoons H_2(g) + S(g) \qquad K_p = ?$$

**15.24.** At 1045 K the partial pressures of an equilibrium mixture of $H_2O$, $H_2$, and $O_2$ are 0.040, 0.0045, and 0.0030 atm, respectively. Calculate the value of the equilibrium constant $K_p$ at 1045 K.

$$2 H_2O(g) \rightleftharpoons 2 H_2(g) + O_2(g)$$

**15.25.** At equilibrium, the concentrations of gaseous $N_2$, $O_2$, and NO in a sealed reaction vessel are $[N_2] = 3.3 \times 10^{-3}$ $M$, $[O_2] = 5.8 \times 10^{-3}$ $M$, and $[NO] = 3.1 \times 10^{-3}$ $M$. What is the value of $K_c$ for the reaction

$$N_2(g) + O_2(g) \rightleftharpoons 2 NO(g)$$

at the temperature of the reaction mixture?

**15.26.** Analyses of an equilibrium mixture of gaseous $N_2O_4$ and $NO_2$ gave the following results: $[NO_2] = 4.2 \times 10^{-3}$ $M$ and $[N_2O_4] = 2.9 \times 10^{-3}$ $M$. What is the value of the equilibrium constant $K_c$ for the following reaction at the temperature of the mixture?

$$2 NO_2(g) \rightleftharpoons N_2O_4(g)$$

**15.27. Hydrogen Production (Step 1)** Hydrogen gas production often begins with the steam–methane reforming reaction

$$CH_4(g) + H_2O(g) \rightleftharpoons CO(g) + 3\,H_2(g) \qquad (15.1)$$

At 1000 K the partial pressures of the gases in an equilibrium mixture are: 0.71 atm $CH_4$, 1.41 atm $H_2O$, 1.00 atm CO, and 3.00 atm $H_2$. What is the value of the $K_p$ of the reaction at 1000 K?

**15.28. Hydrogen Production (Step 2)** When the CO produced in the steam–methane reforming reaction is reacted with more steam at 450 K, the water-gas shift reaction:

$$CO(g) + H_2O(g) \rightleftharpoons CO_2(g) + H_2(g) \qquad (15.2)$$

produces more hydrogen. If the equilibrium partial pressures of the gases in the reactor are 0.35 atm $H_2O$, 0.24 atm CO, 4.47 atm $H_2$, and 4.36 atm $CO_2$, what is the value of $K_p$?

---

**15.29.** The equilibrium constant $K_p$ for the following equilibrium is 32 at 298 K. What is the value of $K_c$ for this same equilibrium at 298 K?

$$A(g) + B(g) \rightleftharpoons AB(g)$$

**\*15.30.** A 100-mL reaction vessel initially contains $2.60 \times 10^{-2}$ mol of NO and $1.30 \times 10^{-2}$ mol of $H_2$. At equilibrium, the concentration of NO in the vessel is 0.161 $M$. At equilibrium the vessel also contains $N_2$, $H_2O$, and $H_2$. What is the value of the equilibrium constant $K_c$ for the following reaction?

$$2\,H_2(g) + 2\,NO(g) \rightleftharpoons 2H_2O(g) + N_2(g)$$

---

**15.31.** At 500°C, the equilibrium constant $K_p$ for the synthesis of ammonia

$$N_2(g) + 3\,H_2(g) \rightleftharpoons 2\,NH_3(g)$$

is $1.45 \times 10^{-5}$. What is the value of $K_c$?

**15.32.** If the value of the equilibrium constant, $K_c$, for the following reaction is $5 \times 10^5$ at 298 K, what is the value of $K_p$ at 298 K?

$$2\,CO(g) + O_2(g) \rightleftharpoons 2\,CO_2(g)$$

---

**15.33.** For which of the following reactions are the values of $K_p$ and $K_c$ the same?
a. $2\,NH_3(g) + 2\,O_2(g) \rightleftharpoons N_2O(g) + 3\,H_2O(g)$
b. $2\,N_2H_4(\ell) + 2\,NO_2(g) \rightleftharpoons 3\,N_2(g) + 4\,H_2O(\ell)$
c. $NH_2F(g) + CaCl_2(s) \rightleftharpoons NH_2Cl(g) + CaClF(s)$

**15.34.** For which of the following reactions are the values of $K_p$ and $K_c$ different?
a. $2\,O_3(g) \rightleftharpoons 3\,O_2(g)$
b. $NH_4NO_3(s) \rightleftharpoons 2\,H_2O(g) + N_2O(g)$
c. $4\,HCl(aq) + MnO_2(s) \rightleftharpoons MnCl_2(aq) + 2\,H_2O(\ell) + Cl_2(g)$

---

**15.35. Bulletproof Glass** Phosgene ($COCl_2$) is used in the manufacture of foam rubber and bulletproof glass. It is formed from carbon monoxide and chlorine in the following reaction:

$$Cl_2(g) + CO(g) \rightleftharpoons COCl_2(g)$$

The value of $K_c$ for the reaction is 5.0 at 327°C. What is the value of $K_p$ at 327°C?

**15.36.** If the value of $K_p$ for the following reaction

$$SO_2(g) + NO_2(g) \rightleftharpoons NO(g) + SO_3(g)$$

is 3.45 at 298 K, what is the value of $K_c$ for the reverse reaction?

## Manipulating Equilibrium Constant Expressions

### CONCEPT REVIEW

**15.37.** Explain why representing the same reaction with different chemical equations, like this:

(1) $2\,N_2(g) + O_2(g) \rightleftharpoons 2\,NO_2(g)$ $\qquad K_{(1)}$
(2) $N_2(g) + \frac{1}{2}\,O_2(g) \rightleftharpoons NO_2(g)$ $\qquad K_{(2)}$

results in different equilibrium constant values ($K_{(1)} \neq K_{(2)}$)

**15.38.** If the value of $K_c$ for the reaction $A(g) \rightleftharpoons B(g)$ is 10, what is the value of $K_c$ for the reaction $B(g) \rightleftharpoons A(g)$?

### PROBLEMS

**15.39.** The equilibrium constant $K_c$ for the reaction

$$I_2(g) + Br_2(g) \rightleftharpoons 2\,IBr(g)$$

is 120 at 425 K. What is the value of $K_c$ for the equilibrium

$$\frac{1}{2}\,I_2(g) + \frac{1}{2}\,Br_2(g) \rightleftharpoons IBr(g)$$

at 425 K?

**15.40.** The equilibrium constant $K_p$ for the synthesis of ammonia:

$$N_2(g) + 3\,H_2(g) \rightleftharpoons 2\,NH_3(g)$$

is $4.3 \times 10^{-3}$ at 300°C. What is the value of $K_p$ for the equilibrium

$$\frac{1}{2}\,N_2(g) + \frac{3}{2}\,H_2(g) \rightleftharpoons NH_3(g)$$

at 300°C?

---

**15.41.** The following reaction is one of the elementary steps in the oxidation of NO:

$$NO(g) + NO_3(g) \rightleftharpoons 2\,NO_2(g)$$

Write an expression for the equilibrium constant $K_c$ for this reaction and for the reverse reaction:

$$2\,NO_2(g) \rightleftharpoons NO(g) + NO_3(g)$$

How are the two $K_c$ expressions related?

**15.42. Making Ammonia** The value of the equilibrium constant $K_p$ for the formation of ammonia,

$$N_2(g) + 3\,H_2(g) \rightleftharpoons 2\,NH_3(g)$$

is $4.5 \times 10^{-5}$ at 450°C. What is the value of $K_p$ for the following reaction?

$$2\,NH_3(g) \rightleftharpoons N_2(g) + 3\,H_2(g)$$

---

**15.43.** The $K_c$ value of the following reaction is $3.0 \times 10^{-4}$ at 298 K:

$$2\,NOBr(g) \rightleftharpoons 2\,NO(g) + Br_2(g)$$

What is the $K_c$ value of the this reaction at 298 K:

$$NOBr(g) \rightleftharpoons NO(g) + \frac{1}{2}\,Br_2(g)$$

**15.44.** How is the value of the equilibrium constant $K_p$ for the reaction

$$2\,H_2O(g) + N_2(g) \rightleftharpoons 2\,H_2(g) + 2\,NO(g)$$

related to the value of $K_p$ for this reaction at the same temperature?

$$H_2O(g) + \tfrac{1}{2}\,N_2(g) \rightleftharpoons H_2(g) + NO(g)$$

---

**15.45.** At a given temperature, the equilibrium constant $K_c$ for the reaction

$$2\,SO_2(g) + O_2(g) \rightleftharpoons 2\,SO_3(g)$$

is $2.4 \times 10^{-3}$. What is the value of the equilibrium constant for each of the following reactions at that temperature?
a. $SO_2(g) + \tfrac{1}{2}\,O_2(g) \rightleftharpoons SO_3(g)$
b. $2\,SO_3(g) \rightleftharpoons 2\,SO_2(g) + O_2(g)$
c. $SO_3(g) \rightleftharpoons SO_2(g) + \tfrac{1}{2}\,O_2(g)$

**15.46.** If the equilibrium constant $K_c$ for the reaction

$$2\,NO(g) + O_2(g) \rightleftharpoons 2\,NO_2(g)$$

is $5 \times 10^{12}$, what is the value of the equilibrium constant of each of the following reactions at the same temperature?
a. $NO(g) + \tfrac{1}{2}\,O_2(g) \rightleftharpoons NO_2(g)$
b. $2\,NO_2(g) \rightleftharpoons 2\,NO(g) + O_2(g)$
c. $NO_2(g) \rightleftharpoons NO(g) + \tfrac{1}{2}\,O_2(g)$

---

**15.47.** Calculate the value of the equilibrium constant $K_p$ at 298 K for the reaction

$$N_2(g) + 2\,O_2(g) \rightleftharpoons 2\,NO_2(g)$$

from the following $K_p$ values at 298 K:

$$N_2(g) + O_2(g) \rightleftharpoons 2\,NO(g) \qquad K_p = 4.4 \times 10^{-31}$$
$$2\,NO(g) + O_2(g) \rightleftharpoons 2\,NO_2(g) \qquad K_p = 2.4 \times 10^{12}$$

**15.48.** Calculate the value of the equilibrium constant $K_p$ at 298 K for the reaction

$$\tfrac{1}{4}\,S_8(s) + 3\,O_2(g) \rightleftharpoons 2\,SO_3(g)$$

from the following $K_p$ values at 298 K:

$$\tfrac{1}{8}\,S_8(s) + O_2(g) \rightleftharpoons SO_2(g) \qquad K_p = 4.0 \times 10^{52}$$
$$2\,SO_2(g) + O_2(g) \rightleftharpoons 2\,SO_3(g) \qquad K_p = 7.8 \times 10^{24}$$

## Equilibrium Constants and Reaction Quotients

**CONCEPT REVIEW**

**15.49.** What is a reaction quotient?
**15.50.** How is an equilibrium constant different from a reaction quotient?
**15.51.** What does it mean when the reaction quotient $Q$ is numerically equal to the equilibrium constant $K$?
**15.52.** Explain how knowing $Q$ and $K$ for an equilibrium system enables you to say whether it is at equilibrium or whether it will shift in one direction or another.

**PROBLEMS**

**15.53.** If the equilibrium constant $K_c$ for the hypothetical reaction $A(g) \rightleftharpoons B(g)$ is 22 at a given temperature, and if $[A] = 0.10\ M$ and $[B] = 2.0\ M$ in a reaction mixture at

that temperature, is the reaction at chemical equilibrium? If not, in which direction will the reaction proceed to reach equilibrium?

**15.54.** The equilibrium constant $K_c$ for the hypothetical reaction

$$2\,C \rightleftharpoons D + E$$

is $3 \times 10^{-3}$. At a particular time, the composition of the reaction mixture is $[C] = [D] = [E] = 5 \times 10^{-4}\ M$. In which direction will the reaction proceed to reach equilibrium?

---

**15.55.** Suppose the value of the equilibrium constant $K_p$ of the following hypothetical reaction

$$A(g) + B(g) \rightleftharpoons C(g)$$

is 1.00 at 300 K. Are either of the following reaction mixtures at chemical equilibrium at 300 K?
a. $P_A = P_B = P_C = 1.0$ atm
b. $[A] = [B] = [C] = 1.0\ M$

**15.56.** In which direction will the following hypothetical reaction proceed to reach equilibrium under the conditions given?

$$A(g) + B(g) \rightleftharpoons C(g) \qquad K_p = 1.00\ \text{at}\ 300\ K$$

a. $P_A = P_C = 1.0$ atm, $P_B = 0.50$ atm
b. $[A] = [B] = [C] = 1.0\ M$

---

**15.57.** If the equilibrium constant $K_c$ for the reaction

$$N_2(g) + O_2(g) \rightleftharpoons 2\,NO(g)$$

is $1.5 \times 10^{-3}$, in which direction will the reaction proceed if the partial pressures of the three gases are all $1.00 \times 10^{-3}$ atm?

**15.58.** At 650 K, the value of the equilibrium constant $K_p$ for the ammonia synthesis reaction

$$N_2(g) + 3\,H_2(g) \rightleftharpoons 2\,NH_3(g)$$

is $4.3 \times 10^{-4}$. If a vessel contains a reaction mixture in which $[N_2] = 0.010\ M$, $[H_2] = 0.030\ M$, and $[NH_3] = 0.00020\ M$, will more ammonia form?

---

**15.59.** Use the information below to determine whether or not a reaction mixture in which the partial pressures of $PCl_3$, $Cl_2$, and $PCl_5$ are 0.20, 0.40, and 0.60 atm, respectively, is at equilibrium at 450 K.

$$PCl_3(g) + Cl_2(g) \rightleftharpoons PCl_5(g) \qquad K_p = 3.8\ \text{at}\ 450\ K$$

If the reaction mixture is not at equilibrium, in which direction does the reaction proceed to achieve equilibrium?

**15.60.** If enough $PCl_3$ were injected into the reaction mixture in the previous problem to double its partial pressure, would the mixture be closer to equilibrium? Explain why or why not.

## Heterogeneous Equilibria

**CONCEPT REVIEW**

**15.61. SO₂ in the Air** Combustion of fossil fuels that contain sulfur is an important source of sulfur dioxide in the atmosphere. Write the $K_p$ expression for the combustion of elemental sulfur:

$$\tfrac{1}{8}\,S_8(s) + O_2(g) \rightleftharpoons SO_2(g)$$

**15.62.** Write $K_p$ expressions for the reactions below that take place during the thermal decomposition of the mineral dolomite (a mixture of calcium and magnesium carbonates):

$$CaCO_3(s) \rightleftharpoons CaO(s) + CO_2(g)$$
$$MgCO_3(s) \rightleftharpoons MgO(s) + CO_2(g)$$

**15.63.** The brown residues that form on the surfaces of plumbing fixtures, such as inside toilet tanks, is the result of the oxidation of more soluble iron(II) compounds with dissolved oxygen, forming less soluble iron(III) compounds. Write the $K_c$ expression for one such reaction:

$$4\,Fe(OH)_2(aq) + O_2(aq) + 2\,H_2O(\ell) \rightleftharpoons 4\,Fe(OH)_3(s)$$

*****15.64.** **Testing Minerals** Write the $K_c$ expression for this reaction, which is used to test whether a white, shiny mineral is marble (calcium carbonate):

$$2\,HCl(aq) + CaCO_3(s) \rightleftharpoons CaCl_2(aq) + H_2O(\ell) + CO_2(g)$$

## Le Châtelier's Principle

**CONCEPT REVIEW**

**15.65.** Does adding reactants to a system at equilibrium increase the value of the equilibrium constant?

**15.66.** Increasing the concentration of a reactant shifts the position of chemical equilibrium toward formation of more products. What effect does adding a reactant have on the rates of the forward and reverse reactions?

**15.67.** **Carbon Monoxide Poisoning** Patients suffering from carbon monoxide poisoning are treated with pure oxygen to remove CO from the hemoglobin (Hb) in their blood. The two relevant equilibria are

$$Hb + 4\,CO(g) \rightleftharpoons Hb(CO)_4$$
$$Hb + 4\,O_2(g) \rightleftharpoons Hb(O_2)_4$$

The value of the equilibrium constant for CO binding to Hb is greater than that for $O_2$. How, then, does this treatment work?

**15.68.** Is the equilibrium constant $K_p$ for the reaction

$$2\,NO_2(g) \rightleftharpoons N_2O_4(g)$$

in air the same in Los Angeles as in Denver if the atmospheric pressure in Denver is lower but the temperature is the same?

**15.69.** Henry's law (Chapter 10) predicts that the solubility of a gas in a liquid increases with its partial pressure. Explain Henry's law in relation to Le Châtelier's principle.

*****15.70.** For the reaction

$$2\,CO(g) + O_2(g) \rightleftharpoons 2\,CO_2(g)$$

why does adding an inert gas such as argon to an equilibrium mixture of CO, $O_2$, and $CO_2$ in a sealed vessel increase the total pressure of the system but not affect the position of the equilibrium?

**PROBLEMS**

**15.71.** Which of the following equilibria will shift toward formation of more products if an equilibrium mixture is compressed into half its volume?
a. $2\,N_2O(g) \rightleftharpoons 2\,N_2(g) + O_2(g)$
b. $2\,CO(g) + O_2(g) \rightleftharpoons 2\,CO_2(g)$
c. $N_2(g) + O_2(g) \rightleftharpoons 2\,NO(g)$
d. $2\,NO(g) + O_2(g) \rightleftharpoons 2\,NO_2(g)$

**15.72.** Which of the following equilibria will shift toward formation of more products if the volume of a reaction mixture at equilibrium increases by a factor of 2?
a. $2\,SO_2(g) + O_2(g) \rightleftharpoons 2\,SO_3(g)$
b. $NO(g) + O_3(g) \rightleftharpoons NO_2(g) + O_2(g)$
c. $2\,N_2O_5(g) \rightleftharpoons 4\,NO_2(g) + O_2(g)$
d. $N_2O_4(g) \rightleftharpoons 2\,NO_2(g)$

**15.73.** What would be the effect of the changes listed on the equilibrium concentrations of reactants and products in the following reaction?

$$2\,O_3(g) \rightleftharpoons 3\,O_2(g)$$

a. $O_3$ is added to the system.
b. $O_2$ is added to the system.
c. The mixture is compressed to one-tenth its initial volume.

**15.74.** How will the changes listed affect the position of the following equilibrium?

$$2\,NO_2(g) \rightleftharpoons NO(g) + NO_3(g)$$

a. The concentration of NO is increased.
b. The concentration of $NO_2$ is increased.
c. The volume of the system is allowed to expand to 5 times its initial value.

**15.75.** How would reducing the partial pressure of $O_2(g)$ affect the position of the equilibrium in the following reaction?

$$2\,SO_2(g) + O_2(g) \rightleftharpoons 2\,SO_3(g)$$

*****15.76.** Ammonia is added to a gaseous reaction mixture containing $H_2$, $Cl_2$, and HCl that is at chemical equilibrium. How will the addition of ammonia affect the relative concentrations of $H_2$, $Cl_2$, and HCl if the equilibrium constant of reaction 2 is much greater than the equilibrium constant of reaction 1?

(1) $\quad H_2(g) + Cl_2(g) \rightleftharpoons 2\,HCl(g)$
(2) $\quad HCl(g) + NH_3(g) \rightleftharpoons NH_4Cl(s)$

**15.77.** In which of the following equilibrium does an increase in temperature produce a shift toward the formation of more product?
a. $PCl_3(g) + Cl_2(g) \rightleftharpoons PCl_5(g) \qquad \Delta H° < 0$
b. $CH_4(g) + H_2O(g) \rightleftharpoons CO(g) + 3\,H_2(g) \quad \Delta H° > 0$
c. $CO(g) + H_2O(g) \rightleftharpoons CO_2(g) + H_2(g) \quad \Delta H° < 0$

**15.78.** In which of the following equilibrium does an increase in temperature produce a shift toward the formation of more product?
a. $N_2(g) + 3\,H_2(g) \rightleftharpoons 2\,NH_3(g) \qquad \Delta H° < 0$
b. $2\,NO_2(g) \rightleftharpoons 2\,NO(g) + O_2(g) \qquad \Delta H° > 0$
c. $C_2H_4(g) + H_2(g) \rightleftharpoons C_2H_6(g) \qquad \Delta H° < 0$

## Calculations Based on *K*

**CONCEPT REVIEW**

**15.79.** Why are calculations based on $K$ often simpler when the value of $K$ is very small?

**15.80.** Could the quadratic equation be used to solve for the equilibrium concentration of $NO_2$ in the following reaction?

$$2\,NO + O_2 \rightarrow 2\,NO_2$$

**PROBLEMS**

**15.81.** For the reaction

$$PCl_5(g) \rightleftharpoons PCl_3(g) + Cl_2(g) \qquad K_p = 23.6 \text{ at } 500 \text{ K}$$

a. Calculate the equilibrium partial pressures of the reactants and products if the initial pressures are $P_{PCl_5} = 0.560$ atm and $P_{PCl_3} = 0.500$ atm.

b. If more chlorine is added after equilibrium is reached, how will the concentrations of $PCl_5$ and $PCl_3$ change?

**15.82.** Enough $NO_2$ gas is injected into a cylindrical vessel to produce a partial pressure, $P_{NO_2}$, of 0.900 atm at 298 K. Calculate the equilibrium partial pressures of $NO_2$ and $N_2O_4$, given

$$2 NO_2(g) \rightleftharpoons N_2O_4(g) \qquad K_p = 4 \text{ at } 298 \text{ K}$$

**15.83.** The value of $K_c$ for the reaction between water vapor and dichlorine monoxide,

$$H_2O(g) + Cl_2O(g) \rightleftharpoons 2 HOCl(g)$$

is 0.0900 at 25°C. Determine the equilibrium concentrations of all three compounds if the starting concentrations of both reactants are 0.00432 $M$ and no HOCl is present.

**15.84.** The value of $K_p$ for the reaction

$$3 H_2(g) + N_2(g) \rightleftharpoons 2 NH_3(g)$$

is $4.3 \times 10^{-4}$ at 648 K. Determine the equilibrium partial pressure of $NH_3$ in a reaction vessel that initially contained 0.900 atm $N_2$ and 0.500 atm $H_2$ at 648 K.

**15.85.** The value of $K_p$ for the reaction

$$NO(g) + \tfrac{1}{2} O_2(g) \rightleftharpoons NO_2(g)$$

is $1.5 \times 10^6$ at 25°C. At equilibrium, what is the ratio of $P_{NO_2}$ to $P_{NO}$ in air at 25°C? Assume that $P_{O_2} = 0.21$ atm and does not change.

**\*15.86. Water Gas** The water–gas reaction is a source of hydrogen. Passing steam over hot carbon produces a mixture of carbon monoxide and hydrogen:

$$H_2O(g) + C(s) \rightleftharpoons CO(g) + H_2(g)$$

The value of $K_c$ for the reaction at 1000°C is $3.0 \times 10^{-2}$.

a. Calculate the equilibrium partial pressures of the products and reactants if $P_{H_2O} = 0.442$ atm and $P_{CO} = 5.0$ atm at the start of the reaction. Assume that the carbon is in excess.

b. Determine the equilibrium partial pressures of the reactants and products after sufficient CO and $H_2$ are added to the equilibrium mixture in part a to initially increase the partial pressures of both gases by 0.075 atm.

**15.87.** The value of $K_p$ for the reaction

$$CO_2(g) + C(s) \rightleftharpoons 2 CO(g)$$

is 1.5 at 700°C. Calculate the equilibrium partial pressures of CO and $CO_2$ if initially $P_{CO_2} = 5.0$ atm and $P_{CO} = 0.0$. Pure graphite is present initially and when equilibrium is achieved.

**15.88. Jupiter's Atmosphere** Ammonium hydrogen sulfide ($NH_4SH$) was detected in the atmosphere of Jupiter subsequent to its collision with the comet Shoemaker–Levy. The equilibrium between ammonia, hydrogen sulfide, and $NH_4SH$ is described by the following equation:

$$NH_4SH(s) \rightleftharpoons NH_3(g) + H_2S(g)$$

The value of $K_p$ for the reaction at 24°C is 0.126. Suppose a sealed flask contains an equilibrium mixture of $NH_4SH$, $NH_3$, and $H_2S$. At equilibrium, the partial pressure of $H_2S$ is 0.355 atm. What is the partial pressure of $NH_3$?

**\*15.89.** A flask containing pure $NO_2$ was heated to 1000 K, a temperature at which the value of $K_p$ for the decomposition of $NO_2$ is 158.

$$2 NO_2(g) \rightleftharpoons 2 NO(g) + O_2(g)$$

The partial pressure of $O_2$ at equilibrium is 0.136 atm.

a. Calculate the partial pressures of NO and $NO_2$.

b. Calculate the total pressure in the flask at equilibrium.

**15.90.** The equilibrium constant $K_p$ of the reaction

$$2 SO_3(g) \rightleftharpoons 2 SO_2(g) + O_2(g)$$

is 7.69 at 830°C. If a vessel at this temperature initially contains pure $SO_3$ and if the partial pressure of $SO_3$ at equilibrium is 0.100 atm, what is the partial pressure of $O_2$ in the flask at equilibrium?

**\*15.91. NO$_x$ Pollution** In a study of the formation of $NO_x$ air pollution, a chamber heated to 2200°C was filled with air (0.79 atm $N_2$, 0.21 atm $O_2$). What are the equilibrium partial pressures of $N_2$, $O_2$, and NO if $K_p = 0.050$ for the following reaction at 2200°C?

$$N_2(g) + O_2(g) \rightleftharpoons 2 NO(g)$$

**\*15.92.** The equilibrium constant $K_p$ for the thermal decomposition of $NO_2$

$$2 NO_2(g) \rightleftharpoons 2 NO(g) + O_2(g)$$

is $6.5 \times 10^{-6}$ at 450°C. If a reaction vessel at this temperature initially contains 0.500 atm $NO_2$, what will be the partial pressures of $NO_2$, NO, and $O_2$ in the vessel when equilibrium has been attained?

**15.93.** The value of $K_c$ for the thermal decomposition of hydrogen sulfide

$$2 H_2S(g) \rightleftharpoons 2 H_2(g) + S_2(g)$$

is $2.2 \times 10^{-4}$ at 1400 K. A sample of gas in which $[H_2S] = 6.00$ $M$ is heated to 1400 K in a sealed high-pressure vessel. After chemical equilibrium has been achieved, what is the value of $[H_2S]$? Assume that no $H_2$ or $S_2$ was present in the original sample.

**15.94. Urban Air** On a very smoggy day, the equilibrium concentration of $NO_2$ in the air over an urban area reaches $2.2 \times 10^{-7}$ $M$. If the temperature of the air is 25°C, what is the concentration of the dimer $N_2O_4$ in the air? Given:

$$N_2O_4(g) \rightleftharpoons 2 NO_2(g) \qquad K_c = 6.1 \times 10^{-3}$$

**\*15.95. Chemical Weapon** Phosgene, $COCl_2$, gained notoriety as a chemical weapon in World War I. Phosgene is produced by the reaction of carbon monoxide with chlorine

$$CO(g) + Cl_2(g) \rightleftharpoons COCl_2(g)$$

The value of $K_c$ for this reaction is 5.0 at 600 K. What are the equilibrium partial pressures of the three gases if a reaction vessel initially contains a mixture of the reactants in which $P_{CO} = P_{Cl_2} = 0.265$ atm and $P_{COCl_2} = 0.000$ atm?

**\*15.96.** At 2000°C, the value of $K_c$ for the reaction

$$2\,CO(g) + O_2(g) \rightleftharpoons 2\,CO_2(g)$$

is 1.0. What is the ratio of [CO] to $[CO_2]$ in an atmosphere in which $[O_2] = 0.0045\ M$?

---

**\*15.97.** The water–gas shift reaction is an important source of hydrogen. The value of $K_c$ for the reaction

$$CO(g) + H_2O(g) \rightleftharpoons CO_2(g) + H_2(g)$$

at 700 K is 5.1. Calculate the equilibrium concentrations of the four gases if the initial concentration of each of them is 0.050 $M$.

**\*15.98.** Sulfur dioxide reacts with $NO_2$, forming $SO_3$ and NO:

$$SO_2(g) + NO_2(g) \rightleftharpoons SO_3(g) + NO(g)$$

If the value of $K_c$ for the reaction is 2.50, what are the equilibrium concentrations of the products if the reaction mixture was initially 0.50 $M$ $SO_2$, 0.50 $M$ $NO_2$, 0.0050 $M$ $SO_3$, and 0.0050 $M$ NO?

## Additional Problems

**\*15.99. CO as a Fuel** Is carbon dioxide a viable source of the fuel CO? Pure carbon dioxide ($P_{CO_2} = 1$ atm) decomposes at high temperatures. For the system

$$2\,CO_2(g) \rightleftharpoons 2\,CO(g) + O_2(g)$$

the percentage of decomposition of $CO_2(g)$ changes with temperature as follows:

| Temperature (K) | Decomposition (%) |
|---|---|
| 1500 | 0.048 |
| 2500 | 17.6 |
| 3000 | 54.8 |

Is the reaction endothermic? Calculate the value of $K_p$ at each temperature and discuss the results. Is the decomposition of $CO_2$ an antidote for global warming?

**15.100.** Ammonia decomposes at high temperatures. In an experiment to explore this behavior, 2.00 moles of gaseous $NH_3$ is sealed in a rigid 1-liter vessel. The vessel is heated at 800 K and some of the $NH_3$ decomposes in the following reaction:

$$2\,NH_3(g) \rightleftharpoons N_2(g) + 3\,H_2(g)$$

The system eventually reaches equilibrium and is found to contain 1.74 moles of $NH_3$. What are the values of $K_p$ and $K_c$ for the decomposition reaction at 800 K?

**\*15.101.** Elements of group 16 form hydrides with the generic formula $H_2X$. When gaseous $H_2X$ is bubbled through a solution containing 0.3 $M$ hydrochloric acid, the solution becomes saturated and $[H_2X] = 0.1\ M$. The following equilibria exist in this solution:

$$H_2X(aq) + H_2O(\ell) \rightleftharpoons HX^-(aq) + H_3O^+(aq) \qquad K_1 = 8.3 \times 10^{-8}$$
$$HX^-(aq) + H_2O(\ell) \rightleftharpoons X^{2-}(aq) + H_3O^+(aq) \qquad K_2 = 1 \times 10^{-14}$$

Calculate the concentration of $X^{2-}$ in the solution.

**\*15.102.** Nitrogen dioxide reacts with $SO_2$ to form $SO_3$ and NO:

$$NO_2(g) + SO_2(g) \rightleftharpoons NO(g) + SO_3(g)$$

An equilibrium mixture is analyzed at a certain temperature and found to contain $[NO_2] = 0.100\ M$, $[SO_2] = 0.300\ M$, $[NO] = 2.00\ M$, and $[SO_3] = 0.600\ M$. At the same temperature, extra $SO_2(g)$ is added to make $[SO_2] = 0.800\ M$. Calculate the composition of the mixture when equilibrium has been reestablished.

**\*15.103. Effective Air Pollution Control** Calcium oxide is used to remove the pollutant $SO_2$ from smokestack gases. The overall reaction is

$$CaO(s) + SO_2(g) + \tfrac{1}{2}O_2(g) \rightleftharpoons CaSO_4(s)$$

The $K_p$ for this reaction is $2.38 \times 10^{73}$.
a. What is $P_{SO_2}$ in equilibrium with air and solid CaO?
b. Consider a sample of gas that contains 100.0 moles of gas. How many molecules of $SO_2$ are in that sample?

**\*15.104. Volcanic Eruptions** During volcanic eruptions, gases as hot as 700°C and rich in $SO_2$ are released into the atmosphere. As air mixes with these gases, the following reaction converts some of the $SO_2$ into $SO_3$:

$$2\,SO_2(g) + O_2(g) \rightleftharpoons 2\,SO_3(g)$$

Calculate the value of $K_p$ for this reaction at 700°C. What is the ratio of $P_{SO_2}$ to $P_{SO_3}$ in equilibrium with $P_{O_2} = 0.21$ atm?

If your instructor assigns problems in **smartwork**, log in at **smartwork.wwnorton.com**.

# 16

# Acid–Base and Solubility Equilibria: Reactions in Soil and Water

**16.1** Acids and Bases: The Brønsted–Lowry Model

**16.2** pH and the Autoionization of Water

**16.3** Calculations Involving pH, $K_a$, and $K_b$

**16.4** Polyprotic Acids

**16.5** Acid Strength and Molecular Structure

**16.6** pH of Salt Solutions

**16.7** The Common-Ion Effect

**16.8** pH Buffers

**16.9** Indicators and Acid–Base Titrations

**16.10** Solubility Equilibria

## Learning Outcomes

**LO1** Identify the Brønsted–Lowry acids and bases and their conjugate bases and acids in chemical reactions
**Sample Exercise 16.1**

**LO2** Calculate, interpret, and interconvert pH and [H⁺] values
**Sample Exercises 16.2, 16.3, 16.6**

**LO3** Relate the strengths of acids and bases to their $K_a$, $K_b$, p$K_a$, and p$K_b$ values and to their percent ionization or dissociation in water
**Sample Exercises 16.4, 16.5**

**LO4** Predict whether a salt is acidic, basic, or neutral
**Sample Exercise 16.7**

**LO5** Calculate the pH values of solutions of weak acids and bases and the salts of their conjugate bases and acids
**Sample Exercises 16.8, 16.9**

**LO6** Explain how pH buffers control pH and calculate the pH of a conjugate acid–base pair
**Sample Exercises 16.10, 16.11, 16.12, 16.13**

**LO7** Interpret the results of an acid–base titration
**Sample Exercise 16.14**

**LO8** Relate the solubility of an ionic compound to its solubility product
**Sample Exercises 16.15, 16.16, 16.17, 16.18**

## A Balancing Act

We saw in Chapter 15 that equilibrium is an important idea in many areas of chemistry, from environmental protection to industrial production. Some of the most vital aspects of equilibria involve acids and bases, a topic we first discussed in Chapter 4. Many chemical and biological reactions depend on maintaining a proper balance between acidic and basic conditions.

For centuries, farmers have added lime (CaO) to their fields to make the soil less acidic. The basic lime neutralizes the organic acids produced when biological matter, such as the plowed-under remnants of last year's crops, decays. Most plants grow best in soil that is neutral—neither acidic nor basic—because the nutrients they need for growth are more available to them under these conditions. This is not universally true; for instance, blueberries actually thrive in acidic soil. However, acid rain—rain containing dissolved oxides of nitrogen and sulfur from fossil fuel combustion, making it acidic—has caused the death of large areas of forest in the northeastern United States and in parts of Europe.

Sometimes soil acidity has less dramatic consequences. For example, the color of hydrangea blossoms is controlled by the availability of aluminum ions ($Al^{3+}$) in the soil. These ions are soluble in acidic solutions, so hydrangeas form blue blossoms in acidic soil. If the soil is neutral or slightly basic, the aluminum ions precipitate as aluminum hydroxide and are no longer available. As a result, the plants lack their blue pigment and produce pink flowers.

For many biological systems, including humans, acid-base balance is vital to good health. Our blood is slightly basic, regulated by the amount of dissolved $CO_2$ in the blood. ($CO_2$ dissolved in water forms carbonic acid.) However, if lung function is impaired by disease, such as emphysema, $CO_2$ concentration in the blood increases, which in extreme cases can lead to coma and death.

**Shades of Purple, Pink, and Blue Flowers** The color of hydrangea flowers depends on the acid content of the soil in which they are grown. ▶

▶❙❙ **CHEMTOUR** Acid Rain

In this chapter we examine the influence of acid–base balance in the environment and in biological systems. We study why changes in this balance occur, what consequences they have, and how they can be regulated. ■

◉◉ **CONNECTION** We introduced acids, bases, and neutralization reactions in Chapter 4, where we defined acids as proton donors and bases as proton acceptors. Neutralization reactions between acids and bases produce water and a salt.

▶❙❙ **CHEMTOUR** Acid–Base Ionization

# 16.1 Acids and Bases: The Brønsted–Lowry Model

In Chapter 4 we defined acids as substances that donate $H^+$ ions and bases as substances that accept $H^+$ ions. These descriptions of acids and bases were developed independently by Danish chemist Johannes Brønsted (1879–1947) and English chemist Thomas Lowry (1874–1936) and published in the same year, 1923. Today these descriptions are known as the **Brønsted–Lowry model** of acids and bases.

## Strong and Weak Acids

As we discussed in Section 4.5, acids in aqueous solution are classified as strong or weak depending on the extent to which they ionize, donating $H^+$ ions to molecules of water and forming hydronium ($H_3O^+$) ions. Strong acids are completely ionized in water; weak acids are only partially ionized. Figure 16.1 shows data on the ionization of two acids with similar chemical structures: nitric acid ($HNO_3$), which is strong, and nitrous acid ($HNO_2$), which is weak. As shown by the bar graphs in Figure 16.1(a), $HNO_3$ in a 0.10 $M$ solution is completely ionized; there are no intact molecules of $HNO_3$ in the solution. Instead, nitrate ions ($NO_3^-$) are present, along with $H^+$ ions that combine with $H_2O$ to form $H_3O^+$ ions:

$$HNO_3(aq) + H_2O(\ell) \rightarrow NO_3^-(aq) + H_3O^+(aq) \quad (16.1)$$
$$\text{(H}^+ \text{ donor)} \quad \text{(H}^+ \text{ acceptor)}$$

In this equation we have highlighted (in red) nitric acid as the $H^+$ ion donor (Brønsted–Lowry acid) and water (in blue) as the $H^+$ ion acceptor (Brønsted–Lowry base). All the molecules of $HNO_3$ in an aqueous solution of the acid are ionized; chemists say that the reaction in Equation 16.1 *goes to completion*. The single reaction arrow in Equation 16.1 conveys this message, indicating that, even though the reaction is reversible, the equilibrium lies very far to the right.

In contrast, only a small percentage of the $HNO_2$ molecules in 0.10 $M$ nitrous acid donate $H^+$ ions to molecules of water to form $NO_2^-$ and $H_3O^+$ ions (Figure 16.1b):

$$HNO_2(aq) + H_2O(\ell) \rightleftharpoons NO_2^-(aq) + H_3O^+(aq) \quad (16.2)$$

At equilibrium, molecules of $HNO_2$ donate $H^+$ ions to molecules of $H_2O$ at the same rate that $H_3O^+$ ions donate $H^+$ ions to $NO_2^-$ ions, reforming $HNO_2$ and $H_2O$. Both the molecular form of the acid and the ions arising from its dissociation coexist in solution.

Two types of strong acids are listed in Table 16.1. First are three binary acids with the generic formula HX, where X is Cl, Br, or I. Next are three *oxoacids* with the generic formula $H_mXO_n$, where X is a nonmetal. Several factors promote the transfer of $H^+$ ions from molecules of these acids (or any acid) to molecules of water. We focus on two of them: (1) the polarity of the bonds to the ionizable

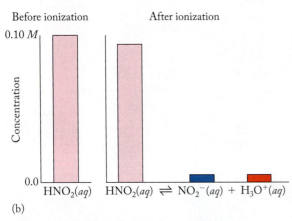

(a)

(b)

**FIGURE 16.1** Difference in degree of ionization for strong and weak acids. (a) The strong acid $HNO_3$ ionizes completely to $H_3O^+$ and $NO_3^-$ ions. (b) The weak acid $HNO_2$ ionizes very little.

| TABLE 16.1 | Strong Acids and Their Ionization Reactions in Water |
|---|---|
| **Strong Acid** | **Reaction in Water** |
| Hydrobromic | $HBr(aq) + H_2O(\ell) \rightarrow Br^-(aq) + H_3O^+(aq)$ |
| Hydrochloric | $HCl(aq) + H_2O(\ell) \rightarrow Cl^-(aq) + H_3O^+(aq)$ |
| Hydroiodic | $HI(aq) + H_2O(\ell) \rightarrow I^-(aq) + H_3O^+(aq)$ |
| Nitric | $HNO_3(aq) + H_2O(\ell) \rightarrow NO_3^-(aq) + H_3O^+(aq)$ |
| Perchloric | $HClO_4(aq) + H_2O(\ell) \rightarrow ClO_4^-(aq) + H_3O^+(aq)$ |
| Sulfuric | $H_2SO_4(aq) + H_2O(\ell) \rightarrow HSO_4^-(aq) + H_3O^+(aq)$ <br> $HSO_4^-(aq) + H_2O(\ell) \rightleftharpoons SO_4^{2-}(aq) + H_3O^+(aq)$ |

**Brønsted−Lowry model** defines acids as $H^+$ ion donors and bases as $H^+$ ion acceptors.

H atoms and (2) the strength of the dipole–dipole interactions between these H atoms and O atoms in molecules of water.

In all strong acids, ionizable H atoms are bonded to atoms of one of these highly electronegative elements: O, Cl, Br, or I. Large differences in electronegativity mean that these bonds are all polar covalent, which sets up strong dipole–dipole interactions between their H atoms and the O atoms of $H_2O$ molecules. These interactions are so strong in aqueous solutions of HCl, for example, that all H—Cl bonds break, and O—H bonds form, producing $Cl^-$ and $H_3O^+$ ions as shown in Figure 16.2.

**CONNECTION** In Chapter 4 we introduced three ways to refer to the hydrogen ions produced by acids in aqueous solution: as the proton [$H^+(aq)$], the hydrogen ion [also $H^+(aq)$], or the hydronium ion [$H_3O^+(aq)$]. The hydronium ion most closely represents the species present in solution.

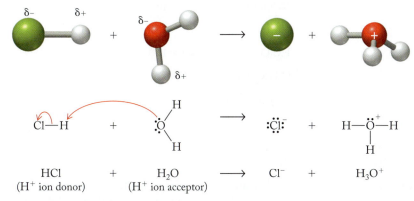

**FIGURE 16.2** When hydrogen chloride dissolves in water, HCl ionizes and forms $Cl^-$ and $H_3O^+$ ions.

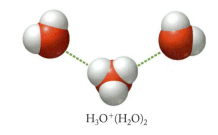

$H_3O^+(H_2O)_2$

We often describe acid–base reactions in terms of the donation and acceptance of $H^+$ ions, but we need to keep in mind that free protons ($H^+$ ions) do not exist in aqueous solutions. Rather they exist as $H_3O^+$ ions. In addition, ion–dipole interactions between water molecules and $H_3O^+$ ions cause $H_2O$ molecules to cluster around central $H_3O^+$ ions, as shown in Figure 16.3. As we describe concentrations of $H^+$ ions in acid–base reactions, remember that we really mean $H_3O^+$ ions with water molecules clustered around them.

As with the gas-phase equilibria we discussed in Chapter 15, the extent to which a reaction takes place before it reaches equilibrium in the aqueous phase is characterized by the value of its equilibrium constant. The equilibrium constants that describe the ionization of acids in water are given the symbol $K_a$. They are concentration-based equilibrium constants, but the subscript "c" is replaced with "a" to indicate that the equilibrium involves the ionization of an *acid*. Table 16.2 lists some common weak acids, their ionization reactions in water, and their $K_a$ values at 25°C. Note that the $K_a$ values of these weak acids are all much less than 1.

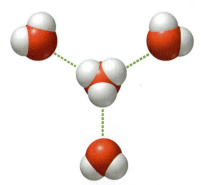

$H_3O^+(H_2O)_3$

**FIGURE 16.3** Water molecules cluster around hydronium ions through hydrogen bonds, forming species with formulas $H_3O^+(H_2O)_2^+$ and $H_3O^+(H_2O)_3^+$.

These small values contrast with the large $K_a$ values of the strong acids in Table 16.1, which are often expressed as "$K_a \gg 1$" to indicate that ionization of the acid is essentially complete in aqueous solutions.

| TABLE 16.2 | Some Common Weak Acids and Their Ionization Reactions in Water at 25°C | |
|---|---|---|
| **Weak Acid** | **Reaction in Water** | $K_a = \dfrac{[A^-][H_3O^+]}{[HA]}$ |
| Acetic | $CH_3COOH(aq) + H_2O(\ell) \rightleftharpoons CH_3COO^-(aq) + H_3O^+(aq)$ | $1.76 \times 10^{-5}$ |
| Formic | $HCOOH(aq) + H_2O(\ell) \rightleftharpoons HCOO^-(aq) + H_3O^+(aq)$ | $1.77 \times 10^{-4}$ |
| Hydrofluoric | $HF(aq) + H_2O(\ell) \rightleftharpoons F^-(aq) + H_3O^+(aq)$ | $6.8 \times 10^{-4}$ |
| Hypochlorous | $HClO(aq) + H_2O(\ell) \rightleftharpoons ClO^-(aq) + H_3O^+(aq)$ | $2.9 \times 10^{-8}$ |
| Nitrous | $HNO_2(aq) + H_2O(\ell) \rightleftharpoons NO_2^-(aq) + H_3O^+(aq)$ | $4.0 \times 10^{-4}$ |

**CONCEPT TEST** ······································································

Rank the following acids in order of decreasing acid strength.

| Weak Acid | $K_a$ |
|---|---|
| $HN_3$ | $1.9 \times 10^{-5}$ |
| $HCOOH$ | $1.77 \times 10^{-4}$ |
| $HF$ | $6.8 \times 10^{-4}$ |
| $HClO$ | $2.9 \times 10^{-8}$ |
| $C_6H_5COOH$ | $6.46 \times 10^{-5}$ |

*(Answers to Concept Tests are in the back of the book.)*
·····································································································

## Conjugate Acid–Base Pairs

In the chemical equilibrium in Equation 16.2, $HNO_2$ functions as a Brønsted–Lowry acid in the forward reaction and the $NO_2^-$ ion functions as a Brønsted–Lowry base in the reverse reaction. The difference between these two species is the $H^+$ ion that a molecule of $HNO_2$ gives away when it forms a $NO_2^-$ ion. Chemists call an acid and a base that are related in this way a **conjugate acid–base pair**. An acid forms its **conjugate base** when it donates a $H^+$ ion, and a base (like $NH_3$) forms its **conjugate acid** ($NH_4^+$) when it accepts a $H^+$ ion. This relationship is expressed in these simple chemical equations:

$$\underset{\text{Acid}}{HNO_2} \rightleftharpoons \underset{\substack{\text{Conjugate}\\\text{base}}}{NO_2^-} + H^+ \qquad \underset{\text{Base}}{NH_3} + H^+ \rightleftharpoons \underset{\substack{\text{Conjugate}\\\text{acid}}}{NH_4^+}$$

**conjugate acid–base pair** a Brønsted–Lowry acid and base differing from each other only by the presence or absence of a $H^+$ ion: acid $\rightleftharpoons$ conjugate base + $H^+$.

**conjugate base** formed when a Brønsted–Lowry acid donates a $H^+$ ion.

**conjugate acid** formed when a Brønsted–Lowry base accepts a $H^+$ ion.

The corresponding general equations for equilibria in aqueous solutions are

$$Acid(aq) + H_2O(\ell) \rightleftharpoons \text{conjugate base}(aq) + H_3O^+(aq) \qquad (16.3)$$
$$Base(aq) + H_2O(\ell) \rightleftharpoons \text{conjugate acid}(aq) + OH^-(aq) \qquad (16.4)$$

Water and the hydronium ion form another conjugate acid–base pair: $H_2O$ is the base and $H_3O^+$ is its conjugate acid in the reaction with nitrous acid; $H_2O$ is the acid and $OH^-$ its conjugate base in the reaction with ammonia.

Identify the conjugate acid–base pairs in the reactions that result when perchloric acid, $HClO_4$, and formic acid, $HCOOH$, dissolve in water.

**Collect and Organize** We are to identify the conjugate acid–base pairs that form in aqueous solutions of $HClO_4$ and $HCOOH$. We know from Table 16.1 that perchloric acid is a strong acid and from Table 16.2 that formic acid is a weak acid.

**Analyze** Acids in aqueous solutions form their conjugate bases by donating $H^+$ ions to molecules of $H_2O$ (Equation 16.3). Therefore, the formulas of their conjugate bases are the formulas of the original acids minus a $H^+$ ion. The formula of perchloric acid has only one H atom in it, which must be the one that it loses as a $H^+$ ion. The formula of formic acid has two H atoms. The ionizable one is bonded to an O atom in the carboxylic acid group.

**Solve** Rewriting Equation 16.3 for aqueous solutions of these two acids, we have

Perchloric acid     $\underset{\text{Acid}}{HClO_4(aq)} + H_2O(\ell) \rightarrow \underset{\text{Conjugate base}}{ClO_4^-(aq)} + H_3O^+(aq)$

Formic acid     $\underset{\text{Acid}}{HCOOH(aq)} + H_2O(\ell) \rightleftharpoons \underset{\text{Conjugate base}}{HCOO^-(aq)} + H_3O^+(aq)$

In both reactions, $H_3O^+$ and $H_2O$ are also a conjugate acid–base pair: $H_2O$ is the base and $H_3O^+$ is its conjugate acid.

**Think About It** A single arrow is used in the $HClO_4$ equation because $HClO_4$ is a strong acid that ionizes completely in water. Equilibrium arrows are used for $HCOOH$ because it is a weak acid and both $HCOOH$ and $HCOO^-$ are likely to be present in solution at equilibrium.

⚙ **Practice Exercise** Identify the conjugate acid–base pairs in the reaction that takes place when the weak organic acid acetic acid ($CH_3COOH$) dissolves in water.

*(Answers to Practice Exercises are in the back of the book.)*

## Strong and Weak Bases

The most common strong bases are hydroxides of group 1 and 2 metals. Table 16.3 shows how these ionic compounds dissociate when they dissolve in water. Their corresponding equilibrium constants all have values much greater than 1 ($K_b \gg 1$, where the "b" subscript indicates that the reactant functions as a *base*). The hydroxide ions produced when these bases dissolve in water are very effective $H^+$ acceptors and so these compounds are strong Brønsted–Lowry bases.

We can also use the Brønsted–Lowry model to explain what happens when a weak base (Table 16.4) dissolves in water. Let's use ammonia as an example. In aqueous solution, $NH_3$ molecules accept $H^+$ ions from molecules of water to form $NH_4^+$ and $OH^-$ ions:

| TABLE 16.3 | Strong Bases and Their Ionization Reactions in Water |
|---|---|
| **Strong Base** | **Reaction in Water** |
| Lithium hydroxide | $LiOH(aq) \rightarrow Li^+(aq) + OH^-(aq)$ |
| Sodium hydroxide | $NaOH(aq) \rightarrow Na^+(aq) + OH^-(aq)$ |
| Potassium hydroxide | $KOH(aq) \rightarrow K^+(aq) + OH^-(aq)$ |
| Calcium hydroxide | $Ca(OH)_2(aq) \rightarrow Ca^{2+}(aq) + 2\,OH^-(aq)$ |
| Barium hydroxide | $Ba(OH)_2(aq) \rightarrow Ba^{2+}(aq) + 2\,OH^-(aq)$ |
| Strontium hydroxide | $Sr(OH)_2(aq) \rightarrow Sr^{2+}(aq) + 2\,OH^-(aq)$ |

$$\underset{\text{Base}}{NH_3(aq)} + H_2O(\ell) \rightleftharpoons \underset{\text{Conjugate acid}}{NH_4^+(aq)} + \underset{\text{Conjugate base}}{OH^-(aq)} \qquad (16.5)$$

Note how the $NH_4^+$ ion is the conjugate acid of $NH_3$, a relationship that may be clearer if we consider the reaction that occurs when an ammonium salt dissolves in water:

$$NH_4^+(aq) + H_2O(\ell) \rightleftharpoons NH_3(aq) + H_3O^+(aq) \qquad (16.6)$$

**leveling effect** the observation that strong acids all have the same strength in water and are completely converted into solutions of $H_3O^+$ ions; strong bases are likewise leveled in water and are completely converted into solutions of $OH^-$ ions.

| TABLE 16.4 | Some Common Weak Bases and Their Ionization Reactions in Water | |
|---|---|---|
| **Weak Base** | **Reaction in Water** | $K_b$ |
| Ammonia | $NH_3(aq) + H_2O(\ell) \rightleftharpoons NH_4^+(aq) + OH^-(aq)$ | $1.76 \times 10^{-5}$ |
| Aniline | $C_6H_5NH_2(aq) + H_2O(\ell) \rightleftharpoons C_6H_5NH_3^+(aq) + OH^-(aq)$ | $4.0 \times 10^{-10}$ |
| Dimethylamine | $(CH_3)_2NH(aq) + H_2O(\ell) \rightleftharpoons (CH_3)_2NH_2^+(aq) + OH^-(aq)$ | $5.9 \times 10^{-4}$ |
| Methylamine | $CH_3NH_2(aq) + H_2O(\ell) \rightleftharpoons CH_3NH_3^+(aq) + OH^-(aq)$ | $4.4 \times 10^{-4}$ |
| Pyridine | $C_5H_5N(aq) + H_2O(\ell) \rightleftharpoons C_5H_5NH^+(aq) + OH^-(aq)$ | $1.7 \times 10^{-9}$ |

Here $NH_4^+$, acting as an acid, donates a $H^+$ ion and thereby forms $NH_3$, its conjugate base. Also note that water molecules are proton donors in Equation 16.5, making water a Brønsted–Lowry acid in this reaction. However, water molecules are proton acceptors in Equation 16.6, making water a Brønsted–Lowry base. We revisit this acid–base duality of water throughout this chapter.

## Relative Strengths of Acids and Bases

We can use the concepts we have developed for describing equilibria to evaluate the relative strengths of acids and bases. To do so, let's revisit the ionization of HCl:

$$HCl(aq) + H_2O(\ell) \rightarrow Cl^-(aq) + H_3O^+(aq)$$

Because HCl is a strong acid, this reaction goes to completion, which means the reverse reaction essentially does not happen at all. This in turn means that a $Cl^-$ ion (the conjugate base of HCl) must be a very weak base because it has no tendency to accept a proton from $H_3O^+$. This contrast in relative strengths applies to all conjugate pairs: strong acids have very weak conjugate bases and strong bases have very weak conjugate acids (Figure 16.4).

In between these extremes are many substances that are weak acids with weak conjugate bases. For example, $HNO_2$ is a weak acid ($K_a = 4.0 \times 10^{-4}$), which means that its conjugate base, $NO_2^-$, is a weak base. This pairing of weakly acidic and weakly basic strengths applies to all conjugate acid–base pairs.

All the strong acids in Figure 16.4 ionize completely in water. The $H_2O$ molecules in their solutions readily accept $H^+$ ions from 100% of the acid molecules. In this context, water is said to *level* the strengths of these acids; they all are equally strong because they cannot be more than 100% ionized. This **leveling effect** means that the conjugate acid of $H_2O$, which is $H_3O^+$, is the strongest $H^+$ donor that can exist in water. An even stronger acid, no matter how much stronger it is, simply donates all its ionizable H atoms to water molecules, forming $H_3O^+$ ions.

On the other hand, weak acids are differentiated by their ability to donate their ionizable H atoms to water molecules. The weak acids higher on the list in Figure 16.4 form more acidic aqueous solutions than acids lower on the list.

A similar pattern is evident in the strengths of bases. The strongest base that can exist in water is the conjugate base of $H_2O$,

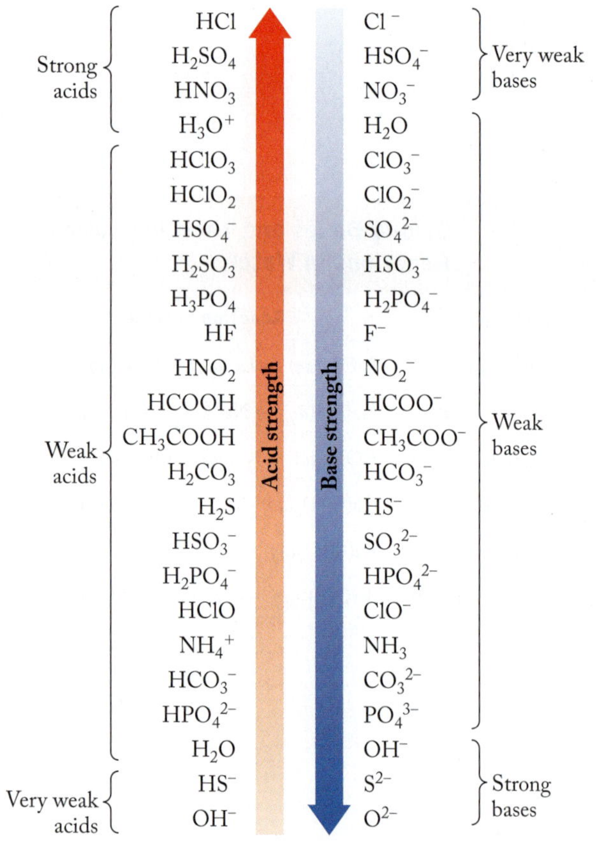

**FIGURE 16.4** Opposing trends characterize the relative strengths of acids and their conjugate bases: the stronger the acid, the weaker its conjugate base. The same is true for bases: the stronger the base, the weaker its conjugate acid.

which is the OH$^-$ ion. Any base that is stronger than OH$^-$ hydrolyzes in water to produce OH$^-$ ions. The oxide ion (O$^{2-}$), for example, is a very strong base, and reacts with water to produce two OH$^-$ ions:

$$O^{2-}(aq) + H_2O(\ell) \rightarrow 2\ OH^-(aq)$$

The strengths of bases weaker than OH$^-$ ions can be differentiated by the fraction of their molecules that accept H$^+$ ions from water molecules in aqueous solutions. Weaker bases are higher on the list in Figure 16.4; stronger bases are lower on the list.

**autoionization** the process that produces equal and very small concentrations of H$_3$O$^+$ and OH$^-$ ions in pure water.

**CONNECTION** Hydrolysis reactions, which are the reactions of substances with water, were introduced in Chapters 3 and 4.

**CONCEPT TEST**

List the following anions in order of decreasing strength as Brønsted–Lowry bases: F$^-$, Cl$^-$, OH$^-$, HCOO$^-$, NO$_2^-$.

## 16.2 pH and the Autoionization of Water

We have seen that the acidity of a solution is directly related to the concentration of H$_3$O$^+$ ions in it. In this section we examine another way to express acidity. To understand this alternative, we first need to understand the **autoionization** of water, which is the process that produces equal and very small concentrations of H$_3$O$^+$ and OH$^-$ ions in pure water:

$$H_2O(\ell) + H_2O(\ell) \rightleftharpoons H_3O^+(aq) + OH^-(aq) \qquad (16.7)$$

One water molecule, acting as an acid, donates a hydrogen ion to another water molecule, which acts as a base (Figure 16.5). The donor H$_2$O forms its conjugate base (OH$^-$), and the acceptor H$_2$O forms its conjugate acid (H$_3$O$^+$). We have already encountered this dual nature of water: molecules of H$_2$O act as H$^+$ ion acceptors in solutions of acidic solutes, and as H$^+$ ion donors in solutions of basic solutes. As we discussed in Chapter 4, any substance that can act as either an acid or a base is said to be amphiprotic. The autoionization of water is an example of amphiprotic behavior.

The equilibrium constant for the autoionization of water shown in Equation 16.7 is written as

$$K_c = \frac{[H_3O^+][OH^-]}{[H_2O][H_2O]} \qquad (16.8)$$

Because water is a pure liquid, its concentration does not appear in the equilibrium constant expression. To further simplify the expression, we substitute [H$^+$] for [H$_3$O$^+$] because these terms represent the same species. This reduces Equation 16.8 to an equilibrium constant expression that is given the symbol $K_w$:

$$K_w = [H^+][OH^-] \qquad (16.9)$$

**CONNECTION** We defined *amphiprotic* compounds in Chapter 4 as having both acidic and basic properties.

**CHEMTOUR** Autoionization of Water

**FIGURE 16.5** The autoionization of water takes place when a proton is transferred from one water molecule to another. Both molecules are converted to ions.

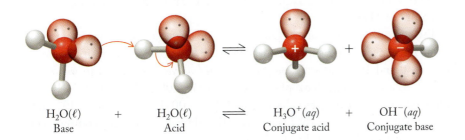

| H$_2$O($\ell$) | + | H$_2$O($\ell$) | $\rightleftharpoons$ | H$_3$O$^+$($aq$) | + | OH$^-$($aq$) |
| Base | | Acid | | Conjugate acid | | Conjugate base |

**pH** the negative logarithm of the hydrogen ion concentration in an aqueous solution.

In pure water at 25°C, $[H^+] = [OH^-] = 1.00 \times 10^{-7}$ $M$. Inserting these values into Equation 16.9 gives

$$K_w = [H^+][OH^-] = (1.00 \times 10^{-7})(1.00 \times 10^{-7}) = 1.00 \times 10^{-14} \quad (16.10)$$

Such a tiny value of $K_w$ confirms that a very tiny fraction of water molecules undergoes autoionization. The reverse of autoionization—the reaction between $[H^+]$ and $[OH^-]$ to produce $H_2O$—has an equilibrium constant of $1/K_w = 1.00 \times 10^{14}$ at 25°C and essentially goes to completion:

$$H^+(aq) + OH^-(aq) \rightleftharpoons H_2O(\ell) \qquad K = 1/K_w = 1.00 \times 10^{14}$$

The value $K_w = 1.00 \times 10^{-14}$ applies to all aqueous solutions at 25°C, not just pure water. (Because this value depends on temperature, throughout this and subsequent discussions, we assume a temperature of 25°C.) Equation 16.10 tells us that an inverse relation exists between $[H^+]$ and $[OH^-]$ in any aqueous sample: as the value of one increases, the value of the other must decrease so that the product of the two is always $1.00 \times 10^{-14}$. A solution in which $[H^+] > [OH^-]$ is acidic, a solution in which $[H^+] < [OH^-]$ is basic, and a solution in which $[H^+] = [OH^-] = 1.00 \times 10^{-7}$ $M$ is neutral (neither acidic nor basic).

The tiny value of $K_w$ means that autoionization of water does not contribute significantly to $[H^+]$ in solutions of most acids or to $[OH^-]$ in solutions of most bases, so we can ignore the contribution of autoionization in most calculations of acid or base strength. However, if acids or bases are extremely weak or if their concentrations are extremely low, $H_2O$ autoionization may need to be taken into account.

## The pH Scale

▶❚❚ **CHEMTOUR** pH Scale

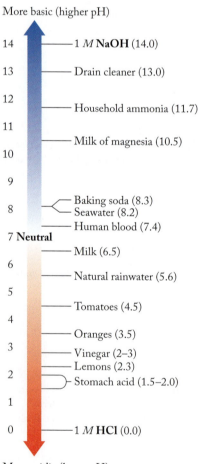

More basic (higher pH)

14 —— 1 $M$ **NaOH** (14.0)

13 —— Drain cleaner (13.0)

12 —— Household ammonia (11.7)

11

—— Milk of magnesia (10.5)

10

9

8 —— Baking soda (8.3)
    —— Seawater (8.2)
    —— Human blood (7.4)

**7 Neutral**

—— Milk (6.5)

6

—— Natural rainwater (5.6)

5

—— Tomatoes (4.5)

4

—— Oranges (3.5)

3

—— Vinegar (2–3)
    —— Lemons (2.3)

2 —— Stomach acid (1.5–2.0)

1

0 —— 1 $M$ **HCl** (0.0)

More acidic (lower pH)

**FIGURE 16.6** The pH scale is a convenient way to express the range of acidic or basic properties of some common materials.

In the early 1900s, scientists developed a device called the *hydrogen electrode* to determine the $[H^+]$ of solutions. The electrical voltage, or *potential*, produced by the hydrogen electrode is a linear function of the logarithm of $[H^+]$. This relation led Danish biochemist Søren Sørensen (1868–1939) to propose a scale for expressing acidity and basicity based on what he termed "the *potential* of the hydrogen ion," abbreviated **pH**. Mathematically, we define pH as the negative logarithm of $[H^+]$:

$$pH = -\log[H^+] \quad (16.11)$$

For example, the pH of a solution in which $[H^+] = 5.0 \times 10^{-3}$ $M$ is

$$pH = -\log(5.0 \times 10^{-3}) = -(-2.30) = 2.30$$

Sørensen's pH scale has several attractive features. Because it is logarithmic, there are no exponents, as are commonly encountered in values of $[H^+]$. The logarithmic scale also means that a change of one pH unit corresponds to a 10-fold change in $[H^+]$, so that a solution with a pH of 5.0 has 10 times the $[H^+]$ of a solution with a pH of 6.0 and is 10 times as acidic. Similarly, a solution with a pH of 12.0 has 1/10 the $[H^+]$, or 10 times the $[OH^-]$, as a solution with a pH of 11.0.

The negative sign in front of the logarithmic term means that most pH values, except for concentrated solutions of strong acids or bases, are positive numbers between 0 and 14. It also means that *large pH values* correspond to *small values of $[H^+]$*. Acidic solutions have pH values less than 7.00 ($[H^+] > 1.00 \times 10^{-7}$ $M$), and basic solutions have pH values greater than 7.00 ($[H^+] < 1.00 \times 10^{-7}$ $M$). A solution with a pH of exactly 7 is neutral. The pH values for some common aqueous solutions are shown in Figure 16.6.

CONCEPT TEST

Match the pH values on the left with the descriptors on the right:

| | |
|---|---|
| 13.77 | strongly acidic |
| 10.03 | weakly acidic |
| 7.00 | weakly basic |
| 4.37 | strongly basic |
| 0.22 | neutral |

CONCEPT TEST

Suppose solution A has a pH of 6.0 and solution B has a pH of 7.0. Which of the following statements about the two solutions is/are true?
a. Solution A is 10 times more acidic than solution B.
b. Solution B is neither acidic nor basic.
c. The concentration of $OH^-$ ions in solution B is 10 times their concentration in solution A.
d. $[OH^-] = [H^+]$ in solution B.
e. The value of $[H^+]$ in solution A is 10 times that in solution B.

A note about expressing pH values to the appropriate number of significant figures is necessary. Because any pH value is the negative logarithm of the hydrogen ion concentration, the first number in the value defines the location of the decimal point in the concentration term. As such, it is not considered when determining the number of significant figures. For example, a hydrogen ion concentration of $2.7 \times 10^{-4}$ has two significant figures in the coefficient of the power of 10. The corresponding pH value with two significant figures is 3.57. The 3 in pH 3.57 is not considered a significant figure because it just means that the $[H^+]$ is between $10^{-3}$ and $10^{-4}$ $M$.

**SAMPLE EXERCISE 16.2**    **Interconverting pH and [H⁺]**    **LO2**

Oxidation of iron(II) sulfide in mine tailings leads to acidic runoff and subsequent environmental damage (Figure 16.7). Is water from the Iron Mountain Mine, with a pH of 1.81, more or less acidic than the water in Lemonade Creek in Yellowstone National Park, where $[H^+] = 6.0 \times 10^{-3}$ $M$?

**Collect and Organize** We are asked to compare the acidities of two solutions. We know the pH of one and the $[H^+]$ of the other.

**Analyze** We can either convert the given pH value into the corresponding $[H^+]$ value, or the given $[H^+]$ value into pH using Equation 16.11:

$$pH = -\log[H^+]$$

The given $[H^+]$ value is between $10^{-2}$ and $10^{-3}$ $M$, which means that the corresponding pH value will have a 2 in front of the decimal place.

**Solve**

(1) pH of Lemonade Creek:
$$pH = -\log(6.0 \times 10^{-3})$$
$$= -(-2.22) = 2.22$$

(2) $[H^+]$ of Iron Mountain Mine runoff:
$$1.81 = -\log[H^+]$$
$$[H^+] = 10^{-1.81} = 1.5 \times 10^{-2} \, M$$

1. The pH of Lemonade Creek (2.22) is higher than the pH of Iron Mountain Mine water (1.81). Therefore, Lemonade Creek water is less acidic.
2. The $[H^+]$ of Iron Mountain Mine water ($1.5 \times 10^{-2}$ $M$) is greater than that of Lemonade Creek ($6.0 \times 10^{-3}$ $M$). Therefore, Iron Mountain Mine water is more acidic.

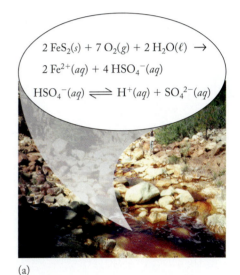

$$2\,FeS_2(s) + 7\,O_2(g) + 2\,H_2O(\ell) \rightarrow$$
$$2\,Fe^{2+}(aq) + 4\,HSO_4^-(aq)$$
$$HSO_4^-(aq) \rightleftharpoons H^+(aq) + SO_4^{2-}(aq)$$

(a)

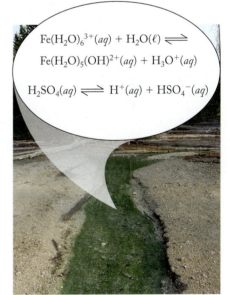

$$Fe(H_2O)_6^{3+}(aq) + H_2O(\ell) \rightleftharpoons$$
$$Fe(H_2O)_5(OH)^{2+}(aq) + H_3O^+(aq)$$
$$H_2SO_4(aq) \rightleftharpoons H^+(aq) + HSO_4^-(aq)$$

(b)

**FIGURE 16.7** Some natural waters have low pH values. (a) The oxidation of iron(II) sulfide in the Iron Mountain Mine leads to solutions of $Fe^{2+}$ $(aq)$, $HSO_4^-(aq)$, and $H^+(aq)$. (b) The mineral content of Lemonade Creek in the West Nymph Creek area of Yellowstone National Park also leads to acidic waters. The green color comes from hardy algae that can tolerate high temperatures and low pH values.

**pOH** the negative logarithm of the hydroxide ion concentration in an aqueous solution.

**Think About It** Our two results agree. Iron Mountain Mine water is more acidic than water from Lemonade Creek according to both calculations. Actually, both solutions have pH values less than 7 and are weakly acidic.

**Practice Exercise** Is solution A with a pH of 9.58 more or less acidic than solution B in which $[H^+] = 4.3 \times 10^{-10}$ $M$?

**CONCEPT TEST**

Is the pH of a 1.00 $M$ solution of a weak acid higher or lower than the pH of a 1.00 $M$ solution of a strong acid?

## pOH

The letter $p$ as used in pH is also used with other symbols to mean *the negative logarithm* of the variable that follows it. For example, just as every aqueous solution has a pH value, it also has a **pOH** value, defined as

$$pOH = -\log[OH^-] \qquad (16.12)$$

We can use Equation 16.10 to relate pOH to pH. We start by taking the negative logarithm of both sides of the equation:

$$K_w = [H^+][OH^-] = 1.00 \times 10^{-14}$$
$$-\log K_w = -\log([H^+][OH^-]) = -\log(1.00 \times 10^{-14})$$
$$pK_w = -(\log[H^+] + \log[OH^-]) = -(-14.00)$$
$$pK_w = pH + pOH = 14.00 \qquad (16.13)$$

Many tables of equilibrium constants list p$K$ values rather than $K$ values because doing so does not require the use of exponential notation and is more convenient. The tables in Appendix 5 of this book contain both sets of values. Use them whenever you need a $K$ or p$K$ value that is not provided in a problem. For example, we can define both p$K_a$ and p$K_b$ values for acetic acid and methylamine, respectively, as well as the other acids and bases in Tables 16.2 and 16.4:

$$pK_a = -\log K_a \qquad\qquad pK_b = -\log K_b$$
$$= -\log(1.76 \times 10^{-5}) \qquad = -\log(4.4 \times 10^{-4})$$
$$= 4.75 \qquad\qquad\qquad = 3.4$$

**CONNECTION** In Chapter 3 we identified nonmetal oxides like $CO_2(g)$, $NO_2(g)$, and $SO_3(g)$ as acid anhydrides, so called because they hydrolyze in water to produce acidic solutions.

**SAMPLE EXERCISE 16.3  Relating [H⁺], [OH⁻], pH, and pOH   LO2**

The carbonic acid that forms when atmospheric $CO_2$ dissolves in rainwater gives rain a normal pH of about 5.6. The pH of acid rain (Figure 16.8), however, can be one or more pH units lower than 5.6. Calculate the values of $[H^+]$, pOH, and $[OH^-]$ in pH 5.6 rainwater and in a sample of acid rain with a pH of 4.3.

**Collect and Organize** We are given pH values of two samples of rainwater and asked to determine the corresponding $[H^+]$, pOH, and $[OH^-]$ values.

**Analyze** We can start with Equation 16.11 to convert pH into $[H^+]$, use Equation 16.13 to convert pH into pOH, and then use Equation 16.12 to convert pOH into $[OH^-]$. Both samples are weakly acidic with pH values below 7. This means, according to Equation 16.13, that the corresponding pOH values must be greater than 7. The acid rain sample has

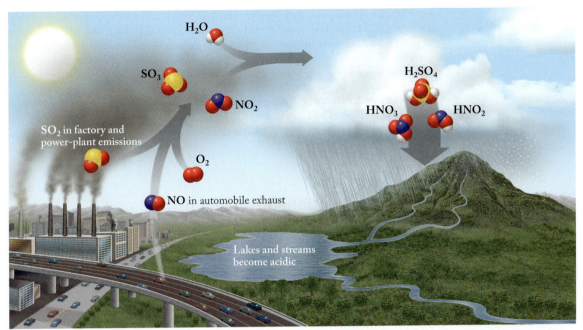

**FIGURE 16.8** Acid rain forms when volatile nonmetal oxides such as NO and $SO_2$ are further oxidized in the atmosphere and dissolve in rain to form nitric ($HNO_3$), nitrous ($HNO_2$), and sulfuric ($H_2SO_4$) acids. These acids ionize, forming $NO_3^-$, $NO_2^-$, and $SO_4^{2-}$ ions, respectively, and the hydronium ($H_3O^+$) ions that make rain acidic.

a lower pH than natural rainwater and should have a higher $[H^+]$ value, which means a smaller $[OH^-]$ value and a higher pOH.

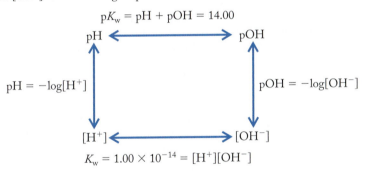

$$pK_w = pH + pOH = 14.00$$

$$pH = -\log[H^+]$$

$$pOH = -\log[OH^-]$$

$$K_w = 1.00 \times 10^{-14} = [H^+][OH^-]$$

**Solve** For $[H^+]$,

$$pH = -\log[H^+]$$
$$[H^+] = 10^{-pH}$$

Normal rain: $[H^+] = 10^{-5.6} = 3 \times 10^{-6}\ M$

Acid rain: $[H^+] = 10^{-4.3} = 5 \times 10^{-5}\ M$

For pOH,

$$pK_w = pH + pOH = 14.00$$
$$pOH = 14.00 - pH$$

Normal rain: $pOH = 14.00 - 5.6 = 8.4$

Acid rain: $pOH = 14.00 - 4.3 = 9.7$

For $[OH^-]$,

$$pOH = -\log[OH^-]$$
$$[OH^-] = 10^{-pOH}$$

Normal rain: $[OH^-] = 10^{-8.4} = 4 \times 10^{-9}\ M$

Acid rain: $[OH^-] = 10^{-9.7} = 2 \times 10^{-10}\ M$

**Think About It** The $[H^+]$ in the acid rain sample is nearly 20 times higher than its concentration in normal rain, and its pH value differs by 1.3 units. These differences make sense because (1) the pH scale is logarithmic, so one unit difference in pH means a 10-fold difference in $[H^+]$ or $[OH^-]$, and (2) pH values decrease as $[H^+]$ increases. The differences in pOH values make sense for the same reasons. All calculated concentration values were rounded off to only one significant figure because each starting pH value had only one significant figure—the single digit after the decimal point.

⚙ **Practice Exercise** What are the values of $[H^+]$ and $[OH^-]$ in household ammonia, an aqueous solution of $NH_3$ that has a pH of 11.7?

■

# 16.3 Calculations Involving pH, $K_a$, and $K_b$

If we know the pH and concentration of a solution of a weakly acidic or basic substance, we can calculate the acid or base equilibrium constant $K_a$ or $K_b$ for that substance. Of more practical importance, if we know the value of $K_a$ or $K_b$, we can calculate the pH of an aqueous solution of a weak acid or weak base.

## Weak Acids

The vast majority of the acids on our planet are weak acids. Among them, as we saw in Section 16.1, is nitrous acid:

$$HNO_2(aq) + H_2O(\ell) \rightleftharpoons NO_2^-(aq) + H_3O^+(aq) \qquad (16.2)$$

Let's begin our quantitative analysis of this equilibrium by writing the equilibrium constant expression for this reaction:

$$K_a = \frac{[NO_2^-][H_3O^+]}{[HNO_2][H_2O]}$$

Water is the solvent in this case and is present in great abundance. Consequently, the concentration of water does not significantly change during the course of reactions in aqueous solutions, so we do not include the $[H_2O]$ term in the equilibrium expression. We also replace $[H_3O^+]$ with $[H^+]$ to simplify the expression. With these changes we have

$$K_a = \frac{[NO_2^-][H^+]}{[HNO_2]}$$

A generic form of this $K_a$ expression that applies to the ionization reaction of any acid (HA):

$$HA(aq) \rightleftharpoons A^-(aq) + H^+(aq) \qquad (16.14)$$

is written

$$K_a = \frac{[A^-][H^+]}{[HA]} \qquad (16.15)$$

We can use Equation 16.15 to calculate the value of the $K_a$ of an unknown weak acid if we know the pH of a solution of the acid and the value of [HA]. Suppose we measure the pH of a 0.100 $M$ solution and find that it is 2.20. To calculate $K_a$ we first convert the pH value to $[H^+]$:

$$pH = -\log[H^+] = 2.20$$
$$[H^+] = 10^{-2.20} = 6.3 \times 10^{-3}\ M$$

Assuming the only source of H⁺ ions is ionization of HA, then [A⁻] must also be $6.3 \times 10^{-3}$ M because the stoichiometry of the reaction is that 1 mole of HA ionizes to form 1 mole of H⁺ and 1 mole of A⁻. If the ionization reaction yields a [H⁺] of $6.3 \times 10^{-3}$ M, then [HA] must have decreased by the same amount. Therefore, at equilibrium:

$$[\text{HA}] = (0.100 - 6.3 \times 10^{-3})\ M = 0.094\ M$$

We substitute the three calculated equilibrium concentrations in the $K_a$ expression:

$$K_a = \frac{[\text{A}^-][\text{H}^+]}{[\text{HA}]} = \frac{(6.3 \times 10^{-3})(6.3 \times 10^{-3})}{(0.094)} = 4.2 \times 10^{-4}$$

The small value of $K_a$ confirms that HA is a weak acid.

The ratio of the concentration of H⁺ ions at equilibrium to the initial concentration of HA represents the **degree of ionization** of HA, which is usually expressed as a percentage of the initial acid concentration. It is also called **percent ionization**. In equation form this relationship is

$$\text{Percent ionization} = \frac{[\text{H}^+]_{\text{equilibrium}}}{[\text{HA}]_{\text{initial}}} \times 100\% \qquad (16.16)$$

Inserting the data from the previous calculation into Equation 16.16 gives

$$\text{Percent ionization} = \frac{6.3 \times 10^{-3}\ \cancel{M}}{0.100\ \cancel{M}} \times 100\% = 6.3\%$$

**CONCEPT TEST**

Describe how the percent ionization of a weak acid is related to the value of its $K_a$.

A plot of percent ionization as a function of initial concentration of nitrous acid is shown in Figure 16.9. This same pattern is observed for all weak acids: the degree to which they ionize increases as their concentration decreases. The following Sample Exercise provides a mathematical perspective on this trend.

> **degree of ionization** the ratio of the quantity of a substance that is ionized to the concentration of the substance before ionization; when expressed as a percentage, called **percent ionization**.

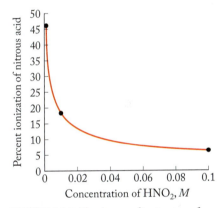

**FIGURE 16.9** The degree of ionization of a weak acid increases with decreasing acid concentration. Here the degree of ionization of nitrous acid increases from about 6% in a 0.100 M solution to 18% in a 0.010 M solution to 46% in a 0.001 M solution.

---

**SAMPLE EXERCISE 16.4**   **Relating pH, $K_a$, and Percent Ionization of a Weak Acid**   **LO3**

The pH of a 1.00 M solution of formic acid (HCOOH), a weak organic acid found in red ants and responsible for the sting of their bite, is 1.88.
a. What is the percent ionization of 1.00 M HCOOH?
b. What is the $K_a$ value of the acid?
c. What is the percent ionization of 0.0100 M HCOOH?

**Collect and Organize** We are asked to determine the $K_a$ value of formic acid and its percent ionization in two solutions of known concentration.

**Analyze** We know the pH of one of the solutions. Equation 16.11 (pH = −log[H⁺]) relates pH to [H⁺]; Equation 16.15 $\left(K_a = \dfrac{[\text{A}^-][\text{H}^+]}{[\text{HA}]}\right)$ is the generic equilibrium constant expression for a weak acid; and Equation 16.16 (Percent ionization = [H⁺]$_{\text{equilibrium}}$/ [HA]$_{\text{initial}}$ × 100%) is the formula for calculating percent ionization. The 1:1:1 stoichiometry of the ionization reaction

$$\text{HCOOH}(aq) \rightleftharpoons \text{HCOO}^-(aq) + \text{H}^+(aq)$$

tells us that in a solution of HCOOH at equilibrium, $[\text{HCOO}^-] = [\text{H}^+]$, and [HCOOH] is equal to the initial concentration of the acid minus the portion of it that ionized, or

$$[\text{HCOOH}]_{\text{equilibrium}} = [\text{HCOOH}]_{\text{initial}} - [\text{H}^+]_{\text{equilibrium}}$$

Inserting these concentration values into the equilibrium constant expression for formic acid based on Equation 16.15:

$$K_a = \frac{[\text{HCOO}^-][\text{H}^+]}{[\text{HCOOH}]}$$

enables us to calculate the value of $K_a$. Once we know $K_a$, we can use the equilibrium constant expression to calculate $[\text{H}^+]$ in any solution of formic acid, and, from $[\text{H}^+]$, the percent ionization of the acid in that solution.

The pH value of the 1.00 $M$ solution is close to 2, which corresponds to $[\text{H}^+] = 10^{-2}\ M$. Therefore the percent ionization of formic acid in this solution should be about 1%. The more dilute solution should be more extensively ionized.

**Solve**

a. The pH of the 1.00 $M$ solution is 1.88. The corresponding $[\text{H}^+]$ is

$$[\text{H}^+] = 10^{-1.88} = 1.32 \times 10^{-2}\ M$$

Inserting this value and the initial concentration of HCOOH in Equation 16.16:

$$\text{Percent ionization} = \frac{[\text{H}^+]_{\text{equilibrium}}}{[\text{HCOOH}]_{\text{initial}}} \times 100\%$$

$$= \frac{1.32 \times 10^{-2}\ M}{1.00\ M} \times 100\% = 1.32\%$$

b. At equilibrium $[\text{HCOO}^-] = [\text{H}^+] = 1.32 \times 10^{-2}\ M$, and the equilibrium concentration of HCOOH is

$$(1.00 - 1.32 \times 10^{-2})\ M = 0.99\ M$$

Inserting these values in the expression for $K_a$:

$$K_a = \frac{[\text{H}^+][\text{HCOO}^-]}{[\text{HCOOH}]} = \frac{(1.32 \times 10^{-2})(1.32 \times 10^{-2})}{(0.99)} = 1.76 \times 10^{-4}$$

c. To calculate $[\text{H}^+]$ in 0.0100 $M$ HCOOH, we use a RICE table, as in the equilibrium calculations in Chapter 15, to solve for $[\text{H}^+]$ at equilibrium. Since it is the unknown in the calculation, we give it the symbol $x$. Filling in the other cells in the RICE table:

| Reaction (R) | HCOOH $\rightleftharpoons$ | HCOO$^-$ + | H$^+$ |
|---|---|---|---|
| | [HCOOH] | [HCOO$^-$] (*M*) | [H$^+$] (*M*) |
| Initial (I) | 0.0100 | 0.0000 | 0.0000 |
| Change (C) | $-x$ | $+x$ | $+x$ |
| Equilibrium (E) | $(0.0100 - x)$ | $x$ | $x$ |

Inserting the equilibrium terms into the $K_a$ expression and using $K_a$ from part b,

$$K_a = \frac{(x)(x)}{(0.0100 - x)} = 1.76 \times 10^{-4}$$

We solve for $x$ using the quadratic equation:

$$x^2 = 1.76 \times 10^{-6} - (1.76 \times 10^{-4})x$$

$$x^2 + (1.76 \times 10^{-4})x - 1.76 \times 10^{-6} = 0$$

$$x = \frac{-1.76 \times 10^{-4} \pm \sqrt{(1.76 \times 10^{-4})^2 - 4(1)(-1.76 \times 10^{-6})}}{2(1)}$$

$$x = 1.24 \times 10^{-3} \quad \text{or} \quad -1.42 \times 10^{-3}$$

The negative value for $x$ has no physical meaning because it gives us a negative concentration value. Therefore, we use only the positive $x$ value. The $x$ value is equal to $[H^+]$ at equilibrium. Therefore, the percent ionization of formic acid in a 0.0100 $M$ solution is

$$\text{Percent ionization} = \frac{[H^+]_{\text{equilibrium}}}{[\text{HCOOH}]_{\text{initial}}} \times 100\%$$

$$= \frac{1.24 \times 10^{-3}}{0.0100} \times 100\% = 12.4\%$$

**Think About It** The solutions to parts a and c agree with our estimates: the percent ionization of HCOOH in the 1.00 $M$ solution was indeed near 1%, and the percent ionization in the 0.0100 $M$ solution was significantly higher: 12.4%. The calculated $K_a$ value also agrees with the tabulated value for formic acid (Appendix 5). Finally, we did not attempt to simplify the calculation of $[H^+]$ in part c by eliminating "$- x$" from the denominator of the equilibrium constant expression. Had we done so, the value of $x$ would have been

$$\frac{(x)(x)}{0.100} \approx 1.76 \times 10^{-4}$$

$$x \approx 1.33 \times 10^{-3} \, M$$

The relative difference between this value and the correct one:

$$\frac{(1.33 \times 10^{-3}) - (1.24 \times 10^{-3})}{1.24 \times 10^{-3}} \times 100\% = 7\%$$

represents a +7% error. In addition, the value of $x$ is 12.0% of the initial [HCOOH], which is greater than the 5% value we used to judge the appropriateness of equilibrium calculation approximations in Chapter 15, so we were correct in not simplifying this calculation.

⊚ **Practice Exercise** The value of $[H^+]$ in a 0.050 $M$ solution of an organic acid is $5.9 \times 10^{-3} \, M$. What is the pH of the solution, the percent ionization of the acid, and its $K_a$ value?

**CONCEPT TEST** • • • • • • • • • • • • • • • • • • • • • • • • • • • • • • • • • • • •

Three weak acids have these $K_a$ values:

| Acid | $K_a$ |
|------|-------|
| A | $3.6 \times 10^{-5}$ |
| B | $4.9 \times 10^{-4}$ |
| C | $9.2 \times 10^{-4}$ |

Which of the three acids is the most extensively ionized in a 0.100 $M$ solution of the acid? Which of the three acids has the lowest percent ionization in a 1.00 $M$ solution of the acid?

## Weak Bases

Now let's consider what happens when a weakly basic compound dissolves in water. As noted in Section 16.1, the weak base ammonia accepts hydrogen ions from water as described by the chemical equation

$$NH_3(aq) + H_2O(\ell) \rightleftharpoons NH_4^+(aq) + OH^-(aq) \qquad (16.17)$$

This reaction is the result of strong intermolecular forces that lead to covalent bonds breaking and new bonds forming, as shown in Figure 16.10. Not all the $NH_3$ molecules in an ammonia solution accept hydrogen ions. Instead, the reaction reaches an equilibrium in which most ammonia molecules are present as

**FIGURE 16.10** (a) Ammonia reacts with water to produce ammonium ions and hydroxide ions in solution. The lone pair of electrons on the nitrogen of $NH_3$ is shared with a transferred $H^+$ ion to make the fourth N—H bond in $NH_4^+$. (b) Ammonia is a weak base, as illustrated by the very small change in the height of the purple $NH_3$ bar at equilibrium and the small amount of hydroxide ion produced, representing the small extent to which the reaction proceeds.

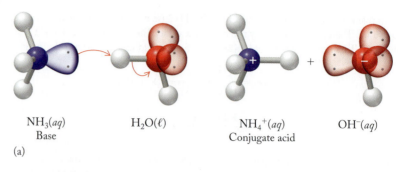

$NH_3(aq)$     $H_2O(\ell)$          $NH_4^+(aq)$          $OH^-(aq)$
Base                              Conjugate acid
(a)

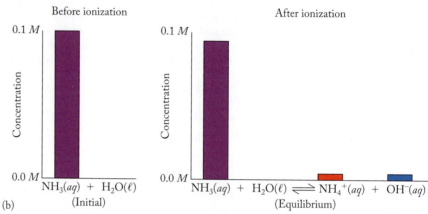

(b)

$NH_3$ rather than $NH_4^+$. The limited strength of ammonia as a base is reflected in its small $K_b$ value at 25°C:

$$K_b = \frac{[NH_4^+][OH^-]}{[NH_3]} = 1.76 \times 10^{-5} \qquad (16.18)$$

Note that no $[H_2O]$ term appears in the expression for $K_b$. As with the ionization of weak acids, the concentration of water does not change significantly in the course of the reaction in Equation 16.17. We can calculate $K_b$ if we know the initial base concentration and the solution pH, or we can work in the other direction and calculate pH from a known $K_b$ value. One additional step is necessary: the unknown in this equilibrium calculation is $[OH^-]$ instead of $[H^+]$, but once we know $[OH^-]$ we can take the negative logarithm of it to calculate pOH and then use that value in Equation 16.13 to calculate pH.

---

**SAMPLE EXERCISE 16.5** **Calculating the pH of a Solution of a Weak Base**   **LO3**

The concentration of $NH_3$ in household ammonia ranges between 50 and 100 g/L, or from about 3 $M$ to almost 6 $M$. What is the pH of a 3.0 $M$ solution of $NH_3$?

**Collect and Organize** We are asked to determine the pH of a 3.0 $M$ solution of ammonia. Equation 16.17 describes the basic behavior of ammonia in aqueous solutions.

**Analyze** The hydrolysis of ammonia produces $OH^-$ ions. We can calculate the equilibrium concentration of $OH^-$ ions using the $K_b$ expression and its value and then convert $[OH^-]$ to pOH and finally to pH. Given the 1:1:1 stoichiometry of $NH_3$, $NH_4^+$, and $OH^-$ in Equation 16.17, $[NH_4^+] = [OH^-]$ at equilibrium, and if that value is $x$, then the change in $[NH_3]$ during the course of the reaction is $-x$. The pH value of a fairly concentrated solution of a base with a $K_b$ value near $10^{-5}$ should be well above 7 but below 14.

**Solve** We begin by setting up a RICE table based on Equation 16.17 and letting $[NH_4^+] = [OH^-] = x$ at equilibrium:

| Reaction | $NH_3 + H_2O \rightleftharpoons$ | $NH_4^+$ + | $OH^-$ |
|---|---|---|---|
| | $[NH_3]$ (M) | $[NH_4^+]$ (M) | $[OH^-]$ (M) |
| Initial | 3.0 | 0.0 | 0.0 |
| Change | $-x$ | $+x$ | $+x$ |
| Equilibrium | $3.0 - x$ | $x$ | $x$ |

Because $K_b$ is small ($1.76 \times 10^{-5}$) relative to the initial concentration of base (3.0 M), we can make the simplifying assumption that $x$ is small compared with 3.0 M, so $3.0 - x \approx 3.0$. With this assumption, our equilibrium constant expression is

$$K_b = \frac{[NH_4^+][OH^-]}{[NH_3]} = \frac{(x)(x)}{(3.0)} = 1.76 \times 10^{-5}$$

Solving for $x$ gives us

$$x = [OH^-] = \sqrt{5.28 \times 10^{-5}} = 7.3 \times 10^{-3} \ M$$

Taking the negative logarithm of $[OH^-]$ to calculate pOH:

$$pOH = -\log[OH^-] = -\log(7.3 \times 10^{-3} \ M) = 2.14$$

We subtract this value from 14.00 to obtain the pH:

$$pH = 14.00 - pOH = 14.00 - 2.14 = 11.86$$

**Think About It** The calculated pH value falls in the range we predicted, given the small $K_b$ value but relatively high initial concentration of ammonia. To check our simplifying assumption, let's compare the value of $x$ to $[NH_3]_{initial}$ (3.0 M):

$$\frac{7.3 \times 10^{-3}}{3.0} = 0.0024 \times 100\% = 0.24\%$$

This small percentage is acceptable, which means our simplifying assumption was justified.

⚙ **Practice Exercise** What is the pH of a 0.200 M solution of methylamine ($CH_3NH_2$, $K_b = 4.4 = 10^{-4}$)?

## pH of Very Dilute Solutions

In the pH calculations thus far, we have not had to consider how much the auto-ionization of water contributes to the concentration of $H^+$ or $OH^-$. Let's now look at one case where we do have to take this into account.

Suppose we want to calculate the pH of $1.00 \times 10^{-8} \ M$ HCl. The acid is completely ionized so that $[H^+] = 1.00 \times 10^{-8}$ and pH (from Equation 16.11) is

$$pH = -\log[H^+] = -\log(1.00 \times 10^{-8}) = 8.000$$

This answer is not reasonable: how could a solution of a strong acid, no matter how dilute, have a weakly basic pH? We would expect the solution to be at least slightly acidic (pH < 7).

To calculate the pH of a solution this dilute, we must consider two sources of $H^+$ ions: ionization of the acid ($1.00 \times 10^{-8} \ M$) and the autoionization of water. Let's use $x$ to represent $[H^+]$ and $[OH^-]$ resulting from autoionization. The $[H^+]$ term in the $K_w$ expression is the sum of $x$ and $1.00 \times 10^{-8} \ M$:

$$K_w = [H^+][OH^-]$$
$$1.00 \times 10^{-14} = (x + 1.00 \times 10^{-8})(x)$$

Rearranging this equation to solve for $x$ gives:

$$x^2 + (1.00 \times 10^{-8}x) - (1.00 \times 10^{-14}) = 0$$

$$x = 9.5 \times 10^{-8} \ M$$

The concentration of hydrogen ion in the solution is therefore

$$[H^+] = (9.5 \times 10^{-8} \ M) + (1.00 \times 10^{-8} \ M) = 10.5 \times 10^{-8} \ M = 1.05 \times 10^{-7} \ M$$

and the pH is

$$pH = -\log(1.05 \times 10^{-7} \ M) = 6.98$$

This value agrees with our prediction that the solution should be slightly acidic.

## 16.4 Polyprotic Acids

Up to this point we have dealt with **monoprotic acids**, which have only one ionizable hydrogen atom per molecule. Acids that contain more than one ionizable hydrogen—such as sulfuric acid ($H_2SO_4$) and phosphoric acid ($H_3PO_4$)—are called **polyprotic acids**. For molecules with two and three ionizable hydrogen atoms, we use the more specific terms *diprotic acids* and *triprotic acids*, respectively. Let's first consider the acidic properties of the strong diprotic acid, sulfuric acid.

### Acid Precipitation

Among the consequences of fossil fuel combustion is the formation of $SO_3$ from sulfur impurities in coal. Sulfur trioxide dissolves in water to give sulfuric acid, the principal contributor to acid rain (see Figure 16.8). Sulfuric acid is a strong acid (Table 16.1) because ionization of the first proton is complete ($K_{a_1} \gg 1$):

$$H_2SO_4(aq) \rightarrow HSO_4^-(aq) + H^+(aq) \qquad (16.19)$$

However, the second ionization step is not complete:

$$HSO_4^-(aq) \rightleftharpoons SO_4^{2-}(aq) + H^+(aq) \qquad K_{a_2} = 1.2 \times 10^{-2} \quad (16.20)$$

Note that these equilibrium constant symbols have an additional subscript, 1 or 2, corresponding to the loss of first one and then a second $H^+$ ion per molecule.

The combination of one complete and one incomplete ionization means that many solutions of $H_2SO_4(aq)$ contain more than 1 mole but less than 2 moles of $H^+(aq)$ for every mole of $H_2SO_4$ dissolved. Let's be more quantitative about this and determine the pH of a 0.100 $M$ solution of $H_2SO_4$. The starting point in this calculation is a solution in which all the $H_2SO_4$ has ionized, as described in Equation 16.19. Therefore, as the second step begins, $[HSO_4^-] = [H^+] = 0.100 \ M$. Ionization of $HSO_4^-$ (Equation 16.20) produces additional $H^+$ ions. To analyze the effect of the second ionization, we set up a RICE table in which $+x$ is the change in $[H^+]$ produced by the second ionization. Given the 1:1:1 stoichiometry of the balanced equation for this step, the change in $[SO_4^{2-}]$ is also $+x$, and the change in $[HSO_4^-]$ is $-x$. Inserting these values in the RICE table and completing the third row:

**monoprotic acid** has one ionizable hydrogen atom per molecule.

**polyprotic acid** has two or more ionizable hydrogen atoms per molecule.

| Reaction | $HSO_4^-$ | $\rightleftharpoons$ | $SO_4^{2-}$ | + | $H^+$ |
|---|---|---|---|---|---|
| | $[HSO_4^-]$ (M) | | $[SO_4^{2-}]$ (M) | | $[H^+]$ (M) |
| Initial | 0.100 | | 0.000 | | 0.100 |
| Change | $-x$ | | $+x$ | | $+x$ |
| Equilibrium | $0.100 - x$ | | $x$ | | $0.100 + x$ |

We insert the equilibrium concentrations in the equilibrium constant expression for $K_{a_2}$:

$$K_{a_2} = \frac{[H^+][SO_4^{2-}]}{[HSO_4^-]} = \frac{(0.100 + x)(x)}{(0.100 - x)} = 1.2 \times 10^{-2}$$

Solving for $x$ using the quadratic equation (or a solver program), we get

$$0.100x + x^2 = 1.2 \times 10^{-3} - (1.2 \times 10^{-2}\, x)$$
$$x^2 + 0.112x - 1.2 \times 10^{-3} = 0$$
$$x = \frac{-0.112 \pm \sqrt{(0.112)^2 - 4(1)(-1.2 \times 10^{-3})}}{2(1)}$$
$$x = +9.85 \times 10^{-3} \text{ or } -0.122$$

The negative value for $x$ has no physical meaning because it gives us a negative $[SO_4^{2-}]$ value. Therefore, we use only the positive $x$ value. At equilibrium,

$$[H^+] = (0.100 + x)\, M = (0.100 + 0.00985)\, M = 0.10985\, M = 0.110\, M$$

The corresponding pH is

$$pH = -\log[H^+] = -\log(0.110\, M) = 0.96$$

As predicted, the value of $[H^+]$ is between one and two times the initial concentration of $H_2SO_4$. The degree of ionization of $HSO_4^-$ is

$$\frac{[SO_4^{2-}]_{equilibrium}}{[HSO_4^-]_{initial}} = \frac{0.010\, M}{0.100\, M} \times 100\% = 10\%$$

Therefore we were correct in our decision not to use the simplifying assumption to avoid solving a quadratic equation.

**CONCEPT TEST** ••••••••••••••••••••••••••••••••••••••••••••••••••••••••

Identify all of the species present in an aqueous solution of phosphoric acid, $H_3PO_4$.

••••••••••••••••••••••••••••••••••••••••••••••••••••••••••••••••••••••••••

## Acidification of the Ocean

As we discussed in Chapters 5 and 9, the combustion of fossil fuels increases the concentration of carbon dioxide, a greenhouse gas, in the atmosphere. Increased $CO_2$ concentrations also have an effect on the acidity of Earth's oceans, because $CO_2$ forms carbonic acid when it dissolves in water (Figure 16.11).

Calculating the pH of a solution of a weak diprotic acid, such as carbonic or sulfurous acid, is actually easier than the above calculation for sulfuric acid. To see why, look closely at the $K_a$ values in Table 16.5. In each of the two pairs of values, $K_{a_2}$ is much smaller than $K_{a_1}$. We can rationalize the difference on the basis

**FIGURE 16.11** Increasing $CO_2$ concentration in the atmosphere means that the partial pressure of $CO_2$ increases. By Henry's law, as $P_{CO_2}$ increases, so does the solubility of $CO_2$ in water, pushing the equilibrium between $CO_2$, hydrogen carbonate ion, and $H^+$ to the right, reducing the pH of the oceans. One of the consequences of a lower pH is the potential for damage to shellfish by eroding their $CaCO_3$ shells.

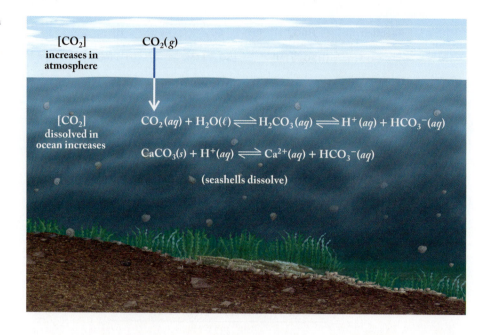

| TABLE 16.5 | Ionization Equilibria for Two Diprotic Acids | |
|---|---|---|
| **Acid** | **Ionization Equilibria** | **$K_a$** |
| Carbonic acid | Step 1: $H_2CO_3(aq) \rightleftharpoons HCO_3^-(aq) + H^+(aq)$ | $K_{a_1} = 4.3 \times 10^{-7}$ |
| | Step 2: $HCO_3^-(aq) \rightleftharpoons CO_3^{2-}(aq) + H^+(aq)$ | $K_{a_2} = 4.7 \times 10^{-11}$ |
| Sulfurous acid | Step 1: $H_2SO_3(aq) \rightleftharpoons HSO_3^-(aq) + H^+(aq)$ | $K_{a_1} = 1.7 \times 10^{-2}$ |
| | Step 2: $HSO_3^-(aq) \rightleftharpoons SO_3^{2-}(aq) + H^+(aq)$ | $K_{a_2} = 6.2 \times 10^{-8}$ |

of electrostatic attractions between oppositely charged ions. The first ionization produces a negatively charged oxoanion—$HCO_3^-$ or $HSO_3^-$. The second ionization requires that a positive ion ($H^+$) dissociate from a negative ion to produce an even more negative oxoanion. Separating oppositely charged ions that are naturally attracted to each other is not a process that we would expect to be favored, and the smaller values for $K_{a_2}$ confirm our expectations. In general, the $K_{a_2}$ of any diprotic acid is less, and often much less, than $K_{a_1}$. The consequence of these large differences is that essentially all of the limited strength of weak polyprotic acids is due to the first ionization reaction.

To see how this separation of ionization steps plays out, let's focus on carbonic acid ($H_2CO_3$), which is present in every drop of rain that falls from the sky. It gets there because the atmosphere is about 0.039% (by volume) $CO_2$. Carbon dioxide is slightly soluble in water, and when it dissolves it forms carbonic acid:

$$CO_2(g) + H_2O(\ell) \rightleftharpoons H_2CO_3(aq)$$

Actually, the $H_2CO_3$ molecule is not stable in aqueous solutions, but we write it as a convenience to show how dissolving $CO_2$ in water produces an acidic solution. The net result of $CO_2$ dissolving in water is

$$CO_2(aq) + H_2O(\ell) \rightleftharpoons HCO_3^-(aq) + H^+(aq)$$

In Sample Exercise 16.6 we use this equilibrium to calculate the natural pH of rainwater.

SAMPLE EXERCISE 16.6   **Calculating the pH of a Solution of a Weak Diprotic Acid**   LO2

What is the pH of rainwater at 25°C in which atmospheric $CO_2$ has dissolved, producing a constant $[H_2CO_3]$ of $1.2 \times 10^{-5}$ $M$?

**Collect and Organize** We are asked to determine the pH of a dilute solution of $H_2CO_3$. There are two ionizable H atoms in $H_2CO_3$; the $K_{a_1}$ and $K_{a_2}$ values are given in Table 16.5. Any $H_2CO_3$ consumed by the reaction is replaced by the dissolution of more $CO_2$, so that $[H_2CO_3]$ remains a constant $1.2 \times 10^{-5}$ $M$.

**Analyze** The large difference between the $K_{a_1}$ and $K_{a_2}$ values indicates that the pH of the solution is controlled by the first ionization equilibrium:

$$H_2CO_3(aq) \rightleftharpoons HCO_3^-(aq) + H^+(aq) \qquad K_{a_1} = 4.3 \times 10^{-7}$$

Because of the small value of $K_{a_1}$ and the small concentration of $H_2CO_3$, we should obtain a pH value that is less than 7 but closer to 7 than to 0.

**Solve** First we set up a RICE table in which $x = [H^+] = [HCO_3^-]$ at equilibrium and the value of $[H_2CO_3]$ at equilibrium is $1.2 \times 10^{-5}$ $M$.

| Reaction | $H_2CO_3$ | $\rightleftharpoons$ | $HCO_3^-$ | + | $H^+$ |
|---|---|---|---|---|---|
| | $[H_2CO_3]$ (M) | | $[HCO_3^-]$ (M) | | $[H^+]$ (M) |
| Initial | $1.2 \times 10^{-5}$ | | 0 | | 0 |
| Change | 0 | | $+x$ | | $+x$ |
| Equilibrium | $1.2 \times 10^{-5}$ | | $x$ | | $x$ |

$$K_{a_1} = \frac{[HCO_3^-][H^+]}{[H_2CO_3]} = \frac{(x)(x)}{1.2 \times 10^{-5}} = 4.3 \times 10^{-7}$$

$$x = [H^+] = 2.3 \times 10^{-6} \, M$$

We take the negative logarithm of $[H^+]$ to calculate pH:

$$pH = -\log[H^+] = -\log(2.3 \times 10^{-6} \, M) = 5.64$$

**Think About It** Carbonic acid is a weak acid, and its concentration here is small, so obtaining a pH value that is only about 1.4 units below neutral pH (7.00) is reasonable.

**Practice Exercise** The pH value in Sample Exercise 16.6 is not far from 7.00, and it raises the question of whether the autoionization of water that produces a $[H^+]$ of $1.00 \times 10^{-7}$ $M$ contributes significantly to $[H^+]$ in the rainwater sample. Recalculate the pH of the rainwater sample taking into account the autoionization of water.

Some acids have three ionizable H atoms per molecule. Two important triprotic acids are phosphoric acid, $H_3PO_4$, and citric acid, the acid responsible for the tart flavor of citrus fruits. Note in Table 16.6 how $K_{a_1} > K_{a_2} > K_{a_3}$ for both acids. This pattern is much like that for the $K_{a_1} > K_{a_2}$ values of diprotic acids and for the same reason: it is more difficult to remove a second $H^+$ ion from the negatively charged ion formed after the first $H^+$ ion is removed, and it is even more difficult to remove a third $H^+$ ion from an ion with a 2− charge.

**CONCEPT TEST**

Do you expect the second or third acid ionization steps in phosphoric acid and citric acid to influence the pH of 0.100 $M$ solutions of either acid?

| TABLE 16.6 | Ionization Equilibria for Two Triprotic Acids | |
|---|---|---|
| **Phosphoric Acid** | | |
| (1) $HO-\overset{\overset{O}{\parallel}}{\underset{\underset{OH}{\mid}}{P}}-OH \rightleftharpoons HO-\overset{\overset{O}{\parallel}}{\underset{\underset{OH}{\mid}}{P}}-O^- + H^+$ | | $K_{a_1} = 6.9 \times 10^{-3}$ |
| (2) $HO-\overset{\overset{O}{\parallel}}{\underset{\underset{OH}{\mid}}{P}}-O^- \rightleftharpoons {}^-O-\overset{\overset{O}{\parallel}}{\underset{\underset{OH}{\mid}}{P}}-O^- + H^+$ | | $K_{a_2} = 6.4 \times 10^{-8}$ |
| (3) ${}^-O-\overset{\overset{O}{\parallel}}{\underset{\underset{OH}{\mid}}{P}}-O^- \rightleftharpoons {}^-O-\overset{\overset{O}{\parallel}}{\underset{\underset{O^-}{\mid}}{P}}-O^- + H^+$ | | $K_{a_3} = 4.8 \times 10^{-13}$ |
| **Citric Acid** | | |
| (1) $HO-\overset{\overset{CH_2COOH}{\mid}}{\underset{\underset{CH_2COOH}{\mid}}{C}}-COOH \rightleftharpoons HO-\overset{\overset{CH_2COO^-}{\mid}}{\underset{\underset{CH_2COOH}{\mid}}{C}}-COOH + H^+$ | | $K_{a_1} = 7.4 \times 10^{-4}$ |
| (2) $HO-\overset{\overset{CH_2COO^-}{\mid}}{\underset{\underset{CH_2COOH}{\mid}}{C}}-COOH \rightleftharpoons HO-\overset{\overset{CH_2COO^-}{\mid}}{\underset{\underset{CH_2COOH}{\mid}}{C}}-COO^- + H^+$ | | $K_{a_2} = 1.7 \times 10^{-5}$ |
| (3) $HO-\overset{\overset{CH_2COO^-}{\mid}}{\underset{\underset{CH_2COOH}{\mid}}{C}}-COO^- \rightleftharpoons HO-\overset{\overset{CH_2COO^-}{\mid}}{\underset{\underset{CH_2COO^-}{\mid}}{C}}-COO^- + H^+$ | | $K_{a_3} = 4.0 \times 10^{-7}$ |

▶‖ **CHEMTOUR** Acid Strength and Molecular Structure

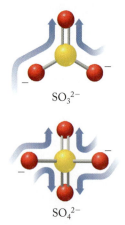

**FIGURE 16.12** Sulfuric acid ($H_2SO_4$) is a stronger acid than sulfurous acid ($H_2SO_3$) because of the greater stability that comes with delocalizing the negative charge of a $SO_4^{2-}$ ion over more atoms (shown by the curved blue arrows).

# 16.5 Acid Strength and Molecular Structure

In Section 16.1, we noted that nitric acid ($HNO_3$) is a strong acid but nitrous acid ($HNO_2$) is weak. Similarly, sulfuric acid ($H_2SO_4$) is a strong acid, but sulfurous acid ($H_2SO_3$) is weak. The reason for these differences in strength lies in subtle differences in molecular structure (Figure 16.12). The ionizable hydrogen atoms in both $H_2SO_3$ and $H_2SO_4$ molecules are bonded to oxygen atoms that are also bonded to the central sulfur atoms. The difference between the two is that the central sulfur atom is also bonded to either one (in $H_2SO_3$) or two (in $H_2SO_4$) other oxygen atoms.

Recall from Section 8.3 that oxygen is the second most electronegative element (after fluorine). This means that oxygen atoms bonded to the central atom of an oxoacid attract electron density toward themselves. The more electron density that is drawn away from the O—H groups, the more spread out (or *delocalized*) is the negative charge on the anion that forms when a $H^+$ ion is lost. Spreading out charge over more atoms has a stabilizing effect on the anion. Thus, $SO_4^{2-}$ ions are more stable than $SO_3^{2-}$ ions, making $H_2SO_4$ a stronger acid than $H_2SO_3$.

This trend of increasing acid strength with increasing numbers of oxygen atoms bonded to the central atom (that is, with increasing oxidation number of the central atom) is true for all oxoacids with the same central atom. The trend is illustrated by the strong acidity of $HNO_3$ and the weak acidity of $HNO_2$, and by the strengths of the oxoacids of chlorine, shown in Figure 16.13.

The strength of an oxoacid is also related to the electron-withdrawing power of the central atom. Consider, for example, the relative strengths of the three hypohalous acids in Figure 16.14. The most electronegative of the three halogen atoms (Cl) has the greatest attraction for the pair of electrons it shares with oxygen. This attraction draws electron density away from hydrogen toward chlorine and toward the oxygen end of the already polar O—H bond. These shifts in electron density make the hypochlorite ($ClO^-$) ions better able to bear a negative charge because the charge is more delocalized. Thus, $HClO(aq)$ is the strongest of the three acids, followed by hypobromous acid [$HBrO(aq)$] and hypoiodous acid [$HIO(aq)$].

**CONCEPT TEST**

Rank the following compounds in order of decreasing acid strength: $H_3PO_4$, $H_3AsO_4$, $H_3SbO_4$, and $H_3BiO_4$.

| Acid | Structure | Oxidation number of Cl | $K_a$ |
|---|---|---|---|
| Hypochlorous HClO | 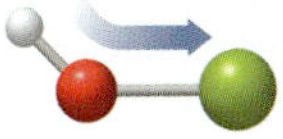 | +1 | $2.9 \times 10^{-8}$ |
| Chlorous HClO$_2$ | 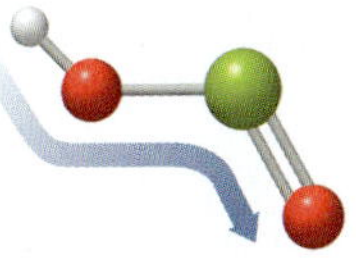 | +3 | $1.1 \times 10^{-2}$ |
| Chloric HClO$_3$ | 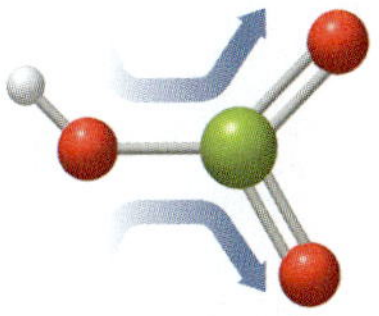 | +5 | ~1 |
| Perchloric HClO$_4$ | 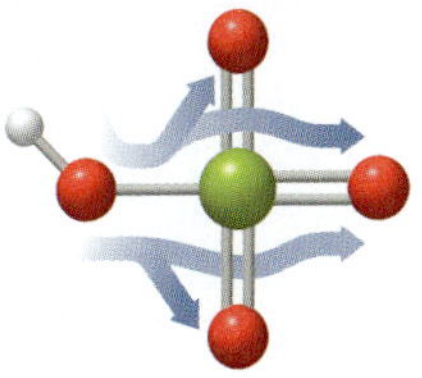 | +7 | Strong acid |

**FIGURE 16.13** In the oxoacids of chlorine, acid strength increases with increasing Cl oxidation number. The higher the oxidation number, the greater the number of O atoms bonded to the Cl. The greater the number of O atoms bonded to Cl, the greater the ability to delocalize the negative charge on the anion created when each acid loses its H.

**CONNECTION** Oxidation numbers were introduced in Chapter 4, Section 4.9.

**CONNECTION** The electronegativities of the elements were given in Chapter 8 (Figure 8.5).

# 16.6 pH of Salt Solutions

Seawater and the freshwater in many rivers and lakes have pH values that range from weakly basic to weakly acidic. How can these waters be more basic than the acidic rainwater (pH ≤ 5.6) that serves, directly or indirectly, as their water supply? The answer is that, when rain soaks into the ground, its pH changes as it flows through soils that contain basic components. To understand the chemical processes that produce neutral or slightly basic groundwater, we first need to examine the acid–base properties of some common ionic compounds present in these waters.

As we discussed in Chapter 4, soluble ionic compounds separate into their component ions when they dissolve in water. For example, a 0.01 $M$ solution of NaCl contains 0.01 $M$ Na$^+$ ions and 0.01 $M$ Cl$^-$ ions. It is also a neutral solution. Neither Na$^+$ ions nor Cl$^-$ ions hydrolyze to form either $H_3O^+$ or $OH^-$ ions. Recall that the Cl$^-$ ion is the conjugate base of a strong acid (HCl). Therefore, the Cl$^-$ ion must be a very weak Brønsted–Lowry base (Figure 16.4).

Now let's consider another sodium salt, NaF. When it dissolves in water it produces F$^-$ ions, which are the conjugate base of the *weak* acid, HF. Therefore, F$^-$ ions should be weakly basic, producing at least some $OH^-$ ions when they dissolve in water:

$$F^-(aq) + H_2O(\ell) \rightleftharpoons HF(aq) + OH^-(aq)$$

Thus, solutions of NaF are weakly basic.

| Acid | Structure | Electronegativity of halogen atom | $K_a$ |
|---|---|---|---|
| Hypochlorous HClO | 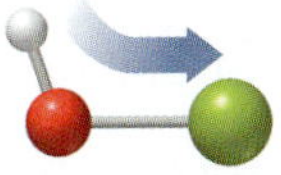 | 3.0 | $2.9 \times 10^{-8}$ |
| Hypobromous HBrO |  | 2.8 | $2.3 \times 10^{-9}$ |
| Hypoiodous HIO | 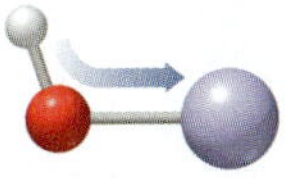 | 2.5 | $2.3 \times 10^{-11}$ |

**FIGURE 16.14** The strengths of these three hypohalous acids are related to the electronegativities of their halogen atoms. The more electronegative the halogen atom, the more it pulls electron density (blue arrows) away from the hydrogen end of the molecule. The less electron density at the H atom, the more easily the H ionizes and the stronger the acid.

If salts that contain the conjugate bases of weak acids can be basic, then it is logical that salts that contain the conjugate acids of weak bases can be acidic. An example of such a salt is $NH_4Cl$. We have seen that the $Cl^-$ ions that are produced when $NH_4Cl$ dissolves have negligible strengths as Brønsted–Lowry bases. However, $NH_4^+$ ions are the conjugate acid of a weak base, $NH_3$. Therefore, $NH_4^+$ ions should be weakly acidic, producing at least some $H_3O^+$ ions:

$$NH_4^+(aq) + H_2O(\ell) \rightleftharpoons NH_3(aq) + H_3O^+(aq)$$

Consequently, solutions of $NH_4Cl$ are weakly acidic.

Table 16.7 summarizes how salts can be acidic, basic, or neutral depending on whether they include cations that are the conjugate acids of weak bases, or anions that are the conjugate bases of weak acids, or both. Note that salts that contain both the conjugate base of a weak acid *and* the conjugate acid of a weak base may be acidic, basic, or neutral, depending on the relative strengths of the acid and base. Ammonium acetate represents the rare example of a salt in which the strengths of the acid (acetic acid) and the base (ammonia) happen to be *exactly the same* [$K_a$ (acetic acid) = $K_b$ (ammonia) = $1.76 \times 10^{-5}$]. As a result, ammonium acetate is a neutral salt.

**TABLE 16.7  Acid–Base Properties of Some Common Salts**

| Anion Is Derived from a | Cation Is Derived from a | pH of Aqueous Solutions | Example |
|---|---|---|---|
| Strong acid | Strong base | 7 | NaCl |
| Strong acid | Weak base | <7 | $NH_4Cl$ |
| Weak acid | Strong base | >7 | NaF |
| Weak acid | Weak base | Depends on relative values of $pK_a$ and $pK_b$ | $pK_a > pK_b$, acidic; $NH_4F$ $pK_b > pK_a$, basic; $NH_4HCO_3$ $pK_a = pK_b$, neutral; $CH_3COONH_4$ |

**CONCEPT TEST**

When ammonium fluoride is heated to about 100°C it decomposes, forming ammonia and ammonium hydrogen fluoride:

$$2\,NH_4F(s) \rightarrow NH_3(g) + (NH_4)HF_2(s)$$

Do you think $(NH_4)HF_2$ is more acidic than $NH_4F$? Why?

**SAMPLE EXERCISE 16.7  Distinguishing Acidic, Basic, and Neutral Salts**  LO4

Is an aqueous solution of NaClO acidic, basic, or neutral?

**Collect and Organize** We are asked whether a solution of NaClO is acidic, basic, or neutral. When this salt dissolves in water, it dissociates into $Na^+$ ions and $ClO^-$ ions.

**Analyze** Sodium ions do not hydrolyze and hence do not produce acidic solutions in water. However, $ClO^-$ ions are the conjugate base of HClO, which is a weak acid (Table 16.2). Therefore, $ClO^-$ ions should partially hydrolyze in water, forming $OH^-$ ions:

$$ClO^-(aq) + H_2O(\ell) \rightleftharpoons HClO(aq) + OH^-(aq)$$

**Solve** Because hydrolysis of $ClO^-$ ions produces $OH^-$ ions, solutions of NaClO are weakly basic.

**Think About It** Any sodium salt that contains an anion that is the conjugate base of a weak acid produces weakly basic aqueous solutions.

⚙ **Practice Exercise** Write a chemical equation for the hydrolysis reaction that explains why an aqueous solution of $K_2SO_4$ is basic.

⬤⬤⬤ ■

Having established that salts can be acidic, basic, or neutral, let's now explore a strategy for calculating the pH values of their aqueous solutions. Our strategy is much like the approach we have taken to calculate the pH values of solutions of weak acids and bases. Let's start with a 0.100 $M$ solution of sodium carbonate, $Na_2CO_3$. (We start with a carbonate salt because the pH of many natural waters—and biological systems—is controlled by the presence of $CO_3^{2-}$ ions and their conjugate acid, $HCO_3^-$ ions.) The $[CO_3^{2-}]$ and $[HCO_3^-]$ values in a solution are linked by the second acid ionization reaction of $H_2CO_3$ (see Table 16.5):

$$HCO_3^-(aq) + H_2O(\ell) \rightleftharpoons CO_3^{2-}(aq) + H_3O^+(aq)$$

$$K_{a_2} = 4.7 \times 10^{-11} \tag{16.21}$$

Carbonate and bicarbonate are also linked by the chemical reaction in which the carbonate ion acts like a Brønsted–Lowry base:

$$CO_3^{2-}(aq) + H_2O(\ell) \rightleftharpoons HCO_3^-(aq) + OH^-(aq) \tag{16.22}$$

None of the tables in this chapter or Appendix 5 contains the $K_b$ value for the reaction in Equation 16.22, so we have to calculate it. We start with the equilibrium constant expression for the reaction:

$$K_{b_1} = \frac{[HCO_3^-][OH^-]}{[CO_3^{2-}]}$$

We have labeled this constant $K_{b_1}$ because it describes the first of two possible hydrolysis reactions that release $OH^-$ ions. In the second, the bicarbonate ion produced in the first reaction also acts like a Brønsted–Lowry base:

$$HCO_3^-(aq) + H_2O(\ell) \rightleftharpoons H_2CO_3(aq) + OH^-(aq)$$

We give the equilibrium constant for this reaction the symbol $K_{b_2}$:

$$K_{b_2} = \frac{[H_2CO_3][OH^-]}{[HCO_3^-]}$$

To calculate the values of $K_{b_1}$ and $K_{b_2}$, we begin with the $K_{a_1}$ and $K_{a_2}$ equilibrium constant expressions for $H_2CO_3$:

$$K_{a_1} = \frac{[HCO_3^-][H^+]}{[H_2CO_3]} = 4.3 \times 10^{-7} \qquad K_{a_2} = \frac{[CO_3^{2-}][H^+]}{[HCO_3^-]} = 4.7 \times 10^{-11}$$

Now let's compare the $K_{b_1}$ expression for the carbonate ion and the $K_{a_2}$ expression for carbonic acid:

$$K_{b_1} = \frac{[HCO_3^-][OH^-]}{[CO_3^{2-}]} \qquad K_{a_2} = \frac{[CO_3^{2-}][H^+]}{[HCO_3^-]} = 4.7 \times 10^{-11}$$

Their similarity becomes more apparent when we write the reciprocal of the $K_{a_2}$ expression:

$$K_{b_1} = \frac{[HCO_3^-][OH^-]}{[CO_3^{2-}]} \qquad \frac{1}{K_{a_2}} = \frac{[HCO_3^-]}{[CO_3^{2-}][H^+]} = \frac{1}{4.7 \times 10^{-11}}$$

**CONNECTION** We noted in Chapter 2 that *bicarbonate* is a more common name for the $HCO_3^-$ ion than *hydrogen carbonate*.

The only difference between the two is the $[OH^-]$ term in the $K_{b_1}$ expression and the $[H^+]$ term in the $1/K_{a_2}$ expression. As we have seen, $[H^+]$ and $[OH^-]$ are linked by $K_w$ (Equation 16.10). Consider what happens when we multiply $1/K_{a_2}$ by $K_w$:

$$\frac{1}{K_{a_2}} \times K_w = \left(\frac{[HCO_3^-]}{[H^+][CO_3^{2-}]}\right)([H^+][OH^-]) = \frac{[HCO_3^-][OH^-]}{[CO_3^{2-}]}$$

The resulting expression is the $K_{b_1}$ expression for the carbonate ion. We know the values of $K_w$ and $K_{a_2}$, so we can calculate $K_{b_1}$:

$$K_{b_1} = \frac{K_w}{K_{a_2}} = \frac{1.0 \times 10^{-14}}{4.7 \times 10^{-11}} = 2.1 \times 10^{-4}$$

The inverse relation between the $K_{a_2}$ of $H_2CO_3$ and the $K_{b_1}$ of $CO_3^{2-}$ also holds for the $K_{a_1}$ of $H_2CO_3$ and the $K_{b_2}$ of $CO_3^{2-}$. The connection between the acidic strength of $H_2CO_3$ and the basic strength of its conjugate base, $HCO_3^-$, is given by

$$K_{b_2} = \frac{K_w}{K_{a_1}} = \frac{1.0 \times 10^{-14}}{4.3 \times 10^{-7}} = 2.3 \times 10^{-8}$$

The $K_w$ connection between $K_a$ and $K_b$ values for carbonic acid equilibria applies to any conjugate acid–base pair:

$$K_b = \frac{K_w}{K_a}$$

$$\text{or} \quad K_a \times K_b = K_w \tag{16.23}$$

Equation 16.23 reinforces the complementary nature of an acid and its conjugate base: as the strength ($K_a$) of the acid increases, the strength ($K_b$) of its conjugate base decreases, and vice versa (see Figure 16.4). In the carbonate and carbonic acid examples, the $CO_3^{2-}$ ion is a stronger base ($K_{b_1} = 2.1 \times 10^{-4}$) than its conjugate acid, the $HCO_3^-$ ion, is an acid ($K_{a_2} = 4.7 \times 10^{-11}$). However, the $HCO_3^-$ ion is a weaker base ($K_{b_2} = 2.3 \times 10^{-8}$) than $H_2CO_3$ is an acid ($K_{a_1} = 4.3 \times 10^{-7}$).

Now that we have a value of $K_{b_1}$ for the carbonate ion, we can calculate the pH of the 0.100 $M$ solution of $Na_2CO_3$. In setting up a RICE table, we assume that the only important source of $OH^-$ is hydrolysis of the carbonate ion and that the autoionization of water does not contribute significantly to $[OH^-]$ at equilibrium. Let $x$ be the equilibrium value of $[OH^-]$. Then $[HCO_3^-]$ also is $x$. The changes in the two must both be $+x$, and the change in $[CO_3^{2-}]$ must be $-x$. Completing the RICE table, we have

| Reaction | $CO_3^{2-}$ | $\rightleftharpoons$ | $HCO_3^-$ | + | $OH^-$ |
|---|---|---|---|---|---|
| | $[CO_3^{2-}]$ (M) | | $[HCO_3^-]$ (M) | | $[OH^-]$ (M) |
| Initial | 0.100 | | 0 | | 0 |
| Change | $-x$ | | $+x$ | | $+x$ |
| Equilibrium | $0.100 - x$ | | $x$ | | $x$ |

Solving for $x$:

$$K_{b_1} = 2.1 \times 10^{-4} = \frac{[HCO_3^-][OH^-]}{[CO_3^{2-}]} = \frac{(x)(x)}{(0.100 - x)}$$

$$x = 4.5 \times 10^{-3} \, M = [OH^-]$$

The calculated [OH⁻] is much greater than [OH⁻] in pure water ($1.0 \times 10^{-7}\ M$), so the assumption that water autoionization is unimportant in this calculation is valid. To calculate pH from [OH⁻], we first calculate pOH:

$$pOH = -\log[OH^-] = -\log(4.5 \times 10^{-3}) = 2.35$$

and then use Equation 16.13 to calculate pH:

$$pH = pK_w - pOH = 14.00 - 2.35 = 11.65$$

A pH of 11.65 is quite basic. If you swam in a pool of water at that pH, you would experience skin irritation and painful burning in your eyes. The hydrolysis of carbonate and silicate minerals is responsible for the alkalinity of hot springs in various parts of the world. For example, Octopus Spring in Yellowstone National Park (Figure 16.15) has a pH close to 9.

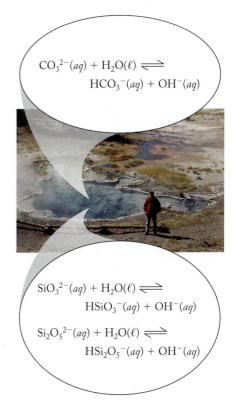

$$CO_3^{2-}(aq) + H_2O(\ell) \rightleftharpoons$$
$$HCO_3^-(aq) + OH^-(aq)$$

$$SiO_3^{2-}(aq) + H_2O(\ell) \rightleftharpoons$$
$$HSiO_3^-(aq) + OH^-(aq)$$

$$Si_2O_5^{2-}(aq) + H_2O(\ell) \rightleftharpoons$$
$$HSi_2O_5^-(aq) + OH^-(aq)$$

**FIGURE 16.15** Octopus Spring in the Lower Geyser Basin of Yellowstone National Park owes its name to its octopus-like shape. It is also an example of the many alkaline hot springs in the area that have waters ranging from pH 8 to 9. Despite the high pH, which has been traced to high concentrations of carbonate and silicate minerals dissolving in water to form hydroxide ion, the spring supports orange and yellow mats of thermophilic bacteria.

**CONCEPT TEST**

In the pH calculations in this chapter we routinely ignore the concentrations of H₃O⁺ and OH⁻ ions produced by the autoionization of water. Suppose calculations of the pH of six different salt solutions produced the results shown in the table. Which, if any, of the calculations should have taken into account the autoionization of water to obtain an accurate result?

| Solution | pH |
|----------|-------|
| A | 2.66 |
| B | 4.12 |
| C | 6.39 |
| D | 7.27 |
| E | 9.10 |
| F | 12.88 |

**SAMPLE EXERCISE 16.8** **Calculating the pH of a Solution of an Acidic Salt** **LO5**

What is the pH of 0.25 $M$ NH₄Cl?

**Collect and Organize** We are to calculate the pH of a solution of NH₄Cl. When NH₄Cl dissolves in water, NH₄⁺ and Cl⁻ ions are released into solution. The NH₄⁺ ion is the conjugate acid of NH₃; the Cl⁻ ion is the conjugate base of HCl.

**Analyze** We have seen that the Cl⁻ ion has negligible strength as a Brønsted–Lowry base, so it does not contribute to the acid–base properties of NH₄Cl. Ammonia is a weak base, which means that its conjugate acid, NH₄⁺, is a weak acid. Therefore, some of the ammonium ions in solution donate H⁺ ions to molecules of water:

$$NH_4^+(aq) + H_2O(\ell) \rightleftharpoons NH_3(aq) + H_3O^+(aq)$$

The $K_a$ value of NH₄⁺ is not given in the problem and is not listed in Appendix 5. However, the $K_b$ value of its conjugate base, NH₃, is given in Table A5.3: $1.76 \times 10^{-5}$. We can calculate the value of $K_a$ from $K_b$ using Equation 16.23.

The $K_b$ value of ammonia is close to $10^{-5}$. The product of $K_a \times K_b$ of a conjugate acid–base pair is $10^{-14}$; therefore, the $K_a$ value of the ammonium ion will be close to $10^{-9}$. Because the ammonium ion is a very weak acid, we can anticipate a pH value that is less than 7 but probably closer to 7 than to 0.

**Solve** The simplified $K_a$ expression for the NH₄⁺ ion is

$$K_a = \frac{[NH_3][H^+]}{[NH_4^+]}$$

Rearranging Equation 16.23 to solve for $K_a$:

$$K_a = \frac{K_w}{K_b} = \frac{1.00 \times 10^{-14}}{1.76 \times 10^{-5}} = 5.68 \times 10^{-10} = \frac{[NH_3][H^+]}{[NH_4^+]}$$

We set up a RICE table in which we make the usual assumptions that the reaction is the only significant source of $H^+$ and that $x = [H^+] = [NH_3]$ at equilibrium:

| Reaction | $NH_4^+$ | $\rightleftharpoons$ | $NH_3$ | $+$ | $H^+$ |
|---|---|---|---|---|---|
| | $[NH_4^+]$ (M) | | $[NH_3]$ (M) | | $[H^+]$ (M) |
| Initial | 0.25 | | 0 | | 0 |
| Change | $-x$ | | $+x$ | | $+x$ |
| Equilibrium | $0.25 - x$ | | $x$ | | $x$ |

$$K_a = 5.68 \times 10^{-10} = \frac{[NH_3][H^+]}{[NH_4^+]} = \frac{(x)(x)}{0.25 - x}$$

Given the very small value of $K_a$, we can make the simplifying assumption that $(0.25\ M - x) \approx 0.25\ M$, which gives us

$$\frac{x^2}{0.25} = 5.68 \times 10^{-10}$$

$$x^2 = 1.42 \times 10^{-10}$$

$$x = 1.19 \times 10^{-5} = [H^+]$$

$$pH = -\log[H^+] = -\log(1.19 \times 10^{-5}) = 4.92$$

**Think About It** This result matches our prediction: the pH of the solution is less than 7, but closer to 7 than to 0. The calculated $[H^+]$ is less than 5% of the initial concentration of $NH_4^+$, so our simplifying assumption was valid.

⚙ **Practice Exercise** What is the pH of a 0.25 $M$ solution of sodium acetate? (*Hint:* The acetate ion is the conjugate base of acetic acid.)

The reactions of carbonate minerals in soils and rocks with acidic groundwater are critically important in mitigating the effects of acidic precipitation. When, for example, rain containing dilute sulfuric acid soaks into soil containing $CaCO_3$, in the form of limestone, marble, or shellfish shells, the acid can be converted into environmentally more benign carbonic acid:

$$CaCO_3(s) + H_2SO_4(aq) \rightleftharpoons CaSO_4(s) + H_2CO_3(aq)$$

or, if enough $CaCO_3$ is available, into calcium sulfate and soluble calcium hydrogen carbonate:

$$2\,CaCO_3(s) + H_2SO_4(aq) \rightleftharpoons CaSO_4(s) + Ca(HCO_3)_2(aq)$$

∞ **CONNECTION** The reaction of acidic groundwater with calcium carbonate and its connection to the formation of limestone caves were described in Chapter 4.

As long as carbonates and other basic substances are present in soils and in the sediments of rivers and lakes, nature has the capacity to neutralize the acid in acid rain and maintain pH in a range that supports aquatic life.

**CONCEPT TEST**

A mineral called dolomite contains a mixture of calcium carbonate and magnesium carbonate. Which can neutralize more acid: $CaCO_3$ or the same mass of $MgCO_3$?

# 16.7 The Common-Ion Effect

Thus far in this chapter we have worked with reaction systems consisting of a single acidic or basic reactant. Natural systems are often more complicated than that: they typically have multiple reactants that can influence pH or resist pH change. Suppose, for example, that a sample of river water contains $1.2 \times 10^{-5} \, M$ $H_2CO_3$ as a result of atmospheric carbon dioxide dissolving in the water. Suppose also that river sediment suspended in the water contains tiny particles of solid calcium carbonate ($CaCO_3$). We have seen how $CaCO_3$ can neutralize strong acids; it can also neutralize weak acids, as shown in the following reaction that produces soluble calcium bicarbonate:

$$CaCO_3(s) + H_2CO_3(aq) \rightleftharpoons Ca(HCO_3)_2(aq)$$

Other acidic substances in the river water could also be neutralized by calcium carbonate, producing additional soluble bicarbonate salts. Suppose that the total concentration of bicarbonate ions in the river water from these reactions is $1.0 \times 10^{-4} \, M$. How would the presence of this much $HCO_3^-$ affect the pH of the water, assuming that the water also contains $1.2 \times 10^{-5} \, M \, H_2CO_3$? To answer this question, we need to keep in mind that $H_2CO_3$ and $HCO_3^-$ represent a conjugate acid–base pair. They are related by the first acid ionization of carbonic acid,

$$H_2CO_3(aq) \rightleftharpoons HCO_3^-(aq) + H^+ (aq) \qquad (16.24)$$

and its equilibrium constant expression:

$$K_{a_1} = \frac{[HCO_3^-][H^+]}{[H_2CO_3]} = 4.3 \times 10^{-7}$$

We can summarize concentrations of reactants and products in the following table:

| Reaction | $H_2CO_3$ | $\rightleftharpoons$ | $H^+$ | $+$ | $HCO_3^-$ |
|---|---|---|---|---|---|
| | $H_2CO_3$ (M) | | $H^+$ (M) | | $HCO_3^-$ (M) |
| Initial | $1.2 \times 10^{-5}$ | | $0$ | | $1.0 \times 10^{-4}$ |
| Change | $-x$ | | $+x$ | | $+x$ |
| Equilibrium | $1.2 \times 10^{-5} - x$ | | $+x$ | | $1.0 \times 10^{-4} + x$ |

Let's insert the given values of $[H_2CO_3]$ and $[HCO_3^-]$ into the $K_{a_1}$ expression, letting $x = [H^+]$:

$$K_{a_1} = \frac{[HCO_3^-][H^+]}{[H_2CO_3]} = \frac{(1.0 \times 10^{-4} + x)(x)}{(1.2 \times 10^{-5} - x)} = 4.3 \times 10^{-7}$$

Given the small value for $K_a$, we assume that $1.0 \times 10^{-4} + x \approx 1.0 \times 10^{-4}$. Since the river water is in contact with $CO_2$ in the atmosphere, the concentration of $H_2CO_3$ will not change and we assume that $1.2 \times 10^{-5} - x = 1.2 \times 10^{-5}$, so the equilibrium constant expression can be written as:

$$K_{a_1} = \frac{(1.0 \times 10^{-4})(x)}{(1.2 \times 10^{-5})} = 4.3 \times 10^{-7}$$

Solving for $x$ gives us:

$$x = 4.3 \times 10^{-7} \times \frac{1.2 \times 10^{-5}}{1.0 \times 10^{-4}} = 5.16 \times 10^{-8} \, M = [H^+]$$

We take the negative logarithm to obtain pH:

$$pH = -\log[H^+] = -\log(5.16 \times 10^{-8} \, M) = 7.29$$

In Sample Exercise 16.6 we calculated that the pH of $1.2 \times 10^{-5} \, M \, H_2CO_3$ is 5.64. When $H_2CO_3$ is the only solute, ionization of $H_2CO_3$ is the only source of

$HCO_3^-$. Dissolution of carbonate minerals provides a second source. The additional $HCO_3^-$ causes the pH of the carbonic acid solution to increase by nearly 2 units. This increase in pH corresponds to a *decrease in [H⁺]* of nearly two orders of magnitude.

Is that the sort of change we should have expected? It does make sense because $[HCO_3^-]$ in equilibrium with $1.2 \times 10^{-5}$ *M* $H_2CO_3$ alone is only $10^{-5.64} = 2.3 \times 10^{-6}$ *M*. By increasing $[HCO_3^-]$ to $1.0 \times 10^{-4}$ *M*, we drive the equilibrium in Equation 16.24 to the left, as predicted by Le Châtelier's principle. The shift to the left lowers $[H^+]$ and raises pH.

This phenomenon illustrates a principle known as the **common-ion effect**: in any ionic equilibrium, a reaction that produces an ion is suppressed when another source of the same ion is added to the system. In the river water sample, ionization of $H_2CO_3$ (Equation 16.24) is suppressed when $HCO_3^-$ from carbonate minerals is added.

We can apply the common-ion effect to any equilibrium involving a weak acid and its conjugate base,

$$\text{Acid}(aq) \rightleftharpoons H^+(aq) + \text{base}(aq)$$

which has the equilibrium constant expression:

$$K_a = \frac{[H^+][\text{base}]}{[\text{acid}]}$$

If we take the negative logarithm of both sides of this expression, we transform $[H^+]$ into pH and $K_a$ into $pK_a$:

$$pK_a = pH - \log\frac{[\text{base}]}{[\text{acid}]}$$

$$pH = pK_a + \log\frac{[\text{base}]}{[\text{acid}]} \qquad (16.25)$$

Equation 16.25 is particularly useful in calculating the pH of a solution in which there are independent sources of both an acid and its conjugate base (or a base and its conjugate acid). It is called the **Henderson–Hasselbalch equation**.

Consider what happens to the logarithmic term in the Henderson–Hasselbalch equation when the concentrations of the acid and base are the same. Then the numerator and denominator in the log term are equal, and the value of the fraction is 1. The log of 1 is 0, and $pH = pK_a$. This equality serves as a handy reference point in an acid–conjugate base system. If the concentration of the basic component is greater than that of the acid, the logarithmic term is greater than zero and $pH > pK_a$. If the concentration of the basic component is less than that of the acid, the logarithmic term is less than zero, and $pH < pK_a$.

Suppose the concentration of the base is 10 times the concentration of the acid; that is, $[\text{base}] = 10[\text{acid}]$. Substituting this equality into Equation 16.25, we have

$$pH = pK_a + \log\frac{10[\text{acid}]}{[\text{acid}]}$$

$$= pK_a + \log 10 = pK_a + 1$$

A 10-fold higher concentration of base produces a pH one unit above the $pK_a$ value. Similarly, if the concentration of the acid component is 10 times that of the base, then $pH = pK_a - 1$.

**common-ion effect** the shift in the position of an equilibrium caused by the addition of an ion taking part in the reaction.

**Henderson–Hasselbalch equation** used to calculate the pH of a solution in which the concentrations of acid and conjugate base are known.

To demonstrate how the Henderson–Hasselbalch equation simplifies pH calculations when we know the concentrations of a weak acid and its conjugate base, let's use that equation to recalculate the pH of our river water sample. Our starting information includes the concentrations of a weak acid, $[H_2CO_3] = 1.2 \times 10^{-5}\ M$, and its conjugate base, $[HCO_3^-] = 1.0 \times 10^{-4}\ M$. They are linked by the equilibrium

$$H_2CO_3(aq) \rightleftharpoons HCO_3^-(aq) + H^+(aq) \qquad K_{a_1} = 4.3 \times 10^{-7}$$

First we take the negative log of $K_{a_1}$:

$$pK_{a_1} = -\log K_{a_1} = -\log(4.3 \times 10^{-7}) = 6.37$$

Then we insert that value and the concentrations into the Henderson–Hasselbalch equation:

$$pH = pK_a + \log\frac{[\text{base}]}{[\text{acid}]}$$

$$= 6.37 + \log\frac{1.0 \times 10^{-4}}{1.2 \times 10^{-5}}$$

$$= 7.29$$

This result matches the one we calculated earlier.

We can also use the Henderson–Hasselbalch equation to calculate the pH of a solution of a weak base and its conjugate acid. An extra step may be involved because we may know the value of the $K_b$ of the base but not of the $K_a$ of its conjugate acid, and only $K_a$ can be used in the Henderson–Hasselbalch equation. However, Equation 16.23 allows us to interconvert $K_a$ and $K_b$ values. Interconverting $pK_a$ and $pK_b$ values is even easier. All we have to do is a logarithmic transformation of Equation 16.23:

$$-\log K_b = -\log(1.00 \times 10^{-14}) - (-\log K_a)$$

$$pK_b = 14.00 - pK_a$$

$$pK_b + pK_a = 14.00 \qquad\qquad (16.26)$$

Thus, converting a $pK_b$ value into the $pK_a$ of its conjugate acid is simply a matter of subtracting the $pK_b$ value from 14.00.

---

**SAMPLE EXERCISE 16.9    Calculating the pH of a Solution of    LO5**
**a Weak Base and Its Conjugate Acid**

Calculate the pH of a solution that is 0.200 $M$ in $NH_3$ and 0.300 $M$ in $NH_4Cl$.

**Collect and Organize** We are to calculate the pH of a solution containing known concentrations of a weak base ($NH_3$) and a salt of its conjugate acid ($NH_4^+$).

**Analyze** We can use the Henderson–Hasselbalch equation to calculate the pH of such a solution from the concentrations of the two components and the $pK_a$ of the acid. Table A5.3 contains $K_b$ and $pK_b$ values of common bases. The $pK_a$ and $pK_b$ values of a conjugate acid–base pair are related by Equation 16.26:

$$pK_b + pK_a = 14.00$$

Our approach involves converting the $pK_b$ value from Appendix 5 (4.75) into $pK_a$. Addition of ammonium ion to a solution of ammonia should produce a solution that is still basic, but not as basic as a solution of only ammonia.

**Solve** Inserting the value of $pK_b$ in Equation 16.26 and solving for $pK_a$:

$$pK_a = 14.00 - pK_b = 14.00 - 4.75 = 9.25$$

**pH buffer** a solution that resists changes in pH when acids or bases are added to it; typically a solution of a weak acid and its conjugate base.

We use this value and the given concentrations of $NH_3$ and $NH_4^+$ in Equation 16.25:

$$pH = pK_a + \log\frac{[\text{base}]}{[\text{acid}]}$$

$$= 9.25 + \log\frac{0.200}{0.300} = 9.07$$

**Think About It** We predicted a result that would be a basic pH, but not as basic as a solution of ammonia alone. To confirm this prediction, we can calculate the pH of 0.200 $M$ $NH_3$ using the approach in Sample Exercise 16.5. The result is a pH of 11.27—over 2 pH units higher and more than 100 times more basic than the solution of ammonia and ammonium chloride in this exercise.

☼ **Practice Exercise** Calculate the pH of a solution that is 0.150 $M$ in benzoic acid and 0.100 $M$ in sodium benzoate.

# 16.8 pH Buffers

The concept of the common-ion effect is applied in a very useful technique for controlling the pH of a solution. A **pH buffer** is a solution that has the capacity to resist pH change by neutralizing small additions of acid or base. Typically it is a solution of a weak acid and its conjugate base. Buffers are an important component of natural water systems and of living organisms. In both settings, the role of a buffer is to maintain pH within a desired range. The internal pH of most living cells is regulated by buffers, including the carbonic acid–bicarbonate system we have been examining. When the pH stabilization provided by these buffers is disturbed in a living system, protein function is impaired, and the health of individual cells or the entire organism may be in jeopardy.

▶❙❙ **CHEMTOUR** Buffers

## An Environmental Buffer

As discussed in the preceding section, the presence of bicarbonate ion in river water gives the water a capacity to resist pH change when, for example, acid rain falls into it. We can use the Henderson–Hasselbalch equation to determine the effect of adding a small amount of a strong acid or base to a buffered solution. Consider, for example, what happens when a quantity of strong acid is added to our model river water in which $[H_2CO_3] = 1.2 \times 10^{-5}$ $M$ and $[HCO_3^-] = 1.0 \times 10^{-4}$ $M$. For reference purposes, we also evaluate the impact of adding the same quantity of acid to pure (unbuffered) pH 7.00 water. We start with 1.00 L each of river water and pure water and add 10.0 mL of $1.0 \times 10^{-3}$ $M$ $HNO_3$ to both.

Adding acid to pure water is an exercise in dilution, a concept introduced in Section 4.3. In this case, 10.0 mL of $1.0 \times 10^{-3}$ $M$ $HNO_3$ is diluted to a final volume of 1.01 L. To calculate the final concentration of $HNO_3$ (and $[H^+]$), we use Equation 4.3:

$$V_{\text{initial}} \times M_{\text{initial}} = V_{\text{diluted}} \times M_{\text{diluted}}$$

Solving for $M_{\text{diluted}}$:

$$M_{\text{diluted}} = [H^+] = \frac{V_{\text{initial}} \times M_{\text{initial}}}{V_{\text{diluted}}}$$

$$= \frac{0.0100 \text{ L} \times (1.0 \times 10^{-3} \text{ } M)}{1.01 \text{ L}} = 9.9 \times 10^{-6} \text{ } M$$

The corresponding pH is $-\log(9.9 \times 10^{-6}\ M) = 5.00$. Thus, the addition of acid dropped the pH of the water by 2.00 pH units.

To calculate the change in pH when the same quantity of acid is added to 1.00 L of pH 7.29 river water, we need to focus on the carbonic acid–bicarbonate equilibrium and the $pK_{a_1}$ of $H_2CO_3$:

$$H_2CO_3(aq) \rightleftharpoons HCO_3^-(aq) + H^+(aq) \qquad K_{a_2} = 4.3 \times 10^{-7} \quad (16.27)$$

Any acid added to an equilibrium mixture of $H_2CO_3$ and $HCO_3^-$ reacts with $HCO_3^-$, producing $H_2CO_3$ as the reaction runs in reverse. The number of moles of $H^+$ added is

$$(0.0100\ \text{L}) \times \left(1.0 \times 10^{-3}\ \frac{\text{mol } H^+}{\text{L}}\right) = 1.0 \times 10^{-5}\ \text{mol } H^+$$

This quantity of $HCO_3^-$ in the river water sample is consumed as the reaction in Equation 16.27 runs in the reverse direction. The number of moles of bicarbonate present initially is

$$\left(1.0 \times 10^{-4}\ \frac{\text{mol}}{\text{L}}\right) \times (1.00\ \text{L}) = 1.0 \times 10^{-4}\ \text{mol}$$

After subtracting the number of moles of $HCO_3^-$ consumed, we have $9 \times 10^{-5}$ mole left in a final volume of 1.01 L. As we have done in previous calculations, we assume that the value of $[H_2CO_3]$ is controlled by the solubility of $CO_2$ in water and does not change from the initial value of $1.2 \times 10^{-5}\ M$. With this information, we can calculate the pH of the river water using the Henderson–Hasselbalch equation:

$$pH = pK_a + \log\frac{[\text{base}]}{[\text{acid}]} = 6.37 + \log\frac{\left(\dfrac{9 \times 10^{-5}\ \text{mol}}{1.01\ \text{L}}\right)}{1.2 \times 10^{-5}\ \dfrac{\text{mol}}{\text{L}}}$$

$$= 7.24$$

The original pH was 7.29. Therefore, the addition of 10.0 mL of strong acid lowered the pH by only 0.05 pH units because of the action of the carbonic acid–hydrogen carbonate buffer. This stands in contrast to a decrease of 2.00 pH units in unbuffered water.

**CONCEPT TEST** ........................................................

Select from the list in Table A5.1 a weak acid that, when mixed with the sodium salt of its conjugate base in approximately equimolar proportions, produces a buffer with a pH of exactly 2.80. Indicate whether the buffer will contain *exactly* the same concentrations of acid and conjugate base, or slightly more acid or base.

........................................................

**SAMPLE EXERCISE 16.10** **Calculating Buffer Response to Addition of Acid or Base** **LO6**

Calculate the change in pH when 1.0 mL of 1.00 $M$ HCl is added to 100 mL of a solution that is 0.100 $M$ in sodium acetate and 0.100 $M$ in acetic acid (Figure 16.16).

**Collect and Organize** We are to determine by how much the pH of a solution of acetic acid and sodium acetate changes when a quantity of strong acid is added to it. This solution functions as a buffer.

**FIGURE 16.16** (a) Adding 1.00 mL of 1.00 $M$ HCl to 100 mL of distilled water leads to a drop in the pH from 7.00 to 2.00. (b) Addition of the same amount of acid, 1.00 mL of 1.00 $M$ HCl, to 100 mL of a buffer containing 0.100 $M$ acetic acid and 0.100 $M$ sodium acetate leads to a much smaller change in pH.

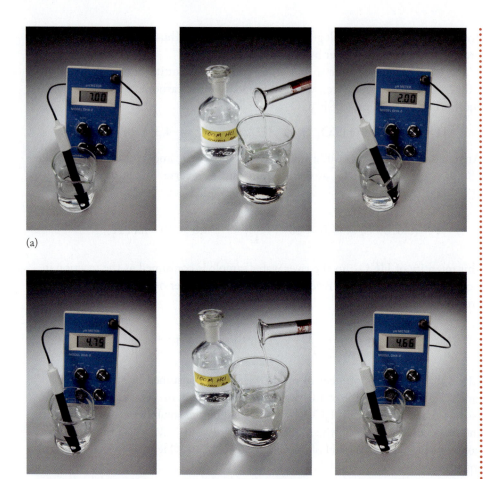

(a)

(b)

**Analyze** The pH of this buffer is controlled by the ionization equilibrium of acetic acid:

$$CH_3COOH(aq) \rightleftharpoons CH_3COO^-(aq) + H^+(aq)$$

The pH of a solution with known concentrations of a conjugate acid–base pair can be calculated using the Henderson–Hasselbalch equation:

$$pH = pK_a + \log\frac{[\text{base}]}{[\text{acid}]}$$

The $pK_a$ value of acetic acid is 4.75 (Appendix 5).

When a strong acid is added to a solution containing acetic acid and an acetate salt, some of the $CH_3COO^-$ ions react with the added $H^+$ ions, forming more $CH_3COOH$ as the ionization reaction runs in reverse. Initially $[CH_3COOH] = [CH_3COO^-]$, which means the log term in the Henderson–Hasselbalch equation is zero and pH = $pK_a$ = 4.75. Addition of a quantity of strong acid that does not consume all the acetate ion should result in a pH that is slightly lower than 4.75.

**Solve** The initial quantities of $CH_3COO^-$ and $CH_3COOH$ in the buffer are both

$$100 \text{ mL} \times \frac{1 \text{ L}}{1000 \text{ mL}} \times \frac{0.100 \text{ mol}}{\text{L}} \times \frac{1000 \text{ mmol}}{1 \text{ mol}} = 10.0 \text{ mmol}$$

The quantity of $H^+$ added

$$1.0 \text{ mL} \times \frac{1 \text{ L}}{1000 \text{ mL}} \times \frac{1.00 \text{ mol}}{\text{L}} \times \frac{1000 \text{ mmol}}{1 \text{ mol}} = 1.00 \text{ mmol}$$

represents the quantity of $CH_3COO^-$ converted into $CH_3COOH$. The impact of this conversion is summarized in the following table:

| Reaction | $CH_3COOH$ | $\rightleftharpoons$ $CH_3COO^- + H^+$ |
|---|---|---|
| | $CH_3COOH$ (mmol) | $CHCOO^-$ (mmol) |
| **Initial** | 10.0 | 10.0 |
| **Change** | +1.0 | −1.0 |
| **Equilibrium** | 11.0 | 9.0 |

Both final quantities are in the same total volume, 101 mL. Therefore we can use the ratio of quantities in lieu of concentration values in Equation 16.25 to calculate pH. Inserting these quantities and the $pK_a$ value of acetic acid in Equation 16.25 gives

$$pH = pK_a + \log\frac{[CH_3COO^-]}{[CH_3COOH]} = 4.75 + \log\frac{9.0}{11.0} = 4.66$$

The change in the pH of the solution after adding the acid is $4.75 - 4.66 = 0.09$ pH units.

**Think About It** The result is reasonable because adding strong acid produces a buffer solution with more acid and less conjugate base, so it has a slightly lower final pH (4.66) than it had initially (4.75).

⚙ **Practice Exercise** Calculate the change in pH when 10.0 mL of a 0.100 $M$ solution of NaOH is added to 1.00 L of a solution that is 1.00 $M$ in sodium acetate and 1.00 $M$ in acetic acid.

---

**SAMPLE EXERCISE 16.11** **Preparing a Buffer Solution with a Given pH** **LO6**

Buffer solutions containing dihydrogen phosphate ($H_2PO_4^-$) and hydrogen phosphate ($HPO_4^{2-}$) are found in the cytoplasm of all cells. What ratio of $HPO_4^{2-}$ to $H_2PO_4^-$ is required to make a buffer with pH = 6.75, the pH of golgi bodies in cancer cells.

**Collect and Organize** We are to determine the ratio of acid to conjugate base that results in a buffer with a prescribed pH of 6.75. From Table A5.1, the $pK_a$ for $H_2PO_4^-$ is 7.21.

**Analyze** The pH of this buffer is controlled by the ionization equilibrium of the dihydrogen phosphate ion:

$$H_2PO_4^-(aq) \rightleftharpoons HPO_4^{2-}(aq) + H^+(aq)$$

The desired pH for the solution is 6.75. The pH of a solution with known concentrations of a conjugate acid–base pair can be calculated using the Henderson–Hasselbalch equation:

$$pH = pK_a + \log\frac{[base]}{[acid]}$$

Since the desired pH is less than the $pK_a$, we predict that [base]/[acid] must be negative and [base]/[acid] < 1.

**Solve** We can rearrange the Henderson–Hasselbalch equation to solve for the ratio of base to acid.

$$\log\frac{[base]}{[acid]} = pH - pK_a$$

Substituting the values for pH and $pK_a$ gives us:

$$\log\frac{[HPO_4^{2-}]}{[H_2PO_4^-]} = pH - pK_a = 6.75 - 7.21 = -0.46$$

Then the ratio of $[HPO_4^{2-}]$ to $[H_2PO_4^-]$ is

$$\frac{[HPO_4^{2-}]}{[H_2PO_4^-]} = 10^{-0.46} = 0.35$$

**Think About It** The calculated ratio of $[HPO_4^{2-}]$ to $[H_2PO_4^-]$ is less than 1, consistent with our prediction. The pH of a buffer equals $pK_a$ when the concentrations of acid and its conjugate base are equal. An excess of weak acid in a buffer will lower the pH, whereas an excess of conjugate base will raise the pH.

**Practice Exercise** What ratio of sodium ascorbate to ascorbic acid is required to make a buffer with pH = 5.25?

## A Physiological Buffer

Buffers are vital components of living systems because most biochemical reactions involved in life-sustaining processes—such as metabolism, respiration, and transmission of nerve impulses—take place only within a narrow pH range. The pH of blood, for instance, needs to be buffered against perturbations caused by the ingestion or internal production of acidic or basic substances. Organisms living in the extreme environments shown in Figures 16.7 and 16.15, in particular, have developed mechanisms for maintaining the pH of their body fluids.

One of the buffer systems the human body relies on to control pH is the same one we have discussed in the context of environmental waters: the carbonic acid–bicarbonate system. This system is intimately tied to respiration. A key feature of pH control in this system is the role of breathing to maintain pH balance by regulating the concentration of dissolved $CO_2$.

Let's consider what may happen to the pH of blood during strenuous exercise. The soccer players in Figure 16.17 expend considerable energy during their game as a series of chemical reactions in their muscle cells convert oxygen and glucose into carbon dioxide and water. If this $CO_2$ is produced in muscle cells more rapidly than it can be transported away from them and to the lungs, there will be an increase in the concentration of $CO_2$ in these cells and also in the blood flowing past them. This increase in dissolved $CO_2$ produces an increase in the concentration of carbonic acid and an increase the concentration of $H^+$ ions as the following equilibria shift to the right:

$$CO_2(aq) + H_2O(\ell) \rightleftharpoons H_2CO_3(aq) \rightleftharpoons HCO_3^-(aq) + H^+(aq)$$

The resulting decrease in blood pH is called *respiratory acidosis*.

We saw at the beginning of this section that the chemical equilibria shown above can effectively buffer the pH of natural waters. These equilibria perform the same function in our bodies, which also have the ability to manipulate the concentrations of dissolved $CO_2$ and $HCO_3^-$ ions to even more tightly control pH. We just saw that vigorous exercise produces additional $CO_2$. One of the jobs of blood is to carry $CO_2$ from cells to the lungs, where it is eliminated as we exhale. During exercise we breathe faster and deeper to take in more $O_2$ *and* to exhale more $CO_2$. Lowering dissolved $CO_2$ concentrations back toward their normal values shifts the above equilibria to the left as predicted by Le Châtelier's principle. This shift decreases $[H^+]$ and helps restore blood pH to its normal range between 7.35 and 7.45.

If lung function is impaired through disease or injury, and the lungs cannot remove enough $CO_2$, respiratory acidosis becomes a serious medical issue. On the other hand, hyperventilation—fast, overly deep breathing—may cause too much $CO_2$ to be exhaled, shifting the above equilibria to the left and causing *respiratory*

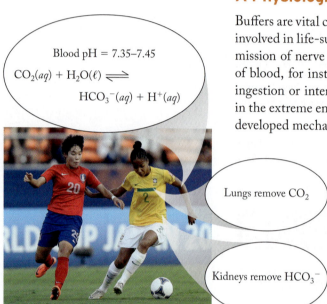

Blood pH = 7.35–7.45
$$CO_2(aq) + H_2O(\ell) \rightleftharpoons$$
$$HCO_3^-(aq) + H^+(aq)$$

Lungs remove $CO_2$

Kidneys remove $HCO_3^-$

**FIGURE 16.17** Strenuous exercise leads to increased concentrations of $CO_2$ in the blood. The body regulates pH by (1) the buffering action of the $CO_2$–$HCO_3^-$ buffer; (2) removing $CO_2$ by exhalation from the lungs; and (3) excretion of bicarbonate ion by the kidneys.

*alkalosis.* One way our bodies cope with alkalosis is by excreting bicarbonate ions through the kidneys. Removing the conjugate base of the carbonic acid/bicarbonate buffer shifts the above equilibria to the right and lowers blood pH into its normal range.

**buffer capacity** the quantity of acid or base that a pH buffer can neutralize while maintaining its pH within a desired range.

**CONCEPT TEST**

The mice used in biomedical experiments are sometimes euthanized by exposing them to high concentrations of $CO_2$ in the air they breathe. Why would this exposure be deadly?

## Buffer Range and Capacity

We have seen how buffers resist pH change when either acid or base is added. When scientists select buffers for particular applications, they need to answer two questions:

1. What is the desired pH range to be maintained? The appropriate buffer is one whose weak acid has a $pK_a$ that is within one pH unit of the desired pH. The Henderson–Hasselbalch equation tells us that over this pH range, the ratio [base]/[acid] varies from 10:1 to 1:10. Expressed another way, if this condition is met:

$$0.1 < \frac{[\text{base}]}{[\text{acid}]} < 10$$

then both components are available to neutralize additions of either acid or base and maintain the desired pH.

2. How much acid or base can the system consume without a large change in pH? The answer to this question defines the **buffer capacity**. A buffer is best able to resist changes in pH when the initial concentrations of acid and conjugate base are comparable to each other and greater than the concentration of acid or base that might be added. The greater the concentration of the buffer components, the greater is the buffer capacity (Figure 16.18).

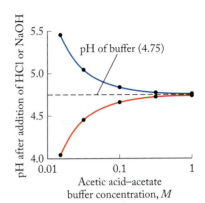

**FIGURE 16.18** When strong acid (red line) or strong base (blue line) is added to a buffer solution, the extent to which the pH changes is inversely proportional to buffer concentration: the higher the concentrations of the buffer components, the smaller the change in pH. In this illustration, 100 mL samples of five solutions that are 0.015, 0.030, 0.100, 0.300, and 1.000 *M* acetic acid and sodium acetate all have an initial pH of 4.75 (dashed line). The graph shows the pH values of these solutions after 1.00 mL of 1.00 *M* HCl or 1.00 *M* NaOH has been added.

**SAMPLE EXERCISE 16.12** **Effect of Concentration on Buffer Capacity** **LO6**

Calculate the final pH of the solution after 0.0100 mol $H^+$ is added to (a) 100 mL of a buffer containing 1.00 *M* acetic acid and 1.00 *M* sodium acetate, and (b) 100 mL of a buffer containing 0.150 *M* acetic acid and 0.150 *M* sodium acetate.

**Collect and Organize** We are asked to calculate the final pH of two buffer solutions after addition of the same quantity of acid (0.0100 mol $H^+$). We are given the composition and concentrations of the buffers. From Table A5.1, the $pK_a$ for $CH_3COOH$ is 4.75.

**Analyze** This exercise follows up on Sample Exercise 16.10. The pH of this buffer is controlled by the ionization equilibrium of acetic acid:

$$CH_3COOH(aq) \rightleftharpoons CH_3COO^-(aq) + H^+(aq)$$

The pH of a solution with known concentrations of a conjugate acid–base pair can be calculated using the Henderson–Hasselbalch equation:

$$pH = pK_a + \log \frac{[\text{base}]}{[\text{acid}]}$$

The two buffer solutions both contain equal concentrations of acid and base, but the starting concentrations are different. We predict that the more concentrated buffer will have a smaller change in pH on addition of acid.

**Solve** The initial quantities of $CH_3COO^-$ and $CH_3COOH$ in buffer (a) are both

$$100 \text{ mL} \times \frac{1 \text{ L}}{1000 \text{ mL}} \times \frac{1.00 \text{ mol}}{\text{L}} = 0.100 \text{ mol}$$

The quantities in buffer (b) are

$$100 \text{ mL} \times \frac{1 \text{ L}}{1000 \text{ mL}} \times \frac{0.150 \text{ mol}}{1 \text{ L}} = 0.0150 \text{ mol}$$

The quantity of $H^+$ added to both buffers is 0.0100 mol, which represents the quantity of $CH_3COO^-$ converted into $CH_3COOH$. The impact of this conversion is summarized in the following RICE table:

| Reaction | $CH_3COOH$ | | $\rightleftharpoons$ | $CH_3COO^- + H^+$ | |
|---|---|---|---|---|---|
| | **BUFFER (a)** | | | **BUFFER (b)** | |
| | $CH_3COOH$ (mol) | $CH_3COO^-$ (mol) | | $CH_3COOH$ (mol) | $CH_3COO^-$ (mol) |
| **Initial** | 0.100 | 0.100 | | 0.0150 | 0.0150 |
| **Change** | +0.0100 | −0.0100 | | +0.0100 | −0.0100 |
| **Equilibrium** | 0.110 | 0.090 | | 0.0250 | 0.0050 |

Inserting these quantities and the $pK_a$ value of acetic acid into the Henderson–Hasselbalch equation for each buffer gives:

Buffer (a) $\quad pH = pK_a + \log \dfrac{[CH_3COO^-]}{[CH_3COOH]} = 4.75 + \log \dfrac{0.090}{0.110} = 4.66$

Buffer (b) $\quad pH = pK_a + \log \dfrac{[CH_3COO^-]}{[CH_3COOH]} = 4.75 + \log \dfrac{0.0050}{0.0250} = 4.05$

**Think About It** The change in the pH of the solution after adding the acid to buffer (a) is 0.09 pH units, whereas the pH change in buffer (b) is 0.70 pH units. As predicted, the buffer with a higher concentration is better able to resist changes in pH when a given amount of acid (or base) is added. This reflects the fact that buffers work by converting conjugate base to acid on addition of $H^+$. Both conjugate base concentrations change by the same amount, but the relative change of the more concentrated buffer is less.

**Practice Exercise** Calculate the change in pH when 0.145 mol $OH^-$ is added to 1.00 L of two buffers: (a) a 1.16 $M$ solution of sodium dihydrogen phosphate containing 1.16 $M$ sodium hydrogen phosphate and (b) a 0.58 $M$ solution of $NaH_2PO_4$ containing 0.58 $M$ $Na_2HPO_4$.

---

**SAMPLE EXERCISE 16.13**    **Effect of [Base]/[Acid] Ratio on Buffer Capacity**    LO6

Calculate the pH of two buffers: (a) 255 mL of 1.00 $M$ formic acid and 0.794 $M$ sodium formate; and (b) 255 mL of 0.332 $M$ HF and 1.00 $M$ sodium fluoride. Calculate the initial pH of each buffer solution and the pH of each after the addition of 0.080 mole of $OH^-$. (Assume the addition of base does not change the volume of buffer significantly.)

**Collect and Organize** We are asked to calculate the initial pH of two buffer solutions containing two different weak acids and their conjugate bases, and then determine their final pH values after addition of equal amounts of base to each. We are given the composition and concentrations of the buffers. From Table A5.1, the $pK_a$ for formic acid ($HCO_2H$) is 3.75 and that for hydrofluoric acid (HF) is 3.17.

**Analyze** The equilibria for these two buffer solutions are described by:

$$HCOOH(aq) \rightleftharpoons HCOO^-(aq) + H^+(aq) \qquad pK_a = 3.75$$
$$HF(aq) \rightleftharpoons F^-(aq) + H^+(aq) \qquad pK_a = 3.17$$

The pH of each solution can be calculated by substituting into the Henderson–Hasselbalch equation:

$$pH = pK_a + \log \frac{[base]}{[acid]} \qquad (16.25)$$

We predict that the buffer with [base]/[acid] closer to 1 will have a smaller change in pH on addition of base.

**Solve** The initial pH of each solution is determined by substitution into Equation 16.25:

$$pH = pK_a + \log \frac{[HCOO^-]}{[HCOOH]} = 3.75 + \log \frac{0.794}{1.00} = 3.65$$

$$pH = pK_a + \log \frac{[F^-]}{[HF]} = 3.17 + \log \frac{1.00}{0.332} = 3.65$$

The initial pH is the same for both solutions. The amounts of acid and conjugate base in both buffers after addition of hydroxide ion are calculated and summarized in the RICE table.

$$255 \text{ mL HCOOH} \times \frac{1.00 \text{ mol}}{1000 \text{ mL}} = 0.255 \text{ mol}$$

$$255 \text{ mL NaHCOO} \times \frac{0.794 \text{ mol}}{1000 \text{ mL}} = 0.202 \text{ mol NaHCOO}$$

$$255 \text{ mL HF} \times \frac{0.332 \text{ mol}}{1000 \text{ mL}} = 0.0847 \text{ mol HF}$$

$$255 \text{ mL NaF} \times \frac{1.00 \text{ mol}}{1000 \text{ mL}} = 0.255 \text{ mol NaF}$$

| Reaction | $HCOOH(aq) \rightleftharpoons HCOO^-(aq) + H^+(aq)$ BUFFER (a) | | $HF(aq) \rightleftharpoons F^-(aq) + H^+(aq)$ BUFFER (b) | |
|---|---|---|---|---|
| | HCOOH (mol) | HCOO⁻ (mol) | HF (mol) | F⁻ (mol) |
| Initial | 0.255 | 0.202 | 0.0847 | 0.255 |
| Change | −0.080 | +0.080 | −0.080 | +0.080 |
| Equilibrium | 0.175 | 0.282 | 0.0047 | 0.335 |

Converting the number of moles into concentrations and inserting them and the $pK_a$ value of each acid into the Henderson–Hasselbalch equation for each buffer:

$$\text{Buffer (a)} \qquad pH = pK_a + \log \frac{[HCOO^-]}{[HCOOH]} = 3.75 + \log \frac{\left(\frac{0.282 \text{ mol}}{0.255 \text{ L}}\right)}{\left(\frac{0.175 \text{ mol}}{0.255 \text{ L}}\right)} = 3.65$$

$$\text{Buffer (b)} \qquad pH = pK_a + \log \frac{[F^-]}{[HF]} = 3.17 + \log \frac{\left(\frac{0.335 \text{ mol}}{0.255 \text{ L}}\right)}{\left(\frac{0.0047 \text{ mol}}{0.255 \text{ L}}\right)} = 5.02$$

The change in the pH of the two solutions after adding base to buffer (a) is 3.96 − 3.65 = 0.31 pH units, whereas the pH change in buffer (b) is 5.02 − 3.65 = 1.37 pH units.

**Think About It** As predicted, addition of OH⁻ to the buffer with [base]/[acid] closer to 1 (the formic acid/sodium formate buffer) leads to a smaller change in pH when a given amount of acid (or base) is added. When a buffer starts with [base]/[acid] closer to 1, the logarithm of [base]/[acid] has a smaller value, even after addition of acid or base, and has a smaller effect on the final pH. In this example, we have exceeded the capacity of buffer (b) to control pH. Buffers normally control pH better than (b) does here.

**pH indicator** a water-soluble weak organic acid that changes color as pH changes.

⚙ **Practice Exercise** Calculate the change in pH when 0.286 mole $H^+$ is added to 1.00 L of two buffers: (a) a 0.584 $M$ solution of pyridine (py) containing 0.500 $M$ pyH$^+$ and (b) a 0.682 $M$ solution of aniline (an) containing 0.964 $M$ anH$^+$.

## 16.9 Indicators and Acid–Base Titrations

Swimming pool operators routinely check pool pH. They often use sodium carbonate to adjust the pH of pools, but to make sure just the right amount is used, they test the pH of the water to ensure that the pool is not too basic. They often use a test kit that includes a **pH indicator** (Figure 16.19), a substance that changes color as pH changes. One such substance is phenol red. It is a weak acid ($pK_a = 7.6$) that is yellow in its un-ionized form (which, for convenience, we assign the generic formula HIn) and violet in its ionized (In$^-$) form. At a pH one unit above the $pK_a$—at pH 8.6—the ratio [In$^-$]/[HIn] is 10:1 according to the Henderson–Hasselbalch equation, and a phenol red solution is violet. At a pH less than 6.6, phenol red is largely un-ionized, and a solution of the indicator is yellow. In the pH range from about 6.8 to 8.6, the color of a phenol red solution changes from yellow to orange to red to violet with increasing pH (Figure 16.20).

As with buffers, different pH indicators are useful in different pH ranges. For example, phenol red is the best choice if the pH values being monitored fall between pH 6.6 and 8.6; one of the other indicators shown in Figure 16.20 would be a better choice in another pH range. Every indicator has a useful pH range defined by its $pK_a \pm 1.0$ pH unit. In addition to their role in determining pH values, indicators are also used to detect the large changes in pH that occur in acid–base titrations.

### Acid–Base Titrations

We discussed titration methods in Section 4.6; here we summarize how they are carried out. Figure 16.21 shows a typical titration apparatus. There are four steps in its use:

1. Accurately transfer a known volume of sample to a flask or beaker.
2. Either add a few drops of an indicator solution to the sample or insert the probe of a pH meter.

(a)

(b)

(c)

**FIGURE 16.19** Many pool test kits include the pH indicator phenol red. A few drops are added to a sample of pool water collected in the tube with the red cap. (a) After a rainstorm, the pH of the pool water is 6.8 (or less), as indicated by the yellow color of the sample. (b) Sodium carbonate is added to the pool to raise the pH. (c) A follow-up test produces a red-orange color, indicating the pH of the pool has been properly adjusted.

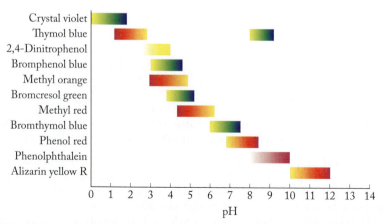

**FIGURE 16.20** A pH indicator is useful within a range of 1 pH unit above and below the $pK_a$ value of the indicator. This array of indicators could be used to determine pH values from 0 to 12.

3. Fill a buret with a solution (the *titrant*) of known concentration of a substance that reacts with a solute (the *analyte*) in the sample.
4. Slowly add titrant to the sample, and monitor the change in pH. The volume of titrant needed to completely consume the analyte is indicated by either a change in indicator color or a rapid change in the pH meter reading.

We use the recorded titrant volume to calculate the concentration of analyte in the sample.

The neutralization titrations in Chapter 4 involved titrating strong acids with strong bases and vice versa. Here we look at another type of titration, called an *alkalinity titration*. The term *alkalinity* is used to indicate the buffer capacity of a solution or sample of natural water against additions of acid. Although the buffers found in natural waters have more species involved than we included in our model of river water, the same carbonic acid–hydrogen carbonate buffer system that we have discussed several times is usually the most important one.

An alkalinity titration is more complex than a strong acid/strong base titration in that an alkalinity titration may have two equivalence points, as we titrate first carbonate ions and then bicarbonate ions. Let's work our way up to dealing with the complexity of an alkalinity titration by starting with titrations of artificial samples containing known concentrations of weak and strong monoprotic acids and *monobasic* bases. (A monobasic base accepts one hydrogen ion per molecule.) In these examples we monitor the change in pH with addition of titrant using a pH electrode.

In the first example, we compare the titration curves of two 20.0 mL samples. One contains 0.100 $M$ NaOH; the other contains 0.100 $M$ NH$_3$. Both are titrated with 0.100 $M$ HCl. The two titration curves are shown in Figure 16.22. The initial pH of the NaOH solution is higher than that of the NH$_3$ solution because NaOH is a strong base.

In the titration of the strong base NaOH with the strong acid HCl, the pH of the sample as acid is added does not change much until it is close to the equivalence point. As the equivalence point is approached, the principal ions present in the reaction mixture are Na$^+$($aq$), Cl$^-$($aq$), and OH$^-$($aq$). Sample pH is determined only by the [OH$^-$] still present. When enough acid has been added to completely consume all the OH$^-$ ions in the sample, the equivalence point is reached. The solution now consists of water and NaCl. The ions in solution (Na$^+$ and Cl$^-$) do not hydrolyze and do not influence pH; therefore, at the equivalence point, pH = 7.00.

The pH of the weak base NH$_3$ changes abruptly with the first few drops of added acid, but then the changes become smaller and the titration curve levels out. In this nearly flat region, additions of acidic titrant are consumed as NH$_3$ combines with H$^+$ ions to form NH$_4^+$ ions. In this region of the titration curve, the sample acts like a pH buffer (a solution of a weak base and its conjugate acid), and as long as there are

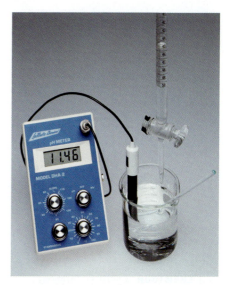

**FIGURE 16.21** A digital pH meter is used to measure pH during a titration.

🔴🔵 **CONNECTION** When we discussed titrations in Chapter 4, we introduced the equivalence point as that point in a titration when just enough standard solution has been added to completely react with all the solute.

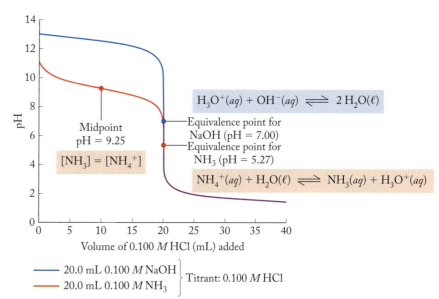

**FIGURE 16.22** The red curve describes a titration of 20.0 mL of 0.100 $M$ NH$_3$ with 0.100 $M$ HCl titrant. The blue curve describes a titration of 20.0 mL of 0.100 $M$ NaOH with 0.100 $M$ HCl titrant. Before the equivalence point, the pH of the weak base (NH$_3$) solution is always lower than the pH of the strong base (NaOH) solution. Once the equivalence point is reached, the two solutions have the same pH as additional titrant is added.

significant concentrations of $NH_3$ and $NH_4^+$ in the sample, the changes in pH with added titrant are small.

When enough titrant has been added to consume all the base in either of the two samples—in other words, at the equivalence point of each titration—pH drops sharply. At the equivalence point, the same volume of acid has been added to both samples, because the volumes of the two samples and their concentrations of base are the same. Whether the base is strong or weak does not affect the volume of titrant needed to reach the equivalence point: only the number of moles of base present determines the number of moles of acid required to neutralize it. Beyond the equivalence point, the curves are identical because the pH is determined only by the amount of HCl added and the total volume.

The pH values at the equivalence points of the two curves, however, are different: 7.00 (neutral) in the strong acid–strong base titration, but 5.27 (slightly acidic) in the strong acid–weak base titration. The neutralization of $NH_3$ with HCl does not produce a neutral solution because the product of the titration reaction is $NH_4Cl$:

$$HCl(aq) + NH_3(aq) \rightarrow NH_4^+(aq) + Cl^-(aq)$$

and, as we saw in Sample Exercise 16.8, ammonium ions are weakly acidic:

$$NH_4^+(aq) + H_2O(\ell) \rightleftharpoons NH_3(aq) + H_3O^+(aq)$$

As a general rule, the pH at the equivalence point in a titration of a weak base with a strong acid is less than 7 because the cation of the salt produced in the neutralization reaction hydrolyzes to produce an acidic solution.

One other important point in the $NH_3$ titration curve lies halfway to the equivalence point. At this *midpoint*, half of the $NH_3$ initially in the sample has been converted into $NH_4^+$ ions. Therefore, $[NH_3] = [NH_4^+]$. Because we know the relative concentrations of an acid and its conjugate base, we can use the Henderson–Hasselbalch equation to determine its pH. Because $[NH_3] = [NH_4^+]$, the logarithmic term is zero and pH = $pK_a$. The $K_a$ of $NH_4^+$ is $5.68 \times 10^{-10}$ (see Sample Exercise 16.8), so

$$pH = -\log(5.68 \times 10^{-10}) = 9.25$$

Just as the concentration of basic compounds in aqueous solution can be determined by titration with known quantities of a strong acid, the concentration of acidic compounds in aqueous solution can be determined by titration with known quantities of a strong base. Figure 16.23 illustrates two such titrations: 20.0 mL of 0.100 $M$ HCl and 20.0 mL of 0.100 $M$ acetic acid ($CH_3COOH$), each titrated with 0.100 $M$ NaOH. The two titration curves differ until the equivalence points are reached, and the pH values are consistently higher for the weak acid. The curves overlap beyond the equivalence points, where pH is controlled only by the increasing concentration of NaOH titrant. The titration curve for acetic acid prior to the equivalence point is nearly flat for the same reason the ammonia curve is nearly flat in Figure 16.22: the partially titrated sample acts like a pH buffer. In this case the buffer consists of a weak acid—the acetic acid in the original sample—and its conjugate base, the acetate ions produced as a result of the titration reaction.

The pH at the equivalence point in the acetic acid titration is 8.73. This value is well above 7, even though it is the point in the titration at which just enough NaOH has been added to exactly neutralize all the acetic acid in the sample. The product of the titration reaction

$$CH_3COOH(aq) + NaOH(aq) \rightarrow CH_3COONa(aq) + H_2O(\ell)$$

▶❚❚ **CHEMTOUR** Acid/Base Titrations

▶❚❚ **CHEMTOUR** Titrations of Weak Acids

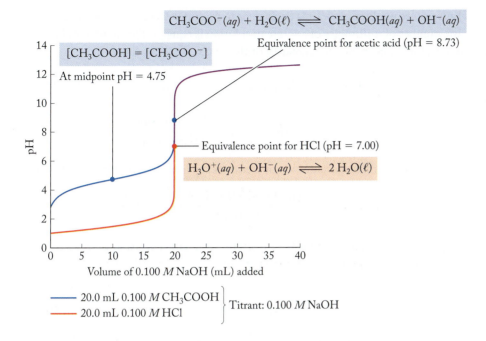

$$CH_3COO^-(aq) + H_2O(\ell) \rightleftharpoons CH_3COOH(aq) + OH^-(aq)$$

Equivalence point for acetic acid (pH = 8.73)

$[CH_3COOH] = [CH_3COO^-]$

At midpoint pH = 4.75

Equivalence point for HCl (pH = 7.00)

$$H_3O^+(aq) + OH^-(aq) \rightleftharpoons 2\,H_2O(\ell)$$

pH

Volume of 0.100 $M$ NaOH (mL) added

— 20.0 mL 0.100 $M$ CH$_3$COOH
— 20.0 mL 0.100 $M$ HCl
Titrant: 0.100 $M$ NaOH

**FIGURE 16.23** The blue curve describes a titration of 20.0 mL of 0.100 $M$ acetic acid (CH$_3$COOH) with 0.100 $M$ NaOH titrant. The red curve describes a titration of 20.0 mL of 0.100 $M$ HCl with 0.100 $M$ NaOH titrant. Before the equivalence point, the pH of the weak acid (acetic acid) solution is always higher than the pH of the strong acid (HCl) solution. After the equivalence point is reached, the two curves overlap.

is a solution of sodium acetate (CH$_3$COONa). Aqueous solutions of sodium acetate are basic because the acetate ion hydrolyzes:

$$CH_3COO^-(aq) + H_2O(\ell) \rightleftharpoons CH_3COOH(aq) + OH^-(aq)$$

As a general rule, the pH at the equivalence point in the titration of a weak acid with a strong base is greater than 7 because the hydrolysis of the anion of the salt produced in the neutralization produces a basic solution.

**CONCEPT TEST** ••••••••••••••••••••••••••••••••••••••••••••••••••••

What is the pH at the midpoint in the titration with a strong base of an aqueous sample that contains an unknown concentration of benzoic acid?

••••••••••••••••••••••••••••••••••••••••••••••••••••••••••••••••••••••

**CONCEPT TEST** ••••••••••••••••••••••••••••••••••••••••••••••••••••

The titration of a weak acid with a weak base is usually not done. Why?

••••••••••••••••••••••••••••••••••••••••••••••••••••••••••••••••••••••

To review the acid–base titration calculation described in Chapter 4, suppose we titrate 10.00 mL of vinegar and find that it requires 16.24 mL of 0.1050 $M$ NaOH to reach the equivalence point. To determine the concentration of acetic acid in the vinegar, we start with the stoichiometry of the balanced equation for the titration reaction:

$$CH_3COOH(aq) + NaOH(aq) \rightarrow CH_3COONa(aq) + H_2O(\ell)$$

One mole of CH$_3$COOH is consumed per mole of NaOH added. Because we are working with volumes in milliliters, a more useful quantity is the millimole (mmol), which is $10^{-3}$ mole. The definition of molarity ($M$) is moles of solute per liter of solution, and we can convert moles to millimoles and liters to milliliters by dividing both by 1000:

$$M = \frac{mol}{L} = \frac{mol/1000}{L/1000} = \frac{mmol}{mL}$$

Thus, moles per liter (mol/L) is equivalent to millimoles per milliliter (mmol/mL). The number of millimoles of a solute in solution is equal to the volume of the solution in milliliters times the molarity of the solute. Thus we can write

$$(V_{CH_3COOH})(M_{CH_3COOH}) = (V_{NaOH})(M_{NaOH}) \qquad (16.28)$$

$$M_{CH_3COOH} = \frac{(V_{NaOH})(M_{NaOH})}{(V_{CH_3COOH})}$$

Then we solve for the concentration of acetic acid by inserting the experimental data:

$$M_{CH_3COOH} = \frac{(16.24 \text{ mL})(0.1050 \ M)}{10.00 \text{ mL}} = 0.1705 \ M$$

We can write an equation for calculating the concentration of any acidic or basic solute from the results of a titration by using a general version of Equation 16.28:

$$V_A M_A = \frac{n_A}{n_B} V_B M_B \qquad (16.29)$$

Here $V_A$ and $M_A$ represent the volume and molarity of the acid, $V_B$ and $M_B$ represent the volume and molarity of the base, and $n_A/n_B$ is the ratio of the moles of acid to moles of base in the balanced chemical equation describing the reaction. For example, in a titration of a solution of sulfuric acid with NaOH, the number of moles of NaOH consumed is twice the number of moles of $H_2SO_4$ consumed:

$$H_2SO_4(aq) + 2 \ NaOH(aq) \rightarrow Na_2SO_4(aq) + 2 \ H_2O(\ell)$$

Therefore,

$$\frac{n_A}{n_B} = \frac{1}{2}$$

and Equation 16.29 becomes

$$(V_{H_2SO_4})(M_{H_2SO_4}) = \tfrac{1}{2}(V_{NaOH})(M_{NaOH})$$

## Alkalinity Titrations

Let's return to the application with which we began this section: determining the alkalinity of a sample of natural water, by which we mean the capacity of the water to neutralize additions of acid. We start with a known volume of the sample and slowly add a strongly acidic titrant from a buret, monitoring the change in pH produced by each drop. If carbonate is present in the sample, the first additions of titrant convert carbonate into bicarbonate:

$$CO_3^{2-}(aq) + H^+(aq) \rightarrow HCO_3^-(aq)$$

This reaction is the first stage of the alkalinity titration. In the second stage, the bicarbonate formed in the first stage, plus any bicarbonate present in the original sample, reacts with additional titrant to form carbonic acid.

$$HCO_3^-(aq) + H^+(aq) \rightarrow H_2CO_3(aq)$$

If more carbonic acid is produced than is soluble (remember, carbonic acid is a solution of $CO_2$ in water), the carbonic acid leaves the solution in the form of bubbles of carbon dioxide:

$$H_2CO_3(aq) \rightarrow H_2O(\ell) + CO_2(g)$$

In the first stage of the alkalinity titration (Figure 16.24), the titration curve has a region in which added acid has little effect on pH. This is a buffering region where $HCO_3^-$ and $CO_3^{2-}$ function as a weak acid–conjugate base pair. When the $CO_3^{2-}$ in the sample is completely consumed, the pH drops sharply, producing the first equivalence point. During the second stage of the titration, the conversion of $HCO_3^-$ ions to $H_2CO_3$ produces a second pH buffer and a plateau of nearly constant pH. Then the $HCO_3^-$ ions are completely consumed and pH drops sharply for a second time, creating a second equivalence point.

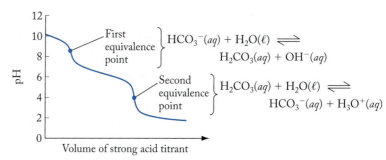

**FIGURE 16.24** The titration curve for an alkalinity titration has two equivalence points. The first marks the complete conversion of carbonate into bicarbonate, and the second marks the conversion of bicarbonate into carbonic acid.

Note that the initial pH of the sample in Figure 16.24 is slightly above 10, which is quite basic and above the pH range tolerated by many species of aquatic life. Such highly basic water may be found in arid regions such as the U.S. Southwest, where rocks containing $CaCO_3$ and other basic compounds are in contact with water. The pH of seawater is about 8.2, which is near the first equivalence point on the alkalinity titration curve in Figure 16.24. This means that the dominant carbonate species in seawater is actually bicarbonate. For this reason, alkalinity titration curves for seawater (and most freshwater samples) have only one equivalence point, coinciding with the pH of the second equivalence point in Figure 16.24.

**CONCEPT TEST**

In the alkalinity titration of a sample that initially contains both $CO_3^{2-}$ and $HCO_3^-$, the volume of titrant required to reach the first equivalence point is less than that required to titrate from the first equivalence point to the second. Why?

We can detect both equivalence points in an alkalinity titration with the appropriate color indicators. Phenol red would not be a good choice because it changes color between pH 6.8 and 8.4. This range is just below the pH of the first equivalence point and well above the pH of the second equivalence point. To detect the first equivalence point, we need an indicator with a $pK_a$ near the pH of the solution at the first equivalence point (8.5). Looking at Figure 16.20, we see that one candidate is phenolphthalein ($pK_a = 8.8$), which is pink in its basic form and colorless at low pH.

To detect the second equivalence point, we could add bromcresol green ($pK_a = 4.6$) after the first equivalence point has been reached. We would not add it before then because its blue-green color in basic solutions would obscure the pink-to-colorless transition of phenolphthalein. We do not need to be concerned about the phenolphthalein obscuring the bromcresol green color change because phenolphthalein is colorless in acidic solutions.

Suppose 50.00 mL of pH 10.0 water from a hot spring is titrated with 0.02075 $M$ HCl. A few drops of phenolphthalein are added at the beginning of the titration, and the solution turns pink. It takes 11.21 mL of titrant to reach the pink-to-clear equivalence point. Then a few drops of bromcresol green are added, and it takes an additional 32.28 mL of titrant before the blue-green color changes to yellow. What are the initial concentrations of carbonate and bicarbonate in the sample?

**Collect and Organize** We are asked to determine the concentrations of two analytes in one sample from the results of a single titration with a monoprotic strong acid, HCl. These determinations are based on the volumes of titrant needed to reach two equivalence points: an initial one at which any $CO_3^{2-}$ in the sample has been converted to $HCO_3^-$:

$$(1) \qquad H^+(aq) + CO_3^{2-}(aq) \rightarrow HCO_3^-(aq)$$

and a second one at which $HCO_3^-$ has been converted to $H_2CO_3$:

$$(2) \qquad H^+(aq) + HCO_3^-(aq) \rightarrow H_2CO_3(aq)$$

**Analyze** The $HCO_3^-$ titrated in reaction 2 includes any $HCO_3^-$ in the original sample plus all the $HCO_3^-$ produced in reaction 1. If there were no $HCO_3^-$ present initially, the volume of titrant needed to react with the $HCO_3^-$ in reaction 2 would be exactly the same as the volume needed to react with $CO_3^{2-}$, 11.21 mL, in reaction 1. However, the volume of titrant required to reach the second equivalence point is much greater: 32.28 mL. The difference between these two volumes, $(32.28 - 11.21) = 21.07$ mL, is the volume of acid required to react with any $HCO_3^-$ that was present in the original sample. Because this difference is nearly twice the volume of titrant required to react with the $CO_3^{2-}$ in the sample, the value of $[HCO_3^-]$ that we calculate for the original sample should be nearly twice its $[CO_3^{2-}]$ value.

The stoichiometry of the reaction tells us that the titrant and carbonate react in a 1:1 mole ratio, so, at the first equivalence point,

$$\text{mol HCl added} = \text{mol } CO_3^{2-} \text{ consumed}$$

This means that the ratio $n_A/n_B$ for this reaction is 1:1. The same is true for the conversion of $HCO_3^-$ into $H_2CO_3$:

$$\text{mol HCl added} = \text{mol } HCO_3^- \text{ consumed}$$

**Solve** We use Equation 16.29 for both calculations:

$$M_{CO_3^{2-}} = \frac{(11.21 \text{ mL})(0.02075\ M)}{50.00 \text{ mL}} = 4.652 \times 10^{-3}\ M$$

$$M_{HCO_3^-} = \frac{(21.07 \text{ mL})(0.02075\ M)}{50.00 \text{ mL}} = 8.744 \times 10^{-3}\ M$$

**Think About It** The titration results confirm that the bicarbonate concentration in the original sample was almost twice the carbonate concentration. To check our two values from the calculations, we can insert the results into the Henderson–Hasselbalch equation and calculate what the initial pH of the sample should have been. To do that, we need the $pK_{a_2}$ value for carbonic acid from Table A5.1: 10.33.

$$pH = pK_a + \log\frac{[\text{base}]}{[\text{acid}]}$$

$$= 10.33 + \log\frac{4.652 \times 10^{-3}\ M}{8.744 \times 10^{-3}\ M}$$

$$= 10.06$$

This calculated pH value agrees with the pH given in the problem statement.

⚙ **Practice Exercise** Your job is to determine the concentration of ammonia in a commercial window cleaner. In the titration of a 25.00 mL sample, the equivalence point is reached after 10.49 mL of 0.155 $M$ HCl has been added.

a. What is the concentration of ammonia in the solution?

b. Which pH indicator in Figure 16.20 would be suitable for this titration?

# 16.10 Solubility Equilibria

In Section 16.1 we noted that the common strong bases are all hydroxides of alkali or alkaline earth elements. One exception is $Mg(OH)_2$, which is a weak base because it has limited solubility in water. Magnesium hydroxide is the active ingredient in a product found in many medicine cabinets: the antacid called *milk of magnesia*. This liquid appears "milky" because it is an aqueous *suspension* (not solution) of solid, white $Mg(OH)_2$. We can express the limited solubility of solid $Mg(OH)_2$ by the following equation:

$$Mg(OH)_2(s) \rightleftharpoons Mg^{2+}(aq) + 2\,OH^-(aq)$$

Because $Mg(OH)_2$ is a solid, its effective concentration does not change as long as some of it is present in the system. As was the case for pure liquids (Section 16.2), pure solids do not appear in the equilibrium constant expression. Therefore, the equilibrium constant for the dissolution of $Mg(OH)_2$ is

$$K_{sp} = [Mg^{2+}][OH^-]^2$$

where $K_{sp}$ represents an equilibrium constant called either the **solubility-product constant** or simply the **solubility product**.

The $K_{sp}$ values of $Mg(OH)_2$ and other slightly soluble compounds are listed in Table A5.4. We can use these values to calculate concentrations of these compounds in aqueous solutions. Two terms are used to describe how much of a solid dissolves in a solvent: *solubility*, which is usually expressed in grams of solute per liter of solution, and *molar solubility*, which is expressed in moles of solute per liter of solution. Let's use the $K_{sp}$ of $Mg(OH)_2$ from Appendix 5 ($5.6 \times 10^{-12}$) to calculate the molar solubility of $Mg(OH)_2$ at 25°C and how many grams of it dissolve in 50.0 mL at 25°C. We start with the $K_{sp}$ expression:

$$K_{sp} = [Mg^{2+}][OH^-]^2 = 5.6 \times 10^{-12}$$

If we let $x$ be the number of moles of $Mg(OH)_2$ that dissolves in 1 L of solution, then $x$ moles of $Mg^{2+}$ ions and $2x$ moles of $OH^-$ ions are produced:

$$K_{sp} = (x)(2x)^2 = (x)(4x^2) = 4x^3 = 5.6 \times 10^{-12}$$
$$x = 1.1 \times 10^{-4}\,M$$

Note that the entire algebraic expression for [$OH^-$], $2x$, is squared in this calculation. Forgetting to square the coefficient is a common mistake. Also note that the molar solubility of $Mg(OH)_2$ is much greater than its solubility product. This difference is true for all sparingly soluble ionic compounds because each $K_{sp}$ is the product of small concentration values multiplied together, producing an even smaller overall $K_{sp}$ value.

Using the molar solubility of $Mg(OH)_2$ that we just calculated ($1.1 \times 10^{-4}$ mol/L), we can determine how many grams of $Mg(OH)_2$ dissolve in 50.0 mL of a saturated

**solubility-product constant** (also called **solubility product, $K_{sp}$**) an equilibrium constant that describes the formation of a saturated solution of a slightly soluble salt.

solution. We first convert molar solubility into an equivalent number of grams of $Mg(OH)_2$ per liter using the molar mass of $Mg(OH)_2$, 58.32 g/mol. Then we convert this value into the mass that dissolves in 50.0 mL of solution:

$$\frac{1.1 \times 10^{-4} \; \cancel{mol}}{\cancel{L}} \times \frac{58.32 \; g}{1 \; \cancel{mol}} \times \frac{1 \; \cancel{L}}{1000 \; \cancel{mL}} \times 50.0 \; \cancel{mL} = 3.2 \times 10^{-4} \; g$$

We can derive a general equation relating the molar solubility of any ionic compound $M_mZ_z$ to its $K_{sp}$ value. We start with the dissolution equilibrium

$$M_mZ_z(s) \rightleftharpoons mM^{n+}(aq) + zZ^{y-}(aq)$$

If $S$ represents the molar solubility of $M_mZ_z$, then there are $(m \times S)$ mol/L $M^{n+}$ ions and $(z \times S)$ mol/L $Z^{y-}$ ions in solution. Inserting these molar concentrations into the equilibrium constant expression for $M_mZ_z$:

$$K_{sp} = [M^{n+}]^m [Z^{y-}]^z = (m \times S)^m (z \times S)^z$$

$$K_{sp} = (m^m z^z) \, S^{(m+z)} \qquad (16.30)$$

Before we get lost in all of these letters, let's apply Equation 16.30 to calculate the molar solubility of calcium phosphate, $Ca_3(PO_4)_2$:

$$K_{sp} = (m^m Z^z) S^{(m+z)} = (3^3 2^2) S^{(3+2)} = 108 \, S^5$$

Note that there are really only two terms in Equation 16.30. The first contains each subscript raised to the same power as its value: every "2" is squared, every "3" is cubed, and so on. The second term is simply molar solubility ($S$) raised to a power equal to the sum of the subscripts. One warning about subscripts and polyatomic ions: use only the subscript outside the parentheses, not the subscripts that are part of the formula of the ion.

Now we can insert the $K_{sp}$ value listed for $Ca_3(PO_4)_2$ in Table A5.4 and solve for $S$:

$$K_{sp} = 2.1 \times 10^{-33} = 108 \, S^5$$

$$S = \sqrt[5]{\frac{2.1 \times 10^{-33}}{108}} = 1.1 \times 10^{-7} \; M$$

**CONCEPT TEST** • • • • • • • • • • • • • • • • • • • • • • • • • • • • • • • • • • • • • • • • • • • •

Use Equation 16.30 to write an equation that relates the $K_{sp}$ value for $Na_2CO_3$ to its molar solubility $S$.

• • • • • • • • • • • • • • • • • • • • • • • • • • • • • • • • • • • • • • • • • • • • • • • • • • • •

**SAMPLE EXERCISE 16.15** **Calculating Molar Solubility from $K_{sp}$** **LO8**

The mineral barite is mostly barium sulfate ($BaSO_4$) and is widely used in industry and in medical imaging of the digestive system. Calculate the molar solubility at 25°C of $BaSO_4$ in (a) pure water and (b) seawater in which the concentration of sulfate ions is 2.8 g/L.

**Collect and Organize** We are to calculate the molar solubility of $BaSO_4$ in both pure water and in seawater that already contains sulfate ions. The $K_{sp}$ value of $BaSO_4$ given in Table A5.4 is $9.1 \times 10^{-11}$.

**Analyze** The dissolution reaction for barium sulfate

$$BaSO_4(s) \rightleftharpoons Ba^{2+}(aq) + SO_4^{2-}(aq)$$

indicates that 1 mole of $Ba^{2+}$ ions and 1 mole of $SO_4^{2-}$ ions form from each mole of $BaSO_4$ that dissolves. If $S$ mol/L of $BaSO_4$ dissolves in pure water, then $[Ba^{2+}] = [SO_4^{2-}] = S$.

However, seawater has a background concentration of $SO_4^{2-}$. According to Le Châtelier's principle and the common-ion effect, the sulfate ion already in seawater should shift the dissolution equilibrium to the left, which means that less $BaSO_4$ should dissolve in seawater than in pure water.

**Solve**

a. In pure water,

$$K_{sp} = [Ba^{2+}][SO_4^{2-}] = (S)(S) = 9.1 \times 10^{-11}$$
$$S^2 = 9.1 \times 10^{-11}$$
$$S = 9.5 \times 10^{-6} \ M$$

b. In seawater, we first need to calculate the value of $[SO_4^{2-}]$ before any $BaSO_4$ dissolves:

$$[SO_4^{2-}]_{initial} = \frac{2.8 \ g}{L} \times \frac{1 \ mol}{96.06 \ g} = \frac{0.029 \ mol}{L}$$

The value of $[SO_4^{2-}]$ at equilibrium is the sum of the background concentration (0.029 mol/L) and the additional $SO_4^{2-}$ ions from the dissolution of $BaSO_4$ (S). Incorporating this value into the $[SO_4^{2-}]$ term in the $K_{sp}$ expression gives

$$K_{sp} = [Ba^{2+}][SO_4^{2-}] = (S)(0.029 + S) = 9.1 \times 10^{-11}$$

Solving for S is simplified if we assume that the $K_{sp}$ of $BaSO_4$ is so small that we can ignore its contribution to the total $SO_4^{2-}$ concentration. This assumption is reasonable because S is likely to be much less than 0.029 $M$ given the limited solubility of $BaSO_4$ in pure water. Therefore,

$$(S)(0.029 + S) \approx (S)(0.029) = 9.1 \times 10^{-11}$$
$$S = 3.1 \times 10^{-9} \ M$$

**Think About It** The calculated molar solubility of $BaSO_4$ in seawater is much less than the initial $[SO_4^{2-}]$ value, so our simplifying assumption was justified. The lower solubility of $BaSO_4$ in seawater is another illustration of the common-ion effect: the dissolution of $BaSO_4$ is suppressed by the $SO_4^{2-}$ ions already present in seawater.

**Practice Exercise** What is the molar solubility of $MgCO_3$, a component of the mineral dolomite, at 25°C?

In the preceding Sample Exercise we saw how the common-ion effect can suppress the solubility of an ionic compound. Other perturbations to solubility equilibria can actually promote solubility, as happens when the anion of the compound is the conjugate base of a weak acid. The molar solubilities of such compounds increase when strong acid is added and pH is lowered. We will see why as we solve the following Sample Exercise.

---

**SAMPLE EXERCISE 16.16** **Calculating the Effect of pH on Solubility** **LO8**

What is the molar solubility of $CaF_2$ at 25°C in (a) pure water and (b) an acidic buffer in which $[H^+]$ is a constant 0.050 $M$?

**Collect and Organize** We are asked to calculate the solubility of $CaF_2$ in both pure water and in an acidic buffer. The dissolution process is described by the equilibrium:

$$(1) \qquad CaF_2(s) \rightleftharpoons Ca^{2+}(aq) + 2 \ F^-(aq)$$

The equilibrium constant expression for the process and the $K_{sp}$ value from Table A5.4 are

$$K_{sp} = [Ca^{2+}][F^-]^2 = 3.9 \times 10^{-11}$$

Calcium fluoride is a basic salt because the $F^-$ ion is the conjugate base of the weak acid HF ($K_a = 6.8 \times 10^{-4}$).

**Analyze** To account for the effect of acid on the solubility of a fluoride salt, we need to consider the chemical equilibrium in which the fluoride ion acts as a Brønsted–Lowry base ($H^+$ ion acceptor):

$$(2) \qquad F^-(aq) + H^+(aq) \rightleftharpoons HF(aq)$$

This reaction is the reverse of the acid ionization reaction:

$$HF(aq) \rightleftharpoons F^-(aq) + H^+(aq)$$

Therefore, the equilibrium constant for reaction 2 is the reciprocal of the $K_a$ of HF:

$$K_2 = \frac{[HF]}{[H^+][F^-]} = \frac{1}{6.8 \times 10^{-4}} = 1.47 \times 10^3$$

As reaction 2 proceeds, $F^-$ ions are consumed, which shifts the equilibrium in reaction 1 to the right, increasing $CaF_2$ solubility. Therefore we can anticipate an increase in the solubility of $CaF_2$ when acid is present.

**Solve**

a. Let $S$ be the molar solubility of $CaF_2$ in pure water. According to the stoichiometry of reaction 1, $[Ca^{2+}] = S$ mol/L and $[F^-] = 2S$ mol/L. Inserting these symbols in the $K_{sp}$ expression:

$$K_{sp} = [Ca^{2+}][F^-]^2 = (S)(2S)^2 = 4S^3$$
$$4S^3 = 3.9 \times 10^{-11}$$
$$S = 2.1 \times 10^{-4} \, M$$

b. In the acidic buffer, $[H^+] = 0.050 \, M$. The $F^-$ and $H^+$ ions combine as in reaction 2. Assuming the buffer pH is constant, $[H^+] = 0.050 \, M$ and the equilibrium constant expression for reaction 2 is

$$K_2 = \frac{[HF]}{[0.050][F^-]} = 1.47 \times 10^3$$

$$\frac{[HF]}{[F^-]} = 73.5$$

$$(3) \qquad [HF] = 73.5[F^-]$$

This calculation tells us that most of the $F^-$ produced when calcium fluoride dissolves is converted into HF. The tiny fraction of all the $F^-$ ions produced by $CaF_2$ that remain as free $F^-$ ions is defined by this ratio:

$$(4) \qquad \frac{[F^-]}{[F^-] + [HF]}$$

Here the numerator is the concentration of free $F^-$ ions at equilibrium and the denominator is the concentration of all the $F^-$ ions produced when $CaF_2$ dissolved. Combining expressions 3 and 4:

$$\frac{[F^-]}{[F^-] + [HF]} = \frac{[F^-]}{[F^-] + 73.5[F^-]} = \frac{[F^-]}{74.5[F^-]} = 0.0134$$

If $S$ is the molar solubility of $CaF_2$ in the acid, $S$ mol/L $Ca^{2+}$ and $2S$ mol/L $F^-$ are produced. However, most of the fluoride ions are converted into HF, and, as we just calculated, the free $F^-$ ion concentration is only 0.0134 of that produced when $CaF_2$ dissolved. Therefore, the $[F^-]$ term in the $K_{sp}$ expression is not $2S$, but only $(0.0134 \times 2S) = 0.0268 \, S$. Inserting this value in the $K_{sp}$ expression,

$$K_{sp} = [Ca^{2+}][F^-]^2 = (S)(0.0268 \, S)^2 = 7.18 \times 10^{-4} \, S^3$$
$$7.18 \times 10^{-4} \, S^3 = 3.9 \times 10^{-11}$$
$$S = 3.8 \times 10^{-3} \, M$$

**Think About It** A comparison of the results from parts a and b reveals that the molar solubility of $CaF_2$ is about 10 times higher in the acidic buffer, as we predicted, because most of the $F^-$ ions produced when $CaF_2$ dissolves in the buffer form HF. This conversion of $F^-$ ions to HF means a product (free $F^-$ ion) is being removed from the dissolution equilibrium. Le Châtelier's principle tells us that the result will be a shift in the position of the equilibrium to the right, in favor of forming product. In this case greater product formation means greater solubility.

⚙ **Practice Exercise** What is the solubility of $ZnCO_3$ at 25°C in a buffer solution with a pH value of 10.33?

■

**CONCEPT TEST**

In part b of Sample Exercise 16.16, we assumed that $[H^+]$ did not change significantly as a result of the reaction $F^-(aq) + H^+(aq) \rightleftharpoons HF(aq)$. If $[H^+]$ had dropped significantly, how would the solubility of $CaF_2$ have been affected?

## $K_{sp}$ and $Q$

In the preceding Sample Exercises we used calculations involving the $K_{sp}$ of slightly soluble salts in cases where the solubility of the salt was either diminished or enhanced. We can also use $K_{sp}$ values to determine if a precipitate will form when two solutions are mixed. In answering such questions, we can use the concept of the reaction quotient $Q$ that we developed in Chapter 15. When applied to the equilibrium governing a slightly soluble salt, $Q$ is sometimes called the *ion product*, a name that describes exactly what $Q$ is in this case: the product of the concentrations of the ions in the precipitate raised to powers equal to their coefficients in the balanced equation. If the calculated $Q$ value is greater than the $K_{sp}$ of a salt ($Q > K_{sp}$), then that salt will precipitate when the solutions are mixed. If $Q < K_{sp}$, no precipitate will form.

**SAMPLE EXERCISE 16.17** **Determining if a Precipitate Forms when Solutions Are Mixed** **LO8**

Lead(II) chloride ($K_{sp} = 1.60 \times 10^{-5}$) is a white pigment used in 15th-century European sculpture. Will $PbCl_2$ precipitate when 275 mL of a 0.134 $M$ solution of $Pb(NO_3)_2$ is added to 125 mL of a 0.0339 $M$ solution of NaCl?

**Collect and Organize** We are asked if $PbCl_2$ will precipitate when two solutions containing $Pb^{2+}$ ions and $Cl^-$ ions are mixed. The process is described by the equilibrium

$$Pb^{2+}(aq) + 2\,Cl^-(aq) \rightleftharpoons PbCl_2(s) \qquad K_{sp} = 1.60 \times 10^{-5}$$

**Analyze** To determine if a precipitate forms, we need to calculate $Q$ and compare its value to $K_{sp}$. If $Q > K_{sp}$, $PbCl_2$ will precipitate; if $Q < K_{sp}$, it will not precipitate. $Q$ has the same form as the equilibrium constant

$$K_{sp} = [Pb^{2+}][Cl^-]^2$$

but the concentrations used in calculating it are the values given for the system, which are probably not equilibrium values.

**Solve** First we must calculate the concentrations of the lead ions and chloride ions in the two solutions immediately on mixing. Remember that mixing two solutions dilutes both, and that the volumes of the solutions (275 mL and 125 mL) may be added together to get the final solution volume (400 mL = 0.400 L).

$$Pb^{2+}(aq): \quad 0.134 \, \frac{mol}{L} \times 0.275 \, L = 0.0369 \, mol$$

$$[Pb^{2+}] = \frac{0.0369 \, mol}{0.400 \, L} = 0.0921 \, M$$

$$Cl^-(aq): \quad 0.0339 \, \frac{mol}{L} \times 0.125 \, L = 0.00424 \, mol;$$

$$[Cl^-] = \frac{0.00424 \, mol}{0.400 \, L} = 0.0106 \, M$$

The value of $Q$ is

$$Q = [Pb^{2+}][Cl^-]^2 = (0.0921)(0.0106)^2 = 1.03 \times 10^{-5}$$

which is smaller than $K_{sp}$, so no precipitate forms.

**Think About It** Lead(II) chloride was on the list of *insoluble* chloride salts in Table 4.5. In this scenario, $PbCl_2$ does not precipitate because the solutions of lead ion and chloride ion are too dilute to provide the concentrations required for the precipitate to form.

⚙ **Practice Exercise** Will calcium fluoride ($K_{sp} = 3.9 \times 10^{-11}$) precipitate when 175 mL of a $4.78 \times 10^{-3} \, M$ solution of $Ca(NO_3)_2$ is added to 135 mL of a $7.35 \times 10^{-3} \, M$ solution of KF?

■ •••••••••••••••••••••••••••••••••••••••••••••••••••••••••

We can use differences in the solubilities of ionic compounds to selectively remove ions from solution. For example, suppose a solution contains $0.10 \, M \, Ca^{2+}$ ion and $0.020 \, M \, Mg^{2+}$ ion. The hydroxide salts of both ions are slightly soluble. Is it possible to remove the $Mg^{2+}$ ions from solution by precipitating them as $Mg(OH)_2$ while leaving the $Ca^{2+}$ ions in solution? Consider the following solubility equilibria and $K_{sp}$ values:

$$Ca(OH)_2 \rightleftharpoons Ca^{2+}(aq) + 2 \, OH^-(aq) \qquad K_{sp} = [Ca^{2+}][OH^-]^2 = 4.7 \times 10^{-6}$$
$$Mg(OH)_2 \rightleftharpoons Mg^{2+}(aq) + 2 \, OH^+(aq) \qquad K_{sp} = [Mg^{2+}][OH^-]^2 = 5.6 \times 10^{-12}$$

Note that $Ca(OH)_2$ has a larger $K_{sp}$ than $Mg(OH)_2$. We can therefore ask: What is the maximum concentration ($x$) of $OH^-$ ions that will *not* cause the $0.10 \, M \, Ca^{2+}$ ion to precipitate? We can calculate that directly from the $K_{sp}$ of calcium hydroxide:

$$K_{sp} = 4.7 \times 10^{-6} = [Ca^{2+}][OH^-]^2 = (0.10)(x)^2$$

$$x = \sqrt{\frac{4.7 \times 10^{-6}}{0.10}} = 6.9 \times 10^{-3} \, M$$

How much of the $Mg^{2+}$ in the solution would precipitate if $[OH^-] = 6.9 \times 10^{-3} \, M$? The concentration $x$ of magnesium ion in the presence of $6.9 \times 10^{-3} \, M$ hydroxide ion may be calculated from its $K_{sp}$:

$$K_{sp} = 5.6 \times 10^{-12} = [Mg^{2+}][OH^-]^2 = (x)(6.9 \times 10^{-3})^2$$

$$x = \frac{5.6 \times 10^{-12}}{(6.9 \times 10^{-3})^2} = 1.2 \times 10^{-7} \, M$$

Because the original solution was $0.020 \, M$ in $Mg^{2+}$ ions, a concentration of $1.2 \times 10^{-7} \, M \, Mg^{2+}$ means that only $[(1.2 \times 10^{-7})/0.020] \times 100\%$, or only $0.00060\%$, of the original amount of $Mg^{2+}$ remains in solution. This corresponds to virtually complete precipitation of $Mg^{2+}$ and successful separation of the solution's calcium ions from its magnesium ions. Indeed, magnesium can be separated from the less abundant calcium ions in seawater by this technique.

**CONCEPT TEST** ....................................................................

In the analysis of using a precipitation reaction to separate magnesium ion from calcium ion in solution, we did not specify the concentration of the hydroxide ion solution used to form the precipitates. Why did we not need this value?

....................................................................

**SAMPLE EXERCISE 16.18** **Separating Ions in Solution** **LO8**

Both lead(II) chloride and lead(II) fluoride are slightly soluble salts. A solution of lead(II) nitrate is added to a solution that is 0.275 $M$ in both $Cl^-(aq)$ and $F^-(aq)$. Can we use this method to separate the two halide ions? If "complete precipitation" is defined as there being less than 0.10% of a particular ion left in solution, is the precipitation of the first salt complete before the second salt begins to precipitate? ($K_{sp}$ $PbCl_2 = 1.6 \times 10^{-5}$; $K_{sp}$ $PbF_2 = 3.2 \times 10^{-8}$.)

**Collect and Organize** We are given a solution that contains two ions that form slightly soluble lead(II) salts and asked if one ion can be completely removed before the second one starts to precipitate when lead(II) ion is added to the solution. We have the $K_{sp}$ values for both salts and the initial concentrations of both ions.

**Analyze** The equilibrium constant expressions for both ions are

$$PbCl_2(s) \rightleftharpoons Pb^{2+}(aq) + 2\,Cl^-(aq) \qquad K_{sp} = [Pb^{2+}][Cl^-]^2 = 1.6 \times 10^{-5}$$
$$PbF_2(s) \rightleftharpoons Pb^{2+}(aq) + 2\,F^-(aq) \qquad K_{sp} = [Pb^{2+}][F^-]^2 = 3.2 \times 10^{-8}$$

In both equilibrium expressions, the concentration of lead ion is raised to the first power and the concentration of the halide ion is raised to the second power, so we can compare the influence of the ion concentrations on the $K_{sp}$ values directly. The $K_{sp}$ of $PbCl_2(s)$ is almost $10^3$ times larger than the $K_{sp}$ of $PbF_2(s)$, so we can assume the $PbF_2(s)$ will precipitate first. We can therefore ask: What is the maximum $Pb^{2+}$ concentration in the solution that will not cause the $PbCl_2(s)$ to precipitate? When we determine that value, we can calculate how much $F^-(aq)$ is in solution under those conditions and determine if the precipitation of $F^-$ is complete.

**Solve** The maximum amount of $Pb^{2+}$ in the solution that will not cause the chloride ion to precipitate is

$$K_{sp} = 1.6 \times 10^{-5} = [Pb^{2+}][Cl^-]^2 = (x)(0.275)^2$$
$$x = \frac{1.6 \times 10^{-5}}{(0.275)^2} = 2.12 \times 10^{-4}\ M$$

The concentration of $F^-(aq)$ in the solution at this concentration of lead(II) ion is

$$K_{sp} = 3.2 \times 10^{-8} = (2.12 \times 10^{-4})(x)^2$$
$$x = \sqrt{\frac{3.2 \times 10^{-8}}{2.12 \times 10^{-4}}} = 0.0123\ M \qquad\qquad (a)$$

The original solution was 0.275 $M$ in $F^-(aq)$, and a concentration of 0.0123 $M$ means that $(0.0123/0.275) \times 100\% = 4.5\%$ of the original amount of $F^-(aq)$ remains in solution. The precipitation of $PbF_2(s)$ is not complete and we could not use this method to separate the two ions.

**Think About It** Determining which salt precipitates first was easy in this example because the concentrations of both ions were the same and both were raised to the same power in the $K_{sp}$ expressions. If the concentrations differ, or if the concentrations of the ions in the $K_{sp}$ expressions are raised to different powers, determining which species precipitates first may have to be determined by calculation rather than simple inspection.

⚙ **Practice Exercise** An aqueous solution of sodium fluoride is slowly added to a water sample that contains barium ion (0.0375 $M$) and calcium ion (0.0667 $M$). Consult Appendix 5 to help you answer these questions:
a. Are both barium fluoride and calcium fluoride slightly soluble salts?
b. Applying the definition of complete precipitation given in Sample Exercise 16.18, determine if $Ba^{2+}$ and $Ca^{2+}$ ions in solution can be completely separated by selective precipitation with $F^-$ ions.

**SAMPLE EXERCISE 16.19** **Integrating Concepts: Determining [CO₂] in Seawater**[1]

Earth's oceans absorb $CO_2$ from the atmosphere, including some of that produced by fossil fuel combustion. As $CO_2$ dissolves in the sea, the following equilibria shift to the right:

$$CO_2(aq) + H_2O(\ell) \rightleftharpoons H_2CO_3(aq) \rightleftharpoons HCO_3^-(aq) + H^+(aq)$$

Many scientists are concerned that these shifts will drive down the pH of Earth's oceans and threaten marine ecosystems. A group of these scientists built an instrument for determining $[CO_2]$ in seawater that measures the intensities of the different colors of light absorbed and emitted by a pH indicator (HIn, $pK_a = 7.75$) and its conjugate base ($In^-$). Suppose that the results of one set of these measurements reveal that the ratio of $[In^-]$ to $[HIn]$ in a sample of seawater is 3.16.

a. What is the pH of the sample?
b. If the results of an alkalinity titration disclose that the concentration of $HCO_3^-$ ions in the sample is 1.05 m$M$, what is the concentration of dissolved $CO_2$ in the sample? Given:

$$H_2CO_3(aq) \rightleftharpoons HCO_3^-(aq) + H^+(aq) \qquad pK_{a_1} = 6.37$$

c. If the temperature of the sample is 25°C, what is the partial pressure of atmospheric $CO_2$ in equilibrium with the sample? Given: the Henry's law coefficient for $CO_2$ is $k_H = 3.5 \times 10^{-2}$ $M$/atm at 25°C.

**Collect and Organize** We are given the $pK_a$ of a pH indicator (HIn) and the ratio of $[In^-]$ to $[HIn]$ in a sample and are asked to calculate the pH of the sample. We are also given the value of $[HCO_3^-]$ in the sample and are asked to calculate $[CO_2(aq)]$. Finally we are asked to calculate the partial pressure of atmospheric $CO_2$ in equilibrium with the sample. The Henderson–Hasselbalch equation (16.25) relates the pH and $pK_a$ of a weak acid to the ratio of the concentrations of the acid and its conjugate base:

$$pH = pK_a + \log\frac{[\text{base}]}{[\text{acid}]} \qquad (16.25)$$

[1]Based on *Analytical Chemistry* (1999) *71*, 154.

**Analyze** The pH of the sample can be calculated directly using Equation 16.25 by inserting the $pK_a$ value of the indicator (HIn), and the ratio of $[In^-]$ to $[HIn]$. We can then use this pH value, the $pK_{a_1}$ value of carbonic acid, and the given $[HCO_3^-]$ value to calculate $[H_2CO_3]$, again using Equation 16.25. As noted on page 798, $[H_2CO_3]$ in these calculations represents the total concentration of carbon dioxide dissolved in an aqueous sample, which consist mostly of molecules of $CO_2$. Finally, we use Henry's law to relate the concentration of $CO_2$ dissolved in the sample to the partial pressure of $CO_2$ in the atmosphere.

**Solve**
a. Calculating the pH of the sample from the indicator data:

$$pH = pK_a + \log\frac{[In^-]}{[HIn]}$$
$$= 7.75 + \log 3.16 = 8.25$$

b. Calculating the concentration of dissolved $CO_2$:

$$pH = pK_a + \log\frac{[HCO_3^-]}{[H_2CO_3]}$$
$$8.25 = 6.37 + \log\frac{1.05 \times 10^{-3}\ M}{x}$$
$$x = 1.38 \times 10^{-5}\ M = [CO_2(aq)]$$

c. Calculating $P_{CO_2}$ using Henry's law:

$$C_{CO_2} = k_H P_{CO_2}$$
$$1.38 \times 10^{-5}\ M = 3.5 \times 10^{-2}\ M/\text{atm}\ P_{CO_2}$$
$$P_{CO_2} = 3.9 \times 10^{-4}\ \text{atm}$$

**Think About It** The calculated partial pressure of $CO_2$ is consistent with the current composition of Earth's atmosphere, which contains about 0.039% $CO_2$ by volume (see page 798). This value translates into 0.039 mol% $CO_2$ or a mole fraction of 0.00039. At an atmospheric pressure of 1.00 atm, this mole fraction value corresponds to a partial pressure of $0.00039 \times 1.00$ atm $= 3.9 \times 10^{-4}$ atm.

# The Chemistry of Two Strong Acids: Sulfuric and Nitric Acids

Sulfuric acid ($H_2SO_4$) and nitric acid ($HNO_3$) are among the most widely used industrial chemicals in the world (Figure 16.25). Sulfuric acid ranks first, with a worldwide production of more than 100 million metric tons each year, about one-third of that coming from North America (U.S. and Canada). Nitric acid production ranks 10th on the North American industrial chemicals list. About 70% of the $H_2SO_4$ and 75% of the $HNO_3$ produced in the United States is used to make fertilizer. The rest is used in a variety of chemical manufacturing processes, including the preparation of synthetic fibers described in Chapter 13.

Pure sulfuric acid is a dense, colorless, oily liquid. When heated, it fumes as it partially decomposes into $H_2O$ and $SO_3$. The residual solution is 98.3% $H_2SO_4$ and 1.7% $H_2O$. This solution, which is 18 $M$ $H_2SO_4$, is the liquid sold as concentrated sulfuric acid. It is very hygroscopic (that is, it absorbs water) and is used as a drying agent and to remove water from many compounds. It can dehydrate sugar, turning the carbohydrate into carbon. Sulfuric acid dissolves in water in a process so exothermic that the solution may boil, which is why concentrated sulfuric acid must be diluted by slowly adding it to cold water. *Never* add water to concentrated sulfuric acid.

The synthesis of sulfuric acid starts with the combustion of sulfur to sulfur dioxide, followed by the oxidation of $SO_2$ to $SO_3$:

$$S_8(s) + 8\,O_2(g) \rightarrow 8\,SO_2(g)$$
$$8\,SO_2(g) + 4\,O_2(g) \rightleftharpoons 8\,SO_3(g)$$

Overall    $S_8(s) + 12\,O_2(g) \rightleftharpoons 8\,SO_3(g)$

As we discussed in Chapter 15, both reactions are equilibrium processes, but the equilibrium constant for the formation of $SO_2$ is large and the reaction essentially goes to completion. The reaction is exothermic but slow at room temperature. Higher temperatures speed the rate of the reaction but decrease the equilibrium concentration of the product. The yield is improved by (1) increasing the pressure of the reactants, (2) using an excess of $O_2$, and (3) harvesting $SO_3$ during the reaction. Vanadium(V) oxide, $V_2O_5$, is used as a catalyst, allowing the reaction to proceed at an acceptable rate at moderate temperatures.

Reacting sulfur trioxide with water yields the final sulfuric acid:

$$SO_3(g) + H_2O(\ell) \rightarrow H_2SO_4(\ell)$$

Note that the steps in the production of sulfuric acid are the same as those that lead to the formation of acid rain in the environment.

The production of nitric acid is linked to the production of ammonia because $NH_3$ is a reactant in the synthesis of $HNO_3$. The controlled, selective oxidation of ammonia to NO and subsequent conversion into nitric acid is known as the Ostwald process. Developing it earned Wilhelm Ostwald (1853–1932) the Nobel Prize in Chemistry in 1909 (Figure 16.26). The three steps in the process are

(1)    $4\,NH_3(g) + 5\,O_2(g) \rightarrow 4\,NO(g) + 6\,H_2O(g)$

(2)    $4\,NO(g) + O_2(g) \rightarrow 2\,NO_2(g)$

(3)    $3\,NO_2(g) + H_2O(\ell) \rightarrow 2\,HNO_3(\ell) + NO(g)$

Note that the oxidation number of nitrogen increases from –3 (in $NH_3$) to +2 (in NO) to +4 (in $NO_2$) to +5 (in $HNO_3$) during the process. A catalyst composed of platinum (or platinum and rhodium) and a reaction temperature of 850°C are needed to achieve a rapid rate of conversion of ammonia into nitric acid.

**FIGURE 16.26** Wilhelm Ostwald was a professor of physical chemistry at Leipzig University in Germany from 1887 until 1906. During that time he mentored several brilliant students. Among them were Jacobus Henricus van 't Hoff, who won the Nobel Prize in Chemistry in 1901, and Svante August Arrhenius, who won the Nobel Prize in Chemistry in 1903.

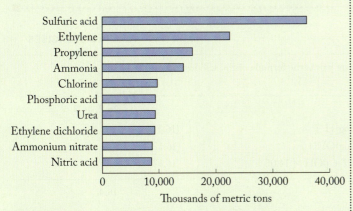

**FIGURE 16.25** The top 10 industrial chemicals produced in the United States and Canada in 2008 included sulfuric acid (1st) and nitric acid (10th). (*Source*: *Chemical and Engineering News*, July 6, 2009, pp. 53, 56.)

## SUMMARY

**Learning Outcome 1** The **Brønsted–Lowry model** of acids and bases defines acids as $H^+$ ion donors and bases as $H^+$ ion acceptors. The $H^+$ ions that acids release combine with water molecules to form hydronium ($H_3O^+$) ions. Most acids are weak acids, which means they ionize only partially in water. When a weak acid HA ionizes, it forms its **conjugate base**, $A^-$. When base B acquires a $H^+$ ion, it forms its **conjugate acid**, $HB^+$. (Section 16.1)

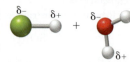

**Learning Outcome 2** In a neutral solution, $[H_3O^+] = [OH^-] = 1.00 \times 10^{-7}$. The **pH** scale is a logarithmic scale for expressing the acidic or basic strength of solutions. Acidic solutions have pH values less than 7; basic solutions have pH values greater than 7. Because pH is the negative logarithm of $H^+$ concentration, the higher the pH, the lower the $H^+$ concentration. An increase in one pH unit represents a decrease in $[H^+]$ to 1/10 of its initial value. (Section 16.2)

**Learning Outcome 3** The values of mass action expressions for acids and bases indicate their relative strengths. The strength of an oxoacid is related to the stability of the anion formed when the acid ionizes. Weak **polyprotic acids** can undergo more than one acid ionization reaction, but the first is the one that usually controls pH. The **percent ionization** of a weak acid HA is the ratio of $[H^+]$ to the initial (total) acid concentration and is usually expressed as a percentage. (Sections 16.3, 16.4, 16.5)

**Learning Outcome 4** A salt solution is acidic if the cation in the salt is the conjugate acid of a weak base and the anion is the conjugate base of a strong acid. A salt solution is basic if the anion in the salt is the conjugate base of a weak acid and the cation is the conjugate acid of a strong base. (Section 16.6)

**Learning Outcome 5** Adding the salt of a weak acid HA to a solution of the acid provides a second source of the conjugate base $A^-$. As predicted by Le Châtelier's principle, the added $A^-$ inhibits the acid ionization reaction, causing pH to rise. Adding a salt of a weak base B to a solution of the base provides a second source of its conjugate acid $HB^+$, which lowers pH. These shifts are examples of the **common-ion effect**. (Section 16.7)

**Learning Outcome 6** A **pH buffer** is a solution that contains either a weak acid and a salt of its conjugate base or a weak base and a salt of its conjugate acid. Buffer solutions resist pH change when acids or bases are added. (Section 16.8)

**Learning Outcome 7** **pH indicators** are used to detect the equivalence points in pH titrations, which are used to determine the concentration of an acid or a base in an aqueous sample. (Section 16.9)

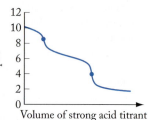

**Learning Outcome 8** The solubility of slightly soluble ionic compounds is described by their $K_{sp}$ or **solubility product**, which is the value of the equilibrium constant for their dissolution. (Section 16.10)

## PROBLEM-SOLVING SUMMARY

| TYPE OF PROBLEM | CONCEPTS AND EQUATIONS | | SAMPLE EXERCISES |
|---|---|---|---|
| **Identifying conjugate acid–base pairs** | The formula of the base in a conjugate pair is the formula of the acid less one $H^+$ ion. | | 16.1 |
| **Interconverting $[H^+]$, $[OH^-]$, pH, and pOH** | Use the following: $$pH = -\log[H^+]$$ $$pOH = -\log[OH^-]$$ $$pK_w = pH + pOH = 14.00$$ | (16.11) (16.12) (16.13) | 16.2, 16.3 |
| **Converting pH into $K_a$ and calculating percent ionization of a weak acid HA** | Use the following: $$[H^+] = 10^{-pH}$$ $$K_a = \frac{[A^-][H^+]}{[HA]}$$ $$\text{Percent ionization} = \frac{[H^+]_{\text{equilibrium}}}{[HA]_{\text{initial}}} \times 100\%$$ | (16.15) (16.16) | 16.4 |
| **Calculating the pH of a solution of weak base B** | Set up a RICE table based on the equilibrium: $$B(aq) + H_2O(\ell) \rightleftharpoons HB^+(aq) + OH^-(aq)$$ Let $x = [OH^-] = [HB^+]$ at equilibrium. Calculate $x$ using $$K_b = \frac{x^2}{[B] - x}$$ | | 16.5 |

| TYPE OF PROBLEM | CONCEPTS AND EQUATIONS | SAMPLE EXERCISES |
|---|---|---|
| | Then use $$pOH = -\log[OH^-]$$ $$pH = 14.00 - pOH$$ | |
| **Calculating the pH of a solution of a weak diprotic acid** | Set up a RICE table based on the $K_{a_1}$ equilibrium: $$H_2A(aq) \rightleftharpoons H^+(aq) + HA^-(aq)$$ Let $x = [H^+] = [HA^-]$ at equilibrium. Calculate $x$ using $$K_{a_1} = \frac{x^2}{[H_2A] - x}$$ Then calculate $pH = -\log[H^+]$. | 16.6 |
| **Distinguishing acidic, basic, and neutral salts** | The cations in acidic salts are the conjugate acids of weak bases. The anions in basic salts are the conjugate bases of weak acids. | 16.7 |
| **Calculating the pH of a solution of an acidic salt** | Assume the salt completely dissociates into $BH^+$ and $X^-$. Set up a RICE table for the equilibrium: $$BH^+(aq) \rightleftharpoons B(aq) + H^+(aq)$$ Let $x = [H^+] = [B]$ at equilibrium. Calculate $x$ using $$K_a = \frac{K_w}{K_b} = \frac{x^2}{[BH^+] - x}$$ Then calculate pH. | 16.8 |
| **Calculating pH of a solution of a weak base and its conjugate acid (or a weak acid and its conjugate base)** | Use the Henderson–Hasselbalch relation: $$pH = pK_a + \log\frac{[base]}{[acid]} \qquad (16.25)$$ | 16.9, 16.10, 16.12 |
| **Preparing a buffer of given pH** | Use the Henderson–Hasselbalch equation: $$pH = pK_a + \log\frac{[base]}{[acid]} \qquad (16.25)$$ | 16.11 |
| **Effect of concentration and [base]:[acid] ratio on buffer capacity** | Use the Henderson–Hasselbalch equation: $$pH = pK_a + \log\frac{[base]}{[acid]} \qquad (16.25)$$ | 16.12, 16.13 |
| **Interpreting results of an alkalinity titration** | Use the relation $$V_A M_A = \frac{n_A}{n_B} V_B M_B \qquad (16.29)$$ where $V_A$ and $M_A$ are the volume and molarity of the acid, $V_B$ and $M_B$ are the volume and molarity of the base, and $n_A$ and $n_B$ are the coefficients of the acid and base, respectively, in the balanced equation. | 16.14 |
| **Calculating molar solubility ($S$) of $M_mZ_z$ from $K_{sp}$** | Use the relation $$K_{sp} = (m^m z^z)S^{(m + z)} \qquad (16.30)$$ | 16.15 |
| **Calculating the effect of pH on the molar solubility ($S$) of MZ** | If $Z^-$ is the conjugate base of a weak acid, calculate the fraction of $Z^-$ that remains as the free ion. Use this fraction as the coefficient of $S$ in the $K_{sp}$ expression. | 16.16 |
| **Determining if a precipitate forms when solutions are mixed and which precipitate forms if more than one is possible** | Compare the ion product $Q$ to $K_{sp}$ to determine if a precipitate will form; use $K_{sp}$ expressions to calculate maximum concentrations of one ion in solution that will not cause another ion to precipitate. | 16.17, 16.18 |

## VISUAL PROBLEMS

*(Answers to boldface end-of-chapter questions and problems are in the back of the book.)*

**16.1.** Which of the lines in Figure P16.1 best represents the dependence of the degree of ionization of acetic acid on its concentration in aqueous solution?

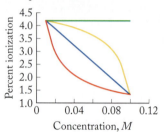

**FIGURE P16.1**

**16.2.** The graph in Figure P16.2 shows the percent ionization of two acids as a function of concentration in water. Which line describes the behavior of $HNO_3$, and which line butanoic acid ($CH_3CH_2CH_2COOH$)?

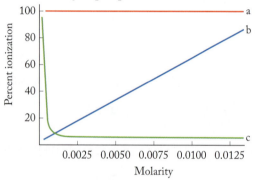

**FIGURE P16.2**

**16.3.** The graph in Figure P16.3 shows the titration curves of a 1 *M* solution of a weak acid with a strong base and a 1 *M* solution of a strong acid with the same base. Which curve is which?

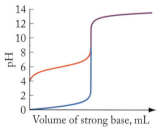

**FIGURE P16.3**

**16.4.** Estimate to within one pH unit the pH of a 0.5 *M* solution of the sodium salt of the weak acid in Problem 16.3.

**16.5.** Suppose you have four color indicators to choose from for detecting the equivalence point of the titration reaction represented by the red curve (upper curve on the left side of the plot) in Figure P16.3. The $pK_a$ values of the four indicators are 3.3, 5.0, 7.0, and 9.0. Which indicator would be the best one to choose?

**16.6.** What is the $pK_a$ value of the weak acid in Figure P16.3?

**16.7.** One of the titration curves in Figure P16.7 represents the titration of an aqueous sample of $Na_2CO_3$ with strong acid; the other represents the titration of an aqueous sample of $NaHCO_3$ with the same acid. Which curve is which?

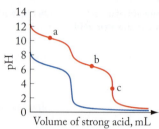

**FIGURE P16.7**

**16.8.** Identify the major species present in solution at points a–c on the red curve in Figure P16.7.

**16.9.** Consider the three beakers in Figure P16.9. Each contains a few drops of the color indicator bromthymol blue, which is yellow in acidic solutions and blue in basic solutions. One beaker contains a solution of ammonium chloride, one contains ammonium acetate, and the third contains sodium acetate. Which beaker contains which salt?

**FIGURE P16.9**

*16.10.** The graphs in Figure P16.10 show the conductivity of a solution as a function of the volume of titrant added. Which of the graphs best represents the titration of (a) a strong acid with a strong base and (b) a weak acid with a strong base?

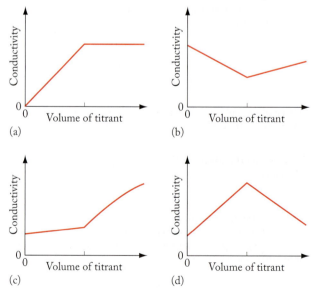

**FIGURE P16.10**

## QUESTIONS AND PROBLEMS ·····························■

## Acids and Bases: The Brønsted–Lowry Model

### CONCEPT REVIEW

**16.11.** In an aqueous solution of HBr, which compound acts as a Brønsted–Lowry acid and which is the Brønsted–Lowry base?

**16.12.** In an aqueous solution of $HClO_4$, which compound acts as a Brønsted-Lowry acid and which is the Brønsted-Lowry base?

**16.13.** In an aqueous solution of NaOH, which species acts as a Brønsted–Lowry acid and which is the Brønsted–Lowry base?

**16.14.** Both NaOH and $Ca(OH)_2$ are strong bases. Does this mean that solutions of the two compounds with the same molarity have the same capacity to neutralize strong acids? Why or why not?

**16.15.** Identify the acids and bases in the following reactions:
  a. $HNO_3(aq) + NaOH(aq) \rightarrow NaNO_3(aq) + H_2O(\ell)$
  b. $CaCO_3(s) + 2\ HCl(aq) \rightarrow CaCl_2(aq) + CO_2(g) + H_2O(\ell)$
  c. $NH_3(aq) + HCN(aq) \rightarrow NH_4CN(aq)$

**16.16.** Identify the acids and bases in the following reactions:
  a. $NH_3(aq) + H_2PO_4^-(aq) \rightleftharpoons NH_4^+(aq) + HPO_4^{2-}(aq)$
  b. $ClO^-(aq) + H_2O(\ell) \rightleftharpoons HClO(aq) + OH^-(aq)$
  c. $(CH_3)_3COH(aq) + H_3O^+(aq) \rightleftharpoons$
  $\qquad\qquad\qquad\qquad (CH_3)_3COH_2^+(aq) + H_2O(\ell)$

**16.17.** Identify the conjugate base of each of the following compounds: $HNO_2$, $HClO$, $H_3PO_4$, and $NH_3$.

**16.18.** Identify the conjugate acid of each of the following species: $NH_3$, $ClO_2^-$, $SO_4^{2-}$, and $OH^-$.

### PROBLEMS

**16.19.** What is the concentration of $H^+$ ions in a 1.50 $M$ solution of $HNO_3$? What is the concentration of $H^+$ ions after adding 10 mL of 0.505 $M$ NaOH to 100 mL of 1.50 $M$ HCl?

**16.20.** What is the concentration of $H^+$ ions in a solution of hydrochloric acid that was prepared by diluting 20.0 mL of concentrated (11.6 $M$) HCl to a final volume of 500 mL?

---

**16.21.** What is the value of $[OH^-]$ in a 0.0800 $M$ solution of $Sr(OH)_2$? What is the concentration of $OH^-$ ions after adding 12 mL of 0.465 $M$ HCl to 100 mL of 0.0800 $M$ $Sr(OH)_2$?

**16.22.** Calcium hydroxide, also known as slaked lime, is the cheapest strong base available and it is used in industrial processes in which low concentrations of base are required. Only 0.16 g of $Ca(OH)_2$ dissolves in 100 mL of water at 25°C. What is the concentration of hydroxide ions in 250 mL of a solution containing the maximum amount of dissolved calcium hydroxide?

---

**16.23.** Describe how you would prepare 2.50 L of a NaOH solution in which $[OH^-]$ = 0.70 $M$, starting with solid NaOH.

**16.24.** How many milliliters of a 1.00 $M$ solution of NaOH do you need to prepare 250 mL of a solution in which $[OH^-]$ = 0.0200 $M$?

## pH and the Autoionization of Water

### CONCEPT REVIEW

**16.25.** Explain why pH values decrease as acidity increases.

**16.26.** Solution A is 100 times more acidic than solution B. What is the difference in the pH values of solution A and solution B?

**16.27.** Under what conditions is the pH of a solution negative?

16.28. Can pOH values ever be less than zero?

**\*16.29.** How does the value of $pK_w$ depend on temperature?

\*16.30. Liquid ammonia at a temperature of 223 K undergoes autoionization. The value of the equilibrium constant for the autoionization of ammonia is considerably less than that of water. Write an equation for the autoionization of ammonia and suggest a reason why the value of K for the process is less than that of water.

### PROBLEMS

**16.31.** Calculate the pH and pOH of the solutions with the following hydrogen ion or hydroxide ion concentrations. Indicate which solutions are acidic, basic, or neutral.
  a. $[H^+] = 3.45 \times 10^{-8}\ M$
  b. $[H^+] = 2.0 \times 10^{-5}\ M$
  c. $[H^+] = 7.0 \times 10^{-8}\ M$
  d. $[OH^-] = 8.56 \times 10^{-4}\ M$

16.32. Calculate the pH and pOH of the solutions with the following hydrogen ion or hydroxide ion concentrations. Indicate which solutions are acidic, basic, or neutral.
  a. $[OH^-] = 1.44 \times 10^{-10}\ M$
  b. $[OH^-] = 6.37 \times 10^{-2}\ M$
  c. $[H^+] = 3.39 \times 10^{-9}\ M$
  d. $[H^+] = 4.92 \times 10^{-3}\ M$

---

**16.33.** Calculate the concentration of the following ions in the solution described:
  a. $[H^+]$ in $8.42 \times 10^{-4}\ M$ NaOH
  b. $[H^+]$ in $3.97 \times 10^{-5}\ M$ $Ca(OH)_2$
  c. $[OH^-]$ in $4.51 \times 10^{-3}\ M$ HCl
  d. $[OH^-]$ in $6.92 \times 10^{-5}\ M$ HCl

16.34. Determine the indicated characteristic of the following solutions:
  a. pH of a solution of pOH 9.47
  b. pH of a solution of pOH 3.86
  c. pOH of a solution of pH 7.91
  d. pOH of a solution of pH 4.23

---

**16.35.** Calculate the pH and pOH of the following solutions:
  a. stomach acid in which [HCl] = 0.155 $M$
  b. 0.00500 $M$ $HNO_3$
  c. a 2:1 mixture of 0.0125 $M$ HCl and 0.0125 $M$ NaOH
  d. a 3:1 mixture of 0.0125 $M$ $H_2SO_4$ and 0.0125 $M$ KOH

16.36. Calculate the pH and pOH of the following solutions:
  a. 0.0450 $M$ NaOH
  b. 0.160 $M$ $Ca(OH)_2$
  c. a 1:1 mixture of 0.0125 $M$ HCl and 0.0125 $M$ $Ca(OH)_2$
  d. a 2:3 mixture of 0.0125 $M$ $HNO_3$ and 0.0125 $M$ KOH

---

**\*16.37.** Calculate the pH of a $1.33 \times 10^{-9}\ M$ solution of LiOH.

**\*16.38.** Calculate the pH of a $6.9 \times 10^{-8}\ M$ solution of HBr.

## Calculations Involving pH, $K_a$, and $K_b$

### CONCEPT REVIEW

**16.39.** One-molar solutions of the following acids are prepared: $CH_3COOH$, $HNO_2$, $HClO$, and $HCl$.
  a. Rank them in order of decreasing $[H^+]$.
  b. Rank them in order of increasing strength as acids (weakest to strongest).

**16.40.** On the basis of the following degree-of-ionization data for 0.100 $M$ solutions, select which acid has the smallest $K_a$.

| Acid | Degree of Ionization (%) |
|---|---|
| $C_6H_5COOH$ | 2.5 |
| HF | 8.5 |
| $HN_3$ | 1.4 |
| $CH_3COOH$ | 1.3 |

**16.41.** A 1.0 $M$ aqueous solution of $NaNO_2$ is a much better conductor of electricity than is a 1.0 $M$ solution of $HNO_2$. Explain why.

**16.42.** Hydrogen chloride and water are molecular compounds, yet a solution of HCl dissolved in $H_2O$ is an excellent conductor of electricity. Explain why.

**16.43.** Hydrofluoric acid is a weak acid. Write the mass action expression for its acid ionization reaction.

**16.44. Early Antiseptic** The use of phenol, also known as carbolic acid, was pioneered in the 19th century by Sir Joseph Lister (after whom Listerine was named) as an antiseptic in surgery. Its formula is $C_6H_5OH$; the red hydrogen atom is ionizable. Write the mass action expression for the acid ionization equilibrium of phenol.

**\*16.45.** The $K_a$ values of weak acids depend on the solvent in which they dissolve. For example, the $K_a$ of alanine in aqueous ethanol is less than its $K_a$ in water.
   a. In which solvent does alanine ionize to the largest degree?
   b. Which is the stronger Brønsted–Lowry base: water or ethanol?

**\*16.46.** The $K_a$ of proline is $2.5 \times 10^{-11}$ in water, $2.8 \times 10^{-11}$ in an aqueous solution that is 28% ethanol, and $1.66 \times 10^{-8}$ in aqueous formaldehyde at 25°C.
   a. In which solvent is proline the strongest acid?
   b. Rank these compounds on the basis of their strengths as Brønsted–Lowry bases: water, ethanol, and formaldehyde.

**16.47.** When methylamine, $CH_3NH_2$, dissolves in water, the resulting solution is slightly basic. Which compound is the Brønsted–Lowry acid and which is the base?

**\*16.48.** When 1,2-diaminoethane, $H_2NCH_2CH_2NH_2$, dissolves in water, the resulting solution is basic. Write the formula of the ionic compound that is formed when hydrochloric acid is added to a solution of 1,2-diaminoethane.

## PROBLEMS

**16.49. Muscle Physiology** During strenuous exercise lactic acid builds up in muscle tissues. In a 1.00 $M$ aqueous solution, 2.94% of lactic acid is ionized. What is the value of its $K_a$?

**16.50. Rancid Butter** The odor of spoiled butter is due in part to butanoic acid, which results from the chemical breakdown of butterfat. A 0.100 $M$ solution of butanoic acid is 1.23% ionized. Calculate the value of $K_a$ for butanoic acid.

**16.51.** At equilibrium, the value of $[H^+]$ in a 0.250 $M$ solution of an unknown acid is $4.07 \times 10^{-3}$ $M$. Determine the degree of ionization and the $K_a$ of this acid.

**16.52.** Nitric acid ($HNO_3$) is a strong acid that is completely ionized in aqueous solutions of concentrations ranging from 1% to 10% (1.5 $M$). However, in more concentrated solutions, part of the nitric acid is present as un-ionized molecules of $HNO_3$. For example, in a 50% solution (7.5 $M$) at 25°C, only 33% of the molecules of $HNO_3$ dissociate into $H^+$ and $NO_3^-$. What is the $K_a$ value of $HNO_3$?

**16.53. Ant Bites** The venom of biting ants contains formic acid, HCOOH, $K_a = 1.8 \times 10^{-4}$ at 25°C. Calculate the pH of a 0.060 $M$ solution of formic acid.

**16.54. Poisonous Plant** Gifblaar is a small South African shrub and one of the most poisonous plants known because it contains fluoroacetic acid. If a 0.480 $M$ solution of fluoroacetic acid has a pH of 1.44, what is the $K_a$ of the acid?

**16.55. Acid Rain** A weather system moving through the American Midwest produced rain with an average pH of 5.02. By the time the system reached New England, the rain it produced had an average pH of 4.66. How much more acidic was the rain falling in New England?

**16.56. Acid Rain II** A newspaper reported that the "level of acidity" in a sample taken from an extensively studied watershed in New Hampshire in February 1998 was "an astounding 200 times lower than the worst measurement" taken in the preceding 23 years. What is this difference expressed in units of pH?

**16.57.** The $K_b$ of aminoethanol, $HOCH_2CH_2NH_2$, is $3.1 \times 10^{-5}$.
   a. Is aminoethanol a stronger or weaker base than ethylamine, $pK_b = 3.36$?
   b. Calculate the pH of a $1.67 \times 10^{-2}$ $M$ solution of aminoethanol.
   c. Calculate the $[OH^-]$ concentration of a $4.25 \times 10^{-4}$ $M$ solution of aminoethanol.

**16.58. Food Dye** Quinoline is a weakly basic liquid used in the manufacture of quinolone yellow, a greenish yellow dye for foods, and also in the production of niacin. Its $pK_b$ is 9.15.
   a. What is the pH of a 0.0752 $M$ solution of quinolone?
   b. What is the hydroxide ion concentration of the solution in part a?

**16.59. Painkillers** Morphine is an effective painkiller but is also highly addictive. Codeine is a popular prescription painkiller because it is much less addictive than morphine. Codeine contains a basic nitrogen atom that can be protonated to give the conjugate acid of codeine.
   a. Calculate the pH of a 0.115 $M$ solution of morphine if its $pK_b = 5.79$.
   b. Calculate the pH of a $3.42 \times 10^{-4}$ $M$ solution of codeine if the $pK_a$ of the conjugate acid is 8.21.

**16.60.** The odor of dead fish is attributed to trimethylamine, $(CH_3)_3N$, one of a series of compounds of nitrogen, carbon, and hydrogen with the general formula $(CH_3)_nNH_{3-n}$ where $n = 0$–3. (The $n = 0$ compound is ammonia.)
   a. The $K_b$ of trimethylamine $[(CH_3)_3N]$ is $6.5 \times 10^{-5}$ at 25°C. Calculate the pH of a $3.00 \times 10^{-4}$ $M$ solution of dimethylamine.
   b. The $K_b$ of methylamine $[(CH_3)NH_2]$ is $4.4 \times 10^{-4}$ at 25°C. Calculate the pH of a $2.88 \times 10^{-3}$ $M$ solution of methylamine.
   \*c. The $K_b$ of dimethylamine $[(CH_3)_2NH]$ is $5.9 \times 10^{-4}$ at 25°C. What concentration of dimethylamine is needed for the solution to have the same pH as the solution in part b?

## Polyprotic Acids

### CONCEPT REVIEW

**16.61.** Why is the $K_{a_2}$ value of phosphoric acid less than its $K_{a_1}$ value but greater than its $K_{a_3}$ value?

**16.62.** In calculating the pH of a 1.0 $M$ solution of sulfurous acid, we can ignore the $H^+$ ions produced by the ionization of the bisulfite ion; however, in calculating the pH of a 1.0 $M$ solution of sulfuric acid, we cannot ignore the $H^+$ ions produced by the ionization of the bisulfate ion. Why?

### PROBLEMS

**16.63.** What is the pH of a 0.300 $M$ solution of $H_2SO_4$?

**16.64.** What is the pH of a 0.150 $M$ solution of sulfurous acid?

**16.65.** Ascorbic acid (vitamin C) is a diprotic acid. What is the pH of a 0.250 $M$ solution of ascorbic acid?

**16.66.** The leaves of the rhubarb plant contain high concentrations of diprotic oxalic acid (HOOCCOOH) and must be removed before the stems are used to make rhubarb pie. What is the pH of a 0.0288 $M$ solution of oxalic acid?

**16.67.** **Addiction to Tobacco** Nicotine is responsible for the addictive properties of tobacco. What is the pH of a $1.00 \times 10^{-3}$ $M$ solution of nicotine?

**16.68.** Pseudoephedrine hydrochloride (Fig. P16.68) is a common ingredient in cough syrups and decongestants. Its $pK_a = 9.22$. What is the pH of a solution that is 0.0295 $M$ in pseudoephedrine hydrochloride?

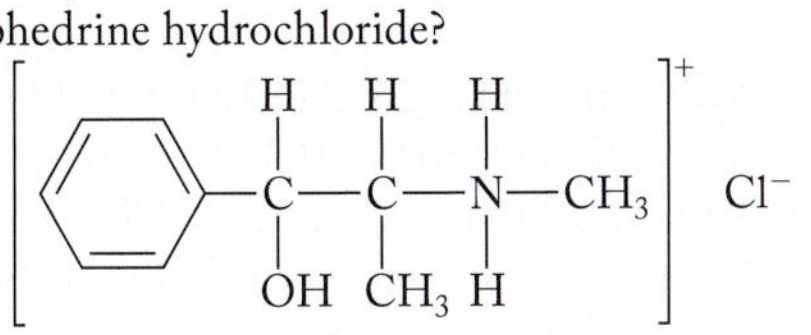

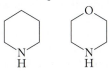

**FIGURE P16.68**

**16.69.** **Malaria Treatment** Quinine occurs naturally in the bark of the cinchona tree. For centuries it was the only treatment for malaria. Calculate the pH of a 0.01050 $M$ solution of quinine in water.

**16.70.** Dozens of pharmaceuticals ranging from cyclizine for motion sickness to Viagra for impotence are derived from the organic compound piperazine, whose structure is shown in Figure P16.70.

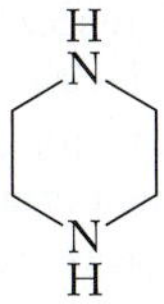

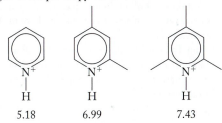

**FIGURE P16.70**

a. Solutions of piperazine are basic ($K_{b_1} = 5.38 \times 10^{-5}$; $K_{b_2} = 2.15 \times 10^{-9}$). What is the pH of a 0.0133 $M$ solution of piperazine?

*b. Draw the structure of the ionic form of piperazine that would be present in stomach acid (about 0.15 $M$ HCl).

## Acid Strength and Molecular Structure

### CONCEPT REVIEW

**16.71.** Explain why the $K_{a_1}$ of $H_2SO_4$ is much greater than the $K_{a_1}$ of $H_2SeO_4$.

**16.72.** Explain why the $K_{a_1}$ of $H_2SO_4$ is much greater than the $K_{a_1}$ of $H_2SO_3$.

**16.73.** Predict which acid in the following pairs of acids is the stronger acid: (a) $H_2SO_3$ or $H_2SeO_3$; (b) $H_2SeO_4$ or $H_2SeO_3$.

**16.74.** Trifluoroacetic acid, $CF_3COOH$, is over $10^4$ times stronger than acetic acid, $CH_3COOH$. Explain why.

*16.75. Explain why the $pK_b$ of ethanolamine, $HOCH_2CH_2NH_2$, is greater than that of ethylamine, $CH_3CH_2NH_2$.

*16.76. Which base in Figure P16.76 do you predict will have the larger $K_b$, piperidine or morpholine?

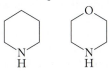

Piperidine    Morpholine

**FIGURE P16.76**

## pH of Salt Solutions

### CONCEPT REVIEW

*16.77. The $pK_a$ values of the conjugate acids of pyridine derivatives shown in Figure P16.77 increase as more methyl groups are added. Do more methyl groups increase or decrease the strength of the parent pyridine bases?

5.18    6.99    7.43

**FIGURE P16.77**

**16.78.** Why is it unnecessary to publish tables of $K_b$ values of the conjugate bases of weak acids whose $K_a$ values are known?

**16.79.** Which of the following salts produces an acidic solution in water: ammonium acetate, ammonium nitrate, or sodium formate?

**16.80.** Which of the following salts produces a basic solution in water: NaF, KCl, $NH_4Cl$?

**16.81.** **Neutralizing the Smell of Fish** Trimethylamine, $(CH_3)_3N$, $K_b = 6.5 \times 10^{-5}$ at 25°C, is a contributor to the "fishy" odor of not-so-fresh seafood. Some people squeeze fresh lemon juice (which contains a high concentration of citric acid) on cooked fish to reduce the fishy odor. Why is this practice effective?

**16.82.** **Nutritional Value of Beets** Beets contain high concentrations of the calcium salt of a dicarboxylic acid with the common name malonic acid and the formula $HOOCCH_2COOH$. Could the presence of the calcium salt of malonic acid affect the pH balance of beets? If so, in which direction? Explain.

### PROBLEMS

**16.83.** If the $K_a$ of the conjugate acid of the artificial sweetener saccharin is $2.1 \times 10^{-11}$, what is the $pK_b$ for saccharin?

**16.84.** If the $K_{a_1}$ value for oxalic acid (HOOCCOOH) is $5.9 \times 10^{-2}$ and the $K_{a_2}$ value is $6.4 \times 10^{-5}$, what are the values of $K_{b_1}$ and $K_{b_2}$ of the oxalate anion (−OOCCOO−)?

**16.85. Dental Health** Sodium fluoride is added to many municipal water supplies to reduce tooth decay. Calculate the pH of a 0.00339 $M$ solution of NaF at 25°C.

**16.86.** Calculate the pH of a $1.25 \times 10^{-2}$ $M$ solution of the decongestant ephedrine hydrochloride if the p$K_b$ of ephedrine (its conjugate base) is 3.86.

## The Common-Ion Effect and pH Buffers

### CONCEPT REVIEW

**16.87.** Why is a solution of sodium acetate and acetic acid a much better pH buffer than is a solution of sodium chloride and hydrochloric acid?

**16.88.** Why does a solution of a weak base and its conjugate acid act as a better buffer than does a solution of the weak base alone?

**16.89.** What effect does adding NaF have on the pH and buffer capacity of an aqueous solution of 1.0 $M$ HF and 0.050 $M$ NaF?

**16.90.** What effect does adding silver nitrate to a $Na_2HPO_4/Na_3PO_4$ buffer have on the pH of the buffer?

**16.91.** Two buffers are prepared with equal concentrations of acid and conjugate base using weak acids with different p$K_a$ values. Do they have the same buffer capacity?

**16.92.** Equal amounts of $H^+$ and $OH^-$ are added to a buffer. What is the effect on the pH of the buffer?

**16.93.** Using Table A5.1, identify a suitable buffer system to maintain a pH of 3.0 in an aqueous solution. Assume the concentrations of acid and conjugate base in the buffer are initially equal.

**16.94.** Using Table A5.1, identify a suitable buffer system to maintain a pH of 9.0 in an aqueous solution. Assume the concentrations of acid and conjugate base in the buffer are initially equal.

### PROBLEMS

**16.95.** Calculate the pH of a buffer that is 0.244 $M$ acetic acid and 0.122 $M$ sodium acetate at 25°C. What is the pH of this mixture at 0°C ($K_a = 1.64 \times 10^{-5}$)?

**16.96.** Calculate the pH of a buffer that is 0.100 $M$ pyridine and 0.275 $M$ pyridinium chloride at 25°C.

**16.97.** Calculate the pH and pOH of 500.0 mL of a phosphate buffer that is 0.225 $M$ $HPO_4^{2-}$ and 0.225 $M$ $PO_4^{3-}$ at 25°C.

**16.98.** Determine the pH and pOH of 0.250 L of a buffer that is 0.0200 $M$ boric acid and 0.0250 $M$ sodium borate at 25°C.

**16.99.** How would you prepare 100 mL of buffer with pH = 3.00 from iodoacetic acid and sodium iodoacetate?

**16.100.** How would you prepare 100 mL of a buffer with pH = 11.0 from dimethylamine (p$K_b$ = 3.2) and dimethylammonium chloride, the conjugate acid of dimethylamine?

**16.101.** Buffers are prepared using 1.0 $M$ solutions of the following: formic acid (p$K_a$ = 3.75), hydrofluoric acid (p$K_a$ = 3.17), and hydrocyanic acid (p$K_a$ = 9.21) and their conjugate bases. Which buffer will have the lowest pH?

**16.102.** Does a buffer prepared from ammonia (p$K_b$ = 4.75) and ammonium chloride have the same pH as one prepared from phenol (p$K_a$ = 9.89) and sodium phenolate (the conjugate base of phenol) if equal concentrations of acid and conjugate base are used?

**16.103.** A 100 mL sample of an acetic acid/sodium acetate buffer is diluted with 100 mL of distilled water. What is the effect of the dilution on the pH of the buffer? What is the effect of the dilution on the buffer capacity?

**16.104.** A co-worker in a biochemistry laboratory suggests to you that a buffer solution with a pH of 8.00 may be prepared by mixing 72.00 mL of 0.200 $M$ $NH_3$ solution with 128.0 mL of 0.200 $M$ $NH_4Cl$ solution.
   a. Is your co-worker correct? What is the pH of the resultant solution?
   b. If the resultant buffer solution were diluted to a volume of 1.00 L, would its pH change?

**16.105.** What is the pH at 25°C of a solution that results from mixing together equal volumes of a 0.05 $M$ solution of ammonia and a 0.025 $M$ solution of hydrochloric acid?

**16.106.** What is the pH at 25°C of a solution that results from mixing together equal volumes of a 0.05 $M$ solution of acetic acid and a 0.025 $M$ solution of sodium hydroxide?

**\*16.107.** The pH of a solution of an unidentified acid and its conjugate base is 3.15. The concentration of the acid is 0.080 $M$ and the concentration of its conjugate base is 0.020 $M$. What is the p$K_a$ of the acid?

**\*16.108.** How much 6.0 $M$ NaOH must be added to 0.500 L of a buffer that is 0.0200 $M$ acetic acid and 0.0250 $M$ sodium acetate to raise the pH to 5.75 at 25°C?

**\*16.109.** Calculate the pH at 25°C of 1.00 L of a buffer that is 0.120 $M$ $HNO_2$ and 0.150 $M$ $NaNO_2$ before and after the addition of 1.00 mL of 12.0 $M$ HCl.

**\*16.110.** An acetic acid–sodium acetate buffer can be prepared by adding sodium acetate to HCl($aq$).
   a. Write an equation for the reaction that occurs when sodium acetate is added to HCl($aq$).
   b. If 10.0 g of $CH_3COONa$ is added to 275 mL of 0.225 $M$ HCl, what is the pH of the resulting buffer?
   c. What is the pH of the solution in part b after the addition of 1.26 g of solid NaOH?

## Indicators and Acid–Base Titrations

### CONCEPT REVIEW

**16.111.** What are the differences between the titration curve of a strong acid titrated with a strong base and that of a weak acid titrated with a strong base?

**16.112.** When carrying out a titration using an acid–base indicator, it is advisable to use as small a quantity of indicator as possible. Why?

**16.113.** Do all titrations of a weak acid with a strong base have the same pH at the equivalence point?

**16.114.** Compare the volume of 0.125 $M$ NaOH solution required to titrate 25.0 mL of a 0.50 $M$ solution of a weak acid to that required for 25.0 mL of a 0.50 $M$ solution of a strong acid. Compare the pH at the equivalence point of both solutions.

### PROBLEMS

**16.115.** A 25.0 mL sample of 0.100 $M$ acetic acid is titrated with 0.125 $M$ NaOH. Calculate the pH at 25°C of the titration mixture after 10.0, 20.0, and 30.0 mL of the base have been added.

**16.116.** A 25.0 mL sample of a 0.100 $M$ solution of aqueous trimethylamine is titrated with a 0.125 $M$ solution of HCl. Calculate the pH of the solution after 10.0, 20.0, and 30.0 mL of acid have been added; $pK_b$ of $(CH_3)_3N = 4.19$ at 25°C.

**16.117.** What is the concentration of ammonia in a solution if 22.35 mL of 0.1145 $M$ HCl is needed to titrate a 100.0 mL sample of the solution?

**16.118.** In an alkalinity titration of a 100.0 mL sample of water from a hot spring, 2.56 mL of a 0.0355 $M$ solution of HCl is needed to reach the first equivalence point (pH = 8.3) and another 10.42 mL is needed to reach the second equivalence point (pH = 4.0). If the alkalinity of the spring water is due only to the presence of carbonate and bicarbonate, what are the concentrations of each?

**16.119.** What volume of 0.0100 $M$ HCl is required to titrate 250 mL of 0.0100 $M$ Na$_2$CO$_3$ to the first equivalence point?

**16.120.** How much 0.0100 $M$ HCl is required to titrate 250 mL of 0.0100 $M$ Na$_2$CO$_3$ and 250 mL of 0.0100 $M$ HCO$_3^-$?

**16.121.** In the titration of a solution of a weak monoprotic acid with a 0.1025 $M$ solution of NaOH, the pH halfway to the equivalence point was 4.44. In the titration of a second solution of the same acid, exactly twice as much of a 0.1025 $M$ solution of NaOH was needed to reach the equivalence point. What was the pH halfway to the equivalence point in this titration?

**16.122.** A 125.0 mg sample of an unknown, monoprotic acid was dissolved in 100.0 mL of distilled water and titrated with a 0.050 $M$ solution of NaOH. The pH of the solution was monitored throughout the titration, and the following data were collected. Determine the $K_a$ of the acid.

| Volume of OH⁻ Added (mL) | pH | Volume of OH⁻ Added (mL) | pH |
|---|---|---|---|
| 0 | 3.09 | 22 | 5.93 |
| 5 | 3.65 | 22.2 | 6.24 |
| 10 | 4.10 | 22.6 | 9.91 |
| 15 | 4.50 | 22.8 | 10.2 |
| 17 | 4.55 | 23 | 10.4 |
| 18 | 4.71 | 24 | 10.8 |
| 19 | 4.94 | 25 | 11.0 |
| 20 | 5.11 | 30 | 11.5 |
| 21 | 5.37 | 40 | 11.8 |

**16.123.** Sketch a titration curve for the titration of 50.0 mL of 0.250 $M$ HNO$_2$ with 1.00 $M$ NaOH. What is the pH at the equivalence point?

**16.124.** Red cabbage juice is a sensitive acid–base indicator; its colors range from red at acidic pH to yellow in alkaline solutions. What color would red cabbage juice have when 25 mL of a 0.10 $M$ solution of acetic acid is titrated with 0.10 $M$ NaOH to its equivalence point?

**16.125.** Sketch a titration curve for the titration of the malaria drug quinine if 40.0 mL of a 0.100 $M$ solution of quinine is titrated with a 0.100 $M$ solution of HCl.

**16.126.** Sketch a titration curve for the titration of 100 mL of $1.25 \times 10^{-2}$ $M$ ascorbic acid with $1.00 \times 10^{-2}$ $M$ NaOH. How many equivalence points should the curve have, and what pH indicator(s) could be used? Refer to Figure 16.20 for colors of indicators.

## Solubility Equilibria

### CONCEPT REVIEW

**16.127.** What is the difference between *molar solubility* and *solubility product*?

**16.128.** Give an example of how the common-ion effect limits the dissolution of a sparingly soluble ionic compound.

**16.129.** Which of the following cations will precipitate first as a carbonate mineral from an equimolar solution of Mg$^{2+}$, Ca$^{2+}$, and Sr$^{2+}$?

**16.130.** If the solubility of a compound increases with increasing temperature, does $K_{sp}$ increase or decrease?

**16.131.** The $K_{sp}$ of strontium sulfate increases from $2.8 \times 10^{-7}$ at 37°C to $3.8 \times 10^{-7}$ at 77°C. Is the dissolution of strontium sulfate endothermic or exothermic?

**16.132.** Identify any of the following solids that are more soluble in acidic solution than in neutral water: CaCl$_2$, Ba(HCO$_3$)$_2$, PbSO$_4$, Cu(OH)$_2$. Explain your choices.

**16.133. Chemistry of Tooth Decay** Tooth enamel is composed of a mineral known as hydroxyapatite with the formula Ca$_5$(PO$_4$)$_3$(OH). Explain why tooth enamel can be eroded by acidic substances released by bacteria growing in the mouth.

**16.134. Fluoride and Dental Hygiene** Fluoride ions in drinking water and toothpaste convert hydroxyapatite in tooth enamel into fluorapatite:

$$Ca_5(PO_4)_3(OH)(s) + F^-(aq) \rightleftharpoons Ca_5(PO_4)_3F(s) + OH^-(aq)$$

Why is fluorapatite less susceptible than hydroxyapatite to erosion by acids?

### PROBLEMS

**16.135.** At a particular temperature the value of [Ba$^{2+}$] in a saturated solution of barium sulfate is $1.04 \times 10^{-5}$ $M$. Starting with this information, calculate the $K_{sp}$ value of barium sulfate at this temperature.

**16.136.** Problem 16.22 said that only 0.16 g of Ca(OH)$_2$ dissolves in 100 mL of water at 25°C. What is the $K_{sp}$ value for calcium hydroxide at that temperature?

**16.137.** What are the equilibrium concentrations of Cu$^+$ and Cl$^-$ in a saturated solution of copper(I) chloride if $K_{sp} = 1.02 \times 10^{-6}$?

**16.138.** What are the equilibrium concentrations of Pb$^{2+}$ and F$^-$ in a saturated solution of lead fluoride if the $K_{sp}$ value of PbF$_2$ is $3.2 \times 10^{-8}$?

**16.139.** What is the solubility of calcite (CaCO$_3$) in grams per milliliter at a temperature at which its $K_{sp} = 9.9 \times 10^{-9}$?

**16.140.** What is the solubility of silver iodide in grams per milliliter at a temperature at which its $K_{sp} = 1.50 \times 10^{-16}$?

**16.141.** What is the pH at 25°C of a saturated solution of silver hydroxide?

**16.142. pH of Milk of Magnesia** What is the pH at 25°C of a saturated solution of magnesium hydroxide (the active ingredient in the antacid milk of magnesia)?

**16.143.** Suppose you have 100 mL of each of the following solutions. In which will the most $CaCO_3$ dissolve? (a) 0.1 $M$ NaCl; (b) 0.1 $M$ $Na_2CO_3$; (c) 0.1 $M$ NaOH; (d) 0.1 $M$ HCl

**16.144.** In which of the following solutions will $CaF_2$ be most soluble? (a) 0.010 $M$ $Ca(NO_3)_2$; (b) 0.01 $M$ NaF; (c) 0.001 $M$ NaF; (d) 0.10 $M$ $Ca(NO_3)_2$

**16.145. Composition of Seawater** The average concentration of sulfate in surface seawater is about 0.028 $M$. The average concentration of $Sr^{2+}$ is $9 \times 10^{-5}$ $M$. If the $K_{sp}$ value of strontium sulfate is $3.4 \times 10^{-7}$, is the concentration of strontium in the sea probably controlled by the insolubility of its sulfate salt?

**16.146. Fertilizing the Sea to Combat Climate Change** Some scientists have proposed adding Fe(III) compounds to large expanses of the open ocean to promote the growth of phytoplankton that would in turn remove $CO_2$ from the atmosphere through photosynthesis. The average pH of open ocean water is 8.1. What is the maximum value of $[Fe^{3+}]$ in seawater if the $K_{sp}$ value of $Fe(OH)_3$ is $1.1 \times 10^{-36}$?

**16.147.** Will calcium fluoride precipitate when 125 mL of 0.375 $M$ $Ca(NO_3)_2$ is added to 245 mL of 0.255 $M$ NaF at 25°C?

**16.148.** Will lead(II) chloride precipitate if 185 mL of 0.025 $M$ sodium chloride is added to 235 mL of 0.165 $M$ lead(II) perchlorate?

**16.149.** A solution is 0.010 $M$ in both $Br^-$ and $SO_4^{2-}$. A 0.250 $M$ solution of lead(II) nitrate is slowly added to it.
  a. Which anion will precipitate first?
  b. What is the concentration in the solution of the first ion when the second one starts to precipitate at 25°C?

**\*16.150.** Solution A is 0.0250 $M$ in $Ag^+$ ion and $Pb^{2+}$ ion. You have access to two other solutions: (B) 0.500 $M$ NaCl and (C) 0.500 $M$ NaBr.
  a. Which would be the better solution to add to separate lead from silver by precipitation? (The better solution is the one that has less lead remaining in solution when the silver begins to precipitate.)
  b. Using the solution you selected in part a, is the separation of the two ions complete? ("Complete" is defined as the point when less than 0.10% of the silver ion is left in the solution when the lead ion begins to precipitate.)

## Additional Problems

**16.151.** Describe the intermolecular forces and changes in bonding that lead to the formation of a basic solution when methylamine ($CH_3NH_2$) dissolves in water.

**16.152.** Describe the chemical reactions of sulfur that begin with the burning of high-sulfur fossil fuel and that end with the reaction between acid rain and building exteriors made of marble ($CaCO_3$).

**16.153.** The value of $K_{a_1}$ of phosphorous acid, $H_3PO_3$, is nearly the same as the $K_{a_1}$ of phosphoric acid, $H_3PO_4$.
  a. Draw the Lewis structure of phosphorous acid.
  b. Identify the ionizable hydrogen atoms in the structure.
  c. Explain why the $K_{a_1}$ values of phosphoric and phosphorous acid are similar.

**16.154.** When silver oxide dissolves in water, the following reaction occurs:

$$Ag_2O(s) + H_2O(\ell) \rightarrow 2\,Ag^+(aq) + 2\,OH^-(aq)$$

If a saturated aqueous solution of silver oxide is $1.6 \times 10^{-4}$ $M$ in hydroxide ion, what is the $K_{sp}$ of silver oxide?

**\*16.155. pH of Baking Soda** A cook dissolves a teaspoon of baking soda ($NaHCO_3$) in a cup of water, then discovers that the recipe calls for a tablespoon, not a teaspoon. So the cook adds two more teaspoons of baking soda to make up the difference. Does the additional baking soda change the pH of the solution? Explain why or why not.

**\*16.156. Antacid Tablets** Antacids contain a variety of bases such as $NaHCO_3$, $MgCO_3$, $CaCO_3$, and $Mg(OH)_2$. Only $NaHCO_3$ has appreciable solubility in water.
  a. Write a net ionic equation for the reaction of each base with aqueous HCl.
  b. Explain how insoluble substances can act as effective antacids.

**\*16.157. pH of Natural Waters I** In a 1985 study of Little Rock Lake in Wisconsin, 400 gallons of 18 $M$ sulfuric acid were added to the lake over six years. The initial pH of the lake was 6.1 and the final pH was 4.7. If none of the acid was consumed in chemical reactions, estimate the volume of the lake.

**16.158. pH of Natural Waters II** Between 1993 and 1995, sodium phosphate was added to Seathwaite Tarn in the English Lake District to increase its pH. Explain why addition of this compound increased pH.

**16.159. Acid–Base Properties of Pharmaceuticals I** Zoloft is a common prescription drug for the treatment of depression. It is sold as a salt of HCl.

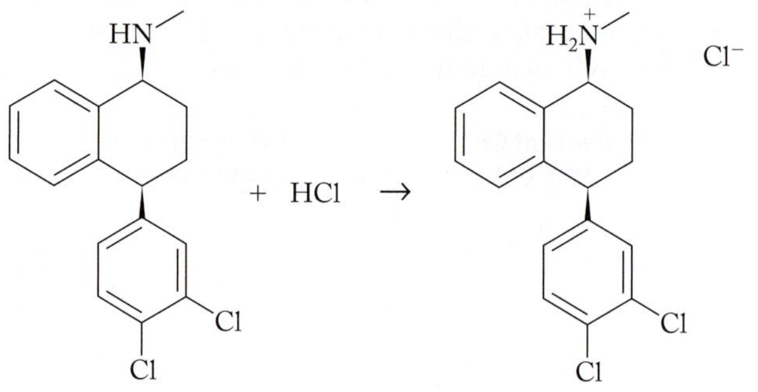

**FIGURE P16.159**

  a. In the reaction shown in Figure P16.159, which structure is that of the acid salt?
  b. When Zoloft dissolves in water, will the resulting solution be acidic or basic?

**16.160. Acid–Base Properties of Pharmaceuticals II** Prozac is a popular antidepressant drug. Its structure is given in Figure P16.160.

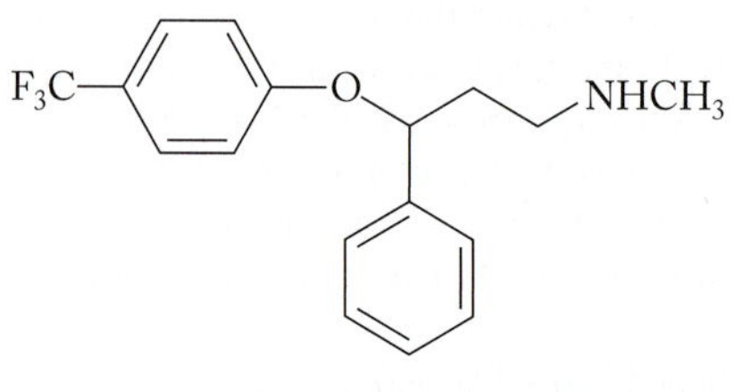

Prozac

**FIGURE P16.160**

a. Is a solution of Prozac in water likely to be slightly basic or slightly acidic? Explain your answer.
b. Prozac is also sold as a salt of HCl. Which atom, N or O, is most likely to react with HCl?
c. Prozac is sold as a salt of HCl because the solubility of the salt in water is higher than Prozac itself. Why is the salt more soluble?

**16.161.** Hydrogen fluoride (HF) behaves as a weak acid in aqueous solution. Two equilibria influence which fluorine-containing species are present in solution.

$$HF(aq) + H_2O(\ell) \rightleftharpoons H_3O^+(aq) + F^-(aq) \qquad K_a = 1.1 \times 10^{-3}$$

$$F^-(aq) + HF(g) \rightleftharpoons HF_2^-(aq) \qquad K = 2.6 \times 10^{-1}$$

a. Is fluoride in pH 7.00 drinking water more likely to be present as $F^-$ or $HF_2^-$?
b. What is the equilibrium constant for this equilibrium?

$$2\,HF(aq) + H_2O(\ell) \rightleftharpoons H_3O^+(aq) + HF_2^-(aq)$$

c. What are the pH and equilibrium concentration of $HF_2^-$ in a 0.150 $M$ solution of HF?

*16.162. Pentafluorocyclopentadiene, which has the structure shown in Figure P16.162, is a strong acid.
a. Draw the conjugate base of $C_5F_5H$.
b. Why is the compound so acidic when most organic acids are weak?

**FIGURE P16.162**

**16.163.** Naproxen (a.k.a. Aleve) is an anti-inflammatory drug used to reduce pain, fever, inflammation, and stiffness caused by conditions such as osteoarthritis and rheumatoid arthritis. Naproxen is an organic acid; its structure is shown in Figure P16.163. Naproxen has limited solubility in water, so it is sold as its sodium salt.
a. Draw the molecular structure of the sodium salt.
b. Should a solution of the salt be acidic or basic? Explain why.
c. Explain why the salt is more soluble than naproxen itself.

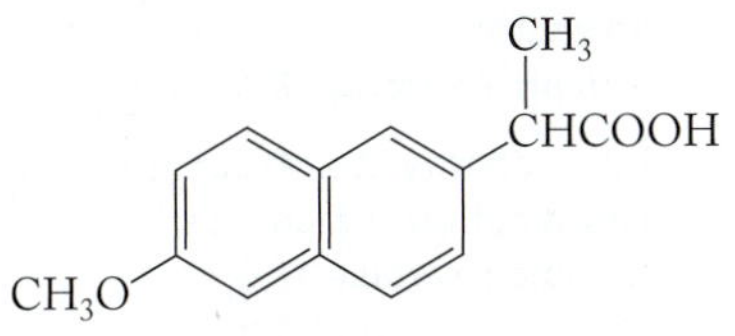

**FIGURE P16.163**

*16.164. **Greenhouse Gases and Ocean pH** Some climate models predict a decrease in the pH of the oceans of 0.77 pH units because of increases in atmospheric carbon dioxide.
a. Explain, by using the appropriate chemical reactions and equilibria, how an increase in atmospheric $CO_2$ could produce a decrease in oceanic pH.

b. How much more acidic (in terms of $[H^+]$) would the oceans be if their pH dropped this much?
c. Oceanographers are concerned about the impact of a drop in oceanic pH on the survival of coral reefs. Why?

## The Chemistry of Two Strong Acids: Sulfuric and Nitric Acids

**16.165.** Complete the following chemical equations with the appropriate product(s).
a. $SO_3(g) + H_2O(\ell) \rightarrow$
b. $3\,NO_2(g) + H_2O(\ell) \rightarrow$
c. $4\,NH_3(g) + 5\,O_2(g) \rightarrow ? + 6\,H_2O(g)$

16.166. In the Ostwald process, which steps should have a higher yield at higher total pressure?

**16.167.** In the Ostwald process, which steps should have higher yields at higher temperature?

16.168. Given $\Delta H_f^\circ$ for $SO_2(g)$ (−296.8 kJ/mol) and $SO_3(g)$ (−395.7 kJ/mol), calculate $\Delta H_{rxn}^\circ$ for the conversion of $SO_2$ to $SO_3$.

**16.169.** Write balanced chemical equations that correspond to $\Delta H_{f,SO_2}^\circ$ and $\Delta H_{f,SO_3}^\circ$. Show how Hess's law can be used to determine $\Delta H^\circ$ for the reaction

$$2\,SO_2(g) + O_2(g) \rightleftharpoons 2\,SO_3(g)$$

*16.170. Reaction of sodium nitrate with sodium oxide at high temperature produces $Na_3NO_4$, which contains the $NO_4^{3-}$ (orthonitrate) anion. However, the corresponding acid, $H_3NO_4$, is unknown. Would you expect $H_3NO_4$ to be a stronger or weaker acid than $HNO_3$?

*16.171. The Henry's law constant for $CO_2$ dissolved in water is $3.5 \times 10^{-2}$ $M$/atm. Do you expect the corresponding constants for $SO_3$ and $NO_2$ to be greater than or less than this value? Explain your answer.

16.172. Write a balanced chemical equation to describe the following reactions of sulfuric acid and nitric acid:
a. Nitric acid reacts with ammonia.
b. Sulfuric acid reacts with ammonia.
c. Sulfuric acid dissolves in water.

*16.173. Thiosulfuric acid, $H_2S_2O_3$, can be prepared by the reaction of $H_2S$ with $HSO_3Cl$:

$$HSO_3Cl(\ell) + H_2S(g) \rightarrow HCl(g) + H_2S_2O_3(\ell)$$

a. Draw a Lewis structure for $H_2S_2O_3$, given that it is isostructural with $H_2SO_4$.
b. Do you expect $H_2S_2O_3$ to be a stronger or weaker acid than $H_2SO_4$? Explain your answer.

*16.174. Sulfuric acid reacts with nitric acid as shown below:

$$HNO_3(aq) + 2\,H_2SO_4(aq) \rightarrow NO_2^+(aq) + H_3O^+(aq) + 2\,HSO_4^-(aq)$$

a. Is the reaction a redox process?
b. Identify the acid, base, conjugate acid, and conjugate base in the reaction. (*Hint*: Draw the Lewis structures for each.)

If your instructor assigns problems in **smartwork**, log in at **smartwork.wwnorton.com**.

# 17

# Metal Ions: Colorful and Essential

## The Company They Keep

**M**any of the metallic elements in the periodic table are essential to the health of most life forms, including us. For example, copper, zinc, and cobalt play key roles in protein function; iron is needed to transport oxygen from our lungs to all the cells of our body; and calcium is important in building strong teeth and bones.

These and other essential metallic elements should be present either in our diets or in the supplements many of us rely on for balanced nutrition. However, the mere presence of these elements is not sufficient—they must be in a form that we can digest. Swallowing an 18 mg steel pellet as if it were an aspirin tablet would not be an effective way for a woman to get her recommended daily allowance of iron. Chewing a gram of calcium metal would be just as ineffective and much less pleasant. If we are to benefit from consuming essential metals in food and nutritional supplements, the metals need to be in compounds, not free-element form, and these compounds must be absorbable by the body.

All the metallic elements essential to human health occur in nature in ionic compounds, but not all ionic forms are absorbed equally well. For example, most of the iron in fish, poultry, and red meat is readily absorbed because it is present in a form called *heme iron*. However, the iron in plants is mostly nonheme and not as readily absorbed. Eating a meal that includes both meat and vegetables improves the absorption of the nonheme iron in the vegetables, as does consuming foods high in vitamin C. All these dietary factors work together at the molecular level to provide us with the nutrients we need to survive.

Metal ions are essential to good health, but the nonmetal ions and molecules that accompany them are equally important because these species promote the solubility of metal ions and make them more chemically reactive and biologically available. These other species influence other properties of metal ions, including the wavelengths of visible light they absorb and the color of their solutions. In this chapter we explore how the chemical environment of metal ions in solids and

**Ocean Blooms** The green and yellow swirls in this satellite image of the Baltic Sea are produced by high concentrations of microscopic algae called phytoplankton. Their color comes from chlorophyll, whose molecules each contain four linked rings of atoms, each containing a nitrogen atom bonded to a central $Mg^{2+}$ ion. ▶

**17.1** Lewis Acids and Bases

**17.2** Complex Ions

**17.3** Complex-Ion Equilibria

**17.4** Naming Complex Ions and Coordination Compounds

**17.5** Hydrated Metal Ions as Acids

**17.6** Polydentate Ligands

**17.7** Ligand Strength and the Chelate Effect

**17.8** Crystal Field Theory

**17.9** Magnetism and Spin States

**17.10** Isomerism in Coordination Compounds

**17.11** Coordination Compounds in Biochemistry

### Learning Outcomes

**LO1** Distinguish between Brønsted–Lowry and Lewis acids and bases
**Sample Exercise 17.1**

**LO2** Use formation constants to calculate the concentrations of free and complexed metal ions in solution
**Sample Exercise 17.2**

**LO3** Name coordination compounds based on their formulas and translate names into formulas
**Sample Exercise 17.3**

**LO4** Identify electron-pair donor groups in potential ligands
**Sample Exercise 17.4**

**LO5** Use crystal field theory to explain why many transition metal compounds and solutions are colored

**LO6** Predict the spin states of metal ions in tetrahedral and octahedral fields produced by ligands with different field strengths
**Sample Exercise 17.5**

**LO7** Identify stereoisomerism in coordination compounds
**Sample Exercise 17.6**

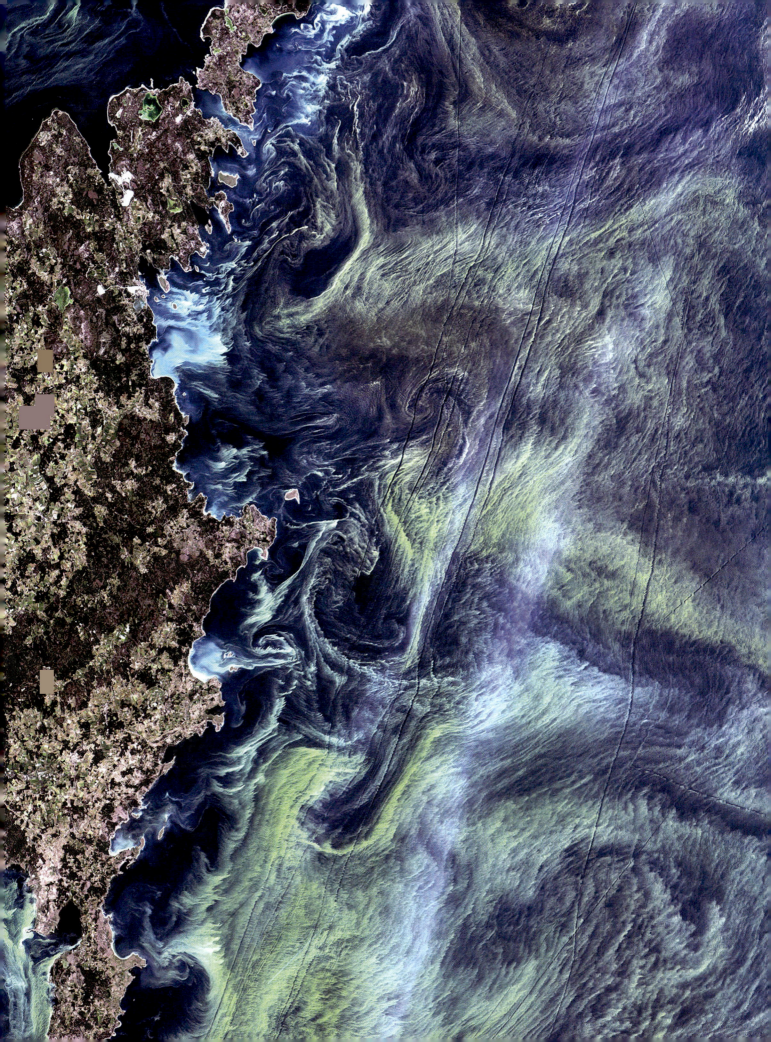

solutions affects their physical, chemical, and biological properties. We answer such questions as why many, but not all, metal compounds have distinctive colors, and how metals play key roles in many biological processes through the formation of complex ions with biomolecules. ■

# 17.1 Lewis Acids and Bases

In this chapter we focus on the interactions between metal ions and the other ions and molecules that surround them in solids and solutions. To understand these interactions, we need to reconsider the definitions of acids and bases we used in Chapter 16. Let's begin by revisiting what happens when ammonia gas dissolves in water:

$$NH_3(g) + H_2O(\ell) \rightleftharpoons NH_4^+(aq) + OH^-(aq)$$

Figure 17.1(a) shows a Brønsted–Lowry interpretation of this reaction: in donating $H^+$ ions to ammonia, $H_2O$ acts as a Brønsted–Lowry acid, and in accepting the protons $NH_3$ acts as a Brønsted–Lowry base.

Another way to view this reaction is illustrated in Figure 17.1(b). Instead of focusing on the transfer of hydrogen ions, consider the two reactants as a donor and an acceptor of a *pair of electrons*. In this view, the N atom in $NH_3$ donates its lone pair of electrons to one of the H atoms in $H_2O$. In the process, one of the H—O bonds in $H_2O$ is broken in such a way that the bonding pair of electrons remains with the O atom. The donated lone pair from the N atom forms a fourth N—H covalent bond. The result is the same as in the Brønsted–Lowry model: a molecule of $NH_3$ bonds to a $H^+$ ion, forming an $NH_4^+$ ion, and a molecule of $H_2O$ loses a $H^+$ ion, becoming a $OH^-$ ion.

**FIGURE 17.1** (a) Brønsted–Lowry view of the reaction between $H_2O$ (proton donor) and $NH_3$ (proton acceptor). (b) Lewis view of the reaction: $H_2O$ acts as a Lewis acid (electron-pair acceptor) and $NH_3$ acts as a Lewis base (electron-pair donor).

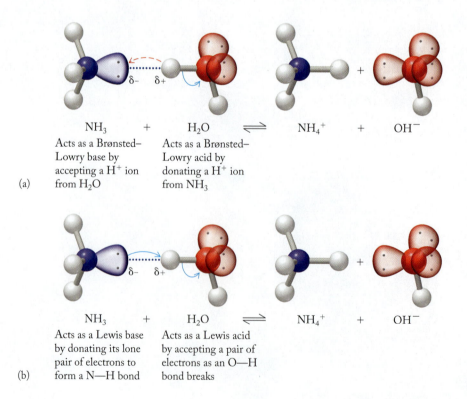

(a)
NH$_3$ + H$_2$O $\rightleftharpoons$ NH$_4^+$ + OH$^-$

Acts as a Brønsted–Lowry base by accepting a H$^+$ ion from H$_2$O

Acts as a Brønsted–Lowry acid by donating a H$^+$ ion from NH$_3$

(b)
NH$_3$ + H$_2$O $\rightleftharpoons$ NH$_4^+$ + OH$^-$

Acts as a Lewis base by donating its lone pair of electrons to form a N—H bond

Acts as a Lewis acid by accepting a pair of electrons as an O—H bond breaks

Viewing this process as the donation and acceptance of an electron pair provides the following basis for defining acids and bases:

➤ A **Lewis base** is a substance that *donates* a lone pair of electrons in a chemical reaction.
➤ A **Lewis acid** is a substance that *accepts* a lone pair of electrons in a chemical reaction.

These definitions are named after their developer, Gilbert N. Lewis, who pioneered research into the nature of chemical bonds (Section 8.2). The Lewis definition of a base is consistent with the Brønsted–Lowry model we explored in Chapter 16 because a substance must be able to donate a pair of electrons if it is to bond with a $H^+$ ion. However, the same parallelism is not true for acids. The Brønsted–Lowry model defines an acid as a hydrogen-ion donor, but the Lewis definition includes species that have no hydrogen ions to donate but that can still accept electrons. One such compound is boron trifluoride, $BF_3$.

With only six valence electrons, the boron atom in $BF_3$ can accept another pair to complete its octet. $NH_3$ is a suitable electron-pair donor, as shown in Figure 17.2. There is no transfer of $H^+$ ions in this reaction, so it is not an acid–base reaction according to the Brønsted–Lowry model. However, $NH_3$ donates a lone pair of electrons and $BF_3$ accepts them, so it is an acid–base reaction according to the broader Lewis model.

Many important Lewis bases are anions, including the halide ions, $OH^-$, and $O^{2-}$. To see how $O^{2-}$ functions as a Lewis base, let's revisit the reaction described in Section 15.6 between $SO_2$ and $CaO$ that is used to reduce $SO_2$ emissions from power stations and smelters:

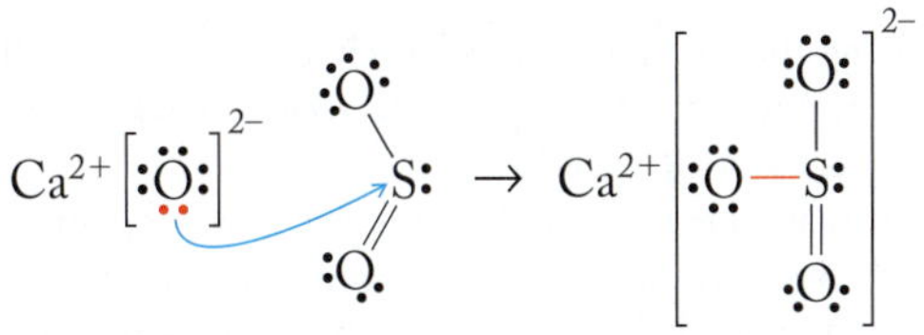

The oxide ion in CaO is the electron-pair donor, so $O^{2-}$ is a Lewis base. Sulfur dioxide is the electron-pair acceptor and is therefore a Lewis acid. Sharing a lone pair of electrons between the oxide ion and the sulfur atom in $SO_2$ results in the formation of a new S—O bond. These bonding changes produce a structure in which the formal charges on the S atom and the double-bonded O atom are both zero and those on the two single-bonded O atoms are −1, giving an overall charge of 2−.

NH₃    +    BF₃    ⟶    H₃N—BF₃

Acts as a Lewis base by donating a pair of electrons to $BF_3$

Acts as a Lewis acid by accepting a pair of electrons from $NH_3$

**Lewis base** a substance that *donates* a lone pair of electrons in a chemical reaction.

**Lewis acid** a substance that *accepts* a lone pair of electrons in a chemical reaction.

**CONNECTION** Lewis's pioneering theories of the nature of covalent bonding were described in Section 8.2.

**CONNECTION** The concept of formal charge and its calculation were described in Section 8.6.

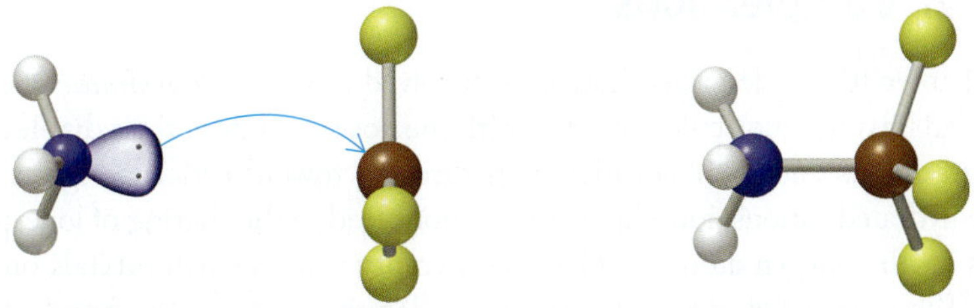

**FIGURE 17.2** In the reaction between $NH_3$ and $BF_3$, $NH_3$ acts as a Lewis base and $BF_3$ acts as a Lewis acid. This is an acid–base reaction in the Lewis sense because we focus on the acceptance and donation of an electron pair, not the transfer of a proton, as in the Brønsted-Lowry system.

**coordinate bond** a covalent bond formed when one anion or molecule donates a pair of electrons to another ion or molecule.

**ligand** a Lewis base bonded to the central metal ion of a complex ion.

**complex ion** an ionic species consisting of a metal ion bonded to one or more Lewis bases.

**inner coordination sphere** the ligands that are bound directly to a metal via coordinate bonds.

---

**SAMPLE EXERCISE 17.1   Identifying Lewis Acids and Bases   LO1**

In the following reaction, which species is a Lewis acid and which is a Lewis base?

$$AlCl_3 + Cl^- \rightarrow AlCl_4^-$$

**Collect and Organize** We are asked to identify which of two reactants is acting as a Lewis acid, meaning an electron-pair acceptor, and which is acting as a Lewis base, meaning an electron-pair donor.

**Analyze** The $Cl^-$ ion has an octet in its valence shell, so is not likely to accept electrons. However, it can *donate* one of its four pairs to form a covalent bond to aluminum. To analyze the capacity of $AlCl_3$ to act as a Lewis base, we need to draw its Lewis structure:

This structure accounts for all the valence electrons ($3 \times 7$ from 3 Cl atoms + 3 from Al = 24) with three Al—Cl single bonds and *no lone pairs*. This leaves Al with 6 valence electrons and thus the capacity to accept one more pair, that is, to act as a Lewis acid.

**Solve** In this reaction, $AlCl_3$ is a Lewis acid and the $Cl^-$ ion is a Lewis base. We can represent the reaction using Lewis structures:

**Think About It** Drawing the Lewis structure of $AlCl_3$ is the key to solving the problem. With its incomplete octet, $AlCl_3$ has the capacity to accept an additional pair of electrons, much like $BF_3$, so it may act as a Lewis acid.

These Lewis structures assume that $AlCl_3$ is a molecular compound and not, like most other metal chlorides, an ionic compound. This assumption is supported by the physical properties of $AlCl_3$ and particularly by the fact that it sublimes at 178°C and 1 atm. Most metal halides do not even melt until heated to temperatures many hundreds of degrees higher than that.

**Practice Exercise** In the following reaction, which reactant is the Lewis acid and which is the Lewis base?

$$CO_2(g) + CaO(s) \rightarrow CaCO_3(s)$$

*(Answers to Practice Exercises are in the back of the book.)*

---

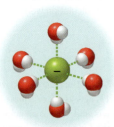

**FIGURE 17.3** In Chapter 10 we discussed how ions in aqueous solutions are surrounded by water molecules oriented with the positive ends of their dipoles (H atoms) directed toward anions and the negative ends (O atoms) directed toward cations. When these O atoms donate a lone pair of electrons to an empty orbital of a cation, a coordinate bond is formed.

# 17.2 Complex Ions

In Chapter 10, we described how ions dissolved in water are *hydrated*; that is, surrounded by water molecules oriented with the positive ends of their dipoles directed toward anions and their negative ends directed toward cations (Figure 17.3). In some hydrated cations, ion–dipole interactions lead to the sharing of lone-pair electrons on the oxygen atoms of $H_2O$ with empty valence-shell orbitals on the cations. These shared electron pairs meet our definition of covalent bonds, but these particular bonds are called *coordinate* covalent bonds, or simply **coordinate bonds**. Such bonds form when either a molecule or an anion donates a lone pair of electrons to an empty valence-shell orbital of an atom, molecule, or cation. Once formed, a coordinate bond is indistinguishable from any other kind of covalent bond.

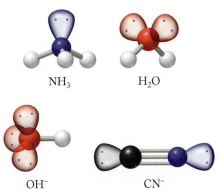

Is the N—B bond that forms between $NH_3$ and $BF_3$ (as shown in Figure 17.2) a coordinate bond? Explain your answer.

*(Answers to Concept Tests are in the back of the book.)*

∙∙∙∙∙∙∙∙∙∙∙∙∙∙∙∙∙∙∙∙∙∙∙∙∙∙∙∙∙∙∙∙∙∙∙∙∙∙∙∙∙∙∙∙∙∙∙∙∙∙∙∙∙∙∙∙∙∙∙∙∙∙∙∙∙∙∙∙∙∙∙∙∙

**FIGURE 17.4** Anions or molecules with lone pairs of electrons have the capacity to donate those electrons to metal cations. When these compounds do this, they are called ligands. All these ligands except $NH_3$ have more than one lone pair, but only one pair at a time can be directed toward a single metal cation.

Molecules or anions (Figure 17.4) that function as Lewis bases and form coordinate bonds with metals are called **ligands**. The resulting species, which are composed of central metal ions and the surrounding ligands, are called **complex ions**, or simply *complexes*. Direct bonding to a central cation means that the ligands in a complex occupy the **inner coordination sphere** of the cation.

To better understand the meaning of these terms and others related to complex formation, let's consider what happens when $Zn(NO_3)_2$ dissolves in a solution containing ammonia. As the salt dissolves, each $Zn^{2+}$ ion is surrounded by four $NH_3$ molecules. Each molecule donates the lone pair of electrons on its N atom to an empty valence orbital of the $Zn^{2+}$ ion, forming a coordinate bond and resulting in a complex with the formula $Zn(NH_3)_4^{2+}$ (Figure 17.5). Note that the charge of the complex is the same as the charge of the $Zn^{2+}$ ion because the ligands are neutral molecules of $NH_3$.

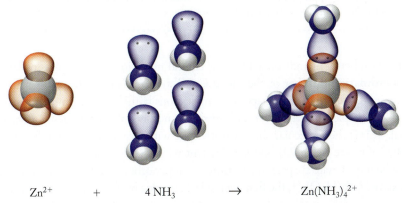

$$Zn^{2+} \quad + \quad 4\,NH_3 \quad \rightarrow \quad Zn(NH_3)_4^{2+}$$

**FIGURE 17.5** The complex ion $Zn(NH_3)_4^{2+}$ is produced when four molecules of ammonia form coordinate bonds with empty $sp^3$ orbitals on a $Zn^{2+}$ ion.

Complex ions have characteristic shapes depending on the number of ligands surrounding the central cation. For example, the four $NH_3$ molecules in $Zn(NH_3)_4^{2+}$ occupy the four corners of a tetrahedron. We can explain its tetrahedral shape this way: each $Zn^{2+}$ ion has the electron configuration $[Ar]3d^{10}$, as shown in the left-hand orbital diagram in Figure 17.6. Its $3d$ orbitals are full, but its $4s$ and $4p$ orbitals are empty. According to valence bond theory (see Section

**FIGURE 17.6** The tetrahedral shape of $Zn(NH_3)_4^{2+}$ is consistent with the formation of a set of $sp^3$ hybrid orbitals in the $Zn^{2+}$ ion's valence shell.

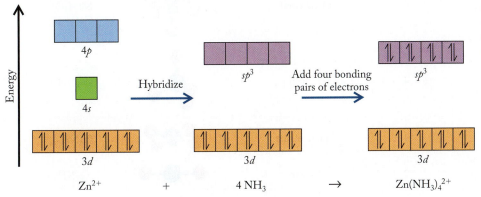

9.4), these orbitals can hybridize, forming four empty $sp^3$ hybrid orbitals. As with all sets of $sp^3$ orbitals, these are directed toward the four corners of a tetrahedron, as shown in the image of the $Zn^{2+}$ ion in Figure 17.5. The four lone pairs of the N atoms from four molecules of $NH_3$ are shared by these $sp^3$ orbitals when the complex forms, yielding a tetrahedral $Zn(NH_3)_4^{2+}$ ion.

The positive charges on the $Zn(NH_3)_4^{2+}$ ions in the solution are balanced by the $NO_3^-$ ions that were released when $Zn(NO_3)_2$ dissolved. The nitrate ions in this solution serve as **counter ions**, a term describing ions of opposite charge that are not part of the inner coordination sphere. These $NO_3^-$ ions are not bonded directly to the $Zn^{2+}$ ions but are electrostatically attracted to them. For this reason, the formula of the solute in this solution is written

$$[Zn(NH_3)_4](NO_3)_2$$

where the brackets separate the formula of the complex ion from the formula of its counter ions. This formula is that of a **coordination compound**—a label that applies to any compound that contains a complex ion.

**CONCEPT TEST** ....................................................

In the coordination compound $Na_3[Fe(CN)_6]$, which ions occupy the inner coordination sphere of the $Fe^{3+}$ ion, and which ions are counter ions?

....................................................

The number of ligands bonded to the central metal ion in a complex ion defines the **coordination number** of the metal ion. Thus $Zn(NH_3)_4^{2+}$ has a coordination number of 4. Table 17.1 lists shapes of complex ions with coordination numbers of 6, 4, and 2. Complex ions that have a coordination number of 6 are almost always octahedral. This is the molecular shape we associate with a central atom that has a steric number of 6 (see Table 9.1) and is consistent with a coordinate bonding pattern in which ligands donate six pairs of electrons to empty orbitals on the central metal ion. For example, $Fe(H_2O)_6^{3+}$, the first octahedral complex listed in Table 17.1, consists of one $Fe^{3+}$ ion with coordinate bonds to the O atoms in six $H_2O$ molecules.

**counter ions** species that provide electrical balance of charges for complex ions in coordination compounds.

**coordination compound** a compound made up of at least one complex ion.

**coordination number** the number of electron pairs surrounding a metal ion in a complex.

**formation constant ($K_f$)** an equilibrium constant describing the formation of a metal complex from a free metal ion and its ligands.

**TABLE 17.1 Common Coordination Numbers and Shapes for Complex Ions**

| Coordination Number | Shape | Structure | Examples |
|---|---|---|---|
| 6 | Octahedral | | $Fe(H_2O)_6^{3+}$ $Ni(H_2O)_6^{2+}$ $Co(H_2O)_6^{3+}$ |
| 4 | Tetrahedral | | $Zn(NH_3)_4^{2+}$ |
| 4 | Square planar | | $Pt(NH_3)_4^{2+}$ |
| 2 | Linear | | $Ag(NH_3)_2^+$ |

Complex ions with a coordination number of 4 may be tetrahedral, but some, such as $Pt(NH_3)_4^{2+}$, are square planar. In square planar complexes, four coordinate bonds are arranged 90° apart in an equatorial plane. A discussion of the reasons a complex with a coordination number of four adopts a tetrahedral or square planar geometry is beyond the scope of this text.

## 17.3 Complex-Ion Equilibria

Let's investigate the formation of complex ions using the mathematical tools we used in Chapters 15 and 16. These tools are appropriate because complex formation processes are reversible, and many of them reach chemical equilibrium rapidly.

We start this investigation with two aqueous solutions, one containing copper(II) sulfate ($CuSO_4$), the other $NH_3$ (Figure 17.7). The $CuSO_4$ solution is robin's-egg blue, the color characteristic of $Cu^{2+}(aq)$ ions, and the ammonia solution is colorless. When the solutions are mixed, the robin's-egg blue turns a dark navy blue, as shown on the right in Figure 17.7. This is the color of tetraamminecopper(II), $Cu(NH_3)_4^{2+}$. The change in color provides visual evidence that the following equilibrium lies far to the right, favoring complex formation:

$$Cu^{2+}(aq) + 4\,NH_3(aq) \rightleftharpoons Cu(NH_3)_4^{2+}(aq)$$

This conclusion is supported by the large equilibrium constant for the reaction:

$$K_f = \frac{[Cu(NH_3)_4^{2+}]}{[Cu^{2+}][NH_3]^4} = 5.0 \times 10^{13}$$

The equilibrium constant $K_f$ is called a **formation constant** because it describes the formation of a complex ion or coordination complex. For the general case in which 1 mole of metal ions ($M^{m+}$) combines with $n$ moles of ligand ($X^{x-}$) to form the complex ion $MX_n^{(m-nx)+}$, the formation constant expression is

$$K_f = \frac{[MX_n^{(m-nx)+}]}{[M^{m+}][X^{x-}]^n}$$

**FIGURE 17.7** The beaker on the left contains a solution of $Cu^{2+}(aq)$, which is a characteristic robin's-egg blue. As a colorless solution of ammonia is added (from the bottle in the middle), the mixture of the two solutions turns dark blue (beaker on the right), which is the color of the $Cu(NH_3)_4^{2+}$ complex ion.

Formation constants can be used to calculate the concentration of complex ions in solution or to calculate the concentration of free, uncomplexed metal ions, $M^{m+}(aq)$, in equilibrium with a given (usually larger) concentration of ligand. Because $K_f$ values are usually very large, equilibrium concentrations of uncomplexed metal ions are usually very small. Therefore, we let $x$ be the concentration of uncomplexed metal ions, just as we let $x$ be the (usually small) equilibrium concentration of $H^+(aq)$ ions in calculating the pH of solutions of weak acids in Chapter 16. However, there $x$ represented the concentration *of a product* in those pH calculations; in complexation calculations, $x$ represents the concentration *of a reactant*. The impact of this difference is illustrated in Sample Exercise 17.2.

**SAMPLE EXERCISE 17.2** **Calculating the Concentration of a Free Metal Ion in Equilibrium with a Complex** **LO2**

Ammonia gas is dissolved in a $1.00 \times 10^{-4}$ $M$ solution of $CuSO_4$, so that initially $[NH_3] = 2.00 \times 10^{-3}$ $M$. Calculate the concentration of $Cu^{2+}(aq)$ ions in the solution after the reaction mixture has come to chemical equilibrium.

**Collect and Organize** The concentration of $CuSO_4$ means that $[Cu^{2+}]$ before complex formation is $1.00 \times 10^{-4}$ $M$. The initial concentration of the ligand ($NH_3$) is $2.00 \times 10^{-3}$ $M$. We know from the text preceding this Sample Exercise that the reaction involves the formation of $Cu(NH_3)_4^{2+}$:

$$Cu^{2+}(aq) + 4\,NH_3(aq) \rightleftharpoons Cu(NH_3)_4^{2+}(aq)$$

Also, the values of $[NH_3]$, $[Cu^{2+}]$, and $[Cu(NH_3)_4^{2+}]$ are related by the formation constant expression

$$K_f = \frac{[Cu(NH_3)_4^{2+}]}{[Cu^{2+}][NH_3]^4} = 5.0 \times 10^{13}$$

**Analyze** The stoichiometric ratio of $NH_3$ to $Cu^{2+}$ is 4:1, but initially $[NH_3]/[Cu^{2+}] = 20$, so there is more than enough $NH_3$ to convert all the $Cu^{2+}$ into $Cu(NH_3)_4^{2+}$. Because $K_f$ is large, we can assume that nearly all the $Cu^{2+}$ ions are converted to complex ions and only a tiny concentration of free $Cu^{2+}$ ions, which we will call $x$, remains at equilibrium:

$$[Cu^{2+}] = x$$

The change in $[Cu^{2+}]$ is equal to the equilibrium value ($x$) less the initial value ($1.00 \times 10^{-4}$ $M$):

$$\Delta\,[Cu^{2+}] = x - 1.00 \times 10^{-4}$$

The decrease in concentration described by this equation is equal in magnitude but opposite in sign to the increase in the concentration of complex ions

$$\Delta\,[Cu(NH_3)_4^{2+}] = -\Delta\,[Cu^{2+}] = 1.00 \times 10^{-4} - x$$

The balanced equation tells us that 4 moles of $NH_3$ are consumed for every mole of $Cu^{2+}$, so the change in ammonia concentration is

$$\Delta\,[NH_3] = 4\,\Delta\,[Cu^{2+}] = 4\,(x - 1.00 \times 10^{-4})$$

We can list these initial values and changes in concentration in a RICE table:

| Reaction (R) | $Cu^{2+}$ | + | $4\,NH_3$ | $\rightleftharpoons$ | $Cu(NH_3)_4^{2+}$ |
|---|---|---|---|---|---|
| | $Cu^{2+}$ (M) | | $NH_3$ (M) | | $Cu(NH_3)_4^{2+}$ (M) |
| Initial (I) | $1.00 \times 10^{-4}$ | | $2.00 \times 10^{-3}$ | | 0 |
| Change (C) | $x - 1.00 \times 10^{-4}$ | | $4\,(x - 1.00 \times 10^{-4})$ | | $(1.00 \times 10^{-4} - x)$ |
| Equilibrium (E) | $x$ | | | | |

Given the large value of $K_f$, we should obtain a $[Cu^{2+}]$ value at equilibrium that is much less than $1.00 \times 10^{-4}$ $M$.

**Solve** First we complete the row of equilibrium concentrations of $NH_3$ and $Cu(NH_3)_4^{2+}$:

| Reaction (R) | $Cu^{2+}$ | + | $4\,NH_3$ | $\rightleftharpoons$ | $Cu(NH_3)_4^{2+}$ |
|---|---|---|---|---|---|
| | $Cu^{2+}$ (M) | | $NH_3$ (M) | | $Cu(NH_3)_4^{2+}$ (M) |
| Initial (I) | $1.00 \times 10^{-4}$ | | $2.00 \times 10^{-3}$ | | 0 |
| Change (C) | $x - 1.00 \times 10^{-4}$ | | $4\,(x - 1.00 \times 10^{-4})$ | | $(1.00 \times 10^{-4} - x)$ |
| Equilibrium (E) | $x$ | | $(1.60 \times 10^{-3}) + 4x$ | | $1.00 \times 10^{-4} - x$ |

Next we make the simplifying assumption that $x$ is much smaller than $1.00 \times 10^{-4}\ M$. If it is, then $4x$ must be much smaller than $1.60 \times 10^{-3}\ M$. Therefore, we can ignore the $x$ terms in the equilibrium values of $[NH_3]$ and $[Cu(NH_3)_4^{2+}]$ and use the simplified values in the $K_f$ expression:

$$K_f = \frac{[Cu(NH_3)_4^{2+}]}{[Cu^{2+}][NH_3]^4}$$

$$= \frac{1.00 \times 10^{-4}}{(x)(1.60 \times 10^{-3})^4} = 5.0 \times 10^{13}$$

$$x = \frac{1.00 \times 10^{-4}}{(1.60 \times 10^{-3})^4(5.0 \times 10^{13})}$$

$$= 3.1 \times 10^{-7}\ M = [Cu^{2+}]$$

**Think About It** This result confirms our simplifying assumption and also our prediction that $[Cu^{2+}]$ at equilibrium is much less than $[Cu^{2+}]$ initially. In fact, more than 99% of the Cu(II) in the solution is present as $Cu(NH_3)_4^{2+}$.

⚙ **Practice Exercise** Calculate the equilibrium concentration of $Ag^+(aq)$ in a solution that is initially $0.100\ M\ AgNO_3$ and $0.800\ M\ NH_3$ after this reaction takes place:

$$Ag^+(aq) + 2\ NH_3(aq) \rightleftharpoons Ag(NH_3)_2^+(aq) \qquad K_f = 1.7 \times 10^7$$

# 17.4 Naming Complex Ions and Coordination Compounds

The names of complex ions and coordination compounds tell us the identity and oxidation state of the central ion, the names and numbers of ligands, the charge (in the case of complex ions), and the identity of counter ions. To convey all this information, we need to follow some simple naming rules.

## Complex Ions with a Positive Charge

1. Start with the identities of the ligand(s). Names of common ligands appear in Table 17.2. If the ion has more than one kind of ligand, list the names alphabetically.
2. In front of the name(s) written in step 1, use the usual prefix(es) to indicate the number of each type of ligand.

| Number of Ligands | Prefix |
|:---:|:---:|
| 2 | Di- |
| 3 | Tri- |
| 4 | Tetra- |
| 5 | Penta- |
| 6 | Hexa- |

3. Write the name of the transition metal ion, with a Roman numeral indicating its oxidation state.

Examples:

| | |
|---|---|
| $Ni(H_2O)_6^{2+}$ | Hexaaquanickel(II) |
| $Co(NH_3)_6^{3+}$ | Hexaamminecobalt(III) |
| $[Cu(NH_3)_4(H_2O)_2]^{2+}$ | Tetraamminediaquacopper(II) |

| | | | | Number of Donor |
| Ligand | Name within Complex Ion | Structure | Charge | Groups |
|---|---|---|---|---|
| Iodide | Iodo | $I^-$ | 1– | 1 |
| Bromide | Bromo | $Br^-$ | 1– | 1 |
| Chloride | Chloro | $Cl^-$ | 1– | 1 |
| Fluoride | Fluoro | $F^-$ | 1– | 1 |
| Nitrate | Nitrato | | 1– | 1 |
| Hydroxide | Hydroxo | $[O{-}H]^-$ | 1– | 1 |
| Water | Aqua | | 0 | 1 |
| Pyridine | Pyridyl | | 0 | 1 |
| Ammonia | Ammine | $NH_3$ | 0 | 1 |
| Ethylenediamine (en) | (same)[a] | | 0 | 2 |
| 2,2'-Bipyridine (bipy) | Bipyridyl | | 0 | 2 |
| 1,10-Phenanthroline (phen) | (same)[a] | | 0 | 2 |
| Cyanide[b] | Cyano | $[C{\equiv}N]^-$ | 1– | 1 |
| Carbon monoxide[b] | Carbonyl | $C{\equiv}O$ | 0 | 1 |

**TABLE 17.2  Names and Structures of Common Ligands**

[a] The names of some electrically neutral ligands in complexes are the same as the names of the molecules.
[b] Carbon atoms are the lone pair donors in these ligands.

It may seem strange having two *a*'s together in the names of the complex ions on the previous page, but it is permitted under current naming rules. In the third name, prefixes are ignored in determining alphabetical order: the order is *ammine* before *aqua*, rather than *di* before *tetra*.

In these three examples, the ligands are all electrically neutral. This makes determining the oxidation state of the central metal ion a simple task because the charge on the complex ion is the same as the charge on the metal ion, which is the oxidation state of the metal. When the ligands include anions, determining the oxidation state of the central metal ion requires us to account for these charges.

## Complex Ions with a Negative Charge

1. Follow the steps for naming positively charged complexes.
2. Add *-ate* to the name of the central metal ion to indicate that the complex ion carries a negative charge (just as we use *-ate* to end the names of oxoanions). For some metals, the base name changes, too. The two most common examples are iron, which becomes *ferrate*, and copper, which becomes *cuprate*.

Examples:

| | |
|---|---|
| $Fe(CN)_6^{3-}$ | Hexacyanoferrate(III) |
| $[Fe(H_2O)(CN)_5]^{3-}$ | Aquapentacyanoferrate(II) |
| $[Al(H_2O)_2(OH)_4]^-$ | Diaquatetrahydroxoaluminate |

In the first two examples we must determine the oxidation state of Fe. We start with the charge on the complex ion and then take into account the charges on the ligand anions to calculate the charge on the metal ion. For example, the overall charge of the aquapentacyanoferrate(II) ion is 3− and it contains five $CN^-$ ions. To reduce the combined charge of 5− from these cyanide ions to an overall charge of 3−, the charge on Fe must be 2+.

**CONCEPT TEST** ......................................................

What is the name of the complex anion with the formula $PtCl_4^{2-}$?

......................................................

## Coordination Compounds

1. If the counter ion of the complex ion is a cation, the cation name goes first, followed by the name of the anionic complex ion.
2. If the counter ion of the complex ion is an anion, the name of the cationic complex ion goes first, followed by the name of the anion. The anion does not have a numeric prefix.

Examples:

| | |
|---|---|
| $[Ni(NH_3)_6]Cl_2$ | Hexaamminenickel(II) chloride |
| $K_3[Fe(CN)_6]$ | Potassium hexacyanoferrate(III) |
| $[Co(NH_3)_5(H_2O)]Br_2$ | Pentaammineaquacobalt(II) bromide |

A key to naming coordination compounds is to recognize from their formulas that they *are* coordination compounds. For help with this, look for formulas that have the atomic symbol of a metallic element and one or more ligands all in brackets, either followed by the symbol of an anion, as in $[Co(NH_3)_5(H_2O)]Br_2$ or preceded by the symbol of a cation, as in $K_3[Fe(CN)_6]$.

**SAMPLE EXERCISE 17.3  Naming Coordination Compounds  LO3**

Name the coordination compounds (a) $Na_4[Co(CN)_6]$ and (b) $[Co(NH_3)_5Cl](NO_3)_2$.

**Collect and Organize** We are asked to write a name for each compound that unambiguously identifies its composition. The formulas of the complex ions appear in brackets in both compounds. Because cobalt, the central metal ion in both, is a transition metal, we express its oxidation state using Roman numerals. The names of common ligands are given in Table 17.2.

**Analyze** It is useful to take an inventory of the ligands and counter ions:

| Compound | Counter Ion | LIGAND | | | |
|---|---|---|---|---|---|
| | | Formula | Name | Number | Prefix |
| $Na_4[Co(CN)_6]$ | $Na^+$ | $CN^-$ | Cyano | 6 | Hexa- |
| $[Co(NH_3)_5Cl](NO_3)_2$ | $NO_3^-$ | $NH_3$ | Ammine | 5 | Penta- |
| | | $Cl^-$ | Chloro | 1 | — |

The oxidation state of each cobalt ion can be calculated by setting the sum of the charges on all the ions in both compounds equal to zero:

a. Ions:     $(4 Na^+$ ions$) + (1$ Co ion$) + (6 CN^-$ ions$)$

Charges:     $4+ \quad + \quad x \quad + \quad 6- \qquad = 0$

$$x = 2+$$

b. Ions:     $(1$ Co ion$) + (1 Cl^-$ ion$) + (2 NO_3^-$ ions$)$

Charges:     $x \quad + \quad 1- \quad + \quad 2- \qquad = 0$

$$x = 3+$$

**Solve**

a. Because the counter ion, *sodium*, is a cation, its name comes first. The complex ion is an anion in this compound. To name it, we begin with the ligand *cyano*, to which we add the prefix *hexa-* and write *hexacyano*. This is followed by the name of the transition metal ion: hexacyano*cobalt*. We then change the ending of the name of the complex ion to *-ate* because it is an anion: hexacyanocobalt*ate*. We then add a Roman numeral to indicate the oxidation state of the cobalt: hexacyanocobaltate(*II*). All together, we arrive at the name: sodium hexacyanocobaltate(II).

b. The complex ion is the cation in this compound, and we begin by naming the ligands directly attached to the metal ion in alphabetical order: *ammine* and *chloro*. We indicate the number (5) of $NH_3$ ligands with the appropriate prefix: *penta*amminechloro. We name the metal next and indicate its oxidation state with a Roman numeral: pentaamminechloro*cobalt(III)*. Finally we name the anionic counter ion: *nitrate*. Putting it all together, we obtain the name: pentaamminechlorocobalt(III) nitrate.

**Think About It** We can check the validity of the proposed names by breaking them down to see if they match up with the starting formulas. For example, the name in part b has a penta (meaning 5) prefix in front of ammine (meaning $NH_3$) followed by chloro (meaning $Cl^-$ ion) before the central ion name, cobalt(III) (meaning $Co^{3+}$). Assembling these symbols into that of a complex ion we have $[Co(NH_3)_5Cl]^{2+}$. The charge of 2+ is the sum of the charge of the central metal ion (3+) and the single chloride ion (1−). Two nitrate ions are needed to balance the charge on the complex ion. Adding them gives us a formula that matches that of the original coordination compound: $[Co(NH_3)_5Cl](NO_3)_2$.

⚙ **Practice Exercise** Identify the ligands and counter ions in (a) $[Zn(NH_3)_4]Cl_2$ and (b) $[Co(NH_3)_4(H_2O)_2](NO_2)_2$, and name each compound.

■

## 17.5 Hydrated Metal Ions as Acids

**CONNECTION** We discussed periodic trends in electronegativity in Chapter 8.

In Chapter 16 we saw that the strength of an oxoacid depends on the electronegativity of its central atom. For example, the relative strengths of the three hypohalous acids HOCl > HOBr > HOI align with the relative electronegativities of their halogen atoms: Cl > Br > I. This alignment makes sense because the more electronegative the halogen, the more it draws electron density away from the oxygen in the polar —OH bond (see Figure 16.13). These shifts make the negative ion formed by dissociation better able to bear a negative charge because of increased delocalization.

A similar shift in electron density occurs in hydrated metal ions having the generic formula $M(H_2O)_6^{n+}$ when $n \geq 2$. The electrons in the O—H bonds of the water molecules surrounding the metal ions are attracted to the positively charged ions. The resulting distortion in electron density increases the likelihood of one of these O—H bonds ionizing and donating a $H^+$ ion to a neighboring molecule of water:

$$M(H_2O)_6^{n+}(aq) + H_2O(\ell) \rightleftharpoons M(H_2O)_5(OH)^{(n-1)+}(aq) + H_3O^+(aq)$$

Figure 17.8 provides a molecular view of this reaction, using $Fe^{3+}(aq)$ as the central ion. Similar reactions allow other hydrated metal ions, particularly those with charges of 3+, to function as Brønsted–Lowry acids. The acidic strengths of several of these ions are indicated by the $K_a$ values in Table 17.3. Note how much stronger the 3+ ions are compared to the 2+ ions, as we would expect given the greater electron-withdrawing power of the more highly charged ions.

Figure 17.8 shows that one of the original six water molecules of hydration is converted into a hydroxide ion as a result of the ionization reaction. This reduces the charge on the complex from 3+ to 2+. The product $Fe(H_2O)_5(OH)^{2+}$ ion can then act as a Brønsted–Lowry acid ($H^+$ donor) in a second acid ionization:

$$Fe(H_2O)_5(OH)^{2+}(aq) + H_2O(\ell) \rightleftharpoons Fe(H_2O)_4(OH)_2^+(aq) + H_3O^+(aq)$$

A third acid ionization is possible, producing solid iron(III) hydroxide:

$$Fe(H_2O)_4(OH)_2^+(aq) + H_2O(\ell) \rightleftharpoons Fe(H_2O)_3(OH)_3(s) + H_3O^+(aq)$$

For simplicity, we usually write the formula of iron(III) hydroxide $Fe(OH)_3(s)$ even though the neutral material contains water of hydration.

Other 3+ cations, including $Cr^{3+}$ and $Al^{3+}$, display similar behavior. However, these two trivalent ions form electrically neutral hydroxide compounds that have an unusual solubility pattern. Whereas $Fe(OH)_3(s)$ and many other metal hydroxides have very limited solubility in basic solutions, $Cr(OH)_3$ and $Al(OH)_3$ are more soluble in strongly basic solutions (pH > 11) than in weakly basic solutions (pH ≈ 8). Why? Because solid $Cr(OH)_3$ and $Al(OH)_3$ may accept additional $OH^-$ ions at high pH, forming soluble anionic complex ions:

$$Cr(OH)_3(s) + OH^-(aq) \rightleftharpoons Cr(OH)_4^-(aq) = Cr(H_2O)_2(OH)_4^-(aq)$$

$$Al(OH)_3(s) + OH^-(aq) \rightleftharpoons Al(OH)_4^-(aq) = Al(H_2O)_2(OH)_4^-(aq)$$

Formation of complex ions explains the solubility of $Cr^{3+}$ and $Al^{3+}$ ions at low and high pH. These ions are soluble in strongly acidic solutions (pH ≈ 2), where

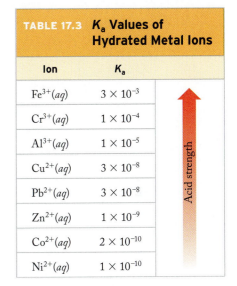

| TABLE 17.3 | $K_a$ Values of Hydrated Metal Ions | |
|---|---|---|
| **Ion** | **$K_a$** | |
| $Fe^{3+}(aq)$ | $3 \times 10^{-3}$ | |
| $Cr^{3+}(aq)$ | $1 \times 10^{-4}$ | |
| $Al^{3+}(aq)$ | $1 \times 10^{-5}$ | |
| $Cu^{2+}(aq)$ | $3 \times 10^{-8}$ | |
| $Pb^{2+}(aq)$ | $3 \times 10^{-8}$ | |
| $Zn^{2+}(aq)$ | $1 \times 10^{-9}$ | |
| $Co^{2+}(aq)$ | $2 \times 10^{-10}$ | |
| $Ni^{2+}(aq)$ | $1 \times 10^{-10}$ | |

Acid strength

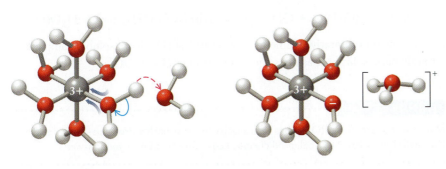

$$Fe(H_2O)_6^{3+}(aq) \quad + \quad H_2O(\ell) \quad \rightleftharpoons \quad Fe(H_2O)_5(OH)^{2+}(aq) \quad + \quad H_3O^+(aq)$$

**FIGURE 17.8** A hydrated $Fe^{3+}$ cation draws electron density away from the water molecules of its inner coordination sphere, which makes it possible for one or more of these molecules to donate a $H^+$ ion to a water molecule outside the sphere. The hydrated $Fe^{3+}$ ion is left with one fewer $H_2O$ molecule and one $OH^-$ ion.

they exist as $Cr(H_2O)_6^{3+}(aq)$ and $Al(H_2O)_6^{3+}(aq)$. In less acidic solutions they exist as positively charged complex ions with generic formulas such as $M(H_2O)_5(OH)^{2+}(aq)$ and $M(H_2O)_4(OH)_2^+(aq)$, and in strongly basic solutions they exist as $M(H_2O)_2(OH)_4^-(aq)$. Zinc hydroxide, $Zn(OH)_2$, is the only other transition metal hydroxide that is soluble at high pH.

Nearly all transition metals share one common characteristic: They exist as $M^{n+}(aq)$ ions only in strongly acidic solutions. They occur as complex ions in which at least one of their ligands is an $OH^-$ ion in aqueous solutions that range from slightly acidic to slightly basic ($3 < pH < 9$). This pH range includes most environmental waters and biological fluids.

We just discussed how little $Fe(OH)_3$ or $Al(OH)_3$ dissolves in pure water. The minuscule solubilities of these compounds are reflected in their $K_{sp}$ values: $1.1 \times 10^{-36}$ and $1.9 \times 10^{-33}$, respectively. However, we must keep in mind that these tiny $K_{sp}$ values apply to equilibrium concentrations of *free metal ions*, that is, the concentrations of $Fe^{3+}(aq)$ and $Al^{3+}(aq)$:

$$K_{sp} = [Fe^{3+}][OH^-]^3 = 1.1 \times 10^{-36}$$
$$K_{sp} = [Al^{3+}][OH^-]^3 = 1.9 \times 10^{-33}$$

Let's consider the implication of the extremely small $K_{sp}$ of $Al(OH)_3$ by answering the question: What is the maximum $[Al^{3+}]$ that can exist in pure water (pH = 7.00)? Solving the $K_{sp}$ expression for $[Al^{3+}]$ and inserting $[OH^-] = 1.0 \times 10^{-7}$, we get

$$[Al^{3+}] = \frac{K_{sp}}{[OH^-]^3} = \frac{1.9 \times 10^{-33}}{(1.0 \times 10^{-7})^3} = 1.9 \times 10^{-12} \, M$$

This very small value for $[Al^{3+}]$ seems to imply that no $Al^{3+}$ salt is soluble in water because the $Al^{3+}$ ions that it releases as it dissolves would immediately precipitate as $Al(OH)_3$. However, that conclusion is not correct. For example, $Al(NO_3)_3$, like all nitrates, is quite soluble in water. This solubility can be explained by the acidic properties of $Al^{3+}(aq)$, or $Al(H_2O)_6^{3+}$. This acidity means that the concentration of the acid $Al(H_2O)_6^{3+}$ at pH = 7.00 is very small for the same reason that the concentration of any weak acid is small compared to the concentration of its conjugate base at a pH above its $pK_a$. Most of the $Al^{3+}$ ions are present in solution as the conjugate base of $Al(H_2O)_6^{3+}$—that is, $Al(H_2O)_5(OH)^{2+}$—or even as the conjugate base of $Al(H_2O)_5(OH)^{2+}$, which is $Al(H_2O)_4(OH)_2^+$. The formation of these complexes means that the overall concentration of all soluble $Al^{3+}$ species is much greater than $[Al(H_2O)_6^{3+}]$ alone. Therefore, the molar solubility of $Al(NO_3)_3$ is actually much greater than $1.9 \times 10^{-12} \, M$. Moreover, its solubility increases if enough strong base is added because soluble $Al(H_2O)_2(OH)_4^-(aq)$ is the dominant $Al^{3+}$ species in alkaline solutions:

$$Al(H_2O)_3(OH)_3(s) + OH^-(aq) \rightleftharpoons Al(H_2O)_2(OH)_4^-(aq) + H_2O(\ell)$$

Among the representative elements, Sn(II) and Sn(IV) form soluble complexes in strongly basic solutions, as does the transition metal ion Cr(III).

**monodentate ligand** a species that forms only a single coordinate bond to a metal ion in a complex.

**polydentate ligand** a species that can form more than one coordinate bond per molecule.

**chelation** the interaction of a metal with a polydentate ligand (chelating agent); pairs of electrons on one molecule of the ligand occupy two or more coordination sites on the central metal.

**CONCEPT TEST**

Is it correct to say that $Al(OH)_3$ has the capacity to act as both a Brønsted–Lowry acid and a Brønsted–Lowry base? Use balanced chemical equations to support your answer.

# 17.6 Polydentate Ligands

The ligands in Figure 17.4 and many of those in Table 17.2 can donate only one pair of electrons to a metal ion. Even atoms with more than one lone pair usually donate only one pair at a time to a given metal ion because the other lone pair or pairs are oriented away from the metal ion. Because these ligands have effectively only one donor group, they are called **monodentate ligands**, which literally means "single-toothed."

Molecules larger than those in Figure 17.4 may be able to donate more than one lone pair of electrons and therefore form more than one coordinate bond to a central metal ion. Ligands in this category are called **polydentate ligands**, or more specifically *bidentate*, *tridentate*, and so on. One group of polydentate ligands is the polyamines, which include the compound ethylenediamine, a bidentate ligand with the structure

$$\text{H}_2\ddot{\text{N}} \qquad \ddot{\text{N}}\text{H}_2$$
$$\text{H}_2\text{C}\!-\!\text{CH}_2$$

Note that the lone pairs on the two –NH$_2$ groups are separated from each other by two –CH$_2$– groups. This combination means that a molecule of ethylenediamine can partially encircle a metal ion so that both lone pairs can bond to the metal ion.

Formation of the ethylenediamine complex of Ni$^{2+}$(*aq*) is shown in Figure 17.9(a). The two bonding orbitals that accept lone pairs of electrons from a molecule of ethylenediamine must be on the same side of the Ni$^{2+}$ ion. However, two more ethylenediamine molecules can bond to other pairs of bonding sites, displacing additional pairs of water molecules and forming a complex in which the Ni$^{2+}$ ion is surrounded by three bidentate ethylenediamine molecules, as shown in Figure 17.9(b).

Each ethylenediamine molecule forms a five-atom ring with the metal ion. If the ring were a perfect pentagon (meaning all bond lengths and bond angles were exactly the same), each of its bond angles would be 108°. These pentagons are not perfect, but each ring's preferred octahedral bond angles of 90° for the N–Ni–N bond and of 107° to 109° for all the other bonds are accommodated with only a little strain on the ideal bond angles.

An even larger ligand, diethylenetriamine (H$_2$NCH$_2$CH$_2$NHCH$_2$CH$_2$NH$_2$), is shown in Figure 17.10(a). The lone pairs of electrons on its three nitrogen atoms give diethylenetriamine the capacity to form three coordinate bonds to a metal ion, meaning this is a *tridentate* ligand (Figure 17.10b).

As you may imagine, larger molecules may have even more atoms per molecule that can bond to a single metal ion. The interaction of a metal ion with a ligand having multiple donor atoms is called **chelation** (pronounced *key-LAY-shun*). The word comes from the Greek *chele*, meaning "claw." The polydentate ligands that take part in these interactions are called *chelating agents*.

Many chelating agents have more than one kind of electron-pair-donating group. One family of such compounds is called *aminocarboxylic acids*. The most important of them is ethylenediaminetetraacetic acid (EDTA), the molecular structure of which is shown in Figure 17.11(a). Note that one molecule of EDTA contains two amine groups and four carboxylic acid groups. When the acid groups release their H$^+$ ions, they form four carboxylate anions, –COO$^-$, in which either of the O atoms can donate a pair of electrons to a central metal ion.

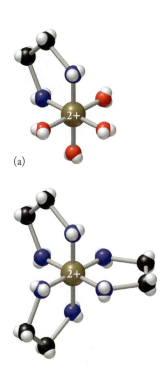

(a)

(b)

**FIGURE 17.9** (a) The bidentate ligand ethylenediamine has two N atoms that can each donate a pair of electrons to the empty orbitals of adjacent octahedral bonding sites on the same Ni$^{2+}$(*aq*) ion (gold sphere), displacing two molecules of H$_2$O. (b) Three ethylenediamine molecules can occupy all six octahedral coordination sites of a Ni$^{2+}$ ion.

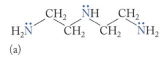

$$\text{H}_2\ddot{\text{N}} \underset{\text{CH}_2}{\overset{\text{CH}_2}{\diagup}} \ddot{\text{N}}\text{H} \underset{\text{CH}_2}{\overset{\text{CH}_2}{\diagdown}} \ddot{\text{N}}\text{H}_2$$

(a)

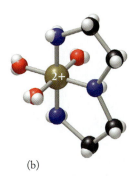

(b)

**FIGURE 17.10** Tridentate chelation. (a) The three amine groups in the tridentate ligand diethylenetriamine are all potential electron-pair donor groups. (b) When these groups donate their lone pairs of electrons to a Ni$^{2+}$(*aq*) ion (gold sphere) in solution, they occupy three of the six coordination sites on the ion.

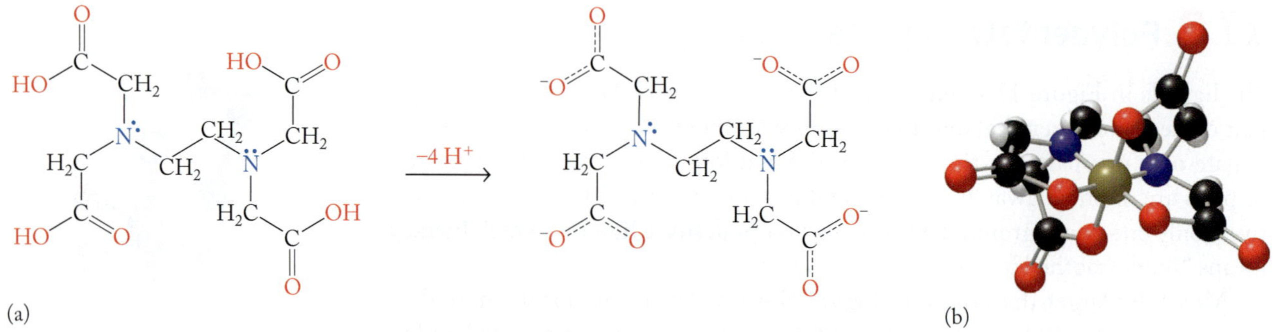

**FIGURE 17.11** (a) In the hexadentate ligand EDTA, the six donor groups are the two amine groups and the four carboxylic acid groups. The acid groups ionize to form coordinate bonding sites. (b) All six Lewis base groups in ionized EDTA can form a coordinate bond with the same metal ion, such as $Co^{3+}$ (the gold sphere) shown here. In the process they form three 5-membered rings.

When O atoms on all four groups do so and the two amine groups do as well, six octahedral bonding sites around the metal ion can be occupied, as shown in Figure 17.11(b).

EDTA forms very stable complex ions and is used as a metal ion *sequestering agent*; that is, a chelating agent that binds metal ions so tightly that they are "sequestered" and prevented from reacting with other substances. For example, EDTA is used as a preservative in many beverages and prepared foods because it sequesters iron, copper, zinc, manganese, and other transition metal ions often present in these foods that can catalyze the degradation of ingredients in the foods. Many foods are fortified with ascorbic acid (vitamin C), which is particularly vulnerable to metal-catalyzed degradation because it is also a polydentate ligand and is more likely to be oxidized when chelated to one of the above metal ions. EDTA effectively shields vitamin C from these ions. We discuss the preferential binding of metal ions to different ligands in the next section.

---

**SAMPLE EXERCISE 17.4**  **Identifying the Potential Electron-Pair Donor Groups in a Molecule**  **LO4**

How many donor groups does this polydentate ligand, nitrilotriacetic acid (NTA), have?

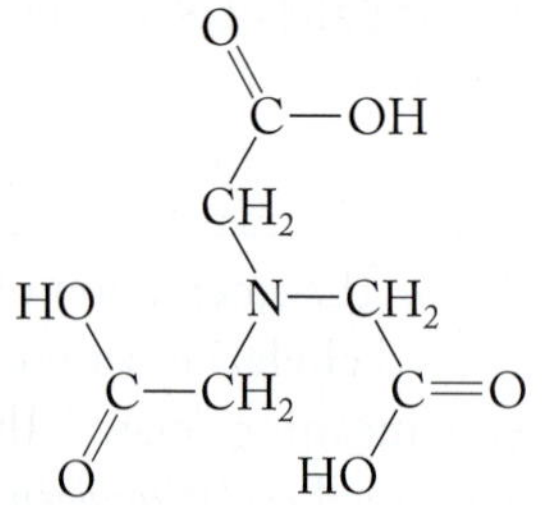

**Collect and Organize** We are asked to examine this molecular structure to find electron pairs that can be donated. The molecule has a nitrogen atom with three single bonds and three carboxylic acid groups.

**Analyze** Because there are three single bonds around the N atom, the atom's fourth $sp^3$ orbital must contain a lone pair of electrons. When all three carboxylic acid groups are ionized, there are three carboxylate groups in the molecule. Each carboxylate group can donate one nonbonding pair of electrons from one of its oxygen atoms to a metal atom. When one of these O atoms or the center N atom simultaneously forms coordinate bonds with a metal ion, a five-atom ring is produced, as in EDTA complexes.

**Solve** The central N atom and an O atom from each of the three carboxylate groups can form a total of four coordinate bonds. Therefore, NTA is potentially a tetradentate ligand with four donor groups.

**Think About It** The tetradentate capacity of NTA is reasonable because, like EDTA, it is an aminocarboxylic acid. It has one fewer amino group and one fewer carboxylic acid group than the hexadentate EDTA.

**Practice Exercise** How many potential donor groups are there in citric acid, a component of citrus fruits and a widely used preservative in the food industry?

## 17.7 Ligand Strength and the Chelate Effect

We have seen that ligands are electron-pair donors, that is, Lewis bases. In this section, let's explore the strengths of several ligands as Lewis bases by considering their affinity for $Ni^{2+}(aq)$ ions. Suppose we dissolve crystals of nickel(II) chloride hexahydrate, $NiCl_2 \cdot 6\ H_2O$, in water. The dot connecting the two halves of the formula and the prefix *hexa* indicate that each $Ni^{2+}$ ion in crystals of nickel(II) chloride is surrounded by six water molecules. Brilliant green crystals of $NiCl_2 \cdot 6\ H_2O$ form green solutions (Figure 17.12). The fact that the solid and its aqueous solution have the same color tells us that the same clustering of water molecules around $Ni^{2+}$ ions occurs in both solid $NiCl_2 \cdot 6\ H_2O$ and aqueous solutions of $Ni^{2+}$ ions. Thus, the $Ni^{2+}$ ion in a solution of nickel(II) chloride is most likely $Ni(H_2O)_6^{2+}$.

**FIGURE 17.12** (a) Solid nickel(II) chloride hexahydrate is green. (b) When it dissolves in water, the resulting solution has the same green color (left), telling us that each $Ni^{2+}$ ion (gold sphere) must be surrounded by $H_2O$ molecules, both in the solid and in the solution. When ammonia gas is bubbled through a solution of $Ni(H_2O)_6^{2+}$, the color changes to blue as $NH_3$ replaces $H_2O$ in the $Ni^{2+}$ ion's inner coordination sphere. When ethylenediamine is added to a solution of $Ni(NH_3)_6^{2+}$, the color turns from blue to purple as the ethylenediamine displaces the ammonia ligands and the $Ni(en)_3^{2+}$ complex forms.

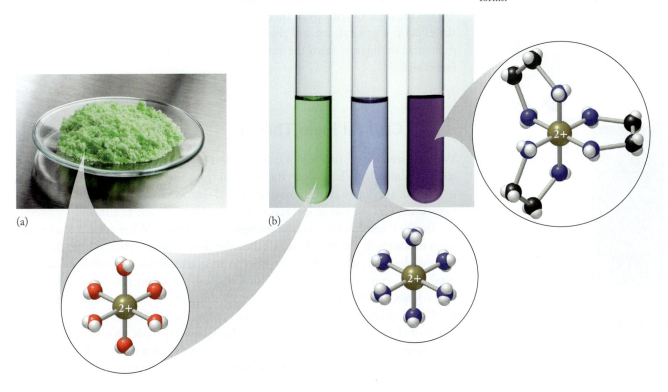

**chelate effect** the greater affinity of metal ions for polydentate ligands than for monodentate ligands.

**crystal field splitting** the separation of a set of *d* orbitals into subsets with different energies as a result of interactions between electrons in those orbitals and lone pairs of electrons in ligands.

**crystal field splitting energy (Δ)** the difference in energy between subsets of *d* orbitals split by interactions in a crystal field.

Now let's bubble colorless ammonia gas through a green solution of $Ni(H_2O)_6^{2+}$ ions. As shown in Figure 17.12(b), the green solution turns blue. The color change means that different ligands are bonded to the $Ni^{2+}$ ions. We may conclude that $NH_3$ molecules have displaced at least some $H_2O$ molecules around the $Ni^{2+}$ ions. If all the molecules of $H_2O$ are displaced, the complex $Ni(NH_3)_6^{2+}$ is formed. The following chemical equation describes this change:

$$Ni(H_2O)_6^{2+}(aq) + 6\ NH_3(g) \rightleftharpoons Ni(NH_3)_6^{2+}(aq) + 6\ H_2O(\ell) \qquad K_f = 5 \times 10^8$$

Keep in mind that the hydrated ion $Ni(H_2O)_6^{2+}$ is often expressed as $Ni^{2+}(aq)$.

This *ligand displacement* reaction illustrates that $Ni^{2+}$ ions have a greater affinity for molecules of $NH_3$ than for molecules of $H_2O$. This affinity is reflected in the large formation constant for the reaction. Many other transition metal ions also have a greater affinity for ammonia than for water. We may conclude that ammonia is inherently a better electron-pair donor and is a stronger Lewis base than water. This conclusion is reasonable because we saw in Chapter 16 that $NH_3$ is a stronger Brønsted–Lowry base than $H_2O$.

Next we add ethylenediamine to the blue solution of $Ni(NH_3)_6^{2+}$ ions. The solution changes color again, from blue to purple (Figure 17.12b), indicating yet another change in the ligands surrounding the $Ni^{2+}$ ions. Molecules of ethylenediamine displace ammonia molecules from the inner coordination sphere of $Ni^{2+}$ ions. This affinity of $Ni^{2+}$ ions for ethylenediamine molecules is reflected in the formation constant for $Ni(en)_3^{2+}$ (where "en" represents ethylenediamine):

$$Ni(H_2O)_6^{2+}(aq) + 3\ en(aq) \rightleftharpoons Ni(en)_3^{2+}(aq) + 6\ H_2O(\ell) \qquad K_f = 1.1 \times 10^{18}$$

This value is more than $10^9$ times the $K_f$ value for $Ni(NH_3)_6^{2+}$.

Why should the affinity of $Ni^{2+}$ ions for ethylenediamine be so much greater than their affinity for ammonia? After all, in both ligands the coordinate bonds are formed by lone pairs of electrons on N atoms. The greater affinity of metal ions for polydentate ligands compared with monodentate ligands is called the **chelate effect**. We will explore the reasons for this effect in Chapter 18 when we discuss the topic of entropy in thermodynamics. For now, simply understand that the effect is a general one and that reactions in which chelating ligands displace monodentate ligands from the inner coordination sphere about a metal ion are favorable.

# 17.8 Crystal Field Theory

We have seen that formation of complex ions can change the color of solutions of transition metals. Why is this? The colors of transition metal compounds and ions in solution are due to transitions of *d*-orbital electrons. Let's explore these transitions using $Cr^{3+}$ as our model transition metal ion (Figure 17.13).

The $Cr^{3+}$ ion has the electron configuration $[Ar]3d^3$. When a $Cr^{3+}$ ion (or any atom or ion) is in the gas phase, all the orbitals in a given subshell have the same energy. However, when a $Cr^{3+}$ ion is in an aqueous solution and surrounded by an octahedral array of water molecules in $Cr(H_2O)_6^{3+}$, the energies of its $3d$ orbitals are no longer all the same. As the six oxygen atoms of the water molecules approach the $Cr^{3+}$ ion during coordinate bond formation, the $3d$ electrons of $Cr^{3+}$ (Figure 17.14a) and the lone pairs of electrons on the ligands repel one another. These repulsions raise the energies of all the *d* orbitals, but to different extents. The $3d_{xy}$, $3d_{xz}$, and $3d_{yz}$ orbitals experience some increase in energy, but the energies of the $3d_{z^2}$ and

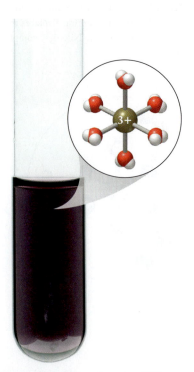

**FIGURE 17.13** When chromium(III) nitrate dissolves in water, the resulting solution has a distinctive violet color due to the presence of $Cr(H_2O)_6^{3+}$ ions.

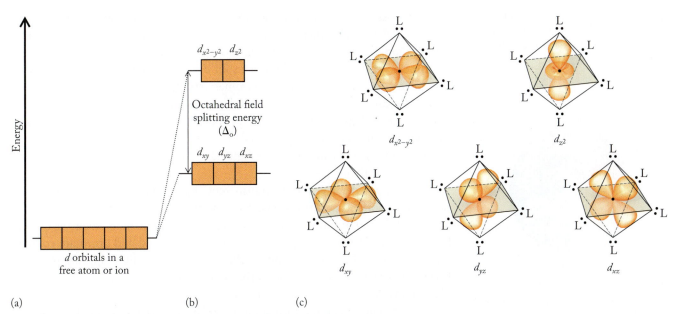

**FIGURE 17.14** Octahedral crystal field splitting. (a) In an atom or ion in the gas phase, all orbitals in a subshell are degenerate, as shown here for the five $3d$ orbitals. (b) When an ion is part of a complex ion in a compound or solution, repulsion between electrons in the ion's $d$ orbitals and ligand electrons raises the energies of the orbitals, as shown here for an octahedral field. (c) The greatest repulsion is experienced by electrons in the $d_{z^2}$ and $d_{x^2-y^2}$ orbitals because the lobes are directed toward the corners of the octahedron and so are closest to the lone pairs on the ligands (L). The lobes of the lower-energy $d_{xy}$, $d_{xz}$, and $d_{yz}$ orbitals are directed toward points that lie between the corners of the octahedron, so electrons in them experience less repulsion.

$3d_{x^2-y^2}$ orbitals increase even more (Figure 17.14b). This happens because the lobes of the $3d_{z^2}$ and $3d_{x^2-y^2}$ orbitals point directly toward the $H_2O$ oxygen atoms at the corners of the octahedron formed by the ligands and are repelled by the electrons on those O atoms (Figure 17.14c). The energies of the $3d_{xy}$, $3d_{xz}$, and $3d_{yz}$ orbitals are not raised as much because the lobes of these three orbitals do not point directly toward the corners of the octahedron, so their electron repulsions are weaker.

This process of changing *degenerate* (equal energy) orbitals to orbitals of different energies is known as **crystal field splitting**, and the difference in energy created by crystal field splitting is called **crystal field splitting energy (Δ)**. The name was originally used to describe splitting of $d$-orbital energies in ionic crystals, but the theory also applies to species in aqueous solutions.

In a $Cr(H_2O)_6^{3+}$ ion, three electrons are distributed among five $3d$ orbitals: three with lower energy than the other two. According to Hund's rule, each of the three electrons should occupy one of the three lower-energy orbitals, leaving the two higher-energy orbitals unoccupied, as shown in Figure 17.15(a). The energy difference between the two subsets of orbitals is denoted by $\Delta_o$, where the subscript indicates that the energy split was caused by an *o*ctahedral array of electron repulsions.

Now let's consider what happens when a $Cr^{3+}$ ion absorbs a photon whose energy is exactly equal to $\Delta_o$. As the photon is absorbed, a $3d$ electron moves from a lower-energy orbital to a higher-energy orbital (Figure 17.15b). The wavelength, $\lambda$, of the absorbed photon is related to the energy difference between the two orbitals—in other words, to the crystal field splitting energy—by Equation 17.1:

▶‖ **CHEMTOUR** Crystal Field Splitting

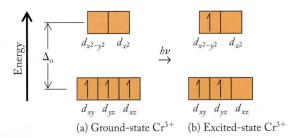

**FIGURE 17.15** A $Cr^{3+}$ ion, $[Ar]3d^3$, in an octahedral field can absorb a photon of light that has energy ($h\nu$) equal to $\Delta_o$. This energy raises a $3d$ electron from one of the lower-energy $d$ orbitals (a) to one of the higher-energy $d$ orbitals (b).

$$E = h\nu = \frac{hc}{\lambda} = \Delta_o \qquad (17.1)$$

**CONNECTION** In Chapter 7, we first used the equation $E = h\nu = hc/\lambda$ in discussing the energy of light in the electromagnetic spectrum.

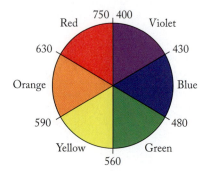

**FIGURE 17.16** A color wheel. Colors on opposite sides of the wheel are complementary to each other. When we look at a solution that absorbs light corresponding to a given color, we see the complementary color. Wavelengths are in nanometers.

As we discussed in Chapter 7, the energy and wavelength of a photon are inversely proportional to each other. Therefore, the larger the crystal field splitting in a complex ion, the shorter the wavelength of the photons the ion absorbs.

The size of the energy gap between split $d$ orbitals often corresponds to radiation in the visible region of the electromagnetic spectrum. This means that the colors of solutions of metal complexes depend on the strengths of metal–ligand interactions. When white light (which contains all colors of visible light) passes through a solution containing complex ions, the ions may absorb energy corresponding to a particular color. The light leaving the solution and reaching our eyes is missing that color.

The color we perceive for any transparent object is not the color it absorbs but rather the color(s) that it transmits. To relate the color of a solution to the wavelengths of light it absorbs, we need to consider complementary colors as defined by a simple color wheel (Figure 17.16). For example, red and green are complementary colors; therefore, a solution that absorbs green light appears red to us because our eyes and brain process the transmitted colors—red, orange, yellow, blue, and violet—as the average of those colors, which is red.

Now we have enough information to explain why the solution of $Cr^{3+}(aq)$ in Figure 17.13 is violet. The radiation absorbed by the electron transition shown in Figure 17.15 happens to be yellow-orange light. Because yellow-orange is the complement of violet, $Cr^{3+}(aq)$ solutions appear to be violet.

Color-averaging also occurs when a substance absorbs more than one color, as many transition metal solutions do. For example, a solution of $Cu(NH_3)_4^{2+}$ ions has a distinctive deep blue color, as we saw in Figure 17.7. The light transmitted by such a solution features an absorption band that spans yellow, orange, and red wavelengths with a minimum transmission of light at 620 nm (Figure 17.17). Our eyes sense the range that is transmitted, which spans violet to blue-green. Then our brain processes this band of transmitted colors and signals to us the average of these colors, a deep navy blue.

In two of the three colored solutions we have examined up to this point—$Ni^{2+}$ complexes and $Cr^{3+}$ complexes—the complex ions are octahedral and involve six ligands. However, in the solution of $Cu(NH_3)_4^{2+}$ the complex ions have only four ligands. This means that the deep blue color of this complex ion is caused by a different type of ligand–metal interaction.

Four ligands around a central metal have either a tetrahedral arrangement or a square planar arrangement (Table 17.1). Square planar geometries tend to be limited to the transition metal ions with nearly filled valence-shell $d$ orbitals, particularly those with $d^8$ or $d^9$ electron configurations. One such ion is $Cu^{2+}$, which has the electron configuration $[Ar]3d^9$. The $Cu(NH_3)_4^{2+}$ complex (Figure 17.17) is square planar, which means that the strongest interactions occur between the

**FIGURE 17.17** The visible light transmitted by a solution of $Cu(NH_3)_4^{2+}$ ions is missing much of the yellow, orange, and red portions of the visible spectrum because of a broad absorption band centered at 620 nm. Our eyes and brain perceive the transmitted colors as navy blue.

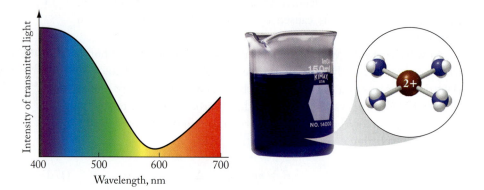

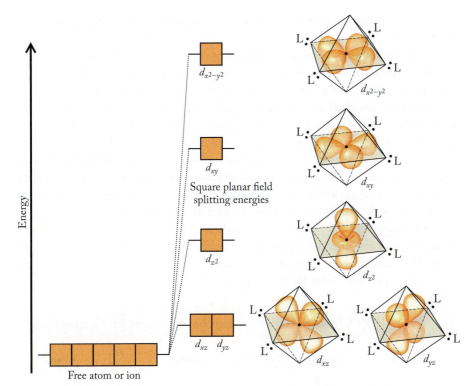

**FIGURE 17.18** Square planar crystal field splitting. The *d* orbitals of a transition metal ion in a square planar field are split into several energy levels, depending on how close the orbital lobes are to the ligand electrons located at the four corners of the square. The $d_{x^2-y^2}$ orbital is raised the most in energy because its lobes are directed right at the four corners of the square.

$3d$ orbitals on $Cu^{2+}$ and the nitrogen atom lone pairs at the four corners of the equatorial plane of the octahedron, as shown in Figure 17.18. In the $Cu^{2+}$ ion, the $3d$ orbital with the strongest interactions, and so the highest energy, is the $d_{x^2-y^2}$ orbital because its lobes are oriented directly at the four corners of the plane. The $d_{xy}$ orbital has slightly less energy because its lobes, though in the $xy$ plane, are directed 45° away from the corners. Electrons in the three $d$ orbitals with lobes out of the $xy$ plane interact even less with the lone pairs of the ligand and thus have even lower energies.

Finally, let's consider the $d$-orbital crystal field splitting that occurs in a tetrahedral complex (Figure 17.19). In this geometry, the greatest electron–electron repulsions are experienced in the $d_{xy}$, $d_{xz}$, and $d_{yz}$ orbitals because the lobes of these orbitals are oriented most directly to the corners of the tetrahedron that are occupied by ligand electron pairs. The two other $d$ orbitals are less affected because their lobes do not point toward the corners. The difference in energy

**FIGURE 17.19** (a) In a tetrahedral complex ion, such as $Zn(NH_3)_4{}^{2+}$, the $d$ orbitals of the metal ion undergo tetrahedral crystal field splitting. (b) The lobes of the higher-energy orbitals—$d_{xy}$, $d_{xz}$, and $d_{yz}$—are closer to the ligands at the four corners of the tetrahedron than the lobes of the lower-energy orbitals are. (One of the four corners of the tetrahedron is hidden in the drawings.)

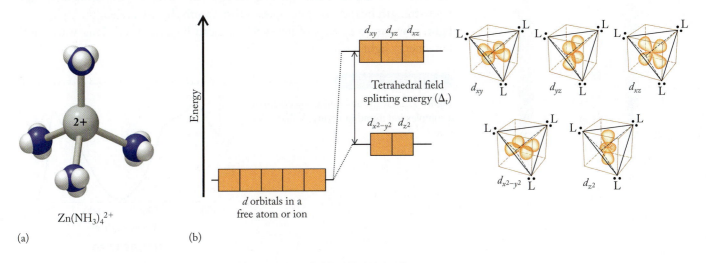

(a)    $Zn(NH_3)_4{}^{2+}$      (b)

**spectrochemical series** a list of ligands rank-ordered by their ability to split the energies of the *d* orbitals of transition metal ions.

between the two subsets of *d* orbitals resulting from the tetrahedral interactions is labeled $\Delta_t$.

Before ending this discussion on the colors of transition metal ions, let's revisit the color changes we saw in Figure 17.12, when first ammonia and then ethylene-diamine were added to a solution of $Ni^{2+}$ ions. Let's think in terms of the colors these solutions *absorb*. A solution of $Ni^{2+}(aq)$ ions is green because it absorbs colors at the red end of the spectrum. Similarly, a solution of $Ni(NH_3)_6^{2+}$ is blue because it absorbs colors opposite blue, which are centered on orange. Also, a solution of $Ni(en)_3^{2+}$ is violet because it absorbs colors centered on the complement of violet, which is yellow. Note how the colors these three solutions absorb are in the sequence red, orange, and yellow. This sequence runs from longest wavelength to shortest, from about 730 nm for red to about 570 nm for yellow. Radiant energy is inversely proportional to wavelength (Equation 17.1); therefore, the ability of these ligands to split the energies of the *d* orbitals of $Ni^{2+}$ ions is $en > NH_3 > H_2O$. Table 17.4 summarizes the observed colors of these $Ni^{2+}$ ion complexes and the meaning we can infer from their colors.

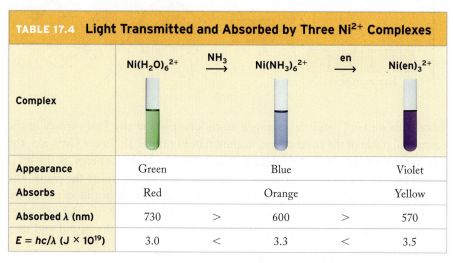

**TABLE 17.4 Light Transmitted and Absorbed by Three $Ni^{2+}$ Complexes**

|  | $Ni(H_2O)_6^{2+}$ | $\xrightarrow{NH_3}$ | $Ni(NH_3)_6^{2+}$ | $\xrightarrow{en}$ | $Ni(en)_3^{2+}$ |
|---|---|---|---|---|---|
| **Complex** |  |  |  |  |  |
| **Appearance** | Green |  | Blue |  | Violet |
| **Absorbs** | Red |  | Orange |  | Yellow |
| **Absorbed λ (nm)** | 730 | > | 600 | > | 570 |
| **$E = hc/\lambda$ (J × $10^{19}$)** | 3.0 | < | 3.3 | < | 3.5 |

Chemists use the parameter *field strength* to describe the relative abilities of ligands to split the energies of *d* orbitals in metal ions, ranking ligands in what is called a **spectrochemical series**. Table 17.5 contains one such series. The ligands at the top of the chart create the strongest repulsions with central-ion electrons, and those at the bottom create the weakest. As the field strength of the ligand increases, the crystal field splitting energy ($\Delta$) increases. Consequently, high-field-strength ligands form complexes that absorb short-wavelength, high-energy light, whereas complexes of low-field-strength ligands absorb long-wavelength, low-energy light.

**TABLE 17.5 Spectrochemical Series of Some Common Ligands**

Field strength ↑ / Orbital splitting ↑

$CN^-$
$NO_2^-$
en
py ≈ $NH_3$
$EDTA^{4-}$
$H_2O$
$OH^-$
$F^-$
$Cl^-$
$Br^-$
$I^-$

**CONCEPT TEST**

Figure 17.20 is the absorption spectrum of a complex ion found in nature. What color is it?

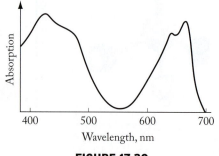

**FIGURE 17.20**

# 17.9 Magnetism and Spin States

In addition to determining the color of transition metal complexes, crystal field splitting can influence their magnetic properties because these properties depend on the number of unpaired electrons in the valence-shell *d* orbitals. The larger this number, the more paramagnetic the ion; that is, the more strongly it is attracted to a magnetic field. For example, an $Fe^{3+}$ ion has up to five *3d* electrons (Figure 17.21a). In an octahedral field, there are two ways to distribute the five *3d* electrons: one way conforms to Hund's rule and has a single electron in each orbital, leaving them all unpaired (Figure 17.21b). However, when $\Delta_o$ is large, as shown in Figure 17.21(c), all five electrons may go into the three lower-energy orbitals. This pattern of electron distribution occurs when the energy of repulsion between two electrons in the same orbital is less than the energy needed to promote an electron to a higher-energy orbital. In this configuration, only one electron is unpaired.

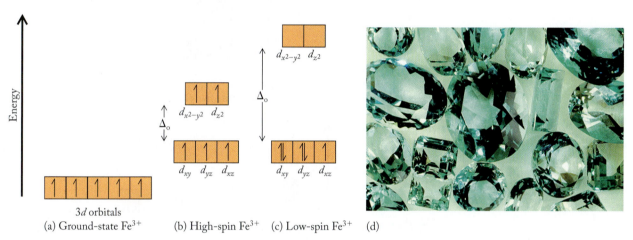

**FIGURE 17.21** Low-spin and high-spin complexes. (a) The ground state of a free $Fe^{3+}$ ion has a degenerate, half-filled set of *3d* orbitals. (b) A weak octahedral field ($\Delta_o$ < electron pairing energy) produces the high-spin state: five unpaired electrons, each in its own orbital. (c) In a strong octahedral field ($\Delta_o$ > electron pairing energy), the energies of the *3d* orbitals are split enough to produce the low-spin state: two sets of paired electrons, one unpaired electron, and two empty higher-energy orbitals. (d) The $Fe^{3+}$ ions in crystals of aquamarine are high-spin.

The configuration with all five electrons unpaired is called the *high-spin state* because the spin on all five electrons is in the same direction, resulting in the maximum magnetic field produced by the spins. The configuration with only one electron unpaired is called the *low-spin state*. Both configurations are paramagnetic because both have at least one unpaired electron. However, the high-spin state is much more paramagnetic.

Not all transition metal ions can have both high-spin and low-spin states. Consider, for example, $Cr^{3+}$ ions in an octahedral field. Because each ion has only three *3d* electrons (Figure 17.15), each electron is unpaired whether the orbital energies are split a lot or only a little. Therefore $Cr^{3+}$ ions have only one spin state. Other metal ions with full or nearly full sets of *d* orbitals have only one spin state because there cannot be more than two unpaired electrons in these orbitals no matter how they are distributed.

**CONNECTION** We introduced the magnetic behavior of matter in Chapter 9 in our discussion of molecular orbital theory.

**CONNECTION** Substances made of atoms, ions, or molecules that contain unpaired electrons are paramagnetic (see Section 9.7).

---

**SAMPLE EXERCISE 17.5** **Predicting Spin States** **LO6**

Determine which of these ions can have high-spin and low-spin configurations when part of an octahedral complex: (a) $Mn^{4+}$; (b) $Mn^{2+}$; (c) $Cu^{2+}$.

**Collect and Organize** To determine whether high-spin and low-spin states are possible, we need to determine the number of *d* electrons in each ion. Then we need to distribute them among sets of *d* orbitals split by an octahedral field to see if it is possible for the ions to have different spin states. Mn and Cu are in groups 7 and 11 of the periodic table, so their atoms have 7 and 11 valence electrons, respectively. Also, in an octahedral field, a set of five *d* orbitals splits into a low-energy subset of three orbitals and a high-energy subset of two orbitals.

**Analyze** The electron configurations of the atoms of the two elements are $[Ar]3d^54s^2$ for Mn and $[Ar]3d^{10}4s^1$ for Cu. When they form cations, the atoms of these transition metals lose their $4s$ electrons first and then their $3d$ electrons. Therefore, the numbers of *d* electrons in the ions are three in $Mn^{4+}$, five in $Mn^{2+}$, and nine in $Cu^{2+}$. It is likely that the ion with the fewest *d* electrons ($Mn^{4+}$) and the one with nearly the most *d* electrons possible ($Cu^{2+}$) will each have only one spin state.

**Solve**
a. $Mn^{4+}$: Putting three electrons into the lowest-energy $3d$ orbitals available and keeping them as unpaired as possible gives this orbital distribution of electrons:

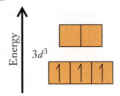

There is no low-spin option for $Mn^{4+}$.
b. $Mn^{2+}$: There are two options for distributing five electrons among the five $3d$ orbitals:

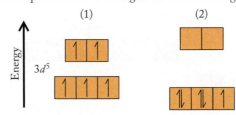

Thus $Mn^{2+}$ can have a high-spin (on the left) or a low-spin (on the right) configuration when part of an octahedral complex.
c. $Cu^{2+}$: The nine $3d$ electrons completely fill the lower-energy orbitals and nearly fill the higher-energy ones. No arrangement is possible other than the one shown, so $Cu^{2+}$ has only one spin state:

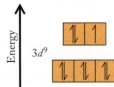

**Think About It** In an octahedral field, metal ions with 4, 5, 6, or 7 *d* electrons can exist in high-spin and low-spin states. Those ions with 3 or fewer *d* electrons have only one spin state, in which all the electrons are unpaired and in the lower-energy set of orbitals. Ions with 8 or more *d* electrons have only one spin state because their lower-energy set of orbitals is completely filled. The magnitude of the crystal field splitting energy, $\Delta_o$, determines which spin state an ion with 4, 5, 6, or 7 *d* electrons occupies.

⚙ **Practice Exercise** Which of the following ions can have high-spin and low-spin configurations when part of an octahedral complex? (a) $V^{4+}$; (b) $Cr^{3+}$; (c) $Ni^{3+}$. Are any of the possible spin states diamagnetic?

As noted earlier, whether a transition metal ion is in a high-spin state or a low-spin state depends on whether less energy is needed to promote an electron to a higher-energy orbital than to overcome the repulsion experienced by two electrons sharing the same lower-energy orbital. Several factors affect the size of $\Delta_o$.

We have already discussed a major one in the context of the spectrochemical series shown in Table 17.5: the different field strengths of different ligands. Because nitrogen-containing molecules are stronger field-splitting ligands than $H_2O$ molecules, hydrated metal ions (small $\Delta_o$) are more likely to be in high-spin states, and metals surrounded by nitrogen-containing ligands (large $\Delta_o$) are more likely to be in low-spin states. Another factor affecting spin state is the oxidation state of the metal ion. The higher the oxidation number (and ionic charge), the stronger the attraction of the electron pairs on the ligands for the ion. Greater attraction leads to more ligand–$d$ orbital interaction and therefore to a larger $\Delta_o$.

Complexes of transition metals in the fifth and sixth rows tend to be low spin because their $4d$ and $5d$ orbitals extend farther out from the nucleus than do $3d$ orbitals. These larger $d$ orbitals overlap more and interact more strongly with the lone pairs of electrons on the ligands, leading to greater crystal field splitting.

Our discussion of high-spin and low-spin states has focused entirely on $d$ orbitals split by octahedral fields. What about spin states in tetrahedral fields? Most tetrahedral complexes are high spin because tetrahedral fields, which are created by only four ligands, are weaker than octahedral fields created by six ligands. Weaker field strength means less $d$-orbital splitting—not enough to offset the energies associated with pairing two electrons in the same orbitals.

**CONCEPT TEST** ........................................................................

Explain the following: (a) $Mn(pyridine)_6^{2+}$ is a high-spin complex ion, but $Mn(CN)_6^{4-}$ is low spin; (b) $Fe(NH_3)_6^{2+}$ is high spin, but $Ru(NH_3)_6^{2+}$ is low spin.

..............................................................................................

# 17.10 Isomerism in Coordination Compounds

We introduced the concepts of structural isomerism and stereoisomerism in Chapter 13. To review:

➤ Structural isomers are compounds that have the same chemical formula but different arrangements of the bonds in their molecules. For example, these two hydrocarbons are structural isomers because they have the same formula ($C_4H_8$) but their C=C double bonds are in different locations:

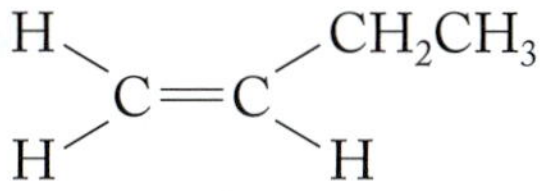

1-Butene 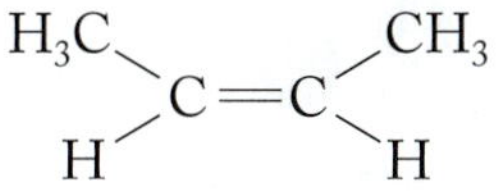                     2-Butene

➤ Stereoisomers are compounds with the same formula *and* the same bonding pattern, but the spatial orientation of the groups connected by those bonds is different. For example, these two hydrocarbons are stereoisomers:

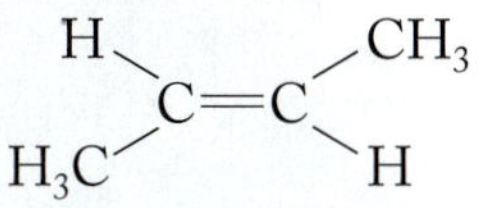

*trans*-2-Butene 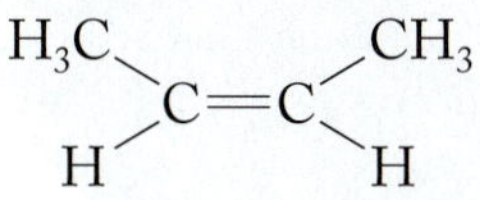              *cis*-2-Butene

In *trans*-2-butene the –$CH_3$ groups are on opposite sides of the C=C double bond, but in *cis*-2-butene they are on the same side. Remember that C=C bonds are rigid, so the ends of molecules cannot rotate around the C=C axis as they can around C—C single bonds.

**CONNECTION** The rigidity of double bonds was described in Section 13.3.

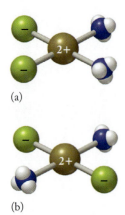

(a)

(b)

**FIGURE 17.22** Two ways to orient the $Cl^-$ ions and $NH_3$ molecules around a $Pt^{2+}$ ion (gold sphere) in the square planar coordination compound $Pt(NH_3)_2Cl_2$. (a) The two members of each pair of ligands are on the same side of the square in *cis*-diamminedichloroplatinum(II). (b) The two members of each pair are at opposite corners in *trans*-diamminedichloroplatinum(II).

# Stereoisomers of Coordination Compounds

Both structural isomerism and stereoisomerism occur in coordination compounds, but stereoisomerism is our focus here. Consider the square planar $Pt^{2+}$ coordination compound $Pt(NH_3)_2Cl_2$. Both molecules of ammonia and the two chloride ions are coordinately bonded to the $Pt^{2+}$ ion, so the name of the compound is diamminedichloroplatinum(II). No counter ions are present because the sum of the charges on the other ions is zero and the complex is neutral.

Two stereoisomers of diamminedichloroplatinum(II) are possible because there are two ways to orient the two pairs of ligands in the square plane of the complex (Figure 17.22): the two members of each pair can be at adjacent corners of the square or at opposite corners. The isomer with the two members of each pair at adjacent corners is called *cis*-diamminedichloroplatinum(II), and the isomer with the pairs at opposite corners is *trans*-diamminedichloroplatinum(II). Note that cis and trans have the same meaning as in the nomenclature of organic compounds.

To illustrate the importance of stereoisomerism, consider this: *cis*-diamminedichloroplatinum(II) is a widely used anticancer drug with the common name *cisplatin*, but the trans isomer is ineffective in fighting cancer. The therapeutic power of cisplatin comes from its structurally specific reactions with DNA. During these reactions, the two Cl atoms of $Pt(NH_3)_2Cl_2$ are replaced by nitrogen-containing bases along a strand of DNA in the nucleus of a cancerous cell. This ability of cisplatin to cross-link the bases distorts the molecular shape of the DNA and prevents it from replicating. Eventually, the cell dies. The trans isomer has a stronger affinity for other cellular proteins and is cleared from the cell before it can inhibit DNA replication.

Stereoisomerism is also possible in octahedral complexes containing more than one type of ligand. For example, there are two possible stereoisomers of $[Co(NH_3)_4Cl_2]Cl$, as shown in Figure 17.23. The two chloro ligands in $[Co(NH_3)_4Cl_2]^+$ are either on the same side of the complex, with a 90° Cl–Co–Cl

**FIGURE 17.23** The two stereoisomers of the coordination compound with the formula $[Co(NH_3)_4Cl_2]Cl$: (a) *cis*-tetraamminedichlorocobalt(III) chloride and (b) *trans*-tetraamminedichlorocobalt(III) chloride.

(a) *cis*-Tetraamminedichlorocobalt(III) chloride

(b) *trans*-Tetraamminedichlorocobalt(III) chloride

bond angle (the cis isomer), or across from each other so that the Cl–Co–Cl bond angle is 180° (the trans isomer). *cis*-Tetraamminedichlorocobalt(III) chloride is violet, and *trans*-tetraamminedichlorocobalt(III) chloride is green.

**SAMPLE EXERCISE 17.6**    **Identifying Stereoisomers**     **LO7**
                                        **of Coordination Compounds**

Sketch the structures and name the stereoisomers of $Ni(NH_3)_4Cl_2$.

**Collect and Organize** We are given the formula of a coordination compound and are to name and draw the structures of its stereoisomers. The $Ni^{2+}$ complexes we have seen so far in the chapter have all been octahedral. Ammonia molecules and $Cl^-$ ions are both monodentate ligands.

**Analyze** The formula contains no brackets, so the $Cl^-$ ions are not counter ions; they must be covalently bonded to the Ni ion. There are two of them, so the charge on Ni must be 2+. There are a total of six ligands, which confirms that the compound is octahedral.

**Solve** There are two ways to orient the chloride ions: opposite each other with a Cl–Ni–Cl bond angle of 180° or on the same side of the octahedron with a Cl–Ni–Cl bond angle of 90°:

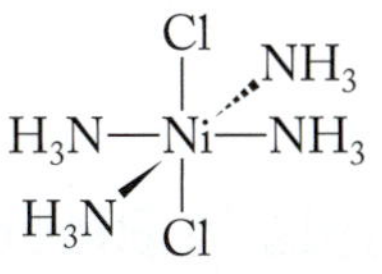

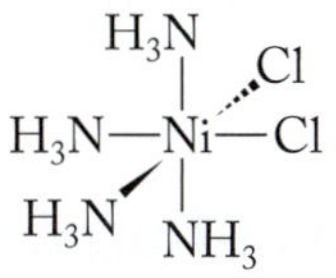

Cl      $NH_3$            $H_3N$     Cl

$H_3N$—Ni—$NH_3$        $H_3N$—Ni—Cl

$H_3N$    Cl            $H_3N$    $NH_3$

The first isomer is *trans*-tetraamminedichloronickel(II); the second is *cis*-tetraamminedichloronickel(II).

**Think About It** You can draw other tetraamminedichloronickel(II) structures that do not look exactly like these two structures. If you flip or rotate them, however, you will see that they match one of the two structures shown above.

 **Practice Exercise** Sketch the stereoisomers of $[CoBr_2(en)(NH_3)_2]^+$.

# Enantiomers

Another kind of stereoisomerism is possible in complex ions and coordination compounds. Consider the octahedral Co(III) complex ion containing two ethylenediamine molecules and two chloride ions. There are two ways to arrange the chloride ions: on adjacent bonding sites in a cis isomer or on opposite sides of the octahedron in a trans isomer. The cis isomer is shown in Figure 17.24.

This complex ion is called *cis*-dichlorobis(ethylenediamine)cobalt(III). Note the prefix *bis-* just before "(ethylenediamine)." In naming complex ions containing a polydentate ligand, we use *bis-* instead of *di-* to indicate that two molecules of the ligand are present and avoid the use of two *di-* prefixes in the same ligand name. The corresponding prefix for three polydentate ligands is *tris-*.

Figure 17.24 shows that *cis*-$Co(en)_2Cl_2^+$ is chiral: it has a mirror image that is not identical to the original. The difference is demonstrated by the fact that there is no way to rotate the mirror image so that its atoms align exactly with those in the original. In other words, the two structures are not *superimposable*. We encountered this phenomenon in Chapter 13 and noted that chemists call such nonsuperimposable stereoisomers *enantiomers*.

**FIGURE 17.24** The complex ion *cis*-dichlorobis(ethylenediamine)cobalt(III) is chiral, which means that its mirror image is not superimposable on the original complex. To illustrate this point, we rotate the mirror image 180° about its vertical axis so that it looks as much like the original as possible. However, note that the top ethylenediamine ligand is located behind the plane of the page in the original but in front of the plane of the page in the rotated mirror image. Thus, the mirror images are not superimposable.

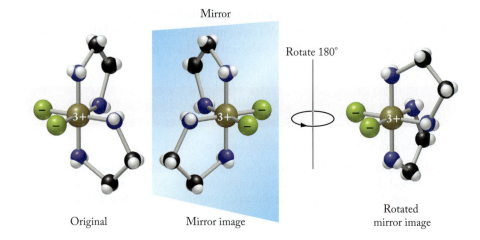

Mirror

Rotate 180°

Original          Mirror image          Rotated mirror image

**CONCEPT TEST** ...........................................................

Would four different ligands arranged in a square planar geometry produce a chiral complex ion? What about four different ligands in a tetrahedral geometry?

.....................................................................................

## 17.11 Coordination Compounds in Biochemistry

At the beginning of the chapter, we noted that metals essential to human health must be present in foods in forms the body can absorb. In this section we explore some biological polydentate ligands that help metal ions participate in processes that are essential to nutrition and good health.

Let's begin with photosynthesis, a chemical process at the base of our food chain. Green plants can harness solar energy because they contain large biomolecules we collectively call *chlorophyll*. All molecules of chlorophyll contain ring-shaped tetradentate ligands called *chlorins* (Figure 17.25a). The structures of chlorins are similar to those of **porphyrins**, another class of tetradentate ligands found in biological systems (Figure 17.25b). Chlorins and porphyrins are members of a larger category of polydentate compounds known as **macrocyclic ligands**. (*Macrocycle* means, literally, "big ring.")

Two of the four nitrogen atoms in porphyrins and chlorins are $sp^3$ hybridized and bound to hydrogen atoms, whereas the other two are $sp^2$ hybridized with no hydrogen atoms. When either ring forms a coordinate bond with a metal ion $M^{n+}$ (Figure 17.25c), the two hydrogen atoms ionize, giving the ring a charge of $2-$ and the complex ion an overall charge of $(n - 2)$. The lone pairs of electrons on the N atoms in the ionized structure are oriented toward the ring center. These

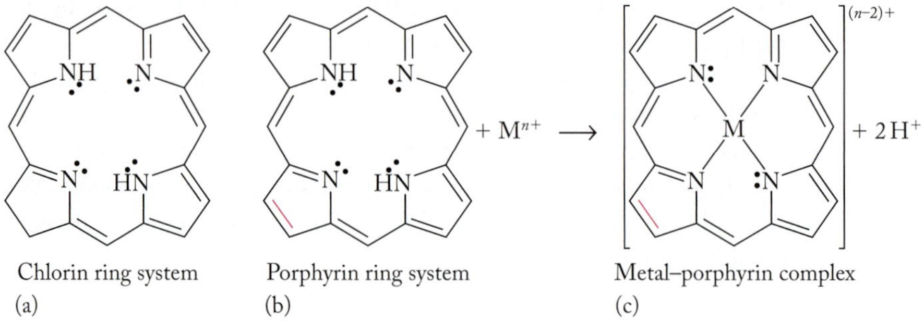

**FIGURE 17.25** (a) Chlorin and (b) porphyrin rings are biologically important tetradentate ligands that have similar core structures. The principal difference is a C=C double bond in the porphyrin structure (shown in red) that is a single bond in chlorin rings. The innermost atoms in each ring are four nitrogen atoms with lone pairs of electrons. (c) All four N atoms form coordinate bonds with a metal ion, as shown in this porphyrin ring.

Chlorin ring system
(a)

Porphyrin ring system
(b)

$+ M^{n+} \longrightarrow$

Metal–porphyrin complex
(c)

$+ 2H^+$

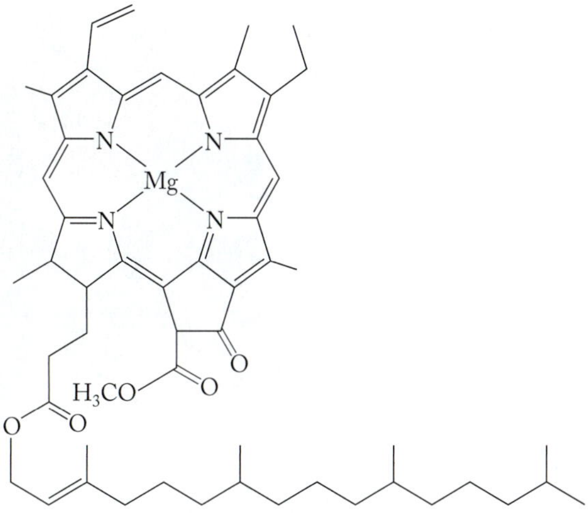

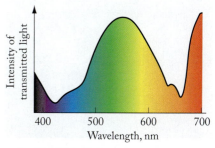

Chlorophyll *a*

**FIGURE 17.26** Chlorophyll absorbs sunlight and initiates a series of reactions that convert sunlight, carbon dioxide, and water into chemical energy stored in the bonds of carbohydrates. All forms of chlorophyll, including chlorophyll *a* shown here, have a $Mg^{2+}$ ion in a chlorin ring as part of their structure.

lone pairs can occupy either the four equatorial coordination sites in an octahedral complex ion or all four coordination sites in a square planar complex ion. In octahedral complex ions, each central metal ion still has its two axial sites available for bonding to other ligands. Depending on the charge of the central ion, the coordination compound may be either ionic (a complex ion) or electrically neutral.

Porphyrin and chlorin rings are widespread in nature and play many biochemical roles. Their chemical and physical properties depend on

1. the identity of the central metal ion
2. the species that occupy the axial coordination sites of octahedral complexes
3. the number and identity of organic groups attached to the outside of the ring

Chlorophyll *a* has a $Mg^{2+}$ ion coordinated at the center of a chlorin ring (Figure 17.26). Delocalized *p* electrons in the conjugated double bonds in and around the ring stabilize it and give it and other plant pigments the ability to absorb wavelengths of red and blue-violet light. Because plant leaves absorb these colors, most of them are green and yellow-green—the colors that are not absorbed (Figure 17.27). In temperate climates, chlorophyll is lost from the leaves of trees such as maples and oaks at the end of each growing season, which reveals the colors of other pigments in the beautiful leaves of autumn (Figure 17.28).

An important porphyrin complex called the *heme* group has a coordinately bonded central $Fe^{2+}$ ion (Figure 17.29). The heme group enables the proteins hemoglobin and myoglobin to transport $O_2$ in the blood and to store $O_2$ in muscle tissues, respectively. The four nitrogen atoms in the porphyrin ring of a heme group occupy equatorial positions in an octahedral complex of an $Fe^{2+}$ ion. Below the ring, a fifth bond is formed between the $Fe^{2+}$ ion and a lone pair of electrons on another nitrogen atom in the protein. The sixth ligand, located above the porphyrin ring, is typically a molecule of $O_2$, as in the oxygenated forms of hemoglobin in blood leaving the lungs.

Each $O_2$ molecule can act as a Lewis base and donate one of its lone pairs of electrons to the iron in heme. This coordinate covalent bond is strong enough to carry oxygen from the lungs to the cells in our bodies but weak enough to break easily when the oxygen reaches a cell. Other ligands of similar size can bind to the $Fe^{2+}$ ion in heme, and problems arise when some of them do. Carbon monoxide

**FIGURE 17.27** Combined transmission spectrum of chlorophyll and other pigments in a typical green leaf. The pigments absorb most of the visible radiation emitted by the sun except yellow-green.

**FIGURE 17.28** The colors of fall in southern Vermont are characterized by the reds, yellows, and golds of leaves that have lost their green pigments and are about to fall.

**porphyrin** a type of tetradentate macrocyclic ligand.

**macrocyclic ligand** a ring containing multiple electron-pair donors that bind to a metal ion.

**FIGURE 17.29** (a) In the porphyrin known as heme, the four nitrogen atoms occupy equatorial positions in an octahedral complex in which the central metal ion is $Fe^{2+}$. (b) Hemoglobin contains four heme groups, each of which is bound to a protein chain. The four protein chains are held together by intermolecular forces.

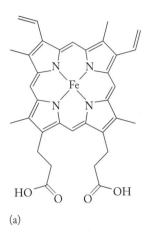

(a)

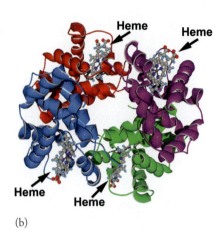

(b)

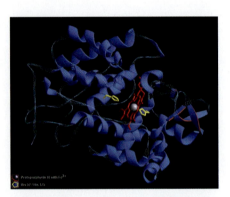

**FIGURE 17.30** The structure of cytochrome proteins, such as cytochrome *c* shown here, includes heme complexes (shown in red) that mediate energy production and redox reactions in living cells. Different cytochromes have different ligands (shown in yellow) occupying the fifth and sixth octahedral coordination sites, and different groups in the protein may be attached to the porphyrin ring.

is such a ligand. Similar in size to $O_2$, a CO molecule easily fits into the sixth binding site. Unfortunately, it binds about 200 times more strongly than $O_2$. If a person breathes air containing carbon monoxide, CO prevents $O_2$ from being taken up by blood flowing through the lungs, causing the symptoms of CO poisoning and, ultimately, death by suffocation.

Proteins called *cytochromes* also contain heme groups (Figure 17.30). Cytochromes mediate oxidation and reduction processes connected with energy production in cells. The heme group conveys electrons as the half-reaction

$$Fe^{3+} + e^- \rightleftharpoons Fe^{2+}$$

rapidly and reversibly consumes or releases electrons needed in the biochemical reactions that sustain life. There are many kinds of cytochrome proteins with different substituents on the porphyrin rings and different axial ligands, each of which influences the function of the complex. This last point has been repeated several times in this chapter: the chemical properties and biological functions of transition metals that are essential to living organisms are linked to their molecular environments and to the formation of stable complex ions with ligands that are strong electron donors; that is, strong Lewis bases.

---

**SAMPLE EXERCISE 17.7  Integrating Concepts: Analysis of an Alloy**

The presence of iron, cobalt, and nickel in homogeneous mixtures of metals called alloys can be detected by a series of chemical tests. The alloy sample is prepared by dissolving it in acid to produce an aqueous solution of $Fe^{2+}$, $Co^{2+}$, and $Ni^{2+}$ ions. After pH adjustment, $H_2S$ gas is bubbled through the solution, producing sulfide precipitates of the three ions.

1. The precipitated mass is collected and treated with hydrochloric acid to dissolve only the most soluble of the three metal sulfides. This solution is retained as solution 1.
2. The precipitate from step 1, containing the two less soluble metal sulfides, is dissolved in nitric acid. This solution is retained as solution 2.
3. Solution 1 is treated with $HNO_3$, which oxidizes the metal ions in solution from their +2 to +3 oxidation states. In the process, $HNO_3$ is converted into NO.
4. A few crystals of $NH_4SCN$ are added to the solution from step 3, producing a blood-red solution indicating the presence of $Fe(NCS)(H_2O)_5^{2+}$ ions. This cyanate ion ($SCN^-$) is written

$NCS^-$ in the complex to indicate that the bond is from the metal ion to the nitrogen atom.
5. $NH_3(aq)$ is added to solution 2 to form ammine complexes of the ions present. The resultant solution is divided into two portions: 5A and 5B.
6. The bidentate ligand dimethylglyoxime (molecular DMG: $C_4H_8N_2O_2$) is added to solution 5A, producing a red precipitate, which is analyzed and determined to have the following percent composition:

Ni: 20.31%;   C: 33.25%;   H: 4.89%;   N: 19.40%.

7. Hydrochloric acid, which converts ammine complexes back to aqua complexes, is added to solution 5B. A few crystals of $NH_4SCN$ are also added, and a blue solution is produced indicating the presence of $Co(SCN)_4^{2-}$, which is a square-planar complex.

Identify the metals involved in each reaction, and write net ionic equations for the reactions involved in these steps. Draw structures of the complex ions that form and identify any possible isomers. The

$K_{sp}$ values for the three sulfides are $6.3 \times 10^{-18}$ for FeS, $4 \times 10^{-21}$ for CoS, and $1 \times 10^{-24}$ for NiS.

**Collect and Organize** A metal sample is dissolved in strong acid, producing a solution containing $Fe^{2+}$, $Co^{2+}$, and $Ni^{2+}$ ions. These ions are precipitated as their sulfides and then selectively redissolved with more acids, including nitric acid, which oxidizes one of the metals to its +3 oxidation state as $HNO_3$ is reduced to NO. The presence of the three metals is confirmed by adding ligands that form complexes whose colors indicate the presence of a particular metal ion. We are asked to write net ionic equations describing the reactions involved in the tests and to draw the structures of the complex ions that are formed.

**Analyze** The $K_{sp}$ values for the binary sulfides of the three metals provide a direct measure of which of them is the most soluble and therefore dissolves in step 1. The reactions of the ion of the most soluble sulfide are described in steps 1, 3, and 4. Writing a balanced net ionic equation for the redox reaction in step 3 will require an analysis of the changes in oxidation states of the reactants and products. The reaction takes place in an acidic solution, so the final steps involving H and O atoms will be done by adding $H_2O$ and $H^+$ ions.

The chemistries of the other two ions are described in steps 2, 5, 6, and 7. In step 6 we are provided elemental composition information and will need to use it to calculate the empirical formula of the coordination compound nickel forms with DMG. From that formula, we can calculate the ratio of DMG ligands per nickel ion. The elemental composition data do not add up to 100%, but note that a value is missing for oxygen, which is present in the molecular formula of DMG. Oxygen's proportion can be calculated by adding up the other percent composition values and subtracting from 100%.

**Solve** Of the metal sulfides in step 1, FeS has the largest $K_{sp}$, so it is the one that dissolves when the sulfide precipitate is treated with hydrochloric acid, producing soluble iron(II) chloride and hydrogen sulfide:

$$FeS(s) + 2\,H^+(aq) \rightarrow Fe^{2+}(aq) + H_2S(g)$$

The collected precipitate from step 1 contains CoS(s) and NiS(s). These solids are dissolved in nitric acid in step 2, producing solutions of cobalt(II) and nickel(II) nitrates and hydrogen sulfide gas:

$$CoS(s) + 2\,H^+(aq) \rightarrow Co^{2+}(aq) + H_2S(g)$$

$$NiS(s) + 2\,H^+(aq) \rightarrow Ni^{2+}(aq) + H_2S(g)$$

Nitric acid oxidizes $Fe^{2+}(aq)$ to $Fe^{3+}(aq)$ in step 3 and forms NO(g) in the process. To write a balanced net ionic equation describing this reaction we start with the known reactants and products:

Unbalanced:   $Fe^{2+}(aq) + NO_3^-(aq) \rightarrow Fe^{3+}(aq) + NO(g)$

Using the half-reaction method from Chapter 4 for a redox reaction taking place in aqueous acid, we separate the two half-reactions:

$$Fe^{2+}(aq) \rightarrow Fe^{3+}(aq)$$

$$NO_3^-(aq) \rightarrow NO(g)$$

then balance them by adding water, hydrogen ions, and electrons:

$$Fe^{2+}(aq) \rightarrow Fe^{3+}(aq) + 1\,e^-$$

$$3\,e^- + 4\,H^+(aq) + NO_3^-(aq) \rightarrow NO(g) + 2\,H_2O(\ell)$$

We balance the number of electrons and then add the two equations:

$$3\,[Fe^{2+}(aq) \rightarrow Fe^{3+}(aq) + 1\,e^-]$$

$$3\,Fe^{2+}(aq) \rightarrow 3\,Fe^{3+}(aq) + \cancel{3\,e^-}$$

$$\cancel{3\,e^-} + 4\,H^+(aq) + NO_3^-(aq) \rightarrow NO(g) + 2\,H_2O(\ell)$$

$$3\,Fe^{2+}(aq) + NO_3^-(aq) + 4\,H^+(aq) \rightarrow$$
$$3\,Fe^{3+}(aq) + NO(g) + 2\,H_2O(\ell)$$

A final check of the charges (9+ on both sides of the reaction arrow) confirms that we have a balanced net ionic equation describing the redox reaction.

In step 4, hydrated $Fe^{3+}$ ions combine with dissolved thiocyanate ions, forming a red solution of $Fe(NCS)(H_2O)_5^{2+}$ complex ions. The net ionic equation describing the complex formation reaction is

$$Fe(H_2O)_6^{3+}(aq) + SCN^-(aq) \rightarrow Fe(NCS)(H_2O)_5^{2+}(aq) + H_2O(\ell)$$

and the structure of the complex ion may be drawn

The added ammonia in step 5 displaces water from the inner coordination sphere and produces the hexaammine complexes of the $Co^{2+}(aq)$ and $Ni^{2+}(aq)$ ions:

$$Co(H_2O)_6^{2+}(aq) + 6\,NH_3(aq) \rightarrow Co(NH_3)_6^{2+}(aq) + 6\,H_2O(\ell)$$

$$Ni(H_2O)_6^{2+}(aq) + 6\,NH_3(aq) \rightarrow Ni(NH_3)_6^{2+}(aq) + 6\,H_2O(\ell)$$

The DMG ligand in step 6 must have a negative charge to form a precipitate (neutral compound) with $Ni^{2+}$, but we do not know what charge it has or how many DMG ligands bind to one $Ni^{2+}$ ion. We can determine the complex's empirical formula after calculating the percent oxygen by summing the other values and subtracting the sum from 100.00%. That value is 22.15%.

| | | |
|---|---|---|
| Ni: | 20.31 g/(58.69 g/mol) = 0.3461 mol | 0.3461/0.3461 ≈ 1 |
| C: | 33.25 g/(12.01 g/mol) = 2.769 mol | 2.769/0.3461 ≈ 8 |
| H: | 4.89 g/(1.008 g/mol) = 4.85 mol | 4.85/0.3461 ≈ 14 |
| N: | 19.40 g/(14.01 g/mol) = 1.385 mol | 1.385/0.3461 ≈ 4 |
| O: | 22.15 g/(16.00 g/mol) = 1.384 mol | 1.384/0.3461 ≈ 4 |

The empirical formula of nickel dimethylglyoxime is $NiC_8H_{14}N_4O_4$, which indicates that there are 2 DMG ligands bonded to each $Ni^{2+}$ ion. However, if the ligands were neutral, 2 DMG would be $2(C_4H_8N_2O_2)$ or $C_8H_{16}N_4O_4$, and not $C_8H_{14}N_4O_4$. The difference of 2 H atoms indicates that each ligand must have lost one $H^+$ ion, giving it a 1− charge. Therefore the molecular formula of the neutral (insoluble) coordination compound is $Ni(C_4H_7N_2O_2)_2$, and the balanced net ionic equation describing its formation in this experiment is

$$Ni(NH_3)_6^{2+}(aq) + 2\,C_4H_8N_2O_2(aq) \rightarrow$$
$$Ni(C_4H_7N_2O_2)_2(s) + 2\,NH_4^+(aq) + 4\,NH_3(aq)$$

DMG is a bidentate ligand, so four sites around the nickel ion are occupied by electron pairs from DMG. A coordination number of

4 could yield either tetrahedral or square planar geometry. However, analyses of the structure of nickel dimethylglyoxime using X-ray diffraction, which we discussed in Chapter 12, have shown that it is square planar:

In step 7, strong acid is added to the hexaamminecobalt(II) complex, which results in regeneration of the hexaaquacobalt(II) complex:

$$Co(NH_3)_6^{2+}(aq) + 6\ H^+(aq) + 6\ H_2O(\ell) \rightarrow$$
$$Co(H_2O)_6^{2+}(aq) + 6\ NH_4^+(aq)$$

Adding $SCN^-$ results in formation of $Co(SCN)_4^{2-}$:

$$Co(H_2O)_6^{2+}(aq) + 4\ SCN^-(aq) \rightarrow Co(SCN)_4^{2-}(aq) + 6\ H_2O(\ell)$$

which has a square planar structure:

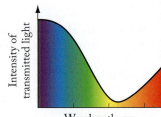

**Think About It** The analysis of the alloy involves precipitation, redox, and complexation in addition to acid–base chemistry. Steps 4 and 6 produce vivid colors and illustrate reactions that are used in qualitative tests for the presence of $Fe^{3+}$ and $Ni^{2+}$ ions in aqueous solution. The formulas of the two thiocyanate complexes indicate that the thiocyanate ligand coordinates to $Fe^{3+}$ ions through the N atom, but it coordinates to $Co^{2+}$ ions through the S atom. When it was first discovered, the nickel–DMG complex was thought to be tetrahedral; it is now known to be among the many 4-coordinate complexes of nickel that are square planar.

## SUMMARY

**Learning Outcome 1** A **Lewis base** is a substance that donates pairs of electrons to a **Lewis acid**, defined as an electron-pair acceptor. The donated electron pair forms a covalent bond. In some Lewis acid–Lewis base reactions, other bonds must break to accommodate the new one. (Section 17.1)

**Learning Outcome 2** The stability of any complex ion is expressed mathematically by its **formation constant ($K_f$)**, which can be used to calculate the equilibrium concentration of free metal ions in a solution of complex ions. (Sections 17.3 and 17.5)

**Learning Outcome 3** The names of **complex ions** and **coordination compounds** provide information about the identities and numbers of **ligands**, the identity and oxidation state of the central metal ion, and the identity and number of counter ions. (Sections 17.2 and 17.4)

**Learning Outcome 4** A **monodentate ligand** donates one pair of electrons in a complex ion; a **polydentate ligand** donates more than one pair in a process called **chelation**. Polydentate ligands are particularly effective at forming complex ions. This phenomenon is called the **chelate effect**. EDTA is a particularly effective sequestering agent, which is a chelating agent that prevents metal ions in solution from reacting with other substances. (Sections 17.6 and 17.7)

**Learning Outcome 5** The colors of transition metals can be explained by the interactions between electrons in different $d$ orbitals and the lone pairs of electrons on surrounding ligands. These interactions create **crystal field splitting** of the energies of the $d$ orbitals. A **spectrochemical series** ranks ligands on the basis of their *field strength* and the wavelengths of electromagnetic radiation absorbed by their complex ions. (Section 17.8)

**Learning Outcome 6** Strong repulsions and large values of crystal field splitting energy can lead to electron pairing in lower-energy orbitals and an electron configuration called a low-spin state. Metals and their ions are less paramagnetic in low-spin states than when their $d$ electrons are evenly distributed across all the $d$ orbitals in the valence shell—a configuration called a high-spin state. (Section 17.9)

**Learning Outcome 7** Complex metal ions containing more than one type of ligand may form stereoisomers. When one type occupies two adjacent corners of a square planar complex, the complex is a cis isomer; when the same ligand occupies opposite corners, it is a trans isomer. (Section 17.10)

## PROBLEM-SOLVING SUMMARY

| TYPE OF PROBLEM | CONCEPTS AND EQUATIONS | SAMPLE EXERCISES |
|---|---|---|
| **Identifying Lewis acids and bases** | Determine which reactant donates a pair of electrons (the Lewis base) and which one accepts them (the Lewis acid). | 17.1 |
| **Calculating the concentration of a free metal ion in equilibrium with a complex** | Set up a RICE table based on formation of the complex. Let $x =$ the concentration of metal ions that *do not* form the complex. If the value of $K_f$ is large (it usually is), assume $x$ is much less than the other concentrations. | 17.2 |
| **Naming coordination compounds** | Follow the naming rules in Section 17.4. | 17.3 |
| **Identifying the potential electron-pair donor groups in a molecule** | Examine the molecular structure of a compound, and find lone pairs that can be donated to a metal. | 17.4 |
| **Predicting spin states** | Sketch a $d$-orbital diagram based on crystal field splitting. Fill the lowest-energy orbitals with the valence electrons. If the number of electron pairs in the diagram is greater than the number of pairs you would have if you distributed the electrons evenly over all five $d$ orbitals, multiple spin states are possible. | 17.5 |
| **Identifying stereoisomers of coordination compounds** | Isomers are possible only if there are at least two types of ligands. If ligands of one type are all on the same side of the complex ion, it is a cis isomer. If ligands of one type are on opposite sides, it is a trans isomer. | 17.6 |

## VISUAL PROBLEMS

*(Answers to boldface end-of-chapter questions and problems are in the back of the book.)*

**17.1.** The chlorides of two of the four highlighted elements in Figure P17.1 are colored. Which ones?

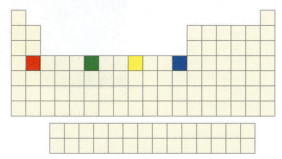

**FIGURE P17.1**

**17.2.** Which of the highlighted transition metals in Figure P17.2 form $M^{2+}$ cations that cannot have high-spin and low-spin states?

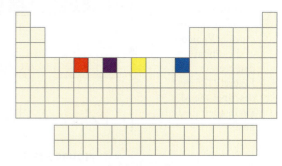

**FIGURE P17.2**

**17.3.** Which of the highlighted transition metals in Figure P17.3 have $M^{2+}$ cations that form colorless tetrahedral complex ions?

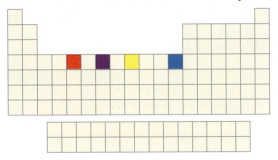

**FIGURE P17.3**

**17.4. Chelation Therapy** The compound with the structure shown in Figure P17.4 has been used to treat people exposed to plutonium, americium, and other actinide metal ions. How many donor groups does the sequestering agent have when the carboxylic acid groups are ionized?

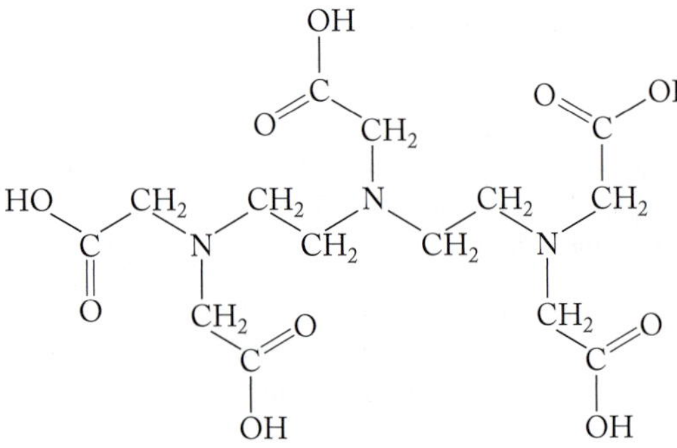

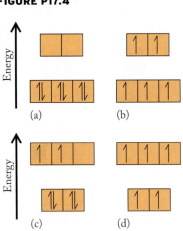

**FIGURE P17.4**

**17.5.** Amethyst has distinctive lavender and purple colors due to the presence of manganese impurities in crystals of silicon dioxide. Which of the orbital diagrams in Figure P17.5 best describes the $Mn^{2+}$ ion in a tetrahedral field?

**FIGURE P17.5**

**17.6.** The orbital diagram of the valence shell of a transition metal ion in octahedral field is shown in Figure 17.6. Which of the following statements about this configuration is(are) true?
  a. It represents a ground state.
  b. It represents an excited state.
  c. It represents a high-spin state.
  d. It represents a low-spin state.
  e. The ion has only one spin state.

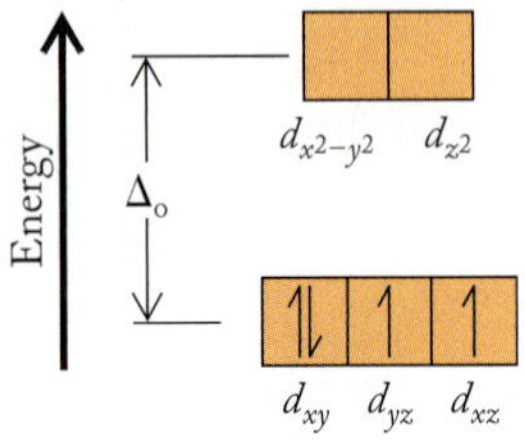

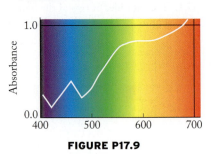

**FIGURE P17.6**

**17.7.** Figure P17.7 shows the absorption spectrum of a solution of $Ti(H_2O)_6^{3+}$. What color is the solution?

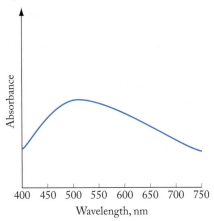

**FIGURE P17.7**

**17.8.** A precious gemstone has the transmittance spectrum shown in Figure P17.8, where the $y$ axis is the percent of light that passes through the gemstone. What color is it?

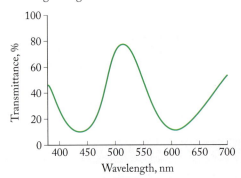

**FIGURE P17.8**

**17.9.** A gemstone has the visible light absorbance spectrum shown in Figure P17.9, where an absorbance value of 0.0 means that no incident light at that wavelength is absorbed and a value of 1.0 means that 90% is absorbed. What is the color of the gemstone?

**FIGURE P17.9**

**17.10.** The three beakers in Figure P17.10 contain solutions of $[CoF_6]^{3-}$, $[Co(NH_3)_6]^{3+}$, and $[Co(CN)_6]^{3-}$. Based on the colors of the three solutions, which of the complex ions is present in each of the beakers?

**FIGURE P17.10**

## QUESTIONS AND PROBLEMS

## Lewis Acids and Bases

### CONCEPT REVIEW

**17.11.** Are all Lewis bases also Brønsted–Lowry bases? Explain why or why not.

17.12. Are all Brønsted–Lowry bases also Lewis bases? Explain why or why not.

**17.13.** Why is $BF_3$ a Lewis acid but not a Brønsted–Lowry acid?

17.14. Sketch the autoionization of water showing how one $H_2O$ molecule acts as a Lewis acid and the other as a Lewis base.

### PROBLEMS

**17.15.** Use Lewis structures to show how electron pairs move and bonds form in the reaction in which trimethylborane, $B(CH_3)_3$, combines with ammonia forming a single product, and identify the Lewis acid and Lewis base.

$$B(CH_3)_3(g) + NH_3(g) \rightarrow \underline{\hspace{2cm}}$$

17.16. Use Lewis structures to show how electron pairs move and bonds form and break in this reaction, and identify the Lewis acid and Lewis base.

$$MgO(s) + CO_2(g) \rightarrow MgCO_3(s)$$

**17.17.** Use Lewis structures to show how electron pairs move and bonds form and break in this reaction, and identify the Lewis acid and Lewis base.

$$SO_2(g) + H_2O(\ell) \rightleftharpoons H_2SO_3(aq)$$

17.18. Use Lewis structures to show how electron pairs move and bonds form and break in this reaction, and identify the Lewis acid and Lewis base.

$$SeO_3(g) + H_2O(\ell) \rightarrow H_2SeO_4(aq)$$

**17.19.** Use Lewis structures to show how electron pairs move and bonds form and break in this reaction, and identify the Lewis acid and Lewis base.

$$B(OH)_3(aq) + H_2O(\ell) \rightleftharpoons B(OH)_4^-(aq) + H^+(aq)$$

*17.20. Use Lewis structures to show how electron pairs move and bonds form and break in this reaction, and identify the Lewis acid and Lewis base. (*Note:* $HSbF_6$ is an ionic compound and one of the strongest Brønsted–Lowry acids known.)

$$SbF_5(s) + HF(g) \rightleftharpoons HSbF_6(s)$$

## Complex Ions

### CONCEPT REVIEW

**17.21.** When $CaCl_2$ dissolves in water, which molecules or ions occupy the inner coordination sphere around the $Ca^{2+}$ ions?

17.22. When $TiCl_3$ dissolves in water, which of the following species are among those nearest the $Ti^{3+}$ ions? (a) other $Ti^{3+}$ ions; (b) $Cl^-$ ions; (c) molecules of $H_2O$ with their O atoms closest to the $Ti^{3+}$ ions; (d) molecules of $H_2O$ with their H atoms closest to the $Ti^{3+}$ ions

**17.23.** When $AgNO_3$ dissolves in a solution of ammonia water, which molecules or ions occupy the inner coordination sphere around the $Ag^+$ ions?

17.24. When $[Ni(NH_3)_6]Cl_2$ dissolves in water, which molecules or ions occupy the inner coordination sphere around the $Ni^{2+}$ ions?

**17.25.** Which ion is the counter ion in the coordination compound $Na_3[Fe(CN)_6]$?

17.26. Which ion is the counter ion in the coordination compound $[Co(NH_3)_4Cl_2]NO_3$?

## Complex-Ion Equilibria

### CONCEPT REVIEW

**17.27. Chelation Therapy** Explain how treating a child suffering from lead poisoning with a chelating agent makes the lead less poisonous. Assume the lead is present in the +2 oxidation state.

17.28. When a solution of NaOH is added to a solution of $CuSO_4$, which is pale blue, a precipitate forms, and the solution above the precipitate is colorless. When ammonia is added to this mixture, the precipitate dissolves, and the solution turns a deep navy blue. Use appropriate chemical equations to explain why the observed changes occur.

*17.29. A lab technician cleaning glassware that contains residues of AgCl washes the glassware with an aqueous solution of ammonia. The AgCl, which is insoluble in water, rapidly dissolves in the ammonia solution. Why?

*17.30. The procedure used in the previous question dissolves AgCl but not AgI. Why?

### PROBLEMS

*Note*: Appendix 5 contains formation constant ($K_f$) values that may be useful in solving the following problems.

**17.31.** A solution is prepared that contains 0.00100 mol $Ni(NO_3)_2$ and 0.500 mol $NH_3$ in a total volume of 1.00 L.
 a. What is the initial concentration of $Ni(NO_3)_2$ in the solution?
 b. What is the concentration of $Ni^{2+}(aq)$ in the solution at equilibrium?

17.32. A 1.00 L solution contains $5.00 \times 10^{-5}$ $M$ $Cu(NO_3)_2$ and $1.00 \times 10^{-3}$ $M$ ethylenediamine. What is the concentration of $Cu^{2+}(aq)$ in the solution?

**17.33.** Suppose a solution contains 1.00 mmol $Co(NO_3)_2$, 0.100 mol $NH_3$, and 0.100 mol of ethylenediamine in a total volume of 0.250 L. What is the concentration of $Co^{2+}(aq)$ in the solution?

*17.34. If 1.00 mL 0.0100 $M$ $AgNO_3$, 1.00 mL 0.100 $M$ NaBr, and 1.00 mL 0.100 $M$ NaCN are diluted to 250 mL with deionized water in a volumetric flask and shaken vigorously, will the contents of the flask be cloudy or clear? Support your answer with the appropriate calculations. (*Hint*: The $K_{sp}$ of AgBr is $5.4 \times 10^{-13}$.)

## Naming Complex Ions and Coordination Compounds

### PROBLEMS

**17.35.** What are the names of the following complex ions?
 a. $Ag(NH_3)_2^+$
 b. $Co(H_2O)_6^{3+}$
 c. $[Fe(NH_3)_5Br]^{2+}$

17.36. What are the names of the following complex ions?
 a. $Cu(NH_3)_2^+$
 b. $Ti(H_2O)_4(OH)_2^{2+}$
 c. $Ni(NH_3)_4(H_2O)_2^{2+}$

**17.37.** What are the names and coordination numbers of the following complex ions?
   a. $CoBr_4^{2-}$
   b. $Zn(H_2O)(OH)_3^-$
   c. $Ni(CN)_5^{3-}$

**17.38.** What are the names and coordination numbers of the following complex ions?
   a. $CoCl_4^{2-}$
   b. $CuI_4^{2-}$
   c. $[Cr(en)(OH)_4]^-$

**17.39.** For the following coordination compounds:
   i. $[Zn(en)_2]SO_4$
   ii. $[Ni(NH_3)_5(H_2O)]Cl_2$
   iii. $K_4Fe(CN)_6$
   a. Name the compound.
   b. Identify the charge on the complex ion, the oxidation state of the metal in the complex ion, and the coordination number of the metal in the complex ion.

**17.40.** For the following coordination compounds:
   i. $(NH_4)_3[Co(CN)_6]$
   ii. $[Co(en)_2Cl](NO_3)_2$
   iii. $[Fe(H_2O)_4(OH)_2]Cl$
   a. Name the compound.
   b. Identify the charge on the complex ion, the oxidation state of the metal in the complex ion, and the coordination number of the metal in the complex ion.

## Hydrated Metal Ions as Acids

### CONCEPT REVIEW

**17.41.** Which, if any, aqueous solutions of the following chloride compounds are acidic? (a) $CaCl_2$; (b) $CrCl_3$; (c) NaCl; (d) $FeCl_2$

**17.42.** If 0.100 $M$ aqueous solutions of each of these compounds were prepared, which one would have the lowest pH? (a) $BaCl_2$; (b) $NiCl_2$; (c) KCl; (d) $TiCl_4$

**17.43.** When ozone is bubbled through an aqueous solution of $Fe^{2+}$ ions, the ions are oxidized to $Fe^{3+}$ ions. How does the oxidation process affect the pH of the solution?

**17.44.** As an aqueous solution of KOH is slowly added to a stirred solution of $AlCl_3$, the mixture becomes cloudy but then clears when more KOH is added.
   a. Explain the chemical changes responsible for the changes in the appearance of the mixture.
   b. Would you expect to observe the same changes if KOH were added to a solution of $FeCl_3$? Explain why or why not.

**17.45.** Chromium(III) hydroxide is amphiprotic. Write chemical equations showing how an aqueous suspension of this compound reacts to the addition of a strong acid and a strong base.

**17.46.** Zinc hydroxide is amphiprotic. Write chemical equations showing how an aqueous suspension of this compound reacts to the addition of a strong acid and a strong base.

**17.47.** **Refining Aluminum** To remove impurities such as calcium and magnesium carbonates and Fe(III) oxides from aluminum ore (which is mostly $Al_2O_3$), the ore is treated with a strongly basic solution. In this treatment, $Al^{3+}$ dissolves but the other metal ions do not. Why?

*17.48. Exactly 1.00 g of $FeCl_3$ is dissolved in each of four 0.500 L samples: 1 $M$ $HNO_3$, 1 $M$ $HNO_2$, 1 $M$ $CH_3COOH$, and pure water. Is the concentration of $Fe(H_2O)_6^{3+}$ ions the same in all four solutions? Explain why or why not.

### PROBLEMS

**17.49.** What is the pH of 0.25 $M$ $Al(NO_3)_3$?
**17.50.** What is the pH of 0.50 $M$ $CrCl_3$?

**17.51.** What is the pH of 0.100 $M$ $Fe(NO_3)_3$?
**17.52.** What is the pH of 1.00 $M$ $Cu(NO_3)_2$?

**17.53.** Sketch the titration curve (pH versus volume of 0.50 $M$ NaOH) for a 25 mL sample of 0.25 $M$ $FeCl_3$.

**17.54.** Sketch the titration curve that results from the addition of 0.50 $M$ NaOH to a sample containing 0.25 $M$ $KFe(SO_4)_2$.

## Polydentate Ligands; Ligand Strength and the Chelate Effect

### CONCEPT REVIEW

**17.55.** What is meant by the term *chelating agent*? What properties make a substance an effective chelating agent?

*17.56. The condensed molecular structures of two compounds that each contain two $-NH_2$ groups are shown in Figure P17.56. The one on the left is ethylenediamine, a bidentate ligand. Does the molecule on the right have the same ability to donate two pairs of electrons to a metal ion? Explain why you think it does or does not.

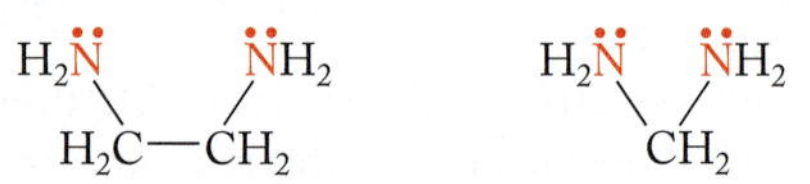

**FIGURE P17.56**

**17.57.** How does the chelating ability of an aminocarboxylic acid vary with changing pH?

*17.58. EDTA, a food preservative, is added to food as the calcium disodium salt, $Na_2[CaEDTA]$, not as the undissociated acid. This salt is actually a coordination compound with a $Ca^{2+}$ ion at the center of a complex ion. Draw a line structure of this compound.

## Crystal Field Theory

### CONCEPT REVIEW

**17.59.** In a square planar crystal field, why is the $d_{xy}$ orbital higher in energy than the $d_{xz}$ and $d_{yz}$ orbitals?

**17.60.** On average, the $d$ orbitals of a transition metal ion in an octahedral field are higher in energy than they are when the ion is in the gas phase. Why?

**17.61.** Aqueous solutions of $Ti^{3+}$ ions are violet, but those of $Ti^{4+}$ are colorless. Why?

**17.62.** Evaporation of an aqueous solution of nickel(II) nitrate yields emerald-green crystals of this compound. When the crystals are placed in a drying oven at 150°C, their green color fades. Why?

### PROBLEMS

**17.63.** Aqueous solutions of one the following complex ions of Cr(III) are violet; solutions of the other are yellow. Which is which? (a) $Cr(H_2O)_6^{3+}$; (b) $Cr(NH_3)_6^{3+}$

**17.64.** Which of the following complex ions should absorb the shortest wavelengths of electromagnetic radiation? (a) $Cu(CN)_4^{2-}$; (b) $Cu(F)_4^{2-}$; (c) $Cu(I)_4^{2-}$; (d) $Cu(Br)_4^{2-}$

**17.65.** The octahedral crystal field splitting energy, $\Delta_o$, of $Co(phen)_3^{3+}$ is $5.21 \times 10^{-19}$ J/ion. What is the color of a solution of this complex ion?

**17.66.** The octahedral crystal field splitting energy, $\Delta_o$, of $Co(CN)_6^{3-}$ is $6.74 \times 10^{-19}$ J/ion. What is the color of a solution of this complex ion?

---

**17.67.** Solutions of $NiCl_4^{2-}$ and $NiBr_4^{2-}$ are different shades of green because they absorb visible light in different regions of the red end of the spectrum. Which complex absorbs the longer wavelengths?

**17.68.** Chromium(III) chloride forms six-coordinate complexes with bipyridine, including $cis$-$[Cr(bipy)_2Cl_2]^+$, which reacts slowly with water to produce two products, $cis$-$[Cr(bipy)_2(H_2O)Cl]^{2+}$ and $cis$-$[Cr(bipy)_2(H_2O)_2]^{3+}$. In which of these complexes should $\Delta_o$ be the largest?

## Magnetism and Spin States

### PROBLEMS

**17.69.** Consider the following cations: $Co^{2+}$; $Cr^{2+}$; $Ni^{2+}$; $Cu^{2+}$.
a. How many $d$ electrons are in each ion?
b. Which cations can have high-spin or low-spin states in octahedral fields produced by ligands with different field strength?

**17.70.** Consider the following cations: $V^{2+}$; $Co^{3+}$; $Fe^{2+}$; $Cr^{3+}$.
a. How many $d$ electrons are in each ion?
b. Which cations can have high-spin or low-spin states in octahedral fields produced by ligands with different field strength?

---

**17.71.** How many unpaired electrons are there in the following transition metal ions in their hexachloro complex ions? $Fe^{2+}$; $Cu^{2+}$; $Co^{2+}$; $Mn^{2+}$

**17.72.** How many unpaired electrons are there in the following transition metal ions in their hexacyano complex ions? $Fe^{3+}$; $Rh^+$; $V^{3+}$; $Mn^{3+}$

---

**17.73.** The manganese minerals pyrolusite, $MnO_2$, and hausmannite, $Mn_3O_4$, contain Mn ions in octahedral holes formed by oxide ions.
a. What are the charges of the Mn ions in each mineral?
b. In which of these compounds could there be high-spin and low-spin Mn ions?

**17.74. Dietary Supplement** Chromium picolinate is an over-the-counter diet aid sold in many pharmacies. The $Cr^{3+}$ ions in this coordination compound are in an octahedral field. Is the compound paramagnetic or diamagnetic?

---

**\*17.75.** One method for refining cobalt involves the formation of the complex ion $CoCl_4^{2-}$. This anion is tetrahedral. Is this complex paramagnetic or diamagnetic?

**\*17.76.** Why is it that $Ni(CN)_4^{2-}$ is diamagnetic, but $NiCl_4^{2-}$ is paramagnetic?

## Isomerism in Coordination Compounds; Coordination Compounds in Biochemistry

### CONCEPT REVIEW

**17.77.** What do the prefixes $cis$- and $trans$- mean in the context of an octahedral complex ion?

**17.78.** What do the prefixes $cis$- and $trans$- mean in the context of a square planar complex?

**17.79.** How many different types of donor groups are required to have stereoisomers of a square planar complex?

**17.80.** With respect to your answer to the previous question, do all square planar complexes with this many different types of donor groups have stereoisomers?

### PROBLEMS

**17.81.** Does $Co(en)(H_2O)_2Cl_2$ have stereoisomers?

**17.82.** Does the complex ion $Fe(en)_3^{3+}$ have stereoisomers?

---

**\*17.83.** Sketch the stereoisomers of the square planar complex ion $CuCl_2Br_2^{2-}$. Name them. Are any of these isomers chiral?

**\*17.84.** Sketch the stereoisomers of the octahedral complex ion $Ni(en)Cl_2(CN)_2^{2-}$. Name them. Are any of these isomers chiral?

## Additional Problems

**17.85. Photographic Film Processing** During the processing of black-and-white photographic film, excess silver(I) halides are removed by washing the film in a bath containing sodium thiosulfate. This treatment is based on the following complexation reaction:

$$Ag^+(aq) + 2\,S_2O_3^{2-}(aq) \rightleftharpoons Ag(S_2O_3)_2^{3-}(aq) \qquad K_f = 5 \times 10^{13}$$

What is the ratio of $[Ag^+]$ to $[Ag(S_2O_3)_2^{3-}]$ in a bath in which $[S_2O_3^{2-}] = 0.250\ M$?

**17.86. Lead Poisoning** Children were once treated for lead poisoning with intravenous injections of EDTA. If the concentration of EDTA in the blood of a patient is $1.0 \times 10^{-7}\ M$ and the formation constant for the complex $[Pb(EDTA)]^{2-}$ is $2.0 \times 10^{18}$, what is the concentration ratio of the free (and potentially toxic) $Pb^{2+}(aq)$ in the blood to the much less toxic $Pb^{2+}$–EDTA complex?

**17.87.** Dissolving cobalt(II) nitrate in water produces a beautiful purple solution. There are three unpaired electrons in this cobalt(II) complex. When cobalt(II) nitrate is dissolved in aqueous ammonia and oxidized with air, the resulting yellow complex has no unpaired electrons. Which cobalt complex has the larger crystal field splitting energy, $\Delta_o$?

**17.88.** A solid compound containing Fe(II) in an octahedral crystal field has four unpaired electrons at 298 K. When the compound is cooled to 80 K, the same sample appears to have no unpaired electrons. How do you explain this change in the compound's properties?

**17.89.** When $Ag_2O$ reacts with peroxodisulfate ($S_2O_8^{2-}$) ion (a powerful oxidizing agent), AgO is produced. Crystallographic and magnetic analyses of AgO suggest that it is not simply Ag(II) oxide, but rather a blend of Ag(I) and Ag(III) in a square planar environment. The $Ag^{2+}$ ion is paramagnetic but AgO, $Ag^+$, and $Ag^{3+}$ are diamagnetic. Explain why.

**17.90.** The iron(II) compound $Fe(bipy)_2(SCN)_2$ is paramagnetic, but the corresponding cyanide compound $Fe(bipy)_2(CN)_2$ is diamagnetic. Why do these two compounds have different magnetic properties?

**17.91.** Aqueous solutions of copper(II)–ammonia complexes are dark blue. Will the color of the series of complexes $Cu(H_2O)_{(6-x)}(NH_3)_x^+$ shift toward shorter or longer wavelengths as the value of $x$ increases from 0 to 6?

If your instructor assigns problems in **smartwork**, log in at **smartwork.wwnorton.com**.

# 18

# Thermodynamics: Spontaneous and Nonspontaneous Reactions and Processes

18.1 Spontaneous Processes

18.2 Thermodynamic Entropy

18.3 Absolute Entropy and the Third Law of Thermodynamics

18.4 Calculating Entropy Changes

18.5 Free Energy

18.6 Temperature and Spontaneity

18.7 Free Energy and Chemical Equilibrium

18.8 Influence of Temperature on Equilibrium Constants

18.9 Driving the Human Engine: Coupled Reactions

18.10 Microstates: A Quantized View of Entropy

## Learning Outcomes

**LO1** Distinguish between spontaneous and nonspontaneous processes using the second law of thermodynamics

**LO2** Predict the signs of entropy changes for chemical reactions and physical processes
**Sample Exercise 18.1**

**LO3** Predict the relative entropies of substances based on their molecular structures
**Sample Exercise 18.2**

**LO4** Calculate entropy changes in chemical reactions using standard molar entropies
**Sample Exercise 18.3**

**LO5** Calculate free-energy changes and standard free-energy changes in chemical reactions
**Sample Exercises 18.4, 18.5**

**LO6** Predict the spontaneity of a chemical reaction as a function of temperature
**Sample Exercise 18.6**

**LO7** Relate the value of the equilibrium constant of a reaction to its change in free energy under standard conditions
**Sample Exercises 18.7, 18.8**

**LO8** Use the van 't Hoff equation to calculate the values of the equilibrium constant of a reaction at different temperatures
**Sample Exercise 18.8**

**LO9** Calculate the net change in free energy of coupled spontaneous and nonspontaneous reactions
**Sample Exercise 18.9**

## The Game of Energy Conversion

Some processes are so familiar we do not even think about them or why they happen. If a car tire is punctured, the air inside rushes out and the tire goes flat; air does not rush back into a punctured tire and reinflate it. Objects made of iron left on the ground or underwater for months or years become clumps of rust; they do not turn back into shiny metal after even more time. A tray of ice cubes accidentally left on a kitchen counter melts into a tray of liquid water. As long as the kitchen remains the same temperature, there is no way the water in the tray will reform to ice cubes.

All of these processes have something in common: they are all spontaneous. This means that they all happen without any ongoing intervention and without work being done on the system. In each case, the reverse process is nonspontaneous—it can only happen as long as energy is continually added. Why are some processes spontaneous and others not? We find one explanation in the second law of thermodynamics and a thermodynamic parameter called entropy.

As we learned in Chapter 5, the first law of thermodynamics tells us that energy cannot be created or destroyed. This means that when we play the game of energy conversion, the best we could do is break even. Furthermore, the second law of thermodynamics dictates that not all of the energy released by a spontaneous reaction, such as burning gasoline in a car engine, is available to do useful work. In other words, in the game of energy conversion, not only can we not win, we *can't* break even.

If energy cannot be destroyed, what happens to the energy that is unavailable to do useful work? The answer is that this energy spreads out, becoming less concentrated over time. This dispersion of energy is called entropy. In this chapter we explore the meaning and some of the impacts of entropy and the second law of thermodynamics on

**Corrosion** Most metal objects in the sea, including the remains of this shipwreck, corrode as a result of spontaneous chemical reactions. Iron metal is converted into iron(III) oxide. ▶

**spontaneous** a process that occurs without outside intervention.

**nonspontaneous** a process that occurs only as long as energy is continually added to the system.

familiar processes. We also explore that component of the energy that is available to do work and how it is related to reaction spontaneity and equilibrium. We will also see how spontaneous reactions can be coupled with nonspontaneous reactions to make multistep chemical and biochemical processes happen. This coupling is important to us because it powers the molecular processes that sustain life. ■

## 18.1 Spontaneous Processes

Before Thomas Edison invented the light bulb in 1879, city streets were often illuminated at night by gas-fueled lamps. Once lit, they burned through the night until their fuel was cut off as dawn approached. Chemical reactions like the combustion of gas-lamp fuel are examples of **spontaneous** reactions: once started, they proceed without outside intervention. The reverse reaction—in this case, converting carbon dioxide and water into fuel and oxygen—is **nonspontaneous**: it cannot happen on its own without the addition of energy from an external source.

The word "spontaneous" can be a bit misleading because many spontaneous reactions do not start all by themselves; they need a little energy boost, such as a spark or external flame to ignite the gas in a gas lamp or the gasoline/air reaction mixture in an automobile engine. However, once initiated, a spontaneous reaction continues without the input of additional energy, as long as reactants are available. It is the self-sustaining nature of the reactions that earn them the label *spontaneous*.

In addition, spontaneous does not necessarily mean rapid. Though combustion reactions certainly are fast, other spontaneous reactions, such as the formation of a layer of rust on an object made of iron

$$4 \, Fe(s) + 3 \, O_2(g) \rightarrow 2 \, Fe_2O_3(s) \qquad \Delta H° = -1648 \text{ kJ}$$

can take a very long time, depending on temperature and the rate at which $O_2$ reaches the iron surface. Spontaneity has nothing to do with kinetics. Still, rust formation does proceed without intervention and meets the definition of thermodynamic spontaneity.

Spontaneous is also *not* a synonym for exothermic, although many scientists once thought so. In the mid-19th century, many chemists thought that exothermic reactions, in which high-enthalpy reactants formed low-enthalpy (more stable) products, should always be spontaneous. It is true that many exothermic reactions, including combustion reactions and metal corrosion, *are* spontaneous. However, some exothermic processes may not be spontaneous, and some endothermic processes may be spontaneous, depending on reaction conditions.

Familiar examples of spontaneous endothermic processes include the phase changes that occur when ice melts and water boils. Both are endothermic processes, but both are spontaneous, depending on the reaction conditions. Consider, also, what happens when baking soda (sodium bicarbonate) is added to room-temperature vinegar (dilute acetic acid), as shown in Figure 18.1. The foaming mixture tells us that a reaction is taking place and the lower-than-room-temperature of the mixture tells us that the reaction is endothermic ($\Delta H > 0$).

Another example is the process that makes instant cold packs cold (Figure 18.2). An instant cold pack has two compartments. One is filled with water; the other contains a water-soluble compound such as ammonium nitrate that has a positive enthalpy of solution ($\Delta H_{soln} > 0$). When the membrane separating the

**CONNECTION** We learned in Chapter 5 that enthalpy change ($\Delta H$) is a thermodynamic quantity that describes energy flow into or out of a system.

▶❚❚ **CHEMTOUR** Dissolution of Ammonium Nitrate

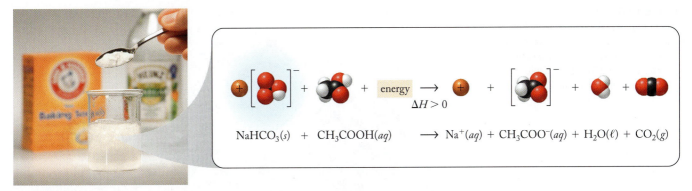

$$NaHCO_3(s) \;+\; CH_3COOH(aq) \;\longrightarrow\; Na^+(aq) \;+\; CH_3COO^-(aq) \;+\; H_2O(\ell) \;+\; CO_2(g)$$

**FIGURE 18.1** Adding baking soda to vinegar produces sodium acetate, water, and carbon dioxide in an endothermic ($\Delta H° = +48.5$ kJ), yet spontaneous, reaction.

two compartments is ruptured, the solid compound ($NH_4NO_3$) mixes with and dissolves in the water. The resulting solution gets very cold, becoming an effective anti-inflammation treatment for bruises and muscle sprains.

Why are endothermic processes such as these spontaneous? The answer to this question lies in something these processes have in common: the particles that make up their products are more spread out and have more freedom of motion than the particles in their starting materials. Consider the molecular changes that accompany ice melting and water boiling (Figure 18.3). The molecules of $H_2O$ in ice occupy fixed positions. Their motion, like that of particles in all crystalline solids, is limited to vibrating in place, not going anywhere. When ice melts, its $H_2O$ molecules become more mobile, acquiring rotational and translational motion, as shown in Figure 18.3. When liquid water evaporates and the molecules enter the gas phase, their rotational and translational motion increases even more. The water vapor molecules are now also able to expand to fill their container, as are all gas particles at atmospheric pressure, which is why gases are compressible, whereas liquids and solids are not.

**CONNECTION** The ion–dipole interactions that promote the solubility of ionic compounds in water were described in Chapter 10.

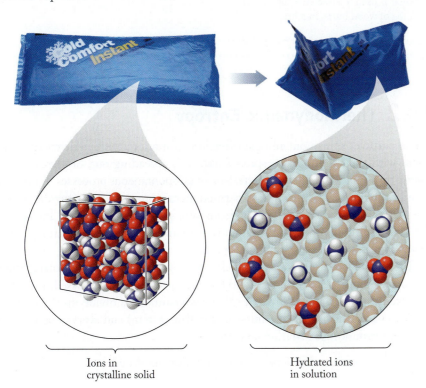

Ions in
crystalline solid

Hydrated ions
in solution

**FIGURE 18.2** Instant cold packs get cold when the water-filled pouch inside is ruptured and the water-soluble compound (e.g., ammonium nitrate) within the pack dissolves. The $NH_4^+$ and $NO_3^-$ ions in solid $NH_4NO_3$ experience increased freedom of motion as they form hydrated $NH_4^+$ and $NO_3^-$ ions in solution. The temperature of the solution drops because the dissolution process is endothermic.

**FIGURE 18.3** Molecules of water have three types of motion: (a) vibrational—which is their only motion in solids; (b) rotational; and (c) translational. They have some rotational and translation motion in the liquid phase, and a lot more of both in the gas phase.

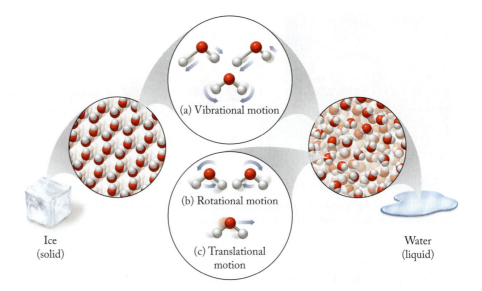

(a) Vibrational motion

(b) Rotational motion

(c) Translational motion

Ice (solid)

Water (liquid)

The particles that make up the reaction mixtures in the two endothermic chemical reactions described above also experience gains in freedom of motion. The bicarbonate ions in the solid reactant ($NaHCO_3$) become liberated as a gaseous product of the reactions ($CO_2$). The instant cold pack gets cold when the solid solute dissolves in water—the particles of the solute acquire more freedom of motion and can move throughout the resulting solution. These particles experience increases in motion similar to those of a melting solid.

**CONCEPT TEST** ······················································

In which of the following processes do particles experience an increase in freedom of motion?
a. A glass of water evaporates.
b. Dew forms overnight on grass and other surfaces.
c. Sugar is used to sweeten a cup of coffee.
d. A log of wood burns in a fireplace.

*(Answers to Concept Tests are in the back of the book.)*
·····················································································

## 18.2 Thermodynamic Entropy

▶❚❚ **CHEMTOUR** Entropy

When particles spread out and gain freedom of motion, the kinetic energy associated with their motion also spreads out. This spreading out, or dispersion, of energy turns out to be a key characteristic of all spontaneous processes. It even has a name: **entropy (S)**. Entropy is a thermodynamic property that provides a measure of the dispersal of energy in a system at a specific temperature. The **second law of thermodynamics** states that entropy of an *isolated* thermodynamic system *always increases* during a spontaneous process.

◉◉ **CONNECTION** In Chapter 5, we defined an isolated thermodynamic system as one that exchanges neither energy nor matter with its surroundings; a closed system exchanges energy but not matter; and an open system exchanges both.

The second law also covers thermodynamic systems that are not isolated (most systems are not isolated; they are either open or closed, as discussed in Section 5.2). Recall from Chapter 5 that in thermodynamics we divide up the universe into two parts: the part we are interested in (the system) and everything else (the system's surroundings). Mathematically,

$$\text{Universe} = \text{System} + \text{Surroundings}$$

Logically, then, the overall change in the entropy of the universe is the sum of the entropy changes experienced by the system and by its surroundings:

$$\Delta S_{univ} = \Delta S_{sys} + \Delta S_{surr} \qquad (18.1)$$

In the case of a spontaneous process in an isolated system, $\Delta S_{sys}$ is greater than zero ($\Delta S_{sys} > 0$) and the entropy of its surroundings is unchanged ($\Delta S_{surr} = 0$). Therefore, according to Equation 18.1, $\Delta S_{univ}$ must also be greater than zero ($\Delta S_{univ} > 0$). The positive value of $\Delta S_{univ}$ is the basis for another way of expressing the second law of thermodynamics that applies to all systems (not just isolated ones): *a spontaneous process produces an increase in the entropy of the universe.*

This version of the second law is built on the assumption that a physical or chemical change in a closed or open thermodynamic system can alter the entropies of both the system and its surroundings. The second law says that a process is spontaneous when one or the other of these entropy changes is greater than zero, so that their sum, $\Delta S_{univ}$, is also greater than zero. The second law provides a thermodynamic requirement for reaction spontaneity as well as a criterion for *non*spontaneity: a process that produces a *decrease* in the entropy of the universe does not occur on its own. Summarizing these relationships:

If $\Delta S_{univ} > 0$, then a process is spontaneous

If $\Delta S_{univ} < 0$, then a process is nonspontaneous

To see how a process affects the entropy of its surroundings, let's focus on a familiar exothermic reaction, the combustion of natural gas (methane):

$$CH_4(g) + 2\,O_2(g) \rightarrow CO_2(g) + 2\,H_2O(\ell) \qquad \Delta H° = -890 \text{ kJ}$$

As written, the reaction consumes three moles of gases and produces two moles of a liquid product and one mole of $CO_2$ gas. Note that there are fewer moles of gas on the product side of the reaction equation. Given the much greater freedom of motion of particles in the gas phase, we can accurately predict that there will be a decrease in entropy of the reaction mixture, which is our thermodynamic system:

$$\Delta S_{sys} < 0$$

However, we know that the reaction is spontaneous, which means

$$\Delta S_{univ} > 0$$

How do we reconcile these opposing inequalities? Equation 18.1 supplies an explanation. If $\Delta S_{univ}$ is greater than zero, then the sum of $\Delta S_{sys}$ and $\Delta S_{surr}$ must also be greater than zero. The fact that $\Delta S_{sys}$ is less than zero simply means that $\Delta S_{surr}$ is not only greater than zero, it must have a large enough positive value to more than offset the negative value of $\Delta S_{sys}$. Expressing this relationship in terms of the absolute values of $\Delta S_{surr}$ and $\Delta S_{sys}$:

$$|\Delta S_{surr}| > |\Delta S_{sys}|$$

Is the combustion of methane likely to produce a large, positive $\Delta S_{surr}$? Absolutely, because the reaction is highly exothermic: combustion of only one mole (16 grams) of methane releases 890 kJ of thermal energy. As energy flows from the system into its surroundings, a dispersion of energy occurs, producing a positive $\Delta S_{surr}$ that more than compensates for the unfavorable (negative) value of $\Delta S_{sys}$.

All exothermic reactions have the capacity to raise the entropy of their surroundings. The more energy ($q$) that flows into the surroundings, or into any collec-

**entropy (S)** a measure of how dispersed the energy in a system is at a specific temperature.

**second law of thermodynamics** the principle that the total entropy of the universe increases in any spontaneous process.

**reversible process** a process that can be run in the reverse direction in such a way that, once the system has been restored to its original state, no net energy has flowed either to the system or to its surroundings.

tion of particles, the greater the dispersion of kinetic energy among the particles and the greater the increase in their entropy ($\Delta S$). However, adding energy to particles that are already hot produces a smaller gain in entropy than adding the same quantity of energy to the same particles at a lower temperature. This inverse relationship between entropy gain and temperature is reflected in Equation 18.2:

$$\Delta S = \frac{q_{rev}}{T} \qquad (18.2)$$

The subscript "rev" means that the heating process is reversible. Theoretically, a **reversible process** happens so slowly that equilibrium is constantly maintained. For example, after an incremental change in the system has occurred in the forward direction, the process can be reversed with a tiny change in process conditions so that the original state of the system can be restored with no net flow of energy into or out of the system. In other words, everything about the system goes back to exactly as it was before the process began. At best, real chemical reactions and physical changes are only approximately reversible, but many are close enough that we can adapt Equation 18.2 to calculate $\Delta S_{sys}$:

$$\Delta S_{sys} = \frac{q_{sys}}{T} \qquad (18.3)$$

For example, it takes 6.01 kJ (or $6.01 \times 10^3$ J) of energy to melt 1.00 mol of ice at 0°C. Assuming the process occurs reversibly, then

$$\Delta S_{sys} = \frac{q_{sys}}{T} = \frac{(1.00 \ \text{mol})(6.01 \times 10^3 \ \text{J/mol})}{273 \ \text{K}} = 22.0 \ \text{J/K}$$

Because energy flows into the ice, $q_{sys}$ is positive, which also makes $\Delta S_{sys}$ positive, as we would expect given the greater freedom of motion of particles in the liquid phase. Note that the units of entropy in this calculation are joules per kelvin. We will continue to use these units in all entropy calculations in this chapter.

Suppose 1.00 mol of ice melts as $6.01 \times 10^3$ J of energy flows into it from room-temperature (22°C or 295 K) surroundings. We can calculate the change in entropy of the surroundings by using another adaptation of Equation 18.2:

$$\Delta S_{surr} = \frac{q_{surr}}{T} = \frac{(1.00 \ \text{mol})(-6.01 \times 10^3 \ \text{J/mol})}{295 \ \text{K}} = -20.4 \ \text{J/K}$$

Note that the sign of $q_{surr}$ is negative because energy flows from the surroundings into the system (ice). Also note that (1) the value of $\Delta S_{surr}$ is less than zero, and (2) the magnitude of the decrease in $\Delta S_{surr}$ is less than the increase in $\Delta S_{sys}$ because the same absolute value of $q$ was divided by a higher temperature to calculate $\Delta S_{surr}$. Therefore, when we sum $\Delta S_{sys}$ and $\Delta S_{surr}$ (Figure 18.4), we get a positive value of $\Delta S_{univ}$:

$$\Delta S_{univ} = \Delta S_{sys} + \Delta S_{surr} = (22.0 - 20.4) \ \text{J/K} = 1.6 \ \text{J/K}$$

The result of this calculation is one example of a general truth about energy flow (that will come as no surprise): energy flows spontaneously into a system that is cooler than its surroundings. Even more generally, energy flows spontaneously from a warm object to an adjacent cooler object. Entropy and the second law of thermodynamics simply provide a mathematical explanation of why.

You may wonder why we did not consider the change in temperature of the surroundings as energy flowed from it into the melting ice. The answer lies in the sheer size of the surroundings (the universe minus a small cube of ice). The temperature of such an enormous mass is not likely to change significantly.

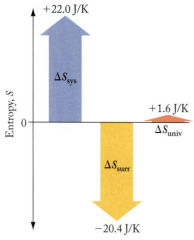

**FIGURE 18.4** If the surroundings experience a decrease in entropy ($\Delta S_{surr} = -20.4$ J/K), the system (ice) must experience an increase in entropy ($\Delta S_{sys} = +22.0$ J/K) that more than offsets the decrease if the process (melting) is spontaneous ($\Delta S_{univ} = +1.6$ J/K).

Now let's consider how entropy changes when the temperature of the surroundings is *lower* than the temperature of the system. Suppose a tray containing 1.00 mol of liquid water at 0°C is placed in a freezer at −10°C. We know the water will freeze. Because the temperature of the surroundings (the freezer) is lower than the temperature of the liquid water, energy spontaneously flows from the water into its surroundings. The net entropy change for this process is

$$\Delta S_{univ} = \Delta S_{sys} + \Delta S_{surr}$$

$$= \frac{(1.00 \text{ mol})(-6.01 \times 10^3 \text{ J/mol})}{273 \text{ K}} + \frac{(1.00 \text{ mol})(+6.01 \times 10^3 \text{ J/mol})}{263 \text{ K}}$$

$$= (-22.0 \text{ J/K}) + (+22.9 \text{ J/K})$$

$$= 0.9 \text{ J/K}$$

Once again the entropy of the universe increases as energy flows spontaneously from the warmer object (liquid water at 0°C) into the colder surroundings (the freezer at −10°C).

**FIGURE 18.5** Water vapor exhaled by this Inuit hunter in the Northwest Territories of Canada was spontaneously deposited on his facial hair as frost—evidence of how cold it was when this photo was taken.

**CONCEPT TEST** ··········································································

Is $\Delta S_{univ}$ greater than, less than, or equal to zero when water vapor exhaled by the Inuit hunter in Figure 18.5 is deposited as crystals of ice on his beard?

··········································································

Entropy-change calculations based on Equation 18.2 assume process reversibility, which, as we have discussed, is an idealized, theoretical concept. In reality, the $\Delta S$ values calculated in this way are *minimum* $\Delta S$ values. When processes take place in the real world, the accompanying changes in entropy are inevitably greater than those calculated using Equation 18.2.

According to the second law, spontaneous processes *always* produce an increase in the entropy of the universe. Table 18.1 shows how the spontaneity of a process depends on the sign and magnitude of $\Delta S_{sys}$ and $\Delta S_{surr}$. Note how processes in which $\Delta S_{sys}$ and $\Delta S_{surr}$ are both greater than zero inevitably result in an increase in $\Delta S_{univ}$, which means they are always spontaneous. On the other hand, processes in which $\Delta S_{sys}$ and $\Delta S_{surr}$ are both less than zero always produce a decrease in $\Delta S_{univ}$, which means they are always nonspontaneous. Between these extremes are four possible combinations of $\Delta S_{sys}$ and $\Delta S_{surr}$ in which these changes have opposite signs. These combinations may or may not produce positive $\Delta S_{univ}$ values, depending on the absolute values of $\Delta S_{sys}$ and $\Delta S_{surr}$. If the larger of the two is the one with the positive value, then $\Delta S_{univ}$ increases, and the process is spontaneous.

| TABLE 18.1 | Spontaneity of Process as a Function of $\Delta S_{sys}$ and $\Delta S_{surr}$ | |
|---|---|---|
| $\Delta S_{sys}$ | $\Delta S_{surr}$ | **Spontaneity of Process** |
| > 0 | > 0 | Always spontaneous |
| < 0 | > 0 | Spontaneous if $|\Delta S_{sys}| < |\Delta S_{surr}|$ |
| | | Nonspontaneous if $|\Delta S_{sys}| > |\Delta S_{surr}|$ |
| > 0 | < 0 | Spontaneous if $|\Delta S_{sys}| > |\Delta S_{surr}|$ |
| | | Nonspontaneous if $|\Delta S_{sys}| < |\Delta S_{surr}|$ |
| < 0 | < 0 | Always nonspontaneous |

**SAMPLE EXERCISE 18.1** **Predicting the Sign of Entropy Change** **LO2**

Predict whether $\Delta S_{sys}$ is greater or less than zero when each of these processes occurs at constant temperature:
a. $H_2O(\ell) \rightarrow H_2O(g)$
b. $NH_3(g) + HCl(g) \rightarrow NH_4Cl(s)$
c. $C_{12}H_{22}O_{11}(s) \xrightarrow{H_2O} C_{12}H_{22}O_{11}(aq)$

**Collect and Organize** We are given equations describing four processes and are to predict the signs of the accompanying entropy changes.

**third law of thermodynamics** the entropy of a perfect crystal is zero at absolute zero.

**standard molar entropy ($S°$)** the absolute entropy of 1 mole of a substance in its standard state.

**Analyze**

a. One mole of liquid water molecules becomes one mole of water vapor molecules, which increases the molecules' freedom of motion.

b. Two moles of gaseous substances form one mole of a solid compound.

c. One mole of a solid dissolves, forming one mole of molecules dispersed in an aqueous solution.

**Solve**

a. $\Delta S_{sys} > 0$ because the kinetic energies of water vapor molecules are more dispersed than they are in the liquid state.

b. $\Delta S_{sys} < 0$ because formation of a solid causes the loss of the translational and rotational motion the gas-phase particles had.

c. $\Delta S_{sys} > 0$ because particles in a solution have translational and rotational motion that particles in crystalline solids do not have.

**Think About It** Entropy increases when solids melt and liquids vaporize because of the increased freedom of motion of their particles and the increased dispersion of their particles' kinetic energies. When gases combine to form a solid, they lose freedom of motion and entropy decreases. When solids dissolve in liquids, however, they gain freedom of motion and experience an increase in entropy.

**Practice Exercise** Predict whether these chemical reactions result in an increase or decrease in the entropy of the system. Assume the reactants and products are at the same temperature and pressure.

a. $CaCO_3(s) + 2\ HCl(aq) \rightarrow CaCl_2(aq) + CO_2(g) + H_2O(\ell)$

b. $NH_3(g) + BF_3(g) \rightarrow NH_3BF_3(s)$

(*Answers to Practice Exercises are in the back of the book.*)

# 18.3 Absolute Entropy and the Third Law of Thermodynamics

We have seen that the entropy of a system depends on temperature. Higher temperatures mean higher particle kinetic energies, which mean the particles have more vibrational, rotational, and translational motion—and more entropy. Conversely, lower temperatures mean that all these quantities are smaller. If we lower the temperature of a substance to absolute zero, in principle all motion should cease. Assuming the substance forms a perfect crystalline solid (Figure 18.6), where each particle is locked in one and only one site within the crystal, then each particle has zero freedom of motion. There is no dispersion of kinetic energy because there is none to disperse. This situation leads to the conclusion that the *absolute* entropy of a perfect crystalline solid is zero at absolute zero (0 K). This conclusion is known as the **third law of thermodynamics**.

Setting a zero point on the entropy scale allows scientists to establish absolute entropy values for pure substances at any temperature. The absolute entropy of a substance is often expressed as its **standard molar entropy ($S°$)**, the entropy of one mole of the substance at 298 K and 1 bar (~1 atm) of pressure in its standard state (Table 18.2). For example, the standard molar entropy of solid NaCl is 72.1 J/(mol · K) at 298 K (Table 18.3). Absolute entropies are determined from careful determinations of the molar heat capacity (or specific heat) of substances as a function of temperature. Our ability to determine absolute entropy values of substances and systems contrasts with the observation in Chapter 5 that we cannot measure absolute enthalpy ($H$), and the best we can do is calculate *changes* in enthalpy, $\Delta H$.

The $S°$ values for liquid water and water vapor in Table 18.3 illustrate a difference in the entropies of liquids and gases that we have discussed before in this

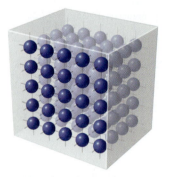

**FIGURE 18.6** A perfect crystal at 0 K has zero entropy because all particles are locked in place and have zero freedom of motion and no kinetic energy.

**CONNECTION** Molar heat capacity and specific heat were defined in Chapter 5.

| TABLE 18.2 | Standard States of Pure Substances and Solutions[a] | |
|---|---|---|
| **Physical State** | **Standard State** | **Pressure**[b] |
| Solid | Pure solid, most stable allotrope of an element | 1 bar |
| Liquid | Pure liquid | 1 bar |
| Gas | Pure gas | 1 bar |
| Solution | 1 $M$ | 1 bar |

[a]The thermodynamic data in Appendices 4–6 and used elsewhere in this book are based on a temperature of 298.15 K (25°C). *Note*: This temperature is not the STP temperature we use for gases (see Chapter 6), which is 273 K.
[b]Since 1982, 1 bar has been the standard pressure for tabulating all thermodynamic data. Prior to 1982, standard pressure was 1 atmosphere (atm) = 1.01325 bar.

| TABLE 18.3 | Selected Standard Molar Entropy Values[a] | | | |
|---|---|---|---|---|
| **Substance** | **$S°$, J/(mol · K)** | **Substance** | **Name** | **$S°$, J/(mol · K)** |
| $Br_2(g)$ | 245.5 | $CH_4(g)$ | Methane | 186.2 |
| $Br_2(\ell)$ | 152.2 | $CH_3CH_3(g)$ | Ethane | 229.5 |
| $C_{diamond}(s)$ | 2.4 | $CH_3OH(g)$ | Methanol | 239.9 |
| $C_{graphite}(s)$ | 5.7 | $CH_3OH(\ell)$ | | 126.8 |
| $CO(g)$ | 197.7 | $CH_3CH_2OH(g)$ | Ethanol | 282.6 |
| $CO_2(g)$ | 213.8 | $CH_3CH_2OH(\ell)$ | | 160.7 |
| $H_2(g)$ | 130.6 | $CH_3CH_2CH_3(g)$ | Propane | 269.9 |
| $N_2(g)$ | 191.5 | $CH_3(CH_2)_2CH_3(g)$ | Butane | 310.0 |
| $O_2(g)$ | 205.0 | $CH_3(CH_2)_2CH_3(\ell)$ | | 231.0 |
| $H_2O(g)$ | 188.8 | $C_6H_6(g)$ | Benzene | 269.2 |
| $H_2O(\ell)$ | 69.9 | $C_6H_6(\ell)$ | | 172.9 |
| $NH_3(g)$ | 192.5 | $C_{12}H_{22}O_{11}(s)$ | Sucrose | 360.2 |

[a]Values for additional substances are given in Appendix 4.

chapter: the molecules in a gas under standard conditions are much more widely dispersed than the molecules in a liquid, and the entropies of the different phases of a given substance at a given temperature follow the order $S_{solid} < S_{liquid} < S_{gas}$.

The entropy changes that occur as one mole of ice at 0 K is heated are shown in Figure 18.7. Note the jump in entropy as the ice melts and the even bigger jump as the liquid water vaporizes. Also note that the lines between the phase changes are curved. The change in entropy, $\Delta S$, with temperature is not linear because, as described by Equation 18.2, heating a substance at a higher temperature produces

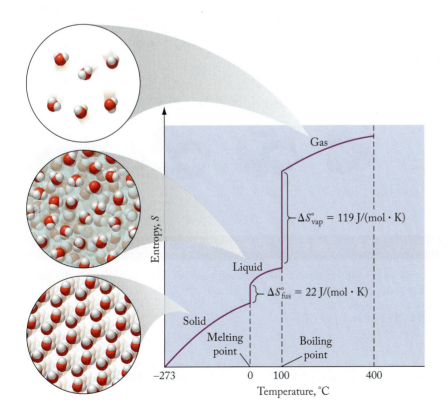

**FIGURE 18.7** Abrupt increases in entropy accompany changes of state, with the greater increase occurring during the transition from liquid to gas.

a smaller entropy increase than adding the same quantity of heat to the same substance at a lower temperature.

Let's summarize the factors that affect entropy change:

1. Entropy increases when temperature increases.
2. Entropy increases when volume increases.[1]
3. Entropy increases when the number of independent particles increases.

In all three cases, entropy increases because each change increases the dispersion of the kinetic energy of a system's particles. We can often make qualitative predictions about entropy changes that accompany chemical reactions based on these three factors, even if we have no thermodynamic data about the reactants and products. For example, when propane burns in air:

$$CH_3CH_2CH_3(g) + 5\ O_2(g) \rightarrow 3\ CO_2(g) + 4\ H_2O(g)$$

Six moles of gaseous reactants combine to form seven moles of gaseous products. The number of moles of gaseous particles increases as the reaction proceeds, so entropy also increases ($\Delta S_{sys} > 0$).

**CONNECTION** In Chapter 6, Avogadro's law told us that the number of moles of gas is directly proportional to the volume occupied by the gas at constant temperature and pressure.

## Entropy and Structure

The data in Table 18.3 contain an important message about the standard molar entropies of substances: they are strongly linked to molecular structure. To see this influence in action, consider the standard molar entropies of the $C_1$ to $C_4$ alkanes in natural gas (Figure 18.8). Note how $S°$ values increase with an increasing number of atoms per molecule. This trend can be explained based on the freedom of motion of the atoms inside their molecules. The more bonds within a molecule, the more opportunities for internal (vibrational) motion, and the greater the standard molar entropy.

Another structural feature that influences entropy is rigidity. The two most common forms of carbon—diamond and graphite—are both polymeric network solids. However, the rigid three-dimensional structure of diamonds gives them much less entropy [$S° = 2.4$ J/(mol · K)] than the less rigid, layered graphite form of carbon [$S° = 5.7$ J/(mol · K)].

**FIGURE 18.8** Among methane, ethane, propane, and butane, standard molar entropies increase as the numbers of atoms and chemical bonds in the molecules increase.

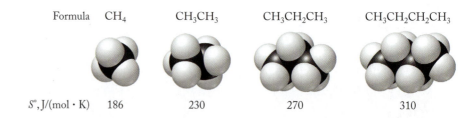

| Formula | $CH_4$ | $CH_3CH_3$ | $CH_3CH_2CH_3$ | $CH_3CH_2CH_2CH_3$ |
|---|---|---|---|---|
| $S°$, J/(mol · K) | 186 | 230 | 270 | 310 |

SAMPLE EXERCISE 18.2 **Comparing Absolute Entropy Values** L03

Select the component in each of the following pairs of compounds that has the greater absolute entropy per mole at a pressure of 1 bar and 298 K.
a. HCl(g), HCl(aq)
b. $CH_3OH(\ell)$, $CH_3CH_2OH(\ell)$

[1]Water is a notable exception to this observation; the molar volume of ice is larger than for liquid water even though the absolute entropy of ice is less than the absolute entropy of liquid water.

c.

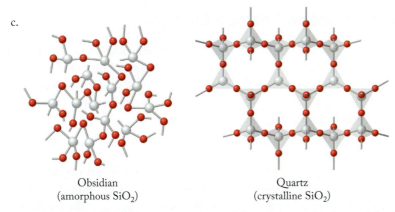

Obsidian
(amorphous $SiO_2$)

Quartz
(crystalline $SiO_2$)

**Collect and Organize** We are given chemical formulas or molecular models of pairs of substances and are to select which component of each pair has the greater absolute entropy per mole under standard conditions; that is, the greater standard molar entropy ($S°$) at 298 K.

**Analyze** Particles in the vapor state have more freedom of motion and entropy than they do in the liquid state, and particles in the liquid state have more freedom of motion and entropy than they do in the solid state. Substances composed of more particles, contributing to more freedom of motion, have more absolute entropy than substances made of fewer particles with less freedom of motion.

**Solve**
a. HCl gas loses freedom of motion when it dissolves in water, so HCl gas has a greater $S°$ value at 298 K.
b. Both compounds are liquids with similar formulas, but one mole of ethanol ($CH_3CH_2OH$) has more atoms and more covalent bonds between atoms than one mole of methanol ($CH_3OH$). Therefore, ethanol has a greater $S°$ value at 298 K.
c. Both forms of $SiO_2$ are solids with extended covalent networks of atoms. However, quartz has a crystalline structure and obsidian has an irregular arrangement of bonds and atoms. The randomness of the obsidian structure gives it a greater $S°$ value at 298 K.

**Think About It** The comparison in part a is complicated by the fact that HCl($aq$) is a strong acid, which means that each molecule produces two ions in solution. However, they are liquid-phase ions and have less combined entropy than half as many gas-phase HCl molecules. (Compare the $S°$ values of HCl($g$), $H^+(aq)$, and $Cl^-(aq)$ in Appendix 4 to see for yourself.)

⚙ **Practice Exercise** Below are ball-and-stick models of four hydrocarbons that each contain six carbon atoms per molecule. Rank these compounds in order of decreasing $S°$ values.

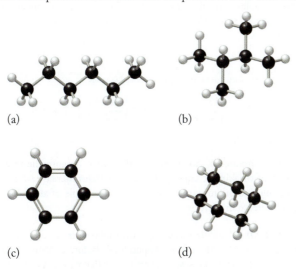

(a)

(b)

(c)

(d)

# 18.4 Calculating Entropy Changes

The entropy of a system (like its enthalpy and internal energy) is a state function, which means that the change in entropy that accompanies a process depends only on the initial and final states of the system, not on the pathway of the process. Therefore the change in entropy experienced by a system is simply the difference between its initial and final absolute entropy levels:

$$\Delta S_{sys} = S_{final} - S_{initial} \qquad (18.4)$$

●○● **CONNECTION** State functions were defined in Chapter 5.

We can adapt Equation 18.4 to calculate the change in entropy that accompanies a chemical reaction under standard conditions, $\Delta S^\circ_{rxn}$, from the difference in the standard molar entropies of moles of reactants, $n_{reactants}$ (the equivalent of $S_{initial}$ in Equation 18.4), and the molar entropies of moles of products, $n_{products}$, (that is, $S_{final}$):

$$\Delta S^\circ_{rxn} = \sum n_{products} S^\circ_{products} - \sum n_{reactants} S^\circ_{reactants} \qquad (18.5)$$

Each individual $S^\circ$ value for a product or reactant is multiplied by the appropriate number of moles from the balanced chemical equation. In other words, just as we saw for $\Delta H^\circ$ in Chapter 5, entropy is an extensive thermodynamic property that depends on the quantities of substances consumed or produced in a reaction. Standard molar entropies of selected substances are listed in Appendix 4.

---

**SAMPLE EXERCISE 18.3  Calculating Entropy Changes**  **LO4**

What is $\Delta S^\circ$ for the dissolution of ammonium nitrate (Figure 18.9, $\Delta H^\circ > 0$) under standard conditions, given the following standard molar entropy values:

$$NH_4NO_3(s) \rightarrow NH_4^+(aq) + NO_3^-(aq)$$

$S^\circ$ [J/(mol · K)]       151.1          113.4          146.4

**Collect and Organize** We are given the standard molar entropy values of a solid ionic compound and its ions in aqueous solution and want to calculate the change in entropy that occurs when the compound dissolves under standard conditions. Entropy changes associated with chemical reactions depend on the entropies of the reactants and products, as described in Equation 18.5.

**Analyze** Equation 18.5 allows us to calculate $\Delta S^\circ$ of the dissolution process from the difference in standard molar entropies of the solid solute and its ions in solution. Dissolving an ionic solid in water increases the freedom of motion of the solute ions, so $\Delta S^\circ$ should be greater than zero.

**Solve**

$$\Delta S^\circ = \sum n_{products} S^\circ_{products} - \sum n_{reactants} S^\circ_{reactants}$$

$$= \left[ 1 \text{ mol} \times \left( \frac{113.4 \text{ J}}{\text{mol} \cdot \text{K}} \right) + 1 \text{ mol} \times \left( \frac{146.4 \text{ J}}{\text{mol} \cdot \text{K}} \right) \right] - 1 \text{ mol} \times \left( \frac{151.1 \text{ J}}{\text{mol} \cdot \text{K}} \right)$$

$$= 108.7 \text{ J/K}$$

**Think About It** As predicted, $\Delta S^\circ$ is greater than zero because the freedom of motion and the dispersion of the kinetic energy of solute particles increase when the solid solute dissolves. The dissolution of ammonium nitrate is an example of an endothermic reaction that results in an increase in entropy.

⚙ **Practice Exercise** Calculate the standard molar entropy change for the combustion of methane gas using $S^\circ$ values from Appendix 4. Before carrying out the calculation, predict whether the entropy of the system increases or decreases. Assume that liquid water is one of the products.

---

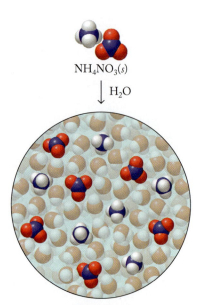

$NH_4NO_3(s)$

↓ $H_2O$

$NH_4NO_3(aq)$

**FIGURE 18.9** Dissolution of solid ammonium nitrate into $NH_4^+$ and $NO_3^-$ ions is endothermic but spontaneous under standard conditions because $\Delta S^\circ$ is greater than zero, which contributes to $\Delta S_{univ} > 0$.

# 18.5 Free Energy

In Sample Exercise 18.3, we used standard molar entropy values to calculate $\Delta S°$. The solute, solvent, and resulting solution together constituted a closed thermodynamic system, which meant that energy could flow into it as its temperature dropped, but matter was not exchanged with its surroundings. Therefore the $\Delta S°$ value that we calculated applied only to the system. There was no evaluation of the change in the entropy of the system's surroundings accompanying the dissolution process, though we might predict that energy flowing from the surroundings into the system would produce a decrease in $S_{surr}$. (Had the dissolution process been exothermic, energy would have flowed from the system into its surroundings, and $\Delta S_{surr}$ would have been greater than zero.)

Thus, the entropy change experienced by the surroundings of any chemical thermodynamic system depends on whether the process occurring in the system is exothermic or endothermic. When heat flows from an exothermic process occurring at constant pressure into a system's surroundings, the quantity of heat is equal in magnitude but opposite in sign to the enthalpy change of the system:

$$q_{surr} = -\Delta H_{sys} \qquad (18.6)$$

When the system hosts an endothermic process, as in Sample Exercise 18.3, the direction of energy flow is reversed, but Equation 18.6 still applies. Assuming these transfers of energy occur reversibly, we can calculate the value of $\Delta S_{surr}$ using this modification of Equation 18.2:

$$\Delta S_{surr} = \frac{q_{surr}}{T} \qquad (18.7)$$

Now let's combine Equations 18.6 and 18.7:

$$\Delta S_{surr} = -\frac{\Delta H_{sys}}{T}$$

We can substitute this expression into Equation 18.1 ($\Delta S_{univ} = \Delta S_{sys} + \Delta S_{surr}$):

$$\Delta S_{univ} = \Delta S_{sys} - \frac{\Delta H_{sys}}{T} \qquad (18.8)$$

The beauty of Equation 18.8 is that it allows us to predict whether or not a process is spontaneous at a particular temperature once we calculate the enthalpy and entropy changes accompanying the process. The downside of Equation 18.8 is that spontaneity relies on the value of a parameter ($\Delta S_{univ}$) that is impossible to determine directly and that has little physical meaning. It would be great if we could substitute a thermodynamic parameter for $\Delta S_{univ}$ that is based only on the system and not the entire universe. Such a parameter exists and is called the change in the system's *free energy*.

In chemistry we focus on a particular kind of free energy called **Gibbs free energy (G)** in honor of American scientist J. Willard Gibbs (1839–1903). Gibbs free energy is the energy released by processes happening at constant temperature and pressure that is available to do useful work.

Like many of the thermodynamic properties we have examined, absolute free energy values of substances are often of less interest than the *changes* in free energy that accompany chemical reactions and other processes. Gibbs proposed that the change in free energy ($\Delta G_{sys}$) of a process occurring at constant temperature and pressure is linked directly to that temperature and $\Delta S_{univ}$:

$$\Delta G_{sys} = -T\,\Delta S_{univ}$$

**CONNECTION** In Chapter 5, we defined a change in enthalpy ($\Delta H$) as the heat gained or lost in a reaction carried out at constant pressure.

▶❚❚ **CHEMTOUR** Gibbs Free Energy

**Gibbs free energy (G)** the maximum energy released by a process occurring at constant temperature and pressure that is available to do useful work.

Because of the $-T$ multiplier, *negative* values of $\Delta G_{sys}$ correspond to *positive* values of $\Delta S_{univ}$. Therefore

- If $\Delta G_{rxn} < 0$, then $\Delta S_{univ} > 0$, and the reaction is spontaneous.
- If $\Delta G_{rxn} > 0$, then $\Delta S_{univ} < 0$, and the reaction is nonspontaneous. Instead, the reaction running in reverse is spontaneous.
- If $\Delta G_{rxn} = 0$, then $\Delta S_{univ} = 0$, and the composition of the reaction mixture does not change with time. In other words, the reaction has reached chemical equilibrium.

We can combine Gibbs' equation with Equation 18.8 by multiplying all of the terms in Equation 18.8 by $-T$:

$$-T\,\Delta S_{univ} = -T\,\Delta S_{sys} + \Delta H_{sys} \tag{18.9}$$

The left side of Equation 18.9 is equal to $\Delta G_{sys}$. Making that substitution and rearranging the terms on the right side gives us

$$\Delta G_{sys} = \Delta H_{sys} - T\,\Delta S_{sys}$$

Since all of the parameters in this equation apply to the system, we typically simplify the equation by eliminating them:

$$\Delta G = \Delta H - T\,\Delta S \tag{18.10}$$

**CONCEPT TEST** ......................................................

(a) Given the thermodynamic data in Table A4.3 in the Appendix, is the conversion of diamond to graphite spontaneous? Explain your answer. (b) If yes, does knowing that the conversion is spontaneous tell you how rapid the conversion is?

........................................................

Equation 18.10 highlights the two thermodynamic driving forces that contribute to a decrease in free energy and to making a process spontaneous:

1. The system experiences an increase in entropy ($\Delta S > 0$).
2. The process is exothermic ($\Delta H < 0$).

One or both of these conditions must be true for a reaction to be spontaneous.

Using Equation 18.10, we can calculate the change in Gibbs free energy of a process if we first calculate the values of $\Delta H$ and $\Delta S$. In the case of a chemical reaction occurring under standard conditions, we can calculate the *standard* change in Gibbs free energy $\Delta G^{\circ}_{rxn}$ by using the following modified version of Equation 18.10:

$$\Delta G^{\circ}_{rxn} = \Delta H^{\circ}_{rxn} - T\,\Delta S^{\circ}_{rxn} \tag{18.11}$$

In Chapter 5 we calculated $\Delta H^{\circ}_{rxn}$ values from the differences in standard enthalpies of formation $\Delta H^{\circ}_{f}$ of products and reactants:

$$\Delta H^{\circ}_{rxn} = \sum n_{products}\,\Delta H^{\circ}_{f,products} - \sum n_{reactants}\,\Delta H^{\circ}_{f,reactants} \tag{5.17}$$

The value of $\Delta S^{\circ}_{rxn}$ can be calculated using Equation 18.5:

$$\Delta S^{\circ}_{rxn} = \sum n_{products}\,S^{\circ}_{products} - \sum n_{reactants}\,S^{\circ}_{reactants} \tag{18.5}$$

We can combine the results of these two calculations in Equation 18.11 to calculate $\Delta G^{\circ}_{rxn}$, as illustrated in Sample Exercise 18.4.

SAMPLE EXERCISE 18.4    **Predicting Reaction Spontaneity**    **LO5**
                       **under Standard Conditions**

Consider the reaction of nitrogen gas and hydrogen gas (Figure 18.10) at 298 K to make ammonia at the same temperature:

$$N_2(g) + 3\,H_2(g) \rightarrow 2\,NH_3(g)$$

a. Before doing any calculations, predict the sign of $\Delta S^\circ_{rxn}$.
b. What is the actual value of $\Delta S^\circ_{rxn}$?
c. What is the value of $\Delta H^\circ_{rxn}$?
d. What is the value of $\Delta G^\circ_{rxn}$ at 298 K?
e. Is the reaction spontaneous at 298 K and 1 bar of pressure?

**Collect and Organize** For a given reaction, we are to predict the sign of the standard entropy change, and then calculate the changes in entropy, enthalpy, and Gibbs free energy. We can then determine the spontaneity of the reaction under standard conditions. Standard molar entropies and standard heats of formation of the reactants and product are in Table 18.3 and Appendix 4.

**Analyze** Figure 18.10 reinforces the point that there are more molecules of gaseous reactants than products in the reaction. We can use Equation 18.5 to calculate entropy changes under standard conditions:

$$\Delta S^\circ_{rxn} = \sum n_{products}S^\circ_{products} - \sum n_{reactants}S^\circ_{reactants} \qquad (18.5)$$

We learned in Chapter 5 how to calculate $\Delta H^\circ_{rxn}$ from standard heats of formation values (in Appendix 4) using this equation:

$$\Delta H^\circ_{rxn} = \sum n_{products}\,\Delta H^\circ_{f,products} - \sum n_{reactants}\,\Delta H^\circ_{f,reactants} \qquad (5.17)$$

The calculated values of $\Delta S^\circ_{rxn}$ and $\Delta H^\circ_{rxn}$ can be combined to calculate $\Delta G^\circ_{rxn}$:

$$\Delta G^\circ_{rxn} = \Delta H^\circ_{rxn} - T\,\Delta S^\circ_{rxn} \qquad (18.11)$$

The sign of $\Delta G^\circ_{rxn}$ will allow us to predict whether or not the reaction is spontaneous under standard conditions ($T = 298$ K). We cannot predict the sign of $\Delta G^\circ_{rxn}$ and whether this reaction is spontaneous without knowing the magnitudes of $\Delta H^\circ_{rxn}$ and $T\Delta S^\circ_{rxn}$. $\Delta H$ values are usually in the kJ range, whereas $T\Delta S$ values are usually in the J range.

**Solve**

a. The number of gas-phase molecules decreases as the reaction proceeds, so $\Delta S^\circ_{rxn}$ is probably less than zero.

b. We use data from Table 18.3 in Equation 18.5 to calculate $\Delta S^\circ_{rxn}$:

$$\Delta S^\circ_{rxn} = \sum n_{products}S^\circ_{products} - \sum n_{reactants}S^\circ_{reactants} = \Delta S^\circ_{sys}$$

$$= \left\{ \left[ 2\ \text{mol} \times \left( \frac{192.5\ \text{J}}{\text{mol} \cdot \text{K}} \right) \right] - \left[ 1\ \text{mol} \times \left( \frac{191.5\ \text{J}}{\text{mol} \cdot \text{K}} \right) + 3\ \text{mol} \times \left( \frac{130.6\ \text{J}}{\text{mol} \cdot \text{K}} \right) \right] \right\}$$

$$= -198.3\ \text{J/K}$$

The entropy change is negative, as predicted in part a.

c. The change in enthalpy that accompanies the reaction under standard conditions is

$$\Delta H^\circ_{rxn} = \sum n_{products}\,\Delta H^\circ_{f,products} - \sum n_{reactants}\,\Delta H^\circ_{f,reactants}$$

$$= \left\{ \left[ 2\ \text{mol} \times \left( \frac{-46.1\ \text{kJ}}{\text{mol}} \right) \right] - \left[ 1\ \text{mol} \times \left( \frac{0.0\ \text{kJ}}{\text{mol}} \right) + 3\ \text{mol} \times \left( \frac{0.0\ \text{kJ}}{\text{mol}} \right) \right] \right\}$$

$$= -92.2\ \text{kJ}$$

d. We insert the values of $\Delta S^\circ_{rxn}$ and $\Delta H^\circ_{rxn}$ calculated in parts a and b into Equation 18.11:

$$\Delta G^\circ_{rxn} = \Delta H^\circ_{rxn} - T\,\Delta S^\circ_{rxn}$$

$$= -92.2\ \text{kJ} - \left[ (298\ \text{K}) \times \left( -198.3\ \frac{\text{J}}{\text{K}} \times \frac{1\ \text{kJ}}{1000\ \text{J}} \right) \right]$$

$$= -33.1\ \text{kJ}$$

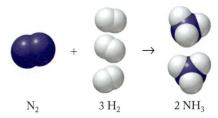

$N_2$      $3\,H_2$      $2\,NH_3$

**FIGURE 18.10** In the synthesis of ammonia, one molecule of nitrogen reacts with three molecules of hydrogen to yield two molecules of ammonia. All substances are gases.

**standard free energy of formation ($\Delta G_f^\circ$)** the change in free energy associated with the formation of 1 mole of a compound in its standard state from its component elements.

e. The decrease in Gibbs free energy tells us that the reaction is spontaneous under standard conditions and $T = 298$ K.

**Think About It** We were unable to determine reaction spontaneity because we did not have a feel for the magnitudes of $\Delta H_{rxn}$ and $\Delta S_{rxn}$. In this case, an unfavorable entropy change ($\Delta S_{rxn}^\circ < 0$) is more than offset by a favorable enthalpy change ($\Delta H_{rxn}^\circ < 0$) so that overall, $\Delta G_{rxn}^\circ < 0$.

 **Practice Exercise** For the reaction $2\,H_2(g) + O_2(g) \rightarrow 2\,H_2O(\ell)$,
a. Predict the sign of the entropy change for the reaction.
b. What is the value of $\Delta S_{rxn}^\circ$?
c. What is the value of $\Delta H_{rxn}^\circ$?
d. Is the reaction spontaneous at 298 K and 1 bar pressure?

**CONCEPT TEST**

The preparation of ammonia from nitrogen and hydrogen in Sample Exercise 18.4 is determined to be spontaneous under standard conditions, yet if we mix the two gases at 298 K, no reaction is observed. Suggest a reason why.

Another way to calculate the change in Gibbs free energy of a reaction under standard conditions is based on another thermodynamic property of substances listed in Appendix 4: **standard free energy of formation ($\Delta G_f^\circ$)**. A compound's $\Delta G_f^\circ$ value is the change in free energy associated with the formation of 1 mole of it in its standard state from its elements in their standard states.

In Chapter 5, we calculated standard heats of reactions ($\Delta H_{rxn}^\circ$) from the difference in the standard heats of formation ($\Delta H_f^\circ$) of their products and reactants. We can also calculate the change in standard free energy of a reaction under standard conditions from the difference in the standard free energies of formation of its products and reactants. As with standard heats of formation, standard free energies of formation of the most stable forms of elements in their standard states are zero. The similarities in the two calculations can be seen from the similar formats of the equations used to calculate $\Delta G_{rxn}^\circ$.

$$\Delta G_{rxn}^\circ = \sum n_{products}\,\Delta G_{f,products}^\circ - \sum n_{reactants}\,\Delta G_{f,reactants}^\circ \quad (18.12)$$

and $\Delta H_{rxn}^\circ$:

$$\Delta H_{rxn}^\circ = \sum n_{products}\,\Delta H_{f,products}^\circ - \sum n_{reactants}\,\Delta H_{f,reactants}^\circ \quad (5.17)$$

Sample Exercise 18.5 illustrates just how similar the two calculations are.

Octane
$\Delta G_f^\circ = 16.3$ kJ/mol

2-Methylheptane
$\Delta G_f^\circ = 11.7$ kJ/mol

3,3-Dimethylhexane
$\Delta G_f^\circ = 12.6$ kJ/mol

**FIGURE 18.11** Molecular structures and $\Delta G_f^\circ$ values of three $C_8H_{18}$ isomers.

**CONCEPT TEST**

Molecular models and standard free energies of formation for three structural isomers with the molecular formula $C_8H_{18}$ are shown in Figure 18.11. All three isomers burn in air, as described by the same chemical equation

$$2\,C_8H_{18}(\ell) + 25\,O_2(g) \rightarrow 16\,CO_2(g) + 18\,H_2O(g)$$

Are the $\Delta G_{rxn}^\circ$ values for the three combustion reactions also the same? Why or why not?

**SAMPLE EXERCISE 18.5** **Calculating $\Delta G^\circ_{rxn}$ Using Appropriate $\Delta G^\circ_f$ Values** **LO6**

Use the appropriate standard free energy of formation values in Appendix 4 to calculate the change in Gibbs free energy as ethanol burns under standard conditions. Assume the reaction proceeds as described by the following chemical equation:

$$CH_3CH_2OH(\ell) + 3\,O_2(g) \rightarrow 2\,CO_2(g) + 3\,H_2O(\ell)$$

**Collect and Organize** We are to calculate the value of $\Delta G^\circ_{rxn}$ for the combustion of ethanol according to Equation 18.12, using the $\Delta G^\circ_f$ values of the reactants and products in the combustion reaction:

| Substance | $CH_3CH_2OH(\ell)$ | $O_2(g)$ | $CO_2(g)$ | $H_2O(\ell)$ |
|---|---|---|---|---|
| $\Delta G^\circ_f$ (kJ/mol) | −174.9 | 0 | −394.4 | −237.2 |

**Analyze** The reaction consumes 1 mole of liquid ethanol and 3 moles of oxygen and produces 2 moles of $CO_2$ gas and 3 moles of liquid $H_2O$. Substituting the above $\Delta G^\circ_f$ values and the appropriate numbers of moles into Equation 18.12 yields the value of $\Delta G^\circ_{rxn}$. The combustion of ethanol, a common additive in gasoline in the United States, is spontaneous, so $\Delta G^\circ_{rxn}$ should be less than zero.

**Solve** We insert the appropriate numbers of moles and $\Delta G^\circ_f$ values into Equation 18.12 and do the math:

$$\Delta G^\circ_{rxn} = \sum n_{products}\,\Delta G^\circ_{f,products} - \sum n_{reactants}\,\Delta G^\circ_{f,reactants}$$

$$= [2\ \text{mol}\ CO_2 \times (-394.4\ \text{kJ/mol}) + 3\ \text{mol}\ H_2O \times (-237.2\ \text{kJ/mol})]$$

$$- 1\ \text{mol}\ CH_3CH_2OH \times (-174.9\ \text{kJ/mol})$$

$$= -1325.5\ \text{kJ}$$

**Think About It** The calculated value represents that part of the total energy released by the combustion of one mole of ethanol under standard conditions that is available to do useful work.

⚙ **Practice Exercise** Use the appropriate standard free energy of formation values in Appendix 4 to calculate the value of $\Delta G^\circ_{rxn}$ for the steam–reforming reaction used to produce $H_2$ gas:

$$CH_4(g) + H_2O(g) \rightarrow CO(g) + 3\,H_2(g)$$

What exactly does "energy available to do useful work" mean? Let's attempt to answer this question, using as our model the internal combustion (gasoline) engines used to power most automobiles.

A combustion reaction is a thermodynamic system that experiences a decrease in internal energy ($\Delta E$) as energy in the form of heat, $q$, flows from it into its surroundings and as it does work, $w$, on its surroundings. These three variables are related by Equation 5.5:

$$\Delta E = q + w \qquad (5.5)$$

From the perspective of the system, all three quantities are less than zero. Internal combustion engines have cooling systems to manage the dissipation of heat, which is wasted energy that does nothing to power the car. The energy that moves the car is derived from the rapid expansion of the gaseous products of combustion in the cylinders of the engine. As Figure 18.12 shows, this expansion pushes down on the piston of a cylinder, increasing the volume of the reaction mixture.

(a)                (b)

**FIGURE 18.12** (a) In a car engine, thermal expansion causes the gases in a cylinder to (b) push down on a piston with a pressure, $P$, represented by the blue arrow. The product of $P$ and the change in volume of the gases, $\Delta V$, is the work done by the expanding gases that propels the car.

The pressure exerted by the reacting gases and their products on the piston, multiplied by the resulting volume change, is $P\,\Delta V$ work (which we discussed in Section 5.2), which propels the car at constant pressure:

$$w = -P\,\Delta V$$

Gibbs free energy is a measure of the *maximum* amount of work that can be done by the energy released during combustion. To see how this theoretical quantity of work compares with the total energy released, let's rearrange Equation 18.10 by isolating the $\Delta H$ term and simplify it by discarding all the "sys" subscripts:

$$\Delta H = \Delta G + T\,\Delta S \qquad (18.13)$$

In this form, the equation that tells us that the enthalpy change that accompanies making and breaking chemical bonds during a chemical reaction may be divided into two parts. One part, $\Delta G$, is the energy that can theoretically be converted into motion and other useful work (like propelling a car or generating electricity for its electrical system). The other part, $T\,\Delta S$, is not usable: it is the portion of energy that spreads out when, for example, hot gases flow out of an automobile exhaust pipe. This part of $\Delta H$ is wasted. Consequently, conversion of chemical energy into useful mechanical energy ($\Delta G$) is never 100% efficient. A portion of $\Delta G$ is also wasted because combustion and the energy conversion happen quickly and, therefore, irreversibly. Maximum efficiency comes with very slow, reversible reaction and energy conversion rates (see Section 18.2), but that is not how automobile engines operate. It turns out that gasoline engines convert only about 30% of the energy produced during combustion into useful work.

## 18.6 Temperature and Spontaneity

Let's revisit the process of ice melting, this time focusing on how the values of the three terms $\Delta H$, $T\,\Delta S$, and $\Delta G$ in Equation 18.10 change as the temperature of a mixture of ice and water increases from $-10°C$ to $+10°C$ (Figure 18.13). As temperature rises over this range, there is little impact on the heat of fusion, $\Delta H$, as shown by the nearly flat green line in Figure 18.13. However, it is only reasonable that increasing the value of $T$ increases the value $T\,\Delta S$ (as shown by the upward slope of the purple line). After all, the $\Delta S$ of melting ice (or any melting solid)[2] has a positive value, so $T\,\Delta S$ must increase as $T$ increases. The green and purple lines intersect at $0°C$, which means $\Delta H$ is equal to $T\,\Delta S$. Put another way, the difference between $\Delta H$ and $T\,\Delta S$ at $0°C$ is zero, and so is $\Delta G$ (remember: $\Delta G = \Delta H - T\,\Delta S$). The temperature at the point in Figure 18.13 where $\Delta G = 0$ defines the melting (or freezing) point, which by definition is the temperature at which the solid melts at the same rate as the liquid freezes. The two phases are in *equilibrium*. At $0°C$, no net change takes place, and both phases coexist.

We also know that ice melts spontaneously above $0°C$, which means that $\Delta G$ for the melting process must be less than zero. The graph in Figure 18.13 shows that indeed it is. Above $0°C$ the value of $T\,\Delta S$ is greater than $\Delta H$. Therefore the difference between them ($\Delta H - T\,\Delta S$) is less than zero and becomes more negative with increasing temperature, as illustrated by the distance from the purple line to the green line in Figure 18.13.

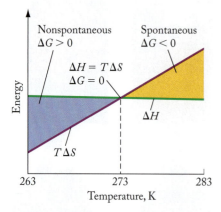

**FIGURE 18.13** Changes in the values of $\Delta H$, the quantity $T\,\Delta S$, and $\Delta G$ for ice melting as temperature increases from $-10°C$ (263 K) to $+10°C$ (283 K).

[2] Helium-3, $^3$He, is an exception to this observation: $\Delta H_{fus,^3He} < 0$.

Below 0°C, ice does not melt spontaneously, which means $\Delta G$ is greater than zero. The graph shows why this is true: at $T < 273$ K (0°C), $T\Delta S$ values (purple line) are less than $\Delta H$ values (green line), which means that $\Delta H - T\Delta S$ (and $\Delta G$) is greater than zero. The positive $\Delta G$ values mean the process is nonspontaneous below 273 K, and ice does not melt below its freezing point. However, the opposite process—liquid water freezing—*is* spontaneous because reversing a process keeps the absolute values but switches the signs of $\Delta H$, $\Delta S$, and $\Delta G$. Therefore, if the melting process is nonspontaneous, then the exothermic freezing process *is* spontaneous at low temperatures.

**CONCEPT TEST** ························································································

A given process is spontaneous at lower temperature and nonspontaneous at higher temperature. What does the graph in Figure 18.13 look like for such a process?

························································································

Table 18.4 summarizes the effects of the signs of $\Delta H°$ and $\Delta S°$ on $\Delta G°$ and on reaction spontaneity.

| TABLE 18.4 | Effects of $\Delta H°$ and $\Delta S°$ on $\Delta G°$ and Spontaneity | | |
|---|---|---|---|
| **$\Delta H°$** | **$\Delta S°$** | **$\Delta G°$** | |
| − | + | Always $< 0$ | Always spontaneous |
| − | − | $< 0$ at lower temperature | Spontaneous at lower temperature |
| + | + | $< 0$ at higher temperature | Spontaneous at higher temperature |
| + | − | Always $> 0$ | Never spontaneous |

Equilibrium also exists for water at 100°C and 1 atm pressure, which are the temperature and pressure at which $\Delta G_{vaporization}$ and $\Delta G_{condensation}$ both equal zero. At 100°C, liquid water vaporizes and water vapor condenses at the same rate; the two phases coexist.

The temperature at which $\Delta G = 0$ for a process can be calculated from Equation 18.10 if we know the values of $\Delta H_{fus}$ and $\Delta S_{fus}$. For example, the values for the fusion of water are

$$H_2O(s) \rightarrow H_2O(\ell) \qquad \Delta H°_{fus} = 6.01 \times 10^3 \text{ J/mol}; \Delta S°_{fus} = 22.0 \text{ J/(mol} \cdot \text{K)}$$

In doing this calculation, we assume that the values of $\Delta H°$ and $\Delta S°$ do not change significantly with small changes in temperature. For this process, as well as for most other physical and chemical processes, this assumption is acceptable. Therefore we can assume that $\Delta G = \Delta H - T\Delta S \approx \Delta H° - T\Delta S°$. Inserting the values of $\Delta H°$ and $\Delta S°$ and using $\Delta G = 0$ for a process at equilibrium, we get

$$\Delta G = (6.01 \times 10^3 \text{ J/mol}) - T[22.0 \text{ J/(mol} \cdot \text{K)}] = 0$$

$$T = \frac{6.01 \times 10^3 \text{ J/mol}}{22.0 \text{ J/(mol} \cdot \text{K)}} = 273 \text{ K} = 0°C$$

This is the familiar value for the melting point of ice.

Like physical processes, chemical reactions can occur with zero change in free energy. As a spontaneous reaction proceeds ($\Delta G < 0$), reactants become products and $\Delta G$ becomes less negative. In fact, $\Delta G$ may reach zero before all

**CONNECTION** In Chapter 10 we discussed the equilibrium between phases of a substance and used phase diagrams to illustrate which physical states are stable at various combinations of temperature and pressure.

the reactants are consumed. In this case, no more products form and no more reactants are consumed. Rather, reactants and products coexist in equilibrium with each other. Reactants still react, and products are still formed, but the reverse reaction, in which products become reactants, proceeds at the same rate as the forward reaction. This is the state we know as chemical equilibrium. No net changes in free energy or the amounts of reactants and products occur once equilibrium is reached.

---

**SAMPLE EXERCISE 18.6** **Relating Reaction Spontaneity to $\Delta H$ and $\Delta S$** **LO7**

A certain chemical reaction is spontaneous at low temperatures but not at high temperatures. Use Equation 18.10 to determine the signs of the enthalpy and entropy changes for this reaction.

**Collect and Organize** We are to determine the signs of $\Delta H$ (enthalpy change) and $\Delta S$ (entropy change) based on the change in spontaneity of a reaction as temperature changes. This means we need to think about how the signs of $\Delta H$ and $\Delta S$ determine how $\Delta G$ varies with temperature.

**Analyze** There are two possible combinations for $\Delta H$ and $\Delta S$ that might lead to spontaneous reactions: $\Delta H > 0$ and $\Delta S > 0$ or $\Delta H < 0$ and $\Delta S < 0$. Only one of these combinations of $\Delta H$ and $\Delta S$ will satisfy the condition that $\Delta G > 0$ at higher temperature and $\Delta G < 0$ at lower temperature. $\Delta H < 0$ and $\Delta S > 0$ always leads to a spontaneous reaction.

**Solve** The importance of $\Delta S$ increases with increasing temperature because the product $T\,\Delta S$ appears in Equation 18.10. The reaction is nonspontaneous at higher temperatures, where the magnitude of $T\,\Delta S$ is more likely to be larger than the magnitude of $\Delta H$. The reaction is spontaneous at low temperatures, however, where the impact of a negative $\Delta S$ value is more than offset by a decrease in enthalpy, a change that favors the reaction. The reaction must have negative $\Delta S$ and negative $\Delta H$ values.

**Think About It** Table 18.4 confirms our prediction that a process that is spontaneous only at low temperatures is one in which there is a decrease in both entropy and enthalpy. Recall that the freezing of water below 0°C is an example of an exothermic reaction ($\Delta H < 0$) that is accompanied by a decrease in entropy ($\Delta S < 0$).

⚙ **Practice Exercise** At high temperatures, ammonia decomposes to nitrogen and hydrogen gases:

$$2\,NH_3(g) \rightarrow N_2(g) + 3\,H_2(g)$$

$\Delta H$ for the reaction is positive and $\Delta S$ is positive. Predict whether the reaction is spontaneous at all temperatures, or only at high temperatures.

---

# 18.7 Free Energy and Chemical Equilibrium

We have just seen that equilibrium exists when the free energy change that accompanies a process is zero. For chemical reactions not at equilibrium, the magnitude of $\Delta G$—how far it is from zero in either a negative or positive direction—indicates how far a system is from its equilibrium position.

In Chapter 15 we discussed how to use the concept of the reaction quotient ($Q$) to determine whether or not a reaction is at chemical equilibrium. When the value of $Q$ is much larger or smaller than the value of the reaction's equilibrium constant, $K$, we know that the reaction is far from chemical equilibrium. For example, we saw in Chapter 15 that when $Q$ is less than $K$, the ratio of product concentrations to reactant concentrations (raised to the appropriate powers) is less than they would be at equilibrium. To reach equilibrium, the reaction proceeds (spontaneously!) in the forward direction until the ratio of product to reactant concentrations matches the value of $K$. In other words, when $Q$ is less than $K$, $\Delta G_{rxn}$ must be less than zero.

On the other hand, when $Q$ is greater than $K$, the ratio of product to reactant concentrations is greater than it would be at equilibrium. To reach equilibrium, the reaction must proceed in reverse, so that products are consumed and reactants form, until the ratio of product to reactant concentrations matches the value of $K$. Therefore, when $Q$ is greater than $K$, the forward reaction is not spontaneous because the *reverse* reaction is spontaneous, which means $\Delta G_{rxn}$ for the forward reaction must be greater than zero.

To summarize these points for any reaction:

1. When $Q = K$, $\Delta G_{rxn} = 0$ and the reaction is at equilibrium.
2. When $Q < K$, $\Delta G_{rxn} < 0$ and the reaction is spontaneous.
3. When $Q > K$, $\Delta G_{rxn} > 0$ and the reaction is nonspontaneous.

The key concept here is that the value of $\Delta G_{rxn}$ for any chemical reaction depends not only on the standard free energy of formation values of its reactants and products—that is, on the value of $\Delta G^\circ_{rxn}$—but also on the concentrations of the reactants and products in the reaction mixture. In other words, the value of $\Delta G_{rxn}$ depends on the value of $Q$. This dependency is expressed in Equation 18.14:

$$\Delta G_{rxn} = \Delta G^\circ_{rxn} + RT \ln Q \qquad (18.14)$$

To illustrate the dependency of $\Delta G_{rxn}$ on $Q$, let's use as our chemical model the decomposition of $N_2O_4$:

$$N_2O_4(g) \rightleftharpoons 2\,NO_2(g)$$

First, we use Equation 18.12 and the $\Delta G^\circ_f$ values for the reactant and product in Appendix 4 to calculate the value of $\Delta G^\circ_{rxn}$:

$$\Delta G^\circ_{rxn} = \sum n_{products} G^\circ_{f,products} - \sum n_{reactants} G^\circ_{f,reactants}$$

$$= 2\text{ mol }(51.3\text{ kJ/mol}) - 1\text{ mol }(97.8\text{ kJ/mol}) = +4.8\text{ kJ}$$

The positive value of $\Delta G^\circ_{rxn}$ indicates that the reaction as written is not spontaneous at $T = 298$ K and under standard conditions, which is when $P_{NO_2} = P_{N_2O_4} = 1$ bar. Consider a reaction vessel at 298 K that initially contains 1 mole of $N_2O_4$ at a partial pressure of 1 bar and no (or hardly any) $NO_2$. Under these conditions, the value of the reaction quotient expressed with partial pressures ($Q_p$) is zero:

$$Q_p = \frac{(P_{NO_2})^2}{P_{N_2O_4}} = \frac{0}{1} = 0$$

Although we do not know the value of $K_p$, it has to be greater than zero. Therefore, $Q_p < K_p$ and the system should respond by spontaneously forming $NO_2$ from $N_2O_4$.

If the reaction in the forward direction is spontaneous, then the change in free energy of the reaction ($\Delta G$) must be less than zero, even though the value of $\Delta G^\circ$ is positive (4.8 kJ/mol). The value of $\Delta G$ *is* negative because the value of $RT \ln Q$ in Equation 18.14 has an increasingly larger negative value as $Q$ approaches zero. Actually, any reversible reaction, regardless of its $\Delta G^\circ$ value, has a negative $\Delta G$ value and is spontaneous when the system has only reactant and no (or practically no) product.

▶‖ **CHEMTOUR** Equilibrium and Thermodynamics

●● **CONNECTION** In Chapter 15 we defined the reaction quotient ($Q$) as the ratio of the concentrations (or partial pressures) of products to reactants in a chemical reaction, each term raised to a power equal to the coefficient of that substance in the balanced chemical equation describing the reaction.

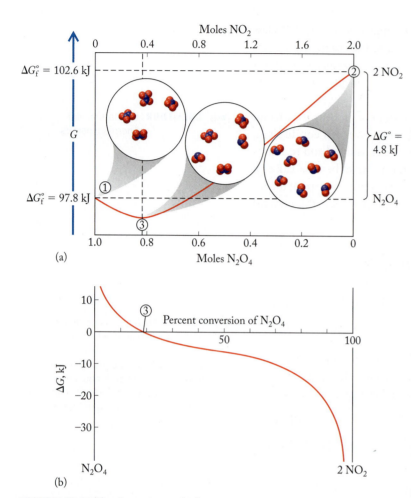

(a)

(b)

**FIGURE 18.14** The change in standard free energy $\Delta G^\circ_{rxn}$ is a constant for a given reaction. (a) Point 3, the point of minimum free energy, defines the composition of a system at equilibrium. At point 1, the system consists of reactants only; at point 2, the system consists of products only. (b) $\Delta G$ is the "distance" from equilibrium in terms of free energy. As a spontaneous reaction proceeds, the composition of the system changes and $\Delta G$ approaches 0 (point 3), at which point no further change in composition occurs.

The opposite situation occurs if we have 2 moles of $NO_2$ in the reaction vessel and essentially no $N_2O_4$. Under these conditions, the denominator of the reaction quotient is nearly zero, making the value of $Q_p$ enormous. Similarly, the $RT \ln Q_p$ term in Equation 18.14 now has a large positive value, guaranteeing that $\Delta G > 0$. This means that the forward reaction is not spontaneous, but the reverse reaction is. We also predict that $NO_2$ in the reaction vessel should combine to form $N_2O_4$, based on the fact that the enormous $Q_p$ must be greater than $K_p$. Therefore, the reaction should run in reverse until $Q_p = K_p$.

Figure 18.14(a) shows how free energy changes as the quantities of $N_2O_4$ and $NO_2$ in the reaction mixture change. The minimum of the curve (point 3) corresponds to a free energy that is lower than that of either pure $N_2O_4$ (point 1) or pure $NO_2$ (point 2). The vertical dashed line, marking the minimum in free energy, crosses the top and bottom axes at values that tell us the reaction mixture at equilibrium contains about 0.83 mol $N_2O_4$ and about 0.34 mol $NO_2$.

The curve in Figure 18.14(b) delivers the same message as in part a. In b, the $y$-axis represents how far $N_2O_4/NO_2$ reaction mixtures of different composition are from equilibrium. Clearly, "mixtures" that are either pure reactant or pure product are furthest from each other, as we discussed above. The reaction curve crosses the $x$-axis ($\Delta G = 0$) at point 3, which defines the composition of the reaction mixture at equilibrium. This value, expressed as the percentage of $N_2O_4$ that has dissociated into $NO_2$, corresponds to the same $N_2O_4/NO_2$ ratio as the minimum in the curve in Figure 18.14(a).

Once a reaction has reached equilibrium, $Q = K$, $\Delta G = 0$, and Equation 18.14 becomes

$$\Delta G_{rxn} = \Delta G^\circ_{rxn} + RT \ln K = 0$$

This can also be written as

$$\Delta G^\circ_{rxn} = -RT \ln K \qquad (18.15)$$

Rearranging Equation 18.15 allows us to calculate the $K$ value for a reaction from its change in standard free energy and absolute temperature. First, we rearrange the terms:

$$\ln K = \frac{-\Delta G^\circ_{rxn}}{RT} \qquad (18.16)$$

Then we take the antilogarithm of both sides:

$$K = e^{-\Delta G^\circ_{rxn}/RT} \qquad (18.17)$$

Equation 18.17 provides the following interpretation of reaction spontaneity under standard conditions: Whenever $\Delta G^\circ$ is negative, the exponent $-\Delta G^\circ/RT$ in Equation 18.17 is positive, and $e^{-\Delta G^\circ/RT} > 1$, making

$K > 1$. Therefore, any reversible reaction with an equilibrium constant greater than 1 is spontaneous under standard conditions, as shown in Figure 18.15(a). This spontaneity has its limits. As reactants are consumed and products are formed, the value of the reaction quotient increases, making the value of $\Delta G$ less negative. When it reaches zero, there is no further change in the composition of the reaction mixture because chemical equilibrium has been achieved.

It follows that a reversible reaction with a less negative value of $\Delta G°$ (Figure 18.15b) is still spontaneous but has a smaller equilibrium constant, so less reactant is consumed and less product is formed before the value of $\Delta G$ reaches 1. Finally, a reaction that has a positive value of $\Delta G°$ (Figure 18.15c) has an equilibrium constant that is less than zero, and is not spontaneous under standard conditions. Instead, the reverse of the reaction is spontaneous.

Let's apply this concept to the equilibrium between $N_2O_4$ and $NO_2$:

$$N_2O_4(g) \rightleftharpoons 2\,NO_2(g) \qquad \Delta G°_{rxn} = 4.8\ kJ$$

We focus on the composition of the reaction mixture of these two gases at equilibrium in Figure 18.14. We start by calculating the value of the exponent in Equation 18.17:

$$-\frac{\Delta G°_{rxn}}{RT} = -\frac{\left(\dfrac{4.8\ \cancel{kJ}}{\cancel{mol}}\right)\left(\dfrac{1000\ \cancel{J}}{1\ \cancel{kJ}}\right)}{\left(\dfrac{8.314\ \cancel{J}}{\cancel{mol}\cdot\cancel{K}}\right)(298\ \cancel{K})} = -1.94$$

Inserting this value into Equation 18.17 gives

$$K = e^{-\Delta G°_{rxn}/RT} = e^{-1.94} = 0.14$$

This result is consistent with the composition of the reaction mixture at equilibrium in Figure 18.14, where $P_{N_2O_4} \approx 0.83$ atm and $P_{NO_2} \approx 0.34$ atm. Inserting these values into the equilibrium constant expression for the reaction gives us an approximate value of $K_p$ that is close to the one we calculated from $\Delta G°_{rxn}$:

$$K_p = \frac{(P_{NO_2})^2}{P_{N_2O_4}} \approx \frac{(0.34)^2}{0.83} = 0.14$$

In this example, a positive $\Delta G°_{rxn}$ value of only few kilojoules corresponds to an equilibrium constant value that is less than one but still greater than 0.1. As a result, the equilibrium reaction mixture contains more reactant than product, but less than an order of magnitude more. Similarly, an equilibrium reaction mixture produced by a reaction with a negative $\Delta G°_{rxn}$ value of only few kilojoules is likely to contain more products than reactants, but with significant quantities of both.

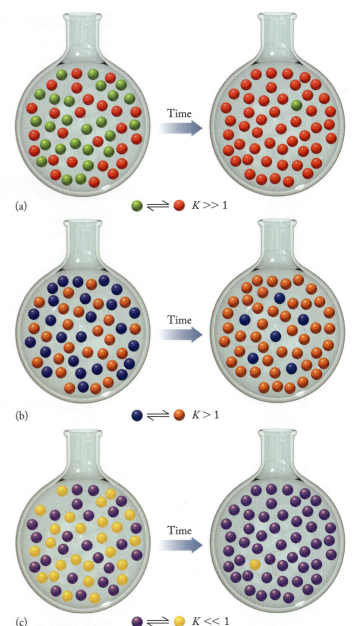

**FIGURE 18.15** The equilibrium constant of a chemical reaction is linked to its $\Delta G°$ value. The three reaction vessels on the left each contain equimolar mixtures of different pairs of gases. The initial partial pressure of each gas is 1 atm. (a) The value of $\Delta G°$ for the formation of "red" gas from "green" gas has a large negative value, which makes $K_p$ much greater than 1. The reaction proceeds in the forward direction, leaving only a little green gas left over. (b) If the value of $\Delta G°$ for the formation of "orange" gas from "blue" gas has a less negative value than in part a, the value of $K_p$ is smaller, though still greater than 1, and there is more orange gas present at equilibrium than blue gas. (c) If the value of $\Delta G°$ for the formation of "purple" gas from "yellow" gas is positive, $K_p$ is less than 1 and the reaction runs in the reverse direction, forming yellow gas from purple gas.

Suppose the $\Delta G°_{rxn}$ value of the hypothetical chemical reaction A $\rightleftharpoons$ B is −3.0 kJ/mol. Which of the following statements about an equilibrium mixture of A and B at 298 K is true?
a. There is only A present.
b. There is only B present.
c. There is an equimolar mixture of A and B present.
d. There is more A than B present.
e. There is more B than A present.

---

**SAMPLE EXERCISE 18.7** **Calculating the Value of K from ΔG°f** **LO8**

Use $\Delta G°_f$ values from Table A4.3 in the Appendix to calculate $\Delta G°_{rxn}$ and the value of $K_p$ for the formation of $NO_2$ from NO and $O_2$ at 298 K:

$$NO(g) + \tfrac{1}{2} O_2(g) \rightleftharpoons NO_2(g)$$

**Collect and Organize** We are to calculate the values of $\Delta G°_{rxn}$ and $K$ for a given reaction, starting with $\Delta G°_f$ values from Table A4.3, which are 51.3 kJ/mol for $NO_2$ and 86.6 kJ/mol for NO. Because $O_2$ gas is the most stable form of the element, its $\Delta G°_f$ value is 0.0 kJ/mol.

**Analyze** We can use Equation 18.12 to calculate $\Delta G°_{rxn}$ and then Equation 18.17 to calculate the value of $K$.

**Solve**

$$\Delta G°_{rxn} = [\Delta G°_f(NO_2)] - [\Delta G°_f(NO) + \tfrac{1}{2}\Delta G°_f(O_2)]$$

$$= [1 \text{ mol } (51.3 \text{ kJ/mol})] - [1 \text{ mol } (86.6 \text{ kJ/mol}) + \tfrac{1}{2} \text{ mol } (0.0 \text{ kJ/mol})]$$

$$= (51.3 - 86.6) \text{ kJ}$$

$$= -35.3 \text{ kJ or } -35,300 \text{ J per mol of } NO_2 \text{ produced}$$

The exponent in Equation 18.17 is

$$-\frac{\Delta G°_{rxn}}{RT} = -\frac{\left(\dfrac{-35,300 \text{ J}}{\text{mol}}\right)}{\left(\dfrac{8.314 \text{ J}}{\text{mol} \cdot \text{K}}\right)(298 \text{ K})} = 14.2$$

The corresponding value of $K_p$ is

$$K_p = e^{-\Delta G°_{rxn}/RT} = e^{14.2} = 2 \times 10^6$$

**Think About It** The exponential relationship between $\Delta G°_{rxn}$ and $K_p$ means that a moderately negative free-energy change of −35.3 kJ/mol corresponds to a very large value of $K_p$: in this case, greater than $10^6$.

⚙ **Practice Exercise** The standard free energy of formation of ammonia at 298 K is −16.5 kJ/mol. What is the value of $K$ for the reaction

$$N_2(g) + 3 H_2(g) \rightleftharpoons 2 NH_3(g)$$

at 298 K?

---

We have yet to explain which kind of equilibrium constant, $K_c$ or $K_p$, is related to $\Delta G°$ by Equation 18.17. The symbol $\Delta G°$ represents a change in free energy under standard conditions. The standard state of a gaseous reactant or product is one in which its *partial pressure* is 1 bar. Thus, the $\Delta G°$ of a reaction *in the gas phase* is linked by Equation 18.17 to its $K_p$ value. However, standard conditions for reactions in solution (the focus of Chapter 16) mean that all dissolved reactants and products are present at a concentration of 1.00 *M*. Thus, the $\Delta G°$ of a reaction *in solution* is related by Equation 18.17 to its $K_c$ value.

# 18.8 Influence of Temperature on Equilibrium Constants

We noted on numerous occasions in Chapter 15 that the value of $K$ changes with temperature. In this section we use the thermodynamics of chemical reactions to explain how and why.

Let's begin by combining a simplified version of Equation 18.11 (leaving out all the "rxn" subscripts):

$$\Delta G° = \Delta H° - T\Delta S°$$

and with a similarly simplified version of Equation 18.16:

$$\ln K = \frac{-\Delta G°}{RT}$$

The result is an equation that relates $K$ to $\Delta H°$ and $\Delta S°$:

$$\ln K = \frac{-\Delta G°}{RT} = -\frac{\Delta H°}{RT} + \frac{T\Delta S°}{RT}$$

$$= -\frac{\Delta H°}{RT} + \frac{\Delta S°}{R} \qquad (18.18)$$

Note how a negative value of $\Delta H°$ or a positive value of $\Delta S°$ contributes to a large value of $K$. These dependencies make sense because negative values of $\Delta H°$ and positive values of $\Delta S°$ are the two factors that contribute to making reactions spontaneous.

Because we are discussing the influence of temperature on $K$, let's identify the factors affected by changes in $T$ in Equation 18.18. The $\Delta S°$ term is not affected, but the influence of $\Delta H°$ does depend on temperature: the higher the temperature, the larger the denominator of the $\Delta H°$ term, and the smaller the influence of a favorable $\Delta H°$; that is, a $\Delta H°$ value that is negative. This temperature dependence makes sense based on Le Châtelier's principle and the notion that energy is a product of exothermic reactions and a reactant in endothermic reactions. Increasing temperature shifts a reaction toward the side opposite the energy term: it promotes endothermic reactions and inhibits exothermic reactions.

If $\Delta H°$ and $\Delta S°$ do not vary much with temperature, then Equation 18.18 predicts that $\ln K$ will be a linear function of $1/T$. Furthermore, we expect a graph of $\ln K$ versus $1/T$ to have a positive slope for an exothermic process and a negative slope for an endothermic process. We can determine $\Delta H°$ from the slope of the line and $\Delta S°$ from its $y$-intercept. Thus, we can calculate fundamental thermodynamic values of a reaction at equilibrium by determining its equilibrium constant at different temperatures.

For example, let's determine the values of $\Delta H°$ and $\Delta S°$ for the reaction

$$2\,CO_2(g) \rightleftharpoons 2\,CO(g) + O_2(g)$$

We start with the $K_p$ values of $2.75 \times 10^{-11}$ at 1500 K, $1.42 \times 10^{-3}$ at 2500 K, and 0.112 at 3000 K. The fact that $K_p$ increases as temperature increases tells us that energy is a reactant and the reaction is endothermic. We can use these data to determine the values of $\Delta H°$ and $\Delta S°$ by plotting $\ln K_p$ values versus $1/T$. To see how this graphical method works, let's rewrite Equation 18.18 so that it fits the form of the equation for a straight line ($y = mx + b$):

$$\ln K = -\frac{\Delta H°_{rxn}}{R}\left(\frac{1}{T}\right) + \frac{\Delta S°_{rxn}}{R} \qquad (18.19)$$

**CONNECTION** In Chapter 5 we defined exothermic reactions as those giving off heat; they have a negative enthalpy change ($\Delta H < 0$). Endothermic reactions absorb heat ($\Delta H > 0$).

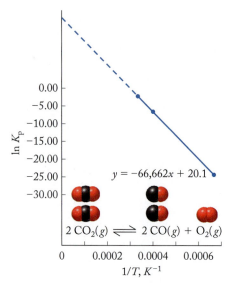

$y = -66,662x + 20.1$

$$2\ CO_2(g) \rightleftharpoons 2\ CO(g) + O_2(g)$$

**FIGURE 18.16** The graph of $\ln K_p$ versus $1/T$ is a straight line. We calculate $\Delta H^\circ_{rxn}$ from the slope and extrapolate the line to the y-intercept, from which we calculate $\Delta S^\circ_{rxn}$.

The graph of $\ln K_p$ versus $1/T$ (Figure 18.16) is indeed a straight line. The slope of this line is $-66,662$ K, which is equal to $-\Delta H^\circ_{rxn}/R$. The corresponding value of $\Delta H^\circ_{rxn}$ is

$$\Delta H^\circ_{rxn} = -\text{slope} \times R$$

$$= -(-66,662\ \text{K})\left(\frac{8.314\ \text{J}}{\text{mol} \cdot \text{K}}\right)$$

$$= 554,228\ \text{J/mol} = 554.2\ \text{kJ/mol}$$

The y-intercept ($\Delta S^\circ_{rxn}/R$) of the graph is 20.1. The corresponding value of $\Delta S^\circ_{rxn}$ is

$$\Delta S^\circ_{rxn} = (20.1)\left(\frac{8.314\ \text{J}}{\text{mol} \cdot \text{K}}\right) = 167\ \text{J/(mol} \cdot \text{K)}$$

This positive entropy change is logical because the forward reaction converts 2 moles of gaseous $CO_2$ into 3 moles of gaseous products.

We can use a modified version of Equation 18.19 to relate the values of $K$ at two different temperatures:

$$\ln\left(\frac{K_2}{K_1}\right) = -\frac{\Delta H^\circ_{rxn}}{R}\left(\frac{1}{T_2} - \frac{1}{T_1}\right) \tag{18.20}$$

Equation 18.20 is called the *van 't Hoff equation* because it was first derived by Jacobus van 't Hoff. It is particularly useful for calculating the value of $K$ at a very high or low temperature if we know what it is at a standard reference temperature (e.g., 298 K). We use such a calculation in the following Sample Exercise.

---

**SAMPLE EXERCISE 18.8**   **Calculating the Value of $K$ at a Specific Temperature**   **LO8, 9**

Use data from Appendix 4 to calculate the equilibrium constant $K_p$ for the exothermic reaction

$$N_2(g) + 3\ H_2(g) \rightleftharpoons 2\ NH_3(g)$$

at 298 K and at 773 K, a typical temperature used in the Haber–Bosch process for synthesizing ammonia.

**Collect and Organize** We are to calculate the $K_p$ value for a reaction at two temperatures. One of the temperatures is 298 K, which is the reference temperature for the standard thermodynamic data in Appendix 4. The $\Delta G^\circ_f$ value of $NH_3$ is $-16.5$ kJ/mol.

**Analyze** We can use Equation 18.17 to calculate the value of $K_p$ at 298 K from the value of $\Delta G^\circ_{rxn}$. Then we can use Equation 18.20 to calculate the value of $K_p$ at 773 K from the value of $K_p$ at 298 K and the value of $\Delta H^\circ_{rxn}$. The value of $\Delta H^\circ_{rxn}$ can be calculated from the enthalpy of formation of $NH_3$ ($-46.1$ kJ/mol). The negative value of $\Delta H^\circ_f$ means that $\Delta H^\circ_{rxn}$ is also negative, which means that the reaction is exothermic. Since heat is a product of the reaction, we can predict that the value of $K_p$ is smaller at 773 K than at 298 K.

**Solve** The value of $\Delta G^\circ_{rxn}$ for the reaction is

$$\frac{2\ \text{mol NH}_3}{\text{mol N}_2} \times \frac{-16.5\ \text{kJ}}{\text{mol NH}_3} \times \frac{1000\ \text{J}}{\text{kJ}} = -33,000\ \frac{\text{J}}{\text{mol N}_2} = -3.30 \times 10^4\ \frac{\text{J}}{\text{mol N}_2}$$

Using this value in Equation 18.17 to calculate the value of $K_p$ gives us

$$K_p = e^{-\Delta G^\circ_{rxn}/RT}$$

$$= e^{\left(\dfrac{-\left(-3.30 \times 10^4\ \dfrac{\text{J}}{\text{mol}}\right)}{8.314\ \dfrac{\text{J}}{\text{mol} \cdot \text{K}} \times 298\ \text{K}}\right)}$$

$$= 6.1 \times 10^5$$

Once we know the value of $K_p$ at 298 K, we can calculate $K_p$ at 773 K by using Equation 18.20. To do this, we must first calculate the value of $\Delta H^\circ_{rxn}$, which is twice the $\Delta H^\circ_f$ of $NH_3$:

$$\Delta H^\circ_{rxn} = \frac{2 \text{ mol NH}_3}{\text{mol N}_2} \times \frac{-46.1 \text{ kJ}}{\text{mol NH}_3} \times \frac{1000 \text{ J}}{\text{kJ}}$$

$$= -92{,}200 \frac{\text{J}}{\text{mol N}_2}, \text{ or } -92.2 \text{ kJ}$$

After substituting $K_1 = 6.1 \times 10^5$, $T_1 = 298$ K, and $T_2 = 773$ K into Equation 18.20, we solve for $K_2$:

$$\ln\left(\frac{K_2}{6.1 \times 10^5}\right) = -\frac{\left(-92{,}200 \dfrac{\text{J}}{\text{mol}}\right)}{8.314 \dfrac{\text{J}}{\text{mol} \cdot \text{K}}}\left(\frac{1}{773 \text{ K}} - \frac{1}{298 \text{ K}}\right)$$

$$\ln\left(\frac{K_2}{6.1 \times 10^5}\right) = -22.87$$

$$\frac{K_2}{6.1 \times 10^5} = e^{-22.87} = 1.2 \times 10^{-10}$$

$$K_2 = 7.3 \times 10^{-5} \text{ at 773 K}$$

**Think About It** The equilibrium constant decreases markedly when the temperature of the reaction is raised from 298 K to 773 K. This decrease fits our prediction for this exothermic reaction. Note that the standard thermodynamic data for the reaction ($\Delta H^\circ_{rxn}$ and $\Delta G^\circ_{rxn}$) are expressed per mole of the reactant with a coefficient of 1 in the balanced chemical equation for $N_2$.

**Practice Exercise** Use data from Appendix 4 to calculate the value of $K_p$ for the reaction

$$2 N_2(g) + O_2(g) \rightleftharpoons 2 N_2O(g)$$

at 298 K and 2000 K.

# 18.9 Driving the Human Engine: Coupled Reactions

The laws of thermodynamics that govern chemical reactions in the laboratory also govern all the chemical reactions that take place in living systems (Figure 18.17). Organisms carry out reactions that release the energy contained in the chemical bonds of food molecules and then use that energy to do work and sustain an array of other essential biological functions. Just like mechanical engines, however, humans and other life forms are far from 100% efficient, which means that life requires a continuous input of energy in terms of the caloric content of the food we eat. Thus, we humans must constantly absorb energy in the form of food and release heat and waste products into our surroundings. Young women have an average daily nutritional need of 2100 Cal; for young men the figure is 2900 Cal. This level of caloric intake provides the energy we need in order to function at all levels, from thinking to getting out of bed in the morning.

In living systems, spontaneous reactions ($\Delta G < 0$) typically involve metabolizing food, whereas nonspontaneous reactions ($\Delta G > 0$) involve building molecules needed by the body. Living systems use the energy from spontaneous reactions to run nonspontaneous reactions; we say that the spontaneous reactions are *coupled* to the nonspontaneous reactions. Part of the study of biochemistry involves deciphering the molecular mechanisms that enable reaction coupling. This topic is covered in Chapter 20, but for now it is sufficient to know that elegant molecular

**CONNECTION** In Chapter 5 we defined 1 Calorie, the "calorie" used in discussing food, as equivalent to 1 kcal.

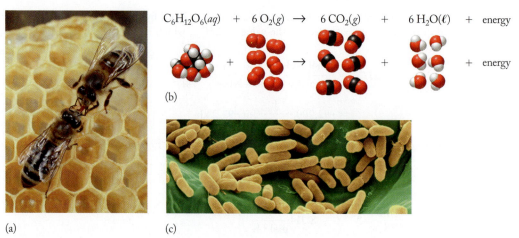

(a)                    (c)

**FIGURE 18.17** The rules of thermodynamics apply to all living systems. (a) Honeybees extract energy from nutrients to support life. The bees store this energy as honey (a mixture containing levulose and dextrose), which is then consumed by the bees, by humans, or by other animals to generate energy. (b) When honey is consumed, heat, water, and carbon dioxide are released, increasing the entropy of the universe. (c) Microscopic organisms—such as the *E. coli* that live in our gastrointestinal tracts—consume nutrients to live and generate heat in the process. Their expenditure of energy also increases the entropy of the universe.

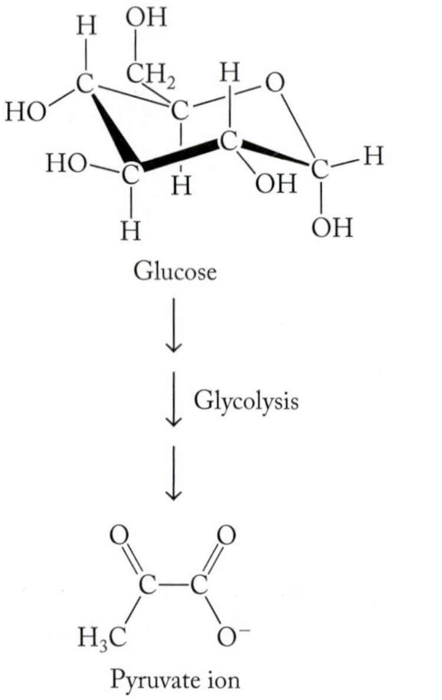

**FIGURE 18.18** In glycolysis, molecules of glucose are converted into twice their number of pyruvate ions.

⊙⊙ **CONNECTION** Photosynthesis and the carbon cycle were introduced in Chapter 3.

**glycolysis** a series of reactions that converts glucose into pyruvate; a major anaerobic (no oxygen required) pathway for the metabolism of glucose in the cells of almost all living organisms.

**phosphorylation** a reaction resulting in the addition of a phosphate group to an organic molecule.

processes have evolved to enable living systems to couple chemical reactions so that the energy obtained from spontaneous reactions can be used to drive the nonspontaneous reactions that maintain life. The metabolic chemical reactions we look at here are *not* presented for you to memorize. Rather, the intent is to aid your understanding of how changes in free energy allow spontaneous reactions to drive nonspontaneous reactions and to illustrate operationally what the phrase "coupled reactions" actually means.

As noted at the beginning of Chapter 5, all the energy contained in the food we eat has sunlight as its ultimate source. Green plants store energy from sunlight in their tissues as molecules such as glucose ($C_6H_{12}O_6$), which they produce from $CO_2$ and $H_2O$ during photosynthesis.

Production of glucose by green plants is a nonspontaneous process, which is why the plants require the energy of sunlight to carry out this reaction. Animals that consume plants use the energy stored in the chemical bonds of glucose and other molecules and release $CO_2$ and $H_2O$ back into the environment. This reaction,

$$C_6H_{12}O_6(s) + 6\ O_2(g) \rightarrow 6\ CO_2(g) + 6\ H_2O(\ell)$$

also releases energy, an event that increases the entropy of the universe, and is, therefore, spontaneous. Thus the processes of life increase the entropy of the universe by converting chemical energy into energy (heat) that flows into the surroundings.

Therefore the free-energy change for the breakdown of glucose must be less than zero, and is actually $\Delta G° = -880$ kJ/mol. This spontaneous process is highly controlled in living systems, however, so that the energy it produces can be directed into the nonspontaneous processes essential to organisms that do not carry out photosynthesis. Figure 18.18 summarizes one portion of glucose metabolism—**glycolysis**—that involves coupled reactions.

In glycolysis, each mole of glucose is converted into two moles of pyruvate ion ($CH_3COCOO^-$). An early step is the conversion of glucose into glucose 6-phosphate (Figure 18.19), an example of a **phosphorylation** reaction. Glucose reacts with the hydrogen phosphate ion ($HPO_4^{2-}$), producing glucose 6-phosphate

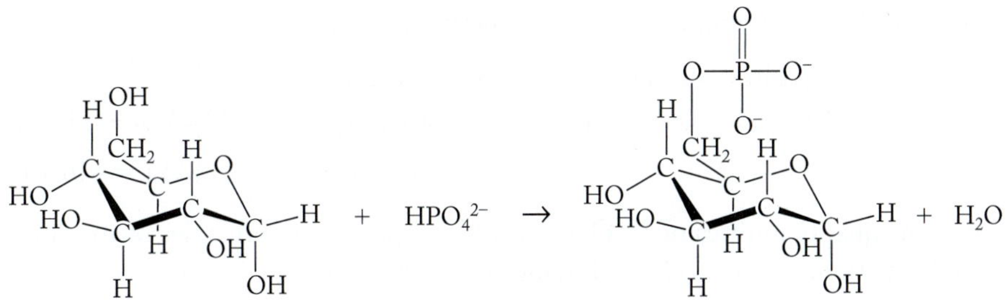

**FIGURE 18.19** The conversion of glucose into glucose 6-phosphate is an early step in glycolysis. This reaction is nonspontaneous, which means energy must be added to make the reaction go: $\Delta G^\circ_{rxn} = 13.8$ kJ/mol.

Glucose($aq$) + HPO$_4^{2-}$($aq$) → Glucose 6-phosphate($aq$) + H$_2$O($\ell$)

and water. This reaction is not spontaneous ($\Delta G^\circ = +13.8$ kJ/mol), and the energy needed to make it happen comes from a compound called adenosine triphosphate (ATP). ATP functions in our cells both as a storehouse of energy and as an energy-transfer agent: it hydrolyzes to adenosine diphosphate (ADP) in a reaction (Figure 18.20) that produces a hydrogen phosphate ion and energy: $\Delta G^\circ_{rxn} = -30.5$ kJ/mol.

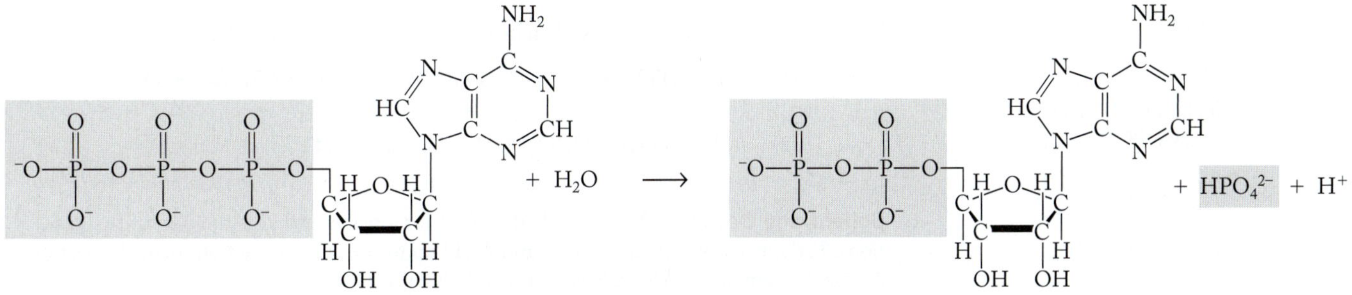

Adenosine triphosphate (ATP$^{4-}$)     Adenosine diphosphate (ADP$^{3-}$)

**FIGURE 18.20** The hydrolysis of ATP to ADP is a spontaneous reaction: $\Delta G^\circ = -30.5$ kJ/mol. The body couples this reaction to nonspontaneous reactions so that the energy released can drive them (see Figure 18.21).

In a living system, spontaneous hydrolysis of ATP consumes water and produces HPO$_4^{2-}$ and H$^+$, whereas the nonspontaneous phosphorylation of glucose consumes HPO$_4^{2-}$ and produces water. The two reactions are coupled: the spontaneous ATP → ADP reaction supplies the energy that drives the nonspontaneous formation of glucose 6-phosphate (Figure 18.21).

This example illustrates another important general point about reactions and $\Delta G^\circ$ values. The $\Delta G^\circ$ values for coupled reactions (and also for sequential reactions) are additive. This is true for any set of reactions, not just those occurring in living systems.

The first two steps in glycolysis are

(1)     ATP$^{4-}$($aq$) + H$_2$O($\ell$) → ADP$^{3-}$($aq$) + HPO$_4^{2-}$($aq$) + H$^+$($aq$)
$$\Delta G^\circ = -30.5 \text{ kJ}$$

(2)     C$_6$H$_{12}$O$_6$($aq$) + HPO$_4^{2-}$($aq$) → C$_6$H$_{11}$O$_6$PO$_3^{2-}$($aq$) + H$_2$O($\ell$)
    Glucose     Hydrogen     Glucose
            phosphate     6-phosphate
$$\Delta G^\circ = 13.8 \text{ kJ}$$

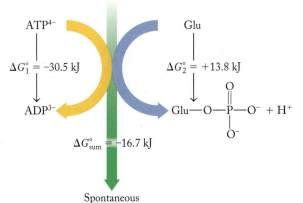

$\Delta G^\circ_1 = -30.5$ kJ    $\Delta G^\circ_2 = +13.8$ kJ

$\Delta G^\circ_{sum} = -16.7$ kJ

Spontaneous

**FIGURE 18.21** The spontaneous hydrolysis of ATP is coupled to the nonspontaneous phosphorylation of glucose. The overall reaction—the sum of the two individual reactions—is spontaneous.

If we add these reactions and their free energies, we get

$$\text{ATP}^{4-}(aq) + \cancel{\text{H}_2\text{O}(\ell)} + \text{C}_6\text{H}_{12}\text{O}_6(aq) + \cancel{\text{HPO}_4^{2-}(aq)} \rightarrow$$
$$\text{ADP}^{3-}(aq) + \cancel{\text{HPO}_4^{2-}(aq)} + \text{C}_6\text{H}_{11}\text{O}_6\text{PO}_3^{2-}(aq) + \cancel{\text{H}_2\text{O}(\ell)} + \text{H}^+(aq)$$
$$\Delta G° = (-30.5 + 13.8)\text{ kJ} = -16.7\text{ kJ}$$

Because equal quantities of $\text{H}_2\text{O}$ and $\text{HPO}_4^{2-}$ appear on both sides of the combined equation, they cancel out, leaving the net reaction

$$\text{C}_6\text{H}_{12}\text{O}_6(aq) + \text{ATP}^{4-}(aq) \rightarrow \text{ADP}^{3-}(aq) + \text{C}_6\text{H}_{11}\text{O}_6\text{PO}_3^{2-}(aq) + \text{H}^+(aq)$$
$$\Delta G° = -16.7\text{ kJ}$$

Since $\Delta G°$ for the net reaction is negative, the reaction is spontaneous.

**CONNECTION** In Chapter 5 we used Hess's law (the enthalpy change of a reaction that is the sum of two or more reactions equals the sum of the enthalpy changes of the constituent reactions) to calculate $\Delta H$. We apply a similar principle here when adding $\Delta G°$ values of coupled reactions.

---

**SAMPLE EXERCISE 18.9    Calculating $\Delta G°$ of Coupled Reactions    LO10**

The body would rapidly run out of ATP if there were not some process for regenerating ATP from ADP. That process is the hydrolysis of 1,3-diphosphoglycerate$^{4-}$ (1,3-DPG$^{4-}$) to 3-phosphoglycerate$^{3-}$ (3-PG$^{3-}$; Figure 18.22):

$$\text{ADP}^{3-} + \text{1,3-diphosphoglycerate}^{4-} \rightarrow \text{3-phosphoglycerate}^{3-} + \text{ATP}^{4-}$$

This hydrolysis is spontaneous. Calculate its $\Delta G°$ value from these values:

(1)    $\text{1,3-DPG}^{4-}(aq) + \text{H}_2\text{O}(\ell) \rightarrow \text{3-PG}^{3-}(aq) + \text{HPO}_4^{2-}(aq) + \text{H}^+(aq)$
$$\Delta G° = -49.0\text{ kJ}$$

(2)    $\text{ADP}^{3-}(aq) + \text{HPO}_4^{2-}(aq) + \text{H}^+(aq) \rightarrow \text{ATP}^{4-}(aq) + \text{H}_2\text{O}(\ell)$
$$\Delta G° = 30.5\text{ kJ}$$

**Collect and Organize** We can calculate $\Delta G°$ for a reaction that is the sum of two reactions. If the reactions in equations 1 and 2 add up to the overall reaction, then the overall $\Delta G°$ is the sum of the $\Delta G°$ values for the individual reactions.

**Analyze** First, we add the reactions described by equations 1 and 2. Assuming that the overall reaction between ADP and 1,3-diphosphoglycerate$^{4-}$ is the sum of the reactions describing the hydrolysis of 1,3-diphosphoglycerate$^{4-}$ and the phosphorylation of ADP, we know that the sum of $\Delta G_1°$ and $\Delta G_2°$ will be less than zero because the overall reaction is spontaneous.

**Solve** We add the reactions in equations 1 and 2 and confirm that they equal the overall reaction:

(1)    $\text{1,3-DPG}^{4-}(aq) + \text{H}_2\text{O}(\ell) \rightarrow \text{3-PG}^{3-}(aq) + \text{HPO}_4^{2-}(aq) + \text{H}^+(aq)$
(2)    $\text{ADP}^{3-}(aq) + \text{HPO}_4^{2-}(aq) + \text{H}^+(aq) \rightarrow \text{ATP}^{4-}(aq) + \text{H}_2\text{O}(\ell)$

---

$$\text{1,3-DPG}^{4-}(aq) + \cancel{\text{H}_2\text{O}(\ell)} + \text{ADP}^{3-}(aq) + \cancel{\text{HPO}_4^{2-}(aq)} + \cancel{\text{H}^+(aq)} \rightarrow$$
$$\text{3-PG}^{3-}(aq) + \cancel{\text{HPO}_4^{2-}(aq)} + \cancel{\text{H}^+(aq)} + \text{ATP}^{4-}(aq) + \cancel{\text{H}_2\text{O}(\ell)}$$

We sum the $\Delta G°$ values for steps 1 and 2 to determine $\Delta G°$ for the overall reaction:

$$\Delta G°_{\text{overall}} = \Delta G_1° + \Delta G_2° = [(-49.0) + (30.5)]\text{ kJ} = -18.5\text{ kJ}$$

**Think About It** The hydrolysis of 1,3-diphosphoglycerate$^{4-}$ provides more than sufficient energy for the conversion of ADP into ATP.

**Practice Exercise** The conversion of glucose into lactic acid drives the phosphorylation of 2 moles of ADP to ATP:

$$\text{C}_6\text{H}_{12}\text{O}_6(aq) + 2\text{ HPO}_4^{2-}(aq) + 2\text{ ADP}^{3-}(aq) + 2\text{ H}^+(aq) \rightarrow$$
Glucose
$$2\text{ CH}_3\text{CH(OH)COOH}(aq) + 2\text{ ATP}^{4-}(aq) + 2\text{ H}_2\text{O}(\ell)$$
Lactic acid
$$\Delta G° = -135\text{ kJ/mol}$$

What is $\Delta G°$ for the conversion of glucose into lactic acid?

$$\text{C}_6\text{H}_{12}\text{O}_6(aq) \rightarrow 2\text{ CH}_3\text{CH(OH)COOH}(aq)$$

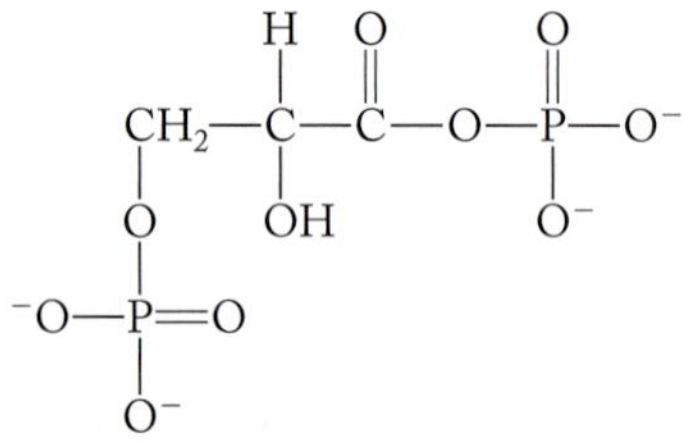

1,3-Diphosphoglycerate$^{4-}$
(1,3-DPG$^{4-}$)

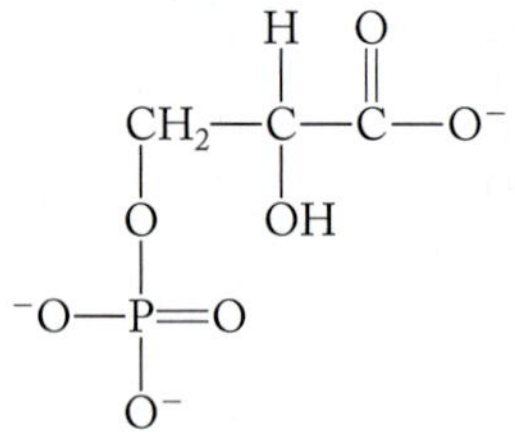

3-Phosphoglycerate$^{3-}$
(3-PG$^{3-}$)

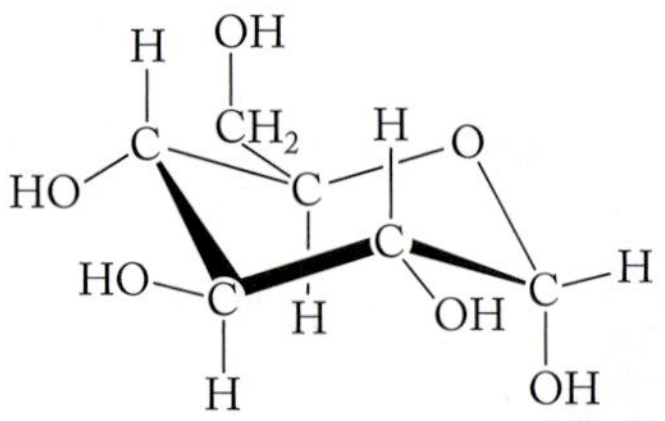

Glucose

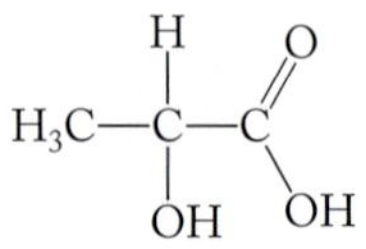

Lactic acid

**FIGURE 18.22** Structures of 1,3-diphosphoglycerate, 3-phosphoglycerate, glucose, and lactic acid.

The ATP produced from the breakdown of glucose (as, for example, via the first reaction in the preceding Practice Exercise) is used to drive nonspontaneous reactions in cells. The metabolism of fats and proteins relies on a series of cycles, all of which involve coupled reactions. The ATP–ADP system is a carrier of chemical energy because ADP requires energy to accept a phosphate group and thus is coupled to reactions that yield energy, whereas ATP donates a phosphate group, releases energy, and is coupled to reactions that require energy. All energy changes in living systems are governed by the first and second laws of thermodynamics, as are all the energy changes in the inanimate world.

## 18.10 Microstates: A Quantized View of Entropy

In this chapter we have used a thermodynamic definition of entropy, describing it as a measure of how energy is distributed throughout a system: the more spread out energy is, the greater the system's entropy is. This thermodynamic view evolved in the middle of the 19th century, largely through the work of German physicist Rudolph Clausius (1822–1888), who was a leader in establishing the field of thermodynamics and who introduced the concept of entropy in 1865.

Clausius' thermodynamic approach views entropy from a macroscopic, system-wide perspective. In much the same way as we use such parameters as pressure, volume, and temperature to describe a quantity of a gaseous substance, change in a system's entropy is defined in terms of the quantity of energy flowing out from or into the system and its temperature (Equation 18.2). However, just as the kinetic molecular theory of gases provides a particle-based explanation of why gases exhibit the macroscopic properties they do, we can also understand entropy changes from the viewpoint of the dispersion of particles and their energies.

In 1877 another German physicist, Ludwig Boltzmann (1844–1906), developed a microscopic view of Clausius' concept of entropy. His theory would eventually incorporate quantum mechanics to explain how energy was dispersed not only throughout a system, but within its molecules. Let's examine Boltzmann's microscopic definition of entropy using two models: a single molecule of $O_2$ and a room full of them.

As we have discussed in this chapter and as shown in Figure 18.23, a molecule of a gas such as $O_2$ is free to undergo three types of motion: (1) *translational motion* as it zips around the space it occupies; (2) *rotational motion* as it spins about imaginary axes perpendicular to the O==O bond, and (3) *vibrational motion* as its bonded atoms move toward and away from each other, like balls on the ends of a spring. All three modes of motion have one thing in common: the greater the thermal energy of the molecule, the greater each type of motion.

In our discussion of quantum mechanics in Chapter 7, we saw that energy is not continuous on the atomic scale. Instead, the energies and motion of atoms and molecules are quantized. At temperatures near room temperature, the differences between the translational energy levels of atoms and molecules in the gas phase are so small that we may consider them a continuum of energy. However, the gaps in vibrational and rotational energy levels are large enough that we must take quantization into account.

To investigate these gaps, let's revisit the change in electrostatic potential energy that occurs when two oxygen atoms approach each other and a covalent bond forms between them (see Figure 8.1). Energy reaches a minimum (Figure 18.24a) when the nuclei of the two atoms are a distance apart that corresponds to the length of the O==O bond, or 121 pm. Figure 18.24(a) also contains additional energy levels represented by horizontal red lines. These are vibrational energy levels. Note that

(a) Translational motion

(b) Rotational motion

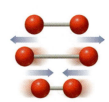
(c) Vibrational motion

**FIGURE 18.23** A diatomic molecule has three fundamental types of motion: (a) translational motion, (b) rotational motion, and (c) vibrational motion.

**FIGURE 18.24** Vibrational and rotational energy levels in a molecule of $O_2$ are quantized. (a) The electrostatic potential energy between two oxygen atoms reaches a minimum when their nuclei are 121 pm apart—the length of an O$=$O bond. (b) Quantized rotational energy levels (represented by the blue horizontal lines) are superimposed on each vibrational energy level, represented by a red horizontal line in part a. (Not all the accessible rotational and vibrational states are shown, and the gaps between them are not to scale.)

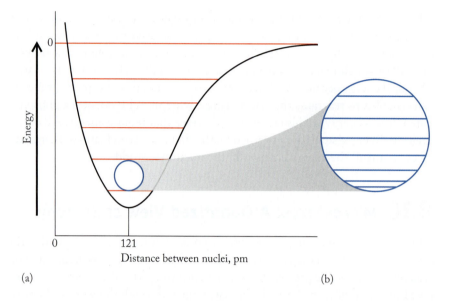

the red lines are longer with greater vibrational energy. Greater length corresponds to greater variation in intermolecular distance; that is, more energetic oscillations of the O$=$O bond occur with increasing vibrational energy.

Superimposed on each vibrational energy level is a set of rotation energy levels, shown for the first vibrational energy level in Figure 18.24(b). Similar sets of rotational energy levels are associated with all the other vibrational energy levels, creating a multitude of energy states that are accessible to every $O_2$ molecule at room temperature.

Now let's focus on one oxygen molecule as it collides with others, sometimes gaining, other times losing, translational kinetic energy. At room temperature and pressure, a single $O_2$ molecule experiences billions of collisions per second, each collision altering its speed and also changing its vibrational and rotational energies. At any instant, the overall, quantized energy of the molecule is a particular combination of translational, vibrational, and rotational energy states. The number of possible combinations that an $O_2$ molecule can have is enormous at room temperature and the number goes up as temperature goes up. In a room full of $O_2$ molecules (our second model), the number of different quantized total-molecule energy states is so large ($e$ raised to a power that is many times Avogadro's number) that it is beyond the capacity of your calculator to compute.

The entirety of all the quantized energy states occupied by all of the atoms in the room at a particular instant is called a **microstate**. Each microstate represents one discrete way for the system to disperse all the energy in the system. Thus, each microstate represents a particular way to distribute the same quantity of energy. The vast number of microstates to which a system has access defines its entropy: the more microstates, the higher its entropy. This linkage between number of accessible microstates ($W$) and entropy ($S$) is defined by a simple mathematical equation developed by Boltzmann:

$$S = k_B \ln W \tag{18.21}$$

Here $k_B$ is the Boltzmann constant, which is equal to the gas constant divided by Avogadro's number:

**microstate** a unique distribution of particles among energy levels.

$$k_B = \frac{R}{N_A} = \frac{8.314 \ \frac{J}{\cancel{mol} \cdot K}}{6.0221 \times 10^{23}/\cancel{mol}} = 1.381 \times 10^{-23} \ J/K$$

Boltzmann's definition of entropy, based on access to energy microstates, is conceptually compatible with the thermodynamic, macroscopic view, based on energy dispersion. After all, each microstate represents one of the multitude of ways that energy can be dispersed in a thermodynamic system. In fact, some facets of entropy are easier to understand using the microstate model. One of them is the concept of absolute entropy, the notion embedded in the third law of thermodynamics that a perfect crystalline solid has zero entropy at a temperature of absolute zero. If the particles of a crystalline solid are perfectly aligned at 0 K, only one distribution of the particles is possible in the space occupied by the crystal. Therefore, the crystal has only one microstate, and, according to Equation 18.21, its entropy is zero:

$$S = k_B (\ln W) = k_B (\ln 1) = 0$$

In this chapter, we addressed the reasons why some reactions and processes are spontaneous while others are not. We have seen that the value of free-energy change ($\Delta G$) determines the spontaneity of a chemical reaction or process. Together, enthalpy and entropy changes for chemical reactions allow us to determine the $\Delta G$ for a chemical reaction or process. The free-energy change represents the maximum amount of work that can be done by the energy associated with a change. For changes carried out in the real world, the maximum amount of work is always less than $\Delta G$.

**CONCEPT TEST** ...........................................................................

Predict which of the following values comes closest to the number of microstates accessible to one mole of liquid water molecules at 25°C: (a) 1; (b) $10^2$; (c) $10^{10}$; (d) $10^{23}$; (e) $\gg 10^{23}$. Use the appropriate $S°$ value in Appendix 4 and Equation 18.21 to determine whether your selection was the best one.

........................................................................................

**SAMPLE EXERCISE 18.10**   **Integrating Concepts: Otzi's Axe and the Chemistry of Copper Refining**

In 1991, two hikers discovered the remains of a caveman frozen in a glacier high in the Ötztal Alps near the Italy/Austria border. Otzi the Iceman, as he came to be known, lived about 5300 years ago. Among his possessions was an axe with a head made of nearly pure (99.7%) copper (Figure 18.25). The purity of the copper, as well as other clues, told scientists that the copper had been skillfully extracted and refined. Though copper does occur in its native elemental state in Earth's crust, ancient craftsmen usually had to extract the metal from copper-containing minerals such as chalcocite ($Cu_2S$). The process required that they find a way to add a lot of energy to a sulfide mineral like $Cu_2S$ because its decomposition reaction has a large, positive $\Delta G°$ value:

$$Cu_2S(s) \rightarrow 2\,Cu(s) + S(s) \qquad \Delta G° = 86.2 \text{ kJ}$$

However, when $Cu_2S$ is roasted in a furnace that provides an ample supply of very hot air, any sulfur produced by the decomposition reaction would be oxidized to $SO_2$ gas in a decidedly spontaneous reaction:

$$S(s) + O_2(g) \rightarrow SO_2(g) \qquad \Delta G° = -300.1 \text{ kJ}$$

a. Can these two reactions be coupled so that the spontaneous second one drives the nonspontaneous first reaction?
b. Is there likely to be an increase in entropy during the overall reaction?

**FIGURE 18.25** Copper axe found alongside Otzi the Iceman, who lived about 5300 years ago.

c. What is the equilibrium constant of the first reaction under standard conditions?
d. How is the equilibrium shifted by the reaction conditions in the furnace?

**Collect and Organize** We are given two chemical reactions and asked to determine whether the free energy released by the spontaneous one can be coupled to and drive the nonspontaneous reaction. We are also to calculate the equilibrium constant of the nonspontaneous reaction and decide how reaction conditions shift the equilibrium state. Equation 18.17 relates the equilibrium constant of a chemical reaction to its $\Delta G°_{rxn}$ value.

**Analyze** Sulfur is a product of the first reaction and a reactant in the second one, so it should be possible to couple the two reactions.

Given the large positive value of $\Delta G^{\circ}_{rxn}$ for the first reaction, its $K$ value should be much less than 1. The second reaction consumes a product (sulfur) of the first reaction, which should shift the equilibrium in the first reaction toward forming more products.

## Solve

a. One mole of sulfur is produced in the first reaction and consumed in the second, so the two reactions can be coupled. The result is an overall reaction with a $\Delta G^{\circ}_{rxn}$ value that is the sum of the first two:

$$Cu_2S(s) \rightarrow 2\,Cu(s) + \cancel{S(s)} \qquad \Delta G^{\circ}_{rxn} = 86.2\ kJ$$
$$+\ \cancel{S(s)} + O_2(g) \rightarrow SO_2(g) \qquad \Delta G^{\circ}_{rxn} = -300.1\ kJ$$
$$\overline{Cu_2S(s) + O_2(g) \rightarrow 2\,Cu(s) + SO_2(g) \qquad \Delta G^{\circ}_{rxn} = -213.9\ kJ}$$

The calculated negative $\Delta G^{\circ}_{rxn}$ value means that the overall reaction is spontaneous.

b. The reactants and products both contain one mole of gas, but $SO_2$ is triatomic and has a larger standard molar entropy (248.4 J/mol · K) than diatomic $O_2$ (205.0 J/mol · K). Also, the other reactant is one mole of a binary ionic solid, but the products include two moles of solid metal. Therefore, based on the $S^{\circ}$ values in Appendix 4, we can safely predict that the system entropy increases in the overall reaction.

c. Using Equation 18.17 to calculate the equilibrium constant for the decomposition reaction under standard conditions (and letting $T = 298\ K$) gives us:

$$K = e^{-\Delta G^{\circ}_{rxn}/RT}$$

First calculate the exponent:

$$-\frac{\Delta G^{\circ}_{rxn}}{RT} = -\frac{86.2\ \dfrac{\cancel{kJ}}{\cancel{mol}} \times 1000\ \dfrac{J}{\cancel{kJ}}}{\left(8.314\ \dfrac{J}{\cancel{mol} \cdot \cancel{K}}\right)(298\ \cancel{K})} = -34.79$$

Then the value of $K$ is

$$K = e^{-34.79} = 7.8 \times 10^{-16}$$

d. As the reaction mixture in the first reaction comes to equilibrium (after forming an insignificant quantity of product), removal of sulfur as $SO_2$ shifts the equilibrium in favor of forming more product, as predicted by Le Châtelier's principle.

**Think About It** As predicted, the nonspontaneous first reaction has a small $K$ value; however, the reaction becomes spontaneous when coupled to the spontaneous second reaction. This shift toward forming more Cu metal is also explained by Le Châtelier's principle and the capacity of the second reaction to remove one of the products (sulfur) of the first.

## SUMMARY

**Learning Outcome 1** **Spontaneous processes** happen on their own without continuous, outside intervention. **Nonspontaneous processes**, which are spontaneous processes in reverse, do not happen on their own. Spontaneous processes may be exothermic or endothermic and are often accompanied by an increase in the freedom of motion of the particles involved in the process. (Section 18.1)

**Learning Outcome 2** **Entropy (S)** is a thermodynamic property that provides a measure of how dispersed the energy is in a system at a given temperature. According to the **second law of thermodynamics**, a spontaneous process is accompanied by an increase in the entropy of an isolated system, or an increase in entropy of the universe for a process occurring in any system.

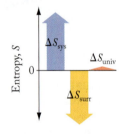

A **reversible process** takes place in very small steps and very slowly so that the system can be restored to its initial state with no net flow of energy to or from its surroundings. (Section 18.2)

**Learning Outcome 3** According to the **third law of thermodynamics**, a perfect crystal of a pure substance has zero entropy at absolute zero. All substances have positive entropies at temperatures above absolute zero. **Standard molar entropies (S°)** are entropy values for substances in their standard states. The entropy of a system increases with increasing molecular complexity and with increasing temperature. (Section 18.3)

**Learning Outcome 4** The entropy change in a reaction under standard conditions can be calculated from the standard entropies of the products and reactants and their coefficients in the balanced chemical equation. (Section 18.4)

**Learning Outcome 5** **Free energy (G)** is defined as the energy available to do useful work. The change in free energy of a process is a state function defining the maximum useful work the system can do on its surroundings. When a process results in a decrease in free energy of a system ($\Delta G < 0$), the process is spontaneous; when $\Delta G > 0$ the process is nonspontaneous. Reversing a process changes the sign of $\Delta G$, and $\Delta G = 0$ for a process at equilibrium. The change in free energy of a reaction under standard conditions can be calculated either from the **standard free energies of formation ($\Delta G^{\circ}_f$)** of the products and reactants or from the enthalpy and entropy changes. (Section 18.5)

**Learning Outcome 6** The temperature range over which a process is spontaneous depends on the relative magnitudes of $\Delta H$ and $\Delta S$. (Section 18.6)

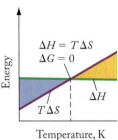

**Learning Outcome 7** Negative values of $\Delta G°$ correspond to $K > 1$ and equilibrium reaction mixtures composed mostly of products. Positive values of $\Delta G°$ correspond to $K < 1$. As spontaneous reactions proceed, the free energy of the reaction mixture increases from an initial negative value and reaches zero when the reaction comes to chemical equilibrium. (Section 18.7)

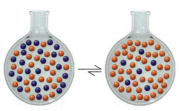

**Learning Outcome 8** Higher reaction temperatures increase the equilibrium constant of an endothermic reaction but decrease the equilibrium constant of an exothermic reaction. The slope of a plot of ln $K$ versus $1/T$ for an equilibrium system is used to determine the standard enthalpy for the equilibrium, and the $y$-intercept of the plot is used to determine the standard entropy change. (Section 18.8)

**Learning Outcome 9** Many important biochemical processes, including **glycolysis** and **phosphorylation**, are made possible by coupled spontaneous and nonspontaneous reactions. The free energy released in the spontaneous processes going on in the body is used to drive nonspontaneous processes. (Section 18.9)

## PROBLEM-SOLVING SUMMARY

| TYPE OF PROBLEM | CONCEPTS AND EQUATIONS | SAMPLE EXERCISES |
|---|---|---|
| **Predicting the sign of entropy change** | Look for the net removal of gas molecules or precipitation of a solute from solution ($\Delta S < 0$ for both). For the reverse processes, $\Delta S > 0$. | 18.1 |
| **Comparing absolute entropy values** | Substances composed of larger, less rigid molecules have more entropy. Among substances with similar molar masses, gases have more entropy than liquids, which have more entropy than solids. | 18.2 |
| **Calculating the entropy change of a chemical reaction** | Use $$\Delta S°_{rxn} = \sum n_{products} S°_{products} - \sum n_{reactants} S°_{reactants} \quad (18.5)$$ where $S°_{products}$ and $S°_{reactants}$ are the standard molar entropies and $n_{products}$ and $n_{reactants}$ are the stoichiometric coefficients for the process. | 18.3 |
| **Predicting reaction spontaneity under standard conditions** | A reaction is spontaneous if $$\Delta S_{univ} = (\Delta S_{sys} + \Delta S_{surr}) > 0$$ | 18.4 |
| **Calculating $\Delta G°_{rxn}$ using appropriate $\Delta G°_f$ values** | Use $$\Delta G°_{rxn} = \sum n_{products} \Delta G°_{f,products} - \sum n_{reactants} \Delta G°_{f,reactants} \quad (18.12)$$ where $\Delta G°_{f,products}$ and $\Delta G°_{f,reactants}$ are the standard molar free energies of formation, and $n_{products}$ and $n_{reactants}$ are the stoichiometric coefficients for the process. | 18.5 |
| **Relating reaction spontaneity to $\Delta H$ and $\Delta S$** | Use $$\Delta G°_{rxn} = \Delta H°_{rxn} - T\Delta S°_{rxn} \quad (18.11)$$ | 18.6 |
| **Calculating the value of $K$ from $\Delta G°_{rxn}$** | $$K = e^{-\Delta G°_{rxn}/RT} \quad (18.17)$$ | 18.7 |
| **Calculating the value of $K$ at a specific temperature** | Use $$\ln\left(\frac{K_2}{K_1}\right) = -\frac{\Delta H°_{rxn}}{R}\left(\frac{1}{T_2} - \frac{1}{T_1}\right) \quad (18.20)$$ Convert $\Delta H°_{rxn}$ from kilojoules per mole to joules per mole to match the units of $R$. | 18.8 |
| **Calculating $\Delta G°$ of coupled reactions** | Free-energy changes are additive. If adding two reactions gives the desired overall reaction, add the free-energy changes of the two reactions to obtain the free-energy change of the overall reaction. | 18.9 |

## VISUAL PROBLEMS

*(Answers to boldface end-of-chapter questions and problems are in the back of the book.)*

**18.1.** Two balloons are inflated at the same temperature to the same volume (Figure P18.1), though it takes more gas to inflate the balloon on the right. In which balloon is the gas under greater internal pressure and in which does the gas have greater entropy?

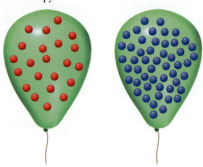

**FIGURE P18.1**

**18.2.** Two cubic containers (Figure P18.2) contain the same quantity of gas at the same temperature. Which cube contains gas with more entropy? If the sample in part b is left unchanged but the sample in part a is cooled so that it condenses, which sample has the higher entropy?

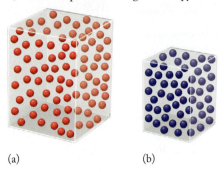

(a)                          (b)

**FIGURE P18.2**

**18.3.** Figure P18.3 shows two connected bulbs that have just been filled with a mixture of two ideal gases: A (red spheres) and B (blue spheres). If the molar mass of A is twice that of B, will the atoms of A eventually fill the bottom bulb and the atoms of B fill the top bulb? Why or why not? Do the bulbs represent an isolated system?

**FIGURE P18.3**

**18.4.** The box on the left represents a mixture of two diatomic gases: $A_2$ (red spheres) and $B_2$ (blue spheres). As a result of the process depicted by the arrow, how do the entropies of $A_2$ and $B_2$ change?

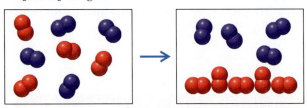

**FIGURE P18.4**

**18.5.** Is the process in Figure P18.4 more likely to be spontaneous at high temperature or low temperature, or is it unaffected by changing temperature?

**18.6.** Figure P18.6 shows the plots of $\Delta H$ and $T\,\Delta S$ for a phase change as a function of temperature.
   a. What is the status of the process at the point where the two lines intersect?
   b. Over what temperature range is the process spontaneous?

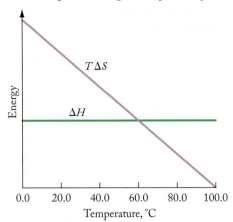

**FIGURE P18.6**

**18.7.** Of the six phase changes (melting, vaporization, condensation, freezing, sublimation, and deposition) which ones have thermodynamic profiles that fit the pattern in Figure P18.6?

18.8. Figure P18.8 presents the $\Delta G_f^\circ$ values of several elements and compounds selected from Appendix 4. Which of the following conversions are spontaneous? (a) $C_6H_6(\ell)$ to $CO_2(g)$ and $H_2O(\ell)$; (b) $CO_2(g)$ to $C_2H_2(g)$; (c) $H_2(g)$ and $O_2(g)$ to $H_2O(\ell)$. Explain your reasoning.

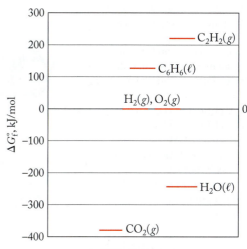

**FIGURE P18.8**

## QUESTIONS AND PROBLEMS

### Spontaneous Processes and Thermodynamic Entropy

#### CONCEPT REVIEW

**18.9.** How is the entropy change that accompanies a reaction related to the entropy change that happens when the reaction runs in reverse?

**18.10.** Identify the following systems as isolated or not isolated, identify the processes as spontaneous or nonspontaneous, and explain your choice.
   a. A photovoltaic cell in a solar panel produces electricity.
   b. Helium gas escapes from a latex party balloon.
   c. A sample of pitchblende (uranium ore) emits alpha particles.

**18.11.** Ice cubes melt in a glass of lemonade, cooling the lemonade from 10.0°C to 0.0°C. If the ice cubes are the system, what are the signs of $\Delta S_{sys}$ and $\Delta S_{surr}$?

**18.12.** Adding sidewalk deicer (calcium chloride) to water causes the temperature of the water to increase. If solid $CaCl_2$ is the system, what are the signs of $\Delta S_{sys}$ and $\Delta S_{surr}$?

#### PROBLEMS

**18.13.** Which of the following combinations of entropy changes for a process are mathematically possible?
   a. $\Delta S_{sys} > 0$, $\Delta S_{surr} > 0$, $\Delta S_{univ} > 0$
   b. $\Delta S_{sys} > 0$, $\Delta S_{surr} < 0$, $\Delta S_{univ} > 0$
   c. $\Delta S_{sys} > 0$, $\Delta S_{surr} > 0$, $\Delta S_{univ} < 0$

**18.14.** Which of the following combinations of entropy changes for a process are mathematically possible?
   a. $\Delta S_{sys} < 0$, $\Delta S_{surr} > 0$, $\Delta S_{univ} > 0$
   b. $\Delta S_{sys} < 0$, $\Delta S_{surr} < 0$, $\Delta S_{univ} > 0$
   c. $\Delta S_{sys} < 0$, $\Delta S_{surr} > 0$, $\Delta S_{univ} < 0$

**18.15.** What are the signs of $\Delta S_{sys}$ and $\Delta S_{univ}$ for the photosynthesis of glucose from carbon dioxide and water?

**18.16.** What are the signs of $\Delta S_{sys}$, $\Delta S_{surr}$, and $\Delta S_{univ}$ for the complete combustion of propane in which the products include water vapor and carbon dioxide?

**18.17.** The nonspontaneous reaction D + E → F decreases the system entropy by 66.0 J/K. What is the maximum value of the entropy change of the surroundings?

**18.18.** The spontaneous reaction A → B + C increases the system entropy by 72.0 J/K. What is the minimum value of the entropy change of the surroundings?

**18.19.** Which of the following ionic solutes experiences the greatest increase in entropy when 0.0100 moles of it dissolves in 1.00 liter of water? (a) $CaCl_2$; (b) NaBr; (c) KCl; (d) $Cr(NO_3)_3$; (e) LiOH

**18.20.** Which of the following molecular solutes experiences an increase in entropy when it dissolves in water? (a) $CO_2(g)$; (b) $HF(g)$; (c) $CH_3OH(\ell)$; (d) $CH_3COOH(\ell)$; (e) $C_{12}H_{22}O_{11}(s)$

### Absolute Entropy and the Third Law of Thermodynamics

#### CONCEPT REVIEW

**18.21.** Which component in each of the following pairs has the greater entropy?
   a. 1 mole of $S_2(g)$ or 1 mole of $S_8(g)$
   b. 1 mole of $S_2(g)$ or 1 mole of $S_8(s)$
   c. 1 mole of $O_2(g)$ or 1 mole of $O_3(g)$
   d. 1 gram of $O_2(g)$ or 1 gram of $O_3(g)$

**18.22.** **Digestion** During digestion, complex carbohydrates decompose into simple sugars. Do the carbohydrates experience an increase or decrease in entropy?

**18.23.** Diamond and the fullerenes are two allotropes of carbon. On the basis of their different structures and properties, predict which has the higher standard molar entropy.

**18.24.** **Superfluids** The 1996 Nobel Prize in Physics was awarded to Douglas Osheroff, Robert Richardson, and David Lee for discovering *superfluidity* (apparently frictionless flow) in $^3$He. When $^3$He is cooled to 2.7 mK, the liquid settles into an *ordered* superfluid state. What is the predicted sign of the entropy change for the conversion of liquid $^3He$ into its superfluid state?

**18.25.** Rank the compounds in each of the following groups in order of increasing standard molar entropy $(S°)$:
   a. $CH_4(g)$, $CF_4(g)$, and $CCl_4(g)$
   b. $CH_2O(g)$, $CH_3CHO(g)$, and $CH_3CH_2CHO(g)$
   c. $HF(g)$, $H_2O(g)$, and $NH_3(g)$

**18.26.** The hydrocarbon $C_5H_{12}$ has three isomers, shown in figure P18.26. Which isomer has the largest molar entropy, $S°$?

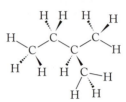
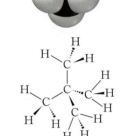

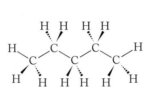

$CH_3$—$CH_2$—$CH_2$—$CH_2$—$CH_3$

Pentane

$CH_3$—$CH_2$—$CH$—$CH_3$
              |
             $CH_3$

2-Methylbutane

$$CH_3—\overset{\overset{\displaystyle CH_3}{|}}{\underset{\underset{\displaystyle CH_3}{|}}{C}}—CH_3$$

2,2-Dimethylpropane

**FIGURE P18.26**

## Calculating Entropy Changes

### CONCEPT REVIEW

**18.27.** Under standard conditions, the products of a reaction have, overall, greater entropy than the reactants. What is the sign of $\Delta S°_{rxn}$?

**18.28.** Do decomposition reactions tend to have $\Delta S°_{rxn}$ values that are greater than zero or less than zero? Why?

**18.29.** Do precipitation reactions tend to have $\Delta S°_{rxn}$ values that are greater than zero or less than zero? Why?

*__18.30.__ Suppose compound A($s$) decomposes to substances B($\ell$) and C($\ell$) at moderately high temperatures. How would running the decomposition reaction at even higher temperatures (above the melting point of A, but still below the boiling points of B and C) affect the value of $\Delta S°_{rxn}$?

### PROBLEMS

**18.31. Smog** Use the standard molar entropies in Appendix 4 to calculate $\Delta S°$ values for each of the following atmospheric reactions that contribute to the formation of photochemical smog.
   a. $N_2(g) + O_2(g) \rightarrow 2\,NO(g)$
   b. $2\,NO(g) + O_2(g) \rightarrow 2\,NO_2(g)$
   c. $NO(g) + \frac{1}{2}O_2(g) \rightarrow NO_2(g)$
   d. $2\,NO_2(g) \rightarrow N_2O_4(g)$

**18.32.** Use the standard molar entropies in Appendix 4 to calculate the $\Delta S°$ value for each of the following reactions of sulfur compounds.
   a. $H_2S(g) + \frac{3}{2}O_2(g) \rightarrow H_2O(g) + SO_2(g)$
   b. $2\,SO_2(g) + O_2(g) \rightarrow 2\,SO_3(g)$
   c. $SO_3(g) + H_2O(\ell) \rightarrow H_2SO_4(aq)$
   d. $S(g) + O_2(g) \rightarrow SO_2(g)$

**18.33.** What is the entropy change to the surroundings when a small, decorative ice sculpture at a temperature of 0°C and weighing 456 g melts on a granite tabletop if the temperature of the granite is 12°C and the process occurs reversibly? Assume a final temperature for the water of 0°C. The heat of fusion of ice is 6.01 kJ/mol.

**18.34.** The decorative ice sculpture described in Problem 18.33 was carved from solid $CO_2$ and held at its sublimation point of −78.5°C. What is the entropy change to the universe if

the $CO_2$ sculpture weighing 389 g sublimes on a granite tabletop if the temperature of the granite is 12°C and the process occurs reversibly? Assume a final temperature of the $CO_2$ vapor is −78.5°C. The heat of sublimation of $CO_2$ is 26.1 kJ/mol.

## Free Energy

### CONCEPT REVIEW

**18.35.** What does the sign of $\Delta G$ tell you about the spontaneity of a process?

**18.36.** What does the sign of $\Delta G$ tell you about the rate of a reaction?

*__18.37.__ Many 19th-century scientists believed that all exothermic reactions were spontaneous. Why did so many of them share this belief?

**18.38.** In which direction does a reaction proceed when its $\Delta G$ value is (a) less than zero; (b) equal to zero; (c) greater than zero?

**18.39.** What are the signs of $\Delta S$, $\Delta H$, and $\Delta G$ for the sublimation of dry ice (solid $CO_2$) at 25°C?

**18.40.** What are the signs of $\Delta S$, $\Delta H$, and $\Delta G$ for the formation of dew on a cool night?

**18.41.** Which of the following processes is(are) spontaneous?
   a. A tornado forms.
   b. A broken cell phone fixes itself.
   c. You get an A in this course.
   d. Hot soup gets cold before it is served.

**18.42.** Which of the following processes is(are) spontaneous?
   a. Wood burns in air.
   b. Water vapor condenses on the sides of a glass of ice tea.
   c. Salt dissolves in water.
   d. Photosynthesis.

## PROBLEMS

**18.43.** Calculate the free-energy change for the dissolution in water of 1 mole of NaBr and 1 mole of NaI at 298 K from the values in the table below.

|  | $\Delta H°_{soln}$ (kJ/mol) | $\Delta S°_{soln}$ J/(mol · K) |
|---|---|---|
| **NaBr** | −1 | 57 |
| **NaI** | −7 | 74 |

**18.44.** The values of $\Delta H°_{rxn}$ and $\Delta S°_{rxn}$ for the reaction

$$2\, NO(g) + O_2(g) \rightarrow 2\, NO_2(g)$$

are −12 kJ and −146 J/K.
 a. Use these values to calculate $\Delta G°_{rxn}$ at 298 K.
 b. Explain why the value of $\Delta G°_{rxn}$ is negative.

---

**18.45.** A mixture of $CO(g)$ and $H_2(g)$ is produced by passing steam over hot charcoal:

$$H_2O(g) + C(s) \rightarrow H_2(g) + CO(g)$$

Calculate the $\Delta G°_{rxn}$ value for the reaction from the appropriate $\Delta G°_f$ data in Appendix 4.

**18.46.** Use the appropriate $\Delta G°_f$ data in Appendix 4 to calculate $\Delta G°_{rxn}$ for the complete combustion of methanol:

$$2\, CH_3OH(g) + 3\, O_2(g) \rightarrow 2\, CO_2(g) + 4\, H_2O(g)$$

---

**18.47.** Determine the value of $\Delta G°$ for the reduction of iron ore with hydrogen gas:

$$Fe_2O_3(s) + 3\, H_2(g) \rightarrow 2\, Fe(s) + 3\, H_2O(g)$$

given the following thermodynamic properties at 25°C:

|  | $Fe_2O_3(s)$ | $H_2(g)$ | $Fe(s)$ | $H_2O(g)$ |
|---|---|---|---|---|
| **$\Delta H°_f$ kJ/mol** | −824.2 | 0 | 0 | −241.8 |
| **$S°$ J/(mol · K)** | 87.4 | 130.6 | 27.3 | 188.8 |

**18.48.** Determine the value of $\Delta G°$ for the reaction of the fuel gas acetylene with hydrogen gas:

$$C_2H_2(g) + 2\, H_2(g) \rightarrow C_2H_6(g)$$

given the following thermodynamic properties at 25°C:

|  | $C_2H_2(g)$ | $H_2(g)$ | $C_2H_6(g)$ |
|---|---|---|---|
| **$\Delta H°_f$ kJ/mol** | 226.7 | 0 | −84.7 |
| **$S°$ J/(mol · K)** | 200.8 | 130.6 | 229.5 |

---

**18.49. Acid Precipitation** Aerosols (fine droplets) of sulfuric acid form in the atmosphere as a result of the reaction below. Use the appropriate $\Delta G°_f$ data in Appendix 4 to calculate $\Delta G°_{rxn}$ for the combination reaction:

$$SO_3(g) + H_2O(g) \rightarrow H_2SO_4(\ell)$$

**18.50.** One source of sulfuric acid aerosols in the atmosphere is combustion of high-sulfur fuels, which releases $SO_2$ gas that then is further oxidized to $SO_3$:

$$2\, SO_2(g) + O_2(g) \rightarrow 2\, SO_3(g)$$

Use the appropriate $\Delta G°_f$ data in Appendix 4 to calculate $\Delta G°_{rxn}$ for this combination reaction. Is it spontaneous under standard conditions?

## Temperature and Spontaneity

### CONCEPT REVIEW

**18.51.** Are exothermic reactions spontaneous only at low temperature? Explain your answer.

**18.52.** Are endothermic reactions never spontaneous at low temperature? Explain your answer.

### PROBLEMS

**18.53.** What is the lowest temperature at which the following reaction (see Problem 18.45) is spontaneous?

$$H_2O(g) + C(s) \rightarrow H_2(g) + CO(g)$$

**18.54.** Deposits of elemental sulfur are often seen near active volcanoes. Their presence there may be due to the following reaction of $SO_2$ with $H_2S$:

$$SO_2(g) + 2\, H_2S(g) \rightarrow \tfrac{3}{8}\, S_8(s) + 2\, H_2O(g)$$

Assuming the values of $\Delta H°_{rxn}$ and $\Delta S°_{rxn}$ do not change appreciably with temperature, over what temperature range is the reaction spontaneous?

---

**18.55.** Use the data in Appendix 4 to calculate $\Delta H°$ and $\Delta S°$ for the vaporization of hydrogen peroxide:

$$H_2O_2(\ell) \rightarrow H_2O_2(g)$$

Assuming that the calculated values are independent of temperature, what is the boiling point of hydrogen peroxide at $P = 1.00$ atm?

**18.56.** Determine the normal melting point of carbon tetrachloride in degrees centigrade given the following data:

$$\Delta H°_{fus} = 2.67 \text{ kJ/mol} \qquad \Delta S°_{fus} = 10.86 \text{ J/(mol · K)}$$

---

**18.57.** Which of the following reactions is spontaneous only at low temperatures; only at high temperatures; at all temperatures?
 a. $2\, NO(g) + O_2(g) \rightarrow 2\, NO_2(g)$
 b. $2\, NH_3(g) + 2\, O_2(g) \rightarrow N_2O(g) + 3\, H_2O(g)$
 c. $NH_4NO_3(s) \rightarrow 2\, H_2O(g) + N_2O(g)$

**18.58.** Which of the following reactions is spontaneous only at low temperatures; only at high temperatures; at all temperatures?
 a. $2\, H_2S(g) + 3\, O_2(g) \rightarrow 2\, H_2O(g) + 2\, SO_2(g)$
 b. $SO_2(g) + H_2O_2(\ell) \rightarrow H_2SO_4(\ell)$
 c. $S(g) + O_2(g) \rightarrow SO_2(g)$

## Free Energy and Chemical Equilibrium

### CONCEPT REVIEW

**18.59.** If the value of $K$ of a reaction is less than 1, what is the sign of $\Delta G°_{rxn}$?

*18.60. The equation $\Delta G° = -RT \ln K$ relates the value of $K_p$, not $K_c$, to $\Delta G°$ for gas-phase reactions. Explain why.

**18.61.** If a reaction mixture contains only reactants and no products, will the reaction proceed in the forward direction even if $\Delta G° > 0$? Explain why or why not.

*18.62. If a gas-phase reaction mixture contains one mole of each reactant and product, is the $\Delta G°_{rxn}$ of the reaction mixture the same as $\Delta H°_{rxn}$? Explain why or why not.

**PROBLEMS**

**18.63.** Which of the following reactions has the largest $K_p$ value at 25°C?
a. $Cl_2(g) + F_2(g) \rightleftharpoons 2\ ClF(g)$  $\Delta G°_{rxn} = 115.4$ kJ
b. $Cl_2(g) + Br_2(g) \rightleftharpoons 2\ ClBr(g)$  $\Delta G°_{rxn} = -2.0$ kJ
c. $Cl_2(g) + I_2(g) \rightleftharpoons 2\ ICl(g)$  $\Delta G°_{rxn} = -27.9$ kJ

**18.64.** Use the appropriate $\Delta G°_f$ value in Appendix 4 to calculate the value of $K_p$ at 298 K for the reaction
$$N_2(g) + 2\ O_2(g) \rightleftharpoons 2\ NO_2(g)$$

**18.65.** Use the appropriate equilibrium constant in Appendix 5 to calculate the value of $\Delta G°_{rxn}$ for the reaction
$$NH_3(g) + H_2O(\ell) \rightleftharpoons NH_4^+(aq) + OH^-(aq)$$

**18.66.** Use the appropriate equilibrium constant in Appendix 5 to calculate the value of $\Delta G°_{rxn}$ for the reaction
$$HClO(aq) + H_2O(\ell) \rightleftharpoons ClO^-(aq) + H_3O^+(aq)$$

## Influence of Temperature on Equilibrium Constants

**CONCEPT REVIEW**

**18.67.** The value of the equilibrium constant of a reaction decreases with increasing temperature. Is the reaction endothermic or exothermic?

**18.68.** The reaction
$$2\ CO(g) + O_2(g) \rightleftharpoons 2\ CO_2(g)$$
is exothermic. Does the value of $K_p$ increase or decrease with increasing temperature?

**18.69.** The value of $K_p$ for the water–gas shift reaction
$$CO(g) + H_2O(g) \rightleftharpoons H_2(g) + CO_2(g)$$
increases as the temperature decreases. Is the reaction exothermic or endothermic?

**18.70.** Does the value of $K_p$ for the reaction
$$CH_4(g) + H_2O(g) \rightleftharpoons 3\ H_2(g) + CO(g)\quad \Delta H° = 206\ kJ$$
increase, decrease, or remain unchanged as temperature increases?

**PROBLEMS**

**18.71. Air Pollution** Automobiles and trucks pollute the air with NO. At 2000°C, $K_c$ for the reaction
$$N_2(g) + O_2(g) \rightleftharpoons 2\ NO(g)$$
is $4.10 \times 10^{-4}$, and $\Delta H° = 180.6$ kJ. What is the value of $K_c$ at 25°C?

**18.72.** At 400 K the value of $K_p$ for the reaction
$$N_2(g) + 3\ H_2(g) \rightleftharpoons 2\ NH_3(g)$$
is 41, and $\Delta H° = -92.2$ kJ. What is the value of $K_p$ at 700 K?

**18.73.** The equilibrium constant for the reaction
$$NO(g) + O_2(g) \rightleftharpoons 2\ NO_2(g)$$
decreases from $1.5 \times 10^5$ at 430°C to 23 at 1000°C. From these data, calculate the value of $\Delta H°$ for the reaction.

**18.74.** The value of $K_c$ for the reaction $A \rightleftharpoons B$ is 0.455 at 50°C and 0.655 at 100°C. Calculate $\Delta H°$ for the reaction.

## Driving the Human Engine: Coupled Reactions

**CONCEPT REVIEW**

**18.75.** Describe the ways in which two chemical reactions must complement each other so that the decrease in free energy of the spontaneous reaction can drive the nonspontaneous reaction.

**18.76.** Why is it important that at least some of the spontaneous steps in glycolysis convert ADP to ATP?

**18.77.** The second step in glycolysis converts glucose 6-phosphate into fructose 6-phosphate (Figure P18.77). Suggest a reason why $\Delta G°$ for this reaction is close to zero.

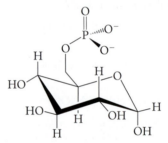

Glucose 6-phosphate     Fructose 6-phosphate

**FIGURE P18.77**

**PROBLEMS**

**18.78.** The methane in natural gas in an important starting material, or feedstock, for producing industrial chemicals, including $H_2$ gas.
a. Use the appropriate $\Delta G°_f$ value(s) from Appendix 4 to calculate $\Delta G°_{rxn}$ for the reaction known as *steam–methane reforming*:
$$CH_4(g) + H_2O(g) \rightarrow CO(g) + 3\ H_2(g)$$
b. To drive this nonspontaneous reaction, the CO that is produced can be oxidized to $CO_2$ using more steam:
$$CO(g) + H_2O(g) \rightarrow CO_2(g) + H_2(g)$$
Use the appropriate $\Delta G°_f$ value(s) from Appendix 4 to calculate $\Delta G°_{rxn}$ for this reaction, which is known as the *water–gas shift reaction*.
c. Combine these two reactions together and write the chemical equation of the overall reaction in which methane and steam combine to produce hydrogen gas and carbon dioxide.
d. Calculate the $\Delta G°_{rxn}$ value of the overall reaction. Is it spontaneous under standard conditions?

**18.79.** In addition to the reactions described in Problem 18.78, methane can, in theory, be used to produce hydrogen gas by a process in which it decomposes into elemental carbon and hydrogen:

$$CH_4(g) \rightarrow C(s) + 2\,H_2(g)$$

and the carbon produced in the first step is then oxidized to $CO_2$:

$$C(s) + O_2(g) \rightarrow CO_2(g)$$

a. Calculate the $\Delta G°_{rxn}$ values of the above two reactions.
b. Write a balanced chemical equation describing the overall reaction obtained by coupling the above two reactions and calculate its $\Delta G°_{rxn}$ value. Is the coupled reaction spontaneous under standard conditions?

**18.80.** Which of the following steps in glycolysis has the largest equilibrium constant?
a. Fructose 1,6-diphosphate $\rightleftharpoons$
   2 glyceraldehyde-3-phosphate
   $$\Delta G°_{rxn} = 24 \text{ kJ}$$
b. 3-Phosphoglycerate $\rightleftharpoons$ 2-phosphoglycerate
   $$\Delta G°_{rxn} = 4.4 \text{ kJ}$$
c. 2-Phosphoglycerate $\rightleftharpoons$ phosphoenolpyruvate
   $$\Delta G°_{rxn} = 1.8 \text{ kJ}$$

**18.81.** The value of $\Delta G°$ for the phosphorylation of glucose in glycolysis is 13.8 kJ/mol. What is the value of the equilibrium constant for the reaction at 298 K?

**18.82.** In glycolysis, the hydrolysis of ATP to ADP drives the phosphorylation of glucose:

$$\text{Glucose} + \text{ATP} \rightleftharpoons \text{ADP} + \text{glucose 6-phosphate}$$
$$\Delta G°_{rxn} = -17.7 \text{ kJ}$$

What is the value of $K_c$ for this reaction at 298 K?

**18.83.** Sucrose enters the series of reactions in glycolysis after the reaction in which it is hydrolyzed, forming glucose and fructose:

$$\text{Sucrose} + H_2O \rightleftharpoons \text{glucose} + \text{fructose}$$
$$K_c = 5.3 \times 10^{12} \text{ at 298 K}$$

What is the value of $\Delta G°_{rxn}$?

## Microstates: A Quantized View of Entropy

### CONCEPT REVIEW

**18.84.** You flip three coins, assigning the values +1 for heads and −1 for tails. Each outcome of the three flips constitutes a microstate. How many different microstates are possible from flipping the three coins? Which value or values for the sums in the microstates are most likely? (*Hint:* The sequence HHT [+1 +1 −1] is one possible outcome, or microstate. Note, however, that this outcome differs from THH [−1 +1 +1], even though the two sequences sum to the same value.)

**18.85.** Imagine you have four identical chairs to arrange on four steps leading up to a stage, one chair on each step. The chairs have numbers on their backs: 1, 2, 3, and 4. How many different microstates for the chairs are possible? (Notice that when viewed from the front, all the microstates look the same. Viewed from the back, you can identify the different microstates because you can distinguish the chairs by their numbers.)

### PROBLEMS

**18.86.** Use the appropriate standard molar entropy value in Appendix 4 to calculate how many microstates are accessible to a single molecule of $N_2$ at 298 K.

**18.87.** Use the appropriate standard molar entropy value in Appendix 4 to calculate how many microstates are accessible to a single molecule of liquid $H_2O$ at 298 K.

## Additional Problems

**18.88.** Methanogenic bacteria convert acetic acid ($CH_3COOH$) into $CO_2(g)$ and $CH_4(g)$.
a. Is this process endothermic or exothermic under standard conditions?
b. Is the reaction spontaneous under standard conditions?

**18.89.** Chlorofluorocarbons (CFCs) are no longer used as refrigerants because they catalyze the decomposition of stratospheric ozone. Trichlorofluoromethane ($CCl_3F$) boils at 23.8°C and its molar heat of vaporization is 24.8 kJ/mol. What is the molar entropy of evaporation of $CCl_3F(\ell)$.

**18.90.** Consider the precipitation reactions described by the following net ionic equations:

$$Mg^{2+}(aq) + 2\,OH^-(aq) \rightarrow Mg(OH)_2(s)$$
$$Ag^+(aq) + Cl^-(aq) \rightarrow AgCl(s)$$

a. Predict the sign of $\Delta S°_{rxn}$ for the reactions.
b. Using the values for $S°$ from Appendix 4, calculate $\Delta S°$ for these reactions.
c. Do your calculations support your prediction?

**\*18.91.** At what temperature is the free-energy change for the following reaction equal to zero?

$$NH_4Cl(s) \rightarrow NH_3(g) + HCl(g)$$

**18.92.** Which of these processes result in an entropy decrease of the system?
a. Diluting hydrochloric acid with water
b. Boiling water
c. $2\,NO(g) + O_2(g) \rightarrow 2\,NO_2(g)$
d. Making ice cubes in the freezer

**\*18.93.** Calculate the standard free-energy change of the following reaction. Is it spontaneous?

$$2\,NO(g) + 2\,H_2(g) \rightarrow N_2(g) + 2\,H_2O(g)$$

**\*18.94.** Estimate the free-energy change of the following reaction at 225°C:

$$C_2H_4(g) + 3\,O_2(g) \rightarrow 2\,CO_2(g) + 2\,H_2O(g)$$

**\*18.95.** Show that hydrogen cyanide (HCN) is a gas at 25°C by estimating its normal boiling point from the following data:

| | $\Delta H^\circ_f$, kJ/mol | $S^\circ_f$, J/mol |
|---|---|---|
| **HCN(ℓ)** | 108.9 | 113 |
| **HCN(g)** | 135.1 | 202 |

**18.96. Making Methanol** The element hydrogen (H$_2$) is not abundant on Earth, but it is a useful reagent in, for example, the potential synthesis of the liquid fuel methanol from gaseous carbon monoxide:

$$2\,H_2(g) + CO(g) \rightarrow CH_3OH(\ell)$$

Under what temperature conditions is this reaction spontaneous?

**18.97. Light Bulb Filaments** Tungsten (W) is the favored metal for light bulb filaments, partly because of its high melting point of 3422°C. The enthalpy of fusion of tungsten is 35.4 kJ/mol. What is its entropy of fusion?

**\*18.98.** Over what temperature range is the reduction of tungsten(VI) oxide by hydrogen to give metallic tungsten and water spontaneous? The standard heat of formation of WO$_3$(s) is −843 kJ/mol, and its standard molar entropy is 76 J/(mol · K).

**18.99.** Two allotropes (A and B) of sulfur interconvert at 369 K and 1 atm pressure:

$$S_8(s, A) \rightarrow S_8(s, B)$$

The enthalpy change in this transition is 297 J/mol. What is the entropy change?

**\*18.100.** Copper forms two oxides, Cu$_2$O and CuO.
  a. Name these oxides.
  b. Predict over what temperature range this reaction

$$Cu_2O(s) \rightarrow CuO(s) + Cu(s)$$

  is spontaneous using the following thermodynamic data:

| | $\Delta H^\circ_f$, kJ/mol | $S^\circ_f$, J/mol |
|---|---|---|
| **Cu$_2$O(s)** | −170.7 | 92.4 |
| **CuO(s)** | −156.1 | 42.6 |

  c. Why is the standard molar entropy of Cu$_2$O(s) larger than that of CuO(s)?

**\*18.101. Lime** Enormous amounts of lime (CaO) are used in steel industry blast furnaces to remove impurities from iron. Lime is made by heating limestone and other solid forms of CaCO$_3$(s). Why is the standard molar entropy of CaCO$_3$(s) higher than that of CaO(s)? At what temperature is the pressure of CO$_2$(g) over CaCO$_3$(s) equal to 1.0 atm?

| | $\Delta H^\circ_f$, kJ/mol | $S^\circ_f$, J/(mol · K) |
|---|---|---|
| **CaCO$_3$(s)** | −1207 | 93 |
| **CaO(s)** | −636 | 40 |
| **CO$_2$(g)** | −394 | 214 |

**\*18.102.** *Trouton's rule* says that the ratio $\Delta H^\circ_{vap}/T_b$ for a liquid is approximately 80 J/K. Here, $\Delta H^\circ_{vap}$ is the molar enthalpy of vaporization of a liquid, and $T_b$ is its normal boiling point.
  a. What idea suggests that $\Delta H^\circ_{vap}/T_b$ for a range of liquids should be approximately constant?
  b. Check Trouton's rule against the data in Figure P18.102. Which liquids deviate from Trouton's rule, and why?

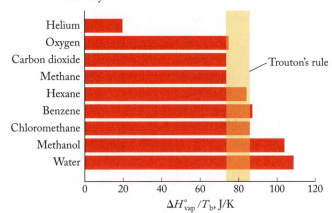

**FIGURE P18.102**

**\*18.103. Melting DNA** When a solution of DNA in water is heated, the DNA double helix separates into two single strands:

1 DNA double helix $\rightleftharpoons$ 2 single strands

  a. What is the sign of $\Delta S$ for the forward process as written?
  b. The DNA double helix re-forms as the system cools. What is the sign of $\Delta S$ for the process by which two single strands re-form the double helix?

c. The melting point of DNA is defined as the temperature at which $\Delta G = 0$. At that temperature, the forward reaction produces two single strands as fast as two single strands recombine to form the double helix. Write an equation that defines the melting temperature ($T$) of DNA in terms of $\Delta H$ and $\Delta S$.

*18.104. **Melting Organic Compounds** When dicarboxylic acids (compounds with two –COOH groups in their structures) melt, they frequently decompose to produce 2 moles of $CO_2$ gas for every 1 mole of dicarboxylic acid melted (shown in Figure P18.104).

   a. What are the signs of $\Delta H$ and $\Delta S$ for the process as written?

   b. Problem 18.103 describes the DNA double helix re-forming when the system cools after melting. Do you think the dicarboxylic acid will re-form when the melted material cools? Why or why not?

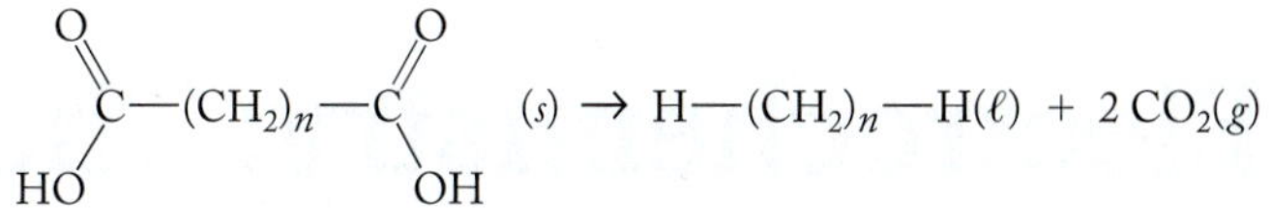

$$\text{(s)} \rightarrow \text{H}-(CH_2)_n-\text{H}(\ell) + 2\,CO_2(g)$$

**FIGURE P18.104**

If your instructor assigns problems in **smartwork**, log in at **smartwork.wwnorton.com**.

# 19

# Electrochemistry: The Quest for Clean Energy

**19.1** Redox Chemistry Revisited

**19.2** Electrochemical Cells

**19.3** Standard Potentials

**19.4** Chemical Energy and Electrical Work

**19.5** A Reference Point: The Standard Hydrogen Electrode

**19.6** The Effect of Concentration on $E_{cell}$

**19.7** Relating Battery Capacity to Quantities of Reactants

**19.8** Corrosion: Unwanted Electrochemical Reactions

**19.9** Electrolytic Cells and Rechargeable Batteries

**19.10** Fuel Cells

### Learning Outcomes

**LO1** Combine the appropriate half-reactions to write net ionic equations of spontaneous redox reactions
**Sample Exercise 19.1**

**LO2** Draw cell diagrams and describe the components of electrochemical cells and their roles in interconverting chemical and electrical energy
**Sample Exercise 19.2**

**LO3** Calculate standard cell potentials from standard reduction potentials
**Sample Exercise 19.3**

**LO4** Interconvert a cell's potential and the change in free energy of the cell reaction
**Sample Exercise 19.4**

**LO5** Use the Nernst equation to calculate cell potentials
**Sample Exercise 19.5**

**LO6** Relate standard cell potentials to the equilibrium constants of the cell reactions
**Sample Exercise 19.6**

**LO7** Use the Faraday constant to relate quantity of charge to changes in the quantities of reactants and products in cell reactions
**Sample Exercises 19.7, 19.8**

## Efficient Locomotion—Batteries Included

In earlier chapters we discussed some of the chemistry involved in air pollution and other problems caused by the burning of fossil fuels. Concern over global climate change, along with political and economic problems associated with the price and availability of gasoline, has spurred development of innovative drive systems for cars that do not use fossil fuels. Hybrid vehicles use combinations of electric motors, powered by rechargeable batteries, and small gasoline engines. All-electric vehicles are powered by banks of high-performance batteries or fuel cells. The challenge for scientists and engineers is to develop lighter-weight, higher-capacity, more reliable batteries and more powerful and less expensive fuel cells.

Batteries convert chemical energy into electrical energy; motors then convert the electrical energy to mechanical energy. These two processes working together tend to be more efficient than a process involving combustion. In addition, a vehicle's electric motor can be used to recharge its batteries. However, the driving range of a battery-powered car is limited by the capacity of its batteries to store and deliver electrical energy. The smaller battery packs in plug-in hybrids (the batteries can be recharged directly from the power grid) provide a driving range of about 50 km (≈ 30 mi). Modern all-electric cars can travel as far as 400 km (250 mi) on a single charge, but the battery packs in these cars account for half of the car's weight.

The chemistry of batteries and fuel cells is an extension of the oxidation–reduction chemistry we studied in Chapter 4. We begin this chapter with a review of some of that material. We then explore the various ways that have been devised to release an electric current from combinations of chemical substances. We will focus on the batteries being developed for electric cars, but we will also look at the batteries used for other modern devices, from laptop computers to cell phones to portable TV and music systems. ■

**An Electrifying Sedan** The 2013 Chevrolet Volt has a 16 kWh lithium–ion battery, giving the Volt an all-electric driving range of about 60 km (≈ 37 mi). A gasoline-powered generator gives the Volt an overall driving range of about 1000 km (≈ 620 mi). ▶

**electrochemistry** the branch of chemistry that examines the transformations between chemical and electrical energy.

# 19.1 Redox Chemistry Revisited

In this chapter, we explore propulsion technologies that are more efficient than internal combustion engines at converting chemical energy into mechanical work and that have the potential to dramatically reduce air pollution. These technologies are based on **electrochemistry**, the branch of chemistry that links chemical reactions to the production or consumption of electrical energy. At the heart of electrochemistry are chemical reactions in which electrons are gained and lost at electrode surfaces. In other words, electrochemistry is based on *red*uction and *ox*idation, or *redox*, chemistry.

The principles of redox reactions were introduced in Section 4.9. Let's briefly review them here:

➤ A redox reaction is the sum of two half-reactions: a reduction half-reaction, in which a reactant gains electrons, and an oxidation half-reaction, in which a reactant loses electrons.
➤ Reduction and oxidation half-reactions happen simultaneously so that the number of electrons gained during reduction exactly matches the number lost during oxidation.
➤ A substance that is easily oxidized is one that readily gives up electrons. This electron-donating power makes the substance an effective reducing agent. In any redox reaction, the *reducing agent is always oxidized.*
➤ A substance that readily accepts electrons and is thereby reduced is an effective oxidizing agent. In any redox reaction, the *oxidizing agent is always reduced.*

(a)          (b)          (c)

**FIGURE 19.1** A strip of zinc is immersed in an aqueous solution of blue copper(II) sulfate. Over time it becomes encrusted with a dark layer of copper as $Cu^{2+}$ ions are reduced to Cu atoms and Zn atoms are oxidized to colorless $Zn^{2+}$ ions.

Let's begin our exploration of electrochemistry with an examination of the redox properties of two neighbors in the periodic table: copper ($Z = 29$) and zinc ($Z = 30$). Elemental zinc has considerable electron-donating power, and $Cu^{2+}$ ions tend to be willing acceptors of donated electrons. These complementary properties are captured in Figure 19.1. When a strip of Zn metal is placed in a solution of $CuSO_4$—that is, a solution of $Cu^{2+}$ ions and $SO_4^{2-}$ ions—Zn atoms spontaneously donate electrons to $Cu^{2+}$ ions, forming $Zn^{2+}$ ions and Cu atoms. The shiny zinc surface turns dark brown as a textured layer of copper metal accumulates on it, and the distinctive blue color of $Cu^{2+}(aq)$ ions fades as these ions acquire electrons and become atoms of copper metal.

In this spontaneous reaction, each mole of Zn atoms that is oxidized *loses* 2 moles of electrons in this oxidation half-reaction:

$$Zn(s) \rightarrow Zn^{2+}(aq) + 2\ e^-$$

Also, every mole of $Cu^{2+}$ ions that is reduced *accepts* 2 moles of electrons in this reduction half-reaction:

$$Cu^{2+}(aq) + 2\ e^- \rightarrow Cu(s)$$

Because the number of moles of electrons lost and gained is the same in the two half-reactions, writing a net ionic equation to describe the overall redox reaction is simply a matter of adding the two half-reactions together:

$$Zn(s) \rightarrow Zn^{2+}(aq) + 2\,e^-$$
$$\underline{Cu^{2+}(aq) + 2\,e^- \rightarrow Cu(s)}$$
$$Zn(s) + Cu^{2+}(aq) + \cancel{2\,e^-} \rightarrow Cu(s) + Zn^{2+}(aq) + \cancel{2\,e^-}$$

Canceling out the equal numbers of electrons gained and lost, we get

$$Zn(s) + Cu^{2+}(aq) \rightarrow Cu(s) + Zn^{2+}(aq) \qquad (19.1)$$

Combining half-reactions is a convenient way to write net ionic equations for redox reactions. In this chapter we use a valuable resource in this equation writing process: a table of common half-reactions in Table A6.1 in Appendix 6. Note that all the half-reactions are written as reduction half-reactions. There is no need for a separate table of *oxidation* half-reactions because any reduction half-reaction can always be reversed to obtain the corresponding oxidation half-reaction. Many of the half-reactions in Table A6.1 are written as if they occur in acidic solutions. We know this because $H^+$ ions are used to balance the number of H atoms in most of the half-reactions in which water is a reactant or product. For example, the reduction of $O_2$ to $H_2O$ is written

$$O_2(g) + 4\,H^+(aq) + 4\,e^- \rightarrow 2\,H_2O(\ell)$$

However, a few half-reactions in Table A6.1 contain $OH^-$ ions, which tells us that they apply to reactions in basic solutions. One of them also involves the reduction of $O_2$:

$$O_2(g) + 2\,H_2O(\ell) + 4\,e^- \rightarrow 4\,OH^-$$

In the following Sample and Practice Exercises you will need to select the appropriate half-reaction that matches the pH conditions of the overall redox reaction.

**⊙⊙ CONNECTION** Keep in mind that hydrogen ions in aqueous solutions are really hydronium ions, $H_3O^+(aq)$, as described in Chapter 4.

---

**SAMPLE EXERCISE 19.1**   **Writing Net Ionic Equations of Redox Reactions by Combining Half-Reactions**   **LO1**

Write a net ionic equation describing the oxidation of $Fe^{2+}(aq)$ by $O_2$ gas dissolved in an acidic solution. (*Hint*: Water is a component of the $O_2$ reduction half-reaction.)

**Collect and Organize**  We find these half-reactions for $O_2$ reduction in Table A6.1:

$$O_2(g) + 4\,H^+(aq) + 4\,e^- \rightarrow 2\,H_2O(\ell)$$
$$O_2(g) + 2\,H_2O(\ell) + 4\,e^- \rightarrow 4\,OH^-(aq)$$

All of the half-reactions in Table A6.1 are reductions, so there is no half-reaction for the oxidation of $Fe^{2+}$ ions. However, $Fe^{2+}$ ions appear as a product in the reduction of $Fe^{3+}$:

$$Fe^{3+}(aq) + e^- \rightarrow Fe^{2+}(aq)$$

We can reverse this reduction half-reaction to make $Fe^{2+}$ the reactant in an oxidation half-reaction:

$$Fe^{2+}(aq) \rightarrow Fe^{3+}(aq) + e^-$$

**Analyze**  The problem specifies acidic conditions, so we should use the $O_2$ half-reaction with $H^+$ rather than $OH^-$ ions in it. In any redox reaction, the number of electrons gained by the substance that is reduced must equal the number lost by the substance that is oxidized. A gain of 4 moles of electrons occurs in the $O_2$ half-reaction and a loss of 1 mole of electrons in the $Fe^{2+}$ half-reaction. Therefore, we need to multiply the $Fe^{2+}$ half-reaction by 4 to obtain a balanced overall redox reaction equation.

**Solve** Multiplying the $Fe^{2+}$ half-reaction by 4 and adding it to the $O_2$ half-reaction containing $H^+$ ions, we get

$$4\,Fe^{2+}(aq) \rightarrow 4\,Fe^{3+}(aq) + 4\,e^-$$
$$O_2(g) + 4\,H^+(aq) + 4\,e^- \rightarrow 2\,H_2O(\ell)$$

$$\overline{4\,Fe^{2+}(aq) + O_2(g) + 4\,H^+(aq) + \cancel{4\,e^-} \rightarrow 4\,Fe^{3+}(aq) + 2\,H_2O(\ell) + \cancel{4\,e^-}}$$

Simplifying gives

$$4\,Fe^{2+}(aq) + O_2(g) + 4\,H^+(aq) \rightarrow 4\,Fe^{3+}(aq) + 2\,H_2O(\ell)$$

**Think About It** We can verify that this is a balanced net ionic equation by confirming that the numbers of Fe, O, and H atoms are the same on the two sides of the reaction arrow, which they are, and that the total electrical charge is the same on the two sides (both are 12+). Note that the result of combining these two half-reactions is not a molecular equation but rather a net ionic equation.

☼ **Practice Exercise** Write a net ionic equation describing the oxidation of $NO_2^-$ to $NO_3^-$ by $O_2$ in a basic solution.

*(Answers to Practice Exercises are in the back of the book.)*

## 19.2 Electrochemical Cells

We have seen that the redox reaction between Zn metal and $Cu^{2+}$ ions is the net result of two separate half-reactions, one involving the oxidation of Zn and the other the reduction of $Cu^{2+}$ ions. Let's now physically separate the two half-reactions using a device called an **electrochemical cell**. Figure 19.2 provides a view of an electrochemical cell based on the $Zn/Cu^{2+}$ reaction. Like nearly all electrochemical cells, it is made of two compartments. One of the compartments in the $Zn/Cu^{2+}$ cell contains a strip of zinc metal immersed in 1.00 $M$ $ZnSO_4$; the other compartment has a strip of copper metal immersed in 1.00 $M$ $CuSO_4$. Sulfate ions are spectator ions in this reaction and are not included in the net ionic equation, though we will see that they do play a role in cell function. The two metal strips serve as the cell's *electrodes*, providing pathways along which the electrons produced and consumed in the two half-reactions flow to and from an external circuit.

As the cell reaction proceeds, oxidation of Zn atoms produces electrons, which travel from the Zn electrode through the external circuit to the surface of the Cu electrode, where they combine with $Cu^{2+}$ ions, forming atoms of Cu metal. In an electrochemical cell the electrode at which the oxidation half-reaction takes place (the zinc electrode in this case) is called the **anode**, and the electrode at which the reduction half-reaction takes place is called the **cathode**.

**FIGURE 19.2** This electrochemical cell consists of two compartments: the one on the left contains a zinc metal anode immersed in a 1.00 $M$ solution of $ZnSO_4$; the one on the right contains a copper metal cathode immersed in a 1.00 $M$ solution of $CuSO_4$. A porous bridge made of either glass or plastic provides an electrical connection through which ions (and their charges) can migrate from one compartment to the other.

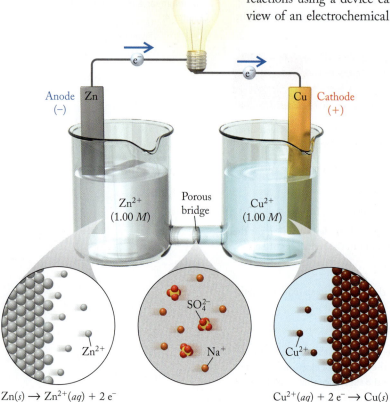

$$Zn(s) \rightarrow Zn^{2+}(aq) + 2\,e^-$$

$$Cu^{2+}(aq) + 2\,e^- \rightarrow Cu(s)$$

You might think that production of $Zn^{2+}$ ions in the left compartment in Figure 19.2 would result in a buildup of positive charge on that side of the cell, and that conversion of $Cu^{2+}$ ions to Cu metal would create an excess of $SO_4^{2-}$ ions and thus a negative charge in the Cu compartment. However, no such charge buildup occurs because the two compartments are connected by a porous bridge and because the solutions surrounding both electrodes contain a *background electrolyte* made of ions that are not involved in either half-reaction. In the $Zn/Cu^{2+}$ cell, $Na_2SO_4$ makes an ideal background electrolyte. Migration of $Na^+$ ions toward the Cu compartment and $SO_4^{2-}$ ions toward the Zn compartment through the permeable bridge as shown in the middle molecular view in Figure 19.2 balances the flow of electrons in the external circuit and eliminates any accumulation of ionic charge in either compartment.

**CONCEPT TEST** ...........................................................

As the cell reaction in Figure 19.2 proceeds, is the increase in mass of the copper strip the same as the decrease in mass of the zinc strip?

*(Answers to Concept Tests are in the back of the book.)*

...........................................................

## Voltaic and Electrolytic Cells

The $Zn/Cu^{2+}$ cell in Figure 19.2 is an example of the kind of electrochemical cell in which a spontaneous cell reaction pumps electrons out from its anode, through an external circuit, and into its cathode. During this process the decrease in free energy of the spontaneous cell reaction is harnessed to do electrical work. A cell that operates in this way is called a **voltaic cell** in honor of Italian physicist Alessandro Volta (1745–1827), who is credited with building the first battery (Figure 19.3). His battery and all of the modern batteries that power familiar electric devices, from cell phones to flashlights to laptop computers, are examples of voltaic cells.

There is a second kind of electrochemical cell that is the operational opposite of a voltaic cell. In this other kind, an external source of electrical energy is applied to the cell and causes a nonspontaneous reaction to occur inside it. In such a cell, electrical energy is converted into chemical energy as low-energy reactants form higher-energy products. A reaction that is driven in this way, by the consumption of electrical energy, is called **electrolysis** and a cell in which electrolysis occurs is called an **electrolytic cell** (Figure 19.4). In electrolytic cells electrons are pumped into the cathodes (making them the negative electrodes), and out from the anodes (making them the positive electrodes). As we will see in Section 19.9, many of the batteries used to power familiar electronic devices are rechargeable, which means that their spontaneous cell reactions can be forced to run in reverse by applying an opposing electrical potential. When this happens, these voltaic cells become electrolytic cells.

## Cell Diagrams

Figure 19.2 provides a view of the physical reality of a $Zn/Cu^{2+}$ electrochemical cell, but we could use a more compact way to represent the components of the cell. A **cell diagram** uses a string of chemical formulas and symbols to show how the components of the cell are connected. A cell diagram does not convey stoichiometry, so any coefficients in the balanced equation for the cell reaction do not appear in the cell diagram.

▶❚❚ **CHEMTOUR** Zinc–Copper Cell

**FIGURE 19.3** Alessandro Volta and his battery, which consisted of a stack of alternating layers of zinc, blotter paper soaked in salt water, and silver.

.......................................................

**electrochemical cell** an apparatus that converts chemical energy into electrical work or electrical work into chemical energy.

**anode** an electrode at which an oxidation half-reaction (loss of electrons) takes place.

**cathode** an electrode at which a reduction half-reaction (gain of electrons) takes place.

**voltaic cell** an electrochemical cell in which chemical energy is transformed into electrical work by a spontaneous cell reaction.

**electrolysis** a process in which electrical energy is used to drive a nonspontaneous chemical reaction.

**electrolytic cell** a device in which an external source of electrical energy does work on a chemical system, turning reactant(s) into higher-energy product(s).

**cell diagram** symbols that show how the components of an electrochemical cell are connected.

.......................................................

**FIGURE 19.4** Voltaic versus electrolytic cells. (a) In a voltaic cell, a spontaneous reaction produces electrical energy and does electrical work on its surroundings, such as lighting a light bulb. (b) In an electrolytic cell, an external supply of electrical energy does work on the chemical system in the cell, driving a nonspontaneous reaction.

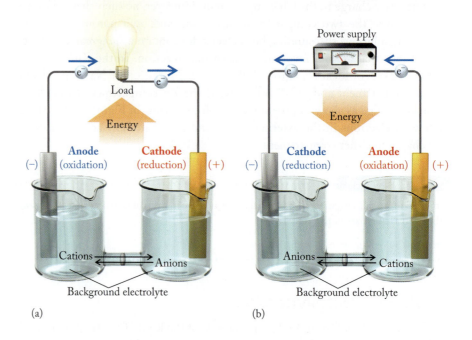

(a)   (b)

When writing a cell diagram, we follow these steps:

1. Write the chemical symbol of the anode at the far left of the diagram, the symbol of the cathode at the far right, and double vertical lines for the connecting bridge halfway between them.
2. Work inward from the electrodes toward the connecting bridge, using vertical lines to indicate phase changes (like that between a solid metal electrode and an aqueous solution). Represent the electrolytes surrounding the electrode using the symbols of the ions or compounds that are changed by the cell reaction. Use commas to separate species in the same phase.
3. If known, use the concentrations of the dissolved species in place of (*aq*) phase symbols, and add the partial pressures of any gases within their (*g*) phase symbols.

Following these steps produces a cell diagram with these components:

anode | anode solution || cathode solution | cathode

In the case of the $Zn/Cu^{2+}$ electrochemical cell in Figure 19.2, the three steps yield the following cell diagram:

(1)   $Zn(s) \dots\dots\dots || \dots\dots\dots Cu(s)$

(2)   $Zn(s) | Zn^{2+}(aq) || Cu^{2+}(aq) | Cu(s)$

(3)   $Zn(s) | Zn^{2+}(1.00\ M) || Cu^{2+}(1.00\ M) | Cu(s)$

**CONCEPT TEST**

Describe in your own words the meaning of the above cell diagram. Start your description with: "The cell consists of a zinc anode, which is oxidized to . . .".

**SAMPLE EXERCISE 19.2** **Diagramming an Electrochemical Cell**  L02

Figure 19.5 depicts an electrochemical cell in which a copper electrode immersed in a 1.00 $M$ solution of $Cu^{2+}$ ions is connected to a silver electrode immersed in a 1.00 $M$ solution of $Ag^+$ ions. Write a balanced chemical equation for this cell reaction and write a cell diagram for this cell.

**Collect and Organize** We have a cell reaction in which electrons spontaneously flow from a copper electrode in contact with a solution of $Cu^{2+}$ ions through an external circuit to a silver electrode in contact with a solution of $Ag^+$ ions. The half-reactions in Table A6.1 involving these metals and ions are

$$Cu^{2+}(aq) + 2\,e^- \rightarrow Cu(s)$$
$$Ag^+(aq) + e^- \rightarrow Ag(s)$$

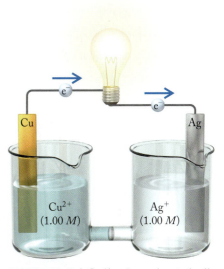

**FIGURE 19.5** A Cu/Ag electrochemical cell.

In a cell diagram, the anode and the species involved in the oxidation half-reaction are on the left and the cathode and the species involved in the reduction half-reaction are on the right. We use single lines to separate phases and a double line to represent the porous bridge separating the two compartments of the cell.

**Analyze** In an electrochemical cell, electrons flow from the anode through an external circuit to the cathode. Therefore, in the cell in Figure 19.5, copper is the anode and silver is the cathode. This means that the Cu reduction half-reaction must run in reverse as an oxidation half-reaction:

$$Cu(s) \rightarrow Cu^{2+}(aq) + 2\,e^-$$

Two moles of electrons are produced in the anode half-reaction, but only 1 mole of electrons is consumed in the cathode half-reaction at the Ag electrode. We therefore need to multiply the silver half-reaction by 2 before combining the two equations.

**Solve** Multiplying the $Ag^+$ half-reaction by 2 and adding it to the Cu half-reaction, we get

$$2\,Ag^+(aq) + 2\,e^- \rightarrow 2\,Ag(s)$$
$$\underline{Cu(s) \rightarrow Cu^{2+}(aq) + 2\,e^-}$$
$$2\,Ag^+(aq) + \cancel{2\,e^-} + Cu(s) \rightarrow 2\,Ag(s) + Cu^{2+}(aq) + \cancel{2\,e^-}$$
$$2\,Ag^+(aq) + Cu(s) \rightarrow 2\,Ag(s) + Cu^{2+}(aq)$$

The equation is balanced, and we are finished with this portion of our task.
    Applying the rules for writing a cell diagram:

1. Anode on the left, cathode on the right, bridge in the middle:

$$Cu(s) \qquad || \qquad Ag(s)$$

2. Adding electrode/solution boundaries and the formulas of the ions produced and consumed in the cell reaction:

$$Cu(s)\,|\,Cu^{2+}(aq)\,||\,Ag^+(aq)\,|\,Ag(s)$$

3. Adding concentration terms:

$$Cu(s)\,|\,Cu^{2+}(1.00\ M)\,||\,Ag^+(1.00\ M)\,|\,Ag(s)$$

**Think About It** To test the validity of the cell diagram, let's translate it into a sentence: *A copper anode is oxidized to aqueous $Cu^{2+}$ ions, which are separated by a porous bridge from an aqueous solution of $Ag^+$ ions that are reduced to Ag atoms at a silver cathode.* This description matches the net ionic equation for the reaction and is consistent with the cell layout and flow of electrons in Figure 19.5.

⚙ **Practice Exercise** Write a balanced chemical equation and the cell diagram for an electrochemical cell that has a copper cathode immersed in a solution of $Cu^{2+}$ ions and an aluminum anode immersed in a solution of $Al^{3+}$ ions.

## 19.3 Standard Potentials

Table A6.1 lists half-reactions in order of a parameter called the **standard reduction potential ($E°$)**. These potentials are expressed in volts (V), in honor of Alessandro Volta. As we noted in Section 19.1, reduction half-reactions can be reversed, making them oxidation half-reactions. Reversing a reduction half-reaction does not affect the magnitude of its standard potential, but does change its sign. In this respect, $E°$ is like other thermodynamic parameters, such as $\Delta G°$, $\Delta H°$, and $\Delta S°$. Recall from Chapters 5 and 18 that reversing a chemical reaction changes the sign of these parameters but does not change their magnitude.

The $E°$ value of a reduction half-reaction is an indication of how likely the half-reaction is to occur. The more positive the value of $E°$, the greater the probability that the reduction half-reaction will couple with an oxidation half-reaction to produce a spontaneous redox reaction. The most positive $E°$ value in Table A6.1 is for the reduction of fluorine:

$$F_2(g) + 2\,e^- \rightarrow 2\,F^-(aq) \qquad E° = +2.866\text{ V}$$

Fluorine's top position means that it is the most easily reduced substance in Table A6.1. It also means that $F_2$ is the strongest oxidizing agent in the table. It is capable of oxidizing any of the substances on the product side of the half-reactions lower in the table.

Substances with very negative $E°$ values at the bottom of Table A6.1 include the major cations in biological systems and environmental waters: $Na^+$, $K^+$, $Mg^{2+}$, and $Ca^{2+}$. Their negative $E°$ values tell us that these ions are not easily reduced to their free metals in aqueous solutions. They are chemically stable (which explains the presence of these cations in nature).

Now let's consider what happens when the half-reactions in Table A6.1 run in reverse. This means that the products of reduction half-reactions become the reactants of oxidation half-reactions. It also means that the half-reaction at the very bottom of the table, with the most negative $E°$ value,

$$Li^+(aq) + e^- \rightarrow Li(s) \qquad E° = -3.05\text{ V}$$

is the one most likely to run in reverse as an oxidation half-reaction:

$$Li(s) \rightarrow Li^+(aq) + e^-$$

Thus, $Li^+$ ions in aqueous solution have little ability to oxidize anything, but Li metal is a very powerful reducing agent that can reduce any of the substances on the reactant side of the half-reactions in Table A6.1.

**CONNECTION** Oxidizing and reducing agents were introduced in Chapter 4.

> **CONCEPT TEST** ...........................................................
>
> Given the positions of the reactants and products in Table A6.1, predict which of the following reactions is/are spontaneous under standard conditions:
>
> a. $Cu(s) + 2\,Fe^{3+}(aq) \rightarrow Cu^{2+}(aq) + 2\,Fe^{2+}(aq)$
> b. $2\,Ag(s) + Zn^{2+}(aq) \rightarrow 2\,Ag^+(aq) + Zn(s)$
> c. $Hg(\ell) + 2\,H^+(aq) \rightarrow Hg^{2+}(aq) + H_2(g)$
>
> ...........................................................

We can use the standard reduction potentials in Table A6.1 to calculate the **standard cell potentials ($E°_{cell}$)** of electrochemical cells. Standard cell potentials, as their name suggests, are related to the chemical (potential) energy stored in a

**CHEMTOUR** Cell Potential

cell and specifically to how forcefully the cell can pump electrons out from their anodes, through external circuits, and into their cathodes.

The difference in the standard reduction potentials of a voltaic cell's cathode and anode define the cell's standard potential; that is,

$$E°_{cell} = E°_{cathode} - E°_{anode} \quad (19.2)$$

To understand why we subtract $E°_{anode}$ from $E°_{cathode}$, keep in mind that $E°_{anode}$, like all standard electrode potentials, is a *reduction* potential. However, the anode is the electrode in a cell where *oxidation* takes place. The half-reaction happening at the anode is essentially a reduction half-reaction running in reverse; that is, an *oxidation half-reaction*. When we reverse a reaction we change the signs of its thermodynamic properties, such as $\Delta H°$, $\Delta S°$, and $\Delta G°$. Standard cell potential is another thermodynamic property, which, as we shall see in Section 19.4, is directly linked to the change in standard free energy of the cell reaction, $\Delta G°_{cell}$. Therefore, it is only reasonable that the sign of $E°_{anode}$ change when the direction of its half-reaction is reversed. That change in sign is the reason for the minus sign in Equation 19.2.

The superscript (°) in Equation 19.2 has its usual thermodynamic meaning—all reactants and products are in their standard states, that is, the concentrations of all dissolved substances are 1 $M$ and the partial pressures of all gases are 1 bar ($\approx$ 1 atm). To obtain a generalized cell potential in which reactants and products are not necessarily in their standard states, we simply refer to the **cell potential** ($E_{cell}$).

Let's use Equation 19.2 to calculate $E°_{cell}$ for the $Zn/Cu^{2+}$ voltaic cell in Figure 19.2. The standard reduction potential of the cathode half-reaction is

$$Cu^{2+}(aq) + 2\,e^- \rightarrow Cu(s) \quad E° = 0.342\text{ V}$$

To obtain the standard potential for the oxidation half-reaction at the zinc anode, we start with the standard reduction potential of $Zn^{2+}$ ions in Table A6.1:

$$Zn^{2+}(aq) + 2\,e^- \rightarrow Zn(s) \quad E° = -0.762\text{ V}$$

Now we use Equation 19.2 to calculate $E°_{cell}$:

$$E°_{cell} = E°_{cathode} - E°_{anode}$$
$$= 0.342 - (-0.762) = 1.104\text{ V}$$

This is the cell potential we would measure if we connected a device called a voltmeter across the two electrodes, as shown in Figure 19.6, at 25°C under standard conditions.

To use Equation 19.2, we need to know which half-reaction occurs at the cathode (reduction) and which occurs at the anode (oxidation). In other words, we need to know which component of the spontaneous cell reaction is more likely to be oxidized and which is more likely to be reduced. As we have seen, this decision can be made based on the data in Table A6.1. In our $Zn/Cu^{2+}$ cell, for instance, the value of $E°$ for the reduction of $Cu^{2+}$ ions to Cu metal is 0.342 V, which is greater than the value of $E°$ for reducing $Zn^{2+}$ ions to Zn metal (−0.762 V). Therefore, in the $Zn/Cu^{2+}$ voltaic cell, $Cu^{2+}$ ions are reduced and Zn metal is oxidized. We can generalize this observation to the cell reaction of any voltaic cell: the half-reaction with the more positive value of $E°$ runs as a reduction and the other one runs in reverse as an oxidation. This way we are assured that $E°_{cell}$ is, overall, a positive value.

**standard reduction potential ($E°$)** the potential of a reduction half-reaction in which all reactants and products are in their standard states at 25°C.

**standard cell potential ($E°_{cell}$)** a measure of how forcefully an electrochemical cell, in which all reactants and products are in their standard states, can pump electrons through an external circuit.

**cell potential ($E_{cell}$)** the energy released when a voltaic cell, in which reactants and products are not necessarily in their standard states, pumps one mole of electrons through an external circuit.

**FIGURE 19.6** A voltmeter displays a cell potential of 1.104 V between a Zn electrode immersed in a 1.00 $M$ solution of $Zn^{2+}$ ions and a Cu electrode immersed in a 1.00 $M$ solution of $Cu^{2+}$ ions.

▶❚❚ CHEMTOUR Alkaline Battery

Cathode cup
Air diffusion layers
Air access hole

Insulator

Zinc
anode

Separator
Anode cup

Porous
carbon/metal
screen

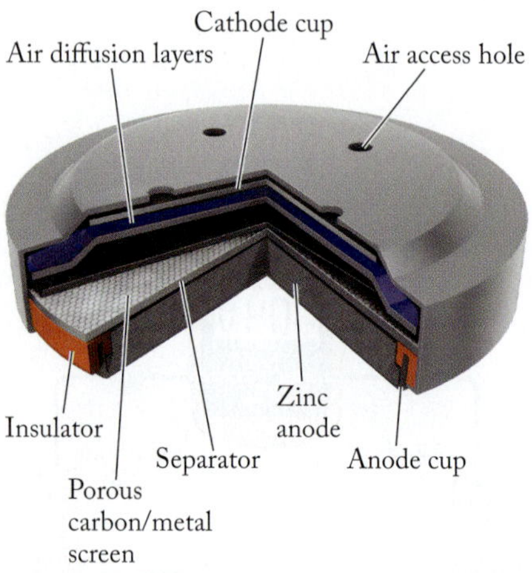

**FIGURE 19.7** Most of the internal volume of a zinc–air battery is occupied by the anode: a slurry of Zn particles in an aqueous solution of KOH, surrounded by a metal cup that serves as the negative terminal of the battery. Oxygen from the air is the reactant at the cathode. Air enters through holes in an inverted metal cup that serves as the positive terminal of the battery. Once inside the battery, air diffuses through layers of gas-permeable plastic film that let air in but keep electrolyte from leaking out. Oxygen in the air is reduced at the porous carbon/metal cathode to OH⁻ ions that migrate toward the anode, where they are consumed in the Zn oxidation half-reaction.

**SAMPLE EXERCISE 19.3**  **Identifying Anode and Cathode Half-Reactions and Calculating the Value of $E°_{cell}$**  **LO3**

The standard reduction potentials of the half-reactions in single-use alkaline batteries are

$$ZnO(s) + H_2O(\ell) + 2\,e^- \rightarrow Zn(s) + 2\,OH^-(aq) \qquad E° = -1.25\ V$$
$$2\,MnO_2(s) + H_2O(\ell) + 2\,e^- \rightarrow Mn_2O_3(s) + 2\,OH^-(aq) \qquad E° = 0.15\ V$$

What is the net ionic equation for the cell reaction and the value of $E°_{cell}$?

**Collect and Organize** We can calculate $E°_{cell}$ using Equation 19.2:

$$E°_{cell} = E°_{cathode} - E°_{anode}$$

However, first we need to decide which half-reaction occurs at the cathode and which at the anode. The equation for a cell reaction is written by combining half-reactions once the loss or gain of electrons in the two half-reactions is balanced.

**Analyze** The $MnO_2$ half-reaction has the more positive $E°$, making it our reduction half-reaction. We must reverse the ZnO half-reaction, turning it into an oxidation half-reaction. The two half-reactions both involve the transfer of two electrons, so we may combine them by simply adding them together.

**Solve** The oxidation half-reaction at the anode is

$$Zn(s) + 2\,OH^-(aq) \rightarrow ZnO(s) + H_2O(\ell) + 2\,e^-$$

The reduction half-reaction at the cathode is

$$2\,MnO_2(s) + H_2O(\ell) + 2\,e^- \rightarrow Mn_2O_3(s) + 2\,OH^-(aq)$$

Combining these half-reactions to obtain the overall cell reaction, we get

$$2\,MnO_2(s) + \cancel{H_2O(\ell)} + Zn(s) + \cancel{2\,OH^-(aq)} + \cancel{2\,e^-} \rightarrow$$
$$Mn_2O_3(s) + \cancel{2\,OH^-(aq)} + ZnO(s) + \cancel{H_2O(\ell)} + \cancel{2\,e^-}$$

Simplifying gives us the net ionic equation for the cell reaction:

$$2\,MnO_2(s) + Zn(s) \rightarrow Mn_2O_3(s) + ZnO(s)$$

The overall $E°_{cell}$ for this reaction is obtained by using Equation 19.2:

$$E°_{cell} = E°_{cathode} - E°_{anode}$$
$$= 0.15\ V - (-1.25\ V) = 1.40\ V$$

**Think About It** The $E°_{cell}$ value is reasonable because the potential of most alkaline batteries is nominally 1.5 V. In this particular cell reaction, the net ionic equation is also the complete molecular equation.

**Practice Exercise** The half-reactions in nicad (nickel–cadmium) batteries are

$$Cd(OH)_2(s) + 2\,e^- \rightarrow Cd(s) + 2\,OH^-(aq) \qquad E° = -0.81\ V$$
$$2\,NiO(OH)(s) + 2\,H_2O(\ell) + 2\,e^- \rightarrow 2\,Ni(OH)_2(s) + 2\,OH^-(aq) \qquad E° = 0.49\ V$$

Write the net ionic equation for the cell reaction and calculate the value of $E°_{cell}$.

Before closing this discussion of standard cell potentials, let's combine two half-reactions in which different numbers of electrons are gained and lost. This combination occurs in a type of battery that has a limitless supply of one of its reactants. It is called the zinc–air battery (Figure 19.7), and it powers devices in which small battery size and low mass are high priorities, such as hearing aids. Most of the internal volume of one of these batteries is occupied by an anode consisting of a paste of zinc particles packed in an aqueous solution of KOH. As in alkaline batteries (Sample Exercise 19.3), the anode half-reaction is

$$Zn(s) + 2\ OH^-(aq) \rightarrow ZnO(s) + H_2O(\ell) + 2\ e^-$$

which is the reverse of the reaction in Table A6.1:

$$ZnO(s) + H_2O(\ell) + 2\ e^- \rightarrow Zn(s) + 2\ OH^-(aq) \qquad E° = -1.25\ V$$

The cathode consists of porous carbon supported by a metal screen. Air diffuses through small holes in the battery and across a layer of Teflon that lets gases pass through but keeps electrolyte from leaking out. As air passes through the cathode, oxygen is reduced to hydroxide ions:

$$O_2(g) + 2\ H_2O(\ell) + 4\ e^- \rightarrow 4\ OH^-(aq) \qquad E°_{cathode} = 0.401\ V$$

To write the overall cell reaction, we need to multiply the oxidation half-reaction by 2 before combining it with the reduction half-reaction:

$$2[Zn(s) + 2\ OH^-(aq) \rightarrow ZnO(s) + H_2O(\ell) + 2\ e^-]$$
$$O_2(g) + 2\ H_2O(\ell) + 4\ e^- \rightarrow 4\ OH^-(aq)$$

$$2\ Zn(s) + 4\ \cancel{OH^-(aq)} + O_2(g) + \cancel{2\ H_2O(\ell)} + \cancel{4\ e^-} \rightarrow$$
$$2\ ZnO(s) + \cancel{2\ H_2O(\ell)} + \cancel{4\ OH^-(aq)} + \cancel{4\ e^-}$$

This simplifies to

$$2\ Zn(s) + O_2(g) \rightarrow 2\ ZnO(s)$$
$$E°_{cell} = E°_{cathode} - E°_{anode} = 0.401\ V - (-1.25\ V) = 1.65\ V$$

Note that when we multiply the anode half-reaction by 2 and add it to the cathode half-reaction, *we do not multiply* the $E°$ of the anode half-reaction by 2. The reason we do not is that $E°$ is an *intensive* property of a half-reaction or a complete cell reaction. It does not change when the quantities of reactants and products change. Thus, a zinc–air battery the size of a pea has the same $E°$ as one the size of a book (like those being developed for electric vehicles). On the other hand, the amount of electrical work a zinc–air battery can do *does* depend on how much zinc is inside it because, as we are about to see, the electrical work that a voltaic cell can do depends on both cell potential *and* the quantity of charge it can deliver at that potential.

## 19.4 Chemical Energy and Electrical Work

When we connect the Zn and Cu electrodes in Figure 19.6 to a digital voltmeter—the Zn electrode to the negative terminal of the meter and the Cu electrode to the positive terminal—the meter reads 1.104 V. These connections tell us that the battery can pump electrons from the Zn electrode through an external circuit to the Cu electrode with a potential of 1.104 V. Under standard conditions this potential is the same as the cell's standard cell potential ($E°_{cell}$); under any other conditions it is simply $E_{cell}$.

Where does the energy come from to pump electrons this forcefully through an external circuit? A hint at the answer comes from the fact that these moving electrons can do electrical work ($w_{elec}$), like lighting a light bulb or turning an electric motor. We saw in Chapter 18 that the ability of a chemical reaction (in this case the Zn/$Cu^{2+}$ cell reaction) to do work (different from expansion or compression) is expressed by the change in free energy ($\Delta G_{cell}$) that accompanies the cell reaction. Recall that when a thermodynamic system does work ($w$) on its surroundings, $w$ is less than zero. Similarly, the change in free energy of the

▶❙❙ **CHEMTOUR** Free Energy

reaction (system) that did this work is also less than zero. Thus, the two quantities would be identical if the efficiency of the energy conversion process were 100%:

$$\Delta G_{cell} = w_{elec} \qquad (19.3)$$

The work done by a voltaic cell on its surroundings is defined as the product of the quantity of electrical charge ($C$) the cell pumps through an external circuit times the cell potential:

$$w_{elec} = -CE_{cell} \qquad (19.4)$$

The negative sign reflects the fact that work done *by* a voltaic cell on its surroundings (the external circuit) corresponds to free energy lost by the cell. Keep in mind that the charge of an electron is 1−, so the passage of one mole of electrons through an external circuit in one direction is matched by the passage of one mole of positive charge ($C$) in the opposite direction. The difference is that electrons are real, and charge is derived from a hypothetical quantity of charge carriers that have the same magnitude, but opposite sign.

Actually, the quantities of electrical charge are not typically expressed in moles but in coulombs (C). As noted in Chapter 2, the magnitude of the charge on a single electron is $1.602 \times 10^{-19}$ coulombs (C). Also note the distinction between italic $C$, a symbol for the variable "charge," and nonitalic C, the abbreviation for the unit "coulomb." The magnitude of electrical charge on 1 mole of electrons is

$$\frac{1.602 \times 10^{-19}\ C}{e^-} \times \frac{6.0221 \times 10^{23}\ e^-}{mol\ e^-} = \frac{9.65 \times 10^4\ C}{mol\ e^-}$$

This quantity of charge, $9.65 \times 10^4$ C/mol $e^-$, is called the **Faraday constant ($F$)** after Michael Faraday (1791–1867), the English chemist and physicist who discovered that redox reactions take place when electrons are transferred from one species to another. The quantity of charge, $C$, flowing through an electrical circuit is the product of the number of moles ($n$) of electrons times the Faraday constant:

$$C = nF \qquad (19.5)$$

Combining Equations 19.4 and 19.5 gives us an equation relating $w_{elec}$ and $E_{cell}$:

$$w_{elec} = -nFE_{cell} \qquad (19.6)$$

If we combine Equations 19.3 and 19.6, we connect the quantity of electrical work a voltaic cell can do on its surroundings with the change in free energy in the cell:

$$\Delta G_{cell} = -nFE_{cell} \qquad (19.7)$$

Perhaps you are wondering how the product on the right side of Equation 19.7 is the equivalent of energy. It is because the units on the right side are

$$\text{moles}\ e^- \times \frac{coulombs}{\text{mole}\ e^-} \times volts = coulombs\text{-}volts$$

A coulomb-volt is the same quantity of energy as a joule:

$$1\ coulomb\text{-}volt = 1\ joule$$
$$1\ C \cdot V = 1\ J$$

The negative sign on the right side of Equation 19.7 tells us that the $E_{cell}$ of any voltaic cell must have a positive value because the sign of $\Delta G_{cell}$ for the spontaneous chemical reaction inside the cell must be negative.

---

**CONNECTION** The sign conventions used for work done *on* a thermodynamic system (+) and the work done *by* the system (−) were explained in Section 5.2.

---

**Faraday constant ($F$)** the magnitude of electrical charge in 1 mole of electrons. Its value to three significant figures is $9.65 \times 10^4$ C/mol $e^-$.

Let's use Equation 19.7 to calculate the change in standard free energy of the $Zn/Cu^{2+}$ cell reaction. We start with the standard cell potential calculated in Section 19.3:

$$E^{\circ}_{cell(Zn/Cu^{2+})} = 1.104 \text{ V}$$

We can convert this standard cell potential into a change in standard free energy ($\Delta G^{\circ}_{cell}$) using Equation 19.7 under standard conditions, so that $\Delta G = \Delta G^{\circ}$ and $\Delta E = \Delta E^{\circ}$:

$$\Delta G^{\circ}_{cell} = -nFE^{\circ}_{cell}$$

$$= -\left(2 \text{ mol e}^- \times \frac{9.65 \times 10^4 \text{ C}}{\text{mol e}^-} \times 1.104 \text{ V}\right) = -2.13 \times 10^5 \text{ C} \cdot \text{V}$$

$$= -2.13 \times 10^5 \text{ J} = -213 \text{ kJ}$$

To put this value in perspective, the $Zn/Cu^{2+}$ reaction produces nearly as much useful energy as the combustion of 1 mole of hydrogen gas:

$$H_2(g) + \tfrac{1}{2} O_2(g) \rightarrow H_2O(g) \qquad \Delta G^{\circ} = -228.6 \text{ kJ}$$

**CONCEPT TEST**

When a rechargeable battery, such as the one used to start a car's engine, is recharged, an external source of electrical power forces the voltaic cell reaction to run in reverse. What are the signs of $\Delta G_{cell}$ and $E_{cell}$ during the recharging process?

---

**SAMPLE EXERCISE 19.4**   **Relating $\Delta G^{\circ}_{cell}$ and $E^{\circ}_{cell}$**   **LO4**

Many of the "button" batteries used in electric watches consist of a Zn anode and a $Ag_2O$ cathode, separated by a membrane soaked in a concentrated solution of KOH (Figure 19.8). At the cathode, $Ag_2O$ is reduced to Ag metal, and at the anode Zn is oxidized to solid $Zn(OH)_2$. Write the net ionic equation for the reaction, and use the appropriate standard reduction potentials in Table A6.1 to calculate the values of $E^{\circ}_{cell}$ and $\Delta G^{\circ}_{cell}$.

**Collect and Organize** We know the reactants and products of the anode and cathode reactions and that the reaction occurs in a basic solution. The following equations should be useful in calculating $E^{\circ}_{cell}$ and $\Delta G^{\circ}_{cell}$ from the appropriate standard potentials:

$$E^{\circ}_{cell} = E^{\circ}_{cathode} - E^{\circ}_{anode}$$
$$\Delta G^{\circ}_{cell} = -nFE^{\circ}_{cell}$$

**Analyze** The half-reaction at the cathode is based on the reduction of $Ag_2O$ to Ag. The appropriate half-reaction in Table A6.1 is

$$Ag_2O(s) + H_2O(\ell) + 2 \text{ e}^- \rightarrow 2 \text{ Ag}(s) + 2 \text{ OH}^-(aq) \qquad E^{\circ}_{cathode} = 0.342 \text{ V}$$

We must reverse the anode oxidation half-reaction to find an entry in Table A6.1 in which $Zn(OH)_2$ is the reactant and Zn is the product:

$$Zn(OH)_2(s) + 2 \text{ e}^- \rightarrow Zn(s) + 2 \text{ OH}^-(aq) \qquad E^{\circ}_{anode} = -1.249 \text{ V}$$

We need to reverse this half-reaction before combining it with the cathode half-reaction. The two half-reactions involve the transfer of the same number of electrons ($n = 2$), so combining them simply means adding them together. The value of $E^{\circ}_{cell}$ is about $[0.35 - (-1.25)]$ or about 1.60 V. This value is half again as large as the $E^{\circ}_{cell}$ of the $Zn/Cu^{2+}$ cell. Therefore, the magnitude of its $\Delta G^{\circ}_{cell}$ value should be half again as large as $-212$ kJ/mol, or about $-300$ kJ/mol.

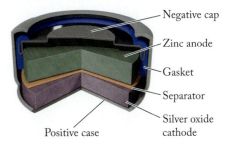

**FIGURE 19.8** Many of the button batteries that power small electronic devices incorporate a Zn anode and a $Ag_2O$ cathode separated by a membrane containing KOH electrolyte.

Negative cap
Zinc anode
Gasket
Separator
Silver oxide cathode
Positive case

**Solve** Reversing the $Zn(OH)_2$ half-reaction and adding it to the $Ag_2O$ half-reaction, we get

$$Zn(s) + 2\ OH^-(aq) \rightarrow Zn(OH)_2(s) + 2\ e^-$$

$$Ag_2O(s) + H_2O(\ell) + 2\ e^- \rightarrow 2\ Ag(s) + 2\ OH^-(aq)$$

$$Ag_2O(s) + H_2O(\ell) + Zn(s) + \cancel{2\ OH^-(aq)} + \cancel{2\ e^-} \rightarrow$$
$$2\ Ag(s) + \cancel{2\ OH^-(aq)} + Zn(OH)_2(s) + \cancel{2\ e^-}$$

This simplifies to

$$Ag_2O(s) + H_2O(\ell) + Zn(s) \rightarrow 2\ Ag(s) + Zn(OH)_2(s)$$

Then we calculate $E^\circ_{cell}$:

$$E^\circ_{cell} = E^\circ_{cathode} - E^\circ_{anode}$$
$$= 0.342\ V - (-1.249\ V)$$
$$= 1.591\ V$$

From this value, we can determine $\Delta G^\circ_{cell}$:

$$\Delta G^\circ_{cell} = -nFE^\circ_{cell}$$
$$= -(2\ \cancel{mol\ e^-} \times 9.65 \times 10^4\ C/\cancel{mol\ e^-} \times 1.591\ V)$$
$$= -3.07 \times 10^5\ C \cdot V = -3.07 \times 10^5\ J = -307\ kJ$$

**Think About It** The positive value of $E^\circ_{cell}$ and negative value of $\Delta G^\circ_{cell}$ are expected because voltaic cell reactions are spontaneous; and the calculated values are close to those we estimated.

⚙ **Practice Exercise** If an alkaline battery produces a cell potential of 1.50 V, what is the value of $\Delta G_{cell}$?

■

Some final thoughts about the $\Delta G^\circ_{cell}$ value calculated in Sample Exercise 19.4: First, it is based on the reaction of 1 mole $Ag_2O$ and 1 mole Zn, which correspond to 232 g $Ag_2O$ and 65 g Zn. The energy stored in a button battery (Figure 19.8), which has a mass of only 1 or 2 grams, would be a tiny fraction of the calculated value. Also, note that no ions appear in the net ionic equation because all the reactants and products in the silver oxide battery reaction are solids, so the net ionic equation and molecular equation are identical.

# 19.5 A Reference Point: The Standard Hydrogen Electrode

We can measure the value of $E_{cell}$ using a voltmeter, but how do we measure the individual electrode potentials of the cathode and anode? The answer to this question is that we arbitrarily assign a value of zero volts to the standard potential for the reduction of hydrogen ions to hydrogen gas:

$$2\ H^+(aq) + 2\ e^- \rightarrow H_2(g) \qquad E^\circ = 0.000\ V \qquad (19.8)$$

**standard hydrogen electrode (SHE)** a reference electrode based on the half-reaction $2\ H^+(aq) + 2\ e^- \rightarrow H_2(g)$ that produces a standard electrode potential of 0.000 V.

An electrode that generates this reference potential, called the **standard hydrogen electrode (SHE)**, consists of a platinum electrode in contact with a solution of a strong acid ($[H^+] = 1.00\ M$) and hydrogen gas at a pressure of 1.00 atm (Figure 19.9). The platinum is not changed by the electrode reaction. Rather, it serves as a chemically inert conveyor of electrons. Electrons are conveyed to the electrode surface if $H^+$ ions are being reduced to hydrogen gas and away from the electrode surface if hydrogen gas is being oxidized to $H^+$ ions. The potential of the SHE is the same for both half-reactions, 0.000 V.

To write the cell diagram for a cell in which the SHE serves as the anode, we represent the SHE half of the cell as follows:

$$Pt \mid H_2(g, 1.00 \text{ atm}) \mid H^+(1.00 \text{ } M) \parallel$$

This indicates that the anode half-reaction involves the oxidation of $H_2$ gas to $H^+$ ions. If the SHE is the cathode, then we diagram its half of the cell this way:

$$\parallel H^+ (1.00 \text{ } M) \mid H_2(g, 1.00 \text{ atm}) \mid Pt$$

This indicates that the cathode half-reaction involves the reduction of $H^+$ ions to $H_2$ gas.

Because the standard reduction (or oxidation) potential of the SHE is 0.000 V, the measured $E^{\circ}_{cell}$ of any voltaic cell in which a SHE is one of the two electrodes— either cathode or anode—is the potential produced by the other electrode. This means that if we attach a voltmeter to the cell, the meter reading is the electrode potential of the other electrode. Suppose, for example, that a voltaic cell consists of a strip of zinc metal immersed in a 1.00 $M$ solution of $Zn^{2+}$ ions in one compartment and a SHE in the other (Figure 19.10a). Also suppose that a voltmeter is connected to the cell so that it measures the potential at which the cell pumps electrons from the zinc electrode to the SHE. This direction of electron flow means that the zinc electrode is the cell's anode and the SHE is the cathode of the cell. At 25°C the meter reads 0.762 V. We know that the value of $E^{\circ}_{cathode}$ is that of the SHE (0.000 V), and that $E^{\circ}_{anode}$ is $E^{\circ}_{Zn}$.

Inserting these values and symbols into Equation 19.2,

$$E^{\circ}_{cell} = E^{\circ}_{cathode} - E^{\circ}_{anode}$$
$$E^{\circ}_{cell} = E^{\circ}_{SHE} - E^{\circ}_{Zn}$$
$$0.762 \text{ V} = 0.000 \text{ V} - E^{\circ}_{Zn}$$
$$E^{\circ}_{Zn} = -0.762 \text{ V}$$

This value is equal to the standard reduction potential of $Zn^{2+}$ in Table A6.1:

$$Zn^{2+}(aq) + 2 \text{ e}^- \rightarrow Zn(s) \qquad E^{\circ} = -0.762 \text{ V}$$

In Figure 19.10(b), the SHE is coupled to a copper electrode immersed in a 1.00 $M$ solution of $Cu^{2+}$ ions. In this cell, electrons flow from the SHE through an external circuit to the copper electrode at a cell potential of 0.342 V at 25°C.

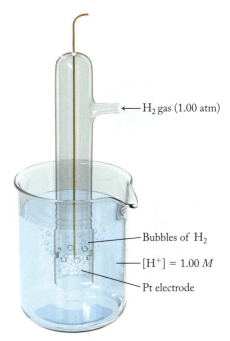

**FIGURE 19.9** The standard hydrogen electrode consists of a platinum electrode immersed in a 1.00 $M$ solution of $H^+(aq)$ and bathed in a stream of pure $H_2$ gas at a pressure of 1.00 atm. Its potential is the same (0.000 V) whether $H^+(aq)$ ions are reduced or $H_2$ gas is oxidized.

— $H_2$ gas (1.00 atm)

— Bubbles of $H_2$

— $[H^+] = 1.00 \text{ } M$

— Pt electrode

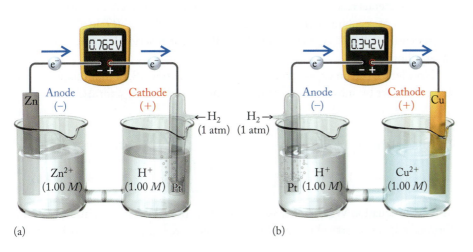

(a)

(b)

**FIGURE 19.10** The standard hydrogen electrode allows us to determine the standard potential of any half-reaction. (a) When coupled to a Zn electrode under standard conditions, the SHE is the cathode ($H^+$ is reduced) and the Zn electrode is the anode. When the SHE is connected to the negative terminal of a voltmeter and the Zn electrode to the positive terminal, the meter measures a cell potential of 0.762 V. (b) When coupled to a Cu electrode under standard conditions, the SHE is the anode ($H_2$ is oxidized), the Cu electrode is the cathode, and the meter measures a cell potential of 0.342 V.

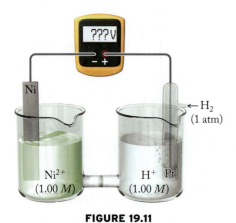

**FIGURE 19.11**

The direction of current flow means that the electrons are consumed at the copper electrode, making it the cathode. The value of $E°$ for the copper half-reaction is calculated as follows:

$$E°_{cell} = E°_{cathode} - E°_{anode}$$

$$= E°_{Cu} - E°_{SHE}$$

$$0.342 \text{ V} = E°_{Cu} - 0.000 \text{ V}$$

$$E°_{Cu} = 0.342 \text{ V}$$

This half-reaction potential matches the value of $E°$ for the reduction of $Cu^{2+}$ to Cu metal in Table A6.1.

**CONCEPT TEST**

A cell consists of a SHE in one compartment and a Ni electrode immersed in a 1.00 $M$ solution of $Ni^{2+}$ ions in the other. If a voltmeter is connected to the electrode as shown in Figure 19.11, what will be the value on the voltmeter's display?

## 19.6 The Effect of Concentration on $E_{cell}$

Reactions stop when one of the reactants is completely consumed. This concept was the basis for our discussion of limiting reactants in Chapter 3. However, a commercial battery usually stops operating at its rated cell potential—1.5 V for a flashlight battery—before its reactants are completely consumed. This happens because the cell potential of a voltaic cell depends on the concentrations of the reactants and products.

### The Nernst Equation

In 1889, German chemist Walther Nernst (1864–1941) derived an expression, now called the **Nernst equation**, that describes how cell potentials depend on reactant and product concentrations. We can reconstruct his derivation starting with Equation 18.14 which relates the change in free energy $\Delta G$ of any reaction to its change in free energy under standard conditions $\Delta G°$:

$$\Delta G = \Delta G° + RT \ln Q \qquad (18.14)$$

As a spontaneous reaction proceeds, concentrations of products increase and concentrations of reactants decrease until the positive value of $RT \ln Q$ offsets the negative value of $\Delta G°$. At that point, $\Delta G = 0$ and the reaction has reached chemical equilibrium.

Now let's write an expression analogous to Equation 18.14 that relates $E_{cell}$ to $E°_{cell}$. We start by substituting $-nFE_{cell}$ for $\Delta G_{cell}$ and $-nFE°_{cell}$ for $\Delta G°_{cell}$:

$$-nFE_{cell} = -nFE°_{cell} + RT \ln Q$$

Dividing all terms by $-nF$ gives

$$E_{cell} = E°_{cell} - \frac{RT \ln Q}{nF} \qquad (19.9)$$

This is the equation Walther Nernst developed in 1889. We can obtain a very useful form of it if we insert values for $R$ [8.314 J/(mol · K)] and $F$ (9.65 × 10⁴ C/mol),

**CONNECTION** In Chapter 18 we discussed the relationship between change in free energy and the reaction quotient $Q$.

**Nernst equation** an equation relating the potential of a cell (or half-cell) reaction to its standard potential ($E°$) and to the concentrations of its reactants and products.

assume $T = 298$ K, and convert the natural logarithm to a base-10 logarithm: ln $Q = 2.303$ log $Q$. With these changes, the Nernst equation becomes

$$E_{cell} = E°_{cell} - \frac{0.0592 \text{ V}}{n} \log Q \qquad (19.10)$$

Equation 19.10 allows us to predict how the potential ($E_{cell}$ in V) of a voltaic cell at 298 K changes as the concentrations of products inside the cell increase and the concentrations of reactants decrease. As they do, $Q$ increases and so does the value of 0.0592 V/$n$ × log $Q$. The negative sign in front of this term in Equation 19.10 means that the value of $E_{cell}$ decreases as reactants are converted into products. Eventually, $E_{cell}$ approaches zero. When it reaches zero, the cell reaction has achieved chemical equilibrium. The cell can no longer pump electrons through an external circuit. In other words, it's dead.

**CONCEPT TEST** • • • • • • • • • • • • • • • • • • • • • • • • • • • • • • • • •

We can also use Equation 19.10 to calculate the potential of a single electrode. Consider the half-reaction at the Ag/Ag$^+$ electrode:

$$\text{Ag}^+(aq) + \text{e}^- \rightarrow \text{Ag}(s) \qquad E° = 0.799 \text{ V}$$

What is the potential of this half-reaction at 25°C when the concentration of Ag$^+$ is 0.100 $M$?

• • • • • • • • • • • • • • • • • • • • • • • • • • • • • • • • • • • • • • • •

Batteries are voltaic cells, so their cell potential should drop with usage. Let's consider how much they drop by focusing on the *lead–acid* battery used to start most car engines. These batteries each contain six electrochemical cells. Their anodes are made of Pb and their cathodes are made of PbO$_2$. Both electrodes are immersed in 4.5 $M$ H$_2$SO$_4$ (Figure 19.12). The value of a fully charged cell is about 2.0 V. The six cells are connected in series so that the operating potential of the battery is the sum of the six cell potentials, or about 12 V.

As the battery discharges, PbO$_2(s)$ is reduced to PbSO$_4(s)$ at the cathodes:

$$\text{PbO}_2(s) + 3\text{ H}^+(aq) + \text{HSO}_4^-(aq) + 2\text{ e}^- \rightarrow \text{PbSO}_4(s) + 2\text{ H}_2\text{O}(\ell)$$
$$E° = 1.685 \text{ V}$$

Also, Pb($s$) is oxidized to PbSO$_4(s)$ at the anodes:

$$\text{Pb}(s) + \text{HSO}_4^-(aq) \rightarrow \text{PbSO}_4(s) + \text{H}^+(aq) + 2\text{ e}^- \qquad E° = -0.356 \text{ V}$$

The reduction half-reaction consumes 2 moles of electrons, and the oxidation half-reaction involves the loss of 2 moles of electrons for each mole of lead.

The net ionic equation for the overall cell reaction is the sum of the two half-reactions:

$$\text{PbO}_2(s) + \text{Pb}(s) + 2\text{ H}^+(aq) + 2\text{ HSO}_4^-(aq) \rightarrow 2\text{ PbSO}_4(s) + 2\text{ H}_2\text{O}(\ell)$$

and the value of $E°_{cell}$ is

$$E°_{cell} = E°_{cathode} - E°_{anode} = 1.685 \text{ V} - (-0.356 \text{ V}) = 2.041 \text{ V}$$

As the battery discharges, the concentration of sulfuric acid decreases, and so does the value of $E_{cell}$ calculated from the Nernst equation:

$$E_{cell} = 2.041 \text{ V} - \frac{0.0592 \text{ V}}{2} \log \frac{1}{[\text{H}^+]^2 [\text{HSO}_4^-]^2}$$

However, the decrease in $E_{cell}$ is very gradual, not falling below 2.0 V until the battery is about 97% discharged, as shown in Figure 19.13. The gradual decrease

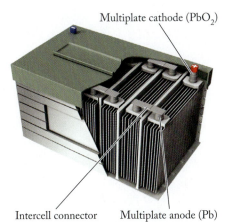

Multiplate cathode (PbO$_2$)

Intercell connector    Multiplate anode (Pb)

**FIGURE 19.12** The lead–acid battery that provides power to start most motor vehicles contains six cells. Each has an anode made of lead and a cathode made of PbO$_2$ immersed in a background electrolyte of 4.5 $M$ H$_2$SO$_4$. The electrodes are formed into plates and held in place by grids made of a lead alloy. The grids connect the cells together in series so that the operating potential of the battery (12.0 V) is the sum of six $E_{cell}$ values (each 2.0 V).

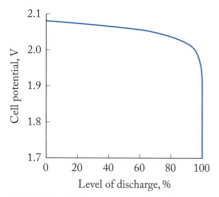

**FIGURE 19.13** The potential of a cell in a lead–acid battery decreases as reactants are converted into products, but the change in potential is small until the battery is nearly completely discharged.

makes sense because of the logarithmic relationship between $Q$ and $E_{cell}$. If, for example, the concentration of sulfuric acid decreased by an order of magnitude, say, from 1.00 $M$ to 0.100 $M$, the value of $E_{cell}$ would decrease by only about 0.1 V—from 2.041 V to

$$E_{cell} = 2.041 \text{ V} - \frac{0.0592 \text{ V}}{2} \log \frac{1}{0.100^2 \times 0.100^2} = 1.982 \text{ V}$$

The logarithmic relationship between $Q$ and $E_{cell}$ means that most batteries can deliver current at a cell potential close to their "design" potential until they are almost completely discharged.

---

**SAMPLE EXERCISE 19.5** **Calculating $E_{cell}$ from $E°_{cell}$ and the Concentrations of Reactants and Products** **LO5**

The standard potential $(E°_{cell})$ of a voltaic cell based on the Zn/Cu²⁺ ion reaction:

$$Zn(s) + Cu^{2+}(aq) \rightarrow Zn^{2+}(aq) + Cu(s)$$

is 1.104 V. What is the value of $E_{cell}$ at 25°C when the concentration of $Cu^{2+}$ is 0.100 $M$ and the concentration of $Zn^{2+}$ is 1.90 $M$?

**Collect and Organize** We are given the standard cell potential and are asked to determine the value of $E_{cell}$ when the concentration of $Cu^{2+}$ is 0.100 $M$ and the concentration of $Zn^{2+}$ is 1.90 $M$. The Nernst equation enables us to calculate $E_{cell}$ values for different concentrations of reactants and products. This equation requires us to work with the reaction quotient $Q$, which we know from Section 15.5 to be the mass action expression for the reaction. Solid copper and zinc are also part of the reaction system, but no terms for pure solids appear in reaction quotients.

**Analyze** The only term in the numerator of the $Q$ expression for this cell reaction is $[Zn^{2+}]$, and the only one in the denominator is $[Cu^{2+}]$. Each $Cu^{2+}$ ion acquires two electrons, and each Zn atom donates two electrons, so the value of $n$ in the Nernst equation is 2. The value of $[Zn^{2+}]$ is greater than $[Cu^{2+}]$, which makes $Q > 1$. The negative sign in front of the $0.0592/n \times \log Q$ term in Equation 19.10 means that the calculated value of $E_{cell}$ should be less than the value of $E°_{cell}$.

**Solve** Substituting the values of $[Zn^{2+}]$ and $[Cu^{2+}]$ in the Nernst equation gives

$$E_{cell} = E°_{cell} - \frac{0.0592 \text{ V} \log Q}{n} = 1.104 \text{ V} - \frac{0.0592 \text{ V}}{2} \log \frac{1.90}{0.100}$$

$$E_{cell} = 1.104 \text{ V} - \frac{0.0592 \text{ V}}{2}(1.279) = 1.066 \text{ V}$$

**Think About It** The calculated $E_{cell}$ value is only 0.038 V less than $E°_{cell}$ because the logarithmic dependence of cell potential on reactant and product concentrations minimizes the impact of changing concentrations.

⚙ **Practice Exercise** The standard cell potential of the zinc–air battery (Figure 19.7) is 1.65 V. If at 25.0°C the partial pressure of oxygen in the air diffusing through its cathode is 0.21 atm, what is the cell potential? Assume the cell reaction is

$$2 Zn(s) + O_2(g) \rightarrow 2 ZnO(s)$$

---

## E° and K

When the cell reaction of a voltaic cell reaches chemical equilibrium, $\Delta G_{cell} = E_{cell} = 0$ and $Q = K$. Therefore, Equation 19.9 becomes

$$0 = E°_{cell} - \frac{RT \ln K}{nF}$$

We can rearrange this equation to

$$\ln K = \frac{nFE^{\circ}_{cell}}{RT}$$

converting the natural logarithm to a base-10 logarithm: $\ln K = 2.303 \log K$

$$\log K = \frac{nFE^{\circ}_{cell}}{2.303RT}$$

At 298 K, this equation simplifies to:

$$\log K = \frac{nFE^{\circ}_{cell}}{0.0592 \text{ V}} \qquad (19.11)$$

We can use Equation 19.11 to calculate the equilibrium constant for any redox reaction, not just those in electrochemical cells. For the more general case, we substitute $E^{\circ}_{rxn}$ for $E^{\circ}_{cell}$:

$$\log K = \frac{nFE^{\circ}_{rxn}}{0.0592 \text{ V}} \qquad (19.12)$$

---

**SAMPLE EXERCISE 19.6** **Calculating $K$ for a Redox Reaction from the Standard Potentials of Its Half-Reactions** **LO6**

Many procedures for determining mercury levels in environmental samples begin by oxidizing the mercury to $Hg^{2+}$ and then reducing the $Hg^{2+}$ to elemental Hg with $Sn^{2+}$. Use the appropriate $E^{\circ}$ values from Table A6.1 to calculate the equilibrium constant at 25°C for the reaction

$$Sn^{2+}(aq) + Hg^{2+}(aq) \rightarrow Sn^{4+}(aq) + Hg(\ell)$$

**Collect and Organize** Equation 19.12 relates the equilibrium constant for any redox reaction to the standard potential $E^{\circ}_{rxn}$. To calculate $E^{\circ}_{rxn}$, we need to combine the appropriate standard reduction potentials. Table A6.1 lists two half-reactions involving our reactants and products:

$$Hg^{2+}(aq) + 2\,e^- \rightarrow Hg(\ell) \qquad E^{\circ} = 0.851 \text{ V}$$
$$Sn^{4+}(aq) + 2\,e^- \rightarrow Sn^{2+}(aq) \qquad E^{\circ} = 0.154 \text{ V}$$

**Analyze** The problem states that $Hg^{2+}$ is reduced by $Sn^{2+}$, so $Sn^{2+}$ is the reducing agent in the reaction, which means that it must be oxidized. Therefore, the second reaction runs in reverse, as an oxidation, and we must subtract its standard potential from that of the mercury half-reaction. The difference in the two half-reaction potentials is about +0.7 V. Therefore the right side of Equation 19.12 will be about $(2 \times 0.7)/0.06 \approx 23$, and the value of $K$ should be about $10^{23}$.

**Solve** We obtain the standard potential for the reaction from a modified version of Equation 19.2:

$$E^{\circ}_{rxn} = E^{\circ}_{Hg} - E^{\circ}_{Sn} = 0.851 \text{ V} - 0.154 \text{ V} = 0.697 \text{ V}$$

Using this value for $E^{\circ}_{rxn}$ in Equation 19.12 and a value of 2 for $n$, we have

$$\log K = \frac{nFE^{\circ}_{rxn}}{0.0592 \text{ V}} = \frac{2(0.697 \text{ V})}{0.0592 \text{ V}} = 23.5$$

$$K = 10^{23.5} = 3.16 \times 10^{23}$$

**Think About It** The calculated value is quite close to what we estimated. Note how a $E^{\circ}_{rxn}$ value of less than 1 V corresponds to a huge equilibrium constant, indicating that the reaction essentially goes to completion. That the reaction goes to completion is one reason it can be reliably used to determine the concentrations of mercury in samples containing $Hg^{2+}$ ions.

⚙ **Practice Exercise** Use the appropriate standard reduction potentials in Table A6.1 to calculate the value of $K$ at 25°C for the reaction

$$5\,Fe^{2+}(aq) + MnO_4^-(aq) + 8\,H^+(aq) \rightarrow 5\,Fe^{3+}(aq) + Mn^{2+}(aq) + 4\,H_2O(\ell)$$

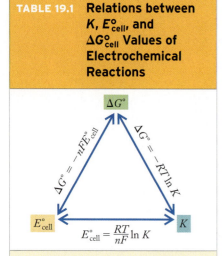

**TABLE 19.1** **Relations between $K$, $E°_{cell}$, and $\Delta G°_{cell}$ Values of Electrochemical Reactions**

| $K$ | $E°_{cell}$ | $\Delta G°_{cell}$ | Favors Formation of |
|-----|-------------|--------------------|---------------------|
| <1 | <0 | >0 | Reactants |
| >1 | >0 | <0 | Products |
| 1 | 0 | 0 | Neither |

Before ending our discussion of how to derive equilibrium constant values at 25°C from $E°_{cell}$ values, we should note that measuring the potential of an electrochemical reaction allows us to calculate equilibrium constant values that may be too large or too small to determine from the equilibrium concentrations of reactants and products. A value of $K$ as large as that calculated in Sample Exercise 19.6 could not be obtained by analyzing the composition of an equilibrium reaction mixture because the concentrations of the reactants would be too small to be determined accurately. Similarly, a cell potential of about −1 V would correspond to a tiny $K$ value and concentrations of products that are too small to be determined quantitatively.

Table 19.1 summarizes how the values of $K$ and $E°_{cell}$ are related to each other and to the change in free energy ($\Delta G°_{cell}$) of a cell reaction under standard conditions. If we know any one of the quantities in Table 19.1, we can calculate the other two. Note how spontaneous electrochemical reactions are those with $E°_{cell}$ values greater than zero and $K$ values greater than 1. The connection between positive cell potential ($E_{cell}$) and reaction spontaneity applies even under nonstandard conditions. Also keep in mind that small positive values of $E°_{cell}$ (only a fraction of a volt, for example) correspond to very large $K$ values and to cell reactions that go nearly to completion.

## 19.7 Relating Battery Capacity to Quantities of Reactants

An important performance characteristic of a battery is its *capacity* to do electrical work; that is, to deliver electrical charge at the designed cell potential. This capacity—the amount of electrical work done—is defined by Equation 19.4,

$$w_{elec} = -CE_{cell}$$

Here $C$ is the quantity of electrical charge delivered in coulombs (C).

Another important unit in electricity is the *ampere* (A), which is the SI base unit of electrical current. An ampere is defined as a current of 1 coulomb per second:

$$1 \text{ ampere} = 1 \text{ coulomb/second}$$

which we can rearrange to

$$1 \text{ coulomb} = 1 \text{ ampere-second}$$

Multiplying both sides of this equation by volts, and recalling that 1 joule of electrical energy is equivalent to 1 coulomb-volt of electrical work, we get

$$1 \text{ (coulomb)(volt)} = 1 \text{ joule} = 1 \text{ (ampere-second)(volt)} \quad (19.13)$$

However, joules are small energy units, so battery capacities are usually expressed in energy units with time intervals longer than seconds. For example, the energy ratings of rechargeable AA batteries (Figure 19.14) are often expressed in ampere-hours at the rated cell potential.

We need even bigger units to express the power and energy capacities of the large battery packs used in hybrid vehicles. They are the *watt* (W), the SI unit of power, and the *kilowatt-hour*, a unit of energy equal to over 3 million joules, as shown in the following unit conversions:

$$1 \text{ watt} = 1 \text{ joule/second}$$
$$1 \text{ kilowatt} = 1000 \text{ W} = 1000 \text{ J/s}$$
$$1 \text{ kilowatt} \cdot \text{hour} = (1000 \text{ W})(1 \text{ hr})$$
$$= (1000 \text{ J/s} \times 60 \text{ s/min} \times 60 \text{ min/hr})(1 \text{ hr})$$
$$= 3.6 \times 10^6 \text{ J}$$

**FIGURE 19.14** The electrical energy rating of these rechargeable nickel–metal hydride AA batteries is 2500 milliampere-hours at 1.2 V.

## Nickel–Metal Hydride Batteries

As we noted at the beginning of this chapter, hybrid vehicles such as the Toyota Prius are powered by combinations of small gasoline engines and electric motors. Electricity for the motors comes from battery packs (Figure 19.15) made of dozens of nickel–metal hydride (NiMH) cells (Figure 19.16). At the cathodes in these cells, NiO(OH) is reduced to Ni(OH)$_2$, and at the anodes, made of one or more transition metals, hydrogen atoms are oxidized to H$^+$ ions. The electrodes are separated by aqueous KOH.

The cathode half-reaction is

$$\text{NiO(OH)}(s) + \text{H}_2\text{O}(\ell) + e^- \rightarrow \text{Ni(OH)}_2(s) + \text{OH}^-(aq) \qquad E° = 1.32 \text{ V}$$

At the anode, hydrogen is present as a *metal hydride*. To write the anode half-reaction, we use the generic formula MH, where M stands for a transition metal or metal alloy that forms a hydride. In a basic background electrolyte, the anode oxidation half-reaction is

$$\text{MH}(s) + \text{OH}^-(aq) \rightarrow \text{M}(s) + \text{H}_2\text{O}(\ell) + e^-$$

The standard potential of this half-reaction depends on the chemical properties of MH, but generally the value is near that of the SHE, or about 0.0 V.

The overall cell reaction from these two half-reactions is

$$\text{MH}(s) + \text{NiO(OH)}(s) \rightarrow \text{M}(s) + \text{Ni(OH)}_2(s)$$

The value of $E°_{cell}$ for the NiMH battery cannot be calculated precisely because we have only an approximate value of $E°_{anode}$. Most NiMH cells are rated at about 1.2 V.

**CONCEPT TEST** ......................................................

In a NiMH battery, what are the oxidation states of (a) Ni in NiO(OH), (b) H in MH, (c) M in MH, and (d) H in H$_2$O?

..........................................................................

**FIGURE 19.15** The 2012 Toyota Prius is powered by a combination of a 73-kW (98-horsepower) gasoline engine and a 60-kW electric motor. Electricity for the motor comes from a nickel–metal hydride battery pack located behind and below the back seat.

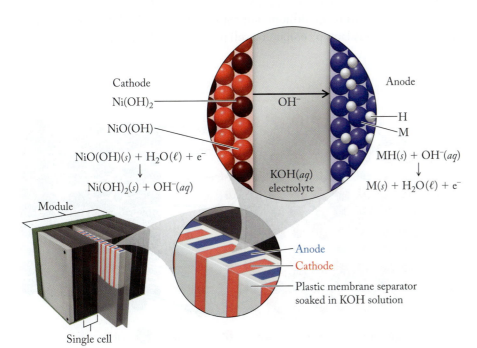

Cathode
Ni(OH)$_2$
NiO(OH)

NiO(OH)($s$) + H$_2$O($\ell$) + e$^-$
↓
Ni(OH)$_2$($s$) + OH$^-$($aq$)

OH$^-$

KOH($aq$)
electrolyte

Anode
H
M

MH($s$) + OH$^-$($aq$)
↓
M($s$) + H$_2$O($\ell$) + e$^-$

Module

Single cell

Anode
Cathode
Plastic membrane separator soaked in KOH solution

**FIGURE 19.16** In the cells of a nickel–metal hydride battery pack, NiO(OH) is reduced to Ni(OH)$_2$ at the cathodes (red plates). The OH$^-$ ions produced by the cathode half-reaction migrate across a KOH-soaked porous membrane and are consumed in the anode half-reaction. At the anodes (blue plates), oxidation of the MH hydrogen atoms produces H$^+$ ions that combine with OH$^-$ ions, forming H$_2$O.

Now let's relate the electrical energy stored in a battery (in other words, its capacity) to the quantities of reactants needed to produce that energy. Consider a rechargeable AA NiMH battery rated to deliver 2.5 ampere-hours of electrical charge at 1.2 V. How much NiO(OH) has to be converted to $Ni(OH)_2$ to deliver this much charge? To answer this question, we need to relate the quantity of charge to a number of moles of electrons, then convert that to an equivalent number of moles of reactant, and finally to a mass of reactant. Recall that an ampere is defined as a coulomb per second, which means the quantity of electrical charge delivered is

$$2.5 \; \text{A} \cdot \text{hr} \times \frac{1 \; \text{C}}{\text{A} \cdot \text{s}} \times \frac{60 \; \text{min}}{1 \; \text{hr}} \times \frac{60 \; \text{s}}{1 \; \text{min}} = 9.0 \times 10^3 \; \text{C}$$

The Faraday constant tells us that 1 mole of charge is equivalent to $9.65 \times 10^4$ C, so the number of moles of charge, which is equal to the number of moles of electrons that flow from the battery, is

$$9.0 \times 10^3 \; \text{C} \left( \frac{1 \; \text{mol e}^-}{9.65 \times 10^4 \; \text{C}} \right) = 0.0933 \; \text{mol e}^-$$

The stoichiometry of the cathode half-reaction tells us that the mole ratio of NiO(OH) to electrons is 1:1. Therefore, the mass of NiO(OH) consumed is

$$0.0933 \; \text{mol e}^- \left( \frac{1 \; \text{mol NiO(OH)}}{1 \; \text{mol e}^-} \right) \left( \frac{91.70 \; \text{g NiO(OH)}}{1 \; \text{mol NiO(OH)}} \right) = 8.6 \; \text{g NiO(OH)}$$

The mass of an AA battery is about 30 g, so this mass for the NiO(OH) is reasonable if we allow for the mass of the anode, background electrolyte, and exterior shell.

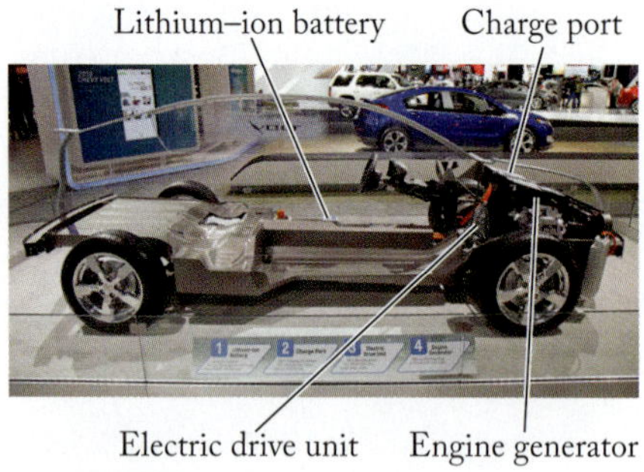

## Lithium–Ion Batteries

The NiMH batteries used in hybrid vehicles do not have the capacity to power them at highway speeds or for extended distances. Nor do these batteries have the energy capacity to power plug-in hybrids such as the Chevrolet Volt on the opening page of this chapter, or all-electric vehicles, such as the Nissan Leaf (Figure 19.17). The electrical power demands of these vehicles require batteries with much greater ratios of energy capacity to battery size. The technology of choice in these applications is the lithium–ion battery (Figure 19.18), the same kind of battery that powers laptop computers, cell phones, and digital cameras.

Lithium–ion battery    Charge port

Electric drive unit    Engine generator

**FIGURE 19.17** The 2013 Nissan Leaf has a 24 kWh battery pack, which gives it an operating range of about 120 km (≈ 75 mi).

**FIGURE 19.18** As this lithium–ion battery discharges, $Li^+$ ions stored in graphite layers of the anode travel to the cathode, which is made of $CoO_2$. During recharging, the direction of ion migration reverses. The crystal structure of the cathode is a cubic closest-packed array of oxide ions in which the $Co^{4+}$ ions occupy half the octahedral holes. $Li^+$ ions move in and out of the remaining holes.

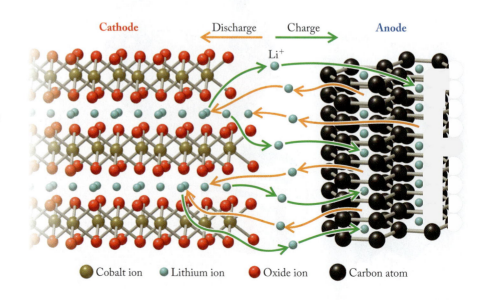

Cathode ← Discharge    Charge → Anode

$Li^+$

● Cobalt ion    ● Lithium ion    ● Oxide ion    ● Carbon atom

In a lithium–ion battery, $Li^+$ ions are stored in a graphite or silicon anode. During discharge, these ions migrate through a nonaqueous electrolyte to a porous cathode. These cathodes are made of transition metal oxides or phosphates that can form stable complexes with $Li^+$ ions. One popular cathode material is cobalt(IV) oxide. Lithium–ion batteries with these cathodes have cell potentials of about 3.6 V (three times that of a NiMH battery). The cell reaction for a lithium–ion battery with this cathode is

$$Li_{1-x}CoO_2(s) + Li_xC_6(s) \rightarrow 6\ C(s) + LiCoO_2(s) \qquad (19.14)$$

In a fully charged cell, $x = 1$, which makes the cathode lithium-free $CoO_2$. As the cell discharges and $Li^+$ ions migrate from the carbon anode to the cobalt oxide cathode, the value of $x$ falls toward zero. To balance this flow of positive charges inside the cell, electrons flow from the anode to the cathode through an external circuit. When fully discharged, the cathode is $LiCoO_2$, and the oxidation number of Co is reduced to +3. The electrodes in a lithium–ion battery may react with oxygen and water, so the background electrolytes (for example, $LiPF_6$) are dissolved in polar organic solvents, such as tetrahydrofuran, ethylene carbonate, or propylene carbonate (Figure 19.19).

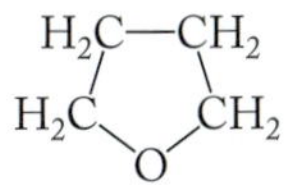

**CONCEPT TEST** ••••••••••••••••••••••••••••••••••••••••••••••••••••••••••••

Which element is oxidized and which is reduced in the $Li^+$ ion cell reaction (Equation 19.14)?

••••••••••••••••••••••••••••••••••••••••••••••••••••••••••••••••••••••••••••••

Tetrahydrofuran

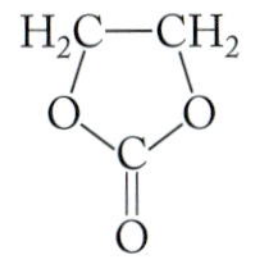

Ethylene carbonate

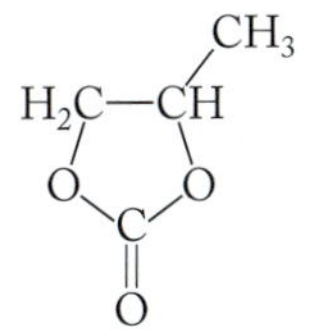

Propylene carbonate

**FIGURE 19.19** Polar organic compounds such as these are the solvents for the background electrolytes in a lithium–ion battery.

---

**SAMPLE EXERCISE 19.7** **Relating Mass of Reactant in an Electrochemical Reaction to Quantity of Electrical Charge** **LO7**

The capacity of the lithium–ion battery in a digital camera is 3.4 W · hr at 3.6 V. How many grams of $Li^+$ ions must migrate from anode to cathode to produce this much electrical energy?

**Collect and Organize** We are asked to relate the electrical energy generated by an electrochemical cell to the mass of the ions involved in generating that energy. We know the cell potential and its capacity in the energy unit watt-hours. Given these starting points and the eventual need to calculate moles and then grams of $Li^+$ ions, the following equivalencies may be useful:

$$1\ watt = 1\ ampere\text{-}volt\ (A \cdot V)$$

$$1\ coulomb\ (C) = 1\ ampere\text{-}second\ (A \cdot s)$$

We may also need to use the Faraday constant, $9.65 \times 10^4$ C/mol $e^-$.

**Analyze** We know the energy capacity of the battery, which is the product of the charge (electrons) it can deliver times the cell potential pumping that charge. This exercise focuses on the quantity of charge, so we need to separate the cell potential's contribution to the energy rating from the charge's contribution. To do that we need to divide the energy rating in watt-hours by the battery's cell potential in volts, and then follow that division with these unit conversions to get to grams of $Li^+$ ions.

$$\boxed{\dfrac{watt \cdot hour}{volt}} \xrightarrow{\dfrac{amp \cdot volt}{watt}} \xrightarrow{\dfrac{coulomb}{amp \cdot second}} \xrightarrow{\dfrac{3600\ second}{hour}} \xrightarrow{\dfrac{1\ mol\ e^-}{9.65 \times 10^4\ C}} \xrightarrow{\dfrac{1\ mol\ Li^+}{1\ mol\ e^-}}$$

$$\xrightarrow{molar\ mass\ Li^+} = \boxed{g\ Li^+}$$

The ratio of the initial values, 3.4 watt-hours and 3.6 volts, is nearly one, so the result of the calculation can be estimated based on the approximate values of the combined conversion factors, or roughly $4000 \times 7/10^5$, or about 0.3.

**Solve** Using the given energy and cell-potential values in the above unit conversion series, we get

$$\frac{3.4 \ \text{W} \cdot \text{hr}}{3.6 \ \text{V}} \times \frac{1 \ \text{A} \cdot \text{V}}{1 \ \text{W}} \times \frac{1 \ \text{C}}{1 \ \text{A} \cdot \text{s}} \times \frac{3600 \ \text{s}}{1 \ \text{hr}} \times \frac{1 \ \text{mol e}^-}{9.65 \times 10^4 \ \text{C}}$$

$$\times \frac{1 \ \text{mol Li}^+}{1 \ \text{mol e}^-} \times \frac{6.941 \ \text{g Li}^+}{1 \ \text{mol Li}^+} = 0.24 \ \text{g Li}$$

**Think About It** The battery that is the subject of this exercise has a mass of about 22 grams, so $Li^+$ ions make up only about 1% of the mass of the battery. This small percentage is not surprising given the masses of the other required components of the cell, including an anode where $Li^+$ ions are surrounded by hexagons of six carbon atoms and a cathode made of $CoO_2$, for example, which has 13 times the molar mass of Li.

⚙ **Practice Exercise** Magnesium metal is produced by passing an electric current through molten $MgCl_2$. The reaction at the cathode is

$$Mg^{2+}(\ell) + 2 \ e^- \rightarrow Mg(s)$$

How many grams of magnesium metal are produced if an average current of 63.7 A flows for 4.50 hr? Assume all of the current is consumed by the half-reaction shown.

# 19.8 Corrosion: Unwanted Electrochemical Reactions

In Chapter 18 we learned about the spontaneous nature of corrosion processes, such as iron rusting. We described the formation of rust in terms of the overall redox reaction:

$$4 \ Fe(s) + 3 \ O_2(g) \rightarrow 2 \ Fe_2O_3(s)$$

In this chapter we have seen how redox reactions are combinations of reduction and oxidation half-reactions. These half-reactions are physically separated in electrochemical cells, and they are often separated in corrosion reactions. In this respect, the chemistry of corrosion is much like the electrochemical reactions in voltaic cells. In fact, we can define **corrosion** as the deterioration of metals due to spontaneous electrochemical reactions. This definition is reflected in several of the factors that promote corrosion:

1. *The presence of water.* Metals that are left out in the rain and snow rust or corrode more rapidly than those that are under cover.
2. *The presence of electrolytes.* Just as electrolytes carry electrical current between anodes and cathodes and facilitate cell reactions, corrosion is much more rapid in, for example, seawater than in freshwater.
3. *Contact between dissimilar metals.* Metals corrode more rapidly when in contact with other metals that are less likely to be oxidized; that is, they have higher reduction potentials.

Let's explore the impact of these factors using a model corrosion problem of a well-known metal structure, the Statue of Liberty (Figure 19.20a). The exterior of the statue is made of sculpted copper sheets. As the statue was built, its exterior copper sheets were attached to and supported by an interior network of iron beams (Figure 19.20b). The French designers of the statue knew these two metals in contact with each other might someday pose a corrosion problem because the two have very different electrochemical properties. The difference is reflected in

**corrosion** a process in which a metal is oxidized by substances in its environment.

(a)

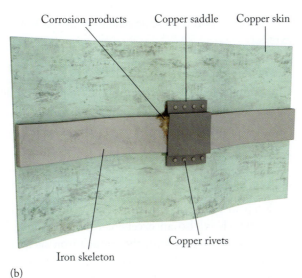

Corrosion products    Copper saddle    Copper skin

Copper rivets

Iron skeleton

(b)

**FIGURE 19.20** (a) The light green patina of the Statue of Liberty is caused by the accumulation of Cu(II) compounds on the surface of the copper sheets that make up its exterior. (b) When the statue was built, these sheets were supported by an iron skeleton that corroded near the points of contact with the sheets and the copper saddles that held the sheets and skeleton together.

the standard reduction potentials of the ions they form when they oxidize, $Fe^{2+}$ and $Cu^{2+}$:

$$Fe^{2+}(aq) + 2\ e^- \rightarrow Fe(s) \qquad E° = -0.447 V$$
$$Cu^{2+}(aq) + 2\ e^- \rightarrow Cu(s) \qquad E° = 0.342\ V$$

As we discussed in Section 19.3, the greater $E°$ of $Cu^{2+}$ means that it is more easily reduced under standard conditions than $Fe^{2+}$, and Fe is more easily oxidized than Cu. We can confirm this by flipping the iron half-reaction and combining it with the copper half-reaction:

$$Cu^{2+}(aq) + 2\ e^- \rightarrow Cu(s)$$
$$\underline{Fe(s) \rightarrow Fe^{2+}(aq) + 2\ e^-}$$
$$Cu^{2+}(aq) + Fe(s) \rightarrow Cu(s) + Fe^{2+}(aq) \qquad (19.15)$$

The combined equation could be that of an electrochemical cell that has a copper cathode, an iron anode, and a positive standard cell potential:

$$E°_{cell} = E°_{cathode} - E°_{anode} = 0.342\ V - (-0.447\ V) = 0.789\ V$$

The positive value of $E°_{cell}$ tells us that under standard conditions, iron in contact with copper will spontaneously oxidize to $Fe^{2+}$ ions as $Cu^{2+}$ ions are reduced.

To suppress the reaction in Equation 19.15, insulators made of asbestos mats soaked in shellac (Figure 19.20b) were used to separate the statue's iron skeleton from its copper exterior. Unfortunately, these insulators did not stand up to the humid, marine environment of New York Harbor. They absorbed water vapor and seawater spray, eventually turning into electrolyte-soaked sponges that actually promoted rather than retarded iron oxidation.

You may be wondering about the source of $Cu^{2+}$ ions that are the oxidizing agents in Equation 19.15. The light green patina of the Statue of Liberty is visual evidence that $Cu^{2+}$ ions are indeed present. They exist on the statue's surface in a mixture of Cu(II) compounds, including $CuCO_3$, $CuSO_4$, and $Cu(OH)_2$. These substances are the products of other corrosion processes involving atmospheric oxidizing agents such as $O_2$ gas:

$$O_2(g) + 4\ H^+(aq) + 4\ e^- \rightarrow 2\ H_2O(\ell)$$

This half-reaction drives the oxidation of Cu atoms to $Cu^{2+}$ ions on the statue's exterior:

$$Cu(s) \rightarrow Cu^{2+}(aq) + 2\ e^-$$

Combining these two half-reactions, we get the following net ionic equation for surface corrosion:

$$O_2(g) + 4\ H^+(aq) + 4\ e^- \rightarrow 2\ H_2O(\ell)$$
$$\underline{2[Cu(s) \rightarrow Cu^{2+}(aq) + 2\ e^-]}$$
$$O_2(g) + 2\ Cu(s) + 4\ H^+(aq) \rightarrow 2\ Cu^{2+}(aq) + 2\ H_2O(\ell) \qquad (19.16)$$

Keep in mind that this reaction occurs on an expansive surface of copper metal, which is also an excellent conductor of electricity and which, unfortunately, was in contact with the statue's iron skeleton through water-logged, ion-rich asbestos mats. This connection meant the reactions in Equations 19.15 and 19.16 were linked together. Let's write an equation describing the combined inside and surface reactions:

$$O_2(g) + 2\ \cancel{Cu(s)} + 4\ H^+(aq) \rightarrow 2\ \cancel{Cu^{2+}(aq)} + 2\ H_2O(\ell)$$
$$\underline{2[\cancel{Cu^{2+}(aq)} + Fe(s) \rightarrow \cancel{Cu(s)} + Fe^{2+}(aq)]}$$
$$2\ Fe(s) + O_2(g) + 4\ H^+(aq) \rightarrow 2\ Fe^{2+}(aq) + 2\ H_2O(\ell) \qquad (19.17)$$

Note how the copper half-reaction has disappeared from the overall reaction. The statue's copper exterior functions as a giant electron delivery system, allowing electrons to flow from iron atoms—as they oxidize to $Fe^{2+}$ inside the statue—to molecules of atmospheric $O_2$ on the exterior surface. The overall effect was a dramatic acceleration of the rate of air-oxidation of the original iron skeleton.

Corrosion of the statue did not end with the conversion of a lot of skeletal iron into $Fe^{2+}$ ions. The final step involved more atmospheric $O_2$ and more moisture and resulted in the conversion of soluble $Fe^{2+}$ ions into hydrous iron(III) oxide:

$$2\ Fe^{2+}(aq) + \tfrac{1}{2}\ O_2(g) + (2 + n)\ H_2O(\ell) \rightarrow Fe_2O_3 \cdot n\ H_2O(s) + 4\ H^+(aq) \quad (19.18)$$

The variable coefficient $n$ is used here to show that $Fe_2O_3$ can incorporate within its solid structure a variable number of water molecules per formula unit—depending on the moisture content of its environment. The common name of hydrous iron(III) oxide is rust!

Let's combine Equations 19.17 and 19.18 to obtain an overall equation describing rust formation:

$$2\ Fe(s) + O_2(g) + \cancel{4\ H^+(aq)} \rightarrow \cancel{2\ Fe^{2+}(aq)} + \cancel{2\ H_2O(\ell)} \qquad (19.17)$$
$$\underline{\cancel{2\ Fe^{2+}(aq)} + \tfrac{1}{2}\ O_2(g) + (2 + n)\ H_2O(\ell) \rightarrow Fe_2O_3 \cdot n\ H_2O(s) + \cancel{4\ H^+(aq)} \quad (19.18)}$$
$$2\ Fe(s) + \tfrac{3}{2}\ O_2(g) + n\ H_2O(\ell) \rightarrow Fe_2O_3 \cdot n\ H_2O(s) \qquad (19.19)$$

The reaction in Equation 19.19 resulted in severe deterioration of the skeletal network that held up the Statue of Liberty, so much so that in the 1980s the iron skeleton had to be replaced with one made of corrosion-resistant stainless steel.

Deterioration of the Statue of Liberty was not an isolated incident. According to one industrial estimate, the direct and indirect cost of corrosion to the U.S. economy in 2012 exceeded $1 trillion. Worldwide, the cost was over $2 trillion. These figures include the cost to repair or replace corroded equipment and structures and to protect them against corrosion. In the latter category is money

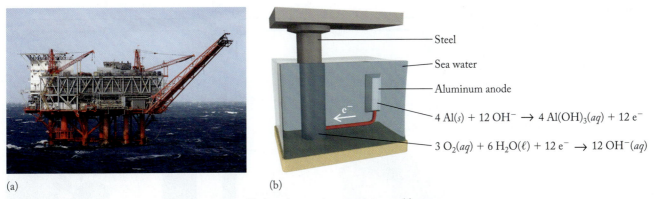

(a)                                                        (b)

**FIGURE 19.21** (a) Ocean-going metal structures are likely to have serious corrosion problems unless (b) they are provided with cathodic protection.

spent on protective coatings, including paint and chemical or electrochemical modification of metal surfaces to make them less reactive.

Another widely used method for inhibiting corrosion involves chemically bonding metal oxide coatings to metal surfaces. For example, a method called bluing is widely used to protect steel tools, gun barrels, wood stoves, and other steel materials. The name comes from the distinctive very dark blue color of a surface layer of magnetite ($Fe_3O_4$), which can be reaction-bonded to steel. Formation of a protective oxide layer is also the mechanism that makes stainless steel resistant to corrosion. Though the main ingredient in all forms of stainless steel is iron, it also contains chromium. Oxidation of surface Cr atoms forms a durable protective layer of $Cr_2O_3$. Similarly, objects made of aluminum are protected by a surface layer of $Al_2O_3$ that inhibits further oxidation and strongly adheres to the underlying metal.

Another way to protect metal structures, especially those that come in contact with seawater (Figure 19.21), involves attaching to them objects made of even more reactive metals. These objects are called *sacrificial* anodes. As their name implies, their role is to form a voltaic cell with the protected structure in which the object is the anode and the structure is the cathode. That way the sacrificial anodes oxidize and the structure does not. This preservation technique is called *cathodic protection*. Many of the sacrificial anodes used in the marine industry are made of zinc or aluminum/magnesium alloys.

$$Zn(s) + 2\ OH^-(aq) \rightarrow ZnO(s) + H_2O(\ell) + 2\ e^-$$
$$Mg(s) + 2\ OH^-(aq) \rightarrow MgO(s) + H_2O(\ell) + 2\ e^-$$
$$2\ Al(s) + 6\ OH^-(aq) \rightarrow Al_2O_3(s) + 3\ H_2O(\ell) + 6\ e^-$$

These materials have the benefit of forming oxide coatings as they oxidize, which partially protect them and slow the rate at which they oxidize further, thereby requiring less frequent replacement.

# 19.9 Electrolytic Cells and Rechargeable Batteries

Lead–acid, NiMH, and lithium–ion batteries are rechargeable, which means that their spontaneous ($\Delta G < 0$) cell reactions that convert chemical energy into electrical work can be forced to run in reverse. Recharging happens when external sources of electrical energy are applied to the batteries. This electrical energy is

**FIGURE 19.22** The lead–acid battery used in many vehicles is based on oxidation of Pb and reduction of $PbO_2$. As the battery discharges (circuit on left), Pb is oxidized to $PbSO_4$ and $PbO_2$ is reduced to $PbSO_4$. When the engine is running (circuit on right), a device called an alternator generates electrical energy that flows into the battery, recharging it as both electrode reactions are reversed: $PbSO_4$ is oxidized to $PbO_2$, and $PbSO_4$ is reduced to Pb.

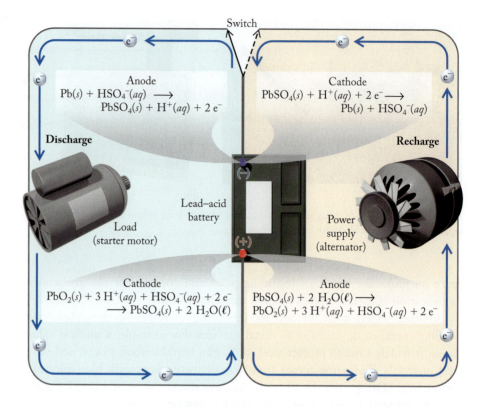

converted into chemical energy as it drives nonspontaneous ($\Delta G > 0$) reverse cell reactions, re-forming reactants from products. To make this possible, the products of the original cell reactions must be substances that either adhere to or are embedded in the electrodes and are available to react with the electrons supplied to the cathodes and drawn away from the anodes by the external power supply.

Figure 19.22 shows the discharge/recharge cycle of a lead–acid battery. Note that electrons flow in one direction when the battery discharges—out of the negative terminal and into the positive terminal—but in the opposite direction when the battery is recharging. Thus, the Pb electrodes, which are connected to the negative battery terminal in Figure 19.22, serve as anodes during discharge but as cathodes during recharge. Any $PbSO_4$ that forms on these Pb electrodes during discharge is reduced back to Pb metal during recharge:

$$PbSO_4(s) + H^+(aq) + 2\ e^- \rightarrow Pb(s) + HSO_4^-(aq)$$

Similarly, the $PbO_2$ electrodes at the positive terminal serve as cathodes during discharge but as anodes during recharge. Any $PbSO_4$ that forms on these $PbO_2$ electrodes during discharge is oxidized back to $PbO_2$ during recharge:

$$PbSO_4(s) + 2\ H_2O(\ell) \rightarrow PbO_2(s) + 3\ H^+(aq) + HSO_4^-(aq) + 2\ e^-$$

---

**SAMPLE EXERCISE 19.8**    **Calculating the Time Required**     **L07**
**to Oxidize a Quantity of Reactant**

If a battery charger for AA NiMH batteries supplies a charging current of 1.00 A, how many minutes does it take to oxidize 0.649 g of $Ni(OH)_2$ to $NiO(OH)$?

**Collect and Organize**   We are asked to calculate the time required for a charging current to oxidize a given mass of $Ni(OH)_2$ to $NiO(OH)$. During the discharge of a NiMH battery, the spontaneous cathode half-reaction is

$$NiO(OH)(s) + H_2O(\ell) + e^- \rightarrow Ni(OH)_2(s) + OH^-(aq) \qquad E° = 1.32\ V$$

**Analyze** During recharging, the spontaneous cathode half-reaction runs in reverse:

$$Ni(OH)_2(s) + OH^-(aq) \rightarrow NiO(OH)(s) + H_2O(\ell) + e^-$$

One mole of electrons is produced for each mole of $Ni(OH)_2$ consumed. Our first steps are to convert 0.649 g of $Ni(OH)_2$ into moles of $Ni(OH)_2$ and then into moles of electrons. We can use the Faraday constant to convert moles of electrons to coulombs of charge. A coulomb is the same as an ampere-second, so dividing by the charging current gives us seconds of charging current. The unit conversions to minutes of charging time are summarized as follows:

$$\boxed{\text{g Ni(OH)}_2} \xrightarrow{\dfrac{1}{\text{molar mass Ni(OH)}_2}} \xrightarrow{\dfrac{1 \text{ mol e}^-}{1 \text{ mol Ni(OH)}_2}} \xrightarrow{\dfrac{9.65 \times 10^4 \text{ C}}{\text{mol e}^-}} \xrightarrow{\dfrac{\text{A} \cdot \text{s}}{\text{C}}} \xrightarrow{\dfrac{1}{\text{A}}} \xrightarrow{\dfrac{1 \text{ min}}{60 \text{ s}}}$$

$$= \text{charging time (min)}$$

**Solve**

$$0.649 \text{ g Ni(OH)}_2 \times \frac{1 \text{ mol Ni(OH)}_2}{92.71 \text{ g Ni(OH)}_2} \times \frac{1 \text{ mol e}^-}{1 \text{ mol Ni(OH)}_2} \times \frac{9.65 \times 10^4 \text{ C}}{1 \text{ mol e}^-}$$

$$\times \frac{1 \text{ A} \cdot \text{s}}{1 \text{ C}} \times \frac{1}{1.00 \text{ A}} \times \frac{1 \text{ min}}{60 \text{ s}} = 11.3 \text{ min}$$

**Think About It** A charging time of 11.3 min may seem short, but the quantity of the $Ni(OH)_2$ to be oxidized (0.649 g) is much less than the total quantity of $Ni(OH)_2$ in a fully charged AA NiMH battery (8.6 g, as calculated in Section 19.7).

**Practice Exercise** Suppose that a car's starter motor draws 230 A of current for 6.0 s to start the car. What mass of Pb is oxidized in the battery to supply this much electricity?

Electrolysis is used in many other processes besides recharging batteries. Electrolytic cells are used to electroplate thin layers of silver, gold, and other metals onto objects, giving these objects the appearance, resistance to corrosion, and other properties of the electroplated metal, but at a fraction of the cost of fabricating the entire object out of the metal (Figure 19.23).

In the chemical industry, electrolysis of molten salts is used to produce highly reactive substances, such as sodium, chlorine, and fluorine; alkali and alkaline earth metals; and aluminum. When NaCl, for instance, is heated to just above its melting point (above 800°C), it becomes an ionic liquid that can conduct electricity. If a sufficiently large potential is applied to carbon electrodes immersed in the molten NaCl, sodium ions are attracted to the negative electrode and are reduced to sodium metal, while chloride ions are attracted to the positive electrode and oxidized to $Cl_2$ gas:

$$2 \text{ Na}^+(\ell) + 2 \text{ Cl}^-(\ell) \rightarrow 2 \text{ Na}(\ell) + \text{Cl}_2(g)$$

A final note about anode and cathode polarity is in order. The reactions in voltaic cells are spontaneous. These cells pump electric current through external circuits and electrical devices with a force equal to their cell potentials. The anode in a voltaic cell is negative because an oxidation half-reaction supplies negatively charged electrons to the device powered by the cell. Electrons flow from the device into the positive battery terminal that is connected to the cathode, where these electrons are consumed in a reduction half-reaction.

The reactions in electrolytic cells are nonspontaneous. They require electrical energy from an external power supply. When the negative terminal of such a power supply is connected to the cathode of the battery, the power supply pumps electrons into the cathode, where they are consumed in reduction half-reactions.

**FIGURE 19.23** The Oscar statuettes given out at the annual Academy Awards are made of an alloy of tin, antimony, and copper that is electroplated with three different materials: copper, nickel silver (a silvery alloy of copper, nickel, and zinc), and finally covered with a layer of 24-karat gold.

Electrons are pumped away from the anode, where they must have been generated in oxidation half-reactions, toward the positive terminal of the power supply. Thus, the cathode of an electrolytic cell is the negative electrode, but the cathode of a voltaic cell is the positive electrode. Similarly, the anode of an electrolytic cell is the positive electrode, but the anode of a voltaic cell is the negative electrode. These "pole reversals" make sense if we keep in mind the fundamental definitions:

➤ Anodes are electrodes where oxidation takes place.
➤ Cathodes are electrodes where reduction takes place.

**CONCEPT TEST** ...................................................................

The electrolysis of molten NaCl produces liquid Na metal at the cathode and $Cl_2$ gas at the anode. However, the electrolysis of an aqueous solution of NaCl produces gases at both cathode and anode. On the basis of the standard potentials in Table A6.1, predict the gas formed at each electrode.

...................................................................

# 19.10 Fuel Cells

▶‖ **CHEMTOUR** Fuel Cell

**Fuel cells** are promising energy conversion devices for many applications, from powering office buildings to cruise ships to electric vehicles. Fuel cells differ from batteries, where energy is stored in electrodes that undergo change during discharge. Fuel cells are energy conversion devices that continue to supply electricity as long as fuel and oxygen are delivered to the anode and cathode, respectively. Ideally fuel cell electrodes are invariant with time, in stark contrast to battery electrodes. From a thermodynamic perspective, most batteries are closed systems and fuel cells are open systems. Chemical energy from the reaction of hydrogen with oxygen is converted directly into electrical energy.

In a typical fuel cell, electrons are supplied to an external circuit by the oxidation of $H_2$ at the anode. Electrons are consumed by the reaction of $O_2$ with protons migrating through the electrolyte to the cathode. The net fuel cell reaction is

$$2\,H_2(g) + O_2(g) \rightarrow 2\,H_2O(\ell)$$

The fuel cells used to power electric vehicles consist of metallic or graphite electrodes separated by a hydrated polymeric material called a *proton-exchange membrane* (PEM). The PEM serves as both an electrolyte and a barrier that prevents crossover and mixing of the fuel and oxidant (Figure 19.24). The electrodes are carbon supported transition-metal catalysts dispersed in a layer upon the PEM. The catalytic layers speed up the electrode half-reactions. Platinum catalysts promote H—H bond breaking during the oxidation of $H_2$ gas to $H^+$ ions at the anode:

$$H_2(g) \rightarrow 2\,H^+(aq) + 2\,e^- \qquad E° = 0.000\ \text{V}$$

At the cathode, a platinum–nickel alloy with the nominal composition $Pt_3Ni$ is particularly effective in catalyzing the formation of free O atoms from $O_2$ molecules that are part of the reduction half-reaction:

$$O_2(g) + 4\,H^+(aq) + 4\,e^- \rightarrow 2\,H_2O(\ell) \qquad E° = +1.229\ \text{V}$$

Hydrogen ions that form at the anode migrate through the PEM to the cathode, where they combine with $O_2$ and electrons from the external circuit. This migration

**fuel cell** a voltaic cell based on the oxidation of a continuously supplied fuel. The reaction is the equivalent of combustion, but chemical energy is converted directly into electrical energy.

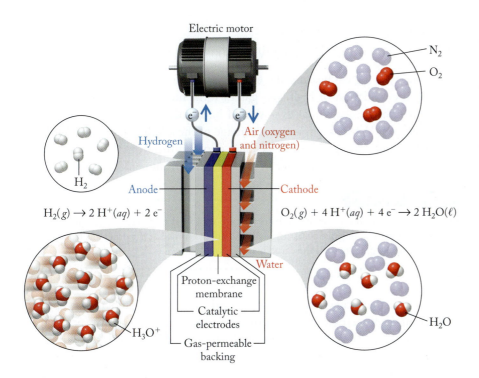

**FIGURE 19.24** Most fuel cells used in vehicles have a proton-exchange membrane between the two halves of the cell. Hydrogen gas diffuses to the anode, and oxygen gas diffuses to the cathode. These electrodes are made of porous material, such as carbon nanofibers, that has a relatively high surface area for a given mass of material. Catalysts on the electrode surfaces also increase the rate of the half-reactions at the anode and the cathode.

of positive charges across the PEM electrolyte is concurrent with the flow of electrons through the device in the external circuit.

A single PEM fuel cell typically delivers useful currents at cell potentials of about 0.8 V. The operating cell voltage is less that the expected $E°_{cell}$ value of 1.23 V because some of the energy of combustion is wasted as heat. When these cells are assembled into fuel cell series *stacks*, they are capable of producing 100 kW of electrical power. That is enough to give a mid-size car, such as the one in Figure 19.25, a top speed of 160 km/hr (99 mi/hr).

PEM fuel cells are well suited for use in vehicles because they are compact, lightweight, and operate at fairly low temperatures of 60 to 80°C. It turns out that the performance of nearly all fuel cells is better at above-ambient temperatures because the rates of the half-reactions increase with temperature as predicted by the Arrhenius equation. The increase in rate results in higher efficiency and more power.

Some fuel cells use basic electrolytes such as concentrated KOH. Pure $O_2$ is supplied to a cathode made of porous graphite containing a nickel catalyst, and $H_2$ gas is supplied to a graphite anode containing nickel(II) oxide. Hydroxide ions formed during $O_2$ reduction at the cathode,

$$O_2(g) + 2\,H_2O(\ell) + 4\,e^- \rightarrow 4\,OH^-(aq) \qquad E° = 0.401\ V$$

migrate through the cell to the anode, where they combine with $H_2$ as it is oxidized to water:

$$H_2(g) + 2\,OH^-(aq) \rightarrow 2\,H_2O(\ell) + 2\,e^-$$

This oxidation half-reaction is the reverse of the following reduction half-reaction in Table A6.1:

$$2\,H_2O(\ell) + 2\,e^- \rightarrow H_2(g) + 2\,OH^-(aq) \qquad E° = -0.828\ V$$

Note that this basic pair of standard electrode potentials yields the same $E°_{cell}$ value:

$$E°_{cell} = 0.401\ V - (-0.828\ V) = 1.229\ V$$

**CONNECTION** The temperature dependence of reaction rates (the Arrhenius equation) was discussed in Chapter 14.

as the acidic pair described earlier:

$$E^\circ_{cell} = 1.229\ V - 0.000\ V = 1.229\ V$$

This equality is logical because the energy, $\Delta G^\circ$, released under standard conditions by the oxidation of hydrogen gas to form liquid water should have only one value, which means that $E^\circ_{cell}$ should have only one value, independent of electrolyte pH. What, then, is the advantage of the alkaline fuel cell? While the thermodynamics of the cell is independent of pH, it turns out that the kinetics of oxygen reduction are much better in an alkaline environment.

**CONCEPT TEST**

During the operation of molten alkali-metal-carbonate fuel cells, carbonate ions are generated at one electrode, migrate across the cell, and are consumed at the other electrode. Do the carbonate ions migrate toward the cathode or the anode?

The same chemical energy that is released in fuel cells could also be obtained by burning hydrogen gas in an internal combustion engine. However, typically only about 20–25% of the chemical energy in the fuel burned in such an engine is converted into mechanical energy; most is lost to the surroundings as heat. In contrast, fuel-cell technologies can convert up to about 80% of the energy released in a fuel-cell redox reaction into electrical energy. The electric motors they power are also about 80% efficient at converting electrical energy into mechanical energy. Thus, the overall conversion efficiency of a fuel-cell powered car is theoretically as high as (80% × 80%) or 64%. Actually, the measured efficiency of the propulsion system of the car in Figure 19.25 is about 60%, which is still more than twice that of an internal combustion engine. In addition, $H_2$-fueled vehicles emit only water vapor; they produce no oxides of nitrogen, no carbon monoxide, and no $CO_2$.

As fuel cells become even more efficient and less expensive, the principal limit on their use in passenger cars will be the availability and cost of hydrogen fuel. Some people worry about the safety of storing hydrogen in a high-pressure tank in a car. Actually, $H_2$ has a higher ignition temperature than gasoline and spreads

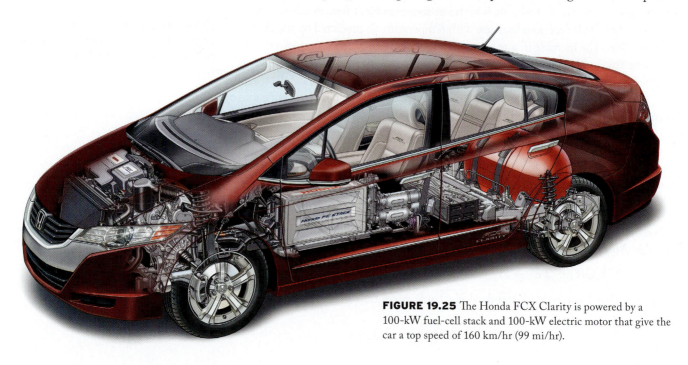

**FIGURE 19.25** The Honda FCX Clarity is powered by a 100-kW fuel-cell stack and 100-kW electric motor that give the car a top speed of 160 km/hr (99 mi/hr).

through the air more quickly, reducing the risk of fire. Still, hydrogen in air burns over a much wider range of concentrations than gasoline, and its flame is almost invisible.

Because of the lack of an extensive hydrogen distribution network, many fuel-cell powered vehicles are buses and fleet vehicles operating from a central location where hydrogen gas is available. With their very large fuel tanks, buses powered by fuel cells have an operating range of 400 km ($\approx$ 250 mi). Longer ranges would be possible if better methods for storing hydrogen gas were available. Currently under development are mobile chemical-processing plants that can extract hydrogen from gasoline, methane, or methanol. On-board production of hydrogen from fossil fuels also generates carbon dioxide as a by-product, so these fuels still produce greenhouse gases. However, the efficiencies of fuel cells and the electric motors they power mean that much less of these fuels will be needed and less $CO_2$ will be emitted.

**CONNECTION** Processes for producing hydrogen gas by reacting methane and other fuels with steam were discussed in Chapter 15.

---

**SAMPLE EXERCISE 19.9    Integrating Concepts: Corrosion at Sea**

In the 18th century, the British Royal Navy began the practice of adding copper sheathing to the bottoms of its sailing vessels to inhibit the growth of barnacles and seaweed that slowed them down. Unfortunately, the first copper sheets used were fastened with iron nails. (a) Explain why using iron nails was not a good idea. (b) Which metal will corrode?

Modern navies continue to encounter shipboard corrosion problems, sometimes with disastrous results. In 1963, the *USS Thresher*, the lead boat in a new class of nuclear submarines, was conducting deep diving tests about 200 miles off the coast of Cape Cod when it suddenly sank. All 129 sailors and civilians on board perished. The cause of the sinking was later determined to be corrosion that occurred where silver solder had been used to connect metal pipes that transported seawater to cool the *Thresher*'s power plant. Metal pipes that carry seawater are made of copper/nickel alloys. (c) Why might silver solder have been considered a logical choice for connecting them? (d) How is the oxidation of silver in seawater influenced by the presence of chloride ions? (e) Use the appropriate half-reactions and standard reduction potentials in Appendix 6 to calculate the cell potential for the air-oxidation of Ag in pH 8.00 seawater in which $[Cl^-] = 0.56$ M. Assume $P_{O_2} = 0.21$ atm. Is the reaction spontaneous?

**Collect and Organize** First we are asked to explain why nailing one metal to another is a bad idea when they will be immersed in seawater and to identify which one will corrode. Then we are to address another corrosion problem involving Ag and Cu/Ni alloys, addressing how their electrochemical properties should have been similar and how the oxidation of Ag in seawater depends on the presence of $Cl^-$ ions. Finally, we are to determine whether Ag spontaneously oxidizes in aerated seawater using half-reactions from Appendix 6 to calculate $E_{cell}$.

**Analyze** The problem with nailing copper sheets with iron nails is similar to the one encountered in the Statue of Liberty (see Figure 19.20), in which the iron skeleton of the statue rusted so badly it had to be replaced.

The logic behind using silver solder to connect pipes made of another metal (or metal alloy) must have been based on their similar electrochemical properties. To determine similarity we need to use the appropriate half-reaction for silver. Appendix 6 lists two

half-reactions: one that is simply $Ag^+(aq) + e^- \rightarrow Ag(s)$ and the other in which the half-reaction occurs in a solution of chloride ions: $AgCl(s) + e^- \rightarrow Ag(s) + Cl^-(aq)$. Given the high $[Cl^-]$ in seawater and the limited solubility of AgCl, we should use the latter to compare electrochemical properties and to determine whether or not the oxidation of Ag is spontaneous.

**Solve**

a. Iron and copper should not be allowed to contact each other in seawater because they are dissimilar metals, as shown by the standard reduction potentials of these half-reactions:

$$Fe^{2+}(aq) + 2\,e^- \rightarrow Fe(s) \qquad E° = -0.447 \text{ V}$$
$$Cu^{2+}(aq) + 2\,e^- \rightarrow Cu(s) \qquad E° = 0.342 \text{ V}$$

When Cu and Fe are in contact with each other in an electrolytic solution, the two become the cathode and anode, respectively, of a voltaic cell:

$$Cu^{2+}(aq) + Fe(s) \rightarrow Cu(s) + Fe^{2+}(aq)$$

This reaction has an $E°_{cell} = E°_{cathode} - E°_{anode} = 0.342$ V $-$ $(-0.447$ V$) = 0.789$ V

b. The spontaneous cell reaction tells us that Fe is the metal that is oxidized (corrodes).

c. Comparing the standard reduction potentials of the appropriate half-reactions involving Ag, Cu, and Ni in Appendix 6:

$$AgCl(s) + e^- \rightarrow Ag(s) + Cl^-(aq) \qquad E° = 0.222 \text{ V}$$
$$Cu^{2+}(aq) + 2\,e^- \rightarrow Cu(s) \qquad E° = 0.342 \text{ V}$$
$$Ni^{2+}(aq) + 2\,e^- \rightarrow Ni(s) \qquad E° = -0.257 \text{ V}$$

We see that $E°$ of the Ag reaction is in between the other two. If the actual electrochemical properties of the alloy are closer to those of Cu than Ni (which, it turns out, they are), then Ag would have been a logical choice to use in connecting pipes made from the alloy.

d. The difference in the potentials of the two silver half-reactions in Appendix 6,

$$AgCl(s) + e^- \rightarrow Ag(s) + Cl^-(aq) \qquad E° = 0.222 \text{ V}$$
$$Ag^+(aq) + e^- \rightarrow Ag(s) \qquad E° = 0.799 \text{ V}$$

indicates that the presence of $Cl^-$ ions has a significant impact on the electrochemical properties of Ag. As Ag atoms are oxidized, they precipitate as AgCl, which keeps the concentration of $Ag^+(aq)$ low. The Nernst equation for the $Ag^+/Ag$ half-reaction is

$$E = E° - 0.0592 \text{ V} \log (1/[Ag^+])$$

A small $[Ag^+]$ value in the denominator means a large value for the log term and an $E$ value for the $Ag^+/Ag$ half-reaction that is much less than its $E°$. A smaller potential for the reduction half-reaction means a greater potential for the oxidation of Ag. Therefore, Ag is more likely to be oxidized in the presence of $Cl^-$ ions.

e. The appropriate half-reactions are

$$O_2(g) + 4 H^+(aq) + 4 e^- \rightarrow 2 H_2O(\ell) \qquad E° = 1.229 \text{ V}$$
$$AgCl(s) + e^- \rightarrow Ag(s) + Cl^-(aq) \qquad E° = 0.222 \text{ V}$$

To assess the spontaneity of the air-oxidation of Ag, we need to flip the second one and multiply it by 4 to balance the loss and gain of electrons:

$$O_2(g) + 4 H^+(aq) + 4 e^- \rightarrow 2 H_2O(\ell)$$
$$4[Ag(s) + Cl^-(aq) \rightarrow AgCl(s) + e^-]$$
$$\overline{O_2(g) + 4 H^+(aq) + 4 Ag(s) + 4 Cl^-(aq) \rightarrow 4 AgCl(s) + 2 H_2O(\ell)}$$

Flipping the silver reaction means that Ag is the anode and the standard cell potential is

$$E°_{cell} = E°_{cathode} - E°_{anode} = 1.229 \text{ V} - 0.222 \text{ V} = 1.007 \text{ V}$$

Using this $E°_{cell}$ value and setting up the Nernst equation for the reaction gives us

$$E_{cell} = E°_{cell} - (0.0592 \text{ V}/4) \log [1/[P_{O_2} [H^+]^4 [Cl^-]^4]$$

We take the antilog of pH = 8.00 and use the $P_{O_2}$ and $[Cl^-]$ values for aerated seawater:

$$E_{cell} = 1.007 \text{ V} - (0.0592 \text{ V}/4) \log [1/(0.21 (1.00 \times 10^{-8})^4 (0.56)^4)]$$
$$= 0.508 \text{ V}$$

The positive cell potential means that the oxidation of silver in seawater should be spontaneous.

**Think About It** The above analysis made it easy to predict why the British Royal Navy should not have used iron nails to fasten copper sheathing to the hulls of their 18th-century wooden sailing ships. We did not come up with the answer for why the silver solder joints failed aboard the *Thresher*. According to modern naval engineering references, silver solder and Cu/Ni alloys should have similar electrochemical properties when immersed in seawater. Unfortunately, similar was apparently not close enough for the 129 souls aboard the *Thresher*.

## SUMMARY

**Learning Outcome 1** **Electrochemistry** is the branch of chemistry that links redox reactions to the production or consumption of electrical energy. Any redox reaction can be broken down into oxidation and reduction half-reactions. (Section 19.1)

**Learning Outcome 2** In an **electrochemical cell**, the oxidation half-reaction occurs at the **anode** and the reduction half-reaction occurs at the **cathode**. Migration of the ions in the cell's electrolyte allows electrical charges to flow between the cathode and anode compartments as electrons flow through an external electrical circuit. A **cell diagram** shows how the components of the cathodic and anodic compartments of the cell are connected. (Section 19.2)

**Learning Outcome 3** The difference in the **standard reduction potentials ($E°$)** of a cell's cathode and anode half-reactions is equal to the **standard cell potential ($E°_{cell}$)**. The value of $E°_{cell}$ is a measure of how forcefully a **voltaic cell**, in which all reactants and products are in their standard states, can pump electrons through an external circuit. All standard cell potentials are referenced to the cell

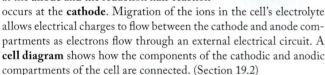

potential of the **standard hydrogen electrode** ($E°_{SHE} = 0.000 \text{ V}$). (Sections 19.3, 19.4, 19.5)

**Learning Outcome 4** A voltaic cell has a positive cell potential ($E_{cell} > 0$) and its cell reaction has a negative change in free energy ($\Delta G_{cell} < 0$). This decrease in free energy in a voltaic cell is available to do work in an external electrical circuit. The **Faraday constant** relates the quantity of electrical charge to the number of moles of electrons and indirectly to the number of moles of reactants. (Section 19.4)

**Learning Outcome 5** The potential of a voltaic cell decreases as reactants turn into products. The **Nernst equation** describes how cell potential changes with concentration changes. (Section 19.6)

**Learning Outcome 6** The potential of a voltaic cell approaches zero as the cell reaction approaches chemical equilibrium, at which point $E_{cell} = 0$ and $Q = K$. We can calculate the equilibrium constant for any redox reaction at 25°C, not just those in electrochemical cells. (Sections 19.6, 19.7, 19.10)

**Learning Outcome 7** The quantities of reactants consumed in a voltaic cell reaction are proportional to the coulombs of electrical charge delivered by the cell. Nickel–metal hydride batteries supply electricity when H atoms are oxidized to $H^+$ ions at the anodes and NiO(OH) is reduced to $Ni(OH)_2$ at the cathodes. In lithium–ion batteries, electricity is produced when $Li^+$ ions stored in graphite anodes migrate toward and are incorporated into transition metal oxide or phosphate cathodes. (Sections 19.7, 19.8, 19.9)

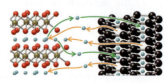

## PROBLEM-SOLVING SUMMARY

| TYPE OF PROBLEM | CONCEPTS AND EQUATIONS | SAMPLE EXERCISES |
|---|---|---|
| **Writing net ionic equations for redox reactions** | Combine the reduction and oxidation half-reactions after balancing the gain and loss of electrons. | 19.1 |
| **Diagramming an electrochemical cell** | Use the format:<br><br>anode \| anode solution \|\| cathode solution \| cathode<br><br>Insert solution concentrations if they are known. | 19.2 |
| **Identifying anode and cathode half-reactions and calculating the value of $E°_{cell}$** | The half-reaction with the more positive standard reduction potential is the cathode half-reaction in a voltaic cell.<br><br>$$E°_{cell} = E°_{cathode} - E°_{anode} \qquad (19.2)$$ | 19.3 |
| **Relating $\Delta G°_{cell}$ and $E°_{cell}$** | $$\Delta G°_{cell} = -nFE°_{cell} \qquad (19.7)$$<br><br>where $n$ is the number of moles of electrons transferred in the cell reaction and $F$ is the Faraday constant, $9.65 \times 10^4$ C/mol e⁻. | 19.4 |
| **Calculating $E_{cell}$ from $E°_{cell}$ and the concentrations of reactants and products** | $E_{cell}$ at 25°C is related to $E°_{cell}$ and the cell reaction quotient by the Nernst equation:<br><br>$$E_{cell} = E°_{cell} - \frac{0.0592\ \text{V}}{n} \log Q \qquad (19.10)$$ | 19.5 |
| **Calculating $K$ for a redox reaction from the standard potentials of its half-reactions** | $E°_{rxn}$ is related to $K$ at 298 K by<br><br>$$\log K = \frac{nFE°_{rxn}}{0.0592} \qquad (19.12)$$ | 19.6 |
| **Relating the mass of a reactant in an electrochemical reaction to a quantity of electrical charge** | Determine the ratio of moles of reactants to moles of electrons transferred; use the Faraday constant to relate coulombs of charge to moles of electrons. | 19.7 |
| **Calculating the time required to oxidize a quantity of reactant** | Use the Faraday constant and the relation between moles of electrons and moles of reactants to describe an electrolytic process. | 19.8 |

## VISUAL PROBLEMS

*(Answers to boldface end-of-chapter questions and problems are in the back of the book.)*

**19.1.** In the voltaic cell shown in Figure P19.1, the greater density of a concentrated solution of $CuSO_4$ allows a less concentrated solution of $ZnSO_4$ solution to be (carefully) layered on top of it. Why is a porous separator not needed in this cell?

**FIGURE P19.1**

**19.2.** In the voltaic cell shown in Figure P19.2, the concentrations of $Cu^{2+}$ and $Cd^{2+}$ are 1.00 $M$. On the basis of the standard potentials in Appendix 6, identify which electrode is the anode and which is the cathode. Indicate the direction of electron flow.

**FIGURE P19.2**

**19.3.** In the voltaic cell shown in Figure P19.3, $[Ag^+] = [H^+] = 1.00\ M$ and $P_{H_2} = 1.00$ atm. Based on the standard potentials in Appendix 6, identify which electrode is the anode and which is the cathode. Indicate the direction of electron flow.

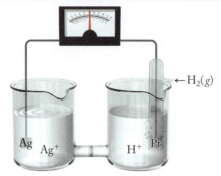

←$H_2(g)$

Ag    $Ag^+$          $H^+$    Pt

**FIGURE P19.3**

**19.4.** In many electrochemical cells the electrodes are metals that carry electrons to and from the cell but are not chemically changed by the cell reaction. Each of the highlighted clusters in the periodic table in Figure P19.4 consists of three metals. Which of the highlighted clusters is best suited to form inert electrodes?

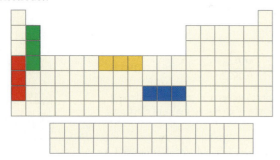

**FIGURE P19.4**

**19.5.** Which of the four curves in Figure P19.5 best represents the dependence of the potential of a lead–acid battery on the concentration of sulfuric acid? Note that the scale of the x-axis is logarithmic.

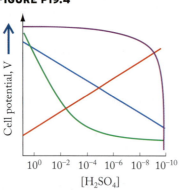

Cell potential, V

$10^0$  $10^{-2}$  $10^{-4}$  $10^{-6}$  $10^{-8}$  $10^{-10}$
$[H_2SO_4]$

**FIGURE P19.5**

**19.6.** Consider the four types of batteries in Figure P19.6. From top to bottom the sizes are AAA, AA, C, and D. The performance of batteries like these is often expressed in units such as (a) volts, (b) watt-hours, or (c) milliampere-hours. Which of the values differ significantly between the four batteries?

**FIGURE P19.6**

**19.7.** The apparatus in Figure P19.7 is used for the electrolysis of water. Hydrogen and oxygen gas are collected in the two inverted burettes. An inert electrode at the bottom of the left burette is connected to the negative terminal of a 6-volt battery; the electrode in the burette on the right is connected to the positive terminal. A small quantity of sulfuric acid is added to speed up the electrolytic reaction.
   a. What are the half-reactions at the left and right electrodes and their standard potentials?
   b. Why does sulfuric acid make the electrolysis reaction go more rapidly?

Overall cell reaction
$H_2O(\ell) \rightarrow H_2(g) + \frac{1}{2} O_2(g)$

**FIGURE P19.7**

**19.8.** An electrolytic apparatus identical to the one shown in Problem 19.7 is used to electrolyze water, but the reaction is speeded up by the addition of sodium carbonate instead of sulfuric acid.
   a. What are the half-reactions and the standard potentials for the electrodes on the left and right?
   b. Why does sodium carbonate make the electrolysis reaction go more rapidly?

*****19.9.** The photo in Figure P19.9 shows a phenomenon known as waterline corrosion. Assuming the oxidizing agent in the corrosion process is $O_2$, propose a reason why metal pilings such as this one tend to corrode the most at the waterline and corrode less at heights above and below the waterline.

**19.10.** Most classic cars, such as the one in Figure 19.10, have chromium-electroplated or *chrome* bumpers.
   a. In the electroplating process is the bumper the anode or the cathode?
   b. How does the presence of a layer of chromium protect the bumper from corroding?

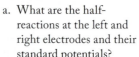

**FIGURE P19.9**

**FIGURE P19.10**

## QUESTIONS AND PROBLEMS

## Redox Chemistry Revisited; Electrochemical Cells

### CONCEPT REVIEW

**19.11.** Regarding the porous separator between the two halves of an electrochemical cell:
   a. Describe how it allows electrical charge to flow between the two half-cells.
   b. Explain why a piece of wire could not perform the same function.

**19.12.** The $Zn/Cu^{2+}$ reactions in Figures 19.1 and 19.2 are the same; however, the reaction in the cell in Figure 19.2 generates electricity, while the reaction in the beaker in Figure 19.1 does not. Why?

### PROBLEMS

**19.13.** Select the appropriate half-reactions from Appendix 6 to write net ionic equations describing the reaction between:
   a. aluminum metal and $Fe^{3+}$ ions in solution that produces dissolved $Al^{3+}$ and $Fe^{2+}$ ions.
   b. $I_2$ and $NO_2^-$ ions in an alkaline solution that produces $I^-$ and $NO_3^-$ ions.
   c. $MnO_4^-$ and $Cr^{3+}$ ions in an acidic solution that produces $Mn^{2+}$ and $Cr_2O_7^{2-}$ ions.

19.14. Select the appropriate half-reactions from Appendix 6 to write net ionic equations describing the reaction between:
   a. tin and $Ag^+$ ions in solution that produces dissolved $Sn^{2+}$ ions and silver metal.
   b. copper and $O_2$ in an acidic solution that produces $Cu^{2+}$ ions.
   c. solid $Cr(OH)_3$ and $O_2$ in a basic solution that produces $CrO_4^{2-}$ ions.

**19.15.** An electrochemical cell with an aqueous electrolyte is based on the reaction between $Ni^{2+}(aq)$ and $Cd(s)$, producing $Ni(s)$ and $Cd^{2+}(aq)$.
   a. Write half-reactions for the anode and cathode.
   b. Write a balanced cell reaction.
   c. Diagram the cell.

19.16. A voltaic cell is based on the reaction between $Cu^{2+}(aq)$ and $Ni(s)$, producing $Cu(s)$ and $Ni^{2+}(aq)$.
   a. Write the anode and cathode half-reactions.
   b. Write a balanced cell reaction.
   c. Diagram the cell.

**19.17.** A voltaic cell with a basic aqueous background electrolyte is based on the oxidation of $Cd(s)$ to $Cd(OH)_2(s)$ and the reduction of $MnO_4^-(aq)$ to $MnO_2(s)$.
   a. Write half-reactions for the cell's anode and cathode.
   b. Write a balanced cell reaction.
   c. Diagram the cell.

19.18. A voltaic cell is based on the reduction of $Ag^+(aq)$ to $Ag(s)$ and the oxidation of $Sn(s)$ to $Sn^{2+}(aq)$.
   a. Write half-reactions for the cell's anode and cathode.
   b. Write a balanced cell reaction.
   c. Diagram the cell.

**19.19. Super Iron Batteries** In 1999, scientists in Israel developed a battery based on the following cell reaction with iron(VI), nicknamed "super iron":

$$2\,K_2FeO_4(aq) + 3\,Zn(s) \rightarrow Fe_2O_3(s) + ZnO(s) + 2\,K_2ZnO_2(aq)$$

   a. Determine the number of electrons transferred in the cell reaction.
   b. What are the oxidation states of the transition metals in the reaction?
   c. Diagram the cell.

19.20. **Aluminum–Air Batteries** In recent years engineers have been working on an aluminum–air battery as an alternative energy source for electric vehicles. The battery consists of an aluminum anode, which is oxidized to solid aluminum hydroxide, immersed in an electrolyte of aqueous KOH. At the cathode oxygen from the air is reduced to hydroxide ions on an inert metal surface. Write the two half-reactions for the battery and diagram the cell. Use the generic $M(s)$ symbol for the metallic cathode material.

## Standard Potentials

### CONCEPT REVIEW

**19.21.** What is the function of platinum in the standard hydrogen electrode?

19.22. Suggest a replacement metal for platinum in the standard hydrogen electrode. Explain why you selected the metal you did.

**19.23.** In some textbooks the formula used to calculate the standard cell is written this way:

$$E^\circ_{cell} = E^\circ_{red}\,(\text{cathode}) + E^\circ_{ox}\,(\text{anode})$$

where the "ox" subscript means the standard potential of the anode half-reaction written as an oxidation. Show how this equation is equivalent to Equation 19.2.

19.24. Suppose there were a scale for expressing electrode potentials in which the standard potential for the reduction of water in base

$$2\,H_2O(\ell) + 2\,e^- \rightarrow H_2(g) + 2\,OH^-(aq)$$

is assigned an $E^\circ$ value of 0.000 V. How would the standard potential values on this new scale differ from those in Appendix 6?

**19.25.** Which is a stronger oxidizer under standard conditions: oxygen or chlorine? Use standard reduction potentials in Appendix 6 to support your answer.

*19.26. Of the group 1 elements Li, K, and Na, which is the strongest reducing agent?

### PROBLEMS

**19.27.** Starting with the appropriate standard free energies of formation in Table A4.3 in Appendix 4, calculate the values of $\Delta G^\circ$ and $E^\circ_{cell}$ of the following reactions:
   a. $2\,Cu^+(aq) \rightarrow Cu^{2+}(aq) + Cu(s)$
   b. $Cu(s) + 2\,Fe^{3+}(aq) \rightarrow Cu^{2+}(aq) + 2\,Fe^{2+}(aq)$

19.28. Starting with the appropriate standard free energies of formation in Appendix 4, calculate the values of $\Delta G^\circ$ and $E^\circ_{cell}$ of the following reactions:
   a. $2\,Na(s) + 2\,H_2O(\ell) \rightarrow 2\,NaOH(aq) + H_2(g)$
   b. $2\,Pb(s) + O_2(g) + 2\,H_2SO_4(aq) \rightarrow 2\,PbSO_4(s) + 2\,H_2O(\ell)$

**19.29.** If a piece of silver is placed in a solution in which $[Ag^+] = [Cu^{2+}] = 1.00\ M$, will the following reaction proceed spontaneously?

$$2\ Ag(s) + Cu^{2+}(aq) \rightarrow 2\ Ag^+(aq) + Cu(s)$$

**19.30.** A piece of cadmium is placed in a solution in which $[Cd^{2+}] = [Sn^{2+}] = 1.00\ M$. Will the following reaction proceed spontaneously?

$$Cd(s) + Sn^{2+}(aq) \rightarrow Cd^{2+}(aq) + Sn(s)$$

**19.31.** Sometimes the anode half-reaction in the zinc–air battery (Figure 19.7) is written with the zincate ion, $Zn(OH)_4{}^{2-}$, as the product. Write a balanced equation for the cell reaction based on this product.

*19.32. Sometimes the cell reaction of nickel–cadmium batteries is written with Cd metal as the anode and solid $NiO_2$ as the cathode. Assuming that the products of the reactions are a solid hydroxide of Cd(II) at the anode and a solid hydroxide of Ni(II) at the cathode, write balanced equations for the cathode and anode half-reactions and the overall cell reaction.

**19.33.** In a voltaic cell similar to the Cu–Zn cell in Figure 19.2, the Cu electrode is replaced with one made of Ni immersed in a solution of $NiSO_4$. Will the standard potential of this Ni–Zn cell be greater than, the same as, or less than 1.10 V?

**19.34.** Suppose the copper half of the Cu–Zn cell in Figure 19.2 were replaced with a silver wire in contact with $1\ M\ Ag^+(aq)$.
  a. What would be the value of $E^\circ_{cell}$?
  b. Which electrode would be the anode?

**19.35.** Starting with standard potentials listed in Appendix 6, calculate the values of $E^\circ_{cell}$ and $\Delta G^\circ$ of the following reactions.
  a. $Cl_2(g) + 2\ Br^-(aq) \rightarrow Br_2(\ell) + 2\ Cl^-(aq)$
  b. $Zn(s) + Ni^{2+}(aq) \rightarrow Zn^{2+}(aq) + Ni(s)$

**19.36.** Voltaic cells based on the following pairs of half-reactions are constructed. For each pair, write a balanced equation for the cell reaction, and identify which half-reaction takes place at each anode and cathode.
  a. $Cd^{2+}(aq) + 2\ e^- \rightarrow Cd(s)$
     $Ag^+(aq) + e^- \rightarrow Ag(s)$
  b. $AgBr(s) + e^- \rightarrow Ag(s) + Br^-(aq)$
     $MnO_2(s) + 4\ H^+(aq) + 2\ e^- \rightarrow Mn^{2+}(aq) + 2\ H_2O(\ell)$
  c. $PtCl_4{}^{2-}(aq) + 2\ e^- \rightarrow Pt(s) + 4\ Cl^-(aq)$
     $AgCl(s) + e^- \rightarrow Ag(s) + Cl^-(aq)$

**19.37.** Under standard conditions is $H_2$ gas capable of reducing:
  a. $Ag^+$ to Ag
  b. $Mg^{2+}$ to Mg
  c. $Cu^{2+}$ to Cu
  d. $Cd^{2+}$ to Cd?

**19.38.** Under standard conditions are $H^+$ ions capable of oxidizing:
  a. Zn to $Zn^{2+}$
  b. $Fe^{2+}$ to $Fe^{3+}$
  c. $Cr(OH)_3$ to $CrO_4{}^{2-}$
  d. Ni to $Ni^{2+}$?

**19.39.** The half-reactions and standard potentials for a nickel–metal hydride battery with a titanium–zirconium anode are as follows:

Cathode:  $NiO(OH)(s) + H_2O(\ell) + e^- \rightarrow Ni(OH)_2(s) + OH^-(aq)$
$$E^\circ = 1.32\ V$$

Anode:  $TiZr_2H(s) + OH^-(aq) \rightarrow TiZr_2(s) + H_2O(\ell) + e^-$
$$E^\circ = 0.00\ V$$

  a. Write the overall cell reaction for this battery.
  b. Calculate the standard cell potential.

*19.40. **Lithium–Ion Batteries** Scientists at the University of Texas, Austin, and at MIT developed lithium–ion batteries that have $LiFePO_4$, which is the composition of the cathode when the battery is fully discharged. Batteries with this cathode are more powerful than those with the same mass with $LiCoO_2$ cathodes. They are also more stable at high temperatures.
  a. What is the formula of the $LiFePO_4$ cathode when the battery is fully charged?
  b. Is Fe oxidized or reduced as the battery discharges?
  c. Is the cell potential of a lithium–ion battery with an iron phosphate cathode likely to differ from one with a cobalt oxide cathode? Explain your answer.

## Chemical Energy and Electrical Work

### PROBLEMS

**19.41.** For many years the 1.50 V batteries used to power flashlights were based on the following cell reaction:

$$Zn(s) + 2\ NH_4Cl(s) + 2\ MnO_2(s) \rightarrow$$
$$Zn(NH_3)_2Cl_2(s) + Mn_2O_3(s) + H_2O(\ell)$$

What is the value of $\Delta G_{cell}$?

**19.42.** **Laptop Battery** The first generation of laptop computers was powered by nickel–cadmium (nicad) batteries, which generated 1.20 V based on the following cell reaction:

$$Cd(s) + 2\ NiO(OH)(s) + 2\ H_2O(\ell) \rightarrow Cd(OH)_2(s) + 2\ Ni(OH)_2(s)$$

What is the value of $\Delta G_{cell}$?

**19.43.** The cells in the nickel–metal hydride battery packs used in many hybrid vehicles produce 1.20 V based on the following cell reaction:

$$MH(s) + NiO(OH)(s) \rightarrow M(s) + Ni(OH)_2(s)$$

What is the value of $\Delta G_{cell}$?

**19.44.** A cell in a lead–acid battery delivers exactly 2.00 V of cell potential based on the following cell reaction:

$$Pb(s) + PbO_2(s) + 2\ H_2SO_4(aq) \rightarrow 2\ PbSO_4(s) + 2\ H_2O(\ell)$$

What is the value of $\Delta G_{cell}$?

## A Reference Point: The Standard Hydrogen Electrode; The Effect of Concentration on $E_{cell}$

### CONCEPT REVIEW

**19.45.** Why does the operating cell potential of most batteries change little until the battery is nearly discharged?

**19.46.** The standard potential of the Cu–Zn cell reaction,

$$Zn(s) + Cu^{2+}(aq) \rightarrow Zn^{2+}(aq) + Cu(s)$$

is 1.10 V. Would the potential of the Cu–Zn cell differ from 1.10 V if the concentrations of both $Cu^{2+}$ and $Zn^{2+}$ were $0.25\ M$?

**PROBLEMS**

**19.47.** Calculate the $E_{cell}$ value at 298 K for the cell based on the reaction

$$Fe^{3+}(aq) + Cu^+(aq) \rightarrow Fe^{2+}(aq) + Cu^{2+}(aq)$$

when $[Fe^{3+}] = [Cu^+] = 1.50 \times 10^{-3}$ $M$ and $[Fe^{2+}] = [Cu^{2+}] = 2.5 \times 10^{-4}$ $M$.

**19.48.** Calculate the $E_{cell}$ value at 298 K for the cell based on the reaction

$$Cu(s) + 2 Ag^+(aq) \rightarrow Cu^{2+}(aq) + 2 Ag(s)$$

when $[Ag^+] = 2.56 \times 10^{-3}$ $M$ and $[Cu^{2+}] = 8.25 \times 10^{-4}$ $M$.

**19.49.** Using the appropriate standard potentials in Appendix 6, determine the equilibrium constant for the following reaction at 298 K:

$$Fe^{3+}(aq) + Cr^{2+}(aq) \rightarrow Fe^{2+}(aq) + Cr^{3+}(aq)$$

**19.50.** Using the appropriate standard potentials in Appendix 6, determine the equilibrium constant at 298 K for the following reaction between $MnO_2$ and $Fe^{2+}$ in acid solution:

$$4 H^+(aq) + MnO_2(s) + 2 Fe^{2+}(aq) \rightarrow$$
$$Mn^{2+}(aq) + 2 Fe^{3+}(aq) + 2 H_2O(\ell)$$

**19.51.** If the potential of a hydrogen electrode based on the half-reaction

$$2 H^+(aq) + 2 e^- \rightarrow H_2(g)$$

is 0.000 V at pH = 0.00, what is the potential of the same electrode at pH = 7.00?

**19.52.** **Glucose Metabolism** The standard potentials for the reduction of nicotinamide adenine dinucleotide ($NAD^+$) and oxaloacetate (reactants in the multistep metabolism of glucose) are as follows:

$$NAD^+(aq) + 2 H^+(aq) + 2 e^- \rightarrow NADH(aq) + H^+(aq)$$
$$E^\circ = -0.320 \text{ V}$$

$$Oxaloacetate^{2-}(aq) + 2 H^+(aq) + 2 e^- \rightarrow malate^{2-}(aq)$$
$$E^\circ = -0.166 \text{ V}$$

a. Calculate the standard potential for the following reaction:

$$Oxaloacetate^{2-}(aq) + NADH(aq) + H^+(aq) \rightarrow$$
$$malate^-(aq) + NAD^+(aq)$$

b. Calculate the equilibrium constant for the reaction at 298 K.

**19.53.** Permanganate ion can oxidize sulfite to sulfate in basic solution as follows:

$$2 MnO_4^-(aq) + 3 SO_3^{2-}(aq) + H_2O(\ell) \rightarrow$$
$$2 MnO_2(s) + 3 SO_4^{2-}(aq) + 2 OH^-(aq)$$

Determine the potential for the reaction ($E_{rxn}$) at 298 K when the concentrations of the reactants and products are as follows: $[MnO_4^-] = 0.250$ $M$, $[SO_3^{2-}] = 0.425$ $M$, $[SO_4^{2-}] = 0.075$ $M$, and $[OH^-] = 0.0200$ $M$. Will the value of $E_{rxn}$ increase or decrease as the reaction proceeds?

*****19.54.** A concentration cell is constructed by using the same half-reaction for both the cathode and anode. What is the value of $E_{cell}$ for a concentration cell that combines copper electrodes in contact with 0.35 $M$ copper(II) nitrate and 0.00075 $M$ copper(II) nitrate solutions? ($E^\circ = 0.3419$ V for Cu/Cu$^{2+}$)

**19.55.** A copper penny dropped into a solution of nitric acid produces a mixture of nitrogen oxides. The following reaction describes the formation of NO, one of the products:

$$3 Cu(s) + 8 H^+(aq) + 2 NO_3^-(aq) \rightarrow$$
$$2 NO(g) + 3 Cu^{2+}(aq) + 4 H_2O(\ell)$$

a. Starting with the appropriate standard potentials in Appendix 6, calculate $E^\circ_{cell}$ for this reaction.
b. Calculate $E^\circ_{cell}$ at 298 K when $[H^+] = 0.500$ $M$, $[NO_3^-] = 0.0550$ $M$, $[Cu^{2+}] = 0.0500$ $M$, and the partial pressure of NO = 0.00250 atm.

**19.56.** Chlorine dioxide ($ClO_2$) is produced by the following reaction of chlorate ($ClO_3^-$) with Cl$^-$ in acid solution:

$$2 ClO_3^-(aq) + 2 Cl^-(aq) + 4 H^+(aq) \rightarrow$$
$$2 ClO_2(g) + Cl_2(g) + 2 H_2O(\ell)$$

a. Determine $E^\circ$ for the reaction.
b. The reaction produces a mixture of gases in the reaction vessel in which $P_{ClO_2} = 2.0$ atm; $P_{Cl_2} = 1.00$ atm. Calculate $[ClO_3^-]$ if, at equilibrium ($T = 298$ K), $[H^+] = [Cl^-] = 10.0$ $M$.

*****19.57.** The oxidation of $NH_4^+$ to $NO_3^-$ in acid solution is described by the following equation:

$$NH_4^+(aq) + 2 O_2(g) \rightarrow NO_3^-(aq) + 2 H^+(aq) + H_2O(\ell)$$

a. Calculate $E^\circ$ for the overall reaction.
b. If the reaction is in equilibrium with air ($P_{O_2} = 0.21$ atm) at pH 7.00, what is the ratio of $[NO_3^-]$ to $[NH_4^+]$ at 298 K?

**19.58.** What is the value of $E^\circ$ for the following reaction?

$$2 AgCl(s) + H_2(g) \rightarrow 2 Ag(s) + 2 HCl(aq)$$

# Relating Battery Capacity to Quantities of Reactants

**CONCEPT REVIEW**

**19.59.** One 12-volt lead–acid battery has a higher ampere-hour rating than another. Which of the following parameters are likely to be different for the two batteries?
a. individual cell potentials
b. anode half-reactions
c. total masses of electrode materials
d. number of cells
e. electrolyte composition
f. combined surface areas of their electrodes

**19.60.** In a voltaic cell based on the Cu–Zn cell reaction

$$Zn(s) + Cu^{2+}(aq) \rightarrow Cu(s) + Zn^{2+}(aq)$$

there is exactly 1 mole of each reactant and product. A second cell based on the Cd–Cu cell reaction

$$Cd(s) + Cu^{2+}(aq) \rightarrow Cu(s) + Cd^{2+}(aq)$$

also has exactly 1 mole of each reactant and product. Which of the following statements about these two cells is true?
a. Their cell potentials are the same.
b. The masses of their electrodes are the same.
c. The quantities of electrical charge that they can produce are the same.
d. The quantities of electrical energy that they can produce are the same.

## PROBLEMS

**19.61.** Which of the following voltaic cells will produce the greater quantity of electrical charge per gram of anode material?

$$Cd(s) + 2\,NiO(OH)(s) + 2\,H_2O(\ell) \rightarrow 2\,Ni(OH)_2(s) + Cd(OH)_2(s)$$

or

$$4\,Al(s) + 3\,O_2(g) + 6\,H_2O(\ell) + 4\,OH^-(aq) \rightarrow 4\,Al(OH)_4^-(aq)$$

**19.62.** Which of the following voltaic cells will produce the greater quantity of electrical charge per gram of anode material?

$$Zn(s) + MnO_2(s) + H_2O(\ell) \rightarrow ZnO(s) + Mn(OH)_2(s)$$

or

$$Li(s) + MnO_2(s) \rightarrow LiMnO_2(s)$$

---

**\*19.63.** Which of the following voltaic cell reactions delivers more electrical energy per gram of anode material at 298 K?

$$Zn(s) + 2\,NiO(OH)(s) + 2\,H_2O(\ell) \rightarrow$$
$$2\,Ni(OH)_2(s) + Zn(OH)_2(s) \qquad E^\circ_{cell} = 1.20\ V$$

or

$$Li(s) + MnO_2(s) \rightarrow LiMnO_2(s) \qquad E^\circ_{cell} = 3.15\ V$$

**\*19.64.** Which of the following voltaic cell reactions delivers more electrical energy per gram of anode material at 298 K?

$$Zn(s) + Ni(OH)_2(s) \rightarrow Zn(OH)_2(s) + Ni(s)$$
$$E^\circ_{cell} = 1.50\ V$$

or

$$2\,Zn(s) + O_2(g) \rightarrow 2\,ZnO(s) \qquad E^\circ_{cell} = 2.08\ V$$

## Corrosion: Unwanted Electrochemical Reactions

### CONCEPT REVIEW

**19.65.** When the iron skeleton of the Statue of Liberty was replaced with stainless steel, the asbestos mats that had separated the skeleton from the copper exterior were replaced with Teflon spacers. Why was Teflon a good choice?

**19.66.** What does a sacrificial anode do to protect a metal structure and why is the process called *cathodic* protection?

**19.67.** The windows of the giant ocean tank at the New England Aquarium (Figure P19.67) are held in place with aluminum frames. What would be a good material to use to make sacrificial anodes for the frames?

**FIGURE P19.67**

**19.68. Corrosion of Copper Pipes** The copper pipes frequently used in household plumbing may corrode and eventually leak. The corrosion reaction is believed to involve the formation of copper(I) chloride:

$$2\,Cu(s) + Cl_2(aq) \rightarrow 2\,CuCl(s)$$

a. Write balanced equations for the half-reactions in this redox reaction.
b. Calculate $E^\circ_{rxn}$ and $\Delta G^\circ_{rxn}$ for the reaction.

## Electrolytic Cells and Rechargeable Batteries

### CONCEPT REVIEW

**19.69.** The positive terminal of a voltaic cell is the cathode. However, the cathode of an electrolytic cell is connected to the negative terminal of a power supply. Explain this difference in polarity.

**19.70.** The anode in an electrochemical cell is defined as the electrode where oxidation takes place. Why is the anode in an electrolytic cell connected to the positive (+) terminal of an external supply, whereas the anode in a voltaic cell battery is connected to the negative (−) terminal?

**19.71.** The salts obtained from the evaporation of seawater can act as a source of halogens, principally $Cl_2$ and $Br_2$, through the electrolysis of the molten alkali metal halides. As the potential of the anode in an electrolytic cell is increased, which of these two halogens forms first?

**19.72.** In the electrolysis described in Problem 19.71, why is it necessary to use molten salts rather than seawater itself?

**19.73. Quantitative Analysis** Electrolysis can be used to determine the concentration of $Cu^{2+}$ in a given volume of solution by electrolyzing the solution in a cell equipped with a platinum cathode. If all of the $Cu^{2+}$ is reduced to Cu metal at the cathode, the increase in mass of the electrode provides a measure of the concentration of $Cu^{2+}$ in the original solution. To ensure the complete (99.99%) removal of the $Cu^{2+}$ from a solution in which $[Cu^{2+}]$ is initially about 1.0 $M$, will the potential of the cathode (versus SHE) have to be more or less negative than 0.34 V (the standard potential for $Cu^{2+} + 2\,e^- \rightarrow Cu$)?

**19.74.** A high school chemistry student wishes to demonstrate how water can be separated into hydrogen and oxygen by electrolysis. She knows that the reaction will proceed more rapidly if an electrolyte is added to the water. She has access to 2.00 $M$ solutions of these compounds: $H_2SO_4$, HBr, NaI, $Na_2SO_4$, and $Na_2CO_3$. Which one(s) should she use? Explain your selection(s).

### PROBLEMS

**19.75.** Suppose the current from a battery is used to electroplate an object with silver. Calculate the mass of silver that would be deposited by a battery that delivers 1.3 A · hr of charge.

**19.76.** A battery charger used to recharge the NiMH batteries used in a digital camera can deliver as much as 0.75 A of current to each battery. If it takes 100 min to recharge one battery, how many grams of $Ni(OH)_2$ are oxidized to NiO(OH)?

---

**19.77.** A NiMH battery containing 4.10 g of NiO(OH) was 75% discharged when it was connected to a charger with an output of 2.00 A at 1.3 V. How long does it take to recharge the battery?

*19.78. How long does it take to deposit a coating of gold 1.00 μm thick on a disk-shaped medallion 2.0 cm in diameter and 3.0 mm thick at a constant current of 45 A? The density of gold is 19.3 g/cm$^3$. The gold solution contains Au(III).

_____

*19.79. **Oxygen Supply in Submarines** Nuclear submarines can stay under water nearly indefinitely because they can produce their own oxygen by the electrolysis of water.
   a. How many liters of $O_2$ at 298 K and 1.00 bar are produced in 1 hr in an electrolytic cell operating at a current of 0.025 A?
   b. Could seawater be used as the source of oxygen in this electrolysis? Explain why or why not.

19.80. In the electrolysis of water, how long will it take to produce $1.00 \times 10^2$ L of $H_2$ at STP (273 K and 1.00 atm) using an electrolytic cell through which the current is 52 mA?

_____

19.81. Calculate the minimum (least negative) cathode potential (versus SHE) needed to begin electroplating nickel from 0.35 $M$ $Ni^{2+}$ onto a piece of iron.

*19.82. What is the minimum (least negative) cathode potential (versus SHE) needed to electroplate silver onto cutlery in a solution of $Ag^+$ and $NH_3$ in which most of the silver ions are present as the complex, $Ag(NH_3)_2{}^+$, and the concentration of $Ag^+(aq)$ is only $2.50 \times 10^{-4}$ $M$?

## Fuel Cells

### CONCEPT REVIEW

19.83. Describe two advantages of hybrid (gasoline engine–electric motor) power systems over all-electric systems based on fuel cells. Describe two disadvantages.

19.84. Describe three factors limiting widespread use of cars powered by fuel cells.

19.85. Methane can serve as the fuel for electric cars powered by fuel cells. Carbon dioxide is a product of the fuel cell reaction. All cars powered by internal combustion engines burning natural gas (mostly methane) produce $CO_2$. Why are electric vehicles powered by fuel cells likely to produce less $CO_2$ per mile?

19.86. To make the refueling of fuel cells easier, several manufacturers offer converters that turn readily available fuels—such as natural gas, propane, and methanol—into $H_2$ for the fuel cells and $CO_2$. Although vehicles with such power systems are not truly "zero emission," they still offer significant environmental benefits over vehicles powered by internal combustion engines. Describe a few of them.

### PROBLEMS

19.87. Fuel cells with molten alkali metal carbonates as electrolytes can use methane as a fuel. The methane is first converted into hydrogen in a two-step process:

$$CH_4(g) + H_2O(g) \rightarrow CO(g) + 3\,H_2(g)$$
$$CO(g) + H_2O(g) \rightarrow H_2(g) + CO_2(g)$$

   a. Assign oxidation numbers to carbon and hydrogen in the reactants and products.

   b. Using the standard free energy of formation values in Table A4.3 in Appendix 4, calculate the standard free-energy changes in the two reactions and the overall $\Delta G°$ for the formation of $H_2 + CO_2$ from methane and steam.

*19.88. A direct methanol fuel cell uses the oxidation of methanol by oxygen to generate electrical energy. The overall reaction, which is given below, has a $\Delta G°$ value of −702.4 kJ/mol of methanol oxidized. What is the standard cell potential for this fuel cell?

$$CH_3OH(\ell) + \tfrac{3}{2}\,O_2 \rightarrow CO_2(g) + 2\,H_2O(\ell)$$

## Additional Problems

*19.89. **Electrolysis of Seawater** Magnesium metal is obtained by the electrolysis of molten $Mg^{2+}$ salts from evaporated seawater.
   a. Would elemental Mg form at the cathode or anode?
   b. Do you think the principal ingredient in sea salt (NaCl) would need to be separated from the $Mg^{2+}$ salts before electrolysis? Explain your answer.
   c. Would electrolysis of an aqueous solution of $MgCl_2$ also produce elemental Mg?
   d. If your answer to part c was no, what would be the products of electrolysis?

*19.90. **Silverware Tarnish** Low concentrations of hydrogen sulfide in air react with silver to form $Ag_2S$, more familiar to us as tarnish. Silver polish contains aluminum metal powder in a basic suspension.
   a. Write a balanced net ionic equation for the redox reaction between $Ag_2S$ and Al metal that produces Ag metal and $Al(OH)_3$.
   b. Calculate $E°$ for the reaction.

19.91. A magnesium battery can be constructed from an anode of magnesium metal and a cathode of molybdenum sulfide, $Mo_3S_4$. The standard reduction potentials of the electrode half-reactions are

$$Mg^{2+}(aq) + 2\,e^- \rightarrow Mg(s) \qquad E° = -2.37\text{ V}$$
$$Mg^{2+}(aq) + Mo_3S_4(s) + 2\,e^- \rightarrow MgMo_3S_4(s) \qquad E° = ?$$

   a. If the standard cell potential for the battery is 1.50 V, what is the value of $E°$ for the reduction of $Mo_3S_4$?
   b. What are the apparent oxidation states of Mo in $Mo_3S_4$ and in $MgMo_3S_4$?
   *c. The electrolyte in the battery contains a complex magnesium salt, $Mg(AlCl_3CH_3)_2$. Why is it necessary to include $Mg^{2+}$ ions in the electrolyte?

*19.92. **Clinical Chemistry** The concentration of $Na^+$ ions in red blood cells (11 m$M$) and in the surrounding plasma (140 m$M$) are quite different. Calculate the electrochemical potential (emf) across the cell membrane as a result of this concentration gradient at 37°C.

If your instructor assigns problems in **smartwork**, log in at **smartwork.wwnorton.com**.

# 20

# Biochemistry: The Compounds of Life

**20.1** The Composition of Proteins

**20.2** Protein Structure and Function

**20.3** Carbohydrates

**20.4** Lipids

**20.5** Nucleotides and Nucleic Acids

**20.6** From Biomolecules to Living Cells

## Learning Outcomes

**LO1** Interpret titration curves for amino acids and use them to define $pK_a$ and $pI$ values
**Sample Exercise 20.1**

**LO2** Name and draw the structures of small peptides and determine the peptide sequence of a polypeptide from its fragments
**Sample Exercises 20.2, 20.3**

**LO3** Describe the four levels of protein structure and how intermolecular forces and covalent bonds stabilize these structures

**LO4** Describe the molecular structures of simple sugars and the types of bonds linking them in polysaccharides; and explain how these compounds are used as energy sources and for energy storage

**LO5** Describe the molecular structure and the physical and chemical properties of lipids
**Sample Exercise 20.4**

**LO6** Describe the structures of DNA and RNA and how they function together to translate genetic information
**Sample Exercise 20.5**

## Function Follows Form

For all the stunning diversity of the biosphere, from single-cell organisms to elephants, whales, and giant redwoods, all life forms consist of substances made from only about 40 or 50 different small molecules. This small number of starting materials can link together to form large molecules and biopolymers that display astonishing variations of structure and function.

In this chapter we explore the composition, structure, and function of four classes of biomolecules: proteins, carbohydrates, lipids, and nucleic acids. Many of these molecules are large and complex, but our knowledge of the behavior of small molecules can serve as a framework for understanding the interactions of these classes of compounds. Each of these four classes performs similar functions in all the life forms in which they occur.

An important aspect of studying biochemical processes is finding out what happens when something goes wrong. It has been estimated that 70% of inherited diseases in people are caused by the absence of particular proteins or some other defect that causes proteins to function improperly. For example, the malformation or the total absence of the protein dystrophin causes a number of debilitating diseases referred to as muscular dystrophy (MD). In mild forms of MD, mis-shapen molecules of dystrophin are unable to build muscle fibers sufficient to function normally. As a result, muscle tissue wastes away and the patient becomes physically disabled. In Duchenne MD, a severe form of the disease, functional dystrophin is absent, and the disability often results in early death. Knowlege of the mechanisms that produce dystrophin has led to promising therapies. Similar approaches based on mechanisms have resulted in the development of new drugs and cell therapies for other inherited and contagious diseases. By learning how biomolecules behave in healthy individuals, scientists have developed new drugs and treatments for many inherited and contagious diseases. ∎

**Similarity, Diversity, Interdependence** The abundant diversity of life on Earth arises from different combinations of 40 or 50 small molecules, and these basic building blocks of life are responsible for the similarity, diversity, and interdependence of all forms of life. ▶

**protein** biological polymer made of amino acids.

**biomolecule** an organic molecule present naturally in a living system.

**amino acid** molecule that contains at least one amine group and one carboxylic acid group; in an *α-amino acid*, the two groups are attached to the same (α) carbon atom.

**essential amino acid** any of the 8 amino acids that make up peptides and proteins but are not synthesized in the human body and must be obtained through the food we eat.

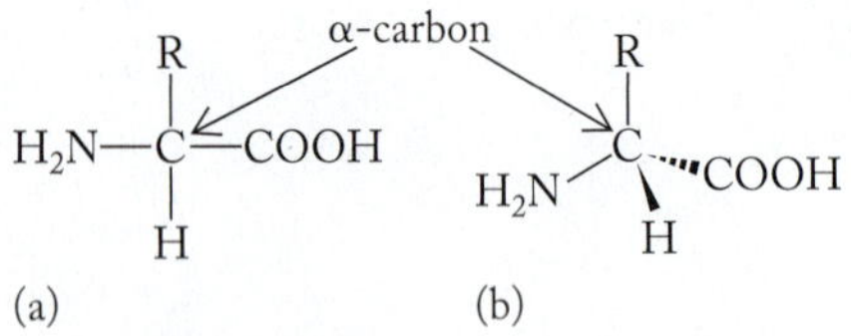

FIGURE 20.1 (a) General structure of an α-amino acid; (b) three-dimensional structure.

**◉◉ CONNECTION** As in Chapter 13, we use the generic symbol R to represent any group of covalently bonded atoms that are connected to the main structure of a molecule via a C—C bond.

**FIGURE 20.2** A meal of red beans and rice provides all the essential amino acids.

# 20.1 The Composition of Proteins

We begin this chapter with a discussion of **proteins**, the most abundant class of **biomolecules** in all animals, including us. Proteins account for about half the mass of the human body that is not water. They are the major component in skin, muscles, cartilage, hair, and nails. Most of the enzymes that catalyze biochemical reactions are proteins, as are the molecules that transport oxygen to our cells and many of the hormones that regulate cell function and growth. Most proteins are large, with molar masses of $10^5$ g/mol or more, and are made from monomers called amino acids. They are examples of *biopolymers*.

## Amino Acids

The molecular structures and biological functions of big biomolecules depend on the identities of their small-molecule building blocks and the sequence in which those small molecules are connected. The molecular building blocks of proteins are **amino acids**, so named because each of them contains at least one amine (–NH$_2$) group and at least one carboxylic acid (–COOH) group. The amino acids in proteins are called α-amino acids because in their structures one carbon atom, called the α-carbon, is bonded to both an –NH$_2$ and a –COOH group (Figure 20.1).

In addition to its single bonds to the –NH$_2$ and –COOH groups, the α-carbon atom in each of the amino acids that make up human proteins is bonded to a hydrogen atom and to one of 20 different *R groups*. The structures of these R groups, often called *side-chain groups*, are highlighted in red in the amino acid structures in Table 20.1. The amino acids are arranged in four categories based on their R groups. In the first category, containing 9 amino acids, the R groups contain mostly carbon and hydrogen atoms, and are nonpolar. The R groups of the remaining 11 amino acids contain at least one *heteroatom* (O, N, or S) bonded to an H atom and are polar. Of these amino acids, two (aspartic acid and glutamic acid) have R groups that contain carboxylic acid functional groups, and three (histidine, lysine, and arginine) contain amine groups, which are weakly basic.

Our bodies can synthesize 12 of the 20 amino acids in Table 20.1, but the other 8 must be present in the food we eat. These 8 are marked with a superscript *b* and are referred to as **essential amino acids**. Most proteins from animal sources, including those in meats, eggs, and dairy products, contain all the essential amino acids needed by the human body, and in close to the correct proportions. These foods are sometimes referred to as *perfect foods* or, more precisely, *complete proteins*. In contrast, most plant proteins from foods such as legumes and vegetables do not contain all the essential amino acids, so vegetarians must be careful to eat a combination of foods that provide all the essential amino acids. A good example is red beans and rice (Figure 20.2), a traditional dish in Latin American cuisine that provides a balance of essential amino acids: rice has all of them but lysine, and beans lack only methionine.

## Chirality

Look carefully at the structure of the amino acid alanine in Figure 20.3. Note how the α-carbon atom is bonded to four different groups. This bonding pattern makes this α-carbon atom a *chiral center* (see Section 13.8). It also means that all the amino acids in Table 20.1 except glycine are chiral compounds: their molecular

**TABLE 20.1   Structures and Abbreviations of the 20 Common Amino Acids[a]**

**Nonpolar R Groups**

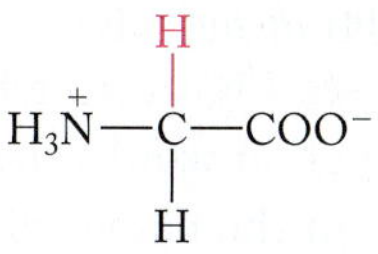

Glycine (Gly; G)

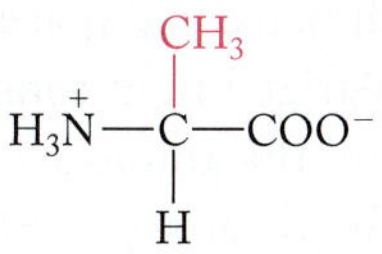

Alanine (Ala; A)

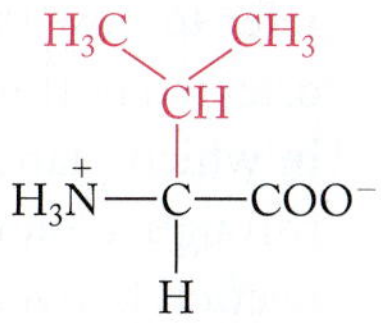

Valine[b] (Val; V)

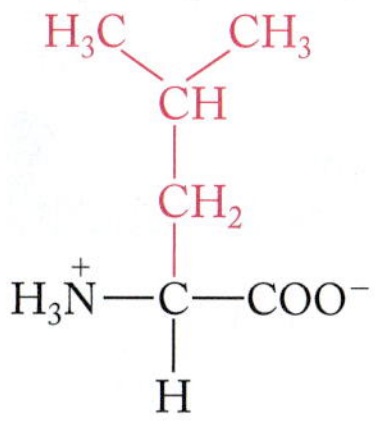

Leucine[b] (Leu; L)

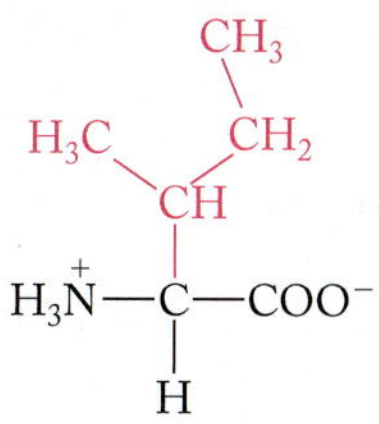

Isoleucine[b] (Ile; I)

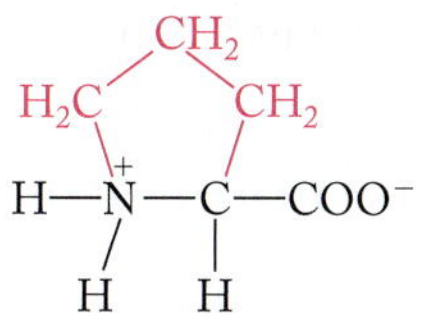

Proline (Pro; P)

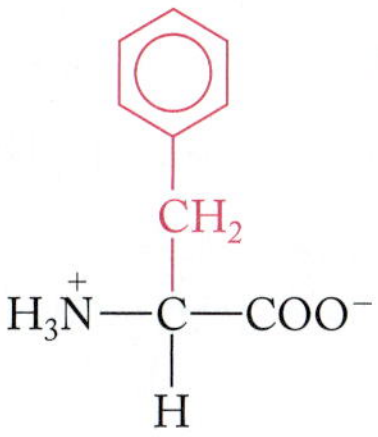

Phenylalanine[b] (Phe; F)

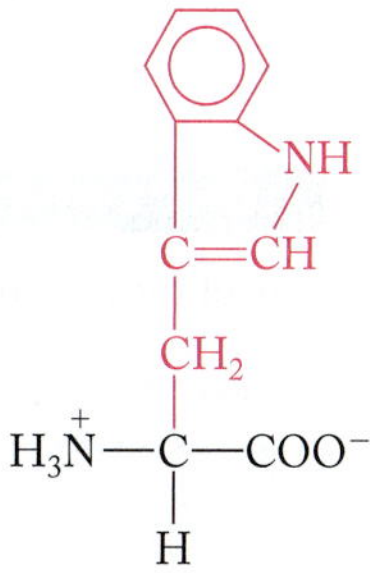

Tryptophan (Trp; W)

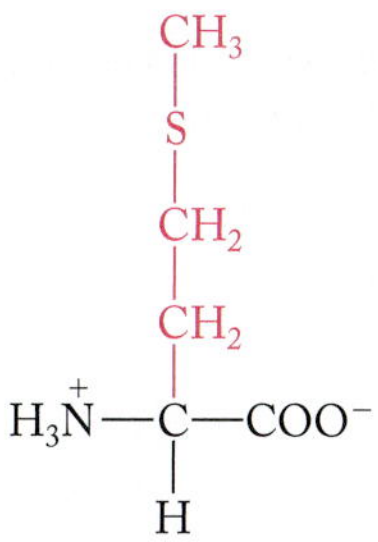

Methionine[b] (Met; M)

**Polar R Groups**

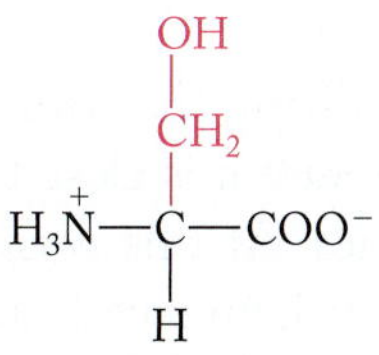

Serine (Ser; S)

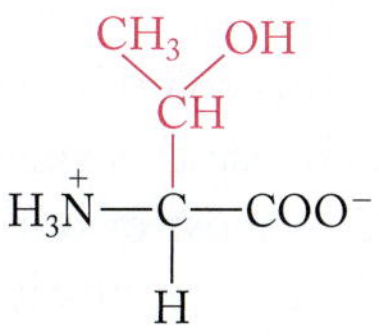

Threonine[b] (Thr; T)

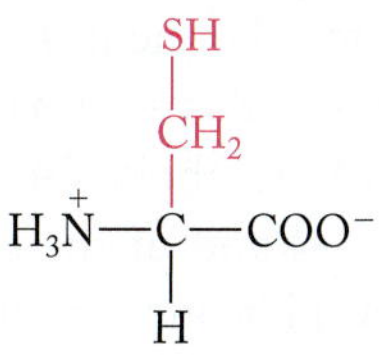

Cysteine (Cys; C)

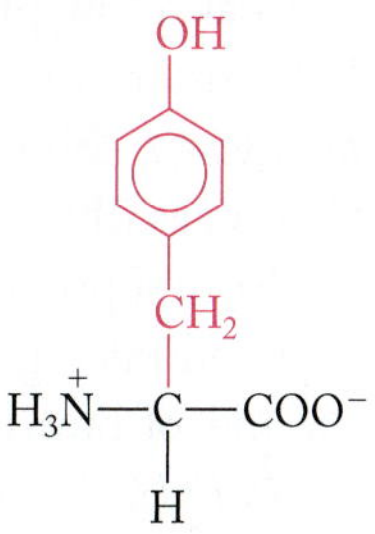

Tyrosine[b] (Tyr; Y)

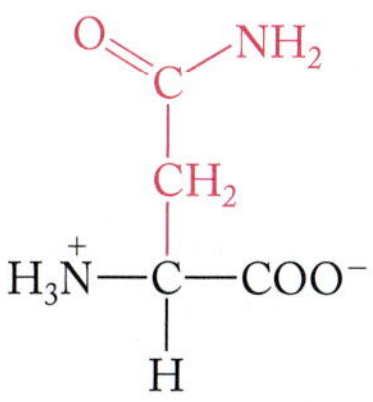

Asparagine (Asn; N)

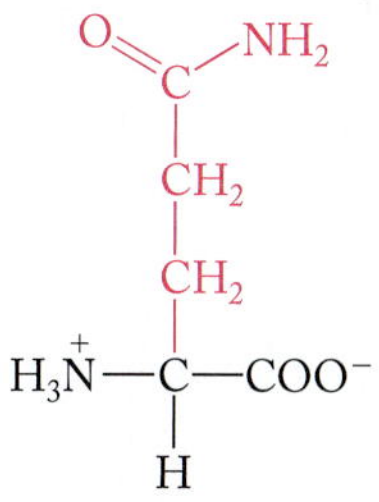

Glutamine (Gln; Q)

**Acid R Groups**

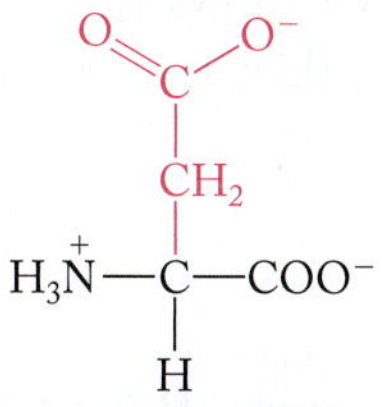

Aspartic acid (Asp; D)

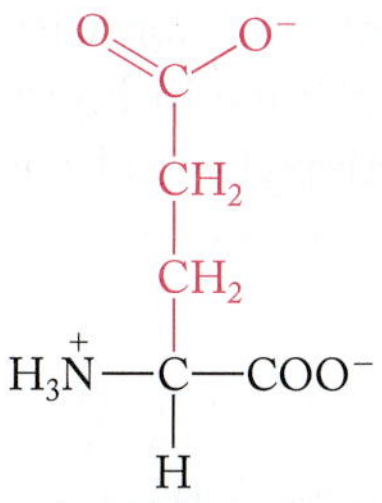

Glutamic acid (Glu; E)

**Basic R Groups**

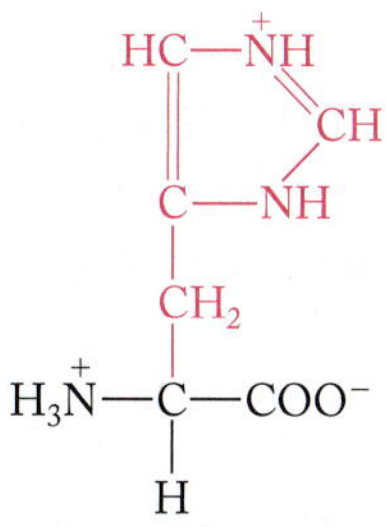

Histidine (His; H)

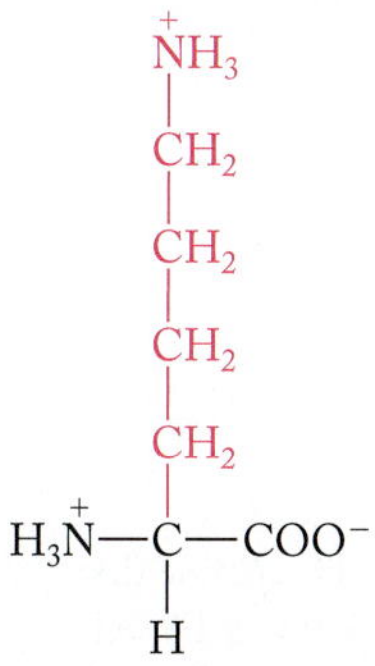

Lysine[b] (Lys; K)

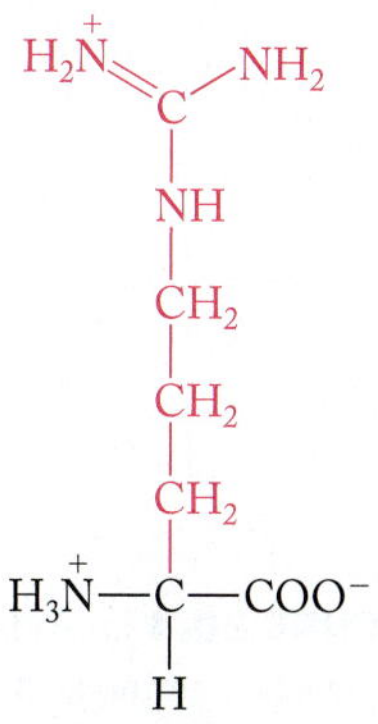

Arginine (Arg; R)

[a]R groups in pink; ionized forms at pH near 7 shown.
[b]The eight essential amino acids for adults (histidine is essential for children).

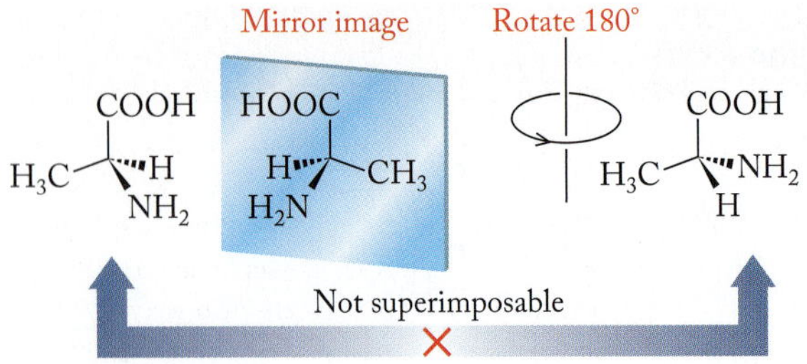

Mirror image    Rotate 180°

Not superimposable

**FIGURE 20.3** Like most α-amino acids, alanine is chiral, which means that its molecular structure is not superimposable on its mirror image. Rotating the mirror image 180° so that the –CH$_3$ and –COOH groups overlap does not produce a structure in which –NH$_2$ and –H overlap.

**◉◉ CONNECTION** Optical isomerism and the concept of chirality were described in Section 13.8.

structures are not superimposable on their mirror images, as illustrated for alanine in Figure 20.3. Instead, each amino acid and its mirror image constitute an enantiomeric pair.

For historical reasons, amino acid enantiomers are designated by the prefixes D- (for *dextro-*, right) and L- (*levo-*, left). These labels refer to how the four groups bonded to each chiral carbon atom are oriented in three-dimensional space. They *do not* refer to the direction in which plane-polarized light rotates (see Figure 13.54) as it passes through a solution of the amino acid. In other words, there is no connection between the D- and L- prefixes in the names of amino acids and the *dextrorotary* and *levorotary* enantiomers that are designated with (+) and (−) signs based on their optical properties. All the chiral amino acids in the proteins in our bodies are L-enantiomers, even though 9 of the 19 are actually dextrorotary.

**CONCEPT TEST** • • • • • • • • • • • • • • • • • • • • • • • • • • • • • • •

One of the 20 amino acids in Table 20.1, glycine, is not chiral. Why?

*(Answers to Concept Tests are in the back of the book.)*

• • • • • • • • • • • • • • • • • • • • • • • • • • • • • • • • • • • • • • • • • • • • •

## Zwitterions

If we dissolve an amino acid in a solution that already contains a strong acid and has a low pH, the carboxylic acid group on each molecule does not ionize. In addition, each amine group, being a weak base, accepts a H$^+$ ion, forming a *protonated* –NH$_3^+$ group. The result is a molecular ion with an overall positive charge, as illustrated for alanine in Figure 20.4(a).

Now suppose we add a strong base to this solution, neutralizing the strong acid and raising the pH to about 7.4. We choose this value because it is close to the pH of human blood and of the fluids in most of our tissues. At this *physiological pH*, the –COOH groups in amino acids are mostly ionized, but nearly all the amine groups are still in the protonated –NH$_3^+$ form because –NH$_3^+$ is the conjugate acid of a weak base and is itself a *very* weak acid. As a result, most of the protonated amine groups still have positive charges and most of the ionized carboxylic acid groups have negative charges. A molecule with this distribution of charges is called a **zwitterion** (literally, a *hybrid ion*) because it has both a positive and a negative functional group (Figure 20.4b). Its net charge is zero.

If we add more base, alanine loses a H$^+$ from its NH$_3^+$ group (we say that it *deprotonates*), and we have a molecular ion with an overall charge of 1− (Figure 20.4c). The titration curve for alanine in Figure 20.5 shows this two-step neutralization process: first the carboxylic acid ionizes, then the protonated amine deprotonates.

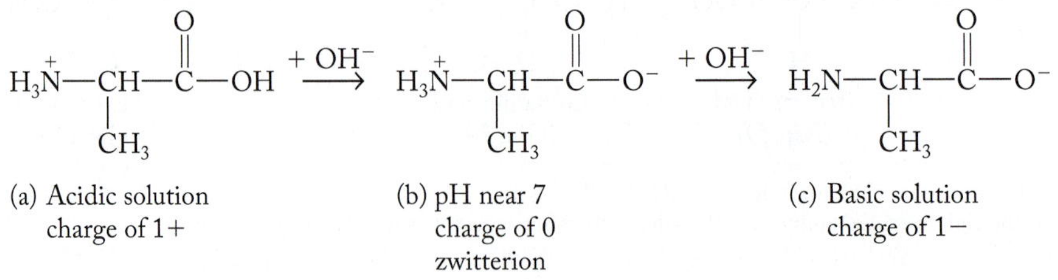

**FIGURE 20.4** (a) At low pH, alanine (like many other amino acids) exists as a 1+ ion. (b) At physiological pH (7), the –COOH group is ionized and the molecule becomes a zwitterion with an overall charge of zero. (c) In basic solutions, the charge decreases to 1− as the –NH$_3^+$ group deprotonates.

(a) Acidic solution charge of 1+

(b) pH near 7 charge of 0 zwitterion

(c) Basic solution charge of 1−

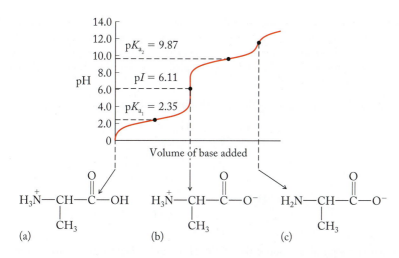

**FIGURE 20.5** The titration curve of alanine resembles that of a weak diprotic acid. The carboxylic acid group ($pK_{a_1} = 2.35$) is neutralized first. The protonated amine group ($pK_{a_2} = 9.87$) is neutralized in the second step of the titration. The dominant forms of alanine present in solution are shown at (a) the starting point, and at (b) and (c), the two equivalence points.

Amino acids with acidic or basic R groups have three-step neutralization processes. For example, an amino acid with a second carboxylic acid group in R will experience the neutralization of both acidic groups and then the deprotonation of the amine. Which acidic group is the stronger; which deprotonates at lowest pH? Think about the structure of the molecule to answer this. The –COOH group bonded to the $\alpha$-carbon atom is only one carbon atom away from an electropositive $NH_3^+$ group. The side-chain –COOH group is at least two carbon atoms away, and the carboxylate ion that it forms when it ionizes is less stabilized by delocalization of its negative charge toward the cationic group. Therefore, the –COOH group bonded to the $\alpha$-carbon atom is the stronger acid and ionizes first.

For an amino acid with a basic R group, the order of ionization in a titration also depends on the relative strengths of the groups as proton donors. The –COOH always ionizes first as base is added; it is the strongest acid in the molecule. The protonated nitrogen groups deprotonate in order of their relative strengths as proton donors.

Each amino acid has a characteristic **pI** value, called the **isoelectric point**, representing the pH at which molecules of the amino acid have, on average, zero charge. The pH at the equivalence point between the two $pK_a$ values is the pI. Its value is the average of the flanking $pK_a$ values. At this pH, the amino acid does not migrate in an electric field. If the pH is more acidic than the pI, the amino acid is positively charged and migrates to the cathode (negative electrode). At pH values more basic than the pI, the amino acid has a net negative charge and migrates toward the anode (positive electrode). This behavior is the basis of several analytical techniques used to characterize and identify amino acids. The pI values of neutral amino acids (Table 20.1) are around pH 7; acidic amino acids have pI values much lower than 7, and basic amino acids, much higher.

**CONNECTION** The inverse relationship between the strengths of acid–base conjugate pairs was discussed in Chapter 16 and illustrated in Figure 16.4.

**SAMPLE EXERCISE 20.1** **Interpreting Acid–Base Titration Curves of Amino Acids** **LO1**

Figure 20.6 shows the titration curve for aspartic acid and its $pK_{a_1}$, $pK_{a_2}$, and $pK_{a_3}$ values. Draw the molecular structures of the principal form of aspartic acid that is in solution at the start of the titration and at each equivalence point. Estimate the pI of aspartic acid.

**Collect and Organize** We are asked to draw the molecular structures of aspartic acid at three characteristic pH values. The molecular structure of aspartic acid is in Table 20.1.

**zwitterion** a molecule that has both positively and negatively charged groups in its structure.

**pI** (also called **isoelectric point**) the pH at which molecules of the amino acid have, on average, zero charge.

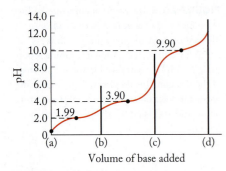

**FIGURE 20.6** Titration curve of aspartic acid. The vertical black lines identify the equivalence points.

**Analyze** Aspartic acid has two carboxylic acid groups and one amine group. At low pH the carboxylic acid groups are not ionized and the amine group is protonated. As pH is raised, the –COOH groups ionize and then, under basic conditions, the protonated amine group releases its proton. The –COOH group bonded to the α-carbon atom should be the stronger acid and ionize first.

**Solve** At (a), the beginning of the titration, the fully protonated 1+ ion is present:

At (b), the first equivalence point, the –COOH group bonded to the α-carbon atom is ionized; the charge on the ion is 0.

At (c), the second equivalence point, the side-chain –COOH group is ionized; the charge on the ion is 1−:

At (d), the third equivalence point in the titration, the amine group is deprotonated, resulting in an ion with a 2− charge:

The p$I$ of aspartic acid is the value midway between p$K_{a_1}$ and p$K_{a_2}$ on Figure 20.6, or (1.99 + 3.90)/2 = 2.94.

**Think About It** The values of p$K_{a_1}$ and p$K_{a_2}$ differ by nearly 2 pH units, which means that the α-COOH group is almost 100 times stronger an acid than the side-chain –COOH group, due to the presence of the α-NH$_3^+$ group.

⚙ **Practice Exercise** Sketch the titration curve for lysine starting at pH = 1.0, and draw the molecular structure of the principal form of lysine that is in solution at each equivalence point.

*(Answers to Practice Exercises are in the back of the book.)*

■ ••••••••••••••••••••

The acid–base characteristics of amino acids are important for functional reasons. In Chapter 14 we mentioned catalysis by proteins called enzymes, and amino acid protonation/deprotonation controls the activity of many catalysts. In addition, the structures of proteins depend in part on hydrogen bonding, which can be disrupted by changes in pH. This process is used in food preparation, when acids like lemon juice are used to "cook" raw fish in dishes like ceviche.

## Peptides

Amino acids bond together to form chainlike molecules with a wide range of sizes. Amino acid *residues* (the name we give to amino acids that are part of a larger molecule) make up each link in the chain. The longest chains, some with molar masses as high as 3 million g/mol, are called proteins, but the shortest chains, called **peptides**, are only a few links long. The smallest peptides contain only two or three amino acid residues. These molecules are called *dipeptides* and *tripeptides*. Peptides up to 20 residues long are called *oligopeptides*. Those made of more than 20 are called *polypeptides*. The size at which a polypeptide becomes a protein is arbitrary but is typically set around 50–75 amino acid residues.

The type of bond linking the amino acids in peptides and proteins is called a **peptide bond** (highlighted in blue in Figure 20.7) or *peptide linkage*. They form when the α-carboxylic acid group of one amino acid condenses with the α-amine group of another, with the loss of a molecule of water. Note that peptide bonds have the same structure as the amide bonds that hold together the monomeric units of synthetic polyamides such as nylon (Chapter 13). This process is not spontaneous and is driven by being coupled to ATP hydrolysis, as we saw in Section 18.9.

**peptide** a compound of two or more amino acids joined by peptide bonds. Small peptides containing up to 20 amino acids are *oligopeptides*; and the term *polypeptide* is used for chains longer than 20 amino acids but shorter than proteins.

**peptide bond** the result of a condensation reaction between the carboxylic acid group of one amino acid and the amine group of another.

▶‖ **CHEMTOUR** Condensation of Biological Polymers

◉◉ **CONNECTION** The formation of amide bonds in reactions between carboxylic acids and amines was described in Chapter 13 and illustrated in Figure 13.48.

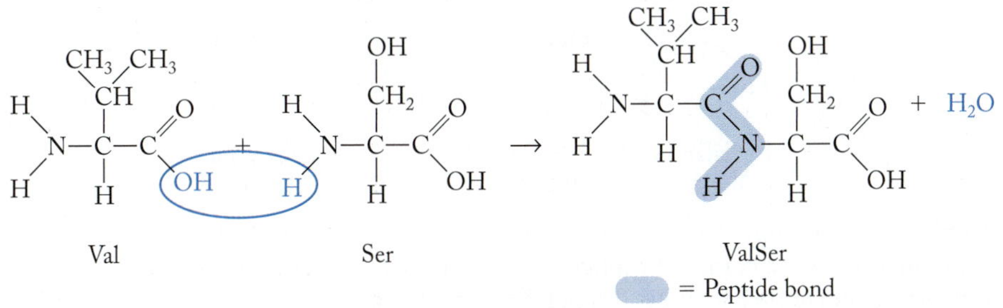

**FIGURE 20.7** When the carboxylic acid group of valine reacts with the amine group of serine (blue oval) to form a peptide bond (blue highlight), the products are the dipeptide valylserine and water.

The convention for drawing the structures of peptides begins by placing the amino acid that has a free α-amine group at the left end of the peptide chain and the amino acid with a free α-carboxylic acid at the right end. The left end is called the *amine* (or *N-*) *terminus* of the peptide and the right end is called the *carboxylic acid* (or *C-*) *terminus*. The name of a peptide is based on the names of its amino acids, starting with the one at the N-terminus. The names of all the amino acids except the one at the C-terminus are changed to end in -*yl*. For example, if a peptide bond forms between the α-COOH group of valine (Val) and the α-NH$_2$ group of serine (Ser) as in Figure 20.7, the dipeptide that is produced is called valylserine, or more typically, using the three-letter or one-letter abbreviations, ValSer or VS.

The artificial sweetener aspartame is the methyl ester of the dipeptide aspartylphenylalanine (Figure 20.8). At pH 7.4, aspartame exists as a zwitterion because the aspartic acid amine group is protonated and the carboxylic acid in its R group is ionized. In contrast, its parent dipeptide has a net charge of

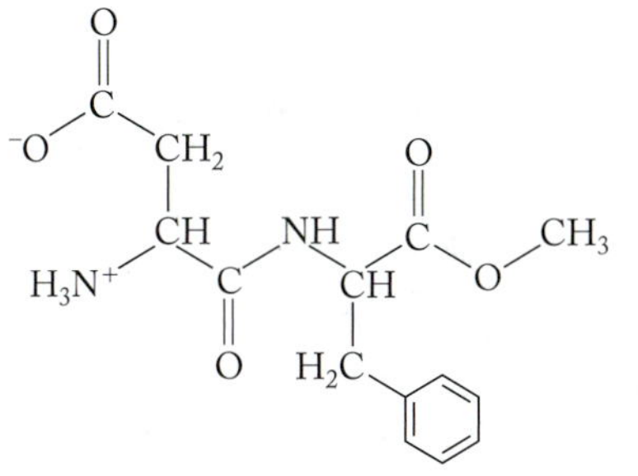

Aspartame

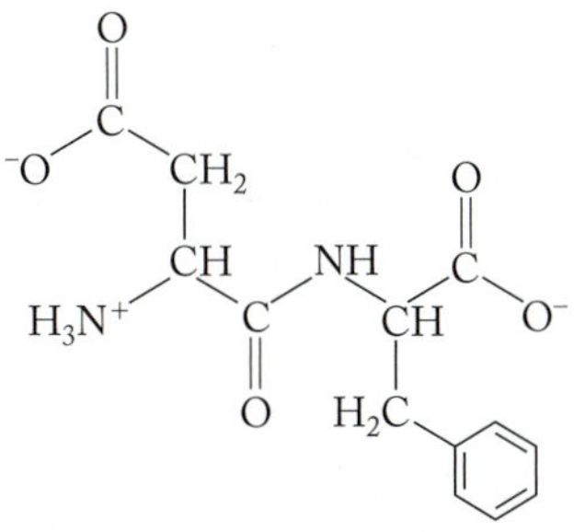

Aspartylphenylalanine

**FIGURE 20.8** Aspartame is the methyl ester of the dipeptide aspartylphenylalanine and so has one fewer –COOH group than its parent dipeptide. Therefore, the overall charge of a molecule of aspartame at pH 7.4 is zero, whereas the charge on aspartylphenylalanine is 1−.

1− because both −COOH groups are ionized at that pH. In an amino acid that has an amine group in its R group, such as lysine or arginine, that amine group is probably protonated at physiological pH, giving the amino acid a net charge of 1+. The overall charge on the peptide at physiological pH is the sum of the positive charges on protonated amine groups and the negative charges of ionized carboxylic acid groups.

---

**SAMPLE EXERCISE 20.2    Drawing and Naming Peptides    LO2**

(a) Name all the dipeptides that can be made by reacting alanine with glycine, and (b) draw their molecular structures in solution at pH 7.4.

**Collect and Organize** We are to draw and name the dipeptides produced when alanine and glycine are linked by a peptide bond. The structures of these amino acids are in Table 20.1.

**Analyze** Two different peptides can be made from two amino acids by changing their sequence from the N-terminus to the C-terminus. For the two dipeptides, the sequences are AlaGly and GlyAla. The R groups in alanine (−CH$_3$) and glycine (−H) have no acidic or basic properties. At pH 7.4, the carboxylic acid terminus −COOH groups should be ionized and the amine terminus −NH$_2$ groups should be protonated.

**Solve**
a. The names of the two peptides with the amino acid sequence GlyAla and AlaGly are glycylalanine and alanylglycine, respectively.
b. At pH 7.4, the −NH$_2$ groups of amino acids are protonated and the −COOH groups are ionized, so the principal forms of these dipeptides are

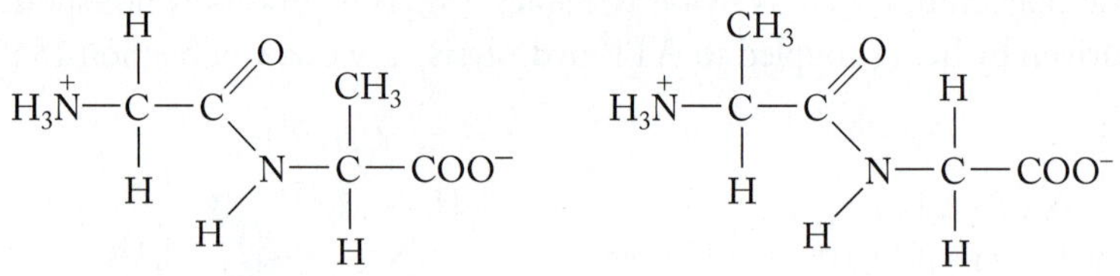

Glycylalanine
(GlyAla)

Alanylglycine
(AlaGly)

**Think About It** Only the N-terminal −NH$_2$ and C-terminal −COOH groups can be protonated or ionized in these dipeptides. Thus, AlaGly and GlyAla are both zwitterionic with net charges of (1+) + (1−) = 0 at pH 7.4.

**Practice Exercise** How many different tripeptides can be synthesized from one molecule of each of three different amino acids Gly, Tyr, and Ser?

---

**CONCEPT TEST**

Rank the following amino acids in order of their net charges at pH 7.4, starting with the most positive net charge: (a) alanine; (b) arginine; (c) aspartic acid.

---

Only two sequences are possible for a peptide containing two different amino acids. Amino acid A and amino acid B can make dipeptide AB or BA. For a peptide made from three different amino acids—A, B, and C—six sequences are possible. For longer peptides, the number of possibilities becomes enormous. The sequence matters in determining the properties of the peptides, and figuring out the amino acid sequence of peptides and proteins is a significant challenge in biochemistry. One of the fundamental experimental methods used to determine sequence involves finding the identities of individual component amino acids, then identifying fragments of the structure containing several amino acids and piecing them together like a puzzle. For example, a tripeptide could be broken down into amino acids A, B, and C. Larger fragments BA and AC could then be identified, which means the overall sequence of the tripeptide must be BAC.

---

**SAMPLE EXERCISE 20.3**    **Determining the Sequence**     **LO2**
                                       **of a Small Peptide**

A small peptide was hydrolyzed completely to its component amino acids: A, B, C, D, E, F, and G. Alternative ways to treat the peptide yielded these larger fragments: BA, CDF, EC, AG, and GEC. What is the sequence of the amino acids in the peptide?

**Collect and Organize** We know the identity of seven amino acids that make up a peptide, and we have several larger fragments.

**Analyze** We need to arrange the larger fragments in a way that reveals the sequence of the larger peptide.

**Solve** Arranging the fragments so that like amino acids are in columns:

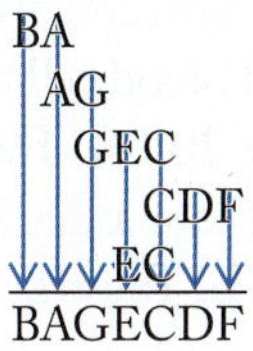

The amino acid sequence of the peptide is BAGECDF.

**Think About It** Longer peptide chains and proteins have many copies of each amino acid, so longer fragments are needed to sequence them.

⚙ **Practice Exercise** The hydrolysis of a peptide yielded the following amino acids: Gly, Ala, Val, Thr, and 2 Leu; and the following larger fragments: LeuThr, AlaLeuLeu, ThrGly, ValAla. What is the sequence of the amino acids in the peptide?

---

# 20.2 Protein Structure and Function

The structure of a protein is crucial to its function. This fact is easiest to appreciate for structural proteins like the collagens, which impart flexibility, strength, and elasticity to skin and tendons. The functions of nonstructural proteins, such as the ability of enzymes to catalyze biochemical reactions, are also closely linked to their structures. These large biomolecules must assume particular three-dimensional conformations to interact with other molecules and to function properly. Table 20.2 lists some of the important functional classifications of proteins and gives examples of each.

## Primary Structure

The **primary (1°) structure** of a protein is the sequence of the amino acids in it, starting with the N-terminus (Figure 20.9a). If two proteins are made up of the same number and type of amino acids but have different amino acid sequences, they are different proteins.

Changing only one amino acid can dramatically alter a protein's function. For example, in hemoglobin the sixth amino acid from the N-terminus of a protein

**primary (1°) structure** the sequence in which amino acid monomers occur in a protein chain.

**TABLE 20.2  Functional Classes of Proteins**

| Class of Protein | Function | Example |
|---|---|---|
| Structural | Gives strength and flexibility to tissues | Collagen: a component of connective tissue |
| Enzymes | Catalyze reactions in cells | Sucrase: promotes conversion of sucrose into simpler sugars |
| Hormones | Regulate processes in organisms | Insulin: helps cells regulate absorption of glucose |
| Signaling | Coordinates the activity of cells | GTPase: regulates intracellular dynamics |
| Immune response | Defense | Antibodies: attack bacteria |
| Storage | Releases energy during metabolism | Ovalbumin: present in eggs |
| Transport | Transports molecules from place to place | Hemoglobin: transports oxygen from lungs to cells |

**FIGURE 20.9** The four levels of protein structure. (a) A protein's primary structure is its amino acid sequence. The green shapes represent different R groups. (b) Secondary structure (here, an α helix) describes the three-dimensional pattern adopted by segments of the protein strand. (c) Tertiary structure is the overall shape of the molecule as segments of it bend and fold. (d) Quaternary structure refers to the overall shape adopted by multiple protein strands that assemble into a single unit.

strand that is 146 amino acids long is normally glutamic acid. In some people, valine substitutes for glutamic acid at this position. This one substitution alters the solubility of the protein and causes the red blood cell to take on a sickle shape instead of the normal, plump disk shape (Figure 20.10). These sickled cells do not pass through capillaries easily and may impede blood circulation. They also break readily and do not last as long as normal blood cells. These factors lead to a diminished capacity of

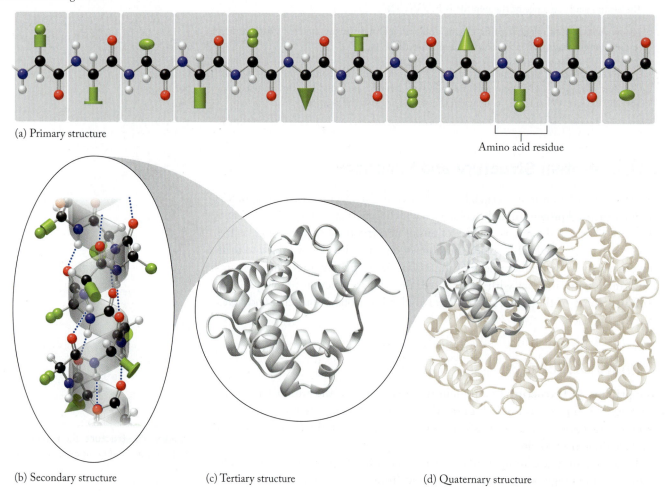

(a) Primary structure

Amino acid residue

(b) Secondary structure          (c) Tertiary structure          (d) Quaternary structure

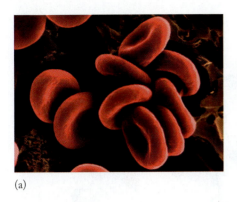

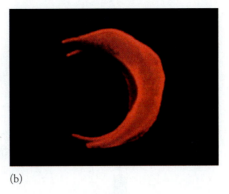

(a)                                (b)

**FIGURE 20.10** (a) Normal red blood cells are plump disks, whereas (b) those in patients suffering from sickle-cell anemia are distorted and incomplete.

the blood to carry oxygen, which is one of the characteristics of the disease called sickle-cell anemia.

Why does switching valine for glutamic acid affect the solubility of hemoglobin? The answer lies in the R groups of these two amino acids (Figure 20.11). The side-chain –COOH group of glutamic acid is ionized at physiological pH, and strong ion–dipole interactions with water molecules enhance the protein's solubility. However, if valine with its nonpolar isopropyl R group replaces glutamic acid, that strong intermolecular interaction with water is lost. The valine creates a hydrophobic patch on the surface of the protein that results in the deoxygenated forms of hemoglobin sticking to each other, producing stiff fibers inside red blood cells. This, in turn, deforms the red blood cell from its usual smooth disk shape (Figure 20.10a) into a sickle shape (Figure 20.10b). The distorted cells are fragile and often rupture, leading to loss of hemoglobin.

Sickle-cell anemia is a debilitating disease, but it provides a survival advantage in regions where malaria is endemic. Having sickle-cell anemia does not protect people from contracting malaria or make them invulnerable to the parasite that causes it. However, children infected with malaria are more likely to survive the illness if they have sickle-cell anemia. Exactly why sickle-cell anemia has this effect is not completely understood, but recent work points toward sickle-hemoglobin increasing the production of carbon monoxide by an enzyme in infected cells; the carbon monoxide protects the host from succumbing to malaria without actually interfering with the parasite.

**CONCEPT TEST** ••••••••••••••••••••••••••••••••••••••••••••••••••••

Which other amino acids besides valine might result in the sickling of the cell when substituted for glutamic acid?

••••••••••••••••••••••••••••••••••••••••••••••••••••••••••••••••••••••

## Secondary Structure

The next level of protein structure, the **secondary (2°) structure**, describes the geometric patterns made by segments of amino acid chains. One common pattern is the **α helix**, a coiled arrangement with the R groups pointing outward (Figure 20.9b). The helical structure is maintained by hydrogen bonds between –NH groups on one part of the chain and C═O groups on amino acids four residues away from them in the sequence. The α helix looks very much like a spring. One group of proteins that are mostly α-helical is the keratins, the proteins in hair and fingernails.

Another common pattern of 2° structure is called a **β-pleated sheet**. These sheets are assemblies of multiple amino acid chains aligned side by side. The pleats

Primary structure

Normal protein:
   Val - His - Leu - Thr - Pro -⎡Glu⎤- Lys - ...
Abnormal protein:
   Val - His - Leu - Thr - Pro -⎣Val⎦- Lys - ...
(a)

$$\begin{array}{c} O \diagdown \ \ \diagup O^- \\ C \\ | \\ CH_2 \\ | \\ CH_2 \\ | \\ H_3\overset{+}{N}-C-COO^- \\ | \\ H \end{array} \qquad \begin{array}{c} H_3C \diagdown \ \ \diagup CH_3 \\ CH \\ | \\ H_3\overset{+}{N}-C-COO^- \\ | \\ H \end{array}$$

   Glutamic acid          Valine
(b)

**FIGURE 20.11** (a) The primary structure of the amine end of a protein in normal hemoglobin and in the abnormal hemoglobin responsible for sickle-cell anemia. (b) The replacement of glutamic acid with its hydrophilic R group by valine with its hydrophobic group is responsible for the disease.

**CONNECTION** We discussed in Chapter 10 how ion-dipole and dipole-dipole interactions are the key to the solubility of solutes in water.

**secondary (2°) structure** the pattern of arrangement of segments of a protein chain.

**α helix** a coil in a protein chain's secondary structure.

**β-pleated sheet** a puckered two-dimensional array of protein strands held together by hydrogen bonds.

**FIGURE 20.12** Each amino acid chain in a β-pleated sheet is folded in a zigzag pattern. Adjacent chains in a sheet are held together by hydrogen bonds (blue dotted lines). The R groups (green structures) extend above and below the sheet, linking it to adjacent sheets via noncovalent intermolecular interactions.

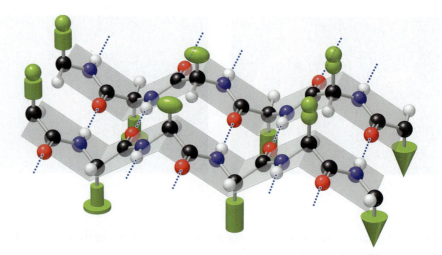

(a) $\vdash CH_2CH_2COO^- \cdots H_3\overset{+}{N}-(CH_2)_4 \dashv$

(b) $\vdash CH_2-\ddot{O}\cdots H-O-CH \dashv$ with H below O, and $CH_3$ below CH

(c) $\vdash CH \overset{CH_3}{\underset{CH_3}{\diagdown}} \; CH_3 \dashv$

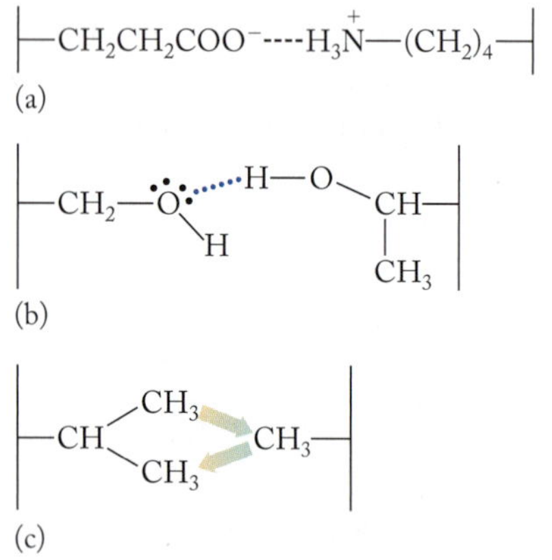

**FIGURE 20.13** Intermolecular interactions that influence the secondary and tertiary structures of proteins include (a) ion–ion interactions between acidic and basic R groups, (b) hydrogen bonding, and (c) van der Waals interactions between nonpolar side chains.

▶❚❚ **CHEMTOUR** Fiber Strength and Elasticity

are caused by the tetrahedral molecular geometries of the atoms along the chains (Figure 20.12). Adjacent chains are linked together by hydrogen bonds, and the collection of side-by-side zigzag chains forms a continuous β-pleated sheet. R groups extend above and below the pleats. Sheets may stack on top of one another like two pieces of corrugated roofing. Stacked sheets are held together by the same interactions that hold all proteins together, including ion–ion, hydrogen-bonding, and van der Waals interactions, depending on the pairs of R groups involved (Figure 20.13).

The proteins that make up strands of silk form thin, planar crystals of β-pleated sheets that are only a few nanometers on a side. Enormous numbers of these crystals form long arrays of sheets stacked together like nanoscale pancakes (Figure 20.14). Hydrogen bonds hold the stacks together and reinforce adjacent sheets. As a result of these interactions, strands of silk are stronger than strands of steel with the same mass; indeed, silk is one of the toughest known materials—natural or synthetic.

Some single-stranded proteins exist as α helices on their own but form β-pleated sheets when they clump together in multistrand aggregates. One consequence of this clumping is the formation of insoluble protein deposits called plaque. Abnormal accumulation of plaque can be a serious health risk: plaque formed by a protein called amyloid β has been linked to the onset of Alzheimer's disease.

If part of a protein is characterized by an irregular or rapidly changing structure, it is said to have a **random coil** 2° structure. The amino acid chain may fold back on itself and around itself, but it has no regular features the way an α helix or a β-pleated sheet does. When proteins *denature*, losing their secondary structure because of heat or change in pH, they may become random coils.

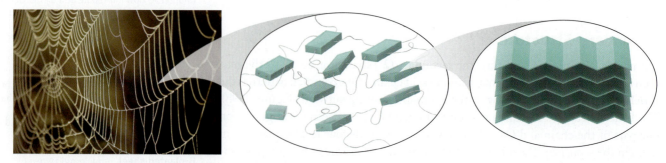

**FIGURE 20.14** The strength of spider silk comes from the flexible cross-linked chains connecting regions of crystalline stacks of β-sheets, shown here as blue-green boxes, which are only a few nanometers in size. Hydrogen bonds make an extremely strong network, and if a hydrogen bond breaks, many more are left that can maintain the material's overall strength.

Large protein molecules may contain all three types of 2° structure. In describing a protein, scientists may indicate the percentage of amino acids involved in each type—for example, 50% α-helical, 30% β-pleated sheet, and 20% random coil. Figure 20.15 shows a model of a protein called carbonic anhydrase, which illustrates all three types.

**CONCEPT TEST**

The aqueous solution of proteins in egg whites turns into a solid mass when eggs are cooked or are dropped into an organic solvent such as acetone.
a. What is the likely secondary structure of the proteins in cooked eggs?
b. Do you think the solidification process occurs as a result of a change in the primary structure of the proteins?

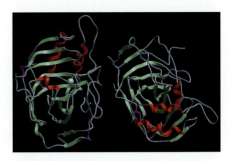

**FIGURE 20.15** The structure of the protein carbonic anhydrase has α-helical regions (red), β-pleated sheet regions (green), and random coil (blue). The light blue sphere in the center is a zinc ion.

## Tertiary and Quaternary Structure

Large proteins have structure beyond the 1° and 2° levels. Their molecules can fold back on themselves as a result of interactions between R groups on amino acids that are considerable distances apart along the protein chain. These interactions may be ion–ion, ion–dipole, or van der Waals forces. They may even involve the formation of covalent bonds. For example, the –SH groups on two cysteine residues may combine to form a disulfide linkage that holds two parts of the protein strand together via an intrastrand –S—S– covalent bond (Figure 20.16). Interactions and reactions such as these determine a protein's **tertiary (3°) structure**, the overall three-dimensional shape of the protein that is key to its biological activity (Figure 20.9c). Because the proteins in living systems exist in an aqueous environment, hydrophobic R groups tend to reside in the interiors of their 3° structures, whereas hydrophilic groups (as we saw in normal hemoglobin) are oriented toward the outside, where they interact with nearby molecules of water. Hydrophobic interactions are the primary force that causes protein folding and compaction, but all the other modes of interaction help a large protein form its unique 3° structure.

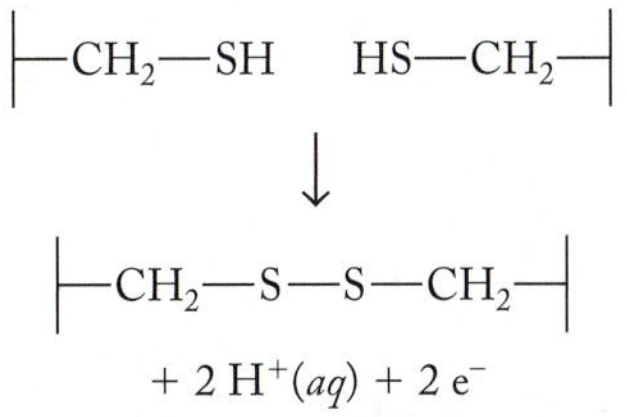

Hemoglobin and some other proteins exhibit an even higher order of structure. One hemoglobin unit (Figure 20.17a) contains four protein strands, each

$$-CH_2-SH \quad HS-CH_2-$$
$$\downarrow$$
$$-CH_2-S-S-CH_2-$$
$$+ 2\,H^+(aq) + 2\,e^-$$

**FIGURE 20.16** The tertiary structure of some proteins is stabilized by intrastrand –S—S– bonds that form between the –SH groups on the side chains of cysteine residues.

**CONNECTION** All the types of intermolecular forces described in Chapter 10 are involved in the *intra*molecular interactions that give proteins their unique 2° and 3° structures.

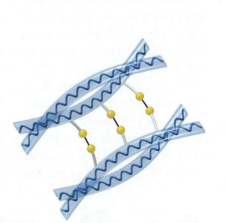

(a) Hemoglobin      (b) Keratin

**FIGURE 20.17** Quaternary structure of proteins. (a) Four protein chains form a single unit in the quaternary structure of hemoglobin. The iron-containing porphyrins are bright green. (b) Pairs of α-helical chains (blue) wound together and linked by interstrand –S—S– bonds (yellow) stabilize the quaternary structure of the keratin in hair and fingernails.

**random coil** an irregular or rapidly changing part of the secondary structure of a protein.

**tertiary (3°) structure** the three-dimensional, biologically active structure of a protein that arises because of interactions between the R groups on its amino acids.

of which enfolds a porphyrin ring containing one $Fe^{2+}$ ion. The combination of four protein strands to make one hemoglobin assembly is an example of **quaternary (4°) structure** (Figure 20.9d). In the 4° structures of keratins (Figure 20.17b), protein strands with α-helical 2° structures coil around each other to make even larger coils. When the protein strands in keratin structures are held together mostly by van der Waals forces, as they are in the keratin in skin tissue, the structures are flexible and elastic. If they are also restrained by many covalent bonds, as in Figure 20.17(b), they produce tissues that are hard and less flexible, like fingernails and the beaks of birds.

## Enzymes: Proteins as Catalysts

We already introduced *enzymes* as biological catalysts in Chapter 14. As described there, the chemical reactions involved with metabolism—both *catabolism* (breaking down of molecules) and *anabolism* (synthesis of complex materials from simple feedstocks)—are mediated in large part by enzymes. Both catabolism and anabolism are organized in sequences of reactions called *metabolic pathways*, and each step in a metabolic pathway is catalyzed by a specific enzyme.

Unlike the inorganic catalysts discussed in Chapter 14, enzymes are highly selective: each catalyzes a particular reaction involving a particular reactant. For example, an enzyme called lactase catalyzes only the reaction by which lactose (the sugar in milk) is broken down during digestion. People who lack this enzyme cannot metabolize this sugar; they are said to be *lactose intolerant*. If they consume dairy products, unmetabolized lactose passes into their large intestines where bacteria ferment it, and unpleasant and painful abdominal disturbances result.

Synthetic reaction pathways catalyzed by enzymes are usually more rapid and involve fewer steps than uncatalyzed pathways to make the same product. The products also tend to be optically pure materials. These advantages can significantly reduce the cost of production of, for example, biological pharmaceuticals. However, the biocatalytic reactions often run best in very dilute solutions, which limits production. A key issue driving interest in such processes is that an enantiomerically pure pharmaceutical is likely to be more potent and produce fewer side effects than a racemic mixture of the same product. As an example, the drug thalidomide is a racemic mixture of two enantiomeric forms: one is very effective in treating morning sickness, but the second is a teratogen (an agent that disturbs the development of a fetus). In perhaps the worst medical tragedy in modern times, more than 10,000 children were born in the late 1950s and early 1960s with serious, frequently fatal, deformities as a result of their mothers taking the racemic drug.

The molecular structure of enzymes contains a region called an **active site** that binds the reactant molecule, called the **substrate**. The action of enzymes was originally explained by a lock-and-key analogy in which the substrate is the key and the active site is the lock (Figure 20.18). The substrate is held in the active site by the same kinds of intermolecular interactions that hold any biomolecules together. Some enzymes become covalently bonded to intermediates in the

**CONNECTION** In Section 14.6 we described heterogeneous catalysts used in automobile exhaust systems. Enzymes are homogeneous catalysts that selectively speed up biochemical reactions.

**CONNECTION** A racemic mixture contains equal proportions of the two enantiomers of a chiral compound, as we discussed in Section 13.8.

**FIGURE 20.18** In the lock-and-key model of enzyme activity, the substrate (orange) fits exactly into the active site of the enzyme (green) that catalyzes a chemical reaction involving the substrate.

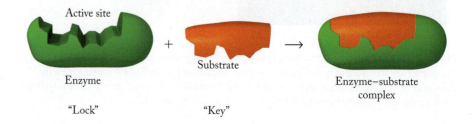

Active site

Enzyme

"Lock"

+ Substrate

"Key"

→ Enzyme–substrate complex

catalytic process. Once in the active site, the substrate is converted into product via a reaction having a lower-energy transition state than it would without the enzyme. The reaction of a substrate S with an enzyme E produces an *enzyme–substrate* (ES) *complex* that decomposes, forming a product P and regenerating the enzyme:

$$E + S \rightleftharpoons ES \rightarrow E + P$$

The lock-and-key analogy does not fully account for enzyme behavior. A more accurate view is provided by the *induced-fit model*, which assumes that the substrate does more than just fit into the existing shape of an active site. This model assumes that as the ES complex forms, the binding site undergoes subtle changes in its shape to more precisely fit the three-dimensional structure of the transition state. The binding energy between the enzyme and the transition state lowers the activation energy barrier, thereby speeding up the reaction. A simplified illustration of such an interaction is shown in Figure 20.19(a).

The induced-fit model helps explain the behavior of compounds called **inhibitors**, which can diminish or destroy the effectiveness of enzymes. An inhibitor may bind to an active site and block it from interacting with the substrate (Figure 20.19b). Alternatively, it may disable the enzyme by preventing it from assuming its active shape. In the latter process the inhibitor may bind to the enzyme at a site other than the active site and from that position prevent the enzyme from achieving its active shape (Figure 20.19c).

**quaternary (4°) structure** the larger structure functioning as a single unit that results when two or more proteins associate.

**active site** the location on an enzyme where a reactive substance binds.

**substrate** the reactant that binds to the active site in an enzyme-catalyzed reaction.

**inhibitor** a compound that diminishes or destroys the ability of an enzyme to catalyze a reaction.

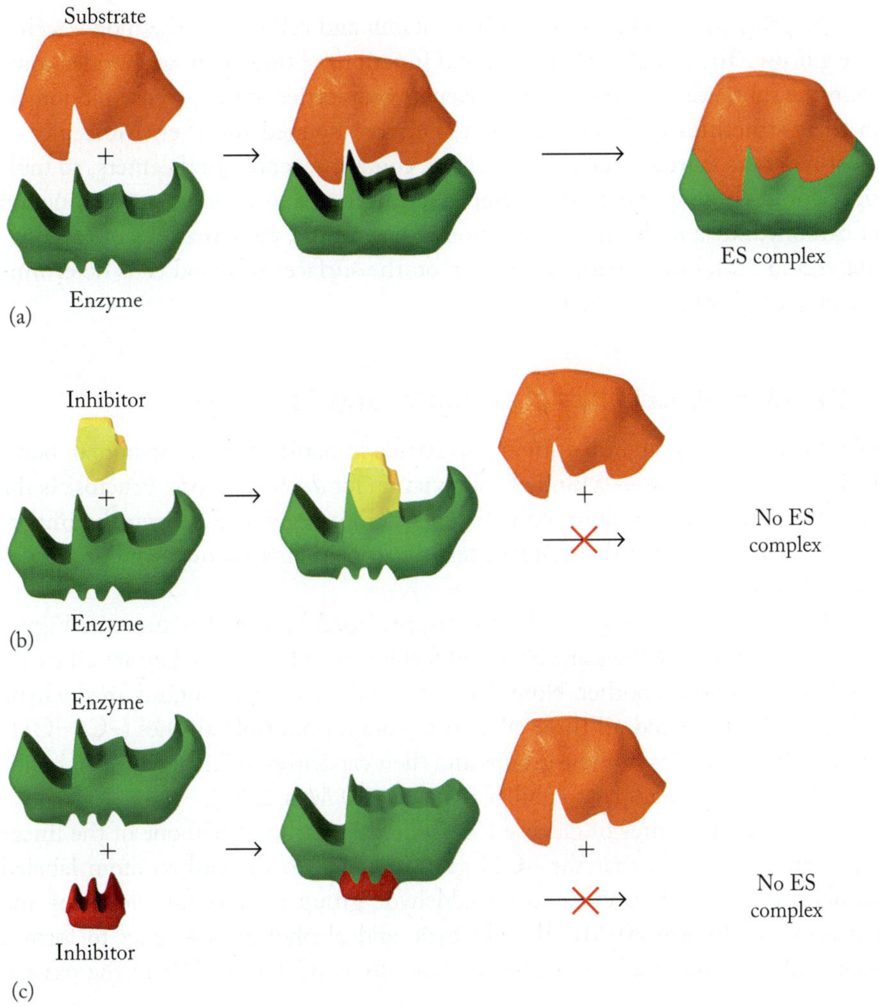

(a) Substrate, Enzyme, ES complex

(b) Inhibitor, Enzyme, No ES complex

(c) Enzyme, Inhibitor, No ES complex

**FIGURE 20.19** (a) The induced-fit model assumes that the shape of the enzyme (green) changes to accommodate the substrate (orange) and form the enzyme–substrate (ES) complex. Inhibitors (yellow and red molecules) may (b) block the enzyme's binding site or (c) cause a change elsewhere in the enzyme's structure that prevents the active site from attaining the shape it needs to form the ES complex.

**carbohydrate** an organic molecule with the generic formula $C_x(H_2O)_y$.

**monosaccharide** a single-sugar unit and the simplest carbohydrate.

**polysaccharide** a polymer of monosaccharides.

Natural enzyme inhibitors play important roles in regulating the rates of reactions that are catalyzed by enzymes. For example, in a multistep reaction pathway, the product of a later step may inhibit an enzyme that catalyzes an earlier reaction. This kind of negative feedback keeps the sequence of reactions from running too quickly and perhaps jeopardizing the health of the organism due to an accumulation of undesirable products or intermediates. Enzyme inhibitors may also be used to fight disease. Powerful drugs have been developed that inhibit enzymes called proteases involved in virus maturation. Several such drugs have been particularly effective in treating HIV.

## 20.3 Carbohydrates

**Carbohydrates** have the generic formula $C_x(H_2O)_y$. This formula gives us a clue about where the name *carbohydrate*, or *hydrate of carbon*, comes from. The smallest carbohydrates are **monosaccharides** (the name means "one sugar"), which bond together to form more complex carbohydrates called **polysaccharides**. Many organisms use monosaccharides as their main energy source but convert them to polysaccharides for the purpose of energy storage. Starch is the most abundant energy-storage polysaccharide in plants. Polysaccharides in the form of cellulose also provide structural support in plants. Plants produce over 100 billion tons of cellulose each year—the woody parts of trees are over 50% cellulose, and cotton is 99% cellulose.

The principal building block of both starch and cellulose is the monosaccharide glucose. The different properties and functions of these polysaccharides come from the subtle differences in the molecular geometries of the glucose monomers in their structures and how the monomers are bonded together. Molecules of most monosaccharides, including glucose, contain several chiral centers, so multiple optical isomers exist. This complexity gives rise to another major function of carbohydrates: molecular recognition. For example, combinations of carbohydrates and proteins, called *glycoproteins*, on the surfaces of blood cells determine the blood type of an individual.

### Molecular Structures of Glucose and Fructose

Glucose is the most abundant monosaccharide in nature and in the human body. It is also called *dextrose*—a kind of abbreviation for *dextro*-glucose. Fructose is the principal monosaccharide in many fruits and root vegetables. Given the importance and abundance of these sugars, let's examine their structures and properties more closely.

Three molecular views of glucose are provided by the structures in Figure 20.20. All three have the same chemical formula ($C_6H_{12}O_6$), so they are all structural isomers of one another. Note that the middle structure contains an aldehyde (–CH=O) group and all three structures contain multiple alcohol (–C—OH) groups. The polarities of these groups and their capacities to form hydrogen bonds give glucose its high molar solubility in water (5.0 $M$ at 25°C).

The cyclic structures of glucose form when the carbon backbone of the linear form curls around so that the –OH group bonded to the carbon atom labeled number 5 (C-5) comes close to the aldehyde group on C-1 (as shown by the red arrow in Figure 20.20). The aldehyde and alcohol groups react to form a six-membered ring made up of five carbon atoms (C-1 to C-5) and the oxygen

**CONNECTION** An aldehyde group (Chapter 13) includes these atoms and bonds:

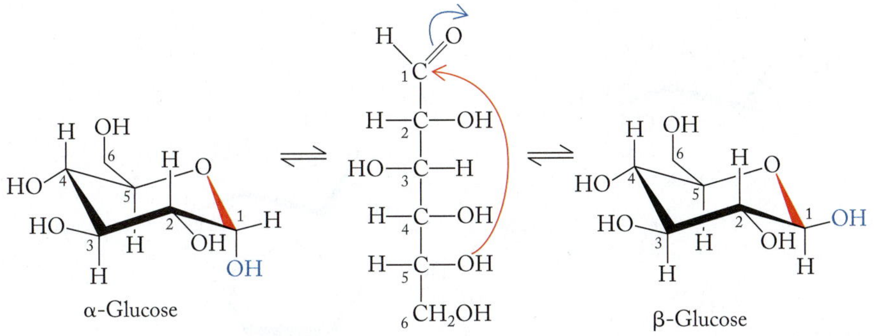

FIGURE 20.20 An equilibrium exists between the linear structure of glucose and the two cyclic forms α-glucose and β-glucose. The difference between the two cyclic structures is the orientation of the −OH group on C-1 (highlighted in blue). The new bond is shown in red in the cyclic structures.

atom of the C-5 alcohol. Note a small but significant difference in the two cyclic structures. In the molecule on the left, called α-glucose, the −OH group on C-1 points down. In the molecule on the right, β-glucose, the C-1 −OH group points up. These two orientations are possible because there are two ways that the carbon chain in the middle structure can form a cyclic structure: the −OH group on C-5 can approach C-1 from either of the two sides of the plane defined by the C-1 carbon atom and the C and H atoms bonded to it. When the C-5 −OH group approaches C-1 as shown in Figure 20.21, the α isomer is produced. Approaching C-1 from the other side yields the β isomer. Notice that in the linear form, C-1 is an aldehyde and is not a chiral center. The cyclization creates a new chiral center in the cyclic molecule.

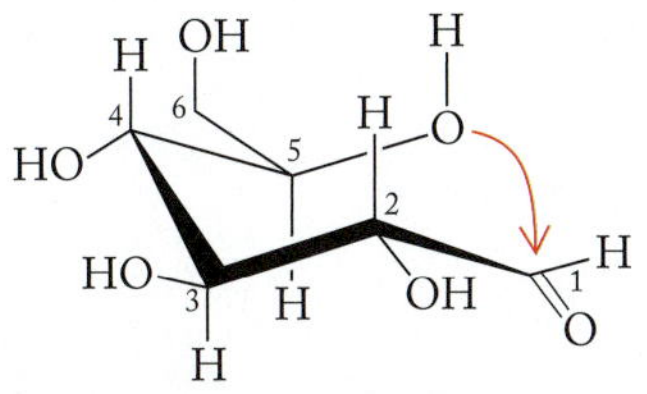

The β form is slightly more stable than the α form and accounts for 64% of glucose molecules in aqueous solution; the α form accounts for the remaining 36%. Both cyclic forms are more stable than the straight-chain form, which exists only as an intermediate between the two cyclic forms. The energy differences are small, however, so glucose molecules in solution are constantly opening and closing in a dynamic structural equilibrium.

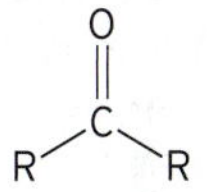

Figure 20.22 shows the structures of the linear and cyclic forms of fructose. Note that the C=O group in the linear form is not on the terminal carbon atom. In other words, fructose is a ketone rather than an aldehyde. It forms a five-membered ring, not a six-membered ring, when the −OH group at C-5 reacts with the carbonyl carbon atom at the C-2 position. The product is a ring with two −CH$_2$OH groups that are either on the same side of the ring (α-fructose) or on opposite sides (β-fructose).

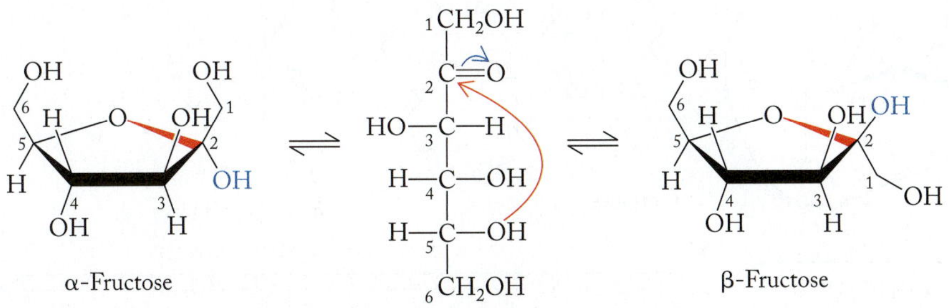

FIGURE 20.21 The −OH group on C-5 may approach the aldehyde at C-1 from either side of the plane defined by the C, H, and O atoms in the aldehyde group. In the approach shown here the product is the α isomer.

⬤◯ **CONNECTION** A ketone group (Chapter 13) contains these atoms and bonds:

FIGURE 20.22 The cyclization of fructose proceeds via the −OH group on C-5 and the ketone on C-2. The new bond is shown in red in the cyclic structures.

## Disaccharides and Polysaccharides

Sucrose, or ordinary table sugar, is a *disaccharide* ("two sugars") that consists of one molecule of α-glucose bonded to one molecule of β-fructose (Figure 20.23). The bond between them forms when the −OH group on the C-1 carbon atom of glucose reacts with the C-2 carbon atom of fructose, producing a C−O−C

**FIGURE 20.23** α-Glucose and β-fructose combine to form a molecule of sucrose, ordinary table sugar. The bond shown in red is the α,β-1,2-glycosidic bond.

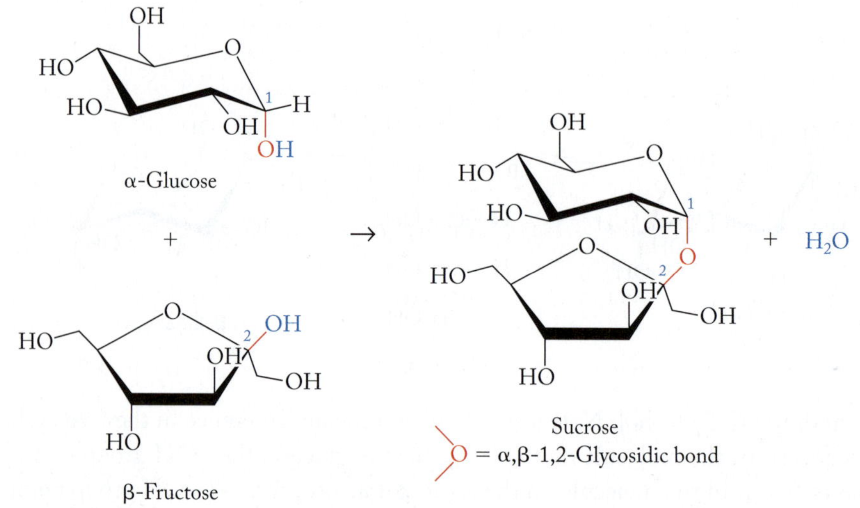

α-Glucose

+

β-Fructose

→

Sucrose

$\displaystyle \diagup\!\!\!\diagdown$ = α,β-1,2-Glycosidic bond

+ $H_2O$

▶‖ CHEMTOUR   Formation of Sucrose

**glycosidic bond** or *glycosidic linkage*. Water is also produced, making this reaction another example of a condensation reaction, as is peptide bond formation. The glycosidic bond in sucrose is called an α,β-1,2 linkage because of the orientations (α and β) of the two –OH groups involved and their positions (C-1 and C-2) in the cyclic structures of the two monosaccharides.

Another important glycosidic bond involves the –OH groups on the C-1 and C-4 carbon atoms of glucose molecules. When glucose molecules link at these positions they can form long-chain polysaccharides. If the starting monomer is α-glucose, the bonds between them are α-1,4-glycosidic linkages, and the product of bond formation is starch (Figure 20.24a). The conversion of α-glucose into starch is an effective way for plants to store energy because formation of the α-1,4 linkage is reversible. With the aid of digestive enzymes, α-1,4 bonds can be

**FIGURE 20.24** (a) Starch is a polysaccharide of α-glucose molecules joined by α-1,4-glycosidic bonds, shown in red. Starch molecules form spirals, much like coiled springs, that pack together to form granules. (b) Cellulose is a polysaccharide of β-glucose molecules joined by β-1,4-glycosidic bonds (in red). Cellulose chains pack together much more tightly to form structural fibers.

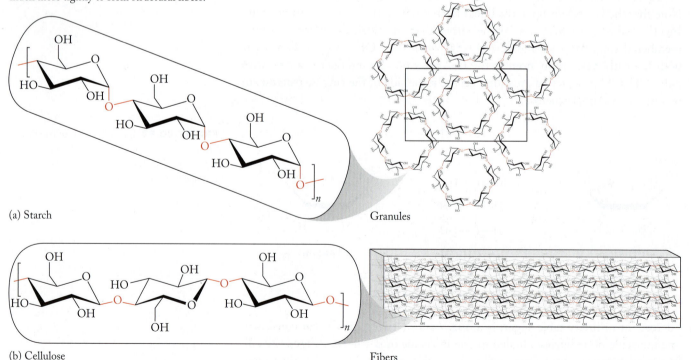

(a) Starch

Granules

(b) Cellulose

Fibers

hydrolyzed and starch converted back into glucose by plants that make the starch, or by animals that eat the plants.

The cellulose that plants synthesize to build stems and other organs has a structure (Figure 20.24b) slightly different from that of starch because the building blocks of cellulose are β-glucose instead of α-glucose, so the monomers are linked by β-1,4-glycosidic bonds. This structural difference is important because it enables starch to coil and make granules for efficient energy storage, whereas cellulose forms structural fibers.

Carbohydrates in plants are a major part of the total organic matter in any given ecological system; that is, they make up much of the system's **biomass**. Humans have used various forms of biomass, including wood and animal dung, as fuel for thousands of years. More recently, we have begun to convert biomass into a liquid *biofuel*, ethanol, for use as a gasoline additive and substitute. Because ethanol contains oxygen, it improves the combustion of a hydrocarbon fuel like gasoline and reduces carbon monoxide emissions. An important industrial application of starch hydrolysis is the conversion of cornstarch into glucose and then, through fermentation, into ethanol:

$$C_6H_{12}O_6(aq) \rightarrow 2\ CH_3CH_2OH(aq) + 2\ CO_2(g)$$

This exothermic reaction provides energy to the yeast cells whose biological processes drive fermentation.

**CONCEPT TEST**

Cellobiose is a disaccharide made from the degradation of cellulose. We cannot digest cellobiose. Which of the two structures in Figure 20.25 represents a molecule of cellobiose?

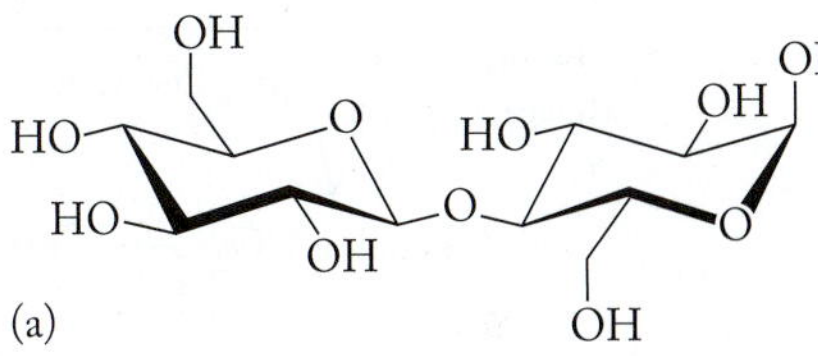

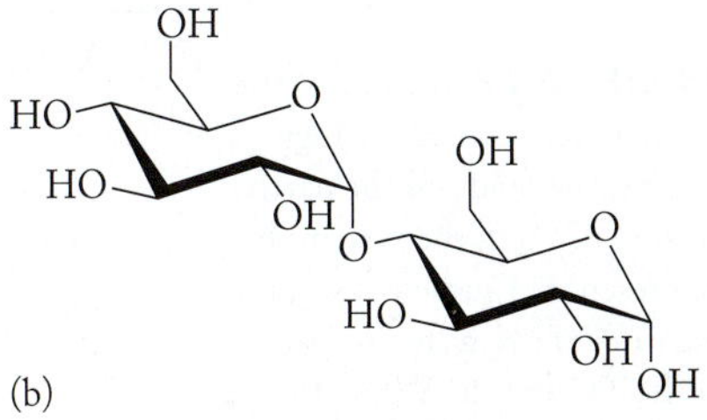

Unlike grazing animals, humans cannot digest cellulose because we do not have microorganisms in our digestive tracts that have enzymes called *cellulases*, which catalyze hydrolysis of β-glycosidic bonds. The challenge of reproducing what the cellulose-eating bacteria do in a laboratory or on an industrial scale is the focus of an enormous research effort as scientists try to develop efficient procedures for converting cellulose to glucose and then to ethanol. This research has focused on more efficient, less energy-intensive ways to break apart cellulose fibers. In addition, scientists are genetically engineering microorganisms like those in cattle stomachs to increase the supply of cellulases.

If this research is successful, it will address several major problems associated with ethanol as a gasoline additive or alternative fuel. First of all, it will lower the cost of production. Today it costs more to produce ethanol from cornstarch than to produce gasoline from crude oil. One reason for this is that most of the mass of a corn plant, or any plant, is cellulose, not starch. Ethanol production from cornstarch is also energy intensive: more than 70% of the energy contained in ethanol is expended in producing it; therefore, the net energy value of ethanol from corn is less than 30%. If ethanol could be produced from cellulose instead of starch, its net energy value could be as high as 80%. Finally, the use of edible cornstarch in fuel production has driven up the price of foods derived from corn, including livestock feed. The impacts of this inflation have been felt worldwide and have been particularly painful in developing countries. The use of agricultural land and consumption of increasingly scarce water resources for ethanol production raise further concerns.

**glycosidic bond** a C—O—C bond between sugar molecules.

**biomass** the sum total of the mass of organic matter in any given ecological system.

(a)

(b)

**FIGURE 20.25**

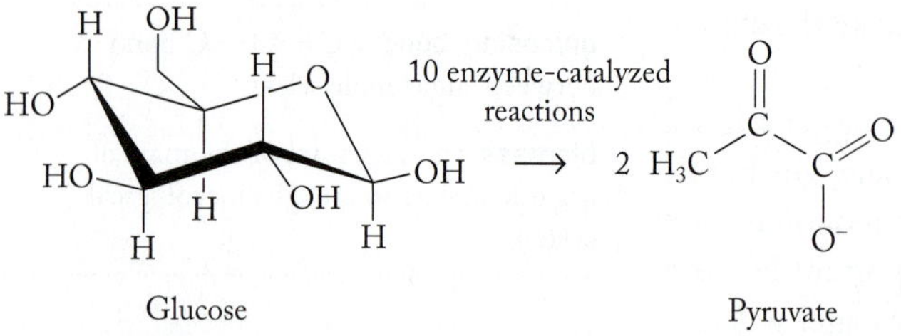

**FIGURE 20.26** In glycolysis, molecules of glucose are broken down to pyruvate ions.

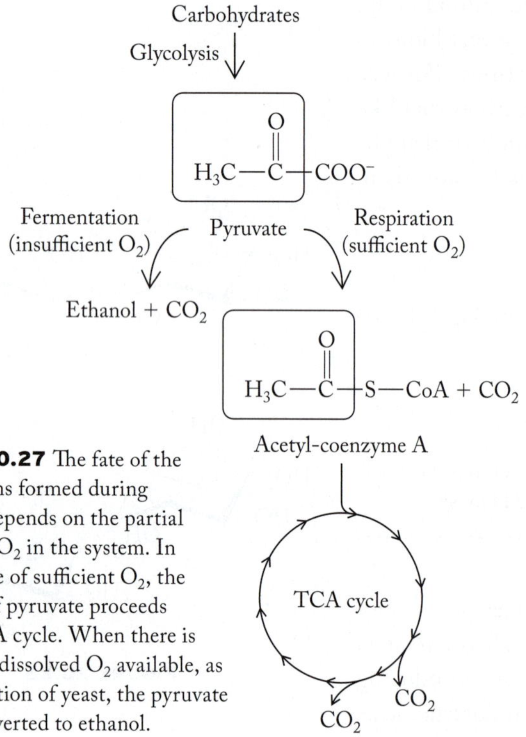

## Glycolysis Revisited

In Chapter 18, we introduced *glycolysis* in a discussion of the thermodynamics of coupled reactions. Most animal cells, including those in our bodies, use glucose as a fuel in a series of reactions, collectively called glycolysis, that oxidize glucose to pyruvate ions (Figure 20.26), the conjugate base of pyruvic acid. Pyruvate sits at a metabolic crossroads and can be converted into different products, depending on the type of cell in which it is generated, the enzymes present, and the availability of oxygen (Figure 20.27). In yeast cells growing under low-oxygen conditions, pyruvate is converted into ethanol and $CO_2$. Another series of reactions, called the **tricarboxylic acid (TCA) cycle** or the *citrate cycle*, occurs in the presence of sufficient dissolved $O_2$ and is fundamental to the conversion of glucose to energy in humans and other animals. A key step prior to the TCA cycle occurs when pyruvate loses $CO_2$ and forms an acetyl group, which then combines with coenzyme A. The resulting product, acetyl-coenzyme A, is a reactant in many biosynthetic pathways, including the production of fats.

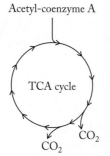

**FIGURE 20.27** The fate of the pyruvate ions formed during glycolysis depends on the partial pressure of $O_2$ in the system. In the presence of sufficient $O_2$, the oxidation of pyruvate proceeds via the TCA cycle. When there is insufficient dissolved $O_2$ available, as in fermentation of yeast, the pyruvate may be converted to ethanol.

## 20.4 Lipids

**Lipids** differ from carbohydrates and proteins in that they are not biopolymers. Lipids are best described by their physical properties rather than by any common structural subunit: lipids do not dissolve in water but are soluble in nonpolar solvents, and they are oily to the touch. Because they are insoluble in water, they are ideal components of cell membranes, which separate the aqueous solutions within cells from the aqueous environments outside them. An important class of lipids called **glycerides** are esters formed between glycerol and long-chain fatty acids (Figure 20.28). The three –OH groups on glycerol allow for mono-, di-, and triglycerides, with the latter being the most abundant. Glycerides account for over 98% of the lipids in the fatty tissues of mammals.

Table 20.3 lists some common fatty acids. The most abundant ones have an even number of carbon atoms because the fatty acids are built from two-carbon subunits. Their biosynthesis begins with the conversion of pyruvate to acetyl-coenzyme A.

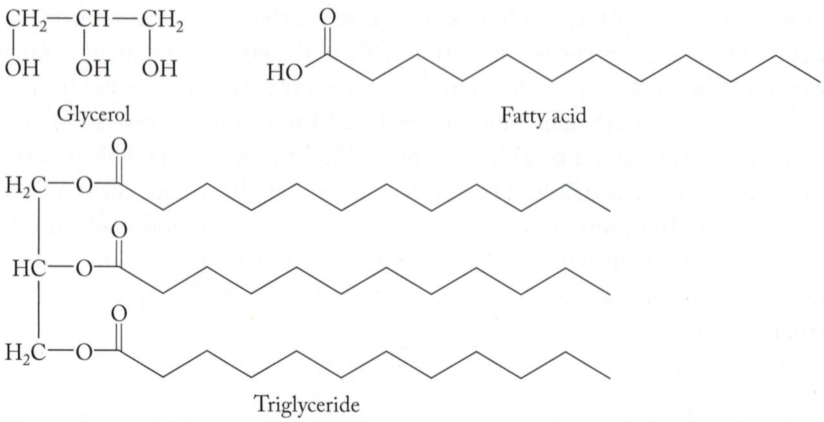

**FIGURE 20.28** Glycerides are esters that form when glycerol combines with fatty acids. When all three –OH groups on glycerol react to form ester bonds, the product is a *tri*glyceride.

| TABLE 20.3 | Names and Structural Formulas of Common Fatty Acids |
|---|---|
| **Common Name (chemical name) (source)** | **Formula** |
| **SATURATED FATTY ACIDS** | |
| Lauric acid (dodecanoic acid) (coconut oil) | $CH_3(CH_2)_{10}COOH$ |
| Myristic acid (tetradecanoic acid) (nutmeg butter) | $CH_3(CH_2)_{12}COOH$ |
| Palmitic acid (hexadecanoic acid) (animal and vegetable fats) | $CH_3(CH_2)_{14}COOH$ |
| Stearic acid (octadecanoic acid) (animal and vegetable fats) | $CH_3(CH_2)_{16}COOH$ |
| **UNSATURATED FATTY ACIDS** | |
| Oleic acid (cis-9-octadecenoic acid) (animal and vegetable fats) | $CH_3(CH_2)_7CH{=}CH(CH_2)_7COOH$ |
| Linoleic acid (cis,cis-9,12-octadecadienoic acid) (linseed oil, cottonseed oil) | $CH_3(CH_2)_4CH{=}CHCH_2CH{=}CH(CH_2)_7COOH$ |
| α-Linolenic acid (cis,cis,cis-9,12,15-octadecatrienoic acid) (linseed oil) | $CH_3CH_2CH{=}CHCH_2CH{=}CHCH_2CH{=}CH(CH_2)_7COOH$ |

Most fatty acids contain between 14 and 22 carbon atoms. As we defined in Chapter 13, saturated organic molecules have no carbon–carbon double or triple bonds. Unsaturated molecules contain carbon–carbon double and/or triple bonds. Fatty acids are either saturated or unsaturated, based on the absence or presence of carbon–carbon multiple bonds. A type of fatty acid much in the news because of their alleged health benefits is the family of omega-3 fatty acids. They are polyunsaturated, meaning they have more than one carbon–carbon multiple bond. The name comes from the location of the first double bond counted from the methyl end of the chain, which is known as the omega (ω) end of the molecule. Omega-3 fatty acids are found in fish oils and some plant oils. α-Linolenic acid (see Table 20.3) is an omega-3 fatty acid.

**SAMPLE EXERCISE 20.4  Identifying Triglycerides  LO5**

How many different triglycerides (including structural isomers and stereoisomers) can be made from glycerol combining with two different fatty acids (X and Y) if each molecule of triglyceride contains at least one molecule of each fatty acid?

**Collect and Organize** We are asked to determine how many different triglycerides can be made from glycerol and two different fatty acids. A triglyceride contains three fatty acid units.

**Analyze** Each fatty acid may bond to one of three –OH groups in glycerol. Each triglyceride has at least one X and one Y residue, which makes two formulas possible: $X_2Y$ and $Y_2X$. Each of these formulas has two structural isomers, depending on whether the single fatty acid in the formula is bonded to the middle carbon or to one of the end carbon atoms. Finally, if a structure has an X on one end carbon atom and a Y on the other, then

**tricarboxylic acid (TCA) cycle** a series of reactions that continue the oxidation of pyruvate formed in glycolysis.

**lipid** a class of water-insoluble, oily organic compounds that are common structural materials in cells.

**glyceride** lipid consisting of esters formed between fatty acids and the alcohol glycerol.

the middle carbon atom is a chiral center, which means that there are two enantiomeric forms of that compound.

**Solve** We can generate four different molecular structures by attaching X and Y in four different sequences to the glycerol –OH groups:

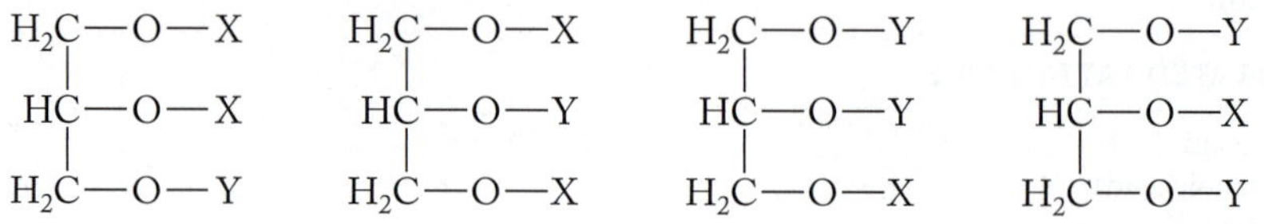

$$
\begin{array}{cccc}
\text{H}_2\text{C}-\text{O}-\text{X} & \text{H}_2\text{C}-\text{O}-\text{X} & \text{H}_2\text{C}-\text{O}-\text{Y} & \text{H}_2\text{C}-\text{O}-\text{Y} \\
| & | & | & | \\
\text{HC}-\text{O}-\text{X} & \text{HC}-\text{O}-\text{Y} & \text{HC}-\text{O}-\text{Y} & \text{HC}-\text{O}-\text{X} \\
| & | & | & | \\
\text{H}_2\text{C}-\text{O}-\text{Y} & \text{H}_2\text{C}-\text{O}-\text{X} & \text{H}_2\text{C}-\text{O}-\text{X} & \text{H}_2\text{C}-\text{O}-\text{Y} \\
\end{array}
$$

$$
\quad\quad (1) \quad\quad\quad\quad (2) \quad\quad\quad\quad (3) \quad\quad\quad\quad (4)
$$

Structures (1) and (3) have chiral centers, so each of these isomers has two enantiomeric forms. Therefore, there are a total of six different triglycerides possible.

**Think About It** The central carbon atoms in structures (1) and (3) are chiral because they are each bonded to a H atom, an O atom, and to two C atoms, which are themselves bonded to different fatty acids: X and Y.

⚙ **Practice Exercise** How many different triglycerides can be made from glycerol and one molecule each of three different fatty acids A, B, and C?

■

## Function and Metabolism of Lipids

Lipids are an important energy source in our diets, providing more energy per gram than carbohydrates or proteins. **Fats** are glycerides composed primarily of saturated fatty acids. They are solids at room temperature because the molecules can pack very tightly together. **Oils** are glycerides composed predominantly of unsaturated fatty acids and are liquids at room temperature, because their chains have kinks in them due to the double bonds, which prevent the chains from packing together as closely as the saturated chains can. Oils can be converted into solid, saturated glycerides by hydrogenation. For example, in the hydrogenation of corn oil, which is a mixture of mostly two unsaturated fatty acids (oleic and linoleic acids), hydrogen is added to convert some or all the $-\text{CH}{=}\text{CH}-$ subunits into $-\text{CH}_2-\text{CH}_2-$ subunits. Hydrogenation converts the oil into a solid at room temperature that is whipped with skim milk, coloring agents, and vitamins to produce the food spread we know as margarine.

The consumption of too much saturated fat is associated with coronary heart disease. One of the purported advantages of the so-called Mediterranean diet is that olive oil, a liquid composed of glycerides containing over 80% oleic acid, is used in cooking rather than animal fats like butter and lard. In addition, this diet tends to be richer in fish and vegetables, both of which contain unsaturated fats. Most animal fats, like the marbling in beef that enhances its flavor, are saturated.

Another problem with hydrogenating vegetable oil arises when the oils are only partially hydrogenated. Partial hydrogenation alters the molecular structure around their remaining C=C double bonds, changing them from their natural cis isomers into trans isomers (Figure 20.29). Unsaturated trans fats like elaidic acid tend to be solids at room temperature because their molecules pack together more uniformly than do molecules of cis unsaturated fatty acids. Consumption of trans fatty acids is associated with increased levels of cholesterol in the blood and other health risks.

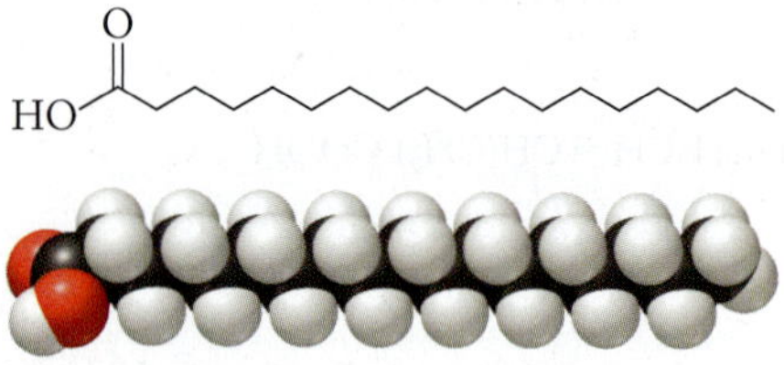

(a) Stearic acid: a saturated fatty acid

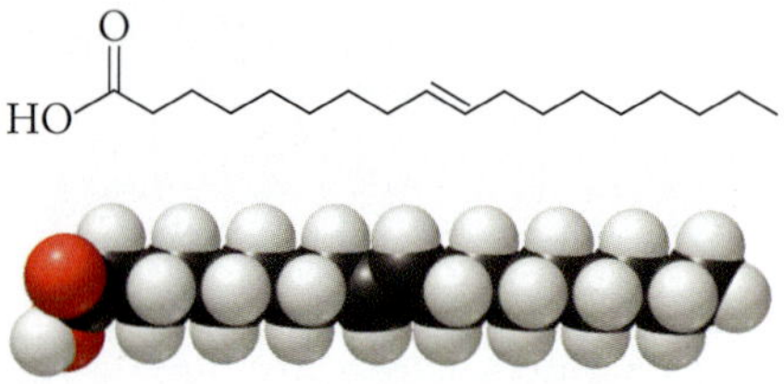

(b) Elaidic acid: a trans unsaturated fatty acid

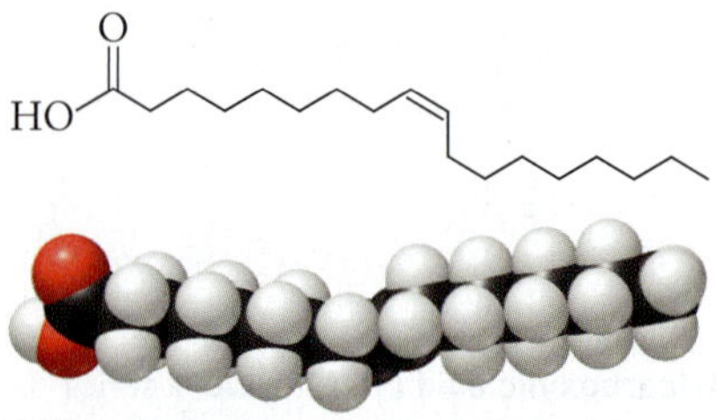

(c) Oleic acid: a cis unsaturated fatty acid

**FIGURE 20.29** Types of fatty acids. (a) Stearic acid, a saturated $C_{18}$ fatty acid. (b) Elaidic acid, an unsaturated $C_{18}$ fatty acid (trans isomer). (c) Oleic acid, an unsaturated $C_{18}$ fatty acid (cis isomer).

The mixture of lipids in the legs of reindeer living close to the Arctic Circle changes as a function of distance from the hoof: The closer to the hoof, the higher the percentage of unsaturated fatty acids. Why would this be an advantage for these reindeer?

The lipids in some prepared foods have been modified to reduce their caloric content but still provide the taste, aroma, and "mouth feel" we associate with lipid-rich foods. The active sites of enzymes that break down natural lipids accommodate triglycerides formed from glycerol and fatty acids. However, chemically modified esters made from the same fatty acids, but attached to an alcohol other than glycerol, cannot be metabolized by these enzymes. Such molecules, if they have the appropriate physical properties and are nontoxic, can be incorporated into foods without adding any calories because they are not metabolized.

Olestra is one such product (Figure 20.30). It is an ester made from long-chain fatty acids and the carbohydrate sucrose. (Remember, sugars have –OH groups and technically are alcohols.) Each of the eight –OH groups in a sucrose molecule reacts with a molecule of fatty acid to make the ester in olestra. The resultant material is used to deep-fry potato chips. Any olestra that remains on the chip does not add calories because it cannot be processed by enzymes that recognize only fatty acid esters on a glycerol scaffold.

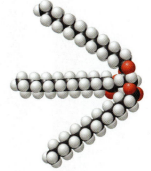

**FIGURE 20.30** Olestra has a very different shape from that of the triglycerides typically metabolized by our bodies. Consequently, it cannot be processed by the enzymes that digest triglycerides.

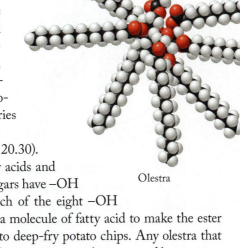

Olestra

Triglyceride

Olestra may be "calorie-free" as a food subject to metabolism in the living system, but how would it compare with a common triglyceride in terms of kilojoules of heat released per mole in a calorimeter experiment?

The enzymes that metabolize triglycerides hydrolyze the esters and release glycerol and the fatty acids that were bonded to it. Glycerol enters the metabolic pathway for glucose. The fatty acids are oxidized in a series of reactions known as β-oxidation: a process that removes two carbon atoms at a time. For example, stearic acid (the saturated $C_{18}$ fatty acid) is transformed into a $C_2$ fragment and the $C_{16}$ acid, palmitic acid. Palmitic acid yields another $C_2$ fragment and myristic acid, and so forth until the fatty acid is completely degraded. Electrons released from this oxidative process eventually are donated to $O_2$. The energy released by this process powers metabolism.

## Other Types of Lipids

Cells contain other types of lipids in addition to triglycerides. One type, **phospholipids** (Figure 20.31a), plays a key role in cell structure. A phospholipid molecule

**fat** solid triglyceride containing primarily saturated fatty acids.

**oil** liquid triglyceride containing primarily unsaturated fatty acids.

**phospholipid** a molecule of glycerol with two fatty acid chains and one polar group containing a phosphate; phospholipids are major constituents of cell membranes.

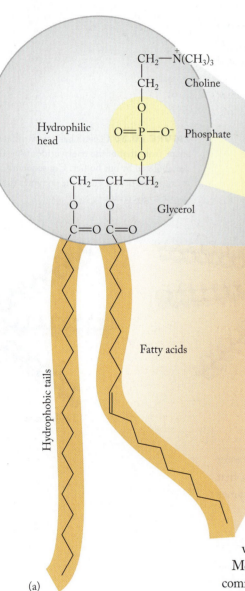

(a)

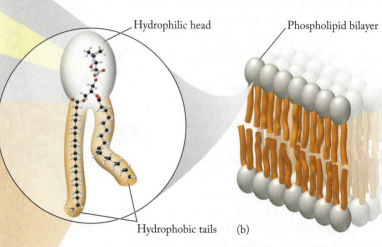

**FIGURE 20.31** Phospholipids are major constituents of cell membranes. (a) The presence of a polar group (here, choline) attached to a phosphate unit on glycerol changes the properties of the resulting diglyceride (here, phosphatidylcholine). (b) In the lipid bilayer that forms cell membranes, phospholipids orient themselves so that the polar groups in one half of the bilayer face the aqueous environment outside the cell while the polar groups in the other half of the bilayer face the aqueous environment of the cell interior. This arrangement leaves the nonpolar fatty acid part of the phospholipids in the interior of the bilayer.

consists of a glycerol molecule bonded to two fatty acid chains and to one phosphate group that is also bonded to polar substituents. The presence of nonpolar fatty acid chains and a polar region in the same molecule makes phospholipids ideal for forming cell membranes. In an aqueous medium, phospholipids form a **lipid bilayer**, a double layer enclosing each cell and isolating its interior from the outside environment. The phospholipid molecules of the bilayer align so that the nonpolar groups interact with each other inside the membrane while the polar groups interact with water molecules outside it (Figure 20.31b). Membranes exist both to isolate the contents of cells and to serve as the locus of communication between processes that occur within and outside the cells.

Cholesterol is a lipid and a key component in the structure of cell membranes. It is also a precursor of the bile acids that aid in digestion and of steroid hormones, which regulate the development of the sex organs and secondary sexual traits, stimulate the biosynthesis of proteins, and regulate the balance of electrolytes in the kidneys. Biosynthesis of cholesterol, like the biosynthesis of fatty acids, begins with the conversion of pyruvate to acetyl-coenzyme A (Figure 20.27). In a healthy person, synthesis and use of cholesterol are tightly regulated to prevent overaccumulation and consequent deposition of cholesterol in coronary arteries. We clearly need cholesterol, but deposition in the arteries can lead to serious coronary disease (Figure 20.32).

**CONCEPT TEST**

A newspaper article contains the wording, "Made primarily by the liver, cholesterol begins with tiny pieces of sugar. . . ." What does this statement mean at the molecular level?

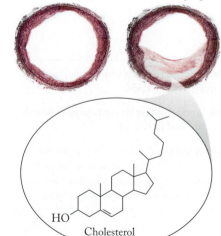

**FIGURE 20.32** Cholesterol deposits called plaques are responsible for restricted blood flow, which results in a variety of sometimes catastrophic cardiovascular problems.

# 20.5 Nucleotides and Nucleic Acids

**Nucleic acids** are our fourth class of biomolecules and third class of biopolymers. We focus on two types: deoxyribonucleic acid (DNA) and ribonucleic acid (RNA). Even though nucleic acids make up only about 1% of a higher organism's mass, they control the metabolic activity of all its cells. The fraction is much higher in yeast and bacteria because those organisms are packed with ribosomes, which are themselves half RNA. DNA carries the genetic blueprint of an organism, and a variety of RNAs use that DNA blueprint to guide the production of proteins.

A nucleic acid is a polymer composed of monomeric units called **nucleotides**. Each nucleotide unit is in turn composed of three subunits: a five-carbon sugar, a phosphate group, and a nitrogen-containing base (Figure 20.33). The phosphate group in each nucleotide is attached to a carbon in the side chain of the sugar called the 5′ carbon atom (the prime number refers to the position of the carbon atom in the sugar molecule). The nitrogen-containing base is attached to the 1′ carbon atom in each sugar molecule. The sugar in Figure 20.33 is called *ribose*, which makes this a nucleotide in a strand of *ribo*nucleic acid, or RNA. If the sugar were *deoxyribose* instead, there would be a H atom in place of the −OH group on the 2′ carbon atom, and the nucleotide would be a building block of *deoxyribo*nucleic acid, or DNA. Because of the ionized phosphate groups, both DNA and RNA are polyanions. Their anionic character is important because it causes them to interact with proteins while not allowing them to pass through cell membranes.

The nitrogen-containing base in Figure 20.33 is called adenine (A). Structural formulas of adenine and the other four bases in nucleic acids—cytosine (C), guanine (G), thymine (T), and uracil (U)—are shown in Figure 20.34. The point of attachment of the sugar–phosphate groups on each base is indicated by −R. In addition to the difference in their sugars, RNA and DNA also differ in one of the bases in their nucleotides: RNA contains A, C, G, and U, but DNA contains A, C, G, and T.

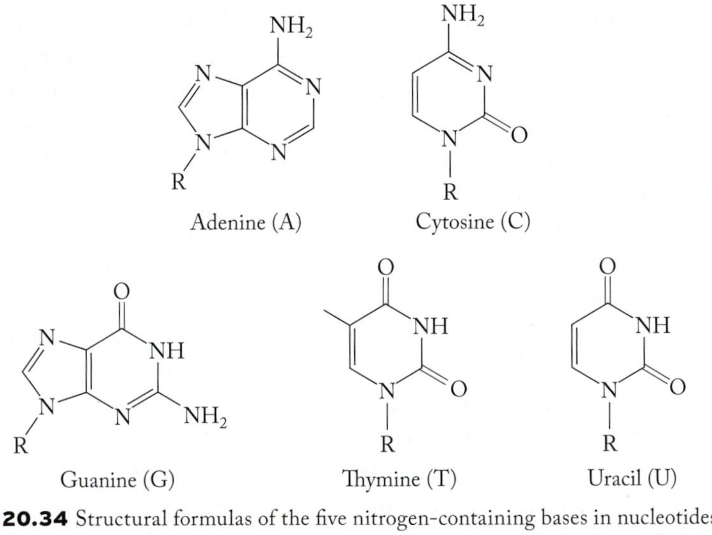

**lipid bilayer** a double layer of molecules whose polar head groups interact with water molecules and whose nonpolar tails interact with each other.

**nucleic acid** one of a family of large molecules, which includes deoxyribonucleic acid (DNA) and ribonucleic acid (RNA), that stores the genetic blueprint of an organism and controls the production of proteins.

**nucleotide** a monomer unit from which nucleic acids are made.

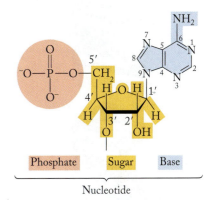

**FIGURE 20.33** A nucleotide consists of a phosphate group and a nitrogen-containing base that are both bonded to a five-carbon sugar.

**FIGURE 20.34** Structural formulas of the five nitrogen-containing bases in nucleotides. The nucleotides in DNA contain A, C, G, and T; those in RNA contain A, C, G, and U. The R group identifies the point of attachment of the sugar residue.

Adenine (A)    Cytosine (C)    Guanine (G)    Thymine (T)    Uracil (U)

In a polymeric strand of nucleic acid, each phosphate is also linked to the 3′ carbon atom in the sugar of the monomer that precedes it in the chain, as shown for a strand of DNA in Figure 20.35. Both DNA and RNA strands are synthesized in the cell from the 5′ to the 3′ direction (downward in Figure 20.35). The structures of DNA and RNA are frequently written using only the single-letter labels of their bases, beginning with the free phosphate group on the 5′ end of the chain and reading toward the free 3′ hydroxyl group at the other terminus.

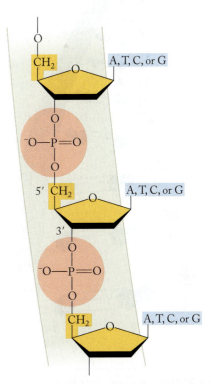

**FIGURE 20.35** The backbone of the polymer chain in DNA consists of alternating sugar units (yellow) and phosphate units (pink). The bases (blue) are attached to the backbone through the C-1′ atom of the sugar unit.

(a)

(b)

**FIGURE 20.36** The nitrogen-containing bases on one strand of DNA pair with the bases on a second strand by hydrogen bonding. (a) Adenine and thymine pair via two hydrogen bonds; guanine and cytosine pair via three hydrogen bonds. (b) DNA as a double helix with the sugar–phosphate backbone on the outside and the base pairs on the inside.

When scientists first isolated DNA and began to analyze its composition, they made a pivotal observation about the abundance of the nitrogen-containing bases. A typical molecule of DNA consists of thousands of nucleotides, and the percentages of the four bases in different samples of DNA can vary over a wide range. However, the percentage of A in a sample always matches the percentage of T. Likewise, the percentage of C always matches that of G. This result makes sense if the bases are paired because a molecule of A can form two hydrogen bonds to a molecule of T, whereas a molecule of C can form three hydrogen bonds with a molecule of G. Therefore, A–T and G–C pairings maximize the number of hydrogen bonds possible (Figure 20.36a). More importantly, the base pairs assembled this way all have the same width and fit together in a regular structure.

The normal structure of DNA has two strands of nucleotides wrapped around each other in a form that is called a *double helix* (Figure 20.36b). The nucleotide backbone is on the outside of the spiraling strands, with hydrogen bonds between the complementary bases keeping the two strands together. Notice also that the base pairs are parallel to each other and perpendicular to the helical axis. The fidelity of this base-pairing—A always with T, and C always with G—gives DNA the ability to copy itself. If a pair of complementary strands is unzipped into two single strands, each strand provides a template on which a new complementary strand can be synthesized via the process called **replication** (Figure 20.37).

During replication, the two strands are separated, forming a structure called the *replication fork*. The fork advances through the DNA as replication proceeds. One of the two resulting strands, the leading strand, is unzipped in the 3′–5′ direction, which allows the new complementary strand to be continuously synthesized in the 5′–3′ direction. The replication of the second strand, called the lagging strand, is more complicated. It proceeds in short fragments, which are then assembled into a continuous strand later in the process.

**replication** the process by which one double-stranded DNA forms two new DNA molecules, each one containing one strand from the original molecule and one new strand.

**SAMPLE EXERCISE 20.5  Using Base Complementarity in DNA          LO6**

If 31.6% of the nucleotides in a sample of DNA are adenine, what are the percentages of cytosine, guanine, and thymine?

**Collect and Organize** We know how much adenine is in a DNA sample and need to calculate the remainder of the nucleotide composition. Nucleotides are paired: A always pairs with T, and C always pairs with G.

**Analyze** Because of base pairing, the percentage of T must equal the percentage of A, and the percentage of C must equal the percentage of G.

**Solve** If A = 31.6%, then T = 31.6%. This leaves $(100 - 2 \times 31.6) = 36.8\%$ left to be equally distributed between G and C. Therefore G = C = 18.4%.

**Think About It** The percentages should total 100%, and they do.

⚙ **Practice Exercise** Indicate the sequence of the complementary strand on the double helix formed by each of these sequences of nucleotides:
a. CGGTATCCGAT
b. TTAAGCCGCTAG

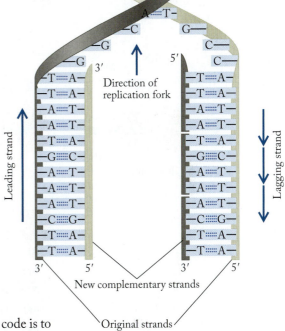

FIGURE 20.37 When DNA replicates, the two strands of a short portion of the double helix are unzipped. The complementary strand to the leading strand is synthesized continuously in the 5'-3' direction. The complementary strand to the lagging strand is synthesized in fragments, which are joined together later to produce a continuous strand.

DNA's double-stranded structure is also the key to its ability to preserve genetic information. The two strands carry the same information, much like an old-fashioned photograph and its negative. Genetic information is duplicated every time a DNA molecule is replicated, a process that is essential whenever a cell divides into two new cells.

## From DNA to New Proteins

Proteins are formed from amino acids in accordance with the *genetic code* contained in the base sequences of DNA strands. The bases A, T, G, and C are the alphabet in this code, and the "words" in the code are three-letter combinations of these four letters, with each word representing a particular amino acid or a signal to begin or end protein synthesis. Using four letters to write three-letter words means there are $4^3 = 64$ combinations possible, more than enough to encode for the 20 amino acids found in cells. The function of the genetic code is to specify the protein's primary structure—the sequence of amino acids in proteins. The flow of genetic information goes from DNA to RNA to proteins, a sequence sometimes called the *central dogma of molecular biology*.

Protein synthesis begins with a process called **transcription** (Figure 20.38a), in which double-stranded DNA unwinds and its genetic information guides the synthesis of a single strand of a molecule called **messenger RNA (mRNA)**. This strand of mRNA has the complementary base sequence of the original DNA. It carries the 3-letter words of that DNA, in the form of three-base sequences called **codons** (Table 20.4), from the nucleus of the cell into the cytoplasm, where the mRNA binds with a cellular structure called a ribosome. Keep in mind that a sequence of A, C, G, and T in the original DNA is transcribed into the following sequence in mRNA:

DNA: ...ACGT...
mRNA: ...UGCA...

At the ribosome the genetic information in the messenger RNA directs the synthesis of particular proteins in a process called **translation**. Another type of RNA, called **transfer RNA (tRNA)**, plays a key role in translation. There are 20 different forms of tRNA in the cell, one for each amino acid. To see how tRNA works, let's look at Figure 20.38(b). The first codon in this piece of an mRNA strand is AUG, which codes for the amino acid methionine (see Table 20.4). In the cytoplasm surrounding the ribosome, molecules of tRNA are reversibly bonded to molecules of every amino acid. The particular tRNA molecules that are bonded

**transcription** the process of copying the information in DNA to RNA.

**messenger RNA (mRNA)** the form of RNA that carries the code for synthesizing proteins from DNA to the site of protein synthesis in a cell.

**codon** a three-nucleotide sequence that codes for a specific amino acid.

**translation** the process of assembling proteins from the information encoded in RNA.

**transfer RNA (tRNA)** the form of the nucleic acid RNA that delivers amino acids, one at a time, to polypeptide chains being assembled by the ribosome–mRNA complex.

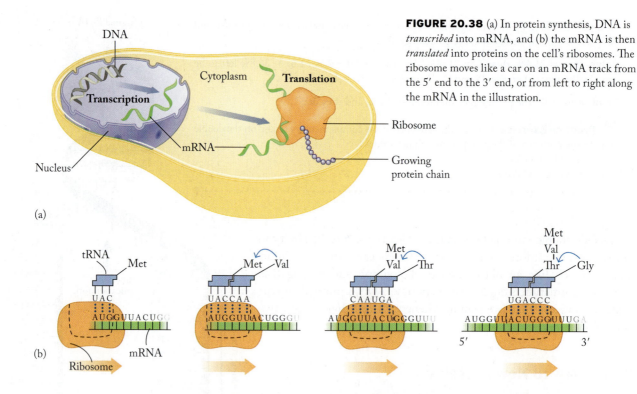

**FIGURE 20.38** (a) In protein synthesis, DNA is *transcribed* into mRNA, and (b) the mRNA is then *translated* into proteins on the cell's ribosomes. The ribosome moves like a car on an mRNA track from the 5′ end to the 3′ end, or from left to right along the mRNA in the illustration.

| TABLE 20.4 | **mRNA Codons** | | |
|---|---|---|---|
| **Ala** | GCU, GCC, GCA, GCG | **Leu** | UUA, UUG, CUU, CUC, CUA, CUG |
| **Arg** | CGU, CGC, CGA, CGG, AGA, AGG | **Lys** | AAA, AAG |
| **Asn** | AAU, AAC | **Met** | AUG |
| **Asp** | GAU, GAC | **Phe** | UUU, UUC |
| **Cys** | UGU, UGC | **Pro** | CCU, CCC, CCA, CCG |
| **Gln** | CAA, CAG | **Ser** | UCU, UCC, UCA, UCG, AGU, AGC |
| **Glu** | GAA, GAG | **Thr** | ACU, ACC, ACA, ACG |
| **Gly** | GGU, GGC, GGA, GGG | **Trp** | UGG |
| **His** | CAU, CAC | **Tyr** | UAU, UAC |
| **Ile** | AUU, AUC, AUA | **Val** | GUU, GUC, GUA, GUG |
| **Start** | AUG | **Stop** | UAG, UGA, UAA |

to methionine also contain the sequence UAC, the complement of AUG, at a site that allows the tRNA to interact with mRNA. As Figure 20.38(b) shows, the segment of mRNA with the AUG codon links with the complementary strand on the tRNA molecule bonded to methionine. In doing so, the methionine is put into a position to unlink from the tRNA and to be the first amino acid residue in the protein being synthesized.

The sequence of events in the preceding paragraph is repeated many times. In this example (Figure 20.38b) the next codon, GUU, links up with a molecule of tRNA that has a CAA binding site and a molecule of valine in tow. In this way valine moves into position to become the next amino acid residue in the protein and to form a peptide bond with the N-terminal methionine. Valine is followed by threonine, which is followed by glycine, and so on until a Stop codon finally signals the end of the translation process.

If a GUU codon attracts valine to the translation site, does a UUG codon do the same thing? Explain your answer.

## 20.6 From Biomolecules to Living Cells

We end this chapter by addressing two fundamental questions about the major classes of biomolecules and their roles in sustaining life: (1) how were they first formed on prebiotic Earth, and (2) how did they assemble into living cells? Experiments conducted in the 1950s at the University of Chicago by Professor Harold Urey (1893–1981) and his student Stanley Miller (1930–2007) showed that amino acids could form from $H_2O$, $CH_4$, $NH_3$, and $H_2$ (Figure 20.39). Although the reactants the two scientists chose are now thought to be different from those present on early Earth, the result still stands: inorganic molecules can react to produce the organic molecules found in living systems.

Evidence also exists that some biomolecules may have reached Earth from extraterrestrial origins. In 2006 the NASA spacecraft *Stardust* returned to Earth with samples collected from the tail of a comet that is believed to have formed at about the same time as the solar system (Figure 20.40). Subsequent analyses disclosed the presence of glycine in the comet. This was not the first experiment to detect amino acids in space. A class of meteorites called carbonaceous chondrites contain isovaline and other amino acids. Interestingly, most of the amino acids from meteorites, which are believed to be fragments of asteroids and comets, are L-enantiomers, the same form that dominates our biosphere. These observations have lead to intriguing suggestions that L-amino acids are somehow "favored" and that life on Earth was "seeded" from elsewhere.

As we have seen in this chapter, RNA is needed to guide the assembly of amino acids into proteins. Since the 1990s, groups of scientists have explored the possibility that strands of RNA may have formed spontaneously from solutions of nucleotides in contact with clay minerals. The crystalline structures of these minerals provide three-dimensional templates that guide the self-assembly of the nucleotides into long strands of RNA. Pools of oligonucleotides made in this way usually contain many chains with random sequences, but some can actually catalyze their own replication. This ability to self-replicate is crucial to life, and the observation that molecules are capable of speeding up their own replication on a clay surface suggests that processes essential to the formation of living cells could have happened spontaneously. Much controversy still exists about these ideas, but the fact that RNA can act as both a source of information and a catalyst is part of the *RNA world hypothesis*. This hypothesis proposes that a world filled with life based on RNA predates the current world of life based on DNA and proteins. The capacity of RNA to both store information like DNA *and* act as a catalyst like an enzyme suggests that RNA alone could have supported cellular or precellular life forms.

Current research is also testing the hypothesis that life on Earth may have evolved near deep-ocean hydrothermal vents. Entire ecosystems have been discovered at these locations since they were first explored in the 1970s. They are sustained by geothermal and chemical energy rather than energy from the Sun. It may be that life actually began in such environments, with hydrothermal energy driving reactions in which inorganic compounds such as carbon dioxide and

**FIGURE 20.39** The apparatus used by Miller (shown) and Urey to simulate the synthesis of amino acids in the atmosphere of early (prebiotic) Earth. Discharges between the tungsten electrodes were meant to provide the sort of energy that might have come from lightning.

**FIGURE 20.40** NASA scientists who analyzed the samples collected by the *Stardust* spacecraft were careful to avoid contaminating it with biological material from Earth. This meant isolating themselves from the sample as they prepared it for analysis.

**FIGURE 20.41** Black clouds of transition metal oxides and sulfides flow into the sea through chimneys like this one at a deep-ocean hydrothermal vent. Some scientists believe that these particles may have guided and catalyzed the formation of the first self-replicating molecules on Earth.

hydrogen sulfide formed small organic compounds. As with the reactions on the surfaces of clay minerals, the synthesis reactions at hydrothermal vents may have been catalyzed and guided by reactants adsorbed on solid compounds such as FeS and $MnO_2$, which pour into the sea in dense black clouds near some vents (Figure 20.41). Among the known products of these reactions are acetate ions ($CH_3COO^-$). Acetate is a key intermediate in many biosynthetic pathways in living organisms. In modern bacteria, the systems that make acetate depend on a catalyst made of iron, nickel, and sulfur that has a structure much like that of particles produced by "black smokers" on the ocean floor.

To take the next step toward forming living cells, large biomolecules must have assembled themselves into even larger structures, such as membranes, that allow cells and structures within them to collect materials and retain them at concentrations different from those in the surrounding medium. Molecules in these assemblies are not necessarily connected by covalent bonds, but rather are held together by the intermolecular interactions we have discussed in this chapter.

## SAMPLE EXERCISE 20.6  Integrating Concepts: Liquid Oil to Solid Fat

Imagine we have a research project to turn soybean oil into a solid for use in a butter substitute. We can accomplish this by hydrogenating the oil. Suppose we are working at the level of a small-scale industrial facility to test the feasibility of this process. A particular sample of soybean oil contains 54% linoleic acid, 24% oleic acid, and 18% palmitic acid, by mass. We need to hydrogenate 10 kg of soybean oil. Our source of hydrogen is the steam-reforming of methane:

$$H_2O(g) + CH_4(g) \rightarrow CO(g) + 3\,H_2(g)$$

a. What volume of methane at 20°C and 1.00 atm of pressure do we need to produce the amount of hydrogen required to completely hydrogenate 10 kg of soybean oil?
b. How many different compounds could result from the hydrogenation of a triglyceride that contained one of each of the three fatty acids in this study: palmitic acid, linoleic acid, and oleic acid?

**Collect and Organize** We need to determine how much methane is needed in the steam-reforming reaction to supply enough hydrogen to react completely with the unsaturated fats present in a 10.0 kg sample of soybean oil. We know the oil contains three fatty acids: linoleic, oleic, and palmitic acid. Table 20.3 contains the molecular structures of all three of these fatty acids.

**Analyze** According to the structures in Table 20.3, palmitic acid is a saturated fatty acid, which means it does not react with hydrogen. Oleic acid has one C=C double bond per molecule and linoleic acid has two, which means one mole of oleic acid combines with one mole of $H_2$, and one mole of linoleic acid combines with two moles of $H_2$. We can use the given weight percentages (24% oleic acid and 54% linoleic acid) to calculate that a 10.0 kg sample contains 2.4 kg oleic acid and 5.4 kg linoleic acid. The structures of these two fatty acids can be used to write their molecular formulas, and from those formulas we can calculate their molar masses. Converting the above masses into grams and

dividing these masses by the molar masses of the two fatty acids yields the moles of each in the sample. We then use the 1:1 and 1:2 hydrogenation reaction stoichiometries to calculate the moles of $H_2$ we need and the stoichiometry of the steam-reforming reaction (1 mol $CH_4$ / 3 mol $H_2$) to convert moles of $H_2$ into moles of $CH_4$. The total moles of $CH_4$ needed along with the given temperature and pressure of the gas are used in the ideal gas law equation to calculate the volume of $CH_4$ needed.

**Solve**
a. Using the condensed structure of the two unsaturated fatty acids to determine their molecular formulas and calculate their molar masses:

| Fatty Acid | Molecular Formula | Molar Mass (g/mol) |
|---|---|---|
| Oleic acid | $C_{18}H_{34}O_2$ | 282.46 |
| Linoleic acid | $C_{18}H_{32}O_2$ | 280.45 |

Converting the masses of these fatty acids in the sample into moles:

$$2.4 \text{ kg oleic acid} \times \frac{1000 \text{ g}}{1 \text{ kg}} \times \frac{1 \text{ mol}}{282.46 \text{ g}} = 8.5 \text{ mol oleic acid}$$

$$5.4 \text{ kg linoleic acid} \times \frac{1000 \text{ g}}{1 \text{ kg}} \times \frac{1 \text{ mol}}{280.45 \text{ g}} = 19 \text{ mol linoleic acid}$$

Calculating the total moles of $CH_4$ needed:

$$8.5 \text{ mol oleic acid} \times \frac{1 \text{ mol } H_2}{1 \text{ mol oleic acid}} \times \frac{1 \text{ mol } CH_4}{3 \text{ mol } H_2}$$

$$= 2.8 \text{ mol } CH_4$$

$$+\ 19 \text{ mol oleic acid} \times \frac{2 \text{ mol } H_2}{1 \text{ mol oleic acid}} \times \frac{1 \text{ mol } CH_4}{3 \text{ mol } H_2}$$

$$= 13 \text{ mol } CH_4$$

Total moles of $CH_4$ = 2.8 + 13 = 16 mol $CH_4$

At a temperature of 20°C and a pressure of 1 atm, 16 moles of $CH_4$ occupies a volume of

$$PV = nRT$$

$$(1.00 \text{ atm})V = 16 \text{ mol} \times \frac{0.08206 \text{ L} \cdot \text{atm}}{\text{K} \cdot \text{mol}} \times 293 \text{ K}$$

$$V = \frac{16 \text{ mol} \times \dfrac{0.08206 \text{ L} \cdot \text{atm}}{\text{K} \cdot \text{mol}} \times 293 \text{ K}}{1.00 \text{ atm}} = 380 \text{ L}$$

Reported to the correct number of significant figures, the reaction requires 380 L of methane.

b. If a single triglyceride contained one of each of the saturated fatty acids that result from this process, it would contain one unit of palmitic acid, which was saturated in the original material and unchanged by the hydrogenation, and two units of stearic acid, the C18 saturated fatty acid that is the product of hydrogenation of oleic acid and linoleic acid. We can represent

this symbolically, letting A = palmitic acid and B = stearic acid. Two structural isomers of triglycerides result:

$$\begin{array}{ccc} CH_2 - CH - CH_2 & \quad & CH_2 - CH - CH_2 \\ | \quad | \quad | & & | \quad | \quad | \\ A \quad B \quad B & & B \quad A \quad B \end{array}$$

The structure on the left also has two enantiomeric forms, because the carbon in the –CH– unit is a chiral center. Therefore the products are two structural isomers, one of which also has two enantiomers (stereoisomers), for a total of three different triglycerides.

**Think About It** The number of moles of methane required seems reasonable based on our estimate. Actual soybean oil probably has several different triglycerides in it that are a combination of saturated and unsaturated fatty acid components, resulting in the observed composition by weight. Natural oils can be quite complex mixtures.

## SUMMARY

**Learning Outcome 1** The acid–base properties of **amino acids** are important for structural and functional reasons. We can define these properties using data derived from titrations. (Section 20.1)

**Learning Outcome 2** The **proteins** and **peptides** in the human body are composed of 20 α-amino acids reversibly linked together by **peptide bonds**. The sequence of amino acids in peptide and protein chains matters in determining their properties. (Section 20.1)

**Learning Outcome 3** The structure of a protein is crucial to its function. It is defined by the sequence of amino acids in the chain, the geometric pattern that segments of a chain adopt, the overall three-dimensional shape of the protein, and any larger structure formed when two or more proteins interact and function as a single unit. (Section 20.2)

**Learning Outcome 4** **Carbohydrates** are produced from $CO_2$ and $H_2O$, and organisms derive energy from glycolysis and the **tricarboxylic acid (TCA) cycle**, the reaction pathways by which glucose is oxidized to $CO_2$ and $H_2O$. **Monosaccharides** are joined through **glycosidic bonds** into **polysaccharides** like starch for

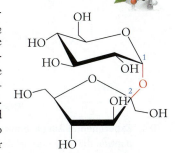

energy storage and cellulose for structural support in plants. (Section 20.3)

**Learning Outcome 5** **Lipids** include important families of molecules, including **glycerides**, **fats**, and **oils**. Some lipids are a major source of energy in our diet, and others play key roles in cell structure. (Section 20.4)

**Learning Outcome 6** The **nucleic acids** DNA and RNA contain an organism's genetic information and control protein synthesis through **transcription** and **translation**. Living cells may have formed as a result of chemical reactions in which inorganic molecules combined to form small organic molecules such as amino acids and **nucleotides**. (Sections 20.5, 20.6)

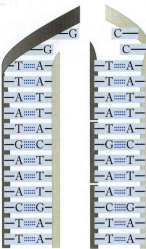

## PROBLEM-SOLVING SUMMARY

| TYPE OF PROBLEM | CONCEPTS AND EQUATIONS | SAMPLE EXERCISES |
|---|---|---|
| **Interpreting acid–base titration curves of amino acids** | At low pH all amino acids have at least two ionizable H atoms, one each from –COOH and $-NH_3^+$. Side-chain carboxylic acid and amine groups may also impart acidic and basic strength to amino acids. | 20.1 |

| TYPE OF PROBLEM | CONCEPTS AND EQUATIONS | SAMPLE EXERCISES |
|---|---|---|
| **Drawing and naming peptides** | Connect the α-amine of one amino acid to the α-carboxylic acid of another with a peptide bond. Starting with the free amine (N-) terminus on the left, name each amino acid residue by changing the ending of the name of the parent amino acid to -*yl* in all but the last (C-terminal) amino acid. | 20.2 |
| **Determining the sequence of a small peptide** | Arrange fragments of the peptide in a way that reveals the sequence of the peptide. | 20.3 |
| **Identifying triglycerides** | The −COOH groups of fatty acids react with the −OH groups in glycerol to form triglycerides and water. | 20.4 |
| **Using base complementarity in DNA** | Identify the base pairs: A pairs with T; G pairs with C. The percentage of T should equal the percentage of A, and the percentage of C should equal the percentage of G. | 20.5 |

## VISUAL PROBLEMS

*(Answers to boldface end-of-chapter questions and problems are in the back of the book.)*

**20.1.** The photochemical reaction of sodium hydrogen phosphite with formaldehyde is shown in Figure P20.1. It may have played a role in the formation of nucleic acids before life existed on Earth. Draw the Lewis structure for the hydrogen phosphite ($HPO_3^{2-}$) ion.

$$Na_2HPO_3(aq) + CH_2O(aq) + 2\,H^+(aq) \xrightarrow{h\nu}$$

**FIGURE P20.1**

**20.2.** The nucleotides in DNA contain the bases with the structures shown in Figure P20.2. Identify the basic functional groups in the structures.

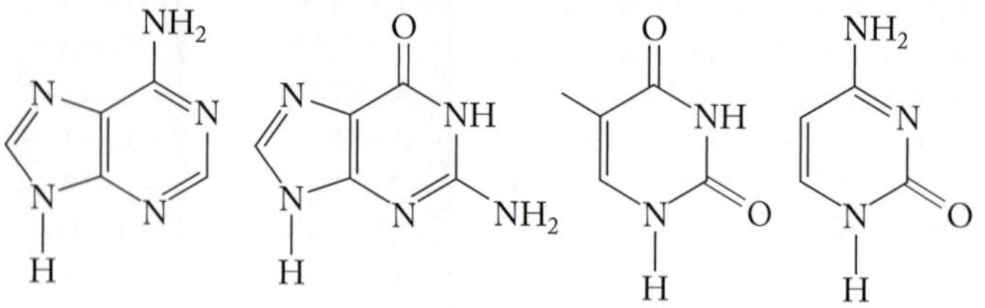

Adenine    Guanine    Thymine    Cytosine

**FIGURE P20.2**

**20.3. Olive Oil** Olive oil contains triglycerides such as those shown in Figure P20.3. Which of the fatty acids in these triglycerides is/are saturated?

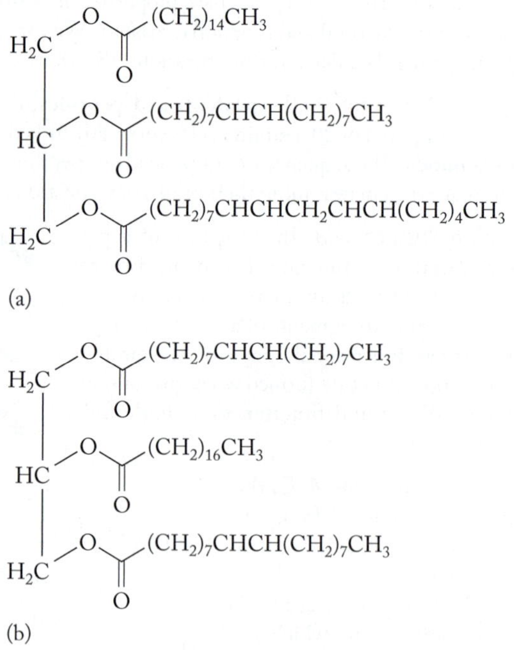

(a)

(b)

**FIGURE P20.3**

**20.4. Experimental Agent** Bentiromide (Figure P20.4) is a peptide that was once evaluated as an agent to monitor the function of the pancreas during therapy. Draw the structures of the amino acids in the structure of bentiromide and indicate if they are α-amino acids.

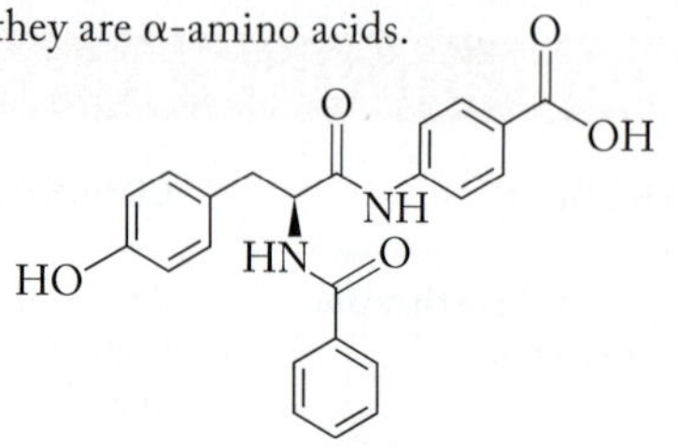

**FIGURE P20.4**

**20.5. Natural Painkillers** The human brain produces polypeptides called *endorphins* that help in controlling pain. The pentapeptide in Figure P20.5 is called enkephalin. Identify the five amino acids that make up enkephalin.

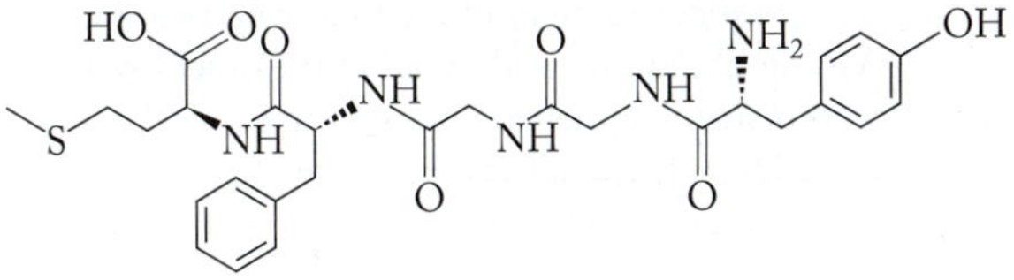

Enkephalin

**FIGURE P20.5**

**20.6. Treating Infections** The major component of the ointment bacitracin is cyclic polypeptide called bacitracin A. It was first isolated in 1943 from a knee scrape from a girl named Margaret Tracy, after whom it is named. Bacitracin is effective topically and is used in ointments for skin, eye, and wound infections. Figure P20.6 shows the structure of bacitracin A. Identify the amino acids found in human proteins that are also part of the structure of bacitracin A.

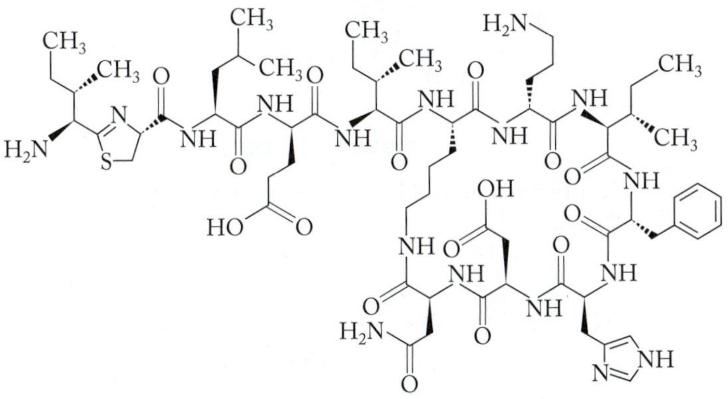

**FIGURE P20.6**

**20.7. Trans Fats** The role of "trans fats" in human health has been extensively debated both in the scientific community and in the popular press. What type of isomerism does the word "trans fat" refer to? Which of the molecules in Figure P20.7 are considered trans fats?

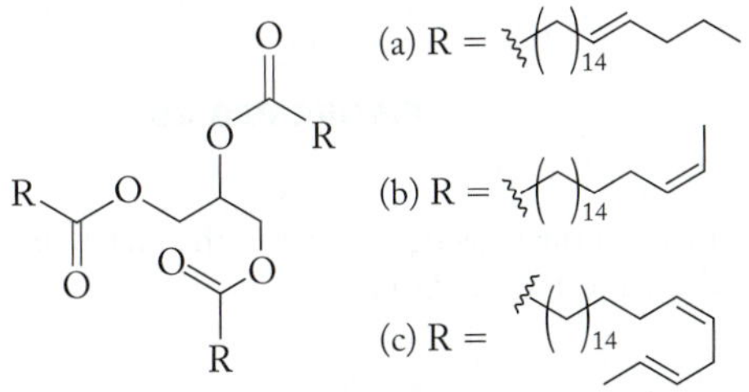

(a) R =
(b) R =
(c) R =

**FIGURE P20.7**

**20.8. Cocoa Butter** Cocoa butter (Figure P20.8) is a key ingredient in chocolate. Cocoa butter is a triglyceride that results from esterification of glycerol with three different fatty acids. Identify the fatty acids produced by hydrolysis of cocoa butter.

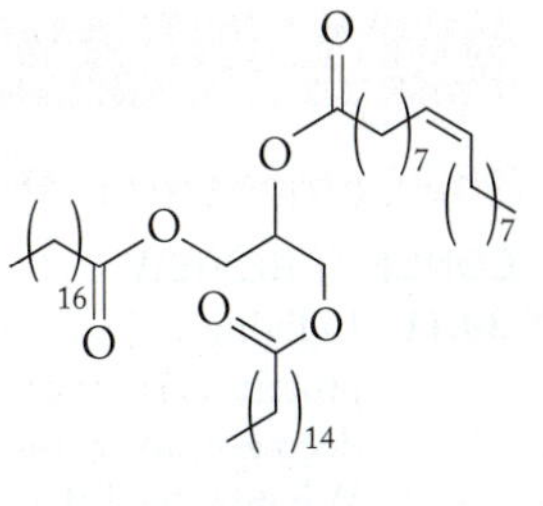

Cocoa butter

**FIGURE P20.8**

**20.9. Sucralose** The molecular structure of the artificial sweetener sucralose (trade name Splenda) is shown in Figure P20.9. Advertising for this product claims that it is made from sugar, implying that it is a natural product. What sugar might it be made from? Comment on the implication that it is a "natural" product.

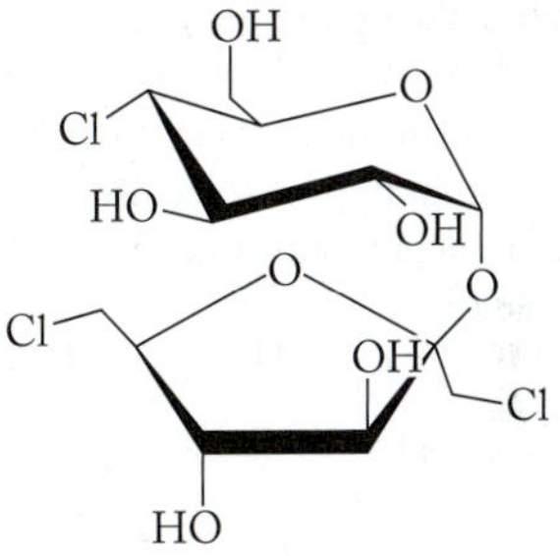

Sucralose

**FIGURE P20.9**

**20.10.** Figure P20.10 contains the titration curve of which of these amino acids: leucine, histidine, or lysine?

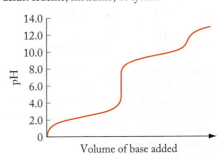

Volume of base added

**FIGURE P20.10**

## The Composition of Proteins

### CONCEPT REVIEW

**20.11.** In living cells, amino acids combine to make peptides and proteins. Are these processes accompanied by increases or decreases in entropy of the reaction system?

**20.12.** What is the difference between a peptide bond and an amide bond?

**20.13.** What does the alpha mean in α-amino acid?

**20.14.** Cystinuria is a metabolic disease that occurs in some breeds of dogs. It is characterized by the presence of kidney stones made of cystine, a dimer formed when two cysteine residues covalently link through an –S—S– bond (Figure 20.16).
  a. Draw the structure of the dipeptide Cys-Cys.
  b. Is cystine a dipeptide? Why or why not?

**20.15.** Meteorites contain more L-amino acids, which are the forms that make up the proteins in our bodies, than D-amino acids. What do the prefixes L- and D- mean?

**20.16.** Do any of the amino acids in Table 20.1 have more than one chiral carbon atom per molecule?

**20.17.** Which of the compounds in Figure P20.17 is not an α-amino acid?

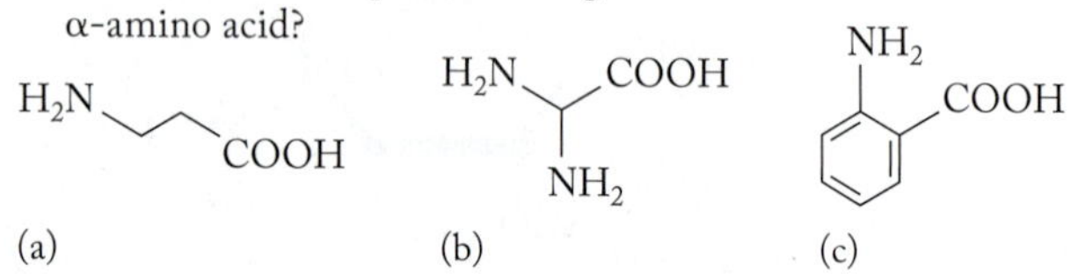

**FIGURE P20.17**

**20.18.** Which of the compounds in Figure P20.18 are α-amino acids?

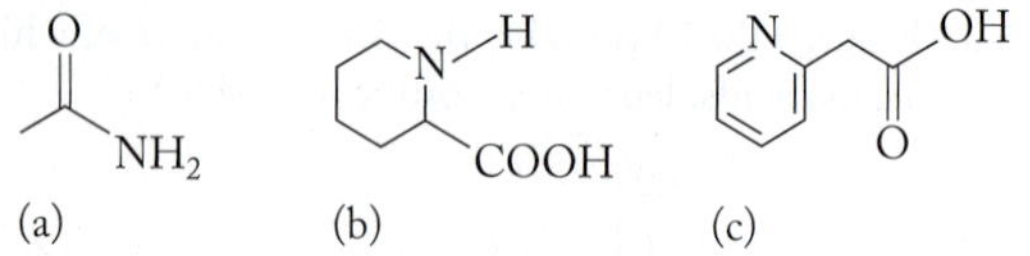

**FIGURE P20.18**

**20.19.** Why do most amino acids exist in the zwitterionic form at physiological pH (pH ≈ 7.4)?

**20.20.** Draw the condensed structural formulas of the amino acid tyrosine that you would expect to predominate in aqueous solution under the following conditions:
  a  in strongly acidic solution
  b. in strongly basic solution
  c. in a solution in which pH = p$I$

## Protein Structure and Function

### CONCEPT REVIEW

**20.21.** When protein strands fold back on themselves in forming stable tertiary structures, lysine residues are often paired up with glutamic acid residues. Why?

**20.22.** Ion–ion interactions are particularly effective at stabilizing tertiary structures of proteins. Suggest a pair of amino acid residues that would be attracted to each other via ion–ion interactions at pH 7.4.

### PROBLEMS

**20.23.** Draw structures and name all possible dipeptides produced from condensation reactions of the following L-amino acids:
  a. alanine + serine
  b. alanine + phenylalanine
  c. alanine + valine

**20.24.** Draw structures of the peptides produced from condensation reactions of the following L-amino acids. Assume the amino acids bond in the order given.
  a. tyrosine + cysteine + threonine
  b. serine + glutamic acid + valine
  c. asparagine + histidine + lysine

**20.25.** Identify the amino acids in the dipeptides shown in Figure P20.25.

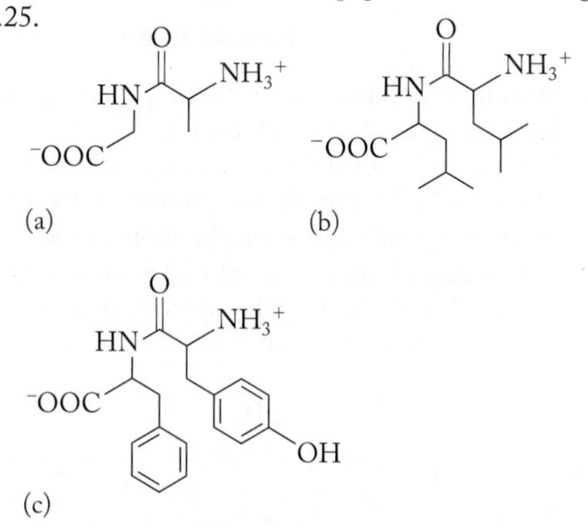

**FIGURE P20.25**

**20.26.** Identify the amino acids in the tripeptides in Figure P20.26.

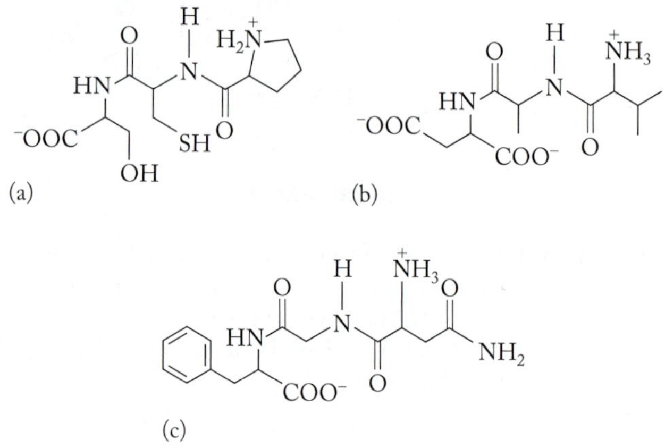

**FIGURE P20.26**

**20.27.** Identify the missing product in the metabolic reaction shown in Figure P20.27.

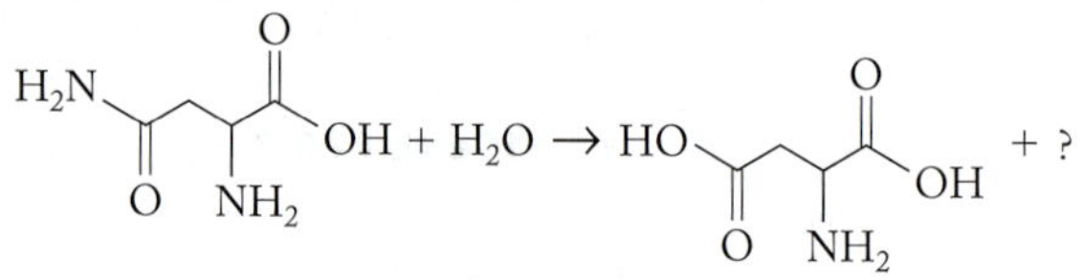

**FIGURE P20.27**

**20.28.** The molecular formula for glycine is $C_2H_5NO_2$. What is the molecular formula of the linear peptide formed when ten glycine molecules are linked together in peptide bonds?

# Carbohydrates

## CONCEPT REVIEW

**20.29.** What are the structural differences between starch and cellulose?

**20.30.** Why is the discovery of enzymes that catalyze cellulose hydrolysis a worthwhile objective?

**20.31.** Is the fuel value (see Chapter 5) of glucose in the linear form the same as that in the cyclic form?

**\*20.32.** Without doing the actual calculation, estimate the fuel values of glucose and starch by considering average bond energies. Do you predict the fuel values of the two substances to be the same or different?

**20.33.** The second step in glycolysis converts glucose 6-phosphate into fructose 6-phosphate. Can you think of a reason why $\Delta G°$ for this reaction is close to zero?

**\*20.34.** Which of the following statements are correct about glycosidic bonds in carbohydrates?
a. The glycosidic bond in maltose is hydrolyzed by people who are lactose intolerant.
b. A glycosidic bond links glucose and fructose together to form sucrose.
c. A glycosidic bond is an ether linkage, but all ether linkages are not glycosidic bonds.

**20.35.** How do we calculate the overall free-energy change of a process consisting of two steps?

**20.36.** During glycolysis a monosaccharide is converted to pyruvate. Do you think this process produces an increase or decrease in the entropy of the system? Explain your answer.

## PROBLEMS

**20.37.** Describe the similarities and differences in the structures of the α and β isomers formed when galactose (Figure P20.37) forms a six-membered ring.

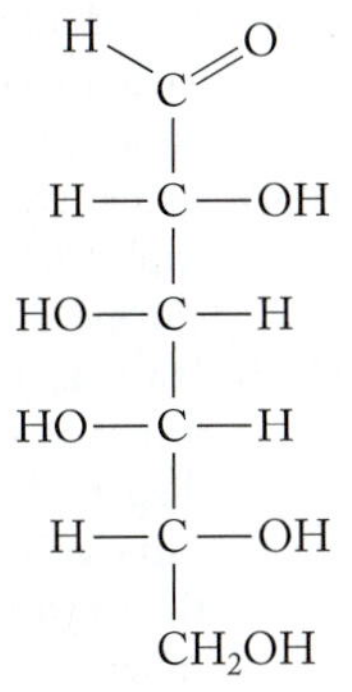

Galactose

**FIGURE P20.37**

**20.38.** Describe the similarities and differences in the structures of the α and β isomers formed when ribose (Figure P20.38) forms a five-membered ring.

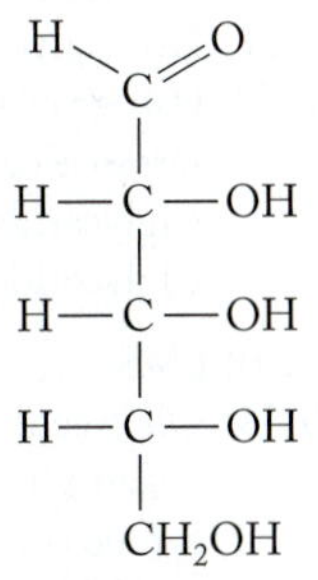

Ribose

**FIGURE P20.38**

**20.39.** Which, if any, of the structures in Figure P20.39 are β isomers of a monosaccharide?

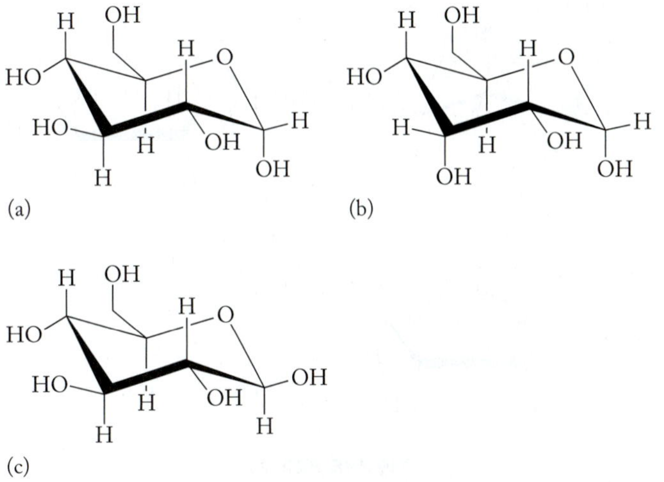

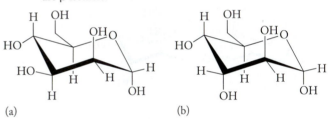

(a)　　　(b)

(c)

**FIGURE P20.39**

**20.40.** Identify which, if any, of the structures in Figure P20.40 are β isomers.

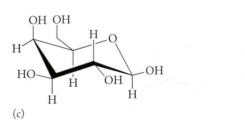

(a)　　　(b)

(c)

**FIGURE P20.40**

**20.41.** Which, if any, of the structures in Figure P20.41 are α isomers?

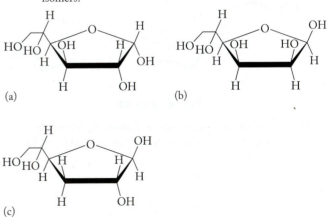

(a)　　　(b)

(c)

**FIGURE P20.41**

**20.42.** Which, if any, of the structures in Figure P20.42 are α isomers?

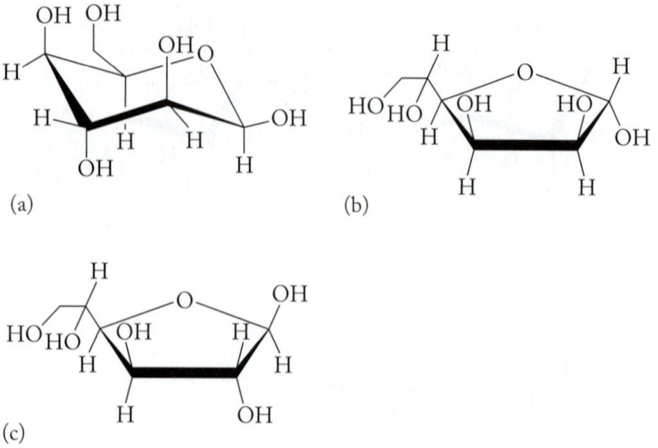

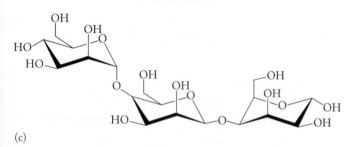

(a)

(b)

(c)

**FIGURE P20.42**

---

**20.43.** Which of the saccharides in Figure P20.43 is digestible by humans?

(a)

(b)

(c)

**FIGURE P20.43**

*20.44. For any of the disaccharides in Problem 20.43 that are not digestible by humans, draw an isomer that would be digested.

---

**20.45.** The structure of the disaccharide maltose appears in Figure P20.45. Hydrolysis of 1 mole of maltose ($\Delta G_f^\circ = -2246.6$ kJ/mol) produces 2 moles of glucose ($\Delta G_f^\circ = -1274.4$ kJ/mol):

$$\text{Maltose} + H_2O \rightarrow 2 \text{ glucose}$$

If the value of $\Delta G_f^\circ$ of water is −285.8 kJ/mol, what is the change in free energy of the hydrolysis reaction?

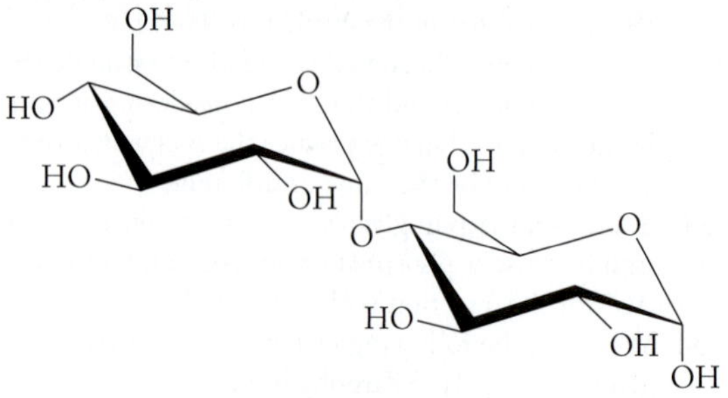

Maltose

**FIGURE P20.45**

**20.46.** If the maltose in Problem 20.45 were replaced by another disaccharide, would you expect the free-energy change for the hydrolysis to be exactly the same or just similar in value? Explain your answer.

## Lipids

### CONCEPT REVIEW

**20.47.** What is the difference between a saturated and an unsaturated fatty acid?

**20.48.** Which of the following lipids would have the lowest energy value in terms of human nutrition: olive oil, margarine, olestra, or butter? Explain your answer.

**20.49.** Some Arctic explorers have eaten sticks of butter on their explorations. Give a nutritional reason for this unusual cuisine.

*20.50. If you agitate a mixture of fatty acids in water, an emulsion forms, in which spherical structures called micelles are dispersed throughout the water. Micelles form when the carboxylic acid groups of the fatty acids face the solvent and their hydrocarbon tails are directed toward the inside of the sphere.
  a. Explain why these structures form with this orientation.
  b. It is sometimes possible to "break" an emulsion, destroying the micelles by adding a strong acid to the mixture. Why would this destroy the micelles?

**20.51.** Do triglycerides have a chiral center? Explain your answer.

*20.52. Using your knowledge of molecular geometry and intermolecular forces, why might polyunsaturated triglycerides be more likely to be liquid than saturated triglycerides?

### PROBLEMS

**20.53.** Oleic acid and α-linolenic acid (Table 20.3) are both unsaturated fats and are liquids at room temperature. They can be converted into solid saturated fats by hydrogenation.
  a. Which would consume the greater amount of hydrogen: 1.0 kg of oleic acid or 0.50 kg of α-linolenic acid?
  b. Could you distinguish between the two fatty acids by determining the identity of their hydrogenation products? Explain your answer.

**20.54.** For each of the pairs of fatty acids in Figure P20.54, indicate whether they are structural isomers, geometric isomers, or unrelated compounds.

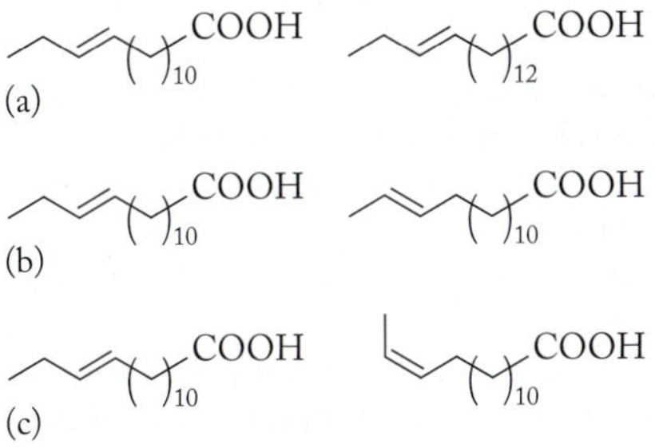

**FIGURE P20.54**

**20.55.** Draw the structures of the three fats formed by reaction of glycerol with (a) octanoic acid ($C_7H_{15}COOH$), (b) decanoic acid ($C_9H_{19}COOH$), and (c) dodecanoic acid ($C_{11}H_{23}COOH$).

**20.56. Oil-Based Paints** Oil-based paints contain linseed oil, a triglyceride formed by esterification of glycerol with linolenic acid (Figure P20.56).
  a. Draw the line structure of linolenic acid.
  *b. Are the double bonds in linolenic acid conjugated?

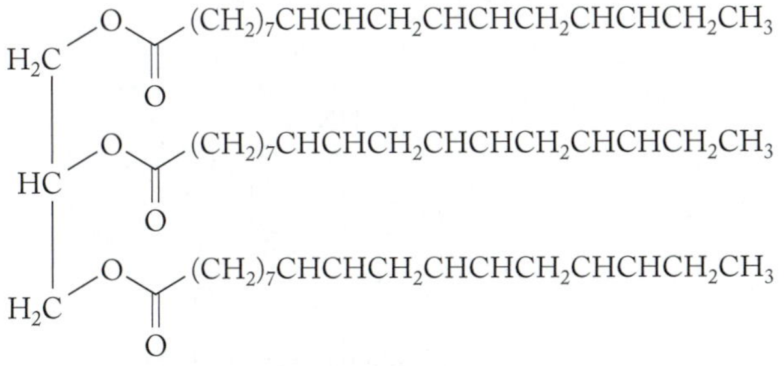

Linseed oil

**FIGURE P20.56**

## Nucleotides and Nucleic Acids

### CONCEPT REVIEW
**20.57.** What are the three kinds of molecular subunits in DNA? Which two form the "backbone" of DNA strands?

**20.58.** Why does a codon consist of a sequence of three, and not two, ribonucleotides?

**20.59.** What kind of intermolecular force holds together the strands of DNA in the double helix configuration?

**20.60.** DNA is a highly charged polyanion. If a solution of DNA is heated, the DNA will separate into individual strands, a process called denaturation. If the salt concentration of the solution is increased, the temperature at which denaturation occurs increases. Suggest a reason why.

### PROBLEMS
**20.61.** Draw the structure of adenosine 5′-monophosphate, one of the four ribonucleotides in a strand of RNA.

**20.62.** Draw the structure of deoxythymidine 5′-monophosphate, one of the four nucleotides in a strand of DNA.

**20.63.** In the replication of DNA, a segment of an original strand has the sequence T-C-G.
  a. What is the sequence of the opposite strand?
  b. Draw the structure of this section of the double helix, clearly showing all hydrogen bonds.

**20.64.** If the sequence of one strand of DNA is 5′ ATTGCCA 3′, what is the sequence (in the 5′ to 3′ direction) of the other strand?

## Additional Problems

**20.65.** Olestra is a calorie-free fat substitute. The core of the olestra molecule (Figure P20.65) is a disaccharide that has reacted with a carboxylic acid; this results in the conversion of hydroxyl groups on the disaccharide into the depicted structure.
  a. What is the name of the disaccharide core of the olestra molecule?
  b. What functional group has replaced the hydroxyl groups on the disaccharide?
  c. What is the formula of the carboxylic acid used to make olestra?

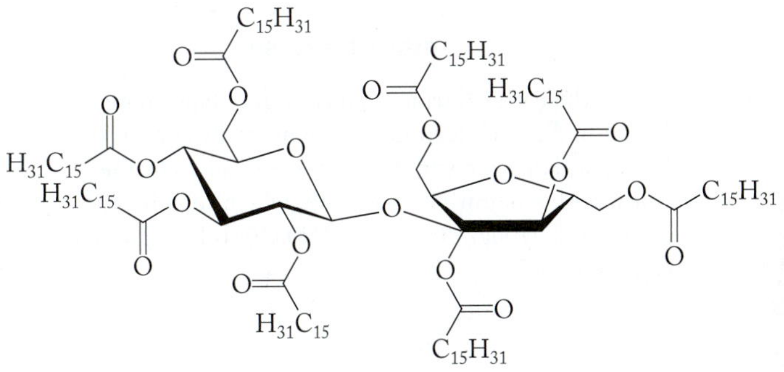

Olestra

**FIGURE P20.65**

**20.66.** When scientists at UC Santa Cruz directed UV radiation at an ice crystal containing methanol, ammonia, and hydrogen cyanide, three amino acids (glycine, alanine, and serine) were detected among the products of photochemical reactions. The formation of these amino acids suggests that they may also be synthesized in comets approaching the Sun (and Earth). Determine the standard free-energy change of the hypothetical formation of glycine in comets, using standard free energies of formation of the reactants and products in this reaction [$\Delta G_f^\circ$ for HCN($g$) is +125 kJ/mol and for solid glycine is −368.4 kJ/mol; other $\Delta G_f^\circ$ values are in Appendix 4].

$$CH_3OH(\ell) + HCN(g) + H_2O(\ell) \rightarrow H_2NCH_2COOH(s) + H_2(g)$$

**20.67.** Homocysteine (Figure P20.67) is formed during the metabolism of amino acids. A mutation in some people's genes leads to high concentrations of homocysteine in the blood and a consequent increase in their risk of heart disease and incidence of bone fractures in old age.
  a. What is the structural difference between homocysteine and cysteine?
  b. Cysteine is a chiral compound. Is homocysteine chiral?

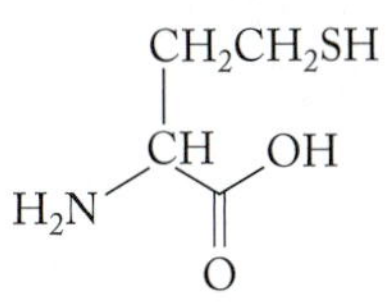

Homocysteine

**FIGURE P20.67**

20.68. Some scientists believe life on Earth can be traced to amino acids and other molecules brought to Earth by comets and meteorites. In 2004, a new class of amino acids called diamino acids (Figure P20.68) were found in the Murchison meteorite.
  a. Which of these diamino acids is not an α-amino acid?
  b. Which of these amino acids is chiral?

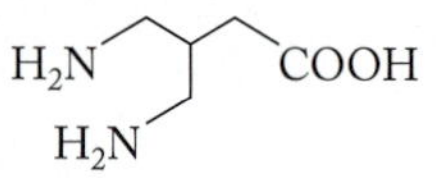

**FIGURE P20.68**

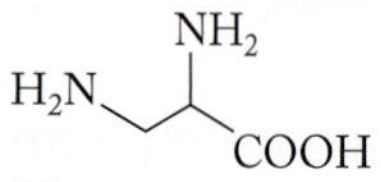

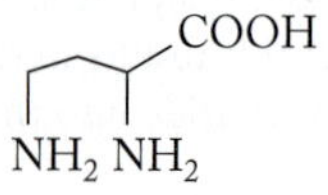

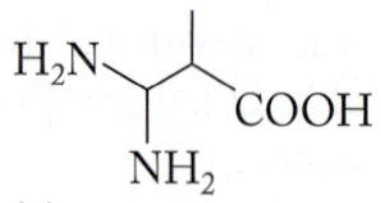

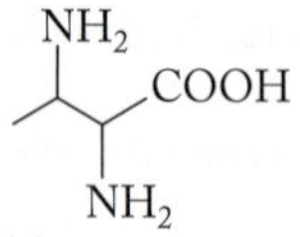

20.69. Ackee, the national fruit of Jamaica, is a staple in many Jamaican diets. Unfortunately, a potentially fatal sickness known as Jamaican vomiting disease is caused by the consumption of unripe ackee fruit, which contains the amino acid hypoglycin (Figure P20.69). Is hypoglycin an α-amino acid?

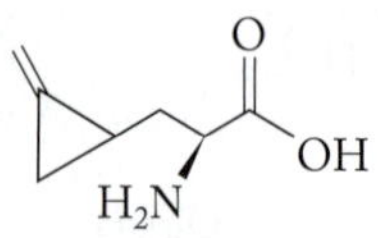

Hypoglycin

**FIGURE P20.69**

*20.70. In Chapter 13, we discussed the rotation of a beam of polarized light when it passes through a solution containing an optically active molecule. Equimolar solutions of α-glucose and β-glucose rotate plane-polarized light by +112° and +18.7°, respectively. If these two solutions are then mixed together and allowed to reach equilibrium, the solution then rotates the polarized light by +53.4°C. Calculate the percent glucose in the α and β forms in this solution.

20.71. Supplements containing BCAA (branched-chain amino acids) have been studied for their effects on athletic performance (in terms of the perception of exertion) and mental fatigue. Current research does not support that they have either of these effects, but they are used medically to slow muscle wasting in bedridden patients. BCAAs are amino acids having aliphatic side chains with a branch, a carbon atom bound to more than two other carbon atoms. Consult Table 20.1 and draw the structures of the BCAAs among the common amino acids.

20.72. In response to specific neural messages, the human hypothalamus may secrete a number of polypeptides, including the tripeptide shown in Figure P20.72.
  a. Sketch the structures of the three amino acids that combine to make this tripeptide.
  b. Which, if any, of the constituent amino acids are among the 20 α-amino acids in proteins?

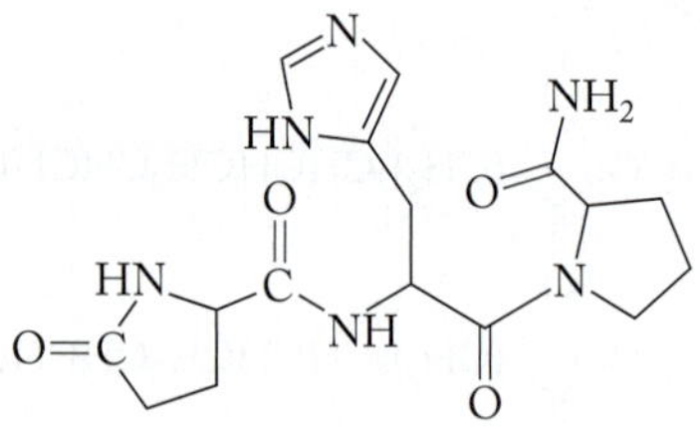

Thyrotropin-releasing factor

**FIGURE P20.72**

**20.73.** Glutathione (Figure P20.73) is an essential molecule in the human body. It acts as an activator for enzymes and protects lipids from oxidation. Which three amino acids combine to make glutathione?

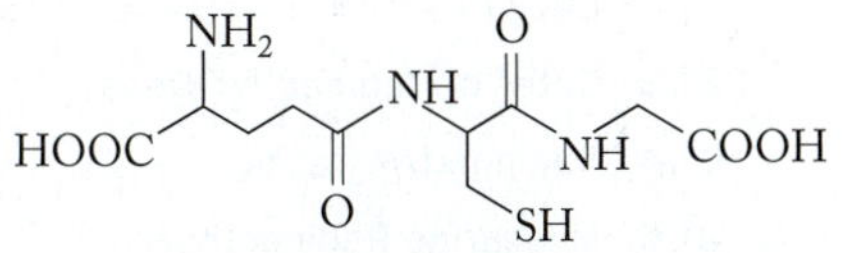

Glutathione

**FIGURE P20.73**

*20.74. Three amino acids—glutamic acid, arginine, and tryptophan—are dissolved in a gel that is buffered at a pH of 5.9. Two electrodes are placed in the gel and an electric current is applied.
   a. Toward which electrode does each amino acid migrate?
   b. Draw the forms of each amino acid present in the gel at a pH of 5.9.

*20.75. Without doing the actual calculation, estimate the fuel values of leucine and isoleucine by considering average bond energies. Should the fuel values of the two amino acids be the same? Actual calorimetric measurements show that isoleucine has a lower fuel value than leucine. Explain why.

20.76. Sucralose (see Figure P20.9) is about 600 times sweeter than sucrose (see Figure 20.23). All substances that taste sweet have functional groups that form hydrogen bonds with "sweetness" receptor sites on the tongue. What does the difference in sweetness between sucralose and sucrose tell you about additional intermolecular interactions between sweet compounds and receptor sites that contribute to their sweet taste?

If your instructor assigns problems in smartwork, log in at **smartwork.wwnorton.com**.

# 21

# Nuclear Chemistry: Applications to Energy and Medicine

21.1 Binding Energy and Nuclear Stability

21.2 Unstable Nuclei and Radioactive Decay

21.3 Rates of Radioactive Decay

21.4 Radiometric Dating

21.5 Measuring Radioactivity

21.6 Biological Effects of Radioactivity

21.7 Medical Applications of Radionuclides

21.8 Nuclear Fission

21.9 Nuclear Fusion and the Quest for Clean Energy

## Learning Outcomes

**LO1** Calculate the binding energy of atomic nuclei

**LO2** Predict the decay modes of radionuclides
**Sample Exercise 21.1**

**LO3** Calculate the quantity of a radionuclide remaining after a defined decay time
**Sample Exercise 21.2**

**LO4** Determine the age of a sample by radiometric dating
**Sample Exercise 21.3**

**LO5** Calculate the level of radioactivity in a sample of a radionuclide
**Sample Exercise 21.4**

**LO6** Describe the dangers of exposure to nuclear radiation and calculate effective radiation doses
**Sample Exercise 21.5**

**LO7** Calculate the energy released in a nuclear reaction from the masses of the products and reactants
**Sample Exercise 21.6**

## The Risks and Benefits of Nuclear Radiation

In Section 2.1 we briefly described how Henri Becquerel's discovery of radioactivity emitted by uranium-containing minerals was critical to unraveling atomic structure in the early 20th century. Ernest Rutherford used $\alpha$ particles from the decay of uranium to discover that the atoms contain dense nuclei composed of protons and neutrons. In 1898 Marie Curie separated and purified two new radioactive elements from uranium—polonium and radium—helping launch a new scientific discipline: radiochemistry.

Interest in the applications of radium grew rapidly. Radium-containing paint was used to create luminous dials for watches and gauges. Ointments containing radium were prescribed as treatments for skin lesions. People were encouraged to visit health spas, such as Saratoga Springs, NY, where the waters contained low concentrations of dissolved radium salts. Proliferation of radium-containing products also led to the discovery that nuclear radiation is dangerous. Young women hired to paint the dials of watches during World War I were instructed to "point" the tips of their brushes by licking them, inadvertently introducing radium into their bodies, where it concentrated in their teeth and bones. Within a few years, many of these young women developed bone cancer and died. Marie Curie herself succumbed to aplastic anemia, caused by years of research with radioactive materials.

In recent years a better understanding of the biological impacts of radiation has led to the development of nuclear methods for diagnosing and treating disease that minimize the hazards of radiation to patients and those who treat them. Nuclear reactors provide doctors and scientists with radionuclides that decay over minutes to hours by predictable pathways. Selective uptake of these nuclides by different organs in the body allows doctors to evaluate organ function and

**Positron Emission Tomography (PET)** In nuclear medicine PET is a ▶ powerful imaging technique based on administering trace concentrations of radioactive nuclides that emit positrons—the antimatter version of electrons. The distribution of these compounds enables physicians to detect abnormal cell activity, for example, in this patient's lymph nodes. The activity seen in the brain is normal.

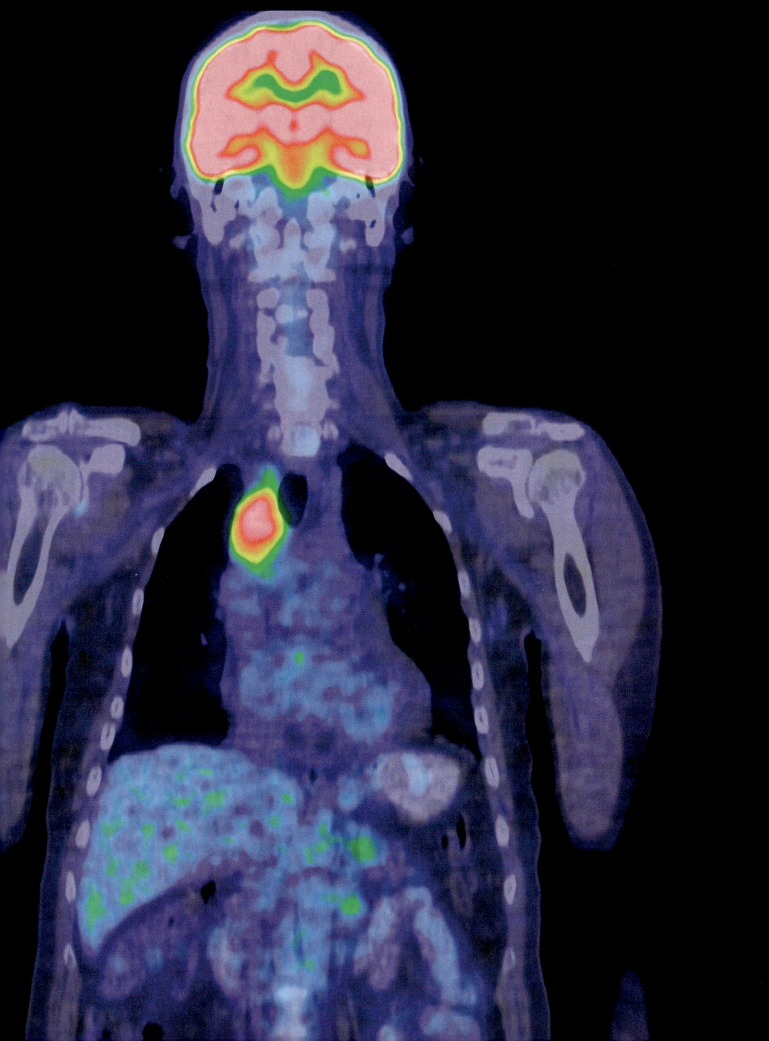

prescribe treatment when function is impaired. Radiation from other nuclides that concentrate in cancerous tissues can be used, often in conjunction with other therapies, to destroy malignant tumors.

In this chapter we examine the origins of nuclear radiation, that is, interactions that take place in the nuclei of atoms. We address why some nuclei are stable and others are not, the nuclear reactions that unstable nuclei undergo, and how the energy released by these reactions can be used to help meet the world's need for electrical power and to diagnose and treat disease. We also address some of the dangers radioactive substances pose to human health and how we can shield ourselves from them. We conclude with an examination of nuclear reactions, similar to those that fuel the Sun, that someday may be used to meet the world's energy needs safely and sustainably. ■

## 21.1 Binding Energy and Nuclear Stability

In Chapter 1 we briefly mentioned the rapid transformation of energy into matter that scientists think followed the Big Bang. The proportions of energy and matter involved in such transformations are described by Einstein's famous equation

$$E = mc^2$$

where $E$ is the amount of energy transformed into matter (or vice versa), $m$ is the mass of that matter, and $c$ is the speed of light. The sheer immensity of the speed of light (squared) means that these transformations are accompanied by enormous changes in energy.

As noted in Chapter 2, primordial nucleosynthesis created an early universe composed mainly of two elements: hydrogen and helium. More massive nuclei, containing up to 26 protons in each nucleus, formed later in the cores of the giant stars that populated the first generation of galaxies. Even more massive nuclei formed during the enormous releases of energy that marked the deaths of these giant stars—events we know as supernovae and have witnessed (from a safe distance) here on Earth (see Figure 2.21).

Why does stellar nucleosynthesis end with the formation of iron ($Z = 26$) nuclei? The answer has to do with the energy inside all nuclei that binds nucleons together. This energy was discovered in the 1930s when scientists determined that the masses of stable nuclei are always less than the sum of the free masses of their nucleons. (The masses and symbols of subatomic particles are listed in Table 21.1.) For example, the total mass of the free nucleons in one $^4$He nucleus—that is, 2 neutrons and 2 protons—is

$$\text{Mass of 2 neutrons} = 2(1.67493 \times 10^{-27} \text{ kg})$$
$$\underline{\text{Mass of 2 protons} = 2(1.67262 \times 10^{-27} \text{ kg})}$$
$$\text{Total mass} = 6.69510 \times 10^{-27} \text{ kg}$$

The difference between this value and the mass of a $^4$He nucleus (from Table 21.1) is

$$\begin{array}{r} 6.69510 \times 10^{-27} \text{ kg} \\ - 6.64465 \times 10^{-27} \text{ kg} \\ \hline 0.05045 \times 10^{-27} \text{ or } 5.045 \times 10^{-29} \text{ kg} \end{array}$$

**TABLE 21.1  Symbols and Masses of Subatomic Particles and Small Nuclei**

| Particle | Symbol | Mass (kg) |
|---|---|---|
| Neutron | $^1_0\text{n}$ | $1.67493 \times 10^{-27}$ |
| Proton | $^1_1\text{p}$ or $^1_1\text{H}$ | $1.67262 \times 10^{-27}$ |
| Electron (β particle) | $^0_{-1}\beta$ or $^0_{-1}\text{e}$ | $9.10939 \times 10^{-31}$ |
| Deuteron | $^2_1\text{D}$ or $^2_1\text{H}$ | $3.34370 \times 10^{-27}$ |
| α Particle | $^4_2\alpha$ or $^4_2\text{He}$ | $6.64465 \times 10^{-27}$ |
| Positron | $^0_1\beta$ | $9.10939 \times 10^{-31}$ |

This difference is called the **mass defect (Δ*m*)** of the nucleus. The energy equivalent to the mass defect of a nucleus is called its **binding energy (BE)** and can be calculated using Einstein's equation:

$$BE = (\Delta m)c^2$$
$$= 5.045 \times 10^{-29} \text{ kg} \times (2.998 \times 10^8 \text{ m/s})^2$$
$$= 4.534 \times 10^{-12} \text{ kg} \cdot \text{(m/s)}^2 = 4.534 \times 10^{-12} \text{ J}$$

This value may seem very small, but it is the binding energy of a single nucleus. Its value per mole of $^4$He is equivalent to billions of kilojoules:

$$\frac{4.534 \times 10^{-12} \text{ J}}{\text{atom}} \times \frac{6.022 \times 10^{23} \text{ atoms}}{\text{mol}} \times \frac{1 \text{ kJ}}{1000 \text{ J}} = 2.3730 \times 10^9 \text{ kJ/mol}$$

For comparison, consider that the bond dissociation energy of $H_2$, 436 kJ/mol, is over 5,000,000 times *smaller* than the binding energy of a mole of He atoms.

Stable nuclei with larger numbers of nucleons than helium-4 have larger binding energies. This makes sense because a nucleus with many protons in close proximity must be held together by an enormous energy to overcome the coulombic repulsion that all of these positively charged particles exert on one another. To make comparisons of nuclear binding energies meaningful for nuclei of different sizes, the energies are usually divided by the number of nucleons in each nucleus.

Expressing binding energy on a per nucleon basis allows us to compare the relative stabilities of different nuclides. When we plot binding energy per nucleon values against atomic number we get the curve in Figure 21.1. Note that these values reach a maximum with $^{56}$Fe. This pattern means that $^{56}$Fe is the most stable nuclide in the universe, which fits what we know about the nuclear processes that synthesize the elements in the cores of giant stars. Recall from our discussion of stellar nuclear synthesis in Chapter 2 that the nuclear furnaces of giant stars are fueled by the energy released when lighter nuclei fuse together to form heavier nuclei, but only when the heavier ones that form are more stable than the lighter ones. Energy is not released if one of the reacting particles is $^{56}$Fe or a heavier nuclide because, according to Figure 21.1, these heavier fusion products are *less stable*, not more stable. Their formation consumes energy instead of producing it. Thus, an aged giant star whose core has turned into iron as a result of multiple nuclear fusion processes has essentially run out of fuel. Without heating from its

**mass defect (Δ*m*)** the difference between the mass of a stable nucleus and the masses of the individual nucleons that comprise it.

**binding energy (BE)** the energy that would be released when free nucleons combine to form the nucleus of an atom. It is also the energy needed to split the nucleus into free nucleons.

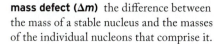 **CONNECTION** As we discussed in Chapter 2, the lighter elements from helium to iron are synthesized by fusion reactions in the cores of giant stars.

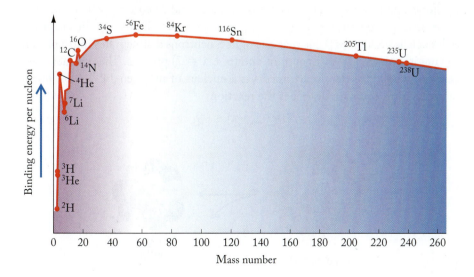

**FIGURE 21.1** The stability of a nucleus is directly proportional to its binding energy per nucleon. For all nuclides up to $^{56}$Fe, fusion reactions lead to products that have greater binding energy per nucleon than the reactants. This means the fusion reactions release energy. However, fusion of nuclei with atomic numbers greater than 26 produces nuclei that have less binding energy per nucleon than the reactants, and thus consumes energy.

**belt of stability** the region on a graph of number of neutrons versus number of protons that includes all stable nuclei.

**radionuclide** an unstable nuclide that undergoes radioactive decay.

**radioactive decay** the spontaneous disintegration of unstable particles accompanied by the release of radiation.

**beta (β) decay** the spontaneous ejection of a β particle by a neutron-rich nucleus.

**nuclear reaction** a process that alters the number of neutrons or protons in the nucleus on an atom.

▶❚❚ **CHEMTOUR** Balancing Nuclear Equations

nuclear furnace, the star's fusion reactions cease and its enormous gravity forces it to collapse into itself. The result is rapid heating as the star's mass is compressed, followed by a gigantic explosion called a supernova.

**CONCEPT TEST**

Why does a negative change in mass when nucleons combine to form a nucleus produce a positive binding energy?

*(Answers to Concept Tests are in the back of the book.)*

## 21.2 Unstable Nuclei and Radioactive Decay

The values of the atomic masses and mass numbers of the elements in the periodic table tell us about the ratios of neutrons to protons in the nuclei of their stable isotopes. The lighter elements have atomic masses that are about twice their atomic numbers and have neutron-to-proton ratios close to unity. For example, $^{12}$C has 6 neutrons and 6 protons, and most oxygen atoms have 8 neutrons and 8 protons.

However, with increasing values of Z, the ratios of neutrons to protons increase. This trend is illustrated in Figure 21.2, where the green dots represent combinations of neutrons and protons that form stable nuclides. The band of green dots runs diagonally through the graph, defining the **belt of stability**. Note how the belt curves upward away from the purple straight line, representing a neutron-to-proton ratio of 1. This curvature shows how the neutron:proton ratios increase from about 1:1 for the lightest stable nuclides to about 1.5:1 for the most massive ones.

The nuclides represented by orange dots in Figure 21.2 are **radionuclides**. They are not stable but instead undergo **radioactive decay**, which is the spontaneous disintegration of radioactive nuclei, accompanied by the release of nuclear radiation. Modes of radioactive decay depend on whether a nuclide is above or below the belt of stability. Those above, such as carbon-14, are *neutron rich* and tend to undergo **β decay**: a **nuclear reaction** in which a neutron disintegrates, producing a proton that remains in the nucleus and a high-speed, high-energy electron, called a β particle, that is emitted from the atom.

For example, a $^{14}$C nucleus contains 6 protons and 8 neutrons, which means its neutron-to-proton ratio is 1.333—a large value for a small nucleus. This neutron-rich nucleus undergoes β decay as one of its 8 neutrons disintegrates into a proton, leaving it with (8 − 1 = 7) neutrons and (6 + 1 = 7) protons. These values mean that the product of the decay process is a nucleus of nitrogen-14. The following radiochemical equation describes the reaction:

$$^{14}_{6}C \rightarrow \,^{14}_{7}N + \,^{0}_{-1}\beta$$

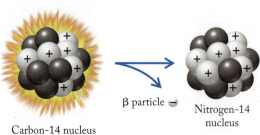

Carbon-14 nucleus    β particle ⊖    Nitrogen-14 nucleus

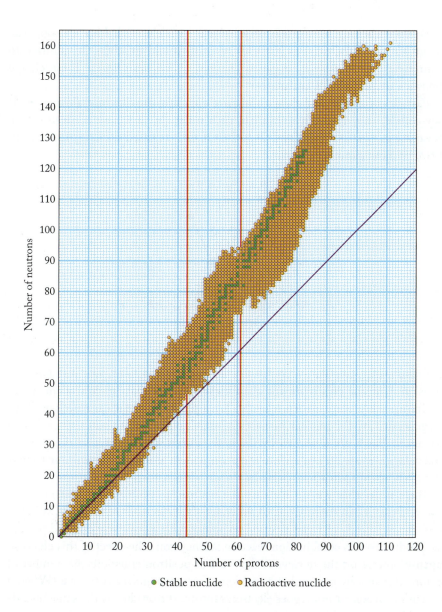

● Stable nuclide  ● Radioactive nuclide

**FIGURE 21.2** The belt of stability. Green dots represent stable combinations of protons and neutrons. Orange dots represent known radioactive (unstable) nuclides. Nuclides that fall along the purple line have equal numbers of neutrons and protons. Note that there are no stable nuclides (no green dots) for $Z = 43$ (technetium) and $Z = 61$ (promethium), as indicated by the two vertical red lines. These elements are the only two among the first 83 that are not found in nature.

Nuclides below the belt of stability are *neutron poor* and undergo decay processes that *increase* their neutron-to-proton ratio. In one of these processes the radioactive nucleus emits a high-velocity particle that has the same mass as an electron, but that has a positive charge. It is called a **positron ($_{1}^{0}\beta$)** and its ejection from a neutron-poor nucleus is called **positron emission**. The net effect of positron emission is the production of a nuclide of a nucleus with one fewer proton and one more neutron, as illustrated in the decay of carbon-11:

$$_{6}^{11}\text{C} \rightarrow {}_{5}^{11}\text{B} + {}_{1}^{0}\beta$$

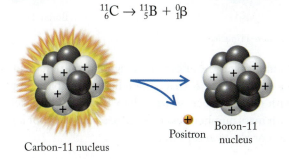

Carbon-11 nucleus        Positron   Boron-11 nucleus

**positron ($_{1}^{0}\beta$)** a particle with the mass of an electron but with a positive charge.

**positron emission** the spontaneous emission of a positron from a neutron-poor nucleus.

Positrons belong to a group of subatomic particles that have the opposite charge but the same mass as particles typically found in atoms. In addition to positrons are protons with negative charges, called *antiprotons*. These charge opposites are particles of **antimatter**.

Particles of matter and their antimatter opposites are like mortal enemies. If they collide, they instantly annihilate each other. In their mutual destruction, they cease to exist as matter, and all of their mass is released as energy in the form of two or more gamma (γ) rays:

⊙⊙ **CONNECTION** Gamma rays are the highest energy form of electromagnetic radiation (see Figure 7.1).

$$\,^0_1\beta + \,^0_{-1}\beta \rightarrow 2\,\gamma$$

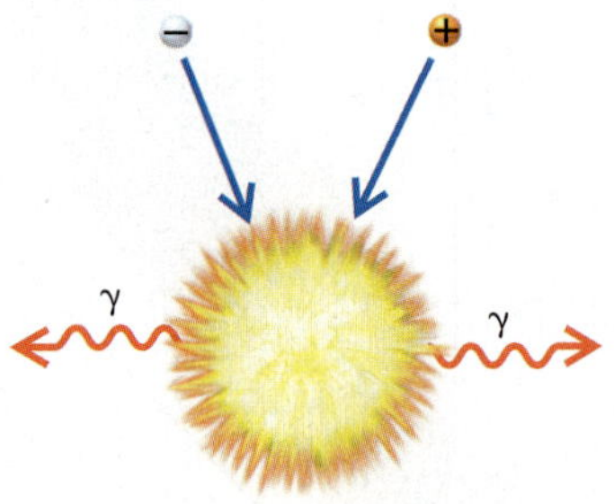

Gamma ray emission accompanies all nuclear reactions, not just positron emission. Gamma rays represent quantities of energy that are equivalent to the loss in mass that occurs when reactants form products in nuclear reactions. They are generated by the nuclear furnaces of stars and permeate outer space. Those that reach Earth are absorbed by the gases in our atmosphere. In the process, molecular gases are broken up into their component atoms, and atomic nuclei may be broken up into subatomic particles.

There is another way to increase the neutron-to-proton ratio of a neutron-poor nucleus: it can capture one of the inner-shell electrons of its atom. When it does, the negatively charged electron combines with a positively charged proton. The product of this combination reaction is a neutron. The effect of this **electron capture** process on the nucleus is the same as positron emission: the number of protons *decreases* by one and the number of neutrons *increases* by one. When a nucleus of carbon-11 undergoes electron capture, the product is the same nuclide formed in positron emission, boron-11:

$$\,^{11}_6C + \,^0_{-1}e \rightarrow \,^{11}_5B$$

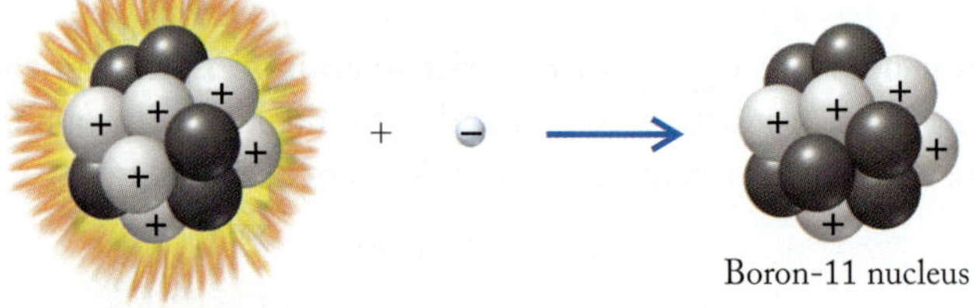

Carbon-11 nucleus

Boron-11 nucleus

**antimatter** particles that are the charge opposites of normal subatomic particles.

**electron capture** a nuclear reaction in which a neutron-poor nucleus draws in one of its surrounding electrons, which transforms a proton in the nucleus into a neutron.

Table 21.2 summarizes the effects of being neutron rich, neutron poor, or neither on isotopes of carbon. Note that carbon has two stable isotopes: $^{12}C$ and $^{13}C$. The isotopes with mass numbers greater than 13 are neutron rich and undergo β decay. Those with mass numbers less than 12 are neutron poor and undergo either positron emission or electron capture.

| TABLE 21.2 | Isotopes of Carbon and Their Radioactive Decay Products | | | | |
|---|---|---|---|---|---|
| Name | Symbol | Mass (amu) | Mode(s) of Decay | Half-Life | Natural Abundance (%) |
| Carbon-10 | $^{10}_{6}C$ | | Positron emission | 19.45 s | |
| Carbon-11 | $^{11}_{6}C$ | | Positron emission, electron capture | 20.3 min | |
| Carbon-12 | $^{12}_{6}C$ | 12.00000 | | (Stable) | 98.89 |
| Carbon-13 | $^{13}_{6}C$ | 13.00335 | | (Stable) | 1.11 |
| Carbon-14 | $^{14}_{6}C$ | | β decay | 5730 yr | |
| Carbon-15 | $^{15}_{6}C$ | | β decay | 2.4 s | |
| Carbon-16 | $^{16}_{6}C$ | | β decay | 0.74 s | |

## SAMPLE EXERCISE 21.1   Predicting the Modes and Products of Radioactive Decay   LO2

Predict the mode of radioactive decay of $^{32}P$, which is one of the most widely used radio-nuclides in biomedical research and treatment. Identify the nuclide that is produced in the decay process.

**Collect and Organize** We are asked to predict the mode of decay of a radionuclide, which depends on whether it is neutron rich or neutron poor. Phosphorus-32 has 17 neutrons and 15 protons per nucleus. Figure 21.2 identifies the stable isotopes of the elements. According to the magnified region of the belt of stability in Figure 21.3, phosphorus ($Z = 15$) has only one stable nuclide, which has 16 neutrons and a mass number of 31.

**Analyze** Phosphorus-32 is represented by the orange dot directly above the $^{31}P$ green dot, which means that it is radioactive and neutron rich. Neutron-rich radioisotopes of lighter elements undergo β decay, so the decay of $^{32}P$ will produce a β particle. We can use a balanced nuclear equation describing the β decay reaction to identify the nuclide that is produced. As we learned in Section 2.7, the sums of the superscripts on the left and right sides must be equal in a balanced nuclear equation, as must the sums of the subscripts.

**Solve** A β particle must be one product of the decay reaction, giving the incomplete nuclear equation:

$$^{32}_{15}P \rightarrow ? + ^{0}_{-1}\beta$$

The missing product must have an atomic number of 16 (so that the subscripts on the right side add up to 15), which makes it an isotope of S. Its mass number must be 32, so the product is sulfur-32:

$$^{32}_{15}P \rightarrow ^{32}_{16}S + ^{0}_{-1}\beta$$

**Think About It** By emitting a β particle, the nucleus increased its number of protons by one and decreased its number of neutrons by one, thereby reducing its neutron "richness" and forming a stable isotope of sulfur.

**Practice Exercise** What is the mode of radioactive decay of $^{28}Al$? Identify the nuclide produced by the decay process.

*(Answers to Practice Exercises are in the back of the book.)*

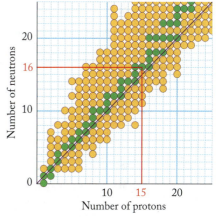

**FIGURE 21.3**

**alpha (α) decay** a nuclear reaction in which an unstable nuclide spontaneously emits an alpha particle.

All known nuclides with more than 83 protons are radioactive. Because there is no stable reference point in the pattern of green dots in Figure 21.2, it is hard to say whether any given $Z > 83$ nuclide is neutron rich or neutron poor. We can make one general statement though: these most massive nuclides tend to undergo either β decay or **alpha (α) decay**. In α decay they produce a nuclide with two fewer protons and two fewer neutrons, as in the case of uranium-238:

$$^{238}_{92}U \rightarrow {}^{234}_{90}Th + {}^{4}_{2}\alpha$$

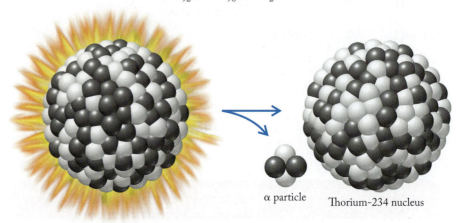

α particle     Thorium-234 nucleus

Uranium-238 nucleus

▶❚❚ **CHEMTOUR**  Radioactive Decay Modes

Figure 21.4 illustrates the changes in atomic number and mass number caused by the various modes of decay.

Among the most massive radionuclides, one radioactive decay process often leads to another in what is referred to as a *radioactive decay series*. Consider, for example, the decay series that begins with the α decay of $^{238}U$ to $^{234}Th$ (Figure 21.5). Thorium has no stable isotopes and undergoes two β decay steps to produce $^{234}U$. In a series of subsequent α decay steps, $^{234}U$ turns into thorium-230, radium-226, radon-222, polonium-218, and finally lead-214. Although some isotopes of lead ($Z = 82$) are stable, $^{214}Pb$ is not one of them. Therefore, the radioactive decay series continues, as shown at the bottom left of Figure 21.5, and does not end until the stable nuclide $^{206}Pb$ is produced.

**FIGURE 21.4** Radioactive decay results in predictable changes in the number of protons and neutrons in a nucleus. In α decay, the nucleus loses 2 neutrons and 2 protons, resulting in a decrease of 2 in atomic number and 4 in mass number. Beta decay leads to an increase of 1 proton at the expense of 1 neutron, so the atomic number increases by 1, but the mass number is unchanged. In positron emission and electron capture, the number of protons decreases by 1 and the number of neutrons increases by 1, so the atomic number decreases by 1, but the mass number remains the same.

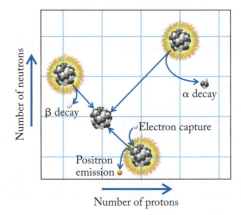

CONCEPT TEST

In the $^{238}U$ radioactive decay series, five α decay steps in a row transform $^{234}U$ into $^{214}Pb$. Given the shape of the belt of stability, why does it make sense that the product of these α decay steps would be a neutron-rich nuclide that undergoes β decay?

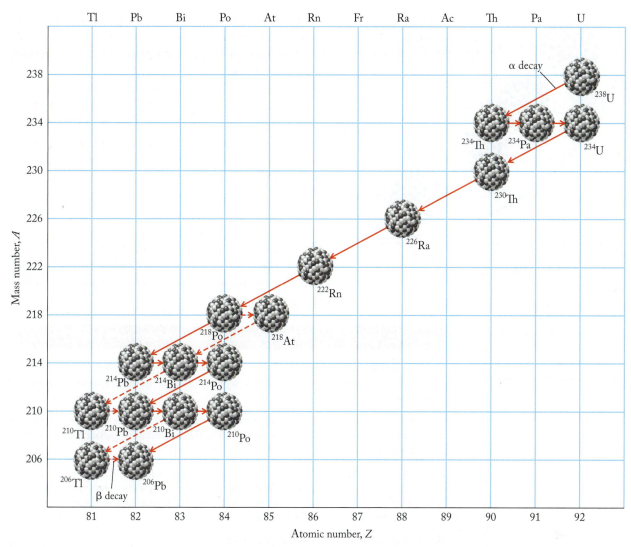

**FIGURE 21.5** Uranium-238 radioactive decay series. The long diagonal arrows represent α decay events; the short horizontal ones represent β decay events. The dashed arrows are alternative pathways representing less than 1% of the decay events in this series. Note that whether decay proceeds by the solid-line pathways or the dashed-line pathways, the end product is always stable lead-206.

# 21.3 Rates of Radioactive Decay

In the preceding section we examined *how* radionuclides undergo radioactive decay; in this section we focus on *how rapidly* they decay. We have defined radioactive decay as the *spontaneous* disintegration of unstable nuclei, but, as with chemical reactions, spontaneous does not necessarily mean rapid. All radioactive decay processes follow first-order kinetics, and each has a characteristic *half-life* ($t_{1/2}$), the time interval during which the quantity of radioactive particles decreases by one half (Figure 21.6); the faster the decay process, the shorter the half-life.

To calculate what fraction of radionuclide remains after decay time $t$, we first convert this time into an equivalent number of half-lives, $n$:

▶❚❚ **CHEMTOUR** Half-Life

◉◉ **CONNECTION** We introduced the concept of *half-life*, $t_{1/2}$, and how its value is inversely proportional to the rate constant, $k$, of a first-order reaction in Chapter 14.

$$n = \frac{t}{t_{1/2}} \tag{21.1}$$

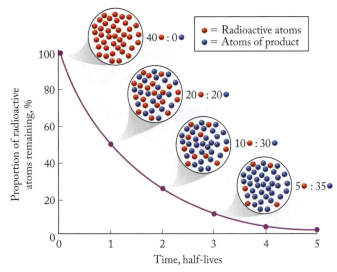

**FIGURE 21.6** Radioactive decay follows first-order kinetics, which means, for example, that if a sample initially contains 40 radioactive atoms, it will contain only one-half that number after a time interval equal to one half-life. Half of the remaining half, or 10 radioactive atoms, remain after two half-lives, and so on.

Once we know the value of $n$, we can calculate the fraction of radioactive nuclei remaining at instant $t$ by writing the ratio of the number of nuclei at instant $t$ ($N_t$) to the number present at $t = 0$ ($N_0$):

$$\frac{N_t}{N_0} = 0.5^n \qquad (21.2)$$

In practice, it is often more convenient to relate quantities of radioactive particles directly to time by combining Equations 21.1 and 21.2:

$$\frac{N_t}{N_0} = 0.5^{t/t_{1/2}} \qquad (21.3)$$

In the following exercises and elsewhere in this chapter we apply Equation 21.3 to various radioactive decay processes. We can do so because all of these processes follow first-order reaction kinetics.

---

**SAMPLE EXERCISE 21.2    Calculations Involving Half-Lives    LO3**

Free neutrons are radioactive with a half-life of 12 minutes, undergoing β decay to a proton and an electron. Starting with a population of $6.6 \times 10^5$ free neutrons, how many remain after 2.0 min?

**Collect and Organize** The problem gives an initial quantity of free neutrons and asks how many are left after 2.0 min. Equation 21.3 relates quantities of radioactive particles to decay times.

**Analyze** In this problem the initial number of neutrons ($N_0$) is $6.6 \times 10^5$, $t = 2.0$ min, $t_{1/2} = 12$ min, and we need to solve for $N_t$. The value of $t$ is only a fraction of $t_{1/2}$; far fewer than half of the initial number of neutrons will have decayed after 2.0 min. In other words, most of the neutrons will still be present.

**Solve** Substituting the data into Equation 21.3 gives

$$\frac{N_t}{6.6 \times 10^5} = 0.5^{2.0 \,\text{min}/12 \,\text{min}}$$

$$N_t = 5.9 \times 10^5$$

**Think About It** The value of $N_t$ is reasonable because, as we predicted, only a small fraction of the initial quantity of free neutrons decayed in two minutes.

⚙ **Practice Exercise** Cesium-131 is a short-lived radionuclide ($t_{1/2} = 9.7$ d) used to treat prostate cancer. How much therapeutic strength does a cesium-131 source lose over 60 days? Express your answer as a percentage of the strength the source had at the beginning of the first day.

---

**radiometric dating** a method for determining the age of an object based on the quantity of a radioactive nuclide and/or the products of its decay that the object contains.

# 21.4 Radiometric Dating

**Radiometric dating** is a term used to describe methods for determining the age of objects based on the tiny concentrations of radionuclides that occur naturally in them and the rates of radioactive decay of these nuclides. The concept originated in the early 1900s in the work of Ernest Rutherford, who had already recognized in his pioneering studies on radioactivity that radioactive decay processes have characteristic half-lives. Rutherford proposed to use this concept to determine the age of rocks and even the age of Earth itself. The basis for his initial attempt was the emission of α particles from uranium ore. He correctly suspected that α particles were part of helium atoms, and he proposed to determine the age of uranium ore samples by determining the concentration of helium gas trapped inside them.

Rutherford's helium method did not yield very accurate results, but it did inspire a young American chemist, Bertram Boltwood (1870–1927), who had determined that the decay of radioactive uranium involves a series of decay events ending with the formation of stable lead (see Figure 21.5). In 1907 Boltwood published the results of dating 43 samples of uranium-containing minerals based on the ratio of lead to uranium in them. The ages he reported spanned hundreds of millions to over a billion years and probably represent the first successful attempt at radiometric dating.

In recent years the development of the mass spectrometer for accurately determining the abundances of individual isotopes of elements, coupled with more accurate half-life values for decay events such as those in Figure 21.5, has allowed scientists to use the ratio of $^{206}Pb$ to $^{238}U$ in geological samples to determine their ages with a precision of about ±1%. Other methods, including one based on the decay of $^{235}U$ to $^{207}Pb$ ($t_{1/2} = 7.0 \times 10^6$ yr), may be used to analyze the same samples, providing independent determinations that mutually ensure more accurate results. These analyses have shown that the oldest rocks on Earth are over 4.0 billion years old and that meteorites that formed as the solar system formed are 4.5 billion years old.

These radiometric methods for dating geological samples yield reliable results only when the sample is a *closed system*, which means that the only loss of the radionuclide is via radioactive decay, and that all of the nuclides produced by the decay processes remain in the sample. In addition, those decay processes must be the *only* source of the product nuclides. For these reasons, scientists must exercise care in selecting the types of samples they subject to radiometric dating analysis. For example, the presence of the mineral zircon ($ZrSiO_4$) in a geological sample is good news for scientists interested in dating it because $U^{4+}$ ions readily substitute for $Zr^{4+}$ ions as crystals of $ZrSiO_4$ solidify from the molten state, but $Pb^{2+}$ ions do not. Therefore, the only source of $^{206}Pb$ and $^{207}Pb$ in a zircon sample should be the decay of $^{238}U$ and $^{235}U$, respectively.

In 1947 American chemist Willard Libby (1908–1980) developed a radiometric dating technique, called **radiocarbon dating**, for determining the age of artifacts from prehistory and early civilizations. The method is based on determining the carbon-14 content of samples derived from plants or the animals that consumed them. Carbon-14 originates in the upper atmosphere, where cosmic rays break apart the nuclei of atoms, forming free protons and neutrons. When one of these neutrons collides with a nitrogen-14 atom, they form an atom of radioactive carbon-14 and a proton:

$$^{14}_{7}N + ^{1}_{0}n \rightarrow ^{14}_{6}C + ^{1}_{1}p$$

Atmospheric carbon-14 combines with oxygen, forming $^{14}CO_2$. The atmospheric concentration of $^{14}CO_2$ amounts to only about $10^{-12}$ of all the molecules of $CO_2$ in the air. These traces of radioactive $CO_2$, along with the stable forms $^{12}CO_2$ and $^{13}CO_2$, are incorporated into the structures of green plants during photosynthesis. The tiny fraction of the plant's mass that is $^{14}C$ gets even tinier after a plant dies,

⊙⊙ **CONNECTION** In Section 2.3 we saw that mass spectrometry can be used to determine the abundances of the isotopes of elements in a sample.

**radiocarbon dating** a method for establishing the age of a carbon-containing object by measuring the amount of radioactive carbon-14 remaining in the object.

or after a part of it stops growing and photosynthesizing, because $^{14}$C undergoes β decay, as we described in Section 21.2:

$$^{14}_{6}C \rightarrow {}^{14}_{7}N + {}^{0}_{-1}\beta$$

The half-life of the decay process is 5730 years.

If we can determine the $^{14}$C content ($N_t$) of an object of historical interest, such as a piece of wood from an ancient building, charcoal from a prehistoric campfire, or papyrus from an early Egyptian scroll, and if we know (or can predict) its $^{14}$C content when the material in it was alive ($N_0$), then we can apply Equation 21.3 to determine its age:

$$\frac{N_t}{N_0} = 0.5^{t/t_{1/2}} \tag{21.3}$$

Predicting the value of $N_0$ is usually done by analyzing samples from growing plants—that is, samples for which the $^{14}$C decay time is zero.

To facilitate radiocarbon dating calculations, let's solve Equation 21.3 for $t$ by first taking the natural log of both sides (keeping in mind that ln 0.5 = −0.693):

$$\ln\frac{N_t}{N_0} = -0.693\,\frac{t}{t_{1/2}}$$

Rearranging the terms to solve for $t$ gives us a useful equation for determining the radiocarbon age $t$:

$$t = -\frac{t_{1/2}}{0.693}\ln\frac{N_t}{N_0} \tag{21.4}$$

---

### SAMPLE EXERCISE 21.3  Radiocarbon Dating                                    LO4

The $^{14}$C content of a wooden harpoon handle found in the remains of an Inuit encampment in western Alaska is 61.9% of the $^{14}$C content of the same type of wood from a recently cut tree. How old is the harpoon?

**Collect and Organize**  We are asked to calculate the age of a sample that contains 61.9% of the $^{14}$C in a modern sample of the same material. The half-life of carbon-14 is 5730 years. Equation 21.4 provides the age $t$ of the artifact if we know the ratio of the $^{14}$C in it today to its initial $^{14}$C content.

**Analyze**  The $^{14}$C content of the modern sample can be used as a surrogate for the initial $^{14}$C content of the artifact. Therefore 61.9% (or 0.619) represents the ratio $N_t/N_0$. This value is greater than 0.5, which means that the age of the sample is less than one half-life (5730 yr).

**Solve**  Substituting into Equation 21.4:

$$t = -\frac{t_{1/2}}{0.693}\ln\frac{N_t}{N_0}$$

$$= -\frac{5730\ \text{yr}}{0.693}\ln(0.619)$$

$$= 3966\ \text{yr} = 3.97 \times 10^3\ \text{yr}$$

**Think About It**  The resulting age is less than one half-life, which is reasonable because it contained more than half the original carbon-14 content. The result is expressed with three significant figures to match that of the starting composition (61.9%).

⚙ **Practice Exercise**  The carbon-14:carbon-12 ratio in papyrus growing along the Nile River today is 1.8 times greater than in a papyrus scroll found near the Great Pyramid at Giza. How old is the scroll?

The accuracy of radiocarbon dating can be checked by determining the $^{14}$C content of the annual growth rings of very old trees, such as the bristlecone pines that grow in the American Southwest (Figure 21.7). When scientists plot the radiocarbon ages of these rings against their actual ages obtained by counting rings starting from the outer growth layer of the tree (representing $t = 0$), they find that the two sets of ages do not agree exactly, as shown in Figure 21.8. There are several reasons for this lack of agreement, including variability in the rate of $^{14}$C production due to changing intensity of the cosmic rays striking Earth's upper atmosphere. To assure accurate $^{14}$C results, scientists must correct for these and other variations, and they must be careful to avoid contaminating ancient samples with modern carbonaceous material. With proper analytical technique, radiocarbon dating results are generally accurate to within ±40 years for samples that are 500–50,000 years old.

**FIGURE 21.7** Radiocarbon dating relies on knowing the atmospheric concentration of carbon-14 over time. Ancient living trees, such as the bristlecone pines in the American Southwest, act as a check of the atmospheric carbon-14 levels over thousands of years. The ages of the rings can be determined by counting them, and their carbon-14 content can be determined by mass spectrometry.

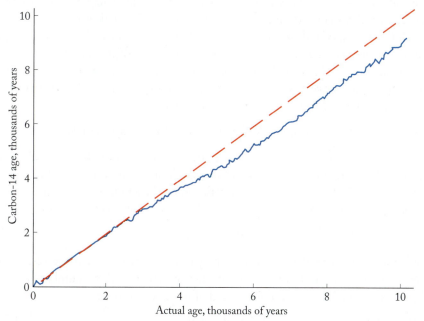

**FIGURE 21.8** Calibration curves for radiocarbon dating allow scientists to accurately calculate the ages of archaeological objects. If the rate of $^{14}$C production in the upper atmosphere were constant, then the age of objects based on their $^{14}$C content would match their actual age—a condition represented by the red dashed line. However, analyses of tree rings, corals, and lake sediments tell us that the rate of $^{14}$C production in the upper atmosphere is variable, so a real plot of $^{14}$C age versus actual age produces the jagged blue line. This plot allows scientists to convert $^{14}$C ages into actual ages.

**CONCEPT TEST** ••••••••••••••••••••••••••••••••••••••••••••••••••••••••••••

How might the increased consumption of fossil fuels over the last century affect the $^{14}$C content of growing plant tissues?

••••••••••••••••••••••••••••••••••••••••••••••••••••••••••••••••••••••••••••

## 21.5 Measuring Radioactivity

French scientist Henri Becquerel (1852–1908) discovered radioactivity in 1896 when he observed that uranium and other substances produce radiation that fogs photographic film. Photographic film is still used to detect radioactivity in the film dosimeter badges worn by people working with radioactive materials to

**scintillation counter** an instrument that determines the level of radioactivity in samples by measuring the intensity of light emitted by phosphors in contact with the samples.

**Geiger counter** a portable device for determining nuclear radiation levels by measuring how much the radiation ionizes the gas in a sealed detector.

**becquerel (Bq)** the SI unit of radioactivity; one becquerel equals one decay event per second.

**curie (Ci)** non-SI unit of radioactivity; $1 \text{ Ci} = 3.70 \times 10^{10}$ decay events per second.

record their exposure to radiation. Detectors called **scintillation counters** use materials called *phosphors* to absorb energy released during radioactive decay. The phosphors then release the absorbed energy as visible light, the intensity of which is a measure of the amount of radiation initially emitted.

Radioactivity can also be measured with many types of **Geiger counters**, which detect the common products of radioactivity—α particles, β particles, and γ rays—based on their abilities to ionize atoms (Figure 21.9). A sealed metal cylinder, filled with gas (usually argon) and a positively charged electrode, has a window that allows α particles, β particles, and γ rays to enter. Once inside the cylinder, these particles ionize argon atoms into $Ar^+$ ions and free electrons. If an electrical potential difference is applied between the cylinder shell and the central electrode, free electrons migrate toward the positive electrode and argon ions migrate toward the negatively charged shell. This ion migration produces a pulse of electrical current whenever radiation enters the cylinder. The current is amplified and read out to a meter and a microphone that makes a clicking sound.

One measure of radioactivity in a sample is the number of decay events per unit time. This parameter is called the radioactivity ($A$) of the sample. The SI unit of radioactivity is the **becquerel (Bq)**, named in honor of Henri Becquerel and equal to one decay event per second. An older radioactivity unit is the **curie (Ci)**, named in honor of Marie and Pierre Curie, where

$$1 \text{ Ci} = 3.70 \times 10^{10} \text{ Bq} = 3.70 \times 10^{10} \text{ decay events/s}$$

Both the becquerel and the curie quantify the *rate* at which a radioactive substance decays, which provides a measure of how much of it is in a sample. As noted in Section 21.3, all decay processes follow first-order kinetics, in which the reaction rate is equal to the rate constant, $k$, times the concentration of a reactant, R:

$$\text{Rate} = k[\text{R}]$$

The same mathematical relationships apply to radioactive decay processes, except that we refer to radioactivity instead of reaction rate and to the number of atoms ($N$) of a radionuclide in a sample instead of its concentration:

$$A = kN \tag{21.5}$$

Because radioactivity is the number of decay events per second, the units of the *decay rate constant ($k$)* are decay events per atom per second.

**FIGURE 21.9** In a Geiger counter, a particle produced by radioactive decay passes through a thin window, usually made of beryllium or a plastic film. Inside the tube, the particle collides with atoms of argon gas and ionizes them. The resulting argon cations migrate toward the negatively charged tube housing, and the electrons migrate toward a positive electrode, creating a pulse of current through the tube. The current pulses are amplified and recorded via a meter and a speaker that produces an audible "click" for each pulse.

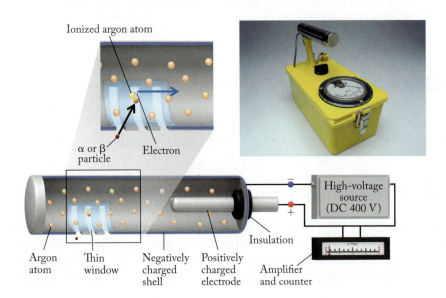

Because radioactive decay is a first-order reaction, the decay rate constant also is related to the half-life of the radionuclide by Equation 14.23:

$$t_{1/2} = \frac{0.693}{k}$$

In the following Sample Exercise, we calculate the radioactivity of a sample based on the quantity and half-life of the radionuclide in the sample.

---

**SAMPLE EXERCISE 21.4    Calculating the Radioactivity    LO5**
**of a Sample**

Radium-223 undergoes β decay with a half-life of 11.4 days. What is the radioactivity of a sample that contains 1.00 μg of $^{223}$Ra? Express your answer in becquerels and in curies.

**Collect and Organize** We are given the half-life and quantity of a radioactive substance and are asked to determine its radioactivity; that is, its rate of decay. The decay rate constant, $k$, is related to half-life by

$$t_{1/2} = \frac{0.693}{k}$$

The radioactivity is the product of the rate constant and the number of atoms of radionuclide in the sample (Equation 21.5, $A = kN$).

**Analyze** Before using Equation 14.23, we must convert the half-life into seconds because both of the radioactivity units we need to calculate are based on decay events per second. To calculate radioactivity, we determine the number of atoms in 1.00 mg of radium. This will likely be a very large number, which should translate into a large number of decay events per second given the relatively short half-life of $^{223}$Ra.

**Solve** The half-life is

$$11.4 \text{ d} \times \frac{24 \text{ hr}}{1 \text{ d}} \times \frac{60 \text{ min}}{1 \text{ hr}} \times \frac{60 \text{ s}}{1 \text{ min}} = 9.85 \times 10^5 \text{ s}$$

Using this value in Equation 14.23 and solving for $k$:

$$k = \frac{0.693}{9.85 \times 10^5 \text{ s}}$$

$$= 7.04 \times 10^{-7} \text{ s}^{-1} = 7.04 \times 10^{-7} \text{ decay events/(atom} \cdot \text{s)}$$

The number of atoms ($N$) of $^{223}$Ra is

$$N = 1.00 \text{ μg} \times \frac{1 \text{ g}}{10^6 \text{ μg}} \times \frac{1 \text{ mol Ra}}{223 \text{ g Ra}} \times \frac{6.022 \times 10^{23} \text{ Ra}}{1 \text{ mol Ra}} = 2.70 \times 10^{15} \text{ atoms Ra}$$

Inserting these values of $k$ and $N$ into Equation 21.5 gives us

$$A = kN = \frac{7.04 \times 10^{-7} \text{ decay events}}{\text{atom} \cdot \text{s}} \times 2.70 \times 10^{15} \text{ atoms Ra} = 1.90 \times 10^9 \text{ decay events/s}$$

Because 1 Bq = 1 decay event/s, the radioactivity of the sample is $1.90 \times 10^9$ Bq. Expressing radioactivity in curies:

$$\frac{1.90 \times 10^9 \text{ decay events}}{\text{s}} \times \frac{1 \text{ Ci}}{3.70 \times 10^{10} \text{ decay events/s}} = 0.0514 \text{ Ci}$$

**Think About It** The large number of decay events per second that we calculated meets our expectation of a high level of radioactivity, given the large number of radioactive atoms in the sample.

⚙ **Practice Exercise** In March 2011, an earthquake and tsunami off the coast of northeast Japan crippled nuclear reactors at a power station in Fukushima, Japan. The resulting explosions and fires released $^{133}$Xe into the atmosphere. Determine the radioactivity in 1.00 μg of this radionuclide ($t_{1/2}$ = 5.25 days) in becquerels and in millicuries.

## 21.6 Biological Effects of Radioactivity

The $\gamma$ rays and many of the $\alpha$ and $\beta$ particles produced by nuclear reactions have more than enough energy to tear chemical bonds apart, producing odd-electron radicals or free electrons and cations. Consequently, these rays and particles are classified as **ionizing radiation**. Other examples include X-rays and short-wavelength ultraviolet rays. The ionization of atoms and molecules in living tissue can lead to radiation sickness, cancer, birth defects, and death. The scientists who first worked with radioactive materials were not aware of these hazards, and some of them suffered for it. Marie Curie died from radiation exposure, and so did her daughter Irène Joliot-Curie, who continued the research program started by her parents.

In medicine, the term *ionizing radiation* is limited to photons and particles that have sufficient energy to remove an electron from water:

$$H_2O(\ell) \xrightarrow{\text{1216 kJ/mol}} H_2O^+(aq) + e^-$$

The logic behind this definition is that the human body is composed largely of water. Therefore water molecules are the most abundant ionizable targets when we are exposed to nuclear radiation. The cation $H_2O^+$ reacts with another water molecule in the body to form a hydronium ion and a hydroxyl free radical:

$$H_2O^+(aq) + H_2O(\ell) \rightarrow H_3O^+(aq) + \cdot OH(aq)$$

The rapid reactions of free radicals with biomolecules can disrupt cell function, sometimes with life-threatening consequences.

Radiation-induced alterations to the biochemical machinery that controls cell growth are most likely to occur in tissues in which cells grow and divide rapidly. One such tissue is bone marrow, where billions of white blood cells are produced each day to fortify the body's immune system. Molecular damage to bone marrow can lead to leukemia: uncontrolled production of nonfunctioning white blood cells that spread throughout the body, crowding out healthy cells. Ionizing radiation can also cause molecular alterations in the genes and chromosomes of sperm and egg cells, increasing the chances of birth defects.

### Radiation Dosage

The biological impact of ionizing radiation depends on how much of it is absorbed by an organism. If the radiation is coming from one radioactive source, then the amount absorbed depends on the radioactivity of the source and the energy of the radiation that is produced per decay event. Tables of radioactive isotopes often include information about their modes of decay and the energies of the particles and gamma rays they emit.

*Absorbed dose* is the quantity of ionizing radiation absorbed by a unit mass of living tissue. The SI unit of absorbed dose is the **gray (Gy)**. One gray is equal to the absorption of 1 J of radiation energy per kilogram of body mass:

$$1\ \text{Gy} = 1\ \text{J/kg}$$

Grays express dosage, but they do not indicate the amount of *tissue damage* caused by that dosage. Different products of nuclear reactions affect living tissue differently. Exposure to 1 Gy of $\gamma$ rays produces about the same amount of tissue damage as exposure to 1 Gy of $\beta$ particles. However, 1 Gy of $\alpha$ particles, which move

**ionizing radiation** high-energy products of radioactive decay that can ionize molecules.

**gray (Gy)** the SI unit of absorbed radiation; 1 Gy = 1 J/kg of tissue.

**relative biological effectiveness (RBE)** a factor that accounts for the differences in physical damage caused by different types of radiation.

**sievert (Sv)** SI unit used to express the amount of biological damage caused by ionizing radiation.

about 10 times slower than β particles but have nearly $10^4$ times the mass, causes up to 20 times as much damage as 1 Gy of γ rays. Neutrons cause 3 to 5 times as much damage. To account for these differences, values of **relative biological effectiveness (RBE)** have been established for the various forms of ionizing radiation (Table 21.3). When absorbed dose in grays is multiplied by an RBE factor, the product is called *effective* dose, a measure of tissue damage. The SI unit of effective dose is the **sievert (Sv)**.

Table 21.4 summarizes the various units used to express quantities of radiation and their biological impact. Two non-SI units are listed that predate their SI counterparts but are still often used. They are *radiation absorbed dose*, or *rad*, which is equivalent to 0.01 Gy, and *rem* for tissue damage, which is an acronym for *roentgen equivalent man*. One rem is the product of one rad of ionization times the appropriate RBE factor. There are 100 rems in 1 Sv.

| TABLE 21.3 | RBE Values of Nuclear Radiation |
|---|---|
| **Radiation** | **RBE** |
| γ Rays | 1.0 |
| β Particles | 1.0–1.5 |
| Neutrons | 3–5 |
| Protons | 10 |
| α Particles | 20 |

**TABLE 21.4  Units for Expressing Quantities of Ionizing Radiation**

| Parameter | SI Unit | Description | Alternative Common Unit | Description |
|---|---|---|---|---|
| Radioactivity | Becquerel (Bq) | 1 decay event/s | Curie (Ci) | $3.70 \times 10^{10}$ decay events/s |
| Ionizing energy absorbed | Gray (Gy) | 1 J/kg of tissue | Rad | 0.01 J/kg of tissue |
| Amount of tissue damage | Sievert (Sv) | 1 Gy × RBE[a] | Rem | 1 rad × RBE[a] |

[a]RBE, relative biological effectiveness.

The RBE of 20 for α particles may lead you to believe that these particles pose the greatest health threat from radioactivity. Not exactly. Alpha particles are so big that they have little penetrating power; they are stopped by a sheet of paper, clothing, or even a layer of dead skin (Figure 21.10). However, if you ingest or inhale an α emitter, tissue damage can be severe because the relatively massive α particles do not have to travel far to cause cell damage. Gamma rays are considered the most dangerous form of radiation emanating from a source outside the body because they have the greatest penetrating power.

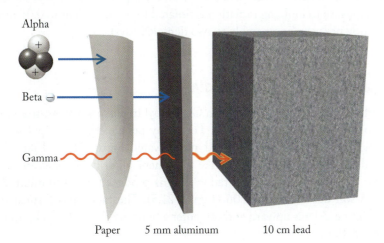

Alpha

Beta

Gamma

Paper     5 mm aluminum     10 cm lead

**FIGURE 21.10** The tissue damage caused by α particles, β particles, and γ rays depends on their ability to penetrate materials that shield the tissues. Alpha particles are stopped by paper or clothing, but are extremely dangerous if formed inside the body because their low penetrating power traps them inside. Stopping gamma rays requires a thick layer of lead or several meters of concrete or soil.

**FIGURE 21.11** The ruins of the nuclear reactor at Chernobyl, Ukraine, that exploded in 1986.

| TABLE 21.5 | **Acute Effects of Single Whole-Body Effective Doses of Ionizing Radiation** |
|---|---|
| **Effective Dose (Sv)** | **Toxic Effect** |
| 0.05–0.25 | No acute effect, possible carcinogenic or mutagenic damage to DNA |
| 0.25–1.0 | Temporary reduction in white blood cell count |
| 1.0–2.0 | Radiation sickness: fatigue, vomiting, diarrhea, impaired immune system |
| 2.0–4.0 | Severe radiation sickness: intestinal bleeding, bone marrow destruction |
| 4.0–10.0 | Death, usually through infection, within weeks |
| >10.0 | Death within hours |

The effects of exposure to different single effective doses of radiation are summarized in Table 21.5. To put these data in perspective, the effective dose from a typical dental X-ray is about 25 μSv, or about 2000 times smaller than the lowest exposure level cited in the table.

Widespread exposure to very high levels of radiation occurred after the 1986 explosion at the Chernobyl nuclear reactor in what is now Ukraine (Figure 21.11). Many plant workers and first responders were exposed to more than 1.0 Sv of radiation. At least 30 of them died in the weeks after the accident. Many of the more than 300,000 workers who cleaned up the area around the reactor exhibited symptoms of radiation sickness, and at least 5 million people in Ukraine, Belarus, and Russia were exposed to fallout in the days following the accident. Studies conducted in the early 1990s uncovered high incidences of thyroid cancer in children in southern Belarus due to $^{131}I$ released in the Chernobyl accident, and children born in the region nearly a decade after the accident had unusually high rates of mutations in their DNA because of their parents' exposure to ionizing radiation. Genetic damage was also widespread among plants and animals living in the region (Figure 21.12).

Radiation exposure was not confined to Ukraine and Belarus. After the accident, a cloud of radioactive material spread rapidly across northern Europe, and within two weeks, increased levels of radioactivity were detected throughout the Northern Hemisphere (Figure 21.13). The accident produced a global increase in human exposure to ionizing radiation estimated to be equivalent to 0.05 mSv per year (50 μSv per year, comparable to two dental X-rays).

(a)

(b)

**FIGURE 21.12** Wildlife surrounding the destroyed nuclear reactor at Chernobyl, Ukraine, was exposed to intense ionizing radiation, which led to deaths and sublethal biological effects such as genetic mutations. One example of the latter is (a) the partially albino barn swallow. (b) A normal swallow has no white feathers directly beneath its beak.

## Evaluating the Risks of Radiation

To put global radiation exposure from Chernobyl in perspective, we need to consider typical annual exposure levels. For many people, the principal source of radiation is radon gas in indoor air and in well water (Figure 21.14). Like all noble gases, radon is chemically inert. Unlike the others, all of its isotopes are radioactive. The most common isotope, radon-222, is produced when uranium-238 in rocks and soil decays to lead-206 (Figure 21.5). The radon gas formed in this decay series percolates upward and can enter a building through cracks and pores in its foundation.

If you breathe radon-contaminated air and then exhale before it decays, no harm is done. However, if radon-222 decays inside the lungs, it emits an α particle that can attack lung tissue. The nuclide produced by the α decay of $^{222}$Rn is radioactive polonium-218, which may become attached to tissue in the respiratory system and undergo a second α decay, forming lead-214:

$$^{222}_{86}\text{Rn} \rightarrow {}^{218}_{84}\text{Po} + {}^{4}_{2}\alpha \qquad t_{1/2} = 3.8 \text{ d}$$

$$^{218}_{84}\text{Po} \rightarrow {}^{214}_{82}\text{Pb} + {}^{4}_{2}\alpha \qquad t_{1/2} = 3.1 \text{ min}$$

As we have seen, α particles are the most damaging product of nuclear decay when formed *inside the body*. How big a threat does radon pose to human health? Concentrations of indoor radon depend on local geology (Figure 21.15) and on how gastight building foundations are. The air in many buildings contains concentrations of radon in the range of 1 pCi per liter of air. How hazardous are such tiny concentrations? There appears to be no simple answer. The U.S. Environmental Protection Agency has established 4 pCi/L as an "action level," meaning that people occupying houses with higher concentrations should take measures to minimize their exposure.

This action level is based on studies of the incidence of lung cancer in workers in uranium mines. These workers are exposed to radon concentrations (and concentrations of other radionuclides) that are much higher than the concentrations in homes and other buildings. However, many scientists believe that people exposed to very low levels of radon for many years are as much at risk as miners exposed to high levels of radiation for shorter periods. Some researchers use a model that assumes a linear relation between radon exposure and incidence of lung cancer. This model is represented by the red line in Figure 21.16, which graphs cancer deaths as a function of radiation absorbed. On the basis of this dose–response model, an estimated 15,000 Americans die of lung cancer each year because of exposure to indoor radon. This number comprises 10% of all lung-cancer fatalities and 30% of those among nonsmokers.

Is this linear model valid? Perhaps—but some scientists believe that there may be a threshold of exposure below which radon poses no significant threat to public health. They advocate an S-shaped dose–response curve, shown by the blue line in

**FIGURE 21.13** Radioactive fallout (shown in pink) from the Chernobyl accident in 1986 was detected throughout the Northern Hemisphere.

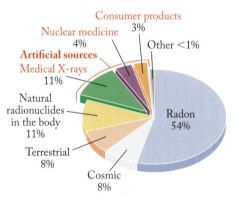

**FIGURE 21.14** Sources of radiation exposure of the U.S. population. On average, a person living in the United States is exposed to 0.0036 Sv of radiation each year. More than 80% of this exposure comes from natural sources, mainly radon in the air and water. Artificial sources account for about 18% of the total exposure.

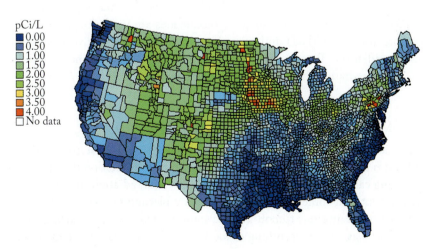

**FIGURE 21.15** Levels of radon gas in soils and rocks across the United States.

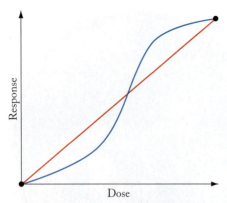

**FIGURE 21.16** The risk of death from radiation-induced cancer may follow one of two models. In the linear response model (red line), risk is directly proportional to the radiation exposure. In the S-shaped model (blue line), risk remains low below a critical threshold and then increases rapidly as the exposure increases. In the S-shaped model, the risk is less than for the linear model at low doses but is higher at higher doses.

Figure 21.16. Notice that at low radiation exposure the risk of death from cancer in the S-shaped curve is much lower than in the linear response model but rises rapidly above a critical value.

---

**SAMPLE EXERCISE 21.5  Calculating Effective Dose    LO6**

It has been estimated that a person living in a home where the air radon concentration is 4.0 pCi/L receives an annual absorbed dose of ionizing radiation equivalent to 0.40 mGy. What is the person's annual effective dose, in millisieverts, from this radon? Use information from Figure 21.14 to compare this annual effective dose of radon with the average annual effective dose of radon estimated for persons living in the United States.

**Collect and Organize** We are given an absorbed radiation dose of 0.40 mGy. Radon isotopes emit α particles, which we know from Table 21.3 have a relative biological effectiveness of 20.

**Analyze** The effective dose caused by an absorbed dose of ionizing radiation is the absorbed dose multiplied by the RBE of the radiation.

**Solve**

$$0.40 \text{ mGy} \times 20 = 8.0 \text{ mSv}$$

Figure 21.14 tells us that the average American is exposed to 3.6 mSv of radiation per year, with 54% of that amount, or 1.9 mSv, from radon. A person living in the home described in the problem has an effective dose slightly more than four times the average value.

**Think About It** The calculated value is more than twice the average annual effective dose of 3.6 mSv from all sources of radiation. The U.S. National Research Council has estimated that a nonsmoker living in air contaminated with 4.0 pCi/L of radon has a 1% chance of dying from lung cancer due to this exposure. The cancer risk for a smoker is close to 5%.

⚙ **Practice Exercise** A dental X-ray for imaging impacted wisdom teeth produces an effective dose of 10 μSv. If a dental X-ray machine emits X-rays with an energy of $6.0 \times 10^{-17}$ J each, how many of these X-rays must be absorbed per kilogram of tissue to produce an effective dose of 10 μSv? Assume the RBE of these X-rays is 1.2.

---

# 21.7 Medical Applications of Radionuclides

Radionuclides are used in both the detection and the treatment of diseases: they are key agents in the respective medical fields of *diagnostic radiology* and *therapeutic radiology*. In diagnostic radiology, radionuclides are used alongside magnetic resonance imaging (MRI) and other imaging systems that involve only nonionizing radiation. Therapeutic radiology, however, is based almost entirely on the ionizing radiation that comes from nuclear processes.

## Therapeutic Radiology

Because ionizing radiation causes the most damage to cells that grow and divide rapidly, it is a powerful tool in the fight *against* cancer. Radiation therapy consists of exposing cancerous tissue to γ radiation. Often the radiation source is external to the patient, but sometimes it is encased in a platinum capsule and surgically implanted in a cancerous tumor. The platinum provides a chemically inert outer layer and acts as a filter, absorbing α and β particles emitted by the radionuclide but allowing γ rays to pass into the tumor.

A nuclide's chemical properties can sometimes be exploited to direct it to a tumor site. For example, most iodine in the body is concentrated in the thyroid gland, so an effective therapy against thyroid cancer starts with ingestion of potassium iodide containing radioactive iodine-123. Other radionuclides used in cancer therapy are listed in Table 21.6.

Surgically inaccessible tumors can be treated with beams of γ rays from a radiation source outside the body. Unfortunately, γ radiation destroys both cancer cells and healthy ones. Thus, patients receiving radiation therapy frequently suffer symptoms of radiation sickness, including nausea and vomiting (the tissues that make up intestinal walls are especially susceptible to radiation-induced damage), and hair loss in the treatment area. To reduce the severity of these side effects, radiologists must carefully control the dosage a patient receives.

| TABLE 21.6 | Some Radionuclides Used in Radiation Therapy | | |
|---|---|---|---|
| Nuclide | Radiation | Half-Life | Treatment |
| $^{32}P$ | β | 14.3 d | Leukemia therapy |
| $^{60}Co$ | β, γ | 5.27 yr | Cancer therapy |
| $^{131}I$ | β | 8.1 d | Thyroid therapy |
| $^{131}Cs$ | γ | 9.7 d | Prostate cancer therapy |
| $^{192}Ir$ | β, γ | 74 d | Coronary disease |

## Diagnostic Radiology

The movement of radionuclides in the body and their accumulation in certain organs provide ways to assess organ function. A tiny quantity of a radioactive isotope is used, together with a much larger amount of a stable isotope of the same element. The radioactive isotope is called a *tracer*, and the stable isotope is the *carrier*. For example, the circulatory system can be imaged by injecting into the blood a solution of sodium chloride containing a trace amount of $^{24}NaCl$. Circulation is monitored by measuring the γ rays emitted by $^{24}Na$ as it decays.

The ideal tracer for medical imaging is one that has a half-life about equal to the length of time required to perform the imaging measurements. It should emit moderate-energy γ rays but no α particles or β particles that might cause tissue damage. Sodium-24 (a γ emitter with a half-life of 15 hours) meets both these criteria. Table 21.7 lists several other radionuclides that are used in medical imaging.

*Positron emission tomography* (PET) is a powerful tool for diagnosing organ and cell function. PET uses short-lived, neutron-poor, positron-emitting radionuclides such as carbon-11, oxygen-15, and fluorine-18. A patient might be administered a solution of glucose in which some of the sugar molecules contain atoms of $^{11}C$, $^{15}O$, or $^{18}F$. The rate at which glucose is metabolized in various regions of the brain is monitored by detecting the γ rays produced by positron–electron annihilations. Unusual patterns in PET images of brains (Figure 21.17) can indicate schizophrenia, manic depression, Alzheimer's disease, damage from strokes, and even nicotine addiction in tobacco smokers.

| TABLE 21.7 | Selected Radionuclides Used for Medical Imaging | | |
|---|---|---|---|
| Nuclide | Radiation | Half-Life (hr) | Use |
| $^{99m}Tc$ | γ | 6.0 | Bones, circulatory system, various organs |
| $^{67}Ga$ | γ | 78 | Tumors in the brain and other organs |
| $^{201}Tl$ | γ | 73 | Coronary arteries, heart muscle |
| $^{123}I$ | γ | 13.2 | Thyroxine production in thyroid gland |

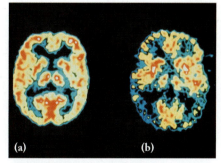

(a)     (b)

**FIGURE 21.17** Positron emission tomography (PET) is used to monitor cell activity in organs such as the brain. (a) Brain function in a healthy person. The red and yellow regions indicate high brain activity; blue and black indicate low activity. (b) Brain function in a patient suffering from Alzheimer's disease.

## 21.8 Nuclear Fission

When an atom of uranium-235 captures a neutron, the nucleus of the unstable product, uranium-236, splits into two lighter nuclei in a process called **nuclear fission**. Several uranium-235 fission reactions can occur, including these three:

$$^{235}_{92}U + {}^{1}_{0}n \rightarrow {}^{141}_{56}Ba + {}^{92}_{36}Kr + 3\,{}^{1}_{0}n$$

$$^{235}_{92}U + {}^{1}_{0}n \rightarrow {}^{137}_{52}Te + {}^{97}_{40}Zr + 2\,{}^{1}_{0}n$$

$$^{235}_{92}U + {}^{1}_{0}n \rightarrow {}^{138}_{55}Cs + {}^{96}_{37}Rb + 2\,{}^{1}_{0}n$$

**nuclear fission** a nuclear reaction in which the nucleus of an element splits into two lighter nuclei. The process is usually accompanied by the release of one or more neutrons and energy.

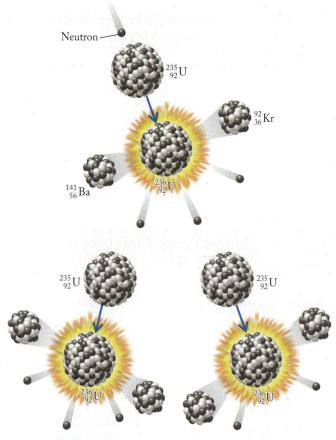

In all these reactions, the sums of the masses of the products are slightly less than the sums of the masses of the reactants. As we observed for nucleosynthesis in Section 21.1, this decrease in mass is released as energy in accordance with Einstein's equation ($E = mc^2$).

These reactions also produce additional neutrons, which can smash into other uranium-235 nuclei and initiate more fission events in a **chain reaction** (Figure 21.18). The reaction proceeds as long as there are enough uranium-235 nuclei present to absorb the neutrons being produced. On average, at least one neutron from each fission event must cause another nucleus to split apart for the chain reaction to be self-sustaining. The quantity of fissionable material needed to assure that every fission event produces another is called the **critical mass**. For uranium-235, the critical mass is about 1 kg of the pure isotope.

Uranium-235 is the most abundant fissionable isotope, but it makes up only 0.72% of the uranium in the principal uranium ore, pitchblende (Figure 21.19a). The uranium in nuclear reactors must be at least 3% to 4% uranium-235, and enrichment to about 85% is needed for nuclear weapons. The most common method for enriching uranium ore involves extracting the uranium in a process that yields a material called yellowcake, which is mostly $U_3O_8$ (Figure 21.19b). This oxide is then converted to $UF_6$, which, despite a molar mass of over 300 grams per mole, is a volatile solid that sublimes at 56°C.

**FIGURE 21.18** Each fission event in the chain reaction of a uranium-235 nucleus begins when the nucleus captures a neutron, forming an unstable uranium-236 nucleus that then splits apart (fissions) in one of several ways. In the first process shown here, the uranium-236 nucleus splits into krypton-92, barium-141, and three neutrons. If, on average, at least one of the three neutrons from each fission event causes the fission of another uranium-235 nucleus, then the process is sustained in a chain reaction.

(a)

(c)

**FIGURE 21.19** Preparing uranium fuel. (a) A piece of pitchblende, source of the uranium fuel for nuclear reactors. Pitchblende ore is ground up and extracted with strong acid. (b) The uranium compounds (mostly $U_3O_8$) obtained from the extract are called yellowcake. (c) Uranium oxides are converted to volatile $UF_6$, which is centrifuged at very high speed to separate $^{235}UF_6$ from $^{238}UF_6$. The less dense and less abundant $^{235}UF_6$ is enriched near the center of the centrifuge cylinder and separated from the heavier $^{238}UF_6$.

(b)

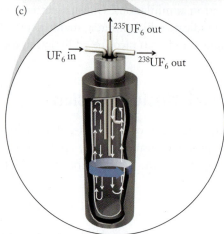

The volatility of this nonpolar molecular compound can be explained by the relatively weak London dispersion forces experienced by its compact, symmetrical molecules (Figure 21.20). Fissionable $^{235}UF_6$ is separated from $^{238}UF_6$ based on their slightly different densities. Elaborate centrifuge systems are used to exploit this difference and speed up the separation (Figure 21.19c).

Harnessing the energy released by nuclear fission to generate electricity began in the middle of the 20th century. In a typical nuclear power plant (Figure 21.21), fuel rods containing 3% to 4% uranium-235 are interspersed with rods of boron or cadmium that control the rate of the chain reaction by absorbing some of the neutrons produced during fission. Pressurized water flows around the fuel and control rods, removing the heat created during fission and transferring it to a steam generator. The water also acts as a moderator, slowing down the neutrons and thereby allowing for their more efficient capture by $^{235}U$ atoms.

In 1952 the first **breeder reactor** was built, so called because in addition to producing energy to make electricity, the reactor makes ("breeds") its own fuel. The reactor starts out with a mixture of plutonium-239 and uranium-238. As the plutonium fissions and the energy from those reactions is collected to produce electricity, some of the neutrons that are produced sustain the fission chain reaction just as in the reactor of Figure 21.21, while others convert the uranium into more plutonium fuel:

$$^{238}_{92}U + {}^{1}_{0}n \rightarrow {}^{239}_{92}U + \gamma \rightarrow {}^{239}_{94}Pu + 2\,{}^{0}_{-1}\beta$$

In less than 10 years of operation, a breeder reactor can make enough plutonium-239 to refuel itself *and* another reactor. Unfortunately, plutonium-239 is a carcinogen and one of the most toxic substances known. Only about half a kilogram is needed to make an atomic bomb, and it has a long half-life: $2.4 \times 10^4$ years.

**FIGURE 21.20** Uranium hexafluoride is a volatile solid that sublimes at only 56°C because it is composed of compact, symmetrical molecules that experience relatively weak London dispersion forces despite their considerable mass.

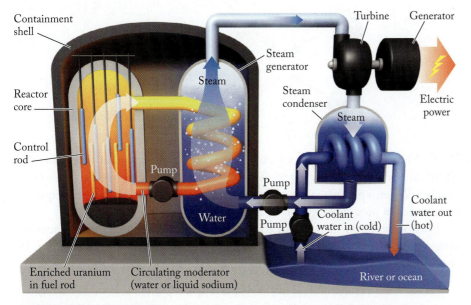

**FIGURE 21.21** A pressurized, water-cooled nuclear power plant uses fuel rods containing uranium enriched to about 4% uranium-235. The fission chain reaction is regulated with control rods and a moderator that is either water or liquid sodium. The moderator slows down the neutrons released by fission so that they can be more efficiently captured by other uranium-235 nuclei. It also transfers the heat produced by the fission reaction to a steam generator. The steam generated by this heat drives a turbine that generates electricity. Nuclear power plants are usually situated near coasts or large rivers so large amounts of coolant water are readily available.

**chain reaction** a self-sustaining series of fission reactions in which the neutrons released when nuclei split apart initiate additional fission events and sustain the reaction.

**critical mass** the minimum quantity of fissionable material needed to sustain a chain reaction.

**breeder reactor** a nuclear reactor in which fissionable material is produced during normal reactor operation.

**nuclear fusion** a nuclear reaction in which subatomic particles or atomic nuclei collide with each other at very high speeds and fuse together, forming more massive nuclei and releasing energy.

▶❚❚ **CHEMTOUR** Fusion of Hydrogen

Understandably, extreme caution and tight security surround the handling of plutonium fuel and the transportation and storage of nuclear wastes containing even small amounts of plutonium. Health and safety matters related to reactor operation and spent-fuel disposal are the principal reasons there are no breeder power stations in the United States, although they have been built in at least seven other countries.

## 21.9 Nuclear Fusion and the Quest for Clean Energy

The energy of the Sun is derived from the high-speed collision and combining of hydrogen nuclei to form helium. This **nuclear fusion** process involves more steps than the process that probably took place during primordial nucleosynthesis (see Chapter 2), when protons (hydrogen nuclei) and neutrons fused together to form deuterons (nuclei of the hydrogen isotope deuterium, D):

$$\,^{1}_{1}\text{H} + \,^{1}_{0}\text{n} \rightarrow \,^{2}_{1}\text{D} \tag{21.6}$$

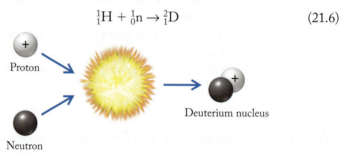
Proton / Neutron / Deuterium nucleus

Once deuterons formed, they also collided with each other and fused together, forming α particles, which are the nuclei of helium-4 atoms:

$$2\,^{2}_{1}\text{D} \rightarrow \,^{4}_{2}\text{He} \tag{21.7}$$

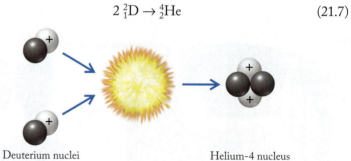
Deuterium nuclei                    Helium-4 nucleus

⊙⊙ **CONNECTION** We learned how to write and balance nuclear equations in Section 2.6.

Recall from Chapter 2 that each subscript in a nuclear equation represents the electrical charge of a particle and each superscript represents the particle's mass number. When a particle is the nucleus of an atom, its electrical charge is equal to the number of protons in it; that is, its atomic number.

Hydrogen fusion in our Sun follows a different path because free neutron concentrations are far lower there than they were in the primordial universe. In the Sun, colliding protons may fuse together to form a deuteron and a positron:

$$2\,^{1}_{1}\text{H} \rightarrow \,^{2}_{1}\text{D} + \,^{0}_{1}\beta \tag{21.8}$$

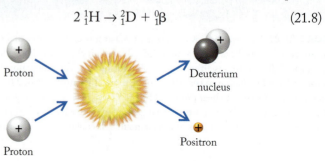
Proton / Proton / Deuterium nucleus / Positron

In the second stage of solar fusion, protons fuse with deuterons to form helium-3 nuclei:

$$^{1}_{1}H + ^{2}_{1}D \rightarrow ^{3}_{2}He \tag{21.9}$$

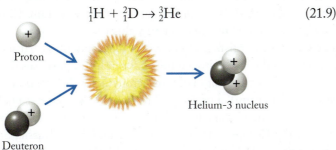

The superscript 3 indicates that the particle has 3 nucleons (2 protons and 1 neutron). Recall from Chapter 2 that any atom with 2 protons in its nucleus is by definition a helium atom, and that atoms of the same element with different numbers of nucleons are called isotopes. Finally, fusion of two helium-3 nuclei produces a helium-4 nucleus and 2 protons:

$$2\,^{3}_{2}He \rightarrow ^{4}_{2}He + 2\,^{1}_{1}H \tag{21.10}$$

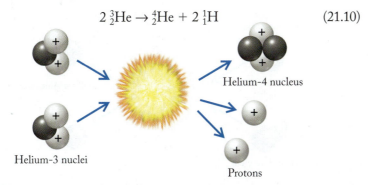

Deuterium and $^{3}He$ nuclei are intermediates in the hydrogen-fusion process because they are made in one step but then consumed in another. To write an overall equation for solar fusion, we combine Equations 21.8, 21.9, and 21.10, multiplying Equations 21.8 and 21.9 by 2 to balance the production and consumption of the intermediate particles:

$$2\,[2\,^{1}_{1}H \rightarrow ^{2}_{1}D + ^{0}_{1}\beta]$$

$$+\,2\,[^{1}_{1}H + ^{2}_{1}D \rightarrow ^{3}_{2}He]$$

$$+\qquad 2\,^{3}_{2}He \rightarrow ^{4}_{2}He + 2\,^{1}_{1}H$$

$$\overline{4\,6\,^{1}_{1}H + 2\,^{2}_{1}D + 2\,^{3}_{2}He \rightarrow 2\,^{2}_{1}D + 2\,^{3}_{2}He + ^{4}_{2}He + 2\,^{0}_{1}\beta + 2\,^{1}_{1}H}$$

This equation reduces to:

$$4\,^{1}_{1}H \rightarrow ^{4}_{2}He + 2\,^{0}_{1}\beta \tag{21.11}$$

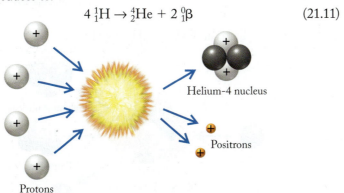

Annihilation reactions between the positrons produced in Equation 21.11 and electrons in the matter surrounding the reactants release considerable energy,

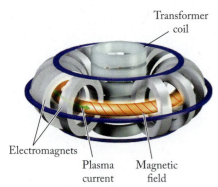

**FIGURE 21.22** A tokamak transmits electrical energy into a toroidal (donut-shaped) chamber containing deuterium and tritium, causing these isotopes of hydrogen to ionize and form a plasma of nuclei and free electrons with a temperature above $10^8$ K. Combinations of electromagnets confine the plasma to the interior of the torus, where collisions between $^2$H and $^3$H nuclei result in their fusing together, forming $^4$He nuclei and free neutrons. The neutrons then collide with Li atoms in the walls of the chamber, initiating additional nuclear reactions that produce more tritium fuel.

but most of the energy from hydrogen fusion comes from the loss in mass as four protons are transformed into an α particle (that is, a helium-4 nucleus) and two positrons (Table 21.1). In Sample Exercise 21.6 we will use Einstein's equation, $E = mc^2$, to calculate how much energy this is.

For decades scientists and engineers have sought to harness the enormous energy released during hydrogen fusion for peaceful purposes. In 2009, construction began on ITER (originally an acronym for International Thermonuclear Experimental Reactor), a project to build the world's largest nuclear fusion reactor. Located at the Cadarache facility in southern France, the project is funded and run by the European Union, India, Japan, China, Russia, South Korea, and the United States. When it is operational, ITER will use a device called a *tokamak* (Figure 21.22) to heat a mixture of deuterium ($^2_1$H) and tritium ($^3_1$H) to temperatures near $1.5 \times 10^8$ K. At such temperatures, all these atoms are ionized, forming an incandescent plasma that is confined by the tokamak's powerful magnets to the center of a donut-shaped tunnel. High-speed collisions between deuterium and tritium nuclei in the plasma produce nuclei of helium-4:

$$^2_1\text{H} + ^3_1\text{H} \rightarrow ^4_2\text{He} + ^1_0\text{n} \qquad (21.12)$$

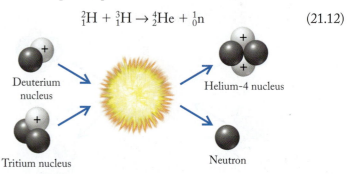

The neutrons produced in the reaction collide with the nuclei of Li atoms in "breeder" blankets surrounding the hydrogen plasma. Two nuclear reactions are initiated by these collisions, depending on which isotope of Li is involved:

$$^1_0\text{n} + ^6_3\text{Li} \rightarrow ^4_2\text{He} + ^3_1\text{H} \qquad (21.13)$$

$$^1_0\text{n} + ^7_3\text{Li} \rightarrow ^4_2\text{He} + ^3_1\text{H} + ^1_0\text{n} \qquad (21.14)$$

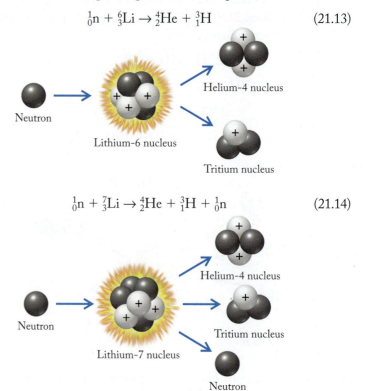

Note that the reactions in Equations 21.13 and 21.14 produce tritium nuclei. In this way the reactions supply more fuel for the primary fusion reaction. The world's supply of deuterium, the other fuel, is enormous (seawater contains about 15 mg of deuterium per kilogram). On the other hand, tritium is not abundant in nature because it is radioactive with a half-life of only 12.3 years. One disadvantage of these reactions is their reliance on lithium during a time when expanding production of lithium-ion batteries (see Section 19.7) is increasing our demand for the element.

**CONCEPT TEST** ....................................................................

Nuclear reactors powered by the energy released by the fission of uranium-235 have been operating since the 1950s, but a reactor powered by the energy released by the fusion of hydrogen has yet to be built. Why is it taking so long to build a fusion reactor?

....................................................................

**SAMPLE EXERCISE 21.6** **Calculating the Energy Released** **LO7**
**in a Nuclear Reaction**

How much energy in joules is released by the overall fusion process in which four protons undergo nuclear fusion, producing an $\alpha$ particle and two positrons (Equation 21.11)?

**Collect and Organize** We are asked to calculate the energy released in the nuclear reaction:

$$4\,{}^{1}_{1}\text{H} \rightarrow {}^{4}_{2}\text{He} + 2\,{}^{0}_{1}\beta \tag{21.11}$$

The energies associated with nuclear reactions are related to differences in the masses of the reactant and product particles and Einstein's equation, $E = mc^2$. The masses of the particles in Table 21.1 are given in kilograms, which is convenient because the relationship between energy and mass is linked to the unit conversion

$$1\,\text{J} = 1\,\text{kg} \cdot (\text{m/s})^2$$

**Analyze** Given the value of the masses in Table 21.1, the difference in mass will probably be less than $10^{-27}$ kg. When multiplied by the square of the speed of light, $(2.998 \times 10^8\,\text{m/s})^2 \approx 10^{17}$, the calculated value of $E$ should be less than $10^{-10}$ J.

**Solve** First we calculate the change in mass:

$$\Delta m = (m_{\alpha\,\text{particle}} + 2\,m_{\text{positron}}) - 4\,m_{\text{proton}}$$
$$= [6.64465 \times 10^{-27} + (2 \times 9.10938 \times 10^{-31})]\,\text{kg} - (4 \times 1.67262 \times 10^{-27})\,\text{kg}$$
$$= -4.40081 \times 10^{-29}\,\text{kg}$$

The energy corresponding to this loss in mass is calculated using Einstein's equation where $m = -4.40081 \times 10^{-29}$ kg:

$$E = mc^2$$
$$= -4.40081 \times 10^{-29}\,\text{kg} \times (2.998 \times 10^8\,\text{m/s})^2$$
$$= -3.955 \times 10^{-12}\,\text{kg} \cdot (\text{m/s})^2 = -3.955 \times 10^{-12}\,\text{J}$$

**Think About It** The decrease in mass translates into energy lost by the reaction system to its surroundings. As we predicted, the absolute value of this energy is less (actually much less) than $10^{-10}$ J, which seems like an awfully small value compared to the world's energy needs. However, this value applies to the formation of a single $\alpha$ particle. If we multiply it by Avogadro's number and convert it to a value in kilojoules per mole, a unit we typically use in thermochemistry, we get

$$\frac{-3.955 \times 10^{-12}\,\text{J}}{\alpha\text{-particle}} \times \frac{6.022 \times 10^{23}\,\alpha\text{-particles}}{\text{mol}} \times \frac{1\,\text{kJ}}{1000\,\text{J}} = -2.382 \times 10^9\,\text{kJ/mol}$$

To put this value in perspective, it is about $10^7$ times the change in free energy from the combustion of one mole of hydrogen gas.

⚙ **Practice Exercise** How much energy is released in the nuclear reaction described by Equation 21.12? Express your answer in kJ/mol. (*Note:* The mass of a tritium nucleus is $5.00827 \times 10^{-27}$ kg.)

## SAMPLE EXERCISE 21.7 Integrating Concepts: Radium Girls and Safety in the Workplace

Radium was discovered by Pierre and Marie Curie in 1898. By 1902, the new element had its first practical use: radium compounds were mixed with zinc sulfide (ZnS) to make paint that glowed in the dark as α particles emitted by the decay of $^{226}$Ra ($t_{1/2}$ = 1600 years) caused ZnS crystals to emit a greenish fluorescence (Figure 21.23). The paint was used to make dials for watches, clocks, and instruments used on naval vessels and, a few years later, in military and civilian airplanes.

By 1914, U.S. companies were making radium-painted dials and employing young women in their late teens and early 20s as dial painters. Soon after their employment, many of the women became very sick, suffering from anemia and other symptoms we now associate with exposure to nuclear radiation. Some developed bone cancer and over one hundred of them, who became known around the world as the Radium Girls, died. Their deaths were linked to the practice of "pointing" the fine paint brushes they used, which meant using their lips to make fine points on the brushes to help them paint the tiny numerals and hands on watch faces (Figure 21.24). In the process, they ingested some of the radioactive paint. Tests later determined that about 20% of the radium ingested was incorporated into their bones, where it attacked bone marrow and caused malignancies known as osteosarcomas.

a. Suggest a reason why radium was concentrated in the victims' bones.
b. Studies of radiation levels and incidence of cancer in over 1000 female dial painters yielded the results in the table below. Which of the two dose–response curves in Figure 21.16 best fits these results?

| Radium Exposure (mg ingested) | Occurrence of Malignancy (% of workers exposed) |
|---|---|
| 1 | 0 |
| 3 | 0 |
| 10 | 0 |
| 30 | 0 |
| 100 | 5 |
| 300 | 53 |
| 1000 | 85 |

c. The green luminescence of radium watch dials began to fade after a few years. Was this loss in luminosity due to decreased radioactivity in the paint? Explain why or why not.
d. In one study, the levels of radioactivity in pocket watches with radium-painted dials were found to be between 0.6 to 1.39 μCi per watch. How many micrograms of $^{226}$Ra produce 1.39 μCi of radioactivity?

**Collect and Organize** We are asked (a) why ingested $^{226}$Ra concentrates in bones; (b) whether malignancy in dial painters was proportional to their exposure to $^{226}$Ra radiation or followed an S-shaped

dose–response curve; (c) why the luminosity of radium-activated paint fades after a few years; and (d) how many micrograms of $^{226}$Ra are needed to produce 1.39 mCi of radiation. One curie (Ci) is equal to $3.70 \times 10^{10}$ decay events/s. The half-life ($t_{1/2}$) of $^{226}$Ra is 1600 years and is related to the first-order rate constant ($k$) of the decay reaction by the equation

$$t_{1/2} = 0.693/k$$

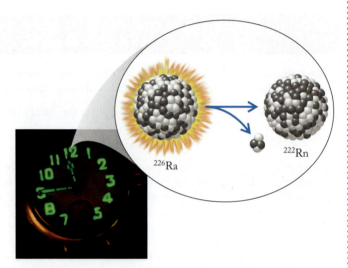

**FIGURE 21.23** During the 20th century, many millions of watches and clocks had dials that glowed in the dark as high-energy α particles emitted by $^{226}$Ra caused crystals of ZnS to fluoresce.

**FIGURE 21.24** This editorial cartoon appeared in Sunday newspapers on February 28, 1926. It portrayed the deadly consequences of young women "pointing" their brushes with their lips as they painted the dials of watches with paint that contained radioactive radium.

The level of radioactivity ($A$) in a sample of radium is equal to the product of the rate constant and the number ($N$) of $^{226}$Ra atoms: $A = kN$.

**Analyze** Radium is a group-2 element and should have chemical and biochemical properties that are similar to those of the other elements in that group, including its association with biological tissues. High concentrations of another group-2 element, calcium, occur in teeth and bones. Worker exposure levels in the above table cover a wide range, but exposure up to nearly 100 mg $^{226}$Ra caused few malignancies, whereas concentrations above 100 mg caused many. The half-life of $^{226}$Ra is 1600 years, so the radioactivity of a sample decreases little over a few years or even over many decades. Relating a half-life expressed in years to a level of radioactivity expressed in a multiple of decay events per second requires converting units of time and then quantities of radioactive atoms to moles and then micrograms.

**Solve**

a. Radium likely accumulates in bones because its chemistry is similar to that of calcium, which means that $^{226}$Ra$^{2+}$ ions are likely to take the place of Ca$^{2+}$ ions in bone tissue.

b. Malignancies did not occur among the dial painters who ingested less than 100 mg of $^{226}$Ra; however, the percentage of the women who suffered malignancies increased sharply with exposure between 100 and 1000 mg. This pattern is described by the S-shaped (blue) curve in Figure 21.16.

c. Given the 1600-year half-life of $^{226}$Ra, the loss of watch dial luminescence was not the result of depleted radioactivity. Rather, it must have been due to less efficient conversion of the energy of radioactive decay into visible light by ZnS crystals.

d. Let's first convert the half-life of $^{226}$Ra into a decay rate constant in units of s$^{-1}$:

$$k = \frac{0.693}{t_{1/2}} = \frac{0.693}{1600 \text{ yr}} \times \frac{1 \text{ yr}}{365.25 \text{ d}} \times \frac{1 \text{ d}}{24 \text{ hr}} \times \frac{1 \text{ hr}}{3600 \text{ s}}$$

$$= 1.372 \times 10^{-11} \text{ s}^{-1}$$

Next, we solve the equation $A = kN$ for $N$, and we use the above rate constant and the radioactivity of the watch to calculate the number of $^{226}$Ra atoms in the dial:

$$N = \frac{A}{k} = \frac{1.39 \text{ μCi}}{1.372 \times 10^{-11} \text{ s}^{-1}} \times \frac{1 \text{ Ci}}{10^6 \text{ μCi}}$$

$$\times \frac{3.70 \times 10^{10} \text{ atoms Ra s}^{-1}}{\text{Ci}} = 3.75 \times 10^{15} \text{ atoms Ra}$$

The corresponding mass in μg is

$$3.75 \times 10^{15} \text{ atoms Ra} \times \frac{1 \text{ mol Ra}}{6.022 \times 10^{23} \text{ atoms Ra}}$$

$$\times \frac{226 \text{ g Ra}}{1 \text{ mol Ra}} \times \frac{10^6 \text{ μg}}{1 \text{ g}} = 1.41 \text{ μg Ra}$$

**Think About It** Did you notice the similarity in the level of radioactivity (1.39 μCi) in the watch dial and the mass of radium (1.41 μg) producing it? This is not a coincidence. When the curie was adopted as the standard unit of radioactivity in the early 20th century, it was chosen to honor the pioneering work of Marie and Pierre Curie, and it was based on what was then believed to be the level of radioactivity in one gram of radium. Newspaper articles published in the 1920s made clear that Marie Curie was deeply troubled by the tragedy of the Radium Girls. Sadly, in 1934 she herself died from aplastic anemia—a disease caused by the inability of bone marrow to produce red blood cells.

## SUMMARY

**Learning Outcome 1** Nuclear chemistry is the study and application of reactions that involve changes in atomic nuclei. The **mass defect ($\Delta m$)** of a nucleus is the difference between its mass and the sum of the masses of its nucleons. **Binding energy (BE)** is the energy released when the nucleons combine to form a nucleus. It is also the energy needed to split the nucleus into its nucleons. Binding energy per nucleon is a measure of the relative stability of a nucleus. When a particle of matter encounters a particle of **antimatter**, both are converted into energy (they annihilate one another), yielding γ rays. (Section 21.1)

**Learning Outcome 2** Stable nuclei have neutron-to-proton ratios that fall within a range of values called the **belt of stability**. Unstable nuclides undergo **radioactive decay**. Neutron-rich nuclides (mass number greater than the average atomic mass) undergo **β decay**; neutron-poor nuclides undergo **positron emission** or **electron capture**. Very large nuclides ($Z > 83$) may undergo β decay or **α decay**. (Section 21.2)

**Learning Outcome 3** Radioactive decay follows first-order kinetics, so the half-life ($t_{1/2}$) of a radionuclide is a characteristic value of the decay process. (Section 21.3)

**Learning Outcome 4** **Radiometric dating** is used to determine the age of an object based on its content of a radionuclide and/or its decay product. **Radiocarbon dating** involves measuring the amount of radioactive carbon-14 that remains in an object derived from plant or animal tissue to calculate the age of the object. The accuracy of the technique relies on calibration of the data with results of radiometric analyses of samples of known age, such as the trunks of trees that have lived for thousands of years and whose age can be confirmed by their growth rings. (Section 21.4)

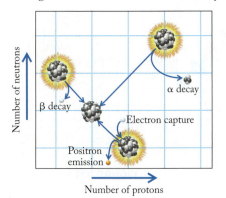

**Learning Outcome 5** **Scintillation counters** and **Geiger counters** are used to measure levels of nuclear radiation. Radioactivity is the number of decay events per unit time. Common units are the **becquerel** (**Bq**; 1 decay event/s) and the **curie** (**Ci**; 1 Ci = $3.70 \times 10^{10}$ Bq). (Section 21.5)

**Learning Outcome 6** Alpha particles, β particles, and γ rays have enough energy to break up molecules into electrons and cations and are examples of **ionizing radiation**, which can damage body tissue and DNA. The quantity of ionizing radiation energy absorbed per kilogram of body mass is called the *absorbed dose* and is expressed in **grays** (**Gy**; 1 Gy = 1.00 J/kg). The effective dose of any type of ionizing radiation is the product of the absorbed dose in grays and the **relative biological effectiveness (RBE)** of the radiation; the unit of effective dose is the **sievert (Sv)**. Alpha particles have a larger RBE than β particles and γ rays but have the least penetrating power of these three types of ionizing radiation. Selected radioactive isotopes are useful as tracers in the human body to map biological activity

and diagnose diseases. Other radioactive isotopes are used to treat cancers. (Sections 21.6, 21.7)

**Learning Outcome 7** Neutron absorption by uranium-235 and a few other massive isotopes may lead to **nuclear fission**, creating lighter nuclei, accompanied by the release of energy that can be harnessed to generate electricity. A **chain reaction** happens when the neutrons released during fission collide with other fissionable nuclei. They require a **critical mass** of a fissionable isotope. A **breeder reactor** is used to make plutonium-239 from uranium-238, while also producing energy to make electricity. **Nuclear fusion** occurs when subatomic particles or atomic nuclei collide with each other and fuse together. Facilities that harness the enormous energy of hydrogen fusion must operate at temperatures over $10^8$ K. (Sections 21.8, 21.9)

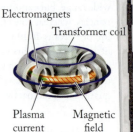

Electromagnets
Transformer coil
Plasma current
Magnetic field

## PROBLEM-SOLVING SUMMARY

| TYPE OF PROBLEM | CONCEPTS AND EQUATIONS | SAMPLE EXERCISES |
|---|---|---|
| **Predicting the modes and products of radioactive decay** | Neutron-rich nuclides tend to undergo β decay; neutron-poor nuclides undergo positron emission or electron capture. | 21.1 |
| **Calculations involving half-lives and radiocarbon dating** | $$\frac{N_t}{N_0} = 0.5^{t/t_{1/2}} \qquad (21.3)$$ $$t = -\frac{t_{1/2}}{0.693}\ln\frac{N_t}{N_0} \qquad (21.4)$$ where $N_t/N_0$ is the ratio of the quantity of radionuclide present in a sample at time $t$ ($N_t$) to the quantity at $t = 0$ ($N_0$) and is the half-life of the radionuclide. | 21.2, 21.3 |
| **Calculating the radioactivity of a sample** | $$A = kN \qquad (21.5)$$ where $$k = \frac{0.693}{t_{1/2}}$$ | 21.4 |
| **Calculating effective dose** | Effective dose = absorbed dose × RBE | 21.5 |
| **Calculating the energy released in a nuclear reaction** | $E = mc^2$ where $m$ is the loss in mass as reactants form products. | 21.6 |

## VISUAL PROBLEMS

*(Answers to boldface end-of-chapter questions and problems are in the back of the book.)*

**21.1.** Which of the highlighted elements in Figure P21.1 currently plays a key role in the controlled fusion of hydrogen?

**21.2.** Exposure to which of the highlighted elements in Figure P21.1 could cause anemia and bone disease?

**21.3.** Which of the highlighted elements in Figure P21.1 is produced by the decay of uranium-238?

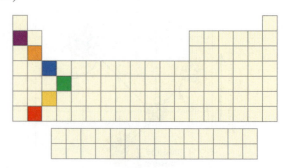

**FIGURE P21.1**

21.4. What radioactive decay processes are represented by the graphs in Figure P21.4?

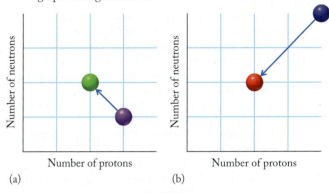

(a)                                    (b)

**FIGURE P21.4**

**21.5.** Which of the graphs in Figure P21.5 illustrates β decay?

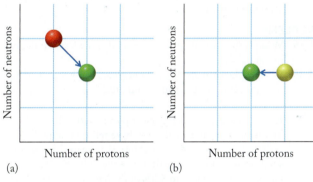

(a)                                    (b)

**FIGURE P21.5**

21.6. Which of the graphs in Figure P21.6 illustrates the overall effect of neutron capture followed by β decay?

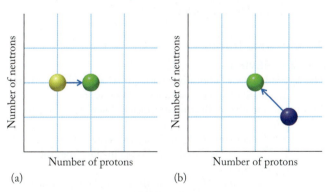

(a)                                    (b)

**FIGURE P21.6**

**21.7.** Which of the curves in Figure P21.7 represents the decay of an isotope that has a half-life of 2.0 days?

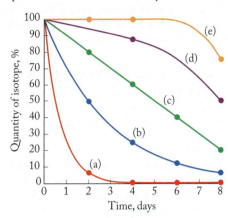

**FIGURE P21.7**

21.8. Which of the curves in Figure P21.7 do(es) not represent a radioactive decay curve?

**21.9.** Which of the models in Figure P21.9 represents fission and which represents fusion?

(1)                                    (2)

**FIGURE P21.9**

21.10. Isotopes in a nuclear decay series emit particles with a positive charge and particles with a negative charge. The two kinds of particles penetrate a column of water as shown in Figure P21.10. Is the "X" particle the positive or the negative one?

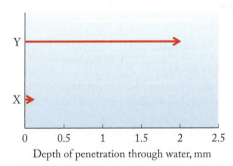

**FIGURE P21.10**

······· QUESTIONS AND PROBLEMS ·····································

## Binding Energy and Nuclear Stability

### CONCEPT REVIEW

**21.11.** What do the terms *mass defect* and *binding energy* mean?

21.12. Why is energy released in a nuclear fusion process when the product is an element preceding iron in the periodic table?

### PROBLEMS

21.13. What is the binding energy of a deuteron?

21.14. What is the binding energy of $^6$Li, which has a nuclear mass of $9.98561 \times 10^{-27}$ kg?

**21.15.** Our Sun is a fairly small star that has barely enough mass to fuse hydrogen to helium. Calculate the binding energy per nucleon of helium-4 based on these masses: $_2^4\text{He}$ (4.00260 amu), $_1^1\text{p}$ (1.00728 amu), and $_0^1\text{n}$ (1.00866 amu).

**21.16.** What is the binding energy per nucleon of $^{12}\text{C}$, the atomic mass of which is 12.00000 amu? (*Note*: Atomic mass includes the mass of 6 electrons.)

## Unstable Nuclei and Radioactive Decay

### CONCEPT REVIEW

**21.17.** How can the belt of stability be used to predict the probable decay mode of an unstable nuclide?

**21.18.** Compare and contrast positron-emission and electron-capture processes.

**21.19.** The ratio of neutrons to protons in stable nuclei increases with increasing atomic number. Use this trend to explain why multiple α decay steps in the $^{238}\text{U}$ decay series are often followed by β decay.

**21.20.** Copper-64 is an unusual radionuclide in that it may undergo β decay, positron emission, or electron capture. What are the products of these decay processes?

### PROBLEMS

**21.21.** Iodine-137 decays to give xenon-137, which decays to give cesium-137. What are the modes of decay in these two reactions?

**21.22.** Write a balanced nuclear equation describing (a) α decay of boron-12; (b) β decay of boron-12; (c) positron emission by manganese-50; (d) electron capture by cadmium-104.

---

**21.23.** If the mass number of an isotope is more than twice the atomic number, is the neutron-to-proton ratio less than, greater than, or equal to 1?

**21.24.** In each of the following pairs of isotopes, select the isotope that has more protons and the isotope that has more neutrons. Also indicate which pairs of isotopes have the same number of neutrons or protons. (a) $^{63}\text{Cu}$ and $^{65}\text{Cu}$; (b) $^{71}\text{Ga}$ and $^{71}\text{Ge}$; (c) $^{39}\text{K}$ and $^{40}\text{Ar}$

---

**21.25.** Aluminum is found on Earth exclusively as $^{27}\text{Al}$. However, $^{26}\text{Al}$ is formed in stars. It decays to $^{26}\text{Mg}$ with a half-life of $7.4 \times 10^5$ years. Write an equation describing the decay of $^{26}\text{Al}$ to $^{26}\text{Mg}$.

**21.26.** Write a balanced nuclear equation describing the β decay of cesium-137, which is produced in nuclear power plants.

---

**21.27.** Predict the modes of decay for the following radioactive isotopes: (a) $^{10}\text{C}$; (b) $^{19}\text{Ne}$; (c) $^{50}\text{Ti}$.

**21.28.** Predict the mode(s) of decay of the following radionuclides: (a) $^{24}\text{Ne}$; (b) $^{38}\text{K}$; (c) $^{45}\text{Ti}$; (d) $^{237}\text{Np}$.

---

**21.29. Elements in a Supernova** The isotopes $^{56}\text{Co}$ and $^{44}\text{Ti}$ were detected in supernova SN 1987A. Predict the decay pathway for these radioactive isotopes.

**21.30.** There are isotopes of nitrogen that have as few as 5 or as many as 11 neutrons in each of their nuclei. Write a balanced nuclear equation describing the decay of the isotope with 11 neutrons.

## Rates of Radioactive Decay

### CONCEPT REVIEW

**\*21.31.** Chlorine has isotopes with atomic numbers from 32 through 39. Two of them, $^{35}\text{Cl}$ and $^{37}\text{Cl}$, are stable.
  a. Which three of the other isotopes emit positrons?
  b. Which three of the other isotopes emit β particles?
  c. Which one of the other isotopes is capable of emitting *either* positrons or β particles?

**\*21.32.** Bromine has isotopes with atomic numbers from 74 through 90. Two of them, $^{79}\text{Br}$ and $^{81}\text{Br}$, are stable.
  a. How many of the others emit positrons or undergo electron capture?
  b. How many of the others emit β particles?
  c. Which one of the other isotopes is capable of emitting *either* positrons or β particles?

**21.33.** What percentage of a sample's original radioactivity remains after two half-lives?

**21.34.** What percentage of a sample's original radioactivity remains after five half-lives?

### PROBLEMS

**21.35.** What is the half-life of a radionuclide if 87.5% of it decays in 6.6 days?

**21.36.** What is the half-life of a radionuclide if only 6.25% of it remains after 8 hours and 20 minutes?

---

**21.37.** Explosions at a disabled nuclear power station in Fukushima, Japan in 2011 may have released more cesium-137 ($t_{1/2} = 30.2$ years) into the ocean than any other single event. How long will it take the radioactivity of this radionuclide to decay to 5.0% of the level released in 2011?

**21.38.** Spent fuel removed from nuclear power stations contains plutonium-239 ($t_{1/2} = 2.41 \times 10^4$ years). How long will it take a sample of this radionuclide to reach a level of radioactivity that is 2.5% of the level it had when it was removed from a reactor?

## Radiometric Dating

### CONCEPT REVIEW

**21.39.** Explain why radiocarbon dating is reliable only for artifacts and fossils younger than about 50,000 years.

**21.40.** Which of the following statements about $^{14}\text{C}$ dating are true?
  a. The amount of $^{14}\text{C}$ in all objects is the same.
  b. Carbon-14 is unstable and is readily lost from the atmosphere.
  c. The ratio of $^{14}\text{C}$ to $^{12}\text{C}$ in the atmosphere is a constant.
  d. Living tissue will absorb $^{12}\text{C}$ but not $^{14}\text{C}$.

**21.41.** Why is $^{40}\text{K}$ dating ($t_{1/2} = 1.28 \times 10^9$ years) useful only for rocks older than 300,000 years?

**21.42.** Where does the $^{14}\text{C}$ found in plants come from?

## PROBLEMS

**21.43. First Humans in South America** Archeologists continue to debate the arrival of the first humans in the Western Hemisphere. Radiocarbon dating of charcoal from a cave in Chile was used to establish the earliest date of human habitation in South America as 8700 years ago. What fraction of the $^{14}C$ initially present remained in the charcoal after 8700 years?

**21.44.** For thousands of years Native Americans living along the north coast of Peru used knotted cotton strands called *quipu* (Figure P21.44) to record financial transactions and governmental actions. A particular quipu sample is 4800 years old. Compared with the fibers of cotton plants growing today, what is the ratio of carbon-14 to carbon-12 in the sample?

**FIGURE P21.44**

**\*21.45.** Figure P21.45 is a close-up of the center of a giant sequoia tree cut down in 1891 in what is now Kings Canyon National Park. It contained 1342 annual growth rings. If samples of the tree were removed for radiocarbon dating today, what would be the difference in $^{14}C/^{12}C$ ratio in the innermost (oldest) ring compared with that ratio in the youngest ring?

**FIGURE P21.45**

**\*21.46.** Geologists who study volcanoes can develop historical profiles of previous eruptions by determining the $^{14}C/^{12}C$ ratios of charred plant remains entrapped in old magma and ash flows. If the uncertainty in determining these ratios is 0.1%, could radiocarbon dating distinguish between debris from the eruptions of Mt. Vesuvius that occurred in the years 472 and 512 ? (*Hint:* Calculate the $^{14}C/^{12}C$ ratios for samples from the two dates.)

**21.47.** Figure P21.47 shows a carved mammoth tusk that was uncovered at an ancient campsite in the Ural Mountains in 2001. The $^{14}C/^{12}C$ ratio in the tusk was only 1.19% of that in modern elephant tusks. How old is the mammoth tusk?

20 cm

**FIGURE P21.47**

**21.48. The Destruction of Jericho** The Bible describes the Exodus as a period of 40 years that began with plagues in Egypt and ended with the destruction of Jericho. Archeologists seeking to establish the exact dates of these events have proposed that the plagues coincided with a huge eruption of the volcano Thera in the Aegean Sea.
   a. Radiocarbon dating suggests that the eruption occurred around 1360 BCE, though other records place the eruption of Thera in the year 1628 BCE. What is the percent difference in the $^{14}C$ decay rate in biological samples from these two dates?
   b. Radiocarbon dating of blackened grains from the site of ancient Jericho provides a date of 1315 BCE ±13 years for the fall of the city. What is the $^{14}C/^{12}C$ ratio in the blackened grains compared with that of grain harvested last year?

## Measuring Radioactivity; Biological Effects of Radioactivity

### CONCEPT REVIEW

**21.49.** What is the difference between a *level* of radioactivity and a *dose* of radioactivity?

**21.50.** What are some of the molecular effects of exposure to radioactivity?

**21.51.** Describe the dangers of exposure to radon-222.

**21.52. Food Safety** Periodic outbreaks of food poisoning from *E. coli* contaminated meat have renewed the debate about irradiation as an effective treatment of food. In one newspaper article on the subject, the following statement appeared: "Irradiating food destroys bacteria by breaking apart their molecular structure." How would you improve or expand on this explanation?

### PROBLEMS

**21.53. Radiation Exposure from Dental X-rays** Dental X-rays expose patients to about 5 μSv of radiation. Given an RBE of 1 for X-rays, how many grays of radiation does 5 μSv represent? For a 50 kg person, how much energy does 5 μSv correspond to?

*21.54. **Radiation Exposure at Chernobyl** Some workers responding to the explosion at the Chernobyl nuclear power plant were exposed to 5 Sv of radiation, resulting in death for many of them. If the exposure was primarily in the form of $\gamma$ rays with an energy of $3.3 \times 10^{-14}$ J and an RBE of 1, how many $\gamma$ rays did an 80 kg person absorb?

*21.55. **Strontium-90 in Milk** In the years immediately following the explosion at the Chernobyl nuclear power plant, the concentration of $^{90}$Sr in cow's milk in southern Europe was slightly elevated. Some samples contained as much as 1.25 Bq/L of $^{90}$Sr radioactivity. The half-life of strontium-90 is 28.8 years.
   a. Write a balanced nuclear equation describing the decay of $^{90}$Sr.
   b. How many atoms of $^{90}$Sr are in a 200 mL glass of milk with 1.25 Bq/L of $^{90}$Sr radioactivity?
   c. Why would strontium-90 be more concentrated in milk than other foods, such as grains, fruits, or vegetables?

*21.56. **Radium Watch Dials** If exactly 1.00 μg of $^{226}$Ra was used to paint the glow-in-the-dark dial of a wristwatch made in 1914, how radioactive is the watch today? Express your answer in microcuries and becquerels. The half-life of $^{226}$Ra is $1.60 \times 10^3$ years.

21.57. In 1999, the U.S. Environmental Protection Agency set a maximum radon level for drinking water at 4.0 pCi per milliliter.
   a. How many decay events occur per second in a milliliter of water for this level of radon radioactivity?
   b. If the above radioactivity were due to decay of $^{222}$Rn ($t_{1/2}$ = 3.8 days), how many $^{222}$Rn atoms would there be in 1.0 mL of water?

21.58. A former Russian spy died from radiation sickness in 2006 after dining at a London restaurant where he apparently ingested polonium-210. The other people at his table did not suffer from radiation sickness, even though they were very near the radioactive tea the victim drank. Why were they not affected?

## Medical Applications of Radionuclides

### CONCEPT REVIEW

21.59. How does the selection of an isotope for radiotherapy relate to (a) its half-life, (b) its mode of decay, and (c) the properties of the products of decay?

21.60. Are the same radioactive isotopes likely to be used for both imaging and cancer treatment? Why or why not?

### PROBLEMS

21.61. Predict the most likely mode of decay for the following isotopes used as imaging agents in nuclear medicine: (a) $^{197}$Hg (kidney); (b) $^{75}$Se (parathyroid gland); (c) $^{18}$F (bone).

21.62. Predict the most likely mode of decay for the following isotopes used as imaging agents in nuclear medicine: (a) $^{133}$Xe (cerebral blood flow); (b) $^{57}$Co (tumor detection); (c) $^{51}$Cr (red blood cell mass); (d) $^{67}$Ga (tumor detection).

21.63. A 1.00 mg sample of $^{192}$Ir was inserted into the artery of a heart patient. After 30 days, 0.756 mg remained. What is the half-life of $^{192}$Ir?

21.64. In a treatment that decreases pain and reduces inflammation of the lining of the knee joint, a sample of dysprosium-165 with a radioactivity of 1100 counts per second was injected into the knee of a patient suffering from rheumatoid arthritis. After 24 hr, the radioactivity had dropped to 1.14 counts per second. Calculate the half-life of $^{165}$Dy.

21.65. **Study of Tourette's Syndrome** Tourette's syndrome is a condition whose symptoms include sudden movements and vocalizations. Iodine isotopes are used in brain imaging of people suffering from Tourette's syndrome. To study the uptake and distribution of iodine in cells, mammalian brain cells in culture were treated with a solution containing $^{131}$I with an initial radioactivity of 108 counts per minute. The cells were removed after 30 days, and the remaining solution was found to have a radioactivity of 4.1 counts per minute. Did the brain cells absorb any $^{131}$I ($t_{1/2}$ = 8.1 days)?

21.66. A patient is administered mercury-197 to evaluate kidney function. Mercury-197 has a half-life of 65 hr. What fraction of an initial dose of mercury-197 remains after 6 days?

21.67. Carbon-11 is an isotope used in positron emission tomography and has a half-life of 20.4 min. How long will it take for 99% of the $^{11}$C injected into a patient to decay?

21.68. Sodium-24 is used to treat leukemia and has a half-life of 15 hr. In a patient injected with a salt solution containing sodium-24, what percentage of the $^{24}$Na remains after 48 hr?

*21.69. **Boron Neutron-Capture Therapy** In boron neutron-capture therapy (BNCT), a patient is given a compound containing $^{10}$B that accumulates inside cancer tumors. Then the tumors are irradiated with neutrons, which are absorbed by $^{10}$B nuclei. The product of neutron capture is an unstable form of $^{11}$B that undergoes $\alpha$ decay to $^7$Li.
   a. Write a balanced nuclear equation for the neutron absorption and $\alpha$ decay process.
   b. Calculate the energy released by each nucleus of boron-10 that captures a neutron and undergoes $\alpha$ decay, given the following masses of the particles in the process: $^{10}$B (10.0129 amu), $^7$Li (7.01600 amu), $^4$He (4.00260 amu), and $^1$n (1.00866 amu).
   c. Why is the formation of a nuclide that undergoes $\alpha$ decay a particularly effective cancer therapy?

21.70. **Balloon Angioplasty and Arteriosclerosis** Balloon angioplasty is a common procedure for unclogging arteries in patients suffering from arteriosclerosis. Iridium-192 therapy is being tested as a treatment to prevent reclogging of the arteries. In the procedure, a thin ribbon containing pellets of $^{192}$Ir is threaded into the artery. The half-life of $^{192}$Ir is 74 days. How long will it take for 99% of the radioactivity from 1.00 mg of $^{192}$Ir to disappear?

## Nuclear Fission

**CONCEPT REVIEW**

**21.71.** How is the rate of energy release controlled in a nuclear reactor?

**21.72.** How does a breeder reactor create fuel and energy at the same time?

**\*21.73.** Why are neutrons always by-products of the fission of most massive nuclides? (*Hint*: Look closely at the neutron-to-proton ratios shown in Figure 21.2.)

**21.74.** Seaborgium (Sg, element 106) is prepared by the bombardment of curium-248 with neon-22, which produces two isotopes, $^{265}$Sg and $^{266}$Sg. Write balanced nuclear reactions for the formation of both isotopes. Are these reactions better described as fusion or fission processes?

**PROBLEMS**

**21.75.** The fission of uranium produces dozens of isotopes. For each of the following fission reactions, determine the identity of the unknown nuclide:
a. $^{235}$U + $^1$n → $^{96}$Zr + ? + 2 $^1$n
b. $^{235}$U + $^1$n → $^{99}$Nb + ? + 4 $^1$n
c. $^{235}$U + $^1$n → $^{90}$Rb + ? + 3 $^1$n

**21.76.** For each of the following fission reactions, determine the identity of the unknown nuclide:
a. $^{235}$U + $^1$n → $^{137}$I + ? + 2 $^1$n
b. $^{235}$U + $^1$n → $^{137}$Cs + ? + 3 $^1$n
c. $^{235}$U + $^1$n → $^{141}$Ce + ? + 2 $^1$n

**21.77.** For each of the following fission reactions, determine the identity of the unknown nuclide:
a. $^{235}$U + $^1$n → $^{131}$I + ? + 2 $^1$n
b. $^{235}$U + $^1$n → $^{103}$Ru + ? + 3 $^1$n
c. $^{235}$U + $^1$n → $^{95}$Zr + ? + 3 $^1$n

**21.78.** For each of the following fission reactions, determine the identity of the unknown nuclide:
a. $^{235}$U + $^1$n → $^{147}$Pm + ? + 2 $^1$n
b. $^{235}$U + $^1$n → $^{94}$Kr + ? + 2 $^1$n
c. $^{235}$U + $^1$n → $^{95}$Sr + ? + 3 $^1$n

## Nuclear Fusion and the Quest for Clean Energy

**CONCEPT REVIEW**

**21.79.** In what ways are the fusion reactions that formed α particles during primordial nucleosynthesis different from those that fuel our Sun today?

**21.80.** How are the fusion reactions that are the basis for power production in the tokamak described in Section 21.9 different from those that power our Sun?

**PROBLEMS**

**21.81.** All of the following fusion reactions produce $^{28}$Si. Calculate the energy released in each reaction from the masses of the isotopes: $^2$H (2.0146 amu), $^4$He (4.00260 amu), $^{10}$B (10.0129 amu), $^{12}$C (12.000 amu), $^{14}$N (14.00307 amu), $^{16}$O (15.99491 amu), $^{24}$Mg (23.98504 amu), $^{28}$Si (27.97693 amu).
a. $^{14}$N + $^{14}$N → $^{28}$Si
b. $^{10}$B + $^{16}$O + $^2$H → $^{28}$Si
c. $^{16}$O + $^{12}$C → $^{28}$Si
d. $^{24}$Mg + $^4$He → $^{28}$Si

**21.82.** All of the following fusion reactions produce $^{32}$S. Calculate the energy released in each reaction from the masses of the isotopes: $^4$He (4.00260 amu), $^6$Li (6.01512 amu), $^{12}$C (12.000 amu), $^{14}$N (14.00307 amu), $^{16}$O (15.99491 amu), $^{24}$Mg (23.98504 amu), $^{28}$Si (27.97693 amu), $^{32}$S (31.97207 amu).
a. $^{16}$O + $^{16}$O → $^{32}$S
b. $^{28}$Si + $^4$He → $^{32}$S
c. $^{14}$N + $^{12}$C + $^6$Li → $^{32}$S
d. $^{24}$Mg + 2 $^4$He → $^{32}$S

**21.83. Tokamak Radiochemistry** How much energy is released per nucleus of tritium produced during the following reactions?
a. $^1_0$n + $^6_3$Li → $^4_2$He + $^3_1$H
b. $^1_0$n + $^7_3$Li → $^4_2$He + $^3_1$H + $^1_0$n

**21.84.** It has been proposed that electrical power production in the future might be based on the fusion of deuterium to helium-4.
a. Write a radiochemical equation describing the reaction (assume that $^4$He is the only product).
b. Calculate how much energy is released during the formation of one mole of $^4$He.

## Additional Problems

**21.85.** Thirty years before the creation of antihydrogen, television producer Gene Roddenberry (1921–1991) proposed to use this form of antimatter to fuel the powerful "warp" engines of the fictional starship *Enterprise*.
a. Why would antihydrogen have been a particularly suitable fuel?
b. Describe the challenges of storing such a fuel on a starship.

**21.86.** Tiny concentrations of radioactive tritium ($^3_1$H) occur naturally in rain and groundwater. The half-life of $^3_1$H is 12 years. Assuming that tiny concentrations of tritium can be determined accurately, could the isotope be used to determine whether a bottle of wine with the year 1969 on its label actually contained wine made from grapes that were grown in 1969? Explain your answer.

**21.87.** The energy released during the fission of $^{235}$U is about $3.2 \times 10^{-11}$ J per atom of the isotope. Compare this quantity of energy with that released by the fusion of four hydrogen atoms to make an atom of helium-4:

$$4\,^1_1\text{H} \rightarrow\,^4_2\text{He} + 2\,^0_1\beta$$

Assume that the positrons are annihilated in collisions with electrons so that the masses of the positrons are converted into energy. In your comparison, express the energies released by the fission and fusion processes in joules per nucleon for $^{235}$U and $^4$He, respectively.

**21.88.** How much energy is required to remove a neutron from the nucleus of an atom of carbon-13 (mass = 13.00335 amu)? (*Hint*: The mass of an atom of carbon-12 is exactly 12.00000 amu.)

**21.89. Smoke Detectors** Americium-241 ($t_{1/2}$ = 433 yr) is used in smoke detectors. The α particles from this isotope ionize nitrogen and oxygen in the air, creating an electric current.

When smoke is present, the current decreases, setting off the alarm.

a. Does a smoke detector bear a closer resemblance to a Geiger counter or to a scintillation counter?

b. How long will it take for the radioactivity of a sample of $^{241}$Am to drop to 1% of its original radioactivity?

c. Why are smoke detectors containing $^{241}$Am safe to handle without protective equipment?

*21.90. **Colorectal Cancer Treatment** Cancer therapy with radioactive rhenium-188 shows promise in patients suffering from colorectal cancer.

a. Write the symbol for rhenium-188 and determine the number of neutrons, protons, and electrons.

b. Are most rhenium isotopes likely to have fewer neutrons than rhenium-188?

c. The half-life of rhenium-188 is 17 hours. If it takes 30 minutes to bind the isotope to an antibody that delivers the rhenium to the tumor, what percentage of the rhenium remains after binding to the antibody?

d. The effectiveness of rhenium-188 is thought to result from penetration of β particles as deep as 8 mm into the tumor. Why wouldn't an α emitter be more effective?

e. Using an appropriate reference text, such as the *CRC Handbook of Chemistry and Physics*, pick out the two most abundant isotopes of rhenium. List their natural abundances and explain why the one that is radioactive decays by the pathway that it does.

21.91. In 2006 an international team of scientists confirmed the synthesis of a total of three atoms of $^{294}_{118}$Uuo in experiments run in 2002 and 2005. They had bombarded a $^{249}$Cf target with $^{48}$Ca nuclei.

a. Write a balanced nuclear equation describing the synthesis of $^{294}_{118}$Uuo.

b. The synthesized isotope of Uuo undergoes α decay ($t_{1/2} = 0.9$ ms). What nuclide is produced by the decay process?

c. The nuclide produced in part b also undergoes α decay ($t_{1/2} = 10$ ms). What nuclide is produced by this decay process?

d. The nuclide produced in part c also undergoes α decay ($t_{1/2} = 0.16$ s). What nuclide is produced by this decay process?

e. If you had to select an element that occurs in nature and that has physical and chemical properties similar to Uuo, which element would it be?

*21.92. Consider the following decay series:

$$A\ (t_{1/2} = 4.5\ s) \rightarrow B\ (t_{1/2} = 15.0\ days) \rightarrow C$$

If we start with $10^6$ atoms of A, how many atoms of A, B, and C are there after 30 days?

21.93. Which element in the following series will be present in the greatest amount after one year?

$$^{214}_{83}\text{Bi} \xrightarrow{\alpha} {}^{210}_{81}\text{Tl} \xrightarrow{\beta} {}^{210}_{82}\text{Pb} \xrightarrow{\beta} {}^{210}_{83}\text{Bi} \longrightarrow$$
$$t_{1/2} = \quad 20\ \text{min} \quad 1.3\ \text{min} \quad 20\ \text{yr} \quad 5\ \text{d}$$

*21.94. **Dating Cave Paintings** Cave paintings in Gua Saleh Cave in Borneo have been dated by measuring the amount of $^{14}$C in calcium carbonate that formed over the pigments used in the paint. The source of the carbonate ion was atmospheric $CO_2$.

a. What is the ratio of the $^{14}$C radioactivity in calcium carbonate that formed 9900 years ago to that in calcium carbonate formed today?

b. The archeologists also used a second method, uranium–thorium dating, to confirm the age of the paintings by measuring trace quantities of these elements present as contaminants in the calcium carbonate. Shown below are two candidates for the U–Th dating method. Which isotope of uranium do you suppose was chosen? Explain your answer.

$$^{235}_{92}\text{U} \rightarrow {}^{231}_{90}\text{Th} \rightarrow {}^{231}_{91}\text{Pa} \rightarrow$$
$$t_{1/2} = \quad 7.04 \times 10^8\ \text{yr} \quad 25.6\ \text{hr} \quad 3.25 \times 10^4\ \text{yr}$$

$$^{234}_{92}\text{U} \rightarrow {}^{230}_{90}\text{Th} \rightarrow {}^{226}_{88}\text{Pa} \rightarrow$$
$$t_{1/2} = \quad 2.44 \times 10^5\ \text{yr} \quad 7.7 \times 10^4\ \text{hr} \quad 1600\ \text{yr}$$

21.95. The synthesis of new elements and specific isotopes of known elements in linear accelerators involves the fusion of smaller nuclei.

a. An isotope of platinum can be prepared from nickel-64 and tin-124. Write a balanced equation for this nuclear reaction. (You may assume that no neutrons are ejected in the fusion reaction.)

b. Substitution of tin-132 for tin-124 increases the rate of the fusion reaction 10 times. Which isotope of Pt is formed in this reaction?

21.96. A sample of drinking water collected from a suburban Boston municipal water system in 2002 contained 0.5 pCi/L of radon. Assume that this level of radioactivity was due to the decay of $^{222}$Rn ($t_{1/2} = 3.8$ days).

a. What was the level of radioactivity (Bq/L) of this nuclide in the sample?

b. How many decay events per hour would occur in 2.5 L of the water?

**21.97. Stone Age Skeletons** The discovery of six skeletons in an Italian cave at the beginning of the 20th century was considered a significant find in Stone Age archaeology. The age of these bones has been debated. The first attempt at radiocarbon dating indicated an age of 15,000 years. Redetermination of the age in 2004 indicated an older age for two bones, between 23,300 and 26,400 years. What is the ratio of $^{14}$C in a sample 15,000 years old to one 25,000 years old?

**\*21.98.** Atmospheric testing of nuclear weapons in the 1950s and 60s produced an increase in the concentration of carbon-14 in the atmosphere. Use one or more balanced nuclear equations to explain how this could have happened.

**21.99. Dating Prehistoric Bones** In 1997 anthropologists uncovered three partial skulls of prehistoric humans in the Ethiopian village of Herto. Based on the amount of $^{40}$Ar in the volcanic ash in which the remains were buried, their age was estimated at between 154,000 and 160,000 years old.

a. $^{40}$Ar is produced by the decay of $^{40}$K ($t_{1/2} = 1.28 \times 10^9$ yr). Propose a decay mechanism for $^{40}$K to $^{40}$Ar.

b. Why did the researchers choose $^{40}$Ar rather than $^{14}$C as the isotope for dating these remains?

**\*21.100. Biblical Archeology** The Old Testament describes the construction of the Siloam Tunnel, used to carry water into Jerusalem under the reign of King Hezekiah (727–698 BCE). An inscription on the tunnel has been interpreted as evidence that the tunnel was not built until 200–100 BCE. $^{14}$C dating (in 2003) indicated a date close to 700 BCE. What is the ratio of $^{14}$C in a wooden object made in 100 BCE to one made from the same kind of wood in 700 BCE?

If your instructor assigns problems in **smartwork**, log in at **smartwork.wwnorton.com**.

# 22

# Life and the Periodic Table

**22.1** The Periodic Table of Life

**22.2** Major Essential Elements

**22.3** Trace and Ultratrace Essential Elements

**22.4** Nonessential Elements

**22.5** Elements for Diagnosis and Therapy

## Learning Outcomes

**LO1** Distinguish between essential and nonessential elements, balance equations, and carry out calculations relevant to their behavior *in vivo*
Sample Exercises 22.1, 22.2

**LO2** Distinguish between major, trace, and ultratrace elements

**LO3** Summarize the pathways for ion transport across cell membranes

**LO4** Explain how radioactive isotopes are used in the diagnosis of disease
Sample Exercises 22.3, 22.4

**LO5** Describe how metal complexes are used as therapeutic compounds
Sample Exercise 22.5

**LO6** Describe how metals and alloys are used in medical devices
Sample Exercise 22.6

## Elements in Our Bodies

Have you ever wondered how many of the elements in the periodic table are in the human body? Or wondered which of them are important to the health of humans?

Roughly one-third of the 90 naturally occurring elements have an identifiable role in human health and in organisms in general. Some—like carbon, hydrogen, oxygen, nitrogen, sulfur, and phosphorus—are the principal constituents of all plants and animals. The alkali metal cations $Na^+$ and $K^+$ act as charge carriers, maintain osmotic pressure, and transmit nerve impulses. The alkaline earth cation $Mg^{2+}$ is important in photosynthesis, and its family member $Ca^{2+}$ forms structural materials such as bones and teeth. Chloride ions balance the charge of $Na^+$ and $K^+$ ions to maintain electrical neutrality in living cells. Other main group elements, such as iodine and selenium, are required in tiny amounts in our bodies, and several transition metal ions, such as $Fe^{2+}$, $Zn^{2+}$, and $Ni^{2+}$, are key components of the enzymes that catalyze biochemical reactions.

Many elements with no known biological function are useful in medicine as either diagnostic tools or therapeutic agents. Some radioactive isotopes act as effective imaging agents for organs and tumors, for example, and others are used to kill cancer cells. Compounds containing platinum, silver, or gold are effective drugs for treating cancer, burns, and arthritis, respectively. Drugs containing lithium ions are used to treat depression, and corrosion-resistant metals such as tantalum are used in artificial joints.

In this chapter we survey the roles of various elements in the human body and their importance to good health. At the same time we call on the knowledge and skills you have acquired in your study of general chemistry to solve problems that link concepts from prior chapters to the central question of this chapter: What are the roles of the elements in the chemistry of life? ■

**MRI** Magnetic resonance imaging is a valuable diagnostic tool in medicine. Gadolinium compounds are used to improve the contrast in this MRI image of the carotid arteries and others in the brain (pink color). ▶

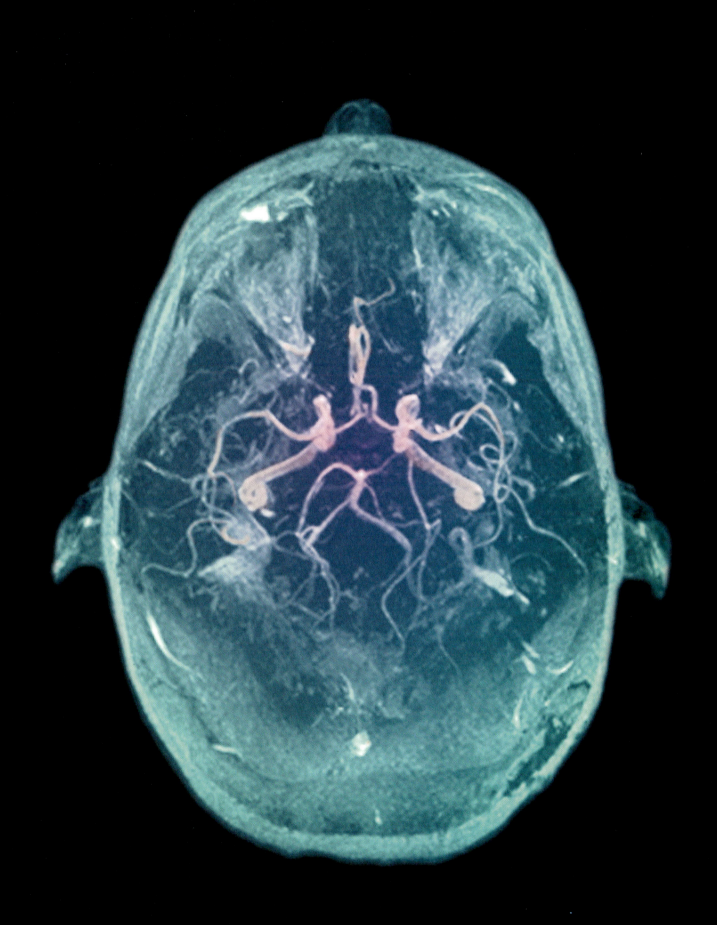

## 22.1 The Periodic Table of Life

The elements found in the human body can be classified as either essential or nonessential to life (Tables 22.1 and 22.2). **Essential elements** are defined as those that have a beneficial physiological function, including those whose absence impairs functioning of the organism. **Nonessential elements** are present in the body but have no known function. In some cases, the presence of a nonessential element has a **stimulatory effect**, which means that the consumption of small amounts of the element affects us via a mechanism that is not yet understood. For example, small amounts of the nonessential element antimony promote growth in some mammals when added to their diets. Nonessential elements, and even toxic elements, are often incorporated into our bodies because their chemical properties are similar to those of an essential element. For example, $Pb^{2+}$ ions are incorporated into teeth and bones because they are similar to $Ca^{2+}$ ions in size and charge. The alkali metal rubidium is the most abundant nonessential element in humans; $Rb^+$ is retained by the body because its size and chemistry are similar to those of $K^+$.

Oxygen in the form of $O_2$ gas is the only element that occurs in the body in elemental form; oxygen is also incorporated into many compounds (e.g., $H_2O$) or ions, such as $HCO_3^-$. When we speak of any other element in the body, we are always referring to an ion or compound containing that element rather than the pure element. For example, when we describe zinc as an essential element, we are referring to zinc ions, $Zn^{2+}$, not zinc metal.

The essential elements are further classified as **major**, **trace**, or **ultratrace essential elements**. Major essential elements are present in gram quantities in the human body and are required in large amounts in our diet. Almost all foods are rich in compounds containing carbon, hydrogen, oxygen, nitrogen, sulfur, and phosphorus. Salt is perhaps the most familiar dietary source of sodium and chloride ions, although both are ubiquitous in food. Vegetables like broccoli and Brussels sprouts and fruits like bananas are rich in potassium. Calcium is found in dairy products and is often added to orange juice as a dietary supplement.

Table 22.3 compares the elemental compositions of the human body, the universe, Earth's crust, and seawater. Note that the composition of our bodies most closely resembles the composition of seawater. The match would be even closer if it were not for the biological processes in the sea that remove essential elements such as nitrogen and phosphorus, and that store others in solid structures like the $CaCO_3$ that makes up corals and mollusk shells.

Our diet should supply us with sufficient quantities of all essential elements. In the United States and Canada, these quantities are called *dietary reference intake* (DRI) values. They are based on the recommendations of the Food and Nutrition Board of the National Academy of Sciences and are frequently updated in response to research. For many essential elements, DRI values have replaced the *recommended dietary allowance* (RDA) values you may be familiar with from labels on food and vitamin packages (Figure 22.1). Table 22.4 compares the DRI and RDA values for several major, trace, and ultratrace essential elements. Among the

### TABLE 22.1 Essential Elements Found in the Human Body

| Major (>1 mg/g body mass) | Trace (1–1000 μg/g body mass) | Ultratrace (<1 μg/g body mass) |
|---|---|---|
| Calcium | Fluorine | Chromium |
| Carbon | Iodine | Cobalt |
| Chlorine | Iron | Copper |
| Hydrogen | Silicon | Manganese |
| Magnesium | Zinc | Molybdenum |
| Nitrogen | | Nickel |
| Oxygen | | Selenium |
| Phosphorus | | Vanadium |
| Potassium | | |
| Sodium | | |
| Sulfur | | |

### TABLE 22.2 Nonessential Elements Found in the Human Body

| Stimulatory | Unknown Role | No Role |
|---|---|---|
| Boron | Antimony | Barium |
| Titanium | Arsenic | Cesium |
| | | Germanium |
| | | Rubidium |
| | | Strontium |

**essential element** element present in tissue, blood, or other body fluids that has a physiological function.

**nonessential element** element present in humans that has no known function.

**stimulatory effect** increased growth or other biological response to the presence of a nonessential element.

| TABLE 22.3 | Comparative Composition*a* of the Universe, Earth's Crust, Seawater, and the Human Body | | | |
|---|---|---|---|---|
| **Element** | **Universe (%)** | **Crust (%)** | **Seawater (%)** | **Human Body (%)** |
| Hydrogen | 91 | 0.22 | 66 | 63 |
| Oxygen | 0.57 | 47 | 33 | 25.5 |
| Carbon | 0.021 | 0.019 | 0.0014 | 9.5 |
| Nitrogen | 0.042 | | | 1.4 |
| Calcium | | 3.5 | 0.006 | 0.31 |
| Phosphorus | | | | 0.22 |
| Chlorine | | | 0.33 | 0.03 |
| Potassium | | 2.5 | 0.006 | 0.06 |
| Sulfur | 0.001 | 0.034 | 0.017 | 0.05 |
| Sodium | | 2.5 | 0.28 | 0.01 |
| Magnesium | 0.002 | 2.2 | 0.033 | 0.01 |
| Helium | 9.1 | | | |
| Silicon | 0.003 | 28 | | |
| Aluminum | | 7.9 | | |
| Neon | 0.003 | | | |
| Iron | 0.002 | 6.2 | | |
| Bromine | | | 0.0005 | |
| Titanium | | 0.46 | | |
| All other elements | < 0.1 | < 0.1 | < 0.1 | < 0.1 |

*a*Compositions are expressed as the percentage of the total number of atoms. Because of rounding, the totals do not equal exactly 100%.

| TABLE 22.4 | Dietary Reference Intakes (DRI) and Recommended Dietary Allowances (RDA) for Selected Essential Elements*a* | |
|---|---|---|
| **Element** | **DRI** | **RDA** |
| Calcium | 1000 mg | 1200 mg |
| Chlorine | 2300 mg | 2300 mg |
| Chromium | 25–35 µg | 35 µg |
| Copper | 900 µg | 900 µg |
| Fluorine | 3–4 mg | 4 mg |
| Iodine | 150 µg | 150 µg |
| Iron | 8–18 mg | 18 mg |
| Magnesium | 420 mg | 320–400 mg |
| Manganese | 1.8–2.3 mg | 2–5 mg |
| Molybdenum | 45 µg | 45 µg |
| Phosphorus | 700 mg | 700 mg |
| Potassium | 4700 mg | 4700 mg |
| Selenium | 55 µg | 55 µg |
| Sodium | 1500 mg | 1500 mg |
| Zinc | 8–11 mg | 11 mg |

*a*DRI and RDA values in mg or µg per day from the U.S. Department of Agriculture (2009) and from the Council on Responsible Nutrition (CRN) for 19- to 30-year-olds.

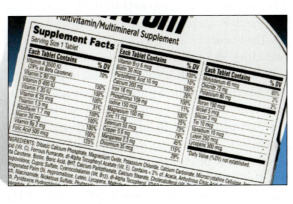

**FIGURE 22.1** The labels on multivitamin supplements may not list DRI or RDA values but rather *% daily values* (DVs). Daily values are based on RDA or DRI values, but there can be inconsistencies, particularly among the ultratrace essential elements.

**major essential element** essential element present in the body in average concentrations greater than 1 mg of element per gram of body mass.

**trace essential element** essential element present in the body in average concentrations between 1 and 1000 µg of element per gram of body mass.

**ultratrace essential element** essential element present in the body in average concentrations less than 1 µg of element per gram of body mass.

major essential elements, DRI/RDA values range from 0.32–0.42 g of magnesium per day to 4.7 g/d of potassium. DRI/RDA values for trace essential elements, including iron and zinc, are in the 10–20 mg/d range. There are DRI/RDA values for some, but not all, of the ultratrace essential elements, ranging from 55 µg/d for selenium and 45 µg/d for molybdenum up to 5 mg/day for manganese.

Some of the elements in the periodic table are toxic. These include radon, beryllium, cadmium, mercury, and lead. As described in Chapter 21, inhaled radon gas poses serious health hazards from α decay taking place inside the body. Beryllium toxicity is most often encountered in industrial settings where beryllium-contaminated dust is inhaled. The $Be^{2+}$ ion replaces $Mg^{2+}$ in the body, where it inhibits $Mg^{2+}$-catalyzed RNA and DNA synthesis in cells.

Prior to the development of nickel–cadmium batteries in the latter part of the 20th century, the toxicity of cadmium compounds was of minimal concern to humans. However, nicad batteries thrown into landfills may introduce $Cd^{2+}$ ions into groundwater. Cadmium(II) ions have an ionic radius (95 pm) close to that of calcium ions (100 pm). Their similarity in size leads to cadmium accumulation in bones, which can weaken them.

Mercury poisoning has a longer history. The Mad Hatter in *Alice in Wonderland* exhibits symptoms that were common among hat makers who used mercury(II) nitrate to make felt easier to work. Mercury(II) ions bind to sulfur-containing amino acids and are readily transported throughout the body. Some of the effects of mercury poisoning include memory loss, tremors, and impaired coordination. Although no longer used in hat manufacture, mercury was used in other ways throughout the 20th century—dental amalgams and thermometers are two familiar applications. Mercury spills from broken thermometers are difficult to clean up because the dense liquid finds its way into cracks and corners. Mercury has a small but significant vapor pressure, so anyone in a room with an open container of Hg metal will inhale its vapor. In our bodies, mercury can be transformed into methylmercury(II) ion, $CH_3Hg^+$, a potent neurotoxin. The methyl group increases the solubility of mercury in the nonpolar portion of cell membranes and allows the $CH_3Hg^+$ to diffuse through the lipid bilayer. In particular, the membranes separating blood from the brain are vulnerable to penetration by $CH_3Hg^+$, which leads to the symptoms of mercury poisoning.

The conversion of mercury into $CH_3Hg^+$ in the environment (*biomethylation*) was responsible for the health problems experienced in the 1950s by people living around Minamata Bay in Japan. Mercury-containing waste from industrial plants was routinely dumped into the bay, and the mercury was converted to $CH_3Hg^+$. Fish accumulated the ion in their tissues. Local residents ate the fish and suffered the effects of mercury poisoning.

The history of lead toxicity dates back thousands of years. Lead compounds, including PbS, $PbCO_3$, and PbCl(OH), were used by the ancient Egyptians in cosmetics. The Romans used lead in pipes for plumbing and in wine carafes and suffered from lead poisoning as a result. In the Middle Ages, lead acetate was used to sweeten wine, a practice with dire consequences. In the 20th century, tetraethyl lead [$(CH_3CH_2)_4Pb$] was used as a gasoline additive to help automobile engines run smoothly. The negative effects of lead on the mental development of children have been well documented, and the use of lead in gasoline in the United States was phased out by 1986. Similarly, paints containing lead(II) carbonate ("white lead") and lead(II) chromate ("chrome yellow") have been banned from residential use in the United States since 1978. The toxicity of lead is traced to its ability to form strong bonds with oxygen and sulfur groups in many enzymes. This interaction inhibits the activity of these enzymes, including those that catalyze hemoglobin synthesis and many other physiological processes.

**CONNECTION** The vapor pressures of liquids were discussed in Chapter 11. Pressure units such as the pascal (Pa) were defined in Chapter 6.

**CONNECTION** The applications of the group 12 elements Zn, Cd, and Hg in electrochemistry and battery technology were discussed in Chapter 19.

**ion channel** group of helical proteins that penetrate cell membranes and allow selective transport of ions.

**CONCEPT TEST**

There are 112 known elements listed in the periodic table. Why do you suppose that none of the elements with atomic numbers 93–112 are essential to life?

*(Answers to Concept Tests are in the back of the book.)*

# 22.2 Major Essential Elements

The eleven elements shown in red in Figure 22.2 and listed in the first column of Table 22.1 are the major essential elements. Together they account for more than 99% of the mass of the human body. Oxygen is the most abundant element by mass, followed by carbon and hydrogen. Although life depends on the presence of elemental oxygen in the form of $O_2$ gas, much of the oxygen in our bodies is combined with hydrogen in water molecules.

The most abundant elements in the human body include seven nonmetals: C, H, O, Cl, S, P, and N. Together they make up most of the mass of the human body and are the building blocks for most of its molecular compounds and the principal polyatomic ions: $HCO_3^-$, $SO_4^{2-}$, and $H_2PO_4^-$, which are dissolved in body fluids. (We discuss the roles of these elements in Chapters 13 and 20, on organic chemistry and biochemistry, respectively.) The average concentrations of four major metals in the human body are listed in Table 22.5. Let's explore some of the roles sodium, potassium, magnesium, calcium, chlorine, and nitrogen play in the biochemistry of the human body. As we do, we will revisit several of the chemical principles discussed in earlier chapters.

## Sodium and Potassium

Regulated concentrations of sodium and potassium ions are crucial to cell function. For example, too much $Na^+$ has been linked to hypertension (high blood pressure). To maintain a constant concentration of these two alkali metal (group 1) ions in body fluids, the ions must be able to move into and out of cells. As noted in Section 20.4, the membrane surrounding a typical cell is a lipid bilayer, with polar groups containing phosphate on the two surfaces of the cell membrane and nonpolar fatty acids oriented toward the middle of the membrane. Direct diffusion of $Na^+$ and $K^+$ through the lipid bilayer is difficult because these polar cations do not dissolve in the nonpolar interior of the membrane.

As Figure 22.3 shows, the cell membrane is pierced by **ion channels**, which are groups of protein complexes that allow selective transport of ions. The ion channels control which ions pass through the membrane, based on the size and charge of the ion as well as the shape of the protein. For example, the protein of the ion channel for potassium ions has its amino acids oriented in such a fashion that favorable ion–dipole interactions occur only for ions with the radius of a $K^+$ ion (138 pm) and not $Na^+$ (102 pm) or any other cation. The sodium ion channel is also selective, excluding $K^+$ and $Ca^{2+}$ even though the radii of $Na^+$ and $Ca^{2+}$ (100 pm) differ by only 2 pm. Another difference between the $Na^+$ and $K^+$ channels is the ability of $H_3O^+$ (hydronium ion, radius 113 pm) to pass through sodium ion channels but not potassium ion channels.

Living organisms also contain oxygen-rich molecules like nonactin (Figure 22.4) that behave as ligands and bond to $Ca^{2+}$, $K^+$, $Na^+$, and $Mg^{2+}$ ions through strong ion–dipole forces. The resulting complex ions consist of a polar, charged alkali metal ion encapsulated in a nonpolar exterior. Because the complex has both a polar portion and a nonpolar portion, it does not require a channel for passage through a cell membrane. Instead, the complex carries its alkali metal cation through both the polar and nonpolar regions of the bilayer, providing an alternative to ion channels for the transport of metal ions.

In addition to ion channels and diffusion via ligands, alkali metal cations can be transported by a third mechanism, one involving $Na^+$–$K^+$ ion pumps. An

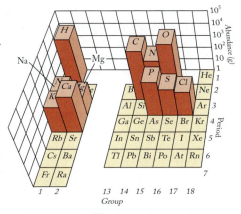

**FIGURE 22.2** The eleven elements shown in red are the major essential elements. Their abundances range from 35 g of magnesium to 46 kg of oxygen in a 70-kg adult human.

| TABLE 22.5 | Average Concentration of Four Metallic Elements in the Human Body | |
| --- | --- | --- |
| **Element** | **mg/g of Body Mass** | |
| Calcium | 15.0 | |
| Potassium | 2.0 | |
| Sodium | 1.5 | |
| Magnesium | 0.5 | |

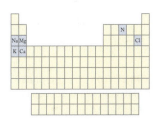

**CONNECTION** We introduced phospholipids in Chapter 20 as consisting of glycerol with two fatty acid chains and a polar region containing a phosphate group.

**CONNECTION** In Chapter 10 we described how alkali metal cations dissolved in water are surrounded by six water molecules. Each water molecule is oriented so that the oxygen atoms point toward the cation. In Chapter 17 we described this interaction as an example of a Lewis acid (cation, electron-pair acceptor) interacting with a Lewis base or ligand (water, electron-pair donor).

**FIGURE 22.3** (a) Cell membranes consist of a bilayer of phospholipids pierced by ion channels. The polar groups of the phospholipids face the aqueous solutions inside and outside the cell, whereas the fatty acids form a nonpolar region within the membrane. (b) An electron micrograph of the membranes separating two adjacent cells.

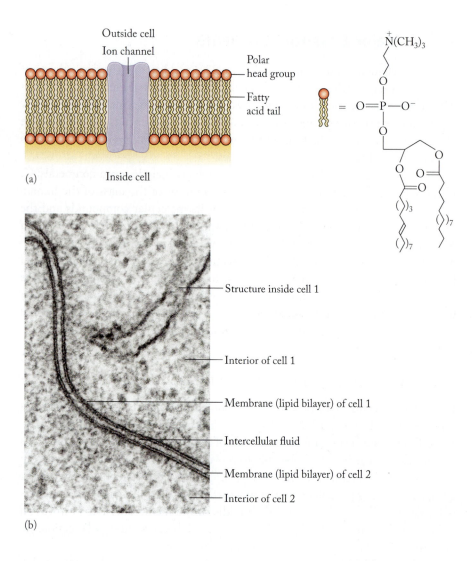

(a)

(b)

Outside cell
Ion channel
Polar head group
Fatty acid tail
Inside cell

Structure inside cell 1
Interior of cell 1
Membrane (lipid bilayer) of cell 1
Intercellular fluid
Membrane (lipid bilayer) of cell 2
Interior of cell 2

**○○ CONNECTION** The role of ATP and ADP in metabolism was described in Chapter 18.

**ion pump** is a system of membrane proteins that exchange ions inside the cell (for example, $Na^+$) with those in the intercellular fluid (for example, $K^+$). Unlike diffusion or transport through ion channels, transport via the $Na^+$–$K^+$ pump requires energy, which is provided by the hydrolysis of ATP to ADP. An example of how the $Na^+$–$K^+$ ion pump works is the response of a nerve cell to touch. Stimulation of the nerve cell causes $Na^+$ to flow into the cell and $K^+$ to flow out via ion channels; this two-way flow of ions produces the nerve impulse. The ion

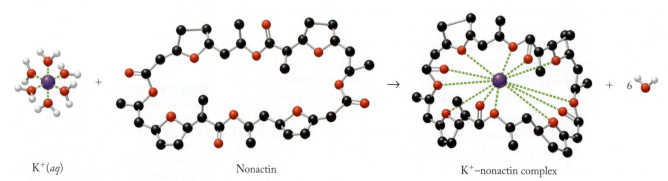

$K^+(aq)$      Nonactin      $K^+$–nonactin complex

**FIGURE 22.4** In living organisms, ligands such as nonactin can form a complex with any one of the four alkali metal and alkaline earth major essential ions and carry the ion directly through a cell membrane. No ion channel is required in this transport pathway.

pump then "recharges" the system by pumping $Na^+$ out of the cell and $K^+$ into the cell so that another impulse can immediately be transmitted along the nerve.

**ion pump** system of membrane proteins that exchange ions inside the cell with those in the intercellular fluid.

**CONCEPT TEST**

Do ion pumps represent spontaneous or nonspontaneous processes?

## Magnesium and Calcium

The biological roles of $Mg^{2+}$ and $Ca^{2+}$ are more varied than those of $Na^+$ and $K^+$. We have mentioned that calcium is a major component of teeth and bones. A prolonged deficiency of calcium can lead to osteoporosis (a disease characterized by low bone density), whereas high concentrations of calcium in muscle cells contribute to cramps. Most kidney stones are made of calcium oxalate or calcium phosphate. Magnesium deficiencies can reduce physical and mental capacity because of the role of $Mg^{2+}$ in the transfer of phosphate groups to and from ATP; slowing this transfer diminishes the amount of energy available to cells. The cellular concentrations of $Mg^{2+}$ and $Ca^{2+}$ are maintained by ion pumps.

Magnesium is a component of chlorophyll, which is one of several molecules used by plants to collect and capture light energy across the visible portion (400 to 700 nm) of the electromagnetic spectrum (Figure 22.5). Chlorophylls from different plants vary slightly in composition, but all of them contain magnesium coordinated to four nitrogen atoms. The presence of magnesium in chlorophyll does not account for the green color of the molecule, nor does it play a direct role in absorption of sunlight. The function of the $Mg^{2+}$ ion is to orient the molecules in positions that allow energy to be transferred to the reaction centers where $H_2O$ is consumed and $O_2$ is produced during photosynthesis. Carotene (and related compounds) is responsible for the orange colors of autumn leaves on deciduous trees when chlorophyll production ceases.

**CONNECTION** The inorganic chemistry of calcium was described in Chapter 4.

**CONNECTION** The stability of complexes formed between metal ions and polydentate ligands, such as the chlorin ring in chlorophyll, was described in Chapter 17.

**CONCEPT TEST**

The colors of metal compounds were discussed in Chapter 17. Most $Mg^{2+}$ compounds are white, not green like chlorophyll. Why?

$Mg^{2+}$ ions play important roles in ATP hydrolysis and ADP phosphorylation. The many $Mg^{2+}$-mediated ATP $\rightarrow$ ADP processes include transferring phosphate to glucose in the conversion of glucose to pyruvate and driving $Na^+$–$K^+$ ion pumps.

To some extent, calcium ions are also capable of mediating ATP hydrolysis, but these ions play other roles in the cell. They are required to trigger muscle contractions, for example—the calcium ions used for this purpose are stored in proteins. Recall that the action of $Na^+$–$K^+$ pumps is responsible for the generation of nerve impulses. One effect of nerve impulses is to trigger the release of $Ca^{2+}$ ions from their storage proteins into the intracellular fluid. In a multistep process, muscle cells contract and relax as calcium ions are released. Once the muscle action is complete, the ions are returned to their storage proteins in a process coupled to $Mg^{2+}$-mediated ATP hydrolysis.

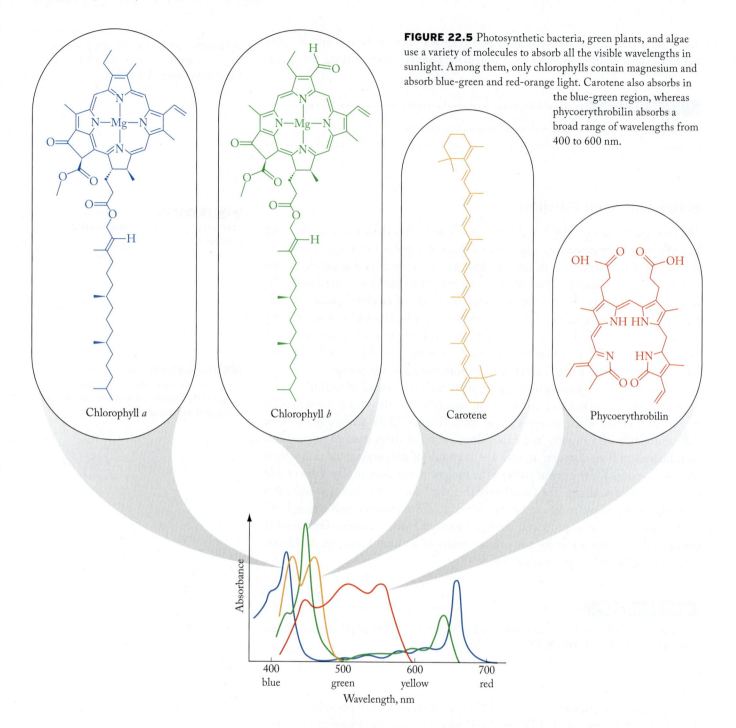

**FIGURE 22.5** Photosynthetic bacteria, green plants, and algae use a variety of molecules to absorb all the visible wavelengths in sunlight. Among them, only chlorophylls contain magnesium and absorb blue-green and red-orange light. Carotene also absorbs in the blue-green region, whereas phycoerythrobilin absorbs a broad range of wavelengths from 400 to 600 nm.

Chlorophyll *a*

Chlorophyll *b*

Carotene

Phycoerythrobilin

Of the four alkali metal and alkaline earth major essential elements, only calcium plays a major role in the formation of teeth and bones. Mammalian bones are a *composite material*, defined as a material containing a mixture of different substances. About 30% of dry bone mass is elastic protein fibers. The remainder of the mass consists of calcium compounds, including the mineral hydroxyapatite, $Ca_5(PO_4)_3(OH)$, which is also a principal component of teeth. Hydroxyapatite crystals are bound to the protein fibers in bone through phosphate groups.

The shells of marine organisms are mostly calcium carbonate ($CaCO_3$) in a matrix of proteins and polysaccharides. Some magnesium is incorporated into the calcium carbonate outer shell of marine organisms that are capable of photosynthesis, such as algae and phytoplankton.

## Chlorine

Of all the halogens, only chlorine (as chloride ion) is present in sufficient quantities to be considered a major essential element in humans. Chloride ions are the most abundant anions in the human body and are involved in many functions. The concentration of chloride ions in the human body (1.5 mg per gram of body mass) is slightly less than one-tenth of the concentration of $Cl^-$ in seawater (19 mg per gram of water) but about 12 times greater than in Earth's crust (0.13 mg per gram of crust). Like the major essential cations, chloride ions are transported into and out of cells primarily via ion channels and ion pumps. To maintain electrical neutrality in a cell, the transport of alkali metal cations is accompanied by the transport of chloride anions. The *cotransport* of $Na^+$ and $Cl^-$ is essential in kidney function, where the ions are reabsorbed by the body rather than eliminated with liquid waste products.

Malfunctioning chloride ion channels are the underlying cause of cystic fibrosis, a lethal genetic disease that causes patients to accumulate mucus in their airways such that breathing becomes difficult. The discovery of high concentrations of $Na^+$ and $Cl^-$ in the sweat of cystic fibrosis patients led to an understanding of the role of chloride ion transport in patients with this disease.

Chloride ions also play a major role in the elimination of $CO_2$ from the body. Because it is nonpolar, carbon dioxide produced during glucose catabolism can pass from muscle cells (for example) into red blood cells, moving easily through the largely nonpolar cell membranes of these cells. Inside the red blood cells, $CO_2$ is converted to bicarbonate ion, $HCO_3^-$. When $HCO_3^-$ is pumped out of the cell, $Cl^-$ enters the cell through an ion channel to maintain charge balance.

Chloride ion concentrations are high in gastric juices because of the presence of hydrochloric acid, which catalyzes digestive processes in the stomach. In response to food in the digestive system, cells tap ATP for the needed energy to pump hydrochloric acid into the stomach.

**CONNECTION** Additional information on the chemistry of chlorine and the other halogens can be found in Chapters 8 and 10.

**CONNECTION** The catabolism of glucose was described in Chapter 20.

**SAMPLE EXERCISE 22.1** **Calculating the Concentration of HCl in Stomach Acid** **LO1**

Acid reflux (sometimes called heartburn, though the heart is not involved) affects many people. It results from acid in the stomach leaking into the esophagus and causing discomfort. Stomach acid is primarily an aqueous solution of HCl. (a) Calculate the molarity of hydrochloric acid in gastric juice that has a pH of 0.80. One treatment for the symptoms of acid reflux is to take an antacid tablet. (b) What volume of gastric juice can be neutralized by a 750-mg tablet of calcium carbonate (a typical size for an over-the-counter antacid)?

**CONNECTION** An introduction to the strengths of acids and the calculation of pH can be found in Chapters 4 and 16.

**Collect and Organize** We are given the pH of a solution and are asked to calculate the concentration of HCl that corresponds to that pH. According to Equation 16.11, $pH = -\log[H^+]$. Hydrochloric acid is a strong acid and ionizes completely to $H^+$ and $Cl^-$ in water:

$$HCl(aq) \rightarrow H^+(aq) + Cl^-(aq)$$

We are also asked to calculate the volume of HCl solution that can be neutralized by a 750-mg tablet of calcium carbonate, $CaCO_3$. We need to write a balanced chemical equation for the neutralization reaction. Recall that in aqueous solution, $H^+(aq)$ is actually present as $H_3O^+(aq)$.

**Analyze** The equation describing the ionization of hydrochloric acid indicates that one mole of $H^+$ ions is formed for every mole of HCl present. The pH of gastric juice falls between 1 and 0, so $[H^+]$ will be between $10^{-1}$ ($= 0.1$) $M$ and $10^0$ ($= 1$) $M$.

The neutralization reaction is

$$CaCO_3(s) + 2\,H^+(aq) \rightarrow Ca^{2+}(aq) + CO_2(g) + H_2O(\ell)$$

This equation indicates that two moles of $H^+$ are consumed for every mole of $CaCO_3$. We are told that the tablet size is typical of an antacid tablet, so common sense leads us to predict that the volume of acid this tablet can neutralize will not be excessively large (greater than 1 L) or small (less than 10 mL): too large a tablet would be a waste of antacid, and too small a tablet would not relieve the symptoms.

**Solve**

a. Substitution into Equation 16.11 gives

$$pH = -\log[H^+] = 0.80$$

We take the antilog of both sides to solve for $[H^+]$:

$$[H^+] = 10^{-0.80} = 0.16\ M\,H^+$$

Therefore, the concentration of HCl is

$$0.16\ M\,H^+ \times \frac{1\ \text{mol HCl}}{1\ \text{mol }H^+} = 0.16\ M\,HCl$$

b. First we calculate the number of moles of $CaCO_3$ present in 750 mg:

$$0.750\ \text{g CaCO}_3 \times \frac{1\ \text{mol CaCO}_3}{100.09\ \text{g CaCO}_3} = 7.49 \times 10^{-3}\ \text{mol CaCO}_3$$

Next we use the stoichiometry of the neutralization reaction to calculate the volume of 0.16 M HCl this quantity of $CaCO_3$ can neutralize:

$$7.49 \times 10^{-3}\ \text{mol CaCO}_3 \times \frac{2\ \text{mol }H^+}{1\ \text{mol CaCO}_3} \times \frac{1\ \text{L}}{0.16\ \text{mol }H^+}$$

$$= 9.36 \times 10^{-2}\ \text{L} = 94\ \text{mL}\ 0.16\ M\,HCl$$

**Think About It** A concentration of 0.16 M seems reasonable because it is indeed within the range of values predicted for a solution with pH < 1. The volume of 0.16 M acid that a 750-mg tablet of $CaCO_3$ can neutralize is also reasonable; 94 mL represents about 3 ounces of gastric juice.

⚙ **Practice Exercise** Calculate the pH of a solution prepared by mixing 10.0 mL of 0.160 M HCl with 15.0 mL of water. How much antacid containing $4.00 \times 10^3$ mg of $Mg(OH)_2$ in 5.00 mL of water is needed to neutralize this volume of acid?

*(Answers to Practice Exercises are in the back of the book.)*

**CONCEPT TEST**

Taking an antacid tablet is often sufficient to treat an occasional case of mild acid reflux. Another remedy is a drug like Prilosec, which inhibits a cell's proton pumps by binding to the site of the pump and disabling it for more than 24 hours. Is the equilibrium constant for the binding of a proton pump inhibitor likely to be less than or greater than 1?

## Nitrogen

Nitrogen is a major essential element found primarily in proteins but also in DNA and RNA. Nitrogen is available in the atmosphere as $N_2$, and soil and water contain nitrate ions, but neither of these forms of nitrogen can be directly incorporated into amino acids, the building blocks of proteins. The biosynthesis of amino acids requires ammonia or ammonium ions. For example, glycine ($NH_2CH_2COOH$), the simplest amino acid, is formed by reaction of $CO_2$ and ammonia in the presence of the appropriate enzyme. Certain bacteria use enzymes called *nitrogenases* to convert $N_2$ to ammonia. Plants convert $NO_3^-$ ions to $NO_2^-$ and then to $NH_3$ by using enzymes called *reductases*. Ultimately, the chemical reactions in these organisms begin a food chain that supplies the essential amino acids for human diets.

---

**SAMPLE EXERCISE 22.2**    **Writing a Balanced Chemical Equation Describing the Reaction of Nitrate Reductases**    **LO2**

Nitrate ion can be reduced to ammonia by enzymes called nitrate reductases. The first step is conversion of nitrate ion to nitrite ion. Assign oxidation numbers to the elements in these ions and write a balanced equation for the following half-reaction in a basic solution:

$$NO_3^-(aq) \rightarrow NO_2^-(aq)$$

**Collect and Organize** We need to assign oxidation numbers based on the guidelines in Section 4.9.

**Analyze** The oxidation numbers of nitrogen and oxygen in a polyatomic ion must add up to the charge on the ion. Oxygen in compounds usually has an oxidation number of −2, whereas the oxidation number of nitrogen is unknown, so we begin by calling it $x$.

**Solve** The oxidation number of nitrogen in $NO_3^-$ is

$$x + 3(-2) = -1$$
$$x = +5$$

The oxidation number of nitrogen in $NO_2^-$ is

$$x + 2(-2) = -1$$
$$x = +3$$

The half-reaction is

$$NO_3^-(aq) \rightarrow NO_2^-(aq)$$

The nitrogen is balanced; we balance oxygen by adding water:

$$NO_3^-(aq) \rightarrow NO_2^-(aq) + H_2O(\ell)$$

Then we balance hydrogen by adding hydrogen ions:

$$2\,H^+(aq) + NO_3^-(aq) \rightarrow NO_2^-(aq) + H_2O(\ell)$$

We balance charge by adding electrons:

$$2\,e^- + 2\,H^+(aq) + NO_3^-(aq) \rightarrow NO_2^-(aq) + H_2O(\ell)$$

We switch to a basic solution by adding the same number of $OH^-$ as we have $H^+$ to both sides of the equation:

$$2\,OH^-(aq) + 2\,e^- + 2\,H^+(aq) + NO_3^-(aq) \rightarrow NO_2^-(aq) + H_2O(\ell) + 2\,OH^-(aq)$$

The hydrogen ions combine with the hydroxide ions to form water, and we cancel the species that are the same on both sides:

$$2\,e^- + \cancel{2}\,H_2O(\ell) + NO_3^-(aq) \rightarrow NO_2^-(aq) + \cancel{H_2O(\ell)} + 2\,OH^-(aq)$$

This gives us a final equation for the reduction half-reaction:

$$2\,e^- + H_2O(\ell) + NO_3^-(aq) \rightarrow NO_2^-(aq) + 2\,OH^-(aq)$$

**Think About It** Assigning oxidation numbers is a convenient way of identifying which element is reduced or oxidized in a half-reaction and of determining how many electrons are gained or lost. The balanced half-reaction confirms that electrons are added to nitrate to reduce it to nitrite ion.

⚙ **Practice Exercise** The reduction half-reaction catalyzed by one type of nitrogenase produces one mole of $H_2(g)$ for every two moles of $NH_4^+(aq)$ under acidic conditions. Write a balanced equation for this half-reaction.

◉◉ **CONNECTION** Another approach, based on balancing the number of O and H atoms with water molecules and $H^+$ ions and then balancing charges with $OH^-$ ions, is described in Chapter 4.

◉◉ **CONNECTION** The inorganic chemistry of nitrogen was discussed in Chapter 6.

In humans and other mammals, excess nitrogen is converted to urea in the liver and excreted via the kidneys. Plants use urea as a source of ammonia by the action of *ureases* via the reaction:

$$\underset{H_2N}{\overset{\overset{\displaystyle O}{\displaystyle \|}}{C}}\underset{NH_2}{} + H_2O \rightarrow 2\,NH_3 + CO_2$$

Unlike reactions catalyzed by nitrogenases and nitrate reductases, the conversion of urea to ammonia and carbon dioxide is not a redox reaction. It is a hydrolysis reaction, similar to the reaction of nonmetal oxides with water described in Chapter 3.

## 22.3 Trace and Ultratrace Essential Elements

Figure 22.6 shows the DRI values of trace (red) and ultratrace (blue) essential elements, including four main group elements (silicon, selenium, fluorine, and iodine) and nine transition elements (chromium, cobalt, copper, iron, manganese, molybdenum, nickel, vanadium, and zinc). Note that eight of the nine transition metals are from the fourth row of the periodic table. Some transition metals in the fifth and sixth periods are also found in enzymes, for example, Mo and W. Compounds of iron, zinc, silicon, and fluorine are present in the body in average concentrations between 1 and 1000 μg of element per gram of body mass and are considered to be trace elements. The remaining elements in Figure 22.6 are considered ultratrace, meaning essential elements present in the body in average concentrations less than 1 μg of element per gram of body mass.

### Trace Essential Main Group Elements

#### Silicon

In mammals, a lack of the trace essential element silicon stunts growth. The presence of silicon as silicic acid [$Si(OH)_4$] is believed to reduce the toxicity of $Al^{3+}$ ions in organisms by precipitating the aluminum as aluminosilicate minerals. Amorphous silica ($SiO_2$) is found in the exoskeletons of diatoms and in the cell membranes of some plants, such as the tips of stinging nettles.

**FIGURE 22.6** The elements shown in red are trace essential elements, and those shown in blue are ultratrace essential elements. The remaining labeled elements are nonessential. The vertical bars show DRI values for these elements in micrograms per day. The value for cobalt reflects the DRI for vitamin $B_{12}$, the principal source of cobalt in our diets.

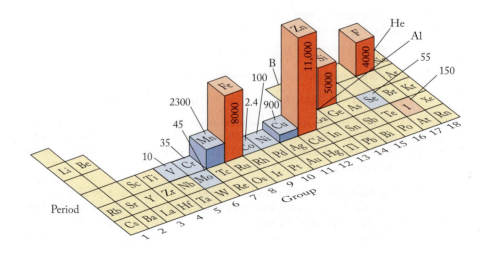

## Fluorine and Iodine

Fluoride ions have significant benefits for dental health. Tooth enamel is composed of the mineral hydroxyapatite, $Ca_5(PO_4)_3(OH)$, which is essentially insoluble in water:

$$Ca_5(PO_4)_3(OH)(s) \rightleftharpoons Ca_5(PO_4)_3^+(aq) + OH^-(aq) \qquad K_{sp} \approx 2.4 \times 10^{-59}$$

When hydroxyapatite comes into contact with weak acids in your mouth, this equilibrium shifts to the right as the acid reacts with the hydroxide ions. This shift effectively increases the solubility of hydroxyapatite, so that your tooth enamel becomes pitted, and dental caries form. This is an example of Le Châtelier's principle. Fluoride ions reduce the likelihood of caries by displacing the $OH^-$ ions in hydroxyapatite to form fluorapatite:

$$Ca_5(PO_4)_3(OH)(s) + F^-(aq) \rightleftharpoons Ca_5(PO_4)_3F(s) + OH^-(aq) \qquad K = 8.48$$

The solubility of fluorapatite is less dependent on pH than is the solubility of hydroxyapatite, so changing tooth enamel to fluorapatite makes your teeth more resistant to decay. This is why toothpaste contains fluoride compounds and why fluoride is added to drinking water in many communities in North America and Europe.

Of all the trace essential elements, iodine may have the best-defined role in human health. The body concentrates iodide ions in the thyroid gland, where they are incorporated into two hormones—thyroxine and 3,5,3'-triiodothyronine (Figure 22.7)—whose role is to regulate energy production and use. The conversion of thyroxine to 3,5,3'-triiodothyronine is catalyzed by selenocysteine-containing proteins. A deficiency of iodine or of either hormone can cause fatigue or feeling cold and can ultimately lead to an enlarged thyroid gland, a condition known as goiter. To help prevent iodine deficiency, table salt sold in the United States and many other countries is "iodized" with a small amount of sodium iodide. An excess of either hormone can cause a person to feel hot and is linked to Graves' disease, an autoimmune disease. The immune system in a patient with Graves' disease attacks the thyroid gland and causes it to overproduce the two hormones.

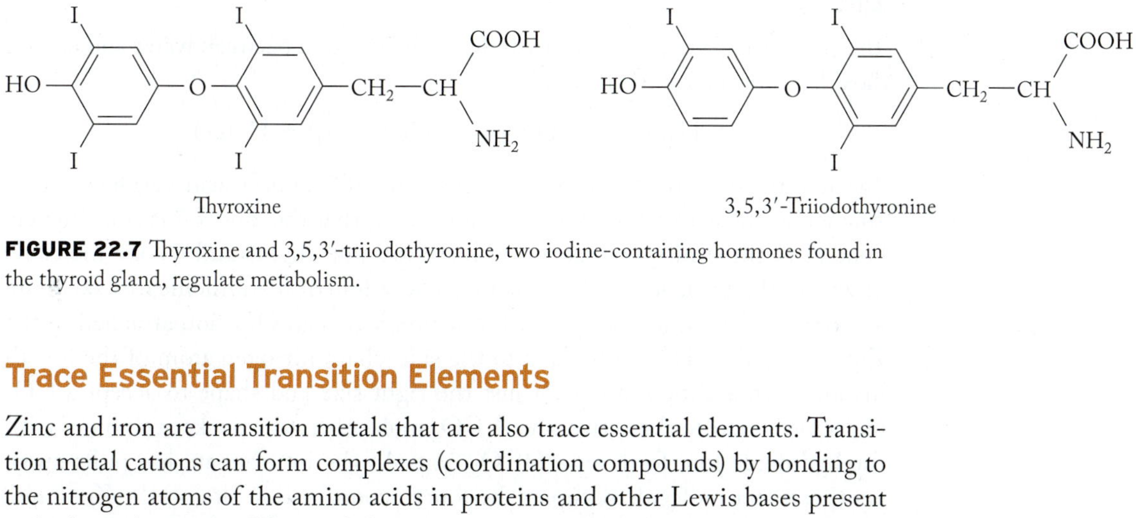

**CONNECTION** Solubility products, or $K_{sp}$, were introduced in Chapter 16. Le Châtelier's principle was discussed in Chapter 15.

**CONNECTION** In Chapter 10 we discussed the toxicity of halogens and some of their roles in medicine.

**FIGURE 22.7** Thyroxine and 3,5,3'-triiodothyronine, two iodine-containing hormones found in the thyroid gland, regulate metabolism.

## Trace Essential Transition Elements

Zinc and iron are transition metals that are also trace essential elements. Transition metal cations can form complexes (coordination compounds) by bonding to the nitrogen atoms of the amino acids in proteins and other Lewis bases present in biological systems. Many of the enzymes in our bodies contain zinc or iron ions; these enzymes, along with others containing transition metal ions, are called *metalloenzymes*. Table 22.6 lists some of the more important ones and the reactions they catalyze.

**TABLE 22.6  Selected Metalloenzymes and Some Reactions and Half-Reactions They Catalyze**

| Metal[a] | Enzyme | Reaction or Half-Reaction Catalyzed |
|---|---|---|
| V(Fe) | Nitrogenase | $N_2 + 10\,H^+ + 8\,e^- \rightarrow H_2 + 2\,NH_4^+$ |
| | Haloperoxidase | $CH_4 + H_2O_2 + Cl^- + H^+ \rightarrow CH_3Cl + 2\,H_2O$ |
| Mo | Nitrate reductase | $NO_3^- + 3\,H^+ + 2\,e^- \rightarrow H_2O + HNO_2$ |
| | Sulfite oxidase | $SO_3^{2-} + H_2O \rightarrow SO_4^{2-} + 2\,e^- + 2\,H^+$ |
| W(Fe) | Formate dehydrogenase | $HCOO^- \rightarrow CO_2 + 2\,e^- + H^+$ |
| Fe | Cytochrome P450 | $R\!-\!H + O_2 + 2\,e^- + 2\,H^+ \rightarrow R\!-\!OH + H_2O$ |
| | Peroxidase | $RCH_2COOH + 2\,H_2O_2 \rightarrow 3\,H_2O + RCHO + CO_2$ |
| Co | Coenzyme $B_{12}$ | Required as coenzyme for many reactions |
| Ni | Urease | (see structure) $+ H_2O \rightarrow 2\,NH_3 + CO_2$ |
| Ni(Fe) | Hydrogenase | $H_2 \rightarrow 2\,H^+ + 2\,e^-$ |
| Cu | $N_2O$ reductase | $N_2O + 2\,e^- + 2\,H^+ \rightarrow N_2 + H_2O$ |
| | Amine oxidase | $CH_3NH_2 + O_2 + H_2O \rightarrow CH_2O + H_2O_2 + NH_3$ |
| Zn | Carbonic anhydrase | $H_2O + CO_2 \rightleftharpoons HCO_3^- + H^+$ |
| | Carboxypeptidase | (see structure) |

[a]The metal in parentheses is also present in the enzyme and is essential to its function.

## Zinc

The enzyme carbonic anhydrase catalyzes the reaction between water and carbon dioxide to form bicarbonate ions:

$$H_2O(\ell) + CO_2(aq) \rightleftharpoons HCO_3^-(aq) + H^+(aq)$$

The α-form of carbonic anhydrase contains 260 amino acid residues and a zinc ion at the active site. Note in Figure 22.8 that the zinc ion is coordinately bonded to three nitrogen atoms on histidine side chains and to one molecule of water. The presence of these ligands and a fourth histidine nearby facilitate ionization of the water molecule. Ionization leaves an $OH^-$ ion attached to the $Zn^{2+}$ ion and an $H^+$ ion bonded to the side-chain nitrogen atom of the fourth histidine. In addition, a pocket just the right size and shape to accept a $CO_2$ molecule is next to the active site. A $CO_2$ molecule in the pocket combines with the hydroxide ion, forming a $HCO_3^-$ ion. As the bicarbonate ion pulls away, another water molecule occupies the fourth coordination site on the $Zn^{2+}$ ion, another $CO_2$ molecule enters the pocket, the histidine is protonated, and the catalytic cycle can be repeated. This reaction is important because it helps eliminate $CO_2$ from cells during respiration and mediates the uptake of $CO_2$ during photosynthesis in some plants.

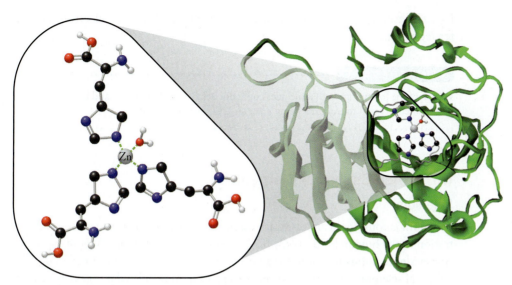

**FIGURE 22.8** The active site of one form of carbonic anhydrase consists of a zinc atom bonded to three histidine molecules and one $H_2O$ molecule. The $OH^-$ ion produced when this $H_2O$ ionizes combines with a molecule of $CO_2$, forming a $HCO_3^-$ ion.

## Iron

In Chapter 20 we discussed the role of iron in hemoglobin's ability to transport oxygen throughout the body. In hemoglobin, $Fe^{2+}$ is coordinated to four nitrogen atoms in the porphyrin ring of a *heme group*. Two additional Lewis bases—water and the amino acid histidine—complete the coordination sphere of $Fe^{2+}$. The heme group is found in several iron-containing enzymes.

Iron-containing peroxidases and catalases are integral to the transfer of oxygen to biomolecules. For example, plants use a fatty acid peroxidase to catalyze the stepwise degradation of fatty acids. One $CH_2$ group at a time is removed from the fatty acid using hydrogen peroxide:

$$\underset{\text{Fatty acid}}{R-CH_2-COOH(aq)} + \underset{\substack{\text{Hydrogen} \\ \text{peroxide}}}{2\,H_2O_2(aq)} \xrightarrow{\substack{\text{fatty acid} \\ \text{peroxidase}}}$$

$$3\,H_2O(\ell) + \underset{\text{Aldehyde}}{R-CHO(aq)} + CO_2(aq)$$

Subsequent oxidation of the aldehyde back to a carboxylic acid yields a new fatty acid with one fewer $CH_2$ group:

$$\underset{\text{Aldehyde}}{2\,R-CHO(aq)} + O_2(aq) \rightarrow \underset{\text{Fatty acid}}{2\,RCOOH(aq)}$$

Iron-containing enzymes called cytochromes catalyze electron transport in photosynthesis and the metabolism of glucose to $CO_2$ and water. The conversion of glucose, $C_6H_{12}O_6$, to $CO_2$ involves the oxidation of carbon. The oxidation number of carbon in $CO_2$ is +4, but the average oxidation number of the carbon atoms in $C_6H_{12}O_6$ is zero.

Iron-containing enzymes also catalyze the reduction of nitrite $(NO_2^-)$ and sulfite $(SO_3^{2-})$ ions:

$$NO_2^-(aq) + 6\,e^- + 8\,H^+(aq) \xrightarrow{\text{nitrite reductase}} NH_4^+(aq) + 2\,H_2O(\ell)$$

$$SO_3^{2-}(aq) + 6\,e^- + 7\,H^+(aq) \xrightarrow{\text{sulfite reductase}} HS^-(aq) + 3\,H_2O(\ell)$$

**CONNECTION** Examples of coordination complexes of transition metals with biologically important ligands can also be found in Chapter 17.

**CONNECTION** The hemoglobin molecule was described in Section 20.2 and shown in Figure 20.17(a).

**CONNECTION** The assignment of oxidation numbers to organic compounds was discussed in Chapter 4.

In each case, the iron in the enzyme is oxidized, providing the electrons needed for reduction.

**CONCEPT TEST** ....................................................................

Which of the following small molecules could not behave as a Lewis base and hence would not be expected to bond to iron in a heme protein? $CO$; $H_2$; $NO_2$; $H_2S$

....................................................................

## Ultratrace Essential Elements

### Selenium

Selenium is considered an ultratrace essential element because its average concentration in the human body is only 0.3 µg per gram of body mass. Mounting scientific evidence points to a need for a minimum daily dose of selenium of 55 µg. The effects of selenium toxicity, however, are apparent in people who ingest more than 500 µg per day. Most of the selenium we need is obtained from selenium-rich produce (garlic, mushrooms, asparagus) or from fish. Selenium occurs in the body as the amino acid selenocysteine (Figure 22.9) and is incorporated into enzymes.

Selenocysteine is an antioxidant. Our bodies need oxygen to survive, yet living in an oxygen-rich atmosphere can lead to the formation of potentially dangerous oxidizing agents in cells. For example, metabolism of fatty acids forms oxidizing agents called alkyl hydroperoxides, which can attack the lipid bilayer of cell membranes. It is believed that aging is related to the inability of the body to inhibit oxidative degradation of tissue. Selenocysteine participates in a series of reactions that result in the decomposition of these alkyl hydroperoxides.

### Molybdenum and Vanadium

Many of the metalloenzymes in Table 22.6 are involved in transformations of nitrogen. Molybdenum-containing reductases are responsible for converting $NO_3^-$ ions to $NO_2^-$ and then to $NH_3$ (Section 22.2). The active site of sulfite oxidase, which converts $SO_3^{2-}$ to $SO_4^{2-}$, also contains molybdenum. Sometimes more than one type of transition metal is found in a metalloenzyme. Both molybdenum and iron are required by xanthine oxidase, an important enzyme along the pathway for degradation of excess nucleic acids (adenine and guanine) to xanthine and then to uric acid for elimination through the kidneys.

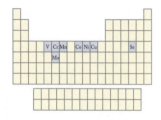

**CONNECTION** The inorganic chemistry of oxygen and sulfur, which like selenium are in group 16 of the periodic table, was discussed in Chapter 9.

**FIGURE 22.9** Selenocysteine is the selenium-containing analog of the amino acid cysteine. Much of the selenium in the human body is found in proteins containing selenocysteine.

Xanthine $+ H_2O + O_2 \xrightarrow[H^+]{\text{xanthine oxidase}}$ Uric acid $+ H_2O_2$

The combination of iron and vanadium is essential to the function of haloperoxidases, a class of enzymes found in some algae, lichens, and fungi that replace C—H bonds with carbon–halogen bonds. The reaction products may function in the defense systems of these organisms.

### Copper

Copper-containing proteins perform several functions in both plants and animals, including oxygen transport in mollusks such as clams and oysters. Copper is also an essential element in the enzymes azurin and plastocyanin, which mediate electron transfer during photosynthesis.

Reactions catalyzed by xanthine oxidase may produce other reactive oxygen species besides $H_2O_2$. One of them is the superoxide ion, $O_2^-$. Superoxide ion is a strong oxidizing agent and must be eliminated to prevent cell damage. The removal of superoxide begins with the action of superoxide dismutases, which convert superoxide to hydrogen peroxide:

$$2\ O_2^-(aq) + 2\ H^+(aq) \xrightarrow{\text{superoxide dismutase}} H_2O_2(aq) + O_2(aq)$$

Researchers have isolated superoxide dismutases containing a variety of transition metals, including a copper–zinc enzyme. The hydrogen peroxide produced in this reaction is decomposed to water and oxygen by iron-containing catalase.

> **CONCEPT TEST**
>
> Why is superoxide ion a good oxidizing agent?

**coenzyme** organic molecule that, like an enzyme, accelerates the rate of biochemical reactions.

## Nickel

Ureases, the enzymes responsible for the conversion of urea to ammonia in plants, contain nickel(II). Nickel is also found with iron in enzymes called *hydrogenases*. Hydrogenases oxidize hydrogen gas to protons:

$$H_2(g) \xrightarrow{\text{hydrogenase}} 2\ H^+(aq) + 2\ e^-$$

Nickel and iron also combine in CO dehydrogenase, an enzyme that catalyzes the formation of acetyl-CoA, which is a key component in the tricarboxylic acid cycle discussed in Section 20.3. Finally, nickel-containing enzymes are found among the enzymes responsible for methane generation by bacteria, as discussed in Section 13.5.

## Cobalt and Coenzymes

Many enzymes require the presence of a **coenzyme**, which is an organic compound that cocatalyzes a biochemical reaction. The coenzyme $B_{12}$ (Figure 22.10) contains the ultratrace essential element cobalt(III) and is a derivative of vitamin $B_{12}$.

The cobalt(III) in coenzyme $B_{12}$ is easily reduced to cobalt(II) and even cobalt(I) in the course of enzyme-catalyzed redox reactions. The change in oxidation state of Co allows for facile transfer of methyl groups, as in the conversion of methionine to homocysteine:

Methionine $\longrightarrow$ Homocysteine

Coenzyme $B_{12}$ is also critical to the function of *mutases*, which are enzymes that catalyze the rearrangement of the skeleton of a molecule, as in the interconversion of glutamate and methylaspartate:

Glutamate $\underset{\text{coenzyme } B_{12}}{\overset{\text{glutamate mutase}}{\rightleftharpoons}}$ Methylaspartate

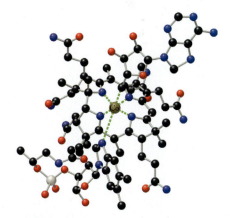

**FIGURE 22.10** Coenzyme $B_{12}$ contains cobalt(III), an ultratrace essential metal.

## Manganese and Photosynthesis

We noted in discussing the chlorophyll structures in Figure 22.5 that the role of the major essential element $Mg^{2+}$ is to orient the chlorophyll molecules in the proper way. Another metal, the ultratrace essential element manganese, also plays a role in the production of oxygen during photosynthesis. To examine that role, let's write an equation for photosynthesis that is slightly different from the equation we are used to seeing. Instead of writing the formula $C_6H_{12}O_6$ for glucose, we use the generic carbohydrate formula $(CH_2O)_n$ so that the coefficient is 1 for all other species in the reaction (rather than 6):

$$H_2O(\ell) + CO_2(g) \rightarrow \tfrac{1}{n}(CH_2O)_n(aq) + O_2(g)$$

Writing the equation in this form makes it easier to see how the overall reaction between water and carbon dioxide involves electron transfer. Photosynthesis is a redox reaction. As usual, the oxidation number for O is −2 in $H_2O$ and $CO_2$, and 0 in $O_2$. This increase in oxidation number means the O is oxidized. The $CO_2$ oxygen atom ends up in the $(CH_2O)_n$ molecule, and the $H_2O$ oxygen atom ends up in the $O_2$. We can write two half-reactions to describe this overall redox reaction:

$$4\,e^- + 4\,H^+ + CO_2 \rightarrow CH_2O + H_2O$$
$$2\,H_2O \rightarrow O_2 + 4\,H^+ + 4\,e^-$$

When the O atom in $H_2O$ is oxidized to $O_2$, what is reduced? If the oxidation number of O is −2 in $CO_2$, then the oxidation number of C must be +4. All of the carbon in $CO_2$ ends up as $CH_2O$, in which the oxidation number of carbon is zero. Thus photosynthesis is accompanied by the reduction of $CO_2$, since the oxidation number of carbon decreases from +4 to 0. The redox process requires metalloenzymes. Manganese-containing biomolecules in which Mn is in the +3 and +4 oxidation states mediate the transfer of electrons from water in photosynthesis. Although the exact structure of these manganese compounds remains undetermined, it is believed that two Mn(III) ions and two Mn(IV) ions are present at the site of $O_2$ production. Recall that copper-containing enzymes are also involved in the series of reactions that make up photosynthesis.

**CONCEPT TEST** ...........................................................

Manganese ions involved in photosynthesis are often surrounded by six Lewis base ligands in an octahedral geometry. Is reduction of a Mn(IV) ion to Mn(III) in an octahedral complex (see Section 17.9) accompanied by a change in spin state of the Mn ion?

..............................................................................

## Chromium

Chromium in the +3 oxidation state is an ultratrace essential element in our diets. It is involved in regulating glucose levels in the blood through a molecule called chromodulin. Chromodulin is a polypeptide incorporating only 4 of the 20 naturally occurring amino acids: glycine, cysteine, glutamic acid, and aspartic acid. Four $Cr^{3+}$ ions are bound to the peptide chain. Cereals and grains contain enough chromium for our daily needs, but certain plants (such as shepherd's purse) concentrate chromium and have been used as herbal remedies in diabetes treatment.

Chromium in the +6 oxidation state, as found in chromate ions ($CrO_4^{2-}$), is acutely toxic and also carcinogenic. Chromate ion enters cells through ion channels that transport $SO_4^{2-}$ ions. Once inside, $CrO_4^{2-}$ is reduced to $Cr^{3+}$, which binds to the phosphate backbone of DNA.

# 22.4 Nonessential Elements

Table 22.2 lists elements that are found in the human body but are classified as nonessential. In this section, we discuss how some of these elements may end up in our bodies, working our way from left to right across the periodic table.

### Rubidium and Cesium

Rubidium is generally regarded as nonessential, yet it is the 15th most abundant element in the body. It is believed that $Rb^+$ is retained by the body because of the similarity of its size and chemistry to that of $K^+$. Like the other cations of group 1, cesium ions ($Cs^+$) are also readily absorbed by the body. Cesium cations have no known function, although they can substitute for $K^+$ and interfere with potassium-dependent functions. In most cases, the concentration of cesium in the environment is low, so exposure to $Cs^+$ is not a health concern. The nuclear accident at Chernobyl in 1986, however, released significant quantities of radioactive $^{137}Cs$ into the environment. The ability of $Cs^+$ to substitute for $K^+$ led to the incorporation of $^{137}Cs^+$ into plants, which rendered crops grown in the immediate area unfit for human consumption because of the radiation hazard posed by this long-lived ($t_{1/2} \approx 30$ yr) β emitter.

### Strontium and Barium

Some single-celled organisms build exoskeletons made with $SrSO_4$ and $BaSO_4$, but the human body appears to have no use for $Sr^{2+}$ and $Ba^{2+}$ ions. These ions do find their way into human bones, where they replace $Ca^{2+}$ ions. At the low concentrations of $Sr^{2+}$ and $Ba^{2+}$ that are typically present in the human body, these elements appear to be benign. However, as in the case of radioactive $^{137}Cs$, incorporation of $^{90}Sr$ ($t_{1/2} = 29$ yr) in bones can lead to leukemia. Atmospheric testing of nuclear weapons over the Pacific Ocean and in sparsely populated regions of the American West in the 1950s released $^{90}Sr$ into the environment. The full extent of the toxic effects of the fallout from these tests did not become apparent for several decades.

### Germanium

It is generally agreed that germanium is a nonessential element and is barely detectable in the human body. The use of bis(carboxyethyl)germanium sesquioxide (Figure 22.11) has been touted as a nutritional supplement, but its efficacy remains controversial.

### Antimony

The role of antimony is also poorly understood. Most antimony compounds are toxic; they cause liver damage. However, ultratrace amounts of antimony may have a stimulatory effect, and selected antimony compounds have been used medically as antiparasitic agents, as discussed in the next section.

**CONCEPT TEST** ·······························

Looking at groups 1, 2, 14, and 15, what periodic trend do you see in the location of the nonessential elements relative to the essential elements in the same group?

**CONNECTION** The biological effects of different types of radiation were described in Chapter 21.

$O_3(GeCH_2CH_2COOH)_2$

**FIGURE 22.11** Bis(carboxyethyl)germanium sesquioxide has been sold as a nutrition supplement, but its benefits are not well established.

## 22.5 Elements for Diagnosis and Therapy

So far we have talked about the biological roles of approximately 30 essential and nonessential elements found in our bodies. Some of these 30 elements are also useful in diagnosing and/or treating diseases, as are some of the 60 other elements in the periodic table (Figure 22.12). In this section, we describe some of the applications of radioactive isotopes in the diagnosis of diseases. We also explore how compounds of essential and nonessential elements have found application in the treatment of a wide range of illnesses.

**FIGURE 22.12** The elements shown in red are used in imaging, those shown in green are used in therapy, and those shown in blue are used in medical devices. Elements with multiple uses are shown in two colors.

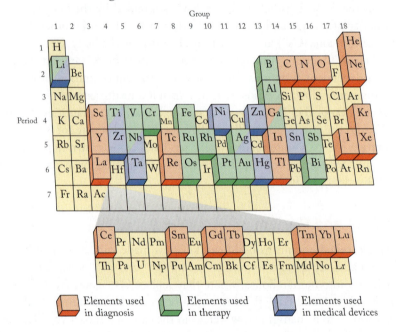

Any diagnostic or therapeutic compound that is injected intravenously must be sufficiently soluble in blood to be delivered to the target. While in transit, the compound must be stable enough not to undergo chemical reactions that result in its precipitation or rapid elimination from the body. Occasionally, the compound can be in the form of a simple salt, but more often a metal ion is introduced as a coordination complex or coordination compound. Ligands used in forming biologically active coordination complexes include amino acids and simple anions like the citrate ion. Chelating ligands like diethylenetriaminepentaacetate (DTPA$^{5-}$; Figure 22.13) are often used in biological applications. A medicinal

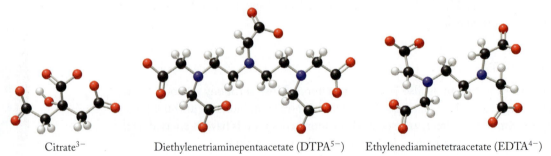

Citrate$^{3-}$     Diethylenetriaminepentaacetate (DTPA$^{5-}$)     Ethylenediaminetetraacetate (EDTA$^{4-}$)

**FIGURE 22.13** Citrate$^{3-}$, diethylenetriaminepentaacetate (DTPA$^{5-}$), and ethylenediaminetetraacetate (EDTA$^{4-}$) are often used as chelating ligands for diagnostic and therapeutic agents based on transition metals. These ions form stable complex ions with 2+ and 3+ metal cations. The solubilities of the complex ions are typically much greater than those of the hydrated ions at physiological pH (7.4).

chemist can also take advantage of substances that occur naturally in the body, such as antibodies, to carry a diagnostic or therapeutic metal ion to its target.

## Diagnostic Applications

Physicians in the 21st century have an array of imaging agents to help in diagnosing disease. Some methods use radionuclides with short half-lives that emit easily detectable gamma rays. Examples cited in Chapter 21 include the use of iodine-131 to image the thyroid gland (Figure 22.14) and of neutron-poor isotopes such as carbon-11 and fluorine-18 for positron emission tomography (PET). Not all imaging depends on radionuclides, however. In magnetic resonance imaging (MRI), for instance, which can diagnose soft-tissue injuries, stable isotopes of gadolinium are used to enhance images.

### Imaging with Radionuclides

The radionuclides used in medicine have short half-lives to limit the patient's exposure to ionizing radiation. If the half-life is too short, however, the nuclide may either decay before it can be administered or not reach the target organ rapidly enough to provide an image. Emission of relatively low-energy γ rays is essential to preventing collateral tissue damage.

Nuclide selection is also governed by the toxicity of both the parent element and the daughter nuclide. The speed at which the imaging agent is eliminated from the body can help mitigate toxic effects. Naturally, the cost and availability of a particular nuclide also factor into its usefulness in the clinical setting.

**CONCEPT TEST** ......................................................

Why is it important to consider the nature of the decay products—α, β, or γ particles, or positrons—when choosing a radionuclide for medical imaging?

......................................................

### Gallium, Indium, and Thallium

Gallium-66, gallium-67, gallium-68, and indium-111 are used as imaging agents for tumors and leukemia. All four nuclides decay by electron capture, and the γ radiation emitted in this nuclear reaction produces the images. In addition, the three gallium isotopes also decay by positron emission, which makes compounds containing these isotopes attractive for positron-emission tomography. The half-lives range from just over 1 hr for gallium-68 to 78 hr for gallium-67. The discovery that indium-111-containing compounds can image a range of cancers has led to the development of the drug Zevalin®, currently used to treat some forms of non-Hodgkin's lymphoma.

The use of the gamma emitter thallium-201 ($t_{1/2}$ = 73 hr) in diagnosing heart disease presents an interesting case for considering the risks and benefits of using a particular isotope in medicine. Although thallium compounds are among the most toxic metal-containing compounds known, the nanogram quantities required for diagnosis pose few, if any, health hazards, meaning that the benefits outweigh the risks.

### Technetium and Rhenium

Technetium and rhenium are transition metals in group 7 of the periodic table, just below manganese. Technetium is unusual in that it does not exist naturally

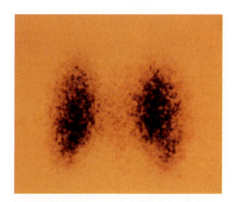

**FIGURE 22.14** Gamma radiation that accompanies the decay of iodine-131 can be used to image the two butterfly-shaped lobes of the thyroid gland.

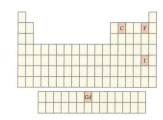

**CONNECTION** Magnetic resonance imaging (MRI) was mentioned in Chapter 3 in our discussion of the use of cryogenic helium for superconducting magnets.

**CONNECTION** Chapter 21 gave a more detailed discussion of nuclear chemistry and nuclear medicine, including an assessment of the effects of different types of radiation on living tissue.

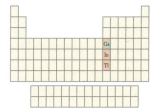

on Earth in easily measurable amounts because it has no stable isotopes; in other words, all technetium isotopes are radioactive. These isotopes can be produced in nuclear reactors for use in medicine. The $^{99m}$Tc used in hospitals for imaging is prepared in technetium generators in which a stable isotope of molybdenum, $^{98}$Mo (23.78% natural abundance), is bombarded with neutrons. Technetium-99 has a half-life greater than 20,000 years; however, when it is produced in a nuclear reactor, its nucleus is in an excited state, called a *metastable* nucleus. The metastable state, designated by adding the letter "m" to the mass number, as in technetium-99m or $^{99m}$Tc, has a half-life of six hours ($t_{1/2}$ of $^{99m}$Tc = 6.0 hr) and decays to the more stable technetium-99 nucleus.

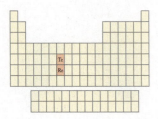

---

**SAMPLE EXERCISE 22.3**  **Identifying Particles in Nuclear Reactions**  **LO4**

Technetium-99m produced from molybdenum-98 decays by $\gamma$ emission followed by $\beta$ decay. Complete the nuclear equations representing this sequence by identifying the missing particles P1, P2, and P3:

$$^{98}_{42}\text{Mo} + \text{P1} \rightarrow {}^{99}_{42}\text{Mo} \rightarrow {}^{99m}_{43}\text{Tc} + \text{P2} \qquad \text{(A)}$$

$$^{99m}_{43}\text{Tc} \rightarrow {}^{99}_{43}\text{Tc} + \gamma \qquad \text{(B)}$$

$$^{99m}_{43}\text{Tc} \rightarrow \text{P3} + {}^{0}_{-1}\beta \qquad \text{(C)}$$

**Collect and Organize** We are provided with three nuclear equations, two of which are missing one or more particles and are not balanced.

**Analyze** When an unstable nucleus decays, it can eject a $\beta$ particle ($^{0}_{-1}\beta$), a positron, ($^{0}_{1}\beta$), or an $\alpha$ particle ($^{4}_{2}\text{He}$) from its nucleus, along with gamma rays. The sum of the electrical charges denoted by the subscripts and the sum of the masses denoted by the superscripts to the left of the reaction arrow must equal the comparable sums to the right of the arrow. To find each missing particle, we must determine its electrical charge and mass. Because Tc and Mo account for most of the mass in Equation A, we predict that the subscripts and superscripts of P1 and P2 will be small. Particle P3, however, will have an atomic number and mass number close to that of Tc.

**Solve** In the first step of Equation A,

$$^{98}_{42}\text{Mo} + \text{P1} \rightarrow {}^{99}_{42}\text{Mo}$$

the atomic number is unchanged, meaning the molybdenum nucleus stays a molybdenum nucleus. The mass changes by one mass unit, from $^{98}$Mo to $^{99}$Mo. The particle with a mass number equal to 1 and a charge of 0 is a neutron ($^{1}_{0}$n) Thus, P1 is a neutron, and we conclude that molybdenum-98 is bombarded by neutrons to form molybdenum-99.

In the second step of Equation A,

$$^{99}_{42}\text{Mo} \rightarrow {}^{99m}_{43}\text{Tc} + \text{P2}$$

the mass number does not change, telling us that P2 must have negligible mass. The atomic number (number of protons) increases from 42 to 43, so the charge on P2 must be −1:

$$42 = 43 + x$$
$$x = -1$$

The particle that has negligible mass and a charge of −1 is a $\beta$ particle ($^{0}_{-1}\beta$). P2 is a $\beta$ particle.

To balance Equation C,

$$^{99}_{43}\text{Tc} \rightarrow \text{P3} + {}^{0}_{-1}\beta$$

P3 must have a mass number of 99. Its charge (that is, the number of protons it contains), determined from the subscripts, must be 44:

$$43 = x + (-1)$$
$$x = 44$$

The element with an atomic number of 44 is ruthenium, so P3 is ruthenium-99, and our sequence is

$$\ce{^{98}_{42}Mo + ^{1}_{0}n -> ^{99}_{42}Mo -> ^{99m}_{43}Tc + ^{0}_{-1}\beta} \tag{A}$$

$$\ce{^{99m}_{43}Tc -> ^{99}_{43}Tc + \gamma} \tag{B}$$

$$\ce{^{99}_{43}Tc -> ^{99}_{44}Ru + ^{0}_{-1}\beta} \tag{C}$$

**Think About It** As predicted, the subscripts (charges) and superscripts (mass numbers) of particles 1 and 2 in Equation A are small. Because $\gamma$ rays have no mass or charge, there is no change in the atomic or mass number of technetium-99m when it emits a $\gamma$ ray in Equation B. We also confirm our prediction that the $\beta$ decay of $^{99}$Tc in Equation C produces an isotope with atomic and mass numbers comparable to those of $^{99}$Tc. Recall that $\beta$ decay leads to an increase of 1 in atomic number (number of protons).

**Practice Exercise** Gamma rays emitted by thallium-201 can be used for cardiac imaging. Thallium-201 decays by electron capture (a proton and an electron form a neutron). Write a balanced nuclear equation for this decay.

Technetium-99m has been widely used as an imaging agent because it has a short half-life and emits low-energy $\gamma$ rays. Patients can be injected with a variety of technetium compounds, depending on the target organ. For imaging the heart, the coordination compounds shown in Figure 22.15 are used.

The ability to deliver radionuclides to a particular organ opens the possibility for selective irradiation of a tumor located in that organ. Therefore, some radionuclides can be used to both image and treat cancers. Rhenium, for example, is being studied as both an imaging and a therapeutic agent for many tumors, including breast, liver, and skin cancer. Rhenium-186 and rhenium-188 undergo $\beta$ decay with half-lives of 3.72 d and 17.0 hr, respectively. Certain tumors have highly selective receptor sites for particular molecules on their surfaces. By including a radioactive rhenium ion in a molecule that binds strongly and specifically to these receptors, physicians can deliver both an imaging agent and a therapeutic agent to the tumor. In principle, the $\beta$ particles destroy the tumor. Several patents have been issued for the use of compounds containing rhenium isotopes, but therapies based on these compounds remain in the experimental stage. One example of a rhenium compound used in these applications is shown in Figure 22.16.

### Scandium, Yttrium, and Lanthanide Elements

Scandium, yttrium, and lanthanum are in group 3 in the periodic table, the first group in the transition metal series. Scandium-46 has been used to image the spleen, and scandium-47 shows promise in diagnosing breast cancer. Among the available radioactive isotopes of yttrium, yttrium-90 has been applied to imaging a variety of tumors, including intestinal, breast, and thyroid cancers. Chelating ligands are used to complex the $^{90}\text{Y}^{3+}$ cation and deliver it to the tumor, where it binds strongly to receptors on the surface of the tumor. Yttrium-90 agents for treatment of non-Hodgkin's lymphoma are currently in clinical trials.

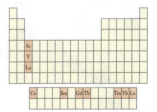

Radioactive isotopes of the lanthanides have seen extensive use in scintigraphic imaging, a procedure that uses a scintillation counter (Section 21.5) to measure the light emitted when radiation strikes a phosphor-coated screen in the instrument. The intensity of the emitted light is translated into a three-dimensional image of the organ of interest, a method similar to the PET technique described in Section

**FIGURE 22.15** (a) Cardiolite® [Tc(CNR)₆] and (b) Myoview [TcO₂(RPCH₂CH₂PR)₂] are used for imaging the heart. (c) Images of a patient's heart after intravenous injection with a technetium-containing drug. The images along the bottom show four measurements of heart function. Radioactive technetium compounds emit gamma rays that allow the blood to be tracked as it is pumped through the heart.

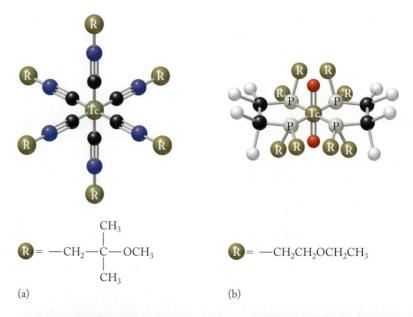

$$R = -CH_2-\overset{\overset{\displaystyle CH_3}{|}}{\underset{\underset{\displaystyle CH_3}{|}}{C}}-OCH_3$$

(a)

$$R = -CH_2CH_2OCH_2CH_3$$

(b)

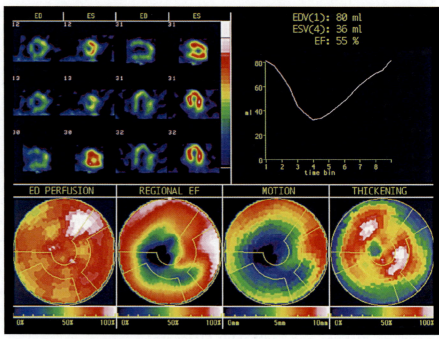

(c)

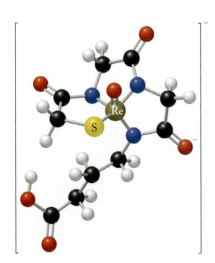

**FIGURE 22.16** This rhenium–mercaptoacetylglycylglycyl-γ-amino acid complex is used to attach $^{186}Re$ or $^{188}Re$ to monoclonal antibodies or peptides.

21.7. Among the nuclides used for this purpose are cerium-141, samarium-153, gadolinium-153, terbium-160, thulium-170, ytterbium-169, and lutetium-177. They have an advantage over gallium isotopes in having lower retention times in blood, thereby reducing the possibility of potentially harmful biological side effects. In some cases, lutetium-177 derivatives have replaced yttrium-90 compounds under investigation as radioactive antitumor agents.

**CONCEPT TEST** • • • • • • • • • • • • • • • • • • • • • • • • • • • • • • • • • • • •

What is wrong with the statement, "A radioactive isotope with twice the half-life of another will allow an image to be collected in half the time"?

• • • • • • • • • • • • • • • • • • • • • • • • • • • • • • • • • • • • • • • • • • • • • •

SAMPLE EXERCISE 22.4 | **Calculating Quantities of Radioactive Isotopes** | LO4

Two isotopes of yttrium, yttrium-86 ($t_{1/2}$ = 14.6 hr) and yttrium-90 ($t_{1/2}$ = 64 hr), have been used in radioimaging and therapy. Which decays faster? If we start with 10.0 mg of each isotope, how much of each remains after 24 hr?

**Collect and Organize** We are given the half-lives of two radionuclides and asked to predict which one will decay faster. We are also asked to calculate how much of 10.0 mg of each isotope remains after 24 hr.

**Analyze** An isotope with a shorter half-life decays faster. Radioactive decay follows first-order kinetics. Quantitatively, the relationship between half-life and amount of material remaining is described by Equation 21.4:

$$\ln\frac{N_t}{N_0} = \frac{-0.693t}{t_{1/2}}$$

Here $N_0$ and $N_t$ refer to the amount of material present initially and the amount at time $t$, respectively. If our prediction for the relative decay rates of the two yttrium isotopes is correct, then more of the isotope with the longer half-life should remain after 24 hr. We need to complete two calculations to determine the amount of each sample present after 24 hr.

**Solve** For the amount of yttrium-86 remaining after 24 hr, we have

$$\ln\frac{N_t}{10.0 \text{ mg}} = \frac{(-0.693)(24 \text{ hr})}{14.6 \text{ hr}} = -1.139$$

Taking the antilog of both sides, we get

$$\frac{N_t}{10.0 \text{ mg}} = 0.320$$

$$N_t = (0.320)(10.0 \text{ mg}) = 3.20 \text{ mg}$$

For yttrium-90,

$$\ln\frac{N_t}{10.0 \text{ mg}} = \frac{(-0.693)(24 \text{ hr})}{64 \text{ hr}} = -0.2599$$

$$\frac{N_t}{10.0 \text{ mg}} = 0.771$$

$$N_t = 7.7 \text{ mg}$$

**Think About It** We predicted that yttrium-86 would decay faster, which means that after 24 hr the quantity of this isotope should be less than the quantity of yttrium-90, and it is.

**Practice Exercise** Two radioactive isotopes of rhenium are used to treat breast, liver, and skin cancer. The half-lives are 89 hr for rhenium-186 and 17 hr for rhenium-188. If we start with 25.0 mg of each isotope, how much of each sample remains after 24 hr?

**CONNECTION** We discussed first-order reactions in Chapter 14 and applied the same equations to the first-order kinetics of radioactive decay in Chapter 21.

## Imaging with Stable Isotopes: Lanthanides, Noble Gases, and MRI

As noted earlier, the quality of an MRI scan (Figure 22.17) can be improved when gadolinium is used as a contrast agent. Prior to the procedure, the patient is injected with a gadolinium compound. Many ligands have been investigated for $Gd^{3+}$ MRI contrast agents. The gadolinium used is a mixture of naturally occurring isotopes of gadolinium. Coordination compounds of other lanthanide ions ($Dy^{3+}$ and $Ho^{3+}$) have been evaluated but do not work as well as those of gadolinium.

So far in this chapter we have had little opportunity to mention the noble gas elements. None of these elements are essential to the human body. Their lack of

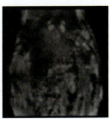

(a) Unenhanced

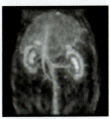

(b) Enhanced

**FIGURE 22.17** MRI scans made (a) without a contrast agent and (b) enhanced by a gadolinium contrast agent.

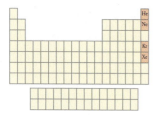

chemical reactivity and ease of introduction into the body by inhalation, however, make selected isotopes of the noble gases, including helium-3, krypton-83, and xenon-129, attractive as agents for enhancing MRI images, particularly of the lungs. Krypton-83 provides greater sensitivity than xenon-129, allowing for better resolution in the images and, in principle, requiring the use of less gas.

Helium-3 has no known side effects and is preferable to xenon-129 for MRI, but it is present in only trace natural abundance. This isotope is obtained from β decay of tritium ($^3$H):

$$^3_1H \rightarrow \,^3_2He + \,^0_{-1}\beta$$

Neon-19 has been used in PET despite its short half-life (17.5 s). A patient positioned in a PET scanner breathes air containing a small amount of this isotope. Positron emission from the neon is recorded, and an image is created. Of the group 18 elements, only argon and radon do not currently have direct medical applications.

## Therapeutic Applications

In this section we examine therapeutic agents that contain metallic elements (including alkali metals), transition elements, and heavier main group elements in addition to carbon, hydrogen, nitrogen, oxygen, and sulfur.

### Lithium, Boron, Aluminum, and Gallium

The similar size of Li$^+$ (76 pm) and Mg$^{2+}$ (72 pm) means that lithium ions can compete with magnesium ions in biological systems. The substitution of lithium for magnesium may account for its toxicity at high concentrations. Nevertheless, lithium carbonate is used to treat bipolar disorder, and other lithium compounds have been used to treat hyperactivity. In all cases, however, the use of lithium-containing drugs must be carefully monitored.

Of the elements of group 13, only boron and aluminum have been detected in humans. The role of boron in our bodies is not fully understood, but this element appears to play a role in nucleic acid synthesis and carbohydrate metabolism. Selected boron compounds appear to concentrate in human brain tumors. This property has opened the door to a treatment known as boron neutron-capture therapy (BNCT). Once a suitable boron compound has been injected and has made its way to a tumor, irradiation of the tumor with low-energy neutrons leads to the nuclear reaction

$$^{10}_5B + \,^1_0n \rightarrow \,^7_3Li + \,^4_2He$$

The α particles generated in the reaction have a short penetration depth but high relative biological effectiveness (RBE), so they can kill the tumor cells without harming surrounding tissue. The identification of compounds suitable for BNCT remains an area of active research.

Aluminum is found in some antacids as aluminum hydroxide, $Al(OH)_3$, or aluminum carbonate, $Al_2(CO_3)_3$. Aluminum sodium sulfate, $AlNa(SO_4)_2 \cdot 12\, H_2O$, is an ingredient in some brands of baking powder. Most of the aluminum in the human body can be traced to these sources. Aluminum is not considered essential to humans, but low-aluminum diets have been observed to harm goats and chickens. High concentrations of aluminum are clearly toxic; the effects are most noticeable in patients with impaired kidney function. The role of aluminum in Alzheimer's disease has been extensively debated but remains unresolved.

**CONNECTION** Tritium is produced in nuclear reactors by bombarding $^6$Li and $^7$Li with high-energy neutrons. The reactions involved are discussed in Chapter 21.

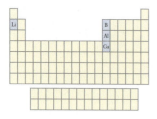

**CONNECTION** The relative biological effectiveness (RBE) of radioactive particles was introduced in Chapter 21.

Simple gallium compounds like gallium(III) nitrate and gallium(III) chloride, either alone or in combination with other drugs, have shown activity on bladder and ovarian cancers. The similar ionic radii of $Ga^{3+}$ (62 pm) and $Fe^{3+}$ (64.5 pm) allow gallium to block DNA synthesis by replacing iron in a protein called transferrin and in other enzymes. Because gallium compounds accumulate in tumors at a higher rate than in healthy tissue, the disruption of DNA synthesis in the tumor cells inhibits tumor growth.

**organometallic compound** a molecule containing direct carbon–metal covalent bonds.

### Antimony and Bismuth

Antimony compounds are generally considered to be toxic. It has been reported, for instance, that exposure of infants to antimony compounds used as fire retardants in mattresses may contribute to sudden infant death syndrome (SIDS). However, this element does appear to have a medical use. Leishmaniasis, an insect-borne disease characterized by the formation of boils or skin lesions, is resistant to most treatments, but patients have been successfully treated with sodium stibogluconate, one of the few applications of antimony compounds in human health.

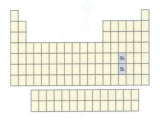

Popular over-the-counter remedies for indigestion, diarrhea, and other gastrointestinal disorders contain bismuth subsalicylate (Figure 22.18). The bismuth in these compounds acts as a mild antibacterial agent that reduces the number of diarrhea-causing bacteria.

### Titanium, Vanadium, and Niobium Cancer Drugs

Compounds of some transition metals demonstrate antitumor activity. Titanium(IV) complexes such as budotitane (Figure 22.19) show promise against colon cancer. Encouraging results against breast, lung, and colon cancers were observed with a series of compounds with formulas $(C_5H_5)_2TiCl_2$, $(C_5H_5)_2VCl$, and $(C_5H_5)_2NbCl_2$. Unfortunately, at therapeutically useful doses the potential for liver damage outweighs the benefits of these compounds. These compounds contain carbon–metal bonds and belong to a class of substances called **organometallic compounds**.

### Vanadium and Chromium

Insulin is a hormone needed for proper glucose metabolism and protein synthesis. People suffering from diabetes either are unable to produce insulin or produce the hormone but are unable to use it effectively. The suggestion that vanadium plays a role in insulin production has prompted investigation of vanadium compounds as diabetes drugs. Encouraging results from animal studies using bis(allixinato) oxovanadium(IV) (Figure 22.20a) have been reported, but the compound is not currently approved for use in humans. Chromium(III) compounds have been shown to lower fasting blood sugar levels and reduce the amount of insulin required by some diabetics (Figure 22.20b).

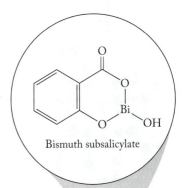

Bismuth subsalicylate

### Iron

The body must regulate the amount of iron in cells in order to produce enough hemoglobin to maintain good health. Deficiencies in iron lead to several diseases broadly classified as anemia. Mild forms of anemia are common among women of child-bearing age and are treated with oral iron supplements. A more serious genetic form of anemia, known as thalassemia, must be treated with blood transfusions that leave patients with too much iron in their blood. To remove the excess iron, these patients are treated with chelating ligands that complex some of the iron and transport it out of the red blood cells and eventually out of the body in

**FIGURE 22.18** Bismuth subsalicylate is found in some antacids.

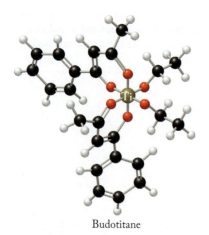

Budotitane

**FIGURE 22.19** Budotitane is a titanium(IV) complex that shows activity against colon cancer.

⊗⊗ **CONNECTION** We encountered enantiomers of metal complexes in Chapter 17, where we saw that *cis*-Co(en)$_2$Cl$_2^+$ has two enantiomers.

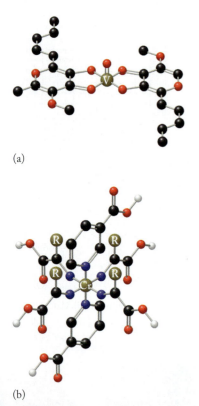

(a)

(b)

**FIGURE 22.20** (a) Vanadium(IV) compounds like bis(allixinato)oxovanadium(IV) can act as insulin mimics. (b) The chromium(III) complex in a glucose tolerance factor contributes to our bodies' ability to regulate insulin levels. The light brown spheres are the side chain R groups of the amino acids in the structure. (To simplify the structures, the hydrogen atoms bonded to carbon atoms are not shown.)

what is known as chelation therapy. Chelation therapy relies on the equilibrium constant for the reaction:

$$FeL(aq) + L'(aq) \rightleftharpoons FeL'(aq) + L(aq)$$

Iron coordinated to ligand L (such as hemoglobin) in the blood is treated with ligand L', which also binds iron. If the formation constant for the complex between iron and L' is greater than for iron and L, then the equilibrium constant for the equation is greater than one, and formation of the iron–L' complex is favored. The ligand L' is chosen so that the complex between iron and L' is eliminated from the body, reducing the concentration of iron and relieving the symptoms caused by thalassemia treatments. Chelation therapy is also the treatment of choice for heavy metal poisoning from toxic metals, including lead or mercury.

The use of chelating ligands to form coordination complexes of iron and other transition metals introduces the possibility of forming optical isomers. Consider the two iron complexes in Figure 22.21; they are nonsuperimposable mirror images of each other and hence are enantiomers. You may wish to build models of the complex and its mirror image to see for yourself that they are not superimposable. One enantiomer of an iron complex often works better than the other in the treatment of diseases like thalassemia.

Mirror

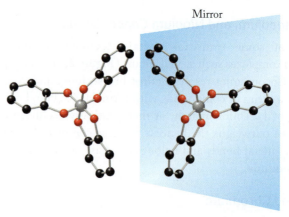

**FIGURE 22.21** Some octahedral transition metal coordination complexes containing polydentate ligands, such as this iron(II) compound, exist as optical isomers.

---

**SAMPLE EXERCISE 22.5** **Calculating the Equilibrium Constant of a Ligand Exchange Reaction** **LO5**

The protein transferrin is involved in the transport of iron into cells. Iron accumulation in the human body has been implicated in diseases such as Parkinson's, Alzheimer's, and thalassemia. The chelating ligand deferoxamine (DFO) is used to treat thalassemia; the complex between iron and DFO is eliminated from the body. Given the formation constants for the complexation of Fe(III) by transferrin (Equation A) and the reaction of deferoxamine with iron (Equation B), calculate the equilibrium constant for the ligand exchange reaction between deferoxamine and transferrin (Equation C).

$$Fe^{3+}(aq) + transferrin(aq) \rightleftharpoons Fe(transferrin)^{3+}(aq) \qquad K_{f,A} = 4.7 \times 10^{20} \quad (A)$$
$$Fe^{3+}(aq) + DFO(aq) \rightleftharpoons Fe(DFO)^{3+}(aq) \qquad K_{f,B} = 4.0 \times 10^{30} \quad (B)$$
$$Fe(transferrin)^{3+}(aq) + DFO(aq) \rightleftharpoons Fe(DFO)^{3+}(aq) + transferrin(aq) \qquad K_C = ? \quad (C)$$

**Collect and Organize** We are given the formation constants of two reactions involving iron and different ligands and are asked to calculate the equilibrium constant for the ligand exchange reaction. You may wish to refer to Chapter 17 for calculations involving formation constants of complexes, as well as to Chapters 15 and 16, where we manipulated equilibrium constants.

**Analyze** Transferrin is a reactant in Equation A and a product in the exchange reaction (C). Therefore, we need to reverse Equation A before adding it to Equation B to obtain Equation C. Reversing a reaction means taking the reciprocal of its equilibrium constant. Combining the reverse of Equation A with Equation B to obtain Equation C means multiplying $1/K_{f,A}$ and $K_{f,B}$ together to obtain $K_C$. The reciprocal of $K_{f,A}$ has a value of about $10^{-20}$. Therefore, the value of $K_C$ should be about $10^{-20} \times 10^{30}$, or about $10^{10}$.

**Solve** Multiplying $1/K_{f,A}$ by $K_{f,B}$ to obtain $K_C$:

$$K_C = \frac{1}{4.7 \times 10^{20}} \times (4.0 \times 10^{30}) = 8.5 \times 10^9$$

**Think About It** The equilibrium constant between the Fe(transferrin) complex and deferoxamine (DFO) is indeed $\sim 10^{10}$ and illustrates why DFO is an effective treatment for thalassemia—the large value for the equilibrium constant means that the equilibrium in Equation C lies toward the products (to the right), removing iron from the blood.

**Practice Exercise** The equilibrium constants for the reactions of penicillamine and methionine with methyl mercury are

$$CH_3Hg^+ + \text{penicillamine} \rightleftharpoons CH_3Hg(\text{penicillamine})^+ \qquad K_f = 6.3 \times 10^{13}$$
$$CH_3Hg^+ + \text{methionine} \rightleftharpoons CH_3Hg(\text{methionine})^+ \qquad K_f = 2.5 \times 10^7$$

Calculate the equilibrium constant for the reaction:

$$CH_3Hg(\text{methionine})^+ + \text{penicillamine} \rightleftharpoons CH_3Hg(\text{penicillamine})^+ + \text{methionine}$$

**CONNECTION** In Chapter 17 we discussed formation constants for metal complexes. Formation constants are equilibrium constants that describe the formation of a metal–ligand complex.

## Platinum Group and Coinage Metals

The period 5 and period 6 elements in groups 8, 9, and 10 (Ru, Rh, Pd, Os, Ir, and Pt) are often referred to as the *platinum group metals*. The group 11 elements of these two periods (Ag and Au) along with Cu are called the *coinage metals*. In this section we explore examples of soluble compounds of these metals used in medications for arthritis, cancer, and other diseases.

The serendipitous discovery in the 1960s that *cis*-diamminedichloroplatinum(II) (cisplatin in Figure 22.22) is effective in treating testicular, ovarian, and other cancers has spurred the development of a host of cancer drugs based on both platinum group metals and coinage metals. The results of this research include a compound known as carboplatin that shows the same activity as cisplatin but has fewer side effects. In addition to platinum compounds, the antitumor activities of complexes of gold, rhodium, ruthenium, and silver have been explored. The effectiveness of all of these drugs lies in their ability to bind to the nitrogen-containing bases in DNA and inhibit cell replication. If their cells cannot divide, tumors cannot grow. The greater toxicity of rhodium compounds relative to those of platinum and ruthenium has limited their clinical application.

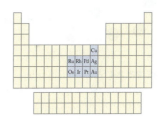

**CONNECTION** The use of the platinum group metals (groups 8–10) as catalysts for inorganic chemical reactions was discussed in Chapter 14.

**FIGURE 22.22** Cisplatin and carboplatin are effective antitumor agents. The ruthenium and rhodium compounds show activity against leukemia but have not yet seen widespread use.

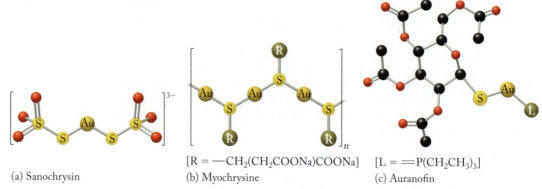

[R = —CH$_2$(CH$_2$COONa)COONa]  [L = =P(CH$_2$CH$_3$)$_3$]

(a) Sanochrysin  (b) Myochrysine  (c) Auranofin

**FIGURE 22.23** (a) Sanochrysin, (b) myochrysine, and (c) auranofin are three gold-containing drugs used to treat arthritis.

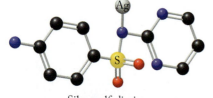

Silver sulfadiazine

**FIGURE 22.24** Silver sulfadiazine is an effective antibiotic when applied to burns. (H atoms are not shown.)

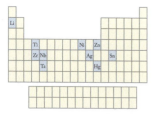

⊙⊙ **CONNECTION** Alloys were described in Chapter 12. The cell potentials of simple voltaic cells were discussed in Chapter 19.

Selected osmium compounds reduce inflammation in joints resulting from arthritis, although the use of such compounds has diminished with the development of other anti-inflammatory agents. The therapeutic effects of aqueous solutions of osmium tetroxide, OsO$_4$, were first investigated in the 1950s. The use of osmium tetroxide was superseded by the use of glucose polymers containing osmium, known as osmarins. The use of osmarins reduces the toxic effects of osmium and illustrates how even toxic metals can be adapted to therapy.

Although the historic use of gold for medicinal purposes dates back millennia, the effective use of gold-containing pharmaceuticals originated with the discovery in the 1920s and 1930s that a gold thiosulfate compound, sanochrysin, alleviates the symptoms of rheumatoid arthritis. The most commonly used gold drugs for the treatment of arthritis today are sold under the names myochrysine and auranofin (Figure 22.23). Myochrysine is injected; auranofin can be taken orally.

Eye drops containing silver salts are used to treat eye infections. Silver sulfadiazine is a broad-spectrum "sulfa" drug used to prevent and treat bacterial and fungal infections (Figure 22.24). It is an active ingredient in creams used to treat thermal and chemical burns.

## Medical Devices and Materials

### Dental Alloys

When the fluoride treatment described in Section 22.3 fails to prevent dental caries, a dentist will clean the cavity and fill it, often with a metal alloy that could contain mercury, silver, tin, copper, and traces of zinc. An alloy containing mercury is called an *amalgam*. The most common dental amalgam contains Ag$_2$Hg$_3$, Ag$_3$Sn, and Sn$_8$Hg. Although mercury and mercury compounds are highly toxic, when combined with silver or tin, the mercury is chemically unreactive.

### Pacemakers

Pacemakers are lifesaving devices implanted in the chests of patients with certain heart conditions. A pacemaker delivers electrical pulses to stimulate the heart to beat at the proper rate. The batteries used in pacemakers must deliver reliable power for long periods because complicated surgery is required to change the batteries. A solid-state zinc/mercury oxide battery (Figure 22.25) meets these stringent requirements and was the battery of choice in the first pacemakers. The zinc container acts as the anode, where oxidation takes place. A steel cathode extends

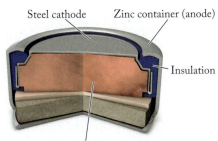

Steel cathode      Zinc container (anode)

Insulation

Paste of HgO, KOH, and Zn(OH)$_2$

**FIGURE 22.25** The Zn/HgO battery used in pacemakers.

into a paste of mercury(II) oxide, potassium hydroxide, and zinc(II) hydroxide. The half-reactions and overall cell reaction are

Anode:    $Zn(s) + 2\,OH^-(aq) \rightarrow ZnO(s) + H_2O(\ell) + 2\,e^-$

Cathode:  $HgO(s) + H_2O(\ell) + 2\,e^- \rightarrow Hg(\ell) + 2\,OH^-(aq)$

Overall:  $Zn(s) + HgO(s) \rightarrow ZnO(s) + Hg(s)$        $E^\circ_{cell} = 1.347\ V$

The hazards associated with mercury, as well as performance issues, have led to the use of lithium/iodine batteries in pacemakers. The cathode is an electrically conductive material prepared by heating iodine with poly(vinylpyridine) (PVP). Lithium metal acts as the anode, and the reactions are

Anode:    $2\,Li(s) \rightarrow 2\,Li^+(s) + 2\,e^-$

Cathode:  $I_2(PVP)(s) + 2\,e^- \rightarrow 2\,I^-(s) + (PVP)(s)$

Overall:  $2\,Li(s) + I_2(PVP)(s) \rightarrow 2\,LiI(s) + (PVP)(s)$        $E^\circ_{cell} = 3.59\ V$

**CONNECTION** In Chapter 19 we learned how to use reduction potentials for half-reactions to calculate overall potentials of reactions.

**CONNECTION** Poly(vinylpyridine) is a polymer. The preparation and properties of synthetic polymers like poly(vinylpyridine) were described in Chapter 13.

---

**SAMPLE EXERCISE 22.6    Calculating the Electrochemical      L06
                          Potential of a Battery**

Hearing aid batteries are based on the zinc/air electrochemical cell. Given the following standard reduction potentials, (a) determine the overall equation for the cell reaction and (b) calculate the standard cell potential.

$Zn(OH)_2(s) + 2\,e^- \rightarrow Zn(s) + 2\,OH^-(aq)$        $E^\circ = -1.249\ V$

$O_2(g) + 2\,H_2O(\ell) + 4\,e^- \rightarrow 4\,OH^-(aq)$        $E^\circ = 0.401\ V$

**Collect and Organize** We are given two half-reactions and their potentials. The cell reaction is the sum of the oxidation and reduction half-reactions, each multiplied by a coefficient to make the number of electrons gained equal the number of electrons lost. The overall electrochemical cell potential is the difference in the potentials of the cathode and anode half-reactions (Equation 19.2). You may wish to review Sections 19.2 and 19.3.

**Analyze** Both half-reactions in the zinc/air battery are written as reductions: that is, the reactant gains electrons. We need to determine which half-reaction actually runs in reverse, that is, as an oxidation at the anode. The electrochemical potential for the overall reaction must be positive ($E^\circ > 0$) in order to have a functioning battery. We also need to determine each half-reaction's coefficient to ensure that the number of moles of electrons gained equals the number of moles of electrons lost.

**Solve**

a. The reduction of $Zn(OH)_2$ has the smaller $E^\circ$. Therefore, the anode half-reaction is the reverse of $Zn(OH)_2$ reduction, that is, oxidation of Zn metal in alkaline solution:

$Zn(s) + 2\,OH^-(aq) \rightarrow Zn(OH)_2(s) + 2\,e^-$

The reduction of $O_2$ to $OH^-$ involves four moles of electrons, so we must multiply the oxidation half-reaction by 2:

$2\,Zn(s) + 4\,OH^-(aq) \rightarrow 2\,Zn(OH)_2(s) + 4\,e^-$

**CONNECTION** In Chapter 13 we learned about the development of high-density polyethylene polymers for use in artificial joints.

(a)

(b)

**FIGURE 22.26** (a) Hemispherical sockets made from tantalum are used in (b) hip replacement.

Adding the two half-reactions gives us

$$2\,Zn(s) + 4\,OH^-(aq) \rightarrow 2\,Zn(OH)_2(s) + 4\,e^-$$
$$O_2(g) + 2\,H_2O(\ell) + 4\,e^- \rightarrow 4\,OH^-(aq)$$

$$2\,Zn(s) + \cancel{4\,OH^-(aq)} + O_2(g) + 2\,H_2O(\ell) + \cancel{4\,e^-} \rightarrow \cancel{4\,OH^-(aq)} + 2\,Zn(OH)_2(s) + \cancel{4\,e^-}$$

The overall reaction is

$$2\,Zn(s) + O_2(g) + 2\,H_2O(\ell) \rightarrow 2\,Zn(OH)_2(s)$$

b. The standard cell potential is the difference in the standard potentials for the half-reactions:

$$E^\circ_{cell} = E^\circ_{cathode} - E^\circ_{anode} = 0.401\text{ V} - (-1.249\text{ V}) = 1.650\text{ V}$$

**Think About It** The standard cell potential, 1.650 V, assumes that $P_{O_2} = 1$ atm. In air $P_{O_2}$ is about 0.21 atm, so the potential of the Zn/air battery is probably less than 1.650 V.

**Practice Exercise** Implantable cardioverter-defibrillators (ICDs) are devices that detect certain types of cardiac arrhythmias and provide a shock directly to the patient's heart, stopping ventricular fibrillation. Lithium silver vanadium oxide batteries are used to power these devices. The equation for the overall reaction is

$$7\,Li(s) + Ag_2V_4O_{11}(s) \rightarrow Li_7Ag_2V_4O_{11}(s) \qquad E^\circ_{cell} = 2.00\text{ V}$$

Using the data in Appendix 6 for the oxidation potential for Li metal, calculate the reduction potential for silver vanadium oxide.

## Surgical Implants

The degeneration of knees and hips that often accompanies aging necessitates their replacement with artificial joints. The biomaterials for these joints must be chemically nonreactive, and they must provide a smooth surface that allows the joint to move. Severely broken bones may need to be reinforced by pins or implants to promote proper healing. Surgeons use staples and sutures to close the cut made during any procedure. All of these applications require strong, nontoxic materials that resist corrosion by body fluids. The chemical properties of the group 4 and 5 elements and their alloys make them particularly useful for biomedical applications.

In Chapter 12 we discussed tantalum as an example of a biocompatible metal. Thin coatings of pure tantalum or niobium, or of alloys of these two metals, are used to make surfaces on artificial joints smooth and abrasion-resistant (Figure 22.26). Pure tantalum and titanium rods can also be used as bone implants in patients suffering from osteonecrosis, a disruption of the blood supply to bones and joints causing weakening of the bone and arthritis.

Shape-memory alloys of titanium and nickel are used to manufacture stents used to keep weakened arteries from collapsing. A piece of wire made from a shape-memory alloy is heated and formed into the desired shape for the stent. Upon cooling, the alloy undergoes a change to a different crystal structure. The structural change allows the stent to be easily bent into a small coil for insertion into an artery. After insertion, the coiled stent is heated, springing back into its original shape as it rapidly reverts to the high-temperature form. At least 15 shape-memory alloys have been identified, ranging from the most commonly used nickel–titanium alloy to alloys containing hafnium, tantalum, titanium, and nickel.

The temperature-dependent structural change in NiTi memory alloys was important in the design of stents; however, not all applications benefit from changes in the crystal structure of an alloy. Substitutional titanium alloys containing 6% aluminum and 4% vanadium are widely used in orthopedic devices precisely because they retain their body-centered cubic structure over a wide temperature range. This property increases the strength of the alloy, an advantage in an orthopedic device. More complex alloys containing Ti, Nb, Ta, and Zr show promise for even better hip joints and dental implants (Figure 22.27).

**CONCEPT TEST**

The table of standard reduction potentials in Appendix 6 does not include values for many of the metals and alloys used in artificial joints and in stents. Do you expect their standard reduction potentials to be greater than or less than that of $Fe^{2+} + 2\,e^- \rightarrow Fe$?

The human body requires about 30 elements to function properly. These elements span the periodic table from group 1 elements like sodium to transition metals like iron to halogens like chlorine. Our bodies need different amounts of these essential elements, ranging from gram to microgram quantities. Through chemical, biochemical, and clinical research, many of the nonessential elements have found application in the diagnosis and treatment of disease.

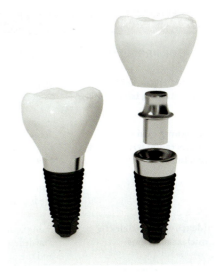

**FIGURE 22.27** Dental implants are often constructed from substitutional titanium alloys containing vanadium and aluminum.

**CONNECTION** The structures of metallic elements and their alloys were introduced in Chapter 12.

**SUMMARY**

**Learning Outcome 1** **Essential elements** have a physiological function in the body. **Nonessential elements** are present in the body but have no known functions. Some may have **stimulatory effects**. (Section 22.1)

**Learning Outcome 2** Essential elements are categorized as **major, trace,** or **ultratrace essential elements** depending on their concentrations in the body. Many enzymes and **coenzymes** contain transition metal ions. Fluoride ions can replace hydroxide ions in hydroxyapatite in teeth and inhibit the formation of cavities. The body concentrates  iodide ions in the thyroid gland where they are incorporated into the hormones that regulate energy production and use. The amino acid selenocysteine is a constituent of proteins that act as catalysts for biochemical reactions. Some elements, such as mercury, cadmium, and lead, have chemical properties that make them highly toxic. (Sections 22.2–22.4)

**Learning Outcome 3** Transport of $Na^+$ and $K^+$ across cell membranes involves **ion pumps** or selective transport through **ion channels** formed by groups of helical proteins. Magnesium ions facilitate the transfer of phosphate groups between ATP and ADP and are found in chlorophyll, the green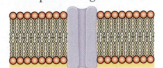

pigment in plants that mediates photosynthesis. Chloride ion is the most abundant anion in the human body, facilitating transport of alkali metal cations and elimination of $CO_2$. Malfunctioning chloride channels are an underlying cause of cystic fibrosis. (Sections 22.2 and 22.3)

**Learning Outcome 4** Radionuclides with short half-lives that emit low-energy γ rays are used in diagnosing disease. The selection of a radionuclide is also governed by the toxicities of the element and its daughter nuclides and the speed at which it is eliminated from the body. (Section 22.5)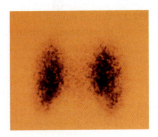

**Learning Outcome 5** Soluble coordination compounds and **organometallic** compounds of the transition metals are used in medications for arthritis, cancer, and other diseases. (Section 22.5)

**Learning Outcome 6** Our understanding of metals and alloys permits us to design batteries that can deliver safe, reliable, and long-lasting power for use in implanted medical devices. Corrosion-resistant metals and alloys are used in dental repair, artificial joints, and stents. (Section 22.5)

## PROBLEM-SOLVING SUMMARY

| TYPE OF PROBLEM | CONCEPTS AND EQUATIONS | SAMPLE EXERCISES |
|---|---|---|
| **Calculating an acid concentration from its pH** | Relate the pH of a solution to the $[H^+]$ by the equation $$pH = -\log[H^+]$$ | 22.1 |
| **Assigning oxidation numbers and writing half-reactions** | Assign oxidation numbers using the guidelines in Chapter 4: Oxidation numbers for pure elements are zero; O.N. for monatomic ions equal the ionic charge; H and O typically have O.N. equal to +1 and −2, respectively. Then use the numbers to determine the number of electrons gained or lost; balance charges with water and $H^+$ ions to balance the numbers of H and O atoms. If in alkaline media, add $OH^-$ ions to both sides of the equation to convert $H^+$ to water. | 22.2 |
| **Identifying particles in nuclear reactions** | Sum the subscripts and superscripts on both sides of the nuclear equation. The difference in the superscripts equals the mass number of the missing particle, and the difference in the subscripts equals the charge on the missing particle. | 22.3 |
| **Calculating quantities of radioactive isotopes** | Use the equation $$\ln\frac{N_t}{N_0} = \frac{-0.693t}{t_{1/2}} \qquad (21.17)$$ where $N_0$ and $N_t$ are the amounts of material present initially and at time $t$, respectively. | 22.4 |
| **Calculating the equilibrium constant of a ligand exchange reaction** | Combine the equilibrium constant expressions for the formation of the two species, after reversing the one that dissociates in the overall reaction, to generate an equivalent equilibrium constant expression for the exchange reaction. | 22.5 |
| **Calculating the electrochemical potential of a battery** | Identify the cathode and anode half-reactions. Then use the equation $$E^\circ_{cell} = E^\circ_{cathode} - E^\circ_{anode} \qquad (19.2)$$ | 22.6 |

## VISUAL PROBLEMS

*(Answers to boldface end-of-chapter questions and problems are in the back of the book.)*

**22.1.** Which part of Figure P22.1 best describes the periodic trend in monatomic cation radii moving up or down a group or across a period in the periodic table? (Arrows point in the direction of increasing radii.)

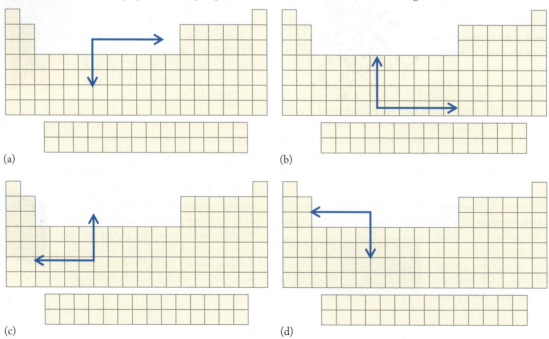

(a)

(b)

(c)

(d)

**FIGURE P22.1**

**22.2.** Which part of Figure P22.1 best describes the periodic trend in monatomic anion radii moving up or down a group or across a period in the periodic table? (Arrows point in the direction of increasing radii.)

**22.3.** Which of the two groups highlighted in the periodic table in Figure P22.3 typically forms ions that have larger radii than the corresponding neutral atoms?

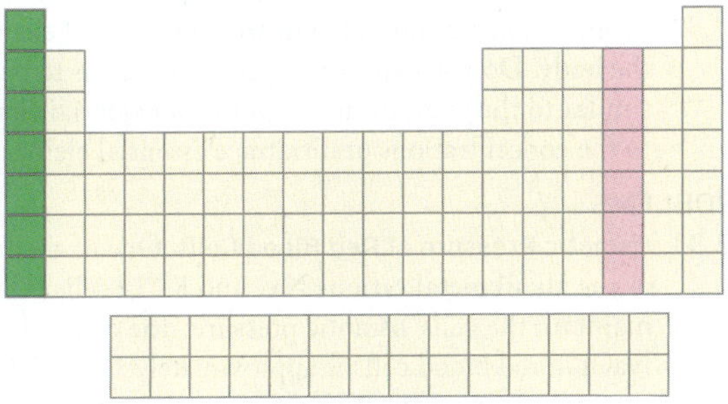

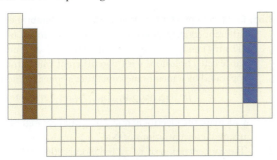

**FIGURE P22.3**

**22.4.** Which of the two groups highlighted in the periodic table in Figure P22.4 typically forms ions that have smaller radii than the corresponding neutral atoms?

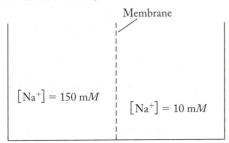

**FIGURE P22.4**

**\*22.5.** As we saw in Chapter 19, the free energy ($\Delta G$) of a reaction is related to the cell potential by the equation $\Delta G = -nFE$. In Figure P22.5, two solutions of $Na^+$ of different concentrations are separated by a semipermeable membrane. Calculate $\Delta G$ for the transport of $Na^+$ from the side with higher concentration to the side with lower concentration. (*Hint*: See Problem 22.35.)

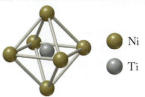

Membrane

$[Na^+] = 150$ m$M$

$[Na^+] = 10$ m$M$

**FIGURE P22.5**

**22.6.** Two solutions of $K^+$ are separated by a semipermeable membrane in Figure P22.6. Calculate $\Delta G$ for the transport of $K^+$ from the side with lower concentration to the side with higher concentration.

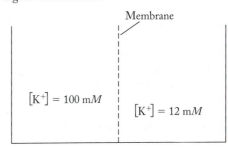

Membrane

$[K^+] = 100$ m$M$

$[K^+] = 12$ m$M$

**FIGURE P22.6**

**22.7.** Describe the molecular geometry around each germanium atom in the structure of the germanium compound shown in Figure P22.7.

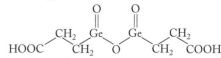

**FIGURE P22.7**

**22.8.** Selenocysteine can exist as two enantiomers (stereoisomers). Identify the atom in Figure P22.8 responsible for the two enantiomers.

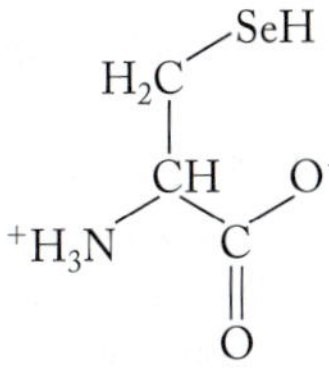

**FIGURE P22.8**

**22.9.** The austenite structure for NiTi is similar to the CsCl structure. Which crystal structure shown in Figure P22.9 is correct for the austenite form of the shape-memory alloy NiTi?

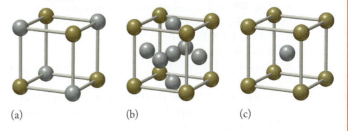

(a)         (b)         (c)

● Ni

● Ti

**FIGURE P22.9**

**\*22.10.** The unit cell for the martensite form of the shape-memory alloy NiTi is shown in Figure P22.10. How many equivalent Ti and Ni atoms are in the unit cell?

● Ni

● Ti

**FIGURE P22.10**

## QUESTIONS AND PROBLEMS

### The Periodic Table of Life

#### CONCEPT REVIEW

**22.11.** What is the difference between an essential element and a nonessential element?

22.12. Are all essential elements major essential elements?

**22.13.** What is the main criterion that distinguishes major, trace, and ultratrace essential elements from one another?

22.14. Should trace essential elements also be considered to be stimulatory?

**22.15.** Why is methylmercury ion, $CH_3Hg^+$, more toxic than mercury metal?

*22.16. Why is $Cd^{2+}$ more likely than $Cr^{2+}$ to replace $Zn^{2+}$ in an enzyme like carbonic anhydrase?

**22.17.** Why is $Be^{2+}$ more likely than $Ca^{2+}$ to displace $Mg^{2+}$ in biomolecules?

22.18. $PbS$, $PbCO_3$, and $PbCl(OH)$ have limited solubility in water. Which of them is(are) more likely to dissolve in acidic solutions?

#### PROBLEMS

**22.19.** Which ion channel must accommodate the larger cation, a potassium or a sodium ion channel?

22.20. Which ion is larger: $Cl^-$ or $I^-$?

---

**22.21.** The concentrations of very dilute solutions are sometimes expressed as parts per million. Express the concentration of each of the following trace and ultratrace essential elements in parts per million:
   a. Copper, 110 mg in 70 kg
   b. Zinc, $3.3 \times 10^{-2}$ g/kg
   c. Iodine, 0.043 g in 100 kg

22.22. In the human body, the concentrations of ultratrace essential elements are even lower than those of trace essential elements and therefore are sometimes expressed in parts per billion. Express the concentrations of each of the following elements in parts per billion:
   a. Cobalt, $4.3 \times 10^{-5}$ g/kg
   b. Boron, 0.014 g/100 kg
   c. Chromium, 5.0 mg/70 kg

---

**22.23.** In the following pairs, which element is more abundant in the human body? (a) silicon or oxygen; (b) iron or oxygen; (c) carbon or aluminum

22.24. In the following pairs, which element is more abundant in the human body? (a) H or Si; (b) Ca or Fe; (c) N or Cr

### Major Essential Elements

#### CONCEPT REVIEW

**22.25. Ion Transport in Cells** Describe three ways in which ions of major essential elements (such as $Na^+$ and $K^+$) enter and exit cells.

22.26. Which transport mechanism for ions requires ATP: diffusion, ion channels, or ion pumps?

**22.27.** Why is it difficult for ions to diffuse across cell membranes?

22.28. Why does $Sr^{2+}$ substitute for $Ca^{2+}$ in bones?

22.29. Which alkali metal ion is $Rb^+$ most likely to substitute for?

22.30. Why don't alkaline earth metal cations substitute for alkali metal cations in cases where the ionic radii are similar?

*22.31. Why might nature have chosen calcium carbonate over calcium sulfate as the major exoskeleton material in shells?

22.32. Bromide ion and fluoride ion are nonessential elements in the body. Do you expect their concentrations to be more similar to the concentrations of major essential elements or to the concentrations of ultratrace essential elements?

#### PROBLEMS

**22.33. Osmotic Pressure of Red Blood Cells** One of the functions of the alkali metal cations $Na^+$ and $K^+$ in cells is to maintain the cells' osmotic pressure. The concentration of $NaCl$ in red blood cells is approximately 11 m$M$. Calculate the osmotic pressure of this solution at body temperature (37°C). (*Hint:* See Equation 11.14.)

22.34. Calculate the osmotic pressure exerted by a 92 m$M$ solution of $KCl$ in a red blood cell at body temperature (37°C). (*Hint:* See Equation 11.14.)

---

*22.35. **Electrochemical Potentials across Cell Membranes** Very different concentrations of $Na^+$ ions exist in red blood cells (11 m$M$) and the blood plasma (160 m$M$) surrounding those cells. Solutions with two different concentrations separated by a membrane constitute a concentration cell.

$$E = E° - (0.0592 \text{ V}/n) \log([Na^+]_{cell}/[Na^+]_{plasma})$$

Calculate the electrochemical potential created by the unequal concentrations of $Na^+$.

22.36. The concentration of $K^+$ in red blood cells is 92 m$M$, and the concentration of $K^+$ in plasma is 10 m$M$. Calculate the electrochemical potential created by the two concentrations of $K^+$.

$$E = E° - (0.0592 \text{ V}/n) \log([K^+]_{plasma}/[K^+]_{cell})$$

---

**22.37.** If the transport of $K^+$ across a cell membrane requires 5 kJ/mol, how many moles of ATP must be hydrolyzed to provide the necessary energy? The hydrolysis of ATP is described by the equation

$$ATP^{4-} + H_2O \rightarrow ADP^{3-} + HPO_4^{2-} + H^+ \qquad \Delta G° = -34.5 \text{ kJ}$$

*22.38. Removing excess $Na^+$ from a cell by an ion pump requires energy. How many moles of ATP must be hydrolyzed to overcome a cell potential of −0.07 V? The hydrolysis of 1 mole of ATP provides 34.5 kJ of energy.

---

**22.39. Plankton Exoskeletons** Exoskeletons of planktonic acantharia contain strontium sulfate. Calculate the solubility in moles per liter of $SrSO_4$ in water at 25°C given that $K_{sp} = 3.4 \times 10^{-7}$.

22.40. Algae in the genus *Closterium* contain structures built from barium sulfate (barite). Calculate the solubility in moles per liter of $BaSO_4$ in water at 25°C given that $K_{sp} = 1.1 \times 10^{-10}$.

## Trace and Ultratrace Essential Elements

### CONCEPT REVIEW

**22.41.** What danger to human health is posed by $^{137}$Cs ($t_{1/2} \approx 30$ yr)?

**22.42.** Why is $^{137}$Cs ($t_{1/2} \approx 30$ yr) considered to be dangerous to human health when naturally occurring $^{40}$K ($t_{1/2} = 1.28 \times 10^6$ yr) is benign?

**22.43.** What are the likely signs of $\Delta S$ and $\Delta G$ for the dissolution of tooth enamel?

**22.44.** Why does fluorapatite resist acid better than hydroxyapatite if both are insoluble in water?

**22.45.** Enzymes are large proteins.
   a. What is the function of enzymes?
   b. Are all proteins enzymes?

**\*22.46.** Why do superoxide ions ($O_2^-$) act as strong oxidizing agents?

**\*22.47.** What effect does an enzyme have on the activation energy of a biochemical reaction?

**\*22.48.** Why might reductases also be described as reducing agents?

**\*22.49.** When a transition metal ion like $Cu^{2+}$ is incorporated into a metalloenzyme, is the formation constant likely to be much greater than one ($K \gg 1$) or much less than one ($K \ll 1$)?

$$Cu^{2+} + \text{protein} \rightleftharpoons \text{metalloenzyme} \qquad K = \frac{[\text{metalloenzyme}]}{[Cu^{2+}][\text{protein}]}$$

**\*22.50.** Carbon monoxide poisoning derives from competitive binding of CO versus $O_2$ to the $Fe^{2+}$ in the coordination sphere of hemoglobin. If CO can be displaced by breathing pure $O_2$, which of the following is likely to be correct under these conditions: $K_{O_2}/K_{CO} < 1$ or $K_{O_2}/K_{CO} > 1$?

**22.51.** What is the likely sign of $\Delta S$ for the reaction in Table 22.6 catalyzed by carboxypeptidase?

**\*22.52.** Why are thyroxine and 3,5,3'-triiodothyronine (Figure 22.7) considered to be amino acids? Why aren't they *essential* amino acids?

### PROBLEMS

**22.53.** What are the products of radioactive decay of $^{137}$Cs? Write a balanced equation for the nuclear decay reaction.

**22.54.** Potassium-40 decays by three pathways: β decay, positron emission, and electron capture. Write balanced equations for each of these processes.

---

**22.55.** Calculate the pH of a $1.00 \times 10^{-3}$ M solution of selenocysteine ($pK_{a_1} = 2.21$, $pK_{a_2} = 5.43$).

**22.56.** Calculate the pH of a $1.00 \times 10^{-3}$ M solution of cysteine ($pK_{a_1} = 1.7$, $pK_{a_2} = 8.3$). Is selenocysteine a stronger acid than cysteine?

---

**22.57.** **Composition of Tooth Enamel** Tooth enamel contains the mineral hydroxyapatite. Hydroxyapatite reacts with fluoride ion in toothpaste to form fluorapatite. The equilibrium constant for the reaction between hydroxyapatite and fluoride ion is $K = 8.48$. Write the equilibrium constant expression for the following reaction. In which direction does the equilibrium lie?

$$Ca_5(PO_4)_3(OH)(s) + F^-(aq) \rightleftharpoons Ca_5(PO_4)_3(F)(s) + OH^-(aq)$$

**\*22.58.** **Effects of Excess Fluoridation on Teeth** Too much fluoride might lead to the formation of calcium fluoride according to the reaction

$$Ca_5(PO_4)_3(OH)(s) + 10\ F^-(aq) \rightleftharpoons \\ 5\ CaF_2(s) + 3\ PO_4^{3-}(aq) + OH^-(aq)$$

Write the equilibrium constant expression for the reaction. Given the $K_{sp}$ values for the following two reactions, calculate $K$ for the reaction between $Ca_5(PO_4)_3(OH)$ and fluoride ion that forms $CaF_2$.

$$Ca_5(PO_4)_3(OH)(s) \rightleftharpoons 5\ Ca^{2+}(aq) + 3\ PO_4^{3-}(aq) + OH^-(aq) \\ K_{sp} = 2.3 \times 10^{-59}$$

$$CaF_2(s) \rightleftharpoons Ca^{2+}(aq) + 2\ F^-(aq) \qquad K_{sp} = 3.9 \times 10^{-11}$$

$$Ca_5(PO_4)_3(OH)(s) + 10\ F^-(aq) \rightleftharpoons \\ 5\ CaF_2(s) + 3\ PO_4^{3-}(aq) + OH^-(aq)$$

---

**22.59.** Tooth enamel is actually a composite material containing both hydroxyapatite and a calcium phosphate, $Ca_8(HPO_4)_2(PO_4)_4 \cdot 6\ H_2O$ ($K_{sp} = 1.1 \times 10^{-47}$).
   a. Is this calcium mineral more or less soluble than hydroxyapatite ($K_{sp} = 2.3 \times 10^{-59}$)?
   b. Calculate the solubility in moles per liter of hydroxyapatite, $Ca_5(PO_4)_3(OH)$, $K_{sp} = 2.3 \times 10^{-59}$ in water at 25°C and pH = 7.0.
   c. What is the solubility of hydroxyapatite at pH = 5.0?

**22.60.** The $K_{sp}$ of actual tooth enamel is reported to be $1 \times 10^{-58}$.
   a. Does this mean that tooth enamel is more soluble than pure hydroxyapatite ($K_{sp} = 2.3 \times 10^{-59}$)?
   b. Does the measured value of $K_{sp}$ for tooth enamel support the idea that tooth enamel is a mixture of hydroxyapatite, $Ca_5(PO_4)_3(OH)$, and a calcium phosphate $Ca_8(HPO_4)_2(PO_4)_4 \cdot 6\ H_2O$ ($K_{sp} = 1.1 \times 10^{-47}$)?
   c. Calculate the solubility in moles per liter of $Ca_8(HPO_4)_2(PO_4)_4 \cdot 6\ H_2O$ ($K_{sp} = 1.1 \times 10^{-47}$) in water at 25°C and pH = 7.0.

---

**22.61.** Some sources give the formula of hydroxyapatite as $Ca_{10}(PO_4)_6(OH)_2$. If the $K_{sp}$ of $Ca_5(PO_4)_3(OH)$ is $2.3 \times 10^{-59}$, what is the $K_{sp}$ of $Ca_{10}(PO_4)_6(OH)_2$?

**22.62.** The same sources mentioned in the previous problem cite the formula of fluorapatite as $Ca_{10}(PO_4)_6F_2$. If the $K_{sp}$ of $Ca_5(PO_4)_3F$ is $3.2 \times 10^{-60}$, what is the $K_{sp}$ of $Ca_{10}(PO_4)_6F_2$?

---

**\*22.63.** The activation energy for the uncatalyzed decomposition of hydrogen peroxide at 20°C is 75.3 kJ/mol. In the presence of the enzyme catalase, the activation energy is reduced to 29.3 kJ/mol. By using the following form of the Arrhenius equation, $RT \ln(k_1/k_2) = E_{a_2} - E_{a_1}$, how much faster is the catalyzed reaction?

**\*22.64.** **Enzymatic Activity of Urease** Urease catalyzes the decomposition of urea to ammonia and carbon dioxide (Figure P22.64). The rate constant for the uncatalyzed reaction at 20°C and pH 8 is $k = 3 \times 10^{-10}$ s$^{-1}$. A urease isolated from the jack bean increases the rate constant to $k = 3 \times 10^4$ s$^{-1}$. By using the $RT \ln(k_1/k_2) = E_{a_2} - E_{a_1}$ form of the Arrhenius equation, calculate the difference between the activation energies.

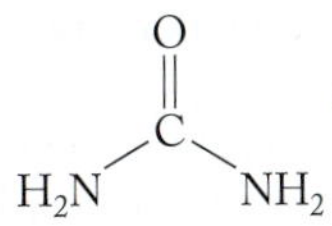

$$\underset{H_2N \qquad NH_2}{\overset{\displaystyle O}{\underset{\displaystyle \|}{C}}}(aq) + H_2O(\ell) \rightarrow 2\ NH_3(aq) + CO_2(g)$$

**FIGURE P22.64**

*22.65. The initial rate of an enzyme-catalyzed reaction depends on the concentration of substrate as shown in Figure P22.65. What is the apparent reaction order at the far right side of the graph?

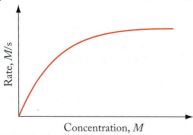

**FIGURE P22.65**

22.66. The rate of an enzyme-catalyzed reaction depends on the concentration of substrate. Which line in Figure P22.66 has the highest reaction rate?

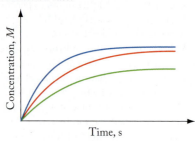

**FIGURE P22.66**

## Elements for Diagnosis and Therapy

### CONCEPT REVIEW

22.67. When choosing an isotope for imaging, why is it important to consider the decay mode of the isotope as well as the half-life?

22.68. Brachiotherapy is a cancer treatment that involves surgical implantation of a small capsule of $^{192}$Ir into a tumor. Iridium-192 lies between two stable isotopes of iridium, $^{193}$Ir and $^{191}$Ir. How does this account for the fact that $^{192}$Ir decays by β-decay, electron capture, and positron emission?

22.69. Why might an α emitter be a good choice for radiation therapy?

*22.70. What advantage does a β emitter have over an α emitter for imaging?

22.71. Gadolinium-153 decays by electron capture. What type of radiation does $^{153}$Gd produce that makes it useful for imaging?

22.72. Gadolinium-153 and samarium-153 both have the same mass number. Why might $^{153}$Gd decay by electron capture whereas $^{153}$Sm decays by emitting β particles?

22.73. How do platinum- and ruthenium-containing drugs fight cancer?

*22.74. Many transition metal complexes are brightly colored. Why might the titanium(IV) compound budotitane be colorless? (*Hint*: See Chapter 17.)

*22.75. Is the glucose tolerance factor that contains chromium(III) paramagnetic or diamagnetic? (*Hint*: See Chapter 17.)

22.76. Why might lithium be an attractive anode material in batteries used in pacemakers?

22.77. Mercury compounds are generally toxic to humans. Why can we use mercury in dental amalgams?

22.78. Tantalum is used in artificial joints. What property of Ta makes it attractive for this purpose?

22.79. Gadolinium(III) ions are used in contrast agents for MRI because they have unpaired electrons. What is the electron configuration for $Gd^{3+}$ and how many unpaired electrons does it have?

22.80. Which of the first-row transition metal ions in Figure 2.17 on page 56 have the most unpaired valence electrons?

### PROBLEMS

22.81. Sodium nitroprusside, $Na_2[Fe(NO)(CN)_5]$, is used in emergency rooms to treat extremely high blood pressure. What is the electron configuration of Fe in this compound? Is this compound likely to be a high spin or low spin complex?

22.82. Manganese(II) chloride was one of the first compounds to be investigated as MRI contrast agents. How many unpaired electrons does Mn have when $MnCl_2$ dissolves in water to form the coordination compound $MnCl_2(H_2O)_4$?

22.83. **PET Imaging with Gallium** A patient is injected with a 5 μ*M* solution of gallium citrate containing $^{68}$Ga ($t_{1/2} = 9.4$ hr) for a PET study. How long is it before the activity of the $^{68}$Ga drops to 5% of its initial value?

22.84. Iodine-123 ($t_{1/2} = 13.3$ h) has replaced iodine-131 ($t_{1/2} = 8.1$ days) for diagnosis of thyroid conditions. How long is it before the activity of $^{123}$I drops to 5% of its initial value?

22.85. The bismuth in over-the-counter antacids is found as $BiO^+$. Draw the Lewis structure for the $BiO^+$ cation.

22.86. Some medicines used in treating depression contain lithium carbonate. Draw the Lewis structure for $Li_2CO_3$.

22.87. The complexation of mercury(II) ion with methionine

$$Hg^{2+} + \text{methionine} \rightleftharpoons Hg(\text{methionine})^{2+}$$

has a formation constant of log $K = 14.2$, whereas the formation constant for the $Hg^{2+}$ complex with penicillamine

$$Hg^{2+} + \text{penicillamine} \rightleftharpoons Hg(\text{penicillamine})^{2+}$$

is log $K = 16.3$. Calculate the equilibrium constant for the reaction

$$Hg(\text{methionine})^{2+} + \text{penicillamine} \rightleftharpoons Hg(\text{penicillamine})^{2+} + \text{methionine}$$

22.88. The complexation of mercury(II) ion with cysteine in aqueous solution

$$Hg^{2+} + \text{cysteine} \rightleftharpoons Hg(\text{cysteine})^{2+}$$

has a formation constant of log $K = 14.2$, whereas the formation constant for the $Hg^{2+}$ complex with glycine

$$Hg^{2+} + \text{glycine} \rightleftharpoons Hg(\text{glycine})^{2+}$$

is log $K = 10.3$. Calculate the equilibrium constant for the reaction

$$Hg(\text{cysteine})^{2+} + \text{glycine} \rightleftharpoons Hg(\text{glycine})^{2+} + \text{cysteine}$$

**22.89.** The equilibrium constant of the reaction

$$CH_3Hg(penicillamine)^+(aq) + cysteine(aq) \rightleftharpoons$$
$$CH_3Hg(cysteine)^+(aq) + penicillamine(aq)$$

is $K = 0.633$. Calculate the equilibrium concentrations of cysteine and penicillamine if we start with a 1.00 $M$ solution of cysteine and a 1.00 m$M$ solution of $CH_3Hg(penicillamine)^+$.

**22.90.** The equilibrium constant of the reaction

$$CH_3Hg(glutathione)^+(aq) + cysteine(aq) \rightleftharpoons$$
$$CH_3Hg(cysteine)^+(aq) + glutathione(aq)$$

is $K = 5.0$. Calculate the equilibrium concentrations of cysteine and glutathione if we start with a 1.20 m$M$ solution of cysteine and a 1.20 m$M$ solution of $CH_3Hg(glutathione)^+$.

**22.91.** Draw the Lewis structure for the citrate ion, using the skeletal drawing in Figure P22.91 as a guide.

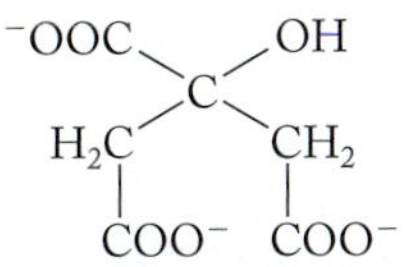

**FIGURE P22.91**

**22.92.** Draw the Lewis structure for the thiosulfate ion, $S_2O_3^{2-}$, which is found in some arthritis drugs containing gold atoms.

**22.93.** Aluminum hydroxide is used in some antacids. Write a balanced net ionic equation for the reaction of aluminum hydroxide with HCl.

**22.94.** Aluminum carbonate is used in some antacids. Write a balanced net ionic equation for the reaction of aluminum carbonate with hydrochloric acid.

**22.95.** A silver/zinc (Ag/Zn) battery has the advantage of being mercury-free. The half-reactions are as follows:

$$ZnO(s) + H_2O(\ell) + 2\,e^- \rightarrow$$
$$Zn(s) + 2\,OH^-(aq) \quad E° = -1.258\ V$$

$$Ag_2O(s) + H_2O(\ell) + 2\,e^- \rightarrow$$
$$2\,Ag(s) + 2\,OH^-(aq) \quad E° = 0.342\ V$$

What is the overall reaction of an electrochemical cell based on these materials? Using the $E°$ values provided, determine the standard cell potential of the Ag/ZnO battery.

**22.96. Power Sources for Pacemakers** Another power source considered for pacemakers is the lithium/copper sulfide battery. Use the half-reactions shown here to determine the overall reaction for an electrochemical cell based on these materials. With the $E°$ values provided, calculate the cell potential of the Li/CuS battery.

$$Li^+(s) + e^- \rightarrow Li(s) \quad E° = -3.05\ V$$
$$CuS(s) + 2\,e^- \rightarrow Cu(s) + S^{2-}(aq) \quad E° = -0.851\ V$$

**22.97.** Iridium-192, used in cancer therapy, crystallizes in a face-centered unit cell with a density of 22.42 g/cm³. Calculate the radius of an iridium atom.

**22.98.** Tantalum metal, used in bone implants, crystallizes in a body-centered unit cell with a density of 16.65 g/cm³. Calculate the radius of a tantalum atom.

**22.99.** Substitutional alloys of titanium containing 6% Al and 4% V by mass are used in orthopedic devices.
  a. What is the composition of the alloy on a mole basis?
  *b. Does the composition of this alloy match a structure based on a body-centered cubic (bcc) arrangement of Ti atoms with two corner atoms substituted by an Al and a V?

**22.100.** Titanium-molybdenum alloys containing 15% Mo by mass offer some strength advantages over T-Al-V alloys in orthopedic applications.
  a. What is the composition of the alloy on a mole basis?
  *b. Does the composition of this alloy match a structure based on a bcc arrangement of Ti atoms with one corner atom substituted by Mo?

If your instructor assigns problems in **smartwork**, log in at **smartwork.wwnorton.com**.

# Mathematical Procedures

## Working with Scientific Notation

Quantities that scientists work with are sometimes very large, such as Earth's mass, and other times very small, such as the mass of an electron. It is easier to work with these values when they are expressed in scientific notation.

The general form of standard scientific notation is a value between 1 and 10 multiplied by 10 raised to an integral power. According to this definition, $598 \times 10^{22}$ kg (Earth's mass) is not in standard scientific notation, but $5.98 \times 10^{24}$ kg is. It is good practice to use and report values in standard scientific notation.

1. **To convert an "ordinary" number to standard scientific notation,** move the decimal point to the left for a large number, or to the right for a small one, so that the decimal point is located after the first nonzero digit.

   A. For example, to express Earth's average density ($5517$ kg/m$^3$) in scientific notation requires moving the decimal point three places to the left. Doing so is the same as dividing the number by 1000, or $10^3$. To keep the value the same we multiply it by $10^3$. So, Earth's density in standard scientific notation is $5.517 \times 10^3$ kg/m$^3$.

   B. If we move the decimal point of a value less than 1 to the right to express it in scientific notation, then the exponent is a negative integer equal to the number of places we moved the decimal point to the right. For example, the value of $R$ used in solving ideal gas law problems is $0.08206$ L $\cdot$ atm/(mol $\cdot$ K). Moving the decimal point two places to the right converts the value of $R$ to scientific notation: $8.206 \times 10^{-2}$ L $\cdot$ atm/(mol $\cdot$ K).

   C. Another value of $R$, $8.314$ J/(mol $\cdot$ K), does not need an exponent, though it could be written $8.314 \times 10^0$ J/(mol $\cdot$ K) because any value raised to the zero power is equal to 1.

2. **For calculations with numbers in scientific notation,** most calculators have a function key for entering the exponents of values expressed in scientific notation. In many calculators it is labeled "E" or "EE" or "Exp." To enter, for example, the speed of light in meters per second, $2.998 \times 10^8$, we enter the value before the exponent, 2.998, followed by the exponent key and then the value of the exponent (8). To enter a value with a negative exponent, use the sign-change key. Sometimes it is labeled "(−)" or "+/−" or "±."

## Working with Logarithms

A logarithm to the base 10 has the following form:

$$\log_{10} x = \log x = p, \text{ where } x = 10^p$$

We usually abbreviate the logarithm function "log" if the logarithm is to the base 10, which means the scale in which $\log 10 = 1$.

A logarithm to the base $e$, called a *natural logarithm*, has the following form:

$$\log_e x = \ln x = q, \text{ where } x = e^q$$

Scientific calculators have (LOG) and (LN) keys, so it is easy to convert a number into its log or ln form. The directions below apply to most calculators.

**Sample Exercise 1** Find the logarithm to the base 10 of 2.247 (log 2.247).

**Solution** In some calculators you enter 2.247 first and then press the (LOG) key. In others, such as the TI 84/89 series, you press the (LOG) key first followed by 2.247 and then the (ENTER) key. Either way, the answer should be 0.3516 (to four significant figures).

**Sample Exercise 2** Find the natural logarithm of 2.247 (ln 2.247).

**Solution** Follow the same procedure as in Sample Exercise 1 except use the $\boxed{\text{LN}}$ key. The answer should be 0.8096. This answer is (0.8096/0.3516) = 2.303 times larger than log value. That is,

$$\ln x = 2.303 \log x$$

**Sample Exercise 3** Find the log of $6.0221 \times 10^{23}$.

**Solution** Following the procedures described above for entering a value with an exponent into your calculator and then taking its log to the base 10, you should obtain the value 23.77974796. We know the original value to five significant figures; the log value should have the same precision, but how do we express it? You might think the log value should be 23.780; however, the 23 to the left of the decimal point reflects the value of the exponent in the original value, and exponents don't count in determining the number of significant figures in a value. Therefore, the value of the logarithm to five significant figures is 23.77975. To understand better why this is so, calculate the log of 6.0221. You should get 0.77975 to five significant figures. Note how the difference between the two log values (23.77975 and 0.77975) is simply the value of the exponent in the first value.

**Combining Logs** The following equations summarize how logarithms of the products or quotients of two or more values are related to the individual logs of those values:

$$\text{logarithm } (ab) = \text{logarithm } a + \text{logarithm } b$$

and

$$\text{logarithm } (a/b) = \text{logarithm } a - \text{logarithm } b$$

## Converting Logarithms into Numbers

If we know the value of log $x$, what is the value of $x$? This question is frequently asked when working with pH, which is the negative log of the concentration of hydrogen ions, $[H^+]$, in solution:

$$\text{pH} = -\log[H^+]$$

Suppose the pH of a solution of a weak acid is 2.50. The concentration of $H^+$ is related to this pH value as follows:

$$2.50 = -\log[H^+]$$

or

$$-2.50 = \log[H^+]$$

To find the value of $[H^+]$, we enter 2.5 into the calculator and press the appropriate key to change its sign to $-2.5$. The next step depends on the type of calculator. If yours has a $\boxed{10^x}$ key (often accessible using a second function key), use it to find the value of $10^{-2.5}$, which is the value we are looking for. The corresponding keystrokes with many graphing calculators are $\boxed{10^x}$, $\boxed{(-)}$, 2.5, $\boxed{\text{ENTER}}$. Some calculators, including the virtual one in many Windows operating systems, have an $\boxed{x^y}$ key. We can use it for this problem by entering 10 and pushing the $\boxed{x^y}$ key, then entering 2.5 and pushing the $\boxed{+/-}$ key followed by the $\boxed{=}$ key. All of these approaches do the same calculation, taking 10 to the $-2.50$ power, and give the same answer, $[H^+] = 3.2 \times 10^{-3}$ to two significant figures. (Remember, pH is a log value, so the digit before the decimal point is not significant.)

## Solving Quadratic Equations

If the terms in an equation can be rearranged so that they take the form

$$ax^2 + bx + c = 0$$

they have the form of a quadratic equation. The value(s) of $x$ can be determined from the values of the coefficients $a$, $b$, and $c$ by using the equation

$$x = \frac{-b \pm \sqrt{b^2 - 4ac}}{2a}$$

For example, if the solution to a problem yields the following expression where $x$ is the concentration of a solute:

$$x^2 + 0.112x - 1.2 \times 10^{-3} = 0$$

Then the value of $x$ can be determined as follows:

$$x = \frac{-b \pm \sqrt{b^2 - 4ac}}{2a}$$

$$= \frac{-0.112 \pm \sqrt{(0.112)^2 - 4(1)(-1.2 \times 10^{-3})}}{2(1)}$$

$$= \frac{-0.112 \pm \sqrt{0.01254 + 0.0048}}{2}$$

$$= \frac{-0.112 \pm 0.132}{2} = +0.010 \text{ or } -0.122$$

In this example, the negative value for $x$ satisfies the equation, but it has no meaning because we cannot have negative concentration values; therefore we use only the $+0.010$ value.

## Expressing Data in Graphical Form

Fitting curves to plots of experimental data is a powerful tool in determining the relationships between variables. Many natural phenomena obey exponential functions. For example, the rate constant ($k$) of a chemical reaction increases exponentially with increasing absolute temperature ($T$). This relationship is described by the Arrhenius equation:

$$k = A e^{-E_a/RT}$$

where $A$ is a constant for a particular reaction (called the frequency factor), $E_a$ is the activation energy of the reaction, and $R$ is the ideal gas constant. Taking the natural logarithms of both sides of the Arrhenius equation gives

$$\ln k = \ln A - \left(\frac{E_a}{RT}\right)$$

This equation fits the general equation of a straight line ($y = mx + b$) if ($\ln k$) is the $y$-variable and ($1/T$) is the $x$-variable. Plotting ($\ln k$) versus ($1/T$) should give a straight line with a slope equal to $-E_a/R$. The slopes of these plots are negative because the activation energies, $E_a$, of chemical reactions are positive. The data for a reaction given in columns 2 and 4 of Table A1.1 are plotted in Figure A1.1. The program that generated the graph also gives us the equation of the straight line that best fits the data. The slope ($-1281$ K) of this line is used to calculate the value of $E_a$:

$$-1281 \text{ K} = -\frac{E_a}{R}$$

$$E_a = -(-1281 \text{ K})[8.314 \text{ J}/(\text{mol} \cdot \text{K})]$$

$$= 10,650 \text{ J/mol} = 10.65 \text{ kJ/mol}$$

| TABLE A1.1 | Rate Constant $k$ as a Function of Temperature $T$ | | |
|---|---|---|---|
| Temperature $T$ (K) | $1/T$ (K$^{-1}$) | Rate Constant $k$ | ln $k$ |
| 500 | 0.0020 | 0.030 | −3.51 |
| 550 | 0.0018 | 0.38 | −0.97 |
| 600 | 0.0017 | 2.9 | 1.06 |
| 650 | 0.0015 | 17 | 2.83 |
| 700 | 0.0014 | 75 | 4.32 |

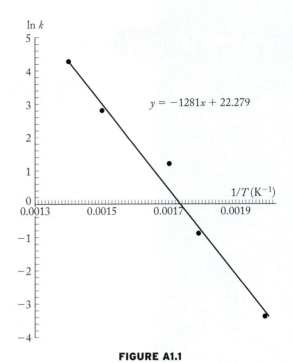

$y = -1281x + 22.279$

**FIGURE A1.1**

# SI Units and Conversion Factors

### TABLE A2.1 Six SI Base Units

| SI Base Quantity | Unit | Symbol |
|---|---|---|
| length | meter | m |
| mass | kilogram | kg |
| time | second | s |
| amount of substance | mole | mol |
| temperature | kelvin | K |
| electric current | ampere | A |

### TABLE A2.2 Some SI–Derived Units

| SI–Derived Quantity | Unit | Symbol | Dimensions |
|---|---|---|---|
| electric charge | coulomb | C | $A \cdot s$ |
| electric potential | volt | V | $J/C$ |
| force | newton | N | $kg \cdot m/s^2$ |
| frequency | hertz | Hz | $s^{-1}$ |
| momentum | newton-second | $N \cdot s$ | $kg \cdot m/s$ |
| power | watt | W | $J/s$ |
| pressure | pascal | Pa | $N/m^2$ |
| radioactivity | becquerel | Bq | $s^{-1}$ |
| speed or velocity | meter per second | m/s | m/s |
| energy | joule (newton-meter) | $J (N \cdot m)$ | $kg \cdot m^2/s^2$ |

### TABLE A2.3 SI Prefixes

| Prefix | Symbol | Multiplier | Prefix | Symbol | Multiplier |
|---|---|---|---|---|---|
| deci | d | $10^{-1}$ | deka | da | $10^1$ |
| centi | c | $10^{-2}$ | hecto | h | $10^2$ |
| milli | m | $10^{-3}$ | kilo | k | $10^3$ |
| micro | $\mu$ | $10^{-6}$ | mega | M | $10^6$ |
| nano | n | $10^{-9}$ | giga | G | $10^9$ |
| pico | p | $10^{-12}$ | tera | T | $10^{12}$ |
| femto | f | $10^{-15}$ | peta | P | $10^{15}$ |
| atto | a | $10^{-18}$ | exa | E | $10^{18}$ |
| zepto | z | $10^{-21}$ | zetta | Z | $10^{21}$ |

## TABLE A2.4    Special Units and Conversion Factors

| Quantity | Unit | Symbol | Conversion[a] |
|---|---|---|---|
| energy[a] | electron-volt | eV | $1 \text{ eV} = 1.6022 \times 10^{-19} \text{ J}$ |
| energy[a] | kilowatt-hour | kWh | $1 \text{ kWh} = 3600 \text{ kJ}$ |
| energy | calorie | cal | $1 \text{ cal} = 4.184 \text{ J}$ |
| mass | pound | lb | $1 \text{ lb} = 453.59 \text{ g}$ |
| mass[a] | atomic mass unit | amu | $1 \text{ amu} = 1.66054 \times 10^{-27} \text{ kg}$ |
| length | angstrom | Å | $1 \text{ Å} = 10^{-8} \text{ cm} = 10^{-10} \text{ m}$ |
| length | inch | in | $1 \text{ in} = 2.54 \text{ cm}$ |
| length | mile | mi | $1 \text{ mi} = 5280 \text{ ft} = 1.6093 \text{ km}$ |
| pressure | atmosphere | atm | $1 \text{ atm} = 1.01325 \times 10^5 \text{ Pa}$ |
| pressure | torr | torr | $1 \text{ torr} = 1/760 \text{ atm}$ |
| temperature | Celsius scale | °C | $T(°C) = T(K) - 273.15$ |
| temperature | Fahrenheit scale | °F | $T(°F) = \frac{9}{5} T(°C) + 32$ |
| volume | liter | L | $1 \text{ L} = 1 \text{ dm}^3 = 10^{-3} \text{ m}^3$ |
| volume | cubic centimeter | cm³, cc | $1 \text{ cm}^3 = 1 \text{ mL} = 10^{-3} \text{ L}$ |
| volume | cubic foot | ft³ | $1 \text{ ft}^3 = 7.4805 \text{ gal}$ |
| volume | gallon (U.S.) | gal | $1 \text{ gal} = 3.785 \text{ L}$ |

[a]From http://physics.nist.gov/cuu/Constants/index.html.

## TABLE A2.5    Physical Constants[a]

| Quantity | Symbol | Value |
|---|---|---|
| acceleration due to gravity (Earth) | $g$ | $9.807 \text{ m/s}^2$ |
| Avogadro's number | $N_A$ | $6.0221 \times 10^{23} \text{ mol}^{-1}$ |
| Bohr radius | $a_0$ | $5.29 \times 10^{-11} \text{ m}$ |
| Boltzmann constant | $k_B$ | $1.3806 \times 10^{-23} \text{ J/K}$ |
| electron charge-to-mass ratio | $-e/m_e$ | $1.7588 \times 10^{11} \text{ C/kg}$ |
| elementary charge | $e$ | $1.602 \times 10^{-19} \text{ C}$ |
| Faraday constant | $F$ | $9.65 \times 10^4 \text{ C/mol}$ |
| mass of an electron | $m_e$ | $9.10938 \times 10^{-31} \text{ kg}$ |
| mass of a neutron | $m_n$ | $1.67493 \times 10^{-27} \text{ kg}$ |
| mass of a proton | $m_p$ | $1.67262 \times 10^{-27} \text{ kg}$ |
| molar volume of ideal gas at 0°C and 1 atm | $V_m$ | $22.4 \text{ L/mol}$ |
| Planck constant | $h$ | $6.626 \times 10^{-34} \text{ J} \cdot \text{s}$ |
| speed of light in vacuum | $c$ | $2.998 \times 10^8 \text{ m/s}$ |
| universal gas constant | $R$ | $8.314 \text{ J/(mol} \cdot \text{K)}$ <br> $0.08206 \text{ L} \cdot \text{atm/(mol} \cdot \text{K)}$ |

[a]From http://physics.nist.gov/cuu/Constants/index.html.

# The Elements and Their Properties

| TABLE A3.1 | Ground-State Electron Configurations, Atomic Radii, and First Ionization Energies of the Elements | | | | |
|---|---|---|---|---|---|
| Element | Symbol | Atomic Number $Z$ | Ground-State Configuration | Atomic Radius (pm) | First Ionization Energy (kJ/mol) |
| hydrogen | H | 1 | $1s^1$ | 37 | 1312.0 |
| helium | He | 2 | $1s^2$ | 32 | 2372.3 |
| lithium | Li | 3 | $[\text{He}]2s^1$ | 152 | 520.2 |
| beryllium | Be | 4 | $[\text{He}]2s^2$ | 112 | 899.5 |
| boron | B | 5 | $[\text{He}]2s^22p^1$ | 88 | 800.6 |
| carbon | C | 6 | $[\text{He}]2s^22p^2$ | 77 | 1086.5 |
| nitrogen | N | 7 | $[\text{He}]2s^22p^3$ | 75 | 1402.3 |
| oxygen | O | 8 | $[\text{He}]2s^22p^4$ | 73 | 1313.9 |
| fluorine | F | 9 | $[\text{He}]2s^22p^5$ | 71 | 1681.0 |
| neon | Ne | 10 | $[\text{He}]2s^22p^6$ | 69 | 2080.7 |
| sodium | Na | 11 | $[\text{Ne}]3s^1$ | 186 | 495.3 |
| magnesium | Mg | 12 | $[\text{Ne}]3s^2$ | 160 | 737.7 |
| aluminum | Al | 13 | $[\text{Ne}]3s^23p^1$ | 143 | 577.5 |
| silicon | Si | 14 | $[\text{Ne}]3s^23p^2$ | 117 | 786.5 |
| phosphorus | P | 15 | $[\text{Ne}]3s^23p^3$ | 110 | 1011.8 |
| sulfur | S | 16 | $[\text{Ne}]3s^23p^4$ | 103 | 999.6 |
| chlorine | Cl | 17 | $[\text{Ne}]3s^23p^5$ | 99 | 1251.2 |
| argon | Ar | 18 | $[\text{Ne}]3s^23p^6$ | 97 | 1520.6 |
| potassium | K | 19 | $[\text{Ar}]4s^1$ | 227 | 418.8 |
| calcium | Ca | 20 | $[\text{Ar}]4s^2$ | 197 | 589.8 |
| scandium | Sc | 21 | $[\text{Ar}]4s^23d^1$ | 162 | 633.1 |
| titanium | Ti | 22 | $[\text{Ar}]4s^23d^2$ | 147 | 658.8 |
| vanadium | V | 23 | $[\text{Ar}]4s^23d^3$ | 135 | 650.9 |
| chromium | Cr | 24 | $[\text{Ar}]4s^13d^5$ | 128 | 652.9 |
| manganese | Mn | 25 | $[\text{Ar}]4s^23d^5$ | 127 | 717.3 |
| iron | Fe | 26 | $[\text{Ar}]4s^23d^6$ | 126 | 762.5 |
| cobalt | Co | 27 | $[\text{Ar}]4s^23d^7$ | 125 | 760.4 |
| nickel | Ni | 28 | $[\text{Ar}]4s^23d^8$ | 124 | 737.1 |
| copper | Cu | 29 | $[\text{Ar}]4s^13d^{10}$ | 128 | 745.5 |
| zinc | Zn | 30 | $[\text{Ar}]4s^23d^{10}$ | 134 | 906.4 |
| gallium | Ga | 31 | $[\text{Ar}]4s^23d^{10}4p^1$ | 135 | 578.8 |
| germanium | Ge | 32 | $[\text{Ar}]4s^23d^{10}4p^2$ | 122 | 762.2 |
| arsenic | As | 33 | $[\text{Ar}]4s^23d^{10}4p^3$ | 121 | 947.0 |
| selenium | Se | 34 | $[\text{Ar}]4s^23d^{10}4p^4$ | 119 | 941.0 |

| TABLE A3.1 | Ground-State Electron Configurations, Atomic Radii, and First Ionization Energies of the Elements *(Continued)* |

| Element | Symbol | Atomic Number $Z$ | Ground-State Configuration | Atomic Radius (pm) | First Ionization Energy (kJ/mol) |
|---|---|---|---|---|---|
| bromine | Br | 35 | [Ar]$4s^2 3d^{10} 4p^5$ | 114 | 1139.9 |
| krypton | Kr | 36 | [Ar]$4s^2 3d^{10} 4p^6$ | 110 | 1350.8 |
| rubidium | Rb | 37 | [Kr]$5s^1$ | 247 | 403.0 |
| strontium | Sr | 38 | [Kr]$5s^2$ | 215 | 549.5 |
| yttrium | Y | 39 | [Kr]$5s^2 4d^1$ | 180 | 599.8 |
| zirconium | Zr | 40 | [Kr]$5s^2 4d^2$ | 160 | 640.1 |
| niobium | Nb | 41 | [Kr]$5s^1 4d^4$ | 146 | 652.1 |
| molybdenum | Mo | 42 | [Kr]$5s^1 4d^5$ | 139 | 684.3 |
| technetium | Tc | 43 | [Kr]$5s^2 4d^5$ | 136 | 702.4 |
| ruthenium | Ru | 44 | [Kr]$5s^1 4d^7$ | 134 | 710.2 |
| rhodium | Rh | 45 | [Kr]$5s^1 4d^8$ | 134 | 719.7 |
| palladium | Pd | 46 | [Kr]$4d^{10}$ | 137 | 804.4 |
| silver | Ag | 47 | [Kr]$5s^1 4d^{10}$ | 144 | 731.0 |
| cadmium | Cd | 48 | [Kr]$5s^2 4d^{10}$ | 151 | 867.8 |
| indium | In | 49 | [Kr]$5s^2 4d^{10} 5p^1$ | 167 | 558.3 |
| tin | Sn | 50 | [Kr]$5s^2 4d^{10} 5p^2$ | 140 | 708.6 |
| antimony | Sb | 51 | [Kr]$5s^2 4d^{10} 5p^3$ | 141 | 833.6 |
| tellurium | Te | 52 | [Kr]$5s^2 4d^{10} 5p^4$ | 143 | 869.3 |
| iodine | I | 53 | [Kr]$5s^2 4d^{10} 5p^5$ | 133 | 1008.4 |
| xenon | Xe | 54 | [Kr]$5s^2 4d^{10} 5p^6$ | 130 | 1170.4 |
| cesium | Cs | 55 | [Xe]$6s^1$ | 265 | 375.7 |
| barium | Ba | 56 | [Xe]$6s^2$ | 222 | 502.9 |
| lanthanum | La | 57 | [Xe]$6s^2 5d^1$ | 187 | 538.1 |
| cerium | Ce | 58 | [Xe]$6s^2 4f^1 5d^1$ | 182 | 534.4 |
| praseodymium | Pr | 59 | [Xe]$6s^2 4f^3$ | 182 | 527.2 |
| neodymium | Nd | 60 | [Xe]$6s^2 4f^4$ | 181 | 533.1 |
| promethium | Pm | 61 | [Xe]$6s^2 4f^5$ | 183 | 535.5 |
| samarium | Sm | 62 | [Xe]$6s^2 4f^6$ | 180 | 544.5 |
| europium | Eu | 63 | [Xe]$6s^2 4f^7$ | 208 | 547.1 |
| gadolinium | Gd | 64 | [Xe]$6s^2 4f^7 5d^1$ | 180 | 593.4 |
| terbium | Tb | 65 | [Xe]$6s^2 4f^9$ | 177 | 565.8 |
| dysprosium | Dy | 66 | [Xe]$6s^2 4f^{10}$ | 178 | 573.0 |
| holmium | Ho | 67 | [Xe]$6s^2 4f^{11}$ | 176 | 581.0 |
| erbium | Er | 68 | [Xe]$6s^2 4f^{12}$ | 176 | 589.3 |
| thulium | Tm | 69 | [Xe]$6s^2 4f^{13}$ | 176 | 596.7 |
| ytterbium | Yb | 70 | [Xe]$6s^2 4f^{14}$ | 193 | 603.4 |
| lutetium | Lu | 71 | [Xe]$6s^2 4f^{14} 5d^1$ | 174 | 523.5 |
| hafnium | Hf | 72 | [Xe]$6s^2 4f^{14} 5d^2$ | 159 | 658.5 |
| tantalum | Ta | 73 | [Xe]$6s^2 4f^{14} 5d^3$ | 146 | 761.3 |
| tungsten | W | 74 | [Xe]$6s^2 4f^{14} 5d^4$ | 139 | 770.0 |
| rhenium | Re | 75 | [Xe]$6s^2 4f^{14} 5d^5$ | 137 | 760.3 |
| osmium | Os | 76 | [Xe]$6s^2 4f^{14} 5d^6$ | 135 | 839.4 |
| iridium | Ir | 77 | [Xe]$6s^2 4f^{14} 5d^7$ | 136 | 878.0 |

*Continued on next page*

| TABLE A3.1 | | Ground-State Electron Configurations, Atomic Radii, and First Ionization Energies of the Elements *(Continued)* | | | |
|---|---|---|---|---|---|
| Element | Symbol | Atomic Number $Z$ | Ground-State Configuration | Atomic Radius (pm) | First Ionization Energy (kJ/mol) |
| platinum | Pt | 78 | $[\text{Xe}]6s^1 4f^{14} 5d^9$ | 139 | 868.4 |
| gold | Au | 79 | $[\text{Xe}]6s^1 4f^{14} 5d^{10}$ | 144 | 890.1 |
| mercury | Hg | 80 | $[\text{Xe}]6s^2 4f^{14} 5d^{10}$ | 151 | 1007.1 |
| thallium | Tl | 81 | $[\text{Xe}]6s^2 4f^{14} 5d^{10} 6p^1$ | 170 | 589.4 |
| lead | Pb | 82 | $[\text{Xe}]6s^2 4f^{14} 5d^{10} 6p^2$ | 154 | 715.6 |
| bismuth | Bi | 83 | $[\text{Xe}]6s^2 4f^{14} 5d^{10} 6p^3$ | 150 | 703.3 |
| polonium | Po | 84 | $[\text{Xe}]6s^2 4f^{14} 5d^{10} 6p^4$ | 167 | 812.1 |
| astatine | At | 85 | $[\text{Xe}]6s^2 4f^{14} 5d^{10} 6p^5$ | 140 | 924.6 |
| radon | Rn | 86 | $[\text{Xe}]6s^2 4f^{14} 5d^{10} 6p^6$ | 145 | 1037.1 |
| francium | Fr | 87 | $[\text{Rn}]7s^1$ | 242 | 380 |
| radium | Ra | 88 | $[\text{Rn}]7s^2$ | 211 | 509.3 |
| actinium | Ac | 89 | $[\text{Rn}]7s^2 6d^1$ | 188 | 499 |
| thorium | Th | 90 | $[\text{Rn}]7s^2 6d^2$ | 179 | 587 |
| protactinium | Pa | 91 | $[\text{Rn}]7s^2 5f^2 6d^1$ | 163 | 568 |
| uranium | U | 92 | $[\text{Rn}]7s^2 5f^3 6d^1$ | 156 | 587 |
| neptunium | Np | 93 | $[\text{Rn}]7s^2 5f^4 6d^1$ | 155 | 597 |
| plutonium | Pu | 94 | $[\text{Rn}]7s^2 5f^6$ | 159 | 585 |
| americium | Am | 95 | $[\text{Rn}]7s^2 5f^7$ | 173 | 578 |
| curium | Cm | 96 | $[\text{Rn}]7s^2 5f^7 6d^1$ | 174 | 581 |
| berkelium | Bk | 97 | $[\text{Rn}]7s^2 5f^9$ | 170 | 601 |
| californium | Cf | 98 | $[\text{Rn}]7s^2 5f^{10}$ | 186 | 608 |
| einsteinium | Es | 99 | $[\text{Rn}]7s^2 5f^{11}$ | 186 | 619 |
| fermium | Fm | 100 | $[\text{Rn}]7s^2 5f^{12}$ | 167 | 627 |
| mendelevium | Md | 101 | $[\text{Rn}]7s^2 5f^{13}$ | 173 | 635 |
| nobelium | No | 102 | $[\text{Rn}]7s^2 5f^{14}$ | 176 | 642 |
| lawrencium | Lr | 103 | $[\text{Rn}]7s^2 5f^{14} 6d^1$ | 161 | — |
| rutherfordium | Rf | 104 | $[\text{Rn}]7s^2 5f^{14} 6d^2$ | 157 | — |
| dubnium | Db | 105 | $[\text{Rn}]7s^2 5f^{14} 6d^3$ | 149 | — |
| seaborgium | Sg | 106 | $[\text{Rn}]7s^2 5f^{14} 6d^4$ | 143 | — |
| bohrium | Bh | 107 | $[\text{Rn}]7s^2 5f^{14} 6d^5$ | 141 | — |
| hassium | Hs | 108 | $[\text{Rn}]7s^2 5f^{14} 6d^6$ | 134 | — |
| meitnerium | Mt | 109 | $[\text{Rn}]7s^2 5f^{14} 6d^7$ | 129 | — |
| darmstadtium | Ds | 110 | $[\text{Rn}]7s^2 5f^{14} 6d^8$ | 128 | — |
| roentgenium | Rg | 111 | $[\text{Rn}]7s^2 5f^{14} 6d^9$ | 121 | — |
| copernicium | Cn | 112 | $[\text{Rn}]7s^2 5f^{14} 6d^{10}$ | 122 | — |
| ununtrium | Uut | 113 | $[\text{Rn}]7s^2 5f^{14} 6d^{10} 7p^1$ | 136 | — |
| flerovium | Fl | 114 | $[\text{Rn}]7s^2 5f^{14} 6d^{10} 7p^2$ | 143 | — |
| ununpentium | Uup | 115 | $[\text{Rn}]7s^2 5f^{14} 6d^{10} 7p^3$ | 162 | — |
| livermorium | Lv | 116 | $[\text{Rn}]7s^2 5f^{14} 6d^{10} 7p^4$ | 175 | — |
| ununseptium | Uus | 117 | $[\text{Rn}]7s^2 5f^{14} 6d^{10} 7p^5$ | 165 | — |
| ununoctium | Uuo | 118 | $[\text{Rn}]7s^2 5f^{14} 6d^{10} 7p^6$ | 157 | — |

| TABLE A3.2 | Miscellaneous Physical Properties of the Elements[a] | | | | |
|---|---|---|---|---|---|

| Element | Symbol | Atomic Number | Physical State[b,c] | Density[d] (g/cm³) | Melting Point (°C) | Boiling Point (°C) |
|---|---|---|---|---|---|---|
| hydrogen | H | 1 | gas | 0.000090 | −259.14 | −252.87 |
| helium | He | 2 | gas | 0.000179 | <−272.2 | −268.93 |
| lithium | Li | 3 | solid | 0.534 | 180.5 | 1347 |
| beryllium | Be | 4 | solid | 1.848 | 1283 | 2484 |
| boron | B | 5 | solid | 2.34 | 2300 | 3650 |
| carbon | C | 6 | solid (gr) | 1.9–2.3 | ~3350 | sublimes |
| nitrogen | N | 7 | gas | 0.00125 | −210.00 | −195.8 |
| oxygen | O | 8 | gas | 0.00143 | −218.8 | −182.95 |
| fluorine | F | 9 | gas | 0.00170 | −219.62 | −188.12 |
| neon | Ne | 10 | gas | 0.00090 | −248.59 | −246.08 |
| sodium | Na | 11 | solid | 0.971 | 97.72 | 883 |
| magnesium | Mg | 12 | solid | 1.738 | 650 | 1090 |
| aluminum | Al | 13 | solid | 2.6989 | 660.32 | 2467 |
| silicon | Si | 14 | solid | 2.33 | 1414 | 2355 |
| phosphorus | P | 15 | solid (wh) | 1.82 | 44.15 | 280 |
| sulfur | S | 16 | solid | 2.07 | 115.21 | 444.60 |
| chlorine | Cl | 17 | gas | 0.00321 | −101.5 | −34.04 |
| argon | Ar | 18 | gas | 0.00178 | −189.3 | −185.9 |
| potassium | K | 19 | solid | 0.862 | 63.28 | 759 |
| calcium | Ca | 20 | solid | 1.55 | 842 | 1484 |
| scandium | Sc | 21 | solid | 2.989 | 1541 | 2380 |
| titanium | Ti | 22 | solid | 4.54 | 1668 | 3287 |
| vanadium | V | 23 | solid | 6.11 | 1910 | 3407 |
| chromium | Cr | 24 | solid | 7.19 | 1857 | 2671 |
| manganese | Mn | 25 | solid | 7.3 | 1246 | 1962 |
| iron | Fe | 26 | solid | 7.874 | 1538 | 2750 |
| cobalt | Co | 27 | solid | 8.9 | 1495 | 2870 |
| nickel | Ni | 28 | solid | 8.902 | 1455 | 2730 |
| copper | Cu | 29 | solid | 8.96 | 1084.6 | 2562 |
| zinc | Zn | 30 | solid | 7.133 | 419.53 | 907 |
| gallium | Ga | 31 | solid | 5.904 | 29.76 | 2403 |
| germanium | Ge | 32 | solid | 5.323 | 938.25 | 2833 |
| arsenic | As | 33 | solid (gy) | 5.727 | 614 | sublimes |
| selenium | Se | 34 | solid (gy) | 4.79 | 221 | 685 |
| bromine | Br | 35 | liquid | 3.12 | −7.2 | 58.78 |
| krypton | Kr | 36 | gas | 0.00373 | −157.36 | −153.22 |
| rubidium | Rb | 37 | solid | 1.532 | 39.31 | 688 |
| strontium | Sr | 38 | solid | 2.64 | 777 | 1382 |
| yttrium | Y | 39 | solid | 4.469 | 1526 | 3336 |
| zirconium | Zr | 40 | solid | 6.506 | 1855 | 4409 |
| niobium | Nb | 41 | solid | 8.57 | 2477 | 4744 |

*Continued on next page*

| TABLE A3.2 | **Miscellaneous Physical Properties of the Elements[a] (Continued)** | | | | | |
|---|---|---|---|---|---|---|
| **Element** | **Symbol** | **Atomic Number** | **Physical State[b,c]** | **Density[d] (g/cm³)** | **Melting Point (°C)** | **Boiling Point (°C)** |
| molybdenum | Mo | 42 | solid | 10.22 | 2623 | 4639 |
| technetium | Tc | 43 | solid | 11.50 | 2157 | 4538 |
| ruthenium | Ru | 44 | solid | 12.41 | 2334 | 3900 |
| rhodium | Rh | 45 | solid | 12.41 | 1964 | 3695 |
| palladium | Pd | 46 | solid | 12.02 | 1555 | 2963 |
| silver | Ag | 47 | solid | 10.50 | 961.78 | 2212 |
| cadmium | Cd | 48 | solid | 8.65 | 321.07 | 767 |
| indium | In | 49 | solid | 7.31 | 156.60 | 2072 |
| tin | Sn | 50 | solid (wh) | 7.31 | 231.9 | 2270 |
| antimony | Sb | 51 | solid | 6.691 | 630.63 | 1750 |
| tellurium | Te | 52 | solid | 6.24 | 449.5 | 998 |
| iodine | I | 53 | solid | 4.93 | 113.7 | 184.4 |
| xenon | Xe | 54 | gas | 0.00589 | −111.75 | −108.0 |
| cesium | Cs | 55 | solid | 1.873 | 28.44 | 671 |
| barium | Ba | 56 | solid | 3.5 | 727 | 1640 |
| lanthanum | La | 57 | solid | 6.145 | 920 | 3455 |
| cerium | Ce | 58 | solid | 6.770 | 799 | 3424 |
| praseodymium | Pr | 59 | solid | 6.773 | 931 | 3510 |
| neodymium | Nd | 60 | solid | 7.008 | 1016 | 3066 |
| promethium | Pm | 61 | solid | 7.264 | 1042 | ~3000 |
| samarium | Sm | 62 | solid | 7.520 | 1072 | 1790 |
| europium | Eu | 63 | solid | 5.244 | 822 | 1596 |
| gadolinium | Gd | 64 | solid | 7.901 | 1314 | 3264 |
| terbium | Tb | 65 | solid | 8.230 | 1359 | 3221 |
| dysprosium | Dy | 66 | solid | 8.551 | 1411 | 2561 |
| holmium | Ho | 67 | solid | 8.795 | 1472 | 2694 |
| erbium | Er | 68 | solid | 9.066 | 1529 | 2862 |
| thulium | Tm | 69 | solid | 9.321 | 1545 | 1946 |
| ytterbium | Yb | 70 | solid | 6.966 | 824 | 1194 |
| lutetium | Lu | 71 | solid | 9.841 | 1663 | 3393 |
| hafnium | Hf | 72 | solid | 13.31 | 2233 | 4603 |
| tantalum | Ta | 73 | solid | 16.654 | 3017 | 5458 |
| tungsten | W | 74 | solid | 19.3 | 3422 | 5660 |
| rhenium | Re | 75 | solid | 21.02 | 3186 | 5596 |
| osmium | Os | 76 | solid | 22.57 | 3033 | 5012 |
| iridium | Ir | 77 | solid | 22.42 | 2446 | 4130 |
| platinum | Pt | 78 | solid | 21.45 | 1768.4 | 3825 |
| gold | Au | 79 | solid | 19.3 | 1064.18 | 2856 |
| mercury | Hg | 80 | liquid | 13.546 | −38.83 | 356.73 |
| thallium | Tl | 81 | solid | 11.85 | 304 | 1473 |
| lead | Pb | 82 | solid | 11.35 | 327.46 | 1749 |
| bismuth | Bi | 83 | solid | 9.747 | 271.4 | 1564 |

| TABLE A3.2 | Miscellaneous Physical Properties of the Elements[a] *(Continued)* | | | | | |
|---|---|---|---|---|---|---|
| Element | Symbol | Atomic Number | Physical State[b,c] | Density[d] (g/cm³) | Melting Point (°C) | Boiling Point (°C) |
| polonium | Po | 84 | solid | 9.32 | 254 | 962 |
| astatine | At | 85 | solid | unknown | 302 | 337 |
| radon | Rn | 86 | gas | 0.00973 | −71 | −61.7 |
| francium | Fr | 87 | solid | unknown | 27 | 677 |
| radium | Ra | 88 | solid | 5 | 700 | 1737 |
| actinium | Ac | 89 | solid | 10.07 | 1051 | ~3200 |
| thorium | Th | 90 | solid | 11.72 | 1750 | 4788 |
| protactinium | Pa | 91 | solid | 15.37 | 1572 | unknown |
| uranium | U | 92 | solid | 19.05 | 1132 | 3818 |

[a]For relative atomic masses and alphabetical listing of the elements, see the flyleaf at the front of this volume.
[b]Normal state at 25°C and 1 atm.
[c]Allotropes: gr = graphite, gy = gray, wh = white.
[d]Liquids and solids at 25°C and 1 atm; gases at 0°C and 1 atm (STP).

| TABLE A3.3 | A Selection of Stable Isotopes[a] | | | | | |
|---|---|---|---|---|---|---|
| Isotope $^{A}X$ | Natural Abundance (%) | Atomic Number Z | Neutron Number N | Mass Number A | Atomic Mass (amu) | Binding Energy per Nucleon (MeV)[b] |
| $^{1}$H | 99.985 | 1 | 0 | 1 | 1.007825 | — |
| $^{2}$H | 0.015 | 1 | 1 | 2 | 2.014000 | 1.160 |
| $^{3}$He | 0.000137 | 2 | 1 | 3 | 3.016030 | 2.572 |
| $^{4}$He | 99.999863 | 2 | 2 | 4 | 4.002603 | 7.075 |
| $^{6}$Li | 7.5 | 3 | 3 | 6 | 6.015121 | 5.333 |
| $^{7}$Li | 92.5 | 3 | 4 | 7 | 7.016003 | 5.606 |
| $^{9}$Be | 100.0 | 4 | 5 | 9 | 9.012182 | 6.463 |
| $^{10}$B | 19.9 | 5 | 5 | 10 | 10.012937 | 6.475 |
| $^{11}$B | 80.1 | 5 | 6 | 11 | 11.009305 | 6.928 |
| $^{12}$C | 98.90 | 6 | 6 | 12 | 12.000000 | 7.680 |
| $^{13}$C | 1.10 | 6 | 7 | 13 | 13.003355 | 7.470 |
| $^{14}$N | 99.634 | 7 | 7 | 14 | 14.003074 | 7.476 |
| $^{15}$N | 0.366 | 7 | 8 | 15 | 15.000108 | 7.699 |
| $^{16}$O | 99.762 | 8 | 8 | 16 | 15.994915 | 7.976 |
| $^{17}$O | 0.038 | 8 | 9 | 17 | 16.999131 | 7.751 |
| $^{18}$O | 0.200 | 8 | 10 | 18 | 17.999160 | 7.767 |
| $^{19}$F | 100.0 | 9 | 10 | 19 | 18.998403 | 7.779 |
| $^{20}$Ne | 90.48 | 10 | 10 | 20 | 19.992435 | 8.032 |
| $^{21}$Ne | 0.27 | 10 | 11 | 21 | 20.993843 | 7.972 |
| $^{22}$Ne | 9.25 | 10 | 12 | 22 | 21.991383 | 8.081 |
| $^{23}$Na | 100.0 | 11 | 12 | 23 | 22.989770 | 8.112 |
| $^{24}$Mg | 78.99 | 12 | 12 | 24 | 23.985042 | 8.261 |
| $^{25}$Mg | 10.00 | 12 | 13 | 25 | 24.985837 | 8.223 |
| $^{26}$Mg | 11.01 | 12 | 14 | 26 | 25.982593 | 8.334 |
| $^{27}$Al | 100.0 | 13 | 14 | 27 | 26.981538 | 8.331 |
| $^{28}$Si | 92.23 | 14 | 14 | 28 | 27.976927 | 8.448 |
| $^{29}$Si | 4.67 | 14 | 15 | 29 | 28.976495 | 8.449 |
| $^{30}$Si | 3.10 | 14 | 16 | 30 | 29.973770 | 8.521 |

*Continued on next page*

**TABLE A3.3    A Selection of Stable Isotopes[a] (Continued)**

| Isotope [A]X | Natural Abundance (%) | Atomic Number Z | Neutron Number N | Mass Number A | Atomic Mass (amu) | Binding Energy per Nucleon (MeV)[b] |
|---|---|---|---|---|---|---|
| [31]P | 100.0 | 15 | 16 | 31 | 30.973761 | 8.481 |
| [32]S | 95.02 | 16 | 16 | 32 | 31.972070 | 8.493 |
| [33]S | 0.75 | 16 | 17 | 33 | 32.971456 | 8.498 |
| [34]S | 4.21 | 16 | 18 | 34 | 33.967866 | 8.584 |
| [36]S | 0.02 | 16 | 20 | 36 | 35.967080 | 8.575 |
| [35]Cl | 75.77 | 17 | 18 | 35 | 34.968852 | 8.520 |
| [37]Cl | 24.23 | 17 | 20 | 37 | 36.965903 | 8.570 |
| [36]Ar | 0.337 | 18 | 18 | 36 | 35.967545 | 8.520 |
| [38]Ar | 0.063 | 18 | 20 | 38 | 37.962732 | 8.614 |
| [40]Ar | 99.600 | 18 | 22 | 40 | 39.962384 | 8.595 |
| [39]K | 93.258 | 19 | 20 | 39 | 38.963707 | 8.557 |
| [41]K | 6.730 | 19 | 22 | 41 | 40.961825 | 8.576 |
| [40]Ca | 96.941 | 20 | 20 | 40 | 39.962591 | 8.551 |
| [42]Ca | 0.647 | 20 | 22 | 42 | 41.958618 | 8.617 |
| [43]Ca | 0.135 | 20 | 23 | 43 | 42.958766 | 8.601 |
| [44]Ca | 2.086 | 20 | 24 | 44 | 43.955480 | 8.658 |
| [46]Ca | 0.004 | 20 | 26 | 46 | 45.953689 | 8.669 |
| [48]Ca | 0.187 | 20 | 28 | 48 | 47.952533 | 8.666 |
| [45]Sc | 100.0 | 21 | 24 | 45 | 44.955910 | 8.619 |
| [46]Ti | 8.0 | 22 | 24 | 46 | 45.952629 | 8.656 |
| [47]Ti | 7.3 | 22 | 25 | 47 | 46.951764 | 8.661 |
| [48]Ti | 73.8 | 22 | 26 | 48 | 47.947947 | 8.723 |
| [49]Ti | 5.5 | 22 | 27 | 49 | 48.947871 | 8.711 |
| [50]Ti | 5.4 | 22 | 28 | 50 | 49.944792 | 8.756 |
| [51]V | 99.750 | 23 | 28 | 51 | 50.943962 | 8.742 |
| [50]Cr | 4.345 | 24 | 26 | 50 | 49.946046 | 8.701 |
| [52]Cr | 83.789 | 24 | 28 | 52 | 51.940509 | 8.776 |
| [53]Cr | 9.501 | 24 | 29 | 53 | 52.940651 | 8.760 |
| [54]Cr | 2.365 | 24 | 30 | 54 | 53.938882 | 8.778 |
| [55]Mn | 100.0 | 25 | 30 | 55 | 54.938049 | 8.765 |
| [54]Fe | 5.9 | 26 | 28 | 54 | 53.939612 | 8.736 |
| [56]Fe | 91.72 | 26 | 30 | 56 | 55.934939 | 8.790 |
| [57]Fe | 2.1 | 26 | 31 | 57 | 56.935396 | 8.770 |
| [58]Fe | 0.28 | 26 | 32 | 58 | 57.933277 | 8.792 |
| [59]Co | 100.0 | 27 | 32 | 59 | 58.933200 | 8.768 |
| [204]Pb | 1.4 | 82 | 122 | 204 | 203.973020 | 7.880 |
| [206]Pb | 24.1 | 82 | 124 | 206 | 205.974440 | 7.875 |
| [207]Pb | 22.1 | 82 | 125 | 207 | 206.975872 | 7.870 |
| [208]Pb | 52.4 | 82 | 126 | 208 | 207.976627 | 7.868 |
| [209]Bi | 100.0 | 83 | 126 | 209 | 208.980380 | 7.848 |

[a]Selection is complete through cobalt-59. Where natural abundances do not add to 100%, the differences are made up by radioactive isotopes with exceedingly long half-lives: potassium-40 (0.0117%, $t_{1/2} = 1.3 \times 10^9$ yr); vanadium-50 (0.250%, $t_{1/2} > 1.4 \times 10^{17}$ yr).

[b]1 MeV (mega electron-volt) = $1.6022 \times 10^{-13}$ J.

## TABLE A3.4  A Selection of Radioactive Isotopes

| Isotope $^{A}X$ | Decay Mode[a] | Half-Life $t_{1/2}$ | Atomic Number Z | Neutron Number N | Mass Number A | Atomic Mass (amu) | Binding Energy per Nucleon (MeV)[b] |
|---|---|---|---|---|---|---|---|
| $^{3}$H | $\beta^{-}$ | 12.3 yr | 1 | 2 | 3 | 3.01605 | 2.827 |
| $^{8}$Be | $\alpha$ | $\sim 7 \times 10^{-17}$ s | 4 | 4 | 8 | 8.005305 | 7.062 |
| $^{14}$C | $\beta^{-}$ | $5.7 \times 10^{3}$ yr | 6 | 8 | 14 | 14.003241 | 7.520 |
| $^{22}$Na | $\beta^{+}$ | 2.6 yr | 11 | 11 | 22 | 21.994434 | 7.916 |
| $^{24}$Na | $\beta^{-}$ | 15.0 hr | 11 | 13 | 24 | 23.990961 | 8.064 |
| $^{32}$P | $\beta^{-}$ | 14.3 d | 15 | 17 | 32 | 31.973907 | 8.464 |
| $^{35}$S | $\beta^{-}$ | 87.2 d | 16 | 19 | 35 | 34.969031 | 8.538 |
| $^{59}$Fe | $\beta^{-}$ | 44.5 d | 26 | 33 | 59 | 58.934877 | 8.755 |
| $^{60}$Co | $\beta^{-}$ | 5.3 yr | 27 | 33 | 60 | 59.933819 | 8.747 |
| $^{90}$Sr | $\beta^{-}$ | 29.1 yr | 38 | 52 | 90 | 89.907738 | 8.696 |
| $^{99}$Tc | $\beta^{-}$ | $2.1 \times 10^{5}$ yr | 43 | 56 | 99 | 98.906524 | 8.611 |
| $^{109}$Cd | EC | 462 d | 48 | 61 | 109 | 108.904953 | 8.539 |
| $^{125}$I | EC | 59.4 d | 53 | 72 | 125 | 124.904620 | 8.450 |
| $^{131}$I | $\beta^{-}$ | 8.04 d | 53 | 78 | 131 | 130.906114 | 8.422 |
| $^{137}$Cs | $\beta^{-}$ | 30.3 yr | 55 | 82 | 137 | 136.907073 | 8.389 |
| $^{222}$Rn | $\alpha$ | 3.82 d | 86 | 136 | 222 | 222.017570 | 7.695 |
| $^{226}$Ra | $\alpha$ | 1600 yr | 88 | 138 | 226 | 226.025402 | 7.662 |
| $^{232}$Th | $\alpha$ | $1.4 \times 10^{10}$ yr | 90 | 142 | 232 | 232.038054 | 7.615 |
| $^{235}$U | $\alpha$ | $7.0 \times 10^{8}$ yr | 92 | 143 | 235 | 235.043924 | 7.591 |
| $^{238}$U | $\alpha$ | $4.5 \times 10^{9}$ yr | 92 | 146 | 238 | 238.050784 | 7.570 |
| $^{239}$Pu | $\alpha$ | $2.4 \times 10^{4}$ yr | 94 | 145 | 239 | 239.052157 | 7.560 |

[a]Modes of decay include alpha emission ($\alpha$), beta emission ($\beta^{-}$), positron emission ($\beta^{+}$), and electron capture (EC).
[b]1 MeV (mega electron-volt) = $1.6022 \times 10^{-13}$ J.

# Chemical Bonds and Thermodynamic Data

| TABLE A4.1 | Average Lengths and Energies of Covalent Bonds | | |
|---|---|---|---|
| Atom | Bond | Bond Length (pm) | Bond Energy (kJ/mol) |
| H | H—H | 75 | 436 |
| | H—F | 92 | 567 |
| | H—Cl | 127 | 431 |
| | H—Br | 141 | 366 |
| | H—I | 161 | 299 |
| C | C—C | 154 | 348 |
| | C=C | 134 | 614 |
| | C≡C | 120 | 839 |
| | C—H | 110 | 413 |
| | C—N | 147 | 293 |
| | C=N | 127 | 615 |
| | C≡N | 116 | 891 |
| | C—O | 143 | 358 |
| | C=O | 123 | 743[a] |
| | C≡O | 113 | 1072 |
| | C—F | 133 | 485 |
| | C—Cl | 177 | 328 |
| | C—Br | 179 | 276 |
| | C—I | 215 | 238 |
| N | N—N | 147 | 163 |
| | N=N | 124 | 418 |
| | N≡N | 110 | 945 |
| | N—H | 104 | 391 |
| | N—O | 136 | 201 |
| | N=O | 122 | 607 |
| | N≡O | 106 | 678 |
| O | O—O | 148 | 146 |
| | O=O | 121 | 498 |
| | O—H | 96 | 463 |
| S | S—O | 151 | 265 |
| | S=O | 143 | 523 |
| | S—S | 204 | 266 |
| | S—H | 134 | 347 |
| F | F—F | 143 | 155 |
| Cl | Cl—Cl | 200 | 243 |
| Br | Br—Br | 228 | 193 |
| I | I—I | 266 | 151 |

[a]The bond energy of C=O in $CO_2$ is 799 kJ/mol.

| TABLE A4.2 | Critical Temperatures ($T_c$) and van der Waals Parameters ($a$, $b$) of Real Gases | | | |
|---|---|---|---|---|
| **Gas**[a] | **Molar Mass (g/mol)** | **$T_c$ (K)** | **$a$ ($L^2 \cdot atm/mol^2$)** | **$b$ (L/mol)** |
| $H_2O$ | 18.015 | 647.14 | 5.46 | 0.0305 |
| $Br_2$ | 159.808 | 588 | 9.75 | 0.0591 |
| $CCl_3F$ | 137.367 | 471.2 | 14.68 | 0.1111 |
| $Cl_2$ | 70.906 | 416.9 | 6.343 | 0.0542 |
| $CO_2$ | 44.010 | 304.14 | 3.59 | 0.0427 |
| Kr | 83.798 | 209.41 | 2.325 | 0.0396 |
| $CH_4$ | 16.043 | 190.53 | 2.25 | 0.0428 |
| $O_2$ | 31.999 | 154.59 | 1.36 | 0.0318 |
| Ar | 39.948 | 150.87 | 1.34 | 0.0322 |
| $F_2$ | 37.997 | 144.13 | 1.171 | 0.0290 |
| CO | 28.010 | 132.91 | 1.45 | 0.0395 |
| $N_2$ | 28.013 | 126.21 | 1.39 | 0.0391 |
| $H_2$ | 2.016 | 32.97 | 0.244 | 0.0266 |
| He | 4.003 | 5.19 | 0.0341 | 0.0237 |

[a]Listed in descending order of critical temperature.

| TABLE A4.3 | Thermodynamic Properties at 25°C | | | |
|---|---|---|---|---|
| **Substance**[a,b] | **Molar Mass (g/mol)** | **$\Delta H_f^\circ$ (kJ/mol)** | **$S^\circ$ [J/(mol · K)]** | **$\Delta G_f^\circ$ (kJ/mol)** |
| **ELEMENTS AND MONATOMIC IONS** | | | | |
| $Ag^+(aq)$ | 107.87 | 105.6 | 72.7 | 77.1 |
| $Ag(g)$ | 107.87 | 284.9 | 173.0 | 246.0 |
| $Ag(s)$ | 107.87 | 0.0 | 42.6 | 0.0 |
| $Al^{3+}(aq)$ | 26.982 | −531 | −321.7 | −485 |
| $Al(g)$ | 26.982 | 330.0 | 164.6 | 289.4 |
| $Al(s)$ | 26.982 | 0.0 | 28.3 | 0.0 |
| $Al(\ell)$ | 26.982 | 10.6 | 39.6 | −1.2 |
| $Ar(g)$ | 39.948 | 0.0 | 154.8 | 0.0 |
| $Au(g)$ | 196.97 | 366.1 | 180.5 | 326.3 |
| $Au(s)$ | 196.97 | 0.0 | 47.4 | 0.0 |
| $B(g)$ | 10.811 | 565.0 | 153.4 | 521.0 |
| $B(s)$ | 10.811 | 0.0 | 5.9 | 0.0 |
| $Ba^{2+}(aq)$ | 137.33 | −537.6 | 9.6 | −560.8 |
| $Ba(g)$ | 137.33 | 180.0 | 170.2 | 146.0 |
| $Ba(s)$ | 137.33 | 0.0 | 62.8 | 0.0 |
| $Be(g)$ | 9.0122 | 324.0 | 136.3 | 286.6 |
| $Be(s)$ | 9.0122 | 0.0 | 9.5 | 0.0 |
| $Br^-(aq)$ | 79.904 | −121.6 | 82.4 | −104.0 |
| $Br(g)$ | 79.904 | 111.9 | 175.0 | 82.4 |
| $Br_2(g)$ | 159.808 | 30.9 | 245.5 | 3.1 |
| $Br_2(\ell)$ | 159.808 | 0.0 | 152.2 | 0.0 |
| $C(g)$ | 12.011 | 716.7 | 158.1 | 671.3 |

*Continued on next page*

**TABLE A4.3  Thermodynamic Properties at 25°C (Continued)**

| Substance[a,b] | Molar Mass (g/mol) | $\Delta H^\circ_f$ (kJ/mol) | $S^\circ$ [J/(mol · K)] | $\Delta G^\circ_f$ (kJ/mol) |
|---|---|---|---|---|
| C(s, diamond) | 12.011 | 1.9 | 2.4 | 2.9 |
| C(s, graphite) | 12.011 | 0.0 | 5.7 | 0.0 |
| $Ca^{2+}(aq)$ | 40.078 | −542.8 | −55.3 | −553.6 |
| Ca(g) | 40.078 | 177.8 | 154.9 | 144.0 |
| Ca(s) | 40.078 | 0.0 | 41.6 | 0.0 |
| $Cl^-(aq)$ | 35.453 | −167.2 | 56.5 | −131.2 |
| Cl(g) | 35.453 | 121.3 | 165.2 | 105.3 |
| $Cl_2(g)$ | 70.906 | 0.0 | 223.0 | 0.0 |
| $Co^{2+}(aq)$ | 58.933 | −58.2 | −113 | −54.4 |
| $Co^{3+}(aq)$ | 58.933 | 92 | −305 | 134 |
| Co(g) | 58.933 | 424.7 | 179.5 | 380.3 |
| Co(s) | 58.933 | 0.0 | 30.0 | 0.0 |
| Cr(g) | 51.996 | 396.6 | 174.5 | 351.8 |
| Cr(s) | 51.996 | 0.0 | 23.8 | 0.0 |
| $Cs^+(aq)$ | 132.91 | −258.3 | 133.1 | −292.0 |
| Cs(g) | 132.91 | 76.5 | 175.6 | 49.6 |
| Cs(s) | 132.91 | 0.0 | 85.2 | 0.0 |
| $Cu^+(aq)$ | 63.546 | 71.7 | 40.6 | 50.0 |
| $Cu^{2+}(aq)$ | 63.546 | 64.8 | −99.6 | 65.5 |
| Cu(g) | 63.546 | 337.4 | 166.4 | 297.7 |
| Cu(s) | 63.546 | 0.0 | 33.2 | 0.0 |
| $F^-(aq)$ | 18.998 | −332.6 | −13.8 | −278.8 |
| F(g) | 18.998 | 79.4 | 158.8 | 62.3 |
| $F_2(g)$ | 37.997 | 0.0 | 202.8 | 0.0 |
| $Fe^{2+}(aq)$ | 55.845 | −89.1 | −137.7 | −78.9 |
| $Fe^{3+}(aq)$ | 55.845 | −48.5 | −315.9 | −4.7 |
| Fe(g) | 55.845 | 416.3 | 180.5 | 370.7 |
| Fe(s) | 55.845 | 0.0 | 27.3 | 0.0 |
| $H^+(aq)$ | 1.0079 | 0.0 | 0.0 | 0.0 |
| H(g) | 1.0079 | 218.0 | 114.7 | 203.3 |
| $H_2(g)$ | 2.0158 | 0.0 | 130.6 | 0.0 |
| He(g) | 4.0026 | 0.0 | 126.2 | 0.0 |
| $Hg_2^{2+}(aq)$ | 401.18 | 172.4 | 84.5 | 153.5 |
| $Hg^{2+}(aq)$ | 200.59 | 171.1 | −32.2 | 164.4 |
| Hg(g) | 200.59 | 61.4 | 175.0 | 31.8 |
| Hg(ℓ) | 200.59 | 0.0 | 75.9 | 0.0 |
| $I^-(aq)$ | 126.90 | −55.2 | 111.3 | −51.6 |
| I(g) | 126.90 | 106.8 | 180.8 | 70.2 |
| $I_2(g)$ | 253.81 | 62.4 | 260.7 | 19.3 |
| $I_2(s)$ | 253.81 | 0.0 | 116.1 | 0.0 |
| $K^+(aq)$ | 39.098 | −252.4 | 102.5 | −283.3 |
| K(g) | 39.098 | 89.0 | 160.3 | 60.5 |

| TABLE A4.3 | **Thermodynamic Properties at 25°C (Continued)** | | | |
|---|---|---|---|---|
| Substance[a,b] | Molar Mass (g/mol) | $\Delta H_f^\circ$ (kJ/mol) | $S^\circ$ [J/(mol · K)] | $\Delta G_f^\circ$ (kJ/mol) |
| K(s) | 39.098 | 0.0 | 64.7 | 0.0 |
| Li$^+$(aq) | 6.941 | −278.5 | 13.4 | −293.3 |
| Li(g) | 6.941 | 159.3 | 138.8 | 126.6 |
| Li$^+$(g) | 6.941 | 685.7 | 133.0 | 648.5 |
| Li(s) | 6.941 | 0.0 | 29.1 | 0.0 |
| Mg$^{2+}$(aq) | 24.305 | −466.9 | −138.1 | −454.8 |
| Mg(g) | 24.305 | 147.1 | 148.6 | 112.5 |
| Mg(s) | 24.305 | 0.0 | 32.7 | 0.0 |
| Mn$^{2+}$(aq) | 54.938 | −220.8 | −73.6 | −228.1 |
| Mn(g) | 54.938 | 280.7 | 173.7 | 238.5 |
| Mn(s) | 54.938 | 0.0 | 32.0 | 0.0 |
| N(g) | 14.007 | 472.7 | 153.3 | 455.5 |
| N$_2$(g) | 28.013 | 0.0 | 191.5 | 0.0 |
| Na$^+$(aq) | 22.990 | −240.1 | 59.0 | −261.9 |
| Na(g) | 22.990 | 107.5 | 153.7 | 77.0 |
| Na$^+$(g) | 22.990 | 609.3 | 148.0 | 574.3 |
| Na(s) | 22.990 | 0.0 | 51.3 | 0.0 |
| Ne(g) | 20.180 | 0.0 | 146.3 | 0.0 |
| Ni$^{2+}$(aq) | 58.693 | −54.0 | −128.9 | −45.6 |
| Ni(g) | 58.693 | 429.7 | 182.2 | 384.5 |
| Ni(s) | 58.693 | 0.0 | 29.9 | 0.0 |
| O(g) | 15.999 | 249.2 | 161.1 | 231.7 |
| O$_2$(g) | 31.999 | 0.0 | 205.0 | 0.0 |
| O$_3$(g) | 47.998 | 142.7 | 238.8 | 163.2 |
| P(g) | 30.974 | 314.6 | 163.1 | 278.3 |
| P$_4$(s, red) | 123.895 | −17.6 | 22.8 | −12.1 |
| P$_4$(s, white) | 123.895 | 0.0 | 41.1 | 0.0 |
| Pb$^{2+}$(aq) | 207.2 | −1.7 | 10.5 | −24.4 |
| Pb(g) | 207.2 | 195.2 | 162.2 | 175.4 |
| Pb(s) | 207.2 | 0.0 | 64.8 | 0.0 |
| Rb$^+$(aq) | 85.468 | −251.2 | 121.5 | −284.0 |
| Rb(g) | 85.468 | 80.9 | 170.1 | 53.1 |
| Rb(s) | 85.468 | 0.0 | 76.8 | 0.0 |
| S(g) | 32.065 | 277.2 | 167.8 | 236.7 |
| S$_8$(g) | 256.520 | 102.3 | 430.2 | 49.1 |
| S$_8$(s) | 256.520 | 0.0 | 32.1 | 0.0 |
| Sc(g) | 44.956 | 377.8 | 174.8 | 336.0 |
| Sc(s) | 44.956 | 0.0 | 34.6 | 0.0 |
| Si(g) | 28.086 | 450.0 | 168.0 | 405.5 |
| Si(s) | 28.086 | 0.0 | 18.8 | 0.0 |
| Sn(g) | 118.71 | 301.2 | 168.5 | 266.2 |
| Sn(s, gray) | 118.71 | −2.1 | 44.1 | 0.1 |
| Sn(s, white) | 118.71 | 0.0 | 51.2 | 0.0 |
| Sr$^{2+}$(aq) | 87.62 | −545.8 | −32.6 | −559.5 |

*Continued on next page*

### TABLE A4.3 Thermodynamic Properties at 25°C (Continued)

| Substance[a,b] | Molar Mass (g/mol) | $\Delta H_f^\circ$ (kJ/mol) | $S^\circ$ [J/(mol · K)] | $\Delta G_f^\circ$ (kJ/mol) |
|---|---|---|---|---|
| Sr(g) | 87.62 | 164.4 | 164.6 | 130.9 |
| Sr(s) | 87.62 | 0.0 | 52.3 | 0.0 |
| Ti(g) | 47.867 | 473.0 | 180.3 | 428.4 |
| Ti(s) | 47.867 | 0.0 | 30.7 | 0.0 |
| V(g) | 50.942 | 514.2 | 182.2 | 468.5 |
| V(s) | 50.942 | 0.0 | 28.9 | 0.0 |
| W(s) | 183.84 | 0.0 | 32.6 | 0.0 |
| $Zn^{2+}(aq)$ | 65.38 | −153.9 | −112.1 | −147.1 |
| Zn(g) | 65.38 | 130.4 | 161.0 | 94.8 |
| Zn(s) | 65.38 | 0.0 | 41.6 | 0.0 |
| **POLYATOMIC IONS** | | | | |
| $CH_3COO^-(aq)$ | 59.045 | −486.0 | 86.6 | −369.3 |
| $CO_3^{2-}(aq)$ | 60.009 | −677.1 | −56.9 | −527.8 |
| $C_2O_4^{2-}(aq)$ | 88.020 | −825.1 | 45.6 | −673.9 |
| $CrO_4^{2-}(aq)$ | 115.994 | −881.2 | 50.2 | −727.8 |
| $Cr_2O_7^{2-}(aq)$ | 215.988 | −1490.3 | 261.9 | −1301.1 |
| $HCOO^-(aq)$ | 45.018 | −425.6 | 92 | −351.0 |
| $HCO_3^-(aq)$ | 61.017 | −692.0 | 91.2 | −586.8 |
| $HSO_4^-(aq)$ | 97.072 | −887.3 | 131.8 | −755.9 |
| $MnO_4^-(aq)$ | 118.936 | −541.4 | 191.2 | −447.2 |
| $NH_4^+(aq)$ | 18.038 | −132.5 | 113.4 | −79.3 |
| $NO_3^-(aq)$ | 62.005 | −205.0 | 146.4 | −108.7 |
| $OH^-(aq)$ | 17.007 | −230.0 | −10.8 | −157.2 |
| $PO_4^{3-}(aq)$ | 94.971 | −1277.4 | −222 | −1018.7 |
| $SO_4^{2-}(aq)$ | 96.064 | −909.3 | 20.1 | −744.5 |
| **INORGANIC COMPOUNDS** | | | | |
| AgCl(s) | 143.32 | −127.1 | 96.2 | −109.8 |
| AgI(s) | 234.77 | −61.8 | 115.5 | −66.2 |
| $AgNO_3(s)$ | 169.87 | −124.4 | 140.9 | −33.4 |
| $Al_2O_3(s)$ | 101.961 | −1675.7 | 50.9 | −1582.3 |
| $B_2H_6(g)$ | 27.669 | 35.0 | 232.0 | 86.6 |
| $B_2O_3(s)$ | 69.622 | −1263.6 | 54.0 | −1184.1 |
| $BaCO_3(s)$ | 197.34 | −1216.3 | 112.1 | −1137.6 |
| $BaSO_4(s)$ | 233.39 | −1473.2 | 132.2 | −1362.2 |
| $CaCO_3(s)$ | 100.087 | −1206.9 | 92.9 | −1128.8 |
| $CaCl_2(s)$ | 110.984 | −795.4 | 108.4 | −748.8 |
| $CaF_2(s)$ | 78.075 | −1228.0 | 68.5 | −1175.6 |
| CaO(s) | 56.077 | −634.9 | 38.1 | −603.3 |
| $Ca(OH)_2(s)$ | 74.093 | −985.2 | 83.4 | −897.5 |
| $CaSO_4(s)$ | 136.142 | −1434.5 | 106.5 | −1322.0 |
| CO(g) | 28.010 | −110.5 | 197.7 | −137.2 |
| $CO_2(g)$ | 44.010 | −393.5 | 213.8 | −394.4 |
| $CO_2(aq)$ | 44.010 | −412.9 | 121.3 | −386.2 |

| TABLE A4.3 | Thermodynamic Properties at 25°C (Continued) | | | |
|---|---|---|---|---|
| Substance[a,b] | Molar Mass (g/mol) | $\Delta H_f^\circ$ (kJ/mol) | $S^\circ$ [J/(mol · K)] | $\Delta G_f^\circ$ (kJ/mol) |
| $CS_2(g)$ | 76.143 | 115.3 | 237.8 | 65.1 |
| $CS_2(\ell)$ | 76.143 | 87.9 | 151.0 | 63.6 |
| $CsCl(s)$ | 168.358 | −443.0 | 101.2 | −414.6 |
| $CuSO_4(s)$ | 159.610 | −771.4 | 109.2 | −662.2 |
| $FeCl_2(s)$ | 126.750 | −341.8 | 118.0 | −302.3 |
| $FeCl_3(s)$ | 162.203 | −399.5 | 142.3 | −334.0 |
| $FeO(s)$ | 71.844 | −271.9 | 60.8 | −255.2 |
| $Fe_2O_3(s)$ | 159.688 | −824.2 | 87.4 | −742.2 |
| $HBr(g)$ | 80.912 | −36.3 | 198.7 | −53.4 |
| $HCl(g)$ | 36.461 | −92.3 | 186.9 | −95.3 |
| $HF(g)$ | 20.006 | −273.3 | 173.8 | −275.4 |
| $HI(g)$ | 127.912 | 26.5 | 206.6 | 1.7 |
| $HNO_3(g)$ | 63.013 | −135.1 | 266.4 | −74.7 |
| $HNO_3(\ell)$ | 63.013 | −174.1 | 155.6 | −80.7 |
| $HNO_3(aq)$ | 63.013 | −206.6 | 146.0 | −110.5 |
| $HgCl_2(s)$ | 271.50 | −224.3 | 146.0 | −178.6 |
| $Hg_2Cl_2(s)$ | 472.09 | −265.4 | 191.6 | −210.7 |
| $H_2O(g)$ | 18.015 | −241.8 | 188.8 | −228.6 |
| $H_2O(\ell)$ | 18.015 | −285.8 | 69.9 | −237.2 |
| $H_2S(g)$ | 34.082 | −20.17 | 205.6 | −33.01 |
| $H_2O_2(g)$ | 34.015 | −136.3 | 232.7 | −105.6 |
| $H_2O_2(\ell)$ | 34.015 | −187.8 | 109.6 | −120.4 |
| $H_2SO_4(\ell)$ | 98.079 | −814.0 | 156.9 | −690.0 |
| $H_2SO_4(aq)$ | 98.079 | −909.2 | 20.1 | −744.5 |
| $KBr(s)$ | 119.002 | −393.8 | 95.9 | −380.7 |
| $KCl(s)$ | 74.551 | −436.5 | 82.6 | −408.5 |
| $KHCO_3(s)$ | 100.115 | −963.2 | 115.5 | −863.6 |
| $K_2CO_3(s)$ | 138.205 | −1151.0 | 155.5 | −1063.5 |
| $LiBr(s)$ | 86.845 | −351.2 | 74.3 | −342.0 |
| $LiCl(s)$ | 42.394 | −408.6 | 59.3 | −384.4 |
| $Li_2CO_3(s)$ | 73.891 | −1215.9 | 90.4 | −1132.1 |
| $MgCl_2(s)$ | 95.211 | −641.3 | 89.6 | −591.8 |
| $Mg(OH)_2(s)$ | 58.320 | −924.5 | 63.2 | −833.5 |
| $MgSO_4(s)$ | 120.369 | −1284.9 | 91.6 | −1170.6 |
| $MnO_2(s)$ | 86.937 | −520.0 | 53.1 | −465.1 |
| $NaCH_3OO(s)$ | 82.034 | −708.8 | 123.0 | −607.2 |
| $NaBr(s)$ | 102.894 | −361.1 | 86.82 | −349.0 |
| $NaCl(s)$ | 58.443 | −411.2 | 72.1 | −384.2 |
| $NaCl(g)$ | 58.443 | −181.4 | 229.8 | −201.3 |
| $Na_2CO_3(s)$ | 105.989 | −1130.7 | 135.0 | −1044.4 |
| $NaHCO_3(s)$ | 84.007 | −950.8 | 101.7 | −851.0 |
| $NaNO_3(s)$ | 84.995 | −467.9 | 116.5 | −367.0 |
| $NaOH(s)$ | 39.997 | −425.6 | 64.5 | −379.5 |

*Continued on next page*

**TABLE A4.3  Thermodynamic Properties at 25°C (Continued)**

| Substance[a,b] | Molar Mass (g/mol) | $\Delta H_f^\circ$ (kJ/mol) | $S^\circ$ [J/(mol · K)] | $\Delta G_f^\circ$ (kJ/mol) |
|---|---|---|---|---|
| $Na_2SO_4(s)$ | 142.043 | −1387.1 | 149.6 | −1270.2 |
| $NF_3(g)$ | 71.002 | −132.1 | 260.8 | −90.6 |
| $NH_3(aq)$ | 17.031 | −80.3 | 111.3 | −26.50 |
| $NH_3(g)$ | 17.031 | −46.1 | 192.5 | −16.5 |
| $NH_4Cl(s)$ | 53.491 | −314.4 | 94.6 | −203.0 |
| $NH_4NO_3(s)$ | 80.043 | −365.6 | 151.1 | −183.9 |
| $N_2H_4(g)$ | 32.045 | 95.35 | 238.5 | 159.4 |
| $N_2H_4(\ell)$ | 32.045 | 50.63 | 121.52 | 149.3 |
| $NiCl_2(s)$ | 129.60 | −305.3 | 97.7 | −259.0 |
| $NiO(s)$ | 74.60 | −239.7 | 38.0 | −211.7 |
| $NO(g)$ | 30.006 | 90.3 | 210.7 | 86.6 |
| $NO_2(g)$ | 46.006 | 33.2 | 240.0 | 51.3 |
| $N_2O(g)$ | 44.013 | 82.1 | 219.9 | 104.2 |
| $N_2O_4(g)$ | 92.011 | 9.2 | 304.2 | 97.8 |
| $NOCl(g)$ | 65.459 | 51.7 | 261.7 | 66.1 |
| $PCl_3(g)$ | 137.33 | −288.07 | 311.7 | −269.6 |
| $PCl_3(\ell)$ | 137.33 | −319.6 | 217 | −272.4 |
| $PF_5(g)$ | 125.96 | −1594.4 | 300.8 | −1520.7 |
| $PH_3(g)$ | 33.998 | 5.4 | 210.2 | 13.4 |
| $PbCl_2(s)$ | 278.1 | −359.4 | 136.0 | −314.1 |
| $PbSO_4(s)$ | 303.3 | −920.0 | 148.5 | −813.0 |
| $SO_2(g)$ | 64.065 | −296.8 | 248.2 | −300.1 |
| $SO_3(g)$ | 80.064 | −395.7 | 256.8 | −371.1 |
| $ZnCl_2(s)$ | 136.30 | −415.1 | 111.5 | −369.4 |
| $ZnO(s)$ | 81.37 | −348.0 | 43.9 | −318.2 |
| $ZnSO_4(s)$ | 161.45 | −982.8 | 110.5 | −871.5 |
| **ORGANIC COMPOUNDS** | | | | |
| $CCl_4(g)$ | 153.823 | −102.9 | 309.7 | −60.6 |
| $CCl_4(\ell)$ | 153.823 | −135.4 | 216.4 | −65.3 |
| $CH_4(g)$ | 16.043 | −74.8 | 186.2 | −50.8 |
| $CH_3COOH(g)$ | 60.053 | −432.8 | 282.5 | −374.5 |
| $CH_3COOH(\ell)$ | 60.053 | −484.5 | 159.8 | −389.9 |
| $CH_3OH(g)$ | 32.042 | −200.7 | 239.9 | −162.0 |
| $CH_3OH(\ell)$ | 32.042 | −238.7 | 126.8 | −166.4 |
| $C_2H_2(g)$ | 26.038 | 226.7 | 200.8 | 209.2 |
| $C_2H_4(g)$ | 28.054 | 52.4 | 219.5 | 68.1 |
| $C_2H_6(g)$ | 30.070 | −84.67 | 229.5 | −32.9 |
| $CH_3CH_2OH(g)$ | 46.069 | −235.1 | 282.6 | −168.6 |
| $CH_3CH_2OH(\ell)$ | 46.069 | −277.7 | 160.7 | −174.9 |
| $CH_3CHO(g)$ | 44.053 | −166 | 266 | −133.7 |
| $C_3H_8(g)$ | 44.097 | −103.8 | 269.9 | −23.5 |
| $CH_3(CH_2)_2CH_3(g)$ | 58.123 | −125.6 | 310.0 | −15.7 |
| $CH_3(CH_2)_2CH_3(\ell)$ | 58.123 | −147.6 | 231.0 | −15.0 |

## TABLE A4.3  Thermodynamic Properties at 25°C (Continued)

| Substance[a,b] | Molar Mass (g/mol) | $\Delta H_f^\circ$ (kJ/mol) | $S^\circ$ [J/(mol · K)] | $\Delta G_f^\circ$ (kJ/mol) |
|---|---|---|---|---|
| $CH_3COCH_3(\ell)$ | 58.079 | −248.4 | 199.8 | −155.6 |
| $CH_3COCH_3(g)$ | 58.079 | −217.1 | 295.3 | −152.7 |
| $CH_3(CH_2)_2CH_2OH(\ell)$ | 74.122 | −327.3 | 225.8 | |
| $(CH_3CH_2)_2O(\ell)$ | 74.122 | −279.6 | 172.4 | |
| $(CH_3CH_2)_2O(g)$ | 74.122 | −252.1 | 342.7 | |
| $(CH_3)_2C\!=\!C(CH_3)_2(\ell)$ | 84.161 | 66.6 | 362.6 | −69.2 |
| $(CH_3)_2NH(\ell)$ | 45.084 | −43.9 | 182.3 | |
| $(CH_3)_2NH(g)$ | 45.084 | −18.5 | 273.1 | |
| $(C_2H_5)_2NH(\ell)$ | 73.138 | −103.3 | | |
| $(C_2H_5)_2NH(g)$ | 73.138 | −71.4 | | |
| $(CH_3)_3N(\ell)$ | 59.111 | −46.0 | 208.5 | |
| $(CH_3)_3N(g)$ | 59.111 | −23.6 | 287.1 | |
| $(CH_3CH_2)_3N(\ell)$ | 101.191 | −134.3 | | |
| $(CH_3CH_2)_3N(g)$ | 101.191 | −95.8 | | |
| $C_6H_6(g)$ | 78.114 | 82.9 | 269.2 | 129.7 |
| $C_6H_6(\ell)$ | 78.114 | 49.0 | 172.9 | 124.5 |
| $C_6H_{12}O_6(s)$ | 180.158 | −1274.4 | 212.1 | −910.1 |
| $CH_3(CH_2)_6CH_3(\ell)$ | 114.231 | −249.9 | 361.1 | 6.4 |
| $CH_3(CH_2)_6CH_3(g)$ | 114.231 | −208.6 | 466.7 | 16.4 |
| $C_{12}H_{22}O_{11}(s)$ | 342.300 | −2221.7 | 360.2 | −1543.8 |
| $HCOOH(\ell)$ | 46.026 | −424.7 | 129.0 | −361.4 |

[a]Substances are arranged alphabetically by chemical formula within each class: (1) elements and monatomic ions; (2) polyatomic ions; (3) inorganic compounds (including CO and $CO_2$); (4) organic compounds (hydrocarbon-based).
[b]Symbols denote standard enthalpy of formation ($\Delta H_f^\circ$), standard third-law entropy ($S^\circ$), and standard Gibbs free energy of formation ($\Delta G_f^\circ$). Entropies in aqueous solution are referred to $S^\circ[H^+(aq)] = 0$, not to absolute zero.

## TABLE A4.4  Vapor Pressure of Water as a Function of Temperature

| T (°C) | P (torr) |
|---|---|
| 0.0 | 4.579 |
| 10.0 | 9.209 |
| 20.0 | 17.535 |
| 25.0 | 23.756 |
| 30.0 | 31.824 |
| 40.0 | 55.324 |
| 60.0 | 149.4 |
| 70.0 | 233.7 |
| 90.0 | 525.8 |
| 100 | 760.0 |
| 105 | 906.0 |

# Equilibrium Constants

| TABLE A5.1 | Ionization Constants of Selected Acids at 25°C | | | |
|---|---|---|---|---|
| **Acid** | **Step** | **Aqueous Equilibrium**[a] | $K_a$ | $pK_a$ |
| acetic | 1 | $CH_3COOH(aq) \rightleftharpoons H^+(aq) + CH_3COO^-(aq)$ | $1.76 \times 10^{-5}$ | 4.75 |
| arsenic | 1 | $H_3AsO_4(aq) \rightleftharpoons H^+(aq) + H_2AsO_4^-(aq)$ | $5.5 \times 10^{-3}$ | 2.26 |
| | 2 | $H_2AsO_4^-(aq) \rightleftharpoons H^+(aq) + HAsO_4^{2-}(aq)$ | $1.7 \times 10^{-7}$ | 6.77 |
| | 3 | $HAsO_4^{2-}(aq) \rightleftharpoons H^+(aq) + AsO_4^{3-}(aq)$ | $5.1 \times 10^{-12}$ | 11.29 |
| ascorbic | 1 | $H_2C_6H_6O_6(aq) \rightleftharpoons H^+(aq) + HC_6H_6O_6^-(aq)$ | $9.1 \times 10^{-5}$ | 4.04 |
| | 2 | $HC_6H_6O_6^-(aq) \rightleftharpoons H^+(aq) + C_6H_6O_6^{2-}(aq)$ | $5 \times 10^{-12}$ | 11.3 |
| benzoic | 1 | $C_6H_5COOH(aq) \rightleftharpoons H^+(aq) + C_6H_5COO^-(aq)$ | $6.25 \times 10^{-5}$ | 4.20 |
| boric | 1 | $H_3BO_3(aq) \rightleftharpoons H^+(aq) + H_2BO_3^-(aq)$ | $5.4 \times 10^{-10}$ | 9.27 |
| | 2 | $H_2BO_3^-(aq) \rightleftharpoons H^+(aq) + HBO_3^{2-}(aq)$ | $<10^{-14}$ | $>14$ |
| bromoacetic | 1 | $CH_2BrCOOH(aq) \rightleftharpoons H^+(aq) + CH_2BrCOO^-(aq)$ | $2.0 \times 10^{-3}$ | 2.70 |
| butanoic | 1 | $CH_3CH_2CH_2COOH(aq) \rightleftharpoons$ $H^+(aq) + CH_3CH_2CH_2COO^-(aq)$ | $1.5 \times 10^{-5}$ | 4.82 |
| carbonic | 1 | $H_2CO_3(aq) \rightleftharpoons H^+(aq) + HCO_3^-(aq)$ | $4.3 \times 10^{-7}$ | 6.37 |
| | 2 | $HCO_3^-(aq) \rightleftharpoons H^+(aq) + CO_3^{2-}(aq)$ | $4.7 \times 10^{-11}$ | 10.33 |
| chloric | 1 | $HClO_3(aq) \rightleftharpoons H^+(aq) + ClO_3^-(aq)$ | $\sim 1$ | $\sim 0$ |
| chloroacetic | 1 | $CH_2ClCOOH(aq) \rightleftharpoons H^+(aq) + CH_2ClCOO^-(aq)$ | $1.4 \times 10^{-3}$ | 2.85 |
| chlorous | 1 | $HClO_2(aq) \rightleftharpoons H^+(aq) + ClO_2^-(aq)$ | $1.1 \times 10^{-2}$ | 1.96 |
| citric | 1 | $HOC(CH_2)_2(COOH)_3(aq) \rightleftharpoons$ $H^+(aq) + HOC(CH_2)_2(COOH)_2COO^-(aq)$ | $7.4 \times 10^{-4}$ | 3.13 |
| | 2 | $HOC(CH_2)_2(COOH)_2COO^-(aq) \rightleftharpoons$ $H^+(aq) + HOC(CH_2)_2(COOH)(COO^-)_2(aq)$ | $1.7 \times 10^{-5}$ | 4.77 |
| | 3 | $HOC(CH_2)_2(COOH)(COO^-)_2(aq) \rightleftharpoons$ $H^+(aq) + HOC(CH_2)_2(COO^-)_3(aq)$ | $4.0 \times 10^{-7}$ | 6.40 |
| dichloroacetic | 1 | $CHCl_2COOH(aq) \rightleftharpoons H^+(aq) + CHCl_2COO^-(aq)$ | $5.5 \times 10^{-2}$ | 1.26 |
| ethanol | 1 | $CH_3CH_2OH(aq) \rightleftharpoons H^+(aq) + CH_3CH_2O^-(aq)$ | $1.3 \times 10^{-16}$ | 15.9 |
| fluoroacetic | 1 | $CH_2FCOOH(aq) \rightleftharpoons H^+(aq) + CH_2FCOO^-(aq)$ | $2.6 \times 10^{-3}$ | 2.59 |
| formic | 1 | $HCOOH(aq) \rightleftharpoons H^+(aq) + HCOO^-(aq)$ | $1.77 \times 10^{-4}$ | 3.75 |
| germanic | 1 | $H_2GeO_3(aq) \rightleftharpoons H^+(aq) + HGeO_3^-(aq)$ | $9.8 \times 10^{-10}$ | 9.01 |
| | 2 | $HGeO_3^-(aq) \rightleftharpoons H^+(aq) + GeO_3^{2-}(aq)$ | $5 \times 10^{-13}$ | 12.3 |
| hydr(o)azoic | 1 | $HN_3(aq) \rightleftharpoons H^+(aq) + N_3^-(aq)$ | $1.9 \times 10^{-5}$ | 4.72 |

**TABLE A5.1** **Ionization Constants of Selected Acids at 25°C (Continued)**

| Acid | Step | Aqueous Equilibrium[a] | $K_a$ | $pK_a$ |
|------|------|------------------------|-------|--------|
| hydrobromic | 1 | $HBr(aq) \rightleftharpoons H^+(aq) + Br^-(aq)$ | $\gg 1$ (strong) | $<0$ |
| hydrochloric | 1 | $HCl(aq) \rightleftharpoons H^+(aq) + Cl^-(aq)$ | $\gg 1$ (strong) | $<0$ |
| hydrocyanic | 1 | $HCN(aq) \rightleftharpoons H^+(aq) + CN^-(aq)$ | $6.2 \times 10^{-10}$ | 9.21 |
| hydrofluoric | 1 | $HF(aq) \rightleftharpoons H^+(aq) + F^-(aq)$ | $6.8 \times 10^{-4}$ | 3.17 |
| hydr(o)iodic | 1 | $HI(aq) \rightleftharpoons H^+(aq) + I^-(aq)$ | $\gg 1$ (strong) | $<0$ |
| hydrosulfuric | 1 | $H_2S(aq) \rightleftharpoons H^+(aq) + HS^-(aq)$ | $8.9 \times 10^{-8}$ | 7.05 |
|  | 2 | $HS^-(aq) \rightleftharpoons H^+(aq) + S^{2-}(aq)$ | $\sim 10^{-19}$ | $\sim 19$ |
| hypobromous | 1 | $HBrO(aq) \rightleftharpoons H^+(aq) + BrO^-(aq)$ | $2.3 \times 10^{-9}$ | 8.64 |
| hypochlorous | 1 | $HClO(aq) \rightleftharpoons H^+(aq) + ClO^-(aq)$ | $2.9 \times 10^{-8}$ | 7.54 |
| hypoiodous | 1 | $HIO(aq) \rightleftharpoons H^+(aq) + IO^-(aq)$ | $2.3 \times 10^{-11}$ | 10.64 |
| iodic | 1 | $HIO_3(aq) \rightleftharpoons H^+(aq) + IO_3^-(aq)$ | $1.7 \times 10^{-1}$ | 0.77 |
| iodoacetic | 1 | $CH_2ICOOH(aq) \rightleftharpoons H^+(aq) + CH_2ICOO^-(aq)$ | $7.6 \times 10^{-4}$ | 3.12 |
| lactic | 1 | $CH_3CHOHCOOH(aq) \rightleftharpoons$ $H^+(aq) + CH_3CHOHCOO^-(aq)$ | $1.4 \times 10^{-4}$ | 3.85 |
| maleic (*cis*-butenedioic) | 1 | $HOOCCH{=}CHCOOH(aq) \rightleftharpoons$ $H^+(aq) + HOOCCH{=}CHCOO^-(aq)$ | $1.2 \times 10^{-2}$ | 1.92 |
|  | 2 | $HOOCCH{=}CHCOO^-(aq) \rightleftharpoons$ $H^+(aq) + {}^-OOCCH{=}CHCOO^-(aq)$ | $4.7 \times 10^{-7}$ | 6.33 |
| malonic | 1 | $HOOCCH_2COOH(aq) \rightleftharpoons$ $H^+(aq) + HOOCCH_2COO^-(aq)$ | $1.5 \times 10^{-3}$ | 2.82 |
|  | 2 | $HOOCCH_2COO^-(aq) \rightleftharpoons$ $H^+(aq) + {}^-OOCCH_2COO^-(aq)$ | $2.0 \times 10^{-6}$ | 5.70 |
| nitric | 1 | $HNO_3(aq) \rightleftharpoons H^+(aq) + NO_3^-(aq)$ | $\gg 1$ (strong) | $<0$ |
| nitrous | 1 | $HNO_2(aq) \rightleftharpoons H^+(aq) + NO_2^-(aq)$ | $4.0 \times 10^{-4}$ | 3.40 |
| oxalic | 1 | $HOOCCOOH(aq) \rightleftharpoons H^+(aq) + HOOCCOO^-(aq)$ | $5.9 \times 10^{-2}$ | 1.23 |
|  | 2 | $HOOCCOO^-(aq) \rightleftharpoons H^+(aq) + {}^-OOCCOO^-(aq)$ | $6.4 \times 10^{-5}$ | 4.19 |
| perchloric | 1 | $HClO_4(aq) \rightleftharpoons H^+(aq) + ClO_4^-(aq)$ | $\gg 1$ (strong) | $<0$ |
| periodic | 1 | $HIO_4(aq) \rightleftharpoons H^+(aq) + IO_4^-(aq)$ | $2.3 \times 10^{-2}$ | 1.64 |
| phenol | 1 | $C_6H_5OH(aq) \rightleftharpoons H^+(aq) + C_6H_5O^-(aq)$ | $1.3 \times 10^{-10}$ | 9.89 |
| phosphoric | 1 | $H_3PO_4(aq) \rightleftharpoons H^+(aq) + H_2PO_4^-(aq)$ | $6.9 \times 10^{-3}$ | 2.16 |
|  | 2 | $H_2PO_4^-(aq) \rightleftharpoons H^+(aq) + HPO_4^{2-}(aq)$ | $6.4 \times 10^{-8}$ | 7.19 |
|  | 3 | $HPO_4^{2-}(aq) \rightleftharpoons H^+(aq) + PO_4^{3-}(aq)$ | $4.8 \times 10^{-13}$ | 12.32 |
| propanoic | 1 | $CH_3CH_2COOH(aq) \rightleftharpoons$ $H^+(aq) + CH_3CH_2COO^-(aq)$ | $1.4 \times 10^{-5}$ | 4.85 |
| pyruvic | 1 | $CH_3C(O)COOH(aq) \rightleftharpoons$ $H^+(aq) + CH_3C(O)COO^-(aq)$ | $2.8 \times 10^{-3}$ | 2.55 |
| sulfuric | 1 | $H_2SO_4(aq) \rightleftharpoons H^+(aq) + HSO_4^-(aq)$ | $\gg 1$ (strong) | $<0$ |
|  | 2 | $HSO_4^-(aq) \rightleftharpoons H^+(aq) + SO_4^{2-}(aq)$ | $1.2 \times 10^{-2}$ | 1.92 |
| sulfurous | 1 | $H_2SO_3(aq) \rightleftharpoons H^+(aq) + HSO_3^-(aq)$ | $1.7 \times 10^{-2}$ | 1.77 |
|  | 2 | $HSO_3^-(aq) \rightleftharpoons H^+(aq) + SO_3^{2-}(aq)$ | $6.2 \times 10^{-8}$ | 7.21 |

*Continued on next page*

**TABLE A5.1  Ionization Constants of Selected Acids at 25°C (Continued)**

| Acid | Step | Aqueous Equilibrium[a] | $K_a$ | $pK_a$ |
|------|------|------------------------|-------|--------|
| thiocyanic | 1 | $HSCN(aq) \rightleftharpoons H^+(aq) + SCN^-(aq)$ | $\gg 1$ (strong) | $<0$ |
| trichloroacetic | 1 | $CCl_3COOH(aq) \rightleftharpoons H^+(aq) + CCl_3COO^-(aq)$ | $2.3 \times 10^{-1}$ | 0.64 |
| trifluoroacetic | 1 | $CF_3COOH(aq) \rightleftharpoons H^+(aq) + CF_3COO^-(aq)$ | $5.9 \times 10^{-1}$ | 0.23 |
| water | 1 | $H_2O(aq) \rightleftharpoons H^+(aq) + OH^-(aq)$ | $1.0 \times 10^{-14}$ | 14.00 |

[a]The formulas of most carboxylic acids are written in an RCOOH format to highlight their molecular structures.

**TABLE A5.2  Acid Ionization Constants of Hydrated Metal Ions at 25°C**

| Free Ion | Hydrated Ion | $K_a$ |
|----------|-------------|-------|
| $Fe^{3+}$ | $Fe(H_2O)_6^{3+}$ | $3 \times 10^{-3}$ |
| $Sn^{2+}$ | $Sn(H_2O)_6^{2+}$ | $4 \times 10^{-4}$ |
| $Cr^{3+}$ | $Cr(H_2O)_6^{3+}$ | $1 \times 10^{-4}$ |
| $Al^{3+}$ | $Al(H_2O)_6^{3+}$ | $1 \times 10^{-5}$ |
| $Cu^{2+}$ | $Cu(H_2O)_6^{2+}$ | $3 \times 10^{-8}$ |
| $Pb^{2+}$ | $Pb(H_2O)_6^{2+}$ | $3 \times 10^{-8}$ |
| $Zn^{2+}$ | $Zn(H_2O)_6^{2+}$ | $1 \times 10^{-9}$ |
| $Co^{2+}$ | $Co(H_2O)_6^{2+}$ | $2 \times 10^{-10}$ |
| $Ni^{2+}$ | $Ni(H_2O)_6^{2+}$ | $1 \times 10^{-10}$ |

**TABLE A5.3  Ionization Constants of Selected Bases at 25°C**

| Base | Aqueous Equilibrium | $K_b$ | $pK_b$ |
|------|--------------------|-------|--------|
| ammonia | $NH_3(aq) + H_2O(\ell) \rightleftharpoons NH_4^+(aq) + OH^-(aq)$ | $1.76 \times 10^{-5}$ | 4.75 |
| aniline | $C_6H_5NH_2(aq) + H_2O(\ell) \rightleftharpoons C_6H_5NH_3^+(aq) + OH^-(aq)$ | $4.0 \times 10^{-10}$ | 9.40 |
| diethylamine | $(CH_3CH_2)_2NH(aq) + H_2O(\ell) \rightleftharpoons (CH_3CH_2)_2NH_2^+(aq) + OH^-(aq)$ | $8.6 \times 10^{-4}$ | 3.07 |
| dimethylamine | $(CH_3)_2NH(aq) + H_2O(\ell) \rightleftharpoons (CH_3)_2NH_2^+(aq) + OH^-(aq)$ | $5.9 \times 10^{-4}$ | 3.23 |
| methylamine | $CH_3NH_2(aq) + H_2O(\ell) \rightleftharpoons CH_3NH_3^+(aq) + OH^-(aq)$ | $4.4 \times 10^{-4}$ | 3.36 |
| nicotine (1) | | $1.0 \times 10^{-6}$ | 6.0 |
| (2) | | $1.3 \times 10^{-11}$ | 10.9 |
| pyridine | $C_5H_5N(aq) + H_2O(\ell) \rightleftharpoons C_5H_5NH^+(aq) + OH^-(aq)$ | $1.7 \times 10^{-9}$ | 8.77 |
| quinine (1) | | $3.3 \times 10^{-6}$ | 5.48 |
| (2) | | $1.4 \times 10^{-10}$ | 9.9 |
| urea | $H_2NCONH_2(aq) + H_2O(\ell) \rightleftharpoons H_2NCONH_3^+(aq) + OH^-(aq)$ | $1.3 \times 10^{-14}$ | 13.9 |

| TABLE A5.4 | Solubility-Product Constants at 25°C | | |
|---|---|---|---|
| **Cation** | **Anion** | **Heterogeneous Equilibrium**[a] | **$K_{sp}$** |
| aluminum | hydroxide | $Al(OH)_3(s) \rightleftharpoons Al^{3+}(aq) + 3\ OH^-(aq)$ | $1.9 \times 10^{-33}$ |
|  | phosphate | $AlPO_4(s) \rightleftharpoons Al^{3+}(aq) + PO_4^{3-}(aq)$ | $9.8 \times 10^{-21}$ |
| barium | carbonate | $BaCO_3(s) \rightleftharpoons Ba^{2+}(aq) + CO_3^{2-}(aq)$ | $2.6 \times 10^{-9}$ |
|  | fluoride | $BaF_2(s) \rightleftharpoons Ba^{2+}(aq) + 2\ F^-(aq)$ | $1.0 \times 10^{-6}$ |
|  | sulfate | $BaSO_4(s) \rightleftharpoons Ba^{2+}(aq) + SO_4^{2-}(aq)$ | $9.1 \times 10^{-11}$ |
| calcium | carbonate | $CaCO_3(s) \rightleftharpoons Ca^{2+}(aq) + CO_3^{2-}(aq)$ | $5.0 \times 10^{-9}$ |
|  | fluoride | $CaF_2(s) \rightleftharpoons Ca^{2+}(aq) + 2\ F^-(aq)$ | $3.9 \times 10^{-11}$ |
|  | hydroxide | $Ca(OH)_2(s) \rightleftharpoons Ca^{2+}(aq) + 2\ OH^-(aq)$ | $4.7 \times 10^{-6}$ |
|  | phosphate | $Ca_3(PO_4)_2(s) \rightleftharpoons 3\ Ca^{2+}(aq) + 2\ PO_4^{3-}(aq)$ | $2.1 \times 10^{-33}$ |
|  | sulfate | $CaSO_4(s) \rightleftharpoons Ca^{2+}(aq) + SO_4^{2-}(aq)$ | $7.1 \times 10^{-5}$ |
| cobalt(II) | carbonate | $CoCO_3(s) \rightleftharpoons Co^{2+}(aq) + CO_3^{2-}(aq)$ | $1.0 \times 10^{-10}$ |
|  | phosphate | $Co_3(PO_4)_2(s) \rightleftharpoons 3\ Co^{2+}(aq) + 2\ PO_4^{3-}(aq)$ | $2.1 \times 10^{-35}$ |
|  | sulfide | $CoS(s) \rightleftharpoons Co^{2+}(aq) + S^{2-}(aq)$ | $4 \times 10^{-21}$ |
| copper(I) | bromide | $CuBr(s) \rightleftharpoons Cu^+(aq) + Br^-(aq)$ | $6.3 \times 10^{-9}$ |
|  | chloride | $CuCl(s) \rightleftharpoons Cu^+(aq) + Cl^-(aq)$ | $1.0 \times 10^{-6}$ |
|  | iodide | $CuI(s) \rightleftharpoons Cu^+(aq) + I^-(aq)$ | $1.3 \times 10^{-12}$ |
| copper(II) | phosphate | $Cu_3(PO_4)_2(s) \rightleftharpoons 3\ Cu^{2+}(aq) + 2\ PO_4^{3-}(aq)$ | $1.4 \times 10^{-37}$ |
|  | hydroxide | $Cu(OH)_2(s) \rightleftharpoons Cu^{2+}(aq) + 2\ OH^-(aq)$ | $4.8 \times 10^{-20}$ |
| iron(II) | carbonate | $FeCO_3(s) \rightleftharpoons Fe^{2+}(aq) + CO_3^{2-}(aq)$ | $3.1 \times 10^{-11}$ |
|  | fluoride | $FeF_2(s) \rightleftharpoons Fe^{2+}(aq) + 2\ F^-(aq)$ | $2.4 \times 10^{-6}$ |
|  | hydroxide | $Fe(OH)_2(s) \rightleftharpoons Fe^{2+}(aq) + 2\ OH^-(aq)$ | $4.9 \times 10^{-17}$ |
|  | sulfide | $FeS(s) \rightleftharpoons Fe^{2+}(aq) + S^{2-}(aq)$ | $6.3 \times 10^{-18}$ |
| lead | bromide | $PbBr_2(s) \rightleftharpoons Pb^{2+}(aq) + 2\ Br^-(aq)$ | $6.6 \times 10^{-6}$ |
|  | carbonate | $PbCO_3(s) \rightleftharpoons Pb^{2+}(aq) + CO_3^{2-}(aq)$ | $1.5 \times 10^{-13}$ |
|  | chloride | $PbCl_2(s) \rightleftharpoons Pb^{2+}(aq) + 2\ Cl^-(aq)$ | $1.6 \times 10^{-5}$ |
|  | fluoride | $PbF_2(s) \rightleftharpoons Pb^{2+}(aq) + 2\ F^-(aq)$ | $3.2 \times 10^{-8}$ |
|  | iodide | $PbI_2(s) \rightleftharpoons Pb^{2+}(aq) + 2\ I^-(aq)$ | $8.5 \times 10^{-9}$ |
|  | sulfate | $PbSO_4(s) \rightleftharpoons Pb^{2+}(aq) + SO_4^{2-}(aq)$ | $1.8 \times 10^{-8}$ |
| lithium | carbonate | $Li_2CO_3(s) \rightleftharpoons 2\ Li^+(aq) + CO_3^{2-}(aq)$ | $8.2 \times 10^{-4}$ |
| magnesium | carbonate | $MgCO_3(s) \rightleftharpoons Mg^{2+}(aq) + CO_3^{2-}(aq)$ | $6.8 \times 10^{-6}$ |
|  | fluoride | $MgF_2(s) \rightleftharpoons Mg^{2+}(aq) + 2\ F^-(aq)$ | $6.5 \times 10^{-9}$ |
|  | hydroxide | $Mg(OH)_2(s) \rightleftharpoons Mg^{2+}(aq) + 2\ OH^-(aq)$ | $5.6 \times 10^{-12}$ |
| manganese(II) | carbonate | $MnCO_3(s) \rightleftharpoons Mn^{2+}(aq) + CO_3^{2-}(aq)$ | $2.2 \times 10^{-11}$ |
|  | hydroxide | $Mn(OH)_2(s) \rightleftharpoons Mn^{2+}(aq) + 2\ OH^-(aq)$ | $5.6 \times 10^{-12}$ |
| mercury(I) | bromide | $Hg_2Br_2(s) \rightleftharpoons Hg_2^{2+}(aq) + 2\ Br^-(aq)$ | $6.4 \times 10^{-23}$ |
|  | carbonate | $Hg_2CO_3(s) \rightleftharpoons Hg_2^{2+}(aq) + CO_3^{2-}(aq)$ | $3.7 \times 10^{-17}$ |
|  | chloride | $Hg_2Cl_2(s) \rightleftharpoons Hg_2^{2+}(aq) + 2\ Cl^-(aq)$ | $1.5 \times 10^{-18}$ |
|  | iodide | $Hg_2I_2(s) \rightleftharpoons Hg_2^{2+}(aq) + 2\ I^-(aq)$ | $5.3 \times 10^{-29}$ |
|  | sulfate | $Hg_2SO_4(s) \rightleftharpoons Hg_2^{2+}(aq) + SO_4^{2-}(aq)$ | $8.0 \times 10^{-7}$ |
| mercury(II) | hydroxide | $Hg(OH)_2(s) \rightleftharpoons Hg^{2+}(aq) + 2\ OH^-(aq)$ | $3.1 \times 10^{-26}$ |
|  | iodide | $HgI_2(s) \rightleftharpoons Hg^{2+}(aq) + 2\ I^-(aq)$ | $2.8 \times 10^{-29}$ |
| nickel(II) | carbonate | $NiCO_3(s) \rightleftharpoons Ni^{2+}(aq) + CO_3^{2-}(aq)$ | $1.4 \times 10^{-7}$ |
|  | phosphate | $Ni_3(PO_4)_2(s) \rightleftharpoons 3\ Ni^{2+}(aq) + 2\ PO_4^{3-}(aq)$ | $4.7 \times 10^{-32}$ |
|  | sulfide | $NiS(s) \rightleftharpoons Ni^{2+}(aq) + S^{2-}(aq)$ | $1 \times 10^{-24}$ |
| silver | bromide | $AgBr(s) \rightleftharpoons Ag^+(aq) + Br^-(aq)$ | $5.4 \times 10^{-13}$ |
|  | carbonate | $Ag_2CO_3(s) \rightleftharpoons 2\ Ag^+(aq) + CO_3^{2-}(aq)$ | $8.5 \times 10^{-12}$ |
|  | chloride | $AgCl(s) \rightleftharpoons Ag^+(aq) + Cl^-(aq)$ | $1.8 \times 10^{-10}$ |
|  | chromate | $Ag_2CrO_4(s) \rightleftharpoons 2\ Ag^+(aq) + CrO_4^{2-}(aq)$ | $1.1 \times 10^{-12}$ |
|  | hydroxide | $AgOH(s) \rightleftharpoons Ag^+(aq) + OH^-(aq)$ | $1.52 \times 10^{-8}$ |
|  | iodide | $AgI(s) \rightleftharpoons Ag^+(aq) + I^-(aq)$ | $8.3 \times 10^{-17}$ |
|  | phosphate | $Ag_3PO_4(s) \rightleftharpoons 3\ Ag^+(aq) + PO_4^{3-}(aq)$ | $8.9 \times 10^{-17}$ |
|  | sulfate | $Ag_2SO_4(s) \rightleftharpoons 2\ Ag^+(aq) + SO_4^{2-}(aq)$ | $1.2 \times 10^{-5}$ |
|  | sulfide | $Ag_2S(s) \rightleftharpoons 2\ Ag^+(aq) + S^{2-}(aq)$ | $1.6 \times 10^{-49}$ |
| strontium | carbonate | $SrCO_3(s) \rightleftharpoons Sr^{2+}(aq) + CO_3^{2-}(aq)$ | $5.6 \times 10^{-10}$ |
|  | fluoride | $SrF_2(s) \rightleftharpoons Sr^{2+}(aq) + 2\ F^-(aq)$ | $4.3 \times 10^{-9}$ |
|  | sulfate | $SrSO_4(s) \rightleftharpoons Sr^{2+}(aq) + SO_4^{2-}(aq)$ | $3.4 \times 10^{-7}$ |
| zinc | carbonate | $ZnCO_3(s) \rightleftharpoons Zn^{2+}(aq) + CO_3^{2-}(aq)$ | $1.2 \times 10^{-10}$ |
|  | hydroxide | $Zn(OH)_2(s) \rightleftharpoons Zn^{2+}(aq) + 2\ OH^-(aq)$ | $3.0 \times 10^{-16}$ |

[a]Equilibrium is between solid phase and aqueous solution.

| TABLE A5.5 | Formation Constants of Complex Ions at 25°C | |
|---|---|---|
| **Complex Ion** | **Aqueous Equilibrium** | **$K_f$** |
| $[Ag(NH_3)_2]^+$ | $Ag^+(aq) + 2\,NH_3(aq) \rightleftharpoons Ag(NH_3)_2^+(aq)$ | $1.7 \times 10^7$ |
| $[AgCl_2]^-$ | $Ag^+(aq) + 2\,Cl^-(aq) \rightleftharpoons AgCl_2^-(aq)$ | $2.5 \times 10^5$ |
| $[Ag(CN)_2]^-$ | $Ag^+(aq) + 2\,CN^-(aq) \rightleftharpoons Ag(CN)_2^-(aq)$ | $1.0 \times 10^{21}$ |
| $[Ag(S_2O_3)_2]^{3-}$ | $Ag^+(aq) + 2\,S_2O_3^{2-}(aq) \rightleftharpoons Ag(S_2O_3)_2^{3-}(aq)$ | $4.7 \times 10^{13}$ |
| $[AlF_6]^{3-}$ | $Al^{3+}(aq) + 6\,F^-(aq) \rightleftharpoons AlF_6^{3-}(aq)$ | $4.0 \times 10^{19}$ |
| $[Al(OH)_4]^-$ | $Al^{3+}(aq) + 4\,OH^-(aq) \rightleftharpoons Al(OH)_4^-(aq)$ | $7.7 \times 10^{33}$ |
| $[Au(CN)_2]^-$ | $Au^+(aq) + 2\,CN^-(aq) \rightleftharpoons Au(CN)_2^-(aq)$ | $2.0 \times 10^{38}$ |
| $[Co(NH_3)_6]^{2+}$ | $Co^{2+}(aq) + 6\,NH_3(aq) \rightleftharpoons Co(NH_3)_6^{2+}(aq)$ | $7.7 \times 10^4$ |
| $[Co(NH_3)_6]^{3+}$ | $Co^{3+}(aq) + 6\,NH_3(aq) \rightleftharpoons Co(NH_3)_6^{3+}(aq)$ | $5.0 \times 10^{31}$ |
| $[Co(en)_3]^{2+}$ | $Co^{2+}(aq) + 3\,en(aq) \rightleftharpoons Co(en)_3^{2+}(aq)$ | $8.7 \times 10^{13}$ |
| $[Co(C_2O_4)_3]^{4-}$ | $Co^{2+}(aq) + 3\,C_2O_4^{2-}(aq) \rightleftharpoons Co(C_2O_4)_3^{4-}(aq)$ | $4.5 \times 10^6$ |
| $[Cu(NH_3)_4]^{2+}$ | $Cu^{2+}(aq) + 4\,NH_3(aq) \rightleftharpoons Cu(NH_3)_4^{2+}(aq)$ | $5.0 \times 10^{13}$ |
| $[Cu(en)_2]^{2+}$ | $Cu^{2+}(aq) + 2\,en(aq) \rightleftharpoons Cu(en)_2^{2+}(aq)$ | $3.2 \times 10^{19}$ |
| $[Cu(CN)_4]^{2-}$ | $Cu^{2+}(aq) + 4\,CN^-(aq) \rightleftharpoons Cu(CN)_4^{2-}(aq)$ | $1.0 \times 10^{25}$ |
| $[Cu(C_2O_4)_2]^{2-}$ | $Cu^{2+}(aq) + 2\,C_2O_4^{2-}(aq) \rightleftharpoons Cu(C_2O_4)_2^{2-}(aq)$ | $1.7 \times 10^{10}$ |
| $[Fe(C_2O_4)_3]^{4-}$ | $Fe^{2+}(aq) + 3\,C_2O_4^{2-}(aq) \rightleftharpoons Fe(C_2O_4)_3^{4-}(aq)$ | $6 \times 10^6$ |
| $[Fe(C_2O_4)_3]^{3-}$ | $Fe^{3+}(aq) + 3\,C_2O_4^{2-}(aq) \rightleftharpoons Fe(C_2O_4)_3^{3-}(aq)$ | $3.3 \times 10^{20}$ |
| $[HgCl_4]^{2-}$ | $Hg^{2+}(aq) + 4\,Cl^-(aq) \rightleftharpoons HgCl_4^{2-}(aq)$ | $1.2 \times 10^{15}$ |
| $[Ni(NH_3)_6]^{2+}$ | $Ni^{2+}(aq) + 6\,NH_3(aq) \rightleftharpoons Ni(NH_3)_6^{2+}(aq)$ | $5.5 \times 10^8$ |
| $[PbCl_4]^{2-}$ | $Pb^{2+}(aq) + 4\,Cl^-(aq) \rightleftharpoons PbCl_4^{2-}(aq)$ | $2.5 \times 10^1$ |
| $[Zn(NH_3)_4]^{2+}$ | $Zn^{2+}(aq) + 4\,NH_3(aq) \rightleftharpoons Zn(NH_3)_4^{2+}(aq)$ | $2.9 \times 10^9$ |
| $[Zn(OH)_4]^{2-}$ | $Zn^{2+}(aq) + 4\,OH^-(aq) \rightleftharpoons Zn(OH)_4^{2-}(aq)$ | $2.8 \times 10^{15}$ |

# Standard Reduction Potentials

| TABLE A6.1 | Standard Reduction Potentials at 25°C | | |
| --- | --- | --- | --- |
| **Half-Reaction** | | **$n$** | **$E°$ (V)** |
| $F_2(g) + 2\,e^- \rightarrow 2\,F^-(aq)$ | | 2 | 2.866 |
| $H_2N_2O_2(s) + 2\,H^+(aq) + 2\,e^- \rightarrow N_2(g) + 2\,H_2O(\ell)$ | | 2 | 2.65 |
| $O(g) + 2\,H^+(aq) + 2\,e^- \rightarrow H_2O(\ell)$ | | 2 | 2.421 |
| $Cu^{3+}(aq) + e^- \rightarrow Cu^{2+}(aq)$ | | 1 | 2.4 |
| $XeO_3(s) + 6\,H^+(aq) + 6\,e^- \rightarrow Xe(g) + 3\,H_2O(\ell)$ | | 6 | 2.10 |
| $O_3(g) + 2\,H^+(aq) + 2\,e^- \rightarrow O_2(g) + H_2O(\ell)$ | | 2 | 2.076 |
| $OH(g) + e^- \rightarrow OH^-(aq)$ | | 1 | 2.02 |
| $Co^{3+}(aq) + e^- \rightarrow Co^{2+}(aq)$ | | 1 | 1.92 |
| $H_2O_2(\ell) + 2\,H^+(aq) + 2\,e^- \rightarrow 2\,H_2O(\ell)$ | | 2 | 1.776 |
| $N_2O(g) + 2\,H^+(aq) + 2\,e^- \rightarrow N_2(g) + H_2O(\ell)$ | | 2 | 1.766 |
| $Ce(OH)^{3+}(aq) + H^+(aq) + e^- \rightarrow Ce^{3+}(aq) + H_2O(\ell)$ | | 1 | 1.70 |
| $Au^+(aq) + e^- \rightarrow Au(s)$ | | 1 | 1.692 |
| $PbO_2(s) + SO_4^{2-}(aq) + 4\,H^+(aq) + 2\,e^- \rightarrow PbSO_4(s) + 2\,H_2O(\ell)$ | | 2 | 1.691 |
| $PbO_2(s) + HSO_4^-(aq) + 3\,H^+(aq) + 2\,e^- \rightarrow PbSO_4(s) + 2\,H_2O(\ell)$ | | 2 | 1.685 |
| $MnO_4^-(aq) + 4\,H^+(aq) + 3\,e^- \rightarrow MnO_2(s) + 2\,H_2O(\ell)$ | | 3 | 1.673 |
| $NiO_2(s) + 4\,H^+(aq) + 2\,e^- \rightarrow Ni^{2+}(aq) + 2\,H_2O(\ell)$ | | 2 | 1.678 |
| $HClO(\ell) + H^+(aq) + e^- \rightarrow \frac{1}{2}\,Cl_2(g) + H_2O(aq)$ | | 1 | 1.63 |
| $Ce^{4+}(aq) + e^- \rightarrow Ce^{3+}(aq)$ | | 1 | 1.61 |
| $Mn^{3+}(aq) + e^- \rightarrow Mn^{2+}(aq)$ | | 1 | 1.542 |
| $MnO_4^-(aq) + 8\,H^+(aq) + 5\,e^- \rightarrow Mn^{2+}(aq) + 4\,H_2O(\ell)$ | | 5 | 1.507 |
| $BrO_3^-(aq) + 6\,H^+(aq) + 5\,e^- \rightarrow \frac{1}{2}\,Br_2(\ell) + 3\,H_2O(\ell)$ | | 5 | 1.52 |
| $ClO_3^-(aq) + 6\,H^+(aq) + 5\,e^- \rightarrow \frac{1}{2}\,Cl_2(g) + 3\,H_2O(\ell)$ | | 5 | 1.47 |
| $PbO_2(s) + 4\,H^+(aq) + 2\,e^- \rightarrow Pb^{2+}(aq) + 2\,H_2O(\ell)$ | | 2 | 1.455 |
| $Au^{3+}(aq) + 3\,e^- \rightarrow Au(s)$ | | 3 | 1.40 |
| $Cl_2(g) + 2\,e^- \rightarrow 2\,Cl^-(aq)$ | | 2 | 1.358 |
| $Cr_2O_7^{2-}(aq) + 14\,H^+(aq) + 6\,e^- \rightarrow 2\,Cr^{3+}(aq) + 7\,H_2O(\ell)$ | | 6 | 1.33 |
| $2\,NiO(OH)(s) + 2\,H_2O(\ell) + 2\,e^- \rightarrow 2\,Ni(OH)_2(s) + 2\,OH^-(aq)$ | | 2 | 1.32 |
| $MnO_2(s) + 4\,H^+(aq) + 2\,e^- \rightarrow Mn^{2+}(aq) + 2\,H_2O(\ell)$ | | 2 | 1.23 |

*Continued on next page*

| TABLE A6.1 Standard Reduction Potentials at 25°C *(Continued)* | | |
|---|---|---|
| Half-Reaction | $n$ | $E°$ (V) |
| $O_2(g) + 4\,H^+(aq) + 4\,e^- \rightarrow 2\,H_2O(\ell)$ | 4 | 1.229 |
| $IO_3^-(aq) + 6\,H^+(aq) + 5\,e^- \rightarrow \frac{1}{2}\,I_2(s) + 3\,H_2O(\ell)$ | 5 | 1.195 |
| $IO_3^-(aq) + 6\,H^+(aq) + 6\,e^- \rightarrow I^-(aq) + 3\,H_2O(\ell)$ | 6 | 1.085 |
| $Br_2(\ell) + 2\,e^- \rightarrow 2\,Br^-(aq)$ | 2 | 1.066 |
| $HNO_2(\ell) + H^+(aq) + e^- \rightarrow NO(g) + H_2O(\ell)$ | 1 | 1.00 |
| $VO_2^+(aq) + 2\,H^+(aq) + e^- \rightarrow VO^{2+}(aq) + H_2O(\ell)$ | 1 | 1.00 |
| $NO_3^-(aq) + 4\,H^+(aq) + 3\,e^- \rightarrow NO(g) + 2\,H_2O(\ell)$ | 3 | 0.96 |
| $2\,Hg^{2+}(aq) + 2\,e^- \rightarrow Hg_2^{2+}(aq)$ | 2 | 0.92 |
| $ClO^-(aq) + H_2O(\ell) + 2\,e^- \rightarrow Cl^-(aq) + 2\,OH^-(aq)$ | 2 | 0.89 |
| $HO_2^-(aq) + H_2O(\ell) + 2\,e^- \rightarrow 3\,OH^-(aq)$ | 2 | 0.88 |
| $Hg^{2+}(aq) + 2\,e^- \rightarrow Hg(\ell)$ | 2 | 0.851 |
| $Ag^+(aq) + e^- \rightarrow Ag(s)$ | 1 | 0.800 |
| $Hg_2^{2+}(aq) + 2\,e^- \rightarrow 2\,Hg(\ell)$ | 2 | 0.797 |
| $Fe^{3+}(aq) + e^- \rightarrow Fe^{2+}(aq)$ | 1 | 0.770 |
| $PtCl_4^{2-}(aq) + 2\,e^- \rightarrow Pt(s) + 4\,Cl^-(aq)$ | 2 | 0.73 |
| $O_2(g) + 2\,H^+(aq) + 2\,e^- \rightarrow H_2O_2(\ell)$ | 2 | 0.68 |
| $MnO_4^-(aq) + 2\,H_2O(\ell) + 3\,e^- \rightarrow MnO_2(s) + 4\,OH^-(aq)$ | 3 | 0.59 |
| $H_3AsO_4(s) + 2\,H^+(aq) + 2\,e^- \rightarrow H_3AsO_3(aq) + H_2O(\ell)$ | 2 | 0.559 |
| $I_2(s) + 2\,e^- \rightarrow 2\,I^-(aq)$ | 2 | 0.536 |
| $Cu^+(aq) + e^- \rightarrow Cu(s)$ | 1 | 0.521 |
| $H_2SO_3(\ell) + 4\,H^+(aq) + 4\,e^- \rightarrow S(s) + 3\,H_2O(\ell)$ | 4 | 0.449 |
| $Ag_2CrO_4(s) + 2\,e^- \rightarrow 2\,Ag(s) + CrO_4^{2-}(aq)$ | 2 | 0.447 |
| $O_2(g) + 2\,H_2O(\ell) + 4\,e^- \rightarrow 4\,OH^-(aq)$ | 4 | 0.401 |
| $Fe(CN)_6^{3-}(aq) + e^- \rightarrow Fe(CN)_6^{4-}(aq)$ | 1 | 0.36 |
| $Ag_2O(s) + H_2O(\ell) + 2\,e^- \rightarrow 2\,Ag(s) + 2\,OH^-(aq)$ | 2 | 0.342 |
| $Cu^{2+}(aq) + 2\,e^- \rightarrow Cu(s)$ | 2 | 0.342 |
| $BiO^+(aq) + 2\,H^+(aq) + 3\,e^- \rightarrow Bi(s) + H_2O(\ell)$ | 3 | 0.32 |
| $AgCl(s) + e^- \rightarrow Ag(s) + Cl^-(aq)$ | 1 | 0.222 |
| $HSO_4^-(aq) + 3\,H^+(aq) + 2\,e^- \rightarrow H_2SO_3(\ell) + H_2O(\ell)$ | 2 | 0.17 |
| $Sn^{4+}(aq) + 2\,e^- \rightarrow Sn^{2+}(aq)$ | 2 | 0.154 |
| $Cu^{2+}(aq) + e^- \rightarrow Cu^+(aq)$ | 1 | 0.153 |
| $2\,MnO_2(s) + H_2O(\ell) + 2\,e^- \rightarrow Mn_2O_3(s) + 2\,OH^-(aq)$ | 2 | 0.15 |
| $S(s) + 2\,H^+(aq) + 2\,e^- \rightarrow H_2S(g)$ | 2 | 0.141 |
| $HgO(s) + H_2O(\ell) + 2\,e^- \rightarrow Hg(\ell) + 2\,OH^-(aq)$ | 2 | 0.0977 |

| TABLE A6.1 | Standard Reduction Potentials at 25°C *(Continued)* | | |
|---|---|---|---|
| **Half-Reaction** | | **n** | **E° (V)** |
| $AgBr(s) + e^- \rightarrow Ag(s) + Br^-(aq)$ | | 1 | 0.095 |
| $Ag(S_2O_3)_2{}^{3-}(aq) + e^- \rightarrow Ag(s) + 2\,S_2O_3{}^{2-}(aq)$ | | 1 | 0.01 |
| $NO_3{}^-(aq) + H_2O(\ell) + 2\,e^- \rightarrow NO_2{}^-(aq) + 2\,OH^-(aq)$ | | 2 | 0.01 |
| $2\,H^+(aq) + 2\,e^- \rightarrow H_2(g)$ | | 2 | 0.000 |
| $Pb^{2+}(aq) + 2\,e^- \rightarrow Pb(s)$ | | 2 | −0.126 |
| $CrO_4{}^{2-}(aq) + 4\,H_2O(\ell) + 3\,e^- \rightarrow Cr(OH)_3(s) + 5\,OH^-(aq)$ | | 3 | −0.13 |
| $Sn^{2+}(aq) + 2\,e^- \rightarrow Sn(s)$ | | 2 | −0.136 |
| $AgI(s) + e^- \rightarrow Ag(s) + I^-(aq)$ | | 1 | −0.152 |
| $CuI(s) + e^- \rightarrow Cu(s) + I^-(aq)$ | | 1 | −0.185 |
| $N_2(g) + 5\,H^+(aq) + 4\,e^- \rightarrow N_2H_5{}^+(aq)$ | | 4 | −0.23 |
| $Ni^{2+}(aq) + 2\,e^- \rightarrow Ni(s)$ | | 2 | −0.257 |
| $PbSO_4(s) + H^+(aq) + 2\,e^- \rightarrow Pb(s) + HSO_4{}^-(aq)$ | | 2 | −0.356 |
| $Co^{2+}(aq) + 2\,e^- \rightarrow Co(s)$ | | 2 | −0.277 |
| $Ag(CN)_2{}^-(aq) + e^- \rightarrow Ag(s) + 2\,CN^-(aq)$ | | 1 | −0.31 |
| $Cd^{2+}(aq) + 2\,e^- \rightarrow Cd(s)$ | | 2 | −0.403 |
| $Cd(OH)_2(s) + 2\,e^- \rightarrow Cd(s) + 2\,OH^-(aq)$ | | 2 | −0.403 |
| $Cr^{3+}(aq) + e^- \rightarrow Cr^{2+}(aq)$ | | 1 | −0.41 |
| $Fe^{2+}(aq) + 2\,e^- \rightarrow Fe(s)$ | | 2 | −0.447 |
| $2\,CO_2(g) + 2\,H^+(aq) + 2\,e^- \rightarrow H_2C_2O_4(s)$ | | 2 | −0.49 |
| $Ni(OH)_2(s) + 2\,e^- \rightarrow Ni(s) + 2\,OH^-(aq)$ | | 2 | −0.72 |
| $Cr^{3+}(aq) + 3\,e^- \rightarrow Cr(s)$ | | 3 | −0.74 |
| $Zn^{2+}(aq) + 2\,e^- \rightarrow Zn(s)$ | | 2 | −0.762 |
| $2\,H_2O(\ell) + 2\,e^- \rightarrow H_2(g) + 2\,OH^-(aq)$ | | 2 | −0.828 |
| $SO_4{}^{2-}(aq) + H_2O(\ell) + 2\,e^- \rightarrow SO_3{}^{2-}(aq) + 2\,OH^-(aq)$ | | 2 | −0.92 |
| $N_2(g) + 4\,H_2O(\ell) + 4\,e^- \rightarrow 4\,OH^-(aq) + N_2H_4(\ell)$ | | 4 | −1.16 |
| $Mn^{2+}(aq) + 2\,e^- \rightarrow Mn(s)$ | | 2 | −1.185 |
| $Zn(OH)_2(s) + 2\,e^- \rightarrow Zn(s) + 2\,OH^-(aq)$ | | 2 | −1.249 |
| $ZnO(s) + H_2O(\ell) + 2\,e^- \rightarrow Zn(s) + 2\,OH^-(aq)$ | | 2 | −1.25 |
| $Al^{3+}(aq) + 3\,e^- \rightarrow Al(s)$ | | 3 | −1.662 |
| $Mg^{2+}(aq) + 2\,e^- \rightarrow Mg(s)$ | | 2 | −2.37 |
| $Na^+(aq) + e^- \rightarrow Na(s)$ | | 1 | −2.71 |
| $Ca^{2+}(aq) + 2\,e^- \rightarrow Ca(s)$ | | 2 | −2.868 |
| $Ba^{2+}(aq) + 2\,e^- \rightarrow Ba(s)$ | | 2 | −2.912 |
| $K^+(aq) + e^- \rightarrow K(s)$ | | 1 | −2.95 |
| $Li^+(aq) + e^- \rightarrow Li(s)$ | | 1 | −3.05 |

# Naming Organic Compounds

While organic chemistry was becoming established as a discipline within chemistry, many compounds were given trivial names that are still commonly used and recognized. We refer to many of these compounds by their nonsystematic names throughout this book, and their names and structures are listed in Table A7.1.

| TABLE A7.1 | Organic Compounds and Their Commonly Used Nonsystematic Names | |
|---|---|---|
| **Name** | **Formula** | **Structure** |
| ethylene | $C_2H_4$ | |
| acetylene | $C_2H_2$ | $HC \equiv CH$ |
| benzene | $C_6H_6$ | |
| toluene | $C_6H_5CH_3$ | |
| ethyl alcohol | $CH_3CH_2OH$ | |
| acetone | $CH_3COCH_3$ | |
| acetic acid | $CH_3COOH$ | |
| formaldehyde | $CH_2O$ | |

The International Union of Pure and Applied Chemistry (IUPAC) has proposed a set of rules for the systematic naming of organic compounds. The basic principles for naming alkanes, alkenes, and alkynes are presented in Chapter 13. These rules are summarized here and extended to include compounds containing other functional groups. When naming compounds or drawing structures based on names, we need to keep in mind that the IUPAC system of nomenclature is based on two fundamental ideas: (1) the name of a compound must indicate how the carbon atoms in the skeleton are bonded together, and (2) the name must identify the location of any functional groups in the molecule.

## Alkanes

Table A7.2 contains the prefixes used for carbon chains ranging in size from $C_1$ to $C_{20}$ and gives the names for compounds consisting of unbranched chains. The name of a compound consists of a prefix identifying the number of carbons in the chain and a suffix defining the type of hydrocarbon. The suffix *-ane* indicates that the compounds are alkanes and that all carbon–carbon bonds are single bonds.

| TABLE A7.2 | Prefixes for Naming Carbon Chains | | | | |
|---|---|---|---|---|---|
| Prefix | Example | Name | Prefix | Example | Name |
| meth | $CH_4$ | methane | undec | $C_{11}H_{24}$ | undecane |
| eth | $C_2H_6$ | ethane | dodec | $C_{12}H_{26}$ | dodecane |
| pro | $C_3H_8$ | propane | tridec | $C_{13}H_{28}$ | tridecane |
| but | $C_4H_{10}$ | butane | tetradec | $C_{14}H_{30}$ | tetradecane |
| pent | $C_5H_{12}$ | pentane | pentadec | $C_{15}H_{32}$ | pentadecane |
| hex | $C_6H_{14}$ | hexane | hexadec | $C_{16}H_{34}$ | hexadecane |
| hept | $C_7H_{16}$ | heptane | heptadec | $C_{17}H_{36}$ | heptadecane |
| oct | $C_8H_{18}$ | octane | octadec | $C_{18}H_{38}$ | octadecane |
| non | $C_9H_{20}$ | nonane | nonadec | $C_{19}H_{40}$ | nonadecane |
| dec | $C_{10}H_{22}$ | decane | eicos | $C_{20}H_{42}$ | eicosane |

## Branched-Chain Alkanes

The alkane drawn here is used to illustrate each step in the naming rules:

$$CH_3$$
$$CH_3CH_2CHCH_2CHCHCH_2CH_2CH_3$$
$$CH_3 \quad CH_2CH_3$$

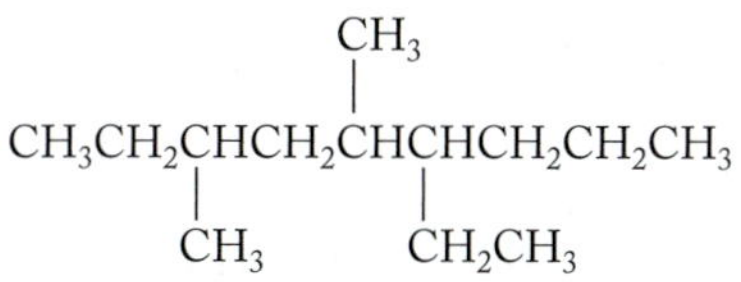

1. **Identify and name the longest continuous carbon chain.**

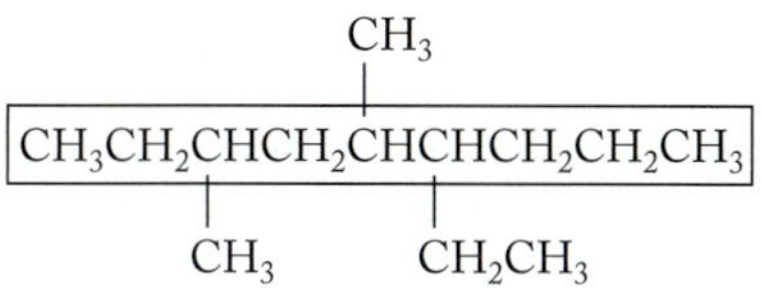

$$CH_3$$
$$\boxed{CH_3CH_2CHCH_2CHCHCH_2CH_2CH_3} \text{ Nonane}$$
$$CH_3 \quad CH_2CH_3$$

2. **Identify the groups attached to this chain and name them.** Names of substituent groups consist of the prefix from Table A7.2 that identifies the length of the group and the suffix *-yl* that identifies it as an alkyl group.

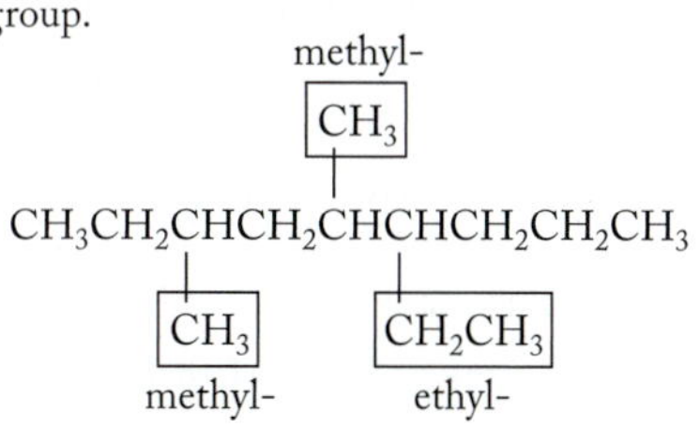

methyl-
$$\boxed{CH_3}$$
$$CH_3CH_2CHCH_2CHCHCH_2CH_2CH_3$$
$$\boxed{CH_3} \quad \boxed{CH_2CH_3}$$
methyl-　　ethyl-

3. **Number the carbon atoms in the longest chain,** starting at the end nearest a substituent group. Doing this identifies the points of attachment of the alkyl groups with the lowest possible numbers.

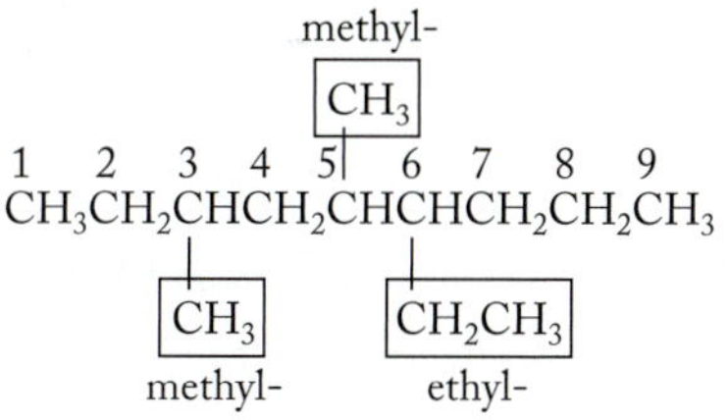

methyl-
$$\boxed{CH_3}$$
$$\begin{array}{ccccccccc} 1 & 2 & 3 & 4 & 5 & 6 & 7 & 8 & 9 \end{array}$$
$$CH_3CH_2CHCH_2CHCHCH_2CH_2CH_3$$
$$\boxed{CH_3} \quad \boxed{CH_2CH_3}$$
methyl-　　ethyl-

4. **Designate the location and identity of each substituent group with a number,** followed by a hyphen, and its name.

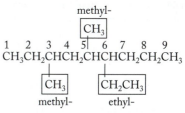

methyl-
$$\boxed{CH_3}$$
$$\begin{array}{ccccccccc} 1 & 2 & 3 & 4 & 5 & 6 & 7 & 8 & 9 \end{array}$$
$$CH_3CH_2CHCH_2CHCHCH_2CH_2CH_3$$
$$\boxed{CH_3} \quad \boxed{CH_2CH_3}$$
methyl-　　ethyl-

3-methyl-, 5-methyl-, 6-ethyl-

5. **Put together the complete name by listing the substituent groups in alphabetical order.** If more than one of a given type of substituent group is present, prefixes *di-*, *tri-*, *tetra-*, and so forth are appended to the names, but these numerical prefixes are not considered when determining the alphabetical order. The name of the last substituent group is written together with the name identifying the longest carbon chain.

$$CH_3$$
$$CH_3CH_2CHCH_2CHCHCH_2CH_2CH_3$$
$$CH_3 \quad CH_2CH_3$$

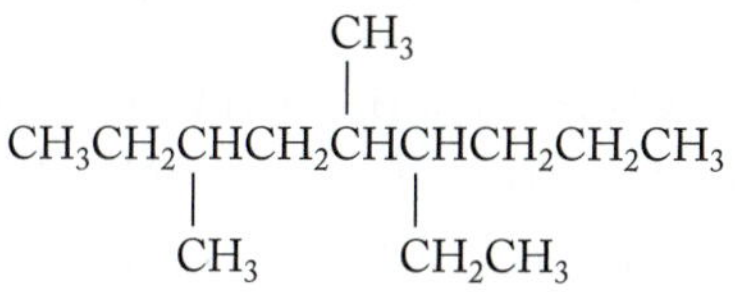

6-Ethyl-3,5-dimethylnonane

## Cycloalkanes

The simplest examples of this class of compounds consist of one unsubstituted ring of carbon atoms. The IUPAC names of these compounds consist of the prefix *cyclo-* followed by the parent name from Table A7.2 to indicate the number of carbon atoms in the ring. As an illustration, the names, formulas, and line structures of the first three cycloalkanes in the homologous series are

$$\triangle \qquad \square \qquad \pentagon$$
$$C_3H_6 \qquad C_4H_8 \qquad C_5H_{10}$$
Cyclopropane　Cyclobutane　Cyclopentane

## Alkenes and Alkynes

Alkenes have carbon–carbon double bonds and alkynes have carbon–carbon triple bonds as functional groups. The names of these types of compounds consist of (1) a parent name that identifies the longest carbon chain that includes the double or triple bond, (2) a suffix that identifies the class of compound, and (3) names of any substituent groups attached to the longest carbon chain. The suffix *-ene* identifies an alkene; *-yne* identifies an alkyne.

The alkene and alkyne drawn here are used to illustrate each step in the naming rules:

$$CH_3 \qquad\qquad CH_3$$
$$CH_3CHCH=CHCH_2CH_2CH_3 \quad CH_3CHC\equiv CCH_2CH_2CH_3$$

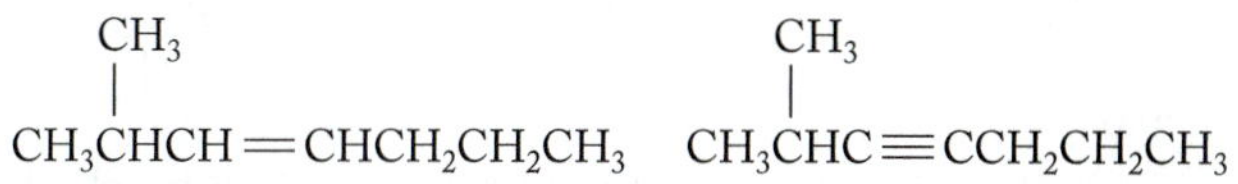

1. **To determine the parent name,** identify the longest chain that contains the unsaturation. Name the parent **compound** with the prefix that defines the number of carbons in that chain and the suffix that identifies the class of compound.

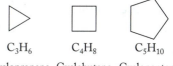

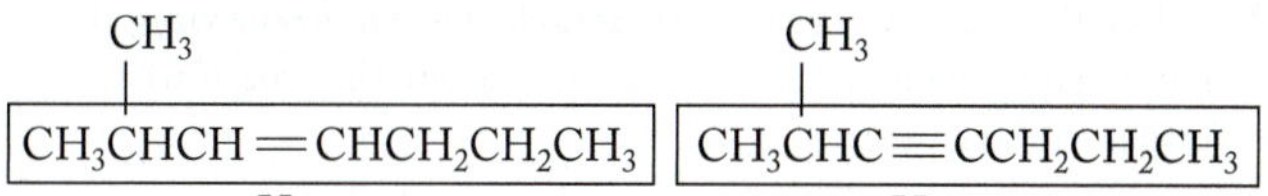

$$CH_3 \qquad\qquad\qquad CH_3$$
$$\boxed{CH_3CHCH=CHCH_2CH_2CH_3} \quad \boxed{CH_3CHC\equiv CCH_2CH_2CH_3}$$
　　　Heptene　　　　　　　　　　　Heptyne

**2. Number the parent chain from the end nearest the unsaturation so that the first carbon in the double or triple bond has the lowest number possible.** (If the unsaturation is in the middle of a chain, the location of any substituent group is used to determine where the numbering starts.) The smaller of the two numbers identifying the carbon atoms involved in the unsaturation is used as the locator of the multiple bond.

$$
\begin{array}{c}
\text{CH}_3\\
\overset{1}{\text{CH}_3}\overset{2}{\text{CHCH}}=\overset{3\ \ \ \ 4\ \ \ 5\ \ \ \ 6\ \ \ 7}{\text{CHCH}_2\text{CH}_2\text{CH}_3}\\
\text{3-Heptene}
\end{array}
\qquad
\begin{array}{c}
\text{CH}_3\\
\overset{1}{\text{CH}_3}\overset{2}{\text{CHC}}\equiv\overset{3\ \ \ \ 4\ 5\ \ \ \ 6\ \ \ 7}{\text{CCH}_2\text{CH}_2\text{CH}_3}\\
\text{3-Heptyne}
\end{array}
$$

**3. Stereoisomers of alkenes are named by writing *cis*- or *trans*- before the number identifying the location of the double bond.** Chapter 19 in the text addresses naming stereoisomers.

**4. The rules for naming substituted alkanes are followed to name and locate any other groups on the chain.**

$$
\begin{array}{c}
\text{CH}_3\\
|\\
\text{CH}_3\text{CHCH}=\text{CHCH}_2\text{CH}_2\text{CH}_3\\
\text{2-Methyl-3-heptene}
\end{array}
\qquad
\begin{array}{c}
\text{CH}_3\\
|\\
\text{CH}_3\text{CHC}\equiv\text{CCH}_2\text{CH}_2\text{CH}_3\\
\text{2-Methyl-3-heptyne}
\end{array}
$$

Halogens attached to an alkane, alkene, or alkyne are named as fluoro- (F–), chloro- (Cl–), bromo- (Br–), or iodo- (I–) and are located by using the same numbering system described for alkyl groups.

## Benzene Derivatives

Naming compounds containing substituted benzene rings is less systematic than naming hydrocarbons. Many compounds have common names that are incorporated into accepted names, but for simple substituted benzene rings, the following rules may be applied.

**1. For monosubstituted benzene rings,** a prefix identifying the group is appended to the parent name benzene:

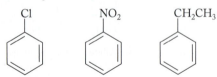

Chlorobenzene    Nitrobenzene    Ethylbenzene

**2. For disubstituted benzene rings,** three isomers are possible. The relative position of the substituent groups is indicated by numbers in IUPAC nomenclature, but the set of prefixes shown are very commonly used as well:

IUPAC:
1,2-Dichlorobenzene    1,3-Dichlorobenzene    1,4-Dichlorobenzene

Common:
*ortho*-Dichlorobenzene    *meta*-Dichlorobenzene    *para*-Dichlorobenzene
*o*-Dichlorobenzene    *m*-Dichlorobenzene    *p*-Dichlorobenzene

**3. When three or more groups are attached to a benzene ring,** the lowest possible numbers are assigned to locate the groups with respect to each other.

1,2,3-Trichlorobenzene    1,2,4-Trichlorobenzene    1,2,3,5-Tetrachlorobenzene
(NOTE: Not 1,3,4-trichlorobenzene; and not 1,3,4,5-tetrachlorobenzene.)

## Hydrocarbons Containing Other Functional Groups

The same basic principles developed for naming alkanes apply to naming hydrocarbons with functional groups other than alkyl groups The name must identify the carbon skeleton, locate the functional group, and contain a suffix that defines the class of compound. The following examples give the suffixes for some common functional groups; when suffixes are used, they replace the final *-e* in the name of the parent alkane. Other functional groups may be identified by including the name of the class of compounds in the name of the molecule.

**Alcohols: Suffix *-ol***

$$
\text{CH}_3\text{CH}_2\text{CH}_2\text{OH}
\qquad
\begin{array}{c}
\text{CH}_3\text{CHCH}_3\\
|\\
\text{OH}
\end{array}
\qquad
\begin{array}{c}
\text{CH}_3\text{CH}_2\text{CH}_2\text{CHCH}_3\\
|\\
\text{OH}
\end{array}
$$

1-Propanol    2-Propanol    2-Pentanol

**Aldehydes: Suffix *-al***

IUPAC:    Methanal    Ethanal
Common:    Formaldehyde    Acetaldehyde

Because the aldehyde group can only be on a terminal carbon, no number is necessary to locate it on the carbon chain.

**Ketones: Suffix *-one***   The location of the carbonyl is given by a number, and the chain is numbered so that the carbonyl carbon has the lowest possible value. Many ketones also have common names generated by identifying the hydrocarbon groups on both sides of the carbonyl group.

IUPAC:    Propan-2-one    Butan-2-one
Common:    Acetone    Methyl ethyl ketone

**Carboxylic Acids: Suffix *-oic acid***   The carboxylic acid group is by definition carbon 1, so no number identifying its location is included in the name.

IUPAC:    Ethanoic acid    *trans*-2-Butenoic acid
Common:    Acetic acid

**Salts of Carboxylic Acids**  Salts are named with the cation first, followed by the anion name of the acid from which -*ic acid* is dropped and the suffix -*ate* is added. The sodium salt of acetic acid is sodium acetate.

Acetic acid          Acetate ion          Sodium acetate

**Esters**  Esters are viewed as derivatives of carboxylic acids. They are named in a manner analogous to that of salts. The alkyl group comes first followed by the name of the carboxylate anion.

Alkyl  Carboxylate          Ethyl acetate

**Amides**  Amides are also derivatives of carboxylic acids. They are named by replacing -*ic acid* (of the common names) or -*oic acid* of the IUPAC names with -*amide*.

Parent acid -amide          Acetamide
(ethanamide)

**Ethers**  Ethers are frequently named by naming the two groups attached to the oxygen and following those names by the word *ether*.

$$CH_3OCH_3 \qquad CH_3CH_2OCH_2CH_3$$

Dimethyl ether          Diethyl ether

**Amines**  Aliphatic amines are usually named by listing the group or groups attached to the nitrogen and then appending -*amine* as a suffix. They may also be named by prefixing *amino-* to the name of the parent chain.

$$H_3C-NH_2 \qquad H_3C-NH \qquad H_2NCH_2CH_2OH$$

Methylamine          Ethylmethylamine          2-Aminoethanol

This brief summary will enable you to understand the names of organic compounds used in this book. IUPAC rules are much more extensive than this and can be applied to all varieties of carbon compounds including those with multiple functional groups. It is important to recognize that the rules of systematic nomenclature do not necessarily lead to a unique name for each compound, but they do always lead to an unambiguous one. Furthermore, common names are still used frequently in organic chemistry because the systematic alternatives do not improve communication. Remember that the main purpose of chemical nomenclature is to identify a chemical species by means of written or spoken words. Anyone who reads or hears the name should be able to deduce the structure and thereby the identity of the compound.

## A

**absolute temperature**    Temperature expressed in kelvins on the absolute (Kelvin) temperature scale, on which 0 K is the lowest possible temperature.

**absolute zero (0 K)**    The zero point on the Kelvin temperature scale; theoretically the lowest temperature possible.

**absorbance**    A measure of the quantity of light absorbed by a sample.

**accuracy**    Agreement between an experimental value and the true value.

**achiral**    Not chiral; describes compounds that can be superimposed on their mirror images.

**acid (Brønsted–Lowry acid)**    A proton donor.

**activated complex**    A species formed in a chemical reaction when molecules have enough energy to react with each other.

**activation energy ($E_a$)**    The minimum energy molecules need to react when they collide.

**active site**    The location on an enzyme where a reactive substance binds.

**activity series**    A qualitative ordering of the oxidizing ability of metals and their cations.

**actual yield**    The amount of product obtained from a chemical reaction, which is often less than the theoretical yield.

**addition polymer**    Macromolecule prepared by adding monomers to a growing polymer chain without the loss of any atoms or molecules.

**addition reaction**    A reaction in which one reactant adds across a multiple bond in another reactant to form one product.

**alcohol**    Organic compound containing the –OH functional group.

**aldehyde**    Organic compound containing a carbonyl group bonded to one R group and one hydrogen; its general formula is RCHO.

**alkali metals**    The elements in group 1 of the periodic table.

**alkaline earth metals**    The elements in group 2 of the periodic table.

**alkane**    Hydrocarbon in which all the bonds are single bonds with the general formula $C_nH_{2n+2}$.

**alkene**    Hydrocarbon containing one or more carbon–carbon double bonds.

**alkyne**    Hydrocarbon containing one or more carbon–carbon triple bonds.

**allotropes**    Different molecular forms of the same element, such as oxygen ($O_2$) and ozone ($O_3$).

**alloy**    A blend of a host metal and one or more other elements, which may or may not be metals, that are added to change the properties of the host metal.

**alpha (α) decay**    A nuclear reaction in which an unstable nuclide spontaneously emits an alpha particle.

**α helix**    A coil in a protein chain's secondary structure.

**alpha (α) particle**    A radioactive emission with a charge of 2+ and a mass equivalent to that of a helium nucleus.

**amide**    Organic compound in which the same carbon atom is single bonded to a nitrogen atom and double bonded to an oxygen atom.

**amine**    Organic compound that contains a group with the general formula $RNH_2$, $R_2NH$, or $R_3N$, where R is any organic subgroup.

**amino acid**    Molecule that contains at least one amine group and one carboxylic acid group; in an *α-amino acid*, the two groups are attached to the same (α) carbon atom.

**Amontons's law**    The pressure of a quantity of gas at constant volume is directly proportional to its absolute temperature.

**amorphous solid**    A solid that lacks long-range order for the atoms, ions, or molecules in its structure.

**amphiprotic**    A substance that can behave as either a proton acceptor or a proton donor.

**amplitude**    The height of the crest or depth of the trough of a wave with respect to the center line of the wave.

**angular momentum quantum number ($\ell$)**    An integer having any value from 0 to $n-1$ that defines the shape of an orbital.

**anion**    A negatively charged particle created when an atom or molecule gains one or more electrons.

**anode**    An electrode at which an oxidation half-reaction (loss of electrons) takes place.

**antibonding orbital**    Term in MO theory describing regions of electron density in a molecule that destabilize the molecule because they decrease the electron density between nuclear centers.

**antimatter** Particles that are the charge opposites of normal sub-atomic particles.

**aromatic compound** A cyclic, planar compound with delocalized π (pi) electrons above and below the plane of the molecule.

**Arrhenius equation** Relates the rate constant of a reaction to absolute temperature ($T$), the activation energy of the reaction ($E_a$), and the frequency factor ($A$).

**atmosphere (1 atm)** A unit of pressure based on Earth's average atmospheric pressure at sea level.

**atmospheric pressure ($P_{atm}$)** The force exerted by the gases surrounding Earth on Earth's surface and on all surfaces of all objects.

**atom** The smallest particle of an element that retains the chemical characteristics of the element.

**atomic absorption spectrum** (also called *dark-line spectrum*) A characteristic series of dark lines produced when free, gaseous atoms are illuminated by an external source of radiation.

**atomic emission spectrum** (also called *bright-line spectrum*) A characteristic series of bright lines produced by high-temperature atoms.

**atomic mass unit (amu)** Unit used to express the relative masses of atoms and subatomic particles; it is exactly 1/12 the mass of one atom of carbon with six protons and six neutrons in its nucleus.

**atomic number ($Z$)** The number of protons in the nucleus of an atom.

**atomic radius** (also called *covalent radius*) Half the distance between identical nuclear centers in a molecule.

**aufbau principle** The method of building electron configurations of atoms by adding one electron at a time as atomic number increases across the rows of the periodic table; each electron goes into the lowest-energy orbital available.

**autoionization** The process that produces equal and very small concentrations of $H_3O^+$ and $OH^-$ ions in pure water.

**average atomic mass** A weighted average of masses of all isotopes of an element, calculated by multiplying the natural abundance of each isotope by its mass in atomic mass units and then summing these products.

**Avogadro's law** The volume of a gas at a given temperature and pressure is proportional to the quantity of the gas.

**Avogadro's number ($N_A$)** The number of carbon atoms in exactly 12 grams of the carbon-12 isotope; $N_A = 6.022 \times 10^{23}$. It is the number of particles in one mole.

## B

**band gap ($E_g$)** The energy gap between the valence and conduction bands.

**band theory** An extension of molecular orbital theory that describes bonding in solids.

**barometer** An instrument that measures atmospheric pressure.

**base (Brønsted–Lowry base)** A proton acceptor.

**becquerel (Bq)** The SI unit of radioactivity; one becquerel equals one decay event per second.

**Beer's Law** Relates the absorbance of a solution ($A$) to concentration ($c$), path length ($b$), and the molar absorptivity ($\varepsilon$) by the equation $A = \varepsilon b c$.

**belt of stability** The region on a graph of number of neutrons versus number of protons that includes all stable nuclei.

**beta (β) decay** The spontaneous ejection of a β particle by a neutron-rich nucleus.

**beta (β) particle** A radioactive emission that is a high-energy electron.

**β-pleated sheet** A puckered two-dimensional array of protein strands held together by hydrogen bonds.

**bimolecular step** A step in a reaction mechanism involving a collision between two molecules.

**binding energy (BE)** The energy that would be released when free nucleons combine to form the nucleus of an atom. It is also the energy needed to split the nucleus into free nucleons.

**biocatalysis** The use of enzymes to catalyze reactions on a large scale; it is becoming especially important in processes that involve chiral materials.

**biomass** The sum total of the mass of organic matter in any given ecological system.

**biomolecule** An organic molecule present naturally in a living system.

**body-centered cubic (bcc) unit cell** A cell with one atom at each of the eight corners of a cube and one at the center of the cell.

**bomb calorimeter** A constant-volume device used to measure the energy released during a combustion reaction.

**bond angle** The angle (in degrees) defined by lines joining the centers of two atoms to the center of a third atom to which they are chemically bonded.

**bond dipole** Separation of electrical charge created when atoms with different electronegativities form a covalent bond.

**bond energy** The energy needed to break 1 mole of a particular covalent bond in a molecule or polyatomic ion in the gas phase.

**bond length** The distance between the nuclear centers of two atoms joined together in a bond.

**bond order** The number of bonds between atoms: 1 for a single bond, 2 for a double bond, and 3 for a triple bond.

**bond polarity** A measure of the extent to which bonding electrons are unequally shared due to differences in electronegativity of the bonded atoms.

**bonding capacity** The number of covalent bonds an atom forms to have an octet of electrons in its valence shell.

**bonding orbital** Term in MO theory describing regions of increased electron density between nuclear centers that serve to hold atoms together in molecules.

**bonding pair** A pair of electrons shared between two atoms.

**Born–Haber cycle** A series of steps with corresponding enthalpy changes that describes the formation of an ionic solid from its constituent elements.

**Boyle's law** The volume of a given amount of gas at constant temperature is inversely proportional to its pressure.

**Bragg equation** Relates the angle of diffraction ($2\theta$) of X-rays to the spacing ($d$) between the layers of ions or atoms in a crystal: $n\lambda = 2d \sin \theta$.

**branched-chain hydrocarbon** An organic molecule in which the chain of carbon atoms is not linear.

**breeder reactor** A nuclear reactor in which fissionable material is produced during normal reactor operation.

**bright-line spectrum** (also called *atomic emission spectrum*) A characteristic series of bright lines produced by high-temperature.

**Brønsted–Lowry model** Defines acids as $H^+$ ion donors and bases as $H^+$ ion acceptors.

**buffer capacity** The quantity of acid or base that a pH buffer can neutralize while maintaining its pH within a desired range.

## C

**calibration curve** Compares a measurable property, such as absorbance, of a solution of unknown concentration to a set of standard samples of known concentration.

**calorie (cal)** The amount of energy necessary to raise the temperature of 1 g of water by 1°C, from 14.5°C to 15.5°C.

**calorimeter**  A device used to measure the absorption or release of energy by a physical change or chemical process.

**calorimeter constant ($C_{calorimeter}$)**  The heat capacity of a calorimeter.

**calorimetry**  The measurement of the quantity of heat transferred during a physical change or chemical process.

**capillary action**  The rise of a liquid in a narrow tube as a result of adhesive forces between the liquid and the tube and cohesive forces within the liquid.

**carbohydrate**  An organic molecule with the generic formula $C_x(H_2O)_y$.

**carbonyl group**  A carbon atom with a double bond to an oxygen atom.

**carboxylic acid**  An organic compound containing the –COOH functional group.

**catalyst**  A substance added to a reaction that increases the rate of the reaction but is not consumed in the process.

**cathode**  An electrode at which a reduction half-reaction (gain of electrons) takes place.

**cathode rays**  Streams of electrons emitted by the cathode in a partially evacuated tube.

**cation**  A positively charged particle created when an atom or molecule loses one or more electrons.

**cell diagram**  Symbols that show how the components of an electrochemical cell are connected.

**cell potential ($E_{cell}$)**  The energy released when a voltaic cell, in which reactants and products are not necessarily in their standard states, pumps one mole of electrons through an external circuit.

**ceramic**  A solid inorganic compound or mixture that has been transformed into a harder, more heat-resistant material by heating.

**chain reaction**  A self-sustaining series of fission reactions in which the neutrons released when nuclei split apart initiate additional fission events and sustain the reaction.

**Charles's law**  The volume of a fixed quantity of gas at constant pressure is directly proportional to its absolute temperature.

**chelate effect**  The greater affinity of metal ions for polydentate ligands than for monodentate ligands.

**chelation**  The interaction of a metal with a polydentate ligand (chelating agent); pairs of electrons on one molecule of the ligand occupy two or more coordination sites on the central metal.

**chemical bond**  The energy that holds two atoms in a molecule together.

**chemical equation**  Notation in which chemical formulas express the identities and their coefficients express the quantities of substances involved in a chemical reaction.

**chemical equilibrium**  A dynamic process in which the concentrations of reactants and products remain constant over time and the rate of a reaction in the forward direction matches its rate in the reverse direction.

**chemical formula**  A notation for representing elements and compounds; consists of the symbols of the constituent elements and subscripts identifying the number of atoms of each element in one molecule.

**chemical kinetics**  The study of the rates of change of concentrations of substances involved in chemical reactions.

**chemical property**  A property of a substance that can be observed only by reacting it to form another substance.

**chemical reaction**  The transformation of one or more substances into different substances.

**chemistry**  The study of the composition, structure, and properties of matter and of the energy consumed or given off when matter undergoes a change.

**chiral**  Compounds having nonsuperimposable mirror images.

**chirality**  Property of a molecule that is not superimposable on its mirror image.

**cis isomer**  (also called Z *isomer*) Molecule with two like groups (such as two R groups or two hydrogen atoms) on the same side of the molecule.

**Clausius–Clapeyron equation**  Relates the vapor pressures of a substance at different temperatures to its heat of vaporization.

**closed system**  A system that exchanges energy but not matter with the surroundings.

**codon**  A three-nucleotide sequence that codes for a specific amino acid.

**coenzyme**  Organic molecule that, like an enzyme, accelerates the rate of biochemical reactions.

**colligative properties**  Characteristics of solutions that depend on the concentration and not the identity of particles dissolved in the solvent.

**combination reaction**  A reaction in which two (or more) substances combine to form one product.

**combined gas law**  (also called *general gas equation*) Relates the pressure, volume, and temperature of a quantity of an ideal gas:

$$\frac{P_1V_1}{T_1} = \frac{P_2V_2}{T_2}$$

**combustion analysis**  A laboratory procedure for determining the composition of a substance by burning it completely in oxygen to produce known compounds whose masses are used to determine the composition of the original material.

**combustion reaction**  A heat-producing reaction between oxygen and another element or compound.

**common-ion effect**  The shift in the position of an equilibrium caused by the addition of an ion taking part in the reaction.

**complex ion**  An ionic species consisting of a metal ion bonded to one or more Lewis bases.

**compound**  A pure substance that is composed of two or more elements bonded together in fixed proportions and that can be broken down into those elements by some chemical process.

**concentration**  The amount of a solute in a particular amount of solvent or solution.

**condensation polymer**  Macromolecule formed by the reaction of monomers yielding a polymer and water or other small molecule as a by-product of the reaction.

**condensation reaction**  Two molecules combining to form a larger molecule and a small molecule (typically water).

**conduction band**  In metals, an unoccupied band higher in energy than a valence band, in which electrons are free to migrate.

**conjugate acid**  Formed when a Brønsted–Lowry base accepts a $H^+$ ion.

**conjugate acid–base pair**  A Brønsted–Lowry acid and base differing from each other only by the presence or absence of a $H^+$ ion: acid $\rightleftharpoons$ conjugate base + $H^+$.

**conjugate base**  Formed when a Brønsted–Lowry acid donates a $H^+$ ion.

**constitutional isomers**  Molecules having the same molecular formula but different arrangements of atoms; they are different compounds and have different chemical and physical properties.

**conversion factor**  A fraction in which the numerator is equivalent to the denominator but is expressed in different units, making the value of the fraction one.

**coordinate bond**  A covalent bond formed when one anion or molecule donates a pair of electrons to another ion or molecule.

**coordination compound**   A compound made up of at least one complex ion.

**coordination number**   The number of electron pairs surrounding a metal ion in a complex.

**copolymer**   A macromolecule formed from the chemical combination of two different monomers.

**core electrons**   Electrons in the filled, inner shells in an atom or ion that are not involved in chemical reactions.

**corrosion**   A process in which a metal is oxidized by substances in its environment.

**counter ions**   Species that provide electrical balance of charges for complex ions in coordination compounds.

**covalent bond**   A bond between two atoms created by sharing one or more pairs of electrons.

**covalent network solid**   A solid consisting of atoms held together by extended arrays of covalent bonds.

**covalent radius**   (also called *atomic radius*) Half the distance between identical nuclear centers in a molecule.

**critical mass**   The minimum quantity of fissionable material needed to sustain a chain reaction.

**critical point**   A specific temperature and pressure at which the liquid and gas phases of a substance have the same density and are indistinguishable from each other.

**critical temperature ($T_c$)**   The temperature below which a material becomes a superconductor.

**crude oil**   A combustible liquid mixture of hydrocarbons and other organic molecules formed under Earth's surface.

**crystal field splitting**   The separation of a set of $d$ orbitals into subsets with different energies as a result of interactions between electrons in those orbitals and lone pairs of electrons in ligands.

**crystal field splitting energy ($\Delta$)**   The difference in energy between subsets of $d$ orbitals split by interactions in a crystal field.

**crystal lattice**   A three-dimensional array of particles (atoms, ions, or molecules) in a crystalline solid.

**crystal structure**   An ordered arrangement in three-dimensional space of the particles (atoms, ions, or molecules) that make up a crystalline solid.

**crystalline solid**   A solid made of an ordered array of atoms, ions, or molecules.

**cubic closest-packed (ccp)**   A crystal structure composed of face-centered cubic unit cells and layers of particles having an *abcabc . . .* stacking pattern.

**curie (Ci)**   Non-SI unit of radioactivity; 1 Ci = $3.70 \times 10^{10}$ decay events per second.

**cycloalkane**   Ring-containing alkane with the general formula $C_nH_{2n}$.

## D

**dalton (Da)**   A unit of mass identical to 1 atomic mass unit.

**Dalton's law of partial pressures**   The total pressure of any mixture of gases equals the sum of the partial pressures of all the gases in the mixture.

**dark-line spectrum**   (also called *atomic absorption spectrum*) A characteristic series of dark lines produced when free, gaseous atoms are illuminated by an external source of radiation.

**degenerate**   Describes orbitals of the same energy.

**degree of ionization**   The ratio of the quantity of a substance that is ionized to the concentration of the substance before ionization; when expressed as a percentage, called *percent ionization*.

**delocalization**   (*adjective*: delocalized) The sharing of bonding pairs of electrons in alternating single and double bonds by three or more atoms in a molecule.

**density ($d$)**   The ratio of the mass ($m$) of an object to its volume ($V$).

**deposition**   Transformation of a vapor (gas) directly into a solid.

**diamagnetic**   Describes a substance with no unpaired electrons that is weakly repelled by a magnetic field.

**diffraction**   Bending of electromagnetic radiation as it passes around an edge of an object or through a narrow opening.

**diffusion**   The spread of one substance (usually a gas or liquid) through another.

**dilution**   The process of lowering the concentration of a solution by adding more solvent.

**dipole moment ($\mu$)**   A measure of the degree to which a molecule aligns itself in a strong electric field; a quantitative expression of the polarity of a molecule.

**dipole–dipole interaction**   An attractive force between polar molecules.

**dispersion force**   (also called *London force*) An intermolecular force between nonpolar molecules caused by the presence of temporary dipoles in the molecules.

**distillation**   A separation technique in which the more *volatile* (more easily vaporized) components of a mixture are vaporized and then condensed, thereby separating them from the less volatile components.

**double bond**   A chemical bond in which two atoms share two pairs of electrons.

## E

**$E$ isomer**   (also called *trans isomer*) Molecule with two like groups (such as two R groups or two hydrogen atoms) on opposite sides of the molecule.

**effective nuclear charge ($Z_{eff}$)**   The attractive force toward the nucleus experienced by an electron in an atom; the positive charge on the nucleus reduced by the extent to which other electrons in the atom shield the electron from the nucleus.

**effusion**   The process by which a gas escapes from its container through a tiny hole into a region of lower pressure.

**electrochemical cell**   An apparatus that converts chemical energy into electrical work or electrical work into chemical energy.

**electrochemistry**   The branch of chemistry that examines the transformations between chemical and electrical energy.

**electrolysis**   A process in which electrical energy is used to drive a nonspontaneous chemical reaction.

**electrolyte**   A substance that dissociates into ions when it dissolves, enhancing the conductivity of the solvent.

**electrolytic cell**   A device in which an external source of electrical energy does work on a chemical system, turning reactant(s) into higher-energy product(s).

**electromagnetic radiation**   Any form of radiant energy in the electromagnetic spectrum.

**electromagnetic spectrum**   A continuous range of radiant energy that includes radio waves, infrared radiation, visible light, ultraviolet radiation, X-rays, and gamma rays.

**electron**   A subatomic particle that has a negative charge and essentially zero mass.

**electron affinity (EA)**   The energy change that occurs when 1 mole of electrons combines with 1 mole of atoms or ions in the gas phase.

**electron capture**   A nuclear reaction in which a neutron-poor nucleus draws in one of its surrounding electrons, which transforms a proton in the nucleus into a neutron.

**electron configuration** The distribution of electrons among the orbitals of an atom or ion.

**electron transition** Movement of an electron between energy levels.

**electronegativity** A relative measure of the ability of an atom to attract electrons in a bond to itself.

**electron-pair geometry** The three-dimensional arrangement of bonding pairs and lone pairs of electrons about a central atom.

**electrostatic potential energy ($E_{el}$)** The energy a particle has because of its electrostatic charge and its position relative to another particle; it is directly proportional to the product of the charges of the particles and inversely proportional to the distance between them.

**element** A pure substance that cannot be separated into simpler substances by any chemical process.

**elementary step** A molecular view of a single process taking place in a chemical reaction.

**empirical formula** A formula showing the smallest whole-number ratio of elements in a compound.

**enantiomer** One of a pair of optical isomers of a compound.

**end point** The point in a titration that is reached when just enough standard solution has been added to cause the indicator to change color.

**endothermic process** One in which energy flows from the surroundings into the system.

**energy profile** A graph showing the changes in potential energy for a reaction as a function of the progress of the reaction from reactants to products.

**energy** The capacity to do work.

**enthalpy ($H$)** The sum of the internal energy and the pressure–volume product of a system; $H = E + PV$.

**enthalpy change ($\Delta H$)** The heat absorbed by an endothermic process or given off by an exothermic process occurring at constant pressure.

**enthalpy of hydration ($\Delta H_{hydration}$)** The energy change when gas-phase ions dissolve in a solvent.

**enthalpy of reaction ($\Delta H_{rxn}$)** The energy absorbed or given off by a chemical reaction; also called *heat of reaction*.

**enthalpy of solution ($\Delta H_{solution}$)** The overall energy change when a solute is dissolved in a solvent.

**entropy ($S$)** A measure of how dispersed the energy in a system is at a specific temperature.

**enzyme** A protein that catalyzes a reaction.

**equilibrium constant ($K$)** The value of the ratio of concentration (or partial pressure) terms in the equilibrium constant expression at a specific temperature.

**equilibrium constant expression** The ratio of the equilibrium concentrations or partial pressures of products to reactants, each term raised to a power equal to the coefficient of that substance in the balanced chemical equation for the reaction.

**equivalence point** The point in a titration at which the number of moles of titrant added is stoichiometrically equal to the number of moles of the substance being analyzed.

**essential amino acid** Any of the 8 amino acids that make up peptides and proteins but are not synthesized in the human body and must be obtained through the food we eat.

**essential element** Element present in tissue, blood, or other body fluids that has a physiological function.

**ester** Organic compound in which the –OH of a carboxylic acid group is replaced by –OR, where R can be any organic group.

**ether** Organic compound with the general formula R—O—R′, where R is any alkyl group or aromatic ring; the two R groups may be different.

**excited state** Any energy state above the ground state in an atom or ion.

**exothermic process** One in which energy flows from a system into its surroundings.

**extensive property** A property that varies with the quantity of the substance present.

## F

**face-centered cubic (fcc) unit cell** An array of closest-packed particles that has one particle at each of the eight corners of a cube and one more at the center of each face of the cube.

**Faraday constant ($F$)** The magnitude of electrical charge in 1 mole of electrons. Its value to three significant figures is $9.65 \times 10^4$ C/mol e$^-$.

**fat** Solid triglyceride containing primarily saturated fatty acids.

**filtration** A process for separating particles suspended in a liquid or a gas from that liquid or gas by passing the mixture through a medium that retains the particles.

**first law of thermodynamics** The energy gained or lost by a system must equal the energy lost or gained by the surroundings.

**food value** The quantity of energy produced when a material consumed by an organism for sustenance is burned completely; it is typically reported in Calories (kilocalories) per gram of food.

**formal charge (FC)** A value calculated for an atom in a molecule or polyatomic ion by determining the difference between the number of valence electrons in the free atom and the sum of lone-pair electrons plus half of the electrons in the atom's bonding pairs.

**formation constant ($K_f$)** An equilibrium constant describing the formation of a metal complex from a free metal ion and its ligands.

**formation reaction** A reaction in which 1 mole of a substance is formed from its component elements in their standard states.

**formula mass** The mass of one formula unit of an ionic compound.

**formula unit** The smallest electrically neutral unit of an ionic compound.

**fractional distillation** A method of separating a mixture of compounds on the basis of their different boiling points.

**Fraunhofer lines** A set of dark lines in the otherwise continuous solar spectrum.

**free radical** An odd-electron molecule with an unpaired electron in its Lewis structure.

**frequency ($\nu$)** The number of crests of a wave that pass a stationary point of reference per second.

**frequency factor ($A$)** The product of the frequency of molecular collisions and a factor that expresses the probability that the orientation of the molecules is appropriate for a reaction to occur.

**fuel cell** A voltaic cell based on the oxidation of a continuously supplied fuel. The reaction is the equivalent of combustion, but chemical energy is converted directly into electrical energy.

**fuel density** The amount of energy released during the complete combustion of 1 liter of a liquid fuel.

**fuel value** The energy released during complete combustion of 1 g of a substance.

**functional group** A structural subunit in organic molecules that imparts characteristic chemical and physical properties.

## G

**gas** A form of matter that has neither definite volume nor shape and that expands to fill its containers; also called *vapor*.

**Geiger counter** A portable device for determining nuclear radiation levels by measuring how much the radiation ionizes the gas in a sealed detector.

**general gas equation** (also called *combined gas law*) Relates the pressure, volume, and temperature of a quantity of an ideal gas:

$$\frac{P_1V_1}{T_1} = \frac{P_2V_2}{T_2}$$

**Gibbs free energy (G)** The maximum energy released by a process occurring at constant temperature and pressure that is available to do useful work.

**glyceride** Lipid consisting of esters formed between fatty acids and the alcohol glycerol.

**glycolysis** A series of reactions that converts glucose into pyruvate; a major anaerobic (no oxygen required) pathway for the metabolism of glucose in the cells of almost all living organisms.

**glycosidic bond** A C—O—C bond between sugar molecules.

**Graham's law of effusion** The rate of effusion of a gas is inversely proportional to the square root of its molar mass.

**gray (Gy)** The SI unit of absorbed radiation; 1 Gy = 1 J/kg of tissue.

**ground state** The most stable, lowest-energy state available to an atom or ion.

**group** All elements in the same column of the periodic table; also called *family*.

# H

**half-life ($t_{1/2}$)** The time in the course of a chemical reaction during which the concentration of a reactant decreases by half.

**half-reaction** One of the two halves of an oxidation–reduction reaction; one half-reaction is the oxidation component, and the other is the reduction component.

**halogens** The elements in group 17 of the periodic table.

**heat capacity ($C_P$)** The quantity of energy needed to raise the temperature of an object 1°C at constant pressure.

**heat** The energy transferred between objects because of a difference in their temperatures.

**Heisenberg uncertainty principle** We cannot determine both the position and the momentum of an electron in an atom at the same time.

**Henderson–Hasselbalch equation** Used to calculate the pH of a solution in which the concentrations of acid and conjugate base are known.

**Henry's law** The concentration of a sparingly soluble, chemically unreactive gas in a liquid is proportional to the partial pressure of the gas.

**hertz (Hz)** The SI unit of frequency, measured in reciprocal seconds: 1 Hz = 1 s$^{-1}$ = 1 cycle per second (cps).

**Hess's law** The standard enthalpy of reaction $\Delta H_{rxn}$ for a reaction that is the sum of two or more reactions is equal to the sum of the $\Delta H_{rxn}$ values of the constituent reactions; also called *Hess's law of constant heat of summation.*

**heteroatom** Any atom other than carbon, hydrogen, or metals in an organic compound.

**heterogeneous catalyst** A catalyst in a different phase from the reactants.

**heterogeneous equilibria** Involve reactants and products in more than one phase.

**heterogeneous mixture** A mixture in which the components are not distributed uniformly, so that the mixture contains distinct regions of different compositions.

**heteropolymer** A polymer made of three or more different monomer units.

**hexagonal closest-packed (hcp)** A crystal lattice in which the layers of atoms or ions in hexagonal unit cells have an *ababab* . . . stacking pattern.

**hexagonal unit cell** An array of closest-packed particles that includes parts of four particles on the top and four on the bottom faces of a hexagonal prism and one particle in a middle layer.

**homogeneous catalyst** A catalyst in the same phase as the reactants.

**homogeneous equilibria** Involve reactants and products in the same phase.

**homogeneous mixture** A mixture in which the components are distributed uniformly throughout and have no visible boundaries or regions.

**homologous series** A set of related organic compounds that differ from one another by the number of common subgroups, such as –CH$_2$–, in their molecular structures.

**homopolymer** A polymer composed of only one kind of monomer unit.

**Hund's rule** The lowest-energy electron configuration of an atom has the maximum number of unpaired electrons, all of which have the same spin, in degenerate orbitals.

**hybrid atomic orbital** In valence bond theory, one of a set of equivalent orbitals about an atom created when specific atomic orbitals are mixed.

**hybridization** In valence bond theory, the mixing of atomic orbitals to generate new sets of orbitals that are then available to form covalent bonds with other atoms.

**hydrocarbons** A class of organic compounds composed of only hydrogen and carbon.

**hydrogen bond** The strongest dipole–dipole interaction. It occurs between a hydrogen atom bonded to a small, highly electronegative element (O, N, F) and an atom of oxygen or nitrogen in another molecule. Molecules of HF also form hydrogen bonds.

**hydrogenation** The reaction of an unsaturated hydrocarbon with hydrogen.

**hydrolysis** The reaction of water with another material. The hydrolysis of nonmetal oxides produces acids

**hydronium ion ($H_3O^+$)** A H$^+$ ion bonded to a molecule of water, $H_2O$; the form in which the hydrogen ion is found in an aqueous solution.

**hydrophilic** A "water-loving" or attractive interaction between a solute and water that promotes water solubility.

**hydrophobic** A "water-fearing" or repulsive interaction between a solute and water that diminishes water solubility.

**hypothesis** A tentative and testable explanation for an observation or a series of observations.

# I

***i* factor** (also called *van 't Hoff factor*) The ratio of the experimentally measured value of a colligative property to the theoretical value expected for that property if the solute were a nonelectrolyte.

**ideal gas** A gas whose behavior is predicted by the linear relations defined by Boyle's, Charles's, Avogadro's, and Amontons's laws.

**ideal gas equation** (also called *ideal gas law*) Relates the pressure, volume, number of moles, and temperature of an ideal gas; expressed as $PV = nRT$, where $R$ is the universal gas constant.

**ideal gas law** (also called *ideal gas equation*) Relates the pressure, volume, number of moles, and temperature of an ideal gas; expressed as $PV = nRT$, where $R$ is the universal gas constant.

**ideal solution** One that obeys Raoult's law.

**induced dipole** (also called *temporary dipole*) The separation of charge produced in an atom or molecule by a momentary uneven distribution of electrons.

**inhibitor** A compound that diminishes or destroys the ability of an enzyme to catalyze a reaction.

**initial rate** The rate of a reaction at $t = 0$, immediately after the reactants are mixed.

**inner coordination sphere** The ligands that are bound directly to a metal via coordinate bonds.

**instantaneous rate** The rate of a reaction at a specific instant during the course of the reaction.

**integrated rate law** A mathematical expression that describes the change in concentration of a reactant in a chemical reaction with time.

**intensive property** A property that is independent of the amount of substance present.

**interference** The interaction of waves that results in either reinforcing their amplitudes (constructive interference) or canceling them out (destructive interference).

**intermediate** A species produced in one step of a reaction and consumed in a subsequent step.

**internal energy** ($E$) The sum of all the kinetic and potential energies of all of the components of a system.

**interstitial alloy** Atoms of the added element occupy the spaces between atoms of the host.

**ion** An atom or group of atoms that has a net positive or negative charge.

**ion channel** Group of helical proteins that penetrate cell membranes and allow selective transport of ions.

**ion exchange** A process by which one ion is displaced by another.

**ion pair** A cluster formed when a cation and an anion associate with each other in solution.

**ion pump** System of membrane proteins that exchange ions inside the cell with those in the intercellular fluid.

**ion–dipole interaction** An attractive force between an ion and a molecule that has a permanent dipole moment.

**ionic bond** A chemical bond that results from the electrostatic attraction between cations and anions.

**ionic character** An estimate of the magnitude of charge separation in a covalent bond.

**ionic compound** A compound composed of positively and negatively charged ions held together by electrostatic attraction.

**ionic radius** Radius derived from the distance between nuclear centers in ionic crystals.

**ionic solid** A solid consisting of monatomic or polyatomic ions held together by ionic bonds.

**ionization energy** (IE) The amount of energy needed to remove 1 mole of electrons from 1 mole of ground-state atoms or ions in the gas phase.

**ionizing radiation** High-energy products of radioactive decay that can ionize molecules.

**isoelectric point** (also called $p$I) The pH at which molecules of the amino acid have, on average, zero charge.

**isoelectronic** Describes atoms or ions that have identical electron configurations.

**isolated system** A system that exchanges neither energy nor matter with the surroundings.

**isotopes** Atoms of an element containing different numbers of neutrons.

**J**

**joule (J)** The SI unit of energy; 4.184 J = 1 cal.

**K**

**Kekulé structure** A structure showing all of the bonds in a covalently bonded molecule using lines but not showing lone pairs on the atoms.

**kelvin (K)** The SI unit of temperature.

**ketone** Organic molecule containing a carbonyl group bonded to two R groups; its general formula is RC(O)R′.

**kinetic energy** (KE) The energy of an object in motion due to its mass ($m$) and its speed ($u$): $\mathrm{KE} = \frac{1}{2}mu^2$.

**kinetic molecular theory** A model that describes the behavior of gases; all equations defining relationships between pressure, volume, temperature, and number of moles of gases can be derived from the theory.

**L**

**lattice energy** ($U$) The enthalpy change that occurs when 1 mole of an ionic compound forms from its free ions in the gas phase.

**law of conservation of energy** Energy cannot be created or destroyed but can be converted from one form into another.

**law of conservation of mass** The sum of the masses of the reactants in a chemical reaction is equal to the sum of the masses of the products.

**law of constant composition** All samples of a particular compound contain the same elements combined in the same proportions.

**law of mass action** The ratio of the concentrations or partial pressures of products to reactants at equilibrium has a characteristic value at a given temperature when each term is raised to a power equal to the coefficient of that substance in the balanced chemical equation for the reaction.

**law of multiple proportions** The ratio of the two masses of one element that react with a given mass of another element to form two different compounds is the ratio of two small whole numbers.

**Le Châtelier's principle** A system at equilibrium responds to a stress in such a way that it relieves that stress.

**leveling effect** The observation that strong acids all have the same strength in water and are completely converted into solutions of $H_3O^+$ ions; strong bases are likewise leveled in water and are completely converted into solutions of $OH^-$ ions.

**Lewis acid** A substance that *accepts* a lone pair of electrons in a chemical reaction.

**Lewis base** A substance that *donates* a lone pair of electrons in a chemical reaction.

**Lewis dot symbol** (also called *Lewis symbol*) The chemical symbol for an atom surrounded by one or more dots representing the valence electrons.

**Lewis structure** A two-dimensional representation of the bonds and lone pairs of valence electrons in a molecule or polyatomic ion.

**Lewis symbol** (also called *Lewis dot symbol*) The chemical symbol for an atom surrounded by one or more dots representing the valence electrons.

**ligand** A Lewis base bonded to the central metal ion of a complex ion.

**limiting reactant** A reactant that is consumed completely in a chemical reaction. The amount of product formed depends on the amount of the limiting reactant available.

**lipid** A class of water-insoluble, oily organic compounds that are common structural materials in cells.

**lipid bilayer** A double layer of molecules whose polar head groups interact with water molecules and whose nonpolar tails interact with each other.

**liquid** A form of matter that occupies a definite volume but flows to assume the shape of its containers.

**London force** (also called *dispersion force*) An intermolecular force between nonpolar molecules caused by the presence of temporary dipoles in the molecules.

**lone pair** A pair of electrons that is not shared.

# M

**macrocyclic ligand** A ring containing multiple electron-pair donors that bind to a metal ion.

**magnetic quantum number ($m_\ell$)** Defines the orientation of an orbital in space; an integer that may have any value from $-\ell$ to $+\ell$, where $\ell$ is the angular momentum quantum number.

**main group elements** (also called *representative elements*) The elements in groups 1, 2, and 13 through 18 of the periodic table.

**major essential element** Essential element present in the body in average concentrations greater than 1 mg of element per gram of body mass.

**manometer** An instrument for measuring the pressure exerted by a gas.

**mass** The property that defines the quantity of matter in an object.

**mass action expression** Equivalent to the equilibrium constant expression, but applied to reaction mixtures that may, or may not, be at equilibrium.

**mass defect ($\Delta m$)** The difference between the mass of a stable nucleus and the masses of the individual nucleons that comprise it.

**mass number ($A$)** The number of nucleons in an atom.

**mass spectrometer** An instrument that separates and counts ions according to their mass.

**mass spectrum** A graph of the data from a mass spectrometer, where $m/z$ ratios of the deflected particles are plotted against the number of particles with a particular mass.

**matter** Anything that has mass and occupies space.

**matter wave** The wave associated with any particle.

**meniscus** The concave or convex surface of a liquid in a small-diameter tube.

**messenger RNA (mRNA)** The form of RNA that carries the code for synthesizing proteins from DNA to the site of protein synthesis in a cell.

**metallic bond** A chemical bond consisting of metal atoms surrounded by a "sea" of shared electrons.

**metallic radius** Half the distance between nuclear centers in the crystal of a metal.

**metalloids** (also called *semimetals*) Elements along the border between metals and nonmetals in the periodic table; they have some metallic and some nonmetallic properties.

**metals** The elements on the left side of the periodic table that are typically shiny solids that conduct heat and electricity well and are malleable and ductile.

**meter** The standard unit of length, named after the Greek *metron*, which means "measure," and equivalent to 39.37 inches.

**methanogenic bacteria** Bacteria using simple organic compounds and hydrogen for energy; their respiration produces methane, carbon dioxide, and water, depending on the compounds they consume.

**methyl group** ($-CH_3$) A structural unit that can make only one bond.

**methylene group** ($-CH_2-$) A structural unit that can make two bonds.

**microstate** A unique distribution of particles among energy levels.

**millimeters of mercury (mmHg)** (also called *torr*) A unit of pressure where 1 atm = 760 torr = 760 mmHg.

**miscible** Capable of being mixed in any proportion (without reacting chemically).

**mixture** A combination of pure substances in variable proportions in which the individual substances retain their chemical identities and can be separated from one another by a physical process.

**molality ($m$)** Concentration expressed as the number of moles of solute per kilogram of solvent.

**molar absorptivity** A measure of how well a compound or ion absorbs light.

**molar heat capacity ($c_P$)** The energy required at constant pressure to raise the temperature of 1 mole of a substance by 1°C.

**molar heat of fusion ($\Delta H_{fus}$)** The energy required to convert 1 mole of a solid substance at its melting point into the liquid state.

**molar heat of vaporization ($\Delta H_{vap}$)** The energy required to convert 1 mole of a liquid substance at its boiling point to the vapor state.

**molar mass ($\mathcal{M}$)** The mass of 1 mole of a substance. The molar mass of an element in grams per mole is numerically equal to that element's average atomic mass in atomic mass units.

**molar volume** Volume occupied by 1 mole of an ideal gas at STP; 22.4 L.

**molarity ($M$)** The concentration of a solution expressed in moles of solute per liter of solution ($M = n/V$).

**mole (mol)** An amount of material (atoms, ions, or molecules) that contains Avogadro's number ($N_A = 6.022 \times 10^{23}$) of particles.

**mole fraction ($X_x$)** The ratio of the number of moles of a component in a mixture to the total number of moles in the mixture.

**molecular compound** A compound composed of atoms held together in molecules by covalent bonds.

**molecular equation** A balanced equation describing a reaction in solution in which the reactants and products are written as undissociated molecules.

**molecular formula** A notation showing the number and type of atoms present in one molecule of a molecular compound.

**molecular geometry** The three-dimensional arrangement of the atoms in a molecule.

**molecular ion ($M^+$)** The peak of highest mass in a mass spectrum; it has the same mass as the molecule from which it came.

**molecular mass** The mass of one molecule of a molecular compound.

**molecular orbital** A region of characteristic shape and energy where electrons in a molecule are delocalized over two or more atoms in a molecule.

**molecular orbital diagram** In MO theory, an energy-level diagram showing the relative energies and electron occupancy of the molecular orbitals for a molecule.

**molecular orbital (MO) theory** A bonding theory based on the mixing of atomic orbitals of similar shapes and energies to form molecular orbitals that extend to two or more atoms.

**molecular recognition** The process by which molecules interact with other molecules in living tissues to produce a biological effect.

**molecular solid** A solid formed by neutral, covalently bonded molecules held together by intermolecular attractive forces.

**molecularity** The number of ions, atoms, or molecules involved in an elementary step in a reaction.

**molecule** A collection of atoms chemically bonded together in characteristic proportions.

**monodentate ligand** A species that forms only a single coordinate bond to a metal ion in a complex.

**monomer** A small molecule that bonds with others like it to form polymers.

**monoprotic acid** Has one ionizable hydrogen atom per molecule.

**monosaccharide** A single-sugar unit and the simplest carbohydrate.

# N

**nanoparticle**   Approximately spherical samples of matter with dimensions less than 100 nanometers ($1 \times 10^{-7}$ m).

**natural abundance**   The proportion of a particular isotope, usually expressed as a percentage, relative to all the isotopes of that element in a natural sample.

**Nernst equation**   An equation relating the potential of a cell (or half-cell) reaction to its standard potential ($E°$) and to the concentrations of its reactants and products.

**net ionic equation**   A balanced equation that describes the actual reaction taking place in aqueous solution; it is obtained by eliminating the spectator ions from the overall ionic equation.

**neutralization reaction**   A reaction that takes place when an acid reacts with a base and produces a solution of a salt in water.

**neutron**   An electrically neutral (uncharged) subatomic particle found in the nucleus of an atom.

**noble gases**   The elements in group 18 of the periodic table.

**node**   A location in a standing wave that experiences no displacement.

**nonelectrolyte**   A substance that does not dissociate into ions and therefore does not enhance the conductivity of water when dissolved.

**nonessential element**   Element present in humans that has no known function.

**nonmetals**   Elements with properties opposite those of metals, including poor conductivity of heat and electricity.

**nonpolar covalent bond**   A bond characterized by an even distribution of charge; electrons in the bonds are shared equally by the two atoms.

**nonspontaneous**   A process that occurs only as long as energy is continually added to the system.

**normal boiling point**   The temperature at which the vapor pressure of a liquid equals 1 atm (760 torr).

**n-type semiconductor**   Semiconductor containing electron-rich dopant atoms that contribute excess electrons.

**nuclear fission**   A nuclear reaction in which the nucleus of an element splits into two lighter nuclei. The process is usually accompanied by the release of one or more neutrons and energy.

**nuclear fusion**   A nuclear reaction in which subatomic particles or atomic nuclei collide with each other at very high speeds and fuse together, forming more massive nuclei and releasing energy.

**nuclear reaction**   A process that alters the number of neutrons or protons in the nucleus of an atom.

**nucleic acid**   One of a family of large molecules, which includes deoxyribonucleic acid (DNA) and ribonucleic acid (RNA), that stores the genetic blueprint of an organism and controls the production of proteins.

**nucleon**   Either a proton or a neutron in a nucleus.

**nucleosynthesis**   The natural formation of nuclei as a result of fusion and other nuclear processes.

**nucleotide**   A monomer unit from which nucleic acids are made.

**nucleus**   (of an atom) The positively charged center of an atom that contains nearly all the atom's mass.

**nuclide**   The nucleus of a specific isotope of an element.

# O

**octahedral**   Molecular geometry about a central atom with a steric number of 6 and no lone pairs of electrons, in which all six sites are equivalent.

**octet rule**   Atoms of main group elements make bonds by gaining, losing, or sharing electrons to achieve a valence shell containing eight electrons, or four electron pairs.

**oil**   Liquid triglyceride containing primarily unsaturated fatty acids.

**open system**   A system that exchanges both energy and matter with the surroundings.

**optical isomers**   Molecules that are not superimposable on their mirror images.

**optically active molecule**   A chiral compound that causes rotation of a beam of plane-polarized light when it passes through a solution of the compound.

**orbital**   A region around the nucleus of an atom where the probability of finding an electron is high; each orbital is defined by the square of the wave function ($\psi^2$) and identified by a unique combination of three quantum numbers.

**orbital diagram**   Depiction of the arrangement of electrons in an atom or ion using boxes to represent orbitals.

**orbital penetration**   The probability that an electron in an outer orbital will be as close to the nucleus as an electron in an inner shell.

**organic chemistry**   The study of carbon-containing compounds, typically those with C—C and C—H bonds.

**organometallic compound**   A molecule containing direct carbon–metal covalent bonds.

**osmosis**   The flow of a fluid through a semipermeable membrane to balance the concentration of solutes in solutions on the two sides of the membrane. The solvent molecules' flow proceeds from the more dilute solution into the more concentrated one.

**osmotic pressure ($\Pi$)**   The pressure applied across a semipermeable membrane to stop the flow of solvent from the compartment containing pure solvent or a less concentrated solution to the compartment containing a more concentrated solution. The osmotic pressure of a solution increases with solute concentration, $M$, and with solution temperature, $T$.

**overall ionic equation**   A balanced equation that shows all the species, both ionic and molecular, present in a reaction occurring in aqueous solution.

**overall reaction order**   The sum of the exponents of the concentration terms in the rate law.

**overlap**   A term in valence bond theory describing bonds arising from two orbitals on different atoms that occupy the same region of space.

**oxidation**   A chemical change in which a species loses electrons; the oxidation number of the species increases.

**oxidation number (O.N.)**   (also called *oxidation state*) A positive or negative number based on the number of electrons an atom gains or loses when it forms an ion, or that it shares when it forms a covalent bond with another element; pure elements have an oxidation number of zero.

**oxidation state**   (also called *oxidation number [O.N.]*) A positive or negative number based on the number of electrons an atom gains or loses when it forms an ion, or that it shares when it forms a covalent bond with another element; pure elements have an oxidation number of zero.

**oxidizing agent**   A substance in a redox reaction that contains the element being reduced.

**oxoanions**   Polyatomic ions that contain oxygen in combination with one or more other elements.

# P

**packing efficiency**   Percentage of the total volume of a unit cell occupied by the spheres.

**paramagnetic**   Describes a substance with unpaired electrons that is attracted to a magnetic field.

**partial pressure**   The contribution to the total pressure made by a component in a mixture of gases.

**partial reaction order**   The order of a reaction with respect to a single reactant.

**Pauli exclusion principle**   No two electrons in an atom can have the same set of four quantum numbers.

**peptide**   A compound of two or more amino acids joined by peptide bonds. Small peptides containing up to 20 amino acids are *oligopeptides*; and the term *polypeptide* is used for chains longer than 20 amino acids but shorter than proteins.

**peptide bond**   The result of a condensation reaction between the carboxylic acid group of one amino acid and the amine group of another.

**percent composition**   The composition of a compound expressed in terms of the percentage by mass of each element in the compound.

**percent ionization**   The ratio of the quantity of a substance that is ionized to the concentration of the substance before ionization, expressed as a percentage.

**percent yield**   The ratio, expressed as a percentage, of the actual yield of a chemical reaction to the theoretical yield.

**period**   A horizontal row in the periodic table.

**periodic table of the elements**   A chart of the elements in order of their atomic numbers and in a pattern based on their physical and chemical properties.

**pH**   The negative logarithm of the hydrogen ion concentration in an aqueous solution.

**pH buffer**   A solution that resists changes in pH when acids or bases are added to it; typically a solution of a weak acid and its conjugate base.

**pH indicator**   A water-soluble weak organic acid that changes color as pH changes.

**phase diagram**   A graphical representation of how the stabilities of the physical states of a substance depend on temperature and pressure.

**phospholipid**   A molecule of glycerol with two fatty acid chains and one polar group containing a phosphate; phospholipids are major constituents of cell membranes.

**phosphorylation**   A reaction resulting in the addition of a phosphate group to an organic molecule.

**photochemical smog**   A mixture of gases formed in the lower atmosphere when sunlight interacts with compounds produced in internal combustion engines and other pollutants.

**photoelectric effect**   The phenomenon of light striking a metal surface and producing an electric current (a flow of electrons).

**photon**   A quantum of electromagnetic radiation.

**physical process**   A transformation of a sample of matter, such as a change in its physical state, that does not alter the chemical identity of any substance in the sample.

**physical property**   A property of a substance that can be observed without changing it into another substance.

**p$I$**   (also called *isoelectric point*) The pH at which molecules of the amino acid have, on average, zero charge.

**pi (π) bond**   A covalent bond in which electron density is greatest around—not along—the bonding axis.

**pi (π) molecular orbitals**   In MO theory, molecular orbitals formed by the mixing of atomic orbitals oriented above and below, or in front of and behind, the bonding axis.

**Planck constant ($h$)**   The proportionality constant between the energy and frequency of electromagnetic radiation expressed in $E = h\nu$; $h = 6.626 \times 10^{-34}$ J · s.

**pOH**   The negative logarithm of the hydroxide ion concentration in an aqueous solution.

**polar covalent bond**   Unequal sharing of bonding pairs of electrons between atoms.

**polarizability**   The relative ease with which the electron cloud in a molecule, ion, or atom can be distorted, inducing a temporary dipole.

**polyatomic ions**   Charged groups of two or more atoms joined together by covalent bonds.

**polydentate ligand**   A species that can form more than one coordinate bond per molecule.

**polymer**   A very large molecule with high molar mass formed by bonding together a large number of small molecules of low molecular mass.

**polyprotic acid**   Has two or more ionizable hydrogen atoms per molecule.

**polysaccharide**   A polymer of monosaccharides.

**porphyrin**   A type of tetradentate macrocyclic ligand.

**positron ($^{0}_{1}\beta$)**   A particle with the mass of an electron but with a positive charge.

**positron emission**   The spontaneous emission of a positron from a neutron-poor nucleus.

**potential energy (PE)**   The energy stored in an object because of its position.

**precipitate**   A solid product formed from a reaction in solution.

**precipitation reaction**   A reaction that produces an insoluble product upon mixing two solutions.

**precision**   The extent to which repeated measurements of the same variable agree.

**pressure ($P$)**   The ratio of force to surface area over which the force is applied.

**pressure–volume ($P$–$V$) work**   The work associated with the expansion or compression of a gas.

**primary (1°) structure**   The sequence in which amino acid monomers occur in a protein chain.

**principal quantum number ($n$)**   A positive integer describing the relative size and energy of an atomic orbital or group of orbitals in an atom.

**product**   A substance formed during a chemical reaction.

**protein**   Biological polymer made of amino acids.

**proton**   A positively charged subatomic particle present in the nucleus of an atom.

**pseudo-first-order**   A reaction in which all the reactants but one are present at such high concentrations that they do not decrease significantly during the course of the reaction, so that reaction rate is controlled by the concentration of the limiting reactant.

**p-type semiconductor**   Semiconductor containing electron-poor dopant atoms that cause a reduction in the number of electrons, which is equivalent to the presence of positively charged holes.

## Q

**quantized**   Having values restricted to whole-number multiples of a specific base value.

**quantum**   (plural *quanta*) The smallest discrete quantity of a particular form of energy.

**quantum mechanics**   (also called *wave mechanics*) A mathematical description of the wavelike behavior of particles on the atomic level.

**quantum number**   A number that specifies the energy and location of an electron in an atom.

**quantum theory**   A model based on the idea that energy is absorbed and emitted in discrete quantities of energy called quanta.

**quarks**   Elementary particles that combine to form neutrons and protons.

**quaternary (4°) structure**   The larger structure functioning as a single unit that results when two or more proteins associate.

## R

**R**   Symbol in a general formula representing an organic group that has one available bond; it is used to indicate the variable part of a molecule so that the focus is placed on the functional group.

**racemic mixture**   A sample containing equal amounts of both optical isomers of a compound.

**radioactive decay**   The spontaneous disintegration of unstable particles accompanied by the release of radiation.

**radioactivity**   The spontaneous emission of high-energy radiation and particles by materials.

**radiocarbon dating**   A method for establishing the age of a carbon-containing object by measuring the amount of radioactive carbon-14 remaining in the object.

**radiometric dating**   A method for determining the age of an object based on the quantity of a radioactive nuclide and/or the products of its decay that the object contains.

**radionuclide**   An unstable nuclide that undergoes radioactive decay.

**random coil**   An irregular or rapidly changing part of the secondary structure of a protein.

**Raoult's law**   The vapor pressure of the solvent in a solution is equal to the vapor pressure of the pure solvent multiplied by the mole fraction of the solvent in the solution.

**rate constant**   The proportionality constant that relates the rate of a reaction to the concentrations of reactants.

**rate law**   An equation that defines the experimentally determined relation between the concentrations of reactants in a chemical reaction and the rate of that reaction.

**rate-determining step**   The slowest step in a multistep chemical reaction.

**reactant**   A substance consumed during a chemical reaction.

**reaction mechanism**   A set of steps that describe how a reaction occurs at the molecular level; the mechanism must be consistent with the rate law for the reaction.

**reaction order**   An experimentally determined number defining the dependence of the reaction rate on the concentration of a reactant.

**reaction quotient ($Q$)**   The numerical value of the mass action expression for *any values* of the concentrations (or partial pressures) of reactants and products; at equilibrium, $Q = K$.

**reaction rate**   How rapidly a reaction occurs; it is related to rates of change in the concentrations of reactants and products over time.

**receptor**   A cavity in a protein molecule that fits a particular molecule.

**recognition**   Process by which molecules in living systems interact with one another.

**reducing agent**   A substance in a redox reaction that contains the element being oxidized.

**reduction**   A chemical change in which a species gains electrons; the oxidation number of the species decreases.

**relative biological effectiveness (RBE)**   A factor that accounts for the differences in physical damage caused by different types of radiation.

**replication**   The process by which one double-stranded DNA forms two new DNA molecules, each one containing one strand from the original molecule and one new strand.

**representative elements**   (also called *main group elements*) The elements in groups 1, 2, and 13 through 18 of the periodic table.

**resonance**   A characteristic of electron distributions when two or more equivalent Lewis structures can be drawn for one compound.

**resonance structure**   One of two or more Lewis structures with the same arrangement of atoms but different arrangements of bonding pairs of electrons.

**reverse osmosis**   A process in which solvent is forced through semipermeable membranes, leaving a more concentrated solution behind.

**reversible process**   A process that can be run in the reverse direction in such a way that, once the system has been restored to its original state, no net energy has flowed either to the system or to its surroundings.

**root-mean-square speed ($u_{rms}$)**   The square root of the average of the squared speeds of all the molecules in a population of gas molecules; a molecule possessing the average kinetic energy moves at this speed.

## S

**salt**   The product of a neutralization reaction; it is made up of the cation of the base in the reaction plus the anion of the acid.

**saturated hydrocarbon**   An alkane.

**saturated solution**   A solution that contains the maximum concentration of a solute possible at a given temperature.

**Schrödinger wave equation**   A description of how the electron matter wave varies with location and time around the nucleus of a hydrogen atom.

**scientific method**   An approach to acquiring knowledge based on observation of phenomena, development of a testable hypothesis, and additional experiments that test the validity of the hypothesis.

**scientific theory (model)**   A general explanation of a widely observed phenomenon that has been extensively tested and validated.

**scintillation counter**   An instrument that determines the level of radioactivity in samples by measuring the intensity of light emitted by phosphors in contact with the samples.

**screening**   (also called *shielding*) The effect when inner-shell electrons prevent outer-shell electrons from experiencing the total nuclear charge.

**second law of thermodynamics**   The principle that the total entropy of the universe increases in any spontaneous process.

**secondary (2°) structure**   The pattern of arrangement of segments of a protein chain.

**seesaw**   Molecular geometry about a central atom with a steric number of 5 and one lone pair of electrons in an equatorial position.

**semiconductor**   A semimetal (metalloid) with electrical conductivity between that of metals and insulators that can be chemically altered to increase its electrical conductivity.

**semimetals**   (also called *metalloids*) Elements along the border between metals and nonmetals in the periodic table; they have some metallic and some nonmetallic properties.

**shielding**   (also called *screening*) The effect when inner-shell electrons prevent outer-shell electrons from experiencing the total nuclear charge.

**sievert (Sv)**   SI unit used to express the amount of biological damage caused by ionizing radiation.

**sigma (σ) bond**   A covalent bond in which the highest electron density lies between the two atoms along the bond axis.

**sigma (σ) molecular orbital**   In MO theory, a molecular orbital in which the greatest electron density is concentrated along an imaginary line drawn through the bonded atom centers.

**significant figures**   All the certain digits in a measured value plus one estimated digit. The greater the number of significant figures, the greater the certainty with which the value is known.

**simple cubic (sc) unit cell**   A cell with atoms or molecules only at the eight corners of a cube.

**single bond** A chemical bond that results when two atoms share one pair of electrons.

**solid** A form of matter that has a definite shape and volume.

**solubility product, $K_{sp}$** (also called *solubility-product constant*) An equilibrium constant that describes the formation of a saturated solution of a slightly soluble salt.

**solubility** The maximum amount of a substance that dissolves in a given quantity of solvent at a given temperature.

**solubility-product constant** (also called *solubility product, $K_{sp}$*) An equilibrium constant that describes the formation of a saturated solution of a slightly soluble salt.

**solute** Any component in a solution other than the solvent. A solution may contain one or more solutes.

**solution** Another name for homogeneous mixture. Solutions are often liquids, but they may also be solids or gases.

**solvent** The most abundant component of a solution in terms of moles.

**$sp$ hybrid orbitals** Two hybrid orbitals on opposite sides of the hybridized atom, formed by mixing one $s$ and one $p$ orbital.

**$sp^2$ hybrid orbitals** Three hybrid orbitals in a trigonal planar orientation, formed by mixing one $s$ and two $p$ atomic orbitals.

**$sp^3$ hybrid orbitals** A set of four hybrid orbitals with a tetrahedral orientation, formed by mixing one $s$ and three $p$ atomic orbitals.

**$sp^3d$ hybrid orbitals** Five equivalent hybrid orbitals with lobes pointing toward the vertices of a trigonal bipyramid, formed by mixing one $s$ orbital, three $p$ orbitals, and one $d$ orbital from the same shell.

**$sp^3d^2$ hybrid orbitals** Six equivalent orbitals that point toward the vertices of an octahedron, formed by mixing one $s$ orbital, three $p$ orbitals, and two $d$ orbitals from the same shell.

**specific heat ($c_s$)** The energy required to raise the temperature of 1 g of a substance 1°C at constant pressure.

**spectator ion** An ion that is present in a reaction vessel when a chemical reaction takes place but is unchanged by the reaction; spectator ions appear in an overall ionic equation but not in a net ionic equation.

**spectrochemical series** A list of ligands rank-ordered by their ability to split the energies of the $d$ orbitals of transition metal ions.

**spectrophotometry** An analytical method that relies on the absorption of light and Beer's law to measure the concentration of a solution.

**sphere of hydration** The cluster of water molecules surrounding an ion in aqueous medium; the general term applied to such a cluster forming in any solvent is *sphere of solvation*.

**spin magnetic quantum number ($m_s$)** Either $+\frac{1}{2}$ or $-\frac{1}{2}$, indicating that the spin orientation of an electron is either up or down.

**spontaneous** A process that occurs without outside intervention.

**square planar** Molecular geometry about a central atom with a steric number of 6 and two lone pairs of electrons that occupy axial sites; the atoms occupy four equatorial positions.

**square pyramidal** Molecular geometry about a central atom with a steric number of 6 and one lone pair of electrons; as typically drawn, the atoms occupy four equatorial and one axial site.

**standard cell potential ($E^\circ_{cell}$)** A measure of how forcefully an electrochemical cell, in which all reactants and products are in their standard states, can pump electrons through an external circuit.

**standard conditions** In thermodynamics: a pressure of 1 bar (~1 atm) and some specified temperature, assumed to be 25°C unless otherwise stated; for solutions, a concentration of 1 $M$ is specified.

**standard enthalpy of formation ($\Delta H^\circ_f$)** The enthalpy change of a formation reaction; also called *standard heat of formation* or *heat of formation*.

**standard enthalpy of reaction ($\Delta H^\circ_{rxn}$)** The energy associated with a reaction that takes place under standard conditions; also called *standard heat of reaction*.

**standard free energy of formation ($\Delta G^\circ_f$)** The change in free energy associated with the formation of 1 mole of a compound in its standard state from its component elements.

**standard hydrogen electrode (SHE)** A reference electrode based on the half-reaction $2\,H^+(aq) + 2\,e^- \rightarrow H_2(g)$ that produces a standard electrode potential of 0.000 V.

**standard molar entropy ($S^\circ$)** The absolute entropy of 1 mole of a substance in its standard state.

**standard reduction potential ($E^\circ$)** The potential of a reduction half-reaction in which all reactants and products are in their standard states at 25°C.

**standard solution** A solution of known concentration used in titrations.

**standard state** The most stable form of a substance under 1 bar pressure and some specified temperature (25°C unless otherwise stated).

**standard temperature and pressure (STP)** 0°C and 1 bar as defined by IUPAC; in the United States, 0°C and 1 atm.

**standing wave** A wave confined to a given space, with a wavelength ($\lambda$) related to the length $L$ of the space by $L = n(\lambda/2)$, where $n$ is a whole number.

**state function** A property of an entity based solely on its chemical or physical state or both, but not on how it achieved that state.

**stereoisomers** Molecules with the same formulas and the same connectivities between their atoms, but with different spatial arrangements of their atoms.

**steric number (SN)** The sum of the number of atoms bonded to a central atom and the number of lone pairs of electrons on the central atom.

**stimulatory effect** Increased growth or other biological response to the presence of a nonessential element.

**stock solution** A concentrated solution of a substance used to prepare solutions of lower concentration.

**stoichiometry** The quantitative relation between reactants and products in a chemical reaction.

**straight-chain alkane** A hydrocarbon in which the carbon atoms are bonded together in one continuous line. Linear alkane chains have a methyl group at each end with methylene groups connecting them.

**strong acid** An acid that completely dissociates into ions in aqueous solution.

**strong base** A base that completely dissociates into ions in aqueous solution.

**strong electrolyte** A substance that dissociates completely into ions when it dissolves in water.

**sublimation** Transformation of a solid directly into a vapor (gas).

**substance** Matter that has a constant composition and cannot be broken down to simpler matter by any physical process; also called *pure substance*.

**substitutional alloy** An alloy in which atoms of the nonhost metal replace host atoms in the crystal lattice.

**substrate** The reactant that binds to the active site in an enzyme-catalyzed reaction.

**superconductor** A material that has zero resistance to the flow of electric current.

**supercritical fluid** A substance at conditions above its critical temperature and pressure, where the liquid and vapor phases are indistinguishable and have some characteristics of both a liquid and a gas.

**supersaturated solution**   A solution that contains more than the maximum quantity of solute predicted to be soluble in a given volume of solution at a given temperature.

**surface tension**   The energy needed to separate the molecules at the surface of a liquid.

**surroundings**   Everything that is not part of the system.

**system**   The part of the universe that is the focus of a thermochemical study.

## T

**temporary dipole**   (also called *induced dipole*) The separation of charge produced in an atom or molecule by a momentary uneven distribution of electrons.

**termolecular step**   A step in a reaction mechanism involving a collision among three molecules.

**tertiary (3°) structure**   The three-dimensional, biologically active structure of a protein that arises because of interactions between the R groups on its amino acids.

**tetrahedral**   Molecular geometry about a central atom with a steric number of 4 and no lone pairs of electrons.

**theoretical yield**   The maximum amount of product possible in a chemical reaction for given quantities of reactants; also called *stoichiometric yield*.

**thermal energy**   The kinetic energy of atoms, ions, and molecules.

**thermal equilibrium**   A condition in which temperature is uniform throughout a material and no energy flows from one point to another.

**thermochemical equation**   The chemical equation of a reaction that includes energy as a reactant or a product.

**thermochemistry**   The study of the relation between chemical reactions and changes in energy.

**thermodynamics**   The study of energy and its transformations.

**third law of thermodynamics**   The entropy of a perfect crystal is zero at absolute zero.

**threshold frequency ($\nu_0$)**   The minimum frequency of light required to produce the photoelectric effect.

**titrant**   The standard solution added to the sample in a titration.

**titration**   An analytical method for determining the concentration of a solute in a sample by reacting the solute with a standard solution of known concentration.

**torr**   (also called *millimeters of mercury [mmHg]*) A unit of pressure, where 1 atm = 760 torr = 760 mmHg.

**trace essential element**   Essential element present in the body in average concentrations between 1 and 1000 µg of element per gram of body mass.

**trans isomer**   (also called E *isomer*) Molecule with two like groups (such as two R groups or two hydrogen atoms) on opposite sides of the molecule.

**transcription**   The process of copying the information in DNA to RNA.

**transfer RNA (tRNA)**   The form of the nucleic acid RNA that delivers amino acids, one at a time, to polypeptide chains being assembled by the ribosome–mRNA complex.

**transition metals**   The elements in groups 3 through 12 of the periodic table.

**transition state**   A high-energy state between reactants and products in a chemical reaction.

**translation**   The process of assembling proteins from the information encoded in RNA.

**tricarboxylic acid (TCA) cycle**   A series of reactions that continue the oxidation of pyruvate formed in glycolysis.

**trigonal bipyramidal**   Molecular geometry about a central atom with a steric number of 5 and no lone pairs of electrons; three atoms occupy equatorial sites and two other atoms occupy axial sites above and below the equatorial plane.

**trigonal planar**   Molecular geometry about a central atom with a steric number of 3 and no lone pairs of electrons.

**triple bond**   A chemical bond in which two atoms share three pairs of electrons.

**triple point**   The temperature and pressure at which all three phases of a substance coexist. Under these conditions, freezing and melting, boiling and liquefaction, and sublimation and deposition all proceed at the same rate.

**T-shaped**   Molecular geometry about a central atom with a steric number of 5 and two lone pairs of electrons that occupy equatorial positions; the atoms occupy two axial sites and one equatorial site.

## U

**ultratrace essential element**   Essential element present in the body in average concentrations less than 1 µg of element per gram of body mass.

**unimolecular step**   A step in a reaction mechanism involving only one molecule on the reactant side.

**unit cell**   The basic repeating unit of the arrangement of atoms, ions, or molecules in a crystalline solid.

**universal gas constant**   The constant $R$ in the ideal gas equation; its value and units depend on the units used for the variables in the equation.

**unsaturated hydrocarbon**   An alkene or alkyne.

## V

**valence band**   A band of orbitals that are filled or partially filled by valence electrons.

**valence bond theory**   A quantum mechanics-based theory of bonding that assumes covalent bonds form when half-filled orbitals on different atoms overlap or occupy the same region in space.

**valence electrons**   Electrons in the outermost occupied shell of an atom having the most influence on the atom's chemical behavior.

**valence shell**   The shell in an atom containing the valence electrons.

**valence-shell electron-pair repulsion theory (VSEPR)**   A model predicting that the arrangement of valence electron pairs around a central atom minimizes their mutual repulsion to produce the lowest-energy orientations.

**van der Waals equation**   An equation that includes experimentally determined factors $a$ and $b$ that quantify the contributions of non-negligible molecular volume and non-negligible intermolecular interactions to the behavior of real gases with respect to changes in $P$, $V$, and $T$.

**van 't Hoff factor**   (also called *i factor*) The ratio of the experimentally measured value of a colligative property to the theoretical value expected for that property if the solute were a nonelectrolyte.

**vapor pressure**   The pressure exerted by a gas at a given temperature in equilibrium with its liquid phase.

**vinyl group**   The subgroup $CH_2{=}CH-$.

**vinyl polymer**   One of the family of polymers formed from monomers containing the subgroup $CH_2{=}CH-$.

**viscosity**   The resistance to flow of a liquid.

**voltaic cell**   An electrochemical cell in which chemical energy is transformed into electrical work by a spontaneous cell reaction.

## W

**wave function ($\psi$)**   A solution to the Schrödinger wave equation.

**wave mechanics**   (also called *quantum mechanics*) A mathematical description of the wavelike behavior of particles on the atomic level.

**wavelength ($\lambda$)**   The distance from crest to crest or trough to trough on a wave.

**wave–particle duality**   Occurs when an object exhibits the properties of both a wave and a particle.

**weak acid**   An acid that only partially dissociates in aqueous solution and so has a limited capacity to donate protons to the medium.

**weak base**   A base that only partially dissociates in aqueous solution and so has a limited capacity to accept protons.

**weak electrolyte**   A substance that only partly dissociates into ions when it dissolves in water.

**work**   A form of energy: the energy required to move an object through a given distance.

**work function ($\Phi$)**   The amount of energy needed to dislodge an electron from the surface of a metal.

## X

**X-ray diffraction (XRD)**   A technique for determining the arrangement of atoms or ions in a crystal by analyzing the pattern that results when X-rays are scattered after bombarding the crystal.

## Z

**Z isomer**   (also called *cis isomer*) Molecule with two like groups (such as two R groups or two hydrogen atoms) on the same side of the molecule.

**zeolite**   Natural crystalline minerals or synthetic materials consisting of three-dimensional networks of channels that contain sodium or other 1+ cations.

**zwitterion**   A molecule that has both positively and negatively charged groups in its structure.

# Answers to Concept Tests and Practice Exercises

## CHAPTER 1

### Concept Tests

p. 6   1:1

p. 8   $CH_2O$

p. 10  Yes. It appears homogeneous, but it is available in several different compositions, as reflected by the octane number; it is now also frequently mixed with ethanol as well as other additives.

p. 12  (b)

p. 15  (b) Less than, because the arrows in Figure 1.14 are shorter for melting ice than for boiling water. Figure 1.13 shows why: fewer intermolecular interactions are broken in melting ice than in boiling water.

p. 17  (b) Decreasing

p. 21  (1) 3 significant figures; (2) 3 significant figures; (4) 4 significant figures

p. 23  Speed = distance/time. Distance, because actual distance traveled will vary depending on whether a car stays closer to the inside or the outside of the oval.

p. 33  Hypothesis, because his explanation had not been thoroughly tested yet.

### Practice Exercises

1.1. Properties (a) and (c) are physical properties; property (d) is a chemical property; property (b) is both physical and chemical.

1.2. a.  The particles in the box on the left represent a gas because the particles are widely spaced and fill the box. The particles in the box on the right represent a solid because the particles are ordered and do not fill the box. The change of state represented is deposition.

 b.  Sublimation

1.3. 1.14

1.4. Statistics (a) and (e) are exact numbers; statistics (b)–(d) have inherent uncertainty.

1.5. $1.9 \times 10^2$; $1.55 \times 10^{-7}$; $1.8179 \times 10^{-1}$

1.6. 0.324 km; $3.24 \times 10^4$ cm

1.7. $1.5 \times 10^2$ cm

1.8. $9 \times 10^{12}$ km/y

1.9. $K_{low} = 40$ K and $K_{high} = 396$ K; $°F_{low} = -387°F$ and $°F_{high} = 253°F$

## CHAPTER 2

### Concept Tests

p. 44  Matter is neutral, so some positive particle must be present to provide electrical neutrality.

p. 50  He left spaces for unknown elements and predicted their properties; when those elements were discovered, his predictions were confirmed.

p. 56  (top) CsN

p. 56  (bottom) $H_2O_2$ and HO

### Practice Exercises

2.1.  a.  $^{56}_{26}Fe$

 b.  $^{15}_{7}N$

 c.  $^{37}_{17}Cl$

 d.  $^{39}_{19}K$

2.2.  (a) 27 p, 33 n; (b) 53 p, 78 n; (c) 77 p, 115 n

2.3. $^{107}Ag = 51.5\%$; $^{109}Ag = 48.5\%$
2.4. a. As, arsenic
 b. Ca, calcium
 c. Hg, mercury
 d. S, sulfur
2.5. 40.0 g
2.6. (a)–(d) are molecular; (e) is ionic.
2.7. a. Tetraphosphorus decoxide
 b. Carbon monoxide
 c. Nitrogen trichloride
 d. $SF_6$
 e. ICl
 f. $Br_2O$
2.8. a. $SrCl_2$
 b. MgO
 c. NaF
 d. $CaBr_2$
2.9. $MnCl_2$ and $MnO_2$
2.10. a. $Sr(NO_3)_2$
 b. $K_2SO_3$
2.11. a. Calcium phosphate
 b. Magnesium perchlorate
 c. Lithium nitrite
 d. Sodium hypochlorite
 e. Potassium permanganate
2.12. a. Hypochlorous acid
 b. Chlorous acid
 c. Carbonic acid

# CHAPTER 3

## Concept Tests

p. 80 Both the mole and a gross represent a specific number of particles independent of their size, mass, or identity.
p. 82 One gram of Ag has more atoms than one gram of Au because Ag has a smaller molar mass.
p. 91 When balancing equations we cannot change the subscripts because doing so changes the identity of the substance.
p. 102 (a) $C_2H_4$ and (b) $C_{20}H_{40}$ have the same empirical formula; (c) $C_2H_2$ and (d) $C_6H_6$ have the same empirical formula.
p. 104 $C_2H_6O_2$ and $C_6H_{12}O_6$ are molecular formulas; $C_3H_8O$ is an empirical formula and could also be a molecular formula.

## Practice Exercises

3.1. $1.5 \times 10^{10}$ atoms
3.2. $6.80 \times 10^{24}$ electrons
3.3. $5.41 \times 10^{-2}$ moles
3.4. 49.2 g
3.5. $CO_2 = 44.01$ g/mol; $O_2 = 32.00$ g/mol; $C_6H_{12}O_6 = 180.16$ g/mol
3.6. $5.00 \times 10^{-3}$ moles; $3.01 \times 10^{-21}$ formula units
3.7. 155 g Pb
3.8.

3.9. (a) $P_4(s) + 5\,O_2(g) \rightarrow P_4O_{10}(s)$;
 (b) $P_4O_{10}(s) + 6\,H_2O(\ell) \rightarrow 4\,H_3PO_4(\ell)$
3.10. $2\,CO(g) + O_2(g) \rightarrow 2\,CO_2(g)$
3.11. $C_3H_8(g) + 5\,O_2(g) \rightarrow 3\,CO_2(g) + 4\,H_2O(\ell)$

3.12. $2\,C_4H_{10}(g) + 13\,O_2(g) \rightarrow 8\,CO_2(g) + 10\,H_2O(\ell)$; 3.03 g $CO_2$ produced
3.13. 321 g $Cu_2S$; 129 g $SO_2$
3.14. 25.99% C; 74.01% F
3.15. $N_4O_3$
3.16. $Li_2CO_3$
3.17. $C_{20}H_{40}$
3.18. $C_5H_8$
3.19. $C_8H_8O_3$
3.20. This fuel–oxygen mixture is rich.
3.21. 3.74 g $C_2H_4N_2$
3.22. 89.9%
3.23. $1.10 \times 10^3$ g $Al_2O_3$; 194 g C

# CHAPTER 4

## Concept Tests

p. 135 (c) Clear cough syrup; (d) Filtered dry air
p. 139 (d) 56,000 n$M$ NaCl is the least concentrated solution.
p. 144 Solvent is added to dilute the stock solution, which decreases the concentration.
p. 147 Equivalent molar concentrations mean that the same number of solute moles are dissolved in a given solution volume. Differences in conductivity are due to the different numbers of ions the solutes make when they dissolve.
p. 149 Drawing (a) describes a weak electrolyte and (c) describes a strong electrolyte.
p. 152 (a) $H_2E^{2-}$; (b) $H_4E$
p. 153 (b) $HSO_4^-$ is amphiprotic.
p. 160 React aqueous lead(II) nitrate with the stoichiometric amount of aqueous potassium dichromate. Stir the mixture for a minute or so; let it stand for 10 min; filter off the yellow $PbCr_2O_7$ precipitate; wash it with water in the filter; and allow the washed solid to air-dry overnight.
p. 168 In the reaction of $Fe_3O_4$ with oxygen, the iron is oxidized, because the product contains more oxygen than the reactant: 3 Fe ions are present for every 4 oxygen ions (1.33 oxygen ions per iron ion) versus 2 Fe ions for every 3 oxygen ions (1.5 oxygen ions per iron ion). If iron is oxidized, the other substance involved in the reaction must be reduced; oxygen is reduced.
p. 171 Reactions (a) and (b) are redox reactions. In reaction (a), $Br_2$ is reduced to $Br^-$, so $Br_2$ is the oxidizing agent, and $Sn^{2+}$ is the reducing agent. In reaction (b), $F_2$ (O.N. is zero) is reduced (O.N. in HF is −1), so $F_2$ is the oxidizing agent. The oxygen atom in $H_2O$ has O.N. = −2, but in $O_2$ O.N. is zero, so $H_2O$ is the reducing agent.

## Practice Exercises

4.1. Well water is 120 times more concentrated in arsenic.
4.2. 1.88 $M$ $MgCl_2$
4.3. 0.109 $M$ KCl
4.4. 4.48 g $NaC_3H_3O_5$
4.5. $V_{initial} = 12.5$ mL
4.6. $1.16 \times 10^{-4}$ $M$
4.7. (a) strong electrolyte; (b) strong acid, strong electrolyte; (c) weak base, weak electrolyte; (d) nonelectrolyte
4.8. a. $H_3PO_4(aq) + 3\,NaOH(aq) \rightarrow 3\,H_2O(\ell) + Na_3PO_4(aq)$
 b. $H_3PO_4(aq) + 3\,Na^+(aq) + 3\,OH^-(aq) \rightarrow 3\,H_2O(\ell) + 3\,Na^+(aq) + PO_4^{3-}(aq)$
 c. $H_3PO_4(aq) + 3\,OH^-(aq) \rightarrow 3\,H_2O(\ell) + PO_4^{3-}(aq)$

4.9. The lemon juice is 0.416 $M$ $C_6H_8O_7$; 100 mL of juice contains 8.00 g $C_6H_8O_7$.

4.10. 0.0987 $M$

4.11. $3\ Ba^{2+}(aq) + 6\ OH^-(aq) + 2\ H^+(aq) + 2\ H_2PO_4^-(aq) \rightarrow$ $Ba_3(PO_4)_2(s) + 6\ H_2O(\ell)$

4.12. a. No precipitate forms.
   b. $Hg_2Cl_2$ precipitates from the mixture.
   c. $Hg_2^{2+}(aq) + 2\ Cl^-(aq) \rightarrow Hg_2Cl_2(s)$

4.13. 0.174 g HgS

4.14. $1.31 \times 10^{-4}\ M\ SO_4^{2-}$

4.15. Yes

4.16. a. +4
   b. +1
   c. +5

4.17. Oxygen is reduced and is the oxidizing agent; $SO_2$ is oxidized and is the reducing agent.

4.18. a. Yes. Because the oxidation numbers for iron and palladium change from reactants to products, this reaction is a redox reaction.
   b. $2\ Fe(s) + 3\ Pd^{2+}(aq) \rightarrow 2\ Fe^{3+}(aq) + 3\ Pd(s)$

4.19. Iron

4.20. Aluminum nitrate and metallic silver
   Oxidation reaction: $Al(s) \rightarrow Al^{3+}(aq) + 3\ e^-$
   Reduction reaction: $1\ e^- + Ag^+(aq) \rightarrow Ag(s)$
   Net ionic equation: $Al(s) + 3\ Ag^+(aq) \rightarrow Al^{3+}(aq) + 3\ Ag(s)$
   Molecular equation: $Al(s) + 3\ AgNO_3(aq) \rightarrow$ $Al(NO_3)_3(aq) + 3\ Ag(s)$

4.21. $5\ H^+(aq) + 3\ HO_2^-(aq) + 2\ MnO_4^-(aq) \rightarrow$ $3\ O_2(g) + 2\ MnO_2(s) + 4\ H_2O(\ell)$

# CHAPTER 5

## Concept Tests

p. 200  No; skier 1 has more PE: $m_1gh = (PE)_{skier\ 1} > (PE)_{skier\ 2} = m_2gh$

p. 201  (top) The PE of skier 2 is less than the PE of skier 1; skier 1 has greater KE:
$$\tfrac{1}{2}\ m_1u^2 = (KE)_{skier\ 1} > (KE)_{skier\ 2} = \tfrac{1}{2}\ m_2u^2$$

p. 201  (bottom) There is far more water in the pool than in the cup. Therefore, the correct answer is "less than," even if the pool temperature is, say, 20°C.

p. 205  (a) open; (b, c) closed, because the bottle and the sandwich wrap can conduct heat; (d) open

p. 220  The mass of the aluminum is much less than the mass of the water and the molar heat capacity of aluminum is much less than that of water.

p. 237  +2219.9 kJ; We are now asked about the reverse of a reaction for which we calculated a negative $\Delta H$; we just need to change the sign on the value.

p. 241  (a) 1 mole $CH_4$; (b) 1 g $H_2$

## Practice Exercises

5.1. About 5% slower

5.2. a. The match is the system, $q < 0$, and the process is exothermic.
   b. The wax is the system, $q < 0$, and the process is exothermic.
   c. The liquid is the system, $q > 0$, and the process is endothermic.

5.3. $\Delta E = 32$ J

5.4. $w = 1.56 \times 10^7$ L · atm

5.5. $1.13 \times 10^3$ g or 1.13 kg

5.6. $\Delta H_{sys}$ is negative; $q_{surr}$ is positive.

5.7. −321 kJ

5.8. 0.0°C

5.9. 51.4°C

5.10. −4.61 kJ/mol

5.11. When 0.500 g of the hydrocarbon mixture is burned, the energy released is 24.6 kJ. When 1.000 g of the hydrocarbon mixture is burned, the energy released is 49.2 kJ.

5.12. Reverse B and then add the three equations.

5.13.

| | |
|---|---|
| $2\ CH_4(g) + 3\ O_2(g) \rightarrow 2\ \text{CO}(g) + 4\ H_2O(g)$ | $\Delta H_{comb} = -1038$ kJ |
| $2\ \text{CO}(g) + O_2(g) \rightarrow 2\ CO_2(g)$ | $\Delta H_{comb} = -566$ kJ |
| $2\ CH_4(g) + 4\ O_2(g) \rightarrow 2\ CO_2(g) + 4\ H_2O(g)$ | $\Delta H_{comb} = -1604$ kJ |

   For 1 mole $CH_4$, $\Delta H_{comb} = -802$ kJ

5.14. a. $Ca(s) + C(s) + \tfrac{3}{2}O_2(g) \rightarrow CaCO_3(s)$
   b. $2\ C(s) + 2\ H_2(g) + O_2(g) \rightarrow CH_3COOH(\ell)$
   c. $K(s) + Mn(s) + 2\ O_2(g) \rightarrow KMnO_4(s)$

5.15. $\Delta H^\circ_{rxn} = -41.2$ kJ

5.16. Recycling reaction: 13 kJ/mol. Refining: −128 kJ/mol. NOTE: It looks like refining consumes about 10 times less energy than recycling. This is not true, because many other processes must be considered. Recycling copper is about 15% more efficient than refining ores.

5.17. Fuel value of kerosene = 41.40 kJ/g; fuel density of kerosene = $3.10 \times 10^4$ kJ/L

5.18. $C_{calorimeter} = 11.2$ kJ/°C

# CHAPTER 6

## Concept Tests

p. 267  (c)

p. 269  The two graphs will have different slopes because the slope depends on $P$. Slope of graph B = $\tfrac{1}{2}$ slope of graph A.

p. 271  (c)

p. 275  b and d

p. 280  Kr

p. 281  (top) (a) i; (b) ii

p. 281  (bottom) (c) Low $T$ and high $P$ give greatest $d$.

p. 284  Because $P = F/A$, the force results from all components combined so one cannot measure the partial pressure since one cannot measure the partial force.

p. 285  Yes

p. 289  Lower

p. 291  $UF_6 < SF_6 < Kr < CO_2 < Ar < H_2$

p. 293  $T$ cancels out

p. 296  $CO_2$, $CH_4$

## Practice Exercises

6.1. $P = 7.71 \times 10^2$ Pa

6.2. At the end of the experiment the mercury levels will be

   $\Delta h = 144$ mmHg

6.3. $V_2 = 10.5$ L

6.4. $V_1/V_2 = 0.622$

6.5. $P_2 = 34$ psi

6.6. $V_2 = 2.10 \times 10^3$ L

6.7. $V = 1.8 \times 10^3$ L

6.8. 111 g

6.9. 259 mL $N_2O$

6.10. The balloon will sink to the floor.

6.11. M = 44.0 g/mol; $CO_2$

6.12. $O_2 = 0.120$, He = 0.880

6.13. $X_{O_2} = 0.042$

6.14. $2.2 \times 10^{-3}$ g $H_2$

6.15. $u_{\text{rms,He}} = 1.37 \times 10^3$ m/s, or 2.65 times faster than $N_2$

6.16. Ar

6.17. $N_2$ behaves more ideally.

# CHAPTER 7

## Concept Tests

p. 320 Higher frequency

p. 321 No change

p. 323 (c)

p. 324 (a) continuous; (b) discrete; (c) continuous; (d) discrete

p. 328 Yes

p. 331 (c) > (a) > (b) > (d)

p. 332 Yes because $He^+$ has only one electron.

p. 334 The vibrations of the strings have wavelengths that are equal to $2l/n$, where $n$ is a whole number.

p. 344 Five

p. 354 A half-filled set of $f$ orbitals is more stable because adding another $f$ electron requires pairing energy.

p. 356 (a) $1s$ > (b) $2s$ > (e) $2p$ > (c) $3s$ > (f) $3p$ > (d) $4s$ > (g) $4p$

p. 360 Because the valence electrons are farther from the nucleus and shielded from it by more inner shell electrons

p. 361 The magnitude of IE and EA both increase with increasing $Z$ across a row (except for group 18). EA values do not display clear trends within groups, whereas IE values decrease with increasing $Z$.

## Practice Exercises

7.1. $\lambda = 3.30$ m

7.2. $E_{450} = 4.4 \times 10^{-19}$ J; $E_{470} = 4.2 \times 10^{-19}$ J

7.3. $\lambda = 2.62 \times 10^{-7}$ m or 262 nm

7.4. 486.3 nm

7.5. Prediction: Less energy is required to remove an electron from the hydrogen atom in the $n = 3$ state than for a hydrogen atom in the $n = 1$ state. Calculated value: $2.420 \times 10^{-19}$ J

7.6. $\lambda = 3.3 \times 10^{-10}$ m

7.7. $\Delta x \geq 7 \times 10^{-11}$ m

7.8. Four

7.9.

| $n$ | $\ell$ | $m_\ell$ | $m_s$ |
|-----|--------|----------|-------|
| 3 | 1 | −1 | $\frac{1}{2}$ |
| 3 | 1 | 0 | $\frac{1}{2}$ |
| 3 | 1 | 1 | $\frac{1}{2}$ |

7.10. Ga: $[Ar]\,3d^{10}4s^24p^1$    As: $[Ar]\,3d^{10}4s^24p^3$

7.11. Co = $[Ar]3d^74s^2$

7.12. $K^+ = [Ar]$; $I^- = [Kr]4d^{10}5s^25p^6 = [Xe]$; $Ba^{2+} = [Xe]$; $Rb^+ = [Kr]$; $O^{2-} = [He]2s^22p^6 = [Ne]$; $Al^{3+} = [Ne]$; $Cl^- = [Ne]3s^23p^6 = [Ar]$. $K^+$ and $Cl^-$ are isoelectronic with Ar.

7.13. $Mn = [Ar]3d^54s^2$; $Mn^{3+} = [Ar]3d^4$; $Mn^{4+} = [Ar]3d^3$

7.14. a. $Li^+ < F^- < Cl^-$
      b. $Al^{3+} < Mg^{2+} < P^{3-}$

7.15. Ne > Ca > Cs

# CHAPTER 8

## Concept Tests

p. 384 (top) Most noble gases do not form covalent bonds.

p. 384 (bottom)

p. 387 No, stretching in $N\equiv N$ or $O=O$ does not result in IR absorptions because the bonds are nonpolar and no change in polarity occurs when they stretch.

p. 394 −1

p. 404 $N_2O > NO_2 > NO$

p. 407 The $O=O$ bond is not as strong as the $N\equiv N$ bond and is more easily broken.

## Practice Exercises

8.1.

8.2.

8.3.

8.4.

8.5. Be—Cl; the bond is considered to be a polar covalent bond.

8.6.

8.7. Resonance forms for $N_3^-$:

Resonance forms for $NO_2^+$:

8.8.

8.9.

8.10.

8.11. Sodium atoms do not have an octet in either structure.

$$Na—\overset{..}{\underset{..}{Cl}}: \qquad Na—\overset{..}{\underset{\overset{|}{:\underset{..}{Cl}—Na}}{Cl}}:$$

8.12. $\Delta H_{rxn} = -79$ kJ

# CHAPTER 9

## Concept Tests

p. 427  Molecular geometry is derived from electron-pair geometry. They are the same if there are no lone pairs but different if there are lone pairs.

p. 431  (b) > (a) > (c)

p. 434  One lone pair on $PH_3$ and two lone pairs on $H_2S$ repel bonding pairs and reduce bond angles.

p. 436  Slightly smaller

p. 438  All the bond dipoles offset each other, as in $CH_4$.

p. 440  Because the difference in electronegativity between H and S is less than between H and O

p. 447  It needs unhybridized $p$ orbitals to form $\pi$ bonds.

p. 451  (a) and (c), because the double bonds are conjugated

p. 453  (b) and (c) are chiral

p. 454  Similar: they are the products of mixing atomic orbitals. Different: MOs are delocalized.

p. 459  Yes, because $O_2$ is paramagnetic and $N_2$ is not.

p. 462  No, the bond orders in the ground states are higher because each has one more electron in bonding orbital and one fewer in an antibonding orbital.

## Practice Exercises

9.1. Tetrahedral:

$$H—\overset{\overset{Cl}{|}}{\underset{\underset{Cl}{|}}{C}}\cdots Cl$$

9.2. The O—S—O angle in $SO_3$ is greater than the O—S—O angle in $SO_2$.

9.3. Tetrahedral; bond angles ~109.5°

9.4. No

9.5. $5p$ and $4p$

9.6. (a) $CCl_4$ and (d) $PH_3$

9.7. $sp^3d^2$

9.8. Each N in diazene is trigonal planar. With one lone pair on each N atom, the molecular geometry around the N atoms is bent with H–N–N bond angles of less than 120°. The N atoms are $sp^2$ hybridized and the molecule is flat. Each N in hydrazine is tetrahedral. With one lone pair on each N atom, the molecular geometry around the N atoms is trigonal pyramidal with H–N–N bond angles of less than 109.5°. The N atoms are $sp^3$ hybridized and the molecule is three-dimensional.

9.9. $H_2^+$ can exist; its bond order is 0.5.

9.10. The bond order increases on the addition of an electron for $Be_2$ to $Be_2^-$, $B_2$ to $B_2^-$, $C_2$ to $C_2^-$, and $Ne_2$ to $Ne_2^-$.

9.11.

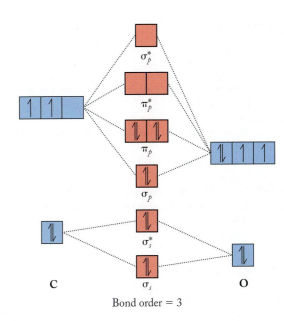

C      $\sigma_s$      O

Bond order = 3

# CHAPTER 10

## Concept Tests

p. 477  (a) LiF < NaF < KF; (b) $CaF_2$ < $CaCl_2$ < $CaBr_2$

p. 481  Acetone

p. 482  (a) $CH_3Br$; (b) $CH_3CH_2OH$; (c) $CH_3NH_2$

p. 486  H < Ne < Ar < Kr

p. 487  Yes. Ion-induced dipole. Slightly stronger than dipole-induced dipole and weaker than dipole–dipole

p. 488  $CCl_4$ is larger and experiences larger dispersion forces.

p. 493  Gasoline

p. 497  Yes, it's possible that increased pressure will cause the ice to melt, based on the slope of the blue line in Figure 10.24 and the footprint of the tires.

p. 498  No

p. 500  No, the adhesive forces are weak.

## Practice Exercises

10.1. $BaO > CaCl_2 > KCl$

10.2. To enter the vapor phase from the liquid phase (to boil), ethylene glycol would need to break two hydrogen bonds compared to isopropanol's one hydrogen bond; therefore, ethylene glycol has a higher boiling point.

10.3. Largest dipole–dipole forces: (a) $H_2NNH_2$; largest dispersion forces: (d) $CH_3CH_2CH_2CH_2CH_3$; lowest boiling point: (c) Ne

10.4. Helium, being smaller and with fewer electrons than nitrogen, is less soluble in blood because it is less polarizable in its interaction with the polar molecules in blood (water).

10.5. Methanol. Both methanol and hexanol have an –OH group that can hydrogen-bond with the oxygen on acetone. Hexanol has a much longer hydrocarbon chain than methanol, and its molecules would interact via dispersion forces, which might cause hexanol to interact more strongly with itself than with acetone and hence cause it to be less soluble in acetone.

10.6. 28.6 kJ/mol

10.7. At 25 atm pressure and –100°C, the sample of $CO_2$ is a solid. As the temperature is increased to about –50°C the solid melts into a liquid, and as the temperature is raised further (to about –20°C) the liquid $CO_2$ boils to form a gas.

# CHAPTER 11

## Concept Tests

p. 515 The vapor pressure of a pure solvent is an intensive property. The vapor pressure of solution is an extensive property.

p. 529 Dimethyl ether

p. 532 (c)

p. 534 The solution is mostly water, which has a density of 1 kg/L.

p. 539 (b)

p. 540 (a) possible; (b) not possible; (c) possible; (d) not possible

p. 543 (top) Toward the KCl

p. 543 (bottom) Doubling the concentration

p. 547 Although more opposing pressure is needed at 50°C than at 20°C, the greater molecular motion would cause osmosis to occur faster.

## Practice Exercises

11.1. $P_{solution} = 0.848$ atm or 644 torr

11.2. $9.5 \times 10^{-5}$ mol/L

11.3. $TiO_2$

11.4. $U = -3792$ kJ

11.5. $U = -670$ kJ/mol

11.6. Distillation of mixture of hexane, heptane, and nonane

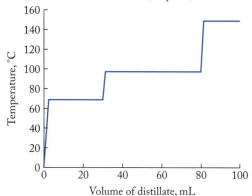

11.7. 1.4

11.8. 0.840 $m$

11.9. 4.4 $m$

11.10. 109.4°C

11.11. $i = 3$; 102.7°C

11.12. −0.12°C

11.13. 28.1 atm

11.14. 3.7 atm

11.15. 27.5 atm

11.16. 180 g/mol

11.17. $6.40 \times 10^4$ g/mol

# CHAPTER 12

## Concept Tests

p. 567 Unit cell is the smallest repeating pattern in a crystal lattice.

p. 572 (top) Yes, because it is homogeneous.

p. 572 (bottom) Both alloys could be either.

p. 575 Overlapping conduction and valence bands, because bands from filled Mg 3$s$ orbitals overlap with empty $p$ orbitals.

p. 582 A group 15 element such as N or P

p. 584 $x = y = 1.2$

## Practice Exercises

12.1. For silver, $d = 10.57$ g/mL, close to the 10.50 g/mL value for the density of silver from Appendix 3; for gold, $d = 19.41$ g/mL, close to the 19.3 g/mL value for the density of gold from Appendix 3.

12.2. Gold forms substitutional alloys with both silver and copper.

12.3. Selenium

12.4. 101 pm; 2.16 g/cm$^3$

12.5. $d = 412$ pm; for $n = 3$, $2\theta = 30.0°$; for $n = 4$, $2\theta = 40.4°$

# CHAPTER 13

## Concept Tests

p. 606 Dispersion forces

p. 608 No, they both consume 2 moles of $H_2$ per mole.

p. 609 $n = 8$

p. 616 The same name means the same compound.

p. 622 (e)

p. 623 (a) No; (b) no

p. 624 Different molecular structures and properties mean LDPE and HDPE must be recycled separately.

p. 626 The two Cl atoms per monomer unit in Saran wrap do not allow the chains to pack tightly; less interaction between molecules makes them more flexible than PVC.

p. 628 The compound is not cyclic.

p. 630 Paraglyine: tertiary; amphetamine: primary; adrenaline: secondary; Benadryl: tertiary

p. 631 Alcohols and ethers are both polar and interact with polar water molecules via hydrogen bonding; hydrocarbons are nonpolar.

p. 635 MTBE > diethyl ether > ethanol > methanol

p. 637 The longer the hydrocarbon portion of the chain, the lower the water solubility.

p. 638 Zingerone: ether, aromatic rings, −OH; carvone: C=C; cinnamaldehyde: aromatic ring, C=C

p. 641

Amine: $\overset{|}{N}$ ; Amide: $\overset{O}{\underset{N<}{\parallel}}$

The nitrogen in the amide is attached to a carbonyl group.

p. 650 The carbon atom bonded to the −OH group is a chiral center. The different orientations of the parts of the molecule on either side of this center result in the enantiomers being analogous to right and left hands. A right hand "matches" another right hand in a handshake. A left hand does not match a right hand. In the same way a right-handed molecule needs a right-handed receptor; a left-handed molecule would not interact correctly with that receptor.

p. 652 If the muscarine came from mushrooms, it would be of one enantiomer and its solution would be optically active. Because the solution did not rotate the plane of polarized light, the muscarine present was a racemic mixture, which had to be the product of a laboratory synthesis. The coroner concluded that the victim was poisoned by someone who had access to synthetic muscarine.

## Practice Exercises

13.1. Yes, we can differentiate between them using hydrogenation reactions. The reaction product, however, is the same for both compounds A and B:

$$CH_3—CH_2—CH_2—CH_2—CH_2—CH_2—CH_3$$

13.2. Hexane:  ; heptane: $CH_3(CH_2)_5CH_3$

13.3. (a) Constitutional isomers; (b) two different compounds; (c) constitutional isomers

13.4. (a) 7 carbon atoms; (b) 9 carbon atoms

13.5.

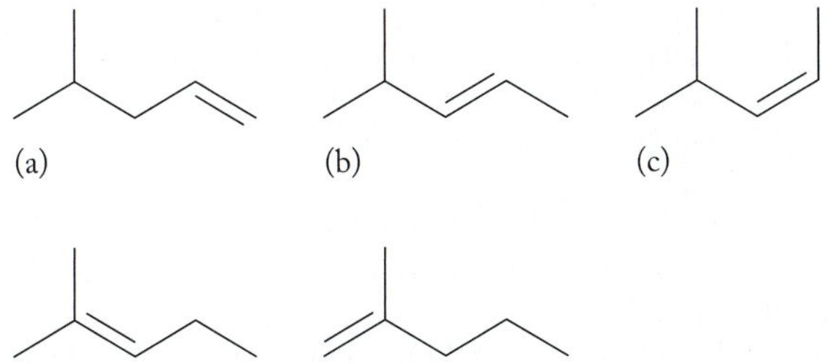

(a)    (b)    (c)

(d)    (e)

The names of the compounds are (a) 4-methyl-1-pentene, (b) *trans*-4-methyl-2-pentene, (c) *cis*-4-methyl-2-pentene, (d) 2-methyl-2-pentene, and (e) 2-methyl-1-pentene.

13.6. The carbon skeleton of the monomer is

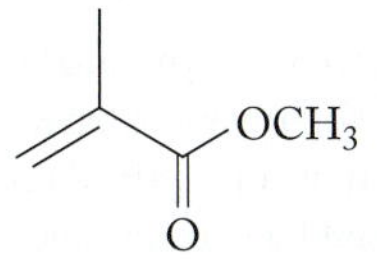

The condensed structure of the monomer is
$H_2C{=}C(CH_3)C(O)OCH_3$

13.7. The heat released from the E-85 fuel is 29% less than that released from the same volume of pure nonane.

13.8. London dispersion and dipole-induced dipole interactions.

13.9.

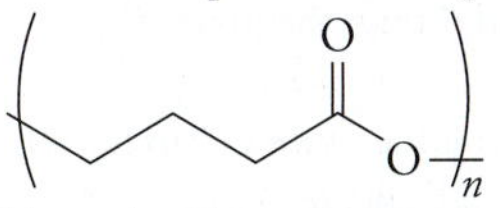

13.10. The polar fibers of cotton and polyester repel very nonpolar greases and oils but attract water molecules, so perspiration wicks out of the gloves to cool the skin.

13.11. The carbon skeleton structures of the monomers are

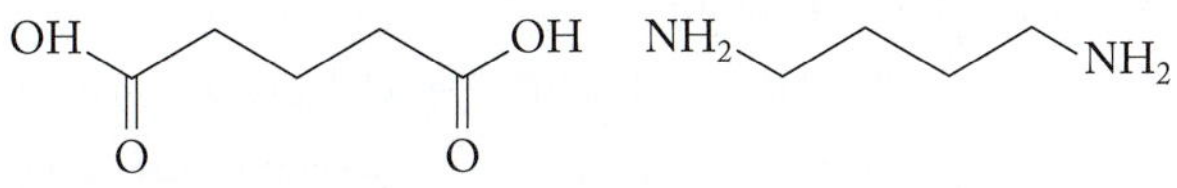

The repeating unit in the polymer is

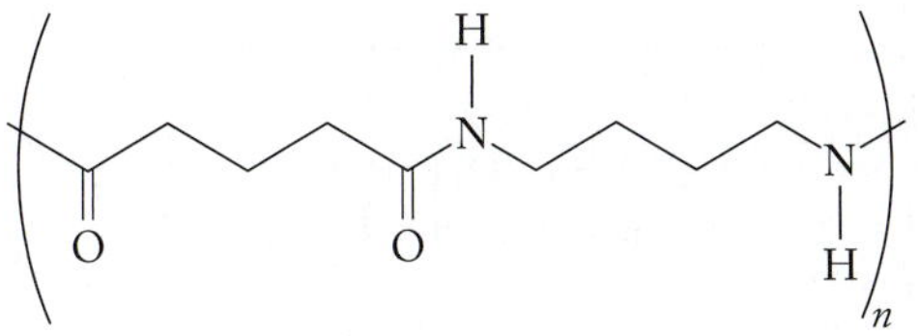

13.12.

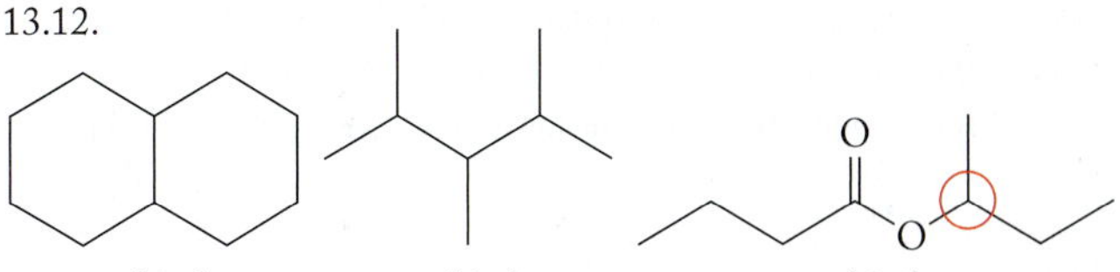

achiral    achiral    chiral

(a)    (b)    (c)

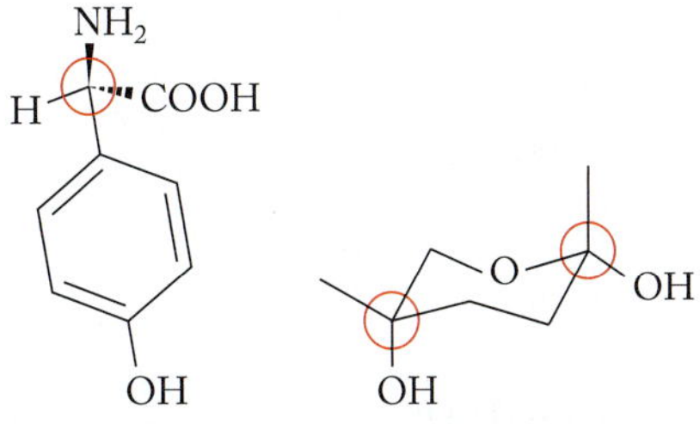

chiral    chiral

(d)    (e)

13.13.

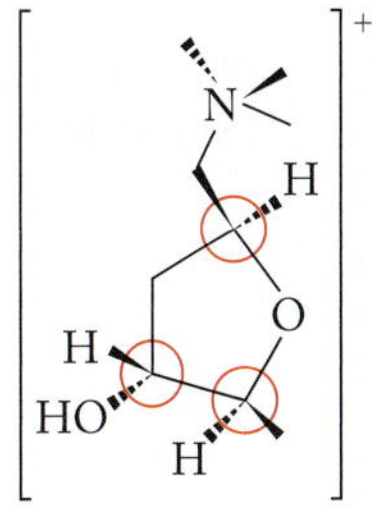

# CHAPTER 14

## Concept Tests

p. 671 Because a negative value would mean the reaction is running in reverse, not as written

p. 676 (d)

p. 680 (a) 1; (b) 0

p. 681 Four: Rate $= k[A][B]^2$; Rate $= k[A]^2[B]$; Rate $= k[A]^3$; Rate $= k[B]^3$

p. 687 Fast

p. 693 Plot $k'$ vs. $[O_3]$; use the slope to determine $k$

p. 695 Molecules are moving faster, so the likelihood of collisions is greater.

p. 699 (c)

p. 701 $n$

p. 704 No

p. 706 Similar rate laws indicate similar reaction mechanisms.

p. 708 (a)

p. 709 Intermediate

## Practice Exercises

14.1. The rate of consumption of CO is twice that of $O_2$. The rate of formation of $CO_2$ (a product) is twice the rate of consumption of $O_2$.

14.2. $\dfrac{\Delta[N_2]}{\Delta t} = 10.8\ M/s$; $\dfrac{\Delta[H_2O]}{\Delta t} = 21.5\ M/s$

14.3. $1.2 \times 10^{-6}\ M/s$

14.4. Rate $= k[NO][NO_3]$; $k = 1.57 \times 10^{10}/(M \cdot s)$

14.5. The decomposition of $H_2O_2$ is first order; $k = 8.30 \times 10^{-4}\ M/s$

14.6. $0.430\ M$

14.7. $k = 2.5 \times 10^{-2}/day$

14.8. This reaction is second order in $[NO_2]$; $k = 0.751/(M \cdot s)$
14.9. The pseudo-first-order rate constant is $k' = 6.11 \times 10^{-4}/\mu s$. The second-order rate constant is $k = 7.2 \times 10^{6}/(M \cdot \mu s)$ or $7.2 \times 10^{12}/(M \cdot s)$.
14.10. No.
14.11. 6.9 kJ
14.12. Rate = $k_{overall}$ $[A]^2[B]^2$
14.13. Because none of the rate laws match the experimental rate law, this proposed mechanism cannot be valid.
14.14. Yes, $NO_2$ acts as a catalyst in this reaction.

# CHAPTER 15

## Concept Tests

p. 737  3.0
p. 739  (b) $[CO_2] = [H_2] > [CO] = [H_2O]$
p. 751  Zone a
p. 754  There is no term for liquid water because its concentration is considered constant.
p. 755  (a) The value of $Q$ decreases. (b) After equilibrium is reached the concentrations of $CO_2$, $H_2O$, and CO will all have decreased and $[H_2]$ will have increased.
p. 758  The number of moles of gaseous reactants and products are the same.

## Practice Exercises

15.1.  $K_c = \dfrac{[CO][H_2]^3}{[CH_4][H_2O]}$        $K_p = \dfrac{(P_{CO})(P_{H_2})^3}{(P_{CH_4})(P_{H_2O})}$
15.2.  $K_c = [CH_3OH]/\{[CO][H_2]^2\} = 290$
15.3.  $K_p = 2.7 \times 10^4$
15.4.  $K_p = 4.5 \times 10^{-12}$
15.5.  $K_{p,reverse} = 2.3 \times 10^2$
15.6.  $K_c = 0.13$
15.7.  $K_{c,overall} = 1.7 \times 10^2$
15.8.  This reaction is not at equilibrium and proceeds to the right.
15.9.  a. $K_p = \dfrac{(P_{CO})^2}{(P_{CO_2})}$

       b. $K_p = \dfrac{(P_{CO})}{(P_{CO_2})(P_{H_2})}$

15.10.  a. When the reaction is cooled and water vapor condenses, one product is removed from the reaction mixture and the equilibrium shifts to the right, forming more $SO_2$.
        b. When $SO_2$ gas dissolves in liquid water as it condenses, products are removed and the equilibrium shifts to the right, forming more products.
        c. When $O_2$ is added, the concentration of one reactant increases and the equilibrium shifts to the right, forming more products.
15.11.  Increasing the pressure shifts the equilibrium in the reaction to the products, the side of the reaction that has the fewest moles of gas.
15.12.  The value of $K$ for the endothermic reaction increases with increasing reaction temperature.
15.13.  $P_{PCl_5} = 1.13$ atm; $P_{PCl_3} = 0.216$ atm; $P_{Cl_2} = 0.216$ atm
15.14.  $P_{Cl_2} = 0.0902$ atm; $P_{Br_2} = 0.0902$ atm; $P_{BrCl} = 0.0196$ atm
15.15.  $P_{ICl} = 0.0044$ atm; $P_{Cl_2} = 0.0044$ atm; $P_{I_2} = 1.36$ atm

# CHAPTER 16

## Concept Tests

p. 782  $HF > HCOOH > C_6H_5COOH > HN_3 > HClO$
p. 785  $OH^- > HCOO^- > NO_2^- > F^- > Cl^-$
p. 787  (top) pH 0.22 = strongly acidic; 4.37 = weakly acidic; 10.03 = weakly basic; 13.77 = strongly basic; 7.00 = neutral
p. 787  (bottom) All are true.
p. 788  Higher
p. 791  The larger its $K_a$, the greater is the percent ionization for a given concentration of HA.
p. 793  Most: C; least: A
p. 797  $H_3PO_4$, $H_2PO_4^-$, $HPO_4^{2-}$, $PO_4^{3-}$, $H_3O^+$, $OH^-$, $H_2O$
p. 799  No
p. 801  $H_3PO_4 > H_3AsO_4 > H_3SbO_4 \approx H_3BiO_4$
p. 802  More acidic because the decomposition reaction released $NH_3$, which is a base. The solid left behind should be more acidic than the reactant.
p. 805  C and D
p. 806  $MgCO_3$, because it has the smaller molar mass
p. 811  Malonic acid; slightly more base.
p. 815  The equilibrium would shift to produce more acid in their blood; they would die of extreme acidosis.
p. 821  (top) pH = $pK_a$ = 4.19
p. 821  (bottom) There would be too small a change in pH at the equivalence point to detect it precisely.
p. 823  Because neutralization of the $CO_3^{2-}$ produces more $HCO_3^-$ which adds to the $HCO_3^-$ present initially in the sample to make the second plateau wider than the first
p. 826  $K_{sp} = 4S^3$
p. 829  If $[H^+]$ decreases, less $F^-$ combines with $H^+$ to form HF. Therefore, the solubility of $CaF_2$ decreases.
p. 831  The value $[OH^-]$ is defined by the initial $[Ca^{2+}]$ and the $K_{sp}$ of $Ca(OH)_2$.

## Practice Exercises

16.1.  $CH_3COOH(aq) + H_2O(\ell) \rightarrow CH_3COO^-(aq) + H_3O^+(aq)$
       acid          base              conjugate base   conjugate acid
16.2.  Less acidic
16.3.  $[H^+] = 2 \times 10^{-12}$ $M$ and $[OH^-] = 5 \times 10^{-3}$ $M$
16.4.  pH = 2.23; percent ionization = 12%; $K_a = 7.9 \times 10^{-4}$
16.5.  pH = 11.97
16.6.  pH = 5.69
16.7.  $SO_4^{2-}(aq) + H_2O(\ell) \rightleftharpoons HSO_4^-(aq) + OH^-(aq)$
16.8.  pH = 9.08
16.9.  pH = 4.01
16.10.  pH = 4.75. There is essentially no change in the pH.
16.11.  About 1.78 parts base to 1 part acid (1.78:1)
16.12.  (a) change in pH of +0.11; (b) change in pH of +0.22
16.13.  (a) change in pH of −0.49; (b) change in pH of −0.50
16.14.  a. Ammonia concentration = 0.0650 $M$
        b. Methyl red (pH at equivalence point = 5.29)
16.15.  $S = 2.6 \times 10^{-3}$ $M$
16.16.  $S = 1.1 \times 10^{-5}$ $M$
16.17.  Yes
16.18.  a. Yes, both $BaF_2$ and $CaF_2$ are slightly soluble.
        b. Yes, $Ba^{2+}$ and $Ca^{2+}$ ions in solution can be completely separated by selective precipitation with $F^-$.

# CHAPTER 17

## Concept Tests

p. 849  Yes, N donates a pair of electrons to form the N—B bond.
p. 850  $CN^-$ ions are in the inner coordination sphere, and the $Na^+$ are counter ions.
p. 855  Tetrachloroplatinate(II)
p. 858  Yes. As a base: $Al(OH)_3(s) \rightleftharpoons Al(OH)_2^+(aq) + OH^-(aq)$; as an acid: $Al(OH)_3(s) + H_2O(\ell) \rightleftharpoons Al(OH)_4^-(aq) + H^+(aq)$
p. 866  Green
p. 869  (a) $CN^-$ is a stronger field ligand than pyridine. (b) $Ru^{2+}$ ions are larger than $Fe^{2+}$ and have a larger $\Delta_o$; their $4d$ electrons interact more with ligand lone pairs than do the $3d$ electrons of $Fe^{2+}$.
p. 872  No for square planar; yes for tetrahedral

## Practice Exercises

17.1.  CaO acts as a Lewis base and $CO_2$ acts as a Lewis acid.
17.2.  $[Ag^+(aq)] = 1.6 \times 10^{-8}\ M$
17.3.

| | | LIGAND | | | | | | | | |
|---|---|---|---|---|---|---|---|---|---|---|
| Compound | Counter Ion | Formula | Name | Number | Prefix | Formula | Name | Number | Prefix | $M^{n+}$ |
| $[Zn(NH_3)_4]Cl_2$ | $Cl^-$ (chloride) | $NH_3$ | ammine | 4 | tetra- | | | | | 2+ |
| $[Co(NH_3)_4 (H_2O)_2](NO_2)_2$ | $NO_2^-$ (nitrite) | $NH_3$ | ammine | 4 | tetra- | $H_2O$ | aqua | 2 | di- | 2+ |

 a.  $[Zn(NH_3)_4]Cl_2$ = tetraamminezinc(II) chloride
 b.  $[Co(NH_3)_4(H_2O)_2](NO_2)_2$ = tetraamminediaquacobalt(II) nitrite

17.4.  Four
17.5.  $Ni^{3+}$ can have either a high-spin or a low-spin configuration; none of the ions is diamagnetic.
17.6.

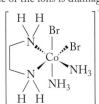

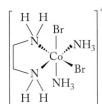

*cis*-Diammine-*cis*-dibromo-ethylenediaminecobalt(III) ion

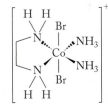

*cis*-Diammine-*trans*-dibromo-ethylenediaminecobalt(III) ion     *trans*-Diammine-*cis*-dibromo-ethylenediaminecobalt(III) ion

# CHAPTER 18

## Concept Tests

p. 886  (a), (c), and (d)
p. 889  $\Delta S_{univ} > 0$
p. 896  (a) Yes, it is spontaneous ($\Delta G < 0$); (b) thermodynamics says nothing about the rate of reaction.
p. 898  (top) The reaction is very slow.
p. 898  (bottom) No. The sum of $\Delta G^\circ_{f,prod}$ are all the same, but the reactants have different $\Delta G^\circ_f$ values, so each reaction will have a different $\Delta G^\circ$ value.
p. 901  The line for $T\Delta S$ has a negative slope.
p. 906  (e)
p. 915  (e)

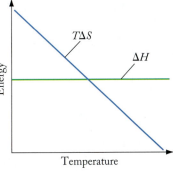

## Practice Exercises

18.1.  (a) increase in entropy; (b) decrease in entropy
18.2.  (a) > (b) > (d) > (c)
18.3.  Prediction: $S_{sys}$ decreases; $\Delta S^\circ_{rxn} = -243$ J/K
18.4.  a.  $\Delta S_{rxn}$ is expected to be negative.
       b.  $\Delta S^\circ_{rxn} = -326$ J/K
       c.  $\Delta H^\circ_{rxn} = -571.6$ kJ/mol
       d.  $\Delta G^\circ_{rxn} = -474.3$ kJ, so the reaction is spontaneous under standard conditions.
18.5.  142.2 kJ/mol
18.6.  The reaction is spontaneous only at high temperatures.
18.7.  780
18.8.  $2.95 \times 10^{-37}$ at 298 K; $9.20 \times 10^{-13}$ at 2000 K
18.9.  $\Delta G^\circ = -196$ kJ

# CHAPTER 19

## Concept Tests

p. 931  No, because the atomic masses of Cu and Zn are different.
p. 932  The cell consists of a zinc anode, which is oxidized to $Zn^{2+}$ ions as $Cu^{2+}$ ions are reduced to Cu metal at the cathode.
p. 934  (a)
p. 939  $\Delta G_{cell} > 0$, positive; $E_{cell} < 0$, negative
p. 942  0.257 V
p. 943  E = 0.740 V
p. 947  (a) +3; (b) −1; (c) +1; (d) +1
p. 949  C is oxidized; Co is reduced.

p. 956 Cathode: $H_2$; anode: $Cl_2$
p. 958 Anode

## Practice Exercises

19.1. $O_2(g) + 2\,NO_2^-(aq) \rightarrow 2\,NO_3^-(aq)$
19.2. The balanced redox reaction is

$$3\,Cu^{2+}(aq) + 2\,Al(s) \rightarrow 3\,Cu(s) + 2\,Al^{3+}(aq)$$

The cell diagram is

$$Al(s)\ |\ Al^{3+}(aq)\ ||\ Cu^{2+}(aq)\ |\ Cu(s)$$

19.3. The net ionic equation is $Cd(s) + 2\,NiO(OH)(s) + 2\,H_2O(\ell) \rightarrow$ $Cd(OH)_2(s) + 2\,Ni(OH)_2(s)$. $E_{cell}° = 1.30$ V.
19.4. $\Delta G_{cell} = -290$ kJ
19.5. $E_{cell} = 1.64$ V
19.6. $K = 1.8 \times 10^{62}$
19.7. 128 g
19.8. 1.5 g

# CHAPTER 20

## Concept Tests

p. 972 Two of the groups on the $sp^3$ C atom are identical.
p. 976 Arg (1+) > Ala (0) > Asp (1–)
p. 979 Those with nonpolar R groups
p. 981 (a) Random coil; (b) no, a change in the 2° structure is involved.
p. 987 Figure 20.25a
p. 991 (top) At low temperatures the unsaturated fatty acids remain liquid.
p. 991 (bottom) Olestra is a much larger molecule with many more C—C and C—H bonds, so on a per mole basis it would give off much more energy.
p. 992 Glycolysis of sugar produces the smaller molecule acetyl-CoA, which is a precursor in the synthesis of cholesterol.
p. 997 No, the direction in which the code is read matters; UUG codes for leucine.

## Practice Exercises

20.1.

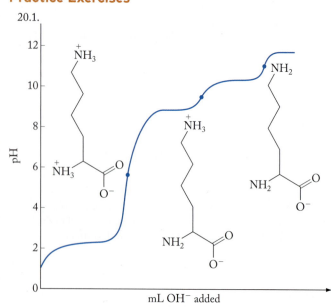

mL $OH^-$ added

20.2. Six
20.3. Val-Ala-Leu-Leu-Thr-Gly

20.4. Six
20.5. (a) GCCATAGGCTA; (b) AATTCGGCGATC

# CHAPTER 21

## Concept Tests

p. 1012 The loss in mass becomes an energy that must be added to separate the nucleons that are bound together.
p. 1016 The neutron-to-proton ratio for stable isotopes increases with atomic number. For heavy isotopes it is about 1.5 to 1. Losing α particles increases this ratio, creating neutron-rich nuclides that undergo β decay.
p. 1021 Increased $CO_2$($^{14}$C depleted) in the air from burned fossil fuels will reduce the $^{14}C/^{12}C$ ratio in the air and in plant tissues.
p. 1035 To fuse nuclei, their coulombic repulsion must be overcome, which requires they collide at very high velocities, requiring very high temperatures. Man-made fusion has not been carried out except in a hydrogen bomb. We cannot make a fusion reactor until the process itself can be safely carried out.

## Practice Exercises

21.1. Beta decay; $^{28}_{14}Si$
21.2. 98.6%
21.3. 4900 years old
21.4. $A = 7.02 \times 10^{12}$ Bq; $A = 190$ mCi
21.5. $1.4 \times 10^{11}$ X-rays
21.6. $1.753 \times 10^9$ kJ/mol

# CHAPTER 22

## Concept Tests

p. 1050 They are not found in nature, and so they can have no natural biological function.
p. 1053 (top) Nonspontaneous—they require energy to pump ions
p. 1053 (bottom) Because $Mg^{2+}$ has no $d$ electrons
p. 1056 $K > 1$ for the drug to be effective
p. 1062 $H_2$ because it has no lone pairs of electrons
p. 1063 Because in $O_2^-$ the oxidation number of O is $-\frac{1}{2}$, it is readily reduced to the more stable 2– ion.
p. 1064 Yes, the spin state changes.
p. 1065 Nonessential elements are generally toward the bottom of the periodic table. They have larger atomic numbers and are less abundant than the essential elements in the same group.
p. 1067 To avoid tissue damage, the nuclide should not emit high-energy α or β particles.
p. 1070 Longer half-life means slower decay and longer time to get the image.
p. 1079 The reduction potentials for these metals and alloys should be less negative than Fe so that the materials are less likely to oxidize than iron.

## Practice Exercises

22.1. pH = 1.19; the volume of $Mg(OH)_2$ solution required to neutralize the acid solution is $5.83 \times 10^{-2}$ mL.
22.2. $N_2(g) + 10\,H^+(aq) + 8\,e^- \rightarrow 2\,NH_4^+(aq) + H_2(g)$
22.3. $^{201}_{81}Tl + ^{0}_{-1}e \rightarrow ^{201}_{80}Hg$
22.4. The amount of $^{186}$Re is 21 mg after 24 hr. The amount of $^{188}$Re is 9.4 mg after 24 hr.
22.5. $K_{overall} = 2.5 \times 10^6$
22.6. $E_{cat}° = 1.05$ V

# Answers to Selected End-of-Chapter Questions and Problems

## CHAPTER 1

1.1.  a.  A pure compound in the gas phase.
    b.  A heterogeneous mixture of blue element atoms and red element atoms: blue atoms are in the gas phase, red spheres are in the liquid phase.

1.3.  b

1.5.  $H_3COH$ or $CH_4O$

1.7.  The sun is an example where matter is being changed into energy through nuclear fusion reactions. Therefore, both students are correct.

1.9.  One chemical property of gold is its resistance to corrosion (oxidation). Gold's physical properties include its density, color, melting temperature, and electrical and thermal conductivity.

1.11.  Add water to the salt–sand mixture to dissolve the salt. Passing the sand–solution mixture through a filter will leave the sand on the filter. The salt can be recovered by evaporating the water from the solution that passed through the filter.

1.13.  (b) Combustion

1.15.  b and d

1.17.  (d) Orange juice (with pulp)

1.19.  We can distinguish between table sugar, water, and oxygen by examining their physical states (sugar is a solid, water is a liquid, and oxygen is a gas) and by their densities, melting points, and boiling points.

1.21.  Density, melting point, thermal and electrical conductivity, and softness (a–d) are all physical properties, whereas tarnishing and reaction with water (e, f) are both chemical properties.

1.23.  a and e

1.25.  a

1.27.  a

1.29.  Sand and water (b) may be separated by filtration.

1.31.  c

1.33.  Extensive properties will change with the size of the sample and therefore cannot be used to identify a substance.

1.35.  To form a hypothesis we need at least one observation, experiment, or idea (from examining nature).

1.37.  Yes.

1.39.  *Theory* in normal conversation is someone's idea or opinion or speculation that can be changed.

1.41.  SI units can be easily converted into a larger or smaller unit by multiplying or dividing by multiples of 10. English units are based on other number multiples and thus are more complicated to manipulate.

1.43.  b (and c, if the zero is *not* significant)

1.45.  b (and c, if the zero is *not* significant)

1.47.  b, c (and d if the zero is significant)

1.49.  (a) 17.4; (b) $7 \times 10^{-14}$; (c) $5.70 \times 10^{-23}$; (d) $3.58 \times 10^{-3}$

1.51.  (a) 7.14 mi/hr; (b) 3.19 m/s

1.53. 1.201 U.S. gallons

1.55. $7.48 \times 10^2$ g

1.57. $9.24 \times 10^3$ mL

1.59. Peter is taller than Paul.

1.61. 1330 cal

1.63. 2.5 mi

1.65. 23 g

1.67. 58.0 cm$^3$

1.69. 73.8 mL

1.71. Yes

1.73. 0.28 cm$^3$

1.75. a. Manufacturer 1 has a range of $0.516 - 0.504 = 0.012$ μm; Manufacturer 2 has a range of $0.514 - 0.512 = 0.002$ μm; and Manufacturer 3 has a range of $0.502 - 0.500 = 0.002$ μm.
   b. Yes, Manufacturers 2 and 3 can justify the claim.
   c. Yes. In the case of Manufacturer 2, where the lines are printed at wider widths than the widths specified, the lines are precise, but inaccurate!

1.77. Yes, $-40°C$ is equal to $-40°F$.

1.79. $-269.0°C$

1.81. 39.2°C

1.83. $-89.2°C$; 183.9 K

1.85. $-38°F$; 230°F

1.87. The $T_c$ for $YBa_2Cu_3O_7$ is already expressed in kelvin, $T_c = 93.0$ K. The $T_c$ of $Nb_3Ge$ converted to K is 23.2 K. The $T_c$ of $HgBa_2CaCu_2O_6$ converted to K is 127.0 K. The superconductor with the highest $T_c$ is $HgBa_2CaCu_2O_6$.

1.89. 0.031 mg/L

1.91. Both mixtures a and b react so that there is neither sodium nor chlorine left over.

1.93. (a) No; (b) 83 plates

1.95. 17 bicycles

1.97. Day 11

1.99. Spring B is stronger.

# CHAPTER 2

2.1. (c) A mixture of $NO_2$ and NO

2.3. b

2.5. a. Chlorine ($Cl_2$) (yellow)
   b. Neon (Ne) (red)
   c. Sodium (Na) (dark blue)

2.7. a. Mg (green) will form MgO.
   b. K (red) will form $K_2O$.
   c. Ti (yellow) will form $TiO_2$.
   d. Al (dark blue) will form $Al_2O_3$.

2.9. Rutherford concluded that the positive (the pudding) charge in the atom could not be spread out in the atom, but must result from a concentration of charge in the center of the atom (the nucleus). Most of the α particles were deflected only slightly or passed directly through the gold foil. So he reasoned that the nucleus must be small compared to the size of the entire atom. The negatively charged electrons do not deflect the α particles, and Rutherford reasoned that the electrons took up the remainder of the space of the atom outside the nucleus.

2.11. The fact that cathode rays were deflected by a magnetic field indicated that the rays were streams of charged particles.

2.13. A *weighted average* takes into account the proportion of each value in the group of values to be averaged.

2.15. Greater than 1

2.17. a and b

2.19. 35.45 amu

2.21. Yes

2.23. 47.95 amu

2.25. Mendeleev knew only the masses of the elements at the time he arranged the elements into his periodic table.

2.27. The order of elements is determined by the number of protons (atomic number), but atomic masses are a weighted average of isotope masses. When Mendeleev put together his periodic table, he was unaware of the presence of subatomic particles. Because of this, he did not know that the atomic masses he used to sort the elements were a weighted average of the naturally occurring isotopes of a given element. Elements with a large natural abundance of heavier isotopes exhibit a higher average atomic mass than might be predicted by atomic number alone, hence creating a different order.

2.29.

| Atom | Mass Number | Atomic Number = Number of Protons | Number of Neutrons = Mass Number − Atomic Number | Number of Electrons = Number of Protons − the Charge |
|---|---|---|---|---|
| (a) $^{14}C$ | 14 | 6 | 8 | 6 |
| (b) $^{59}Fe$ | 59 | 26 | 33 | 26 |
| (c) $^{90}Sr$ | 90 | 38 | 52 | 38 |
| (d) $^{210}Pb$ | 210 | 82 | 128 | 82 |

2.31.

| Symbol | $^{23}Na$ | $^{89}Y$ | $^{118}Sn$ | $^{197}Au$ |
|---|---|---|---|---|
| Number of Protons | 11 | 39 | 50 | 79 |
| Number of Neutrons | 12 | 50 | 68 | 118 |
| Number of Electrons | 11 | 39 | 50 | 79 |
| Mass Number | 23 | 89 | 118 | 197 |

2.33.

| Symbol | $^{37}Cl^-$ | $^{23}Na^+$ | $^{81}Br^-$ | $^{226}Ra^{2+}$ |
|---|---|---|---|---|
| Number of Protons | 17 | 11 | 35 | 88 |
| Number of Neutrons | 20 | 12 | 46 | 138 |
| Number of Electrons | 18 | 10 | 36 | 86 |
| Mass Number | 37 | 23 | 81 | 226 |

2.35. (c) Be

2.37. (c) $S^{2-}$

2.39. (a) $S^{2-}$; (b) $P^{3-}$; and (d) $Ca^{2+}$

2.41. (b) Br

2.43. $F^-$

2.45. $Cl_2$

2.47. Na

2.49. NaCl and $Na_2SO_4$; KCl and $K_2SO_4$; $CaCl_2$ and $CaSO_4$; $MgCl_2$ and $MgSO_4$

2.51. Dalton's atomic theory states that, because atoms are indivisible, the ratio of the atoms (elements) in a compound is a ratio of whole numbers. Thus, in water, the ratio of volumes of hydrogen to oxygen is 2:1, a whole-number ratio, because the atoms in water are in the ratio of 2:1.

2.53. Molecular compounds are composed of non-metals; ionic compounds are composed of a cation derived from a metal or the $NH_4^+$ cation, and an anion, which is derived from a nonmetal or polyatomic anion.

2.55. 1.5

2.57. 7.5 g

2.59. $NaCl$, $MgCl_2$, $CaCl_2$, $KCl$, $SrCl_2$; $Na_2SO_4$, $MgSO_4$, $CaSO_4$, $K_2SO_4$, $SrSO_4$

2.61. Magnesium hydroxide.

2.63. (a) hydrogen bromide; (b) carbon disulfide; (c) nitrogen trifluoride; (d) phosphorus pentafluoride

2.65. a and d are molecules; b and c consist of ions.

2.67. (a) 4 atoms; (b) 5 atoms; (c) 8 atoms; (d) 15 atoms

2.69. $XO_2^{2-}$

2.71. Roman numerals indicate the charge on the transition metal cation.

2.73.   a.  $NO_3$, nitrogen trioxide
     b.  $N_2O_5$, dinitrogen pentoxide
     c.  $N_2O_4$, dinitrogen tetroxide
     d.  $NO_2$, nitrogen dioxide
     e.  $N_2O_3$, dinitrogen trioxide
     f.  NO, nitrogen monoxide
     g.  $N_2O$, dinitrogen monoxide
     h.  $N_4O$, tetranitrogen monoxide

2.75.   a.  $Na_2S$, sodium sulfide
     b.  $SrCl_2$, strontium chloride
     c.  $Al_2O_3$, aluminum oxide
     d.  LiH, lithium hydride

2.77. (b) $LiSO_4$ is incorrect (should be $Li_2SO_4$).

2.79. (a) $BrO^-$; (b) $SO_4^{2-}$; (c) $IO_3^-$; (d) $NO_2^-$

2.81. (a) potassium carbonate; (b) sodium cyanide; (c) lithium hydrogen carbonate; (d) calcium hypochlorite

2.83. (a) hydrofluoric acid; (b) bromic acid; (c) $H_3PO_4$; (d) $HNO_2$

2.85. (a) sodium oxide; (b) sodium sulfide; (c) sodium sulfate; (d) sodium nitrate; (e) sodium nitrite

2.87. (a) $K_2S$; (b) $K_2Se$; (c) $Rb_2SO_4$; (d) $RbNO_2$; (e) $MgSO_4$

2.89. $Na_2SO_3$

2.91. (a) cobalt(II) oxide; (b) cobalt(III) oxide; (c) cobalt(IV) oxide

2.93. (a) manganese(II) sulfide; (b) vanadium(II) nitride; (c) chromium(III) sulfate; (d) cobalt(II) nitrate; (e) iron(III) oxide

2.95. (d) Vanadium(IV) oxide is incorrect (should be $VO_2$).

2.97. Chemistry is the study of the composition, structure, properties, and reactivity of matter. Cosmology is the study of the history, structure, and dynamics of the universe. A few of the ways that these two sciences are related might be: (1) Because the universe is composed of matter and the study of matter is chemistry, the study of the universe is really chemistry; (2) the changing universe is driven by chemical and atomic or nuclear reactions, which are also studied in chemistry; (3) cosmology often asks what the universe (including stars, black holes, etc.) is made of at the atomic level.

2.99. Because quarks combine to make up the three particles that are important to the properties and reactivity of atoms: protons, neutrons, and electrons.

2.101. The density of the universe is decreasing.

2.103. The electron is twice as hard to remove from a helium atom compared to a hydrogen atom because it is being held by a nucleus of 2+ charge rather than one of 1+ charge.

2.105. $^{21}_{10}Ne + ^4_2\alpha \rightarrow ^1_0n + ^{24}_{12}Mg$

2.107.   a.  Three

     b.  The spots would all be in different locations, one at the center of the screen, one closer to the negatively charged plate, and one closer to the positively charged plate. The spot closest to the negatively charged plate would be from $\alpha$ particles, the spot closest to the positively charged plate would be from $\beta$ particles, and the one at the center would be from $\gamma$ rays.

2.109. (a) 12 H atoms for every 1 He atom. (b) There is more helium present now. (c) Stars consume hydrogen and produce helium. (d) Look at the elemental compositions of older galaxies and compare them to our own galaxy.

2.111. (a) $Cr_2O_3$, chromium(III) oxide and NiO, nickel(II) oxide. (b) $Cr^{3+}$ and $Ni^{2+}$.

2.113. group V

2.115.   a.  magnesia = magnesium oxide, MgO
     b.  Epsom salt = magnesium sulfate, $MgSO_4$
     c.  K-Dur = potassium chloride, KCl
     d.  lime = calcium oxide, CaO
     e.  baking soda = sodium bicarbonate, $NaHCO_3$
     f.  caustic soda = sodium hydroxide, NaOH
     g.  muriatic acid = hydrogen chloride solution, HCl(*aq*)
     h.  zirconia = zirconium dioxide, $ZrO_2$

2.117. (a) $S_2O_3^{2-}$; (b) $Na_2S_2O_3$

2.119. 49.59%

2.121. $^{25}Mg$ 10.00%, and $^{26}Mg$ 11.01%

2.123. (a) $^{24}_{12}Mg^{2+}$; (b) $^{121}_{51}Sb^{3+}$; (c) $^{84}_{36}Kr$

2.125. Radium will adopt a 2+ charge to form $Ra^{2+}$ ions. It will likely be malleable, relatively dense, conduct heat and electric current, and melt at a fairly high temperature.

| | Melting Point (°C) | | Melting Point (°C) |
|---|---|---|---|
| $CaCl_2$ | 772 | CaO | 2572 |
| $SrCl_2$ | 874 | SrO | 2531 |
| $BaCl_2$ | 962 | BaO | 1923 |
| $RaCl_2$ | (950 to 1050) | RaO | (1700 to 2000) |

2.127. Despite being heavier (on average), argon contains 18 protons, whereas potassium contains 19 protons. Since the modern periodic table is organized by increasing atomic number, argon is placed before potassium.

## CHAPTER 3

3.1.   a.  $4X(g) + 4Y(g) \rightarrow 4XY(g)$
     b.  $4X(g) + 4Y(g) \rightarrow 4XY(s)$
     c.  $4X(g) + 4Y(g) \rightarrow 2XY_2(g) + 2X(g)$
     d.  $4X_2(g) + 4Y_2(g) \rightarrow 8XY(g)$

3.3. b

3.5. a and c are the same ($NO_2$); b and d are the same ($N_2O$).

3.7.   b. $2SO_2 + O_2 \rightarrow 2SO_3$
     d. $CS_2 + 3O_2 \rightarrow CO_2 + 2SO_2$

3.9. $Fe(\ell)$

3.11. Less than

3.13. It is too small a unit to express the very large number of atoms, ions, or molecules present in laboratory quantities such as a mole.

3.15. No, the molar mass of a substance does not directly correlate to the number of atoms in a molecular compound. The statement would be true only if the two compounds were composed of the same elements.

3.17. (a) $7.3 \times 10^{-10}$ mol Ne; (b) $7.0 \times 10^{-11}$ mol $CH_4$; (c) $4.2 \times 10^{-12}$ mol $O_3$; (d) $8.1 \times 10^{-15}$ mol $NO_2$

3.19. (a) $7.53 \times 10^{22}$ Ti atoms; (b) $7.53 \times 10^{22}$ Ti atoms; (c) $1.51 \times 10^{23}$ Ti atoms; (d) $2.26 \times 10^{23}$ Ti atoms

3.21. (a) Both contain the same; (b) $N_2O_4$; (c) $CO_2$

3.23. (a) 3.00 mol; (b) 4.50 mol; (c) 1.50 mol

3.25. 41.63 mol

3.27. $7.0 \times 10^{19}$ Ir atoms

3.29. (a) 1 mol; (b) 2 mol; (c) 1 mol; (d) 3 mol

3.31. (a) 64.06 g/mol; (b) 48.00 g/mol; (c) 44.01 g/mol; (d) 108.02 g/mol

3.33. (a) 152.16 g/mol; (b) 164.22 g/mol; (c) 148.22 g/mol; (d) 132.17 g/mol

3.35. (a) NO; (b) $CO_2$; (c) $O_2$

3.37. 0.752 mol $SiO_2$

3.39. 10.3 g

3.41. Diamond

3.43. No

3.45. No

3.47.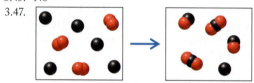

3.49. a. $3\,FeSiO_3(s) + 4\,H_2O(\ell) \rightarrow Fe_3Si_2O_5(OH)_4(s) + H_4SiO_4(aq)$
b. $Fe_2SiO_4(s) + 2\,CO_2(g) + 2\,H_2O(\ell) \rightarrow 2\,FeCO_3(s) + H_4SiO_4(aq)$
c. $Fe_3Si_2O_5(OH)_4(s) + 3\,CO_2(g) + 2\,H_2O(\ell) \rightarrow 3\,FeCO_3(s) + 2\,H_4SiO_4(aq)$

3.51. a. $N_2(g) + O_2(g) \rightarrow 2\,NO(g)$
b. $2\,NO(g) + O_2(g) \rightarrow 2\,NO_2(g)$
c. $NO(g) + NO_3(g) \rightarrow 2\,NO_2(g)$
d. $2\,N_2(g) + O_2(g) \rightarrow 2\,N_2O(g)$

3.53. a. $N_2O_5(g) + Na(s) \rightarrow NaNO_3(s) + NO_2(g)$
b. $N_2O_4(g) + H_2O(\ell) \rightarrow HNO_3(aq) + HNO_2(aq)$
c. $3\,NO(g) \rightarrow N_2O(g) + NO_2(g)$
d. $2\,C_2H_2(g) + 5\,O_2(g) \rightarrow 4\,CO_2(g) + 2\,H_2O(g)$

3.55. Yes

3.57. (a) $4.5 \times 10^{11}$ mol C; (b) $2.0 \times 10^{10}$ kg $CO_2$

3.59. a. $2\,NaHCO_3(s) \rightarrow CO_2(g) + H_2O(g) + Na_2CO_3(s)$
b. 6.55 g $CO_2$

3.61. 1.17 kg

3.63. 1.5 metric tons $SO_2$

3.65. $2\,C_8H_{18} + 25\,O_2 \rightarrow 16\,CO_2 + 18\,H_2O$
$C_2H_6O + 3\,O_2 \rightarrow 2\,CO_2 + 3\,H_2O$
Octane produces more $CO_2$.

3.67. 346 g Cu

3.69. An empirical formula shows the lowest whole-number ratio of atoms in a substance. A molecular formula shows the actual numbers of each kind of atom that compose one molecule of the substance.

3.71. No

3.73. Yes, all of these have the same empirical and molecular formulas.

3.75. No

3.77. a. 74.19% Na, 25.81% O
b. 57.48% Na, 40.00% O, 2.52% H
c. 27.37% Na, 1.20% H, 14.30% C, 57.13% O
d. 43.38% Na, 11.33% C, 45.28% O

3.79. Pyrene, $C_{16}H_{10}$, has the greatest percent carbon by mass. The empirical formulas differ.

3.81. $CH_4$

3.83. $Ti_6Al_4V$

3.85. (a) MgO; (b) $2\,Mg(s) + O_2(g) \rightarrow 2\,MgO(s)$

3.87. $Mg_3Si_2H_4O_9$

3.89. Empirical formula: CHN. Molecular formula: $C_5H_5N_5$.

3.91. The excess of oxygen is required in combustion analysis to ensure the complete reaction of the hydrogen and carbon to form water and carbon dioxide.

3.93. Yes

3.95. $NO_2$

3.97. The empirical formula is $C_2H_3$. The molecular formula is $C_{20}H_{30}$.

3.99. $C_{10}H_{18}O$

3.101. $C_9H_8O_4$

3.103. (c) Less than the sum of the masses of Fe and S to start

3.105. Reactions do not always go to completion because the reaction may be slow or may have, for a portion of the reaction, yielded different products than expected.

3.107. 3 cups

3.109. $NH_3(g) + HCl(g) \rightarrow NH_4Cl(s)$; 0.7 g $NH_3$

3.111. a. $2\,PbO(s) + 2\,NaCl(aq) + H_2O(\ell) + CO_2(g) \rightarrow Pb_2Cl_2CO_3(s) + 2\,NaOH(aq)$
b. 12.2 g
c. 22.3%

3.113. 59%

3.115. (a) $C_6H_{12}O_6(aq) \rightarrow 2\,C_2H_5OH(\ell) + 2\,CO_2(g)$; (b) 77.1%

3.117. (a) 36.09%; (b) 36.07%

3.119. (a) 870 g; (b) 1020 g; (c) 1400 g

3.121. (a) $a = 1$, $b = 3$, charge on U is 6+; (b) $c = 3$, $d = 8$, charge on U is 5.33+; (c) $x = 2$, $y = 2$, $z = 6$

3.123. (a) $5.838 \times 10^{20}$ molecules of $C_{13}H_{18}O_2$; (b) $3.008 \times 10^{21}$ molecules of $CaCO_3$; (c) $9 \times 10^{18}$ molecules of $C_{16}H_{19}N_2Cl$

3.125. a. $Mn_2O_3$ is manganese(III) oxide and $MnO_2$ is manganese(IV) oxide.
b. % Mn in $Mn_2O_3$ = 69.60%; % Mn in $MnO_2$ = 63.19%.
c. These compounds contain the same elements, but in different atom ratios.

3.127. (a) 0.966 g; (b) $C_{10}H_{20}O_{10}$

3.129. $1 \times 10^{-8}$ mol

3.131. 55 mol ethanol

3.133. Re

3.135. 82.4%

3.137. a. 6.0 metric tons
b. $2\,SO_2(g) + 2\,H_2O(g) + O_2(g) \rightarrow 2\,H_2SO_4(\ell)$
c. 9.2 metric tons

3.139. 3.06 g $H_2SO_4$

3.141. $Mg_2SiO_4$

3.143. (a) $3.05 \times 10^3$ g $KO_2$; (b) More $Na_2O_2$; (c) $KO_2$

# CHAPTER 4

4.1. Yellow

4.3. a. Cl (purple)
b. S (orange)
c. N (green)
d. P (blue)

4.5. a and c are strong electrolytes; c is a strong acid; b is a weak acid; b is a weak electrolyte; d is a nonelectrolyte.

4.7. Hydronium and nitrate ions will remain in solution.

4.9. The solvent is usually the liquid component of the solution. If the solvent and solute are both liquids or both solids, the solvent is that component present in the greatest amount.

4.11. $1.00\ M$

4.13. a. $5.6\ M\ BaCl_2$
   b. $1.00\ M\ Na_2CO_3$
   c. $1.30\ M\ C_6H_{12}O_6$
   d. $5.92\ M\ KNO_3$

4.15. a. $0.14\ M\ Na^+$
   b. $0.11\ M\ Cl^-$
   c. $0.096\ M\ SO_4^{2-}$
   d. $0.20\ M\ Ca^{2+}$

4.17. a. $11.7$ g NaCl
   b. $4.99$ g $CuSO_4$
   c. $6.41$ g $CH_3OH$

4.19. $2.72$ g

4.21. a. $9.6 \times 10^{-3}$ mol
   b. $7.80 \times 10^{-4}$ mol
   c. $8.8 \times 10^{-2}$ mol
   d. $4.22$ mol

4.23. Orchard sample: $3.4 \times 10^{-4}$ mmol/L, 0.12 ppm
   Residential area sample: $5.6 \times 10^{-5}$ mmol/L, 0.020 ppm
   After storm sample: $3.2 \times 10^{-2}$ mmol/L, 11 ppm

4.25. $1.15 \times 10^{-6}$ mg/kg $NF_3$ per kg air

4.27. $4.4 \times 10^{-5}\ M$

4.29. a. The final concentration after diluting will be $1.81 \times 10^{-2}\ M\ Na^+$.
   b. The final concentration after diluting will be $2.7 \times 10^{-1}$ m$M$ LiCl.
   c. The final concentration after diluting will be $1.28 \times 10^{-2}$ m$M$ $Zn^{2+}$.

4.31. $58.63$ mL

4.33. $1.95\ M$

4.35. $12.3$ mL

4.37. $A = 0.75$

4.39. Table salt produces $Na^+$ and $Cl^-$ ions in solution when it dissolves. Sugar does not dissociate into ions because it is not a salt. Ions are required to conduct electricity.

4.41. The lack of ions in methanol means that the liquid is nonconductive. Molten NaOH, however, has freely moving $Na^+$ and $OH^-$ ions, which can conduct electricity.

4.43. In order of decreasing conductivity, $1.0\ M\ Na_2SO_4$ (c) $> 1.2\ M$ KCl (b) $> 1.0\ M$ NaCl (a) $> 0.75\ M$ LiCl (d).

4.45. (a) $0.025\ M$; (b) $0.050\ M$; (c) $0.075\ M$

4.47. (b) $1\ M\ CaCl_2$

4.49. Acid

4.51. Strong acids include $HCl$, $HNO_3$, $HClO_4$, $H_2SO_4$, $HI$, $HBr$; weak acids include $CH_3COOH$, $HCOOH$, $HF$, $H_3PO_4$.

4.53. Base

4.55. Strong bases include NaOH, KOH, CsOH, LiOH, RbOH, $Ba(OH)_2$, $Sr(OH)_2$, $Ca(OH)_2$; weak bases include $NH_3$, $CH_3NH_2$, $C_5H_5N$.

4.57. a. Ionic and net ionic equation: $H^+(aq) + HSO_4^{2-}(aq) + Ca^{2+}(aq) + 2\ OH^-(aq) \rightarrow CaSO_4(s) + 2\ H_2O(\ell)$
   The acid is $H_2SO_4$; the base is $Ca(OH)_2$.
   b. Ionic and net ionic equation: $PbCO_3(s) + H^+(aq) + HSO_4^{2-}(aq) \rightarrow PbSO_4(s) + CO_2(g) + H_2O(\ell)$
   $PbCO_3$ is the base; sulfuric acid is the acid.

   c. Ionic equation: $Ca(OH)_2(s) + 2\ CH_3COOH(aq) \rightarrow Ca^{2+}(aq) + 2\ CH_3COO^-(aq) + 2\ H_2O(\ell)$
   Calcium is a spectator ion. $Ca(OH)_2$ is the base; $CH_3COOH$ is the acid.
   Net ionic equation: $OH^-(aq) + CH_3COOH(aq) \rightarrow CH_3COO^-(aq) + H_2O(\ell)$

4.59. a. Molecular equation:
   $Mg(OH)_2(s) + H_2SO_4(aq) \rightarrow MgSO_4(aq) + 2\ H_2O(\ell)$
   Net ionic equation:
   $Mg(OH)_2(s) + H^+(aq) + HSO_4^-(aq) \rightarrow Mg^{2+}(aq) + 2\ H_2O(\ell) + SO_4^{2-}(aq)$
   b. Molecular equation:
   $MgCO_3(s) + 2\ HCl(aq) \rightarrow MgCl_2(aq) + H_2CO_3(aq)$
   Net ionic equation:
   $MgCO_3(s) + 2\ H^+(aq) \rightarrow Mg^{2+}(aq) + H_2O(\ell) + CO_2(g)$
   c. Molecular equation: $NH_3(g) + HCl(g) \rightarrow NH_4Cl(s)$
   This is also the net ionic equation.
   d. Molecular equation:
   $SO_3(g) + 2\ NaOH(aq) \rightarrow Na_2SO_4(aq) + H_2O(\ell)$
   Net ionic equation:
   $SO_3(g) + 2\ OH^-(aq) \rightarrow SO_4^{2-}(aq) + H_2O(\ell)$

4.61. $PbCO_3(s) + 2\ H^+(aq) \rightarrow Pb^{2+}(aq) + CO_2(g) + H_2O(\ell)$;
   $Pb(OH)_2(s) + 2\ H^+(aq) \rightarrow Pb^{2+}(aq) + 2\ H_2O(\ell)$

4.63. (a) $5.00$ mL; (b) $31.5$ mL; (c) $215$ mL

4.65. $500$ mL

4.67. $280$ mL

4.69. A saturated solution contains the maximum concentration of a solute. A supersaturated solution *temporarily* contains *more* than the maximum concentration of a solute at a given temperature.

4.71. $CaCO_3$

4.73. A saturated solution may not be a concentrated solution if the solute is only sparingly or slightly soluble in the solution. In that case, the solution is a saturated dilute solution.

4.75. (a) Barium sulfate is insoluble; (e) lead hydroxide is insoluble; (f) calcium phosphate is insoluble.

4.77. a. Balanced reaction:
   $Pb(NO_3)_2(aq) + Na_2SO_4(aq) \rightarrow PbSO_4(s) + 2\ NaNO_3(aq)$
   Net ionic equation:
   $Pb^{2+}(aq) + SO_4^{2-}(aq) \rightarrow PbSO_4(s)$
   b. No precipitation reaction occurs.
   c. Balanced reaction:
   $FeCl_2(aq) + Na_2S(aq) \rightarrow FeS(s) + 2\ NaCl\ (aq)$
   Net ionic equation: $Fe^{2+}(aq) + S^{2-}(aq) \rightarrow FeS(s)$
   d. Balanced reaction:
   $MgSO_4(aq) + BaCl_2(aq) \rightarrow MgCl_2(aq) + BaSO_4(s)$
   Net ionic equation: $Ba^{2+}(aq) + SO_4^{2-}(aq) \rightarrow BaSO_4(s)$

4.79. $2.11 \times 10^{-2}$ g

4.81. $5.4 \times 10^{-2}$ g

4.83. $130$ kg

4.85. a. $[Na^+]$ and $[NO_3^-]$ remain the same, while $[Ag^+]$ and $[Cl^-]$ decrease.
   b. $[Na^+]$ and $[Cl^-]$ remain the same, while $[H^+]$ and $[OH^-]$ decrease.
   c. All ionic concentrations remain the same.

4.87. To deionize water, cations such as $Na^+$ and $Ca^{2+}$ are exchanged for $H^+$ at cation-exchange sites. Anions such as $Cl^-$ and $SO_4^{2-}$ are exchanged for $OH^-$ at the anion-exchange sites. The released ions ($H^+$ and $OH^-$) at these sites combine to form $H_2O$.

4.89. To deionize water, the cation at the cation-exchange site must be $H^+$ and the anion at the anion-exchange site must be $OH^-$. When these combine, they form $H_2O$.

4.91. The number of electrons gained or lost is directly related to the change in oxidation number of a species.

4.93. (a) −1; (b) +1; (c) −2; (d) −3

4.95. Silver

4.97. $Na^+ + e^- \rightarrow Na(s)$
$2\,Cl^- \rightarrow Cl_2(g) + 2\,e^-$

4.99. (a) +1; (b) +5; (c) +7

4.101. a. $2\,e^- + Br_2(\ell) \rightarrow 2\,Br^-(aq)$; reduction
b. $Pb(s) + 2\,Cl^-(aq) \rightarrow PbCl_2(s) + 2\,e^-$; oxidation
c. $2\,e^- + O_3(g) + 2\,H^+(aq) \rightarrow O_2(g) + H_2O(\ell)$; reduction
d. $2\,H_2SO_3(aq) + H^+(aq) + 2\,e^- \rightarrow HS_2O_4^-(aq) + 2\,H_2O\ (\ell)$; reduction

4.103. a.

| Reactants | Products |
|---|---|
| $SiO_2$: Si = +4, O = −2 | $Fe_2SiO_4$: Fe = +2, Si = +4, O = −2 |
| $Fe_3O_4$: Fe = +8/3, O = −2 | $O_2$: O = 0 |
| Oxygen is oxidized ($O^{2-}$ to $O_2$) and iron is reduced from an average of +8/3 to +2. | |

b.

| Reactants | Products |
|---|---|
| $SiO_2$: Si = +4, O = −2 | $Fe_2SiO_4$: Fe = +2, Si = +4, O = −2 |
| Fe: Fe = 0 | |
| $O_2$: O = 0 | |
| Iron is oxidized ($Fe^0$ to $Fe^{2+}$) and oxygen is reduced ($O_2$ to $O^{2-}$). | |

c.

| Reactants | Products |
|---|---|
| FeO: Fe = +2, O = −2 | $Fe(OH)_3$: Fe = +3, O = −2, H = +1 |
| $O_2$: O = 0 | |
| $H_2O$: H = +1, O = −2 | |
| Iron is oxidized ($Fe^{2+}$ to $Fe^{3+}$) and oxygen is reduced ($O_2$ to $O^{2-}$). | |

4.105. a. $O_2(aq) + 4\,FeCO_3(s) \rightarrow 2\,Fe_2O_3(s) + 4\,CO_2(g)$
b. $O_2(aq) + 6\,FeCO_3(s) \rightarrow 2\,Fe_3O_4(s) + 6\,CO_2(g)$
c. $O_2(aq) + 4\,Fe_3O_4(s) \rightarrow 6\,Fe_2O_3(s)$

4.107. $NH_4^+(aq) + 2\,O_2(g) \rightarrow NO_3^-(aq) + 2\,H^+(aq) + H_2O(\ell)$

4.109. a.

| Reactants | Products |
|---|---|
| $HCrO_4^-$: H = +1, Cr = +6, O = −2 | $Cr_2O_3$: Cr = +3, O = −2 |
| $H_2S$: H = +1, S = −2 | $SO_4^{2-}$: S = +6, O = −2 |

b. $3\,H_2S(aq) + 8\,HCrO_4^-(aq) + 2\,H^+ \rightarrow$
$3\,SO_4^{2-}(aq) + 4\,Cr_2O_3(s) + 8\,H_2O(\ell)$
c. 3

4.111. $2\,Fe(OH)_2^+(aq) + Mn^{2+}(aq) \rightarrow$
$2\,Fe^{2+}(aq) + 2\,H_2O(\ell) + MnO_2(s)$

4.113. a. $8\,H_2O(\ell) + 2\,MnO_4^-(aq) + 7\,S^{2-}(aq) \rightarrow$
$2\,MnS(s) + 5\,S(s) + 16\,OH^-(aq)$
b. $H_2O(\ell) + 2\,MnO_4^-(aq) + 3\,CN^-(aq) \rightarrow$
$2\,MnO_2(s) + 3\,CNO^-(aq) + 2\,OH^-(aq)$
c. $H_2O(\ell) + 2\,MnO_4^-(aq) + 3\,SO_3^{2-}(aq) \rightarrow$
$2\,MnO_2(s) + 3\,SO_4^{2-}(aq) + 2\,OH^-(aq)$
d. $2\,H_2O(\ell) + 4\,Ag(s) + 8\,CN^-(aq) + O_2(g) \rightarrow$
$4\,Ag(CN)_2^-(aq) + 4\,OH^-(aq)$

4.115. Zinc and aluminum.

4.117. Vanadium is placed below aluminum, and scandium is placed above aluminum on the activity series. We could use magnesium to test vanadium's position. If magnesium is oxidized by $Sc^{3+}$ ions, scandium must lie between aluminum and magnesium.

4.119. a. $Cr_2O_7^{2-}(aq) + 14\,H^+(aq) + 6\,Fe^{2+}(aq) \rightarrow$
$2\,Cr^{3+}(aq) + 7\,H_2O(\ell) + 6\,Fe^{3+}(aq)$
b. 0.123 $M$

4.121. $7.98 \times 10^{-4}\ M\ SO_4^{2-}$

4.123. (a) 11.7 $M$; (b) 42.7 mL; (c) 1.72 kg

4.125. a. $2\,OH^-(aq) + 2\,H_2O(\ell) + 3\,S_2O_4^{2-}(aq) + 2\,CrO_4^{2-}(aq) \rightarrow$
$6\,SO_3^{2-}(aq) + 2\,Cr(OH)_3(s)$
b. Sulfur is oxidized; chromium is reduced.
c. Oxidizing agent = $CrO_4^{2-}$; reducing agent = $S_2O_4^{2-}$.
d. 38.7 g

4.127. a. Balanced equation: $2\,Ag(s) + H_2S(g) \rightarrow Ag_2S(s) + H_2(g)$
Oxidation numbers:
Ag = 0           $Ag_2S$: Ag = +1, S = −2
$H_2S$: H = +1, S = −2     $H_2$: H = 0
One mole of electrons is transferred per mole of Ag.
b. $3\,Ag_2S(s) + 12\,H_2O(\ell) + 4\,Al(s) \rightarrow$
$6\,Ag(s) + 3\,H_2S(g) + 3\,H_2(g) + 4\,Al(OH)_3(s)$

4.129. a. Balanced equation:
$HC_2H_3O_2(aq) + KOH(aq) \rightarrow H_2O(\ell) + KC_2H_3O_2(aq)$
Net ionic equation:
$HC_2H_3O_2(aq) + OH^-(aq) \rightarrow H_2O(\ell) + C_2H_3O_2^-(aq)$
b. Balanced equation:
$Na_2CO_3(aq) + CaCl_2(aq) \rightarrow CaCO_3(s) + 2\,NaCl(aq)$
Net ionic equation: $CO_3^{2-}(aq) + Ca^{2+}(aq) \rightarrow CaCO_3(s)$
c. Balanced equation:
$CaO(s) + H_2O(\ell) \rightarrow Ca(OH)_2(aq)$
Net ionic equation:
$CaO(s) + H_2O(\ell) \rightarrow Ca^{2+}(aq) + 2\,OH^-(aq)$

4.131. (a) $NaClO_4$, $NH_4ClO_4$; (b) 427 kg; (c) $2.80 \times 10^{10}$ gal; (d) The MA lab

4.133. a. $3\,CH_2O \rightarrow CO_2 + C_2H_5OH$
b. $C_2H_5OH + O_2 \rightarrow HC_2H_3O_2 + H_2O$
c. $CH_2O$: C = 0
$CO_2$: C = +4
$C_2H_5OH$: C = −4 average of both carbon atoms. Oxidation number on each carbon = −2
$HC_2H_3O_2$: C = 0
d. 66.7 g acetic acid

4.135. a. $H^+(aq) + HSO_4^-(aq) + Ba^{2+}(aq) + 2\,OH^-(aq) \rightarrow$
$BaSO_4(s) + 2\,H_2O(\ell)$
b. Graph c

4.137. $HC_2H_3O_2(aq) + OH^-(aq) \rightarrow H_2O(\ell) + C_2H_3O_2^-(aq)$

4.139. a. Oxidation: $O_2^-(aq) \rightarrow O_2(aq) + 1\,e^-$
Reduction: $2\,H^+(aq) + O_2^-(aq) + 1\,e^- \rightarrow H_2O_2(aq)$
b. $2\,H^+(aq) + 2\,O_2^-(aq) \rightarrow H_2O_2(aq) + O_2(aq)$

4.141. a. The first reaction is a redox reaction; eight electrons are transferred.
b. $2\,H^+(aq) + SO_4^{2-}(aq) + CaCO_3(s) \rightarrow$
$CaSO_4(s) + H_2O(\ell) + CO_2(g)$
c. $SO_4^{2-}(aq) + CaCO_3(s) \rightarrow CaSO_4(s) + CO_3^{2-}(aq)$

4.143. c and d

# CHAPTER 5

5.1. At 35 ft above street level KE = 350 J; just before hitting the street KE = 150 J

5.3. a.

　b. The piston is higher in the cylinder.
　c. Yes
　d. The system did work on the surroundings.
5.5. a. A closed system
　b. The internal energy of the system will increase.
　c. No
5.7. a. Because the heat of formation of an element in its standard state is defined as zero
　b. Because the enthalpy of its formation is positive
　c. By subtracting the sum of the enthalpies of formation of the reactants (multiplied by the number of moles in the balanced equation for each) from the sum of enthalpies of formation of the products (again multiplied by their molar amounts from the balanced equation)
5.9. Energy makes work possible.
5.11. The value of a state function is independent of the path; only the initial and final values are important.
5.13. a. The potential energy in a battery consists of the energy stored in the molecules, which is released when they react via a redox reaction.
　b. The potential energy in a gallon of gasoline consists of the chemical bonds in the fuel that release heat, and do work via expansion as the fuel is combusted.
　c. The potential energy of the crest of a wave is due to its position above the ground.
5.15. The system is that part of the universe that we are interested in. The surroundings are everything else, extending to the entire universe.
5.17. The sign of $\Delta E$ depends on the magnitude of $q$ and $w$.
5.19. a. Exothermic
　b. Endothermic
　c. Exothermic
5.21. Energy is absorbed as heat from the surroundings. Thus, $q$ increases, and therefore $\Delta E$ increases.
5.23. $w = -0.500 \text{ L} \cdot \text{atm} = -50.7 \text{ J}$
5.25. a. 80.0 J
　b. 9.2 kJ
　c. −940. J
5.27. b
5.29. a. 127 J of work is done.
　b. $\Delta E = -2.47 \text{ kJ}$
5.31. A change in enthalpy is the sum of the change of internal energy and the product of the system's pressure and change in volume.
5.33. If the system transfers energy to the surroundings, its energy will be less after the process than at the start of the process.
5.35. Negative
5.37. Positive
5.39. Negative
5.41. Specific heat is specified for a gram of the substance. Heat capacity does not take into account how much of a substance there is; it is defined for a given object.

5.43. No
5.45. Water's high heat capacity relative to air means that water carries away more energy from the engine for every degree Celsius rise in temperature, so water is a better choice for cooling automobile engines.
5.47. 29.3 kJ
5.49.

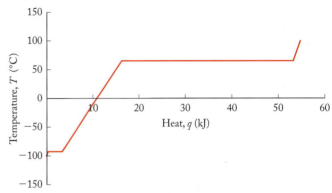

5.51. 886 g $H_2O$
5.53. Gold has the highest temperature.
5.55. −47.5°C
5.57. Because we need to know how much energy (generated or absorbed by the system) is required to change the temperature of the surroundings (the calorimeter) in order to calculate the heat capacity or final temperature of the system in an experiment
5.59. Yes
5.61. $NH_4NO_3$ provides the greatest temperature drop.
5.63. 8.044 kJ/°C
5.65. −5129 kJ/mol
5.67. 23.29°C
5.69. a. $Mg(s) + 2 H^+(aq) \rightarrow H_2(g) + Mg^{2+}(aq)$
　b. $\Delta H_{rxn} = -44.4 \text{ kJ/mol}$
5.71. 0.145 g
5.73. When we apply Hess's law, all the energy is accounted for in the reaction; energy is neither created nor destroyed when using Hess's law.
5.75. Reverse the direction of the second reaction, and add the first reaction to obtain the third reaction (after canceling species common to both sides).
$$CH_4(g) + NH_3(g) \rightarrow HCN(g) + 3 H_2(g)$$
5.77. $\Delta H_{rxn} = 54.08 \text{ kJ}$
5.79. If we write out the chemical equations for the $\Delta H_f°$ for the reactants and products, these formation reactions will add up to the overall reaction.
5.81. No
5.83. a and d
5.85. −164.9 kJ
5.87. −35.9 kJ
5.89. −7198 kJ
5.91. Reverse the direction of the combustion reaction (flip the sign of $\Delta H_{comb}$), and add the reaction describing the formation of $CO_2(g)$ from pure $C(s)$ and $O_2(g)$ to obtain the reaction for the formation of $CO(g)$.
$$\Delta H_f°(CO) = \Delta H_f°(CO_2) - \Delta H_{comb}(CO)$$
5.93. 12.6 kJ
5.95. The energy per gram a fuel releases on burning
5.97. The fuel value (kJ/g) is obtained by dividing the molar heat of combustion (kJ/mol) by the molar mass (mol/g).

5.99. Gasoline has a higher fuel value.

5.101. a. 48.99 kJ/g
   b. $4.90 \times 10^4$ kJ
   c. 5.97 g

5.103. Diethyl ether has a fuel value of 36.781 kJ/g and a fuel density of $2.624 \times 10^4$ kJ/L. Both of these values are lower than those of diesel.

5.105. 12.9 kJ

5.107. $3.54 \times 10^4$ mol

5.109. a. −92.2 kJ
   b. 46.1 kJ

5.111. 58.1 g

5.113. 33.3°C

5.115. a. $6\, FeO(s) + O_2(g) \rightarrow 2\, Fe_3O_4(s)$
   b. $\Delta H_{rxn} = -636$ kJ

5.117. a. The reaction is exothermic.
   b. Bonds must be broken, which requires more energy than breaking intermolecular forces.
   c. 793 kJ of heat produced

5.119. The balanced chemical equation for the reaction of iron(II) oxide with oxygen to form iron(III) oxide is
   $4\, FeO(s) + O_2(g) \rightarrow 2\, Fe_2O_3(s)$   $\Delta H_{rxn} = -560.8$ kJ

5.121. Under these conditions, $\Delta H_{vap} = 40$ kJ/mol.

5.123. a. Exothermic.
   b. −471.8 kJ
   c. −882.3 kJ
   d. Reaction would be less exothermic.

5.125. $\Delta H_3 = (-\Delta H_1) + (-\Delta H_2)$

5.127. $-1.39 \times 10^3$ kJ

5.129. −1234.7 kJ

5.131. 38.5°C

5.133.

| Element | $\mathcal{M}$ (g/mol) | $c_s$[J/(g · °C)] | $\mathcal{M} \cdot c_s$ |
|---|---|---|---|
| Bismuth | 210.8 | 0.120 | 25.3 |
| Lead | 207.2 | 0.123 | 25.5 |
| Gold | 197.0 | 0.125 | 24.6 |
| Platinum | 195.1 | 0.130 | 25.3 |
| Tin | 118.7 | 0.215 | 25.5 |
| Silver | 108.6 | 0.233 | 25.3 |
| Zinc | 65.38 | 0.388 | 25.4 |
| Copper | 63.5 | 0.397 | 25.2 |
| Cobalt/nickel | 58.4 | 0.433 | 25.3 |
| Iron | 55.8 | 0.460 | 25.7 |
| Sulfur | 32.1 | 0.788 | 25.3 |
| Average value in column 4 | | | 25.3 |

5.135. First, heat the piece of metal by immersing a dry test tube containing the metal into a boiling water bath long enough for the metal to come to thermal equilibrium. The temperature of the boiling water will be the $T_i$ of the metal. After recording $T_i$ of the water in the Styrofoam cup, quickly transfer the metal from the test tube to the water in the cup. Record the highest temperature of the water as it heats up from the transfer of energy from the hot metal to the water. This temperature will be $T_f$ for both the metal and the water. Because all the energy lost by the metal is gained by the water:

$$-q_{lost,metal} = q_{gained,water}$$
$$-mc_{s,metal}(T_f - T_{i,metal}) = nc_{P,water}(T_f - T_{i,water})$$

where the mass ($m$) of the metal, $T_f$, $T_{i,water}$, $T_{i,metal}$, moles ($n$) of the water, and the specific heat of water [75.3 J/(mol · °C)] are all known. This leaves $c_{s,metal}$ as the only unknown in the equation.

5.137. The reaction is endothermic, $\Delta H_{comb} = 68$ kJ

5.139. a. For $H_2(g)$, the fuel density is 10.78 kJ/L, whereas for $H_2(\ell)$, the fuel density is 8492 kJ/L.
   b. 
   $H_3NBH_3(g) \rightarrow NH_3(g) + BH_3(g)$    $\Delta H^\circ_{rxn} = 102.2$ kJ
   $H_3NBH_3(g) \rightarrow H_2(g) + H_2NBH_2(g)$    $\Delta H^\circ_{rxn} = -28.40$ kJ
   $H_2NBH_2(g) \rightarrow H_2(g) + HNBH(g)$    $\Delta H^\circ_{rxn} = 123.4$ kJ
   c. 76.6 kg

5.141. Exothermic

5.143. 1.8 mL

# CHAPTER 6

6.1. Barometer a

6.3. a

6.5. c

6.7. Line 1

6.9. Line 2

6.11. The molar mass of helium is 4 g/mol. The line on the graph in Figure P6.10 for this gas goes below line 1. The molar mass of NO is 30 g/mol, so the line on the graph for this gas goes above line 2. All lines will converge at $P = 0$ and $d = 0$.

6.13. b

6.15. Curve 1 is $SO_2$. Curve 2 could be $CO_2$ or $C_3H_8$.

6.17. $Br_2$ (orange)

6.19. c

6.21. a

6.23. Force is the product of the mass of an object and the acceleration due to gravity. Pressure uses force in its definition: It is the force an object exerts over a given area.

6.25. 760 torr = 1 atmosphere

6.27. The ethanol barometer

6.29. A sharpened blade has a smaller area over which the force is distributed compared to a dull blade.

6.31. As we go up in altitude, the overlying mass of the atmosphere above us decreases, so the pressure also decreases.

6.33. $3.9 \times 10^3$ Pa

6.35. a. 0.020 atm
   b. 0.739 atm

6.37. a. 814.6 mmHg
   b. 1.072 atm
   c. 1086 mbar

6.39. a. 91 atm
   b. $CO_2$ is heavier than $N_2$, so the atmosphere on Venus has a greater mass than that on Earth.

6.41. The higher the temperature, the faster the gas molecules move. The faster they move, the more often they collide with the walls of the container and the greater the force with which gas molecules hit the walls. Both of these result in increased pressure as temperature is raised.

6.43. The balloonist should decrease the temperature.

6.45. The gas pressure increases.

6.47. a. 2.00 atm
   b. 0.664 atm
   c. 1.67 atm

6.49. 2.30 atm; 13.0 m

6.51.

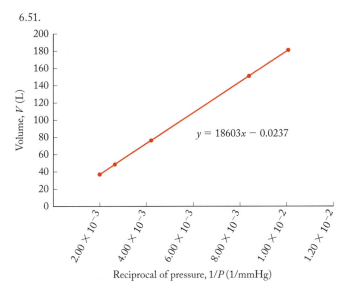

Yes, the graph is exactly the same for the same number of moles of argon gas.

6.53.

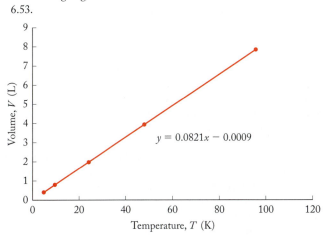

If the amount of gas were halved, the graph would still be linear, but the slope of the line would be halved.

6.55. 596 K or 323°C

6.57. a. 4.27 L
b. 2.02 L
c. 2.41 L

6.59. −4.22 kJ

6.61. b

6.63. a. No change
b. Decrease to 1/4 the original volume
c. Increase of 17%

6.65. 144 L

6.67. 6.6 atm

6.69. STP is defined as 1 atm and 0°C (273 K); $V = 22.4$ L

6.71. The product of the number of moles of gas in the sample, the temperature, and the gas constant

6.73. 0.67 mol

6.75. 1.50 atm

6.77. a. $1.8 \times 10^3$ L
b. $1.8 \times 10^5$ L
c. The balloon will break under both sets of conditions.

6.79. a. 0.0419 mol/hr
b. 10.7 g

6.81. $1.5 \times 10^3$ g $(NH_4)_2Cr_2O_7$ and $1.2 \times 10^2$ g $KClO_3$

6.83. 715 g

6.85. The densities of different gases are not necessarily the same for a particular temperature and pressure.

6.87. Density (a) increases with increasing pressure and (b) increases with decreasing temperature.

6.89. a. 9.08 g/L
b. In the basement

6.91. $SO_2$

6.93. CO

6.95. The pressure that a particular gas individually contributes to the total pressure

6.97. Sample c

6.99. $X_{H_2} = 0.20$, $P_{total} = 2.46$ atm, $P_{N_2} = 1.7$ atm, $P_{H_2} = 0.49$ atm, $P_{CH_4} = 0.25$ atm

6.101. a. $4\,NH_3(g) + 7\,O_2(g) \rightarrow 4\,NO_2(g) + 6\,H_2O(g)$
b. The ratio of $NH_3:O_2$ is 4:7

6.103. a. 0.0190 mol
b. No. There will be fewer moles of $O_2$ if collected over ethanol.

6.105. a. Greater than
b. Lower than
c. Greater than
d. Greater than

6.107. 1.7 times more

6.109. 680 mmHg

6.111. 25%

6.113. The speed of a molecule in a gas that has the average kinetic energy of all the molecules of the sample

6.115. a. As the molar mass increases, $u_{rms}$ decreases
b. As temperature increases, the $u_{rms}$ increases

6.117. To determine the molar mass of an unknown gas, measure the rate of its effusion $(r_x)$ relative to the rate of effusion of a known gas $(r_y)$. Since we know the molar mass of the known gas, $\mathcal{M}_y$, we can use the equation to solve for the unknown $\mathcal{M}_x$.

6.119. Diffusion is the spread of one substance into another. Effusion is the escape of a gas from its container through a tiny hole. The gas is escaping from a region of higher pressure to one of lower pressure.

6.121. The rank order in terms of increasing root-mean-square speed is $SO_2 < NO_2 < CO_2$.

6.123. Gas C

6.125.

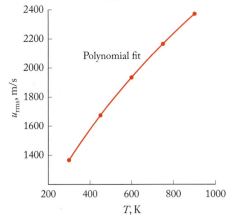

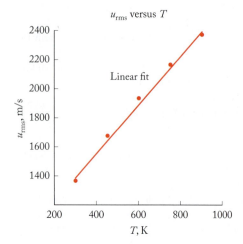

$u_{rms}$ versus $T$

Linear fit

6.127. 0.711
6.129. No, the molar masses are nearly identical.
6.131. 18.2 g/mol
6.133. a. $r(^{12}CO_2)/r(^{13}CO_2) = 1.01$
       b. $^{12}CO_2$ diffuses faster than $^{13}CO_2$.
6.135. The smaller balloon contains hydrogen.
6.137. At low temperatures the gas particles move more slowly, and
       their collisions become inelastic; they stick together due to the
       weak attractive forces between them. The particles, there-
       fore, do not act separately to contribute to the pressure in the
       container, and the pressure is lower than would be expected by
       the ideal gas law. Also, the gas particles take up real volume
       in the container and as the pressure increases, the volume of
       the particles takes up a greater volume of the free space in the
       container. This has the effect of raising the pressure–volume
       product above what we would expect from the ideal-gas law (in
       a plot of $PV/RT$ versus $P$).
6.139. Since $b$ is a measure of the volume that the gas particles occupy,
       $b$ increases as the sizes of the particles increase.
6.141. $H_2$
6.143. a. $P = 910$ atm
       b. $P = 476$ atm
6.145. 126 L
6.147. 27.3 L
6.149. 21.1 atm
6.151. 3.31 L
6.153. 0.25 m/s
6.155. 18 kg
6.157. Xe
6.159. a. $NH_3(g) + HCl(g) \rightarrow NH_4Cl(s)$
       b. The ring of $NH_4Cl$ should appear closer to the end with
          HCl because its molar mass is greater and it diffuses more
          slowly along the tube compared to $NH_3$.
       c. 0.594 m
6.161. $1.38 \times 10^5$ L
6.163. $4.5 \times 10^3$ psi (within the safety limit)
6.165. 26.7 g
6.167. 5.56 atm
6.169. 137 g/mol
6.171. $u_{rms} = \sqrt{\dfrac{3P}{d}}$
6.173. 2.7 atm
6.175. 38.4 L $N_2$, 192 L $CO_2$, 1.74 g/L
6.177. 382 L

6.179. 54.9 L
6.181. a. $4\,NH_3(g) + 5\,O_2(g) \rightarrow 4\,NO(g) + 6\,H_2O(g)$
       b. $NH_3(g) + HNO_3(\ell) \rightarrow NH_4NO_3(s)$
6.183. a. $NH_4NO_2(s) \rightarrow N_2(g) + 2\,H_2O(g)$
       b. Yes
6.185. Sodium is a highly reactive element. It reacts quickly with
       moist air to form NaOH and $H_2$. NaOH is caustic, since it is
       a strong base, and the $H_2$ produced could form an explosive
       mixture with oxygen in the air.
6.187. −642.2 kJ

# CHAPTER 7

7.1. a. Purple (Na), red (Cr), and orange (Au)
     b. Blue (Ne)
     c. Orange (Au)
     d. Red (Cr)
     e. Blue (Ne) and green (Cl)
7.3. Green (Cl)
7.5. Blue (Rb), green (Sr), and orange (Y)
7.7. Orange $(Y^{3+})$ < green $(Sr^{2+})$ < blue $(Rb^+)$ < gray $(I^-)$ < red $(Te^{2-})$
7.9. Blue (Rb)
7.11. Absorption: (d), (e), and (f); emission: (a), (b), and (c)
7.13. e
7.15. (a) Na; (b) K; (c) $Na^+$
7.17. a and b
7.19. a
7.21. All these forms of light have perpendicular, oscillating electric
      and magnetic fields that travel together through space.
7.23. The lead shield must be used to protect the parts of our bodies
      that might be unnecessarily exposed to X-rays (but do not need
      to be imaged). Lead is a very high density metal with many
      electrons that interact with X-rays and can absorb nearly all the
      X-rays before they can reach our bodies.
7.25. X-rays and ultraviolet radiation
7.27. $4.87 \times 10^{14}$ s$^{-1}$
7.29. (a) 2.88 m; (b) 2.95 m; (c) 2.98 m
7.31. (a) The radio station has the lower frequency.
7.33. 8.3 min
7.35. The hydrogen absorption spectrum consists of dark lines at
      wavelengths specific to hydrogen. The emission spectrum has
      bright lines on a dark background with the lines appearing at
      exactly the same wavelengths as the dark lines in the absorp-
      tion spectrum.
7.37. In the study of atomic spectra, each element was found to have
      a distinctive and unique set of absorption and emission lines.
      The location of dark Fraunhofer lines in spectra of sunlight
      correspond to the location of bright lines in emission spectra
      of elements, thus allowing us to deduce the Sun's elemental
      composition.
7.39. A quantum is the smallest indivisible amount of radiant energy
      that an atom can absorb or emit. A photon is the smallest indi-
      visible packet or particle of light energy.
7.41. Provided the incoming light has enough energy to cause the
      emission of electrons to occur in the first place, increasing the
      intensity of the light causes more electrons to be emitted as a
      function of time.
7.43. b
7.45. Potassium; $8.04 \times 10^5$ m/s

7.47. No, light with a wavelength of 500 nm does not have enough energy to liberate electrons.

7.49. $3.17 \times 10^{18}$ photons/s

7.51. Because the single electron interacts only with the proton in the nucleus; there are no other electrons to repel it.

7.53. It is the difference between $n$ levels that determines emission energy.

7.55. a

7.57. No

7.59. At $n = 7$, the wavelength of the electron's transition ($n = 7$ to $n = 2$) has moved out of the visible region.

7.61. 1875 nm; infrared

7.63. (a) Decreases; (b) no

7.65. 72.9 nm

7.67. In the de Broglie equation, $\lambda$ is the wavelength the particle, of mass $m$, exhibits as it travels at speed $v$, where $h$ is Planck's constant. This equation states that any moving particle has wavelike properties because a wavelength can be calculated through the equation, and the wavelength of the particle is inversely related to its momentum (mass multiplied by velocity).

7.69. No

7.71. (a) 10.8 nm; (b) 0.180 nm; (c) $1.24 \times 10^{-27}$ nm; (d) $3.68 \times 10^{-54}$ nm

7.73. c

7.75. (a) $1.77 \times 10^3$ m/s; (b) 0.0565 nm

7.77. $\Delta x \geq 1.3 \times 10^{-13}$ m

7.79. The Bohr model orbit showed the quantized nature of the electron in the atom as a particle moving around the nucleus in concentric orbits. In quantum theory, an orbital is a region of space where the probability of finding the electron is high. The electron is not viewed as a particle, but as a wave, and it is not confined to a clearly defined orbit; rather, we refer to the probability of the electron being at various locations around the nucleus.

7.81. Three: $n$, $\ell$, and $m_\ell$.

7.83. (a) 1; (b) 4; (c) 9; (d) 16; (e) 25

7.85. 3, 2, 1, 0

7.87. (a) 2; (b) 6; (c) 10; (d) 2

7.89. b

7.91. Degenerate orbitals have the same energy and are indistinguishable from each other.

7.93. As we start from an argon core of electrons, we move to potassium and calcium, which are located in the $s$ block on the periodic table. It is not until Sc, Ti, V, etc., that we begin to put electrons into the $3d$ shell.

7.95. Electrons may be promoted to many different orbitals with energy higher than the ground state; each of these corresponds to an excited state with a different energy. Because the ground state is the lowest possible energy state, each atom has only one.

7.97. (c) $3s$ < (a) $3d$ < (d) $4p$ < (b) $5p$

7.99. Li: [He]$2s^1$        Na$^+$: [He]$2s^2 2p^6$ or [Ne]
Li$^+$: $1s^2$ or [He]        Mg$^{2+}$: [He]$2s^2 2p^6$ or [Ne]
Ca: [Ar]$4s^2$        Al$^{3+}$: [He]$2s^2 2p^6$ or [Ne]
F$^-$: [He]$2s^2 2p^6$ or [Ne]

7.101. K: [Ar]$4s^1$        Ba: [Xe]$6s^2$
K$^+$: [Ar]        Ti$^{4+}$: [Ne]$3s^2 3p^6$ or [Ar]
S$^{2-}$: [Ar] or [Ne]$3s^2 3p^6$        Al: [Ne]$3s^2 3p^1$
N: [He]$2s^2 2p^3$

7.103. (a) 3; (b) 2; (c) 0; (d) 0

7.105. Ti; two unpaired electrons

7.107. Cl$^-$; no unpaired electrons

7.109. No

7.111. Al$^{3+}$, N$^{3-}$, Mg$^{2+}$, and Cs$^+$

7.113. a and d

7.115. a. B: $1s^2 2s^2 2p^1$ or [He] $2s^2 2p^1$
O: $1s^2 2s^2 2p^4$ or [He] $2s^2 2p^4$
b. H (+1), B (+3), O(−2)
H$^+$: $1s^0$
B$^{3+}$: $1s^2 1s^2$ or [He]
O$^{2-}$: $1s^2 2s^2 2p^6$ or [Ne]

7.117. $5p$; yes

7.119. If electrons do not repel each other as much in Na$^+$ as they do in Na, they will have lower energy and be, on average, closer to the nucleus, resulting in a smaller size. When electrons are added to an atom (Cl), the e$^-$−e$^-$ repulsion increases, so the electrons have higher energy and they will be, on average, farther from the nucleus, thereby creating a larger-size species (Cl$^-$).

7.121. Rb. The size of atoms increases down a group because electrons have been added to higher $n$ levels.

7.123. a. As the atomic number increases down a group, electrons are added to higher $n$ levels, leading to a decrease in ionization energy.
b. As the atomic number increases across a period, the effective nuclear charge increases. This means that the ionization energy increases across a period of elements.

7.125. Fluorine, with a higher nuclear charge, exerts a higher $Z_{eff}$ on the $2p$ electrons than boron does, resulting in a higher ionization energy.

7.127. Sr

7.129. a. $-1.11 \times 10^{-26}$ J
b. 17.9 m
c. A radio telescope

7.131. a. Cu: $1s^2 2s^2 2p^6 3s^2 3p^6 4s^1 3d^{10}$ or [Ar] $4s^1 3d^{10}$
b. The difference in the population of the 3d orbital:
Cu$^+$: $1s^2 2s^2 2p^6 3s^2 3p^6 4s^0 3d^{10}$ or [Ar] $4s^0 3d^{10}$
Cu$^{2+}$: $1s^2 2s^2 2p^6 3s^2 3p^6 4s^0 3d^9$ or [Ar] $4s^0 3d^9$

7.133. a. Silver forms a 1+ ion through the loss of a high-energy $5s$ electron. Palladium ([Kr]$4d^8 5s^2$) and cadmium ([Kr]$4d^{10} 5s^2$) each lose two $5s$ electrons to form 2+ cations.
b. The heavier group 13 elements may form 1+ cations through the loss of one $np$ electron and form 3+ cations through the loss of the $np$ electron and the two $ns$ electrons.
c. The heavier group 14 elements may form 2+ cations through the loss of the two $np$ electrons and form 4+ cations through the loss of both $np$ electrons and the two $ns$ electrons. The group 4 elements may lose the two $ns$ electrons to form 2+ cations and may lose both the $ns$ electrons and the two $(n-1)d$ electrons to form 4+ cations.

7.135. a. Cl$^-$: [Ne]$3s^2 3p^6$ or [Ar]
Cl: [Ne]$3s^2 3p^5$
Ag: [Kr]$4d^{10} 5s^1$
Ag$^+$: [Kr]$4d^{10}$
b. An excited state occurs when an electron occupies a higher energy orbital. The electron is not in its lowest energy state.
c. More energy is needed to remove an electron from Cl$^-$ compared to Br$^-$ because the electron removed from Cl$^-$ is at a lower $n$ (principal quantum number) level and is held more tightly by the nucleus.
d. If AgBr were used in place of AgCl, longer wavelength (lower energy) light would remove the electron. The AgBr sunglasses would darken perhaps in the infrared region.

7.137. a.  $Sn^{2+}$: $[Kr]4d^{10}5s^2$
   $Sn^{4+}$: $[Kr]4d^{10}$
   $Mg^{2+}$: $[He]2s^22p^6$ or $[Ne]$

   b.  Cadmium has the same electron configuration as $Sn^{2+}$, and neon has the same electron configuration as $Mg^{2+}$.

   c.  $Cd^{2+}$

7.139. a.  Ne, 5.76; Ar, 6.76

   b.  The outermost electron in argon is a $3p$ electron, which is shielded by the electrons in the $n = 2$ level (10 electrons) and the $n = 1$ level (2 electrons), whereas the outermost electron in neon is a $2p$ electron, which is shielded only by the electrons in the $n = 1$ level (2 electrons).

7.141.  When we think of the electron as a wave, we can envision the node between the two lobes as a wave of zero amplitude and the $p$ orbital as a standing wave.

7.143.  No, this wavelength is in the infrared portion of the electromagnetic spectrum.

7.145.  The heavier noble gases are easier to ionize (IE decreases down a group in the periodic table) and therefore can combine with oxygen and fluorine.

7.147.  The high velocity of helium atoms means that once helium atoms are released into the atmosphere, they can escape Earth's gravitational pull. Therefore, Earth's atmosphere contains very little helium.

# CHAPTER 8

8.1.  a.  Group 1 (red)
   b.  Group 14 (blue)
   c.  Group 16 (purple)

8.3.  $Mg^{2+}$

8.5.  Group 14 (blue, carbon)

8.7.  Lithium (red) and fluorine (lilac)

8.9.  b

8.11.  The arrangement of the atoms in two of the structures is S—O—S and in the other two structures it is S—S—O. Because the arrangement of atoms differs, they are not resonance structures. Also, for each arrangement, the structures do not show a different arrangement of electrons on the atoms; the only difference is that the bonds are drawn bent, not straight. The "bent form" and "linear form" are not resonance forms of each other if the numbers of lone pairs and bonding pairs of electrons on each atom are the same.

8.13.  a

8.15.  Fluorine (purple) and oxygen (light blue)

8.17.

8.19.  Yes, for hydrogen and helium

8.21.  Yes

8.23.  In the diatomic molecule XY shown here

   Lewis counts 6 e$^-$ in 3 lone pairs on both X and Y. He also counts the 2 e$^-$ shared between X and Y separately (2 e$^-$ for X

and 2 e$^-$ for Y). However, there are not 4 e$^-$ being shared, only 2 e$^-$. It seems that the Lewis counting scheme counts the shared electrons twice.

8.25.  For the H—O—H bonding pattern, the oxygen of the central atom forms bonds to the two hydrogen atoms. This uses 4 of the 8 e$^-$, leaving 4 e$^-$ left over for the 2 lone pairs. Each hydrogen atom has a duet of electrons, so the lone pairs reside on oxygen and form an octet on oxygen.

For H—H—O bonding, the two covalent bonds again use 4 of the 8 e$^-$, leaving 4 e$^-$ for 2 lone pairs. If these are placed on the oxygen atom as shown here,

oxygen does not complete its octet and the central hydrogen atom has 4 e$^-$, not a duet. This structure would violate the Lewis structure formalism.

8.27.  Li·  ·Mg·  ·Äl·

8.29.  The corrected Lewis symbols are:

8.31.  I$^-$ and $Ca^{2+}$ have a complete valence-shell octet.

8.33.  a.  $[X\cdot]^+$
   b.  $[X]^{3+}$

8.35.  (a) 8; (b) 8; (c) 8; (d) 10

8.37.  a.  :C≡O:

   b.

   c.

   d.  $[:C≡N:]^-$

8.39.  Groups 1, 13, 15, and 17

8.41.  a.

   b.

   c.

8.43.

**8.45.** :Cl̈—Cl̈—Ö:

$$\left[\ :\ddot{O}—\overset{\displaystyle :\ddot{O}:}{\overset{|}{Cl}}—\ddot{O}:\ \right]^{-}$$

**8.47.** If there is an electronegativity difference of 2.0 or greater, the bond between the atoms is ionic; below 2.0, the bond is covalent.

**8.49.** The size of an atom is the result of the number of electrons and the pull of the nucleus on those electrons. The higher the nuclear charge, the stronger the pull on the electrons within a given valence shell. This is why the size of atoms generally decreases across a period. A small atom will form a shorter bond with another atom, and the electrons in the bond will feel a strong pull from the nucleus of a smaller atom since the bonding electrons will be "closer" to the nucleus. This stronger pull is the reason for the higher electronegativity for smaller atoms.

**8.51.** A polar covalent bond is one in which the electrons are shared, but not equally, by the atoms.

**8.53.** The polar bonds and the atoms with the greater electronegativity (underlined) are C̲—Se, C—O̲, N̲—H, and C̲—H.

**8.55.** CsI

**8.57.** Like the panes of glass in a greenhouse, the greenhouse gases in the atmosphere are transparent to visible light. The visible light warms the surface of the earth and is reemitted as infrared (lower energy) light. The greenhouse gases absorb the infrared light, accumulating and holding heat in the atmosphere just as the panes of glass hold heat inside a greenhouse.

**8.59.** The N—O bond would be expected to absorb IR radiation on stretching.

**8.61.** CO does absorb IR radiation because stretching its linear bond gives rise to a fluctuating electric field.

**8.63.** Infrared radiation has longer wavelengths and lower energy than UV radiation, thus it causes chemical bonds only to stretch and bend, but not to break.

**8.65.** More

**8.67.** Resonance occurs when two or more valid Lewis structures may be drawn for a molecular species. The true structure of the species is a hybrid of the structures drawn.

**8.69.** A molecule or ion shows resonance when there is more than one correct Lewis structure; that is, when the electrons in the correct Lewis structure may be distributed in more than one way. Often, when the central atom has both a single and a double bond, resonance is possible.

**8.71.** Either N—O bond in the $NO_2$ structure could be double-bonded and the formal charges for each structure are identical, so there is more than one correct Lewis structure, and $NO_2$ will exhibit resonance.

:Ö⁻—N̈⁺¹=Ö⁺⁰ ⟷ ⁺⁰Ö=N̈⁺¹—Ö⁻:

The resonance forms of $CO_2$ show that one is dominant (the one in which all formal charges are zero) and so the other forms contribute little to the true structure of $CO_2$.

:Ö⁰=C⁰=Ö⁰: ⟷ :O⁺¹≡C⁰—Ö⁻¹: ⟷ :Ö⁻¹—C⁰≡O⁺¹:

**8.73.**

H—C≡N—Ö: ⟷ H—C̈—N≡O: ⟷ H—C̈=N̈=O:

**8.75.** $N_2O_2$:

Ö=N̈—N̈=Ö ⟷ :O≡N—N̈—Ö: ⟷ Ö=N̈=N̈—Ö: ⟷

:Ö—N≡N—Ö: ⟷ :Ö—N̈—N≡O: ⟷ :Ö—N̈=N=Ö:

$N_2O_3$:

:Ö=N̈—N(=Ö)(—Ö:) ⟷ :Ö=N̈—N(—Ö)(—Ö:) ⟷

:Ö=N=N(—Ö)(—Ö:) ⟷ :O≡N—N̈(—Ö)(—Ö:)

**8.77.** a.   H—Ö—Ö—H

:F̈—Ö—Ö—F̈:

b. The presence of an O=O double bond shortens the bond length.

:F̈—Ö⁺—=Ö:   :F̈:⁻

**8.79.** :Cl̈—S̈e—N̈⁻—S̈⁺=Ö:

:Cl̈—S̈e—N̈=S⁺—Ö⁻:

:Cl̈—S̈e=N̈—S̈—Ö⁻:

:Cl̈⁺=S̈e—N̈⁻—S⁺—Ö⁻:

:Cl̈—S̈e—N̈=S̈=Ö:

**8.81.** The best possible structure for a molecule, judging by formal charges, is the structure in which the formal charges are minimized and the negative formal charges are on the most electronegative atoms in the structure.

**8.83.** No

**8.85.**  H⁰—N̈⁺¹≡C̈⁻¹:          H⁰—C⁰≡N⁰:

The formal charges are zero for all the atoms in HCN, whereas in HNC the carbon atom, with a lower electronegativity than N, has a −1 formal charge.

**8.87.**  (H₂)N̈⁺¹=C⁰=N̈⁻¹: ⟷ (H₂)N⁰—C⁰≡N⁰:

The preferred structure is the one with the C triple bonded to N.

**8.89.**

N—S⁺(—N⁻)(ring with S—As) ⟷ N(—S)(—N⁻)(ring with S⁺—As) ⟷

N⁻(—S)(—N⁻)(ring with S=As⁺) ⟷ N⁻(—S)(—N)(ring with S—As⁺) ⟷ N⁻(—S⁺)(—N)(ring with S—As)

**8.91.** $:\overset{-2}{N}-\overset{+2}{O}\equiv\overset{0}{N}: \longleftrightarrow :\overset{-1}{N}=\overset{+2}{O}=\overset{-1}{N}: \longleftrightarrow :\overset{0}{N}\equiv\overset{+2}{O}-\overset{-2}{N}:$

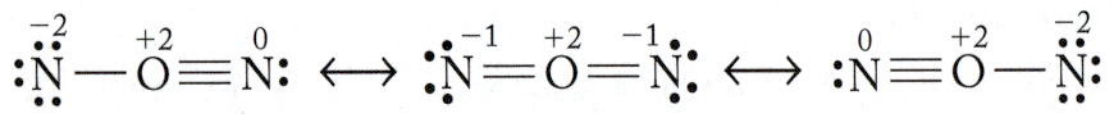

Since oxygen is more electronegative than nitrogen, none of these structures is likely to be stable because the formal charge on O is positive.

**8.93.** Yes

**8.95.** In order for the atom to accommodate more than 8 e$^-$ in covalently bonded molecules, it would require the use of orbitals beyond $s$ and $p$. The $d$ orbitals are not available to the small elements in the second period.

**8.97.** (a) $SF_6$, (b) $SF_5$, and (c) $SF_4$

**8.99.**
  a. 6
  b. 8
  c. 12
  d. 12

**8.101.**

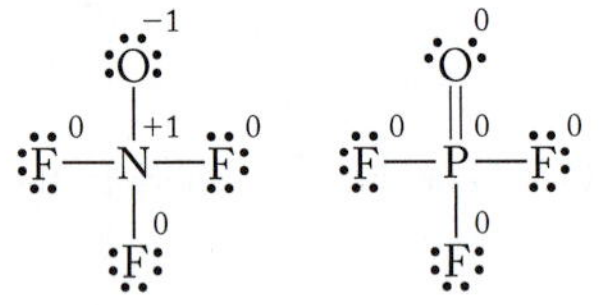

In $POF_3$ there is a double bond and no formal charges; in $NOF_3$ there are only single bonds, and formal charges are present on N and O.

**8.103.**

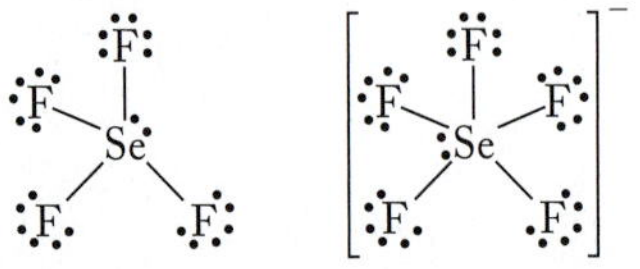

In both structures Se has more than 8 valence electrons.

**8.105.**

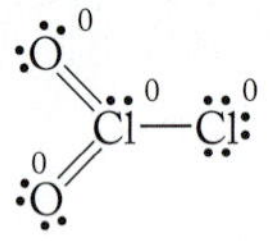

The central chlorine atom has an expanded octet.

**8.107.** (c) $ClO_4$, (d) $ClO_3$, and (e) $ClO_2$

**8.109.** d

**8.111.** The formation of a dimer completes the octet on Al.

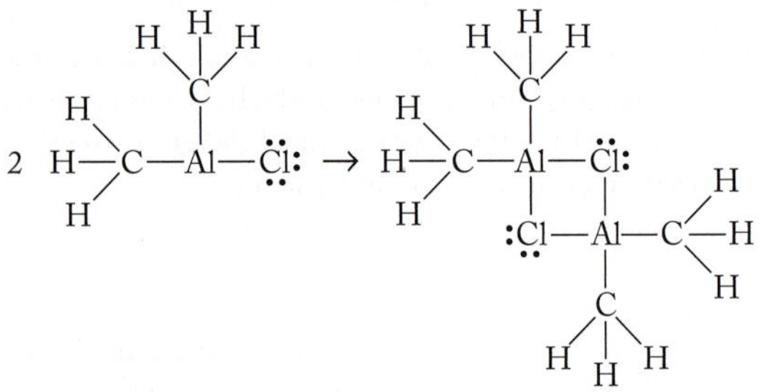

**8.113.** The ionic form is less favorable because of the presence of formal charges on S and F (no formal charges are required with an expanded octet on S).

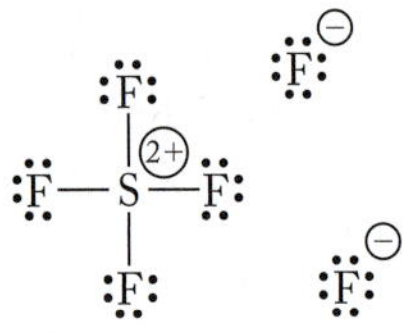

**8.115.** No

**8.117.** The nitrogen–oxygen bond in $N_2O_4$ has a bond order of 1.5 due to four equivalent resonance forms:

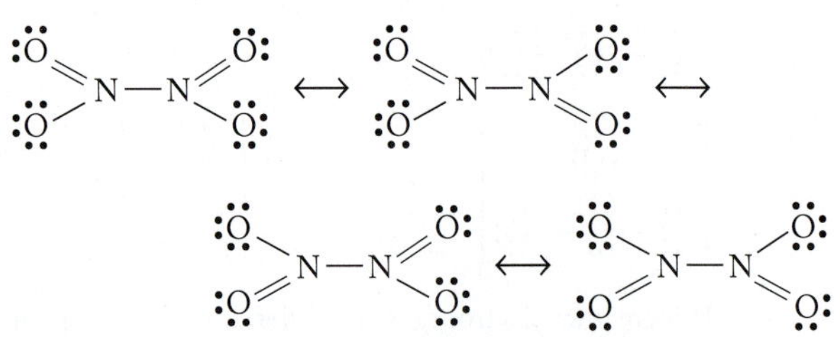

The nitrogen–oxygen bond in $N_2O$ has a bond order of 1.5 due to resonance between three resonance forms (although the last resonance structure shown does not significantly contribute to the structure of the molecule because of the high formal charges):

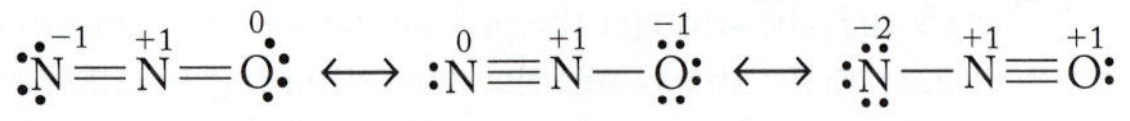

$:\overset{-1}{N}=\overset{+1}{N}=\overset{0}{O}: \longleftrightarrow :\overset{0}{N}\equiv\overset{+1}{N}-\overset{-1}{O}: \longleftrightarrow :\overset{-2}{N}-\overset{+1}{N}\equiv\overset{+1}{O}:$

Therefore, owing to resonance, $N_2O_4$ and $N_2O$ are expected to have nearly equal bond lengths.

**8.119.**
  a. $NO^+ < NO_2^- < NO_3^-$
  b. $NO_3^- < NO_2^- < NO^+$

**8.121.** No

**8.123.** We must account for all the bonds that break and all the bonds that form in the reaction. To do this, we must have a balanced chemical reaction.

**8.125.** If the compounds are in the solid or liquid phase, interactions between molecules may slightly change the bond energy for a given bond.

**8.127.**
  a. 862 kJ
  b. 98 kJ
  c. 93 kJ

**8.129.** 1068 kJ/mol

**8.131.** The incomplete combustion reaction releases 278 kJ less than the complete combustion reaction.

**8.133.** −667 kJ

**8.135.** 552 kJ/mol

**8.137.** $:O=C=C=C=O:$

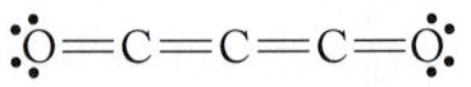

Both carbon–oxygen bonds are equal.

**8.139.** (a) $\cdot Be\cdot$ (b) $\cdot Al\cdot$ (c) $\cdot \dot{C}\cdot$ (d) $He:$

**8.141.** $:\overset{0}{S}=\overset{0}{C}=\overset{0}{S}: \quad :\overset{0}{S}=\overset{+2}{S}=\overset{-2}{C}:$

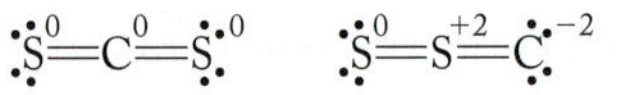

The preferred structure for carbon disulfide is when C is the central atom.

**8.143.** a.

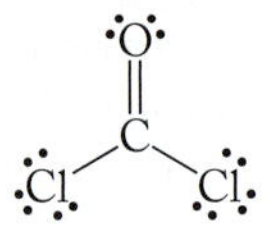

b.

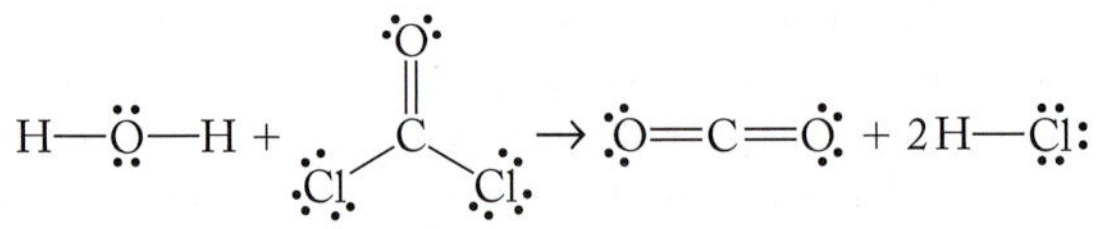

**8.145.**

a. 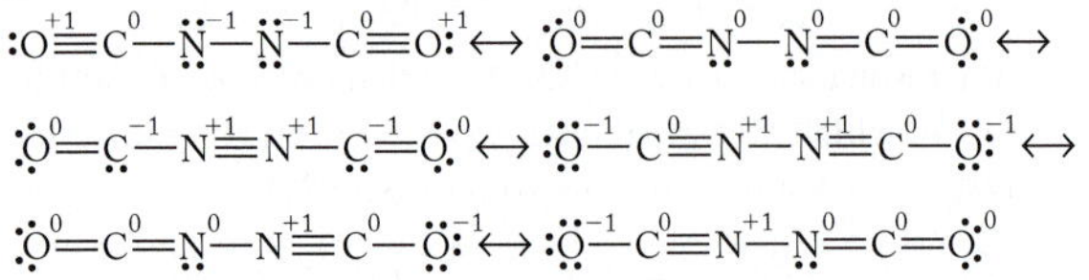 $:O\equiv\overset{+1}{C}-\overset{0}{N}-\overset{-1}{N}-\overset{-1}{C}\equiv O: \longleftrightarrow :\overset{0}{O}=\overset{0}{C}=\overset{0}{N}-\overset{0}{N}=\overset{0}{C}=\overset{0}{O}: \longleftrightarrow$

$:\overset{0}{O}=\overset{0}{C}-\overset{+1}{N}\equiv\overset{+1}{N}-\overset{-1}{C}=\overset{0}{O}: \longleftrightarrow :\overset{-1}{O}-\overset{0}{C}\equiv\overset{+1}{N}-\overset{+1}{N}\equiv\overset{0}{C}-\overset{-1}{O}: \longleftrightarrow$

$:\overset{0}{O}=\overset{0}{C}=\overset{0}{N}-\overset{+1}{N}\equiv\overset{0}{C}-\overset{-1}{O}: \longleftrightarrow :\overset{-1}{O}-\overset{0}{C}\equiv\overset{+1}{N}-\overset{0}{N}=\overset{0}{C}=\overset{0}{O}:$

b. 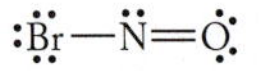 $:\overset{\cdot\cdot}{Br}-\overset{\cdot\cdot}{N}=\overset{\cdot\cdot}{O}:$

$:O\equiv\overset{+1}{C}-\overset{0}{N}=\overset{0}{N}-\overset{-1}{O}: \longleftrightarrow :\overset{0}{O}=\overset{0}{C}-\overset{-1}{N}=\overset{0}{N}=\overset{+1}{O}: \longleftrightarrow$

$:\overset{-1}{O}-\overset{-1}{C}=\overset{0}{N}-\overset{+1}{N}\equiv O:$

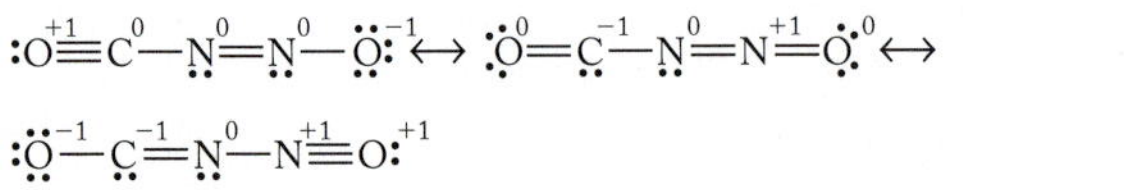

c.　$:O\equiv\overset{+1}{C}-\overset{-1}{N}-\overset{-1}{C}-\overset{-1}{N}-\overset{0}{C}\equiv O:^{+1} \leftrightarrow :\overset{0}{O}=\overset{0}{C}=\overset{0}{N}-\overset{0}{C}-\overset{0}{N}=\overset{0}{C}=\overset{0}{O}: \leftrightarrow$

$:\overset{0}{O}=\overset{0}{C}=\overset{0}{N}-\overset{+1}{C}=\overset{0}{N}=\overset{0}{C}=\overset{0}{O}:^{-1} \leftrightarrow :\overset{-1}{O}-\overset{0}{C}\equiv\overset{+1}{N}-\overset{+1}{C}-\overset{0}{N}\equiv\overset{0}{C}-\overset{-1}{O}: \leftrightarrow$

$:\overset{-1}{O}-\overset{0}{C}\equiv\overset{+1}{N}-\overset{0}{C}-\overset{0}{N}=\overset{0}{C}=\overset{0}{O}: \leftrightarrow :\overset{0}{O}=\overset{0}{C}=\overset{0}{N}-\overset{0}{C}-\overset{+1}{N}\equiv\overset{0}{C}-\overset{-1}{O}:$

8.147. For $Cl_2O_6$ with a Cl—Cl bond

(Lewis structure: O=Cl—Cl=O with O atoms, all formal charges 0)

For $Cl_2O_6$ with a Cl—O—Cl bond

(Lewis structure with Cl—O—Cl bridge, all formal charges 0)

For $ClO_2$

$\overset{0}{O}=\overset{0}{Cl}=\overset{0}{O}$

8.149. a. $\cdot C\equiv N:$

The more likely structure for cyanogen is the one that contains the C—C bond.

b. Because oxalic acid has a C—C bond, we would expect that oxalic acid retained this C—C bond from the cyanogen from which it formed in reaction with water. This supports the predicted structure for cyanogen from formal charge analysis.

8.151.

(Lewis structure: N triple/double bonded to C, F—S—F with F below, all formal charges 0)

8.153.

$\left[\begin{array}{c} \text{(octahedral Te with 5 F and 1 O}^{-1}\text{, Te}^{-1}\text{)} \end{array}\right]^{2-}$

8.155. (a) 2; (b) 4; (c) 1; (d) 5; (e) 8

8.157. a. If we can distinguish atoms by electron density using this technique, we would be able to distinguish X—A—A from A—X—A.

b. Electron diffraction cannot distinguish among resonance forms. Remember that resonance forms are not "real"—the molecule does not fluctuate between the resonance forms, but is rather a hybrid. If the resonance forms for A—X—A shown are all equally weighted (none is more preferred than another), we would expect the average X—A bond to be a double bond, as in A=X=A.

8.159. $\left[\begin{array}{c} H \\ | \\ H-N-H \\ | \\ H \end{array}\right]^{+}$ $\left[:\overset{..}{S}-H\right]^{-}$

There cannot be a nitrogen–sulfur covalent bond because the nitrogen atom in $NH_4^+$ has a complete octet through its bonding with hydrogen and it cannot expand its octet because it is a second period element.

8.161. a, b.

$:\overset{0}{N}\equiv\overset{+1}{N}-\overset{0}{N}=\overset{-1}{N}: \leftrightarrow :\overset{-1}{N}=\overset{+1}{N}=\overset{+1}{N}=\overset{-1}{N}: \leftrightarrow :\overset{-1}{N}=\overset{0}{N}-\overset{+1}{N}\equiv\overset{0}{N}:$

The middle structure has the most nonzero formal charges across three bond lengths, so this one is least preferred. The first and last resonance structures are preferred and are indistinguishable from each other.

c. (ring structures of four N atoms)

$\begin{array}{c} \overset{0}{N}-\overset{0}{N} \\ || \quad || \\ \overset{0}{N}-\overset{0}{N} \end{array} \leftrightarrow \begin{array}{c} \overset{0}{N}=\overset{0}{N} \\ | \quad | \\ \overset{0}{N}=\overset{0}{N} \end{array}$

8.163.

$\begin{array}{c} :\overset{0}{Cl}: \\ | \\ :\overset{0}{F}-Al-\overset{0}{Cl}: \\ \quad \overset{0}{} \end{array}$

8.165. b and c

8.167. a, b.

$\left[:\overset{-2}{N}-\overset{0}{N}=\overset{0}{N}-\overset{+1}{N}\equiv\overset{0}{N}:\right]^{-} \leftrightarrow \left[:\overset{-2}{N}-\overset{0}{N}=\overset{+1}{N}=\overset{+1}{N}=\overset{-1}{N}:\right]^{-} \leftrightarrow$

$\left[:\overset{-2}{N}-\overset{+1}{N}\equiv\overset{+1}{N}-\overset{0}{N}=\overset{-1}{N}:\right]^{-} \leftrightarrow \left[:\overset{-1}{N}=\overset{0}{N}-\overset{+1}{N}\equiv\overset{+1}{N}-\overset{-2}{N}:\right]^{-} \leftrightarrow$

$\left[:\overset{0}{N}\equiv\overset{+1}{N}-\overset{0}{N}=\overset{0}{N}-\overset{-2}{N}:\right]^{-} \leftrightarrow \left[:\overset{-1}{N}=\overset{0}{N}-\overset{0}{N}=\overset{+1}{N}=\overset{-1}{N}:\right]^{-} \leftrightarrow$

$\left[:\overset{-1}{N}=\overset{+1}{N}=\overset{0}{N}-\overset{0}{N}=\overset{-1}{N}:\right]^{-} \leftrightarrow \left[:\overset{0}{N}\equiv\overset{+1}{N}-\overset{-1}{N}-\overset{0}{N}=\overset{-1}{N}:\right]^{-} \leftrightarrow$

$\left[:\overset{-1}{N}=\overset{0}{N}-\overset{-1}{N}-\overset{+1}{N}\equiv\overset{0}{N}:\right]^{-}$

The structures that contribute most have the lowest formal charges (last four structures shown).

c. $N_3^-$ has the Lewis structures

$\left[:\overset{-2}{N}-\overset{+1}{N}\equiv\overset{0}{N}:\right]^{+} \leftrightarrow \left[:\overset{-1}{N}=\overset{+1}{N}=\overset{-1}{N}:\right]^{-} \leftrightarrow \left[:\overset{0}{N}\equiv\overset{+1}{N}-\overset{-2}{N}:\right]^{-}$

From these resonance structures we see that each bond is predicted to be (on average) of double bond character in $N_3^-$. Therefore, in $N_5^-$ there are two longer N—N bonds (single) than in $N_3^-$. $N_3^-$ has the higher average bond order.

8.169.

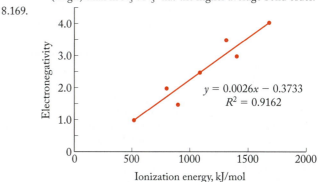

$y = 0.0026x - 0.3733$
$R^2 = 0.9162$

The plot is linear. Using the equation for the best-fit line, where $x =$ the ionization energy of neon, gives a value of $y$ (electronegativity) of neon: $y = 5.0$.

8.171. a. Isoelectronic means that the two species have the same number of electrons.

b–d.

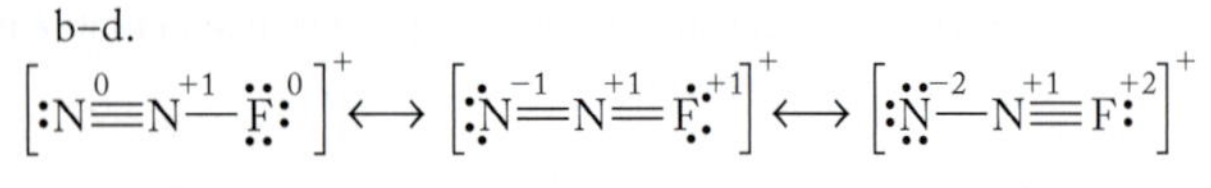

$$\left[:N\equiv\overset{+1}{N}-\overset{0}{\ddot{F}}:\right]^{+} \longleftrightarrow \left[:\overset{-1}{N}=\overset{+1}{N}=\overset{+1}{\ddot{F}}:\right]^{+} \longleftrightarrow \left[:\overset{-2}{\ddot{N}}-\overset{+1}{N}\equiv\overset{+2}{F}:\right]^{+}$$

The central nitrogen atom in all the resonance structures always carries a +1 formal charge. The first resonance form shown is preferred because it has the minimal formal charges on the atoms.

e. Yes, fluorine could be the central atom in the molecule, but this would place significant positive formal charge on the fluorine atom (the most electronegative element). These structures are unlikely:

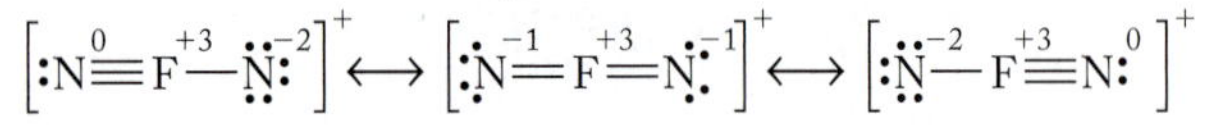

$$\left[:N\equiv\overset{+3}{F}-\overset{-2}{\ddot{N}}:\right]^{+} \longleftrightarrow \left[:\overset{-1}{N}=\overset{+3}{F}=\overset{-1}{N}:\right]^{+} \longleftrightarrow \left[:\overset{-2}{\ddot{N}}-\overset{+3}{F}\equiv\overset{0}{N}:\right]^{+}$$

8.173. $5\ F_2(g) + 5\ H_2O(\ell) \rightarrow 8\ HF(g) + O_2(g) + H_2O_2(\ell) + OF_2(g)$

a. $F_2$ is the oxidizing agent. $H_2O$ is the reducing agent.

b. :F̈—F̈:     Ö (H H)     H—F̈:

    Ö=Ö     Ö—Ö (H H)     Ö (F̈ F̈)

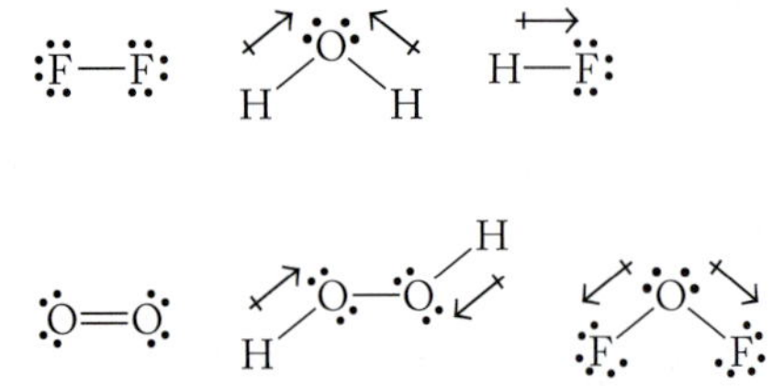

8.175.

F, F, C=C, F, F

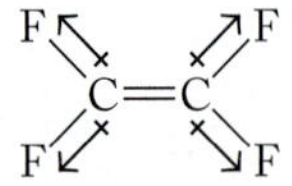

This molecule is nonpolar overall because the individual bond dipoles are equal in magnitude and, as vectors, they cancel each other out.

# CHAPTER 9

9.1. Yes

9.3. $N_2F_2$ and NCCN are planar; there are no delocalized $\pi$ electrons in any of these molecules.

9.5. More

9.7. The axial-F–Re–axial-F bond angle is 180°. The axial-F–Re–equatorial-F angle is 90°. The equatorial-F–Re–equatorial-F bonds are all 72°.

9.9. Because the electrons take up most of the space in the atom and because the nucleus is located in the center of the electron cloud, the electron clouds repel each other before the nuclei get close enough to interact.

9.11. Because both have three atoms bonded with no lone pairs on the central atom.

9.13. Because the lone pair feels attraction from only one nucleus, it is less confined than bonding pairs and therefore occupies more space around the central N atom in ammonia.

9.15. The seesaw geometry has only two lone-pair–bond-pair interactions at 90° (compared to trigonal pyramidal's three), so it has lower energy.

9.17. (b) octahedral < (c) tetrahedral < (a) trigonal planar

9.19. Trigonal bipyramidal, seesaw, T-shaped, octahedral, square pyramidal, and square planar

9.21. Tetrahedral

9.23. Pentagonal pyramidal and distorted octahedral

9.25. (a) tetrahedral; (b) trigonal pyramidal; (c) bent; (d) tetrahedral

9.27. a. 109.5°
   b. <109.5°
   c. <120°
   d. 90°, 180°

9.29. (a) tetrahedral; (b) tetrahedral; (c) trigonal planar; (d) linear

9.31. $O_3$, $SO_2$, and $S_2O$ are bent (120°); $N_2O$ and $CO_2$ are linear.

9.33. The structure on the right contributes most.

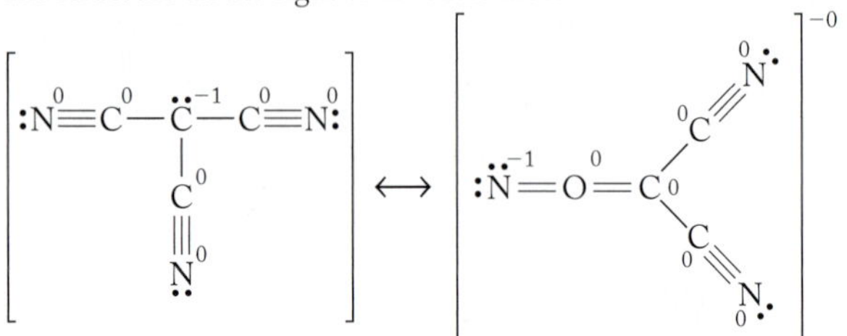

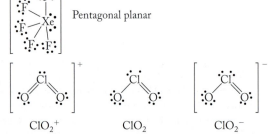

9.35. Silicon may exceed its octet, changing the geometry around the nitrogen atom to trigonal planar.

9.37.

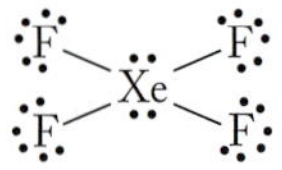

F̈ F̈ (Xe) F̈ F̈     Square planar

[F̈ F̈ (Xe) F̈ F̈ F̈]⁻     Pentagonal planar

9.39.

[Cl (Ö Ö)]⁺     Cl (Ö Ö)     [Cl (Ö Ö)]⁻

$ClO_2^{+}$          $ClO_2$          $ClO_2^{-}$

9.41. A polar bond is only between two atoms in a molecule. Molecular polarity takes into account all the individual bond polarities and the geometry of the molecule. A polar molecule has a permanent, measurable dipole moment.

9.43. Yes

9.45. Polar molecules are (b) $CHCl_3$, (d) $H_2S$, and (e) $SO_2$. Nonpolar molecules are (a) $CCl_4$ and (c) $CO_2$.

9.47. All of the molecules (a–c) are polar.

9.49. a. i
   b. i
   c. ii

9.51. (a) tetrahedral; (b) nonpolar

9.53. Sigma bonds are formed when two electrons are shared between two atoms along a straight line connecting the two atoms. Pi bonds are also formed when two electrons are shared between two atoms, but the overlapping electron density occurs in a plane above and below the straight line connecting the two atoms. Sigma bonds may be formed from the overlap of any type of orbital, whereas $\pi$ bonds are never formed from an overlap of $s$-orbitals. Most commonly $\pi$ bonds are formed from overlapping $p$-orbitals, although other combinations of orbitals are possible.

9.55. Yes

9.57. a. $sp$ hybridized
   b. $sp^2$ hybridized
   c. $sp$ hybridized
   d. $sp^2$ hybridized
   e. $sp^2$ hybridized

9.59. Both Lewis structures have three resonance forms: one with two N=N bonds and two with an N≡N and an N—N bond. For both molecules, the central nitrogen atoms are $sp$ hybridized.

9.61. The hybridizations are: $SF_2$ ($sp^3$); $SF_4$ ($sp^3d$); $SF_6$ ($sp^3d^2$)

9.63. The N—O bond is a single bond.

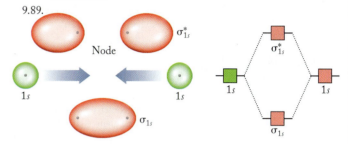

9.65.

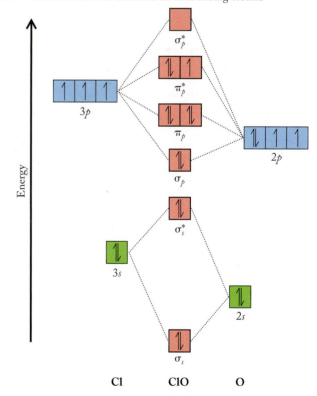

Tetrahedral molecular geometry. At first glance this would mean that the hybridization would be assigned as $sp^3$. However, notice that the Cl forms three π bonds to three of the oxygen atoms. This requires that three of the $p$ orbitals on Cl not be involved in the hybridization so that it can form parallel π bonds. Therefore, Cl must use low-lying $d$ orbitals in place of the $p$ orbitals for $sd^3$ hybridization to form the four σ bonds to oxygen.

9.67. H—Ar—F:     $sp^3d$ hybridized

9.69. a.

tetrahedral, $sp^3$ hybridized

b.

c. Trigonal bipyramidal

9.71. Yes

9.73. Yes, in resonance structures the electron distribution is blurred across all the resonance forms, which, in essence, defines the delocalization of electrons.

9.75. b and c

9.77. Each carbon in $C_6H_6$ has SN = 3 and $sp^2$ hybridization with trigonal planar geometry. Each carbon in $C_4H_8$ has SN = 4 and $sp^3$ hybridization with tetrahedral geometry. The double-bonded carbon atoms in $C_4H_6$ have SN = 3 and $sp^2$ hybridization with trigonal planar geometry. The single-bonded carbon atoms in $C_4H_6$ have SN = 4 and $sp^3$ hybridization with tetrahedral geometry.

9.79. All three carbon atoms are trigonal planar. The nitrogen and oxygen atoms both adopt a bent (<109.5°) geometry, and the sulfur atom is tetrahedral. The ring C—O bond is an overlap of

$C_{sp^2}/O_{sp^3}$, whereas the C=O bond is an overlap of $C_{sp^2}/O_{sp^2}$ and $C_{2p}/O_{2p}$. The C—N bond is formed by an overlap of $C_{sp^2}/N_{sp^3}$. The extra electron on N is located in an $sp^3$ hybrid orbital.

9.81. a and c

9.83. No

9.85. No, because the overlap of 1s and 2s orbitals is not as efficient as 1s–1s or 2s–2s overlaps. The match in size and energy is prohibitive.

9.87. If a molecule or ion has an even number of valence electrons distributed over an odd number of orbitals, or if the highest occupied molecular orbital is a π orbital, paramagnetic species could result.

9.89.

9.91. $N_2^+$ BO = 2.5
$O_2^+$ BO = 2.5
$C_2^+$ BO = 1.5
$Br_2^{2-}$ BO = 0
All species with nonzero bond order ($N_2^+$, $O_2^+$, and $C_2^+$) are expected to exist.

9.93. The species with one or more unpaired electrons are (a) $N_2^+$, (b) $O_2^+$, (c) $C_2^{2+}$, (e) $O_2^-$, (g) $N_2^{2-}$, and (h) $F_2^+$.

9.95. The odd electron is located in an antibonding orbital.

9.97. a and b

9.99. No

**9.101.**

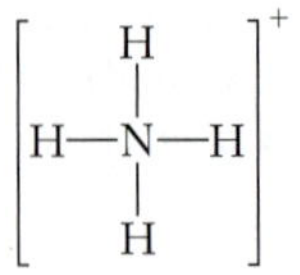

Tetrahedral

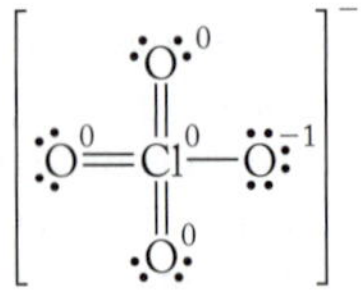

Tetrahedral

**9.103.**

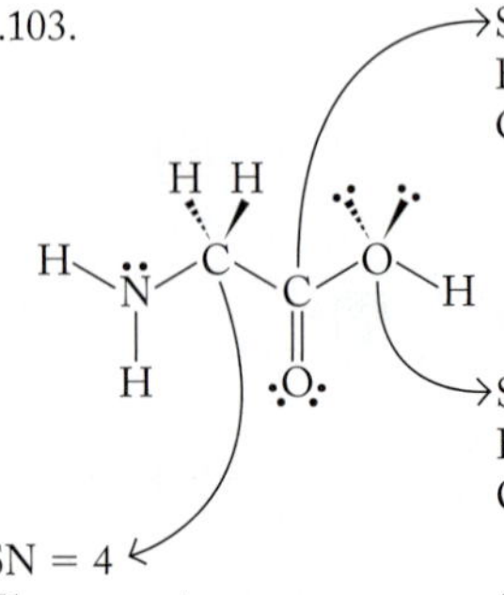

SN = 3
Electron-pair geometry = trigonal planar
O—C—O bond angle = 120°

SN = 4
Electron-pair geometry = tetrahedral
C—O—H bond angle = 109.5°

SN = 4
Electron-pair geometry = tetrahedral
N—C—C bond angle = 109.5°

**9.105.** (a) No, neither of the two molecules is linear. (b) They do not have the same dipole moments.

**9.107.** a. 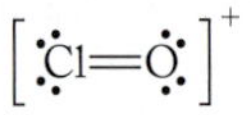 $\left[ \ddot{C}l{=}\dot{\ddot{O}} \right]^{+}$

b. BO = 2

**9.109.**

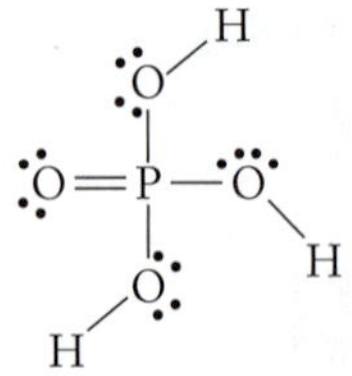

Molecular geometry around P = tetrahedral

**9.111.** $:B—B{=}C{=}\ddot{O}:$  Both B atoms have incomplete octets.
Molecular geometry = linear

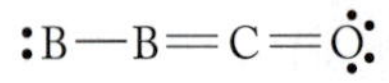

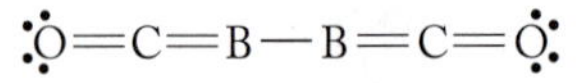

$\ddot{O}{=}C{=}B—B{=}C{=}\ddot{O}:$  Both B atoms have incomplete octets.
Molecular geometry = linear

**9.113.**

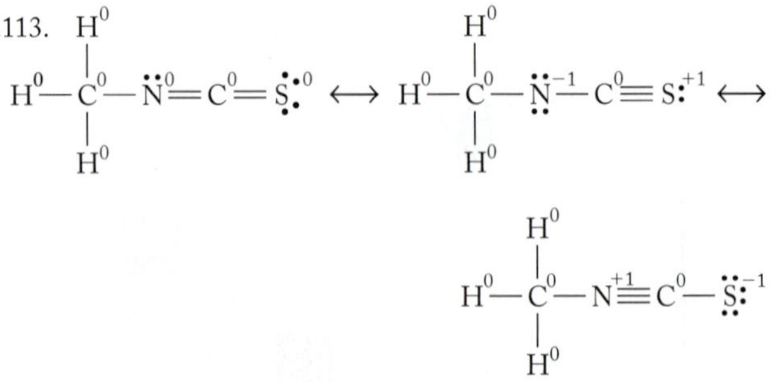

The resonance structure in which all formal charges are zero contributes the most to the bonding. The methyl (CH₃) carbon, SN = 4, so this carbon is tetrahedral. At the isothiocyanate (NCS) carbon, SN = 2, so the molecular geometry at this carbon is linear.

**9.115.** Yes

**9.117.** Bond order = 0

**9.119.** $N_2O_5$ and $N_2O_3$

**9.121.** This species is diamagnetic.

**9.123.** $sp^3$

**9.125.** This molecule is polar because, although the oxygen–oxygen bonds themselves are nonpolar, the lone pair has its own "pull" on the electrons in the molecule. Also, the $\pi$ bonds between the oxygen atoms place slightly more electron density on the terminal O atoms, making the "nonpolar O—O bond" polar.

**9.127.** a. Alliin:

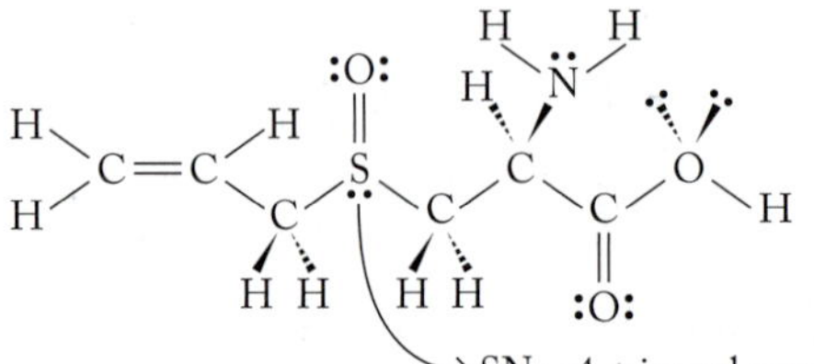

SN = 4, trigonal pyramidal molecular geometry

Allicin:

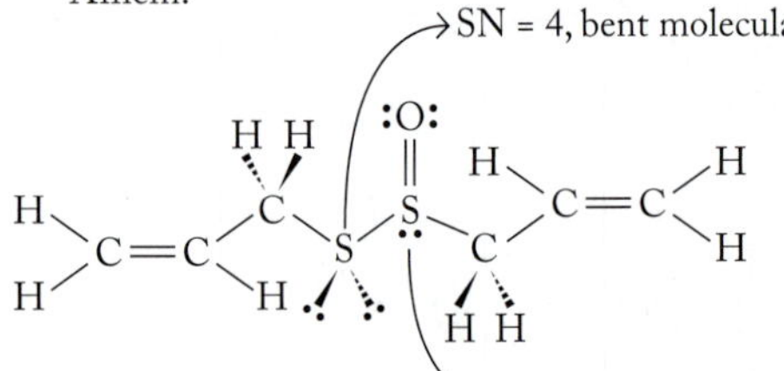

SN = 4, bent molecular geometry

SN = 4, trigonal pyramidal molecular geometry

b. The C—S—S bond angle is predicted to be the same as the S bond angles in $H_2S$, $CH_3SH$, and $(CH_3)_2S$.

## CHAPTER 10

**10.1.** KF

**10.3.** $XH_3$

**10.5.** The graph on the left has stronger intermolecular forces.

**10.7.** Increases

**10.9.** c

**10.11.** The ion–ion bond in $CaSO_4$ is stronger than in NaCl because of the higher charges on the cation and anion. For $CaSO_4$, this is greater than the ion–dipole interactions that would occur when $Ca^{2+}$ and $SO_4^{2-}$ dissolve, so $CaSO_4$ is not very soluble in water. NaCl has a lower ion–ion bond strength and its ion–dipole interactions with water are strong, so it dissolves in water.

**10.13.** $CsBr < KBr < SrBr_2$

**10.15.** The water molecule is oriented around $Cl^-$ so as to point the partially positive hydrogen atoms toward the $Cl^-$ ion.

**10.17.** Because of the full positive or negative charge on the ion, the ion–dipole interaction is stronger than the dipole–dipole interaction.

**10.19.** The charge buildup on H (partially positive) and the electronegative element (partially negative) means that the X—H bond is polar. It is still a dipole–dipole interaction except that its strength is noticeably higher than other dipole–dipole interactions.

**10.21.** $Cl^-$

**10.23.** $CH_3F$ is a polar molecule and therefore has stronger intermolecular forces than the nonpolar molecules of $CH_4$, which have only the weak dispersion forces. As it takes more energy to overcome strong intermolecular forces, $CH_3F$ has a higher melting point than $CH_4$.

10.25. The H atoms in fluoromethane are bound to the relatively low electronegativity C atom. Therefore, the carbon–hydrogen bond is not polar enough to exhibit hydrogen bonding. In HF, however, the hydrogen atom is bound to the extremely electronegative fluorine atom, so hydrogen bonding is expected.

10.27. b

10.29. a and d

10.31. Dispersion forces

10.33. Because real gases do have a volume, and when cooled they have less kinetic energy to overcome their intermolecular forces, reducing elasticity of collisions.

10.35. The greater dispersion forces of $CH_2Cl_2$ add to the dipole–dipole interactions to give stronger intermolecular forces between the $CH_2Cl_2$ molecules compared to those of $CH_2F_2$ molecules. Also, the molar mass of $CH_2Cl_2$ is higher than that of $CH_2F_2$, so it takes more energy to vaporize.

10.37. (a) $CCl_4$; (b) $C_3H_8$

10.39. $PCl_5$ is a solid at room temperature; $PCl_3$ is a liquid. Despite being nonpolar, $PCl_5$ has a higher molar mass and a larger surface area, so the dispersion forces for $PCl_5$ outweigh the weaker dispersion forces and dipole–dipole interactions for $PCl_3$.

10.41. Miscible solutes and solvents dissolve completely in each other; an insoluble solute does not dissolve at all.

10.43. 1,1-dichloroethane is soluble in water and 1,2-dichloroethane is not.

10.45. Hydrophilic substances dissolve in water. Hydrophobic substances do not dissolve, or are immiscible, in water.

10.47.  a.  $CHCl_3$
      b.  $CH_3OH$
      c.  NaF
      d.  $BaF_2$

10.49. a–d

10.51. b

10.53. d

10.55. $SO_3$

10.57. b

10.59. Yes

10.61. (a) $CH_3CH_2OH$ < (b) $CH_3OCH_3$ < (c) $CH_3CH_2CH_3$

10.63. 41.0 kJ/mol

10.65. In sublimation the solid does not first liquefy before evaporating; the solid goes directly into the gas phase. In evaporation a liquid becomes a gas.

10.67. When the pressure above a liquid is decreased, the temperature required to boil (i.e., enter the gas phase) is also decreased. Similarly for sublimation, as the pressure is decreased, the temperature required for particles to enter the vapor phase is decreased. The relationship between the pressure and temperature required for liquid/solid phase changes differs depending on the density of the solid and liquid phases. Transitions from the more dense phase (frequently a solid) to the less dense phase (frequently a liquid) occur at a lower temperature when the pressure is decreased. In all cases, the opposite is true if the pressure is increased.

10.69. (a) solid phase; (b) gas phase

10.71. Yes

10.73. Approximately 55°C to 60°C

10.75. Ethylene glycol has stronger intermolecular forces. We can tell this from the higher normal boiling point.

10.77. Reduce the temperature from 25°C to 0.01°C, then reduce the pressure from 1 atm to 0.006 atm.

10.79. The water vaporizes from liquid to gas.

10.81. −57°C

10.83. (a) liquid; (b) gas; (c) gas

10.85. A needle floats on water but not on methanol because of the high surface tension of water. This is because water can hydrogen bond through two O—H bonds with other water molecules, whereas methanol has only one O—H bond through which to form strong hydrogen bonds.

10.87. The expansion of water in the pipes on freezing may create sufficient pressure on the wall of the pipes to cause them to burst.

10.89. The cohesive forces within mercury are stronger than the adhesive forces of the mercury to the glass.

10.91. Molecules in the bulk liquid are "pulled" by all the other liquid molecules surrounding them and they are, therefore, "suspended" in the bulk liquid. Molecules on the surface of a liquid, however, are only pulled by molecules under and beside them, creating a tight film of molecules on the surface.

10.93. As temperature increases, the surface tension decreases because the molecular "film" on the surface of the liquid has fewer molecules held together by tight intermolecular forces. Likewise, the viscosity decreases as the temperature increases because molecules have more energy to readily break the intermolecular forces to enable them to slide past each other more freely.

10.95. Water

10.97. Liquid B is expected to have the higher surface tension and higher viscosity because its higher melting point is indicative of stronger intermolecular forces.

10.99. The sublimation point of ice will increase with increasing pressure. Higher external pressures mean that a molecule of ice will require a higher temperature to overcome intermolecular forces and enter the vapor phase.

10.101. Although the dispersion forces between methanol molecules are greater than those between water molecules (because methanol has more electrons and greater molar mass), water can form two hydrogen bonds compared to methanol's one hydrogen bond. This greater number of stronger interactions between water molecules raises the boiling point of water above that of methanol.

10.103. Increases

10.105. Contracts

10.107. Water in the solid (ice) form has a lower density than in the liquid form due to the open lattice formed by extensive hydrogen bonding between the water molecules.

10.109. b

10.111. $NH_3$: hydrogen bonding, dipole–dipole, London dispersion
      $SO_2$: dipole–dipole, London dispersion
      $C_2F_6$: London dispersion only

10.113. Both $Br_2$ and $I_2$ are nonpolar, and so only have London dispersion forces.

10.115. Astatine (At)

10.117. $NH_2Cl$ is trigonal pyramidal. It should be soluble in water because of its ability to form hydrogen bonds.

10.119.  a.  hypochlorite = $ClO^-$
      b.  chlorite = $ClO_2^-$
      c.  chlorate = $ClO_3^-$
      d.  perchlorate = $ClO_4^-$

10.121. $3\,OH^-(aq) + IO_3^-(aq) + 3\,HSO_3^-(aq) \rightarrow$
      $I^-(aq) + 3\,SO_4^{2-}(aq) + 3\,H_2O(\ell)$

# CHAPTER 11

11.1. b

11.3. 2

11.5. a. It is a mixture.
  b. (i) Three components
  (ii) 20 mL:50 mL:30 mL
  (iii) 80°C, 95°C, and 130°C

11.7. Solution A must be the more concentrated solution because solvent flows through the membrane from the least to the most concentrated side.

11.9. A nonvolatile solute is a compound that dissolves into a solvent and does not, under conditions to maintain the solution, enter appreciably into the gas phase.

11.11. When the average kinetic energy of the liquid molecules increases, more of the molecules can escape the liquid phase and enter the gas phase. An increase in the molecules in the gas phase causes an increase the vapor pressure.

11.13. The vapor pressure of pure ethanol is greater than the vapor pressure of the sugar in ethanol solution. The rate of evaporation will be higher for the pure ethanol. On the other hand, the rate of condensation in the sugar solution beaker is higher than the rate of evaporation, so eventually the sugar solution beaker will overflow.

11.15. Mole fraction of water in solution: $X = 0.70$; vapor pressure of solution at 25°C: $P_{solution} = 17$ torr

11.17. $P_{soln} = X_{solvent} \cdot P°_{solvent}$
$X_{solvent} + X_{solute} = 1$
$X_{solvent} = 1 - X_{solute}$
$P_{soln} = (1 - X_{solute}) \cdot P°_{solvent}$
$P_{soln} = P°_{solvent} - P°_{solvent} \cdot (X_{solute})$
$P°_{solvent} \cdot (X_{solute}) = P°_{solvent} - P_{soln}$
$X_{solute} = \dfrac{(P°_{solvent} - P_{soln})}{P°_{solvent}}$

11.19. At low temperatures, the solubility of a gas increases because fewer gas molecules dissolved in the solvent have sufficient (kinetic) energy to overcome the intermolecular forces between the solvent molecules. At low temperatures the gas molecules are "trapped" in the solvent and cannot "escape" into the gas phase.

11.21. No, the Henry's law constant for air is not simply the sum of $k_H$ for $N_2$ and $O_2$.

$$C_{air} = k_{H,air}P_{air} = C_{O_2} + C_{N_2}$$
$$k_{H,air}P_{air} = k_{H,O_2}P_{O_2} + k_{H,N_2}P_{N_2}$$
$$= k_{H,O_2}(0.22P_{air}) + k_{H,N_2}(0.78P_{air})$$
$$= 0.22k_{H,O_2}P_{air} + 0.78k_{H,N_2}P_{air}$$
$$= (0.22k_{H,O_2} + 0.78k_{H,N_2})P_{air}$$
$$k_{H,air} = 0.22k_{H,O_2} + 0.78k_{H,N_2}$$

11.23. $3.7 \times 10^{-2}$ mol/(L · atm)

11.25. a. $2.7 \times 10^{-3}$ M
  b. $2.3 \times 10^{-2}$ M

11.27. At 10°C, $k_H = 1.6 \times 10^{-3}$ mol/(L · atm)
  At 20°C, $k_H = 1.3 \times 10^{-3}$ mol/(L · atm)
  At 30°C, $k_H = 1.1 \times 10^{-3}$ mol/(L · atm)

11.29. When a gaseous ion, $M^+$ or $X^-$, dissolves in a water, new ion–dipole interactions are formed between the water molecules and the ions. In order for these interactions to form, some water–water (dipole–dipole) interactions must be broken. Both of these terms combine to give the enthalpy of hydration.

11.31. $S^{2-}$ has stronger ion–ion interactions than $NO_3^-$, so sulfides are mostly insoluble.

11.33. Melting point decreases as the atomic number of X increases.

11.35. a

11.37. −723 kJ/mol

11.39. −690. kJ/mol

11.41. The components of crude oil can be separated by fractional distillation, which uses differences in boiling points of the compounds.

11.43. $C_5H_{12}$

11.45. 60 torr

11.47. a. 23 styrene to 27 ethylbenzene
  b. 90°C

11.49. Molarity is the moles of the solute in one liter of solution. Molality is the moles of solute in one kilogram of solvent.

11.51. The difference between molarity and molality will increase.

11.53. While a molecular compound will not dissociate, an ionic compound may dissociate in the solvent. This dissociation may yield two or more particles in solution from each dissolved solute particle. This results in greater changes in the melting and boiling points, compared to those of a molecular solute that does not dissociate.

11.55. The theoretical value of $i$ for $CH_3OH$ is 1 because methanol is molecular and does not dissociate in a solvent such as water. NaBr has a theoretical value of $i = 2$ because it dissociates into two particles upon dissolution ($Na^+$ and $Br^-$). $K_2SO_4$ has a theoretical value of $i = 3$ because it dissociates into three particles upon dissolution (2 $K^+$ and $SO_4^{2-}$).

11.57. A semipermeable membrane is a boundary between two solutions through which some molecules may pass but others cannot. Usually, small molecules may pass through but large molecules are blocked.

11.59. Solvent flows across a semipermeable membrane from the more dilute solution side to the more concentrated solution side to balance the concentration of solutes on both sides of the membrane.

11.61. Reverse osmosis transfers solvent across a semipermeable membrane from a region of higher solute concentration to a region of lower solute concentration. Because reverse osmosis goes against the natural flow of solvent across the membrane, the key component needed is a pump to apply pressure to the more concentrated side of the membrane. Other components needed include a containment system, piping to introduce and remove the solutions, and a tough semipermeable membrane that can withstand the high pressures needed.

11.63. The dressing will pull water out of the salad greens by osmosis, leading to wilted salad.

11.65. a. 0.21 m
  b. 0.572 m
  c. 0.440 m

11.67. a. 749 g
  b. 31.7 g
  c. 104.6 g

11.69. $6.5 \times 10^{-5}$ m $NH_3$; $8.7 \times 10^{-6}$ m $NO_2^-$; $2.195 \times 10^{-2}$ m $NO_3^-$

11.71. 3.81°C

11.73. $2.52 \times 10^{-2}$ m

11.75. −1.89°C

11.77. 0.5 m $CaCl_2$

11.79. 0.0100 m $Mg(NO_3)_2$

11.81. (a) 0.06 m $FeCl_3$ < (b) 0.10 m $MgCl_2$ < (c) 0.20 m KCl

11.83. a. From side A to side B
b. From side B to side A
c. From side A to side B

11.85. a. 57.5 atm
b. 0.682 atm
c. 52.9 atm
d. 46.5 atm

11.87. a. $2.75 \times 10^{-2} \ M$
b. $1.11 \times 10^{-3} \ M$
c. $1.00 \times 10^{-2} \ M$

11.89. False. The molarity of the NaCl solution would be greater (by 1.5 times) than the molarity of $CaCl_2$.

11.91. a. Osmotic pressure increases
b. Freezing point decreases
c. Boiling point increases

11.93. 94.1 g/mol

11.95. Molar mass = 164 g/mol. The molecular formula of eugenol is $C_{10}H_{12}O_2$.

11.97. a

11.99. Yes

11.101. For 0.0935 $m$ $NH_4Cl$, $i = 1.85$
For 0.0378 $m$ $(NH_4)_2SO_4$, $i = 2.46$

11.103. 2.3 atm

11.105. 4270 g/mol

11.107. $[Na^+]_{saline} = [Na^+]_{RL} + [K^+]_{RL} + [Ca^{2+}]_{RL}$
$[Cl^-]_{saline} = [Cl^-]_{RL} + [C_3H_5O_3^-]_{RL}$

# CHAPTER 12

12.1. b and d are crystalline; a and c are amorphous.

12.3. The chemical formula is $A_4B_4$ or AB.

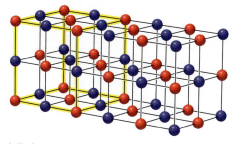

12.5. $A_4B_2C$

12.7. $Li_2S$

12.9. Cs (blue) and Sr (purple)

12.11. $MgB_2$

12.13. Cubic closest-packed structures have an *abcabc* . . . pattern and hexagonal closest-packed structures have an *abab* . . . pattern.

12.15. Body-centered cubic

12.17. These structural forms are not allotropes because iron is not molecular.

12.19. 104.2 pm

12.21. 513 pm

12.23. c

12.25. 0.963 g/cm³

12.27. 4.50 g/cm³

12.29. In a substitutional alloy, some of the atoms of one lattice are replaced with atoms of another element. One example of a substitutional alloy is bronze, in which up to 30% of the Cu atoms in a copper lattice have been replaced with tin atoms. The tin atoms in this alloy are randomly distributed throughout the lattice, occupying any lattice position normally occupied by a copper atom. An interstitial alloy contains solutes in the spaces (or "holes") between atoms when the solvent lattice forms. One example of an interstitial alloy is austenite, an alloy containing carbon in the octahedral holes of the iron fcc lattice.

12.31. The edge length of the unit cell with Ni at the center will be shorter.

12.33. Increase

12.35. Yes, both are correct.

12.37. a. 90 pm
b. Substitutional alloy

12.39. Substitutional alloy

12.41. $V_2C$

12.43. 5.78 g/cm³

12.45. a. A substitutional alloy with a bcc lattice
b. $CuZn_3$

12.47. In the sea-of-electrons model, electrons in metallic bonds are highly delocalized, forming a "sea" of mobile electrons that move freely among all the atoms in a metallic solid. The application of electrical potential to gold causes its mobile valence electrons to move toward the positive potential through this delocalized "sea."

12.49. Ionic bonds are stronger than metallic bonds.

12.51. Yes

12.53. Groups 2 and 12

12.55. Doping a metalloid generates a semiconductor with a smaller bandgap than the starting material. n-Type semiconductors have "extra" electrons higher in energy than the valence band for the undoped material. p-Type semiconductors have "extra" holes lower in energy than the conduction band for the undoped material. In both types of doped semiconductor the effect is similar; the band gap is lowered, and electrons flow with a smaller energy barrier.

12.57. Group 14

12.59. a. n-Type
b.

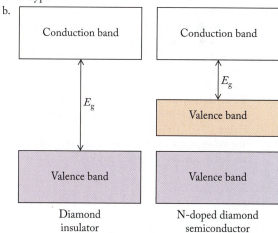

Diamond insulator / N-doped diamond semiconductor

c. $4.68 \times 10^{-19}$ J

12.61. The wavelength is shorter.

12.63. $K^+$ is large and so does not fit well into the octahedral holes of the fcc lattice.

12.65. The radius of $Cl^-$ is 181 pm and the radius of $Cs^+$ is 170 pm, so their radii are very similar. The $Cs^+$ ion at the center of Figure P12.65 occupies the center of the cubic cell, so CsCl could be viewed as a body-centered cubic structure when taking into account the ions' slight difference in size. However, if we look at the ions as roughly equal in size, the unit cell becomes two interpenetrating simple cubic unit cells.

12.67. The rock salt structure would not provide the correct ratio of ions for $CaCl_2$.

12.69. Yes, the alloy is more dense.

12.71. Less than

12.73. $MgFe_2O_4$

12.75. (a) octahedral; (b) half

12.77. a. The rock salt structure may be thought of as an fcc lattice of anions with cations in the octahedral holes. Alternately, the rock salt structure is somewhat like two interpenetrating fcc lattices. In contrast, the cesium chloride structure may be thought of as a simple cubic lattice of anions with cations in the cubic holes. Alternately, the cesium chloride structure is somewhat like two interpenetrating simple cubic lattices.
   b. Both structures have the same ratio of $Mg^{2+}$ to $Se^{2-}$ ions (1:1).

12.79. 5.25 g/cm$^3$

12.81. 421 pm

12.83. The hybridization of the phosphorus atom in black phosphorus is $sp^3$, with bond angles of 102°, giving a puckered ring. In graphite, the carbon atoms are $sp^2$ hybridized, with bond angles of 120°, which gives graphite a flat geometry.

12.85. a. 1.708 g/cm$^3$
   b. 498.6 pm

12.87. 3.81 g/cm$^3$

12.89. Ductility, electrical and thermal conductivity, and malleability describe metals. Ceramics are electrical and thermal insulators and are brittle.

12.91. $Mg_3(Si_2O_5)(OH)_4$

12.93. $2\ KAlSi_3O_8(s) + 2\ H_2O(\ell) + CO_2(g) \rightarrow Al_2(Si_2O_5)(OH)_4(s) + 4\ SiO_2(s) + K_2CO_3(aq)$; this is not a redox reaction.

12.95. a. $3\ CaAl_2Si_2O_8(s) \rightarrow Ca_3Al_2(SiO_4)_3(s) + 2\ Al_2SiO_5(s) + SiO_2(s)$
   b. In anorthite, the silicate anion is $Si_2O_8{}^{8-}$.
      In grossular, the silicate anion is $SiO_4{}^{4-}$.
      In kyanite, the silicate anion is $SiO_5{}^{6-}$.

12.97. Cubic holes can accommodate $Ba^{2+}$. Octahedral holes can accommodate $Ti^{4+}$.

12.99. An amorphous solid has no regular, repeating lattice to diffract X-rays.

12.101. X-rays have wavelengths about the same size as the distance between atoms in crystals. Microwaves have wavelengths too long to be diffracted by crystal lattices.

12.103. If a crystallographer uses a shorter λ wavelength, the data set can be collected over a smaller scanning range.

12.105. Halite

12.107. The values of $n$ are 2 ($\theta = 6.99°$) and 3 ($\theta = 10.62°$). The average lattice spacing is $d = 582$ pm.

12.109. 4.76°

12.111. $XYZ_3$

12.113. a. bcc
   b. Two phase changes, hcp → bcc → hex

12.115. a. $2.3 \times 10^5$ unit cells/particle
   b. $9.1 \times 10^5$ Ag atoms
   c. $8.4 \times 10^{12}$ Ag particles

12.117. a. 7.53 g/cm$^3$
   b. 3.42 g/cm$^3$
   c. 3.36 g/cm$^3$

12.119. AuZn

12.121. Substitutional alloys

12.123. A cluster of three simple cubic unit cells, with an Al atom in the center of one of them, consistent with the formula ($Cu_3Al$).

12.125. a. $SiO_2(s) + C(s) \rightarrow CO_2(g) + Si(s)$
   b. $Si(s) + 2\ Cl_2(g) \rightarrow SiCl_4(\ell)$
   c. $Ge(g) + 2\ Br_2(\ell) \rightarrow GeBr_4(s)$

12.127. a. 598 pm
   b. 299 pm
   c. In deoxidizing the iron, silicon is oxidized to $Si^{4+}$, and any iron cations (such as $Fe^{3+}$) are reduced to Fe.

# CHAPTER 13

13.1. a. One degree of unsaturation
   b. Two degrees of unsaturation
   c. No degrees of unsaturation
   d. Three degrees of unsaturation

13.3. Pine oil and oil of celery

13.5. a.

trans                              cis

   b.

(both isomers equivalent)

   c.

trans                              cis

13.7. a and b

13.9. Benzyl acetate contains an ester group; carvone contains two alkene groups and a ketone group; cinnamaldehyde contains an alkene and an aldehyde group.

13.11.

$$\left[ \begin{array}{c} CH_3 \\ | \\ -Si-O- \\ | \\ CH_3 \end{array} \right]_n$$

13.13. For *cis*-polyisoprene:

For *trans*-polyisoprene:

13.15. The functional groups present are alcohol, ester, and alkene.

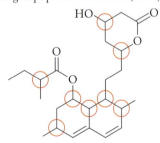

13.17. An $sp$ hybridized carbon atom can form two double bonds or one triple bond; an $sp^2$ hybridized carbon atom can form one double bond; an $sp^3$ hybridized carbon atom can form only single bonds.

13.19. No

13.21. Some examples from Chapter 9 are:
Acrolein (alkene + aldehyde) in Section 9.5
Spearmint and caraway (alkene + ketone) in Figure 9.38

13.23. 3565 monomers

13.25. Yes

13.27. $sp^3$

13.29. The structure of cyclohexane shows that C atoms are $sp^3$ hybridized with bond angles of 109.5°. It cannot be a planar molecule.

13.31. No

13.33. No

13.35.

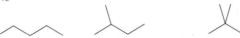

Pentane          2-Methylbutane          2,2-Dimethylpropane

13.37. (b) 2,3,4-trimethylpentane; (c) 3-ethyl-2-methylpentane; (d) 3-ethyl-3-methylpentane; (e) 3,4-dimethylhexane.

13.39. a. $C_8H_{18}$
b. $C_9H_{20}$
c. $C_8H_{18}$
d. $C_8H_{18}$
e. $C_9H_{20}$

13.41. −124 kJ

13.43. $C_3H_8 < C_8H_{16} < C_{14}H_{30}$

13.45. Structural isomers have different connectivity of the atoms; geometric isomers have the same connectivity of the atoms but a different spatial arrangement.

13.47. No

13.49. When the double bond is "terminal" (occurs at the end or beginning of the carbon chain) there are three like groups (H) so no cis and trans isomers are possible.

13.51. The C=C double bond outside of the ring does not show cis–trans isomerism because there are not two dissimilar groups on the terminal carbon atom. The C=C double bond in the ring of carbon atoms is cis in the structure of carvone. This bond cannot be trans or the ring of 6 carbon atoms would not be possible.

13.53. Ethylene has a C=C bond with which HBr is reactive, but polyethylene has only saturated C—C bonds that do not react with HBr.

13.55. a is trans, $E$; b is cis, $Z$

13.57. a. 681.2 kJ, endothermic
b. $CH_4$ is oxidized

13.59.

13.61. In benzene, each C atom is $sp^2$ hybridized with bond angles of 120°. This geometry at each of the carbon atoms in the ring makes benzene a planar molecule.

13.63. Tetramethylbenzene has three structural isomers; pentamethylbenzene has no structural isomers.

13.65. Yes

13.67.

13.69. Fuel value for 1 mole benzene = 41.83 kJ/g
Fuel value for 3 moles ethylene = 50.30 kJ/g
1 mole benzene has a lower energy content than 3 moles ethylene.

13.71. Methylamine has a smaller nonpolar hydrocarbon chain compared to n-butylamine, so it is more soluble in water.

13.73.

HO    NH₂ 1° amine
N 2° amine
H

NH₂ 1° amine

Serotonin          Amphetamine

13.75. −138.7 kJ

13.77. The more oxygenated the fuel, the lower the fuel value.

13.79. Ethers have lower boiling points compared to alcohols because they have weaker dipole–dipole forces compared to the alcohols, which have hydrogen bonding between the molecules.

13.81. Evaporation of ethanol from the skin is an endothermic process (phase change from liquid to vapor). The heat transfers from the skin to the ethanol so the skin feels cold.

13.83. a and d are alcohols, b and c are ethers; (b) < (c) < (d) < (a)

13.85. Fuel value for diethyl ether = 36.74 kJ/g
Fuel value for n-butanol = 36.10 kJ/g
Diethyl ether has a slightly higher fuel value.

13.87. Fuel value for methanol = 22.67 kJ/g
Fuel value for ethanol = 29.67 kJ/g
Yes, the answer supports the prediction made in Problem 13.78.

13.89. Both carboxylic acids and aldehydes have polar functional groups. Carboxylic acids, however, are more soluble in water because they form strong hydrogen bonds with water.

13.91. Yes

13.93. No

13.95. Structure a, because all of the formal charges are zero

13.97. An amide includes a carbonyl (C=O) as part of its functional group in addition to the −NH₂ group.

13.99. a, b, and d

13.101. b

13.103.

The plot of C:H ratio versus number of C atoms for aldehydes correlates exactly to that of alkenes and poorly to that of alkanes.

13.105. a. Pineapples

Acetic acid          Butanol

+ H₂O

b. Bananas

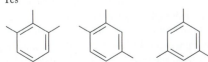

2-Methylbutanoic acid          Ethanol

+ H₂O

c. Apples

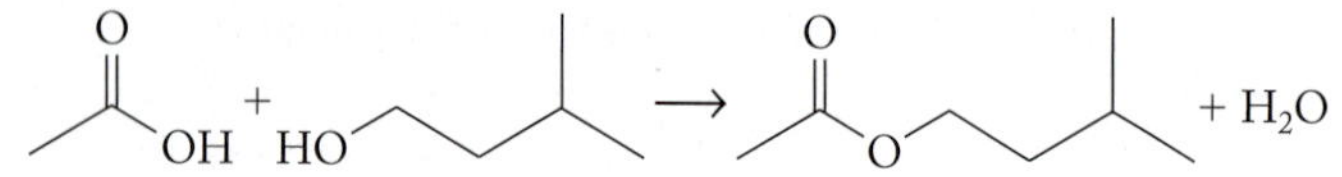

Acetic acid     3-Methylbutanol

13.107.

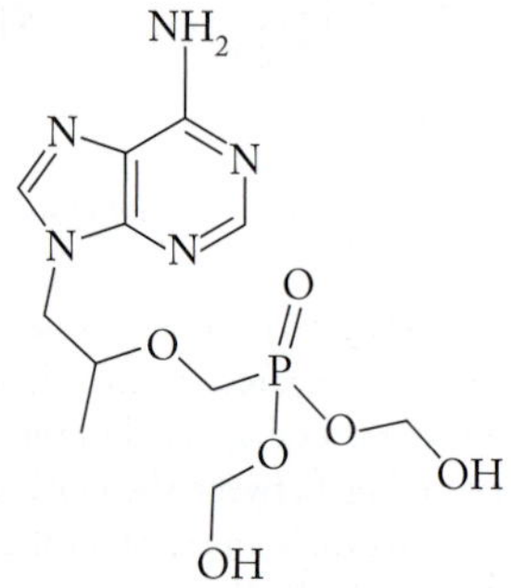

13.109. Fuel value for formaldehyde = 19.00 kJ/g
Fuel value for formic acid = 6.531 kJ/g
Formaldehyde has a significantly higher fuel value than formic acid.

13.111. For reaction 1, $\Delta H^{\circ}_{rxn}$ = 17.5 kJ
For reaction 2, $\Delta H^{\circ}_{rxn}$ = 312.1 kJ

13.113. (a) 4; (b) 6; (c) 8

13.115. a. Condensation; methanol
b. Because of the presence of the six-membered ring, Kodel might be better able to accept nonpolar organic dyes.

13.117. No; enantiomer and optically active can describe the same chiral molecule, but achiral cannot.

13.119. Homogeneous

13.121. Glycine has no chiral carbon centers.

13.123. $sp^3$

13.125. a, c, and d

13.127. a

13.129.

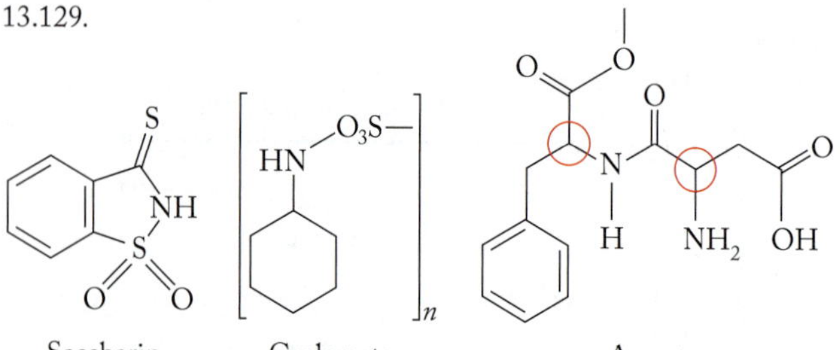

Saccharin     Cyclamate     Aspartame

13.131. a. Yes; 2 chiral centers.
b. Yes, the *cis* isomer is shown below:

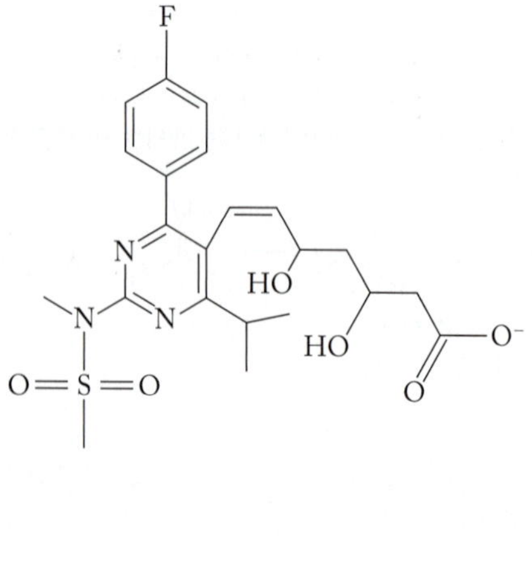

13.133. Nicotine's highlighted N atom is in a tertiary amine group; Valium's highlighted N atom is in an amide group.

13.135. 2.52 g methanol; 3.45 g carbon dioxide

13.137. Compound A is diethyl ether; compound B is butanol

13.139. a. 12
b. Alkene and aldehyde
c. Stereoisomers (cis and trans) and structural isomers

13.141. Yes

13.143.

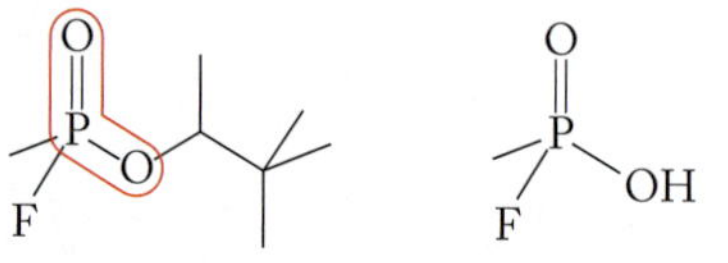

Parent acid

The formula for the parent acid is $PF(CH_3)(OH)$.

13.145. Amide groups

13.147. a.

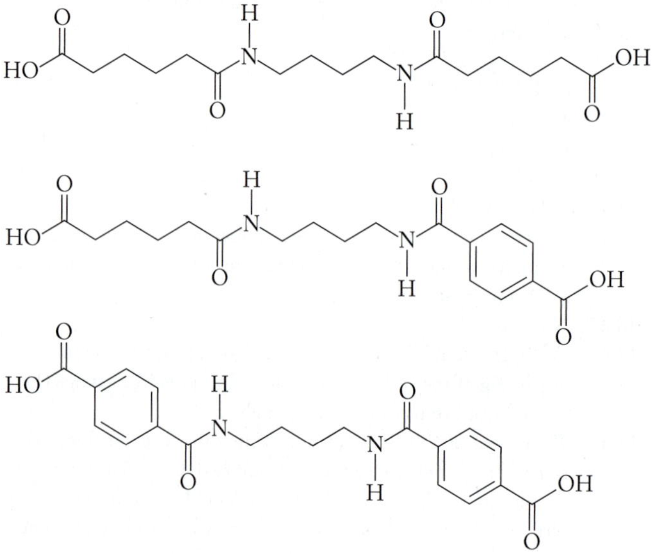

b. 1 mole adipic acid : 1 mole terephthalic acid : 2 moles putrescine

13.149. a.

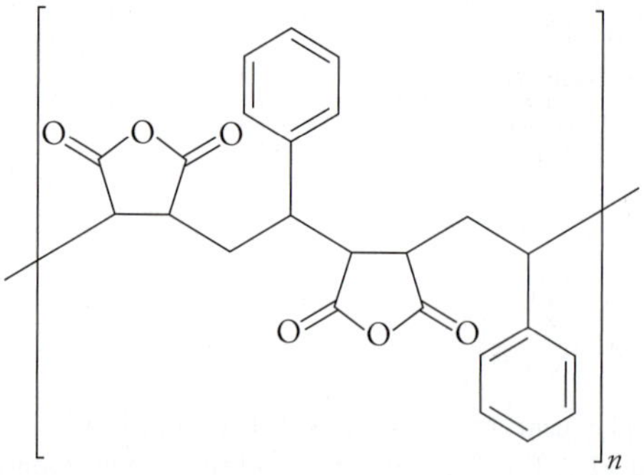

b. It would be more hydrophilic and less rigid.

13.151. a. $R_2SiCl_2(aq) + H_2O(\ell) \rightarrow R_2SiCl(OH)(aq) + HCl(aq)$
$R_2SiCl(OH)(aq) + (HO)SiClR_2(aq) \rightarrow$
$R_2ClSi-O-SiClR_2(aq) + H_2O(\ell)$
b. The side chains (R groups) are nonpolar.

13.153. The heat of combustion determined using experimental means is different from that calculated from average bond energies because the bond energy of a particular bond depends on the structure of the rest of the molecule.

13.155. a. Ether, ester, alkane
   b. Alcohol, ketone, carboxylic acid, ester, alkane
   c. Ether, alkene (both *cis*), alkane

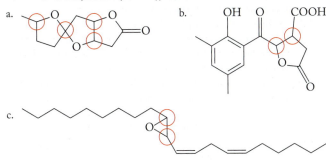

# CHAPTER 14

14.1. The $[N_2O]$ is represented by the green line and $[O_2]$ is represented by the red line.

14.3. b

14.5. a. Graph a
   b. Graph b

14.7. c

14.9. b

14.11. a. 3*
   b. a
   c. b

14.13. Nitrogen (light blue)

14.15. Palladium (blue) and platinum (orange)

14.17. The presence of $NO_2$ in the atmosphere and ample sunlight allows the O atoms to react with $O_2$ to generate $O_3$. The reactant $NO_2$ is present in the atmosphere due to automobile exhausts, which build up during the day. The higher concentration of $O_3$ occurs later in the day, when $[NO_2]$ has increased and the sunlight becomes stronger as midday approaches.

14.19. During the evening rush hour, the sunlight (and UV radiation) is less intense, so the photochemical breakdown of $NO_2$ does not occur to as great an extent as after the morning rush hour.

14.21. O atoms are more reactive due to the incomplete octet and unpaired electrons on each atom.

14.23. a. $2\,N_2(g) + O_2(g) \rightarrow 2\,N_2O(g)$
   b. $2\,N_2(g) + 5\,O_2(g) \rightarrow 2\,N_2O_5(g)$

14.25. The average rate is the rate averaged over a defined time interval, whereas the instantaneous rate is the rate at a specific moment.

14.27. As the reaction proceeds, the rate of change in concentration of the product will decrease.

14.29. a. $\dfrac{\Delta[N_2]}{\Delta t} = \dfrac{\Delta[O_2]}{\Delta t}$

   b. $\dfrac{\Delta[N_2]}{\Delta t} = -\dfrac{1}{2}\dfrac{\Delta[NO]}{\Delta t}$

14.31. a. Rate $= -\dfrac{\Delta[F_2]}{\Delta t} = +\dfrac{\Delta[HOF]}{\Delta t} = +\dfrac{\Delta[HF]}{\Delta t}$

   b. Rate $= -\dfrac{1}{3}\dfrac{\Delta[HCl]}{\Delta t} = +\dfrac{\Delta[H_2]}{\Delta t}$

   c. Rate $= -\dfrac{1}{4}\dfrac{\Delta[NH_3]}{\Delta t} = -\dfrac{1}{3}\dfrac{\Delta[O_2]}{\Delta t} = +\dfrac{1}{2}\dfrac{\Delta[N_2]}{\Delta t} = +\dfrac{1}{6}\dfrac{\Delta[H_2O]}{\Delta t}$

14.33. $1.0 \times 10^{-3}\ M/s$

14.35. a. Rate $= \dfrac{\Delta[CO_2]}{\Delta t} = -\dfrac{2}{3}\dfrac{\Delta[CO]}{\Delta t}$

   b. Rate $= \dfrac{\Delta[COS]}{\Delta t} = -\dfrac{\Delta[SO_2]}{\Delta t}$

   c. Rate $= \dfrac{\Delta[CO]}{\Delta t} = 3\dfrac{\Delta[SO_2]}{\Delta t}$

14.37. a. $1.2 \times 10^7\ M/s$
   b. $2.9 \times 10^4\ M/s$

14.39. Between 0 and 100 μs: $1.4 \times 10^{-5}\ M/\mu s$
   Between 200 and 300 μs: $5.5 \times 10^{-6}\ M/\mu s$

14.41. For the change in concentration of ClO versus time we obtain the following plot:

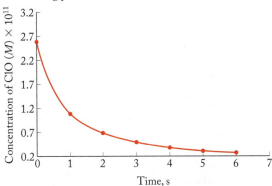

The instantaneous rate at 1 s is $8.28 \times 10^{10}$ molecules/(cm$^3$ · s). For the change in concentration of $Cl_2O_2$ versus time we obtain the following plot:

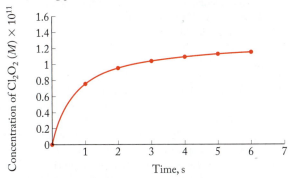

The instantaneous rate at 1 s is $4.13 \times 10^{10}$ molecules/(cm$^3$ · s).

14.43. The concentration of reactants decreases over time, so the likelihood of collisions decreases, and the rate of reaction decreases.

14.45. Yes

14.47. The half-life will be halved.

14.49. a. First order in both A and B, and second order overall
   b. Second order in A, first order in B, and third order overall
   c. First order in A, third order in B, and fourth order overall

14.51. a. Rate $= k[O][NO_2]$; $k$ units $= M^{-1}s^{-1}$
   b. Rate $= k[NO]^2[Cl_2]$; $k$ units $= M^{-2}s^{-1}$
   c. Rate $= k[CHCl_3][Cl_2]^{1/2}$; $k$ units $= M^{-1/2}s^{-1}$
   d. Rate $= k[O_3]^2[O]^{-1}$; $k$ units $= s^{-1}$

14.53. a. Rate $= k[BrO]$
   b. Rate $= k[BrO]^2$
   c. Rate $= k[BrO]$
   d. Rate $= k[BrO]^0 = k$

14.55. No. The same behavior would be observed if the reaction were second order in one reactant, and zero order in the other.

14.57. a. Rate = $k[NO_2][O_3]$
b. $4.9 \times 10^{-11}$ $M/s$
c. $4.9 \times 10^{-11}$ $M/s$
d. The rate doubles.

14.59. c

14.61. Rate $= -\dfrac{1}{2}\dfrac{\Delta[N_2O_5]}{\Delta t} = +\dfrac{1}{4}\dfrac{\Delta[NO_2]}{\Delta t} = +\dfrac{\Delta[O_2]}{\Delta t} = 3.4 \times 10^{-5}$ s$^{-1}$

14.63. Rate = $k[ClO_2][OH^-]$; $k = 14$ $M^{-1}s^{-1}$

14.65. Rate = $k[NO]^2[H_2]$; $k = 6.32$ $M^{-2}s^{-1}$

14.67. a. $-3.56 \times 10^{-4}$ $M \cdot min^{-1}$
b. $-3.56 \times 10^{-4}$ $M \cdot min^{-1}$

14.69. 0.293 $M$

14.71. a. $k = 8.31 \times 10^{-5}$ s$^{-1}$
b. 0.616 $M$

14.73. a. Rate = $k[N_2O]$
b. 4

14.75. a. Rate = $k[^{32}P]$
b. 0.0485 day$^{-1}$
c. 14.3 days

14.77. $k = 5.40 \times 10^{-12}$ cm$^3$ molecules$^{-1}$ s$^{-1}$; $t_{1/2} = 0.712$ s

14.79. Rate = $k[C_{12}H_{22}O_{11}][H_2O] = k'[C_{12}H_{22}O_{11}]$;
$k' = 6.21 \times 10^{-5}$ s$^{-1}$

14.81. Rate depends on concentration and activation energy. Concentrations of reactants vary widely, and some reactions are catalyzed (e.g., by enzymes or metals) whereas others are not.

14.83. An increase in temperature increases the frequency and the kinetic energy at which the reactants collide. This speeds up the reaction. The order of the reaction is unaffected.

14.85. The reaction with the larger activation energy (150 kJ/mol).

14.87. $E_a = 17.1$ kJ/mol; $A = 1.002$

14.89. a. $E_a = 314$ kJ/mol
b. $A = 5.03 \times 10^{10}$
c. $k = 1.06 \times 10^{-44}$ $M^{-1/2}s^{-1}$

14.91. $E_a = 39.1$ kJ/mol; $A = 1.27 \times 10^{12}$

14.93. Yes, though it does not have to. If a catalyst or intermediate is present, the reaction could proceed in two or more steps.

14.95. Pseudo-first-order kinetics occurs when one of the reactants is in sufficiently high concentration that its concentration does not change appreciably over the course of the reaction.

14.97. It could occur in one step, though it is possible that there is a fast second step that does not affect the overall rate of reaction.

14.99. a. Rate = $k[SO_2Cl_2]$; unimolecular
b. Rate = $k[NO_2][CO]$; bimolecular
c. Rate = $k[NO_2]^2$; bimolecular

14.101. $N_2O_5(g) + O(g) \rightarrow 2\ NO_2(g) + O_2(g)$

14.103. The second step

14.105. The first step

14.107. Thermal decomposition: b or c
Photochemical decomposition: a

14.109. Yes

14.111. Yes

14.113. Because the catalyst itself is not involved in the rate-limiting step

14.115. NO is the catalyst.

14.117. The reaction of $O_3$ with Cl has the larger rate constant.

14.119. When the concentration of a reactant ($O_2$ for the combustion reaction) increases, the rate of combustion increases.

14.121. The bodily reactions that use $O_2$ are slower at colder temperatures.

14.123. Yes, we could use other times, not just $t = 0$, as long as the rate of the reverse reaction is still much slower than the forward reaction. Concentrations at time $t$ would be used in place of

initial concentrations, subtracting these values from those at time $t + m$, where $m$ is the time interval being investigated.

14.125. In a plot of $1/[X] - 1/[X]_0$ divided by $t - t_0$, the slope of the line corresponds to $k$, the reaction rate constant.

14.127. For an elementary step to take place, there must be some involvement from a molecular or atomic species. Therefore, there cannot be *no* dependence (or zero order) of the reactant in an elementary step of a reaction mechanism.

14.129. The rate of consumption of $O_3$ is the same as the rate of formation of $N_2O_5$ and $O_2$ and one-half the rate of consumption of $NO_2$.

14.131. $k = 3.6 \times 10^{-4}$ s$^{-1}$; Rate $= (3.6 \times 10^{-4}$ s$^{-1})[N_2O_5]$

14.133. a. Yes
b. $E_a = 62.5$ kJ/mol
c. Rate $= 1.2 \times 10^{-12}$ $M/s$
d. At 10°C (283 K), $k = 21$ $M^{-1}s^{-1}$; at 35°C (308 K), $k = 1.8 \times 10^2$ $M^{-1}s^{-1}$

14.135. a. Rate = $k[Na(H_2O)_6{}^+]$
b. Neither

14.137. a. Rate = $k[NO][ONOO^-]$; $k = 1.30 \times 10^{-3}$ $M^{-1}s^{-1}$
b. Preferred

c. $-55$ kJ

14.139. a. Second order
b. No

14.141. a. Rate = $k[NH_2][NO]$
b. $1.2 \times 10^9$ $M^{-1}$ s$^{-1}$

# CHAPTER 15

15.1. a. $K_c = \dfrac{[A_3B]^2}{[A_2]^3[B_2]}$
b. $K_c = 1.0$

15.3. (a) A $\rightleftharpoons$ B; (b) $K_c = 2.0$

15.5. The reaction is endothermic. As temperature increases, $K$ increases, indicating that more products form at higher temperatures.

15.7. A system is at equilibrium when the rate of the forward reaction equals the rate of the reverse reaction.

15.9. No, because at 20 μs the concentrations of A and B are still changing.

15.11. $K = \dfrac{k_{forward}}{k_{reverse}}$

15.13.

| Molar Mass | Compound | How Present |
|---|---|---|
| 28 | $^{14}N_2$ | Originally present |
| 29 | $^{15}N^{14}N$ | From decomposition of $^{15}N^{14}NO$ |
| 30 | $^{15}N_2$ | From decomposition of $^{15}N_2O$ |
| 32 | $O_2$ | Originally present |
| 44 | $^{14}N_2O$ | From combination of $^{14}N_2$ and $O_2$ |
| 45 | $^{15}N^{14}NO$ | From combination of $^{15}N^{14}N$ and $O_2$ |
| 46 | $^{15}N_2O$ | Originally present |

15.15. 0.333

15.17. When $\Delta n = 0$; when the number of moles of gaseous products equals the number of moles of gaseous reactants

15.19. a. $K_c = \dfrac{[C_6H_6]}{[C_2H_4][H_2]}$ and $K_p = \dfrac{(P_{C_6H_6})}{(P_{C_2H_4})(P_{H_2})}$

   b. $K_c = \dfrac{[SO_3]^2}{[SO_2]^2[O_2]}$ and $K_p = \dfrac{(P_{SO_3})^2}{(P_{SO_2})^2(P_{O_2})}$

15.21. 0.50

15.23. 0.068

15.25. 0.50

15.27. 27

15.29. 780

15.31. 0.0583

15.33. a and c

15.35. 0.10

15.37. When scaling the coefficients of a reaction up or down, the new value of the equilibrium constant is the first $K$ raised to the power of the scaling constant.

15.39. 11.0

15.41. $K_{c,forward} = \dfrac{[NO_2]^2}{[NO][NO_3]}$; $K_{c,reverse} = \dfrac{[NO][NO_3]}{[NO_2]^2}$;

   $K_{c,reverse} = \dfrac{1}{K_{c,forward}}$

15.43. $K_c = 1.7 \times 10^{-2}$

15.45. a. 0.049
   b. 420
   c. 20

15.47. $1.1 \times 10^{-18}$

15.49. The reaction quotient $Q$ is the ratio of the concentrations or partial pressures of the products of a reaction, raised to their stoichiometric coefficients, to the concentrations of reactants, raised to their stoichiometric coefficients. The reaction quotient has the same form as the equilibrium constant $K$ expression, but the reaction is not necessarily at equilibrium.

15.51. The system is at equilibrium.

15.53. No, $Q < K$ so the reaction proceeds to the right to reach equilibrium.

15.55. Mixture (a) is at equilibrium.

15.57. $Q > K$, so the reaction will proceed to the left.

15.59. The reaction is not at equilibrium. It will proceed to the left (reactant side) to approach equilibrium.

15.61. $K_p = \dfrac{(P_{SO_2})}{(P_{O_2})}$

15.63. $K_c = \dfrac{1}{[Fe(OH)_2]^4[O_2]}$

15.65. No

15.67. As the concentration of $O_2$ increases, the reaction shifts to the right and the CO on the hemoglobin is displaced.

15.69. According to Le Châtelier's principle, an increase in the partial pressure (or concentration) of $O_2$ above the water shifts the equilibrium to the right so that more oxygen becomes dissolved in the water. This is consistent with Henry's law.

15.71. b and d

15.73. a. Increasing the concentration of the reactant, $O_3$, shifts the equilibrium to the right, increasing the concentration of the product, $O_2$.
   b. Increasing the concentration of the product, $O_2$, shifts the equilibrium to the left, increasing the concentration of the reactant, $O_3$.

   c. Decreasing the volume of the reaction to 1/10 its original volume shifts the equilibrium to the left, increasing the concentration of the reactant, $O_3$.

15.75. The equilibrium shifts to the left.

15.77. b

15.79. When $K$ is small, the amount of reactants that are transformed into products may be so small that at equilibrium the concentrations of the reactants are approximately equal to the initial concentrations. This means that we can make an approximation in the $K$ expression to make our calculations easier.

15.81. a. $P_{PCl_5} = 0.024$ atm, $P_{PCl_3} = 1.036$ atm, $P_{Cl_2} = 0.536$ atm
   b. The concentration of $PCl_3$ decreases, and the concentration of $PCl_5$ increases.

15.83. $[H_2O] = [Cl_2O] = 3.76 \times 10^{-3}$ $M$; $[HOCl] = 1.13 \times 10^{-3}$ $M$

15.85. $6.9 \times 10^5$ to 1

15.87. $P_{CO} = 2.4$ atm, $P_{CO_2} = 3.8$ atm

15.89. a. $P_{NO} = 0.272$ atm; $P_{NO_2} = 7.98 \times 10^{-3}$ $M$
   b. $P_T = 0.416$ atm

15.91. $P_{O_2} = 0.17$ atm, $P_{N_2} = 0.75$ atm, $P_{NO} = 0.080$ atm

15.93. 5.75 $M$

15.95. $P_{CO} = P_{Cl_2} = 0.258$ atm, $P_{COCl_2} = 0.00680$ atm

15.97. $[CO] = [H_2O] = 0.031$ $M$, $[CO_2] = [H_2] = 0.069$ $M$

15.99. Yes. The reaction is endothermic:
   At 1500 K, $K_p = 5.5 \times 10^{-11}$
   At 2500 K, $K_p = 4.0 \times 10^{-3}$
   At 3000 K, $K_p = 0.40$
   This reaction does not favor products even at very high temperature, so this is not a viable source of CO and is not a remedy to decrease $CO_2$ as a contributor to global warming. Also, the process produces poisonous CO gas.

15.101. $9 \times 10^{-22}$ $M$

15.103. a. $P_{SO_2} = 9.2 \times 10^{-74}$ atm
   b. $2.63 \times 10^{-47}$ molecules

## CHAPTER 16

16.1. Red line

16.3. The blue titration curve represents the titration of a 1 $M$ solution of strong acid. The red titration curve represents the titration of a 1 $M$ solution of weak acid.

16.5. The indicator with a $pK_a$ of 9.0

16.7. The red titration curve represents the titration of $Na_2CO_3$; the blue titration curve represents the titration of $NaHCO_3$.

16.9. $NH_4Cl$ is dissolved in the yellow solution; $NaC_2H_3O_2$ is dissolved in the blue solution; and $NH_4C_2H_3O_2$ is dissolved in the light green solution.

16.11. HBr is the acid; $H_2O$ is the base.

16.13. $OH^-$ is the base; $H_2O$ is the acid.

16.15. a. $HNO_3$ is the acid; NaOH is the base.
   b. HCl is the acid; $CaCO_3$ is the base.
   c. HCN is the acid; $NH_3$ is the base.

16.17. $NO_2^-$; $OCl^-$; $H_2PO_4^-$; $NH_2^-$

16.19. 1.50 $M$ $H^+$ before; 1.32 $M$ $H^+$ after

16.21. 0.160 $M$ before addition of HCl, 0.093 $M$ after addition.

16.23. Dissolve 70.0 g of NaOH($s$) in water and dilute to a total volume of 2.50 L.

16.25. Because the pH function is a $-\log$ function, as $[H^+]$ increases, the value of $-\log[H^+]$ decreases.

16.27. When $[H^+]$ is greater than 1 $M$

16.29. $pK_w$ decreases as temperature increases

16.31. a. pH = 7.462; pOH = 6.538; basic
b. pH = 4.70; pOH = 9.30; acidic
c. pH = 7.15; pOH = 6.85; basic
d. pH = 10.932; pOH = 3.068; basic

16.33. a. $1.19 \times 10^{-11}$ M
b. $1.26 \times 10^{-10}$ M
c. $2.22 \times 10^{-12}$ M
d. $1.45 \times 10^{-10}$ M

16.35. a. pH = 0.810; pOH = 13.190
b. pH = 2.301; pOH = 11.699
c. pH = 1.903; pOH = 12.097
d. pH = 1.200; pOH = 12.800

16.37. 7.006

16.39. a. $HCl > HNO_2 > CH_3COOH > HClO$
b. $HClO < CH_3COOH < HNO_2 < HCl$

16.41. $NaNO_2$ is soluble in water, separating into $Na^+$ and $NO_2^-$ ions, each at 1.0 $M$ concentration, for a total ion concentration of 2.0 $M$. $HNO_2$, however, only weakly dissociates in water and so produces just slightly greater than 1.0 $M$ ions in solution. Therefore, $NaNO_2$, with more dissolved ions in solution, is a better conductor of electricity.

16.43. $K_a = \dfrac{[H^+][F^-]}{[HF]}$

16.45. a. Water
b. Water

16.47. $H_2O$ is the acid; $CH_3NH_2$ is the base.

16.49. $8.91 \times 10^{-4}$

16.51. 1.63%; $K_a = 6.74 \times 10^{-5}$

16.53. 2.49

16.55. 2.3 times

16.57. a. Aminoethanol is a weaker base than ethylamine.
b. 10.857
c. $1.00 \times 10^{-4}$

16.59. a. 10.635
b. 9.356

16.61. With each successive ionization, it becomes more difficult to remove $H^+$ because the species becomes more negatively charged.

16.63. 0.51

16.65. 2.80

16.67. 9.50

16.69. 10.27

16.71. Sulfur is more electronegative than selenium. The higher electronegativity on the sulfur atom stabilizes the anion $HSO_4^-$ more than the anion $HSeO_4^-$.

16.73. a. $H_2SO_3$
b. $H_2SeO_4$

16.75. Ethanolamine has a larger $pK_b$ (is a weaker base) than ethylamine. The oxygen atom in ethanolamine pulls electron density away from the nitrogen lone pair, rendering the nitrogen less basic than in ethylamine.

16.77. Increase

16.79. Ammonium nitrate

16.81. The citric acid in the lemon juice neutralizes the volatile trimethylamine to make a nonvolatile dissolved salt.

16.83. 3.32

16.85. 7.35

16.87. A solution of acetic acid and acetate ions can neutralize additions of acid or base. However, a solution of HCl and NaCl has no acid-neutralizing power because the $Cl^-$ ion is too weak a base.

16.89. Adding NaF will increase the buffer pH and the buffer capacity.

16.91. Yes

16.93. Iodoacetic acid/sodium iodoacetate

16.95. At 25°C, pH = 4.453; at 0°C, pH = 4.484

16.97. pH = 12.34; pOH = 1.65

16.99. Mix 15.8 g (0.0759 mol) of sodium iodoacetate and 18.6 g (0.100 mol) of iodoacetic acid, and add water to a volume of 100 mL.

16.101. The hydrofluoric acid/fluoride buffer will have the lowest pH.

16.103. Dilution will have no effect on buffer pH, but will decrease the buffer capacity.

16.105. 9.25

16.107. 3.75

16.109. pH = 3.50 before adding HCl; pH = 3.42 after adding HCl

16.111. The weak acid titration curve has an initial pH that is higher (less acidic) than that of an equimolar solution of a strong acid (lower pH, more acidic). The pH at the equivalence point in the titration of a strong acid is 7.00 whereas the pH at the equivalence point for a weak acid is basic.

16.113. No

16.115. After 10.0 mL of $OH^-$ has been added, pH = 4.754; after 20.0 mL of $OH^-$ has been added, pH = 8.750; after 30.0 mL of $OH^-$ has been added, pH = 12.356

16.117. 0.02559 $M$

16.119. 250 mL

16.121. 4.44

16.123.

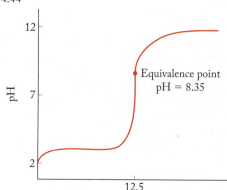

16.125.

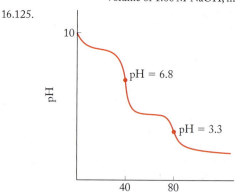

16.127. Molar solubility is the quantity (moles) of substance that dissolves in a liter of solution. The solubility product is the equilibrium constant for the dissolution of a substance.

16.129. $Sr^{2+}$

16.131. Endothermic

16.133. Acidic substances react with the $OH^-$ released on dissolution of hydroxyapatite. The equilibrium is shifted to the right, dissolving more hydroxyapatite.

16.135. $1.08 \times 10^{-10}$

16.137. $[Cu^+] = [Cl^-] = 1.01 \times 10^{-3}\ M$

16.139. $9.96 \times 10^{-6}\ g/mL$

16.141. 10.091

16.143. d

16.145. No

16.147. Yes

16.149. a. $SO_4^{2-}$
  b. $1.34 \times 10^{-4}\ M$

16.151. The hydrogen bonds between some of the water molecules must break and re-form around the species $CH_3NH_2$. Also, the amine hydrolyzes and forms $CH_3NH_3^+$ and $OH^-$; resulting in ion–dipole forces between these ions and the surrounding water molecules.

16.153. a, b.

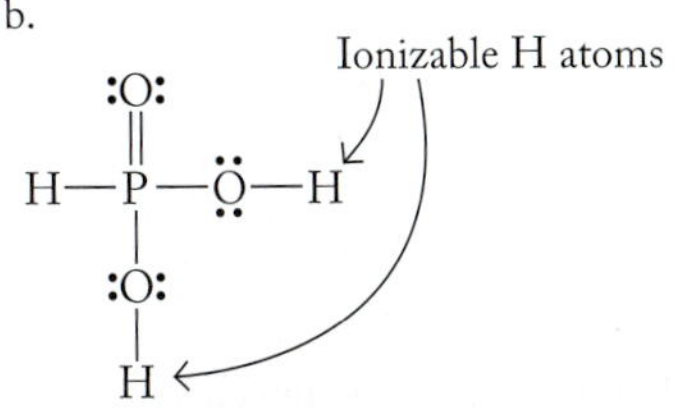

  c. Phosphoric acid has a similar structure with a single $P=O$ bond and ionizable H atoms also bonded to oxygen atoms.

16.155. Subsequent additions of $HCO_3^-$ react with water to form bicarbonate's conjugate acid ($H_2CO_3$) and its conjugate base ($CO_3^{2-}$) in the same proportions as the first addition, so pH does not change.

16.157. $1.4 \times 10^9\ L$

16.159. a. The acid salt form has an $H^+$ ion on the amine ($R_2NH$) group with $Cl^-$ as a counter ion. This structure is shown on the right of Figure P16.159.
  b. Acidic

16.161. a. Because HF is weak, the $[F^-]$ is low compared to that of water and so HF reacts with $H_2O$ to form $F^-$ as the major anionic species.
  b. $K_{overall} = 2.9 \times 10^{-4}$
  c. pH = 1.91; $[HF_2^-]_{eq} = 4.40 \times 10^{-4}\ M$

16.163. a.

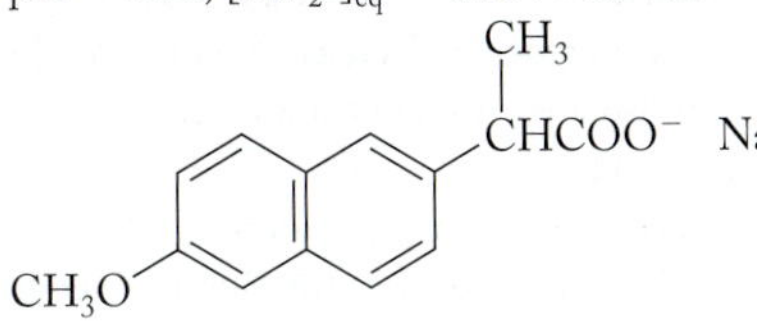

  b. Basic

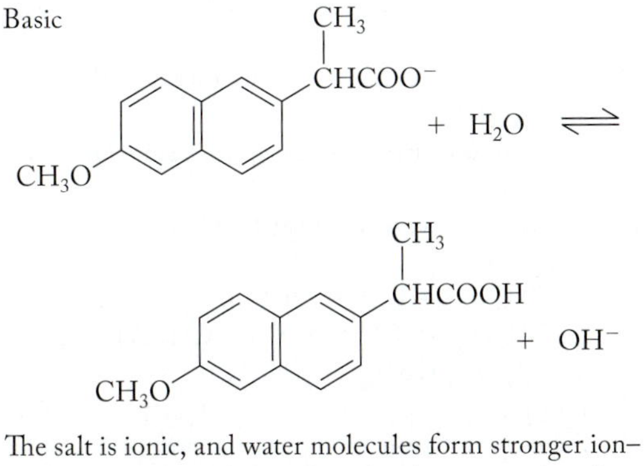

  c. The salt is ionic, and water molecules form stronger ion–dipole forces around the salt molecule compared to the dipole-induced dipole forces between the neutral molecule and water.

16.165. a. $SO_3(g) + H_2O(\ell) \rightarrow H_2SO_4(\ell)$
  b. $3\ NO_2(g) + H_2O(\ell) \rightarrow 2\ HNO_3(\ell) + NO(g)$
  c. $4\ NH_3(g) + 5\ O_2(g) \rightarrow 4\ NO(g) + 6\ H_2O(g)$

16.167. None

16.169. The balanced formation reactions for $SO_2$ and $SO_3$ are
$$\tfrac{1}{8} S_8(s) + O_2(g) \rightarrow SO_2(g) \qquad \Delta H^\circ_{f,SO_2}$$
$$\tfrac{1}{8} S_8(s) + \tfrac{3}{2} O_2(g) \rightarrow SO_3(g) \qquad \Delta H^\circ_{f,SO_3}$$
$$\Delta H^\circ_{rxn} = 2\Delta^\circ_{f,SO_3} - 2\Delta^\circ_{f,SO_2}$$

16.171. The corresponding constant for $NO_2$ should be greater than the Henry's law constant for $CO_2$ because $NO_2$ is polar, whereas $CO_2$ is nonpolar. When $SO_3$ gas comes in contact with water, it rapidly reacts with it instead of simply dissolving in it, as $SO_3$ and $H_2O$ combine to form sulfuric acid. $SO_3$ does not have a measurable Henry's law constant.

16.173. a.

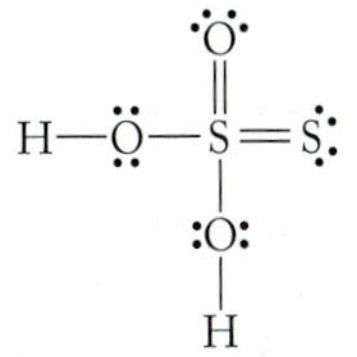

  b. When a less electronegative sulfur atom replaces an oxygen atom in the acid, the acidity decreases. Therefore, $H_2S_2O_3$ is a weaker acid than $H_2SO_4$.

# CHAPTER 17

17.1. Chromium (green) and cobalt (yellow)

17.3. Zinc (blue)

17.5. d

17.7. Red

17.9. Purple/blue

17.11. No. Some Lewis bases donate an electron pair, but do not accept a proton while doing so.

17.13. $BF_3$ can accept electron pairs but has no H atoms to donate to be a Brønsted–Lowry acid.

17.15. $B(CH_3)_3$ is the Lewis acid, $NH_3$ is the Lewis base.

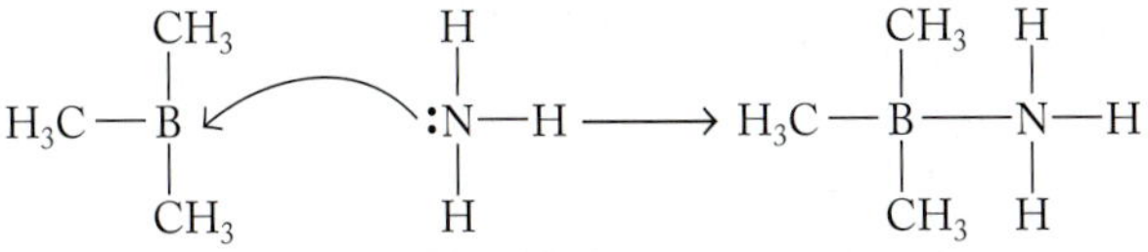

17.17. The sulfur atom in $SO_2$ and the hydrogen atom in $H_2O$ are acting as Lewis acids; the oxygen atoms in $SO_2$ and $H_2O$ are acting as Lewis bases.

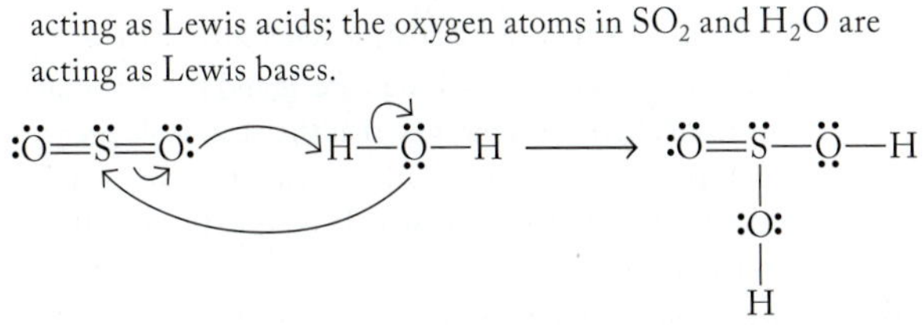

17.19.

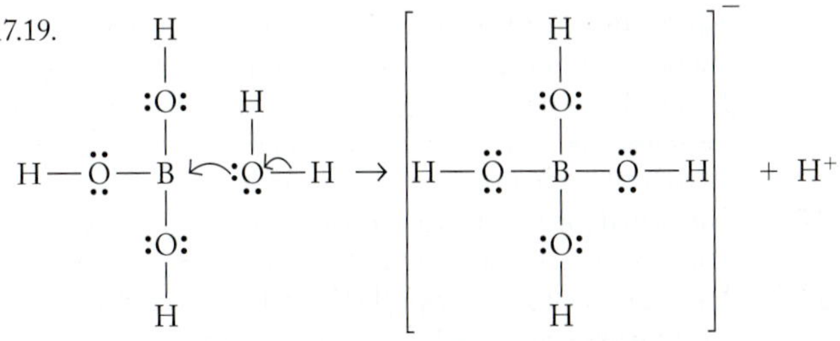

$B(OH)_3$ is the Lewis acid, and $H_2O$ is the Lewis base.

17.21. Water

17.23. Ammonia

17.25. $Na^+$

17.27. The $Pb^{2+}$ ion is complexed by the EDTA (a chelating agent), forming a compound that is less toxic than $Pb^{2+}$ on its own.

17.29. $Ag^+$ forms a soluble complex with $NH_3$, removing $Ag^+$ from solution and shifting the equilibrium for the dissolution of AgCl to the right.

17.31. (a) $4.0 \times 10^{-3}\ M$; (b) $6.3 \times 10^{-10}\ M$

17.33. $2.48 \times 10^{-13}\ M$

17.35. a. Diamminesilver(I)
  b. Hexaaquacobalt(III)
  c. Pentaamminebromoiron(III)

17.37. a. Tetrabromocolbaltate(II)
  b. Aquatrihydroxozincate(II)
  c. Pentacyanonickelate(II)

17.39. (i) Ethylenediaminezinc(II) sulfate; 2+ charge on ion; $Zn^{2+}$; CN = 4.
  (ii) Pentaammineaquanickel(II) chloride; 2+ charge on ion; $Ni^{2+}$; CN = 6.
  (iii) Potassium hexacyanoferrate(II); 4– charge on ion; $Fe^{2+}$; CN = 6.

17.41. b and d

17.43. The solution will become more acidic.

17.45. In basic solution: $Cr(OH)_3(s) + OH^-(aq) \rightleftharpoons Cr(OH)_4^-(aq)$
  In acidic solution: $Cr(OH)_3(s) + 3\,H^+(aq) \rightleftharpoons Cr^{3+}(aq) + 3\,H_2O(\ell)$

17.47. $Al(OH)_3$ reacts with $OH^-$ in solution to form soluble $Al(OH)_4^-$. The other ions do not form this type of soluble complex ion.

17.49. 2.80

17.51. 1.80

17.53.

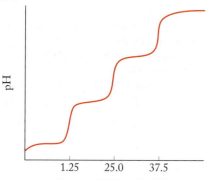

17.55. A chelating agent is a multidentate ligand that separates metal ions from other substances so that they can no longer react. Properties that make a chelating agent effective include strong bonds formed between the metal and the ligand and large formation constants.

17.57. As pH increases the chelating ability increases because $OH^-$ removes the H on the carboxylic acid groups, providing an additional site for binding to the metal cation.

17.59. The repulsions due to the ligands in a square-planar crystal field are highest for the $d_{xy}$ orbital, so it is higher in energy because this orbital lies in the plane of the ligands.

17.61. Compounds of $Ti^{4+}$ are colorless because $Ti^{4+}$ has no electrons in its $d$ orbitals, so no $d$–$d$ transitions can occur.

17.63. The yellow solution contains (b) $Cr(NH_3)_6^{3+}$. The violet solution contains (a) $Cr(H_2O)_6^{3+}$.

17.65. Colorless

17.67. $NiCl_4^{2-}$

17.69. a. $Co^{3+}$ has 7 $d$ electrons; $Cr^{2+}$ has 4 $d$ electrons; $Ni^{2+}$ has 8 $d$ electrons; and $Cu^{2+}$ has 9 $d$ electrons.
  b. Both $Co^{2+}$ and $Cr^{2+}$ may have high-spin and low-spin states.

17.71. $Fe^{2+}$ has 4 unpaired electrons. $Cu^{2+}$ has 1 unpaired electron. $Co^{2+}$ has 3 unpaired electrons. $Mn^{2+}$ has 5 unpaired electrons.

17.73. a. $Mn^{4+}$ in $MnO_2$; 2 $Mn^{3+}$ and 1 $Mn^{2+}$ in $Mn_3O_4$
  b. Both low-spin and high-spin configurations are possible in $Mn_3O_4$ ($d^4$ and $d^5$) but not in $MnO_2$ ($d^3$).

17.75. Paramagnetic

17.77. For an octahedral geometry, *cis–* means that two ligands are side by side and have a 90° bond angle between them. Ligands that are *trans–* to each other have a 180° bond angle between them.

17.79. At least two different ligands

17.81. Yes

17.83.
$$\left[\begin{array}{c} \text{Br} \quad \text{Cl} \\ \text{Cu} \\ \text{Br} \quad \text{Cl} \end{array}\right]^{2-} \qquad \left[\begin{array}{c} \text{Cl} \quad \text{Br} \\ \text{Cu} \\ \text{Br} \quad \text{Cl} \end{array}\right]^{2-}$$
  Cis          Trans

  cis– and trans– dibromodichlorocuprate(II). No, neither isomer is chiral.

17.85. $3.2 \times 10^{-13}$

17.87. The yellow complex containing $Co^{3+}$ in aqueous ammonia has the larger $\Delta_o$.

17.89. $Ag^{2+}$ has 9 $d$ electrons, leaving an unpaired electron in the $d_{x^2-y^2}$ orbital to make it paramagnetic. $Ag^{3+}$ has 8 $d$ electrons and $Ag^+$ has 10 $d$ electrons. Both have all electrons paired, so those silver ions are diamagnetic.

17.91. Toward longer wavelengths

# CHAPTER 18

18.1. The balloon on the right has a greater pressure and entropy.

18.3. No. This is a low probability because each gas would then be confined to a smaller volume and would have more order, so this change would involve a decrease in entropy. The bulbs are one system, not isolated from one another.

18.5. At low temperature

18.7. Condensation, freezing, and deposition

18.9. They are equal in magnitude, but the sign is reversed.

18.11. $\Delta S_{sys}$ is positive; $\Delta S_{surr}$ is negative.

18.13. a and b

18.15. $\Delta S_{sys}$ is negative, $\Delta S_{universe}$ is negative, and the reaction is non-spontaneous.

18.17. $\Delta S_{surr}$ must be less (more negative) than $+66.0$ J/K.

18.19. (d) $Cr(NO_3)_3$

18.21. (a) $S_8(g)$; (b) $S_2(g)$; (c) $O_3(g)$; (d) $O_2(g)$

18.23. Fullerenes

18.25. a. $CH_4(g) < CF_4(g) < CCl_4(g)$
  b. $CH_2O(g) < CH_3CHO(g) < CH_3CH_2CHO(g)$
  c. $HF(g) < H_2O(g) < NH_3(g)$

18.27. Positive

18.29. Solids have a lower entropy (are a more ordered phase) than particles dissolved in aqueous solution. As a result of the precipitation reaction, fewer particles are present in any phase of the reaction. Both of these lead to $\Delta S_{rxn}$ being less than zero.

18.31. (a) 24.9 J/K; (b) −146.4 J/K; (c) −73.2 J/K; (d) −175.8 J/K

18.33. −534 J/(mol · K)

18.35. Spontaneous processes have negative $\Delta G$ values.

18.37. In general, exothermic reactions have a corresponding increase in entropy. Both of these lead to a spontaneous reaction, though now we realize that it is possible for an exothermic reaction to be spontaneous at low temperatures if disfavored by entropy.

18.39. $\Delta S$ is positive; $\Delta H$ is positive; $\Delta G$ is negative.

18.41. d

18.43. For NaBr, −18 kJ/mol; for NaI, −29 kJ/mol

18.45. +91.4 kJ

18.47. +56.5 kJ

18.49. −81.7 kJ

18.51. No, if $\Delta S_{rxn}$ were positive, the exothermic reaction would be spontaneous at all temperatures.

18.53. 981.3 K

18.55. $\Delta H° = 44.0$ kJ; $\Delta S° = 118.9$ J/K or 0.1189 kJ/K; 370.1 K or 96.9°C

18.57. (a) low temperatures; (b) low temperatures; (c) all temperatures

18.59. Positive

18.61. Yes, the reaction will proceed in the forward direction. When $\Delta G° < 0$ then ln $K$ of the reaction is positive, giving $K > 1$. The reaction favors the formation of products, so the reaction will shift to the right.

18.63. c

18.65. +27.124 kJ/mol

18.67. Exothermic

18.69. Exothermic

18.71. $1.3 \times 10^{-31}$

18.73. −115 kJ/mol

18.75. Some of the products of one reaction must be reactants of the second reaction, and the overall value of $\Delta G$ when the reactions are summed must be negative.

18.77. The bond arrangements are only slightly different between the two structures, so the enthalpies and entropies of the product and reactant are very close in value.

18.79. a. +50.8 kJ and −394.4 kJ

b. $CH_4(g) + O_2(g) \rightarrow 2 H_2(g) + CO_2(g)$

$\Delta G_{rxn} = -343.6$ kJ; the reaction is spontaneous as written.

18.81. $3.81 \times 10^{-3}$

18.83. −73 kJ/mol

18.85. 24

18.87. $4.47 \times 10^3$

18.89. 83.5 J/(mol · K)

18.91. $T = 618$ K

18.93. −630.4 kJ; spontaneous

18.95. $T_b \approx 294$ K. Since the boiling point of HCN is below room temperature (298 K), it will be a gas under standard conditions.

18.97. 9.58 J/(mol · K)

18.99. 0.805 J/(mol · K)

18.101. There are more atoms in $CaCO_3$ than in CaO, so the $S°$ is more positive; 1099 K

18.103. a. Positive

b. Negative

c. $T = \Delta H/\Delta S$

## CHAPTER 19

19.1. Because of the careful layering, each half-cell has its metal in contact with its cation solution. The solutions are not mixing, but nevertheless the layers allow the ions, needed to balance the charge in each half-cell, to pass through.

19.3. Ag is the cathode; Pt in the SHE is the anode; electrons flow from the SHE to Ag.

19.5. Blue line

19.7. a. $2 H_2O(\ell) + 2 e^- \rightarrow H_2(g) + 2 OH^-(aq)$

$E°_{cathode} = -0.8277$ V

$2 H_2O(\ell) \rightarrow O_2(g) + 4 H^+(aq) + 4 e^-$

$E°_{anode} = 1.229$ V

b. Because it increases the conductivity of the solution.

19.9. Water at the post provides a medium for the redox reaction to occur. At the air/water interface, both $O_2$ (the oxidizing agent) and water (the solvent) are present. Above the waterline, $O_2$ is present, but not the solvent. Underwater, the solvent is present, but the concentration of dissolved $O_2$ in water is low. When both species are present, the metal is oxidized to yield $M^{n+}$ ions and $n$ electrons.

19.11. a. The separator allows nonreactive ions to pass through, maintaining electrical neutrality and completing the electric circuit.

b. A wire would allow electrons to pass, but not other ions. Since electrons cannot travel through the solution, a wire would not complete the circuit.

19.13. a. $Fe^{3+}(aq) + 1 e^- \rightarrow Fe^{2+}(aq)$

$Al^{3+}(aq) + 3 e^- \rightarrow Al(s)$

$Al(s) + 3 Fe^{2+}(aq) \rightarrow Al^{3+}(aq) + 3 Fe^{3+}(aq)$

b. $I_2(aq) + 2 e^- \rightarrow 2 I^-(aq)$

$NO_3^-(aq) + H_2O(\ell) + 2 e^- \rightarrow NO_2^-(aq) + 2 OH^-(aq)$

$NO_2^-(aq) + 2 OH^-(aq) + I_2(s) \rightarrow NO_3^-(aq) + H_2O(\ell) + 2 I^-(aq)$

c. $MnO_4^-(aq) + 8 H^+(aq) + 5 e^- \rightarrow Mn^{2+}(aq) + 4 H_2O(\ell)$

$Cr_2O_7^{2-}(aq) + 14 H^+(aq) + 6 e^- \rightarrow 2 Cr^{3+}(aq) + 7 H_2O(\ell)$

$6 MnO_4^-(aq) + 10 Cr^{3+}(aq) + 11 H_2O(\ell) \rightarrow$
$6 Mn^{2+}(aq) + 5 Cr_2O_7^{2-}(aq) + 22 H^+(aq)$

19.15. a. $Ni^{2+}(aq) + 2 e^- \rightarrow Ni(s)$  cathode

$Cd(s) \rightarrow Cd^{2+}(aq) + 2 e^-$  anode

b. $Ni^{2+}(aq) + Cd(s) \rightarrow Cd^{2+}(aq) + Ni(s)$

c. $Cd(s) | Cd^{2+}(aq) || Ni^{2+}(aq) | Ni(s)$

19.17. a. $MnO_4^-(aq) + 2 H_2O(\ell) + 3 e^- \rightarrow$
$MnO_2(s) + 4 OH^-(aq)$  cathode

$Cd(s) + 2 OH^-(aq) \rightarrow Cd(OH)_2(s) + 2 e^-$  anode

b. $2 MnO_4^-(aq) + 4 H_2O(\ell) + 3 Cd(s) \rightarrow$
$2 MnO_2(s) + 3 Cd(OH)_2(s) + 2 OH^-(aq)$

c. $Cd(s) | Cd(OH)_2(s) || MnO_4^-(aq) | MnO_2(s) | Pt(s)$

19.19. a. Six

b. $FeO_4^{2-}$ has $Fe^{6+}$; $Fe_2O_3$ has $Fe^{3+}$; Zn has $Zn^0$; ZnO and $ZnO_2^{2-}$ have $Zn^{2+}$

c. $Zn(s) | ZnO(s) | ZnO_2^{2-}(aq) || FeO_4^{2-}(aq) | Fe_2O_3(s) | Pt(s)$

19.21. The platinum electrode transfers electrons to the half-cell; it is inert and not involved in the reaction.

19.23. Because $E°_{ox}$ (anode) $= -E°_{red}$, we can substitute $-E°_{red}$ (anode) for $E°_{ox}$ (anode) in the expression

$$E°_{cell} = E°_{red} \text{(cathode)} + E°_{ox} \text{(anode)}$$

to obtain

$$E°_{cell} = E°_{red} \text{(cathode)} + [E°_{red} \text{(anode)}]$$
$$= E°_{red} \text{(cathode)} - E°_{red} \text{(anode)}$$

This is equal to the expression for $E°_{cell}$ in Equation 19.2.

19.25. $Cl_2$ ($E°_{red}$ = 1.3583 V) is a better oxidizing agent than $O_2$ ($E°_{red}$ = 1.229 V). The higher the reduction potential, the more powerful the oxidizing agent, and the more readily it is itself reduced.

19.27. a. $\Delta G°$ = −34.5 kJ; $\Delta E°_{cell}$ = 0.358 V
b. $\Delta G°$ = 82.9 kJ; $\Delta E°_{cell}$ = 0.430 V

19.29. No

19.31. $O_2(g) + 2\,H_2O(\ell) + 2\,Zn(s) + 4\,OH^-(aq) \rightarrow 2\,Zn(OH)_4{}^{2-}(aq)$

19.33. Less than 1.10 V

19.35. a. $\Delta E°_{cell}$ = 0.292 V; $\Delta G°$ = −56.4 kJ
b. $\Delta E°_{cell}$ = 0.548 V; $\Delta G°$ = −97.4 kJ

19.37. a and c

19.39. a. $NiO(OH)(s) + TiZr_2H(s) \rightarrow TiZr_2(s) + Ni(OH)_2(s)$
b. 1.32 V

19.41. −290 kJ

19.43. −116 kJ

19.45. Voltage of a battery (a voltaic cell) is governed by the Nernst equation:

$$E_{cell} = E°_{cell} - \frac{RT}{nF}\ln Q$$

As a battery discharges, the value of $Q$, the reaction quotient, changes:

$$Q = \frac{[\text{products}]^x}{[\text{reactants}]^y}$$

At the start of the reaction, $Q$ is very small because [reactants] >> [products]. As the reaction proceeds, [products] grows, and $Q$ increases but does not increase significantly until significant amounts of products form, that is, when the battery is nearly discharged.

19.47. 0.617 V

19.49. $8.56 \times 10^{19}$

19.51. −0.414 V

19.53. $E_{rxn}$ = 1.55 V; $E_{rxn}$ will decrease

19.55. a. 0.62 V; b. 0.66 V

19.57. a. 0.349 V; b. $6.41 \times 10^{59}$:1

19.59. c and f

19.61. Al–$O_2$

19.63. Li–$MnO_2$

19.65. Teflon is a polymeric material composed of poly(tetrafluoroethylene). Since no metals or ionizable ions are present in teflon (the C–F bond is extremely stable), no redox activity is expected. Teflon is also water repellent, which keeps the metal structure drier that it would be with asbestos pads.

19.67. The material used must be more readily oxidized than aluminum. Magnesium is a stable, nonreactive metal that would serve as a good sacrificial anode for aluminum.

19.69. In a voltaic cell, the electrons are produced at the anode so a negative (−) charge builds up there; in an electrolytic cell, electrons are being forced onto the cathode so the cathode builds up a negative (−) charge. The flow of electrons in the outside circuit is reversed in an electrolytic cell compared to the flow in a voltaic cell.

19.71. $Br_2$

19.73. More negative

19.75. 5.2 g

19.77. 27.0 minutes

19.79. (a) $5.78 \times 10^{-3}$ L; (b) no, some toxic $Cl_2$ and $Br_2$ would also be produced, rendering the oxygen unsafe to breathe.

19.81. −0.270 V

19.83. A hybrid vehicle uses a relatively inexpensive fuel (gasoline) in the internal combustion engine and has good fuel economy, but still gives off emissions. A fuel-cell vehicle does not give off emissions (the reaction produces $H_2O$) but requires a more expensive and explosive fuel (hydrogen); moreover, current battery technologies incorporate materials that are still very expensive and bulky.

19.85. Electric engines are more efficient by converting more of the energy into motion instead of losing it as heat. The fuel cell will thus require less fuel than a combustion engine, so less $CO_2$ is produced by driving the same distance.

19.87. a. $\overset{-4}{C}\overset{+1}{H_4}(g) + \overset{+1}{H_2}O(g) \rightarrow \overset{+2}{C}O(g) + 3\,\overset{0}{H_2}(g)$
$\overset{+2}{C}O(g) + \overset{+1}{H_2}O(g) \rightarrow \overset{+1}{H_2}(g) + \overset{+4}{C}O_2(g)$
b. For the reaction of $CH_4$ with $H_2O$, $\Delta G°_{rxn}$ = 142.2 kJ. For the reaction of CO with $H_2O$, $\Delta G°_{rxn}$ = −28.6 kJ. For the overall reaction, $\Delta G°_{overall} = \Delta G°_{rxn_1} + \Delta G°_{rxn_2}$ = 113.6 kJ.

19.89. a. Cathode
b. No. $Mg^{2+}$, with a higher positive charge, has a lower (less negative) reduction potential than $Na^+$.
c. No
d. $H_2$ and $O_2$

19.91. a. −0.87 V
b. $Mo_3S_4$: Mo = +2.67; $MgMo^3S_4$: Mo = +2
c. $Mg^{2+}$ is added to the electrolyte to better carry the charge in the cell. This cation is produced at the anode and consumed at the cathode.

## CHAPTER 20

20.1.

20.3. (a) Palmitic acid; (b) stearic acid

20.5. Tyrosine, glycine, glycine, phenylalanine, and methionine

20.7. Trans fats exhibit geometric isomerism around the C=C bond where similar groups on the two carbon atoms are situated on opposite sides of the double bond. Structures a and c contain trans fats.

20.9. Sucrose. The difference in the structures is that in sucralose, three –OH groups on sucrose have been replaced by Cl atoms. Being derived from sucrose implies that the sugar is natural, but the presence of Cl atoms on sugars is not natural.

20.11. Decreases

20.13. The "α" refers to the single carbon atom in amino acids to which both −NH₂ and −COOH groups are bonded.

20.15. D- and L- refer to how the four groups on a chiral carbon are oriented.

20.17. a and c

20.19. Most amino acids are zwitterions at pH ≈ 7.4 because the amino group will be protonated and the carboxylic acid group will be deprotonated, giving $RC\overset{+}{N}H_3COO^-$.

20.21. Lysine contains two amino groups, one of which is on a long carbon tail. This can react with the carboxylic acid on the carbon tail of glutamic acid to form a salt bridge.

20.23.  a.  AlaSer (AS)

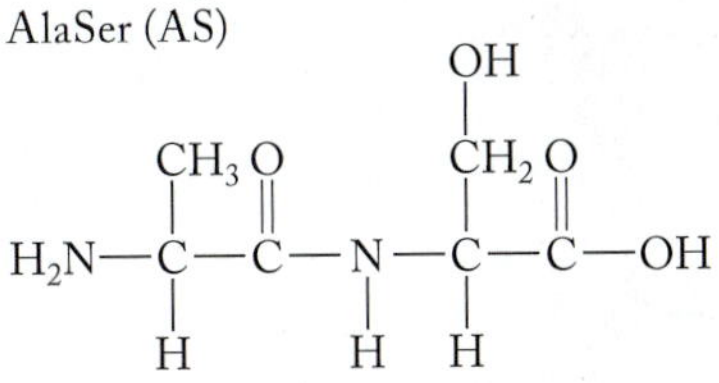

  b.  AlaPhe (AF)

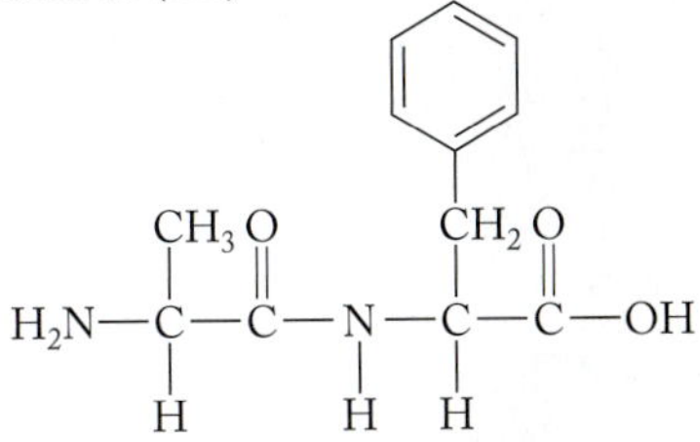

  c.  AlaVal (AV)

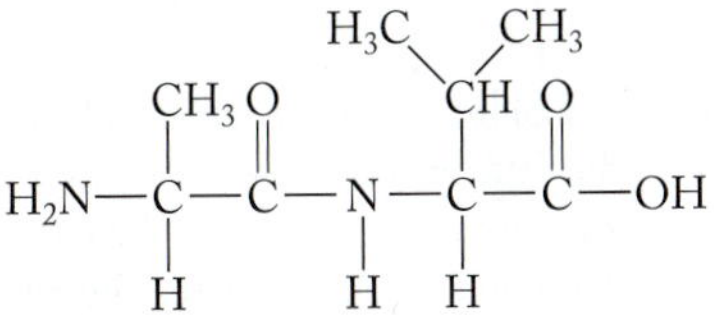

20.25.  a.  Alanine + glycine
  b.  Leucine + leucine
  c.  Tyrosine + phenylalanine

20.27. NH₃

20.29. Starch has α-glycosidic bonds, and cellulose has β-glycosidic bonds. Starch coils into granules, and cellulose forms linear molecules.

20.31. No

20.33. The bonding in fructose and glucose is nearly the same.

20.35. To calculate the free-energy change for a two-step process we need only to sum the individual ΔG values for each reaction.

20.37. The position of the hydroxy group on carbon 1 differs between the α and β forms of galactose. The relative positions of the hydroxy groups on carbons 2, 3, and 4 are the same on both isomers.

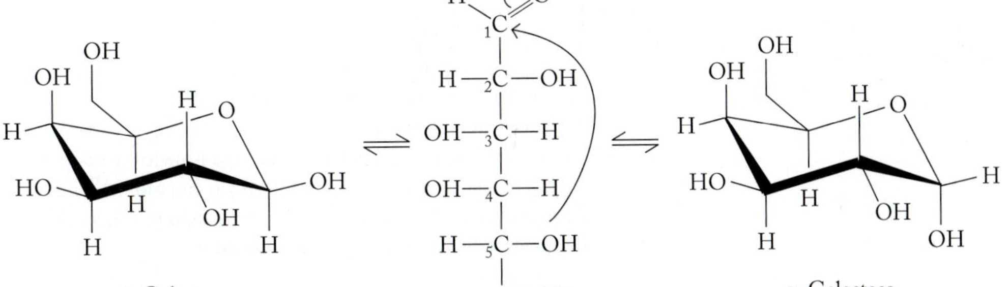

β-Galactose                    α-Galactose

20.39. c

20.41. a

20.43. b

20.45. −16.4 kJ

20.47. Saturated fatty acids have all C—C single bonds in their structure; unsaturated fatty acids have C=C double bonds.

20.49. Fatty acids have a high fuel value (see Problem 20.48) and eating sticks of butter affords Arctic explorers with more energy per gram of food compared to carbohydrates or proteins.

20.51. If the two fatty acids linked to the glycerol at C-1 and C-3 are different, the triglyceride has a chiral center.

20.53.  a.  0.50 kg of α-linolenic acid would consume more hydrogen.
  b.  No. The hydrogenation product of both species is stearic acid.

20.55.

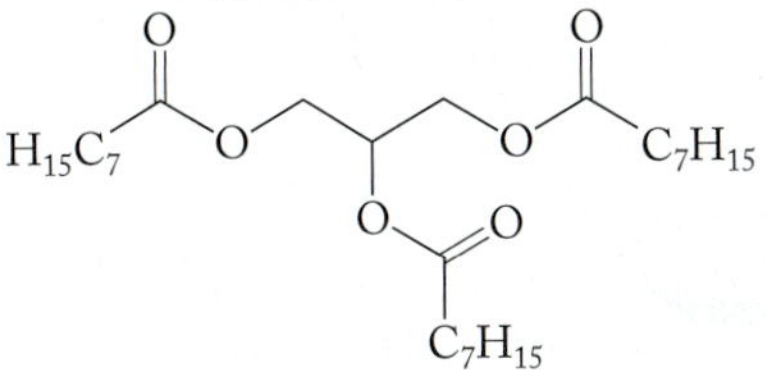

(a) Glycerol with octanoic acid

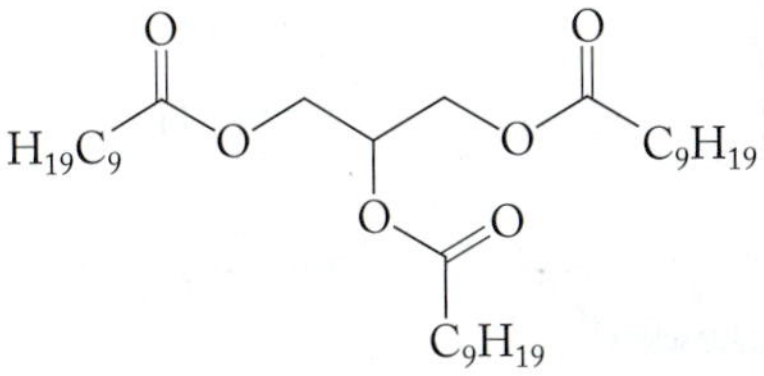

(b) Glycerol with decanoic acid

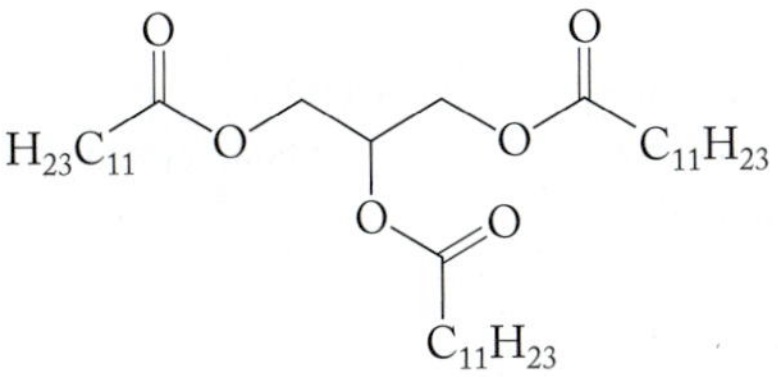

(c) Glycerol with dodecanoic acid

20.57. Phosphate group, a five-carbon sugar, and a nitrogen base; the "backbone" of DNA is composed of alternating sugar residues and phosphate groups.

20.59. Hydrogen bonds

20.61.

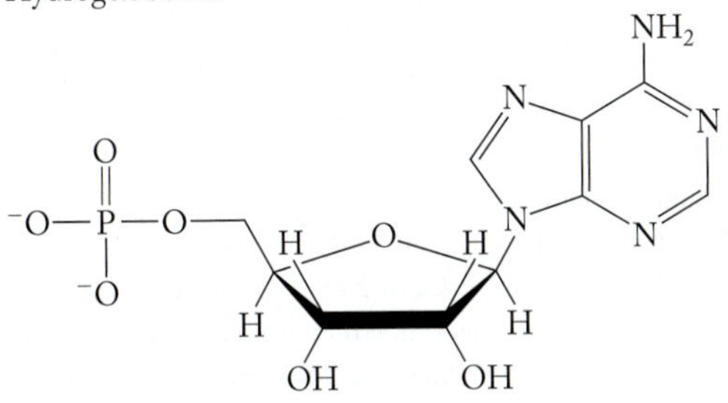

20.63. a. A-G-C

b.

20.65. (a) Sucrose; (b) esters; (c) $C_{15}H_{31}COOH$

20.67. a. There is an extra $-CH_2-$ group in homocysteine's sulfur-containing side chain.

b. Yes

20.69. Yes

20.71.

Valine          Leucine          Isoleucine

20.73. Glutamic acid, cysteine, and glycine

20.75. Yes. Because there is no difference in the number of C—C, C—H, C=O, C—O, or N—H bonds between the two compounds, we expect, based on average bond energies, that the fuel values of leucine and isoleucine should be identical. Isoleucine might have a lower fuel value because the $CH_3$ group is closer to the COOH and $NH_2$ groups, and this difference in shape must contribute to the slightly different fuel values.

# CHAPTER 21

21.1. Purple (lithium)

21.3. Red (radium)

21.5. a

21.7. (b) Blue line

21.9. Model 1 represents fission; model 2 represents fusion.

21.11. The *mass defect* is the difference between the mass of the nucleus of an isotope and the sum of the masses of the individual nuclear particles that make up that isotope. The *binding energy* is the energy released when individual nucleons combine to form the nucleus of an isotope.

21.13. $3.46 \times 10^{-13}$ J

21.15. $1.09 \times 10^{-12}$ J/nucleon

21.17. If the nuclide lies in the belt of stability (green dots on the plot in Figure 21.3), it is not radioactive and is stable. If it lies above the belt of stability, then it is neutron-rich and tends to undergo β decay to increase the number of protons and reduce the number of neutrons in its nucleus. If it lies below the belt of stability, it is neutron-poor and tends to undergo positron emission or electron capture to increase the number of neutrons and reduce the number of protons in its nucleus.

21.19. Alpha decay increases the neutron-to-proton ratio to produce less stable isotopes that can then be made more stable through β emission, decreasing the neutron-to-proton ratio.

21.21. Both of these processes are β decays.

21.23. Greater than 1

21.25. $^{26}_{13}Al \rightarrow ^{0}_{1}\beta + ^{26}_{12}Mg$

21.27. a. Electron capture or positron emission

b. Electron capture or positron emission

c. This isotope is stable.

21.29. $^{56}$Co has 27 protons and 29 neutrons and is neutron-poor; it may undergo electron capture or positron emission. $^{44}$Ti has 22 protons and 22 neutrons and is neutron-poor; it may undergo electron capture or positron emission.

21.31. (a) $^{32}$Cl, $^{33}$Cl, and $^{34}$Cl; (b) $^{38}$Cl and $^{39}$Cl; (c) $^{36}$Cl

21.33. 25%

21.35. 2.2 days

21.37. 131 years

21.39. After 8.726 half-lives, the ratio of $^{14}$C present to that originally in an artifact is $N_t/N_0 = 0.50^{8.726} = 0.00236$ or 0.236%. This is too little to detect.

21.41. After 0.00023 half-lives the ratio of $^{40}$K present to that originally in a sample is $N_t/N_0 = 0.50^{0.00023} = 0.9998$ or 99.98%. This level is just when we can detect the difference in amounts of $^{40}$K.

21.43. 35%

21.45. The oldest ring would have a 15% lower $^{14}$C/$^{12}$C ratio.

21.47. 36,640 y

21.49. The level of radioactivity is the amount of radioactive particles present in a given instant of time, which is equal to the rate constant times the number of atoms. The dose is the accumulation of exposure over a length of time.

21.51. When radon-222 decays to polonium-218 while in the lungs, the $^{218}$Po, a reactive solid that is chemically similar to oxygen, lodges in the lung tissue where it continues to emit α radiation. Alpha radiation is one of the most damaging kinds of radiation when in contact with biological tissues. The result of exposure to high levels of radon is an increased risk for lung cancer.

21.53. 5 µSv = 5 µGy; 250 µJ

21.55. a. $^{90}_{38}$Sr → $^{0}_{-1}$β + $^{90}_{39}$Y

b. $3.28 \times 10^8$

c. Strontium-90 is found in milk and not other foods because it is chemically similar to calcium, and milk is rich in calcium.

21.57. (a) 0.15 decays/s; (b) $7.0 \times 10^4$

21.59. a. The half-life should be long enough to effect treatment of the cancerous cells but not so long as to cause damage to healthy tissues.

b. Because α radiation does not penetrate far beyond a tumor, the α decay mode is best.

c. Products should be nonradioactive, if possible, or have short half-lives and be able to be flushed from the body by normal cellular and biological processes.

21.61. a. Positron emission or electron capture

b. Positron emission or electron capture

c. Positron emission or electron capture

21.63. 74.3 days

21.65. Yes

21.67. 136 min

21.69. a. $^{10}_{5}$B + $^{1}_{0}$n → $^{7}_{3}$Li + $^{4}_{2}$α

b. $4.43 \times 10^{-13}$ J

c. Alpha particles have a high RBE, and they do not penetrate very far into healthy tissue if the radionuclide is placed inside a tumor.

21.71. Control rods made of boron or cadmium are used to absorb the excess neutrons to control the rate of energy release.

21.73. The neutron-to-proton ratio for heavy nuclei is high and when the nuclide undergoes fission to form smaller nuclides, it must emit neutrons because the fission products require a lower neutron-to-proton ratio for stability.

21.75. (a) $^{138}_{52}$Te; (b) $^{133}_{51}$Sb; (c) $^{143}_{55}$Cs

21.77. (a) $^{103}_{39}$Y; (b) $^{130}_{48}$Cd; (c) $^{138}_{52}$Te

21.79. Today's Sun has fewer neutrons than the primordial Sun. The primordial Sun formed α particles via the fusion of a proton and a neutron:

$^{1}_{1}$p + $^{1}_{0}$n → $^{2}_{1}$H

$^{2}_{1}$H + $^{2}_{1}$H → $^{4}_{2}$He

Today's Sun forms α particles via the fusion of two protons:

$^{1}_{1}$p + $^{1}_{1}$p → $^{2}_{1}$H + $^{0}_{1}$β

$^{2}_{1}$H + $^{1}_{1}$p → $^{3}_{2}$He

$^{3}_{2}$He + $^{3}_{2}$He → $^{4}_{2}$He + 2 $^{1}_{1}$p

21.81. a. $4.37 \times 10^{-12}$ J

b. $6.80 \times 10^{-12}$ J

c. $2.69 \times 10^{-12}$ J

d. $1.60 \times 10^{-12}$ J

21.83. (a) $7.67 \times 10^{-13}$ J; (b) $-3.96 \times 10^{-13}$ J

21.85. a. Besides releasing a large amount of energy to power the starship *Enterprise*, hydrogen is an abundant fuel in the universe and therefore could easily react with any antihydrogen produced.

b. If any of the antimatter fuel came into contact with conventional matter (such as the ship, the crew, the warp engine), the two would annihilate one another, releasing a large quantity of energy in the form of an explosion.

21.87. The energy released in the fusion reaction is $\Delta E = 9.91 \times 10^{-13}$ J/nucleon. The energy released in the fission reaction is $\Delta E = 1.4 \times 10^{-13}$ J/nucleon. On a per nucleon basis, the fusion reaction generates more energy.

21.89. a. Geiger counter

b. 2877 years

c. $^{241}$Am is an α emitter and α particles do not travel more than a few inches in air and cannot penetrate the first layer of skin.

21.91. a. $^{249}_{98}$Cf + $^{48}_{20}$Ca → $^{294}_{118}$Uuo + 3 $^{1}_{0}$n

b. $^{290}_{116}$Uuh

c. $^{286}_{114}$Uuq

d. $^{282}_{112}$Cn

e. Because $^{294}$Uuo is a member of the noble gas family, it has chemical and physical properties similar to naturally occurring radon.

21.93. $^{210}$Pb

21.95. (a) $^{64}_{28}$Ni + $^{124}_{50}$Sn → $^{188}_{78}$Pt; (b) $^{196}_{78}$Pt

21.97. 3.35

21.99. a. $^{40}_{19}$K → $^{40}_{18}$Ar + $^{0}_{1}$β

b. Because the half-life of $^{40}$K is so much longer than that of $^{14}$C

# CHAPTER 22

22.1. d

22.3. Group 16 (pink)

22.5. −6.72 kJ

22.7. Both have trigonal planar geometry

22.9. c

22.11. Without an essential element, biological processes that rely on that element would shut down or deteriorate. If a nonessential element is missing, there would not be severe deleterious effects.

22.13. The main criterion that distinguishes major, trace, and ultra-trace essential elements from one another is their concentration in the body. Ultratrace elements are present in less than micro-gram quantities, trace elements are present in microgram to milligram quantities, and major elements are present in greater than milligram quantities.

22.15. The methyl group on $CH_3Hg^+$ is relatively nonpolar and this helps the cation be soluble in nonpolar environments. Its charge, on the other hand allows it to be soluble in polar environments. Mercury metal, however, is neutral and has no nonpolar substituents, and so it is less soluble in biological systems.

22.17. The difference in behavior must be due to size. $Ca^{2+}$ is too large to fit into the biomolecules where $Mg^{2+}$ is important, whereas $Be^{2+}$ is closer in size to $Mg^{2+}$.

22.19. $K^+$

22.21. (a) 1.6 ppm; (b) 33 ppm; (c) 0.43 ppm

22.23. (a) Oxygen; (b) oxygen; (c) carbon

22.25. Osmosis, ion channels, and ion pumps

22.27. The nonpolar, hydrophobic interior of the cell membrane makes it difficult to transport polar, hydrophilic ions through the cell membrane.

22.29. Potassium

22.31. The greater insolubility of $CaCO_3$ compared to $CaSO_4$ makes calcium carbonate a better structural material. Also, the partial pressure of $CO_2$ in the atmosphere is higher than $SO_3$, so the carbonate solubility equilibrium is shifted more to the left by Le Châtelier's principle than is the sulfate equilibrium.

22.33. 0.56 atm

22.35. 0.0688 V

22.37. 0.15 mol

22.39. $5.8 \times 10^{-4} M$

22.41. $^{137}Cs^+$ may substitute for $K^+$ in cells; as a β emitter with a relatively long half-life, it may cause cancer.

22.43. $\Delta S$ is likely positive, and $\Delta G$ is likely negative.

22.45. (a) To catalyze (lower the $E_a$ of) biological processes; (b) no

22.47. Lowers the activation energy

22.49. Much greater than 1

22.51. Positive

22.53. $^{137}_{55}Cs \rightarrow {}^{0}_{-1}\beta + {}^{137}_{56}Ba$

22.55. 3.06

22.57. $K = [OH^-]/[F^-] = 8.48$; the equilibrium lies to the right.

22.59. a. $Ca_8(HPO_4)_2(PO_4)_4 \cdot 6\,H_2O$ is more soluble than hydroxyapatite.
b. $8.5 \times 10^{-8} M$
c. $1.5 \times 10^{-7} M$

22.61. $5.3 \times 10^{-118}$

22.63. $1.59 \times 10^8$ times faster.

22.65. Zero order.

22.67. We must consider the decay mode to ensure that α emitters are not used inside the body. γ emitters are preferred because γ radiation has a low RBE and can easily escape the body, minimizing damage to tissue and organs.

22.69. Low tissue penetration and high RBE

22.71. γ radiation

22.73. By binding to the nitrogen atoms in DNA to stop the division of cells

22.75. Paramagnetic

22.77. The amalgam renders the mercury insoluble and chemically unreactive.

22.79. $Gd^{3+}$ has seven unpaired electrons, and the configuration [Xe] $6s^0\,4f^7$.

22.81. [Ar] $3d^6\,3d^6$; low spin.

22.83. 41 hr

22.85. $[:Bi\equiv O:]^+$

22.87. 127

22.89. $[\text{penicillamine}]_{eq} = 4.43 \times 10^{-4} M$;
$[\text{cysteine}]_{eq} = 5.6 \times 10^{-4} M$

22.91.

22.93. $Al(OH)_3(s) + 3\,H^+(aq) \rightarrow 3\,H_2O(\ell) + Al^{3+}(aq)$

22.95. $Zn(s) + Ag_2O(s) \rightarrow ZnO(s) + 2\,Ag(s)$; $E° = 1.600$ V

22.97. 136.4 pm

22.99. (a) The ratio of Al:V:Ti is approximately 3:1:24; (b) no

## Chapter 1

Pages 2–3: NASA, ESA, J. Hester, A. Loll (ASU); p. 4: (all): Courtesy NASA/JPL; p. 5: (a): Photographer's Choice/Punchstock; (b): Daniel Smith/Corbis; (c): NRH Photography; (d): NRH Photography; p. 6: Lester V. Bergman/Corbis; p. 7 (left): (a): IBM/Science Source/Photo Researchers; (b): Phototake Inc/Alamy; p. 9 (top left): (a): NASA's Goddard Space Flight Center/USGS; (b): Courtesy Markus Geisen/NHMPL; p. 9 (center right): © 2009 Richard Megna/Fundamental Photographs; p. 9 (bottom left): © 2009 Richard Megna/Fundamental Photographs; p. 10 (top right): Aquacone courtesy Solar Solutions, Inc.; p. 14: (a): (solid) Owen Franken/Corbis; (b): (liquid) David Wrobel/Visuals Unlimited; (c): (gas) Larry Stepanowicz/Visuals Unlimited; p. 16: © Todd Jason Baker; p. 20: Courtesy of A&D Weighing, San Jose, CA. www.andweighing.com; p. 22: ©2009 Richard Megna/Fundamental Photographs; p. 23: aeropix/Alamy; p. 27 (top): Crown ©/The Royal Collection © 2007, Her Majesty Queen Elizabeth II; p. 27 (bottom): Courtesy Chris Joosen/White Mountain National Forest; p. 31 (top): Courtesy Special Collections, Princeton University Library; p. 31 (bottom): (a): Jim Pickerell/Stock Collection Blue/Alamy; (b): © Pixac | Dreamstime.com; (c): Andrew Holt/Photographer's Choice/Getty Images; (d): ShutterStock; (e): Digital Vision/Getty Images; p. 32 (top): AP Images; p. 32 (bottom): Courtesy of NASA/WMAP Science Team; p. 33 (left): (both): Courtesy of NASA/JPL-Caltech/Univ. of Arizona; p. 34 (left): Aquacone courtesy Solar Solutions, Inc.; p. 34 (top right): David Wrobel/Visuals Unlimited; p. 37: (a): Car Culture/Corbis; (b): AP Images.

## Chapter 2

Pages 40–41: Laurent/Science Source/Photo Researchers; p. 42: The Royal Institution, London/Bridgeman Art Library; p. 44: Courtesy Jacob Lewis Bourjaily; p. 55: Dirk Wiersma/Science Photo Library; p. 64: Courtesy of NASA/HST/J. Morse/K. Davidson; p. 65 (top): Courtesy of NASA/CXC/SAO; p. 65 (bottom): Vaz A and Griffiths M. Parathyroid Imaging and Localization Using SPECT/CT: Initial Results. J Nucl Med Technol. 2011; 39(5): 195–200. Figure 8; p. 66: Courtesy of NASA/CXC/SAO; p. 72: Taxi/Getty Images; p. 73: Harry Taylor/Getty Images.

## Chapter 3

Pages 74–75: MARK THIESSEN/National Geographic Stock; p. 77: Courtesy of Austin Post, USGS/CVO/Glaciology Project; p. 78: Richard Megna/Fundamental Photographs; p. 79: icefront/iStockphoto; p. 80 (top left): © Zach Holmes/Alamy; p. 80 (bottom): © 2009 Richard Megna/Fundamental Photographs; p. 81: moodboard/Corbis; p. 82: Kei Shooting/Shutterstock; p. 87: © Roger Ressmeyer/CORBIS; p. 92: vovan13/iStockphoto; p. 97: © Davids1993 | Dreamstime.com; p. 98: Gilbert S Grant/Photo Researchers RM/Getty Images; p. 99 (top): vupulepe/Shutterstock; p. 99 (bottom): michaeljung/Shutterstock; p. 102: (a) Biophoto Associates/Science Source/Photo Researchers; (b) Krzysztof Stanek | Dreamstime.com; (c) © FocusTechnology/Alamy; p. 118: AP Images; p. 119: (a): Peter Saloutos/Corbis; (b) AP Photo; p. 120 Richard Megna/Fundamental Photographs; p. 124: Courtesy of Purest Colloids, Inc., www.purestcolloids.com; p. 130: Smithsonian Institution/Corbis.

## Chapter 4

Pages 132–133: Bill Curtsinger/Getty Images; p. 134 (a): NASA; (b): Time Life Pictures/Getty Images; (c): Steve Schmeissner/Photo Researchers, Inc.; p. 134 (bottom) Courtesy of NASA/JPL-Caltech/MSSS and PSI; p. 135 (a):Colin Anderson/Getty Images/Brand X; (b): Nuridsany et Perennou/Science Source; (c): Shutterstock; p. 136: (a): Richard Megna/Fundamental Photographs; (b): Dr. E. R. Degginger/www.color-pic.com; p. 139: Leigh Smith Images/Alamy; p. 141: Richard Megna/Fundamental Photographs, NYC; p. 143: (all): © 1996 Richard Megna, Fundamental Photographs, NYC; p. 145 (top): Photo courtesy of Thermo Fisher Scientific; p. 145 (bottom): Courtesy of The University of North Carolina at Pembroke; p. 147: (all): © 2009 Richard Megna, Fundamental Photographs, NYC; p. 151: Richard Thom/Visuals Unlimited; p. 152: (both): ©TopFoto/The Image Works; p. 154: (both): Richard Megna/Fundamental Photographs; p. 156: Dirk

Degginger/Photo Researchers; p. 563: (a): © Goruppa | Dreamstime. com; (b): © Raja Rc | Dreamstime.com; (c): Purest Colloids, Inc., www. purestcolloids.com, P 609-267-2112; p. 564: (a): Dorling Kindersley/Getty Images; (b): Karlene Lowell Schwartz; p. 565: © 2013 Paul Silverman - Fundamental Photographs; p. 571: Kris Mercer/Alamy; p. 571: (a): Photo courtesy of Keren Peled (www.kelkajewelry.com); (b): Wikimedia Commons/Walters Art Museum; p. 572: Philippe Plailly/Photo Researchers, Inc.; p. 577: Richard Megna/Fundamental Photographs, NYC; p. 578: Andrew Silver/U.S. Geological Survey.; p. 579: Courtesy of www.VassichkoMinerals.com © 2007 JWV; p. 581: (a): Jon Stokes/Photo Researchers; (b): Shutterstock; (c): Andre Geim & Kostya Novoselov/Science Source; (d): © Dennis Kunkel Microscopy, Inc./Visuals Unlimited/Corbis; p. 583: Jose Manuel Sachis Calvete/Corbis; p. 583: E. R. Degginger/Photo Researchers; p. 584: Arthur Hill/Visuals Unlimited, Inc.; p. 585: Ashley Cooper/Alamy; p. 586: © Tom Pantages; p. 590: Malcolm Fielding, The BOC Group plc/Science Source; p. 591: Richard Megna/Fundamental Photographs, NYC; p. 597: Smithsonian Institution, Washington, D.C./Bridgeman Art Library; p. 597: Courtesy Ledtronics, Inc.; p. 598: Beinecke Rare Book and Manuscript Library, Yale University; p. 600: From Phase Diagrams of the Elements by David A. Young, p. 94, Fig. 7.7. Copyright © 1991, The Regents of the University of California. Reprinted by permission of the University of California Press.; p. 601: © Greenshoots Communications/Alamy.

## Chapter 13

Pages 602–603: Courtesy of Takao Someya, University of Tokyo; p. 606: Bernd Vogel/Solus-Veer/Corbis; p. 617: Dr. Keith Wheeler/Photo Researchers; p. 618: (a): Eric and David Hosking/Corbis; (b): Cabannes/PhotoCuisine/Corbis; (c): Digital Vision/Jupiterimages; p. 624: Tek Image/Photo Researchers; p. 624: Jeffrey Hamilton/Getty Images; p. 625: bag: Emily Spence/Lexington Herald-Leader/MCT/Newscom; rug: Fotosearch; chairs: The HON Company; rope: studiomode/Alamy; p. 628: Polystyrene cup, Photo by Luiscarlosrubino: http://creativecommons.org/licenses/by-sa/3.0/; p. 629: Stephen Stickler/Getty Images; p. 629: © 2010 Richard Megna/Fundamental Photographs NYC; p. 642: Barry Slaven/The Medical File/Peter Arnold; p. 643: Creatas/Jupiterimages; p. 646: Courtesy DuPont; p. 651: Phil Degginger/www.color-pic.com; p. 652: Phil Dotson/Science Photo Researchers, Inc.; p. 655: Peppermint: John Madere/Corbis; camphor tree: Douglas Peebles/Corbis.

## Chapter 14

Pages 666–667: Chris Detrick/Salt Lake Tribune; p. 668: Courtesy of Bob Burkhart; p. 670: (a): © Kateleigh | Dreamstime.com; (b): © 2005 Richard Megna/Fundamental Photographs; (c): Charles D Winters/Getty Images; p. 709: NASA.GSFC/Scientific Visualization Studio; p. 715: Courtesy of Platinum Today (www.platinum.matthey.com); p. 715: Yoshitazu Tsuno/AFP/Getty Images; p. 723: Lawrence Migdale/Photo Researchers; p. 728: Andrew Lambert Photography/Science Photo Library/Photo Researchers, Inc.

## Chapter 15

Pages 732–733: Courtesy of PotashCorp-Aurora; p. 736: Robert Landau/Corbis; p. 736: © 1993 Richard Megna, Fundamental Photographs, NYC; p. 752: (a): Stan Pritchard/Alamy; (b): Robert Jones/Alamy; p. 753:

Thermo Scientific Model 48000/Courtesy of Thermofisher; p. 755: © 2010 Richard Megna/ Fundamental Photographs NYC; p. 760: Richard Megna/Fundamental Photographs, NYC; p. 769: Robert Jones/Alamy.

## Chapter 16

Pages 778–779: Purestock/Alamy; p. 787: (a): Courtesy of EPA and CH2M; (b): The American Southwest; p. 805: Courtesy of Todd M. Lowe; p. 812: (all): © 2013 Richard Megna, Fundamental Photographs, NYC; p. 814: Photo by Ian Walton - FIFA/FIFA via Getty Images; p. 818: (all): Larry Stepanowicz/Visuals Unlimited; p. 819: © 1994 Richard Megna, Fundamental Photographs, NYC; p. 833: Hulton-Deutsch Collection/Corbis; p. 836: © 1995 Richard Megna/Fundamental Photographs.

## Chapter 17

Pages 844–845: NASA's Goddard Space Flight Center/USGS; p. 851: Richard Megna, Fundamental Photographs, NYC; p. 861: © 1994 Richard Megna, Fundamental Photographs, NYC; p. 862: © 1997 Richard Megna, Fundamental Photographs, NYC; p. 864: © 1990 Richard Megna, Fundamental Photographs, NYC; p. 866: © 1994 Richard Megna, Fundamental Photographs, NYC; p. 867: Vaughan Fleming/Photo Researchers, Inc.; p. 870: Phil Degginger/www.color-pic.com; p. 873: Terry W. Eggers/Corbis; p. 874: European Bioinformatics Institute (Protein Data Bank).

## Chapter 18

Pages 882–883: iStockphoto; p. 885: Phil Degginger/Alamy; p. 885: © 1987 Richard Megna, Fundamental Photographs, NYC; p. 889: B. & C. Alexander/Photo Researchers, Inc.; p. 910: (a): Hans Reinhard/zefa/Corbis; (c): SCIMAT/Science Photo Library/Photo Researchers, Inc.; p. 915: South Tyrol Museum of Archaeology.

## Chapter 19

Pages 926–927: 2013 Chevrolet Volt, photographed by John Voelcker for GreenCarReports.com; p. 928: (a): Charles D. Winters/Science Source; (b): Phil Degginger/www.color-pic.com; (c): Charles D. Winters/Science Source; p. 946: Tom Gilbert; p. 948: © Ambient Images Inc./Alamy; p. 951: Istockphoto; p. 953: © Richard Goldberg | Dreamstime.com; p. 955: © John Gress/Reuters/Corbis; p. 958: Courtesy of American Honda Motor Company Inc.; p. 960: Phil Degginger/www.color-pic.com; p. 962: Alix/Photo Researchers, Inc; p. 962: © Glenda Powers | Dreamstime.com; p. 962: Shutterstock; p. 966: © Franz-Marc Frei/Corbis.

## Chapter 20

Pages 968–969: Shutterstock; p. 970: Valueline/Punchstock; p. 979: (a): Dennis Kunkel Microscopy, Inc./Visuals Unlimited, Inc.; (b): Omikron/Photo Researchers, Inc.; p. 980: © Kokodrill | Dreamstime.com; p. 981: Courtesy of N.I.S.T.; p. 981: Chemical Design/Science Photo Library/Photo Researchers; p. 997: Roger Ressmeyer/Corbis; p. 997: NASA; p. 998: © W.R. Normak, courtesy USGS.

## Chapter 21

Pages 1008–1009: © Centre Jean Perrin/ISM/PhototakeUSA.com; p. 1021: Tom Bean/Corbis; p. 1022: Photodisc/Alamy; p. 1026: Igor Kostin/Corbis Sygma; p. 1026: © 2006 T. A. Mousseau and A. P. Moller; p. 1027: DOE Photo; p. 1029: Dr. Robert Friedland/Photo Researchers, Inc.; p. 1030: (a): Astrid & Hanns-Frieder Michler/Science Photo Library/Photo Researchers, Inc.; (b): Tom Tracey Photography/Alamy; p. 1031: Argonne National Laboratory, US DOE Office of Environmental Management; p. 1036: © 1990 Richard Megna/Fundamental Photographs; p. 1036: The American Weekly, 1926; p. 1037: Tom Bean/Corbis; p. 1041: Mireille Vautier/Alamy; p. 1041: © blickwinkel/Alamy; p. 1041: Reprinted by permission from Macmillan Publishers Ltd.: "Human Presence in the European Arctic Nearly 40,000 Years Ago," Nature 413 (September 6, 2001): 64–67, Figure 4—mammoth tusk showing human markings, © 2001.

## Chapter 22

Pages 1046–1047: ISM/Phototake, Inc.; p. 1049: (both): Phil Degginger/www.color-pic.com; p. 1052: Don W. Fawcett/Photo Researchers, Inc.; p. 1065: Phil Degginger/www.color-pic.com; p. 1067: SIU/Visuals Unlimited, Inc.; p. 1070: Zephry/Science Photo Library/Photo Researchers, Inc.; p. 1071: From "Small molecular gadolinium (III) complexes as MRI contrast agents for diagnostic imaging," Chan Kannie Wai-Yan and Wong wing-Tak, Coordination Chemistry Reviews, Sept., 2007, Elsevier B.V., Copyright Clearance Center; p. 1073: Phil Degginger/www.color-pic.com; p. 1078: (a): Southern Illinois University/Science Source; (b): David Pedre/Getty Images; p. 1079: Sergii Dmytriienko/Getty Images.

NOTE: Material in figures or tables is indicated by italic page numbers. Footnotes are indicated by n after the page number.

## A

absolute temperature, 268
   *See also* Kelvin temperature scale
absolute zero (0 K), 30, 268, 269, 270
absorbance (*A*), 144–46, *1054*
absorbed dose, 1024–25, 1028
*Acantharia*, 135
accuracy, 18, 24, 25–26
acetaldehyde, 635–36, 669
acetamide, *640*
acetic acid, 132
   hydrogen bonding, *482*
   metabolism by methanogenic bacteria, 639
   reaction with sodium bicarbonate, 884, *885*
   structure, *639, 640*
   titration, 155, 820–22
   as weak electrolyte, 147
   *See also* vinegar
acetone
   dipole moment, 481, 529
   as solvent, *9*, 638
   structure, *481, 529, 532, 638*
acetyl-coenzyme A, 988, 992, 1063
acetylene
   enthalpy of formation, *233*, 234
   Lewis structure, 381
   pi (π) bonds, *445*, 446
   production from alkanes, 619
   *sp* hybrid orbitals, 445–46
   use in welding, 618
achiral molecules, 453, 647
acid rain
   environmental effects, 90, 778
   formation of $SO_3$ from $SO_2$ and $O_2$, 77, 90, 740, 747
   neutralization by carbonates and bicarbonates, 806, 810–11
   from nonmetal oxides, 151, *183*, 778, *790*
   pH and pOH, 788–90
   reaction with calcium carbonate, 151, *152*, 163
acid–base chemistry, 148–57, 778–843
   acid ionization equilibrium constant ($K_a$), definition, 781
   acid–base properties of common salts, *802*
   alkalinity titrations, 819, 822–25, 832
   autoionization of water, 784, 785–86, 795
   base ionization equilibrium constant ($K_b$), definition, 783
   Brønsted–Lowry acids, 148, 149, 153, 780–82, 846–47
   Brønsted–Lowry bases, 148, 149, 153, 783–84, 846–47

Brønsted–Lowry model of acids and bases, 780–85
   calculations with pH and $K_a$, 790–93
   calculations with pH and $K_b$, 793–95
   common-ion effect, 806–10, 827
   conjugate acid–base pairs, 782–83
   dynamic equilibrium, 152–53
   hydrated metal ions as acids, 856–58
   $K_a$ values of hydrated metal ions, *857*
   leveling effect of water, 784
   Lewis acids, 846–48
   Lewis bases, 846–48
   net ionic equations, 150–51, 156–57
   neutralization reactions, 148, 150–51, 153–57
   overall ionic equations, 150–51, 156
   proton transfer, 148–53, *785*
   relationship between $K_a$ and $K_b$, 803–4
   relative strengths of acids and bases, 784–85
   strong acids, ionization reactions with water, *781*, 784
   strong bases, ionization reactions with water, *783*
   titrations, 153–57, 818–25
   very dilute solutions, 795–96
   water, 148–57, 782, 785–86, 846
   weak acids, ionization reactions with water, *782*
   weak bases, ionization reactions with water, *784*
   *See also* acids; bases; pH; pH buffers; solutions; titrations
acid–base indicators. *See* pH indicators
acids, 846–48
   acid ionization equilibrium constant ($K_a$), definition, 781
   acid strength and molecular structure, 800–801
   binary acids, naming, 62
   Brønsted–Lowry acids, 148, 149, 153, 780–82, 846–47
   calculations with pH and $K_a$, 790–93
   conjugate acid–base pairs, 782–83
   definition, 148
   diprotic acids, 152, 796, 797–99
   hydrated metal ions as acids, 856–58
   hypohalous acids, acid strength, 801, 856–57
   ionization of weak acids, 780
   $K_a$ values of hydrated metal ions, *857*
   Lewis acids, 846–48
   monoprotic acids, 152, 796
   oxoacids, acid strength, 781, 801
   oxoacids, naming, *61*, 62

acids *(cont.)*
  polyprotic acids, 152, 796–800
  relative strengths of acids and bases, 784–85
  strong acids, 152, 153, 780–81
  strong acids, ionization reactions with water, 781, 784
  triprotic acids, 796, 799, *800*
  volatile nonmetal oxides and their acids, 151, *152*
  weak acids, 152, 153, 780, 781–82, 784, 790–93
  weak acids, ionization reactions with water, *782*
  *See also* acid–base chemistry; amino acids; pH
acrolein, 450, 638
actinides, *51, 52,* 350, 353–54
activated complexes, 696, *697,* 700–701
activation energy ($E_a$)
  and Arrhenius equation, 694–99
  calculation from rate constants, 698–99
  and catalysts, 708, 761
  definition, 694
  and reaction mechanisms, 701
active sites
  enzymes, 712, 982–83, 991, 1060, *1061,* 1062
  molecular recognition, 424, 449
activity series for metals, 175–77
actual yield, 114–17
addition polymers, 623, 625–26, 635, *647*
addition reactions, definition, 620
adenine, *483,* 994
adenosine diphosphate (ADP), 911–13
adenosine triphosphate (ATP) hydrolysis, 911–13
adipic acid, *644*
adrenaline, 630
aerogel, *4*
alanine, 970, *971, 972, 973,* 976
albuterol, 651
alcohols, 630–33
  definition, 631
  functional group, 604, *605*
  naming alcohols, 631
  polymers of alcohols and ethers, 635–37
  solubilities of homologous series, 490–91, *631*
  *See also* specific types
aldehydes, overview, *605,* 637, 638
alkali metals, *51, 52*
alkaline earth metals, *51, 52*
alkanes, 606–18
  constitutional (structural) isomers, 611–13
  cycloalkanes, 616–17
  definition, 606, 607
  drawing alkane structures, 609–11
  functional group, *605,* 606
  melting points and boiling points, 608–9
  naming alkanes, 608, 613–16
  physical properties and structure, *605,* 608–9
  prefixes for naming, *608*
  sources and uses, 617–18
  straight-chain alkanes, 608–9
alkenes, 618–26
  addition reactions, 620

chemical reactivities, 619
cis and trans isomers, 620–23
definition, 606, 607, 618
electron distributions in π orbitals, 619–20
functional group, *605,* 606
hydrogenation, 607–8
isomers, 620–21
melting points and boiling points, *619*
naming alkenes, 622–23
polymers of alkenes, 623–26
stereoisomers, 620, 621
alkynes, 618–26
  addition reactions, 620
  chemical reactivities, 619
  definition, 606, 607, 618
  electron distributions in π orbitals, 619–20
  functional group, *605,* 606
  hydrogenation, 607–8
  isomers, 620
  melting points and boiling points, *619*
  naming alkynes, 622
allotropes, definition, 387
alloys, 560
  aluminum alloys, 560
  analysis of metals in, 874–75
  brass, 560
  bronze, 571–72, 573
  definition, 560, 571, 572
  dental alloys, 1050, 1076
  gold, 560, 571
  heterogeneous alloys, 571
  homogeneous alloys, 571
  interstitial alloys, 572–74
  nitinol, 571–72
  octahedral holes, 572–73
  shape-memory alloys, 571–72, 1078–79
  sterling silver, 573–74
  substitutional alloys, 570, 571–72, 573–74
  superconductors, 585
  tetrahedral holes, 572–73
  *See also* metals; mixtures; steel
alpha (α) particles
  alpha (α) decay, 1016, *1017,* 1027
  in boron neutron-capture therapy (BNCT), 1072
  definition, 44
  discovery, 44
  electric charge, 44
  mass, 44, *1010*
  relative biological effectiveness (RBE), 1072
  Rutherford gold foil experiment, 44–45
α helix, *978, 979, 980, 981*
alumina ($Al_2O_3$), 238, 584
aluminosilicates, 583–84, *585*
aluminum
  alloys, 560
  aluminum compounds in antacids, 1072
  electron configuration, 348
  energy cost of producing, 238–39
  energy cost of recycling, 239
  Hall–Héroult process, 238–39

health effects, 1073
hydrated aluminum ions, 857–58
molar heat capacity ($c_p$), *215,* 219, 222–24
in semiconductors, 577
specific heat ($c_s$), determination of, 222–24
aluminum chloride, 401, 403, 408, 848
Alzheimer's disease, 980, 1029, 1072, 1074
amalgams, 1050, 1076
*Amanita muscaria, 653*
amides, *605,* 637, 640–41
  *See also* polyamides
amines, *605,* 629–30
amino acids, 970–72
  biosynthesis, 1056
  α-carbon, 970, 973, 974
  chirality, 970, 972
  D- *(dextro-)* and L- *(levo-)* prefixes, 972
  definition, 970
  essential amino acids, 970
  isoelectric point (pI), 973, 974
  messenger RNA (mRNA) codons, 995–96
  in meteorites, 74
  (+) and (−) optical rotation, 972
  primary (1°) structure of proteins, 977–79
  R groups (side-chain groups), 970, *971*
  structures, *971*
  synthesis from inorganic materials, 74
  titration curves, 972–74
  transfer RNA (tRNA) and amino acids, 995–96
  zwitterions, 972–75, 976
  *See also* peptides; proteins
aminoacetonitrile, 463
ammonia
  hydrogen bonding, 481, *482,* 490
  reaction with boron trifluoride, 847
  reaction with methane, 87–88
  solubility, 489–90
ammonia ($NH_3$)
  as Brønsted–Lowry or Lewis base, 846
  calculations with pH and $K_b$, 793–95
  commercial uses, 114
  conversion to urea, 1058, 1063
  enthalpy of formation, *233,* 234
  entropy and spontaneity of synthesis, 897–98
  formation reaction, 233–34, 672–73
  from Haber–Bosch process, 118, 301, 760–61, 768
  Lewis structure, 379, 433
  melting and boiling points, *499*
  molecular geometry, 432–33
  nitric acid production, by Ostwald process, 768
  reaction rates (synthesis reaction), 672–73
  shifting equilibrium in synthesis, 755–56, *757,* 759
  $sp^3$ hybrid orbitals, 442–43
  synthesis, calculation of $K_p$, 908–9
  titration curve, 819–20
  as weak base, 153
ammonium chloride, 802, 805–6, 809–10
ammonium nitrate dissolution (cold packs), 884–85, 894

Amontons, Guillaume, 272
Amontons's law, 271–73, *275*, 289, *290*
ampere (A), definition, 946
amphetamine, 630
amphiprotic substances, 153, 785
amplitude (waves), 319
amyloid β, 980
anabolism, 982
anions, definition, 55
anodes, 42, *43*, 930, 931, 954, 955–56
anthracene, *452*, *628*
antifluorite structure, 580
antimatter, definition, 1014
antimony, 1048, 1065, 1073
aqueous solutions, definition, 77, 135
   *See also* solutions
arginine, 970, *971*, 976
argon, 347, 348, 359–60, 362–63
aromatic compounds, 626–29
   definition, 450, 451, 627
   functional group, *605*
   isomers, 627–28
   polycyclic aromatic hydrocarbons (PAHs),
      451, *452*, 628
   polymers containing aromatic rings, 628–29
   *See also* benzene
Arrhenius, Svante August, 695, *833*
Arrhenius equation, 694–99
arsenic, 138, 562, 577
artificial joints, 1046, 1078–79
asbestos, 410, 583, *584*
ascorbic acid (vitamin C), 844, 860
asparagine, *971*
"asparagus urine" odor, 465
aspartame, 975
aspartic acid, 970, *971*, 973–74, 975
aspartylphenylalanine, 975–76
aspirin, 20, 640, 693, 732
astatine, *88*, 504
Aston, Francis W., 46–47, 49, 106, 363
astrobiology, 74
atmosphere (Earth)
   composition, 260
   gases in Earth's early atmosphere, 74, 76–77,
      87, 111
   hydrogen as nonpolluting fuel, 118
   mass, *260*, 262
   pollutant transport rate constants, 687
   pollution from vehicle exhaust, 666–69
   thickness, 260
atmospheric pressure
   altitude effect, 261–62
   definition, 209, 260, 261
   and gravity, 209, 260, 262–63
   isobars (pressure), 262
   measurement, 263–65
   millibars (mbar), 262
   millimeters of mercury (mmHg), 261, *262*
   pascal (Pa), definition, 261–62
   and pressure–volume (P–V) work, 209
   standard atmosphere (atm), 261, *262*

torr, 261, *262*
units, 261–62
*See also* pressure (P)
atomic absorption spectra, 322–23, 331
atomic emission spectra, 322, *323*, 331
atomic mass, 46, 48–50
atomic mass units (amu), 46
atomic models
   Bohr model of hydrogen, 329–32
   nuclear model, 42–46
   Rutherford, 44–45, *46*, 329
   Thomson's plum-pudding model, 43, 44, *45*
atomic number (Z)
   definition, 47
   with element symbol, 47, 63
   and periodic table, 47–48, 50
   *See also* nuclear charge
atomic orbitals
   aufbau principle, 346, 347, 349, 350
   condensed orbital diagrams, 348
   *d* orbitals, sizes and shapes, 345–46
   *d* orbitals, spin states in complex ions, 867–69
   *d* orbitals, transitions in complex ions, 862–66
   definition, 336, 337
   degenerate orbitals, 347, 457
   electron configuration of atoms, 346–51
   Hund's rule, 347
   orbital diagrams, 346–47, *348*
   orbital penetration, 354–55
   *p* orbitals, sizes and shape, 345–46, *441*
   and periodic table, 345–51
   quantum numbers, 337–41
   *s* orbitals, sizes and shape, 343–44
   sizes and shapes, 343–45
   subshells, letters and quantum numbers, 338–41
   *See also* electron configuration; molecular
      orbital (MO) theory; wave mechanics
atomic radius, 354, *355*, 356, *363*, 579–80
atoms
   definition, 6
   sizes, 354–57, *363*, 477, *478*
   *See also* atomic models
ATP (adenosine triphosphate) hydrolysis,
   911–13
aufbau principle, 346, 347, 349, 350
auranofin, 1076
auroras (northern lights), 453–54, 462
austenite (iron), 572–73
automobile air bags, 277, 279, 301
average atomic mass, 48–49
Avogadro, Amedeo, 77, 271, 288, 289
Avogadro's law, 270–71, 274, *275*, 279, 288, 289
Avogadro's number ($N_A$), 78–79, 80, *82*, 83–84, 85
azides, 277, 279, 301

## B

Bacon, Francis, 16
ball-and-stick models, 8
balloons
   helium, 119, 362

helium balloons, 211
*Hindenburg*, 362
hot-air balloons, 209–10, 265, 281, 299
permeable skin, 293
and P–V work, 209–10, 211
*Spirit of Freedom* balloon, 211
weather balloons, 265
Balmer, Johann, 328–29, 339
Balmer equation, 328
band gap ($E_g$), 574, 576–77, 582
band theory, 574–75, 584–85
bar (pressure), definition, 232, 233, *262*,
   274, 275
Bardeen, John, 590
barium, 349
barium sulfate, *157*, 159–60, 826–27, 1065
barium sulfide, 82
barograph, 263, *264*
barometers, 260–61, 263, *264*
bases
   base ionization equilibrium constant ($K_b$),
      definition, 783
   Brønsted–Lowry bases, 148, 149, 153,
      783–84, 846–47
   calculations with pH and $K_b$, 793–95
   conjugate acid–base pairs, 782–83
   definition, 148
   Lewis bases, 846–48
   ligands as Lewis bases, 848, 849, *860*,
      861–62, 873–74
   names and formulas of basic minerals, *153*
   pOH, 788–90
   relative strengths of acids and bases,
      784–85
   strong bases, 153, 783
   strong bases, ionization reactions with
      water, *783*
   weak bases, 153, 783–84, 793–95
   weak bases, ionization reactions with
      water, *784*
   *See also* acid–base chemistry; pH
batteries, 926
   AA batteries, *946*, 948, 954–55
   alkaline batteries, half-cell reactions, 936
   alkaline batteries, standard cell potential, 936
   capacity related to quantities of reactants,
      946–50
   hearing aid batteries, 1077
   hybrid vehicles, 926, 946–50
   invention by Volta, *931*
   lead–acid automobile batteries, 132, 134, 591,
      943–44, 954
   lithium silver vanadium oxide batteries, 1078
   lithium/iodine batteries, 1077
   lithium–ion batteries, 948–50
   nickel–cadmium (nicad) batteries, 936, 1050
   nickel–metal hydride (NiMH) batteries,
      947–48, 954–55
   pacemaker batteries, 1076–77
   rechargeable batteries, 926, 931, 946, 948,
      953–56

batteries *(cont.)*
  sulfuric acid solution in automobile batteries, 132, 134
  zinc–air batteries, 936–37, 944, 1077–78
  Zn/Ag button batteries, 939–40
  Zn/HgO batteries, 1076–77
  *See also* electrochemical cells
bauxite, 238
BCNU, 503
Becquerel, Henri, 44, 1008, 1021, 1022
becquerel (Bq), 1022
Beer's law, 144–46
belt of stability, 1012–13, 1015
Benadryl, 629, 630, 634
bent or angular molecular geometry, 432, 433, *434, 448*
benzene
  bonding and resonance structures, 391, 451, 627
  delocalized π electrons, 626–27
  freezing point depression constant ($K_f$), 548, *549*
  reactivity, 627
  solubility, 491–92
benzo*[a]*pyrene, *452*
benzoic acid, 228, 229
beryllium, 347, 360, 376, 1049–50
beryllium chloride, 401
beta (β) decay, 1012, 1014–18, 1020, 1023, 1068–69, 1072
beta (β) particles, 44
  *See also* electrons
β-pleated sheet, 979–80, *981*
Big Bang theory
  cosmic microwave background radiation, 31–32
  interstellar space temperature, 28–29, 31, *32*
  leftover warmth, 29
  origins of theory, 17
  primordial nucleosynthesis, 63–64, 1032
  timeline for energy and matter transformation, *63*
  *See also* universe
binding energy (BE), 1010–12
biocatalysis, 713
biochemistry, 968–1007
  *See also* carbohydrates; lipids; nucleic acids; proteins
biomass, 95, 502, 630, 639, 987
biomolecules
  coordination compounds in, 872–74
  definition, 970
  extraterrestrial origins, 74, 997
  nitrogen in, 300
  synthesis from inorganic materials, 74, 997–98
  *See also* carbohydrates; lipids; nucleic acids; proteins
bis(methylthio)methane, 465
bismuth subsalicylate, 1073
body-centered cubic (bcc) unit cell, *565*, 567, 568–70, 572–73
body-centered cubic packing, *567*

Bohr, Niels, 316, 329, 331, 332, 336, *341*
Bohr model of hydrogen, 329–32
boiling point
  boiling point elevation, 534–35, *536*, 537–39, 549
  boiling point elevation constant ($K_b$), 535, *549*
  common hydrocarbons and alcohols, 483–84, *486*, 487–88
  compounds with mass similar to water, *499*
  and dispersion forces, 485–88
  and hydrogen bonds, 481–82, 483–84, 488
  normal boiling point, definition, 492
  and vapor pressure, 492–93
Boltwood, Bertram, 1019
Boltzmann, Ludwig, 913
Boltzmann constant, 914
Boltzmann distribution, 915
Boltzmann equation, 914–15
bomb calorimeters, 227–29, 243
bond angles, definition, 427
  *See also* molecular geometry
bond dipole, 437
bond order, 404, 406, 456, 458–60, 461–62
bond polarity, definition, 382
  *See also* polar bonds
bonding and bonds. *See* chemical bonds
bonding capacity, definition, 376
Born, Max, 337
Born–Haber cycle, 522–25
boron
  biological role, 1072
  boron neutron-capture therapy (BNCT), 1072
  boron trifluoride, 400–401, 429, 847
  boron-11, *1013*, 1014
  electron configuration, 347
  exceptions to octet rule, 376
  in nuclear reactors, 1031
Bosch, Carl, 760
Boyle, Robert, 266–67, 270
Boyle's law, 265–68, 271, 274, *279*, 288
Bragg equation, 587
branched-chain hydrocarbons, 611–12, 614–16
brass, 560
Brattain, Walter H., 590
breeder reactors, 1031
bright-line spectra, 322
bromine, 88, 504, 505
bromochlorofluoromethane (CHBrClF), 452–53
Brønsted, Johannes, 780
Brønsted–Lowry acids, overview, 148, 149, 153, 780–82, 846–47
Brønsted–Lowry bases, overview, 148, 149, 153, 783–84, 846–47
Brønsted–Lowry model of acids and bases, 780–85
bronze, 571–72, 573
Bryce Canyon National Park, 177
buckminsterfullerene (buckyballs), 246, *581*, 582
budotitane, 1073, *1074*
Bunsen, Robert Wilhelm, 322
bupropion, 651

butane, 97, 101, 115, *487, 611, 892*
butanethiol, 465
butyric acid, 640

### C

cadmium
  nickel–cadmium (nicad) batteries, 936, 1050
  in nuclear reactors, 1031
  toxicity, 1048, 1050
calcium, 182–83, 1053–54
  DRI/RDA value, *1049*
  electron configuration, 348
  as essential element, 183, 844, 1046, 1048, 1053–54
calcium carbonate, 182
  acid–base chemistry, 151, *152*, 265, 551, 806, 807
  antacid tablets, 84
  diffraction of light, 182
  in oyster shells, 265
  reaction with acid rain, 151, *152*, 163
  reaction with carbonic acid, 151
calcium fluoride, 161–62, 478–79, 524–25, 578–79, 827–29
calcium hydroxide, 141–42, 182, 830
calcium hypochlorite, 182
calcium oxide. *See* lime (quicklime; CaO)
calcium silicate ($CaSiO_3$), 182
calcium sulfate (gypsum), 182, 476, 477
calibration curves, 144, 145, *1021*
Calorie (Cal, nutritional Calorie), 208, 209, 242, 243–44
calorie (cal) definition, 208, 209
calorimeters, 215, 222, 223, 226–29, 243–44
calorimetry
  coffee-cup calorimetry, 226–27
  constant-volume calorimetry, 227–28
  definition, 222, 223
  food value, 242–44
  fuel value, 240–42
  measuring calorimeter constants, 227–29
  molar heat capacity determination, 222–24
  specific heat determination, 222–25
camphor, 116–17
capillary action, 500
capillin, 618, *619*
caraway, 424, 452–53
carbohydrates, 984–88
  cellobiose, 987
  cellulose, 246, 639, 984, *986*, 987
  chirality, 452
  deoxyribose, 994
  disaccharides, 985–86, 987
  formulas, 984
  fructose, 104, 984, 985
  glycosidic bonds, 985–87
  hydrogen bonds, 984
  monosaccharides, 984
  polysaccharides, 984, 986–87
  ribose, 994

carbohydrates *(cont.)*
  starch, 984, 986–87
  *See also* glucose; sucrose (table sugar)
carbon, 246
  atomic radius, 573
  buckminsterfullerene (buckyballs), 246, *581*, 582
  carbon-11 electron capture, 1014
  carbon-11 positron emission, 1013
  carbon-14 β decay, 1012
  combustion, 95–96
  crystalline forms, 246
  diamonds, 26–27, 246, 580–82, 892
  electron configuration, 347
  as essential element, 118, 1046, 1048
  fullerenes, 246, 580, *581*, 582
  graphene, *581*, 582
  isotopes and radioactive decay products, *1015*
  nanotubes, *581*, 582
  radiocarbon dating with carbon-14, 1019–21
  reduced form in carbohydrates and fossil fuels, 246
  stellar nucleosynthesis, 64
  *See also* graphite
carbon cycle, 94–98, 246
carbon dioxide
  absorption by NaOH, *107*, 108
  absorption of infrared radiation, 372–74
  bond angles, 426
  bond length, 404
  in cans of soda, 758–59
  density, 280–81
  determining concentration in seawater, 832
  in Earth's early atmosphere, 76
  effect on physiological pH, 814–15
  from fossil fuel combustion, 95–97, 280
  as greenhouse gas, 372–74, 618
  hydrolysis by carbonic anhydrase, 712, 1060, *1061*
  increase in atmospheric concentration, 280, 372
  Lewis structures compared to ball-and-stick model, *426*
  molecular structure, *5*, *54*, 429
  phase diagram, *497*
  polar bonds in nonpolar molecule, 437
  reduction in photosynthesis, 1064
  released in volcanic eruptions, 76
  resonance structures based on formal charge, 394
  sublimation, 497
  thermodynamics of decomposition reaction, 907–8
  valence bond theory model, 446
  vibrating bonds and greenhouse effect, 386
  from volcanoes, 280
carbon monoxide
  in atmosphere, 92
  in automobile exhaust, 710–11
  binding to hemoglobin, 873–74
  bond length, 404

combustion, 92, 230–31
  enthalpy of combustion, 230–31
  methanol synthesis, 631
  molecular structure, *54*
  reaction with nitrogen monoxide, 690, 693–94
  *See also* water–gas shift reaction
carbon tetrachloride
  empirical formula, 100, 101
  molecular geometry, 429–30
  molecular structure, *101*, *490*
  solubility, 489–90
carbon tetrafluoride, 437
carbonate minerals, 246, 806, 807
  *See also* calcium carbonate; limestone
carbonic acid
  calculations with pH and $K_a$, 797–99, 807–9, 810–11
  chemical weathering, 151
  common-ion effect, 807–9
  as environmental pH buffer, 810–11
  from partially neutralized acid rain, 806
  as physiological pH buffer, 814–15
carbonic anhydrase, 981, 1060, *1061*
carbon-skeleton structures of organic molecules, *609*, 610–11, 612–13
carbonyl compounds, 637–38
  *See also* aldehydes; ketones
carbonyl group, definition, 637, 638
carboplatin, 1075
carboxylate anion, 165, 639
carboxylic acids, 638–39
  carboxylic acid group, 637, 638
  definition, 639
  functional group, 604, *605*, 637
  hydrogen bonding, *482*, 638–39
  pharmaceuticals, 640
  *See also* amides; esters; polyamides; polyesters
η-Carinae (Eta Carinae), *64*
Carlsbad Caverns, *151*
carotene, 325, 1053, *1054*
carvone, 638, 651
catabolism, 982
catalysts, 706–13
  and activation energy ($E_a$), 708, 761
  and chemical equilibrium, 760–61
  definition, 706
  enzymes as protein catalysts, 970, 975, 982–84
  fuel cells, 715, 956, 957
  heterogeneous catalysts, 709, 713
  homogeneous catalysts, 708–9
  metal catalysts in catalytic converters, 710–11
  and the ozone layer, 707–10
  and reaction rate, 706
catalytic converters
  catalysts, 710–11
  development, 666, 669
  effect on air quality, 666, 669
  location in automobiles, *711*
  reduction of NO emissions, 669, 710

cathode rays, 42–43, 46
cathode-ray tubes (CRTs), 42–43, 44n2
cathodes, 42, *43*, 930, 931, 954, 955–56
cathodic protection, 953
cations, definition, 55
cell membranes
  cholesterol in, 992
  function, 988, 992
  ion channels, 1050, 1051–52, 1055, 1064
  ion pumps, 1051–53, 1055
  lipid bilayers, 992, 993, 1050, 1051, *1052*, 1062
  osmosis, 541–42, 543, 544–45
  phospholipids in, 991–92, 1051, *1052*
  structure, 992, *1052*
cellobiose, 987
cellulases, 987
cellulose, 246, 639, 984, *986*, 987
Celsius temperature scale, 29–30
cembrene A, 108
ceramics, 582–86
  aluminosilicates in, 583–84, *585*
  band theory, 584–85
  definition, 582
  earthenware, 583
  ionic silicates, 583–84
  mullite ($Al_6Si_2O_{13}$), 584
  polymorphs of silica, 583
  production from clay, 582, 584
  superconductors, 584–86
cesium, 1018, 1065
Chadwick, James, 46
chalcogens, 464–65
chalk, 246
Chandra X-ray Observatory, *65*
changes of state. *See* phase changes
charcoal, combustion, 230, 277–78
Charles, Jacques, 268, 269, 271
Charles's law, 268–70, 271, 274, *275*, 289, *290*
chelate effect, 862
chelating agents, 858, 859–60, 1066, 1069, 1073–74
  *See also* polydentate ligands
chelation, 858, 859, 861–62
chelation therapy, 1074
Chemical Abstracts Service (CAS), 653
chemical bonds, 372–424
  bond energy (bond strength), 374, 404–7
  bond length, 374, 403–4
  bond order, 404, 406, 456, 458–60, 461–62
  coordinate bonds, definition, 848
  definition, 8
  delocalized electrons in bonds, 451–52
  double bond, definition, 379
  metallic bonds, 375
  single bond, definition, 377
  triple bond, definition, 380
  types of bonds, comparison, *376*
  vibrating bonds and greenhouse effect, 386–87
  *See also* covalent bonds; ionic bonds; polar bonds; resonance; valence bond theory

chemical equations, 77, 87–92
  balancing, 77, 87–92
  coefficients, 77, 85, 87–92
  definition, 6, 7–8
  molecular equations, 150
  and moles, 85–86
  net ionic equations, 929–30
  net ionic equations, definition, 150
  nuclear equations, 64
  overall ionic equations, definition, 150
  symbols, 77
chemical equilibrium, 732–77
  and catalysts, 760–61
  complex ions, 851–53
  definition, 734
  dynamics of chemical equilibrium, 732, 734–37
  effects of adding or removing products or reactants, 755–56
  effects of pressure changes, 756–59
  effects of temperature changes, 759–60
  effects of volume changes, 756–57
  endothermic and exothermic processes, 759–60
  and free energy (G), 902–6
  gas-phase equilibria, 736, 756–59
  heterogeneous equilibria, 752–54
  homogeneous equilibria, 752
  solubility equilibria, 825–31
  solubility-product constant ($K_{sp}$), 825–32
  See also dynamic equilibrium; equilibrium constants (K); Le Châtelier's principle; physical equilibrium
chemical formulas, 53–55
  coordination compounds, 849
  definition, 6, 7
  formula unit, definition, 55, 82
  structural formulas, 8
  See also empirical formulas; molecular formulas
chemical kinetics, 666–731
  definition, 669
  See also reaction rates
chemical properties, 12–13
chemical reactions, definition, 6
chemical reactions, overview, 76–77
  See also specific types
chemical weathering, 151, 152, 177
chemistry, definition, 4
Chernobyl, Ukraine, nuclear reactor explosion, 1026, 1027, 1065
Chevrolet Volt, 926–27, 948
chirality, 452–53, 647–52
  achiral molecules, 453, 647
  amino acids, 970, 972
  chiral carbon atoms, 453
  chiral centers, 647–48, 649–50, 661
  chiral complex ions and coordination compounds, 871–72
  chiral compounds, definition, 647
  mirror images, 424, 452–53, 647–48, 649–50

  and molecular recognition, 452–53, 650–51, 652–53
  in nature, 650–52
  optical isomers, 452–53, 647–51
  optically active molecules, 648–49
  pharmaceuticals, 650, 651, 982
  proteins, 452
  racemic mixtures, 651
  See also enantiomers
chlor–alkali process, 505
chloramines, 505
chloride ion
  acid–base chemistry, 784, 801
  analysis by precipitation reactions, 161, 162–63
  concentration in seawater and drinking water, 137, 138
  as essential element, 1046, 1048, 1055–56
  ion formation, 55
chlorine, 504–5, 1055–56
  diatomic element, 88, 354
  DRI/RDA value, 1049
  electron affinity, 360
  as essential element, 1046, 1048, 1055–56
  free radicals, 411
  ion formation, 55
  oxoanions of chlorine and their acids, 61
chlorine monoxide, 411, 689–90, 692, 698–99, 708–9
chlorins, 872–73
chlorofluorocarbons (CFCs), 411, 707–8
chloroform
  dipole moment ($\mu$), 439
  Lewis structure, 378–79
chlorophyll
  chlorins, 872–73
  $Mg^{2+}$ ion in chlorophyll a, 844–45, 873, 1053, 1054, 1064
  phytoplankton, 9, 844–45
  spectra of chlorophyll and other pigments, 325, 873, 1054
  structure, 873, 1054
  See also photosynthesis
cholesterol, 990, 992
chromium
  alloys, 560
  chromium(III) hydroxide, 857
  diabetes drugs, 1073, 1074
  DRI/RDA value, 1049
  electron configuration, 349, 862–63
  as essential ultratrace element, 1048, 1064
  hydrated chromium ions, 857–58, 862–64
  spin states of complex ions, 867
  toxicity of chromate ions, 1064
chromodulin, 1064
chrysotile (asbestos), 583, 584
cinnamaldehyde, 638
cisplatin [cis-diamminedichloroplatinum(II)], 870, 1075
citrate ion, 1066
citric acid, 155, 799, 800
Clausius, Rudolph, 913

Clausius–Clapeyron equation, 493–94, 495
clay and clay minerals, 582, 584, 997
climate change, 95, 118, 218, 237, 280
clusters (crystalline solids), 581, 582, 583
COAST (collect, organize, analyze, solve, and think about the answer), 10–11
cobalt, as essential ultratrace element, 844, 1048, 1063
cobalt(II) chloride solution, 759–60
coccolithophores, 9
codons, 995–96
coefficients in chemical equations, 77, 85, 87–92
coenzyme $B_{12}$, 1063
coinage group metals, 1075–76
  See also copper; gold; silver
colligative properties, 532–50
  boiling point elevation, 534–35, 536, 537–39, 549
  boiling point elevation constant ($K_b$), 535, 549
  definition, 514, 515
  freezing point depression, 535–37, 538–40, 548–49
  freezing point depression constants ($K_f$), 536, 540, 549
  ideal solutions, 514, 515
  molality (m), 533–34
  molar mass measurement, 548–50
  osmotic pressure ($\Pi$), 541–46, 549, 1046
  reverse osmosis, 546–47
  van 't Hoff factor (i factor), 537–41
  See also vapor pressure
color and vision, 864
colors of metal compounds
  crystal field splitting, 862, 863–66
  crystal field splitting energy ($\Delta$), 862, 863–66, 867–69
  crystal field theory, 862–66
  $Cu^{2+}(aq)$, characteristic color, 851
  ethylenediamine complex of $Ni^{2+}$, 861–62, 866
  nickel(II) chloride hexahydrate, 861–62, 866
  $Ni(NH_3)_6^{2+}$, 861–62, 866
  spectrochemical series of common ligands, 866, 869
  tetraamminecopper(II), $Cu(NH_3)_4^{2+}$ complex, 851–53, 864–65
  vision and color, 864
  See also complex ions (complexes); metals
combination reactions, 77
combined gas law, 274
combustion analysis, 106–10
combustion reactions, 92–94
  carbon, 95–96
  carbon monoxide, 92, 230–31
  combustion analysis, 106–10
  definition, 12, 92
  enthalpy of combustion ($\Delta H_{comb}$), 225, 228–29, 230–33, 241, 632
  ethane, 93–94
  fossil fuels, 95–97, 246, 618

combustion reactions *(cont.)*
hydrocarbons, 92–94
lean mixture, 113
methane, *74–75*, 92–93, 95–96, 167–68, 170–71
rich mixture, 113
*See also* oxidation–reduction reactions
comets, *4*
common-ion effect, 806–10, 827
complex ions (complexes), 848–81
in biomolecules, 872–74
coordination numbers, 850
counter ions, 850
*d* orbitals and spin states, 867–69
*d* orbitals and transitions, 862–66
diagnostic and therapeutic agents, 1066–67
enantiomers, 871–72
equilibria, 851–53
formation constant ($K_f$), 852
Hund's rule, 863, 867
hybrid orbitals, 849–51
hydrated metal ions as acids, 856–58
inner coordination sphere, 848, 849
introduction and definitions, 848–51
$K_a$ values of hydrated metal ions, *857*
magnetism and spin states, 867–69
naming, 853–56
shapes and geometries, 849–51, 863, 864–65
*See also* colors of metal compounds; coordination compounds; ligands; metals
compounds, definition, 6
*See also* coordination compounds; ionic compounds; molecular compounds
concentration
definition, 136
determining concentration, 144–46
dilute and concentrated solutions, 136
dilutions, 142–46
effect on cell potential ($E_{cell}$), 942–46
effect on reaction rates, 677–94
molality (*m*), 533–34
molarity (*M*), definition and unit conversions, 138–40
molarity (*M*) and density, 140–41
parts per billion (ppb), 137
parts per million (ppm), 137, 138
preparing solutions of known concentration, 141–42, 143
units, in solutions, 136–42
*See also* solutions
condensation polymers, 642, 643, 646, *647*
condensation reactions, 640, *640*, 641, 643
condensed structures of organic molecules, 609–11, 612–13
conduction bands, 574–75, 576–77, 584–85
conservation of energy, law of, 200, 201, 208
conservation of mass, law of, 77, 86
constant composition, law of, 6
constructive interference, 321
conversion factors, *19*, 26–28

coordinate bonds, definition, 848
coordination compounds
in biomolecules, 872–74
chemical formulas, 849
chirality, 871–72
definition, 850
enantiomers, 871–72
naming, 855–56, 871
stereoisomers of, 870–71
*See also* complex ions (complexes)
coordination numbers, 850
copolymers, 636, 637, 642
copper
alloys, 560, 571–72
atomic radius, 573
band theory, 574–75
chalcocite, copper(I) sulfide, 98, 915–16
characteristic color of $Cu^{2+}(aq)$, 851
corrosion, 951–52
Cu/Ag electrochemical cell, 933
cuprite, copper(I) oxide, 97–98
DRI/RDA value, *1049*
electron configuration, 349
as essential ultratrace element, 844, *1048*, 1063–64
malleability, 571, 572
melting point, 240
metallic bond and arrangement of atoms, *571*
molar heat capacity ($c_P$), 240
molar heat of fusion ($\Delta H_{fus}$), 240
reaction with silver nitrate, 172–74
redox reaction with zinc, 928–29
refining, 915–16
spin states of complex ions, 867–68
tetraamminecopper(II), $Cu(NH_3)_4^{2+}$ complex, 851–53, 864–65
X-ray diffraction (XRD) analysis, 588
$Zn/Cu^{2+}$ electrochemical cell, 930–31, 932, 935, 937, 939, 944
corn oil, 990
corrosion, 560, *882–83*, 884, 950–53, 959–60
Cosmic Background Explorer (COBE) satellite, 32
cosmic microwave background radiation, 31–32
cosmology, definition, 2
cotton, 246, 643–44, 984
coulomb (C), definition, 43, *46*, 938
coulombic interactions, 202–3, 477–79
Coulomb's Law, 202
counter ions, 850
coupled reactions, 909–13
covalent bonds
bond energy (bond strength), 404–7
bond length, 403–4
bond order, 404, 406
covalent radius, 354
definition, 54
ionic character of bonds, 383, 403
nonpolar covalent bonds, 383
polar covalent bonds, 382–86
in polyatomic ions, 60

types of bonds, comparison, *376*
*See also* chemical bonds; Lewis structures; molecular compounds
covalent network solids, 580–82, 583, 590, 892
covalent radius, 354, 363
Crab Nebula, *2–3*
critical point in phase diagrams, 497
critical temperature ($T_c$) for superconductors, 584, 585
crucibles, *753*
crude oil
alkanes from, 617–18
alkenes in, 618, 626
definition, 527
nitrogen under pressure used in oil deposits, 300, *301*
separation by distillation, 527, 618
volatile compounds in, 527
cryogens
helium, 119
nitrogen, 300
oxygen (LOX), 411
cryolite ($Na_3AlF_6$), 238
crystal field theory, 862–66
crystal field splitting, 862, 863–66, 867–69
crystal field splitting energy ($\Delta$), 862, 863–66, 867–69
magnetism and spin states, 867–69
crystal structure
antifluorite structure, 580
body-centered cubic packing, *567*
covalent network solids, 580–82, 583, 892
crystal lattices, overview, 563–64
crystalline nonmetals, 580–82
crystalline solids, definition, 562, 563
cubic closest-packed (ccp) structures, *565*, 566, 567
cubic packing, *565*, 566–67
definition, 564
diamond structure, 580, 591
face-centered cubic (fcc) unit cell, *565*, 566, 567–68, 569–70, 573–74, 578–79
fluorite structure, 579–80
halite structure, 101
hexagonal closest-packed (hcp) structures, 564, *565*
hexagonal unit cell, 564, *565*, 567
ionic solids, 577–80
octahedral holes, 572–73, 577–78
packing efficiency, 566, 567, 570–71
rock salt structure, 578, 580, 589
simple cubic (sc) unit cell, *565*, 566–67
sphalerite structure, 578
square packing, 566
stacking patterns, overview, 563–64, *567*
tetrahedral holes, 572–73, 578–79, 589
unit cells of metals in periodic table, *566*
X-ray diffraction (XRD), 586–88
crystalline solids, definition, 562, 563
cubic closest-packed (ccp) structures, *565*, 566, 567

cubic packing, *565*, 566–67
Curie, Marie and Pierre, 44, 1008, 1022, 1024, 1036
curie (Ci), 1022
*Curiosity* Martian rover, 33, 134
Curl, Robert, 246
cycles per second (cps), 319
cycloalkanes, 616–17
cyclobutane, 617
cyclohexane, 617
cyclopentane, 617
cyclopropane, 617
cysteine, *971*, 981, 1062
cystic fibrosis, 1055
cytochromes, 874, 1061
cytosine, *483*, 994

**D**

da Vinci, Leonardo, 167
Dacron, 643–44
Dalton, John, 46, 53–54, 288
dalton (Da), 46
Dalton's atomic theory, 288
Dalton's law of partial pressures, 282, 283–87, 288–89, 298
dark-line spectra, 322–23
de Broglie, Louis, 332, 333–34, 335, 337
de Broglie wavelengths, 332–35
deferoxamine (DFO), 1074–75
delocalization of electrons, 375, 451–52
density
    air, 280, 282–83
    calculation, 12, 21–22, 23
    calculation from unit cells, 579–80, 589
    definition, 12
    gas density, 260, 280–83
    gold, 21–22, 23
    molarity (*M*) and density, 140–41
    water, density changes with temperature, 501, 502
dental amalgams, 1050, 1076
dental fillings, 79
Denver, smog, *668*
deoxyribonucleic acid (DNA)
    double helix, *452*, 482, *483*, 994–95
    effect of polycyclic aromatic compounds, 451, *452*
    genetic code, 995
    hydrogen bonds, 482, *483*, 994
    intercalation, 451, *452*
    nitrogen in, 300
    replication, 994–95
    structure, *40–41*, 993–95
    transcription, 995, *996*
    *See also* nucleic acids; ribonucleic acid (RNA)
deoxyribose, 994
deposition, definition, 14, 15, 495
desiccants, 182
destructive interference, *320*, 321

deuterium ($^2$H or D), 63, 118–19, 1032, 1033, 1034, 1035
    *See also* hydrogen
deuterons, 63, *1010*, 1032–33
dextrose. *See* glucose
diagnostic applications of elements, 1046, 1066–72
    diagnostic radiology, 1029
    imaging with radionuclides, 1008, 1029, 1067–72
    imaging with stable isotopes, 1071–72
    *See also specific elements*
diamagnetism, 458, 459–60
1,4-diaminobenzene, *645*
diamonds, 26–27, 246, 580–82, 892
diatomic elements, 7, 88, 354
diazene ($N_2H_2$), 450
diazotrophs, 300
dibromochloromethane ($CHBr_2Cl$), 452–53
*cis*-dichlorobis(ethylenediamine)cobalt(III), 871, *872*
Dicke, Robert, 31
Dieng Plateau (Indonesia), 280, 281
dietary reference intake (DRI) values, 1048, *1049*, 1058
diethyl ether, 493, 634–36, 636
diethylenetriamine, 859
diethylenetriaminepentaacetate ($DTPA^{5-}$), 1066
diffraction, 182, 320–21, 327, 586–88
diffusion, 294–95, 1051–52
dilutions, 142–46
dimensional analysis, 26–28
dimethyl disulfide, *465*
dimethyl ether, 464–65, 481, 483–84, 529
dimethyl sulfide, 465
dimethyl sulfoxide, *465*
dimethylbenzene (xylene), 628
2,2-dimethylpropane, 486–87
dinitrogen monoxide (nitrous oxide), 391–94, 395, 686
    percent composition, 102
dinitrogen pentoxide, 91–92, 151, 684–86
dipole moment ($\mu$), 438–40
dipole–dipole interactions, 481
diprotic acids, 796, 797–99
disaccharides, 985–86, 987
dispersion forces (London forces), 484–88
distillation
    crude oil separation, 527, 618
    definition, 8
    energy transfer, 206–7
    fractional distillation, 363, 528–31
    gasoline, 527
    laboratory setup, *207, 528*
    octane, 528–30
    seawater, 9
DNA. *See* deoxyribonucleic acid (DNA)
dolomite [$CaMg(CO_3)_2$], 246
double bond, definition, 379
drinking water standards, 136, 138
drugs. *See* pharmaceuticals

Duchenne muscular dystrophy, 968
dynamic equilibrium
    dynamics of chemical equilibrium, 732, 734–37
    glucose and fructose cyclization, 985
    liquid–vapor equilibrium, 529–30
    saturated solutions, *164*
    solutions of weak electrolytes, 152–53
    sulfur dioxide with sulfur trioxide, 732–34
    *See also* chemical equilibrium; physical equilibrium
dystrophin, 968

**E**

Earth
    age, 76
    crust, 74, 76, 134, 157, *1049*
    early atmosphere, 74, 76–77, *87*, 111
    inner and outer core, 76
    mantle, 76
    surface area, *260*
    "water planet," 134
effective nuclear charge ($Z_{eff}$), 356, 357, 358, 359
effusion, 292, 293–95
eggs (chicken), 550–51
eicosene, 106, 548–49
Einstein, Albert
    on de Broglie wavelengths, 335
    mass–energy formula, 17, 332, 1010, 1030, 1034, 1035
    photoelectric effect, 326, 327
    on quantum mechanics, 341–42
elaidic acid, 990
electric vehicles, 926, *927*
    Chevrolet Volt, *926–27*, 948
    driving range, 926, *948*
    fuel cells, 926
    lithium–ion batteries, *927*, 948–50
    Nissan Leaf, 948
    *See also* hybrid vehicles
electrochemical cells
    anodes, 930, 931, 954, 955–56
    bridge, *930*, 931
    cathodes, 930, 931, 954, 955–56
    cell diagrams, 931–32
    cell potentials ($E_{cell}$), 935
    chemical free energy ($\Delta G_{cell}$), 937–40, 946
    concentration effect on cell potential ($E_{cell}$), 942–46
    definition, 931
    electrical work ($w_{elec}$), 937–38, 946
    electrodes, 930
    identifying anode and cathode half-reactions, 935–36
    illustrations, *930, 932, 933, 935*
    Nernst equation, 942–44
    standard cell potentials ($E°_{cell}$), 934–37, 1077–78
    standard cell potentials ($E°_{cell}$), *K*, and $\Delta G°$, 944–46

electrochemical cells *(cont.)*
   standard reduction potentials (*E*°), 934–37
   Zn/Cu²⁺ electrochemical cell, 930–31, 932, 935, 937, 939, 944
   *See also* batteries; voltaic cells
electrochemistry, 926–67
   definition, 928
   *See also* electrochemical cells; oxidation–reduction reactions
electrodes, 146
   anodes, 42, *43*, 930, 931, 954, 955–56
   cathodes, 42, *43*, 930, 931, 954, 955–56
   standard hydrogen electrode (SHE), 940–42
   voltaic cells *vs.* electrolytic cells, 954, 955–56
   *See also* electrochemical cells
electrolysis, 118, 505, 931, 955–56
electrolytes
   overview, 146–47, 152
   strong electrolytes, 146, 147, 149, 152
   weak electrolytes, 146, 147, 149, 152
electrolytic cells
   comparison to voltaic cells, 931, *932*, 954, 955–56
   definition, 931
   electrolysis, 505, 931, 955–56
   lead–acid automobile batteries, 132, 134, 943–44, 954
   nickel–metal hydride (NiMH) batteries, 947–48, 954–55
   rechargeable batteries, 926, 931, 946, 948, 953–56
   and rechargeable batteries, 953–56
   *See also* batteries; electrochemical cells
electromagnetic radiation
   definition, 318, 319
   diffraction, 182, 320–21, 327
   electromagnetic spectrum, *31*, *318*
   frequency (ν), 318, 319, 361
   gamma rays, 319, 335, 1014, 1024–25, 1028
   from hot objects, *316–17*, 323, *324*
   infrared (IR) radiation, 320, 323
   ionizing radiation, 1024–29
   radio waves, *318*, 319
   refraction, 320
   speed of light (*c*), 319
   ultraviolet (UV) radiation, *318*, 320
   wave properties, 318–20, 327
   wavelength (λ), 318, 319, 361
   wave–particle duality, 327–28
   *See also* light; waves
electromagnetic spectrum, *31*, *318*, 319
electron affinities (EA), 360–61
electron capture, 1014, *1016*
electron configuration
   condensed electron configurations, 346–51
   condensed orbital diagrams, 348
   core electrons, definition, 346, 347
   definition, 346, 347
   degenerate orbitals, 347, 457
   energy levels in multielectron atoms, *349*

Hund's rule, 347, 457
ions of main group elements, 351–52
orbital diagrams, 346–47, *348*
orbital filling order in atoms, 346–51
transition metal cations, 353–54
valence electrons, definition, 347
*See also* atomic orbitals
electronegativity
   and bond polarity, 383, 384–86
   halogens, *384*
   and ionization energy, 383–84
   Pauling electronegativity scale, 383
electron-pair geometry, 427
   *See also* valence-shell electron-pair repulsion theory (VSEPR)
electrons
   annihilation reactions with positrons, 1014, 1029, 1033
   beta (β) particles, discovery, 44
   de Broglie wavelengths, 332–35
   definition, 43
   discovery, 42
   electric charge, 43, *46*, 938
   electron transitions, 330, 331–32
   energy level diagram, *331*
   energy levels (*n*), 329–32
   excited states, 330
   ground state, 330
   mass, 43, *46*, *1010*
   mass-to-charge ratio (*m/e*), 42, 44
   properties, *46*, 333–35
   spin, 340–41
   wave properties, 333–35, 344
electroplating, 955
electrostatic potential energy (*E*ₑₗ), 202–3, 374–75, 477–79
elementary steps, 700–10, 713
elements
   belt of stability, 1012–13, 1015
   definition, 5
   diagnostic applications, 1008, 1029, 1046, 1066–72
   diatomic elements, 7, 88, 354
   in Earth's crust, 74, *1049*
   essential elements, definition, 1048
   essential elements, list, *1048*
   free elements in nature, 6, 12, 563, 915
   in human body, *1049*, *1051*
   Lewis symbols of main group elements, *377*
   life and the periodic table, 1046–85
   main group elements or representative elements, *51*, 52
   major essential elements, 1046, 1048, 1050–58
   nonessential elements, 1048, 1065
   in seawater, *137*, 535, *1049*
   stable (nonradioactive) elements, 374
   stimulatory effect of nonessential elements, 1048, 1065
   in the sun, 65
   synthesis of, 62–65, 1010–11, 1032

therapeutic applications, 504, 1028–29, 1046, 1066–67, 1072–76
trace essential elements, 1046, 1048, 1058–62
ultratrace essential elements, 1048, 1058, 1062–64
in universe, *1049*
*See also* periodic table; *specific elements*
empirical formulas
   from combustion analysis, 106–10
   definition, 55, 100
   and molecular formulas, 101, 103–6, 107–10
   from percent composition, 98–103, 181
   *See also* chemical formulas
enantiomers, 648–51
   amino acids, 972, 997
   complex ions, 1074
   complex ions and coordination compounds, 871–72
   definition, 648, 649
   and drug activity, 650, 651
   enzyme selectivity, 982
   in nature, 650–52
   *See also* chirality
end point, definition, 154, 155
endothermic processes, 205–8, 759–60, 884
   *See also* heat (*q*)
energy, 198–204
   definition, 4
   internal energy (*E*), 208, 209–10, 212, 217, 228, 233
   law of conservation of energy, 200, 201, 208
   mass–energy formula, 17, 332, 1010, 1030, 1034, 1035
   transformations, 196, 200
   units, 208, 209, 211
   *See also* kinetic energy (KE); potential energy (PE); work
energy efficiency, 900, 958–59
energy levels (*n*), 329–32
   *See also* quantum theory
energy profiles of reactions, *695*, 696, *697*, 701
energy transfer
   and changes of state, 198, 205–8
   closed systems, 204–5, 205
   definition, 198
   open systems, 204–5, 205
   in phase changes, 198, 205–8
   sign convention, 206, 210
   and temperature changes, 198, 205
enthalpy (*H*), 211–14
   and bond energies, 405–7
   Born–Haber cycle, 522–25
   definition, 212
   effects of Δ*H*, Δ*S*, and *T* on free energy, 895–98, 900–902
   enthalpy change (Δ*H*), definition, 211
   enthalpy change (Δ*H*) in phase changes, 212–13
   enthalpy of combustion (Δ*H*comb), 225, 228–29, 230–33, 241, 632

enthalpy (*H*) (cont.)
  enthalpy of hydration ($\Delta H_{hydration}$), 520–21, 525–27
  enthalpy of reaction ($\Delta H_{rxn}$), 696, 897
  enthalpy of reaction ($\Delta H_{rxn}$), determining, 225–27
  enthalpy of solution ($\Delta H_{solution}$), 520–21, 526–27
  Hess's law, 229–33, 237–38
  molar heat of fusion ($\Delta H_{fus}$), 212, 216, 217, 218
  molar heat of vaporization ($\Delta H_{vap}$), 216–17, 494
  sign convention, 212, 213–14, 231–32, 236
  standard enthalpies for selected compounds, *233*
  standard enthalpy of formation ($\Delta H_f^\circ$), 232, 233–35, 237–38, 240, 244–45
  standard enthalpy of reaction ($\Delta H_{rxn}^\circ$), 230, 232, 233–38
  as state function, 212, 231, 236
  *See also* heat (*q*)
entropy (*S*), 886–94
  absolute entropy, 890–93, 894, 915
  ammonia synthesis, 897–98
  ammonium nitrate dissolution (cold packs), 894
  changes as ice is heated, 888, 891, 900–901
  in chemical reactions, 894
  definition, 882, 886, 887
  and heat, 888–89
  microstates and entropy changes ($\Delta S$), 913–15
  and molecular structure, 892–93
  reversible processes, 888–89, 895
  second law of thermodynamics, 882, 886–87, 888–89, 913
  spontaneity as function of $\Delta S_{sys}$ and $\Delta S_{surr}$, *888*, 889–90, 895
  and spontaneous processes, 886–87, 888–89, 894–96
  standard entropy of selected substances, *891*
  standard molar entropy (*S*°), 890–91
  as state function, 894
  thermodynamic entropy, 886–90
  third law of thermodynamics, 890–92, 915
enzymes, 982–84
  active sites, 712, 982–83, 991, *1061, 1062*
  as catalysts, 712–13, 970, 975, 982–84
  definition, 712
  enzyme–substrate (E–S) complexes, 712, 982–83
  induced-fit model of activity, 983
  inhibitors, 983
  kinetics, 713
  lock-and-key model of activity, 982–83
  and metabolic pathways, 712
  metalloenzymes and their reactions, *1060*
  selectivity for enantiomers, 982
  substrates, 712, 982–83
  turnover number, 712
  *See also specific enzymes*
equilibrium. *See* chemical equilibrium; dynamic equilibrium; physical equilibrium
equilibrium constants (*K*)
  acid ionization equilibrium constant ($K_a$), definition, 781
  autoionization of water ($K_w$), 785–86
  base ionization equilibrium constant ($K_b$), definition, 783
  calculations based on *K*, 761–68
  combined reactions, overall *K* values, 748–50
  definition, 737
  effects of temperature, 907–9
  for equation multiplied by a number, 746–48
  equilibrium constant expressions, 737–42, 753–54
  formation constant ($K_f$), 852
  and free energy (*G*), 904–7
  heterogeneous equilibria, 753–54
  law of mass action, 738
  mass action expression, 738, 750–52, 755
  and molar concentrations ($K_c$), 739
  and partial pressures ($K_p$), 739
  reaction quotients (*Q*), 750–52, 829–31, 903–4
  relationship between $K_a$ and $K_b$, 803–4
  relationship between $K_c$ and $K_p$, 739, 742–45
  relationship to *E*° and $\Delta G$° in electrochemical cells, *946*
  for reverse reactions, 745–47
  RICE tables (Reaction, Initial values, Changes, Equilibrium values), 761–67
  solubility-product constant ($K_{sp}$), 825–32
  standard cell potentials ($E_{cell}^\circ$) and *K*, 944–46
  and thermodynamics, 739
  van 't Hoff equation, 908
  very large or small values of *K*, 737, 742
  *See also* chemical equilibrium; reaction quotients (*Q*)
equivalence point, 154, 155
*Escherichia coli*, 134, *910*
essential amino acids, 970
essential elements for biological systems
  definition, 1048
  dietary reference intake (DRI) values, 1048, *1049*, 1058
  list of essential elements, *1048*
  major essential elements, 1046, 1048, 1050–58
  nonessential elements, 1048, 1065
  percent daily values (DV), *1049*
  recommended dietary allowance (RDA) values, 1048, *1049*
  stimulatory effect of nonessential elements, 1048, 1065
  trace essential elements, 1046, 1048, 1058–62
  ultratrace essential elements, 1048, 1058, 1062–64
  *See also specific elements*
esterification reactions, 640, 641–42
esters, *605, 637,* 640–41
  *See also* polyesters
estuaries, 143–44

ethane, 93–94, 438, 449–50, *892*
ethanethiol, 464
ethanol, 464, 632–33
  in alcoholic beverages, 626, 632
  as biofuel, 987
  boiling point, 483–84, 493
  enthalpy of combustion, 632
  fuel value, 632–33, 987
  hydrogen bonds, 483–84, 491, 493
  miscibility in water, *489*, 491, *631*
  molecular mass, 201
  molecular structure, *202*, 532
  as nonelectrolyte, 146–47
  production and synthesis, 632, 633
  structure, *8*
  use in gasoline, *631*, 632–33, 987
ethers, 630–31, 634–35
  diethyl ether, 634–35, 636
  dimethyl ether, 464–65, 481, 483–84, 529
  functional group, *605*
  methyl *tert*-butyl ether (MTBE), 635, 687
  polymers of alcohols and ethers, 635–37
ethyl butyrate, 640
ethyl group, 614
ethylene (ethene)
  enthalpy of formation, *233*, 234
  melting point, 606
  polymerization, 606, 623–24
  and ripening process, *424–25*, 449
  structure, 449–50, 451
ethylene glycol, *484*
  in antifreeze, 104, 512, 514, 516
  boiling point, *484*, 493
  solubility, 491–92
  structure, *484*
  vapor pressure of solutions, 515–16
ethylenediamine, 859, *861*, 862, 866, 871
ethylenediaminetetraacetic acid (EDTA), 152, 859–61, *1066*
eugenol, 109
EVAL (ethylene and vinyl acetate copolymer), 636
exact (countable) numbers, 22, 24–25
excited state, 330
exothermic processes, 205–8, 759–60, 884
  *See also* heat (*q*)
extensive properties, 12

# F

face-centered cubic (fcc) unit cell, *565*, 566, 567–68, 569–70, 573–74, 578–79
Fahrenheit temperature scale, 29–30
Faraday, Michael, 938
Faraday's constant (*F*), 938, 948, 949, 955
fatty acids, 989–92
  β-oxidation, 991
  elaidic acid, 990
  hydrogenation, 118, 990, 998–99
  linoleic acid, 990, 998–99
  linolenic acid, 989
  myristic acid, 991

fatty acids *(cont.)*
names and formulas, *989*
oleic acid, 990, 998–99
omega-3 fatty acids, 989
palmitic acid, 991, 998–99
polyunsaturated fatty acids, 989
saturated fatty acids, *989*, 990, 998–99
stearic acid, *990*, 991, 999
trans fats, 118, 990
in triglycerides, *988*, 989–90
unsaturated fatty acids, *989*, 990, 998–99
*See also* lipids
femtochemistry, 700
fermentation, 632, 633, 634, 987, 988
ferrite, 573
filtration, 8–9
fireworks, 350, 361, 505, 525
first law of thermodynamics, 209–10, 214, 222–23, 882, 913
first-order reactions, rate laws, 681, 683–86
fluorapatite, 1059
fluoride for dental caries prevention, 410, 1059, 1076
fluoride ion in seawater, 137
fluorine, 410–11, 504, 1059
diatomic element, 88
DRI/RDA value, *1049*
as essential trace element, *1048*, 1059
reaction with water, 410
standard reduction potential ($E°_{red}$), 934
as strong oxidizing agent, 410, 934
fluorite (calcium fluoride), 578–79
fluorite structure, 579–80
fluorocarbons, 410
5-fluorouracil (5-FU), 503
Fluosol-DA ($C_{10}F_{18}$), 99
food energy, 909
food value, 242–44
forces, intermolecular, 474–512
adhesive, 499
capillary action, 500
cohesive, 499
combinations of intermolecular forces, 490–92
coulombic interactions, 202–3, 477–79
dipole–dipole interactions, definition, 481
dispersion forces (London forces), 484–88
electrostatic potential energy ($E_{el}$), 202–3, 374–75, 477–79
hydrogen bonds, overview, 481–84
ion–dipole interactions, 480, 481, 488–89
ion–ion interactions in salts, 477–79
and melting and boiling points, 476
relative strengths, *485*
strength of ion–molecule interactions, 479–80
surface tension, 499–501
van der Waals forces, 981–82
viscosity, 501
formal charge, 391–95
formaldehyde
bond angles, 431
C–H bond length, 404
dipole moment, 439–40
molecular geometry, 430–31
pi ($\pi$) bonding, 444
$sp^2$ hybrid orbitals, 443–44
structure, *8*, *104*, 380–81, 430–31, *638*
formation reactions, 232, 233–35, 237
formic acid, 639, 783, 791–93
formula mass, 82–85
formula unit, definition, 55, 82
fossil fuels
and climate change, 95, 118
combustion, 95–97, 246, 618
formation, *95*
natural gas, use as fuel, 92–94, 95, 96
reduced carbon in, 246
source of methane, 236
use in hydrogen production, 236
fractional distillation, 363, 528–31
Fraunhofer, Joseph von, 321–22
Fraunhofer lines, 321–23
free energy (*G*), 895–907
ammonia synthesis, 897–98
calculating free energy changes, 896–99
and chemical equilibrium, 902–6
chemical free energy ($\Delta G_{cell}$), 937–40
coupled reactions, 909–13
effects of $\Delta H$, $\Delta S$, and $T$, 895–98, 900–902
and equilibrium constants (*K*), 904–7
Gibbs free energy, definition, 895–96
glycolysis, 910–13
in life processes, 909–13
and mechanical work, 895, 899–900
and reaction quotient (*Q*), 902–4, 942, 944
relation to *K* and *E*° in electrochemical cells, *946*
in reversible processes, 903, 905
standard free energy of formation ($\Delta G°_f$), 898–99
standard free energy of reaction ($\Delta G°_{rxn}$), 898–99
temperature and spontaneity, 900–902
thermodynamics of melting ice, 900–901
*See also* spontaneous processes
free radicals
chlorine atoms, 411
chlorine monoxide, 411
definition, 396
hydroxyl, 669, 1024
nitric oxide (NO), 396
and radioactivity, 1024
freezing point depression, 535–37, 538–40, 548–49
freezing point depression constants ($K_f$), 536, 540, *549*
frequency factor (*A*), 694, 695, *697*
frequency ($\nu$), 318, 319, 361
fructose, 104, 984, 985
fuel cells, 956–59
basic electrolytes in, 957–58
catalysts, 715, 956, 957
definition, 956
in electric vehicles, 926
energy efficiency, 958–59
in Honda FCX Clarity, *958*
hydrogen as fuel, 118, 235, 956, 958–59
proton-exchange membrane (PEM), 956–57
stacks, 957, *958*
fuels
fuel density, 241–42
fuel values, 118, 240–42, 632–33
melting points and boiling points of straight chain alkanes, *609*
*See also* fossil fuels
Fukushima, Japan, nuclear reactor explosions and fire, 1023
Fuller, R. Buckminster, 196, 246, 582
fullerenes, 246, 580, *581*, 582
functional groups (organic chemistry), 604–5, *634*
*See also* specific types
fusion reactions. *See* nuclear fusion

**G**

gadolinium, *1046–47*, 1067, 1070, 1071
gallium, 562, 576, 1067, 1070, 1073
gallium arsenide, 562, 576–77
gallium chloride, 408–9
gamma rays, 319, 335, 1014, 1024–25, 1028
gas density, 260, 280–83
gas laws, 265–77
Amontons's law, 271–73, *275*, 289, *290*
Avogadro's law, 270–71, 274, *275*, 279, 288, 289
Boyle's law, 265–68, 271, 274, *275*, 288, 298–99
Charles's law, 268–70, 271, 274, *275*, 289, *290*
Dalton's law of partial pressures, 282, 283–87, 288–89, 298
gas density, 260, 280–83
general gas equation (combined gas law), 274
Graham's law of effusion, 292, 293–95
ideal gas equation, 273, 277
ideal gas law, 273–77, 295
ideal gas law and stoichiometry, 277–83, 286–87
molar volume, 274, 275–76, 278
standard temperature and pressure (STP), 274, 275
universal gas constant (*R*), 273, 291
van der Waals equation, 297–98
gases, 258–315
characteristic properties, 13
in chemical reactions, 277–83
in Earth's early atmosphere, 74, 76–77, *87*, 111
gas phase, 260
ideal gases, definition, 273
importance of, 258–60
properties, 260
real gases, deviation from ideality, 295–96
solubility in water, 517–19, 551

gases *(cont.)*
  viewed at atomic level, 14
  *See also* kinetic molecular theory of gases
gasoline
  distillation from crude oil, 527
  ethanol additive, 630, 632–33, 987
  fuel value and fuel density, 241–42
  hydrocarbons in, 527, *609*, 618, 626
  reformulated gasoline, 630–31
Geiger, Hans, 44
Geiger counters, 44, 1022
general gas equation, 274
genetic code, 995
geothermal vents, 997–98
germanium, 577, 580, 590–91, 1065
giant stars, 64, 74, 1010, 1011
Gibbs, Willard, 895
Gibbs free energy. *See* free energy (*G*)
glass, 583, 590, 591, 1058
global warming. *See* climate change
glucose
  aldehyde group, 984
  α-glucose structure, 985
  β-glucose structure, 985
  in cellulose, 246
  D5W solution (dextrose), 545–46
  empirical formula, 104
  enthalpy of combustion, 225
  food value, 243–44
  formula, 94, 104, 110
  glycolysis, 910–13, 988
  molecular structure, 984–85
  phosphorylation to glucose-6-phosphate,
    910–13
  from photosynthesis, 83, 94, 110, 243,
    246, 910
  and respiration, 94, 112
  in starch, 984, 986–87
  *See also* carbohydrates
glutamate, *1063*
glutamic acid, 970, *971*, 978, 979
glutamine, *971*
glycerides, 988–91
glycerol, 516, 988, 989–90, 991, 992
glycine
  biosynthesis, 1056
  in comets, 997
  from Miller/Urey experiment, 87, 113–14
  nonchiral molecule, 970, 972
  in peptides, 976, 996
  precursor molelcules, 87, 463
  structure, 463, *971*
glycolaldehyde, 103–4
glycolic acid, 642
glycolysis, 910–11, 910–13, 988
glycoproteins, 984
glycosidic bonds, 985–87
gold
  alloys, 560, 571
  in antitumor drugs, 1075
  in arthritis drugs, 1046, 1076

atomic structure, *46*
crystal structure, 564
density, 21–22, 23
malleability, 563, 571
nanoparticles, 560, 563
properties, 12, 13, 560, 563
Rutherford gold foil experiment, 44–45
Goudsmit, Samuel, 340
Graham, Thomas, 293
Graham's law of effusion, 292, 294–95
Grand Canyon, 134
grapefruit aroma, 465
graphene, *581*, 582
graphite, 246, 580–82
  anodes in lithium–ion batteries, *948*, 949
  standard molar entropy (*S°*), 892
  standard state of carbon, 233
  structure and properties, 246, 580–82, 892
  *See also* carbon
Graves' disease, 1059
gravitational potential energy (PE), 199–200
gravity and atmospheric pressure, 210, 260,
    262–63
gray (Gy), 1024–25
Great Salt Lake, 140–41
greenhouse effect, 372–74, 386–87
  *See also* climate change
greenhouse gases, 372–74
  absorption of infrared radiation, 372
  carbon dioxide, 372–74, 618
  carbon dioxide from fossil fuel combustion,
    95–97, 280
  dinitrogen monoxide, 686
  increase in atmospheric CO2, 280, 372
  methane, 639
  sulfur hexafluoride, 397
guanine, *483*, 994
Guldberg, Cato, 738
gypsum (calcium sulfate), 182

**H**

Haber, Fritz, 301, 760
Haber–Bosch process, 114–15, 118, 301, 760–61,
    768
half-lives (*t*₁/₂) of chemical reactions, 686–87,
    690, 694
half-lives (*t*₁/₂) (radioactive decay), 1017–18,
    1020, 1023, 1029, 1031, 1035, 1036–37
half-reactions for balancing redox reactions,
    172–74, 176–77, 178, 179, 928–30
halides, 162
  binary acids, as strong acids, 781
  binary acids, naming, 62
  bond polarity of hydrogen halides, 384–86
halite structure, 101
Hall, Charles M., 238
Hall, Julia, *238*
Hall–Héroult process, 238–39
halogens, 504–5
  as diatomic elements, *88*, 354

electron affinities, 360
electronegativity, *384*
molar masses and boiling points, 485, *486*
oxoanions, 505
in periodic table, *51*, 52
physical properties, *504*
haloperoxidases, 1062
halothane (C2HBrClF3), 410
hard water, 165–66
heat (*q*)
  from chemical reactions, 196
  from combustion of hydrogen, 203
  definition, 198, 206
  endothermic processes, 205–8, 884
  exothermic processes, 205–8, 884
  sign convention, 206, 243
  *See also* enthalpy (*H*)
heat capacity (*C*P), 215
  *See also* molar heat capacity (*c*P); specific heat (*c*s)
heat (enthalpy) of reaction. *See under* enthalpy (*H*)
heat transfer, 198, 209, 219–22, 223, 225
  entropy, 888
  in phase changes, 205–8
  sign convention, 206, 210, 243
  *See also* calorimetry; energy transfer
heating curves, 214–22
heavy water (D2O), 118–19
Heisenberg, Werner, 316, 335–36
Heisenberg uncertainty principle, 335–37, 354
helium, 118–19, 362
  absorption spectrum, *323*
  abundance, *5*, 118
  alpha (α) particles, 44
  atomic and molar mass, 80
  balloons, 119, 362
  binding energy (BE), 1010–11
  as cryogen, 119
  density, *280*
  discovery, 52, 362
  electron configuration, 346
  in gas deposits, 119
  helium-3, 1033, 1072
  ionization energy, 357
  mass defect (Δ*m*), 1010–11
  molecular orbital theory, 455–56
  primordial nucleosynthesis, 63–64, 64,
    1010, 1032
  *Spirit of Freedom* balloon, 211
  from stellar nuclear fusion, 64, 1033–34
  uses, 119
hematite (Fe2O3), 77, 84–85, 101, 168, 177
heme group
  heme iron, 844, 873–74, 1061
  O2 binding, 873–74
  porphyrin, 873–74, *981*, 982, 1061
  structure, *874*
hemoglobin
  heme group, 873–74
  oxygen binding, *512–13*, 519
  oxygen transport, *512–13*, 519, 873–74
  primary (1°) structure, 977–78, *979*

hemoglobin *(cont.)*
  quaternary (4°) structure, *874, 978,* 981–82
  secondary (2°) structure, *978*
  and sickle-cell anemia, 977–79
  tertiary (3°) structure, *978,* 981
Henderson–Hasselbalch equation, 808–9,
    810–13, 815–17, 818, 820, 824, 832
Henry, William, 518
Henry's law, 517–19, 551
Henry's law constants, 518, *519*
HEPA (high-efficiency particulate air) filters, 9
heptane, 528–31
2-heptanone, 638
Héroult, Paul Louis-Toussaint, 238
hertz (Hz), 318, 319
Hess's law, 229–33, 237–38
heteroatoms, 629
heterogeneous equilibria, 752–54
heterogeneous mixtures, 5
heteropolymers, 636, 637
hexagonal closest-packed (hcp) structures, 564, *565*
hexagonal unit cell, 564, *565,* 567
hexamethylenediamine, *644*
1,3,5-hexatriene, *628*
hexene, 622–23
HFC-134a ($C_2F_4H_2$), 99
HFC-143a ($C_2F_3H_3$), 100–101
high-density polyethylene (HDPE), 623–24
high-efficiency particulate air (HEPA) filters, 9
*Hindenburg,* 362
histidine, 970, *971,* 1061
homocysteine, 1063
homogeneous equilibria, 752
homogeneous mixtures, 5
homologous series, definition, 608, 609
homopolymers, 623
Honda FCX Clarity, *958*
honey, *910*
honeybees, *910*
hot-air balloons, 209–10, 265, 269, 281, 299
Hoyle, Fred, 17
Hund's rule
  atomic orbitals, 347
  complex ions (complexes), 863, 867
  molecular orbital (MO) theory, 457
hybrid orbitals
  in complex ions, 849–51
  *sp* hybrid orbitals, 445–47
  $sp^2$ hybrid orbitals, 443–45
  $sp^3$ hybrid orbitals, 442–43, 849–50
  $sp^3d$ hybrid orbitals, 447–48
  $sp^3d^2$ hybrid orbitals, 447
  summary of hybridization schemes and the
    orientations of orbitals, *448*
hybrid vehicles, 926
  batteries, 926, 946–50
  driving range, 926
  plug-in hybrids, 926
  Toyota Prius, 947
  *See also* electric vehicles
hydrangeas, 778, *779*

hydrated metal ions. *See* complex ions
    (complexes); metals
hydration, sphere of, 480, 481
hydrazine, 172, 301, 450
hydrides, 464, 481–82
hydrobromic acid, 62
hydrocarbons
  branched-chain hydrocarbons, 611–12, 614–16
  combustion analysis, 106–8
  combustion reactions, 92–94
  definition, 92, 606–7
  fuel density, 241–42
  fuel values, 241–42
  hydrogenation of unsaturated hydrocarbons,
    606, 607, 608
  saturated hydrocarbons, 606, 607
  unsaturated hydrocarbons, 606, 607
  *See also specific types*
hydrochloric acid
  hydronium ion formation, 148, *781,* 784
  neutralization reaction, 150
  reaction with water, 148
  in stomach acid, 1055–56
  *See also* hydrogen chloride
hydrofluoric acid
  acid–base chemistry, 801, 816–17
  hydrogen bonding, 490
  precipitation reactions, 161–62
  *See also* hydrogen fluoride
hydrogen, 118–19
  absorption spectrum, *323,* 331
  abundance, 118
  Balmer equation, 328
  Bohr model of hydrogen, 329–32
  bond energy (bond strength), 374, 1011
  bond length, 374, *375*
  commercial uses, 118–19
  deuterium ($^2H$ or D), 63, 118–19, 1032, 1033,
    1034, 1035
  diatomic element, 7, 88
  from electrolysis of water, 118
  electron configuration, 346
  electrostatic potential energy ($E_{el}$) of H–H
    bond, 374, *375*
  emission spectrum, 328–29, 331, 337, 339–40
  energy level diagram, *331*
  as essential element, 118, 1046, 1048
  exceptions to octet rule, 375, 376
  freezing point, 12–13
  as fuel, 115, 118
  in Haber–Bosch process, 118
  heat energy from combustion, 203
  hydrogen-powered vehicles, *203*
  ionization energy, 331–32, 357
  molecular ions ($H_2^-$), 456
  molecular orbitals of $H_2$, 455, 456
  orbital overlap and σ bond formation,
    440–41
  primordial nucleosynthesis, 63, 64
  rocket fuel, *203*
  Rydberg equation, 328–29

  from steam–methane reforming, 118, *230,*
    631, 734
  stellar fusion, 64, 1032–33, 1033–34
  tritium ($^3H$), 118–19, 1034–35, 1036, 1072
  use in fuel cells, 118, 235, 956, 958–59
  from water–gas shift reaction, 115–16, *230*
hydrogen bonds, 481–84
  amides, 640
  amines, 630
  and boiling point, 481–82, 483–84, 488
  carbohydrates, 984
  carboxylic acids, 639–40
  DNA, 482, *483,* 994
  in Kevlar, 646
  poly(ethylene glycol) (PEG), 637
  polymers, 644, 646
  proteins, 975, 979–80
  surface tension, 499
hydrogen bromide, *385*
hydrogen chloride
  atomic orbitals and covalent bond
    formation, 441
  bond polarity, 383
  enthalpy of formation reaction ($\Delta H_{rxn}$), 407–8
  formation of $Cl^-$ and $H_3O^+$ in water, 148,
    *781,* 784
  polarity, 384, *385*
  standard heat of formation, 407
  in stratosphere, 709
  *See also* hydrochloric acid
hydrogen fluoride
  dipole moment ($\mu$), *438*
  hydrogen bonding, 481, *482,* 490
  melting and boiling points, *499*
  polarity, 384, *385*
  solubility, 489–90
  *See also* hydrofluoric acid
hydrogen iodide, *385,* 761–63
hydrogen peroxide
  decomposition, *670,* 685
  fatty acid degradation, 1061
  Lewis structure, 377
  molecular structure, *56, 403*
  and superoxide dismutases, 1063
  and xanthine oxidase, 1062, 1063
hydrogen phosphate ion, in glycolysis, 911–12
hydrogen selenide ($H_2Se$), 464
hydrogen sulfate ion, 152
hydrogen sulfide, 434, 440, 464, 465
hydrogen telluride ($H_2Te$), 464
hydrogenases, 1063
hydrogenation, 607, 990
hydroiodic acid, 62
hydrolysis, definition, 150, 151
hydronium ion
  clusters of $H_2O$ molecules around hydronium
    ions, *781*
  as conjugate acid of water, 782
  formation from protons in water, 148, 490,
    780, 781, 784
  ion channels, 1051

hydrophilic interactions, 491
hydrophobic interactions, 491–92
hydrothermal vents, 132, 997–98
hydroxide ions
  Brønsted–Lowry bases, 148, 783, 784–85, 857–58, 1059
  neutralization reactions, 150–51
  in oxidation–reduction equations, 936, 937, 956–57
  pOH, 788–90
  pOH and pH, 788–89, 794, 795
hydroxyapatite, 1059
hydroxyl radicals, 669, 1024
hypertension, 1051
hypertonic solutions, 541, 545
hypochlorites, 182, 505, 801, 802–3
hypotheses, 17
hypotonic solution, 541, 545
hypoxia, 518

**I**

ibuprofen, 640
ice
  density, *501*
  macroscopic and molecular-level views, *217*
  molar heat capacity ($c_P$), 215, 220
  molar heat of fusion ($\Delta H_{fus}$), 212, 216, 217, 218
  thermodynamics of melting ice, 882, 884–85, 888, 891, 900–901
  *See also* water
Iceman (Otzi), 915
ideal gas law, 273–77, 295
  and stoichiometry, 277–83, 286–87
  ideal gas equation, 273, 277
  ideal gas law equilibrium constants, 742–45
ideal gases
  definition, 273
  deviations from ideality, 295–96
  molar mass and gas density, 282–83
ideal solutions, 514, 515
implantable cardioverter-defibrillators (ICDs), 1078
indium, 1067
infrared (IR) radiation
  in electromagnetic spectrum, *31*, *318*
  emission from warm objects, 31, 325
  hydrogen emission lines, 328, *331*
  vibrating bonds and greenhouse effect, 386–87
  wavelengths, 320
inhibitors, 983
inner coordination sphere, definition, 849
insoluble compounds, definition, 159
intensive properties, 12, 140
interference (waves), *320*, 321
intermediates, 700–701, 703, 708
intermetallic compounds, 571
intermolecular forces. *See* forces between ions and molecules

internal energy ($E$), 208, 209–10, 212, 217, 228, 233
interstellar dust, 2, 4, 628
International Union of Pure and Applied Chemistry (IUPAC), 275, 613
iodine, 504–5, 1059
  DRI/RDA value, *1049*
  as essential trace element, 504, 1046, *1048*, 1059
  [123]I as imaging agent, 504–5
  [123]I for thyroid cancer therapy, 1028–29
  [125]I as imaging agent, 504–5
  [131]I and health risks, 1026
  [131]I as imaging agent, 1067
  [131]I for thyroid cancer therapy, 504, *1029*
  lithium/iodine batteries, 1077
  oxidation–reduction reactions, 174
  thyroid hormones, 505, 1059
  toxicity, 504
iodine monochloride (ICl), decomposition, 768
ion channels, 1050, 1051–52, 1055, 1064
ion exchange, 165–67
ion pairs, 539, 541
ion pumps, 1051–53, 1055
ion–dipole interactions, 480, 481, 488–89
ionic bonds
  and bond polarity, 383, 385–86
  definition, 374
  types of bonds, comparison, *376*
  valence electron transfer, 383
  *See also* chemical bonds
ionic character of bonds, 383, 403
ionic compounds
  binary ionic compounds, overview, 55–57
  Born–Haber cycle, 522–25
  crystalline structure of ionic solids, 577–80
  definition, 54, 55
  energy changes during dissolution of, 520–21
  formula mass of, 82, 83
  ionic solids, definition, 562
  lattice energy ($U$), 521–27
  Lewis structures, 381–82
  making insoluble salts, 157–60
  naming binary ionic compounds, 58–60
  naming binary transition metal compounds, 59–60
  naming polyatomic ion compounds, 60–62
  solubility rules, 157–58
  valence electron transfer, 383
ionic radius, 354, 355, *356*
ion–ion interactions, potential energy, 477–79
ionization energy (IE), 357–60
  definition, 332, 357
  and electronegativity, 383–84
  first 10 elements, *359*
  hydrogen atom, 331–32, 357
  and nuclear charge, 358–59, 383
  and periodic table, 358–60, 383, *384*
  successive ionization energies of elements, 359
ionizing radiation, 1024–29

ions
  charges, from periodic table, 55–56
  definition, 47
  dissolved ions in living cells, 132
  electron configuration, main group elements, 351–52
  electron configuration, transition metals, 353–55
  interactions in salts, 477–79
  ion pairs in solutions, 539, 541
  ionic radius, 354, 355, *356*, 477–79
  molecular ions, 106, 107, 109, 110, 453–54, 456, 460, 462
  monatomic ions, 55–56
  oxoanions, 60–62
  in positive-ray analyzer, 47
  *See also* complex ions (complexes); polyatomic ions
iridium, 715
iron
  anemia treatment, 1073–74
  atomic radius, 573
  austenite structure, 572–73
  chelation therapy, 1073–74
  corrosion, 560, *882–83*, 950–52, 959, 960
  in cytochromes, 874, 1061
  DRI/RDA value, *1049*
  electron configuration of cations, 353
  enantiomers of complex ions, 1074
  energy cost of recycling, 239–40
  in enzymes, 1059, 1061–62, 1063
  as essential trace element, 844, 1046, *1048*, 1061–62
  [56]Fe nuclear stability, 64, 1010, 1011
  ferrite structure, 573
  heme iron, 844, 873–74, 1061
  ligand exchange reactions, 1074–75
  net ionic equation of oxidation–reduction reactions, 929–30
  oxidation–reduction reactions, 168, 929–30
  spin states of complex ions, 867
  stellar nucleosynthesis, 64, 1010, 1011
  *See also* steel
iron compounds
  enantiomers of complex ions, 1074
  $Fe(H_2O)_6^{3+}$ complex ion, 850, 857
  hematite, $Fe_2O_3$, 77, 84, 101, 177
  iron ore, reduction to iron, 168
  iron(II) sulfide, 787–88
  iron(III) hydroxide, 857–58
  iron(III) oxide (rust), 177, *882–83*, 884
  iron(III) sulfate, $Fe_2(SO_4)_3$, 77
  magnetite, $Fe_3O_4$, 84–85, 168
  oxidation–reduction reactions, 177–80
  in soil, *177*
  wustite, FeO, 101
isobars (pressure), 262
isobutane, *487*
isoelectric point (p$I$), 973, 974
isoleucine, *971*

isomers
    alkanes, 611–13, 614–16
    alkenes, 620–21
    alkynes, 620
    aromatic compounds, 627–28
    branched-chain hydrocarbons, 611–12, 614–16
    cis isomers, 620, 621, 869
    constitutional (structural) isomers, definition, 869
    constitutional (structural) isomers, overview, 611–13
    of coordination compounds, 870–71
    definition, 452
    $E$ isomers, 620, 621
    optical isomers, 452–53, 647–51, 984
    stereoisomers, definition, 620, 621, 647, 869
    trans isomers, 620, 621, 869
    $Z$ isomers, 620, 621
    *See also* enantiomers
isopropanol, *484*, *487*
isotonic solutions, 541, 545
isotopes
    and average atomic mass, 49
    carbon isotopes and radioactive decay products, *1015*
    definition, 47, 1033
    discovery, 46–47
    natural abundance, 49
    notation and format, 47
    radioactive isotopes, 65–66
    *See also* nuclides

**J**

Joliot-Curie, Irène, 1024
joule (J) definition, 208, 209, 946

**K**

Kalkwasser solution, 141–42
kaolinite, 584, *585*
Kekulé, August, 609
Kekulé structures, 609, 629
Kelvin temperature scale, 29–30
    absolute zero (0 K), 30, 268, 269, 270
    gas laws, 268–70
    *See also* temperature
keratins, 979, *981*, 982
kerosene, 242
ketones, *605*, 637, 638
Kevlar, 645–46
kinetic energy (KE)
    conversions between potential and kinetic energy, 200
    definition, 200, 201
    formula, 200, 201
    gases, 288, 289, 290–93
    at molecular level, 201–4, 288, 289, 290–93

photoelectrons, 326
    and temperature, 201, *202*, 204, 207–8, 288, 289, 290–93
    thermal energy, 201
kinetic molecular theory of gases
    assumptions, 288, 292, 295, 296
    average speed ($u_{avg}$) of molecules, 201, 272, 290, 292
    definition, 288
    distribution of speeds of molecules, 290–91
    elastic collisions, 288
    molecular speeds and kinetic energy, 290–93
    most probable speed ($u_m$) of molecules, 290–91
    root-mean-square speed ($u_{rms}$) of molecules, 291–94
    solubility of gases in water, 517–18
kinetics, chemical. *See* chemical kinetics; reaction rates
Kirchhoff, Gustav Robert, 322
Kroto, Harold, 246
krypton, 347, 363, 1072

**L**

lactase, 982
lactic acid, 642, 912
lactose, 982
lactose intolerance, 982
lanthanides
    electron configuration, 350, 353–54
    gadolinium, *1046–47*, 1067, 1070, 1071
    magnetic resonance imaging, 1071
    in periodic table, *51*, *52*, 350
    scintigraphic imaging, 1069–70
lanthanum, 52, 1069
lasers, 342–43
lattice energy ($U$) of ionic compounds, 521–27
law of conservation of energy, 200, 201, 208
law of conservation of mass, 77, 86
law of constant composition, 6
law of mass action, 738
law of multiple proportions, 53–54
Le Châtelier, Henri Louis, 754–55
Le Châtelier's principle, 754–61
    acid–base equilibria, 808, 814
    adding or removing products or reactants, 755–56
    pressure changes, 756–59
    solubility equilibria, 827, 829, 1059
    temperature changes, 759–60
    volume changes, 756–57
    *See also* chemical equilibrium
lead, 591, 1050
    crystal structure, 580
    electron configuration, 350
    lead acetate, 1050
    lead-206, 1016, *1017*, 1026
    lead-214, 1016, 1027

lead–acid automobile batteries, 591, 943–44, 954
leaded crystal, 590
lead(II) carbonate, 1050
lead(II) chloride, 591, 829–30, 831
lead(II) chromate, 1050
lead(II) dichromate, 160
lead(II) fluoride, 831
lead(II) iodide, 158–59
lead(II) nitrate, 158–59
lead(IV) oxide, 591, 943
    solubility and precipitation of compounds, 158–59, 160, 163, 829–30
    tetraethyl lead [$(CH_3CH_2)_4Pb$], 1050
    toxicity, 1049, 1050, 1074
lean mixture (fuel combustion), 113
legumes, 300, *301*
Lemaître, Georges-Henri, 17
leucine, *971*
Lewis, Gilbert N., 375–76, 382, 847
Lewis acids, overview, 846–48
Lewis bases, overview, 846–48
Lewis structures, 375–82
    atoms with more than an octet, 397–400
    bonding capacity, definition, 376
    bonding pairs, 377
    comparison to ball-and-stick model, *426*
    definition, 377
    double bonds, 379–81
    electron-deficient compounds, 376–77
    formal charge, 391–95
    guidelines for drawing, 377–79, 384
    ionic compounds, 381–82
    Kekulé structures, 609, 629
    Lewis symbols (Lewis dot symbols), 376–77
    lone pairs, 377
    odd-electron molecules, 395–97
    organic molecules, 609–10
    resonance structures (resonance hybrids), 388–91
    single bonds, 377
    steps for drawing, 377–79, 387
    triple bonds, 379–81
    *See also* covalent bonds; octet rule
Libby, Willard, 1019
life
    diversity of life, *968–69*
    entropy changes ($\Delta S$) in life processes, 910
    free energy changes ($\Delta G$) in life processes, 909–13
    origins of life, 74, 132, 997–98
    and the periodic table, 1046–85
    water and aquatic life, *132–33*, 502
ligands
    chelate effect, 862
    chelating agents, 858, 859–60, 1066–67, 1069, 1073–74
    chelation, 858, 859, 861–62
    deferoxamine (DFO), 1074–75

ligands (cont.)
  definition, 848, 849
  diagnostic and therapeutic agents, 1066–67
  diethylenetriamine, 859
  diethylenetriaminepentaacetate
    (DTPA$^{5-}$), 1066
  displacement reactions, 861, 862
  electron-pair donor groups, 846, 847–48, 859,
    861–62
  ethylenediamine, 859, 861, 862, 866, 871
  ethylenediaminetetraacetic acid (EDTA),
    859–61, 1066
  iron in ligand exchange reactions, 1074–75
  Lewis bases, 848, 849, 860, 861–62
  macrocyclic ligands, 872–74
  monodentate ligands, 858, 859, 862, 871
  names and structures, 854, 871
  nitrilotriacetic acid (NTA), 860–61
  polydentate ligands, 858, 859–61, 862, 871,
    872–74
  spectrochemical series of common ligands,
    866, 869
  strengths of ligands, 861–62
  See also complex ions (complexes)
light
  absorbance (A), 144–46, 1054
  diffraction, 182, 320–21, 327
  particle properties, 327–28
  photons, 325
  plane-polarized light, 452, 648–49, 650,
    651–52
  refraction, 320
  speed of light (c), 17, 28, 319
  visible spectrum, 318, 323
  wave properties, 318–20, 327
  See also electromagnetic radiation
lime (quicklime; CaO)
  ion–ion attractive forces, 478–79
  limelights, 182
  from limestone, 752–53
  production, 182
  scrubbing sulfur oxides from exhaust gases,
    182–83, 752, 847
  in steelmaking, 182
  use to "sweeten" soil, 752, 778
limestone
  decomposition to lime, 182, 752–53
  limestone kilns, 752
  stalagmites in Carlsbad Caverns, 151
  See also calcium carbonate
limiting reactants, 110–16, 163
limonene, 108, 647–48
linear molecular geometry, 427–28, 429, 436,
    445–47, 448
linoleic acid, 990, 998–99
lipid bilayers, 992, 993, 1050, 1051, 1052, 1062
lipids, 988–92
  coupled reactions in fat metabolism, 913
  definition and properties, 988, 989
  fats, 990, 991

  glycerides, 988–91
  hydrogenated oils, 118, 990, 998–99
  lipid bilayers, 992, 993, 1050, 1051, 1052, 1062
  metabolism, 913, 990–91
  oils, 118
  oils, definition, 990, 991
  phospholipids, 991–92, 1051, 1052
  trans fats, 118, 990
  triglycerides, 988, 989–90, 991
  See also fatty acids
liquefaction, 495
liquids
  characteristic properties, 13
  comparison of liquids, gases, and solids, 260
  miscibility of liquids, 491–92
  surface tension, 499–500
  vapor pressure of pure liquids, 492–94
  viewed at atomic level, 14
  viscosity, 501
  See also solutions
lithium
  drugs for depression, 1046, 1072
  electron configuration, 346
  in fireworks, 361
  ionization energy, 359
  lithium silver vanadium oxide batteries, 1078
  lithium/iodine batteries, 1077
  lithium–ion batteries, 948–50
  standard reduction potential ($E°_{red}$), 934
  as strong reducing agent, 934
  toxicity, 1072
lithium aluminum hydride, 402
lithium carbonate, 83–84, 361
lithium chloride, crystal structure, 579–80
lithium perchlorate, 538–39
London, Fritz, 484
London forces, 484–88
longleaf pine (Pinus palustris), 98
Los Angeles, smog, 736
low-density polyethylene (LDPE), 623–24
Lowry, Thomas, 780
lycopene, 105
Lyman, Theodore, 328
lysine, 970, 971, 976

## M

macrocyclic ligands, 872–74
magnesium, 1053–54
  in chlorophyll, 844–45, 873, 1053, 1054, 1064
  DRI/RDA value, 1049
  electron affinity, 360
  as essential element, 1046, 1048, 1049,
    1053–54
  ionization energy, 357
  from $MgCl_2$ electrolysis, 950
  valence band, 575
magnesium hydroxide, 825–26, 830
magnesium oxide, 521, 525, 953
magnetic resonance imaging (MRI)

  contrast agents, 1046–47, 1067, 1071
  helium use, 119
  medical diagnosis, 1046–47
  superconducting magnets, 119, 585
magnetite ($Fe_3O_4$), 84–85, 135, 168
malleability
  copper, 571, 572
  definition, 12, 52
  gold, 12, 13, 563, 571
  metals, 52, 563, 571
  polymers, 606
maltose, 231–32
manganese
  DRI/RDA value, 1049
  as essential ultratrace element, 1048, 1064
  and photosynthesis, 1064
  spin states of complex ions, 867–68
manometers, 263–65
margarine, 118, 990
Mars, 33
Mars, surface water, 134
Mars Reconnaissance Orbiter, 33
Marsden, Ernest, 44
Martian meteorites, 134
Martian rovers, 33, 134
mass
  definition, 4
  mass–energy formula, 17, 332, 1010, 1030,
    1034, 1035
  molar mass, 80–82, 548–50
  molecular mass and formula mass, 82–85,
    105, 107, 109
  mole–mass–particle conversions, summary,
    82–83
mass action expressions, 738, 750–52, 755
  See also equilibrium constants (K); reaction
    quotients (Q)
mass defect (Δm), 1010–11
mass number (A), 47
mass spectrometer, 47, 106, 107, 363
mass spectrometry
  and atomic mass, 47
  mass spectrum, definition, 106, 107
  mass/charge (m/z) values, 106, 107
  molecular formula determination, 106, 107,
    109, 110
  molecular ions, 106, 107, 109, 110
  and molecular mass, 106, 107, 109
  positive-ray analyzer and, 47, 106, 363
  radiometric dating, 1019, 1021
mass–energy formula, 17, 334, 1010, 1030,
    1034, 1035
matter
  atomic view, 6–8
  classes of, 4–6
  definition, 4
  matter waves, 333–35, 337, 344
  properties, 12–13
maximum contaminant levels (MCLs), 136n1
Maxwell, James Clerk, 319, 324

measurement
    importance in science, 18
    precision and accuracy, 25
    SI units, 18–19
    significant figures, in calculations, 21–25
    significant figures, overview, 19–21
    unit conversions and dimensional analysis,
        26–28, 33
medical application of radionuclides,
    1028–29
    *See also* radiation biology; radionuclides
medical devices and materials, 1076–79
medications. *See* pharmaceuticals
Mediterranean diet, 990
Meissner effect, 586
melting points
    compounds with mass similar to water, *499*
    and lattice energy, 521–22
    *See also* freezing point depression
Mendeleev, Dmitri, 50, 52, 65
meniscus, 22, 499–500, 503, 591
menthol (oil of mint), 633
mercury, 1050
    dental amalgams, 1050, 1076
    emission and absorption spectra, *323*
    meniscus, 499, 500
    methylation, 1050
    in Minamata Bay, Japan, 1050
    reduction by $Sn^{2+}$, 945
    threshold frequency ($\nu_0$), 326
    toxicity, 1049, 1050
    work function ($\Phi$), 326
    Zn/HgO batteries, 1076–77
mercury(II) sulfide, synthesis, 162
metabolism, 982, 990–91
metallic bonds, 375, 574–75
    band theory, 574–75, 584
    delocalized electrons in bonds, 375
    sea-of-electrons model, 375, 574
    valence bands, 575
    valence bond theory, 574
metallic radius, 354
metalloids or semimetals, *51*, 52
    *See also* semiconductors
metals
    activity series for metals, 175–77
    characteristics, 562–63
    crystalline solids, 562, 563–71
    as essential nutrients, 844
    hydrated metal ions as acids, 856–58
    $K_a$ values of hydrated metal ions, *857*
    in periodic table, *51*, 52
    specific heat ($c_s$), 222–25
    stacking patterns, overview, 563–64, *567*
    unit cells of metals in periodic table, *566*
    *See also* alloys; colors of metal compounds;
        complex ions (complexes); metallic bonds;
        transition metals; *specific elements*
meteorite ALH840001, *134*
meter, 19, 20

methane
    in atmosphere, 92
    bond angles, 426, 433, 441, 442
    bond length, 404
    combustion, bond energies ($\Delta H$), 405–6
    combustion reactions, *74–75*, 92–93, 96–97,
        167–68, 170–71
    enthalpy of combustion to make CO, 229–31
    enthalpy of combustion to make $CO_2$, 225
    as greenhouse gas, 639
    hydrogen from steam–methane reforming,
        88–99, 118, *230*, 631, 734
    Lewis structures compared to ball-and-stick
        model, *426*
    melting and boiling points, *499*
    molecular structure, 432–33, *892*
    production by bacteria, 630, 639
    production from renewable resources,
        235–36
    reaction with ammonia, 87–88
    reaction with water, 88–89
    sources and uses, 630
    $sp^3$ hybrid orbitals, 442
    standard molar entropy ($S°$), *892*
methanethiol, 464, 465
methanogenic bacteria, 639
methanol
    conversion to methane by bacteria, 639
    fuel value, 631–32
    miscibility in water, *489*, *631*
    properties, 464, 631–32
    sources and synthesis, 631
*Methanosarcina*, 630
methionine, 970, *971*, 995–96, 1063, 1075
methyl groups, 608, 609
methyl *tert*-butyl ether (MTBE), 635, 687
methylaspartate, *1063*
2-methylbutane, 486–87
methylene groups, 608, 609
metric system. *See* SI units (*Système International
    d'Unités*)
micromolarity ($\mu M$), 139
microstates, 913–15
microwaves, *318*
Miller, Stanley, 74, 87–88, 113, 997
Miller/Urey experiment, 74, 87–88, 113
millibars (mbar), 262
Millikan, Robert, 42–43
millimeters of mercury (mmHg), definition,
    261, *262*
millimolarity (m$M$), 139
Minamata Bay, Japan, 1050
mineral oil, 618
mirror images, 452–53, 647–48, 649–50
miscibility of liquids, 491–92
miscible, definition, 260, 261
mixtures, 5–6, 8–10
    *See also* alloys; solutions
molality ($m$), 533–34
molar absorptivity ($\varepsilon$), definition, 144, 145

molar heat capacity ($c_P$), 215–22
    and absolute entropy, 890
    aluminum, *215*, 219, 222–24
    copper, 240
    definition, 215
    experimental determination, 222–24
    heat equation, 215
    ice, 215
    of selected substances, *215*
    water, 215, 216, 218, 219
    water vapor, *215*, 216
    *See also* heat capacity ($C_P$); specific heat ($c_s$)
molar heat of fusion ($\Delta H_{fus}$), 212, 216,
    217, 218
molar heat of vaporization ($\Delta H_{vap}$), 216–17, 494
molar mass ($\mathcal{M}$), 80–82, 282–83, 548–50
molar volume, 274, 275–76, 278
molarity ($M$), 138–42
mole (mol), 77–87
    Avogadro's number ($N_A$), 78–79, 80, *82*,
        83–84, 85
    and chemical equations, 85–86
    definition, 78, 79
    molar mass, overview, 80–82
    molar volume, 274, 275–76, 278
    molecular mass and formula mass, 82–85
    mole–mass–particle conversions, summary,
        82–83
    mole–particle conversions, 78–80
    *See also* stoichiometry
mole fraction ($X$)
    definition, 282, 283
    Henry's law, 519
    and partial pressure, 283–86, 832
    Raoult's law, 514, 515–16, 530–31
mole ratio
    combustion analysis, 108, 109
    empirical *vs.* molecular formulas, 104–5, 105,
        108, 109
    limiting reactant calculations, 110–16
    percent composition calculations, 100–103
    ratio of the coefficients, 85
    stoichiometric calculations, 96
    *See also* stoichiometry
molecular compounds
    binary molecular compounds, overview,
        54–55, 56–57
    definition, 54
    naming binary molecular compounds, 57–58
    naming prefixes, *57*
    *See also* covalent bonds
molecular formulas
    definition, 54, 55
    and empirical formulas, 101, 103–6, 107–10
    from mass spectrometry, 106, 107, 109
molecular geometry, 424–73
    bent or angular molecular geometry, 432, 433,
        *434*, *448*
    bond angles, definition, 427
    chirality, 452–53, 647–52

molecular geometry *(cont.)*
definition, 427
electron-pair geometries, summary, *435*
handedness, 424, 452–53
hybrid orbitals in complex ions, 849–51
Lewis structures compared to ball-and-stick model, *426*
linear molecular geometry, 427–28, 429, 445–47, *448*
octahedral molecular geometry, 428, 429, 430, 447, *448*
seesaw (disphenoidal) molecular geometry, 434, 436, *448*, 449
shape and interactions with large molecules, 449–51
*sp* hybrid orbitals, 445–47
*sp*$^2$ hybrid orbitals, 443–45
*sp*$^3$ hybrid orbitals, 442–43, 849–50
*sp*$^3$*d* hybrid orbitals, 447–48
*sp*$^3$*d*$^2$ hybrid orbitals, 447
square planar molecular geometry, 436, 437, *448*
square pyramidal molecular geometry, 436, 437, *448*
steric number, definition, 427
tetrahedral molecular geometry, 428, 429–30, 442–43, *448*
trigonal bipyramidal molecular geometry, 428, 429, 430, 447–48
trigonal planar molecular geometry, 428, 429, 431, 443–45, *448*
trigonal pyramidal molecular geometry, 433
T-shaped molecular geometry, 434, *448*
*See also* molecular orbital (MO) theory; valence bond theory; valence-shell electron-pair repulsion theory (VSEPR)
molecular ions, 106, 107, 453–54, 456, 460, 462
molecular mass, 82–85, 106
molecular motion, types of, *208*
molecular orbital (MO) theory, 453–63
antibonding orbital, definition, 454
auroras (northern lights), 453–54, 462
band theory, 574–75
and bond order, 456, 458–60, 461–62
bonding orbital, definition, 454
definition, 453
diamagnetism, 458, 459–60
helium, 455–56
heteronuclear diatomic molecules, 460–62
homonuclear diatomic molecules, 456–60
Hund's rule, 457
hydrogen molecular ions (H$_2^-$), 456
hydrogen molecules (H$_2$), 455, 456
light emission by molecules, 453–54
magnetic behavior of molecules, 453, 459–60
molecular ions, 453–54, 456, 460, 462
molecular orbital diagrams, 454, 455–58, *459*, 460–62
molecular orbitals, definition, 454
nitrogen molecular ions (N$_2^-$), 462

nitrogen molecules (N$_2$), 457–59
nitrogen monoxide, 461–62
oxygen molecules (O$_2$), 457–59
paramagnetism, 458, 459–60
pi ($\pi$) molecular orbitals, 457
sigma ($\sigma$) molecular orbitals, 454, 455
valence bands, 575
*See also* atomic orbitals; molecular geometry
molecular recognition
active sites, 424, 449
chirality, 452–53, 650–51, 652–53
definition, 49, 449, 648, 649
and molecular shape, 424, 449–51
receptors, 449–50, 452, 465, 648, 649, 650–51
molecular solids, 562–63
molecules, definition, 6, 7
molecules in space, 463
Molina, Mario, 707
molybdenum, *1048*, 1049, 1062, 1068
monodentate ligands, 858, 859, 862, 871
monomers, definition, *139*, 606
monoprotic acids, 796
monosaccharides, 984
Montreal Protocol, 411, 707
Mount St. Helens, 76, *77*
Mount Washington, New Hampshire, 27–28
Mt. Everest, 33
Mt. Sharp (Mars), 33
mullite (Al$_6$Si$_2$O$_{13}$), 584
multiple proportions, law of, 53–54
multivitamin package labels, 1048, *1049*
muscarine, 653
muscular dystrophy (MD), 968
mutases, 1063
myochrysine, 1076
myoglobin, 873
myristic acid, 991

## N

naming compounds, 57–62
alkanes, 613–16
alkenes, 622–23
alkynes, 622
binary acids, 62
binary ionic compounds, 58–60
binary molecular compounds, 57–58
binary transition metal compounds, 59–60
complex ions with negative charge, 855
complex ions with positive charge, 853–54
coordination compounds, 855–56, 871
ligands, *854*, 871
oxoacids, *61*, 62
oxoanions, 60–62
peptides, 975, 976
polyatomic ions, 60–62
prefixes for alkanes, *608*
prefixes for molecular compounds, 57
nanomolarity (n*M*), 139
nanoparticles, 560, 563

nanotubes, *581*, 582
naphthalene, *452*, *628*
naproxen, 640
natural abundance (isotopes), 49
natural gas
composition, 92, 225, 237, 527, 617
ethanethiol use as odorant, 464
helium in gas deposits, 119
use as fuel, 92–94, 95, 96
*See also* methane
neon, 362–63
atomic absorption spectrum, *323*
average atomic mass calculation, 49
electron configuration, 347
emission spectrum, 316
gas discharge tubes, 362, *363*
isotopes, 46–47, 49
neon-19, 1072
stellar nucleosynthesis, *64*
Nernst, Walther, 942
Nernst equation, 942–44
neutrons, 45–46
definition, 45
half-life of free neutrons, 1018
nucleosynthesis, 63
properties, *1010*
symbol, 63
nickel
alloys, 560, 571–72, 1078–79
color of Ni$^{2+}$(*aq*), 861–62, 866
electron configuration, 353
as essential ultratrace element, 1046, *1048*, 1063
ethylenediamine complex of Ni$^{2+}$, 859, *861*, 862, 866
fuel cell catalyst, 956, 957
light transmitted and absorbed by Ni$^{2+}$ complexes, *866*
nickel–cadmium (nicad) batteries, 936, 1050
nickel(II) chloride hexahydrate, 861–62, *866*, 875
nickel–metal hydride (NiMH) batteries, 947–48, 954–55
Ni(NH$_3$)$_6^{2+}$ complex ions, 861–62, 866, 875
NiTi (nitinol) shape-memory alloy, 571–72, 1078–79
tetraamminedichloronickel(II) geometric isomers, 871
niobium, 1073, 1078
Nissan Leaf, 948
nitinol, 571–72
nitrates
ammonium nitrate dissolution (cold packs), 894
commercial uses, 300–301
in Earth's crust, 300
resonance structures, 390–91
silver nitrate, 161, 172–74
nitric acid, 833
from ammonia, by Ostwald process, 768, 833
complete ionization in water, 780

nitric acid *(cont.)*
   formation, 151, 237–38, 301, 833
   as strong acid, 152
nitrides, 301
nitrilotriacetic acid (NTA), 860–61
nitrogen, 300–301, 1056–58
   in biomolecules, 300
   as cryogen, 300
   diamagnetic molecules, 453, *459*, 460
   diatomic element, 88, 91, 354
   electron configuration, 347
   as essential element, 300, 1046, 1048, 1056–58
   metabolism, 1056–58
   molecular ions ($N_2^-$), 462
   nitrogen fixation, 300, *301*
   pi ($\pi$) bonds, *446*
   *sp* hybrid orbitals, 446
   use in crude oil deposits, 300, *301*
nitrogen cycle, *300*
nitrogen dioxide
   in Earth's early atmosphere, 76
   equilibrium of synthesis and decomposition,
      746–47, 748–49, 906
   equilibrium with $N_2O_4$, 736–37, 740–42,
      745–46, 756–57, 903–5
   formation, 86–87
   photochemical decomposition, *668*, 669, 736
   properties, 301
   reaction mechanism for decomposition,
      700–703, 705–6
   resonance, 396–97
   structure, 396–97
   synthesis, 395, 673–74
   synthesis, rate constant, 681
   synthesis, reaction order, 678–79, 680–81
   synthesis, reaction rates, 673–74, 675–77,
      *678*, *682*
   thermal decomposition, 688–89, 690, 700
nitrogen fixation, 300, *301*
nitrogen monoxide (nitric oxide)
   calculations based on *K*, 764–65, 766–67
   in Earth's early atmosphere, 76
   free radicals, 396
   molecular orbital diagram, 461–62
   in photochemical smog, 395, 668–69
   reaction with carbon monoxide, 690, 693–94
   reaction with ozone, 669, 679–80, 690–92,
      695–98
   removal by catalytic converters, 669
   structure, *54*, 395–96
   synthesis, 395, 668, 671–72, 674–75
   synthesis, equilibrium, 751–52, 764–65,
      766–67
   synthesis, rate law and rate constants, 681–82
   synthesis, reaction rates, 671–72, 674–75, *682*
nitrogen oxides ($NO_x$)
   dinitrogen monoxide, 391–94, 395, 686
   dinitrogen pentoxide, 91–92, 151, 684–86
   in photochemical smog, 668
   *See also* nitrogen dioxide; nitrogen monoxide

nitrogenases, 1056, 1057–58
nitrosonium ion ($NO^+$), 204
nitrous acid
   calculations with pH and $K_a$, 790–91
   incomplete ionization in water, 780
   $NO_2^-$ conjugate base, 782, 784
Nobel Prizes
   Arrhenius, Svante August, *833*
   Bardeen, John, 590
   Bednorz, J. G., 595
   Brattain, Walter H., 590
   Curl, Robert, 246
   de Broglie, Louis, 335
   Einstein, Albert, 327
   Kroto, Harold, 246
   Lee, David, 919
   Müller, K. A., 595
   Osheroff, Douglas, 919
   Ostwald, Wilhelm, 833
   Penzias, Arno A., 31
   Planck, Max, *324*
   Richardson, Robert, 919
   Rutherford, Ernest, *44*
   Shockley, William B., 590
   Smalley, Richard, 246
   Urey, Harold, 74
   van 't Hoff, Jacobus Henricus, *833*
   Wilson, Robert W., 31
   Zewail, Ahmed H., 700n1
noble gases, 362–63
   electron configuration, 347–48, 363
   fractional distillation from air, 363
   ionization energies, 358, 360
   magnetic resonance imaging, 1071–72
   molecular masses and boiling points, 485, *486*
   in periodic table, 52
   *See also individual elements*
nodes (waves), 334, 335, 344, 345
nomenclature. *See* naming compounds
nonane, 529–30, 632–33
nonelectrolytes, 146–47
nonmetals
   clusters, 573, *581*, 582, 583
   covalent network solids, 580–82, 583,
      590, 892
   nonmetal oxides, 151, *152*
   in periodic table, *51*, 52
   properties, 52
   structure of crystalline nonmetals, 580–82
nonpolar covalent bonds, 383
nonspontaneous processes, definition, 882, 884
nuclear binding energies, 1010–12
nuclear charge
   and atomic size, 354, 356, 357
   effective nuclear charge ($Z_{eff}$), 356, 357,
      358, 359
   and ionization energy, 358–59, 383
   and shielding, 355–56, 359, 485
nuclear chemistry, 1008–45
   annihilation reactions, 1014, 1029, 1033

belt of stability, 1012–13, 1015
binding energies (BE), 1010–12
chain reactions, 1030–31
critical mass, 1030, 1031
mass defect ($\Delta m$), 1010–11
mass number *vs.* binding energy per
   nucleon, *1011*
nuclear equations, 64
nuclear fission, 1029–32
nuclear reactors, 118–19, 1008, 1026,
   1030–32, 1035
nuclear stability of $^{56}$Fe, 64, 1010, 1011
predicting modes and products of radioactive
   decay, 1015
symbols and masses of subatomic particles
   and small nuclei, *1010*
*See also* nuclear fusion; nuclides; radiation
   biology; radioactivity and radioactive decay
nuclear fission, 1029–32
nuclear fusion, 1032–36
   definition, 1032
   energy calculations, 1035–36
   positrons from, *1032*, 1033–34, 1035
   primordial nucleosynthesis, 63–64, 1010, 1032
   stellar hydrogen fusion, 64
   stellar nucleosynthesis, 64–65, 1010, 1033–34
   tokamak hydrogen fusion, 1033
   *See also* nuclear chemistry
nuclear power plants, 1031
nuclear reaction, definition, 1012
nuclear reactors, 118–19, 1008, 1026,
   1030–32, 1035
nucleic acids, 993–97
   adenine, *483*, 994
   cytosine, *483*, 994
   definition, 993
   double helix, 452, 482, *483*, 994–95
   guanine, *483*, 994
   nitrogen-containing bases, 994–95
   nucleotides, 993–95, 997
   replication, 994–95, *998*
   structure, 993–95
   thymine, *483*, 994
   uracil, 994
   *See also* deoxyribonucleic acid (DNA);
      ribonucleic acid (RNA)
nucleons, definition, 47
nucleosynthesis, 62–65
   definition, 63
   primordial nucleosynthesis, 63–64,
      1010, 1032
   stellar nucleosynthesis, 64–65, 1010, 1011,
      1033–34
   supernovas, 65
nucleotides, 993–95, 997
nucleus, discovery, 45
nuclides
   belt of stability, 1012–13, 1015
   beta ($\beta$) decay in neutron-rich nuclides, 1012,
      1014, 1015, 1068–69

nuclides *(cont.)*
  binding energy *vs.* mass number, *1011*
  definition, 47
  positron emission in neutron-poor nuclides, 1013, 1014, *1016*, 1029
  radionuclides, 1008, 1012, 1016, 1017–19, 1022–23, 1028–29, 1067–72
  symbols, 47–48, 63
  *See also* isotopes; nuclear chemistry
nylon, *644*, 645

## O

obsidian (volcanic glass), *562*, 583, 1058
  *See also* silica (silicon dioxide)
oceans, 132
  *See also* seawater
octahedral holes, 572–73, 577–78
octahedral molecular geometry, 428, 429, 430, 447, *448*
octane
  boiling point, 493, 494
  distillation, 528–30
  fuel value, 632
  molar heat of vaporization ($\Delta H_{vap}$), 494
  octane ratings, 626, 630
  solubility, *489*, 490–91
  standard free energy of formation of isomers, *898*
  vapor pressure, 494
octanol, *490*, 491–92, 501, 502–3
octet rule
  atoms with more than an octet, 397–400
  definition, 375
  exceptions, 375, 376–77, 395–403
  ionic compounds, 381–82
  and odd-electron molecules, 395–97
  *See also* Lewis structures
oleic acid, 990, 998–99
olestra, 991
oligomers, 605
oligopeptides, 975
olive oil, 990
optical isomers, 452–53, 647–51, 984
  *See also* chirality; isomers
optically active molecules, 648–49, 972
  *See also* chirality
orbital penetration, 354–55
orbitals. *See* atomic orbitals; molecular orbital (MO) theory
organic chemistry, 602–65
  carbon-skeleton structures of organic molecules, *609*, 610–11, 612–13
  condensed structures of organic molecules, 609–11, 612
  definition, 602, 604, 605
  drawing organic molecules, 609–11
  functional groups, 604–5
  *See also* carbon; *specific compounds*
organometallic compounds, 1073

origins of life
  biomolecule synthesis from inorganic materials, 74, 997–98
  extraterrestrial origins, 997
  hydrothermal or geothermal vents, 132, 997–98
  RNA world hypothesis, 997
osmium, 715, 1076
osmosis
  cell membranes, 541–42, 543, 544–45
  definition, 541
  and molar mass calculations, 549–50
  osmotic pressure ($\Pi$), 541–46, 549, 1046
  reverse osmosis, 546–47
  reverse osmotic pressure, 546–47
  semipermeable membranes, 541–45, 546–47, 550–51
Ostwald, Wilhelm, 833
Ostwald process, 768, 833
Otzi the Iceman, 915
oxidation number (O.N. or oxidation state), 168–70
oxidation–reduction reactions, 167–81, 928–30
  activity series for metals, 175–77
  electron transfer, 168, 170–72
  half-reactions for balancing redox reactions, 172–74, 176–77, 178, 179, 928–30
  iron compounds, 177–80
  methane, 167–68
  in nature, 177–80
  net ionic equations, 173–74, 176–77, 929–30
  oxidation, definition, 168, 169, 928
  oxidizing agents, 171–72, 928
  pH and half-reactions, 929
  reducing agents, 171–72, 928
  reduction, definition, 168, 169, 928
  standard potentials of half-reactions and $K$, 945
  standard reduction potentials ($E°$), 934–37
  *See also* combustion reactions; electrochemistry
oxide ion as Lewis base, 847
oxidizing agents, 171–72, 928
oxoacids
  acid strength of oxoacids of chlorine, 801
  naming, *61*, 62
  strength of hypohalous acids, 801, 856–57
  strong acids and their reactions with water, *781*
  volatile nonmetal oxides and their acids, 151, *152*
oxoanions, 60–62
oxygen, 410–11, 464
  atoms, in photochemical smog, 668–69
  bond lengths, 388, 403
  from chlorates, 277, 278, 286
  comparison to sulfur, 464
  diatomic element, 7, 88, 91, 354
  as essential element, 167, 387, 410, 1046, 1048
  freezing point, 13

liquid oxygen (LOX) as cryogen, 411
  microstates, 914
  molecular motion, 913–14
  in oxidation–reduction reactions, 167
  paramagnetic molecules, 453, *459*, 460
  partial pressure in blood, 518–19
  solubility in water, 474, 489–90, 512
  stellar nucleosynthesis, *64*
  superoxide ion ($O_2^-$), 1063
ozone
  bond lengths, 388, 403
  decomposition in the presence of NO, 709–10
  effect of chlorofluorocarbons (CFCs), 411, 707–8
  environmental damage in lower atmosphere, 387
  Lewis structures, 387–88
  from lightning, 387
  ozone hole over Antarctica, 668, 689, 706–10
  photochemical decomposition rates, 683–84, 707
  in photochemical smog, 668, 669
  production in upper atmosphere, 411
  reaction with chlorine monoxide, 411, 692, 708–9
  reaction with nitrogen monoxide, 679–80, 690–92, 695–98
  resonance, 388–89, 403
  ultraviolet shielding in upper atmosphere, 387
  *See also* oxygen

## P

pacemakers, 1076–77
packing efficiency, 566, 567, 570–71
palladium, *562*, 567, 711, 715
palmitic acid, 991, 998–99
panaxytriol, 618, *619*
paraffins, 608
  *See also* alkanes
paramagnetism, 458, 459–60
pargyline, 618, *619*, 629, *630*
partial pressure
  equilibrium constants ($K_p$), 739
  and mole fraction ($X$), 283–86, 832
  oxygen in blood, 518–19
  and solubility (Henry's law), 517–19, 551
partial pressures
  Dalton's law, 282, 283–87, 288–89, 298
particle properties of light, 327–28
partition coefficient determination, 502–3
Pascal, Blaise, 261
pascal (Pa), definition, 261–62
Paschen, Friedrich, 328
Pauli, Wolfgang, 340, *341*
Pauli exclusion principle, 340–41
Pauling, Linus, 383, 440
Pauling electronegativity scale, 383
pearls, 182
pentane, 486–87, 491–92, *609*

pentanol, *490*, 491
pentene, *620*, *621*, 622
Penzias, Arno A., 31, *32*
peptide bonds, 975, 996
peptides, 975–77
    *See also* amino acids; proteins
percent composition
    carbon tetrachloride, 100
    from chemical formulas, 99
    definition, 98
    empirical formulas from percent composition,
        98–103, 181
    HFC-134a ($C_2F_4H_2$), 99
    HFC-143a ($C_2F_3H_3$), 100–101
    methane, 98–99
    nitrous oxide, 102
    pinene, 98–99
    spodumene, 102–3
percent yield, 114–16
perchlorates, 505, 538–39
perchloric acid, 783
periodic table, 50–53, 345–51
    actinides, *51*, *52*, 350
    alkali metals, *51*, *52*
    alkaline earth metals, *51*, *52*
    and atomic number, 47–48, 50, *51*
    and atomic orbitals, 345–51
    atomic radii, *355*
    and charges on ions, 55–56
    columns (groups or families), 51
    definition, 47
    electron affinities (EA), 360
    groups or families, 51–52
    halogens, *51*, 52
    ionization energies of elements, 358–60,
        383, *384*
    lanthanides, *51*, *52*, 350
    Lewis symbols of main group elements, *377*
    and life, 1046–85
    main group elements or representative
        elements, *51*, *52*
    Mendeleev, 50, 52, 53, 65
    metalloids or semimetals, *51*, *52*
    metals, *51*, *52*
    modern periodic table, 5, 47, 50–52
    noble gases, 52
    nonmetals, *51*, *52*
    Pauling electronegativity scale, *383*
    periods (rows), 50, 51
    and properties of the elements, 40, 50
    transition metals, *51*, *52*, 349–50
    unit cells of metals, *566*
peroxyacetyl nitrate, 668–69
petroleum. *See* crude oil; hydrocarbons
pH
    autoionization of water, 785–86, 795
    calculations with pH and $K_a$, 790–93
    calculations with pH and $K_b$, 793–95
    common-ion effect, 806–10
    definition, 786

effect on solubility, 827–29
Henderson–Hasselbalch equation, 808–9,
    810–13, 815–17, 818, 820, 824, 832
isoelectric point (p*I*), 973, 974
pH scale, 786–88
and pOH, 788–89
salt solutions, 801–6
very dilute solutions, 795–96
*See also* acid–base chemistry; acids; bases; pH
    buffers; pH indicators
pH buffers
    buffer capacity, 815–17
    buffer range, 810, 814–15
    carbonic acid as physiological pH buffer,
        814–15
    definition, 810
    environmental buffers, 810–11
    Henderson–Hasselbalch equation, 810–13,
        815–17
    physiological buffers, 813, 814–15
    response to acid or base addition, 810–14
    *See also* acid–base chemistry; pH
pH indicators
    definition, 818
    phenol red, 818, 823
    phenolphthalein, 154, 823, 824
    swimming pool test kits, *818*
    in titrations, 154, 818–19, 823–25
    useful ranges of indicators, *818*
    *See also* acid–base chemistry; pH
pH meters, 818–19
pharmaceuticals
    alcohols in, 633
    alkanes, 618
    alkynes, 618, *619*
    amines, 629, 630
    calcium carbonate, 182
    carboxylic acids, 640
    chromium compounds, 1073, *1074*
    enzyme inhibitors, 984
    esters, 640
    fluorine compounds, 410
    gold compounds, 1046, 1076
    lithium compounds, 1046, 1072
    partition coefficient determination, 502–3
    platinum compounds, 870, 1046, 1075
    poly(ethylene glycol) (PEG) attachments, 636
    shelf-stability, 181
    silver compounds, 1046, 1075, *1076*
    vanadium compounds, 1073, *1074*
phase changes
    changes of state, 14–15
    critical point, 497
    deposition, 14, 15, 495, *889*
    energy transfer, 198, 205–8
    enthalpy change ($\Delta H$), 212–13
    equilibrium lines of phase diagram,
        495–96, 498
    liquefaction, 495
    molar heat of fusion ($\Delta H_{fus}$), 212, 216, 217, 218

molar heat of vaporization ($\Delta H_{vap}$),
    216–17, 494
phase diagrams, 494–98, 532–33
sublimation, 14, 15, 495
supercritical fluids, 497
supercritical region of phase diagram,
    495, 497
*See also* states of matter
phase diagrams, 494–98, 532–33
phenanthrene, *452*, *628*
phenol red, 818, 823
phenolphthalein, 154, 823, 824
phenylalanine, *971*
pheromones, 106
phosgene ($COCl_2$), decomposition, 767–68
phosphate ion, Lewis structure, 399–400
phosphine ($PH_3$), 434
phospholipids, 991–92, 1051, *1052*
phosphorescence, 42, *43*
phosphoric acid, 640–41, 796, 799, *800*
phosphorus
    as dopant, 576
    DRI/RDA value, *1049*
    as essential element, 1046, 1048
    phosphorus pentachloride, 447
    phosphorus pentafluoride, 430
    phosphorus-32, 1015
    semiconductor doping, 576
phosphorus pentachloride production, 763–64
phosphorylation, definition, 910
photochemical smog, 666–69, 675, 713–14, 736
photoelectric effect, 316, 326–27
photoelectrons, 326, 328
photons, 325, 333, 361
photosynthesis
    and carbon cycle, 94, *95*, 246
    carbon dioxide reduction, 1064
    glucose production, 10, 94, 110, 243, 246
    magnesium, *844–45*, 873, 1046, 1053–54
    and manganese, 1064
    reactions, 94, 110
    *See also* chlorophyll
phototubes, *326*
phycoerythrobilin, *1053*
physical equilibrium
    equilibrium lines of phase diagrams,
        495–96, 498
    liquid–vapor equilibrium, 529–30
    thermal equilibrium, 198, *222*, 223, 224
    *See also* chemical equilibrium; dynamic
        equilibrium
physical processes, definition, 4, 5
physical properties, 12, 13
physiological saline, 144, 545
phytoplankton, 8–9, *844–45*
p*I* (isoelectric point), 973, 974
picomolarity (p*M*), 139
pinene, 98–99
pinene (pine oil), *618*
pitchblende, 44, 65–66, 1030

Planck, Max, 324–25, 326, 328
Planck's constant, 325, 331, 333, 335
plane-polarized light, 452, 648–49, 650, 651–52
plaques (amyloid β protein), 980
plaques (cholesterol), *992*
platinum, 715
    ammonia conversion to nitric acid, 833
    average atomic mass calculation, 49–50
    cancer drugs, 870, 1046, 1075
    carboplatin, 1075
    in catalytic converters, 711, 715
    *cis*-diamminedichloroplatinum(II) (cisplatin),
        870, 1075
    fuel cell catalyst, 956
    isotopes, *49*
    kilogram mass standard, 19
    *trans*-diamminedichloroplatinum(II), 870
platinum group metals (PGMs), 715, 1075–76
    *See also individual elements*
plutonium, 118, 1031–32
polar bonds
    bond dipole, 437
    bond polarity, definition, 382
    dipole moment ($\mu$), 438–40
    electronegativity and bond polarity, 383,
        384–86
    partial charges, 383, 386
    and polar molecules, 437–40
    symbols and chemical notation, 382–83, 385
    vibrating bonds and greenhouse effect, 386–87
polar covalent bonds, 382–86
polar molecules
    dipole–dipole interactions, definition, 481
    hydrogen bonds, 481–84
    ion–dipole interactions, 480, 481, 488–89
    and polar bonds, 437–40
    polarity and solubility, 488–92
    sphere of hydration, 480, 481
    sphere of solvation, 480, 481
    temporary dipoles (induced dipoles), 484–85
polarizability, 485–86, *487*
polarized light (plane-polarized light), 452,
    648–49, 650, 651–52
polonium, 567, 1008, 1016, 1027
polyamides, 644–46
polyamines, 859
polyatomic ions, 60–62
    estimated radii, *477*
    formal charges in, 392, 393
    in ionic solids, 562, 577
    Lewis structures, 377, 399–400
    naming, 60–62
    resonance structures, 390–91
    *See also* ions
polycyclic aromatic hydrocarbons (PAHs), 451,
    *452*, 628
polydentate ligands, 858, 859–61, 862, 871,
    872–74
    *See also* chelating agents; ligands

poly(dimethylsiloxane) or PDMS, 591
polyesters, 641–44, 645
poly(ethylene glycol) (PEG), 636–37
poly(ethylene oxide) (PEO), 636
polyethylene (PE), 606, 623–24, 625
poly(ethylene terephthalate) (PETE), 635,
    636, 642
polyisoprene, 625
polymers, 605–6, *647*
    addition polymers, 623, 625–26, 635, *647*
    from alcohols and ethers, 635–37
    of alkenes, 623–26
    alternating copolymers, 636
    aromatic rings in polymers, 628–29
    block copolymers, 636
    cellulose, 246, 639, 984, *986*, 987
    condensation polymers, 642, 643, 646, *647*
    copolymers, 636, 637, 642
    cotton, 246, 643–44, 984
    EVAL (ethylene and vinyl acetate
        copolymer), 636
    heteropolymers, 636, 637
    high-density polyethylene (HDPE),
        623–24
    homopolymers, 623
    Kevlar, 645–46
    low-density polyethylene (LDPE), 623–24
    poly(1,1-dichloroethylene) (Saran wrap), 625
    polyamides, 644–46
    polyesters, 641–44, 645
    poly(ethylene glycol) (PEG), 636–37
    poly(ethylene oxide) (PEO), 636
    polyethylene (PE), 606, 623–24, 625
    poly(ethylene terephthalate) (PETE), 635,
        636, 642
    polyisoprene, 625
    polymerized butadiene, 625
    polypropylene, 625–26
    polystyrene (PS), *628*, 629, 646
    polytetrafluoroethylene (Teflon), 410,
        606, 625
    poly(vinyl acetate) (PVAC), *635*, 636
    poly(vinyl alcohol) (PVAL), 635, 636
    properties, 606
    random copolymers, 636
    starch, 984, 986–87
    ultrahigh molecular weight polyethylene
        (UHMWPE), 624
    vinyl polymers, 626, 627
polypeptides, 975, 1064
polypropylene, 625–26
polyprotic acids, 796–800
polysaccharides, 984, 986–87
    cellulose, 246, 639, 984, *986*, 987
    starch, 984, 986–87
polystyrene (PS), *628*, 629, 646
poly(vinyl acetate) (PVAC), *635*, 636
poly(vinyl alcohol) (PVAL), 636
polyvinyl chloride (PVC), 139, 606, 626

porphyrins, 872–74, *981*, 982, 1061
positive-ray analyzer, 47, 106, 363
positron-emission tomography (PET), *1008–9*,
    1029, 1067, 1072
positrons, 1013–14
    annihilation reactions, 1014, 1029, 1033, *1033*
    as antimatter version of electrons, 1009, 1014
    from hydrogen fusion, *1032*, 1033–34, 1035
    positron emission, 1013, 1014, *1016*, 1029
    symbol and mass, *1010*, 1013
potassium, 1051–53
    biological transport, 1051–53
    DRI/RDA value, 1049
    electron configuration, 348
    as essential element, 1046, 1048, 1051–53
potential energy (PE)
    conversions between potential and kinetic
        energy, 200
    definition, 198, 199
    electrostatic potential energy ($E_{el}$), 202–3,
        374–75, 477–79
    formula, 199, *200*
    and gravity, 199–200
    at molecular level, 201, 202–3
    as state function, 199–200
    and work, 199–200
precipitation reactions, 157–65
    analysis of metals in alloys, 874–75
    insoluble salts, 157–60
    net ionic equations, 159–60, 161, 162
    overall ionic equations, 159–60, 161, 162
    precipitates, definition, 157
    solubility rules, 157–58
    used in analysis, 161–63
precision, 25
pressure ($P$)
    Amontons's law, 271–73, *275*, 289, *290*
    bar, definition, 232, 233, *262*, 275
    Dalton's law of partial pressures, 282, 283–87,
        288–89, 298
    definition, 260, 261
    effect on chemical equilibrium, 756–59
    measurement, 263–65
    millimeters of mercury (mmHg), definition,
        261, *262*
    pascal (Pa), definition, 261–62
    pressure–volume ($P$–$V$) work, 209–11,
        899–900
    pressure–volume relationship of gases, 260,
        265–68, 274, *275*, 288–89
    standard atmosphere (atm), 261, *262*
    standard temperature and pressure (STP),
        274, 275
    torr, definition, 261, *262*
    units, 261–62
    *See also* atmospheric pressure
primary amines, 629
problem solving guidelines, 10–11
products, definition, 77

proline, *971*
promethium, *1013*
propane
    combustion reactions, 113
    enthalpy of combustion, 225, 236–37, *242*
    fuel value, 241
    in natural gas, 93
    standard molar entropy ($S°$), *892*
    structure, *485*, *607*, 610–11, *892*
propanol, *487*
propene, *607*, 626
propylene, 713–14
propyne, *607*
proteins, 970–84
    α helix, *978*, *979*, *980*, *981*
    amine (or N-) terminus, 975
    amino acids, 970–72
    β-pleated sheet, 979–80, *981*
    carboxylic acid (or C-) terminus, 975
    complete proteins, 970
    complex ions in, 873–74
    coupled reactions in protein metabolism, 913
    definition, 970
    denaturation, 980
    determining sequence of small
        peptides, 977
    disulfide (S–S) linkages, 981
    enzymes as protein catalysts, 970, 975,
        982–84
    functional classes, *978*
    hydrogen bonding, 975, 979–80
    levels of protein structure, summary, *978*
    oligopeptides, 975
    peptide bonds, 975, 996
    peptides, 975–77
    polypeptides, 975
    primary (1°) structure, 977–79
    quaternary (4°) structure, *978*,
        981–82
    random coil, 980, 981
    receptors, 648, 649, 650–51
    secondary (2°) structure, *978*,
        979–81
    synthesis, 995–96
    tertiary (3°) structure, *978*, 981
    *See also* amino acids; peptides
proton-exchange membrane (PEM),
    956–57
protons, 45–46
    definition, 45
    nucleosynthesis, 63
    properties, 1003
pseudo-first-order reactions, 690–93
pyrites, 153
pyruvate ion
    conversion to acetyl-coenzyme A, 988,
        989, 992
    conversion to ethanol, 988
    from glycolysis, 910, 988

    oxidation in tricarboxylic acid cycle, 988
    structure, *910*, 988

**Q**

quantized, definition, 324, 325
quantized light, 324
quantum, definition, 324, 325
quantum mechanics. *See* wave mechanics
quantum numbers
    angular momentum quantum number ($\ell$), 338
    definition, 330, 335, 337
    magnetic quantum number ($m_\ell$), *338*
    and orbitals, 338–41
    principal quantum number ($n$), 336, 337, 354
    Schrödinger wave equation, 336, 337–41
    shells and quantum numbers, 338–41
    spin magnetic quantum number ($m_s$), 340–41
    subshells, letters and quantum numbers,
        338–41
quantum theory, 323–28
    Balmer equation, 328
    Bohr model of hydrogen, 329–32
    de Broglie wavelengths, 332–35
    definition, 324, 325
    electron transitions, 330, 331–32
    energy level diagram, *331*
    energy levels ($n$), 329–32
    energy of photons, 325, 361
    excited states, 330
    ground state, 330
    Heisenberg uncertainty principle, 335–37, 354
    history, 316–18, 324, 341, *342*
    matter waves, 333–35, 337, 344
    multielectron atoms, *349*
    photoelectric effect, 316, 326–27
    photons, 325
    Planck, Max, 324–25, 326, 328
    Rydberg equation, 328–29
    Schrödinger wave equation, 336, 337–41
    standing waves, 334–35, 344, 345
    variable lifetimes of excited states, 342
    wave–particle duality, 327–28
    *See also* wave mechanics
quarks, 63
quartz, 562, *565*, 583, *587*, 590
    *See also* silica (silicon dioxide)
quinidine, 651
quinine, 651

**R**

R (symbol for functional groups), 604, 605
racemic mixtures, 651
radiation. *See* electromagnetic radiation;
    radioactivity and radioactive decay
radiation biology, 1008–10, 1024–29
    acute effects of ionizing radiation, *1026*
    boron neutron-capture therapy
        (BNCT), 1072

    Chernobyl, Ukraine, nuclear reactor
        explosion, 1026, *1027*, 1065
    $^{137}$Cs hazards, 1065
    diagnostic radiology, 1029, 1067–72
    effective dose, 1025–26, 1028
    evaluating risks of radiation, 1026–28,
        1036–37
    Fukushima, Japan, nuclear reactor explosions
        and fire, 1023
    gamma ray treatment of cancer, 1028
    genetic mutation from radiation
        exposure, 1026
    $^{123}$I as imaging agent, 504–5
    $^{125}$I as imaging agent, 504–5
    $^{123}$I for thyroid cancer therapy, 504, 1028–29
    $^{131}$I for thyroid cancer therapy, 504, *1029*
    $^{131}$I health risks, 1026
    $^{131}$I as imaging agent, 1067
    imaging with radionuclides, 1008, 1029,
        1067–72
    ionizing radiation, 1024–29
    measuring radioactivity, 1021–23
    radiation dosage, 1024–26
    radiation dose from dental X-rays, 1026, 1028
    radioactive fallout, 1026, *1027*
    radionuclides used in radiation therapy, 504,
        1010, 1028–29
    radon exposure, 1026–28
    sources of radiation exposure, 1026–27
    $^{90}$Sr hazards, 1065
    therapeutic radiology, 504, 1028–29
    tracers for medical imaging, 1029, 1067–72
    units for expressing quantities of ionizing
        radiation, *1025*
    *See also* nuclear chemistry
radio waves, *318*, 319
radioactivity and radioactive decay
    absorbed dose, 1024–25, 1028
    acute effects of ionizing radiation, *1026*
    alpha (α) decay, 1016, *1017*, 1027
    becquerel (Bq), 1022
    belt of stability, 1012–13, 1015
    beta (β) decay, 1012, 1014–18, 1020, 1023,
        1068–69, 1072
    biological effects, 1024–29
    carbon isotopes and radioactive decay
        products, *1015*
    Chernobyl, Ukraine, nuclear reactor
        explosion, 1026, *1027*, 1065
    curie (Ci), 1022
    decay rate constant, 1022–23
    definitions, 44, 1012, 1017
    effective dose, 1025–26, 1028
    electron capture, 1014, *1016*
    evaluating risks of radiation, 1026–28,
        1036–37
    first-order kinetics, 1017–18, 1022–23, 1036
    Fukushima, Japan, nuclear reactor explosions
        and fire, 1023

radioactivity and radioactive decay *(cont.)*
  Geiger counters, 1022
  genetic mutation from radiation exposure, 1026
  gray (Gy), 1024–25
  half-lives ($t_{1/2}$), 1017–18, 1020, 1023, 1029, 1031, 1035, 1036–37
  ionizing radiation, 1024–29
  measuring radioactivity, 1021–23
  positron emission, 1013, 1014, *1016*, 1029
  predicting modes and products of decay, 1015
  radiation dosage, 1024–26
  radioactive decay series, 1016, *1017*
  radioactive fallout, 1026, *1027*
  radioactive isotopes, 65–66
  relative biological effectiveness (RBE), 1024, 1025, 1028, 1072
  sievert (Sv), 1024, 1025
  units for expressing quantities of ionizing radiation, *1025*
  *See also* nuclear chemistry; radiation biology
radiometric dating, 1018–21
radionuclides, 1008, 1012, 1016, 1017–19, 1022–23, 1028–29, 1067–72
  *See also* nuclides; radiation biology; radioactivity and radioactive decay
radium, 1008, 1016, 1023, 1036–37
radon, 347, 363, 1016, 1026–28, 1049
Ramsay, William, 362
random coil protein structure, 980, 981
Raoult's law, 514, 515–16, 530–32
rate constants ($k$), overview, 679–82
rate laws
  definition, 680
  first-order reactions, 683–86
  integrated rate laws, definition, 684
  overview, 680–82
  pseudo-first-order reactions, 690–93
  and reaction mechanisms, 702–6
  second-order reactions, 681, 688–90
  zero-order reactions, 693–94, 706
  *See also* reaction rates
rate-determining steps, 702–5, 706
reactants, definition, 77
reaction mechanisms, 699–706
  bimolecular steps, 701, *703*, 704
  definition, 700, 701
  elementary steps, 700–701, 702–4
  intermediates, definition, 700, 701
  molecularity of elementary steps, 701, 702
  and rate laws, 702–6
  rate-determining steps, 702–5, 706
  termolecular (three-molecule) steps, 701, 704
  transition states, 701
  unimolecular steps, 701
  zero-order reactions, 706
reaction quotients ($Q$), 750–52, 829–31, 902–4, 942, 944
  *See also* mass action expressions

reaction rates, 669–99
  activation energy ($E_a$), 695
  Arrhenius equation, 694–99
  average rate, 674–75, 676n1
  and catalytic converters, 669
  concentration effect on rates, 677–94
  definition, 669
  and energy profiles of reactions, *695*, 696, *697*, 701
  experimentally determined rates, 673–74
  half-lives ($t_{1/2}$) of reactions, 686–87, 690, 694
  initial rate, 677–78
  integrated rate laws, definition, 684
  and molecular orientation, 695
  overall reaction order, 680–81, 682–83
  partial reaction order, 680
  and photochemical smog, 669, 713–14
  pseudo-first-order reactions, 690–93
  rate constants ($k$), overview, 679–82
  rate laws, definition, 680
  rate laws, first-order reactions, 683–86
  rate laws, overview, 680–82
  rate laws, second-order reactions, 681, 688–90
  rate-determining steps, 702–5, 706
  rates, 674–77
  reaction order, 677–82
  temperature effect on rates, 694–99
  zero-order reactions, 693–94, 706
  *See also* chemical kinetics
recognition. *See* molecular recognition
recommended dietary allowance (RDA) values, 1048, *1049*
red blood cells, *512–13*
  oxygen binding, *512–13*
  and partial pressure of $O_2$, 519
  semipermeable membrane and osmotic pressure, *541*, 543, 544–45
  in sickle-cell anemia, 978–79
redox reactions. *See* oxidation–reduction reactions
reducing agents, 171–72, 928
reductases, 1057–58, 1062
reduction, 168
  *See also* oxidation–reduction reactions
refraction, 320
relative biological effectiveness (RBE), 1024, 1025, 1028, 1072
resonance, 387–91
  benzene, 391, 451, 627
  carbon dioxide, 394
  definition, 388
  dinitrogen monoxide, 391–94, 395
  formal charge, 391–95
  nitrates, 390–91
  ozone, 388–89, 403
  polyatomic ions, 390–91
  resonance structures (resonance hybrids), 388–89
  sulfur trioxide, 389–90

respiration, 94, *95*, 112, 231, 814, 1060
reversible processes
  definition, 888
  efficiency, 900
  entropy, 888–89, 895
  free energy change, 903, 905
  spontaneous processes, 903, 905
rhenium, 1067, 1069, *1070*, 1071
rhodium, 711, 715, 833, 1075
ribonucleic acid (RNA), 993, 994, 995–96
  codons, 995–96
  DNA transcription, 995, *996*
  messenger RNA (mRNA), 995–96
  nitrogen in, 300
  and protein synthesis, 993, 995–96
  transfer RNA (tRNA), 995–96
  translation, 995–96
  *See also* deoxyribonucleic acid (DNA); nucleic acids
ribose, 994
ribosomes, 993, 995, *996*
RICE tables (Reaction, Initial values, Changes, Equilibrium values), 761–67
rich mixture (fuel combustion), 113
Ringer's lactate, 142
RNA. *See* ribonucleic acid (RNA)
robotic hand, *602–3*
rock candy, 164
rock cycle, 132
rock salt. *See* sodium chloride (table salt)
rock salt structure, 578, 580, 589
Röntgen, Wilhelm Conrad, 44
root-mean-square speed ($u_{rms}$) of molecules, 291–94
rose gold, 571
rotational motion, *208*, 885, *886*, 890, 913–14
rounding off, 21–23
Rowland, Sherwood, 707
rubidium, 1048, 1065
ruthenium, 715, 1069, 1075
Rutherford, Ernest, *44*
  atomic model, 44–45, *46*, 329, 1008
  beta (β) particles, 44
  gold foil experiment, 44–45
  radiometric dating, 1018–19
Rydberg, Johannes Robert, 328–29

**S**

Sagan, Carl, 42
Salt Lake City, smog, *666–67*
salt (table salt). *See* sodium chloride (table salt)
salts
  acid–base properties of common salts, *802*
  crystalline structure of ionic solids, 577–80
  definition, 148
  formation in acid–base reactions, 148

salts *(cont.)*
    ion–ion interactions in, 477–79
    pH of solutions, 801–6
Saran wrap [poly(1,1-dichloroethylene)], 625
saturated fatty acids, *989*, 990, 998–99
saturated hydrocarbons, 606, 607
    *See also* alkanes
saturated solutions, 164–65
    *See also* solubility; solutions
scandium, 349, 353, 1069
scanning tunneling microscope (STM), 7
Schrödinger, Erwin, 337
Schrödinger wave equation, 336, 337–41
scientific method, 2, 16–17
scientific theories (models), 16, 17
scintillation counters, 1022, 1069
screening (shielding) of nuclear charge, 355–56, 359, 485
scuba diving, 260, 268, 284–85, 286, 288, 298–99
seawater
    acidification of the ocean, 797–800
    chloride ion concentration, *137*, 138
    concentrations of major ions, *137*, 535, *542*, 543, *1049*
    determining $CO_2$ concentration in, 832
    distillation, 9, *10*
    evaporation and salt precipitation, 476, *505*
    filtration, 8
    fluoride ion concentration, 137
    freezing of, 537
    and osmotic pressure, 541–45
    phytoplankton in, 8–9, *844–45*
    reverse osmosis desalination, 546–47
    sodium chloride concentration, 533, 537
    sodium ion concentration, *137*, 141
    toxic effects, 541–42
    unchanging composition, 132
    vapor pressure of water and seawater, 514–15
    *See also* water
second law of thermodynamics, 882, 886–87, 888–89, 913
secondary amines, 629
second-order reactions, rate laws, 681, 688–90
seesaw (disphenoidal) molecular geometry, 434, 436, *448*, 449
selenium, 1062
    DRI/RDA value, 1049, *1049*, 1062
    as essential ultratrace element, 1046, *1048*, 1062
    selenium hydride ($H_2Se$), 464
    selenium tetrafluoride, 448
    selenocysteine, 1062
    toxicity, 1062
selinene (oil of celery), *618*
semiconductors, 575–77
    acceptor bands, 576, *577*
    band gap ($E_g$), 574, 576–77, 582
    conduction bands, 576–77

donor level, *576*
doping and dopants, 576–77, 584
    "holes," 576, *577*
    n-type semiconductors, 576–77
    preparation from silicon, 590–91
    p-type semiconductors, 576–77
    valence bands, 575–77
semimetals (metalloids), *51*, 52
semipermeable membranes, 541–45, 546–47, 550–51
serine, *971*, 975
shakudo gold, 571
shape-memory alloys, 571–72, 1078
shielding (screening) of nuclear charge, 355–56, 359, 485
Shockley, William B., 590
SI units (*Système International d'Unités*), 18–19
    base units, 19
    conversion factors, *19*, 26–28
    dimensional analysis, 26–28
    prefixes, 18–19
    temperature conversions, 30, 33
    unit conversions, 26–28, 33
    units for expressing quantities of ionizing radiation, *1025*
sickle-cell anemia, 977–79
sievert (Sv), 1024, 1025
significant figures, 19–25
silane ($SiH_4$), 434
silica (silicon dioxide), 583
    amorphous forms (volcanic glass or obsidian), *562*, 583, 1058
    covalent network solid, 583, 590
    polymorphs, 583
silicates, 583–84, *585*, 590–91
silicon, 590–91
    atoms, 7
    band theory and semiconducting, 575–77
    crystal structure, 580
    doping and dopants, 576
    as essential trace element, *1048*, 1058
    in microelectronics, 7, 560, 590
silicones (organosilicon compounds), 591
silk, 980
silver
    Ag/Ag$^+$ electrode half-reaction, 943
    in antitumor drugs, 1075
    atomic radius, 573
    for burn treatment, 1046, 1076
    corrosion, 959–60
    Cu/Ag electrochemical cell, 933
    in dental amalgams, 1076
    electron configuration, 353
    isotopes, 50
    lithium silver vanadium oxide batteries, 1078
    sterling silver, 573–74
    for treating infections, 1075
    Zn/Ag button batteries, 939–40
silver chloride, solubility, 161, 162–63

silver nitrate, 161, 162, 172–73
silver sulfadiazine, 1076
simple cubic (sc) unit cell, *565*, 566–67
single bond, definition, 377
skunk odor, 465
slaked lime [$Ca(OH)_2$]. *See* calcium hydroxide
smog, 666–69, 675, 713–14, 736
sodium, 1051–53
    biological transport, 1051–53
    DRI/RDA value, *1049*
    electron configuration, 348
    emission spectrum, 319–20, 322
    as essential element, 1046, 1048, 1051–53
    ion formation, 55
    sodium-24, 1029
sodium acetate, *165*
sodium azide, 277, 279, 301
sodium bicarbonate (sodium hydrogen carbonate), 61–62, *135*, 156, 803–4, 884, *885*
sodium carbonate, 803–5, 818
sodium chloride (table salt)
    analysis by precipitation, 161
    Born–Haber cycle for formation, 522–24
    concentration in seawater, 533, 537
    crystals, 55, *101*, 480, *562*, 577–78
    electrolysis, 505, 955
    electrolyte solutions, 146, *147*
    ion–ion interactions, 476, 477
    lattice energy, 522–23
    octahedral holes, 577–78
    saline (salt) solutions in living cells, 132
    sodium ion in Great Salt Lake or seawater, *137*, 140–41
    sphere of hydration, 480
    unit cell, 577–78
sodium fluoride, 478–79, 522, 524–25, 527, 801
    *See also* hydrofluoric acid
sodium hydroxide standard solution, 153–54
sodium hypochlorite as weak base, 802–3
soil chemistry
    clay minerals and clay, 582, 584
    groundwater, acid–base chemistry, 801, 803–5, 806
    lime use to "sweeten" soil, *752*, 778
    nutrient bioavailability and soil acidity, 778
    soil color from mineral content, 177
*Sojourner* Martian rover, *33*
solar still, *10*
solar system, 74
solar wind, 454
solids, 560–601
    amorphous solids, definition, 562
    characteristic properties, 13
    clusters, *581*, 582, 583
    covalent network solids, 580–82, 583, 590, 892
    crystal lattices, overview, 563–64
    crystalline solids, definition, 562, 563

solids (cont.)
  crystalline structure of ionic solids, 577–80
  cubic closest-packed (ccp) structures, 565, 566, 567
  hexagonal closest-packed (hcp) structures, 564, 565
  ionic solids, crystalline structure, 577–80
  ionic solids, definition, 562
  lattice energy (U) of ionic compounds, 521–27
  molecular solids, 562–63
  stacking patterns, overview, 563–64, 567
  viewed at atomic level, 14
  See also specific types
solubility
  combinations of intermolecular forces, 490–92
  common-ion effect, 827
  definition, 164, 165
  of gases in water, 517–19, 551
  insoluble compounds, definition, 159
  ion–dipole interactions and solubility, 488–89
  miscibility of liquids, 491–92
  nonpolar solutes, 489–90
  partition coefficient determination, 502–3
  pH effect on solubility, 827–29
  polarity and solubility, 488–92
  saturated solutions, 164–65
  solubility equilibria, 825–31
  solubility rules for ionic compounds, 157–58
  solubility-product constant ($K_{sp}$), 825–32
  solute–solute interactions, 488–89
  solvent–solute interactions, 488–89, 490
  solvent–solvent interactions, 488–89, 490
  supersaturated solutions, 164–65
solubility-product constant ($K_{sp}$), 825–32
solutes, definition, 135, 136
solutions, 132–95, 512–59
  aqueous solutions, definition, 77, 135
  in biological systems, 132, 512
  concentration units, 136–42
  D5W solution (dextrose), 545–46
  definition, 5, 135
  dilute and concentrated solutions, 136
  dilutions, 142–46
  electrolytes, 146–47
  energy changes during dissolution of ionic compounds, 520–21, 520–27
  enthalpy of solution ($\Delta H_{solution}$), 520–21, 526–27
  ideal solutions, 514, 515
  ion pairs, 539, 541
  nonelectrolytes, 146–47
  phase diagram of water and solution, 532–33
  physiological saline, 144, 545
  preparing solutions of known concentration, 141–42, 143
  saturated solutions, 164–65
  solid solutions, 135
  solutes, definition, 135, 136

standard solution, definition, 153
stock solutions, 142–44
strong electrolytes, 146, 147, 149, 152
supersaturated solutions, 164–65
vapor pressure, with nonvolatile solutes, 514–16
vapor pressure, with volatile solute mixtures, 527–32
weak electrolytes, 146, 147, 149, 152
See also colligative properties; concentration; precipitation reactions; saturated solutions
solvation, sphere of, 480, 481
solvent, definition, 135, 136
Sørensen, Søren, 786
space shuttle, 505
space-filling models, 7, 8
spearmint, 424, 452–53, 638, 651
specific heat ($c_s$), 215, 222–25
See also heat capacity ($C_P$); molar heat capacity ($c_P$)
spectator ions, 150, 930
spectra
  atomic absorption spectra, 322–23, 331
  atomic emission spectra, 322, 323, 331
  chlorophyll and other pigments, 873, 1054
  doublets, 340–41
  electromagnetic spectrum, 31, 318
  Fraunhofer lines, 321–23
  hydrogen energy level diagram, 331
  neon emission spectra, 316
  visible spectrum, 318, 323
spectrochemical series of common ligands, 866, 869
spectrophotometry, 144–46
speed of light ($c$), 17, 28, 319
sphalerite (ZnS), 578
sphere of hydration, 480, 481
sphere of solvation, 480, 481
Spirit Martian rover, 33
Spirit of Freedom balloon, 211, 212
Splenda (sweetener), 504
spodumene, 102–3
spontaneous processes
  ammonia synthesis, 897–98
  ammonium nitrate dissolution (cold packs), 894
  definition, 882, 884
  effects of $\Delta H$, $\Delta S$, and $T$, 895–98, 900–902
  and entropy, 886–87, 888–89, 893–96
  exothermic processes, relationship, 884
  as function of $\Delta S_{sys}$ and $\Delta S_{surr}$, 888, 889–90, 895
  reversibility, 903, 905
  temperature and spontaneity, 900–902
  thermodynamics of melting ice, 882, 884–85, 888, 891, 900–901
  See also free energy ($G$)
spring bloom, 502
square packing, 566

square planar molecular geometry, 436, 437, 448
square pyramidal molecular geometry, 436, 437, 448
stalactites and stalagmites, 151
standard conditions, 232, 233
standard hydrogen electrode (SHE), 940–42
standard solution, definition, 153
standard states
  of elements, 233, 234, 237, 522
  of substances and solutions, 891, 935
standard temperature and pressure (STP), 274, 275
standing waves, 334–35, 344, 345, 454, 455
starch, 984, 986–87
Stardust probe, 4, 997
state functions
  definition, 198, 199–200, 208, 231, 236
  enthalpy, 212, 231, 236
  entropy, 894
  internal energy, 208, 212
  potential energy, 199–200
states of matter, 13–15
  changes of state, 14–15
  kinetic energy differences, 476
  standard states of elements, 233, 234, 237, 522
  standard states of substances and solutions, 891, 935
  symbols, 77
  See also gases; liquids; phase changes; solids
Statue of Liberty, 950–53
steam–methane reforming
  enthalpy changes, 230, 236
  equilibrium dynamics, 734–35
  hydrogen gas production, 118, 631, 734
  in methanol production, 631
stearic acid, 990, 991, 999
steel
  bluing, 953
  carbon steel, 560, 573
  as interstitial alloy, 572, 573
  stainless steel, 560, 952–53
  steel alloys, 560, 561, 573
  steelmaking, 182, 572–73
  See also alloys; iron
stellar nucleosynthesis, 64–65, 1010, 1011, 1033–34
stents, 1078–79
steric number, 427
  See also molecular geometry
stock solutions, 142–44
stoichiometry, 85–117
  actual yields vs. theoretical yields, 114–17
  calculating product mass from reactant mass, 96–97
  and carbon cycle, 94–98
  definition, 86
  fossil fuel combustion, 95–97
  and ideal gas law, 277–83, 286–87
  limiting reactants, 110–16

stoichiometry *(cont.)*
   percent yield, 114–16
   theoretical (stoichiometric) yield, 112–17
   *See also* mole (mol); mole ratio
strong electrolytes, 146, 147, 149, 152
strontium, 350, 1065
strontium sulfate, 135
styrene, *629*
subatomic particles
   history and discovery, 42–43, 45–46
   properties, *46*
   quarks, 63
   symbols and masses, *1010*
   *See also* electrons; neutrons; positrons; protons
sublimation, 14, 15, 495
substances, definition, 4
substrates, 712, 982–83
sucrose (table sugar), 985–86
   aqueous solutions, *136*, 163
   food value, 244
   molecular structure, *986*
   rock candy, 164
   *See also* carbohydrates
sulfate ion, Lewis structure, 398–99
sulfur, 464–65, 1046, 1048
sulfur dioxide
   in Earth's early atmosphere, 76
   equilibrium with $SO_3$, 732–34, 740, 747
   molecular structure, *54*, 431–32
   reaction with atmospheric oxygen, 710
   reaction with calcium oxide (lime), 182–83, 847
   released in volcanic eruptions, 76
   resonance structures, 431–32
sulfur hexafluoride, 397, 400, 430, 447
sulfur tetrafluoride, 436–37
sulfur trioxide
   in Earth's early atmosphere, 76
   equilibrium with $SO_2$, 740, 747
   formation, 90–91
   molecular structure, *54*
   reaction with water, 77, 86, 110–12, 151, 833
   resonance structures, 389–90
sulfuric acid, 833
   in acid rain, *789*, 796–97
   calculations with $K_a$, 796–97
   comparison to sulfurous acid, 800–801
   dissociation, 152
   formation from pyrites, 153
   formation from $SO_3$, 77, 86, 110–12, 151, 732, 833
   industrial production, 732–34, 833
   ionization, 152, 796–97
   Lewis structure, 399
   neutralization reaction, 150–51, 153–54
   as strong acid, 152
   titration, 153–54, 822
sulfurous acid, 797–98, 800–801
sun, 65, *196–97*

sunlight
   as energy source for Earth, *196–97*, 910
   energy transformations, 196
   Fraunhofer lines, 321–23
superconductors, 119, 584–86
supercritical fluids, 497
supercritical region of phase diagram, 495, 497
supergiant stars, 64
supernovas, 65, 1010, 1012
superoxide dismutases, 1063
superoxide ion ($O_2^-$), 1063
supersaturated solutions, 164–65
surface tension, 498–500
systems
   closed systems, 204, 205
   definition, 204
   energy transfer, 204–5
   isolated systems, 204–5
   open systems, 204, 205
   pressure–volume ($P$–$V$) work, 209–11, 899–900
   surroundings, 204

## T

tantalum, 1046, 1078
taxol, 116–17
technetium, 65–66, *1013*, 1067–69, *1070*
Teflon (polytetrafluoroethylene), 410, 606, 625
temperature
   absolute zero (0 K), 30, 268, 269, 270
   Amontons's law, 271–73, *275*, 289, *290*
   Celsius scale, 29–30
   Charles's law, 268–70, 271, 274, *275*, *289*, *290*
   conversion between scales, 33
   effect on chemical equilibrium, 759–60
   effect on reaction rates, 694–99
   Fahrenheit scale, 29–30
   interstellar space temperature, 28–29, 31, *32*
   Kelvin scale, 29–30
   and kinetic energy, 201, *202*, 204, 695
   and molecular motion, 272, 290
   standard temperature and pressure (STP), 274, 275
temporary dipoles (induced dipoles), 484–85
terephthalic acid (benzene-1,4-dicarboxylic acid), *645*
terpineol (oil of turpentine), 633
tertiary amines, 629
tetraamminedichlorocobalt(III) chloride, 870–71
tetrafluoroethylene, 410
tetrahedral holes, 572–73, 578–79, 589
tetrahedral molecular geometry, 428, 429–30, 442–43, *448*
thalassemia, 1073–75
thalidomide, 982
thallium, 1067, 1069

theoretical yield, 112–17
therapeutic applications of elements, 504, 1028–29, 1046, 1066–67, 1072–76
thermal energy, definition, 201
thermal equilibrium, 198, *222*, 223, 224
thermochemical equations, definition, 198
thermochemistry, 196–257
   *See also* enthalpy ($H$); heat ($q$); heat capacity ($C_P$)
thermocline, 502
thermodynamics, 882–925
   definition, 198
   and equilibrium constants ($K$), 739
   first law of thermodynamics, 209–10, 214, 222–23, 882, 913
   second law of thermodynamics, 882, 886–87, 888–89, 913
   third law of thermodynamics, 890–92, 915
   *See also* enthalpy ($H$); entropy ($S$); free energy ($G$); spontaneous processes
thin layer chromatography (TLC), 9
third law of thermodynamics, 890–92, 915
Thomson, Joseph John, 42, 43, 44, *45*, 46
thorite, 589
thorium, 589, 1016
threonine, *971*, 996
*Thresher* (nuclear submarine), 959–60
threshold frequency ($\nu_0$), 326–28
thymine, *483*, 994
thyroid, imaging with technetium, *65*, 66
tin, 591
   atomic radius, 572
   in bronze, 571, 572
   crystal structure, 580, 591
   dental alloys, 1076
   mercury reduction by $Sn^{2+}$, 945
   oxidation–reduction reactions, 174
titanium
   alloys, 560, 571, 1078–79
   cancer drugs, 1073, *1074*
   crystal structure, 564, *565*, 567
   electron configuration, 349, 353
   medical uses, 560, 564, 571, 1073, *1074*, 1078–79
   shape-memory alloys, 1078–79
   surgical implants, 1078–79
titrants, 153, 819–20, *821*, 822–23, 824
titrations, 153–57, 818–25
   acetic acid (vinegar), 155, 820–22
   acid–base neutralization reactions, 153–57
   acid–base titrations, 153–57, 818–25
   alkalinity titrations, 819, 822–25, 832
   amino acids, 972–74
   ammonia, 819–20
   definition, 153
   end point, 154, 155
   equivalence point, 154, 155, 813, 819–21, 824–25
   pH indicators in, 818–19, 823–25

titrations (cont.)
    shelf-stability of drugs, 181
    sulfuric acid, 153–54, 822
tokamak, 1034
toluene, 627–28
torr, definition, 261, 262
Torricelli, Evangelista, 261
Toyota Prius, 947
tracers for medical imaging, 1029, 1067–72
trans fats, 118, 990
transferrin, 1073, 1074–75
transition metals
    actinides, 51, 52, 350, 353–54
    electron configuration of atoms, 349–50, 351
    electron configuration of cations, 353–55
    lanthanides, 51, 52, 350, 353–54
    naming binary transition metal compounds,
        59–60
    in periodic table, 51, 52, 349–50
    trace essential transition elements, 1059–62
    See also crystal field theory; metals; specific
        elements
transition states, 695, 701, 712
translational motion, 208, 885, 886, 890, 913–14
tricarboxylic acid cycle (TCA cycle, citrate
    cycle), 988, 989, 1063
trichlorofluoromethane, 439
triglycerides, 988, 989–90, 991
trigonal bipyramidal molecular geometry, 428,
    429, 430, 447–48
trigonal planar molecular geometry, 428, 429,
    431, 443–45, 448
triple bond, definition, 380
triple point, 496, 497–98, 532
triprotic acids, 796, 799, 800
tritium ($^3$H), 118–19, 1034–35, 1036, 1072
    See also hydrogen
tryptophan, 971
T-shaped molecular geometry, 434, 448
Tsou, Peter, 4
tungsten, 563
Tycho supernova, 65
tyrosine, 971

U
Uhlenbeck, George, 340
ultrahigh molecular weight polyethylene
    (UHMWPE), 624
ultraviolet (UV) radiation, 31, 318, 320
unit cells
    and atomic radius, 579–80
    body-centered cubic (bcc) unit cell, 565, 567,
        568–70, 572–73
    definition, 564
    and density calculations, 579–80, 589
    dimensions, 567–71
    face-centered cubic (fcc) unit cell, 565, 566,
        567–68, 569–70, 573–74, 578–79

hexagonal unit cell, 564, 565, 567
in interstitial alloys, 572–74
ionic solids, 577–80
of metals in periodic table, 566
packing efficiency, 566, 567, 570–71
simple cubic (sc) unit cell, 565, 566–67
summary, 567
See also crystal structure
unit factor method (dimensional analysis), 26–28
universal gas constant ($R$), 273, 291, 542–43
universe
    composition, 1049
    expansion, 16–17, 29
    origins, 2, 16–17
    timeline for energy and matter
        transformation, 63
    See also Big Bang theory
unsaturated fatty acids, 989, 990, 998–99
unsaturated hydrocarbons, 606, 607
    See also alkenes; alkynes
uracil, 994
uranium
    enrichment, 1030–31
    in pitchblende, 44, 1030
    preparing uranium fuel, 1030
    radiometric dating, 1018–19
    $UF_6$ gas, 1030–31
    uranium-235 critical mass, 1030
    uranium-235 fission, 1029–30
    uranium-238 alpha decay, 1016
    uranium-238 radioactive decay series, 1017
    yellowcake ($U_3O_8$), 1030
urea, 604, 1058, 1063
ureases, 1058, 1063
Urey, Harold, 74, 87–88, 113, 997

V
valence bands, 574, 575–77
valence bond theory, 440–48, 574
    benzene, 627
    hybrid atomic orbitals, definition, 441
    hybridization, definition, 441
    molecular overlap, 440–41
    pi ($\pi$) bonds, 444
    sigma ($\sigma$) bonds, definition, 441
    $sp$ hybrid orbitals, 445–47
    $sp^2$ hybrid orbitals, 443–45
    $sp^3$ hybrid orbitals, 442–43
    $sp^3d$ hybrid orbitals, 447–48
    $sp^3d^2$ hybrid orbitals, 447
    summary of hybridization schemes and the
        orientations of orbitals, 448
    See also molecular geometry
valence electrons, definition, 347
valence-shell electron-pair repulsion theory
    (VSEPR), 427–37
    central atoms with lone pairs, 431–37
    central atoms with no lone pairs, 427–31

definition, 427
electron-pair geometries and molecular
    geometries, summary, 435
electron-pair geometry, definition, 427
linear electron-pair geometry, 427, 429
octahedral electron-pair geometry, 428, 429,
    430, 436
steric number, definition, 427
tetrahedral electron-pair geometry, 428, 429,
    430, 432–33, 434
trigonal bipyramidal electron-pair geometry,
    428, 429, 430, 434–36
trigonal planar electron-pair geometry,
    428, 429
and valence bond theory, 440, 441, 443, 444,
    445, 449–51
See also molecular geometry
valine, 971, 975, 978, 979, 996
van der Waals constants, 297
van der Waals equation, 297–98
van der Waals forces, 981–82
    See also dipole–dipole interactions; dispersion
        forces (London forces); hydrogen bonds
van 't Hoff, Jacobus Henricus, 538, 695, 833, 908
van 't Hoff equation, 908
van 't Hoff factor ($i$ factor), 537–41
vanadium
    bis(allixinato)oxovanadium(IV), 1073, 1074
    in cancer drugs, 1073, 1074
    diabetes drugs, 1073
    electron configuration, 349
    as essential ultratrace element, 1048, 1062
    lithium silver vanadium oxide batteries, 1078
    in orthopedic devices, 1079
    vanadium(V) oxide, 833
vanillin, 110
vapor pressure, 492–94, 514–16
    and boiling point, 492–93
    Clausius–Clapeyron equation, 493–94, 495
    definition, 492
    mixtures of volatile solutes, 527–32
    and nonvolatile solutes, 514–16
    of pure liquids, 492–94
    Raoult's law, 514, 515–16, 530–32
    and temperature, 492–93
    See also colligative properties
vermillion, 162
vibrational motion, 208, 386, 886, 890, 892,
    913–14
vinegar
    commercial production, 639
    as mixture (solution), 5, 132, 155, 639
    reaction with sodium bicarbonate, 884, 885
    titration, 155
    See also acetic acid
vinyl acetate, 635, 636
vinyl chloride, 139–40
vinyl group, 626, 627
vinyl polymers, 626, 627

viscosity, 501
vitamin B$_{12}$, *1058*, 1063
vitamin C (ascorbic acid), 844, 860
vitamin package labels, 1048, *1049*
volatile organic compounds (VOCs), 668–69
volatile sulfur compounds (VSCs), 465
volatility, 8, 9, 493–94
Volta, Alessandro, *931*
voltaic cells
    chemical energy and electrical work,
        937–40
    comparison to electrolytic cells, 931, *932*, 954,
        955–56
    Cu/Ag voltaic cell, 933
    definition, 931
    effect of chemical concentration, 942–46
    fuel cells, 956–59
    standard cell potential, 935
    and standard hydrogen electrode (SHE),
        940–42
    Zn/Cu$_2^+$ voltaic cell, 930–31, 932, 935, 937,
        939, 944
    *See also* electrochemical cells
voltmeters, 935
volume
    molar volume, 274, 275–76, 278
    pressure–volume ($P$–$V$) work, 209–11,
        899–900
    pressure–volume relationship of gases, 260,
        265–68, 274, 275, 288–89
    temperature–volume relationship of
        gases, 260
VSEPR. *See* valence-shell electron-pair repulsion
    theory (VSEPR)

**W**

Waage, Peter, 738
water
    acid–base chemistry, 148–57, 782,
        785–86, 846
    alkalinity titrations, 819, 822–25, 832
    amount on Earth's surface, 134
    and aquatic life, *132–33*, 474, 502, 512
    atomic view of formation, 7–8, *110*
    attractive forces between molecules, 217
    autoionization, 785–86, 795
    capillary action, 500
    changes of state, 14–15
    density changes with temperature, 501,
        502, *502*
    dipole moment, 440
    drinking water standards, 136, 138
    equilibrium constant for autoionization ($K_w$),
        785–86
    as essential compound for life, 118, 132, 180,
        464, 474–76, 502
    freezing point, 12, *499*
    hard water, 165–66

heating curves, 214–22
hydrogen bonding, 481, *482*
law of constant composition, 6
leveling effect on strong acids and bases, 784
Lewis structure, 433
macroscopic and molecular-level views, *217*
on Mars, 134
melting and boiling point, *499*
meniscus, 22, 499–500, 503, 591
molar heat capacity ($c_P$), *215*, 216, 218, 219
molar heat of fusion ($\Delta H_{fus}$), 212, 216,
    217, 218
molar heat of vaporization ($\Delta H_{vap}$), 216–17
molecular structure, *54*, *110*, 432–33
phase changes and energy transfer,
    205–8, *209*
phase diagrams, *496*, *498*, 532–33
polar bonds and molecule, 437–38
properties, 474–76, 499–502
reaction with fluorine, 410
$sp^3$ hybrid orbitals, 443
specific heat, 223
surface tension, 499–500
triple point, *496*, 497–98, *532*
viscosity, 501
water softening, 165–66
    *See also* ice; seawater; water vapor
water softening, 165–66
water vapor
    absorption by Mg(ClO$_4$)$_2$, 106–7
    deposition, *889*
    Earth's early atmosphere, 76
    in Earth's early atmosphere, 76
    macroscopic and molecular-level views, *217*
    molar heat capacity ($c_P$), *215*, 216
    released in volcanic eruptions, 77
    *See also* water
water–gas shift reaction
    calculations based on $K$, 765–66
    effect of adding or removing products or
        reactants, 755
    enthalpy changes, 230
    equilibrium constant expression, 737–38, 739
    equilibrium dynamics, 734–35
    hydrogen generation, 115–16, *230*
    initial and equilibrium concentrations, *738*
watt (W), definition, 946
wave functions ($\psi$), 336, 337, 345
wave mechanics (quantum mechanics)
    angular momentum quantum number ($\ell$), 338
    definition, 336
    lack of determinacy, 342–43
    magnetic quantum number ($m_\ell$), *338*
    orbital sizes and shapes, 343–45
    orbital ($\psi^2$), definition, 336, 339
    orbitals and quantum numbers, 338–41
    Pauli exclusion principle, 340–41
    principal quantum number ($n$), 336, 337, 354
    quantum number, definition, 330, 335, 337

Schrödinger wave equation, 336, 337–41
shells and quantum numbers, 338–41
spin magnetic quantum number ($m_s$),
    340–41
subshells, letters and quantum numbers,
    338–41
wave functions ($\psi$), 336, 337, 345
    *See also* quantum theory
wavelength ($\lambda$), 318, 319, 361
wave–particle duality, 327–28
waves
    amplitude, 319
    behavior, 320–21
    circular waves, 321, 334–35
    diffraction, 182, 320–21, 327
    frequency ($\nu$), 318, 319
    fundamentals, 334
    harmonics, 334
    interference, *320*, 321
    matter waves, 333–35, 337, 344
    nodes, 334, 335, 344, 345
    ocean waves, *474–75*
    properties, 319–20
    refraction, 320
    standing waves, 333–35, 344, 345
    wavelength ($\lambda$), 318, 319
    *See also* electromagnetic radiation
weak electrolytes, 146, 147, 149, 152
weak-link principle, 21, 22, 23
Wilson, Robert W., 31, *32*
Wollaston, William Hyde, 321
work
    definition, 198
    efficiency, 900
    electrical work ($w_{elec}$), 937–38, 946
    formula, 198
    and free energy, 895, 899–900
    and potential energy, 199
    pressure–volume ($P$–$V$) work, 209–11,
        899–900
    sign convention, 210
    units, 208, 209, 211
    *See also* energy
work function ($\Phi$), 326, 327–28, 332
wustite (FeO), 101

**X**

xanthine oxidase, 1062, 1063
xenon, 347, 363, 1072
X-ray diffraction (XRD), 586–88, 589, 592
X-rays
    discovery, 44, 44n2
    in electromagnetic spectrum, *31*, *318*
    ionizing radiation, 42, 1024
    radiation dose from dental X-rays, 1026, 1028
    wavelength, 586
    X-ray astronomy, *65*
xylene (dimethylbenzene), 628

## Y

yttrium, *585*, 1069, 1070
yttrium–barium–copper (YBC) oxide
    superconductors, *585*

## Z

zeolites, 165–67
zero-order reactions, 693–94, 706
Zewail, Ahmed H., 700n1

zinc
    conduction band, 575
    DRI/RDA value, *1049*
    as essential trace element, 844, 1046, 1059–60
    redox reaction with copper, 928–29
    sphalerite (ZnS), 578
    zinc–air batteries, 936–37, 944, 1077–78
    Zn/Ag button batteries, 939–40
    Zn/Cu$^{2+}$ electrochemical cell, 930–31, 932,
        935, 937, 939, 944

Zn/HgO batteries, 1076–77
Zn(NH$_3$)$_4$$^{2+}$ complex ion, 849–50, *865*
zingerone, 638
zingiberene (oil of ginger), *618*
zircon (ZrSiO$_4$), 1019
zwitterions, 972–75, 976

# ATOMIC COLOR PALETTE

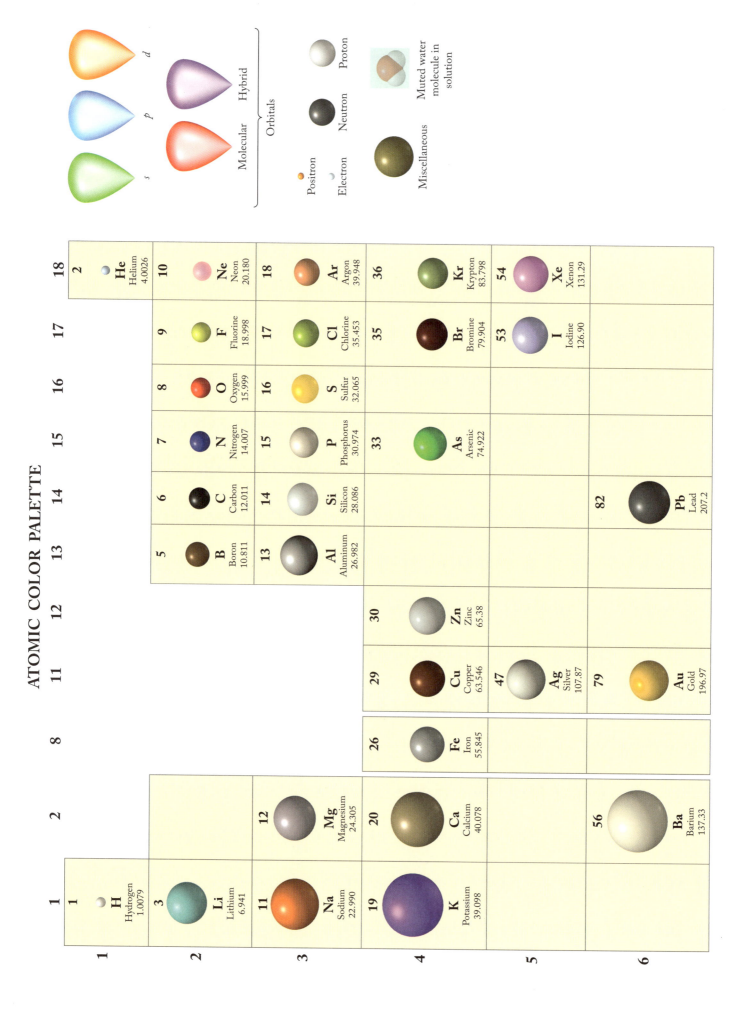

## Special Units and Conversion Factors

| Quantity | Unit | Symbol | Conversion |
|---|---|---|---|
| energy[a] | electron-volt | eV | $1\text{ eV} = 1.6022 \times 10^{-19}\text{ J}$ |
| energy[a] | kilowatt-hour | kWh | $1\text{ kWh} = 3600\text{ kJ}$ |
| energy | calorie | cal | $1\text{ cal} = 4.184\text{ J}$ |
| mass | pound | lb | $1\text{ lb} = 453.59\text{ g}$ |
| mass[a] | atomic mass unit | amu | $1\text{ amu} = 1.66054 \times 10^{-27}\text{ kg}$ |
| length | angstrom | Å | $1\text{ Å} = 10^{-8}\text{ cm} = 10^{-10}\text{ m}$ |
| length | inch | in | $1\text{ in} = 2.54\text{ cm}$ |
| length | mile | mi | $1\text{ mi} = 5280\text{ ft} = 1.6093\text{ km}$ |
| pressure | atmosphere | atm | $1\text{ atm} = 1.01325 \times 10^{5}\text{ Pa}$ |
| pressure | torr | torr | $1\text{ torr} = 1/760\text{ atm}$ |
| temperature | Celsius scale | °C | $T(°\text{C}) = T(\text{K}) - 273.15$ |
| temperature | Fahrenheit scale | °F | $T(°\text{F}) = \frac{9}{5}T(°\text{C}) + 32$ |
| volume | liter | L | $1\text{ L} = 1\text{ dm}^3 = 10^{-3}\text{ m}^3$ |
| volume | cubic centimeter | cm³, cc | $1\text{ cm}^3 = 1\text{ mL} = 10^{-3}\text{ L}$ |
| volume | cubic foot | ft³ | $1\text{ ft}^3 = 7.4805\text{ gal}$ |
| volume | gallon (U.S.) | gal | $1\text{ gal} = 3.785\text{ L}$ |

[a]From http://physics.nist.gov/cuu/constants/index.html.

## Physical Constants[a]

| Quantity | Symbol | Value |
|---|---|---|
| Avogadro's number | $N_A$ | $6.0221 \times 10^{23}\text{ mol}^{-1}$ |
| Bohr radius | $a_0$ | $5.29 \times 10^{-11}\text{ m}$ |
| Boltzmann constant | $k_B$ | $1.3806 \times 10^{-23}\text{ J/K}$ |
| electron charge-to-mass ratio | $-e/m_e$ | $1.7588 \times 10^{11}\text{ C/kg}$ |
| elementary charge | $e$ | $1.602 \times 10^{-19}\text{ C}$ |
| Faraday constant | $F$ | $9.65 \times 10^{4}\text{ C/mol}$ |
| mass of an electron | $m_e$ | $9.10938 \times 10^{-31}\text{ kg}$ |
| mass of a neutron | $m_n$ | $1.67493 \times 10^{-27}\text{ kg}$ |
| mass of a proton | $m_p$ | $1.67262 \times 10^{-27}\text{ kg}$ |
| molar volume of ideal gas at 0°C and 1 atm | $V_m$ | $22.4\text{ L/mol}$ |
| Planck constant | $h$ | $6.626 \times 10^{-34}\text{ J} \cdot \text{s}$ |
| speed of light in vacuum | $c$ | $2.998 \times 10^{8}\text{ m/s}$ |
| universal gas constant | $R$ | $8.314\text{ J/(mol} \cdot \text{K)}$ <br> $0.08206\text{ L} \cdot \text{atm/(mol} \cdot \text{K)}$ |

[a]From http://physics.nist.gov/cuu/constants/index.html.